CONCISE ENCYCLOPEDIA OF

MATERIALS FOR ENERGY SYSTEMS

Elsevier Science and Technology Homepage - http://www.elsevierdirect.com
Consult the Elsevier homepage for full catalogue information on all books, major reference works, journals, electronic products and services.

Elsevier Titles of Related Interest

K.H.J. BUSCHOW
The Encyclopedia of Materials: Science and Technology (2001)
Major Reference Work
ISBN-13: 978-0-08-043152-9

R. CAHN
Concise Encyclopedia of Materials Characterization (2004)
ISBN-13: 978-0-08-044547-2

K.H.J. BUSCHOW
Concise Encyclopedia of Magnetic and Superconducting Materials (2005)
ISBN-13: 978-0-08-044586-1

A. MORTENSON
Concise Encyclopedia of Composite Materials (2006)
ISBN-13: 978-0-08-045126-8

J.W. MARTIN
Concise Encyclopedia of Structure of Materials (2006)
ISBN-13: 978-0-08-045127-5

J.W. MARTIN
Concise Encyclopedia of Mechanical Properties of Materials (2006)
ISBN-13: 978-0-08-046525-8

J.W. MARTIN
Concise Encyclopedia of Chemical Properties of Materials (2007)
ISBN-13: 978-0-08-046526-5

J.W. MARTIN
Concise Encyclopedia of Properties of Materials Surfaces and Interfaces (2008)
ISBN-13: 978-0-08-054811-1

CONCISE ENCYCLOPEDIA OF

MATERIALS FOR ENERGY SYSTEMS

Editor

JOHN W. MARTIN

ELSEVIER

Amsterdam—Boston—Heidelberg—London—New York—Oxford
Paris—San Diego—San Francisco—Singapore—Sydney—Tokyo

Elsevier Ltd., The Boulevard, Langford Lane, Kidlington, Oxford, OX5 1GB, UK

Material in the work originally appeared in the *Encyclopedia of Materials Science and Technology*, edited by Jürgen Buschow, Robert Cahn, Mert Flemings, Bernhard Ilschner, Ed Kramer and Subbash Mahajan (Elsevier, B.V. 2001).

First edition 2009

A catalogue record for this book is available from the British Library

ISBN 978-0-08-096496-6

09 10 11 12 13 10 9 8 7 6 5 4 3 2 1

This book is printed on acid-free paper
Printed and bound in the UK

CONTENTS

M

N

O

P

R

S

T

EDITOR'S PREFACE

The Concise Encyclopedia of Materials for Energy Systems draws together in a single volume a range of articles which appear in the *Encyclopedia of Materials Science and Technology (EMSAT)*. This very substantial work first appeared in 2001, the editors-in-chief being Jürgen Buschow, Robert Cahn, Mert Flemings, Bernhard Ilschner, Ed Kramer and Subbash Mahajan; it has been kept up-to-date by its editors-in-chief in the form of periodic updates loaded on to the internet version of EMSAT.

The range of energy systems we have considered is widely inclusive. It covers thermal and gas turbines, renewable power, fuel cells (low and high temperature), nuclear power generation (fission and fusion) and the transmission and storage of the energy produced. We have included a consideration of the hydrogen economy, as well as the materials employed in the design of batteries, and in energy conversion in a range of devices and actuators.

Over 100 articles have been selected which discuss the applications of materials to these energy systems, with probably the greatest emphasis being placed upon the generation of nuclear energy. The whole assembly provides the reader with an up-to-date understanding of the applications to energy systems of the entire range of materials of engineering importance, namely metals, ceramics and polymeric materials.

John W. Martin
Editor

ALPHABETICAL LIST OF ARTICLES

A

Actuator Materials for Small-scale Devices

E. Quandt

The fabrication of a microsystem requires the processing of a large variety of different materials in order to realize their different electronic, sensor, and actuator components. The fabrication processes of these materials should be compatible with microsystem technologies, i.e., with silicon micromachining. One possibility for the realization of sensors and actuators in microsystems is the use of "smart" materials that directly transduce electrical, magnetic, or thermal energy into mechanical energy or vice versa. The related physical effects are the piezoeffect, magnetostriction, or the shape memory effect. In general, thin film technologies are an attractive approach to integrate these materials into microsystems as this approach has almost no materials limitations, offers easy down-scaling into the micrometer range by a cost-effective manufacturing technology, avoids assembly and interconnection processes, allows the protection of the microsystem by protective layers (Quandt and Holleck 1998), and provides the possibility to design new materials as, for example, metastable phases, multilayers or gradient layers. Physical vapor deposition (PVD) techniques, in particular magnetron sputtering, offer the opportunity of tailoring the constitution of the thin films. Additionally, the interaction with the substrate enables fabrication of highly textured films which, in the case of piezoelectric materials, show improved properties compared to conventional polycrystalline bulk materials. Furthermore, it can be expected that the integrated sensing features of smart film actuators—e.g., the resistive effects present in all three classes of materials—will provide means to realize "intelligent" actuators by using thin film "smart" materials.

Thin film actuators in the case of piezoelectric or magnetostrictive materials are always used as compounds with micromachined substrates working as bending transducers, whereas in the case of shape memory thin films both free-standing thin film actuators and compounds with metallic or silicon substrates are realized. Since free-standing shape memory films have to be trained for the two-way effect to introduce internal stresses into the material, which is quite complicated in small dimensions, the film substrate compounds represent the generally more important approach for all smart film materials. In this case the materials development target has to be modified compared to bulk actuator materials. Whereas in bulk materials the strain has to be maximized, the resulting bending stress of a thin film substrate compound is proportional to the strain and the Young's modulus of the film. As a consequence, this stress of the thin film actuator material should be as large as possible. In the following sub-sections the main results for magnetostrictive, piezoelectric, and shape memory thin films for actuators will be discussed in view of their main application areas.

1. *Magnetostrictive Films*

The development of room temperature giant magnetostrictive bulk and thin film materials is based on the rare earth Fe_2 Laves phase, while at low temperatures the highest strains are found in hexagonal Tb_xDy_{1-x} and body-centered $Tb_xDy_{1-x}Zn$ alloys. The giant magnetostrictive materials were optimized in terms of their magnetostriction to magnetic anisotropy ratio in order to attain large strains at reasonable magnetic fields (see *Magnetostrictive Materials*). Different approaches have been taken based on the Laves phases $(Tb,Dy)(Fe,Co)_2$ with positive or $Sm(Fe,Co)_2$ with negative magnetostriction. The fabrication of these rare earth based materials is restricted to PVD methods, the most prominent being magnetron sputtering.

Because the magnetic saturation field is proportional to the magnetic anisotropy constant and inversely proportional to the saturation magnetization, its reduction is achieved either by decreasing the magnetic anisotropy constant and/or by increasing the saturation magnetization. To reduce the macroscopic anisotropy, amorphous rare earth–iron thin film materials have been fabricated in a wide composition range. It was found that increasing the rare earth content compared to the Laves phase composition results in the highest low-field magnetostriction in amorphous thin films. For amorphous TbFe films with an inplane magnetic easy axis an important improvement of the low-field magnetostriction and the hysteresis can be observed in comparison to nanocrystalline $(Tb,Dy)Fe_2$ films.

Assuming the same local environment in the amorphous compared to the crystalline state, a further approach to lower the remaining anisotropy is by Tb/Dy substitution, which was investigated over a wide composition range for rare earth–iron or –cobalt thin film materials. This approach leads to a further reduction in the magnetic saturation field, but the lower Curie temperature due to the dysprosium alloying significantly reduces the saturation magnetostriction, resulting in negligible gain in low-field magnetostrictive strain of these ternary films.

The second route to lower the macroscopic anisotropy is by increasing the saturation magnetization. The magnetostrictive thin films discussed so far have rather low magnetizations, owing to their ferrimagnetic nature. Unfortunately, an increase of the saturation magnetization using amorphous homogeneous rare earth–transition metal materials is impossible, because the compositions of interest contain 30–45at.% rare earth elements. The rare earth moments dominate and thus any decrease will further reduce the magnetization, while an increase of their content results in a lowering of the Curie temperature and consequently a lowering of the room temperature magnetization. However, using multilayers, it is possible to develop new composite materials which show properties that overcome these limitations. One layer is composed of the giant magnetostrictive amorphous $Tb_{0.4}Fe_{0.6}$ alloy, while the other is magnetically very soft and has a very high magnetization as well as considerable magnetostriction itself ($Fe_{0.5}Co_{0.5}$). By fabricating these layers with thicknesses smaller than the ferromagnetic exchange length, domain wall formation at the interfaces is prevented and the magnetic properties of such an exchange-coupled multilayer system are determined by the average of those of each individual layer. The comparison of the magnetostriction of this multilayer with the state-of-the-art amorphous and nanocrystalline single-layer materials demonstrates that these novel multilayers give the highest magnetostrictions at very low fields with a saturation field of only 20 mT (Fig. 1).

The most important advantages of magnetostrictive actuators are the possibility of wireless operation and the anisotropic behavior of the magnetostriction. The former is especially attractive for ultrasonic motors and for valves or pumps operating in harsh environments or on rotating or moving parts. The anisotropy of the magnetostriction can be employed for the realization of a two-dimensional scanner using a simple cantilever geometry in combination with the excitation of bending and torsion modes.

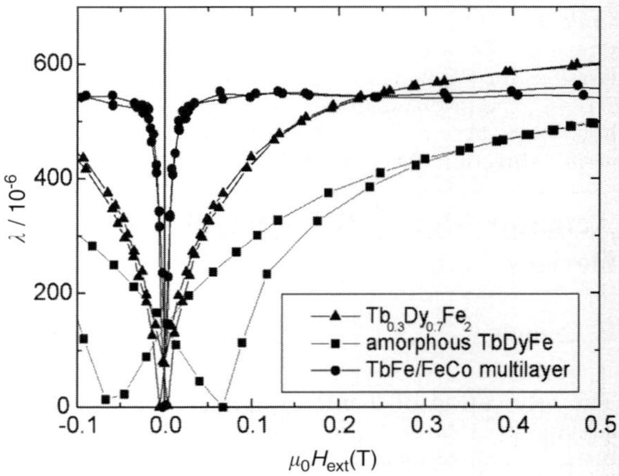

Figure 1
Magnetostriction (λ) versus external magnetic field (H_{ext}) of different thin film materials: nanocrystalline $Tb_{0.3}Dy_{0.7}Fe_2$ single layer, amorphous TbDyFe single layer, and TbFe/FeCo multilayer.

2. *Piezoelectric Films*

The development of piezoelectric thin films for actuators is related either to wurtzite-type structure materials (e.g., ZnO, AlN) or to ferroelectric perovskite materials (e.g., $Pb(Zr,Ti)O_3$). The general merits of piezoelectric thin film actuators are their low power consumption and their driving voltages compatible with transistor–transistor logic (TTL) signal levels. The drawbacks are their high fabrication temperatures, especially in the case of ferroelectric thin films, their comparably low sputtering rates, and the high precision of stoichiometry control required. In the development of piezoelectric thin film materials, high piezoelectric strains, breakdown field strengths, electrical resistivity, Young's modulus, long-time stability and fatigue strength, and in the case of the ferroelectric materials high Curie temperature, are further important targets. The wurtzite-type piezoelectric films can be fabricated by chemical or physical vapor deposition (CVD or PVD) methods, the latter being mainly sputtering or laser-ablation routes. Even more fabrication methods exist in the case of ferroelectric thin film materials; the most prominently used are CVD, PVD and sol–gel methods.

Although their piezoelectric constants are one or two orders of magnitude lower than in $Pb(Zr,Ti)O_3$ (PZT), wurtzite-type thin films are of interest for many applications, predominantly because of their simpler fabrication, mainly owing to lower fabrication temperatures, their hysteresis-free response, and their higher temperature range of use. Furthermore, their much higher Young's moduli compensate in part for their lower piezoelectric constants in the case of bending-type thin film actuators. In order to obtain high piezoelectric constants, highly c-axis-textured wurtzite-type films are required having low resistivity perpendicular to the substrate. It has been found that the microstructure and the texturing depend critically on the deposition conditions. By

Table 1
Measured piezoelectric and dielectric coefficients. $d_{33,f}$ describes the strain x_3 in the normal direction ($x_3 = d_{33,f} E_3$), $e_{31,f}$ the in-plane strain ($\sigma_{1,2} = e_{31,f} E_3$) of thin film materials, where E_3 is the applied electric field in the normal direction (Muralt 1997).

	AlN	PZT
$d_{33,f}$	3.9	90 ... 200
$e_{31,f}$	−0.9	−4 ... −6
ε_{33}	11	1100 ... 1300

adjusting the sputtering conditions, highly c-axis-oriented films can be obtained. Although ZnO shows the higher piezoelectric strain, AlN films are very promising owing to their higher resistivity and their better compatibility with microelectronics batch fabrication processes.

Ferroelectric thin films having a perovskite structure are of wide interest mostly because of their high dielectric constant and high remanence, e.g., as capacitors for dynamic random access memories (DRAMs). The high piezoelectricity in PZT materials leads to further interest in these materials for applications as microsystem actuators, especially as bulk PZT is widely used for high precision or ultrasonic actuators. In general, PZT films can be fabricated either by a single-step process with substrate temperatures of approx. 550 °C or by a two-step process using postdeposition crystallization for the amorphous as-deposited films. Employing this two-step process, crystallization in the perovskite structure was found at temperatures between 550 and 600 °C, whereas the texture of the films depends strongly on the bottom electrode and the substrate being used. Owing to the inherent film stress and its texture the unpoled polycrystalline PZT films being fabricated by sputtering already exhibit a polarization and a piezoelectric response which reach approximately half the values of poled or single crystal thin film materials.

The piezoelectric data of both classes of materials are compared in Table 1, which demonstrates the special potential of PZT films provided that all fabrication issues are solved. The most important features of piezoelectric thin film actuators are the low voltage, low power, high-frequency operation being compatible to TTL signal levels. Consequently, high frequency applications such as surface acoustic wave (SAW) devices, ultrasonic motors, and resonant sensors for mechanical quantities are the domain of piezoelectric thin film actuators.

3. *Shape Memory Thin Films*

The development of shape memory thin films is based on TiNi, which is superior to other potential shape memory materials with respect to attainable strains and stresses. Owing to the low transformation temperatures of binary TiNi films, the materials research is concentrated on Ni/Pd substitution for an increase in transformation temperature and Ni/Cu substitution for a reduction in transformation hysteresis. Furthermore, it is known that TiNiCu films are much less sensitive in their transformation temperatures to variations in the titanium content. Further interest is related to the training procedures as well as to the actuator design to achieve the two-way behavior in either free-standing films or in combination with a substrate.

It has been found that Ni/Pd substitution is very effective in adjusting the transformation temperatures between room temperature and temperatures above 500 °C, whereas the measured austenite finish temperatures of sputtered films are in good agreement with the reported data of bulk materials (Fig. 2) (Quandt *et al.* 1996).

By Ni/Cu substitution it is possible to reduce the transformation hysteresis. In the case of the ternary system Ti–Ni–Cu this substitution affects almost only the martensite finish temperatures, which are raised by increasing the copper content. The martensite and austenite finish temperature (M_f, A_f, respectively), as well as the hysteresis A_f–M_f are shown in Fig. 3 as a function of the copper content while the titanium content was kept constant. The results show that Ni/Cu substitution of about 20% leads to a reduction of the hysteresis from 70 K to approx. 20 K for free-standing, stress-free thin-film materials, which is not further reduced by higher copper contents.

Compounds of substrates and shape memory films are an attractive approach to realize actuators that

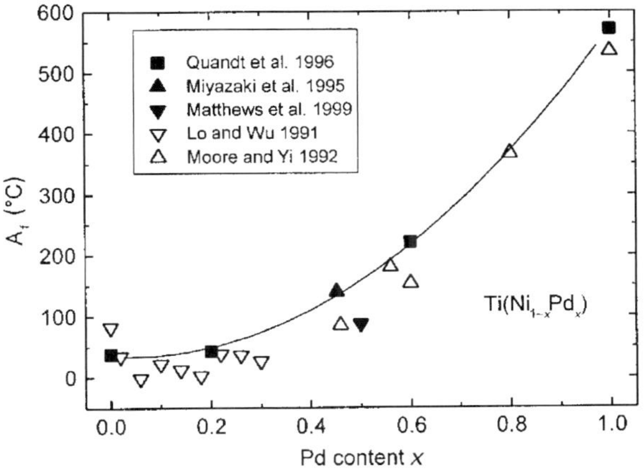

Figure 2
Austenite finish temperature A_f of Ti(Ni$_{1-x}$Pd$_x$) bulk (open symbols) and thin film (closed symbols) materials, as a function of palladium content (after Quandt and Holleck 1998).

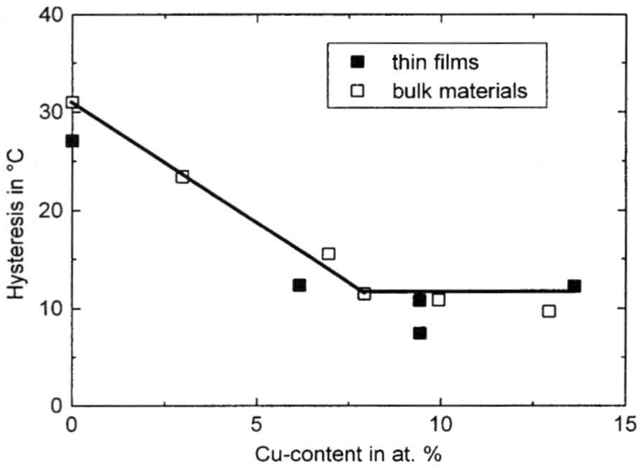

Figure 3
Hysteresis of Ti(Ni,Cu) bulk (open symbols) and thin film (closed symbols) materials (after Miyazaki *et al.* 1996).

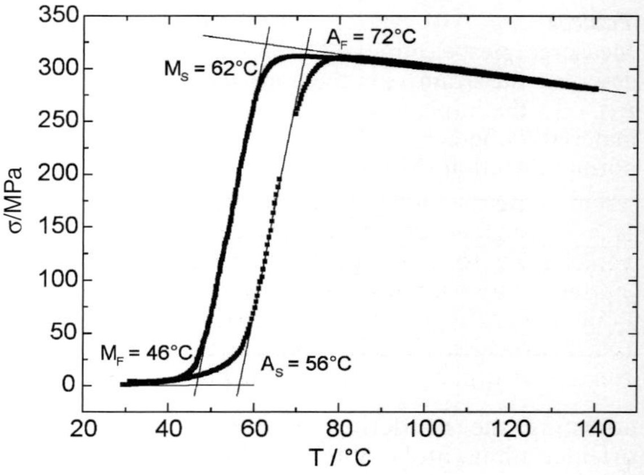

Figure 4
Stress–temperature curve of a TiNiCu shape memory film on molybdenum. A_s and A_f denote the austenite start and finish temperature; M_s and M_f are the martensite start and finish temperature.

do not require a special training treatment, which is necessary in the case of free-standing films. The internal stress resulting in a two-way behavior is introduced after the crystallization upon cooling, owing to different thermal expansion coefficients of substrate and shape memory film. In the case of lower thermal expansion coefficients of the substrate (e.g., silicon, molybdenum) the shape memory films are under tensile stress while higher thermal expansion coefficients of the substrate (e.g. $Fe_{72}Cr_{18}Ni_{10}$) result in compressive film stress. Below the martensite start temperature these stresses vanish, owing to the formation of selected martensite variants corresponding to the lowest free energy of the system. Upon heating, the shape memory effect leads to a restoration of the original stress state. Thus high stresses exceeding 400 MPa are restored in a narrow temperature range defined by the transformation hysteresis of the shape memory material (Fig. 4). Using these shape memory compounds it is possible to fabricate shape memory bending transducers in a fully batch technology by sputtering the metallic substrate layer in combination with sacrificial layers. This two-way actuator does not require any training procedures.

The unique feature of shape memory actuators is the very high energy density, which is considered to be of special importance, owing to the small volumes of actuators in microelectromechanical systems (MEMS). Applications realized so far make use of this feature, e.g., in the case of microvalves to be employed in pneumatics or hydraulics.

See also: Magnetostrictive Materials

Bibliography

Miyazaki S, Hashinaga T, Ishida A 1996 Martensitic transformations in sputter-deposited Ti–Ni–Cu shape memory alloy films. *Thin Solid Films* **281–2**, 364–7

Muralt P 1997 Piezoelectric thin films for MEMS. *Integr. Ferroelect.* **17**, 297–307

Quandt E, Halene C, Holleck H, Feit K, Kohl M, Schloßmacher P, Skokan A, Skrobanek K D 1996 Sputter deposition of TiNi, TiNiPd, and TiPd films displaying the two way shape memory effect. *Sensors Actuat. A.* **53**, 433–8

Quandt E, Holleck H 1998 Transducer and protective PVD-films for applications in microsystem components. *Microsyst. Technol.* **26**, 49–58

Winzek B, Quandt E 1999 Shape memory Ti-Ni-X films (X = Cu, Pd) under constraint. *Z. Metallkd.* **90**, 796–802

Advanced Automotive Powertrains: Materials

J. Newman

To increase the energy efficiency of automobiles there is a substantial effort to develop a hybrid electric vehicle, whereby energy is stored in a battery and subsequently used for acceleration. One power plant being developed for hybrid electric vehicles is the fuel cell. This article describes the materials requirements for the unique components of such vehicles, the fuel cell and the battery.

The concept of the hybrid electric vehicle (HEV) is to provide a means of load leveling so that the primary power plant, be it an internal combustion engine or a fuel cell, can operate at its optimum average conditions to give a good combination of high-energy efficiency and low emissions. At the same time, the primary power plant can be reduced in size, because it produces only the average power, and this further increases the overall energy efficiency of the vehicle. With an energy storage system in place, one can explore the option of recovering braking energy and reducing the overall energy consumption further.

The cost and complexity are increased for an HEV compared to a conventional vehicle. An overall optimization in the face of inexpensive energy has not favored the HEV. The development of HEVs and of fuel cells for automotive applications will permit knowledgeable optimization of vehicle configurations in the face of increased energy prices.

1. Fuel Cells

In its simplest form, the fuel cell consists of a positive electrode, where oxygen is reduced, a negative electrode, where hydrogen is oxidized, an ionically conducting material between these electrodes, and a bipolar separator plate with good electronic conductivity. The mere enumeration of the components of the fuel cell gives a list of the materials with problems.

The use of hydrogen and oxygen (really air) generally means that the vehicle needs a fuel processor to convert a hydrocarbon fuel or methanol into a hydrogen stream, since transport of compressed hydrogen involves a weight penalty and a safety issue. An attractive alternative is a direct methanol fuel cell in which methanol is oxidized at the negative electrode.

The bipolar separator plate needs to be impervious to oxygen and hydrogen and needs to be resistant to both oxidation and reduction. Machined graphite has been an expensive and fragile standby. Molded, electronically conductive plastic is being developed as a cheaper and more durable alternative, and carbon papers can be used to provide flow channels for the gases. Metals have also been tried.

While fuel cells for other applications have been or are being developed with ionic conduction provided by phosphoric acid, potassium hydroxide, molten carbonates, or solid oxides, automotive applications are focused on ion-exchange polymers in the proton (H^+) form consisting of a polyperfluoroethylene backbone with fluorinated side chains terminated in a sulfonic acid group. These provide excellent chemical durability (the membrane needs to withstand the reducing environment at the negative electrode and the oxidizing environment at the positive electrode) and have high proton conductivity. Some effort has been devoted to developing cheaper, less durable membranes, but it may prove mainly to stimulate lower-cost, high-volume production of the conventional, perfluorinated membranes.

The conventional membranes have tolerable leakage rates of hydrogen and oxygen, but methanol can cross from the negative to the positive electrode relatively easily. Significant efforts have been devoted to developing membranes with low permeation rates for methanol to improve the prospects of direct methanol fuel cells. The membranes also require resilience, flexibility, and creep resistance.

The principal component of the electrodes is the catalyst, platinum for oxygen and hydrogen electrodes and a platinum–ruthenium alloy for the methanol electrode. While the quest for better catalysts goes on, it should be understood that the search has already spanned more than a century. The catalysts, including the alternatives sought, are expensive, and substantial effort has been devoted to methods for producing high surface area catalysts and positioning them optimally.

The catalyst needs to touch the separator material for ionic transfer to the counterelectrode and be connected electronically to the separator plate so that electric current can flow to or from the adjacent cell. A catalyst loading of $2\,mgcm^{-2}$ of separator area (on each electrode) might be considered a high value, while $0.1\,mgcm^{-2}$ (on each electrode) can yield good results under favorable conditions. Thus, the details of putting the catalyst on the separator, perhaps partially embedding it in the separator material while maintaining electronic contact, are critical.

Compromising the reduction of the catalyst loading is poisoning of the catalyst by carbon monoxide or sulfur. The fuel cell can be made somewhat more CO tolerant by adding a few percent of oxygen to the fuel, with a consequent loss of energy efficiency. From the mechanism of methanol oxidation, it can be expected that a good methanol catalyst will also be more tolerant to CO in a conventional system.

The catalyst layer may be only $5\,\mu m$ thick. There is customarily a gas diffusion layer, perhaps made of carbon cloth or fibers, between the catalyst layer and the adjacent bipolar separator plate. This layer may be perhaps 20 μm thick and facilitates diffusion of hydrogen or oxygen under the ribs of the bipolar separator plate. It needs to be electronically conducting. The membrane separator may be 125 μm thick, and the whole pattern of components repeats in a "bipolar fashion" from bipolar separator plate, fuel gas diffusion layer, anode catalyst layer, membrane separator, cathode catalyst layer, air gas diffusion layer, and back to the bipolar separator plate.

Since air would be used as the oxidant, the mole fraction of oxygen would be less than 0.21. The fuel stream from a fuel processor (reformer) would probably have a mole fraction of hydrogen less than 0.5. Operation at atmospheric pressure is desirable since compressors to reach ~4 atm would constitute additional, bulky components. Methanol would probably be fed as an aqueous liquid at about 1 mol L^{-1} (10 vol.%).

Carrying hydrogen fuel on a vehicle has disadvantages for reasons of fuel infrastructure, safety, and weight. Storage of hydrogen in a hydride, as well as gas and liquid forms, is being explored. Other possibilities include using a fuel processor or reformer to convert a fuel to a gaseous mixture of hydrogen and carbon dioxide with smaller amounts of water and carbon monoxide. Reforming a fuel followed by a water–gas shift reaction to convert CO into CO_2 should be a straightforward chemical process. However, putting such a chemical plant on a vehicle pushes the limits of the technology and compromises the overall energy efficiency. To reform gasoline to the desired mixture of H_2 and CO_2 requires several catalysts and an operating temperature of 700 °C. This temperature need be only 400 °C if the fuel is methanol (for which the fuel infrastructure is absent). In either case, the reformer temperature is substantially higher than that of the fuel cell (about 80 °C), and this thermal mismatch compromises energy efficiency, which is already reduced by any reforming process.

Because the fuel cell has limited tolerance to CO and even less to sulfur, a compromise needs to be reached in improving the purity of the reformer effluent, increasing the fuel cell tolerance, and overall energy efficiency.

To summarize:

- Four components of the fuel cell tend to be costly: the two catalysts, the membrane separator, and the bipolar separator plate. Costs are being reduced, but the materials requirements for all components are stringent.
- The fuel processor or reformer is also a challenge because it must produce high-purity hydrogen with good energy efficiency and low cost, weight, and volume.
- The direct methanol fuel cell is the most attractive alternative to the hydrogen fuel cell. Critical materials issues here are a catalyst for the methanol electrode and a membrane with reduced permeability to methanol.
- The overall efficiency of a fuel cell might be comparable to an efficient diesel engine in an HEV, but the emissions characteristics would be better.

2. *Batteries*

Moderately high power is required for the battery or energy storage device in an HEV, compared to that needed for an all-electric vehicle (EV). This is gauged from the time of discharge; the HEV battery may be designed for an acceleration discharge pulse of 18 s and a regenerative braking charge pulse of 10 s, both contained within an overall 1 min cycle that approximates urban driving. The EV battery may need to be designed for a 30 s discharge pulse but an overall discharge period of 3 h. The power-to-energy ratio is substantially higher for the HEV (but not as high as that for starting an internal combustion engine).

The specifications for energy storage do not specify the battery chemistry; consequently, a wide variety of chemistries compete in the marketplace. The automotive application also places restriction on cost and weight.

One battery system emerging for automotive applications uses a nickel oxyhydroxide electrode (like that in the nickel cadmium battery) for the positive electrode and a metal hydride, such as alloys of lanthanum and other rare-earth metals (nickel, manganese, titanium, iron, zirconium, and other metals added to achieve special beneficial properties), for the negative electrode, giving a specific energy about twice that of the lead–acid battery but at a substantially higher cost. With an aqueous potassium hydroxide electrolyte, the system works by alternatively storing protons (H^+) inside the active materials of the electrodes. The attainable potential is limited by the aqueous electrolyte, but this also permits overcharge protection by allowing oxygen generated at the positive electrode to be consumed at the negative electrode. Overall energy efficiency is essentially compromised by this oxygen transport and also by an inherent hysteresis in the potential of the positive electrode. Corrosion of the metal hydride can be a life-limiting process.

Lithium and lithium ion batteries have generated hopes for a revolution with EVs and HEVs. Lithium ion batteries work much like the nickel–metal hydride batteries discussed above, except that it is lithium ions that are alternately stored in the active materials of the two electrodes. The cell potential can be 3–4 V. The electrodes are intercalation materials, usually oxides (such as lithium, cobalt, nickel, and manganese oxides) for the positive and carbon for the negative but with sulfides also being possible. The use of a metallic lithium negative electrode provides hope for a truly high-energy battery, but dendrite formation on recharge can seriously shorten life and contribute to safety problems. Polymer electrolytes, such as a lithium salt dissolved in poly(ethylene oxide), are used to address the safety issue, although oxidative instability of the polymer limits the range of possible positive electrode materials. To compensate for the inherently low conductivity of the polymer electrolyte, the batteries are operated at about 85°C. Alloys and intermetallic compounds are other possibilities for the negative electrode.

Diffusion of lithium (or of hydrogen in the nickel–metal hydride system) within the intercalation material can be a rate-limiting process, with particle size of the active materials being a compromise among rate, resistance, and side reactions. The extreme electrode potentials mean that solvent and electrolyte decomposition is a problem, and reaction products can attack the active materials themselves. Furthermore, the reaction of the solvent (and impurities and other components) can lead to covering layers that can both protect the underlying active material and lead to a significant increase in resistance. These side reactions are major contributors to end of life. In addition, nonaqueous and, particularly, polymeric solvents make ohmic resistance, solubility, and transport of lithium ions major concerns.

Other materials issues in batteries include corrosion of current collectors, use of conductive fillers, and contact among active materials and conducting phases. Any high-energy battery involves putting strong oxidizing and reducing agents in close proximity, and safety concerns are of prime importance. This requires careful attention to the thermal stability of the materials and to strategies for dealing with shorts and thermal runaway. At best, the energy density will be small compared to systems where oxygen from the air can be used as the oxidant. This may severely limit the application of an EV, but HEVs, with a smaller energy storage requirement, are better suited to develop a compromise among cost, life, and performance.

Amorphous Semiconductor Solar Cells

S. Guha

Amorphous silicon (a-Si) alloy has received a great deal of attention as a low-cost material for solar cells that convert sunlight into electricity. Amorphous semiconductors absorb sunlight very efficiently because of the inherent disorder, and only a very thin (<500 nm) film is needed to complete the solar cell structure. The material cost for the fabrication of solar cells is thus very low. Innovative manufacturing processes have also been developed to reduce production cost, and a-Si alloy technology is now being extensively used for the production of solar cells and panels for a variety of applications (Guha *et al.* 1998). The challenge has been to increase the light-to-electricity conversion efficiency.

a-Si alloy is a very intriguing material. The lack of long-range order and the presence of dangling bonds and other defects give rise to states in the mobility gap that hinder carrier transport. In addition, exposure to light creates metastable states in the gap that degrade material and cell performance. An enormous effort (Von Roedern *et al.* 1996) has gone into obtaining a better understanding of these phenomena. In this article, we shall discuss the progress in our understanding of these materials and devices that has resulted in the achievement of high and stable cell and module efficiency. We shall also discuss future directions.

1. *Cell Structure*

The simplest single-junction cell structure consists of *n i p* layers deposited on stainless steel coated with a back reflector (Fig. 1). The top of the cell is covered with a transparent conducting oxide that serves as an antireflective coating. The intrinsic (*i*) layer is an a-Si alloy with an optical gap of about 1.7 eV. The solar spectrum consists of photons covering a wide range of energies from 0.6 to 2.5 eV, and this simple structure with a-Si alloy as the intrinsic layer cannot capture all these photons efficiently. As we shall discuss later, the single-junction cells also suffer severely from light-induced degradation. A multijunction structure is usually used to improve stability and spectral response. Although many different configurations are used (Fig. 1), the highest efficiency has been achieved with a triple-junction structure in which the component cells have different optical gaps. In this structure (Fig. 1), the top cell uses a-Si alloy with an optical gap of 1.8 eV to capture the blue photons. The middle cell uses an amorphous silicon-germanium (a-SiGe) alloy with a germanium content of 10–20%. This alloy has an optical gap of about 1.6 eV and captures the green photons. The bottom cell uses an a-SiGe alloy with about 40% germanium to reduce the gap to about 1.4 eV so as to capture the red photons. Light that is not absorbed is reflected from the back-reflector. Contacts between the component cells and from the cell to the conducting oxides are facilitated by the use of doped *n* and *p* layers. In order to improve the efficiency, the challenge has been to improve the quality of all the different thin-film layers. We should mention that alternative cell structures have also been extensively used where a glass superstrate is used (Arya *et al.* 1998). In this structure, the light enters through the superstrate, which is coated with a transparent conducting oxide. The cell structure is deposited onto the oxide layer, and the back reflector is deposited last.

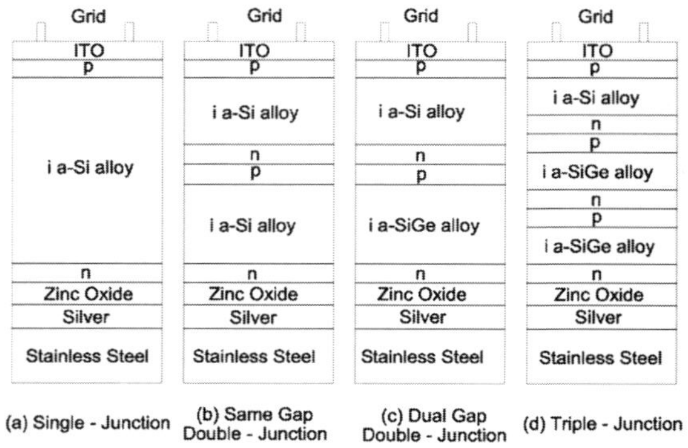

Figure 1
Schematic diagram of different cell structures.

2. *Material Research*

a-Si alloy is usually grown by a plasma-enhanced chemical vapor deposition process in which silane is dissociated in an RF plasma. The deposition conditions determine the different species in the plasma that contribute to the growth process (Matsuda 1996). The quality of the material is determined by the plasma chemistry as well as by the growth kinetics. Extensive work has been carried out to find the optimum deposition parameters for obtaining the best film quality.

The plasma contains many different ions and neutral radicals generated by electron–molecule collisions which, in turn, undergo various secondary reactions as they move to the substrate. It is now generally accepted that the highest quality material is grown from the SiH_3 radicals that have large surface mobility to find energetically favorable sites. Other radicals such as SiH_2 and higher silanes have a high sticking coefficient and can be incorporated readily onto the growing surface. The presence of these and other heavy radicals in the plasma leads to films with poorer microstructure.

The best quality films are grown under the following conditions: pressure, 0.1–1 torr; electrode separation, 4–6 cm; temperature, 150–300 °C; and power density, 10–100 mW cm^{-2}. This range of deposition conditions ensures avoidance of gas-phase polymerization in the plasma and the allowance of adequate time for the impinging species on the growing surface to find energetically favorable sites. A wide variety of gas mixtures has been explored to improve material quality, and dilution of silane with hydrogen has been the most successful.

Since the first observation (Guha *et al.* 1981) of improvement of stability of film properties after light exposure in films grown with hydrogen dilution, the technique has been used extensively for growing both a-Si and a-SiGe alloys. It is generally believed that increasing the hydrogen dilution improves the hydrogen coverage of the growing surface. The impinging SiH_3 species, therefore, diffuse over a larger distance before being incorporated in the film, resulting in an improved structure. In fact, hydrogen dilution beyond a certain limit can lead to even better order, resulting in the formation of microcrystalline material. Hydrogen is an etchant, and may be responsible for elimination of weak bonds at the growing surface.

The best quality material is grown at a hydrogen dilution just below the threshold of the amorphous-to-microcrystalline transition (Yang and Guha 1999). The material has an improved order, as confirmed by Raman, transmission electron microscopy, and x-ray diffraction studies. The improved order of this protocrystalline or "on the edge" (OTE) material is considered to be responsible for the observed superior stability against light exposure.

The recognition that the material grown at the onset of crystallization has superior properties and is ideal for making high-efficiency devices has created new challenges in the optimization of the material. The critical hydrogen dilution necessary to grow the material depends on temperature, pressure, electrode geometry, substrate, and germanium content of the film. Order also improves as the thickness increases and the quality of the material becomes more inhomogeneous in the direction of growth. In fact, a "phase diagram" has been proposed (Koh *et al.* 1998) where, for a given dilution, the material changes from amorphous to microcrystalline at a certain thickness. We should mention that high-quality material showing improved order has also been obtained using deposition conditions that form silicon clusters in the plasma (Roca i Cabarrocas 1998).

So far, we have discussed the optimization methods for a-Si alloys only. The same approach of hydrogen dilution has been successful for growing other alloys such as a-SiC and a-SiGe. In general, however, the introduction of any other element substituting for silicon causes a deterioration of transport properties. Some of the problems are related to a change in structure; others relate to the creation of new defects due to compositional disorder. The latter, for example, causes a widening of the conduction bandtails and reduction in electron drift mobility when germanium or carbon is incorporated in a-Si alloys. a-SiC alloy was considered to be a good candidate for the top cell because of its wider bandgap, but in spite of many years of effort, the material quality and stability are poor. Most researchers use a-Si alloy deposited at a low temperature (to increase the bandgap) in the top cell.

a-SiGe alloys have met with more success and are used in both the middle and the bottom cells. The most successful approach for growing high-quality a-SiGe alloy is by using a dilute mixture of disilane and germane in hydrogen. Germane has a much higher dissociation rate than silane in a plasma and, for a given power density needed to break up silane, there may be plasma polymerization of germane. Use of disilane rather than germane alleviates this problem. Although significant improvement in the quality of a-SiGe alloy has taken place over the years, the minority carrier transport is still poorer than that of a-Si alloy.

Part of the problem comes from the compositional disorder of the alloy that contains both silicon and germanium, which makes the conduction bandtail wider. Moreover, the GeH_3 molecule is heavy and this reduces the adatom mobility at the growing surface, resulting in a structure with large microvoid density. In order to overcome the poorer transport of minority carriers in this material, innovative bandgap profiling approaches have been developed in which the germanium content varies as a function of the thickness of the *i* layer so as to develop an internal field to help carrier collection.

3. Light-induced Degradation

Many properties of a-Si alloys undergo metastable changes on exposure to light. Annealing at temperatures around 150 °C for an hour restores the original values. These metastable changes in the material properties have a pronounced effect on solar-cell performance. Since the 1970s, extensive work has been carried out (see, for example, Fritzsche 1997) both to obtain a fundamental understanding of light-induced degradation and to develop materials and devices that will be immune to this degradation. It is generally agreed that energy released from the recombination of excess electron–hole pairs generated by illumination breaks a weak bond, creating a pair of dangling-bond defects. To prevent the two neighboring broken bonds from recombining, a hydrogen atom moves in from an adjacent Si–H bond. The defect pair may get separated further by additional bond switching. In the annealing process, the hydrogen atoms revert back to their original positions.

There are still many open questions regarding the universal applicability of the above model. No spatial correlations between dangling bond defects and between those and hydrogen have been observed. Moreover, the observation that annealing of defects created at low temperature (4 K) starts at 150 K, where hydrogen diffusion is negligible, suggests that the bond-breaking model together with hydrogen diffusion is not adequate to explain all the different features of this phenomenon. A model for light-induced degradation has been proposed (Branz 1999) which postulates the formation of a metastable complex containing two Si–H bonds. Light exposure generates the electron–hole pairs that recombine through multi-phonon excitation of Si–H vibrations.

The recombination breaks the Si–H bonds and promotes hydrogen to a transport state. When this mobile hydrogen in the form of Si–H and dangling-bond pairs meets another pair of the same kind, the two dangling bonds annihilate each other, leaving behind a metastable complex of two Si–H bonds. The model has been successful in explaining the kinetics of many of the defect creation and annealing behaviors. To explain the kinetics of defect creation at low temperatures, enhanced hydrogen diffusion due to carrier capture effects has been postulated; further studies will be necessary to explain the annealing behavior in the dark at low temperatures when, ordinarily, hydrogen diffusion will be negligible.

Several groups have shown that light exposure causes structural changes in the material. A hydrogen flip model has been proposed to explain the structural changes. Since recent experiments indicate that the light-induced defects are created a finite time after light exposure, it will be interesting to see whether there is a similar time delay with these structural changes.

How does one improve stability? As we have discussed earlier, hydrogen dilution has been the most successful in obtaining a more stable material, giving rise to improvement in cell stability. Further discussion on cell stability is given below.

4. *Solar Cell Research*

In order to make an efficient solar cell structure (Schropp and Zeman 1998), in addition to the high quality of the *i* layers, it is necessary to optimize the doped *p*- and *n*-type layers and the back reflector. The doped layers need to have high conductivity so that they can, first, provide low-resistance contacts between themselves and also with the external world; and, second, increase the built-in potential in the cell to obtain a higher open-circuit voltage. The layers must be transparent to reduce unnecessary optical losses. For the *p*-type material, a-Si alloy is quite absorbing, and two approaches have been used to alleviate the problem. One is the use of a-SiC alloy which is transparent but in which the conductivity is low.

The other is the use of a microcrystalline *p* layer that satisfies both the conductivity and optical absorption requirement. High built-in potentials have been achieved using this layer, and low tunnel-junction losses have been demonstrated. Superior tunnel-junction performance has also been demonstrated using a multilayer structure involving both microcrystalline *n* and *p* layers. This has resulted in further reduction of electrical losses leading to an improvement in efficiency.

The back-reflector should be characterized by a high reflectivity. It should also be textured so that light is reflected at an angle that is large enough to facilitate total internal reflection. There is, however, a loss associated with reflection from a textured surface, and the texture needs to be optimum in order to facilitate multiple reflection, and yet not to lose reflectivity. A bilayer of silver and zinc oxide deposited at a temperature between 200 and 350°C has been found to be optimal.

The highest initial and stable active-area cell efficiencies for the different structures (Fig. 1) as obtained by United Solar are shown in Table 1. The stable efficiency values represent the highest reported in the literature for the different structures. We can see from Table 1 that the highest initial cell efficiency obtained using a single-junction structure is 10.6%. By increasing the thickness of the intrinsic layer, an initial efficiency of 12% has been reported. These cells, however, degrade much more after light exposure since, in thicker cells, the carriers have to move a longer distance before they reach the electrodes. In the multijunction structures, the individual cells are thin, and hence one can get high efficiency with good stability. As one goes from the single-junction to the same-bandgap, double-junction structure, the stable efficiency improves from 9.3% to 10.1%. Incorporation of germanium in the bottom cell improves the spectral response and the efficiency is increased to 12.4%. Finally, use of a triple-junction, triple-bandgap structure results in an efficiency of 13%.

5. *Alternative Technologies and Future Directions*

Several other approaches have been used to fabricate cells using alternative materials and deposition techniques. Microcrystalline silicon is a promising candidate to replace the a-SiGe alloy in a multijunction cell since it absorbs the infrared photons efficiently. The conventional microcrystalline material contains many grain boundaries that hinder the transport of carriers.

The Neuchatel group (Shah *et al.* 1997) has used very high frequency (VHF) for making high-efficiency solar cells using microcrystalline silicon. There

Table 1
Highest stable cell efficiencies (active area $\sim 0.25\,cm^2$) at United Solar for different junction configurations.

		J_{sc} (mA cm^{-2})	V_{oc} (V)	FF	η (%)	Deg. (%)
a-Si:H	Initial	14.65	0.992	0.730	10.6	
single junction	Stable	14.36	0.965	0.672	9.3	12.3
a-Si:H/a-Si:H	Initial	7.9	1.89	0.76	11.4	
same gap	Stable	7.9	1.83	0.70	10.1	11.4
double junction						
a-Si:H/a-SiGe:H	Initial	11.04	1.762	0.738	14.4	
dual gap	Stable	10.68	1.713	0.676	12.4	13.9
double junction						
a-Si:H/a-SiGe:H/	Initial	8.57	2.357	0.723	14.6	
a-SiGe:H	Stable	8.27	2.294	0.684	13.0	11.0
triple junction						

are now many efforts all over the world that have resulted in the achievement of a stable 11.5% efficiency using a multijunction structure consisting of amorphous and microcrystalline silicon. The above results are very encouraging and open up new research areas. The major issue that needs to be addressed before this technology is commercialized is the improvement of the deposition rate. In contrast to the case of a-SiGe alloy for the bottom cell where an *i*-layer thickness of 100 nm is adequate, use of a microcrystalline *i* layer needs a thickness that is 30 times greater. Unless the deposition rate is increased 30 times, the throughput will be low. Extensive work is in progress to maintain high efficiency at a higher deposition rate.

Where does one go from here? Future research is focusing on two important areas. One is to maintain the same level of efficiency at a higher deposition rate. This will help to improve the throughput of production machines, thereby reducing cost. The other approach is to improve efficiency at the same lower deposition rate.

The deposition rate can be increased easily by applying higher power density to the plasma. However, this results in the formation of higher silanes in the plasma, which have a high sticking coefficient at the growing surface, and results in poor microstructure. Even if one can eliminate plasma polymerization, a high deposition rate will reduce the time available for the species to find energetically favorable sites on the growing surface. It is therefore important to separate these two effects by using deposition conditions with different plasma chemistry. Several new deposition methods have been developed to fabricate solar cells at high deposition rates. These include VHF, microwave, hot-wire, and gas-jet methods. Although preliminary results have demonstrated cells with reasonably high efficiencies, these are still lower than cells made at lower deposition rates.

The progress in our understanding of a-Si alloys since the early 1990s has been spectacular, and this has resulted in significant improvement of cell and module efficiency. Many challenges still remain (Wagner *et al.* 2000). In order to improve efficiency at the same deposition rate, the correlation between improved order and light-induced degradation needs to be investigated further. Improvement in the quality of the low-bandgap material will also pay rich dividends. One approach is to obtain a-SiGe alloys with improved order as was achieved in the case of a-Si alloys. Lowering the bandgap of a-Si alloys with less hydrogen incorporation is another interesting alternative. We should note that the theoretical efficiency achievable from a multijunction structure is 71%, and there is an immense opportunity for improvement. The low material cost, however, makes this technology commercially attractive even at a modest 8–10% efficiency.

6. *Product Applications and Market*

The first use of a-Si alloy solar cells was for indoor applications. The spectral response of a-Si alloys to fluorescent light is excellent, and the cells have been in use for powering calculators since the 1980s. The improvement of efficiency with multijunction structures led several manufacturers to construct new production facilities to address the outdoor terrestrial market.

The total annual production capacity in 1998 was about 40 MW with sales of about 19 MW, representing about 12% of the total world market. A wide variety of products is being manufactured, addressing both the off-grid and grid-connected markets. There are two advantages of a-Si alloy products that are attracting customer attention. The bandgap of a-Si alloy is higher than its crystalline counterpart, and the power output is much less sensitive to temperature. For the same power rating, therefore, a-Si alloys provide more power under actual outdoor conditions, where the cell temperature could be as high as 60°C. a-Si alloy solar cells can also be deposited on flexible substrates such as stainless steel, and the products are lightweight, flexible, and rugged. Novel products like photovoltaic shingles have been developed that can easily be integrated with the roof, and are finding enthusiastic market response. There is another market that has been relatively unexplored for this technology and is gaining importance. The lightweight a-Si alloy products are radiation hard, which offers a tremendous advantage for space applications. Specific weight exceeding 2000 $W kg^{-1}$ has been demonstrated using a-Si alloys deposited on Kapton® substrates. Launching cost is a key economic issue for satellites, and the light weight, radiation hardness, and superior high-temperature performance of these products are opening new markets in space.

Acknowledgments

The author is grateful to J. Yang, A. Banerjee, S. R. Ovshinsky, and H. Fritzsche for discussions, and to Virginia Trudeau for preparation of the manuscript. This work was supported in part by NREL under Subcontract No. ZAK-8-17619-09.

Bibliography

Arya R R, Carlson D E, Chen L F, Ganguly G, He M, Lin G, Middya R, Wood G, Newton J, Bennett M, Jackson F, Willing F 1998 R&D issues in scale-up and manufacturing of amorphous silicon tandem modules. In: *Proc. AIP Conf. 462*, Denver, CO, pp. 94—9

Branz H M 1999 Hydrogen collision model: quantitative description of metastability in amorphous silicon. *Phys. Rev. B* **59**, 5498–512

Fritzsche H 1997 Search for explaining the Staebler–Wronski effect. In: Wagner S, Hack M, Schiff E A, Schropp R, Shimizu I (eds.) *Proc. Mater. Res. Soc. Symp.*, San Francisco, CA, 19–30

Guha S, Narasimhan K L, Pietruszko S M 1981 On light-induced effects in amorphous hydrogenated silicon. *J. Appl. Phys.* **52**, 859–60

Guha S, Yang J, Banerjee A, Hoffman K, Call J 1998 Manufacturing issues for large volume production of amorphous silicon alloy photovoltaic modules. In: Al-Jassim M, Thornton J P, Gee J M (eds.) *Proc. AIP Conf. 462*, Denver, CO, pp. 88–93

Koh J H, Lee Y H, Fujiwara H, Wronski C R, Collins R W 1998 Optimization of hydrogenated amorphous silicon p-i-n solar cells with two-step i layers guided by real-time spectroscopic ellipsometry. *Appl. Phys. Lett.* **73**, 1526–28

Matsuda A 1996 Control of plasma and surface conditions for low defect density a-Si:H at high growth rates. In: *Conf. Record 25th IEEE Photovoltaic Specialists Conf. 1996*. Washington, DC, pp. 1029–34.

Roca i Cabarrocas P 1998 Plasma deposition of silicon clusters: a way to produce silicon thin films with medium-range order? In: Schropp R, Branz H M, Hack M, Shimizu I, Wagner S (eds.) *Proc. Mater. Res. Soc. Symp.*, San Francisco, CA, pp. 855–65

Schropp R EI, Zeman M 1998 *Amorphous and Microcrystalline Silicon Solar Cells*. Kluwer, Boston

Shah A, Meier J, Torres P, Kroll U, Fischer D, Beck N, Wyrsch N, Keppner H 1997 Recent progress on microcrystalline solar cells. In: *Conf. Record 26th IEEE Photovoltaic Specialists Conf. 1997*, Anaheim, CA, pp. 569–74

Von Roedern B, Zweibel K, Schiff E, Cohen J D, Wagner S, Hegedus S S, Peterson T 1996 Progress report on the amorphous silicon teaming activities. In: Witt C E, Al-Jassim M, Gee J M (eds.) *Proc. AIP Conf. 394*, Lakewood, CO, pp. 3–12

Wagner S, Carlson D E, Branz H M 2000 Amorphous and microcrystalline silicon solar cells. *J. Electrochem. Soc.* (in press)

Yang J, Guha S 1999 Amorphous silicon alloy materials and solar cells near the threshold of microcrystallinity. In: Branz H M, Collins R W, Okamoto H, Guha S, Schropp R (eds.) *Proc. Mater. Res. Soc. Symp.*, San Francisco, CA, pp. 239–50

Automotive Engine Materials

R. C. McCune and G. A. Weber

The automobile in the broadest sense, including passenger cars, trucks, and other vehicles, is fundamentally defined by the incorporation of a self-contained power source that converts energy into mechanical forces thereby propelling the vehicle. This has traditionally been accomplished with an internal-combustion engine that converts stored energy from a fuel into mechanical torque capable of rotating a set of wheels through a transmission and driveline. Major engineering and materials advances for internal combustion engines during the twentieth century focused on improvements in efficiency and reductions in emissions, through such technologies as fuel injection, catalytic converters, lightweight materials, and electronic engine controls. At the turn of the twenty-first century, automobile manufacturers are developing alternative power sources employing either purely electric propulsion or hybrid schemes which combine the traditional internal combustion engine with electric motors, generators, batteries, fuel cells, or other forms of energy storage and conversion. The forces behind such dramatic change after over 100 years of reliance on the internal combustion engine reflect mounting concerns over both future availability of fossil fuels (e.g., coal, gasoline, diesel, kerosene, natural gas, etc.) as well as alleged long-term climatic changes brought about, in part, by accumulation of so-called greenhouse gases including CO_2, a normal combustion product of carbonaceous fuels. While earlier concerns about automobile engine emissions (typically hydrocarbons, ozone, carbon monoxide, and oxides of nitrogen) and their effects on the atmosphere were addressed by such innovations as catalytic converters and electronic engine controls, the unavoidable generation of CO_2 as a normal combustion byproduct of carbonaceous fuels is a systemic issue requiring totally different automotive propulsion technologies and materials. This article first reviews typical and advanced materials used in present-day internal combustion engines and second, considers some of the materials issues envisioned for such future powertrains.

1. Powertrain Configurations

Figures 1(a)–(d) illustrate several common automobile powertrain configurations in use or under development at the turn of the twenty-first century. Figure 1(a) illustrates a traditional configuration in

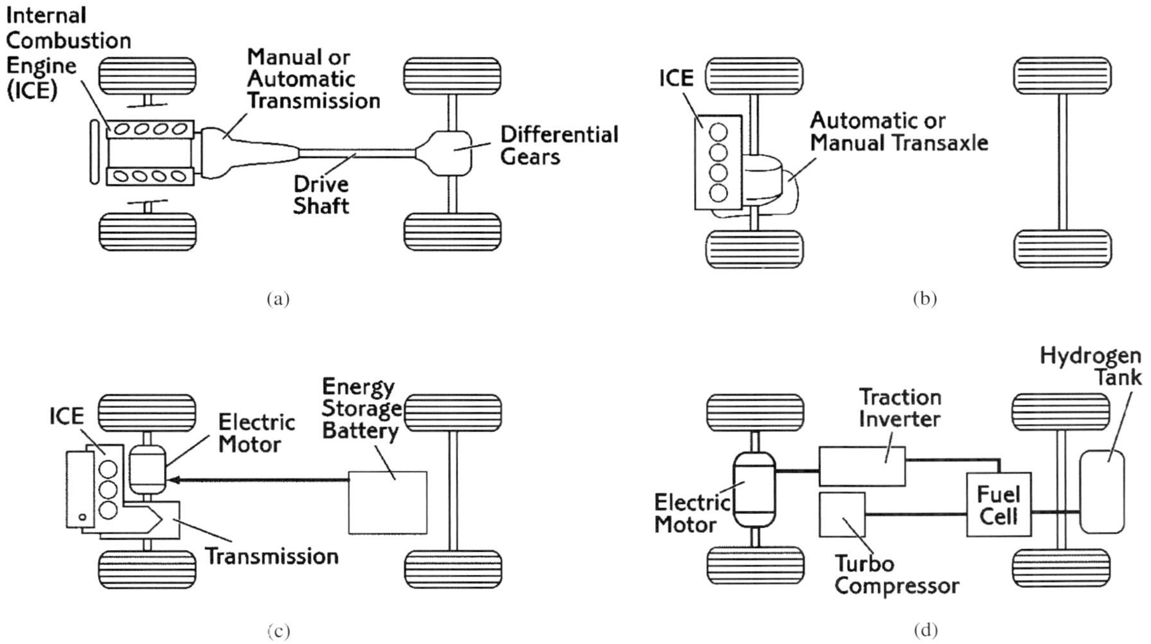

Figure 1
Powertrain architectures for current and next-generation vehicles: (a) historic internal combustion engine drive line configuration; (b) front wheel drive powertrain; (c) parallel hybrid electric vehicle powertrain; (d) fuel-cell powered electric vehicle.

which the internal-combustion engine is coupled in-line with a transmission package through either direct mechanical connection using a clutching arrangement with manual gearbox or through a fluidically-coupled "automatic" transmission incorporating a torque converter which permits smooth power flow between engine and transmission. In both arrangements, a drive shaft delivers torque from the transmission through a differential gear set to the rear wheels. Figure 1(b) illustrates a transversely mounted internal combusiton engine coupled through a trans-axle to the front wheels of the vehicle. The transaxle may be either a manual gearbox or fluid-based automatic transmission, incorporating the differential gearing feature that permits slight variations in wheel velocity during cornering. So-called front-wheel drive systems also incorporate continuous velocity (CV) joint connectors between driveline and wheels for this purpose. Figure 1(c) illustrates a "parallel" hybrid powertrain arrangement in which a conventional internal combustion engine can provide direct power to a driveline, or alternately a storage array such as a battery can be used to deliver power through a motor set to the wheels of the vehicle. The mechanical engine and electric motor may thus operate in parallel to power the wheels of the vehicle. In a "series" hybrid arrangement, the internal combustion or other engine (i.e., a turbine) is used solely to operate an electrical generator that then powers a motor set to the vehicle wheels. Electric vehicles do not generally utilize separate transmissions, although gear sets are employed for transfer of torque to the wheels from the electric motor. Figure 1(d) illustrates an electric vehicle utilizing a fuel cell to provide power to an electric motor. Many electric vehicles incorporate storage devices to recover a fraction of the kinetic energy of the vehicle, which can be converted when the vehicle decelerates from its driving speed. Typical storage devices include flywheels, electrical capacitors, and batteries.

2. *Materials for the Internal Combustion Engine*

2.1 Power Cylinder

Figure 2 illustrates the fundamental components of a typical internal-combustion engine power cylinder and ancillary mechanisms. Operational details of internal combustion engines have been reviewed at length elsewhere (Taylor 1968, Heywood 1988). The fundamental principle of operation is a combustion process occurring near the point of greatest compression of the air and fuel mixture by the piston in the cylinder (top dead center). The expanding gases then force the piston to the point of greatest defined cylinder volume (bottom dead center), resulting in a power stroke, whereby the force of combustion does work on the crankshaft through the connecting rod which attaches the piston to the crankshaft. The

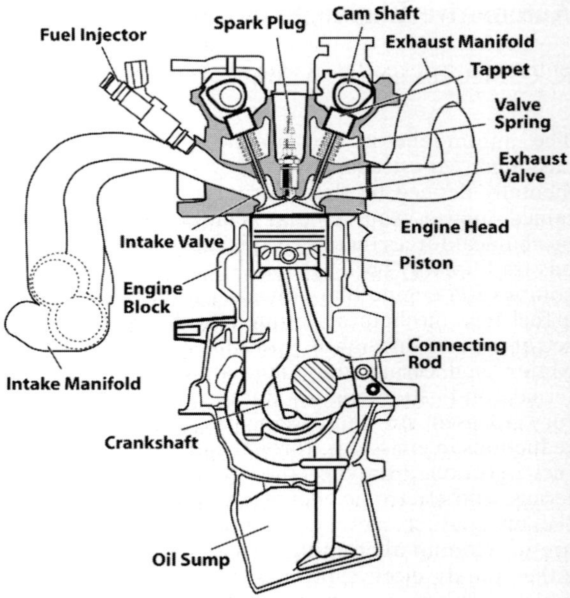

Figure 2
Cross-section schematization of a typical modern overhead cam, spark-ignited internal combustion engine.

expanded gases are exhausted when the piston, on its continuing stroke, forces them out of the cylinder, typically through an arrangement of valves. The valvetrain, in both spark-ignited (SI) and compression-ignited (CI) or diesel engines, controls the flow of air and fuel into the cylinder, seals the cylinder, and manages outward flow of exhaust gases. In general, CI engines operate under more severe cylinder pressure peaks during both compression and expansion, placing much greater demands on materials of construction. SI and CI engines are considered four-cycle, whereas in two-cycle engines, found in many motorcycles and smaller automobiles, there is not a valvetrain; the inlet and exhaust gases flow into the cylinder being effected by ports in the cylinder wall, which are open or closed in sequence with the passage of the piston.

The principal materials systems in the power-cylinder portion of the engine are the cylinder block, engine head, connecting rod, piston, crankshaft, and valvetrain. Naturally, an operating engine will incorporate hundreds of ancillary parts including manifolds, fuel handling system, bearings, fasteners, timing chains or belts, and flywheel, etc. However, the above items fundamentally dictate the manufacturing protocol for the engine. Depending on its design, the valvetrain may incorporate differing materials technologies employed to achieve such objectives as reduced weight, extreme durability, or high strength. New technologies for the valvetrain include mechanisms for variable valve timing (VVT) and electrically or hydraulically actuated valves which eliminate the need for a separate camshaft and drive arrangement.

2.2 Cylinder Block and Head

The pre-eminent material of cylinder block and head construction since the inception of the internal combustion engine has been gray cast iron. The basic specifications for the currently accepted versions of gray iron are given by industry handbook (SAE Handbook 2000). Gray cast iron is produced to its near final shapeby sand casting, permitting incorporation of appropriate coolant channels, attachment bosses, and air/fuel passageways of the cylinder head, through use of sand cores. Gray iron provides a durable, machinable, and recyclable material for the block and head application with a nearly ideal running surface for the piston and ring pack of the power cylinder. The material is provided with topographic features for oil retention introduced by honing of the cylinder wall surface, as well as excellent anti-scuff properties due to the intrinsic graphitic component. In the cylinder head, the gray iron can provide for the valve seating surface and valve guide surfaces, often without further modification. Durability of the parent gray iron for the valve seating surface is provided either through use of inserts or induction hardening to produce a tempered martensitic structure (Bisaro 1973). Concerns regarding engine weight have led over the years to an increasing use of aluminum for automotive cylinder blocks and heads, such that current worldwide usage of aluminum in this application approaches 50% of all blocks and cylinders (Anonymous 1990). For the running surface of the cylinder bore, provision must be made for wear and scuff resistance through use of inserts (often gray iron) or coating processes (McCune 1995). Particularly wear resisting grades of aluminum, such as AA-390 (Jorstad 1983) permit direct action of the piston rings on the aluminum surface, although special precautions must be taken in surface finishing and piston coating to avoid scuffing. Aluminum for engine heads, using such sand or permanent mold casting alloys as 319, 333 and 356, requires either use of inserts or coating processes for wear surfaces such as valve seats (Strong and Liang 1997) and valve guides. Block and head construction for compression ignition (e.g., diesel) engines has traditionally relied on gray iron due to the greater strength demands of the system, however, more recently, aluminum is finding use in diesel applications where weight reduction is the primary consideration (Keppeler *et al.* 1998). Compacted graphite iron is an emerging cylinder block material having a higher specific strength than conventional gray iron, and with improved fatigue strength and acoustic properties relative to aluminum, thereby allowing cast iron blocks of reduced thickness and mass to be constructed (Tholl 1996).

2.3 Crankshaft

The crankshaft, supported by the journal bearings and crankcase in in-line and V-configurations, transmits the force of combustion, via the pistons and connecting rods, through the transmission to the driveline, or in hybrid configurations to an electrical generator/storage system. The crankshaft receives multiple points of torque from the various cylinders of the engine block and is thus subject to both bending moments as well as torsional loading, requiring materials of both high strength and fatigue endurance. For engines under significant loads, the traditional crankshaft material is forged, alloyed steel, including microalloyed grades containing small additions of vanadium, titanium, or niobium, which permit development of superior surface strength, resistance to fatigue, and surface durability in the bearing contacts (Hashimoto *et al.* 1982). In recent years, smaller, lower-power internal combustion engines have exploited the use of nodular cast iron, having both economic implications for manufacture and excellent vibration damping capability (Nallicheri *et al.* 1991). The journal surfaces of crankshafts are usually induction hardened to resist scuffing wear on cold-start conditions when journal lubrication may be minimal.

2.4 Connecting Rods

The connecting rod provides the mechanical linkage between piston and crankshaft and must exhibit properties of high strength, low inertial mass, and uniformity of mass with the other connecting rods attached to the crankshaft. Coupling to the crankshaft is achieved through solid journal bearing half shells inserted on both sides of the split large end. Coupling to the piston is usually through a hardened steel gudgeon pin through the piston body. In recent years, excellent connecting rod fits at the split large end have been achieved through controlled fracture of the connecting rod large end (Olaniran and Stickels 1993). The connecting rod large ends are often threaded directly such that the split portion may be attached during assembly with the crankshaft. Materials for connecting rods have included powder metallurgy steels, which are formed into an initial shape then forged to near final dimension, as well as medium carbon steels, which develop superior strength either through separate heat treating processes or by controlled cooling following the forging step. Racing engines may utilize titanium alloys such as Ti–6Al–4V for connecting rods in order to achieve a high ratio of strength to mass of the part.

Table 1
Typical powertrain materials.

Component	Material	Common Designation or Specification	UNS Designation	Typical Composition (wt%)
Engine Block, Head	Gray Cast Iron	SAE J 431 G3000	F10006	bal. Fe, 31.–3.4C,1.9–2.3Si, 0.6–0.9Mn, 0.15P, 0.15S
	Cast Aluminum	SAE J 452 319	A03190	bal. Al, 6.5–8.0Si, 3.0–4.0Cu, 0.20–0.35Mg, 0.40 max Fe
Crankshaft	High Strength Nodular Cast Iron	SAE J 434 D5506	F33800	bal. Fe, 3.2–4.1C, 1.8–3.0Si, 0.1–1.0Mn, 015–.1P, 005–.035S
	Forged Microalloyed Steel	SAE J 403 1536 + nitrogen	G15360	bal. Fe, 0.36–0.40C, 1.35–1.45Mn, 0.50–0.65Si, 0.013–0.020N, 0.08–0.12V
Camshaft	Chilled Cast Iron	SAE J 431 G4000d	F10008	bal. Fe, 3.10–3.90C, 1.50–2.50Si, 0.60–0.90Mn, 0.80 max Mo, 1.0 max Cu
	Tubular Assy – P/M Lobe	Tube: SAE J403 1026	G10260	bal. Fe, 0.22–.28C, 0.15–3.0Si, 0.30–.60Mn, 0.04 max P, 0.05 max S
		Lobes: Sintered Cr-Ni-Mo Steel	N/A	bal. Fe, 1.90–3.30C, 6.5–9.5Cr, 1.0–2.5Ni, 1.0–3.0Mo
Pistons	Hypereutectic Al–Si Alloy	SAE J 452 B390		bal. Al, 14–18Si, 1.9–3.6Cu, 0.5–1.0Mg, 0.45 max Mn, 0.35–0.80Ni
Connecting Rods	Forged P/M Alloy	ASTM B848 P/F-11C40	N/A	bal. Fe, 0.40–0.55C, 1.8–2.2Cu, 0.30–0.50Mn, 0.15 max S
	Forged Steel	SAE J1249 1151	G11510	bal. Fe, 0.48–.55C, 0.15–.30Si, 0.70–1.0Mn, 0.04 max P, 0.05 max S
Valves				
Intake	Plain Carbon Steel	SAE J1268 1541H	H15410	bal. Fe, 0.45–0.45C, 1.25–1.75Mn, 0.04 max P, 0.05 max S.
		SAE J775 Sil 1	S65007	bal. Fe, 0.40–.50C, 3.0–3.5Si, 0.20–.60Mn, 8.0–9.0Cr
Exhaust	Austenitic Stainless Steel	SAE J775 21–2N	S63012	bal. Fe, 0.5–.6C, 7.0–10.0Mn, 0.25 max Si, 19.25–21.5Cr, 1.5–2.75Ni, 0.2–0.4N
		SAE J775 21–4N	S63008	bal. Fe, 0.48–0.58C, 8–10Mn, 0.25 max Si, 20–22Cr, 3.25–4.5Ni, 0.35–0.50N
	Nickel-base superalloy	Inconel 751	N07751	min. 70Ni, 14.0–17.0Cr, 5.0–9.0Fe, 0.9–1.5Al, 0.7–1.2Cb, 1.0 max Mn
				2.0–2.6Ti, 0.1 max C, 0.50 max Cu, .01 max S, 0.5 max Si
Valve Covers	Die Cast Aluminum	SAE J452 380	A03800	bal. Al, 7.5–9.5Si, 2 max Fe, 3.0–4.0Cu, 0.5 max Mn, 0.1 max Mg, 0.5 max Ni, 3 max Zn, 0.35 max Sn

Exhaust Manifold	Cast High Si–Mo Ductile Iron	SAE J434 D4512 + Si + Mo		bal. Fe, 3.0–3.8C, 3.5–5.0Si, 0.70 max Mn, 0.50–1.25Mo
	Cast Austenitic Stainless Steel	ASTM A 297 Type HH	J93503	bal. Fe, 0.40–0.50C, 24–27Cr, 12–15Ni, 0.80–1.25 Si, 0.75–1.10Mn, 1.70–2.10Nb
	Tubular Ferritic Stainless Steel	ASTM A268 TP409	S40900	bal. Fe, 0.08C, 1.0Si, 1.0Mn, 10.5–11.7Cr, 0.5Ni, 0.48–.75Ti
Intake Manifold	Cast Aluminum Nylon 6,6	SAE J452 319	A03190	bal. Al, 6.5–8.0Si, 3.0–4.0Cu, 0.20–0.35Mg, 0.40 max Fe
Catalyst Monolith	Cordierite Ceramic			
Catalyst Cannister	Ferritic Stainless Steel	ASTM A268 TP409	S40900	bal. Fe, 0.08C, 1.0Si, 1.0Mn, 10.5–11.7Cr, 0.5Ni, 0.48–.75Ti
Transmission Gears	Alloy Steel	SAE J404 8620	G86200	bal. Fe, 0.18–.23C, 0.4–0.6Cr, 0.7–0.9Mn, 0.15–0.25Mo, 0.4–0.70Ni, .035 max P, 040 max S, 0.15–0.35Si
		SAE J404 5120	G51200	bal. Fe, 0.17–0.22C, 0.7–0.9Cr, 0.7–0.9Mn, 0.035 max P, 0.040 max S, 0.15–0.35Si
Transmission Shafts	Alloy Steel	SAE J404 4027	G40270	bal. Fe, 0.25–0.30C, 0.7–0.9Mn, 0.20–0.30Mo, 035 max P, 0.040 max S, 0.15–0.35 Si
Transmission Housing	Die Cast Aluminum	SAE J452 380	A03800	bal. Al, 7.5–9.5Si, 2 max Fe, 3.0–4.0Cu, 0.5 max Mn, 0.1 max Mg, 0.5 max Ni, 3 max Zn, 0.35 max Sn

2.5 Pistons

Automotive engine pistons are most frequently made from aluminum alloys, usually with high silicon content to improve stiffness while reducing overall density, and may be cast or forged. Many diesel and high power density engines require pistons of alloy cast iron such as nickel-containing grades, due to the high temperatures encountered in the combustion chamber. Pistons may have unique engineered shapes to the combustion facing surface, as well as surface treatments to resist wear and ring sticking in the ring grooves, and mitigation of thermal effects on the piston crown. Friction-reducing coatings, including graphite or molybdenum disulfide, are applied on the piston skirt surfaces which may contact the cylinder walls under certain operating conditions. Hypereutectic Al–Si alloys such as AA390, used for the cylinder block and corresponding bore surface, require a coating such as electrolytic iron on the piston skirt to avoid scuffing under low lubrication conditions.

2.6 Valvetrain Components

In four-cycle SI and CI engines, the valvetrain provides for the mechanical action which synchronizes the flows of air, fuel, and exhausts with the intake, compression, power, and exhaust strokes of the piston in cylinder. A number of valvetrain actuation mechanisms have been devised during the history of the internal combustion engine and are described elsewhere (Taylor 1968). These systems consist fundamentally of the engine valves, camshaft, and intervening mechanisms that typically include tappets,followers, springs, retainers, and drive mechanisms. Variable valve timing (VVT) using mechanical, electromechanical or hydraulic mechanisms has been under development to allow greater control of engine emissions and performance (Higashi and Ueda 1994). Ultimately, highly computer-controlled internal-combustion engines may employ camless systems to permit control of valve operation without mechanical linkage to components such as the crankshaft (Schechter and Levin 1996).

(a) Valve materials. Valves may be made of plain carbon alloy steels in the case of intake valves, however, higher hot-strength, corrosion, and wear resistance are required for exhaust valves. Materials of choice are heat-treatable Fe–Cr–Si (Silchrome) alloys or austenitic stainless steels (e.g., designation 21-4N). Current typical materials are included in Table 1 (Tunnecliffe 1997). Performance engines may employ high strength materials such as titanium alloys (Allison *et al.* 1987) or even ceramic valves fabricated of silicon nitride, these latter approaches being utilized to achieve weight and friction reduction of the valvetrain. Valves for extreme durability conditions may be constructed of nickel-based superalloys such as Nimonic or Inconel grades. For severe operating conditions such as found in diesel and gaseous-fuel engines, a facing may be required for the valve seating surface. This is often accomplished with cobalt- or nickel-based hardfacing alloys.

(b) Camshaft. The cams and camshaft provide actuation for the valve motion in concert with the motion of the pistons and crankshaft, usually by a drive connection to the crankshaft, through direct gearing, chains, or belt drives, which assures mechanical synchronicity. Camshafts with integral cam lobes may be formed from white or chilled cast iron or forged steel. Surface treatment of the cam lobe itself is often required since the cam/tappet contact is severe at the points of maximum valve spring compression. Recently, composite camshafts have been developed wherein the individual cam lobes are assembled onto a tubular steel shaft (Nakamura *et al.* 1996). The cam lobes are typically fabricated by a powder metallurgy process using low alloy steels, and can incorporate grades of wear-resistant steel containing chromium and carbon at the contact surface.

(c) Tappet materials. Tappets provide for the mechanical contact between valve tip and cam. In the simplest formulation of a direct-acting mechanical "bucket" tappet as shown in Fig. 2, the tappet may be formed from steel, although optimally using lightweight materials such as forged aluminum (Ezaki *et al.* 1990), with either a separate steel shim or surfaced layer to contact the cam. Aluminum alloy bucket tappets require wear resisting surface coatings to resist scuffing wear when aluminum is also used as the engine head material.

2.7 Air and Fuel Handling

The schematized modern internal combustion engine shown in Fig. 2, illustrates the use of an advanced air intake manifold, in which the complex channels for directing air into the individual cylinders are formed by molding the part using Nylon 6,6 (Palmer *et al.* 1996), with use of a vibrational welding process to join two halves of the manifold. This manufacturing technique allows for weight reduction and more complex manifolds than could be fabricated from aluminum die or sand casting, which are alternate methods. The ability to have extraordinary control over the introduction of fuel into the intake manifold is presently achieved through the use of fuel injectors, which are electromagnetically controlled valves operating under pressurization of the fuel. The injector is a critical part of diesel or compression ignition engines, and is being advanced in extreme pressure versions for direct cylinder injection in gasoline direct-injection (GDI) or direct-injection spark-ignition (DISI) engines. Injectors employ advanced, heat treatable stainless steels such as the 440 series for internal parts, and often use advanced physical-vapor deposition coatings such as titanium nitride (TiN) to minimize corrosion, erosion, or sub-component seizure.

2.8 Exhaust System

Various cast irons have traditionally been used in the construction of exhaust manifolds. Gray iron was used for many years, but was replaced by compacted graphite iron and then by ductile iron as increasing exhaust gas temperatures required increased high-temperature strength. Ductile iron with significant additions of silicon and molybdenum is the current material of choice for normal-duty applications. Austenitic ductile irons with large nickel additions have been replaced in heavy-duty applications by cast stainless steels. When space restrictions require complex designs, exhaust manifolds are often fabricated from stainless steel tubing of various grades, including 409, 439, and 304 (UNS S40900, S43035, and S30400), which are also used in other portions of the exhaust system where high-temperature strength and corrosion resistance are required. These "fabricated" manifolds allow for weight reduction and reduce the thermal inertia of the manifold, which permits faster "light-off" of the catalyst used for emissions control.

The catalytic converter (Jacoby 1999) is one of the most critical architectural features of advanced gasoline and diesel internal combustion engines developed in the twentieth century. A typical converter package, shown in Fig. 3, consists of a stainless steel canister with an interior ceramic monolith or honeycomb (usually made of cordierite—magnesium-aluminum silicate) onto which the active catalyst material has been dispersed in the form of a "washcoat," i.e., a ceramic powder mixture with active catalyst particles, usually of the platinum group metals, including rhodium and palladium. The washcoat, typically composed of γ-alumina, may include other functional materials, such as cerium oxide or zirconia, which act to control local oxygen levels as well as retard sintering of the particulate media which can render the catalyst ineffective. Metallic support materials made of corrugated layers of high chromium, aluminum stainless steel (Fe–Cr–Al–Y) have also been employed, where the passivated steel surface retains the washcoat dispersion. A further feature of metallic support materials is that they may be preheated by passing an electric current through the metal assembly, allowing for a rapid "light-off" of the catalyst.

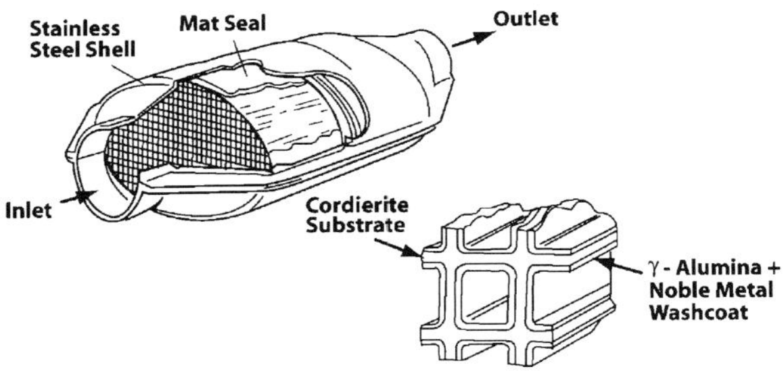

Figure 3
Schematization of catalytic converter assembly and detail of ceramic monolith structure (inset).

3. *Materials for Transmissions*

3.1 Conventional Manual and Automatic Transmission Gear Materials

The transmission package, whether manual or automatic relies on a train of gears, shafts, and clutches to transmit power developed at the engine crankshaft through to the remainder of the driveline, ultimately to be deposited at the vehicle wheels. The principal materials technologies of both transmission types are for the gears, bearings, shafts, and housing bodies. The overall package is usually determined by the torque transmitting capability of the gear sets which are usually fabricated from high-strength steels, usually incorporating surface treatments such as carburizing, nitriding, or carbonitriding to improve surface strength and reduce both contact fatigue and tooth bending fatigue. Typical gear steels in use for passenger cars would include AISI grades 8620, 5120, and 4027. Shafts are typically plain carbon steels such as grades 1040, 1050M, and 1540, which are amenable to surface hardening by induction heating. Transmission housings are often formed of large aluminum die castings, typically of type 380.

3.2 Continuously Variable Transmissions

In the search for improving fuel economy, automobile manufacturers have developed continuously variable transmissions (CVTs), in which distinctive gear pairings are replaced by drive arrangements where the torque delivered to the wheels varies continuously over the vehicle speed range. Principal current designs utilize either belt-driven sheaves or so-called traction drives which rely on contact between rollers and planar elements. Both types of CVT require extraordinary material durability and finishing in the contact regions. Materials developments for CVTs will focus on improved materials for severe contacts, surface treatments, and advanced lubricants.

4. *Materials for Hybrid and Electric Vehicles*

As research and development intensify in the field of vehicle powertrains, the emergence of electrically based systems predominates. Increased use of electric drivetrains, based upon motor driven wheels with various combinations of hybrid or chemical power, is expected. Materials challenges will shift from mechanically based systems to electrical and electronic devices. In addition to typical materials of construction for electrical machines incorporating copper windings, soft iron cores, and hard magnets (Nd–Fe–B), there will be increased demands for high-current capacity electrical switching devices and commensurate packaging and cooling capabilities.

5. *Materials for Fuel Cells*

One of the most intriguing technologies being considered for automotive propulsion at the turn of the twenty-first century is the fuel cell (Oei 1997). Capable of extracting the chemical energy from the reaction of hydrogen and oxygen to form water, the fuel cell, itself, can be considered a zero emission device. A variety of fuel cell architectures exist, but the device of current interest for automotive propulsion is based on the proton exchange membrane or PEM. Figure 4 shows the architecture of an assembled fuel cell plate wherein the membrane electrode assembly (MEA) is contained within a package that also provides a delivery structure for hydrogen gas, oxygen or air, and water removal as a byproduct of the reaction.

Materials challenges for the PEM fuel cell include the polymer membrane, currently based on fluorinated polymers such as Dupont Nafion, and catalytic deposition of platinum adjacent to the membrane to promote either the dissociation of hydrogen gas or water formation. Channeled plates for gas delivery are produced from conductive, nonreactive, dense materials such as machined graphite or stainless steel. Mass manufacturing, packaging, weight and cost reduction, and efficiency improvements of PEM-based fuel cells are some of the challenges facing automobile manufacturers. Additional materials challenges are presented if the hydrogen source for the PEM fuel cell is produced on-board from either gasoline or methanol using reforming technologies, which require further catalysts, air compressors and gas containment piping.

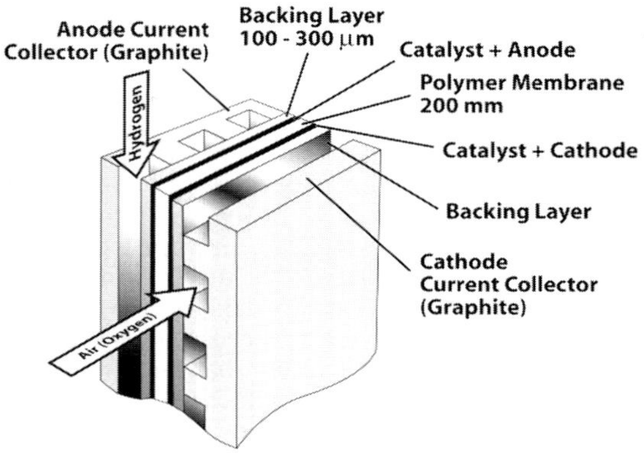

Figure 4
Schematization of idealized membrane electrode assembly (MEA) and backing plates for a polymer electrolyte membrane (PEM) fuel cell.

Bibliography

Allison J E, Sherman A M, Bapna M R 1987 Titanium in engine valve systems. *J. Met.* **39**, 15–8

Anon 1990 *Automotive News Market Data Book.* Crain Communications, Detroit, MI

Bankovic D R, Yeager D A 1994 Powder forging makes a better conn rod. *Adv. Mater. Proc.* **146**, 26–30

Bisaro R E 1973 Hardening valve seats by induction. *Metal Prog.* **104**, 82–8

Ezaki S, Masuda M, Fujita H, Hayashi S, Tereshima Y, Motosugi K 1990 Aluminum valve lifter for Toyota new V-8 engine. SAE Paper 900450. Society of Automotive Engineers, Warrendale, PA

Hashimoto H, Serino Y, Aoyama Y, Hashimoto K 1982 Non heat-treated vanadium alloyed steel crankshaft. SAE Paper 820125, Society of Automotive Engineers, Warrendale, PA

Heywood J B 1988 *Internal Combustion Engine Fundamentals.* McGraw-Hill, New York

Higashi H, Ueda K 1994 Technical trends and an outlook in the future of variable valve train. *J. Soc. Automotive Engrs. Jpn.* **48**, 6–15

Jacoby M 1999 Getting auto exhausts to pristine. *Chem. Eng. News* **77** (4), 36–45

Jorstad J L 1983 Reynolds 390 engine technology. SAE Paper 830010, Society of Automotive Engineers, Warrendale, PA

Keppeler S, Schulte H, Exker H J 1998 The technical ramifications of downsizing HSDI diesel technology to the 300 cc displacement class. SAE Special Publication SP-379. Society of Automotive Engineers, Warrendale, PA, SAE Paper 981916

McCune R C 1995 Thermal spraying of cylinder bore surfaces for aluminum engine blocks. *Weld. J.* **74**, 41–7

Nakamura Y, Egami Y, Shimizu 1996 Development of an assembled camshaft by mechanical bonding. Special Publication SP-1138. Society of Automotive Engineers, Warrendale, PA, SAE Paper 960302

Nallicheri N V, Clark J P, Field III F R 1991 Material alternatives for the automotive crankshaft: a competitive assessment based on manufacturing economics. SAE Paper 910139, Society of Automotive Engineers, Warrendale, PA

Oei D G 1997 Fuel cell engines for vehicles. *Automotive Eng.* **105** (2), 115–7

Olaniran M A, Stickels C A 1993 Separation of forged steel conn rods and caps by fracture splitting. SAE Paper 930033, Society of Automotive Engineers, Warrendale, PA

Palmer R J, Norrey C, Hayashi R, Tres D 1996 Meeting present and future global materials requirements for nylon air intake manifolds. SAE Paper 960144. Society of Automotive Engineers, Warrendale, PA

SAE Handbook 2001 Society of Automotive Engineers, Warrendale, PA

Schechter M, Levin M 1996 Camless engine. SAE Paper 960581, Society of Automotive Engineers, Warrendale, PA

Strong G, Liang X 1997 A review of valve seat insert material properties required for success. In: Bolton H A, Larson J M (eds.) *Valvetrain System Design and Materials.* ASM International, Materials Park, OH, pp. 121–8

Taylor C F 1968 *The Internal Combustion Engine in Theory and Practice.* MIT Press, Cambridge, MA

Tholl M 1996 Practical experience with passenger car engine blocks produced in high-quality compacted graphite iron. SAE Special Publication SP-1138. Society of Automotive Engineers, Warrendale, PA, SAE Paper 960297

Tunnecliffe T 1997 The development of passenger car valves in the United States. In: Bolton H A, Larson J M (eds.) *Valvetrain System Design and Materials.* ASM International, Materials Park, OH, pp. 99–105

B

Batteries: Glassy Electrolytes

J.-L. Souquet and M. Duclot

Many new secondary battery systems have been studied since the 1980s for a wide range of applications. At the present time, the so-called lithium-ion batteries are widely used in portable devices. In the near future, an extension of lithium battery applications towards higher energy supplies (e.g., for electric vehicles) or lower energy consuming systems (e.g., for smart cards and implantable medical devices) is expected.

Conventional lithium battery systems contain a liquid electrolyte which is generally a concentrated organic solution of a lithium salt. The liquid state offers good contact with the solid electrodes and an ionic conductivity occurs (which is slightly temperature dependent) allowing stable performance in the temperature range of -20 to $+60°C$. However, lithium cations, and also their associated counter anions and solvent molecules, are mobile and give rise to the formation of passivation layers at the electrode/electrolyte interface.

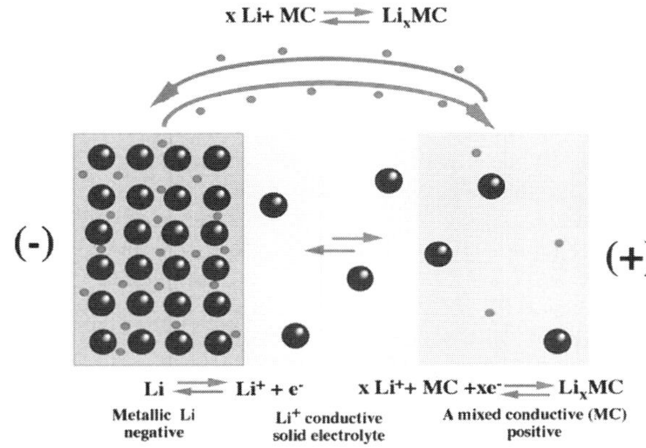

Figure 1
Charge and discharge electrochemical reactions in a lithium solid-state battery. The negative electrode is metallic lithium or an alloy, the positive is a mixed conductive (MC) material in which the ionic current of the electrolyte extends into the electrode and is gradually replaced by the electronic current at the terminals. Large and small circles represent Li^+ ions and electrons, respectively.

The use of a solid electrolyte may overcome this difficulty. In a lithium crystalline or glassy solid electrolyte, the only mobile species is the Li^+ cation because its associated anion is immobilized in a crystalline lattice or in a rigid macromolecular network. Examples include LiI, lithium solid electrolytes of NASICON-type (sodium super ionic conductor) structure, and inorganic oxide or sulfide glasses. Their use in all-solid-state batteries is described in previous review articles (Julien 1994, Jones and Akridge 1995a, b, Kondo 1998). The blocking of the anion and solvent prevents the detrimental passivation of the electrodes, subsequently enlarging the practical redox stability domain of the solid electrolyte far above the range fixed by thermodynamics. Under these conditions, the electrolyte can coexist with redox couples which are highly reducing at the negative electrode and highly oxidizing at the positive electrode.

Basically, a solid-state battery, schematically illustrated in Fig. 1, corresponds to an electrochemical chain characterized by the continuity of the cationic transport from one electrode to the other. During discharge this chain enables the transfer of lithium from a high chemical potential (anode) to a lower chemical potential (cathode). The electrical energy, EF, which is the product of the potential E (volts) and the Faraday constant F, liberated while discharging is equal to the change in lithium free energy $\Delta\bar{G}_{\mathrm{Li}}$ associated with this transfer:

$$EF = -\Delta\bar{G}_{\mathrm{Li}} \tag{1}$$

Practically, part of this electrical energy is dissipated by the Joule effect in the electrolyte and at each positive and negative interface. Optimizing the available energy requires minimization of the electrolyte resistance and a decrease of the current density at the interfaces, which can be achieved by increasing the contact areas.

For these reasons, glassy materials as solid electrolytes are of special interest: several glass compositions exhibit good lithium conductivities (1×10^{-4} to 1×10^{-3} S cm^{-1} at room temperature) (Rao and Ganguli 1995). In addition, thin-film technologies allow the manufacture of solid-state batteries with thin electrolyte and large electrochemically active interfaces. In many cases, these techniques lead to the preferential formation of amorphous phases for the electrolyte or mixed conductive electrodes.

1. Lithium Glassy Electrolytes

Ionically conducting glasses generally involve three compounds: a network former, a network modifier, and a doping salt. The absence of structural constraints allows their relative content to be easily modified to optimize electrical and electrochemical properties.

Network formers are covalent oxides or sulfides such as SiO_2, P_2O_5, B_2O_3, SiS_2, P_2S_5, etc. They form strongly cross-linked macromolecular chains that facilitate glass formation by quenching from a liquid or vapor phase.

When Li_2O or Li_2S are added to the network former it chemically reacts breaking the oxygen or sulfide bridges. This chemical reaction decreases the average length of the macromolecular chains. Increasing the amount of modifier makes the glass less stable and decreases the glass transition temperature.

Ionic salts (e.g., LiI and Li_2SO_4) are often added to a glassy matrix since such an addition significantly enhances the ionic conduction by several orders of magnitude compared to that of a glassy matrix made of only network formers and network modifiers. It has been shown that these doping salts do not chemically react with the network former (Börjesson *et al.* 1989) and their dissolution only occurs by electrostatic interactions. It has also been shown that the associated anion has a very low transference number (Robinel *et al.* 1983).

A lithium-ion conducting glass usually contains one or two network formers, one network modifier, and sometimes one doping salt. Ionic transport measurements, as well as cyclic voltammetry, show that the Li^+ transference number is unity. Below the glass transition temperature the ionic conductivity σ_+ obeys the Arrhenius law expressed by:

$$\sigma_+ T = A_+ \exp\left(-\frac{E_\sigma}{RT}\right) \tag{2}$$

where A_+ is the pre-exponential factor, E_σ the ionic conductivity activation energy, and T the absolute temperature. For all glass compositions, the A_+ values are almost constant ($\sim 1\times 10^5\,S\,cm^{-1}$) and the wide range of conductivities observed at room temperature is the result of the dependence of E_σ on composition, typically varying from 0.35 eV to 0.70 eV. The ionic transport mechanism could be attributed to the formation and migration of interstitial cationic pairs (Souquet 1995).

The chemical compositions of some Li^+ glassy electrolytes used in lithium batteries and their conductivities at room temperature are reported in Table 1.

2. Mixed Conductive Glasses as Electrode Materials

Oxide glasses containing transition metal oxides are known to exhibit electronic conductivity. Because pure transition metal oxides do not easily form a glass alone, the stability of glassy phases is obtained by the addition of a conventional network former such as SiO_2, P_2O_5, or TeO_2. They may also incorporate in their structure intercalated alkali metals. In the case of lithium, the Li^+ cations are dispersed in the voids of the glassy matrix and the associated electron is located in the *d* orbitals of the transition metals. Due to the disordered structure, the electron displacement is generally interpreted by the small-polaron model (Austin and Mott 1969).

The intercalation reaction of lithium in an electronically conducting glass has been thermodynamically and kinetically studied in several V_2O_5-containing glasses, for instance in the V_2O_5–P_2O_5, V_2O_5–B_2O_3, and V_2O_5–P_2O_5–B_2O_3 systems and pure V_2O_5 glass (Pagnier *et al.* 1983, Jourdaine *et al.* 1986, Sakurai *et al.* 1987, Lebrun *et al.* 1990, Machida *et al.* 1990, Gavrilyuk 1997).

The monotonic variation of $\Delta\bar{G}_{Li}$ observed with lithium content in glassy $Li_xV_2O_5$ leads to the supposition that the inserted lithium ions are distributed among energetically nonequivalent sites in the glassy matrix. By an impulse technique or impedance spectroscopy it has been possible to determine the lithium diffusion coefficient $\bar{D}_{Li}$ for V_2O_5-containing glasses and the usual values at room temperature lie between $1\times 10^{-8}\,cm^2\,s^{-1}$ and $1\times 10^{-15}\,cm^2\,s^{-1}$. For pure amorphous V_2O_5 films, $\bar{D}_{Li}$ exhibits a smooth and continuous decrease with increasing *x* (McGraw *et al.* 2000).

Table 1
Chemical compositions of some Li^+ glassy electrolytes and their conductivities at room temperature.

Glass composition	$\sigma_{25^\circ C}$ ($S\,cm^{-1}$)	Reference
$0.45LiI–0.37Li_2S–0.18P_2S_5$	1.0×10^{-3}	Malugani and Robert 1980
$0.6Li_2O–0.3B_2O_3–0.08SiO_2$	1.2×10^{-6}	Levasseur *et al.* 1983
$0.45LiI–0.34Li_2S–0.02Li_2O–0.18P_2O_5$	5.0×10^{-4}	Akridge and Vourlis 1988
$0.01Li_3PO_4–0.63Li_2S–0.36SiS_2$	2.0×10^{-3}	Aotani *et al.* 1994
$Li_{3.3}PO_{3.8}N_{0.22}$	3.3×10^{-6}	Xiaohua *et al.* 1997

The cyclability of glassy cathode materials has been tested: a better stability in the specific capacity is observed for glassy V_2O_5 compared to crystalline V_2O_5 (Sakurai *et al.* 1987).

Most mixed conductive glasses currently contain V_2O_5 as the active component and are studied as positive electrode materials. Some compositions may also be of interest as negative electrodes. This is the case for amorphous Fe_3O_4 deposited as a thin layer by radio-frequency (RF) sputtering (Sarradin *et al.* 1996) or glasses in the $SnO–B_2O_3–P_2O_5$ system obtained by quenching (Idota *et al.* 1997) or mechanochemical synthesis (Morimoto *et al.* 1999). For these glasses it has been shown that during the first discharge SnO is reduced to metallic tin and lithium forms Li_2O. Metallic tin is dispersed as clusters in a $Li_2O–B_2O_3–P_2O_5$ glass. The tin reacts with further lithium to the composition limit $Li_{4.4}Sn$ which corresponds to a specific anodic capacity of 991vmAh g^{-1}, and during charge Li is removed from the lithium–tin alloy. During cycling, however, the tin cluster size increases and simultaneously a serious loss in capacity is observed (Courtney *et al.* 1999). Similar behavior has been observed in thin films of amorphous silicon–tin oxynitride anode deposited by RF-magnetron sputtering. While charging, this compound with the composition $SiSn_{0.87}O_{1.2}N_{1.72}$ is reduced to Li_xSn and Li_xSi dispersed in an amorphous matrix of Li, Sn, and Si oxynitrides (Neudecker *et al.* 1999).

3. *Glassy Electrolyte Lithium Coin Cells*

The first attempt to produce a primary lithium battery using a glassy electrolyte was made in 1981 by the French SAFT Company (Malugani *et al.* 1983) using the pressed powder of a glass with the composition $0.29LiI–0.47Li_2S–0.24P_2S_5$ ($0.55LiI–0.45Li_4P_2S_7$), a lithium anode, and a composite cathode (63% TiS_2, 7% graphite, and 30% glass electrolyte). An alternative battery consisted of $Bi_4B_2O_9$ as the cathode, leading to the irreversible formation of Bi, Li_2O, and B_2O_3.

A similar glassy electrolyte with a higher content of LiI ($0.71LiI–0.29Li_4P_2S_7$) is used in a commercially marketed solid-state battery by Energizer Company (Akridge and Vourlis 1986, 1988, 1990). The battery is manufactured by first pressing a composite cathode (50–85% TiS_2 and the rest with glassy electrolyte) and then pressing a thin layer of pure electrolyte onto the cathode. Finally, a Li metallic disk is added and the complete stack is isostatically pressed and sealed in a coin cell package. These batteries can be discharged at a current density up to $300\,\mu A\ cm^{-2}$, showing good pulse performance and operating from $-40\,°C$ to $+150\,°C$. Later some of the sulfur atoms were substituted with oxygen and the glassy electrolyte $0.71LiI–0.29Li_4P_2O_{0.125}S_{6.875}$ was synthesized and used. This allowed a good reversibility at 12% depth of discharge.

More recently, Matsushita Battery Co, Japan, has produced solid-state batteries using a sulfide-based glass as the electrolyte with a composition of $0.01Li_3PO_4–0.63Li_2S–0.36SiS_2$ obtained by twin roller quenching. The doping of the sulfide glass by a small amount of Li_3PO_4 increases the ionic conductivity (up to $1 \times 10^{-3}\,S\ cm^{-1}$ at room temperature) as well as the electrochemical stability (Aotani *et al.* 1994). The absence of I^- in the glass structure which may be slowly oxidized, allows the use of highly oxidizing positive electrode materials.

A solid-state lithium battery was fabricated with $Li_{1-x}CoO_2$ as the positive and metallic indium as the negative electrode, allowing the total reaction:

$$LiCoO_2 + In \leftrightarrow Li_{1-x}CoO_2 + InLi_x \tag{3}$$

Coin-type cells of 20 mm in diameter and 1.6 mm in thickness were assembled having 15 mAh capacity and $104\,mWh\,cm^{-3}$ energy density. The charge–discharge efficiencies were almost 100% and no degradation was found after 100 cycles. A lithium "rocking chair" cell was also constructed using $Li_{4/3+x}Ti_{5/3}O_4$ as the negative electrode, showing excellent cyclability (Takada *et al.* 1996). Such a cell is represented in Fig. 2 (Iwamoto *et al.* 1999).

For all the solid-state lithium cells described above the glassy electrolyte is ground to a fine powder and pressed into pellets of about $200\,\mu m$ thick between the two electrodes. Compared to the bulk glass, the pellet's conductivity is about one order of magnitude lower and this technique is hence restricted to highly conductive sulfide glasses which cannot be handled in air because of their extreme sensitivity to oxygen and moisture.

In addition, this technique does not allow good and stable interfacial contacts between the different solid components of the cell. The volume changes that accompany the intercalation–deintercalation processes or lithium dissolution lead to an irreversible decrease of the efficiency of the electrochemically active interface (Menetrier *et al.* 1992). This is

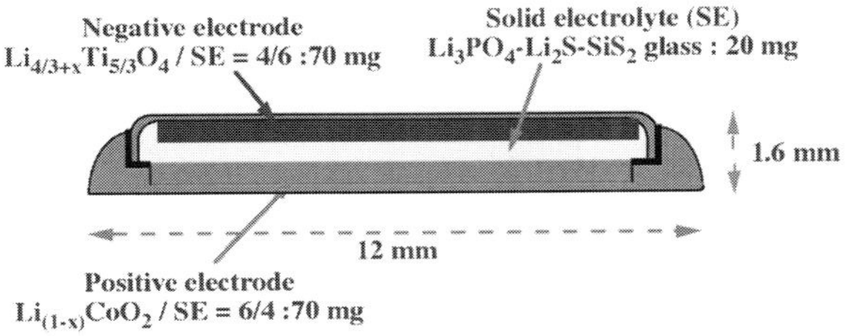

Figure 2
Cross-section view of the 1216 size solid-state battery (after Iwamoto *et al.* 1999).

probably the reason why no cycling tests reported on solid-state coin cells exceed 20% for depth of discharge. To overcome these two inconveniences, i.e., loss in ionic conductivity and poor interfacial contacts, glassy polymer composite electrolytes have been recently investigated (Cho *et al.* 1998).

4. Thin-film Lithium Batteries with Glassy Electrolytes

The miniaturization of electronic devices has resulted in very low current and power requirements for many applications (e.g., smart cards, complementary metal oxide semiconductor (CMOS) backup, implantable medical devices and microelectromechanical systems (MEMS)). This has enabled the use of thin-film rechargeable microbatteries as power sources for these devices.

As pointed out by (Jones and Akridge 1995a,b) some of the advantages offered by thin-film microbatteries include: (i) they are manufactured by the same techniques as currently used in the microelectronics industry, (ii) very low thickness allows the use of oxide glasses as the electrolyte and though these are less conductive they are easier to handle than the sulfide glasses, and (iii) the successive deposition processes allow large and good electrochemical active interfaces to be produced.

Interestingly, the use of new techniques, like sputtering, allowed the synthesis of new and unexpected compositions for the electrolytes (e.g., LiPON glass) or the positive electrode (e.g., amorphous titanium oxysulfide).

The first thin-film microbatteries were designed in the 1980s (Kanehori *et al.* 1983, Meunier *et al.* 1989, Creus *et al.* 1989) and three projects have presently reached a pre-industrial level.

The first is a joint project of Bordeaux University, France (Levasseur and Vinatier 1998, Terrat *et al.* 1998) with HEF (Hydromécanique et Frottements, a French company specializing in thin-film technologies). The positive electrode is an amorphous titanium or molybdenum oxysulfide ($TiO_{0.2}S_{1.4}$ or $MoO_{1.6}S_{1.8}$) film deposited from TiS_2 or MoS_2 targets. The presence of oxygen in the sputtered films is due to traces of oxygen in the sputtering chamber and it has been found that the partial substitution of sulfur by oxygen improves the electrochemical behavior. A glassy oxide-based electrolyte film with a similar thickness is then deposited by the same technique. Finally, the negative lithium electrode (4–5 μm in thickness) is obtained by vacuum evaporation. A hydrophobic polyimide (Capton) film protects the battery from moisture (Fig. 3(a)).

The thickness of the thin-film battery is $<10\,\mu m$ and the final product, including the air-tight packaging is <0.1 mm thick. The cells have a surface area of $\sim 2\,cm^2$ and can deliver a maximum current density of $\sim 200\,\mu A\ cm^{-2}$ for 10 minutes at an average potential of 2.0 V. More than 10000 charge–discharge cycles have been performed.

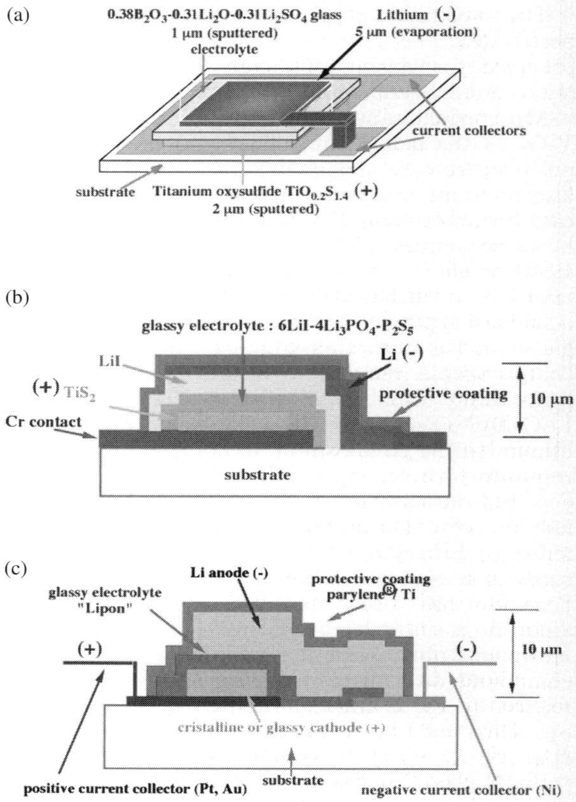

Figure 3
Microbatteries with glassy thin-film electrolytes. (a) Thin-film Li/glassy electrolyte/TiS_2 microbattery (Levasseur and Vinatier 1998), (b) cross-section of the Eveready Battery Company microbattery ($10\,cm^2$, Jones and Akridge 1995), and (c) cross-section of the Oak Ridge National Laboratory thin-film battery (Bates 1997).

A similar project has been developed at the Eveready Battery Company (Jones and Akridge 1993). Thin-film solid-state Li/TiS_2 microbatteries (Fig. 3(b)) are fabricated by sputtered depositions of the metal contacts, with TiS_2 as the cathode and an oxysulfide glass as the electrolyte. This electrolyte is sputter deposited from a target with the composition $6LiI–4Li_3PO_4–P_2S_5$. The conductivity of the glassy thin films is $2 \times 10^{-5}\,S\ cm^{-1}$. It has been found that the metallic lithium anode could react with this electrolyte forming a high-resistance layer, probably Li_2S. To avoid direct contact between lithium and the electrolyte, an intermediate LiI layer is deposited by vacuum evaporation. The total ionic conductivity of this two-layer electrolyte (glassy electrolyte + LiI) is 2×10^{-6} S cm^{-1}.

Typically, these microbatteries have a total thickness around 10 μm including the protective package and an open-circuit voltage (OCV) between 2.4 V and 2.5 V. Their capacity range lies between 5 μAh cm^{-2} and 100 μAh cm^{-2}, depending on the amount of active materials. They can routinely sustain more than 1000 cycles between 1.4 V and 2.8 V with greater than 90% cathode utilization, at current densities as high as 300 μA cm^{-2}. They cycle at temperatures as low as − 10 °C at a current density of 100 μA cm^{-2} and are capable of supplying a pulsed current of several mA cm^{-2}. Microbatteries as large as 10 cm^2 have been fabricated and cycled over 1000 times at close to 100% cathode efficiency (Jones and Akridge 1996).

A third example is the solid-state rechargeable lithium battery developed at the Oak Ridge National Laboratory (Bates *et al.* 1995), shown in Fig. 3(c). This battery consists of amorphous or crystalline cathodes of oxide lithium intercalation compounds ($LiCoO_2$, $LiMn_2O_4$ or V_2O_5) and a lithium metal anode separated by a glassy electrolyte. The first step is the fabrication of a sputtered film of the cathode material, eventually annealed to crystallize the amorphous phase for better Li^+ diffusion. Then, an amorphous lithium phosphorus oxynitride is deposited by RF-magnetron sputtering of Li_3PO_4 in N_2. This material with a composition of Li_{3+x}-$PO_{4-x}N_x$ ($x \sim 0.2$) is commonly denominated as "LiPON glass." It has an Li^+ conductivity of about 1×10^{-6} S cm^{-1} at room temperature and is stable in contact with lithium and oxidizing cathode materials such as CoO_2. The anode is deposited over the electrolyte by thermal evaporation of lithium metal and the battery is then sealed with a multi-layer coating of polyphenyl ethylene (Parylene) and titanium metal. The performance of the battery is determined by the cathode material which is $Li_xV_2O_5$, $Li_xMn_2O_4$, or Li_xCoO_2. For high-rate applications crystalline $LiCoO_2$ has been found to be the best choice. With a 1 $cm^2 \times 2$ μm layer of $LiCoO_2$, the battery can deliver about 100 μAh between 4.2 V and 3 V at a continuous rate of 1 mA cm^{-2}. Li/LiPON glass/$LiCoO_2$ batteries have been cycled over 40 000 times at room temperature with less than a 5% total capacity loss (Bates 1997).

Recently, the same research team (Neudecker *et al.* 2000) has proposed an improvement of the manufacturing technology by forming the lithium anode *in situ* by lithium deposition on a copper layer from a $LiCoO_2$ positive electrode layer.

5. Concluding Remarks

This article on glassy materials for lithium batteries is a survey of our present knowledge of the electrochemical properties of these materials and of the present state of the art of their working in all-solid-state lithium batteries. We have intentionally restricted the device descriptions to those studied at the pre-industrial level.

Considering the performances, it seems that coin cells realized by pressing together electrolyte powders and electrode materials remain less efficient than lithium-ion batteries using a liquid electrolyte. Their applications will be restricted as primary batteries operating in the temperature range of 0–100 °C since conventional organic solutions of lithium salts allow a deeper charge and discharge cycles at lower temperatures.

However, thin-film solid-state batteries have no conventional competitor and are an attractive possibility for electrical energy supply for smart credit cards, electronic labels, and applications on a very large industrial scale. This market is very large since 4 billion smart cards are estimated to have been produced in the year 2000.

Bibliography

Akridge J R, Vourlis H 1986 Solid state batteries using vitreous solid electrolytes. *Solid State Ionics* **18–19**, 1082–107

Akridge J R, Vourlis H 1988 Performance of Li/TiS_2 solid state batteries using phosphorous chalcogenide network former glasses as solid electrolytes. *Solid State Ionics* **28–30**, 841–6

Akridge J R, Vourlis H 1990 Rechargeable solid state batteries. In: Akridge J R, Balkanski M (eds.) *Solid State Microbatteries*. NATO ASI Series 217. Plenum, New York, pp. 353–70

Aotani N, Iwamoto K, Takada K, Kondo S 1994 Synthesis and electrochemical properties of lithium ion conductive glass Li_3PO_4–Li_2S–SiS_2. *Solid State Ionics* **68**, 35–9

Austin I G, Mott N F 1969 Polarons in crystalline and non-crystalline materials. *Adv. Phys.* **18**, 41–102

Bates J B 1997 Thin-film rechargeable lithium batteries. *Electron. Engin.* **69**, 63–4

Bates J B, Dudney N J, Lubben D C, Gruzalski G R, Kwak B S, Xiaohua Y, Zuhr R A 1995 Thin-film rechargeable lithium batteries. *J. Power Sources* **54**, 58–62

Börjesson L, Torell L M, Dahlborg U, Howells W S 1989 Brillouin scattering and neutron diffraction in ion-conducting glasses. *Philos. Mag. B* **59**, 105–23

Cho J, Kim G, Lim H, Liu M 1998 Electrochemical properties of GeS_2-based glass-polymer composite electrolytes for lithium-ion batteries. *J. Electrochem. Soc.* **145**, 1949–52

Courtney I A, McKinnon W R, Dahn J R 1999 On the aggregation of tin in SnO composite glasses caused by the reversible reaction with lithium. *J. Electrochem. Soc.* **146**, 59–68

Creus R, Sarradin J, Astier R, Pradel A, Ribes M 1989 The use of ionic and mixed conductive glasses in microbatteries. *Mater. Sci. Eng. B* **3**, 109–12

Gavrilyuk V I 1997 Thermodynamics of lithium intercalation in glassy vanadium pentoxide. *Tech. Phys.* **42**, 121–3

Idota Y, Kubota T, Matsuliji A, Maekawa Y, Miyasaka T 1997 Tin-based amorphous oxide: a high-capacity lithium-ion-storage material. *Science* **276**, 1395–7

Iwamoto K, Takada K, Fujino M, Kondo S 1999 Performance of coin type solid state battery with $Li_{4/3}Ti_{5/3}O_4$ as negative electrode material. *Electrochemistry* **67**, 151–4
Jones S D, Akridge J R 1993 A thin-film solid-state microbattery. *J. Power Sources* **43–44**, 505–13
Jones S D, Akridge J R 1995a Lithium glassy electrolyte batteries. In: Munshi M Z A (ed.) *Handbook of Solid State Batteries and Capacitors*. World Scientific, Singapore
Jones S D, Akridge J R 1995b Development and performance of a rechargeable thin-film solid-state microbattery. *J. Power Sources* **54**, 63–7
Jones S D, Akridge J R 1996 A microfabricated solid-state secondary Li battery. *Solid State Ionics* **86–88**, 1291–4
Jourdaine L, Bonnat M, Souquet J-L 1986 Electrochemical measurement of the diffusion coefficient of silver in thin-films of V_2O_5–P_2O_5 glass. *Solid State Ionics* **18–19**, 461–6
Julien C 1994 Thin film technology and microbatteries. In: Pistoia G (ed.) *Lithium Batteries, New Materials, Developments and Perspectives*. Industrial Chemistry Library, Vol. 5, Elsevier, Amsterdam
Kanehori K, Matsumoto K, Miyauchi K, Kudo J 1983 Thin film solid electrolyte and its application to secondary lithium cell. *Solid State Ionics* **9–10**, 1445–8
Kondo S 1998 All solid-state lithium secondary battery with highly ion conductive glassy electrolyte. In: Wakihora M, Yamamoto O (eds.) *Lithium-ion batteries*. Wiley-VCH, New York
Lebrun N, Levy M, Souquet J-L 1990 Electronic conductivity in glasses of the TeO_2–V_2O_5–MoO_3 system. *Solid State Ionics* **40–41**, 718–22
Levasseur A, Kbala M, Hagenmuller P, Couturier G, Danto Y 1983 Elaboration and characterization of lithium conducting thin-film glasses. *Solid State Ionics* **9–10**, 1439–44
Levasseur A, Vinatier P 1998 Microbatteries: an original concept with new materials for micropower devices. In: Chowdari B V R (ed.) *Solid State Ionics: Science and Technology*. World Scientific, Singapore
Machida N, Fuchida R, Minami T 1990 Fe_2O_3–V_2O_5 glasses as cathode in lithium batteries. *J. Electrochem. Soc.* **137**, 1315–6
Malugani J-P, Fahys B, Mercier R, Robert G, Duchange J-P, Baudry S, Broussely M, Gabano J-P 1983 De nouveaux verres conducteurs par l'ion lithium et leurs applications dans des générateurs électrochimiques. *Solid State Ionics* **9–10**, 659–66
Malugani J-P, Robert G 1980 Preparation and electrical properties of the $0.37Li_2S$–P_2S_5–$0.45LiI$ glass. *Solid State Ionics* **1**, 519–23
McGraw J M, Bahn C S, Parilla P A, Perkings J D, Ready D W, Ginley D S 2000 Li-ion diffusion measurements in crystalline and amorphous V_2O_5 thin-film battery cathodes. New Materials for Batteries and Fuel Cells. *Mater. Res. Soc. Symp. Proc.* **575**, 103–8
Menetrier M, Delmas C, Levasseur A 1992 On the behavior of intercalation compounds in solid state batteries. *Mater. Sci. Eng. B* **15**, 101–4
Meunier G, Dormoy R, Levasseur A 1989 New positive-electrode materials for lithium thin film secondary batteries. *Mater. Sci. Eng. B* **3**, 19–23
Morimoto H, Nakai M, Tatsumisago, Minami T 1999 Mechanochemical synthesis and anode properties of SnO-based amorphous materials. *J. Electrochem. Soc.* **146**, 3970–3973
Neudecker B J, Dudney N J, Bates J B 2000 "Lithium-free" thin-film battery with in situ plated Li anode. *J. Electrochem. Soc* **147**, 517–23
Neudecker B J, Zuhr R A, Bates J B 1999 Lithium silicon tin oxynitride (Li_ySiTON): high-performance anode in thin-film lithium-ion batteries for microelectronics. *J. Power Sources* **81–82**, 27–32
Pagnier T, Fouletier M, Souquet J-L 1983 Electrochemical properties of phosphate based semi-conductive glasses. *Solid State Ionics* **9–10**, 649–54
Rao K J, Ganguli M 1995 Lithium ion conducting glasses. In: Munshi M Z A (ed.) *Handbook of Solid State Batteries and Capacitors*. World Scientific, Singapore
Robinel E, Koné A, Duclot M, Souquet J-L 1983 Silver sulfide based glasses (II). Electrochemical properties of GeS_2–AgS–AgI glasses: transference number measurement and redox stability range. *J. Non-cryst. Solids* **57**, 59–70
Sakurai Y, Okada S, Yamaki J, Okada T 1987 Electrochemical of amorphous V_2O_5–P_2O_5 cathodes for lithium secondary batteries. *J. Power Sources* **20**, 173–7
Sarradin J, Guessous A, Ribes M 1996 Synthesis and characterization of lithium intercalation electrodes based on iron oxide thin films. *J. Power Sources* **62**, 149–54
Souquet J-L 1995 Ionic transport in glassy electrolytes. In: Bruce P G (ed.) *Solid State Electrochemistry*. Cambridge University Press, Cambridge
Takada K, Aotani N, Iwamoto K, Kondo S 1996 Solid state lithium battery with oxysulfide glass. *Solid State Ionics* **8 6–88**, 877–82
Terrat J-P, Martin M, Levasseur A, Meunier G, Vinatier P 1998 Les microgénérateurs: micropiles et microbatteries. *Techniques de l'Ingénieur—Génie Electrique D* **3342**, 4–10
Xiaohua Y, Bates J B, Jellison G E, Hart F X 1997 A stable thin-film lithium electrolyte: lithium phosphorus oxynitride. *J. Electrochem. Soc.* **144**, 524–32

Batteries, Rechargeable

L. Zhang

As electrical and electronic devices become increasingly essential parts of modern society, we are ever more dependent on our sources of electrical power. Batteries are one of the few practical methods of storing electrical energy. As such, they are vital components in electrical and electronic devices ranging from cellular phones to satellites in space. Recent advances in battery technology, both in new battery types and in improvements to existing batteries, have fueled a surge in battery applications. As battery applications become more diverse and more critical to system operation, it is especially important that system designers and users understand the fundamentals of battery function.

Batteries can be roughly divided into primary and secondary batteries. Primary batteries, such as the very popular alkaline–manganese batteries, cannot be electrically charged. However, secondary batteries (rechargeable batteries) can be electrically charged, which offers savings in costs and resources. The market demand obviously is a strong driving force for the research and development of rechargeable batteries.

The rapid development of and demand for portable cordless consumer electronics such as video cameras, shavers, power tools, cellular phones, laptop computers, and, in particular, electric vehicles (EV), and hybrid electric vehicles (HEV) demand high-performance power sources; that is, compact, high-energy, high-power, long-lasting, and maintenance-free rechargeable batteries. Nickel–cadmium (NiCd) has been a very important power source since the 1960s. Since 1980, the variety of practical batteries has increased dramatically. Recently, the market for nickel–metal hydride (NiMH) and lithium ion batteries (Li-Ion) has been developed and has grown remarkably.

A battery is a device that converts the chemical energy contained in its active materials directly into electrical energy by means of an electrochemical oxidation-reduction (redox) reaction. "Battery," which is a term often used in practical application and manufacturing, is an assembly of one or more cells connected in series or parallel (or a combination), depending on the requirements of the output voltage and capacity. "Cell" is the basic electrochemical unit consisting of major components and undergoing essential chemical reactions to generate electrical energy for use. At least two reaction partners undergo a chemical process during reaction. The energy of this reaction is available as electric current at a defined voltage and time. A characteristic feature of an electrochemical cell is that the electrochemical processes at the electrodes generate the electronic current, which is the movement of electrons in the external circuit. In contrast to the electronic current, the charge is transported between the positive and the negative electrode in the electrolyte by ions.

In this section, the fundamental aspects, properties, performance, and applications of rechargeable batteries, including NiCd, NiMH, and Li-Ion systems will be discussed. In this review, the terms "battery" and "cell" are generally used interchangeably.

1. Nickel–Cadmium Batteries

The NiCd cell has a unique set of desirable physical and electrical characteristics. Although NiCd cells can be designed and manufactured in different forms and a variety of sizes, the basic chemistry remains the same. The NiCd cell is an electrochemical system in which the active materials contained in the electrodes change in oxidation state without any deterioration in physical state. These active materials are present as solids that are insoluble in the alkaline electrolyte. Unlike many other systems, in the NiCd system, its charge and discharge reactions do not require the transfer of material from one electrode to the other. The electrodes are long lasting, since the active materials in them are not consumed during operation or storage.

1.1 Charge/Discharge Reactions

In the NiCd cell, nickel oxyhydroxide, NiOOH, is the active material in the charged positive electrode. During discharge it is reduced to the lower valence state, nickel hydroxide, $Ni(OH)_2$, by accepting electrons from the external circuit:

$$2NiOOH + 2H_2O + 2e^- \rightarrow 2Ni(OH)_2 + 2OH^-$$

Cadmium metal is the active material in the charged negative electrode. During discharge, the cadmium is oxidized to cadmium hydroxide, $Cd(OH)_2$, and releases electrons to the external circuit:

$$Cd + 2OH^- \rightarrow Cd(OH)_2 + 2e^-$$

The net reaction occurring in the potassium hydroxide (KOH) electrolyte during discharging can be expressed as follows:

$$Cd + 2H_2O + 2NiOOH \rightarrow 2Ni(OH)_2 + Cd(OH)_2$$

while during charging, these reactions reverse directions.

Sealed cells are today the dominant form of NiCd battery, which can provide convenient, clean, reliable, and maintenance-free services. Sealed NiCd cells are usually made in cylindrical shapes. It uses a wound plate, sealed construction with a nickel-plated steel can as the negative terminal, and a metallic cover as the positive terminal. The cell cover is an assembly that includes a high-pressure safety vent mechanism and insulating ring. Each electrode, which is a continuous conductive strip containing active material, is isolated from the other electrode by a separator; a nonconducting, porous, fibrous, polymeric material. The two electrodes and their accompanying separators are wound together into a roll configuration. This roll is then inserted into a can. Figure 1 shows a cutaway of a typical sealed cell (Gates Energy Products 1989). In Fig. 2, discharge curves of a sealed NiCd cell at different rates at room temperature are displayed. It can be seen that the cell is capable of delivering energy at very rapid rates.

1.2 Recombination Reactions in the Sealed Operation

The sealed cells normally operate at internal pressures well below the vent pressure because the gas evolved during charging is readily recombined. To accomplish this feature, the capacity of the negative electrode must be greater than that of the positive one (the so-called positive limited design). As a result, the positive electrode achieves full charge and emits oxygen before the negative electrode is fully charged and emits hydrogen, which cannot be readily recombined. The electrolyte solution (KOH) must be uniformly distributed by the separator, as a thin film across the surface of the two electrodes. Oxygen gas must be free to pass between the electrodes. There must be sufficient open area in and around the electrolyte and separator for oxygen to diffuse efficiently from the positive electrode to the negative electrode.

In charging a sealed NiCd cell, the positive electrode will reach full charge before the negative electrode. At this stage, additional charging causes the positive electrode potential to rise until all the incoming current is oxidizing hydroxyl ions and generating oxygen gas at the positive electrode:

$$2OH^- \rightarrow 1/2O_2 + H_2O + 2e^-$$

The oxygen generated at the positive electrode diffuses rapidly to the negative electrode where it is reduced back to hydroxyl ions:

$$1/2O_2 + H_2O + 2e^- \rightarrow 2OH^-$$

Thus, in overcharge, all the current generates oxygen that is subsequently recombined. The process described above is called oxygen recombination. The oxygen pressure initially increases but stabilizes at a low equilibrium pressure determined by the cell design, the ambient temperature, and the charge rate. It is important to remember that the recombination of oxygen is an exothermic reaction. When the cell is in overcharge, essentially all of the energy in the current coming in is converted to heat. With proper matching of the charger to the battery and attention to battery location and heat dissipation, the battery system can be designed to reach a thermal steady state.

1.3 Cathode Active Material—Nickel Hydroxides

Nickel hydroxides have been used as the active material in the positive electrodes of several alkaline batteries, including the NiCd system, for over a century. Because of the commercial importance of NiCd batteries and, in particular, of NiMH batteries,

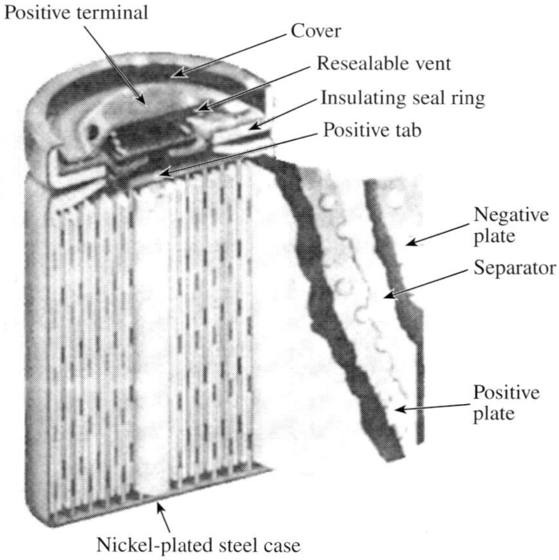

Figure 1
Cutaway of a typical cylindrical NiCd cell (courtesy of Gates Energy Products 1989).

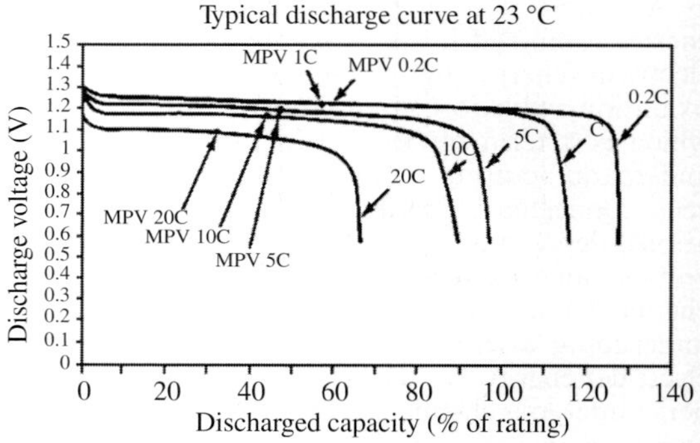

Figure 2
Discharge curves of a sealed NiCd cell at different rates at room temperature.

the nickel oxyhydroxide materials continue to attract much attention. The most significant advances in the understanding of the overall reaction of the nickel hydroxide electrodes was made by Bode and his co-workers (McBreen 1999). It was established that both the discharged material ($Ni(OH)_2$), and the charged material (NiOOH), could exist in two forms. One form of $Ni(OH)_2$, which was designated as β-$Ni(OH)_2$, is anhydrous and has a layered brucite ($Mg(OH)_2$) structure as shown in Fig. 3. The other form, α-$Ni(OH)_2$, is hydrated and has intercalated water between brucite-like layers. Oxidation of β-$Ni(OH)_2$ on charge produces β-NiOOH, and oxidation of α-$Ni(OH)_2$ produces γ-NiOOH. Discharge of β-NiOOH yields β-$Ni(OH)_2$ and discharge of γ-NiOOH yields α-$Ni(OH)_2$. The two reaction schemes are often referred to as the β/β and the α/γ cycles. On discharge, the α-$Ni(OH)_2$ can dehydrate and recrystallize in the concentrated alkaline electrolyte to form β-$Ni(OH)_2$. Also, β-$Ni(OH)_2$ can be converted to γ-NiOOH when the electrode is overcharged. Their overall reaction scheme is shown schematically in Fig. 4. It is also found that the transformation of β-$Ni(OH)_2$ to β-NiOOH experiences some degree of volume expansion, while when the β-$Ni(OH)_2$ converts to γ-NiOOH, its volume increases drastically.

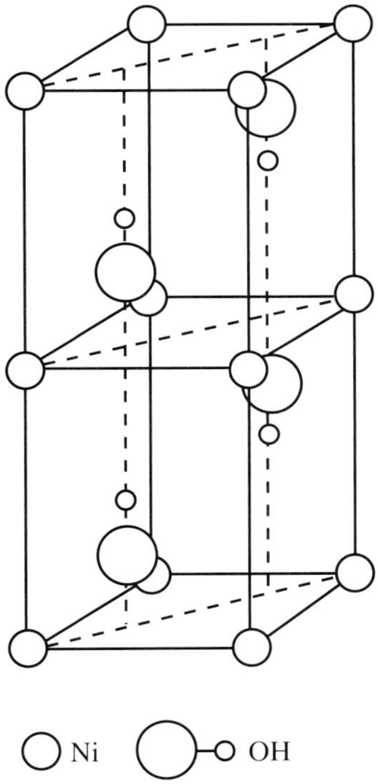

Figure 3
The brucite structure of $Ni(OH)_2$ showing the orientation of the O–H bonds (after McBreen 1999).

Therefore, conventional nickel hydroxide electrodes are designed to operate on the β/β cycle in order to accommodate the volume change that occurs during cycling, since the β/β cycle results in less volume increase. During the β/β transformation, the theoretical capacity of the β-$Ni(OH)_2$ is $289\,mAhg^{-1}$. The discharged phase of the positive electrode, β-$Ni(OH)_2$, is a poor electrical conductor. The conductivity of β-NiOOH is more than five orders of magnitude higher than that of β-$Ni(OH)_2$. As a result, there is difficulty in the charge/discharge processes, particularly the discharge, because the resistive discharged products cause the electrode not to be discharged at a useful rate. Operation on the β/β cycle is ensured by the use of a combination of additives such as cobalt, cobalt oxide, and zinc, as well as by the control of the electrolyte composition.

In the NiCd cell, the nickel hydroxide electrodes can be made using either sintered or foam substrate materials into which the active material is deposited. In the case of sintered electrodes, a layer of high-surface-area and dendritic-type nickel powder is coated on a steel foil strip (up to 100 μm thick). The coated strip is passed through a high-temperature furnace having a reducing atmosphere, and the dendritic projections of the nickel powder fuse together forming a strong, highly porous substrate material known as nickel plaque. Nickel hydroxide, the active material, is impregnated into the pores of the nickel plaque either by a chemical impregnation process or by an electrochemical technique. The sintered electrodes offer excellent high-rate discharge capabilities.

For the foam electrode, the paste of powdered nickel hydroxide along with other additives is mechanically pasted, compressed, or sprayed into the pores of the foam substrate. The larger amount of active material, which is accommodated by the larger pores and higher pore volume of the foam, results in positive plates whose energy density is 15–20% higher than that of sintered electrodes. However, the large open structure of the nickel foam substrate may cause variations of charge distribution. To improve charge-distribution variations resulting from the foam's open structure, conductive additives such as cobalt, cobalt oxide and/or cobalt hydroxide are mixed with the active $Ni(OH)_2$ material. A spherical, high-capacity $Ni(OH)_2$ powder (Ohta *et al.* 1994) has

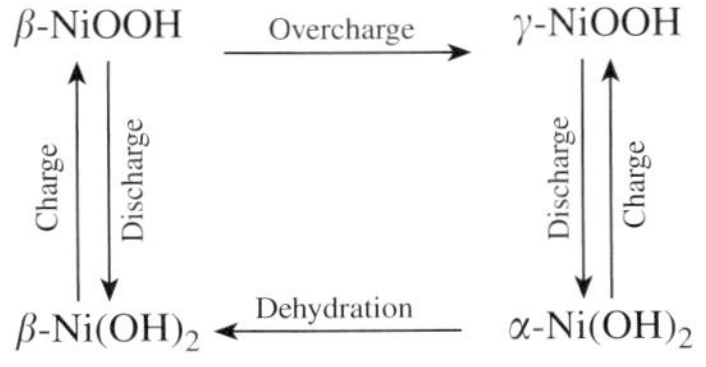

Figure 4
Reaction scheme of nickel hydroxide electrodes.

been developed to increase the energy density of the foam-type positive electrodes. In addition, additives such as cobalt, zinc, and/or cadmium by coprecipitation can reduce the electrode swelling which results from phase transformation during charge and discharge cycling. Nickel electrode swelling is believed to be the primary cause of electrode failure. High-performance nickel foam with fine structure and high porosity has become the norm. Thin, chemically stable separators are being developed to reduce interelectrode spacing. In turn, the volume in the cell available for active materials is increased.

1.4 Anode Active Materials—Cadmium

The theoretical capacity of cadmium metal is $480\,mAhg^{-1}$. However, cadmium is not usually applied as a metal to form a battery anode. The cadmium electrode may be formed starting with a mixed cadmium hydroxide, and/or cadmium oxide and a certain amount of cadmium powder. Two types of cadmium electrode are also widely used. One is the pasted cadmium electrode. Pressing the paste of mixed cadmium with a binder and additives on a nickel-plated steel foil substrate makes the pasted electrode. Due to high hydrogen over-voltage of cadmium in the caustic electrolyte, no amalgamation is needed. Another type of cadmium electrode is the sintered one, incorporating the paste of mixed cadmium material with binder and other additives in the pore system of a sintered nickel plate, which then undergoes a sintering process. The sintered cadmium electrodes provide the capability of high discharge rates when compared to pasted electrodes.

1.5 Separators

Separators generally serve two primary functions: (i) keeping the positive electrode physically apart from the negative in order to prevent any electronic current passing between them, and (ii) permitting an ionic current with the least possible hindrance. To meet the two opposing requirements, the separator materials must be thin, porous, mechanically durable, chemically stable, and electrically resistant.

Two kinds of fabric materials are widely used as separators for NiCd batteries: polyamide ("nylon") and polyolefin, which can be polypropylene (PP), or polyethylene (PE), or a combination. In the case of sealed batteries, these fabric materials have proven themselves. As discussed above, the working principle of the sealed batteries is based on internal oxygen consumption. During the charging step, the positive electrodes reach their fully charged state earlier and start to evolve oxygen, which migrates through voids in the electrolyte to the negative electrode to discharge cadmium and form water. Therefore, the separator has to be permeable to gaseous oxygen. This is achieved by the separator pores being of a specific minimum size and not all of them being filled with electrolyte, thus leaving some gas channels. Due to their pore-size distribution, these fabric materials can simultaneously absorb sufficient electrolyte and allow oxygen transfer.

For longer cycle life and better charge retention, in particular for higher-temperature application (up to 60°C), polypropylene fleeces are preferred since they offer better chemical stability and do not contribute to electrolyte carbonation. However, polypropylene has to be pretreated by fluorination, or by coating and crosslinking with hydrophilic substances (e.g., polyacrylic acid) on the surface of the fiber to improve electrolyte absorption.

Cylindrical cells are manufactured automatically at very high speed, in which a layer of separator material is spirally wound, each with two electrodes. High energy density requires the separator to be very thin (0.05–0.3 mm). All these make the separator mechanical strength an important criterion. The melt-blown polypropylene fleeces (Bohnstedt 1999) can be made with very thin fiber and provide low-cost hydrophilization, giving attractive properties such as small pore size, and excellent tensile performance for use in highly automated assembly processes. In this aspect, the production of melt-blown polypropylene fleeces with their excellent tensile properties offers an interesting option.

1.6 Applications

As one of the most popular alkaline secondary batteries, NiCd is available in several cell designs and in a wide range of sizes. Cells with different design and internal construction are provided for the needs of various applications in different environment conditions. The vented pocket-type cells are used in heavy-duty industrial applications, such as mining vehicles, railway signaling, materials-handling trucks, diesel engine starting, and emergency or standby power. This type of battery is very rugged and can withstand both electrical and mechanical abuses, has very long life and requires little maintenance except an occasional topping up with water. The sintered-plate construction has higher energy density that provides high discharge rates and low temperature performance, which are used in applications requiring lighter weight and superior performance, such as aircraft starting, communication and electronics equipment. The clean, maintenance-free sealed cells are available in prismatic, button, and cylindrical configurations and are used in consumer applications such as wireless phones, cellular phones, lighting, shavers, radios, etc., and small industrial applications. The high-rate discharge capability makes them ideal batteries for use in devices such as power tools and appliances, which demand high power. The sealed NiCd battery is one of many candidates for electrical-vehicle applications.

The advantages of NiCd batteries are the flat discharge curve, extremely good cycle life, very high discharge rate, and low cost. On the other hand, the electrochemical equivalent (480 Ahg^{-1}) is one of the lowest for all metallic anodes and the open circuit voltage of 1.35 V for the NiCd cell is not favorable for many applications. Additionally, the use of cadmium should be restricted for environmental reasons. One of the main advantages of NiMH batteries over NiCd batteries is the environmental aspect. The replacement of NiCd by NiMH, which offers better performance and is more environment-friendly, is a strong trend.

2. *Nickel—Metal Hydride Batteries*

The NiMH battery is a viable alternative to NiCd, which has been widely used in portable electronics since the 1960s. The 30–50% higher energy density, nontoxic, and environmentally friendly constituents, as well as plentiful raw materials, make the NiMH superior to the NiCd battery. Since 1980, extensive application-oriented research and development work has been conducted on metal hydride (MH) alloys, MH electrodes, and NiMH batteries (see *Hydrogen–Metal Systems: Basic Properties (1)* and *Hydrogen–Metal Systems: Basic Properties (2)*). NiMH rechargeable batteries have become increasingly popular since they were commercialized in 1990. This is the most successful application of hydrogen absorption materials and hydride technology. Research and development work on the MH materials and NiMH battery has been extensively reviewed (Sakai 1995, Notten 1995, Zhang 1997).

2.1 Charge and Discharge Reactions

The high-energy-density NiMH cell is a combination of NiCd technology and the advanced metal hydride materials. In other words, when MH alloys are used to replace cadmium as the active material in the negative electrode, a NiCd cell becomes a NiMH cell. The concept of a NiMH cell consisting of nickel electrode (+) and AB_5 metal-hydride electrode (−) is schematically represented in Fig. 5. The electrodes are insulated electrically from each other by a separator that is usually fibers made from either polyamide or polyolefin in the form of nonwoven fabrics. An alkaline solution, KOH, is the electrolyte, which provides ionic conductivity between the two electrodes. During charging, at the positive electrode the nickel hydroxide, $Ni(OH)_2$, is oxidized to nickel oxyhydroxide, NiOOH. All of these are the same as in the NiCd cell discussed in the previous chapter. However, at the negative electrode, the metal hydride alloy forms hydride by absorbing hydrogen which is

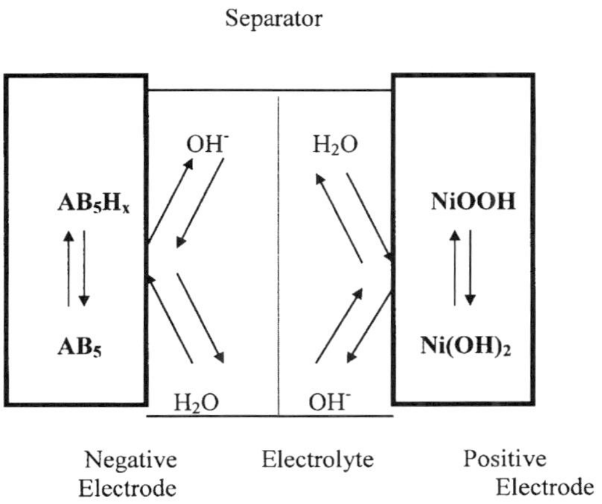

Figure 5
Schematic representation of charge (↑) and discharge (↓) reactions of a rechargeable NiMH cell.

generated from water electrolysis. The electrochemical reactions during the charge and discharge cycle can be expressed using the following equations:

MH electrode (−):

$$M + xH_2O + xe^- \leftrightarrow MH_x + xOH^-$$

Ni electrode (+):

$$Ni(OH)_2 + OH^- \leftrightarrow NiOOH + H_2O + e^-$$

Overall reaction:

$$xNi(OH)_2 + M \leftrightarrow xNiOOH + MH_x$$

A key consideration underlying the development of a sealed NiMH cell is to prevent the buildup of hazardous gas pressure, due to oxygen and hydrogen evolution which results in the loss of electrolyte. This is achieved by balancing the quantity of active materials of the positive and the negative electrodes. In the NiMH cell, the MH electrode is designed to have higher capacity than the $Ni(OH)_2$ counter. The excess MH materials are called "charge reserve." Ideally, the negative electrode should never be fully charged. However, a small amount of overcharge is necessary to ensure that the cell is fully charged. In principle, hydrogen gas is generated as the negative electrode is absorbed by hydride alloy via gas-phase reaction, while the oxygen generated at the positive electrode penetrates through the separator and diffuses to the MH electrode. The oxygen is reduced at the MH/electrolyte interface to form hydroxyl ions. This complicated and multistep process is called oxygen recombination, which is similar to that in NiCd batteries discussed in the previous chapter.

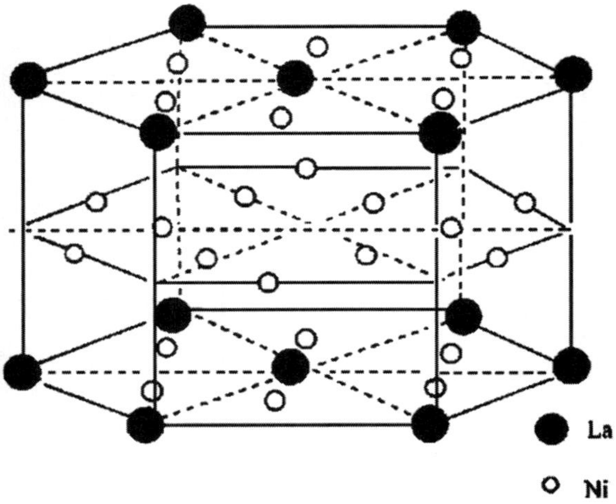

Figure 6
Schematic of hexagonal lattice of $LaNi_5$ intermetallic compound.

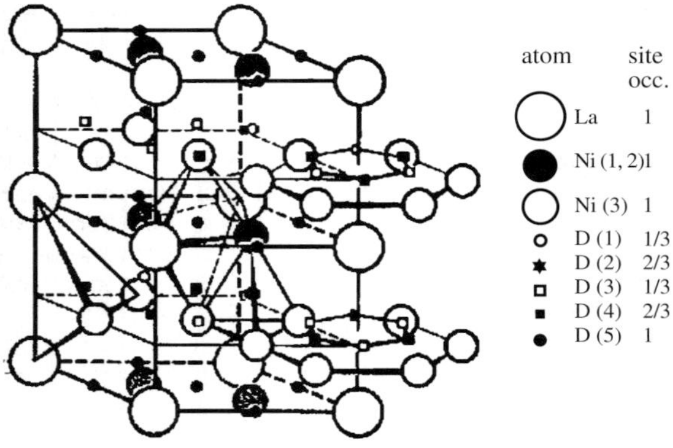

Figure 7
Crystal structure of $LaNi_5D_7$.

2.2 Anode: AB_5-type Metal Hydride Electrodes

There are many intermetallic compounds that exhibit the ability to reversibly absorb large amounts of hydrogen, among which a number of MH alloys have been evaluated for battery applications (van Rijswick 1978), such as the $LaNi_5$, $La_{0.8}Nd_{0.2}Ni_{2.5}Co_{2.4}Si_{0.1}$, $Mm(NiCoMnAl)_5$ system, the TiNi, Ti_2Ni, $TiMn_{0.5}$, Ti-Zr-based system, and the Zr-Mn-Ni-based system. The research work on MH materials has concentrated mainly on rare earth-based AB_5 systems, which are derived from the $LaNi_5$ intermetallics and are considered the best materials for NiMH cells. Another system that has been extensively studied is traditionally called AB_2, which actually includes Ti-

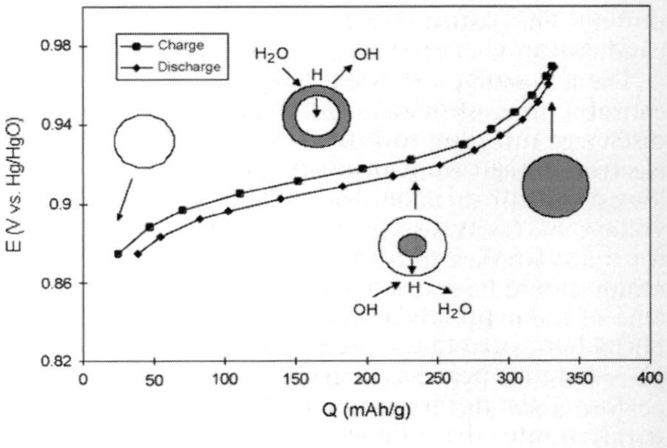

Figure 8
Electrochemical charge and discharge processes of a typical AB_5 alloy in 8 N KOH at 25°C. The MH alloy is subjected to phase transformation between α and β.

Zr-Ni-based multiphase alloys and the Zr-Mn-Ni-based AB_2-type Laves phase alloys.

$LaNi_5$ is the starting compound for the AB_5-type alloys, which has the hexagonal $CaCu_5$ structure, space group *P6/mmm*, as shown in Fig. 6. It can easily form hydride, $LaNi_5H_7$, with a theoretical capacity of $372\,mAhg^{-1}$ or $2600\,mAhcm^{-3}$. The crystal structure of the hydride can be illustrated, as shown in Fig. 7, by $LaNi_5D_7$. However, for $LaNi_5H_7$, its equilibrium pressure is higher than one atmosphere at room temperature ($P_{abs} = 2.3$ atm, $P_{des} = 1.6$ atm at 25°C), making it difficult to charge. In addition, its capacity was found to decline drastically, following an exponential pattern, during repeated electrochemical charging and discharging, which was ascribed to the decomposition of $LaNi_5$ caused by the corrosive action of the electrolyte. Since the 1970s, approaches have been applied to modify $LaNi_5$ for electrochemical use.

The charge and discharge processes of a typical commercialized AB_5-type MH electrode in KOH are illustrated in Fig. 8. The curves are the electrochemical equivalent of the PCIs (pressure-composition isotherms) of gas phase reaction, while a graphic description of phase transformation is displayed. Actually, there is a thermodynamic correlation between the equilibrium pressure measured in the gas phase reaction and the electrode potential measured in an electrochemical cell (Notten 1995):

$$E = -0.932 - 0.029 \log P_{H2}$$

where E is in volts, and P_{H2} is in atm. Usually, the MH alloy must be activated prior to testing. Therefore, the starting alloy would be an α-phase which is a solid solution of hydrogen. Charging a MH electrode consists of electrolysis of water on the MH surface, generating hydrogen atoms and hydroxyl ions. This step is the charge transfer process. The generated hydrogen atoms are adsorbed on the MH surface. Ideally, the adsorbed hydrogen subsequently dissolves in the alloy to form a metal hydride β-phase, therefore a growing product layer of the $\beta\text{-}AB_5H_x$ advances inward from the surface. On discharging, hydrogen atoms transport to the MH surface to react with hydroxyl ions forming water, and the hydride decomposes to α-phase alloy. Again, a growing α-phase layer proceeds inward from the surface. It should be noted that there is a competitive reaction in the charging process, in parallel with hydrogen diffusion: if the diffusion rate is slower than the charge transfer rate, the atomic hydrogen may combine to form molecular hydrogen resulting in the build-up of the internal gas pressure.

Another characteristic of the hydride alloy is that the particles will be broken down during cycling due to repeated lattice expansion/shrinkage, and ultimately to ~3 μm in diameter, along with which, a part of the active material is corroded and the cell internal impedance increases. The charge and discharge mechanism is described in the following equations, from which it can be seen that each absorbed or desorbed hydrogen atom corresponds to the storage or release of one electron:

Charge transfer reaction:

$$H_2O + e^- \rightarrow H_{ad} + OH^-$$

Hydrogen diffusion (MH formation):

$$H_{ad} \rightarrow H_{bulk}$$

Hydrogen combination:

$$2H_{ad} \rightarrow H_2$$

From the above discussion, it can be concluded that a good MH alloy for electrochemical application must meet the following properties for its optimum use in MH electrodes:

(i) Capacity for high energy density;
(ii) Chemical stability for long cycle life;
(iii) Mechanical stability to reduce the decrepitation rate;
(iv) Rapid charge and discharge kinetics for high rate capability;
(v) Rapid activation rate to allow freshly prepared electrodes to quickly respond;
(vi) Low cost.

MH electrode characteristics strongly depend on the chemical composition of the intermetallic compounds. A considerable improvement was realized (Willems 1984) when lanthanum was partially substituted by neodymium, and nickel by cobalt and aluminum or silicon. The substitution not only adjusted the equilibrium pressure to below 1 atm at room temperature, but also significantly improved cycle life. The cobalt substitution has the special effect of reducing the lattice expansion during charging, dramatically decreasing the lattice expansion in the c-axis direction, and in turn reducing the pulverization rate and improving cycle life. Lattice expansion along the c-axis has more influence on cracking due to the layer structure of the alloy. Furthermore misch metal (Mm, a mixture of rare earths including lanthanum, cerium, neodymium, and praseodymium) was used to replace pure lanthanum to reduce its cost. However, the smaller size in atomic radii of cerium, neodymium, and praseodymium compared to that of lanthanum results in the so-called "rare earth contraction effect," decreasing the cell volume and causing an increase in the hydrogen equilibrium pressure. Doping with manganese, in addition to cobalt and aluminum, could effectively reduce the equilibrium pressure without a decrease in capacity. The synergistic effects by alloying also showed improvement in corrosion resistance, attributed to modifications in both chemical and mechanical stability. Therefore, multicomponent AB_5 systems, with formulation $Mm(NiCoMnAl)_5$, were developed and considered the most suitable MH electrode.

For an AB_5 compound, its element distribution, i.e., the alloy's microstructure, plays a key role on the cell cycle life. The microstructure here involves phase distribution, grain size, grain shape, element segregation, grain boundary, etc. If the melted alloy was cooled at high rate using a plate-like casting, a clear columnar structure with a smaller grain size (20–30 μm) formed, and showed an increase in cycle life. This is explained by the columnar structure, with the c-axis in parallel to the cooling plane, having less lattice strain and less pulverization rate. In contrast, manganese facilitates nucleation, therefore the addition of more manganese resulted in an equiaxed structure and, in turn, high lattice strain. The lattice strain in the alloy facilitates the cracking and high decrepitation rate during cycling, thus the alloy displays shorter cycle life. Melt-spinning and gas atomization can produce alloys with small grain size (<10 μm), and better homogeneity. Such a fine microstructure makes the protective surface layer, for example, formed by aluminum-containing compound, to be more effective during cycling which reduces the decrepitating rate and improves cycle life. Increasing the cooling rate followed by post-heat treatment can eliminate the formation of a second phase, modify element segregation along the grain boundary and within the crystallite, remove lattice defects such as dislocations, and thus further improve cycle life. Amorphous alloys, generated either by the thin-film technique or by prolonged mechanical grinding, the $LaNi_5$ or Mm $(NiCoMnAl)_5$ alloys showed less decrepitating rate and longer cycle life, but also resulted in a decrease in the discharge capacity.

The alloy surface activity is an important property directly associated with electrochemical kinetics such as the rate of charge transfer reaction (water electrolysis), and the oxygen recombination reaction. Usually, the surface activity is evaluated by its exchange current density, i_0, which can be obtained from the Tafel relationship expressed as follows:

$$\eta = a - b \log i, \ a = 2.3RT/\beta F \log i_0$$

where η is the overpotential, i is the current density under the overpotential, η. i_0 is the exchange current density usually expressed in $\mathrm{A\,cm^{-2}}$. However, for MH alloys, i_0 is expressed using $\mathrm{A\,g^{-1}}$ due to its dynamic characteristics in particle-size distribution.

Figure 9 is a Tafel plot of a representative $Mm(NiCoMnAl)_5$ alloy at a state of charge of 35%. On the left side of the y-axis, it is a charge process (cathodic reaction), while on the right side, a discharge process (anodic reaction). The peak point ($i = 0$) is the equilibrium potential, $E_{(i=0)}$, of 0.92 V against Hg/HgO.

From this plot it can be seen that there is no linearity behavior between −0.91 V and −0.7 V, indicating that there is no Tafel relationship in the anodic process. A similar behavior was also observed in the cathodic process. This observation suggests that for the multicomponent AB_5 system, it is not appropriate to use the Tafel equation to obtain its exchange oxides density. Instead, we can use the linear polarization technique within a small range around the equilibrium potential ($E_{(i=0)}$) ($\eta = \pm 10\,\mathrm{mV}$). The non-Tafel behavior is also an indication that the surface activity, and in turn the charge transfer reaction, is strongly affected by mass transport. This may involve the diffusion of hydrogen in the solid alloy, OH^- and H_2O in the electrolyte (van Rijswick 1978).

It is generally believed that the multicomponent AB_5 electrochemical hydrogen reaction has good kinetic properties; however, the element substitution resulted in a substantial decrease of the surface activity from that of $LaNi_5$ compound. The surface reactivity relies on two factors: effective reaction area, and catalytic functioning. It was reported that the rate capability of MH electrodes made of $MmNi_{3.6}Co_{0.7}Mn_{0.4}Al_{0.3}$ could be modified significantly by the addition of metal oxide powders, such as CuO, CoO, Y_2O_3, and $Y(OH)_3$.

Double-phase AB_5-type alloys reported by Notten (1995) are capable of increasing the rate capability of MH electrodes. It can be considered as an alternative approach to improving the charge transfer reaction by the use of a catalyst. When additional nickel and molybdenum are put into a "standard" alloy, $La_{0.8}Nd_{0.2}Ni_{2.5}Co_{2.4}Si_{0.1}$, the alloy will become a nonstoichiometric $AB_{5.5}$ material. When a highly electroactive compound $MoCo_3$ was generated as a second phase in the $AB_{5.5}$ alloy (e.g., $La_{0.8}$

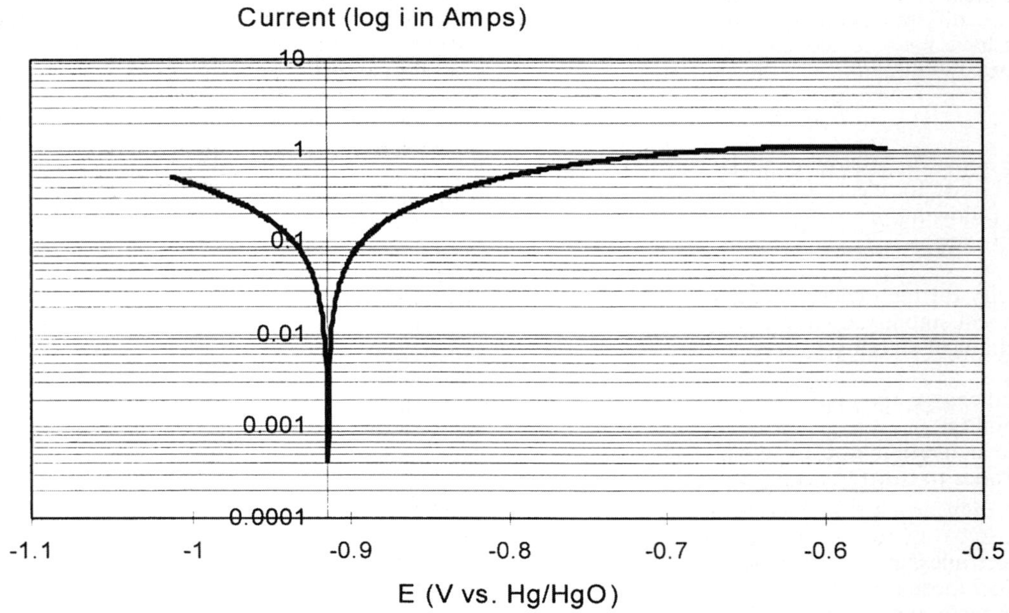

Figure 9
Tafel plot of a typical AB_5 alloy 35% charged. The current density is expressed by amps per 250 mg MH alloy sample ($0.5\,\mathrm{mV\,s^{-1}}$).

$Nd_{0.2}Ni_{2.9}Mo_{0.1}Co_{2.4}Si_{0.1}$), the alloy's discharge efficiency was significantly improved from $190\,mAg^{-1}$ to $600\,mAg^{-1}$. The enhanced surface reactivity was ascribed to the excellent catalytic activity of $MoCo_3$ or $NiCo_3$ decorating the exterior of the MH particles. The introduction of a second phase, such as $MmCo_4B$ and MmB_4, is also reported to improve discharge efficiency.

Fluorine treatment was initiated a few years ago. It was reported that the fluorine treatment forms a condensed LaF_3 layer on the top surface to protect the MH alloy from being corroded. The most striking progress is that nickel particles could be implanted in the RF_3 layer simultaneously during the fluorination process so as to significantly enhance the reactivity, as well as the activation and the corrosion protection, of MH materials (see *Hydrogen–Metal Systems: Electrochemical Reactions (Fundamentals and Applications)*; *Hydrogen–Metal Systems: Fluorinated Metal Hydrides*; *Hydrogen–Metal Systems: Technological and Engineering Aspects*).

2.3 *Development of MH Electrode and NiMH Cells*

To optimize the use of MH as active material in the NiMH cells, a great effort has been made to develop MH electrode technology. The aim of the MH electrode technology is to make MH electrodes possess an appropriate energy capacity, suitable geometric dimensions, good mechanical integrity, satisfactory electrochemical functioning, and long cycle life. At present, two different techniques for preparing MH electrodes have been developed and are widely used in the battery industry: (i) the sintering method (Fetcenko *et al.* 1990) for Ti-Zr-V-based alloy by pressing the powder materials without additive on a nickel-mesh sheet, followed by a sintering process; and (ii) the pasting method (Kinoshita *et al.* 1996) by extruding the substrate through a slurry containing MH powder, conductive powders such as carbon, nickel, cobalt, etc., and other additives to improve the electrode conductivity and surface reactivity. The pasting technology is more complicated because the properties of the numerous additives have to be optimized to obtain the desired electrode properties. The binder materials, such as polytetrafluoroethylene (PTFE), polyvinyl alcohol (PVA), silicon rubber, SEBS rubber, etc. have been characterized for use in the MH electrode to obtain satisfactory mechanical strength; additives like carbon black will form a solid–gas interface to accelerate the oxygen recombination; a hydrophilic agent may generate a solid–liquid interface to enhance the water electrolysis, etc.

A significant improvement in electrode technology, for both MH and $Ni(OH)_2$, allows improvement in the packing factor of the active materials without sacrificing the electrochemical efficiency, and in turn improves the NiMH energy density (see Table 1). Since 1991, NiMH batteries have become an important part, with growing potential, of rechargeable batteries and a well-developed new industrial domain, including manufacturing of raw materials and component parts around the NiMH batteries, has been established.

2.4 *Challenges and Opportunities*

Rapidly developing portable electronics demands even higher-performance batteries. Efforts to increase MH alloy capacity are a first step to meet the market demand for higher energy-density MH cells. Either discovering new alloys or further improving the existing AB_5 can achieve this. Theoretically, if the alloy capacity increases by 30% (i.e., to about $3000\,mAhcm^{-3}$), the cell energy density will increase approximately 10% by adjusting the amount of active materials in both negative and positive electrodes. This increase would be without changing the cell design. Essentially, alloy chemistry determines the alloy hydrogen capacity. Metallurgical processing can improve the alloy capacity to some extent.

Obviously, price is always a major concern, and a low-cost battery is desirable. Improvement of cost performance of NiMH batteries depends heavily on the MH materials' cost. In 1990, the AB_5 alloy's price was about \$40 kg^{-1}. Since then, the material price has gradually decreased. At present (2000), the alloy price is in the range of \$15–20 kg^{-1}, a reduction of almost 60%. The major cause of the price reduction is worldwide industrialization. The demand for large-quantity alloys, standardization in misch metal composition and MH formulation, well-developed MH manufacturing technology, and stabilized raw material resources all contributed to the production efficiency, which makes the operating costs decrease dramatically. However, it is believed that further decreases in the material cost may mostly rely on modification of the alloy's chemical formula, based on a typical AB_5 formulation, $MmNi_{3.55}Co_{0.75}Mn_{0.4}Al_{0.3}$.

Intensive attempts in research and development on the AB_5-type alloys, with new chemistries to reduce materials cost, are ongoing around the world. Misch metal, nickel, and cobalt are major components in the AB_5 alloys in terms of functioning role and cost.

Table 1
Examples to show the progress of a NiMH cell capacity (in mAh).

Size	1991	1998
AA	1000	1000–1500
4/3A	2500	3000–4000

Reduction of one or all of the components will reduce the cost, but some technical approaches must be taken for performance compensation. To eliminate or partly reduce the cobalt content is a first attempt in this aspect. The major problem is poor cycle life, since cobalt is a key element in maintaining corrosion resistance.

Corrosion is an issue that needs to be resolved to improve NiMH performance under severe service environments. Alloys with high corrosion resistance, when subjected to electrochemical cycling, are desirable for improved performance. From the thermodynamic aspect, corrosion of AB_5 alloys in KOH under operating conditions is favorable. The free energy change of $LaNi_5$ oxidation, $\Delta G = -472$ $kJmol^{-1}$, is almost twice the enthalpy change of $LaNi_5$ formation ($-127\,kJmol^{-1}$), giving the alloy a strong driving force for oxidation in KOH solution. However, it has already been proven that it would not be possible to eliminate the causation, but it is technically possible to reduce the corrosion rate.

The improvement of corrosion rate can be directly transferred to longer cycle life. Particularly when the cell serves under high temperature, its corrosion becomes more severe. The corrosion rate improvement is also directly related to developing high-capacity and low-cost alloys, as seen from previous sections of this paper. Taking advantage of the durable MH electrode, we can reduce the amount of excess MH materials that are used for charge reserve, making space for more active materials in the positive electrode to increase the cell energy density (assuming the corrosion improvement allows reduction of the excess MH alloy, which is for charge reserve).

The alloy chemistry is a determining factor in improving the alloy corrosion resistance. Aluminum substitution for nickel can improve the alloy's cycle life, probably due to its surface protection effect. The partial substitution of cobalt for nickel results in considerable improvement. Composition modification for the A side, such as the cerium element, significantly improves the chemical stability and in turn improves cycle life. This is ascribed to cerium's unique mixed $+3/+4$ valence state in the alloy, and the formation of a stable protective oxide layer on top of the MH particles.

Voltage performance is important for cell applications. Figure 10 is a set of discharge profiles of a typical AB_5 electrode in a half-cell at different rates. The voltage was measured against a Hg/HgO reference electrode. From the curves, it can be seen at 5C rate ($1.5\,Ag^{-1}$) that the electrode could deliver power with a midpoint voltage of about 0.56 V. Figure 11 displays the electrode overpotentials (voltage drop) at various discharge rates. The overpotentials are obtained here using the voltage differences between the middle-point voltage of each discharge profile and the equilibrium potential, $E_{(i=0)}$, of the same electrode at 50% DOD. A cell possessing high-rate

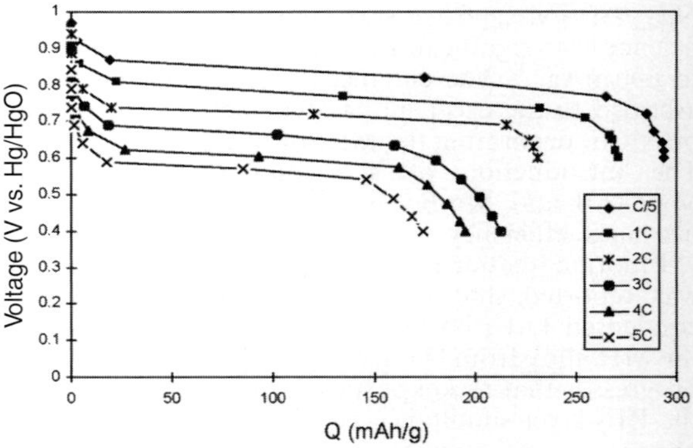

Figure 10
Discharge profiles of a typical AB_5 electrode at different discharge rates at room temperature. The test was conducted in a half cell. The voltage was measured against Hg/HgO reference electrode.

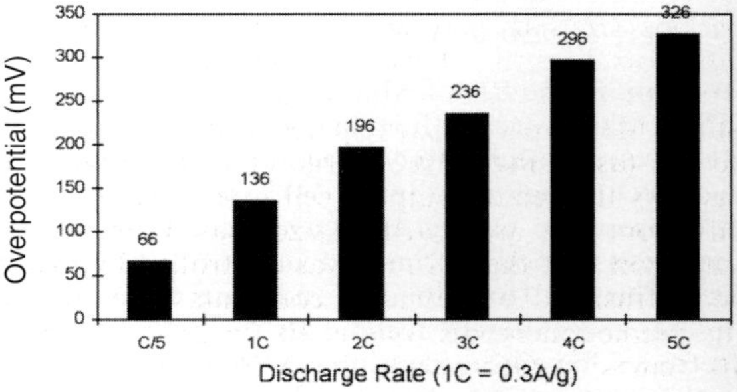

Figure 11
The overpotentials of the MH electrode at different discharge rates.

capability can deliver power at high rate, or under normal operation, can benefit from less voltage drop. High-rate capability is directly related to the electrode kinetics which can be caused by resistance resulting from three sources: charge-transfer reaction, mass-transport barrier, and internal contact resistance.

Charge-transfer reaction rates are governed by surface reactivity, which may be modified by the addition of oxides or a second phase. The mass-transport resistance may result from hydrogen diffusion in the bulk alloy, or from the diffusion of OH^- and H_2O through the electrolyte, particularly at higher rate. The hydrogen diffusion rate is an intrinsic property of MH materials and is related to enthalpy of formation. MH alloy's diffusion constant may be in a range from 10^{-8} to $10^{-12}\,cm^2s^{-1}$ (Willems 1984, van Rijswick 1978), depending on the nature of the alloys and the measuring technique. For

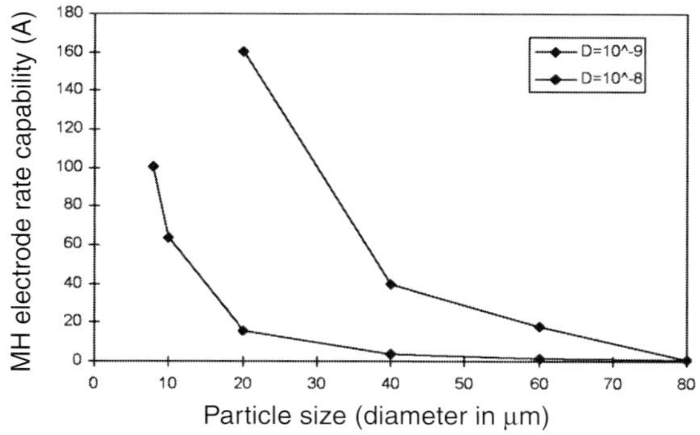

Figure 12
Prediction correlation between the particle size and the electrode rate capability, assuming the reaction is controlled by hydrogen diffusion.

a given alloy, its hydrogen diffusion constant is a fixed number at a certain temperature. However, the particle size can play a role due to the change of diffusion distance. If we take the MH as spherical particles, for a cell containing 10 g of MH, only 70% of the MH capacity participates in the discharge reaction (assuming 30% charge reserve). Figure 12 displays the relationship of cell rate capability with diffusion rate and particle size, based on the assumption that the cell rate was controlled by hydrogen diffusion. Two diffusion constants were taken in the prediction model: one is $10^{-9}\,cm^2s^{-1}$, the other $10^{-8}\,cm^2\,s^{-1}$. Of course, the MH particle will be broken down with continuous cycling. However, at least the rate capability on the early charge/discharge cycling can be enhanced by using smaller particles. To obtain a rate of 40 A, if the diffusion constant is $10^{-9}\,cm^2s^{-1}$, the particle size should be no larger than 15 μm, while for $10^{-8}\,cm^2s^{-1}$, 40 μm particles are enough. In addition, the smaller particle size also provides larger surface area which definitely increases reactivity.

Contact resistance is related to many factors. Various authors systematically analyzed the electrode reaction using electrochemical impedance spectroscopy. It was found that the contact resistance consisted of three parts: electrolyte resistance, contact resistance between the substrate and MH alloy, and contact resistance between the MH particles. The contact resistance between the MH and substrate and between the particles significantly increased with cycling, which is obviously related to the alloy degradation products of the oxides. The structure and thickness of the electrode, the alloy corrosion resistance, the kinds of additives, and the nature of binder materials all contribute to the resistance. They suggested that copper coating could significantly improve the electrode contact resistance.

2.5 *Applications*

The NiMH battery can be designed in a variety of forms, such as button cells, prismatic cells, and cylindrical cells, and in different sizes. The characteristics of the NiMH battery present opportunities for use over a wide range, and it will become one of the leading rechargeable battery systems.

Because of the same voltage value as the NiCd battery, all devices using NiCd can adopt NiMH as their power sources. As a result, the NiMH battery is increasingly used in a wide range of consumer electronic devices, such as cellular phones, camcorders, shavers, transceivers, computers, and other portable applications. Another niche market suitable for NiMH battery is power tools, which require high power discharge capability over a wide temperature range.

The NiMH battery has attractive energy density, high power capability, and good cycle life, all of which make the NiMH a competitive choice for EV and HEV applications, which may become a very important market for rechargeable batteries in the first decade of the twenty-first century. The commercialization of HEV cars by Toyota, Honda who use NiMH batteries, the desire for improving environments, and the concerns of fossil fuel resources have fueled a worldwide surge in the development of various battery-assisted vehicle applications. A major issue for users of portable electronics, EV and HEV applications, is the estimate of the battery's state of charge (SOC), which can translate useful information necessary for managing the battery system, such as how much energy is stored in the battery, how much runtime is left before the need for recharging, what recharge or discharge rate can be applied, etc. Therefore, a kind of "fuel gauge" has been expected, and a variety of schemes for measuring battery SOC have been suggested. In general, experience with NiMH cells indicates that, due to the flatness of the voltage plateau under normal conditions, voltage sensing may not be used to accurately determine the SOC. However, coulometry is a good technique for sensing the SOC. With careful initial calibration, appropriate compensation for environmental conditions, sophisticated charge-flow tracking, and estimation of self-discharge losses, predictions of SOC with moderate accuracy can be obtained.

The high-rate discharge and fast recharge capability of the NiMH battery also make it a candidate to combine with fuel cells, solar cells, and other batteries or internal combustion engines to handle peak loads or provide power when the prime power source is not adequate or not available.

3. *Lithium Ion Rechargeable Batteries*

Rapid development of new technologies, such as portable consumer electronics and electric vehicles, has generated the need for batteries that provide both high energy density and high power capability. Considering thermodynamic reasons, lithium shows a chemical and electrochemical behavior which favors it for selection as an anode material to be used in high-energy and high-power batteries.

Lithium metal has a very large theoretical capacity of $3860\,mAhg^{-1}$, in contrast to the value of $480\,mAhg^{-1}$ for cadmium and $372\,mAhg^{-1}$ for AB_5 metal-hydride alloy (Murphy and Caride 1979). Lithium is much more electropositive than hydrogen, with high standard potential (Li: $-3.01\,V$; Cd: $-0.4\,V$) which allows the realization of significantly higher voltage. In addition, lithium possesses a low density and hence low electrochemical equivalent weight (Li: $0.259\,gAh^{-1}$; Cd: $2.10\,gAh^{-1}$; AB_5: $3.30\,gAh^{-1}$).

During the past decades lithium rechargeable batteries, using lithium metal as the anode and intercalation compounds as the cathode, have been intensively studied. The difficulties associated with the use of metallic lithium stem from the changes that occur after repetitive charge–discharge cycling and its reactivity with the electrolyte. The thermal stability of lithium metal foil in many organic electrolytes is good, with minimal exothermic reaction occurring up to temperatures near the melting point of lithium (180°C). However, during recharge, lithium is electroplated onto the metallic lithium electrode, and forms a more porous deposit with a larger surface area than the original metal, which significantly increases its reactivity and lowers the thermal stability limit of the system. Therefore, the cells become increasingly sensitive to abuse as they are cycled. Another disadvantage is the short cycle life. This happens because lithium is not thermodynamically stable in the organic electrolytes and the surface of the lithium is covered with a film of the reaction products between the lithium and the electrolyte. Every time the lithium is stripped and replated during the discharge and charge, a new lithium surface is exposed and then passivated with a new film, consuming lithium. In order to obtain a reasonable cycle life, a three- to five-fold excess of lithium is required, which increases safety challenges. To overcome these problems, a safer approach is to replace lithium metal with a lithium intercalation compound, usually a carbon one (Lazzari and Scrosati 1980). In 1990, the so-called "rocking-chair" or "lithium-ion" (Li-Ion) battery was put on the market by Sony Energetic Inc. (Nishi 1998). In the Li-Ion battery, lithium ions swing between the carbon anode and the layered, highly oxidizing cathode through an organic liquid electrolyte dissolving inorganic lithium salt, like a rocking chair rocking from side to side. The principal concept is based on the intercalation reaction and is rather different from conventional secondary batteries which are based on chemical reactions.

A well known typical Li-Ion cell, in which a lithiated carbon intercalation material (Li_xC_6) is used as the negative electrode, a lithiated transition metal intercalation compound (lithium cobaltite, $LiCoO_2$) is used as the positive active material, and an aprotic organic solution (lithium salt $LiPF_6$ dissolved in mixed propylene carbonate (PC) -ethylene carbonate (EC) solvent) as the electrolyte, is expressed as follows (Scrosati and Megahed 1993):

$$Li_xC_6/LiPF_6\ \text{in PC-EC}/Li_{1-x}CoO_2$$

The reactions at the electrodes and the overall cell reaction can be represented by the following equations:

$$\text{Positive}: LiCoO_2 \leftrightarrow Li_{1-x}CoO_2 + xLi^+ + xe^-$$

$$\text{Negative}: C + xLi^+ + xe^- \leftrightarrow Li_xC_6$$

$$\text{Overall}: LiCoO_2 + C \leftrightarrow Li_xC_6 + Li_{1-x}CoO_2$$

Lithium ions move back and forth between the positive and negative electrodes during charge and discharge. At the negative electrode, the electrochemical process is the uptake of lithium ions during charge and their release during discharge, rather than lithium plating and stripping. As metallic lithium is not present in the cell, Li-Ion cells are less chemically reactive and have a longer cycle life than the cells containing metallic lithium. The reaction mechanism of the Li-Ion cell is shown graphically in Fig. 13.

Lithiated carbon is air-sensitive. In practice, the Li-Ion cell is manufactured in a fully discharged state. The fabrication procedure is based on the use of a lithium-rich intercalation compound as the cathode (positive electrode). The cell is assembled by coupling this lithium-rich or lithium source cathode compound with a lithium-accepting anode (negative electrode). The cell is "activated" by charging, which transfers lithium ions from the cathode to the anode which is thus lithiated. Most powdered intercalation compounds are pyrophoric when loaded with lithium. However, lithium will not deintercalate to react with air or moisture if it is sufficiently bound in the intercalation compound. Lithium intercalation compounds such as $Li_{1-x}CoO_2$ $(0<x<0.5)$, $LiNiO_2$ $(0<x<0.5)$, and $LiMn_2O_4$ $(0<x<0.85)$, are generally air stable when the chemical potential is less than $-3.6\,eV$.

The specific energy density of a Li-Ion battery with high discharge voltage (3.6 V) is nearly twice as high as the NiCd battery, and about 60% higher than NiMH batteries. In addition, the Li-Ion battery has demonstrated excellent cycle life and much-improved intrinsic safety.

3.1 *Anode Active Materials—Carbon*

Carbon materials can reversibly accept and donate significant amounts of lithium without affecting their mechanical and electrical properties. The lithiated

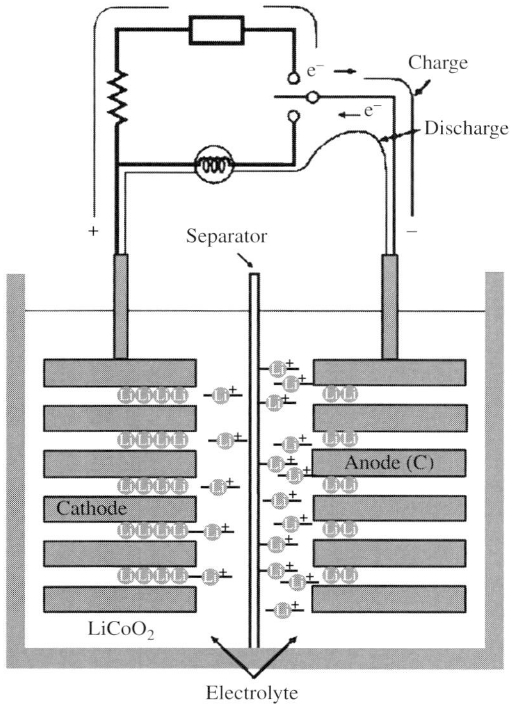

Figure 13
Schematic diagram of the principles of a Li-Ion rechargeable cell (after Nishi 1998).

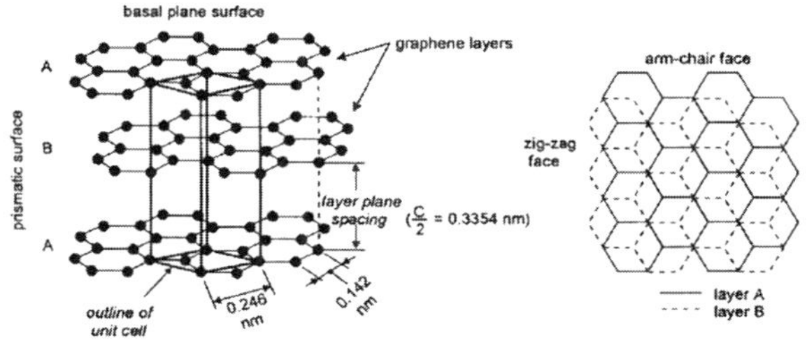

Figure 14
Left: schematic drawing of the crystal structure of hexagonal graphite showing the AB graphene layer stacking sequence and the unit cell. Right: view perpendicular to the basal plane of hexagonal graphite (after Winter and Besenhard 1999).

carbon also has a Fermi energy only about 0.5 eV below that of lithium metal. Therefore, in the Li-Ion cell, carbon is used for the anode instead of metallic lithium, and thus the electrochemical cell will have almost the same open-circuit voltage as one made with metallic lithium.

The carbon anode is a key component of the Li-Ion battery, and various carbon materials ranging from graphite to amorphous carbon have been proposed. Graphite has a layered structure. The graphite can be reduced by electrochemical intercalation with lithium in an aprotic organic electrolyte containing lithium salts to form LiC_6 with a capacity of $372\,mAhg^{-1}$. The formed LiC_6 can be electrochemically oxidized by lithium deintercalation. Graphite is a passable material as an anode; however, some drawbacks have been observed, including limitation of capacity, anode bulging, and poor cyclability. From the viewpoint of materials science as well as battery applications, carbon attracts much interest because of its variety of structures. It is the structure that plays the most important role in electrochemical performance and makes the diverse characteristics of the carbon.

Carbons bind themselves together by sp^3, sp^2, and sp hybrid orbitals, forming many kinds of organic compounds. However, the carbon materials discussed here are mainly built up by sp^2 bonding (Tran *et al.* 1995). Repeating sp^2 bonding forms a large network of 6-ring structures leading to a two-dimensional graphene sheet. The stacking order of these graphene sheets is important; van der Waals forces bind the sheets together forming layered crystalline structures. The ideal stacking order of graphite has two patterns: the carbon positioned in ABAB pattern has hexagonal symmetry with space group $P6_3/mmc$, while the ABCABC pattern has rhombohedral symmetry with $R3m$. Figure 14 schematically displays the crystal structure of hexagonal graphite showing the AB graphene layer stacking sequence and the unit cell. Natural graphite usually comprises these two crystal structures, but the rhombohedral phase is below 3–4%. For graphite, the weak van der Waals force between layers enables the planes to slide easily. Practical carbon materials generally have varying degrees of stacking faults resulting in carbon atoms deviating from regular positions, and periodic stacking is no longer maintained. In general, carbon materials capable of lithium intercalation can be roughly divided into graphitic carbon and nongraphitic carbon. Graphitic carbons are carbonaceous materials of layered structure but with some structural defects (turbostratic orientation). The graphitic carbons are commonly called "graphite" regardless of stacking order, since perfectly stacked graphite crystals are practically not available.

For nongraphitic carbonaceous materials, carbon atoms are arranged in a planar hexagonal network but without the crystallographic order in the C-direction of the graphite structure. Nongraphitic carbons are mostly prepared by pyrolysis of organic polymer or hydrocarbon precursors (Pierson 1993). It is also common to separate the nongraphitic carbon into "graphitizing" and "nongraphitizing" carbons. Graphitizing carbons can be easily converted to graphite, which has an orderly layered structure, by calcinations at temperatures as high as 3000°C, and lithium can intercalate only into the spacings between the layers to form LiC_6. Graphitizing carbon, which is also called "soft carbon", has a slightly disordered structure. On the other hand, nongraphitizing carbon materials consist of randomly oriented small crystallites, and this type of carbon cannot be converted to graphite by calcinations even at 3000°C, which is the reason it is called "hard carbon." We will see the difference between the "soft carbon" and the "hard carbon" in the following section.

3.2 Charge/Discharge Mechanisms of Carbon Anodes

The reversible insertion of mobile lithium guests into the structure of a solid carbon host is commonly referred to as "intercalation." In general, the designation of an insertion reaction by the term intercalation involves the condition that the host matrix units mostly retain their integrity with respect to composition and structure during the intercalation and deintercalation processes. Electrochemical intercalation reactions are confined to coupled electron/ion transfer ("mixed condition") reactions. For lithium intercalation into a carbon electrode from an appropriate Li^+-containing electrolyte, this means that due to the cathodic reduction of the carbon, the lithium guest ions penetrate into the carbon host, forming a lithium/carbon intercalation compound, as illustrated in Fig. 15. The corresponding negative charges are accepted into the carbon host lattice. The reversibility of the intercalation reaction can be checked by subsequent anodic oxidation of LiC_n, i.e., by removal or deintercalation of Li^+:

$$Li_xC_n \leftrightarrow xLi^+ + xe^- + C_n$$

The host/guest charge transfer, although not complete, usually leads to an increased electron density in the conductivity band of carbon and thus to an increase of the in-plane electronic conductivity by several orders of magnitude. Therefore, the carbon intercalation compounds are occasionally named "synthetic metals."

It was concluded that soft carbons always have at least two mechanisms to accommodate lithium (Maubachi *et al.* 1995). In the potential range from 0 V to 0.25 V, lithium is intercalated into graphitic parts and the capacity can be described as follows:

$$C_{0.25V} = 372 \times P_1 \ (\text{mAh g}^{-1})$$

where P_1 is a fractional ratio of perfect graphite stacking. The maximum capacity is $372\,\text{mAhg}^{-1}$ when $P_1 = 1$ for graphite. In the potential region above 0.25 V, the intercalation mechanism changes and depends on heat-treatment temperature. At lower heat-treatment temperatures, e.g., at 1000°C, the soft carbon electrode shows capacity larger than

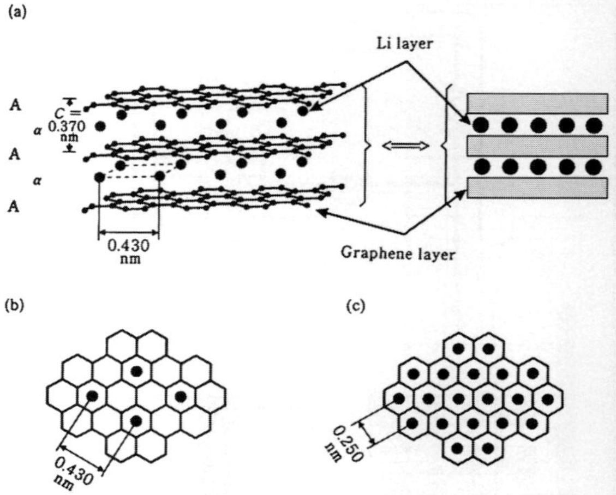

Figure 15
(a) Left: schematic showing graphite AA-layer attacking and lithium intercalate αα interlayer ordering. Right: showing simplified schematic representation. (b) In-layer ordering model of LiC_6. (c) In-layer ordering model of LiC_2 (after Winter and Besenhard 1999).

$372\,\text{mAhg}^{-1}$. This is attributed to the existence of micropores or "cavities" formed between crystallites. It is proposed that lithium intercalates into the graphite layers and the cavities at the same time during charge; in discharge, lithium is first deintercalated from the layers, and then extracted from cavities. This mechanism is illustrated in Fig. 16. In the case of hard carbon, lithium can be electrochemically charged into ultramicropores (with a diameter of 0.7 nm to 0.8 nm) as well as the layers of crystallites themselves. The ultramicropores are probably able to trap lithium in clusters therefore resulting in higher capacity, up to $600\,\text{mAh g}^{-1}$, which exceeds the stoichiometric capacity of LiC_6 (Ishikawa *et al.* 1994). The hard carbons are made up of single graphene sheets which are arranged like a "house of cards" (Xue and Dahn 1995). On charging, intercalated lithium is adsorbed onto both sides of each of the graphene sheets, leading to high lithium capacity. Discharge curve profiles of Li-Ion cells are illustrated in Fig. 17. The sloping discharge profile is characteristic of the Li-Ion cells using hard carbon materials, compared to the flatter discharge profile of the cells using graphite.

In addition, it is observed that the hard carbon is very stable during charge/discharge cycles and exhibits excellent cycle performance. During charge and discharge cycles, graphite and soft carbon repeatedly expand and shrink. This repeated expanding and shrinking between layers, resulting in rapid capacity fade during cycling, damages the graphite structure. Furthermore, this expansion causes deformation of electrodes. In contrast, hard carbon has broader

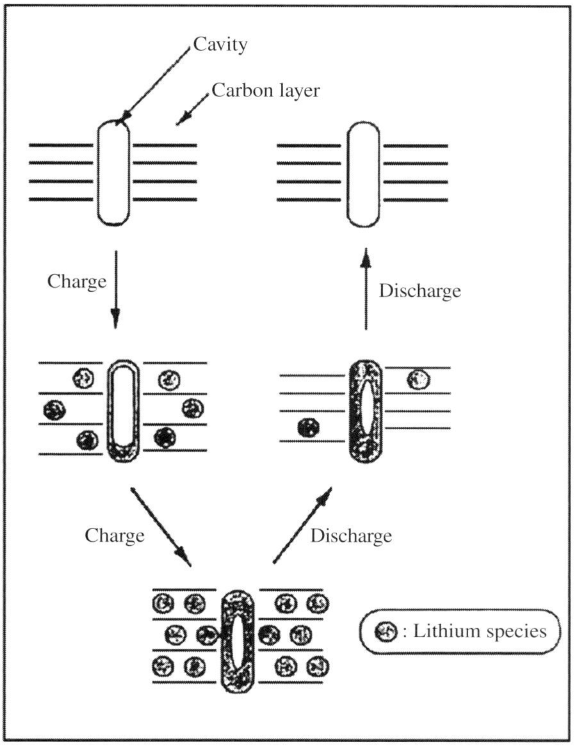

Figure 16
Schematic illustration of the charge–discharge mechanism of a carbon anode involving cavities (after Maubuchi *et al.* 1995).

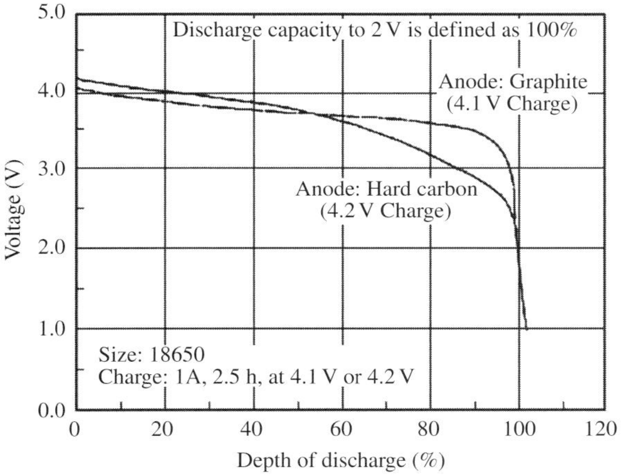

Figure 17
Discharge curve profiles of Li-Ion cells with graphite and hard carbon anodes (after Nishi 1998).

spacing between layers (over 0.372 nm vs. 0.335 nm for graphite) and suffers less damage during the cycling. Hard carbon therefore becomes a powerful candidate as a practical anode material.

3.3 Surface Reactions of a Carbon Anode

During the first electrochemical intercalation of lithium into the carbon, some lithium is irreversibly consumed and a significant amount of capacity cannot be recovered in the following discharge. This irreversible capacity, which depends on the electrolyte solution and the type of carbon material, can be explained on the basis of the reduction of electrolyte solution and the formation of a passivation film at the Li_xC interface.

The anode/electrolyte interphase, which is named the "solid-electrolyte interphase" (SEI) (Winter and Besenhard 1999), plays a key role in Li-Ion batteries. It determines the safety, power capability, lower-temperature performance, faradaic efficiency on charge, the cycle life, and the irreversible capacity loss in the first charge cycle. The irreversible decomposition of organic solvents is one of the major problems regarding the carbon anode, which results in irreversible losses. A passivation surface layer is always formed in any electrolyte system and the characteristics of the formed film greatly affect electrode behavior. Prior to lithium intercalation, charge current was consumed forming a surface passivation film. In the case of PC electrolyte, the film produced by cathodic decomposition of PC appears to be quite porous and can not cover the surface well, therefore solvated lithium ions easily intercalate from the uncovered part at a potential positive to lithium deposition ($\sim$0.8 V vs. Li/Li^+). This causes exfoliation of graphite sheets and damages the reversible behavior of the anode. If the electrolyte consists of EC and EC-based solvents, the film is formed uniformly over the surface, completely screening the graphite from the electrolyte; thus further decomposition is prevented and the capacity loss is limited to a negligible level. In addition, this surface film is a good lithium conductor and lithium ions can migrate through it without solvent molecules. It can be seen that the stability of the solvents determines the electrochemical behavior of a graphite anode. The long-term stability of the anode clearly depends on the property of the protective surface film.

Many efforts have been made to improve anode performance of carbon materials. It is important to reduce the irreversible decomposition of electrolyte and enhance the reversible capacity. Texture control can, to some degree, change the electrical conductivity of the electrode, the active surface area, the porosity and the structure of the interface between electrode and electrolyte, and therefore improve the electrochemical performance of the carbon anode (Peled 1979). The effects of doping with phosphorus, boron, silicon, and nitrogen have been studied and it was found that such doping may improve capacity, coulombic efficiency, and cycle life (Imanish *et al.* 1993). For operation under heavier loads, further improvement of the carbon anodes is essential. There is strong interest in developing new anode materials to replace the carbon for better performance. A lithium cell with the trademark Stalion announced by Fujifilm Celltec Co. has attracted significant interest (Weydanz *et al.* 1994), because it provides a higher energy density, as shown in Fig. 18, a promising cycle life, and a higher power density. The improvement is achieved by replacement of the carbon anode with an amorphous tin-based composite oxide (TCO or ATCO).

The specific energy capacity of the Li-Ion systems depends largely on the type of carbon materials, the lithium intercalation efficiency, and the irreversible capacity associated with the first charge process. It has been found that coke-type carbon, having physical properties such as ash content $<0.1\%$, surface area $<10\,m^2g^{-1}$, true density $<2.15\,gcm^{-3}$, and interlayer spacing $>3.45\,A$, is suitable for Li-Ion systems. These types of carbon materials can provide about $185\,mAhg^{-1}$ capacity ($\sim Li_{0.5}C_6$). By controlling the temperature of the heat treatment, carbon materials having specific properties such as density and interlayer spacing can be prepared.

3.4 Cathode Active Materials—Lithiated Oxides

There is a wide range of choice for materials that can be selected for the positive electrode of Li-Ion rechargeable batteries. However, a number of factors have to be considered in choosing the intercalation compound, such as reversibility of the intercalation reaction, electronic conductivity, free energy of reaction with lithium, cell voltage window, variation of the voltage with the state of charge, and availability and cost of the compound.

Compared with lithium metal batteries, the cathode materials suitable for Li-Ion batteries are highly oxidizing compounds which may compensate for the loss in cell potential at the anode (Pistoia 1994). The best cathodes for Li-Ion rechargeable batteries are those where there is little bonding and structural modification of the active materials during the charge–discharge reaction. Candidate cathode materials include layered compounds with the general formula $LiMO_2$ (Thackery 1999), such as $LiCoO_2$, $LiNiO_2$, and $LiNi_{1-x}Co_xO_2$. In the ideal layered $LiMO_2$ structure, the Li^+ and the M^{3+} ions occupy octahedral sites in alternate layers between cubic close-packed oxygen layers as shown in Fig. 19. The layered metal oxide framework provides a two-dimensional interstitial space, which allows easy removal of the lithium ions. The intercalation and deintercalation of the lithium ion for the layered structure compounds are represented by the following equation:

$$LiMO_2 \leftrightarrow Li_{1-x}MO_2 + xLi^+ + xe^-$$

When the $Li/LiMO_2$ cell is cycled over the limited composition range $0<x<0.5$ in $Li_{1-x}MO_2$, rechargeability and charge retention are good. However, the rechargeable capacity fades rapidly for deep charge and discharge cycles, i.e., for $x>0.5$.

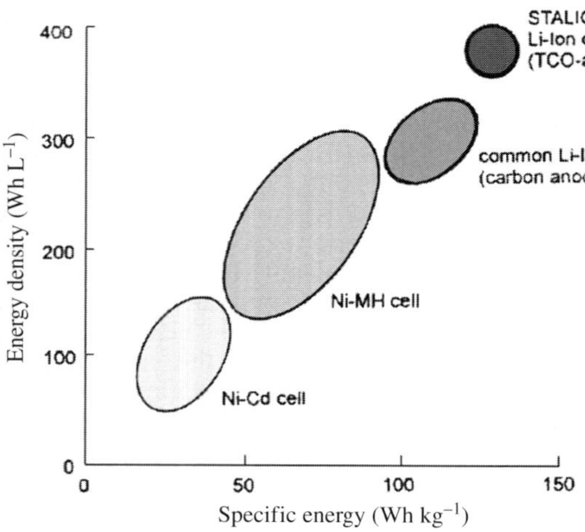

Figure 18
Energy density comparison for NiCd, NiMH, and Li-Ion rechargeable batteries. TCO: amorphous tin-based composed oxide (after Weydanz *et al.* 1994).

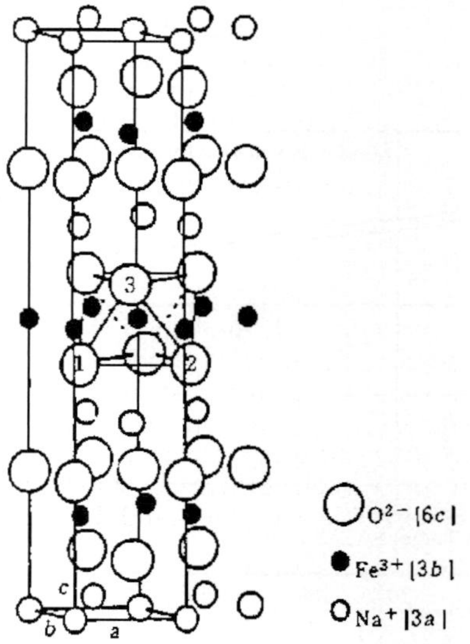

Figure 19
The structure of layered α-$NaFeO_2$, prototype of $LiCoO_2$, and $LiNiO_2$ (after Thackery 1999).

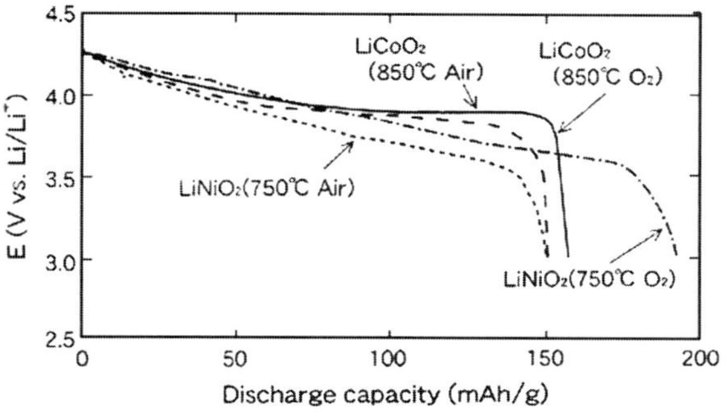

Figure 20
Discharge curve profiles of Li-Ion cells with layered $LiMO_2$ cathodes.

Discharge curve profiles of lithium ion cells with layered $LiMO_2$ cathodes are displayed in Fig. 20.

The compound $LiCoO_2$ is an attractive positive electrode because it has a stable structure which is easy to prepare with the ideal layered configuration. Therefore, practically, the maximum capacity of $LiCoO_2$ is about 125 mAh g^{-1} based on weight of $LiCoO_2$. $LiNiO_2$ and $LiNi_{1-x}Co_xO_2$ have rechargeable capacities more than 150 mAhg^{-1}. It is difficult to prepare large and reproducible $LiNiO_2$ batches with the ideal layered structure. It was also observed that extensively delithiated (charged) $Li_{1-x}NiO_2$ electrodes are less stable than $Li_{1-x}CoO_2$ electrodes and its thermal behavior is a threat to safety (Dahn *et al.* 1991). The drawbacks of the $LiNiO_2$ electrodes can be partly overcome by using cobalt-substituted $LiNi_{1-x}Co_xO_2$ or aluminum-substituted $LiAl_{0.25}Ni_{0.75}O_2$. A very promising cathode material is the three-dimensional compound which is represented by $LiMn_2O_4$ (Thackery 1999). Figure 21 displays the spinel framework structure. The intercalation and deintercalation of the lithium ion for the spinel structure compounds is expressed as follows:

$$LiMn_2O_4 \leftrightarrow Li_{1-x}Mn_2O_4 + xLi^+ + x\bar{e}$$

The reversible value of x is as high as 0.85 and its reversible capacity is 135 mAhg^{-1}. The manganese-based materials offer the following advantages: better structural stability, lower cost resulting from the natural abundance of manganese, and lower toxicity. For practical applications, considering the high cost of cobalt or nickel, $LiMn_2O_4$ spinel compounds are more desirable cathode candidates for Li-Ion batteries.

3.5 *Electrolytes*

The magnitude of the open-circuit voltage along with other performance issues (such as theoretical energy density, high rate capability, etc.) is constrained not only by the electrochemical potentials (the Fermi energy) of the anode reductant (μ_A) and the cathode oxidant (μ_C), but also by the chemistry of the electrolyte. For example, it is the energy gap E_g between the highest occupied molecular orbital and the lowest unoccupied molecular orbital of a liquid electrolyte that determines the thermodynamic stability. The selection of electrolyte for the Li-Ion batteries is critical.

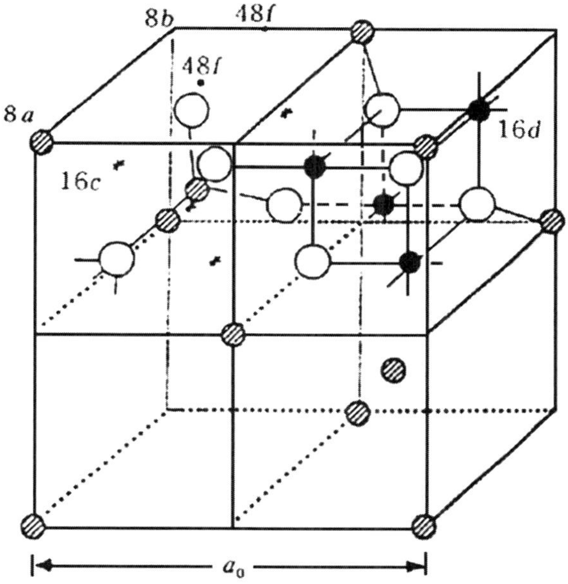

Figure 21
The spinel structure of a $LiMn_2O_4$-type compound (after Thackery 1999).

The "electrolyte" here actually consists of a solvent and a lithium salt. Nonaqueous organic electrolyte solutions are widely used in commercial Li-Ion batteries. The electrolyte solution must possess a wide potential window (0–5 V), low vapor pressure, good ionic conductivity, and compatibility with other cell components. The solvents should be aprotic so as to be stable at fairly low potentials, and so that they do not react with metallic lithium, but should have high polarity in order to dissolve lithium salts, giving high ionic conductivity. Low values of both melting point and boiling point of the solvent are preferred, since they directly relate to the operating temperatures of the battery system and reflect the physical chemistry of the solvent such as the molecular structure and intermolecular forces. The relative permittivity and viscosity of the solvent are the most important properties that determine the ionic conductivity of the electrolyte solution (Li and Currie 1997). Lithium salts with monovalent anions are preferred for the battery electrolyte because of the higher extent of dissociation of the salt and higher mobility of the resulting ions.

The most popular aprotic organic electrolyte solutions are made by dissolving very ionic lithium salts such as $LiClO_4$, $LiBF_4$, $LiAsF_6$, or $LiPF_6$ in empirically optimized mixtures of propylene carbonate (PC), ethylene carbonate (EC), or dimethyl carbonate (DMC). Voltages approaching 4.5 V have been sustained in an electrolyte of $LiPF_6$ in a 1:2 DMC:EC mixture. However, these organic liquid electrolytes generally have conductivities of $\sim 10^{-2}\,S\,cm^{-1}$ at most, about two orders of magnitude lower than aqueous electrolytes ($8\times 10^{-1}\,S\,cm^{-1}$ of H_2SO_4 for lead-acid batteries, $5\times 10^{-1}\,S\,cm^{-1}$ of KOH for alkaline systems). Therefore, many efforts have been attempted to improve the ionic conductivity and other performance of the organic electrolyte solutions.

Polymer electrolyte is an alternative to the liquid electrolyte, and is formed by incorporating lithium salts into polymer matrices, followed by a casting procedure to obtain a thin film. By proper control of the conditions of the synthesis, it is possible to make membranes with a thickness between 20 μm and 100 μm. A thinner polymer film (5 μm) with acceptable conductivity, adequate mechanical strength, and electrochemical stability, is an ideal material as an electrolyte. The polymer membranes can be used as both electrolyte and separator. These electrolytes are less reactive with lithium and should enhance the battery's safety, eliminate the possibility of electrolyte leakage, and provide good adhesion to the electrode, thus ensuring good interfacial contact. Batteries with polymer electrolytes can be placed in metallized plastic bags allowing the construction of batteries with customized shapes. However, due to the slow ionic transport within the polymer structure, these solid electrolytes have ionic conductivities of below $10^{-4}\,S\,cm^{-1}$ at 20 °C, much lower than the liquid one, and, practically, the battery requires higher temperatures in order to obtain enough conductivity for acceptable performance.

Another class of polymer electrolytes called "gel" electrolytes are developed by the addition of a polymer or fumed silica into the liquid electrolyte or by cross linking of a dissolved monomer (Abraham *et al.* 1995). Conductivities as high as $10^{-3}\,S\,cm^{-1}$ at 20 °C were achieved. However, their chemical stability and reactivity with lithium should be improved for practical applications.

3.6 Separators for Li-Ion Batteries

All the requirements for separators used in alkaline storage batteries, such as mechanical strength and chemical stability, should be provided for Li-Ion batteries. The separator should prevent migration of particles of ~10 μm between electrodes, so the effective pore size should be less than 1 μm. During winding, small pieces of electrode material may come off and be forced into the separator by winding tension. The separator must not be punctured; otherwise the battery will be shorted. Generally, a Li-Ion battery is used at a C rate, and it may require a current density of $0.5–3\,mA\,cm^{-2}$ depending on cell design. For high-rate applications, the current density will be much higher. The electrical resistance of the separator should not limit battery performance under any circumstances. It is also regulated (Underwriters Laboratories 1993) that the battery be able to withstand a short circuit without fire or explosion. A positive temperature coefficient (PTC) device (called a Polyswitch) has been used for external short-circuit protection. The PTC device is placed in series inside the cell. Its resistance increases by orders of magnitude at high currents and resulting high temperatures, and suddenly jumps to infinite value at a certain temperature level (so-called trip temperature; generally ~100°C). As soon as the temperature reaches the trip value at external shortage, the short-circuit current is stopped due to the infinite electrical resistance, preventing thermal runaway of the cell. However, in the case of an internal short, e.g., if the positive tab comes loose and connects to the interior of the negative metal can, the PTC will not respond quickly enough. During overcharging, the PTC may not be activated at low current. In these cases, it is an advantage that a separator provide safety functions and prevent thermal runaway by fusing itself. In fact, the separator could act as a fuse. That is, the impedance of the separator increases by two or three orders of magnitude due to an increase in cell temperature, and melts down so that it closes its pores and effectively stops the current flow between the electrodes.

It is found (Yu *et al.* 1994) that a trilayer structure of PP/PE/PP Celgard microporous membranes offers unique characteristics for Li-Ion batteries. In addition to its exceptional puncture strength, the low-melting PE layer (135 °C) can act as a thermal fuse, while the higher-melting PP (165 °C) layers provide physical integrity. The fuse layer melts and loses porosity at elevated temperatures; the other layers do not melt and continue to provide mechanical integrity after the fuse layer has melted.

3.7 Applications

Performance characteristics of NiCd, NiMH, and Li-Ion systems are summarized in Table 2 as well as in Fig. 18. Li-Ion batteries offer advantages in high energy density, high voltage, high power capability

Table 2
Comparison of NiCd, NiMH, and Li-Ion battery systems.

	Energy density		Power			
System	$Whkg^{-1}$	WhL^{-1}	Wkg^{-1}	WL^{-1}	Cycle life[b]	Self-discharge
Ni-Cd	50	90	250	600	>1200	30[c]
NiMH[a]	65	160	200	475	>1200	30[c]
Li-Ion[a]	110	165	750	1600	>600	10[d]

[a]Estimated from Anderman et al.(2000); [b] At room temperature 100% DOD to 80% of initial capacity; [c] At room temperature for 30 days; [d] At room temperature for 90 days.

and good charge retention, and low self-discharge rate. As an advanced rechargeable battery system, Li-Ion cells can be designed in any of the typical sealed-cell constructions: coin, spirally wound cylindrical, or prismatic configurations, and have a wide range of applications. Most of the developments have concentrated on the smaller cells for portable applications. Particularly for cellular phones and laptop computers, Li-Ion batteries become predominant power sources. Large-size cells have also been aggressively developed for applications requiring high energy density and high power capability, such as electrical vehicles (EV), and hybrid electrical vehicles (HEV).

Compared with NiCd and NiMH systems, a Li-Ion cell displays a sloping discharge curve profile. In addition, the discharge rate of a Li-Ion cell cannot be as high as that of NiCd and NiMH systems because, at high-rate discharge, diffusion steps become critical in the intercalation processes. Figure 22 displays the diffusion coefficient (Guyomont and Tarascon 1992) for the petroleum coke (Li_xC_6), showing that the value of the diffusion coefficient of the lithium ion in the carbon anode varies with the lithiation content at ambient temperature, which is responsible for the sloppy discharge curves. While in graphite, its diffusion coefficient is about 10^{-11} cm^2s^{-1} at ambient temperature, it is about two orders of magnitude lower than the petroleum coke. The graphite anode cells usually have a higher capacity at low discharge rates than the hard carbon cells, but they may lose this advantage at higher current drains. The diffusion coefficients of lithium ion (D_{Li+}) in the lithiated metal oxides are in a range from 2×10^{-7} cm^2s^{-1} to 10^{-11} cm^2s^{-1} at room temperature, depending on the nature of the oxide, and also the measuring techniques. The diffusion steps play a significant role in the mechanistic aspects of the intercalation process. On the other hand, aprotic organic electrolytes must be used because of the reactivity of lithium in aqueous electrolytes, but their conductivities are poor. Therefore, these intrinsic properties may limit the high-rate capability of Li-Ion rechargeable batteries.

In high-power applications, the relative low-rate capability during continuous discharge is a drawback. However, in addition to the intrinsic structural and chemical features of the active materials, the battery performance is also strongly affected by engineering/processing factors. It is desirable that the technical conception of the electrodes should provide maximum utilization of the reversible capacity and the high-power performance, which means the lithium intercalation/deintercalation reaction (charge/discharge reaction) allows high current densities. Thin-layer electrodes that are made from small particle-size graphite may achieve the high-rate capability of a carbon electrode. Furthermore, as the rate-limiting process is diffusion polarization, the maximum current can be very high on pulse discharges as well as on pulse charge. Therefore, the Li-Ion battery can be capable of handling the peak power requirements encountered in high-power applications such as HEV.

For high-energy and high-power applications, large-size Li-Ion cells are demanded. It has to be noted that the processing control becomes more important. A significant effect of inhomogeneous current density distribution, caused by ohmic drops along the electrode, difference in the morphology of

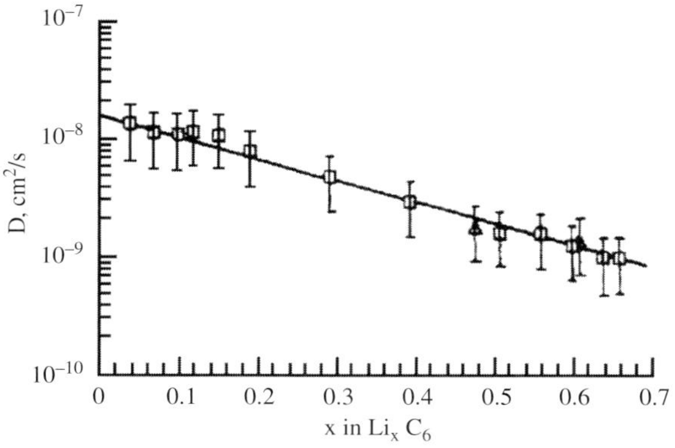

Figure 22
Diffusion coefficient of lithium ion in petroleum coke carbon varying with lithium content (after Guyomont and Tarascon 1992).

the active materials, variations of separator properties, or the distance between the electrodes should be addressed and well controlled. Variation in current density is paralleled by variation in electrolyte concentration, which affects the potential of lithium intercalation or deposition.

For Li-Ion batteries, overdischarging should be avoided as it may cause a performance problem, due to internal short-circuiting resulting from the anode current collector. It is also necessary to control the charging process. The charging voltage is a key parameter and should be limited during charging. For example, for $LiCoO_2$ cells, it is about 4.2 V; for $LiNiO_2$ cells, it is about 4.0 V. Higher charge voltages may cause decomposition of the electrolyte at the positive electrode with a rise in the internal pressure and the plating of lithium on the anode surface.

Li-Ion batteries have relatively shorter cycle life compared to NiCd and NiMH batteries. This may be attributed to structure damage resulted from lattice expansion and shrinkage as well as to the continuous lithium deposition on the anode during repetitive charge/discharge cycling. Similarly to all other rechargeable battery systems, limiting the charge/discharge range during operation to a small window of its entire capacity will enhance the Li-Ion cell durability. Nissan commercialized HEV (Tino) was equipped with Li-Ion batteries in 2000, which indicates that Li-Ion battery technology is a promising choice for HEV applications.

Proper design of the battery pack is important to assure optimum, reliable, and safe operation. It should be noted that the performance of a cell in a battery could be significantly different from that of an individual cell. The cells cannot be manufactured identically. When the cells encounter a somewhat different environment in the battery pack, the behavior of each cell may be different. During cycling, the cells in the pack can become imbalanced and have different voltages. This could result in poor performance or safety problems. Good battery technology provides opportunities to make the best use of cells.

Among the battery technologies, an important requirement is thermal control and management. The heat generated during charge and discharge, especially under high-rate operation, must be dissipated effectively to maintain the battery within a safe temperature range. To achieve this management and control, many control devices are installed in the battery pack to monitor the temperature, such as thermistor (for ΔT, $\Delta T/\Delta t$ control); thermostat (for TCO control); thermal fuse (for protection against thermal runaway); and positive temperature coefficient (PTC, for current as well as temperature control) device.

Recent advances in battery energy control have incorporated the use of microprocessor-based controllers within the battery to manage both the charge and discharge. Chips are designed for monitoring and controlling the battery and are being incorporated into the battery pack, creating the so-called "smart battery," into the battery-using equipment, or into the battery charger. Some of the features include:

(i) A capacity indicator, commonly known as a "fuel gauge," to estimate the remaining battery capacity by factoring in such variables as the discharge rate and time, temperature, self-discharge, charge rate and battery history.
(ii) Charge control. The microprocessor can monitor the battery during charge by controlling the voltage, charge rate, and other termination parameters, such as t, $\Delta T/\Delta t$, $\Delta V/\Delta t$, to cut off the charge or switch to a lower charge rate; or switch from one charge method to another. The constant-current method, constant-voltage method, and pulse charging can all be controlled by the microchips.
(iii) Discharge control, which is provided to control rate, voltage, cell equalization, and temperature management.

Furthermore, a foundation of battery normalization, modeling and control techniques (Usuda and Kayano 1997, Wiegman and Vandenput 1998) has been developed and presented for high-energy, high-power EV, and for high-power and charge-sustaining HEV applications. The system controls the discharge current and power output based on battery state of charge (SOC) and vehicle demand, manages cell balancing, and regulates the regeneration charge process, as well as thermal management. The sophisticated battery modeling technology maintains energy balance by both monitoring and regulating the SOC in predetermined limits (e.g., 0.4–0.8), in order to make the best use of the battery in terms of availability of different functions, battery life, charge and discharge efficiency, etc.

4. *Conclusion*

The current status of rechargeable batteries, including the NiCd system, the NiMH system, and the Li-Ion system, has been reviewed. The market demand is obviously a strong driving force for the research and development of advanced energy storage technology. Scientific breakthroughs are always the pioneer to create new materials, new technology, and new rechargeable battery products.

Since 1990, the interest in electrochemical technology and rechargeable batteries, in particular heightened by the recent emphasis on battery development for portable electronics, energy storage, and electric vehicles, has resulted in significant improvements in the NiCd system, and created new technology such as NiMH and Li-Ion rechargeable batteries. These three systems are currently the major players for mobile electrical energy resources. The development of rechargeable batteries with higher energy density, higher power capability, better reliability, and lower cost to meet the market needs remains a persistent challenge.

For many practical applications, in particular for telecommunications, electric vehicles, and hybrid electric vehicles, the technology of battery management, including thermal management and energy management, has attracted great efforts and obtained great successes, both in electronic controllers and in battery modeling.

Electrochemical technology and rechargeable batteries are areas with very promising futures and technical challenges, and provide a wide range of opportunities for materials science, electrochemistry, and engineering.

Bibliography

Abraham K M, Alamgir M, Hoffman D K 1995 Polymer electrolytes reinforced by Celgard membranes. *J. Electrochem. Soc.* **142**, 683–6

Anderman M, Kalhammer F, MacArthur D 2000 *Advanced batteries for electric vehicles: an assessment of performance, cost, and availability*. Submitted for publication

Bohnstedt W 1999 In: Besenhard J O (ed.) *Handbook of Battery Materials*. Wiley-VCH, New York, pp. 245–92

Dahn J R, von Sacken U, Jukow M R, Ai-Janaby H 1991 Rechargeable $LiNiO_2$ /carbon cells. *J. Electrochem. Soc.* **138**, 2207–11

Fetcenko M A, Venkatesan S, Dhar S 1990 Ni-MH batteries for practical applications. *Progr. Batteries Solar Cells* **9**, 2532–2258

Gates Energy Products 1989 *Sealed Rechargeable Batteries: Application Manual*. Gates Energy Products, Inc

Guyomont D, Tarascon J M 1992 Lithium metal-free rechargeable $LiMn_2O_4$/carbon cells: their understanding and optimization. *J. Electrochem. Soc.* **139**, 937–41

Imanish N, Kashiwagi H, Ichikawa T, Takeda Y, Yamamoto O, Inagaki M 1993 Charge–discharge characteristics of mesophase-pitch-based carbon fibers for lithium cells. *J. Electrochem. Soc.* **140**, 315–20

Ishikawa M, Sonobe N, Chuman H, Iwasaki T 1994 In: 35th *Battery Symp*. Japan, Nagoya, Ext. Abstr. 9.49

Kinoshita K, Okawa T, Takano T, Hiroshima T 1996 US Patent #5527638

Lazzari M, Scrosati B 1980 A cyclable lithium organic electrolyte cell based on two intercalation electrodes. *J. Electrochem. Soc.* **127**, 773

Li W, Currie J C 1997 Morphology effects on the electrochemical performance of $LiNi_{1-x}Co_xO_2$. *J. Electrochem. Soc.* **144**, 2773–9

Li Z P, Higuchi E, Liu B H, Suda S 1999 Electrochemical properties and characteristics of a fluorinated AB2-alloy. *J. Alloys Compds.* **293–5**, 593–600

Maubuchi A, Tokumitsu K, Fujimoto H, Kasuh K 1995 Charge–discharge characteristics of mesocarbon microbeads heat treated at different temperatures. *J. Electrochem. Soc.* **142**, 1041–6

McBreen J 1999 In: Besenhard J O (ed.) *Handbook of Battery Materials*. Wiley-VCH, New York

Murphy D W, Caride J N 1979 Low-voltage behavior of lithium/metal dichalogenide topochemical cells. *J. Electrochem. Soc.* **126**, 349

Nishi Y 1998 Performance of the first lithium ion battery and its process technology. In: Wikihara W, Yamamoto O (eds.) *Lithium Ion Batteries*. Wiley-VCH, New York, pp. 181–98

Notten P H L 1995 Rechargeable nickel–metal hydride batteries: a successful new concept. In: Grandjean F (ed.) *Interstitial Intermetallic Alloys*. pp. 151–95

Ohta K, Ikoma M, Hamada S, Matsuka H, Morishita N 1994 Nickel hydroxide: Improvement of charge efficiency at high temperature. In: Bennett P D, Sakai T (eds.) *Proc. Symp. Hydrogen and Metal Hydride Batteries*. The Electrochemical Society, Pennington, NJ, pp. 296–302

Peled E 1979 The electrochemical behavior of alkali and alkaline earth metals in nonaqueous battery systems—the solid electrolyte interface model. *J. Electrochem. Soc.* **126**, 2047

Pierson H O (ed.) 1993 *Handbook of Carbon, Graphite, Diamond and Fullerenes*. Noyes, Park Ridge, NJ, p. 43

Pistoia G (ed.) 1994 *Lithium Batteries, New Materials, Developments and Perspectives: Industrial Chemistry Library*, Vol. 5. Elsevier, Amsterdam

Sakai T 1995 Rare earth intermetallics for metal-hydride batteries. In: Gschneider K A (ed.) *Handbook on the Physics and Chemistry of Rare Earths*, Vol. 21. pp. 133–78

Scrosati B and Megahed S (eds.) 1993 *Electrochemical Society Short Course, New Orleans*.

Thackery M M 1999 Materials for alkali metal batteries. In: Besenhard J O (ed.) *Handbook of Battery Materials*. Wiley-VCH, New York, p. 293

Tran T D, Feikert J H, Song X, Kinoshita K 1995 Commercial carbonaceous materials as lithium intercalation anodes. *J. Electrochem. Soc.* **142**, 3297–302

Underwriters Laboratories 1993 Outline of investigation for household and commercial batteries. *Subject 2054*

Usuda M, Kayano M 1997 Process for detecting remaining capacity of battery. US patent No. 5658682

van Rijswick M H J 1978 Metal hydride electrodes for electrochemical energy storage. In: Andresen A F, Maeland A J (eds.) *Hydrides for Energy Storage*. Pergamon, Oxford, pp. 261–71

Weydanz W J, Way B M, van Buuren T, Dahn J R 1994 Behavior of nitrogen-substituted carbon (N_zC_{1-z}) in Li/$Li(N_zC_{1-z})_6$ cells. *J. Electrochem. Soc.* **141**, 900–7

Wiegman H L, Vandenput A J A 1998 *Battery state control techniques for charge-sustaining applications. SAE Technical Paper Series* No 981129

Willems J J G 1984 Metal hydride electrodes stability of $LaNi_5$-related compounds. *Philips J. Res.* **39**, (Suppl. No. 1)

Winter M, Besenhard J O 1999 Lithiated carbon. In: Besenhard J O (ed.) *Handbook of Battery Materials*. Wiley-VCH, New York, p. 406

Xue J S, Dahn J R 1995 Dramatic effect of oxidation on lithium insertion in carbon made from epoxy resins. *J. Electrochem. Soc.* **142**, 3668

Yu W C, Callahan R W, Dwiggins C F, Fisher H M, Geiger M W, Schell W J 1994 In: *North America Membrane Society Conference*. Breckenridge, CO

Zhang L Y 1997 Rare earth-based metal hydrides and NiMH rechargeable batteries. In: Bautista R G, Bounds C O, Ellis T W, Kilbourn B T (eds.) *Rare Earths: Science, Technology and Applications III*. pp. 235–57

Battery Carbons

S. Flandrois

From the very beginning of battery manufacturing in the second half of the nineteenth century, carbon materials have proved to be very useful as constituents of many systems because of their high electrical conductivity and their relative electrochemical stability. Although thermodynamics indicates a restricted domain of carbon stability in potential–pH diagrams, oxidation and reduction reactions do not generally occur under normal conditions of temperature and pressure. Thus, carbon rods have been used, and are still used, as current collectors in Leclanché Zn/MnO_2 cells. As oxides and hydroxides, used as the cathode of many types of battery, are not good electronic conductors, they have to be mixed with a conducting powder.

For this purpose, graphite and carbon blacks have found universal application. For example, nickel hydroxide (in some alkaline secondary batteries) and mercuric oxide (in Mallory primary batteries) are mixed with 10–20% by weight of graphite flakes. In manganese dioxide cathodes of Leclanché cells, certain carbon blacks, such as acetylene black, are preferred, because their chain-like structure (the so-called "primary structure") gives them a high absorptive power for electrolyte, in addition to conductivity enhancement. In zinc/air primary batteries, derived from the zinc/manganese dioxide cell, the central MnO_2/carbon black paste is replaced by a gas diffusion electrode in the form of a cylinder of porous activated carbon connected to an external air supply through its top surface. Porous carbons or carbon felts are also used as current collectors in Li/$SOCl_2$ primary batteries or as supports for noble metal catalysts in fuel cells.

These few examples show that the field of carbon application to batteries is broad. In addition, it has expanded tremendously with the development of the so-called lithium ion batteries which will constitute the main topic of this article. Lithium metal combines the lowest standard potential ($-3\cdot045$ V vs. NHE) with a low atomic weight ($M=6\cdot94$) to give an exceptionally high theoretical specific capacity of $3860\,\mathrm{m\,Ah\,g^{-1}}$. In combination with a cathode material, generally a transition metal oxide, and using a suitable nonaqueous electrolyte, battery voltages of about 4 V are obtained. Primary lithium batteries have been fabricated on this basis and commercialized since the end of the 1960s. However, the development of rechargeable lithium batteries was prevented because of the problem of dendrite formation during charging, resulting in poor cycling efficiency and cell shorting. The problem was solved in 1990 when Sony announced the commercialization of lithium ion batteries, where lithium metal is replaced by a carbon host structure that can reversibly absorb and release lithium ions at low electrochemical potentials.

The smart solution of electrode morphology stabilization by using a host structure makes these batteries belong to the so-called "rocking-chair" battery family, where ions spontaneously exchange from one intercalation structure to another. In the case of carbon-based lithium ion batteries, the lithiated carbon is a powerful reducing agent (negative electrode) whereas a metal oxide constitutes the positive electrode. This concept was proposed as early as 1980 by Armand (1980). At that time, carbons were known to intercalate lithium by a chemical route, the maximum amount being achieved with graphite (LiC_6), but early attempts to use graphite as a negative host structure in lithium ion batteries failed because of strong reactivity with the electrolyte. Other less crystalline materials (heat-treated cokes) were found to be less sensitive to the electrolyte, but presented still modest performances. Then, highly disordered materials (non-graphitizing carbons and low-temperature carbons) were shown to exhibit different electrochemical behavior due to their differences in structure, texture, and heteroatom content. Finally, crystalline materials regained interest when it was demonstrated that they could be used in electrolytes with suitable solvent composition, mainly based on ethylene carbonate (EC). This article outlines the main achievements toward the comprehension and the improvement of carbon performances for lithium ion reversible absorption.

1. *Main Properties of Carbon Electrodes*

1.1 *Specific Energy*

The replacement of metallic lithium by a lithiated carbon results in a decrease in energy density because of the presence of the host structure. The basic requirement is the highest possible amount of lithium ions reversibly intercalated into the carbon matrix at a potential close to that of metallic lithium. This property depends on the carbon host structure. Owing to the importance of this subject for this article, it will be presented in Sects. 2–5 with separate treatment of the main carbon families.

1.2 *Reversibility*

All carbon materials experience irreversible side reactions during the first electrochemical absorption of lithium ions. These reactions consume lithium and have to be minimized, since they lead to a capacity loss (the so-called "irreversible capacity"). A well-known source of irreversibility is electrolyte instability at low potential, which yields insoluble products. No electrolyte has been found that withstands the low electrochemical potential of metallic lithium or lithiated carbons.

As the potential of the negative electrode lowers, the electrolyte is reduced until the formation of an electronically insulating (passivating) layer on the carbon particles, often called SEI (solid electrolyte interphase), allows further lithium exchange. This process is clearly visible on the galvanostatic curves as an irreversible plateau at a potential close to 0·8V vs. Li^+/Li (Fig. 1). A major contribution to the knowledge of the layer composition and electrolyte reduction mechanism came from the work of Aurbach and co-workers (Aurbach *et al.* 1997). They showed that cyclic carbonates like propylene carbonate (PC) or ethylene carbonate (EC) are the best suited solvents, as they yield double lithium alkyl carbonates with a good passivating ability.

A direct way to minimize the irreversible capacity associated with the formation of the passivating layer is to decrease the exchange surface area. As the use of carbon powders is mandatory to increase the kinetics of lithium exchange to an acceptable level, a compromise solution is a specific surface area of a few square meters per gram with particle sizes of a few micrometers. For this purpose, materials with a special particle morphology, such as carbon spheres grown from mesophase or carbon fibers, can be attractive. However, as shown in Sects. 2–5, other sources of irreversible losses in the first cycle can exist, such as exfoliation or lithium reaction with heteroatoms and structural defects. Obviously, they depend on the structure, texture, and composition of the carbon.

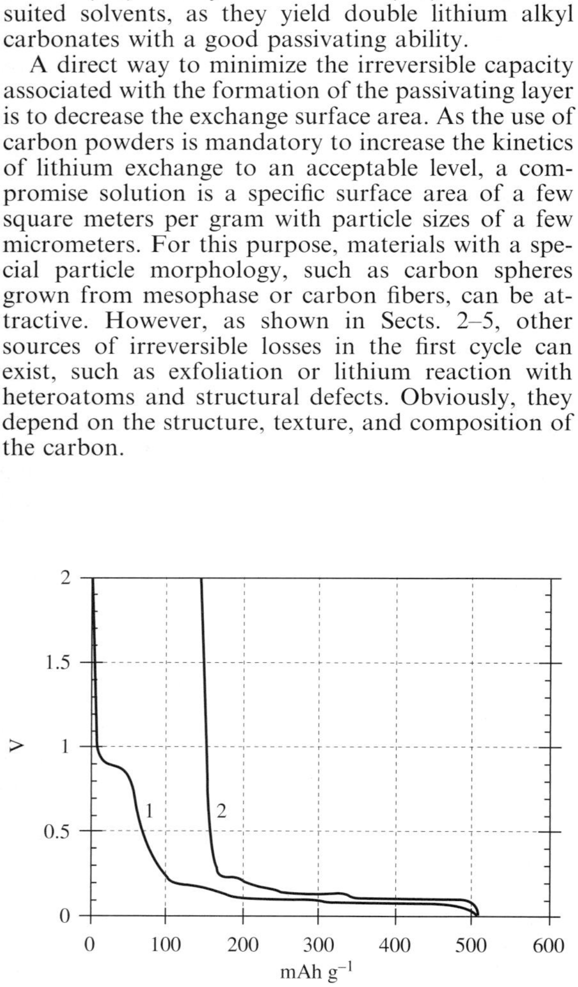

Figure 1
First discharge (1) and charge (2) curves of a lithium/natural graphite cell. The plateau at 0·8–0·9 V on the discharge curve is characteristic of passivating layer formation. The other plateaus at potentials below 0·3 V are signatures of transitions between different stages (see Sect. 2·2).

2. *Lithium Intercalation into Graphites*

2.1 *Irreversible Capacity and Exfoliation*

Highly crystalline carbons often exhibit a large irreversible capacity regardless of their surface area. This occurs especially with PC-containing electrolytes and results from exfoliation of graphene layers. In this case, before the complete formation of the passivating layer, solvated lithium ions are intercalated. Further reduction causes the decomposition of the solvent molecules between the graphene layers, with production of gaseous compounds like propylene which exfoliate the graphite matrix. The newly created surface area then needs extra passivation, resulting in a dramatic increase in the irreversible capacity (from a few tens up to 1000 m Ah g^{-1} of carbon in some cases).

As crystalline materials are highly desirable to achieve high energy densities, improvement of their electrochemical behavior has been the subject of many studies. It has been shown (Ohzuku *et al.* 1993, Besenhard *et al.* 1995) that solvent co-intercalation, and thus exfoliation, does not generally occur in EC based electrolytes. The difference in behavior of such close molecules as EC and PC is not fully understood as yet. However, the beneficial effect of EC against exfoliation is not observed for all graphites.

A better way to prevent exfoliation is the use of graphite grades with a high percentage of rhombohedral phase. The most common structure of graphite is the hexagonal Bernal structure, where the carbon layers are arranged in the …ABAB… sequence with a shift of the B layers with respect to the A layers. Another crystalline form of graphite, thermodynamically metastable, can exist with a rhombohedral structure of …ABCABC… stacking sequence. This form is never pure. It is always mixed with the hexagonal form in variable amounts, which can be increased up to 30–40% of rhombohedral phase by a mechanical treatment such as grinding. Although graphites with different rhombohedral phase content are indistinguishable in terms of reversible lithium capacity, it is not the same for the irreversible capacity, as shown by Simon and co-workers (Simon *et al.* 1998). When the rhombohedral phase content is higher than 30% graphite exfoliation does not occur, even with electrolytes containing a high percentage of PC, a solvent otherwise interesting for its physical properties. From Raman spectroscopy experiments it has been shown that structure defects, localized at grain boundaries between rhombohedral and hexagonal domains, hinder the layer opening necessary for the intercalation of bulky solvated lithium ions.

2.2 *Reversible Capacity and Phase Diagram*

In the conditions where unsolvated lithium ions are electrochemically intercalated into graphite, and once the passivating layer has been formed, reversible intercalation of lithium takes place. The theoretical capacity of 372 m Ah g^{-1} (LiC_6) is generally approached if the charge–discharge regime is slow enough (rate typically lower than $C/20$, where C is the capacity of the cell). The potential curves exhibit several reversible plateaus in the range 0–0·25 V vs. Li^+/Li. As is well known, the existence of potential plateaus corresponds to transitions between different single phases and the end points of the plateaus give the composition range of biphasic domains. In the case of graphite intercalation compounds (GICs), the single phases are obviously compounds of defined stage. Indeed, one of the most important characteristics of GICs is the staging phenomenon, which corresponds to a periodical arrangement of intercalated layers within the graphite layer matrix. GICs are thus classified by a stage index denoting the number of graphite layers that separate two successive intercalated layers. The electrochemical lithium–carbon phase diagram has been determined using *ex situ* or *in situ* x-ray diffraction and Raman spectroscopy. Analysis of the data leads to the following conclusions. Transitions between stage 1 (LiC_6) and stage 2 (LiC_{12}), stage 2 and a dilute lattice-gas disordered stage 2 called stage 2L (LiC_{18}), stage 2L and stage 3, and stage 4 and a dilute stage 1 occur at about 0·09, 0·12, 0·14, and 0·20 V, respectively. The compositions of stage 3 and stage 4 compounds are not always well defined. In addition, the transition between these two stages seems to be continuous with a progressive variation of the cell voltage, which is not understood as yet. For low lithium content, the intercalated lithium is randomly distributed throughout the graphite host in a dilute stage 1 phase.

3. *Lithium Intercalation into Graphitizable Carbons*

Graphitizable ("soft") carbons are carbon materials whose structure evolves progressively toward the graphite structure when they are heat treated at high temperatures, up to 3000 °C. They are composed of more or less misoriented crystallites whose size and crystalline order increase with the heat-treatment temperature (HTT). At the beginning of the graphitization step (HTT ≈ 1200–1300 °C), the size of the crystallites is of the order of 5 nm, both parallel and perpendicular to the layers (L_a and L_c, respectively), with an average interplanar distance d_{002} close to 0·344 nm. The crystallite size is about 10 nm at HTT ≈ 2000 °C and reaches several tens of nanometers with higher HTTs, while d_{002} approaches the graphite value (0·3354 nm). Typical soft carbons are graphitizable cokes, ex-mesophase fibers, vapor-grown fibers, or mesocarbon microbeads (MCMB). All these materials, heat treated in the temperature domain where graphitization occurs (1300–3000 °C), exhibit common features for lithium electrochemical intercalation. First, the irreversible losses in the first cycle decrease with HTT, except at the end of the graphitization process. This is a consequence of a decrease in active surface area resulting from the structural reorganization (elimination of microporosity). As expected, the effect is more important at the beginning of the graphitization process (HTT ≈ 1300–2000 °C). In highly graphitized samples (HTT ≈ 2800 °C and above), some exfoliation occurs, as in graphite, producing an increase in the irreversible losses. Second, the reversible capacity values are lower than for graphite and exhibit a minimum for samples treated at about 2000 °C. In addition, this HTT value delimits two domains with different potential behaviors. For soft carbons heat treated above this temperature, the cell voltage exhibits the progressive appearance of plateaus characteristic of diphasic domains, whereas for carbons heat treated at 2000 °C and below, the potential varies continuously, as expected for monophasic systems.

Thus, the results point to the existence of two domains of electrochemical behavior according to whether the heat-treatment temperature is higher or lower than 2000 °C. In fact, it is well known that this value of HTT is a critical value in the graphitization process. It corresponds to the beginning of the three-dimensional ordering, which is evidenced by the appearance of modulations on the (hk) bands of the X-ray spectra. The increase of reversible capacities at HTTs above 2000 °C can therefore be understood as resulting from increased crystallinity. On the other hand, the capacity increase for lower HTTs is unexpected. Several explanations have been proposed, based on Franklin's or Mering's graphitization models (Flandrois and Simon 1999).

4. *Lithium Insertion into Low-temperature Carbons and Nongraphitizing Carbons*

The pyrolytic conversion of organic compounds to carbon residues at temperatures below 1000 °C occurs by a complex chemistry, which involves a great number of parallel and sequential reactions and depends on the nature of the starting materials. Nevertheless, after the primary step of carbonization (up to 500–600 °C), all carbonaceous materials are made up of similar elemental bricks with different relative arrangements (Oberlin 1989). The elemental unit, or basic structural unit (BSU), contains planar aromatic structures consisting of about 10–20 rings, which are piled up more or less parallel in groups of two to four. In the case of soft carbons, by increasing HTT the BSUs form distorted columns which coalesce during the graphitization step. In non-graphitizing ("hard") carbons, the presence of cross-linking groups, especially when the precursor contains oxygen, hinders the formation of columns and their coalescence. Whatever the graphitization degree they can reach, carbonaceous materials heat-treated between 500 and 1000°C exhibit common features: low crystallinity, large amount of micro- or nanoporosity, and presence of heteroatoms left from the organic precursor, essentially hydrogen, nitrogen, oxygen, and sulfur, depending on the precursor.

Whereas it is justifiable to talk about lithium "intercalation" in the case of soft carbons during the graphitization step (i.e., lithium introduction into the interspace of carbon layers), in the case of hard carbons and low-temperature carbons where the BSU size is 1–1·5 nm, the lithium storage cannot be restricted to the interspace between aromatic molecules. Clearly, the low crystallinity, the internal porosity, and the presence of heteroatoms and functional groups have a strong influence on the reversible and irreversible capacities. The electrochemical behavior of a large number of carbonaceous materials pyrolyzed at temperatures between 500 and 1100°C has been investigated. The main features are: (i) high reversible capacities, which can be more than twice the value of graphite; (ii) high irreversible capacities in the first cycle (up to several hundred $\mathrm{m\,Ah\,g^{-1}}$); and (iii) large hysteresis in the potential curves, i.e., lithium ions are inserted near 0 V and removed at about 1V. The hysteresis in the potential profile has been correlated to the residual hydrogen content (Zheng *et al.* 1996). Heat treatments above 1000°C eliminate hydrogen from the carbon, but the reversible capacity also decreases. Some hard carbons heat treated between 1000 and 1200°C can maintain reversible capacities higher than $400\,\mathrm{m\,Ah\,g^{-1}}$. In addition they develop a pronounced low potential plateau (below 0·1V) and little hysteresis. If their irreversible capacity could be reduced, as it has been done, for example, through the chemical vapor deposition (CVD) of a carbonaceous layer on the surface of a carbon prepared from sucrose (Buiel and Dahn 1998), such carbon materials would be of great interest.

5. *Lithium Intercalation into Doped Carbons*

Carbons, and especially low-temperature carbons, are not pure carbon. They usually contain from a few parts per million up to a few per cent of foreign atoms. However, it is possible to introduce foreign atoms into the carbon lattice intentionally, a process known as doping. Doping methods include: (i) co-deposition by CVD of carbon and foreign atoms (e.g., boron, nitrogen), (ii) pyrolysis of organic molecules containing foreign atoms (e.g., nitrogen, oxygen, silicon), and (iii) chemical treatment of the carbon (e.g., boron, phosphorus, oxygen, halogens). One of the main problems is the actual location of the dopant atoms in the carbon network. They can be substituted to carbon in aromatic rings, located in interstitial sites, or, most often, chemically bound to the edges of the carbonaceous macromolecules.

In most cases, the effect of doping on lithium electrochemical insertion is an increase in the reversible capacities, up to values of the order of 600–700 $\mathrm{m\,Ah\,g^{-1}}$. However, the potential profiles generally develop large hysteresis and large irreversible capacities. At the present time, unquestionable improvement in performance has been obtained in two cases: mild oxidation of graphite (Peled *et al.* 1996) and boron doping (Way and Dahn 1994).

6. *Concluding Remarks and Trends*

The use of carbons as a lithium reservoir in rechargeable batteries has led to what can now be considered as the major breakthrough in new practical battery systems in the second half of the twentieth century. A number of studies have been devoted to the search for alternatives to the carbon electrode (e.g., tin oxides, vanadates, transition metal nitrides). Although these compounds may display large reversible capacities (up to $900\,\mathrm{m\,Ah\,g^{-1}}$), they also show large irreversible capacities and the potential vs. lithium is markedly increased, which would result in a significant battery voltage drop.

Therefore, the use of carbons seems to be well established, at least in the medium term. Within this class of materials, development is still moving rapidly. The choice of a particular grade of carbon is always a compromise, based on consideration of electrode processing and costs. After the initial interest in non-crystalline carbons, the general trend among battery manufacturers has been to use crystalline carbons. Because of their good performance and low price, natural or synthetic graphites, especially those enriched in rhombohedral phase, seem to be preferred by many manufacturers. Nevertheless, the versatility of the element carbon gives hope for future improvements.

Bibliography

Armand M B 1980 In: Murphy D W, Broadhead J, Steele B C H (eds.) *Materials for Advanced Batteries.* Plenum, New York

Aurbach D, Zaban A, Ein-Eli Y, Weissman I, Chusid O, Markowsky B, Levi M, Levi E, Schechter A, Granot E 1997 Recent studies on the correlation between surface chemistry, morphology, three-dimensional structures and performance of Li and Li-C intercalation anodes in several important electrolyte systems. *J. Power Sources.* **68**, 91–8

Besenhard J O, Winter M, Yang J, Biberacher W 1995 Filming mechanism of lithium–carbon anodes in organic and inorganic electrolytes. *J. Power Sources.* **54**, 228–31

Buiel E, Dahn J R 1998 Reduction of the irreversible capacity in hard carbon anode materials prepared from sucrose for Li-ion batteries. *J. Electrochem. Soc.* **145**, 1977–81

Dahn J R, Zheng T, Liu Y, Xue J S 1995 Three mechanisms for lithium insertion in carbonaceous materials. *Science* **270**, 590–4

Flandrois S, Simon B 1999 Carbon materials for lithium-ion rechargeable batteries. *Carbon* **37**, 165–80
Oberlin A 1989 In: Thrower P A (ed.) *Chemistry and Physics of Carbon*, Vol. 22Dekker, New York
Ohzuku T, Iwakoshi Y, Sawai K 1993 Formation of lithium–graphite intercalation compounds in nonaqueous electrolytes and their application as a negative electrode for a lithium ion (shuttlecock) cell. *J. Electrochem. Soc.* **140**, 2490–8
Peled E, Menachem C, Bar-Tow D, Melman A 1996 Improved graphite anode for lithium-ion batteries. *J. Electrochem. Soc.* **143**, L4–7
Pistoia G (ed.) 1994 *Lithium Battery, New Materials and Perspectives*. Elsevier, Amsterdam
Simon B, Flandrois S, Fevrier-Bouvier A, Biensan P 1998 Hexagonal vs rhombohedral graphite: the effect of crystal structure on the electrochemical intercalation of lithium ions. *Mol. Cryst. Liq. Cryst.* **310**, 333–40
Way B M , Dahn J R 1994 The effect of boron substitution in carbon on the intercalation of lithium. *J. Electrochem. Soc.* **141**, 907–12
Zheng T, McKinnon W R , Dahn J R 1996 Hysteresis during lithium insertion in hydrogen containing carbons. *J. Electrochem. Soc.* **143**, 2137–45

Burnable Poisons in Nuclear Fuels

K. W. Hesketh

Burnable poisons or burnable absorbers perform a very valuable function in nuclear fuels. They are materials such as boron, gadolinium, or erbium, which in their unirradiated state have a high propensity for neutron absorption, but which transform into ones that absorb very little after capturing a neutron. By this means the neutron multiplication factor can be controlled in the reactor throughout the lifetime of the fuel with less reliance on insertion of neutron-absorbing control rods or other control mechanisms.

The first application of burnable poisons was actually for a submarine reactor. It was discovered that control rods would be insufficient on their own to control the neutron multiplication factor for the specified refueling cycle time of the core and burnable poisons were used for supplementary control. Today, burnable poisons are used extensively in commercial light water reactors (LWRs), comprising western pressurized water reactors (PWRs) and the Russian equivalent (denoted VVERs), and boiling water reactors (BWRs). LWRs were by far the most numerous of the approximately 440 commercial reactors in operation in 2001. The UK's advanced gas-cooled reactors (AGRs) also rely heavily on burnable poisons. The other reactor types in commercial operation (Candu, Russian RBMK reactors, and the UK Magnox reactors) do not normally use burnable poisons.

1. Control of Neutron Multiplication Factor in Nuclear Reactors

A nuclear reactor operating at a steady power has a neutron multiplication factor of precisely 1.0. In this state each fission event yields, on average, one further fission reaction to maintain the chain reaction. If the multiplication factor is <1.0, the nuclear chain reaction will die away quickly, while if >1.0, the chain reaction will build up exponentially. A definition of the neutron multiplication factor, usually denoted k, is:

$$k = (\text{no. of neutrons born in generation } j)/(\text{no. of neutrons born in generation } j-1)$$

In a reactor operating in the steady state, it is possible to maintain k extraordinarily close to 1.0 using the inherent negative feedback effects, so that the power output precisely matches the amount of thermal energy extracted from the reactor; thermal power production from the reactor in excess of that removed by the turbine generator raises the temperature of the reactor, and temperature feedback effects reduce the multiplication factor to establish a new balance between reactor output and thermal load.

Some reactors are operated on-load, meaning that fuel can be inserted during full power operation. In such a reactor, it is theoretically possible to start the reactor with just sufficient fissile material (such as ^{235}U) that k just marginally exceeds 1.0 at full power. As the fissile material is depleted (burns up), fresh fuel is added to compensate and a corresponding quantity of spent fuel discharged.

However, on-load refueling is not practical in all reactors; many are designed such that access to the fuel is only possible with the reactor shut down and such reactors operate with batch refueling. This means that the reactor is shut down periodically (typically every 12–24 months) and a fraction of the fuel replaced (with an equivalent amount of the highest burnup fuel discharged). In between refueling, the reactor operates with no change in the fissile inventory except for those changes due to fuel depletion and the buildup of fission products and transmutation elements from neutron reactions. The next refueling occurs when the fissile content is depleted to the point where, with all control absorbers removed from the core, the reactor is just capable of maintaining $k = 1.0$ at full power.

The concept of reactivity is useful to note at this point, which is a dimensionless quantity usually measured in units of percent mill (pcm) or 10^{-5}. It is defined by:

$$\rho = 10^5[(k-1)/k] \text{ pcm}$$

A reactor which is just critical has a reactivity of one and is in steady-state operation. Positive reactivity corresponds to a reactor where the neutron flux and power output are increasing with time, while, conversely, negative reactivity corresponds with decreasing flux and power. Control rod insertion is said to cause a decrease in reactivity, while withdrawing them is said to add reactivity.

The upper curve in Fig. 1 illustrates the variation of k with irradiation for a typical LWR fuel assembly with an initial ^{235}U enrichment of 4.45 wt.% and no burnable poison. At low irradiation k is around 1.3, while by the time the fuel has reached its discharge burnup of ~ 55 MWd kg^{-1} (1 MWdv kg^{-1} is a convenient unit of thermal energy output for nuclear fuel equivalent to 8.64×10^{10} J kg^{-1}), k for the assembly is less than one. Thus by the time the fuel is ready for discharge it is incapable of sustaining a chain reaction by itself. However, the core always contains a mix of fresh fuel in its first cycle of irradiation along with highly irradiated fuel and the fresh fuel effectively donates excess neutrons to the high irradiation fuel such that k for the core as a whole is >1.0.

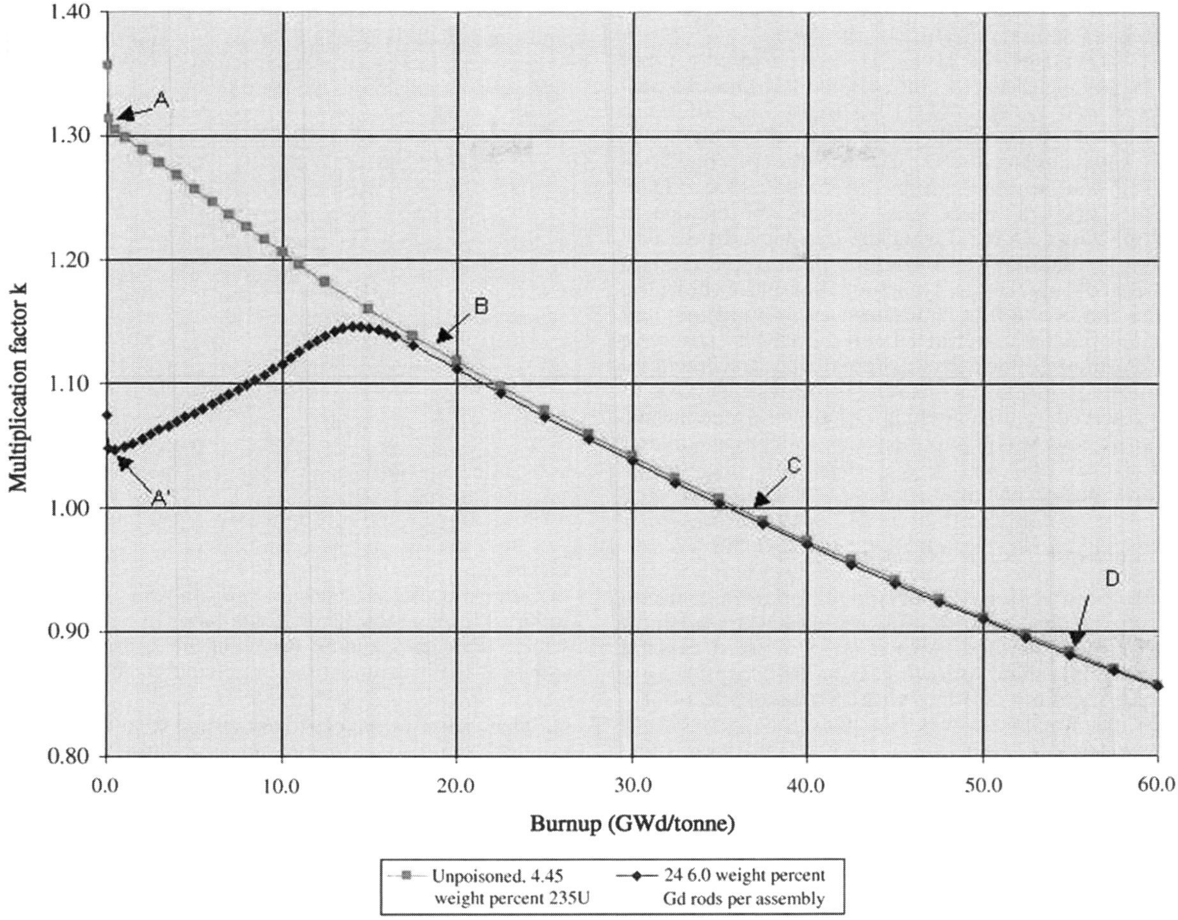

Figure 1
Variation of multiplication factor in a PWR with irradiation: 4.45% enriched PWR assembly with no burnable poison and with 24 poisoned fuel rods containing 6% Gd_2O_3.

Batch refueling causes a problem, however, because if the core is capable of attaining $k = 1.0$ at the end of the cycle, then by implication the uncontrolled k must exceed 1.0 at earlier times. In principle, of course, control rods can be inserted to control the excess multiplication factor, but in practice there are limitations. First, the control rods have a limited effectiveness and more might be required than are practical within the engineering constraints. Second, inserting control rods distorts the spatial distribution of neutron flux and power within the core, causing localized power peaking that might easily exceed the acceptable performance limits of the fuel. Some means is, therefore, needed to supplement the control rods.

PWRs are to an extent able to achieve this with boron dissolved in the coolant as boric acid. But soluble poison has limitations as well, as will be seen later, and soluble absorber is not an option for most other reactor types. Burnable poisons are used to fill this reactivity control gap. The lower curve in Fig. 1 illustrates how a burnable poison (in this case 24 rods per assembly consisting of 6% Gd_2O_3 in UO_2 enriched to 4.45%) might be used to control the excess reactivity of fresh fuel by holding down the multiplication factor in low burnup fuel assemblies. Consider for example a three-batch fueling scheme in which the core is made up of equal quantities of fuel assemblies in their first, second, and third irradiation cycles. At the start of the fuel cycle the first cycle assemblies will have zero burnup, corresponding to points A and A′, unpoisoned and poisoned, respectively. Similarly, the second and third cycle assemblies will correspond to points B and C. The multiplication factor of the unpoisoned core will be $k_1 = [k(\mathrm{A}) + k(\mathrm{B}) + k(\mathrm{C})]/3$ at the start of the cycle.

At the end of the cycle the three assembly types will have burned up to points B, C, and D and k will be $k_2 = [k(\mathrm{B}) + k(\mathrm{C}) + k(\mathrm{D})]/3 = 1.0$. The difference in reactivity between the start and end of the cycle is $k_1 - k_2 = [k(\mathrm{A}) - k(\mathrm{D})]/3$. For the poisoned assembly the corresponding difference is $k_1' - k_2' = [k(\mathrm{A}') - k(\mathrm{D})]/3$. Since $k(\mathrm{A}') < k(\mathrm{A})$, the change in multiplication factor during the cycle is considerably reduced. Ideally, the burnable poison material is depleted during the fuel cycle such that there is none left at the end of the first irradiation cycle indicated; any residual poison material left at the end of the cycle means that the cycle will end prematurely or will require an increased initial ^{235}U loading if the cycle length is to be preserved, both of which equate to an economic penalty. The small gap between the poisoned and unpoisoned curves in Fig. 1 represents just such a residual poison penalty.

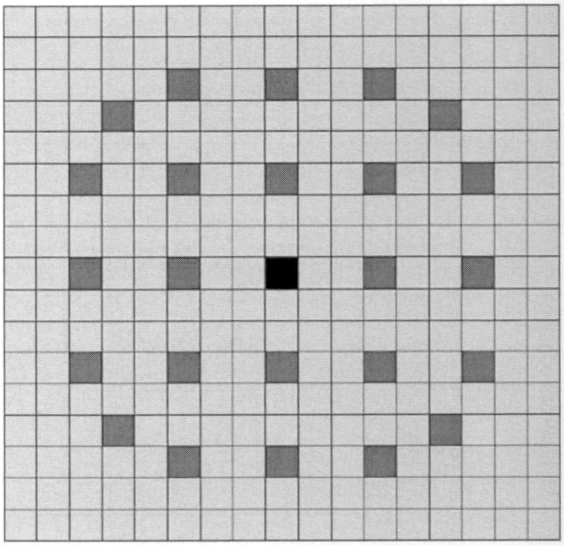

Figure 2
Locations of discrete burnable poison rods (dark gray squares) in the control rod guide tubes of a PWR assembly. Pale gray squares represent fuel rods. The black square is the instrument tube (see colour plate section).

2. *Burnable Poison Materials*

There are three burnable poison materials in common use: boron, gadolinium, and erbium. They all have one or more isotopes with a large neutron capture cross-section that, following a neutron capture, are transformed into isotopes with small cross-sections.

2.1 Boron

Natural boron consists of the two isotopes ^{10}B and ^{11}B at 19.9% and 80.1% natural abundance, respectively. ^{10}B has a thermal neutron capture cross-section (meaning the neutron absorption cross-section for a neutron moving with 0.0256 eV kinetic energy) of 3837 barns (Emsley 1991) (where a barn is a convenient unit of area $10^{-28}\,\mathrm{m}^2$). Following a neutron capture, it is transmuted to ^{7}Li and ^{4}He, both of which have very low neutron absorption cross-sections. For some applications, for example where it is difficult to incorporate a sufficient quantity of ^{10}B as natural boron, isotopically enriched ^{10}B is available commercially. The neutron capture cross-section of ^{10}B decreases smoothly with increasing neutron energy, so that slow-moving (thermal) neutrons are the most likely to be absorbed.

Early applications of boron burnable poisons in PWRs incorporated natural boron in borosilicate glass rods that fitted into the control rod guide tubes of assemblies not positioned under a control rod location, as illustrated in Fig. 2. Such a design is called a discrete burnable poison rod. It has the advantage that the burnable poison material does not need to be incorporated in the fuel and can therefore be manufactured separately. Also, the precise number of rods can be fine-tuned at the last minute before the fuel assembly is loaded in the core. Such discrete poison rods are now seldom used because of the fact that the stainless steel outer clad and other structural components continue to absorb neutrons after all the ^{10}B is depleted; this represents a parasitic absorption penalty that costs the utility because a higher initial enrichment of ^{235}U is needed to compensate. Another disadvantage is that discrete poison rods constitute an additional source of intermediate level waste.

An alternative to discrete boron burnable poison rods is to incorporate a boron compound with the fuel rods themselves. The so-called integral fuel burnable absorber (IFBA) is used in many commercial PWRs in the USA. It consists of a thin coating of zirconium diboride (ZrB_2) which is applied to the surface of some of the UO_2 fuel pellets in the fuel assembly. Since zirconium has a very low capture cross-section and there is no structural material, IFBA has practically a zero residual absorption penalty. A disadvantage is that the application of the coating adds an extra process in fuel manufacture and since it is hygroscopic, manufacturing needs to take place under a dry atmosphere in glove boxes. Figure 3 illustrates a typical distribution of IFBA rods in a PWR assembly.

2.2 Gadolinium

Gadolinium is now the most commonly used burnable poison in commercial reactors. Natural gadolinium consists of the isotopes 152, 154, 155, 156, 157, 158, and 160 with abundances of 0.2%, 2.18%, 14.8%, 20.47%, 15.65%, 24.84%, and 21.86%, respectively

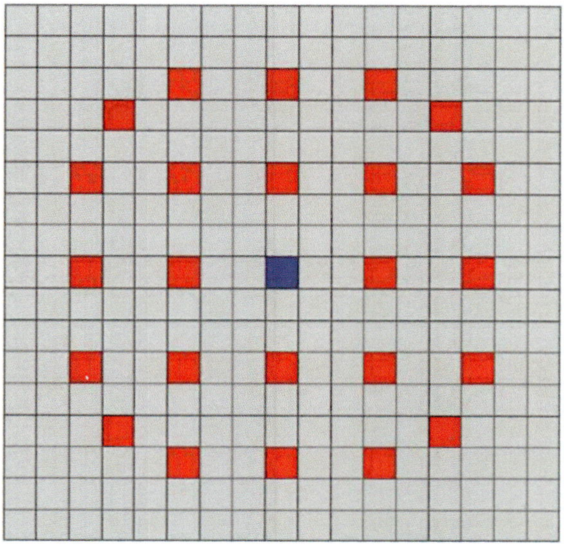

Plate 1
Locations of discrete burnable poison rods (red squares) in the control rod guide tubes of a PWR assembly. Gray squares represent fuel rods. The blue square is the instrument tube (For B/W version, see page 58).

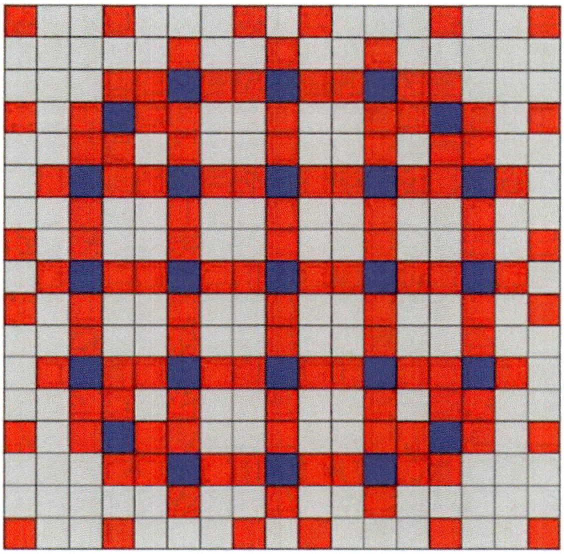

Plate 2
Typical distribution of integral fuel burnable absorber (ZrB_2) rods (red squares) in a PWR assembly. Pale gray squares represent fuel rods. The central blue square is the instrument tube, the other blue are control rod guide tubes (For B/W version, see page 59).

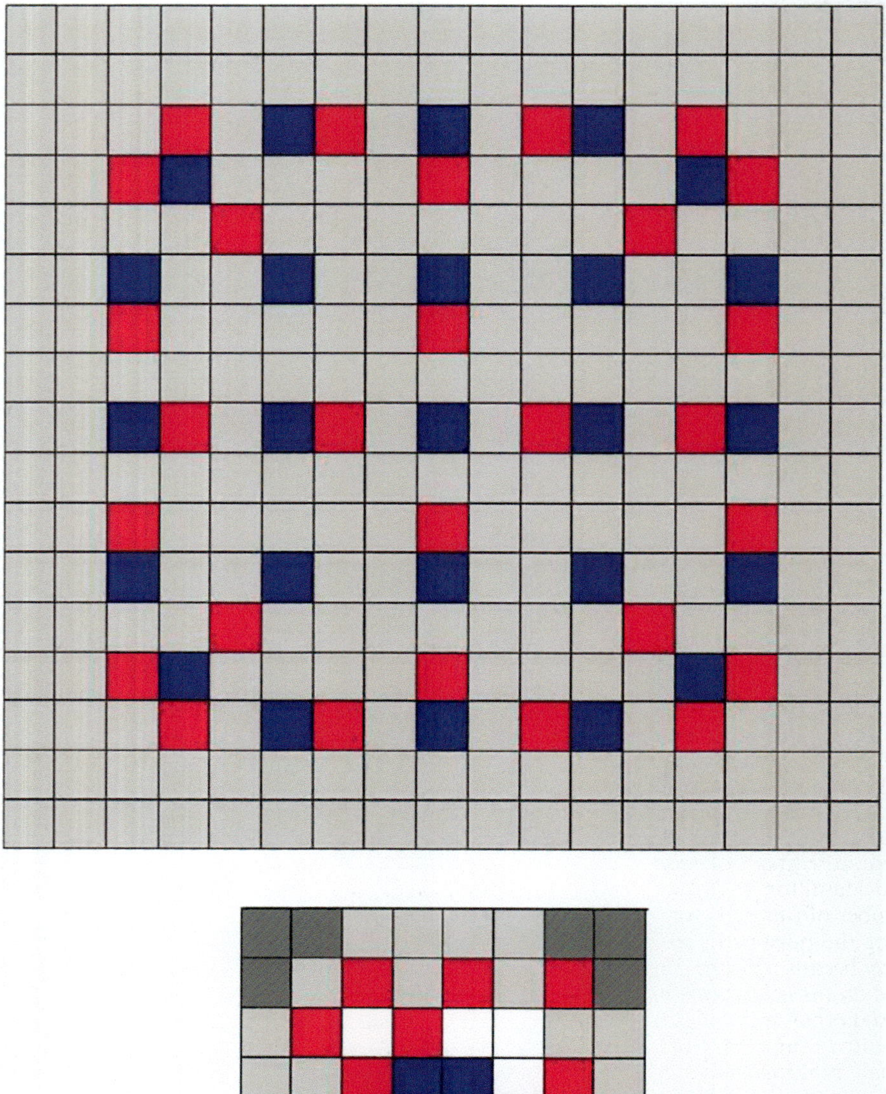

(A)

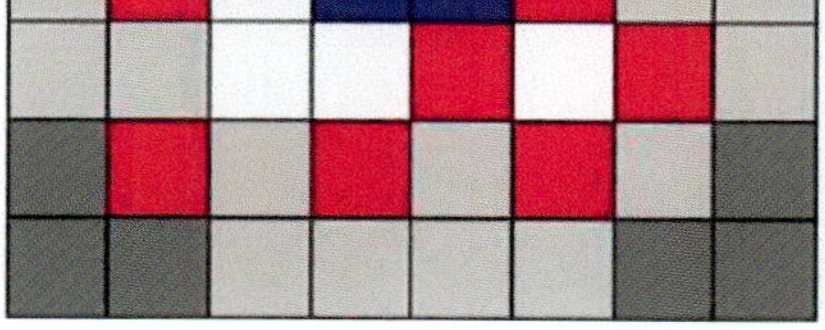

(B)

Plate 3
Typical distribution of Gd_2O_3/UO_2 burnable poison rods (pink squares) in PWR and BWR fuel assemblies. In the PWR assembly, gray squares represent unpoisoned fuel rods. The central blue square is the instrument tube, the other blue squares being control rod guide tubes. In the BWR assembly, the pink squares represent Gd_2O_3/UO_2 rods, the gray squares represent unpoisoned fuel rods and the blue squares represent water channels (For B/W version, see page 60).

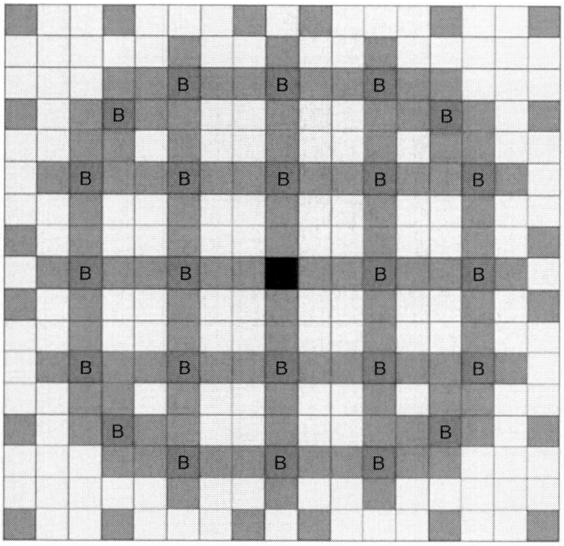

Figure 3
Typical distribution of integral fuel burnable absorber (ZrB_2) rods (dark gray squares) in a PWR assembly. Pale gray squares represent fuel rods. The central square is the instrument tube, the squares marked B are control rod guide tubes (see colour plate section).

(Emsley 1991). Of these, ^{155}Gd and ^{157}Gd have extraordinarily high thermal neutron capture cross-sections of 61 000 and 254 000 barns respectively. This means that gadolinia can severely suppress neutron multiplication at very low concentrations. However, gadolinia is usually concentrated spatially rather than dispersed to take advantage of spatial self-shielding (Duderstadt and Hamilton 1976); when concentrated into a small number of fuel rods for example, thermal neutrons entering the poisoned fuel rod are absorbed near the surface because of the high cross-section. ^{155}Gd and ^{157}Gd atoms near the center of the fuel rod see virtually zero thermal neutron flux and are thereby shielded from neutron captures. Self-shielding thereby limits the initial multiplication effect of lumped gadolinium and extends the depletion time for ^{155}Gd and ^{157}Gd. Self-shielding has the effect that the initial reaction rate is proportional to the surface area, while the time to deplete the ^{155}Gd and ^{157}Gd is proportional to the ratio of surface area to volume. It allows the nuclear designer flexibility in controlling both the initial absorption effect and the depletion rate independently.

Gadolinium is usually incorporated in fuel assemblies in the form of gadolinia/urania (Gd_2O_3/UO_2) fuel pellets, with properties only slightly different to those of conventional UO_2 pellets and a virtually identical manufacturing route. Figure 4 shows typical distributions of Gd_2O_3/UO_2 rods in PWR and BWR fuel assemblies. Gadolinia rods offer the nuclear designer flexibility in choosing the optimal combination of initial reactivity worth and poison depletion rate. There is a residual absorption penalty, though less than that for discrete burnable poison rods, that could, in theory, be reduced by using gadolinium enriched in ^{157}Gd. Figure 5 shows the radial distribution of ^{157}Gd atoms as a function of time. The self-shielding effect is evident in that the depletion proceeds from the outside in shells, usually likened to peeling an onion. ^{155}Gd behaves similarly, though with an extended timescale because of its smaller cross-section.

Gadolinia is used routinely in the UK's AGRs, but in a different form to that used in LWRs. AGR fuel elements incorporate discrete absorber cables containing gadolinia Gd_2O_3 powder in a stainless steel tube. These are located on a support structure within the outer graphite sleeve, with provision for up to three cables at each of the bottom, middle, and top of the element, giving a maximum of nine cables per element. The complete AGR fuel assembly, called a stringer, consists of eight fuel elements stacked on top of one another and at the interfaces between elements the absence of fuel causes the neutron flux and thermal power production to peak. The gadolinia cables are used to counter this deleterious axial heterogeneity.

2.3 *Erbium*

Erbium is a less commonly used burnable poison material, but has been used in some commercial PWRs and is potentially useful for specialized applications. Natural erbium consists of the isotopes 162, 164, 166, 167, 168, and 170 with abundances of 0.14%, 1.61%, 33.6%, 22.95%, 26.8%, and 14.9%, respectively (Emsley 1991). The thermal neutron capture cross-sections are modest, however, ^{167}Er having the highest thermal cross-section of 700 barns and the others all <20 barns, which makes erbium characteristically slow to deplete in the reactor and causes it to have a high residual absorption penalty. It is normally deployed in the form of erbia/urania (Er_2O_3/UO_2) fuel pellets in selected fuel rods.

Unlike boron, erbium acts as a resonant absorber, meaning that there are resonance peaks in the absorption cross-section plotted against incident neutron kinetic energy (Fig. 6). Erbium is, therefore, able to mimic the negative fuel temperature feedback mechanism due to resonance absorptions in ^{238}U, which could theoretically be helpful in nonuranic fuels currently being considered to irradiate and destroy the minor actinides Np, Am, and Cm.

3. *Burnable Poison Materials' Property Requirements*

Any burnable poison material needs to meet a number of important property requirements if it is to

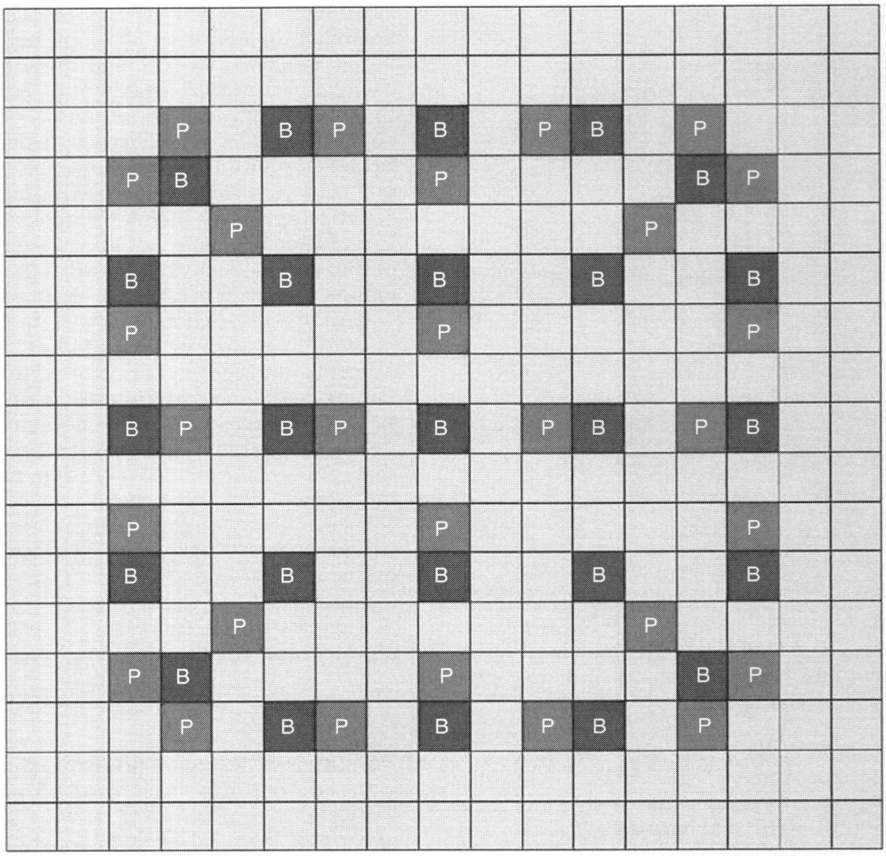

(A)

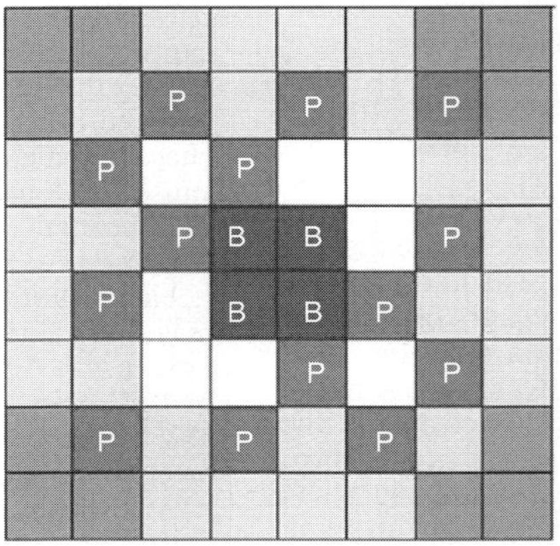

(B)

Figure 4
Typical distribution of Gd_2O_3/UO_2 burnable poison rods (P squares) in PWR and BWR fuel assemblies. In the PWR assembly, gray squares represent unpoisoned fuel rods. The central B square is the instrument tube, the B squares being control rod guide tubes. In the BWR assembly, the P squares represent Gd_2O_3/UO_2 rods, the gray squares represent unpoisoned fuel rods and the B squares represent water channels (see colour plate section).

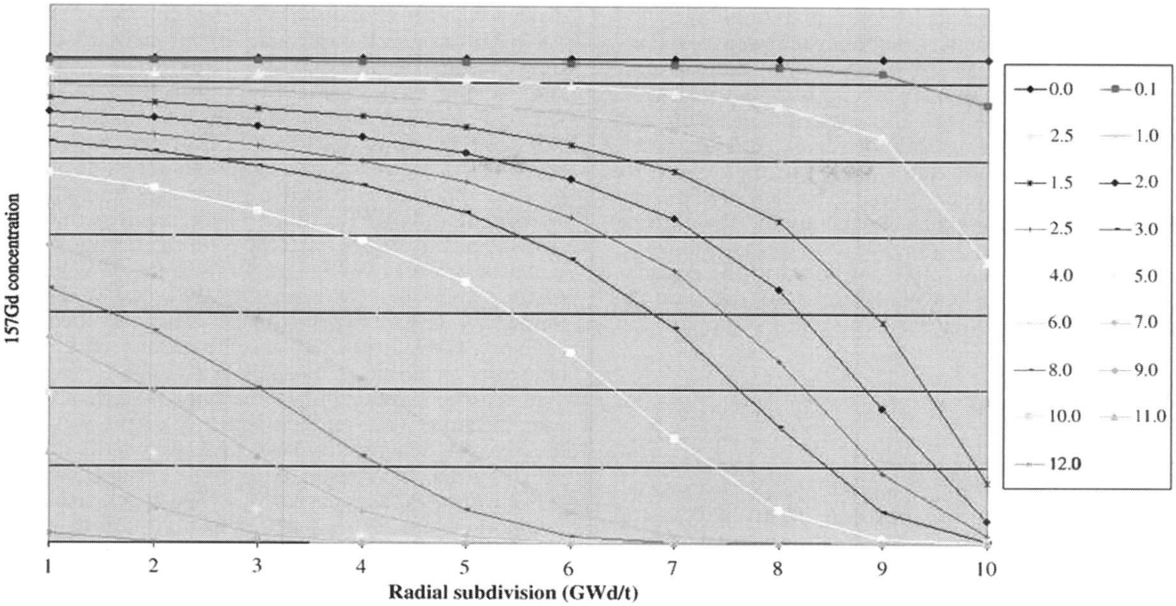

Figure 5
Radial distribution of ^{157}Gd versus burnup across a gadolinia poisoned fuel rod (subdivision 10 represents the outside of the fuel pellet).

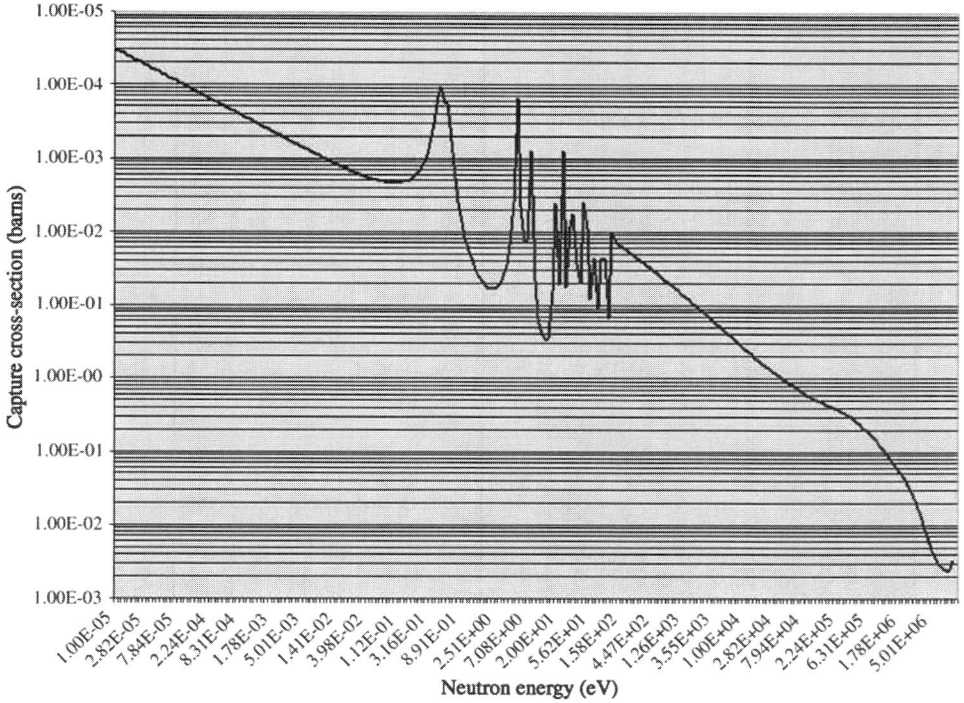

Figure 6
Neutron capture cross-section of ^{167}Er versus neutron energy.

be used in nuclear fuel. Such materials must be physically stable under the high-temperature conditions, they must not be susceptible to corrosion in the sometimes harsh chemical environment, and they must be compatible with the other materials present in nuclear fuel, such as UO_2, Zircaloy, stainless steel, and the coolant materials water, steam, or gas, as appropriate to the reactor type. The burnable poison materials must also behave well under the intense neutron irradiation and gaseous releases must be accommodated. This latter consideration is particularly important for boron, because the neutron absorption leads to helium production through the reaction:

$$^{10}B + n \rightarrow {}^{7}Li + {}^{4}He$$

If used in discrete poison rods, there must be sufficient free volume to accommodate the helium release without overpressurization. With IFBA, the helium released accumulates in the plenum volume provided to accommodate the gaseous fission products and adds to the end-of-life internal pressure.

Additionally, burnable poisons must be designed so that there is no possibility that the poison material could melt or slump out of the active core in an accident scenario or in any way interfere with the insertion of control rods. The migration of absorber material away from the active core would add reactivity during the accident scenario that could potentially worsen the consequences of an accident.

4. *Effect of Burnable Poisons on Fuel and Core Operating Parameters*

Burnable poisons affect the operating characteristics of the core by reducing the excess reactivity control requirements, reducing power peaking in fresh fuel assemblies, and by modifying the reactivity feedback coefficients. BWRs have the most stringent requirements for reactivity control, since in the absence of burnable poison an excessive amount of control rod insertion would be required, especially if the reactor is operating with long fuel cycles (18 months or more). Control rod insertion in power operation is undesirable as it distorts the radial and axial neutron flux distributions, increasing the core power peaking factor (the ratio of peak power to average power in the core). Burnable poisons allow a much more uniform flux distribution and a lower peaking factor. A further deleterious effect of control rod insertion is that fuel rods near the control rods operate at reduced powers and can be subject to a rapid power ramp when the control rods are removed. This can lead to fuel failures which burnable poisons help to avoid.

PWRs have a smaller reactivity control requirement than BWRs because of the use of boron dissolved in the coolant. PWRs typically operate with 1000–1500 ppm of boric acid in the coolant at the start of a fuel cycle, ramping gradually to zero by the end of the cycle. This can control as much as 1.5×10^4 pcm of excess reactivity. Even this is insufficient for long fuel cycles, however, and more boric acid cannot be added indefinitely due to the need to maintain a negative moderator temperature coefficient and also for other reasons such as the need to control the acidity of the coolant carefully.

An increase in water temperature in a PWR normally reduces reactivity because the water is also the moderator and the fuel assembly design is such that reducing moderation reduces the density of thermal neutrons available to propagate the chain reaction. The presence of boron has the opposite effect, however, because when the temperature of the water increases, the density decreases and with it the density of ^{10}B atoms decreases. At a high enough concentration of boric acid (usually $\sim$1500 ppm), the effect of changing absorption outweighs that of moderation and the moderator temperature coefficient becomes positive. This sets an upper limit to the reactivity hold-down that is achievable with soluble boron and burnable poisons are now routinely used to supplement soluble boron.

The presence of burnable poisons in a fuel assembly needs to be accounted for in the thermal-hydraulic and fuel mechanical performance design assessments. For discrete burnable poisons it is important to ensure that there is sufficient cooling to remove the heat production that accompanies neutron captures in the poison material. Where the poison material is incorporated in the fuel rod the generation of helium with zirconium diboride needs to be accounted for in the calculation of fuel rod internal pressure. Gadolinia also has an effect on internal rod pressure, even though there is no gas generation from the poison material. The effect is indirect and caused by the reduction in thermal conductivity of Gd_2O_3/UO_2 compared with pure UO_2. This can lead to higher fuel pellet temperatures and increased fission gas release and the usual design fix is to slightly reduce the initial ^{235}U enrichment of the gadolinia rods relative to the unpoisoned fuel rods.

5. *Burnable Poison Engineering Design Process and Economics*

The engineering design process starts with the utility defining its preferred refueling interval (usually between 12 and 24 months depending on the grid demand profile) and the average load factor expected. This in turn determines the excess reactivity at the beginning of the cycle and the difference between this and the reactivity hold-down capability of the control rods and soluble poison (if applicable) defines the reactivity hold-down requirement for burnable poisons. The nuclear designer must then set the number and location of burnable poisons to meet this requirement. The next step is to adjust the burnable poison loading in each location such that the poison material is nearly completely depleted by the end of the cycle. For IFBA poisons, this may involve choosing to use enriched ^{10}B, while for gadolinia it will involve choosing the optimum initial concentrations (usually in the range 4–8% gadolinia by weight). This is usually an iterative process carried out as part of the core design; it involves finding the fuel assembly loading pattern that best meets the design and safety constraints. Clearly a compromise is required between the nuclear designer's ideal, which might demand different numbers of poison rods in each assembly and different initial poison concentrations, and the manufacturing constraints, which demand as simple a solution as possible.

The initial choice of burnable poison is usually dictated by the manufacturing capabilities of the fuel supplier. The IFBA design requires a major capital investment in the fuel manufacturing plant, while gadolinia rods usually have to be manufactured in a dedicated facility to avoid inadvertent contamination of the main fuel production lines (where even a few ppm of gadolinia contamination can cause fuel to go out of specification). These capital costs and the associated operational costs are passed on to the utility as part of the fuel fabrication price and appear as an indirect cost to the utility. Depending on the burnable poison type, there may also be a residual absorption penalty. For a given cycle length this will increase the initial enrichment requirement of the fuel, which the utility sees as a direct cost.

Though large in absolute terms, relative to the overall generating costs of a commercial reactor these costs are not so important. The benefit to the utility comes through being able to extend the fuel cycle length as much as possible. For a fixed refueling and maintenance outage time, the overall electrical output can be increased if the time between refueling outages is extended. Since fuel costs typically only represent around 20% of the overall generating cost (with capital and finance accounting for about 60%), the penalty on the fuel cost is easily outweighed by even a modest increase in output.

6. *Future Developments*

Utility requirements are tending to become ever more demanding as they attempt to extend cycle lengths further and further and at the same time demand increased fuel discharge burnup. If this trend continues, it is likely that burnable poisons will be used even more extensively and with ever increasing sophistication (which usually means more heterogeneous distribution of the poison material radially and axially within the fuel assemblies). The use of enriched ^{10}B, already available commercially, is likely to increase and there may also arise a demand for isotopically enriched gadolinia and erbia. These developments are likely to increase the pressure on fuel manufacturers to develop automated fabrication lines that can deal more easily with more complicated burnable poison loadings and on the developers of nuclear design codes to accommodate very heterogeneous burnable poison loadings more easily through a higher level of automation. Since there are only around 90 natural elements in the Periodic Table, the number of candidate burnable poison materials are virtually exhausted and it is unlikely that any new ones will emerge.

See also: Nuclear Reactors: Moderator and Reflector Materials; Nuclear Reactor Materials; Nuclear Reactors: Coolant Materials

Bibliography

Chart of the Nuclides. 1998 Forschungszentrum Karlsruhe.
Duderstadt J J, Hamilton L J 1976 *Nuclear Reactor Analysis*. Wiley, New York, pp. 551–4
Emsley J 1991 *The Elements*, 2nd edn. Oxford University Press, Oxford
Graves H W Jr. 1979 *Nuclear Fuel Management*. Wiley, New York, Sect. 5.5

C

Composites for Sensors and Actuators

C. Boller

Composite materials by nature have the advantage of allowing the combination of significantly different types of material. This has principally been used to integrate reinforcing fibers with a passive load-bearing functionality, which carry load along selected and optimized paths. This allows one mainly to avoid fatigue problems and to optimize strength and weight of structural components. Composites can, however, also integrate reinforcing elements that feature other functionalities than load-bearing ones, such as sensing and/or actuation. Resulting materials and structures are known as *adaptive, intelligent*, or *smart*.

Smart materials and structures—briefly expressed—are materials and structures with built-in sensing and/or actuation functions, which contain sensors and actuators linked via a controller. The level of intimacy of coupling between the material and its active elements can range from a classical mechanical coupling of two structures (such as is found in many mechanical structures) down to the level of molecular design. As such, "smart" composites represent a convenient intermediate level of coupling, which allows one to conveniently design and develop a smart material to be applied in a smart structure.

The reasons for developing smart materials and structures are various. These range from a desire to sense the material condition in service (e.g., damage, loads, shape, temperature, pressure, etc.) to design materials capable of adaptive actuation (e.g., changing deformation, shape, stiffness, inner stresses, temperature, color, etc.). Controllers may be built into the composite itself, but so far these have been situated outside the material in the vast majority of cases. The matrix material of such composites is mainly polymer resin; however, metal or concrete matrices have also been considered.

Electromagnetic composites represent another type of smart composite allowing one to modify dielectric permittivity, magnetic permeability, and electric resistivity, which is used for stealth applications. These composites are a combination of multilayered structures, or feature dispersed particles in a matrix. The particles used for passive absorbers are typically graphite powder/fibers, aluminum particles, iron whiskers, or nickel flakes. Active absorbers only absorb electromagnetic energy through integrated semiconductor diodes or other electronic devices when being triggered externally. Further embedded materials can be photosensitive (e.g., fiber optics), pyrosensitive (e.g., silver iodide, Ni–Cr filaments), or voltage/current sensitive arrays of elements, which can be designed for either narrow- or broad-band frequency selection.

With respect to different frequency bands (radio frequency to microwaves), further active roles for "smart" materials can be designed. Examples include phantom materials (absorbing composites simulating human bone tissue and body fluid), electromagnetic shielding (lossy dielectric layers), quasioptic composites consisting of rare-earth materials and ferrites, composite dielectrics (which are two-phase mixtures of low- and high-permittivity components such as polystyrene and barium titanate), chiral composites which include particulates of asymmetric shape allowing the cross-coupling of electric and magnetic effects, and electrochemical composites such as porous carbon impregnated with hydrophobic agents used for fuel cells.

Techniques related to nanoscale materials and composites are atomic and molecular layer manipulation with a focus on semiconductors.

1. Sensor, Actuator, and Control Options

1.1 Sensor Options

Components already existing in a classical structural composite may be used as sensors. An example is carbon fibers, of which variations in the electrical resistance can be used for monitoring damage such as delaminations, or matrix or fiber cracking. A wider application can be seen with fiber optical sensors being attached to, or integrated into, conventional structural composites. Polyimide-clad Bragg grating sensors are the most popular type of such composites. Another option is to make use of piezoelectric materials, available as both ceramics and polymers. These emit a voltage when subjected to stress, and can be shaped to any size and form and adapted onto, or integrated into, the composite. Furthermore, piezoelectric elements can additionally work as actuators when a voltage is applied. Finally, the relationship between electrical resistance and strain of shape memory alloys (SMAs) is another option allowing one to use a classical actuation element as a sensor as well.

1.2 Actuator Options

Piezoelectric materials, usually ceramics (e.g., lead–zirconate–titanate, PZT), are highly favored as actuators owing to their high precision and very rapid response; however, piezoelectric actuators can only work under dynamic loading conditions. Patches or even fibers can be integrated into, or bonded onto, the composite. Deflections achieved are relatively small and may be used for influencing structural damping behavior. Larger deflections need to be achieved through leverage mechanisms. Brittleness of the piezoelectric ceramics, manifested as low allowable strain, must be considered in these applications. Magnetostrictive materials have similar or even slightly better properties as actuators when compared to piezoelectric materials; here the electrical field is replaced by a magnetic field.

SMAs allow one to recover large deformations as a result of a transformation process between martensite and austenite and work at frequencies of about 3 Hz at maximum. Their change in Young's modulus by a factor of three between martensite and austenite allows one to influence damping properties. Transformation is, however, mainly achieved at temperatures less than 60°C, which limits their applications.

The shape memory effect is also observed for specific classes of polymers (e.g., polyurethane of the polyester polyole series) but polymeric materials produce about two orders of magnitude less in force when compared to more conventional metallic SMAs. Density is much lower (0.9–1.1 gcm^{-3} vs. 6–8 gcm^{-3}) and deformation significantly higher (250–800% vs. 6–8%). More information on these materials can be obtained from the literature (e.g., Otsuka and Wayman 1998).

Other types of polymeric composite worth mentioning are ionic polymer–metal composites (IPMCs). These are mainly used as artificial muscles based on charged polyelectrolyte membranes such as collagen or fibrous protein, poly(acrylic acid) (PAA), and poly(vinyl chloride) (PVC). IPMCs show large deformation ($>10\%$) in the presence of low applied voltage (4–7 V) and operate best in a humid environment, but can also be made as self-contained encapsulated actuators to operate in dry environments. Their density is similar to that of the polymers mentioned above but the reaction time is of the order of a microsecond to a second, and thus these are faster. More details are given in Shahinpoor *et al.* (1998).

Electrorheological (ER) fluids are used in order to influence the damping characteristics of materials. These consist of a dielectric liquid (e.g., oil), polarizable particles (e.g., cornstarch), and water as an activator. Through application of an electrical field their rheological properties (e.g., viscosity) are changed. In a zero-field environment the particles are randomly distributed. In an electrical field, however, the particles line up according to the field lines, which then leads to the more rigid behavior of the fluid.

1.3 Control Options

The classical procedures of feedforward and feedback control are used for smart structures as well. Owing to the possibly large number of sensors and actuators to be handled, artificial neural networks have gained broad interest as a means for pattern recognition and novelty detection. Genetic algorithms and wavelet transforms are further procedures that allow for automated determination of optimum solutions and advanced sensor signal processing, respectively.

2. *Design and Manufacturing*

2.1 General Considerations

A major consideration is the compatibility between the sensing/actuation and the host materials. The stiffness ratio between these different materials is a significant parameter. If the actuation or sensing material is too soft, then it will be difficult to transfer the actuation to the host material for actuation, or to the sensing material for sensing designs. A value for this stiffness ratio between 0.5 and 2 is a good design rule of thumb. Another important issue is the quality of the interface between the different materials, because this must have the ability to transfer significant shear stresses. Otherwise the sensing/actuation elements work rather as artificial delaminations. Poor interface stress transfer can also result from chemical incompatibility, or from increased temperature generated by heating of the actuation element.

Size, shape, and quantity of the sensing/actuation elements need to be considered with regard to any notch effects these elements may cause when being integrated into the composite (Warkentin and Crawley 1991). Depending on the application, a large number of widely dispersed small elements may be more useful than one bulky element placed at one location only. The location and thus the number of sensing/actuation elements also very much depend on the geometry and function of the component and structure considered. Optimum design of the smart composite can therefore often very much rely on global component design.

Smart composites are not manufactured on an industrial scale basis, nor are such materials yet qualified according to industrial standards. The vast majority of activity on "smart" composites and structures is still at a research and development stage. There are, however, practical applications for which research and development of such composites has achieved relatively advanced levels, and in which their use is generating significant interest. A few of these cases are described below.

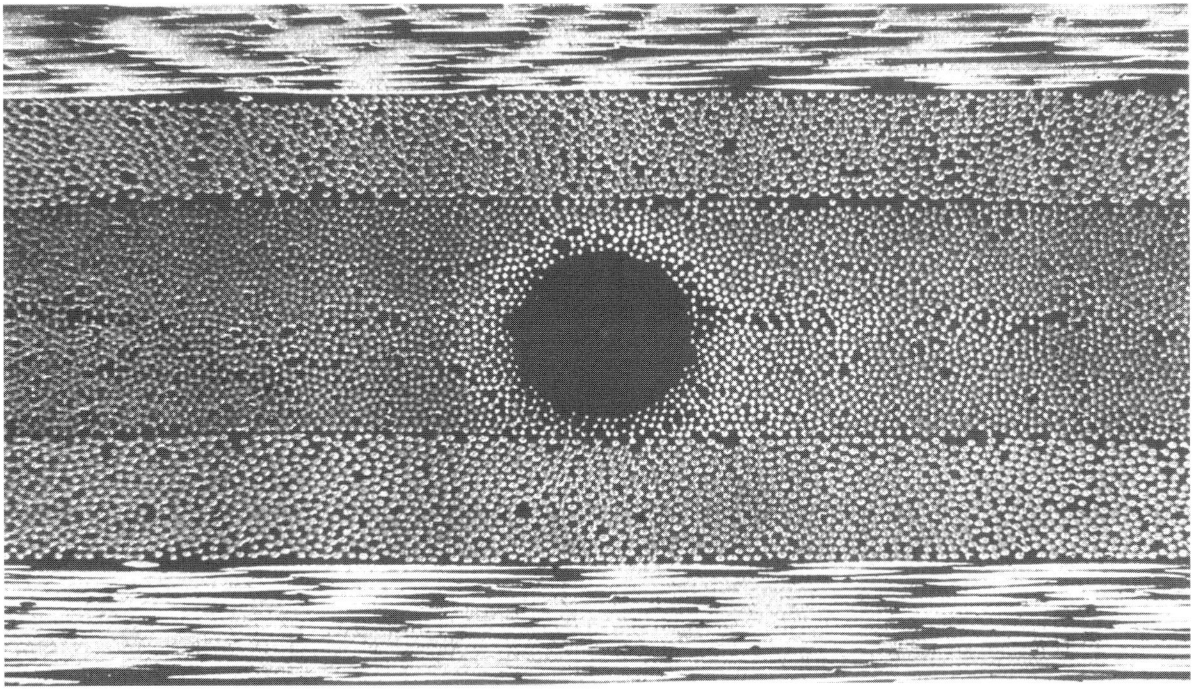

Figure 1
Fiber optic sensor in a quasi-isotropic laminate (courtesy of M. Trutzel, DaimlerChrysler R&T).

2.2 Fiber Optics

Optical fibers are among the most frequently used sensing elements in "smart composites." These are preferably embedded in the composite (Fig. 1), because this generates a more reliable sensor signal output; however, there are techniques that allow surface attachment while maintaining sensor output quality. The smaller the diameter of the optical fiber, the better is its integration into the composite, with diameters $\leqslant 125\,\mu m$ being ideal. Larger diameters are feasible; however, one needs to take into account the formation of resin-rich zones around the fiber. These zones are especially likely if the optical fibers are embedded at an angle to the adjacent ply directions. Very thin optical fibers may, however, lead to positioning problems during processing owing to electrostatic forces. Cleaning of the optical fibers with acetone helps to improve adhesion significantly. One advantage of the use of optical fibers as sensors is that structural integrity is not compromised, since the volume fraction of optical fibers usually required is small compared to the volume fraction of the reinforcing fibers. Another advantage of fiber optic sensors is their very low weight and volume and their immunity to electromagnetic interference.

Critical locations of fracture in the resulting composites are the intersections where the optical fiber comes out of the material, as well as optical fiber terminations and connections. A solution to this makes use of locally cocured synthetic rubber implants that are put around the fiber at these intersections. Adjacent to these are connectors, which are conventional optical fiber plugs incorporating a ceramic precision ferrule. Special techniques have been developed to integrate these connectors into the composite. Results of some initial studies are reported in Green *et al.* (2000).

2.3 Piezoelectric Materials

Piezoelectric sensors are kept small (a few millimeters in diameter and less than 1 mm thick). Examples are "smart layers" (http://www.acellent.net) made of dielectric films with an embedded network of distributed piezoceramic elements, manufactured by a process similar to a circuit printing technique. The layer can either be integrated into the prepreg composite, or it can be bonded onto its surface without compromising the strength of the composite.

The use of piezoelectric elements as actuators (Fig. 2(a)) requires elements of relatively large size, such as $10\,mm \times 4\,mm \times 0.1\,mm$. Special cutouts have to be prepared in the prepreg material in order to integrate such actuation devices. Considerable care is required regarding the selection of the

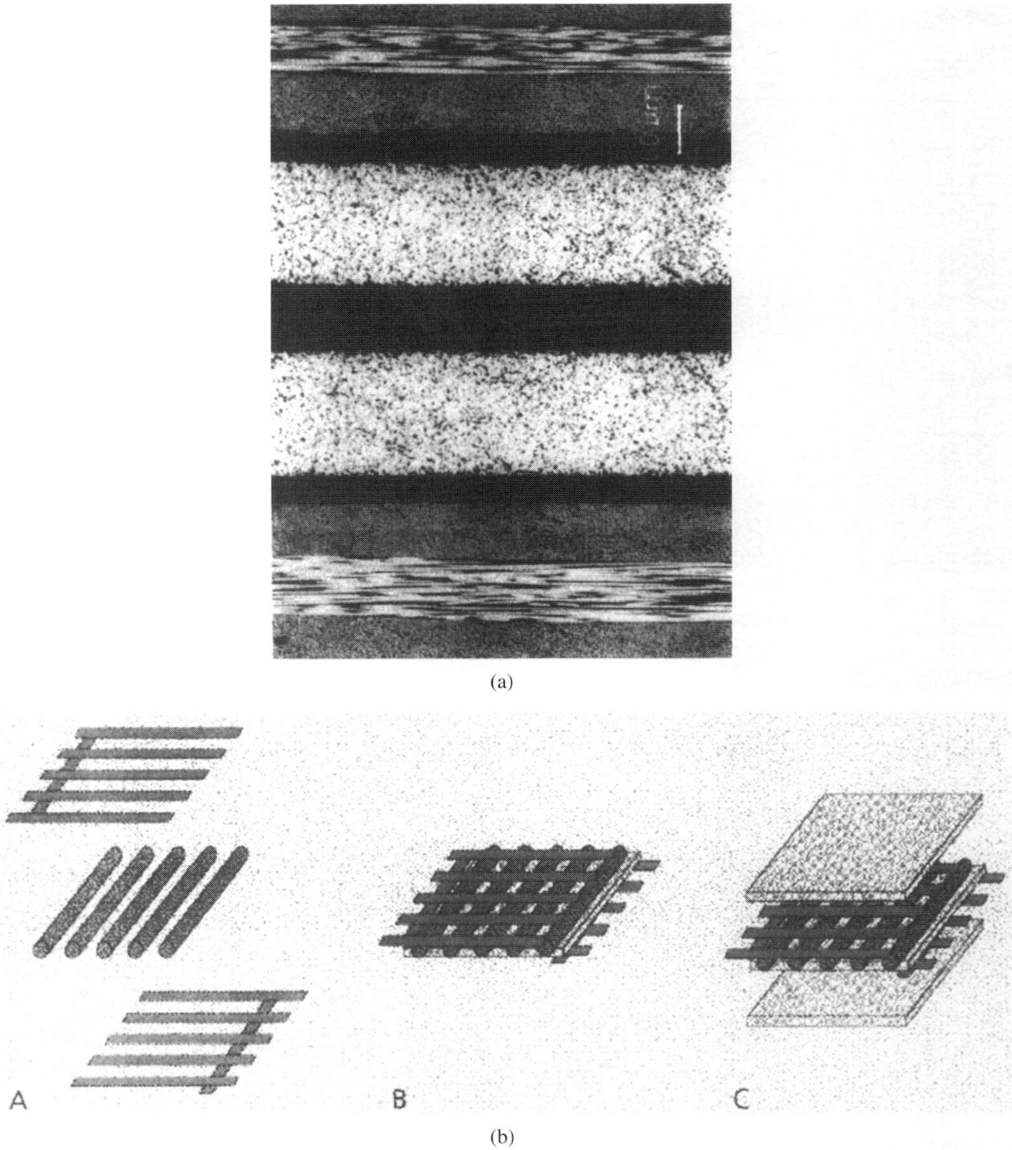

Figure 2
(a) Piezoelectric wafers laminated into a composite (courtesy of H. Zaglauer, U. Herold-Schmidt, F. Dürr, DaimlerChrysler R&T). (b) Piezoelectric fiber concept for integration into composites (courtesy of H. Schönecker *et al.*, Fraunhofer-Institute, IKTS, Dresden, Germany).

appropriate bonding technique to avoid any detrimental effect to the composite. A compressive prestress on the piezoelectric element can help to improve its performance. When integrating the piezoelectric elements care is also required in order to limit the influence of heat being generated by these actuators when operating at higher frequencies and possibly as an actuator cluster. Special conductive layers for heat transfer help to alleviate this problem.

Piezoelectric fibers (Fig. 2(b)) can be integrated like conventional fibers and allow one to avoid any cutouts in the prepregs. A problem, however, exists in producing the required electrical field; this has been solved by developing an interdigitated electrode pattern that is put on the top and bottom of a layer of piezoelectric fibers. This new type of prepreg can then be included into the stacking sequence during composite production.

Other types of piezoelectric elements used as actuators are 1–3 piezoelectric actuator composites. This type of composite consists of a series of piezoelectric rods surrounded by a soft matrix. The rods are manufactured by an injection molding process and are directed vertically to the plane of the composite, which is also the direction of polarization. This allows one to make the best use of sensing and actuation capabilities of the piezoelectric elements, as long as deformations are required in the thickness direction of a plate.

To overcome the problem of high brittleness of piezoelectric ceramics, piezoelectric rubbers and paints have been developed based on the principle that a piezoelectric material is milled into a very fine particulate-sized powder, mixed to a rubber or epoxy resin, and then polarized to obtain again the piezoelectric effect. This allows one easily to shape these types of sensors and actuators. However, a reduction in the sensing and actuation effect is observed when compared to the original piezoelectric material.

2.4 SMAs

State-of-the-art integration of SMA elements—mainly wires with films as an option—into a composite material requires a frame with a comb at its ends where the SMA wire is continuously looped around (Fig. 3). The ends of the frame in the SMA fiber direction are then pulled, which allows one to produce a desired prestress and prestrain on the wires. The whole system is then stacked with the other prepregs and cured, producing a SMA reinforced composite. Appropriate surface treatment of the wires plays a major role in allowing shear stresses to be transferred; however, this should not compromise strength. Techniques of weaving SMA wires into prepregs and even designing complete preforms are a further step being explored. An important issue is that of heat generated by the SMA when going through larger deformations and higher frequencies, and which needs to be

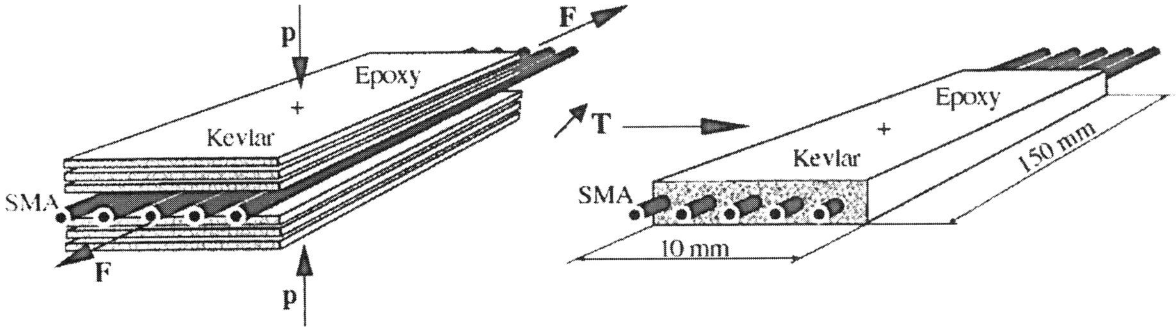

Figure 3
Manufacturing of SMA-reinforced composites. Frame for prestraining SMA fibers (top), stacking and curing process (bottom) (courtesy of V. Michaud, R. Gotthardt *et al.*, EPFL Lausanne, Switzerland).

transferred from the interface between the SMA and host materials.

SMAs may also be plasma sprayed or sputtered as very thin coatings of arbitrary shape on either the composite surface or the prepreg surface prior to composite consolidation. These developments are, however, still at a very early stage. Another novel approach is the integration of SMA particles into a metal matrix, which leads to an arrest effect on crack propagation.

2.5 ER Fluids

ER fluids by definition can only be filled in cavities between composite plates and rubber seals along the thickness axis. CFRP plates are advantageous because they can be used directly as electrodes owing to their electrical conductivity.

2.6 Electrochromatic Effects

Electrochromatic windows allow one to change light transmission as a result of applied voltage. The window consists of a sandwiched cross-section including two transparent electrodes, two electrochromatic layers, and an electrolytic polymer layer. A small current of 1–2 V imposed by the electrodes on the system allows an exchange of ions between the two electrochromatic layers owing to a chemical process and leads to darkening of these layers. Reversing the direction of the current allows one to reverse the darkening process.

2.7 Manufacturing Methods

Manufacturing of composites for sensors and actuators requires care because the sensors, actuators, and wiring must not be affected by mechanical, thermal, or chemical reactions during processing. This is with respect to their properties as well as location within the composite. Adequate care may be achieved through individual hand laying and bonding; however, this is the most costly process. Initial research activities are under way to integrate active fibers such as the ones made of SMA or piezoelectric materials using automatic methods. Automated fiber and tape laying are alternatives but these techniques still need to be adapted to the specifics of these new composites. One approach is a filament winding/tape laying technique with directed energy heating through a laser to synthesize a composite structure layer by layer. Finally, coating techniques represent another promising area where homogeneous material qualities have, however, to be guaranteed. Other low-cost manufacturing processes that have been proposed include extrusion processes for polymer materials designed such that optical fibers or films of piezoelectric materials can be included in the material.

3. Applications

3.1 Sensing

Monitoring of loads is a big issue for determining used structural life in components. The sensing elements that are mainly considered are fiber optics and piezoelectric elements, the latter being useful for vibrational loads of frequencies >10Hz. A related issue is damage monitoring, which includes detection of barely visible delaminations, or of localized fiber and matrix fracture resulting from overloads such as impacts. Here fiber optic sensors are mainly used for monitoring damage-inducing loads while the piezoelectric elements can even be used as actuators by sending into the material a chosen (e.g., ultrasonic) signal which is then recorded by another sensor. Nondestructive testing thus becomes an integral part of the composite material. Another option specific for carbon fiber-reinforced composites is to monitor electrical resistance. This can be achieved by simply contacting the carbon fibers with an adequate current generating and sensing device across the composite. Fiber optics and piezoelectric sensors are also well suited for monitoring the curing process during composite processing. Finally, one can mention electromagnetic sensing, which can be conducted using specially designed composites, and is classically known for radars in aerospace and general defense applications, but is becoming increasingly popular with ground traffic control as well.

3.2 Actuation

Apart from introducing ultrasonic signals, shape control, vibration damping, and electromagnetic antennas are further significant applications of "smart" composites. Shape control originally comes from a desire to adapt the shape of fluid dynamic profiles according to different operational conditions by changing their camber or twisting rotorblades. Shape control can, however, also be desired in order to generate or control the grips of arms such as those used for robots. Applications can also be seen in the medical field, such as stents or valves. The actuators mainly considered are ones that make use of SMAs and, for some specific applications such as microrobots with very low forces, also shape memory polymers. For vibration damping applications, improved passive damping can be achieved by integrating SMAs into the composite, with the side effect that the dynamic properties of the SMA can be changed by heating. SMA-reinforced composites are also used because of their improved impact resistance when compared to the standard composite. Active damping can be added to other functionalities by using piezoelectric elements and/or ER fluids. Much effort is focused on active constrained layer damping using sandwiched structures consisting of the parent structure, a viscoelastic layer, a constraining layer, and a PZT actuator. The actuator is then able to distribute more efficiently its damping energy through the constrained layer over a wider area. Specific applications are seen with turboengines as well as in aircraft generating noise. 1–3 piezoelectric composites represent another solution used as hydrophones for acoustic damping. Actuation is also seen with electromagnetic wave-absorbing composites for antennas, which are used to send out well-defined signals but can be converted to materials that are electromechanically invisible with regard to stealth applications.

4. *Conclusion*

Although composites with sensors and actuators are generally not available "off the shelf," their design as well as their potential applications are already well understood to a large extent. Full-scale experimental verification in an application-oriented environment will allow one to identify and hopefully close any remaining gaps. Specification, qualification, and manufacturing procedures can then be established that may have an impact on further improvement of available sensor and actuator materials.

Bibliography

Green A K, Zaidman M, Shafir E, Tur M, Gali S 2000 Infrastructure development for incorporating fibre-optic sensors in composite materials. *Smart Mater. Struct* **9**, 316–21

Otsuka K and Wayman C M (eds.) 1998 *Shape Memory Materials*. Cambridge University Press, Cambridge

Shahinpoor M, Bar-Cohen Y, Simpson J O, Smith J 1998 Ionic polymer–metal composites (IPMCs) as biomimetic sensors, actuators and artificial muscles—a review. *Smart Mater. Struct.* **7**, R15–30

Warkentin D J, Crawley E F 1991 Embedded electronics for intelligent structures In: *Proc. AIAA Structures, Structural Dynamics and Materials Conf.* AIAA-91-1084-CP. AIAA, Baltimore, MD

Creep Design and Life Extension of Large-scale Plant

D. J. Gooch

Materials used for high-temperature components in large-scale power plants are required to operate for many years under severe conditions of temperature and stress. Variations in detailed chemical composition and heat treatment allowed within specifications result in a wide data band for the creep properties of any particular alloy. Designs must be conservative and are based on laboratory-generated data, together with appropriate safety factors. Consequently, at the end of its nominal design life a component may still be capable of many years further safe operation. Sometimes this may be accomplished simply by a re-appraisal of the original design together with knowledge of past and future operating conditions. However, in many cases a more detailed metallurgical assessment is required. The principles underlying many of the techniques for remaining life assessment have been reviewed by Bhadeshia *et al.* (1998). This article is concerned with the practical application of these techniques, with particular reference to power generation plants.

1. Creep Design

The design of high temperature pressure containing components is governed by national and international standards. Generally, allowable stresses are based upon mean 100 000 h creep rupture data with appropriate safety factors to account for data scatter and departures from design conditions. The safety factor for a life of 100 000 h for a straight pipe is usually 1.5, with higher factors being applied for longer design lives.

Additional design factors for specific components include allowances for wall thinning during bending of pipework, and ligament efficiencies for the areas between penetrations in headers.

National design standards may not exist for non-pressure-containing components, which are then designed to company standards; examples include rotors and blades in steam turbines. Furthermore, these components may be designed using creep strain rather than creep rupture criteria if dimensional stability is critical. Design lives of up to 250 000 h are common for rotors, compared with 100 000 h in the 1960s.

Creep rupture data used in design are generated through long-term tests in air on specimens of typically 5–10 mm diameter. The question then arises as to the effect of oxidation on these specimens compared with a thick section component. If it is assumed that the oxide spalls immediately when it is formed, then theoretical correction factors of >5 on rupture life may be obtained. In practice, oxide will be load bearing for a time before it cracks under accumulated creep strain and true correction factors will be lower. Only limited data are available but these suggest that extensions of design life by up to 50% could be possible.

2. Life Extension

Continued operation beyond design life requires assessment not only of the condition of the component concerned but also of the costs and practicalities of replacement. Table 1 lists the major high-temperature components in a steam power plant together with the time required for "cold" ordering and installing replacements.

A typical life extension program will follow the three-stage approach pioneered in the UK (Gooch and Townsend 1987).

- Stage 1 is essentially a recalculation of the original design life taking into account the actual operating conditions and component dimensions.
- Stage 2 will generally involve access to the component for condition monitoring procedures, which may be performed *in situ*, e.g., installation of additional thermocouples or direct measurement of strain.
- Stage 3 will usually require the removal of samples and the application of techniques, which are not yet considered to be routine procedures, e.g., microstructural techniques.

Life extension issues for specific components will be considered in more detail.

3. Boiler Headers

3.1 Life-limiting Mechanisms

Boiler headers in units constructed during the 1960s and 1970s are thick section cylindrical pressure vessels made from steels such as 1CrMo, 2CrMo, or CrMoV and contain numerous welded stubs for the tube attachments. These headers enable operation at

Table 1
Components with potential for life extension programs.

Component	Lead time (yr)	Outage (weeks)
Boiler drums and headers	2	4–9
High-pressure pipework	1	9–15
Pipework valves	1	2
Turbine valve chests	2	6–9
Turbine casings	3	12
Turbine rotors	3	9

steam temperatures up to 565°C. Headers in newer plants designed for higher-temperature operation and, increasingly, replacement headers in older plants, are made from high chromium martensitic steels such as 9CrMoVNb. High strength austenitic steels such as Esshete 1250 are also used for replacements. There have been extensive life extension programs for the older stations but those with the high chromium steels have not yet reached that stage.

The major areas, excluding welds, which require addressing are: (i) thermal fatigue cracking of the ligaments on the internal bore surface; and (ii) creep life exhaustion, particularly of the ligaments between stub penetrations.

The issue of ligament thermal fatigue cracking is, strictly speaking, one of design and operation rather than design life exhaustion. It is due to the generation of high thermal stresses during transient operation and it has resulted in the replacement of headers before their design life has been reached. Solutions have included the use of the higher creep strength 9CrMoVNb class of materials, which permits thinner sections to be used with consequent reduction in thermally induced stresses.

3.2 *Temperature Measurement*

The role of boiler headers as collection points for steam from smaller components operating across the boiler gas stream means that they are particularly susceptible to variations in temperature along their length. They are also susceptible to transient temperature overshoots during cyclic operation. The temperature dependence of creep of low alloy steels is such that a 10°C increase results in a halving of the creep life. The effect of pressure variation is a second-order effect. Consequently much effort has been expended in monitoring operating temperatures in order more accurately to calculate creep life exhaustion. Attempts have also been made to assess effective operating temperatures by monitoring changes in the detailed precipitate composition or distribution. For example, the increase in chromium concentration of the cementite precipitate as a function of time and temperature has been proposed as a means of measuring the cumulative effective operating temperature. However, such techniques have received limited practical application because of the degree of uncertainty associated with the measurements. The use of reference material "implants" for the same purpose has also been proposed.

3.3 *Microstructural Techniques*

Microstructural degradation, as indicated by the change in precipitate spacing during creep, has been studied as a potential means of direct measurement of the remaining life of headers. Computer models based on carbide coarsening kinetics and the movement of dislocations between carbides have been written for this purpose and calibrated using extensive creep testing programs (Askins 1988). It has also been suggested that hardness may be used as a measure of the precipitate spacing to avoid the need for electron microscope examinations. However, the degree of uncertainty associated with the predictions has prevented these finding practical application.

Creep cavitation detection using plastic replication techniques is one of the most widely used life assessment procedures for headers. Originally devised as a qualitative technique to determine future inspection intervals, it has been variously modeled for quantitative life prediction (Shammas 1987) but remains a largely qualitative technique, again because of the scatter in the experimental data. It also suffers from being a surface measurement technique, whereas the degree of creep damage varies through the header wall.

3.4 *Creep Testing*

The most reliable method of life prediction for headers is that of direct creep testing of ex-service material. Procedures have been developed for producing and testing creep specimens from blanks of material, which may be as small as 13 mm long and 3 mm diameter. Samples of this size may be removed from a component using special jigs leaving a depression, which may be removed by grinding (Gooch 1988). However, more usually, full sized specimens are obtained by trepanning a through thickness plug of material from the header and welding a "blind stub" over the resultant hole.

Accelerated creep testing needs to be employed to obtain life predictions within an acceptable timescale; the two available alternatives are to increase stress or temperature. Increasing stress has the advantage that tests may be performed in air rather than an inert atmosphere. However, increasing stress may result in a change in the creep mechanism such that extrapolation of short-term data to service stress levels becomes uncertain. Nevertheless tests of this nature may be used to "position" the specific cast of steel within the scatter band for that grade. In many cases this may be sufficient to underwrite the continued operation of the component and is often the preferred route.

Changes in creep mechanism are less likely using accelerated temperature because the structural changes influencing creep deformation, such as carbide coarsening, are accelerated to a similar degree to the creep process, both being diffusion controlled.

3.5 *Structural Analysis*

Assessment of the remaining life of complex components and structures may require the use of sophisticated techniques such as finite element analysis to determine stresses and temperatures at critical points. Alternatively the overall behavior of a component may be described by use of a *reference stress*, defined as that stress which would fail a simple tensile creep specimen at the same temperature in the same time as the component (Ainsworth *et al.* 1987). When the temperature varies with time a reference temperature can also be defined at which the comparative specimen test should be performed. The reference stress and temperature may be determined by theoretical and/or experimental work involving computer modeling and model or full-size component tests.

4. High-pressure Steam Pipework

4.1 Life-limiting Mechanisms

High-pressure pipework systems are generally constructed from low alloy steels such as 2CrMo and CrMoV or, in the case of some continental European plants from the martensitic 12CrMoV. In the most modern plants with advanced steam conditions the 9CrMoVNb grades are being used. Welds, particularly in the CrMoV grade, have historically been the most common failure location. Ultrasonic and magnetic particle examinations for crack detection combined with surface replication for cavity detection have been the staple inspection tools. Although attempts have been made to apply fracture mechanic techniques to predict the growth of defects in welds, in practice statistical approaches based on accumulated plant experience have provided the most reliable and cost-effective procedures.

Welds can generally be repaired and, from a life exhaustion perspective, the most critical areas are bends, which are prone to wall thinning and ovality during production and may also experience high local system stresses. These arise from pipe self weight and thermal expansion rather than internal pressure. In principle they should be low at operating temperature, but in practice they are difficult to quantify. Straight sections of pipe are also examined to determine the general condition of complete systems.

4.2 Assessment Techniques

Techniques used in practice include the following:

(i) direct strain measurement on bends or straights, either at outages using creep pips previously welded onto the pipes, or continuously using high-temperature capacitance gauges;
(ii) cavitation assessment by plastic replication—in particular this is used for bend examination; and
(iii) sample removal for creep testing.

Other parameters that have been investigated, but not developed to the point of plant application, include (Eggeler and Hurst 1994):

(iv) ultrasound velocity attenuation—shown to be more sensitive to cavitation detection than conventional ultrasonic techniques;
(v) electrical resistance—may have potential for near surface cavity detection; and
(vi) micromagnetic parameters—shown not to correlate with cavitation.

5. High-temperature Rotors

5.1 Life-limiting Mechanisms

High-temperature rotors are usually made from low alloy forgings, in particular 1CrMoV, although high chromium martensitic forgings are available for higher temperatures. The critical regions for damage accumulation are the bore (if present), the blade root fixings, balance holes, and stress-concentrating features, such as stress relief grooves, on the outer surface (Carlton *et al.* 1987).

Modern high-temperature rotors are generally not bored, as manufacturing and inspection procedures are sufficiently advanced to avoid unacceptable defects. However, many older rotors contain bores and these are often the positions of highest creep strain accumulation. The potential failure mode that has to be guarded against is the initiation and growth of intergranular creep cracks to a size that could result in brittle fracture, either at cold start-up, due to thermal and centrifugal stresses, or during over-speed testing. The initiation of creep cracks may occur either by exhaustion of creep ductility or by growth of pre-existing defects. The creep ductility of 1CrMoV forgings is extremely varied, being dependent on factors such as creep strength, austenitizing temperature, quenching medium, severity of tempering, major and residual element levels, steel making practice, etc. Long-term studies have shown that cavitation may be detected at strains of $<1\%$ in low ductility forgings. Life assessment procedures for rotor bores are therefore based on limiting strain criteria.

Potential failure modes at blade root fixings are either creep crack initiation and growth or shear failure due to creep strain accumulation. Blade detachment does not have the same potential catastrophic consequences as a rotor burst and, consequently, a creep rupture rather than a limiting strain criterion may be used for assessment.

On some designs the position of maximum creep damage is at the balance holes in the disc or at the transition region between disk and rotor body. These locations are more accessible for nondestructive testing than the bore or blade fixings but must still be included in life assessments. Limiting creep strain criteria are generally applied but the possibility of thermal fatigue crack initiation and growth must also be considered.

5.2 *Dimensional Measurements*

In the UK many rotors were provided with means for taking diametral measurements during outages. These were used to monitor strain accumulation during service. Theoretical calculations show that, after approximately 10000 h service the ratio of the strain rates at different radial positions is constant. This enables bore strains to be calculated from diametral strains and strains calculated in this way are consistent with those predicted from design calculations.

5.3 *Creep Testing*

In the case of a rotor the scope for removal of samples for creep testing is limited and must be determined by stress analysis and economics. Nevertheless, this remains one of the most effective contributions to life assessment, being suitable for direct life estimation or for positioning within the scatter band when material from critical regions is unavailable. Miniature or standard specimens may be used.

5.4 *Cavitation Assessment*

The use of cavitation assessment for rotor life prediction is complicated by the inaccessibility of some critical regions, particularly the bore. However, remotely operated sampling devices, which may be positioned in the bore, are available for this purpose. A further complication is the variability of ductility between forgings, which influences the relationship between cavitation and remaining life. High ductility forgings do not exhibit cavitation until a late stage and consequently it has been shown that the technique should not be used quantitatively on rotors with a grain size less than 25 microns. For larger grain sizes a linear relationship between maximum remaining life and cavity density, as measured by the "A" parameter (see *Creep Life Prediction*), has been demonstrated for a range of rotor forgings.

5.5 *Hardness Measurements*

The use of hardness to indicate historical effective operating temperature and residual creep life has received particular attention in relation to high-temperature rotors. It has been suggested that hardness can be correlated with rupture life for a class of rotors. However, other work on a range of rotors suggests that such correlations will vary from forging to forging and that no universal correlation exists (Strang *et al.* 1994). An extensive study of hardness changes during creep testing and thermal aging of a range of 1CrMoV rotors indicated that average service temperatures could be estimated to an accuracy of $\pm 20\,^{\circ}C$ using a Larson–Miller relationship between hardness, time, and temperature. In practice this degree of accuracy is only adequate for use in conjunction with other assessment techniques.

6. *High-temperature Bolts*

High-temperature bolts are used to secure valve chests and turbine casings; typical materials are 1CrMoVNbTi, 12CrMoVNb, and Nimonic 80A. Expended life is calculated from the service hours and the number of tightenings the bolt has received. Bolts are tightened when cold to a predetermined strain (typically 0.15%), which allows for differential expansion effects during start-up and stress relaxation during operation. Potential failure modes are: (i) excessive relaxation resulting in steam leakage; (ii) fracture originating from creep cracks formed at thread roots due to bending stresses generated by temperature gradients; or (iii) fracture from stress corrosion cracks (Nimonic 80A only) formed off load. A policy of replacement is generally pursued because of its ease and relatively low cost and the potential consequences of failure.

7. *Future Steam Plants*

Most life extension programs have been performed on components constructed from low alloy steels, being the predominant materials used in plant coming to the end of design life. However, higher chromium martensitic steels are increasingly used in new plants and there are programs to develop them further for use at steam temperatures up to 625 °C. Future studies must therefore address these.

8. *Gas Turbines*

Life extension studies have concentrated on steam power plant. However, much of the power plant currently being commissioned in countries with access to natural gas is combined cycle gas turbine (CCGT) plant. The critical components in the hot gas path section of the gas turbines are manufactured from nickel base alloys and are subject to extremely onerous operating conditions, with turbine inlet temperatures exceeding 1300 °C and metal temperatures exceeding 900 °C. Major overhauls are conducted at 25 000 h intervals with the expectation of major component replacements at 50 000 h intervals. Replacement costs are high, for example a row of the latest single crystal blades may cost up to $1 million. There is, therefore, a great incentive to develop similar life extension procedures to those for steam plant. However, the task is difficult because of the complexity of the temperature and stress profiles involved. Furthermore, the nickel base alloys exhibit fewer quantifiable degradation phenomena than the low alloy steels. An exception to this is the coarsening of the gamma prime precipitate with time and temperature, which is being developed as a measurement of operating temperature. Changes in the power-generating industry have also resulted in less willingness to invest resources in detailed scientific investigation. Consequently it is unlikely that gas turbine plant life extension will receive the same detailed study as that given to steam turbine plant.

9. *Concluding Remarks*

A life extension program involves the application of a range of techniques tailored to the component in question. A typical program would follow a three-stage approach, in which increasing complexity is of necessity accompanied by increasing uncertainty. Direct measurement or updated calculation of stress, strain, and temperature, accompanied by systematic nondestructive testing, including surface replication for indications of cavitation, remain the most reliable means on which to base an extension. Removal of samples for creep testing may also provide a sufficient degree of confidence, particularly if forward planning allows the degree of test acceleration to be minimized. Techniques based on quantitative characterization of microstructural degradation have proven to be insufficiently accurate to be used in isolation although they may provide valuable supportive information.

Bibliography

Ainsworth R A, Chell G G, Coleman M C, Goodall I W, Gooch D J, Haigh J R, Kimmins S T, Neate G J 1987 CEGB assessment procedure for defects in plant operating in the creep range. *Fatigue Fract. Eng. Mater. Struct.* **10**, 115–27

Askins M C 1988 Creep life assessment by measurement of interparticle spacing In: *Proc. Int. Conf. Life Extension and Assessment.* Netherlands Institute voor Lastechniek, The Hague, Netherlands

Bhadeshia H K D H, Strang A, Gooch D J 1998 Ferritic power plant steels: remanent life assessment and the approach to equilibrium. *Int. Mater. Rev* **43**, 45–69

Carlton R G, Gooch D J, Hawkes E M, Williams H D 1987 The CEGB approach to the determination of the remanent life of high temperature rotors In: *Proc. Int. Conf. Refurbishment and Life Extension of Steam Plants*, London, October 1997. Institute of Mechanical Engineers, London, pp. 45–52

Eggeler G F, Hurst R C 1994 Development and validation of models for the lifetime assessment of steam power plant components In: *Proc. Int. Conf. Materials for Advanced Power Engineering.* Kluwer, Liege, Belgium, pp. 109–32

Gooch D J 1988 Plant life extension—remanent life assessment of high-temperature components. *Power Eng. J* **2**, 323–31

Gooch D J, Townsend R D 1987 CEGB remanent life assessment procedures In: Proc. EPRI Conf. Life Assessment and Extension of Fossil Fuel Plants (Washington DC). Pergamon, Oxford, pp. 927–43

Shammas M S 1987 Remanent life assessment of ferritic weld heat-affected zones by a metallographic measurement of cavitation damage—the "A" parameter In: *Proc. Int. Conf. Refurbishment and Life Extension of Steam Plant.* Institute of Mechanical Engineers, London, pp. 289–95

Strang A, Beech S M, Gooch D J 1994 Remanent life evaluation of 1CrMoV steam turbine rotors In: *Proc. Int. Conf. Materials for Advanced Power Engineering.* Kluwer, Liege, Belgium, pp. 549–60

Strang A and Gooch D J (eds.) 1997 *Microstructural Development and Stability in High Chromium Ferritic Power Plant Steels.* Institute of Materials, London

Creep-resistant Materials for Steam Turbines

S. R. Holdsworth

Large steam turbines used for power generation typically comprise arrangements of up to three types of turbine module referred to as high pressure (HP), intermediate pressure (IP), and low pressure (LP) (e.g., Fig. 1). High pressure, high temperature steam is fed into the inlet stage of the HP turbine to expand through a series of stages of increasing blade height to the exhaust. Each stage consists of a row of stationary blades attached to the casing and a row of moving blades attached to the turbine rotor. The momentum gained by the steam due to the pressure drop across each stage is converted into torque on the rotor through its action on the moving blades. In traditional fossil fuel units, the HP exhaust steam is fed into the inlet of the IP turbine, following reheat in certain designs, and expands through a sequence of stages, as for the HP turbine. IP exhaust steam is fed into the inlet stage(s) of the LP turbine(s) and the exhaust steam directed to a condenser.

For many years, the steam inlet–exhaust temperatures of HP, IP, and LP turbines were typically of the order of 550–400 °C, 550–400 °C and 350–70 °C, respectively. Hence, a key material property of critical components in HP/IP turbines is resistance to creep deformation and rupture.

1. *Material Requirements for HP/IP Steam Turbines*

Large steam turbines are typically designed to operate for 250 000 h. The creep rupture strength data used in their design are usually for 100 000 h (e.g., Fig. 2). Materials for high temperature casing bolts are also required to have adequate relaxation strength to ensure joint tightness and thereby prevent steam leakage between full-inspection outages. Relaxation strength data for 30 000 h are typically used in the design of such fasteners (e.g., Fig. 3).

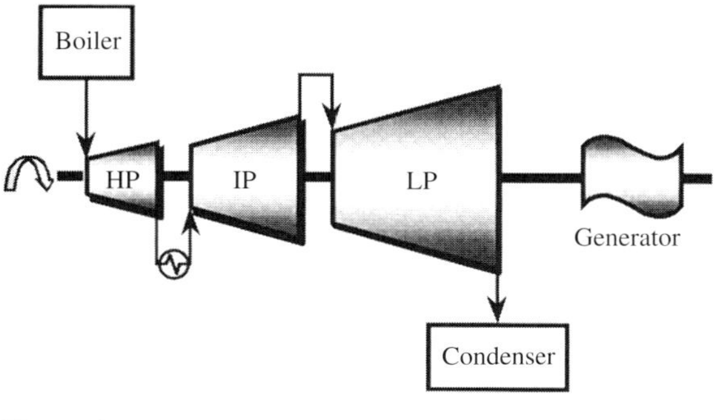

Figure 1
Schematic of a large steam turbine arrangement.

In addition to the requirement for a high resistance to creep deformation and rupture, HP/IP turbine materials need to exhibit an adequate level of rupture ductility and fracture toughness. Rupture ductility is an important property because of its strong influence on high temperature notch tolerance, defect integrity, and creep-fatigue resistance. Steels with uniaxial rupture elongations of $\geqslant 10\%$ are typically notch strengthening, exhibit good creep crack initiation and growth properties, and are less prone to ductility exhaustion arising from thermal fatigue. Adequate fracture toughness is required for defect tolerance, in particular during cold starts. For a number of the steels reviewed below, the metallurgical condition giving rise to high creep strength can also be responsible for low rupture ductility and fracture toughness. The balance of these properties is an important consideration in the specification of HP/IP turbine materials.

2. *Materials*

Since the 1920s, but mainly during the 1980s and 1990s, the development of high temperature creep resistant materials for steam turbines has been strongly influenced by the requirement for further improvements in turbine efficiency and performance. Turbine efficiency is directly related to the steam inlet temperature, and hence the driving force for higher temperature materials (Table 1). High creep strength enables component section sizes to be kept to a minimum and turbine start-up times to be reduced without additional risk of thermal fatigue. However, for large turbines, a limiting factor continues to be the ability to manufacture the large homogeneous forgings required for certain critical components from materials with inherently high strength at high temperatures.

The creep strength of the alloys used to manufacture turbine components is determined to some extent by the microstructure and grain size, but largely by the presence of uniform distributions of fine second phase particles (obstacles to dislocation movement) which are stable for long times at high temperatures. Three categories of creep resistant material are covered in the following sections, i.e., low alloy steels, 9–12Cr martensitic stainless steels and nickel-base alloys (Table 2).

2.1 *Low Alloy Steels*

The present generation of low alloy creep resistant steels originated from carbon steel, initially by the addition of 0.5–1.0% molybdenum (together with up to 2.25% chromium or 0.25% vanadium) to facilitate

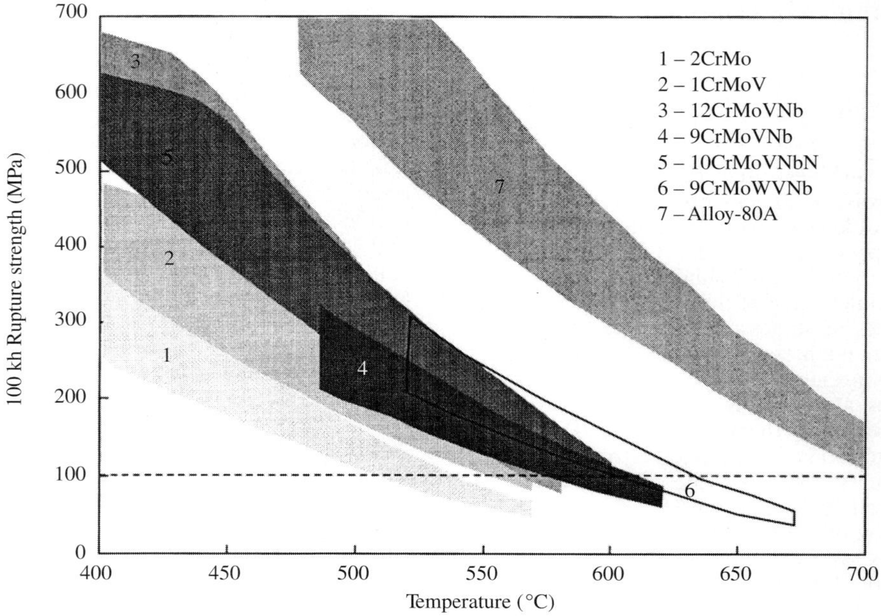

Figure 2
Comparison of 100 000 h creep rupture strengths for steam turbine materials (*the 100 MPa comparison line represents a target 100 kh rupture strength requirement for rotors*).

Table 1
Steam turbine components requiring creep resistant materials.

Inlet steam conditions	180 bar/540°C	240 bar/565°C	300 bar/600°C
Rotor	1CrMoV, 2CrNiMoWV	1CrMoV, 2CrNiMoWV	10CrMoVNb
Blades	12CrMoV	12CrMoVNb, Alloy-80A	11CrMoVNb, Alloy-80A
Casings	$2\frac{1}{4}$CrMo, $\frac{1}{2}$CrMoV, 1CrMoV	$\frac{1}{2}$CrMoV, 1CrMoV, 9CrMoVNb	9CrMoVNb
Casing bolts	1CrMoVTiB, 12CrMoVNb	1CrMoVTiB, 12CrMoVNb, Alloy-80A	11CrMoVNb, Alloy-80A
Pipework	$2\frac{1}{4}$CrMo, $\frac{1}{2}$CrMoV, 1CrMoV	$\frac{1}{2}$CrMoV, 1CrMoV, 9CrMoVNb	9CrMoVNb

formation of the fine carbide precipitates required for long-term creep resistance.

$2\frac{1}{4}$CrMo derives its creep strength from a fine dispersion of $(CrMo)_2C$ particles usually in a microstructure of tempered bainite. It is used as a casing and pipe material (Table 1). The 100 000 h creep rupture strength is at the bottom end of the spectrum shown in Fig. 2, but the steel exhibits good rupture ductility and high fracture toughness.

In low alloy steels, higher long time creep strength is obtained by the addition of vanadium to form fine dispersions of V_4C_3 particles in tempered bainite microstuctures, and various compositions are adopted for casings and pipework, e.g., $\frac{1}{2}$CrMoV and 1CrMoV (Fig. 2). For many years the universally adopted HP/IP rotor material has been 1CrMoV steel. Different turbine manufacturers adopt subtly different compositional balances and heat treatment practices within a broadly similar specification to achieve the optimum balance of creep strength, rupture ductility, and fracture toughness to meet the requirements of their individual design philosophies, e.g., Norton and Strang 1969). Optimized 1CrMoV has a significant temperature advantage over $2\frac{1}{4}$CrMo

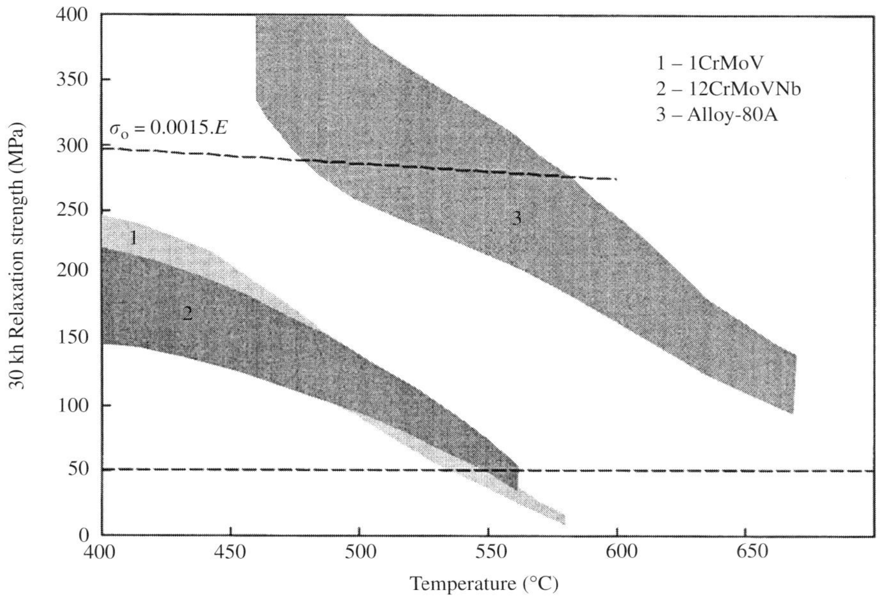

Figure 3
Comparison of 30,000 h relaxation strengths for steam turbine bolting alloys (for 0.15% tightening strain).

(Fig. 2), but at the expense of fracture toughness. An advance is the development of a 2NiCrMoWV rotor steel which has almost the same creep-rupture properties as 1CrMoV, but superior toughness.

A derivative of the 1CrMoV rotor steel is the 1CrMoVTiB alloy developed in the UK for bolting applications (Fig. 3).

2.2 *Martensitic Stainless Steels*

Various grades of 12Cr steel have been commonly used in large steam turbines for a number of years, in particular for high temperature blade and bolting applications, e.g., 12CrMoV and 12CrMoVNb (Wickens *et al.* 1980). A common feature of the alloys employed is a compositional balance to ensure a microstructure of fully tempered martensite with no residual ferrite (responsible for low fracture toughness). The traditional 12Cr steels rely on a fine distribution of $(CrMoV)_2(CN)$ particles for their long-term creep strength. These alloys offer a small temperature advantage over 1CrMoV (Figs. 2 and 3).

The martensitic stainless steels have been the focus of developments for (i) 600°C and (ii) 650°C steam inlet applications (Vanstone and Thornton 1997). The basis for these is the US 9CrMoVNb (Steel-91) and Japanese 10CrMoVNb (TMK1) alloys developed in the 1980s. Notably, both alloys have lower niobium and lower carbon than the earlier generation 12CrMoVNb steel (Table 2). Optimized variants of the 9CrMoVNb alloy are now used for pipework and casing applications at temperatures up to 600°C and at lower temperatures where their higher strength allows them to be used in thinner section than the conventional low alloy materials (Fig. 2). Rotors manufactured from TMK1 and optimized variants with and without tungsten additions are now in service at temperatures up to 600°C (e.g., 10CrMoVNb in Table 2).

For steam inlet temperatures of up to 650°C, alloy development activity is focusing on higher tungsten or boron-containing alloys with 12% chromium for the higher oxidation resitance needed at these higher temperatures. 9–12Cr alloys with such additions have been investigated with mixed success. 9CrMoVWN (Fig. 2) represents a successful development with measured creep properties out to 50 000 h. In Japan, a 12Cr alloy with 2.5% tungsten and 2.5% cobalt (to counter balance the delta ferrite forming effect of high W) shows promise on the basis of short-term properties, but the material's long-term stability is yet to be proven.

2.3 *Nickel-base Alloys*

Apart from the use of Alloy-80A for some blade and bolt applications (Table 1), usually when a specific design has imposed a size limitation on the component, nickel-base alloys have not been widely used as steam turbine materials for inlet temperatures of <565°C. The high creep rupture and relaxation strengths of such alloys usually far exceed

Table 2
Examples of typical compositions of creep resistant steam turbine materials.

Material	C	Mn	N	Al	Co	Cr	Mo	W	Nb	Ni	Ti	V
$2\frac{1}{4}$CrMo	0.12	0.6		0.02		2.2	1.0					
$\frac{1}{2}$CrMoV	0.14	0.6				0.5	0.5					0.26
1CrMoV	0.25	0.7		<0.01		0.9	0.8			0.6		0.33
1CrMoVTiB	0.20	0.5		0.04		1.1	1.0			0.1	0.11	0.70
2CrMoNiWN	0.22	0.7		<0.01		2.1	0.8	0.65		0.7		0.30
12CrMoV	0.15	0.6	0.03	<0.01		12	0.5		0.10	0.5		0.17
12CrMoVNb	0.15	0.8	0.06	<0.01		12	0.6		0.30	0.7		0.30
9CrMoVNb	0.10	0.5	0.05	<0.01		9	1.0		0.10	0.4		0.21
10CrMoVNb	0.12	0.5	0.05	<0.01		10	1.5		0.06	0.6		0.20
9CrMoVWNb	0.10	0.5	0.05	<0.01		9	0.5	1.80	0.06	0.1		0.20
Alloy-80A	0.07			1.40	1.0	20				bal	2.26	

requirements (Figs. 2 and 3) and do not warrant the high costs of such materials. For higher steam inlet temperatures, the use of Alloy-80A is more widespread (Table 1). The creep strength of Alloy-80A is primarily derived from a stable precipitate dispersion of $Ni_3(AlTi)$ γ' particles.

A characteristic feature of Alloy-80A (and other nickel base alloys), below ~550°C, is negative creep (as reflected by the relaxation strength values in excess of the initial loading stress line in Fig. 3). The behavior is due to an ordering reaction and has important practical implications for bolts manufactured from such alloys which contract in this temperature range and may over tighten, leading to failure (Beech *et al.* 1987).

3. The Future

Developments for the future are aimed at the provision of materials for HP turbines with inlet temperatures approaching 700°C (Vanstone and Thornton 1997). There is no shortage of candidate alloys for critical components at these higher temperatures since 700°C is relatively modest in terms of gas turbine technology (the rupture properties for Alloy-80A in Fig. 2 represent a lower bound of the possibilities for such alloys). The challenge is manufacturing components in these high strength high temperature alloys in the large sizes required for steam turbines (gas turbine parts are relatively small).

Bibliography

Beech S M, Holdsworth S R, Mellor H G, Miller D A, Nath B 1987 An assessment of Alloy-80A as a high temperature bolting material for advanced steam conditions. *Proc. Conf. Adv. in Materials Technology for Fossil Power Plants.* ASM/EPRI, Chicago, IL, pp. 403–10

Norton J F, Strang A 1969 Improvements of creep and rupture properties of large 1%CrMoV steam turbine rotor forgings. *JISI* **207**, 193–203

Vanstone R W, Thornton D V 1997 New materials for advanced steam turbines. *Proc. Conf. Adv. Steam Plants.* IMechE, London, C522/007, pp. 87–98

Wickens A, Strang A, Oakes G 1980 High temperature properties of creep resistant 12Cr steam turbine blading steels. *Proc. Conf. Engineering Aspects of Creep.* IMechE, Sheffield, UK, C217, pp. 11–8

D

Direct Borohydride Fuel Cell

S. Suda

Borohydrides (BH_4^-, tetrahydroborates) bonding with Li, K, or Na are classified as chemical complex hydrides with high hydrogen content. For instance, sodium borohydride ($NaBH_4$) contains 10.6 mass% of hydrogen and readily liberates hydrogen by hydrolysis. It has been widely utilized as a convenient source of hydrogen in various chemical industries. Recently, it has been developed as a source of hydrogen in the polymer electrode membrane fuel cell (PEMFC; see *Fuel Cells: Materials*). The hydrolysis,

$$BH_4^- + 2H_2O = BO_2^- + 4H_2 \quad (1)$$

proceeds by applying such catalysts as Pt, Ru, Mg_2Ni, and Ni.

Based on the liberation of hydrogen by hydrolysis, a direct electrochemical utilization of the borohydride has been attracting interest for developing a new type of fuel cell, which is called the "direct borohydride fuel cell (DBFC)." It offers a theoretical potential of 1.64 V, which is 33% higher than that of PEMFC. The negative electrode potential is −1.24 V and it is −0.412 V more negative than in conventional PEMFC.

In an aqueous NaOH solution:

$$BH_4^- + 8OH^- \rightarrow BO_2^- + 6H_2O + 8e \quad E^0 = -1.24\,V \quad (2)$$

$$H_2 + 2OH^- \rightarrow 2H_2O + 2e \quad E^0 = -0.828\,V \quad (3)$$

In Eqn. (2), the BH_4^- ion releases eight electrons, offering a theoretical energy density of 9.3 Wh g^{-1}-$NaBH_4$ in a DBFC. Due to the high energy density, BH_4^-, in an aqueous alkaline solution, is an attractive liquid-type fuel source in a DBFC.

In this section, electrochemical principles of a DBFC in comparison with those of PEMFC are introduced briefly.

1. *Fundamentals of DBFC*

In principle, the electrochemical reaction of a DBFC takes place in an aqueous alkaline solution diluted by NaOH in order to maintain the long-term stability to hydrolysis. The electrode and cell reactions can be expressed as follows:

$$\text{Anode}: BH_4^- + 8OH^- \rightarrow B(OH)_4^- + 4H_2O + 8e^- \quad E_a^0 = -1.24\,V \quad (4)$$

$$\text{Cathode}: 2O_2 + 4H_2O + 8e^- \rightarrow 8OH^- \quad E_c^0 = 0.40\,V \quad (5)$$

$$\text{Over-all reaction}: BH_4^- + 2O_2 \rightarrow B(OH)_4^- \quad E^0 = 1.64\,V \quad (6)$$

A sodium ion (Na^+) plays an important role as a charge carrier in a DBFC. It exists in an aqueous solution of $NaBH_4$ and NaOH on the anode side, and transfers from the anode to the cathode through a polymer electrolyte membrane as a charge carrier, and then, in the NaOH solution, on the cathode side. The principle of the DBFC is schematically illustrated in Fig. 1.

It should be noted that one BH_4^- ion generates eight electrons, and the electromotive force (EMF) yields 1.64 V, which is 1.33 times higher than that of PEMFC and 1.35 times higher than that of DMFC.

Unlike other fuel cells, the DBFC is an electrochemical system that directly applies the "protide (H^-)," which exists in BH_4^- as a state of hydrogen, to release two electrons per protide and generates eight electrons through a reaction between BH_4^- and OH^- in an aqueous alkaline solution at the anode surface, as shown in Eqn. (4). For a detailed description of the protide state, see "Bibliography" (Suda *et al.* 2004).

Table 1 shows a brief comparison between a DBFC and a PEMFC. It is noted that the theoretical energy conversion efficiency of a DBFC is 0.91 and this is higher than 0.83 of a PEMFC.

2. *Practical Aspects of DBFC*

2.1 *Features of DBFC*

A DBFC does not apply gaseous hydrogen as a PEMFC does, but it applies the protide (H^-, hydride ion) in the BH_4^- ion to generate electrons directly at the contact surface of the anode catalyst (Eqn. (2)). The most significant feature of a DBFC is the ease with which large amounts of hydrogen can be stored

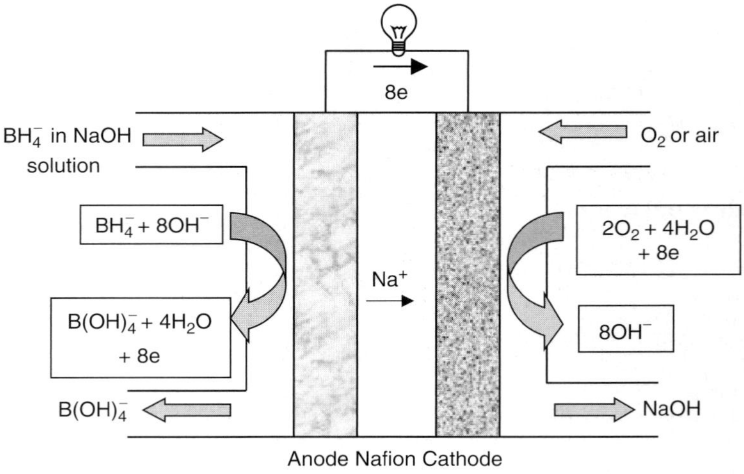

Figure 1
Schematic diagram of the DBFC.

Table 1
A comparison of thermodynamics for DBFC and PEMFC.

DBFC	PEMFC
Cell reaction	Cell reaction
$BH_4^-(l) + 2O_2(g) = B(OH)_4^-(l)$	$H_2(g) + 1/2\ O_2(g) = H_2O(l)$
$\Delta H^0 = -1392.26\,kJ\,mol^{-1}$	$\Delta H^0 = -285.83\,kJ\,mol^{-1}$
$\Delta G^0 = -1267.56\,kJ\,mol^{-1}$	$\Delta G^0 = -237.13\,kJ\,mol^{-1}$
EMF: $E^0 = -\Delta G^0/nF = 1.64\,V$	EMF: $E^0 = -\Delta G^0/nF = 1.23\,V$
Theoretical energy conversion efficiency	Theoretical energy conversion efficiency
$\eta = \Delta G^0/\Delta H^0 = 0.91$	$\eta = \Delta G^0/\Delta H^0 = 0.83$

in a liquid form under ambient conditions. Hydrogen is carried as an aqueous solution in compact and portable tanks, cartridges, and canisters. The electrochemical reaction proceeds under low temperature conditions and can occur even at ambient temperatures. Na^+ is the charge carrier in a DBFC whereas protons (H^+) act as the charge carrier in a PEMFC. As a few molecules of water accompany the Na^+ ion through the membrane from the anode to the cathode, it easily forms a boundary layer of aqueous NaOH at the surface of the cathode which hinders an effective contact with O_2 in the air, thus reducing the power density. The NaOH solution so generated must be removed as quickly as possible and returned to the anode for recycling.

In a PEMFC, H_2 gas is first absorbed physically at the Pt surface and then chemically absorbed at the surface region. Second, the absorbed H_2 is transformed to monoatomic hydrogen (H^0: protium). Finally, H^0 is transformed to proton and electron ($H^0 \rightarrow H^+ + e^-$) when the electrochemical reaction is initiated.

The anode reaction ($BH_4^- \rightarrow B(OH)_4^-$) occurs only at the extreme surface of the electrode material, as shown in Fig. 1. This should be clearly distinguished from the proton (H^+) as the charge carrier in a PEMFC. Noble metals and alloys conveniently applied in a PEMFC for converting H_2 to H^+ are not required in a DBFC. In this sense, the anode material in a DBFC does not necessitate the functions required for hydrogenation and dehydrogenation reactions between hydrogen and metals/alloys.

2.2 Anode and Cathode Materials

Under practical experimental conditions, the observed potential is less negative than the theoretical standard (–1.24 V) given in Eqn. (2). For example, 1 M $NaBH_4$ in 6 N NaOH solution shows a potential of about –1.05 V versus SHE against a nickel electrode and –0.95 V versus SHE against Pd or Pt electrode (Liu *et al.* 2004). The lower potential is due to the partial hydrolysis of BH_4^-, to generate gaseous hydrogen, as shown in Eqn. (1). The electronegative characteristics of a DBFC are apparently reduced according to the degree of hydrolysis. Because of the competition between the hydrolysis and electrochemical reaction, the ideal eight-electron reaction shown in Eqn. (2) is hardly realized in practice during development of DBFCs.

Noble metals, alloys, and their carbon composites have been used conveniently as anode catalysts in PEMFCs and DMFCs. On the other hand, in DBFCs some inexpensive materials can be employed when they exhibit strong chemical and electrochemical durability under long-term alkaline conditions. Various electrode materials used in Ni–MH batteries and alkaline fuel cells (AFCs) are available for DBFCs when replacing noble metal catalysts such as Pt, Pd, and Ru. Recent efforts to replace such noble materials by low-cost materials have attempted to use surface-treated Zr-based Laves phase alloys (Li *et al.* 2003a).

The major problem to be addressed in the DBFC anode development is to suppress the generation of gaseous hydrogen by hydrolysis that results in polarization loss. The Coulombic efficiency on a porous Ni electrode was reduced to 50%. Kim prepared the anode using a Pt catalyst and Nafion®, and found that the cell performance was significantly influenced by the anode composition (Kim *et al.* 2004). It is uncertain from the experimental standpoint to confirm the report that an efficiency of 90% was attained by an Au electrode (Amendola *et al.* 1999). Studies showed that the electrochemical efficiency of DBFCs depends strongly on the anode performance and BH_4^- concentration (Liu *et al.* 2004).

The anode material should have a larger specific surface area at the contact site of the BH_4^- ions to enhance the reaction in Eqn. (4) and to minimize the generation of gaseous hydrogen. Development of anode materials is significantly important for avoiding partial hydrolysis that leads to a loss of the high potential capability of DBFCs.

Since the reaction at the DBFC and the AFC cathodes is the same, catalysts used in AFCs may be applicable in DBFCs as well. A carbon-supported Ag electrode was reported to perform similarly as the Pt electrode (Liu *et al.* 2005). MnO_2 was also reported as a cathode catalyst for DBFCs (Feng *et al.* 2005, Verma *et al.* 2005).

2.3 Recent Achievements

In recent years, the DBFC has been attracting technological interest for the development of more practical and commercially more viable fuel cells, especially for such applications as portable power generators and battery chargers for mobile electronic devices.

A series of extensive efforts in the development of DBFCs has been reported (Li *et al.* 2003a, 2003b, 2004, Liu *et al.* 2003, 2004, 2005). In DBFCs, Zr-based Laves phase alloys as the anode catalyst and a carbon-supported Pt as the cathode catalyst have been applied in combination with a conventional cationic membrane to form a membrane electrode assembly (MEA) that is considerably durable against Na^+-based BH_4^- solutions. A power density of 290 mW cm^{-2} at 60 °C was achieved in a single cell, as shown in Fig. 2. A five-cell stack with an effective area of 67 cm^2 generated 110 W constantly at 60 °C after starting from room temperature conditions without air humidification at the cathode. Figure 3 shows a picture of an experimental model of a 200 W DBFC stack composed of 10 cells.

3. Problems for DBFC

The cell performance of a DBFC, particularly of an MEA, has been greatly improved in recent years. However, there are still some technical difficulties to be solved. First, it is imperative to find a solution for suppressing the generation of H_2 gas on the anode surface. The Coulombic efficiency of DBFCs can be

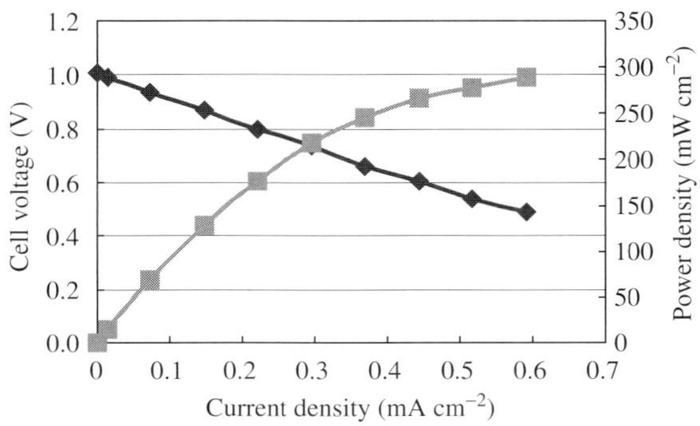

Figure 2
Performance of a single DBFC cell. Experimental details: effective electrode area: 67 cm^2; fuel: 10 wt.% $NaBH_4$ in 20 wt.% NaOH; fuel flow rate: 0.15 l min^{-1}; air flow rate: 5 l min^{-1} (60 °C humidification); and operation temperature: 60 °C.

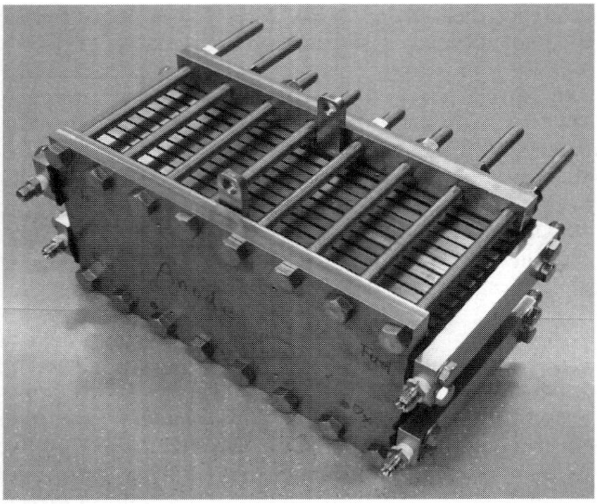

Figure 3
The photo showing a 200 W DBFC stack.

increased only by an improvement of the surface properties and characteristics of the anode materials.

The second problem is related to the BH_4^- concentration as the aqueous solution of $NaBH_4$ and NaOH. As the energy density of a DBFC is basically determined by the BH_4^- concentration, it is desirable to increase its concentration as far as possible. However, $NaBO_2 \cdot 4H_2O$ (sodium metaborate), the oxidized product of BH_4^- after releasing electrons, has a relatively low solubility in such a solution. Precipitation of crystalline material in the DBFC system, especially in anode structures, will lead to a fatal decrease of the electric capacity and loss of cell life. A higher BH_4^- concentration as a H^- source was also found to impose negative effects on the cell performance. The optimum concentration is found in the range between 5 wt.% and 15 wt.% $NaBH_4$.

Another common but serious problem of such fuel cells as DBFCs and DMFCs, which apply diluted liquid fuel, is the volume occupied by the fuel tanks, cartridges, or canisters. It has been pointed out that, in practice, not only liquid-sourced fuel cells such as DBFCs and DMFCs but also H_2-sourced PEMFCs can hardly replace secondary batteries such as the Ni–MH battery and the Li ion battery because of their overall volume-based energy density. DBFCs and DMFCs are not suitable as alternative power sources for these secondary batteries.

See also: Fuel Cells: Electrochemical Reactions; Fuel Cells: Materials

Bibliography

Amendola S, Onnerud P, Kelly M, Petillo P, Sharp-Goldman S, Binder M 1999 A novel high power density borohydride-air cell. *J. Power Sources* **84** (1), 130–3

Feng R X, Dong H, Wang Y D, Ai X P, Cao Y L, Yang H X 2005 A simple and high efficient direct borohydride fuel cell with MnO_2-catalyzed cathode. *Electrochem. Commun.* **7**, 449–52

Indig M E, Snyder R N 1962 Sodium borohydride, an interesting anodic fuel. *J. Electrochem. Soc.* **109**, 1104–6

Kim J H, Kim H S, Kang Y M, Song M S, Rajendran S, Han S C, Jung D H, Lee J Y 2004 Carbon-supported and unsupported Pt anode for direct borohydride liquid fuel cells. *J. Electrochem. Soc.* **151**, 1039–43

Li Z P, Liu B H, Arai K, Suda S 2003a A fuel cell development for using borohydrides as the fuel. *J. Electrochem. Soc.* **150**(7), A868–72

Li Z P, Liu B H, Morigasaki N, Suda S 2003b Preparation of sodium borohydride by the reaction of MgH_2 with dehydrated borax through ball milling at room temperature. *J. Alloys Compd.* **349** (1), 232–6

Li Z P, Liu B H, Arai K, Asaba K, Suda S 2004 Evaluation of alkaline borohydride solutions as the fuel for fuel cell. *J. Power Sources* **126**, 28–33

Liu B H, Li Z P, Suda S 2003 Anodic oxidation of alkali borohydrides catalyzed by nickel. *J. Electrochem. Soc.* **150** (3), A398–402

Liu B H, Li Z P, Suda S 2004 Electrocatalysts for the anodic oxidation of borohydrides. *Electrochem. Acta* **49**, 3097–105

Liu B H, Li Z P, Arai K, Suda S 2005 Performance improvement of a micro borohydride fuel cell operating at ambient conditions. *Electrochem. Acta* **50**, 3719–25

Suda S, Iwase Y, Morigasaki, Li Z P 2004 Metal-hydrogen complex compounds as hydrogen storage materials. *Proc. Adv. Mater. Energy Convers. II, TMS* 123–33

Verma A, Jha A K, Basu S 2005 Manganese dioxide as a cathode catalyst for a direct alcohol or sodium borohydride fuel cell with a flowing alkaline electrolyte. *J. Power Sources* **141**, 30–4

DUPIC Fuel Cycle

Myung Seung Yang and R. Gadsby

The concept of the direct use of spent pressurized water reactor (PWR) fuel in the Canadian deuterium–uranium (CANDU) reactor is based upon dry processing technology to manufacture CANDU fuel from spent PWR fuel material, without separating fissile materials and stable fission products from the spent fuel. The concept was proposed and termed the direct use of spent PWR fuel in CANDU reactors (DUPIC) fuel cycle in a joint development program involving the Korea Atomic Energy Research Institute (KAERI), Atomic Energy Canada Limited (AECL), and the US Department of State in 1991.

The DUPIC fuel cycle builds upon the CANDU reactor fuel cycle flexibility and the synergism between PWR and CANDU reactors. Spent PWR fuel typically contains 0.9 wt.% fissile uranium and 0.6 wt.% fissile plutonium, which exceeds the natural uranium fissile content of 0.711 wt.%. The neutron economy of a CANDU reactor is sufficient to allow DUPIC fuel to be used in a CANDU reactor, which was originally designed for natural uranium fuel, even if the neutron-absorbing fission products originating from the spent PWR fuel remain in the DUPIC fuel. This concept of DUPIC fuel is similar to the Atomics International reduction oxidation (AIROX) process, which was developed in the mid-1950s in the USA.

There are two conventional nuclear fuel cycles: (i) the once-through direct disposal fuel cycle, in which the spent PWR fuel is not recycled but treated for final disposal, and (ii) the recycling fuel cycle, in which the fissile elements such as uranium and plutonium are separated from the fission products by wet processes, the plutonium is refabricated into mixed oxide (MOX) fuel for PWR or fast breeder reactor (FBR), and the recovered uranium is re-enriched for use in PWRs. The DUPIC fuel cycle is suggested as an alternative. In the DUPIC fuel cycle, the spent PWR fuel, having a nominal burn-up of 35 MW d $kgHE^{-1}$ (megawatt days per kilogram of heavy elements), would be processed into DUPIC fuel for an additional burn-up of about 15 MW d $kgHE^{-1}$ (twice the burn-up of natural uranium CANDU fuel) through irradiation in a CANDU reactor.

1. *Benefits of the DUPIC Fuel Cycle*

The DUPIC fuel cycle offers several benefits to countries with PWR and CANDU reactors. The benefits include the additional energy extracted from the fuel in a CANDU reactor, efficient natural uranium utilization, and a significant reduction in spent fuel arising through a proliferation-resistant process of reusing the spent fuel.

1.1 Proliferation Resistance

The proliferation risk in the back end of the fuel cycle has long been a topic of polemics in the international nuclear community. A major attraction of the DUPIC fuel cycle is the high degree of proliferation resistance. Several features of the DUPIC processes significantly enhance the proliferation resistance, relative to fuel cycles employing separated plutonium. First, only dry thermal and mechanical processes are employed, from the initial stage of disassembling spent PWR fuel to the final fabrication of the DUPIC fuel. Second, there is no purposeful separation of fissile isotopes and fission products, and with no selective separation the plutonium concentration remains dilute throughout the entire fabrication process. Third, all stages of the fabrication process are highly radioactive, making physical access to the material, and its removal, extremely difficult. Fourth, the processing facility is entirely self-contained, and therefore there is no transport of intermediate products—only spent PWR fuel as an input and finished DUPIC fuel as the final product. With these inherent self-protecting attributes of the DUPIC process, the proliferation barriers that are present in spent fuel are also present in the whole DUPIC fuel cycle.

1.2 Savings in Uranium Resources

The DUPIC fuel cycle reduces the amount of mined uranium required to produce a given amount of electrical energy. For an equilibrium system in which the PWR-to-CANDU reactor ratio of electrical generation is such that the spent PWR fuel provides all the fuelling needs of the CANDU reactors, the DUPIC fuel cycle could reduce the natural uranium requirements by about 25% compared with direct disposal in a two-reactor open system, in which CANDU reactors are fuelled with natural uranium and PWR reactors are fuelled with enriched uranium, assuming nominal PWR and CANDU–DUPIC burn-ups of 35 MW d $kgHE^{-1}$ and 15 MW d $kgHE^{-1}$, respectively.

1.3 Reduced Quantity of Spent Fuel

Approximately 40–60% more energy can be derived from the PWR fuel by burning it again as DUPIC fuel. In an equilibrium system of PWR and CANDU reactors, this results in a threefold reduction in the quantity of spent fuel arising per unit electricity, compared with direct disposal in a two-reactor open system. Thus, reusing spent PWR fuel as DUPIC fuel may offer an option for reducing the spent fuel management problem. This reduction in the total quantity of spent fuel is attractive because it can also reduce the handling, storage, and transportation activities prior to final disposal.

2. *The DUPIC Fuel Manufacturing Process*

The key feature of the DUPIC fuel manufacturing process is the oxidation and reduction of oxide fuel (OREOX) process. During this dry process, the spent PWR fuel material is sequentially oxidized and reduced. The processes of oxidation and reduction cause crystallographic transformations in the pellet matrix, causing the pellets to break into to a fine powder. Once the powder is conditioned to improve its sinterability, the rest of the DUPIC fuel fabrication process is similar to the the conventional CANDU fuel manufacturing processes that utilize the powder–pellet route, as depicted in Fig. 1. Because the fuel material remains highly radioactive throughout the process, all processing must be done in hot cells.

The fabrication process begins with the disassembly of the PWR fuel assembly to extract the fuel rods. The extracted rods are declad using a mechanical/thermal method. In the oxidative decladding process, for example, a slit rod is subjected to oxidation at about 450°C in air. This causes the enclosed fuel pellets to swell, opening up the cladding, and releasing the fuel material. The fuel material is then subjected to thermal cycles of oxidation and reduction at about 450°C in air and about

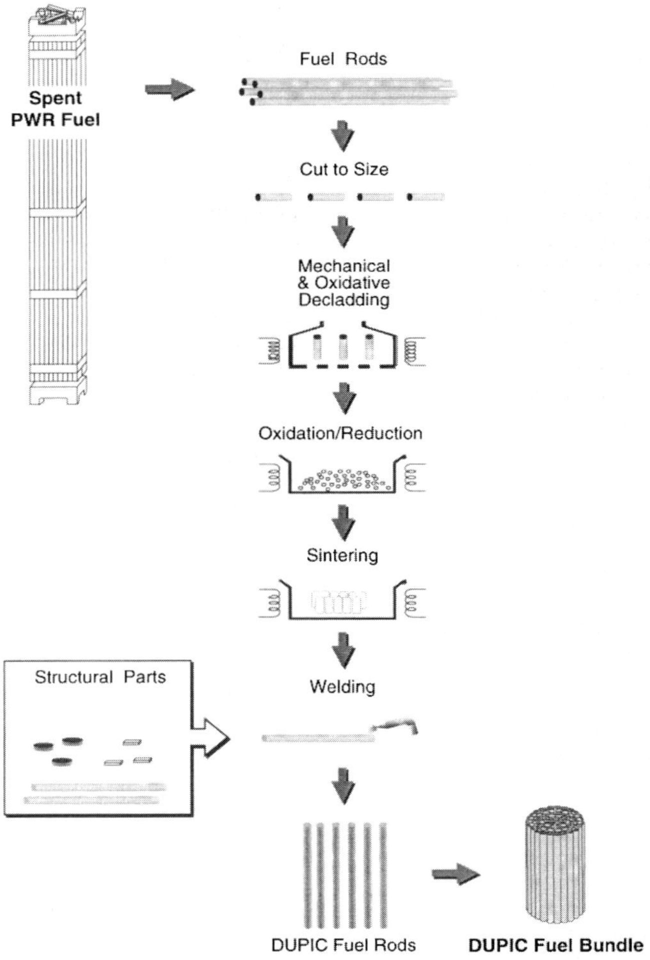

Figure 1
DUPIC fuel fabrication process.

700°C in an argon–4% hydrogen atmosphere, respectively, to break the fuel into a fine powder. Subsequent powder conditioning, such as milling, improves the sinterability of the powder. From this point on, fabrication processes such as precompaction, granulation, mixing with lubricants, final compaction, sintering, and grinding are similar to conventional CANDU fuel fabrication. Careful controls on the sintering and dry centerless grinding of sintered pellet are required due to the relatively poor processing characteristics of the powder.

Once the DUPIC fuel pellets are prepared, they are loaded into the sheaths. The sheaths are prepared in an inactive working area used for attaching various appendages and applying a graphite coating inside the tube. The seal welding of end cap following pellet loading is done in a hot cell using tungsten inert gas welding or optical fiber transmission laser welding. The assembling of the DUPIC fuel bundle is also performed in a hot cell.

The volatile fission products such as cesium, iodine, and ruthenium released during the thermal process are trapped by a ceramic filter, and gaseous products such as krypton and xenon are cryogenically trapped.

Besides the fabrication technology, other technological challenges lie in the operation and maintenance of the fabrication equipment. The equipment design must be simple for easy remote operation, and modular for easy repair and replacement. Electronic components should be separated from the mechanical parts of the equipment and, if possible, located in an inactive area to avoid degradation due to the high radiation fields found in hot cells. Quality control measures to manufacture nuclear fuel to satisfy the tight nuclear fuel specifications must also be established.

The OREOX process is potentially a much simpler process than conventional wet chemical reprocessing. The DUPIC process does not involve the separation of sensitive materials, and the final fuel product is fabricated in the same facility in which the spent fuel is processed. In contrast, in conventional wet reprocessing, the spent fuel is separated into uranium, plutonium, fission products, and actinides, and then the MOX fuel is fabricated remotely in a different processing facility. Hence, DUPIC fuel processing and fabrication offer the potential of being less expensive than conventional reprocessing and MOX fuel fabrication.

3. *DUPIC Fuel Performance*

A premise in the DUPIC fuel cycle development is to make the maximum use of existing CANDU reactors to burn DUPIC fuel. The inherent features of the CANDU reactor, such as excellent neutron economy and on-power refuelling, result in an unsurpassed degree of fuel cycle flexibility, enabling DUPIC fuel to be utilized safely and economically in existing reactors. Detailed reactor physics assessments, such as maximum channel and bundle power, channel power peaking factor, power distribution, and the reactor regulating system, have shown that the DUPIC fuel can be burned in current CANDU-6 reactors without significant modifications. However, a detailed assessment of the handling of radioactive DUPIC fuel in the reactor fuelling system will be required.

From a fuel performance viewpoint, a major characteristic of DUPIC fuel is its initial content of fission products as impurities. As a result, the thermal conductivity is expected to be lower than for fresh fuel, and the centerline temperature of DUPIC fuel pellets would be somewhat higher than in standard CANDU fuel. This may have a negative impact on the fuel performance behavior such as fission gas release, swelling, and thermal expansion of the pellets. Also, the burn-up of DUPIC fuel is twice that of natural uranium CANDU fuel. Hence, the behavior of the high-burn-up DUPIC fuel needs to be further evaluated through irradiation tests in research reactors.

See also: Heavy Water Reactor Fuel Design and Performance; Light Water Reactor Fuel Design and Performance; Nuclear Fuel Cycles; Nuclear Reactor Fuel Fabrication (Including Quality Control)

Bibliography

Choi H B, Rhee B W, Park H S 1997 Physics study on direct use of spent pressurized water reactor fuel in CANDU (DUPIC). *Nucl. Sci. Eng.* **126**, 80–93

Lee J S, Song K C, Yang M S, Chun K S, Rhee B W, Hong J S, Park H S, Rim C S, Keil H 1993 Research and development program of KAERI for DUPIC. *Proc. Int. Conf. Future Nuclear Systems: Emerging Fuel Cycle and Waste Disposal Options. Global '93*, Seattle, WA, pp. 733–9

Yang M S, Kim B G, Song K W, Bae K K, Kim S S, Kim W K, Jung I H, Park H S 1997 Characteristics of DUPIC fuel fabrication technology. *Proc. Int. Conf. Future Nuclear Systems: Challenge towards Second Nuclear Era with Advanced Fuel Cycles. Global '97*, Yokohama, Japan, pp. 535–7

E

Electrochemical Light Emitting Cells from Semiconducting Polymers

A. J. Heeger and J. Gao

Polymer light-emitting electrochemical cells (LECs) provide an alternative approach to electroluminescence (EL) from semiconducting polymers (Pei *et al.* 1995). Unlike polymer light-emitting diodes (LEDs), which are based on the undoped, intrinsic properties of semiconducting polymers, the polymer LEC is a solid-state device in which the semiconducting polymer is doped *in situ*. The active material of an LEC is a phase-separated mixture of luminescent polymer and polymer electrolyte (Cao *et al.* 1996). As a result, LEC materials are both electrically and ionically conductive. When a large enough voltage bias is applied to an LEC, a dynamic p–i–n junction is formed *in situ* by electrochemical doping, thereby facilitating injection of electrons and holes, respectively, into the n- and p-type regions. Light is emitted from an LEC through the radiative recombination of electrons and holes in the undoped, intrinsic region.

The p–i–n junction nature of LEC operation offers significant advantages compared to polymer LEDs, especially when the dynamic junction is stabilized. More importantly, an LEC is the first semiconducting polymer device that is based on a doping-induced homojunction. In terms of operating mechanism, an LEC is more like an inorganic p–n junction LED than is a polymer LED. This article outlines the operating mechanism and device characteristics of LECs. In addition, frozen-junction LECs, which offer many of the combined advantages of LECs and LEDs, are introduced.

1. *LEC Device Configuration*

An LEC can be made in both sandwich and surface cell configurations, similar to a polymer LED. Figure 1 shows the sandwich cell configuration of an LEC. The active material of an LEC is a mixture of two components: a luminescent, semiconducting polymer and a polymer electrolyte. The semiconducting polymer is chosen from the class of soluble conjugated polymers; typical examples are the soluble derivatives of poly(phenylene vinylene) (PPV). Molecular structures of typical LEC materials are shown in Fig. 1. The polymer electrolyte is a solid-state ionic conductor formed by the dissolution of a salt (e.g., $LiCF_3SO_3$) in certain ion-solvating polymers (e.g., polyethylene oxide, PEO).

The processing of a sandwich LEC includes three steps. First, a blend solution is made by dissolving the semiconducting polymer and polymer electrolyte in a common solvent (e.g., cyclohexanone). The blend solution is then spin-cast onto a half-coated ITO glass substrate to form a polymer thin film. Finally, top electrodes (e.g., aluminum) are thermally evaporated on top of the polymer film through a shadow mask. A surface cell LEC can be made by spin-casting an LEC blend film onto a substrate with prepatterned electrodes. LECs are fabricated by these relatively simple processing steps, an advantage which is a more general feature of polymer semiconducting devices (Heeger 1998).

2. *LEC Operating Mechanism*

The operating mechanism of an LEC is shown schematically in Fig. 2. At zero bias, the semiconducting polymer is in the neutral, undoped form. At voltages approximately equal to E_g/e (typically 2–3 V), where E_g is the energy gap of the semiconducting polymer, an electrochemical redox reaction is initiated, and the solvated free ions in the polymer electrolyte redistribute to compensate the injected electronic charges. As a result, the polymer blend film is p-type doped near the anode, n-type doped near the cathode,

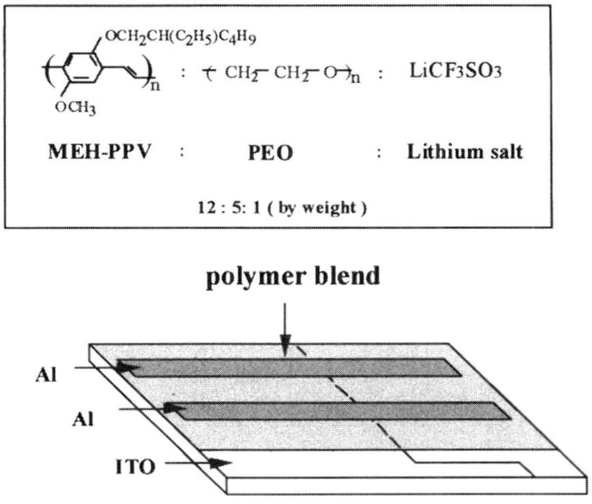

Figure 1
LEC device configuration and molecular structures.

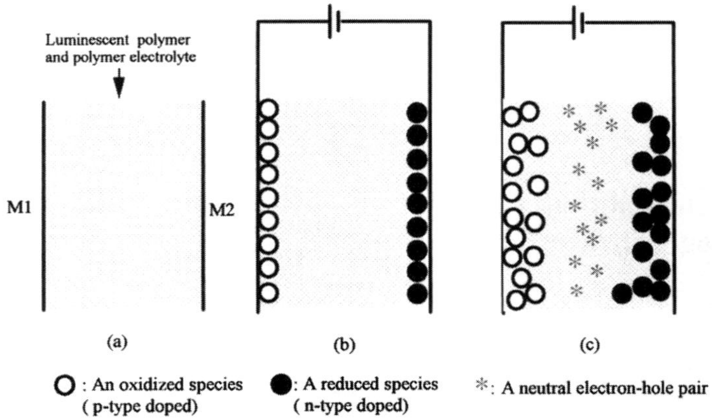

Figure 2
Schematic diagram of LEC operation. The two metal electrodes are labeled M1 and M2. (a) Cell before an external voltage bias is applied. (b) Doping of opposite sides begins as a large enough bias is applied. (c) Final doping profile and radiative decay.

and a dynamic p–i–n junction is created *in situ*. Injected electrons and holes recombine radiatively in the insulating (i) region where the light is emitted. The emission spectrum of an LEC is typically the same as that of an LED made with the same luminescent polymer.

The electrochemical doping in LECs is reversible, and the p–i–n junction formed *in situ* is dynamic. When the external bias is removed, the LEC relaxes by discharging, and the semiconducting polymer returns to the undoped neutral form. In addition, changing the polarity of the applied bias can reverse the polarity of the p–i–n junction. Thus, polymer LECs show bipolar current–voltage (*I–V*) and light emission–voltage (*L–V*) characteristics, as shown in Fig. 3.

In early publications on LECs, the doping-induced homojunction was described as a p–n junction (Pei *et al.* 1995, 1996). However, since the redox occurs near the anode and cathode, and since the ions in LECs are mobile, a graded p–n junction or a p–i–n junction, is formed (Manzanares *et al.* 1998). Throughout this article, therefore, we will refer to the doping-induced junction in LECs as a p–i–n junction.

3. *Direct Confirmation of Electrochemical Junction Formation in LEC Operation*

As discussed above, the LEC operating mechanism involves the formation of a dynamic p–i–n junction by *in situ* electrochemical doping. This basic picture of LEC operation has been directly confirmed by a series of delicate experiments.

Optical beam-induced current (OBIC) microscopy has been used to image the structure of the junction in an LEC (Dick *et al.* 1996). In OBIC measurements, the electric field profile across the LEC structure is measured by scanning a focused laser beam across a surface cell LEC. The surface cell LEC is prepared with a pair of gold strips as the contact electrodes, separated by a gap of approximately 20 µm. When the excitation beam is within the junction, photogenerated electrons and holes are separated by the built-in electric field, resulting in an external photovoltaic signal. When the photoexcited region is confined either to the p-type side or to the n-type side, there will be no external signal. From the measured OBIC photovoltage vs. position characteristics, the built-in electric field is found to be zero near the two contacts and to have a maximum in the i-region; the peak value of the built-in electric field and the spatial width of the field distribution change with the bias voltage. Spatially resolved photoluminescence (PL) and EL have also been measured on the same device. The built-in electric field profile measured by OBIC microscopy is consistent with the PL and EL profiles. Thus, the OBIC data directly confirm the electrochemical doping and junction formation in an operating LEC. Figure 4 shows the EL from a surface cell LEC.

The electrochemical junction formation in an operating LEC has also been confirmed by a.c. impedance and electroabsorption (EA) measurements (Li *et al.* 1998, Campbell *et al.* 1998, Gao *et al.* 1999). Both a polymer LEC and a polymer LED can be modeled as a parallel RC circuit in series with a contact resistance.

In a polymer LED, the semiconductor layer is fully depleted as a result of the low carrier concentration in the undoped polymer film. Consequently, the capacitance is nearly independent of bias voltage, although a trapping effect has been observed (Campbell *et al.* 1995). In addition, the EA signal at the second harmonic frequency of the applied a.c. bias is independent of the applied d.c. bias.

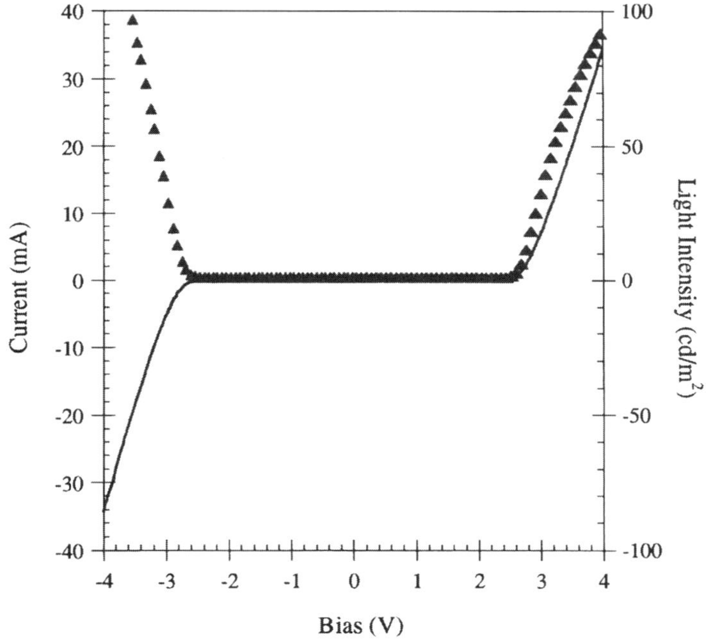

Figure 3
Bipolar current–voltage and light emission–voltage characteristics of a dynamic-junction LEC.

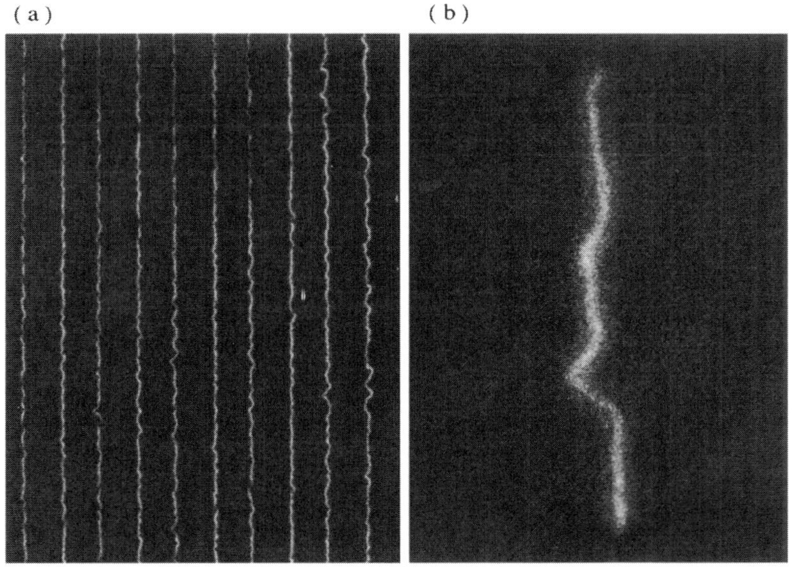

Figure 4
Photograph of a polymer LEC in a surface cell configuration with interdigitated gold electrodes (dark lines); the spacing between electrodes is 15 μm. Yellow-green light emission can be seen when a voltage bias of 4 V is applied; the width of the junction is about 1–3 μm.

In a polymer LEC, however, both the EA signals and the capacitance have been observed to increase sharply above the redox threshold voltage. The increase in capacitance and EA signal confirms the formation of a dynamic p–i–n junction, whose width (the width of the intrinsic region) decreases as the applied d.c. bias increases, as demonstrated by OBIC measurements.

4. *LECs vs. LEDs*

Polymer LEDs can be modeled as metal–insulator–metal (MIM) structures; they are field-driven devices in which the electron and hole currents are limited by carrier injection at the metal–semiconductor contact (Parker 1995). As a result, polymer LEDs are characterized by sensitivity to electrode materials and device thickness. Owing to the p–i–n junction nature of LEC operation, polymer LECs possess some advantages compared to polymer LEDs:

(i) *High efficiency*. Since the doped semiconducting polymer has low resistivity and tends to form ohmic contacts with metal electrodes, more nearly balanced electron and hole injection is expected. Thus, in LECs higher quantum efficiencies can be achieved than in LEDs made with same luminescent polymer. External quantum efficiencies of 2–4% have been demonstrated.

(ii) *Low operating voltages*. The light turn-on voltage of an LEC is roughly equal to E_g/e (2–3 V). Red, green, blue, and broadband white emission from polymer LECs has been demonstrated with low operating voltages (<4 V).

(iii) *Insensitivity to electrode materials*. The charge carrier injection in an LEC is not sensitive to the work function of the electrode material. Air-stable metals such as aluminum are used for electron injection in LECs (in contrast to polymer LEDs where reactive low work function metals are required for good electron injection). Moreover, polymer LECs can be made in the surface cell configuration with symmetric contact electrodes (e.g., Au–polymer–Au).

(iv) *Insensitivity to film thickness*. LECs are voltage-driven devices. The applied bias voltage drops mostly in the more resistive undoped intrinsic region, which only occupies a small portion of the total LEC device thickness. As a result, the device characteristics of LECs are not sensitive to device thickness. Surface cell LECs (see Fig. 4) have been demonstrated with an operating voltage of only few volts. For surface cell LEDs with similar dimensions, however, the operating voltage is few hundred volts. In addition, thicker polymer films avoid pinholes and reduce device leakage in sandwich LECs.

These advantages of polymer LECs result from electrochemical doping and p–i–n junction formation in LEC operation. However, unlike an inorganic p–n junction, the p–i–n junction in an LEC is dynamic, i.e., it relaxes by discharging when the externally applied voltage is removed. A new operation requires that the junction to be rebuilt before the maximum light output can be achieved. As a result, LECs also have serious disadvantages:

(i) *Slow response*. The light emission turn-on of an LEC is slow because seconds or even minutes are required for the redistribution of ions during the creation of the p–i–n junction.

(ii) *Poor stability at high operating voltage*. When the applied bias exceeds the electrochemical stability window of the system (semiconducting polymer, electrolyte, and metal electrodes), the device degrades owing to "overdoping." As a result, the operating voltage must be limited to below ~4 V to avoid device failure. Thus, high brightness cannot be achieved in dynamic-junction LECs.

(iii) *Hysteresis behavior*. The cyclic *I–V* and *V–L* curves of polymer LECs exhibit hysteresis. The hysteresis is caused by the interplay of the slow ionic processes and the fast electronic processes involved in LEC operation.

(iv) *Bipolar light emission and I–V characteristics*. The bipolar characteristics of LECs will introduce cross-talk in column–row addressable displays.

Consequently, dynamic-junction LECs cannot be used in any serious display applications. In order to solve the problems of dynamic-junction LECs, a stabilized junction is needed so that after the junction is formed only fast electronic processes are involved in the operation. The dynamic p–i–n junction in an LEC can be stabilized by the "frozen-junction" approach, as discussed below.

5. *Ion Transport in Polymer Electrolytes*

Ion transport in polymer electrolytes takes place only in the amorphous, elastomeric phase of the ion–polymer complexes. In PEO–Li^+ electrolytes, the motion of Li^+ ions is enabled by the simultaneous making and breaking of bonds with the oxygen atoms of the PEO chain, and is assisted by segmental motion of the PEO chain. Therefore, the ion transport in polymer electrolytes is thermally activated and is related to the glass transition phenomena. The temperature dependence of the ionic conductivity can be expressed by the Vogel–Tamman–Fulcher (VTF) equation:

$$\sigma = AT^{-1/2} \exp[-E_a/(T - T_0)] \tag{1}$$

where A is a constant, T is temperature, T_0 is the "ideal" glass transition temperature (about 50°C below the actual glass transition temperature, T_g), and E_a is the activation energy. The strong temperature dependence of ionic conductivity is a general feature.

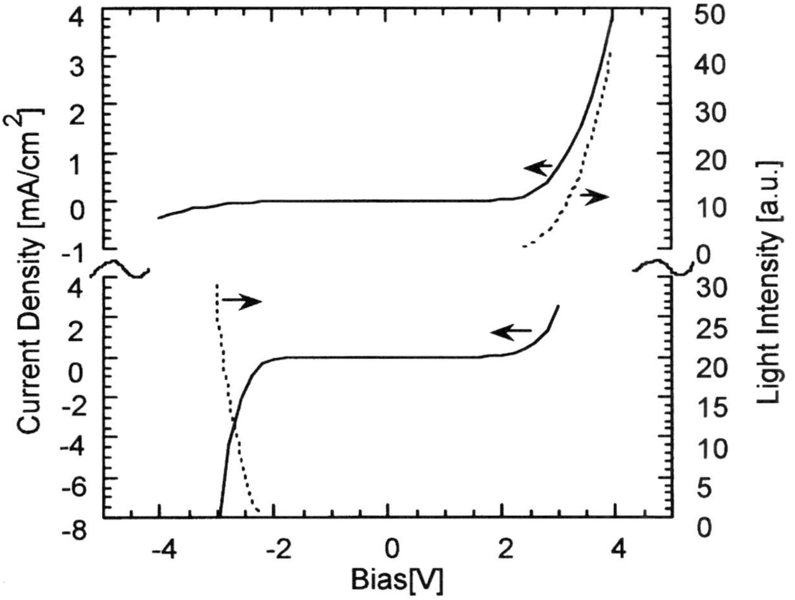

Figure 5
Current–voltage and light emission–voltage characteristics of a frozen-junction LEC. The LEC is prebiased at +4 V (upper curves) and −3 V (lower curves).

At temperatures below T_g, the ion mobility is negligible because the chain segmental motion is quenched.

This "freeze-out" of the ion mobility allows the stabilization of the dynamic p–i–n junction in LECs by lowering the operating temperature. At temperatures below T_g of the ion transport polymer (about 208 K for PEO), the ions will be immobilized owing to the quenching of the polymer chain segmental motion. Both the doping profile and the p–i–n junction will be stabilized, giving "frozen-junction" LECs.

6. *Frozen-junction LECs*

In order to make a frozen-junction LEC, a voltage bias (less than 4 V) is applied to the LEC at room temperature to initiate the electrochemical redox and to build the p–i–n junction. The LEC is then cooled to 100 K under bias in a cryostat. The voltage bias applied to the LEC at room temperature and during the cooling process is called the prebias. Without a prebias, the cooled device is highly resistive and nonemitting, because no junction has been created. Frozen-junction LECs based on MEH-PPV:PEO:Li were first demonstrated at 100 K (Gao *et al.* 1997). The unique characteristics of these devices distinguish them from both polymer LEDs and dynamic-junction LECs. Figure 5 shows the *I–V* and *L–V* characteristics of a frozen-junction LEC at 100 K. The LEC exhibits both rectifying diode behavior and unipolar light emission, with polarity the same as that used to create the junction. The light turn-on voltage is about 2 V, typical for an LEC fabricated with MEH-PPV as the light-emitting polymer. These data are all obtained from the same device. After testing the +4 V prebiased cell at 100 K, the LEC is heated to 300 K. A cooling process to 100 K is repeated with an applied voltage of −3 V. The *I–V* and *L–V* characteristics shown in Fig. 5 for a prebiased LEC are stable and repeatable. However, when the cooling process is carried out without a prebias, the device has a much lower current level and no light emission. The rectification ratios of these frozen-junction LECs are quite small compared to MEH-PPV LEDs. It is also found that the rectification ratio increases as the operating temperature increases (below 200 K). A frozen-junction LEC can be regarded as a diode in series with a resistor. The resistance of the resistor arises mostly from the undoped region, and it becomes dominant at low temperatures, giving rise to reduced rectification ratio.

Frozen-junction LECs have much faster response than dynamic-junction LECs. Figure 6 compares the temporal response of LEC light emission between a frozen-junction LEC and a dynamic-junction LEC. A 50% duty cycle square wave voltage signal is applied to the LEC. While it takes seconds or even minutes for the dynamic-junction LEC to achieve maximum light emission, the light turn-on time of the frozen-junction LEC is found to be only about 40 μs. In fact

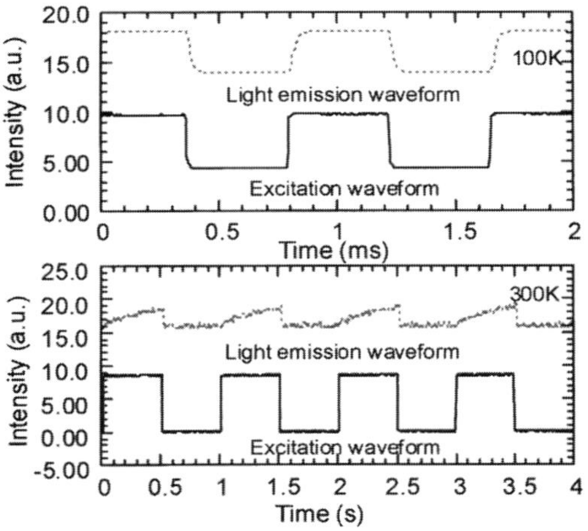

Figure 6
Temporal response of LEC light emission. Top: frozen-junction LEC operated at 100 K; bottom: dynamic-junction LEC operated at 300 K.

these data are limited by the detector response; even faster response (678 ns) is revealed when a fast photodiode is used to detect the light emission, about 10^6 times faster than any dynamic-junction LEC. The fast response of frozen-junction LECs implies that only electronic processes are involved in the operating mechanism.

Because the electrochemistry is shut off in the frozen-junction device, frozen-junction LECs are free of hysteresis and stable at high driving voltages. In fact, frozen-junction LECs have overcome all the drawbacks of dynamic-junction LECs listed in Sect. *4*. Frozen-junction LECs offer the combined advantages of polymer LEDs and LECs, such as fast response, unipolar light emission, and high stability.

The only remaining problem is the operating temperature. Detailed studies have shown that, for frozen-junction LECs based on MEH-PPV:PEO:Li, the highest operating temperature is around 200 K, close to the T_g of PEO. When poly(ethylene adipate) (PEA) is used as the ion transport polymer, frozen-junction LECs have been demonstrated with operating temperatures as high as 273 K. The ultimate goal of making room temperature frozen-junction LECs can be achieved by using polymer electrolyte with T_g higher than room temperature. Frozen-junction behavior has been observed in LECs using crown ether as the electrolyte, although the junction lasts only a few days.

7. Conclusions

LECs are novel polymer light-emitting devices, fundamentally distinct from polymer LEDs. The operation of an LEC involves the formation of a dynamic p–i–n junction by *in situ* electrochemical doping. The dynamic junction in an LEC can be frozen, or stabilized by lowering the temperature to below the T_g of the polymer electrolyte. Frozen-junction LECs offer combined advantages of polymer LEDs and LECs. The future goal of LEC research is to make frozen-junction LECs that can be operated at high temperatures (e.g., room temperature). This can be done by developing suitable polymer electrolytes with high T_g or by cross-linking the ion transport polymer after junction formation.

A frozen-junction LEC is the first polymer device whose operation is based on a stable, doping-induced homojunction. In addition, the idea of a frozen junction can be used in other device applications. For example, polymer frozen-junction photovoltaic cells have been demonstrated with a unique polarity-reversible photovoltaic response (Gao *et al.* 1998).

See also: Light Emitting Diodes from Molecular and Polymer Semiconductors

Bibliography

Campbell I H, Smith D L, Ferraris J P 1995 Electrical impedance measurements of polymer light-emitting diodes. *Appl. Phys. Lett.* **66**, 3030–2

Campbell I H, Smith D L, Neef C J, Ferraris J P 1998 Capacitance measurements of junction formation in polymer light-emitting electrochemical cells. *Appl. Phys. Lett.* **72**, 2565–7

Cao Y, Yu G, Heeger A J, Yang C Y 1996 Efficient, fast response light-emitting electrochemical cells: electroluminescent and solid electrolyte polymers with interpenetrating network morphology. *Appl. Phys. Lett.* **68**, 3218–20

Dick D J, Heeger A J, Yang Y, Pei Q 1996 Imaging the structure of the p–n junction in polymer light-emitting electrochemical cells. *Adv. Mater.* **8**, 985–7

Gao J, Heeger A J, Campbell I H, Smith D L 1999 Direct observation of junction formation in polymer light-emitting electrochemical cells. *Phys. Rev. B: Rap. Commun.* **59**, R2482–5

Gao J, Yu G, Heeger A J 1997 Polymer light-emitting electrochemical cells with frozen p–i–n junction. *Appl. Phys. Lett.* **71**, 1293–5

Gao J, Yu G, Heeger A J 1998 Polymer p–i–n junction photovoltaic cells. *Adv. Mater.* **10**, 692–5

Heeger A J 1998 Light emission from semiconducting polymers: light-emitting diodes, light-emitting electrochemical cells, lasers and white light for the future. *Solid State Commun.* **107**, 673–9

Li Y, Gao J, Yu G, Cao Y, Heeger A J 1998 AC impedance of polymer light-emitting electrochemical cells and light-emitting diodes: a comparative study. *Chem. Phys. Lett.* **287**, 83–8

Manzanares J A, Reiss H, Heeger A J 1998 Polymer light-emitting electrochemical cells: a theoretical study of junction formation under steady-state conditions. *J. Phys. Chem. B* **102**, 4327–36

Parker I D 1995 Carrier tunneling and device characteristics in polymer light-emitting diodes. *J. Appl. Phys.* **75**, 1656–6

Pei Q, Yang Y, Yu G, Zhang C, Heeger A J 1996 Polymer light-emitting electrochemical cells: in situ formation of a light-emitting p–n junction. *J. Am. Chem. Soc.* **118**, 3922–9

Pei Q, Yu G, Zhang C, Yang Y, Heeger A J 1995 Polymer light-emitting electrochemical cells. *Science* **269**, 1086–8

Electroluminescent Phosphors

M. Leskelä, W. -M. Li, and M. Ritala

Electroluminescence (EL) is a phenomenon where light is generated by an electric field. The phenomenon is applied in display devices, which can be divided into two categories: (i) low field devices which are better known as light emitting diodes (LED), where the light is generated by electron–hole pair recombination at a *p–n* junction, and (ii) high field devices where the impact of the high energy electrons excites the luminescent centers in the phosphor material and the EL emission is obtained upon the relaxation of the excited state of the luminescent center (Ono 1995). The electric field in the high field devices is in the order of $10^6\,\mathrm{Vcm^{-1}}$. The high electric field devices can further be classified as a.c. or d.c. operated devices and, based on the type of the phosphor material, as powder or thin film devices. From the practical point of view the a.c. operated thin film electroluminescent (TFEL) devices are the most important ones and they are the topic of this article.

The development of TFEL devices began in the early 1970s. By 2000, TFEL devices had a solid position in the monochrome flat panel display market in demanding applications where the advantages of TFEL devices are appreciated. These advantages include: fully solid structure, which is not sensitive to vibrations, temperature variations, and dusty environments; wide viewing angle; high contrast even in ambient light; fast response allowing video rate; high resolution; and long lifetime. A wide selection of monochrome devices based on the yellow emission of ZnS:Mn has been commercially available since the early 1980s. Multicolor devices have also been on the market since 1993 but the commercialization of full color devices was delayed because of the lack of an efficient blue phosphor. At the time of writing, the most promising approach to the full color devices is that using two phosphors with wide emission bands (ZnS:Mn/SrS:Ce or SrS:Cu), resulting in white emission. The different colors are obtained by color filters (Fig. 1(b)) (Törnqvist 1997).

1. Device Structure

A TFEL device has a MISIM (metal–insulator–semiconductor–insulator–metal) structure where the luminescent layer, typically 0.5–1 μm thick, is sandwiched between two dielectric layers (0.2–0.4 μm). The MISIM structure is completed by transparent conducting ITO (indium–tin–oxide) column electrodes and metallic row electrodes (Fig. 1). The thin film package is grown on a glass substrate. Viewing in the traditional devices, as shown in Fig. 1(a), is through the glass substrate but an inverted structure (Fig. 1(b)) is also applicable. In the inverted structure the substrate may be opaque and the upper electrode is transparent, to allow viewing through it. The light emitting area is fixed by the crossing of the two perpendicular electrodes.

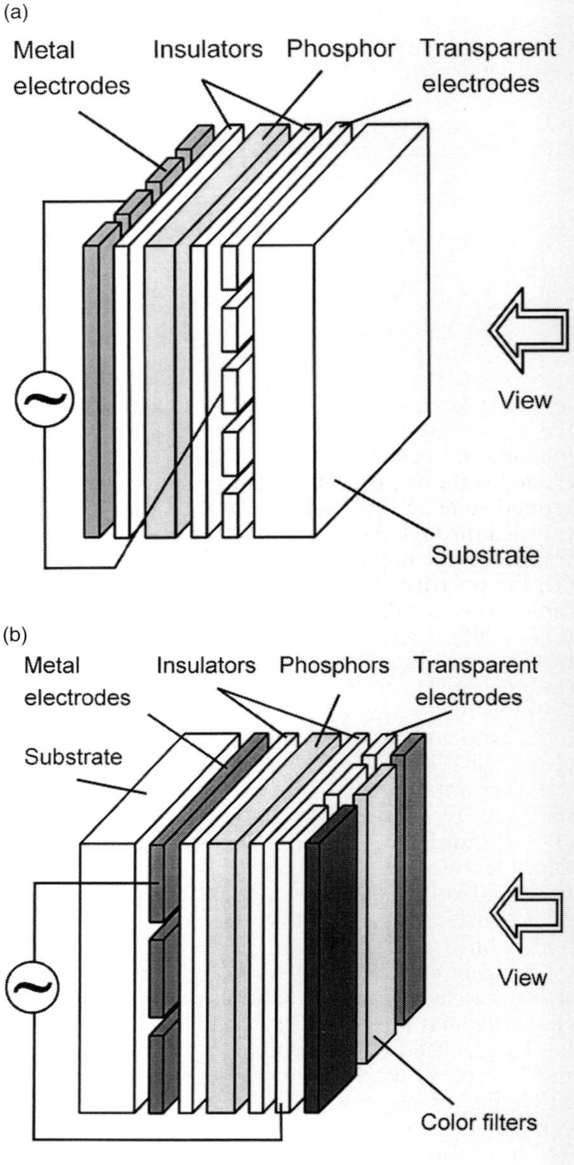

Figure 1
Cross-section of a monochrome ACTFEL device with a traditional structure (a) and a multicolor device based on broad band phosphors and color filters with an inverted structure (b).

The small (0.5–2 in, approx. 1.25–5 cm) high-resolution head-mounted displays (AMEL) have in the late 1990s gained particular interest. These devices are realized by active matrix technology where the active matrix circuitry is made on silicon-on-insulator (SOI) wafers allowing preparation of high voltage transistors on a small area. The films grown on the circuitry are continuous and because the underlying surface is nonplanar, conformality is required from the deposition. The active matrix addressing allows the use of high frequencies (5 kHz), which give high brightness, and in the case of the ZnS:Mn/SrS:Ce phosphor layers improves white emission by enhancing the bluish-green emission of SrS:Ce compared with the normal 60 Hz (King 1996).

2. *EL Mechanism*

The EL emission process can be divided into the following steps: (i) once the voltage is high enough, above the so-called threshold voltage, electrons are injected into the phosphor layer from the interface states between the phosphor and insulator; (ii) the injected electrons gain energy in the high field; (iii) the high-energy electrons (hot electrons) excite the luminescent centers by impact excitation; (iv) the excited state of the luminescent center relaxes to the ground state and EL emission is obtained; (v) having traveled through the phosphor layer, the electrons are trapped at the opposite phosphor/insulator interface; (vi) the polarity of the electrodes is reversed and the same process takes place in the opposite direction (Ono 1995). Impact excitation is the widely accepted mechanism for ZnS-based phosphor materials (Mueller 1984). For alkaline earth metal sulfides, CaS:Eu for example, another excitation mechanism—field induced ionization—has been proposed (Crandall 1987).

Important characteristics of an EL device are luminance (L), luminous efficiency (η), current density (I) and transferred charge (ΔQ). They all are voltage dependent and show a threshold (Fig. 2). Below the threshold voltage (V_{th}), light emission is almost zero. At V_{th}, which is also known as the turn-on voltage and defined as the voltage where $L = 1\,\mathrm{cd\,m^{-2}}$, the injection of electrons into the phosphor layer begins to be in significant amounts. Above V_{th} all the characteristics show a steep increase and a saturation level is reached at higher voltages. A steep increase in the L–V curve within a 20–50 V range is desired because the on-off modulation voltage of the device cannot be large. Efficiency also increases above V_{th} and has its maximum just at the steepest point of the L–V curve. Luminance is, of course, also dependent on the frequency of the a.c. driving voltage. In standard reports the luminance values are given for a frequency of 60 Hz and a voltage 40 V above the threshold (L_{40}).

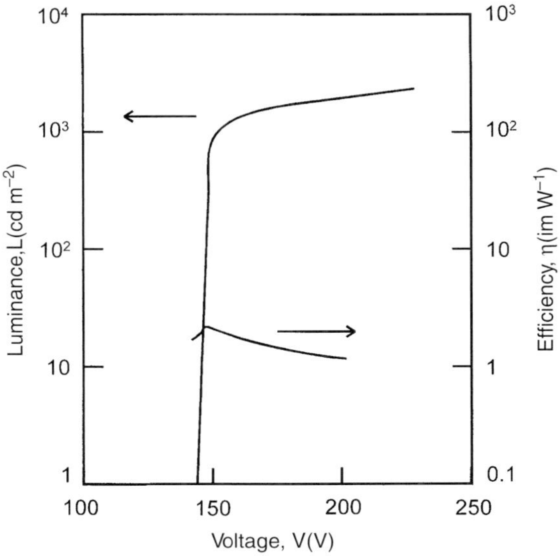

Figure 2
Typical luminance (L) and efficiency (η) versus voltage curves of a ZnS:Mn ACTFEL device.

EL devices should be very stable so that the threshold voltage does not change and the L–V curve should not soften (a decrease of the slope) within a time interval of thousands of hours. Often, changes take place in as-prepared devices but then they stabilize. Therefore, devices are pre-aged before use.

3. *Preparation of Thin Film EL Devices*

Sputtering, evaporation, and atomic layer epitaxy (ALE) are used in industrial fabrication of the films needed in TFEL devices. All the methods, when optimized, produce phosphor films with rather similar EL performances as has been exemplified by many studies on ZnS:Mn and SrS:Ce (Ono 1990, 1995, Törnqvist 1997). Detailed comparisons between the different methods in depositing dielectric layers have not been carried out. The methods that can be used to deposit several thin films, for example, dielectric and phosphor films, in one process are the most attractive.

Fabrication of the so-called hybrid EL devices does not rely only on thin film deposition techniques; the metal and the first dielectric layers may be deposited by thick film screen printing followed by high temperature sintering and sol–gel planarization of the dielectric surface (Wu 1996). Fabrication costs can be lowered this way but at the expense of resolution.

4. Dielectrics and Conducting Films

The quality of the dielectric layers is of utmost importance for the performance of the whole device. High dielectric strength, transparency, and pinhole freeness are required from the dielectrics. Not only high dielectric strength but also the maximum charge density which can be stored in the dielectric layer, defined as $\varepsilon_0 \varepsilon_r E_{BD}$ where ε_r is the relative dielectric constant and E_{BD} the breakdown field, is of importance. The performances of different dielectric materials do not differ very much because the materials having high dielectric constant have low E_{BD} and vice versa. The best results have been obtained with perovskites and oxide multilayers (Ono 1995, Rack and Holloway 1998, Leskelä *et al.* 1999).

ITO (In_2O_3:Sn) is by far the most used transparent conductor material in TFEL devices. A 300–500 nm thick ITO film typically exhibits a resistivity around $10^{-4}\,\Omega$cm, sheet resistance around $5\,\Omega$ and transparency of 90% (Chopra *et al.* 1983).

In the traditional TFEL devices aluminum is used as the metallic conductor whereas in inverted structures, molybdenum and tungsten are the common choices. The desired characteristics for the metal electrodes are low resistivity, good adhesion, no migration at high electric fields, and patternability. In the inverted structures, where the metal electrode is deposited before the other films, the metal must have good thermal stability and thermal expansion coefficient close to that of the substrate. The optical reflectance of the metal electrode has a twofold effect. High reflectance, like that of aluminum, ensures high luminance but results in poor contrast requiring the use of contrast enhancers. Low reflectance, like that of molybdenum, means lower luminance but good contrast for the inverted structures (Törnqvist 1992).

5. Phosphor Materials

Requirements for the ACTFEL phosphor matrix (host) materials are: (i) a band gap large enough for visible emission from the luminescent center to be possible; (ii) the ability to withstand high electric fields without electric breakdown; (iii) to behave like an insulator below the threshold voltage; (iv) to be able to be doped with the activator, giving bright emission at the desired wavelength; (v) toleration of annealing temperatures up to 600 °C; (vi) to provide a medium for the efficient transport of high energy electrons (>2 eV); and (vii) to enable the material to be deposited in thin film form. The luminescent centers must fulfill the following requirements: (i) they should emit the desired light; (ii) the cross-section for the impact excitation must be large; (iii) it must be possible to incorporate them properly into the host lattice; and (iv) they must be stable in the high electric fields used (Ono 1995).

The host materials which fulfill the above requirements are mostly large band gap II–VI compounds (ZnS, ZnSe, CaS, SrS). Ternary sulfides ($CaGa_2S_4$, $SrGa_2S_4$), some oxides ($Zn_2Si_{1-x}Ge_xO_4$, $ZnGa_2O_4$, Ga_2O_3) and fluorides have been used as host materials but because of their high band gap (>4 eV) they suffer from low capability of transporting high enough current densities of hot electrons. The dopant ions are typically transition metal (manganese, copper, chromium) or rare earth metal (cerium, praseodymium, europium, terbium) ions. The luminescent centers are isolated and their energy levels locate deep in the band gap. ZnS-based phosphors known to be efficient in a cathode ray tube (CRT), such as ZnS:Ag, usually rely on donor–acceptor luminescence with shallow levels and thus have poor performance in EL.

For an EL device, luminance (cd m^{-2}), efficiency (lm W^{-1}) and aging are the most important properties of the phosphors. In the following the different phosphor materials are described and Table 1 presents the current ACTFEL phosphor materials and their performances.

5.1 ZnS:Mn

Yellow-emitting ZnS:Mn is by far the most studied and used phosphor material in TFEL devices. Monochrome yellow–black panels based on this material were commercialized in the early 1980s and a wide variety of different displays are now available. Companies use different techniques for depositing the films for the devices but the performances of the displays are practically the same: luminance 300–500 cd m^{-2} at 60 Hz, efficiency 3–6 lm W^{-1}, and lifetime well beyond 300 00 hours.

ZnS has two crystal structures, cubic and hexagonal. When processed at low temperatures the cubic structure is formed, whereas at temperatures above 400°C the hexagonal is dominant (Mikami *et al.* 1991). The EL emission peak position of ZnS:Mn is different in the two structures: 585 nm in the cubic and 580 nm in the hexagonal form. Manganese locates in ZnS at the zinc site and the solid solubility of manganese is high because MnS and ZnS have the same crystal structure. The optimum manganese concentration for EL is around 0.8–1 mol.%. The decay time of Mn EL is in the order of a millisecond.

The basic physics of the ZnS:Mn devices has been under discussion for decades. The impact excitation is a widely accepted mechanism for manganese emission. Transport of high energy electrons has been approached in two ways: Monte Carlo simulations based on the band structure (Müller 1984,

Table 1
The most important ACTFEL phosphors and their EL perfomance.

EL phosphors	Luminance L_{40} at 60 Hz ($cd\,m^{-2}$)	Efficiency ($lm\,W^{-1}$)	CIE*x*	CIE*y*	Emission color
ZnS:Mn	300	5	0.53	0.47	yellow
ZnS:Mn/red filter	70	0.8	0.65	0.35	red
ZnS:Mn/green filter	80		0.45	0.55	green
ZnS:TbOF	100–120	1.3–1.7	0.3	0.6	green
ZnS:Sm,Cl	12	0.08	0.64	0.35	red
ZnS:Tm,F	0.2	<0.01	0.11	0.09	blue
SrS:Ce	100	0.8–1.6	0.3	0.5	bluish-green
SrS:Ce/filter	10		0.13	0.18	blue
SrS:Ce,Mn,Ag,Cl	170	1.5	0.26	0.48	green-blue
CaS:Eu	12	0.05	0.68	0.31	deep red
SrS:Cu	34	0.24	0.17	0.27	blue
SrS:Cu,Ga,Ag	35	0.24	0.16	0.21	blue
CaS:Pb	80		0.16	0.12	blue
$SrGa_2S_4$:Ce	5	0.02	0.15	0.1	blue
$CaGa_2S_4$:Ce	10	0.04	0.15	0.19	blue
$ZnGa_2O_4$:Mn	10	0.2	0.08	0.68	green
$ZnSi_{0.5}Ge_{0.5}O_4$:Mn	~30	~0.2	0.27	0.67	green
ZnS:Mn/SrS:Ce	470	1.5	0.44	0.48	yellow–white
ZnS:Mn/SrS:Cu	160	0.62	0.4	0.42	white

Bhattacharyya *et al.* 1993) and mean-free path of high-energy electrons with energy loss by phonons (lucky drift model) (Bringuier 1994).

The aging of ZnS:Mn EL devices has been a subject of intensive studies. Both positive and negative shifts of the threshold voltage have been observed (Mikami *et al.* 1992). The origin of the shifts is poorly understood and changes in the behavior have been observed when different materials and deposition techniques have been employed. This indicates that the threshold voltage is dependent in a complex way on impurities, morphology and interfacial states in the structure. The *L–V* curves of the ALE deposited dielectric ZnS:Mn structures show softening, the origin of which is also not fully understood (Abu-Dayah *et al.* 1993).

Because of the broad emission band and the superior EL properties, ZnS:Mn can also be used as a red and green phosphor by using color filters. Thus, a multicolor device based only on ZnS:Mn phosphor is possible (Haaranen *et al.* 1992). In fact filtered ZnS:Mn is the best red EL phosphor known so far (Table 1). The green emission of ZnS:Mn can be enhanced by replacing zinc partially by magnesium. The enlarged band gap and the modified crystal field in $Zn_{1-x}Mg_xS$:Mn shifts the emission to shorter wavelengths (Mikami *et al.* 1996).

5.2 SrS:Ce

Strontium sulfide doped with cerium was discovered in the early 1980s and during the 1990s it was widely studied as a possible blue EL phosphor. Its emission (Fig. 3) is, however, bluish-green, having a maximum at 480 nm and a shoulder at 535 nm (CIE color coordinates $x=0.3$, $y=0.5$, see Fig. 4). The brightness 50–100 cdm^{-2} and efficiency (0.5–1 lmW^{-1}) are reasonable but if a filter is used to get pure blue emission, 80–90% of the intensity is lost. The optimum cerium concentration is very low (0.1–0.2 mol.%) because of the allowed transition involved in the cerium luminescence. In ions having the allowed transition the energy transfer (Ce→Ce) instead of radiative emission easily takes place, causing concentration quenching because of the increased probability of losing excitation energy by lattice defects. Higher cerium concentrations also cause a green shift in the emission.

A lot of effort has been put into improving EL performance, viz. brightness, color purity, and

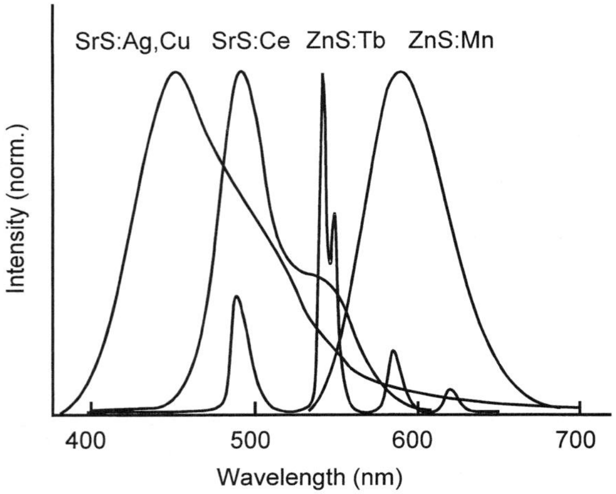

Figure 3
Normalized EL emission curves of the dominating phosphors: ZnS:Mn, ZnS:Tb, SrS:Ce, SrS:Ag,Cu.

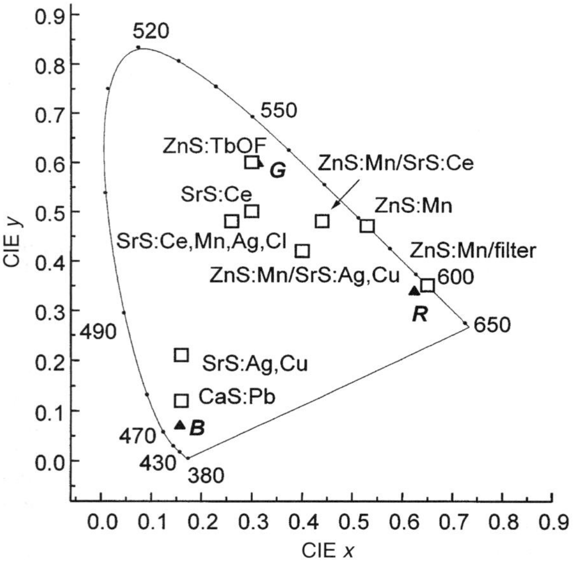

Figure 4
CIE color coordinates of the currently dominant TFEL phosphors. The coordinates of CRT phosphors (*B*, *G*, *R*) are marked for comparison.

stability of the SrS:Ce devices. Attempts have been made to improve the material by codoping, minimizing defects, modifying the lattice, filling the possible vacancies, using flux agents, and annealing. Problems arise from the charge mismatch between Ce^{3+} and Sr^{2+} resulting in vacancies and other defects (Godlewski and Leskelä 1994, Leskelä 1998).

Good crystallinity of the material is important for a good EL. This has been verified by increasing the deposition temperature, using alkali halide flux to improve the crystallization and using postdeposition annealing. Increasing the annealing temperature above 700°C for the ALE deposited films causes a blue shift originating from the increased occupancy of cerium ions at the regular octahedral strontium sites (Li *et al.* 1999a). Also, the much better brightness of SrS:Ce powders compared to films confirms the effect of crystallinity, because the powders prepared at high temperatures (1000°C) are very well crystallized (Hüttl *et al.* 1995).

SrS:Ce films have been codoped with several ions, for example F^-, Cl^-, Li^+, Na^+, K^+, Zn^{2+}, Mn^{2+}, and Cu^{2+}, to compensate for the charge mismatch between the Sr^{2+} and Ce^{3+} ions and to fill vacancies. The correlation between the codopants and EL properties is not simple, since often upon codoping other properties, for example crystallinity, may change. The experiments with ion implantation codoping have shown that incorporation of F^- and K^+ ions increases the brightness and causes a blue shift (Leskelä *et al.* 1999). Zinc and manganese coevaporation has a remarkable effect on the EL performance of SrS:Ce and they are expected to fill strontium vacancies in the phosphor (Hüttl *et al.* 1996). The best SrS:Ce based materials are, however, those codoped with silver, which is expected to act as a charge compensator (Velthaus *et al.* 1997). Sulfur vacancies are fatal for the EL and therefore the films made by PVD methods (sputtering, evaporation) can be improved by additional sulfur evaporation to ensure the correct 1:1 stoichiometry (Onisawa *et al.* 1988).

SrS:Ce films are hygroscopic and their EL properties used to suffer from aging. Now the best films, when crystalline and stoichiometric, are rather stable.

*5.3 SrS:*Cu and SrS:Ag,*Cu*

SrS:Cu thin films with good EL properties were reported for the first time in 1997 (Sun *et al.* 1997). Their emission is a broad band between 390 and 620 nm peaking at about 470 nm and therefore the hue of the EL is blue ($x = 0.15$, $y = 0.24$). The brightness ($> 30\,cd\,m^{-2}$) and efficiency ($0.24\,lm\,W^{-1}$) are better than those of the filtered SrS:Ce and Ce doped thiogallates (Table 1). The achievement of good blue EL requires postdeposition annealing at high temperatures (750–800 °C), which requires substrates that withstand such high temperatures and thus makes the devices expensive. The emission in SrS:Cu originates from isolated Cu^+ centers (Yamashita 1991). It is also possible to obtain pure green EL

(520 nm) from SrS:Cu (Li *et al.* 1999b). The origin of the green emission is unknown but postulates on different copper sites and copper pairs have been presented.

Codoping of SrS:Cu with silver makes the blue phosphor even better (Sun 1998). The role of silver has been studied but it is still unknown. SrS:Ag shows weak PL but no EL at room temperature (Troppenz *et al.* 1998). However, SrS:Ag,Cu films show a blue shift compared to SrS:Cu and the excitation spectrum contains new maxima at 281 and 295 nm suggesting the participation of The EL emission of SrS:Ag,Cu samples also gives an indication that the presence of both silver and copper are crucial for the blue EL.

5.4 *Rare Earth Doped ZnS Phosphors*

Rare earth (RE) doped ZnS EL films have been studied since the late 1960s because by using the rare earth ions it is possible to achieve all the colors. The problem is the oxidation state mismatch between RE^{3+} and Zn^{2+} and the large size of the rare earth ions. The rare earth ions have often been introduced as halides (REF_3, $LnCl_3$) into the evaporated or sputtered films. In the final device, after being treated at high temperatures upon preparation or annealing, the RE/halide ratio approaches 1. Improvement of EL performances has been attempted using co-dopants of various kinds (Leskelä and Niinistö 1992, Godlewski and Leskelä 1994, Leskelä 1998, 1999).

Although almost all the rare earth ions have been studied as dopants in ZnS, only the terbium-doped samples have shown EL properties adequate for practical applications. The best results have been obtained in films with TbOF as a luminescent center (Ono 1995). The luminescence spectrum of ZnS:Tb is dominated by the green transition from the $^5D_4$ level at 550 nm. The brightness ($> 100\,cd\,m^{-2}$), efficiency ($> 1\,lmW^{-1}$) and stability are sufficient for applications in multi- and full-color devices using the patterned phosphor concept.

The optimum content of terbium in ZnS is around few mol.%. Because of the large size of the Tb^{3+} ion only a very small amount can replace zinc in the ZnS lattice. At practical concentrations, and because in the physically deposited films terbium is introduced as a halide molecule, it is possible that a separate luminescent center is formed. In the ALE deposited films, EXAFS measurements have shown the presence of a Tb_2O_2S type luminescent center (Charreire *et al.* 1992). The formation of a "defect" type of center can be confirmed by the decreasing crystallinity of the ZnS films with increasing terbium content.

Samarium has been studied as a dopant in red ZnS phosphor (Kobayashi *et al.* 1985). The chromaticity of its emission is good but obviously, as in all samarium-doped phosphors, the efficient luminescence of Sm^{3+} suffers from a strong cross-relaxation quenching process. Thulium could be a potential dopant for blue EL but unfortunately in the EL spectrum, transitions in the IR region dominate. ZnS:Pr^{3+} and ZnS:Ho^{3+} give white emission because they have several emission lines in the visible area. Their brightness has remained, however, on an impractically low level. In addition, praseodymium shows only two lines (490 and 660 nm), thus hindering the fabrication of a full-color device.

5.5 *Thiogallates*

Rare earth doped thiogallates have been studied as CRT phosphors since the early 1970s. They were introduced to TFEL devices in the early 1990s. The most interesting phosphor was that doped with Ce^{3+}, which shows a deep blue emission (Tuenge 1992). $CaGa_2S_4$:Ce has better brightness ($10\,cd\,m^{-2}$) but inferior color purity compared to $SrGa_2S_4$:Ce, which emits at 445 nm. Problems with the thiogallates are the high annealing temperature ($> 700\,^\circ C$) needed for obtaining crystalline films and the large band gap, which makes electron injection and transport difficult. The brightness values achieved originally were promising but they have not been improved since 1995. Thiogallates have been used in prototype EL devices based on three different phosphors (ZnS:Tb green, ZnS:Mn filtered red) and dual substrates (Barrow *et al.* 1994).

5.6 *Other Alkaline Earth Sulfide Phosphors*

Research on rare earth doped alkaline earth sulfides other than SrS:Ce has been rather limited, CaS:Eu and SrS:Pr being the most often studied. EL emission of all the rare earth ions in the SrS matrix has been demonstrated and small differences to ZnS:RE^{3+} have been seen. However, the EL performance of SrS:RE phosphors is far behind that required for practical applications (Okamoto and Nakazawa 1995).

Eu^{2+} together with Ce^{3+} are the only rare earth ions where emission is from the 5*d* orbital to the 4*f* orbital, thus making the emission color strongly dependent on the host matrix and the crystal field. CaS:Eu^{2+} shows a deep red luminescence with a peak at 650 nm but luminance has remained at 10–12 $cd\,m^{-2}$ despite trials to codope, improve the crystallinity and change the oxide dielectrics to nitrides (Ono 1993, 1995, 1997, Leskelä *et al.* 1999).

SrS:Pr gives white EL emission which is considerably more bright ($30\,cd\,m^{-2}$) and efficient ($0.1\,lm\,W^{-1}$) than that of ZnS: Pr but still too low for applications (Tanaka *et al.* 1988).

Lead doped alkaline earth sulfides have also been studied as blue EL phosphors (Nykänen *et al.* 1992). The blue emission comes from lead pairs and CaS:Pb shows better blue color than the blue–green SrS:Pb. It has been possible to obtain from CaS:Pb EL devices a relatively high luminance ($\sim 80\,cd\,m^{-2}$) with pure blue color ($x=0.16$, $y=0.12$) (Yun *et al.* 1998). A further improvement on EL performance can be expected.

5.7 *Oxide Phosphors*

A wide variety of oxide phosphors are known from other applications but, due to the high processing temperatures and their insulating character, oxide phosphors are not much used for EL. Devices employing oxide phosphors have a single thick insulating layer beneath the phosphor and the structures are completed by metallic back and transparent front electrodes (i.e., hybrid structure). The best results have been obtained with green-emitting manganese doped zinc silicates and gallates (Table 1). Very promising brightness up to several hundred $cd\,m^{-2}$ at 60 Hz and efficiency of $2.5\,lm\,W^{-1}$ have been reported (Minami 1998). Rare earth doped gallium oxides and alkaline earth gallates have also shown good EL performances, even when prepared as amorphous films at low temperatures (200 °C) (Kitai *et al.* 1997). The good results and the stability of the oxide phosphors certainly encourage further studies of these materials.

6. *Conclusions*

The development of TFEL phosphors can be summarized as, in the 1970s stable ZnS:Mn devices were obtained and the basic physics was studied; in the 1980s ZnS:Mn based devices were commercialized and rare earth doped ZnS films were developed; and in the 1990s multicolor devices were introduced into the market and blue phosphors were intensively studied.

TFEL devices have found their position in high quality flat panel displays, but still the need of a better blue phosphor limits the use of TFEL devices. Phosphor development has been slower than expected because of the few groups working in the field and the complexity of the EL phenomenon, which prohibits the drawing of simple conclusions from experimental results. Steady progress in phosphor development in the 1990s with SrS:Ce and later with SrS:Cu shows, however, that commercialization of full color TFEL devices is possible. The oxide phosphors form another interesting and promising group of luminescent materials and they might quickly be developed to a commercial level for hybrid structure TFEL devices.

It must be pointed out here that in TFEL devices not only phosphors are important but also the whole conductor–insulator–phosphor–insulator–conductor structure. Progress in dielectrics and understanding of the interface properties are of utmost importance for the development of bright, efficient full color TFEL devices with long lives.

Bibliography

Abu-Dayah A, Wager J F, Kobayashi S 1993 Electrical characterization of atomic layer epitaxy ZnS:Mn alternating current thin film electroluminescent devices subject to various waveforms. *J. Appl. Phys.* **74**, 5575–81

Barrow W A, Coovert R C, Dickey E, Flegal T, Fullman M, King C N, Laakso C 1994 A high contrast, full color, 320.256 line TFEL display In: *Conf. Records Int. Display Research Conf.* Society for Information Displays, New York, pp. 448–51

Bhattacharyya K, Goodnick S M, Wager J F 1993 Monte Carlo simulation of electron transport in alternating-current thin-film electroluminescent devices. *J. Appl. Phys.* **73**, 3390–5

Bringuier E 1994 Tentative anatomy of ZnS-type electroluminescence. *J. Appl. Phys.* **75**, 4291–312

Charreire Y, Marbeuf A, Tourillon G, Leskelä M, Niinistö L, Nykänen E, Soininen P, Tolonen O 1992 EXAFS study of terbium activated zinc sulfide thin films. *J. Electrochem. Soc.* **139**, 619–21

Chopra K L, Major S, Pandya K D 1983 Transparent conductors—a status review. *Thin Solid Films* **102**, 1–46

Crandall R S 1987 Mechanism of electroluminescence in alkaline-earth sulfides. *Appl. Phys. Lett.* **50**, 551–3

Godlewski M, Leskelä M 1994 Excitation and recombination processes during electroluminescence of rare-earth activated materials. *Crit. Rev. Solid State Mater.* **19**, 199–239

Haaranen J, Törnqvist R, Koponen J, Pitkänen T, Surma-aho M, Barrow B, Laakso C 1992 A 9-in. diagonal high-contrast multicolor TFEL display. *Soc. Inform. Displays Digest* **23**, 348–51

Hüttl B, Troppenz U, Venghaus H, Mauch R H, Kreissl J, Garcia A, Fouassier C, Benalloul P, Barthou C, Benoit J, Gendron F, Ronda C 1995 Luminescence yield of SrS:Ce,Na powders. *Mater. Sci. Forum* **182/184**, 263–6

Hüttl B, Velthaus K O, Troppenz U, Herrmann R, Mauch R H 1996 SrS:Ce,Mn,Cl—a novel efficient EL phosphor. *J. Cryst. Growth* **159**, 943–6

King C N 1996 Electroluminescent displays. *J. Vac. Sci. Technol.* **14**, 1729–35

Kitai A, Xiao T, Liu G, Li J H 1997 Doped amorphous and crystalline gallium oxides and alkaline-earth gallates as EL materials. *Soc. Inform. Displays Digest* **28**, 419–22

Kobayashi H, Tanaka S, Shanker V, Shiiki M, Kunou T, Mita J, Sasakura H 1985 Multicolor electroluminescent ZnS thin films doped with rare earth fluorides. *Phys. Stat. Solidi A* **88**, 713–20

Leskelä M 1998 Rare earths in electroluminescent and field emission phosphors. *J. Alloys Comp.* **275/277**, 702–8

Leskelä M, Li W-M, Ritala M 1999 Materials in thin-film electroluminescent devices. In: Mueller G O (ed.) *Semiconductor and Semimetals, Vol. 64, Electroluminescence*, Academic Press, San Francisco, pp. 413–87

Leskelä M, Niinistö L 1992 Applications of rare earth in full-colour EL displays. *Mater. Chem. Phys.* **31**, 7–1

Li W-M, Ritala M, Leskelä M, Lappalainen R, Soininen E, Niinistö L, Barthou C, Benalloul P, Benoit J 1999a Improved blue luminescence in Ag-codoped SrS:Ce thin films made by atomic layer epitaxy and ion implantation. *Appl. Phys. Lett.* **74**, 2298–300

Li W-M, Ritala M, Leskelä M, Niinistö L, Soininen E, Sun S-S, Tong W, Summers C J 1999b Photo- and electroluminescence of SrS:Cu and SrS:Ag,Cu,Ga thin films. *J. Appl. Phys.* 5017–25

Mikami A, Terada K, Okibayashi K, Tanaka K, Yoshida M, Nakajima S 1991 Chemical vapor deposition of ZnS:Mn electroluminescent films in a low-pressure halogen transport system. *J. Cryst. Growth* **110**, 381–94

Mikami A, Terada K, Okibayashi K, Tanaka K, Yoshida M, Nakajima S 1992 Aging characteristics of ZnS:Mn electroluminescent films grown by a chemical vapor deposition technique. *J. Appl. Phys.* **72**, 773–82

Mikami A, Yashima L, Kajikawa F 1996 New developments in ZnS type EL displays. In: Mauch R H, Gumlich H-E (eds.) *Inorganic and Organic Electroluminescence*, Wissenschaftliche und Technik, Berlin, pp. 369–74

Minami T 1998 Oxide phosphor thin film electroluminescent devices using thick insulating ceramic sheets In: *4th Int. Conf. Science and Technology of Display Phosphors. Extended Abstracts*, Society for Information Displays, Bend, OR, pp. 195–8

Müller G O 1984 Basics of electron-impact-excited luminescence devices. *Phys. Status Solidi A* **81**, 597–608

Nykänen E, Lehto S, Leskelä M, Niinistö L, Soininen P 1992 Blue electroluminescence in Pb^{2+} doped CaS and SrS thin films In: *Proc. 6th Int. Workshop Electroluminescence.*, Cinco Puntos Press, El Paso, TX, pp. 199–203

Okamoto S, Nakazawa E 1995 Transient emission mechanism in thin-film electroluminescent devices with rare-earth-ion-activated SrS phosphor layers. *Jpn. J. Appl. Phys.* **34**, 521–6

Onisawa K, Fuyama M, Taguchi K, Tamura K, Ono Y A 1988 Luminance improvement of blue-emitting SrS:Ce El cell by controlling vacuum conditions with sulfur addition. *J. Electrochem. Soc.* **135**, 2631–4

Ono Y A 1990 Different techniques employed in deposition of TFEL devices. *Acta Polytechn. Scand. Ser. Appl. Phys.* **170**, 41–8

Ono Y A 1993 Electroluminescence. In: Trigg G L (ed.) *Encyclopedia of Applied Physics*, VCH, New York, pp. 295–326

Ono Y A 1995 *Electroluminescent Displays* . World Science, Singapore

Ono Y A 1997 Materials for full-color electroluminescent displays. *Annu. Rev. Mater. Sci.* **27**, 283–303

Rack P D, Holloway P H 1998 The structures, device physics and materials properties of thin film electroluminescent devices. *Mater. Sci. Eng. Rep.* **21**, 171–219

Sun S-S 1998 Blue emitting SrS:Ag,Cu TFEL devices In: *4th Int. Conf. Science and Technology of Display Phosphors.* Extended Abstracts., Society for Information Displays, Bend, OR, pp. 183–6

Sun S-S, Dickey E, Kane J, Yocom P N 1997 A bright and efficient new blue TFEL phosphor In: *Conf. Rec. Int. Display Research Conference.* Society for Information Displays, Toronto, Canada, pp. 301–4

Sun S-S, Tuenge R T, Kane J, Liang M 1994 Electroluminescence and photoluminescence of cerium-activated alkaline earth thiogallate thin films and devices. *J. Electrochem. Soc.* **141**, 2877–83

Tanaka S, Yoshiyama H, Nishiura J, Ohshio S, Kawakami H, Kobayashi H 1988 Bright white-light electroluminescent devices with new phosphor thin film based on SrS. *Proc. Soc. Inform. Displays* **29**, pp. 305–10

Törnqvist R 1992 Thin-film electroluminescence high performance without complexity. *Displays* **13**, 81–8

Törnqvist R 1997 TFEL "color by white." *Soc. Inform. Displays Digest* **28**, 855–8

Troppenz U, Hüttle B, Storz U, Kratzert P, Velthaus K-O, Sun S-S, Tuenge D 1998 Photoluminescence and electroluminescence studies on Cu and Ag doped SrS ACTFEL devices In: *4th Int. Conf. Science and Technology of Display Phosphors. Extended Abstracts.* Society for Information Displays, Bend, OR, pp. 187–90

Tuenge R T 1992 Current status of color TFEL phosphors In: *Proc. 6th Int. Workshop Electroluminescence*, Cinco Puntos Press, El Paso, TX, pp. 173–8

Velthaus K O, Hüttl B, Troppenz U, Herrmann R, Mauch R H 1997 New deposition process for very blue and bright SrS:Ce,Cl TFEL devices. *Soc. Inform. Displays Digest* **28**, 411–41

Wu X 1996 Hybrid EL displays. In: Mauch R H, Gumlich H-E (eds.) *Inorganic and Organic Electroluminescence.* Wissenschaftliche und Technik, Berlin, pp. 285–9

Yamashita N 1991 Photoluminescence properties of Cu^{+} centers in MgS, CaS, SrS and BaS. *Jpn. J. Appl. Phys.* **30**, 3335–40

Yun S J, Kim Y S, Park S K, Cho K I 1998 Blue-emitting Pb-doped calcium sulfide electroluminescent devices grown using tetraethyl lead as Pb-precursor In: *Conf. Records Int. Display Research Conf.* Society for Information Displays, Seoul, Korea, CD-ROM

Electrostrictive Ceramics for Sonar Projectors

S. M. Pilgrim

Accompanying the first successful submarine efforts was an urge to communicate underwater and to find various underwater objects and features. Since water has excellent acoustic properties, early workers soon focused on the use of acoustic signals for these uses. Mimicking the natural performance of cetaceans, this artificial acoustic projection was named SOund NAvigation and Ranging (SONAR). Similar to the more common air acoustic systems, an underwater acoustic system contains three major parts: a projector (analogous to a speaker or voice); a hydrophone (analogous to a microphone or ear); and a processor (analogous to a computer or human brain). In order to have a successful system all three components must be present. The hydrophone and projector components act to interconvert, or transduce, electrical and mechanical energy and vibrations. Since oscillating mechanical energy results in a vibration, a transducer can be considered to interconvert electrical and acoustic energies. Those devices that are electrically driven to provide a projected acoustic wave are called "projectors."

In order to function effectively, the projector must embody a process or property that interconverts the desired energies. The earliest systems used single-crystal piezoelectrics of sodium potassium tartrate tetrahydrate (Rochelle salt) and quartz for this function. Although these projectors saw limited use during World War I, they could convert only a small portion (<10%) of the electrical energy into acoustic energy. During the 1920s and 1930s understanding of projector design improved and electrical models of projectors were developed (Van Dyke 1928, Mason 1981). Thus, World War II began with improved projectors, Helmholtz and Tonpilz (German for "mushroom"), with enhanced performance. These devices formed the basis of ASDIC and sonar throughout the war period. While some of these transducers used improved materials like ammonium dihydrogen phosphate (ADP), performance remained low.

After World War II the development of barium titanate and lead zirconate titanate piezoceramics provided the impetus for major improvements in sonar performance. Unlike previous materials, these piezoelectric ceramics could be easily formed into a variety of shapes since they are processed from compressed and sintered powders. The availability of plates, disks, and rings enabled construction of improved transducer designs. Matched with the improved materials, this provided a great improvement in performance. Specifically, the coupling efficiency increased to nearly 50% (the percentage of input electrical power converted to output acoustic power scales linearly with the coupling efficiency).

While lead zirconate titanate remains the commonest material in transducers, there is limited use of a variety of single crystals for calibration and special use (lithium sulfate ($LiSO_4 \cdot H_2O$) and tourmaline). Several additional ceramics are used for specialized purposes, e.g., lead titanate and lead metaniobate ($PbNb_2O_6$). In all cases the primary effect is piezoelectric—a strain is produced as a linear response to either an induced polarization or an applied electric field. Beginning in the late 1970s, interest arose in a centric class of materials with a different transduction mechanism—electrostriction.

1. *Basic Effect*

"Electrostriction" in a material is commonly defined as a mechanical deformation with an applied electrical stimulus. Thus, electrostriction is the fundamental electromechanical coupling in all materials—*everything* is electrostrictive (Devonshire 1954). Since all materials possess atoms, ions, molecules, or domains that are either polarized or polarizable, application of an electrical excitation (field or induced polarization) distorts the charge distribution, which is then coupled to distortion of the actual dimensions. In short, elastic strain in the material results from distortion of the bond lengths, bond angles, electron distribution functions, or electric dipoles with an applied electric stress. This anisotropic effect is described in mathematical tensor form (Herbert 1982, Jaffe *et al.* 1971):

$$S_{ij} = Q_{ijkl} D_k D_l \tag{1}$$

where S_{ij} are the strain components (unitless), D_l are the dielectric displacements (Cm^{-2}), and Q_{ijkl} are the components of the electrostrictive coefficient ($C^2\,m^{-4}$). Since the materials are anisotropic, multiple coefficients are needed to define the properties. Specifically, the transverse properties differ from the longitudinal values. Typical behavior is illustrated in Fig. 1 where the maximum lateral contraction is 300 ppm, i.e., microstrain. Although the material and effect are anisotropic, the equations can often be simplified by dropping the tensor subscripts. Thus, for materials that are also piezoelectric Eqn. (1) becomes

$$S = gD + QD^2 \tag{2}$$

where there is a linear contribution to the strain from the piezoelectric g coefficients ($Vm\,N^{-1}$). Under practical operating conditions (~ 0.1–$5\,MVm^{-1}$) for piezoelectrics, the quadratic component is sufficiently small to be neglected or results in an acceptably low level of harmonic generation. Under practical

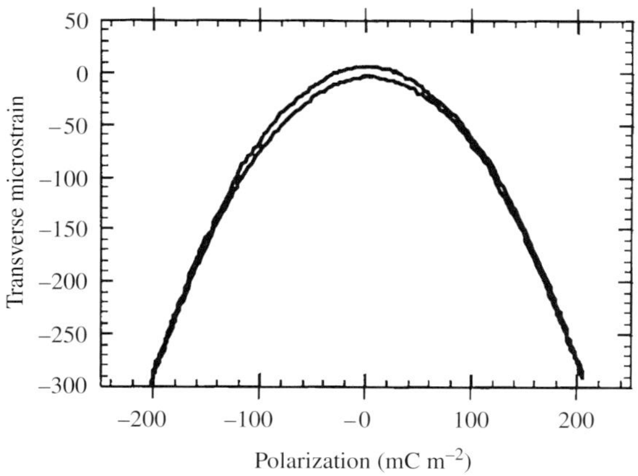

Figure 1
Transverse strain as a function of polarization for a PMN-based electrostrictor (1 MV m^{-1} applied field at 1 Hz) (reproduced by permission of the Institute of Electrical and Electronic Engineers from *IEEE Trans. Ultrason. Ferroelec. Freq. Control*, 1999, **47**, 861–76).

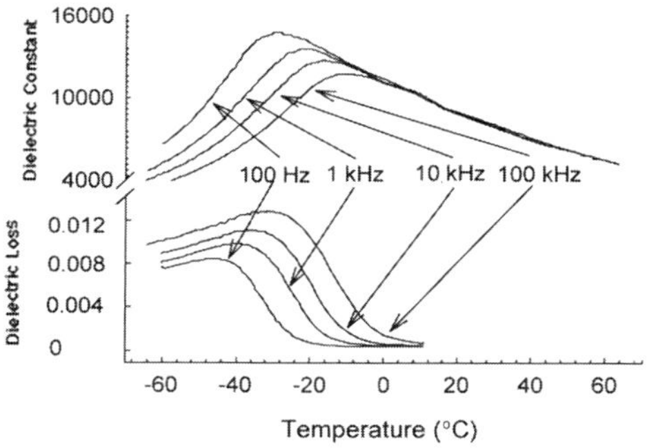

Figure 2
Weak-field dielectric response of a pure PMN electrostrictor showing the frequency dispersion and decrease in permittivity with increasing frequency and the "waterfall" in loss near the transition (reproduced by permission of the American Ceramic Society from *J. Am. Ceram. Soc.* 1999, **82** (11), 3075–9).

operating conditions for electrostrictors, the linear component is sufficiently small to be neglected.

The ideal electroactive material for sonar application would have a:

- very large dielectric displacement (D);
- linear electrical response for ease of use;
- large coefficient between strain and dielectric displacement;
- high breakdown field; and
- stable response over a wide range of environmental conditions.

These criteria may be extended and modified (Wilson 1985). While no material meets all the criteria, electrostrictive ceramics based on the relaxor ferroelectric lead magnesium niobate ($Pb(Mg_{1/3},Nb_{2/3})O_3$ or PMN) display impressive performance. As ferroelectrics they possess a paraelectric high-temperature phase and a piezoelectric low-temperature phase. However, their behavior is significantly different from other ferroelectrics in three important ways: the peak in the permittivity vs. temperature curve is frequency dependent, broad, and diffuse (Fig. 2) (Cross 1994). This provides a broad response range; however, it implies that the performance is temperature dependent.

2. *Structure and Physical Properties*

PMN is a perovskite ceramic (ABO_3 structure type). Commonly formed by mixed oxide synthesis with a columbite precursor by otherwise conventional ceramic processing routes, it can be viewed as a face centered array of oxygen anions and A-site (primarily lead with barium, strontium, lanthanum dopants) with interstitial B-site cations (primarily magnesium and niobium with titanium, zirconium, iron dopants). Depending on the dopants the materials range from pale yellow to cream to brown in color. A deleterious pyrochlore phase is sometimes noted during synthesis—this is usually bright yellow and degrades the electrical, mechanical, and electromechanical properties. Basic PMN is typically modified with various added elements to improve the properties. These elements are most commonly shown as equivalent perovskite additions (primarily $PbTiO_3$ (PT), $SrTiO_3$ (ST), $BaTiO_3$ (BT)) or as site dopants (La^{3+}, Fe^{2+}, etc.). PT is used to increase the transition temperature (T_{max}, material indicator of the phase transition and peak in Fig. 2). Small additions of PT (<15 mol%) provide electrostrictive materials with use temperatures at or below 25 °C. With the addition of sufficient PT (nominally 35 mol%) the material provides its best performance in a piezoelectric mode. Either mode may be used for transducers. Both BT and ST are used to broaden the response behavior with respect to temperature and improve the reproducibility of processing. The site dopants, less than ~3 mol%, are used to modify electromechanical response for specific applications.

The properties of PMN-based materials are very similar to those of the piezoceramic for mechanical and thermal properties. The primary differences are in the electromechanical response of the materials. These arise from the proximity of the use temperature and the transition temperature (T_{max}), the basic differences in mechanism, and the state of PMN development. Regardless of dopants, PMNs are poor in tension, of high density (8.1 gcm^{-3}), and thermally stable. Characteristic strengths range from 45 MPa to 66 MPa with fracture toughness values near

0.64 MPam$^{1/2}$. They have a heat capacity of ~0.35 Jg^{-1}C^{-1} and low thermal expansion coefficients (~2.3 ppm°C^{-1}).

3. Material Properties

From a practical point of view the parameters and properties that are most important to the actual application of an electrostrictor are those related to the frequency and temperature dependence of the energy use, energy conversion, and induced strain. For actual material or device selection these important parameters should be quantified under use conditions. One *must* remember that the relaxors do not show a linear response with applied field, nor are their properties independent of frequency.

Typical properties used to characterize electrostrictive materials for quality control include density, phase purity, weak-field peak permittivity, weak-field dielectric loss, temperature of maximum permittivity (T_{max}), and diffuseness (δ). These are important to the response of the electrostrictive material, but may not be indicative of the actual performance of the material in a transducer.

3.1 Weak-field Properties

Weak-field or dielectric properties, relative permittivity, dielectric loss, and T_{max} are determined directly from automated measurements of dielectric behavior as functions of temperature and frequency (see Fig. 2). These are frequently termed "Curie" runs by analogy to piezoceramics and are typically conducted while cooling the material at 1–4 °Cmin^{-1} in an automated system. Values corresponding to the important points on the curve are then determined (T_{max} is typically reported at 1 kHz). Diffuseness is a derived dielectric parameter that provides a measure of the breadth of the dielectric relaxation (Smolenskii 1970):

$$\varepsilon_{max} = \varepsilon \exp\left[\frac{(T - T_{max})^2}{2\delta^2}\right] \quad (3)$$

Variation in T_{max} occurs with composition and frequency. For each mole percent addition of ST or BT T_{max} varies by 8–12°C; for each 10-fold increase in frequency T_{max} increases by 3–5°C. In general, the weak-field dielectric properties can be scaled to the desired frequency of use by considering $T - T_{max@frequency}$ as the independent variable.

3.2 High-field Properties

Weak-field properties are ideal for quality control during manufacturing and as necessary checks on the performance of devices, but are insufficient to characterize or predict the performance at high electric fields. This requires measurement of polarization and strain as functions of applied electric field—typically at low frequency (~0.1–500 Hz) for applied fields of 1 MV m^{-1} at various temperatures. The salient features of such measurements are the peak polarization, peak strain, and the average hysteresis. Average hysteresis denotes the average separation of the increasing and decreasing arms of the loop normalized by the maximum response. Thus, the product of H_{ave} for polarization and the maximum field excitation is directly proportional to the energy loss on cycling. H_{ave} is defined for both y-variables, polarization, and strain:

$$H_{ave} = \frac{\sum_{i=1}^{n}(y_i^{dec} - y_i^{inc})}{n} \cdot \frac{100}{y_{max}} \quad (4)$$

Another way to view the electromechanical performance is by determining the induced strain as a function of electric drive for various temperatures. For better visualization, the peak strain values and the corresponding hysteresis values can be combined to provide a materials selection plot. This performance of strain with temperature may also be a function of frequency, although it is difficult to separate the effect of frequency from the effect of heating—both result in behavior as shown in Fig. 3.

In an engineering sense, the decrease in strain or an increase of 50% in the average polarization or strain hysteresis between two successive 5°C temperature steps has been used to set the lower limit of use. A

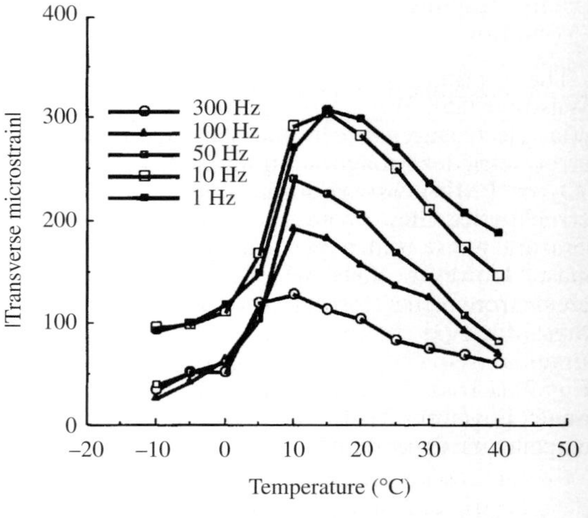

Figure 3
Frequency and temperature dependence of the magnitude of transverse strain in a PMN ceramic at 1 MV m^{-1} (reproduced by permission of Technomic from "Proc. 3rd Int. Workshop on Transducers for Sonics and Ultrasonics," 1992).

second limitation at low temperatures arises from the increase in the dielectric loss. The upper use temperature is set by the reduction in strain below the acceptable threshold or by the onset of an increase in dielectric loss from conductivity (analogous to that found in traditional piezoelectrics). In approximate terms, the important electromechanical parameters are given in Table 1.

Note that as T_{max} approaches the use temperature the temperature dependence of the listed properties will increase. It will also be more difficult to use the properties at higher electric fields without field-induced changes. The electrostrictive compositions are operated very near their T_{max} values. This results in improved strain and hysteresis properties; however, it also restricts optimal response to specific temperature and stress regions. However, regardless of specific composition, the electrostrictive ceramics show remarkable unity of response when properties are referenced to T_{max} as determined at the frequency of use.

3.3 *Electromechanical Coupling*

Determination of electromechanical coupling in electrostrictors is based on the same principles as for piezoelectrics (IEEE 1988). However, the frequency dependence, field dependence, and the need for a bias field make it more difficult to determine experimentally the appropriate coefficients. Direct application of the IEEE equations results in an underestimation of the coupling coefficient. This arises from neglecting the electrostrictive terms and also from the frequency dependence of the properties. Application of the underlying constitutive model yields the characteristic coupling for electrostrictors (a monotonic increase with saturation near 0.5). It can be extended to predict the stress and modulus of the materials. Coupling may also be determined directly from the energy balance or by *in situ* quasistatic measurements.

In addition to the temperature and expected compositional dependence, both the weak- and high-field behavior of electrostrictive ceramics may be time dependent (Schulze and Ogino 1988). These "aging" effects occur in both the high- and weak-field properties. In both cases, there is a reduction in the polarization response as a function of the logarithm of time. This results in the formation of a saddle point in the weak-field response and a constriction near the origin in the high-field polarization response. Although the specific origin is not known, aging is more prevalent in materials with grain sizes greater than ~5 μm and those containing aliovalent dopants.

4. *Transducer Performance*

Under unipolar drive, e.g., bias conditions, the electrostrictive ceramics mimic the familiar behavior of piezoceramics as transducer drive materials. There is some variation in the electrical response of the materials with bias field and amplitude. However, this effect is quantifiable and does not lead to the generation of undesirable harmonics. There is also some variation in modulus (~20% decrease) and resonance frequency with bias field. However, transducer testing has revealed that a Class IV single-shell flextensional shell using an electrostrictive PMN can outperform an equivalent piezoceramic transducer by

Table 1
Electromechanical properties of electrostrictive PMN ceramics and conventional piezoceramics.

Property	Notation	Electrostrictive PMN	Piezoceramic
T_{max} or Curie temperature (°C)		25–40	325
Longitudinal strain (ppm at 1 MV m^{-1})		1000–1200 ($Q_{11}P_1$)	285 (d_{33})
Transverse strain (ppm at 1 MVm^{-1})		− 400 ($Q_{21}P_1$)	− 122 (d_{31})
Relative dielectric constant (RT at 1 kHz)	K_3^T	8000–25000	1300
Dissipation factor (%)	DF; tan δ	1–3	0.4
Mechanical Q	Q_m	100	500
Coupling	k_{33}	0.55	0.70
Coupling	k_{31}	0.25	− 0.33
Hysteresis (strain/field)		<3%	Not reported
Modulus (MPa at 0.5 MVm^{-1})	Y_{33}^E	70–90	66
Frequency constant (kHz in)	N_t	~50	2200
Piezoelectric constant (mVmN^{-1})	g_{33}	Minimal	24.9
Piezoelectric constant (mVmN^{-1})	g_{31}	Minimal	− 10.6
Electrostrictive constant (m^{4}C^{-2})	Q_{1122}	-0.65×10^{-3}	Not applicable
Electrostrictive constant (m^{4}C^{-2})	Q_{1111}	1.7×10^{-3}	Not applicable

Reproduced by permission of the Institute of Electrical and Electronic Engineers from *IEEE Trans. Ultrason. Ferroelec. Freq. Control*, 1999, 47, 861–76.

>5 dB SPL (sound pressure level) across the band 1–4 kHz (McCollum *et al.* 1992a). Other types of transducers may also be constructed with electrostrictive drive elements (McCollum *et al.* 1992b). These generally exploit increased energy density and/or the lower Q_m of electrostrictive materials to enable broadband, high-power, directional transducers.

5. *Summary*

Electrostrictive ceramics have demonstrated potential and promise as improved sonar transducer materials. They broaden the range of design options by providing a mix of properties unavailable in traditional piezoelectrics. Under bias conditions, electrostrictive ceramics mimic the familiar behavior of piezoelectrics (IEEE 1988). The electrostrictors produce strain at a doubled frequency; however, the materials are usually run with an applied d.c. bias that results in unipolar excitation. In a transducer, the amplitude of the harmonics will be reduced through the action of the shell or radiator. In practice, an optimum bias level must be selected for each material/transducer pair to ensure that harmonic distortion is below the acceptable threshold. Since the deviations from linearity are comparable to existing piezoelectrics this has not proved to be a major problem.

Typical properties are: 1000 microstrain across the interval 0–1 MVm^{-1} with a coupling of 0.5; high-field electrical impedance about five times that of traditional piezoceramics; smooth and repeatable property variations with temperature, frequency, and prestress; and rapid and nonrecoverable failure in tension. Tension within the ceramic can be avoided by initially prestressing the ceramic and by proper transducer design. Additional strain, and hence acoustic output, can be achieved by using higher drive fields; however, this approach is restricted by corona discharge and reliability issues within the transducer.

Despite the similarities, the user must remember that electrostrictors are not direct replacements for traditional piezoceramics. They are distinct materials with different electromechanical behavior. This difference is most obvious in the electromechanical response behavior of the electrostrictive materials with changes in bias field, drive level, frequency, and prestress. Despite the innovative nature of electrostrictive transducers, most have demonstrated improved performance when compared to traditional piezoceramic transducers for specific applications.

See also: Piezoelectric Thin Films for MEMS

Bibliography

Berlincourt D A 1968 Piezoelectric and ferroelectric energy conversion. *IEEE Trans. Sonics Ultrason* **SU-15** (2), 89–97

Berlincourt D A, Curran D R, Jaffe H 1964 Piezoelectric and piezomagnetic materials and their function in transducers. In: Mason W P (ed.) *Physical Acoustics*, Vol. IA. Academic Press, New York

Bobber R J 1988 *Underwater Electroacoustic Measurements*. Peninsula, Los Altos, CA

Cross L E 1994 Relaxor ferroelectrics: an overview. *Ferroelectrics* **151**, 305–20

Devonshire A F 1954 Theory of ferroelectrics. *Phil. Mag* **3** (10), 85–130

Herbert J M 1982 *Ferroelectric Transducers and Sensors*. Gordon and Breach, New York

IEEE 1988 *IEEE Standard on Piezoelectricity*. Institute of Electrical and Electronics Engineers Ultrasonics, Ferroelectrics, and Frequency Control Society, ANSI/IEEE Std 176-1987

Jaffe B, Cook W R, Jaffe H 1971 *Piezoelectric Ceramics*. Academic Press, New York

Jona F, Shirane G 1993 Ferroelectric Crystals. Dover, New York

Lines M E, Glass A M 1977 *Principles and Applications of Ferroelectrics and Related Materials*. Oxford University Press, Oxford

Mason W P 1981 Piezoelectricity, its history and applications. *J. Acoust. Soc. Am* **70**, 1561–6

McCollum M D, Hamonic B F, Wilson O B 1992a Electrostrictive Sonar drivers for flextensional transducers. In: *Proc. 3rd Int. Workshop on Transducers for Sonics and Ultrasonics*. Technomic, Lancaster, PA

McCollum M D, Hamonic B F, Wilson O B 1992b Innovative approaches for generating high-power, low-frequency sound. In: *Proc. 3rd Int. Workshop on Transducers for Sonics and Ultrasonics*. Technomic, Lancaster, PA

Pilgrim S M 1999 Electrostrictive ceramics for low frequency active transducers. *IEEE Trans. Ultrason. Ferroelec. Freq. Control* **47**, 861–76

Schulze W A, Ogino K 1988 Review of literature on aging of dielectrics. *Ferroelectrics* **87**, 361–77

Smolenskii G A 1970 Physical phenomena in ferroelectrics with diffused phase transition In: *Proc. 2nd Int. Meeting on Ferroelectrics*. Physical Society of Japan, pp. 26–36

Stansfield D 1990 *Underwater Electroacoustic Transducers*. Bath University Press and Institute of Acoustics, UK

Uchino K 1986 Electrostrictive actuators: materials and applications. *Ceram. Bull* **65** (4), 647–52

Urick R J 1975 *Principles of Underwater Sound*. McGraw-Hill, New York

Van Dyke K S 1928 The piezoelectric resonator and its equivalent network. *Proc. IRE* **16**, 742–64

Wilson O B 1985 *An Introduction to the Theory and Design of Sonar Transducers*. **64** (4), US Government Printing Office

F

Fast Breeder Reactors: Fuels

P. Millet

Fast neutron reactors have been mainly studied with a plutonium-breeding objective (creation of fissile atoms, mainly ^{239}Pu, from fertile ^{238}U by neutron capture). More than three decades of prototype and demonstration fast breeder reactor (FBR) operation have provided a wealth of information on core and fuel behavior in Europe, the USA, Japan, Russia, and India, and have allowed commercial reactors to be developed in several countries.

A FBR core comprises nuclear fuel subassemblies, boron carbide control rods, and a uranium blanket (to enable the reactor to "breed" plutonium). The cooling is usually by liquid sodium.

Fuel pins and subassemblies must be designed to meet the economic objectives of future commercial reactors and ensure load-following without shutdowns owing to fuel-related problems. This has to be done in accordance with the requirements of different steps in the whole fuel cycle: fabrication, in-pile performance, out-of-pile handling, and reprocessing. Safety requirements corresponding to potential incidents are also taken into account at the design stage.

1. Fuel Subassembly—Design and Irradiation Performance

1.1 Design

The fuel in commercial fast reactors is generally a uranium/plutonium dioxide mix with a plutonium content of about 15–20%. However, other ceramic (such as nitride or carbide) or metallic (such as UPuZr alloy) materials have also been tested.

A fuel pin consists of a stack of cylindrical fissile pellets placed in a sealed cladding tube; there has been some limited experience with pins directly filled with vibro-compacted powders (see *Oxide Fuels, Nonaqueous Reprocessing of*). The sealed cladding forms the first barrier against possible dispersion of fissile material and fission products. A large plenum allows all fission gases released from the fuel to be trapped.

Pins are grouped in bundles placed inside a hexagonal wrapper tube. Fuel pins are separated from one to another to ensure adequate coolant flow by a helical spacer wire or by spacer grids similar to the ones found in pressurized water reactors (PWRs).

Steels (austenitic or ferritic) and nickel alloys are considered as materials for structural components (cladding, wrapper tube, wire or grids, etc.).

As an example, Fig. 1 shows the French Superphenix (SPX) subassembly design.

1.2 Irradiation

The main specific characteristics are the following:

(i) high neutron flux ($\sim 6 \times 10^{15}$ n cm^{-2} s^{-1} at the center of the SPX core) with sufficient energy to cause atom displacement;
(ii) high specific power rating and high heat flux ($\sim$400–450 W cm^{-1}); and
(iii) high cladding temperature (from $\sim$400 °C at the bottom to $\sim$600–650 °C at the top of the fissile column).

When subjected to fast neutron flux, metallic structures tend to swell and become brittle. As the neutron energy exceeds the atom displacement threshold, each initial impact under irradiation results in a chain reaction of interatomic shocks, the importance of which depends on the neutron flux and energy spectrum.

As a consequence of these shocks many atoms are displaced from their equilibrium positions. The resulting defects (vacancies and interstitials) have a tendency to move to restore a more stable state. They

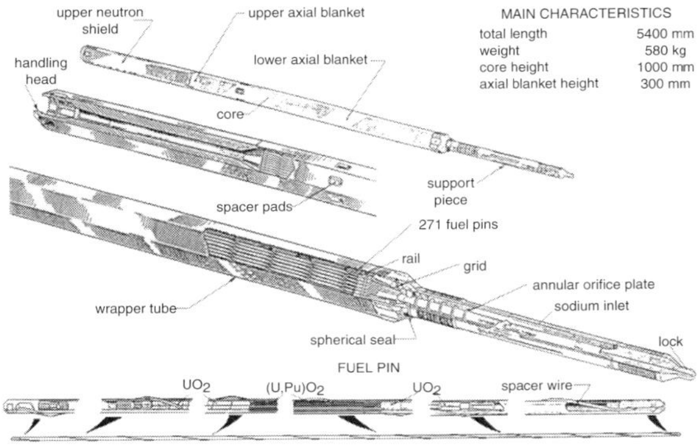

Figure 1
Superphenix fuel assembly.

either diffuse towards existing defects and disappear, or collect to form more extensive defects.

The motion of irradiation-induced defects, their annealing, coalescence, or trapping, are the fundamental mechanisms explaining the major phenomena encountered in metallic structures and fuels, namely embrittlement, swelling, and irradiation creep (see *Nuclear Reactor Materials: Irradiation Effects*).

1.3 Performance

The main characteristic is the burn-up (τ, expressed in at.%), which corresponds to the percentage of heavy atoms (uranium and plutonium) undergoing fission during the period considered. The energy released during this period per unit mass of fuel is often expressed in $MWd\,t^{-1}$ of oxide. The relationship between these two characteristics is 1 at.% $\sim$8500 $MWd\,t^{-1}$ oxide.

Steel damage can be expressed by the neutron flux integrated during irradiation (expressed in $n\,cm^{-2}$). The number of displacement per atom (dpa), a unit that characterizes the damage created in the material, is more often chosen.

A burn-up value of $\sim$20 at.% (corresponding to $\sim$170–190 dpa) is the target for commercial fast reactors.

2. *Fuel Pin In-pile Behavior*

The discussion will first be focused on mixed oxide pins. The pellet "smear density" is about 85% covering annular pellets with high oxide density ($\sim$95% of theoretical density) or solid pellets with a lower oxide density ($\sim$92% of theoretical density). The filling of cladding by fuel can be quantified by smear density. Smear density value is given by smear density (fuel/cladding section) plus measured/theoretical oxide densities. Pellet fabrication is based on powder metallurgy processes; basically UO_2 and PuO_2 powders are blended and pressed, then pellets are sintered (see *Nuclear Reactor Fuel Fabrication (Including Quality Control)*). After irradiation, reprocessing is based on the "PUREX" process: dissolution of fuel in nitric acid, followed by extraction with tributyl phosphate (TBP).

Cold-worked austenitic steels (type AISI 316 or 15 Cr/15 Ni with possible addition of titanium) are the alloys mostly used as cladding material. Ferritic steel or nickel-based alloys have also been tested.

Each pin is filled with helium gas.

Based on many years of intensive multinational development, a large amount of experience on highly irradiated mixed oxide fuel pins has been gained (several tens of thousands of pins with $\tau > 15$ at.%). Figure 2 shows, with a schematic cross-section, the main phenomena affecting fuel pin behavior during irradiation, and also shows the evolution of temperature at the center of the pellet.

2.1 Beginning of Life (From Loading to 3–4 at.%)

Owing to poor fuel–cladding heat transfer and oxide conductivity values, high temperatures and high radial thermal gradients are observed at the very beginning of life: typically 2000–2200 °C is the temperature reached at the center of the pellet. Then fuel microstructure and geometry evolve significantly (migration of the porosity towards the center of the pellet, radial redistribution of plutonium by an evaporation–condensation mechanism, closure of the fabrication gap, creation of a central hole for the initially solid pellet, etc.) and explains the important fuel temperature decrease shown at the beginning of life in Fig. 2.

2.2 Behavior at Moderate Burn-Up (From 3–4 to 7–8 at.%)

The fuel–cladding gap is nearly closed at about 3–4 at.%. The residual gap of a few micrometers fills gradually with a mixture of gases that come from plutonium fission (xenon, krypton, etc.), which pollute the initial helium decreasing its thermal conductivity and leads to the central temperature reincrease. When irradiation reaches 7–8 at.%, the gas present in this gap is replaced by a mixture of compounds formed from fission products such as molybdenum, cesium, etc. This last phenomenon contributes to limiting the fuel pellet temperature when the cladding begins to swell.

2.3 Behavior at High Burn-Up (>12 at.%)

The presence of an ever-growing quantity of fission products plays an important role in fuel pin behavior. At 12 at.%, in a SPX-type pin, the volume of fission gases created is roughly 750 cm^3 with a release rate from the oxide always very high (80–100% of the volume of gas created). These fission gases generate stresses on the cladding, even if a large plenum is provided for in the design. Concerning the solid fission products, their chemical state in the oxide depends on the oxygen potential of the fuel, a potential that in turn varies with irradiation.

Towards 12 at.% (corresponding to $\sim$100 dpa), evolution of the cladding becomes noticeable in terms of geometry, corrosion, and stresses.

(a) Deformation. Cladding geometry evolves under the influence of two mechanisms: swelling and irradiation creep.

- Swelling of steels induces, at least initially, an isotropic strain. The increase in diameter is equal to the increase of length. The main

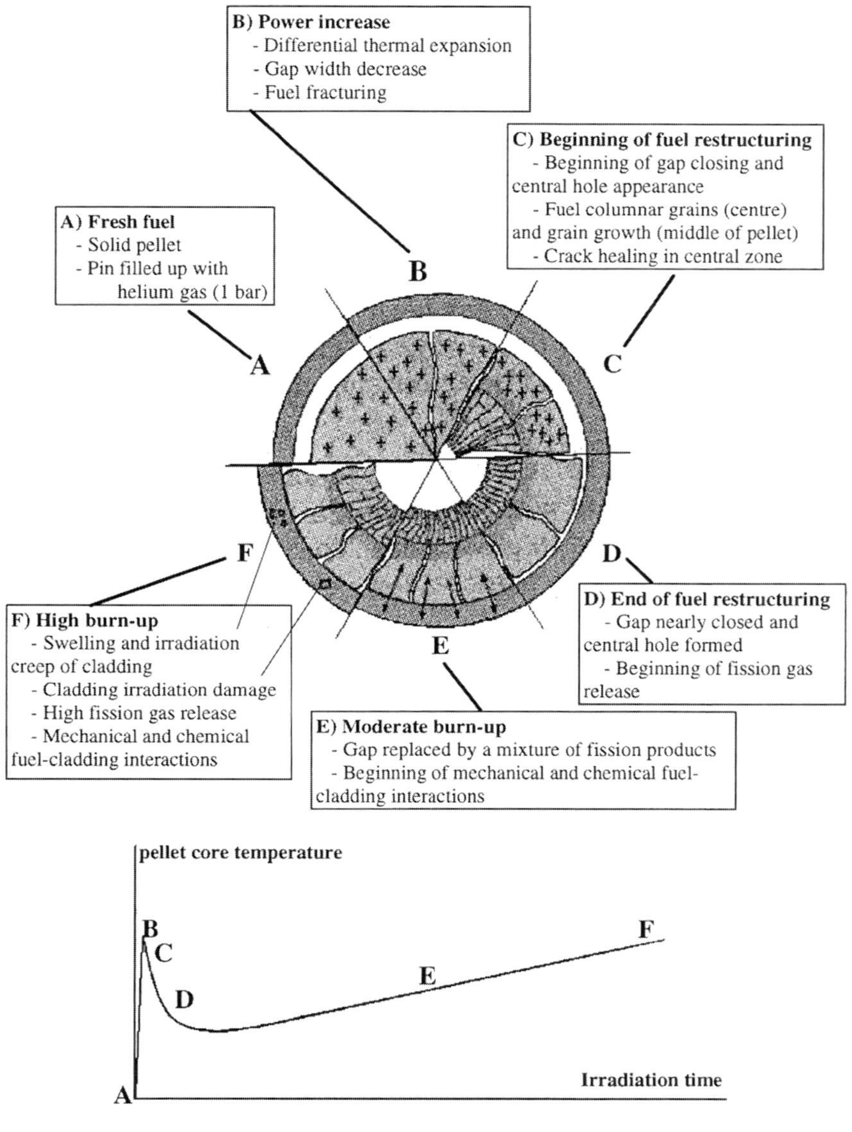

Figure 2
Fuel pin behavior. The main phenomena and related pellet temperature evolution.

parameters that control swelling are the temperature and the dose.

- Irradiation creep strain mainly depends on dose and stress. This mechanism is practically independent of temperature and does not participate in the damage of the material even at high strain values (a few percent).

As irradiation continues, these two phenomena can induce significant cladding deformation with consequences on cladding embrittlement (owing to swelling) and reduction of coolant flow (owing to diametrical deformation).

Cladding deformation is the phenomenon limiting fuel pin performance, and many studies aiming to improve swelling resistance have been performed.

- Optimization (composition, thermomechanical treatments) of austenitic steels has been performed and a dose of about 160–180 dpa can be considered with these types of steels (see *Fast Reactor Cores, Austenitic Steels for*).

- Ferritic or ferritic–martensitic steels (9Cr type) present an excellent resistance to swelling, but their poor thermal creep resistance towards 600–650°C does not allow their use unless at lower peak cladding temperature. A new class of ferritic steels strengthened by oxide dispersion (ODS steels) combining good swelling and thermal creep resistance seems very promising; irradiation of such pins is in progress (see *Fast Reactor Cores, Ferritic–Martensitic Steels for*).
- Nickel alloys also combine good resistance to swelling and thermal creep but the significant irradiation embrittlement observed (low ductility above 500 °C) makes them unattractive as cladding material.

(b) Corrosion. The austenitic steels show good compatibility with sodium. External cladding corrosion does not affect more than 20–30 μm in the hottest parts of the cladding. Conversely, internal cladding corrosion can reach significant depths (up to 100–150 μm). Affecting the upper part of the fissile column, this fuel–cladding chemical interaction (FCCI) leads to a loss of cladding thickness with progressive burn-up. Without it being insignificant, the influence of parameters such as fuel linear power or cladding temperature is less obvious (probably because the operating conditions are, in all cases, above those required for the development of this corrosion).

Mechanisms explaining these forms of corrosion are based on the role of caesium telluride (Cs_2Te) created by combination of caesium and tellurium fission products. The volatility of this compound allows it to migrate from its production site to the fuel–cladding interface where, in the presence of oxygen supplied by the fuel, it can react with steel constituents according to reactions of type

$$Cs_2Te + Cr + 2O_2 \Leftrightarrow Cs_2CrO_4 + Te$$

This reaction is controlled by the oxygen potential of the mixed oxide and depends on local fuel surface temperature and burn-up.

Cladding deformation plays an unfavorable role in FCCI (explained by the increase of surface oxide temperature). So, the swelling-resistant materials (nickel alloys or ferritic steels) show a better behavior towards FCCI than austenitic steels.

Studies of remedies are mainly focused on the incorporation of metallic getters (titanium, vanadium, etc.) inside the pin to trap oxygen required for the decomposition of Cs_2Te. However, these solutions have the drawback of lowering the O/M ratio of the oxide leading to a degradation of the fuel thermal behavior.

*(c) Fuel–Cladding Mechanical Interaction (*FCMI*).* Fuel swelling under irradiation is roughly 0.62% per at.%. The resultant increase in the outside diameter of the pellet can be greater than that of cladding on all or part of the fissile column. Owing to stress relief by irradiation creep, the stress induced in cladding by this mechanical interaction is relatively low during stable operating conditions. During operating transients, additional high stress can be induced by differential fuel–cladding expansion because high strain rates prevent stress relief.

2.4 Cladding Failures

The fundamental point is the incompatibility of $(U,Pu)O_2$ mixed oxide with sodium over a very wide range of temperatures. Following penetration of the cladding, sodium entry in the pin results in an oxide–sodium reaction. The sodium uranoplutonate produced, which has a lower density than that of the initial oxide, increases the stress level of the cladding and can induce an increase in the size of the initial leak or the occurrence of a secondary crack. In the final stage of this evolution, the fissile material can leave the pin, a situation that has to be avoided. In order to prevent this, the evolution of cladding failure is monitored by delayed neutron detection (DND), i.e., detection of delayed neutrons released by the fission products that escape from the pin, thus allowing the subassembly to be removed from the core. Accumulated experience has demonstrated that breaches are benign events and has provided the database for reactor operators to define their own criteria for managing such events.

2.5 Behavior under Off-normal Conditions

Severe irradiation conditions can result from hypothetical accidents such as sudden control rod withdrawal, partial or total loss of coolant flow, earthquake, etc. These transient conditions will induce a degradation of the pin and subassembly behavior that could be significant: for example, increase of the thermal regime of the oxide and additional stresses on the cladding. Each situation of this type must be carefully investigated both from the point of view of the straightforward consequences (cladding failure, fuel melting, etc.) and of transmitted events that could result (e.g., dry-out of coolant channels). Although the key parameters differ from one event to another, two points have to be emphasized:

- The importance of the state of the pin or subassembly at the time of the accident ("T0" state). The pin characteristics evolve significantly during irradiation, so it is important to deal with all the T0 states and to be able to characterize each of them completely (stresses in

the cladding, oxide temperature, fission gas behavior, etc.).
- The transient kinetics. From a fraction of a second to several minutes (and more rarely several hours), this kinetics is much faster than that of power variations during normal operating conditions.

In order to describe the pin and subassembly behavior under these off-normal conditions, specific codes have been developed as also have specific tests aimed at precise determination of the influence of evolution rate on mechanisms and data. Moreover, large experimental programs aiming to test pins or bundles under off-normal irradiation conditions have been conducted in experimental reactors.

2.6 *Carbide and Nitride Fuel Pins*

Compared to oxide, nitride or carbide fuel exhibits attractive properties such as good thermal conductivity, excellent compatibility with sodium, and high heavy metal atom density.

With a design very similar to oxide fuel pins, carbide, carbonitride, and nitride fuel pins have been fabricated and irradiated. The main observations are:

- swelling kinetics of these ceramics under irradiation is higher than that of the oxide, thus requiring a limited smear density in the cladding; assuming a swelling rate of roughly 1.1–1.6% per at.%, it is necessary to limit the smear density to 75–78% (instead of 85% for oxide) in order to reach a burn-up of about 20 at.%;
- a low gas release from these fuels; and
- a lack of cladding chemical interaction with nitride fuels; conversely, cladding carburization is often observed with carbide fuels.

While carbide fuels are no longer studied (because of cladding carburization and poor chemical compatibility with oxygen and moisture leading to difficulties during the fuel cycle), the irradiation results provide favorable expectations for the performance of nitride fuels. The performance of nitride fuels has to be confirmed at high burn-up levels with results on a larger statistical scale. In addition, several other issues have to be considered, including fabrication of ^{15}N-enriched fuels (to avoid production of radioactive ^{14}C) and transient performances (with specific studies on high-temperature nitride dissociation).

2.7 *Metallic Fuel Pins*

UPu alloy cannot be used because of the existence of a eutectic with the iron (from the cladding) appearing at temperatures close to operating conditions. UPuZr alloy (with a zirconium content of roughly 10%) pushes back this eutectic to 750°C and allows for the design of metallic fuel pins. In parallel, a specific fuel cycle has been developed associating reprocessing by pyrometallurgy and fuel fabrication in the same unit.

The fuel pin design must take into account different factors, the most important ones being:

- fast swelling kinetics of the fuel (at least at the beginning of life); a large fuel–cladding gap is required in order to limit fuel smear density; and
- excellent fuel thermal conductivity, but a relatively low melting temperature; a significant safety margin with regard to melting is obtained by the presence of sodium (allowing improved heat transfer) between fuel and cladding.

Irradiations under normal or incidental (cladding rupture, overpower, etc.) conditions have shown the reliable performance of metallic fuel and its potential to achieve high burn-up in prototypical elements. Qualification on a larger scale of the in-pile behavior and of the associated cycle is again necessary.

3. *Subassembly Behavior*

Like the cladding, under the effect of neutron flux the hexagonal wrapper tube undergoes change. Geometrical changes, induced by swelling and irradiation creep (under the effect of dynamic sodium pressure), concern length, across-flat dimension, and, for subassemblies placed in the periphery of the core, bowing (differential swelling induced by differential flux and temperature between opposite faces of the wrapper tube). Moreover, embrittlement owing to irradiation or temperature has to be mastered in order to avoid rupture in the reactor or during the back-end of the cycle (handling, washing, etc.). Cold-worked austenitic steels (with composition closed to that of the cladding) were used first but are now frequently replaced by more swelling-resistant ferritic–martensitic steels (9Cr type), in so far as the peak wrapper temperature, lower than that of cladding, allows avoidance of thermal creep problems.

Differential swelling between cladding, wrapper tube, and spacing system (spacer wire or grids) can induce additional stresses and deformations on these components and/or generate coolant flow perturbations in the bundle. These phenomena have been studied and have particularly shown that wires or grid materials have to be chosen and specified in order to minimize differential swelling.

4. *Conclusions*

Supported by a large experimental base (irradiation tests, fabrication, and reprocessing), mixed oxide can be considered as the reference solution for fast breeder fuel. Alternative solutions based on nitride (with a similar fuel cycle) or metallic (with a specific fuel cycle) fuels have also been tested and have given favorable results. These, however, have to be confirmed on a large scale.

Conversely to the objective of plutonium breeding, fast reactors can also be considered as plutonium burners. For this objective, the pin design has to be modified. The first stage consists of the suppression of the fertile parts (uranium blankets) and the second stage of the modification of the fuel composition by increasing the plutonium/uranium ratio. In this way, plutonium fuel without uranium (plutonium dispersed in a ceramic or metallic inert matrix) can be studied.

See also: Heavy Water Reactor Fuel Design and Performance; Inert Matrix Fuels; Light Water Reactor Fuel Design and Performance; Polymer Light Emitting Diodes, Inkjet Printing of

Bibliography

Cecchi P F, Jones R B, Millet P, Norris D, Pay A, Tobbe H 1986 Life limiting features and technical problems to reach high burn-up in fast reactors In: *Int. Conf. on Reliable Fuels for Liquid Metal Reactors*. Tucson meeting 1986

International Atomic Energy Agency 1999 *Status of Liquid Metal Cooled Fast Reactor Technology*. IAEA-TECDOC-1083. IAEA, Vienna

Lallement R, Mikailoff H, Mustelier J P, Villeneuve J 1999 Fast breeder reactor fuel design principles and performances. *Nucl. Energy* **28** (1), 41–9

Millet P 1999 Fast reactor fuel subassembly. In: Bailly H, Ménessier D, Prunier C (eds.) *The Nuclear Fuel of Pressurized Water Reactors and Fast Reactor Design and Behaviour*. Lavoisier, Paris

Fast Reactor Cores, Austenitic Steels for

S. L. Mannan and P. V. Sivaprasad

Austenitic stainless steels have been employed worldwide as core structural materials in fast breeder reactors (FBRs) because they exhibit good mechanical properties at high temperatures and compatibility with liquid sodium and ceramic fuels of uranium and plutonium. Materials in a fast reactor core are subjected to a very demanding environment of high fast-neutron flux ($\sim 10^{15}\,\mathrm{ncm^{-2}s^{-1}}$) coupled with high temperatures (523–973 K). One major consequence of the high flux of fast neutrons is the occurrence of very high levels of radiation damage in the core structural materials. This damage, characterized in terms of the average number of times that an individual atom is displaced from its lattice site (displacements per atom, dpa) can be as large as 100–200 dpa for some core components. Under such radiation conditions and high temperatures, austenitic stainless steels undergo significant changes in phase stability and consequent microstructural alterations, which induce dimensional changes and affect mechanical properties. Four major phenomena that have an impact on the performance of core components are phase stability, void swelling, irradiation creep, and neutron-induced changes in mechanical properties. These phenomena are intimately linked and it has been shown (Garner 1993) that void swelling depends sensitively on the evolution of phases in austenitic stainless steels and is the dominant determinant of both irradiation creep behavior and mechanical properties. Alterations in chemical composition and microstructure, which influence void swelling, also exert a strong effect on irradiation creep.

Microstructural alterations are the result of atomic displacements owing to collision of energetic neutrons with atomic nuclei. The basic defects produced are vacancies and interstitials, with typical displacement rates of the order of $10^{-6}\,\mathrm{dpas^{-1}}$. High supersaturations of these point defects along with helium produced by (n, α) reactions play an important role in the radiation-induced microstructural evolution. Interactions between various basic defects, depending on temperature of irradiation and the evolving microstructure of austenitic stainless steels, determine their agglomeration into extended defects such as dislocation loops, voids, and helium bubbles. Precipitates play an important role in the evolution of microstructure during irradiation both via interaction of precipitate–matrix interfaces with radiation-induced point defects and removal of certain solutes from the matrix which otherwise interact strongly with vacancies and interstitials and influence their mobility. Phase stability is an important aspect of irradiated austenitic stainless steels and this is discussed first. This is followed by a discussion of void swelling, irradiation creep, and neutron-induced changes in mechanical properties. The influence of chemical composition owing to variations in major and minor alloying elements in austenitic steels and prior thermomechanical history on irradiation effects are examined and approaches followed in optimizing composition and microstructure for the development of austenitic stainless steels resistant to irradiation effects are highlighted.

Engineering consequences of irradiation-induced changes such as void swelling, irradiation creep, and embrittlement are very important as these determine permissible life of core components and hence have an impact on the achievable burn-up and consequent economic viability of FBRs. The gradients in neutron flux and temperature that exist across or along a component which in turn determine evolution of microstructure and macroscopic properties exert a very strong influence on the component performance. This is discussed in the final section.

1. Phase Stability under Irradiation

The chemical compositions of major austenitic stainless steels used in FBRs are given in Table 1. Generally, these can be considered as variants of AISI type 316 stainless steel. A variety of phases precipitate during thermal aging of these steels depending on chemical composition, prior thermomechanical treatment, and service history (Weiss and Stickler 1972, Harries 1981). These include carbides ($M_{23}C_6$, M_6C) and various topologically close-packed phases like sigma (Fe–Cr), Laves (Fe_2Mo), and χ (Fe–Cr–Mo). In steels containing titanium and/or niobium, MC carbides also form during aging. The formation of these phases is either enhanced, i.e., they are produced more rapidly, more abundantly, and at lower temperatures or retarded, i.e., they are produced less rapidly, less abundantly, and at higher temperatures, by irradiation. M_6C is generally enhanced and $M_{23}C_6$ is retarded owing to radiation (Russel 1984, Maziasz 1993, Ukai *et al.* 1999). The compositions of some of the phases get altered and some phases that do not precipitate during thermal exposure are also formed during irradiation. These latter are known as radiation-induced phases and the phases in this category include Ni_3Si (γ'), G-phase ($M_6Ni_{16}Si_7$) silicides, and needle-shaped phosphides.

Radiation-induced microstructural alterations occur owing to the presence of large supersaturation of vacancies and interstitials. Solutes that can bind with either type of point defect tend to flow down any microstructurally induced gradient of that defect, providing a new mechanism of solute segregation called solute drag. This is particularly important for binding of smaller solute atoms with interstitials. Another mechanism of radiation-induced segregation

Table 1
Typical chemical compositions (wt.%) of various austenitic stainless steels used worldwide as in-core materials in FBRs.

Alloy	Cr	Ni	Mo	Mn	Si	C	Ti	Nb	P	Fe
Type 316	18.0	13.0	2.60	1.90	0.80	0.050	0.05			Balance
US D9	13.8	15.2	1.50	1.74	0.92	0.052	0.23		0.003	Balance
316Ti	17.1	14.1	2.75	1.50	0.49	0.049	0.34		0.015	Balance
15-15Ti	14.7	14.7	1.15	1.6	0.43	0.096	0.43		0.007	Balance
Si-modified 15-15Ti	14.9	14.8	1.46	1.5	0.95	0.085	0.5		0.007	Balance
PNC 316	16.0	14.0	2.50	1.80	0.80	0.055	0.10	0.08	0.028	Balance
PNC 1520	15.0	20.0	2.5	1.90	0.80	0.060	0.25	0.11	0.025	Balance
1.4970	15.1	15.1	1.26	1.3	0.49	0.088	0.48		0.004	Balance
FV548	16.5	11.8	1.44	1.14	0.35	0.11		0.92		Balance

is the inverse Kirkendall effect, whereby differences in elemental diffusivities (D_{Fe}, D_{Ni}, D_{Cr}) via vacancy exchange lead to segregation of the slowest diffusing species at the bottom of sink-induced vacancy gradients. This mechanism is particularly effective in segregating nickel in austenitic steels at all sinks that annihilate vacancies. This segregation arises because the elemental diffusivities of various alloy species are different, with $D_{Cr} > D_{Fe} > D_{Ni}$. The formation of γ'- and G-phase precipitates removes elements like nickel and silicon which are important for suppressing void swelling.

The onset of void swelling, therefore, often coincides with the formation of nickel- and silicon-rich phases. Because of their influence on effective vacancy diffusivity, these elements enhance absorption of point defects at sinks and reduce supersaturation. Phosphorus has an even stronger effect in reducing supersaturation of vacancies. Formation of fine precipitates such as phosphides and TiC also enhances recombination of point defects and delay void swelling. Segregation effects eventually predominate and cause dissolution of phosphides and precipitation of nickel- and silicon-rich phases with the consequent acceleration of void nucleation and growth. It has been demonstrated (Shibahara *et al.* 1996) that the incubation period for swelling is prolonged by extending the duration of the neutron exposure required to cause phosphide dissolution. Titanium additions lead to the formation of more stable phosphides. The effects of fine precipitates have also been attributed to dispersion of helium concentration to subcritical levels so that nucleation of voids is delayed (Lee and Mansur 1992).

2. *Void Swelling*

Generally, voids form in austenitic stainless steels in the temperature range 673–973 K in FBR irradiation. Voids result from the accumulation of vacancies at cavities stabilized by helium produced by (n, α) reactions (Maziasz and McHargue 1987, Mansur 1986). The formation and growth of voids is sensitive to nearly all metallurgical variables like chemical composition and thermomechanical history and irradiation parameters like fluence, dose rate, and irradiation temperature. The fluence dependence of swelling can be described as a low swelling transient period followed by an acceleration to a regime of steady-state swelling (Fig. 1). The influence of most of the metallurgical parameters is seen in the duration of the transient regime before and during the void nucleation phase. Microchemical evolution is the dominant factor during the transient regime. The steady-state swelling regime has proved to be reasonably insensitive to these variables with the swelling rate of most austenitic alloys being ~1% per dpa over a wide range of irradiation temperatures. The influence of major alloying elements on swelling

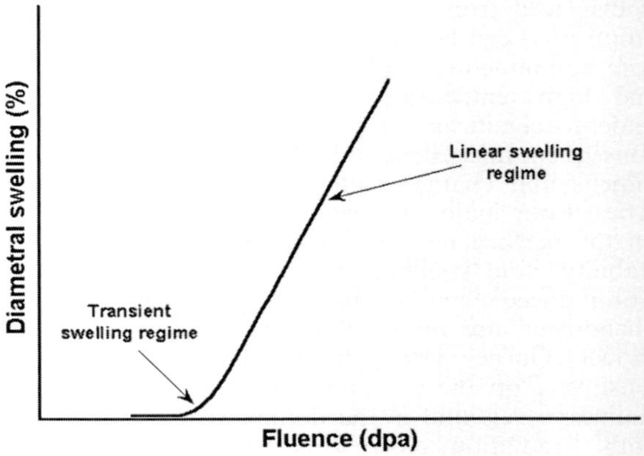

Figure 1
Schematic of different regimes in the variation of swelling with fluence (irradiation temperature ~600–950 K).

behavior of austenitic stainless steels is discussed, followed by the effects of minor elements. The role of thermomechanical history in influencing void swelling is also examined.

2.1 *Effects of Major Alloying Elements*

A variety of investigations on model Fe–Cr–Ni ternary alloys as well as on austenitic stainless steels have established the role of nickel and chromium additions in void swelling behavior of austenitic steels. At any given temperature and dose level swelling first decreases strongly with increasing nickel content, then reaches a minimum value at some intermediate nickel level (~35–45%), and thereafter increases relatively slowly at higher nickel levels. The dependence of neutron-induced swelling on nickel content has been shown to arise almost exclusively from the duration of the transient regime that precedes the onset of steady-state swelling of ~1% per dpa (Garner 1984). The influence of chromium has been found to be monotonic, with increasing chromium causing higher swelling at all nickel levels. The presence of nickel in solution is suggested to enhance the effective vacancy diffusion coefficient (Garner and Kumar 1987) and therefore to lead to reduced void nucleation rate via reduction of vacancy supersaturation. Radiation-induced segregation of nickel at sinks in the microstructure reduces its concentration in the matrix and makes the matrix susceptible to increased swelling with increasing dose. The presence of minor alloying elements in the commercial austenitic steels has been found to be very effective in extending the transient regime in comparison to Fe–Cr–Ni model alloys. Increasing nickel and reducing chromium would also increase stability of the γ phase and prevent precipitation of undesirable phases that remove certain solutes responsible for conferring void swelling resistance to these steels. The trend in development of radiation-resistant modifications of 300 series austenitic stainless steels has been to increase the nickel content and decrease the chromium concentration. Type 316 stainless steel has been employed most extensively in early cores of FBRs and its radiation-resistant variants followed this trend as in the case of D9, 15-15Ti steels, and Japanese PNC varieties (Table 1).

2.2 *Role of Other Alloying Additions*

The influence of a variety of solute additions to type 316 stainless steel on void swelling has been studied and it has been found that certain solutes like titanium, silicon, phosphorus, niobium, boron, and carbon play a dominant role in improving void swelling resistance. Further, their influence is primarily to extend the duration of the transient regime. These solutes act either singly or in a synergistic manner to influence the evolution of microstructure that is resistant to void swelling.

Carbon, in solution as well as in the form of fine precipitates, has been found beneficial in reducing swelling. The presence of fine carbides of titanium and niobium provides recombination sites for point defects and hence reduces their supersaturation and consequently void swelling. Another effect of increasing carbon content is to shift the peak in swelling to higher temperatures for both solution-annealed and cold-worked type 316 stainless steel.

The addition of titanium has been found to be very effective in reducing void swelling. Titanium-stabilized austenitic steels (316Ti and 15-15Ti) have, therefore, been chosen as reference structural materials of the fuel subassemblies of the French FBRs Phenix and Superphenix. Titanium is also an important alloying addition in alloy D9 developed in the USA. These materials are used in the cold-worked condition to minimize dimensional changes in the fuel subassemblies principally owing to void swelling and creep during irradiation. Figure 2 compares the performances of a few clad materials and clearly brings out the gain achieved when moving from unstabilized 316 to 316Ti, 15-15Ti, and the silicon-modified version of 15-15Ti steel. The important difference between these four alloys shown in Fig. 2 is owing to large differences in the incubation dose for swelling. These doses are 45 dpa for CW 316, 95 dpa for CW 316Ti and more than 100 dpa for 15-15Ti and its silicon modification.

Fine TiC precipitates play a crucial role in reducing void swelling. The Ti/C ratio has been shown to play an important role in determining irradiation behavior of austenitic steels. Maximum swelling resistance has been obtained when the Ti/C ratio is

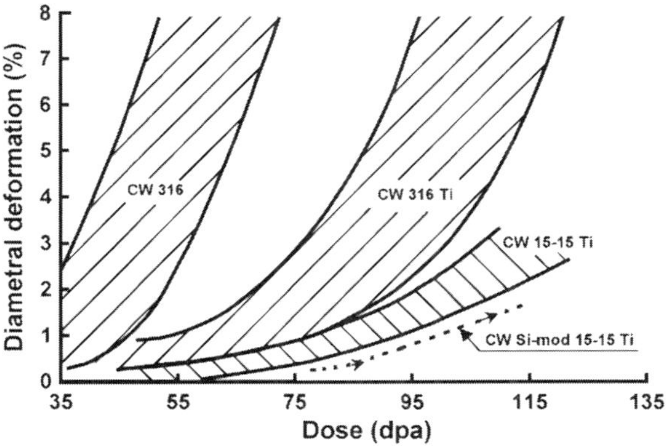

Figure 2
Comparison of different cladding materials. Variation of the maximum diametrical deformation of pin with dose; irradiation temperature 670–870 K (after Seran *et al.* 1992).

below the stoichiometric composition, i.e., when the material is understabilized, i.e., for a titanium content of less than four times the carbon content in weight (Herschbach *et al.* 1993). Carbon, in the absence of titanium, is captured in large precipitates of $M_{23}C_6$ and is not available for vacancy trapping responsible for lowering of supersaturation by recombination processes. The fine TiC particles redissolve by recoil dissolution and continue to contribute to trapping so long as the steel is understabilized. The role of niobium is similar to that of titanium where the presence of fine NbC precipitates confers void swelling resistance to the austenitic steels, particularly in the temperature regime of their precipitation.

Silicon improves void swelling resistance of austenitic stainless steels and is therefore included as an alloying addition in most of the void swelling-resistant stainless steels developed in different countries (Table 1). Silicon enhances vacancy diffusion and thereby reduces supersaturation of vacancies by their relatively fast migration to sinks.

Phosphorus additions have been found to be extremely beneficial for improving the void swelling resistance of stabilized stainless steels (Dubuisson *et al.* 1992). The influence of phosphorus is more pronounced when other solutes like titanium, silicon, and niobium are also present. Addition of phosphorus increases the swelling resistance of cold-worked, titanium-stabilized 316 stainless steel by extending the incubation period beyond 100 dpa (Fig. 3). The improvement in swelling resistance by phosphorus addition arises owing to a decrease in void density especially when a uniform distribution of needle-shaped phosphides appears in the microstructure. In titanium-stabilized steels, the phosphides are FeTiP, whereas titanium is replaced by

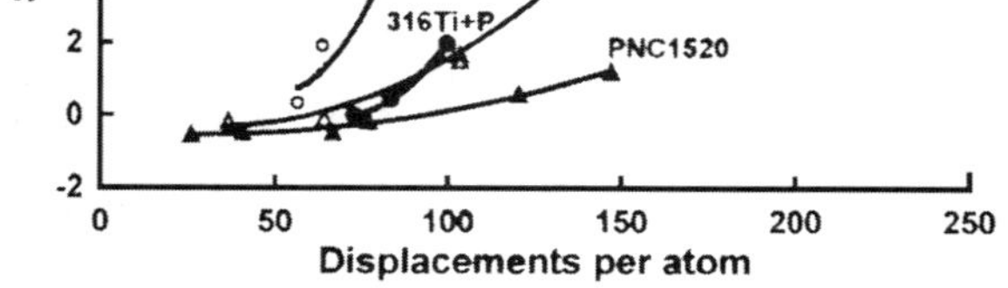

Figure 3
Dose dependence of swelling in 16–20% cold-worked advanced austenitic stainless steels (316Ti and 316Ti + phosphorus irradiated at 728 K and PNC 1520 irradiated at 678 K).

niobium in the phosphides of the multistabilized steel, leaving titanium in solution that further inhibits swelling. Mechanisms suggested that contribute to void swelling resistance of these steels include reduction in vacancy supersaturation because of a strong binding between phosphorus in solution and point defects, phosphorus acting as a trap for helium atoms, enhanced point defect recombination at the matrix–precipitate interfaces, stabilization of dislocation network, reduction in radiation-induced solute segregation, and prevention of the formation of nickel- and silicon-rich phases such as the G phase.

2.3 High Nickel Austenitic Stainless Steels

The Japanese FBR development program aims at developing high nickel advanced austenitic stainless steels for neutron exposures to about 150 dpa. The major effort has been directed towards improvement in creep rupture strength and swelling resistance. PNC 316 (Table 1) is the reference FBR core structural material. Further improvements in creep strength and swelling resistance have been achieved by increasing the nickel and titanium contents (PNC 1520, Table 1), while keeping other elements and cold-work levels the same as in PNC 316 (Shibahara *et al.* 1996). The remarkable extension of the transient regime of swelling in high nickel advanced austenitic stainless steels is indicated in Fig. 3.

Unacceptably high levels of void swelling (beyond 50 dpa) can occur in type 316 stainless steel (Figs. 2 and 3) in the temperature range 670–870 K unless the chemical composition is suitably modified to effect significant extension of the low swelling transient regime. The existence of fine, stable precipitates is necessary to prevent void swelling. Therefore, the composition of austenitic stainless steels should be so controlled as to produce optimum precipitation during irradiation. There are synergistic effects of various alloying additions that have been used in the development of swelling-resistant austenitic stainless steels in various countries. The modifications that improve void swelling resistance of these steels also lead to better creep strength.

2.4 Effect of Thermomechanical Treatment

Since dislocations act as sinks for point defects, their density is expected to influence the population of point defects. The degree of cold work that influences dislocation density is an important parameter determining the void swelling behavior of austenitic stainless steels. Further, it is seen that phase stability and scale of precipitation play an important role in void swelling behavior. Cold work also influences swelling by providing sites, namely dislocations, for nucleation of precipitates as well as for trapping of helium bubbles. Since dislocations also serve as sinks for point defects and thereby generate point defect gradients in their vicinity, they also participate in the radiation-induced segregation processes with consequent effects on the phase stability of these alloys. The degree of swelling suppression for a given level of cold work is, thus, dependent on alloy composition as well as irradiation parameters. Cold work is also specified with a view to enhancing creep strength.

Annealed austenitic steels have dislocation densities of the order of $10^8\,cm^{-2}$ while in the cold-worked condition, dislocation densities are in the range 10^{11}–$10^{12}\,cm^{-2}$. The initial dislocation density is not maintained during irradiation, however. There is a tendency for the dislocation microstructure to evolve towards an equilibrium level with a density of network dislocations of the order of $6.3 \times 10^{10}\,cm^{-2}$. This involves an order of magnitude reduction in dislocation density of cold-worked steels and an increase in that of annealed steels. The presence of high dislocation density leads to suppression of swelling in cold-worked austenitic stainless steels. The effect of cold work is primarily by extension of the transient regime or incubation period of swelling (Appleby *et al.* 1977). The stability of the dislocation substructure induced by the cold work plays an important role in increasing swelling resistance. The titanium addition to 316 stainless steel and increased titanium content in PNC 1520 stabilize the dislocation structure by fine MC precipitation. The precipitation of fine MC precipitates is more prevalent in cold-worked material compared to solution-annealed material.

The temperature of final solution annealing before cold working has also been found to be an important variable affecting swelling behavior. The temperature of annealing determines the distribution and availability of elements such as carbon, phosphorus, and titanium in solution as a result of dissolution of precipitates. In general, the higher the annealing temperature, the greater the availability of such solutes and the better is the swelling resistance. Aging prior to irradiation also influences swelling behavior of austenitic stainless steels. As these alloys are unstable, aging leads to precipitation of various phases like $M_{23}C_6$ carbides and intermetallics which are deleterious to swelling resistance as they remove certain alloying elements which through their interaction with point defects enhance recombination with consequent reduction in swelling.

2.5 *Influence of Environmental Variables*

Apart from irradiation dose, which has a strong influence on void swelling, it is important to appreciate that there are many other environmental variables that influence swelling. These include displacement rate ($dpa\,s^{-1}$), temperature, and stress. These parameters are not constant and exhibit spatial and temporal variations. Because of the variability of these parameters in the reactor core, there are variations in the changes in dimensions both along the length of the fuel elements as well as from one subassembly to another depending on their position in the core. These variations have, in fact, a more significant effect on the permissible life of the fuel elements than the absolute values of the dimensional changes.

Increase in displacement rate generally leads to extension of the transient regime. The most important effect of temperature is on the duration of the transient regime via its influence on microchemical evolution of the alloy matrix. Swelling is reduced both at low temperatures as well as at high temperatures. Recombination of vacancies and interstitials is enhanced at low temperatures while at very high temperatures vacancy emission from voids begins to overwhelm the inflow of excess vacancies. Swelling is therefore more prevalent at intermediate temperatures. Another important variable that influences swelling is the presence of stress. This is important, as in fuel pins stress increases with the increase in burn-up owing to fission gas pressure and this can enhance swelling. The presence of stress has been found to shorten the transient regime.

3. *Irradiation Creep*

Creep is an important deformation mechanism at temperatures greater than about half the melting point of the material in kelvin. The operating conditions (i.e., temperature in the range 673–973 K) in the case of FBRs encompass the temperature regime for austenitic stainless steels where creep strength is an important design consideration. The choice of austenitic stainless steels worldwide for cladding and subassembly wrappers of fast reactors is based on their good creep resistance under FBR operating conditions. However, under irradiation, creep rates at temperatures below the thermal creep regime have been found to be comparable to those of typical thermal creep conditions (Grabova *et al.* 1986). The degree of enhancement of creep is greatest for lower irradiation temperatures, decreasing as the thermal creep regime is approached.

Irradiation creep is primarily due to the presence of both interstitial atoms and vacancies at very large supersaturation levels. Measurements of activation energy associated with irradiation creep reveal that it is the interstitials that are involved in the process rather than the vacancies. Stress-induced preferential absorption (SIPA) of interstitials at dislocations is considered the dominant mechanism of irradiation creep (see Nuclear Reactor Materials: Irradiation Effects).

Irradiation creep in a fast reactor environment has two major components. One is swelling-driven and the other is unrelated to void swelling. The effective strain rate per unit stress and dpa, B, can be written as

$$B = B_o + D\dot{S}$$

where B_o is the creep compliance, D is the creep–swelling coupling coefficient, and $\dot{S}$ is the instantaneous volumetric swelling rate per dpa. B_o shows very weak dependence on temperature between 673 K and 823 K for most of the austenitic stainless steels with its observed value of the order of 10^{-6} $MPa^{-1}dpa^{-1}$ and $D \sim 0.6 \times 10^{-2}\,MPa^{-1}$ (Ehrlich 1981, Garner *et al.* 1999). An important consequence of the direct coupling between irradiation creep and swelling is that strategies adopted to produce swelling-resistant materials also effectively reduce irradiation creep in these materials.

4. *Irradiation-induced Embrittlement*

Fast neutron irradiation strongly affects the mechanical properties of austenitic alloys. One area of major concern is the loss of ductility in these alloys. The embrittlement is dependent on the temperature of irradiation and testing, prior microstructural condition, and neutron flux. The temperature of irradiation is a very important variable as it determines the stability, distribution, and morphology of the defect structures such as dislocation loops, network dislocations, voids, and helium bubbles. At low irradiation and testing temperatures (<673 K), irradiation hardening and consequent reduced capacity of the material to work harden leads to embrittlement (see Nuclear Reactor Materials: Irradiation Effects). At higher irradiation and testing temperatures ($\gtrsim 823$ K), the reduction in ductility is attributed to helium embrittlement (Ullmaier and Trinkaus 1992). Helium is practically insoluble in steels and precipitates as helium bubbles. These bubbles are swept to grain boundaries by dislocations.

The stress-induced growth of helium gas bubbles on grain boundaries leads to reduced ductility. These bubbles grow on grain boundaries normal to the applied stress by stress-directed diffusion of vacancies. Coalescence of adjacent bubbles eventually takes place, possibly by ductile tearing of the ligaments between the cavities, producing cracks and intergranular failure. Many approaches have been examined for alleviating the problem of radiation-induced high-temperature embrittlement in austenitic alloys. These include grain refinement, pinning of helium bubbles at particles within the grains, and formation of precipitates at the grain boundaries to reduce their effectiveness as sources of vacancies. Formation of fine MC carbides in the matrix leads to a high dispersion of helium bubbles and improves postirradiation ductility in titanium-modified austenitic alloys.

At temperatures and doses where void swelling is significant, it exerts a strong influence on mechanical properties (Fissolo *et al.* 1990, Seran *et al.* 1992). It is observed that the austenitic stainless steels such as cold-worked silicon-modified 15-15Ti developed for improved resistance to void swelling also exhibit better postirradiation ductility when compared to cold-worked 316Ti. The correlation between ductility and swelling can be attributed to either the direct effect of voids in promoting brittle fracture or to the indirect effect of radiation-induced segregation owing to the presence of voids. Radiation-induced segregation could lead to instability of the austenite matrix owing to nickel segregation at void surfaces facilitating transformation to martensite, which would induce embrittlement. As there is a direct correlation between swelling and embrittlement, measures to improve swelling resistance of austenitic stainless steels also enhance resistance to embrittlement.

5. *Engineering Consequences of Irradiation Effects*

Void swelling, irradiation creep, and embrittlement caused by fast neutron exposure of core structural materials are important phenomena that determine the permissible life of fuel elements in the core of FBRs. The objective of materials development is to increase the life of fuel elements in the core with a view to increasing the burn-up. To understand how these irradiation phenomena influence fuel subassembly behavior, it is important to visualize the FBR core fuel elements. A FBR core consists of hexagonal wrapper tubes that contain thin-walled fuel pins (Fig. 4). Fuel pins are separated from one another to ensure adequate coolant flow by a helical spacer wire attached to the end plugs of each pin. Deformations of various components of the subassemblies can occur owing to void swelling, thermal creep, and irradiation creep. Swelling and irradiation creep are very sensitive to environmental variables and temporal and spatial gradients that exist throughout the core. Limits have to be specified on deformations in view of the brittle nature of irradiated austenitic stainless steels as well as to avoid interactions between the core components. The latter aspect is very important in permitting insertion and removal of core components during refueling using only moderate loads.

Differential swelling can occur because of gradients in flux and temperature in the core. Deformations owing to differential swelling between the fuel pin and spacer wire should be limited. If the spacer wire deforms more than the pin, the coolant flow can be reduced and consequently the fuel pins overheat slightly. When the pin deformation is higher than that of the spacer wire, there is a mechanical interaction between the wire and the pin and the fuel pin

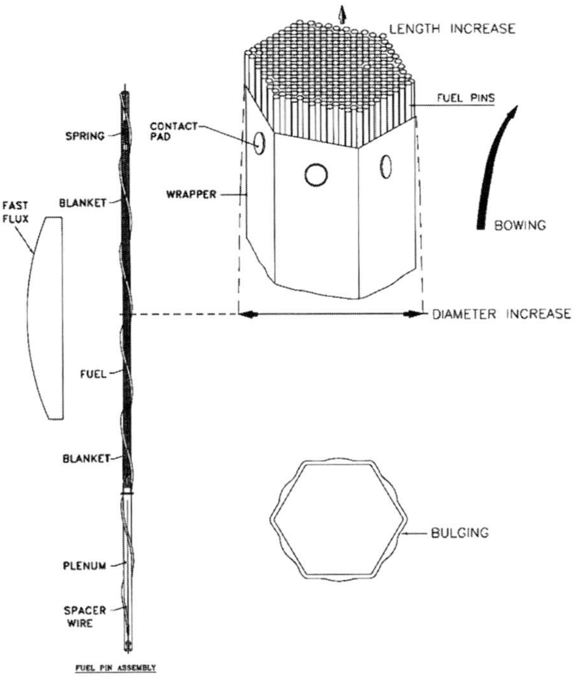

Figure 4
Schematic of fuel subassembly showing the cut out of fuel pins, bulging, and bowing.

undergoes a helical deformation. If both fuel pin and spacer wire swell excessively, there are interactions between the pins and between the fuel pins and the wrapper. Wrapper deformation should be limited otherwise interactions between wrappers lead to excessive loads for fuel handling. Irradiation-induced void swelling and creep can produce different types of wrapper deformation. At the center of the core, subassemblies are expected to remain straight with an elongation and an increase of distance across flats. But at the periphery, subassemblies may bow outwards owing to differential void swelling on the opposite faces of the wrappers as a consequence of the neutron flux gradient. The combination of void swelling and creep induced by internal sodium pressure produces dilation and rounding of wrapper faces.

6. *Conclusion*

Core materials in FBR environments are subjected to very hostile conditions of high, fast neutron flux and elevated temperatures. Austenitic stainless steels undergo phase alterations, dimensional changes, and modifications in mechanical properties owing to this environment. Void swelling, irradiation creep, and embrittlement are important phenomena that determine the permissible life of fuel subassemblies in the core. These in turn depend on various metallurgical and environmental factors. It is important, therefore, to fix precise specifications with narrow limits on chemical composition to avoid differences in deformations arising out of heat-to-heat variations for different core components. Void swelling has been found to be the dominant issue and strategies to combat void swelling in austenitic stainless steels via compositional and thermomechanical modifications have also been very effective in reducing irradiation creep and embrittlement. Improved versions of austenitic stainless steels can be used safely for exposures up to 100–120 dpa levels.

However, there is a necessity to increase the burn-up further and reduce fuel cycle cost. For this purpose, the wrappers may have to be changed to ferritic steels (see *Fast Reactor Cores, Ferritic–Martensitic Steels for*), which swell much less than the austenitic steels. However, these steels have poor high-temperature mechanical properties and are not suitable for fuel clad applications where creep is important. There are developments in the area of dispersion-strengthened ferritic steels that have adequate high-temperature mechanical properties. These materials, however, are more difficult to fabricate into tubes and welding is an issue to be resolved. Another consideration is the increase in ductile to brittle transition temperature in ferritic steels, which might pose some problems in handling of fuel elements after irradiation.

Nickel alloys have also been considered for core applications in FBRs in view of their high resistance to swelling and creep. These alloys, however, are prone to helium embrittlement. Therefore, improved versions of austenitic stainless steels will continue to be the major core materials for FBRs in the short term.

See also: Fusion Conditions; Radiation Damage Theory

Bibliography

Appleby W K, Bloom E E, Flinn J E, Garner F A 1977 Swelling in neutron-irradiated 300-Series stainless steels In: *Proc. Int. Conf. Radiation Effects in Breeder Reactor Structural Materials*.. Metallurgical Society of AIME Scottsdale, NY, pp. 509–27

Dubuisson P, Maillard A, Delalande C, Gilbon D, Seran J L 1992 The effect of phosphorus on the radiation induced microstructure and stabilized austenitic stainless steels. In: Stoller R E, Kumar A S, Gelles D S (eds.) *Effects of Radiation on Materials: 15th Int. Symp., ASTM STP 1125*. American Society for Testing and Materials, Philadelphia, PA, pp. 995–1014

Ehrlich K 1981 Irradiation creep and interrelation with swelling in austenitic stainless steels. *J. Nucl. Mater.* **100**, 149–66

Fissolo A, Cauvin R, Hugot J-P, Levy V 1990 Influence of swelling on irradiated CW titanium modified 316 embrittlement In: *Effects of Radiation on Materials: 14th Int. Symp. ASTM STP 1040*. American Society for Testing and Materials, Philadelphia, PA, pp. 700–13

Garner F A 1984 Recent insights on the swelling and creep of irradiated austenitic alloys. *J. Nucl. Mater.* **122/123**, 459–71

Garner F A 1993 Irradiation performance of cladding and structural steels in liquid metal reactors. In: Frost B R T (ed.) *Nuclear Materials: Part 1. Materials Science and Technology A Comprehensive Treatment*. VCH Publishers, Vol. 10. pp. 420–557

Garner F A, Kumar A S 1987 The influence of both major and minor elements composition on void swelling in austenitic steels. In: Garner F A, Packan N H, Kumar A S (eds.) *Radiation-Induced Changes in Microstructure: 13th Int. Symp. (part I), ASTM STP 955*. American Society for Testing and Materials, Philadelphia, PA, pp. 289–314

Garner F A, Toloczko M B, Munro B, Adaway S, Standring J 1999 Comparison of irradiation creep and swelling of an austenitic alloy irradiated in FFTF and PFR. In: Nanstad R K, Hamilton M L, Garner F A, Kumar A S (eds.) *Effects of Radiation on Materials: 18th Int. Symp., ASTM STP 1325*. American Society for Testing and Materials, Philadelphia, PA, pp. 713–24

Grabova R B, Goncharenko Yu D, Gorbatov V K 1986 The effect of reactor radiation on the dislocation structure and creep characteristics of the (high-nickel) alloy 03Kh20N45M4BCh. *Fiz. Met. Metalloved.* **61** (6), 1164–9

Harries D R 1981 Physical metallurgy of iron–chromium–nickel austenitic steels In: *Proc. Conf. Mechanical Behavior And Nuclear Applications of Stainless Steel at Elevated Temperatures*. Metals Society, London, pp. 1–14

Herschbach K, Schneider W, Ehrlich K 1993 Effects of minor alloying elements upon swelling and in-pile creep in model plain Fe–15Cr–15Ni stainless steels and in commercial DIN 1.4970 alloys. *J. Nucl. Mater.* **203**, 233–48

Lee E H, Mansur L K 1992 Relationships between phase stability and void swelling in Fe–Cr–Ni alloys during irradiation. *Metall. Trans.* **23A**, 1977–86

Mansur L K 1986 Swelling in irradiated metals and alloys. In: Bever M B (ed.) *Encyclopedia of Materials Science and Engineering*. Pergamon, Oxford, Vol. 6, pp. 4834–8

Maziasz P J 1993 Overview of microstructural evolution in neutron-irradiated austenitic stainless steels. *J. Nucl. Mater.* **205**, 118–45

Maziasz P J, McHargue C J 1987 Microstructural evolution in annealed austenitic steels during neutron irradiation. *Int. Mater. Rev.* **32** (4), 190–219

Russel K C 1984 Phase stability under irradiation. *Prog. Mater. Sci.* **28**, 229–434

Sekimura N, Hamada K, Ishino S 1994 The observation and interpretation of microstructural evolution and swelling in austenitic steels under irradiation. In: Kumar A S, Gelles D S, Nanstad R, Little E A (eds.) *Effects of Radiation on Materials: 16th Int. Symp., ASTM STP 1175*. American Society for Testing and Materials, Philadelphia, PA, pp. 992–1000

Seran J L, Levy V, Dubuisson P, Gilbon D, Maillard A, Fissolo A, Touron H, Cauvin R, Chalony A, Le Boulbin E 1992 Behavior under neutron irradiation of the 15-15Ti and EM10 steels used as standard materials of the Phenix fuel subassembly. In: Stoller R E, Kumar A S, Gelles D S (eds.) *Effects of Radiation on Materials:15th Int. Symp., ASTM STP 1125*. American Society for Testing and Materials, Philadelphia, PA, pp. 1209–33

Shibahara I, Akasaka N, Onose S 1996 Effects of metallurgical variables on swelling of modified 316 and higher Ni austenitic stainless steels. In: Gelles D S, Nanstad R, Kumar A S, Little E A (eds.) *Effects of Radiation on Materials: 17th Int. Symp., ASTM STP 1270*. American Society for Testing and Materials, Philadelphia, PA, pp. 858–73

Ukai S, Akasaka N, Hattori K, Onose S 1999 The effects of phase stability on void swelling in P, Ti-modified 316 stainless steels during neutron irradiation. In: Nanstad R K, Hamilton M L, Garner F A, Kumar A S (eds.) *Effects of Radiation on Materials: 18th Int. Symp., ASTM STP 1325*. American Society for Testing and Materials, Philadelphia, PA, pp. 808–21

Ullmaier H, Trinkaus H 1992 Helium in metals: effect on material properties. *Mater. Sci. Forum* **97–99**, 451–72

Weiss S, Stickler 1972 Phase instabilities during high temperature exposure of 316 austenitic stainless steels. *Metall. Trans.* **3**, 851–66

Fast Reactor Cores, Ferritic–Martensitic Steels for

J. L. Séran, J. C. Brachet, and A. Alamo

The successful operation of fast reactors is dependent to a large extent on the good performance of metallic core structures—fuel cladding and wrapper (see *Fast Breeder Reactors: Fuels, Fast Reactor Cores, Austenitic Steels for*)—that experience a temperature range around 300–700°C and high fast-neutron irradiation damage levels (commonly expressed in dpa, i.e., the number of times an atom is displaced from its lattice site).

The maximum doses that structural materials can sustain are around 100 dpa. But, in order to ensure the economy of fast reactor systems, the metallic structures of subassemblies should enable the fuel to reach the high burn-ups required and thus withstand doses above 200 dpa. Owing to poor swelling resistance, this goal is recognized as unreachable by austenitic steels used as standard structural materials for core components. However, ferritic–martensitic steels are highly resistant to void swelling and are generally considered as the best alternative material solution for long-term applications.

The purpose of this article is first to introduce the physical metallurgy and the out-of-pile properties of these materials and second to compare their behavior under irradiation and their postirradiation mechanical properties with those of austenitic steels.

1. *Materials*

The main candidate materials developed in Western countries for in-core applications are HT9 (12Cr–1MoWVNb) in the USA (Straalsund and Johnson 1981) and, in Europe (Bagley *et al.* 1987), EM10 (9Cr–1Mo) and EM12 (9Cr–2MoVNb) in France, FV448 (10Cr–0.7MoWVNb) in the UK, and 1.4914 (10.5Cr–0.6MoVNb) in Germany.

Optimization of these steels aims to obtain fully martensitic materials, which appear as the most promising candidates. Typical heat treatments applied to martensitic materials, i.e., austenitization (normalization) and tempering, make it possible to obtain a wide range of mechanical properties by modifying the heat treatment conditions (temperature, time, cooling rate) (Klueh and Alexander 1998). Thus, high strength in the range 20–650°C can be reached by increasing the austenitization temperature ($T \geqslant 1100$°C), in particular for vanadium-doped steels. In contrast, a better toughness level is favored by the fine grain size structure obtained by austenitization treatments performed at lower temperature ($T < 1100$°C). Tempering at lower temperatures and/or the application of final cold working can also enhance strength.

2. *Physical Metallurgy*

Owing to the occurrence of significant microstructural evolution during thermal treatments of ferritic–martensitic steels, it is important to take into account the physical metallurgy of these materials. This includes ferrite (α) and/or martensite $\leftrightarrow$ austenite (γ) phase transformations and precipitation reactions of carbides and/or nitrides. Three main aspects are considered below.

2.1 *Austenitization Treatments*

These thermal treatments are generally performed at temperatures ranging from 950°C up to 1200°C to obtain homogenization of the γ matrix by the dissolution of carbides and nitrides. Small additions of vanadium and/or niobium promote finer γ grain sizes up to ~1100°C. This behavior can be related to the pinning of grain boundaries by (V,Nb) carbonitrides.

2.2 *Phase Transformation on Cooling*

When initiated from the austenite phase, continuous cooling transformations (usually described in CCT diagrams) in these kinds of steels (Finkler and Schirra 1996) lead to martensite and/or ferrite phases, depending on the cooling rate. Figure 1 shows that the critical cooling rates (R_{cc}) to obtain either 100% martensite or 100% ferrite depend on the prior austenitic grain size (d_γ) according to the following relationship: $R_{cc} \sim 1/(d_\gamma)^m$, where $m \sim 1$ (Brachet *et al.* 1999).

2.3 *Tempering Treatments*

During tempering of the martensite, different types of precipitation reactions occur. For temperatures below ~450°C, few microstructural changes take

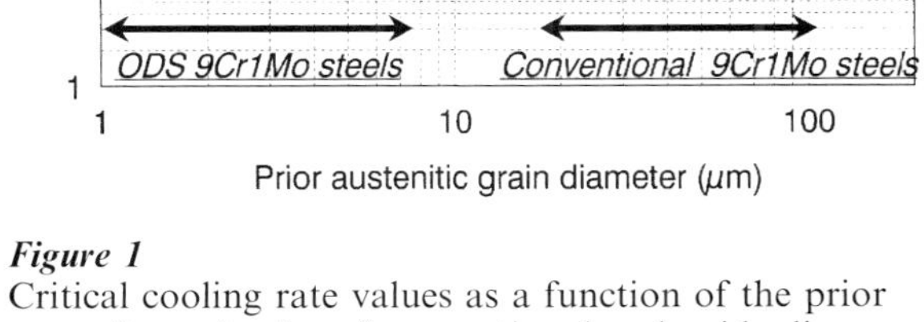

Figure 1
Critical cooling rate values as a function of the prior austenite grain size of conventional and oxide dispersion-strengthened 9Cr1Mo(Nb,V) steels.

place. For temperatures close to ~500°C, a secondary hardening phenomenon is observed, owing to the early precipitation stages of intragranular chromium-rich carbonitrides of the M_2X type. For temperatures ranging from ~600°C up to ~800°C, a heavy precipitation of chromium-rich $M_{23}C_6$ carbides and a recovery of the martensitic matrix occur with, consequently, an important softening effect (Pickering and Vassiliou 1980).

3. *Out-of-pile Properties: Thermal Aging Effects*

The optimization of the in-service behavior of these steels needs to consider not only the best compromise of initial properties, but also their evolution during thermal aging and under irradiation. In general, thermally aged 9–12Cr/0.5–1.0Mo (W, Nb,V) steels behave differently in the low and high temperature range because of different microstructural evolutions (Brachet *et al.* 1995, Alamo *et al.* 1996):

T < *450 °C*. Aging induces an increase of the tensile strength along with a moderate degradation of impact properties, an increase of the ductile–brittle transition temperature (DBTT), and a decrease of the upper shelf energy (USE). The magnitude of the induced hardening and changes in DBTT and USE are correlated to the chromium content of steels. 9Cr steels display the highest stability of properties in this range. No modification of the distribution of carbides and carbonitrides is detected. In the case of steels containing more than about 11% chromium, experiments using small-angle neutron scattering techniques have shown the occurrence of chromium-rich α' clusters of about 1 nm size, which makes them prone to hardening and embrittlement in this temperature range.

T > *450 °C*. The main effects observed are on the impact properties. The DBTT shift and decrease of USE are related to the precipitation of intermetallic Laves phase of $Fe_2(Mo,W)$ type (Senior *et al.* 1988), which occurs as a brittle film on boundaries of prior austenite grains and martensite laths and promotes the intergranular fracture mode.

Consequently, increasing contents of molybdenum and tungsten are responsible for the embrittlement detected in this range of aging temperature.

Besides the compositional effects, another important factor governing the aging behavior is related to the initial metallurgical condition. Figure 2 shows the evolution of the DBTT after aging of 9Cr–1MoVNb steel produced with different initial metallurgical conditions. Compared to the conventional tempered condition, materials with a final cold work (~10%) present more stable mechanical properties (lower DBTT shift and USE decrease) and a best combination of strength and ductility after aging (Alamo *et al.* 1996).

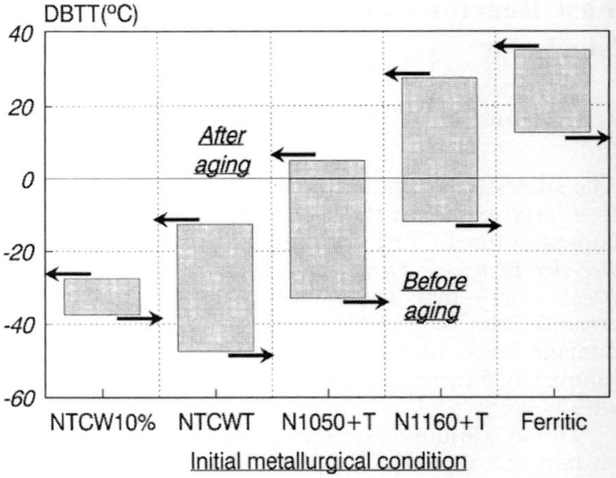

Figure 2
Evolution of the DBTT after aging for 10 000 hours at 550°C of 9Cr1MoNbV (T91) steel as a function of the initial metallurgical condition: NTCW10% = normalized + tempered + cold worked 10%; NTCWT = normalized + tempered + cold worked 10% + tempered; N1050 + T = normalized at 1050°C + tempered (at 760°C); N1160 + T = normalized at 1160°C + tempered (at 760°C); Ferritic = fully ferritic structure obtained from isothermal decomposition of austenite at 670°C.

4. *Postirradiation Properties*

Metallic structural materials of the fast fuel subassembly are subjected to an irradiation-controlled aging of about 15 000 hours to reach a dose of 100 dpa. Besides the thermal aging effects described above, neutron irradiation can induce specific microstructural changes and degradation of mechanical properties, as well as the appearance of new phenomena like swelling and irradiation creep (see *Nuclear Reactor Materials: Irradiation Effects*).

The direct data from fuel subassembly performances in the main fast reactors (Phénix, FFTF, PFR, BN600) and all results from the different experimental programs confirm the high resistance to swelling and irradiation creep of the b.c.c. ferritic–martensitic steels (Garner and Gelles 1990, Séran *et al.* 1992) and, at least up to a dose of around 200 dpa, the swelling resistance of these steels does not appear to be the life-limiting factor of fuel subassemblies.

In fact, the main problem arising for this class of materials is the degradation of impact properties observed for irradiation temperatures lower than 500°C. The dose dependence of this embrittlement seems to exhibit a saturation for doses higher than about 30 dpa (Powell *et al.* 1986, Séran *et al.* 1992, Klueh and Alexander 1998), but this fact must be proved for very high doses (~200 dpa). Degradation

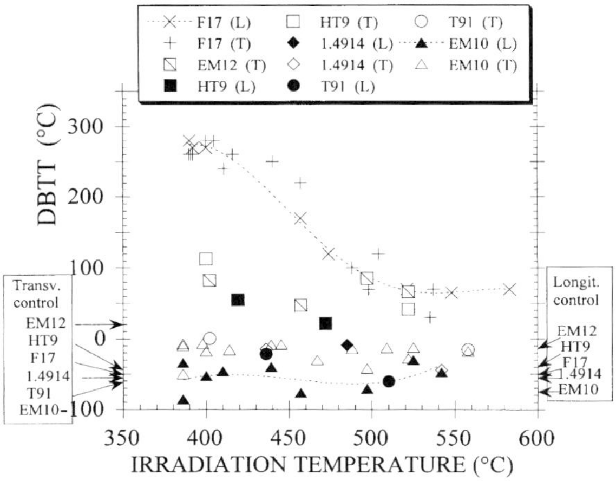

Figure 3
Evolution of the DBTT with irradiation temperature determined on ferritic–martensitic steels irradiated in the Phénix reactor with doses in the range 70–110 dpa. Charpy specimens are obtained with the LT (L) and TL (T) orientations. EM10, EM12, T91, and 1.4914 are 9–10Cr steels, EM12 is duplex ferritic–martensitic, and EM10, T91, and 1.4914 are fully martensitic. HT 9 is a 12Cr and F17 is a 17Cr.

of toughness is generally correlated with an irradiation-controlled hardening owing to the formation of dislocation loops and, depending on the alloy chemistry, to the precipitation of embrittling phases (Maziasz and Klueh 1992, Dubuisson *et al.* 1993).

As shown in Fig. 3 (Séran *et al.* 1994), in this low irradiation temperature range, the major parameter controlling this embrittlement seems to be the chromium content: the lower the chromium concentration in the 9–17 wt.% range investigated, the lower is the embrittlement. One can observe from Fig. 3 that only the 9–11Cr fully martensitic steels studied here (EM10, T91, and 1.4914) exhibit postirradiation DBTT values lower than room temperature and thus can be used without any problem in the normal operating conditions of a fast reactor. In the higher temperature range (>500°C), postirradiation properties are essentially controlled by thermal aging effects.

5. *Conclusion*

Ferritic–martensitic steels are potential candidates for long-life fast reactor core applications because they exhibit an excellent microstructural stability and swelling resistance with adequate mechanical properties after prolonged service in irradiation environments. In this class of materials, fully martensitic steels having about 9% chromium content are the most promising. One can consider that they are nearly qualified for long-term wrapper applications or for any highly irradiated structure of reactors operating at temperatures lower than about 600°C. These materials constitute a basis for research and development on reduced activation martensitic variants where the qualification is in progress in the frame of fusion technology studies (see *Fusion Reactors: Low-activation Structural Materials*).

In the case of fuel cladding that must operate at ~700°C, further effort must be made to develop a ferritic–martensitic steel resistant to thermal creep. One promising way (Alamo *et al.* 1992, Ukai *et al.* 1998) is to develop ferritic–martensitic steels strengthened by an oxide dispersion. At this stage of development, the main problem that needs to be solved is the demonstration of the industrial fabrication feasibility for these kinds of advanced materials.

See also: Radiation Damage Theory; Nuclear Reactors: Pressure Vessel Steels

Bibliography

Alamo A, Brachet J C, Castaing A, Foucher C 1996 Effects of the initial metallurgical condition on the microstructure and mechanical properties of 9Cr–1Mo and 9Cr–1MoVNb thermal aged steels. In: Liaw P K, Viswanathan R, Murthy K L, Frear D, Simonen E P (eds.) *Proc. Int. Conf. Microstructures and Mechanical Properties of Aging Materials II.* Minerals, Metals and Materials Society pp. 121–34

Alamo A, Regle H, Bechade J L 1992 Effects of processing on textures and tensile properties of oxide dispersion strengthened ferritic alloys obtained by mechanical alloying In: Capus J, German R M (eds.) *Proc. 1992 Powder Metallurgy World Congress of the Metal Powder Industries Federation.* American Powder Metallurgy Institute Vol. 7, 169–182

Bagley K Q, Little E A, Lévy V, Alamo A, Ehrlich K, Anderko K, Calzabini A 1987 European development of ferritic–martensitic steels for fast reactor wrapper applications *Proc. Int. Conf. Materials for Nuclear Reactor Core Applications.* British Nuclear Energy Society, London, Vol. 2, pp. 37–46

Brachet J C, Castaing A, Foucher C 1995 Effects of the chemical composition on the behaviour of 9–12Cr/0.5–1Mo/WNbV thermal aged steels *Proc. Int. Symp. Materials Ageing and Component Life Extension.* Engineering Materials Advisory Services, UK, pp. 75–87

Brachet J C, Lambard V, Alamo A 1999 Studies of phase transformations occurring in conventional, low activation and ODS 7–11%Cr–(Mo,W,V,Nb,Ta) martensitic steels. *Proc. Int. Conf. Solid–Solid Phase Transformations* **'99**, Kyoto, Japan

Dubuisson P, Gilbon D, Séran J L 1993 Microstructural evolution of ferritic–artensitic steels irradiated in the fast breeder reactor Phénix. *J. Nucl. Mater.* **205**, 178–89

Finkler H, Schirra M 1996 Transformation behaviour of the high temperature martensitic steels with 8–14% chromium. *Steel Res.* **67**, 328–42

Garner F A, Gelles D S 1990 Neutron-induced swelling of commercial alloys at very high exposures. In: Packan N H, Stoller R E, Kumar A S (eds.) *Proc. 14th Int. Symp. Effects of Radiation on Materials*. ASTM STP 1046. American Society for Testing and Materials, Philadelphia, Vol. 2, pp. 673–83

Klueh R L, Alexander D J 1998 Heat treatment effects on impact toughness of 9Cr–1MoVNb and 12Cr–1MoVW steels irradiated to 100 dpa. *J. Nucl. Mater.* **258–63**, 1269–74

Maziasz P J, Klueh R L 1992 Comparison of the effects of long-term thermal aging and HFIR irradiation on the microstructural evolution of 9Cr1MoVNb steel. In: Stoller R E, Kumar A S, Gelles D S (eds.) *Proc. 15th Int. Symp. Effects of Radiation on Materials ASTM STP 1125*. American Society for Testing and Materials, Philadelphia, pp. 1135–56

Pickering F B, Vassiliou A D 1980 Effect of austenitizing temperature on constitution, transformation and tempering of 9Cr–1Mo. *Met. Technol.* **7** (Oct.), 409–13

Powell R W, Johnson G D, Hamilton M L, Garner F A 1986 LMR cladding and duct materials development *Int. Conf. Reliable Fuels for Liquid Metal Reactors*. American Nuclear Society, La Grange Park, GA, pp. 4.17–4.

Senior B A, Noble F W, Eyre B L 1988 The effect of ageing on the ductility of 9Cr–1Mo steel. *Acta Metall.* **36** (7), 1855–62

Séran J L, Alamo A, Maillard A, Touron H, Brachet J C, Dubuisson P, Rabouille O 1994 Pre- and post-irradiation mechanical properties of ferritic–martensitic steels for fusion applications: EM10 base metal and EM10/EM10 welds. *J. Nucl. Mater.* **212–15**, 588–93

Séran J L, Lévy V, Dubuisson P, Gilbon D, Maillard A, Fissolo A, Touron H, Cauvin R, Chalony A, Le Boulbin E 1992 Behavior under neutron irradiation of the 15-15Ti and EM10 steels used as standard materials of the Phénix fuel subassembly. In: Stoller R E, Kumar A S, Gelles D S (eds.) *Proc. 15th Int. Symp. Effects of Radiation on Materials*. ASTM STP 1125. American Society for Testing and Materials, Philadelphia, pp. 1209–33

Straalsund J L, Johnson G D 1981 Alloy development for cladding and duct applications. *Trans. Am. Nucl. Soc.* **38**, 263

Ukai S, Nishida T, Okuda T, Yoshitake T 1998 R&D of oxide dispersion strengthened ferritic–martensitic steels for FBR. *J. Nucl. Mater.* **258–63**, 1745–9

Fuel Cells: Electrochemical Reactions

H. Nagamoto

The electrochemical devices that convert chemical energy directly to electrical energy are called cells. Fuel cells are devices that can continuously generate electricity by continuously supplying a fuel and an oxidant. Fuel cells chemically combine the molecules of a fuel and oxidant without burning.

When hydrogen is used as a fuel no carbon dioxide is produced, which makes fuel cells extremely attractive. Fuel cells have several attractive characteristics: (i) high efficiency in energy conversion; (ii) low chemical and acoustic pollution; (iii) flexibility in fuel; (iv) capability of cogeneration; (v) compactness in shape; and (vi) fast load response.

The first fuel cell was built in 1839 by Sir William Grove, a Welsh judge and scientist. Serious interest in the fuel cell as a practical generator did not begin until the 1960s, when the American space program chose fuel cells to supply power for the Gemini and Apollo spacecrafts. Fuel cells still provide electricity and water for the space shuttle. In the 1990s fuel cells were loaded on to electric vehicles in place of batteries.

1. *Principles*

All fuel cells have the same basic operating principle. A fuel cell consists of two electrodes sandwiched around an electrolyte. In polymer electrolyte fuel cells (PEFCs) an oxidant, oxygen or air, passes over the cathode and a fuel, hydrogen, over the anode, generating electricity, water, and heat, as shown in Fig. 1.

The hydrogen molecule is dissociatively adsorbed over the catalytic anode to form hydrogen atoms Eqn. (1). The hydrogen atom splits into a proton and an electron (Eqn. (2)), which take different paths to the cathode.

$$H_2 \Leftrightarrow 2H \tag{1}$$

$$H \Leftrightarrow H^+ + e^- \tag{2}$$

The proton passes through the electrolyte. The electrons create a separate current that can be utilized before they return to the cathode to be reunited with the proton and oxygen in a molecule of water.

$$H^+ + \tfrac{1}{2}O_2 + e^- \Leftrightarrow H_2O \tag{3}$$

Then the overall reaction is therefore

$$H_2 + \tfrac{1}{2}O_2 \Leftrightarrow H_2O \tag{4}$$

The open circuit voltage, E, is given by Nernst equation:

$$E = \frac{RT}{2F}\ln\left[\frac{P_{H_2}(\text{anode})}{P_{H_2}(\text{cathode})}\right] \tag{5}$$

where R is the gas constant, T the cell temperature, F the Faraday constant, and P_{H_2}(anode) is the hydrogen partial pressure over the anode.

When the reaction expressed by Eqn. (4) over the cathode is in equilibrium, E is given by

$$E = E^0 + \frac{RT}{2F}\ln\left[\frac{P_{H_2}(\text{anode})\sqrt{P_{O_2}(\text{cathode})}}{P_{H_2O}(\text{cathode})}\right] \tag{6}$$

where P_{O_2} and $P_{H_2}O$ are the partial pressures of oxygen and water, respectively, and E^0 is expressed by using the standard Gibbs energy change of Eqn. (4) ΔG^0 as

$$E^0 = -\frac{\Delta G^0}{2F} \tag{7}$$

Eqn. (7) shows that ΔG^0 can be changed to electrical energy. The overall reaction given by Eqn. (4) equates to the combustion of hydrogen which accompanies the evolution of reaction heat Q.

In the case of chemical reactions the reaction heat Q is identical to the negative value of the standard enthalpy change ΔH^0. The relation between ΔG^0 and ΔH^0 is given by

$$\Delta H^0 = \Delta G^0 + T\Delta S^0 \tag{8}$$

where ΔS^0 is the standard entropy change. The second term implies the heat emitted to the environment.

Fuel cells continuously provide electricity as long as there is a supply of fuel and oxygen, where fuel

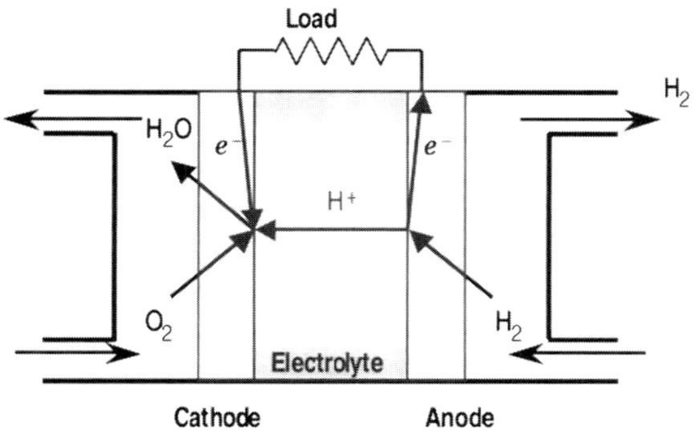

Figure 1
Principles of a fuel cell.

cells are different from dry batteries (see *Batteries, Rechargeable*).

2. Theoretical and Technical Background

The cell voltage V is given by

$$V = E - IR - |\eta \text{ (anode)} - |\eta \text{ (cathode)} \quad (9)$$

where I is the current passing through the cell, R is the electrical resistance of the cell, and η is the overpotential which causes a voltage loss.

The voltage loss can be kept low by reducing R by using a thin electrolyte of high ionic conductivity, or reducing the overpotentials by using electrodes of high catalytic activity for electrochemical reactions and by optimizing the electrode structure.

A fuel cell system that includes a "fuel reformer" can use the hydrogen from any hydrocarbon fuel, from natural gas to methanol, and even gasoline. Since the fuel cell relies on chemistry and not combustion, emissions from this type of a system would still be much smaller than emissions from the cleanest fuel combustion processes.

3. Thermodynamics and Electrochemistry

The performance of fuel cells is usually expressed by the electrical efficiency, which is defined as the ratio of the electrical power output to the chemical energy of the fuel. The electrical power is theoretically identical to the negative value of the standard Gibbs energy change $\Delta G^0$of the overall reaction (Eqn. (4)). The chemical energy of the fuel is represented by the standard enthalpy change ΔH^0 of the overall reaction, which is equal to the negative value of the overall reaction heat, $-Q$. Thus, the theoretical efficiency η_{th} is given by

$$\eta_{th} = \frac{-\Delta G^0}{-\Delta H^0} = 1 - T\frac{\Delta S^0}{\Delta H^0} \quad (10)$$

In Eqn. (4) η_{th} decreases with an increase in temperature since $(\Delta S^0/\Delta H^0)$ is positive and has only a small dependence on the temperature. For example, η_{th} for Eqn. (4) is 0.945 at 298 K and 0.776 at 1000 K.

The practical efficiency, η, which is defined by

$$\eta = \frac{PW}{Q} = \frac{VI}{Q} = \frac{(E - IR - /\eta \text{ (anode)}/ - /\eta \text{ (cathode)}/)I}{Q} \quad (11)$$

where PW is the electrical power obtained as the product of the cell voltage V and the output current I. When one takes up the overall reaction where liquid water is produced, Q includes the condensation heat of gaseous water in addition to the standard enthalpy change for the reaction producing gaseous water. The electrical efficiency in this case is called the efficiency based on high heat value (HHV).

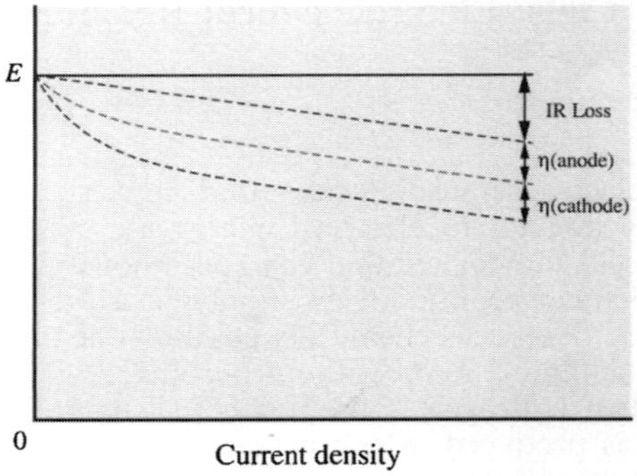

Figure 2
Cell voltage and voltage losses.

The practical efficiency is affected by the cell voltage losses shown in Eqn. (9), the dependence of which on the current is shown in Fig. 2. IR loss (Ohmic loss) is caused by the resistances of the electrolyte, the electrodes, and the interconnect between unit cells. The IR loss may be reduced by selecting materials of high electrical conductivity. The generation of overpotential is concerned with two rate processes: the reaction process; and the mass transfer process. The activation overpotential is caused by the small reaction rate of the charge transfer process over the electrode such as Eqn. (2) or Eqn. (3). In other words, the activation overpotential enhances the charge transfer rate at the expense of electrical efficiency.

The small rate of a mass transfer process such as the diffusion process causes the concentration overpotential. Based on Fick's law, the mass transfer conducted by diffusion requires the concentration difference:

$$J = D\frac{\Delta C}{L} \quad (12)$$

where J is the diffusion flux, D is the diffusion coefficient,L is the diffusion length, and ΔC is the concentration difference of a reactant. This concentration difference leads to a drop in the chemical potential of the reactant, resulting in a potential drop, i.e., the concentration overpotential.

4. Three-phase Boundary

The electrode reactions for fuel cells are considered to take place at the site called the three-phase boundary (TPB). The TBP is defined as the boundary where a dissociated hydrogen atom releases an electron going

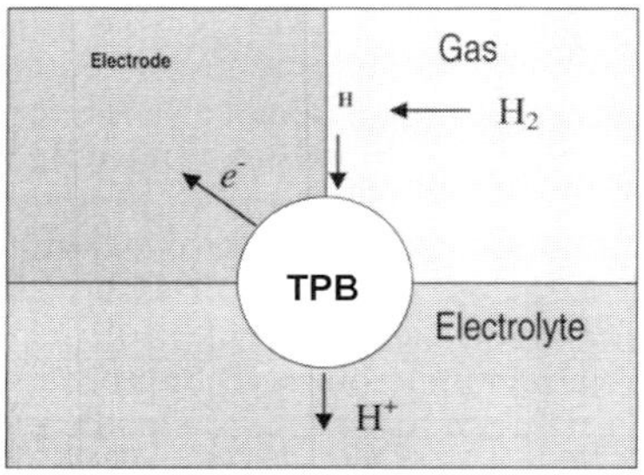

Figure 3
Sketch of a three-phase boundary (TPB).

through the electrode and simultaneously dissolves into the electrolyte for the cathode, or the boundary where ionized hydrogen reacts with oxygen from the gas phase and an electron from the electrode for the anode. Figure 3 shows the TPB for the anode.

One of the key technologies to fabricate successful fuel cells depends on the construction of many TPBs.

5. *Types of Fuel Cells*

There are four primary types of fuel cells that are based on the electrolyte employed: (i) phosphoric acid fuel cell (PAFC); (ii) molten carbonate fuel cell (MCFC); (iii) solid oxide fuel cell (SOFC); and (iv) proton exchange membrane fuel cell (PEFC). Others include the alkaline fuel cell (AFC) and the direct methanol fuel cell (DMFC).

5.1 Phosphoric Acid Fuel Cell

The PAFC is the most mature fuel cell technology in terms of system development and commercialization activities. It has been under development for more than 20 years and has received a total worldwide investment in development and demonstration of the technology in excess of $500 million. The PAFC was selected for substantial development a number of years ago because of the belief that among the low-temperature fuel cells it was the only technology that showed relative tolerance for reformed hydrocarbon fuels and thus could have widespread applicability in the short term.

The PAFC uses liquid phosphoric acid as the electrolyte. The phosphoric acid is contained in a polytetrafluoroethylene (PTFE) bonded silicon carbide matrix. The small pore structure of this matrix preferentially keeps the acid in place through capillary action. Some acid may be entrained in the fuel or oxidant streams and addition of acid may be required after many hours of operation. Platinum-catalyzed, porous carbon electrodes are used on both the fuel (anode) and oxidant (cathode) sides of the electrolyte.

Fuel and oxidant gases are supplied to the backs of the porous electrodes by parallel grooves formed in carbon or carbon–composite plates. These plates are electrically conductive and conduct electrons from the anode to the cathode of the adjacent cell. In most designs, the plates are "bipolar" in that they have grooves on both sides—one side supplies fuel to the anode of one cell while the other side supplies air or oxygen to the cathode of the adjacent cell.

The byproduct water is removed as steam on the cathode (air or oxygen) side of each cell by flowing excess oxidant past the backs of the electrodes. This water removal procedure requires that the system be operated at temperatures around 190°C. At lower temperatures the product water will dissolve in the electrolyte and not be removed as steam. At approximately 210°C the phosphoric acid begins to decompose.

Excess heat is removed from the fuel cell stack by providing carbon plates containing cooling channels every few cells. Either air or a liquid coolant, such as water, can be passed through these channels to remove excess heat.

The PAFC reactions that occur are the same as the PEFC ones:

$$\text{Anode}: \; H_2 \rightarrow 2H^+ + 2e^-$$

$$\text{Cathode}: \; \tfrac{1}{2}O_2 + 2H^+ + 2e^- \rightarrow H_2O$$

PAFC power plant designs show electrical efficiencies in the range from 36% HHV to 42% HHV. The higher efficiency designs operate with pressurized reactants but this requires more components and tends to cost more. PAFC power plants supply usable thermal energy at an efficiency of 37–41% HHV. A portion of the thermal energy can be supplied at temperatures of ~120°C to ~150°C; however, the majority of the thermal energy is supplied at ~65°C. The PAFC has a power density of 1.7–1.9 KW m^{-2} of active cell area.

5.2 Molten Carbonate Fuel Cell

The MCFC evolved from work in the 1960s aimed at producing a fuel cell that would operate directly on coal. While direct operation on coal seems less likely today, operation on coal-derived fuel gases or natural gas is viable.

The MCFC uses a molten carbonate salt mixture as its electrolyte. The composition of the electrolyte varies, but usually consists of lithium carbonate, Li_2CO_3, and potassium carbonate, K_2CO_3. At the operating temperature of about 650°C the salt mixture is liquid and a good ionic conductor. The electrolyte is suspended in a porous, insulating and chemically inert ceramic ($LiAlO_2$) matrix.

The reactions that occur in the MCFC are:

$$\text{Anode}: \; H_2 + CO_3^{2-} \rightarrow H_2O + CO_2 + 2e^-$$

$$\text{or } CO + CO_3^{2-} \rightarrow 2CO_2 + 2e^-$$

$$\text{Cathode}: \; \tfrac{1}{2}O_2 + CO_2 + 2e^- \rightarrow CO_3^{2-}$$

The anode process involves a reaction between hydrogen and carbonate ions (CO_3^{2-}) from the electrolyte that produces water and carbon dioxide (CO_2) while releasing electrons to the anode. The cathode process combines oxygen and CO_2 from the oxidant stream with electrons from the cathode to produce carbonate ions, which enter the electrolyte. The need for CO_2 in the oxidant stream requires a system for collecting CO_2 from the anode exhaust and mixing it with the cathode feed stream.

As the operating temperature increases, the theoretical operating voltage for a fuel cell decreases and with it the maximum theoretical fuel efficiency. On the other hand, increasing the operating temperature increases the rate of the electrochemical reaction and thus the current which can be obtained at a given voltage. The net effect for the MCFC is that the real operating voltage is higher than the operating voltage for the PAFC at the same current density.

The higher operating voltage of the MCFC means that more power is available at a higher fuel efficiency from a MCFC than from a PAFC of the same electrode area. As size and cost scale roughly with electrode area, this suggests that a MCFC should be smaller and less expensive than a "comparable" PAFC.

The MCFC also produces excess heat at a temperature that is high enough to yield high-pressure steam which may be fed to a turbine to generate additional electricity. In combined cycle operation, electrical efficiencies in excess of 60% HHV have been suggested for mature MCFC systems.

The MCFC operates between 600°C and 650°C which is necessary to achieve sufficient conductivity of the electrolyte. To maintain this operating temperature, a higher volume of air is passed through the cathode for cooling purposes.

As mentioned above, the high operating temperature of the MCFC offers the possibility that it could operate directly on gaseous hydrocarbon fuels such as natural gas. The natural gas would be reformed to produce hydrogen within the fuel cell itself.

The need for CO_2 in the oxidant stream requires that CO_2 from the spent anode gas be collected and mixed with the incoming air stream. Before this can be done, any residual hydrogen in the spent fuel stream must be burned. Future systems may incorporate membrane separators to remove the hydrogen for recirculation back to the fuel stream.

At cell operating temperatures of 650°C noble metal catalysts are not required. The anode is a highly porous sintered nickel powder alloyed with chromium to prevent agglomeration and creep at the operating temperatures. The cathode is a porous nickel oxide material doped with lithium. Significant technology has been developed to provide electrode structures that position the electrolyte with respect to the electrodes and maintain that position while allowing for some electrolyte boil-off during operation. The electrolyte boil-off has an insignificant impact on cell stack life. A more significant factor of life expectancy has to do with corrosion of the cathode.

The MCFC operating temperature is about 650°C. At this temperature the salt mixture is liquid and is a good conductor. The cell performance is sensitive to operating temperature and a change in cell temperature from 650°C to 600°C results in a drop in the cell voltage of almost 15%. This reduction is due to increased ionic and electrical resistance and a reduction in electrode kinetics.

5.3 Solid Oxide Fuel Cell

The SOFC uses a ceramic, solid-phase electrolyte that reduces corrosion considerations and eliminates the electrolyte management problems associated with liquid electrolyte fuel cells. To achieve adequate ionic conductivity in such a ceramic, however, the system must operate at about 1000°C. At that temperature, internal reforming of carbonaceous fuels should be possible, and the waste heat from such a device could be easily utilized by conventional thermal electricity generating plants to yield excellent fuel efficiency.

The SOFC is based on the use of a solid ceramic as the electrolyte. The preferred material, dense yttria-stabilized zirconia (YSZ), is an excellent conductor of negatively charged oxygen (oxide) ions at high temperatures. The SOFC is a solid-state device and shares certain properties and fabrication techniques with semiconductor devices. The anode is a porous nickel/zirconia cermet while the cathode is magnesium-doped lanthanum manganite. The Westinghouse cell design constructs the fuel cell around a porous zirconia support tube through which air is supplied to the cathode which is deposited on the outside of the tube. A layer of electrolyte is then deposited on the outside of the cathode and finally a layer of anode is deposited over the electrolyte. A number of cells is connected together by high-temperature semiconductor contacts.

In operation, hydrogen or carbon monoxide (CO) in the fuel stream reacts with oxide ions (O^{2-}) from the electrolyte to produce water or CO_2 and to deposit electrons into the anode. The electrons pass outside the fuel cell, through the load, and back to the cathode where oxygen from air receives the electrons and is converted into oxide ions which are injected into the electrolyte. It is significant that the

SOFC can use CO as well as hydrogen as its direct fuel.

The reactions that occur in a SOFC include:

$$\text{Anode}: H_2 + O^{2-} \rightarrow H_2O + 2e^-$$

$$\text{or } CO + O^{2-} \rightarrow CO_2 + 2e^-$$

$$\text{Cathode}: \tfrac{1}{2}O_2 + 2e^- \rightarrow O^{2-}$$

In development cells and small stacks the SOFC has demonstrated 0.6 V per unit cell at about 250 mA cm^{-2}. Lifetimes in excess of 30000 hours for single cells have been demonstrated as have a number of heat/cool cycles. Unpressurized SOFCs deliver fuel to electric efficiencies in the range of 45% HHV. Argonne National Laboratories suggests that pressurized systems could yield fuel efficiencies of 60% HHV. Bottoming cycles, using the high-temperature waste heat from the fuel cell, could add another few percent to the fuel efficiency of the SOFC system.

The SOFC operates at approximately 1000 °C. Temperature management is achieved through maintenance of the correct volume of the air stream into the cell. As mentioned above, the high operating temperature of the SOFC offers the possibility of internal reforming. As in the MCFC, CO does not act as a poison and can be used directly as a fuel. The SOFC is also the most tolerant of any fuel cell type to sulfur, tolerating several orders of magnitude more sulfur than other fuel cells.

The anode consists of metallic nickel and a YSZ skeleton which serves to inhibit sintering of the metal particles and to provide a thermal expansion coefficient comparable with those of the other cell materials. The anode structure is fabricated with a porosity of 20–40% to facilitate mass transport of reactant and product gases. The strontium-doped lanthanum manganite ($La_{1-x}Sr_xMnO_{3-\delta}$, $x = 0.10 - 0.15$) that is most commonly used for the cathode material is a p-type conductor. Similar to the anode, the cathode is a porous structure that must permit rapid mass transport of reactant and product gases.

The 1000 °C operating temperature of the SOFC requires a significant start-up time. The cell performance is very sensitive to operating temperature. A 10% drop in temperature results in ~12% drop in cell performance due to the increase in internal resistance to the flow of oxygen ions. The high temperature also demands that the system include significant thermal shielding to protect personnel and to retain heat. While such requirements are acceptable in a utility application, they are not consistent with the demands of most transportation applications nor do they lend themselves to small, portable or transportable applications.

5.4 Proton Exchange Membrane Fuel Cell

The PEFC offers a power density an order of magnitude higher than that of any other fuel cell system, with the exception of the advanced aerospace AFC, which has comparable performance. The PEFC can operate on reformed hydrocarbon fuels, with pretreatment, and on air. The use of a solid polymer electrolyte eliminates the corrosion and safety concerns associated with liquid electrolyte fuel cells. Its low operating temperature provides instant start-up and requires no thermal shielding to protect personnel. Advances in performance and design offer the possibility of a lower cost than any other fuel cell system.

The PEFC uses as its electrolyte a polymer membrane. This membrane is an electronic insulator but an excellent conductor of hydrogen ions (see *Polymeric Membranes for Gas Separation*). The materials used to date consist of a fluorocarbon polymer backbone, similar to Teflon, to which are attached sulfonic acid groups. The acid molecules are fixed to the polymer and cannot "leak" out, but the protons on these acid groups are free to migrate through the membrane. With the solid polymer electrolyte, electrolyte loss is not an issue with regard to stack life.

The electrolyte membrane looks rather like a thick sheet of food wrap and can be handled easily and safely. The anode and cathode are prepared by applying a small amount of platinum black to one surface of a thin sheet of porous, graphitized paper which has previously been wet-proofed with Teflon. The electrolyte is then sandwiched between the anode and cathode, and the three components are sealed together under heat and pressure to produce a single membrane/electrode assembly. This assembly, which is the heart of the fuel cell, is less than 1 mm thick.

The anode and cathode are contacted on the backside by flow field plates made of graphite in which channels have been formed. The ridges between the channels make electrical contact with the backs of the electrodes and conduct the current to the external circuit. The channels supply fuel to the anode and oxidant to the cathode.

The electrode reactions in the PEFC are analogous to those in the PAFC:

$$\text{Anode}: H_2 \rightarrow 2H^+ + 2e^-$$

$$\text{Cathode}: \tfrac{1}{2}O_2 + 2H^+ + 2e^- \rightarrow H_2O$$

Hydrogen from the fuel gas stream is consumed at the anode, yielding electrons to the anode and producing hydrogen ions which enter the electrolyte. At the cathode, oxygen combines with electrons from the cathode and hydrogen ions from the electrolyte to produce water. The water does not dissolve in the electrolyte and is, instead, rejected from the back of the cathode into the oxidant gas stream. As the

PEFC operates at about 80°C, the water is produced as liquid water and is carried out of the fuel cell by excess oxidant flow.

PEFC performance has improved over the last several years. Current densities of 0.915 A cm^{-2} are achieved at 0.7 V cell^{-1} on hydrogen and oxygen at 0.45 MPa, and over 0.54 A cm^{-2} is obtained on air at the same pressure.

Early in 1987 Ballard Power Systems received an experimental membrane electrolyte from Dow Chemical. When placed in the improved Ballard hardware, the new Dow membrane produced 4.3 A cm^{-2} at 0.5 V cell^{-1} on hydrogen and oxygen at 0.45 MPa. This represents a power density of 2.15 W cm^{-2}. The performance of a six-cell Ballard stack containing the Dow membrane is comparable with the performance on Nafion. At 0.7 V cell^{-1} the Ballard/Dow experimental technology produced 2.15 A cm^{-2} on hydrogen/oxygen at 0.45 MPa and over 1.08 A cm^{-2} on hydrogen/air at 0.45 MPa.

The power density described above is roughly a factor of 10 greater than that observed for the other fuel cell systems. This represents a potential for a significant reduction in stack size and cost over that possible for other systems. In 5 KW production fuel cell stacks, Ballard is achieving 0.7 V at 0.70 A cm^{-2} on hydrogen on air at 0.31 MPa. The stack dimensions are $25 \times 25 \times 42.4$ cm, yielding a stack-only power density of over 0.19 W cm^{-3}. Power densities approaching 0.5 W cm^{-3} are certainly feasible.

A near-term system, including fuel and oxidant controls, cooling, and product water removal which operates on hydrogen and air at 0.31 MPa will provide 0.044 W cm^{-3} and 88 W Kg^{-1}.

At the fuel cell stack level, electrical efficiency is determined by the voltage at which the fuel cell is operated. Thus, a PAFC, a MCFC, or a PEFC operating at 0.7 V cell^{-1} will be operating at the same stack efficiency. The PEFC has the advantage that, at a given stack efficiency (cell voltage), it will deliver more current and thus more power for a given size fuel cell. Equally, the PEFC can provide the same power at a given stack efficiency from a smaller, less costly stack.

When hydrocarbon fuels are to be used or air is to be used as the oxidant, higher-temperature fuel cells—the MCFC, the SOFC, and to some extent the PAFC—have an efficiency advantage over the PEFC at the system level. The higher-temperature waste heat of these systems can be used to drive air compressors, to assist in the reforming of hydrocarbon fuels, and, in the case of the MCFC and SOFC, to produce steam for thermal electric generation or other thermal load.

The PEFC can be operated using either air or liquid cooling. For those applications requiring a compact power generator, liquid cooling will be the method of choice. This will also be the case if the excess heat is to be used for space or hot water heating in residential or utility cogeneration applications.

In addition to pure hydrogen, the PEFC can also operate on reformed hydrocarbon fuels without removal or recirculation of the by-product CO_2. The traces of CO produced during the reforming process must be converted to CO_2 by a selective oxidation process before the fuel gas enters the fuel cell. This is a simple catalytic process which can easily be integrated into a fuel supply system.

The PEFC can operate on air. As is true with all fuel cells, performance is improved by pressurizing the air. In any application, there will be a tradeoff between the energy and financial cost associated with compressing air to higher pressures and improved performance. Pressures above 0.31 MPa are not likely to be advantageous for most applications.

Because the PEFC uses a solid electrolyte, a significant pressure differential can be maintained across the electrolyte. This allows the operation of the PEFC with low-pressure fuel and higher-pressure air, if desired, to optimize performance.

The PEFC uses platinum at both the anode and cathode. At the end of the NASA program typical platinum loadings were about 4.4 mg cm^{-2} for each electrode. At the cell performance then being obtained, this represented a very significant cost penalty compared with the PAFC.

With the present performance of the PEFC production units, even at 4.4 mg cm^{-2}, platinum requirements would be about 17 g KW^{-1} for operation on hydrogen and air. Significant reductions in platinum loading are possible. Work at Los Alamos National Laboratory and Texas A&M University has demonstrated techniques to reduce the platinum loading to about 0.31 mg cm^{-2} while maintaining performance. This represents a requirement of about 1.2 g KW^{-1} on air or about 0.6 g KW^{-1} on oxygen. Further reductions in platinum loading and further improvements in PEFC performance can reasonably be expected to reduce platinum requirements to 1.0 g KW^{-1} or about \$2 KW^{-1}.

Since the PEFC operates at low temperatures and does not contain a liquid electrolyte, catalyst migration and recrystallization are not problems. Cell operating lifetimes in excess of 50 000 hours were demonstrated for the PEFC during the NASA program. The PEFC typically operates at 70–85°C. About 50% of maximum power is available immediately at room temperature. Full operating power is available within about three minutes under normal conditions. The low temperature of operation also reduces or eliminates the need for thermal insulation to protect personnel or other equipment. The excess heat can be used for space heating or for residential hot water, but it is not hot enough for generating steam for fuel reforming or for utility bottoming cycles.

5.5 Other Fuel Cells

(a) Alkaline fuel cell. The AFC is based on the Bacon cell, a 6 kW cell which was demonstrated in 1959 by a British engineer, Francis Thomas Bacon. NASA then developed it for use with the space program. An alkaline solution of potassium hydroxide (25–50%) is used as the electrolyte and it is contained in a matrix of asbestos or a composite material of potassium titanate and polymer. A platinum catalyst was usually used for anode and cathode.

The reactions that occur in the AFC are:

$$\text{Anode}: \ H_2 + 2OH^- \rightarrow 2H_2O + 2e^-$$

$$\text{Cathode}: \ \tfrac{1}{2}O_2 + 2H_2O + 2e^- \rightarrow 2OH^-$$

These cells can achieve power generating efficiencies of up to 70 % HHV. Until recently they were too costly for commercial applications, but several companies are examining ways to reduce costs and improve operating flexibility.

(b) Direct methanol fuel cell. The DMFC is a relatively new member of the fuel cell family. It is similar to the polymer electrolyte membrane (PEM) cell in that they both use a polymer membrane as the electrolyte. However, in the DMFC the anode catalyst itself draws the hydrogen from the liquid methanol, eliminating the need for a fuel reformer. The reactions that occur in the DMFC are:

$$\text{Anode}: \ CH_3OH + H_2O \rightarrow CO_2 + 6H^+ + 6e^-$$

$$\text{Cathode}: \ \tfrac{3}{2}O_2 + 6H^+ + 6e^- \rightarrow 3H_2O$$

Efficiencies of about 40% are expected with this type of fuel cell, which would typically operate at a temperature of 40–90°C. Higher efficiencies are achieved at higher temperatures.

The DMFC is a regenerative fuel cell, which would be attractive as a closed-loop form of power generation. Water is separated into hydrogen and oxygen by a solar-powered electrolyzer. The hydrogen and oxygen are fed into the fuel cell which generates electricity, heat, and water. The water is then recirculated back to the solar-powered electrolyzer and the process begins again.

6. *Technological Problems*

6.1 Phosphoric Acid Fuel Cell

The PAFC is the most mature fuel cell, but its price is not sufficiently low for it to be widespread. There are three reasons for the higher price: (i) complicated controlling system; (ii) high cost of cell construction; and (iii) high cost of cell maintenance.

The PAFC is controlled by a system requiring many monitoring devices and a complicated method for cell control. A simplified control system would be desirable.

The cell construction cost is pushed up by the cost of carbon materials and platinum used for the electrode and by the fairly low power density that causes an increase in the number of unit cells for practical use. The platinum catalyst loaded on carbon black, which has been reduced to a large extent, is easily affected by carbon monoxide. Accordingly, CO concentration in the reformed hydrocarbon or methanol should be lowered as far as possible, or the anode catalyst should be made more resistant to poisons such as CO. To increase the power density, the TPB should be increased by improving the electrode–electrolyte interface structure.

The operating temperature of PAFC is about 40°C higher than the melting point of pure phosphoric acid, 158°C. Therefore, phosphoric acid steadily evaporates through the pores of electrodes, which means that a supply of additional phosphoric acid is necessary.

6.2 Molten Carbonate Fuel Cell

The MCFC is operated at temperatures above 600°C, which leads not only to the direct utilization of reformed gas, but also to increased corrosion of anode materials by the molten carbonates and the dissolution of NiO at the cathode.

$$NiO + CO_2 \rightarrow Ni^{2+} + CO^{2-}{}_3$$

$$Ni^{2+} + CO^{2-}{}_3 + H_2 \rightarrow Ni\downarrow + H_2O + CO_2$$

Since the latter reaction above causes a short-circuit of the cell, the dissolution of NiO should be suppressed.

6.3 Solid Oxide Fuel Cell

The SOFC is usually operated at 900°C or above, as shown in Table 1. Such a high temperature promotes even reactions between solids, changing their properties, e.g., from an electrically conductive solid to insulating one. Careful selection of materials for a SOFC is therefore necessary. On the other hand, low reactivity tends to cause an insufficient seal between the different materials, so an appropriate sealant is necessary. Solid oxides are resistant to the oxidative environment at elevated temperatures, but it is not feasible to fabricate them into the various forms required.

6.4 Proton Exchange Membrane Fuel Cell

The materials and structure of electrodes for the PEFC are similar to those for the PAFC. Therefore, PEFC shares the technological problems for

Table 1
Comparison of various types of fuel cells.

	PAFC	MCFC	SOFC	PEFC	AFC
Electrolyte	Phosphoric acid H_3PO_4	Molten carbonate salt K_2CO_3, Na_2CO_3	Solid oxide YSZ	Polymer electrolyte	Alkaline solution
Operating temperature (°C)	190	650	1000	80	80
Fuels	H/Reformate	H_2/CO/Reformate	H_2/CO/CH_4/ Reformate	H_2/Reformate	H_2/Hydrazine
Reforming	External	External/Internal	External/ Internal	External	External
Oxidant	O_2/Air	CO_2/O_2/Air	O_2/Air	O_2/Air	O_2/Air
Efficiency (%) (HHV)	40–50	45–55	50–60	40–50	40–70

electrodes with PAFC: weakness against CO poisoning, and low conductivity of electrodes. The former problem is worse for the PEFC than for the PAFC since the PEFC works at a lower temperature. The PEM requires water vapor to provide ionic conductivity, so a constant supply of vapor to the PEFC is required for stable operation. Since a pressurized atmosphere raises the PEFC performance, it is important to make the cell structure strong enough for the high pressure.

6.5 *Alkaline Fuel Cell*

The biggest drawback of AFC is that only hydrogen can be used as a fuel. The reformed hydrocarbon or methanol involves carbon monoxide or carbon dioxide rather than hydrogen. In particular, carbon dioxide contained in the fuel reacts with the alkaline solution to produce carbonates and therefore reduces the concentration of hydroxide ion.

7. *Power Generation Systems*

7.1 *Power Plants*

The system for a fuel cell power station is shown schematically in Fig. 4. The power generating part is made of stacks of fuel cell units which are connected in parallel or in series to increase the output voltage and current. Pressurized air and fuel gas, which is abundant in hydrogen, are introduced separately into the fuel cell stack.

Fuel is fed to the pretreatment apparatus in gaseous form: natural gas is fed as it is, coal is gasified, and oil is cracked to lower hydrocarbons. These gases are pretreated by the exit gas and the poisonous impurities such as sulfur and chlorine are removed. The hydrocarbons are then converted to a hydrogen-enriched gas mixture by steam reforming, water gas shift reaction, or other reactions. In the case of PAFC or PEFC, carbon monoxide should be less than 100 ppm or 10 ppm, respectively, to avoid poisoning the electrodes.

A post-treatment is applied to the exit gas from the fuel cell stack. Heat is removed from the exit gas for high efficiency of heat utilization, the methods of utilization being dependent on the heat level. The temperature of the exit gas from the fuel cell stack is lower than 100°C for AFC and PEFC, and its low-level heat cannot be used efficiently. The heat of the exhaust gas from the PAFC and MCFC produces hot water or steam. In the case of the SOFC, the high-level heat can be converted to electricity through bottoming cycles by using a gas turbine generator or a steam turbine generator, and, moreover, hot water and steam can be produced by the heat of the exit gas.

The operation of the fuel cell stack is controlled by the control system. The generation rate of electricity is controlled to meet demand, but the maximum generation rate does not always reach the level of

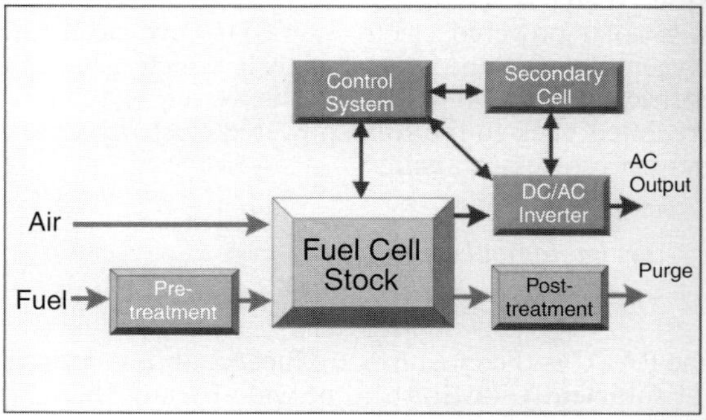

Figure 4
Fuel cell power generation system.

demand. Some generated electricity is saved in a secondary cell (battery) at night as the demand falls, and the saved electricity is released at peak demand during the daytime. The electricity generated at the fuel cell stack, which is d.c., is converted to a.c. and distributed.

7.2 On-site Power Generation and Storage Systems

Distributed power generation reduces the capital investment and improves the overall conversion efficiency of fuel to end-use electricity by reducing transmission losses. In high-growth or remotely located areas, distributed power generation could reduce or eliminate transmission and distribution problems by reducing the need for new capacity or siting new lines. Fuel cells are well suited to these applications

7.3 Cogeneration Systems

A cogeneration system is one that provides electrical power and heat simultaneously. The chemical energy of fuel is converted partly to electrical power, as mentioned previously, and the other part is converted to heat. Even if the electrical efficiency is as low as 40%, by using the heat produced the overall efficiency can be raised to 80%.

In the PAFC the operation temperature is as high as 200°C, which can be used to produce heated water and water steam at about 120–150°C. In the case of the PEFC, development for domestic application has started, where hot water as hot as 60°C can be distributed to the kitchen, bath, and so on.

One of the problems for cogeneration is how much electrical power should be produced and how much hot water should be available. As mentioned before, a maximum power output is associated with each fuel cell. The maximum quantity of hot water produced is given by the difference between Q and PW, but the practical quantity of hot water is determined by the capability of the heat exchanger. The demand for hot water or electricity varies depending on time of the day and on the season of the year, and the variations of the two demands are not always identical. Therefore, there could be a shortage of hot water, for instance, in winter when more heat is required than electricity, or in summer when much less heat is needed than electricity. It is therefore necessary to balance the two outputs to meet the demand.

8. Summary

Fuel cells could be power generators with a higher efficiency than any system produced so far, and cells could dramatically reduce urban air pollution and decrease the consumption of oil, natural gas, and coal. In addition, fuel cells running on hydrogen derived from a renewable source will produce nothing but water vapor. The future of fuel cells is promising, but there still remain several obstacles to bringing them into practical use. Eventually, fuel cells will be used as not only stationary power plants, but also as electrochemical engines for vehicles, power plants for homes, and batteries in portable electronics by down-sizing.

Fuel Cells: Materials

N. P. Brandon

Fuel cells are devices for directly converting the chemical energy of a fuel into electrical energy in a constant temperature process, and are seen as an emerging clean, high-efficiency power source. Potential applications range from battery replacement in consumer goods and portable computers, through residential scale combined heat and power (CHP), to distributed energy generation. However, the key problem limiting the significant commercial uptake of fuel cells is cost, and it is cost reduction, together with the need to demonstrate high levels of reliability and durability, that are the primary concerns of fuel cell developers today.

1. Introduction to Fuel Cells and Fuel Cell Materials

A fuel cell comprises an electrolyte, which is an ionic conductor, and two electrodes (the negative anode and positive cathode) where electrochemical reaction occurs (see *Fuel Cells: Electrochemical Reactions*). The characteristics of the electrolyte are such that it should display a high ionic conductivity, negligible electronic conductivity, zero porosity, adequate mechanical properties, and stability under operating conditions. The nature of the electrolyte distinguishes the different classes of fuel cells, discussed further in Sect. 2. The nature of the ion transfer varies between the different types of cells, but the following discussion for the polymer electrolyte membrane fuel cell (PEMFC) is representative.

In the PEMFC, hydrogen is fed to the anode of the cell where it splits into a proton and electron: the former passing through the electrolyte and the latter forced around an external circuit where it drives a load. The proton and electron combine with oxygen from the air at the cathode, producing pure water and a small amount of heat. The overall reaction is thus

$$H_2 + 0.5O_2 \Leftrightarrow H_2O \qquad (1)$$

Fuel cells largely operate on gaseous fuels and oxidants (essentially air). As such the electrodes must be porous to enable gas transport, display high electronic conductivity, and must be good electrocatalysts for either fuel oxidation (anode) or oxygen reduction (cathode). Electrodes are also often designed such that they exhibit ionic conductivity, either inherently, or via the use of composite materials formed from both electronic and ionic conductors.

Figure 1 illustrates the shape of the current–voltage characteristics that would be expected from a typical fuel cell.

When no current is being drawn from the fuel cell, the cell voltage is at a maximum, the open circuit voltage (E). As current is drawn from the cell, additional irreversible losses result in a decrease in the cell voltage (E_{cell}), according to

$$E_{cell} = E - iR - \eta_a - \eta_c \qquad (2)$$

where iR refers to ohmic losses within electrodes, interconnects, and current take-offs due to the finite resistance of the materials used, η_a refers to the overpotential at the anode, reflecting losses due to both electrode kinetics and mass transport limitations, and η_c refers to the equivalent overpotential at the cathode. Different loss terms are reflected in different regions of the current–voltage curve, with the initial fall in voltage reflecting electrode kinetics, the central linear region being dominated by iR losses, before mass transport limitations dominate at high current densities. Materials selection and microstructure play a vital role in minimizing these losses, and therefore in maximizing fuel cell performance. To help guide materials selection, it is useful to think in terms of the area specific resistance of the cell. For example, given that the open circuit voltage typically falls ~ 1 V, and assuming a target power density ≈ 0.5 W cm^{-2} at a cell voltage of 0.65 V, then the target area specific resistance for the fuel cell would be (1–0.65)/0.5 V (A cm^{-2})$^{-1}$, or 0.7 Ω cm^2. Generally, the required ASR falls between 0.25 Ω cm^2 and 1.0 Ω cm^2, the value being dependent on the application, and the fuel cell type.

In order to achieve a useful output power, individual cells are connected together in a stack. This is achieved using an interconnect or bipolar plate, which joins the anode of one cell to the cathode of the next. The interconnect also separates and often distributes

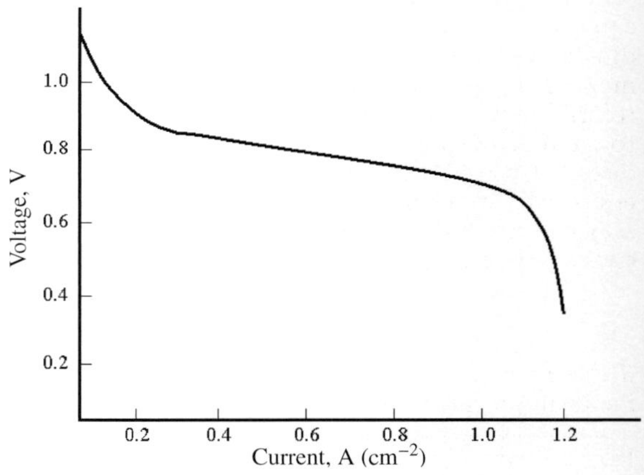

Figure 1
Schematic illustration of a typical fuel cell current–voltage curve.

the fuel and oxidant. The key requirements of this material are a high electronic conductivity, negligible ionic conductivity, zero porosity, and stability in both the fuel and air environment under operating conditions. The interconnect frequently provides mechanical support for the electrolyte, so its mechanical properties are also important, together with cost and ease of manufacture.

The fuel cell stack is then incorporated into a system, which meets the demand of a specific application. Fuel cell systems can include a range of balance of plant (BoP), depending on the fuel used, the application, and the fuel cell type.

Further reading on fuel cell technology can be found in Minh and Takahashi (1995), Kordesch and Simader (1996), and Larminie and Dicks (2000).

2. *Fuel Cell Types*

There are five main classes of fuel cell, each with differing characteristics, and differing advantages and disadvantages. The five types are summarized in Table 1, each taking the name of the electrolyte used in its fabrication.

2.1 Alkaline Fuel Cells

The alkaline fuel cell (AFC) has a long history in the space program. It is still used in the space shuttle in an expensive guise, producing power for the onboard systems by combining the pure hydrogen and oxygen stored in the rocket-fueling system.

The electrolyte is concentrated (85 wt.%) potassium hydroxide (KOH) in AFCs operated at 250 °C, or less concentrated (35–50 wt.%) KOH for operation below 120 °C. The electrolyte is retained in a porous matrix (typically asbestos), and electrocatalysts include nickel and noble metals. The main difficulties with this fuel cell type are: (i) carbon monoxide, always found in hydrogen produced by reforming hydrocarbon or alcohol fuels, is a poison to the precious metal electrocatalysts, and (ii) carbon dioxide (in either fuel or air) will react with the KOH electrolyte to form potassium carbonate. As such, applications are essentially limited to those where either pure oxygen or hydrogen can be used.

2.2 Polymer Electrolyte Fuel Cells

PEMFCs have high power density, rapid startup, and low-temperature operation (~80–120 °C), and so are ideal for use in applications such as transport and battery replacement. The electrolyte used is a proton conducting polymer. This is typically a perfluorinated polymer . The catalytically active layer sits adjacent to the membrane, supported on a PTFE-treated carbon paper which acts as current collector and gas diffusion layer. For operation on pure hydrogen, platinum is the most active catalyst, but alloys of platinum and ruthenium are used when higher levels of carbon monoxide are present (CO is a poison in all low-temperature fuel cells). Water management in the membrane and electrodes is critical for efficient performance, because the membrane must be hydrated, whilst avoiding flooding of the electrode pores with water.

2.3 Phosphoric Acid Fuel Cells

Phosphoric acid fuel cells (PAFCs) operate at 150–220 °C, using a 100% phosphoric acid electrolyte retained in a silicon carbide matrix. Generally, platinum electrocatalysts are used in both anode and cathode. However, it has not been possible to reduce the cost of this technology below around £3000 kWe^{-1}, and the leading developer of this fuel cell type has recently ceased to manufacture the product.

2.4 Molten Carbonate Fuel Cells

The electrolyte in the molten carbonate fuel cell (MCFC) is usually a combination of alkali carbonates, retained in a ceramic $LiAlO_2$ matrix. The temperature of operation is 600–700 °C, where the alkali carbonates form a highly conducting molten salt, with carbonate ions providing the means for ionic conduction. The increased temperature of operation means that precious metal electrocatalysts are not needed, and generally nickel anodes and nickel oxide cathodes are used. A design constraint with the MCFC is the need for CO_2 recirculation, meaning that it is difficult to operate below the 100 kWe scale. Avoiding degradation of the electrodes by reaction with the corrosive molten salt electrolyte is also a challenge.

Table 1
Summary of the five main fuel cell types.

Fuel cell type	Alkaline	Polymer	Phosphoric acid	Molten carbonate	Solid oxide
Acronym	AFC	PEMFC	PAFC	MCFC	SOFC
Operating temperature	60–250 °C	80–120 °C	150–220 °C	600–700 °C	500–1000 °C

2.5 Solid Oxide Fuel Cells

Solid oxide fuel cells (SOFCs) operate at elevated temperatures, generally above 800°C for the high-temperature variant, and in the range 500–800°C for intermediate temperature SOFCs (IT-SOFCs). The electrolyte is a dense ceramic, usually yttria stabilized zirconia (YSZ), which is an oxide ion conductor at elevated temperatures. The cathode is typically a perovskite material such as strontium-doped lanthanum manganite, often mixed with the YSZ in the form of a composite. The anode is a cermet of nickel and YSZ. SOFC geometry is either tubular or planar: the former offers advantages in terms of thermal shock resistance and ease of sealing, whereas the latter offers advantages in terms of power density and ease of cell manufacture.

2.6 Summary

Following this brief review of all the fuel cell types, the following sections will focus on the developments in materials for the two leading fuel cell technologies, namely PEMFC (the leading technology for portable and transportation applications) and SOFC (the leading technology for stationary applications).

3. Developments in Polymer Fuel Cell Materials

3.1 Electrolyte

Proton conducting membranes form the core of PEMFCs. The following membrane properties need to be optimized for fuel cell applications: (a) fast proton transport, (b) good mechanical, chemical, and thermal properties requiring the selection of a suitable polymer backbone, and possibly membrane reinforcement, (c) low gas permeability, and (d) low levels of swelling.

There is significant interaction between these properties and the type of backbone polymer, the degree of sulfonation, and nano-phase separation into hydrophilic and hydrophobic domains (e.g., high levels of sulfonation typically lead to high conductivity but also to a high level of swelling). A number of materials are under consideration in an attempt to meet these conflicting requirements.

Classically, Nafion™ type perfluorinated polymer membranes, which have a PTFE-like backbone and relatively low equivalent weight, are used as the electrolyte. However, the cost of these materials remains high, around US\$56 m^{-2} in large volumes (Little 2000). The upper temperature limit of operation of Nafion is generally regarded to lie ~140°C. Considerable effort is being applied to the development of lower cost, usually fluorine-free, hydrocarbon-based membrane materials, and Jones and Roziere (2001) have, for example, recently reviewed developments in membranes based on polybenzimidazole and polyetherketone.

3.2 Electrodes

A catalytically active layer sits adjacent to the electrolyte membrane within both the anode and cathode of a PEMFC. The layer is supported on a PTFE-treated carbon paper which acts as a current collector and gas-diffusion layer, thus enabling the transport of gas, water, and electrons. For operation with pure hydrogen and air, platinum is the most active electrocatalyst. To reduce cost, nano-particles of platinum on a carbon support have been developed, and on-going development of stable nano-structures continues to be an important aspect of fuel cell development.

However, a key aspect of electrocatalyst development relates to the development of anode catalysts for operation on reformed hydrogen from methanol, natural gas, or gasoline. This is because, while gas-diffusion electrodes with a loading of 0.1–0.2 mg cm^{-2} of dispersed platinum on carbon show very small polarization losses when operating on pure H_2, these losses are raised to unacceptable values when even small amounts of CO are present (10–100 ppm). This is always the case in fuel mixtures originating from reformed natural gas or alcohols, unless stringent measures are taken to reduce the CO content to low levels during fuel processing. Many attempts have been made to understand the mechanism of CO poisoning using a large variety of experimental approaches. It is generally proposed that CO poisoning occurs because of a strong adsorption of CO on the catalyst surface, which blocks the hydrogen adsorption step.

There are two main strategies to overcome the CO poisoning problem in PEMFCs. The first approach is to operate the PEMFC at higher temperatures, which helps suppress the adverse impact of CO. However, this requires the development of new high-temperature membranes, in itself a significant materials challenge, and one discussed by Kreuer (2001).

The second approach requires the development of new CO tolerant electrocatalysts. Most work in this area has concentrated on Pt–M, where M is usually a transition metal, e.g., Mo, Fe, Ni, Co, and Mo.

3.3 Bipolar Plates

Early bipolar plates in PEMFCs were made of machined graphite. This is not a technologically viable option, due to both cost and weight considerations, with the result that other materials are under development, primarily metals and carbon–polymer composites.

Frequently, 316 stainless steel has been the alloy of choice for metallic bipolar plates. However, studies have shown that, because of the formation of an oxide layer with high resistivity, and poisoning of the membrane electrode assembly (MEA) by corrosion products, the use of stainless steel is only possible with protective coatings.

Carbon–polymer composites involve the hot molding of a carbon or graphite filler in a thermosetting or thermoplastic matrix. There is a trade-off between mechanical strength and electrical conductivity. Some typical conductivity values are ~1000 S m^{-1} for polymer/graphite materials, well above that of the membrane, though still below the conductivity values of 5.3×10^8 S m^{-1} for iron based alloys, and 2.4×10^8 S m^{-1} for titanium (Steele and Heinzel 2001).

4. *Developments in SOFC Materials*

Kawada and Yokokawa (1997) and Badwal (2001) have reviewed the rationale behind SOFC materials selection, often based on the result of empirical development programs. Although similar compositions are usually selected, the specific manufacturing process chosen, the stack configuration, and the temperature of operation all influence materials selection.

4.1 *Electrolyte*

For tubular SOFCs the cathode is often used as the support substrate, and at present relatively expensive processes such as electrochemical vapor deposition (EVD), and plasma spraying, are used to deposit the YSZ electrolyte film (typically 40 μm thick). However, cost targets are difficult to achieve with these techniques and a variety of slurry processes are under active development (see *Zirconia Electrolyte-based Solid Oxide Fuel Cells*).

Planar SOFC configurations were originally based around self-supported (~150 μm thick) YSZ electrolytes which required operating temperatures in excess of 950°C to avoid excessive resistive losses associated with the electrolyte component. Operation at 950°C introduced significant problems, including stability of the electrolyte/cathode interface, selection of the bi-polar plate material, and the optimal material composition for the seals that are necessary in planar SOFC stacks. This has led to an increased focus on IT-SOFCs operating in the temperature range of 500–800°C. Assuming that the electrolyte component should not contribute more than $0.15\,\Omega\,cm^2$ to the total cell area specific resistance, then for a thick film thickness (L) of 15 μm the associated specific ionic conductivity (σ) value of the electrolyte should exceed 10^{-2} S cm^{-1} ($\sigma = L/ASR = 0.0015/0.15$). The ionic conductivity of YSZ attains this target value at 700°C, and for $Ce_{0.9}Gd_{0.1}O_{1.95}$ (CGO) the relevant temperature is 500°C. The use of a thick film electrolyte requires this component to be supported on an appropriate substrate. As the substrate is the principal structural component in these cells, it is necessary to optimize the conflicting requirements of mechanical strength and gaseous permeability. Most development work on planar IT-SOFC systems has involved YSZ electrolyte thick films, and so far most groups have used anode (Ni–YSZ) substrates, which allows the electrolyte powder to be densified ~1400°C.

Although YSZ is still the favored electrolyte material for SOFCs, alternative electrolyte materials are being considered. The use of ceria-based electrolytes such as CGO should in principle, allow the cell operating temperature to be lowered to 500°C. Steele (2000) has shown that at such temperatures the reduction of Ce^{4+} ions to Ce^{3+} in the anode compartment is minimized, and can be neglected under typical cell operating conditions.

Another electrolyte, doped $LaGaO_3$ (LSGM), is also attracting much attention for IT-SOFC applications. Although its conductivity is slightly smaller than CGO at 500°C, its ionic domain is wider and it could be more appropriate to use this electrolyte at temperatures ~600°C, where the reduction of Ce^{4+} in CGO is becoming significant. It has been difficult to fabricate pure single-phase ceramic electrolytes, and second phases such as $SrLaGa_3O_7$ and $La_4Ga_2O_9$ are often detected in the grain boundaries.

4.2 *Electrodes*

Typically the anode material used in SOFC prototype demonstration units has been the Ni–YSZ cermet. As the fuel requirement increasingly focuses on hydrocarbon fuels with lower steam content, then the specification for the anode alters. Increasingly, discussions regarding anode materials are highlighting the direct oxidation of hydrocarbon fuels at the anode, for which the traditional Ni-based cermets are unsuitable owing to carbon deposition. Sulfur tolerance is also becoming important, as developers seek to run on fuels such as diesel, gasoline, and biofuels.

The established cathode for SOFCs operating at temperatures above 800°C is strontium-doped lanthanum manganite (LSM). This material has demonstrated high levels of durability within the Siemens Westinghouse tubular SOFC. To improve performance, it is often mixed with YSZ to form a composite cathode. This increases the number of triple-phase contact points between the ionic phase (YSZ), the electronic phase (LSM), and pore space, which are all necessary for oxygen reduction to occur. However, as developers continue to drive down the operating temperature of SOFCs, then the activity of the LSM–YSZ material becomes an increasing constraint. Doped lanthanum ferrites have been shown to be a promising class of materials at temperatures down to 650°C.

4.3 *Interconnect*

High-temperature SOFCs have largely used doped lanthanum chromite as the interconnect material. However, one of the drivers to lower operating temperature is the ability to use lower cost metallic bipolar plates as the interconnect component. Initially a Cr–5%Fe (1%Y_2O_3) oxide dispersion-strengthened (ODS) alloy was developed for 950°C operation, but now most prototype stacks incorporate a ferritic stainless steel as the bipolar plate because of the low ($12.5 \times 10^{-6}\,K^{-1}$) thermal expansion coefficient of these alloys. At present the optimal composition of the alloy remains uncertain due to conflicting requirements for the oxide coating. An appropriate protective film has to be formed to withstand the oxidizing environment of the cathode compartment, and the high steam content developed in the anode compartment. In addition, the oxide layer has to exhibit sufficient electronic conductivity at the operating temperature to ensure that contact area specific resistances are below $10\,m\Omega\,cm^2$.

Another problem associated with the ferritic steels is the transport of Cr gaseous species (e.g., CrO_3, $CrO_2(OH)_2$, $Cr(OH)_3$) into the porous cathode and anode components and subsequent formation of deleterious phases such as $MnCr_2O_4$ and $NiCr_2O_4$ which can severely degrade the electrode performance. At present many developers are reducing the Cr-activity by coating the steel with films of doped $LaCrO_3$ using plasma spraying techniques, but a reduction in operating temperature to below 700°C would probably be the optimal solution.

5. Summary

It is recognized that the capital costs of prototype fuel cell systems are presently too high (\$3000–10 000 kWe^{-1}) compared to commercial cost targets of order \$50 kWe^{-1} (for transport applications) to some \$500 kWe^{-1} (for stationary applications). Much of this cost reduction can only be realized by utilizing cheaper materials and/or manufacturing techniques, or by developing materials which enable system simplification and therefore cost reduction. Thus, materials development and selection remains a key challenge to the commercialization of fuel cell technology.

Bibliography

Badwal S P S 2001 Stability of solid oxide fuel cell components. *Solid State Ionics* **143**, 39–46

Jones D J, Roziere J 2001 Recent advances in the functionalisation of polybenzimidazole and polyetherketones for fuel cell applications. *J. Membrane Sci.* **185**, 41–58

Kawada T, Yokokawa H 1997 Material selection for fuel cells. *Key Eng. Mater.* **125–126**, 187–248

Kordesch K, Simader G 1996 *Fuel Cells and Their Applications.* VCH, Weinheim, Germany

Kreuer K D 2001 On the development of proton conducting polymer membranes for hydrogen and methanol fuel cells. *J. Membrane Sci.* **185**, 29–39

Larminie J, Dicks A 2000 *Fuel Cell Systems Explained.* Wiley, UK

Little A D 2000 Cost analysis of fuel cell systems for transportation: baseline cost estimate. SFAA DE-SCO2-98EE50526, DOE Final report

Minh N Q, Takahashi T 1995 *Science and Technology of Ceramic Fuel Cells.* Elsevier, Amsterdam

Steele B C H 2000 Appraisal of CGO electrolytes for IT-SOFC operating at 500°C. *Solid State Ionics* **129**, 95–110

Steele B C H, Heinzel A 2001 Materials for fuel cell technologies. *Nature* **414**, 345–52

Fuel Cells: Microsystems

H. Yu and C. Hebling

"Fuel cell," is a term often used for an electrochemical device, which converts the chemical energy of the fuel into electrical energy directly. The cell is the basic electrochemical compartment or unit where the electrochemical process takes place.

The chemical processes by which the electric current is generated are electrochemical, rather than thermochemical. Since no combustion reaction is involved, fuel cells do not produce any of the undesirable products normally associated with the oxidation of fossil fuels in conventional energy conversion systems. As a result, fuel cells do not emit SO_2, NO_x, or particulate matter. Without moving components inside the fuel cells, there is no noise during fuel cells operation. Fuel cells are environmental-friendly devices. The first fuel cell was demonstrated by William Grove in 1839. Tremendous interest in the potential use of fuel cell systems has arisen in the 1990s because of the high-energy conversion efficiency of fuel cells. To date, fuel cells are entering more and more into daily life. Especially the micro-fuel cells are the hopeful successor of today's battery; although there are still some obstacles to be overcome before fuel cell commercialization.

Generally, fuel cells are classified by their electrolyte: proton exchange membrane fuel cell (PEMFC), direct methanol fuel cell (DMFC) in which a polymer membrane serves as the electrolyte, alkaline fuel cell (AFC) in which the electrolyte is a base such as potassium hydroxide (KOH), phosphoric acid fuel cell (PAFC) in which the electrolyte is phosphoric acid (H_3PO_4), molten carbonate fuel cell (MCFC) in which molten carbonate, (Li, K)CO_3, serves as the electrolyte transporting carbonate ions, and solid oxide fuel cell (SOFC) in which the electrolyte is a solid such as yttria (Y_2O_3)-stabilized zirconia (ZrO_2) that transports oxygen ions.

The operation temperature is another way to classify fuel cells. Low-temperature fuel cells have an operation temperature lower than 100°C. This type includes the PEMFC, DMFC, and the alkaline fuel cell. The medium temperature fuel cells have an operating temperature between 100°C and 300°C. This type includes the Bacon alkaline fuel cell and the PAFC. High-temperature fuel cells have an operating temperature between 600°C and 1000°C and include the MCFC and the SOFC. But with the advances in various fuel cells, there is a trend that the operating temperatures of fuel cell change a lot. For example, to avoid poisoning the PEMFC with CO, researchers are trying to raise the operating temperature over 100°C. A higher operating temperature would also be advantageous for the DMFC. Meanwhile, researchers are making an effort to lower the operating temperature of SOFC to simplify the system.

Among the various types of fuel cells, the PEMFC and the DMFC operate in a relatively low temperature range, meaning a quick start. They are suitable for portable applications, for example, laptops, camcorders, personal digital assistants (PDA), mobile phones, etc. With the effort made by research and development institutions and product manufacturers all over the world, micro-fuel cells have a bright future. Many manufacturers are working vehemently to produce microfuel cells for the replacement of rechargeable batteries in portable electronics. This is expected to be realized in the near future.

In this section, fundamentals, materials, and application of microfuel cells will be described. The principles and the development trends of the key materials in microfuel cells, such as electrocatalyst, electrode, membrane, gas diffusion layer, reformer, or hydrogen storage materials, are also discussed for both PEMFC (Fig. 1) and DMFC.

1. Fundamentals

In a fuel cell, the reactant and oxidant are fed in and an electrochemical reaction takes place at the membrane electrode assembly (MEA), where electricity is generated. A fuel cell consists of reactant gas

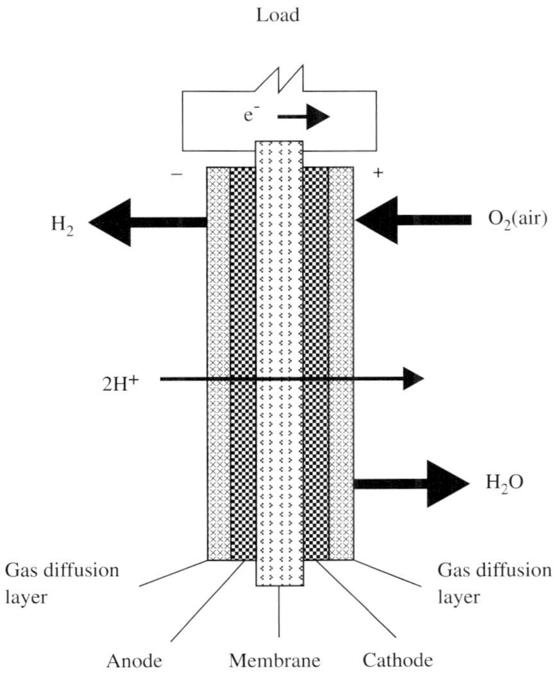

Figure 1
Principle of PEMFC.

suppliers (gas diffusion layer), electrochemical active sites (MEA) as well as of the fuel supplier out of fuel cell (reformer or hydrogen storage).

The basic electrochemical reactions which take place in a proton exchange membrane fuel cell are as follows:

(i) At the anode side

$$H_2 \rightarrow 2H^+ + 2e^-, \quad \varphi^o = 0\ \text{V}$$

(ii) At the cathode side

$$\frac{1}{2}O_2 + 2H^+ + 2e^- \rightarrow H_2O, \quad \varphi^o = 1.229\ \text{V}$$

(iii) Total reaction

$$H_2 + \frac{1}{2}O_2 \rightarrow H_2O, \quad E^o = 1.229\ \text{V}$$

1.1 The Process in a Fuel Cell

At the anode in a typical fuel cell design, the hydrogen is broken down into two components: the hydrogen nucleus (also known as a proton) and an electron. The electron represents the electric current and is transferred from the anode to the cathode through the electric circuit. The proton migrates from the anode to the cathode through the electrolyte. At the cathode, the proton reacts with the electron coming from the load and with the oxygen molecule. They combine together to form water, which is the only chemical product of the fuel cell.

The electrochemical reactions that take place at the electrodes need to be facilitated by an electrocatalyst, platinum being the most common electrocatalyst in PEMFC. At the two electrodes, the electrochemical reactions occur on the catalyst surface where the electrolyte and fuel (or oxidant) meet. This interface is often referred to as the triple or three-phase interface, since in some fuel cell designs, the fuel (or oxidizer) is gas, the electrolyte is liquid, and the electrode/catalyst surface is solid. The state of this three-phase interface plays a critical role in determining the electrochemical performance of a fuel cell; hence, this three-phase interface has been the focus of much research in improving fuel cells.

1.2 Faraday's Law

The relationship between fuel and oxidant consumption and the electricity generated in a fuel cell follows Faraday's law (founded by Faraday in 1833). In the equation for Faraday's first law, the consumption of fuel and oxidant is proportional to the output electricity Q as given below

$$\Delta m = k_e \cdot Q = k_e \cdot I \cdot t$$

where Δm is the consumption, Q the electricity, k_e the proportionality factor which represents the mass of chemicals required to generate one mole electrons. It is called the electrochemical equivalent. When the electricity unit is in coulomb, $k_e^H = 104 \times 10^{-5}$ g A^{-1} s^{-1}, $k_e^O = 8.29 \times 10^{-5}$ g A^{-1} s^{-1}.

Faraday's second law is given by

$$\frac{m_1}{m_2} = \frac{M_1/z_1}{M_2/z_2}$$

where the M/z is the molar mass of an ion equivalent. Faraday's 2nd law shows that the ratio of the masses of the material converted at the two electrodes is equal to the ratio of the molar masses of the ion equivalents (Hamann *et al.* 1998). For example, in the fuel cell when 1 M fuel and 1 M oxidant react, 96485 C of electricity is generated. For the hydrogen/oxygen fuel cell, this is 1.008 g of oxygen and 8.000 g of hydrogen.

1.3 DMFC Electrochemical Reaction

For the DMFC, the cathode reaction is the same as for the PEMFC cathode, oxygen electrochemical reduction is

$$\frac{3}{2}O_2 + 6H^+ + 6e^- \rightarrow 3H_2O, \quad \varphi^o = 1.229\ \text{V}$$

While at the DMFC anode, the CH_3OH electrochemical oxidation reaction is

$$CH_3OH + H_2O \rightarrow CO_2\uparrow + 6H^+ + 6e^-, \quad \varphi^o = 0.046\ \text{V}$$

The total reaction is

$$CH_3OH + \frac{3}{2}O_2 \rightarrow CO_2\uparrow + 2H_2O, \quad E^o = 1.183\ \text{V}$$

Methanol is a liquid fuel, which can be oxidized electrochemically in the presence of water. Methanol has a high energy density (3450 Wh kg and 3000 Wh l in a 1:1 molar ratio with water). It is easy to handle and is completely miscible with water. It can also be produced from a variety of sources and is available in pharmacies at a low price. However, methanol is toxic and the crossover of methanol through the polymer electrolyte membrane results in a poisoning of the cathode catalyst by the formation of a mixed potential. Another problem is the slow anode electrocatalysis of the methanol, which requires significantly more noble metal catalyst than the hydrogen PEMFC. As a result, the power density of the DMFC stack at room temperature is currently at least a factor of 5 smaller than that of hydrogen-fed PEMFC.

1.4 The Efficiency of Fuel Cell

The thermal efficiency of the fuel cell can be defined as the percentage of useful electrical energy produced relative to the heat that would have been obtained through the combustion of the fuel (enthalpy of Combustion):

$$\eta = \frac{\text{Useful energy}}{\Delta H}$$

In the ideal case of an electrochemical converter, the change in Gibbs free energy ΔG is available as useful electrical energy at the temperature of the conversion. The ideal thermodynamic efficiency of a fuel cell, operating irreversibly, is

$$\eta = \frac{\Delta G}{\Delta H}$$

When ΔH is calculated, high heat value (HHV) is based on liquid water as product, and low heat value (LHV) is based on gaseous water as product. In practice, the efficiency of the fuel cell can be expressed as the percentage ratio of operating cell voltage relative to the ideal cell voltage. The actual operating voltage is less than the ideal voltage because of the irreversible losses associated with the fuel cell electrochemistry. The thermodynamic potential is defined as $E = \frac{\Delta G}{nF}$, and the voltage efficiency is defined as $f_V = V/E = V/(\Delta G/nF)$, which is often used in practice.

2. Polarization

When a fuel cell is operating, the potential will deviate from its equivalent value. This is called polarization. Polarization is the energy loss between the fuel cell static status ($i = 0$) and the working status ($i > 0$), which includes activation (Tafel) polarization, concentration polarization, and ohmic polarization. Figure 2 illustrates the ideal and actual voltage of the low-temperature hydrogen/oxygen fuel cells.

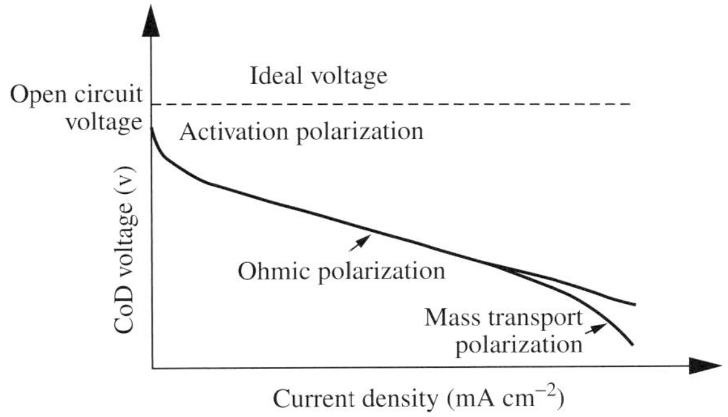

Figure 2
Ideal and actual voltage–current curves of a low-temperature fuel cell.

2.1 Activation Polarization

When any electrochemical reaction that includes at least one process to accept or lose electrons occurs, activation polarization takes place. The potential difference between the thermodynamic rest potential and the onset of a "visible current" is termed the activation polarization. It could be described by a semi-empirical equation, called the Tafel equation, given below:

$$\eta_{\text{act}} = \frac{RT}{\alpha\eta F} \ln \frac{i}{i_o}$$

where α is the electron transfer coefficient of the reaction at the electrode and i_o is the exchange current density.

The usual form of the Tafel equation that can be easily expressed by a Tafel plot is

$$\eta_{\text{act}} = a + b \log i$$

where $a = (-2.3RT/\alpha\eta F) \log i_o$ and $b = 2.3RT/\alpha\eta F$, where b is the Tafel slope.

Processes involving absorption of reactant species, transfer of electrons across the double layer, desorption of product species, and the nature of the electrode surface all contribute to activation polarization.

2.2 Ohmic Polarization

Ohmic loss occurs because of resistance to the flow of ions in the electrolyte and resistance to flow of electrons through the electrode materials. Both the electrolyte and the fuel cell electrodes follow Ohm's law. The ohmic loss is given by

$$\eta_{\text{ohm}} = iR$$

where i is the fuel cell current and R is the total fuel cell resistance, which includes the electronic, ionic, and contact resistance.

2.3 Concentration Polarization

Chemical migration and reaction that lead to a change in the concentrations of the reactant and the product at the electrode, and thus a change in the electrode potential is called concentration polarization. Only when the electrochemical reaction is the speed-determination step for the entire process, can the concentration polarization be ignored.

Since a reactant is consumed at the triple interface in electrochemical reaction, there is a loss of potential due to the inability of the surrounding material to maintain the initial concentration of the bulk fluid. Thus, a concentration gradient is formed. Several processes may contribute to concentration polarization: (i) slow diffusion in the gas phase in the electrode pores, (ii) solution/dissolution of reactants/products into/out of the electrolyte, or (iii) diffusion of reactants/products through the electrolyte to/from the electrochemical reaction site. At practical current densities, slow transport of reactants/products to/from the electrochemical reaction site is a major contributor to concentration polarization. The transport of reactant and products is mainly classified in three ways: convection, molecular diffusion, and electro-migration.

(*a*) *Convection.* The particle moves with the fluid flow. The density difference caused by difference in temperature, as well as the fluid flow caused by stirring and pressure differences result in convection. The convection flow is given by

$$N_{\mathrm{convection},i} = \vec{V}C_i = (V_x + V_y + V_z)C_i$$

(*b*) *Molecular diffusion.* The gradient of chemical potential is approximately the concentration gradient. The flow caused by molecular diffusion is

$$N_{\mathrm{diff},i} = -D_i\nabla C_i = -D_i\left(\vec{i}\frac{\partial}{\partial x} + \vec{j}\frac{\partial}{\partial y} + \vec{k}\frac{\partial}{\partial z}\right)C_i$$

(*c*) *Electro-migration.* The migration of charged particle is caused by the potential gradient.

$$N_{\mathrm{electromigration}} = \pm Eu_i^0 C_i = \pm(E_x + E_y + E_z)u_i^0 C_i$$

where E_x, E_y, E_z is the electric field intensity in the direction of x, y, z, respectively, u_i^0 is defined as the particle flow rate in the equivalent electric field. When all three types of migration are considered, it follows that:

$$N_i = VC_i - D_i\nabla C_i \pm Eu_i^0 C_i$$

for $O + ne^- \rightarrow R$

If it is assumed that there is no convection and electro-migration, the diffusion process of oxidant "O" to the active site is the control step. Therefore, the other steps could be treated as in equilibrium. It is assumed that the thickness of the gas diffusion layer is δ. The concentration in the gas diffusion layer shows linearity. The bulk concentration of "O" is C^{o} and the concentration at the surface is C^{S}. Then

$$N_o = D\frac{C^o - C^{\mathrm{S}}}{\delta}$$

The electrochemical reaction rate is given by

$$i = \frac{N_o}{nF} = \frac{D}{nF}\frac{C^{\mathrm{o}} - C^{\mathrm{S}}}{\delta}$$

When $C^{\mathrm{S}} \rightarrow 0$, the corresponding electrochemical velocity i is the limiting current density, i_{d}. Then

$$i_{\mathrm{d}} = \frac{D}{nF}\frac{C^{\mathrm{o}}}{\delta}$$

Upon substituting i_d in i, we get

$$C^{\mathrm{S}} = C^{\mathrm{o}}\left(1 - \frac{i}{i_{\mathrm{d}}}\right)$$

The concentration polarization is

$$\eta_{\mathrm{conc}} = \frac{RT}{nF}\ln\left(1 - \frac{i}{i_{\mathrm{d}}}\right)$$

It is known that the small i means small concentration polarization.

3. *Comparison with Batteries*

A fuel cell, although having components and characteristic similar to those of a typical battery, differs in several aspects. Since the battery is an energy storage device, the maximum energy available is determined by the amount of chemical reactant stored within the battery itself. The battery will cease to produce electrical energy when the chemical reactants are consumed (i.e., discharged). In a secondary battery, the reactants are regenerated by recharge, which involves putting energy into the battery from an external source. Unlike a battery, a fuel cell does not run down or require recharging. It will generate energy in the form of electricity and heat as long as fuel is supplied. It can be instantly charged by feeding in fuel. Furthermore, there is a negligible self-discharge. The electrodes in fuel cells do not take part in the fuel cell reaction but only act as reaction sites. Therefore, the cycle times of fuel cells are much longer than that of batteries.

The difference between PEMFC and DMFC and batteries depends on the fuel storage form. The comparison for batteries and fuel cells of the volumetric and gravimetric energy density strongly depends on the system design and material used. Therefore, it is difficult to make a general comparison of the energy or power density between battery and fuel cell systems, but the advantage of fuel cells over batteries is clear. Provided there is a sufficient supply of fuel for the fuel cell, electrical power can be generated continually and indefinitely. Fuel cells are noiseless, nonpolluting, and vibration-free. Fuel cells offer much the same advantages as batteries, especially for small appliances such as cell phones and laptops. With their high-energy density and environmental-friendly characteristic, fuel cells are natural candidates for use as battery replacements in the near future.

4. *Electrocatalyst*

Electrocatalysis takes place at the interface of the electrode (electrocatalyst) and the electrolyte and is a branch of multi-phase catalysis. The main characteristic of electrocatalysis is that the reaction rate depends on the electrocatalyst activity and the double layer as well as on the electrolyte. From electrode process dynamic equation, the reaction rate is

$$i = i_o[e^{\alpha nF\eta/RT} - e^{-(1-\alpha)nF\eta/RT}]$$

There are different ways to accelerate the electrochemical reaction. One is to increase the exchange current density i_o, which increases the activity of electrocatalyst. The other is to decrease polarization η, which depends on the electrode structure. Up to now, the most effective electrocatalyst is platinum both for PEMFC and DMFC. For DMFC anode and reformat PEMFC anode, PtRu is the best. Despite the high electrocatalysis activity, the electrocatalyst or the catalyst support should be a good electrical conductor. Furthermore, the electrocatalyst should have a high corrosion resistance at the fuel cell operating conditions. From the cost and the resource point of view, it is required to reduce the precious metals loading. The electrocatalyst should also be highly dispersed on the catalyst support.

4.1 Platinum as the Electrocatalyst for Oxygen Reduction

Platinum is the first choice of the low-temperature fuel cell electrocatalyst. For hydrogen oxidation, the exchange current density is 0.1–100 mA cm^{-2}. When the working current density of hydrogen oxidation at gas diffusion electrode is several hundred mA cm^{-2}, the polarization is only 1–20 mV, whereby it almost functions as the reversible electrode. But for the oxygen reduction, the exchange current density is only several μA cm^{-2}. At room temperature, it is even as low as 10^{-10} A cm^{-2}. The main activation polarization occurs at the cathode side.

The mechanism of oxygen reduction reaction (ORR) is extremely complex and a number of factors contribute to a lowering of the catalytic activity on Pt. The ORR mechanism is explained as the direct four-electron reduction (Eqn. (1)). To maximize the electrical efficiency of the PEMFC, it is required to break a strong O–O bond early in the reaction (Eqn (1)). Additionally the open circuit voltage (OCV) is between 1.0 and 1.20 V, lower than the thermodynamic potential for oxygen reduction on Pt. In an acid environment, Pt may be oxidized by H_2O (Eqn. (2)) under fuel cell oxygen electrode condition. Reaction in Eqn. (1) takes place at the surface of Pt–O. At a stationary state, the OCV is the mix potential of (1) and (2). It is determined by the electrocatalyst composition, preparation process, etc. Up to now no existing electrocatalyst could avoid the open circuit polarization. The diffusion of hydrogen across the membrane also contributes to OCV. The fuel cell efficiency would improve tremendously if the open circuit polarization could be eliminated.

$$O_2 + 4H^+ + 4e^- \rightarrow 2H_2O \quad E^o_{25^\circ C} = 1.23\ V\ (vs.\ NHE) \tag{1}$$

$$Pt + H_2O \rightarrow Pt\text{-}O + 2H^+ + 2e^- \quad E^o_{25^\circ C} = 0.88\ V\ (vs.\ NHE) \tag{2}$$

$$O_2 + 2H^+ + 2e^- \rightarrow H_2O_2 \quad E^o_{25^\circ C} = 0.68\ V\ (vs.\ NHE) \tag{3}$$

Another possible ORR mechanism is that the ORR follows a serial pathway through the intermediate formation of absorbed H_2O_2 (Eqn. (3)). Some recent findings support the formation of peroxide mechanism (Schmidt and Gasteiger 2003). The formation of peroxide can also explain the degradation of the polymer membrane.

4.2 Platinum Alloy as the Electrocatalyst for Oxygen Reduction

As the cathode polarization takes a great part of PEMFC performance loss, many researchers have been trying to find a way to reduce the cathode polarization. Despite the good initial activity, pure Pt catalysts have been shown to deactivate with time due to Pt particle agglomeration. The limited platinum resource is another reason to investigate the replacement of pure platinum as the ORR catalyst. Efforts have been made to find more active and stable oxygen reduction catalysts, which led to the discovery of the ability to make carbon supported Pt alloys with high surface area and higher activity for oxygen reduction than Pt itself.

The Pt–M alloy, where M is a transition metal, could reach a higher exchange current density of ORR. Possible reasons why the alloy has a higher electrocatalysis activity than platinum have been proposed by Appleby and Foulkes (1993); the measured improvement in the stability to sintering, surface roughening due to removal of some base metals, which increases the Pt surface area, preferential crystal orientation, a more favorable Pt–Pt interatomic distance, electronic effects, and oxygen adsorption differences due to modified anion and water adsorption. But the disadvantage is that if the metal leaches from the cathode into the proton exchange membrane, the conductivity of the membrane will be affected. The base metal ions occupy the sulfonic acid groups in the proton-conducting electrolyte normally available for proton conduction. Only the more stable PtCr, PtZr, and PtTi base metal alloys are not leached into the acidic membrane electrolyte. The PtFe, PtMn, and PtNi alloys deposited Fe, Mn and Ni, respectively, in the membrane and in the anode of the MEA (Ralph and Hogarth 2002). This severely restricts the type of Pt alloys that can be used in the PEMFC. Few Pt alloys have been adapted for the practical fuel cell applications.

4.3 Electrocatalyst at PEMFC Anode

When reformat serves as the fuel of the PEMFC, even ppm CO content in the fuel may lead to the platinum poisoning. The development of the CO tolerance electrocatalyst is currently very active. The most successful electrocatalyst is PtRu/C. There are two explanations of CO tolerance mechanism of Pt–Ru. One is the popular bifunction mechanism. Ru-$(OH)_{ads}$ is formed based on Ru and the absorbed CO is oxidized by Ru-$(OH)_{ads}$.

$$M-(CO)_{ads}+Ru-(OH)_{ads}\rightarrow CO_2+H^++e^-$$

in which M represents Pt or Ru.

The other explanation is that Ru enters the Pt crystal lattice and changes the CO absorbtion status, facilitating CO oxidation. It has been proved that the best composition of Pt and Ru is 1:1. Recently, some transition metals have been investigated as alloys with PtRu for high CO tolerance, for example, W, Mo, Sn, etc. Meanwhile, some composite supports have demonstrated advantages over the traditional active carbon.

4.4 Effect of Electrocatalyst Size

The electrocatalyst particle size also has an effect on the electrochemical reaction. At the anode of the PEMFC, the oxidation of hydrogen is not a structure-sensitive reaction. The higher the dispersion of the platinum, the higher the activity of the electrocatalyst will be. The platinum loading of PEMFC anode could be decreased to $<0.1\,mg\,cm^{-2}$, but the electro-reduction mechanism of oxygen in the PEMFC cathode is sensitive to the structure of the electrocatalyst. The platinum lattice should have a certain size (e.g., 4–5 nm) in order to have high catalysis activity.

4.5 Electrocatalyst for Methanol Oxidation

In DMFC, the methanol has a dramatic poisoning effect on the conventional electrocatalyst Pt. The reaction that takes place on the DMFC anode is

$$CH_3OH+H_2O\rightarrow CO_2+6H^++6e^-$$

The possible reaction mechanism (Hamnett 1997) is

$$CH_3OH+Pt\rightarrow Pt-CH_2OH+H^++e^-$$

$$Pt-CH_2OH\rightarrow Pt-CHOH+H^++e^-$$

$$Pt-CHOH\rightarrow Pt-COH+H^++e^-$$

$$Pt-COH\rightarrow Pt-CO+H^++e^-$$

$$Pt-CO+M-H_2O\rightarrow Pt-M+CO_2+2H^++2e^-$$

The intermediates of methanol electrooxidation are absorbed on the surface of the electrocatalyst and thus poisoning the electrocatalyst. Modifying the Pt environment by alloying Pt with some other metals is the method most used in the electrooxidation of methanol. The major issue that limits DMFC in practical applications is its insufficient performance. The methanol oxidation on Pt–M alloy remains the main topic of DMFC research work in recent years (Wasmus and Kuver 1999). Although some ternary (PtRu-M) and even quaternary alloys (Choi *et al.* 2002, Jusys *et al.* 2002) have much higher activity, PtRu retains its most important role in DMFC anode electrocatalyst, especially in the commercialization. It is believed that in PtRu alloys Ru is more easily electrooxidized than pure Pt and forms Ru–OH_{ads} at lower potentials. The –OH_{ads} species spill over onto the neighboring Pt sites where they react with –CO_{ads}. This occurs at lower potentials than with pure Pt. There are some published works describing how to optimize PtRu composition for the DMFC anode. (Dinh *et al.* 2000, Lizcano-Valbuena *et al.* 2002)

Unsupported PtRu alloy blacks are the most widely used electrocatalysts employed at the DMFC anode, primarily because they provide very high Pt loadings (2–10 mg Pt cm^{-2}) to maximize the platinum surface area in the electrode. Such electrodes are very active for methanol electrooxidation but are too expensive for most commercial applications.

Some researchers are investigating novel electrocatalysts for DMFC. Carbonate Tungsten (WC) is very stable both in an acid and alkaline environment. It also has a good electrical conductivity. Moreover, the electric structure is similar to that of platinum. Especially in H_2SO_4, WC is active for hydrogen oxidation, and its CO tolerance is higher than that of platinum. When platinum is highly dispersed on WC, it shows good performance both for hydrogen oxidation and oxygen reduction. But unfortunately, the preparation cost is expensive than that of Pt/C because WC comes from WO_3 at 600–700 °C. Tungsten bronzes, M_xTO_3, are another type of promising electrocatalysts for DMFC, where M is an alkaline metal and T is a transition metal. It is reported that the highest electrical conductivity of Na_xWO_3 is $2.5\times10^4\,\Omega^{-1}\,cm^{-1}$; and it is very stable in acid. It shows activity for oxygen reduction. For the DMFC cathode, since the methanol comes from the anode side through the membrane by crossover, the electrocatalyst at the cathode side needs high methanol resistance. Recent research has shown that Pt/Ni, Pt/

Co, and Pt/Fe have higher tolerance toward methanol than Pt (Drillet *et al.* 2001)

4.6 Electrocatalyst Supports

At the beginning of fuel cell use in Gemini in the 1960s, Pt loading was several mg cm^{-2}. Until the end of 1980s with Pt black in Ballard's Mark V, the Pt loading was 2–4 mg cm^{-2}. The disadvantage was obvious: high cost but low Pt utilization. The development of carbon-supported Pt as oxygen reduction catalyst began in the 1970s in PAFC by the United Technologies Corporation. Using carbon black as a catalyst support makes Pt highly dispersed to maximize Pt active surface area. To reduce the electrocatalyst loading and to achieve a higher catalyst surface area, the electrocatalyst is supported by carbon black, for example, XC-72C is the electrocatalyst support most used, with 30 nm as the average particle size. Its specific surface area is 250 $m^2 g^{-1}$. Supported by active carbon, Pt loading could be 0.1–0.5 mg cm^{-2} without a dramatic drop in fuel cell performance.

Discovered in 1991 (Iijima 1991), carbon nanotube (CNT) has special characteristics, which will benefit fuel cells. These are high mechanical strength and high thermal stability. Furthermore, nanometer-scale spaces inside and among the CNTs provide large gas adsorption capacities and the possibility to increase the catalyst utilization. As efficient carriers of the Pt particles in the fuel cell reaction (Qin *et al.* 2000), the single-wall carbon nanohorn (SWNH) is likely to be used in fuel cells within a few years. The CNTs used for the experimental manufacturing are named nanohorns due to their irregular horn-like shape. Nanohorns have the same graphitic carbon atom structure as normal CNTs. The main characteristic of the carbon nanohorns is that when many nanohorns groups get together, an aggregate (a secondary particle) of about 100 nm is created. When the carbon nanohorn is used as an electrode material for a fuel cell, the advantage is not only the extremely large surface area, but also the ease with which the gas and liquid can permeate to the inside. Compared with normal nanotubes, nanohorns are easily prepared with high purity and it is expected to become a low-cost raw material.

Nowadays, more and more developers begin the activity of CNT applications in fuel cell. In 2001, NEC announced that they developed the tiny PEMFC with carbon nanohorn as the electrocatalyst support. It is reported by Carbon Nanotechnologies, Inc. that they have developed a PEMFC anode with SWNT/Pt. The catalyst loading on the anode side is only 7.61 μg cm^{-2} of Pt, but it could achieve double current density as compared with an E-TEK^{SM} anode with 400 μg cm^{-2} Pt. In November 2003, it was announced that a research team at the University of California at Riverside successfully developed a fuel cell using of multiwalled carbon nanotubes (MWCNT) as a platinum support. The Pacific Fuel Cell Corp. has tested nano-material in PEMFC electrode (Kazuhide and Kawamura 2004). They store hydrogen in nano-material at anode side and simplify the fuel cell system. It has been reported that CNTs serve as the platinum catalyst support at DMFC cathode (Li *et al.* 2004). Higher activity of the oxygen reduction and better performance of the DMFC have been achieved. Honeywell also reported that when CNTs sit at the interface of gas diffusion layer (GDL) and catalyst layer, the fuel cell will have a stable performance after thousand hours operation.

From the progress on using carbon nano-material as the electrocatalyst support, the advantages both in the fuel cell performance and in the bright future for commercialization could be seen. It is obvious that carbon nano-material will play an extremely important role in fuel cell development in the near future.

5. Electrode

High specific surface area, high limiting current density, and stable reaction interface are the basic requirements for fuel cell electrodes. Porous gas diffusion electrodes meet these requirements. Generally, porous gas diffusion electrodes in PEMFC and DMFC include the GDL and the catalyst layer.

5.1 Catalyst Layer

Till date, there are three kinds of catalyst layers. The first type of catalyst layer is a hydrophobic electrode made of PTFE and Pt/C with 20–50 μm thickness. When the membrane electrode assembly is prepared, ion exchange resin is also imbibed to form proton passage. The second type of catalyst layer is a thin hydrophilic electrode, <5 μm thick. It is made up of proton exchange resin and Pt/C electrocatalyst. Because there is no hydrophobic agent added, the reactants, for example, H_2 and O_2, will diffuse through either the hydrophilic resin or water to the active site. The third type of catalyst layer is made by the sputtering method. The sputtered layer is only several dozen nanometers thick. The sputtered method could result in higher catalyst utilization.

5.2 Gas Diffusion Layer

Generally, a gas diffusion layer consists of carbon cloth or carbon paper, which is fabricated from carbon fiber, for example, Toray® and SGL®, etc. GDL, with 100–300 μm thickness, collects current as well as transports reactants and products. GDL also serves as a membrane supporter, especially for coated membrane, so sufficient mechanical strength is also required. For gas transport, some GDLs are wet-proofed by polytetrafluoroethylene (PTFE) to form hydrophobic passages. In some applications, microporous layers are prepared for GDLs to protect the polymer membranes and improve the interface between electrode and membrane. The most common electrode membrane assembly (MEA) for PEMFC is the catalyst-coated membrane, for example, Nafion® MEA from DuPont, PRIMEA® from Gore. Concerning the CO poisoning in PEMFC and methanol oxidation in DMFC, composite electrodes have been proposed to deal with the hydrogen, CO, and methanol oxidation.

6. *Membrane*

The proton exchange membrane is another key component in PEMFC and DMFC. When the fuel cell was utilized for space use in the 1960s, the PEM materials were cross-linked, sulfonated polystyrenes. Sulfonated fluorocarbon membranes such as Nafion®, which was introduced by DuPont in the early 1970s, then sulfonated fluorocarbon membranes, not only Nafion®, but also Aciplex®, Flemion®, and Dow®, were used in PEMFC. The chemical structure of sulfonated fluorocarbon membrane is shown in Fig. 3. in which,

Nafion® : $x = 6 - 10,\ y = 1,\ z = 1,\ n = 2$;

Aciplex® : $x = 6 - 8,\ y = 0 - 1,$
$z = 1\ (0 - 2),\ n = 2 - 5$;

Flemion® : $x = 6 - 10,\ y = 1,\ z = 1,\ n = 2$; and

Dow® : $x = 3 - 10,\ y = 1,\ z = 0,\ n = 2$

Equivalent weight (EW) represents the weight (g) of 1 molar sulfonate group. The smaller the EW is, the larger the electrical conductivity will be. The mechanical strength, however, will become low. EW of Nafion® is mostly between 1000 g mol^{-1} and 1200 g mol^{-1} and EW of Dow® is between 800 g mol^{-1} and 850 g mol^{-1}.

The most acceptable structure of perfluorinated sulfonated membrane is the Cluster-Network model (Hsu and Gierke 1983), as shown in Fig. 4.

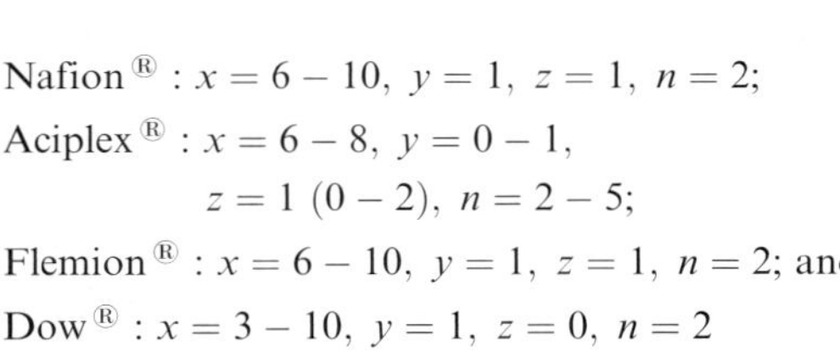

Figure 3
Chemical structure of sulfonated fluorocarbon membrane.

According to this model, the polymer ions and the absorbed water exist in spherical domains that are separated from the PTFE matrix. The hydrophobic perfluorinated chain forms the crystal hydrophobic zone and the sulfonic groups absorb water to form counter-ions. The diameter of the cluster is 4.0 nm and the clusters are connected by short narrow channels of 1 nm diameter. In this model, the counter-ions, the fixed sites, and water phase separate from the hydrophobic fluorocarbon matrix into approximately spherical domains interconnected by short narrow channels. The sulfonic acid in the ionomer is not diluted by the product water. The acidity of the ionomer is described as EW of the ionomer. With increasing EW, the diameter of clusters, the number of sulfonic groups, and the water molecules on each sulfonic group will decrease. But the distance between clusters, crystallinity, and the polymer rigidity will increase.

Although fluoride sulfonic acid membrane is widely used in PEMFC stacks, there are three main disadvantages. First , the manufacturing process of perfluorinated sulfonate resin is very complex and may cause environmental pollution. Second, the electron conductivity of the membrane is strongly affected by the water content. The membrane performance will follow the water content change. Third, the dimensional size of the membrane is not stable enough.

To serve as the solid electrolyte in fuel cells, the proton exchange membrane should have high proton conductivity (0.1 S cm^{-1}), good mechanical, chemical and thermal strength. From the efficiency point of view, the membrane should be kept from gas and methanol permeability. Different kinds of membranes have been investigated as new potential candidates in PEMFC and DMFC.

Partially fluorinated ionomer membranes, for example BAM3G®, have been developed to reduce the cost of the membrane. The single fuel cell test lasted for 15000 h. Its lifetime, however, is less than that of the perfluorinated ionomer membrane because it is

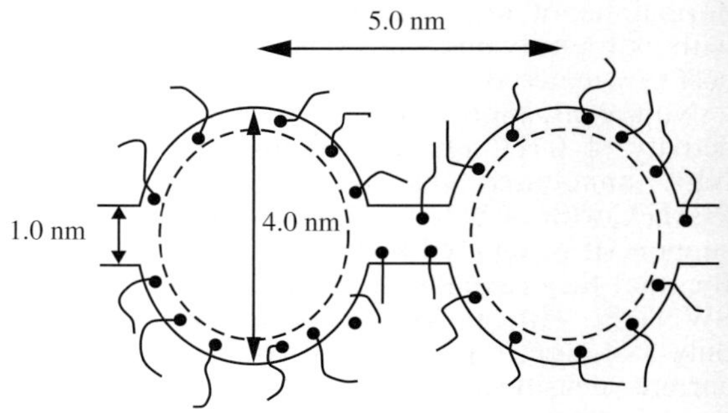

Figure 4
Cluster network model of perfluorinated ionomers.

brittle and degrades easily. Poly(ether ether ketone) PEEK, polyether sulfones (PES), and polysulfones (PS) such as polyimide (PI) have excellent stability and mechanical strength. Unfortunately, the longest duration of such a proton exchange membrane is only several thousand hours. To speed up market entry of PEMFC and DMFC, self-humidification and/or medium temperature membranes would be advantageous.

6.1 *Self-humidification Membrane*

For portable applications, the membrane should maintain a good performance under low relative humidity fuel cell operation. Self-humidification membrane is required. Platinum catalyst is added into the membrane to catalyze the hydrogen and oxygen diffusion through the membrane from the anode and cathode side. As a result, water is produced within the membrane (Watanabe 1994). Recently, the development of self-humidification membrane is combined with medium temperature membrane research.

6.2 *Medium Temperature Membrane*

Increasing the operating temperature to 100–200°C will increase the fuel cell system's efficiency, improve the electrochemical reaction activity, and potentially reduce the size, cost, and complexity of the fuel cell system. In the DMFC, higher temperatures (130–200°C) will improve the kinetics of the methanol oxidation reaction at the anode, which significantly enhances the efficiency of the fuel cell. Meanwhile, the water and heat management system will be simplified. Several approaches have been attempted to prevent the loss of water from the ionic regions (pores) of the membrane. If this is achieved, the conductivity of the ionomer membranes could be maintained even at temperatures near and above the normal boiling point of water.

One of the approaches is incorporating hydrophilic, inorganic material into ionomer membranes to increase the binding energy of water. The efficacy of these hydrophilic additives has been demonstrated when heteropolyacids are in Nafion®. The hybrid organic/inorganic nano-composite membranes consisting of SiO_2/poly(ethylene oxide) (PEO) doped with acidic molecules, such as monododecyl phosphate (MDP, acidic surfactant molecule) or phosphotungstic acid (PWA, heteropolyacid), show good proton conductivities above 100°C.

Silicon oxide/Nafion composite membranes have been studied for operation in hydrogen/oxygen PEMFCs from 80°C to 140°C (Adjemian *et al.* 2002). The silicon oxide improved the water retention of the composite membranes, increasing proton conductivity at elevated temperatures. Furthermore, silicon oxide-modified membranes are more robust than the control membranes—unmodified Nafion®115 and recast Nafion®, which degrades after high operational temperature and thermal cycling.

Hybrid nanostructured organic/inorganic composite membranes represent a novel trend toward minimizing humidification requirements. Some membranes, which could operate at low humidification operation are prepared by impregnating acids, sulfonic acid, phosphoric acid, or phosphoric acid into membranes such as Nafion®, polybenzeimidazole, or other composite membranes. These composite membranes can operate over 100°C under ambient pressure. But the liquid acid electrolyte might be lost during long-term operation. Solid acids show the advantages as the additives in composite membrane synthesis. For example, a composite membrane based on Nafion and solid proton conductor, Nafion–teflon–phosphotungstic Acid (NTPTA), can obtain a high fuel cell performance without humidifier over 100°C. Nafion–teflon–zirconiumhydrogen phosphate (NTZP) and Nafion–zirconiumhydrogen phosphate (NZP) also show good medium temperature performance (Lin *et al.* 2003).

Another approach is to use a nonaqueous, low volatility solvent to replace water as the proton carrier within the ionomer membrane. It has been demonstrated in membranes with phosphoric acid, imida-zole, butyl methyl imidazolium triflate, and butyl methyl imidazolium tetrafluoroborate. The rationale is that other liquid solvents can perform the function of water in proton conduction but with improved physical characteristic (i.e., low volatility). DuPont has carried out some preliminary studies with doped Nafion membranes containing molten acidic salts such as 1-butyl and 3-methyl imidazolium trifluoromethane sulfonate (BMITf). This non-volatile proton-conducting material produces membranes with high proton conductivities of $\sim 0.1\,\mathrm{S\,cm^{-1}}$ at 180°C, even under anhydrous conditions.

While the above two approaches rely on an acidic membrane such as Nafion® and a liquid solvent for proton transport, the third approach is based on a solid state material that conducts protons in the absence of a liquid solvent, such as phosphoric acid doped PBI. The immersion of PBI film in aqueous phosphoric acid leads to an increase in both its conductivity and thermal stability. Conductivities in the range $5\times10^{-3}-2\times10^{-2}\,\mathrm{S\,cm^{-1}}$ at 130°C and even $5\times10^{-2}\,\mathrm{S\,cm^{-1}}$ at 190°C have been reported. In addition, solid oxide conductors have been employed for many years in high temperature fuel cells, but work is just beginning on low-temperature solid-state proton conductors for fuel cells.

The recent research work on membranes has shown a trend that medium temperatures with low humidification will promise a bright future for PEMFC and DMFC applications. In the long-term, a complete novel invention, for example, polyphosphazene proton exchange membranes (Pintauro and Tang 2002), is expected to decrease the membrane cost and enhance the stability at 150–200°C.

7. Hydrogen and Reformer

Fuel storage is another crucial issue in the fuel cell application. It is restricted by several factors, such as the volumetric and gravimetric energy density, the weight and lifetime of the fuel container, the conversion efficiency of the chemical energy to electricity, the safety of the fuel, and fuel container during refueling and transportation and finally the fuel distribution infrastructure. For the portable fuel cell application, both hydrogen and methanol are very attractive. Hydrogen has the higher potential in terms of power density whereas the methanol fuel cell is advantageous from the energy density point of view.

7.1 Hydrogen

As hydrogen is the lightest element, it escapes easily from the storage tank. This makes hydrogen storage difficult. Nowadays, the most common hydrogen storage methods are compressed gaseous hydrogen at high pressure, liquid hydrogen at −253°C, metal hydride, chemical hydride, etc. Recently, carbon nano material seems to be another candidate for hydrogen storage, as it may result in a higher weight efficiency. In portable microfuel cell applications, compressed hydrogen, the metal hydride, chemical hydride, and carbon nanotube are believed to be of more practical use.

7.2 Compressed Hydrogen

Compressed hydrogen is the most common way for fuel cell hydrogen storage. Hydrogen compressed up to 70 MPa is now available for fuel cell applications.

7.3 Metal Hydride

Metal hydrides, which can be charged and discharged reversibly, are metallic alloys (see *Hydrogen–Metal Systems: Basic Properties (1)*; and *Hydrogen–Metal Systems: Basic Properties (2)*). Hydrogen reacts at elevated temperatures with many transition metals and their alloys to form hydrides. Metal hydrides can absorb large amount of hydrogen at a constant pressure because of phase transition. Metal hydride storage reaction is

$$M + 0.5 \times H_2 \leftrightarrow MH_x + \text{heat}$$

The composition of the alloy determines the temperature and pressure levels for desorption and adsorption process. Some metal hydrides absorb and desorb hydrogen at ambient temperature and close to atmospheric pressure, which is an essential factor for portable applications. All the reversible hydrides working around ambient temperature and atmospheric pressure consist of transition metals. Because some metal powder might be generated due to the metal hydride during the hydrogen desorption, a filter is required. Container and valves also contribute to the weight and size of the system. The gravimetric hydrogen density is limited to <3 mass%. The typical storage density of these hydride materials is 1.3–1.5% in weight of hydrogen. Metallic hydrides could reach a volumetric hydrogen density of $115\,\text{kg}\,\text{m}^{-3}$, e.g., $LaNi_5$. The highest volumetric hydrogen density reported (Züttel 2003) is $150\,\text{kg}\,\text{m}^{-3}$ in Mg_2FeH_6 and $Al(BH_4)_3$.

Metal hydrides promise a low-pressure, compact, and moderately inexpensive storage system, and the technology for making and processing them is well developed. To increase the hydrogen storage capacity at room temperature still remains a challenge, e.g., explore the properties of the lightweight metal hydrides.

7.4 Complex Hydride

Another choice of hydrogen storage is using complex hydride. The main difference between the complex and metallic hydrides is the transition to an ionic or covalent compound upon hydrogen absorption. The hydrogen in the complex hydrides is often located in the corners of a tetrahedron with B or Al in the center. The negative charge of the anion, $[BH_4]^-$ and $[AlH_4]^-$, is compensated by a cation, for example, Li or Na. The hydride complexes of borane, the tetrahydroborates $M(BH_4)$, and the tetrahydroaluminates $M(AlH_4)$ are nonflammable and stable in air. They only decompose at elevated temperatures, often above the melting point of the complex, the byproduct in hydrogen is water vapor. Hydrogen could be generated even at 0°C.

$$NaBH_4 + 2H_2O \rightarrow NaBO_2 + 4H_2 + \text{Heat}$$

In 1996, adsorption and desorption pressure-concentration isotherms for catalyzed $NaAlH_4$ at temperatures of 180°C and 210°C (Bogdanovic and Schwickardi 1997) were discovered. Such a chemical hydride could have a high H_2 storage efficiency.

$$NaAlH_4 \leftrightarrow \frac{1}{3}Na_3AlH_6 + \frac{2}{3}Al + H_2 \ (3.7\ \text{wt.\%}\ H_2)$$

$$Na_3AlH_6 \leftrightarrow NaH + Al + \frac{3}{2}H_2 \ (3.0\ \text{wt.\%}\ H_2)$$

The problem of the system is that the reaction kinetics is rather slow. But recently, the kinetics have been improved by catalyst introduction procedures and particle size control. The Ti-catalyzed system reversibly absorbs and desorbs hydrogen of up to 4.2 wt.%. Since the waste makes the concentration of chemical hydrides decrease during the hydrogen generation and since a spent fuel collection system is required, the waste treatment of the chemical hydrides remains an important issue, especially in micro-fuel cell applications.

7.5 Carbon Nanotube

Material with a large specific surface area, like activated or nano-structure carbon, is suitable to be the substrate for physisorption. This property is due to the ability of the material to exist in a very fine powdered form with a highly porous structure and the existence of particular interactions between carbon atoms and gas molecules. In microporous solids with capillaries, which have a diameter less than a few molecules, the potential fields from opposite walls overlap so that the attractive force acting upon adsorbed molecules is increased compared with that on a flat carbon surface. The investigation of the hydrogen–CNT interaction is an important topic in this research area.

The adsorption of hydrogen in carbonaceous materials corresponds to the amount of hydrogen adsorption that takes place near the carbon surface solid. It is only due to the physical forces—Van der Waals interactions—that carbon atoms exert on hydrogen molecules, therefore, it is called physisorption. As a result of van der Waals forces, both SWCN and MWNT self-assemble to form bundles, typically 10–100 nm in diameter. In a bundled sample, there are several sites where hydrogen could be adsorbed: inside the tubes, in the space between the tubes, or only on the outer surface of the bundle. Graphite nano-fibers (GNFs) exhibit hydrogen adsorption properties. It is reported that more than 50 wt.% hydrogen storage in GNFs at room temperature (Chambers *et al.* 1998) could be realized.

However, complex problems lie in the carbon nano-structure materials' reproducibility. Researchers have not yet clearly distinguished between chemisorption and physisorption. Chemisorption (the covalent bonding of hydrogen) would require high temperature and high energy for hydrogen release, whereas any practical fuel cell application would require low adsorption and desorption energies. Therefore, there is still a lot to do before carbon nano-structure materials can be used for hydrogen storage applications, for example, challenges to mass-produce controlled nanostructures at reasonable cost.

7.6 Microreformer

Since challenges still exist for hydrogen storage applications, liquid fuel provides another choice. Hydrogen-rich gas could be generated by microreformers. Generally, a reformer system consists of fuel preheater, fuel reformer, gas clean-up device, heat exchanger, and combustor. The chemical reaction, taking place in the reformer, could be typically accomplished by steam reforming, partial oxidation, or autothermal reforming. Small-scale reformers offer the use of hydrocarbons with their high-energy density at reasonable fuel costs. The striking feature is the worldwide availability of liquid fuels like gasoline, diesel, or butane. Unfortunately, methanol is the only fuel, which can be converted at intermediate temperatures of 250–300°C into a hydrogen-rich gas mixture. For the other fuels, higher temperatures above 700°C are necessary. The methanol steam reformer has the advantages of a simple design and no NO_x exhaust at low temperatures, but requires a preheater.

$$CH_3OH + H_2O \rightarrow CO_2 + 3H_2$$
$$CH_3OH \rightarrow CO + 2H_2$$

At present, the main efforts on microreformers are to select appropriate catalyst, increasing the fuel convert efficiency, and to integrate the reformer into the fuel cell system to achieve a high-power density. For example, Motorola Lab has developed a methanol reformer for portable fuel cell systems. Liquid methanol fed into a steam reformer produces fuel for a high-temperature fuel cell operating at 150°C. With micro-channel technique, the reformer is integrated with the heat exchanger, a compact system. Pacific Northwest National Laboratory has also developed an integrated fuel processor. The methanol conversion efficiency is 87–90%. In Lawrence Livermore National Laboratory (LLNL), CuO and ZnO on Al_2O_3 have been investigated and methanol conversion of 80 to 96% at 250 to 300°C has been reached. The selectivity of the catalysts should be optimized with respect to the suppression of CO formation in order to reduce the size and weight of the gas purification unit.

8. System Design and Manufacturing

Based on the key materials of fuel cell, compact system design is expected for micro-power system. The fuel and oxidant supplying components (micropumps and fans), water management, heat recovery and integration, as well as flow sensors and electronic control units, d.c.–d.c. converter, etc., should be highly integrated and optimized. The weight power density and volumetric power density depend not only on the material used but also on the design of the system. As a requirement of the portable fuel cell system, the manufacturing technique, for example, microelectronic manufacturing system (MEMS), is an essential factor in the compact system. The manufacture of fuel cell components, such as bipolar plates and reformers, benefits when the well-developed electronic device manufacturing technique is used. Print circuit technique, silicon chip, ceramic, etc. have been used in micro-fuel cell system. As porous silicon contains millions of microscopic tunnels, it can maintain its energy-producing capabilities far longer than flat proton exchange membrane (PEM) plates. The existing silicon chip-making technologies could make fuel cell cheaper. Silicon-based strategy is possible to make extremely small fuel cells with a large capacity. Different prototypes of portable fuel cells, ranging from 1 kW to several watts, have been developed, for example, Ballard Nexa, Fraunhofer Power Box, fuel cell powered camcorder, etc. Some institutions, such as Smart Fuel Cell, Casio, Toshiba, NEC, MTI, Samsung, Motorola Lab, Pacific Northwest National Laboratory, Manhattan Scientifics, etc., are working on the portable fuel cell commercialization. The portable fuel cell application is almost ready to enter the market of the small power sources, e.g., laptop, camcorder, cell phone, PDA, etc.

9. Summary

As compact power sources, fuel cells have already shown their advantages. From the material point of view, the commercialization of PEMFC and DMFC is being accelerated by the following developments: high dispersed electrocatalyst, novel electrocatalyst with high activity for oxygen reduction and for high CO and methanol tolerance, carbon nano-material, novel membranes with higher operation temperature and low humidity requirements. The manufacturing technique for mass production is expected in the near future.

See also: Fuel Cells: Electrochemical Reactions; Fuel Cells: Materials; Hydrogen–Metal Systems: Applications of Gas–Solid Reactions; Hydrogen–Metal Systems: Technological and Engineering Aspects.

Bibliography

Adjemian K T, Lee S J, Srinivasan S, Benziger J, Bocarsly A B 2002 Silicon oxide Nafion composite membranes for Proton-exchange membrane fuel cell operation at 80–140°C. *J. Electrochem. Soc.* **149**, A256–61

Appleby J, Foulkes F R 1993 *Fuel Cell Handbook*. Krieger Publishing Company, Malabar, Florida

Bogdanovic B, Schwickardi M 1997 Tri-doped alkali metal aluminium hydrides as potential novel reversible hydrogen storage materials. *J. Alloys Compd.* **253–254**, 1–9

Chambers A, Park C, Terry R, Barker K, Rodriguez N 1998 Hydrogen storage in graphite nanofibers. *J. Phys. Chem. B* **102**, 4253–6

Choi W C, Kim J D, Woo S I 2002 Quaternary Pt-based electrocatalyst for methanol oxidation by combinatorial electrochemistry. *Catal. Today* **74**, 235–40

Dinh H N, Ren X, Garzon F H, Zelenay P, Gottesfeld S 2000 Electrocatalysis in direct methanol fuel cells*in situ* probing of PtRu anode catalyst surfaces. *J. Electroanal. Chem.* **491**, 222–33

Drillet J F, Friedmann A E, Schmidt V M 2001 *Electrochemical Society Meeting Abstracts*. San Francisco, CA, No. 361

Hamnett A 1997 Mechanism and electrocatalysis in the direct methanol fuel cell. *Catal. Today* **38**, 445–57

Hamann C H, Hamnett A, Vielstich W 1998 In: *Electrochemistry*. Wiley-VCHP, Weinheim, p. 7

Hsu W Y, Gierke T D 1983 Ion transport and clustering in Nafion perfluorinated membranes. *J. Membr. Sci.* **13**, 307–26

Iijima S 1991 Helical microtubules of graphitic carbon. *Nature* **354**, 56–8

Jusys Z, Schmidt T J, Dubau L, Lasch K, Garche J, Behm R J 2002 Activity of $PtRuMeO_x$ (Me = W, Mo or V) catalysts towards methanol oxidation and their characterization. *J. Power Sources* **105**, 297–304

Kazuhide O, Kawamura 2004 *US Patent No.* 6 706 431

Li W, Liang C, Zhou W, Qiu J, Li H, Sun G, Xin Q 2004 Homogeneous and controllable Pt particles deposited on multi-wall carbon nanotubes as cathode catalyst for direct methanol fuel cells. *Carbon* **42**, 436–9

Lin J C, Kunz H R, Fenton J M 2003 Membrane/electrode additives for low-humidification operation. In: Vielstich W, Gasteiger H A, Lamm A (eds.) *Handbook of Fuel Cells – Fundamentals, Technology and Applications*. Wiley, England, Chap. 36, Vol. 3, pp. 456–63

Lizcano-Valbuena W H, Paganin V A, Gonzalez E R 2002 Methanol electro-oxidation on gas diffusion electrodes prepared with Pt-Ru/C catalysts. *Electrochim. Acta* **47**, 3715–22

Pintauro P N, Tang H 2002 *US Patent No.* 6 365 294

Qin L C, Zhao X, Hirahara K, Miyamoto Y, Ando Y, Iijima S 2000 Materials sciencethe smallest carbon nanotube. *Nature* **408**, 50

Ralph T R, Hogarth M P 2002 Catalysis for low temperature fuel cells. Part Ithe cathode challenges. *Platinum Metals Rev.* **46**, 3–14

Schmidt T J, Gasteiger H A 2003 Rotating thin-film method for supported catalysts. In: Vielstich W, Gasteiger H A, Lamm A (eds.) *Handbook of Fuel Cells-fundamentals, Technology and Applications*. John Wiley & Sons, Ltd, England, Chap. 22, Vol. 2, p. 328

Wasmus S, Kuver A 1999 Methanol oxidation and direct methanol fuel cellsa selective review. *J. Electroanal. Chem.* **461**, 14–31

Watanabe M 1994 *US Patent No.* 5 472 799

Züttel A 2003 Materials for hydrogen storage. *Mater. Today* 24–33

Fusion Conditions: Radiation Damage Correlation

N. M. Ghoniem

Most materials experience significant mechanical and physical property changes when irradiated by energetic particles. In a fusion reactor, neutrons (the majority having energy in the 14 MeV range) will impart substantial energy to lattice atoms, thus initiating collision cascades and subcascades. The initial energy of the primary knock-on atom (PKA) is typically six to seven orders of magnitude above its thermal value. This local high-energy density is quickly shared amongst neighboring atoms over a short period ($\sim$10 ps), resulting in permanent damage. A fraction of the PKA energy is first consumed in heating the electron system, while the remainder of the deposited energy produces freely migrating defects (FMDs). These defects result in substantial modifications to the microstructure of the material and hence its properties. In neutron-irradiated materials, damage initially manifests itself as vacancies, interstitials, and their clusters in metals and alloys. In ordered intermetallics and covalently bonded multi-species materials, additional forms of damage include the formation of anti-site defects, disordering, and amorphous zones. Nuclear reactions result in transmutations of lattice atoms, where internal helium and hydrogen gases can cause additional degradation mechanisms.

Material testing is performed in existing fission reactors to produce a database, which can be used to qualify their utilization in future fusion energy systems. Measures of material damage (such as swelling, irradiation creep, loss of ductility, etc.) can be obtained from fission neutron test data if appropriate correlations are used. Such correlations are usually based on the accumulated fluence of energetic neutrons or, more precisely, on accumulated displacement damage. Damage measures used in such correlations are discussed here.

1. *Neutron Flux, Spectrum, and Fluence*

Neutron interactions are described in terms of the energy, flux, spectrum, and total fluence. The current is defined as the number of neutrons crossing a unit area perpendicular to the neutron direction per unit time. This concept is reasonable for applications where neutrons predominantly travel in one direction (e.g., shielding). In a fission reactor, the analogous concept is the neutron flux, which is defined as the number of neutrons crossing a sphere of unit projected area per unit time. When the flux is restricted to neutrons with energy in the range E to $(E+\mathrm{d}E)$, we obtain the differential energy flux: $\Phi(E)\mathrm{d}E=$ the number of neutrons with energies in the range $(E, \mathrm{d}E)$ crossing a unit sphere per unit time. The total neutron flux is simply obtained by integration over the entire energy range.

The differential energy flux is also referred to as the neutron spectrum, because it is represented as a distribution of the differential flux as a function of neutron energy. In an LWR fission reactor, the neutron spectrum is "soft," while it is "hard" in a fast-breeder reactor, since many of the neutrons have energies in excess of 1–2 MeV. Under fusion reactor conditions, neutrons are produced at 14.1 MeV in a D–T fusion event, and after several scattering collisions, a "very hard" spectrum is established. Therefore, neutrons that interact with structural materials in a fusion reactor are expected to cause a higher level of damage as compared to (correlated with) lower energy neutrons in a fission reactor. Damage correlations are not simply the ratio of neutron energies in fission and fusion systems, but one must account for the mechanisms of damage production by energetic neutrons. The total neutron flux integrated over a period is referred to as the neutron fluence, which is a crude measure of material damage accumulation.

2. *Correlations for Defect Production and Retention*

Kinchin and Pease (1955) developed the first model for calculating the number of displaced atoms in a solid. No knowledge of the atomic structure is required in this model, and the number of displaced atoms, ν, by a PKA of energy E_{PKA} is readily calculated once the displacement threshold energy, E_{d}, is known. Thus:

$$\nu_{\mathrm{KP}}=\frac{E_{\mathrm{PKA}}}{2E_{\mathrm{d}}}$$

Brinkman (1954), on the other hand, was the first to propose the displacement spike concept, in which the atomic structure was considered. Subsequently, Seeger (1958) shed considerable light on Brinkman's model and proposed the ejection of interstitials from the cascade along low-index crystallographic directions in replacement collision sequences (RCSs), and the formation of a depleted zone at the cascade center. No single interatomic potential function is capable of representing the wide range of energies of displaced atoms. One must therefore resort to a description based on composite potential functions. In the high-energy range (i.e., at close separation between nuclei), the Coulomb potential is sufficient. That is:

$$V(r)=\frac{Z_1Z_2e^2}{r}$$

where Z_1 and Z_2 are the atomic numbers of interacting atoms, r is their separation distance, and e the electronic charge. As the energy decreases, the situation becomes more complex due to the interaction of overlapping electron shells. Atomic collisions above approximately 100–200 eV can be satisfactorily described by a binary collision approximation (BCA), in which the interatomic potential is a screened form of the pure coulomb. Below this energy range, several atoms can interact simultaneously, because of the local electronic structure. Since many important features can occur in the low-energy regime, such as displacement threshold reactions, defect stability, and clustering, etc., it is imperative to include many-body interactions in defect production calculations. As a result, estimations of defect production are based on the BCA for the gross features, and on detailed molecular dynamics (MD) for a more quantitative analysis. The concept of displacements per atom (dpa) is based on the BCA, and utilizes a modified form of the Kinchin–Pease estimate for the number of displaced atoms per collision.

The form of the interatomic potential determines energy transfers during atomic collisions. In BCA calculations, the potential at separations somewhat smaller than equilibrium is assumed to be represented by the Born–Mayer potential:

$$V(r) = A\exp\left(-\frac{r}{\rho}\right)$$

where A and ρ are constants. As the separation between the two atoms decreases, a screened Coulomb form is used. The inverse power law potential $V(r) = \mathrm{Ar}^{-s}$, $s = 2,3$ is often used. In this case, exact analytical expressions for the energy transfer cross-section can be readily found, and efficiently used in BCA calculations. The energy transfer cross-section in a collision between the energetic atom (energy E) and the struck atom (emerging with energy T) is analytically given by Chou and Ghoniem (1987) as: $\sigma(E,T)\,\mathrm{d}T = C_s E^{-s} T^{-1-s} \mathrm{d}T$, where C_s is a constant appropriate for the power s. Matching procedures must be used to ensure smooth transitions in energy transfer during binary BCA energy loss calculations.

Alternatively, exponential screening functions can be used to describe the influence of outer shell electrons on diminishing the interaction between nuclei. Ziegler *et al.* (1985) have successfully fitted many atomic scattering experiments with the potential:

$$V(r) = \frac{Z_1 Z_2 e^2}{r} \sum_{k=1}^{4} C_k e^{-b_k r/a_0}$$

where C_k and b_k are fitting constants, and a_0 is a screening length. Stability and final configuration of displaced atoms are properties that are critically dependent on accurate representation of the interatomic potential at energies just around the threshold for displacement (e.g., tens of electronvolts) down to thermal energies (e.g., fractions of electronvolts). Empirical potentials, which are based on equilibrium properties of the solid, are strictly accurate in the fraction of electronvolt range. Atomic collisions treated via interatomic potentials lead to energy loss from the PKA and redistribution to target atoms participating in the collision cascade. However, in between such collisions, energy is continuously lost from energetic atoms to the electron system in the lattice. The rate of this loss, per target atom per unit area, S_e, is described by the Lindhard formula at low energy (below 25 keV/amu):

$$S_e = 8\pi e^2 a_0 \frac{Z_1^{7/6} Z_2}{Z} \frac{v}{v_0}$$

where $a_0 = 0.0529$ nm, Z_1 and Z_2 are the charge number for projectile and target, with $Z^{2/3} = Z_1^{2/3} + Z_2^{2/3}$, v is the velocity of the projectile, and v_0 a characteristic electronic speed.

In the high-energy regime (i.e., PKA energy greater than 200 keV/amu), the Bethe-Bloch stopping power is used:

$$S_e = \frac{2\pi Z^2_{\mathrm{eff}} e^4 (M_1/m_e)}{E} \ln\left(\frac{4E}{(M_1/m_e)\bar{I}}\right) \quad (1)$$

where Z_{eff} is an effective charge, M_1 is the projectile mass, m_e the electron mass, $\bar{I}$ the average ionization energy of the target atom, and E the PKA energy. In the intermediate energy range of 20–200 keV/amu, the electronic conductance (inverse of S_e) as a linear combination of conductances calculated from the Lindhard and the Bethe-Bloch equations above is used. For polyatomic materials, such as SiC, Al_2O_3, etc., Bragg's additivity rule is used in accordance with the stoichiometry of the compound.

The concept of "displacements per atom," or simply dpa, has been extensively used as a measure of damage accumulation in irradiated materials. We outline here the procedure for dpa calculations in a multi-species material for generality (Huang and Ghoniem 1993). Assume that the material is composed of two species, type A and B (e.g., SiC). If a PKA of type A travels a distance $\mathrm{d}x$, it will eventually result in a number of displacements of type A atoms (ν_{AA}), and of type B (ν_{BB}), and similarly for type B PKA. Each element of the matrix ν_{ij} (i, $j = 1$, 2) is governed by a conservation equation:

$$\nu_{ij}(E)=\sum_k\int_0^{\Lambda_{ik}E}\{[\Gamma(E-T-E_{jd}/\Lambda_{ij})\nu_{ij}(E-T)$$
$$+\Gamma(T-E_{kd})(\delta_{kj}+\nu_{kj}(T))]N_k\sigma_{ik}(E,T)\}\mathrm{d}x\,\mathrm{d}T$$
$$+\int_0^{T_{em}}\nu_{ij}(E-T_e)N_e\sigma_{ie}(E,T)\mathrm{d}x\,\mathrm{d}T$$
$$+\left(1-\sum_k N_k\sigma_{ik}(E)\mathrm{d}x-N_e\sigma_{ie}(E)\mathrm{d}x\right)\nu_{ij}(E) \quad (2)$$

In Eqn. (2), $\Gamma(x)$ is the unit step function, N_i the atomic density of species i, $\sigma_{ij}(E)$ the total scattering cross-section between i and j, $\Lambda_{ij}=4m_im_j/(m_i+m_j)^2$ is the energy transfer efficiency between masses m_i and m_j, and δ_{ij} is the Kronecker delta. The first integral represents the number of displacements ν_{ij} conserved over the probability of a PKA of type i colliding with target atoms k ($k=1, 2$) during its travel a distance dx. It has two terms; the displacements caused by the PKA and those caused by the secondary knock-on. The second integral represents additional displacements, conserved over the probability of PKA collisions with electrons. The last term in parentheses on the right is for conservation of displacements over the probability of collisions with neither atoms nor electrons. Solution of Eqn. (2) depends on assumptions regarding the energy transfer cross-section (hence the interatomic potential), as well as the form of electronic energy loss. For the case of a single-element material, with no electronic energy loss, and a hard-sphere interatomic potential, the Kinchin–Pease estimate, ν_{KP}, is obtained. If a "softer potential" is used, for example, the inverse power law for which the differential energy transfer cross-section is given by Eqn. (1), this number is reduced by a factor $K=0.5$–1.0, depending on the potential. An acceptable value for K is 0.8, and is due to Norgett *et al.* (1974)—the NRT standard. In addition, when the energy loss to electrons is accounted for, the number of displaced atoms is reduced even further. It takes the form:

$$\nu_{NRT}(E)=K\xi(E)\nu_{KP}(E) \quad (3)$$

where the factor K is defined as the *displacement efficiency*, and the function $\xi(E)$ is defined as the *damage efficiency*. Both are less than unity.

When the neutron flux described by a differential spectrum $\phi(E_n)\mathrm{d}E_n$, the differential energy transfer cross-section from a neutron, energy E_n, to a PKA, energy E, is $\sigma_n(E_n, E)\mathrm{d}E$. The *displacement cross-section* σ_d can be conveniently defined as:

$$\sigma_d(E_n)=\int_{E_d}^{\Lambda E_n}\sigma_n(E_n, E)\nu(E)\mathrm{d}E$$

The displacement rate, in units of displacements per atom (dpa), is obtained by integration over the neutron spectrum:

$$R_d=\int_{E_d/\Lambda}^{\infty}\sigma_d(E_n)\phi(E_n)\mathrm{d}E_n \quad (4)$$

Current damage correlations for material behavior in one type of neutron spectrum (e.g., fission) to another (e.g., fusion) are based on Eqn. (4). Figure 1 shows calculated dpa cross-sections for iron, carbon (in SiC), silicon (in SiC), and SiC. The cross-sections are up to several thousand barns, and are very dependent on the neutron energy. Because of a smaller E_d for carbon in SiC, the dpa cross-section for carbon is about four times higher than silicon for fusion neutrons (Huang and Ghoniem 1993).

The damage calculation procedures outlined above are used as a first approximation to correlating fission and fusion neutron effects. The actual value of ν, obviously the basis of dpa, is somewhat uncertain because the theory ignores immediate restructuring of the lattice. As much as 99% restoration of displacements can occur over several time scales. To obtain more details on the nature of displacement damage for correlations of fission and fusion neutron effects, the Monte Carlo (MC) and the molecular dynamics (MD) methods are utilized. The gross features of large-scale cascades can be studied by the MC technique, and involve the use of the previous energy loss and cross-section equations to define probabilities of interactions and magnitudes of energy transfer. These computer simulations are based on the BCA; the primary examples are the computer codes MARLOWE (Oen and Robinson 1964), TRIM (Biersack and Haggmark 1980), and TRIPOS (Chou

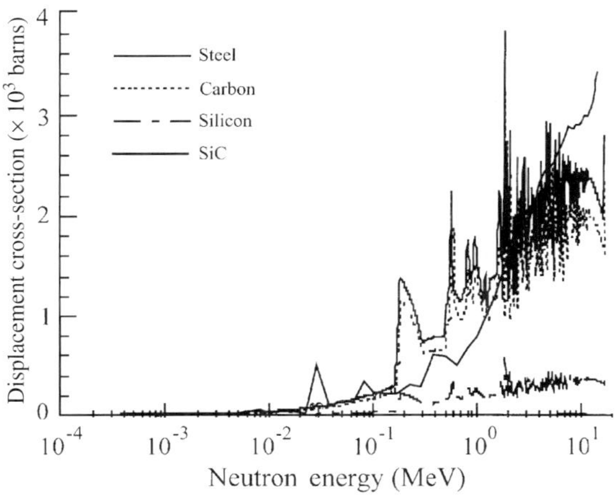

Figure 1
Displacement cross-sections as a function of neutron energy for Si (in SiC), C (in SiC), SiC, and Fe [10].

and Ghoniem 1986). The power of MC cascade analysis stems from the fact that large crystal volumes can be simulated, allowing study of high-energy interactions. When an assumed recombination distance of 5.75 lattice parameters is invoked in subsequent annealing simulations, the surviving defect fraction is shown to be consistent with resistivity measurements for copper (Heinisch 1981). MD simulations are based on direct solution for the equations of motion (EOM) for a large number of atoms. In metals, the production mechanism of stable Frenkel pairs has been established to be the replacement collision sequence (RCS). Atoms receiving energy just above the displacement threshold initiate a supersonic RCS, which terminates by displacing one atom at the end of the chain. Consequently, energetic atoms directed along principal cubic directions in the f.c.c lattice (i.e., $\langle 110 \rangle$ and $\langle 100 \rangle$) are easier to displace because of the focusing effect. In open directions, such as $\langle 111 \rangle$, an energetic atom emanating from a corner must squeeze through a triangle made up of face atoms. In copper, E_d along $\langle 110 \rangle$ or $\langle 100 \rangle$ is of the order of 25 eV, while it is almost twice as large close to the $\langle 111 \rangle$ direction. Similar results have been recently obtained for iron. MD calculations indicate that the lattice temperature has a significant effect on the length of the RCS in copper, where it decreases from 6.5 lattice constants to only three at temperatures below melting. In covalently bonded materials, such as SiC, the displacement process depends on whether the PKA is a carbon or a silicon atom. Displacements of carbon or silicon atoms occur at low energies in directions close to the [111] gaps, while they require substantially more energy to be displaced along bond directions (i.e., $[11\bar{1}]$). If an energetic atom is ejected along the direction of its neighbor, it will be reflected and then will be displaced through the opposite [111] gap. Stable displaced atoms (interstitials) then reside on $\langle 111 \rangle$ planes within the neighboring primitive cell, known as hexagonal (or H) sites. In contrast to metals, it appears that displacement damage in covalent materials does not require the large separation of Frenkel pairs via the RCS mechanism. The low displacement threshold of carbon as compared to silicon indicates that displaced atoms will predominantly be of the carbon type. If a silicon atom is ejected through the [111] gap bounded by a triplet of carbon atoms, it will displace the entire triplet on its way. Instead of RCSs emanating from a molten core, atomic displacements in covalent materials are produced in close proximity (within a neighboring primitive cell) because of a strong covalent bond. In b.c.c. metals, such as iron and vanadium, the general behavior of the cascade is similar to the f.c.c. case, with some additional complications. MD studies by Soneda and de la Rubia (1997) of cascades in b.c.c. iron showed that the general dependence of the fraction of surviving defects on PKA energy is similar to f.c.c. metals (see Fig. 2). However, no vacancy clusters were produced at the cascade core of either iron or vanadium, in sharp contrast to studies of copper and gold. Also, it is found that only small interstitial loops nucleate within the cascade volume. Once nucleated, di- and tri-interstitial loops glide very quickly along $\langle 111 \rangle$ directions. The activation energy for their migration is estimated as 0.2, 0.067, and 0.17 eV for mono-, di- and tri-interstitials, respectively.

The process of damage accumulation involves the following: (i) intercascade interactions, (ii) intra-cascade interactions, and (iii) diffusional migration of FMD. The last item, however, can conveniently be included with microstructure evolution processes, and will therefore not be dealt with here. Within one single cascade, the fraction of surviving defects (SD) is strongly dependent on the PKA energy. This fraction, χ, will be defined as the *defect production efficiency* by the equation: $\nu_{SD} = \chi K \xi(E) \nu_{KP}$. Figure 2 shows the results of several MD simulations of cascades for χ, up to 20 keV, together with experimental resistivity change data. These results indicate that even for one cascade, χ is on the order of 20–30% of the NRT value, for cascade energies above a few kiloelectronvolts. It is to be noted that the concept of dpa as a measure of damage accumulation has to be coupled with more detailed mechanistic studies of microstructure evolution to provide a basis for fusion–fission correlations.

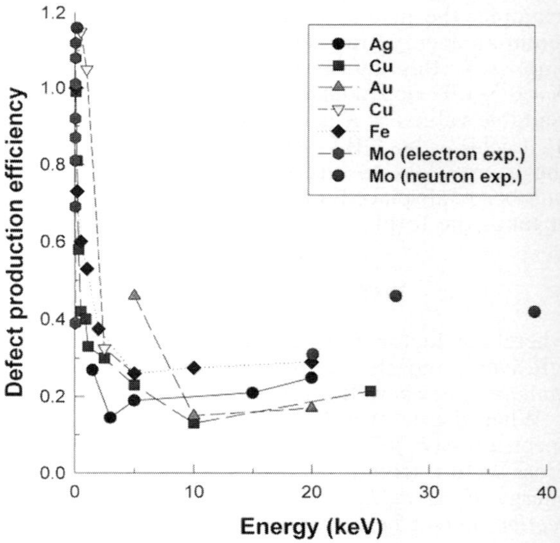

Figure 2
Defect production efficiency as a function of PKA energy—MD and experimental.

3. Nuclear Transmutation and Gas Effects

Nuclear reactions with lattice atoms in structural materials result in the transmutation of a fraction of the original elements and the production of new isotopes. Thus, the composition of the material is altered as new elements are formed. The rate of nuclear transmutation is higher in a fusion reactor, as compared to a fission reactor, because many of the reactions require a threshold energy level. The most significant nuclear reactions, which lead to property changes, are those leading to the formation of helium and hydrogen. To gauge the effects of helium on damage accumulation, the ratio of generated helium (appm) to the displacement damage dose (dpa)—He appm/dpa ratio—is usually employed as a correlation parameter.

Considering steel, as an example, we find that the sources of helium are dependent on the neutron spectrum. In a thermal flux spectrum, the main source of helium is neutron reaction with B^{10} (since boron in steel is mostly found at grain boundaries, the generated helium can readily result in the formation of grain boundary bubbles). Once B^{10} is burned out, helium generation will be determined by a two-step nuclear reaction with Ni^{58}. It is therefore characteristic to have a very rapid rate of helium production from B^{10}, followed by a steady increase in its concentration by neutron reactions with Ni^{58}. After one year of continuous irradiation in a typical light water reactor (LWR), the helium concentration reaches 30 appm. However, in a mixed spectrum reactor (e.g., HFIR at Oak Ridge National Laboratory), the helium concentration is 2000 appm, while the displacement damage dose is only 30 dpa.

In a harder neutron spectrum (e.g., a fast-breeder reactor), high energy neutrons produce helium (and hydrogen) in all elements of the structural material. After one year of neutron irradiation in a fast-breeder reactor (e.g., EBR-II at Argonne National Laboratory), helium concentration reaches 60 appm, while the displacement damage dose is about 50 dpa. In a Tokamak-type fusion reactor, it is expected that the displacement damage dose is about 20–30 dpa/year, while the helium-to-dpa ratio is ~15–20 appm He/dpa (1Ghoniem and Takata 1982). In correlating a material microstructure irradiated in a fission reactor to that in a fusion reactor, the two parameters of dpa and the appm-He-to-dpa ratio must be simultaneously used. Additional modeling is necessary to determine how close the microstructure of one facility is expected to be similar to the other. A combination of modeling techniques is generally used, such as MD and MC for atomic-scale defect behavior, and Rate Theory for the global features of the microstructure.

Bibliography

Biersack J P, Haggmark L G 1980 A Monte Carlo computer program for the transport of energetic ions in amorphous targets. *Nucl. Instrum. Methods* **174** (1–2), 257

Brinkman J A 1954 On the nature of radiation damage in metals. *J. Appl. Phys* **25** (8), 961

Chou S P, Ghoniem N M 1986 Collisional aspects of preferential sputtering using the Monte Carlo method. *J. Nucl. Mater.* **141–3**, 216–20

Chou S P, Ghoniem N M 1987 Applications of the Monte Carlo code TRIPOS to surface and bulk ion transport problems. *Nucl. Instrum. Methods Phys. Res.* **B28**, 175–84

Ghoniem N M, Takata M L 1982 A rate theory of swelling induced by helium and displacement damage in fusion reactor structural materials. *J. Nucl. Mater* **105** (2–3), 276–92

Heinisch H L 1981 Effects of temperature in binary collision simulations of high-energy displacement cascades. *J. Nucl. Mater* **108** (1–2), 62

Huang H, Ghoniem N M 1993 Neutron displacement damage cross-sections for SiC. *J. Nucl. Mater* **199**, 221–30

Kinchin G H, Pease R S 1955 The displacement of atoms in solids by radiation. *Rep. Progr. Phys* **18**, 1

Norgett M J, Robinson M T, Torrens I M 1975 A proposed method of calculating displacement dose rates. *Nucl. Eng. Des* **33** (1), 50

Oen O S, Robinson M T 1964 Monte Carlo range calculations for a Tomas-Fermi potential. *J. Appl. Phys* **35** (8), 2515

Seeger A 1958 Proc. 2nd United Nations Int. Conf. Peaceful Uses of Atomic Energy, Geneva, 6 250. United Nations, New York

Ziegler J F, Biersack J P, Littmark U 1985 In: *The Stopping and Range of Ions in Solids*. Pergamon, New York, p. 48

Fusion Reactors (Magnetically Confined)—Tokamaks: Materials

S. Ishino

Like many other power generating systems, fusion reactors are also composed of various components, as shown in Fig. 1 for a tokamak machine. Specific objectives or requirements are imposed on each of these components. Materials used for each component are selected to accomplish the objectives of the component throughout the service life of the component. In this context, service conditions must be clearly identified and performance modeling must be carried out to show that the materials can be used for the whole period of the target lifetime. This is, in reality, an ideal case because materials databases are not sufficient, particularly with regard to materials behavior under fusion neutron irradiations.

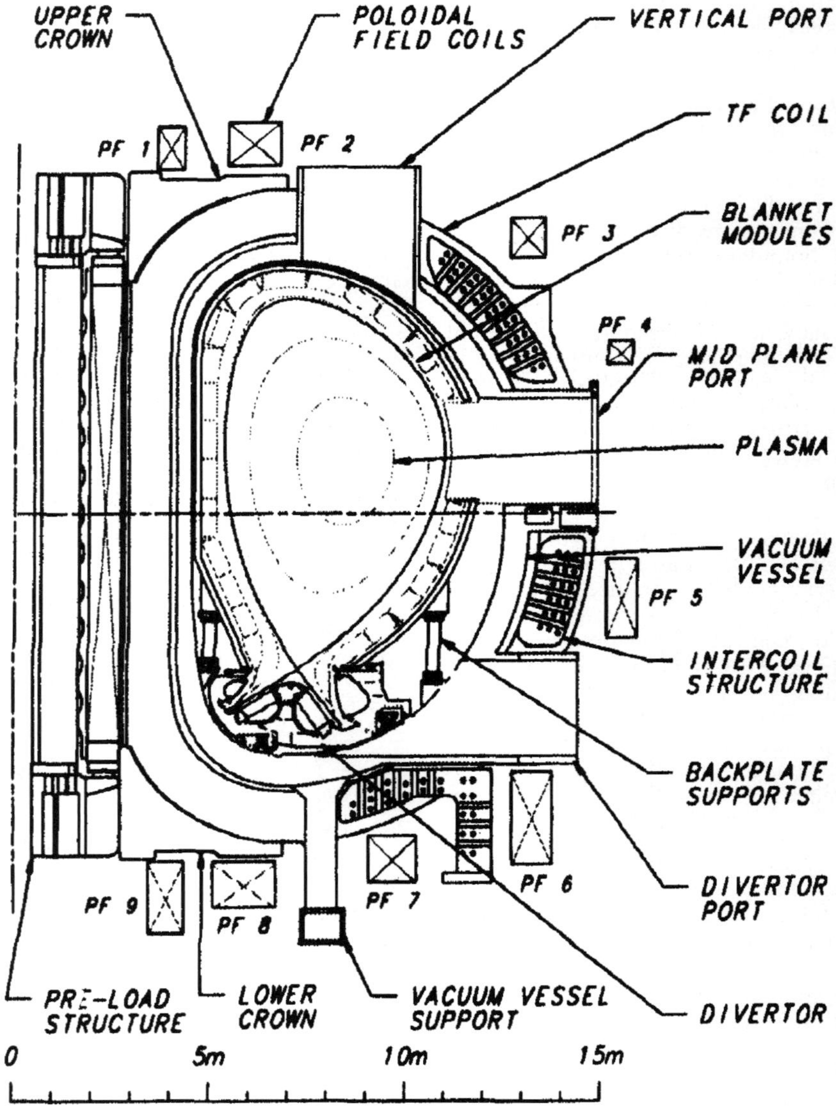

Figure 1
Cross-section of the ITER Tokamak.

1. Components of a Magnetically Confined Fusion Reactor and the Materials

Brief descriptions of each component of a tokamak will be given below.

1.1 Plasma–Fusion Fuel

Plasma in a fusion reactor corresponds to a nuclear fuel in a fission reactor. There are a number of nuclear fusion reactions among which the D–T reaction,

$$^{2}D + {}^{3}T = {}^{4}He(3.5\ MeV) + {}^{1}n(14.1\ MeV) \quad (1)$$

is regarded as closest to energy application. In this article, descriptions will be confined to a tokamak machine using the D–T reaction.

1.2 Plasma Facing and High Heat Flux Component Materials

First wall, divertor, limiter, armor, etc. are facing the plasma, typically with kinetic temperature exceeding 10^8 K (~ 10 keV) and a high heat load is expected in these components. The wall facing the plasma is not a load-bearing member of the reactor. Replacement after some period of service is expected. The first wall interacts directly with the plasma particles coming out through a scrape-off layer and various plasma-wall interactions occur. The divertor is a component to neutralize escaped plasma particles and "fusion product" helium, and to lead them to an exhaust system.

1.3 Blanket

The blanket is an important component on which the following three major missions are imposed.

(a) Tritium breeding. Tritium is an isotope of hydrogen that does not exist in the environment because it is a weak beta emitter with a half-life of 12.3 yr. Therefore, D–T fusion reactors must breed more tritium than they consume. Lithium and its compounds are used for tritium breeding.

(b) Energy conversion. Most of the energy liberated by the D–T fusion nuclear reaction is in the form of kinetic energy of neutrons (14.1 MeV per reaction). The energy of alpha particles will be used to heat the plasma. The kinetic energy of neutrons is recovered in the blanket region as heat.

(c) Shielding. Various radiations and heat must be shielded. In particular, the superconducting magnet, which must be used in a magnetic confinement fusion power reactor, is used at cryogenic temperatures and the superconducting materials are sensitively degraded by neutron exposure.

1.4 First Wall and Blanket Structures

These are the most important parts in fusion reactors because they are placed in the severest environment near the plasma and yet are permanent or semi-permanent components. They are the life-limiting component of a fusion reactor. A more detailed account will be given later.

1.5 Coolant

Selection of coolant materials is important because it dictates the selection of structural and blanket materials from the viewpoint of compatibility, tritium breeding ratio, temperature, and other factors. Pressurized water, helium gas, liquid metals and alloys, and molten salts have been considered as candidates.

1.6 Superconducting Magnet

Magnetic confinement power reactors must use a superconducting magnet to eliminate the ohmic loss occurring in normal conductors. The magnet comprises coils and support structures.

2. First Wall and Blanket Structural Materials

2.1 Importance of Structural Materials

First wall and blanket structures are permanent components of a fusion reactor. Their positions are close to the plasma, receiving high flux of high-energy neutrons and high heat load. The conditions are severest among all the components in the reactor. Therefore, it is a key issue of fusion energy development to develop suitable structural materials. The major requirements imposed on the structural materials for fusion power reactors are:

(i) superior radiation resistance, in particular, to high-fluence high-energy neutrons;
(ii) minimal induced radioactivity; and
(iii) excellent mechanical, physical, and chemical properties throughout their lifetime with high reliability.

Materials research is essentially a proactive exercise, because fusion reactor design must be based on the existing database of materials. For example, 316 stainless steel is adopted as a major structural material for the ITER (the next large research fusion reactor) because 316 stainless steel is an established material with a relatively ample database. Current emphasis of the development of structural materials for fusion reactors is aimed at those for a power demonstration reactor (DEMO), which should come after ITER.

All the three requirements listed above are related to neutron irradiation. One of the most important parameters determining the operating condition of the neutron-interacting fusion materials is neutron wall loading. This is defined as the total energy of neutrons coming from the plasma to the unit area of the first wall and is expressed usually in MWm^{-2}.

This is determined by the fusion reaction rate in a plasma volume and the surface area of the first wall. 1 MW/m^2 corresponds to incoming primary neutrons of 4.43×10^{17} (14 MeV) nm^{-2}s^{-1}. This should not be taken as a neutron flux nor energy deposition at the first wall because the neutron flux contains neutrons back-scattered from the blanket region in addition to those coming from the plasma. Usually, neutron flux values become an order of magnitude larger, with one-quarter to one-third of the total flux coming from the plasma. For a fixed value of the wall loading, the flux varies according to the details of the design such as selection, amount, and configuration of materials outside the plasma region. Cumulative neutron wall loading is a measure of neutron fluence.

2.2 *Characteristic Features of Radiation Damage in Fusion Structural Materials*

There are several characteristic features in radiation effects in fusion structural materials; they are:

(i) D–T fusion reaction produces monoenergetic 14 MeV neutrons. Consequently, displacement damage is produced by high-energy primary knock-on atoms (PKAs). Moreover, a much higher nuclear transmutation rate is expected than in the case of fission reactor neutrons. Nuclear transmutation reactions can be categorized into two parts: one is gas production, particularly helium and hydrogen; the other is production of solid transmutants, sometimes resulting in alloying, and sometimes causing concern regarding induced radioactivity.

(ii) To realize economical fusion power generation, it is imperative for the structural materials to withstand high-energy neutron irradiation to a very high fluence. For austenitic stainless steels, 1 MW · yr m^{-2} of cumulative wall loading corresponds roughly to a displacement damage of 10 dpa and helium accumulation of 150 at.ppm, where dpa means displacements per atom (see *Radiation Damage Theory*). For several fusion devices, end-of-life displacement damage levels, helium and hydrogen production in at.ppm are summarized in Table 1 (Ishino 1996). Several hundred dpa may be envisaged as a goal fluence for future power reactors. This is a very hard task to tackle, because there are no intense neutron sources to introduce such high levels of irradiation with fusion-reactor spectrum neutrons within a reasonable irradiation time. Various surrogate means including fission reactors, ion accelerators, and high-voltage electron microscopes (HVEMs) have been utilized, but it is very difficult to predict material behavior under fusion neutron irradiation with high accuracy.

2.3 *Important Property Changes in Fusion Structural Materials*

Neutron irradiation to high fluence will lead to changes in the properties of the structural materials. Properties of greatest concern are swelling, creep, helium embrittlement, and thermal conductivity degradation (see *Nuclear Reactor Materials: Irradiation Effects*). In the case of fusion neutron irradiation, it is not easy to carry out irradiation experiments with the fusion-relevant He/dpa ratios. Various techniques as shown in Table 2 (Ishino and Sekimura 1990) have been applied. Differences between these techniques and fusion neutron irradiations are sometimes significant (see *Fusion Conditions: Radiation Damage Correlation*). A high-energy high-flux intense neutron source is necessary to verify the applicability of these techniques.

2.4 *Effect of Nuclear Transmutation—Changes in Chemical Constituents and Induced Radioactivity*

In order to assess the effects of nuclear transmutation, nuclear databases including various nuclear

Table 1
Damage parameters for an austenitic stainless steel first wall in various fusion reactors.

Fusion device	Integrated neutron wall loading (MW · yr m^{-2})	Displacement damage level (dpa)	Helium production	Hydrogen production (at.ppm)
FER	0.3	3	45	150
NET	1	10	150	500
ITER[a]				
Basic performance phase (BPP)	0.5–1	5–10	75–150	250–500
Enhanced performance phase (EPP)	1–3	10–50	150–450	500–2500
DEMO	10	100	1500	5000
Commercial reactor	40	400	6000	20 000
cf				
Maximum achieved with RTNS-II (1×10^{23} nm^{-2} 14 MeV)	10^{-3} (equiv.)	10^{-2}	0.2	0.7

[a]In the newest design of ITER these values are reduced roughly by a factor of 3.

Table 2
Techniques to simulate the effect of helium due to fusion neutron irradiation.

I. Utilization of fission reactors
1. Use of Ni double reactions for Ni-bearing alloys: $^{58}Ni\ (n,\gamma)^{59}Ni\ (n,\alpha)^{56}Fe$
   (a) Spectral tailoring technique
   (b) Isotope tailoring technique
   (c) Injector foil technique
2. Use of Cu triple reactions: $^{63}Cu(n,\gamma)^{64}Cu \xrightarrow{\beta} {}^{64}Zn(n,\gamma)^{65}Zn(n,\alpha)^{62}Ni$
3. Use of Fe double reactions: $^{54}Fe\ (n,\gamma)^{55}Fe\ (n,\alpha)^{52}Cr$
4. Boron (n, α) reaction: $^{10}B\ (n,\alpha)^{7}Li$
   (a) $^{10}B/^{11}B$ isotopic tailoring
   (b) Continuous ^{10}B supply technique
5. Use of tritium decay: $^{3}T \xrightarrow{\beta(12.3y)} {}^{3}He$
   (a) Static method
   (b) Dynamic helium charging experiment (DHCE).
6. Use of triple fission of uranium

II. Fission reactor coupled with charged particle accelerator

III. Utilization of ion accelerators
1. Helium pre-implantation
2. Dual-beam irradiation
3. Triple-beam irradiation to study the effect of both helium and hydrogen
4. Light ion irradiation giving rise to similar He/dpa ratio

reaction cross-sections such as elastic: (n, n), inelastic: (n, n′γ), absorption: (n, γ), charged particle producing reactions: (n, p), (n, α) etc., multiple particle producing reactions;(n, n′p), (n, n′α), (n, 2n), (n, 3n), etc., nuclear energy levels and so on must be established.

Because of the complexity of nuclear reactions, materials for the neutron interactive components are vulnerable to conversion into different materials. For example, pure vanadium will change to V–Cr–Ti alloys, pure copper to Cu–Ni–Zn alloys, tantalum to tungsten, then to rhenium, and further to osmium. Together with radiation-induced solute segregation, the matrix composition will change after prolonged irradiations. One of the advantages of fusion energy is that there are no strongly radioactive fission products or long-lived α-emitting transuranium elements. However, the total mass of neutron-interactive structural components of a reactor is quite large, several thousand tons for power reactors. Therefore, low activation structural materials are necessary to take full advantage of relative cleanliness of fusion reactors (*Fusion Reactors: Low Activation Materials*).

2.5 *Candidate Materials for Fusion Reactor Structural Components*

In this section, four kinds of materials: austenitic stainless steels, ferritic/martensitic steels, vanadium alloys, and SiC/SiC composites will be briefly mentioned.

(i) *Austenitic stainless steels.* AISI 316 stainless steels have been used as a cladding material of fast breeder reactors because of their high strength and excellent compatibility with sodium . Largely because of the availability of an ample irradiation database, the material has been adopted as a major structural material for ITER. There are, however, several differences between fast reactor and fusion reactor conditions. First, the helium production rate in austenitic stainless steels in a fusion environment is typically 15 at.ppm/dpa, roughly two orders of magnitude larger than in the fast reactor core conditions. High helium production rate in stainless steels has been simulated by various means such as: isotope tailoring, spectral tailoring, and dual or triple beam irradiations (see *Fusion Conditions: Radiation Damage Correlation*).

For ITER conditions, the steel will be used below ~300°C. Irradiation database at this temperature range for stainless steels to a high fluence is astonishingly scarce. Note that for irradiation temperatures below 300°C, uniform elongation becomes very low and a larger irradiation creep rate occurs at lower irradiation temperatures (Rowcliffe *et al.* 1991).

(ii) *Low activation ferritic/martensitic steels.* Fe–Cr–Mo ferritic/martensitic steels have been used for heat exchangers and core structural materials for fast breeder reactors because of low void swelling rate and reasonable high-temperature strength. For fusion applications, one wt.% of molybdenum may be replaced by 2 wt.% tungsten. Also from the systematic study of irradiation response on ductility, the optimum Cr content has been set around 9% (Kohyama *et al.* 1996, Hishinuma *et al.* 1998) (see also *Fast Reactor Cores, Ferritic Martensitic Steels for*).

High-temperature strength in ferritic/martensitic steels is inferior to that in austenitic steels. To overcome this, applying oxide (Y_2O_3) dispersion strengthening with ferritic/martensitic steel matrix is under consideration as a means to enhance high-temperature strength considerably.

(iii) *Vanadium alloy.* Vanadium has excellent nuclear properties as a base metal of low activation materials (see *Fusion Reactors: Low Activation Materials*). Vanadium alloys have been developed since the 1960s, first for fast breeder cladding, then for fusion. In particular, vanadium in combination with liquid lithium is excellent in view of the tritium breeding ratio as well as of corrosion and compatibility and this combination is presumed to be taken as an operating condition of vanadium components. Vanadium alloys have higher high-temperature strength than ferritic alloys and lower swelling rate, but their sensitivity to irradiation embrittlement and large hydrogen inventory are a few examples of the weak points of vanadium alloys. Minimal ductile-brittle transition temperature (DBTT) is obtained for vanadium alloys containing 0–4 wt.% Cr and 4–5 wt.% Ti (Loomis *et al.* 1994). V–4Cr–4Ti alloy and the alloys modified by adding Si, Al, Y, and so on have been proposed and are being examined. It should also be noted that the properties of vanadium alloys are modified strongly by interstitial impurities; oxygen, carbon, nitrogen, and hydrogen.

(iv) *SiC/SiC composites.* Both silicon and carbon are low activation elements and if they are pure enough, radioactivity decays one month after reactor shutdown to a level that could allow contact maintenance and repair. Long term radioactivity is due to ^{26}Al (half life: 7.2×10^5 yr) from ^{28}Si, but the threshold energy of the reaction is high so that the activation is significant only in the initial layer of the first wall (Fenici *et al.* 1998). The material can be used up to about 1000°C. SiC/SiC composite has high potential but is still regarded as an advanced material for a next-generation machine and a number of issues need to be clarified. Areas requiring further development are: (i) production of high-purity radiation- resistant SiC fibers and fabrication technology for low porosity composites; (ii) joining and large-scale production technology; (iii) evaluation and design methodology; and (iv) irradiation database, particularly degradation of thermal conductivity, creep resistance, and fracture toughness under high helium generation rate of 1500 at.ppm/MW · yrm^{-2}.

3. *Plasma Facing and High Heat Flux Component Materials*

3.1 *Plasma-materials Interactions*

(*a*) *Parameters controlling the plasma-materials interactions.* Plasma facing components interact with plasma and various surface phenomena occur. The phenomena are important not only from erosion of the wall material but also from the plasma side. As a result of the interactions, impurities from the materials in the plasma will enhance radiation loss by Bremsstrahlung and resulting cooling of the plasma will be detrimental to the plasma confinement. Change of reflectivity of the wall will change the energy balance of the plasma. Inventory, release, and permeation of fusion fuel, i.e., hydrogen isotopes, will modify the material balance of the plasma. These phenomena are very complex and cannot be treated from material side alone. Important parameters defining the problem may be grouped into the following four items:

(i) surface bombardment parameters: kind of incident particles, their energy spectrum, flux, angle of incidence, etc.;

(ii) surface characteristics: structure and morphology, chemistry of surface, i.e., surface chemical composition, structure, and characteristics of surface oxide, presence and characteristics of adsorbed atoms, defects on and near surface, spatial variation of chemical composition, etc.;

(iii) characteristics of emitted particles: kind of particles emitted from surface, intensity, energy, angular distribution, charge states, etc.; and

(iv) instrumental characteristics: details of the fusion device, edge plasma characteristics, temperature profiles, magnetic and electric field, etc.

As major phenomena of plasma materials interactions, physical sputtering, chemical erosion, radiation-enhanced sublimation (Hofer and Roth 1996, Davis and Haasz 1997), and hydrogen recycling are briefly described.

(*b*) *Physical sputtering.* Sputtering is a phenomenon in which constituent atoms of materials are lost from the surface by bombardment of charged particles, neutrals, or neutrons with energies exceeding certain threshold values. One of the mechanisms is knocking out of surface atoms by atomic collisions, called physical sputtering. The number of atoms emitted per incident particle is called sputtering ratio or sputtering yield. Physical sputtering is a phenomenon quite similar to radiation damage. The sputtering yield is correlated well with the nuclear stopping cross-section at the surface. Theoretical treatment based on elastic collisions with threshold energy of sputtering has been successful to predict the sputtering yield except at low energy region near the threshold (Davis and Haasz 1997). In a real tokamak machine, the sputtered atoms may hit the wall again, causing self-sputtering.

(*c*) *Chemical erosion.* Erosion of wall material by chemical reaction is an important phenomenon in plasma materials interaction and is often called chemical sputtering. Typical example of chemical sputtering occurs in carbon materials bombarded by hydrogen or oxygen ions or atoms resulting in erosion by emitting methane or carbon monoxide. Methane yield has a peak around 800 K for ~1 keV hydrogen ions in graphite, but for lower energy hydrogen ions the yield curve as a function of temperature broadens and the peak temperature shifts to lower temperatures. This is ascribed to the formation of C_2H_x and C_3H_y hydrocarbons for lower hydrogen energies. Unlike in physical sputtering, no threshold energy for the chemical sputtering seems to exist.

(*d*) *Radiation-enhanced sublimation* (*RES*). In the case of plasma disruption, the wall temperature rises abruptly. Wall material is lost by thermal sublimation, which is called disruption erosion. In graphite, thermal sublimation occurs above 2000 K, but under ion irradiation, sublimation occurs at much lower temperatures, approx. 1200 K. This phenomenon is called radiation-enhanced sublimation (RES), which has been known since the early 1980s. The basic mechanism is thought to be due to displacement of atoms, particularly to interstitial atoms created in a cascade.

Erosion phenomena are complicated because the surface is modified by redeposition, surface segregation, and so on, which again modify the condition of incoming particles.

(*e*) *Hydrogen recycling*. Hydrogen ions and atoms hitting the wall will produce various phenomena: reflection, adsorption, absorption, reemission, desorption, chemical sputtering, and so on. Absorbed hydrogen will diffuse into the material, trapped by defects and impurities, and permeate to the opposite surface. These phenomena are important in fuel balance, tritium inventory, and radiation safety. These processes have been modeled, but again modification of the surface by deposition makes quantitative estimation difficult.

3.2 *High Heat Flux Component Materials*

Low atomic number (Z) materials, particularly carbon materials such as graphite and carbon fiber composites (CFC), have been used for plasma-facing high heat flux components such as the first wall, limiter, and divertor. However, since a high erosion rate due to chemical sputtering and radiation enhanced sublimation has become evident, high-Z refractory metals such as tungsten as well as low-Z, such as beryllium and boron carbide, are to be used. Though tungsten is a high-Z element and inclusion in plasma as an impurity is deleterious, its erosion rate under ion bombardment is far less than for carbon and the threshold energy for physical sputtering is high. Therefore, if the plasma edge temperature is low enough, the erosion rate may be negligible (Hirooka 1998).

3.3 *Heat Sink Materials for High Heat Flux Components*

First wall, limiter and divertor will be used in a high heat flux condition depending on the mode of operation. For example, in ITER with average heat loading of 0.25 MW m^{-2}, startup heat flux at the limiter will become 8 MW m^{-2} and in some part of the divertor during off-normal transient, the heat load will go up to ~20 MW m^{-2} (in the case of vertical disruption event (VDE), up to 60 MW m^{-2}) (Davis and Kalinin 1998). Refractory materials as given in Sect. 3.2. are necessary for the plasma facing wall, but in order to remove the heat efficiently, materials with high strength and high thermal conductivity are required for the heat sink. High-strength high-conductivity copper alloys are to be used in ITER for the purpose. Currently, for primary wall and limiter, dispersion, strengthened copper alloys such as Cu–Al_2O_3 with Al content of 0.15, 0.25, 0.60 wt.% and for divertor, precipitation-hardened copper alloys such as Cu–Cr, Cu–Cr–Zr, Cu–Ni–Be, and Cu–Ni–Cr–Si are among the candidates.

Temperature ranges for these materials may be limited to below 250–300 °C, where substantial softening occurs. Degradation of thermal conductivity due to radiation damage and nuclear transmutants, nickel and zinc, must also be considered.

4. *Blanket Materials*

4.1 *Tritium Breeding Materials*

D–T fusion reactors must breed more tritium than they consume. Lithium and its compounds are used for tritium breeding. Tritium production reactions are

$$^{6}\text{Li} + \text{n} \rightarrow \text{T} + {}^{4}\text{He} + 4.8\ \text{MeV} \tag{2}$$

$$^{7}\text{Li} + \text{n} \rightarrow \text{n}' + \text{T} + {}^{4}\text{He} - 2.5\ \text{MeV} \tag{3}$$

Combination of tritium breeding materials with neutron multipliers, coolant and structural materials is important. The tritium breeding materials are divided into two, solid blanket and liquid blanket materials. Major tritium breeding materials with their characteristics are listed in Table 3 (Tanaka 1997). Solid breeding materials are lithium-containing ceramics, whereas liquid breeding materials are lithium metal, Li–Pb alloy and molten salt, "Flibe." Important issues are tritium recovery characteristics, breeding ratio, and corrosion and compatibility.

Tritium recovery characteristics for liquid breeding materials can be evaluated by Sieverts' constant, K_s. Lithium has a large K_s value so that recovery of tritium becomes difficult and tritium inventory becomes large. For Li–Pb and Flibe, tritium permeation and leakage from the coolant circuit are the major issues. Both lead and beryllium are regarded as neutron multiplier elements.

Table 3
Major tritium breeding materials and their properties (Tanaka 1997).

	Li	$Li_{17}Pb_{83}$	FLIBE:Li_2BeF_4	Li_2O	$LiAlO_2$	Li_2SiO_3	Li_4SiO_4	Li_2ZrO_3	Li_2TiO_3
Density, gcm^{-3}	0.48	9.4	2.0	2.02	2.6	2.52	2.28	4.15	2.0
Li density, gcm^{-3}	0.48	6.4×10^{-2}	0.28	0.94	0.27	0.36	0.54	0.33	0.25
Melting point, °C	180	235	363	1423	1610	1200	1250	1615	1535
Breeding property	good	good	need neutron mulipier	good	need neutron multiplier	need neutron multiplier	need neutron multiplier	Zr is neutron multiplier	need neutron mulipier
Chemical reactivity	react violently with air and water	react with water	almost no reaction	react with water	almost no reaction	almost no reaction	small reactivity	almost no reaction	almost no reaction
Tritium recovery	difficult high solubility	easy high tritium pressure	easy high tritium pressure		easy needs H_2 addition to sweep gas				
Low temperature limit, °C	180(a)	235(a)	363(a)	320(b)	450(b)	400(b)	390(b)	320(b)	300?(b)
High temperature limit, °C	~500(PCA)(c) ~600(HT-9) >750(V15Cr5Ti)	~420(PCA)(c) 460(HT-9) >750(V15Cr5Ti)	~600(SS)(c) 800(Ni alloy)	800(d)	970(e)	700(e)	730(e)	970(e)	920(e)

[a]melting point, [b]from tritium diffusion, [c]from compatability with structural materials, [d]from mass transfer of LiOH, LiOT, [e]from sintering.

For solid breeding materials, a mixture of He and H_2 is used as a sweep gas for tritium recovery. The process of release of tritium from a lithium ceramic matrix into the sweep gas is similar to fission gas release from nuclear fuel in a fission reactor. In order to reduce tritium inventory in the blanket, the tritium concentration in the breeding materials must be as low as several at.ppm or less. Tritium dwell time in the breeding materials must be short, say several days. To enhance tritium recovery, isotopic dilution is applied, adding a small amount of H_2 (~0.1%) in the sweep gas.

4.2 Neutron Multiplier

In order to attain a tritium breeding ratio greater than one, the selection of the combination of breeding materials, coolant, and structural materials is important (Song *et al.* 1989). For a certain combination, tritium breeding can only be achieved by use of neutron multipliers. The neutron multiplier must have large (n, 2n) cross-sections for neutrons with energies below 14 MeV, small capture cross-sections and large atomic density. According to these criteria, ^7Li, ^9Be, Zr, Pb, Bi are selected as candidate multiplier elements. Among these, beryllium has excellent characteristics as a neutron multiplier. Cross-section of total (n, 2n) reaction of ^9Be for 14 MeV neutrons is 484 mb (millibarn $= 10^{-31} m^2$), that of (n, t) reaction is 19 mb, (n, α) 9.2 mb and capture cross-section is only 3.2 mb. $^9Be(n, \alpha)\,^6He \rightarrow {}^6Li$ reaction also leads to tritium production. Beryllium may be used as pebbles, in a block shape, or in a pin shape.

5. Superconducting Magnet Materials

Tokamak power reactors must use superconducting magnets to eliminate the ohmic loss occurring in normal conductors. Several tokamak machines with superconducting toroidal magnets have been built; T-7 (NbTi, 1978), TRIAM-1M (Nb_3Sn, 1988), T-15 (Nb_3Sn, 1988), Tore Supra (NbTi, 1989). Large Helical Device (LHD, NbTi/Cu, 1998), which is not of a tokamak type, uses superconducting coils for both toroidal (helical) and poloidal magnets. Here, superconductors and year of operation are shown in parentheses. In future tokamak machines, superconducting magnets will be applied to both toroidal and poloidal field coils. The magnet comprises coils and support structures. The coils are wound with conductors comprising superconducting materials, stabilizers, insulators, and structures. Since the stored energy in the active magnet is huge, becoming of the order of several tens of gigajoules for ITER-class tokamak machines, strong support structures are required.

Superconducting materials being used or planned to be used are of metallic type, alloys such as Nb–Ti and A-15 type compounds such as Nb_3Sn. Materials of the latter type have superior properties. Highest critical field of 40 T at 4.2 K has been achieved in Nb_3 (Al, Ge). To enhance the stability of performance, superconducting materials are usually fabricated in compacted strands of thin multifilamentary wires surrounded by stabilizers such as copper or aluminum. Various types of conductors have been designed, depending on the method of cooling and other factors.

A-15 type superconductors are more sensitive to radiation than alloy superconductors. To limit the lifetime fluence, fast neutron flux at the magnet must be below $10^{13}\,m^{-2}s^{-1}$. To avoid thermal conductivity degradation, radiation damage for copper stabilizers must be below 5×10^{-4} dpa. Various high-temperature oxide superconductors have been discovered since 1986. Among them, $YBa_2Cu_3O_x$ (YBCO) having a T_c well above liquid nitrogen temperature has been applied to a high-current transport device of a superconducting magnet (Mito *et al.* 1998).

6. Functional (Special Purpose) Materials

Electric insulators, rf windows, optical fibers, and other diagnostic materials are a few examples of functional materials of a tokamak fusion reactor. Since most of them are ceramic materials or glass, to be discussed in other chapters, these subjects will not be discussed here.

Bibliography

Davis J W, Haasz A A 1997 Impurity release from low-Z materials under light particle bombardment. *J. Nucl. Mater.* **241–3**, 37–51

Davis J W, Kalinin G M 1998 Material properties and design requirements for copper alloys used in ITER. *J. Nucl. Mater.* **258–63**, 323–8

Fenici P, Frias Rebelo A J, Jones R H, Kohyama A, Snead L L 1998 Current status of SiC/SiC composites R&D. *J. Nucl. Mater.* **258–63**, 215–25

Hirooka Y 1998 Effects of plasma impurities on materials erosion and mixing in a steady state magnetic fusion device—a comparison between beryllium and tungsten. *J. Nucl. Mater.* **258–63**, 1045–9

Hishinuma A, Kohyama A, Klueh R L, Gelles D S, Dietz W, Ehrlich K 1998 Current status and future R&D for reduced-activation ferritic/martensitic steels. *J. Nucl. Mater.* **258–63**, 193–204

Hofer W O and Roth J (eds.) 1996 *Physical Processes of the Interactions of Fusion Plasma with Solids*. Academic Press, Amsterdam

Ishino S 1996 Implications of fundamental radiation damage studies in the research and development of materials for a fusion reactor. *J. Nucl. Mater.* **239**, 24–33

Ishino S, Sekimura N 1990 Role of charged particle irradiations in the study of radiation damage correlation. *J. Nucl. Mater.* **174**, 158–67

Kohyama A, Hishinuma A, Gelles D S, Klueh R L, Dietz W, Ehrlich K 1996 Low activation ferritic and martensitic steels for fusion application. *J. Nucl. Mater.* **233–7**, 138–47

Loomis B A, Chung H M, Nowicki L J, Smith D L 1994 Effects of neutron irradiation and hydrogen on ductile-brittle transition temperatures of V–Cr–Ti alloys. *J. Nucl. Mater* **212–5**, 799–803

Mito T, Maehata K, Mizokami M, Ishibashi K, Takeo M, Iwamoto A, Hirano N, Shintomi T, Kimura K, Sawamura M, Yamada S, Satoh S, Motojima O 1998 Development of high-T_c superconducting current feeders for a large-scale superconducting coil system. *J. Nucl. Mater* **258–63**, 1940–5

Rowcliffe A F, Hishinuma A, Grossbeck M L, Jitsukawa S 1991 Radiation effects at fusion reactor He:dpa ratios: Overview of US/Japan spectrally tailored experiments. *J. Nucl. Mater.* **179–81**, 125–9

Smith D L, Mattas R L, Billone M C 1994 Fusion reactor materials. In: Cahn R W, Haasen P, Kramer E J, (eds.) Frost B R T, (Vol. ed.) *Materials Science and Technology*, VCH, Weinheim, Germany, Vol. 10B

Song M, Youssef M Z, Abdou M A, Raffray A R 1989 Impact of neutronics on the selection of solid breeder and multiplier materials and configurations. *Fusion Eng. Design* **10**, 47–55

Tanaka S 1997 Tritium behavior in blanket systems. *J. Plasma Fusion Res* **73**, 1333–40

For comprehensive references of original and review papers, see proceedings of "*Topical Meeting on Fusion Reactor Materials*" and of "*International Conference on Fusion Reactor Materials*", all published in a conference volume of *J. Nucl. Mater.* A few thousand papers are included altogether in a series of these Proceedings, being the largest source of references in the fusion materials area. Volume numbers and total pages of these proceedings are: *J. Nucl. Mater.* **85–6** (1979) pp. 1–1214 (1st Topical Meeting), **103–4** (1981) pp.1–1625 (2nd Topical Meeting), **122–3** (1984) pp.1–1645 (3rd Topical Meeting), **133–4** (1985) pp. 1–932 (ICFRM-1), **141–3** (1986) pp. 1–1153 (ICFRM-2), **155–7** (1988) pp. 1–1381 (ICFRM-3), **179–81** (1991) pp. 1–1274 (ICFRM-4), 191–4 (1992) pp. 1–1506 (ICFRM-5), **212–5** (1994) pp. 1–1802 (ICFRM-6), 233–7 (1996) pp. 1–1692 (ICFRM-7), **258–63** (1998) pp. 1–2192, and **271–2** (1999) pp. 1–612 (ICFRM-8), **283–7** (2000) pp. 1–1556 (ICFRM-9). For plasma wall interactions, proceedings of a series of conferences on "*Plasma-Surface Interactions in Controlled Fusion Devices*" are comprehensive sources of references. These proceedings are also published in a conference volume of *J. Nucl. Mater.* **111–2** (1982), **128–9** (1984), **145–7** (1987), **162–4** (1989), **176–7** (1990), **196–8** (1992), **220–2** (1995), 241–3 (1997), **266–9** (1999), **290–3** (2001) pp. 1–1295.

Fusion Reactors: Low-activation Structural Materials

F. Abe and T. Noda

In the selection of low-activation materials for the first wall of a fusion reactor (see *Fusion Reactors (Magnetically Confined)–Tokamaks: Materials*, niobium, molybdenum, nickel, and copper, for which the cooling time of surface dose rate is longer than 100 years, are unfavorable (Okada *et al.* 1989). Based on the examination of the induced activity, the following materials are recommended. For the short-term operation of $<1\,\mathrm{MWy\,m^{-2}}$, where $1\,\mathrm{MWy\,m^{-2}}$ corresponds to a displacement damage level of about 10 displacements per atom (dpa), aluminum alloys, carbon, SiC, Al_2O_3, vanadium alloys are the candidate materials. For long-term operation of $>10\,\mathrm{MWym^{-2}}$, the materials are selected from the viewpoint of waste disposal rather than from the surface dose rate. This is due to the difficulty in controlling the impurity concentration at a level of ppm or less, and iron alloys such as Cr–W–V ferritic and Cr–Mn austenitic steels, vanadium alloys, titanium alloys and ceramic composites such as SiC/SiC become preferable. Figure 1 shows the decay of surface dose rate of induced activity of 9Cr–2W ferritic and 10Cr–30Mn austenitic steels, V–4Cr–4Ti alloy, and SiC/SiC composite after reactor shutdown, together with that of the unfavorable materials such as Mod.9Cr-1Mo steel and type 316 stainless steel. In Fig. 1, the induced activity is calculated using a neutron spectrum at the first wall position of a fusion experimental reactor (FER), which was designed by the Japan Atomic Energy Research Institute. It is also assumed that these materials were irradiated to $10\,\mathrm{MWy\,m^{-2}}$. The decay time, which is defined as the time needed for the dose rate to be lowered to a safety level of $25\,\mu\mathrm{Sv\,h^{-1}}$ from the material with an activation volume of $1\,\mathrm{m^2}$ in area and 1 cm in thickness—which corresponds to $3\times10^{-10}\,\mathrm{Svh^{-1}\,g^{-1}}$ in the case of steel—is evaluated to be 10–20 years for the 9Cr–2W ferritic and 10Cr–30Mn austenitic steels, and about 10^5 years for Mod.9Cr–1Mo steel and type 316 stainless steel. The γ-ray intensity of high-purity SiC/SiC composite decreases by six orders of magnitude in a day after shutdown, making it the best among the materials from the waste disposal viewpoint (Noda *et al.* 1994).

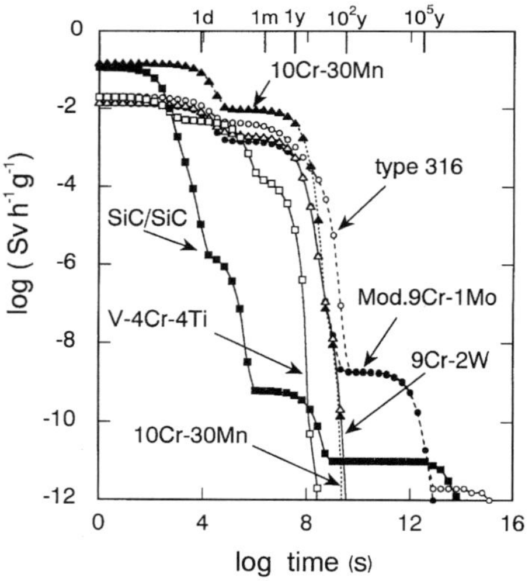

Figure 1
Decay behavior of γ-ray intensity of the materials after irradiation of $10\,\mathrm{MWy\,m^{-2}}$.

1. Reduced-activation Ferritic Steels

From the middle of the 1980s, the development of reduced-activation ferritic and austenitic steels has been pursued in the USA, Europe, and Japan. Calculations on the decay behavior of γ-ray intensity indicate that molybdenum, niobium, nickel, copper, and nitrogen in conventional steel must be eliminated or minimized. Therefore, the development of reduced-activation ferritic steels involves the replacement of molybdenum and niobium in conventional Cr–Mo ferritic steels by tungsten and tantalum, respectively. At present, Cr–W–V reduced-activation ferritic/martensitic steels are very promising as the first wall and blanket structural materials.

The Japanese and US programs for the development of reduced-activation ferritic steels have included both the 2.25Cr steels having bainitic microstructure and the 9–12Cr steels having martensitic microstructure, while the European programs have concentrated on the 9–12Cr martensitic steels. The bainitic 2.25Cr steels are easier to weld and require simpler post-weld heat treatments than the high-chromium martensitic steels, but they are more susceptible to irradiation hardening and embrittlement. In the unirradiated condition, the reduced-activation (9–12)Cr–W–V–Ta martensitic steels have higher tensile and creep-rupture strengths and superior fracture toughness characteristics, shown by lower ductile-to-brittle transition temperature and by higher upper shelf energy compared to those of the conventional (9–12)Cr–Mo–V–Nb steels, as shown in Fig. 2 (Abe *et al.* 1994). Oxide dispersion strengthened (ODS) ferritic/martensitic steels can also be expected to be candidate materials because of their high strength properties at elevated temperature.

Ductile-to-brittle transition temperature (DBTT) has been a major concern because the DBTT of ferritic/martensitic steels increases rapidly with irradiation to above room temperature. Of the reduced-activation Cr–W–V steels with chromium

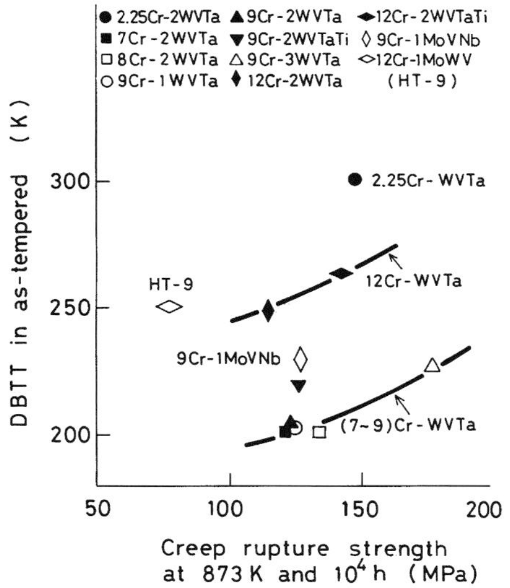

Figure 2
Relationship between DBTT and creep rupture strength for reduced-activation ferritic/martensitic steel.

concentrations varying from 2.25% to 12%, the 9Cr steels showed the greatest resistance to irradiation embrittlement; the increase in DBTT due to irradiation to 7 dpa at 638 K in the Fast Flux Test Facility is 68 K for the 9Cr–2W–0.25V steel, while it is only 4 K for the 9Cr–2W–0.25V–0.07Ta steel (Klueh *et al.* 1992). These increases in DBTT are low compared to increases of over 100°C for the 2.25Cr–2W–0.25V and 12Cr–2W–0.25V steels. The results are encouraging in that they indicate that the effect of irradiation on toughness can be favorably affected by changing the compositions.

Based on a database for the reduced-activation ferritic/martensitic steels, the design window for the steels in the water coolant system of a fusion reactor has been determined as shown in Fig. 3 (Hishinuma *et al.* 1998). The lower temperature is limited by irradiation-induced DBTT increase. In determining the upper temperature limit, softening or loss of creep strength during irradiation are key factors. Thermal fatigue may also be an important factor. Although it has been a concern whether the ferritic/martensitic steels can be used for structural components of fusion devices because of the concern for the influence of magnetic materials on plasma behavior, it has been shown that plasma discharge should be possible with a ferromagnetic vacuum vessel in Tokamak devices.

2. Reduced-activation Austenitic Steels

Proposals for reduced-activation austenitic steels involve the replacement of nickel in conventional Ni–Cr austenitic stainless steels by manganese. The Cr–Mn austenitic steels are readily fabricable and have comparable or superior mechanical properties such as tensile, creep-rupture and fatigue strengths to those of type 316 stainless steel in unirradiated condition. Figure 4 shows the creep-rupture data for the 10Cr–30Mn–2W–0.2V–0.2Ti–0.15C steel at 873 K, together with those for prime candidate alloy (PCA) which is a titanium-modified 316 steel, A-25MN (14Cr–25Mn–0.2Ti–0.25C), type 304 steel and Mod.9Cr–1Mo ferritic steel (Abe *et al.* 1992). The creep-rupture strength of the 10Cr–30Mn–2W–0.2V–0.2Ti–0.15C steel is evaluated to be 260 MPa at 873 K

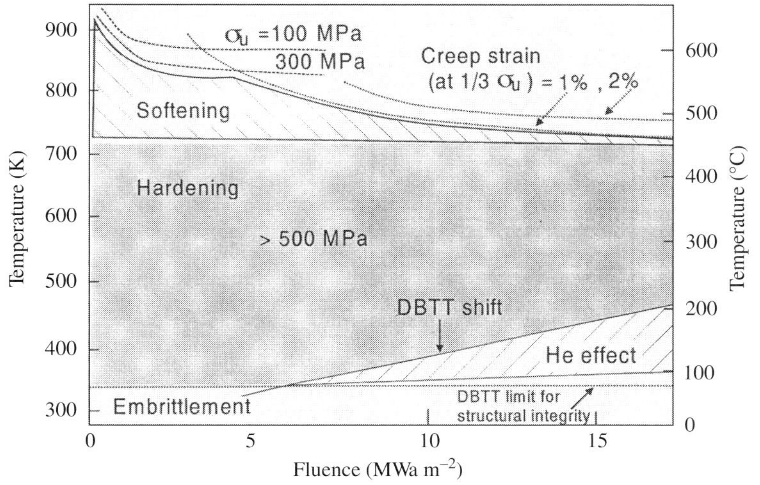

Figure 3
Schematic design window for reduced-activation ferritic/martensitic steels in a Tokamak-type fusion reactor.

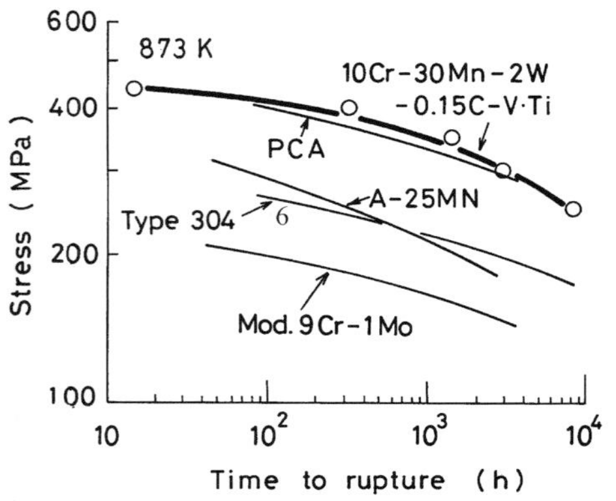

Figure 4
Creep-rupture data for the reduced-activation austenitic steels 10Cr–30Mn–2W–0.15C–VTi and A-25MN at 873 K, comparing with conventional austenitic steels.

and 10 000 h, which is as high as that of PCA and is superior to that of A-25MN, type 304, and Mod.9Cr–1Mo steels. The ultimate tensile strength of the 10Cr–30Mn–2W–0.2V–0.2Ti–0.15C steel is 420 MPa at 873 K, which is slightly lower than that of 12Cr–20Mn–2W–0.1V–0.25C (about 500 MPa) developed in the USA because of the lower carbon content. Thus, the research in progress has been brought to a satisfactory conclusion.

In contrast, the high-manganese steels are unstable under both thermal and irradiation conditions, causing a decomposition of austenite matrix into sigma phase and a new austenite phase having different composition, and have a tendency to exhibit irradiation-enhanced aqueous corrosion. Safety considerations associated with the high decay heat and volatility of manganese are the most serious for the high-manganese austenitic steels as the first wall of a fusion reactor.

3. Low-activation Vanadium Alloys

Vanadium alloys based on V–Cr–Ti are attractive candidates for structural applications in a fusion reactor because of their low activation properties, high thermal stress factor, good strength at elevated temperatures, and good compatibility with liquid lithium (Zinkle *et al.* 1998). One of the often cited disadvantages associated with vanadium alloys is the lack of large industrial infrastructure. Although essentially most data on vanadium alloys were generated on small laboratory heats with a typical mass of 1–50 kg, two different V–4Cr–4Ti alloy heats with masses of 0.5 and 1.2 tons have recently been fabricated. Recent work has shown that unirradiated V–(3–6)%Cr–(3–6)%Ti alloys have favorable mechanical properties and can be fabricated into tubes and other practical shapes. Considerable advances have also been made on joining techniques for vanadium alloys, including gas tungsten arc, electron beam, laser, and inertia welds. Systematic studies have also been conducted to investigate the effects of oxygen and oxidation on the microstructure and tensile properties of V–(4–5)%Cr–(4–5)%Ti alloys. The post-exposure tensile strength of 1 mm thick specimens at 293 and 773 K are not significantly affected by exposure to a range of oxygen pressure (10^{-4}–10^5 Pa) for 250–275 vh. However, significant reductions in total and uniform elongation are observed, particularly in room-temperature tensile test. Exposure to oxygen at 773 K is also observed to increase the sensitivity to hydrogen embrittlement.

The irradiation-induced increase in yield stress (irradiation hardening) of V–(4–5)%Cr–(4–5)%Ti alloys is particularly high at temperatures below 673 K. A pronounced decrease in the irradiation hardening occurs for temperatures above about 673 K, although the irradiated yield stress at 873 K is still about 60% higher than the unirradiated value. The specimens irradiated at temperatures below 673 K exhibit pronounced flow localization caused by dislocation channeling, with uniform elongation less than 0.2%. The uniform elongation slowly increases at temperatures above 673 K. The DBTT for V–4Cr–4Ti specimens is far below room temperature in unirradiated conditions, which increases to 373–473 and about 703 K after irradiation to 0.5 and 5 dpa, respectively, resulting from irradiation hardening.

4. Low-activation SiC/SiC

Silicon carbide (SiC)-based ceramic matrix composites (CMCs) such as SiC/SiC are considered as one of the candidates for the first wall and blanket structural materials of future fusion reactors because of the low activation properties of SiC and dimensional stability under irradiation for temperatures up to 1273 K (Noda *et al.* 1994, Fenici *et al.* 1998). Design projects for future fusion power reactors have been carried out considering SiC/SiC composites as a structural material, attempting both to establish priorities for the development of properties in available materials and to address requirements for component fabrication and assembly. Availability of small-diameter fibers, with moderate high-temperature stability, has allowed the production of CMCs with relatively predictable thermomechanical behavior, overcoming the brittle behavior associated with monolithic ceramics.

At present, chemical vapor infiltration (CVI) is the most extensively studied, because CVI has the advantages of high purity, minimizing damage to fibers, and making near-net-shape products. In the CVI process, Nicalon SiC is generally used for the

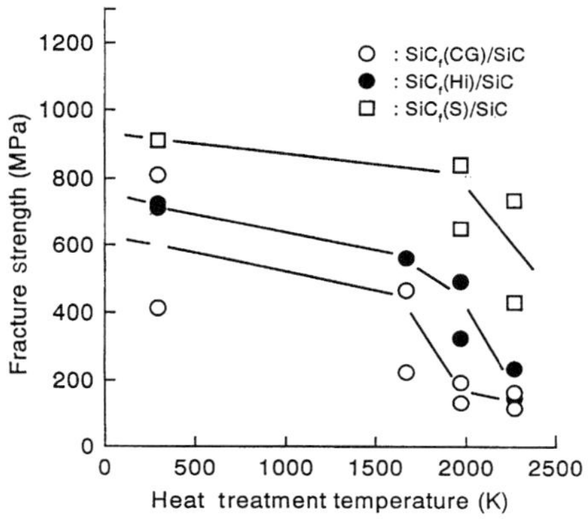

Figure 5
Fracture strength of SiC/SiC composites as a function of heat treatment temperature.

preform because it is a continuous fiber with high flexibility. Figure 5 shows the fracture strength in bending tests for three types of SiC/SiC composites prepared by CVI, as a function of heat treatment temperature (Araki *et al.* 1998). The specimens were heated for 1 h in a vacuum at each temperature. Nicalon commercial grade fiber (SiC_f(CG)), Hi-Nicalon SiC fiber (SiC_f(HI)), and Hi-Nicalon Type-S fiber (SiC_f(S)) with oxygen concentrations of 11.7, 0.5, and 0.2%, respectively, were used as preforms. The strength is in the sequence SiC_f(S)/SiC > SiC_f(HI)/SiC > SiC_f(CG)/SiC. The strength abruptly decreases with temperatures above 1700 K for both the SiC_f(CG)/SiC and SiC_f(CG)/SiC composites, while the SiC_f(S)/SiC composite keeps its strength well even at around 2000 K. The fracture toughness of both the SiC_f(CG)/SiC and SiC_f(CG)/SiC composites as received is 10–27 MPam$^{1/2}$ which is about one order of magnitude higher than that of monolithic SiC. The SiC_f/SiC composites prepared with stoichiometric SiC fibers show high strength and thermal stability in vacuum at temperatures up to 2200 K. The mechanical behavior of the SiC_f(CG)/SiC composite following neutron irradiation has also been investigated by bending tests. Degradation of the mechanical properties continues up to doses of 5 dpa, after which only minimal losses occur up to 80 dpa.

The development of irradiation-resistant CMCs emerges as the main objective for future studies. Enhanced irradiation resistance will depend on the availability of stoichiometric SiC and on avoiding the presence of glassy/amorphous phases. The progressive availability of advanced fibers with high thermomechanical stability will require deep engagement of the industry in the technological development of CMCs (Fenici *et al.* 1998).

See also: Fusion Reactors (Magnetically Confined)–Tokamaks: Materials; Fusion Conditions: Radiation Damage Correlation; Nuclear Reactor Materials: Irradiation Effects; Radiation Damage Theory

Bibliography

Abe F, Noda T, Araki H, Nakazawa S 1992 High temperature strength of simple and solute-modified 10Cr–30Mn austenitic steels. *J. Nucl. Mater.* **191–4**, 668–71

Abe F, Noda T, Araki H, Okada M 1994 Development of reduced-activation martensitic 9Cr steels for fusion reactors. *J. Nucl. Sci. Technol.* **31**, 279–92

Araki H, Suzuki H, Yang W, Sato S, Noda T 1998 Effect of high temperature heat treatment in vacuum on microstructure and bending properties of SiC_f/SiC composites prepared by CVI. *J. Nucl. Mater.* **258–63**, 1540–5

Fenici P, Frias Rebelo A J, Jones R H, Kohyama A, Snead L L 1998 Current status of SiC/SiC composites R&D. *J. Nucl. Mater.* **258–63**, 215–25

Hishinuma A, Kohyama A, Klueh R L, Gelles D S, Dietz W, Ehrlich K 1998 Current status and future R&D for reduced-activation ferritic/martensitic steels. *J. Nucl. Mater.* **258–63**, 193–204

Klueh R L, Alexander D J, Maziasz P J 1992 Impact behavior of reduced-activation ferritic steels irradiated in the fast flux test facility. *J. Nucl. Mater.* **186**, 185–95

Noda T, Araki H, Suzuki H, Abe F, Okada M 1994 Impurities and evaluation of induced activity of SiC composites prepared with chemical vapor infiltration. *J. Nucl. Sci. Technol.* **31**, 1059–65

Okada M, Noda T, Abe F 1989 On the development of low-activation materials for fusion reactors. *J. Nucl. Mater.* **169**, 249–56

Zinkle S J, Matsui H, Smith D L, Rowcliffe A F, Van Osch E, Abe K, Kazakov V A 1998 Research and development on vanadium alloys for fusion applications. *J. Nucl. Mater.* **258–63**, 205–14

G

GaN-based Optoelectronic Devices

D. P. Bour

The development of highly efficient and visible light emitting diodes (LEDs) from nitride semiconductors has had a tremendous impact on many important systems technologies. For example, the efficiency of InGaN green LEDs (like those used in traffic signals) now exceeds that of traditional incandescent lamps; and it is rapidly approaching that of fluorescent lamps. Furthermore, blue and green nitride LEDs are now the basis of bright, full-color displays, when combined with existing red LEDs (fabricated from arsenide or phosphide semiconductors). In this application, the efficiency and color purity of the LEDs permit a very broad gamut of colors to be mixed, spanning a substantial portion of all perceived colors. Moreover, since white light can be generated through such color mixing, or by combining UV-blue LEDs with luminescence-conversion phosphors, LEDs are now also being considered for general illumination and white-lighting.

Similarly, lasers of these primary colors may also be incorporated in full-color film printers and projection displays. However, the primary motivation for developing inexpensive, compact nitride semiconductor laser diodes is optical data storage, where a short wavelength translates into a small focused spot size, as required for maximizing the density and transfer rate of stored data. Currently available DVD–ROM systems use red (650nm) semiconductor lasers to increase the storage density compared to traditional near-infrared (780nm) CD systems. Converting these systems to violet lasers ($\lambda \sim 400$nm) would dramatically enhance performance, leading to capacity > 10 Gbyte for a single DVD disk. High-resolution printing enjoys a similar advantage from short-wavelength lasers.

The wavelength ranges spanned by LEDs and laser diodes fabricated from various compound semiconductor material systems are shown in Fig. 1 Nitride laser diodes and LEDs were pioneered by Dr. Nakamura and co-workers at Nichia Chemical Industries, Japan (Nakamura *et al.* 1995, 1996a, 1996b, 1998, 1999, 2000, Nakamura and Fasol 1997). Prior to this, the only blue LEDs available were made of silicon carbide; but these are only about 1% as bright as the newer nitride devices. Likewise, the shortest wavelength laser diodes available were fabricated from II–VI compound semiconductors. Although these offered excellent performance for green wavelengths, they suffered rapid degradation during operation, with lifetime limited to a few hundred hours. However, since nitride lasers were invented, blue laser lifetimes exceeding 10000h have been demonstrated. These performance characteristics are suitable for their incorporation into DVD–ROM systems; and still higher power nitride lasers are evolving, for DVD-recordable systems, projection displays, and high-speed, high-resolution laser printers. The range of available wavelengths is also being extended beyond the near UV or violet, for applications such as displays and spectroscopy.

1. The Nitride Material System

The properties of group III nitride semiconductors were first investigated in the 1970s. Among the most influential of these early investigators was Jacques Pankove, of the RCA Laboratories (Princeton, NJ, USA), who recognized that the large, direct bandgap of nitride semiconductors made them ideal candidates for visible and ultraviolet light emitters. At RCA, scientists conducted epitaxial GaN film depositions on sapphire substrates, by the chloride vapor transport technique (Pankove *et al.* 1971). They were successful at achieving material capable of efficient blue and UV luminescence, and made metal-insulator-semiconductor LEDs from their GaN layers.

Further progress toward GaN semiconductor devices ensued in the 1980s, when Professors Isamu Akasaki and Hiroshi Amano proved that the growth

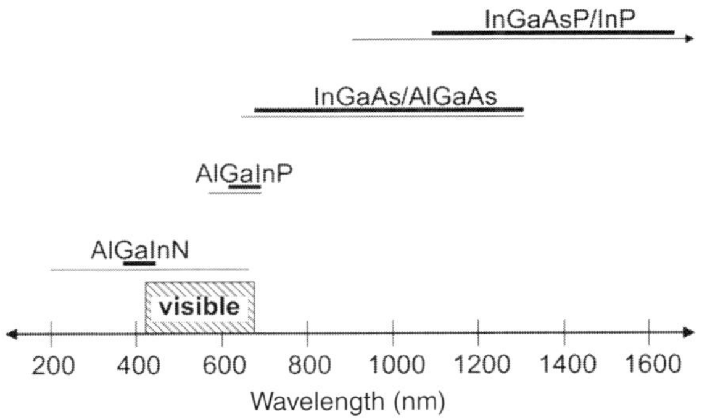

Figure 1
The wavelength range spanned by LEDs (narrow lines) and laser diodes (bold lines), for various compound semiconductor material systems.

of smooth GaN films was possible, using metalorganic chemical vapor deposition (MOCVD). In particular, they incorporated a very thin, amorphous buffer layer of AlN at the sapphire/GaN epitaxial interface (Akasaki and Amano 1994). Akasaki and Amano were also the first to demonstrate *p*-type doping of GaN films, using magnesium as an acceptor species. Subsequent to establishing the capability of *p*-type doping, they proceeded to first demonstrate blue GaN and ultraviolet AlGaN *p–n* junction LEDs. Their persistent efforts were thus the basis for the subsequent great success with nitride laser diodes and LEDs.

An additional key element of nitride laser diode development was InGaN alloy growth. In 1993, using InGaN heterostructures, Nichia introduced blue LEDs, followed shortly thereafter by green diodes (Nakamura and Mukai 1992). Finally, in 1995 and 1996, respectively, Shuji Nakamura achieved pulsed operation, and then continuous-wave (c.w.) operation of nitride laser diodes (Nakamura *et al.* 1995, 1996a, 1996b, Nakamura and Fasol 1997). Since then, this research group has dominated nitride laser diode and LED development, becoming the first to manufacture large volumes of these devices (Nakamura *et al.* 1998, 1999, 2000).

The nitride semiconductors have the hexagonal (wurtzite) crystal structure. The general properties of the binary nitride compound semiconductors InN, GaN, and AlN are summarized in Table 1. All of these nitride binaries are direct bandgap semiconductors, therefore, capable of supporting efficient radiative recombination. In particular, the bandgap energy of the InGaN alloy spans the entire visible spectrum, from red ($\lambda \sim 650$nm) for InN, into the near-UV ($\lambda \sim 365$nm) for GaN, as shown in Fig. 2 This demonstrates a plot of the bandgap energies as a function for the "a" lattice parameter (i.e., in the basal plane) of their hexagonal unit cell. The target wavelength for nitride laser diodes for DVD applications is 395–415nm. This corresponds to a moderate indium alloy content of about 10% ($In_{0.1}Ga_{0.9}N$). On the other hand, the InGaN active region of nitride LEDs may incorporate much higher In content for emission in the blue (20–30% In), green (40–50% In), or yellow spectrum.

AlGaInN lasers and LEDs are most commonly grown by MOCVD on sapphire (Al_2O_3) substrates. In order to achieve smooth films, and to accommodate the large lattice mismatch between the nitride semiconductor films and the sapphire substrate, growth commences with a thin, amorphous buffer layer of either GaN or AlN (Akasaki and Amano 1994, Nakamura and Fasol 1997). Upon heating, this layer undergoes a solid-phase crystallization, forming a dense concentration of crystallites, each in epitaxial alignment with respect to the sapphire substrate. These crystallites then serve as seed crystals, upon which thick, epitaxial films may be nucleated. To promote two-dimensional film growth, the deposition temperature is typically very high (900–1100°C).

Semiconductor LEDs and laser diodes are single-crystal, thin-film devices, for which a common crystal lattice structure underlies the entire multiple-layer heterostructure. This crystal structure is established by the seed crystal upon which the layers are epitaxially deposited. Typically, this seed crystal is a polished, 2-or 3-inch diameter wafer sliced from a large, single-crystal boule. Among the requirements for an ideal seed crystal-substrate are: (i) an appropriate lattice structure for epitaxial deposition; and (ii) chemical inertness at high temperature and in an ambient of reactive gases.

Ideally, nitride lasers would be deposited on bulk GaN substrate wafers, to minimize lattice mismatch. Indeed, in the case of arsenide and phosphide semiconductors (such as those used in infrared or red

Table 1
Basic physical properties of the nitride semiconductors, along with those of common substrate materials (sapphire and silicon carbide).

		InN	GaN	AlN	Al_2O_3	6H-SiC
Bandgap energy (eV)		1.9	3.4	6.2		3.0
Lattice parameters (Å)	a	3.55	3.19	3.11	4.76	3.08
	c	5.76	5.18	4.98	12.99	15.12
Thermal expansion	a	4	5.6	4.2	7.5	4.2
(10^{-6} K^{-1})	c	3	3.2	5.3	8.5	4.7
Refractive index ($\lambda = 400$nm)		2.8	2.6	2.2	1.8	2.8
m_{hh}^*/m_o			1			
m_e^*/m_o		0.11	0.2			
Thermal conductivity ($Wcm^{-1}K^{-1}$)		0.8	1.3	2	0.5	4.9
Valence band offset (eV, w.r.t. GaN)		0.3		0.7		

LEDs and laser diodes), a perfectly lattice-matched substrate wafer is essential. Otherwise, if a lattice-mismatched substrate were used, the epitaxial films would be heavily laden with dislocations that would render the material optically dead. For ideal nitride semiconductor device substrates, however, true bulk GaN crystals have been very difficult to produce, because of the extremely high nitrogen pressures involved in their growth. Thus, as shown in Fig. 2, there is not (yet) a commonly available single-crystal substrate wafer for nitride epitaxy. The most common substrates are sapphire (Al_2O_3) and silicon carbide (SiC), with lattice mismatches of 14.0 and 3.5%, respectively. Despite its poor lattice match, sapphire has thus become the most common substrate for nitride laser and LED epitaxy, primarily because of its low cost and refractory nature. While c-face (basal plane, or (0001)-orientation) sapphire is most often used, excellent device performance has also been obtained for growth on a-face sapphire substrates.

As a consequence of the lattice mismatch between GaN and sapphire, GaN films exhibit a very large density of dislocations, of density 1×10^{8}–$1\times10^{10}\,cm^{-2}$. Transmission electron microscopy (TEM) and x-ray diffraction have revealed that GaN grows epitaxially with a (0001)-oriented columnar structure. To accommodate lattice mismatch, there are slight twists or tilts between the columns; and the boundaries between the columns correspond to the dislocations. This is illustrated by Fig. 3, a TEM micrograph of a nitride LED, showing the high density of threading dislocations that originate at the GaN/sapphire interface and thread through the entire film and heterostructure (Ponce and Bour 1997).

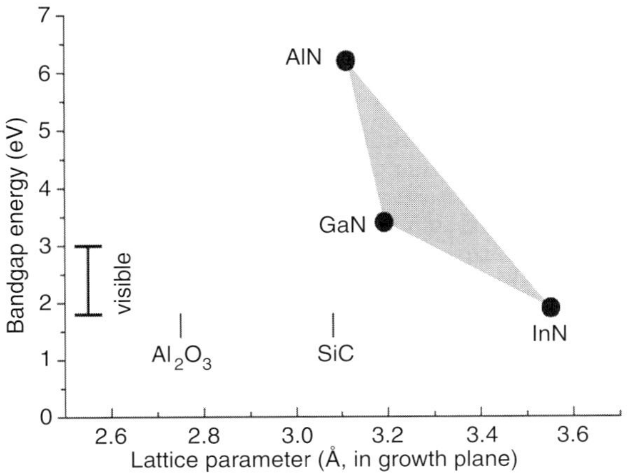

Figure 2
Bandgap energy vs. lattice parameter, for the nitride semiconductor material system. The lattice parameter of the two most common substrates, sapphire and silicon-carbide are also indicated.

The electrical character of these extended defects is still being debated. Compared to other III–V LED materials, the nitrides are remarkably insensitive to dislocations. While the device lifetimes and radiative efficiencies of other materials are seriously degraded by such a high defect density, the internal quantum efficiency of InGaN LEDs remains peculiarly high. This benign nature of dislocations in GaN is still a mystery. Nevertheless, there is also some contrary evidence that dislocations are responsible for a state within the bandgap that gives rise to nonradiative recombination e.g., performance improvements have been reported, for devices grown on low-defect-density material. Likewise, the luminescence efficiency of InGaN is superior to that of GaN, perhaps as a consequence of carrier localization induced by potential fluctuations, which suppresses defect-related recombination (Deguchi *et al.* 1999). Altogether, their unusually high luminescence efficiency, in spite of their high defect density, is a very unusual aspect of the nitrides.

2. AlGaInN Light-emitting Diodes

The layer structure comprising the simplest nitride heterostructure LEDs includes an InGaN quantum well (QW) active region (either a single QW or multiple QWs), surrounded by GaN (or AlGaN). Furthermore, the GaN is doped with magnesium (for *p*-type conductivity) on one side of the active region, and silicon (for *n*-type conductivity) on the opposite side. Thus, the QW active region lies precisely at the *p–n* junction of a diode. When forward biased, electrons are injected across the junction from the n-side, while holes are injected across the junction from the p-side. However, these injected electrons and holes are energetically confined in the InGaN QW(s), because of their lower bandgap energy with respect to GaN. The subsequent recombination of electrons and holes in the InGaN produces visible light whose photon energy equals the bandgap energy of the InGaN. Consequently, the LED color may be controlled by adjusting the composition of the InGaN active region.

This mechanism of LED operation may be understood with the energy-band diagram shown in Fig. 4. This shows how the conduction band energy (E_c), valence band energy (E_v) and Fermi energy (E_f) vary through the device structure. On the left is the band structure at zero bias, while on the right is the case for a positive-voltage bias (+3V, applied to the p-side of the diode). In this situation, the band edges have flattened, as the bias overcomes the field associated with the p–n junction. It is also apparent that the injected carriers are energetically confined in the InGaN, where they eventually recombine. At the junction, the applied voltage ($V_{applied}$) corresponds to the difference between the quasi-fermi levels defining the electron and hole distributions.

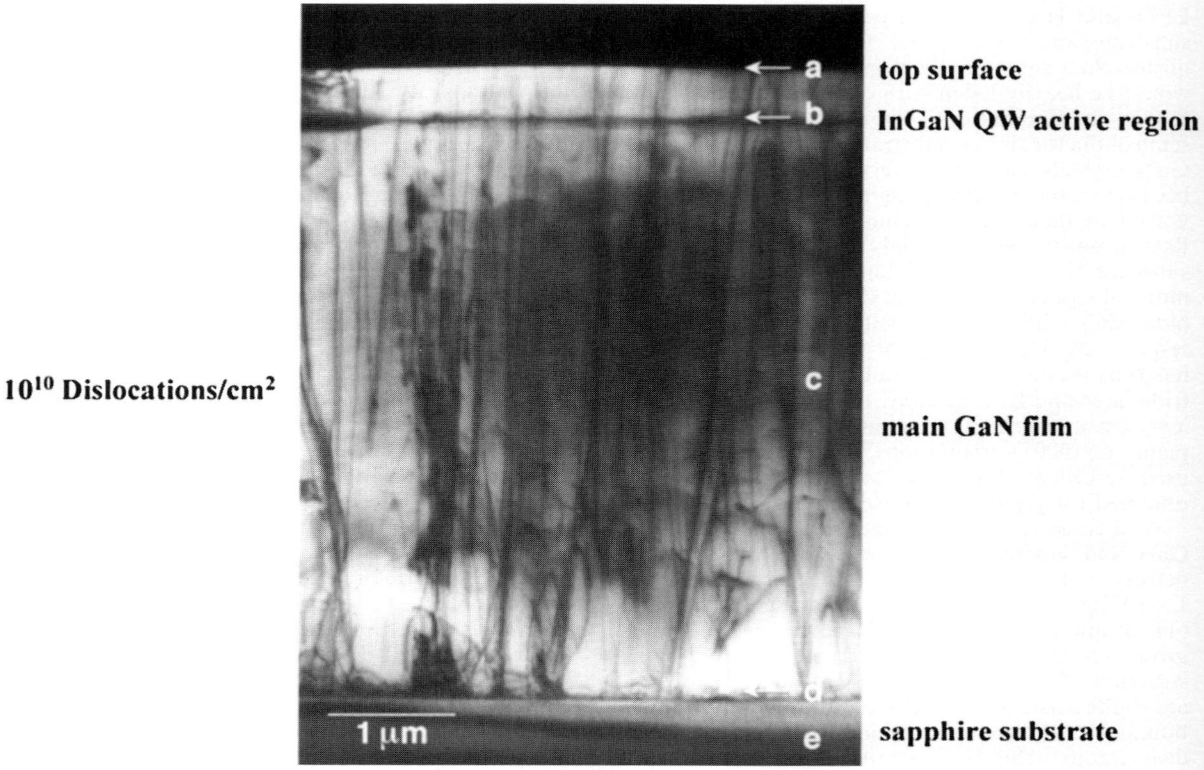

Figure 3
TEM of a nitride LED, showing the highly dislocated structure.

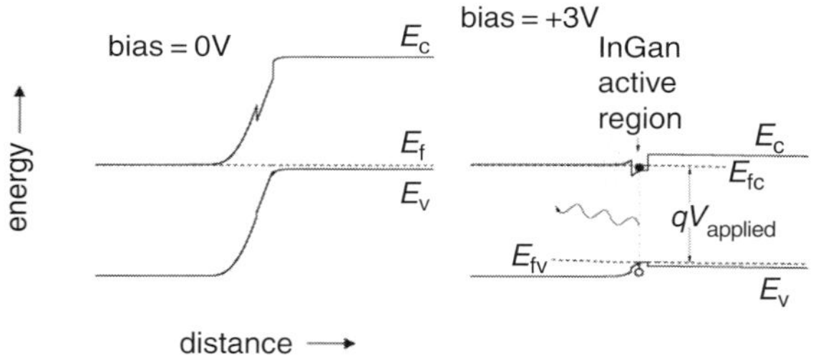

Figure 4
Energy band diagram illustrating the operation of a typical heterostructure LED.

A nitride LED chip is shown schematically in Fig. 5. Typically, the chip is 200–500µm square. The top surface is p-type material, which is contacted with a semi-transparent metal. A very thin nickel-gold metal is often used for this purpose. Because the *p*-type layers in an InGaN LED are fairly resistive, such a metal film is necessary to spread the current uniformly across the chip area. The *n*-contact is established after etching down to expose the n-type material, at one corner of the chip. This chip is mounted in a reflector cup that redirects the edge emission upwards. This assembly is subsequently embedded in a transparent epoxy dome that defines the emission pattern.

In early investigations of InGaN QWs and LEDs, a variety of intriguing characteristics were identified. Specifically, a sizable spectral shift was detected, between the electroluminescence emission wavelength and the absorption edge wavelength (Takeuchi *et al.* 1998, Chichibu *et al.* 1998, 1999). Typically, the emission and absorption-esdge energies are expected to approximately coincide, representing the bandgap energy of the active region. However, in InGaN LEDs the emission energy was appreciably lower than the absorption edge energy, by several hundred meV. This energy difference, often labeled a "Stokes shift," was first understood to be evidence for InGaN alloy segregation. Such a shift indicated that InGaN does not grow as an ideal, homogeneous alloy, but rather undergoes some compositional segregation during growth. More recently, the understanding of this complicated behavior has evolved further, with

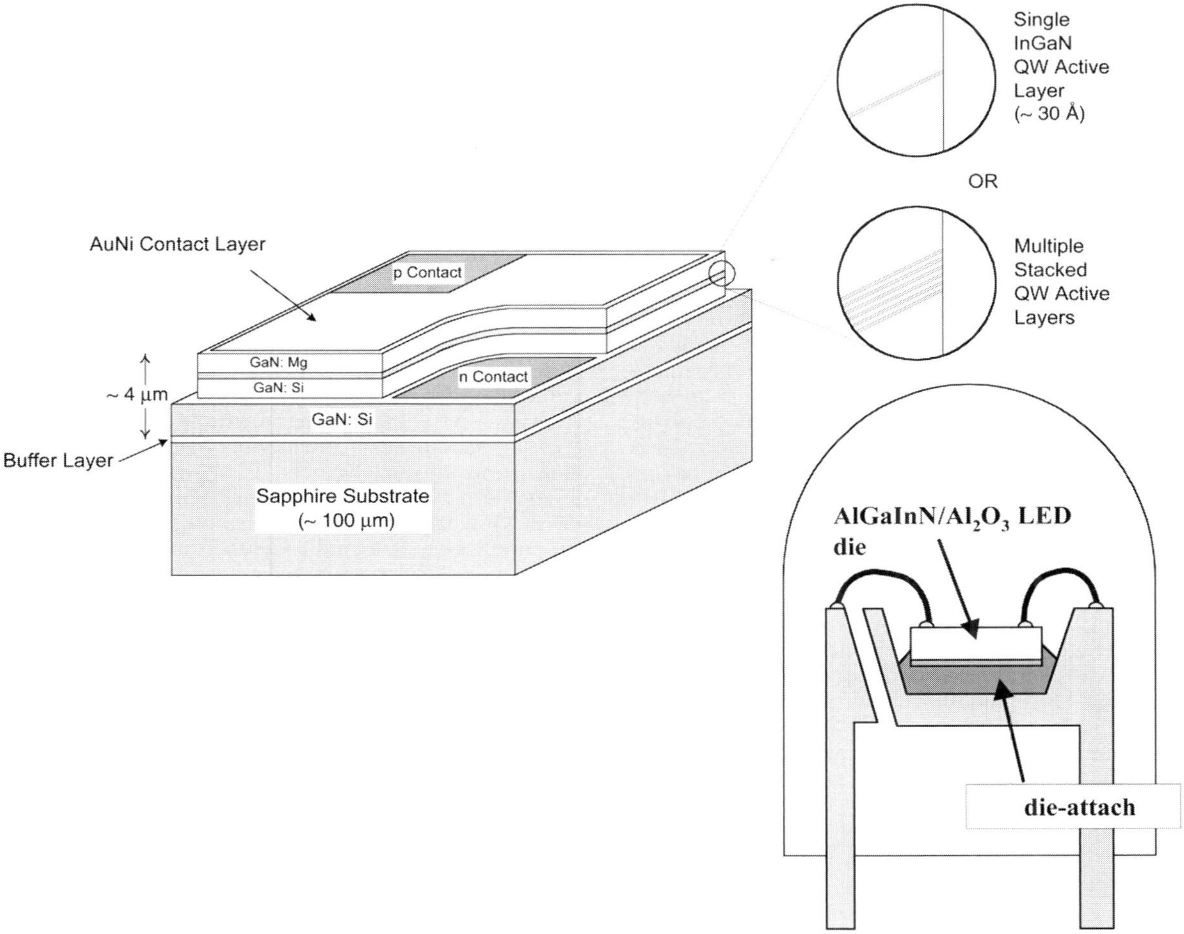

Figure 5
Schematic diagram of a nitride LED chip. The QW active region is shown expanded in the inset.

the recognition that polarization fields (involving both spontaneous and/or piezoelectric components) play a dominant role.

InGaN alloy segregation offers a convenient explanation for the Stokes shift observed for InGaN emitters. First, InGaN alloy segregation is not unexpected, as instabilities often occur on mixing of binary materials with great differences in lattice parameters. Therefore, as a result of the alloy's thermodynamic instability, InGaN would be likely to separate into regions of different alloy content, producing an inhomogeneous composition. Volumes of high-In-content material would correspond to quantum dots, because of their lower bandgap energy and small volume, which offer quantum confinement in three dimensions. Injected carriers would thus be energetically confined to these regions of local potential minima. Consequently, at low injected carrier density, the electroluminescence would represent recombination of carriers localized to these regions. On the other hand, because their volume is so small, these bandtail states would not contribute significantly to the absorption spectrum. Hence, the observed Stokes shift between emission and absorption. Furthermore, the shift would be greater for high-In-content alloys, simply because the magnitude of the composition fluctuations would be greater. For instance, the emission energy may be lower than the absorption edge by ~500meV for green LEDs with $In_{0.4}Ga_{0.6}N$ active regions, compared to only ~100meV for violet-emitting strusctures with $In_{0.1}Ga_{0.9}N$ active regions (Chichibu *et al.* 1999).

A further observation which was initially explained by alloy fluctuations was the relatively strong wavelength shift (with respect to current) observed for nitride LEDs. In this case, the emission wavelength at low currents was red-shifted, corresponding to recombination of carriers localized at potential minima. At increasingly high current densities, the emission wavelength became shorter, as a result of these localized states becoming filled with carriers. This bandfilling effect was especially pronounced for high-In-content QWs.

Another consequence of alloy fluctuations is broadening of the emission spectra, as shown in Fig. 6. To achieve longer wavelengths, the In content of the QW must be increased. However, the absolute magnitude of the composition inhomogeneity is also increased in this situation, producing the broadened spectra shown in Fig. 6. Such broad emission is an undesirable characteristic for the material comprising a laser active region, because it translates into a similarly broad gain spectrum, since generally, a narrow electroluminescence emission spectrum transforms into a narrow gain spectrum. The peak gain is, therefore, greater than that available from a structure whose luminescence spectrum is broader. This points out the importance of material with a characteristic narrow emission linewidth for laser active regions. As the In content is increased to achieve longer wavelength lasers, the increased spectral broadening suggests that threshold currents will likewise become higher.

Despite these problems resulting from InGaN alloy fluctuations, there is substantial evidence that these alloy fluctuations also perform an extremely important role in improving the performance of nitride lasers and LEDs. Namely, because the nitrides are highly dislocated materials, the potential fluctuations may suppress nonradiative recombination via states associated with defects such as dislocations, thereby yielding greater radiative recombination efficiency (Deguchi 1999). This mechanism is illustrated in Fig. 7, which compares band diagrams for a homogeneous alloy (a) and an alloy exhibiting potential fluctuations (b). Dislocations are represented by a nonradiative midgap state. For a homogeneous alloy, injected carriers have extended wavefunctions, and are, therefore, free to recombine through the dislocation-related states. In contrast, potential fluctuations lead to carrier localization. As long as the carrier localization occurs on a scale with dimension less than the spacing between dislocations, the nonradiative recombination will be suppressed.

There exists much experimental evidence that potential fluctuations in InGaN play a beneficial role by localizing injected carriers. First, the radiative efficiency of InGaN QWs is far greater than for devices with GaN active regions. Furthermore, there is no apparent benefit to reducing the dislocation density (e.g., by lateral epitaxial overgrowth) of InGaN QW LEDs. This is not the case for UV GaN QW LEDs, however, for which potential fluctuations cannot exist. This implies that dislocation-related recombination is suppressed in InGaN, presumably by alloy potential fluctuations. These observations, therefore, indicate that localization performs an important role in improving the radiative efficiency of InGaN alloys. Presumably, this benefit accrues from suppression of rapid, nonradiative recombination via extended crystal defects.

In laser diodes, however, where the carrier densities associated with lasing are 1–2 orders of magnitude greater than in LEDs, the localized (bandtail) states may be completely filled with injected carriers, so that many of the injected carriers are no longer localized. Consequently, recombination via dislocation states again becomes important in InGaN

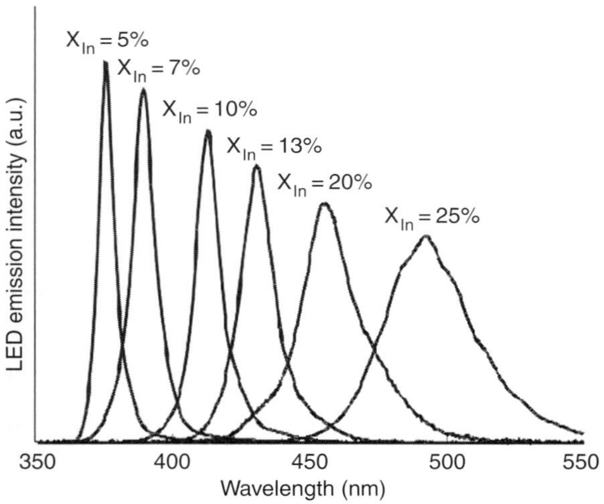

Figure 6
Electroluminescence spectra of InGaN QW LEDs with various indium-alloy-content in the QW, showing spectral broadening at long wavelengths (high In).

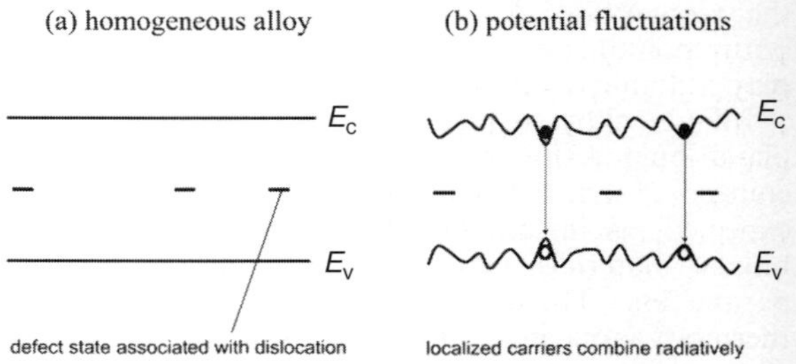

Figure 7
Energy band diagrams comparing the case of a perfectly homogeneous alloy (a), with a segregated alloy (b), for which the alloy composition variations translate into potential fluctuations that confine (localize) electrons and holes.

multiple quantum well (MQW) lasers. Thus, low-defect-density materials (e.g., obtained by lateral epitaxial overgrowth) are particularly important for improving the performance of nitride laser diodes; but not necessarily such a great benefit for nitride LEDs.

Compared to LEDs fabricated from other materials, the strong color tuning (with current and QW thickness) and the Stokes shift are peculiar, and can be explained qualitatively by the presence of alloy segregation. These interesting behaviors relate to the fundamental mechanisms of spontaneous emission, which is still the subject of much investigation. However, it has been more recently recognized that much of the complex behavior of InGaN LEDs arises to a significant degree from the piezoelectric field in InGaN (Chichibu *et al.* 1998, Takeuchi *et al.* s1998). The piezoelectric field arises from the severe lattice mismatch between GaN and InGaN, such that the InGaN experiences a large compressive strain. The group-III nitrides have particularly large piezoelectric constants, so that strained layers develop an appreciable internal electrostatic field. This is especially true for the InGaN QW used for LED or laser diode active regions, for which fields on the order of MV cm^{-1} are possible. Consequently, there exists strong evidence that piezoelectric fields are partially responsible for these spectral shifts. Nevertheless, alloy fluctuations still play a role, as they are the most likely source of the spectral broadening and high radiative efficiency in III-nitride emitters.

The band structure of an InGaN/GaN QW is shown schematically in Fig. 8. The piezoelectric field introduces a severe tilt to the QW's band edges. As a result, the electron and hole wavefunctions (represented by the shaded regions in Fig. 8(a)) are concentrated on opposite sides of the QW. The spatial overlap between these wavefunctions is, therefore, reduced by the piezoelectric field, leading to a proportional reduction in the transition matrix element associated with spontaneous and stimulated emission. As the well thickness increases, therefore, the spatial overlap between the electron and hole wavefunctions decreases. This impacts the recombination dynamics in InGaN QWs. For instance, the radiative lifetime becomes longer for thick QWs. In the presence of a nonradiative recombination pathway (such as through a midgap state associated with dislocations), the radiative efficiency is also decreased by the piezoelectric-field-induced separation of electrons and holes.

A further consequence of the piezoelectric field is a red-shift of the emission wavelength, since the ground-state transition energy is lowered by the field. Furthermore, the band-to-band absorption at the emission wavelength is also weak, because of the poor overlap between electron and hole wavefunctions. A strong absorption edge still exists, but it is displaced to a higher energy (relative to the ground-state transition) by the presence of the field. Thus, the piezoelectric field offers a consistent explanation of the Stokes-like shift between absorption and emission. However, injected charge will screen the piezoelectric field. Subsequently, the emission wavelength demonstrates a shift to shorter wavelength, as the QW carrier density is increased. This effect is illustrated in (b) and (c) of Fig. 8. The short-wavelength shift, associated with screening the piezoelectric field via charge injection, may, therefore, appear similar to a bandfilling effect. In very-high-In-content QW LEDs, this shift can be quite pronounced, leading to drastic color changes with the variation of injection current.

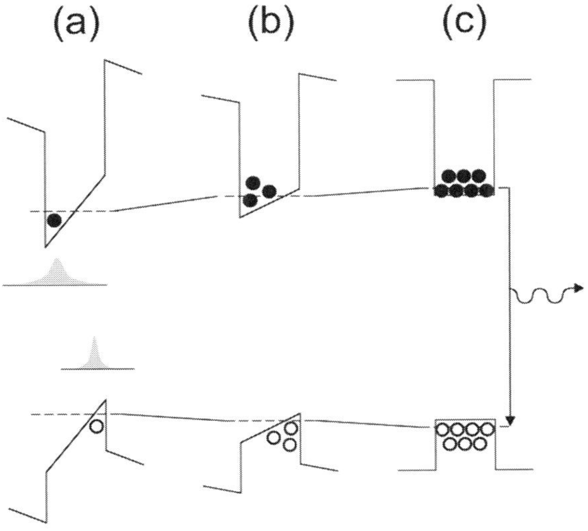

Figure 8
Energy band diagram of an InGaN QW, showing the effect of a very high piezoelectric field. At low injection current (a), the recombination is red-shifted and inefficient, because of the displacement between the electron and hole distributions. As the injected carrier density is increased (b and c), the piezoelectric field is screened, so that the recombination becomes more efficient and it shifts to a shorter wavelength.

3. *AlGaInN-based Laser Diodes*

A typical nitride laser structure is shown schematically in Fig. 9. Nitride lasers differ from more common III–V lasers in a number of important respects. First, they are most commonly deposited on sapphire substrates. Consequently, because sapphire is insulating, a lateral n-contact is required. In contrast, arsenide and phosphide laser diodes have the *n*-electrode on the backside of their conducting substrate, representing a vertical contact scheme. To ensure a low-resistance electron-current path, the nitride laser

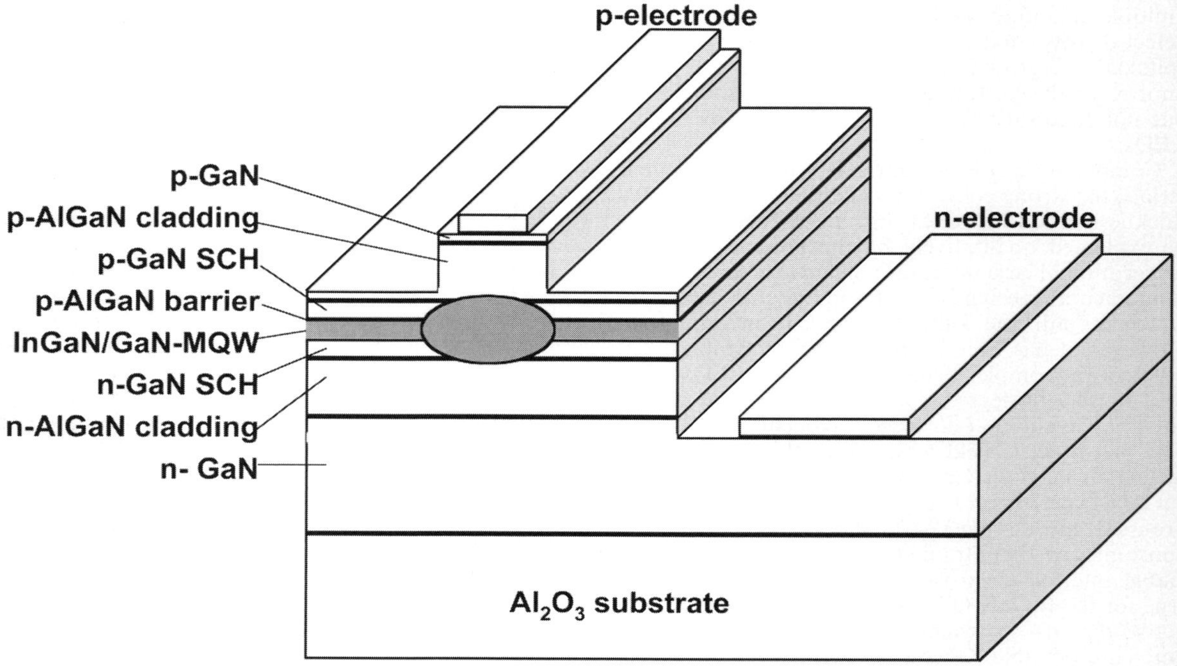

Figure 9
Typical nitride laser heterostructure grown on a sapphire substrate.

heterostructure is, therefore, deposited over a thick (several μm), *n*-type Si-doped GaN layer. Alternatively, a vertical contact structure may be realized if nitride lasers are deposited on SiC substrates, which may be made conductive.

AlGaN layers comprise the cladding layers of the nitride laser. Unlike the case of other III–V lasers, these claddings layers are not lattice matched with respect to the surrounding (largely GaN) structure. Instead, they experience biaxial tension, and so are prone to cracking. Thus, the thickness and Al content of the AlGaN cladding layers is limited.

In optimized nitride laser structures (such as described by Nakamura *et al.* 1996a, 1996b, 1998, 1999, 2000), the cladding layer composition is typically $Al_{0.07}Ga_{0.93}N$. For this composition, the thickness of bulk AlGaN cladding layers is limited to about 0.5–0.7 μm. Unfortunately, this does not offer sufficient optical confinement. In this case, the transverse beam pattern exhibits a strong modulation due to operation in a high-order mode of the entire epitaxial layer structure. It was not until thicker (~1.0 μm), short-period, superlattice cladding layers were developed, that the transverse beam profile became sufficiently fundamental-mode-like (Nakamura *et al.* 1999). With the superlattice replacing bulk AlGaN, the cracking threshold was increased nearly twofold, thereby permitting thicker cladding layers for more complete mode confinement.

The nitride laser active region contains multiple InGaN QWs, with the optimum number of QWs determined by the threshold gain requirement. In the first demonstration of a nitride lasers, Nakumura and colleagues incorporated 26 QWs, an unusually large number (Nakamura *et al.* 1996a). Since then, as material quality has improved and optical loss has been reduced, these structures have evolved to include only 1–4 QWs (Nakamura *et al.* 1998, 1999, 2000). The barriers between the wells are most often silicon-doped InGaN of low In content; but GaN barriers are also possible. The QWs are embedded within a GaN waveguiding region, or separate confinement heterostructure (SCH). The thickness of the *n*- and *p*-waveguides are adjusted so that there is maximum spatial overlap (Γ) between the optical mode and the QW gain. For typical structures with $Al_{0.07}Ga_{0.93}N$ cladding layers, and 1–5 $In_{0.1}Ga_{0.9}N$ QWs, the optimum waveguide layer thickness are ~0.1μm.

Nitride laser mirrors have been fabricated by either dry-etching or cleaving. Etching techniques employed include chemically assisted ion-beam etching (CAIBE), reactive-ion etching (RIE), or focused-ion-beam etching (FIBE). Of these FIBE produces the most ideal surface; but it is a slow technique that is not suitable for production. With respect to cleaving, alternative substrates such as GaN or SiC offer a tremendous advantage. Of the many polytypes of SiC, both 6H and 4H are available in bulk, single-crystal wafers and have been demonstrated as viable substrates for nitride lasers with cleaved mirrors. Similarly, pseudo-bulk GaN wafers have been prepared by HVPE growth of very thick GaN layers on sapphire substrates, followed by substrate removal via either polishing or UV-laser liftoff. For nitride lasers grown on either SiC or GaN, cleaving mirror facets is straightforward, and analogous to the well-developed procedures applied to laser diodes fabricated from other materials. Likewise, these alternative substrates are also much better thermal conductors than sapphire, thereby enabling c.w., high-power, and long-lived operation of nitride laser diodes.

Most optical systems demand lasers that emit in a single spatial mode. While the transverse spatial mode is governed by the epitaxial layer structure, lateral mode control depends upon the subsequent lithographic definition of a waveguide. Most often, this is accomplished with a ridge waveguide structure (Fig. 9), whereby a lateral index step is created by etching close to the active region, outside the stripe. For the proper combination of index step (i.e., etch depth) and stripe width, single-mode operation is possible. In practice, very narrow ridges (1–4 μm) are preferable, since this results in low threshold current and also promotes lateral heat dissipation.

The c.w. light-current and voltage-current characteristics of a 2 μm × 300 μm ridge waveguide nitride laser diode with reflection-coated mirrors are shown in Fig. 10. For c.w. operation, thermal management is essential, due to the heating associated with the high operating current density (~ 10 kA cm^{-2}). Heat dissipation may be facilitated by soldering the laser chip to a heatsink, preferably in a junction-down configuration. However, junction-down mounting is difficult, so often the diodes are mounted junction-side up, after thinning the sapphire substrate to ~ 100μm thickness. In terms of heat dissipation, the high thermal conductivity of SiC offers a tremendous advantage over sapphire.

Apparent in Fig. 10 is a pulsed threshold current of 47mA; but the c.w. threshold is 60mA, due to the heating that accompanies continuous current injection. Likewise, the differential quantum efficiency is lower for c.w. operation, and the c.w. power output exhibits a thermal rollover. The threshold voltage is 7.3V. Aside from the 3.1V junction voltage, the next largest component of the overall diode voltage at threshold arises from the *p*-contact and *p*-layer's bulk resistance.

The best performance characteristics of nitride lasers as excellent as those of red or infrared lasers. For instance, threshold currents as low as 16 mA,

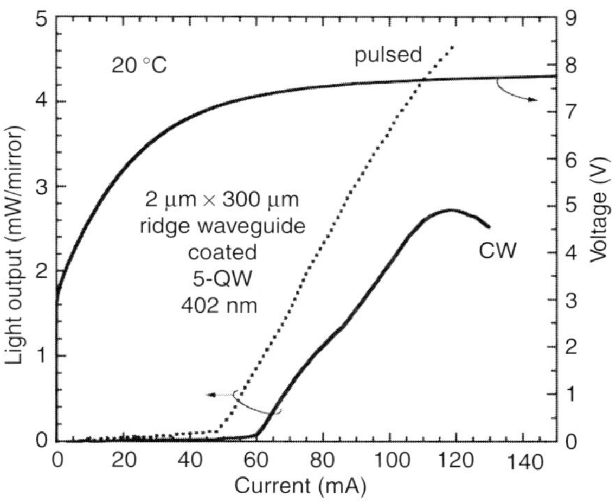

Figure 10
Room temperature, c.w. light-current and voltage-current characteristics of a 2μm × 300μm ridge waveguide nitride laser diode with reflection-coated mirrors. The dotted curve shows the pulsed light output–current curve for the same diode.

threshold current densities as low as 1.2 kA cm^{-2}, threshold voltage as low as 4.3 V, and lifetime exceeding 10 000 h have been achieved (Nakamura *et al.* 1998, 1999, 2000). One especially remarkable achievement is a c.w. power output of 420 mW, from a 3 μm wide ridge waveguide nitride laser diode (Nakamura *et al.* 1998). This power output corresponds to a power density exceeding 10 MW cm^{-2} at the uncoated mirror. The ability to survive such high power densities attests to the robust nature of nitride semiconductors.

The performance of nitride laser diodes is limited by the highly dislocated nature of nitride semiconductor films grown on sapphire or SiC substrates. For instance, dislocations appear to be responsible for nonradiative recombination, thereby reducing the internal quantum efficiency. Cathodoluminescence of GaN films has indicated that there exists an "optically dead zone" surrounding dislocations, in which nonradiative recombination dominates. Thus, as the dislocation density increases, the overall quantum efficiency must decrease accordingly, since the cumulative "dead" volume would increase. Similarly, dislocations act as scattering centers, increasing optical loss in the laser cavity. Finally, it is also expected that the presence of dislocations will decrease the longevity of nitride lasers, either directly (by moving to form dark-line-defects, as occurs in AlGaAs lasers; or by serving as routes for electromigration of the p-contact metal); or indirectly (by degrading the performance, so that the operating current and the temperature are higher than they would otherwise be).

Thus, among the many key developments contributing to the performance improvements of nitride lasers was reducing the dislocation density, through growth over epitaxially-laterally-overgrown (ELOG) GaN substrates. In general, such lateral overgrowth processes incorporate dielectric stripes which interrupt the dislocations, and prevent them from threading through the material overlying the dielectric mask, as shown by the sequence in Fig. 11. In ELOG, or equivalently lateral epitaxial overgrowth (LEO), an initial layer of GaN is first grown over the sapphire substrate (Hiramatsu *et al.* s1999). A dielectric mask is then deposited over this layer, and patterned into narrow stripes, e.g., a typical structure uses 100–200 nm SiO_2 mask, patterned into 8 μm stripes on 11 μm centers. This structure is then returned to the MOCVD reactor, for selective growth of GaN. If the stripes are oriented in the proper direction $\langle 1\bar{1}00 \rangle$, and the growth conditions are chosen properly, the lateral growth rate will be enhanced, perhaps by a factor of 2–4 over the vertical growth rate. Above the dielectric mask, growth is selective and, therefore, purely lateral; and if a thick enough layer is deposited, the laterally-growing material will coalesce over the center of each SiO_2 stripe.

This coalescence plane is structurally poor, consisting of voids and grain boundaries. Otherwise, however, the dislocation density over the mask stripe is decreased substantially, because defects are prevented from propagating by the mask. Consequently,

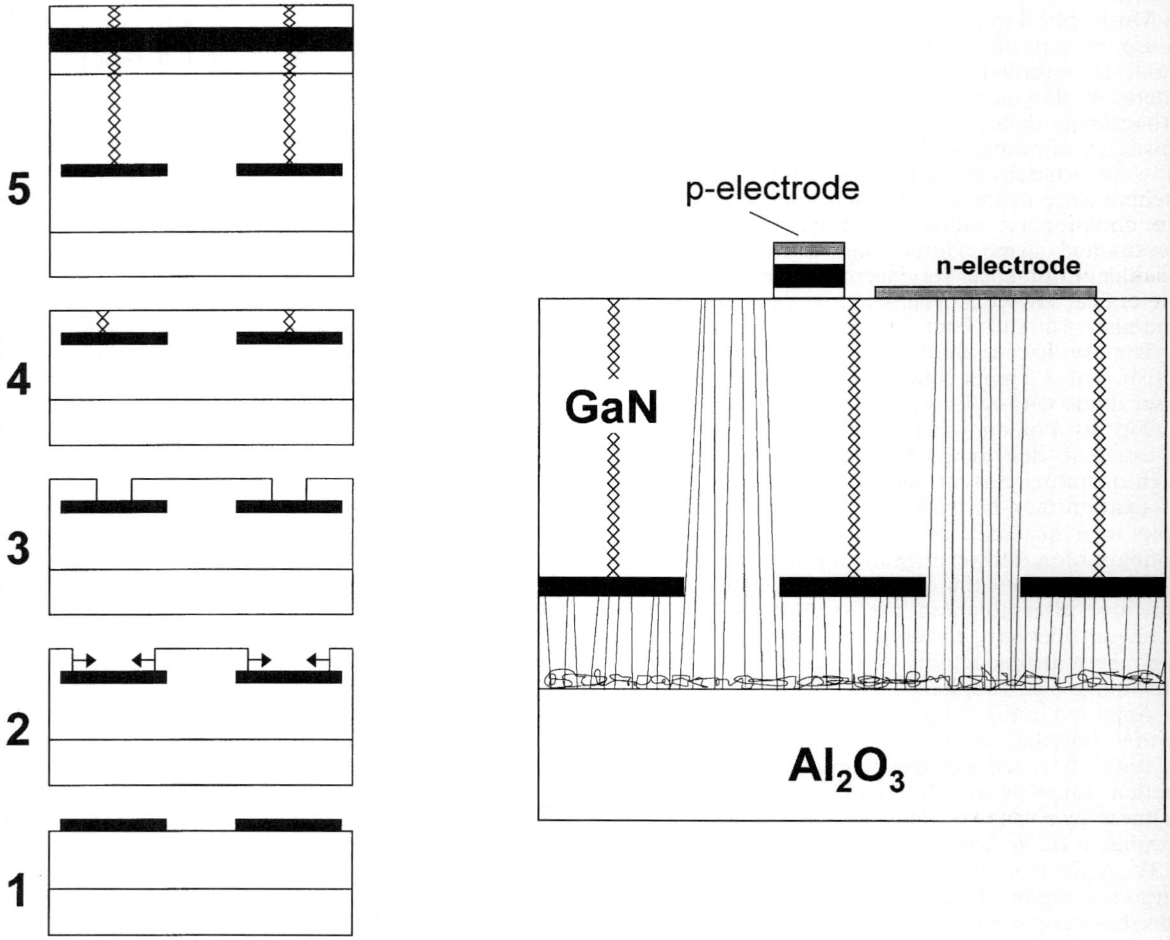

Figure 11
Schematic illustration of the LEO or ELOG sequence, for producing low-defect-density nitride semiconductor films on mismatched substrates.

for the 8μm dielectric mask stripe, there exists a region ~4 μm wide above each half of the mask stripe, that is of excellent structural quality. This width is sufficient for 2 or 3μm ridge waveguide laser fabrication. This technique drastically reduces the dislocation density over a limited area, thereby reducing both the scattering loss and nonradiative recombination. Consequently, the threshold current, quantum efficiency, and lifetime have all been improved by employing LEO to reduce the dislocation density. Lifetimes exceeding 10 000 h, for 2 mW c.w. operations at 20 °C, have also been reported for nitride lasers deposited on ELOG wafers or quasi-bulk GaN (Nakamura *et al.* 1999). Likewise, lifetimes of several thousand hours have also been demonstrated at higher power (5–30 mW) and higher temperature (up to 50 °C), through defect reduction by ELOG.

The wavelength range for nitride lasers is largely limited by the properties of the InGaN alloy active region. If the In alloy content is too high, phase separation occurs. The phase separation (segregation) may also be enhanced by the presence of dislocations. The resultant composition fluctuations lead to broadening of both the spontaneous emission and the optical gain spectra, for lower peak gain. Despite this, c.w. operation at wavelengths as long as 450nm has been accomplished; but performance characteristics are optimized at the shorter wavelength (and correspondingly lower In content) range of 400–410 nm (Nakamura *et al.* 2000). This is an ideal wavelength for optical storage; but blue and green wavelength operation is still being pursued for applications such as projection displays. At shorter wavelengths (< 390 nm), thresholds increase partly because of reduced carrier confinement. In addition, it also appears that some minimum In-alloy-content is required, in order to localize injected carriers. As stated previously, carrier localization is important in these dislocated materials, in order to suppress the nonradiative recombination associated with defect states.

The theoretical understanding of nitride laser operation is still rather immature. There appears to be a rich variety of influences, however, which are relatively unimportant in red and infrared lasers, but which become far more significant in the nitrides. For instance, since the carrier effective masses are high, the density-of-states is similarly great, and the threshold carrier density is accordingly higher. Therefore, many-body-effects such as Coulomb enhancement of the gain and spontaneous emission become important. Likewise, piezoelectric fields and alloy segregation in the InGaN QW active region are also important to consider (Peng *et al.* 1999). In addition, the multitude of dislocations threading through nitride device structures are also a unique feature whose importance is still undergoing investigation.

See also: Light Emitting Diodes

Bibliography

Akasaki I, Amano H 1994 Widegap column-III nitride semiconductors for UV/blue light-emitting devices. *J. Electrochem. Soc.* **141**, 2266–71

Chichibu S F, Abare A C, Mack M P, Minsky M S, Deguchi T, Cohen D, Kozodoy P, Fleischer S B, Keller S, Speck J S, Bowers J E, Hu E, Mishra U K, Coldren L A, DenBaars S P, Wada K, Sota T, Nakamura S 1999 Optical properties of InGaN quantum wells. *Mater. Sci. Eng.* **B59**, 298–303

Chichibu S F, Abare A C, Minsky M S, Keller S, Fleischer S B P, Sota T 1998 Effective band gap inhomogeneity and pizoelectric field in InGaN/GaN multiquantum well structures. *Appl. Phys. Lett.* **73**, 2006–8

Deguchi T, Torii K, Shimada K, Sota T, Matsuo R, Sugiyama M, Setoguchi A, Chichibu S, Nakamura S 1999 Optical properties of an InGaN active layer in ultraviolet light emitting diode. *Jpn. J. Appl. Phys.* **38**, L975–7

Hiramatsu K, Matsushima H, Shibata T, Kawaguchi Y, Sawaki N 1999 Selective area growth and epitaxial lateral overgrowth of GaN by metalorganic vapor phase epitaxy and hydride vapor phase epitaxy. *Mater. Sci. Eng.* **B59**, 104–11

Nakamura S, Fasol G 1997 The Blue Laser Diode. Springer, Berlin

Nakamura S, Mukai T 1992 High quality InGaN films grown on GaN films. *Jpn. J. Appl. Phys.* **31**, L1457–59

Nakamura S, Senoh M, Iwasa N, Nagahama S 1995 High-brightness InGaN blue, green, and yellow light-emitting diodes with quantum well structures. *Jpn. J. Appl. Phys.* **34**, L797–9L1332–5

Nakamura S, Senoh M, Nagahama S, Iwasa N, Matsushita T, Mukai T 2000 Blue InGaN-based laser diodes with an emission wavelength of 450nm. *Appl. Phys. Lett.* **76**, 22–4

Nakamura S, Senoh M, Nagahama S, Iwasa N, Yamada T, Matsushita T, Kiyoku H, Sugimoto Y 1996a InGaN-based multi-quantum-well-structure laser diodes. *Jpn. J. Appl. Phys.* **35**, L74–6

Nakamura S, Senoh M, Nagahama S, Iwasa N, Yamada T, Matsushita T, Kiyoku H, Sugimoto Y 1996b Room-temperature continuous-wave operation of InGaN multi-quantum-well-structure laser diodes. *Appl. Phys. Lett.* **69**, 4056–8

Nakamura S, Senoh M, Nagahama S, Iwasa N, Yamada T, Matsushita T, Kiyoku H, Sugimoto Y, Kozaki T, Umemoto H, Sano M, Chocho K 1998 Violet InGaN/GaN/AlGaN-based laser diodes with an output power of 420mW. *Jpn. J. Appl. Phys.* **37**, L627–9

Nakamura S, Senoh M, Nagahama S, Matsushita T, Kiyoku H, Sugimoto Y, Kozaki T, Umeoto H, Sano M, Mukai T 1999 Violet InGaN/GaN/AlGaN-based laser diodes operable at 50°C with a fundamental transverse mode. *Jpn. J. Appl. Phys.* **38**, L226–9

Pankove J I, Miller E A, Berkeyhauser J E 1971 GaN electroluminescent diodes. *RCA Rev.* **32**, 383–92

Peng L-H, Chuang C-W, Lou L-H 1999 Piezoelectric effects in the optical properties of strained InGaN quantum wells. *Appl. Phys. Lett.* **74**, 795–7

Ponce F A, Bour D P 1997 Nitride-based semiconductors for blue and green light-emitting devices. *Nature.* **386**, 351–9

Takeuchi T, Wetzel C, Yamaguchi S, Sakai H, Amano H, Akasaki I, Kaneko Y, Nakagawa S, Yamaoka Y, Yamada N 1998 Determination of piezoelectric fields in strained GaInN quantum wells using the quantum confined Stark effect. *Appl. Phys. Lett.* **73**, 1691–3

H

Heavy Water Reactor Fuel Design and Performance

F. C. Iglesias and D. H. Barber

The reliability of modern heavy water reactor fuel provides a suitable carrier for numerous fuel cycles. Given the flexibility permitted by heavy water reactor design, advanced fuel cycles can be adopted to meet the national priorities of those countries with heavy water reactors.

1. Background

"Heavy water" and "heavy hydrogen" were discovered by Harold C. Urey and co-workers in 1931. Heavy water is the oxide of deuterium, an isotope of hydrogen with one proton and one neutron in its nucleus. At the time of these discoveries, the neutron had not been discovered, thus the name "heavy hydrogen" was used for what is now called deuterium. The name "heavy water" for water containing a large proportion of deuterium oxide (D_2O) has stuck. Its chemical and thermodynamic properties are similar to those of regular "light" water.

The nuclear properties of heavy water are quite different from those of light water. In a thermal reactor, fission produces fast neutrons that must be slowed down (moderated) to increase the probability of inducing further fission. The material that does the slowing down is called a moderator. Two properties are essential in a good moderator. It must contain atoms that are light enough that elastic scattering slows down the fission neutrons in a small number of collisions and it must not absorb too many neutrons whilst doing so (Foster and Wright 1977) (see *Nuclear Reactors: Moderator and Reflector Materials*). The probability of thermal neutron capture in heavy water is much smaller than in light water (Choppin and Rydberg 1980, Appendix N). This property with its moderating ability makes it possible to create a heavy water moderated nuclear reactor that can be fueled with natural uranium. High purity graphite has a similar combination of properties and was used as the moderator for the world's first nuclear reactor, CP-1, at the University of Chicago. The first nuclear reactor outside the USA was the Canadian ZEEP (Zero Energy Experimental Pile) heavy water natural uranium reactor at Chalk River.

The natural uranium fuel cycle means that there is no requirement for uranium enrichment, which in the early years of nuclear power was usually related to nuclear weapons programs. Countries with natural uranium reactors but without uranium enrichment capability do not become dependent on a foreign supply of enriched uranium. The technology needed for heavy water production was available before World War II in countries such as Norway and Canada, due to the low cost of hydroelectric power used in the electrolytic D_2O production process. More recently, commercial heavy water has been separated by a variety of isotopic exchange techniques. The higher cost of the heavy water moderator (compared to light water) meant that high capacity factors were desirable for heavy water reactors, thus on-power refueling has been a feature of heavy water power reactors. The larger size of a heavy water reactor allowed the adoption of a pressure tube design that avoided the need to manufacture a large pressure vessel, allowing countries lacking the industrial base to produce large pressure vessels to produce nuclear reactors. This design also means that reactivity control devices do not need to be associated with the fuel assembly, thus simplifying fuel assembly design.

Three types of heavy water power reactor have been built. The CANadian Deuterium Uranium Pressurized Heavy Water Reactor (CANDU-PHWR) is a successful commercial design. Power reactors of this design have been built in Canada, India, Pakistan, South Korea, Argentina, and Romania. Two units are under construction in the People's Republic of China. The fuel assemblies are contained in circular horizontal pressure tubes. Depending on the specific design, there are up to 13 fuel bundles in each pressure tube. The pressure tube is isolated from the moderator by an annular gas-filled space between the pressure tube and a calandria tube. The annulus gas serves as thermal insulation between the hot coolant and much cooler moderator. The moderator is contained in a large horizontal calandria vessel. In more recent CANDU designs there is a limited degree of coolant boiling in some high-power channels. The heavy water coolant is pumped through the pressure tubes, where it picks up heat from the fuel, to the steam generators that transfer the heat from the heavy water to light water producing steam that drives a turbo-generator (Fig. 1, Barber and Nott 1989). The coolant passes from the steam generator to the pumps and then

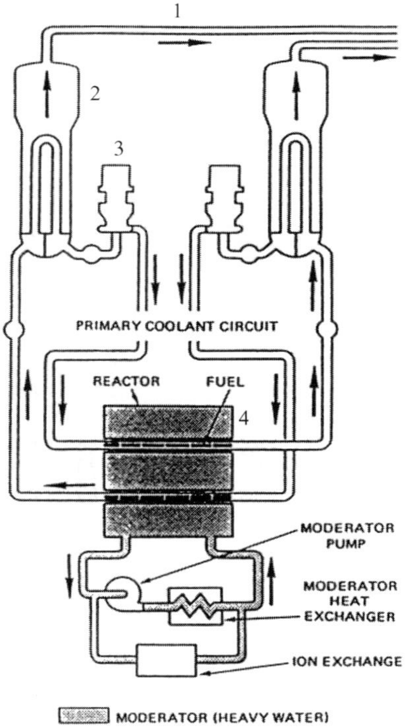

Figure 1
CANDU primary heat transport: 1. light water steam to turbogenerator, 2. steam generators, 3. coolant pumps, 4. pressure tube.

back into the core. The horizontal design and on-power refueling enable the reactor to be fueled bidirectionally. In this way, excess reactivity of fresh fuel can be partially balanced by decreased reactivity of irradiated fuel at the outlet end of an adjacent channel.

Three steam generating heavy water reactors (SGHWR) have been built. These were Gentilly 1 CANDU-BLW in Quebec, Canada; Winfrith SGHWR in Great Britain; and Fugen ATR in Japan (Ikeda *et al.* 1986). They used a heavy water moderator, but light water coolant. The coolant boils in the core to produce steam to drive the turbine directly without using separate steam generators.

The Siemens-designed Atucha 1 has operated in Argentina since 1974. It has a large pressure vessel with 6 m long vertical fuel assemblies similar to pressurized water reactor (PWR) assemblies (Adelfang 1999). The pressure vessel has a large number of penetrations that allow for on-power refueling. The reactor is a pressurized heavy water reactor with steam generators. A second reactor of this type (Atucha II) is under construction.

2. *CANDU Fuel Design and Irradiation Performance*

A single fuel rod (or fuel pin) is made of two materials: Zircaloy-4 for all structural components and uranium dioxide (UO_2) as fissionable material. The UO_2 is incorporated as sintered polycrystalline pellets surrounded by a Zircaloy-4 sheath closed by two end-caps. The additions of spacers and/or bearing pads complete the assembly. A fuel assembly is composed of several single fuel rods. These fuel rods are precisely located and maintained in a predetermined geometry by the use of end-plates, support structures, separators and/or spacer pads. In the CANDU-PHWR design (Hastings 1983), the number of fuel rods in the bundle has increased from seven in early assemblies for the Nuclear Power Demonstration reactor (NPD), to 19 for later NPD fuel and the Douglas Point commercial prototype, 28 for the Pickering reactors, 37 for Bruce, Darlington, and early CANDU-6 reactors, and 43 for the CANFLEX fuel (Gibb *et al.* 1999) tested in 1999 at the CANDU-6 plant at Point Lepreau NGS in New Brunswick, Canada. Figure 2 shows a Bruce fuel assembly in use today.

In general, the UO_2 constitutes about 90% of the total weight of a fuel assembly. The prominent design features of a heavy water reactor fuel are:

- fuel designed for on-power refueling and associated fuel handling operations;
- high-density natural UO_2 pellets, which ensure dimensional stability—important for the remote, on-power handling;
- use of Zircaloy-4 for fuel rod sheaths and other fuel assembly components for neutron economy—neutron economy is a key requirement for a heavy water natural uranium reactor;
- high integrity of fuel rod end-cap welding to prevent the release of fission products during irradiation;
- excellent tolerance of fuel rods to power changes during on-power refueling—this can be accomplished by using stress-corrosion barriers between the fuel pellets and the sheath or by improved fueling operational practices.

Since the first CANDU reactor operation in 1971 until 1999, more than a million fuel assemblies have been irradiated in all PHW reactors. The fuel performance depends on the reactor design. The general fuel performance in PHW reactors has been very good (Adelfang 1999, Ganguly 1999, Judah 1996, Lau *et al.* 1999), e.g., operating Canadian CANDU plants had an average fuel assembly defect rate $<0.02\%$ for 1997–1998 (Lau *et al.* 1999).

Early CANDU fuel showed some sensitivity to the power ramps during refueling. This condition was solved through a combination of modified refueling

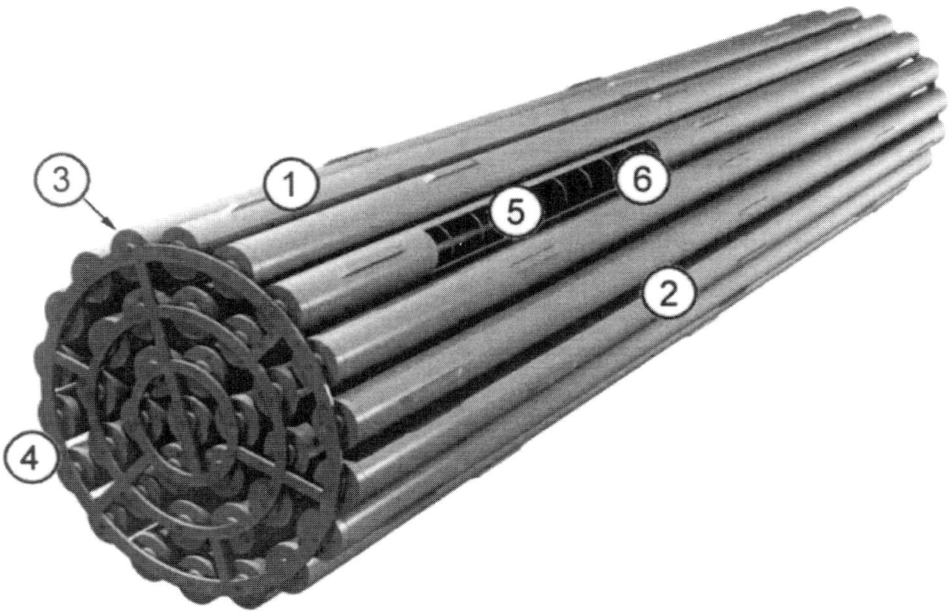

Figure 2
CANDU fuel assembly (Bruce design): 1. Zircaloy bearing pads, 2. Zircaloy fuel sheath, 3. Zircaloy end cap, 4. Zircaloy end support plate, 5. uranium dioxide pellets, 6. CANLUB graphite interface.

sequences and the addition of graphite CANLUB barrier between the pellets and the sheath. Currently, the analysis of fuel defects occurring in PHW reactors has shown that sheath debris fretting is now the main causes of these events (see Sect. 3).

3. *Sheath Failure Mechanisms*

The fuel rod sheath is the primary barrier to prevent the release of fission products to the coolant. If this barrier fails, the radioactive fission products leak into the reactor coolant. The use of ion exchange in the coolant purification system reduces radioactivity of ionic species (e.g., iodine) in the coolant. However, as the dynamics of fission product release, radioactive decay, and ion exchange may still lead to an increased steady-state activity in the coolant, maintenance of reactor components becomes more difficult due to the increased radiation emitted by the reactor cooling system.

In heavy water reactors, due to the on-power refueling activities, the removal of failed fuel from the reactor is a standard operation that does not required reactor shutdown. The detection of defected fuels in the core is performed through the analysis of reactor coolant activity. Once it is known that fuel failures are present through the detection of increased radioiodine activity in the coolant, the determination of the reactor channel that contained the failed fuel assembly can be performed by the use of several localization methods (Lipsett and Palmer 1967). The most common among them is the measurement of the coolant activity at the outlet of each channel by scanning individual channel piping or the use of multiplex valves. Also, some CANDU reactors have a system based on the measurement of delayed neutrons at the channel outlets. The delayed neutrons are neutrons emitted by certain fission products up to about a minute after the original fission event. After the channel containing the failed fuel is localized, refueling of such a channel is planned and performed. This practice maintains the reactor coolant activity to very low levels without the need of power reductions.

The most common mechanisms for fuel failure in heavy water reactors are, in order of decreasing frequency:

- sheath penetrations caused by fretting by small debris in the reactor coolant;
- stress corrosion cracking (*SCC*) *of the fuel sheaths caused by concurrent power change induced stress and chemical conditioning of fuel during irradiation;*
- hydride cracking of the sheath by the dissolution of hydrogen and precipitation of zirconium hydrides (*this mechanism normally occurs after sheath failure by one of the previous mechanisms*)*;*
- fabrication flaws.

Currently, the heavy water reactor plants have reported very low defect rates due to the high quality standards implemented by the fuel manufacturers.

4. *Fission Product Release in the Event of Sheath Failure*

Under normal operating conditions, the fuel sheath provides an effective barrier between the fuel and the reactor coolant. The transport of fission products within the fuel is of interest primarily as an initial condition for the analysis of scenarios involving sheath failure. These scenarios could be in-service fuel failures, or postulated transients. Within the fuel itself, fission products form in the UO_2 grains due to the fission process. The energetic fission products formed directly by fission may have enough energy to escape from the fuel grain by recoil. Fission products near the surface of the fuel grains may be struck by an energetic fission product and acquire sufficient energy to be knocked out of the grain. These two mechanisms (recoil and knockout) release only a small fraction of the fission product inventory from the grains under normal conditions. At the operating temperatures of nuclear fuel, solid phase diffusion of fission products within the grains to the grain boundaries can be significant, particularly for long-lived nuclides.

Solutions to the diffusion equation for both short-lived and long-lived nuclides were determined by Booth and Rymer (1958). Since radioactive decay competes with diffusion to remove a given fission product, releases are greater for longer-lived fission products. At fuel operating temperatures, the fuel grains grow. The growth is at the expense of other fuel grains. As the shrinking grains are depleted of UO_2, those fission products that are not soluble in the matrix will be left on the grain boundaries.

The fission products released from the fuel grains initially accumulate at the grain boundaries in the form of gas bubbles, or as solid or liquid inclusions. Given a sufficient inventory of fission gas, the bubbles can interlink to form tunnels that release the gases to the gap between the fuel pellets or between the fuel pellets and the fuel sheath. Under certain heating conditions microcracking of the fuel can lead to the process of grain boundary separation in which the microcracks offer a release path from the grain boundary to the gap.

With the sheath intact, the releases of fission products to the pellet–sheath gap do not pose any hazard to the public or to plant staff. However, in case of a fuel sheath failure the gap gas can seep out through the failure. The amount of these less-volatile fission products that is released depends on the amount of that fission product available on the fuel surfaces, the chemical environment on the fuel surfaces, and on the vapor pressure of the fission product under these conditions.

Under shutdown conditions, the fuel may cool to a temperature at which liquid water enters the sheath of failed fuel. In this case, soluble fission products may be dissolved from the gap and fuel surfaces. The fission products released by this mechanism are those that were not mobile enough in the gap gas to be released on-power. This mechanism is observable for iodine after the shutdown of a reactor with defected fuel. The iodine concentration in the coolant may rise after reactor shutdown until the removal rate by purification catches up with the increased low temperature release rate. This phenomenon of iodine spiking has been described in the literature (Lewis *et al.* 1997).

Under some upset conditions, hot fuel sheaths may experience rapid cooling which can affect the sheath integrity. If the sheath were to fail under these conditions, the soluble fission products on the fuel surfaces become available for release by dissolution. The extent of this "rewet" release depends on available pellet–sheath gap and fuel surface inventories.

5. *Fuel Degradation After Failure*

In case of sheath failure, the UO_2 fuel is no longer completely isolated from the external chemical environment in the reactor coolant. At temperatures below 1000°C, the affinity of the UO_2 for oxygen is higher than that of the sheath. Consequently, a punctured sheath is not an effective barrier or sink for oxygen. The UO_2 exposed to the steam-filled fuel gap, will oxidize to higher oxides until equilibrium is established between the uranium oxide and partial pressure of oxygen in the pellet-sheath gap (Lewis *et al.* 1993, Lewis 1999). If the formed oxide is higher that U_4O_9, considerable increase in volume occurs inducing fuel rod deformation that could result in fuel sheath rupture. In addition, the progression to higher oxides increases the release of fission products as demonstrated by experiments with fuel rods (Hastings *et al.* 1985, Lewis *et al.* 1990) and bare fuel pellets (McCracken 1985).

6. *Advanced Fuel Cycles*

Heavy water reactors have the flexibility to use a number of fuel cycles. This capability is the driver for advanced fuel cycles with large economical or environmental benefits (Choi *et al.* 1995, Adelfang 1999, Alvarez *et al.* 1999). Between them the most promising options are:

- use of slightly enriched uranium;
- use of spent PWR fuel and mixed-oxide fuel cycles;
- destruction of military plutonium and other actinide wastes; and
- use of thorium.

The underlying goals are to develop fuels and fuel cycles to improve the performance of existing fuel designs, to reduce capital and operational costs in new plants, to enhance resource utilization, to reduce waste handling volumes, to enhance reactor safety, and/or to decrease the excess of weapons stockpiles and actinide wastes.

7. *Concluding Remarks*

This article has reviewed the historical background of heavy water power reactors. Fuel bundle design and irradiation performance under normal operating conditions have been described. Fuel sheath failure mechanisms, fuel degradation, and fission product release have been described. Finally, an overview of current projects and future directions in advanced fuel has been presented. Heavy water reactor fuel is of a robust design and has the flexibility to be modified to suit the national priorities of those countries with heavy water reactors.

See also: Nuclear Fuel Cycles; Nuclear Reactor Fuel Fabrication (Including Quality Control); Mixed Oxide (MOX) Fuel; Light Water Reactor Fuel Design and Performance

Bibliography

Adelfang P 1999 Nuclear fuel cycle activities in Argentina. In: Sejnoha R (ed.) *CANDU: Fuel for the Next Millennium, Proc. 6th Int. Conf. on CANDU Fuel.* Canadian Nuclear Society, Toronto, Canada, Vol. 2, pp. 40–51

Alvarez L, Casario J, Fink J, Perez R, Higa M 1999 SEU Fuel in Atucha-1 NPP, a valuable experience for a CANDU-6 core enrichment. In: Sejnoha R (ed.) *CANDU: Fuel for the Next Millennium, Proc. 6th Int. Conf. CANDU Fuel.* Canadian Nuclear Society, Toronto, Canada, Vol. 2, pp. 206–14

Barber D H, Nott B R 1989 Kinetic imbalance in mixed bed ion exchange In: *Water Chemistry of Nuclear Reactor Systems 5.* British Nuclear Energy Society, London, pp. 257–62

Booth A H, Rymer G T 1958 Determination of the Diffusion Constants of Fission Xenon in UO_2 Crystals and Sintered Compacts. Report AECL-692. Atomic Energy of Canada Ltd., Chalk River

Choi H, Rhee B W, Park H 1995 Comparison of refueling schemes for DUPIC core. In: Cox D (ed.) *Proc. 4th Int. Conf. CANDU Fuel.* Canadian Nuclear Society, Toronto, Canada, Vol. 2, pp. 4A-59–4A-68

Choppin G R, Rydberg J 1980 *Nuclear Chemistry: Theory and Application.* Pergamon, Oxford

Foster A R, Wright R L 1977 *Basic Nuclear Engineering.* Allen and Bacon, Boston, (4th edition 1983)

Ganguly C 1999 PHWR fuel experience in India. In: Sejnoha R (ed.) *CANDU: Fuel for the Next Millennium, Proc. 6th Int. Conf. CANDU Fuel.* Canadian Nuclear Society, Toronto, Canada, Vol. 2, pp. 19–32

Gibb R A, Sancton R W, Steed R G, Reid P J, Bullerwell J 1999 CANFLEX demonstration irradiation at Point Lepreau: a status update. In: Sejnoha R (ed.) *CANDU: Fuel for the Next Millennium, Proc. 6th Int. Conf. CANDU Fuel.* Canadian Nuclear Society, Toronto, Canada, Vol. 2, pp. 91–102

Hastings I J (ed.) 1983 Invited Presentations IAEA/CNEA International Seminar on Heavy Water Fuel Technology. AECL Report AECL-MISC-250, Atomic Energy of Canada Ltd., Chalk River

Hastings I J, Mizzan E, Kelm J R, Moeller R E, Novak J 1985 Postirradiation behavior of naturally and artificially defected UO_2 fuel elements at 250°C in air. *Nucl. Technol* **68**, 418–22

Ikeda H, Hayashi T, Iijima T 1996 International Atomic Energy Agency, Vienna 401–13

Judah J 1996 Overview of fueling, fuel issues and fuel performance at Ontario Hydro. In: *Proc. 27th Ann. Int. Utility Nuclear Fuel Performance Conf.* Idaho Falls (unpublished paper)

Lau J H K, Inch W W R, Cox D S, Steed R G, Kohn E, Macici N N 1999 Canadian CANDU fuel development program and recent fuel operating experience. In: Sejnoha R (ed.) *CANDU: Fuel for the Next Millennium, Proc. 6th Int. Conf. CANDU Fuel.* Canadian Nuclear Society, Toronto, Canada, Vol. 2, pp. 66–77

Lewis B J 1999 An overview of defective fuel analysis in CANDU and light water reactors. In: Sejnoha R (ed.) *CANDU: Fuel for the Next Millennium, Proc. 6th Int. Conf. CANDU Fuel.* Canadian Nuclear Society, Toronto, Canada, Vol. 1, pp. 403–27

Lewis B J, Iglesias F C, Cox D S, Gheorghiu E 1990 A Model of fission gas release and fuel oxidation behavior for defected UO_2 fuel elements. *Nucl. Technol* **92**, 353–62

Lewis B J, Iglesias F C, Postma A K, Steininger D A 1997 Iodine spiking model for pressurized water reactors. *J. Nucl. Mat* **244**, 153–67

Lewis B J, MacDonald R D, Ivanoff N V, Iglesias F C 1993 Fuel performance and fission product studies for defected fuel elements. *Nucl. Technol* **103**, 220–45

Lipsett J J, Palmer J F 1967 Locating Fuel Failures by Fission Product Deposition in CANDU-PHW Reactors. AECL Report AECL-2786, Atomic Energy of Canada Ltd., Chalk River

McCracken D R 1985 Oxidation of UO_2 at 400 to 1000°C in Air and its Relevance to Fission Product Release. AECL Report AECL-642, Atomic Energy of Canada Ltd., Chalk River

Hot Cells, Glove Boxes, and Shielded Facilities

K. V. Kasiviswanathan

Facilities required for the safe handling of radioactive materials are generally classified as fume hoods, glove boxes, and shielded/hot cell facilities. It is necessary to protect personnel both from bodily intake of radiotoxic material and from exposure to penetrating radiation. Hence, these facilities are provided with physical and dynamic barriers, which establish the containment necessary to provide adequate radiological protection to operating personnel and prevent release of radioactivity to the environment. Low-level beta- and gamma-emitting materials are handled inside fume hoods, whereas glove boxes, which have highly leak-tight containment, are used for handling alpha-emitting or highly reactive materials Both fume hoods and glove boxes are also suitable for handling toxic materials, although for highly toxic materials glove boxes are needed. Hot cells are shielded facilities used for handling of materials that emit penetrating gamma radiation. Table 1 summarizes the materials of construction generally used, and the ventilation and negative pressure requirements. These facilities are housed in a laboratory building, which is divided into different zones, isolated from each other by airlocks to avoid any spread of contamination. Details can be obtained from IAEA (1969) and Ferguson *et al.* (1964).

1. *Fume Hoods*

Figure 1 shows a typical series of fume hoods. The viewing panel at the front, made of transparent plastic or toughened laminated glass, can be moved up and down to facilitate material movement. The size of a fume hood varies from a fraction of a cubic meter to one large enough to allow a person to work inside it (walk-in fume hood). Fume hoods are

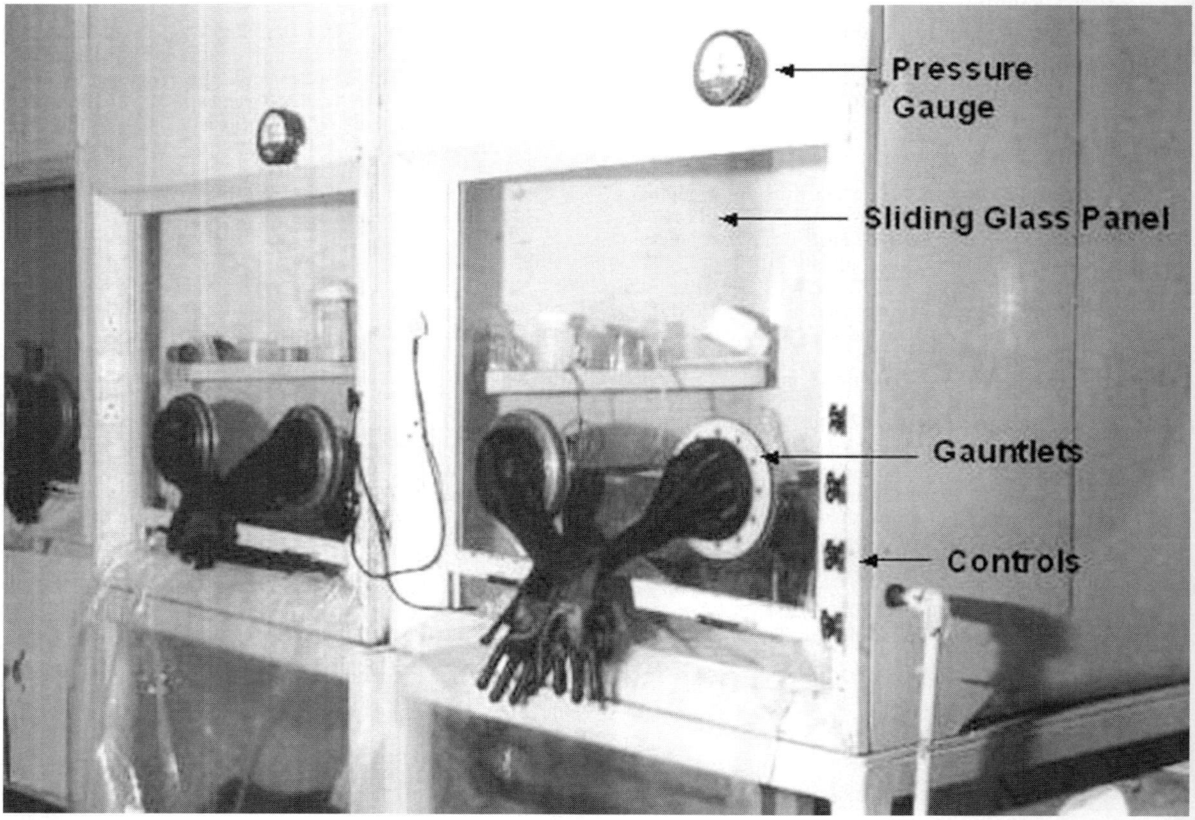

Figure 1
A series of fume hoods.

Table 1
Materials of construction generally used, and ventilation and negative pressure requirements of special enclosures.

Enclosure	Material of construction	Activity handled	Differential pressure (negative)	Ventilation rate
Fume hood	Steel, stainless steel, PVC, reinforced resin-bonded laminates	Low-level β and γ, acids, corrosive chemicals	0.01–0.02 m of water gauge	$0.5\,ms^{-1}$ across a slash opening of 0.3 m
Glove box	Stainless steel	α and low-level β and γ, highly reactive materials	0.01–0.02 m of water gauge	15–20 air changes per hour
Hot cell	Concrete, lead, stainless steel, steel	Very high-level α, β, γ	0.015–0.025 m of water gauge	>20 air changes per hour

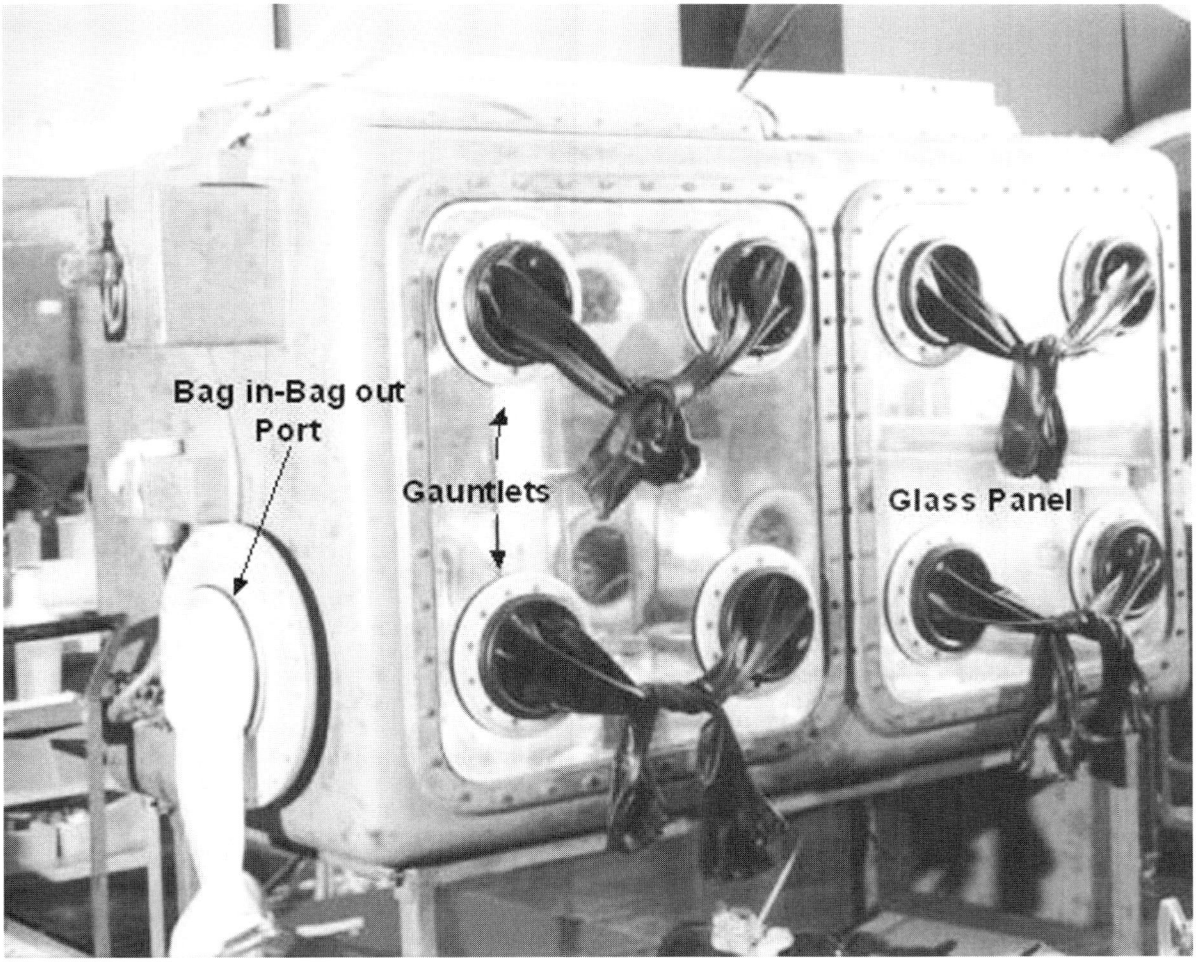

Figure 2
A typical glove box.

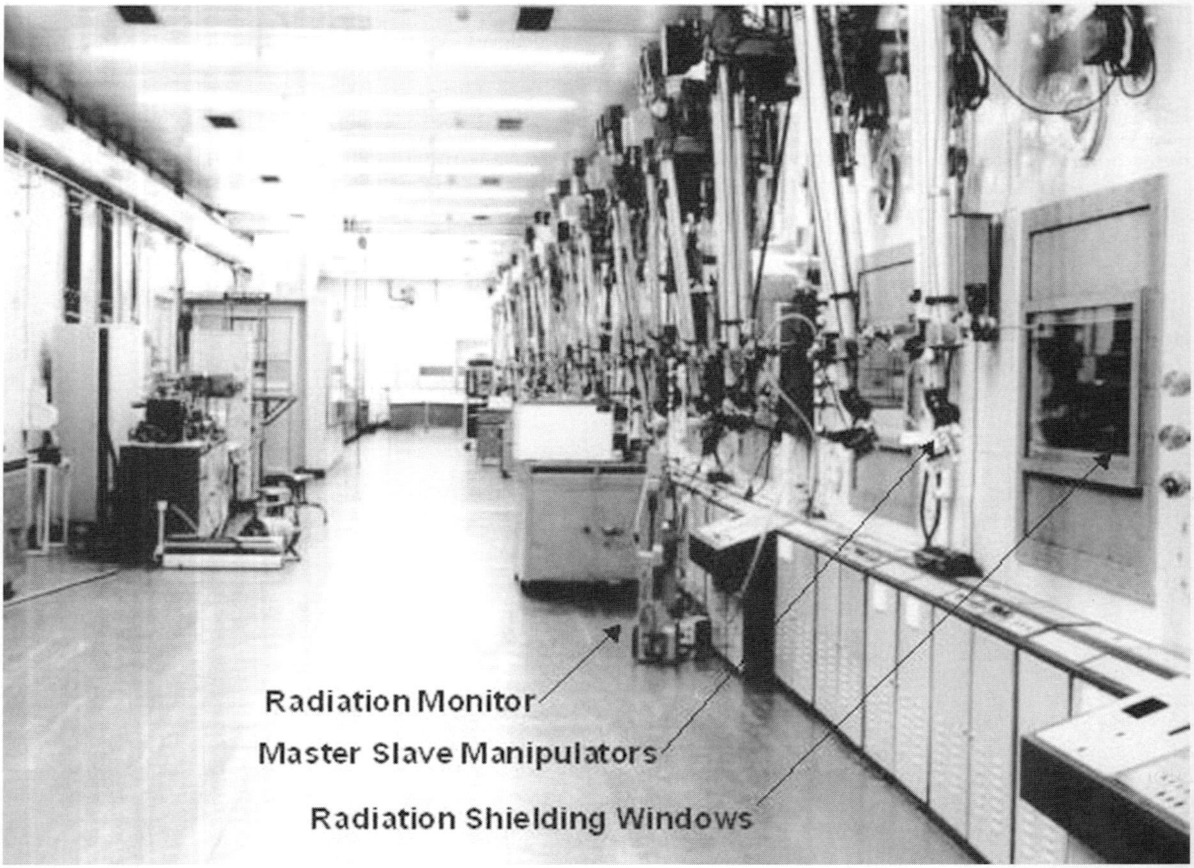

Figure 3
View of the operating area of a hot cell facility at the Indira Gandhi Center for Atomic Research, Kalpakkam, India.

provided with gauntlets (gloves) for handling of toxic or radioactive materials.

Suitable apparatus is mounted inside the fume hood depending on the specific work. Water, gas, power, illumination, and other services are provided with controls outside to facilitate easy maintenance. Dust and vapor are prevented from escaping by maintaining a negative pressure inside the hood and a continuous flow of air through openings via a ventilation system. The ventilation system is provided with a high-efficiency particulate air (HEPA) filter to prevent hazardous particles from escaping to the environment. The very idea of a fume hood is that of a dynamic barrier of air which contains hazardous substance from escaping to the surroundings.

2. *Glove Boxes*

A typical glove box is shown in Fig. 2. Glove boxes are widely used in radiochemical and radiometallurgical laboratories and in production plants. The glove box is provided with toughened laminated safety glass panels, at least on two sides of the box for adequate visibility, and gauntlets for handling of materials.

The ventilation system provided for glove boxes ensures an adequate number of air/inert gas changes in the box, as required by the material being handled, and maintains adequate negative pressure inside. The supply and exhaust air or gas is filtered using a prefilter and HEPA filter assembly. Materials are "bagged in" or "bagged out" using a PVC bag attached to a port. Glove boxes are provided with systems to tackle any possible emergency arising from glove rupture, fire, overpressurization, criticality, etc. The main function of a glove box is to isolate the contaminants from the environment by confining them to the enclosed volume. Design considerations are available in ASTM (1995).

3. *Shielded/Hot Cell Facilities*

Hot cells are classified as concrete or lead cells depending on the material used for shielding the gamma radiation. They generally have a leak-tight stainless steel enclosure mounted permanently inside, which acts as primary physical containment, and are shielded by high-density concrete or lead, which is constructed either adjacent to the cell or as an integral part of it. The secondary dynamic containment is achieved by keeping the hot cell under negative pressure with respect to its surroundings. Figure 3 shows the operating area of an alpha, beta, and gamma hot cell facility operating with inert atmosphere.

Hot cells are employed for the examination of irradiated fuel and cladding materials, reprocessing of spent fuel, metallurgical research, processes for radioisotope separation, and other high-level radiation research work. Several cells, approximately the size of a room, are constructed modularly, with each module having its own design features. Design and radiation shielding considerations are discussed by Ring (1970).

Operations inside hot cells are carried out using mechanical and electrical remote handling systems. The most common variety of remote handling system is the master slave manipulator (MSM), which is provided with radiation-resistant flexible booting to achieve leak-tightness. Cranes and power manipulators are used for heavy material handling inside the cells. Sealed penetrations are provided in the cell wall for service lines, cables, etc. Viewing systems used in hot cells are radiation shielding glass windows, periscopes, CCD cameras, etc. Illumination is provided using sodium/mercury vapor lamps.

Air or inert gas ventilation is selected according to the chemical nature of the materials being handled. Ventilation systems, with elaborate control and data-logging facilities, are provided with HEPA filters on the supply and exhaust lines. Provisions are made to tackle any emergency arising from accidental criticality, booting rupture, positive pressurization, power failure, etc.

4. *Future Trends*

Robotic and automated devices are finding increasing applications in radioactive material handling owing to their increased reliability and high productivity, resulting in reduced risk of radiation exposure to operators. Equipment design is highly modular in construction for easy remote maintenance using robotic devices, employing advanced materials with high radiation resistance for increased life. American Nuclear Society (1984–94) is suggested for further reading.

Bibliography

American Nuclear Society 1984–94 *Proc. 32nd–41st Conf. on Remote Systems Technology*. American Nuclear Society, La Grange Park, IL

ASTM 1995 Design criteria for plutonium glove boxes, C852-93. *Annual Book of Standards*. American Society for Testing and Materials Philadelphia, Vol. 12

Ferguson K R, Doe W B, Goertz R C 1964 Remote handling of radioactive materials. Reactor Handbook—Engineering. Interscience, New York, Vol. IV

IAEA 1969 *Manual on Safety Aspects of the Design and Equipment of Hot Laboratories*, Safety Series no. 30. International Atomic Energy Agency, Vienna

Ring Jr. F 1970 Design and shielding of hot cells for research. In: Jaeger R A, (*et al*). (eds.) *Engineering Compendium on Radiation Shielding*. Springer-Verlag, New York, Vol. III Chap. 10.5

Hydrogen–Metal Systems: Applications of Gas–Solid Reactions

S. Suda

Conversion of molecular hydrogen to monatomic hydrogen requires higher temperatures than 3500°C in the thermal decomposition. However, molecular hydrogen is easily decomposed to monatomic hydrogen under moderate temperature and pressure conditions if it were absorbed in a hydride-forming material.

The monatomic hydrogen (called *protium*) can be placed in the crystalline lattice and desorbed reversibly back to molecular hydrogen. Electrochemical conversion of proton to protium also takes place reversibly by the electrochemical charging and discharging. Hydrogen can exist even as the minus ion in the specific hydrogen-metal compounds such as, for example, MgH_2 and Mg_2NiH_4. Today, such kinds of hydrogen-light metal complexes as $NaAlH_4$, $NaBH_4$, and KBH_4 have been developed as the reversible hydrogen storage materials.

Hydrogen-metal systems have a variety of potentials in technological and engineering applications because of the reversible nature of hydrogen uptake and release reactions and the amount of protium sited in the crystalline lattice.

This article will focus on the applications of the reversibility of hydrogen-metal systems and their thermodynamic properties and characteristics. The extreme reactivity of protium has been intended to apply as the catalysts for chemical reduction. The most advanced hydrogen storage materials such as Al-and B-based hydrogen metal complexes will be introduced as the hydrogen fuel source.

Technological and engineering developments of the hydrogen-metal systems since the 1970s have been extended to a variety of applications but only a few practical devices have been successfully put on the commercial market to date. It is the major purpose of this article to analyze the reasons why hydrogen-metal systems have not yet reached to the practical stages of applications (Suda and Sandrock 1994). Irreversible hydrogen to metal reactions such as the hydrogen embrittlement are not included.

1. Basic Properties and Characteristics of Hydrogen–Metal Systems in View of Technological and Engineering Applications

1.1 Monatomic Hydrogen (Protium, H°)

The terms of absorption and desorption do not correctly represent the hydrogen uptake and release phenomena of hydrogen-metal systems, but these terms have been conveniently used instead of hydrogenation and dehydrogenation or hydriding and dehydriding, respectively. These terms are sometimes abbreviated to H/D reactions, H/D processes, and H/D cycle.

The molecular hydrogen is "absorbed" in hydride-forming metals and alloys in the crystalline lattice as monatomic hydrogen (protium) to form metal hydrides and it is "desorbed" from the crystalline lattice to return to molecular hydrogen and these H/D reactions are reversible. The protium sited in the crystalline lattice is the source of hydrogen stored as given in Eqn. (1):

$$H_2 \leftrightarrow 2H^\circ \text{ (protium in crystalline lattice)} \quad (1)$$

1.2 Available Hydrogen Content

Technological and engineering applications of the gas–solid are based on the reversibility of H/D reactions as shown in Eqn. (2), but the amount of reversibly transferred in any engineering devices is limited to a part of protium stored in metal hydrides as given in Eqn. (2):

$$\text{In principle}: M + (x/2)H_2 \leftrightarrow M\overset{\circ}{H}_x \quad (2)$$

$$\text{In actuality}: M\overset{\circ}{H}_x \leftrightarrow M\overset{\circ}{H}_{(x-y)} + (y/2)H_2 \quad (3)$$

where x and y are defined as *absolute hydrogen content* and *available hydrogen content*, respectively.

The relations of x, y, and $y-x$ are illustrated in a schematic diagram of the pressure-composition isotherm in Fig. 1. The $x-y$ in Eqn. (2) is the amount of protium trapped in the crystalline lattice during H/D processes and the part is not available and only the y part can be applicable. Accordingly, it is not practical to consider that all of the x part of hydrogen contents as given in Eqn. (2) might be available. In any practical applications of gas–solid reactions, the applicable amounts of hydrogen are limited within the available part of hydrogen expressed by available hydrogen content as given as $(y/2)H_2$ in Eqn. (3).

It should be noted that these $x-y$ and y parts are not the fixed value but vary significantly depending on the dynamic conditions (see *Hydrogen–Metal Systems: Technological and Engineering Aspects*). The available hydrogen content, y, is obtained only under equilibrium states and it varies as a function of the hydrogen transfer kinetics and the thermal transfer characteristics of metal hydride beds during dynamic conditions as indicated as actual hydrogen content available.

1.3 Actual Enthalpy Changes of H/D Reactions

The H/D reactions in a metal hydride system are reversible in principle but irreversible in practice. Thermal energy must be exchanged between the system and the surroundings during H/D processes. Due to the sensible heat of a specific hydrogen storage device, the thermal energy that is extracted from the system is smaller than that transferred from the

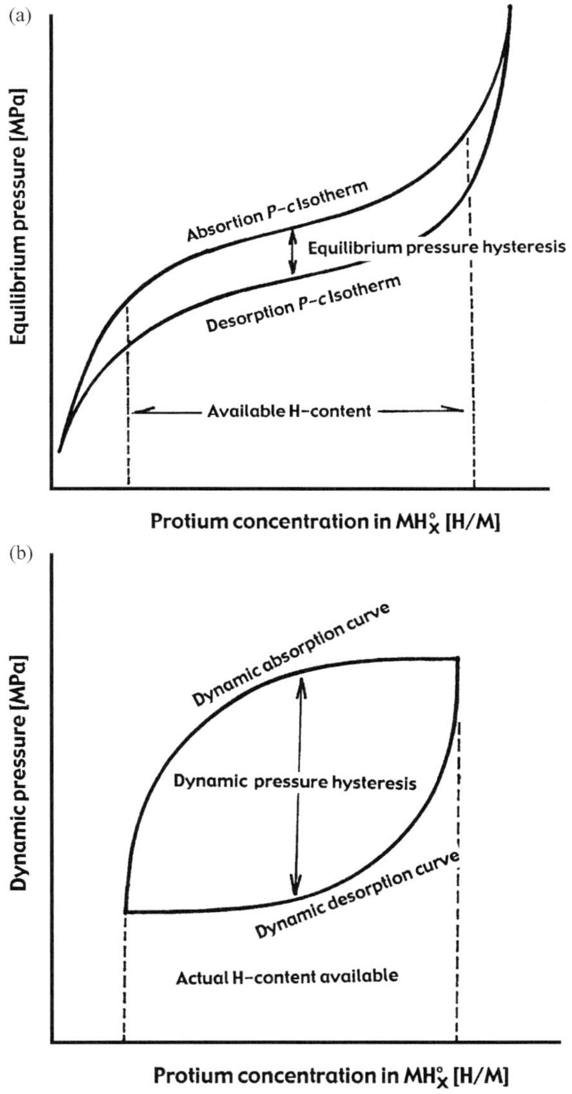

Figure 1
(a) Schematic diagram of equilibrium absorption and desorption P–c–T relations. (b) Schematic diagram of dynamic absorption and desorption P–c–T relations.

surroundings. The thermal energy from and to the surroundings is not equivalent to the reversible heats of H/D reactions. The difference is contributed to the sensible heat retained within a hydrogen storage unit (container structures and metal hydride beds). The sensible heat is defined as the thermal mass ($Q_{mass} = mC_p\Delta T$, total weight multiplied by heat capacity of a storage unit and temperature difference).

Extra thermal energy that compensates the thermal mass is required in any engineering device in order to transfer a desired amount of hydrogen from and to a hydrogen storage unit such as hydrogen storage reservoirs and tanks, and heat exchangers for thermodynamic devices.

2. *Actual Hydrogen Content Available in Engineering Applications*

The actual hydrogen capacity is limited within the value of ($y/2$) where y is defined as available hydrogen content in Eqn. (3). The available hydrogen content is significantly reduced to smaller values due to the dynamic nature of mass and heat transfer characteristics of a storage unit. Under the dynamic conditions, the P–c–T relation depends greatly on the thermal energy transferred between the system and the surroundings and it changes as a function of time (see *Hydrogen–Metal Systems: Technological and Engineering Aspects*). The experimentally measured dynamic P–c–T relations are illustrated for the Mg_2Ni hydride in Fig. 2 (Friedlmeier 1997). The thermal energy transferred is limited by the thermal mass of a storage unit.

At present, the maximum available hydrogen content is less than 1.0 wt.% which is equivalent to 11.21 at STP (1 kg-H_2 in 100 kg container). Furthermore, the actual hydrogen content available is estimated less than 0.8 wt.%. This is the major reason why hydrogen-metal systems as metal hydrides have not been succeeded in practical applications in the past. It is, therefore, not practical to use the metal hydride as the hydrogen storage material not only from the viewpoints of weight penalty but also from the materials properties and characteristics viewpoints.

Much attention has been focused on the development of high hydrogen capacity metal hydrides over 3 mass%. It should be noted that the targeted value by material scientists and metallurgists is not the weight ratio to the total weight of a storage unit plus material, but the hydrogen capacity as the total mass percent protium contained in the crystalline lattice.

From engineering viewpoints, materials that may store at least 5 wt.% per overall weight of a storage unit under moderate pressure and temperature conditions must be developed. Furthermore, the available hydrogen content under dynamic conditions (the actual hydrogen content available) should be higher than 3 wt.%.

New innovative research has been recently undertaken with regard to the hydrogen-light metal systems such as $NaAlH_4$, KBH_4, and $NaBH_4$. These materials have been well known as hydrogen donors in reduction reactions of organic compounds and they form the hydrogen-metal complexes that can generate hydrogen in the approximate ranges of 5–10 mass%.

$NaAlH_4$ has been known as the highly reactive material, especially in contact with moisture and water to generate hydrogen gas violently. The catalytically enhanced dissociation/re-association reactions to

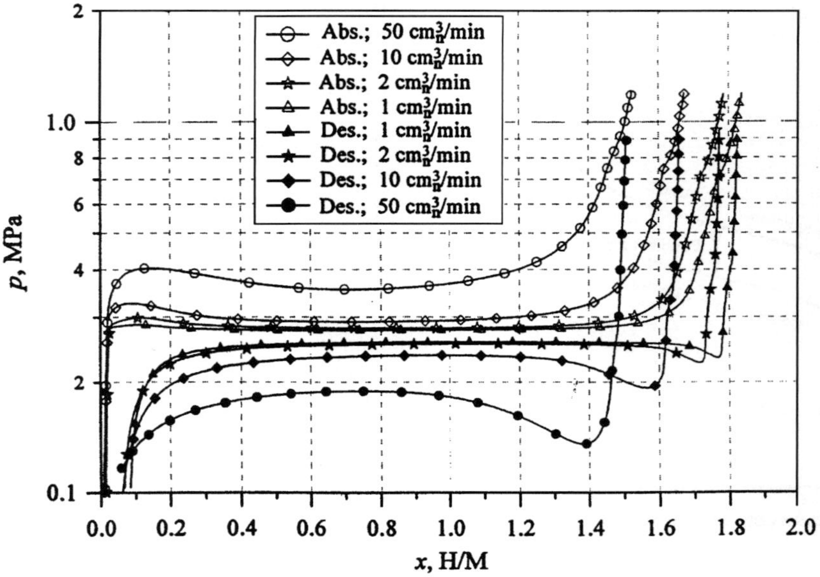

Figure 2
Dynamic absorption and desorption P–c relations of Mg_2Ni-hydride at 316°C as a function of hydrogen flow rates.

generate hydrogen by a stepwise thermal decomposition as shown in Eqn. (4) have been developed by Bogdanovic and Schwickardi (1997) and Jensen *et al.* (1999).

$$NaAlH_4 \leftrightarrow 1/3Na_3AlH_6 + 2/3Al + H_2 \qquad (4)$$

$NaAlH_4$ has been developed extensively as a hydrogen storage material (Yamashita *et al.* 2000, Sandrock *et al.* 2000, Gross 2000a, 2000b, Lewndowski *et al.* 2000, Thomas 2000). The reversibility after the stepwise decomposition has been experimentally demonstrated although the process needs to be developed to be more practical.

It is well known that KBH_4 and $NaBH_4$ can be stabilized in alkaline solutions such as aqueous KOH and NaOH. Hydrogen can be generated by catalytic hydrolysis of the solution. The reaction has been attracting engineering interest from the viewpoint of hydrogen capacity of these alkaline-stabilized hydrogen-metal complex solutions. The hydrolysis reaction is expressed in Eqn. (5):

$$BH^-{}_4 + 2H_2O \rightarrow 4H_2 + BO^-{}_2 \quad \text{(in alkaline solutions)} \qquad (5)$$

In Eqn. (5), 2 mol of water decompose to generate 2 mol of H_2 where the BH_4^- complex ion generates 2 mol of H_2 to result in 4 mol of H_2. Hydrogen gas generated in this reaction is 10.8 mass% in its stoichiometry. However, it should be noticed that the available hydrogen content is greatly dependent on the BH_4^- concentration and its solubility limit in an alkaline solution.

A series of metals and alloys such as cobalt, nickel, and Mg_2Ni work as the catalysts for hydrolysis rather than precious metals such as palladium, platinum, and rubidium (Suda 2000a, 2000b, Suda *et al.* 2000a, 2000b). Borohydrides such as KBH_4 and $NaBH_4$ can be reversibly produced by a chemical process by using the used-fuel as the form of BO_2^- in an alkaline solution that is put in a drying process to prepare potassium metaborate tetrahydrate ($KBO_2 \cdot 4H_2O$) and sodium metaborate tetrahydrate ($NaBO_2 \cdot 4H_2O$). Sodium tetraborate decahydrate ($Na_2B_4O_7 \cdot 10H_2O$) is also used as the starting material for producing sodium metaborate tetrahydrate ($NaBO_2 \cdot 4H_2O$) and it is in turn reduced to $NaBH_4$.

Light metal-complex hydrides composed of lithium, potassium, or sodium with aluminum or boron will be developed in the future as hydrogen storage materials and will be applied in various engineering devices such as the large capacity hydrogen storage units, fuel cells, and residential appliances.

3. *Applications*

3.1 *Hydrogen Storage*

It is not practical to apply metal hydrides as the hydrogen storage material, particularly for large-scale applications such as stationary and transportation. No practical application has been developed commercially with respect to hydrogen storage devices except for small-scale laboratory use.

There are three different types of existing hydrogen storage units; (i) high-pressure cylinders, (ii) liquid hydrogen reservoirs, and (iii) metal hydride containers. However, the available hydrogen content is restricted in each of these units by (i) pressure level and weight of cylinder, (ii) insulation level of vessel and boil-off ratio, and (iii) available hydrogen content of metal hydrides and weight of a container.

High-pressure cylinders that contain hydrogen gas compressed over 30–50 MPa have been developed in order to supply hydrogen gas to proton-exchange-membrane fuel cell (PEMFC) cars. From technological and engineering viewpoints, the high-pressure cylinder is considered the best option among the three hydrogen storage units presently used for PEMFC cars, particularly in view of its high discharge capability during hydrogen release under controlled conditions. Metal hydrides as hydrogen storage materials will not be used for the large capacity storage applications due to their weight penalty and high materials cost as well as the poor available hydrogen content.

International co-operative projects such as Task-12 and Task-17 (International Energy Agency, IEA) have been conducted in order to develop hydrogen storage materials to at least 5 wt.% H_2 recoverable at lower than 100 °C in Task-12 (Sandrock 1996, 1997, 1998) and 80 °C (Task-17 to be started on 2001) under 0.1 MPa and the rate of hydrogen release suitable for practical applications. A Japanese project, Protium New Function in Sub-Nano Materials, was conducted during 1998–2001 to develop hydrogen storage materials with the capacity over 3 wt.% (Okada 1998, 1999, 2000). However, it should be noted that even after succeeding in developing such materials, it is necessary to evaluate the practical applicability with regards to the overall weight efficiency. From practical standpoints, it is desirable to have at least 5 wt.% available hydrogen content and 3 wt.% availability under dynamic states in any engineering devices, particularly in hydrogen storage applications.

3.2 *Thermodynamic Devices*

Enthalpy changes during H/D reactions are from -30 to $-40\,\mathrm{kJmol^{-1}}\ H_2$. The enthalpy value changes slightly depending on the H° concentration in MH_x. The average value for a specific metal hydride can be obtained as the P–T slope of van't Hoff plots.

The heats of exothermic and endothermic reactions have been considered applicable in the thermodynamic devices such as CFC- and HCFC-free air conditioner units, refrigerators, and heat pumps when the reactions are cycled between the coupled reactor systems. The principles of the thermodynamic cycle based on the reversed Carnot principle are illustrated in Fig. 3 for the heat pump and Fig. 4 for the refrigeration or air conditioning cycle.

The cyclic operation can be achieved by preparing two sets of coupled heat exchanger system.

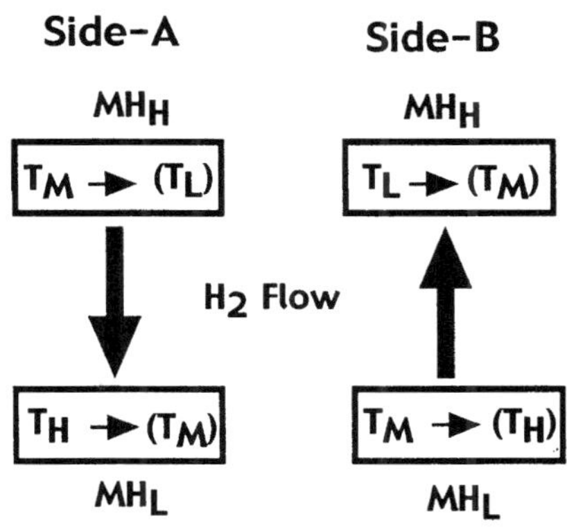

Figure 3
Basic concept of a heat pump (batch-wise hydrogen flow in two sets of a coupled heat exchanger system is shown. The hydrogen flows are periodically switched back and forth both in side A and side B in order to maintain the semicontinuous cyclic operation).

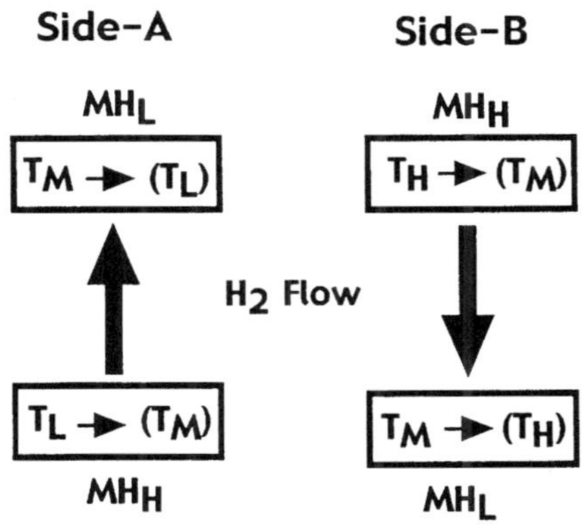

Figure 4
Basic concept of a refrigerator (batch-wise hydrogen flow in two sets of a coupled heat exchanger system is shown. The hydrogen flows are periodically switched back and forth both in side A and side B in order to maintain the semicontinuous cyclic operation).

In a coupled heat exchanger system, two metal hydrides with different hydriding/dehydriding P–c–T relations are chosen. In each set, hydrogen is transferred between the coupled heat exchanger system in a back and forth manner. The hydrogen flow is switched to the opposite direction between a coupled heat exchanger system periodically after a certain amount of hydrogen has been transferred from one side to the other. Each set is operated in the opposite direction to maintain the constant hydrogen flow to realize a semi-continuous and cyclic operation.

The cyclic system is composed of two sets of coupled heat exchangers that are operated periodically in the batch-wise manner. This can be understood as a type of the reversed Carnot cycle that generates heats as a heat pump and extracts heat from a closed space to the surroundings as with an air conditioning unit or a refrigerator.

The thermal energy sources are of three different temperature levels: (i) T_H as heat source or heat to be generated, (ii) T_M as heat source or sink, or (iii) T_L as low temperature source to be generated or heat sink. Any heat source, e.g., waste heat, solar thermal energy, or low-pressure steams, is used as the heat source at T_M. The heat pump cycle and refrigeration cycle are illustrated in Figs. 5 and 6, respectively.

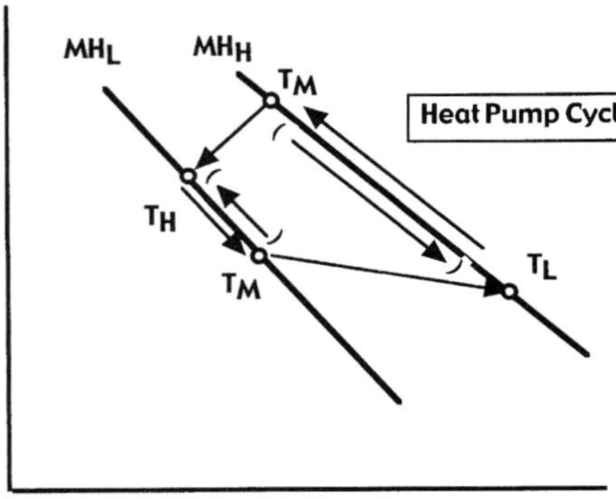

Figure 5
Principles of a heat pump cycle (van't Hoff plots on P–T relations).

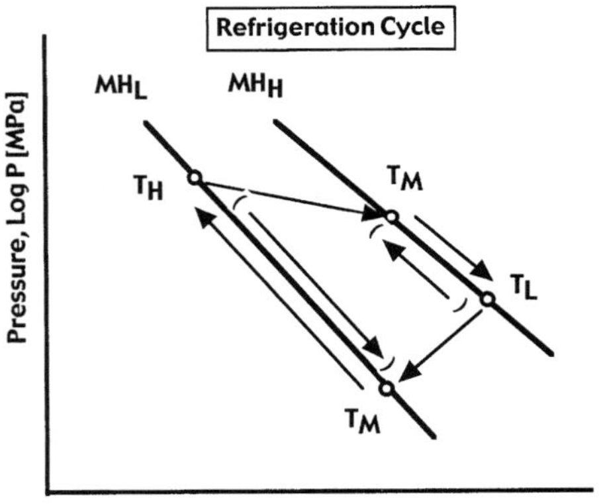

Figure 6
Principles of refrigeration cycle (van't Hoff plots on P–T relations). Heat source T_H (waste heats, solar heat), Heat sink T_M (ambient air, water), Heat extracted T_L (refrigeration, air conditioning).

In Fig. 5, ambient air is applied as the heat sink at T_L. The thermal energy at T_M is upgraded to T_H by the exothermic reaction when hydrogen is transferred between a coupled heat exchanger system. The cyclic direction is anticlockwise in the heat pump cycle. In a refrigeration cycle, hydrogen is transferred in a clockwise manner where T_H is used as the heat source to release heats at T_M before and after generating low temperature source at T_L (see Fig. 6).

Heat exchangers where the *exothermic* reaction proceeds act as the hydrogen acceptor and conversely, heat exchangers where the *endothermic* reaction proceeds act as the hydrogen donor. The pressure level of hydrogen acceptor must be lower than that of hydrogen donor as the pressure difference is the driving force of hydrogen transfer. The driving force of any cyclic devices is the pressure difference in a coupled heat exchanger system with one in the high-pressure side and another in the low-pressure side. The pressure difference is largely dependent on (i) rate of hydrogen transfer, (ii) dynamic P–c–T behaviors of a coupled metal hydride system, and (iii) heat transfer characteristics of heat exchangers. Accordingly, it is not possible to design such devices without knowing the above listed factors that are only available by the collected experimental data on (i) effective thermal conductivity, (ii) overall heat capacity of a heat exchanger, and (iii) dynamic P–c–T relations between the coupled heat exchanger systems.

A typical dynamic P–T relations during the heat pump operation is illustrated in Fig. 7 and the thermodynamic efficiency of the cyclic devices is described in the article *Hydrogen–Metal Systems: Technological and Engineering Aspects*. Nagel and coworkers (1986a, 1986b, 1989) have worked extensively on the dynamic behaviors of the hydrogen-metal hydride systems in a refrigerator.

Various R&D work has been conducted in the past for developing such devices. A double-stage heat pump was developed for the recovery of industrial waste-steam. This might be the most successful illustration of this kind of thermodynamic application where the low thermodynamic efficiency of single-stage heat pumps is improved significantly by introducing the three sets of coupled heat exchangers (Suda *et al.* 1991).

However, no commercial devices have been developed due to (i) low available hydrogen content, (ii) heavy weight-heat exchanger, (iii) poor effective thermal conductivity, (iv) large thermal mass, and (v) dynamic P–c–T behaviors of the batch-wise H/D cycles. Thermodynamic efficiency of cyclic devices is inevitably reduced due to these five shortcomings, and it has been estimated that any cyclic devices may not be commercialized even if metal hydrides with high available hydrogen content (over 5 wt.%) were developed in the future.

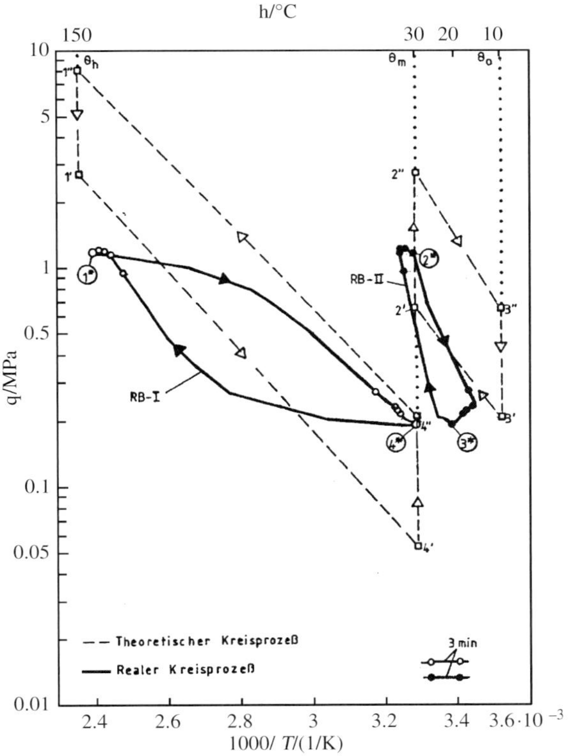

Figure 7
An illustration of actual P–T behavior of a coupled metal hydride system in a refrigerator with $MmNi_{4.0}Fe_{1.0}$–hydride (high-pressure side) and $LaNi_{4.65}Al_{0.3}$-hydride (low-pressure side) (Suda *et al.* 1991).

3.3 *Optical Switches*

Yttrium forms a hydride as YH_x ($0<x<3.0$). This metal hydride behaves as a metal in $0<x<2.1$ and as an isolator (nonconductive) in $2.6<x<3.0$, and it exhibits transition states between metal and metal hydrides as a function of H° concentration as protium in the metal lattice.

Light can penetrate through the thin layer of $Y\mathring{H}_x$ when it forms metal hydride in the range of $x=2.6$–3.0 but cannot penetrated it in the range of $x=0$–2.1. The degree of clearness changes as a function of H° concentration in yttrium hydride. This phenomenon was discovered by Huiberts *et al.* (1996) and has been considered applicable as optical sensors switchable between visible and invisible (transparent and opaque) states (see Fig. 8).

The phenomenon is caused by the energy gap of electrons (~1.5 eV) between the nonconductive and conductive states of yttrium metal hydride. For preparing such an optical switch, a thick yttrium layer is deposited on a clear glass substrate first and then a thin layer of Pd is deposited over the Y-layer in order to protect from oxidation. Then the plate is put in a hydrogen reactor to form hydride under ambient temperature conditions.

Changing the hydrogen pressure by 0.2–0.3 MPa can control the H° concentration in $Y\mathring{H}_x$. The degree of clearness between complete opacity and transparency is freely changeable by changing H° concentration. However, no practical devices have been developed to date.

3.4 *Fuel Cell*

Fuel cell is a typical application of gas–solid reactions between molecular hydrogen and the catalytic electrode to generate electric power through electrochemical combustion (see *Fuel Cells: Electrochemical Reactions*).

Among various fuel cells, the proton-exchange-membrane fuel cell (PEMFC) has been extensively developed for automobile applications. It applies gaseous hydrogen to the anode (negative electrode) by decomposing it to monatomic hydrogen (protium) at the palladium surface to separate into proton (H^+) and electron (e^-). Then, the proton is delivered through the proton exchange membrane to the cathode (positive electrode) to form water with hydroxide ion (OH^-) in alkaline solutions such as 6N–KOH. The transfer processes of hydrogen starting from molecular hydrogen to proton in the course of generating electron and water are shown below.

$$H_2 \rightarrow 2H^\circ \text{ (gas to protium at anode surface)}$$

$$2H^\circ \rightarrow 2H^+ + 2e^- \text{ (protium to proton and electron in anode)}$$

$$2H^+ + 2OH^- \rightarrow 2H_2O \text{ (proton to water at cathode)}$$

$$H_2 + 2OH^- \rightarrow 2H_2O \text{ (overall)}$$

The major theoretical features of PEMFC are: (i) low operating temperature and pressure, (ii) high efficiency, (iii) compact design, and (iv) simple configuration. But the major technical and engineering problems are: (i) high cost of palladium electrodes used both in anode and cathode, (ii) high cost of proton exchange membrane, (iii) low efficiency, (iv) high temperature and pressure operating conditions, (v) hydrogen sources not commercially available to date. The most serious problem is the hydrogen source producible in the commercial scale with reasonable costs.

Metal hydrides have been developed as one of the options for on-board hydrogen storage for PEMFC cars. However, it has been difficult to overcome several requirements under the extreme driving

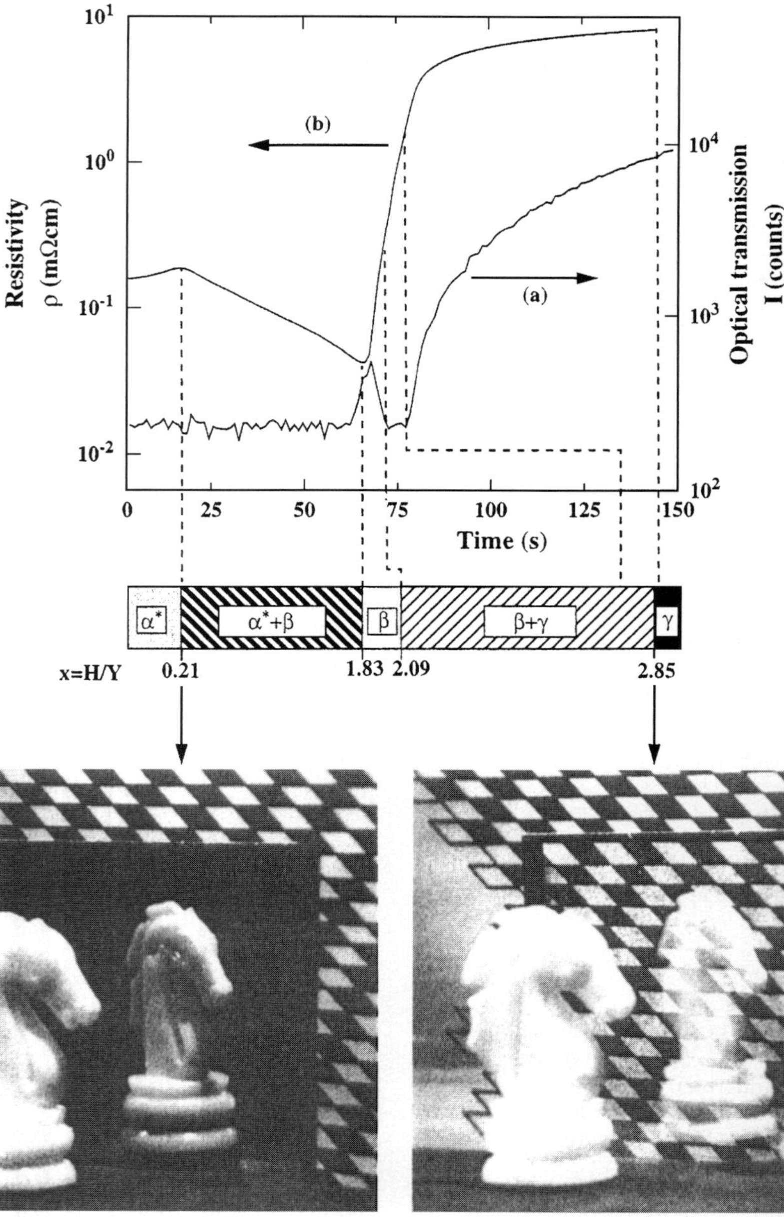

Figure 8
Optical properties of YH_x-hydride as illustrated by the opacity and transparency (Griessen *et al.* 1997). An illustration of the drastic optical changes that take place during hydrogen loading, two photographs of the film are shown (Huiberts *et al.* 1996). The appearance of the film in the visible part of the spectrum remains essentially unchanged as long as $x < 2$ in YH_x. The knight that is placed in front of the film is perfectly reflected and the film is opaque for the light of a lamp placed behind the film and the transparent chess board pattern. However, as soon as the γ-trihydride phase is formed, the transparency of the film increases within a few seconds and the chess board pattern becomes visible throughout the photograph. At the same time the mirror picture of the knight is much weaker. The residual reflection originates mainly from the protective Pd cap-layer.

conditions such as (i) start-up under low ambient temperature, (ii) response to acceleration and high load drives, and (iii) thermal energy sources available for releasing hydrogen. So far, the high-pressure hydrogen has been considered as the most usable hydrogen source among several options such as liquid hydrogen, metal hydrides, methanol, and gasoline for PEMFC car applications.

The hydrogen source must be developed from the following viewpoints; (i) hydrogen generation kinetics, (ii) recyclable hydrogen resources, (iii) environmental issues, (iv) cost and economic infrastructure, and (v) technical and engineering infrastructure for fueling and delivery systems.

3.5 *Sensing Devices*

The equilibrium pressure of metal hydrides varies as a function of protium (H°) concentration and temperature. As the pressure–temperature relation can be expressed by a straight line in a van't Hoff plot, the pressure changes as a function of temperature have been expected to be applicable for sensing devices such as temperature sensors, pressure actuators, and limiting switches. However, the pressure change is significantly dependent on the dynamic P–c–T relations that are influenced by the overall heat transmission characteristics of sensing devices; (i) packing density, (ii) thermal mass (total weight multiplied by the heat capacity of a container), (iii) thermal conductivity, and (iv) wall thickness. It should be remembered that the accuracy and sensitivity of a sensing device is strongly reliant on the heat transmission characteristics and the dynamic P–c–T relation, particularly the dynamic hysteresis.

Power as the boundary work exerted in a pressure actuator is not high enough compared with oil-pressure devices due to the damping effect of the pressure–volume (P–V) relation of gaseous hydrogen. Pressure hysteresis between H/D processes is the source of errors and the slow thermal kinetics causes the inaccuracy. To date, no device has been developed commercially that is available in the market.

3.6 *Catalysts for Hydrogen Reduction*

Monatomic hydrogen stored in the crystalline lattice exhibits strong reactivity against materials that are in contact with the surface or physically adsorbed at the surface of metal hydride particles both in gas–solid reactions ($H_2 \leftrightarrow 2H^{\circ}$) and electrochemical reactions ($H^+ \leftrightarrow H^{\circ} + e^-$).

A typical example of such chemical reductions is the synthesis of pure methane from CO_2 adsorbed on the fluorinated metal hydride surface;

$$CO_2 + 8H^{\circ} \rightarrow CH_4 + 2H_2O \qquad (6)$$

The reaction in Eqn. (6) proceeds under moderate temperature and pressure conditions by switching the operating temperature and pressure conditions periodically between hydriding and dehydriding processes. It had been developed on a pilot-plant scale for recovering CO_2 to synthesize methane (CH_4) as a fuel (Suda *et al.* 1997). In the process, the produced methane and steam easily leave the fluorinated metal hydride surface to be replaced with CO_2 molecules approaching consecutively to the fluorinated surface. No poisoning by the adsorbed CO_2 has been observed during a long-term operation.

Fluorinated Mg_2NiH_4 has been found to be effective as a catalyst for liberating hydrogen from alkaline-stabilized BH_4^--containing solutions as the hydrogen storage material (Suda in press). The reaction can be illustrated as Eqn. (7)

$$BH_4^- + 2H_2O \rightarrow 4H_2\uparrow + BO_2^- \qquad (7)$$

3.7 *The HDDR Process*

The HDDR process has been developed for the Nd–Fe–B magnet system that utilizes a series of four transformations, i.e., hydrogenation, decomposition, desorption, and recombination (Takeshita and Nakamura 1989, Harris 1990).

When the $Nd_2Fe_{14}B$ intermetallic compound is heated under a constant hydrogen pressure at 0.1 MPa at 850–950 °C, it firstly absorbs hydrogen to form $Nd_2Fe_{14}BH_x$ (hydrogenation), and decomposes to form fine grains of NdH_2 hydride, Fe_2B, and α-Fe (decomposition). Then the NdH_2 hydride desorbs hydrogen (desorption) as the temperature is raised and recombines with Fe_2B and α-Fe to form the original $Nd_2Fe_{14}B$ intermetallic compound (recombination). These processes are applied as a useful method for texture control of Nd–Fe–B magnet powder and are briefly illustrated in Fig. 9 (Takeshita 1995, see *Magnets: HDDR Processed*).

The ternary Nd–Fe–B magnet powder produced by the HDDR process is magnetically isotropic but it possesses a high coercivity force (>10 kOe). The coercivity force is one of the most important magnetic properties of a permanent magnet which is strongly dependent on the microstructure of the magnetic materials. The coercivity increases as the particle size decreases until the particle size of a magnetic single domain is reached. When the particle size decreases beyond the magnetic single domain, it causes a decrease in coercivity because of the appearance of behavior as a single paramagnetic material with a giant magnetic moment (formation of supermagnetic system). The desorption temperature and time are considered important for enhancing the coercivity.

The HDDR process has been applied for improving the isotropic nature of Nd–Fe–B magnet powders that has the magnetic energy less than 25% of an anisotropic magnet. The substitution of iron with cobalt and addition of zirconium, hafnium,

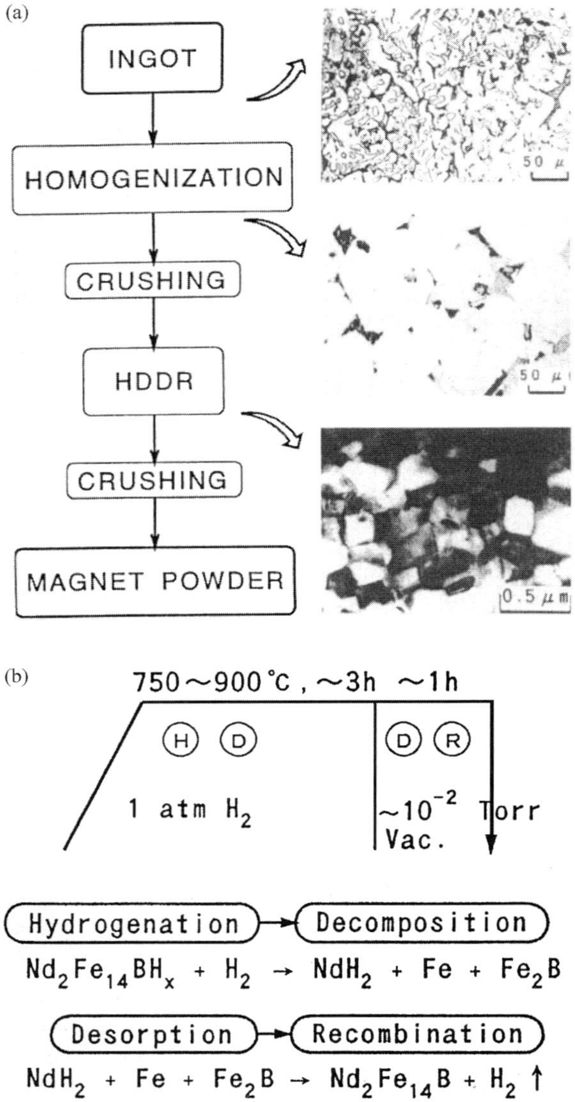

Figure 9
(a) Production flow-chart of the HDDR process for the Nd–Fe–B system and the microstructures at individual steps (Takeshita 1995). (b) HDDR process parameters for the production of Nd–Fe–B magnet powders (Takeshita 1995).

niobium, gallium, and other metals is effective for enhancing the magnetic anisotropy in the HDDR magnet powders of the Nd–Fe–B-based magnet systems. Takeshita (1995) presented an overview on the HDDR process. Okada *et al.* (1995) studied the macrostructural evolutions during HDDR phenomena in $Sm_2Fe_{17}N_x$ compounds from the processing viewpoints. The procedure is illustrated in Fig. 10.

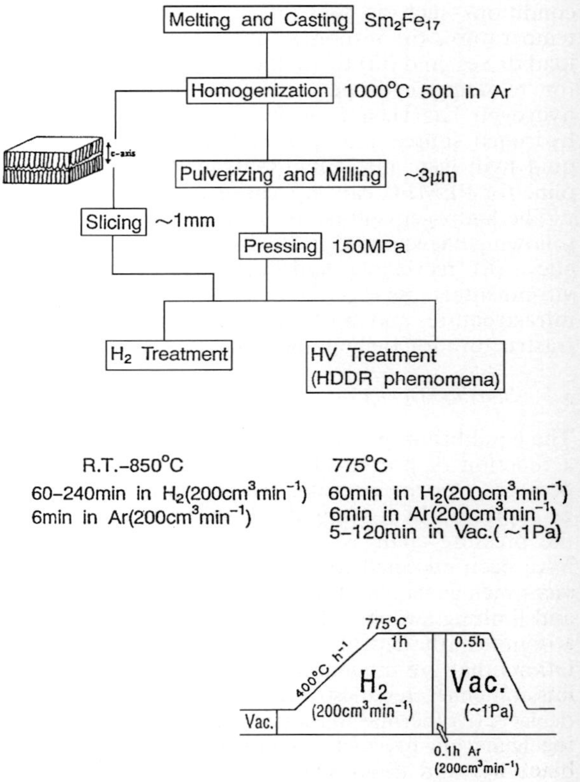

Figure 10
Procedures used for the production of Sm–Fe–N magnet powders (Okada *et al.* 1995).

3.8 Superplasticity

Materials with high elongation for high strength can be obtained a H/D process other than the HDDR process. The H/D process composed of (i) hydrogenation, (ii) heat treatment and hot working, and (iii) dehydrogenation in vacuum at elevated temperature has been found possible to bring the grain size down to 1–3 μm. By changing the above three conditions, the grain size can reduced down to 0.3–0.5 μm.

Superplasticity was observed in the $(\alpha+\beta)$Ti–6Al–4V by the H/D process and the elongation over 6000% has been attained at a testing temperature of 1073 K and a strain rate of $1\times10^{-3}\,s^{-1}$ (Yoshimura 1999a). No practical application has been developed at present.

3.9 Separation and Purification

The metal hydride surface is easily poisoned and deactivated by contact with air, moisture, CO, NO_2, SO_2, H_2S, and other materials. This is caused by the extreme reactivity of the activated surface of metal hydride powders. It should be remembered that the metal hydride surface is reactive in the gas–solid system not only with hydrogen molecules but also other materials.

The only possibility for the separation and purification is to develop the surface that does not exhibit reactivity against impurities (Wang *et al.* 1995). A fluorination method has been developed for this aim (see *Hydrogen–Metal Systems: Fluorinated Metal Hydrides*). However, the protective surface has been known to be destroyed rapidly by repeating H/D cycles and to lose the hydrogen capacity drastically due to the appearance of fresh surfaces by the pulverization of particles. Presently, no practical separation and purification process has been developed except the small laboratory devices. It is unsuitable to apply the H/D processes for purifying hydrogen in the industrial scale, particularly under the existence of H_2S and CO.

3.10 *Hydrogen Compressors*

The pure hydrogen gas can be easily compressed by applying a thermodynamic cycle if there are the enough heat source available. The maximum pressure attained by the compression is dependent on the P–T relation of a hydrogen-metal hydride system and is limited by the critical point of the selected system.

However, it should be noted that the amount of high-pressure hydrogen obtained by this type of a batch-wise operation is limited considerably by the (i) amount of hydrogen-absorbing material, (ii) switchable numbers of H/D cycles per unit time, and (iii) heat transfer characteristics of pressure vessels.

It is estimated that the installation and operating costs must be higher than the conventional compressors. The compression system is only available for the pure hydrogen gas but not for the hydrogen gas with air, moisture, and other gases. No industrial application has been reported.

4. *Concluding Remarks*

Features of the hydrogen-metal systems can be placed in the reversible H/D reactions that accompany the transfers of endothermic and exothermic thermal energy. The transition of hydrogen between molecular and protium states and between the proton and protium states attracts engineering interest both in gas–solid reactions and electrochemical reactions.

Recent innovative findings are the optical properties of yttrium hydride as optical switches and the magnetic properties of Nd–B–Fe that apply the HDDR process in which hydrogen has important roles for improving magnetism and coercivity. Superplasticity of Ti-based alloys was reached over 8000 times in its elongation by the H/D process.

However, there are inherent shortcomings in the hydrogen-metal systems that make any engineering devices unpractical. It is concluded that the conventional hydrogen-metal systems are not suitable for hydrogen storage applications for both stationary and mobile uses. The thermodynamic devices such as heat pumps and refrigerators will not be realized commercially because the large quantity of the thermal mass retained by the metal hydride bed that reduces the thermodynamic efficiency and the cost-effectiveness.

It is suggested that the extremely reactive nature of protium will be applied in various areas of engineering chemistry and electrochemistry. The scopes for searching hydrogen storage materials should be extended to the light-metal hydrides and hydrogen-metal complexes both as the gas–solid and gas–liquid systems.

See also: Hydrogen–Metal Systems: Basic Properties (1); Hydrogen–Metal Systems: Hydride Forming Alloys Properties and Characteristics, Database Information)

Bibliography

Bogdanovic B, Schwickardi M 1997 Ti-doped alkali metal aluminum hydrides as potential novel reversible hydrogen storage materials. *J. Alloys Comp.* **253–4**, 1–9

Friedlmeier G 1997 *Charakterisierung von Hochtemperatur-metallhydriden auf Magnesium-basis.* Fortschr.-Ber VDI Reihe5, VDI Verlag, Dusseldorf, Germany

Griessen R, Huiberts J N, Kremers M, Gogh A T M, Koeman N J, Dekker J P, Notten P H L 1997 Yttrium and lanthanum hydride films with switchable optical properties. *J. Alloys Comp.* **253–4**, 44–50

Gross K J, Jensen C, Takara S, Thomas G J 2000a Catalyzed alanates for hydrogen storage. In: *Proc. Int. Symp. Metal-Hydrogen Systems.* Noosa, Australia, p. 287

Gross K J, Thomas G 2000b Real time X-ray diffraction of catalyzed alanates. In: *Proc. Int. Symp. Metal-Hydrogen Systems.* Noosa, Queensland, Australia, p. 173

Harris I R 1990 *Proc. 11th Workshop on Rare-Earth Magnets and their Applications.* Pittsburgh, PA, pp. 21–24.

Huiberts J N, Griessen R, Rector J H, Wijngaarden R J, Dekker J P, de Groot D G, Koeman N J 1996 *Nature.* **380**, 231

Jensen C M, Zidan R, Mariels N, Hee A, Hagen C 1999 Advanced titanium doping of sodium aluminum hydride: segue to a practical hydrogen storage material? *Int. J. Hydrogen Energy* **24**, 461–5

Lewandowski B, Seidl T, Takara S, Sun D, Jensen C M 2000 Disassociation/reassociation kinetics of catalytically enhanced $NaAlH_4$. In: *Proc. Int. Symp. Metal-Hydrogen Systems.* Noosa, Queensland, Australia, p. 288

Nagel M 1989 *Untersuchung des Betriebsverhaltens einer periodisch mit Metall-hydriden arbeitenden Kaltemaschine.* Fortschr.-Ber VDI Reihe 19, VDI Verlag, Dusseldorf, Germany

Nagel M, Komazaki Y, Matsubara Y, Suda S 1986a Dynamic behaviour of paired metal hydrides II: analytical survey of the experimental results. *J. Less-Common Met.* **123**, 47–58

Nagel M, Komazaki Y, Suda S 1986b Dynamic behavior of paired metal hydrides I: Experimental method and results. *J. Less-Common Met.* **120**, 45–53

Nakahigashi J, Yoshimura H 2000 Superplasticity of fine-grained titanium alloys obtained through protium treatment. *J. Jpn. Soc. Mech. Eng.* (in Japanese). **60**, in press

Okada M 1998 *Progress Report on the New Protium Function in Sub-Nano Lattice Matters (in Japanese). Grant-in-Aid for Scientific Research on Priority Areas (A)*

Okada M 1999 *Progress Report on the New Protium Function in Sub-Nano Lattice Matters (in Japanese). Grant-in-Aid for Scientific Research on Priority Areas (A)*

Okada M 2000 *Progress Report on the New Protium Function in Sub-Nano Lattice Matters (in Japanese). Grant-in-Aid for Scientific Research on Priority Areas (A) (in press)*

Okada M, Saito M, Nakamura H, Sugimoto S, Homma M 1995 Macrostructural evolution during HDDR phenomena in $Sm_2Fe_{17}N_x$ compounds. *J. Alloys Comp.* **231**, 60–5

Sandrock G 1996 Task 12-hydrogen storage in metal hydrides. *IEA Agreement on the Production and Utilization of Hydrogen Annual Report*, pp. 63–6

Sandrock G 1997 Task 12-metal hydrides and carbon for hydrogen storage. *IEA Agreement on the Production and Utilization of Hydrogen Annual Report*, pp. 41–65

Sandrock G 1998 Task 12-metal hydrides and carbon for hydrogen storage. *IEA Agreement on the Production and Utilization of Hydrogen 1998 Annual Report*, pp. 29–38

Sandrock G 1999 Task 12-Metal hydrides and carbon for hydrogen storage. *IEA Agreement on the Production and Utilization of Hydrogen 1999 Annual Report.* (in press)

Sandrock G, Gross K, Thomas G, Jensen C, Takara S 2000 Engineering considerations in the use of catalyzed sodium alanates for hydrogen storage. In: *Proc. Int. Symp. Metal-Hydrogen Systems.* Noosa, Queensland, Australia, p. 278

Suda S 1999 Catalytic generation of hydrogen from metal hydride solutions. In: *1st Int. Symp New Protium Function in Materials*, Osaka, Japan, p. 17

Suda S 2000a Metal-hydrogen complex ions as high H-capacity storage materials and their application to fuel cells. In: *Proc. 10th Canadian Hydrogen Conf. Quebec*, Canada, p. 36

Suda S 2000b Jpn. pat.

Suda S, Iwata K, Sun Y M, Komazaki Y, Liu F J 1997 A pressure–temperature swing process for the methanation of carbon oxides using fluorinated hydriding alloys. *J. Alloys Comp.* **253–4**, 668–72

Suda S, Komazaki Y, Narasaki H, Uchida M 1991 Development of a double-stage heat pump: Experimental and analytical surveys. *J. Less-Common Met.* **172–4**, 1092–110

Suda S, Sandrock G 1994 Three decades of intermetallic hydrides—What happened to the applications? *Z. f. Physik. Chem.* **183**, 149–56

Suda S, Sun Y M, Liu B H, Zhou Y, Arai K, Iwata K, Uchida M, Li Z P 2000 Aqueous hydrogen-metal systems for hydrogen storage. In: *Proc. Int. Symp. Metal-Hydrogen Systems.* Noosa, Queensland, Australia, p. 284

Suda S, Sun Y M, Liu B H, Zhou Y, Morimitsu S, Arai K, Tsukamoto N, Uchida M, Candra Y 2001 Catalytic generation of hydrogen by applying fluorinated-metal hydrides as catalysts. *J. Appl. Phys. A.* (in press)

Suda S, Sun Y M, Uchida M, Liu B H, Morimitsu S, Arai K, Zhou Y,Tsukamoto N, Candra Y, Li Z P 2000a Fluorinated metal hydrides for the catalytic hydrolysis of metal-hydrogen complexes. *In: 2nd Korean Inst. Metals Materials—Japanese Inst. Metals Joint Symp. Hydrogen Absorbing Materials*, Seoul, Korea, pp. 19–28

Takeshita T 1995 Some applications of hydrogenation–decomposition–desorption–recombination-(HDDR) and hydrogen–decrepitation (HD) in metals processing. *J. Alloys Comp.* **231**, 51–9

Takeshita T, Nakamura R 1989 In: *Proc. 10th Int. Workshop on Rare Earth Magnets and Their Applications.* Kyoto, Japan, p. 551

Thomas G J, Gross K J, Yang N Y C 2000 Microstructural characterization of catalyzed $NaAlH_4$. In: *Proc. Int. Symp. Metal-Hydrogen Systems.* Noosa, Queensland, Australia, p. 289

Wang X L, Iwata K, Suda S 1995 Hydrogen purification using fluorinated $LaNi_{4.7}Al_{0.3}$ alloy. *J. Alloys Comp.* **231**, 860–4

Yamashita I, Zidan R A, Takara S, Jensen C M 2000 Thermodynamic properties of catalytically enhanced sodium aluminum hydride. In: *Proc. Int. Symp. Metal-Hydrogen Systems.* Noosa, Queensland, Australia, p. 29

Yoshimura H, Nakahigashi J 1999a Fine-grained structures and superplasticity of protium-treated (a-b) type titanium alloys. In: *Proc. 1st Int. Symp. New Protium Function in Materials*, Osaka, Japan, p. 9

Yoshimura H, Nakahigashi J 1999b Tensile and impact properties of mesoscopic-grained $\alpha+\beta$-type titanium alloys obtained through hydrogen treatments. *J. Alloys Comp.* **293–5**, 858–61

Hydrogen–Metal Systems: Basic Properties (1)

K. Tanaka and O. Yoshinari

Hydrogen-storage alloys have attracted much attention in recent years because they are going to act as key materials in the forthcoming age of hydrogen energy. They can absorb and desorb hydrogen gas quickly and almost reversibly. A hydrogen molecule (H_2) sticking on a metal surface is dissociated into two hydrogen atoms at the surface, which then diffuse into the inside of the metal independently. The amount of dissolved hydrogen in equilibrium with the surrounding hydrogen gas increases with increasing gas pressure. If the hydrogen concentration reaches a certain limit, a hydride precipitates in the metal. Hydrogen is mainly stored in the form of hydrides in hydrogen-storage alloys. The absorption and desorption of hydrogen in the alloys can be controlled by adjusting the external hydrogen pressure and the temperature. This means that the amount of stored hydrogen in an alloy is essentially determined by the thermodynamics of the hydrogen–metal (H–M) system. However, for a practical use of hydrogen-storage alloys, subsidiary knowledge concerning the reaction kinetics with hydrogen gas, possible deviations from the thermodynamic relations, and deterioration of the storage properties are required.

In this article several fundamental aspects of the H–M system, such as the thermodynamics, energetics of metal hydrides, equilibrium process of absorption and desorption, surface properties, reaction kinetics, and alloy degradation, are briefly reviewed.

1. *Thermodynamics of Hydrogen–Metal Systems*

1.1 Phase Diagram of the Pd–H System

The Pd–H system has been one of the most intensively studied systems to date. This is because palladium manifests a higher surface reactivity with hydrogen and a simpler thermodynamic behavior than any other ordinary metal. In the following, the thermodynamics of the H–M system is reviewed by making special reference to the Pd–H system. Figure 1 shows the phase diagram of the Pd–H system. In the primary solid solution (α-phase), hydrogen atoms randomly occupy the interstitial sites of the host metal. At temperatures below 293°C, a new phase (α') is formed in a higher concentration region, and it coexists with the α-phase in the intermediate concentration region. The higher concentration phases beyond the primary solid solution phase are generally called hydrides. In the case of the Pd–H system, the α'-phase is apparently similar to the α-phase; both phases commonly adopt the f.c.c. structure but have quite different hydrogen concentrations.

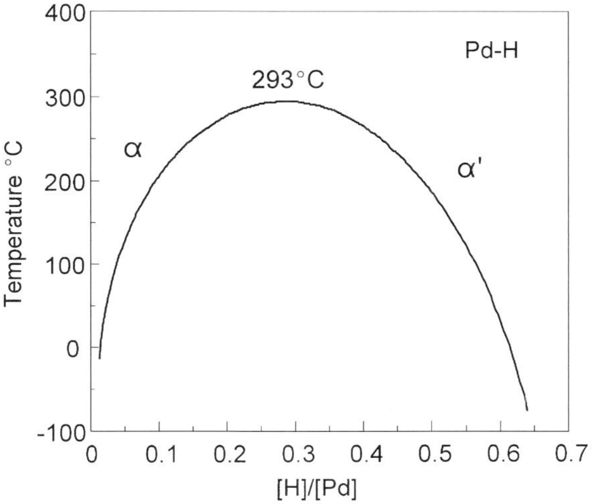

Figure 1
Phase diagram of the Pd–H system (after Massalski 1990).

1.2 Solid Solution of Hydrogen

The thermal equilibrium between hydrogen in the solid solution phase (α-phase) and the external hydrogen gas H_2 is given by

$$\mathrm{M} + \tfrac{1}{2}c\mathrm{H}_2 \leftrightarrow \mathrm{MH}_c \qquad (1)$$

where c is the hydrogen concentration of the α-phase defined as the ratio of the molar number of H atoms to that of metal (M) atoms, i.e., $c = [\mathrm{H}]/[\mathrm{M}]$. The chemical potential of hydrogen in the α-phase can then be written as

$$\mu^\alpha = h^\alpha - Ts^\alpha + RT\ln\frac{c}{r-c} \qquad (2)$$

where r is the maximum hydrogen concentration in the α-phase, h^α is the partial molar enthalpy and s^α is the nonconfigurational (vibrational) part of the partial molar entropy in the α-phase. The third term originates from the configurational entropy of hydrogen atoms. The zero of energy is taken at the energy of a single H atom in vacuum. Alternatively, the chemical potential of H_2 gas is obtained as

$$\mu^g = RT\ln\frac{p}{p_0(T)} - E_\mathrm{d} \qquad (3)$$

where p is the gas pressure and E_d is the heat of dissociation of molar H_2 molecules (432.1 kJmol^{-1}), i.e., $-E_\mathrm{d}$ is the molar enthalpy of the hydrogen molecules. The quantity $p_0(T)$ is independent of the pressure but dependent on the temperature as $T^{7/2}$. Since the equilibrium condition is given by $\mu^{\alpha}=(1/2)\mu^{g}$, the following relation holds:

$$\frac{c}{r-c}=\left(\frac{p}{p_0(T)}\right)^{1/2}\exp\left(\frac{s^{\alpha}}{R}\right)\exp\left(-\frac{E_d/2-h^{\alpha}}{RT}\right) \quad (4)$$

When the hydrogen concentration is small ($c \ll r$), c is approximately given by

$$c \approx K_S^{-1}\sqrt{p} \quad (5)$$

where

$$\ln K_S = \tfrac{1}{2}\ln p_0(T) - \frac{s^{\alpha}}{R} + \frac{E_d/2-h^{\alpha}}{RT} \quad (6)$$

Equation (5) is known as the Sieverts' law. The Sieverts' constant K_S cannot be given by a simple function of temperature since it involves the temperature-dependent term, $p_0(T)$. However, this term is usually neglected and the following relation is assumed:

$$\ln K_S = -\frac{\Delta S_S}{R} + \frac{\Delta H_S}{RT} \quad (7)$$

The parameters ΔS_S and ΔH_S are called the *solution entropy* and the *solution enthalpy*, respectively. It should be noted that ΔH_S, which can be obtained from the Arrhenius-type plot of experimentally obtained K_s versus T relations, implicitly involves a contribution from the temperature dependence of $p_0(T)$, and, therefore, becomes slightly more negative than $E_d/2-h^{\alpha}$.

Representative values of ΔS_S and ΔH_S for several metals obtained experimentally are given in Table 1. The negative value of ΔH_S means that the hydrogen solution is an exothermic reaction, and metals with large negative ΔH_S have large affinities for hydrogen.

Table 1
Solution enthalpy (ΔH_S) and entropy (ΔS_S) of hydrogen in metals.

Metal	ΔH_S (kJmol^{-1}H)	$\Delta S_S/R$ molH
Mg	+21	−4
Al	+67	−6
Y	−82	−6
La (f.c.c.)	−80	−8
Ce (f.c.c.)	−74	−7
Ti (h.c.p.)	−53	−7
Zr (h.c.p.)	−63	−6
V	−27	−8
Nb	−34	−8
Cr	+58	−5
Fe (b.c.c.)	+29	−6
Ni	+16	−6
Pd	−10	−7
Cu	+42	−6

Source: Fukai 1993, p. 7.

1.3 Phase Separation

The phase separation observed in the Pd–H system below 293°C is discussed in the following. It is known that the solution enthalpy, ΔH_S, becomes more negative with increasing hydrogen concentration. This means that the hydrogen atoms go into more stable sites at high hydrogen concentrations. This phenomenon is understood in terms of an attractive elastic interaction between hydrogen atoms. The host metal lattice is expanded by hydrogenation and, hence, the elastic energy for the hydrogen introduction is gradually reduced. Now, if h^{α} in Eqn. (2) changes with c as $h^{\alpha}=h^{0}-uc$ due to the elastic interaction, the chemical potential is also influenced by the interaction, as shown in Fig. 2. Since the configurational entropy term is dominant at high temperatures, μ^{α} monotonically increases with c. However, below a critical temperature T_c, it tends to decrease in the middle concentration range. The α solid solution phase is unstable in this range and decomposes into two phases with a lower concentration

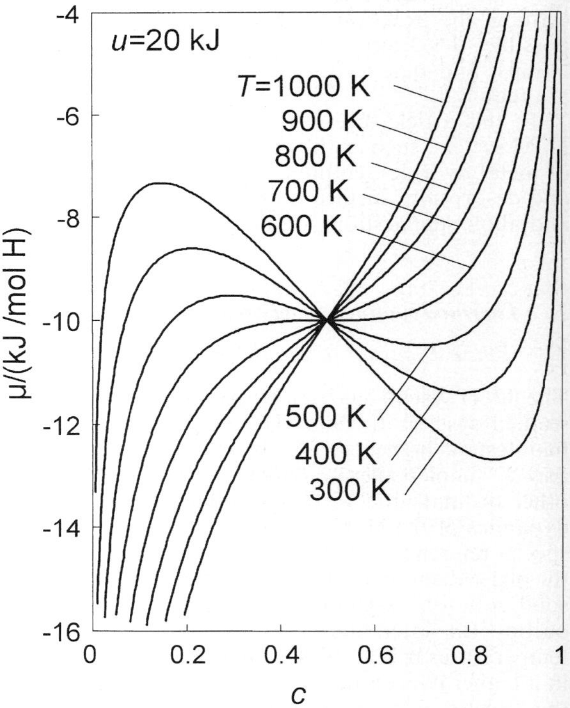

Figure 2
Chemical potential of the α-phase. The parameters in Eqn. (2) are set as follows; $h^{\alpha}=h^{0}-uc$ ($h^{0}=0$, $u=20$ kJ), $r=1$, $s^{\alpha}=0$.

c_1 and a higher concentration c_2. This change is called *spinodal decomposition*. The two concentrations, c_1 and c_2, are given by the so-called *Maxwell's equal-area construction*:

$$\mu^{\alpha}(c_1) = \mu^{\alpha}(c_2) = \frac{1}{c_2 - c_1}\int_{c_1}^{c_2} \mu^{\alpha}(c)\mathrm{d}c \equiv \mu(\mathrm{const}) \qquad (8)$$

The phase diagram of the phase-separation type is, therefore, obtained by calculating the terminal concentrations c_1 and c_2 that satisfy this equation at various temperatures below T_c. The critical temperature for the spinodal decomposition is calculated as $T_c = ur/4R$, and the critical concentration for the decomposition as $c_c = r/2$.

1.4 p–c–T Diagram

At a constant temperature T, the hydrogen concentration c varies with the hydrogen gas pressure p under the thermal equilibrium. The relationship between p and c at the temperature is called a *p–c–T* diagram (or a *p–c* isotherm). In the solid solution region p increases according to the Sieverts' law (Eqn. (5)). As c attains c_1 where the phase separation $\alpha \rightarrow \alpha + \alpha'$ begins, p takes a constant value. In the two phase region, $c_1 < c < c_2$, the molar ratio of the two phases changes with c while maintaining the respective concentrations c_1 and c_2.

The pressure increases again when the α-phase is totally transformed into the α'-phase. This change of the equilibrium pressure with hydrogen concentration is explained according to Gibbs' phase rule. Taking the number of components as C, the number of phases as P, and the degree of freedom as F, a relationship $F = C - P + 2$ is established among them. F indicates the number of intensive variables that can be changed externally. The phase rule for the H–M system is summarized in Table 2. The components of the system are always fixed to M and H, i.e., $C = 2$. In the α-phase region, $P = 2$ (α-phase + gas) resulting in $F = 2$. This means that when p and T are settled the remaining variables are automatically determined. In the two-phase region, the number of phases increases, i.e., $P = 3$, leading to $F = 1$. Here, only one variable can be chosen freely, causing the other to be fixed. Thus, for a given temperature $T < T_c$, a constant pressure p (plateau pressure) results and, furthermore, the compositions (c_1 and c_2) are fixed in this region. In the α'-phase region, the freedom recovers to $F = 2$, and p increases again with c.

Table 2
Application of Gibbs' phase rule to the H–M system.

	α-phase (Solid solution)	Coexistence of α- and α'-phases	α'-phase (Hydride)
C	2	2	2
P	2	3	2
F	2	1	2

1.5 Phase Diagrams of Other H–M Systems

Although the phase diagram of the Pd–H system is of a very simple phase-separation type, many other H–M systems are more complicated. In these systems, ordered phases usually appear at low temperatures where hydrogen atoms preferentially occupy particular interstitial sites that are configured regularly. Two examples of this type are shown in Figs. 3 and 4.

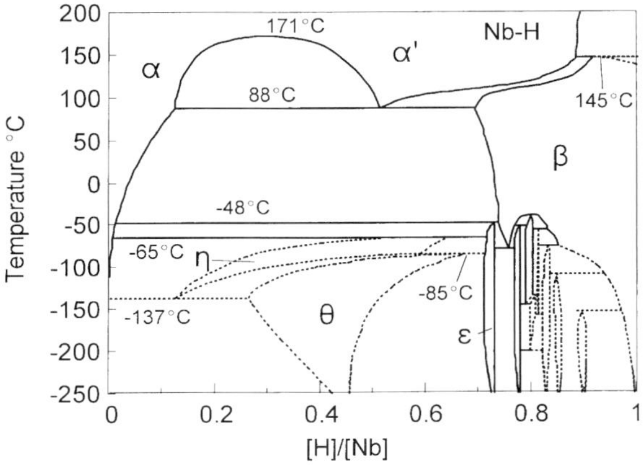

Figure 3
Phase diagram of the Nb–H system (after Massalski 1990).

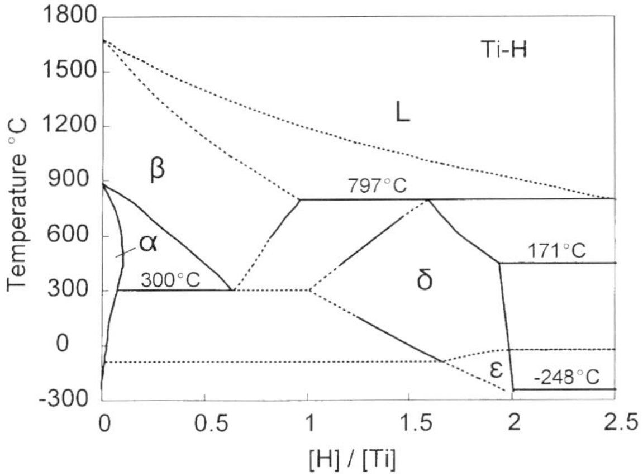

Figure 4
Phase diagram of the Ti–H system (after Massalski 1990).

In the Nb–H system, a phase separation similar to Pd–H takes place in the high temperature region, but at lower temperatures several ordered phases appear. In the Ti–H system, hydrogen atoms occupy the interstitial sites in the host h.c.p. lattice (α-phase) while a b.c.c. solid solution phase (β-phase) appears in a higher concentration range. This is because the hydrogenation stabilizes the b.c.c. phase (the high-temperature phase of pure titanium) relative to the α-phase. This effect causes the Ti–H system to show a phase diagram of the eutectic type.

1.6 Formation of Hydrides

As can be seen from the phase diagrams of the H–M systems, a hydride phase generally precipitates in the solid solution α-phase at appropriate temperatures. The hydride phase is designated as the β-phase. The reaction forming the hydride of concentration c_β from the α-phase of concentration c_α is expressed as

$$\frac{2}{c_\beta - c_\alpha}MH_{c_\alpha} + H_2 \leftrightarrow \frac{2}{c_\beta - c_\alpha}MH_{c_\beta} + Q$$

The formation enthalpy of the reaction, $\Delta H_{\alpha\rightarrow\beta}$, is taken as the value per mole H_2 gas. The reaction is exothermic when $\Delta H_{\alpha\rightarrow\beta}<0$; that is, associated with the reaction, a heat $Q=-\Delta H_{\alpha\rightarrow\beta}>0$ is released. Taking the formation energies of the α-phase and the β-phase as ΔH_α and ΔH_β, respectively, $\Delta H_{\alpha\rightarrow\beta}$ is given by

$$(c_\beta - c_\alpha)\Delta H_{\alpha\rightarrow\beta} = c_\beta \Delta H_\beta - c_\alpha \Delta H_\alpha \quad (10)$$

If $c_\beta \gg c_\alpha$, it follows that $\Delta H_{\alpha\rightarrow\beta} \sim \Delta H_\beta$. Alternatively, the formation entropy $\Delta S_{\alpha\rightarrow\beta}$ is given as

$$\Delta S_{\alpha\rightarrow\beta} = \frac{2}{c_\beta - c_\alpha}(S_\beta - S_\alpha) - SH_2^0 + R\ln\frac{p}{p^0} \quad (11)$$

where S_α and S_β are the entropies of the α and β-phases, respectively, $S^0_{H_2}$is the entropy of hydrogen gas at standard pressure ($p^0 = 1$ atm). When this reaction is in equilibrium condition, the following relation holds:

$$\Delta H_{\alpha\rightarrow\beta} = T\Delta S_{\alpha\rightarrow\beta} \quad (12)$$

Accordingly,

$$\ln\frac{p}{p^0} = \frac{\Delta H_{\alpha\rightarrow\beta}}{RT} - \frac{\Delta S^0_{\alpha-\beta}}{R} \quad (13)$$

The quantity $\Delta S^0_{\alpha\rightarrow\beta}$ is the standard entropy change at $p=p^0$, and is given by

$$\Delta S^0{}_{\alpha\rightarrow\beta} = \frac{2}{c_\beta - c_\alpha}(S_\beta - S_\alpha) - S^0{}_{H_2} \quad (14)$$

Thus, the formation enthalpy and the standard formation entropy can be experimentally obtained from Arrhenius-type plots of the equilibrium (plateau) pressures p against various measuring temperatures T. This plot is conventionally called a van't Hoff plot (see Fig. 5). The experimentally obtained values of the formation enthalpy and the standard formation entropy are given in Table 3. The values of $\Delta S^0_{\alpha\rightarrow\beta}/R$ are roughly constant regardless of the H–M systems. This is because $\Delta S^0_{\alpha-\beta}$ includes the loss of the entropy of H_2 gas, $-S^0_{H_2}$, which is considerably larger in magnitude than the entropy change between the α- and β-phases.

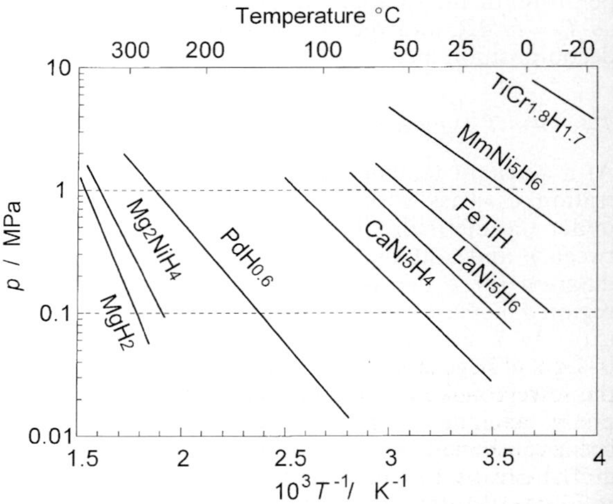

Figure 5
van't Hoff plot of various H–M systems (after Sandrock *et al.* 1992).

Table 3
Enthalpy and entropy of formation of hydrides.

System	$H_{\alpha\rightarrow\beta}$ ($kJmol^{-1}H_2$)	$S_{\alpha\rightarrow\beta}/R$ $(molH_2)^{-1}$
Mg–MgH_2	−74	−16
Ca–CaH_2	−182	−16.8
Y–YH_2	−226	−17.4
La–LaH_2	−208	−18.2
Ce–CeH_2	−206	−17.8
Ti (h.c.p.)–TiH_2	−130	−12
V–$VH_{0.5}$	−72	−13
Nb–$NbH_{0.65}$	−92	−16
Ni–NiH	−58	−14
Pd–$PdH_{0.5}$	−40	−10
Mg_2Ni–Mg_2NiH_4	−64	−14.4
TiFe–TiFeH	−28	−12.4
$LaNi_5$–$LaNi_5H_4$	−32	−13

Source: Fukai 1993, p. 31.

2. *Semi-empirical Calculation of the Formation Enthalpy of the Hydride*

If the equilibrium pressure p in the two-phase region is given, the absorption or desorption temperature can be roughly estimated from the formation enthalpy of a hydride using Eqn. (13), since $\Delta S^0_{\alpha-\beta}$ is almost constant. Therefore, from the viewpoint of application of hydrogen storage alloys, it is very important to determine the alloy composition with a suitable value of formation energy. Griessen and Driessen (1984) suggested a semi-empirical method to obtain the formation enthalpy from the electronic band structures of host metals and alloys. According to this method, the formation enthalpy, $\Delta H_{\alpha\to\beta}$, of a binary hydride MH_{c_β} is given by

$$\Delta H_{\alpha\to\beta} = \frac{n_s}{2}(\alpha \Delta E + \beta) \qquad (15)$$

where $\alpha = 29.62\ \mathrm{kJeV^{-1}mol^{-1}H}$ and $\beta = 135.0\ \mathrm{kJmol^{-1}H}$ are empirical parameters, and ΔE is defined as

$$\Delta E \equiv E_F - E_s \qquad (16)$$

In Eqn. (15), n_s is the number of electrons per atom in the lowest *s*-like conduction band of the host metal and is taken as $n_s = 1$ for the alkali metals and $n_s \sim 2$ for the others. E_s is the energy for which the integrated density of states of the host metal is equal to $n_s/2$, and E_F is the Fermi energy that corresponds to the energy where the integrated density of states reaches the number of valence electrons of the metal. In Fig. 6, E_s and E_F are indicated for Pd. The calculated values of the formation enthalpies of binary hydrides are compared with experimental values in Fig. 7. The nearly linear correlation between them suggests that the formation enthalpy of a binary hydride may be approximately obtained if the band structure of the host metal is given.

The heat of formation of a ternary hydride $A_{x_A}B_{1-x_A}H_{c_\beta}$ is also obtained by extending the method to binary alloys. Since the band structures of binary alloys in general have not been calculated, the band structure of a binary A–B alloy has been constructed from the respective band structures of pure metals A and B with an approximation called the coherent-potential method. The formation enthalpies of ternary hydrides have been calculated in the same manner as the binary hydrides. As shown in Fig. 8, the calculation qualitatively reproduces the relative trend of the experimental values.

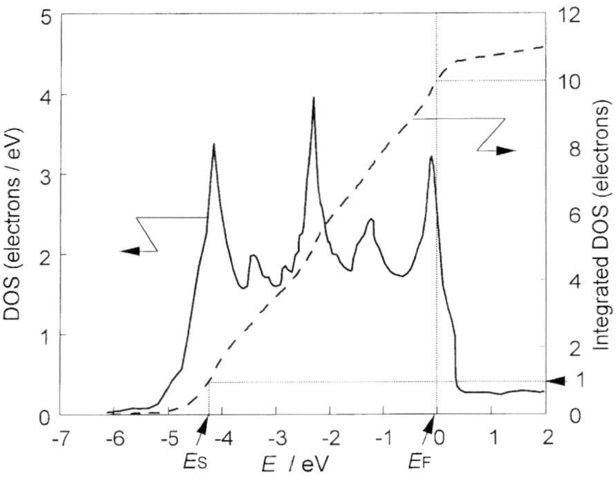

Figure 6
Electronic density of states of Pd. Positions of E_s and E_F are indicated (after Moruzzi *et al.* 1978).

3. *Hysteresis*

The hydrogen pressure for the formation of a hydride is generally different from that for the decomposition of the hydride. Figure 9 shows a schematic *p–c–T* diagram for the hydrogen absorption and desorption processes. Such a difference is also found in the van't Hoff plots; the slope of the plot for the desorption process is larger than that for the absorption process. This means that an excess energy is needed for the hydrogenation. The specific volume of a hydride is larger than that of the host metal and, therefore, the matrix around a hydride is subjected to a large internal stress. If the stress exceeds the elastic limit of

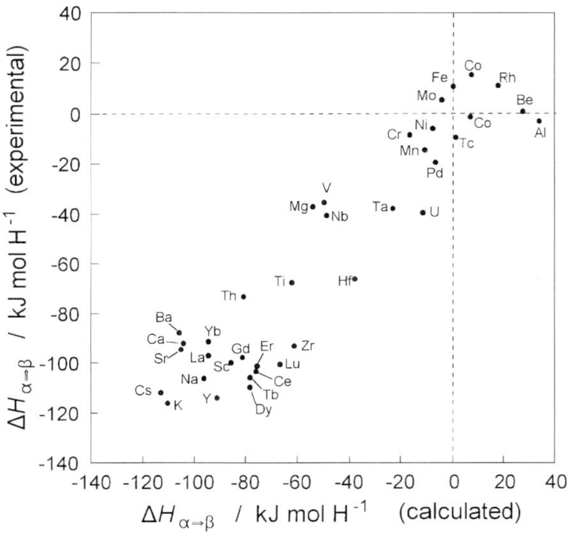

Figure 7
Relation between the experimental and calculated formation enthalpies by the semi-empirical band structure model (after Griessen and Riesterer 1992).

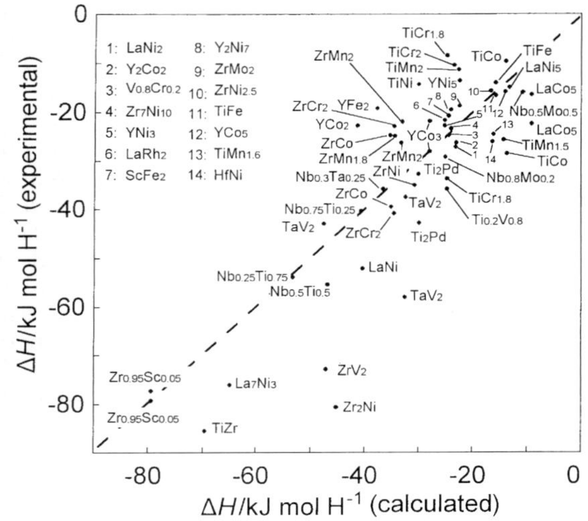

Figure 8
Relation between the experimental and calculated formation enthalpies of ternary hydrides by the semi-empirical band structure model (after Griessen and Reisterer 1992).

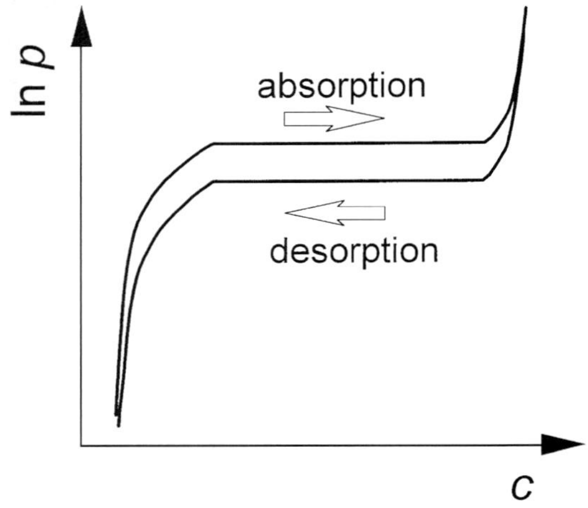

Figure 9
Schematic *p–c–T* diagram for hydrogen absorption and desorption.

the host metal, dislocations are generated around the hydrides. Alternatively, in the dehydrogenation process, the dislocations remain almost unchanged.

The mechanical loss associated with the irreversible plastic deformation during the hydrogenation–dehydrogenation cycle is the origin of the hysteresis. Generally, the magnitude of the hysteresis is larger for a hydride with larger specific volume. However, in

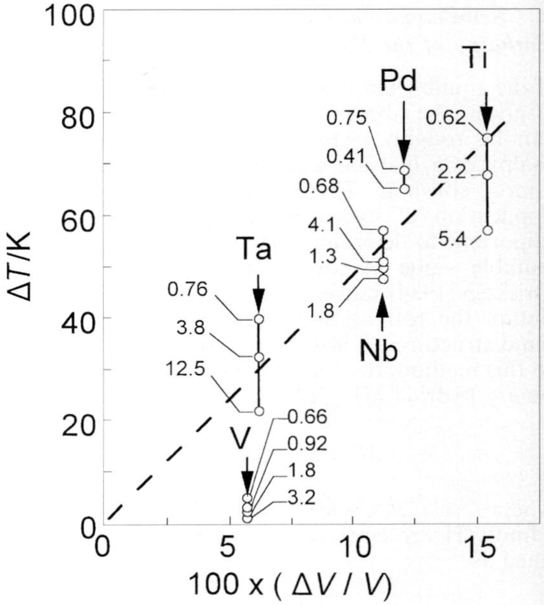

Figure 10
Thermal hysteresis with the precipitation and dissolution of hydrides (Yoshinari and Koiwa 1987). The magnitude of the thermal hysteresis is plotted as a function of the relative difference in specific volumes of metals and metal hydrides, $\Delta V/V$. The number for each data point indicates the hydrogen concentration in at.%.

certain H–M systems, hydrides precipitate in the form of plates and, therefore, the internal stress in the matrix is substantially reduced. The hysteresis is reduced in this case. The V–H system is the typical example of the latter case. Figure 10 summarizes the relationship between the volume change and the thermal hysteresis associated with the hydride precipitation and dissolution in several H–M systems. The hydride of the V–H system (V_2H) has a b.c.t. structure with large axial ratio ($c/a \sim 1.1$) and the elastic energy associated with the hydride precipitation can be reduced by selecting the crystallographic orientation of the hydride platelets.

4. *Surface Properties*

The hydrogen absorption process on a clean metal surface may be represented as a 1D potential model, as shown in Fig. 11. This process can be divided into several subprocesses as follows:

(i) *Outside of metal.* The states of a hydrogen molecule H_2 and two dissociated hydrogen atoms 2H are shown by two separated curves. At a place far from the surface, the difference between the two

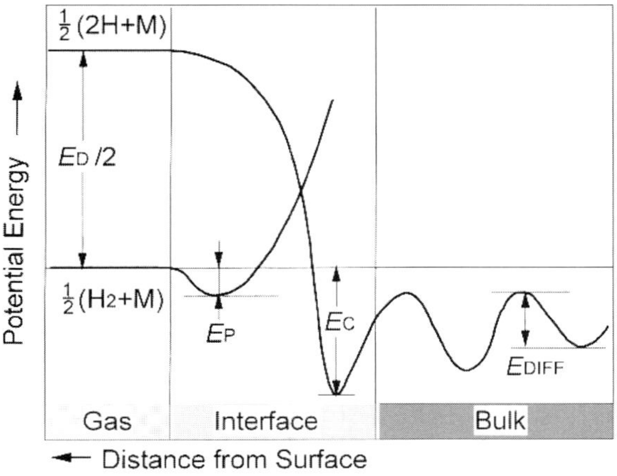

Figure 11
Potential energy curves for a hydrogen molecule across a clean metal surface (after Schlapbach 1992).

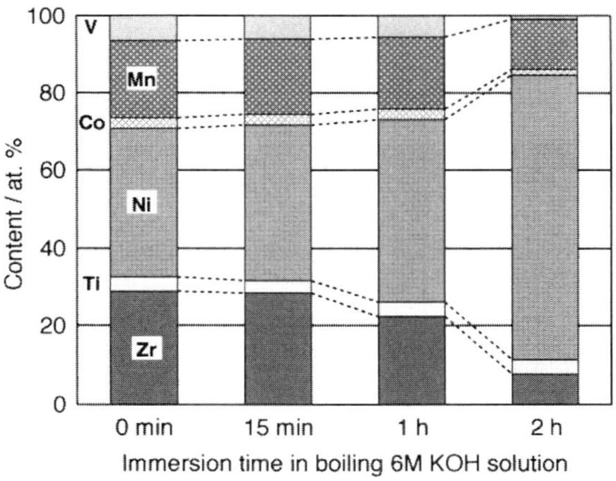

Figure 12
Potential energy contours for a Ni_2–H_2 cluster as a function of the H–H distance, D, and the distance between Ni_2 and H_2 pairs, Z (Harris and Anderson 1985). AU refers to the atomic unit where 1 AU = 0.053 nm and corresponds to the Bohr radius of the hydrogen atom.

curves equals the heat of dissociation $E_D/2 = 218\,kJmol^{-1}H$

(ii) *Physisorption*. The minimum of the "$H_2 + M$" curve corresponds to the physisorption of the H_2 molecule. The physisorption energy, E_P, is about $10\,kJmol^{-1}H$

(iii) *Chemisorption*. The curves "$H_2 + M$" and "$2H + M$" intersect, beyond which the latter shows a deep minimum of chemisorption state. The heat of chemisorption, E_C is about $50\,kJmol^{-1}H$

(iv) *Surface migration*. The adsorbed hydrogen atoms migrate on the metal surface (not shown in the Fig. 11). They interact with one another and form a surface phase. In further steps a hydrogen atom finds an appropriate entrance point and penetrates into the inside of the metal

(v) *Diffusion*. Hydrogen atoms migrate in the metal by a diffusion process. The energy of the hydrogen site E_S is different for different metals. If $E_S < 0$, the reaction of hydrogen dissolution is exothermic

The absorption and desorption of hydrogen are reversible processes, i.e., the desorption occurs tracing the reverse course (v)→(i).

The situation of the chemisorption process can be more readily understood by observing the potential energy against the distance D between H–H atoms of a H_2 molecule. Figure 12 is a result of the calculation of the potential energy for a Ni_2–H_2 cluster, where equi-potential lines are drawn as a function of D and Z, the distance between the Ni–Ni cluster and the H_2 molecule. Although this potential is calculated for a small cluster, it may reflect to some extent the situation of a H_2 molecule in the vicinity of a nickel surface. The potential is independent of Z when $Z > 5$ AU and it takes a minimum at $D = 1.4$ AU, i.e., the interatomic distance of a free H_2 molecule. The H–H distance increases rapidly for $Z < 4$ AU, and the potential takes a negative minimum at $D = 3.5$ AU and $Z = 2.6$ AU. This minimum corresponds the chemisorbed state.

The process of hydrogen absorption becomes complicated when the surface is covered with oxides or certain segregates of alloy components.

5. *Kinetics of Hydride Formation*

The kinetics of the formation process of hydrides is considerably complicated. For practical use of hydrogen-storage alloys, however, it is important to ascertain the rate-determining step among many steps influencing the process, such as gas transfer, transmission from surface to bulk, diffusion, nucleation and growth of the hydrides, etc.

Usually, the amount of hydrides formed is experimentally obtained as a function of time as illustrated in Fig. 13. When the time dependence is sigmoidal (Fig. 13(a)), the nucleation or the growth is considered to be the rate-determining process. Alternatively, if the reaction rate is decreasing with time (Fig. 13(b)), a process other than the nucleation and growth determines the rate. It is important to note that the rate is much affected by the surface condition of the material. This is greatly changed by the pretreatment before hydrogenation, the initialization (first hydrogenation) condition and the post-contamination, etc.

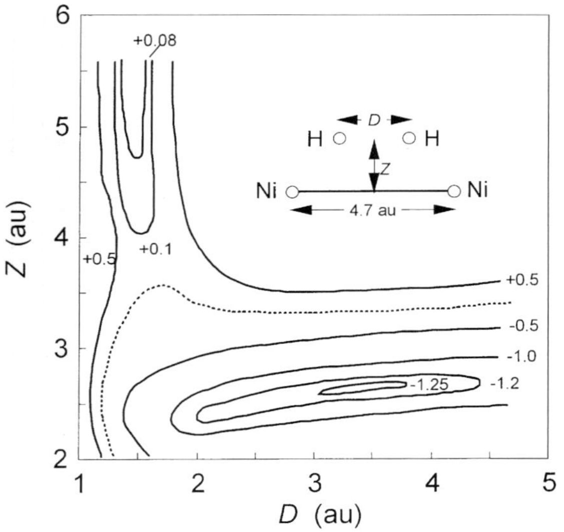

Figure 13
Time dependence of the hydride formation (after Gérard and Ono 1992). The transformation rate, ξ, is defined as $\xi(t)=$(amount of hydride at time t)/(amount of hydride at equilibrium condition).

The heat transfer is also an important factor that affects the hydride-forming kinetics. For the exothermic reaction as is the case of hydrogen-storage alloys, the released heat acts to suppress the hydrogenation. Since this effect becomes more important as the amount of material increases, the elimination of the heat is indispensable for a practical use of the hydrogen storage alloys.

6. *Disproportionation Reaction*

In hydriding reactions of alloys containing more than two metal species, decomposition of the host alloys sometimes occurs. For an A-B alloy, where A refers to a hydride former and B a nonformer, the normal hydride-forming reaction is expressed as

$$\frac{2}{c_\beta - c_\alpha} A_{x_A} B_{1-x_A} H_{c_\alpha} + H_2 \leftrightarrow \frac{2}{c_\beta - c_\alpha} A_{x_A} B_{1-x_A} H_{c_\beta} + Q \quad (17)$$

However, by repeating the hydriding–dehydriding cycle, the following decomposition reaction is occasionally induced:

$$\frac{2}{c'_\beta x_A} A_{x_A} B_{1-x_A} + H_2 \rightarrow \frac{2}{c'_\beta} AH_{c'_\beta} + \frac{2(1-x_A)}{c'_\beta x_A} B + Q' \quad (18)$$

This is called a disproportionation reaction and it proceeds as an irreversible process. This reaction effectively reduces the hydrogen-storage capacity of the alloy and leads to its degradation. Most of the hydrogen-storage alloys tend to be deteriorated through this reaction. However, they can usually be recovered by annealing in vacuum at appropriate temperatures.

See also: Hydrogen–Metal Systems: Applications of Gas–Solid Reactions; Hydrogen–Metal Systems: Electrochemical Reactions (Fundamentals and Applications); Hydrogen–Metal Systems: Technological and Engineering Aspects

Bibliography

Fukai Y 1993 *The Metal Hydrogen Systems*. Springer, Berlin
Gérard N, Ono S 1992 Hydride formation and decomposition kinetics. In: Schlapbach L (ed.) *Hydrogen in Intermetallic Compound II*. Springer, Berlin, Chap. 4, pp. 165–95
Griessen R, Driessen A 1984 Heat of formation and band structure of binary and ternary metal hydrides. *Phys. Rev. B* **30**, 4372–81
Griessen R, Riesterer T 1992 Heat of formation model. In: Schlapbach L (ed.) *Hydrogen in Intermetallic Compound II*. Springer, Berlin, Chap. 6, pp. 219–84
Harris J, Anderson S 1985 H_2 dissociation at metal surfaces. *Phys. Rev. Lett.* **55**, 1583–6
Massalski T B 1990 *Binary Alloy Phase Diagrams*, 2nd edn., ASM International, Materials Park, OH
Moruzzi V L, Janak J F, Williams A R 1978 *Calculated Electronic Properties of Metals*. Pergamon, New York
Sandrock G, Suda S, Schlapbach L 1992 Applications. In: Schlapbach L (ed.) *Hydrogen in Intermetallic Compound II*. Springer, Berlin, Chap. 5, pp. 197–258
Schlapbach L 1992 Surface properties and activation. In: Schlapbach L (ed.) *Hydrogen in Intermetallic Compound II*. Springer, Berlin, Chap. 2, pp. 15–95
Yoshinari O, Koiwa K 1987 Low-frequency internal friction study of Pd–H alloys. *J. Phys. F* **17**, 59–74

Hydrogen–Metal Systems: Electrochemical Reactions (Fundamentals and Applications)

C. Iwakura, H. Inoue, and S. Nohara

Hydrogen storage alloys have not yet been put to practical use on a large scale, except for the nickel–metal hydride battery (nickel–hydrogen battery) in which metal hydride is used as the active material of the negative electrode. In relation to the nickel–metal hydride battery, electrochemical characteristics of hydrogen storage alloys have been extensively investigated from both fundamental and practical viewpoints.

Further information can be found in a short review concerning electrochemical utilization of metal hydrides by Bittner and Badcock (1983), a detailed review of various types of secondary batteries and fuel cells using hydrogen storage alloy by Iwakura (1985), and more comprehensive reviews by Iwakura *et al.* (1991a, 1991b) and Sakai *et al.* (1995). This article describes several selected terms for electrochemical characterization of hydrogen storage alloys with special emphasis placed on nickel–metal hydride batteries.

1. Electrochemical Properties and Characteristics of Metal Hydrides

Half-cell reactions at positive and negative electrodes in nickel–metal hydride batteries and the overall reaction are represented in the following equations.

$$(\text{Positive})\text{Ni(OH)}_2 + \text{OH}^- = \text{NiOOH} + \text{H}_2\text{O} + \text{e}^- \quad (1)$$

$$(\text{Negative})\text{M} + \text{H}_2\text{O} + \text{e}^- = \text{MH} + \text{OH}^- \quad (2)$$

$$(\text{Overall})\text{Ni(OH)}_2 + \text{M} = \text{NiOOH} + \text{MH} \quad (3)$$

where M and MH denote a hydrogen storage alloy and the metal hydride, respectively. During charging the positive electrode, nickel hydroxide ($Ni(OH)_2$), is oxidized to nickel oxyhydroxide (NiOOH) (Eqn. (1)) while at the negative electrode metal hydride (MH) is formed by water electrolysis (Eqn. (2)). As a result, the overall reaction is represented by a transfer of hydrogen between M and $Ni(OH)_2$ (Eqn. (3)).

The charging process at the negative electrode is represented by the following equations in more detail:

$$\text{H}_2\text{O} + \text{e}^- = \text{H}_{\text{ad,s}} + \text{OH}^-(\text{charge transfer}) \quad (4)$$

$$2\text{H}_{\text{ad,s}} = \text{H}_2(\text{hydrogen evolution}) \quad (5)$$

$$\text{H}_{\text{ad,s}} + \text{H}_2\text{O} + \text{e}^- = \text{H}_2 + \text{OH}^-(\text{hydrogen evolution}) \quad (6)$$

$$\text{H}_{\text{ad,s}} = \text{H}_{\text{ab,s}}(\text{hydrogen absorption}) \quad (7)$$

$$\text{H}_{\text{ab,s}} = \text{H}_{\text{ab,b}}(\text{hydrogen diffusion}) \quad (8)$$

where $H_{ad,s}$, $H_{ab,s}$, and $H_{ab,b}$ represent adsorbed hydrogen on the alloy surface, absorbed hydrogen near the alloy surface, and absorbed hydrogen in the bulk alloy, respectively.

First, $H_{ad,s}$ is produced on the alloy surface by a charge transfer reaction (Eqn. (4)) known as the Volmer reaction. Then the atomic hydrogen is absorbed into the alloy (Eqn. (7)) in addition to hydrogen evolution due to a coupling of the atomic hydrogen, known as the Tafel reaction (Eqn. (5)), or an additional charge transfer of the atomic hydrogen known as the Heyrovsky reaction (Eqn. (6)). Equations (5) and (6) are side reactions in the charging process. The absorbed hydrogen diffuses from the surface to the bulk (Eqn. (8)) before hydride formation.

In general, battery performances such as high-rate capability, discharge efficiency, initial activation, discharge capacity, and charge retention (self-discharge rate) are greatly influenced by the electrocatalytic activity of hydrogen storage alloys for the charge transfer reaction (Eqn. (4)) and diffusivity of hydrogen in the alloys. Exchange current density and Tafel slope, b, are measures of the former while the diffusion coefficient (D) is a measure of the latter.

1.1 Exchange Current Density

At equilibrium the apparent current density in Eqn. (4) is zero, but the current density in the backward direction of Eqn. (4) is equal to that in the forward direction. The balanced current density is expressed in terms of exchange current density, i_0. In other words, the i_0 value is denoted as the current density at zero overvoltage. The i_0 value is generally determined from extrapolation of the linear logarithmic current density ($\log i_0$) – overvoltage (η) curves, called the Tafel line, at high values of $|\eta|$ (ca. >70 mV) to the ($\log i_0$) axis. Alternatively, the i_0 value can be determined using the following equation induced from the Butler–Volmer equation in the region of small overvoltage ($|\eta| < 5$ mV):

$$i = (i_0 RT/F)\eta \quad (9)$$

where R, F, and T are the gas constant, Faraday constant, and absolute temperature, respectively. Experimentally the i_0 value is determined by inserting the slope of the linear micropolarization curve in the region of small η into Eqn. (9). The i_0 value of hydrogen storage alloy electrodes has also been determined by using this equation. It has been reported that the discharge efficiency of $La_{0.8}Nd_{0.2}Ni_xCo_{2.4}Si_{0.1}$ ($2.5 \leqslant x \leqslant 3.0$) electrodes depends on the i_0 value or the electrocatalytic activity.

Microencapsulation of alloys with metals such as Cu or Ni is a useful method for improving the i_0 value. The alloy was first encapsulated with a copper layer by electroless plating. Subsequently it was discovered that microencapsulation of the alloy with Cu, Ni–P, and Ni–B and addition of metal oxide powders, such as RuO_2, Co_3O_4, CoO, etc., or surface modification of the alloy with an alkaline solution containing KH_2PO_2 or KBH_4 as a reducing agent remarkably increased the i_0 value or the electrocatalytic activity, greatly increasing the high-rate dischargeability.

Similar results were obtained for encapsulation with Ni–P or Ni alkali treatment with KBH_4. Surface treatment of the Laves-phase alloy electrode with a boiling 6 M KOH solution greatly increased the i_0 value even before activation, leading to rapid activation. By sintering $TiMnV_{0.5}$ alloy with nickel powder *in vacuo*, it was found that the i_0 value increased with the content of Ni due to an increase in electroactive TiNi formed during the sintering. The annealing of $ZrV_{0.7}Mn_{0.5}Ni_{1.2}$ alloy *in vacuo* increased the i_0 value due to increased Ni content in the main phase resulting from the disappearance of the second phase, leading to an improvement in rate capability.

Logarithmic current density–overvoltage curves of $LaNi_{5-x}Al_x$ and $LaNi_4Cu$ electrodes were derived in terms of diffusion theory, and parameters such as the i_0 value, transfer coefficient, and activation energy of diffusion were estimated.The relationship between logarithmic current density and overvoltage in charge transfer and diffusion processes was analyzed for the $TiMn_{1.5}H_x$ electrode. Using this derived equation, the i_0 value and symmetry factor were determined for $Ni_{64}Zr_{36}$ alloy. These factors were also evaluated for different $LaNi_5$-type alloys. Theoretical equations for the dependence of equilibrium potential and exchange current density on the surface hydrogen concentration were derived by assuming that hydrogen diffusion and charge transfer at the surface controlled the discharge process, and discharge curves were simulated.

The variation of the i_0 value with the charge state simulated theoretically agreed well with experimental data. From this simulation, a parameter showing the interaction between hydrogen in the alloy and unhydrided metal lattice was determined and it was found to alter significantly the shape of the predicted discharge curves. In the case of hydrogen-absorbing Ti–Ni alloys and electrocatalytic Mo–Co, Mo–Ni, and Zr–Ni alloys, intermetallic phases and alloys between two pure constituents obeyed typical volcano plots for both hydridic and electrocatalytic activity in accordance with the changes in the electronic density of states and well-defined crystal structures.

The exchange current densities for Volmer and Tafel reactions (i_{0V}, i_{0T}) were evaluated from extrapolation of Tafel lines for these reactions measured by a current interrupter method. It was found that acceleration of the Volmer reaction by surface modification with RuO_2 or Co_3O_4 powder or an increase in the B-site components in alloys such as $Mm(Ni_{3.6}Mn_{0.4}Al_{0.3}Co_{0.7})_{1.12}$ (Mm: mischmetal) was much greater than that of the Tafel reaction, resulting in the enrichment of adsorbed hydrogen or higher charging efficiency. It was also found, using the same method, that the Tafel reaction on the Laves-phase alloy electrodes was effectively suppressed with increasing unit cell volume of the alloys.

1.2 Tafel Slope

The Tafel slope is generally determined from a slope of the log $i_0 - \eta$ curves in the region $|\eta|$ ca. >70 mV and it is used as a measure of electrocatalytic activity as well as the i_0 value. The Tafel slope can also be used for elucidating the mechanism of electrode reactions. For example, hydrogen electrode reactions are composed of three elemental reactions as follows:

$$H_2O + e^- = H_{ad} + OH^- \text{(Volmer reaction)} \quad (10)$$

$$H_{ad} + H_{ad} = H_2 \text{(Tafel reaction)} \quad (11)$$

$$H_{ad} + H_2O + e^- = H_2 + OH^- \text{(Heyrovsky reaction)} \quad (12)$$

The hydrogen evolution reaction proceeds according to the Volmer–Tafel mechanism or Volmer–Heyrovsky mechanism. From a large number of investigations into the mechanism of the hydrogen electrode reactions, Tafel slopes for Volmer, Tafel, and Heyrovsky reactions have been experimentally and theoretically evaluated at approximately 120, 30, and 40 mV decade^{-1}, respectively. The hydrogen electrode reactions at a Pd electrode were investigated in detail by measuring the potential decay immediately after interruption of the applied current. The total overvoltage was successfully divided into two components corresponding to the Tafel and Volmer reactions and it was found that the reactions proceeded via the Volmer–Tafel mechanism. A current interrupter method was applied to various hydrogen storage alloy electrodes and it has been reported that the Volmer reaction was accelerated by mixing the alloy with RuO_2 or Co_3O_4 powder or by an increase in unit-cell volume of the alloy.

1.3 Diffusion Coefficient

The diffusion coefficient (D) of hydrogen in hydrogen storage alloys was evaluated by nuclear magnetic resonance and quasi-elastic neutron scattering techniques in 1970s, but the reported D values of hydrogen in $LaNi_5$ varied from $10^{-6}\,cm^2s^{-1}$ to $10^{-9}\,cm^2\,s^{-1}$ even when using the same method. Electrochemical evaluation of D has hardly been carried out except for wire-type or foil-type electrodes of hydrogen storage metals such as Pd and Ni. The battery performance of metal hydride electrodes is determined by diffusion of hydrogen in the alloy as well as by charge transfer reactions on the alloy surfaces. Values of D determined electrochemically for various alloys are summarized in Table 1.

Table 1
Diffusion coefficient of hydrogen evaluated by various electrochemical methods.

Alloy	Method	D ($cm^2 s^{-1}$)	Reference
$Ni_{64}Zr_{36}$	Chronopotentiometry	$(2.12-4.86)\times10^{-10}$	Iwakura *et al.* (1995)
$MmNi_{4.2}Al_{0.5}M_{0.3}$ (M = Cr, Mn, Fe, Co, Ni)	Potential step	$(1.62-3.24)\times10^{-8}$	Matsuoka *et al.* (1993)
$MmNi_{4.0-x}Mn_{0.75}Al_{0.25}Co_x$ ($x=0-0.6$)	Potential step	$(3.18-7.49)\times10^{-8}$	Notten and Hokkeling (1991)
$LaNi_5$	Potential step	4.3×10^{-8}	Ura *et al.* (1995)
$LaNi_{4.27}Sn_{0.24}$	Constant current discharge	6.75×10^{-11}	Iwakura *et al.* (1998a)
$LaNi_{4.25}Al_{0.75}$	Constant potential and current discharge	$(2.97-3.30)\times10^{-11}$	Tamura *et al.* (1983)
$Mg_{1.9}Y_{0.1}Ni_{0.9}Al_{0.1}$- 5 wt.% MO (MO = Ag_2O, Fe_2O_3, MoO_3, RuO_2, V_2O_5)	Constant potential discharge	$(1.28-1.37)\times10^{-11}$	Wakao and Yonemura (1983)
$LaNi_{4.27}Sn_{0.24}$	AC impedance spectrometry	$(1.67-3.85)\times10^{-11}$	Witham *et al.* (1997)

Chronopotentiometry was used for evaluating D in $Ni_{64}Zr_{36}$ ribbons and it was found that it depended strongly on hydrogen concentration, which could be accounted for in terms of the site distribution theory. Iwakura *et al.* (1995) evaluated D of α-hydrogen in $MmNi_{4.2}Al_{0.5}M_{0.3}$ (M = Cr, Mn, Fe, Co, Ni) alloys by the potential step method. They found that the Einstein diffusion coefficient of hydrogen increased with decreasing atomic radius of the foreign metals except for Mn, and this order is in good agreement with that of the enthalpy change of hydride formation. In addition, Iwakura *et al.* (1998a) reported that diffusion of α-hydrogen in $MmNi_{4.0-x}Mn_{0.75}Al_{0.25}Co_x$ decreased with increasing cobalt content because of the intensive attractive interaction between the absorbed hydrogen.

Ura *et al.* (1995) evaluated D by applying the potential step method to Pd and $LaNi_5$ microelectrodes which were constructed as a single particle in contact with a carbon fiber filament. The diffusion coefficient of hydrogen in palladized $LaNi_{4.25}Al_{0.75}$ and $LaNi_{4.27}Sn_{0.24}$was evaluated with constant-potential and constant-current discharge techniques by assuming spherical diffusion of hydrogen in alloy particles. According to this analysis, the diffusion coefficient of hydrogen in $Mg_{1.9}Y_{0.1}Ni_{0.9}Al_{0.1}$ ball-milled with various metal oxides was also evaluated and it was found that hydrogen diffusion behavior in the alloys hardly changed with modification by oxide additions. Moreover, the a.c. impedance method (see Sect. 11) was used for evaluating the diffusion coefficient of hydrogen in $LaNi_{4.27}Sn_{0.24}$.

2. *Theoretical Capacity of Hydrogen Storage Alloys*

Hydrogen storage alloys basically consist of elements A (La, Ti, Zr, Mg, etc.) and B (Ni, Co, Fe, Mn, etc.) whose hydrides are thermodynamically stable and unstable, respectively, and accordingly they are classified into intermetallic compounds of several types such as AB_5, AB_2, AB, and A_2B. The alloys of the solid-solution type such as V–Ti having a b.c.c. lattice are also well known.

If the composition of a metal hydride which is produced by absorption of hydrogen in a hydrogen storage alloy is known, the theoretical capacity (C_H) of the hydrogen storage alloy is calculated according to Faraday's law as follows:

$$C_H(\mathrm{mAh\,g^{-1}}) = xF/(3.6M) \tag{13}$$

where x, F, and M are the number of absorbed hydrogen atoms, the Faraday constant, and the molecular weight of the hydrogen storage alloy, respectively. For example, the theoretical capacity for $LaNi_5$ alloy is calculated as 372 mAh g^{-1} using Eqn. (13) because the metal hydride $LaNi_5H_6$ is formed by absorption of hydrogen, i.e., $x=6$.

The hydrogen contents of typical metal hydrides and their theoretical capacities are listed in Table 2. The capacities of the AB_2, AB, and A_2B type alloys are much larger than that for the $LaNi_5$ alloy. With increasing ratio of A to B the capacity has a tendency to increase. Nevertheless, there are also unfavorable tendencies such as a decrease in the rate of reaction with hydrogen and an increase in the reaction temperature. The V-based solid solutions have a markedly large theoretical capacity, although only half of the hydride phase VH_2 (γ phase) is available because the hydride phase VH (β phase) is very stable.

The electrochemical properties of these alloys have been improved by many methods such as substitution with foreign metals and surface modification, and the AB_5-type Mm-based and AB_2-type Zr-based multicomponent alloys have been mainly used as negative

Table 2
Hydrogen contents in typical metal hydrides and theoretical capacities (C_H).

Composition	Metal hydride	Hydrogen content		C_H (mAh g^{-1})
		(H/M)	(wt.%)	
AB_5	$LaNi_5H_6$	1.00	1.38	372
	$CaNi_5H_6$	1.00	1.78	482
AB_2	$Ti_{1.2}Mn_{1.8}H_3$	1.00	1.90	514
	$ZrMn_2H_3$	1.00	1.48	400
	$ZrV_2H_{4.5}$	1.50	2.30	625
AB	$TiFeH_2$	1.00	1.91	517
A_2B	Mg_2NiH_4	1.33	3.62	999
A	MgH_2	2.00	7.60	2205
	VH_2	2.00	3.82	1053

electrodes in commercialized nickel–metal hydride batteries. The Mg-based and V-based alloys have also received attention due to their markedly large theoretical capacities, and many efforts have been made to improve their electrochemical properties.

3. *Evaluation of Hydrogen Storage Alloys as Negative Electrode Materials*

For evaluating the applicability of various hydrogen storage alloys to negative electrodes in nickel–metal hydride batteries, the following properties are often characterized.

3.1 *Initial Activation*

An initial activation is generally performed by repeating charge–discharge cycles until constant discharge capacity is obtained. Ideally, the maximum discharge capacity should be obtained at the first cycle without any initial activation. However, when an alloy contains easily oxidizable constituent elements such as Ti or Zr a number of charge–discharge cycles is needed for activation of the alloy electrode due to the hindrance of hydrogen absorption and desorption by the formed passivation film. Therefore, in this case the suppression of the passivation is a key point.

3.2 *Discharge Capacity*

Discharge capacity is limited by the amount of absorbed hydrogen calculated from a pressure composition isotherm. A higher discharge capacity is desirable, but a higher utilization ratio or higher ratio of discharge capacity to charge capacity is also important. Since the discharge capacity is influenced by the crystal structure and unit-cell volume of alloys, equilibrium hydrogen pressure, rates of charge transfer on the alloy surface, hydrogen diffusion in the alloys, etc., detailed discussion of crystallographic and thermodynamic data is useful to improve the discharge capacity.

3.3 *Rate Capability (High-rate Dischargeability and High-rate Chargeability)*

High-rate chargeability and high-rate dischargeability are measures for evaluating rapid chargeability of batteries and suitability for batteries requiring high power such as electric tools and electric vehicles. Therefore, higher-rate capability (high-rate dischargeability and chargeability) is desirable. The electrocatalytic activity of the alloy surface and diffusibility of hydrogen in the alloy bulk are important for improving the rate capability.

3.4 *Charge/Discharge Cycle Life*

Charge/discharge cycle life is a measure of the durability of batteries; therefore, a longer cycle life is desirable. The cycle life is shortened by dissolution and passivation of the constituent elements, and pulverization of the alloy particles and their suppression is important.

3.5 *Self-discharge Rate (Charge Retention)*

Self-discharge is phenomenologically expressed as the decrease in battery capacity while the fully charged battery remains under open-circuit. Self-discharge rate or charge (capacity) retention is generally defined as the ratio of the discharge capacity after the battery has been allowed to stand for a certain period to the original discharge capacity and it is used as a measure of the self-discharge. In addition, the self-discharge rate is also defined as follows:

$$\text{self-dischargerate}(\%) = [1 - 2C_2/(C_1 + C_3)] \times 100 \quad (14)$$

where C_1 is the discharge capacity measured after activation by repeated charge–discharge cycling, C_2 is the discharge capacity measured after storage under open-circuit conditions for a given period of time, and C_3 is the discharge capacity measured in the charge–discharge cycle immediately after the measurement of C_2.

Ikoma *et al.* (1996) suggested the following five reasons as possible factors that would generally affect self-discharge in nickel–metal hydride batteries:

(i) self-decomposition of positive active material (NiOOH);
(ii) shuttle reactions of nitrate ions in the positive electrode;
(iii) reduction of a NiOOH positive electrode by hydrogen gas coming from the metal hydride negative electrode or oxidation of the negative electrode by oxygen gas coming from the positive electrode;
(iv) shuttle reactions of organic impurities dissolved in an alkaline electrolyte solution from a separator;
(v) shuttle reactions of metal ions dissolved in the alkaline electrolyte solution from the hydrogen storage alloy.

Ikoma *et al.* (1996) found that when conventional polyamide was used as a separator, ammonia (and amines such as hexamethylenediamine and adipate) produced by hydrolysis of the separator in an alkaline electrolyte solution participated in the following shuttle reactions:

(Positive electrode)

$$NH_4OH + 6NiOOH + OH^- \rightarrow 6Ni(OH)_2 + NO_2^- \quad (15)$$

$$NO_2^- + 2NiOOH + H_2O \rightarrow 2Ni(OH)_2 + NO_3^- \quad (16)$$

(Negative electrode)

$$NO_3^- + MH_x \rightarrow NO_2^- + MH_{x-2} + H_2O \quad (17)$$

$$NO_2^- + MH_x \rightarrow NH_4OH + MH_{x-6} + OH^- \quad (18)$$

It was concluded that the shuttle reactions between NO_3^- or NO_2^- and NH_4OH (factor (iv)) accelerate the self-discharge of both electrodes, which is similar to the role of NO_3^- in the nickel–cadmium battery. In addition, they demonstrated the superiority of non-woven fabric made of sulfonated polypropylene with hydrophilic properties (H-PP) over a polyamide (PA) separator. The capacity retention of the former greatly exceeded that of the latter separator for nickel–cadmium batteries, as shown in Fig. 1. In addition, they also suggested the possibility of factor (v) occurring. It has been reported that the self-discharge of the fully-charged metal hydride electrode in an open cell originated from the desorption of hydrogen from the metal hydride electrode and deterioration of the alloy. Microencapsulation of the

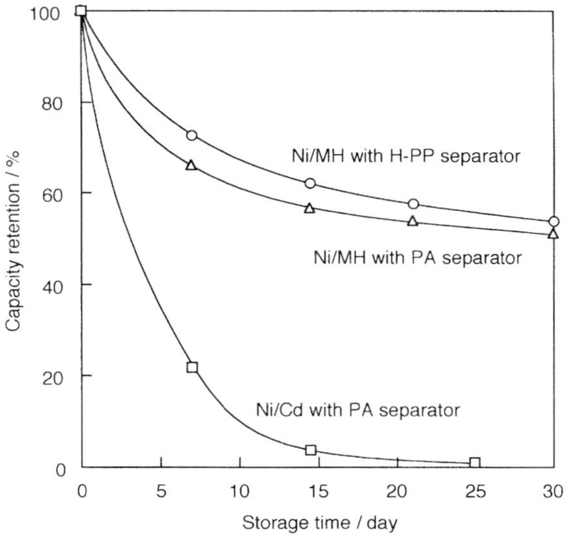

Figure 1
Effect of separator material on self-discharge of Ni/MH and conventional Ni/Cd batteries (after Ikoma *et al.* 1996).

alloy with Cu proved to be effective in reducing the self-discharge. It was also found that the addition of small mounts of Zr or Al to the alloy influenced the charge retention because the surface layer worked as a diffusion barrier against hydrogen release. The removal of lattice strain or increased flatness of the pressure plateau improved the charge retention for $Mm(NiCoMnAl)_5$ alloys.

Sealed all-solid-state nickel–metal hydride batteries made from alkaline solid polymer electrolyte based on poly(ethylene oxide), KOH, and a trace of water showed good self-discharge characteristics.

4. *Electrochemical Pressure–Composition Isotherms*

The relation between the equilibrium potential $E_{eq}(H_2O/H)$ and equilibrium hydrogen pressure $P(H_2)$ of a metal hydride electrode can be obtained via following process (Iwakura *et al.* 1988). The electrochemical hydriding/dehydriding reactions of the metal hydride electrode proceed as

$$H_2O + e^- = H + OH^- \quad (19)$$

$$(1/x)M + H = (1/x)MH_x \quad (20)$$

where H represents hydrogen dissolved in the metal or metal hydride phase. Since the equilibrium potential $E_{eq}(H_2O/H)$ is determined by the reaction of

Eqn. (19), it can be expressed according to the Nernst equation by

$$E_{eq}(H_2O/H) = E^0(H_2O/H) + (RT/F)\ln[a(H_2O)/a(H)a(OH^-)] \quad (21)$$

where $E^0(H_2O/H)$ is the standard electrode potential of the H_2O/H couple and $a(H_2O)$, $a(H)$, and $a(OH^-)$ are the activities of water, hydrogen atom, and hydroxide ion, respectively. Equation (21) indicates that $E_{eq}(H_2O/H)$ depends on $a(H)$.

Between $E^0(H_2O/H)$ of reaction (Eqn. (19)), $E^0(H_2O/H_2)$ of the hydrogen electrode reaction in alkaline solution (Eqn. (22)), and the equilibrium constant K of the hydrogen dissociation reaction in hydrogen gas (Eqn. (23)), the relation given in Eqn. (24) holds:

$$H_2O + e^- = (1/2)H_2 + OH^- \qquad \Delta G^0 = -FE^0(H_2O/H_2) \quad (22)$$

$$(1/2)H_2 = H \quad \Delta G^0 = -RT\ln K \quad (23)$$

$$E^0(H_2O/H) = E^0(H_2O/H_2) + (RT/F)\ln K = E^0(H_2O/H_2) + (RT/F)\ln[a(H)/a(H_2)^{1/2}] \quad (24)$$

Therefore, Eqn. (21) can be transformed into Eqn. (25):

$$E_{eq}(H_2O/H) = E^0(H_2O/H_2) + (RT/F)\ln[a(H_2O)/a(H_2)^{1/2}a(OH^-)] \quad (25)$$

Since the reaction and $E_{eq}(HgO/Hg)$ of a HgO/Hg electrode are expressed by Eqns. (26) and (27), respectively, the relation of the value for $E_{eq}(H_2O/H)$ to the HgO/Hg reference electrode is expressed by the fugacity coefficient $\gamma(H_2)$ and the equilibrium hydrogen pressure $P(H_2)$ as in Eqn. (28):

$$HgO + H_2O + 2e^- = Hg + 2OH^- \quad (26)$$

$$E_{eq}(HgO/Hg) = E^0(HgO/Hg) + (RT/2F)\ln[a(H_2O)/a(OH^-)^2] \quad (27)$$

$$\begin{aligned} E_{eq}(H_2O/H) - E_{eq}(HgO/Hg) &= [E^0(H_2O/H) - E^0(HgO/Hg)] + (RT/2F)\ln[/a(H_2O)/a(H_2)] \\ &= [E^0(H_2O/H) - E^0(HgO/Hg)] + (RT/2F)\ln[a(H_2O)/\gamma(H_2)P(H_2)] \end{aligned} \quad (28)$$

Wakao and Yonemura (1983) reported that the plateau pressures obtained from the electrochemical measurement of $LaNi_{5-x}M_x$ (M=Cu, Al) electrodes showed excellent agreement with those obtained by the gas method (Sieverts' method). Iwakura *et al.* (1988) calculated the equilibrium pressure according to Eqn. (28)s using the standard potential, the activity of water, and the fugacity coefficient in a concentrated KOH solution as a function of temperature and solution concentration. For example, the equilibrium potential at 20°C and 6 M KOH under 1 atm was expressed by the following equation:

$$E_{eq}(H_2O/H) - E_{eq}(HgO/Hg) = -0.9324 - 0.0291[\log P(H_2)] \quad (29)$$

For $LaNi_{4.5}M_{0.5}$ and $LaNi_4M$ (M=Cu, Cr, Mn, Al) electrodes, the change of $E_{eq}(H_2O/H)$ with quantity of electric charge during charge and discharge was investigated, and it was confirmed that these experimental $E_{eq}(H_2O/H)$ values agreed well with the theoretical values calculated from $P(H_2)$ obtained by the gas method using Eqn. (29), as shown in Fig. 2.

Electrochemical pressure–composition–temperature isotherms (P–C–T curves) were obtained at different temperatures for various alloys to evaluate the charge and discharge pressures of the electrodes. As can be seen in Fig. 3, the electrochemical P–C–T curves for $Mm(NiMnAlCo)_5$ alloy electrodes correspond closely to those measured by Sieverts' method above a hydrogen pressure of 0.01 MPa. Furthermore, they can be measured to much lower pressure ranges than with Sieverts' method.

The electrochemical method is useful for obtaining P–C–T curves of metal hydride electrodes over a wide pressure range with a small amount of sample (at least several mg). A pressure change of one order corresponds to a change in potential of 29 mV according to Eqn. (29). If the $E_{eq}(H_2O/H)$ value changes from −0.93V to −0.64V vs. HgO/Hg, $P(H_2)$ changes from

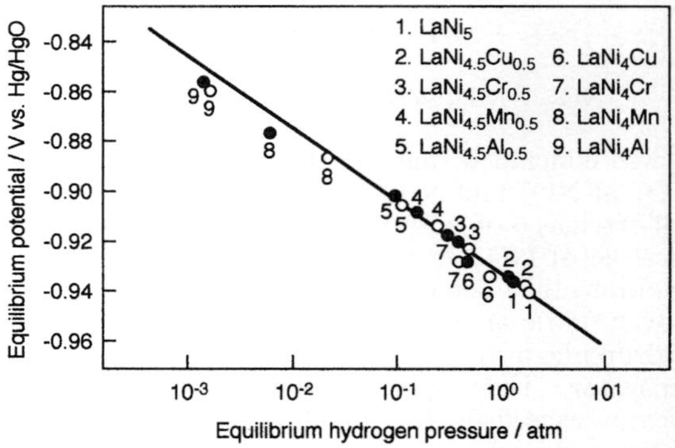

Figure 2
Plot of equilibrium potential against equilibrium hydrogen pressure in each plateau region for $LaNi_{4.5}M_{0.5}$ and $LaNi_4M$ negative electrodes: ○ charge, hydrogen absorption; • discharge, hydrogen desorption; straight line: theoretical line represented by Eqn. (29) (after Iwakura *et al.* 1988).

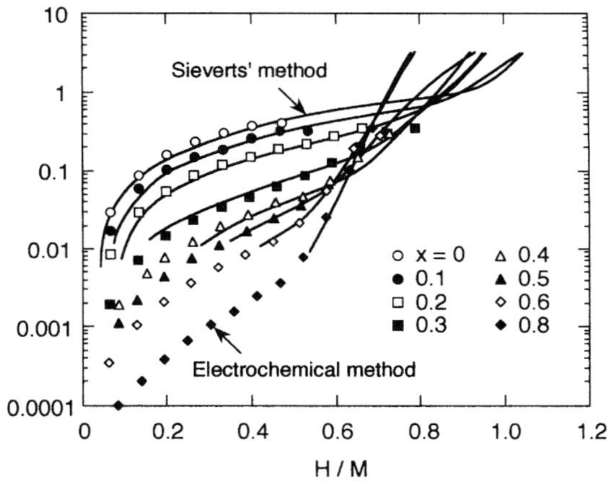

Figure 3
Pressure–composition isotherms for hydrogen absorption in $MmNi_{3.9-x}Mn_{0.4}Al_xCo_{0.7}$ alloy–hydrogen systems at 25 °C measured by Sieverts' method (solid line) and electrochemical method (symbols) (after Senoh *et al.* 2000).

1 atm to 10^{-10} atm. This means that it is not necessary to control exactly the alloy plateau pressure, as is generally the case for conventional applications such as storage tanks, chemical heat-pumps, and purification systems. This allows a wider choice of alloys for nickel–metal hydride batteries.

However, some control of the hydrogen pressure is still important for electrode alloys. If $P(H_2)$ is higher than the cell pressure the charging efficiency of hydrogen decreases, leading to lower storage capacity and hydrogen gas evolution. The influence of the cell pressure and temperature on the discharge capacity of $LaNi_{5-x}M_x$ ($x = 0, 0.5, 1$, M = Cu, Cr, Al, Mn) alloy electrodes was investigated and the observations were explained on the basis of the P–C–T curves of the alloy–hydrogen system (Iwakura *et al.* 1988). The $P(H_2)$ value should be controlled, depending on the purpose of the battery, e.g., high-temperature use, low-temperature use, and high-power use. For ordinary batteries the $P(H_2)$ value should be 0.5–5 atm at 45 °C. The importance of suitable control of the plateau pressure for obtaining high charging efficiency was confirmed for $MmNi_{4.0}Mn_{0.2}Al_{0.2}Co_{0.6}$ by changing the rare earth composition (La, Ce, Nd) in the Mm component.

5. Electrochemical Performance of Metal Hydride Electrodes

The electrochemical performance of metal hydride negative electrodes for use in nickel–metal hydride batteries depends on discharge capacity, rate capability (high-rate dischargeability or high-rate chargeability), initial activation, charge/discharge cycle life, self-discharge rate (charge and capacity retention), kinetics of charge transfer reaction on the alloy surface, and hydrogen diffusion in the alloy. Mischmetal-based alloys such as $MmNi_{3.55}Co_{0.75}Mn_{0.4}Al_{0.3}$ used as a negative electrode material for a commercial nickel–metal hydride battery show good electrochemical performance. However, the electrochemical performance of the alloy is poorer under severe conditions such as temperatures lower than 0 °C or higher than 40 °C. Furthermore, the replacement of expensive Co by the other cheaper elements is recommended for reducing battery costs. In this section the improvement of electrochemical performance at higher or lower temperatures and the development of low-Co or Co-free metal hydride electrodes are introduced.

5.1 Temperature Dependence of the Electrochemical Performance of Metal Hydride Negative Electrodes

Since nickel–metal hydride batteries are used under a variety of circumstances it is important for them to operate satisfactorily over a wide range of temperatures. In general, when the temperature is raised the initial activation rate and high-rate dischargeability of the metal hydride electrodes increase, but the cycle life shortens and the self-discharge rate increases. The discharge capacity tends to decrease both at low and high temperatures compared with near room temperature.

It was shown in the literature that $LaNi_{4.5}Al_{0.5}$ and $LaNi_{4.25}Al_{0.75}$ alloys maintained the discharge capacity at 30 °C and even at 45 °C. It was also found that charge–discharge performance of $LaNi_{4.7}Al_{0.3}$ alloy at low temperatures was improved by encapsulation with Cu. Those Zr–Ti–Cr–Mn–Co–V–$NiMm_{0.05}$ alloys with C14-hexagonal structure showed good dischargeability (290–340 mAh g^{-1}) and relatively good rate capability even at low temperatures.

The temperature effect of the electrochemical performance of a crystalline Mg_2Ni alloy prepared by induction melting was investigated. Although the alloy showed poor charge–discharge characteristics at 30 °C due to the extraordinary stability of its hydride, the discharge capacity of the alloy increased greatly at 70 °C. This was the first case of electrochemical desorption of most of the hydrogen absorbed in the Mg_2Ni alloy. This was deduced from charge–discharge curves as well as measured P–C–T curves.

5.2 Electrochemical Performance of Low-Co or Co-free Metal Hydride Electrodes

Cobalt is an essential element for a long charge–discharge cycle life due to the reduction in the lattice expansion of the alloys upon hydriding (Willems 1984), as shown in Table 3. It leads to some ductility and to good conductive networks on the alloy surface

Table 3
Initial storage capacity ($C_t(0)$), amount of absorbed hydrogen (x), and stability factor (S_{400}) (after Willems 1984).

Compound		$C_t(0)$ (mAh g^{-1})	x (H/AB_5)	S^a_{400}
$LaNi_{5-z}Co_z$	$z=0$	372	6	0.12
	1	372	6	0.25
	2	342	5.5	0.30
	2.5	324	5.2	0.45
	3	292	4.7	0.61
	3.3	273	4.4	0.64
$LaNi_{5-z}Co_zAl_{0.1}$	$z=0$	355	5.8	0.16
	1	365	5.9	0.30
	2	315	5.1	0.59
	2.5	313	5.1	0.82
	3	289	4.7	0.94
$LaNi_{5-z}Co_zSi_{0.1}$	$z=1$	350	5.7	0.33
	2	310	5.0	0.65
	2.5	296	4.8	0.81
	3	280	4.6	0.96
$La_{0.8}Nd_{0.2}Ni_3Co_2$		302	4.9	0.48
$La_{0.7}Nd_{0.3}Ni_{2.5}Co_{2.4}Al_{0.1}$		293	4.8	0.87
$La_{0.8}Nd_{0.2}Ni_{2.5}Co_{2.4}Si_{0.1}$		293	4.7	0.88
$La_{0.7}Nd_{0.2}Ti_{0.1}Ni_{2.5}Co_2Al_{0.5}$		254	3.9	0.90

$S^a_{400} = C_t(400)/C_t(0)$, $C_t(400)$: discharge capacity after 400 cycles.

due to the dissolution of Co and subsequent reprecipitation on the alloys. Therefore, Co has been considered to be indispensable for preparing stable AB_5-type alloys. However, since it is an expensive element, the preparation of low-Co or, preferably, Co-free alloys with battery performances that are comparable with conventional Co-containing alloys is an important research and development subject.

Nonstoichiometric single phase alloys in which Ni was partially replaced by Cu were prepared by the melting of raw materials followed by annealing. It was found that the cycle life was strongly affected by the degree of nonstoichiometry and chemical composition and that $LaNi_{4.4}Cu$ was an optimum alloy in terms of both discharge capacity and long-term cycle stability. Such an improvement was attributed to a crystal structure change with a miscible gap between the α and β phases, resulting in a decrease in the unit-cell volume change in the α–β phase transformation.

However, the above idea was not discovered while attempting to prepare Co-free alloys. Two kinds of Co-free alloys were specifically prepared by another group. First, Co-free AB_5-type alloys containing silicon were prepared, of which $Lm_{0.5}Mm_{0.5}Ni_{4.2}Mn_{0.2}Al_{0.3}Si_{0.3}$ (Lm = La-rich mischmetal) showed a maximum capacity of 270 mAh g^{-1} and good cyclability. Second, iron-containing AB_5-type alloys were prepared, of which $MmNi_{3.6}Fe_{0.7}Al_{0.3}Mn_{0.4}$ exhibited a maximum capacity of 280 mAh g^{-1} and good cyclability. The alloys containing both Si and Fe such as $LmNi_{4.3}Si_{0.1}Fe_{0.6}$ and $LmNi_{4.0}Al_{0.3}Si_{0.1}Fe_{0.6}$ were also found to show a relatively high discharge capacity and good cycle performance.

It has been reported that among various Co-free and low-Co AB_5-type alloys, $LmNi_{3.3}Al_{0.7}Cu_{0.8}Co_{0.2}$ is the most appropriate alloy in terms of the discharge capacity, cyclability, high-rate dischargeability, and cost.

A low-Co $MmNi_{3.65}Co_{0.22}Mn_{0.36}Al_{0.27}Cr_{0.2}Cu_{0.2}Si_{0.1}$ alloy with a discharge capacity of 273 mAh g^{-1} and good cyclability was also prepared. It was suggested that a secondary phase network that formed between the grain boundaries of the main phase as a result of the combined replacement by Cr, Cu, and Si was responsible for the improved cycle life. In addition, it was found that the annealing did not improve the cyclability of the low-Co alloy. The Co-free $MmNi_{3.60}Mn_{0.36}Al_{0.54}Cu_{0.4}Si_{0.1}$ alloy showed a more satisfactory cyclability, comparable to that of the commercial $MmNi_{3.55}Co_{0.75}Mn_{0.4}Al_{0.3}$ alloy with high Co content, but its discharge capacity was low (254 mAh g^{-1}).

Moreover, it has been reported in the literature that Co-free $La_{0.8}Y_{0.2}Ni_{4.7}Sn_{0.3}$ and $La_{0.9}Ni_{5.05}Sn_{0.35}Pb_{0.04}$ alloys showed a relatively high discharge capacity and long cycle life. A series of Co-free

Table 4
Composition of second phase analyzed by x-ray diffraction patterns: discharge capacity at two rates (C_1: 200 mA g^{-1}, C_2: 50 mA g^{-1}, 1st cycle) and their ratio for $MmNi_{3.2}CoAl_{0.2}Mn_{0.6}X_{0.09}$. (after Nogami et al. 1993).

Element X	Composition of second phase	C_1 (mAh g^{-1})	C_2 (mAh g^{-1})	$C_1/C_2 \times 100$ (%)
Mo	Mo–Ni–Co	255	294	86.6
B	$MmCo_4B$	239	286	83.6
Ta	$TaNi_3$, Ta_2Co	239	286	83.6
W	W	223	301	74.2
Zr	$ZrCo_2$	200	292	68.4
no additive		142	295	48.2

nonstoichiometric $La(Ni, Sn)_{5+x}$ alloys was also prepared, and an optimum composition in terms of the discharge capacity and cycle life was $LaNi_{4.84}Sn_{0.32}$.

6. *Initial Activation Characteristics*

For most of the hydrogen storage alloy electrodes, electrochemical activation is necessary before the discharge capacity reaches its maximum value. During the activation cycles, the alloy particles are cracked by hydriding/dehydriding volume changes, e.g., 15–25% for $LaNi_5$, causing the appearance of many new fresh surfaces and diffusion of electrolyte solution inside the particles. The increase in the reaction surface area results in increases in discharge capacity and high-rate dischargeability with increasing cycle number. Moreover, if a porous and Ni-rich layer on the alloy surface is formed by the dissolution of some elements into the alkaline solution during the cycles it works as a good electric conductor, catalyst, and protective layer, leading to an improvement in electrochemical properties. In order to reduce the initial activation cycles but still benefit from the good electrochemical properties of the early cycles, promotion of cracking of the alloy particles and formation of the electrocatalytic surface layer are essential. Many methods such as the addition of foreign metals and chemical treatment of the alloy surface have been used.

Nogami *et al.* (1993) reported that second-phase formation on the grain boundaries prompted by the addition of B, W, Ta, and Mo improved the high-rate dischargeability of $Mm(NiCoMnAl)_5$-based alloys in the first cycle (Table 4). The second-phase formation facilitates the cracking of the alloy particles since the expansion rates of the two phases caused by hydriding are different. In the case of boron addition, the second phase ($MmCo_4B$) on the grain boundaries causes a rapid increase in the specific surface area of the alloy during initial activation cycles. Moreover, the second phases formed by addition of the metals are expected to exist on the new fresh surfaces after cracking and work as electrocatalysts, such as in the case of $MoCo_3$, reported by Notten and Hokkeling (1991).

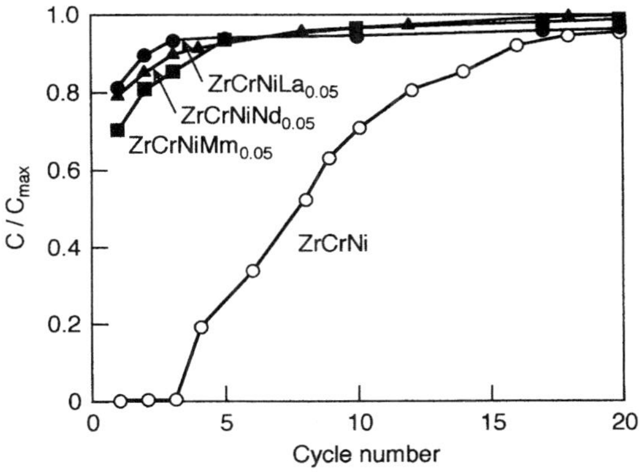

Figure 4
Activation profiles of ZrCrNi and $ZrCrNiM_{0.05}$ (M = La, Nd, Mm) alloy electrodes at room temperature (after Kim and Lee 1992).

Since the Laves-phase alloys require far more activation cycles than AB_5-type Mm-based alloys, additions of some metals and compounds have been carried out to lower the activation cycles. Kim and Lee (1992) improved the activation behavior of ZrCrNi by adding light rare-earth metals such as La, Nd, or Mm (Fig. 4). Activation of $ZrCrNiLa_{0.05}$ alloy electrodes required about five cycles, whereas without La about 20 cycles were required. In the $ZrCrNiLa_{0.05}$ alloys spherical particles of a La-rich phase were distributed in the alloy matrix and fine cracks were induced around these particles. These were expanded by hydrogen absorption and thus cracking of the whole alloy particles was facilitated.

The influence of mixing with several La compounds such as $La(OH)_3$, La_2O_3, and LaF_3 on the properties of $ZrMn_{0.5}V_{0.2}Cr_{0.2}Co_{0.1}Ni_{1.1}$ alloy electrodes was investigated. In the case of $La(OH)_3$ the alloy pulverized even in the first cycle with high charge efficiency, causing ultimately a high discharge capacity due to the earlier cycles. It has also been reported that the initial activation steps of AB_2-type alloys such as ZrCrNi, ZrCrFe, and $Zr(Ni_{0.6}Mn_{0.15}Cr_{0.1}V_{0.15})_2$ are completely eliminated by ball-milling with a small amount (typically 5%) of amorphous Mg_xNi_{1-x} ($x = 0.4$, 0.5) alloy. The amorphous Mg–Ni alloy was found to be an effective electrocatalyst in all the tested alloys.

The pretreatment of the alloy electrodes with KOH or fluoride-containing solution is one of the most effective methods for the activation of hydrogen storage alloy electrodes. The alkaline etching treatment for $Mm(NiCoMn)_5$-based alloy electrodes was carried out by Ogawa *et al.* (1988). A porous nickel-rich layer was formed on the alloy surface by the dissolution of cobalt, leading to improvements in electrochemical properties. For $Zr(V_{0.33}Ni_{0.67})_{2.4}$ alloy electrodes, preliminary long-time (5 days) or high-temperature (45°C) immersion in an alkaline electrolyte was found to decrease significantly the number of activation cycles. An anodic oxidation treatment at −0.6 V (vs. Hg/HgO) for 3 min was also effective in improving the activation behavior of the alloy electrodes. Züttel *et al.* (1994) carried out a pretreatment of $Zr(V_{0.25}Ni_{0.75})_2$ alloys with HF, NH_4F, and KOH solutions in order to dissolve the surface oxides that inhibit or slow down the absorption rate of hydrogen and minimize the activation cycles. The samples pretreated with a 6 M KOH solution at 80°C and a 1 M HF solution at 25°C showed an improved initial activation behavior. Analyses using x-ray photoelectron spectroscopy (XPS) indicated that increasing Ni content in the top layer by etching of the surface oxides led to a decrease in the number of activation cycles (Fig. 5). It has been reported that fluoride treatment of AB_2-type (Ti, Zr) $(Mn, Cr, Ni)_2$ alloys substituted with La improved the initial activation behavior and storability in terms of long-term air and water exposure, which was attributed to the high affinity for hydrogen and the protective role of LaF_3 formed on the alloy surface. For AB_2-type Zr–Ti–Ni–Cr(Co)–Mn–V alloys, a hot-charging treatment in 6 M KOH solution at 80°C and alkali treatment in a boiling 6 M KOH solution were very effective in obtaining the high discharge capacity from first cycle (Fig. 6). Dissolution of oxides such as ZrO_2 and metals such as Co, Mn, and V from the alloy surface caused a nickel-rich surface layer.

It has been reported that $Zr(V_{0.25}Ni_{0.75})_2$ alloy electrodes using nickel powder as compacting material showed the fastest activation compared with powders of copper, nickel, cobalt, gold, and platinum on carbon. This indicates that nickel as a binder also has a high catalytic effect on the activation of $Zr(V_{0.25}Ni_{0.75})_2$ alloys. Investigations of the initial activation mechanism of hydrogen storage alloys

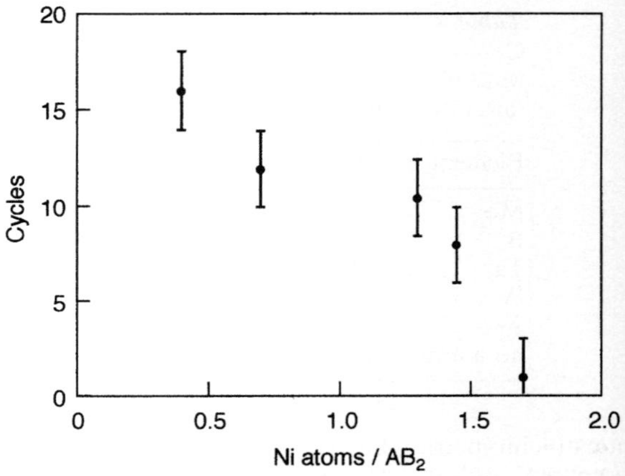

Figure 5
Number of cycles needed to reach the discharge capacity of 200 mAh g^{-1} (60%) as a function of the nickel content in the top layer (30 Å) (after Züttel *et al.* 1994).

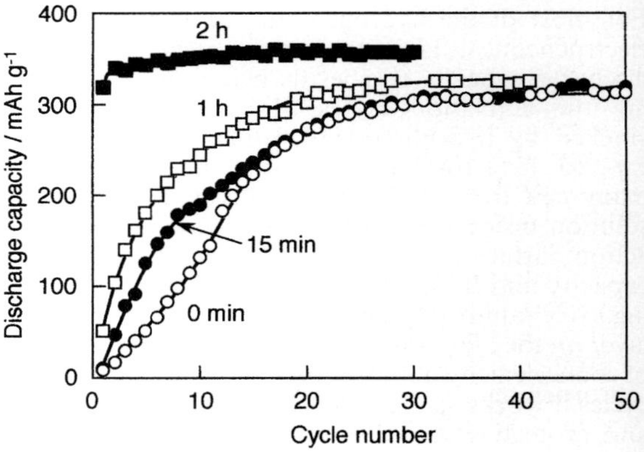

Figure 6
Activation profiles of $Zr_{0.9}Ti_{0.1}Ni_{1.1}Co_{0.1}Mn_{0.6}V_{0.2}$ alloy electrodes treated by immersing them in a boiling 6 M KOH solution for different periods of time (after Choi *et al.* 1999).

produced by gas atomization were also made and compared with conventional melting methods.

7. *Discharge Capacity*

The discharge capacity of metal hydride electrodes in nickel–metal hydride batteries is generally evaluated as the amount of charge passing during discharge to the cut-off potential at a given discharge current density, as shown in Fig. 7. The discharge capacity greatly depends on discharge conditions such as the discharge current density and temperature, unit cell

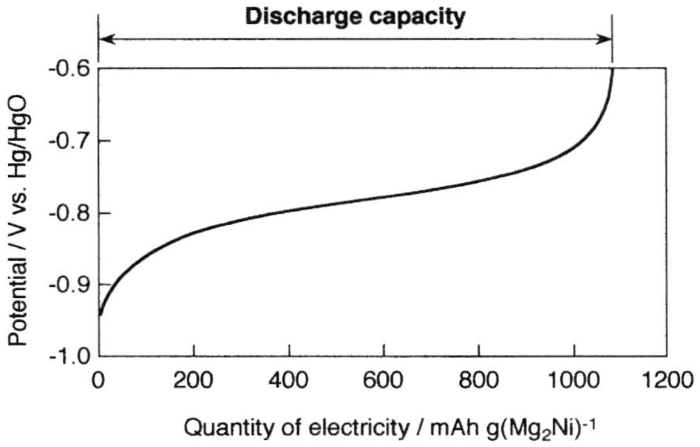

Figure 7
Typical discharge curve of a Mg_2Ni–Ni composite electrode (after Iwakura *et al.* 1998b).

volume, equilibrium hydrogen pressure, and enthalpy of hydride formation. In general, however, the discharge capacity of commercial nickel–metal hydride batteries is controlled by that of the positive electrode.

If the composition of a metal hydride produced by absorption of hydrogen in a hydrogen storage alloy is known, the discharge capacity is calculated using Eqn. (13). Even if the composition of the metal hydrides is unknown, the discharge capacity can be calculated using the amount of absorbed hydrogen evaluated from the measured pressure–composition isotherms (P–C–T curves) (Ogawa *et al.* 1988). By assuming that the hydrogen concentration at 0.5 MPa ($(H/M)_{0.5}$) equals the maximum amount of absorbed hydrogen, the maximum discharge capacity is expressed as follows:

$$C_H(\text{mAh g}^{-1}) = [(H/M)_{0.5} - (H/M)_{0.01}]F/(3.6M) \quad (30)$$

where $(H/M)_{0.5}$ and $(H/M)_{0.01}$ represent atomic ratio of hydrogen to metal at 0.5 and 0.01 MPa, respectively. This is based on the equation expressing the relationship between equilibrium hydrogen pressure and equilibrium potential of the negative electrode (Iwakura *et al.* 1988). Calculated discharge capacities for various hydrogen storage alloys is shown in Table 2. AB_5- and AB_2- type hydrogen storage alloys, which can be called alloys of the first and second generations for negative electrode materials, are actively investigated from a viewpoint of practical use. In contrast, Mg- and V-based alloys with higher discharge capacity such as Mg_2Ni and $TiV_3Ni_{0.5}$ are thought to be promising candidates for third-generation negative electrode materials due to the large amount of absorbed hydrogen, as seen from Table 2. However, they desorb very little hydrogen because of their very low equilibrium hydrogen pressures.

To increase the discharge capacity of these alloys, partial substitution of foreign metals for constituent metals and deviations from the stoichiometric composition have conventionally been used. In addition, surface modification such as microencapsulation of the alloy with Ni or Cu, addition of metal oxides or carbon materials, immersion in alkaline solution with and without a reducing agent, hot-charging, and so on also improved the discharge capacity as well as other battery performance parameters such as high-rate dischargeability and initial activation.

New methods for desorbing hydrogen in very stable hydrides such as Mg-based alloys at room temperature have been reported. It was found that a pellet electrode of $Mg_{50}Ni_{50}$ prepared by mechanical alloying had a high discharge capacity of ~500 mAh g^{-1} at a discharge current density of 20 mA g^{-1}. Surface modification of MgNi prepared by mechanical alloying with graphite using a planetary mill greatly increased the discharge capacity of the ball-milled MgNi paste electrode from ~370 mAh g^{-1} to ~510 mAh $g(MgNi)^{-1}$ due to surface modification of MgNi with graphite. A Mg_2Ni–Ni composite prepared by ball-milling crystalline Mg_2Ni alloy with Ni powder (mole ratio of Mg to Ni = 1) showed the maximum discharge capacity of 750 mAh $g(Mg_2Ni)^{-1}$ [485 mAh $g(Mg_2Ni + 100\text{ mol.\% Ni})^{-1}$].

It has been also reported that similar results were obtained for a Mg_2Ni–Ni composite except that the prepared composite had a uniform amorphous structure (Iwakura *et al.* 1998b). The prepared composite exhibited an extremely high discharge capacity of 1082 mAh $g(Mg_2Ni)^{-1}$ [636 mAh $g(Mg_2Ni + 70\text{ wt.\% Ni})^{-1}$] at 30°C, which exceeded the theoretically calculated value of 999 mAh $g(Mg_2Ni)^{-1}$. A large increase in the discharge capacity by ball-milling of $Mg_{1.9}Al_{0.1}Ni_{0.8}Co_{0.1}Mn_{0.1}$ with Ni powder for an appropriate milling time was also demonstrated. It was found that nanocrystalline $Mg_2Ni_{0.9}Cu_{0.1}$ prepared by mechanical alloying showed a discharge capacity much higher than that of the polycrystalline alloy. Moreover, it was found that nanocrystalline $Mg_{1.95}Y_{0.05}Ni_{0.92}Al_{0.08}$ prepared by a powder metallurgical technique followed by mechanical milling with Ni powder under an argon atmosphere greatly improved the discharge capacity. It has also been reported that $La_{0.7}Mg_{0.3}Ni_{2.8}Co_{0.5}$ in new La–Mg–Ni-type alloys showed a large discharge capacity (410 mAh g^{-1}) and good cycle performance.

8. *High-rate Dischargeability*

From a practical viewpoint high-rate dischargeability (rate capability) is one of the most important properties to optimize. The rate of the charge–discharge reaction of metal hydride electrodes is expected to depend on the following factors:

(i) catalytic activity for the charge transfer reaction; diffusivity of hydrogen from the surface to the bulk alloy;
(ii) diffusivity of OH^- ion and H_2O to/from reaction surface;
(iii) surface conductivity of the alloy particles;
(iv) effective surface area per unit weight of alloy or per unit area of electrode which is always changing by decrepitation, surface oxidation, and dissolution of component elements;
(v) contact resistance between alloy and current collector.

The potential difference between the equilibrium and dynamic potentials, i.e., polarization, includes the total resistance resulting from the above factors. Each contribution from contact resistance, charge transfer resistance, and diffusion resistance can be estimated separately by impedance analysis described in Sect. 11.

In order to improve electronic conductivity for providing electrons to the alloy surface, various preparation methods of electrodes have been tried since Willems *et al.* (1984) made electrodes by pressing a mixture of alloy and copper powder in the weight ratio of 1:4. Various kinds of conductive powders such as Cu, Ni, Co, Fe, and carbon and matrices such as nickel, foam, and fiber have been used by many researchers to obtain longer cycle life and better rate capability for the metal hydride electrodes and sealed cells. In addition, various binding and/or pasting polymers such as poly(tetrafluoroethylene) (PTFE) and poly(vinyl alcohol) (PVA) were used.

Second-phase formation in the grain boundaries is sometimes effective in improving the high-rate dischargeability. Notten and Hokkeling (1991) found that the rate capability of metal hydride electrodes increased when second phases such as electrocatalytic elements (Ni–Co) or a compound ($MoCo_3$) were present in the $La_{0.8}Nd_{0.2}Ni_{2.5}Co_{2.4}Si_{0.1}$ alloy. The second phase of Ni–Co solid solution was formed on the grain boundaries by increasing the nickel content ($La_{0.8}Nd_{0.2}Ni_{3.0}Co_{2.4}Si_{0.1}$). Moreover, when molybdenum was introduced into the alloy ($La_{0.8}Nd_{0.2}Ni_{2.9}Mo_{0.1}Co_{2.4}Si_{0.1}$), a desired second phase of $MoCo_3$ was formed. These two-phase alloys showed a much higher rate capability than the original single-phase alloy with increasing exchange current density, as shown in Fig. 8. They ascribed the improvement in the rate capability to enhanced catalytic activity on the alloy surface due to a linear relationship between the rate capability and exchange current density.

Second-phase formation by adding high-melting point elements such as Zr, W, Ta, B, or Mo into the

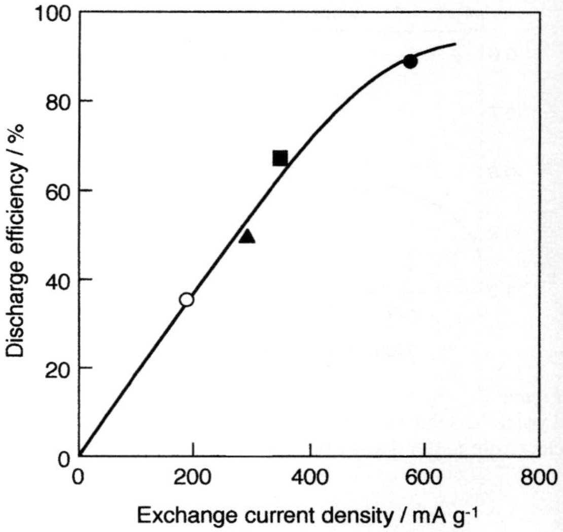

Figure 8
The discharge efficiency under the extreme conditions at 0°C and at a high discharge current of $840\,mA\,g^{-1}$ as a function of the exchange current density for the standard alloy (○), the pentary $AB_{5.5}$ alloys (▲, ■), and the Mo-containing $AB_{5.5}$ alloys (•) (after Notten and Hokkeling 1991).

$Mm(NiCoMnAl)_5$-based alloys has been also reported to be effective in improving the rate capability, as shown in Table 4. These second phases facilitate the cracking of the alloys, increasing rapidly the specific surface area after the first activation cycle. The cycle life was the longest for boron addition. The second phase by boron addition was identified as a $MmCo_4B$ phase. With increasing boron content the capacity decreased while rate capability increased, leading to the suitable composition $MmNi_{3.2}CoAl_{0.2}Mn_{0.6}B_{0.03}$.

It was pointed out in the literature that when the surface of a $MmNi_{3.6}Mn_{0.4}Al_{0.3}Co_{0.7}$ electrode was modified by an electroless copper or nickel coating, the exchange current density increased with increasing rate capability of the metal hydride electrode. The same effect was found when the electrolyte solution (6 M KOH) contained a reducing agent such as 0.08 M H_3PO_2 or 0.001–0.1 M KBH_4. Hydrogen produced by the reducing agent could reduce the surface nickel oxide, forming a nickel-rich surface layer which could provide the good surface conductivity and high catalytic activity. Some of the hydrogen was stored in the alloy. It was suggested that when the discharge reaction is limited by surface reaction, a linear dependence of rate capability on the exchange current density would be expected. The parabolic relationship in Fig. 9 suggests that

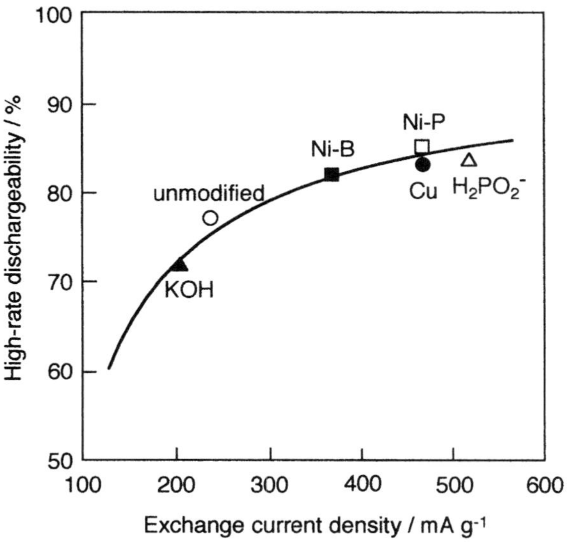

Figure 9
Relationship between high-rate dischargeability and exchange current density: ○ original alloy; • copper plating; □ nickel plating containing phosphorus; ■ nickel plating containing boron;▲ alkaline etching treatment in 6 M KOH at 80°C; □ reducing treatment at 80°C using 0.08 M reducing agent (H_3PO_2) contained in electrolyte solution. The high-rate dischargeability is defined as a capacity ratio of 1786 mA g^{-1} discharge to 178.6 mA g^{-1} discharge (after Iwakura *et al.* 1992).

hydrogen diffusivity becomes progressively a rate-determining step with increasing exchange current density.

Modification with some metal oxides is also effective to improve the high-rate dischargeability. It has been reported that mixing $MmNi_{3.6}Mn_{0.4}Al_{0.3}Co_{0.7}$ alloy powder with RuO_2 or Co_3O_4 powder improved the rate capability with increasing exchange current density. This was explained on the basis of improved electrocatalytic activity and reduced contact resistance as well as an increase in the effective surface area. $L_mNi_{3.45}(CoMnTi)_{1.55}$ alloy was modified by mixing with Bi_2O_3 and the discharge capacity and high-rate dischargeability increased strongly. It was suggested that a network of metallic bismuth formed on the alloy surface due to the reduction of Bi_2O_3 during the charging reaction, resulting in an improvement in the electrical conductivity and electrocatalytic activity.

For AB_2-type Laves-phase alloy electrodes, many efforts have been made to improve the high-rate dischargeability. The effects of compacting materials (copper, nickel, cobalt, gold, and platinum on carbon) on the electrode performance such as cycle life and high-rate dischargeability of $ZrV_{0.5}Ni_{1.5}$ alloy electrodes were reported. The electrochemical characteristics, thermodynamic properties, and crystal structure of Zr–Ti–V–Ni–Mn–M (M = Co, Cr, Fe) alloys were investigated and it was found that the alloy which had an enthalpy of hydride formation of approximately $-36\,kJmol^{-1}$ to $-42\,kJmol^{-1}$ of hydrogen showed a good high-rate dischargeability. Surface modification by electroless Ni–P and Ni coatings was effective for $Ti_{0.35}Zr_{0.65}Ni_{1.2}V_{0.6}Mn_{0.2}$ alloy electrodes.

It was also found that the $Zr_{0.9}Ti_{0.1}(Mn_{0.7}V_{0.5}Ni_{1.2})_{0.92}$ alloy, having an optimum alloy composition, showed a large discharge capacity (392 mA h g^{-1} at 0.25 C) and a good rate capability equaling that of the commercialized AB_5-type alloys. Annealing of the $ZrMn_{0.7}V_{0.5}Ni_{1.2}$ alloy at 1000°C caused disappearance of the second phase (Zr–Ni phase) which existed in the as-cast alloy and improved the high-rate dischargeability with increasing exchange current density. This was concluded to be due to an increase in the nickel content in the matrix resulting from homogenization of the alloy, which could act as a good electrocatalyst.

The high-rate dischargeability for Mg-based alloy electrodes is generally lower than those for AB_5- and AB_2-type alloys because of the low reactivity with hydrogen at room temperature and the formation of a $Mg(OH)_2$ layer on the alloy surface in alkaline solution. It has been reported that the microencapsulation of $Mg_{1.9}Y_{0.1}Ni_{0.9}Al_{0.1}$ alloy by electroless Ni–P, Ni–Pd–P, and Ni–B alloy coating increased the discharge capacity, high-rate dischargeability, and cycle life. It was suggested from cyclic voltammograms and impedance spectra that the microencapsulation greatly improved the electrocatalytic activity for the hydrogen electrode reaction and reduced the overpotential of the charge–discharge reactions.

9. *Charge–Discharge Cycle Life*

Charge–discharge cycle life or cyclability of metal hydride electrodes is expressed as the degree of retention of the discharge capacity during repeated charge–discharge cycles. In general, the charge–discharge cycle life shortens due to the formation of hydroxides and oxides on the fresh surface generated by pulverization of the alloy due to dissolution of the alloy components into the electrolyte solution. The charge–discharge cycle life may be improved by the following strategies:

(i) partial replacement of alloy components by foreign metals such as cobalt;
(ii) amorphization of the alloy;
(iii) microencapsulation of the alloy surface with foreign metals such as copper or nickel;
(iv) others.

9.1 Partial Substitution of Alloy Components by Foreign Metals

Hydrogen storage alloys are pulverized during charge–discharge processes due to repeated lattice expansion and shrinkage as a result of hydrogen absorption and desorption. This is a serious problem leading to deterioration of the nickel–metal hydride batteries. In order to suppress such pulverization, preparation of alloys with larger lattice volume is important because of their smaller volume expansion in hydrogen absorption. Thus, alloy components have been partially replaced by foreign metals such as cobalt. It was found that the lattice volume of $LaNi_{5-x}Co_x$ $(0 \leqslant x \leqslant 3.3)$ alloys increased with x, leading to an improvement in charge–discharge cycle life (Willems 1984). In particular, the lattice expansion with hydrogen absorption was suppressed in the direction of the c-axis, which, it was suggested, brought about high resistance to pulverization. Similar results were obtained by some other groups. It was found that the partial substitution of Ni by Al or Si improved the charge–discharge cycle life. Partial substitution of Sn or Ge for Ni in $LaNi_5$ improved the cycle life but the initial discharge capacity decreased with increasing Sn or Ge content.

Partial substitution of Ce for La and Sn and Co for Ni in a $LaNi_5$ alloy has been reported to be effective for longer cycle life due to the suppression of oxygen reduction at the electrode surface. The coexistence of three elements retained the discharge capacity in addition to the long cycle life. A small amount of Ce was also effective for retarding corrosion regardless of the lattice expansion due to hydrogen absorption. This was ascribed to the formation of a protective CeO_2 film. Moreover, it was reported that $ZrCrNiLa_{0.05}$ showed good durability because of the protection of the alloy by enrichment of Ni and $La(OH)_3$ at the alloy surface.

The decrepitation rate decreased with Vickers hardness, leading to a longer charge–discharge cycle life. For $LaNi_{5-x}M_x$, the decrepitation rate and Vickers hardness decreased in the order M = Mn, Cu, Cr, Al, Co, and the cycle life increased in the same order. The addition of Co gives the alloys some ductility in addition to the smaller lattice expansion described above, which is responsible for the longer cycle life.

Partial substitution of La and Ni in $LaNi_5$ by a small amount of Zr and Al also improved the cycle life of the alloys. This was explained by the formation of a surface oxide film by the added Zr and Al which effectively protected the alloys from further corrosion.

9.2 Amorphization of the Alloy

Amorphous alloys generally show a long cycle life in comparison with their crystalline counterparts, but they show lower discharge capacity. It was found that amorphous MNi_2 (M = La, Ce, Pr) alloys showed low discharge capacities of less than $100\,mAh\,g^{-1}$ at room temperature, but they had a long cycle life. Investigations of the effect of the solid ification rate on the charge–discharge cycle life showed that control of the solidification rate by the linear velocity of the copper wheel in melt-spinning influenced grain size and crystal structure. High solidification rates led to alloys with smaller grains containing nanocrystalline and amorphous phases, resulting in good cyclability. An amorphous $Mg_{0.9}V_{0.1}Ni$ alloy showed a longer charge–discharge cycle life in comparison with a MgNi alloy since the V component in the former suppressed formation of $Mg(OH)_2$ on the alloy surface during the charge–discharge cycling in alkaline solution. In this case, the amorphous MgNi and $Mg_{0.9}V_{0.1}Ni$ alloys were prepared from powders of the components by mechanical alloying under an argon atmosphere using a planetary mill.

9.3 Microencapsulation of Alloy Surface with Foreign Metals

A very small amount of platinum black was painted on $LaNi_5$ electrode surface to protect the alloy from oxidation by oxygen evolved at the positive electrode. Mixing of $LaNi_5$ and its ternary alloys with a large amount of copper powder increased the charge–discharge cycle life. However, these methods seem to be undesirable from a viewpoint of energy density and cost.

A microencapsulated $LaNi_5$ alloy powder covered with an electroless copper layer was used as the negative electrode material in a nickel–metal hydride battery. This caused an increase in electrical and thermal conductivity of the alloy, an improvement in the oxygen resistance of the alloy, and suppression of disintegration of the alloy due to repeated hydrogen absorption and desorption. In addition, the Cu microencapsulation led to a four-fold increase in the charge–discharge cycle life and an improvement in the charge–discharge characteristics at high rates and low temperatures. Subsequently microencapsulation of various alloys with Ni and Co in addition to Cu was carried out to improve the cycle life.

9.4 Others

Ikoma *et al.* (1999) found that the cycle life of sealed AA-size batteries was greatly improved, as shown in Fig. 10, when $MmNi_{4.3-x}Mn_{0.4}Al_{0.3}Co_x$ alloys were treated by immersing them in 7.2 M KOH solution at 80°C.

It has been also reported that nonstoichiometric almost single-phase $AB_{5.5}$ alloys prepared by melt-spinning showed a long charge–discharge cycle life due to the suppression of pulverization relating to the anisotropically deformed crystal lattice.

The surface of a crystalline Mg_2Ni alloy prepared by induction melting was modified by ball-milling

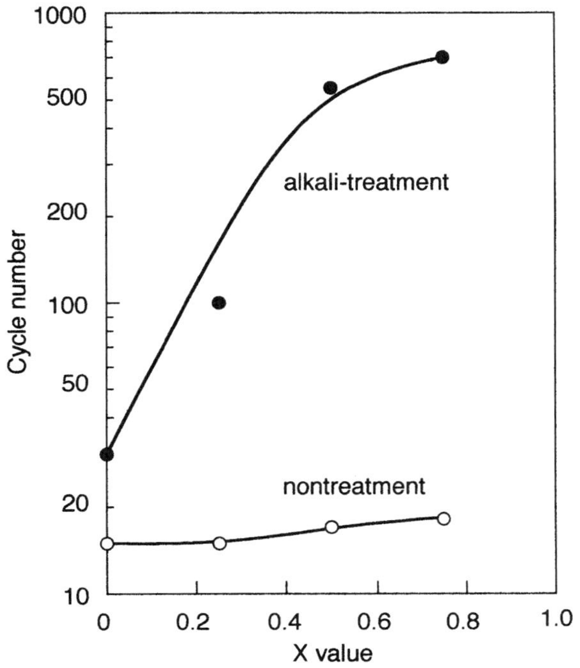

Figure 10
Influence of alkali treatment of alloy powders and Co content on cycle life/amount of Co substitution for $MmNi_{4.3-x}Mn_{0.4}Al_{0.3}Co_x$ alloy (after Ikoma *et al.* 1999).

with 3 wt.% Co powder. The surface-modified alloy electrode showed excellent corrosion resistance, cyclability, and high-rate dischargeability. Moreover, it was found that nanocrystalline Mg_2Ni-type hydrogen storage alloys prepared by powder metallurgical techniques and subsequent mechanical milling with nanosized nickel powder under an argon atmosphere showed excellent cycle life, high discharge capacity, and good high-rate dischargeability.

10. Dissolution Behaviors of Alloy Constituents

The chemical stability of alloys in alkaline solutions is related to three types of behavior for the constituent elements:

(i) immunity when the metal is thermodynamically stable against oxidation;
(ii) corrosion when the metal continuously oxidizes and dissolves in the solution;
(iii) passivation when the oxides or hydroxides are insoluble and form a protective layer.

The behavior changes with both the potential and the pH of the solution. In the potential range over which hydrogen is charged and discharged in the KOH solution, for example, the lanthanum is oxidized (although the solubility of the hydroxide is extremely low), while nickel is immune up to about 150 mV above the hydrogen potential after which it is oxidized to hydroxide, causing passivation. Cobalt is immune up to about 100 mV above the hydrogen potential, but above that the oxidation product dissolves in the KOH solution. Manganese continuously corrodes because its oxides are soluble as manganite ions. The effect of alloying results in a regularly packed structure of several kinds of atoms. Selective oxidation, especially dissolution of the less noble metals, leaves the more noble metal in the surface layer, and these determine the electrochemical and corrosive properties of the alloys.

Ogawa *et al.* (1988) have reported that alkaline etching treatment improved the cycle life of $Mm(NiCoMn)_5$-based alloys (Fig. 10). Cobalt is dissolved from the alloy surface during the alkaline etching treatment before nickel oxidation occurs. As a result, a porous nickel-rich layer is formed on the alloy surface, which is expected to be a good electrical conductor, catalyst, and protective layer. About 25 wt.% of the total Mn in $MmNi_{4.2}Mn_{0.6}Al_{0.2}$ was also found to be dissolved from the alloy and precipitated on the alloy surface as oxide after 200 electrochemical cycles. With increasing Co content in $Mm(NiCoMnAl)_5$ the dissolving ratio of manganese was greatly decreased, suppressing the decrease in manganese content in the alloy during cycles. Furthermore, cobalt could be dissolved and then precipitated easily during the charging process, giving good conductive networks on the surface layer.

A detailed and systematic surface analysis for $LaNi_5$-based alloy electrodes before and after cycles was carried out by using XPS (Meli *et al.* 1994). The partial replacement of Ni by Al and Si also caused higher enrichment of nickel in the metallic state after 30 cycles on the surface layer (Fig. 11). The added aluminum and silicon were not detected down to 50 Å and 10 Å, respectively, showing that they are readily soluble in KOH solution and do not take part in the formation of the surface oxide layer. It was suggested that the metallic nickel layer enriched on the surface produces the protective character on the alloy surface. The alloy surface of $MmNi_{3.5}Al_{0.8}Co_{0.7}$ after 50 and 200 cycles was also investigated by XPS and a model of the surface film was proposed. Aluminum was not present in the outer porous region which mainly consisted of oxidized rare-earth metals, although it existed as oxides in the inner dense region. Moreover, the model showed the existence of a continuous Ni–Co network connecting the metallic substrate with the outer region of the film. This surface film could protect the substrate against oxidation and provide electrical conductivity, retaining the good electrochemical properties.

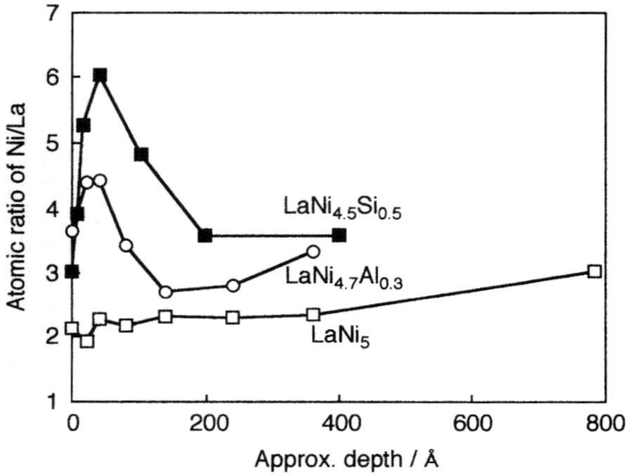

Figure 11
XPS depth profile of $LaNi_{4.5}Si_{0.5}$, $LaNi_{4.7}Al_{0.3}$, and $LaNi_5$ powder electrodes after 30 cycles in 6 M KOH. Atomic ratio of Ni/La is shown as a function of approximate sputter depth (2 keV Ar^+; about 4 Å min^{-1}) (after Meli *et al.* 1994).

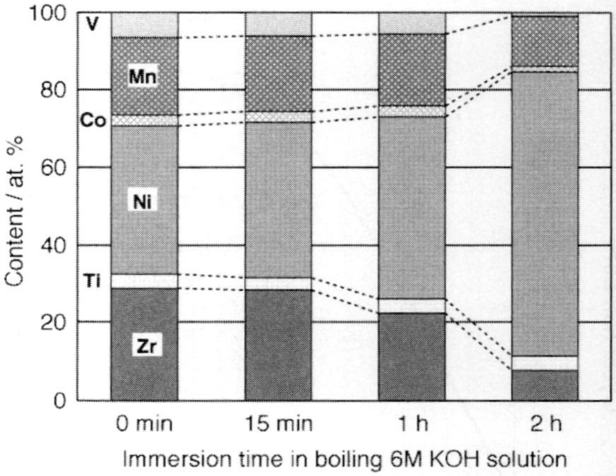

Figure 12
Surface composition change of $Zr_{0.9}Ti_{0.1}Ni_{1.1}Co_{0.1}Mn_{0.6}V_{0.2}$ alloy electrodes treated by immersing them in a boiling 6 M KOH solution for different periods of time: 0 min, 15 min, 1 h, and 2 h (after Choi *et al.* 1999).

The effects of dissolution of the constituent elements in AB_2-type alloy electrodes have also been investigated by several groups. For a $ZrV_{0.4}Ni_{1.6}$ electrode, the surface vanadium, which was present as an oxide before cycling, had completely disappeared after 30 cycles. On the other hand, the nickel was enriched and the zirconium remained as a ZrO_2 on the alloy surface. It was assumed that the dissolution of vanadium during the activation cycles increased the nickel content in the top surface layer, leading to promotion of electrochemical reactions and an increase in discharge capacity with increasing cycle number. A hot-charging treatment improved the activation property and the rate capability of a $Zr_{0.7}Ti_{0.3}Cr_{0.3}Mn_{0.3}V_{0.4}Ni$ alloy electrode. By charging the electrode in hot 6 M KOH solution, the formation of cracks and nickel enrichment on the surface by the dissolution of the surface oxide layer occurred simultaneously, leading to full activation after the first cycle.

The improvement in activation behavior of a $Zr_{0.9}Ti_{0.1}Ni_{1.1}Co_{0.1}Mn_{0.6}V_{0.2}$ alloy electrode by alkali treatment in a boiling 6 M KOH solution has been reported. With increasing immersion time in the boiling solution, the nickel content in the alloy surface increased remarkably, while the contents of zirconium, cobalt, manganese, and vanadium decreased, as shown in Fig. 12. It was suggested that Zr which existed as ZrO_2 before the treatment also dissolved from the alloy surface. During the alkali treatment, hydrogen was also found to be absorbed in the alloy, while selective dissolution of the aggressive components occurred.

Such a phenomenon was explained by a local mechanism. Since the redox potentials of the aggressive components such as Zr and V are more negative than the potental of a H_2O/H couple, it can be assumed that the overall reaction proceeds as follows:

$$M_{alloy} + H_2O_{solution} \rightarrow M^+_{solution} + H_{alloy} + OH^-_{solution} \quad (31)$$

where M denotes the aggressive components which can dissolve easily in the boiling alkaline solution.

11. Impedance Spectrometry of Metal Hydride Electrodes

Impedance spectrometry is an electrochemical method to observe the way in which a system follows perturbation at a steady state. Impedance is evaluated from the response of the current to the varying frequency of the imposed alternating voltage over a wide range and it is separated into real and imaginary components. The relationship obtained between both components is represented as a Cole–Cole (Nyquist) plot. The Cole–Cole plot is generally simulated with an equilibrium circuit composed of resistance, capacitance, and sometimes inductance, in which the electrode/electrolyte solution interphase is virtually structured. Therefore, impedance spectrometry can be used for analyzing the state of the electrode surface and the kinetics of the metal hydride electrode.

Kuriyama *et al.* (1992) applied impedance spectrometry to the analysis of the deterioration process

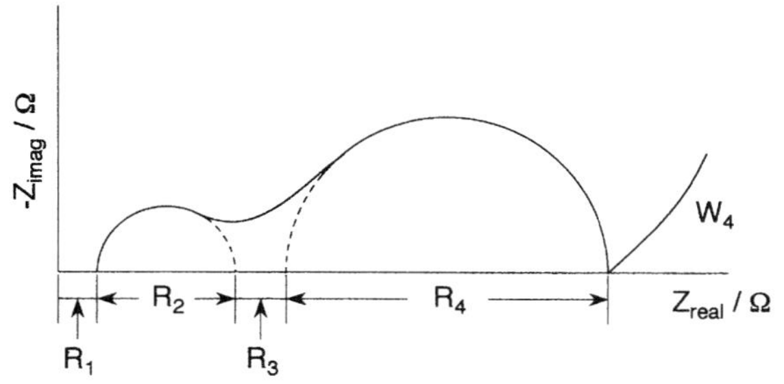

(a) Typical Cole-Cole plot.

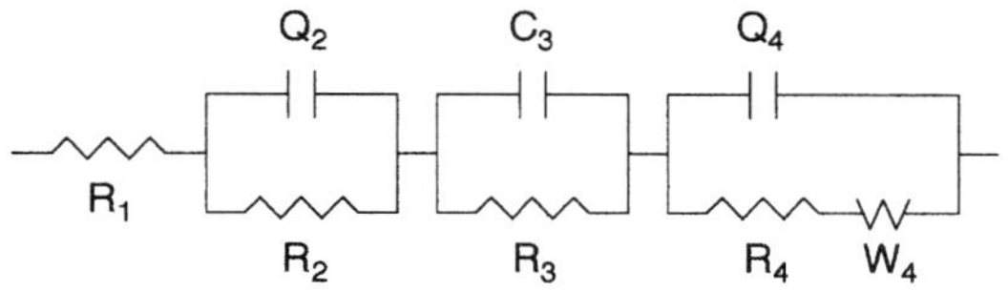

(b) An equivalent circuit.

R_1 : Electrolyte resistance
R_2 : Contact resistance between current collectors and alloy particles
R_3 : Contact resistance between alloy particles
R_4 : Reaction resistance on alloy surface
Q_2 : Capacitance between current collectors and alloy particles
C_3 : Capacitance between alloy particles
Q_4 : Double layer capacitance on the surface
W_4 : Warburg impedance

Figure 13
(a) Typical Cole–Cole plot and (b) an equivalent circuit proposed by Kuriyama *et al.* (after Kuriyama *et al.* 1992).

of metal hydride electrodes. They simulated Cole–Cole plots for some metal hydride electrodes using the equivalent circuit shown in Fig. 13 in which all resistances and capacitances were identified. As a result, they clarified the difference of the deterioration mechanism between Cu-coated and bare $MmNi_{3.5}Co_{0.7}Al_{0.8}$ alloy electrodes. In addition, they used a reaction resistance to represent the charge transfer resistance in the following reaction:

$$M + H_2O + e^- = MH_{ad} + OH^- \qquad (32)$$

and considered an activation enthalpy of the charge transfer reaction to explain the difference in battery performance between mischmetal-based and titanium-based alloy electrodes. The other resistance components, except for the charge transfer resistance, were influenced by the preparation of the pellet electrode due to a well-extended network of a mixture of binder and copper powder.

Impedance spectrometry was used to gain a fundamental understanding of the chemical and physical processes that limited the application of metal hydrides. A $LaNi_5$ electrode was used and a mathematical model was fitted to the experimental data that incorporated the effects of water reduction, adsorption of hydrogen into metal, and diffusion of hydrogen through the metal ingot.

The effect of depth of discharge on charge transfer resistance and on the Warburg coefficient evaluated from a.c. impedance spectra of a Mg_2Ni-type alloy paste electrode was investigated. It was found that both parameters increased with the depth of discharge when this depth was less than 40%. However, when it was over 40% the Warburg coefficient showed a higher increase rate than the charge transfer resistance, suggesting that the rate-determining step of the dehydriding reaction changes with the depth of discharge from a mixed-controlled step in the initial stage of the discharge process to a diffusion-controlled step in the last stage.

The charge transfer resistance for various alloy electrodes was evaluated using the equilibrium circuit in Fig. 13 in terms of electrocatalytic activity. Impedances measured in the frequency range of 0.6–239 Hz were also used for prediction of the state of charge.

Hydrogen absorption and evolution were distinguished by introducing the fraction of absorbed hydrogen into the equilibrium circuit in Fig. 13. According to the curve fitting, the hydrogen absorption of $LaNi_5$ changed at a depth of charge of ca. 60%. In addition, the hydrogen absorption was strongly influenced by the Ni cluster on the alloy surface and the diffusibility of hydrogen in the alloy.

Zhang *et al.* (1995) considered not only the charge transfer reaction (Eqn. (32)) but also the following hydrogen transfer reaction from surface to bulk in their mathematical model for the equilibrium circuit:

$$MH_{ad,\ surface} = MH_{ab,\ bulk} \qquad (33)$$

The equilibrium circuits for planar (ingot) and porous electrodes are shown in Fig. 14. For the porous electrode, the three arcs in the Nyquist plot were assigned to charge transfer at the electrode/electrolyte solution interface and hydrogen transfer from surface to bulk for the low-frequency region, the contact impedance between the alloy particles in the electrode for the middle-frequency region, and the contact impedance between the current collector and the sample (alloy + binder) for the high-frequency region, respectively. It was found that the impedance of metal hydrides is mainly determined by the faradaic process in the low-frequency region and there was a good correlation between the rate

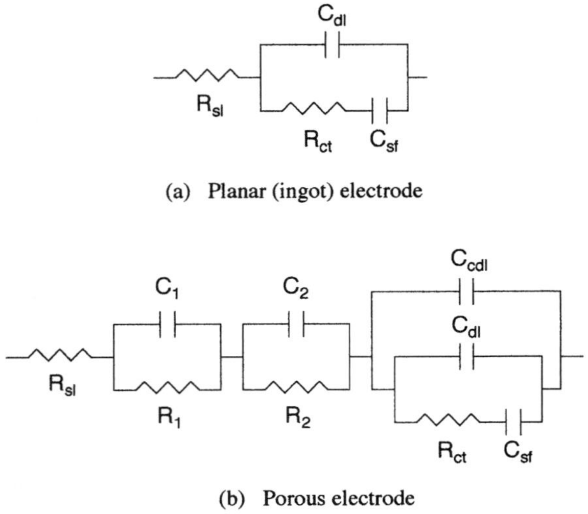

R_{sl} : Electrolyte resistance
R_1 : Contact resistance between current collectors and sample (alloy + binder) particles
R_2 : Contact resistance between alloy particles
R_{ct} : Charge-transfer resistance at the electrode/solution interface
C_{dl} : Double layer capacitance of active alloy particles
C_{cdl}: Double layer capacitance of carbon additives
C_{sf} : Pseudo-capacitance of the MH electrode due to the surface reaction (Eqn. 2)
C_1 : Capacitance between current collectors and sample (alloy + binder) particles
C_2 : Capacitance between alloy particles

Figure 14
Equivalent circuits proposed by Zhang *et al.* for (a) planar (ingot) and (b) porous electrodes (after Zhang *et al.* 1995).

capability and the charge–discharge reaction rate at the alloy surface.

A new mathematical model for the equilibrium circuit was developed in which a surface transition from adsorbed sites on the electrode surface to absorbed sites in a near-surface region (Eqn. (34)) is followed by diffusion of the absorbed hydrogen from the near surface into the bulk (Eqn. (35)). Sometimes transformation from the α phase to the β phase (Eqn. (36)) was considered in addition to a charge transfer reaction (Eqn. (32)).

$$MH_{ad} = MH_{ab,\ surface} \quad (34)$$

$$MH_{ab,\ surface} = MH_{ab,\ bulk,\ \alpha} \quad (35)$$

$$MH_{ab,\ bulk,\ \alpha} = MH_{ab,\ bulk,\ \beta} \quad (36)$$

The simulated Cole–Cole plot showed that the three arcs can be assigned to the charge transfer reaction at the electrode/electrolyte solution interface for the higher-frequency range, the hydrogen transfer between absorbed and adsorbed states for the middle-frequency range, and the diffusion of absorbed hydrogen through the two-phase alloy in the lower-frequency range, respectively. From a large number of model simulations, the state of discharge and the fraction of α phase was found to play an important role in Cole–Cole plots of metal–hydride electrodes.

The diffusion coefficient of hydrogen in $LaNi_{4.27}Sn_{0.24}$ alloy was evaluated by a.c. impedance spectrometry at various states of discharge. The theoretical model for the simulation consisted of the charge transfer reaction (Eqn. (32)) and a semi-infinite or infinite spherical diffusion of hydrogen. The simulated diffusion coefficient affected not only the magnitude of the impedance but also the transition from semi-infinite to finite diffusion control. The diffusion coefficient calculated from the slope of the Cole–Cole plot in the transition region increased with the state of charge, which was attributed to phase transformation at the low state of charge.

See also: Hydrogen–Metal Systems: Applications of Gas–Solid Reactions; Hydrogen–Metal Systems: Basic Properties (1); Hydrogen–Metal Systems: Hydride Forming Alloys Properties and Characteristics, Database Information)

Bibliography

Bittner H F, Badcock C C 1983 Electrochemical utilization of metal hydrides. *J. Electrochem. Soc.* **130**, 193C–8C

Choi W K, Yamataka K, Zhang S G, Inoue H, Iwakura C 1999 Effects of surface treatment with boiling alkaline solution on electrochemical and physicochemical properties of the $Zr_{0.9}Ti_{0.1}Ni_{1.1}Co_{0.1}Mn_{0.6}V_{0.2}$ alloy electrode. *J. Electrochem. Soc.* **146**, 46–8

Ikoma M, Hoshina Y, Matsumoto I, Iwakura C 1996 Self-discharge mechanism of sealed-type nickel/metal-hydride battery. *J. Electrochem. Soc.* **143**, 1904–7

Ikoma M, Komori K, Kaida S, Iwakura C 1999 Effect of alkali-treatment of hydrogen storage alloy on the degradation of Ni/MH batteries. *J. Alloys Comp.* **284**, 92–8

Iwakura C 1985 Batteries using hydrogen storage alloys. In: Yoshizawa S (ed.) *Technologies and Economics of Fuel Cells*. CMC, Tokyo, pp. 178–208

Iwakura C, Asaoka T, Yoneyama H, Sakai T, Ishikawa H, Oguro K 1988 Electrochemical characteristics of $LaNi_5$ system hydrogen-absorbing alloys as negative electrode materials for nickel–hydrogen batteries. *Nippon Kagaku Kaishi* (*J. Chem. Soc. Jpn.*) 1482–8

Iwakura C, Fukuda K, Senoh H, Inoue H, Matsuoka M, Yamamoto Y 1998a Electrochemical characterization of

$MmNi_{4.0-x}Mn_{0.75}Al_{0.25}Co_x$ electrodes as a function of cobalt content. *Electrochim. Acta.* **43**, 2041–6

Iwakura C, Inoue H, Zhang S G, Nohara S 1998b Hydriding and electrochemical characteristics of a homogeneous amorphous Mg_2Ni–Ni composite. *J. Alloys Comp.* **270**, 142–4

Iwakura C, Matsuoka M 1991a Application of hydrogen storage alloys to battery-related fields: nickel–hydrogen batteries. *Progr. Batteries Battery Mater.* **10**, 81–114

Iwakura C, Matsuoka M, Asai K, Kohno T 1992 Surface modification of metal hydride negative electrodes and their charge/discharge performance. *J. Power Sources.* **38**, 335–43

Iwakura C, Matsuoka M, Sakai T, Ishikawa H 1991b Present status of research and development of nickel–hydride batteries. Parts I and II. *Kagaku to Kogyo.* **65**, 248–54, 346–50

Iwakura C, Oura T, Inoue H, Matsuoka M, Yamamoto Y 1995 Effect of alloy composition on hydrogen diffusion in the AB_5-type hydrogen storage alloys. *J. Electroanal. Chem.* **398**, 37–41

Kim S R, Lee J Y 1992 Activation behavior of $ZrCrNiM_{0.05}$ metal hydride electrodes (M = La, Mm (misch metal), Nd). *J. Alloys Comp.* **185**, L1–4

Kuriyama N, Sakai T, Miyamura H, Uehara I, Ishikawa H, Iwasaki T 1992 Electrochemical impedance spectra and deterioration mechanism of metal hydride electrodes. *J. Electrochem. Soc.* **139**, L72–3

Matsuoka M, Asai K, Fukumoto Y, Iwakura C 1993 Surface modification of hydrogen storage alloy electrode with an alkaline solution containing potassium borohydride as a reducing agent. *Electrochim. Acta.* **38**, 659–62

Meli F, Züttel A, Schlapbach L 1994 Effect of silicon on the properties of AB_5-based alloys for battery electrode application. *Z. Phys. Chem.* **183**, 371–7

Nogami M, Tadokoro M, Kimoto M, Chikano Y, Ise T, Furukawa N 1993 Development of high performance hydrogen absorbing alloys for batteries. I. A study on nonstoichiometric hydrogen-absorbing alloys. II. A study on hydrogen absorbing alloys forming second phase with additive elements. *Denki Kagaku (presently Electrochemistry)* **61**, 1088–93; 1094–102

Notten P H L, Hokkeling P 1991 Double-phase hydride forming compounds: a new class of highly electrocatalytic materials. *J. Electrochem. Soc.* **138**, 1877–85

Ogawa H, Ikoma M, Kawano H, Matsumoto I 1988 Metal hydride electrode for high energy density sealed nickel–metal hydride battery. *Power Sources* **12**, 393–410

Sakai T, Matsuoka M, Iwakura C 1995 Rare earth intermetallics for metal–hydrogen batteries. In: Gschneidner K A Jr., Eyring L (eds.) *Handbook on the Physics and Chemistry of Rare Earths.* Vol. 21. Elsevier Science, Amsterdam, pp. 133–78

Senoh H, Morimoto K, Inoue H, Iwakura C, Notten P H L 2000 Relationship between equilibrium hydrogen pressure and exchange current for the hydrogen electrode reaction at $MmNi_{3.9-x}Mn_{0.4}Al_xCo_{0.7}$ alloy electrodes. *J. Electrochem. Soc.* **147**, 2451–5

Tamura H, Iwakura C, Kitamura T 1983 Hydrogen evolution at $LaNi_5$-type alloy electrodes. *J. Less-common Met.* **89**, 567–74

Ura H, Nishina T, Uchida I 1995 Electrochemical measurements of single particles of Pd and $LaNi_5$ with a microelectrode technique. *J. Electroanal. Chem.* **396**, 169–73

Wakao S, Yonemura Y 1983 Anodic polarization behaviour of hydride–deuteride electrodes. *J. Less-common Met.* **89**, 481–8

Willems J J G 1984 Metal hydride electrodes stability of $LaNi_5$-related compounds. *Philips J. Res.* **39**, 1–94

Witham C, Hightower A, Fultz B, Ratnakumar B V, Bowman R C Jr 1997 Electrochemical properties of $LaNi_{5-x}Ge_x$ alloys in Ni–MH batteries. *J. Electrochem. Soc.* **144**, 3758–64

Zhang W, Kumar M P S, Srinivasan S, Ploehn H J 1995 AC impedance studies on metal hydride electrodes. *J. Electrochem. Soc.* **142**, 2935–43

Züttel A, Meli F, Schlapbach L 1994 Effects of pretreatment on the activation behavior of $Zr(V_{0.25}Ni_{0.75})_2$ metal hydride electrodes in alkaline solution. *J. Alloys Comp.* **209**, 99–105

Hydrogen–Metal Systems: Fluorinated Metal Hydrides

Z. P. Li

The thermodynamic properties of a hydrogen storage alloy are mainly determined by its composition and structure, while the activation rate and kinetic properties of the alloy are dependent on its surface activity and hydrogen diffusion rate in the alloy. In practical uses, surfaces of the hydrogen storage alloy are usually contaminated when exposed to air so that its thermodynamic properties are difficult to reach up to theoretical values at the beginning of the hydrogenation. Some alloys are easily poisoned by impurities in hydrogen gas and lose their hydrogen sorption capability because of the existing surface barrier caused by impurities. This surface barrier to the hydrogenation process is called the *passive surface layer*. This layer significantly influences the activation property and kinetic characteristics of the hydrogen storage alloy because the alloy is hydrogenated or dehydrogenated through a surface reaction. If the alloy is easily pulverized to generate fresh surfaces through hydriding and dehydriding cycling or its passive surface layer is loose, the alloy (like AB_5) usually shows a good activation property and quick kinetic characteristics of hydrogenation and dehydrogenation. If the passive surface layer is too loose to block off impurities attacking to the alloy, the alloy will show a poor durability when being cycled by hydriding and dehydriding in gas solid system or by charging and discharging in alkaline electrolyte. If the passive surface layer is condensed and difficult to crack through hydriding and dehydriding cycling, hydrogen is difficult to penetrate through the passive layer to approach to catalytic sites where the hydrogenation or dehydrogenation reaction occurs, thus the alloy will show a poor activation property and slow kinetic characteristics. A_2B alloys, e.g., Mg_2Ni, are typical examples. They cannot bring the hydrogen sorption capability into full play under mild conditions especially in alkaline electrolytes because of the oxide layers existing on alloy surfaces that lack catalytic sites for charge and discharge. For Mg_2Ni, the condensed oxide film and the leanness in catalytic sites on the alloy surfaces are not the only reasons for its poor hydrogenation kinetics as the low hydrogen diffusion rate in the alloy bulk is another fatal shortcoming.

The effects of the surface layer on hydrogenation properties are strongly dependent on the surface composition and structure, e.g., La_2O_3 on $LaNi_5$ alloy surfaces, ZrO_2 on zirconium-based AB_2 alloy surfaces, and MgO on magnesium-based A_2B alloy surfaces. Their effects on the conversion of molecular hydrogen or ionic hydrogen to atomic hydrogen at alloy surfaces are different. La_2O_3 surface layer is porous and shows little retardation of the activation but ZrO_2 and MgO are condensed and show severe retardation of the activation. In order to improve the hydrogenation properties of the hydrogen storage alloy, it is necessary to replace the original passive surface layer to a highly reactive surface layer by some surface modifications. How to design a suitable surface structure for certain hydrogen storage alloys to meet a special requirement for practical uses and how to construct this desired surface structure are the key points to develop surface treatment techniques. Table 1 enumerates the requirements related to the surface state of a hydrogen storage alloy and the desired surface structure.

From Table 1 it can be concluded that the surface layer of a desired hydrogen storage alloy with a good activation property, quick kinetic characteristics, and long durability should be a protective surface layer in which highly dispersed catalytic sites exist. To improve the durability of the alloy without a sacrifice in reactivity, this protective surface layer should allow hydrogen to pass through but hinder impurities from invading. The alloy should yield a specific surface area as large as possible. In order to approach this desired hydrogen storage alloy, many surface treatment techniques have been developed such as chemical plating, hot alkaline treatment, acid treatment, and fluorination treatment. It was found that the fluorination treatment technique is a useful weapon to overcome many shortcomings of hydrogen storage alloys. Fluorination-treated surfaces exhibit significantly improved characteristics in regard to the hydrogen uptakes and the protection against impurities and electrolyte solution. In addition, highly catalytic metal nickel can be chemically reduced on surfaces of alloy particles by some fluorination

Table 1
The requirements related to the surface state of a hydrogen storage alloy and the desired surface structure.

Requirement for a hydrogen storage alloy	Desired surface structure
Good activation property	Highly dispersed catalytic sites
	Large specific surface area
	Antipoisonous catalytic sites
	Few oxides existing
Quick kinetic characteristics	Highly dispersed catalytic sites
	Large specific surface area
Long durability	Protective layer against impurities or electrolyte
	Antipoisonous catalytic sites

treatments. A summary presentation about fluorination techniques has been published, that elucidated the basic conception of fluorination treatment techniques (Suda *et al.* 1998). Here, the principle of fluorination treatments and treatment effects on improving the hydrogenation and dehydrogenation properties are summarized to give some examples how to construct a desired surface layer by using fluorination treatment technologies.

1. Principle of the Fluorination Treatment

The so-called fluorination treatment is a method to treat alloy powders by fluorine-containing aqueous solutions. It is known that fluorine is chemically the most reactive of all the elements and combines directly at ordinary or elevated temperatures with all the elements other than nitrogen, oxygen, and the lighter noble gases due to its strongest electronegativity. Fluorine can easily form complex ions with many transition metal elements such as zirconium, titanium, and manganese whose complex ions with fluorine easily dissolve into aqueous treatment solutions. Alternatively, F^- ions react with magnesium, calcium, and lanthanides to form insoluble fluorides on surfaces of hydrogen storage alloys. Fluorination treatment techniques make full use of these advantages to modify the surfaces of the hydrogen storage alloys in accordance with the alloy composition.

It is possible to produce fine alloy powders with a large specific surface area and store them safely in air or water for long time without surface activity degradation for hydrogen uptake by using fluorination treatments. The activation and kinetic problems related to the alloy surface can be resolved by removing surface oxides, increasing specific surface area and creating more catalytic sites in alloy surface layers. The stability and tolerance to impurities can be improved by forming a protective fluoride layer on alloy surfaces. For example, a protective surface layer (LaF_3) on $LaNi_{4.7}Al_{0.3}$ alloy surface can improve the durability against the CO and KOH aqueous solution. Removing ZrO_2 and chemically depositing catalytic metallic nickel on zirconium-based AB_2 alloy surfaces improves the conversion capability of ionic hydrogen to atomic hydrogen at the surfaces.

A desired surface structure of a hydrogen storage alloy can be constructed by a certain fluorination treatment according to the requirements for the practical use. Four typical types of fluorinated alloy surface structures are illustrated in Fig. 1, which are the fluoride surface layer, the compact metallic nickel layer, the composite surface layer with fluorides and metallic nickel, and the porous metallic nickel layer. Fig. 1(a) shows the magnesium fluoride formed on magnesium-based hydrogen storage alloy surface, Fig. 1(b) shows the nickel film formed on the alloy surface, Fig. 1(c) illustrates a composite structure of fluorides and nickel clusters, and Fig. 1(d) shows the nickel network formed on zirconium-based AB_2 alloy surfaces.

The surface structure deeply depends on the treatment conditions and the composition of alloys to be treated. To obtain the desire surface structure, the treatment solution, method, and conditions should be carefully determined according to the alloy composition. The fluorination treatment mechanism is different when using different treatment solutions to different alloys. Each treatment mechanism will be described in details at the part of influence of the treatment solution.

2. Influence of Fluorination Treatments on the Surface Structure of the Hydrogen Storage Alloy

All the fluorination treatments are designed to remove the original oxide layer of the hydrogen storage alloy then form a desired surface layer with respect to the special requirement for a certain application. There are two ways to remove the surface oxide. One is converting oxides into fluorides, which often occurs in fluorination treatments of AB_5 and magnesium-based alloys. For example, the La_2O_3 surface layer of $LaNi_{4.7}Al_{0.3}$ alloy can be converted into a LaF_3 fluoride layer through the following reaction:

$$La_2O_3 + 6F^- + 3H_2O \rightarrow 2LaF_3 + 6OH^- \quad (1)$$

The other is dissolving oxides into the treatment solution, which usually occurs in the fluorination treatments of AB_2 and AB alloys whose constituent elements react with F^- ions to form soluble complex ions in aqueous solution. For instance, the ZrO_2 surface layer of the zirconium-based AB_2 alloy is dissolved in the treatment solution by forming zirconium fluorine complex ions $[ZrF_6]^{2-}$ by the following reaction:

$$ZrO_2 + 6F^- + 2H_2O \rightarrow [ZrF_6]^{2-} + 4OH^- \quad (2)$$

There are many factors that influence the treatment effects on the hydrogenation properties. The key point of the fluorination treatment techniques is how to make up the treatment solution to form a desired surface layer. Other points are the treatment temperature, the treatment time, and the pH range of the treatment solution. The hydrogen storage alloys must be modified by a suitable treatment solution under suitable conditions, otherwise the treatment effect will be negative.

2.1 Influence of the Treatment Solution

The fluorination treatment solution is mainly composed of three parts. The first part is the fluorine resource such as fluorhydric acid (HF) and KF.

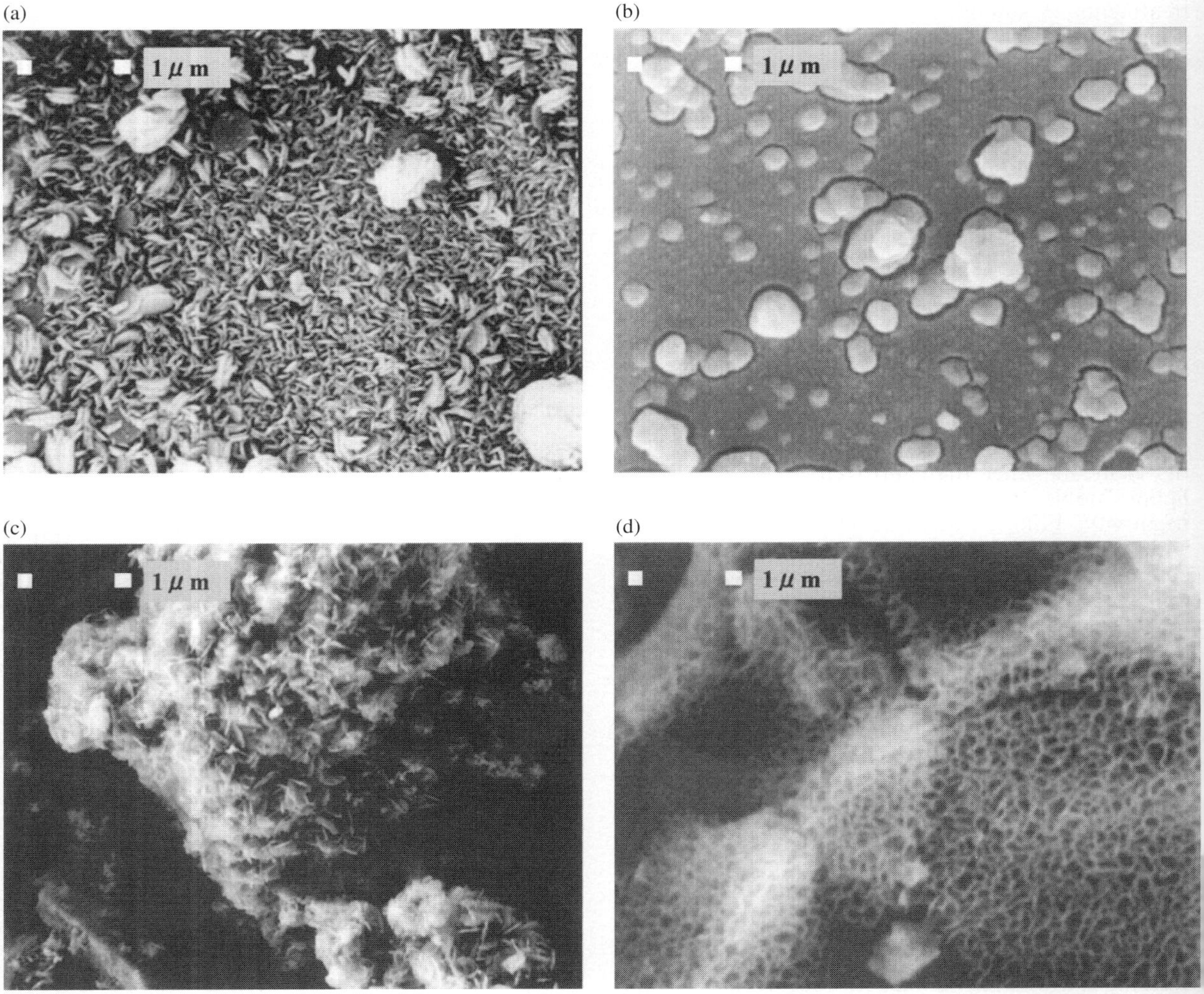

Figure 1
Typical surface structures of fluorinated alloys: (a) fluoride surface layer, (b) compact nickel surface layer, (c) composite nickel–fluoride surface layer, (d) porous nickel surface layer.

Sometimes a dilute HF solution is sufficient to treat hydrogen storage alloys. It is possible to control the pH value by diluting or adding KF to the treatment solution. The second part is the catalyst resource such as NiF_2, or $CoCl_2$. It is known that metallic nickel and cobalt are suitable for hydrogenation in addition to noble metals. The third part is a reducing agent such as sodium phosphinate. Sometimes, in order to stabilize transition metal ions in treatment solutions, a ligand such as sodium citrate is needed.

(*a*) F-1 *solution.* The simplest fluorination treatment solution is a simple KF and HF aqueous solution, 1 l of treatment solution containing 6 g KF and 0.6 ml of 48% HF. This treatment solution is nominated as F-1 solution. The treatment using F-1 solution is called F-1 treatment. During treating the hydrogen storage alloys, F^- ions are supplied constantly by the flowing stepwise equilibrium relations:

$$HF^-{}_2 \leftrightarrow HF + F^- \qquad (3)$$

$$HF \leftrightarrow H^+ + F^- \qquad (4)$$

In the F-1 treatment, a large amount of Ni dissolves into the treatment solution while fluorides are formed on the alloy surfaces. The F-1-treated alloys show a very large specific surface area due to the formation of the fluoride layer on the alloy surfaces. Extensive studies have been done on the reaction kinetics of F-1 treated $LaNi_{4.7}Al_{0.3}$ and very positive results have been obtained.

(*b*) *F-2 solution.* In order to minimize the nickel dissolution from the alloy surface to the treatment solution, a small amount of NiF_2 is added to the F-1 solution reaching its saturation limit at a certain temperature. This solution is called F-2 solution and the treatment using this solution is denoted as F-2 treatment.

(*c*) *F-3 solution.* An $HF–KF–NiF_2–NaH_2PO_2–Na_3C_6H_5O_7$ (denoted as F-3 solution) treatment solution is made up to chemically deposit catalytic nickel on the alloy surface in order to improve the electrochemical properties. One liter of the F-3 solution contains 6 g KF, 0.6 ml of 48% HF, 8 g NiF_2, 8 g NaH_2PO_2, and 3 g $Na_3C_6H_5O_7$. The treatment using this solution is denoted as F-3 treatment. The reaction mechanism of the nickel implantation can be described by the following reaction:

$$2H_2PO^-{}_2 + 2H_2O + [Ni^{2+}C_6H_5O_7^{3-}] = Ni^0 + 2HPO_3^{2-} + 4H^+ + 2H^0 + C_6H_5O_7^{3-} \quad (5)$$

where Ni^0 denotes metallic nickel chemically reduced on the alloy surfaces, and H^0 denotes monatomic hydrogen (protium) absorbed in the alloy lattice. In the F-3 solution, $C_6H_5O_7^{3-}$ is used as a ligand to preventing the precipitation of the metallic nickel out of the alloy surfaces, and the $H_2PO_2^-$ serves as the reducing agent.

The surface structure and composition of the hydrogen storage alloy after fluorination treatment are significantly influenced by the treatment condition composition. For example, when a magnesium-based alloy is treated by the F-1 solution, a fluoride layer is formed on alloy surfaces and some nickel dissolves into the treatment solution as shown in Fig. 1(a). When an AB_2 alloy is treated by the F-2 solution at higher temperature, nickel ions in the treatment solution are chemically reduced by the constituent elements of the alloy like zirconium to form a porous nickel layer, as shown in Fig. 1(d), instead of constituent nickel dissolution at lower treatment temperature. When nickel ions are chemically reduced by a reducing agent, nickel grows on the metal basis. The deposited nickel forms the sites where nickel can grow easily. Therefore, a compact nickel layer will be formed like chemical coating as shown in Fig. 1(b). If an F-1 treated AB_5 alloy is treated by an F-3 solution again, a composite layer of metallic nickel and fluoride will be created on the alloy surfaces as shown in Fig. 1(c).

2.2 *Influence of the Treatment Temperature*

The constituent nickel of the hydrogen storage alloy shows a tendency to dissolve into the treatment solution when the alloy is treated at a low temperature because the fluorination treatment solutions are weak acid aqueous solutions. However, if the alloy is treated at a higher temperature, the nickel ions in the treatment solution can be chemically deposited to the alloy surfaces through chemically reduction by the metallic constituent elements or via reducing agents based on the following chemical replacement reaction or Eqn. (5):

$$0.5(x-y)\,[NiF_6]^{4-} + M(\text{in alloy}) + (3y-2x)F^- \rightarrow [MF_x]^{y-} + 0.5(x-y)Ni^0 \quad (6)$$

where M stands for the constituent elements whose chemical reduction capability is higher than that of nickel. It is an oxidation–reduction reaction.

The reduction capability of the constituent elements is determined by their oxidation potentials. Generally the constituent A-side elements in the alloy formula AB_x ($x = 0.5, 1, 2,$ or 5) possess a higher chemical reduction capability than metallic nickel so that the A-side elements usually serve as the reducing agents, such as lanthanum, calcium, magnesium, zirconium, etc. Some B-side elements like manganese, vanadium, and aluminum are also able to chemically reduce the nickel ions. The sites occupied by the constituent elements with higher reduction capability are first to be replaced by the chemically reduced nickel if the alloy is treated by an F-2 solution at a higher treatment temperature. As a result, a porous nickel surface layer can be formed as shown in Fig. 1(d). This treatment using an F-2 solution at high temperature is referred to as F-5 treatment.

If the alloy is treated by an F-3 solution at low temperature (room temperature), the nickel ions cannot be chemically reduced though the nickel dissolution can be minimized due to the addition of the reducing agent. The metallic nickel can only be chemically reduced at a higher temperature such as 70°C. The nickel precipitation has nothing to do with the constituent elements of the alloy when using F-3 solutions. Every F-3-treated hydrogen storage alloy shows a similar compacted nickel layer formed on the alloy surfaces.

(*a*) *Influence of the alloy composition.* Except for F-3 treatment, the surface structure and composition after fluorination treatments are strongly dependent on the alloy composition. An LaF_3 or CaF_2 fluoride layer in which metallic nickel clusters exist can be formed in the case of AB_5 alloys such as $LaNi_5$ or $CaNi_5$. Similar to the AB_5 alloys, the magnesium-based alloys generate a protective and reactive MgF_2 layer on their surfaces for hydrogen uptake, but the AB_2 alloys cannot form insoluble fluoride layers on the alloy surfaces. If an AB_2 alloy is doped with some fluoride formation elements like lanthanum, fluorides can also form when an lanthanum-rich phase appears on the alloy surface during fluorination treatments.

(*b*) *Combination with other treatment methods.* The fluorination treatment can not only be applied alone but can also be combined with other treatments (including other fluorination treatments) to improve the over all properties of the hydrogen storage alloys. For example, an F-1-treated alloy can be treated again by an F-3 solution to form a nickel-enriched fluoride layer on alloy surfaces with a very high specific surface area. The fluorination treatment combining the F-2 treatment with the alloy pulverization through a planetary ball mill which successfully increases the specific surface area and catalytic activity of the AB_2 alloy. It is denoted as F-9 treatment.

Even though the F-9 treatment is conducted at room temperature, the nickel ions can be deposited on the alloy surfaces by the mechanical energy provided by the ball mill. The mechanical energy produces nickel implants on the fresh surfaces generated during the particle pulverization instead of nickel dissolution from the alloy like in the F-2 treatment at low temperature. If the F-9-treated AB_2 alloy was heated to 70°C again with the treatment solution containing nickel ions, more metallic nickel can be deposited on the alloy surfaces. This treatment can be considered as a combination of the F-9 treatment with the F-5 treatment.

3. *Applications of the Fluorination Treatments*

Generally, fluorination treatments can successfully modify the alloy surfaces by increasing the specific surface area and catalytic activity. The functions of the surface modification through the fluorination treatments can be attributed to (i) removing the passive oxide layer from surfaces of hydrogen storage alloys, (ii) increasing the specific surface area, (iii) creating a protective layer that works as selective passage for hydrogen and keeps hydrogen storage alloys from contamination by other materials such as air and moisture, and (iv) depositing the catalytic metals onto the alloy surface layer.

The hydrogen storage alloys usually are classified as AB_5, AB_2, AB, and A_2B type alloys. In practical use, it is used to classify as the rare earth-based, zirconium-based, titanium-based, and magnesium-based alloy. Here some typical alloys are selected to describe the fluorination treatment effects in accordance with the shortcomings of each series of alloys. For instance, RNi_5-type alloys (R = rare-earth element) are selected as the example for AB_5 alloys, zirconium-based alloys as the example of AB_2 alloys, and magnesium-based alloys as the example of A_2B alloys.

3.1 *AB_5-type Alloys*

It is known that AB_5 alloys are composed of lanthanides, calcium, or other elements such as yttrium, or zirconium (A-side elements) and many transition elements or aluminum (B-side elements). They generally have a hexagonal crystal structure (prototype $CaCu_5$). The AB_5 alloys are easily activated at mild conditions. They are pulverized to fine powders by few hydriding and dehydriding cycles. The fine powders are pyrophoric if suddenly exposed to air. A small amount of O_2, H_2O, CH_4, or CO_2 existing in hydrogen gas will lead to decrease the hydrogen storage capability and reaction kinetics of the alloy. CO is a strong poison for the hydrogenation of the alloy. When they are used as electrode materials, the alloys are corroded to form hydroxides during charge–discharge cycling in alkaline aqueous solutions. The formed lanthanide or calcium hydroxides usually are porous which cannot prevent the alloy from being further corroded.

With the aim of decreasing the pyrophoricity of the AB_5 alloy and improving its impurity resistance, an F-1 solution was applied to the alloy to remove the oxide then to create a protective fluoride layer on the alloy surface. By the fluorination treatment of the $LaNi_{4.7}Al_{0.3}$ alloy, a LaF_3 layer is formed on the surfaces of the particles. Metallic nickel precipitates in a sublayer below the LaF_3 layer. The fluorinated $LaNi_{4.7}Al_{0.3}$ shows a high resistance against the air and water as shown in Fig. 2(b) and (c) (Liu and Suda 1995) and an excellent protection against CO poisoning as shown in Fig. 3(a) and (b) (Wang *et al.* 1995). Due to the excellent protective capability of the fluoride layer on the alloy surface against the impurities such as CO, CO_2, CH_4, and H_2O, the fluorinated $LaNi_{4.7}Al_{0.3}$ alloy was successfully applied as a catalyst for methane synthesis without being influenced by the reactants. The products as shown in Fig. 4 (Suda *et al.* 1997). Carbon dioxide was effectively converted into methane by using the fluorinated alloy with higher specific surface area.

This fluoride layer cannot only effectively protect the catalytic sublayer against impurities even against the extremely poisonous carbon monoxide but also improve the cycling durability of the $LaNi_{4.7}Al_{0.3}$ electrode in an alkaline aqueous electrolyte as shown in Fig. 5 (Li *et al.* 1994). However, it has been found that the fluorination treatment decreased the discharge capacity because of the fluoride formed on the surface that decreased the electrical conductivity of the electrode. The decrease in electrical conductivity has no influence on the gas–solid reaction, but it really is a big problem when the fluorination treatment is applied to treat the electrode materials. In order to overcome this shortcoming, an $HF–KF–NiF_2–NaH_2PO_2–Na_3C_6H_5O_7$ (F-3 solution) treatment solution can be applied for increasing the metallic Ni content in surface layer of the alloy ($LmNi_{4.0}Co_{0.4}Al_{0.3}Mn_{0.3}$) used in electrochemical system. It was found that this fluorination technique considerably improved the durability and initial activation properties of the AB_5-type electrodes as shown in Fig. 6 (Sakashita *et al.* 1997).

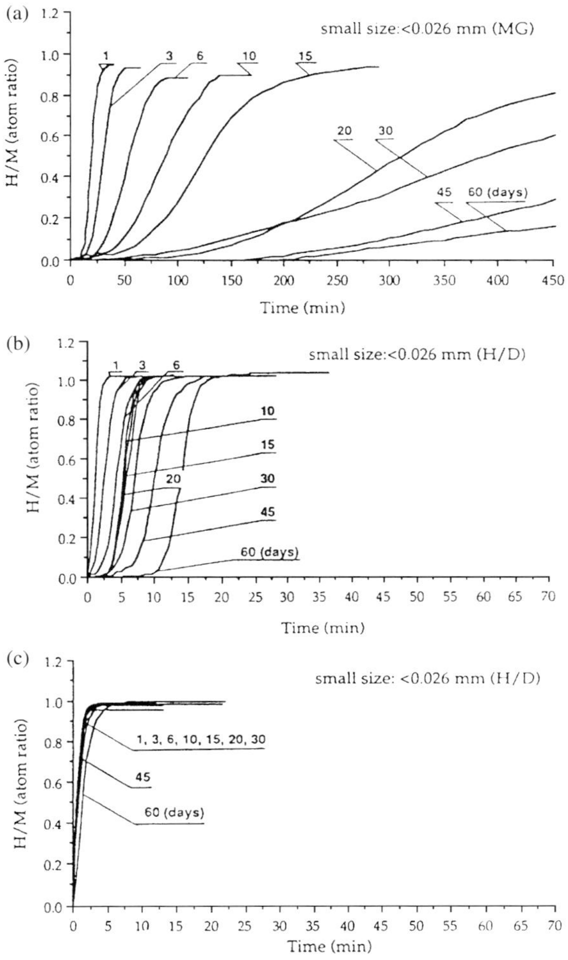

Figure 2
Resistance of the fluorinated $LaNi_{4.7}Al_{0.3}$ to air and water: (a) air-exposed, untreated; (b) air-exposed, fluorine treated; (c) water-exposed, fluorine treated.

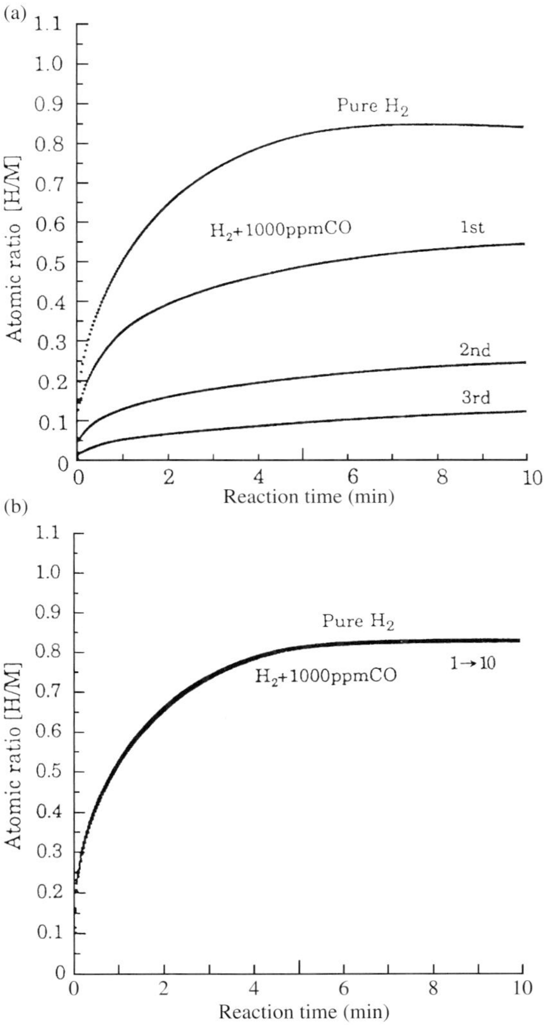

Figure 3
Protection capability of fluorinated $LaNi_{4.7}Al_{0.3}$ from CO poisoning: (a) untreated, (b) fluorine treated.

It is possible to produce a surface layer on AB_5 alloy with a nickel–fluoride composite by a duplicated fluorination technique. The duplicated fluorination treatment consists of two processes (A and B). In process A, the pH of the treatment solution is controlled by a CH_3COONa–CH_3COOH buffer solution to increasing the specific surface area of the alloy. In process B, Ni^{2+} ions are chemically reduced to metallic nickel by adding the reducing agent NaH_2PO_2 as shown in Eqn. (5). The specific surface structure formed during the treatment results in a significant improvement in the initial hydrogen uptake of the $LaNi_{4.7}Al_{0.3}$ alloy in the gas–solid reaction by extending the treatment time in process B. The electrochemical lifecycle was improved by increasing treatment time in process A due to the formation of fluoride at the particle surface as shown in Fig. 7 (Sun *et al.* 1999).

From the research results with AB_5 alloys, it can be concluded that fluorination treatments eliminate surface oxides, increase the specific surface area, decrease the pyrophoricity of the alloys, improve their resistance to impurities and air or water, and increase the anticorrosion capability of the AB_5 electrodes against alkaline electrolyte by choosing suitable fluorination treatment conditions.

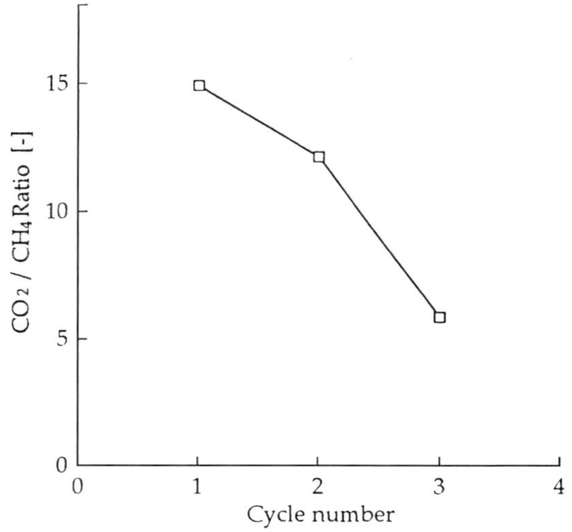

Figure 4
Methane formation from CO_2 and H_2 using fluorine-treated $LaNi_{4.7}Al_{0.3}$ during three *P–T* swings.

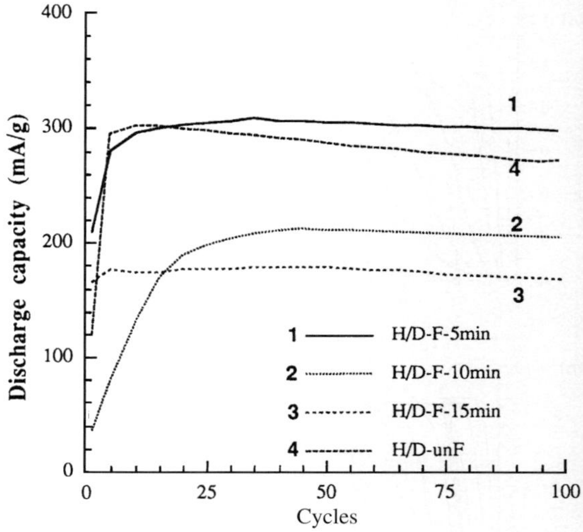

Figure 6
Effect of fluorination time on electrochemical cycling behavior of $LmNi_{4.0}Co_{0.4}Al_{0.3}Mn_{0.3}$ alloy under discharge current of $150\,mAg^{-1}$ at 20°C.

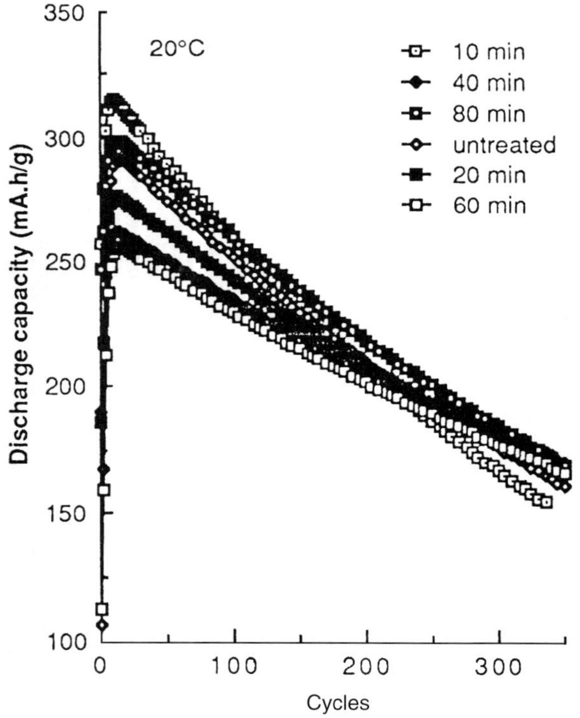

Figure 5
The fluorine-treatment effect on the durability of $LaNi_{4.7}Al_{0.3}$ electrodes.

3.2 AB_2-type Alloys

The A-side elements in AB_2 alloys are often from the group IVA (mainly zirconium and titanium). The B-side elements can be a variety of transition metals such as nickel, chromium, and manganese. The AB_2 alloys are largely based on two relative Laves phase crystal structures, i.e., hexagonal (prototype $MgZn_2$, C14) and cubic (prototype $MgCu_2$, C15). AB_2 alloys usually have a higher hydrogen storage capacity but are more difficult to activate than AB_5 alloys. Their electrochemical kinetic properties are poor compared with AB_5 alloys though there is hardly any difference in gas–solid reaction between the two. The slow activation and poor kinetic properties in electrochemical reactions are caused by the dense oxide layer formed on the alloy surfaces and the intrinsic nickel lean surface layer of the AB_2 alloy. Compared with AB_5 alloys, AB_2 alloys usually show a quick capacity degradation and quick increase in reaction impedance during charge–discharge cycling. It is caused by the surface passivation. This means that the fresh alloy surface after activation is oxidized again to form a condensed oxide layer. The alloy surface after charge–discharge cycling is back to the original surface state before activation. Therefore, removing the original oxide layer, avoiding formation of the condensed oxide layer again during cycling, and increasing the metallic nickel content in surface layer is the central task for the surface treatment of AB_2 alloys. Some fluorination treatments have been developed around this central task.

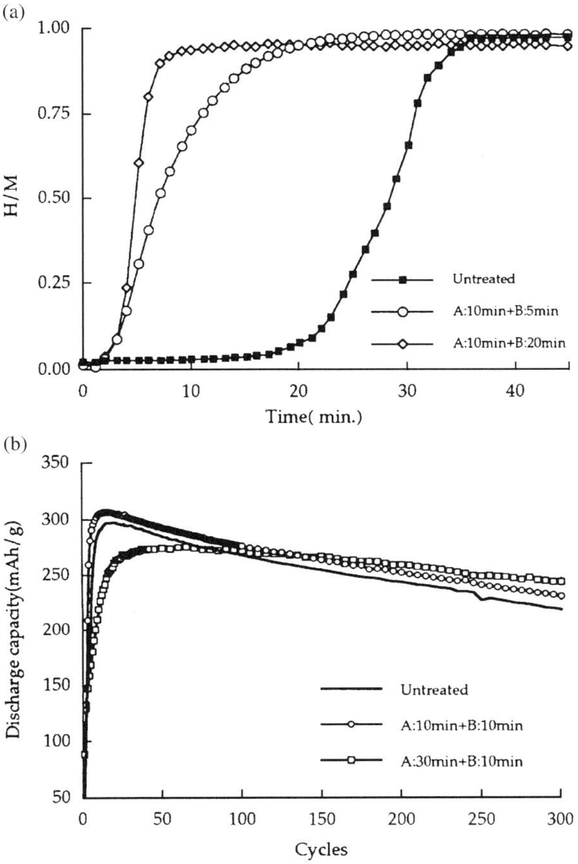

Figure 7
The initial activation of fluorine-treated AB_5 alloys in (a) gas–solid reaction and (b) electrochemical reaction and the cycle life of their electrodes.

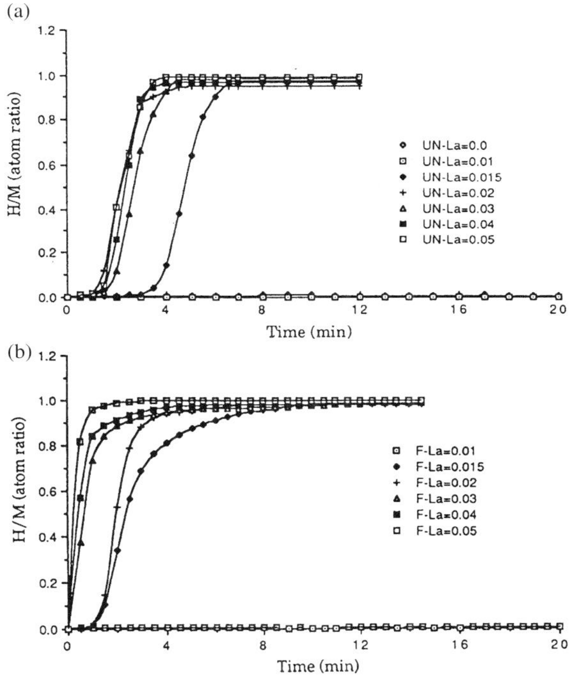

Figure 8
The effect of La content on the initial activation characteristics of freshly crushed $Ti_{0.5-x/2}Zr_{0.5-x/2}La_xMn_{0.8}Cr_{0.8}Ni_{0.4}$ alloys ($P_i = 1.0$ MPa, $T_s = 40°C$): (a) untreated, (b) fluorine treated.

For example, to crack the condensed oxide surface layer, a small amount of lanthanum can be added to the AB_2 alloy ($Ti_{0.5}Zr_{0.5}Mn_{0.8}Cr_{0.8}Ni_{0.4}$). It has been reported that the fluorination treatment of AB_2 alloy containing a small amount of lanthanum could dramatically improve the initial activation as shown in Fig. 8 (Liu and Suda 1995). This is because the lanthanum cannot be substituted into the AB_2 phase but forms lanthanum-rich particles that serve as catalytic sites for hydrogen uptake. The fluoride formed on the lanthanum-rich particles after the fluorination treatment keeps these catalytic sites active. No significant deterioration of the surface reactivity was observed for F-treated lanthanum-incorporated AB_2 alloy even after 6 months exposure to air and water.

An F-treatment in which the alloy powders are first cleaned in a 0.02N HCl solution, and then immersed in a dilute HF solution for 20–60 minutes, has successfully improved the electrochemical activation property of as-atomized alloy powders. It was reported that fluorine ions preferably dissolved the oxide layer on the powder surface, and formed channels that connect the surface and nickel-rich layers inside the atomized powders (Huang *et al.* 2000). The fluorine-treatment enhanced the rate capability of alloy electrodes and improved its long-term antioxidation ability.

An excess stoichiometric AB_2 alloy with lanthanum incorporation ($Zr_{0.9}Ti_{0.1}V_{0.2}Mn_{0.6}Ni_{1.3}La_{0.05}$) was step-treated with an F-2 solution at 70°C then with F-1 solutions at room temperature (Gao *et al.* 1999). A fine network structure of both the lanthanum fluoride and the nickel-rich layer was formed on the alloy surfaces. The specific surface area of the fluorinated particles was increased by a factor of about 20 compared to the untreated one, i.e., from $0.27\,m^2\,g^{-1}$ to $5.31\,m^2\,g^{-1}$. This increase in specific surface area can be contributed to the formation of lanthanum fluoride and to nickel implantation on the alloy surfaces. After fluorination, (i) the initial discharge capacity reached to about 97% of its maximum discharge capacity, (ii) the surface reaction activity was significantly improved after the first charge, and (iii) the high rate discharge capability was greatly increased (Fig. 9).

In order to eliminate the oxide surface layer and to increase the metallic Ni content on the alloy surface layer, an HF–KF–NiF_2 (F-2 solution) and HF–KF–NiF_2–NaH_2PO_2–$Na_3C_6H_5O_7$ (F-3 solution) treatment solution were applied for improving the activation property of the AB_2 alloy ($Zr_{0.9}Ti_{0.1}V_{0.2}Mn_{0.6}Co_{0.1}Ni_{1.1}$) used in electrochemical systems. One liter of the F-2 treatment solution contains 6 g KF, 0.6 ml of 48%HF, and 8 g NiF_2. One liter of the F-3 treatment solution contains 6 g KF, 0.6 ml of 48%HF, 8 g NiF_2, 8 g NaH_2PO_2, and 3 g $Na_3C_6H_5O_7$. During the fluorination treatments of the AB_2 alloys, metallic nickel was deposited not only by the reducing agents but also by the constituent zirconium and manganese that dissolved from the alloy surfaces to the treatment solution. When using an F-2 solution, the oxide layer on the alloy surface was removed by reaction (2) and the nickel reduction reactions are shown in following reactions:

$$Zr + 2[NiF_6]^{4-} = [ZrF_6]^{2-} + 2Ni^0 + 6F^- \quad (7)$$

$$Mn + [NiF_6]^{4-} = [MnF_6]^{4-} + Ni^0 \quad (8)$$

When using an F-3 solution, the oxide layer on the alloy surface was removed by reaction (2) and the nickel reduction reactions proceeded via reaction (5). Different surface layers on the AB_2 alloys were formed due to the different nickel chemical reduction mechanisms. The fluorination treatment using an F-2 solution created a porous nickel layer on the alloy surface but the treatment using an F-3 solution created a compact nickel layer. The specific surface area of the treated alloy was increased by the constituent element dissolution and the porous nickel layer formation. The fluorination treatment using an F-2 solution further decreased the reaction impedance and improved the surface catalysis to increase the exchange current density compared with the treatment using an F-3 solution. The electrochemical activation and rate discharge capacity were further improved as shown Fig. 10 (Li *et al.* 1999).

A zirconium-based AB_2 ($Zr_{0.9}Ti_{0.1}V_{0.2}Mn_{0.6}Co_{0.1}Ni_{1.1}$) alloy was treated by ball-milling the alloy powder in an NiF_2 and HF–KF solution through a planetary ball mill at room temperature. It was found that the electrode of the treated alloy exhibited excellent electrochemical performance. Its activation property and the high-rate discharge capacity were found to be satisfactory compared with the conventional AB_5-based electrodes. The fluorinated AB_2 electrode demonstrated a significant decrease of the reaction resistance measured by an impedance spectrometry (Fig. 11). The fluorine-treated alloy was found to have a large specific surface area and to form a highly nickel-covered surface layer (Liu *et al.* 1999).

A porous metallic nickel layer can be formed on the alloy surfaces when the nickel ions are chemically reduced by the constituent elements of the alloy through an F-5 treatment. The porous nickel layer improved the electrochemical properties of AB_2 alloys compared with a compact nickel layer. As a result, batteries using fluorinated alloys with porous metallic nickel layers on the alloy surfaces showed better discharge performances (Li *et al.* 2000).

It was reported that the sub-C batteries using fluorinated AB_2 alloys showed high discharge capacity (2600 mAh) and good activation properties. Their rate discharge capacity reached up to 2.3 Ah at 4 C (10.4 A) as shown in Fig. 12. The battery sustained over 400 cycles at a 0.2 C charge discharge rate.

The fluorination treatment cannot only be used to improve the activation properties and reaction kinetics of the AB_2 alloys, but also to avoid alloy contamination from the air or water. The fluorinated lanthanum-incorporated AB_2 alloys show a very strong resistance to air and water which has been introduced in fluorination treatment effects of AB_5-type alloys as shown in Fig. 13.

From the research results on AB_2 alloys mentioned above, it can be concluded that fluorination treatment successfully improve the storability and activation properties of the AB_2 alloys. They also decrease the reaction impedance and increase the high rate discharge capabilities of the their electrodes in alkaline electrolyte.

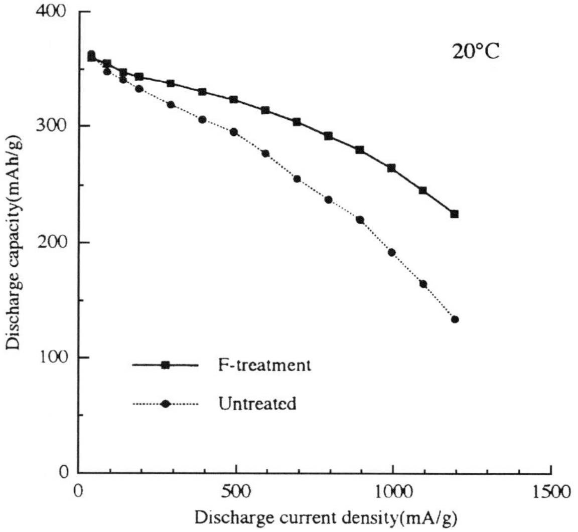

Figure 9
The discharge current dependency of discharge capacity of the F-treated and untreated AB_2 electrodes after full activation.

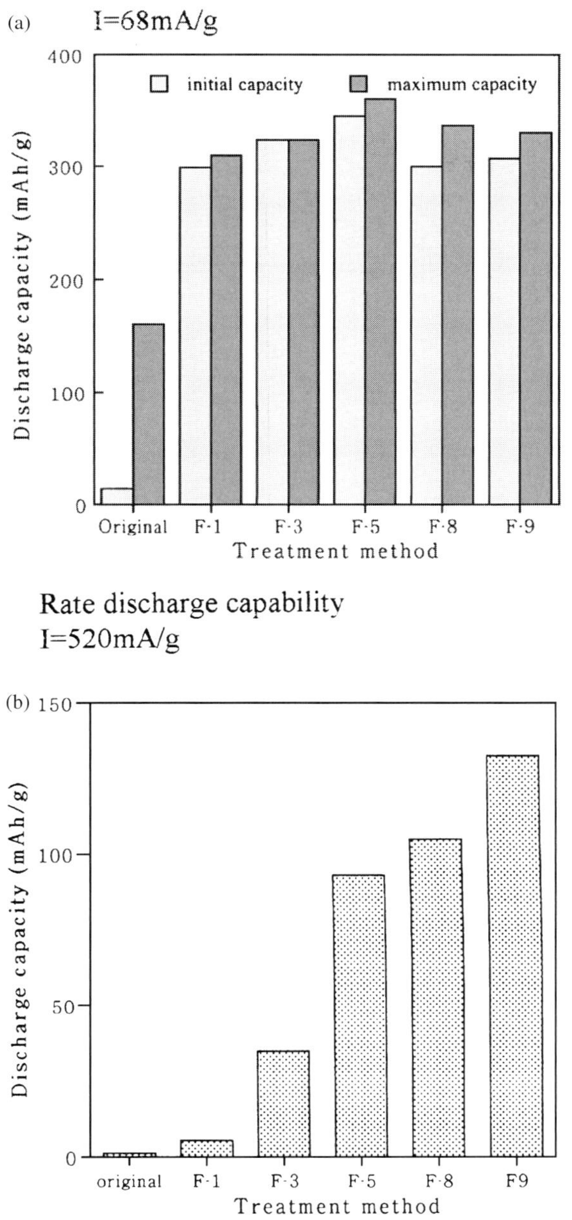

Figure 10
(a) The initial discharge capacities and (b) the rate discharge capabilities of fluorinated AB_2 alloys.

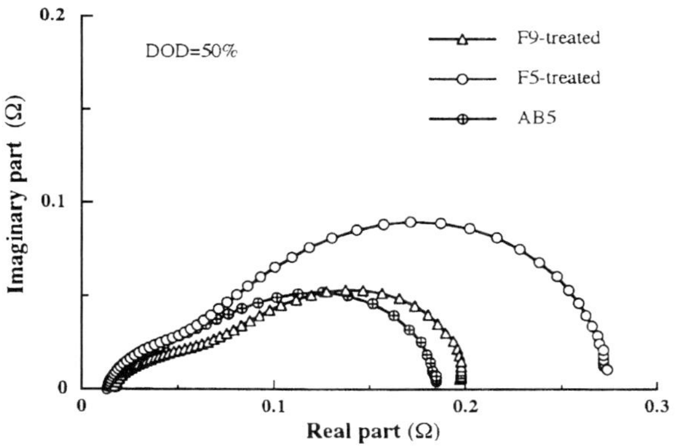

Figure 11
Impedance spectra of the fluorinated AB_2 electrodes at DOD = 50%.

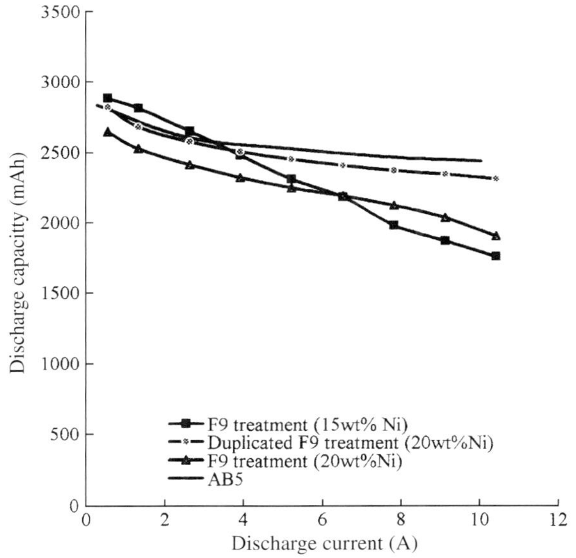

Figure 12
The rate discharge capabilities of sub-C sealed batteries using fluorinated AB_2 alloys.

3.3 A_2B-type Alloys

Various crystal structures are possible for A_2B-type alloys. A-side elements are group IIA and IVA elements such as magnesium, calcium, zirconium, and titanium. Mg_2Ni is a typical A_2B alloy that has been extensively researched because of its high hydrogen storage capacity though its high hydrogen desorption temperature and very slow hydrogen sorption and desorption kinetics are not satisfactory to meet application requirements. The fluorination treatment cannot change the hydrogen desorption temperature, but it is possible to improve the activation and decrease the hydrogen absorption temperature which are related to the reaction kinetics.

An HF–KF (F-1 solution) solution has been successfully used to treat the magnesium-based alloys such as the amorphous alloy $MgLa_{1-x}Ni_x$. A highly

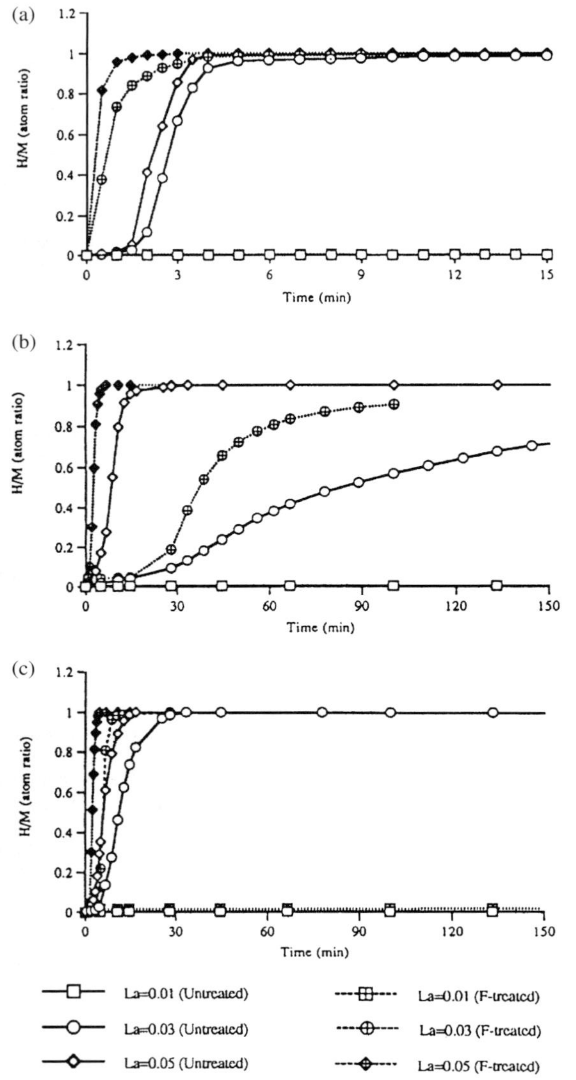

Figure 13
Effect of lanthanum content and storage method on the initial activation of $Ti_{0.5-x/2}Zr_{0.5-x/2}La_xMn_{0.8}Cr_{0.8}Ni_{0.4}$ with and without fluorine-treatment: (a) 1 day air exposure, (b) 6 months air exposure, (c) 6 months water exposure.

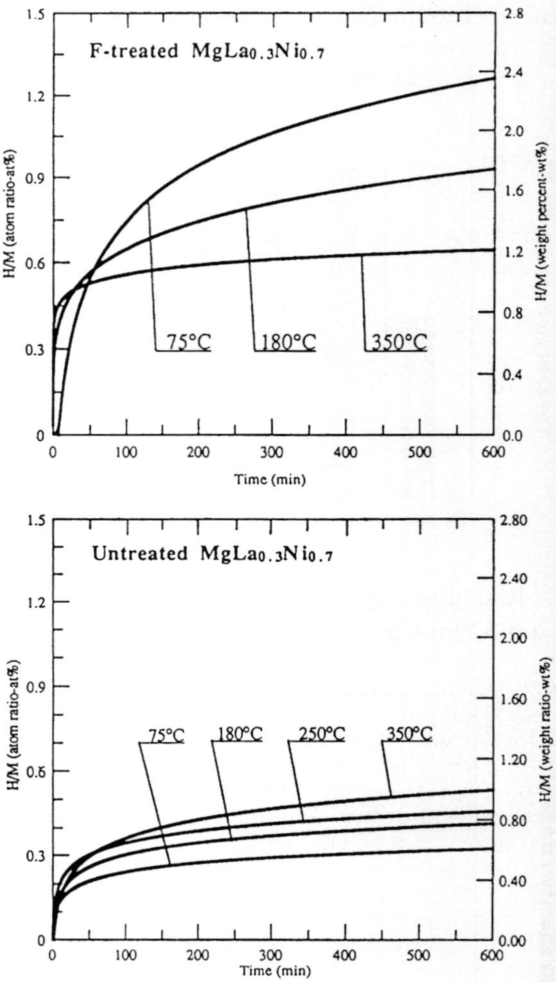

Figure 14
Initial activation characteristics of (a) fluorine-treated and (b) untreated $MgLa_{0.3}Ni_{0.7}$ amorphous alloy at different temperatures under 1.0 MPa H_2 pressure.

active Mg–La-fluorinated surface was created during a fluorination treatment for hydrogen uptake. This fluoride layer improved the initial activation of the alloys at moderate conditions. The maximum hydrogen storage capacity of the fluorinated $MgLa_{0.3}Ni_{0.7}$ during 10 h activation at 75 °C reached ~2.4 wt.%, which is nearly three-fold that of the untreated one (Fig. 14, Liu and Suda 1995). The low hydrogenation temperature prevented the amorphous alloy from crystallization to maintain the amorphous structure.

Yang has suggested using a 0.06 MNH_4F solution to treat $Mg_2Ni_{0.75}M_{0.25}$ (M = Ti, Cr, Mn, Fe, Co, Ni, Cu, or Zn) alloys (Yang *et al.* 2000). It was reported that the hydrogen desorption capability of the magnesium-based alloy reached the highest value after several hydrogen adsorption–desorption cycles. A study of the surface state of the alloys showed that a metallic nickel-rich layer was formed beneath a honeycomb MgF_2 layer. This layer was believed to have a high activity with a nickel concentration

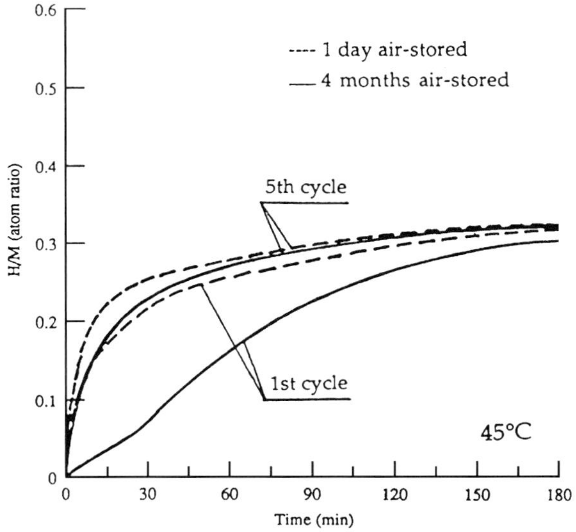

Figure 15
Effect of air exposure period on the initial H_2 absorption of fluorine-treated Mg_2Ni at 45°C under 1.0 MPa H_2 pressure.

approaching 70 at.%, which is much higher than 25 at.% of that in the alloy bulk.

Like many fluorinated AB_5 and lanthanum-incorporated AB_2 alloys, the fluoride layer formed on A_2B alloy surfaces protected the alloys from air and water contamination. For example, it was reported that there is no hydrogenation incubation time needed for the fluorinated Mg_2Ni alloy after 4 months exposure to air as shown in Fig. 15 (Liu and Suda 1995).

Though research work on A_2B alloy is less than that on AB_5 and AB_2 alloys, the fluorination treatments have shown encouraging results for magnesium-based alloys by decreasing the hydrogenation temperature and improving the resistance against air.

4. *Concluding Remarks*

Fluorination treatment is a surface modification method by means of which a desired surface structure of a certain hydrogen storage alloy can be constructed according to the requirements for the practical use. The hydrogenation and dehydrogenation performances of the hydrogen storage alloys in gas–solid reactions or in alkaline electrolytes can be improved by a surface modification through fluorination treatments. It is necessary to point out that the fluorination treatments are not omnipotent. They can only resolve the kinetic problems related to the alloy surface but cannot improve the thermodynamic properties of the hydrogen storage alloys. A proper selection of the treatment solution and treatment method as well as of the treatment conditions is very important to obtain the desired treatment effects.

The general features of the fluorination treatment techniques can be summarized by the following: (i) nonpyrophoricity when in air, (ii) treatment ability in air and water, (iii) removal of oxide layers, (iv) fluoride formation, (v) enlarged specific surface area, (vi) catalyst implantation to surface layers, (vii) stable surface against impurities and alkaline electrolyte, (viii) extra-ordinary rapid activation characteristics, (ix) hydrogen capacity closer to the theoretical value, and (x) low surface reaction impedance.

See also: Hydrogen–Metal Systems: Basic Properties (1); Hydrogen–Metal Systems: Hydride Forming Alloys (Properties and Characteristics, Database Information)

Bibliography

Gao X P, Sun Y M, Higuchi E, Toyoda E, Suda S 1999 Electrochemical properties and characteristics of the fluorinated $Zr_{0.9}Ti_{0.1}V_{0.2}Mn_{0.6}Ni_{1.3}La_{0.05}$ electrode. *J. Alloys Comp.* **293–295**, 707–11

Huang S S, Chuang H J, Chan S L I 2000 Effects of fluorination on the hydriding and electrochemical properties of a gas-atomized Zr-based hydrogen storage alloy. In: Gary E, Sholl C (eds.) *Abstr. Proc. Int. Symp. Metal–Hydrogen Systems*. Noosa, Australia, p. 51

Li Z P, Higuchi E, Liu B H, Suda S 1999 Electrochemical properties and characteristics of a fluorinated AB_2-alloy. *J. Alloys Comp.* **293–295**, 593–600

Li Z P, Liu B H, Hitaka K, Suda S 2000 Effects of surface structure of fluorinated AB_2 alloys on their electrode and battery performances. In: Gary E, Sholl C (eds.) *Abstr. Proc. Int. Symp. on Metal–Hydrogen Systems*. Noosa, Australia, p. 269

Li Z P, Yan D Y, Suda S 1994 The effects of F-treatment level on the electrochemical behavior of $LaNi_{4.7}Al_{0.3}$. *Trans. Mater. Res. Soc. Jpn.* **18B**, 1209–12

Liu B H, Li Z P, Higuchi E, Suda S 1999 Improvement of the electrochemical properties of Zr-based AB_2 alloys by an advanced fluorination technique. *J. Alloys Comp.* **293–5**, 702–6

Liu F J, Sandrock G, Suda S 1994 Effects of F-treatment on the incubation time of $LaNi_{4.7}Al_{0.3}$ alloy after long-time air exposure. *Trans. Mater. Res. Soc. Jpn.* **18B**, 1237–40

Liu F J, Suda S 1995a F-treatment effect on hydriding properties of the La-substituted AB_2 compound (Ti, Zr)(Mn, Cr, $Ni)_2$. *J. Alloys Comp.* **231**, 666–9

Liu F J, Suda S 1995b F-treatment effect on the initial activation characteristics of Mg–La–Ni amorphous alloy. *J. Alloys Comp.* **231**, 696–701

Liu F J, Suda S 1995c A method for improving the long-term storability of hydriding alloys by air/water exposure. *J. Alloys Comp.* **231**, 411–6

Sakashita M, Li Z P, Suda S 1997 Fluorination mechanism and its effects on the electrochemical properties of metal hydrides. *J. Alloys Comp.* **253–4**, 500–5

Suda S, Iwata K, Sun Y M, Komazaki Y, Liu F J 1997 A pressure–temperature swing process for the methanation of carbon oxides using fluorinated hydriding alloys. *J. Alloys Comp.* **253–254**, 668–72

Suda S, Li Z P, Sun Y M, Liu B H, Gao X P 1998 Surface passivation of metal hydrides for their applications. *Mater. Res. Soc. Symp. Proc.* **513**, 25–36

Sun Y M, Gao X P, Araya N, Higuchi E, Suda S 1999 A duplication fluorination technique for hydrogen storage alloys. *J. Alloys Comp.* **293–5**, 364–8

Wang X L, Iwata K, Suda S 1995 Effects of carbon monoxide on the hydriding reactions of the untreated and fluorinated $LaNi_{4.7}Al_{0.3}$ alloys. *J. Alloys Comp.* **231**, 829–34

Wang X L, Suda S 1995 Stability and tolerance to impurities of the fluorinated surface of hydrogen-absorbing alloys. *J. Alloys Comp.* **227**, 58–62

Yang H, Yuan H, Yang E, Ji J, Zhou Z, Zhang Y 2000 Characteristics of $Mg_2Ni_{0.75}M_{0.25}$ (M = Ti, Cr, Mn, Fe, Co, Ni, Cu and Zn) alloys after surface treatment. In: Gary E, Sholl C (eds.) *Abstr. Proc. Int. Symp. on Metal–Hydrogen Systems*. Noosa, Australia, p. 155

Hydrogen–Metal Systems: Hydride Forming Alloys (Properties and Characteristics, Database Information)

B. -H. Liu

Although many elements in the periodic table can react with hydrogen to form hydrides, most of them are not suitable for hydrogen storage as they are not able to absorb and desorb hydrogen readily and reversibly at moderate temperatures and pressures. If the practically favorable temperature and pressure ranges are set as 0–100 °C and 0.01–10 MPa, only VH_2 and $PdH_{0.6}$ are within these ranges and MgH_2 is near the ranges according to the Van't Hoff plots shown in Fig. 1. Despite the fact that the palladium–hydrogen system is the most extensively studied element–hydrogen system, it is of no practical interest as a hydrogen storage material because palladium is too expensive and its hydride has a low hydrogen content. Thus the hydriding properties of palladium will not be discussed here. The properties of V–H and Mg–H will be described later.

The discovery of hydride formation of some intermetallic compounds such as Mg_2Ni, TiFe, and $LaNi_5$ has opened a new scope for hydrogen storage. Because of their excellent hydriding and dehydriding properties under mild temperatures and hydrogen pressures, intermetallic hydrides immediately attracted much research interest. They have become the main focus of hydrogen storage materials research.

Hydrogen storage intermetallic compounds are usually classified into several groups according to their structure types such as AB_5, AB_3, AB_2, AB, and A_2B. A represents strong hydride-forming elements such as rare-earth metals, calcium, magnesium, zirconium, and titanium, and B represents transition metals such as nickel, cobalt, iron, manganese, and chromium. In this discussion hydrogen storage materials are classified into the following categories: (i) AB_5 alloys; (ii) AB_2 alloys; (iii) AB alloys; (iv) magnesium and magnesium-based alloys; (v) vanadium and vanadium-based b.c.c. alloys; (vi) disordered alloys; (vii) alloy composites; and (viii) complex hydrides.

A brief review of the characteristics and properties of hydride-forming alloys including their crystal structure, *P–C–T* relationship, Van't Hoff relationship, activation property, kinetic property, gas impurity effects, cyclic stability, metallurgy, and preparation methods is presented here with emphasis on gas reaction properties. The review is mainly based on a number of review articles (e.g., Sandrock 1997, 1999, Tamura 1998), and a detailed reference list is not included here. More detailed information can be found in review articles (e.g., Sandrock 1997, 1999, Tamura 1998) and references therein, and in the available database. The online database of metal hydrides can be found at http://hydpark.ca.sandia.gov.

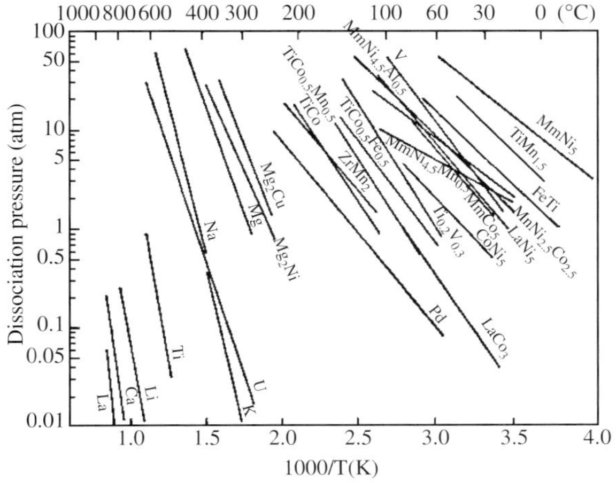

Figure 1
Van't Hoff plots of different metal (alloy)–H systems.

1. *AB_5 Alloys*

Since the discovery of hydride formation of $LaNi_5$ in the Philips Laboratory, AB_5 alloys have attracted a lot of research interest as hydrogen storage materials. Rare earth-based AB_5 alloys are extensively used as the anode material in the nickel metal hydride battery, a substitute for the nickel cadmium battery.

In the formula AB_5, A represents one or more rare-earth elements or calcium and B represents transition metals such as nickel, cobalt, or copper. They have a hexagonal $CaCu_5$-type structure.

1.1 Crystal Structure and Thermodynamic Properties

(*a*) *$LaNi_5$ and RNi_5 (R = rare-earth metal).* $LaNi_5$ is the prototype of the AB_5 hydride-forming alloy family. Its lattice parameters are $a = 0.5017$ nm and $c = 0.3982$ nm. On hydrogenation, a solid solution $LaNi_5H_x$ phase (α phase) forms first. When the hydrogen content is further increased, a hydride $LaNi_5H_6$ phase (β phase) forms. $LaNiH_6$ has a hydrogen content of 1.3 wt.% and its lattice parameters increase to $a = 0.5440$ nm and $c = 0.4310$ nm. Although the structure of $LaNi_5H_6$ remains uncertain, it is well accepted that hydrogen atoms mainly occupy interstices surrounded by La_2Ni_2 or (La_2Ni_4) and $LaNi_3$ (Yvon and Fischer 1988).

It has also been reported that a third phase, $LaNi_5H_3$ (γ phase), appears under certain conditions.

Figure 2 shows the *P–C–T* curves of $LaNi_5$ at different temperatures. The plateau pressure is about 0.2–0.3 MPa under ambient conditions. Hydrogen

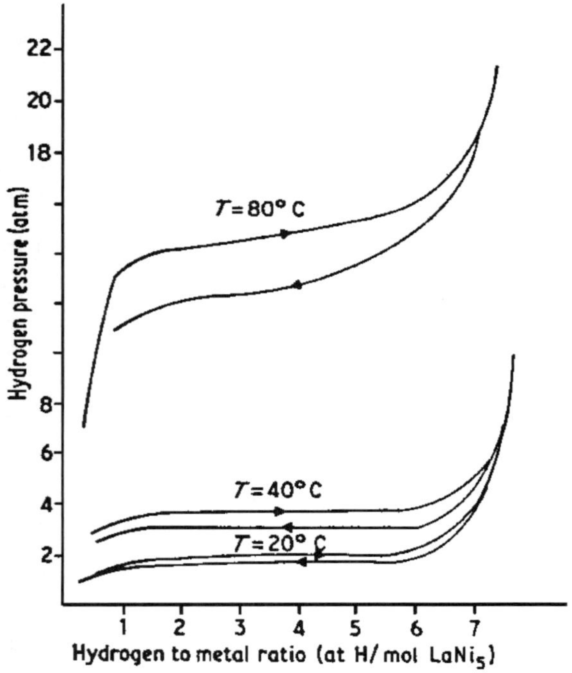

Figure 2
P–C isotherms of $LaNi_5$.

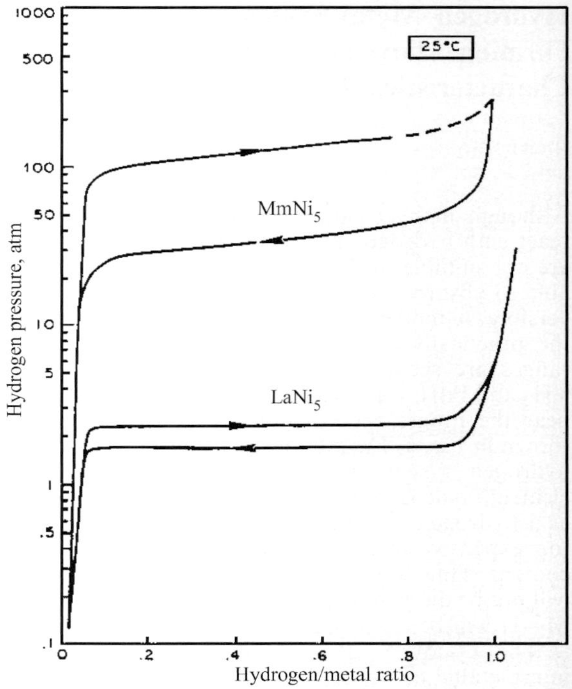

Figure 3
Hydrogen absorption–desorption loops at 25°C for $LaNi_5$ and $MmNi_5$.

can be readily absorbed and desorbed by changing the temperature or hydrogen pressure. The enthalpy change of the reaction is $-30\,kJ\,mol^{-1}\,H_2$. As can be seen in Fig. 2, the plateau of $LaNi_5$ is rather flat and well defined with a hysteresis factor of 0.11.

Other rare-earth metals also form AB_5-type alloys with nickel. These RNi_5 alloys show variable hydrogen storage capacities and plateau pressures. Among them, $CeNi_5$ shows a large hysteresis effect between absorption and desorption.

(*b*) *$MmNi_5$*. Considering that pure rare-earth metals are very expensive, some pioneer researchers used an unrefined rare-earth mixture (mischmetal, Mm) to replace lanthanum. The composition of Mm depends on mineral sources. Usually Mm represents a cerium-rich mischmetal with a typical composition (wt.%) of 48–50% cerium, 32–34% lanthanum, 13–14% neodymium, 4–5% praseodymium, and 1.5% other rare-earth elements. Lanthanum-rich mischmetal is usually denoted as Lm or Ml. A typical composition of lanthanum-rich mischmetal is 48% lanthanum, 25% cerium, 6% praseodymium, 21% neodymium and 0.3% others.

As shown in Fig. 3, the substitution of Mm for lanthanum does not change the hydrogen storage capacity, but results in a considerable increase of the plateau pressure and hysteresis, resembling the effect of cerium substitution for lanthanum. Fortunately, the undesirable effects can be eliminated by the substitution of B in AB_5, which is described below.

(*c*) *$CaNi_5$*. $CaNi_5$ not only has the same crystal structure as $LaNi_5$, but also has similar lattice parameters. However, $CaNi_5$ forms three hydride phases with increasing hydrogen content. Thus its *P–C* isotherms show three plateaus. A hydrogen capacity of 1.9 wt.% can be achieved if the hydrogen pressure is higher than the third plateau pressure. $CaNi_5$ is also an attractive hydrogen storage alloy from an economic point of view because calcium is cheaper than rare-earth metals.

(*d*) *Substituted AB_5 alloys*. Both A and B of AB_5 can be partially or fully replaced by other metals to form multicomponent alloys without changing the $CaCu_5$-type structure. The excellent compatibility of the AB_5 structure makes it possible to tailor alloy properties by alloy substitution for versatile applications.

For A, the rare-earth metals can substitute for each other. The hydriding properties of the rare-earth mixture alloys depend on the portion of one specific element and the properties of the parent RNi_5 alloys. For example, the substitution of cerium, praseodymium, and neodymium for lanthanum will result in an increase of the plateau pressure because $CeNi_5$, $PrNi_5$, and $NdNi_5$ have higher plateau pressures than $LaNi_5$.

Also, rare-earth metals and calcium can substitute for each other. Figure 4 shows the desorption isotherms of $Mm_{1-x}Ca_xNi_5$. The replacement of Mm by calcium lowers the plateau pressure and decreases the hysteresis, but increases the slope of the plateau.

It has also been reported that Mm can be partially replaced by other metals such as titanium, zirconium, boron, copper, etc.

The nickel atoms in ANi_5 can also be partially replaced by other metals such as cobalt, manganese, aluminum, chromium, iron, copper, tin, silicon, boron, etc. Figure 5 shows the effects of substitution of cobalt for nickel. In cobalt-substituted alloys, a second plateau appears. Also, the pressure of the first plateau and the hydrogen capacity are decreased. Cobalt substitution is unnecessary in gas storage applications but is essential in electrochemical applications because it greatly improves the cyclic stability of alloy electrodes by reducing the volume change during hydrogenation (Willems 1984).

Manganese and aluminum are also frequently used as substitution metals. Manganese substitution for nickel lowers the high plateau pressure and the hysteresis of $MmNi_5$ without decreasing the capacity. As shown in Fig. 6, aluminum substitution leads to a

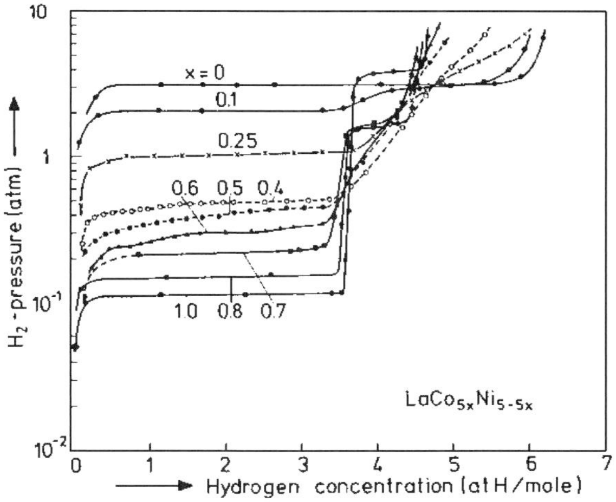

Figure 5
P–C isotherms for various $LaCo_{5x}Ni_{5-5x}$–H systems obtained by desorption at 40°C.

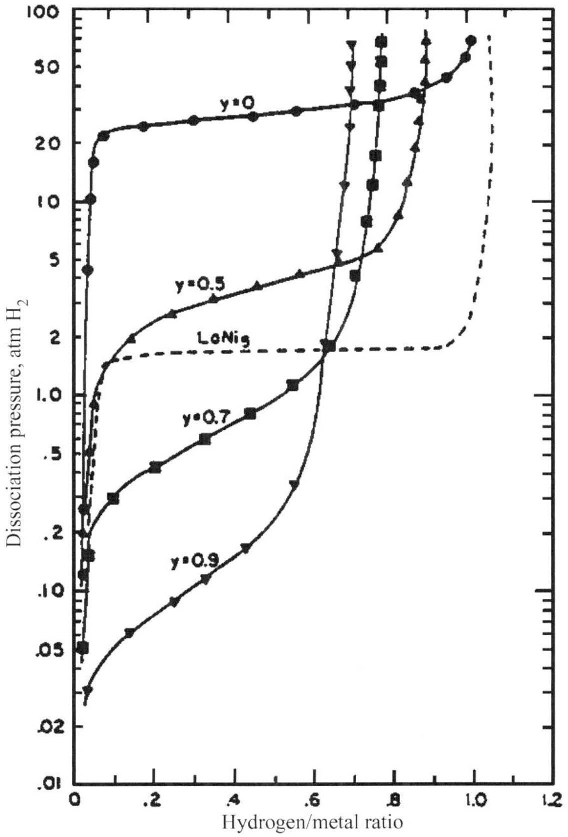

Figure 6
$MmNi_{5-y}Al_y$–H isotherms at 25°C.

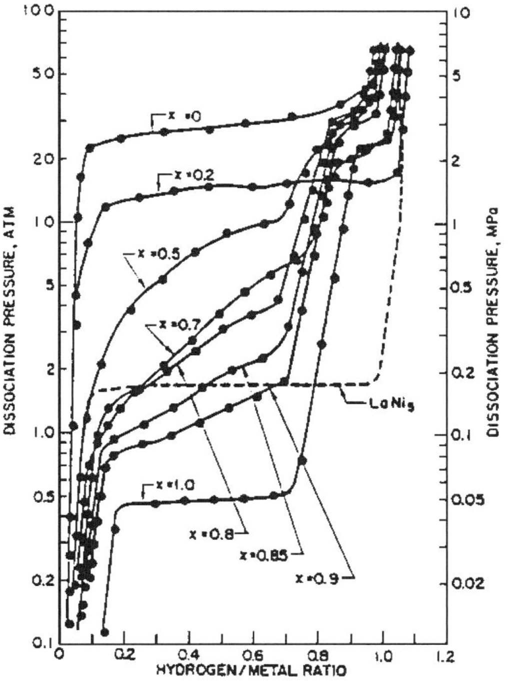

Figure 4
Dissociation isotherms of $Mm_{1-x}Ca_xNi_5$–H at 25°C.

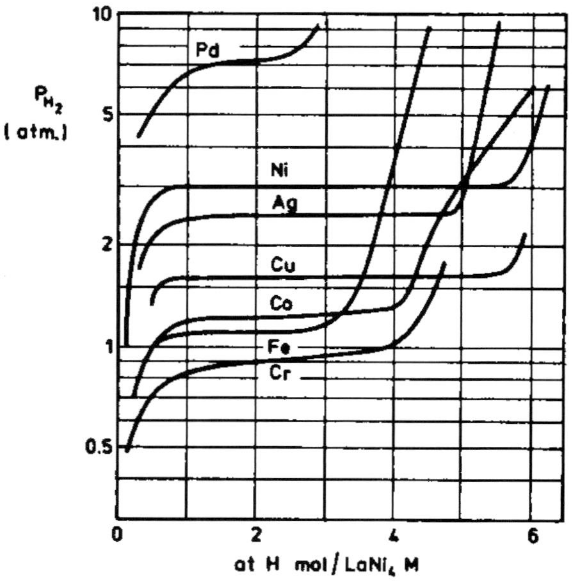

Figure 7
Desorption isotherms at 40°C for various compounds of $LaNi_4M$.

lower plateau pressure and hysteresis with a smaller capacity. The effects of substitution of some other metals are shown in Fig. 7. In most cases, the substituted alloys show lower plateau pressures and smaller capacities.

However, it has been reported that $CaNi_5$ is much less tolerant to substitution of B than is $LaNi_5$ (Jensen and Bjerrum 1999).

By alloy substitution, the hydriding properties of AB_5 alloys can be easily adjusted for various applications, making them a very versatile alloy system.

1.2 Activation Properties

On the whole, AB_5 alloys are easily activated compared with other hydrogen storage intermetallic compounds. Air-exposed alloys can be easily activated by evacuating the reactor and then introducing H_2 at a pressure higher than the absorption pressure at room temperature. The incubation time varies from seconds to hours depending on the degree of oxidation of the surfaces. The incubation time can be shortened or eliminated by heating the alloy in a vacuum or H_2 atmosphere or by surface pretreatments such as F-treatments (see *Hydrogen–Metal Systems: Fluorinated Metal Hydrides*). Several hydriding–dehydriding cycles are necessary before a reproducible *P–C–T* relationship and stable kinetics are obtained.

The easy activation of AB_5 alloys is mainly attributed to the surface segregation that usually takes place in an O_2-containing atmosphere. In the case of $LaNi_5$, the surface segregates to La_2O_3 (or $La(OH)_3$) and metallic nickel clusters. La_2O_3 or $La(OH)_3$ is somewhat transparent to H_2 and the metallic nickel clusters are catalytically active to the chemical dissociation of H_2, resulting in a highly reactive surface.

During the activation process, alloy particles crack and automatically disintegrate into fine powder with a final size of several micrometers owing to a large volume expansion on hydrogenation. The alloy pulverization results in a great increase of fresh new surfaces, which substantially enhance hydriding and dehydriding kinetics. Highly dense defects are also introduced into the crystal structure during the initial hydriding process. The defects not only release the lattice strain during hydrogenation, but also facilitate hydrogen diffusion.

1.3 Intrinsic Kinetics

Although there have been numerous studies of the hydriding and dehydriding kinetics of AB_5 alloys, they offer different experimental results. This originates from the fact that the reaction kinetics is sensitive to some extrinsic factors, especially the heat transfer of the reaction bed. Because of a large thermal effect accompanying the hydriding and dehydriding reaction along with the poor thermal conductivity of metal powder, the isothermal condition is usually difficult to maintain especially during the initial part of the reaction. Thus many kinetics studies do not necessarily reflect the intrinsic kinetics because they fail to consider the heat transfer effect.

Despite the discrepancies of experimental data, it is recognized that the intrinsic hydriding and dehydriding kinetics of AB_5 alloys is good enough and practical reaction rates are mostly limited by the heat transfer.

1.4 Gas Impurity Effects and Cyclic Stability

The cyclic performance of hydriding alloys depends on two factors: extrinsic impurity effects and intrinsic alloy disproportionation.

Gaseous impurities in hydrogen affect the hydriding and dehydriding properties of alloys in four different ways: poisoning, retardation, reaction, and innocuous.

As illustrated in Fig. 8, poisoning gas impurities such as H_2S and CO result in a rapid decrease in hydrogen capacity with cycling. In general, a monolayer coverage of such impurity molecules is enough to deactivate an alloy particle. The poisoned sample can be reactivated by cycling and/or heating in pure H_2 with little permanent loss in capacity because the damage is only localized to the surface. A retardation effect such as the effect of NH_3 is due to the milder and more homogeneous impurity–alloy interaction

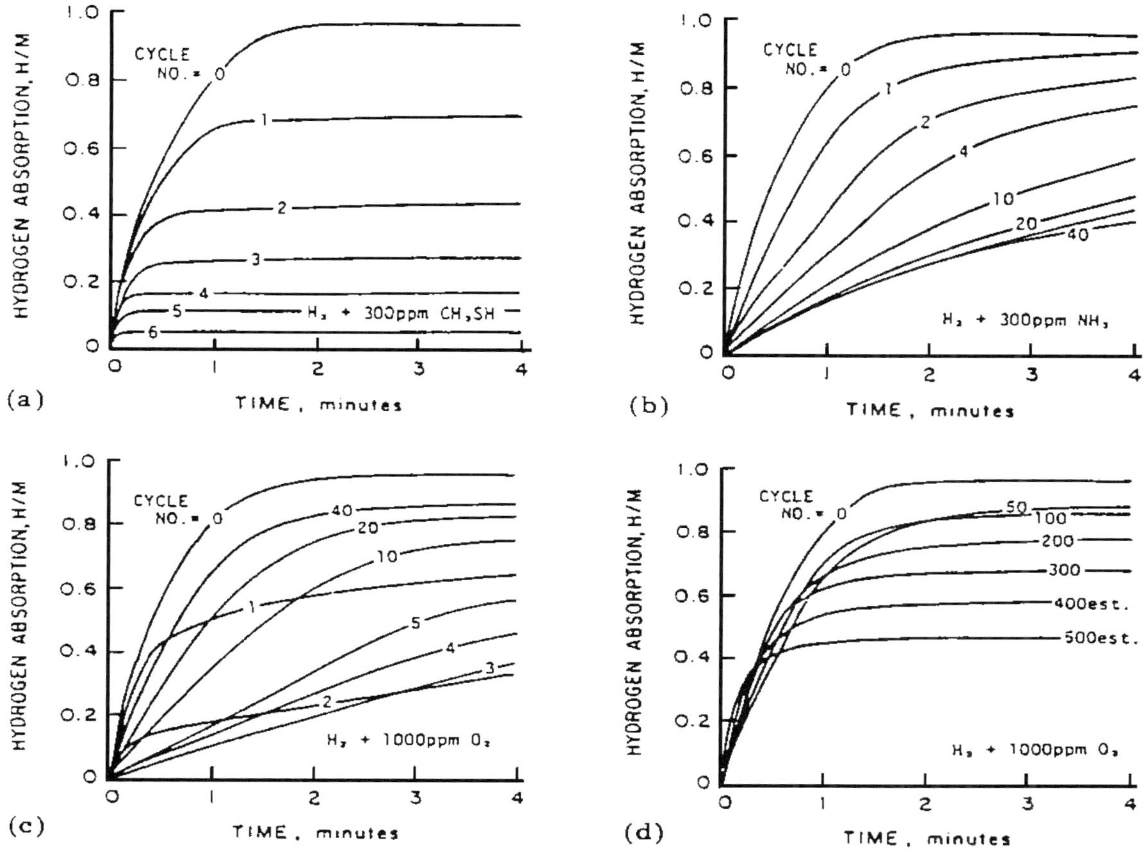

Figure 8
Different types of $LaNi_5$–impurity interactions: (a) poisoning; (b) retardation; (c) retardation–recovery; (d) reaction (25°C and 3.4 atm H_2 pressure).

characterized by a reduction in absorption kinetics without significant loss in capacity. A short-term $LaNi_5$–O_2 interaction is classified as a retardation–recovery effect because the kinetics and capacity substantially recover after further cycling, despite the fact that they show considerable decrease in the first few cycles. A long-term $LaNi_5$–O_2 interaction is a reaction process where the alloy sample is gradually consumed by a bulk corrosion reaction that transforms lanthanum into La_2O_3 or $La(OH)_3$, resulting in an irreversible capacity loss. Some gases such as N_2 and CH_4 are regarded as innocuous, as they do not induce capacity or kinetics loss.

Compared with other families of intermetallic compounds, rare earth-based AB_5 alloys demonstrate high resistance to gaseous impurities, especially to H_2O and O_2. They have been successfully used in some gas separation or purification applications.

Long-term performances of hydride-forming alloys are not only influenced by extrinsic gas impurity effects, but also depend on their inherent stability during extended cycling. Alloys may be subject to a disproportionation reaction because of the metastable nature of their hydrides. For example, the following disproportionation reaction of $LaNi_5$ is more thermodynamically favored than the familiar hydrogenation reaction:

$$LaNi_5 + 3H_2 \rightarrow LaNi_5H_6 \text{ (hydrogenation reaction)}$$

$$LaNi_5 + H_2 \rightarrow LaH_3 + 5Ni \text{(disproportionation reaction)}$$

Although the disproportionation reaction does not proceed easily at room temperature, it does show effects after extended cycling or at high temperatures. On cycling $LaNi_5$ from 30°C to 300°C, it is found that half of the reversible storage capacity is lost after about 750 cycles. Decreasing the cycling temperature to 150°C increases cycle life to an estimated 7500 cycles. Cycling of $LaNi_5$ not only induces capacity loss, but also a splitting of the reaction plateau.

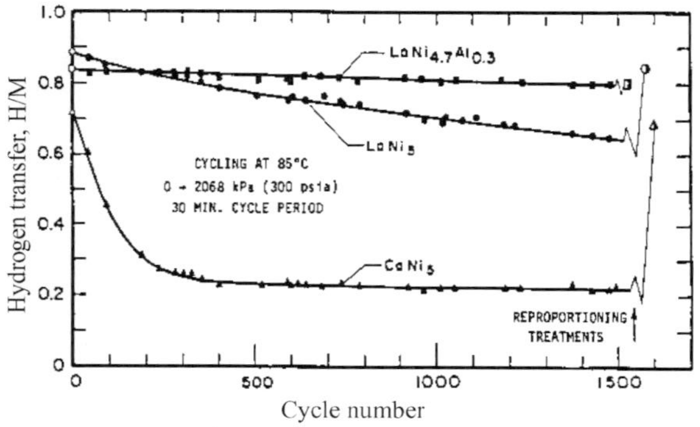

Figure 9
Effect of 85°C pressure cycling on the reversible hydrogen capacity of three AB_5 alloys.

Cyclic stability of $LaNi_5$ can be improved by alloy substitution. $LaNi_{4.7}Al_{0.3}$ exhibits higher stability than $LaNi_5$ when cycled at 85°C for 1500 cycles, as shown in Fig. 9. It has also been reported that $LaNi_{4.8}Sn_{0.2}$ demonstrates higher resistance to disproportionation than $LaNi_5$ by a factor of 20. However, these alloys still show considerable capacity degradation during long-term thermal cycling (Friedlmeier *et al.* 1995). For example, $LaNi_{4.7}Al_{0.3}$ shows a degradation rate of 1.2% 10^{-4} H/M per cycle when cycled between 100°C and 205°C, while $LaNi_{4.85}Sn_{0.15}$ has a degradation rate of 3.8% 10^{-5} H/M per cycle when cycled between 170°C and 45°C. The degradation mechanism of these AB_5 alloys is related to the disproportionation of the reversible hydrides into more stable hydrides and metallic nickel.

$CaNi_5$ shows a rapid degradation on cycling, as can be seen in Fig. 9. The low cyclic stability of $CaNi_5$ limits its application to a large extent although it is considered as a promising low-cost hydrogen storage material.

Fortunately, alloys degraded by disproportionation can be regenerated simply by heating them to 500°C in a vacuum.

1.5 Preparation and Metallurgy

As the hydriding properties of an intermetallic compound are directly determined by its composition and microstructure, careful control of the whole preparation process and thereafter a full chemical and structural characterization are necessary.

AB_5 alloys are usually prepared by vacuum induction melting (VIM) or arc melting of the constituent metals. Starting materials with a purity of more than 99% are highly recommended. For VIM, care should be taken in selecting a crucible in case the melt reacts with the crucible and gets contaminated. After melting, the common casting methods are usually used. However, it has been reported that alloys prepared by gas atomization, melt spinning, or other rapid solidification methods exhibit better properties than those obtained by conventional casting methods.

On a laboratory scale, AB_5 alloys are usually prepared by arc melting in a water-cooled copper hearth under an argon atmosphere. However, precautions should be taken in considering severe evaporation of some elements such as manganese.

After melting, an annealing treatment in vacuum or inert gas atmosphere is usually required to eliminate phase or compositional inhomogeneity. The annealing temperature should be carefully determined not only to ensure sufficient diffusion but also to avoid excessive oxidation of the alloy.

Other methods like the reduction–diffusion method have also been reported for the preparation of AB_5 alloys.

Because of the high vapor pressure of calcium, $CaNi_5$-based alloys need air or argon blanket melting.

2. *AB_2 Alloys*

AB_2 represents the intermetallic compounds with a Laves phase structure. There are three types of Laves phase structures: C14 ($MgZn_2$ type, hexagonal), C15 ($MgCu_2$ type, cubic), and C36 ($MgNi_2$ type, hexagonal). There are only tetrahedral interstices formed in the C14 and C15 Laves phase structures.

Although numerous AB_2 alloys are capable of forming hydrides, titanium- and zirconium-based AB_2 alloys are the most important in terms of their attractive hydriding properties.

2.1 Structural and Thermodynamic Properties

(*a*) *Binary AB_2 alloys.* Zirconium-based AB_2 Laves phase alloys include ZrM_2 (M = V, Cr, Mn, Fe, Co, Mo). The *P–C–T* relationships of $ZrCr_2$–H_2, ZrV_2–H_2, and $ZrMo_2$–H_2 systems have been studied between 0°C and 900°C at hydrogen pressures of 10^{-4}–760 torr. ZrV_2, a C15 structure alloy, can absorb a large amount of hydrogen at low temperatures. However, it forms a hydrogen solid solution rather than a separate hydride phase within the temperature and pressure range. $ZrCr_2$ has the C14 structure at low temperatures but changes to C15 around 900°C. It also forms a solid solution of hydrogen at temperatures above 0°C, but there is an increasing tendency for phase separation at lower temperatures. $ZrMo_2$ forms a solid solution with low hydrogen capacity and low thermodynamic stability. No hysteresis effect is observed for all three systems. $ZrMn_2$, a C14 Laves phase, has a wide homogeneity range from $ZrMn_{1.8}$ to $ZrMn_{3.8}$. Increase of the stoichiometry leads to a rapid rise of plateau pressure. As the hydrogen capacities of

ZrM_2 alloys change as $ZrV_2 > ZrCr_2 > ZrMn_2 > ZrMo_2 > ZrFe_2 > ZrCo_2$, it has been suggested that Laves phases with lower electron concentration may accommodate more hydrogen than those with higher electron concentration.

While hydrides of zirconium-based AB_2 alloys have relatively high stability, titanium-based AB_2 alloys demonstrate low hydride stability. $TiCr_2$ has a C14 structure at high temperatures and C15 at low temperatures. At room temperature, the homogeneity range of C15 is actually between $TiCr_{1.71}$ and $TiCr_{1.92}$. At temperatures as low as $-78°C$, two plateaus are observed for $TiCr_{1.8}$ with little or no hysteresis. The plateau pressure of $TiCr_2$ is too high for room temperature storage, but may be used in low-temperature applications. Stoichiometric $TiMn_2$ (C14 structure) cannot be hydrogenated under conventional conditions, but nonstoichiometric $TiMn_{1.5}$ alloy readily absorbs and desorbs hydrogen but with a large hysteresis.

Although the hydriding properties of binary AB_2 alloys may not be as good as $LaNi_5$, they can be enhanced by alloy substitution. Substituted AB_2 alloys form a large family, like the AB_5 family.

(b) Ternary and multicomponent AB_2 alloys. Although hydrides of some zirconium-based Laves phase alloys such as ZrV_2, $ZrCr_2$, and $ZrMn_2$ have high capacities, they are too stable for hydrogen storage at conventional conditions. Partial replacement of chromium, vanadium, or manganese by other metals such as iron, cobalt, nickel, aluminum, or copper has been shown to decrease hydride stability.

By partially substituting iron for chromium, it is found that the hydride stability can be decreased to desirable levels. $Zr(Fe_xCr_{1-x})_2$ alloys have predominantly C14 structure. A distinct hydride phase with the same crystal structure as the parent compound is formed after hydrogen absorption. Among the intermetallics in this series, $Zr(Fe_{0.75}Cr_{0.25})_2$ exhibits the best overall hydrogen storage properties with a hydrogen capacity similar to that of $LaNi_5$ and even better activation and kinetic properties, except that its plateau is sloping.

$Zr(Fe_xMn_{1-x})_2$ ($x = 0-0.8$) alloys have the C14 structure when $x<0.7$ and the C15 structure when $x>0.7$. Ternary Zr–Mn–Fe alloys including those with nonstoichiometric compositions reveal interesting hydriding properties.

Zirconium-based AB_2 alloys are of interest as the negative electrode material in the Ni–MH battery. To electrochemically charge and discharge hydrogen, nickel is a necessary constituent in alloys. Thus ternary alloys such as Zr–Cr–Ni, Zr–Mn–Ni, and Zr–V–Ni have been extensively studied. Their gaseous absorption properties are sometimes mentioned along with their electrochemical properties.

Partial substitution of zirconium by titanium is another effective way to decrease the hydride stability.

The hydridng properties of Ti–Cr–Mn and Ti–Cr–V ternary alloys have been studied in much detail. For $Ti_{1+x}Cr_{2-y}Mn_y$ ($x = 0.1-0.3$, $y = 0-1.0$) alloys, an increase of titanium leads to an increase in the total capacity but a decrease in the usable capacity. However, an increase of chromium results in a decrease in hydrogen capacity as well as in hysteresis, and increase in plateau pressure. For $Ti_xCr_{2-y}V_y$ ($x = 1.1-1.3$, $y = 0.5-1.0$) alloys, the total capacity is increased by increasing the amount of vanadium, but the usable capacity is less than the total capacity because some part of the hydrogen cannot be released.

Numerous AB_2 alloys have been designed by multiple substitutions into both the A and B sites to serve for different purposes. For these multicomponent AB_2 alloys, the A component usually includes titanium, zirconium, or hafnium. The B component includes many 3d transition elements such as manganese, chromium, vanadium, nickel, cobalt, iron, and copper. These alloys may consist of only a single phase. But in many cases they consist of several phases including the Laves phases, a b.c.c. phase, and other phases.

One example of a commercialized AB_2 alloy is $Ti_{0.98}Zr_{0.02}V_{0.43}Fe_{0.09}Cr_{0.05}Mn_{1.5}$. Also, alloys such as $Ti_{0.9}Zr_{0.1}Mn_{1.4}V_{0.2}Cr_{0.4}$ have been developed. As shown in Fig. 10, these alloys have hydrogen capacities as high as 1.9–2.1 wt.% with moderate plateau pressures. They are suitable for gaseous storage of hydrogen. Numerous zirconium-based AB_2 materials have been developed for electrochemical storage of hydrogen.

2.2 *Activation Properties*

In general, there is no special difficulty in activation of AB_2 alloys. Most alloys in the $Zr(Co_xM_{1-x})_2$ and $Zr(Fe_xM_{1-x})_2$ ($M = V$, Cr, Mn; $0<x<1$) series absorb hydrogen immediately and no activation process is necessary. The$Zr(Fe_xCr_{1-x})_2$ alloys except $ZrCr_2$ and $Zr(Fe_{0.25}\ Cr_{0.75})_2$ are activated quite readily in the first absorption cycle, while $ZrCr_2$ and $Zr(Fe_{0.25}\ Cr_{0.75})_2$ need to be heated to 300–350°C for activation.

This suggests that the ease of activation depends on compositions for AB_2 alloys. Usually alloys with high manganese content can easily be activated, while alloys with high chromium content may need to be heated before activation. For example, Ti–Mn-based alloys are reported to be activated very easily at room temperature without heating.

All these results suggest that most AB_2 alloys can easily be activated, with a few exceptions that need to be heated.

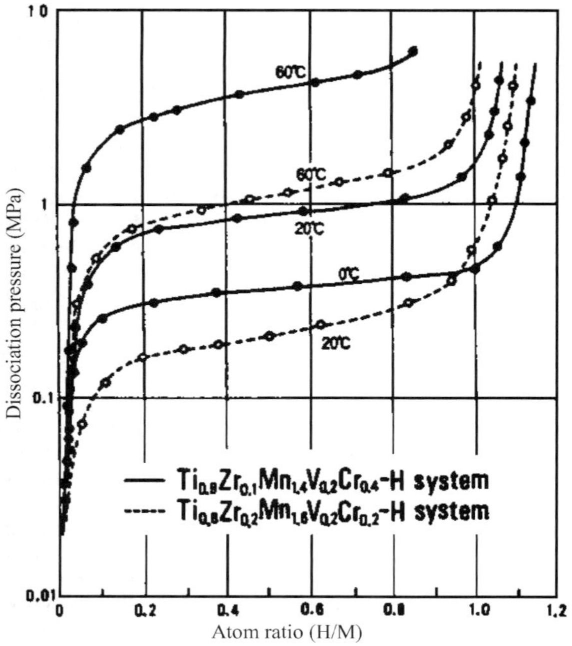

Figure 10
P–C desorption isotherms for two multicomponent AB_2 alloys.

2.3 Intrinsic Kinetics

It has been reported that AB_2 alloys have excellent gaseous absorption kinetics, sometimes even better than AB_5 alloys. The good hydriding kinetics of $ZrMn_x$ has been ascribed to its relatively thin oxide coating and manganese-enriched surface. The absorption rates of $Zr(Fe_{0.75}Cr_{0.25})_2$ and $LaNi_5$ have been compared under the same experimental conditions. The results show that $Zr(Fe_{0.75}Cr_{0.25})_2$ has a higher absorption rate than $LaNi_5$.

The quick kinetics of $Zr(Fe_{0.75}Cr_{0.25})_2$ is at least partially due to the smaller particle size after certain cycles. Experimental results demonstrate that the average particle size is stable at about 9 μm for $Zr(Fe_{0.75}Cr_{0.25})_2$, and 23 μm for $LaNi_5$ after 10 cycles. Ti–Mn-based AB_2 alloys also exhibit high kinetics. It has been reported that the intrinsic kinetics of $Ti_{0.98}Zr_{0.02}V_{0.43}Fe_{0.09}Cr_{0.05}Mn_{1.5}$ is faster than the speed of heat transfer. All these results indicate that AB_2 alloys may provide excellent hydriding and dehydriding kinetics.

2.4 Gas Impurity Effects and Cyclic Stability

Although no systematic investigations of gas impurity effects have been performed in AB_2 alloys as has been done for AB_5 alloys, some results suggest that they may be more sensitive to gaseous impurities than AB_5 alloys. Some data are available for the gas impurity effects in the commercial alloy $Ti_{0.98}Zr_{0.02}V_{0.43}Fe_{0.09}Cr_{0.05}Mn_{1.5}$. When the gas impurities CO, CO_2, and O_2 are at a level of 50 ppm, the alloy is subject to 54%, 27%, and 11% capacity losses, respectively, after 100 cycles. Fortunately, these capacity losses can be recovered by cycling in pure H_2 without heating.

Experiments on cyclic stability of $Ti_{0.98}Zr_{0.02}V_{0.43}Fe_{0.09}Cr_{0.05}Mn_{1.5}$ show that this alloy has excellent stability. A 5% capacity loss of this alloy has been reported after 2000 cycles between 20°C and 80°C. Friedlmeier *et al.* (1995) reported that this alloy does not show any measurable degradation after 4.24×10^4 thermal cycles between 45°C and 120°C. These results show that this AB_2 alloy has a strong resistance to disproportionation.

Cycle tests on $TiMn_{1.5}$ show that the capacity loss is not more than 30% after 10^4 cycles. Substituted Ti–Mn alloys such as $Ti_{0.8}Zr_{0.2}Mn_{1.8}Mo_{0.2}$ and $Ti_{0.9}Zr_{0.1}Mn_{1.4}V_{0.2}Cr_{0.4}$ also show 20–30% capacity losses after 10^4 cycles. Also, $Ti_{1.2}Cr_{1.9}Mn_{0.1}$ is reported to have a capacity reduction of 25% after 2×10^4 cycles between 20°C and 60°C. Few cycling data are available for zirconium-based alloys.

2.5 Preparation and Metallurgy

AB_2 Laves phase alloys in small amounts (less than 100 g) can be easily prepared by arc melting, induction melting, or levitation melting of pure constituent metals under an inert atmosphere. Special care should be taken when preparing alloys with high amounts of manganese because manganese has a high vapor pressure. An annealing treatment is usually essential to ensure a homogeneous structure and to eliminate undesired second phases. During the annealing treatment, the alloy samples can be wrapped in an envelope of a getter material like zirconium to avoid excessive oxidation.

Fabrication of AB_2 alloys on an industrial scale is more difficult because of the high melting points and high reactivity of most constituent metals in AB_2 alloys. As low oxygen levels are difficult to achieve in a one-stage VIM process, a two-stage process has been developed for the production of $TiMn_2$-based alloys like $Ti_{0.98}Zr_{0.02}V_{0.43}Fe_{0.09}Cr_{0.05}Mn_{1.5}$. A fraction of the manganese is first melted with ferrovanadium to make an oxygen-poor MnVFe alloy. The ingot is crushed and then remelted together with the rest of the manganese and with titanium and zirconium. Small amounts of mischmetal are also added to deoxidize the melt (Percheron-Guegan and Welter 1988).

3. *AB Alloys*

Many AB-type intermetallic compounds with CsCl-type crystal structures are able to form hydrides.

Among them, TiFe(FeTi) and related alloys have received most attention because of the low cost and high abundance of the raw materials. They are considered as prime candidates for safe and economical storage of hydrogen.

3.1 Structural and Thermodynamic Properties

(*a*) *Binary AB alloys.* The hydriding and dehydriding properties of TiFe have been studied in detail. There are two intermetallic compounds in the Ti–Fe phase diagram: TiFe and $TiFe_2$. $TiFe_2$ does not react with hydrogen under normal temperature and pressure conditions. TiFe reacts directly and reversibly with hydrogen to form, stepwise, two hydrides: $TiFeH_{1.04}$ (β phase) and $TiFeH_{1.95}$ (γ phase). The reactions are as follows:

$$2.13TiFeH_{0.1} + H_2 = 2.13TiFeH_{1.04}$$

$$2.13TiFeH_{1.04} + H_2 = 2.20TiFeH_{1.95}$$

The *P–C–T* characteristics of TiFe are shown in Fig. 11, which clearly shows two plateaus and a large hysteresis.

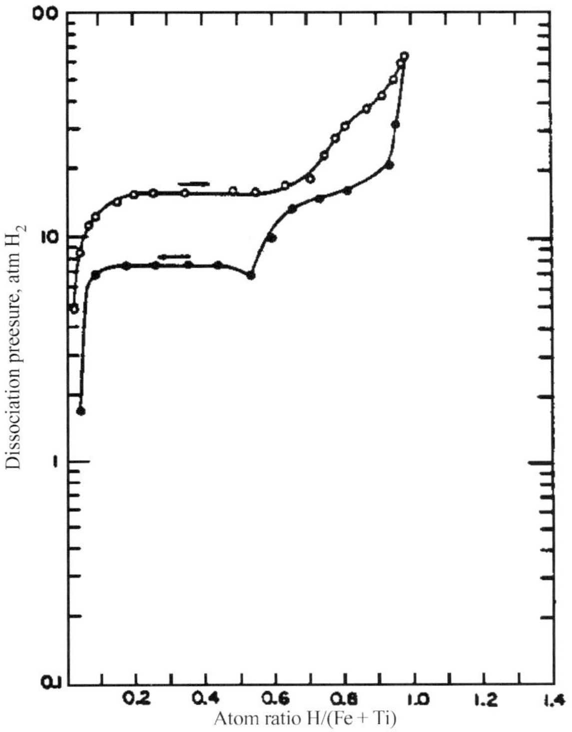

Figure 11
Hydrogen absorption–desorption loop at 40°C for TiFe–H.

The hydridng property of TiFe is highly dependent on the exact ratio of titanium to iron. This property is also influenced by contaminants such as oxygen introduced during the metallurgy process.

Several other AB alloys such as TiCo and ZrNi also react with hydrogen to form hydrides. TiCo forms $TiCoH_{1.4}$ that has a higher stability than the hydride of TiFe. Its *P–C* isotherms also show two plateaus.

ZrNi reacts with hydrogen easily and reversibly to form intermetallic hydrides. The hydride $ZrNiH_3$ has an intermediate stability between that of ZrH_2 and NiH. Its *P–C* isotherms show a flat plateau but with a large hysteresis. The stability of $ZrNiH_3$ is too high to be of practical interest.

(*b*) *Substituted AB alloys.* To modify the activation or other properties, TiFe is partially substituted by other transition elements such as manganese, chromium, vanadium, cobalt, nickel, molybdenum, and copper. As shown in Fig. 12, all the substitutions lead to a decrease in plateau pressure but to an increase in plateau slope. The high stability of the hydrides results from the increases in lattice parameters of the corresponding ternary alloys.

As an example, $TiFe_{1-x}Mn_x$ ($x = 0.1 - 0.3$) can be easily activated at room temperature. The hydrogen capacity is not reduced and the hysteresis effect is improved after the manganese substitution. The plateau can also be flattened by an annealing treatment.

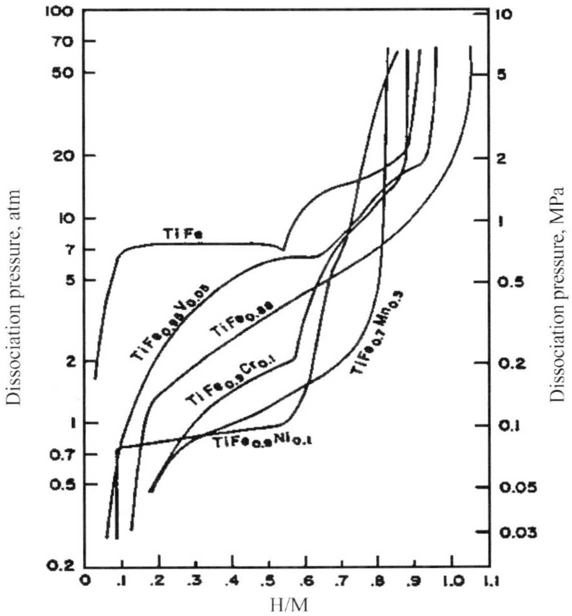

Figure 12
Desorption isotherms (40°C) for partially substituted AB compounds.

Some TiFe-based quaternary alloys have been prepared and their hydriding properties examined. $TiFe_{0.8}Mn_yA_z$ (A = Zr, Al) can easily be activated. Experiments on $TiFe_{1-x}Ni_yA_z$ (A = Al, Co, Cr, La, Mn, Mo, Nb, V, Zr) alloys show that niobium, vanadium, and zirconium substitutions result in increases in total storage capacity but the usable capacity is not increased because some absorbed hydrogen cannot be released.

The hydriding properties of partially substituted TiCo alloy have also been investigated.

3.2 Activation Properties

Unlike most AB_5 or AB_2 alloys, TiFe cannot be activated by only exposure to H_2 at room temperature. It needs to be heated to more than 300°C before the introduction of high-pressure H_2. The reason is that the oxide layer present on the TiFe surface prevents hydrogen from dissociating and penetrating through it. The surface of the activated TiFe contains segregated titanium oxide and metallic iron clusters. It is suggested that the iron clusters serve as the catalyst for chemical dissociation of H_2.

It has also been found that the activation property is strongly influenced by second phases. $Fe_7Ti_{10}O_3$ is the main second phase in the oxygen-contaminated TiFe alloy. It is found that the alloy of high oxygen content is more easily activated than that of low oxygen content. Also, it is found that second phases like $TiFe_2$ or an excess titanium solid solution dramatically increase the activation rate. Even insoluble mischmetal addition significantly aids the activation. However, in most cases second phases decrease the storage capacity of the TiFe alloy.

The activation property can also be improved by alloy substitutions such as manganese, nickel, or aluminum. Most ternary alloys like $TiFe_{1-x}Mn_x$ can be activated at room temperature.

TiCo and its related alloys are easier to activate than TiFe.

3.3 Intrinsic Kinetics

Once activated, TiFe or TiFe-based alloys also have fast hydriding and dehydriding kinetics, although the intrinsic kinetics of TiFe may not be as fast as those of AB_5 or AB_2 alloys.

After activation, the surface of TiFe is composed of titanium oxide and metallic iron clusters. The metallic iron clusters act as a catalyst for the chemical dissociation of H_2 and facilitate the kinetics. Although the diffusion coefficient of hydrogen in bulk TiFe alloy is low, the phase boundaries between hydride phases seem to be extremely mobile, thus the diffusion of hydrogen is not the rate-controlling step. The intrinsic kinetics of TiFe is still rapid enough compared to practical heat transfer rates.

The pulverization behavior of TiFe is slightly different from that of AB_5 or AB_2 alloys. It is reported that the TiFe alloy is not pulverized by hydriding. In fact, the activated alloy particles are internally highly cracked with a specific surface area of $0.5\,m^2\,g^{-1}$, although the fragments are somewhat resistant to detachment. This highly cracked structure ensures rapid reaction kinetics of TiFe.

3.4 Gas Impurity Effects and Cyclic Stability

TiFe and related alloys are found to be more sensitive to gas impurities than AB_5 alloys. Systematic studies of gas impurity effects on TiFe and $TiFe_{0.85}Mn_{0.15}$ show that both alloys demonstrate a steady decrease in hydrogen capacity when they are cycled in O_2- or H_2O-containing H_2, as demonstrated in Fig. 13. Thus the interaction of TiFe-related alloys with O_2 or H_2O is described as being of a retardation type.

It has been reported that TiFe and related alloys are very resistant to disproportionation. There is no deterioration observed after TiFe is subject to about 1.3×10^4 hydriding–dehydriding cycles between −7°C and 110°C. Also, there is no evidence of the presence of any TiH_2 and iron in the cycled alloy. The thermal cycling test on $TiFe_{0.85}Mn_{0.15}$ also shows that there is no capacity loss after 3×10^4 cycles. However, experiments show that the upper plateau pressure increases with cycling although the lower plateau remains unchanged.

3.5 Preparation and Metallurgy

A detailed study has been made of the metallurgy and preparation of TiFe and TiFe-related alloys. A variety of melting techniques have been used to prepare the alloy, including consumable and nonconsumable electrode inert gas arc melting, VIM, and air induction melting. The effects of contaminants such as oxygen, nitrogen, carbon, and silicon on the hydriding and dehydridng properties of the alloys have been studied in detail.

From the Ti–Fe binary phase diagram, the TiFe phase exists over a narrow composition region. To avoid the appearance of the secondary phases that will decrease the overall capacity, the composition should be controlled within the range 49.5–52 at.% titanium.

Much care should be taken to minimize the oxygen contamination during melting because it will significantly decrease the hydrogen storage capacity. Oxygen comes not only from raw materials and the melting atmosphere, but may also come from the crucible. Thus, it is necessary to start with high-purity raw materials and a suitable crucible. For VIM, a graphite crucible gives better results than the commonly used oxide crucibles because of the high reactivity of liquid FeTi with the oxide crucibles, which results in not only a high oxygen contamination of the melt but also in poor life of the crucibles.

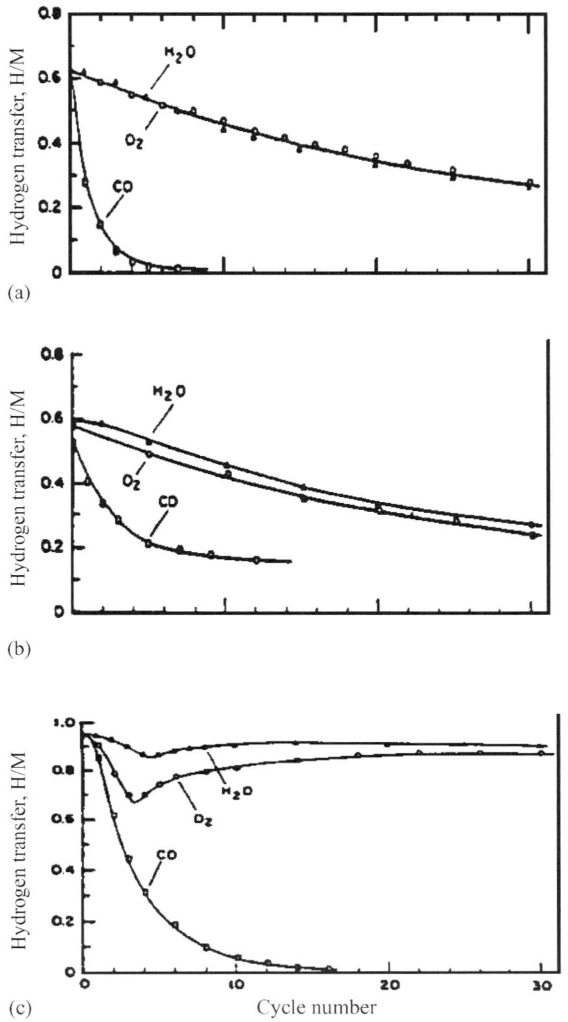

Figure 13
Effect of 300 ppm O_2, H_2O, and CO on the 25°C reversible hydrogen capacity of (a) TiFe, (b) $TiFe_{0.85}Mn_{0.15}$, and (c) $LaNi_5$.

Technologies for commercial production of TiFe and related alloys have been developed successfully. Not only vacuum arc melting by a consumable electrode, but also VIM and even air induction melting can give satisfactory results. TiFe has been successfully prepared in an air induction furnace by using mischmetal to deoxidize the melt.

4. Magnesium-based Alloys

Magnesium has long been considered as a potential hydrogen storage medium because it reacts with hydrogen to form MgH_2, a hydride containing

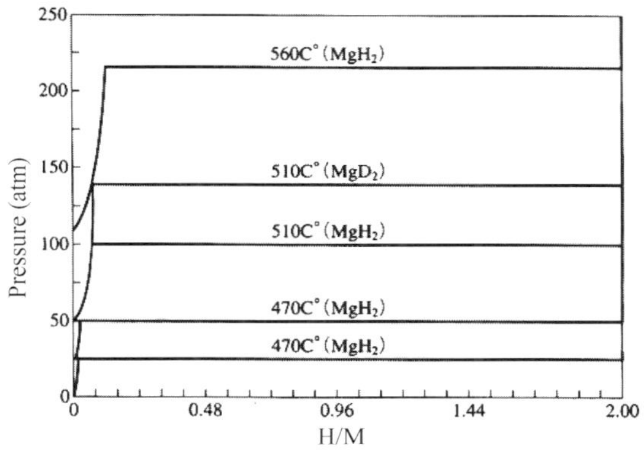

Figure 14
Decomposition isotherms of MgH_2 and MgD_2.

7.6 wt.% hydrogen. However, magnesium poses some problems when it is used for hydrogen storage. First, its hydride is thermodynamically too stable to desorb hydrogen at moderate temperatures. According to the *P–C–T* relationship shown in Fig. 14, the dissociation plateau pressure can reach 1 atm at 285°C. Second, its hydriding and dehydriding kinetics is poor because its surface acts as a barrier for hydrogen dissociation and hydrogen transfer and the MgH_2 formed prevents hydrogen from diffusing through. As a result, magnesium can absorb and desorb hydrogen only at temperatures well above 300°C, making it impractical for use as a hydrogen storage material.

Some magnesium-based intermetallic compounds like Mg_2Ni, Mg_2Cu, and La_2Mg_{17} offer better hydriding and dehydriding properties.

4.1 Structural and Thermodynamic Properties

(*a*) *Mg–Ni*. Magnesium and nickel form two intermetallic compounds Mg_2Ni and $MgNi_2$. Mg_2Ni has a hexagonal structure. It reacts with hydrogen to form Mg_2NiH_4, a stoichiometric hydride containing 3.6 wt.% hydrogen. The *P–C* isotherms are shown in Fig. 15. The dissociation plateau pressure reaches 1 atm at 252°C, indicating that Mg_2NiH_4 is only slightly unstable compared to MgH_2. However, the hydriding and dehydriding kinetics is significantly improved.

Mg_2NiH_4 has some special aspects to its structure. It shows a structure change around 245°C. The high-temperature phase has a CaF_2 cubic structure in which one nickel atom and four hydrogen atoms form a $[NiH_4]^{4-}$ complex.

Mg–Ni alloys containing both magnesium and Mg_2Ni may be more attractive as hydrogen storage materials because they combine the merits of magnesium and Mg_2Ni in having high hydrogen storage capacity and good kinetics.

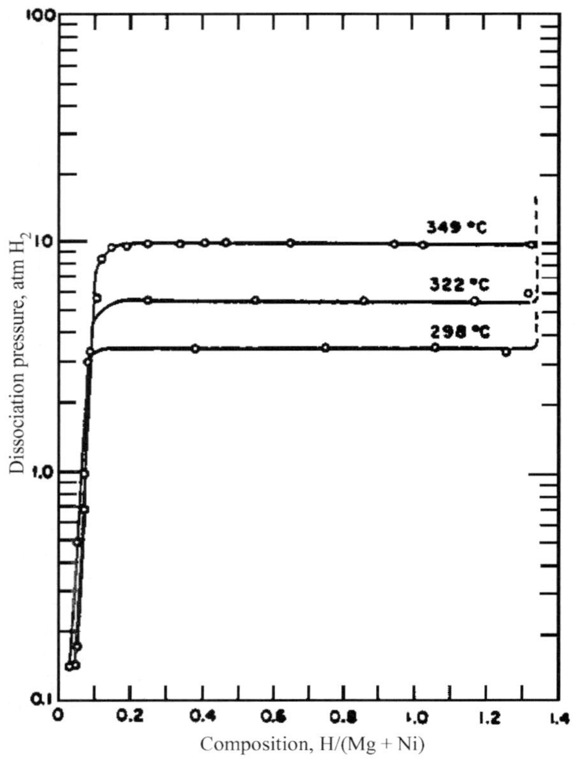

Figure 15
P–C isotherms for the Mg_2Ni–H system.

(*b*) *Mg–Cu.* Mg_2Cu reacts with hydrogen as follows:

$$2Mg_2Cu + 3H_2 = 3MgH_2 + MgCu_2$$

In this reaction, not a ternary hydride but two disproportionation products are formed. Although this system is a little more unstable, the hydrogen storage capacity is largely decreased to only 2.6 wt.%. Hence it is of less practical interest. But alloys with magnesium and Mg_2Cu two-phase structures have higher capacity and are more suitable for hydrogen storage purposes in which Mg_2Cu can act as a catalyst for the hydrogenation of magnesium.

(*c*) *Magnesium–rare earth alloys.* Some intermetallic compounds of magnesium and rare-earth metals such as $LnMg_{12}$ (Ln = La, Ce, Mm), Ln_2Mg_{17} (Ln = La, Ce), and Ln_5Mg_{41} (Ln = Ce) have good hydrogen storage properties. These alloys have specific characteristics in their way of reacting with hydrogen. The first hydrogenation of these intermetallics results in an irreversible decomposition of the alloys:

$$Ln_xMg_y + \tfrac{1}{2}(3x + 2y)H_2 \rightarrow xLnH_3 + yMgH_2$$

The hydriding and dehydriding cycles proceed as follows:

$$XLnH_3 + yMgH_2 \rightarrow XLnH_z(z < 3) + yMg + \tfrac{1}{2}(3x + 2y -$$

The thermodynamic stability of the above reaction is close to that of the $Mg–H_2$ system. However, the hydriding and dehydriding kinetics of these intermetallics are strongly enhanced by the presence of the rare-earth hydrides. After activation, they have a significant absorption rate even at room temperature. Because these alloys can absorb a large amount of hydrogen, they are promising hydrogen storage materials.

(*d*) *Other binary magnesium alloys.* Numerous magnesium alloys have been screened for possible hydrogen absorption, including $Mg_{17}Ba$, Mg_3Cd, Mg_3Sb_2, MgSn, MgZn, Mg_2Pb, Mg_2Ca, Mg_2Sn, Mg_2Si, Mg–Li, and Mg–Al alloys. Studies of Mg–Al alloys show that the hydrogenation leads to a decomposition into MgH_2 and aluminum. The low reaction rates exclude them from practical applications.

A series of magnesium alloys can be prepared by small additions of metals such as silver, aluminum, cadmium, indium, lead, zinc, yttrium, tin, calcium, manganese, bismuth, cobalt, silicon, and antimony. Among them, the Mg–Y alloys have impressive properties.

Some ternary hydrides such as Mg_2CoH_5 and Mg_2FeH_6 can be synthesized, although there exist no such alloys as Mg_2Co and Mg_2Fe.

(*e*) *Substituted magnesium alloys.* Magnesium-based intermetallic compounds such as Mg_2Ni and Mg_2Cu are not very tolerant to substitution. It is reported that nickel in Mg_2Ni cannot be successfully replaced with iron and cobalt. However, Mg_2Ni can be partially substituted by vanadium, chromium, iron, cobalt, copper, and zinc. The substituted alloys $Mg_2Ni_{0.75}M_{0.25}$ (M = V, Cr, Fe, Co, Zn) react with hydrogen to form quaternary hydrides while the copper-substituted alloy $Mg_2Ni_{0.75}M_{0.25}$ decomposes on hydrogenation like Mg_2Cu. The copper substitution leads to a small decrease in hydride stability. Unfortunately, all these substitutions result in losses of hydrogen storage capacity.

Attempts have been made to partially replace magnesium in $CeMg_{12}$ by other 3*d* transition metals like vanadium, chromium, manganese, iron, cobalt, nickel, copper, and zinc to form $CeMg_{11}M$. It is found that no pseudobinary alloys are formed except for the zinc-containing alloy. In most cases, two- or three-phase structures are obtained. La_2Mg_{17} is also reported to be partially substituted by nickel (for magnesium) and calcium (for lanthanum). The substituted alloys show enhanced hydriding and dehydriding kinetics.

Much effort has been spent in developing magnesium-based alloys for hydrogen storage in vehicular applications. The alloys based on Mg–Li and Mg–Al systems do not have impressive hydriding properties. It is found that a Mg–Ni–Cu–Y quaternary alloy has the best performance.

Amorphous and nanocrystalline magnesium-based alloys prepared by mechanical alloying are of considerable interest (see Sect. 6).

4.2 *Activation Properties*

Air-exposed magnesium and magnesium alloys can only be activated at elevated temperatures. Pure magnesium has to be heated to temperatures higher than 350°C to initiate hydrogen absorption. This is because the oxide layer existing on the surface inhibits the chemical dissociation of hydrogen molecules and/or blocks the transfer of hydrogen into the bulk. At high temperatures, when the oxide film on magnesium cracks and hydrogen can directly make contact with metal atoms, a hydride phase nucleates. At subsequent cycles, the hydrogen absorption rate will increase but the capacity will decrease because an outer hydride layer is formed and prevents hydrogen diffusion.

The activation of magnesium alloys such as Mg_2Ni is easier than that of pure magnesium because of the surface segregation in which magnesium is preferentially oxidized and leaves the other element, e.g., nickel, in the metallic state. However, activation of Mg_2Ni and hydrogen absorption are still difficult at room temperature.

4.3 *Intrinsic Kinetics*

The hydriding and dehydriding kinetics of pure magnesium strongly depends on the reaction temperature. To obtain appreciable absorption and desorption rates, it needs to be heated to temperatures higher than 300°C even after the activation. Although there is debate as to the kinetics of the Mg–H_2 reaction, it is usually regarded as a nucleation and growth process. Initially, the hydriding rate is controlled by the chemical dissociation or nucleation. After a hydride layer is formed, the hydriding rate is controlled by the hydrogen diffusion through the hydride layer.

The hydriding and dehydriding kinetics are strongly improved by alloying of magnesium with other metals such as nickel, copper, or rare-earth metals. In alloys such as Mg_2Ni and Mg_2Cu, metallic nickel and copper clusters on the subsurface act as catalysts for the chemical dissociation of hydrogen, resulting in fast kinetics. Mg_2Ni or Mg_2Cu also acts as the catalyst for the hydriding of magnesium in Mg–Mg_2Ni or Mg–Mg_2Cu alloys.

The magnesium–rare earth alloys have good hydriding properties even at room temperature. This is surely related to the hydrides of the rare-earth metals resulting from the disproportionation reaction.

4.4 *Gas Impurity Effects and Cyclic Stability*

Effects of gas impurities such as CO_2, CH_4, H_2O, and N_2 on Mg–Ni alloys have been investigated. It is reported that CH_4 and N_2 have little effect on the hydriding and dehydriding reaction of Mg–Ni, while H_2O has no effect on Mg_2Ni but a small inhibition on the Mg–5 wt.% Ni alloy. CO_2 reacts with H_2 to form CH_4 and H_2O owing to the catalysis of the alloy. Thus CO_2 has an inhibition effect on the hydriding of both Mg_2Ni and the Mg–5 wt.% Ni alloy, but has no effect on the dehydriding reaction.

The hydriding and dehydriding cycling test on pure magnesium shows that the capacity decreases by 50–60% after 500 cycles. The capacity decrease is attributed to loss of magnesium through the formation of MgO and to the sintering of agglomerates.

The Mg–5 wt.% Ni alloy has an abrupt capacity decrease within 100 cycles and then a slow decrease until 470 cycles. The *P–C–T* measurements show that the alloy does not degrade at all. The capacity loss is ascribed to sintering of the hydride bed because the alloy shows only slow capacity loss when the reactor is designed to prevent sintering.

The cycling test on Mg–2 wt.% Ni by Friedlmeier *et al.* (1995) shows that the reversible capacity decreases by 36% after 4300 cycles, but the *P–C–T* measurements show that the original capacity is obtained after the cycling. Thus it is also suggested that the capacity reduction was due to the slower kinetics that may be caused by sintering of the sample.

However, Mg_2Ni shows almost no capacity decrease within 1500 cycles, suggesting that Mg_2Ni is rather stable with respect to cycling.

4.5 *Preparation and Metallurgy*

The low melting temperature and high vapor pressure of magnesium cause difficulties in the preparation of magnesium alloys. The commonly used arc melting method is not suitable for preparation of magnesium alloys. Although magnesium alloys can be prepared by induction melting, a special technique is required to control the composition and to avoid contamination.

Some special preparation methods have been developed for magnesium alloys. Among them, mechanical alloying is accepted as an effective preparation method. Magnesium alloys prepared by mechanical alloying exhibit unique structural and hydriding properties compared to those prepared by conventional melting methods.

5. Vanadium and Vanadium-based B.c.c. Alloys

Vanadium forms two hydrides: VH (β phase) and VH_2 (γ phase). Thus the *P–C* isotherms of vanadium show two plateaus. The *β* phase hydride is too stable to be of practical interest. The plateau pressure corresponding to the *β–γ* reaction is about 0.5 MPa at 313 K, as shown in Fig. 16. The reaction accompanies a hydrogen capacity change of 1.9 wt.%. The large hydrogen storage capacity and suitable plateau pressure of vanadium make it an attractive hydrogen storage medium. However, investigations have shown that the use of pure vanadium as a hydrogen storage material is not practical.

First, its hydriding properties such as hydrogen capacity and plateau pressure are very sensitive to impurities such as boron, carbon, nitrogen, oxygen, and metallic elements. An amount of 1 wt.% oxygen will result in a 1/10 hydrogen storage capacity. But pure vanadium is very expensive. Second, vanadium is difficult to activate and also sensitive to gas impurities. Third, its hydriding properties show considerable degradation with cycling even though no disproportionation or segregation is possible. It has been reported that the effective storage capacity decreases by 20% and the hysteresis factor increases from 1.8 to 2.7 after 1000 thermal cycles (Cantrell and Bowman 1999). Also, a significant change in morphology is observed. The sharp irregular particles transform to a sponge-like appearance accompanied by an increase of the specific surface area from $0.35\,m^2\,g^{-1}$ to $15.5\,m^2\,g^{-1}$. Therefore, pure vanadium is not a suitable hydrogen storage material.

The hydriding and dehydriding properties have been investigated for various binary vanadium alloys containing small amounts of boron, carbon, cobalt, chromium, iron, nickel, niobium, titanium, zirconium, tantalum, tungsten, germanium, tin, silicon, etc. One study focused on V–Ti-based alloys including V–Ti–M (M = Fe, Cr, Mn, Ge). For the $(V_{0.9}Ti_{0.1})_{1-x}Fe_x$ (x = 0–0.075), it was reported that the alloy with $x = 0.05$ had the best properties.

Alloys based on V–Ti b.c.c. solid solutions are of considerable interest. These alloys usually have multiphase structures consisting of Laves phases besides the b.c.c. phase. It is reported that some multiphase Ti–V–Mn and Ti–V–Cr alloys have high hydrogen absorption capacities and also good desorption properties. Because of the presence of Laves phases, the activation and kinetic property of these alloys are enhanced. However, their cyclic stability and sensitivity to gas impurities have not been reported.

Alloys based on V–Ti–Ni are also being developed for electrochemical storage of hydrogen.

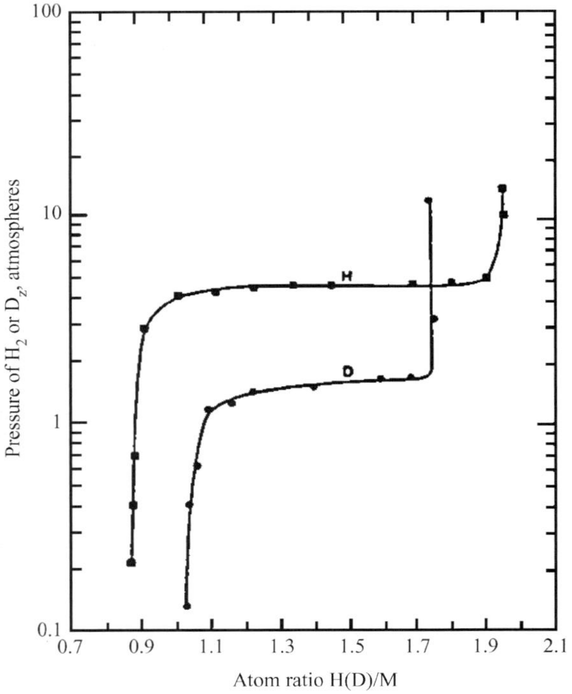

Figure 16
Desorption isotherms (40 °C) for the V–H and V–D systems.

6. Disordered Alloys

In the above, all the alloys described are prepared by conventional melting techniques and are crystalline. When disorder is introduced into alloys, their hydriding properties change.

6.1 Amorphous Alloys

Figure 17 shows the *P–C* isotherms of amorphous $Zr_{50}Ni_{50}$ and crystalline ZrNi. As can be seen, the hydriding properties of amorphous alloys show the following characteristics:

(i) There is no plateau in the *P–C* isotherms of amorphous alloys.
(ii) Amorphous alloys usually show lower hydrogen storage capacity than the corresponding crystalline alloys. But some amorphous alloys such as Ti–Cu and Pd–Zr show a larger capacity than the crystalline ones.
(iii) Compared with crystalline alloys, amorphous alloys show some resistance to pulverization.

Because of their metastable nature, amorphous alloys will crystallize at elevated temperatures. It is found that hydriding aids the crystallization by decreasing the transformation temperature.

Amorphous alloys can be prepared by rapid solidification, or by mechanical alloying or grinding. Amorphous alloys prepared by mechanical alloying

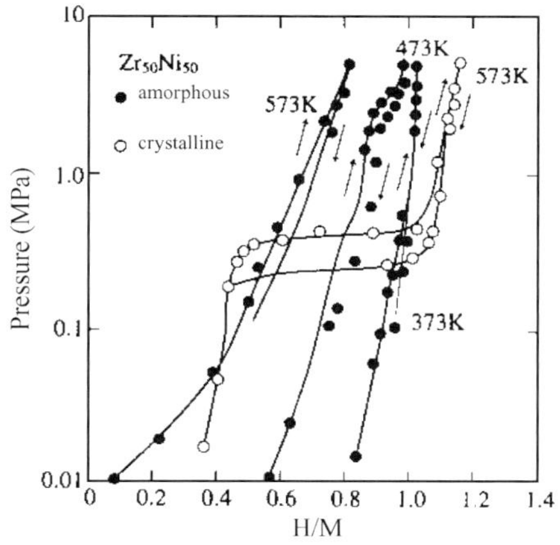

Figure 17
P–C isotherms of amorphous $Zr_{50}Ni_{50}$ and crystalline ZrNi.

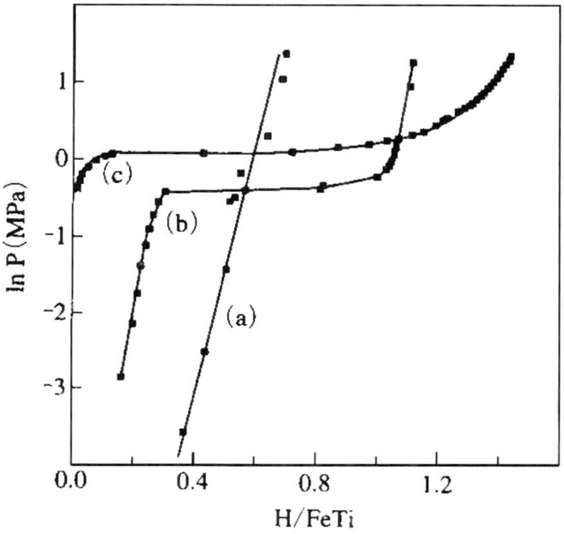

Figure 18
Room temperature hydrogen *P–C* isotherms for TiFe: (a) amorphous; (b) nanocrystalline; (c) crystalline.

are of considerable interest because they exhibit excellent kinetic properties.

In summary, amorphous alloys are not favored for gas storage applications because of the hydriding characteristics described above.

6.2 *Nanocrystalline Alloys*

Nanocrystalline alloys are of significant interest because of their unique hydrogenation properties. Nanocrystalline alloys commonly have a grain size less than 50 nm. Therefore, grain boundary regions take up a large volume fraction. Their hydriding properties are different from those of crystalline and amorphous alloys both in thermodynamic and kinetic aspects. As shown in Fig. 18 and from published results the following can be concluded:

(i) the hydrogen solubility in nanocrystalline alloys is increased while the plateau region is shortened;
(ii) the plateau pressure of nanocrystalline alloys is different from those of crystalline ones; and
(iii) the activation and kinetic properties are drastically improved for nanocrystalline alloys prepared by mechanical alloying.

Nanocrystalline magnesium-based alloys have attracted research interest not only for the improvement in mild or even room temperature hydriding kinetics, but also for the possibility of lowering the hydrogen desorption temperature. Magnesium-based alloys prepared by mechanical alloying exhibit significantly improved hydriding kinetics under moderate temperatures. Unfortunately, the thermodynamic stability of the hydrides is not significantly lowered.

From a practical point of view, a major concern for nanocrystalline alloys is their stability over prolonged hydriding–dehydriding cycles.

7. *Alloy Composites*

Composite materials for hydrogen storage are mixtures of more than two kinds of materials, at least one of which is a hydride-forming material. In these composites, different component materials serve for different purposes during the hydrogenation process. For example, in a Mg_2Ni/Pd nanostructure composite prepared by ball milling, palladium acts as catalyst for hydrogen dissociation and Mg_2Ni acts as a hydrogen absorber.

Numerous composites based on magnesium or magnesium-based alloys have been prepared in an attempt to improve the hydriding properties, including Mg/$LaNi_5$, Mg/$Ti_{0.6}Zr_{0.4}Mn_{0.8}CrCu_{0.2}$, Mg/$ZrFe_{1.4}Cr_{0.6}$, Mg/Ni, Mg/graphite, and Mg_2Ni/Pd.

Some composites are prepared to solve engineering problems such as hydride container expansion and heat transfer resulting from alloy pulverization. The microencapsulation of hydrogen storage alloys by chemically plated copper or nickel not only enhances heat transfer and avoids alloy pulverization in gas storage applications, but also improves electric conductivity and corrosion resistance in electrochemical applications.

8. *Complex Hydrides*

Complex hydrides can be classified into two groups: transition metal complex hydrides and nontransition metal complex hydrides.

A typical transition metal complex hydride is Mg_2NiH_4. Mg_2CoH_5 and Mg_2FeH_6 are also such complexes. These complex hydrides have very high hydrogen content but release of hydrogen is difficult because of very strong hydrogen–metal bonding.

Nontransition metal complex hydrides have the general formula $M(M'H_4)_n$, where M is a metal typically in groups IA and IIA (M = Li, Na, K, Mg) and M′ is a element in group IIIB (M′ = B, Al). These complex hydrides contain very high hydrogen contents.

Sodium borohydride, $NaBH_4$, is the most common alkali borohydride. It reacts with water as follows:

$$NaBH_4 + 2H_2O = NaBO_2 + 4H_2$$

In this reaction, about 10.8 wt.% of reactants change to H_2 gas. Thus it is considered as a convenient hydrogen source with high hydrogen density. However, the reaction is not reversible.

$NaAlH_4$ is representative of alkali aluminum hydrides. It contains 7.5 wt.% hydrogen. $NaAlH_4$ can be synthesized directly from sodium, aluminum, and hydrogen under high pressure and temperature:

$$Na + Al + 2H_2 = NaAlH_4$$

$NaAlH_4$ decomposes in a two-step process:

$$NaAlH_4 \rightarrow 1/3Na_3AlH_6 + 2/3H_2 \rightarrow NaH + Al + 3/2H_2$$

The above reaction is reversible under suitable conditions. Titanium-doped $NaAlH_4$ shows two plateaus in the *P–C* isotherms shown in Fig. 19.

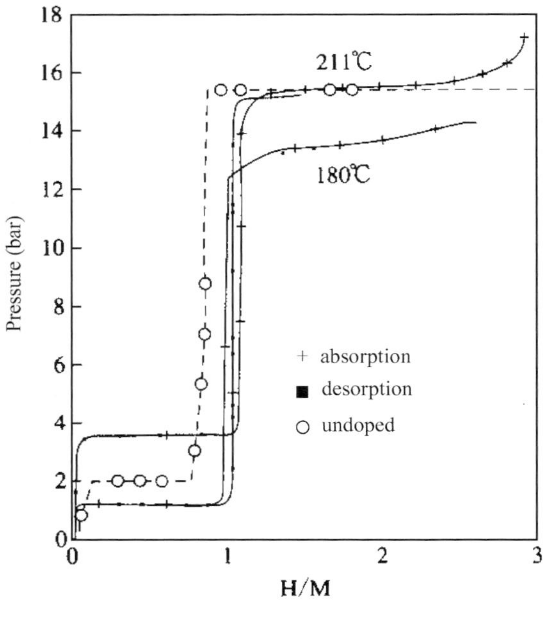

Figure 19
P–C isotherms for titanium-doped $NaAlH_4$.

The theoretical hydrogen storage capacity for $NaAlH_4$ in the above reaction is 5.6 wt.% and that for Na_3AlH_6 is 2.96 wt.%.

Relatively high temperatures (>150°C) and high hydrogen pressures (6–15 MPa) are required for the reaction, and the kinetics is rather slow even with catalysts. Also they show some instability in cycling. Efforts to improve hydriding and dehydriding properties of $NaAlH_4$ and Na_3AlH_6 are continuing (Jensen and Bjerrum 1999, Zaluska *et al.* 2000). It has been reported that the hydriding and dehydriding kinetics of $NaAlH_4$ and Na_3AlH_6 are strongly enhanced by mechanical grinding and the addition of carbon (Zaluska *et al.* 2000).

9. *Conclusions*

Metal hydrides have been developed rapidly since the 1970s and they are expected to play a more important role in the future "hydrogen economy". The problem with the existing hydride-forming alloys is their low reversible capacities at or near room temperature conditions. Intermetallic compounds like AB_5 and AB_2 have excellent hydriding and dehydriding properties, but their hydrogen storage capacities are limited to 1.5–2.0 wt.%. Some alloys such as magnesium-based alloys have high hydrogen storage capacity, but they are not able to absorb and desorb hydrogen readily at room temperature because of the high stability of their hydrides and slow kinetics. Other metal–hydrogen systems like hydride complexes are promising but still have their problems. Breakthroughs are being anticipated in the development of metal hydrides.

See also: Hydrogen–Metal Systems: Basic Properties (1); Hydrogen–Metal Systems: Technological and Engineering Aspects; Hydrogen–Metal Systems: Applications of Gas–Solid Reactions

Bibliography

Cantrell J S, Bowman R C Jr. 1999 Phase composition and the effect of thermal cycling for VH_x, $V_{0.995}C_{0.005}H_x$ and $V_{0.975}Zr_{0.020}C_{0.005}H_x$. *J. Alloys Compounds* **293–295**, 156–60

Friedlmeier G, Manthey A, Wanner M, Groll M 1995 Cyclic stability of various application-relevant metal hydrides. *J. Less-Common Metals* **231**, 880–7

Jensen O J, Bjerrum N J 1999 Systematic B-metal substitution in $CaNi_5$. *J. Less-Common Metals* **293–295**, 185–9

Percheron-Guegan A, Welter J M 1988 Preparation of intermetallics and hydrides. In: Schlapbach L (ed.) *Hydrogen in Intermetallic Compounds: I. Electronic, Thermodynamic, and Crystallographic Properties, Preparation.* Springer-Verlag, Berlin

Sandrock G D 1997 State-of-the-art review of hydrogen storage in reversible metal hydrides for military fuel cell applications. *Final Report for the US Office of Naval Research*

Sandrock G D 1999 A panoramic overview of hydrogen storage alloys from a gas reaction point of view. *J. Alloys Compounds* **293–295**, 877–88
Selvam P, Viswanathan B, Swamy C S, Srinivasan V 1986 Magnesium and magnesium alloy hydrides. *Int. J. Hydrogen Energy* **11**, 169–92
Tamura H (ed.) 1998 *Hydrogen Storage Alloys—Fundamentals and Frontier Technologies*. NTS, Japan
Willems J J G 1984 Metal hydride electrodes: stability of $LaNi_5$-related compounds *Philips J. Res.* **39**(suppl. 1), 1–94
Yvon K, Fischer P 1988 Crystal and magnetic structures of ternary metal hydrides: a comprehensive review. In: Schlapbach L (ed.) *Hydrogen in Intermetallic Compounds: I. Electronic, Thermodynamic, and Crystallographic Properties, Preparation*. Springer-Verlag, Berlin
Zaluska A, Zaluski L, Strom-Olsen J O 2000 Sodium alanates for reversible hydrogen storage. *J. Alloys Compounds* **298**, 125–34

Hydrogen–Metal Systems: Technological and Engineering Aspects

S. Suda

Many metals, alloys, and intermetallic compounds can "*absorb*" reversibly large quantities of hydrogen gas (see *Hydrogen–Metal Systems: Hydride Forming Alloys (Properties and Characteristics, Database Information)*). Several of these hydrogen-metal systems have been considered for technical applications.

The inherent shortcomings of materials will first be described, and then the technological and engineering problems will be introduced in connection with the nonequilibrium (dynamic) behaviors of the hydrogen–metal systems (see *Hydrogen–Metal Systems: Applications of Gas–Solid Reactions*).

1. Shortcomings of Hydrogen-metal Systems—Metal Hydrides: Properties and Characteristics

1.1 Initial Activation Characteristics

All hydrogen absorbing materials require pretreatment procedures that are repeated between high hydrogen pressure and evacuated conditions at high temperature. This procedure is called the initial activation and it has long been introduced in order to initiate hydrogenation reaction for hydrogen uptakes. The equilibrium pressure-composition-temperature (P-c-T) relation can be reached after several activation treatments.

Such a procedure is necessary to remove the surface oxide layer that acts as the resistance for hydrogen uptake and its kinetics. The isothermal pressure-composition relation (P-c isotherm) varies until it reaches to the equilibrium state as a function of hydrogenation/dehydrogenation or hydriding/dehydriding (H/D) cycles. It was found that the initial pressure level in P-c isotherms is reduced significantly by generating lattice defects that trap hydrogen in solid solution as monatomic hydrogen (Inui *et al.* 2000, Yamamoto *et al.* 2000), i.e., protium (denoted H°).

It should be noted that a high-pressure vessel is required during initial activation because the pressure level is excessively high compared with the pressure when an equilibrium state is reached. The pressure difference before and after reaching the equilibrium state generates technological and engineering difficulties in designing lightweight hydrogen storage units such as tanks and reservoirs as hydrogen storage units and heat exchangers for heat pumps and refrigerators. The most serious disadvantage is the weight penalty that lowers thermodynamic efficiency and cost-effectiveness as stated in Sect. 3.1.

Several pretreatment procedures have been proposed for avoiding high-pressure conditions during initial activation by washing powder-state alloys in an acid or alkaline solution. However, P-c-T relations are easily destabilized which results in increasing pressure levels and surface oxidization to reduce the H-capacity after the repeated H/D cycles.

A procedure called the *fluorination treatment* (F-treatment) in a KF/HF solution (see article *Hydrogen–Metal Systems: Fluorinated Metal Hydrides*) has been developed for solving such problems. With an F-treatment the oxide layer is removed to form a fluoride, and it protects the surface from poisoning caused by moisture, air, and impurities such as CO, CO_2, and unsaturated hydrocarbons during H/D cycles.

1.2 Surface Sensitivity

Any hydriding alloys once hydrogenated exhibit an extremely sensitive nature at the surface not only against hydrogen but also against impurities such as O_2, CO, CO_2, NO_2, SO_2, unsaturated organic compounds, moisture, water, air, and N_2. The available hydrogen content is greatly influenced by such contaminants and results in a loss of hydrogen capacity. It should be remembered that the metal hydride surface exhibits strong reactivity not only to hydrogen but also to the above mentioned materials.

From these facts, metal hydrides are applicable in principle only with the pure hydrogen for long-term uses. However, reactivity can be retained as far as the reactive surface is protected by surface treatment techniques. A surface treatment by fluorination is the typical example of special techniques for protecting the surface functions without loosing the available hydrogen content for long-term uses (see article *Hydrogen–Metal Systems: Fluorinated Metal Hydrides*). It is essentially important to protect the surface by applying such treatments.

The highly reactive nature of the surface, in turn, is suitable to apply metal hydrides as hydrogen donors in chemical reduction of such materials listed above. Alternatively, hydrogen gas containing saturated hydrocarbons can be separated and purified by repeated H/D cycles in a separation process, although no industrial process has been developed yet.

1.3 Pulverization and Volume Expansion

During H/D cycles, metal hydride particles reduce the particle size considerably and increase the specific surface area. Pulverization is inherent in any hydrogen-metal system and it accompanies volume expansion and contraction during H/D cycles. The volume expansion is caused by the increase of the lattice parameters.

From engineering viewpoints, the volume expansion and contraction generate unfavorable design problems resulting in (i) severe mechanical stress and damage, (ii) restricted packing density of metal hydride powders, (iii) decreased effective thermal conductivity of powder beds, and (iv) decreased overall hydrogen capacity per volume of container.

The phenomenon can not be overcome technically, but the performance can be improved by incorporating materials with buffer functions such as porous or sintered aluminum or nickel metals with high thermal conductivity. During long-term H/D cycles, some of hydriding alloys segregate or disproportionate to materials that change the properties and characteristics to decrease the hydrogen capacity. Such phenomena are called *segregation* or *disproportionation* depending on the nature of decomposition of hydrogen-metal systems.

1.4 Enthalpy Changes

The hydrogenation and dehydrogenation is accompanied by enthalpy changes that can be obtained as the slopes of van't Hoff plots in *P-T* relations at a given hydrogen concentration in metal hydrides. In general, these enthalpy changes are conveniently called heats of reaction, i.e., exothermic in hydrogenation and endothermic in dehydrogenation. The heats of reaction are discussed in detail in *Hydrogen–Metal Systems: Basic Properties (1)*.

In hydrogen storage devices, the exothermic heat generated during hydrogen uptake must be released to the surroundings, and the endothermic heat required for releasing hydrogen from the system must be extracted from the surroundings. The heats of reaction are unfavorable from the engineering viewpoint because the hydrogen storage unit requires some provisions for the heat transfer. The heat transfer structure results in larger weight of the storage unit and increases the impracticality of hydrogen-metal systems.

Furthermore, it is important to install such a heat transfer structure in the thermodynamic devices that apply the heats of reaction in the reversed Carnot cycles both in refrigeration and heat pump cycles. Such a heat transfer structure increases the overall heat capacity and the sensible heat loss that can not be delivered in an effective manner as the thermal energy to be applied. The magnitude of such heat losses is measured as the thermal mass, mass ($Q_{th} = mC_P\Delta T$, the thermal energy required for heating or cooling a storage unit between two different temperature levels).

In order to maintain a desired hydrogen flow when hydrogen is absorbed or desorbed, heat must be transferred into or released from the powder-state metal hydride bed in any applications. For this reason, hydrogen storage by metal hydrides, in principle, is not practical when a high heat transfer rate is required or when such a heat source or a heat sink as ambient air is not available. It is also not practical to apply metal hydrides where an extremely high discharge rate is required because of the relatively slow dehydrogenation kinetics of metal hydrides under moderate temperature and pressure conditions.

1.5 Effective Thermal Conductivity

The thermal conductivity of powder-state beds is called the *effective thermal conductivity*. The value of the effective thermal conductivity pertaining to the unhydrided condition can be estimated from a simple model for metallic powder although it differs considerably from the hydrided and dehydrided values.

It should be taken into consideration that the effective thermal conductivity changes greatly depending on whether the bed is hydrided or dehydrided. It is determined by knowing (i) the average particle size or specific surface area, (ii) the packing density or void fraction, and (iii) whether it is hydrided or unhydrided.

The effective thermal conductivity is, in general, extremely small for the hydrided power bed in the average magnitude of $0.5\,Wm^{-1}K^{-1}$ and $1.0\,Wm^{-1}K^{-1}$ at maximum. Such a small value results in the poor heat transmission of a packed bed and it leads to essential requirements in any engineering device to improve the effective thermal conductivity from the heat transmission viewpoints. The mass transfer rate of hydrogen, i.e., reaction kinetics, is regulated by the heat transfer rate of the powder beds.

Fine powders of pure copper or nickel have been found to be effective when they are mixed with the metal hydride powders. Improvement of the effective thermal conductivity should be taken into account together with the volume expansion of the metal hydride bed.

There has been some work on improving the effective thermal conductivity. Al- or Ni-based foamed metals have been used effectively to increase the effective thermal conductivity up to $14.0\,Wm^{-1}K^{-1}$ (Suda and Komazaki 1991, Fig. 1).

1.6 Heat Capacity and Thermal Mass

The heat capacity of the metal hydride bed with a fine particle size is quite different from that of the bulk material. It is dependent on the overall characteristics of metal hydride beds, i.e., the overall heat capacity, which depends on the (i) average particle size, (ii) packing density and void fraction, (iii) structure and nature of the materials of hydrogen storage units, and (iv) levels of hydrogenation. Therefore, it is not easy to estimate the value either from the metallic

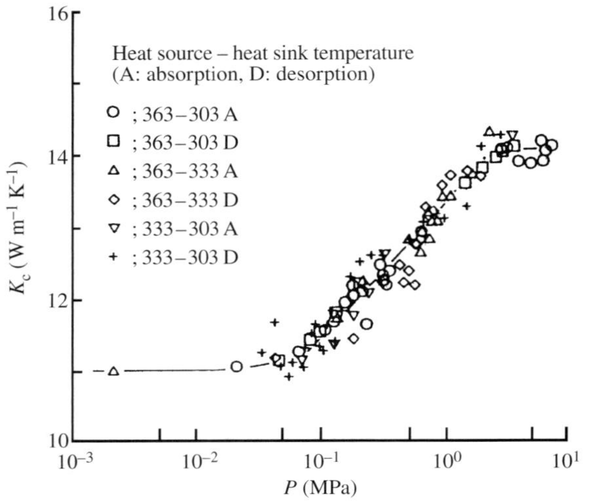

Figure 1
Effective thermal conductivity of $MmNi_{4.52}Al_{0.47}$-hydride bed (H/D cycled 20 times, void fraction 0.353) packed in the Cu-punched metal sheets (void fraction 0.94) (after Suda and Komazaki 1991).

powder before hydrogenation or from the bulk properties.

However, the overall heat capacity of a hydrogen storage unit such as tanks, reservoirs, and heat exchangers is obtained by a simple experiment. The amounts of heat transferred per weight of a system can be measured by recording temperature changes during heating or cooling as a function of time elapsed and it is converted to heat capacity ($Wkg^{-1}K^{-1}$ or $Wkg^{-1}{}^{\circ}C^{-1}$). It is also possible to calculate the thermal mass ($Q_{th} = mC_P\Delta T$, the thermal energy required for heating or cooling a storage unit between two different temperature levels.)

The overall heat capacity becomes the predominant factor in thermodynamic devices such as heat pumps, refrigerators, and air-conditioner units in which very high rates of hydrogen transfer between a coupled reactor system is required. In such applications, it is important to reduce the overall heat capacity as much as possible. The magnitude of the heat capacity value is related to the sensible heat loss of an engineering device that influences directly the thermodynamic efficiency and cost-effectiveness of a device.

It is understandable that the reasons why thermodynamic devices that apply enthalpy changes during exothermic and endothermic reactions have not been successful for heating and cooling in the past. One of the major reasons is the difficulty to extract the heats of reaction from the heat exchangers effectively. It is necessary to reduce the overall weight of heat exchangers in order to increase the thermodynamic efficiency.

1.7 Isothermal Pressure Hysteresis

This phenomenon occurs in the equilibrium pressure-composition relations between absorption and desorption at a given isothermal condition. In most metal hydride systems, there exists pressure deviation between hydrogenation (absorption) and dehydrogenation (desorption) isotherms. The deviation is called *pressure hysteresis*.

The hysteresis has not yet been clearly explained by crystallography and it has been subjected to detailed studies. It should be noted that hysteresis is a phenomenon that occurs under equilibrium states and the pressure deviation is enlarged significantly under dynamic and nonequilibrium (transitional) conditions.

The equilibrium P-c-T relations can be used only as a measure to select a proper metal hydride or a set of metal hydrides in a specific engineering application. The P-c isotherm has, in principle, a flat and constant pressure region called the *plateau* pressure. However, it is actually not flat but sloped for the most metal hydrides and for the wide ranges of hydrogen concentration.

1.8 Dynamic P-c-T Relations

The equilibrium P-c-T relations cannot be used directly in any practical applications in which no equilibrium state is reached during H/D processes. The system pressure is always higher during hydrogenation (absorption) and lower during dehydrogenation (desorption) than the pressure under equilibrium states. It varies greatly dependent on the mass and heat transfer characteristics of a hydrogen storage unit.

The actual pressure exerted during hydrogen transfer is called *dynamic* pressure, and the corresponding P-c-T relation is called *dynamic* P-c-T relation. The degree of pressure hysteresis is enlarged during hydrogen transfer in H/D processes. The smaller the temperature difference and the faster the hydrogen transfer rate, the larger is the pressure difference between H/D processes. Depending on the conditions as listed below, the dynamic desorption pressure becomes higher than the dynamic absorption pressure in a coupled metal hydride system under some operating conditions. These conditions include (i) hydrogen transfer rates, (ii) heat transfer rates, and (iii) equilibrium P-c-T relations of a coupled set of metal hydrides. It is desirable from engineering viewpoints that the hydrogen transfer rate is maintained constant, not depending on the other operating conditions.

The P-c relations under dynamic conditions (transitional conditions) were measured for the fluorinated La-incorporated $Zr_{0.9}Ti_{0.1}V_{0.2}Mn_{0.5}Co_{0.1}Ni_{1.1}$–hydride as a function of time after starting the H/D reactions (Suda *et al.* 1999). The results showed quite

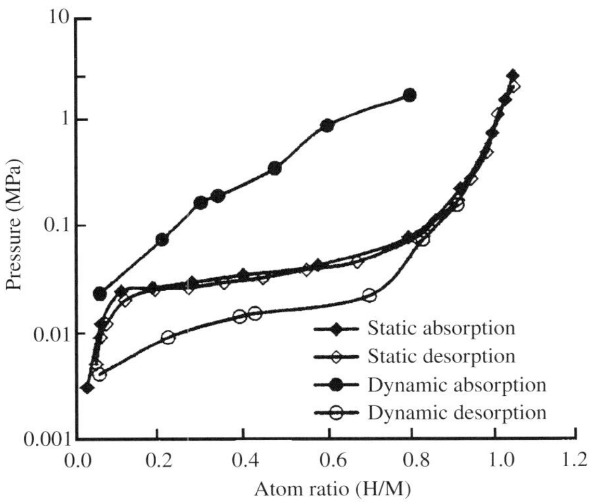

Figure 2
Dynamic P-c-T taken for the fluorinated La-incorporated $Zr_{0.9}Ti_{0.1}V_{0.2}Mn_{0.5}Co_{0.1}Ni_{1.1}$-hydride as a function of time after starting the H/D reactions (after Suda *et al.* 1999).

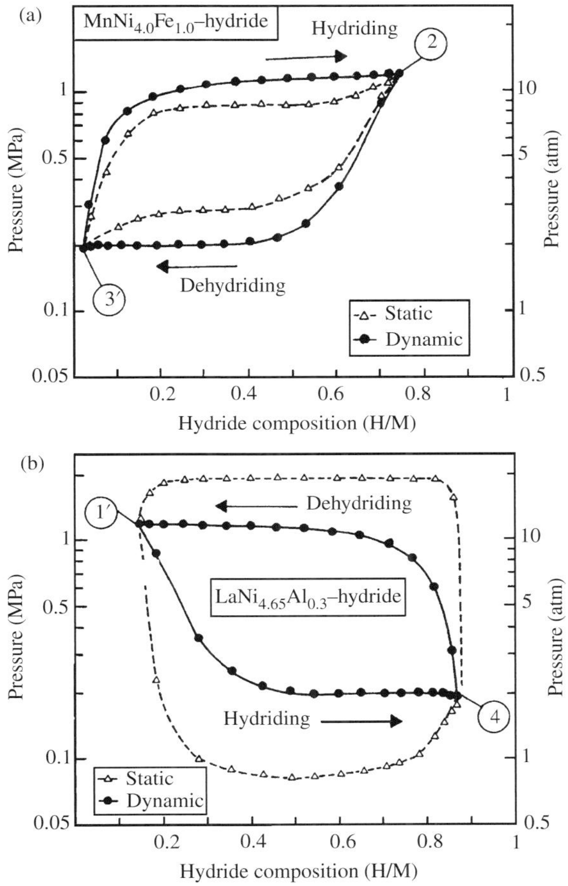

Figure 3
(a) Dynamic P-c relations of $Mn_{4.0}Fe_{1.0}$–hydride in a coupled heat exchanger system composed of $MnNi_{4.0}Fe_{1.0}$–hydride (high-pressure side) and $LaNi_{4.65}Al_{0.3}$–hydride (low-pressure side). (b) Dynamic P-c relations of $LaNi_{4.65}Al_{0.3}$–hydride in a coupled heat exchanger system composed of $MnNi_{4.0}Fe_{1.0}$–hydride (high-pressure side) and $LaNi_{4.65}Al_{0.3}$–hydride (low-pressure side).

different behaviors from the data taken under equilibrium (static) conditions as illustrated in Fig. 2.

The dynamic P-c relations of a coupled heat exchanger system have been studied extensively by Nagel *et al.* (1986a, 1986b). In a coupled heat exchanger system composed of $MnNi_{4.0}Fe_{1.0}$–hydride (high-pressure side) and $LaNi_{4.65}Al_{0.3}$–hydride (low-pressure side), the dynamic P-c behaviors were measured by varying the H/D cycle time between 10-15 minutes. (Fig. 3). Under such a transitional condition, the P-c relations exhibited completely different characteristics from the equilibrium relations. The temperature profiles of the coupled metal hydride system is illustrated in Fig. 4.

The dynamic P-c-T relation is not the thermodynamic property but the transitional and engineering characteristics that are greatly dependent on operating conditions. Accordingly, it is important to know experimentally the dynamic behavior of a metal hydride system in designing engineering devices and processes.

2. *Technical Problems Caused by the Properties and Characteristics of Hydrogen-metal Systems*

2.1 *High-pressure Requirements*

Most hydriding alloys require high pressure and evacuated conditions during initial activation procedures. This means that any hydrogen storage devices must be designed in consideration of the pressure tolerability at the initial activation. As a result, hydrogen storage units such as containers, tanks, reservoirs, and heat exchangers become heavier and increase the weight penalty to reduce thermal and thermodynamic efficiencies.

For solving such problems, particle surfaces must be treated in order to reduce the initial pressure down to a proper level by such techniques as KOH and fluorination treatments before putting metal hydride powders into hydrogen storage units. A hydrogenation treatment to generate lattice defects works effectively for reducing the initial pressure level (Inui et al. 1999). However, as the hydrogenated surface

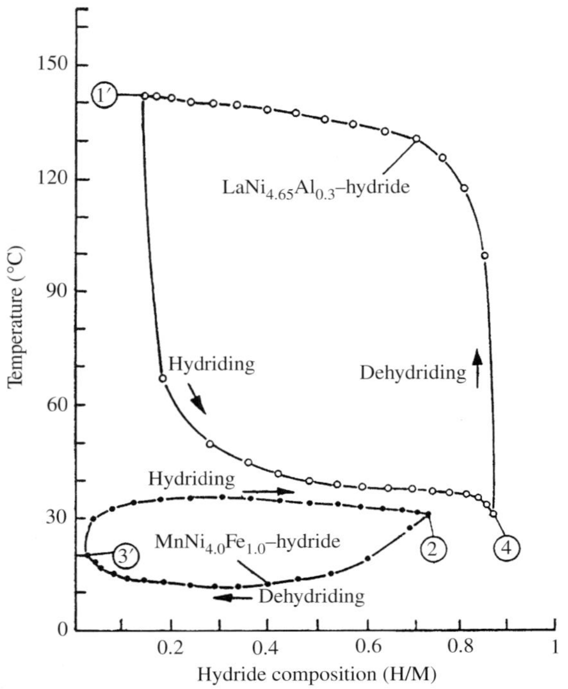

Figure 4
Temperature profiles of the coupled metal hydride system measured in the powder bed.

exhibits strong sensitivity to air and moisture, it is necessary to deal with the hydrogenated material under an inactive atmosphere for avoiding any unexpected accident caused by pyrophoricity, The hydrogenation treatment prior to applications requires the expensive and tedious processes.

2.2 Durability and Life

Impurities contained in hydrogen gas reduce the hydrogen capacity during long-term uses, i.e., repeated H/D cycles. The capacity is also reduced in contact with air, moisture, and water. It frequently happens that the entire hydrogen capacity is lost by contact with CO, H_2S, NO_2, and other gases.

The most typical poisoning phenomenon is the oxidation of surface in gas-solid reactions. The segregation or disproportionation might be expected in some hydriding alloys during long-term H/D cycles. The hydrogen capacity is also lost by disintegration of alloy phases. The durability and life of an engineering device are considerably dependent on the purity of hydrogen gas and the disintegration characteristics of hydrogen-metal systems.

2.3 Volume Expansion–Contraction During H/D Cycles

Volume expansion and contraction are caused by changes of the cell volumes, i.e., changes of lattice parameters during H/D cycles. The volume difference between "hydrogenated and dehydrogenated" conditions sometimes reaches values of more than 30%. The volume change not only causes a loss of heat transfer capability of metal hydride beds but also a fatal mechanical damage of hydrogen storage units.

The only way to avoid the mechanical destruction of a storage unit is (i) to maintain the void fraction within a limited range over 50% (or packing density less than 50%), (ii) to set a storage unit in a horizontal position, and (iii) to charge powders in aluminum- or nickel-porous or sintered materials that has the buffer function against the mechanical stress.

2.4 Heat Transmission Characteristics of the Packed Bed

Thermal energy to be transferred during H/D cycles is largely dependent on the packing density (or void fraction) of a metal hydride bed. The packing density changes according to the degree of expansion/contraction due to hydrogenation and dehydrogenation, and accordingly on the hydrogen concentration in metal hydride.

It should be noted that the amount of thermal energy to be transferred in any engineering device is not determined by the enthalpy change calculated from the equilibrium states and it can only be estimated by knowing the dynamic P-T relations. Some amount of thermal energy is not transferred between the wall of a hydrogen storage unit depending on the magnitude of the thermal mass of a unit.

2.5 Overall Heat Capacity and Thermal Mass

The overall heat capacity of a system is given as the combined heat capacities of packed bed and container itself. The packing density of a powder bed is limited to less than 50% due to the volume expansion/contraction of metal hydride powder beds and the P-T relations during initial activation regulate the weight of hydrogen storage units. This causes the increase of sensible heat retained in a container as the thermal mass without exchanging thermal energy between the system and the surroundings. As a result, thermodynamic efficiency is limited by reduced thermal energy to be transferred.

2.6 Batch Types of H/D Reactions

The amount of hydrogen to be transferred is inherently limited by the nature of H/D reactions. These reactions are composed of batch-wise processes and these processes are maintained as semi-continuous cycle in an engineering device. The dynamic P-c-T relation limits hydrogen to be transferred as a function of pressure difference between the desorption- and absorption sides or the system and the surroundings. Because of the heat that is required for heating and cooling the hydrogen storage unit itself, a large amount of the heat of the reaction is retained without transferring between the system and the surroundings. It reduces the amount of thermal energy to be exchanged between the system and the surroundings, and as a result, it reduces the thermodynamic efficiency of a system.

3. Engineering Problems

3.1 Hydrogen Storage Units (Containers, Tanks and Reservoirs) and Heat Exchangers

In a dynamic condition, the amount of available hydrogen in a storage unit is considerably smaller than the available hydrogen content of a metal hydride under equilibrium states. This is the reason to reduce the practicality of metal hydride systems. At present, the best weight ratio between the amount of available hydrogen and the storage unit (metal hydride + structure materials) is less than 0.8 wt.% in rare-earth-based materials and less than 1.0 wt.% in Laves phase materials.

However there are two remaining problems, i.e., (i) low available hydrogen content of metal hydrides under dynamic conditions, and (ii) heavy weight of storage units. It has been estimated that applying metal hydrides to hydrogen storage devices will not be practical nor economical, even if hydrogen capacities over 5.0 wt.% could be attained in the future, without improving the initial activation characteristics, in particular, without reducing the pressure levels during initial activation.

3.2 Thermodynamic Efficiency of Engineering Devices

In a hydrogen storage unit, thermal energy is required in both cases of discharging and charging hydrogen from/to the unit. The rate of hydrogen transfer is directly regulated by the rate of thermal energy transfer. The quality of the surrounding thermal energy sources, as well as the heat transmission characteristics of a unit during thermal energy intake and release, determines the thermodynamic efficiency.

It is not adequate to use thermal efficiency based on a reversible process or cycle such as $\eta_{th} = 1 - T_L/T_H$ where η_{th} denotes the Carnot thermal efficiency of a reversible system that works between high temperature, T_H, and low temperature, T_L. The thermal efficiency is always less than unity and it is difficult to evaluate the efficiency of a device by comparing the small differences. It does not make any sense when the thermal efficiency or the coefficient of performance (COP) stated below is applied to different devices operated under different temperature levels.

In the case of thermodynamic cycles such as refrigerators, air-conditioner units, and heat pumps, the COP has usually been introduced by supposing a reversed Carnot cycle, i.e., $COP_R = 1 - 1/(T_H/T_L - 1)$ for a refrigerator and $COP_{HP} = 1/(1 - T_L/T_H)$ for a heat pump where $COP_{HP} = COP_R + 1$.

However, the COP is strongly dependent on the temperature difference between high and low thermal energy sources, i.e., thermal energy source and sink, and particularly, it depends on the individual temperature levels of thermal energy sources. The COP based on the reversed Carnot cycle should not be used as the measure of thermodynamic efficiency of a given device because it is always larger than unity and is not applicable for evaluating efficiencies under different temperature conditions.

In order to evaluate the efficiency from COP, the second law thermodynamic efficiency, η_{II}, as the ratio between the ideal (reversible) and actual conditions should be introduced: $\eta_{II} = COP_{actual}/COP_{ideal}$. The COP_{actual} as a measure of efficiency under actual operation conditions is evaluated by taking account of the dynamic behaviors of a given cycle. These include (i) dynamic P-c-T relation, (ii) overall heat transfer characteristics (effective thermal conductivity + heat capacity of metal hydride beds), (iii) operating temperature conditions (heat source and heat sink), and (iv) time intervals required per H/D batch-wise reactions.

At present, there are only a limited number of reports available that describe the thermodynamic efficiency of engineering devices, particularly from the viewpoint of the second thermodynamic law. The second law efficiency obtained from the experimental data on a double-stage heat pump is shown in Fig. 5 (Suda *et al.* 1991). The actual COP_{actual} was 0.21 in comparison with the theoretical COP_{ideal} of 0.36.

It has been pointed out that thermodynamic applications of H/D cycles are not considered practical because of the low available hydrogen content under dynamic conditions and the high sensitive heat loss that is retained in metal hydride beds, i.e., the thermal mass.

No practical example of this kind of applications has appeared commercially in the market yet.

3.3 Cost-effectiveness

Due to the high cost of metal hydrides and low thermodynamic efficiency, engineering devices should be considered impractical to develop at present in view of cost-competitiveness with conventional refrigerators, air-conditioner units, and heat pumps.

The low cost-effectiveness is caused by the poor available hydrogen content and the poor transferable

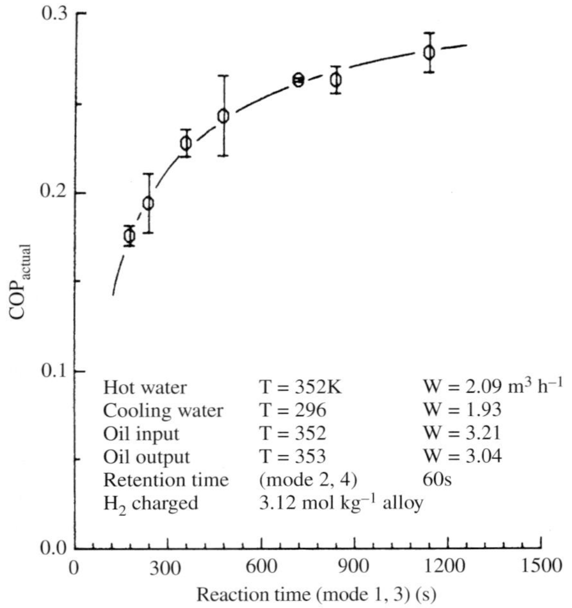

Figure 5
Actual COP of a double-stage heat pump based on the second thermodynamic law (after Suda *et al.* 1991).

thermal energy during dynamic operation in order to bring the batch-wise processes to semi-continuous operations. The cost of hydriding alloys should be evaluated by the amounts of H_2 available per unit weight and preferably by the amount of H_2 transferable during H/D cycles. The average price available in the world market is in the range of US\$25–35 per kg in 2000 but US\$50–65 per g H_2 available in an engineering device.

4. Concluding Remarks

The inherent shortcomings of hydrogen-metal systems have been reviewed from the technological and engineering viewpoints. The most unpractical point is found in the dynamic nature during hydrogenation and dehydrogenation. The dynamic behaviors are entirely different from the equilibrium properties and characteristics, in particular, are the differences in the mass and heat transfer characteristics.

The hydrogen-metal system seems to exhibit far reaching stages in the practical applications. It is difficult, but essentially important, to increase the available hydrogen capacity under dynamic conditions. More basic research on the engineering properties and characteristics under dynamic conditions are necessary with regards to (i) surface treatments for initial activation characteristics and protection from poisoning materials, (ii) mass and heat transfer capability with higher effective thermal conductivity and lower overall heat capacity, and (iii) conversion of the batch-wise H/D processes to the semi-continuous cycle by minimizing the heat losses during switching the H/D reactions.

See also: Hydrogen–Metal Systems: Applications of GasSolid Reactions; Hydrogen–Metal Systems: Electrochemical Reactions (Fundamentals and Applications); Hydrogen–Metal Systems: Basic Properties (1)

Bibliography

Inui H, Yamamoto T, Hirota M, Yamaguchi M 2000 Lattice defects introduced during hydrogen absorption–desorption cycles and their effects on *PCT* characteristics in some intermetallic compounds In: *Proc. Int. Symp. Metal-Hydrogen Systems*. Noosa, Queensland, Australia, p. 12

Inui T, Yamamoto T, Yamaguchi M 1999 Influence of lattice defects on protium absorption pressures in $LaNi_5$ In: *1st Int. Symp. on New Protium Function in Materials*, Osaka, Japan. p. 40

Nagel M, Komazaki Y, Matsubara Y, Suda S 1986a Dynamic behaviour of paired metal hydrides II: analytical survey of the experimental results. *J. Less-Common Met.* **123**, 47–58

Nagel M, Komazaki Y, Suda S 1986b Dynamic behaviour of paired metal hydrides I: experimental method and results. *J. Less-Common Met.* **120**, 45–53

Suda S, Imai M, Uchida M, Komazaki Y, Higuchi E 1999 Dynamic P-c-T relations of the La-incorporated/fluorinated AB_2 hydriding alloys. *J. Alloys Comp.* **293–5**, 391–5

Suda S, Komazaki Y 1991 The effective thermal conductivity of a metal hydride bed packed in a multiple-waved sheet metal structure. *J. Less-Common Met.* **172–4**, 1130–7

Suda S, Komazaki Y, Narasaki H, Uchida M 1991 Development of a double-stage heat pump: experimental and analytical surveys. *J. Less-Common Met.* **172–4**, 1092–110

Yamamoto T, Inui H, Yamaguchi M 2000 Effects of ternary additions on defect generation during hydrogen absorption-desorption cycles in $LaNi_5$ In: *Proc. Int. Symp. Metal-Hydrogen Systems*. Noosa, Queensland, Australia, p. 15

I

Inert Matrix Fuels

N. Chauvin and M. Pelletier

"Inert-matrix fuels" are fuels that do not contain a "fertile isotope" so that no conversion of fertile to fissile material takes place and only the initially present fissile atoms are burnt (see *Nuclear Fuel Cycles*). Their introduction has been envisaged, with the current interest in employing fast "breeders" as fast "burners," mainly for the utilization of the accumulated plutonium without "breeding" more plutonium. Thus in the plutonium cycle, inert matrix fuels are those that are free from uranium-238, and an inert material that does not contain a fissionable or fertile isotope is used as the matrix with a plutonium-bearing material.

The main interest in uranium-free fuels, inert-matrix fuel (IMF), is in reaching higher burn-up for fissile atoms because of the better behavior of IMFs under irradiation than that of standard fuels. The fissile atom and support matrix can either form a solid solution or a composite fuel such as a ceramic–ceramic or ceramic–metal material.

IMFs are generally thought useful for reasons including the following:

- improving the fuel management costs of standard reactors by increasing the average discharge burn-up while guaranteeing improved safety performance;
- burning/using plutonium (Akie *et al.* 1998, Delpech *et al.* 1997); and
- transmuting minor actinides (neptunium, americium, and curium) and thus reducing the radiotoxic inventory of high-level wastes (Tommasi *et al.* 1997).

1. *The Choice*

1.1 *Criteria for Selection*

Possible ceramic or metallic candidate materials can be chosen using the following criteria:

- The inert matrix must be transparent to neutrons: its neutron cross-section must therefore be low and there should be no generation of activation products. Some metals do not fulfill the first requirement in a thermal neutron spectrum. Nitrides must be enriched in ^{15}N in order to avoid ^{14}C formation.
- Specific physical properties are required, such as high melting temperature. In the case of a phase transformation, the temperature must be at a value well above the operating temperature of the fuel. However, there is some leeway in establishing the thermal conductivity requirements since some scenarios involve very different linear heating rates.
- The feasibility of the fuel cycle must be demonstrated: fabrication processing, cost, handling, reprocessing or storage.
- Compatibility with neighboring materials is also of concern: cladding, coolant, fissile phase, and fission products. Indeed new phases formed as a result of chemical reactions between the matrix and the fissile phase may lower the fuel melting temperature.
- The last criterion concerns the behavior under irradiation. Three main areas are of concern: the behavior of materials with respect to irradiation damage, and to the thermomechanical and thermochemical behaviors.

1.2 *Behavior under Irradiation*

In a reactor the fuel is submitted to three different sources of damage: fast neutrons, fission fragments, and alpha decay products (alpha particle + heavy recoil atom). These energetic particles produce damage through electronic and atomic interaction, as there is energy deposition in the material (Matzke 1982) (see ionization or atomic displacement in Table 1). The consequences of a collision depend on the target material and the incident particle is characterized by its mass, electric charge, and energy. The mean free path of a particle in a solid compound and the dose in displacements per atom (dpa) determine the effect of a given collision.

The total number of defects depends on the crystallography of the target: the fluorite structure is known to be more suitable than hexagonal or monoclinic structures. Defects depend on the threshold displacement energy of each atom in the matrix (Zinkle and Kinoshita 1997). However, defect recovery may occur by spatial events or thermal effects. The responses to radiation damage are significant changes in bulk properties: changes of lattice parameter or phase changes including amorphization or swelling (Wiss and Matzke 1999) and evolution of

Table 1
Properties of different damage sources in ceramic nuclear fuels (Matzke 1988).

Damage source	Energy (keV)	Fraction of total energy lost in nuclear stopping	Range in UO_2 (μm)	Number of defects (dpa)
Alpha particle	~5000	~0.01	~12	~200
Recoil atom from alpha decay	~95	~0.90	~0.025	~1200
Fission, median light fission product	95 000	~0.03	~9	~40 000
Fission, median heavy fission product	67 000	~0.05	~7	~60 000

thermal and mechanical properties (Hobbs *et al.* 1994).

Solids respond differently to radiation. As little change in physical properties as possible is desirable. The exact understanding of radiation damage is far from being complete. However, there are in-pile experimental and analytical means, such as instrumented irradiations in reactors or more controlled implantation experiments, where separated effects of each kind of damage can be evaluated. For example, zirconia (ZrO_2) has been studied under ion irradiation and it seems to be a very promising matrix (Sickafus *et al.* 1999) since in the same time $MgAl_2O_4$ becomes amorphous under irradiation by heavy ions or fission products.

The second issue concerning IMF behavior is the thermomechanical behavior. A large margin to melting and limited cracking are encouraging signs if they are not greatly dependent on damage and burn-up (fission products or fission gas inclusions).

In thermodynamic studies, all oxides are characterized by their oxygen potential, which controls their stability. During irradiation, the degree of oxidation of the fission product compounds generally contributes to increase this property. Implantation of large quantities of fission products in an IMF may lead to an enhancement of its oxygen potential which may be at the origin of a noticeable cladding oxidation. Carbide and nitride compounds are often less thermodynamically stable than oxides. At high operating temperature, nitride compounds may be partially decomposed in nitrogen and the associated metal. A transfer of nitrogen inducing a chemical fuel–cladding interaction such as nitriding of the cladding material may then take place.

2. *Optimization of IMFs*

IMFs may be optimized in order to improve specific behavioral features. These can either pertain to their stability at extremely high radiation doses or to their capacity to allow a high production of fission gases. Achieving high burn-ups while maintaining safety requirements remains the driving force for developing IMFs. The fuel may be modified in two different ways (microstructure and macrostructure), in order to improve the design of IMFs.

2.1 Microstructure Design

The choice of the microstructure is based on phenomena occurring both in the fissile phase and in the support phase. The aim is always to reach an extremely high proportion of burnt actinides, which implies a radical change in the chemical composition of the fuel. It is not clear whether binary americium or plutonium oxides are stable at very high burn-up, so cubic or pyrochloric phases (such as $Am_2Zr_2O_7$) look to be promising.

This phase is called the "host phase" and its geometrical dispersion in the inert matrix may be controlled in terms of the shape and size of inclusions. Usually the host phase is made up of small spherical particles. In this instance, however, irradiation damage due to fission fragments or alpha decay will be distributed more or less homogeneously throughout the fuel. The overlap of damage zones often results in a severe deterioration of the matrix properties.

Since fission fragment or alpha particle tracks usually lie in the 10–20 μm range, large inclusions are prepared so as to concentrate radiation effects in a small shell. This is particularly true for fuels irradiated over long periods (Fig. 1).

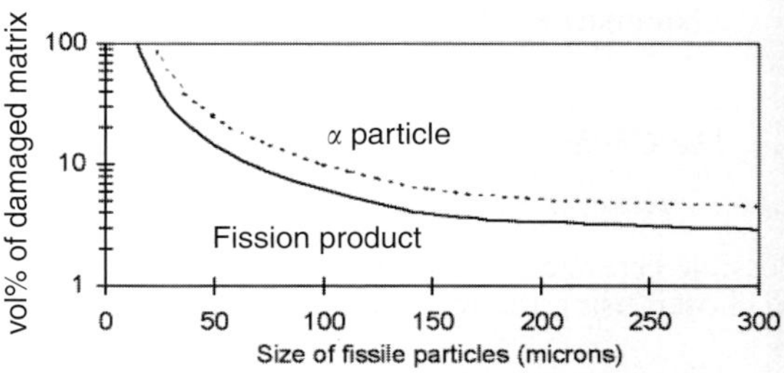

Figure 1
Volume of inert matrix damaged by recoil of fission products or alpha particles. The IMF is $MgAl_2O_4 + 20\,wt.\%\ AmO_x$.

Table 2
Fission gas and helium production in various fuels.

Reactor type	Fuel composition	Burn-up (% FIMA[a])	Gas production[b] ($cm^3 \cdot g^{-1}$ of actinide oxide)	
			Fission gases	Helium
PWR	UO_2	6.2	1.8	0.025
PWR	MOX	5.0	1.3	0.1
FNR	$(UPu)O_2$	17.5	3.6	0.15
Transmutation in fast reactor	IMF: MgO + 20 wt. % AmO_x	**85**	**20.6**	**36**

[a]FIMA = fissile initial metal atom; here refers to heavy metals (actinides) only. [b]At 298.15 k and 1 bar.

A gap between the host phase (particles) and the matrix may exist. Host phase particles are also sometimes coated in order to prevent swelling or to ensure fission gas retention. For the transmutation of americium the total quantity of gases (fission gases and helium) produced will be in excess of that for standard UO_2 fuels (Table 2). Similarly, in plutonium-based matrix fuels, high burn-ups will inevitably lead to the production of a large volume of fission gases. The fuel microstructure must therefore be designed so as to cope with this situation (Chauvin *et al.* 1999).

2.2 *Macrostructure Design*

Macrostructure design is more linked to fabrication processes, gas retention/release, and reactor conditions (Matzke 1986). For example, fission gases may be insoluble in the matrix and the pressure in the fuel increases with burn-up leading to swelling and failure.

There are various concepts used: pellets filled in a cladding, annular pellets, an annular fuel design with inner and outer cladding, coated particles, vibro-compacting fuel, etc. All the existing concepts are not detailed here because they are numerous and many are applied for specific objectives.

In all cases, the IMF and the design of the whole fuel element will depend on the reactor type: the conventional pressurized water reactor (PWR), the fast neutron reactor (FNR), or even the accelerator driven system (ADS).

Calculations are needed to optimize all the design parameters, and some concepts have to be tested in reactors to validate previous assumptions.

3. *Conclusion*

The use of an inert matrix as a support for fissile materials must be considered as part of an entire fuel cycle; for example, rock-like matrices would be optimal for ultimate deep geological formation storage.

In some more innovative cases, IMFs have been investigated where the matrix is not so inert but contributes to a specific purpose through the addition of burnable poisons (erbium, gadolinium), or some moderated materials which thermalize the neutron spectrum to increase transmutation.

The usefulness of IMFs has been demonstrated and the great number of materials and concepts must serve the objectives dedicated to these specific fuels.

The wide range of IMFs and the large number of degrees of freedom for fuel element concepts make experimental validation challenging; international collaboration is required to meet these challenges.

See also: Nuclear Fuel Cycles

Bibliography

Akie H, Yamashira T, Nitani N, Kimura H, Takano H, Muromura T 1998 Disposition of excess plutonium by ROX-LWR system. *AEN-NEA Conf. on Advanced Reactors with Innovative Fuels*. Villingen, Switzerland

Chauvin N, Konings R J M, Matzke Hj 1999 Optimisation of inert matrix fuel concepts for americium transmutation. *J.Nucl. Mater.* **274**, 105–11

Delpech M, Salvatores M, Tommasi J, Zaetta A 1997 Review of systems for plutonium utilisation. *Global '97 Conf.* Yokohama, Japan

Hobbs L W, Clinard F W Jr., Zinkle S J, Ewing R C 1994 Radiation effects in ceramics. *J. Nucl. Mater.* **216**, 291–321

Matzke Hj 1982 Radiation damage in crystalline insulators, oxides and ceramic nuclear fuels. *Radiation Effects* **64**, 3–33

Matzke Hj 1986. *Science of Advanced LMFBR Fuels*. North-Holland, Amsterdam, pp. 75–118

Matzke Hj 1988 Radiation damage effects in nuclear materials. *Nucl. Instrum. Methods Phys. Res.* **B32**, 455–70

Sickafus K E, Hanrahan R J Jr, McClellan K J, Mitchell J N, Wetteland C J, Butt D P, Chodak P, Ramsey K B, Blair T H, Chidester K, Matzke H, Yasuda K, Verrall R A, Yu N 1999 A burn and bury option for the destruction and disposal of plutonium using a cubic zirconia matrix as both fuel-form and waste-form. *Am. Ceram. Soc. Bull.* **78**, 69–74

Tommasi J, Delpech M, Huang S L, Puill A, Rome M, Harislur A, Majournat B, Daudin L 1997 Heterogeneous recycling of americium in thermal and fast reactors. *Global '97 Conf.* Yokohama, Japan

White D W, Beard A P, Willis A H 1957 Irradiation behavior of dispersion fuels. USA EC TID 7546

Wiss T, Matzke Hj 1999 Heavy ion induced damage in $MgAl_2O_4$, an inert matrix candidate for the transmutation of minor actinides. *Radiation Measurements* **31**, 507–14

Zinkle S J, Kinoshita C 1997 Defect production in ceramics. *J Nucl. Mater.* **251**, 200–17

In-reactor Creep Testing: Techniques

P. V. Kumar, S. Murugan, and P. Rodriguez

Creep is the slow, time-dependent strain that occurs in a material at a constant stress (or load). Irradiation creep is the additional strain that occurs as a result of the combined effect of irradiation and stress in components. The strain occurring in the absence of irradiation is referred to as *thermal creep* in this article. Irradiation creep is the sum of both irradiation-enhanced creep (thermal creep increase due to irradiation) and irradiation-induced creep. Performance of fuel, clad tube, and other in-core components of a nuclear reactor can be affected by the dimensional changes caused by creep. Hence, it becomes necessary to determine the amount of creep expected to be suffered by a component during its lifetime in a reactor. This will enable the designer to assess the useful life of the component in the reactor and design suitable components after taking irradiation creep into consideration.

1. *Factors to be Considered in Designing the Tests*

As the cost of in-reactor testing far exceeds that of out-of-pile testing, planning and design are of great importance for a successful experiment. The size and shape of the experimental facility usually dictate miniaturization and exacting construction. The measurement of irradiation creep in an operating nuclear reactor poses several problems and challenges. Maintaining constant conditions of temperature, load/stress, and neutron flux over sufficiently long periods of time at the sample location is difficult. Gamma-ray heating is an important factor in the design of an elevated temperature experiment. It may be so great that cooling of the specimen will be required to maintain the test temperature. In contrast, auxiliary heating may be needed if the nuclear heating does not produce sufficient heat to maintain the required test temperature. Accurate measurement of the creep strain by remote techniques is a challenge.

Other considerations to be taken into account in planning in-reactor experiments include: availability of space for the experimental setup in the reactor, reactivity considerations due to presence of the experimental setup inside the reactor, safety implications of these for smooth operation of the reactor, and radiation damage to the experimental equipment. The contribution of thermal effects and swelling of materials under irradiation have to be taken into account while quantifying the effect of irradiation on creep. Often, separate experiments are needed in the laboratory to determine the thermal effects without irradiation.

2. *Basic Requirements*

The measurement of creep in a reactor environment requires the following: A stressing system to maintain the test specimen under the desired conditions of stress, means to maintain and determine the temperature of the test specimen during the experiment, a measurement system to monitor the change in dimensions of the test specimen (either on-line or outside the reactor), and means to determine the neutron fluence seen by the test specimen.

3. *Types of In-reactor Creep Tests*

In-reactor creep tests can be of two types: (i) instrumented tests for on-line monitoring of test conditions and the creep rate, or (ii) noninstrumented tests where the evaluation of the specimen is done outside the reactor and test parameters are derived from reactor operating conditions. In instrumented experiments, temperature of specimen, load exerted on the specimen, extension of the specimen, etc. are monitored using suitable instruments that are positioned in close proximity to the specimen under investigation Gilbert *et al.* (1972). The selection of the instruments used should take into account the effects of radiation inside the reactor on their performance. Instrumented experiments are complicated to design and are expensive. Very few instrumented experiments can be carried out at a time due to limited test positions and space available in reactors. Noninstrumented experiments are comparatively less expensive and enable larger numbers of specimens to be tested at a time. However, in noninstrumented experiments, control over experimental variables is absent and accuracy of data regarding the variables is lower than in instrumented experiments.

4. *Temperature Control*

The temperature of a test specimen can be maintained at the desired level by (i) electric heaters in conjunction with flowing gas streams around the sample, (ii) control of the flow rate and composition of the gas surrounding the specimen which is being heated due to nuclear heating, (iii) heat pipes designed to suit the experimental setup (Fig. 1), and (iv) an orifice to control the flow rate of the reactor coolant around the specimen.

5. *Temperature Measurement*

Temperature of the test specimen is determined in several ways. In the case of instrumented experiments, thermocouples are used to monitor the specimen temperature. Chromel–alumel thermocouples are used for temperatures below 1000°C, platinum–platinum–rhodium thermocouples are used for

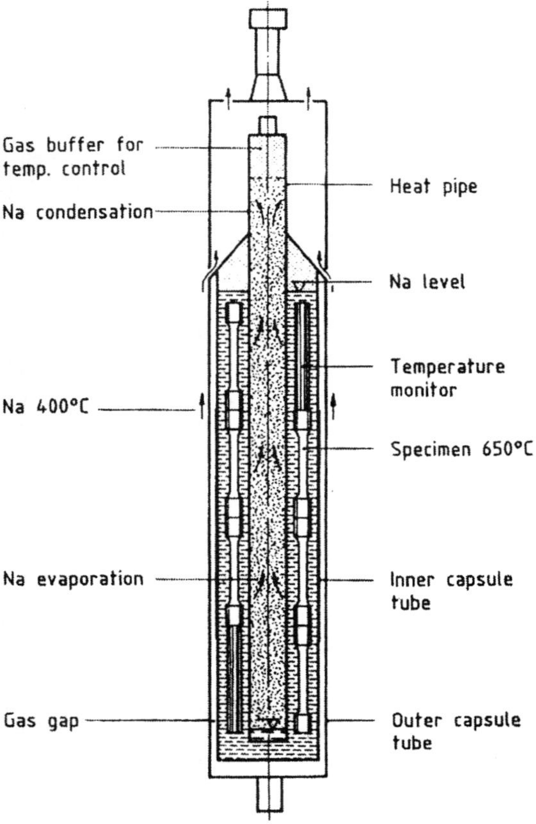

Figure 1
Irradiation capsule with heat pipe (reproduced by permission of Elsevier Science from *J. Nucl. Mater.*, 1986, **139**, 156–7).

temperatures below 1300°C, and tungsten–rhenium alloy thermocouples are used for temperatures below 2200°C. In the case of instrumented experiments on fuels, a rhenium sleeve is used to protect the thermocouple end portion inserted into annular fuel pellets. Temperature measurement accuracy is about 5°C in these cases. In the case of noninstrumented experiments, peak temperature sensors such as temperature expansion detectors (TED), Templugs, and melt wires are used to record the maximum temperature seen by the test specimen. These are loaded into the test capsule along with the test specimens and examined on completion of irradiation of the specimens.

The TED works on the principle of permanent plastic deformation of a tube due to expansion of a fluid inside the tube. The extent of deformation is calibrated against the peak temperature seen by the TED. Accuracy of temperature measurement using TED is ±5°C. Templugs use the change in hardness of the sensor on exposure to temperature as a measure of the temperature. En52 steel (UK standard) is one of the materials used in making Templugs. Sensors made of SiC are also used as indicators of the *average* temperature seen by the specimen. These work on the principle of change in lattice parameter due to the temperature in the reactor. Melt wires are made of a set of eutectic alloys designed to melt at temperatures separated by a small margin (e.g., 10°C). The examination of the wires after discharge from the reactor will show the peak temperature seen by the specimen. The effective temperature of the test specimen can be calculated based on the capsule geometry and dimensions, its position in the reactor core, rate of flow of coolant around the capsule, inlet and outlet temperatures of the reactor coolant, nuclear heating of the material of the specimen and the capsule, and the linear heat-generation rate of the surrounding fuel rods. The accuracy of temperature prediction by such a method is ±20°C.

6. *Measurement of Extension of Specimens*

The extension of the specimen can be measured online using (i) extensometers based on a system of levers, bellows, and linear voltage transducers (accuracy of measurement can be ±2 μm) or optical transducers based on the Moire fringe technique (accuracy of measurement can be ±0.5 μm), (ii) resonant cavity type of strain measuring system (accuracy of measurement about ±2 μm on 25 mm), (iii) gas flow gauges (these have the advantage that their performance is not affected by radiation and hence can be located in close proximity to the specimen. Sensitivity of the gauge can be 0.25 μm). For noninstrumented experiments, out-of-pile measurements are carried out intermittently by x-ray or neutron radiography Markgraf *et al.* (1995), measuring jigs or laser interferometry-based measuring apparatus (accuracy of measurement about ±0.5 μm). The use of radiographic measurement techniques may not be as accurate as on-line measurement methods, but it simplifies the experiment as the measuring equipment need not be built into the experiment and hence is cheaper. It also allows re-irradiation of the same sample for accumulating progressively higher neutron fluences without the need to open the capsules to carry out physical measurements and then reassemble them for further irradiation.

7. *Measurement of Fluence*

The neutron fluence seen by the specimens is measured using activation foils, which are placed along with the samples in the reactor. The selection of suitable foils with appropriate half-life and neutron energy spectrum sensitivity is an important factor in the design of the experiments. Iron, niobium, nickel, and copper are some of the materials used as foils for determining the fluence accumulated by the test specimens. Travelling in-core ionization chambers can be used for flux measurements during the experiments, if such a facility is available in the reactor.

8. Creep Test Techniques

Creep test techniques employed can be broadly grouped as experiments using (i) stress relaxation methods, (ii) pressurized tube samples, (iii) samples under constant applied load, (iv) measurements on actual reactor components and experimental components, or (v) charged particles to simulate reactor environment (these are not in-reactor experiments).

9. Stress Relaxation Tests

In the stress relaxation type of tests, samples are loaded to some initial stress and are placed in the reactor. The relaxation of applied stress occurs through deformation of the sample. This deformation is monitored to determine the remaining stress after brief spells of irradiation in the reactor. Stress relaxation and creep are related analytically by considering stress relaxation to be equivalent to creep under a stress, which decreases continuously to maintain constant strain. The samples used in relaxation tests can be in the form of loaded springs that are irradiated and, after a selected period, the remaining stress in the springs is measured by loading them and measuring the deflection. In this case, the samples are loaded nearly uniformly in pure shear. In another type of experiment, strips of the material to be tested are pre-irradiated to different levels of fluence and then are bent around a mandrel and held in position by a fixture (Fig. 2). The assembly is then irradiated for a further period, after which the fixture is released and the residual curvature of the strips, which is an indication of the creep strain suffered, is measured.

In a third type of stress relaxation experiment, samples in the form of beams subjected to uniform or four-point bending are loaded in tubes with loading mechanisms and irradiated Chin *et al.* (1978). The residual deformation of the beams on unloading after irradiation is measured and is an indication of the creep strain. The three-point bending method is also used with springs to exert the central load. Parabolic cantilever beams and cantilever beams with uniformly varying cross-section have been used to generate constant stresses over the sections of the specimens. In another method of testing, tubes with slits in which wedges are inserted are used to measure the stress relaxation (Fig. 3). The force required to pull the wedge out is an indication of the stress relaxation.

10. Tests Using Pressurized Tubes

In experiments using pressurized tube samples, the tubes are made from the material to be investigated and the dimensions are so selected that the stresses

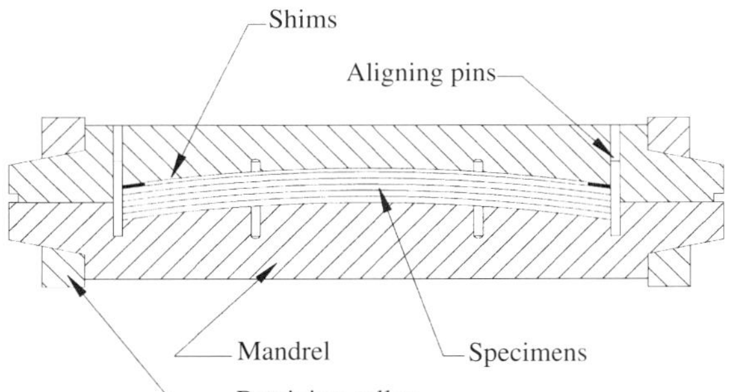

Figure 2
Beam stressing mandrel. Five beams are held at constant radius of curvature within the cavity (reproduced by permission of Elsevier Science fromJ. *Nucl. Mater.*, 1977, **64**, 254–64).

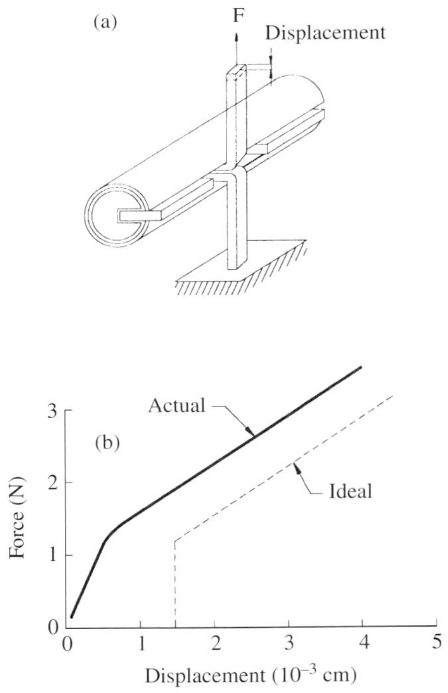

Figure 3
Slit tube measurement (a) schematic of the slit tube measuring device, (b) determination of "clamping force" from a displacement curve (reproduced by permission of Elsevier Science from *J. Nucl. Mater.*, 1978, **74**, 260–6).

induced in the tube walls, on pressurization of the tube, are of the same magnitude as the working stresses in the reactor components. The loading is biaxial and is more representative of the actual loading expected in service. This method allows the use of a large number of specimens with varying conditions to generate the creep data for the material. The tubes are usually less than 100 mm long and of small diameter, and will fit easily into the limited space available in the experiment carriers, which are loaded into the reactor. Welded end caps close the ends of the tubes, which are pressurized using an inert gas such as helium (Fig. 4). A calculated amount of water can also be placed into the tubes instead of gas, so that at the reactor operating temperature the water is converted to superheated steam and exerts the desired pressure. The tubes can be removed several times from the reactor, measured for dimensional changes, and returned to the reactor for further irradiation. Tubes with no pressure are also included in the experiments to determine the contribution of swelling/growth to the dimensional changes. Disadvantages of small, pressurized tube tests are that the diametral strain measurements are intermittent and likely to be inaccurate, control over the temperature of the specimens during test is limited, and the metallurgical variables are often different from those of large reactor components. Multiaxial loading of the tube is possible using a special design of tube with a central rod connected to one end cap, the other end of the rod being connected to bellows, which are in turn connected to the other end cap of the tube (Fig. 5). Different stress states are attained in this design by varying the pressure in the tube and bellows. Tests can also be done using long tubes made by joining a number of specimen tubes together by brazing or welding and inserting the tube into a test position in a reactor. The tube is pressurized using flowing water or other medium at the required temperature. At periodic intervals, the reactor is shut

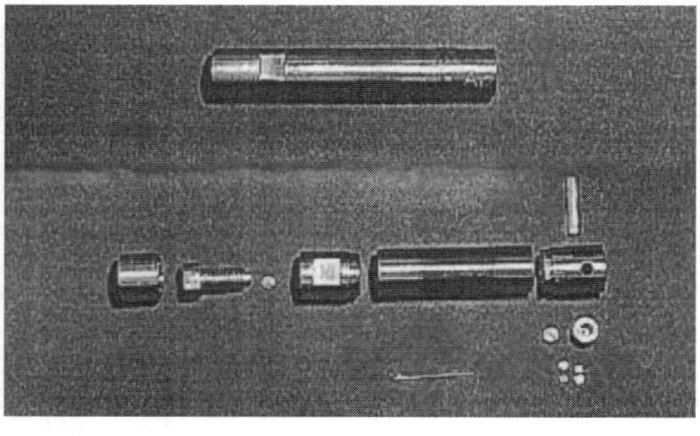

Figure 4
Pressurised capsule used for irradiation creep testing in a fast breeder test reactor (courtesy of P. V. Kumar).

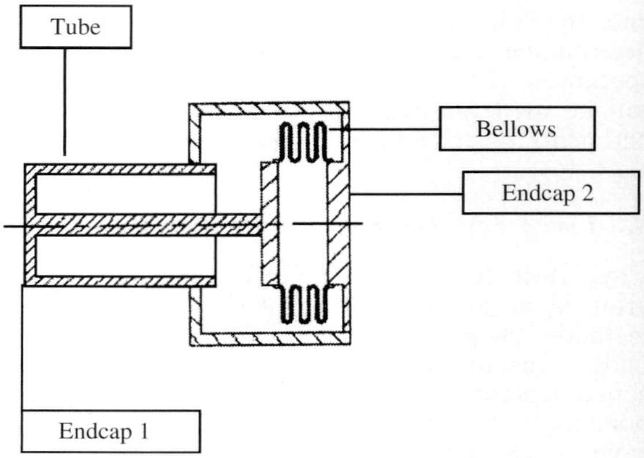

Figure 5
Schematic of a multiaxially loaded tube specimen.

down, the tube is emptied of the pressurizing medium, and the variation in internal diameter of the tube is gauged and recorded. In these experiments, the walls of samples can be made of varying thicknesses to generate different levels of stresses in the tube walls.

11. Tests Using Samples Under Constant Applied Load

Here, helical springs, pre-machined tensile creep test samples, and compression specimens are used. With helical spring samples, the applied stress is nearly pure shear. The extension of the springs, which are loaded by dead weights, is periodically evaluated by x-ray radiography of the capsules containing the specimens during reactor shutdowns. The design of the helix produces a strain magnification that allows precise measurements. Spring specimens are small in size, and simultaneous testing of a larger number of springs is possible. Disadvantages of this type of specimen are that the specimen form and metallurgical condition differ from those of the reactor components and there is limited control on the specimen temperature during the test.

Uniaxial stress is applied in the case of pre-machined samples and these are usually tensile creep tests. The tensile stress is applied either by a pneumatic bellows system subjected to external pressure, or by screw and nut systems in conjunction with feedback devices and load cells. Advantages of this type of test include precise control of stress and temperature even during reactor shutdown, which enables measurement of thermal creep during the shutdown period. Also, the specimens can be machined from the actual reactor component to represent the material accurately. However, only a few tests can be carried out at a time because of space requirements, complexity, and cost.

12. Measurements on Actual Reactor Components

Measurements on actual reactor components and experimental components provide data based on the actual operating conditions in a reactor. Pre- and post-irradiation strain measurements of pressurized heavy water reactor (PHWR) and steam generating heavy water reactor (SGHWR) pressure tubes and fast reactor (Rapsodie-France) subassembly wrappers have enabled irradiation creep deformations to be estimated. Fuel and clad material experiments are often carried out in the reactor using experimental fuel pins with the necessary instrumentation to monitor the changes in the specimen. Data generated from actual reactor components is directly applicable to reactor design. Disadvantages of this type of data generation are: limited control over experimental variables; only intermittent measurement of strain is possible and accuracy of measurements is limited; investigation of material variables is not facilitated; and irradiation creep component of measured strain has to be estimated from total strain using other data or models.

13. Simulated Experiments

Irradiation creep data can be obtained from simulation experiments in beams of accelerated particles. In these experiments, thin specimens stressed to a desired state are exposed to beams of electrons, protons, neutrons, deuterons, or alpha particles in an accelerator or other source. The test conditions for the specimen can be varied widely and also during a test. Strain can be measured continuously. Great care is required to minimize thermal transients in the specimen, as small temperature changes can result in large strain errors. Temperature control of specimens is quite complicated in these experiments; however, data regarding the stress, temperature, and fluence is known more precisely than in in-reactor experiments. These types of experiments are carried out to study the physical mechanisms of irradiation creep and other radiation effects and to screen candidate materials for reactor application. These experiments are generally of short duration and hence restricted to investigation of primary and transient irradiation creep.

14. Concluding Remarks

Various aspects involved in the in-reactor creep testing are briefly discussed in this article. Analyses have shown that steady-state creep rates determined by continuous and intermittent strain measurements are equivalent. Interrelation of creep data obtained by different test methods and applied stress systems is a complicated process and no general agreement exists among researchers.

See also: Heavy Water Reactor Fuel Design and Performance; Nuclear Reactor Materials: Irradiation Effects; Nuclear Applications: Zirconium Alloys; Fast Reactor Cores, Austenitic Steels for; Light Water Reactor Fuel Design and Performance

Bibliography

Chin B A, Straalsund J L 1978 The effect of a temperature change on the in-reactor creep of 20% cold worked AISI 316 stainless steel. *J. Nucl. Mater.* **74**, 260–6

Chin B A, Straalsund J L, Wire G L 1978 Irradiation creep of previously irradiated 304 stainless steel in a fast flux environment. *J. Nucl. Mater.* **75**, 274–81

Coghlan W A 1986 Review of recent irradiation creep results. *Int. Mater. Rev.* **31**(6), 245–57

Davis J S (ed.) 1972 Irradiation embrittlement and creep in fuel cladding and core components. In: *Proc. Conf. British Nuclear Energy Society*. British Nuclear Energy Society, London

Deverall J E, Watson H E 1973 Temperature control of irradiation experiments with gas-controlled heat pipes In: *Proc. Topical Meeting Irradiation Experimentation in Fast Reactors*. American Nuclear Society, USA, pp. 267–78

Gilbert E R, Chin B A 1981 In-reactor creep measurements. *Nucl. Technol.* **52** (2), 273–83

Gilbert E R, Kaulitz D C, Holmes J J, Claudson T T 1972 Fast reactor induced creep in 20% cold worked type 316 stainless steel In: *Proc. Conf. Irradiation Embrittlement and Creep in Fuel Cladding and Core Components*. British Nuclear Energy Society, London, pp. 239–51

Hall M M Jr 1978 Multiaxial in-reactor creep in the presence of significant swelling. *Trans. Am. Nucl. Soc.* **28**, 146–7

Journal of Nuclear Materials 1977 Issue devoted to irradiation creep testing. 65

Markgraf J F W, Tartaglia G P, Tsotridis G 1995 Irradiation devices and irradiation programmes at the high flux reactor (HFR) Petten for the investigation of irradiation behavior of stainless steels. *Nucl. Eng. Des.* **59**, 81–9

von der Hardt P, Rottger H (eds.) 1982 *Irradiation Technology: Proc. Int. Topical Meeting*, Grenoble, France. Reidel, Dordrecht, The Netherlands

Wire G L, Straalsund J L 1977 Irradiation-induced stress relaxation of previously irradiated 304 stainless steel in a fast flux environment. *J. Nucl. Mater.* **64**, 254–64

J

Jet Engine Materials

R. E. Schafrik and J. C. Williams

Aeronautical turbine engines have continually improved since their invention in the 1930s and their first application in aircraft in the 1940s. These engines were originally developed to power fighter aircraft, but were quickly adapted for transport aircraft propulsion in the 1950s. Today the commercial market for jet engines dwarfs the military market. Consequently, new constraints such as cost have a major influence on new product offerings and on the direction of the aircraft engine industry. New military fighter engine applications still demand improved performance, but other factors such as low radar and IR signatures have become quite important (Sect. 7).

Modern commercial jet engines, such as those that power the Boeing 777, benefit from improvements made in performance, durability, and reliability. Fig. 1 shows one of these engines, the GE90, in cut-away view. In this case, performance is defined by several parameters: thrust per unit engine weight, fuel efficiency, engine operability, engine control, exhaust emissions, and noise level. Better durability means longer times between necessary overhaul or repair, and slower deterioration of performance during operation. Reliability equates to fewer instances of unanticipated failures and greatly reduced in-flight shut-downs. Advances in structural materials, including processing, and in manufacturing methods for critical components, have facilitated many of these improvements and will continue to do so in the future. This review highlights ongoing changes in materials and their processing (M&P) that will advance the state-of-the-art of jet engine technology.

The NASA-funded Engine Propulsion Materials (EPM) Program, which supported the High Speed Civil Transport (HSCT) Program was terminated in 1999. Even though the HSCT airplane, designed to cruise at Mach 2.4–2.6, was not built, materials advancements made by the team of GE Aircraft Engines, Pratt & Whitney, and NASA Glenn Research Center fostered the next generation of turbine disk material, the next generation of single crystal, high-pressure turbine superalloy, and SiC/SiC ceramic matrix composites among other achievements.

Material advancements are necessarily driven by the overarching needs of the airplane designers, who in turn reflect the needs of the airplane operators, i.e., commercial airlines or the military. Since requirements differ somewhat between commercial and military users, the materials focus can vary by customer. The needs of military users are discussed in more detail below. Commercial users now dominate the product

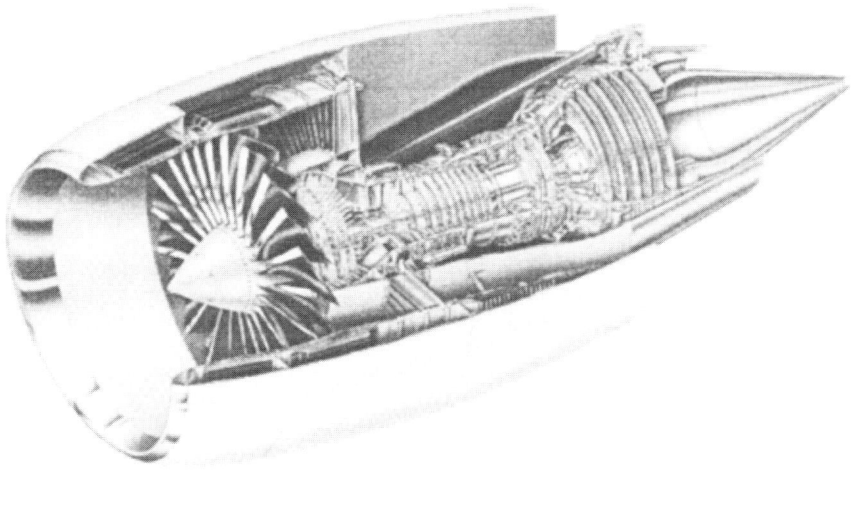

Figure 1
Cut-away view of the GE90 engine.

creation process and are very cost-sensitive. Their applications thus favor M&P that is inherently lower total cost while providing a high degree of durability and reliability, even if the material does not allow the absolute ultimate in performance.

Furthermore, commercial airplanes are subject to increasingly strict environmental rules governing engine emissions and noise levels which place additional constraints on these products. These regulations are gradually being extended to military airplanes as well. Many of the materials and process advances made during the 1990s benefited both military and commercial products. For example, the continued drive to increase operating temperatures to achieve increased performance and lower fuel consumption is common to both military and commercial applications. This article therefore focuses mainly on the M&P aspects of achieving these results.

There are M&P requirements for military engines that can be significantly different than for commercial engines, although the gap is narrowing. These military specific needs originate from the requirement for global deployment and wide-ranging missions. Engine characteristics required to ensure mission effectiveness and crew safety in combat include: low observable radar and infrared signatures, high reliability in severe operating environments (e.g., the deck of an aircraft carrier), and high performance during maneuvering. These needs have led to M&P developed to meet military requirements; over time, these advanced materials find their way into commercial engine applications as the technology becomes better understood, more affordable, and older material solutions fall short of requirements. The military services have also begun to accept commercial specifications, and have instituted development programs aimed at reducing cost and schedule risk by favoring M&P technologies already proven in commercial applications.

A modern, high by-pass ratio jet engine consists of two principal sections: the high-pressure system and the low-pressure system. The high-pressure system consists of the high-pressure compressor, the combustor, and the high-pressure turbine. The low-pressure system consists of the fan, the low-pressure compressor (also called the booster), and the low-pressure turbine.

Materials and processes used in the production of a jet engine vary considerably depending on the operating temperatures and stresses; these parameters also vary between sections of the engine. Fig. 2 shows a cross-section through a large commercial, high by-pass ratio turbo-fan engine, the CF6–80C2, which powers wide-body aircraft. The generic classes of materials used in each section are shown.

M&P highlights consisting of a description of the current state-of-the art and the expected trends for future products are discussed below for each section of the engine. Unique military requirements also are identified where these exist.

1. Fan and Booster Sections

The fan of a transport engine provides a significant fraction of the total thrust; the exact fraction depends on the by-pass ratio. The by-pass ratio can be simply defined as the ratio of the mass of the air that passes through the fan duct to that which passes through the core. The fan air is lower velocity air, which has better momentum coupling to the airplane and also encapsulates the high-velocity core air and reduces the noise. Consequently, high by-pass ratio engines are quieter and more fuel-efficient.

The challenge in fan design and construction is to reduce the mass of the fan section so that the benefit of increased by-pass ratio is not offset by increased engine weight. The use of higher strength titanium alloys for the rotor disk has helped in this regard, but processing higher strength alloys to achieve acceptable fracture toughness and fatigue crack growth properties requires substantial process development and manufacturing costs. The other essential requirement of all rotors (also called disks) in a jet engine is the use of design and materials technology to achieve an absolute minimum probability of catastrophic fracture during service. This is necessary because it is impractical to design a jet engine capable of containing all the pieces of a rotor if it fractures in service. It is this overarching requirement that creates the need for retaining the good fracture toughness and fatigue growth characteristics of the higher strength alloys. Another possible source of premature rotor fracture is the presence of a melt-related defect in the disk material. The detailed nature of these defects differ for titanium and nickel-based alloys, but their potential presence is the cause of the extraordinary care taken during melting and processing of so-called rotor grade materials in fracture-critical applications. Improved melting practices such as cold hearth melting of titanium alloys, and triple melting of titanium and nickel-based alloys, coupled with improved inspection methods such as multi-zone ultrasonic inspection, have significantly decreased the likelihood of such a catastrophic event during service.

Although nickel-based alloys are not generally used in the fan and booster, it is helpful to mention improved melt and inspection practices for both alloy classes together, as these are related to the integrity of all rotors in the engine. Because of the consequences of an uncontained failure, jet engine rotors must be designed and constructed using the best design practices and highest integrity materials.

It also has been necessary to change the construction and materials for the fan blades in very high by-pass ratio engines. Today, engines on the Boeing 777 use blades made from either carbon fiber-reinforced and toughened polymer composites or superplastically formed and diffusion-bonded titanium. One of the most critical requirements for the fan blade is that it must be capable of withstanding

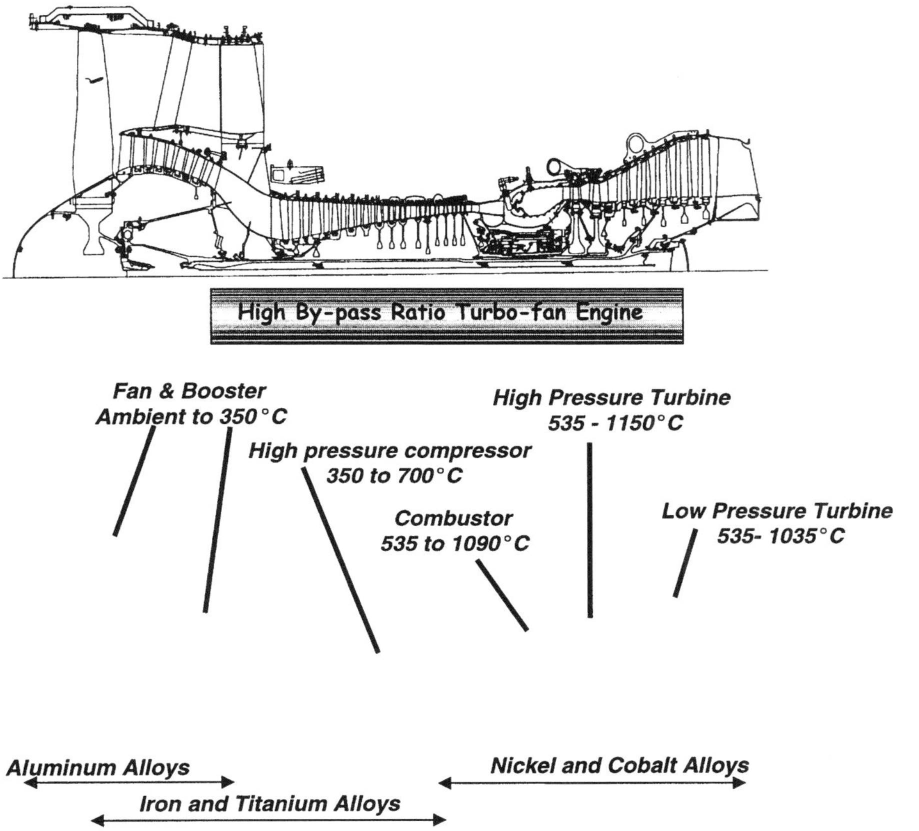

Figure 2
Cross-section through a high by-pass ratio turbo-fan engine, the CF6–80C2.

the impact of a large bird ingested during flight. Both types of blade construction mentioned above have been certified and are used in commercial service. An additional requirement for the fan is the containment of a liberated blade within the engine case and nacelle. This requires reinforcement of the fan case to prevent penetration by the liberated blade, and any fragments from adjacent blades that may be created as it attempts to exit the rotor. Containment becomes more difficult as the fan size and blade mass increase. The most common approach is to over-wrap the fan case with a lightweight, strong fiber such as Kevlar. This approach has proved effective and is currently used on most Boeing 747/Airbus 330 and Boeing 777 class engines.

There has been a concerted effort to improve the aerodynamic efficiency of all the airfoils in the engine from the fan to the low-pressure turbine. The use of three-dimensional solutions to the Euler equations for compressible fluid flow has enabled step improvement in airfoil efficiency. The resulting airfoil designs have new and quite different airfoil shapes, which pose new manufacturing challenges. Except for the largest fan blades, forging has been the common production method for fan, booster, and high-pressure compressor airfoils. The new complex shapes are considerably more difficult to make by forging. Progress in precision investment casting may result in the use of this method instead of forging, especially for nickel-based alloys used at the rear (hottest) section of the high-pressure compressor. Cast airfoils would also allow use of higher temperature capability nickel-based alloys than can routinely be forged.

2. *High-pressure Compressor*

In addition to cast airfoils, the other change in the high-pressure compressor has been higher operating temperatures. This has necessitated use of higher temperature capability materials. In a large transport engine, the high-pressure compressor consists of 10–14 consecutive rotating stages. The front stages are made from titanium alloys and the final stages from nickel-based alloys. The transition depends on the compression ratio and the compressor exit gas temperature, which can range from 500–700°C, depending on the engine design. Unlike the turbine section discussed in Sect. 4, there is no provision to cool the components in the compressor. Thus, these gas temperatures are identical to the material temperatures in the gas path during engine operation. Consequently, there has been a need to introduce higher temperature capability alloys to keep pace with the increased operating temperatures of more fuel-efficient engines. For the titanium alloy stages, this has led to the use of improved processing and higher temperature capability alloys such as Ti–6Al–2Sn–4Zr–2Mo(+Si) and IMI-834. For the nickel-based stages, powder metallurgy disk alloys such as IN-100 and R-88DT have been introduced. The introduction of powder metallurgy disks increases the cost and creates the need for further understanding of processing effects on properties and materials behavior. Powder metallurgy disks have better uniformity of structure and composition; this is an additional benefit if they are processed to achieve coarser grains for improved creep resistance. Fig. 3 shows the benefits of coarse grain processing on creep strength. While the introduction of powder metallurgy nickel disks has been successful, there is general agreement that a higher temperature alloy produced by conventional ingot metallurgy would offer a significant cost benefit.

3. *Combustor*

The combustor can be thought of as a small pressure vessel as the outer portion of the combustor (the combustor case) must withstand both pressure and temperature. The levels of these parameters necessitate the use of high-temperature nickel or cobalt-based alloys. The combustor consists of this case, a liner that shields the case from the direct flame, and fuel nozzles that spray the fuel into the combustion zone. There are numerous other individual parts in a typical combustor, but its functionality is determined by these three major components. Combustor design changed considerably during the 1990s in response to requirements for lower emissions and higher fuel economy. Several designs now dominate, each with its unique materials requirements. Perhaps the most significant change in materials has been the incorporation of improved cobalt-based alloys for the liner, the replacement of a fabricated case by an investment casting that is subjected to hot isostatic pressing (HIP) to assure mechanical properties, and the use of cast pieces made from nickel-based turbine blade alloys.

The temperatures of combustion gases may exceed 1500°C, considerably hotter than the incipient melting temperature of metal parts in contact with these gases. Successful operation depends on supplying cooling air to the combustor liner: air from the high-pressure compressor is bled into the combustor and impinged on the liner to reduce part temperatures as dictated by strength and oxidation resistance requirements. This cooling air is unavoidably mixed with the fuel–air mixture and thus detracts from the ability to control the exact fuel–air ratio desired for emission control. It would therefore be desirable to use material for the combustor liner that has a higher temperature capability than current nickel and cobalt alloys such as Hastelloy X and Haynes 188. Ceramic

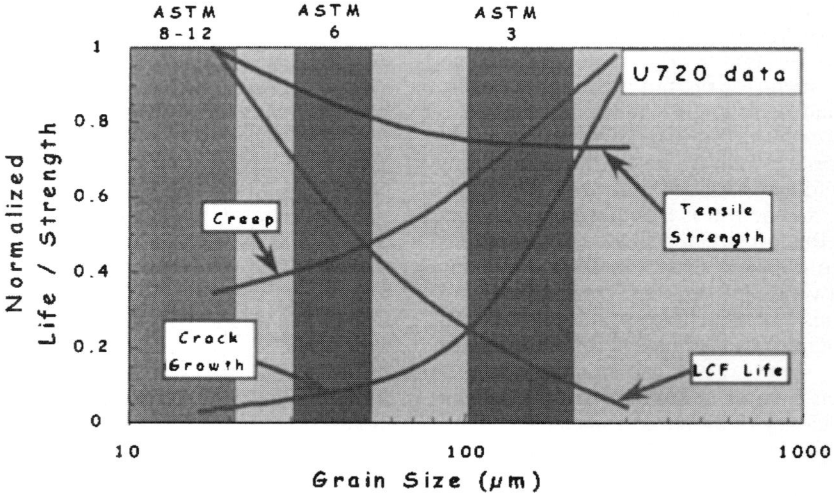

Figure 3
Benefits of coarse grain processing on creep strength.

composites with SiC matrix and SiC fiber reinforcement have the promise of fulfilling this need, but are currently not mature enough for introduction. The improvements required before these materials can be introduced include: manufacturing capability, reproducibility of properties, cost, and development of a suitable protective coating for use in the mildly reducing environment of the combustor.

4. *High-pressure Turbine*

In many regards, the high-pressure turbine is the heart of a jet engine. The temperature capability of the turbine and its efficiency determine, to a significant degree, the performance characteristics of the engine. For example, more thrust can be obtained from an existing engine if the turbine inlet temperature is increased. Thus there is an incentive to continue to raise the turbine inlet gas temperature.

The turbine inlet gas temperatures of modern engines are typically in the range 1350–1450°C. The airfoil materials used for the high-pressure turbine are cast nickel alloys that are directionally solidified to create a single crystal structure. Since the inception of single crystal turbine airfoils in the late 1970s, three additional generations of alloys have been developed—each having a higher temperature capability, but also being more difficult to cast as single crystal blades. The temperature capability of these new blade alloys is shown in Fig. 4. Even the most advanced turbine airfoils cannot operate at these high temperatures for extended time periods. However, metal temperatures can be maintained at satisfactory levels through a combination of improved cooling designs and thermal barrier coatings (TBCs).

Just as in the case of the combustor liner, compressor bleed air cools the high-pressure turbine airfoils. This cooling air can be thought of as "overhead", inasmuch as it is air pumped by the engine that does not directly contribute to the propulsion force of the engine. The cooling schemes employed for turbine airfoils are conceptually simple but difficult to execute. Cooling of the inner surface of the airfoil is achieved through convection—intricately cast serpentine passageways form internal cooling circuits for cool compressor bleed air. In addition, a thin film of cooling air is formed over the outer surface of the airfoil by compressor bleed air pumped from the hollow interior of the cast airfoil through an array of small, shaped holes extending to the external surface of the airfoil. The details of the cooling hole array depend on the particular application, but these holes are usually drilled through the wall using a high-energy, solid state laser.

In addition, a TBC layer is applied to the airfoil to form an insulating layer between the metallic component and the hot gases of combustion. This allows constant metal operating temperatures at reduced cooling air use or increased gas temperatures at constant metal temperature and cooling flow. In

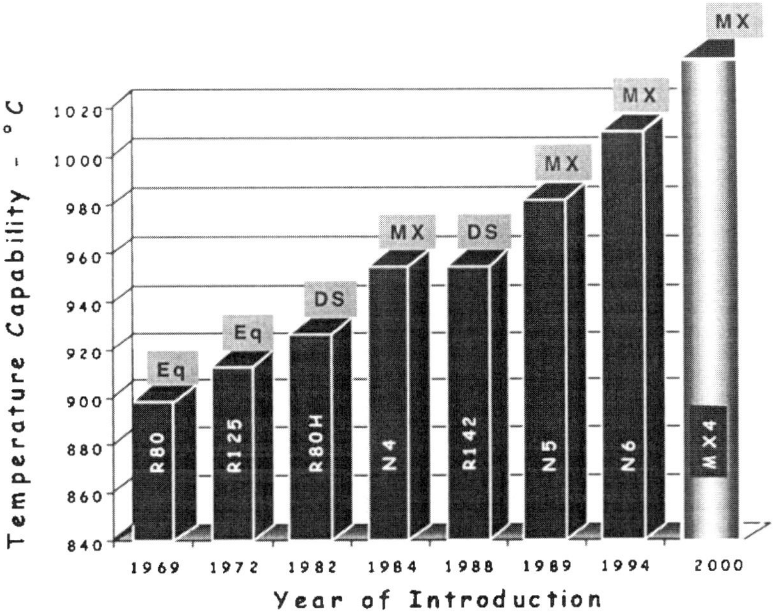

Figure 4
Temperature capability of new blade alloys introduced for use in high-pressure turbines.

either case, use of these coatings enables improved performance in terms of fuel consumption or thrust. TBCs are ~125μm thick ceramic coatings typically consisting of yttrium-stabilized zirconia (YSZ), i.e, ZrO_2 containing 5% to 10% Y_2O_3.

Ceramic TBCs do not provide environmental resistance to the superalloys. An additional metallic layer is therefore placed between the substrate and the TBC. This coating may be an overlay alloy, such as MCrAlY, where M is an element(s) selected from the group Fe, Ni and Co, or a diffusion nickel aluminide, or a platinum aluminide. As an added requirement, these coatings must serve as an interface layer between the TBC and the airfoil alloy.

These coatings are applied by a variety of processes. These are determined by the airfoil configuration, coating composition and thickness, cost, and required quality level. Methods include pack cementation, chemical vapor deposition, plasma spray, low-pressure plasma spray, physical vapor deposition, and sputtering.

5. *Low-pressure Turbine*

The increased temperatures in the low-pressure turbine and the increase in size of the low-pressure turbine due to increasing fan size (by-pass ratio) have created some new requirements. Among these are the need for higher temperature airfoil materials and lighter weight materials with similar temperature capability to those materials currently in use. Modern transport engines typically use solid (uncooled) cast nickel-based alloy airfoils. Depending on the temperature requirement, these can be conventionally cast (equi-axed grains), directionally solidified (elongated grains), or single crystals. The processing benefit obtained through controlled solidification to produce elongated grains or monocrystals is also illustrated in Fig. 3.

Because of cost, there is a strong preference for conventionally cast airfoils. There is a new class of material based on titanium and aluminum known as Gamma TiAl. Airfoils cast from this material have a temperature capability approximately equivalent to conventionally cast nickel-based alloys, but have about half their weight. The total weight benefit obtained from introduction of two stages of TiAl airfoils into the low-pressure turbine of a B777 class engine is of the order of more than 100 kg. By engine industry standards, this is a very large and attractive number. The barriers to the introduction of TiAl are primarily cost, but this becomes acceptable once a significant volume of these airfoils has been produced. There are some minor technical concerns related to the low ductility of this material, but these have been largely addressed by design engineers and demonstrated by successful factory engine testing.

6. *Augmentor (Military Only)*

Military fighter engines achieve higher thrust-to-weight ratios than transport engines by using an augmentor (also called an afterburner). The augmentor section, which is part of the exhaust nozzle, injects additional fuel in the augmentor. This fuel is ignited to produce additional propulsive force (thrust). The augmentor nozzle also has a variable geometry that employs movable flaps and seals to create a variable nozzle area. When the augmentor is lit, the flame impinges directly on these flaps and seals. This creates a need for flap and seal materials with very high temperature capability. The fairly recent use of cast pieces for flaps and seals has allowed longer lifetimes to be achieved between replacement, but the actual times are still quite short. A higher temperature capability material could allow longer life and lower maintenance costs. Ceramic matrix composites (CMCs) have been shown to be excellent candidates for augmentor flap and seal applications. CMC manufacturing technology has advanced appreciably for this application; the increase in durability offsets a cost increase for this new material system. Consequently, the latest fighter engines are successfully using ceramic composite flaps and seals.

7. *Other Special Military Requirements*

Another requirement for military engines that does not exist for commercial engines is the need to reduce the radar signature from the inlet and the IR

Table 1
Typical mechanical properties of selected titanium alloys.

Alloy	Tensile properties at 20°C			
	UTS (MPa)	YS (MPa)	Elongation (%)	RA (%)
Ti–6Al–4V	910	840	10	20
Ti–6Al–2Mo–4Zr–2Sn (+Si)	895	825	10	25
Ti–5.8Al–4Sn–3.5Zr–0.7Nb–0.5Mo–0.3Si (IMI-834)	950	1050	12	20
Ti–6Al–2Sn–4Zr–6Mo	1090–1240	1020–1090	6–14	13–42
Ti–5Al–2Sn–2Zr–4Mo–4Cr (Ti–17)	1105–1240	1035–1075	8–15	20–45

Table 2
Mechanical properties of combustor alloys.

	Tensile properties at 20°C		
Alloy	UTS (MPa)	YS (MPa)	Elongation (%)
Hastelloy X	970	464	62
HA 188	785	360	43

Table 3
Mechanical properties of disk alloys.

	Tensile properties at 20°C			
Alloy	UTS (MPa)	YS (MPa)	Elongation (%)	RA (%)
IN718	1435	1185	21	
Waspaloy	1275	795	25	
Udimet 720	1635	1215	18	17
MERL 76/In 100	1585	1025	20	20
R95	1650	1300	18	21
R88DT	1550	1150	21	NA

signature from the exhaust nozzle. The well-known stealth fighter and bomber aircraft have demonstrated the feasibility of achieving lower signatures through a combined use of special materials and design methods. Material possibilities include radar-absorbing and low emissivity coatings. Coatings offer flexibility in the design of the underlying structure, but durability can be an issue. Materials possessing the intrinsic capability to absorb radar or have low emissivity offer attractive solutions. Work is continuing on these technologies.

8. *Typical Material Properties*

Some typical properties of titanium and nickel-based alloys are shown for reference in Table 1, Table 2 and Table 3. More complete data on these and other alloys can be found in many of the references in the bibliography.

9. *Summary*

This article has addressed the reasons for changing materials and process requirements in jet engines and has outlined the nature of these changes. Detailed accounts of advances in the metallurgy and processing of titanium and nickel alloys can be found in the references in the bibliography.

There has been a major shift in emphasis in materials technology since the mid-1980s. The major changes are:

- A transition from military focus to commercial focus in terms of establishing a market for new materials technologies.
- Commercial jet engines are developed with the intent of achieving performance parity at lowest cost.
- The strong driving force to achieve higher temperatures continues because it enables better fuel economy and longer range—this offers significant challenges to the materials community as nickel-based superalloys reach their inherent physical limitations.
- Powder metallurgy disk alloys are now commonly used in commercial engines to allow higher operating temperatures and better fuel consumption characteristics; even higher temperature disks will be required in the future.
- Thermal barrier coatings are commonly used on high-pressure turbine blades to extend their life and reduce the amount of cooling air required.
- High by-pass ratio fans with high aerodynamic efficiency have necessitated the introduction of both complex-shaped polymer composite and hollow titanium fan blades.
- Advances in process technologies are as important as material advancements. For instance, the need to significantly reduce uncontained rotor failures was satisfied by the introduction of new and improved melting technologies for both nickel and titanium disk alloys, as well as developing more capable non-destructive inspection technology to minimize the possibility of using materials in rotors that contain melt defects.
- The need to reduce emissions has changed the design and construction of combustors and created the need for better liner materials. Ceramic matrix composites are attractive in this regard, but require additional development for introduction.
- The need to reduce inlet and exhaust signatures in military engines has created a special set of materials technologies.

Bibliography

Antolovich, Stusrud, MacKay, Anton, Khan, Kissinger, and Klarstrom (eds.) 1992 *Superalloys 1992*. The Minerals, Metals & Materials Society (TMS), Warrendale, PA

Boyer, Welsch, Collings (eds.) 1994 *Materials Property Handbook–Titanium Alloys*. American Society for Metals, Materials Park, OH

Bradley E F (ed.) 1988 *Superalloys: A Technical Guide*. American Society for Metals, Materials Park, OH

History Office, Aeronautical Systems Division 1986 An Encounter Between the Jet Engine Inventors, Sir Frank Whittle and Dr. Hans von Ohain, 3–4 May 1978. Wright-Patterson Air Force Base, OH

Kissinger, Deye, Anton, Cetel, Nathal, Pollock, and Woodford (eds.) 1996 *Superalloys 1996*. The Minerals, Metals & Materials Society (TMS), Warrendale, PA

Lee J, Schafrik R E (eds.) 1998 Modern manufacturing. In: Kreith F (ed.) *CRC Handbook of Mechanical Engineering*. CRC Press, Boca Raton, FL

Mattingly J D, Heiser W H, Daley D H 1987 *Aircraft Engine Design*. American Institute of Aeronautics and Astronautics, Washington, DC

Nathal, Darolia, Liu, Martin, Miracle, Wagner, and Yamaguchi (eds.) 1997 *Structural Intermetallics 1997*. The Minerals, Metals & Materials Society (TMS), Warrendale, PA

National Research Council 1996 *Coatings for High-temperature Structural Materials: Trends and Opportunities*. National Academy Press, Washington, DC

National Research Council 1998 *Visionary Manufacturing Challenges for 2020*. National Academy Press, Washington, DC

Sims C T and Hagel W C (eds.) 1972 *The Superalloys*. J Wiley, New York

Wilson D G 1984 *The Design of High-efficiency Turbomachinery and Gas Turbines*. MIT Press, Cambridge, MA

L

Light Conversion: Organized Lanthanide-containing Molecular Systems

J.-C. G. Bünzli and C. Piguet

Modern technology relies heavily on information, data transfer, and signaling processes. One of the most convenient vectors to convey signals is light because it travels almost instantaneously and can easily reach regions of a complex edifice that are not accessible to particular or molecular messengers. Moreover, light carries a substantial amount of energy which can be transmitted to the material, inducing a specific triggering or switching effect. Since light is easily detected by very sensitive devices and techniques, signaling or analytical processes based on the emission of light are among the most sensitive known: under special conditions, the sensitivity of a single molecule detection has been frequently achieved.

The lanthanides, or 4f elements, which extend from lanthanum (La, atomic number 57) through lutetium (Lu, 71) have always played a prominent role in lighting and light conversion technologies. Indeed, the first industrial use of a rare-earth-containing material—the rare earth group of elements comprises the lanthanides plus scandium (Sc, 21) and yttrium (Y, 39)—dates back to 1891 when the Austrian entrepreneur Karl Auer von Welsbach reported his patent for cheap incandescent mantles composed of 99% thorium oxide and 1% cerium oxide. It is said that until 1935, when electric light bulbs became ubiquitous, five billion Auer mantles had been sold. Just after the turn of the twentieth century, the French chemist G. Urbain discovered a powerful red-emitting luminescent material—yttrium oxide doped with a few percents of trivalent europium, $Y_2O_3 \cdot Eu^{III}$—with a brilliant red–orange emission which is still used today in fluorescent lamps and television and display screens (see *Rare Earth Phosphors: Fundamentals and Applications*). In the early 1960s laser action (**l**ight **a**mplification by **s**timulated **e**mission of **r**adiation) was discovered and neodymium gave birth to one of the most widely used optical devices, the YAG laser (yttrium aluminum garnet, $YAlO_3 \cdot Nd^{III}$).

Present-day technologies see the development of many lanthanide-containing materials with light emitting and/or conversion properties, ranging from amplifiers for fiber-optic communication to electroluminescent devices (*Electroluminescent Phosphors*), x-ray intensifying screens (*Medical Imaging: Cerium-activated Scintillators*) or materials for bioassays (*Bioassays: Luminescent Materials*). Most of these applications imply an energy conversion step in which the luminescent material transforms electrical energy, high energy radiation (x-rays or electron beams, for instance) or UV light into visible light. In some instances, two infrared photons are absorbed and the resulting energy is also transformed into visible light (up-converters). All these converting materials are named "phosphors" when they are inorganic compounds (mainly oxides or halides) while no specific nomenclature exists for molecular converters using organic receptors, although "organic phosphors" is sometimes used in the field of electroluminescent displays. The denominations "luminescent probes" or "luminescent labels" tend to prevail in biological and medical assays. A large proportion of light-converting materials belongs to the class of inorganic phosphors (see Fig. 1) but there is a

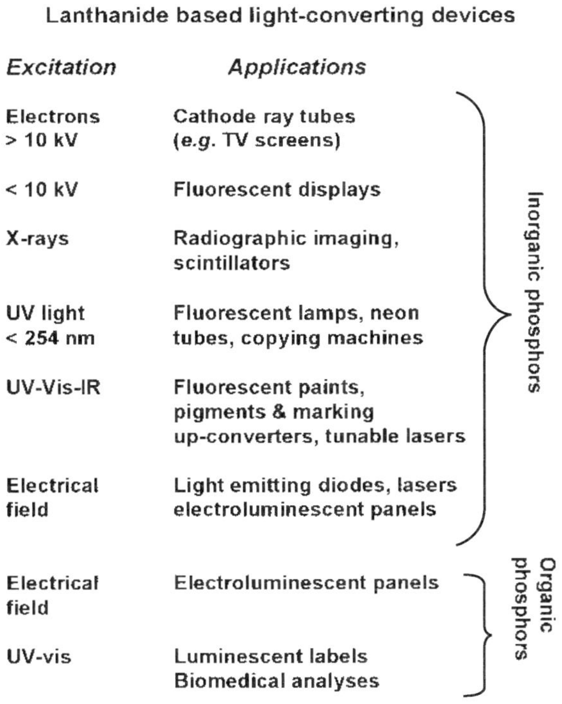

Figure 1
Classification and main uses of lanthanide-based light-converting devices.

growing interest in molecular devices built from carefully tailored organic ligands.

This is mainly due to the emergence of time-resolved fluoroimmunoassays in the mid 1980s and to other biological or medical applications, such as DNA- and RNA-labeling, luminescent responsive systems for *in vivo* trace analysis, or cancer phototherapy. On the other hand, organic ligands for the selective complexation of lanthanide ions have also come under the scrutiny of developers of high-performance optical materials and systems (tunable lasers, light-emitting diodes, electrooptical switches). This article outlines the principles to be followed for the efficient design of organic receptors able to encapsulate trivalent lanthanide ions and to enhance their photophysical properties, particularly their luminescence in view of their use as analytical labels. It also presents selected examples of highly luminescent systems.

1. Exciting Trivalent Lanthanide Ions

1.1 4f Transitions

Most of the electronic transitions of the trivalent lanthanide (Ln^{III}) ions involve solely a redistribution of electrons within the 4f subshell. Electric dipole selection rules forbid such transitions but these rules may be relaxed by several mechanisms. An important one is the coupling with vibrational states, where a molecular vibration temporarily changes the geometric arrangement around the metal ion and, therefore, its symmetry. The strength of the *vibronic* coupling of *vib*rational and electr*onic* parts of the total wavefunction depends on the intensity of the interaction between the metal orbitals and the surrounding ligands. Since Ln^{III} ions have [Xe]$4f^n$ electronic configurations in which the 4f electrons are shielded from external interactions by the $5s^25p^6$ filled subshells, the vibronic interaction remains quite weak for these ions.

The immediate consequence is that the f–f transitions are extremely narrow (atom-like) and faint, the molar absorption coefficients being usually smaller than $1M^{-1}cm^{-1}$. Therefore, even if many lanthanide-containing compounds display a good quantum yield (the ratio between the number of emitted and absorbed photons), direct excitation of the Ln^{III} ions rarely yields highly luminescent materials. Indirect excitation (sensitization) therefore prevails and is achieved either by the lattice in inorganic materials or by the ligands attached to the metal ion in molecular compounds (Fig. 2). Three steps are involved in both cases: (i) light harvesting by the host or ligands; (ii) energy transfer onto the metal ion; and (iii) metal-centered emission of visible light.

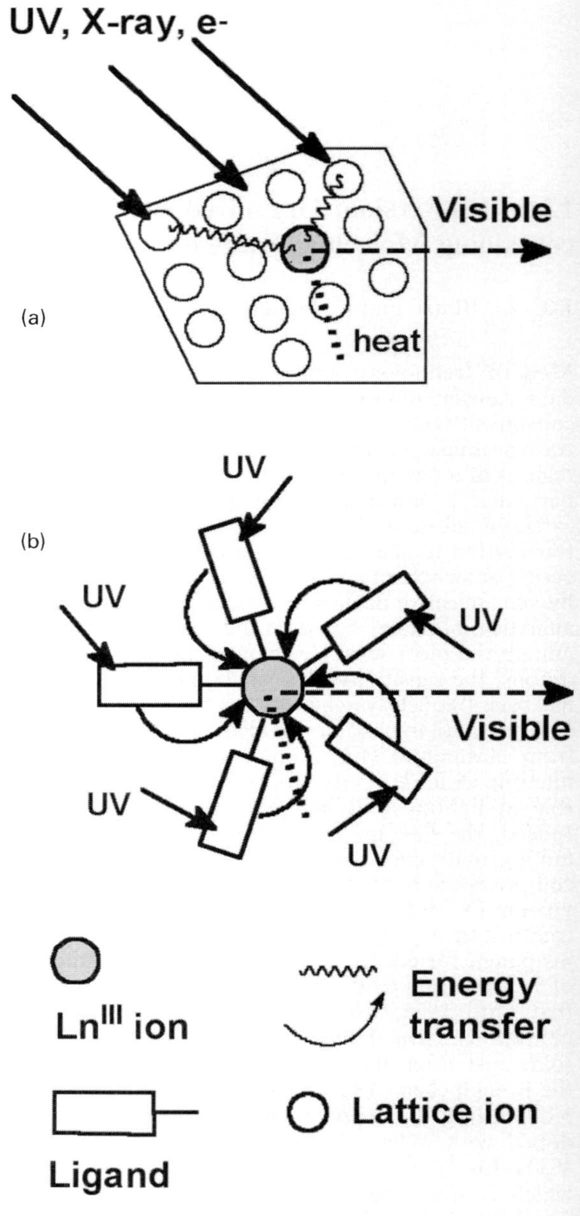

Figure 2
Light conversion by lanthanide-containing devices based on sensitization of the Ln^{III} ions: (a) by the lattice of an inorganic material; (b) by the ligands in a molecular edifice. The energy of nonradiative de-excitation is usually transformed into heat via internal vibrations in the sample.

1.2 Other Transitions

Some lanthanide ions, such as divalent Ln^{II} or trivalent Ce^{III} and Tb^{III}, display $4f^n \rightarrow 4f^{n-1}5d^1$ transitions which are broader and somewhat more intense. In addition, the ions which can be either reduced (e.g., Sm^{III}, Eu^{III}, Yb^{III}) or oxidized (Ce^{III}, Tb^{III}) present, respectively, ligand-to-metal (LMCT) or metal-to-ligand (MLCT) charge-transfer bands with molar absorption coefficients around 10^2–$10^3\,M^{-1}\,cm^{-1}$. Some of these transitions are used in light-converting devices, particularly the blue 5d–4f emission of Eu^{II}, but they will not be specifically discussed in this article.

1.3 Luminescent Properties of the Trivalent Ions

According to modern rules (Bünzli 1989), the luminescence of the Ln^{III} ion is a phosphorescence process since the excited and ground states have different spin quantum numbers. Due to the spin rearrangement, the emission is slower than fluorescence which occurs between levels having the same spin state. However, the latter designation is often used as a generic term and many books and articles describe lanthanide ion fluorescence. The intensity of the lanthanide-centered luminescence depends on the chemical (inorganic or organic) environment of the cations, as experienced by Sir William Crookes as early as the mid 1880s while studying the spectroscopy of yttrium, samarium, and other rare earths with the help of the vacuum discharge tube bearing his name. Symmetry is particularly important and, in the case of Eu^{III} luminescence for instance, point groups of symmetry with inversion centers will result in electric dipole transitions (usually the most intense in the spectra) being forbidden. Nonradiative deactivation processes have also to be avoided.

One important deactivation path goes through dissipation of the excited state energy by vibrational modes. Therefore, the larger the gap ΔE between the excited and ground state, the less likely such a deactivation path will be important. As a consequence, Eu^{III} ($\Delta E \approx 12\,000\,cm^{-1}$), Gd^{III} ($\Delta E \approx 32\,000\,cm^{-1}$), and Tb^{III} ($\Delta E \approx 14\,000\,cm^{-1}$) are expected to be the most luminescent Ln^{III} ions, which is the case, although Gd^{III} is very difficult to excite due to the high energy of its excited state. Other luminescent ions include Nd^{III} ($\Delta E \approx 5\,000\,cm^{-1}$), Sm^{III}, Dy^{III} ($\Delta E \approx 8\,000\,cm^{-1}$), Er^{III}, Tm^{III} ($\Delta E \approx 6\,000\,cm^{-1}$), and Yb^{III} ($\Delta E \approx 10\,000\,cm^{-1}$). The main incentives for using lanthanide-centered luminescence are the easily recognizable line-like emission spectra of these ions, the long lifetimes of the excited levels which range between microseconds and milliseconds, and the large apparent Stokes shifts (i.e., the energy difference between the excitation and emission bands) which often result in a good separation of the metal-centered and ligand- or medium-centered emission.

1.4 Antenna Effect

Turning to coordination compounds, in which a central metal ion is surrounded by one or several ligands, irradiation with UV light may result in three types of emissive processes: (i) fluorescence and phosphorescence from the ligand itself, perturbed by the presence of the metal ion; (ii) luminescence from direct excitation of the metal ion, perturbed by the binding of the ligands; and (iii) metal-centered emission due to nonradiative intramolecular energy transfer from the ligand excited states to the cation (Fig. 2(b)). The last mechanism was discovered by Weissman (1942) when exciting appropriate functional groups of various chelates and is termed the "antenna effect". Due to the long lifetime of the Ln^{III} excited levels, it is generally accepted that the ligand-to-metal energy transfer occurs mainly via the lowest triplet state of the ligands through a Dexter-type exchange mechanism, although several other processes may also take place (*vide infra*).

1.5 Requirements for Designing a Chemical Environment for a Luminescent Ln^{III} Label

The more important energy transfer, emission, and degradation processes that occur in a lanthanide-containing edifice are depicted in Fig. 3. The overall luminescence efficiency of the metal-centered luminescence is influenced by the energy-harvesting properties of the ligand, the population rate of the $^3\pi\pi^*$ state, k_{isc}, the rate of the ligand-to-metal energy transfer, k_t, and the various rates of radiative and nonradiative de-excitation of the singlet, triplet, and Ln^{III} excited states. In addition to easy de-excitation of the Ln^{III} excited states by high-energy vibrations, quenching through low-lying ligand-to-metal charge-transfer states or photoinduced electron transfer from the excited singlet states of the ligand may also occur, especially for Eu^{III}, while Tb^{III} is easily amenable to back-transfer processes where energy

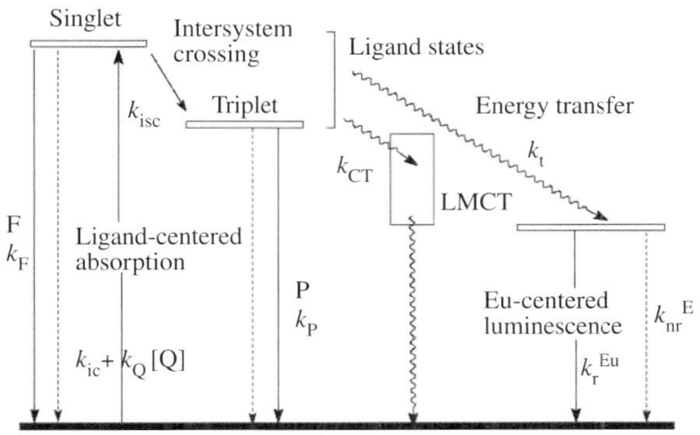

Figure 3
Main energy migration paths in a lanthanide-based light-converting device (F: fluorescence, P: phosphorescence, Q: quencher, LMCT: ligand-to-metal charge-transfer, ic: internal conversion, isc: intersystem crossing, r: radiative, nr: nonradiative).

migrates back from the metal ion to the ligands. As a consequence, several stringent requirements must be met to design a favorable protective environment for the Ln^{III} ions.

At this stage, one has to realize that the overlap integral between the 4f and ligand wavefunctions will influence both k_{isc} and the yield of the ligand-to-metal energy transfer. A somewhat enhanced intersystem crossing is usually observed in presence of a 4f metal ion because of the large spin–orbit coupling of the latter. On the other hand, the overlap integral between the 4f and $^3\pi\pi^*$ states is very sensitive to slight variations either in the position or intensity of the broad $^3\pi\pi^*$ emission since the 4f absorption spectrum is composed of very narrow lines. This means that prediction of energy transfer yields is extremely difficult. Some rationalizations have however been made. For instance, the quantum yields of the metal-centered luminescence upon ligand excitation has been determined for a series of similar compounds derived from substituted calixarenes: a sizable energy transfer is observed when the energy difference ΔE between the 0–0 transition of the $^3\pi\pi^*$ state and the luminescent Eu^{III} or Tb^{III} ^5D_J state is around $3500 cm^{-1}$. For Tb^{III} probes, phonon-assisted back transfer ($^5D_4 \rightarrow {}^3\pi\pi^*$) occurs easily, especially when $\Delta E < 1500 cm^{-1}$, while low-lying LMCT states have to be avoided for Eu^{III} probes (Steemers *et al.* 1995). These have to be considered as guidelines rather than rules since numerous other factors affect the quantum yield.

The decay of the metal-ion excited state is governed both by the radiative rate constant k_r and by nonradiative processes described by k_{nr}. For an isolated molecule, i.e., in the absence of concentration and of collisional intermolecular quenching processes, the nonradiative relaxation between various J states occurs mainly by interaction of the electronic levels of the Ln^{III} ions with the vibrational modes of their chemical environment. The efficiency of these processes is a function of the energy gap ΔE discussed in Section 2.3 and of the energy of the oscillator. High-energy vibrations, such as O–H, N–H, or C–H stretching modes, are therefore particularly efficient in deactivating excited Ln^{III} states. Comparing the experimental rate constant of deactivation in both hydrogenated (e.g., H_2O) and deuterated (e.g., D_2O) solvents allows a determination of the number of high-energy oscillators directly bound to the metal ion; closely diffusing oscillators in the second coordination sphere of the metal ion may also contribute to this nonradiative process (Beeby *et al.* 1999). As stated above, the metal-ion excited state may also be deactivated by thermally activated transfer processes on to states of other electronic configurations (e.g., LMCT for Eu^{III}). Taking into account the various deactivation processes, the observed decay rate constant (obtained from the measured lifetime τ^{Ln}) can be written as:

$$k_{obs}^{Ln} = (\tau^{Ln})^{-1} = k_r^{Ln} + k_{nr}^{Ln}(\mathrm{vibr}) + k_{ct}(T)$$

If the thermally activated process does not play any role at low temperature, $k_{ct}(T)$ may simply be estimated from the measurement of the lifetime at low (77 or, better, 10 k) and room temperature:

$$k_{ct}(T) = (\tau_{10\,K}^{Ln})^{-1} - (\tau_{298\,K}^{Ln})^{-1}$$

Keeping in mind all these energy transfer and energy degradation processes, a convenient receptor for a lanthanide-based luminescent material or analytical sensor should meet the following requirements:

Chemical features

(i) Provides high thermodynamic stability.
(ii) Ensures kinetic inertness (long-term stability).
(iii) Fulfils the coordination requirement of the Ln^{III} ions.
(iv) Does not bring high-energy oscillators (OH, NH, CH groups) in the vicinity of the metal ion and guarantees a rigid coordination environment.
(v) Can be easily synthesized and derivatized.

Spectroscopic features

(i) Bears chromophores with large molar absorption coefficients.
(ii) Has an efficient intersystem crossing (that is, the population of the $^3\pi\pi^*$ state is large).
(iii) Possesses a triplet state with a 0–0 transition ca. $3500 cm^{-1}$ above the metal excited state.
(iv) Does not induce low-lying LMCT or MLCT states.
(v) Minimizes other nonradiative processes (i.e., maximizes the quantum yield of the metal-centered luminescence).

Biological features (whenever relevant)

(i) Features chemical substituents that can be coupled easily and mildly with biological material.
(ii) Warrants low nonspecific binding to biological molecules (i.e., no affinity to surfaces or proteins).
(iii) Does not affect biological activity.
(iv) The probe must be water-soluble.
(v) In case of *in vivo* use, the probe must be easily excreted.

1.6 Alternative means of Excitation

The sensitization mechanism described above

$$h\nu + {}^1\pi\pi \rightarrow {}^1\pi\pi^* \rightarrow {}^3\pi\pi^* \rightarrow Ln^* \rightarrow Ln + h\nu' \qquad (1)$$

is the most commonly encountered and has been first evidenced for the β-diketonate $[Ln(dbm)_3]$ complexes (Buono-Core *et al.* 1990). It also leads to extremely efficient energy transfer, as demonstrated in the case of benzoate and cinnamate complexes (Fig. 4).

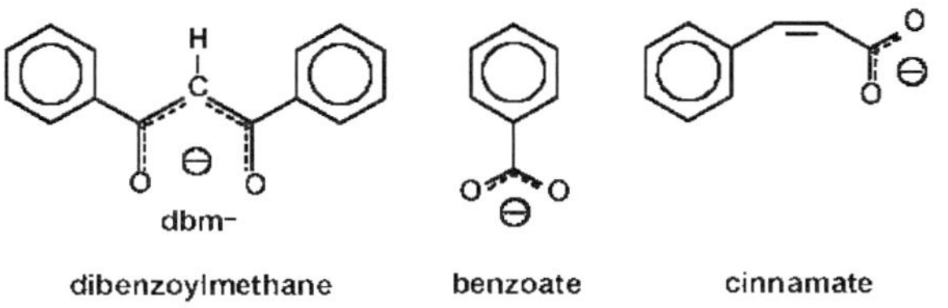

Figure 4
Structure of the ligands.

Benzoate sensitizes Tb^{III} with a quantum yield (254nm excitation, solid state) of 1.0 while the europium complex has a much lower quantum yield (0.2) due both to the presence of a LMCT state and probably to some back-transfer onto the ligand. For cinnamate, the situation is inverted: the ligand triplet state lies below the $Tb(^5D_4)$ level so that no luminescence from this ion is observed while the quantum yield of the europium compound amounts to 0.8 (Bredol *et al.* 1991).

In addition to this mechanism, several other intramolecular energy migration paths have been evidenced. One of them is the direct transfer of energy from the ligand excited singlet state:

$$h\nu + {}^1\pi\pi \rightarrow {}^1\pi\pi^* \rightarrow \mathrm{Ln}^* \rightarrow \mathrm{Ln} + h\nu' \qquad (2)$$

This mechanism explains why sometimes the $Tb(^5D_4)$ luminescence can be excited despite the fact that $E(^3\pi\pi^*)$ is lower in energy than $E(^5D_4)$. Another mechanism involves a more complicated back and forth energy exchange between the ligand and the metal ion, the energy of the singlet state being transferred on a high-energy excited state of the metal ion, shifted to the ligand triplet state and finally retransferred onto the metal ion:

$$h\nu + {}^1\pi\pi \rightarrow {}^1\pi\pi^* \rightarrow \mathrm{Ln}^*(1) \rightarrow {}^3\pi\pi^* \rightarrow \mathrm{Ln}^*(2) \rightarrow \mathrm{Ln} + h\nu' \qquad (3)$$

Finally, intermolecular energy transfer, especially in solution, can be an important sensitization process. An excited $^3\pi\pi^*$ state from a nonluminescent Ln^{III} (e.g., Y, La) complex transfers energy onto the triplet state of a luminescent Ln^{III} (e.g., Eu, Tb) complex (Fig. 5). Direct excitation onto the Ln^{III} excited state does not occur, as demonstrated by adding the luminescent ion to solutions of kinetically inert nonluminescent complexes. This type of excitation considerably enhances the Ln^{III} ion luminescence and several analytical assays take advantage of this.

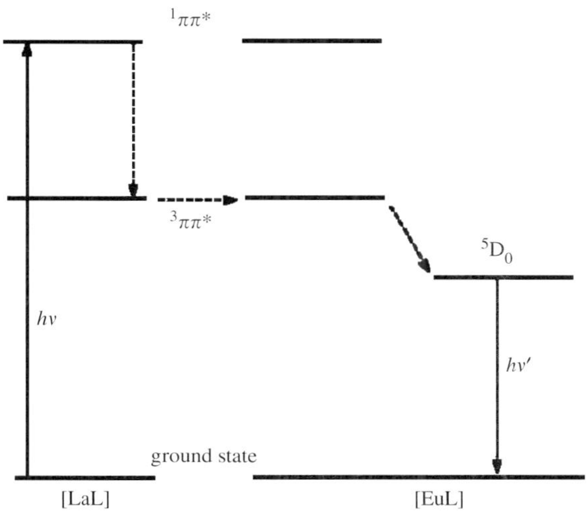

Figure 5
Intermolecular sensitization of an Eu^{III} complex.

2. *Fundamental Properties of Ln^{III} Ions*

The lanthanide elements are associated with the progressive filling of the 4f-electron subshell. The energy and the spatial extension of the 4f wavefunctions drop suddenly at the beginning of the series and the maximum of the 4f wavefunctions lies inside the $5s^25p^6$ closed subshells, which provides an efficient shielding of these orbitals towards external interactions (Bünzli 1998). Therefore the trivalent lanthanide ions with electronic configuration $[Xe]4f^n$ are spherical ions with no preferential geometric requirements, interacting essentially in an electrostatic manner with their chemical environment. As a consequence ligand–field effects are small, in the order of $500\,cm^{-1}$ ($6\,kJ\,mol^{-1}$), contrary to what is observed with d-transition metal complexes where these effects extend commonly from 15 000 to 25 000 cm^{-1} (180–300 $kJ\,mol^{-1}$). This results in atom-like spectra and the coordination chemistry of the Ln^{III} ions, and associated Sc^{III} and Y^{III} ions, is therefore subtle, with geometrical arrangements of the ligands essentially determined by steric effects among them.

2.1 *Ionic Radii and Coordination Numbers*

One characteristic property of Ln^{III} ions is their large ionic radius which decreases slightly and monotonically between La^{III} and Lu^{III}. It is interesting to note that the difference $r_i(La)$–$r_i(Lu)$ amounts to only 18pm (for coordination number CN = 9, Fig. 6), a number only marginally larger than the difference between, for instance, Ca^{II} ($r_i = 118$pm) and Sr^{II} ($r_i = 131$pm). This situation renders difficult the tailoring of an organic receptor specifically designed to encapsulate one particular Ln^{III} ion, especially if other Ln^{III} ions are present. On the other hand, the ionic radius largely depends upon the coordination number (i.e., the number of donor atoms of the receptor chemically bound to the metal ion,), with an

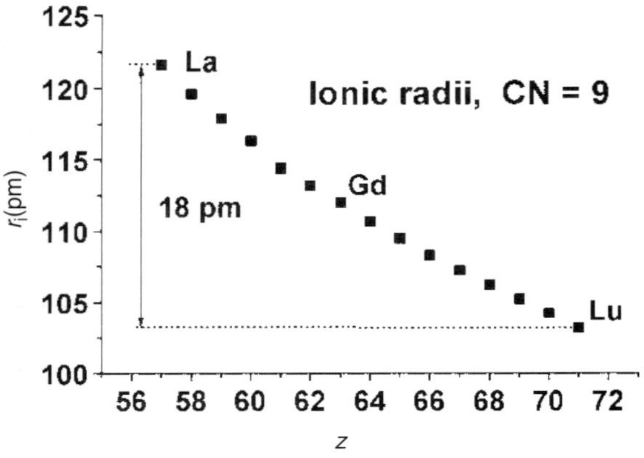

Figure 6
Monotonic decrease in the ionic radius of Ln^{III} ions, for a coordination number of nine.

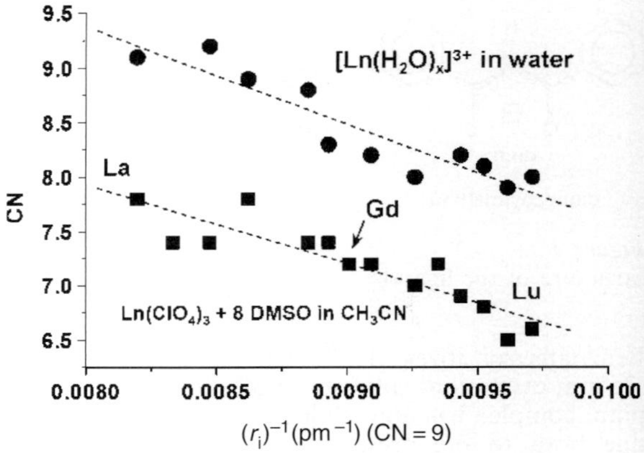

Figure 7
Coordination numbers of Ln^{III} ions in aqueous solution (•) and in anhydrous organic solution (■) vs. the inverse ionic radius for a coordination number of nine.

increase of about 30pm between CN = 6 and CN = 12. As a consequence, Ln^{III} ions easily adapt to many different coordination environments and, as a matter of fact, every coordination number between three and 12 has been found in the solid state. In solution, Ln^{III} ions have the tendency to adopt large coordination numbers.

Investigation of lanthanide perchlorates and nitrates in acetonitrile in the presence of a strong donor such as dimethylsulfoxide in the ratio $R = [DMSO]_t/[Ln^{III}]_t = 8$ has revealed a decrease in CN of about 1.5 units in going from La^{III} (mean CN ≈ 8) to Lu^{III} (mean CN ≈ 6.5) and, moreover, an increase in the mean CN of a specific ion by about 1.5 to 2 units upon addition of a larger quantity of dimethylsulfoxide ($R = 15$). This exemplifies the ability of the Ln^{III} ions to expand their coordination shell when required (Bünzli and Milicic-Tang 1995). In general, fractional values of CNs are obtained, pointing to the presence of equilibria between species with different coordination numbers. In water, the same trend is observed (Fig. 7). After decades of work involving x-ray and neutron diffraction techniques, it is now firmly established that larger Ln^{III} ions have CN = 9 while smaller ions have CN = 8; the intermediate part of the series (Eu–Dy) displays intermediate values, reflecting again the presence of an equilibrium between two differently solvated species (Rizkalla and Choppin 1991).

2.2 *Hydrolysis and Hydration Enthalpies*

One factor to take into account while attempting to complex Ln^{III} ions in water is their large hydrolysis constants, with $*\beta_{11}$ values ranging between −9 (La) and −8 (Lu). When the pH is greater than 6.5–7, the formation of highly insoluble hydroxides competes with complexation

$$*\beta_{l,\,m} = \frac{|[Ln_m(OH)_l(H_2O)_{n-l}^{(3m-l)+}| \cdot |H^+|^l}{|[Ln(H_2O)_n]^{3+}|^m}$$

The complexation process of a hydrated (solvated) ion can be decomposed into two steps. The first involves the desolvation of both the metal ion and the ligand and is highly enthalpy-unfavorable since ΔH_{hydr} values are very large for the Ln^{III} ions. They decrease from −3326 kJmol^{-1} for La^{III} to −3759 kJmol^{-1} for Lu^{III}, compared with −1585 kJ mol^{-1} for the otherwise chemically similar Ca^{II} ion. As a consequence, the enthalpy-favorable second step in which the two desolvated species combine to form the complex cannot compensate the dehydration enthalpy and most of the complexes are entropy-stabilized. This "compensation model" predicts a linear relationship between ΔH_r and ΔS_r, which is effectively often observed. Moreover, it is responsible for the increase in the formation constants with decreasing size of the Ln^{III} ions, as commonly seen for complexes with acyclic ligands.

3. *Inserting the Functional Ln^{III} Ions into Chemical Edifices*

Since the mid-1980s, the development of organic thin-film luminescent materials, as well as the increasing use of luminescent reporter groups attached to antibodies for performing immunoassays and for luminescent mapping of biological molecules (Bünzli 1989, Piguet and Bünzli 1999, Hemmilä *et al.* 1995), has prompted a large development in the quest for adequate organic receptors able to encapsulate Ln^{III}

ions and to enhance their luminescent properties. Given the above-listed requirements, the challenge is daunting and the next sections give some clues for solving this problem using the concepts of metallosupramolecular chemistry.

3.1 Coordination Compounds

The first luminescent lanthanide-containing complexes studied were conventional coordination compounds characterized by ion–dipole bonds formed between an organic (generally polydentate) ligand and the Ln^{III} ions, with no special interaction between the ligand strands. Among these, β-diketonates play a special role because they are good sensitizers of the Ln^{III} ions and because they were extensively studied after the initial work of Weissman (1942) on the antenna effect, in particular for demonstrating the energy migration processes outlined above. Since metallosupramolecular chemistry offers more controlled ways of designing an adequate metal ion environment, we are concentrating on this aspect rather than listing the numerous ligands that have been used for synthesizing luminescent complexes.

3.2 Rigid Receptors

In view of the spherical nature of 4f ions, inorganic chemists have tried to resort to the *lock and key* principle (Fig. 8) put forward by Emil Fischer in 1894 (Lehn 1995) for the selective complexation of these ions, as has been done for s-block ions: a monocyclic (coronand) or polycyclic (cryptand) receptor (Fig. 9) is engineered, which bears enough donor atoms to fulfil most of the coordination requirements of the metal ion and the cavity of which perfectly matches the ionic diameter of the metal ion.

Application of this principle to the selective complexation of 4f ions has however been somewhat deceptive, because adding one $-CH_2-CH_2-X-$ ($X=O$, N) group enlarges the mean cavity of a coronand by about 50 to 70pm, a difference greater than the ionic diameter variation between La^{III} and Lu^{III}. For instance, a stability difference of $\Delta\log K$(La–Lu) ≈ 0.5–1.5 only has been observed for crown ethers and $\Delta\log K < 1$ is usually found between complexes with two successive macrocycles (Bünzli 1987). In other words, the size of the coronand cavity cannot be tuned finely enough to accommodate a given Ln^{III} ion.

Moving to tridimensional receptors does not solve the problem: for instance, [2.2.1] cryptates display almost the same stability in water, around log $K=6.5$, for the entire Ln^{III} series. The reasons are that the cavity is slightly too small for lighter Ln^{III} ions and slightly too large for the heavier ions while the side chains grafted onto the N bridgehead atoms are somewhat flexible and adapt the cavity to the actual size of the metal ion. Introducing more rigidity into these receptors, e.g., by using bipyridyl groups, may help, as exemplified by the stable and luminescent $[Eu(bpy.bpy.bpy)]^{3+}$ cryptate used in homogeneous fluoroimmunoassays (Mathis 1998).

3.3 Designing Induced Cavities

The idea is to apply the *induced fit* concept and to introduce some flexibility into the organic ligand, now called a predisposed receptor, which allows it to wrap around the metal ion and to form an induced cavity specifically adapted to the size of the metal ion. Several families of flexible macrocyclic receptors have been developed.

(*a*) *Large flexible macrocycles.* Since crystal structure determinations pointed to many middle-sized macrocycles displaying some flexibility, an obvious way of introducing more versatility into the receptors was to resort to large macrocycles, especially for the synthesis of dimetallic edifices. Interest in the latter rose because of potential applications: (i) as luminescent probes, since two different luminescent centers can be helpful in imaging techniques; (ii) as precursors for doped materials where metal

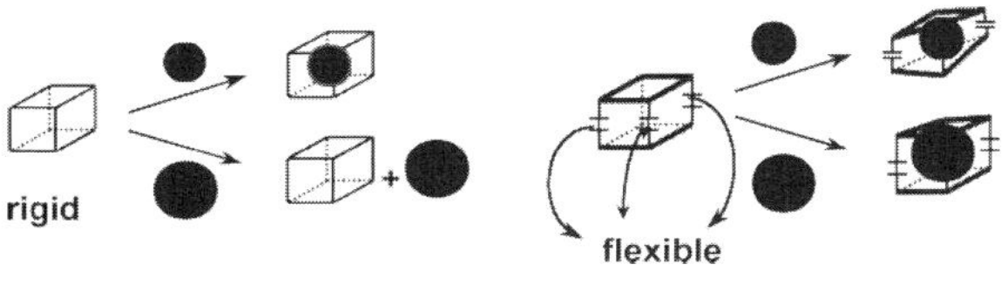

Figure 8
Schematic illustration of the lock and key (left) and induced fit concepts (right) for the complexation of spherical ions.

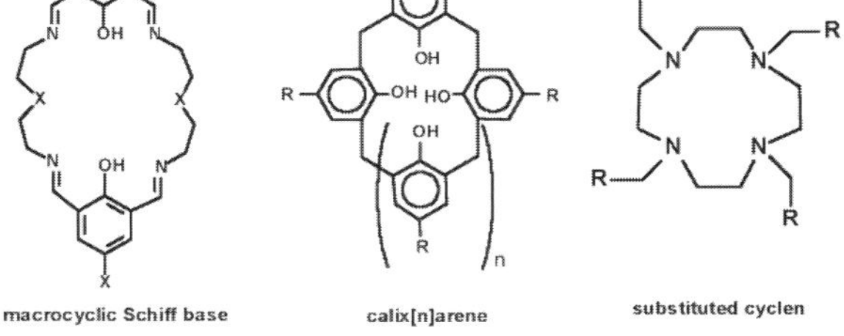

Figure 9
Cyclic and polycyclic receptors.

centers must be implemented at a fixed distance; and (iii) as analytical sensors, for instance for *in vivo* mapping of Ca^{II} concentration. Several classes of large macrocycles have been developed, including Schiff bases, cryptands, and calixarenes, particularly calix[8]arenes. For example, it has been shown that photophysical properties of dimetallic calix[8]arene complexes can be easily tuned by minor modification of the substituents on the upper rim, leading to selective sensitization of either Eu^{III} or Tb^{III} (Bünzli *et al.* 2000). It is noteworthy that the formation of the dimetallic edifices usually requires a large reorganization of the ligand conformation, a factor which is energetically unfavorable.

(*b*) *Ligands bearing pendant arms.* To minimize reorganization energy, coordinating pendant arms have been grafted on to small macrocycles such as derivatized cyclens or calix[*n*]arenes (usually $n = 4, 5$). In the first case, the coordination cavity is achieved by the four nitrogen donor atoms of the ring and by additional (usually four) donor atoms located on the dangling arms. Large stabilities are obtained: log*K* for the $[Ln(dota)]^-$ complexes ($R = CO_2^-$, Fig. 9) in water are in the range 23–26, that is, 17–20 orders of magnitude bigger than the stability of the [2.2.1] cryptates. Several luminescent probes and analytical sensors based on these systems are presently being developed. On the other hand, the phenoxy donor groups of calix[4]- and calix[5]arenes do not necessarily participate in the binding of Ln^{III} ions, depending on which rim is substituted, but these ligands also induce highly stable edifices.

(*c*) *Podands.* When the design of an edifice with given functionalities requires the presence of bidentate or tridentate arms, their grafting onto a relatively large macrocyclic platform is not always optimum, so podands build from a smaller anchor are more efficient. Ligands with four arms are built from small aromatic rings, while tri-armed receptors are simply engineered from a single atom, usually nitrogen, boron, or a transition metal ion. One example of such a podand is tris[3-(2-pyridyl)pyrazolyl]hydroborate, which yields 1:2 complexes in which the Ln^{III} ions lie in a 12-coordination site with a symmetry close to I_h. One advantage of podands is that the number of donor atoms can be changed easily by varying both the number of the arms and their denticity. On the other hand, these ligands are less predisposed than the macrocycles fitted with pendant arms, so that the correct orientation of the arms upon complexation may require more conformational work. A remedy to this drawback is the use of weak noncovalent interactions, such as hydrogen bonding (Renaud *et al.* 1999), to position the arms so that the donor atoms are organized in the required orientation before complexation.

(*d*) *Self-assembly processes.* Molecular recognition, which is at the heart of life processes, is often based on self-assembly, an action in which carefully tailored particles spontaneously assemble around the selected guest (Lehn 1995). Application of these processes in coordination chemistry is relatively recent but has produced fascinating di- and tridimensional functional architectures. The key to self-assembly in coordination chemistry is that the formation of the host cavity not only relies on the strong ion–dipole bonds to organize the ligands strands spatially but also on the presence of programmed weak noncovalent interactions between these strands. A series of monotopic and ditopic ligands has been synthesized which self-assemble with Ln^{III} ions to yield monometallic and dimetallic 3d–4f and 4f–4f edifices with predetermined luminescent or magnetic properties. As an example, work on triple-stranded dimetallic 4f–4f helicates is briefly summarized. Ligand L^A (Piguet *et al.* 1993), L^B (Martin *et al.* 1998) and L^C (Elhabiri *et al.* 1999) were designed to form dimetallic edifices both in the solid state and in solution (Fig. 10). The self-assembly process yields dimetallic helicates with large stability constants, on the order of $\log\beta_{23} = 25$ (L^A, L^B in acetonitrile) or even 50 (L^C in water). The distance between the two metal ions is around 0.8–0.9nm and the pitch of the helix around 2.4–2.5nm. The diethylamido groups were introduced in L^B because helicates with L^A are sensitive to hydrolysis and the europium-containing compound is amenable to luminescence quenching by a low-lying LMCT state. Despite this process, the directional intermolecular Tb^{III}-to-Eu^{III} energy transfer process could be studied and was shown to follow a dipole–dipole mechanism. The introduction of the terminal carboxylic acid functions leads to stable helicates in water, meaning that the self-assembly process easily overcomes the

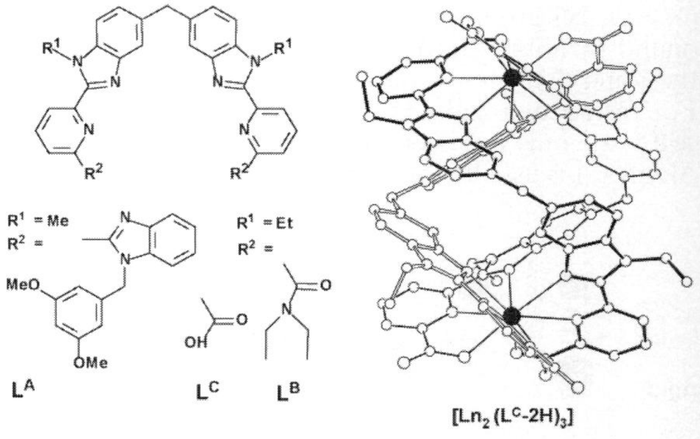

Figure 10
Dimetallic helicate obtained by self-assembly processes between $Ln(ClO_4)_3$ and ligands L^A, L^B (in acetonitrile), and L^C (in water).

unfavorable dehydration enthalpy. Moreover, the Ln^{III} ions are well protected from water interaction, resulting in quite long lifetimes of the excited $Eu(^5D_0)$ and $Tb(^5D_4)$ levels (2.1 and 2.5ms, respectively). The quantum yield of the metal-centered luminescence is sizable for the europium helicate (6%), but small for the terbium edifice because of a back transfer process. This new class of lanthanide carboxylates are promising compounds for the development of water-soluble luminescent stains and simple modifications of L^C should enable easy coupling with biological material, provide a better antenna effect, and lead to even more stable edifices.

4. *Selected Lanthanide-containing Luminescent Systems with Organic Ligands*

4.1 Schiff Base Complexes

Archer *et al.* (1998) demonstrated that Eu^{III} complexes derived from Schiff bases follow the same energy transfer mechanism (Eqn. (1) above) as the β-diketonates, as shown by the correlation between the energy of the ligand triplet state and the quantum yield of the metal-centered luminescence (Fig. 11). This work also illustrates that the triplet state energy alone does not determine the overall quantum yield. Indeed, compounds with a 1:2 stoichiometry, $Na[EuL_2]$, exhibit a quantum yield which is 60–70% larger compared with the one measured for $[Eu_2L_3]$, despite $E(^3\pi\pi^*)$ being the same for pairs of compounds with the same ligand (as measured on the Y^{III} analogues at 77K) and despite the coordination number (eight) being the same. In fact, half of the Eu^{III} ions in the 2:3 compounds attain this coordination number by completing their coordination sphere with water molecules, which explains the lower quantum yield.

Furthermore, these systems demonstrate an additional interesting feature. Two molecules sketched in Fig. 11 (R=X=Y=H) can be linked together by a methylene bridge to yield a dinucleating ligand H_4L'. Upon reaction with Ln^{III} ions, $[NaLnL']_n$ polyelectrolytes are formed, with mean n values around 21–22000 (Chen and Archer 1996). The Eu^{III} polyelectrolyte is quite luminescent ($Q^{Eu}=0.44$ under the conditions of Fig. 11) but its quantum yield can be boosted by the introduction of Y^{III} ions. For Eu:Y = 4:1, the quantum yield reaches 0.74 because excitation energy residing on the ligand triplet state of Y^{III} centers migrates on to Eu^{III} centers (cf. analogy with intermolecular energy transfer processes).

4.2 Systems with Ligands Based on Salicylic Acid Derivatives

Salicylamide and 2-hydroxyisophthalamide moieties are also good complexing groups for trivalent lanthanide ions, as well as good chromophores. These functionalities can be introduced either into podands or cryptands (Fig. 12). The resulting ligands yield highly stable complexes, especially at physiological pH since the hydroxy proton has a PK_a value of around 6–6.5. In addition, these groups are good chromophores and provide an excellent sensitization of several Ln^{III} ions, particularly Tb^{III}. For instance, the quantum yield of $[Eu(L^2)_2]^+$ amounts to 0.10 in water and it reaches 0.45 for the Tb^{III} analogue. On the other hand, the encapsulating ligand L^3 produces an entity $[Tb(L^3)]^+$ with a quantum yield as large as 0.60 in water (Petoud *et al.* 1999). It is foreseen that derivatives of these podands and cryptands will have applications in homogeneous fluoroimmunoassays.

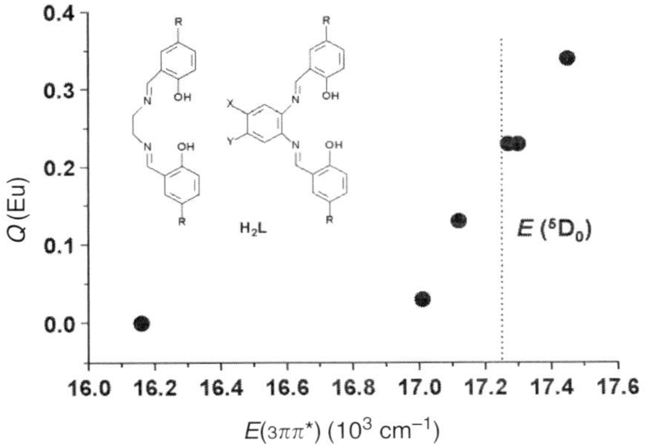

Figure 11
Quantum yield of the Eu^{III}-centered luminescence upon ligand excitation in $[Eu_2L_3]$ Schiff base complexes 10^{-3}–10^{-4}M in frozen dimethyl formamide (77K) as a function of the energy of the ligand triplet state.

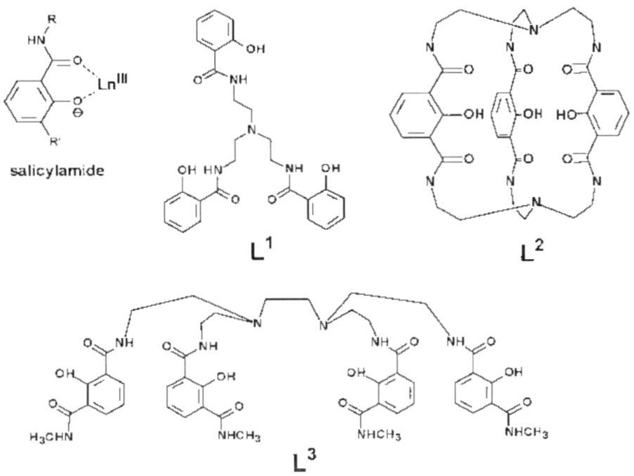

Figure 12
Podands and cryptate based on salicylamide and 2-hydroxyisophthalamide.

4.3 Systems with Ligands Based on Macrocycles

Interesting systems have been investigated which take advantage of the encapsulation of Ln^{III} ions either by cryptands or by macrocycles fitted with pendant arms (Fig. 9, Fig. 13). In the first category, we note that the $[Eu(bpy.bpy.bpy)]^{3+}$ cryptate, written $[Eu \subset (bpy.bpy.bpy)]^{3+}$ to point to the inclusion of the metal ion into the macrocycle, has a deceptively low quantum yield in water (0.02), mainly because the complexed species contain 2–3 water molecules directly bound to the metal ion (one water molecule contributes about $1 ms^{-1}$ to k_{nr}, cf. Fig. 3). The $[Tb \subset (bpy.bpy.bpy)]^{3+}$ cryptate also displays a low quantum yield (0.03) because, in addition to quenching by OH oscillators (here one water molecule contributes about $0.25 ms^{-1}$ to k_{nr}), the energy of the ligand triplet state is only $1200 cm^{-1}$ above $Tb(^5D_4)$ and an efficient thermally-activated back transfer process takes place. Modification of the cryptand design by introducing the N-oxide groups improved considerably the photophysical properties of the corresponding supramolecular assemblies (Fig. 13): the quantum yields of $[Eu \subset (\mathbf{1})]^{3+}$ and $[Eu \subset (\mathbf{2})]^{3+}$ amount to 0.15 and 0.20, respectively. However, Tb^{III} luminescence cannot be sensitized because the ligand triplet state lies at too low energy, a reason why the branched macrocycle **3** was developed: $[Tb(\mathbf{3})]^{3+}$ is highly luminescent with $Q^{Tb} = 0.38$ (Sabbatini *et al.* 1993).

Other interesting receptors include derivatized calixarenes. For instance, the calix[4]arene fitted with bipyridine pendant arms **4** complexes Eu^{III} and the good energy transfer process between the bipyridine moieties and the metal ion leads to a quantum yield of 0.15 in acetonitrile (Ulrich *et al.* 1997). On the other hand, $[Eu \subset (\mathbf{5})]^{3+}$ is poorly luminescent in water while $[Tb \subset (\mathbf{5})]^{3+}$ has a quantum yield equal to 0.2 (Sabbatini *et al.* 1990). The series of derivatized calixarenes **6**–**8** was synthesized to study the influence of the triplet state energy on the ligand-to-metal energy transfer process (Steemers *et al.* 1995). The ligand containing the triphenylene group was found to be more efficient in sensitizing both Eu^{III} and Tb^{III} in water.

4.4 Luminescent Complexes for Chemical Sensors

Many molecular sensors and switches contain luminescent building blocks which respond to a given stimulus. For instance, the structure at the top of Fig. 14 displays a smooth luminescence enhancement upon the complexation of potassium ions into the end macrocycles and can therefore be used as analytical sensor for this otherwise spectroscopically silent ion (de Silva *et al.* 1999). The quantum yield of the Eu^{III}–K^I system in methanol is as high as 0.47, whereas the Eu^{III} complex alone has $Q^{Ln} \approx 0.05$ only: the complexation of the alkali cation suppresses the photoinduced electron transfer from the azacrown nitrogen atom to the luminophore.

The neutral complex between the ligand at the bottom of Fig. 14 and Eu^{III} (R=H) displays a metal-centered luminescence which is pH-dependent. A substantial luminescence enhancement is observed between pH 1.5 and 6.8 (a factor of about 500 is reported) which arises from large variations in the photophysical properties of the sensitizing phenanthridinium chromophore. The maximum of absorption shifts towards longer wavelength and the molar absorption coefficient increases a great deal upon protonation of the tertiary amine, so that the Eu-luminescence can be turned on by excitation at 380nm, where the unprotonated chromophore does not absorb. When R=Me, the luminescence of the

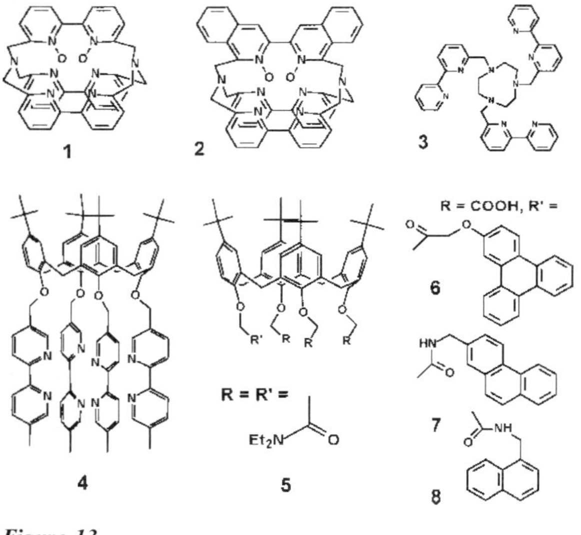

Figure 13
Macrocyclic receptors.

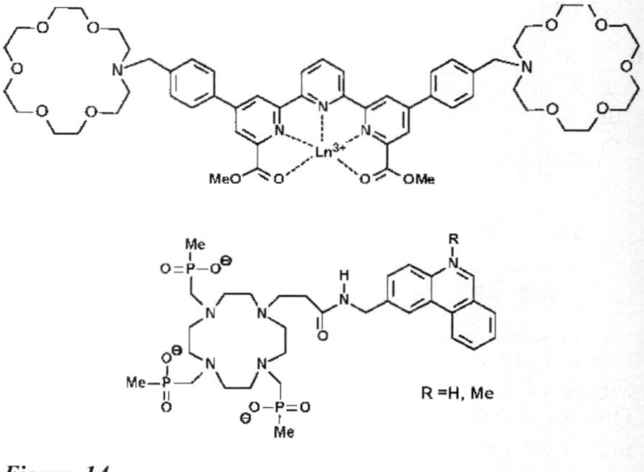

Figure 14
Luminescent responsive systems.

resulting Eu complex is selectively quenched by chloride ions and the system may be used for the quantitative analysis of the latter (Parker 2000).

5. *Perspectives*

The few examples discussed above show the large range of applications spanned by lanthanide-containing organic luminescent stains. During the last years, there has been considerable improvement in mastering the various parameters influencing the metal-centered luminescence and, therefore, the design of suitable receptors can be more conveniently planned. Although one important limitation in these systems lies in the inner-filter effect, there is no doubt that efficient analytical systems will be developed based on luminescent lanthanide-containing chemosensors. The main applications to be foreseen are in the field of biomedical analyses, luminescence mapping of biological molecules, and material sciences. Light-converting systems will also benefit from these technologies; one interesting example is the use of Eu^{III}, Dy^{III}, and Tb^{III} tris(picolinato) complexes to produce luminescent invisible inks used in bar codes (Auslander and Berson 1996).

New systems able to produce near-infrared light upon UV or visible excitation with reasonable yields have been tested and they offer interesting novel possibilities for the design of multiplex analytical assays, for imaging purposes, or for materials with specific photophysical properties. A precursor of the last is a fluorescein-bearing derivatized calix[4]arene (Fig. 15) which sensitizes the infrared luminescence of both Nd^{III} and Er^{III} at 1.06 and 1.54μm, respectively (Wolbers *et al.* 1998). These lines have a potential for use in the design of optical amplifiers for telecommunications and the organic ligands (especially the fluorescein attachment) have much larger molar absorption coefficients than inorganic matrices, which is an advantage for pumping more energy into the resulting edifices.

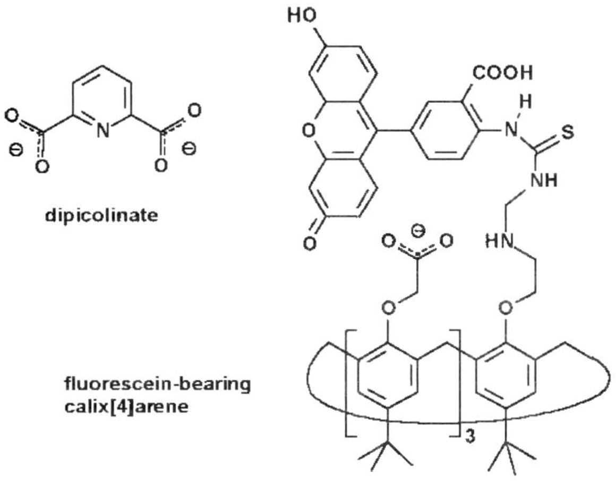

Figure 15
Formulae of the ligands cited in Sect. 5

Bibliography

Archer R D, Chen H Y, Thompson L C 1998 Synthesis, characterization, and luminescence of europium(III) Schiff base complexes. *Inorg. Chem.* **37**, 2089–95

Auslander J D Berson W 1996 Bar codes using luminescent invisible inks. US Pat. 94-347740 (US 5 552 971)

Beeby A, Clarkson I M, Dickins R S, Faulkner S, Parker D, Roayle L, de Sousa A S, William J A G, Woods M 1999 Non-radiative deactivation of the excited states of europium, terbium and ytterbium complexes by proximate energy-matched OH, NH and CH oscillators: an improved luminescence method for establishing solution hydration states. *J. Chem. Soc., Perkin Trans.* **2**, 493–503

Bredol M, Kynast U, Ronda C 1991 Designing luminescent materials. *Adv. Mater* **3**, 361–7

Bünzli J-C G 1987 Complexes with synthetic ionophores. In: Gschneidner K A Jr., Eyring L (eds.) *Handbook on the Physics and Chemistry of Rare Earths*, Elsevier, Amsterdam, Vol. 9 Chap. 60

Bünzli J-C G 1989 Luminescent probes. In: Bünzli J-C G, Choppin G R (eds.) *Lanthanide Probes in Life, Chemical and Earth Sciences: Theory and Practice*. Elsevier, Amsterdam, Chap. 7

Bünzli J-C G 1998 Coordination chemistry of the trivalent lanthanide ions: an introductory overview. In: Saez-Puche R, Caro P (eds.) *Rare Earths*. Editorial Complutense, Madrid, pp. 223–59

Bünzli J-C G, Ihringer F, Besançon F 2000 Bimetallic supramolecular edifices with lanthanides. In: Lumetta G J, Rogers R D, Gopalan A (eds.) *Calixarenes for Separations, ACS Symposium Series 757*. American Chemical Society, Washington, DC, Chap. 14

Bünzli J-C G, Milicic-Tang A 1995 Solvation and anion interaction in organic solvents. In: Gschneidner K A Jr., Eyring L (eds.) *Handbook on the Physics and Chemistry of Rare Earths*. Elsevier, Amsterdam, Vol. 21, Chap. 103

Buono-Core G E, Li H, Marciniak B 1990 Quenching of excited states by lanthanide ions and chelates in solution. *Coord. Chem. Rev.* **99**, 55–87

Chen H Y, Archer R D 1996 Synthesis and characterization of linear luminescent Schiff-base polyelectrolytes with europium(III) in the backbone. *Macromolecules* **29**, 1957–64

de Silva A P, Fox D B, Huxley A J M, McClenaghan N D, Roiron J 1999 Metal complexes as components of luminescent signaling systems. *Coord. Chem. Rev.* **185**, 297–306

Elhabiri M, Scopelliti R, Bünzli J-C G, Piguet C 1999 Lanthanide helicates self-assembled in water: a new class of highly stable and luminescent dimetallic carboxylates. *J. Am. Chem. Soc.* **121**, 10747–62

Hemmilä I Ståhlberg T Mottram P 1995 Bioanalytical Applications of Labelling Technologies, 2nd edn. Wallac Oy, Turku, Finland

Lehn J-M 1995 *Supramolecular Chemistry: Concepts and Perspectives*. VCH, Weinheim, Germany

Martin N, Bünzli J-C G, McKee V, Piguet C, Hopfgartner G 1998 Self-assembled dinuclear lanthanide helicates:

substantial luminescence enhancement upon replacing terminal benzimidazole groups by carboxamide binding units. *Inorg. Chem.* **37**, 577–89

Mathis G 1998 Biological applications of rare earth cryptates. In: Saez-Puche R, Caro P (eds.) *Rare Earths.* Editorial Complutense, Madrid, pp. 285–97

Parker D 2000 Luminescent lanthonide sensors for pO_2 and selected anions. *Coord. Chem. Rev.* **205**, 109–30

Petoud S, Cohen S M, Xu J, Raymond K N, Bünzli J-C G 1999 New highly luminescent and water-soluble lanthanide complexes for homogeneous time-resolved fluoroimmunoassays based on salicylamide-type ligands In: *Abstr. 6th Int. Conf. Methods and Applications of Fluorescence Spectroscopy.* Paris, France

Piguet C, Bünzli J-C G 1999 Mono- and polymetallic lanthanide-containing functional assemblies: a field between tradition and novelty. *Chem. Soc. Rev.* **28**, 347–58

Piguet C, Bünzli J-C G, Bernardinelli G, Hopfgartner G, Williams A F 1993 Self-assembly and photophysical properties of lanthanide dinuclear triple-helical complexes. *J. Am. Chem. Soc.* **115**, 8197–206

Renaud F, Piguet C, Bernardinelli G, Bünzli J-C G, Hopfgartner G 1999 Nine-coordinate lanthanide podates with predetermined structural and electronic properties: facial organization of unsymmetrical tridentate binding units by a protonated covalent tripod. *J. Am. Chem. Soc.* **121**, 9326–42

Rizkalla E N, Choppin G R 1991 Hydration and hydrolysis of lanthanides. In: Gschneidner K A Jr., Eyring L (eds.) *Handbook on the Physics and Chemistry of Rare Earths.* Elsevier, Amsterdam, Vol.15, Chap. 103

Sabbatini N, Guardigli M, Lehn J M 1993 Luminescent lanthanide complexes as photochemical supramolecular devices. *Coord. Chem. Rev.* **123**, 201–28

Sabbatini N, Guardigli M, Mecati A, Balzani V, Ungaro R, Ghidini E, Casnati A, Pocchini A 1990 Encapsulation of lanthanide ions in calixarene receptors. A strongly luminescent terbium(3+) complex. *J. Chem. Soc., Chem. Commun.* 878–9

Steemers F J, Verboom W, Reinhoudt D N, Vandertol E B, Verhoeven J W 1995 New sensitizer-modified calix[4]arenes enabling near-UV excitation of complexed luminescent lanthanide ions. *J. Am. Chem. Soc.* **117**, 9408–14

Ulrich G, Ziessel R, Manet I, Guardigli M, Sabbatini N, Fraternali F, Wipf G 1997 Calix[4]arene podands and barrelands incorporating 2,2'-bipyridine moieties and their lanthanide complexes: luminescence properties. *Chem. Eur. J.* **3**, 1815–22

Weissman S I 1942 Intramolecular energy transfer. The luminescence of complexes of europium β-diketonate complexes. *J. Chem. Phys.* **10**, 214–7

Wolbers M P O, van Veggel F C J M, Peters F G A, van Beelen E S E, Hofstraat J W, Geurts F A J, Reinhoudt D N 1998 Sensitized near-infrared emission from Nd^{3+} and Er^{3+} complexes of fluorescein-bearing calix[4]arene cages. *Chem. Eur. J.* **4**, 772–80

Light Emitting Diodes

R. J. Roedel

Light-emitting diodes (LEDs) are pn-junction devices designed to emit optical radiation when operated under forward electrical bias. In LEDs, the optical radiation is generated by spontaneous recombination (as opposed to the stimulated recombination in a laser), and is generally characterized as an incoherent light emission process with a relatively large linewidth encompassing many optical modes. The dominant emission energy is close to the bandgap of the semiconductor used to fabricate the LED, so the wavelength of an LED can be tailored by the choice of the semiconductor material employed. With bandgap engineering LEDs can be constructed to operate in infrared or visible wavelength regimes.

Light-emitting diodes are widely used in many optoelectronic applications. As emitters of infrared light, they are employed in signaling systems (remote control devices), in integrated optoelectronic structures (opto-isolators), and in optical communication systems. They are also extensively used as indicator lamps, display elements, and sensors when manufactured to emit visible light. Directions taken in LED research and development, such as superradiant and short wavelength LEDs, demonstrate that they will remain a valuable component in photonics systems.

1. *Basic Operating Principles*

The light-emitting diode is a transducer—it converts input electrical energy into output optical energy in the visible or infrared portion of the spectrum—through the physical process of electroluminescence. When a forward bias is applied to a pn-junction, majority carrier electrons are injected from the n-side of the device into the p-side, whereupon they become excess minority carriers. Similarly, majority holes from the p-side become excess minority carriers when injected into the n-side. This is depicted schematically in Fig. 1. These injected minority carriers have a density considerably higher than their equilibrium value, so thermodynamic forces come into play to reduce their number.

Specifically, the excess minority carriers find and recombine with majority carriers on both sides of the junction, and in so doing, liberate their combined kinetic and potential energies. The energy in a single electron–hole recombination event may be used to create a photon (radiative recombination) or may be dissipated in some other manner in the lattice (nonradiative recombination). The ratio of the radiative recombination rate to the total recombination rate is the internal quantum efficiency of the LED. Since the potential energy difference between the electron and

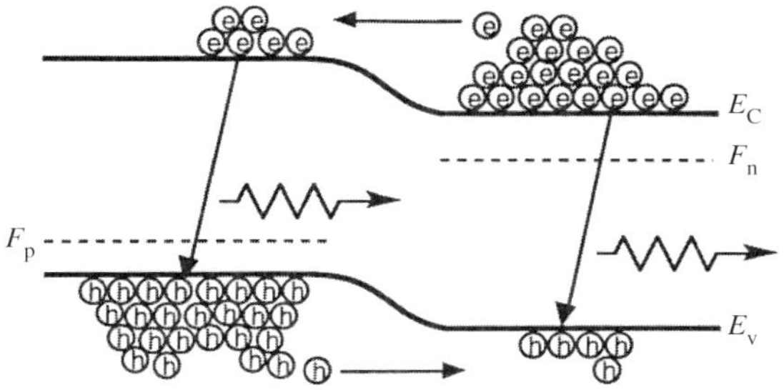

Figure 1
Injection of electrons and holes in a forward-biased pn-junction leading to spontaneous emission of photons.

the hole is the bandgap energy, and the carrier kinetic energies are small (typically of the order of kT), the photon energy is close to the bandgap energy. Since the bandgap of a semiconductor depends upon the specific material, the wavelength of an LED is largely determined by the choice of semiconductor material. Table 1 lists some of the semiconductor materials currently used for LED fabrication.

It should be noted that in the electron–hole pair to photon conversion process, momentum must be conserved as well as energy. Since photons in the energy range of 1–2 eV have almost no momentum, either the electron–hole pair must have nearly zero net momentum or some other process that relaxes the momentum selection rule must come into play. As shown in Table 1, light-emitting diodes have been made from both direct and indirect semiconductors. Direct semiconductors have the conduction band minimum and valence band maximum located at the $k=0$ point in the Brillouin zone so electron–hole pairs have negligible momentum. On the other hand, indirect semiconductors have a substantial difference in crystal momentum between the conduction band minimum and valence band maximum, so radiative recombination of an electron–hole pair requires phonon interaction or some other process to conserve momentum. This reduces the radiative transition rate essentially to zero. However, the introduction of an isoelectronic impurity into an indirect semiconductor lattice (such as nitrogen in GaP) produces a site at which electrons and holes can first form a stable exciton (at room temperature) and then recombine radiatively. The localized nature of the isoelectronic impurity state produces a spread in k-space that relaxes the momentum selection rule and permits radiative transitions to proceed—but at much lower efficiency than in a direct semiconductor.

The basic operating characteristics of an LED, constructed from a direct bandgap semiconductor, can be simply modeled with rate equations based on the continuity equation. Letting n and p represent the

Table 1
Semiconductor material used for visible and IR LED fabrication.

Wavelength (nm)	Active layer material	Type	External quantum efficiency (%)
470 (blue)	ZnS:Al	Indirect	<0.1
450–530 (blue/green)	InGaN/GaN	Direct	>5
565 (green)	GaP:N	Indirect	<0.2
590 (yellow)	GaAsP:N	Indirect	<0.3
590–620 (yellow/orange)	InGaAlP	Direct	>5
610 (orange)	GaAsP:N	Indirect	<0.3
630 (orange/red)	GaAsP:N	Indirect	<0.3
650 (red)	GaAsP	Direct	<0.5
680 (red)	AlGaAs	Direct	>5
700 (red)	GaP:Zn,O	Indirect	<1
850	AlGaAs	Direct	>5
860	GaAs	Direct	>2
1300	InGaAsP	Direct	>1
1550	InGaAsP	Direct	>1

electron and hole densities,

$$n = n_o + \Delta n \quad (1)$$

$$p = p_o + \Delta p \quad (2)$$

where Δn and Δp are the excess carrier concentrations, the rate equation for electrons (on the p-side) is:

$$\frac{dn}{dt} = g_n - r_n \quad (3)$$

and for holes (on the n-side)

$$\frac{dp}{dt} = g_p - r_p \quad (4)$$

where g and r are the generation and recombination rates, respectively. The generation rate for electrons, for example, can be given by:

$$g_n = g_{th} + g_{n,inj} = g_{th} + \frac{J_n}{qL_n} \quad (5)$$

The first term is the thermal generation rate and the second term describes the generation of excess electrons produced by current injection within a minority carrier diffusion length of the junction. An expression for the recombination rate must include terms for both radiative and nonradiative recombination paths. One way of viewing this problem is to consider the physical situation depicted in Fig. 2, in which the nonradiative recombination is modeled as occurring through a trapping center located within the energy gap. We assume that the trap is neutral when unoccupied by an electron and negative after it is has captured an electron, so the trap densities are expressed as:

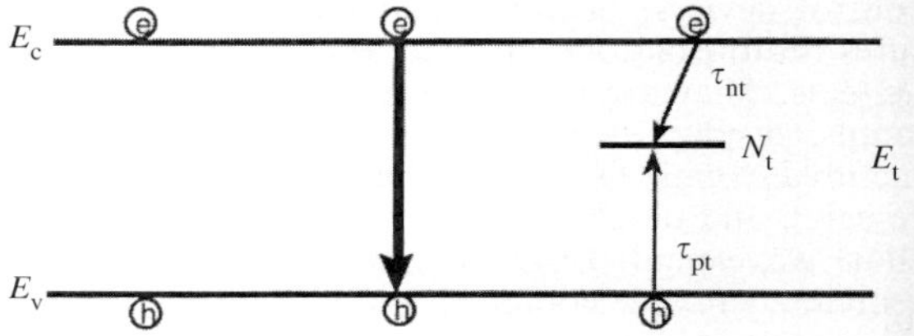

Figure 2
Competing recombination processes in a light-emitting diode.

$$N_t = N_t^o + N_t^- \quad (6)$$

The electron and hole capture times are given by:

$$\tau_{nt} = \frac{1}{\sigma_n v_n N_t} \quad \tau_{pt} = \frac{1}{\sigma_p v_p N_t} \quad (7)$$

where σ and v are the capture cross-sections and thermal velocities, respectively. The total recombination rates can be written as:

$$r_n = B_{sp}np + \frac{n}{\tau_{nt}}\left(\frac{N_t^o}{N_t}\right) \quad (8)$$

$$r_p = B_{sp}np + \frac{p}{\tau_{pt}}\left(\frac{N_t^-}{N_t}\right) \quad (9)$$

In Eqns. (8) and (9), the first term represents the radiative recombination process and B is the spontaneous recombination coefficient, while the second term corresponds to the nonradiative recombination

Table 2
Spontaneous recombination coefficient for a variety of semiconductors.

Semiconductor	Type	B_{sp} (cm^3s^{-1})
GaAs	direct	7.2×10^{-10}
GaSb	direct	2.4×10^{-10}
InP	direct	12.6×10^{-10}
InAs	direct	0.9×10^{-10}
InSb	direct	0.5×10^{-10}
Si	indirect	0.2×10^{-10}
Ge	indirect	5.3×10^{-10}
GaP	indirect	5.4×10^{-10}

process in which an electron is first captured by an empty trap, followed by hole capture at the occupied trap. The spontaneous recombination coefficients for several semiconductors are listed in Table 2. In Eqn. (8), the ratio in parentheses is the fraction of traps that are unoccupied, and in Eqn. (9), the ratio is the fraction of traps occupied by an electron. Substituting Eqns. (8) and (9) into Eqns. (3) and (4), and looking for steady state solutions, we have:

$$\frac{dn}{dt} = 0 = -B_{sp}np + g_{n,\,inj} + \frac{n}{\tau_{nt}}\left(\frac{N_t^- - N_t}{N_t}\right) \quad (10)$$

and

$$\frac{dp}{dt} = 0 = -B_{sp}np + g_{p,\,inj} - \frac{p}{\tau_{pt}}\left(\frac{N_t^-}{N_t}\right) \quad (11)$$

In Eqn. (10), Eqn. (4) was used to eliminate the density of neutral traps. Consider the case in which the electron current density is substantially larger than the hole current density so that the radiative recombination occurs predominantly on the p-side. In this case, the majority hole density remains close to the equilibrium value, p_o. Equations (10) and (11) can easily be solved for the electron density:

$$n = n_\alpha + \left[n_\alpha^2 + \frac{\tau_{nt}\, g_n}{\tau_{pt} B_{sp}}\right]1/2 \quad (12)$$

where:

$$n_\alpha = \frac{g_p \tau_{pt} - p_o(B_{sp}p_o\tau_{nt} + 1)}{2B_{sp}p_o\tau_{nt}} \quad (13)$$

The results of this analysis can be used to examine how the light output from an LED varies with excitation. Equation (12) can be used to compute the quantity:

$$\frac{B_{sp}np_o}{g_n}$$

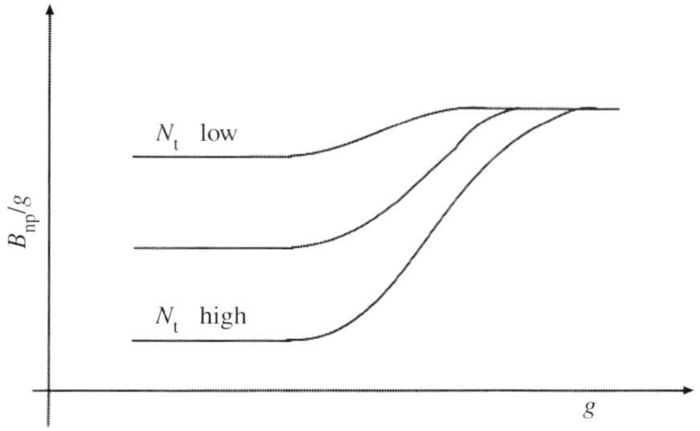

Figure 3
Normalized optical output versus electrical input for a direct bandgap LED.

which is the ratio of the photon generation rate to the excess hole generation rate, or the ratio of optical output to electrical input. This quantity is plotted versus hole generation rate in Fig. 3. The interpretation of this figure is the following: at low excitation rates, nonradiative recombination is the dominant recombination process, with the optical output dropping significantly as the number of trapping centers increases. However, at higher excitation rates, radiative recombination eventually overwhelms the nonradiative process when the density of injected minority carriers exceeds the density of available trapping sites. Also note that the plateaus on the plot correspond to operational regimes in which the optical output is actually directly proportional to the electrical input. It is evident that LEDs require a low density of defects and other trapping centers for high performance, or high current densities will be needed to overcome their deleterious effect.

The internal quantum efficiency of an LED can also be calculated by examining Eqn. (10). Substituting Eqns. (1) and (2) into the expression for the spontaneous recombination rate, we obtain:

$$\begin{aligned} r &= B_{sp}(n_o + \Delta n)(p_o + \Delta p) \\ &= B_{sp}\, n_o\, p_o + B_{sp}\, \Delta n(n_o + p_o + \Delta n) \end{aligned} \quad (14)$$

The first term equals the thermal generation rate, while the second term corresponds to the radiation produced in excess of normal blackbody radiation. If the excess carrier concentration ($\Delta n = \Delta p$) is small compared to the equilibrium concentrations of electrons or holes, this is the low injection condition, and the spontaneous recombination rate is linearly proportional to the excess carrier concentration:

$$r_{sp} \approx B_{sp}\Delta n(n_o + p_o) \quad (15)$$

On the other hand, under high injection, Δn exceeds the background concentration, and the recombination rate becomes proportional to the square of the excess carrier concentration:

$$r_{sp} \approx B_{sp} \Delta n^2 \tag{16}$$

Under low injection, the rate equation can then be written as:

$$\frac{dn}{dt} = \frac{d\Delta n}{dt} = -B_{sp}\Delta n(p_o + \Delta n) + g_{n,inj} + \frac{\Delta n}{\tau_{nt}}\left(\frac{N_t^-}{N_t} - 1\right) \tag{17}$$

In Eqn. (17), it was assumed that

$$p_o, \Delta n \gg n_o \tag{18}$$

so it can be rewritten as:

$$\frac{d\Delta n}{dt} = g_{n,inj} - \frac{\Delta n}{\tau_r} - \frac{\Delta n}{\tau_{nr}} = g_{n,inj} - \frac{\Delta n}{\tau_n} \tag{19}$$

The quantity τ_r is the radiative lifetime:

$$\tau_r = \frac{1}{B_{sp}(p_o + \Delta n)} \tag{20}$$

and is a constant only under low injection conditions. The quantity τ_{nr} is the nonradiative lifetime:

$$\tau_{nr} = \tau_{nt}\left(\frac{N_t}{N_t^o}\right) = \tau_{nt}\left(\frac{N_t}{N_t - N_t^-}\right) \tag{21}$$

and must be modified when different kinds of trapping or recombination centers are present. Finally, the quantity τ_n is the overall electron lifetime:

$$\frac{1}{\tau_n} = \frac{1}{\tau_r} + \frac{1}{\tau_{nr}} \tag{22}$$

The internal quantum efficiency of an LED is defined as:

$$\eta_{int} = \frac{\dfrac{\Delta n}{\tau_r}}{\dfrac{\Delta n}{\tau_n}} = \frac{1}{1 + \dfrac{\tau_r}{\tau_{nr}}} \tag{23}$$

or as stated previously, the ratio of the radiative recombination rate to the total recombination rate. Another way of phrasing this is that the internal quantum efficiency is the number of photons produced per electron crossing the pn-junction. At steady state, the excess electron concentration on the p-side of the diode is found by setting Eqn. (19) to zero:

$$\Delta n = g_{n,inj}\tau_n = \frac{J_n \tau_n}{qL_n} = \frac{J_n \tau_r}{qL_n}\eta_{int} \tag{24}$$

The photon generation rate at steady state is then given by:

$$\frac{\Delta n}{\tau_r} = \frac{J_n}{qL_n}\eta_{int} \tag{25}$$

It is clear that the LED engineer must strive to make the internal quantum efficiency as close to unity as possible.

The rate equations can also be used to examine LED kinetics. Consider the case of an LED that had reached steady state, but had the current abruptly terminated at $t=0$. The decay of the electroluminescence can be found by writing Eqn. (19) as:

$$\frac{d\Delta n}{dt} = -\frac{\Delta n}{\tau_r} \tag{26}$$

In the case of low injection, the radiative lifetime is a constant given by:

$$\tau_r = \frac{1}{B_{sp}\, p_o} \tag{27}$$

so the solution follows conventional exponential decay:

$$\Delta n(t) = \frac{J_n \tau_n}{qL_n}\exp\left[-\frac{t}{\tau_r}\right] \tag{28}$$

On the other hand, if the current flow satisfies high injection conditions, the rate equation becomes nonlinear:

$$\frac{d\Delta n}{dt} = -B_{sp}\Delta n^2 \tag{29}$$

The solution is found to be:

$$\Delta n(t) = \frac{J_n \tau_n}{qL_n}\left[\frac{1}{1 + \dfrac{J_n \tau_n}{qL_n}B_{sp}t}\right] \tag{30}$$

The current decay is hyperbolic, at first. Eventually, the excess carrier concentration will decline until low injection conditions become valid and exponential decay resumes.

For additional LED device theory, the interested reader should consult the texts by Williams and Hall (1978), Saleh and Teich (1991), and Peyghambarian *et al.*(1993). An analysis of the kinetics in a GaP LED is found in the paper by Dishman *et al.* (1970).

2. *Device Structures and Characteristics*

The analysis so far has been confined to the internal processes within a light-emitting diode. However, an LED must actually emit the light to be useful and there are several physical factors that work against the efficient emission of radiation. These factors will be described in this section.

2.1 *Internal Loss Mechanisms*

A useful quantity to describe the operation of an LED is the external quantum efficiency, which can be defined as the ratio of the optical power emitted to the electrical input power, or the number of photons extracted from the device per electron crossing the junction. There is a relation between the internal and external quantum efficiencies:

$$\eta_{ext} = f_{Fresnel}\, f_{reflection}\, f_{absorption}\, \eta_{int} \tag{31}$$

where the modifying coefficients are the three most important factors that reduce the photon emission probability.

Semiconductor materials typically have high refractive indices, so the semiconductor–air interface is a dielectric discontinuity that will cause substantial electromagnetic reflection. The transmission coefficient for normal incidence is given by the Fresnel equation:

$$T_N = f_{Fresnel} = \frac{4n}{(n+1)^2} \tag{32}$$

where n is the semiconductor refractive index. In the case of GaP, for example, where $n = 3.29$, the Fresnel factor is:

$$f_{Fresnel} = \frac{4 \cdot 3.29}{(3.29+1)^2} = 0.72 \tag{33}$$

There is a second reflection loss associated with the semiconductor–air interface. From Snell's Law, light will be totally internally reflected when the incidence angle is greater than the critical angle given by:

$$\theta_{cr} = \sin^{-1}\left(\frac{1}{n}\right) \tag{34}$$

For GaP, the critical angle is approximately 17.7°. Since photons are emitted isotropically from the pn-junction region, only a small fraction will strike the LED surface inside the critical angle. This fraction can be estimated by evaluating these integrals:

$$f_{reflection} = \frac{1}{2}\,\frac{\int_0^{2\pi}\int_0^{\theta_{cr}} \sin\theta \, d\theta \, d\phi}{\int_0^{2\pi}\int_0^{\pi/2} \sin\theta \, d\theta \, d\phi} \tag{35}$$

The factor of 1/2 is used to calculate the emission from, say, the top surface only. This fraction is easily shown to be:

$$f_{reflection} = \frac{1-\cos\theta_{cr}}{2} = \sin^2\frac{\theta_{cr}}{2} \approx \frac{1}{4n^2} \tag{36}$$

For GaP, this factor is approximately

$$f_{reflection} \approx \frac{1}{4n^2} = 0.023 \tag{37}$$

The product of the two reflection factors is given by (and evaluated for GaP):

$$f_{Fresnel}\, f_{reflection} = \frac{4n}{(n+1)^2}\,\frac{1}{4n^2} = \frac{1}{n(n+1)^2} = 0.017 \tag{38}$$

The final factor is reabsorption of the photons. Following the treatment of Hakki (1971), consider an LED with a planar pn-junction located at a distance z_j from the surface, as shown in Fig. 4. Photons generated at a distance z from the junction will contribute to the overall optical spectral intensity:

$$I(E,z) = \frac{I_o}{L_n} g(E) \exp\left(-\frac{z}{L_n}\right) \tag{39}$$

where I_o is the total optical irradiance and $g(E)$ is the normalized spectral distribution:

$$\int_0^{\infty} g(E)\, dE = 1 \tag{40}$$

and

$$\int_0^{\infty}\int_0^{\infty} I(E,z)\, dE\, dz = I_o \tag{41}$$

The light intensity reaching the surface can be expressed as:

$$I_s(E) = \int_0^{z_j} I(E,z) P(E,z) dz \tag{42}$$

where $P(E,z)$ is the probability that a photon of energy E emitted at a distance z from the junction will reach the surface:

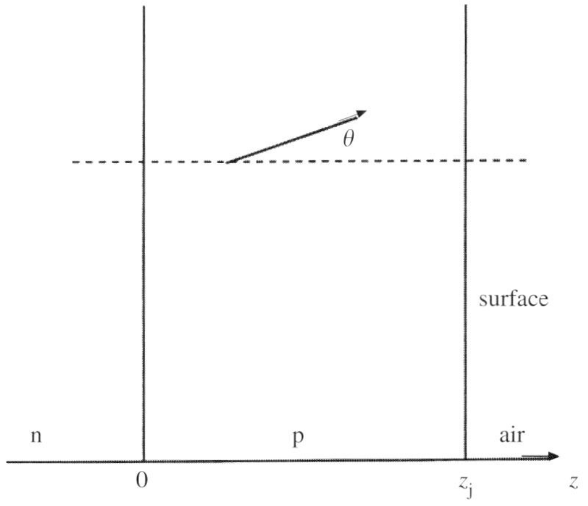

Figure 4
Photon generation and propagation in an LED.

$$P(E,z) = \frac{1}{2}\exp[-\alpha'(E)(z_j - z)] \quad (43)$$

As shown by Hakki, the quantity α' is approximately 2.2 times the value of the nominal absorption coefficient, accounting for off-axis photons. The integral in Eqn. (42) is found to be:

$$I_s(E) = \frac{I_o g(E)}{2}\frac{\exp(-\alpha' z_j) - \exp(-z_j/L_n)}{1-\alpha' L_n} \quad (44)$$

Therefore the loss component for self-absorption is given by:

$$f_{\text{absorption}} = \int_0^\infty \frac{\exp(-\alpha' z_j) - \exp(-z_j/L_n)}{2(1-\alpha' L_n)} g(E)\mathrm{d}E \quad (45)$$

Note that in Eqn. (45), there are two interesting limiting cases. The integral goes to zero as the junction depth goes to zero—there is no light emission because there is no radiative recombination volume. The integral also goes to zero as the junction depth goes to infinity—there is no light reaching the surface because there is complete reabsorption.

The integral in Eqn. (45) can be evaluated only when the energy dependence of the spectral distribution and absorption coefficient is specified. For illustration purposes, consider the case of a spectral distribution given by the delta function (i.e., zero linewidth). Consequently, the absorption factor is given by:

$$f_{\text{absorption}} = \frac{\exp(-\alpha' z_j) - \exp(-z_j/L_n)}{2(1-\alpha' L_n)} \quad (46)$$

This function is plotted in Fig. 5 and it shows that there is an optimal junction depth to limit the impact of self-absorption. This depth is given by:

$$z_{j,\text{opt}} = \frac{L_n}{\alpha' L_n - 1}\ln(\alpha' L_n) \quad (47)$$

which is approximately 0.7 μm for a semiconductor in which the diffusion length is 2 microns and the absorption coefficient is $10^4\,\text{cm}^{-1}$, yielding a value of the absorption factor of roughly 0.10. In other words, there can be substantial reabsorption of radiation and subsequent emission losses when the photon energy is close to the bandgap energy.

2.2 *Strategies to Reduce the Internal Loss Factors*

In short, the three loss factors can severely limit the optical output of an LED unless steps are taken to reduce their influence. Some strategies that have been employed to mitigate their effects are the following.

To reduce the impact of Fresnel reflection, antireflection films, or index-matching coatings can be applied to the LED. Anti-reflection films can reduce the Fresnel reflection to a negligible amount. However, anti-reflection films are usually multi-layer structures with tight thickness tolerances, adding substantially to the cost of LED manufacture. Since LEDs must be packaged in a glass or plastic enclosure, a better strategy is to make the packaging material have an index of refraction as close as possible to the value:

$$n_{\text{package}} = \sqrt{(n_{sc}n_{air})} \cong \sqrt{n_{sc}} \quad (48)$$

To reduce the effect of total internal reflection, it is necessary to eliminate or reduce the planarity of the devices. To this end, hemispherical LEDs have been fashioned, as shown in Fig. 6. If the pn-junction is confined to a small volume near the center of the hemisphere, almost every photon moves in a radial path, thus striking the surface at a nearly normal incidence angle. The manufacture of this kind of structure is complicated, but for indicator and display LEDs, it is still occasionally employed.

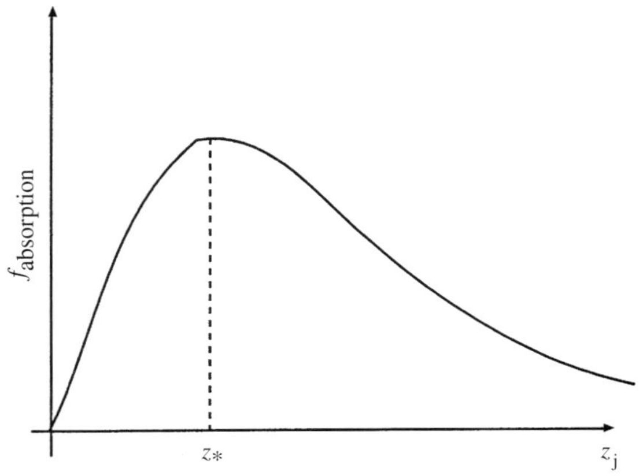

Figure 5
Absorption losses in an LED as a function of junction depth.

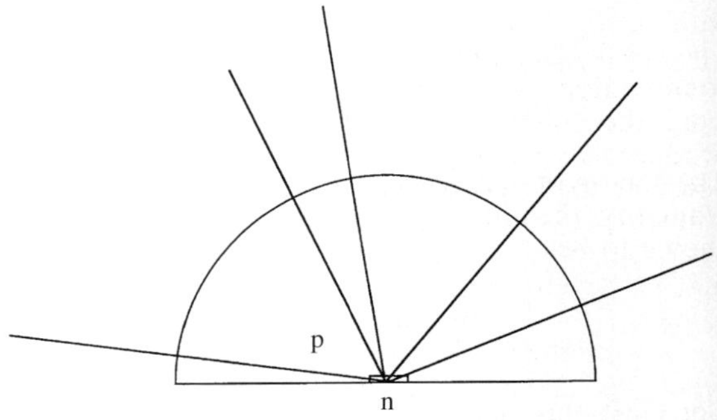

Figure 6
A hemispherical LED structure, designed to minimize internal reflection losses.

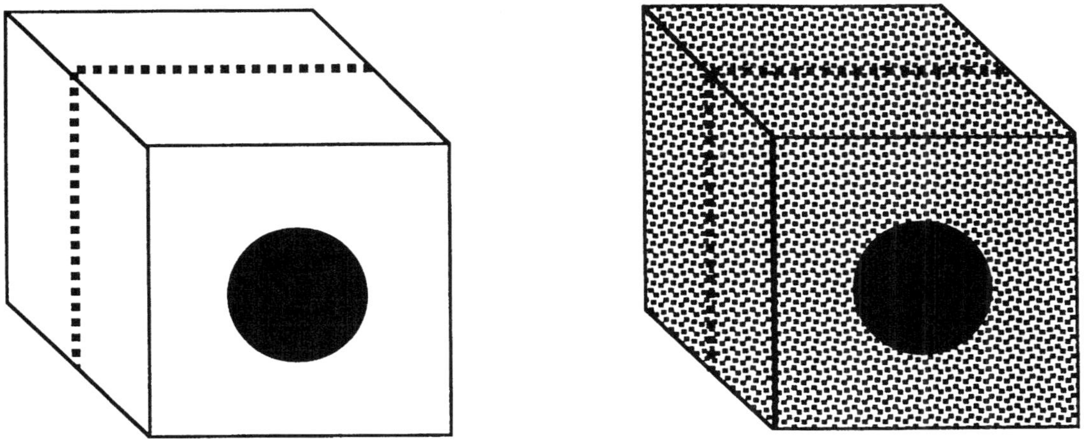

Figure 7
The LED on the left has smooth planar surfaces, while the one on the right has surfaces roughened by acid etching, and has considerably higher external quantum efficiency than the smooth LED.

An alternate and simpler approach to reduce internal reflection is to roughen the surface of the LED by chemical etching, as depicted in Fig. 7. A GaAs LED designed to emit radiation in all directions (as in an opto-isolator) is vastly improved by etching the LED in an acid bath to produce a variety of randomly oriented facets on the nominally smooth and planar surfaces. These facets provide greater opportunity for the photons to strike the surface at an angle smaller than the critical angle. This can improve the external quantum efficiency by more than a factor of two.

The reduction of internal absorption is difficult when the device employs direct band-to-band radiative recombination. Certain LED structures use a band-to-impurity recombination process that reduces the photon energy below the bandgap, and the semiconductor then appears nearly transparent to the generated radiation. Both GaAs and GaAlAs, when doped with silicon, have been used to fabricate LEDs with external quantum efficiencies approaching 40%. The basic principle of the band-to-impurity recombination process is shown in Fig. 8. Electrons from the conduction band move into the silicon-produced impurity band states and recombine with holes at an energy 100 meV below the bandgap. Unfortunately, the use of deep centers for radiative recombination produces a recombination process that has a long time constant, so high efficiency is gained at the expense of speed of operation.

For further information about LED emission processes, the interested reader should consult the texts by Bergh and Dean (1976), Kressel and Butler (1977), Jones (1987), and Wilson and Hawkes (1989).

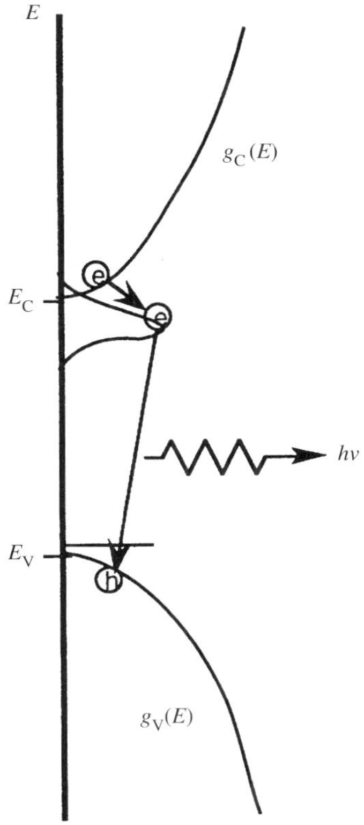

Figure 8
Emission process in a GaAs:Si LED.

3. *Device Design and Current Flow Considerations*

The current flow in a light-emitting diode has an enormous impact on the photon generation and emission pattern. As shown in Eqn. (5), the local generation rate is determined by the local current density. If an LED were to be constructed with full surface metallization on each side of the pn-junction, the current density through the junction would be uniform, as would the photon generation. But full surface metallization blocks photon passage, so generally at least one of the ohmic contacts has patterning to allow optical emission. However, the patterning causes the current flow to be nonuniform. As depicted in Fig. 9, current from a small area contact through a junction to a full surface contact may remain roughly collimated until it crosses the junction, or it may first move laterally before crossing the junction. As shown by Joyce and Wemple (1970), the flow pattern is determined by the junction depth, the ratio of upper to lower contact radii, and the resistivity of the upper semiconductor layer. As expected, current tends to flow in the path of least resistance, so if the upper layer has high resistivity, the current crowds under the contact, and if the upper layer has low resistivity, the current will spread laterally. This phenomenon can be exploited in LED design. By constructing the LED to have significant current crowding, the local current density under the restricting contact can be much larger than the geometrically average current. This type of LED can then generate a bright, centered beam of light suitable for coupling into an optical fiber. On the other hand, an LED without current crowding will produce a more uniform light emission pattern and is appropriate for an LED to be used as an indicator bulb.

There are many commercial LEDs that have a simple composition, such as the two layer GaP visible LEDs and the amphoterically doped GaAs:Si and GaAlAs:Si LEDs for opto-isolator applications. However, the LEDs to be used in high-speed optical communication links or as super bright indicator lamps require a more sophisticated and complicated structure. The normal embodiment of a high performance LED is very similar to that for a semiconductor laser, employing a double heterostructure as shown in Fig. 10. The device construction includes a buffer layer to isolate the substrate, two large bandgap layers straddling a smaller bandgap active layer to form the double heterostructure, a heavily doped low bandgap contact layer, a patterned dielectric layer to define the metal/semiconductor contact areas, and ohmic contact metallization. This kind of structure has been used with great success in the GaAs/GaAlAs, the InP/InGaAsP, the InGaAlP/GaAs, and the GaInAlN/GaN material systems.

The main advantage of the double heterostructure device is the high efficiency the structure naturally provides. As shown in Fig. 11, both holes and electrons are injected into and contained by the low bandgap active layer. The close confinement increases the recombination probability dramatically. In fact, the generation rate can now be written as:

$$g_{\mathrm{n,inj}} = \frac{J_{\mathrm{n}}}{q d_{\mathrm{active}}} \tag{49}$$

where d_{active} is the active layer thickness. This also causes the apparent recombination lifetime to become:

$$\tau_{\mathrm{r}} = \frac{q d_{\mathrm{active}} \Delta n}{J_{\mathrm{n}} \eta_{\mathrm{int}}} \approx \sqrt{\left(\frac{q d_{\mathrm{active}}}{B_{\mathrm{sp}} J_{\mathrm{n}} \eta_{\mathrm{int}}}\right)} \tag{50}$$

after modifying Eqn. (25) with the insertion of Eqn. (20) for the case of high injection. In other words, the higher the current density and the smaller the active layer thickness, the smaller the radiative lifetime, or the higher the device speed.

For additional current LED device configurations and design strategies, the interested reader should consult the texts by Bhattacharya (1997) and Fukuda (1999).

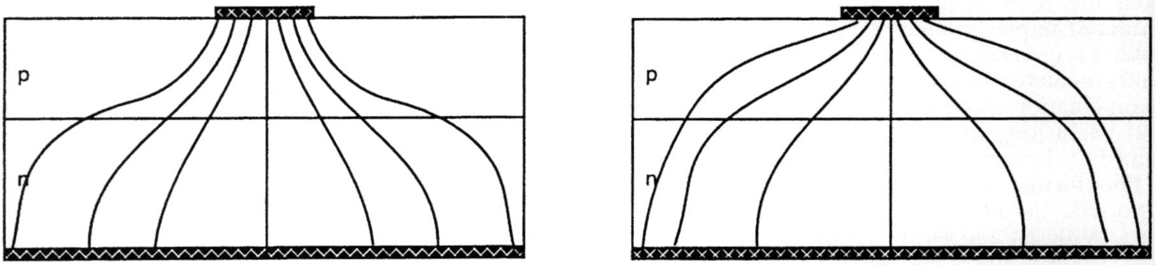

Figure 9
Possible current flow patterns in a typical LED structure.

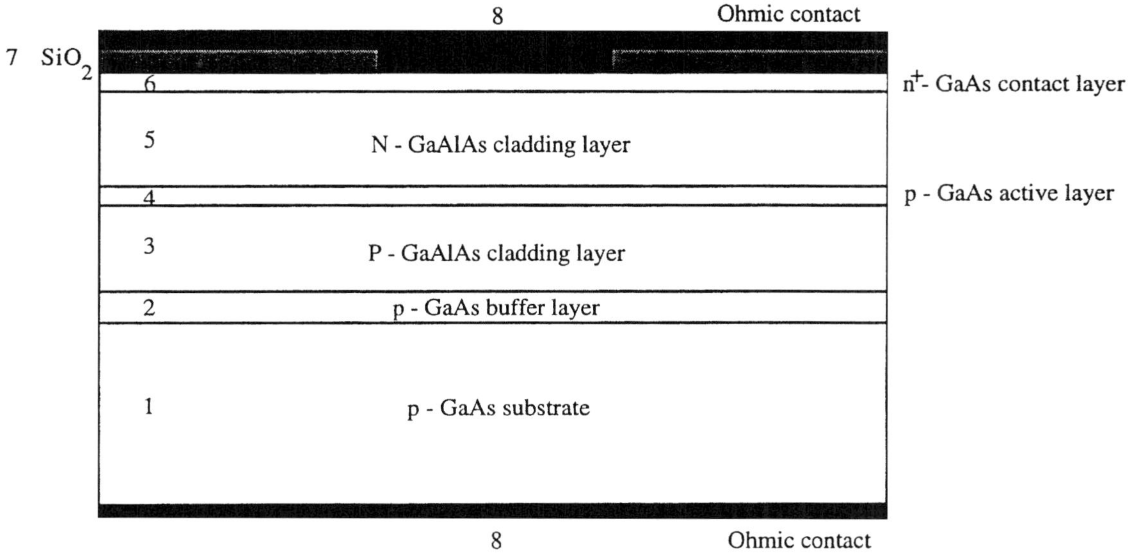

Figure 10
Cross-section of a typical GaAlAs/GaAs double heterostructure LED.

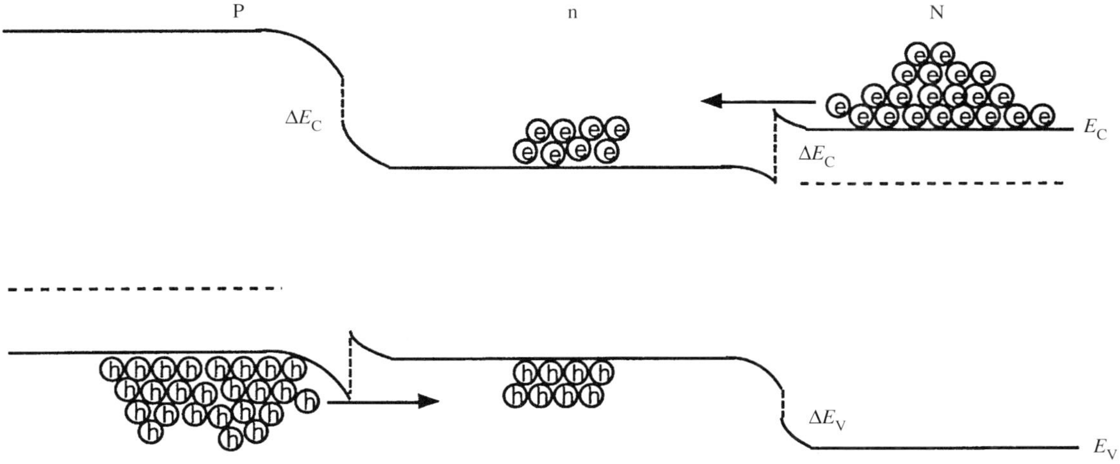

Figure 11
Energy band diagram of a double heterostructure LED with forward bias.

4. *Concluding Remarks*

Although most modern optical systems employ semiconductor laser devices because of the combination of high power, high speed, and recently reduced cost, there remain many areas where light-emitting diodes are the optical source of choice. These applications include: low cost, solid-state indicator lamps; super bright displays; opto-isolator and coupled pair structures; short haul optical communication systems; and systems that require high reliability and/or low power consumption. We can expect additional refinements in device structure and materials to continue.

See also: GaN-based Optoelectronic Devices

Bibliography

Bergh A A, Dean P J 1976 *Light Emitting Diodes*. Clarendon Press, Oxford

Bhattacharya P 1997 *Semiconductor Optoelectronic Devices*. Prentice-Hall, Englewood Cliffs, NJ

Dishman J M, DiDomenico M Jr., Caruso R 1970 Luminescence and minority carrier recombination in p-type GaP(Zn,O). *Phys. Rev. B* **2**, 1988–2009

Fukuda M 1999 *Optical Semiconductor Devices*. Wiley, New York

Hakki B W 1971 Theory of luminescent efficiency of ternary semiconductors. *J. Appl. Phys.* **42**, 4981–5015

Jones K A 1987 *Introduction to Optical Electronics*. Harper and Row, New York

Joyce W B, Wemple S H 1970 Steady-state junction current distributions in thin resistive films on semiconductor junctions. *J. Appl. Phys.* **41**, 3818–30

Kressel H, Butler J K 1977 *Semiconductor Lasers and Heterojunction LEDs*. Academic Press, New York

Peyghambarian N, Koch S W, Mysyrowicz A 1993 *Introduction to Semiconductor Optics*. Prentice-Hall, Englewood Cliffs, NJ

Saleh B E A, Teich M C 1991 *Fundamentals of Photonics*. Wiley, New York

Williams E W, Hall R 1978 *Luminescence and the Light Emitting Diode*. Pergamon, New York

Wilson J, Hawkes J F B 1989 *Optoelectronics, an Introduction*. Prentice-Hall, New York

Light Emitting Diodes from Molecular and Polymer Semiconductors

N. Tessler

Organic light emitting diodes (LEDs) have progressed in the past few years to an almost mature level. High quality devices are being reported by various groups showing high power efficiencies with a corresponding external efficiency that exceeds 5% (Cao *et al.* 1999, O'Brien *et al.* 1999, Ho *et al.* 2000). This progress is largely attributed to the improved purity of the starting material and the understanding of its sensitivity to various processing stages. The main drive for their use is the potential low cost which in the case of polymers is associated with the solution processability of such polymers. The other attribute is the coverage of the entire visible spectrum making them suitable for display applications. The progress in the field is due to contributions from a variety of disciplines ranging from chemistry through physics to device engineering. This article will discuss some fundamental properties of conjugated materials which are relevant to light emitting devices such as LEDs.

This article will cover more polymer-based devices compared to those made of small molecules. The figures used in this work are largely taken from data acquired by the author and his colleagues. However, the rapid progress in this field is due to a large number of research groups, as the text will show (Bradley 1996, Sheats *et al.* 1996, Heeger 1998, Friend *et al.* 1999, Tessler 1999, Cacialli 2000).

1. *Organic Semiconductors*

The term organic semiconductors is very general and this article will be concerned with the family of conjugated polymers which were largely developed in the 1990s mainly for light emitting purposes, with the color display being in mind.

A sample of the materials that can be used for LEDs is shown in Fig. 1. The chemical structure follows the convention that each line corresponds to a chemical bond. The bonds are between carbon atoms unless shown otherwise. Although the structures are projected onto the plane of the paper it should be noted that the real structure is 3D and not necessarily planar. The structures shown include the polymer poly(*p*-phenylenevinylene) (PPV, Burroughes *et al.* 1990) that is a prototypical polymer, which serves as a basis for a wide range of chemical derivatives as poly(2-methoxy-5-(2′-ethyl-hexyloxy)-*p*-phenylenevinylene) (MEH-PPV, Braun and Heeger 1991) or CN-PPV (Cyano substituted PPV derivative). Other types of polymers, which exhibit more than 50% photoluminescence efficiency, are the polyfluorenes (PFO). For the polymers, only the

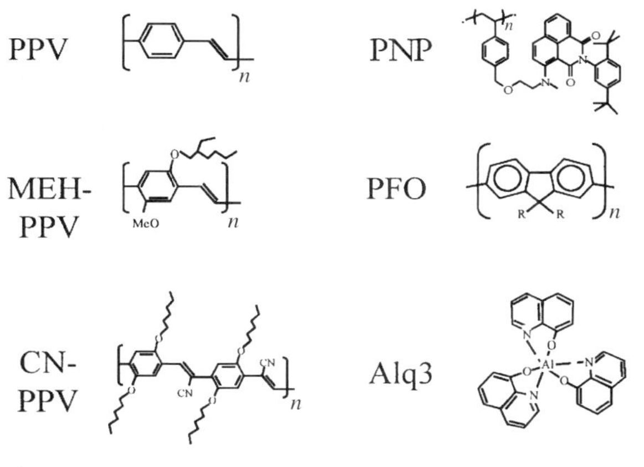

Figure 1
Chemical structure of a sample of luminescent organic materials.

basic unit (monomer) is sketched and the "$()_n$" means that this structure repeats itself many times to form the polymer chain. Note that in all structures the atoms are connected by at least one bond. This bond, called a sigma bond, is strong and it creates the skeleton of the polymer. The drawn structures also include a second bond which is alternating between every two atoms. This alternating bond is a convention of representing the fact that there is conjugation in the structure, i.e., there is an electron cloud that is shared between the atoms and which links them through an additional, less strong, bond. These electrons which are shared and are not so localized are called the π electrons and are those that contribute to the charge transport (conductivity). If the monomer structure repeats itself without interruptions (chemical or physical) there could be an electronic continuity over more than one unit cell along the chain. The distance between interruptions is then called the *conjugation length*, which has a similar effect to a width of a quantum well in a "particle in a box" picture. Namely, the shorter the conjugation length, the shorter would be the corresponding emission wavelength (it will be of higher energy). The other type of organic materials are the sublimed small molecules that are represented here by Alq3 (Tang and VanSlyke 1987). Most of the examples (figures) are polymer based, but the small molecule organics are also potentially important and many scientific and technological achievements have been reported through the use of sublimed molecules (see references). We expect that the physical properties and attributes that will be presented for polymer materials can be extended to include the small organic molecule-based devices.

As stated above, one of the attractive properties of organic materials is the ability to cover the entire visible spectrum and in the above example PFO is blue, PPV and Alq3 are green, and CN-PPV has an emission extending to the infra-red (see Fig. 2). The spectra shown in Fig. 2 for the blue and green polymers exhibit the typical vibronic structure which appears as a shoulder in the emission spectrum. These shoulders appear due to strong coupling between the emission and the vibration of the polymer chain. Nonuniformity in the local environment or the material physical structure throughout the film (as conjugation length, intermolecular coupling, etc.) may give rise to spatial fluctuations in the emission wavelength which appear as an inhomogeneous broadening of the emission spectrum as depicted by the width of the vibration lines. In the extreme case, as for the red polymer, this vibration structure is completely smeared out by the broadening.

The strong coupling between the excitations (charged or neutral) and the physical structure makes them polaronic in nature. The polaronic nature means that the charges modify (polarize) the environment they reside in (hence polarons), which results in self-trapping. In a polymer the change of environment would be a change in bond lengths and/or twists of parts of the chain. The emitting species are commonly believed to be excitons. The exciton is an electron-polaron which is correlated with a hole-polaron where the dissociation energy is typically much larger than the thermal energy (kT), at room temperature (i.e., they are bound together). In this case, the luminescence is the radioactive decay of the exciton. The existence of excitons may have a large effect on the maximum available electroluminescence (EL) efficiency, namely under voltage or current excitation. The correlation between the electron and the hole imply that their spins are correlated and the total spin of the exciton can take either the value of zero or one. The spin zero is associated with only one state and is called a *singlet* while the spin one has three available states and hence is called a *triplet*. Since the triplet carries momentum it is very unlikely that it will undergo radioactive decay (see exceptions in the following section). In current injection schemes there is no correlation between the injected electrons and holes and hence some of the electron-hole recombination (exciton formation) events will result in singlet excitons and some in triplet excitons.

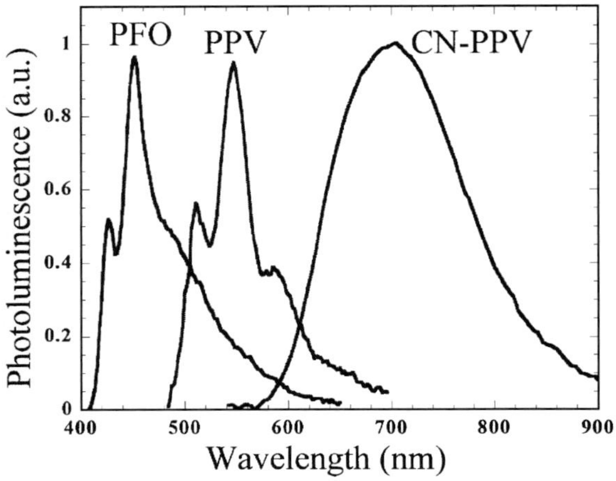

Figure 2
Photoluminescence spectrum of a sample of polymers covering the visible spectrum.

If the only argument would be that of spin statistics then it would be expected that when two oppositely charged polarons recombine they may form also nonemitting triplet excitons which may take up to 75% of the totally generated excitons. Therefore there could be a maximum of 25% of singlet (emissive) excitons generated which would set the theoretical maximum of EL efficiency at 25%. Conversely, it could be that due to the nature of the triplet state the generation of triplet excitons is a more difficult process (Shuai *et al.* 2000) and thus the balance is shifted towards singlet excitons and higher possible efficiency values. The 25% argument was recently supported by a paper probing the singlet and triplet excitons in a small molecule device (Baldo *et al.* 1999). Other recent papers, analyzing the efficiency of polymer LEDs, suggest that the probability for generating singlet excitons is beyond the 25% limit (Cao *et al.* 1999, Ho *et al.* 2000). It is too early to judge which result would be more wide spread and how or if such a result would depend on the specific material structure under study.

2. LEDs

One of the most common structures used for studying light-emitting-polymer device-physics is the LED. Since the device is being used to extract the polymer physical properties (hence device-physics), it is important to realize the relation between the measured properties of the device and the intrinsic properties of the polymer making the device, i.e., understanding the effect of the device structure on properties as internal quantum efficiency, emission spectrum, and more is still an issue. As the following paragraphs will show, organic devices are not like ones based on inorganic semiconductors and there is a rather large flexibility in the choice of material or device-structure combination.

Figure 3 shows a schematic description of an LED. It consists of two electrodes where at least one is transparent (as ITO) with emissive organic material in between. The top panel describes a single layer structure and the bottom panel a multilayer one. Also shown in Fig. 3 are the interfaces which are sometimes modified to improve the diode properties (injection/leakage). To provide good hole injection and stability the ITO is often coated/replaced by a doped

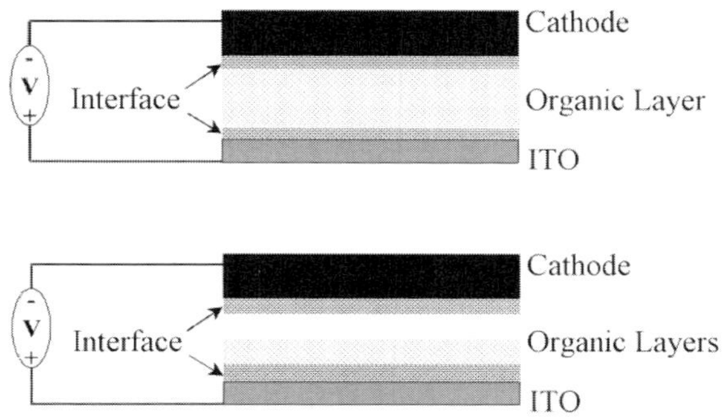

Figure 3
Schematic description of a single (top) and double (bottom) layer LED.

polymer. For electron injection, low work function metals such as calcium or lithium which are covered by aluminum to improve environmental stability are often used. Since the organic layer is nominally intrinsic (undoped) the charge transport is generally space charge limited. In cases where the energy barrier between the contact and the organic layer is large ($\geqslant 0.4\,\mathrm{eV}$) the injected charge density is relatively low so that the space charge is not significant. In this case the operation is considered injection limited.

The first electroluminescent devices were composed of a single organic layer in between two injecting electrodes (Burroughes *et al.* 1990, Braun and Heeger 1991, Tang and VanSlyke 1987, Pope *et al.* 1963). This single-layer functions both as charge transport layer and a charge recombination (and emission) layer. Due to this dual function the carriers that are injected from one electrode may be transported to the other electrode without recombining with the other carrier type. When a current flows directly from one electrode to the other it is a "leakage current" that reduces the diode efficiency. Another possible drawback of using a single layer is that both charge-transport and light-emission properties need to be optimized in the same material. Despite the above it was demonstrated, for some materials, that an optimized choice of electrodes or interfacial layers can enable highly efficient light emission from "single" layer structures (Cao *et al.* 1999, Ho *et al.* 2000).

It is well known that inorganic LEDs and laser diodes (LDs) performances have significantly improved with the introduction of heterostructure. In that case, the discontinuity in the electronic band can be viewed as an effective electric field ($E = \partial V/\partial Z$) which confines electrons and holes, in space, thus improving the probability of charge recombination and light emission ($e^- + h^+ \rightarrow$ photon). In the case of organic device structures, the heterostructure also show an improved performance over the single layer LED (Adachi *et al.* 1990, Brown *et al.* 1992). However, in the context of electronic transport in organic LEDs the term "band" is misleading. In general, carriers are moving in an amorphous molecular structure with discrete states (Van der Auweraer *et al.* 1994) rather than in a crystal with a continuum of states (band of states). Moreover, the transport properties are not governed solely by the position of the highest-occupied and lowest-unoccupied molecular orbital (HOMO and LUMO) levels but also by various microscopic details of the organic material structure. These details can, for example, change the mobility of holes and/or electrons by orders of magnitude. Hence, it is more appropriate to discuss discontinuity in the electronic transport rather than a discontinuity in electronic bands.

The other option for enhanced flexibility in the material choice or design is the use of mixtures or blends instead of multilayers. One approach is to create a solid solution where the bulk of the material (the matrix) is chosen for its transport properties and the dissolved entities (dopants) are chosen for their light emission properties and/or good matching to the injecting electrodes (Kido *et al.* 1992, Baldo *et al.* 1998, Cao *et al.* 1999). The second approach is to create a material that is made of several building units each intended to perform a different function. An example is the use of copolymers or block copolymers (Chen *et al.* 1999).

Finally, it is also important to note a new class of LEDs which is based on phosphorescence (triplet emission) rather then on singlet emission (Baldo *et al.* 1998, Cleave *et al.* 1999). In this scheme a phosphorescent dopant is added to the organic matrix which acts as an emission center. Such a dopant is the platinum octaethylporphyrin (PtOEP) where the heavy platinum atom enhances the spin-orbit coupling which facilitates the transfer of excitations from the singlet to the triplet state as well as the efficiency of light emission from the triplet state (25–50% in this case). It has been shown that it is possible to channel both singlet (Förster transfer) and triplet (Dexter transfer) exciton, which may be created in the host, to the dopant and thus recover the "lost" triplets which would otherwise decay nonradioactively in the host. Figure 4 illustrates the states that could be formed in the host and the possible transfer of both singlet and triplet energy to a dopant. As was mentioned above, up to 75% of the excitation in charge injection schemes (LEDs) could be of the triplet-state so recovering them could be significant in terms of efficiency. In this context it should also be mentioned that efficiency could also improve if the dopant acts as a recombination center where the triplet energy results in efficient radiation. In this case one can also consider doping with complexes as the lanthanide complexes (McGehee *et al.* 1999) where the lanthanide is sensitized via the triplet state of the complex.

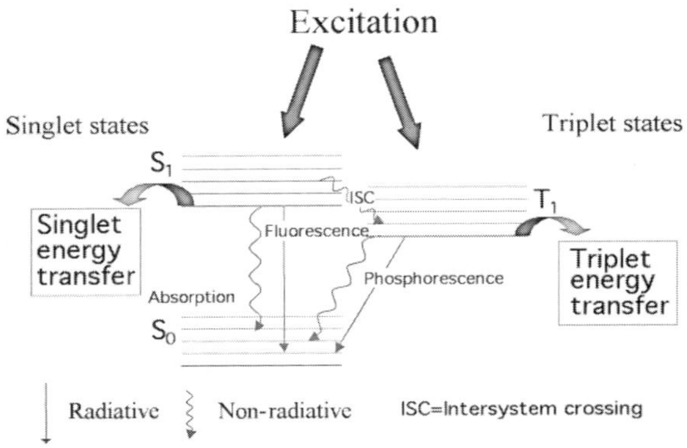

Figure 4
Schematic description of the singlet exciton and triplet exciton states which could be formed in current excitation. The figure also shows the possible internal energy transfer between singlet and triplet states as well as energy transfer to other molecules

3. Concluding Remarks

As was shown in the above sections, organic materials are rather flexible and they allow for a range of device and material configurations. While at the early stages this freedom made it hard to reproduce results (especially across the globe), the progress in the understanding of the factors governing the material/device properties makes it possible to use this freedom as a true advantage. Today, high brightness LEDs are being reported with external efficiencies (defined as the number of photons emitted from the structure divided by the number of injected charge carriers) exceeding 5% (Cao *et al.* 1999, O'Brien *et al.* 1999, Ho *et al.* 2000). Lately, there is an accelerated progress in this area and it can be partly attributed to the serious interest demonstrated by commercial bodies which see the organic display as a promising first product. In this context it is also important to mention the evolution in new device fabrication techniques as printing (Hebner *et al.* 1998) or soft lithography (Duffy *et al.* 1999). It is expected that these and other methods will play an important role in future applications such as the integrated optoelectronic device (Sirringhaus *et al.* 1998, Dodabalapur *et al.* 1998) or electronic paper.

Bibliography

Adachi C, Tsutsui T, Saito S 1990 Confinement of charge carriers and molecular excitons within 5-nm-thick emitter layer in organic electroluminescent devices with a double heterostructure. *Appl. Phys. Lett.* **57**, 531–3

Baldo M A, O'Brien D F, Thompson M E, Forrest S R 1999 Excitonic singlet–triplet ratio in a semiconducting organic thin film. *Phys. Rev. B. Condens. Matter* **60**, 14422–28

Baldo M A, O'Brien D F, You Y, Shoustikov A, Sibley S, Thompson M E, Forrest S R 1998 Highly efficient phosphorescent emission from organic electroluminescent devices. *Nature* **395**, 151–4

Bradley D 1996 Electroluminescent polymers: materials, physics and device engineering. *Current Opinion in Solid State & Mater. Sci.* **1**, 789–97

Braun D, Heeger A J 1991 Visible light emission from semiconducting polymer diodes. *Appl. Phys. Lett.* **58**, 1982–4

Brown A R, Bradley D D C, Burroughes J H, Friend R H, Greenham N C, Burn P L, Holmes A B, Kraft A 1992 Poly(p-phenylenevinylene) light emitting diodes: Enhanced electroluminescent efficiency through charge carrier confinement. *Appl. Phys. Lett.* **61**, 2793–5

Burroughes J H, Bradley D D C, Brown A R, Marks R N, Mackay K, Friend R H, Burns P L, Holmes A B 1990 Light-emitting diodes based on conjugated polymers. *Nature* **347**, 539–41

Cacialli F 2000 Organic semiconductors for the new millennium. *Phil. Trans. R. Soc. Lond. A* **358**, 173–92

Cao Y, Parker I D, Yu G, Zhang C, Heeger A J 1999 Improved quantum efficiency for electroluminescence in semiconducting polymers. *Nature* **397**, 414–7

Chen J P, Markiewicz D, Lee V Y, Klaerner G, Miller R D, Scott J C 1999 Improved efficiencies of light-emitting diodes through incorporation of charge transporting components in tri-block polymers. *Synth. Met.* **107**, 203–7

Cleave V, Yahioglu G, LeBarny P, Friend R H, Tessler N 1999 Harvesting singlet and triplet energy in polymer LEDs. *Advanced Materials* **11**, 285–8

Dodabalapur A, Bao Z, Makhija A, Laquindanum J G, Raju V R, Feng Y, Katz H E, Rogers J 1998 Organic smart pixels. *Appl. Phys. Lett.* **73**, 142–4

Duffy D C, Jackman R J, Vaeth K M, Jensen K F, Whitesides G M 1999 Electroluminescence materials with feature sizes as small as 5 μm using elastomeric membranes as masks for dry lift-off. *Adv. Mater.* **11**, 546–53

Friend R H, Gymer R W, Holmes A B, Burroughes J H, Marks R N, Taliani C, Bradley D D C, Dossantos D A, Bredas J L, Logdlund M, Salaneck W R 1999 electroluminescence in conjugated polymers. *Nature* **397**, 121–8

Greenham N C, Friend R H 1995 Semiconductor device physics of conjugated polymers. In: Ehrenreich H, Spaepen F A (eds.) *Solid State Physics*. Academic Press, New York, Vol. 49, pp. 2–150

Hebner T R, Wu C C, Marcy D, Lu M H, Sturm J C 1998 Ink-jet printing of doped polymers for organic light emitting devices. *Appl. Phys. Lett.* **72**, 519–21

Heeger A J 1998 Light emitting diodes, light emitting electrochemical cells, lasers and white light for the future. *Solid State Communications* **107**, 673–9

Ho P K H, Kim J-S, Burroughes J H, Becker H, Li S F Y, Brown T M, Cacialli F, Friend R H 2000 Molecular-scale interface engineering for polymer light-emitting diodes. *Nature* **404**, 481–4

Kido J, Kohda M, Okuyama K, Nagai K 1992 Organic electroluminescent devices based on molecularly doped polymers. *Appl. Phys. Lett.* **61**, 761–3

McGehee M D, Bergstedt T, Zhang C, Saab A P, O'Regan M B, Bazan G C, Srdanov V I, Heeger A J 1999 Narrow bandwidth luminescence from blends with energy transfer from semiconducting conjugated polymers to europium complexes. *Adv. Mater.* **11**, 1349–54

O'Brien D F, Baldo M A, Thompson M E, Forrest S R 1999 Improved energy transfer in electrophorescent devices. *Appl. Phys. Lett.* **74**, 442–4

Pope M, Kallmann H, Magnante P 1963 Electroluminescence in organic crystals. *J. Chem. Phys.* **38**, 2042–3

Sheats J R, Antoniadis H, Hueschen M, Leonard W, Miller J, Moon R, Roitman D, Stocking A 1996 Organic electroluminescent devices. *Science* **273**, 884–8

Shuai Z, Beljonne D, Silbey R J, Bredas J L 2000 Singlet and triplet exciton formation rates in conjugated polymer light-emitting diodes. *Phys. Rev. Lett.* **84**, 131–4

Sirringhaus H, Tessler N, Friend R H 1998 Integrated optoelectronic devices based on conjugated polymers. *Science* **280**, 1741–3

Tang C W, VanSlyke S A 1987 Organic electroluminescent diodes. *Appl. Phys. Lett.* **51**, 913–5

Tessler N 1999 Lasers based on semiconducting organic materials. *Adv. Mater.* **11**, 363–70

Van der Auweraer M, de Schryver F C, Borsenberger P M, Bassler H 1994 Disorder in charge transport in doped polymers. *Adv. Mater.* **6**, 199–213

Light Water Reactor Fuel Design and Performance

D. R. Olander

Many nuclear reactor types have been designed and developed, and some have had brief commercial lives. The light-water reactor (LWR) has survived as the major provider (85%) of nuclear-generated electricity in the world. This article covers two aspects of LWR fuel and associated hardware: their design and performance. Three basic components are emphasized. The first is the ceramic *fuel pellet* and the second is the thin-walled metal tube called *cladding* that hermetically contains the fuel. Taken together, these two components constitute a *fuel element*, sometimes called a *fuel rod*. The third component is the supporting hardware that holds 100–200 fuel elements into a unit that provides adequate space for water coolant flow during operation and sufficient rigidity for handling, loading and unloading, and transportation to and from the reactor. This package of fuel rods is called a *fuel assembly*. (See *Nuclear Reactor Fuel Fabrication (Including Quality Control)*).

The LWR uses ordinary water as the coolant to remove heat generated by fission and as the moderator to slow down neutrons to energies at which the fission process is efficient. Like the internal combustion engine, the basic design of the LWR fuel element has not changed since its introduction in the early 1950s. It consists of a stack of pellets containing fissionable ^{235}U and ^{239}Pu held in a thin-walled tube called cladding. The fuel is ceramic UO_2 with ^{235}U enrichments up to 5%. The Canadian heavy-water reactor called CANDU uses D_2O as the moderator so that natural uranium in the form of UO_2 can be employed as the fuel (see *Heavy Water Reactor Fuel Design and Performance*).

1. *Fuel Pellets*

The standard fuel pellet is a solid cylinder of polycrystalline UO_2 1cm or less in diameter. The two variants of the LWR, the *boiling water reactor* (BWR) and the *pressurized water reactor* (PWR), use slightly different pellet sizes, that for the BWR being larger. The pellet height in both fuels is between 1 cm and 2 cm.

There are strict limits on the impurity elements in UO_2. It is particularly important to maintain concentrations of fluorine and water in the low tens of ppm in order to avoid internal corrosion of the cladding. Oxygen in excess of the 2:1 ratio of stoichiometric UO_2 is also to be avoided. The excess oxygen in UO_{2+x} (even when $x \approx 0.01$) is corrosive towards the cladding. Hyperstoichiometry also produces undesirable property changes in the fuel. Excess oxygen reduces the thermal conductivity, causing the fuel temperature to increase, and enhances the mobility of the fission products, making their release easier.

The microstructure of fuel pellets is also closely controlled. The usual pellet fabrication process produces polycrystals with grain diameters of about 8 μm. The porosity of the pellets consists almost exclusively of closed spherical cavities with negligible open porosity. Porosity is expressed as a percentage of theoretical density (TD). Most fuels are 95–96% TD. Retention of a few percent of porosity is desirable. The pores serve as sinks for fission gases and lessen their release. Swelling of the fuel owing to solid fission products is also reduced by filling in the internal voidage.

The size of the pores is as important as the total porosity. In operation, very fine pores (<1 μm) are eliminated by a process called radiation densification. In essence, the pores are converted to their component vacancies, which are then removed by sinks such as dislocations and grain boundaries. If not mitigated, this process causes an initial rapid reduction of the pellet diameter and concomitant collapse of the cladding on the shrunken fuel.

Over time, the standard cylindrical fuel pellet design has been modified in numerous ways, as shown in Fig. 1. The standard pellet is shown in Fig. 1(a). During operation, the temperature distribution in the pellet causes its shape to change to the "hourglass" shape shown in Fig. 1(b). The edges of the top and bottom pellet surfaces deform the cladding with high local strains and stresses, risking perforation. Cladding subjected to this type of deformation resembles a stalk of bamboo. To avoid this type of cladding deformation, and to minimize chipping during manufacturing and handling, the fuel pellets are chamfered as shown in Fig. 1(c). Additional void space to accommodate fission product swelling and axial thermal expansion of hot pellet centers is provided by "dishing" the top and bottom pellet surfaces.

In a solid pellet, fission heat generation and removal creates differences in temperature between the centerline and the pellet surface as large as 1000°C. This is undesirable for a number of reasons, the principal one being release of fission gases from the hot center. To reduce the centerline temperatures at the same linear power (i.e., power produced per unit fuel height), the designs shown in Figs. 1(d) and (e) have been developed. Both move the heat source closer to the heat sink (the coolant) than in the standard fuel pellet. In Fig. 1(d), the outer radial zone of the pellet is more highly enriched in ^{235}U than the inner zone, thus reducing heat production near the center and lowering the centerline temperature.

In Fig. 1(e), the center is removed entirely, resulting in an annular pellet. To produce the same power, this pellet must contain a higher enrichment than the standard solid cylindrical pellet. The example shown in Fig. 2 compares the temperature distributions in solid and annular designs that produce the same linear power. The diameter of the central hole is one-fifth of

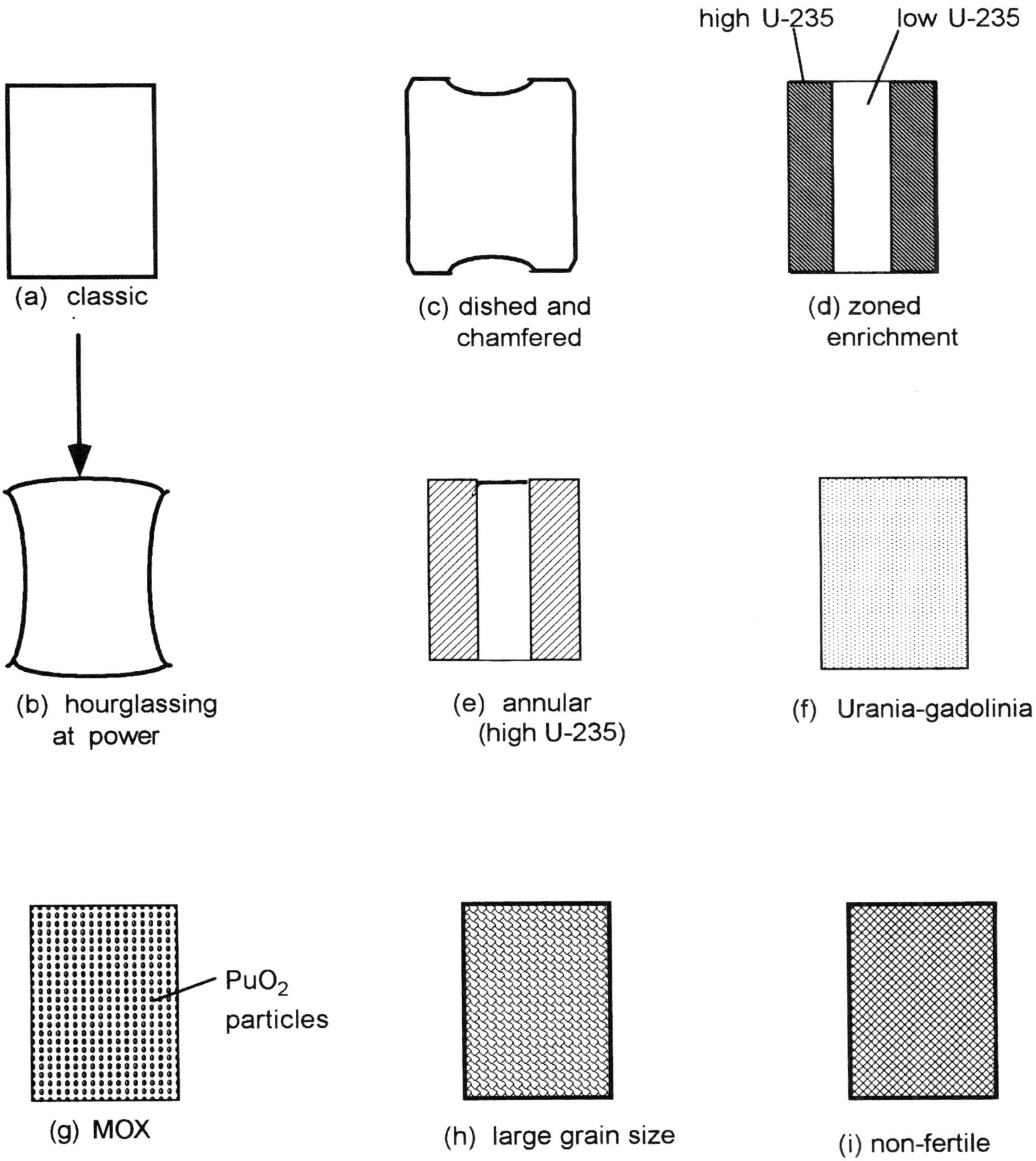

Figure 1
Ceramic oxide pellet designs.

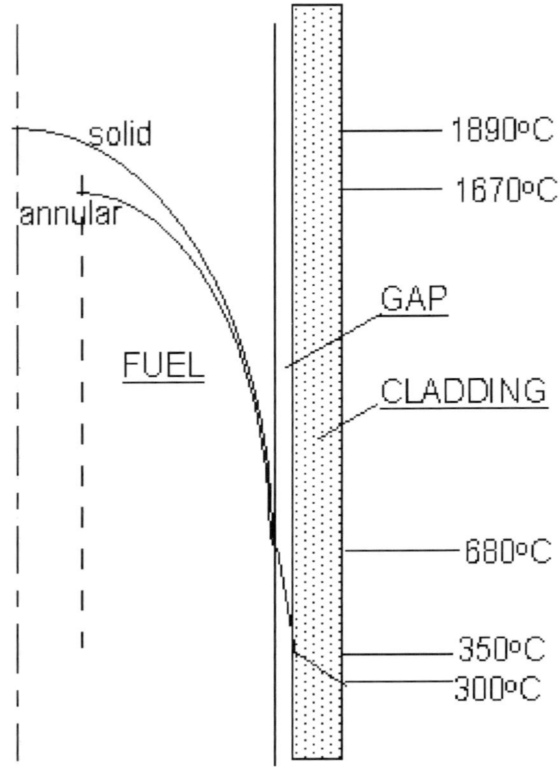

Figure 2
Radial temperature distributions in solid and annular fuel pellets. Linear power: 350 W cm^{-1}; gap thickness: 50 μm; fuel diameter: 10 mm; control hole diameter in annular pellet: 2 mm; gap gas: 20% xenon, 80% helium.

the pellet diameter. The annular pellet requires only 4% higher ^{235}U enrichment, but reduces the maximum fuel temperature by over 200 °C.

In two designs, urania is mixed with other oxides for reasons related to the nuclear processes. In Fig. 1(f), the additive is Gd_2O_3, where the gadolinium acts as a burnable poison that allows a longer irradiation time without excessive reactivity held in control rods at beginning-of-life. In Fig. 1(g), UO_2 is mixed with PuO_2 to produce what is called a mixed oxide or MOX. The plutonia is in the form of small particles in a matrix of UO_2. Since plutonium is the principal fissionable nuclide, these particles are hotter than the rest of the fuel. Despite this nonuniformity of temperature, MOX fuel behaves very similarly to standard UO_2 fuel in reactors.

In Fig. 1(h), the microstructure of the fuel is purposely altered to increase the grain size from 8 μm to as much as 40 μm by using oxides such as Nb_2O_5 as a fuel additive. Niobium promotes grain growth during the sintering process. The objective of the large grains is to reduce fission product release by increasing the length of the diffusion path from the grain interior to the grain boundary.

So-called "nonfertile" or "inert matrix" fuel, depicted in Fig. 1(i), is a response to the desire to burn excess plutonium without simultaneously producing this element from a ^{238}U component, as is the case with MOX fuel. Nonfertile fuels contain PuO_2 dissolved in a matrix such as $(Zr,Eb)O_2$. For additional details about this fuel type see *Inert Matrix Fuels*.

2. Cladding

Standard LWR fuel cladding is an alloy composed of 98% zirconium, ~ 1.5% tin, and the balance being iron, nickel, and chromium. Its commercial name is Zircaloy, or Zry for short. Zry comes in two varieties, Zry-2 and Zry-4. These two differ by minor changes in the concentrations of the transition metals (Zry-4 has no nickel). PWRs use Zry-4 while BWRs employ Zry-2 as cladding and either Zry-2 or Zry-4 for the duct or channel holding the fuel-element bundle. The transition metals iron and chromium appear in the microstructure as precipitates of the intermetallic compounds $Zr_2(Ni,Fe)$ and $Zr(Cr,Fe)_2$. These minor alloying elements enhance corrosion resistance to an extent dependent on the size of the precipitates.

The original purpose of tin in Zry was to mitigate the loss of corrosion resistance caused by pickup of impurity nitrogen during fabrication of the alloy. Zry production methods now offer far better control of gaseous impurities than early fabrication technology, so the original reason for adding tin is moot. Since tin is a substitutional impurity, its function is to increase the strength of Zry over that of pure zirconium.

The mechanical properties of the cladding can be controlled by the heat treatment during fabrication. "Annealed" Zry is softer than "stress-relieved" material. Because of the combination of the anisotropy of the crystal structure of zirconium and the orientation of the grains in the tubing produced by the extrusion process, Zry cladding has properties that are different in the radial and axial (longitudinal) directions. The grains or crystallites are preferentially oriented with the *c*-axis in the hexagonal structure pointing in the radial direction. Mechanical properties such as yield strength are different in the radial and axial directions in the tubing. Point defect (vacancy and interstitial) diffusion coefficients are greater in the longitudinal direction than in the radial direction because of the more open crystal structure in directions perpendicular to the *c*-axis. This consequence of anisotropy is believed to be related to the phenomenon of growth of Zry that is caused by displacement damage in a neutron flux (see *Nuclear Reactor Materials: Irradiation Effects*; *Nuclear Applications: Zirconium Alloys*).

In order to eliminate waterside corrosion of cladding as a life-limiting process, modified Zircaloys have been developed. Under PWR water conditions, reducing the tin content of Zry to 1% and adding a comparable concentration of niobium produces an alloy with a corrosion rate several times lower than that of the standard Zry. Modified low-tin alloys can be used either as a complete replacement for the Zry tube wall or as a thin outer skin on standard Zry substrate tubing.

The older BWR cladding fabrication method produced $Zr_2(Ni,Fe)$ and $Zr(Cr,Fe)_2$ second-phase particles as large as a few micrometers in diameter. These are large enough to provide an electrical short circuit in the ZrO_2 scale that grows around them during in-reactor operation. This loss of the insulating properties of the oxide is believed to accelerate the corrosion rate. As a remedy, rapid cooling of the Zry ingot during fabrication (called beta quenching) produces a much finer intermetallic precipitate population. This distribution of the transition metals maintains the strength of the Zry but improves its corrosion resistance, possibly by short-circuiting charge-transport paths between the scale surface and the metal–oxide interface.

Standard Zry is sufficiently hard to be susceptible to stress-corrosion cracking (SCC) when the pellet–cladding gap closes and the swelling fuel generates substantial tensile hoop stresses in the cladding. In the presence of an aggressive fission product (iodine is most often cited), a brittle crack can be initiated and penetrate the cladding wall. This is a primary cladding failure that admits steam into the rod with the accompanying potential for a more serious secondary failure due to hydriding. The combination of conditions (susceptible metal, tensile stress, and chemical enhancer) that produces the primary defect is termed *pellet–cladding interaction* (PCI).

As is well known in other instances of SCC, suppression of any one of the three necessary conditions cited above eliminates the potential for SCC. The response of BWR fuel vendors has been to line the inside surface of the cladding with a soft metal that is not susceptible to brittle cracking. Sponge zirconium (essentially pure zirconium) has been used for this purpose. This remedy did indeed reduce the incidence of PCI but, as is often the case with many *ad hoc* solutions, another problem arose as a result. The high-purity Zry liner corrodes in steam more rapidly than Zry, and hence produces more hydrogen. In the event of cladding penetration by an unrelated mechanism (e.g., debris fretting), more H_2 is produced and the risk of a massive secondary hydriding failure is increased. In response to this unforeseen consequence, several fixes were introduced. In one, a second inner-surface liner of ordinary Zry is extruded on the sponge zirconium liner in order to restore corrosion resistance. In other modifications, the sponge zirconium is alloyed with iron or tin to improve oxidation resistance without affecting the softness that provides immunity from PCI.

The thickness of the cladding is a design parameter dictated by conflicting performance requirements. The effect of loss of metal by inner or outer corrosion is less important for thick cladding than for thin-walled tubing. For the same external loading, the stress produced in the cladding, either external from the coolant or internal from fission-product swelling of the fuel, is reduced in proportion to the wall thickness. However, thick cladding is detrimental to neutron economy, increases the volume of the reactor core, raises fuel temperature, and costs more than thin-walled tubing. The compromise between these competing demands has produced wall thicknesses of ~0.9 mm in BWRs and ~0.7 mm in PWRs.

3. *Fuel Elements*

Inserting a stack of UO_2 pellets into a length of cladding, placing a spring on the pellet stack for mechanical stability, and welding Zry caps on the ends of the tube produces the fuel element shown schematically in Fig. 3 (see *Nuclear Reactor Fuel Fabrication (Including Quality Control)*). The function of the fuel element is to hold fissile material in a mechanically stable, coolable geometry, to retain any fission products that are released from the fuel, and to prevent the coolant from making contact with the UO_2 pellets. The water contains enough dissolved O_2 to oxidize UO_2 to friable U_4O_9, which can be swept into the coolant (see *Nuclear Reactors: Coolant Materials*). The height of the fuel column is approximately 4m and the diameter of the fuel element ranges from 9mm to 11mm, PWR fuel being smaller than BWR elements.

The finer points of the design of the fuel rod are directed in large part towards mitigating the deleterious effects of fission gases (chiefly xenon) that are released from the fuel during operation. Fission gas collects mainly in the two void spaces in the rod, namely the radial gap between the fuel and the cladding and the empty space above the fuel stack, which is called the plenum.

The thickness of the gap in the as-fabricated fuel element is a prime design specification. The gap is filled with a gas, which is a very effective thermal insulator (even compared to ceramic oxides). To alleviate this effect, the rod is sealed while filled with helium, chosen because of its inertness and high thermal conductivity. Slightly diluted with fission gas, a rod operated at a linear heat rating of 350 W cm^{-1} with a gap thickness of 50 μm would support a 330°C temperature drop across the gap (see Fig. 2). Since the temperature of the inner cladding wall is fixed by the heat flux, the temperature difference across the gap directly adds to that in the fuel. The increased fuel temperature experienced by rods with thick gaps leads to numerous undesirable consequences, the most serious of which is increased fission-product release.

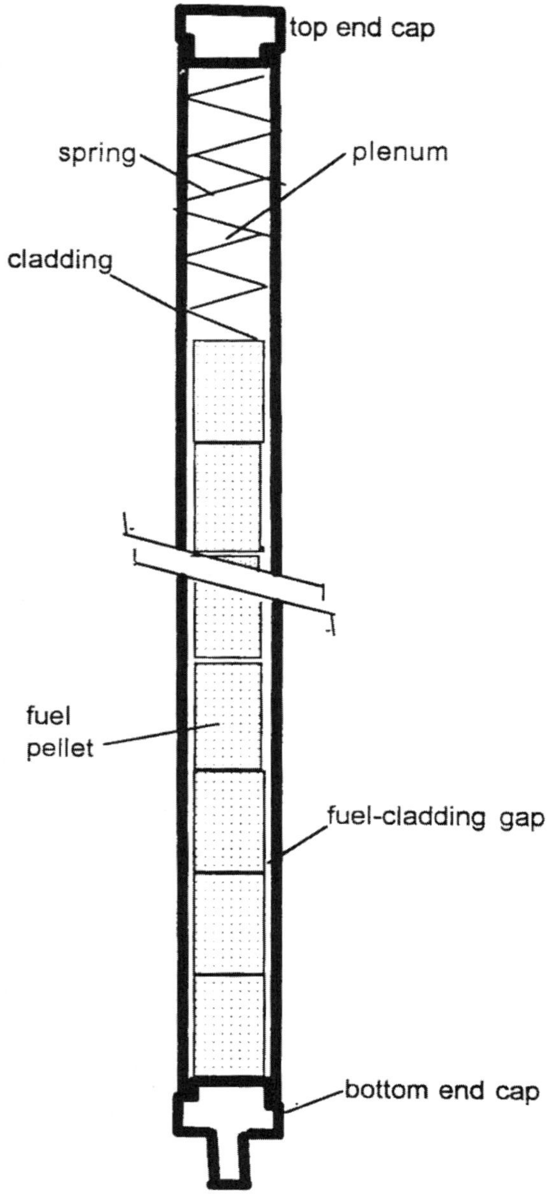

Figure 3
LWR fuel element.

The thickness of the fuel–cladding gap varies in a complex manner during irradiation. When first brought to power from a cold condition, the gap partially closes owing to differential thermal expansion of the pellet and the Zry tube. On a short time scale (months), fine porosity in the fuel is removed by radiation densification, which shrinks the fuel and opens the gap. At a slower rate, the gap is reduced in size by fission-product swelling of the fuel in concert with creep collapse of the cladding. The latter occurs because the external (coolant) pressure outside the rod is higher than the internal pressure due in part to the helium fill gas and in part to the released fission gases.

By design, the combination of the above processes leads to gap closure after 1–2 years at power. For this to occur, the initial cold gap radial thickness is specified in the range of 50–100 μm. However, contact of fuel and cladding does not assure elimination of the thermal resistance of the gap. Cladding creep collapse does not preserve the original circular shape of the tube but instead produces an oval shape that only partially makes contact with the fuel pellet. Moreover, even at locations of solid–solid contact, roughness of the metal and oxide surfaces and contamination of the interface by released fission products prevents tight binding; occasionally the thin ZrO_2 corrosion layer on the inner cladding surface reacts with the adjacent fuel to form a bonding interface of $(U,Zr)O_2$. Although the thermal resistance of the closed gap is less than that of an open gap, it is not zero and, moreover, is difficult to model mathematically.

Once fuel–cladding hard contact has been made, fission-product-induced swelling of the fuel produces hoop stresses in the cladding and strains the metal. If the cladding has become embrittled by radiation damage or accumulation of corrosion-product hydrogen, a rapid power rise can result in a through-wall crack. The fuel element has failed and releases a burst of volatile fission products (cesium, iodine, and xenon) to the coolant. More significantly, this PCI defect admits steam to the rod interior. Corrosion of the inner cladding surface produces hydrogen that can lead to massive hydride blisters or axial cracking of the cladding that forces reactor shutdown.

The temperature drop across the gap depends on the thermal conductivity of the gas as well as on gap thickness. This is the reason for the choice of helium as the filling gas. However, instead of sealing fuel rods at 1atm helium pressure, pressures as high as 20atm are used. The justification for introducing this added complexity to the fabrication process has to do with the thermal conductivity of the gas in the gap, but in an indirect way.

According to the kinetic theory of gases, the thermal conductivity is independent of pressure. The reason for high-pressure helium filling is to mitigate the reduction in thermal conductivity when fission-product xenon (which has a thermal conductivity 1/20 that of helium) released from the fuel mixes with the initial charge of helium in the gap. Because the gap is very thin, axial diffusion of xenon to the large helium reservoir in the plenum at the top of the rod is slow. If the initial helium filling pressure is low (say 1atm), xenon released from the fuel substantially increases the temperature difference across the gap and consequently raises the fuel temperature by the same amount. Since fission gas release increases exponentially with temperature, more xenon enters the gap and further reduces the thermal conductivity of the gas mixture. This autofeedback process causes a local hot spot in the fuel and nearly completely strips fission gas from the central zone of the affected region. Increasing the initial helium filling pressure to the 10–20 atm range has been found to eliminate this problem by decreasing the effect of xenon release on gap heat transfer.

In addition to the fuel–cladding gap, the other feature of the fuel element, shown in Fig. 3, that is directed exclusively to minimizing potentially harmful effects of fission gas release is the plenum. This seemingly wasted space (which can be as much as 20 cm long) is intended to provide sufficient volume for collecting released fission gases (xenon and krypton) without overpressurizing the rod interior. Internal pressurization does not threaten cladding integrity during operation because the coolant pressure (150 atm in PWRs, 70 atm in BWRs) is generally larger than the internal rod pressure. If the opposite is true, as could occur at high burnups in BWRs, the pressure differences causes the cladding to creep away from the fuel. Such displacement increases the gap width, with the attendant degradation of rod heat transfer. This phenomenon is called cladding liftoff, and is the reverse of cladding creepdown.

Even if the internal rod pressure is lower than the coolant pressure during operation, the reverse is true when the reactor is shut down, e.g., for refueling. With an external pressure of 1 atm, the tensile hoop stress produced by excessive internal rod pressure could cause brittle fracture of the cold cladding. Heavily hydrided cladding or fuel elements that are improperly handled during defueling or transport are particularly susceptible to this type of damage.

4. *Fuel Assemblies*

For the purposes of handling during transportation, loading/unloading, and storage, and for maintaining mechanical stability and proper spacing in a rapidly flowing coolant, fuel elements are packaged into units known as fuel assemblies (see *Nuclear Reactor Fuel Fabrication (Including Quality Control)*). Figure 4 shows fuel assemblies for two types of LWRs, the BWR assembly (top) and the PWR version (bottom).

The PWR assembly contains 17 fuel elements along each of its four sides, and is referred to as a 17×17 assembly. The overall assembly shape is square, as is the pitch of the rods forming the bundle. The square dimension of the assembly is 25–30 cm. Some 200 fuel assemblies are packed into the core of a 1000 MW reactor.

The external holes on the top-end fitting of the PWR assembly are guide tubes, or thimbles, in which an array of control rods can be moved in and out from above. The control rod material is an Ag–In–Cd alloy chosen for the high neutron absorption cross-section of the component elements and for the slow rate at which this absorptivity changes with time. The guide thimbles also hold gadolinia burnable poison rods and neutron sources for reactor startup.

One of the most important components of the fuel assembly is the series of equally spaced grids that hold the fuel rods in place. The top two of these appear as straps around the fuel bundle in Fig. 4, but they actually resemble egg cartons with holes in the bottom of each cup. They are made either of Zircaloy or of the high-nickel alloy Inconel. The transverse spacing of the fuel rods in the grids and hence in the assembly as a whole is as small as possible consistent with avoiding excessive pressure drop in the upward-flowing coolant. The vertical spacing of the grids along the fuel bundle is chosen to minimize vibration of the rods produced by the flowing coolant.

The vertical sides of the holes through which the fuel rods pass are formed into springs in a leaf or "I" shape, or a dimple. These are the contact points with the fuel rod, and the force that the grid exerts on the fuel rods must be set with great care. Too large a force causes cladding deformation, usually in the form of bowing. Too small a force allows the fuel rods to rattle about in their guide hole. The grid-to-rod contact is a major source of cladding failures. These failures are due to rubbing, or fretting, of the cladding against either the grid proper or against a small piece of tramp metal carried by the coolant and caught between the grid and rod. The relative movement of the grid and the rod cause the debris to rub against the cladding wall. Fretting defects are through-wall openings of submillimeter size. However, they permit steam to enter the rod and allow volatile fission products to leak into the coolant.

Grid-to-rod fretting arises from rod vibration generated by the high coolant velocity through the grid structure. This type of cladding wear can be minimized by clever design of the coolant passages through the grid that avoids excessive turbulence. The type of cladding wear due to trapped foreign objects is called *debris fretting*. It can be alleviated by preventing small metal particles from entering the assembly at the bottom. The bottom end-fitting (not shown in Fig. 4) contains holes to receive the bottom end caps of the fuel elements and a set of smaller holes that allow coolant to enter but filter out most of the debris that might be carried in from other parts of the primary circuit.

The fuel assembly shown in the upper part of Fig. 4 is a 10×10 BWR design. Although BWR rods are larger than those of the PWR, the lateral dimension of the BWR assembly is smaller. The central positions in some assemblies contain no fuel rods or empty, perforated cladding tubes (called "water rods"). These spaces are filled by liquid water instead of a two-phase mixture, thereby enhancing neutron moderation and increasing the fission rate in the remaining rods in the assembly. Like the PWR assembly, the BWR assembly contains a series of spacer grids, but these are not visible in Fig. 4. Cladding failure by debris fretting occurs in BWRs, but grid-to-rod fretting is not observed because the coolant is mostly steam rather than liquid water.

Also not shown in Fig. 4 is the square box into which the BWR fuel bundle fits. The walls of this box are ~3 mm thick and are made of Zircaloy. The box is called either a channel or a duct. There is no PWR analog of the BWR channel. Although channels serve to separate coolant of different steam quality and provide mechanical stability to the assembly, the principal reason for their use is connected with the method used for controlling nuclear reactivity in BWRs. Instead of control rods of the same diameter as the fuel elements that are inserted into the fuel assembly, the BWR control material (boron carbide) is contained in four cruciform blades of the same lateral dimension as the assembly. The four blades fit between four assemblies and make contact with the channel walls at a few dimples on the blades. The control blade assembly is moved into and out of the core from the bottom of the reactor pressure vessel.

5. *Fuel Performance*

The discussion up to now has been centered on the design issues of fuel elements and fuel assemblies. Some consideration of performance goals is necessary in order to justify design choices. In the following sections, the emphasis is on fuel performance. Performance is affected by imposed operating conditions as well as by design. A design is intended for a particular environment. If the latter changes, the design must be modified in order to maintain acceptable economic and regulatory performance under the new conditions.

For many years, LWRs have operated under rather static requirements: assembly-average burnups of ~30 MWd kg^{-1} for BWRs and 35–40 MWd kg^{-1} for PWRs (burnup is a measure of the degree of fission of the fuel; the unit MWd kg^{-1} refers to the fission energy delivered per unit mass of uranium; in terms of the fractional burnup, 10 MWd kg^{-1} is equivalent to fission of 1% of the initial uranium); thermal output per unit height of fuel (also called the linear heat rating) of 35 kW m^{-1} for PWRs and 30 kW m^{-1} for BWRs; and operating cycles (the time between shutdowns for refueling) of one year. Recent economic pressure from gas-fired electricity production has forced nuclear utilities to demand more of the fuel: average burnups approaching 60 MWd kg^{-1}

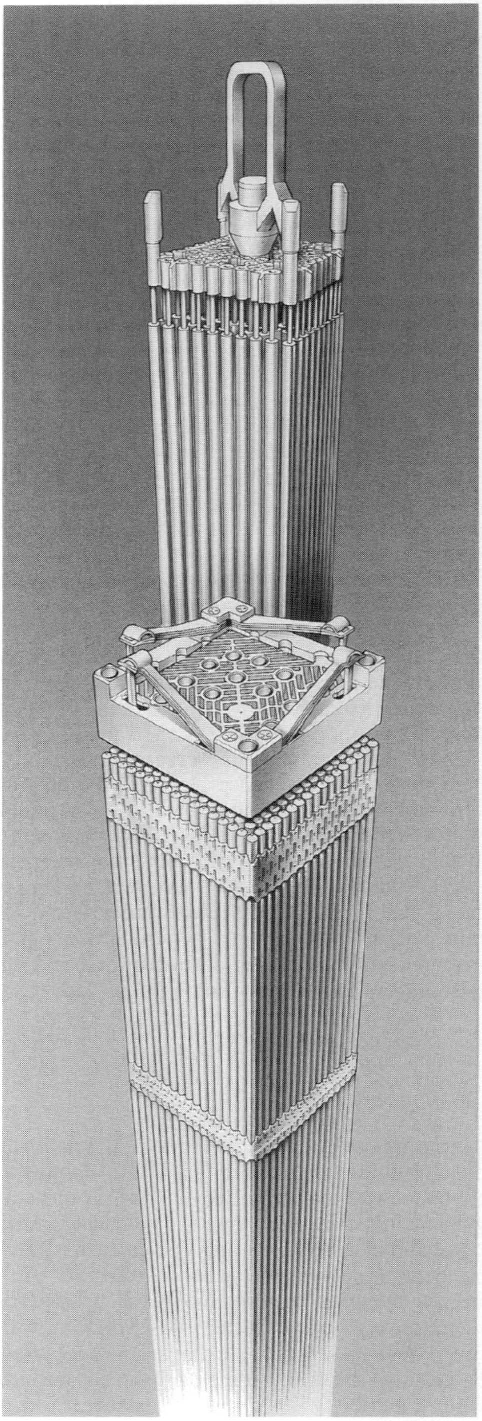

Figure 4
LWR fuel assemblies: BWR (top); PWR (bottom) (courtesy of Siemens Power Corp.).

for PWRs and 50 MWd kg^{-1} for BWRs; linear heat rating increases of 10–15% for PWRs; and operating cycles of 1.5–2 years.

Performance is measured on three levels. The first and broadest is the plant *capacity factor*. This measure is the ratio of the actual electrical energy delivered by the plant in a year to that which could have been produced had the plant operated at full power for the entire year. The capacity factor can never reach 100% if measured over a long time period because of the need to periodically shut down and refuel the reactor. After a stagnant period until 1985 during which the average capacity factor of US nuclear power plants hovered around 60%, improved performance, based mainly on better handling and resolution of fuel-related materials problems, led to capacity factors approaching 80% by the end of the twentieth century. Part of this improvement has been due to the movement by reactor operators to longer operating cycles (i.e., time between refueling outages).

Most of the other sources of reduced capacity factor are related to fuel failures that either necessitate an unplanned reactor outage to replace the damaged assembly or force the operator to use power maneuvering that does not stress the fuel but which results in lower energy production. Forced outages are due either to conditions that violate safety regulations or to fuel failures that excessively contaminate the primary coolant with fission products or fuel particles.

Two categories of cladding breaches are observed. The first and least serious is opening of a sub-millimeter-size hole in the cladding due to one of a number of causes. The main source is fretting at grid locations from either debris or vibration (the latter only in PWRs). In the same category are holes created by several mechanisms associated with waterside corrosion. Thick ZrO_2 scales on PWR cladding can spall and expose fresh metal. Deposits on the cladding consisting of oxides of the structural metals (iron, chromium, nickel, cobalt) originating from the ex-core primary coolant system can cause cladding temperature increases that accelerate corrosion. Stress-related mechanisms such as PCI (in BWRs) and cladding creep collapse (in PWRs) have been implicated in creation of pinhole leaks.

The small holes produced by the mechanisms listed above are termed *primary defects*, or simply *failures*. They are immediately recognizable by an increase in fission product radioactivity in the primary coolant. Failures of this type do not generally require shutdown of the reactor because the activity level in the coolant usually falls below the technical specifications that would trigger an outage. Additionally, the failure rates are low. Approximately 1–2 in every 10^5 fuel rods discharged suffer a primary defect. This translates into roughly 100 failed fuel elements per year removed from the more than 100 reactors operating in the USA.

The principal concern arising from a fuel failure of the type described above is the possibility that it will lead to the far more serious second category of cladding breach. The steam that enters the rod via the primary defect is responsible for this mode of internal corrosion. Zirconium is one of the few elements that react strongly with both oxygen (as either O_2 or H_2O) and hydrogen. The steam reacts with the cladding inner wall and in so doing releases H_2. Because the gap between the fuel and the cladding is at most a few tens of micrometers thick, replenishment of the reacted steam by diffusion from the primary defect is slow. As a result, locations in the gap far removed from the primary defect become steam-starved as H_2 replaces H_2O. When the H_2/H_2O ratio exceeds ~ 1000, the normally hydrogen-impervious ZrO_2 scale on the inner cladding surface loses its protective quality. Hydrogen is then capable of attacking the cladding and producing zirconium hydride. Severe cladding damage occurs as blisters of hydride form and drop off, exposing large areas of fuel directly to the coolant. Alternatively, if massive hydriding has occurred but has not broken through the cladding wall, a reactor power ramp that causes the fuel to stress the cladding can produce meter-long axial cracks as the brittle hydride fractures.

A fuel rod that has experienced this form of secondary damage is termed *degraded*. The presence of a degraded fuel rod in the core nearly always requires immediate reactor shutdown because of the large quantities of fission products and fuel introduced into the coolant. The problem confronting reactor operators (particularly of BWRs) is that not all primary defects lead to massive secondary degradation by hydriding. In the few instances that this type of degradation has occurred, the time interval between the opening of the primary defect and secondary degradation ranges from days to years. The reactor operator needs to make the painful decision either to shut down the reactor and remove the leaking assembly, thereby incurring significant costs for replacement power, or to continue operating and hope that the failed rod does not degrade before the next scheduled outage. A third option is to suppress power in the region of the failed rod by inserting nearby control rods. Reduced power lowers the stress on the rod and enhances the likelihood that it survives until the next regularly scheduled outage.

6. *Irradiation Growth and Bowing*

Other aspects of performance are associated with dimensional changes of the fuel rods in PWRs and the channels in BWRs. These result from irradiation growth, a phenomenon that can be likened to creep without applied stress. Growth used in this context is characterized by elongation of a metal component in the axial direction, compensated by sufficient shrinkage in the lateral directions to maintain constant solid volume. Irradiation growth requires three ingredients: a metal with an anisotropic crystal structure; preferred orientation of the grains, also known as texture; and production of displacements (vacancy–interstitial pairs) by high-energy neutrons.

The unit cell of the α-zirconium crystal structure is a hexagonal prism. The top and bottom faces are called basal planes and the six sides are termed prism planes. Mechanical properties such as yield strength and atomic properties such as self-diffusivity differ significantly in the directions perpendicular to these two planes. The direction perpendicular to the basal plane is called the c-axis and the direction perpendicular to one of the prism planes is the a-direction.

The step in the fabrication process (see *Nuclear Reactor Fuel Fabrication (Including Quality Control)*; *Nuclear Applications: Zirconium Alloys*) that produces tubing from an ingot results in the c-axes of the grains preferentially pointing in the radial direction of the tube. The longitudinal and circumferential directions are mainly perpendicular to prism planes. Atoms displaced by fast-neutron collisions and surviving recombination with vacancies tend to condense selectively as disks (or loops) on these prism planes. Because of the preferred orientation of the crystallites, the macroscopic consequence is growth of the entire tube in the axial direction accompanied by slight reduction in the wall thickness.

LWR cladding tubes may grow several centimeters before discharge, and the assembly design must provide space at the top end-cap to accommodate this deformation. If growth is uneven circumferentially, the long cladding tubes may lose straightness, or *bow*. This type of deformation constricts coolant flow, impairs heat transfer from the cladding to the water, and may cause local overheating of the fuel. If the guide thimbles bow, vertical movement of the control rods they contain may be impaired. Both of these consequences of rod bowing are significant safety issues.

LWR rod bowing can also arise from purely mechanical sources, independent of the neutron flux. Bowing can result from relaxation of internal stresses in the tubing or uneven forces exerted on the tube by the springs in the spacer grids.

BWR channels also change shape during irradiation. Two types of channel geometry changes occur, bulging and bowing. Bulging is ballooning of the channel due to the higher coolant pressure inside the channel than in the stagnant liquid between channels. The elevated internal pressure provides the vertical driving force for moving coolant up the fuel bundle. The lateral pressure differential across the channel wall is greatest at the bottom and decreases to zero at the top of the assembly. Deformation due to the stresses in the channel arising from the pressure difference takes place by a combination of thermal and irradiation creep, the latter being a consequence of atom displacements by the fast-neutron flux.

Bowing of the channels is driven by the radial variation of the fast-neutron flux in the core. The channel wall closest to the core axis experiences a higher neutron flux than the opposing face. Since irradiation growth of Zircaloy increases with the fast-neutron dose (the fast-neutron dose is the integral of the fast flux over time, and is also termed the fluence; dose can also be expressed as displacements per atom, dpa; see *Radiation Damage Theory*), the centermost face elongates more than the outer face. As a result of this uneven growth, the channels bow outward, away from the core center.

The distortions caused by both bulging and bowing act to close the interchannel gap and could impede motion of the control elements. Stuck control blades have occurred in the operation of BWRs. PWR control rods have also experienced a sticking in their guide tubes due to Zry growth and hydride swelling.

7. *Waterside Corrosion*

The term waterside corrosion refers to the chemical attack of the exterior surface of the cladding by the water coolant. Corrosion of Zry proceeds by the reaction:

$$Zr + 2H_2O \rightarrow ZrO_2 + 2H_2(\text{or } 4H \text{ in metal})$$

As indicated by this reaction, some of the hydrogen reaction product (about 15%) is absorbed by the substrate metal instead of returning to the coolant as H_2. The absorbed hydrogen is potentially damaging to the cladding because at low temperatures it forms the hydride ZrH_2, which is very brittle and can cause cladding failure. Hydrogen also migrates readily in temperature gradients in the cladding, concentrating in the coldest spots where it increases the risk of fracture.

Up to a thickness of about 2 μm, the oxide is dense and protective, with the kinetics controlled by atomic transport in the solid oxide. The other elementary steps comprising the overall reaction, namely decomposition of H_2O that produces oxygen ions in the surface layer, conversion of the substrate metal to Zr^{4+} at the metal–oxide interface, and migration of electrons from there to the surface, are rapid compared with the oxygen transport process. At the metal–oxide interface, newly formed ZrO_2 crystallites are very small ($\sim$5 nm diameter), which favors oxygen transport in the grain boundaries of the oxide rather than through the lattice of the grains.

The process described above would be expected to yield parabolic kinetics, in which the oxide scale thickness, L, grows as the square root of time, t. Experimentally, however, this early stage of Zry oxidation instead follows the so-called cubic rate law, for which $L \propto t^{1/3}$. Many theories for this nonclassical corrosion behavior have been advanced (see *Nuclear Applications: Zirconium Alloys*).

A transition from cubic to linear kinetics occurs when the scale thickness reaches about 2 μm. At this thickness, open porosity develops in the oxide above the barrier layer. The porosity may be augmented by stress-induced vertical cracks. The pore/crack network permits direct and easy access of water to the top of the barrier layer, through which the rate-controlling oxygen diffusion process takes place. Since the barrier layer thickness does not change with time, the corrosion rate is constant in time, or the kinetics are linear.

7.1 PWR

Under PWR water conditions (subcooled liquid, about 300 °C; see *Nuclear Reactors: Coolant Materials*), the ZrO_2 reaction product forms on the Zry substrate as a uniform scale that grows with time in a complex manner. Of particular concern is the acceleration of the corrosion rate from the constant value in the post-transition region.

Since the number of fissions is approximately proportional to the time that the fuel rod is exposed to the hot coolant, corrosion kinetics are commonly expressed as oxide scale thickness (or weight gain per unit area) as a function of fuel burnup. Standard Zry is satisfactorily resistant to oxidation by high-pressure, 300 °C water or steam to a burnup of at least 30 MWd kg^{-1}U (or ~3% consumption of the initial uranium). For higher burnups, the corrosion rate accelerates, and the rate is described by $L \propto (BU)^n$, where BU is the burnup and $n > 1$. As indicated in Sect. 2, the most effective way to mitigate accelerated corrosion is modification of the minor element concentrations in the zirconium alloy or control of the size of the iron-, nickel-, and chromium-bearing precipitates in the cladding fabrication process. A number of explanations of the accelerated corrosion phenomenon have been advanced with the hope that additional remedies will be uncovered.

An easily understandable source of corrosion acceleration is called thermal feedback. The corrosion rate is highly temperature dependent (activation energy ~100 kJ mol^{-1}). With a heat flux passing through the oxide from the fuel, the temperature of the metal–oxide interface is as much as 10 °C higher than at the surface in contact with the coolant. The rate of corrosion is controlled by the temperature of the metal–oxide interface because that is the location of the barrier layer through which oxygen diffusion occurs. As the scale thickens, the temperature at the metal–oxide interface rises. This causes an acceleration of the corrosion rate compared to the linear rate characteristic of thin scales.

A second heat transfer resistance that forms on top of the oxide scale bears the inelegant name "crud" (US slang for an unpleasant material; also claimed to be the acronym for Chalk River Unidentified Deposit, after the Canadian nuclear laboratory). This deposit consists principally of the oxides of iron and chromium but may contain lesser quantities of cobalt as well. Transition metals are present as aqueous ions in the coolant as a result of corrosion of steel piping and other components in the primary circuit. Under LWR water conditions, iron and chromium exhibit retrograde solubility, which means that the saturation concentrations of these species decrease with increasing temperatures. Being the hottest surfaces encountered by the coolant, these species deposit on the cladding in the form of the oxides Fe_2O_3 and $(Fe,Cr)_3O_4$. The crud deposits are often highly porous, and because they contain a high water content, have thermal conductivities even lower than solid oxides. Since the crud layer supports the rod heat flux, it also has a temperature gradient that contributes to the rise of the temperature of the metal–oxide interface. As a result, crud deposition enhances waterside corrosion of the cladding.

In addition to accelerating corrosion, crud deposits have two other features that are detrimental to system performance. The cobalt that the crud contains is converted by the neutron flux to highly radioactive ^{60}Co. Some of the activated cobalt returns to the coolant and is deposited in ex-core components where it contributes to the radiation dose received by maintenance personnel.

The crud layer also has the capacity to sequester boron which, as boric acid, is added to PWR coolant for nuclear reactivity control. The boron content of the crud can reach levels that significantly alter the distribution of neutron absorption in the core. The result is a perturbation of the neutron flux called the *axial offset anomaly*.

Were boric acid the only additive to PWR water, unacceptable corrosion and crud formation would occur because of the low pH. To provide an acceptably basic solution, lithium hydroxide is added to raise the pH to 6.9 (at 300 °C the pH of neutral water is 5.5). Even higher pH would be desirable for minimization of corrosion in the primary circuit and for more effective prevention of crud deposition on the fuel rods. However, at a pH of ~7.3, LiOH accelerates corrosion of the cladding. This phenomenon is believed to be due to lithium, high concentrations of which increase the porosity of the oxide scale. The increased porosity permits more rapid ingress of water to the metal–oxide interface thereby enhancing corrosion.

The narrow range of lithium concentrations needed to avoid corrosion of both primary circuit components and cladding poses a difficulty in extending the burnup of PWR fuel. To permit significant increase in discharge burnup, the enrichment of the fuel must be raised. At beginning of life, the higher enrichment requires greater negative reactivity, either in control rods or in boron in the coolant. If the boron concentration is increased to provide the necessary neutronic control, the LiOH concentration must also be increased to maintain the proper water pH. However, the higher Li^+ concentration enhances cladding corrosion, which may then set the burnup limit.

Hydrogen arises from two distinct sources that have different effects on the cladding. The first source is gaseous H_2 that is added to the feed water in both BWRs and PWRs. This dissolved H_2 directly reduces corrosion of stainless steel components by lowering the electrochemical potential of the water, mainly by scavenging radiolytically produced O_2 and H_2O_2. However, dissolved H_2 does not appreciably diminish Zircaloy corrosion.

The second source is corrosion-product hydrogen released by water decomposition on the oxide surface. The underlying metal absorbs a small fraction (15–30%) from this source. In addition to embrittling the metal, hydrogen in the metal appears to accelerate corrosion. Because of the thermal diffusion properties of the H/Zr system, hydrogen concentrates in the coldest part of the cladding, which is the metal–oxide interface. The enhanced hydrogen concentration at this location is, in part at least, responsible for accelerated waterside corrosion that is observed in high-burnup fuel.

Extension of the burnup of fissile material in fuel elements clad with Zry tubing is limited by corrosion of the exterior cladding surface. When the ZrO_2 scale thickness reaches $\sim 100\,\mu m$, it begins to flake off due to stresses induced by the 50% volume expansion of zirconium upon conversion to the oxide. The bared metal surface is cooler than the adjacent metal still protected by oxide, so hydrogen collects at the flaked-off spots, which become regions of rapid degradation.

7.2 BWR

BWR coolant conditions are less aggressive towards Zry than PWR water, chiefly because the cladding outer surface temperature is 10–20°C lower in BWRs than in PWRs. The enhancement of the corrosion rate with burnup observed on PWR cladding does not occur on BWR cladding. Patches of oxide called *nodules* develop on the uniform corrosion layer on BWR cladding. However, unless significant copper deposition on the oxide has occurred during operation, the oxide nodules do not jeopardize cladding performance.

8. Fission Product Behavior

Fission products affect the fuel in four principal ways, only one of which is desirable. The beneficial effect is the deposition of the energy of nuclear fission in the UO_2 lattice, from which the energy is conducted as heat to the coolant. The undesirable features are swelling of the fuel, release of mobile fission products from the fuel to the void spaces in the rod, and chemical attack of the Zircaloy cladding. (The intense radioactivity of the fission products is also an unwanted property, but is more of biological concern than an influence on material and mechanical properties.)

Fission products number in the hundreds, but the vast majority quickly decay into stable or long-lived species. For dealing with their effect on material and chemical properties of the fuel, efficiency is achieved by grouping the fission products with similar physical and chemical properties into "pseudo-elements" with the average characteristics of the group. As an example, the stable (or long-lived) nuclides in the decay chains at masses 133, 135, and 137 are isotopes of cesium. These nuclides, together with the minor alkali metal rubidium, are collectively categorized as cesium.

The fraction of fissions that produce elements having the same mass number is called the mass chain yield. When combined according to chemical similarity, the group yield and the physical states in the fuel are shown in Table 1.

The sum of the yields of the five groups is 1.96; all other fission product elements (tellurium, iodine, antimony, silver, etc.) together have yields that sum to less than 0.04. Roughly one half of the fission products, those with stable III and IV oxidation states, replace uranium in the fluorite crystal structure. The remaining fission products have zero or

Table 1
Yields and physicochemical states of fission product groups.

Group	Elements in group	Yield	State in fuel[a]
Stable III and IV oxidation states	Zr, Mo, rare earths	1.07	Substitutional cations in UO_2 lattice
Noble metals	Ru, Tc, Rh, Pd	0.26	Metallic precipitate
Alkali metals	Cs, Rb	0.23	Oxide or compound with UO_2
Alkaline earths	Ba, Sr	0.15	Dissolved in fuel at low burnup; oxide precipitate at high burnup
Rare gases	Xe, Kr	0.25	Inter- or intragranular bubbles

[a] At low temperatures, all fission products exist as substitutional or interstitial atoms or ions in the fuel lattice. The states listed form only at temperatures high enough for atomic mobility ($\sim >1000°C$).

very limited solubilities in UO_2 and form separate phases in the fuel microstructure.

Some small-yield fission products, tellurium and iodine in particular, are important because they are volatile and escape more easily from the core during a severe accident than the refractory species. Even in normal operation, iodine, along with cesium, reach the fuel–cladding gap and are released to the coolant in the event of a cladding breach. The elements that dissolve in the fuel or form second oxide or metallic phases are nonvolatile and remain with the fuel even under severe accident conditions.

The chemical states of the fission products indicated in Table 1 are consistent with the thermochemistry of the elements in each group, with perhaps one exception. The susceptibility of Zircaloy to the form of SCC called PCI was indicated in *Sect. 2*. The chemically aggressive fission product that is believed to be responsible for PCI is iodine. Laboratory experiments do indeed show that elemental iodine is very effective in causing SCC of Zircaloy. Thermochemically, however, all of the iodine in a fuel rod should be combined with cesium as CsI, which is not an SCC agent for Zircaloy. Radiation chemistry/chemical kinetic theoretical calculations claim to show that the intense fission fragment flux in the fuel–cladding gap sustains a nonequilibrium pressure of elemental iodine sufficient to cause SCC of the cladding. This argument has an *ad hoc* flavor because there is no evidence that radiation affects the chemical states of any of the other fission products. The minor fission product cadmium is known to cause liquid-metal embrittlement of Zry. Since cadmium is a volatile element, it may be involved in PCI.

An important feature of the fission products is the swelling of the fuel that they cause. Expansion rather than contraction is expected because fission produces two atoms from one. If all fission products occupied the same volume as the uranium from which they originated, the fractional swelling would be equal to the fractional burnup, or the fraction of the initial uranium atoms that have undergone fission. However, the swelling coefficient due to the solid fission products is less than the fractional burnup for two reasons. First, the fission gases are treated separately because of the vast density difference between gaseous and condensed matter. Second, the atomic volumes of the solid fission products depend on their phase and chemical form: the noble metals have an average atomic volume only one-third that of uranium in UO_2; barium in the oxide precipitate and cesium in the cesium fuel compound are nearly twice as large as uranium in UO_2. When the atomic volumes of the fission products are weighted with their yields and summed, the solid fission-product swelling coefficient is about 2/3 of the burnup expressed in atom percent.

The group that is of most concern regarding fuel-rod integrity is the fission gases, 90% of which consist of the stable isotopes of xenon. Their gaseous nature permits them to escape from the fuel more readily than the solid fission products. Once released, the gaseous fission products internally pressurize the fuel rod, which can be life-limiting.

Like other fission products, the rare gases are born as individual atoms lodged in the fuel matrix. At low temperatures ($< \sim 1000\,^\circ C$), gas-atom mobility is too small for the formation of bubbles. The swelling of the fuel that the gas atoms cause is basically the same as that of the solid fission products. Similarly, the low diffusivity of the fission gases prevents them from reaching free surfaces and escaping from the fuel. The only release mechanism at low temperatures is direct recoil of fission fragments created at a depth beneath a free surface equal to their range in UO_2 (about 7 μm).

At temperatures above about 1000 °C, the radiation-produced point defects (in particular vacancies) acquire sufficient mobility to jump randomly in the lattice and to form clusters around gas-atom fission products. Xenon atoms are so large that they cannot move in the crystal without first associating with two oxygen ion vacancies and one uranium ion vacancy. This unit is capable of migrating in the crystal structure of the fuel grain.

The mobile gas atoms take part in a number of microscopic processes, which have the macroscopic consequences of swelling the fuel or escaping from the fuel. At low burnup ($< \sim 1\ \mathrm{MWd\,kg^{-1}U}$), the gas atoms behave as isolated entities in the fuel; lattice diffusion delivers gas atoms to the grain boundaries, where they are trapped and do not escape from the pellet.

As the burnup increases, the higher gas concentration results in an ever-higher probability of random convergence of two gas atoms on neighboring sites in the lattice or on grain boundaries. Such an event produces a stable diatom, which is the nucleus of a gas bubble. The nucleation process continues until a newly produced gas atom has an equal chance of being absorbed by an existing nucleus (or bubble) as it does of encountering another gas atom and forming a new nucleus. The attachment of migrating gas atoms to existing clusters (bubble nuclei or larger bubbles) is termed *growth*, and quickly supplants nucleation as the dominant process once the above condition is met. During the growth period, the number density of gas bubbles remains substantially constant but their size increases by collection of diffusing gas atoms.

The rare gases are thermodynamically insoluble in the fuel. In the absence of a radiation field, gas atoms in bubbles never redissolve into the surrounding solid. However, fission tracks intersect gas bubbles in the fuel, and occasionally a fission fragment collides with a gas atom and imparts sufficient energy to it to drive the gas atom back into the surrounding solid. This process is called *re-solution*, and reduces the efficiency of the bubbles as gas-atom traps.

At high temperatures, as might be encountered in a reactor overpower transient, gas bubbles acquire mobility as a whole by virtue of the random migration of the UO_2 molecules on their solid surfaces. If two moving bubbles happen to meet, they coalesce into a single bubble with a volume greater than the sum of the volumes of the initial pair. This decrease in gas density as the bubbles enlarge, whether by single-atom growth or by coalescence, is due to the effect of the surface tension acting on the spherical surface of the enclosing solid. The effective negative pressure produced by surface tension is inversely proportional to bubble radius. As the bubble grows in size the gas pressure, and hence the gas density, decreases.

The combined effects of diffusional trapping and resolution make the intragranular bubbles temporary traps for the migrating single gas atoms. This effect is manifested as a reduction of the apparent diffusion coefficient of the gas atoms. However, most gas atoms produced by fission eventually reach the grain boundaries of the fuel, which are thermodynamically black sinks for the gas (and for other fission products).

The processes described above for intragranular gas bubbles (i.e., nucleation, growth, resolution, and coalescence) also occur in the two-dimensional confines of the grain boundaries. At high burnup, the accumulated intergranular gas bubbles coalesce by growing into each other. The porosity of the grain boundaries and the grain edges become large enough to permit extensive interconnection of tunnels and establishment of routes to external fuel surfaces. At this stage, gas release occurs by hydrodynamic flow through the porous medium (Darcy flow).

Once the fission gases have precipitated into bubbles, the swelling they cause is much larger than that due to the same number of solid fission products; even though the pressure in the bubble can be quite high (approaching 1000 bar for 10 nm diameter cavities), the gas density is still much less than that of a solid. Fission gases constitute only 1/8 of the total fission products, but provide over half of the fuel swelling.

Many computer codes have attempted to calculate the swelling and release of fission gases by first-principles modeling of the elementary process described earlier in this section. However, these computations are not sufficiently accurate for practical predictions. The reasons include insufficient quantitative understanding of the physical processes (e.g., bubble nucleation), inability to model satisfactorily a process (e.g., fuel cracking), and poorly known physical properties (e.g., fission gas diffusivities).

As a result of the limited success of detailed mechanistic descriptions of fission gas behavior, semiempirical methods for handling fission gas release have been developed. The most widely used of these is the Booth model. In this description, the fuel is represented as an assembly of equivalent spheres from which fission gases escape by diffusion. Once gas atoms arrive at the periphery, their release to the fuel rod is assumed (in the Booth model) to be instantaneous. The basic process of intragranular diffusion is explicitly retained in the Booth model, and all of the other complications of the release process are lumped into a single parameter, the diameter of the equivalent sphere. The Booth spheres are larger than the grains (to account for gas trapping by bubbles) but smaller than the fuel pellet dimension (to represent release-enhancing processes such as escape via cracks, grain boundary diffusion, and flow in open porosity).

A major attraction of the Booth model is the simplification of the complex gas transport process to a single temperature-dependent parameter, the ratio of the gas diffusivity in the grain to the square of the radius of the Booth sphere. The former may be reduced by the intragranular bubble-trapping processes described earlier and augmented by radiation-enhanced diffusion. The latter is dependent on the fuel microstructure and its evolution with burnup.

9. *Fuel Microstructure Evolution with Burnup*

Conventional LWR fuel is polycrystalline with an average grain diameter of 8–10 μm. During irradiation the original features of the solid are changed by thermal stresses, new fission product phases, grain growth, thermal gradient restructuring, and a unique structure that develops at high burnup. Most of the new microstructures were first observed in fuel pellets for the liquid metal fast breeder reactor (LMFBR) (see *Fast Breeder Reactors: Fuels*). Because of the high operating temperature of this design, porosity migration up the temperature gradient creates a hole in the pellet center that is surrounded by columnar grains left behind by the moving pores. Following this structure is a radial annulus of equiaxed grains enlarged by grain growth. The cool outer periphery of the LMFBR fuel pellet retains the original features of fresh fuel.

LWR fuel pellets are larger in diameter than LMFBR pellets and operate at a power density about one-fourth of the latter. Consequently, LWR fuel temperatures are too low (a maximum of about 1400°C) to permit the thermal-gradient-driven porosity movement that is such a distinctive feature of LMFBR fuel. Nonetheless, the temperature gradient in LWR fuel is steep enough to generate thermal stresses that exceed the fracture stress of UO_2 over a significant fraction of the pellet radius. This causes radial cracks to be produced because the largest thermal stress component is azimuthal. The pellet splits into six or seven pie-shaped wedges immediately upon ascension to power. Rather than impeding heat transfer, the wedges move radially outward and partially close the fuel–cladding gap. The reduction of this thermal resistance lowers fuel temperatures.

One of the most striking microstructural features of irradiated LWR fuel is the population of lenticular bubbles on grain faces and tunnels on grain edges. These cavities develop at high burnup and contain fission gas at high pressure. They grow with time until the network becomes dense enough to interlink and provide a tortuous flow path to the pellet exterior. Following purging of the gas content, the bubbles and tunnels shrink and close under the driving force of UO_2 surface tension. Once sealed from the open space in the fuel rod, the cavities resume trapping of fission gas arriving from the grains until interlinkage again occurs and venting takes place.

Solid fission product precipitate phases are also observed in high-burnup fuel. The noble metals listed in Table 1, perhaps including molybdenum, form micrometer-size particles often associated with intragranular fission gas bubbles. Ternary oxides such as $BaZrO_3$, $CsUO_4$, and $CsMoO_4$ can form, the last two only at high oxygen potentials (i.e., oxidized fuel).

Despite long irradiation times at elevated temperatures, grain growth does not always occur because grain boundary mobility is reduced by fission products. Whether or not extensive grain growth takes place depends on the temperature–burnup history of the fuel. If the fuel is operated at high temperatures early in life before fission product concentrations have built up, grain growth can take place in the hot fuel center. Conversely, long initial low-temperature irradiation produces impurity pinning that locks grain boundaries in any subsequent high-temperature operation.

The general trend towards increasing LWR fuel burnup to $>70\,\mathrm{MWd\,kg^{-1}U}$ has revealed the development of a hitherto unknown microstructural change. This new feature was initially called the *rim effect* because it begins at the periphery of the fuel pellet. As irradiation proceeds, the new microstructure advances radially inward. The realization that the new microstructure developed at a certain burnup and was not restricted to the pellet surface led to renaming the phenomenon more descriptively as the *high burnup structure* (HBS).

The HBS is characterized by complete loss of the original grains ($\sim 10\,\mu m$ diameter) and their replacement by a much larger number of small grains with diameters ranging from $0.2\,\mu m$ to $0.6\,\mu m$. In addition, the solid lattice is denuded of fission gas, all of which is collected in a population of bubbles whose diameters range from $1\,\mu m$ to $2\,\mu m$. The relative sizes of the bubbles and the grains in the HBS are the inverse of that characteristic of low burnup. In the latter, the bubbles are smaller than the grains and the bubble surface is that of a single grain (for intragranular bubbles) or of two adjacent grain surfaces (for intergranular bubbles). The surface of the bubbles in the HBS, however, consists of 50–100 individual subgrains.

The reason that the HBS begins at the pellet outer surface is that the threshold burnup is first attained here. The neutron absorption cross-section of ^{238}U has strong resonances in the epithermal energy region. Fission neutrons are thermalized by hydrogen atoms in the coolant so that thermal and epithermal neutrons enter the fuel pellet at its outer surface. The high epithermal cross-section of ^{238}U strips the entering flux of this energy group and in the process produces ^{239}U. This nuclide decays to ^{239}Pu (via ^{239}Np), which undergoes fission in the thermal flux. As a result of these nuclear processes, the burnup at the outer edge of the pellet is about twice that in the pellet center. It is therefore natural that the threshold burnup for HBS formation is first reached at the pellet periphery.

Although the condition for HBS formation is couched in terms of burnup, it is more likely that more fundamental radiation processes are responsible. Stopping of fission fragments creates copious quantities of point defects, which are vacancies and interstitials of both uranium and oxygen. A small fraction of these survive recombination. The remaining interstitials coalesce into disks that are converted to dislocation loops. These loops are mobile, and when sufficiently numerous (i.e., at high burnup), they organize themselves into dislocation walls that effectively are new grain boundaries. These tiny embryo grains then grow (recrystallize), obliterating the original grain boundaries and collecting the fission gases in the volume they sweep out.

How the large bubbles form is more problematic. At the temperatures of the pellet periphery ($\sim 400\,^{\circ}C$), the vacancy diffusion coefficient is very low. The gas in the bubbles is pressurized beyond the equilibrium value ($p_{eq} = 2\gamma/R$, where γ is the surface tension of UO_2 and R is the bubble radius). Nonetheless, it is a gas and not a solid, and so must contain vacancies to provide the low density of a gas. Since the vacancy mobility at the temperature of the pellet surface is very low, vacancies most probably cannot cause the bubble to grow by a diffusion process. Rather, an athermal cavity expansion mechanism such as loop punching (the high gas pressure continuously generating circular dislocations at the bubble surface) may be responsible for expansion.

Although the detailed mechanism of formation of the HBS is uncertain, several effects on fuel rod performance are expected. The most important is the propensity for release of the gas that has been moved from the single-atom state in the fuel lattice to the overpressurized bubbles. There does not appear to be significant release of fission gas from the HBS during normal operation. However, it is possible that the rapid temperature rise during a reactivity insertion accident (RIA) could shatter the HBS and place sufficient stress on the cladding to cause failure.

The development of porosity in the HBS locally degrades the thermal conductivity of the fuel. The resulting increase in fuel centerline temperature can reach 100°C. The porosity of the HBS can also contribute to a sizable increase in fuel swelling at high burnup.

10. Conclusions

The LWR fuel rod has evolved over its lifetime into a rugged and reliable vehicle for conversion of fission energy into thermal energy. New performance requirements include extended burnup, higher linear heat rating, and longer intervals between refueling. In view of the plethora of performance limitations (waterside corrosion, fission gas internal pressure, hydriding of punctured cladding, etc.), the ability of the UO_2–Zry classic fuel element to withstand the more demanding environments is not assured.

See also: Nuclear Reactor Fuel Fabrication (Including Quality Control)

Bibliography

Bailly H, Menessier D, Prunier C 1999 *The Nuclear Fuel of Pressurized Water Reactors and Fast Neutron Reactors: Design and Behavior*. Lavoisier, Paris

American Nuclear Society 1997 *Proc. 1997 Int. Topical Meeting on LWR Fuel Performance*. American Nuclear Society, Portland, OR

Olander D R 1976 *Fundamental Aspects of Nuclear Reactor Fuel Elements, TID-26711-P1*. National Technical Information Services, US Department of Commerce, Springfield, VA

Roberts J T A R 1981 *Structural Materials in Nuclear Power Systems*. Plenum, New York

M

Magnetic Fields: Thermoelectric Power

K. Yagasaki and A. T. Burkov

The coefficient of the thermoelectric power (S) measured in variable magnetic fields (B) is called magnetothermopower. This physical quantity can provide very valuable complementary information concerning details of electronic band structure. In comparison with the magnetoresistivity, the magnetothermopower is not sensitive to a uniform change of mobility of conduction electrons by the Lorentz force. The magnetothermopower of metallic materials basically depends on a relative change of the mobility of conduction electrons with different energy near to the Fermi surface.

The effect of magnetic fields on thermopower is important also from a practical point of view. Thermocouples are widely used as temperature sensors. Therefore, in low-temperature thermometry in equipment with strong magnetic fields, such as research cryostats with superconducting magnets, a knowledge of the sensitivity of a thermocouple to the magnetic field is essential. Another significant and expanding field for the application of thermoelectric phenomena is thermoelectric heat–energy conversion and thermoelectric cooling (see *Thermoelectric Devices: Refrigeration and Power Generations With No Moving Parts*). It is known that the thermoelectric efficiency of some materials can be considerably enhanced in a magnetic field (examples are bismuth-based alloys which are used for low-temperature refrigeration).

1. Theoretical Aspects of Magnetothermopower

The thermopower, or Seebeck, coefficient S is defined as the coefficient in the relation: $\mathbf{E} = S\nabla T$ between the temperature gradient ∇T and the electrical field $\mathbf{E}$ in a conducting material, where the net current through the sample is zero. There are two other thermoelectric phenomena: the Peltier heat (characterized by the coefficient Π) and the Thomson heat (characterized by the coefficient μ_T), which can be observed under different conditions. However, these three coefficients S, Π, and μ_T are not independent; they are connected to each other by the Thomson relations. For an isotropic medium, without a magnetic field, the relations read: $\Pi = TS$ and $\mu_T = T\mathrm{d}S/\mathrm{d}T$ (Barnard 1972). In a general case, S, Π, and μ_T are tensors of second rank, which can have as many as nine independent components. The most general form of the Thomson relation is given by the expression: $\Pi_{ik}(B) = TS_{ki}(-B)$ (Landau and Lifshitz 1977).

If there is, in addition to the temperature gradient, an external magnetic field applied to the sample, the thermoelectric field can be written as: $\mathbf{E} = S\nabla T + N\mathbf{B} \times \nabla T$ (Abrikosov 1972). The second term in this equation is the so called Nernst–Ettingshausen (N–E) effect, with N the Nernst–Ettingshausen coefficient. However, apart from the N–E effect, the coefficient S can also be dependent on the magnetic field. The change of S in an external magnetic field is called magnetothermopower: $\Delta S = S(B, T) - S(0, T)$. In the weak field limit, this dependence is a second-order correction of the thermoelectric field $\mathbf{E}$ in the presence of a magnetic field. With respect to experimental conditions, two different situations are possible, depending on the mutual orientation of the magnetic field and the temperature gradient: the magnetic field parallel to the temperature gradient, and the magnetic field perpendicular to the temperature gradient. One distinguishes, respectively, longitudinal magnetothermopower ΔS_L and transversal magnetothermopower ΔS_T. In the case of anisotropic materials, special care should be taken to separate ΔS_T and the influence of the Nernst–Ettingshausen effect. According to the mechanism of the change of thermopower in a magnetic field, the effect can be subdivided into two main categories. In the first category, the effect is connected to the orbital motion of the conduction electrons in a magnetic field caused by the Lorentz force. In the second category, the effect is related to a change in the electronic structure and/or scattering mechanism in a magnetic field.

With regard to the Lorentz force-driven magnetothermopower, two different regimes are distinguishable, depending on the strength of the magnetic field. This field strength can be characterized by the product $\omega_c\tau$, where $\omega_c = eB/m^*c$ is the classical cyclotron frequency, τ is the average time between collisions of a conduction electron with scattering centers, m^* is the effective electron mass, and c is the speed of light. The low-field regime corresponds to $\omega_c\tau \ll 1$, while $\omega_c\tau \gg 1$ is the condition for the high-field range. From a theoretical point of view, the high-field case has been much better understood than the low-field one. In the high-field regime, the scattering processes become unimportant, and galvano- and thermomagnetic phenomena are determined by the topological properties of the Fermi surface. In the high-field limit

$\omega_c \tau \gg 1$, the longitudinal magnetothermopower ΔS_L is independent of the magnetic field for any Fermi surface geometry.

The transversal magnetothermopower ΔS_T is proportional to B for closed electron orbits in compensated metals, and is independent of the field in other cases (Blatt *et al.* 1976). A special case of the high-field regime is realized under the additional condition $\hbar\omega_c > k_B T$ ($k_B = 1.3806568 \times 10^{-23}\,\mathrm{J\,K^{-1}}$ is the Boltzmann constant, $\hbar = 1.05457266 \times 10^{-34}\,\mathrm{Js}$ is the Planck constant). Under this condition, Landau quantization of the orbital motion of the conduction electrons in the plane, normal to the magnetic field, becomes important and all electronic properties of metals, including thermopower, show an oscillatory dependence on B^{-1} with a period ΔB^{-1}. The period depends on the extreme cross-sectional area A_e of the Fermi surface in a plane normal to the magnetic field: $\Delta B^{-1} = (2\pi e/\hbar c)(1/A_e)$ (Landau and Lifshitz 1977). These oscillations can best be observed in a very pure single crystal at low temperatures: $T < (\hbar e/m^* c k_B) B \approx 1.34B$, where B is the field in teslas.

In the low-field regime, the scattering mechanism is essential and this makes a theoretical description extremely difficult. Within this limit, the Lorentz force driven magnetothermopower is expected to be proportional to B^2 (Abrikosov 1972, Landau and Lifshitz 1977, Ri *et al.* 1994).

The magnetothermopower due to a change of electronic structure can be dominant in non-simple metals, magnetic or nearly magnetic materials, strongly correlated systems, and in conductors with magnetic impurities. The effect essentially depends on details of the electronic structure of a particular system. No general theory has been developed up to now. The influence of a magnetic ordering on the thermopower has been discussed and investigated both theoretically and experimentally . On a qualitative level, the variation of thermopower of itinerant ferromagnets in the vicinity of Curie temperature has been explained by Mott, based on his *s–d* scattering model (Blatt *et al.* 1976). A summary of theoretical and experimental results can be found in Blatt *et al.* 1976, and in Yagasaki *et al.* 1996).

A somewhat special field represents the thermomagnetic phenomena in superconductors. This field has experienced a burst of theoretical and experimental activities in connection with the discovery of high-temperature superconductors. Large thermomagnetic effects were observed in the mixed state of high-T_c materials. These effects are related to the magnetic flux flow driven by a temperature gradient. For a review, see Huebener (1979) and Ri *et al.* (1994).

A review of the theory of galvanomagnetic and thermomagnetic effects in metals and of experimental results prior to 1957 is given in Jan (1957).

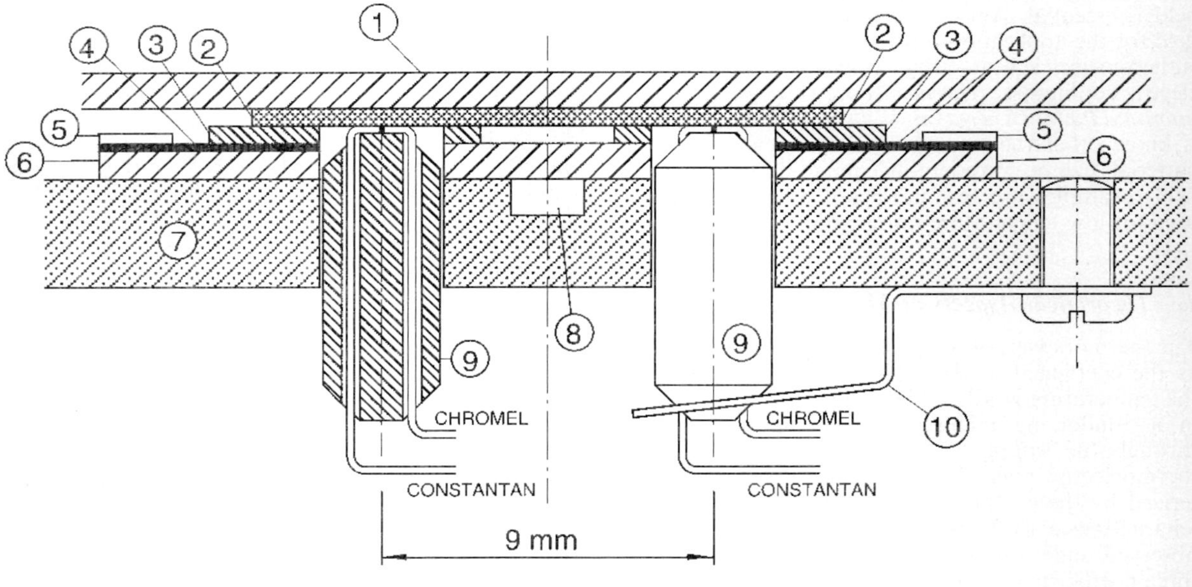

Figure 1
A part of the sample holder for thermopower measurement in a magnetic field. (1) and (6) sample-pressing and sample-supporting plates (glass epoxy); (2) sample; (3) AlN plates; (4) copper layers; (5) gradient heater; (7) heat sink (copper-beryllium bronze); (8) temperature sensor; 9) Al_2O_3 tubes holding thermocouples; (10) spring (copper–beryllium bronze).

2. *Details of the Measurement*

Compared to measurements of the thermopower without a magnetic field, magnetothermopower measurements require several additional considerations:

(i) In the high-field limit, thermopower can be anisotropic even for crystals of cubic symmetry.
(ii) Contributions from galvano- and thermomagnetic effects, different from thermopower, should be carefully eliminated (the Nernst–Ettingshausen effect particularly).
(iii) Calibration of temperature sensors and reference electrodes in a magnetic field is important.
(iv) Care should be taken to avoid a magnetic field gradient.

Several different designs have been introduced for measurements of magnetothermopower. As an example, we mention the apparatus described by Resel *et al.* (1996). With this setup, the thermopower can be measured in magnetic fields from 0 T to 17 T in the temperature range from 3 K to 300 K. The main part of the sample holder is shown in Fig. 1. A special modulation technique for the temperature gradient is used in this setup to eliminate spurious voltages. Two gradient heaters working in opposite directions are utilized to generate an alternating temperature gradient in the sample. The total heating power produced by the two heaters is constant. Therefore, the average temperature of the sample is not affected by the gradient modulation. In order to avoid the N–E effect contribution, a parallel alignment of the temperature gradient and magnetic field direction is chosen. The temperature difference across the sample is measured with the help of two Chromel–Constantan thermocouples.

The sensitivity of this thermocouple is independent of the magnetic field within an error of 5%, in the field range from 0 T to 17 T at temperatures from 3 K to 300 K. For the reference electrode a Chromel wire is used, since the thermopower of this alloy has a comparatively weak and simple dependence on the external magnetic field and the temperature. However, for precise measurements of magnetothermopower of metals, where the effect is small, it is necessary to make corrections in the thermopower of Chromel, so that $S_{Ch}(B, T) = S_{Ch}(0, T) + \Delta S(B, T)$. It was found that $\Delta S(B, T)$ can be approximated by an empirical function $f(B, T)$ with an accuracy of $0.1\,\mu VK^{-1}$. This function reads:

$$f(B, T) = aT\exp(bT^2) + cT^2\exp(dT)$$

where

$$a = \frac{2.1\ 10^{-2}B}{-51.4 + 49.1\exp(-2.91 \times 10^{-2}B)}$$

$$b = -2.2 \times 10^{-4} - 1.02 \times 10^{-5}\exp(0.187B)$$

$$c = -4.18 \times 10^{-5}B$$

$$d = -7.47 \times 10^{-5}B^2 + 2.33 \times 10^{-3}B - 4.38 \times 10^{-2}$$

(where B is the magnetic field in teslas).

3. *Experimental Results*

Most systematic and comprehensive investigations of the magnetothermopower have been performed for simple monovalent and divalent metals as well as for noble metals. It was found that in many cases $\Delta S = S(B, T) - S(0, T)$ is large at low temperatures. In noble and divalent metals, a large contribution to ΔS is referred to as an increase of the phonon drag thermopower in a magnetic field. A summary of these studies is given in Blatt *et al.* (1976). A comprehensive investigation of ΔS in the high-field regime was made for silver (Stanley 1974). Some results for single silver crystals are presented in Fig. 2.

The quantum oscillations of the magnetothermopower and the Peltier coefficient were measured for bismuth, tin, antimony, aluminum, and zinc single

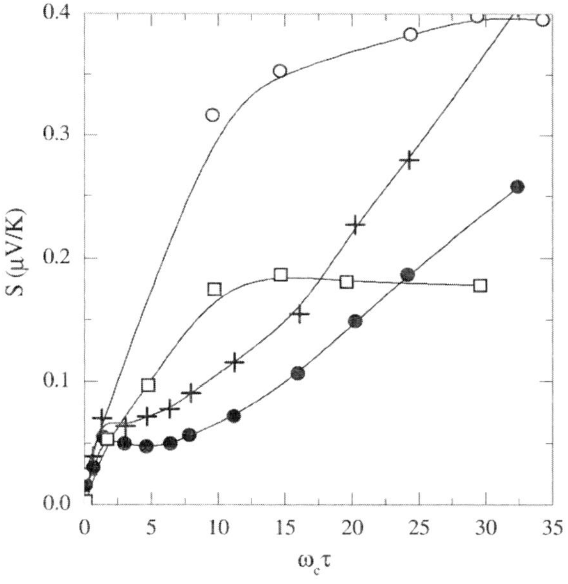

Figure 2
Magnetothermopower of silver against a magnetic field (Stanley 1974). Results for single crystal specimens of silver of different purity and crystallographic orientation are shown. Thermopower was measured with the longitudinal orientation of the magnetic field with respect to the temperature gradient. □: RRR = 1380, along [741] crystallographic direction, $T = 2.5$ K; ○: the same sample, $T = 4.4$ K; •: RRR = 2280, along [111], $T = 4.9$ K; and +: the same sample at $T = 6.2$ K. RRR $\equiv \rho(300)/\rho(4.2)$.

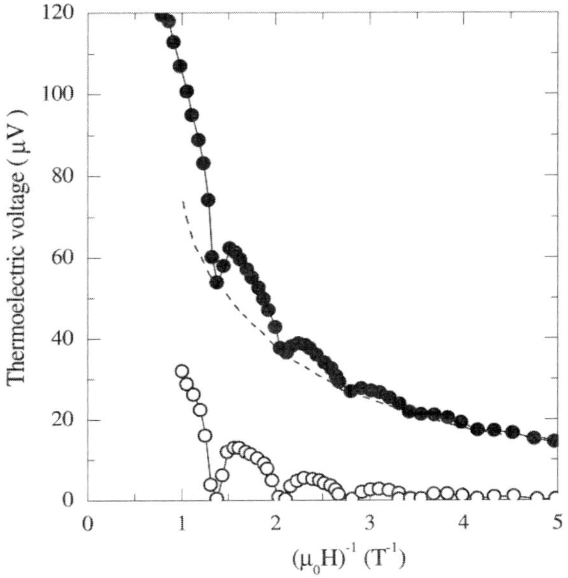

Figure 3
Quantum oscillations of the thermoelectric voltage versus the inverse magnetic field for Bi are shown. ●: total thermoelectric voltage; ○: oscillating part (broken line shows the envelope curve).

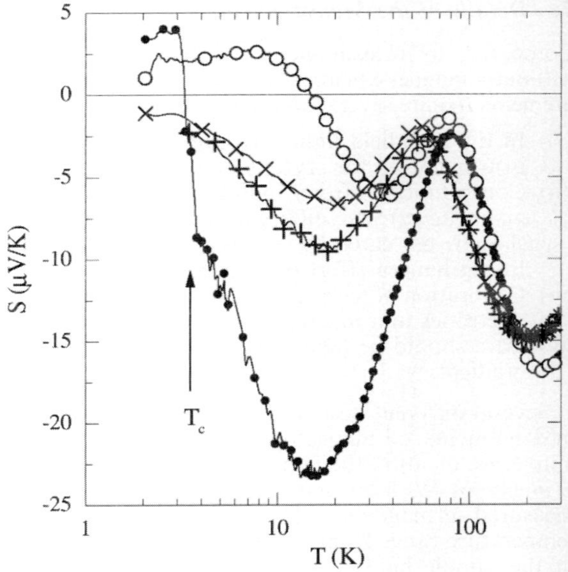

Figure 5
Thermopower of a nearly magnetic YCo_2 and of a ferromagnet $TmCo_2$ against temperature. ●: $TmCo_2$, $\mu_0H = 0\,T$; ○: $TmCo_2$, $\mu_0H = 15\,T$; +: YCo_2, $\mu_0H = 0\,T$; ×: YCo_2, $\mu_0H = 15\,T$. The arrow indicates Curie temperature of $TmCo_2$.

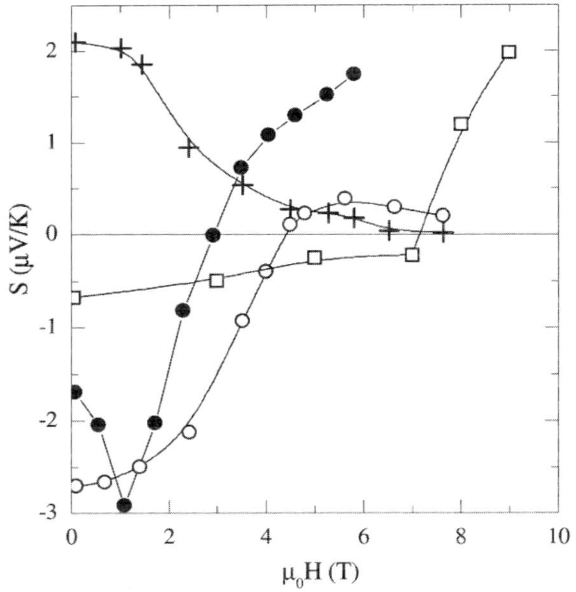

Figure 4
Thermopower of heavy-fermion compounds against a magnetic field (Marcenat *et al.* 1990). ○: $CeAl_2$ at $T = 0.58\,K$; ●: $CeAl_3$ at $T = 0.5\,K$; +: $(La\text{Ce})Al_2$ at $T = 0.58\,K$; □: $CeRu_2Si_2$ at $T = 1.5\,K$.

crystals. The results of these studies are summarized in Blatt *et al.* (1976). The quantum oscillations of the thermoelectric voltage versus the inverse magnetic field at $T = 4.2\,K$ for bismuth are presented in Fig. 3.

Very little has been done with respect to the magnetothermopower of magnetic materials and intermetallic compounds. Some results on ΔS of heavy-fermion compounds can be found in Marcenat (1990): see Fig. 4. Results for magnetic and nearly magnetic compounds are shown in Fig. 5.

Applications of thermomagnetic effects for thermoelectric energy conversion form a separate and large field. Semimetals and heavily doped semiconductors have mainly been used in these applications. Theoretical basics and a review of important experimental results can be found in Hartman and Honig (1967).

Bibliography

Abrikosov A A 1972 *Introduction to the Theory of Normal Metals*. Academic Press, New York

Barnard R D 1972 *Thermoelectricity in Metals and Alloys*. Taylor and Francis, London

Blatt F J, Schroeder P A, Foiles C L 1976 *Thermoelectric Power of Metals*. Plenum, New York

Foiles C L 1985 Thermopower of Pure Metals. In: Hellwege K-H, Olsen J L (eds.) *Landolt-Börnstein. Numerical Data and Functional Relationships in Science and Technology: new series, Group III–Crystal and Solid State Physics*. Vol. 15. *Metals: Electronic Transport Phenomena*. Subvolume b. Springer, Berlin, pp. 48–104

Hartman T C, Honig J M 1967 *Thermoelectric and Thermomagnetic Effects and Applications*. McGraw-Hill, New York

Huebener R P 1979 *Magnetic Flux Structures in Superconductors*. Springer, Berlin

Jan J-P 1957 Galvanomagnetic and Thermomagnetic Effects in Metals. In: Seitz F, Turnbull D 1957 (eds.) *Solid State Physics*. *Vol. 5*. Academic Press, New York, pp. 1–96

Landau L D, Lifshitz E M 1977 In: *Course of Theoretical Physics: Quantum Mechanics*. 3rd edn., Vol. 3. Pergamon, Oxford

Marcenat C, Jaccard D, Sierro J, Flouquet J, Onuki Y, Komatsubahara T 1990 Extended transport measurements on high-purity CeB_6. *Journal of Low Temperature Physics* **78**, 261–85

Resel R, Gratz E, Burkov A T, Nakama T, Higa M, Yagasaki K 1996 Thermopower measurements in magnetic fields up to 17 tesla using the toggled heating method. *Rev. Sci. Instrum.* **67**, 1970–5

Ri H-C, Gross R, Gollnik F, Beck A, Huebener R P 1994 Nernst, Seebeck, and Hall effects in the mixed state of $YBa_2Cu_3O_{7-\delta}$ and $Bi_2Sr_2CaCu_2O_{8+x}$ thin films: a comparative study. *Phys. Rev. B* **50**, 3312–29

Stanley D J 1974 Low-temperature magnetothermopower of silver. *Proc. R. Soc. London A* **339**, 97–117

Yagasaki K, Nakama T, Higa M, Sakai E, Burkov A T, Gratz E, Resel R 1996 Thermopower of $GdAl_2$ with see-saw heating system. *J. Phys. Soc. Jpn.* **65**, Suppl. B, 181–7

Magnetic Materials: Domestic Applications

S. J. Collocott

Hard magnetic materials or permanent magnets, and soft magnetic materials are fundamental to the technological success of many devices and products, in particular electric machines and related devices (see Fig. 1). In modern industrialized western economies the electric motor is ubiquitous and consumes 40–45% of all electrical energy generated. Most of this energy is consumed by large industrial electric drives, but nonetheless an increasing proportion is consumed by small electric motors used in domestic appliances, for example, in refrigerators which operate continuously. Regardless of the motor application there is pressure to improve electric motor efficiency as people and governments become more aware of the consequences of global warming, due to the emission of the greenhouse gas, CO_2, from large coal-fired power stations.

Electric motors are present in many household products (for the purposes of this section "domestic" is defined as any product that is found in the home). Products containing electric motors can be grouped according to their domestic application area and a list, by no means exhaustive, includes: laundry—washing machine and clothes dryer; kitchen–dishwasher, microwave oven, fan-forced convection oven, refrigerator, food blender, electric knife, garbage disposal unit, mixmaster, etc; bathroom—hairdryer, electric tooth brush, hair clippers, electric shaver, etc; entertainment—videorecorder, CD player, hi-fi system, camcorder, walkman, camera, etc; home study—personal computer, printer, photocopier, facsimile machine, etc; general household—vacuum cleaner, exhaust fans, ceiling fans, swimming pool and spa pumps, air-conditioner, fans in gas and electric heating systems, etc; and garage—electric drills, saws, routers and grinders, electric screwdriver, electric lawn mower, edge clipper, garden shears, garage door/gate opener, etc.

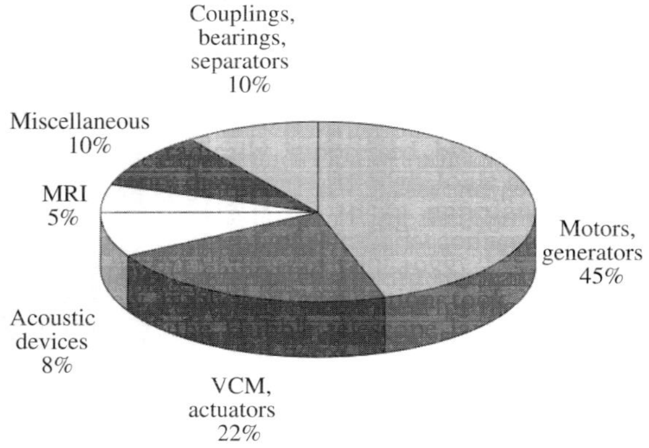

Figure 1
Expected application of $Nd_2Fe_{14}B$ in Europe for 1995 (Krebs 1995).

Some products or appliances may contain as many as four or five individual motors, for example, a camcorder, and where there is a motor there is likely to be power and control electronics, for example, a switch-mode power supply that incorporates magnetic components. Not included in domestic appliances is the family automobile, which may contain up to 40 or even more electric motors. The use of magnetic materials is not just confined to motors, as they are used in speakers in televisions and hi-fi systems, holding devices, water-meters, watt-hour meters and fluorescent lighting ballasts.

1. Motor Technology

A conventional brushed d.c. electric motor can be viewed as having three key components: a field system, which usually consists of a winding around teeth on a stator yoke; an armature, consisting of a series of coils wound in slots on a rotor; and a commutator, for making the electrical connection to the armature coils (Hamdi 1994). The stator and armature laminations are made from a soft magnetic material, usually low-loss silicon steel laminations, and are part of the magnetic circuit which carries and guides the magnetic flux in the machine. In a permanent magnet machine the field system uses permanent magnets rather than producing the field by passing current through coils. The replacement of the wound field system by permanent magnets results in a significant improvement in motor efficiency, as no electrical energy is used in field production. The motor can then be turned inside-out with the field magnets being attached to the rotor, a multiphase armature winding on the stator, and the commutator replaced by electronic switching devices which switch the current through the stator windings (see Fig. 2). This configuration has the advantage of improved thermal performance, as improved cooling is obtained by having the high-current stator winding in good thermal contact with the motor case. Motors of this type are known as permanent magnet brushless d.c. motors or just brushless d.c. motors (Hughes 1993, Gieras and Wing 1997) (see *Rotary and Linear Machines*).

Modern permanent magnet brushless d.c. motors have the advantages of high torque or power per unit volume, improved efficiency, better dynamic response (due to the low inertia of the rotor), reduced volume, reduced weight, mechanical simplicity and reliability (as there are no brushes or commutator), and quiet operation (Hanitsch 1989). These improvements did

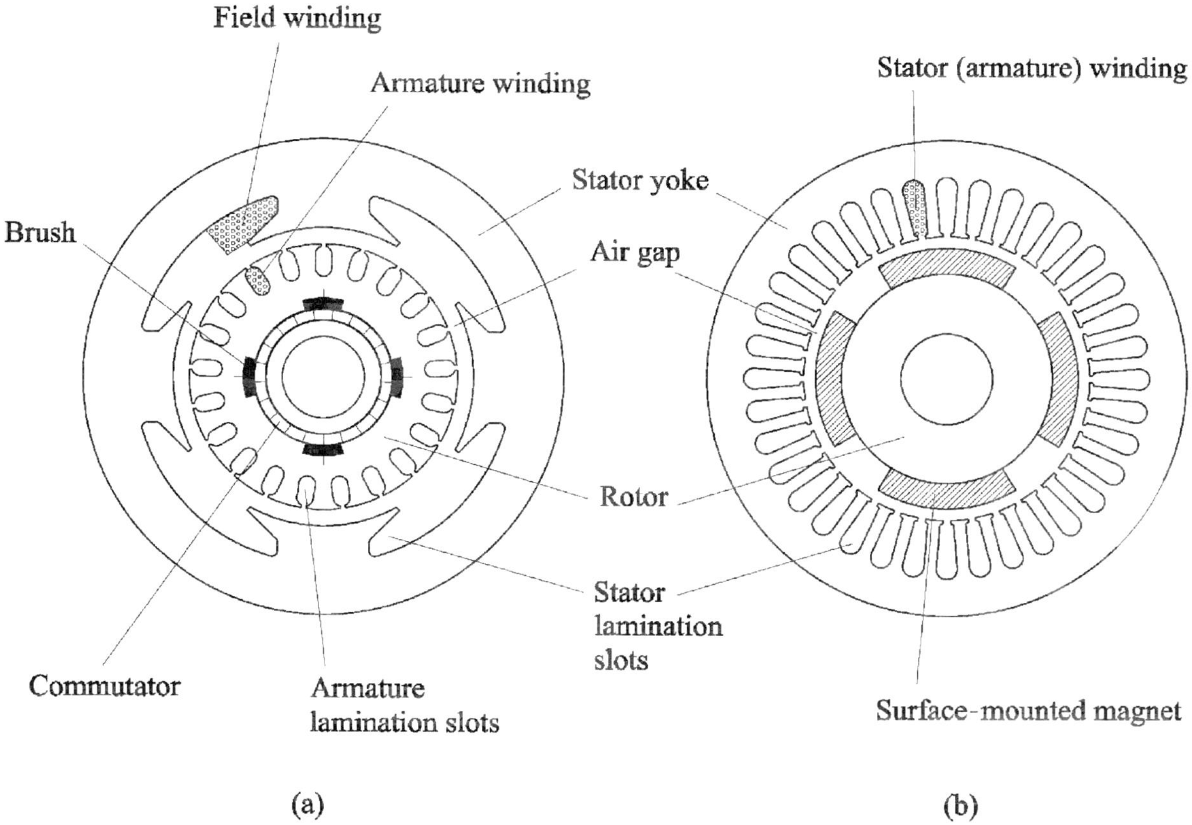

Figure 2
Schematic showing (a) a wound field brushed DC motor and (b) a brushless permanent magnet motor.

not happen overnight, but are the result of advances in materials science over the past 30 years, particularly the development of rare-earth permanent magnets (Sm_2Co_{17} and $Nd_2Fe_{14}B$) , and new soft magnetic materials, such as amorphous or glassy metals . In parallel with these developments in materials science there have been major advances in magnetic circuit design, using computational methods based on finite-element techniques, and a revolution in semiconductors, resulting in low-cost microprocessors and digital signal processors for motor control, and high-power switching devices.

Losses in electric motors have three main sources:

- copper losses or I^2R losses, which are due to the finite electrical resistance of the copper wire used in the windings (in a brushed machine this also includes losses from brush contact resistance);
- magnetic or iron losses from both magnetic hysteresis losses in the soft magnetic components and eddy currents; and
- friction and windage losses from bearing friction, brushes, if present, and the rotor peripheral speed (Hamdi 1994).

The design of a high-efficiency motor requires minimization of the losses, and may be achieved by the use of novel motor topologies, good magnetic circuit design, and judicious choice of materials, for example, low-loss electrical steels. Innovative motor designs may not use any iron at all; an example is an axial flux in-wheel motor developed for a solar powered car that has an efficiency of 98% (Ramsden 1998).

Motors used in domestic applications range in power from a fraction of a watt, or a few watts, to about 2 kW, have an output torque in the range from a few mNm to about 10 Nm, an efficiency of 60 to 70%, and an operational speed from a few rpm to 25 000 or 30 000 rpm. No one motor type satisfies all these power, torque, and speed requirements. A variety of motor types is used, each with performance characteristics to match the application. For example, a vacuum cleaner motor is required to operate at high speed, and may be only used for an hour per week. With this intermittent operation there is no great need for high reliability or high efficiency; rather, manufacturers aim for a motor of minimum cost. The situation is somewhat different with a washing machine. Here the motor has two extremes of operation; it is required to operate at low speed with high torque for the wash cycle and high speed with low torque during the spin cycle, to remove water from the clothes. On the other hand, for a refrigerator that operates continuously or a battery-powered hand-held drill there is an advantage in having a motor of high efficiency as this will minimize energy consumption, and for the drill it will give a longer period of use between battery charges.

Generally most domestic appliances are low-cost items, and the motor is usually chosen with this in mind. A common motor type used in intermittently operated hand-held appliances is the brushed universal motor (it has the same configuration as a brushed d.c. motor). Universal motors are made in a variety of sizes, are very low cost, and have the advantage of a simple controller for variable speed operation, though they are noisy at high speeds. Universal motors are used in hair-dryers, food blenders, vacuum cleaners, electric drills, saws and routers, mixmasters, etc. Induction motors tend to be favored in constant-speed continuous operation situations, for low noise and for their low-cost: for example, in fans, compressors in air-conditioners and refrigerators, clothes dryers, pool and spa pumps, etc. Permanent magnet motors may use either hard-ferrite or rare-earth permanent magnets (very rarely AlNiCo), with the former often used in small pumps in washing machines and dishwashers, and in compact low-noise fans and the latter where the features of high efficiency and high power/torque per unit volume are required, e.g., video-recorders, computer disk drives, CD players, camcorders, cameras, facsimile machines, battery-powered screw drivers, and drills. The situation with washing machines is interesting and reflects market diversity. In Europe frontloading washing machines dominate and tend to use brushed universal motors, while in the USA front-loaders may use a switched reluctance motor. For toploading machines, which are popular in the USA and Australia, both induction motors and direct-drive ferrite permanent magnet motors are used.

In domestic applications hard ferrite magnets dominate, with only limited use of rare-earth permanent magnet motors. More widespread use of rare-earth permanent magnet motors will occur when some, or all, of their advantages of compactness, high efficiency, high torque/power per unit volume, and improved dynamic response are required by the particular product. Domestic applications are highly cost-sensitive, and manufacturers are constantly seeking the lowest cost solution, which inhibits the take-up of motors using rare-earth permanent magnets as they are often of higher cost compared with competing motor types. This is changing as rare-earth magnet prices decrease, motor numbers increase, and the technical advantages of rare-earth permanent motors in particular applications become more apparent.

2. *Soft Magnetic Materials*

In an electrical machine, the magnetic losses or core losses are due to the soft magnetic material operating in a time-varying magnetic field. Core losses are made up of eddy current losses that arise from circulating electrical currents induced in conducting materials (the electrical steel used in the stator and copper used in the motor windings) and magnetic hysteresis losses which are proportional to the enclosed area of the B–H loop of the particular material. The characteristics required of the soft magnetic material are:

- high saturation induction to minimize weight and volume of the iron parts;
- high permeability for design of a low reluctance magnetic circuit;
- low coercivity to minimize hysteresis losses; and
- high resistivity to minimize eddy current losses (minimization of eddy currents is also achieved by using thin laminations that are covered by a thin insulating layer to provide interlaminar insulation).

These desirable magnetic properties are not found in one material, as there are many factors that affect the magnetic properties. The magnetic properties of a material may be tailored by the addition of chemical elements, mechanical working of the material, and heat treatment. The characteristics of a range of soft magnetic materials are shown in Fig. 3 and Table 1

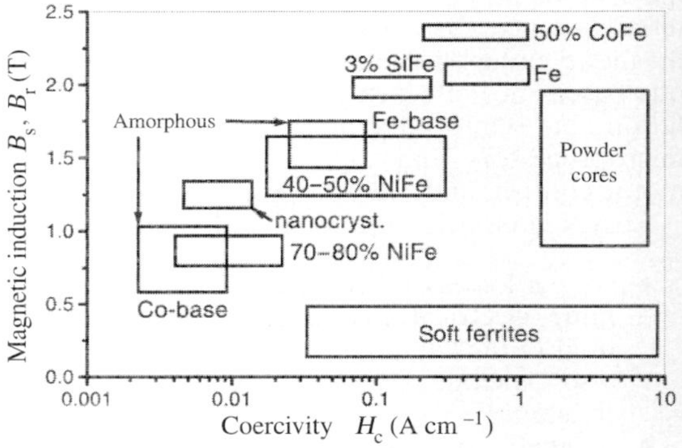

Figure 3
Typical property ranges of soft magnetic materials.

Table 1
Properties of some common soft magnetic materials.

Grade	Thickness (mm)	Nominal silicon content (%)	Resistivity (μΩ m)	Core loss at 1.5 T and 50 Hz (W kg^{-1})
Low-loss Si–Fe steel, nonoriented				
Transil 300-35-A5	0.35	2.9	0.48	2.95
Losil 400-50-A5	0.50	2.4	0.44	3.60
Losil 800-65-A5	0.65	1.3	0.29	6.5
Low-loss Si–Fe steel, oriented				
Unisil 35M6	0.35	3.1	0.48	1.00
Permendur 49 (Co49/Fe49/V2)	0.35	NA	0.40	3.0
METGLAS 2605SA1	0.025	NA	1.3	0.16[a]
SMC (iron powder)	NA	NA	NA	10

[a]Measured at 1.4 T.

(for reference data see Warlimont 1990 and Clegg *et al.* 1993).

2.1 Silicon–Iron Electrical Steels

The most common soft magnetic material used in electrical machines is silicon–iron electrical steel in the form of thin laminations. The addition of silicon to soft iron results in a significant decrease in coercivity, a slight decrease in saturation magnetization, and an increase in resistivity. There are two distinct categories of electrical steels: nonoriented isotropic products with silicon contents in the range 1–3.5%, and grain-oriented anisotropic products which contain 2.9–3.15% Si (+Al). Silicon levels above 3.5% result in the steel becoming very brittle and hard to work. Oriented steels have the lowest core loss (see Table 1) and are used in transformers, whereas nonoriented steel is used in electrical machines as there are varying flux directions.

Domestic appliances are dominated by low-cost universal and induction motors, with efficiencies of 60–70%. The quest for low cost results in motors that use lower-quality nonoriented electrical steels and thicker laminations (typically 0.65 mm), the latter resulting in a reduction in the number of stamping operations necessary to produce the stator lamination stack.

2.2 Iron–Cobalt Alloys

Iron–cobalt alloys (such as Permendur 49, 49Co–49Fe–2 V) have the highest saturation magnetization of all the soft magnetic materials (see Fig. 3). They have the disadvantage of a much higher cost than silicon–iron steels, but are used in applications where it is advantageous to reduce size and weight. Iron–cobalt alloys are used in aircraft generators, some 400 Hz motors, and active magnetic bearings.

2.3 Amorphous or Glassy Metals

Amorphous metals are made from alloys whose constituents may include Fe, Ni, and Co and a metalloid or glass former such as silicon, boron, or carbon. A typical amorphous metal, offered for sale by AlliedSignal Inc, USA, is METGLAS 2826, which has the composition $Fe_{40}Ni_{40}P_{14}B_6$. Amorphous metals are very thin, of the order of 0.025 to 0.04 mm. They have the lowest coercivity of all the soft magnetic materials and a higher resistivity than electrical steels, resulting in a material with the lowest core loss (see Table 1).

There is a number of issues regarding the use of amorphous metals in electrical machines. The very desirable feature of low coercivity is counter-balanced by the lower saturation magnetization, ≈1.5 T compared with ≈2 T in silicon–iron steels. The cutting and forming of amorphous metal laminations is expensive due to increased tool wear from their hardness (over C-80 Rockwell), and being very thin there is a greater number of stamping operations, and the material does not stack as well. Amorphous metals have been used in electrical motors where high efficiency is all-important (>94%), but it is in distribution transformers where the material is proving most beneficial (DeCristofaro 1998).

2.4 Soft Magnetic Composites

Soft magnetic composite (SMC) materials are made from bonded iron powders. The iron powder is coated with an insulating layer and pressed into a solid material using a die before final heat treatment to cure the bond. It is not a powder metallurgy process where sintering is used. SMC materials are rather new and have a number of advantages. The material is isotropic, enabling the design of magnetic circuits that have three-dimensional flux paths, and since there is an insulating layer between the iron powder particles, losses due to eddy currents are minimized.

The material can be used at frequencies up to 100 k Hz, and the cross-over point for core loss for it and 3% silicon steel (0.5 mm thick) is about 400 Hz, where at 1.5 T the loss for both materials is about 90 W kg^{-1}. The main advantage of the material may be in low-cost manufacture, as parts can be made to net shape, using highly-automated mass production methods. The main disadvantage of the material is its high core loss at 50 Hz when compared with silicon–iron electrical steels. Nonetheless, a number of prototype transverse flux and claw-pole electrical machines have been made using SMCs (Jack 1998).

2.5 *Soft Ferrites*

The magnetically soft ferrites have the general chemical formula $MeFe_2O_4$, where Me is most commonly MnZn (MnZn ferrite) or NiZn (NiZn ferrite), and have the cubic structure of the mineral spinel. The properties of the soft ferrite may be further adjusted by additions of Mg, Cu, or Co, or a mixture of these. The soft ferrites have the general characteristics of low coercivity, high permeability, and high resistivity.

The high resistivity results in negligible losses due to eddy currents, enabling the material to be used at high frequencies ($\gg$1 MHz). Generally the NiZn ferrites have a higher resistivity than the MnZn ferrites, but a lower saturation induction. The soft ferrites are manufactured in a range of core shapes, rings, pots, and E and C cores, for use in inductors and transformers. Their use is primarily in compact electronic power supplies and telecommunication components. Their relatively low saturation induction makes them unsuitable for use in power or high-flux transformers.

Another class of soft ferrites, the iron garnets with the general formula $R_3Fe_5O_{12}$, where R is a rare-earth, usually yttrium, is used in specialized microwave devices.

3. *Hard Magnetic Materials*

The role of hard magnetic materials, in particular high-energy permanent magnets, in electric motors is replacement of the field winding. In any application, including motors, there is a multitude of factors that require consideration to ensure that the optimum magnet material is selected (Coey 1996a, Kirchmayr 1996). Of importance are:

- magnetic parameters: remanence, intrinsic coercivity, and magnetic energy product, BH_{max};
- temperature stability of the magnetic parameters;
- ease of magnetizing the material;
- ease of forming the magnet into the desired shape;
- environmental factors, such as resistance to corrosion; and
- cost (cost can be treated in a variety of ways, and includes material cost, cost per unit of magnetic energy product, and the cost of forming the material into the desired shape and its magnetization).

Selection of the permanent magnet material impacts on the motor design; for example, a motor using a hard ferrite has narrow stator teeth, compared to the wider teeth in a machine using rare-earth permanent magnets, where the magnetic flux density is much higher. It is also desirable for motors that the magnet material has a linear *B–H* curve in the second, or demagnetizing, quadrant of the hysteresis loop. Both hard ferrite and rare-earth magnets have this characteristic, whereas AlNiCo differs by exhibiting a highly nonlinear demagnetization behavior. For the motor designer rare-earth permanent magnets offer the advantages of a high BH_{max}, which reduces the amount of magnet material required, and high-intrinsic coercivity, which is important if the motor is to withstand high armature currents, which could lead to demagnetization of the magnets. Typical properties for a range of permanent magnet materials are given in Tables 2 and 3. For detailed information on magnet types see McCaig and Clegg (1987) and Coey (1996b)).

The use of any permanent magnet material in a particular application is not only governed by the magnetic, physical, and chemical parameters, but also by the cost of the material. Applications and products are driven by markets, and the success or failure of a product in a particular market is driven by cost. The world market for various magnet types is given in Table 4 (Ormerod and Constantinides 1997), and it shows that the market is dominated by the low-cost hard ferrites. Included in Table 4 is the price per kilogram for various magnet materials. These prices are at best estimates, and are based on information from a variety of sources, including Abraham (1995) and material presented at the 1999 Gorham/Intertech NdFeB '99 Meeting (Gorham/Intertech Consulting, Portland, ME, USA).

It is difficult to give accurate pricing, as this information is often kept confidential by manufacturers, can be very dependent on quantity and complexity of the magnet shape, and subject to market volatility. For example, the advent of low-cost $Nd_2Fe_{14}B$ from the Far-East has had a considerable impact. Nonetheless, Table 4 does give a guide to the relative prices. In terms of cost per unit of energy, both sintered hard ferrite and sintered $Nd_2Fe_{14}B$ are approaching parity (hard ferrite of 27 kJm^{-3}, \$2.50–3.00 kg^{-1}, $\approx$\$0.10 $(kJm^{-3})^{-1}$ and $Nd_2Fe_{14}B$ of 250 kJm^{-3}, \$30 kg^{-1}, $\approx$\$0.12 $(kJm^{-3})^{-1}$). In recent

Table 2
Typical properties of some common permanent magnet materials.

Material	Remanence (T)	Intrinsic coercivity (kAm^{-1})	Energy product (kJm^{-3})
Sintered $Nd_2Fe_{14}B$	1–1.4	3200–1000	190–380
Sintered Sm_2Co_{17}	1.04–1.12	2070–800	200–240
Sintered $SmCo_5$	0.90–1.01	2400–1500	160–200
Anisotropic bonded HDDR $Nd_2Fe_{14}B$	0.81–0.87	915–1154	123
Isotropic plastic bonded $Nd_2Fe_{14}B$	0.4–0.7	1000–600	30–76
Sintered anisotropic AlNiCo	0.72–1.26	1920–610	20–44
Sintered isotropic AlNiCo	0.62–0.84	1190–125	4–18
Hard ferrite, anisotropic	0.36–0.40	180–270	25–31
Hard ferrite plastic bonded anisotropic	0.22–0.30	240–190	15–18
Hard ferrite isotropic	0.22–0.28	230–300	8.5–10
Hard ferrite plastic bonded isotropic	0.1–0.15	180–230	2–4

Table 3
Thermal characteristics for some permanent magnet materials (α is the reversible temperature coefficient of the remanence and β is the reversible temperature coefficient of the intrinsic coercivity).

Material	α ($\%K^{-1}$)	β ($\%K^{-1}$)	Max. operating temperature (°C)	Curie temperature (°C)
Sintered $Nd_2Fe_{14}B$	− 0.13	− (0.5–0.65)	200	310
Sintered $SmCo_5$	− 0.045	− (0.2–0.3)	250	720
Sintered Sm_2Co_{17}	− 0.03	− 0.19	350	825
AlNiCo	− (0.01–0.02)	0.01–0.03	500	830
Ferrites	− 0.2	0.3–0.4	300	450

Table 4
World permanent magnet market in 1994 and current prices.

Material	Market 1994 (US$ Millions)	Price ($\$USkg^{-1}$)
Bonded $Nd_2Fe_{14}B$	194	30–100
Sintered $Nd_2Fe_{14}B$	603	30–150
Bonded ferrite	673	2.50–6
Bonded rare-earth cobalt	30	230
Sintered ferrite	1244	2.50–5
Sintered rare-earth cobalt	218	250–300
Other (including AlNiCo)	248	50–80[a]
Total	3210	

[a]For cast/sintered AlNiCo.

years, both sintered and bonded $Nd_2Fe_{14}B$ have decreased in price, and it appears this trend will continue as market share increases and $Nd_2Fe_{14}B$ becomes cost-effective in new applications and replaces existing magnet materials in current applications.

3.1 *Rare-earth Cobalt Magnets*

Sintered and polymer-bonded magnets based on both the $SmCo_5$ and Sm_2Co_{17} alloy composition are available from a variety of manufacturers. The main advantage of Sm–Co-based magnets is their high

Curie temperature, making them suitable for use in applications at elevated temperatures, but they suffer from the disadvantage of high cost. The sintered material tends to be somewhat harder and more brittle than $Nd_2Fe_{14}B$, but it has excellent corrosion resistance. Sintered Sm–Co magnets are used in applications where arduous conditions apply, for example, in motors that are used in oil-wells, high-temperature automotive sensors, and microwave switches in telecommunication satellites.

3.2 *Neodymium–Iron–Boron* ($Nd_2Fe_{14}B$) *Magnets*

$Nd_2Fe_{14}B$ magnets are commercially available in a number of forms—anisotropic sintered and anisotropic and isotropic powders for polymer bonded magnets. Isotropic powders are prepared using a rapid solidification method and anisotropic powders by the HDDR (hydrogenation, desorption, disproportionation, recombination) process (Kirchmayr 1996). Projections suggest a cost of $30 kg^{-1} for isotropic powder by 2001. Polymer-bonded magnets are formed by either compression or injection molding, and binders include nylon 6/6. The great advantage of polymer-bonded magnets is that complicated shapes and geometries can be prepared to near-net or net shape using very low-cost production processes.

For applications requiring magnets of the highest possible energy product, sintered $Nd_2Fe_{14}B$ is the magnet of choice . The thermal stability of sintered $Nd_2Fe_{14}B$ magnets has improved, with operating temperatures of 200 °C now attainable. Similar advances have been made in corrosion resistance with the addition of cobalt and other alloying components. Sintered magnets find uses in applications that require the best possible performance, for example, servomotors in machine tools.

Polymer-bonded magnets are the most rapidly growing section of the permanent magnet market. The reason for this rapid expansion is the demand for magnets for computer disk drives. Currently more than 50% of rare-earth magnets produced in Japan are used in computer disk drives, and the world market for computer disk drives is expected to exceed 250 million units by 2002. There is also considerable demand for polymer-bonded magnets for small motors in video recorders, camcorders, printers, facsimile machines, office automation, automotive, and portable drills (Ormerod and Constantinides 1997).

3.3 *Aluminum–Nickel–Cobalt (AlNiCo) Magnets*

AlNiCo magnets are manufactured by a powder metallurgy process or by casting, and are available in both isotropic and anisotropic forms. They tend to be expensive due to the cost of the raw materials, but they have good corrosion resistance and a very respectable remanence, 1.26 T, when compared with the value of 1–1.4 T for $Nd_2Fe_{14}B$. The major disadvantage of AlNiCo is its low coercivity, which creates problems in handling and its use in electric motors. For example, removal of AlNiCo magnets from the magnetic circuit in a small electric motor often results in their partial, or complete, self-demagnetization. A consequence of this is that it is usually necessary to magnetize AlNiCo magnets *in situ*. The advantage that AlNiCo has compared with other commonly used permanent magnets is its excellent temperature stability. It has a reversible temperature coefficient of remanence of -0.01 to $-0.02\%K^{-1}$, compared with $-0.13\%K^{-1}$ for $Nd_2Fe_{14}B$. The excellent temperature stability of AlNiCo makes it the magnet of choice for watt-hour meters, ammeters, and voltmeters. Additionally, AlNiCo magnets are frequently used in small servomotors in aircraft and military hardware .

3.4 *Hard Ferrite Magnets*

In terms of the global permanent magnet market hard ferrite magnets still dominate. Ferrite magnets have the stoichiometry $BaFe_{12}O_{19}$ or $SrFe_{12}O_{19}$, the latter being increasingly used due to its higher coercivity. They are very cheap, as the raw materials from which they are made are abundant. A further advantage is their resistance to many chemicals, and being an oxide they are ideal for use in damp or wet environments. Isotropic and anisotropic sintered magnets, and polymer-bonded magnets formed from the milled powder, are widely available and used extensively. The temperature coefficient of the intrinsic coercivity is positive for the hard ferrites, in contrast to rare-earth permanent magnets, and consequently there is a risk of demagnetization if the hard ferrite magnet is exposed to too low a temperature. Hard ferrite magnets are found in applications ranging from door seals in refrigerators to small electric motors used in dishwashers and washing machines.

4. *Selected Applications*

4.1 *Computer Disk Drives*

Computer disk drives are a prime example of where performance enhancements are traceable to improvements in permanent magnets. In a fixed disk drive permanent magnets are used in the spindle motor, in the actuator that drives the read/write heads (generally termed the voice coil motor (VCM)), and the latch assembly (see Fig. 4). Disk drives nowadays have a 2.5, 3, or 3.5 inch form factor, as emphasis has switched from miniaturization to making the storage capacity of the drive as large as possible.

The availability of high-remanence rare-earth magnets has enabled the disk drive designer to reduce dramatically the size of the VCM, as only a small volume of magnet material is required, and to decrease access times. Disk drives of $>$1–2 GB capacity use a high-grade sintered $Nd_2Fe_{14}B$, showing typically a remanence of 1.4 T, an intrinsic coercivity of 1,190 kAm^{-1}, and an energy product of 360 kJm^{-3}. One or two magnets may be employed, with a total weight of 15.5 g, and return poles are made of low-carbon steel. Spindle motors may use sintered or polymer-bonded material, with the latter often favored as it can be easily molded to shape. Spindle motor speed is important as the higher the speed the shorter the seek time.

High-capacity drives found in file-servers have a spindle speed of 10000 rpm, while drives used in desktops and notebooks have speeds of 7200 rpm and 5400 rpm, respectively. Removable cartridge drives of 100 to 250 MB capacity use linear motors and polymer-bonded $Nd_2Fe_{14}B$, to keep their cost down. On the horizon are compact disk drives for use in digital cameras. IBM is building a 1-inch 340 MB hard drive for that application that uses $Nd_2Fe_{14}B$ magnets.

4.2 *White Goods*

The domestic appliances most often associated with motors are refrigerators, washing machines, and dish washers. On the whole these products appear to have changed little over the years, but there is increasing pressure to improve their energy efficiency. In the USA the Department of Energy mandated a 30% reduction in energy use for refrigerators manufactured after July 2001. An energy-efficient refrigerator has been developed in the USA that uses 1.16 $kWhd^{-1}$, and part of this efficiency improvement comes from the use of DC brushless motors (Braham 1997).

A permanent magnet brushless motor of innovative design is used in a top-loading washing machine made by Fisher & Paykel, New Zealand, and is shown in Fig. 5. The motor is of an axially compact design, approximately 270 mm diameter and 65 mm long, and uses an outer rotor with hard ferrite permanent magnets. The agitator is coupled directly to the motor. This simplifies the design of the washing machine, as the gearbox or belt-drive reduction is eliminated. Fisher & Paykel also make a low-profile pump motor for use in dishwashers (Fig. 6). Again,

Figure 4
Hard disk drive unit; the VCM actuator can be seen in the top left hand corner.

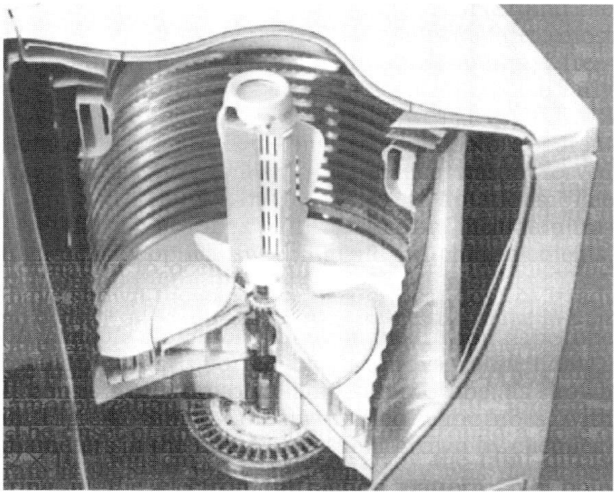

Figure 5
Cutaway showing direct-drive permanent magnet washing machine motor. (Photograph courtesy of Fisher & Paykel, New Zealand).

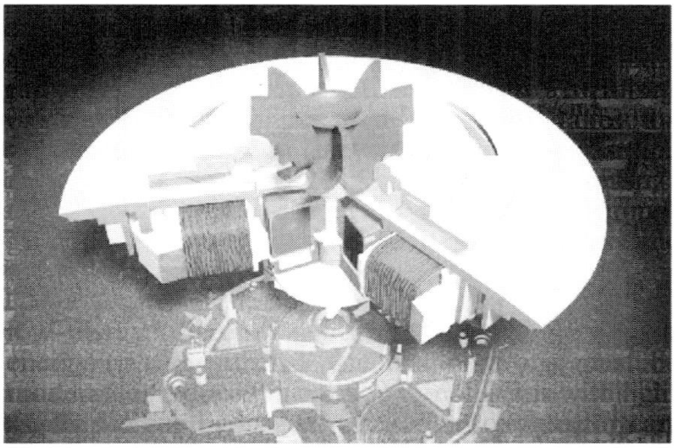

Figure 6
View of a hard-ferrite permanent magnet motor used in dishwashers. (Photograph courtesy of Fisher & Paykel, New Zealand).

this is a hard ferrite brushless permanent magnet motor. The magnets are located on the rotor, and the rotor runs in contact with the water detergent mix of the dishwasher. This is possible due to the excellent resistance of ferrite to chemical attack.

5. The Future

Hard and soft magnetic materials will continue to be fundamental to many domestic applications, and as improvements are made in their physical, chemical, and magnetic properties there will be corresponding product improvements. In the future one can foresee the "plastic" motor which will be constructed from an injection or compression-molded soft magnetic composite stator and polymer-bonded magnets, bonded directly on the rotor. The great attraction of such a motor will be its excellent performance and low cost, as its key components can be manufactured using low-cost plastic molding methods.

Bibliography

Abraham T 1995 Magnets and magnetic materials: a technical economic analysis. *Int. J. Powder Metall.* **31**, 133–6

Braham J 1997 Energy efficiency sends appliance designers into a spin. *Machine Des.* 40–6

Clegg A G, Beckley P, Snelling E C, Major R V 1993 Magnetic materials. In: Jones G R, Laughton M A, Say M G (eds.) *Electrical Engineers Reference Book, Fifteenth Edition.* Butterworth Heinemann, Oxford, UK, Chap. 14

Coey J M D 1996a Industrial applications of permanent magnetism. *Phys. Scr.* **T66**, 60–9

Coey J M D 1996b *Rare-earth Iron Permanent Magnets.* Clarendon Press, Oxford, UK

DeCristofaro N 1998 Amorphous metals in electric-power distribution applications. *MRS Bull.* (**May**). 50–6

Gieras J F, Wing M 1997 *Permanent Magnet Motor Technology.* Marcel Dekker, New York

Hamdi E S 1994 *Design of Small Electrical Machines.* Wiley, Chichester, UK

Hanitsch R 1989 Electromagnetic machines with Nd–Fe–B magnets. *J. Magn. Magn. Mater.* **80**, 119–30

Hughes A 1993 *Electric Motors and Drives.* Newnes, Oxford, UK

Jack A G 1998 Experience with the use of soft magnetic composites in electrical machines. In: Ertan B, Üctu Y (eds.) *Proc. Int. Conf. on Electrical Machines.* Middle East Technical University, Ankara, Turkey, pp. 1441–8

Kirchmayr H R 1996 Permanent magnets and hard magnetic materials. *J. Phys. D: Appl. Phys.* **29**, 2763–78

Krebs J 1995 Hot pressed and die-upset forged magnets. *Magnews* 19–21

McCaig M, Clegg A G 1987 *Permanent Magnets in Theory and Practice.* 2nd edn. Pentech Press, London

Ormerod J, Constantinides S 1997 Bonded permanent magnets: current status and future opportunities. *J. Appl. Phys.* **81**, 4816–20

Ramsden V S 1998 Application of rare-earth magnets in high-performance electric machines. In: Schultz L, Müller K-H (eds.) *Proc. 15th Int. Workshop on Rare-earth Magnets and their Applications.* Werkstofi-Informationsgesellschaft mbH, Frankfurt, Germany, pp. 623–42

Warlimont H 1990 Magnetic property assessment: between physical maximum and economic optimum. *IEEE Trans. Magn.* **26**, 1313–21

Magnetic Micro-actuators (MAGMAS)

G. Reyne, J. Delamare, and O. Cugat

Magnetic micro-actuators and systems (MAGMAS; also called "MAG-MEMS," (micro-electro-mechanical systems)) present outstanding performance for powerful integrated conversion from electrical to mechanical energy. MAGMAS, due to available forces and energies, belong to the recently appearing "Power-MEMS" family. Besides, the diversity of structures and the range of possible applications, including distance and remote-controlled actuation, explains the many recent publications and the large scope of research in this field. The topics addressed range from power energy microsources, to micromotors, microactuators, microswitches, levitation and remote-controlled magnetic micro-objects. The corresponding devices comprise optical, electrical, radiofrequency (RF), biomedical, and fluid dynamics applications.

MEMS are presently diversifying towards different principles of actuation, either electrostatic, magnetic, piezoelectric, or thermal, or actuation based on shape memory alloys. Each principle presents specific properties so as to cover a wider range of possibilities and applications. MAGMAS' development is recent, rapid, and promising as most specific and limiting technological problems are progressively being solved. It combines magnetic materials now available on silicon: microcoils, soft magnetic materials, micromagnets, magnetostrictive thick layers, etc., with either an integrated or remote-control field resulting in numerous possibilities. MAGMAS deserve more credence and attention from MEMS designers, and materials scientists should continue to develop better thick-film patterned permanent magnets compatible with microsystem technologies.

Stress will be put on downscaling laws. Due to both new available materials and the impressive increase of current density in microcoils, it is demonstrated that downscaling actually results in more effective electromagnetic converters. That magnetic interactions actually improve on downscaling is quite new and revolutionary.

1. A Brief History of MEMS and Magnetism

Developments in microelectronics led to MEMS, where the first actuations of thin beams employed were electrostatic, aluminum conductors and electrodes being standard. Thermal bimorph actuation came later. Development of other actuation principles (electromagnetism, piezoelectricity, magnetostriction, shape memory, etc.) has been slower because basic components and specific materials are not standard.

Early on, it was felt strongly that electromagnetism would not be very relevant for MEMS. Based on the simplified interactions between conductors in air or between a conductor in air and a piece of iron, but no effective electromagnetic power conversion, downscaling is catastrophic! This is even more so when drastic changes of essential parameters such as gradients or current densities were not considered when downscaling! Obviously, many recently published "downscaling laws" still favor electrostatics considering electromagnetism in its less appropriate and less downscalable configuration. These theories did not consider direct and homothetic scale reduction of the familiar electrical machines in daily use.

Generally, the perspectives for downscaling of different types of actuation for MEMS were neutral or pessimistic for magnetism, whereas a rigorous electrical engineering approach led to very positive conclusions (Jufer 1994). In the late 1980s, the first teams designed and promoted magnetic micro-actuators (Wagner and Benecke 1990, Guckel *et al.* 1991, Benecke 1994). They wisely remarked that permanent magnets are vital to magnetic actuation but that unfortunately their integration needs to be mastered (Guckel 1996). This is still a bottleneck. Many other works deserve to be read (Nami *et al.* 1996, Busch-Vishniac 1992). More details on MAGMAS, principles, materials, and applications have been collected in the book of Cugat (2002). The reader is also referred to the Feynman (1992, 1993) articles for perspectives in the early 1990s.

2. Magnetic Interactions and Downscaling at Constant Current Density δ

For simple energy considerations, electrical engineering clearly indicates that macroscopic devices (electrical engines and actuators that dominate the macroscopic world) are mainly composed of ferromagnetic materials (either soft magnetic materials or permanent magnets). Hence, the focus is on laws of scale reduction applied to such structures.

2.1 Magnetic Field of Magnets

A magnet of volume v_1 and magnetic polarization J_1 generates a scalar potential V at any point P located at a distance r from the magnet. The magnetic field H is the local gradient of the scalar potential V:

$$V(P) = \frac{v_1}{4\pi\mu_0}\frac{\boldsymbol{J}\cdot\boldsymbol{r}}{r^3}, \qquad H = \nabla V \qquad (1)$$

Homothetic miniaturization (all dimensions reduced by the same factor $k = 10, 100, 1000, \ldots$) of a given magnet (homogeneous material and constant polarization J_1) generates a constant magnetic field H, whereas the scalar potential $V(P)$ is divided by k. The relative geometry and magnitude of the field map around a magnet remain unchanged after scale reduction (Fig. 1). A direct and remarkable consequence is that the field gradients are multiplied by k. This has a tremendous impact on the interaction forces.

2.2 *Interactions between Magnets and Current-carrying Conductors*

Intuitively, one considers the forces of magnets on current-carrying conductors. Homothetical downscaling of the magnetic field maps due to the magnets leaves the Lorentz forces on conductors unchanged.

Obviously it is equivalent to consider the forces exerted by the conductors on the magnets. The Biot–Savart law states that the magnetic field H created in point P by a conductor of length dl and cross-sectional area S, carrying a current density δ (current $I = \delta \cdot S$), is

$$\boldsymbol{H}(P) = \frac{1}{4\pi} \frac{\delta \cdot \boldsymbol{S} \cdot \mathrm{d}\boldsymbol{l} \times \boldsymbol{r}}{r^3} \tag{2}$$

Hence, scale reduction by a factor of $1/k$ decreases the fields H due to the conductors by a factor of k. The magnetic force F on the magnet derives from their mutual magnetic interaction energy W_i:

$$W_i = -\boldsymbol{J} \cdot v \cdot \boldsymbol{H}, \qquad \boldsymbol{F} = -\nabla W_i \tag{3}$$

Scale reduction by a factor k reduces the energy W_i by a factor $1/k^4$, F by a factor $1/k^3$, and the force-to-weight (or force-to-volume) ratio does not change.

2.3 *Torque on a Magnet Due to External Field*

The torque experienced by a magnet of polarization J and volume v immersed in a homogeneous field H is proportional to M and H (where the magnetic moment equals $M = vJ$). During the homothetic $1/k$ reduction, both H and J remain constant, thus the torque-to-volume ratio also remains constant.

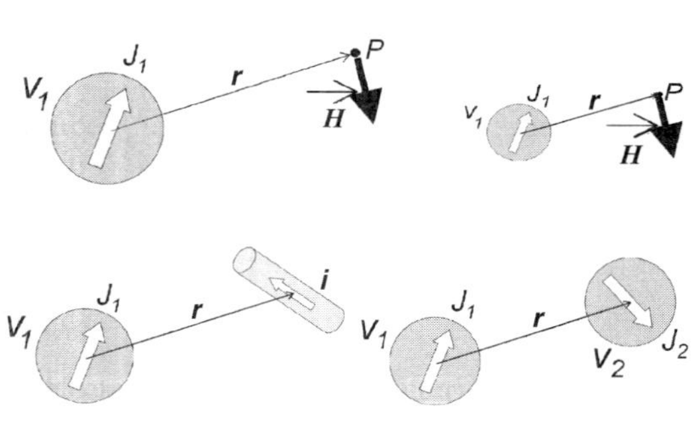

Figure 1
Top: Magnet—magnetic field unchanged with scale. Bottom: Interactions—magnet–current and magnet–magnet.

2.4 *Magnet–Magnet and Magnet–Iron Interactions*

Once again, the magnetic field of magnets remains unchanged upon downscaling. Hence, for magnet–magnet or magnet–iron interactions, Eqn. (3) shows a decrease of W_i by a factor $1/k^3$, F by a factor $1/k^2$ and an increase by k of the force-to-volume ratio.

Not only are macroscopic magnetic systems the most performing electric-to-mechanical converters but their scale reduction by $1/k$ can increase the force-to-weight ratio by k, allowing MAGMAS to provide large forces exactly as an ant can carry much heavier charges relative to its own weight than can an elephant.

2.5 *Scale-reduction Laws on Magnetic Interactions*

Similar calculations can be done for the magnetic interactions involving soft ferromagnetic materials combined to currents, and time-variation-induced currents (Cugat 2002, see chapter 2); see Fig. 2.

For a constant current density δ, the following conclusions can be drawn:

- The main magnetic interactions benefit from scale reduction.
- The most efficient magnetic interactions in devices of small dimensions involve permanent magnets.
- The forces between two magnets or between a magnet and a soft iron core increase by a factor k upon scale reduction by a factor $1/k$.
- The forces between a magnet and current-carrying conductors, when effective on a macroscopic scale, remain as effective once miniaturized.
- Inductive effects are to be avoided.

3. *Permanent Magnets for MAGMAS*

The application of MAGMAS requires the integration of small but high-quality magnets. Material

Reduction factor $1/k$	Magnet	Current	Iron	Induction $E = d\Phi/dt$
Magnet	$\times k$		$\times k$	$\times 1/k$
Current		$\times 1/k$	$\times 1/k$	$\times 1/k^2$

Figure 2
Effect of scale reduction $1/k$ on volume force for basic magnetic interactions, for a constant current density.

researchers are addressing this bottleneck and new results are presently being reported. Large availability of thick high-quality magnets in microtechnology will boost the performance of electrical-to-mechanical conversion and, more generally, of MEMS.

3.1 Micromagnet Fabrication Techniques

The most common and most powerful micromagnets currently used in MAGMAS are those individually micromachined from bulk rare-earth permanent magnets (REPMs) such as Nd–Fe–B or Sm–Co magnets , generally using machining by spark cutting. However, this method is hardly compatible with full integration or batch fabrication.

Many other ways to obtain micromagnets for MAGMAS are being developed . These methods all provide good results but each suffers from a moderate or major drawback (Cugat 2002, chapter 5). Some techniques are well adapted to microfabrication (electroplating of Co–Pt, screen printing of bonded powders) but the resulting magnetic properties are relatively poor compared to bulk REPMs (Zangari *et al.* 1996). Other techniques giving excellent magnetic properties are sputtering (Linetsky and Kornilov 1995), pulsed laser deposition (Cadieu *et al.* 1998), low-pressure plasma spraying (Rieger *et al.* 2000), or direct sintering (Yamashita *et al.* 2002). But either the thickness of the deposited layer is too thin, or it is difficult to adapt the process to microtechnology and batch fabrication (high deposition temperature, chemical pollution, slow deposition rate, small deposition surface). An important aspect currently emerging is the patterning of thick magnet films (Budde and Gatzen 2002). Optimal magnetic orientation of micromagnets is also studied (Kruusing 2002). The perspectives of achievements with low-temperature electroplated high-quality REPMs are discussed by Schwartz *et al.* (1998).

The large impact of the expected results justifies the increasing number of research efforts made in this field. However, perfect candidates for cheap, compatible, and fast integration of thick patterned layers of good-quality micromagnets remain to be found. Recently, technologies have improved rapidly and breakthroughs in this critical field are likely.

3.2 Equivalence between a Permanent Magnet and a Current-carrying Coil

A magnet is equivalent to a current-carrying coil (current density i, radius r) whose circumferential current density ($i/2\pi r$) remains constant (Amperian current model). This means that the equivalent volume current density increases by k when the magnet (or the equivalent coil) is downscaled by $1/k$.

4. Current Density in Microcoils

Magnetic sensors use microcoils and MAGMAS require them for variable fields and forces. Microcoils of various shapes and sizes can stand very high current densities. In macro-applications, the limitation to 3–5 A mm^{-2} is mainly due to thermal overheating. In micro-applications, thermal downscaling laws allow densities of 10^3–10^4 A mm^{-2}. While Joule conduction losses are volume related, the heat flow (for cooling) is proportional to the cross-section and inversely proportional to the thickness the heat flows through. Hence, heating scales as k^3, whereas heat flow dissipated by conduction scales as k. For the same temperature, Joule losses can increase by $k^3/k = k^2$. These Joule losses being proportional to the square of the current density, the current density can increase as k. Hence microcoil downscaling by $1/k$ can be done with the current density increasing by a factor of k. This is equivalent to a constant surface current density or, in other words, magnets and actual microcoils downscale in the same way. As forces and energies are proportional either to that current density or to the square of that density, this deeply boosts scale reduction laws.

The small volume-to-surface ratio of microcoils (planar geometry) and the direct contact with good heat-conducting substrates (e.g., Si), allows even higher DC currents, not to speak of pulsed currents (Cugat 2002, chapters 3 and 4, MacKay *et al.* 2000). However, it cannot be avoided that such increases in current densities and the concomitant increases of thermal losses in conductors can reduce both the efficiency and the autonomy (if used on batteries), and needs adapted heat sinks. These side-effects are practical limiting parameters for real design of MAGMAS.

5. Optimal Downscaling Laws ($\delta \sim k$)

With good magnets and current densities increasing as k, the downscaling laws can become significantly modified (Fig. 3).

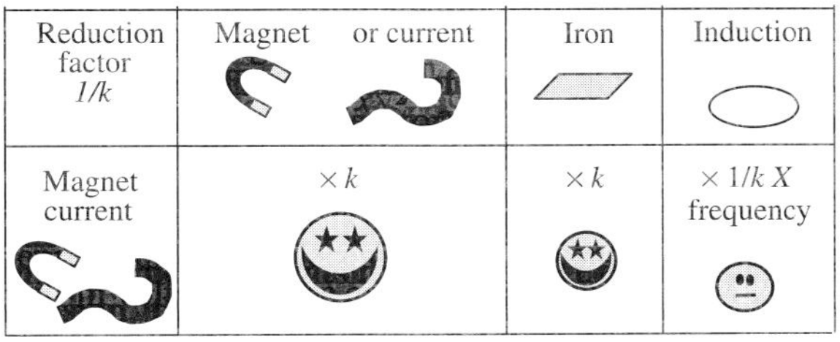

Figure 3
Effect of scale reduction $1/k$ on basic magnetic interactions, with current density increasing as k (constant temperature).

Table 1
Optimal downscaling laws.

Downscaling	$1/k$
Volume, mass, or gravity force	$1/k^3$
Heat flow dissipated by conduction	$1/k$
Electrostatic forces	$1/k^2$
Current density (constant temperature)	k
Most magnetic interactions ($1/k^2$)	$1/k^2$
Force/volume for most magnetic interactions	k

One can draw the following outstanding conclusions (see Table 1):

- All magnetic interactions benefit from scale reduction.
- Magnet–magnet or magnet–iron core forces increase as k.
- Current volume forces increase the same way and possibly faster (pulses). It depends on the specific characteristics of the application and the associated side-effects stated earlier.
- Magnetic forces downscale as electrostatic forces. Here one has to bear in mind that magnetic forces largely dominate in the macroscopic world.
- Inductive effects could benefit from the increase of frequencies when downscaling. The smaller the object, the faster are thermal, mechanical, or electrical effects. However, size reduction increases resistances and decreases inductances, thus handicapping inductive effects.

6. *Additional Characteristics of MAGMAS*

In addition to high energy densities and forces, electromagnetic interactions present other advantages, but also drawbacks for the actuation of microscopic systems (Reyne 2002).

6.1 *Permanent Forces, Bistability, and Suspensions*

Permanent magnets provide constant magnetic fields. Hence, simple latching or bistability is achievable without supply. This ensures energy savings or guarantees excellent safety in case of power failure (Fisher *et al.* 2001, Ruan *et al.* 2001). Bistability combines particularly well with pulsed-switching currents. Such permanent forces can also be implemented into passive magnetic suspensions/bearings, providing an elegant solution to the problem of friction in MEMS (Klöpzig 1998, Cugat *et al.* 1996).

6.2 *Long-range, Remote, or Wireless Actuation*

- Magnetic fields and gradients can be effective over long distances relative to the size of MEMS. This allows integrated large-throw and/or wide-angular actuators.
- Contactless magnetic interaction allows remote actuation through sealed interfaces allowing wireless actuation or vacuum packaging of resonant systems for high Q factors (no air viscosity dampening). Furthermore, remote actuation through sealed interfaces makes magnetic actuators very well suited to harsh environment (e.g., ABS sensors).
- External magnets or macrocoils can be added to power the microsystem thus avoiding integration of supply and most side-effects. It considerably simplifies fabrication. This characteristic is unique and deserves to be used for medical applications with implanted wireless MAGMAS actuated by distance external fields.

6.3 *Possible Future Actuation Modes for MAGMAS*

Major drawbacks of magnetic actuation are side-effects of Joule losses in conductors. One might overcome this problem by exploring new ways to modify magnetic fields:

- thermal demagnetization of a thermomagnetic material,
- magnetic "reprogramming" of semi-hard materials by demagnetization and remagnetization,
- strain-induced modulation of the magnetization of a magnetostrictive material by hybridization with a voltage-actuated piezoelectric element, and superconducting films can be used for levitation and high-temperature superconductivity may be a future possible solution for magnetizing MAGMAS for specific applications (space, cryogenic systems, etc.).

6.4 *Power Supplies, Control, and Coolers*

Electrostatic actuators use high voltages and low currents, whereas MAGMAS require high currents and low voltages. Current pulses, if used, need faster command and control. Hence, final performances strongly rely on working conditions and development of appropriate integrated supplies and coolers.

7. *Applications*

Magnetic sensors are already well established in commercial products (HDD read-heads, fluxgates, transmission coils, ABS sensors, etc.). However, few MAGMAS have, so far, reached industrial production. Laboratory-developed prototypes include radio frequency μ-switches for mobile phones, read/write heads and μ-positionners, matrixes of optical μ-commutators for fiber optic networks, μ-motors for noninvasive surgery and μ-robotics, μ-pumps, and μ-valves for lab-on-chip and μ-fluidic devices, electrical

μ-generators for autonomous power supplies, μ-mirrors for adaptive optics, magnetic suspensions for hard disk drives, etc.

8. Conclusion

Electromagnetic interactions deserve a larger interest from the MEMS community. MAGMAS offer large forces, large strokes, remote or distance control, bistability, and high robustness, all these properties offering great opportunities for new devices in many applications. The large current densities available at small scales, together with the development of magnetic thick films, specific supplies, and coolers should soon greatly promote the development of MAGMAS.

Bibliography

Benecke W 1994 Scaling behavior of microactuators. *Proc. ACTUATOR* **94**, 19–24Bremen

Budde T, Gatzen H H 2002 Patterned sputter deposited Sm-Co films for MEMS applications. *J. Magn. Magn. Mater.* **April**

Busch-Vishniac I J 1992 The case for magnetically-driven micro-actuators. *Sensors & Actuators* **A33**, 207–20

Cadieu F J, Rani R, Quian X R, Chen L 1998 High coercivity Sm-Co based films made by pulsed laser deposition. *J. Appl. Phys.* **83–11**, 6247–9

Cugat O, Fernandez V, Roy D, Reyne G, Delamare J 1996 Miniature permanent magnet bearings: application to planar micromotors. *Proc. Europe-Asia Conference MECATRONICS* **96**

Cugat O 2002 Microactionneurs electromagnétiques – MAGMAS. *Hermès-Science/Lavoisier ISBN 2-7462-0449-5*, 340pp

Feynman R P 1992 & 1993 There's plenty of room at the bottom. 1959 Infinitesimal machinery. 1983 reprinted *J. MEMS* **1**(1), 60–66 & *J. MEMS* **2**, 4–14

Fisher K, *et al.* 2001 A latching bistable optical fiber switch combining LIGA technology with micro-machined permanent magnets. *Transducers* **01**

Guckel H, et al. 1991 Fabrication and testing of the planar magnetic micromotor. *J. Micromech. Microeng.* **1**, 135–8

Guckel H 1996 Progress in electromagnetic actuators. *Proc. ACTUATOR* **96**, Bremen, 45–8

Jufer M 1994 Size limits and characteristic influence of electromagnetic actuators. *Proc. 4th International Conference on New Actuators*, 390–3

Klöpzig M 1998 A novel linear micromachined electromagnetic actuator including magnetic suspension. *Proc. ACTUATOR* **98**, 548–51

Kruusing A 2002 Actuators with permanent magnets having variable in space orientation of magnetization. *Sens. & Act.* **A101**, 168–74

Linetsky Y L, Kornilov N V 1995 Structure and magnetic properties of sputtered Nd–Fe–B alloys. *J. Mat. Eng. and Performance* **4–2**, 188–95

MacKay K, Bonfirm M, Givord D, Fontaine A 2000 50 T pulsed magnetic fields in microcoils. *J. Appl. Phys.* **87–4**, 1996–2002

Nami Z, Ahn C, Allen M G 1996 An energy-based design criterion for magnetic micro-actuators. *J. Micromech. Microeng.* **6**, 337–44

Reyne, G., 2002. Electromagnetic actuation for MOEMS, examples, advantages and drawbacks of MAGMAS. *MMM Journal of Magnetism and Magnetic Materials, Elsevier* **242–245**, Part II, 1119–25

Rieger G, Wecker J, Rodewald W, Sattler W, Bach Fr-W, Duda T, Unterberg W 2000 Nd–Fe–B permanent magnet (thick films) produced by a vacuum-plasma-spraying process. *J. Appl. Phys.* **87–9**, 5329–31

Ruan M, Shen J, Wheeler C B 2001 Latching micromagnetic relays. *J. MEMS* **10–4**, 511–7

Schwartz M, et al. 1998 Thin film alloy electrodeposits of transition rare-earth metals from aqueous media. *Electrochemical Society Proc.* **98–20**, 646–59

Wagner B, Benecke W 1990 Magnetically driven micro-actuators: design considerations In: *Microsystem Technologies*. Springer Verlag, Vol. 90, p. 838

Yamashita F, et al. 2002 Preparation of thick-film Nd–Fe–B magnets by direct Joule heating. *Proc. REM* **XVII**, 668–74

Zangari G, Bucher P, Lecis N, Cavallotti P L, Callegaro L. Puppin E 1996 Magnetic properties of electroplated Co-Pt films. *J. Mag. Mag. Mat.* **157/158**, 256–7

Magnetic Refrigerator: Design and Performance

C. B. Zimm

Introduction

Magnetic refrigeration (MR) technology is based upon the magnetocaloric effect (MCE), an intrinsic property of magnetic materials near their magnetic ordering temperature (for ferromagnetic materials this is the Curie temperature, T_c). In the case of ferromagnetic materials, such as gadolinium, the MCE is the reduction in entropy as the magnetic moments of the atoms are aligned upon application of a magnetic field and the increase of entropy when the magnetic moments become randomly oriented on removing the field. Under isothermal conditions, the MCE is the isothermal entropy change, ΔS_T, and is observed as the release or absorption of heat. Under adiabatic conditions, the MCE is observed as an increase or decrease in the material's temperature (adiabatic temperature change, ΔT_{ad}). Gadolinium, a typical magnetocaloric material (MCM), has a maximum ΔT_{ad} of 2.5 °C at the Curie temperature in a 1 T field. The Curie temperature, and thus the temperature of the peak ΔT_{ad}, may be moved by adjusting the MCM composition.

The response of an MCM to a magnetic field is similar to the response of a gas to compression and expansion. Below 4 K, the MCE is relatively large compared to the temperature span of interest, and thus a very simple MR cycle called adiabatic demagnetization refrigeration (ADR) has been used for many years. The ADR uses a fixed MCM, a slowly charged and discharged superconducting magnet, and direct thermal contact via solid thermal conduction in the high-field (warm) and low-field (cold) states of the MCM to carry small (less than 1 W) thermal loads. The subject of this article are the design refinements required for carrying larger cooling loads at higher temperatures, where the MCE is not large enough to directly attain the thermal span and the thermal loads are too large to be carried by direct thermal contact.

1. The AMR Cycle

The MCE above 10 K is relatively small, and the refrigerant in MR is a solid. Design of MR for use above 10 K is dominated by these two limitations. The small MCE means that regeneration or recuperation must be used to obtain a temperature span greater than the MCE. The small MCE also means that high magnetic fields (1 T or more) must be used, the MCM should be used in a concentrated form, and the volume of heat-transfer apparatus inside the expensive high-field region of the magnet must be minimal. The solid nature of the MCM means that an auxiliary heat-transfer fluid must be used (which also allows for regeneration). In addition, either the MCM or the magnet must be physically moved with respect to one another. An alternative of charging and discharging an electromagnet used in the ADR is too inefficient for the high magnetic field and rapid field charging needed to do useful cooling above 20 K. The active magnetic regeneration (AMR) cycle deals with the limitations of MR above 10 K in a simple manner, and is the only implementation found so far to simultaneously produce both useful temperature span and cooling power above 20 K.

The AMR cycle is operated as shown in Fig. 1 for a steady-state condition. Central in each figure is a bed of MCM that is porous to allow fluid flow. To the left of the bed is a cold heat exchanger (HEX), which in this example is at 5 °C, and to the right of the bed is a hot HEX at 21 °C. In Fig. 1(a), the initial temperature profile is for the bed in its demagnetized state in zero magnetic field (dashed line). When a magnetic field is applied to the refrigerant, each particle in the bed warms because of the MCE to form the final magnetized bed temperature profile (solid line). The amount each particle warms is equal to the adiabatic temperature change upon magnetization at the initial temperature of the particle, reduced by the effect of the heat capacity of the fluid in the pores between the particles. Next, the cold fluid flows through the bed from the cold end to the hot end (Fig. 1(b)). The bed is cooled by the fluid, lowering the temperature profile across the bed, and the fluid in turn is warmed by the bed, emerging at a temperature close to the temperature of the bed at the warm end. This temperature is higher than 21 °C, so heat is removed from the fluid at the hot heat sink as the fluid flows through the hot HEX. After the fluid flow is stopped, the magnetic field is removed, cooling the bed by the MCE (Fig. 1(c)). The refrigeration cycle is completed by forcing the fluid to flow from the hot to the cold end of the bed (Fig. 1(d)). The fluid is cooled by the bed, emerging at a temperature below 5 °C, and removes heat from the cold sink as the fluid passes through the cold HEX. The heat flow from cold to hot in this cycle is driven by the net mechanical work input needed to move the magnet with respect to the magnetocaloric bed. The higher magnetization of the cold bed matrix results in a greater force to remove the magnet from the bed than to initially cover the warmer, lower-magnetization, bed by the magnet.

The AMR cycle overcomes the difficulties of the small MCE and solid MCM. The regeneration in the AMR cycle can increase the temperature lift to much more than that of the MCE alone. The large heat-transfer surface of the porous AMR bed and the heat-transfer fluid flow efficiently couple the MCE of the solid via the HEXs to the cold load and hot heat sink. Each section of the AMR bed sees only a

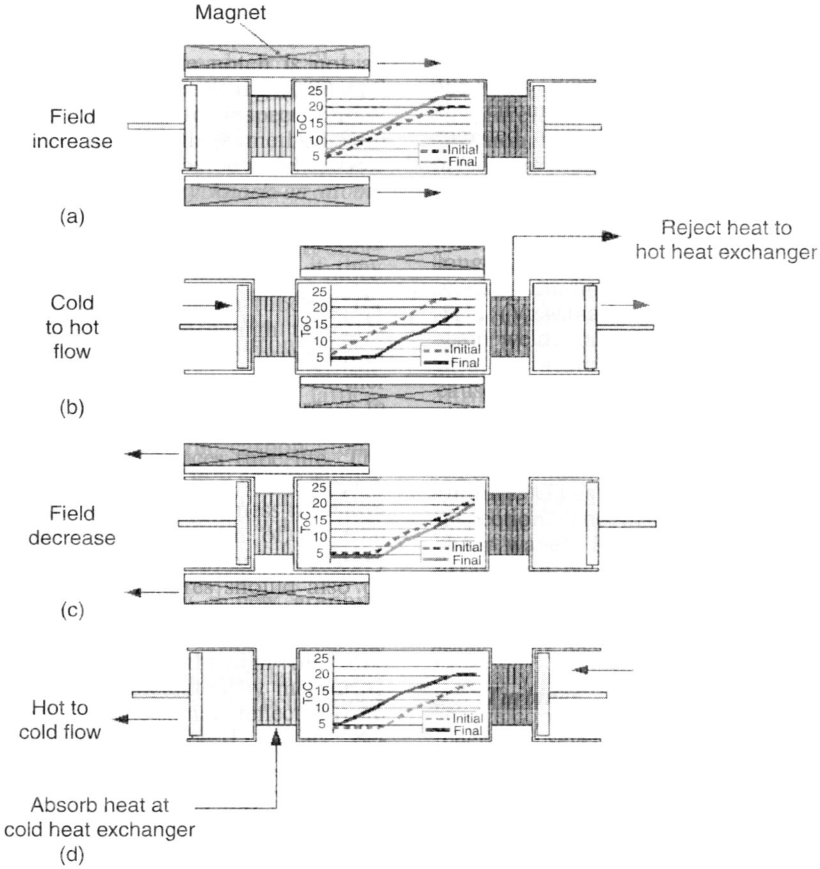

Figure 1
The AMR cycle. Reproduced from Zimm C, Sternberg A, Russek S, King K 2003 Magnetic refrigeration for shipboard centralized and distributed chilled water. In: *Proceedings of the Marine Environmental Engineering Technology Symposium* (on CD-ROM), Arlington VA, Aug. 22, 2003, Session A3.

portion of the total temperature span, and hence the bed may be made into layers, each layer containing a magnetic material composition with properties optimized for the operating temperature of that layer.

2. *Alternatives to the AMR Cycle*

The most direct way to implement an MR cycle similar to the vapor cycle would be to use an MCM suspended in a ferrofluid that is pumped into and out of the magnetic field. Unfortunately, ferrofluids that retain fluidity when entering the high-field region have maximum MCM loadings of no more than a few percent. Such a dilute MCM would have too small of an MCE to do useful cooling.

Another alternative to the AMR cycle is to use a fluid column regenerator. G. V. Brown demonstrated such a near-room-temperature regenerative MR with a 47 K span in 1976 (Brown 1976). In Brown's refrigerator, there is a porous bed of MCM much as in the AMR, but there is almost no temperature gradient within the MCM. The magnetized MCM serves to warm a column of fluid that contains the full temperature gradient of the MR, and then the demagnetized MCM cools the column of fluid. The problem with this design is that the process must be done very slowly in order to not churn the column of fluid and destroy the temperature gradient. Cooling power and efficiency were thus low in this type of device.

Recuperation has been tried as an alternative to regeneration. Recuperation in wheel-type designs requires a rotating wheel of MCM with counter-rotating flow of heat transfer fluid in the regions of changing field. The design requires complex flow control structures to maintain the correct fluid flow.

Fluid bypass flow has prevented this design from working well.

3. *Minimizing Losses in the AMR*

The physical form of the MCM and the dimensions of the AMR bed should be chosen to minimize entropy generation while maintaining a high cooling power. This involves a compromise between three main loss mechanisms: finite heat transfer, fluid pumping loss, and longitudinal thermal conduction. The loss due to finite heat transfer between the fluid and the solid MCM is reduced by increasing the wetted surface area of the MCM, increasing the Nusselt number Nu, and increasing the thermal conductivity of the fluid. In contrast, the fluid pumping losses are reduced by decreasing the friction factor and increasing the hydraulic diameter, which unfortunately tends to decrease Nu and reduce wetted area. The fluid pumping losses are also reduced by using a low-viscosity fluid. The thermal conduction from the hot to the cold sides of the bed is reduced by reducing the thermal conductivity of the fluid and the solid MCM, which unfortunately increases the heat-transfer loss, and by increasing the length of the bed, which unfortunately increases the fluid pumping losses. It is advisable to construct an analytical model of the AMR bed in order to properly choose bed dimensions and MCM physical form to balance these three main losses. Several accounts of such models are in the literature (DeGregoria 1992, Engelbrecht *et al.* 2007). The latter article also reviews recent developments in MR.

In addition to the above three main losses, careful bed design and construction can prevent problems from fluid flow nonuniformity, MCM deterioration due to cyclic magnetic forces, and eddy current heating.

Choice of the heat-transfer fluid is an important consideration for the AMR. The fluid should have a low viscosity and a high volumetric heat capacity to reduce pumping losses. The fluid should also have a reasonably high thermal conductivity to reduce heat-transfer losses. Near room temperature, water (perhaps with antifreeze) is the best choice. At lower temperatures, helium is the best gas to use if the system can be made completely leak tight.

Design of components in the fluid circuit external to the AMR bed is also important. The reciprocating fluid flow needed for heat transfer can be generated by a reciprocating displacer or by a continuously running pump and valves. In either case, precisely phased fluid flow should be delivered to the AMR beds. Friction in the pump and valves or displacer, and the pressure drop and heat losses in the connecting piping should be considered. Placing the pump on the hot side of the bed reduces the impact of its losses on cooling power. The hot and cold HEX should be well coupled to the AMR beds and must be sized to efficiently handle the fluid flow and heat loads.

Magnet design is critical. The magnet circuit needs to produce uniform high-field and low-field regions. The cooling power and span will be proportional to the difference in magnetic fields. If the field regions are to move, this should be done with minimal generation of eddy currents. Use of fiber-reinforced plastic AMR bed housings will reduce eddy currents in both the moving bed and moving magnet cases.

4. *Performance of AMR Magnetic Refrigerators*

The first AMR device to achieve a useful temperature span was built in the 1980s by the Navy Laboratory at Annapolis (Green *et al.* 1990). The MCM was 0.73 kg of gadolinium metal wound in a spiral ribbon. The heat-transfer fluid was air, and the magnetic field was cycled by charging of a superconducting magnet. Efficiency was not reported, but the long time required for charging the magnet and the high pressure drop of a gaseous heat-transfer fluid limited the cooling power.

A small cryogenic AMR used helium gas for heat transfer and a moving superconducting magnet to allow operation at higher frequency than that possible with the charging of a superconducting magnet (Wang *et al.* 1995). As a result, a remarkable 40% of Carnot efficiency over a 20 K span was obtained. This efficiency, however, did not include the work needed to cool the superconducting magnet.

Performance improved when two magnetic refrigerators were built that used water as a heat-transfer fluid (Zimm *et al.* 1998, Hirano *et al.* 2002). Each of these refrigerators attained cooling power exceeding 100 W over a 20 °C span. These refrigerators used superconducting magnets; however, this makes for low overall efficiency when accounting for the work needed to cool the superconducting magnet.

An effective implementation of the AMR cycle at small to moderate loads can be done by efficiently cycling matrices of MCM between low and high magnetic fields produced by a permanent magnet that does not consume power to maintain its magnetic field. To attain high efficiency, work from the material entering the high-field region should be recovered to assist in removing material exiting the high-field region, which is simply done when the MCM is arranged in a wheel, and the MCM or the magnet moves in a circular motion. Simultaneously, hot fluid must be efficiently transferred out of the hot side of the beds in the high-field region to a hot HEX and cold fluid transferred from the cold side of the beds in the low-field region to a cold HEX. An example of a device that implements these functions well is a rotary AMR bed, permanent-magnet-based, magnetic refrigerator (RBMR) built in 2001, as described by Zimm *et al.* (2005) and shown in Fig. 2.

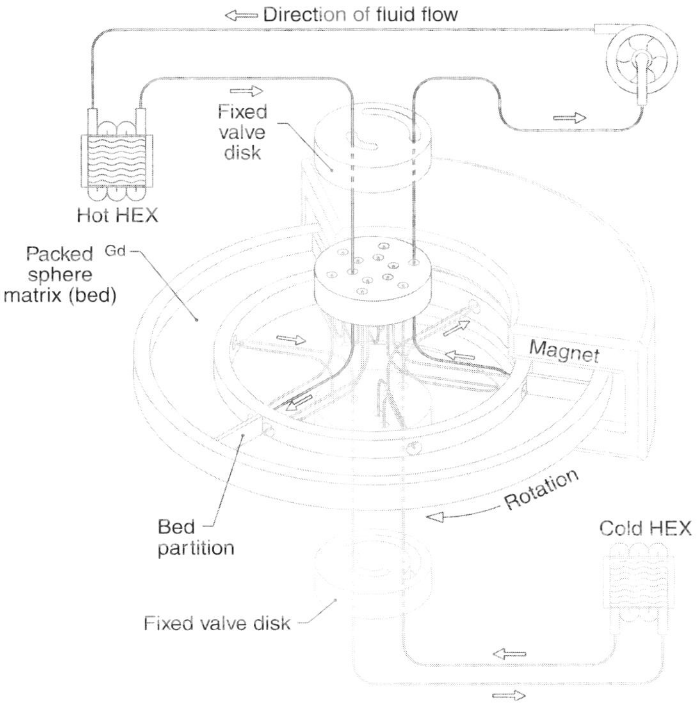

Figure 2
Rotary AMR using a permanent magnet. Reproduced from Zimm C B, Boeder A, Chell J, Sternberg A, Fujita A, Fujieda S, Fukamichi K 2005 Design and performance of a permanent magnet rotary refrigerator. *Int. J. Refrig.* **29**, 1302–6, with permission from Elsevier.

The wheel in Fig. 2 is divided into three sectors, each packed with porous MCM. Radial pipes carry water from the cold HEX into the middle of each sector and remove fluid to the hot HEX from the ends of each sector while the sector is in the high-field gap of the magnet. Fluid is carried in the opposite direction when the sector is the low-field area outside the magnet, thus producing reciprocating circumferential fluid flow in each sector.

The radial pipes from the hot ends of the sectors connect to ports in a rotating part of a disk valve above the wheel. The fixed portion of the upper valve is connected to the hot HEX and a pump. The radial pipes from the cold middle of each sector connect to the rotating part of a disk valve below the wheel. The fixed portion of the lower valve is connected to the cold HEX. The ports in the rotating and fixed parts of the disk valves interact to correctly switch fluid flow as the wheel rotates. The fluid flow in the beds reverses in direction, but the piping and valve switching ensures unidirectional flow in the pipes and HEXs, avoiding the dead volume problem that constrains the efficiency of Stirling and thermoacoustic refrigerators. The valve arrangement and flow circuit allow smooth operation to 4 Hz, with low noise and vibration. A cooling power of 15 W was obtained at a 20 °C span, using 160 g of Gd and Gd alloy MCM arranged in two layers.

5. *Material Properties*

The MCM for an AMR should be chosen so the MCE is increasing slowly with temperature over the span of the MR. In the absence of losses, the ideal MCE should be proportional to absolute temperature (Cross *et al.* 1988). Cooling power will be proportional to the MCE, so a material should be chosen with a high concentration of magnetic spins and a strong magnetic transition in the operating region of the MR. Materials with a first-order magnetic transition have the largest MCE. An AMR using multiple layers of a series of first-order transition MCMs, each layer made to operate in the region around the first-order phase transition, should attain a large cooling power if the hysteresis losses of the first-order transition are not excessive. A review of recent advances in MCM has been published by Gschneidner Jr. *et al.* (2005).

6. *Conclusions*

AMR technology has been advancing steadily, although it is not quite ready for commercial application. However, the history of MR includes many unsuccessful devices. Designers can reduce the chance of repeating past mistakes by reviewing past papers and patents. In addition, one should keep the fundamental characteristics of MR in mind: charge–discharge high-field magnets are very lossy except at cryogenic temperatures, so either the material or magnets must move and heat exchange must be done at the high-field and low-field points; the MCE is small, so a high field and MCMs with high concentration of spins are necessary; MCMs are solids, so an auxiliary heat-exchange fluid is required with excellent thermal contact to the MCM. Obtaining high efficiency requires properly balancing the competing losses in the AMR using a detailed model of the AMR beds. Constructing a good model requires a lot of effort due to the coupled hydraulic, thermal, and magnetic interactions of the AMR.

Bibliography

Brown G V 1976 Magnetic heat pumping near room temperature. *J. Appl. Phys.* **47**, 3673–80

Cross C R, Barclay J A, DeGregoria A J, Jaeger S R, Johnson J W 1988 Optimal temperature–entropy curves for magnetic refrigeration. *Adv. Cryog. Eng.* **33**, 767–75

DeGregoria A J 1992 Modeling the active magnetic regenerator. *Adv. Cryog. Eng.* **41**, 867–9

Engelbrecht K L, Nellis G F, Klein S A, Zimm C B 2007 Recent developments in room temperature active magnetic regenerative refrigerators. *HVAC&R Res.* **13**, 525–42

Green G, Chafe J, Stevens J, Humphrey J 1990 A gadolinium–terbium active regenerator. *Adv. Cryog. Eng.* **35**, 1165–74

Gschneidner K A Jr., Pecharsky V K, Tsokol A O 2005 Recent developments in magnetocaloric materials. *Rept. Prog. Phys.* **68**, 1479–539

Hirano N, Nagaya S, Takahasahi M, Kuriyama T, Ito K, Nomura S 2002 Development of magnetic refrigeration for room temperature applications. *Adv. Cryog. Eng.* **47**, 1027–34

Wang A A, Johnson J W, Niemi R W, Sternberg A, Zimm C B 1995 Experimental results of an efficient active magnetic regenerator refrigerator. In: Ross R G (ed.) *Proceedings of the 8th International Cryocooler Conference*. Plenum, New York, p. 665

Zimm C B, Boeder A, Chell J, Sternberg A, Fujita A, Fujieda S, Fukamichi K 2005 Design and performance of a permanent magnet rotary refrigerator. In: *First IIF-IIR International Conference on Magnetic Refrigeration at Room Temperature*, Montreux, Switzerland, Sep. 27–30.

Zimm C B, Boeder A, Chell J, Sternberg A, Fujita A, Fujieda S, Fukamichi K 2005 Design and performance of a permanent magnet rotary refrigerator. *Int. J. Refrig.* **29**, 1302–6

Zimm C B, Jastrab A G, Sternberg A, Pecharsky V K, Gschneidner K A Jr., Osborne M, Anderson I 1998 Description and performance of a near-room temperature magnetic refrigerator. *Adv. Cryog. Eng.* **43**, 1759–66

Zimm C, Sternberg A, Russek S, King K 2003 Magnetic refrigeration for shipboard centralized and distributed chilled water. In: *Proceedings of the Marine Environmental Engineering Technology Symposium* (on CD-ROM), Arlington VA, Aug. 22, 2003, Session A3

Magnetostrictive Materials

J. C. Peuzin

Magnetostriction is the term given to the mechanical deformation induced in a solid substance by the modification of its magnetic state. Inverse magnetostriction refers to the modification of the magnetic state by an applied stress.

These phenomena are significant only in substances which are magnetically ordered or at least close to the onset of a magnetic order, and have practical importance only in those materials which exhibit a spontaneous magnetization, i.e., ferro- or ferrimagnetic materials. Only the latter will be considered in this article.

1. *Spontaneous Magnetostriction*

Generally speaking, spontaneous magnetostriction is defined as that part of strain which is due to the onset of a magnetic order in a material (du Tremolet de Lacheisserie 1993).

A phenomenological approach consists of treating the onset of spontaneous magnetization as a small perturbation of a hypothetical nonmagnetic state. At a given temperature, the perturbation is entirely defined by the magnetization direction cosines $\alpha_1,\alpha_2,\alpha_3$. The spontaneous magnetostriction tensor **S** is then expanded as a function of this perturbation. In a convenient frame (Nye 1957), allowing for time reversal invariance, and using the usual Einstein convention on repeated indices, this gives:

$$S_{ij} = Q_{ijkl}\alpha_k\alpha_l + \text{even higher order terms}(i,j = 1,2,3) \quad (1)$$

The tensor **Q** and the higher-order tensors are expected to be temperature-dependent. Consideration of the first term only is sufficient to account for the properties of most materials, so that only the form of **Q** will be considered in the following. Since **Q** is a polar tensor of order 4, it follows the same symmetry rules as the elastic tensors (Nye 1957). Two cases are worth considering: cubic crystals and isotropic materials. In the following, emphasis will be placed on cubic symmetry since it contains the isotropic situation as a special case.

In cubic crystals, the convenient reference frame is obviously the one that coincides with crystallographic axes.

There are only 15 nonzero components of **Q** out of which only three are independent, namely

$$Q_{1111} = Q_{2222} = Q_{3333} = Q_{11}$$

$$Q_{1122} = Q_{2211} = Q_{1133} = Q_{3311} = Q_{2233} = Q_{3322} = Q_{12}$$

$$Q_{2323} = Q_{3232} = Q_{1313} = Q_{3131} = Q_{1212} = Q_{2121} = Q_{44} \quad (2)$$

so that the components of spontaneous magnetostriction can now be written as:

$$S_{11} = (Q_{11} - Q_{12})\alpha_1^2 + Q_{12}; \quad S_{22} = (Q_{11} - Q_{12})\alpha_2^2 + Q_{12};$$
$$S_{33} = (Q_{11} - Q_{12})\alpha_3^2 + Q_{12}$$
$$S_{23} = S_{32} = Q_{44}\alpha_2\alpha_3; \quad S_{13} = S_{31} = Q_{44}\alpha_1\alpha_3;$$
$$S_{12} = S_{21} = Q_{44}\alpha_1\alpha_2 \quad (3)$$

In isotropic material the reference frame can be chosen arbitrarily. The relevant **Q** components and the spontaneous magnetostriction are immediately derived from Eqns. (2) and (3), respectively, by expressing the isotropy relation, $Q_{44}=Q_{11}-Q_{12}$.

2. *Volume Magnetostriction*

The relative volume expansion of a fully saturated sample with respect to the nonmagnetic reference state is $\Delta V/V = S_{11} + S_{22} + S_{33}$, which, from Eqn. (3), turns out to be equal to $Q_{11} + 2Q_{12}$ whatever the orientation $\alpha_1,\alpha_2,\alpha_3$. It can be concluded that in cubic crystals (and therefore in isotropic materials) volume magnetostriction is actually independent of the technical magnetization state and equal to $Q_{11} + 2Q_{12}$.

3. *Joule Magnetostriction*

Joule magnetostriction is by definition the deformation of a magnetic material which is induced at a fixed temperature by those technical magnetization processes that involve a rotation of the spontaneous magnetization.

It is interesting from an experimental viewpoint to redefine the zero deformation reference state as the isotropic demagnetized state instead of the hypothetical nonmagnetic state. By this, it is meant a multidomain state with an isotropic distribution of the magnetization orientation and, of course, vanishing average (or technical) magnetization. The strain components as measured with respect to this new reference state (called for convenience spontaneous Joule deformations) are then:

$$S_{11} = (Q_{11} - Q_{12})(\alpha_1^2 - 1/3);$$
$$S_{22} = (Q_{11} - Q_{12})(\alpha_2^2 - 1/3);$$
$$S_{33} = (Q_{11} - Q_{12})(\alpha_3^2 - 1/3)$$
$$S_{23} = S_{32} = Q_{44}\alpha_2\alpha_3; \quad S_{13} = S_{31} = Q_{44}\alpha_1\alpha_3;$$
$$S_{12} = S_{21} = Q_{44}\alpha_1\alpha_2 \quad (4)$$

Joule magnetostriction in a cubic crystal may thus be characterized by only two parameters, namely $(Q_{11} - Q_{12})$ and Q_{44}. They can be determined, at least in principle, by measuring the relative elongation $\lambda(\alpha_1,\alpha_2,\alpha_3)$ along various magnetization directions when the sample is brought from the isotropic demagnetized state to the saturated state. Two particular orientations are worth considering, ⟨100⟩ and ⟨111⟩: a simple calculation gives

$$\lambda_{100} = 2/3(Q_{11} - Q_{12}); \quad \lambda_{111} = 2/3Q_{44}$$

so that the expression of the spontaneous Joule deformations can be rewritten as:

$$S_1^0 = 3/2\lambda_{100}(\alpha_1^2 - 1/3); \quad S_2^0 = 3/2\lambda_{100}(\alpha_2^2 - 1/3); \quad S_3^0 = 3/2\lambda_{100}(\alpha_3^2 - 1/3); \quad S_4^0 = 3\lambda_{111}\alpha_2\alpha_3; \quad S_5^0 = 3\lambda_{111}\alpha_1\alpha_3; \quad S_6^0 = 3\lambda_{111}\alpha_1\alpha_3 \qquad (5)$$

Here we have used the contracted technical notation for strains (Nye 1957): $S_{11} = S_1$, $S_{22} = S_2$, $S_{33} = S_3$, $S_{23} + S_{32} = S_4$, $S_{13} + S_{31} = S_5$, $S_{12} + S_{21} = S_6$. In practice, due to the difficulty of actual realization of an isotropic demagnetized state, the preferred method of measurement relies on the strain variations caused by controlled modifications of the orientation of fully saturated states (du Tremolet de Lacheisserie 1993).

In isotropic materials, Joule magnetostriction is characterized by only one parameter since $\lambda_{100} = \lambda_{111} = \lambda_s$ where λ_s is called the saturation magnetostriction.

4. *Free Energy Density and Other Magnetostrictive Effects*

The knowledge of equations of state is essential to most applications of a material. This is related to the knowledge of its free energy density function F. The free energy density function of a single domain cubic magnetostrictive crystal at a given temperature can be easily calculated for a given elastic and saturated magnetic state specified by stress components $T_{11} = T_1$, $T_{22} = T_2$, $T_{33} = T_3$, $T_{23} = T_{32} = T_4$, $T_{13} = T_{31} = T_5$, $T_{12} = T_{21} = T_6$ and magnetization direction cosines α_1, α_2, α_3. Calculation can be made in two different ways. First the crystal is saturated in the specified directions α_1, α_2, α_3 allowing it to deform freely (zero applied stress), so that it takes the spontaneous Joule deformations (Eqn. (5)). Then suitable increasing stresses are applied to the sample in order to bring it reversibly into its specified elastic state while blocking the magnetization. The result is:

$$F = K_0^T(\alpha_1, \alpha_2, \alpha_3) + 1/2s_{11}[T_1^2 + T_2^2 + T_3^2] + s_{12}[T_2T_3 + T_1T_3 + T_1T_2] + 1/2s_{44}[T_4^2 + T_5^2 + T_6^2] \qquad (6)$$

where $K_0^T(\alpha_1, \alpha_2, \alpha_3)$ is the zero stress anisotropy energy of the crystal (defined to within a constant) and the s_{ij} are the elastic compliances at blocked magnetization.

Alternatively, the stress **T** can be applied in the demagnetized state, and then the magnetization is brought to state $\alpha_1, \alpha_2, \alpha_3$ which still produces the spontaneous deformation (Eqn. (5)). Therefore, the total work supplied to the material is:

$$F = 1/2s_{11}[T_1^2 + T_2^2 + T_3^2] + s_{12}[T_2T_3 + T_1T_3 + T_1T_2] + 1/2s_{44}[T_4^2 + T_5^2 + T_6^2] + K^T{}_T(\alpha_1, \alpha_2, \alpha_3) + 3/2\lambda_{100}[T_1(\alpha_1^2 - 1/3) + T_2(\alpha_2^2 - 1/3) + T_3(\alpha_3^2 - 1/3)] + 3\lambda_{111}[T_4\alpha_2\alpha_3 + T_5\alpha_1\alpha_3 + T_6\alpha_1\alpha_2] \qquad (7)$$

A comparison of Eqns. (6) and (7) shows that the anisotropy energy density under stress **T** is given by

$$K^T{}_T(\alpha_1, \alpha_2, \alpha_3) = \boldsymbol{K}_0^T(\alpha_1, \alpha_2, \alpha_3) + 3/2\lambda_{100}[T_1(\alpha_1^2 - 1/3) + T_2(\alpha_2^2 - 1/3) + T_3(\alpha_3^2 - 1/3)] - 3\lambda_{111}[T_4\alpha_2\alpha_3 + T_5\alpha_1\alpha_3 + T_6\alpha_1\alpha_2] \qquad (8)$$

Finally, noticing that $T_i = c_{ij}(S_j - S_j^0)$ where i, j = 1 – 6, and putting $-3/2\ (c_{11} - c_{12})\lambda_{100} = B_1$ and $-3c_{44}\lambda_{111} = B_2$, it is easy to obtain from Eqns. (5) and (6):

$$F = K_0^S(\alpha_1, \alpha_2, \alpha_3) + 1/2C_{11}[S_1^2 + S_2^2 + S_3^2] + c_{12}[S_2S_3 + S_1S_3 + S_1S_2] + 1/2c_{44}[S_4^2 + S_5^2 + S_6^2] + B_1[S_1(\alpha_1^2 - 1/3) + S_2(\alpha_2^2 - 1/3) + S_3(\alpha_3^2 - 1/3)] + B_2[S_4\alpha_2\alpha_3 + S_5\alpha_1\alpha_3 + S_6\alpha_1\alpha_2] \qquad (9)$$

where the c_{ij} are the elastic stiffnesses, and $K_0^S(\alpha_1, \alpha_2, \alpha_3)$ is the anisotropy energy at zero strain which, for symmetry reasons, has the same form as K_0^T. B_1 and B_2 are called magnetoelastic coupling coefficients.

For an isotropic material, since $\lambda_{100} = \lambda_{111} = \lambda_s$ and $c_{11} - c_{12} = 2c_{44}$, Eqn. (9) reduces to

$$F = K_0^S(\alpha_1, \alpha_2, \alpha_3) + 1/2c_{11}[S_1^2 + S_2^2 + S_3^2] + c_{12}[S_2S_3 + S_1S_3 + S_1S_2] + 1/2c_{44}[S_4^2 + S_5^2 + S_6^2] + B[S_1(\alpha_1^2 - 1/3) + S_2(\alpha_2^2 - 1/3) + S_3(\alpha_3^2 - 1/3) + S_4\alpha_2\alpha_3 + S_5\alpha_1\alpha_3 + S_6\alpha_1\alpha_2] \qquad (10)$$

where $B = -3\lambda_s c_{44}$ is the unique magnetoelastic coupling coefficient.

4.1 *Inverse Magnetostrictive Effect*

Equation (8) shows that the application of a stress on a magnetostrictive sample induces an additional anisotropy. This is the actual mechanism of the inverse effect since in general this added anisotropy will produce a reorientation of the magnetization.

4.2 ΔE Effect

The effective elastic constants of a magnetostrictive material are generally modified by the application of a static magnetic field (ΔE effect). The reason is that the effective elastic constants are different depending on whether the magnetization is free to rotate (at zero d.c. field) or bound to a given direction (at high d.c. field). Calculation of the ΔE effect is easy only in simple domain configurations. A particularly simple example is that of a single domain cubic crystal bearing an anisotropy of the usual form $K_0^S(\alpha_1,\alpha_2,\alpha_3)=-K_1[\alpha_2^2\alpha_3^2+\alpha_1^2\alpha_3^2+\alpha_1^2\alpha_2^2]$ with $K_1>0$, magnetized along 100 by a static field H. Starting from Eqn. (9), the differential shear stiffness can be given by

$$\partial T_4/\partial S_4=\partial T_5/\partial S_5=c_{44}-B_2^2/(K_1+1/2\,\mu_0 M_s H)$$

As expected, it tends to the blocked magnetization value at high field and decreases when the applied field is reduced. A measure of the ΔE effect is the relative variation controlled by the applied field, i.e., $B_2^2/(K_1c_{44})=k_{15}$ (Nye 1957). The interest in this parameter and the reason for this notation will be made clear in the next section.

4.3 Nonlinear elasticity

In correlation with the ΔE effect, strong nonlinear elastic effects are also observed in high coupling factor materials. The source of this elastic nonlinearity lies in the nonlinearity of the magnetic response itself. Some calculations can be made for simple single-domain configurations but they will not be developed here.

5. Magnetostriction of Unsaturated States: Piezomagnetic Formulation

Below saturation, the magnetostrictive response of polycrystalline and amorphous materials involves domain processes which can be quite complicated. The usual method, at least for application purposes, is to characterize the material properties only around the relevant bias point as far as possible by phenomenological linear and reversible equations of states.

In general, both a bias field and a bias uniaxial stress are applied to the material along the same direction. The material has then uniaxial symmetry around the bias direction which is taken as axis three of a rectangular frame. In this frame the linearized properties are therefore described by matrix relations that are analogous to those of piezoelectric ceramics (Nye 1957). Taking, for example, differential field **h**, and differential stress **T** as independent variables, the equations of state are written as

$$S=s\mathbf{T}+\mathbf{d}*\mathbf{h};\, b=\mathbf{dT}+g\mathbf{h}(12)$$

Here, taking advantage of the contracted notation, **T** and the differential strain **S** are six-component vectors whereas **b**, the differential induction, and **h** are three-component vectors. **d** is thus a three line, six columns rectangular matrix. **d*** is actually the transposition of **d**, a consequence of the reversibility assumption.

The form of the piezomagnetic matrix **d** is the same as its piezoelectric analogue, i.e., there are only five nonzero components out of which only three are independent, namely $d_{31}=d_{32}$, $d_{24}=d_{15}$, and d_{33}.

In general in transducer configurations, stress and strain are collinear single-component vectors, as are the field and induction. Vector Eqn. (12) then reduces to a system of two scalar linear equations (no summation on repeated subscripts!):

$$\mathbf{S}_i=\mathbf{s}_{ii}\mathbf{T}_i+\mathbf{d}_{il}\mathbf{h}_l;\, \mathbf{b}_l=\mathbf{d}_{il}\mathbf{T}_i+\mu_{11}\mathbf{h}_l$$

It is convenient to define a dimensionless coefficient $k_{il}=(d_{il}^2/s_{ii}\,\mu_{11})^{1/2}$ called the magnetomechanical coupling factor. The reason is that its square actually measures the ratio of output to input energy in a static reversible conversion process. By a conversion process, it can define a transformation of magnetic energy into mechanical energy or vice versa. Three useful magnetomechanical coupling factors may be defined in uniaxial magnetostrictors, namely:

$$k_{33}=(d_{33}^2/\mu_{33}s_{33})^{1/2};$$
$$k_{31}=k_{32}=(d_{31}^2/\mu_{33}s_{11})^{1/2},\quad \text{and}$$
$$k_{15}=k_{24}=(d_{15}^2 s_{44})^{1/2}$$

6. Microscopic Origin of Magnetostriction

Within the model of localized electron ferromagnetizm, two microscopic mechanisms are the origin of magnetostriction: exchange interaction between spins, and interaction of a given spin with its local electric field via the spin–orbit coupling. Although it may provide some good clues even in the case of transition metals and alloys, the localized electron model is inadequate for these materials and must be replaced by the band magnetism model. In this model magnetostriction appears as an interaction between the strain and the spin-dependent band structure (du Tremolet de Lacheisserie 1993).

6.1 Spin–Spin Interaction

The dependence of the isotropic Heisenberg exchange on distance between neighboring spins leads to a volume magnetostriction which is generally negative and therefore leads to an enhancement of the lattice thermal expansion. This is the "normal" situation, observed, for example, in pure nickel. The strong positive volume magnetostriction observed in the so-called INVAR alloys is a band magnetism effect and will not be discussed here.

In addition to the isotropic Heisenberg contribution, the exchange interaction also includes, in general, anisotropic terms of increasing order which depend both on the distance between interacting spins and on the orientation of the bond with respect to spin direction. The orientation dependence is the origin of the magnetic anisotropy energy. Both the distance dependence and the orientation dependence are the origin of Joule magnetostriction.

6.2 Crystal Field Interaction

Ions with a nonspherical electron distribution interact with the nonuniform electric field produced by their neighbors—the so-called crystal field. This interaction can be expanded in terms of orbital multipolar moments and associated field spatial derivatives. For example, the lowest moment is quadrupolar and it is coupled to the first field derivatives (the field gradient tensor). In any case this electrostatic energy depends on the orientation of the electron orbitals with respect to the crystal field and hence the crystal axes . In magnetic ions the magnetic moment is either tied to the orbitals or coupled to them by spin–orbit interaction. The final result is that there is a contribution to the free energy of the crystal that depends on the orientation of the spontaneous magnetization with respect to the crystal axes, i.e., an anisotropy energy. The dependence of this anisotropy energy on (anisotropic) strain leads to the Joule magnetostriction. It must be emphasized here that high magnetostriction does not necessarily imply large anisotropy since a function may be zero at a given value of the variable while its derivative is not.

7. *Practical Magnetostrictive Materials and their Applications*

From a practical point of view it is worth distinguishing four main families of magnetostrictive materials: (i) the so-called INVAR alloys; (ii) the high coupling factor alloys; (iii) the high-saturation magnetostriction alloys; and, finally (iv) the ferrite crystals or ceramics which, in contrast to the previous metallic materials, are almost perfect insulators.

7.1 INVAR Alloys

It was discovered by Guillaume in 1896 that in some alloys of nickel and iron the volume magnetostriction is positive and its temperature dependence just compensates for the normal thermal expansion of the alloy. The result is a metallic alloy bearing a very weak thermal expansion coefficient over a wide temperature range . The applications of these INVAR alloys include very large structures such as methane tankers as well as the most tiny parts of metrology, together with very specific electronic components, such as the masks for high-resolution color TV (du Tremolet de Lacheisserie 1993). A drawback of these materials is that they exhibit a very large forced volume magnetostriction, $\delta V/V\delta H \cong 10^{-10}$ m A^{-1}, preventing their use in a high magnetic field environment.

7.2 High Coupling Factor Alloys

This category includes magnetically soft polycrystalline and amorphous alloys based on iron, nickel, and cobalt. Their saturation magnetostriction is rather low ($\lambda_s < 50 \times 10^{-6}$) and their high coupling factor is rather a consequence of their very low anisotropy.

The most remarkable and useful example of this category of materials is an amorphous alloy known as Metglass 2605 SC with the composition $Fe_{81}B_{13.5}Si_{3.5}C_2$. Due to its fabrication technique (Kavesh 1977), this alloy is available only in the form of a thin ribbon (around 25 μm thick). Table 1 lists some of its relevant characteristics. The last four parameters may be considered as "soft" since they depend strongly on postdeposition treatments which modify all the properties that are related to anisotropy. It is well known that intrinsic anisotropy in amorphous materials is actually a random function of the spatial position with a correlation length typically less than 1 μm. Extrinsic large-scale anisotropy can, however, be induced by appropriate treatments.

Typically, annealing at 642 K for 10 min under a saturating magnetic field parallel to the ribbon plane induces a uniaxial anisotropy with an easy axis parallel to the applied field. The value of the induced anisotropy field is in the order of 50A m^{-1}. Such a field-annealed Metglass ribbon has uniaxial symmetry and, as explained in Sect. 5, it may be characterized by differential piezomagnetic parameters around a given bias point. It is usual to take the easy axis as the z or "3" axis. Table 1 refers to a field-annealed Metglass alloy in the demagnetized state under zero field and zero stress, unless otherwise specified.

Remarkable points in Table 1 are: (i) an extremely high coupling factor, (ii) a giant ΔE effect (from 25 GPa at zero field up to 200 GPa at high bias field, i.e., at blocked magnetization); and (iii) an extreme sensitivity of permeability to applied stress. However, this is associated with a rather low saturation magnetostriction. All these characteristics make the Metglass alloy a very good candidate for low-power transducer applications (du Tremolet de Lacheisserie 1993, du Tremolet de Lacheisserie *et al.* 1998).

Force transducers make use of the stress dependence of permeability, and more generally of the

Table 1
Some relevant characteristics of Metglass 2605SC.

Young's modulus at blocked magnetization Y_{33}^{M} (GPa)	200
Curie temperature (K)	643
Spontaneous induction at 300 K (T)	1.61
Saturation magnetostriction at 300 K, 10^{-6}	30
Relative permeability, μ_{33}	20 000
Relative permeability μ_{33} under stress bias ($T_{33} = 1$ MPa)	80 000
Static magnetostrictivity d_{33} (nm A^{-1})	1000
Magnetomechanical coupling factor k_{33} (under 50A m^{-1} bias field)	0.97
Young's modulus at free magnetization Y_{33}^{H} (GPa)	25

Table 2
Main characteristics of Terfenol ($TbFe_2$) and Terfenol D ($Tb_{0.3}Dy_{0.7}Fe_2$). Curie temperature T_c, spontaneous induction $\mu_0 M_s$, saturation magnetostriction λ_s, magnetomechanical coupling factor k_{33}, and static magnetostrictivity d_{33}.

Material	T_c (K)	$\mu_0 M_s$ (T)	$10^6\ \lambda_s$	k_{33} max	d_{33} max(nm A^{-1})
$TbFe_2$	698	1.1	1750	0.35	
$Tb_{0.3}Dy_{0.7}Fe_2$	653	1.0	1100	0.6	57

modification of magnetic response by stress. A significant advantage of these transducers is that since magnetic measurements do not require physical contact with the material, it is quite easy to sense forces acting on moving parts. As an example, consider the noncontact measurements of the torque transmitted by a rotating shaft. A Metglass ribbon is wrapped around the shaft and is submitted to the shearing forces associated with the torque. The ribbon permeability is measured by a fixed coil arrangement placed close to the moving ribbon surface. Permeability must be measured along one of the principal stress directions, in this case at 45° off the shaft axis.

7.3 *High-saturation Magnetostriction Alloys*

Giant magnetostrictive effects were discovered in 1975 in the intermetallic cubic compound $TbFe_2$, known as Terfenol (Clark 1980). This alloy exhibits magnetostrictive deformations which are 50 to 100 times larger than the high coupling factor materials. Such a giant magnetostrictive effect can be clearly explained by the localized electron model as a crystal field effect on the rare-earth site. Unfortunately this is obtained at the price of a relatively large anisotropy. Therefore, high driving fields are necessary to generate the magnetostrictive deformation.

The improved alloy $Tb_{0.3}Dy_{0.7}Fe_2$, known as Terfenol D, still has a very large magnetostriction but a much lower anisotropy. This is because the main contributions to anisotropy, due to terbium and dysprosium, have opposite signs and nearly compensate for each other at this composition, whereas the respective contributions to magnetostriction do not. Table 2 lists some useful room-temperature characteristics of Terfenol and Terfenol D. Even the improved material remains relatively hard and despite the very large saturation magnetostriction, the magnetomechanical coupling factors are definitely lower than those of low-saturation magnetostriction but very soft alloys.

Progress in giant magnetostriction materials has been directed towards the fabrication of highly textured materials and even single crystals (Clark 1986). Potential applications of giant magnetostriction alloys include large displacement transducers and high-power ultrasonic generation. However for static displacement generation, piezoelectric transducers based on PZT ceramics are still the best choice mainly because they do not require a permanent excitation power.

In contrast, Terfenol D is a better candidate than PZT for high-power ultrasonic generation (Claeyssen *et al.* 1988) because it is not brittle and it generates a higher deformation and higher power density. A noticeable disadvantage of these alloys is their metallic electrical conductivity which, due to the so-called skin effect, prevents their use at high frequency.

Giant magnetostriction alloys were also prepared as thin films, with the idea that new specific routes to a vanishing anisotropy could be followed in view of applications to low driving field microactuators (du Tremolet de Lacheisserie *et al.* 1998). Thin films of amorphous $Tb_{1-x}Dy_x(Co_{0.55}Fe_{0.45})_y$ were deposited by Duc *et al.* (1996) using d.c. sputtering. Due to their amorphous structure these films are expected to have a vanishing large-scale magnetic anisotropy together with an isotropic, and still giant, magnetostriction. In practice they always bear some induced anisotropy of various origins. It is preferable in most cases intentionally to induce a homogeneous in-plane uniaxial anisotropy rather than to rely on uncontrolled (and often spatially inhomogeneous) parasitic effects.

This is done both by controlling the sputtering conditions (Quandt *et al.* 1994) and by applying a bias field, either during the deposition itself or during a postdeposition treatment (typically 10 min to 1 hour at 150–300 °C under a 10 000 A m^{-1} field). Films are deposited on substrate plates (silicon or pyrex) of known elastic properties thereby constituting a magnetoelastic bimorph (Klokholm 1976). If the magnetostrictive layer is much thinner than the substrate plate, the magnetostrictive stress at zero deformation, rather than the magnetostrictive deformation at zero stress, can be directly determined (du Tremolet de Lacheisserie and Peuzin 1994). This is done by measuring the curvature and torsion of the magnetoelastic bimorph induced by a known applied field. Magnetostrictive deformation at zero stress can then be calculated by making reasonable estimations of the layer elastic modulus.

For example, in $Tb_{0.32}(Fe_{0.45}Co_{0.55})_{0.68}$, a fairly well-defined easy axis can be induced by an annealing at 250 °C. The elongation, $\lambda(H)$, produced along the in-plane hard direction is then determined as a function of the field H applied along the same direction. Giant magnetostriction is observed with $\lambda(H) = 0.8 \times 10^{-3}$ at $\mu_0 H = 0.3\,T$ and still $\lambda(H) = 0.3 \times 10^{-3}$ at $\mu_0 H = 0.1\,T$. However, these driving fields are still high for microsystem applications. This is due in part to the ferrimagnetic character of these alloys which leads to a rather low spontaneous magnetization ($\mu_0 M_s \cong 0.3\,T$) and therefore to a high anisotropy field.

In order to improve further the low field magnetostrictive response, a periodic laminar nanocomposite was fabricated by repeated stacking of a weakly magnetostrictive, but high magnetization, layer of polycrystalline FeCo and a highly magnetostrictive amorphous TbCo layer (Betz 1997). Typical layer thicknesses were about 6 nm and up to hundreds of layers could be stacked by using sequential sputter deposition. As a consequence of the strong exchange coupling between the FeCo and TbCo layers the obtained films behaved as a homogeneous ferromagnetic alloy combining relatively high magnetization and giant magnetostriction. Deformation of up to 0.4×10^{-3} was obtained in a field of only 20 mT. These thin films of giant magnetostriction alloys have given rise to the development of various types of micromechanical devices (du Tremolet de Lacheisserie *et al.* 1998) among which are a micropump (Quandt 1997), a torsional microactuator which is claimed to be free of thermal drift (Betz *et al.* 1996) and a two-dimensional optical microscanner (Garnier *et al.* 1999). These devices are, however, still at the laboratory level.

7.4 Magnetostrictive Insulators: Ferrites

Two families of magnetoelastic ferrites are worth mentioning from the practical point of view: the spinel ferrites of general formula MFe_2O_4 where M is a metal of the first transition series; and the garnets of general formula $R_3Fe_5O_{12}$ where R is a rare-earth metal or yttrium.

Magnetoelastic spinel ferrites such as $NiFe_2O_4$ and $Ni_{0.98}Co_{0.02}Fe_2O_4$ have a resistivity in excess of $10^4\,\Omega$.m, a spontaneous induction of 0.33 T, and a Curie temperature of 863 K. Their spontaneous magnetostriction is comparable with that of high coupling factor materials but since they are somewhat less soft, their magnetomechanical coupling factors are modest (0.2–0.38). These materials face a difficult competition with piezoelectric ceramics.

The garnets, especially $Y_3Fe_5O_{12}$ (YIG), have found many applications in general microwave devices, mostly because they are almost perfect insulators and bear remarkable magnetodynamic properties (von Aulock 1965). Because of their excellent acoustic properties, and despite their rather weak magnetostriction (λ_{100} and λ_{111} are in the order of 10^{-6}), YIG single crystals have been exploited, at least at the laboratory level, in magnetoelastic signal processing devices taking advantage of such effects as magnetic field control of acoustic propagation and coupling between magnetostatic waves and acoustic waves (Auld 1968, 1971).

Bibliography

Auld B A 1968 *Nonlinear Interaction of Spin Waves and Elastic Waves: Advances in Microwaves.* Academic Press, New York

Auld B A 1971 Magnetostatic and magnetoelastic wave propagation in solids. *Appl. Solid State Sci.* **2**, 1–106

Betz J 1997 Magnetostriction géante de couches minces et microactionneurs magnétostrictifs pour des technologies intégrées. Thesis, University of Grenoble, France

Betz J, Mackay K, Peuzin J, Halstrup B, Lhermet N 1996 Torsion based, drift free magnetostrictive microactuator In: *Proc. Actuator '96.* Axon Technology Consult GmbH, Bremen, Germany, pp. 283–6

Claeyssen F, Boucher D, Pohlenz C 1988 Application of magnetostrictive rare earth iron alloys to sonar transducers In: *Proc. Undersea Defence Technology '88.* Microwave Exh. and Pub. Ltd. Ed., London, p. 711

Clark A E 1980 Magnetostrictive rare earth—Fe_2 compounds. In: Wohlfarth E P (ed.) *Ferromagnetic Materials.* North Holland, Amsterdam

Clark A E 1986 Magnetostriction in twinned [112] crystals of $Tb_{0.27}Dy_{0.73}Fe_2$. *IEEE Trans. Magn.* **MAG-22**, 973–5

du Tremolet de Lacheisserie E 1993 *Magnetostriction. Theory and Applications of Magnetoelasticity.* CRC Press, Boca Raton, FL du Tremolet de Lacheisserie E, Mackay K, Betz J, Peuzin J C 1998 From bulk to film magnetostrictive actuators. *J. Alloy. Comp.* 685–91

du Tremolet de Lacheisserie E, Peuzin J C 1994 Magnetostriction and internal stresses in thin films: the cantilever method revisited. *J. Magn. Magn. Mater.* **136**, 189–96

Duc N H, Mackay K, Betz J, Givord D 1996 Giant magnetostriction in amorphous $(Tb_xDy_{1-x})(Fe_{0.45}Co_{0.55})_y$ films. *J. Appl. Phys.* **79**, 973–7

Garnier A, Bourouina T, Fujita H, Hiramoto T, Orsier E, Peuzin J C 1999 Magnetic actuation of bending and torsional vibrations for 2D-optical-scanner application In: *Proc. 10th Int. Conf. on Solid State Sensors and Actuators*. Sendai, Japan

Kavesh S 1977 In: Leany H S, Gilman J J (eds.) *Metallic glasses*. American Society of Metals, Metals Park, OH

Klokholm E 1976 The measurement of magnetostriction in ferromagnetic thin films. *IEEE Trans. Magn.* **12**, 819–21

Nye J F 1957 *Physical Properties of Crystals*. Clarendon Press, Oxford

Quandt E 1997 Giant magnetostrictive thin film materials and applications. *J. Alloy. Comp.* **258**, 126–32

Quandt E, Gerlach B, Seemann K 1994 Preparation and applications of magnetostrictive thin films. *J. Appl. Phys.* **76**, 7000–2

von Aulock W 1965 Handbook of Microwave Ferrite Materials. Academic Press, New York

Metal Amides: New Hydrogen Storage Systems

W. Luo and P. Chen

Recently, research on hydrogen storage in solids for vehicular applications has been focused on materials which can absorb and release hydrogen reversibly with hydrogen capacity of 6 wt.% or more. The hydrides of metals, alloys, or intermetallic compounds have hydrogen capacity of 3.5 wt.% or less (see *Hydrogen–Metal Systems: Basic Properties (1)*). In the late 1990s, catalyzed alanates were found to have a theoretical hydrogen capacity up to 5.6 wt.%, that is, much higher than metallic hydrides, but still below the desirable value, 6 wt.%.

Recently, Chen *et al.* have introduced a new type of hydride material for hydrogen storage, the Li–N–H system, that is, a mixture of $LiNH_2$ and LiH, which has a theoretical hydrogen capacity of 11 wt.%. A two-step reaction is involved in its hydrogen uptake and release:

$$LiNH_2 + 2LiH \Leftrightarrow Li_2NH + LiH + H_2 \Leftrightarrow Li_3N + 2H_2 \quad (1)$$

The stored H-capacity in this system is distributed equally in the two steps in Eqn. (1). Figure 1 shows thermo-gravimetric results for H_2 uptake/release for this system. It can be seen that more than 9 wt.% of hydrogen can be absorbed reversibly.

The operating temperature, corresponding to 1.0 bar equilibrium desorption pressure, is above 250 °C for the Li–N–H system, which is too high for fuel cell applications. Partial substitution of the Li in this system by Mg forms an Li–Mg–N–H system and reduces the operating temperature significantly. Figure 2 shows the absorption/desorption isotherms at 180, 200, and 220 °C for the $(Mg(NH_2)_2 + 2LiH)$ system with isotherms for the $(LiNH_2 + LiH)$ system at 280 °C included for comparison. The H-release/uptake of $(Mg(NH_2)_2 + 2LiH)$ is shown by Eqn. (2), which corresponds to an H-capacity of 5.4 wt.%:

$$Mg(NH_2)_2 + 2LiH \Leftrightarrow Li_2Mg(NH)_2 + 2H_2 \quad (2)$$

Figure 3 shows van't Hoff plots for the Li–N–H and Li–Mg–N–H systems. It can be seen that the Mg substitution destabilizes the Li–N–H system, allowing this material to be able to deliver H_2 at more desirable pressures and temperatures.

It was found that the mixture of $(2LiNH_2 + MgH_2)$ converts into $(Mg(NH_2)_2 + 2LiH)$ before its initial hydrogen release. Equation (3) corresponds to hydrogen uptake/release for a mixture of $(2LiNH_2 + MgH_2)$ with the first step as an irreversible structure rearrangement and the second step as reversible hydrogen release/uptake:

$$2LiNH_2 + MgH_2 \Rightarrow Mg(NH_2)_2 + 2LiH \Leftrightarrow Li_2Mg(NH)_2 + 2H_2 \quad (3)$$

Higher capacities can be achieved by varying the ratio of amide/hydride as shown below. The imide dehydrogenates partially for Eqn. (4) and fully for Eqn. (5):

$$3Mg(NH_2)_2 + 8LiH \Leftrightarrow Mg_3N_2 + 4Li_2NH + 8H_2 \ (6.9\,\text{wt.\%}) \quad (4)$$

$$3Mg(NH_2)_2 + 12LiH \Leftrightarrow Mg_3N_2 + 4Li_3N + 12H_2 \ (9.1\,\text{wt.\%}) \quad (5)$$

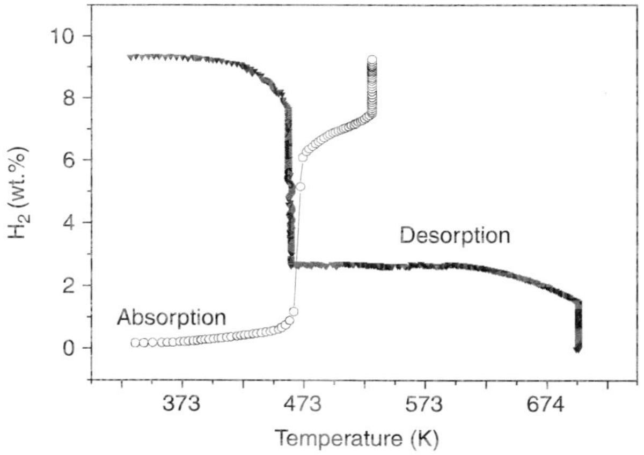

Figure 1
Temperature dependences of hydrogen absorption and desorption over an Li_3N sample.

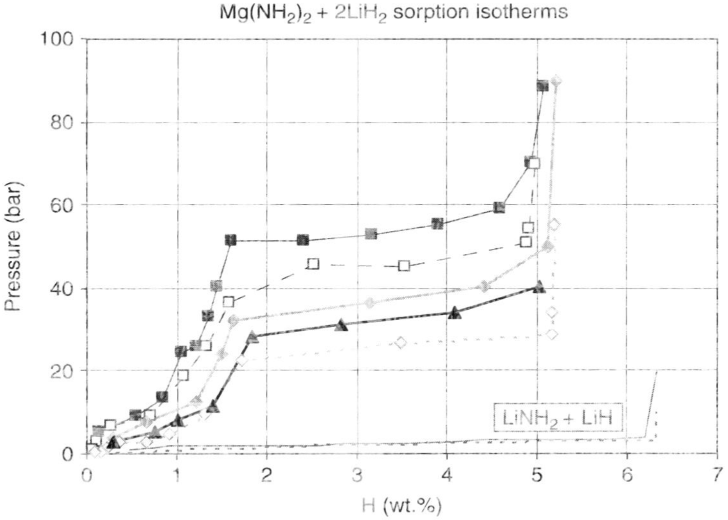

Figure 2
Sorption isotherms of ($2LiNH_2 + MgH_2$). Filled symbols denote absorption and open ones denote desorption. The squares are at 220 °C; the diamonds at 200 °C; the triangles at 180 °C. Isotherms for ($LiNH_2 + LiH$) at 280 °C are included for comparison.

The dehydrogenation of (imide/hydride ⇒ nitride) takes place at a higher temperature than that for (amide/hydride ⇒ imide).

The $LiNH_2$ decomposes itself at a temperature above 300 °C, releasing only NH_3. LiH decomposes at a temperature above 650 °C, releasing H_2. When these two species are mixed, however, the mixture decomposes at 200 °C or below, releasing H_2 only. In other words, $LiNH_2$ destabilizes LiH. This property of the Li–N–H system has received significant attention. The "opposite charge attraction" mechanism proposes that the destabilization results from the

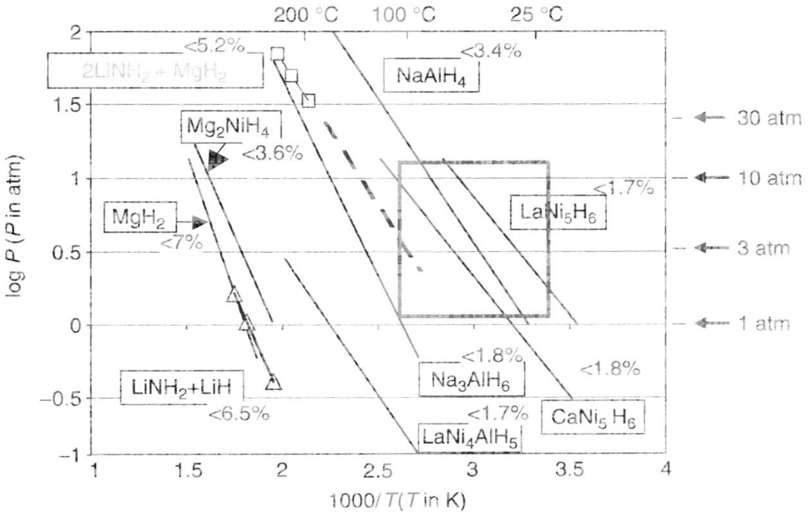

Figure 3
van't Hoff plots for Li–N–H and Li–Mg–N–H systems, with other common reversible metal hydrides included for comparison. The square red box indicates the most desirable operating temperature and pressure ranges for fuel cell applications. The numbers attached to the materials names are the H-capacity values (in wt.%) for the materials.

attraction between opposite charges, which facilitates hydrogen release, as shown in the following diagram:

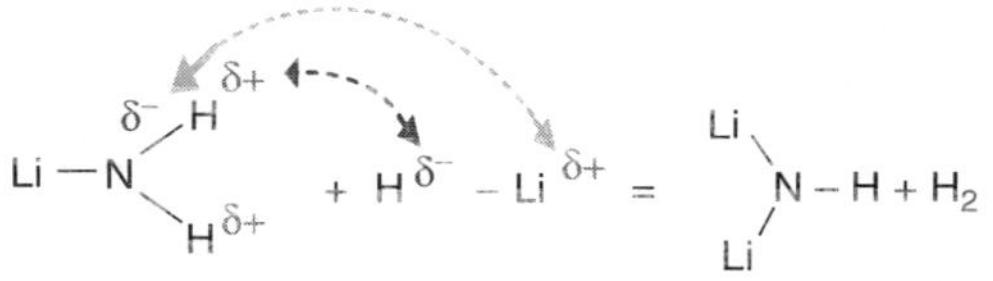

The "opposite charge attraction" mechanism is applicable to different combinations. For example, $LiBH_4$ does not decompose to release H_2 below 350 °C; however, when mixed with $LiNH_2$, it releases H_2 below 250 °C. While the "opposite charge attraction" mechanism destabilizes hydrides, the rehydrogenation becomes more difficult or even thermodynamically unfavorable. It is very challenging to destabilize a system while maintaining its reversibility. So far, the $Mg(NH_2)_2$–LiH is the most promising combination that can deliver hydrogen for vehicular applications.

Currently, the amide-containing materials are the most promising hydrogen storage materials for vehicular applications even though there are still many obstacles to be overcome before being employed.

See also: Hydrogen–Metal Systems: Applications of Gas–Solid Reactions; Hydrogen–Metal Systems: Electrochemical Reactions (Fundamentals and Applications); Hydrogen–Metal Systems: Technological and Engineering Aspects

Bibliography

Bogdanovic B, Bohmhammel K, Christ B, Reiser A, Schlichte K, Vehlen R, Wolf U 1999 Thermodynamic investigation of the magnesium–hydrogen system. *J. Alloys Compd.* **282**, 84–92

Bogdanovic B, Schwickardi M 1997 Ti-doped alkali metal aluminum hydrides as potential noval reversible hydrogen storage materials. *J. Alloys Compd.* **253–4**, 1–9

Chen P, Xiong Z, Luo J, Lin J, Tan K L 2002 Interaction of hydrogen with metal nitride and imides. *Nature* **420** (21), 302–4

Chen P, Xiong Z, Luo J, Lin J, Tan K L 2003 Interaction between lithium amide and lithium hydride. *J. Phys. Chem. B* **107** (39), 10967–70

Grochala W, Edwards P 2004 Thermal decomposition of the non-interstitial hydrides for the storage and production of hydrogen. *Chem. Rev.* **104**, 1283–315

Kojima Y, Matsumoto M, Kawai Y, Haga T, Ohba N, Miwa K, Towata S, Nakamori Y, Orimo S 2006 Hydrogen absorption and desorption by the Li–Al–N–H system. *J. Phys. Chem. B* **110**, 9632–6

Leng H, Ichikawa T, Isobe S, Hino S, Hanada N, Fujii H 2004 Hydrogen absorption properties of Li–Mg–N–H system. *J. Alloys Compd.* **400**, 245–8

Leng H, Ichikawa T, Hino S, Nakagawa T, Fujii H 2005 Mechanism of hydrogenation reaction in the Li–Mg–N–H system. *J. Phys. Chem. B* **109**, 10744–48

Luo W 2004 ($LiNH_2$–MgH_2): A viable hydrogen storage system. *J. Alloys Compd.* **381**, 284–7

Luo W, Ronnebro E 2005 Towards a viable hydrogen storage system for transportation application. *J. Alloys Compd.* **404–6**, 392–5

Luo W, Sickfoose S 2005 Thermodynamic and structural characterization of the Mg–Li–N–H hydrogen storage system. *J. Alloys Compd.* **407**, 274–81

Nakamori Y, Kitabara G, Ninomiya A, Aoki M, Noritake T, Towata S, Orimo S 2005 Guidelines for developing amide-based hydrogen storage materials. *Mater. Trans.* **46** (9), 2093–7

Pinkerton F E, Martin M S, Meisner G P, Balogh M P 2006 Improved hydrogen release from $LiB_{0.33}N_{0.67}H_{2.67}$ with noble metal additions. *J. Phys. Chem. B* **110**, 7967–74

Schlapbach L, Züttel A 2001 Hydrogen-storage materials for mobile applications. *Nature* **414**, 354–8

Xiong Z, Wu G, Hu J, Chen P 2004 Ternary imide for hydrogen storage. *Adv. Mater.* **16** (17), 1522–5

Micro Devices and Micro Systems, Materials for

S. M. Spearing

Microsystems, also known as MEMS (microelectromechanical systems), have the potential to become a pervasive technology for sensors and actuators, and possibly also for power generation and distributed sensing and actuating systems. The majority of MEMS currently in production utilize materials and processes that have been developed for complementary metal oxide semiconductor (CMOS) devices used in the microelectronics industry. This materials set has been extended slightly by the introduction of metals and polymers and the associated processes used in first-level packaging for microelectronic devices, such as electrodeposition of metals and the fabrication of polymer–metal multilayers. However, these MEMS technologies have yet to mature sufficiently to be considered commercially viable. It is apparent that the set of materials available to create microsystems has been largely defined by the available processes that can create mechanical and electrical elements at the required scales (typically 1–1000 μm) and with the necessary geometric resolution (typically 50 nm–1 μm). The limitation on the available processes has hitherto denied the MEMS designer the ability to pursue materials selection based solely on the combinations of material properties required to optimize the design.

This article reviews the current status of materials within and outside the existing MEMS materials set and assesses the likelihood of the translation of new materials technologies into viable MEMS products. Readers might also wish to refer to more detailed references. In particular Senturia (2000) provides an excellent reference for the overall design process for microsystems and Madou (1997) provides an equally comprehensive reference on microfabrication processes. Spearing (2000) provides a more detailed description of materials issues in MEMS.

1. System Drivers

Before discussing the material-specific issues in micro-systems it is important to understand the requirements of materials in the context of the system as a whole. There are two key components to this: (i) the effects of the devices' small scale on their design and the consequent impact on materials, and (ii) the issue of material selection for MEMS.

1.1 Scaling

Scale effects enter into the design of MEMS, and thus impinge on materials issues, via several distinct routes. In this context it is useful to subdivide scaling effects into three categories: quasi-fundamental, mechanism-dependent and extrinsic (or indirect).

(a) Quasi-fundamental scaling. Truly fundamental scaling laws can be obtained by dimensional analysis alone, whereas quasi-fundamental involve assumptions slightly beyond this. For instance, it is usually assumed in scaling arguments that mass scales with volume and thus length raised to the third power. However, this is only strictly true so long as it is reasonable to assume that density is a scale independent property—which is generally a good assumption for solid matter down to the micrometer scale. Perhaps the most important quasi-fundamental scaling law that applies to MEMS is the cube-square scaling that relates volumetric scale dependencies to quantities that scale with area. The most important structural example is that inertial forces depend on the acceleration and mass, and thence volume, whereas the stress that results scales with the cross-sectional area. Thus for a given material strength the acceleration that a structure can withstand increases linearly with decreasing length scale—providing a rationale for microfabricated accelerometers. Table 1 summarizes the quasi-fundamental scaling of physical parameters down to the range of sizes of interest to MEMS designers.

1.2 Mechanism-dependent Scaling

Quasi-fundamental scaling generally assumes that physical parameters or material properties remain independent of scale. However, these assumptions break down when the length scale of the device approaches the characteristic length scale of the mechanisms that control the property of interest. For mechanical elements, material properties such as the thermoelastic constants, density, and conductivities are essentially scale independent in MEMS; however, the scaling of strength-related properties are affected by mechanisms at scales which are characteristic of MEMS devices.

For ductile metals there is an extensive literature on the effect of scale on strength, which is summarized by Arzt (1998). The introduction of constraints, such as hardening particles or grain boundaries or the surfaces and interfaces bounding a thin film act to restrict dislocation formation and motion, resulting in very high strengths at small scales. For MEMS devices this increase in strength can have both positive and negative consequences, i.e., it allows for an increased force transmission capability but it can result in much higher residual stresses than would be normal at large scales.

Other commonly encountered mechanism-dependent scaling effects include the effect of increased surface area to volume ratio at small scales.

This has the effect of increasing surface diffusion as a mechanism in thermally activated deformation processes such as creep. It is also worth noting that the concept of toughness as a material property for ductile materials ceases to be meaningful at small scales since the toughness is largely determined by the plastic dissipation, which is controlled by the structural dimensions rather than any intrinsic material scale.

For brittle materials the strength is governed by the maximum flaw size, typically at the surface. Simple statistical scaling arguments, suggest that the probability of finding a flaw of a given size decreases with the volume (or area) of material under load. Thus mechanical elements with small characteristic dimensions would be expected to be inherently stronger at small scales. In addition to this statistical scaling, extrinsic scaling factors also play a role in defining the strength of brittle materials. The use of microfabrication processes such as etching or deposition, and single-crystal substrates with low defect densities allows the creation of structures with very fine surfaces and, therefore, very high strengths. This favorable scaling of strength may be enabling for MEMS with very high mechanical power densities.

1.3 Extrinsic (Indirect) Scaling

Some of the most important effects of scale on MEMS design or performance cannot be attributed to a single physical factor. For instance the processing routes that are capable of achieving the necessary dimensional tolerances are restrictive in the shapes that they can achieve. Thus MEMS devices generally consist of electromechanical elements which are planar, consist of stacked layers, or are cylindrical shapes. This restriction often drives MEMS into regions of the design space that would never be considered in macroscopic elements. Similarly the high value placed on the functionality offered by MEMS, as for micro-electronic devices, allows the use of materials which would never be used in macroscale applications.

Extrinsic scaling factors often have several quasi-fundamental origins. For instance, the restriction on geometrical shapes can be traced back to the processing routes that are available. In turn, the use of

Table 1
Key quasi-fundamental length-scale dependencies of design parameters for MEMS.

	Parameter	Scaling	Comments
1	Length (L)	L^1	Fundamental
2	Area (A)	L^2	Fundamental
3	Volume (V)	L^3	Fundamental
4	Surface area/volume (A/V)	L^1	Fundamental
5	Mass (M)	L^3	Assumes scale-independent density
6	Inertial force (F)	L^3	Scales with mass
7	Maximum inertial stress (σ)	L^{-1}	For fixed acceleration spring-mass system
8	Maximum centrifugal stress (σ)	L^0	For fixed rim speed on rotating disk
9	Power (W)	L^2	Assumes (5)
10	Power/volume (W/V)	L^{-1}	Assumes (5), also power/mass
11	Structural natural frequency (ω)	L^{-1}	Given scale-independent modulus and density
12	Characteristic diffusion time (t)	$L^{1/2}$	For fixed diffusion coefficient
13	Electrostatic force (F_{El})	L^2	Assumes scale-independent permittivity, dielectric breakdown voltage
14	Maximum acceleration due to electrostatic force (a_{El})	L^{-1}	Given (13) and (5)
15a	Magnetic force (electromagnet) (F_{emag})	L^4	Assumes scale-independent maximum current density
15b	Magnetic force (permanent magnet) (F_{pmag})	L^3	Assumes scale-independent magnetic strength
16	Piezoelectric force (F_{piezo})	L^2	Assumes scale-independent piezomechanical constants, breakdown voltage
17	Thermal losses (Q_{Th})	L^2	For scale-independent heat transfer coefficient
18	Thermal shock resistance (ΔT_{crit})	$\sim L^{-1}$	Scales with Biot number
19	Surface tension	L^1	Assumes scale independence

Material performance indices for mechanical elements applied to micro-fabricatable materials, where possible data from microfabricated structures are used. Note the fracture strength data are the most subject to variation and the maximum allowable temperature will vary with the requirements of the application.

these processing routes, such as deposition, etching, and diffusion-limited doping, can be traced back to the scaling argument that processes which primarily act on surfaces are more economically attractive at small scales due to the cube-square scaling of volume to surface area.

By way of demonstrating how such scaling factors combine it is interesting to consider why electrostatic actuation predominates at small scales for prime movers rather than electromagnetic actuation at larger scales. Quasi-fundamental arguments suggest that the electrostatic force which can be generated between two surfaces held at a fixed potential difference scales with the second power of a characteristic length. In contrast, the force that can be generated by an electromagnet operating at a fixed current density scales with the fourth power of the length (the force due to a permanent magnet made of a particular material scales with the volume).

Thus, given that the mass acted on by the force scales as the third power of length, the acceleration achievable by electrostatic actuation benefits from cube-square scaling whereas electromagnetic actuation is either neutral with scale or actually deteriorates at small scales. In addition to this quasi-fundamental scaling an important mechanism dependence is introduced. The breakdown voltage of an air gap at the macroscale is of the order of $2\times10^{6}\,\mathrm{Vm^{-1}}$. However, reduction of the air gap to dimensions close to the mean free path of air molecules ($\sim 1\,\mu\mathrm{m}$) increases the breakdown voltage to $1\times10^{8}\,\mathrm{Vm^{-1}}$. This increase of nearly two orders of magnitude in breakdown voltage results in a proportional increase in the scaling of the actuation force. Finally, electromagnets require 3D electrical windings, which are not readily achieved via microfabrication, which further tilts the balance in favor of electrostatic actuation.

2. *Material Selection*

Regarding the choice of materials for MEMS, it is useful to divide the materials used into those which are used for mechanical elements and those which are used as transducers. The latter category includes piezoactive materials, shape memory alloys, and materials which can act as chemical and biological sensors. Transducer materials in microdevices tend to be very application-specific and are described later in this article. Ashby (1992) provides a general methodology for materials selection for mechanical elements in engineering systems and these materials selection principles apply equally to MEMS devices.

Three basic mechanical elements that are commonly used in MEMS devices are diaphragms for pressure sensors, beam elements capable of operating at high frequencies for accelerometers and gyroscopes, and rotating disks for pumps and power-producing turbines. The performance indices for these applications are

$$\frac{\sigma_f^{3/2}}{E},\ \frac{E}{\rho},\ \frac{\sigma_f}{\rho}$$

respectively, where σ_f is the fracture strength, E is Young's modulus, and ρ is the density. In addition, the maximum operating temperature T_{max} is a concern if the microsystem is intended for elevated temperature operation (for instance in the case of micro-gas turbine engines or high-temperature pressure sensors). Table 2 summarizes the performance of several microfabricatable materials with respect to these performance indices. It is clear that silicon, either in single-crystal or polycrystalline form, is a very attractive material for high-strength applications such as pressure sensors and turbomachinery.

Table 2
Material performance indices for musical elements applied to microfabricatable materials; where possible data from microfabricated structures are used.

Material	Density, ρ ($\mathrm{kgm^{-3}}$)	Modulus, E (GPa)	Fracture strength, σ_f (MPa)	Maximum temperature, T_{max} (K)	E/ρ ($\mathrm{GNkgm^{-1}}$)	σ_f/ρ ($\mathrm{MNkgm^{-1}}$)	$\frac{\sigma^{3/2}{}_f}{E}$ ($\sqrt{}$(MPa))
Silicon	2330	129–187	4000	825–1000	72	1.7	1.5
Silicon oxide	2200	69	1000	800	36	0.45	0.43
Quartz	2650	72	1000	1000	27	0.38	0.43
Silicon nitride	3200	270	1000	1300	84	0.31	0.12
Nickel	8900	207	500	1000	23	0.06	0.54
Aluminum	2710	69	300	400	25	0.11	0.75
Aluminum oxide	3970	393	2000	1500	99	0.50	0.228
Silicon carbide	3300	430	2000	2000	130	0.303	0.208
Diamond	3510	1035	1000	950	295	0.28	0.31

The fracture strength data are the most subject to variation and the maximum allowable temperature will vary with the requirements of the application.

However, for devices in which specific stiffness is critical, diamond, silicon carbide, and aluminum oxide may offer significant performance enhancements. In applications where elevated temperature is a concern, silicon carbide, silicon nitride, and aluminum oxide may be candidate materials. The list of materials which are potential alternatives to silicon is rather short, reflecting the advantages enjoyed by materials in the CMOS materials set in addition to the wealth of micro-fabrication experience. It is significant that these alternative materials for mechanical elements are all ceramics, as opposed to metals or polymers. At the scales of most MEMS devices, there is little or no favorable scaling of material properties for metals, whereas for ceramics, strength-related properties can be significantly superior at small scales.

3. *Fabrication Routes*

Three fabrication routes account for the vast majority of existing MEMS devices; surface micromachining, bulk micromachining, and electroforming molding processes. Since the materials used in MEMS are to a great extent defined by these manufacturing processes it is worth briefly reviewing these processes. Far more comprehensive descriptions of these processes are available elsewhere (e.g., Madou 1997).

3.1 Surface Micromachining

Surface micromachining (e.g., Bustillo *et al.* 1998) has evolved directly from the CMOS processes used to fabricate very large-scale integration (VLSI) devices. These devices consist of thin deposited layers of conductors, insulators, and semiconductors and passivation layers on doped silicon wafer substrates. In VLSI devices the layers are deposited, patterned, and etched to yield highly integrated electronic devices with very small feature sizes. In surface micromachined MEMS the layers are patterned and etched to yield electromechanical elements or are used as sacrificial layers to allow motion of the mechanical layers. The use of CMOS compatible processes and materials permits a high degree of integration of mechanical devices with the electronics required for control, signal processing, and power distribution. Commercial examples of highly integrated surface micromachined devices include microaccelerometer chips for controlling automobile airbag deployment and mirror arrays for portable projectors. A micrograph of such a device is shown in Fig. 1. However, surface micromachining is typically limited to layers of thicknesses $<5\,\mu m$ which restricts the ability to create devices which can deliver significant mechanical forces or power levels (see Table 1), or to define channels or cavities for fluidic, chemical, or biological applications.

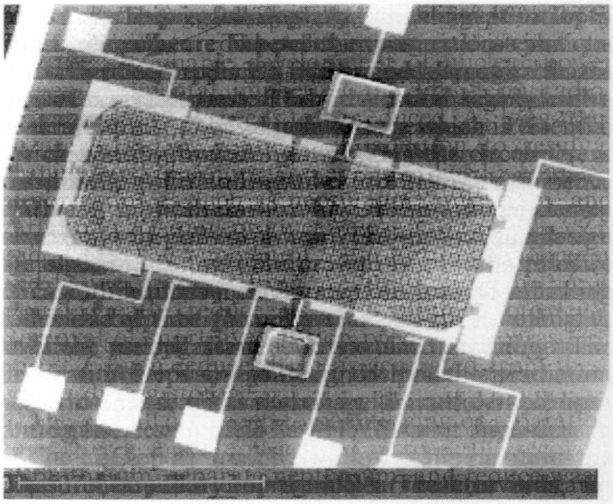

Figure 1
A micromachined accelerometer created by deep-reactive ion etching and surface micromachining (courtesy of J. Borenstein, Draper Laboratory).

3.2 Bulk Micromachining

Bulk micromachining (e.g., Kovacs *et al.* 1998) involves etching features directly into silicon wafers or other substrates. Typically, if integrated electrical function is required the microelectronic elements are created using CMOS processes on the top side of the silicon wafer. Then bulk micromachining commences from the other side of the wafer to yield mechanical elements (such as thin diaphragms or beams on the top side of the wafer) or passages for fluid flow. This strategy has been used for many years to create small pressure sensors, and more recently it has been extended to enable microfluidic applications and even microturbomachinery. The use of the silicon substrate as the basis for mechanical elements of devices permits larger, and particularly deeper, features to be used than in surface micromachining. This is an important consideration in MEMS where higher mechanical power or force levels are desired, or in applications involving fluids, in which large losses would be associated with flow through the smaller channels that could be realized by surface micromachining. Figure 2 shows a micrograph of a deep-etched, microengine rotor, a component of an integrated microgas turbine engine.

3.3 LIGA

The third prevalent manufacturing process used for MEMS is the creation of the mechanical elements of the device by deposition of material into a microfabricated mold (e.g., Mohr 1998). The most widespread such process is "LIGA" (the acronym stems

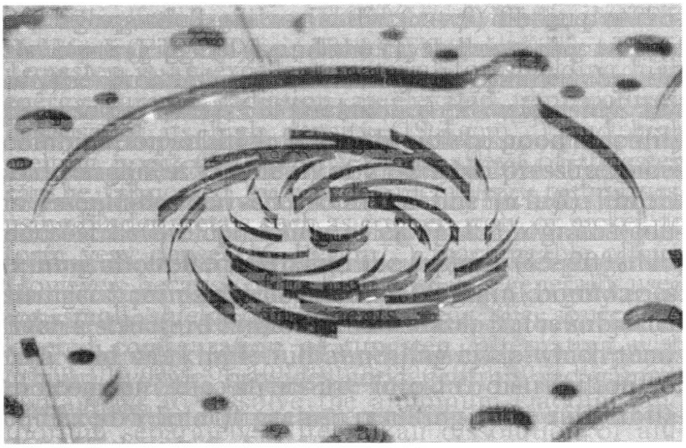

Figure 2
Deep-reactive ion etched microgas turbine rotor, showing the compressor. The rotor diameter is 8 mm and the blades are 240 μm tall (courtesy of J. Protz, A. Ayon, and X. Zhang, MIT MicroEngine Project).

from the German expressions for the major process steps: Lithography, Galvanoformung (electroforming) and Abformung (molding)). The basic process consists of creating a polymer mold by lithography (often x-ray lithography to create high aspect ratio structures) and then electroplating metal into the mold cavities. Although a wealth of experience has been accumulated for metal plating, there is a significant need to refine the capabilities of this technology so as to be able to achieve the very precise control of dimensions and material properties required for MEMS applications.

The conventional classification of microfabrication into three principal fabrication routes is somewhat crude, and there are examples of hybrid approaches emerging that challenge the restrictions described above. Examples include the use of sacrificial LIGA (SLIGA) on silicon wafers, and the use of wafer bonding to combine microfabrication routes on a single device (Schmidt 1998). It can also be argued that wafer bonding itself represents a distinct microfabrication route in its own right inasmuch as it offers the ability to transfer functionality that would traditionally be achieved through first-level packaging to the microfabricated device itself.

4. *The MEMS Materials Set*

The three fabrication routes described above have hitherto largely defined the materials set available to MEMS designers. One of the keys to achieving the high level of reliability and low unit cost of microelectronic devices is that a limited set of materials are used and their composition is very carefully controlled to ensure reproducible performance. The principal materials used in VLSI devices include: doped single- crystal silicon wafers as the semiconductor substrate and deposited layers of polycrystalline silicon ("polysilicon") for resistive elements; aluminum (and now copper) as the principal conductor; and silicon oxide, silicon nitride, and titanium nitride for electrical insulation and passivation/protection, respectively. This limited selection of materials has also formed the basis for the vast majority of surface micromachined and bulk micromachined MEMS. Silicon, polysilicon, and silicon nitride are generally used for mechanical elements, aluminum as the electrical conductor for power and signal transmission, and silicon oxide as a sacrificial layer to allow the release of moving or deforming mechanical elements. The restriction to this set of materials ensures compatibility with the processes used to create the microelectronic elements and, therefore, permits a high degree of integration on a single chip. As was noted in the discussion of materials selection these materials are also attractive based on purely mechanical considerations. Although these materials are of low toughness, the high strength of brittle materials at small scales increases the available strain levels and reduces the susceptibility to damage and fracture that prevents the use of these materials for macroscale devices.

Notwithstanding the utility of CMOS materials for mechanical elements, the silicon microelectronics material set is somewhat restrictive if the full potential for MEMS is to be realized and a wide range of other materials are being explored. Broadly the opportunities for expansion of the MEMS material set can be divided into materials which enable higher performance mechanical elements and those required for transducer elements which permit power or signal conversion from one physical domain to another.

4.1 *Materials for Mechanical Elements*

The breadth of materials potentially available to MEMS designers is rapidly expanding. It is to be expected that the range of materials should ultimately exceed that available to the designer of devices at the macroscale. The use of materials in very small quantities largely eliminates constraints of cost and availability that dominate material selection for macroscale devices. However, the ability to introduce new materials is still somewhat restricted by the need to maintain compatibility with existing processes and microfabrication tools. This is particularly true in cases where MEMS fabrication shares tools that are used for VLSI/CMOS processes which require very low levels of material contamination.

As noted in the previous section on materials selection the most desirable candidate materials for increasing the performance of mechanical elements are ceramics. The question of the feasibility of these candidate materials reduces to one of whether economical fabrication routes can be developed which can yield the required geometrical shapes at the appropriate scales and with the necessary tolerances. Just as for silicon-based MEMS, the processing routes available for ceramic materials can be divided into additive and subtractive processes. Some progress has been made for some of these candidate materials using each of these approaches.

Silicon carbide is available in relatively large single- crystal wafers. However, SiC is relatively unreactive and so etch rates are typically one to two orders of magnitude slower than can be obtained in silicon. The two most promising routes for etching single-crystal silicon carbide are reactive ion etching and photochemically assisted etching. Beheim and co-workers at NASA's Glenn Research Center have achieved etches of several hundred micrometers depth using an inductively coupled plasma etch. This process has also been able to etch features with aspect ratios of higher than six. Etch rates of the order of 0.3 μmmin^{-1} have been achieved, which allows bulk micro-machining as a practical proposition. Examples of the resulting structures are shown in Fig. 3.

Photochemically assisted etching (Tuller and Mlack 1998) of silicon carbide has yielded higher etch rates than plasma etching, but is less capable of creating high aspect ratio structures. However, it has the advantage that doped *n*- or *p*-type epitaxial layers can be used to create etch stops or to act as masks. This permits sealed cavities to be created which enable devices such as pressure sensors to be considered. It also can be used to create surface micromachined structures, such as thin suspended beams. An example of an SiC MEMS structure produced by photochemically assisted etching is shown in Fig. 4.

The major barrier to the creation of commercially attractive single-crystal silicon carbide microsystems is the high cost of the wafers. This has restricted their consideration to small highly specialized devices. As yet large single-crystal wafers are not yet available for diamond and silicon nitride and although they are available for aluminum oxide, there are no viable etches available.

Additive approaches to creating ceramic microsystems are applicable to a wider set of materials. Deposition processes are available for the whole range of ceramics and chemical vapor deposition is particularly widely used. Silicon nitride films produced by plasma-enhanced CVD or low-pressure CVD are already well established as part of the CMOS material set and are widely used in MEMS. Recent developments have shown that MEMS structures can be produced in amorphous diamond Sullivan *et al.* (2001) and an example of a interdigitated comb drive is shown in Fig. 5. Both diamond and silicon nitride are currently restricted to thin films of the order of 1 μm in thickness.

For larger structural components, with dimensions of several millimeters, and thicknesses of hundreds of micrometers, the greatest progress has been made using CVD silicon carbide in micromolding processes, somewhat analogous to the LIGA process used to create metallic MEMS components. Molds are created by the etching and wafer bonding of silicon wafers. These molds are then placed within a CVD reactor and filled with SiC. The molds are then etched away, typically using a wet etch. An example of a part produced by such a process is the microturbine rotor shown in Fig. 6. However, it is worth

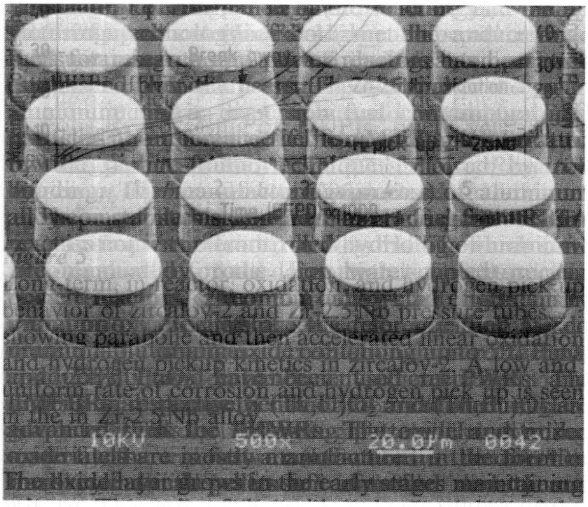

Figure 3
An array of deep-reactive ion etched features in single-crystal silicon carbide (courtesy of G. Beheim, NASA Glenn Research Center).

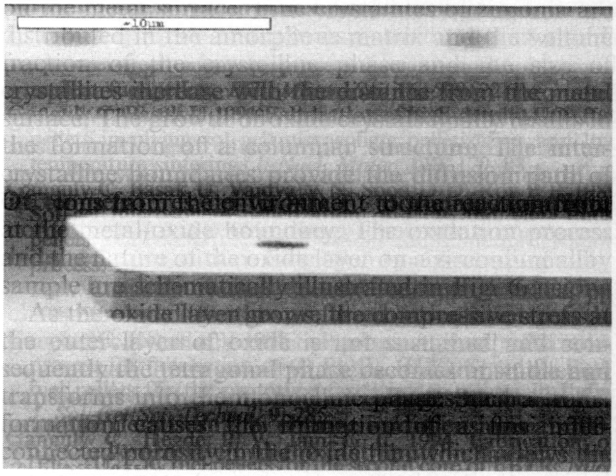

Figure 4
SiC cantilever created by photoassisted etching (courtesy of R. Mlack, Boston Microsystems).

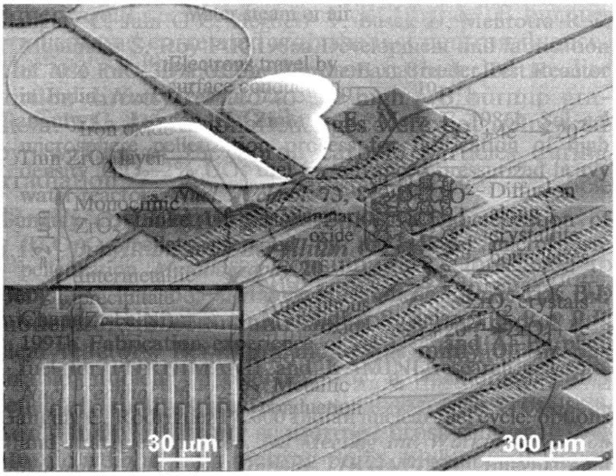

Figure 5
Amorphous diamond thin-film comb drives (courtesy of J. Sullivan, Sandia National Laboratory).

Figure 6
SiC microturbine rotor, created by chemical vapor deposition into a deep reactive ion-etched Si mold, which is subsequently removed by etching. The rotor is 4 mm in diameter (courtesy of K. Lohner, MIT MicroEngine Project, R. Shinavski, HyperTherm Inc).

noting that this approach is not yet capable of yielding parts with the necessary dimensional tolerances for high precision MEMS, although it may be feasible as a means to create parts with reduced requirements on dimensional tolerances.

Although CVD and other processes are available for aluminum oxide and other oxide ceramics, at the time of writing none have been applied to create MEMS devices. Key challenges in the use of deposition processes to create MEMS components are the need to control residual stresses that arise during the deposition. Also, micromolding processes often require planarization (lapping and chemical–mechanical polishing) in order to create complete structures. These processes are not well developed for the hard materials of interest. Furthermore, there are no proven bonding technologies available. It is anticipated that these will remain active areas of research and technology development for the foreseeable future.

4.2 Materials for Transducer Elements

MEMS sensors and actuators require means of converting mechanical inputs to electrical outputs and vice versa. As noted earlier the most common transduction principle is that of electrostatics, in which case capacitance changes are used to measure displacements in pressure sensors and accelerometers or electrostatic forces are used to cause displacements in actuators such as comb drives or micromotors.

For macroscale devices electromagnetic forces are the dominant means of converting electrical power to mechanical; however, there has been much less use of this principle at the microscale. As noted earlier this stems from the favorable scaling and ease of implementation of electrostatic operation, and the relative difficulties of microfabricating coils for inductors and motors. Recent success has been demonstrated in using electrodeposition to create micromachined permanent magnets which offer some promise for electromagnetic devices. Given the capability for microfabricating electromagnetic devices it can be shown that their performance would be broadly comparable to those of electrostatic devices, and probably superior at the larger end of the MEMS size range.

Piezoelectric materials are capable of very high energy and power densities at small scales. The high frequency of operation inherent to MEMS devices matches well with the high frequency capability of piezoelectric materials and the favorable scaling of strength at small scales overcomes some of the limitations encountered in using piezoceramics for macroscale devices. The most commonly used piezomaterials in MEMS devices are lead zirconate titanate (PZT), zinc oxide (ZnO), and aluminum nitride (AlN). These are typically deposited as thin films by sputtering or, in some cases, sol–gel deposition onto silicon micromachined elements.

In addition, molding techniques can be used in conjunction with conventional slurry processing of ceramics to create somewhat larger piezoelements suitable for integration with LIGA fabrication processes. At the time of writing the performance of thin-film piezoceramics (and shape memory alloys) is somewhat inferior to those fabricated by more conventional, macroscale processing routes. In addition, it is worth noting that silicon itself is a piezoresistor and this property has been used in many pressure sensors. Other phenomena have also been employed to create MEMS transducers including shape memory alloys and magnetostrictive materials.

Although the earliest MEMS generally addressed electrical–mechanical energy transduction, more recent advances in biological and chemical sensors require specialty materials to permit detection of specific biological or chemical agents. In addition devices for chemical and biological synthesis or combustors for power production may require catalysts. Transition metals such as platinum or palladium are often used for this purpose in macroscale chemical plants and these metals can be deposited in thin-film form, suitable for integration with other microfabrication processes.

There are clearly considerable opportunities for expanding and improving the materials for transducer elements in MEMS, although it is difficult to generalize, since the choice of material tends to be very application specific.

5. *Concluding Remarks*

It is clear that materials science and technology is a key for the growth of microsystems as a commercial sector. The existing material set is largely derived from that used in microelectronics. However, given the drive towards devices which require higher operating temperatures and for resonant devices with higher natural frequencies there are opportunities to extend it, particularly for ceramic materials. The process technologies available to create structures from the most promising candidate materials (silicon carbide, silicon nitride, and diamond) are being developed but are relative immature compared to those available for the CMOS materials set that is used for the great majority of commercial MEMS.

See also: Actuator Materials for Small-scale Devices

Bibliography

Arzt E 1998 Overview no. 130—Size effects in materials due to microstructural and dimensional constraints: a comparative review. *Acta Mater* **46**, 5611–26

Ashby M F 1992 *Materials Selection in Mechanical Design*. Pergamon, Oxford, UK

Bustillo J M, Howe R T, Muller R S 1998 Surface micromachining for micro-electromechanical systems. *Proc. IEEE* **86**, 1552–74

Kovacs G T A, Maluf N I, Petersen K E 1998 Bulk micromachining of silicon. *Proc. IEEE* **86**, 1536–51

Madou M 1997 *Fundamentals of Microfabrication*. CRC Press, Boca Raton, FL

Mohr J 1998 LIGA—A technology for fabricating microstructures and microsystems. *Sensors Mater.* **10**, 363–73

Schmidt M A 1998 Wafer-to-wafer bonding for microstructure formation. *Proc. IEEE.* **86**, 1575–85

Senturia S D 2000 *Microsystem Design*. Kluwer, Boston

Spearing S M 2000 Materials issues in MEMS. *Acta Mater* **48**, 179–96

Sullivan J P, Friedmann T A, de Boer M P, LaVan D A, Hohlfelder R J, Ashby C I H, Dugger M T, Mitchell M, Dunn R G, Magerkurth A J 2001 In: de Boer M P, Judy M, Kahn H, Spearing S M (eds.) *Microelectromechanical Systems (MEMS) devices III*. Materials Research Society Symposium Proceedings, Vol. 657, Warrendale, PA

Tong Q Y, Gosele U 1998 *Semiconductor Wafer Bonding: Science and Technology*. Wiley, New York

Tuller H L, Mlcak R 2000 Advanced sensor technology based upon oxide thin film-MEMS integration. *J. Electroceram.* **4**, 415–25

Mixed Oxide (MOX) Fuel

H. S. Kamath and D. S. C. Purushotham

During the 1960s with the first commercial use of nuclear energy it was thought that uranium would soon become scarce. For nuclear energy to make effective contributions it was felt necessary that the plutonium produced during nuclear power generation be recovered (by reprocessing) and recycled, preferably in fast reactors. The early work on development of plutonium-bearing fuels clearly established the superiority of uranium–plutonium mixed oxide (MOX) for fast reactors. The proportion of plutonium in the mixed oxide used in fast power reactors is usually 20–30 wt.%. The fuel used in prototype fast reactors such as Phoenix, PFR, MONJU, FFTF, as well as in the commercial demonstration fast reactor Superphoenix is this type of fuel. The delays in the introduction of commercial fast reactors since the 1980s have resulted in increasing use of mixed oxide of UO_2 with ~4% PuO_2 composition in light water reactors (LWRs). Although "mixed oxide" in general terms encompasses both thermal and fast reactor fuels, the term MOX is popularly used to refer to UO_2 with ~4% PuO_2 fuels used in LWRs. In this article, we cover MOX fuels belonging to both types of reactors although a stronger emphasis is placed on thermal reactor MOX because of the interest shown in it.

1. *Physical Characteristics*

UO_2 and PuO_2 both have f.c.c. (CaF_2) crystal structures and exhibit complete solid solubility over the entire composition range at and near an oxygen to metal ratio (O/M) = 2. However, beyond UO_2–30% PuO_2 composition, two-phase regions (of f.c.c. and b.c.c. or b.c.t.) have been reported at O/M values of less than 1.94.

A review of the thermal conductivity of MOX fuels indicates that the presence of PuO_2 in UO_2 slightly decreases the thermal conductivity. The thermal resistivity is represented by $1/K = A + BT$, where A is the phonon scattering (the presence of plutonium increases A) and BT is the so-called intrinsic thermal resistivity arising from phonon–phonon interactions which increases with temperature, T. Laser flash methods are used for thermal conductivity measurements and the decrease is typically 4% for 10% Pu/U + Pu. With burn-up, the conductivity of MOX fuel degrades further owing to irradiation damage and fission products in dynamic solution, the phenomenon being similar to that in UO_2.

The O/M ratio is an important variable that affects thermal conductivity, melting point, and the lattice parameter. In fast-reactor MOX fuels , the O/M ratio is important as it affects plutonium migration behavior, and fission product attack on clad as well sodium fuel compatibility. The O/M generally specified for fast-reactor MOX is between 1.95 and 1.98, whereas it is close to 2.00 for thermal-reactor MOX.

The thermal expansion coefficient of MOX (LWR) is about 1% higher (Haas 1993) and the melting point 20 °C lower compared to UO_2. The melting point and the lattice parameter follow Vegard's law at O/M = 2 for both thermal and fast-reactor MOX. There are reports that O/M affects melting point, particularly in the case of fast-reactor MOX. An O/M corresponding to an O/Pu ratio of about 1.75 is reported to have always the highest melting point owing to high thermal stability seen when trivalent and tetravalent plutonium atoms are in equal proportion. Some of the important physical properties of UO_2, PuO_2, UO_2–4% PuO_2, and UO_2–20% PuO_2 are given in Table 1.

The thermal creep, both for UO_2 and MOX, is generally modeled as the sum of a linear term proportional to applied stress (diffusional creep) and a term proportional to stress to the power of 4.5 (dislocation climb). The irradiation creep is the sum of both irradiation-enhanced creep (thermal creep increased owing to irradiation) and irradiation-induced creep (temperature independent and occurs only in the presence of irradiation) (see *Nuclear Reactor Materials: Irradiation Effects*). MOX fuels are reported to show increased in-pile creep (Palmer *et al.* 1999), resulting in their improved transient behavior.

Table 1
Physical properties of MOX fuel (unirradiated).

No.	Parameter	UO_2	PuO_2	UO_2–4% PuO_2	UO_2–20% PuO_2
1	Theoretical density (g cm^{-3})	10.96	11.46	10.98	11.04
2	Latticeparameter(Å)	5.47	5.396	5.467	5.456
3	Melting point (°C)	2840 ! 40	2400 ! 40	2820	2810
4	Crystal structure	Fluorite: SC, O^{2-}; f.c.c., M^{4+}			
5	Coeff. of linear expansion (10^{-6} °C^{-1})	10.1	10.9	10.15	10.1
6	Thermal diffusivity at 95% Th D (cm^{2}s^{-1})	0.022 at 1600 °C	0.023 at 1000 °C		0.021

2. *MOX Fuel Design*

MOX thermal reactor fuel is now being used essentially in LWRs designed for low enriched uranium (LEU) fuels and hence has to be mechanically, neutronically, and operationally interchangeable with uranium fuels. MOX fuel must perform at least as well as its uranium counterpart (see *Light Water Reactor Fuel Design and Performance*). This is ensured by using almost the same hardware, design, and operating limits and also by limiting the loading of MOX to one-third of the core so that the reactor can be operated within the same licensed safety parameters. The high resonance capture of MOX fuel in a thermal neutron spectrum leads to hardening of the neutron spectrum and this is overcome by introducing a larger number of water rods in MOX fuel assemblies (FAs) compared to uranium FAs. Further, for reasons of economy, fabrication, transport, and storage, "all-Pu" MOX assemblies are preferred over "island-Pu" MOX assemblies in LWRs (Schlosser 1993).

MOX fuel has been the cornerstone of fast-reactor fuel because of its excellent burn-up potential. The life-limiting phenomenon in the case of fast reactors comes from the clad (swelling) (see *Nuclear Reactor Materials: Irradiation Effects, Fast Breeder Reactors: Fuel*) rather than from the fuel matrix. The advanced MOX fuel design for fast reactors uses annular fuel pellets, a lower O/M, and advanced cladding (D-9, HT-9) alloys. The design burn-up in advanced fast-reactor MOX fuel exceeds 200 GW d ton^{-1}.

3. *MOX Fuel Fabrication*

Four countries, Belgium, France, India, and the UK, are engaged in the manufacture of MOX fuels for LWRs. Japan has two MOX fabrication plants—one manufacturing MOX fuel for the Advanced Thermal Reactor (ATR) and the other for its fast reactor (MONJU). The Russian Federation also has two MOX plants—one for pellet fuel and the other for Vibro-fuel for experimental irradiation in fast reactors.

Almost all MOX fabrication plants have adopted fabrication flowsheets based on mechanical mixing/milling techniques. The micronized master blend (MIMAS) developed (Hass 1993) by Belgonucleaire (BN) is used by BN and France. The UK (Macdonald 1994) and India (Kamath *et al.* 1998) use modern attritor technology, which produces micro-homogenous MOX that is likely to have superior performance at high burn-up (Walker *et al.* 1996).

The techniques used in MOX fuel fabrication essentially are powder metallurgy techniques involving cold pressing, sintering, centerless grinding, encapsulation, welding, and fuel rod assembly (see *Nuclear Reactor Fuel Fabrication (Including Quality Control)*). The advances that are taking place in fuel manufacture are mainly with respect to fuel performance, simplification of process steps, high level of automation to reduce operator dose, and inert matrix-based fuels for weapon-plutonium disposal.

4. *MOX Use and Performance*

As at 1999, 35 commercial LWRs were loaded with MOX fuels in five countries up to a maximum of 30% of the core. Several prototype fast reactors, including Superphoenix (1200 MW(e)), have been operated with MOX fuel. More than 85 tons of plutonium and 1000 tons of MOX have been irradiated in thermal and fast reactors.

The scale of MOX fuel fabrication and use is an order of magnitude less than that of UO_2, but its performance is much the same as UO_2. MOX fuel designers are now confident that MOX fuel can be designed to perform to the same operational and performance criteria as uranium fuels (up to 45 $MWdkg^{-1}$). In fact, there is now enough evidence to indicate that MOX fuel performs better than UO_2 with respect to pellet-clad interaction (PCI) (Haas 1989). It is seen that threshold power for PCI failures for MOX fuel is higher compared to UO_2, as derived from the ramp experimental data from Patten and Halden experiments. The plutonium migration noticed in high-power fast-reactor MOX near stoichiometric ratio is absent in thermal reactors.

The only performance-related issue of some concern in the case of MOX appears to be its high fission gas release. This seems to be particularly true for heterogeneous as opposed to homogeneous MOX fuel (Trotabas 1994). The increased internal rod pressure is partly contributed to by enhanced helium generation and release in the case of MOX compared to UO_2 fuels (Kamimura 1999).

Most of the countries using MOX fuel in LWRs have not yet finalized their options for the spent MOX fuel. The storage or disposal of spent MOX fuel entails similar problems to that of spent UO_2 fuel, with somewhat enhanced neutronic dose and increased radiotoxicity owing to the presence of minor actinides. Studies on "Multirecycling MOX" and 100% MOX in advanced reactor concepts are also underway as well as studies on one or two MOX recycles in LWRs to be followed by multicycling in FBRs.

5. *Economics of MOX Fuel*

The MOX option is not necessarily decided only by economics, as it is also an important part of nuclear resource management and a method for reduction of radiotoxicity in the spent fuel. Nevertheless, some cost studies (OECD-NEA 1994) have indicated only a marginal (10%) advantage in costs for the once-through option compared to the recycle option (see *Nuclear Fuel Cycles*). This margin is too small to be taken seriously in view of the uncertainties in the estimates. The once-through option also suffers from the big disadvantage of uncertainty in finalization of site selection for final disposal and licensability of the technology. The studies seem to indicate that MOX fuel fabrication may be three to four times costlier than for UO_2, but this factor is likely to be reduced as MOX fuel fabrication grows in scale and understanding. With utilities having already invested in and committed to reprocessing (zero-value plutonium), the contained cost (which includes nuclear material and fabrication) of MOX is not very different from that of UO_2 fuel. As burn-up of the fuel increases, utilities may find MOX to be economically more attractive than UO_2 as, unlike in LEU, the increased plutonium enrichment does not lead to higher costs in the case of MOX fuel.

See also: Heavy Water Reactor Fuel Design and Performance; Nuclear Reactor Fuel Fabrication (Including Quality Control)

Bibliography

Haas D 1989 MOX fuel in-pile behaviour up to 60 000 MWd/T. *Nucl. Eur.* **1–2**, 14–6

Haas D 1993 In-pile behaviour of MOX with particular emphasis on MIMAS fuel. *Nucl. Technol* **102**, 47–53

Haas D, Vanderghegnst 1994 Mixed oxide fuel fabrication technology and experience at BN CFCa plants and further developments for Melox plant. *Nucl. Technol.* **106**, 60–81

Izutsu S, Sasagawa M, Masayama H, Suzuki T 1999 Progress of fuel MOX core design in ABWR In: *Int. Symp. on MOX Fuel Technologies*. IAEA-SM-358/27. IAEA, Vienna, pp. 362–7

Kamath H S, Majumdar S, Purushotham D S C 1998 *Developments in MOX pellet fabrication technology: Indian experience. IAEA-TECDOC-1036*. IAEA, Vienna, pp. 103–12

Kamimura K 1999 Helium generation and release in MOX fuels In: *Int. Symp. on MOX Fuel Cycle Technologies*. IAEA-SM-358. IAEA, Vienna, Vienna, pp. 263–70

Lippens M, Basselier J 1998 Comparative thermal behaviour of MOX and UO_2 fuels In: *Int. Seminar on Thermal Performance of LWR Fuel*. OECD/NEA, Cadarache. pp. 243–6

Macdonald G 1994 The MOX demonstration facility—the stepping stone to commercial MOX production. *Nuclear Energy* **33** (3), 173–8

Muromura T 1995 Pu rock like fuel integrated R&D—unconventional options for Pu disposition. IAEA TECDOC-840, pp. 253–61

OECD-Nuclear Energy Agency 1994 Economics of the Nuclear Fuel Cycle. OECD, Paris

OECD-Nuclear Energy Agency 1997 Management of Separated Pu: The Technical Options. OECD, Paris

Palmer I, Rossiter G, White R J 1999 Development and validation of enigma code for MOX fuel performance modeling. IAEA-SM-358/20, pp. 271–81

Schlosser G J 1993 Experience in PWR and BWR mixed oxide fuel management. *Nucl. Technol.* **102**, 54–67

Stoll W 1996 *Concept of Advanced MOX Fuel Fabrication Technology—Pu Futures*. Topical conference on Pu and actimides, Los Alamos, CA, 8.1–8.4

Trotabas M 1994 MOX experience in French power reactors In: *Int. Topical Meeting on LWR Fuel Performance*, West Palm Beach, FL pp. 718–23

Walker C T, Goll W, Matsumura T 1996 Effect of inhomogeneity on the level of fission gas and Cs release from OCOM MOX fuel during irradiation. *J. Nucl. Mater.* **228**, 8–17

Molecular Motors

J.-P. Sauvage and L. Raehm

Many "molecular motors" or "machines" play an essential role in biology. These systems are multicomponent assemblies undergoing large-amplitude geometrical changes or leading to the locomotion of one of the components, either under the action of an external stimulus (pH change, redox process, light pulse, etc.) or of a chemical gradient. Many examples of proteins are known which undergo important shape modifications (folding–defolding) after a signal has been sent to the protein. Real molecular motors consist of several units among which some parts are considered as motionless and some others move continuously while a "fuel" is "burnt," which translates, in biological systems, as ATP being hydrolyzed. Linear motors such as myosin–actin (from skeletal muscle) have been known for many years and our understanding of them has made considerable progress. By contrast, few rotary motors have been identified and investigated. The most spectacular example is that of ATPase, whose rotational motion was postulated long ago but which only now can be directly observed. ATPase consists of a "stator" with six proteins ($\alpha_3\beta_3$) assembled to form a cylinder-like unit, a seventh rod-like protein (γ) being threaded in the hollow part of the $\alpha_3\beta_3$ unit to form the "rotor" (Fig. 1). The γ subunit rotates inside the $\alpha_3\beta_3$ assembly with concomitant hydrolysis of ATP or ATP synthesis when a pH gradient is present.

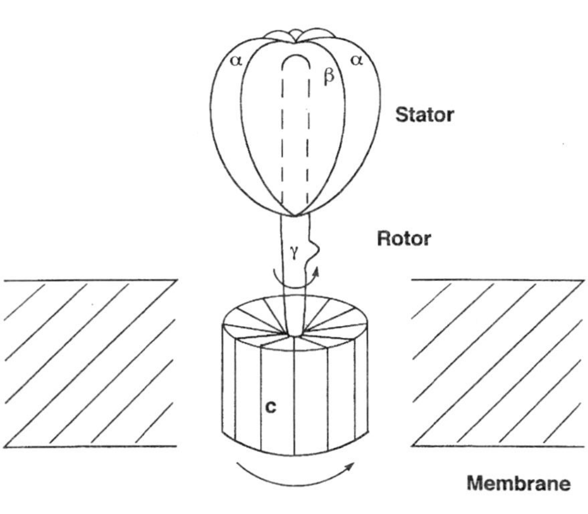

Figure 1
Schematic representation of ATP synthase, a biological rotary motor. γ (and c) are mobile whereas the aggregate $\alpha_3\beta_3$ is fixed to the membrane and constitutes the stator of the motor.

The number of synthetic molecular assemblies whose dynamic behavior is reminiscent of biological motors is very limited. If a ring is threaded onto a rod, it can either rotate around the rod axis or undergo a translation movement. Similarly, in catenanes (i.e., compounds consisting of two interlocking rings), a ring can glide at will and spin within another ring. A new dimension can be added by moving components of these molecular systems at will, under the action of a photochemical or electrochemical signal. In fact, compounds containing interlocking rings or rings threaded onto an acyclic fragment are the ideal precursors to molecular machines, i.e., multicomponent systems for which selected parts can be set in motion while the other fragments are motionless. As represented schematically in Fig. 2, these topologies can provide the working parts of machines or motors.

Molecules whose shapes are modified by sending a signal to the system are of course numerous.

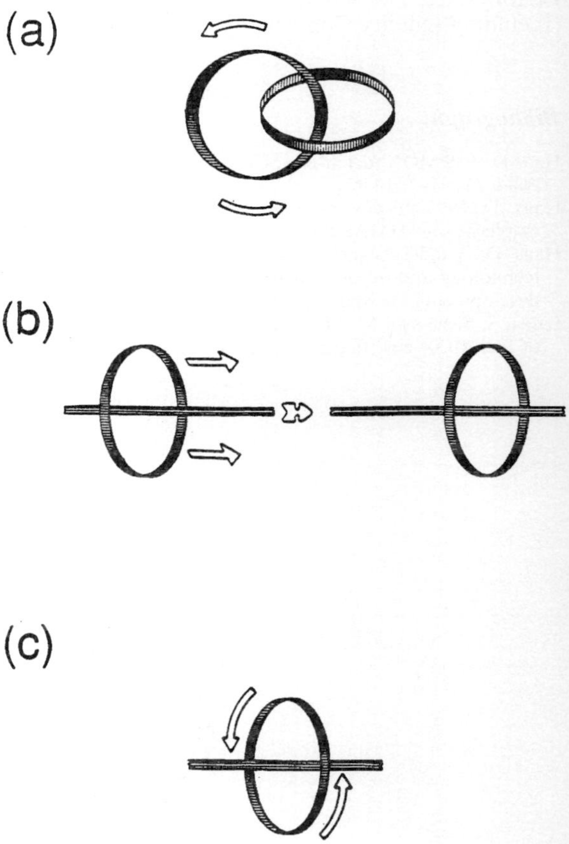

Figure 2
Molecules in motion. (a) Gliding of one ring around another; (b) translation of a ring along its axis; (c) ring rotating around its axis.

However, molecular machines tend to refer to *large-amplitude* motions leading to real translocation of some parts of the compound, reversibility being an essential feature of the system. Several examples of such compounds have been studied using threaded and interlocked molecules based either on acceptor–donor and hydrogen-bonded complexes or on transition metal complexes. In purely organic systems, interesting processes have been observed, such as dethreading of an acyclic component from a ring under the action of light or translation of a ring between two distinct positions of a thread induced by a redox signal. A very large family of interlocking and threaded compounds constructed on acceptor–donor aromatic complexes have been created. This remarkable synthetic work has been sustained by very elegant photochemical and electrochemical studies. These demonstrate that molecular movements can be induced in such systems, either by irradiating the compounds with visible light in the presence of other additional reagents or by using electrochemical reactions (Fig. 3).

One route that has been developed is based on copper complexes. Copper(I) is generally low-coordinated (coordination number $\leqslant 4$) whereas copper(II) is preferably square planar or higher coordinated (coordination number of 5 or 6). By reducing or oxidizing the copper center, from a situation corresponding to a stable complex, the system is set out of equilibrium. The relaxation process of the compound implies a large-amplitude motion that will bring the system back to its new equilibrium position. An example of a molecular motor is a catenane containing two different interlocking rings. Gliding of a ring around another ring can be chemically or electrochemically triggered and the motion can be followed either by electrochemical or spectroscopic techniques. Translation of a molecular ring along its axis is also observed in a rotaxane, i.e., one or more macrocycles encircling a single dumbbell component that has two large blocking groups or stoppers at the ends of the linear thread. The principle of the motor in the case of a ring rotating around its axis is shown in Fig. 4.

Molecular motors constitute an emerging field of research, at the interface with many areas of molecular sciences. For instance, the concepts discussed in this article can certainly be generalized to

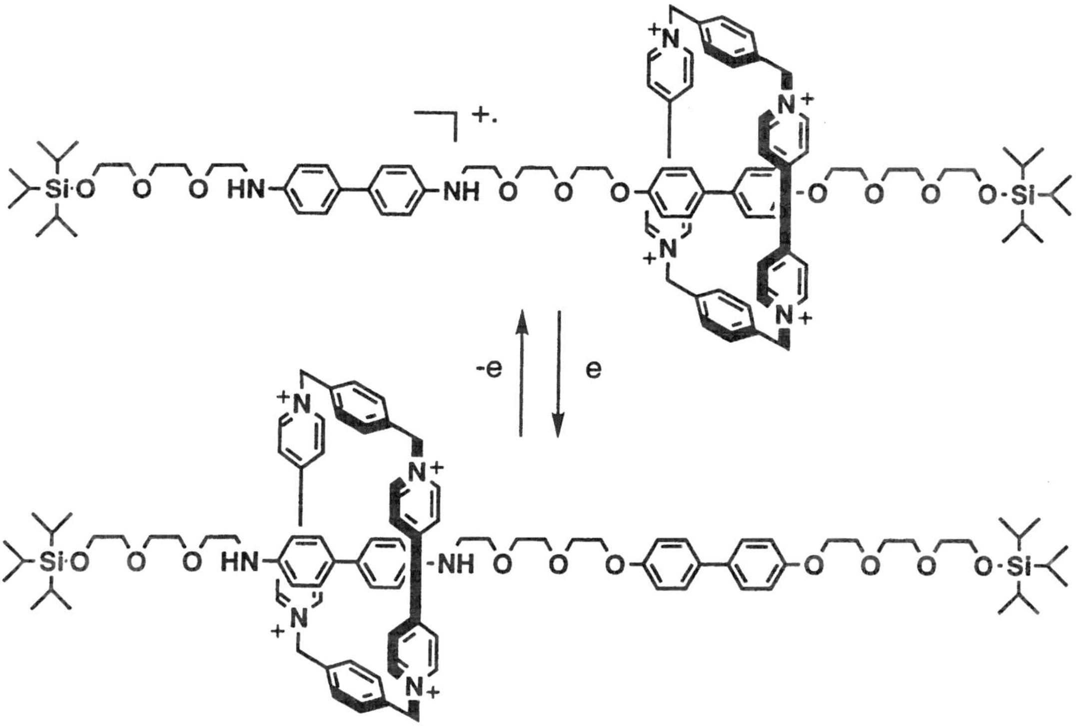

Figure 3
A switchable rotaxane based on acceptor–donor complexes. Before oxidation, the electron acceptor (ring) interacts preferentially with the benzidine nucleus (donor). After electrochemical oxidation of the latter, the ring is shifted towards the biphenol group. The process is reversible.

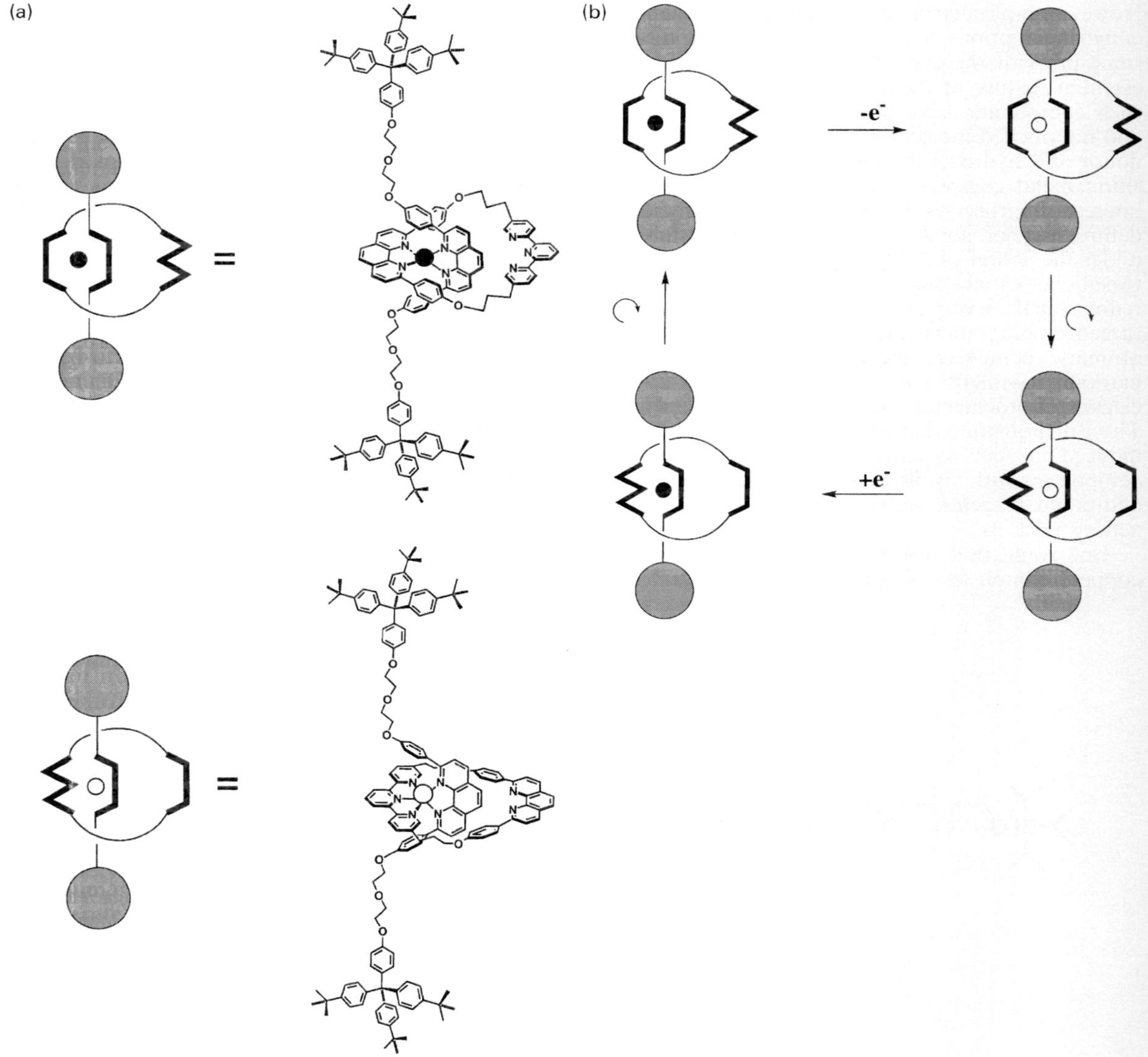

Figure 4
Principle of the electrochemically induced molecular motions in a copper complex rotaxane. The stable four-coordinate monovalent complex is oxidized to an intermediate tetrahedral divalent species. This compound undergoes a rearrangement giving a stable five-coordinate Cu^{II} complex. On reduction, the five-coordinate monovalent state is formed as transient. Finally, the latter undergoes a reorganization process that regenerates the starting complex (● = Cu^{I}, ○ = Cu^{II}).

organized assemblies of molecules (liquid crystals) or to molecular components attached to an electrode surface. The overall properties of the system (now a "device"), such as liquid-crystal character, ability to carry electrons or transport molecules, etc., can be switched by applying a given signal. The variety of impulses used is large: redox (as in the present examples), photonic, heat or pressure change, magnetic field variation, pH change, chemical signal (recognition of a molecule), electric field, etc. The effect obtained is also multifarious and can lead to dramatic modifications of the bulk properties of systems. Since a simple signal can change the shape and the volume of a compound and its assemblies, fascinating features related to mechanics (contraction or stretching) can be imagined, reminiscent of biological systems such as muscles or other biological molecular machines.

Bibliography

Balzani V, Gómez-López M, Stoddart J F 1998 Molecular machines. *Acc. Chem. Res.* **31**, 405–14

Bissele R A, Còrdova E, Kaifer A E, Stoddart J F 1994 A chemically and electrochemically switchable molecular shuttle. *Nature* **369**, 133–7

Collin J-P, Gaviña P, Sauvage J-P 1997 Electrochemically induced molecular motions in copper-complexed threaded systems: from the unstoppered compound to the semi-rotaxane and the fully blocked rotaxane. *New. J. Chem.* **21**, 525–8

Davies A P 1999 Synthetic molecular motors. *Nature* **401**, 120–1

Howard J 1997 Molecular motors: structural adaptations to cellular functions. *Nature* **389**, 561–7

Kelly T R, De Silva H, Silva R A 1999 Unidirectional rotary motion in a molecular system. *Nature* **401**, 150–2

Koumura N, Zijlstra R W J, van Delden R A, Harada N, Feringa B L 1999 Light-driven monodirectional molecular rotor. *Nature* **401**, 152–5

Livoreil A, Dietrich-Buchecker C, Sauvage J-P 1994 Electrochemically triggered swinging of a [2]-catenate. *J. Am. Chem. Soc.* **116**, 9399–400

Raehm L, Kern M, Sauvage J-P 1999 A transition-metal containing rotaxane in motion: electrochemically induced pirouetting of the ring on the threaded dumbbell. *Chem. Eur. J.* **5**, 3310–7

Sauvage J-P 1998 Transition metal-containing rotaxanes and catenanes in motion: toward molecular machines and motors. *Acc. Chem. Res.* **31**, 611–9

Stryer L 1981 *Biochemistry*, 2nd edn., Freeman, San Francisco

Zelikovich L, Libman J, Shanzer A 1995 Molecular redox switches based on chemical triggering of iron translocation in triple-stranded helical complexes. *Nature* **374**, 790–2

Multilayer Ceramic Actuators

K. Uchino

Piezoelectric and electrostrictive ceramic materials have become key components in smart actuator/sensor systems for use as precision positioners, miniature ultrasonic motors, and adaptive mechanical dampers. In particular, multilayer structures have been widely investigated in order to improve their reliability and to expand their applications. Developments in the USA, Japan, and Europe are compared.

1. *Trends in Ceramic Actuators*

Piezoelectric actuators are forming a new field midway between electronic and structural ceramics (Uchino 1993, 1994, 1995b, 1996). Application fields are classified into three categories: positioners, motors, and vibration suppressors. The manufacturing precision of optical instruments such as lasers and cameras, and the positioning accuracy for fabricating semiconductor chips, which are adjusted using solid-state actuators, is of the order of 0.1 μm. Regarding conventional electromagnetic motors, tiny motors smaller than 1 cm are often required in office or factory automation equipment, and are rather difficult to produce with sufficient energy efficiency. Ultrasonic motors whose efficiency is insensitive to size are superior in the minimotor area. Vibration suppression in space structures and military vehicles using piezoelectric actuators is also a promising technology.

Multilayer structures are mainly used for practical applications because of their low drive voltage, high energy density, quick response, and long lifetime (Yoshikawa and Shrout 1993, Uchino 1995a). Figure 1 illustrates multilayer, multimorph, and multilayer–moonie structures, which will be covered in this article. Investigations have focused on the improvement of reliability and durability in multilayer actuators.

This article reviews the investigations of device structures, reliability issues, and applications of multilayer actuators, comparing the developments in the USA, Japan, and Europe.

2. *Multilayer Structures*

Two preparation processes are possible for multilayer ceramic devices: one is a cut-and-bond method and the other is a tape-casting method. The tape-casting method requires expensive fabrication facilities and sophisticated techniques, but is suitable for the mass-production of thousands of pieces per month. Tape casting also provides thin dielectric layers, leading to low drive voltages of 40–100 V (Dibbern 1995, Takada *et al.* 1994).

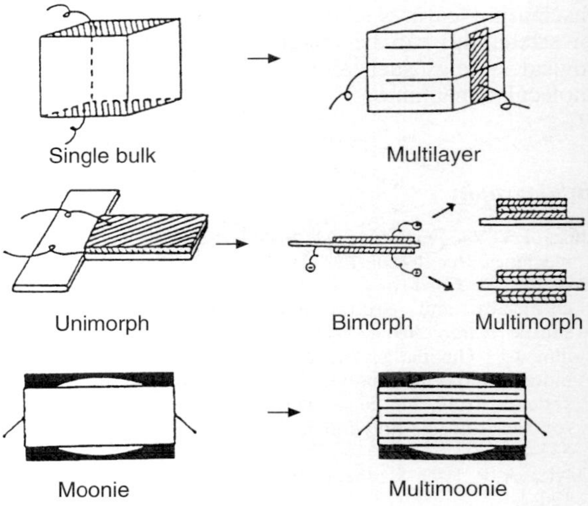

Figure 1
Examples of multilayer, multimorph, and multilayer–moonie structures.

A multilayer actuator with interdigital internal electrodes has been developed by Tokin (Ohashi *et al.* 1993a, 1993b). In contrast to the conventional electrode configuration in Fig. 1, line electrodes are printed on piezoelectric ceramic green sheets, and are stacked in such a way that alternating electrode lines are displaced by one-half pitch (see Fig. 2). This actuator generates motions at right angles to the stacking direction using the longitudinal piezoelectric effect. Long ceramic actuators up to 74 mm in length are manufactured.

A three-dimensional positioning actuator with a stacked structure has been proposed by PI Ceramic (Fig. 3), in which shear strain is used to generate x and y displacements (Bauer and Moller 1995).

Composite actuator structures called "moonies" and "cymbals" have been developed at Penn State University to provide characteristics intermediate between the multilayer and bimorph actuators. These transducers exhibit an order of magnitude larger displacement than the multilayer, and much larger generative force with quicker response than the bimorph (Sugawara *et al.* 1992, Onitsuka *et al.* 1995). The device consists of a thin multilayer piezoelectric element and two metal plates with narrow moon-shaped cavities bonded together as shown in Fig. 1. A moonie 5 mm × 5 mm × 2.5 mm in size can generate a 20 μm displacement under 60 V, eight times as large as the displacement of a multilayer of the same size (Goto *et al.* 1992). This new compact actuator has been used to make a miniaturized laser beam scanner (Goto *et al.* 1992). Moonie/cymbal characteristics have been investigated for various constituent materials and sizes (Tressler *et al.* 1994, Dogan *et al.* 1994).

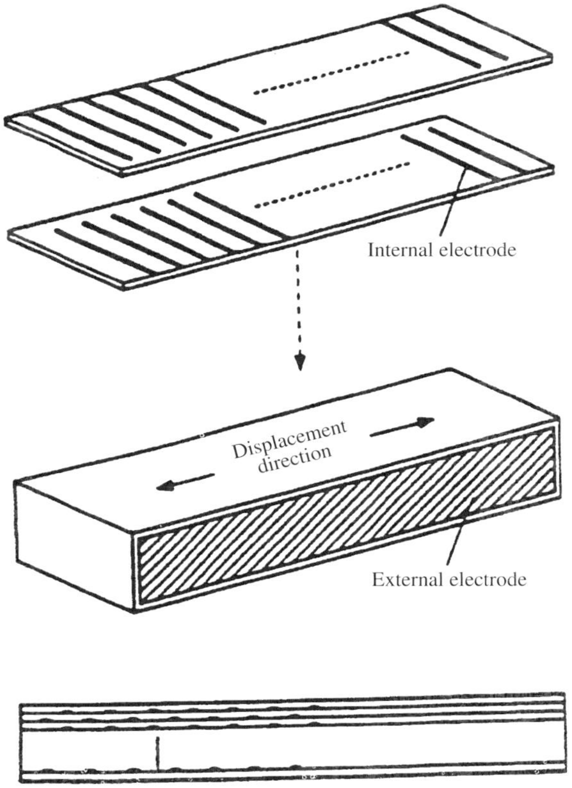

Figure 2
Structure of an internal interdigital electrode actuator.

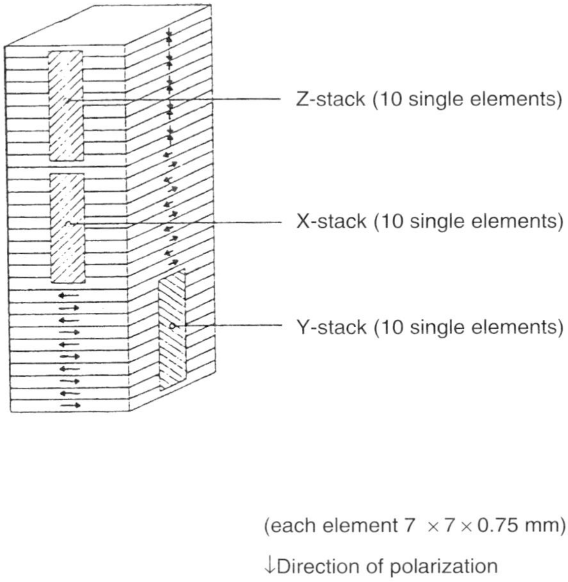

Figure 3
Three-dimensional controllable multilayer piezoelectric actuator.

3. *Actuator Materials*

Actuator materials are classified into three categories: piezoelectric, electrostrictive, and phase-change materials. Modified lead zirconate titanate (PZT, $Pb(Zr,Ti)O_3$) ceramics are the leading materials for piezoelectric applications. The PLZT ($(Pb,La)(Zr,Ti)O_3$) 7/62/38 compound is one such composition (Furuta and Uchino 1986). The strain curve is shown in Fig. 4(a) (left). When the applied field is small, the induced strain, x, is nearly proportional to the field, E ($x = dE$, where d is called the piezoelectric constant). As the field becomes larger (i.e., greater than about $1\,kVcm^{-1}$), however, the strain curve deviates from this linear trend and significant hysteresis is exhibited owing to polarization reorientation. This sometimes limits the usage of such materials in actuator applications that require nonhysteretic response.

An interesting new family of actuators has been fabricated in Germany from a barium stannate titanate system ($Ba(Sn,Ti)O_3$) (von Cieminski and Beige 1991). The useful property of $Ba(Sn_{0.15}Ti_{0.85})O_3$ is its unusual strain curve, in which the domain reorientation occurs only at low fields, and there is then a long linear range at higher fields (Fig. 4(a) (right)); i.e., the coercive field is unusually small. Moreover, this system is particularly intriguing since it contains no lead ions, an essential feature as ecological concerns grow in the future.

The second category of actuators is based on electrostriction as in PMN ($Pb(Mg_{1/3}Nb_{2/3})O_3$)-based ceramics, developed in the USA. Although a second-order phenomenon of electromechanical coupling ($x = ME^2$, where M is the electrostrictive constant), it can be extraordinarily large (more than 0.1%) (Cross *et al.* 1980). An attractive feature of these materials is the near absence of hysteresis (Fig. 4(b)). The superiority of PMN over PZT has been demonstrated in a scanning tunneling microscope (STM) (Uchino 1988). The PMN actuator can provide extremely small distortion of the image even when the probe is scanned in the opposite direction.

The third category is based on phase-change-related strains, i.e. polarization induced by switching from an antiferroelectric to a ferroelectric state (Uchino and Nomura 1983). Figure 4(c) shows the field-induced strain curves taken for the lead zirconate stannate ($Pb_{0.99}Nb_{0.02}((Zr_xSn_{1-x})_{1-y}Ti_y)_{0.98}O_3$)-based system. The longitudinally induced strain reaches more than 0.3%, which is much larger than that expected in normal piezostrictors or electrostrictors. A rectangular hysteresis in Fig. 4(c) (left), referred to as a "digital displacement transducer" because of the two on/off strain states, is interesting. Moreover, this field-induced transition exhibits a shape memory effect in

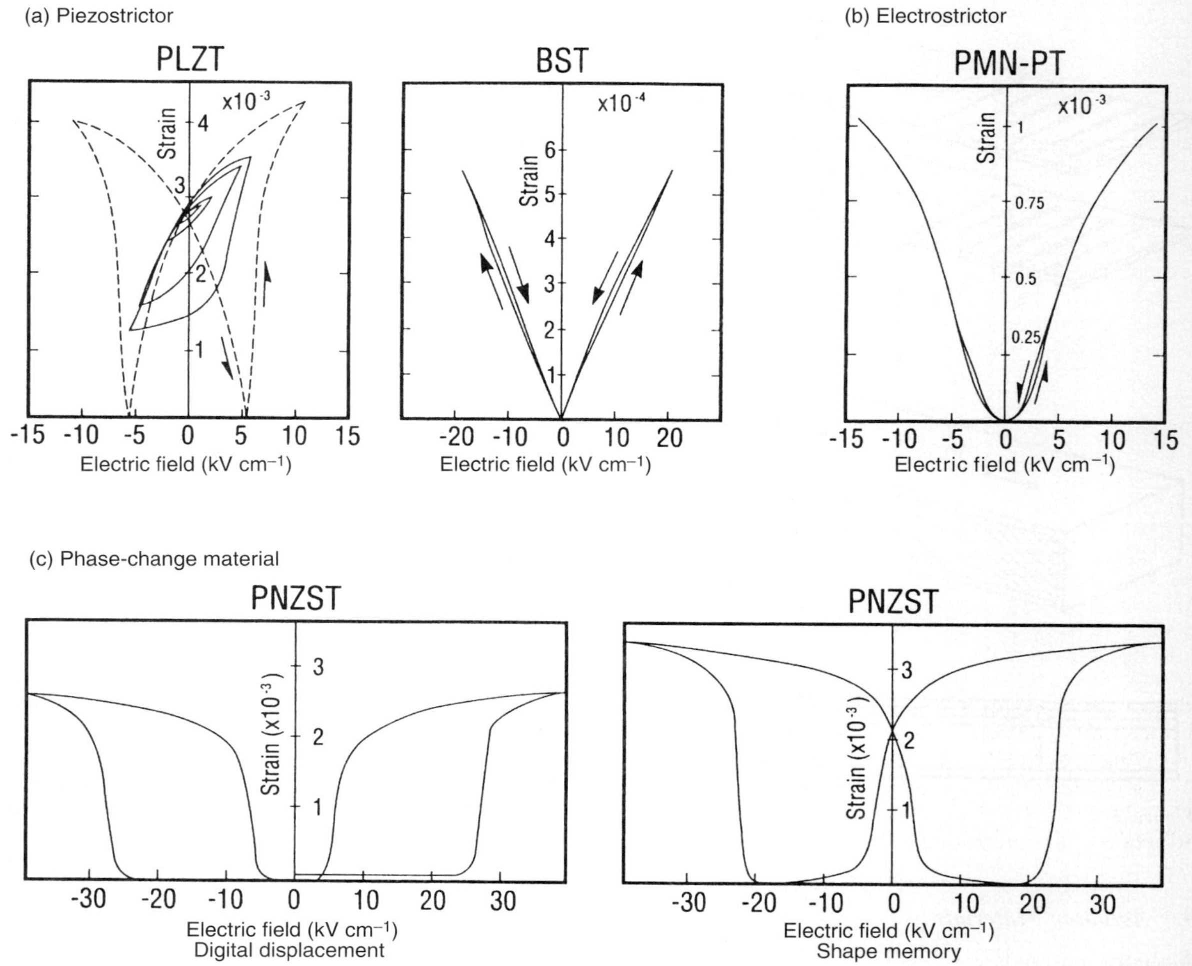

Figure 4
Electric field-induced strains in ceramics: (a) piezoelectric $(Pb,La)(Zr,Ti)O_3$ and $Ba(Sn,Ti)O_3$; (b) electrostrictive $Pb(Mg_{1/3}Nb_{2/3},Ti)O_3$; (c) phase-change material $Pb(Zr,Sn,Ti)O_3$.

appropriate compositions (Fig. 4(c) (right)). Once the ferroelectric phase has been induced, the material "memorizes" its ferroelectric state even under zero-field conditions, although it can be erased with the application of a small reverse bias field (Furuta *et al.* 1992). This shape memory ceramic is used in energy-saving actuators.

4. *Reliability of Multilayer Actuators*

As application fields expand, the reliability and durability issues of multilayer actuators become increasingly important. The reliability of ceramic actuators depends on a number of complex factors, which can be divided into three major categories: reliability of the ceramic itself, reliability of the device design, and the drive technique.

Compositional changes of actuator ceramics and the effect of doping are primary issues used in stabilizing the temperature and stress dependence of the induced strains. A multilayer piezoactuator for use at high temperatures (150°C) has been developed by Hitachi Metal, using Sb_2O_3-doped $(Pb,Sr)(Zr,Ti)O_3$ ceramics (Watanabe *et al.* 1993). Systematic data on uniaxial stress dependence of piezoelectric characteristics have been collected on various navy PZT materials (Zhang *et al.* 1996). Grain size and porosity control of the ceramics are also important in controlling the reproducibility of actuators (Okada *et al.* 1993). Aging phenomena, especially the degradation of strain response, are, in general, strongly dependent on the applied electric field as well as on temperature, humidity, and mechanical bias stress (Sakai *et al.* 1992).

The device design strongly affects its durability and lifetime. Silver electrode metal tends to migrate into the piezoceramic under a high electric field in high humidity. Silver–palladium alloys suppress this behavior effectively. Resistive coatings of the device should also be taken into account (Takada 1992). To overcome electrode delamination, improved adhesion can be realized by using a mesh-type electrode or an electrode material with mixed metal and ceramic (the matrix composition) powders. Pure ceramic electrode materials have also been developed using semiconductive perovskite oxides (barium titanate-based PTCR ceramics) (Abe *et al.* 1986). The lifetime characteristics of a multilayer actuator with applied d.c. or unipolar a.c. voltage at various temperatures (Nagata and Kinoshita 1994, 1995b) and at various humidities (Nagata and Kinoshita 1995a) have been investigated. The relationship between the logarithm of the lifetime and the reciprocal of absolute temperature shows linear characteristics similar to Arrhenius-type ones. Nevertheless, the degradation mechanism remains a critical problem.

In multilayer actuators, reduction of the tensile stress concentration around the internal electrode edge of the conventional interdigital configuration is the central problem. Regarding the destruction mechanism of multilayer ceramic actuators, systematic data collection and analysis have led to considerable progress (Furuta and Uchino 1993, 1994, Aburatani *et al.* 1994, Wang and Singh 1995, Cao and Evans 1994, Schneider *et al.* 1994, Suo 1993, Hao *et al.* 1996). Two typical crack patterns are generated in a conventional interdigital multilayer device: one is a Y-shaped crack located on the edge of an internal electrode, and the other is a vertical crack located in a layer adjacent to the top or bottom inactive layer, connecting a pair of internal electrodes.

To overcome this crack problem, three electrode configurations have been proposed, as illustrated in Fig. 5: plate-through, interdigital and slit, and interdigital and float electrode types. The "float electrode" type is an especially promising design, which can be fabricated using almost the same process as the conventional multilayer actuator, and leads to much longer lifetimes (Aburatani *et al.* 1995). An empirical rule, "the thinner the layer, the tougher the device" (Aburatani *et al.* 1994, Furuta and Uchino 1994), is also very intriguing, and will be more theoretically investigated in the near future.

Failure detection or lifetime prediction methods are expected to increase remarkably the reliability of multilayer actuators. Acoustic emission and surface potential monitoring are promising methods (Uchino and Aburatani 1994). A modified multilayer actuator has been developed at Pennsylvania State University containing a strain gauge as an internal electrode (Sano *et al.* 1993). This internal strain gauge electrode can detect the crack initiation sensitively and monitor the field-induced strain (Aburatani and Uchino 1996).

Regarding drive techniques for ceramic actuators, pulse drive and a.c. drive require special attention; the vibration overshoot associated with a sharp-rise step/pulse voltage causes a large tensile force, leading to delamination of the multistacked structure, while long-term application of a.c. voltage generates considerable heat. A special pulse drive technique using a mechanical bias stress is required in the first case, and heat generation can be suppressed by changing the device design. An analytical approach to the heat generation mechanism in multilayer actuators has been reported, indicating the importance of larger surface area (Zheng *et al.* 1996). Heat generation in piezoelectric ceramics is mainly attributed to the P–E hysteresis loss under large electric field drive (Hirose *et al.* 1995b, Takahashi *et al.* 1995). For ultrasonic motors, antiresonance drive is preferable

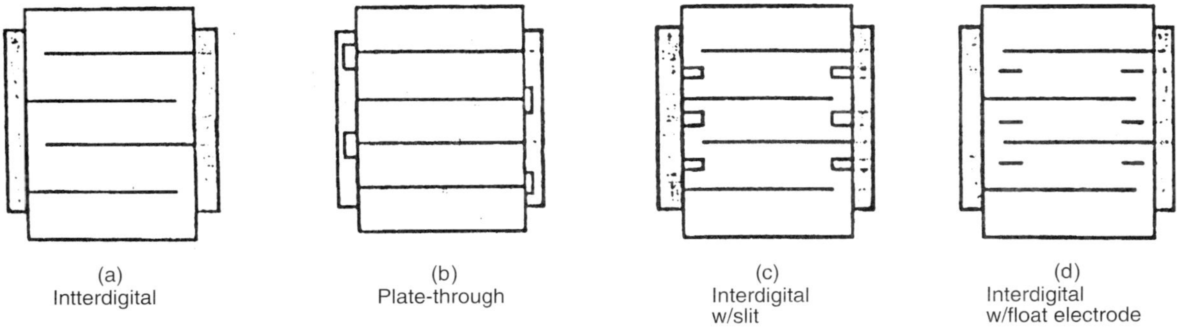

Figure 5
Various internal electrode configurations in multilayer actuators: (a) interdigital, (b) plate-through, (c) slit-insert, and (d) float electrode.

Table 1
Ceramic actuator developments in the USA, Japan, and Europe.

	USA	Japan	Europe
Target	Military-oriented products	Mass consumer products	Laboratory equipment
Category	Vibration suppressor	Minimotor, positioner	Minimotor, positioner, vibration suppressor
Application field	Space structure, military vehicles	Office equipment, cameras, precision machines, automobiles	Laboratory stage-stepper, aircraft, automobiles, hydraulic systems
Actuator size	Up-sizing (30 cm)	Down-sizing (1 cm)	Intermediate size (10 cm)
Major manufacturer	AVX/Kyocera, Morgan Matroc, Itek Opt. Systems, Burleigh, AlliedSignal	Tokin Corp., NEC, Hitachi Metal, Mitsui-Sekka, Canon, Seiko Instruments	Philips, Siemens, Hoechst Ceram Tec., Ferroperm, Physik Instrumente

to resonance drive because of higher efficiency and lower heat generation for the same vibration level (Hirose *et al.* 1995a).

5. *Applications of Multilayer Actuators*

Table 1 compares a variety of ceramic actuator developments in the USA, Japan, and Europe. Additional details are described below.

5.1 USA

The principal target is military-oriented applications such as vibration suppression in space structures and military vehicles. Substantial increase in the size of the actuators is required for these purposes.

A typical example is found in the aircraft wing proposed by NASA (Heeg 1993). A piezoelectric actuator is installed near the support of the wing, allowing immediate suppression of unwanted mechanical vibrations. Several publications have reported on damper and noise cancellation applications (Mather and Tran 1993, Agrawal and Bang 1993).

Passive dampers constitute another important application of piezoelectrics, where mechanical noise vibration is radically suppressed by the converted electric energy dissipation through Joule heat when a suitable resistance, equal to an impedance of the piezoelectric element of $1/\omega C$, is connected to the piezoelement (Uchino and Ishii 1988).

A widely publicized application took place with the repair of the Hubble telescope launched by the Space Shuttle. Multilayer PMN electrostrictive actuators corrected the image by adjusting the phase of the incident light wave (Fig. 6) (Wada 1993). PMN electrostrictors provided superior adjustment of the telescope image because of negligible strain hysteresis.

5.2 Japan

Japanese industries seek to develop mass-produced consumer products, especially minimotors and micropositioners, aiming at applications such as office equipment and cameras/video cameras. Tiny actuators smaller than 1 cm^3 are the main focus in these products.

A dot matrix printer was the first widely commercialized product using multilayer ceramic actuators (Yano *et al.* 1984). Each character formed by such a printer is composed of a 24×24 dot matrix in which a printing ribbon is impacted by a multiwire array. The printing element is composed of a multilayer piezoelectric device with a sophisticated hinge lever magnification mechanism. The magnification by a factor of 30 results in an amplified displacement of 0.5 mm and an energy transfer efficiency greater than 50%. A modified impact printer head has been developed by Tokin, using new interdigital internal electrode-type actuators, which has lowered production costs (Ono and Fuda 1994). A color ink-jet printer has also been commercialized by Seiko Epson, using multilayer piezoactuators (Yonekubo 1995).

Automotive applications by Toyota Motor have been increasing. Multilayer actuators have been introduced to an electronically controlled suspension (Fukami *et al.* 1994) and a pilot injection system for diesel engines has been developed (Abe *et al.* 1994).

Efforts have been made to develop high-power ultrasonic vibrators as replacements for conventional electromagnetic motors. The ultrasonic motor is characterized by "low speed and high torque," which is contrasted with the "high speed and low torque" of electromagnetic motors. After the invention of π-shaped linear motors (Tohda *et al.* 1989), various modifications have been reported (Funakubo *et al.* 1995, Saigoh *et al.* 1995). The Mitsui Petrochemical model is of particular interest because the motor body is composed of only one component prepared by a co-firing method, as illustrated in Fig. 7 (Saigoh

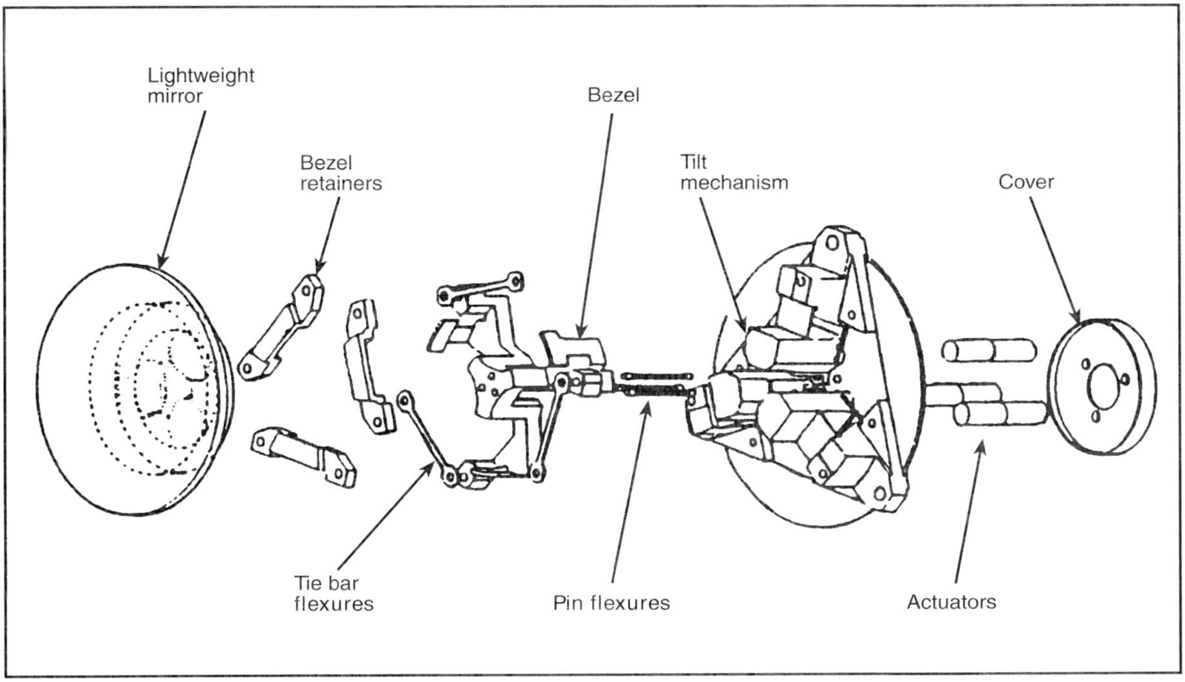

Figure 6
Articulating fold mirror using PMN multilayer actuators.

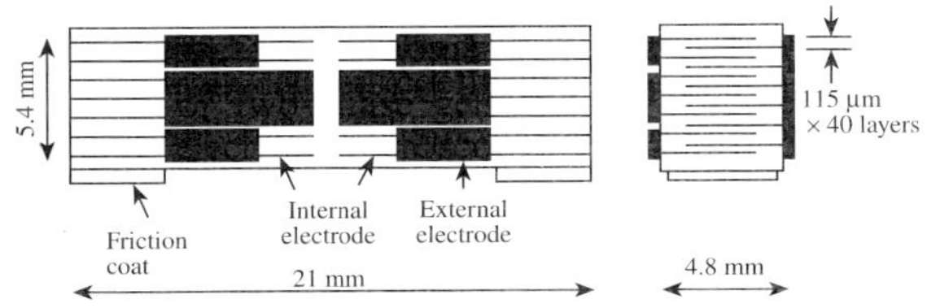

Figure 7
Inchworm using multilayer piezoelectric actuators.

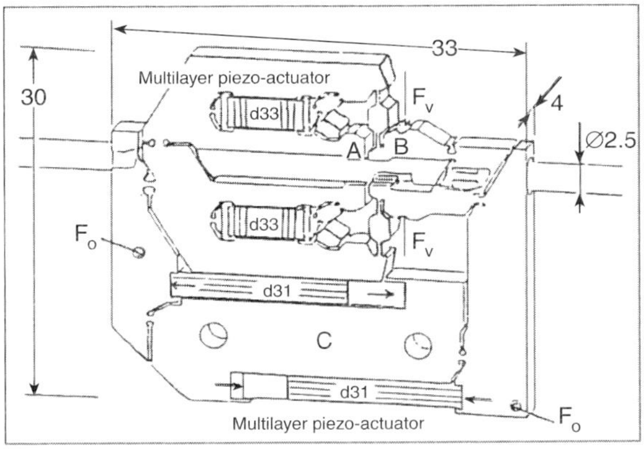

Figure 8
Monolithic multilayer piezoelectric linear motor.

et al. 1995). A maximum speed of $200\,mms^{-1}$ and a maximum thrust of 60 gf (0.6 N) has been reported for this motor.

Camera motors use a traveling elastic wave induced by a thin piezoelectric ring. A ring-type slider in contact with the "rippled" surface of the elastic body bonded onto the piezoelectric can be driven in both directions by exchanging the sine and cosine voltage inputs. Another advantage is its thin design, making it suitable for installation in cameras as an automatic focusing device. Nearly 80% of the exchange lenses in Canon's EOS camera series have been replaced by ultrasonic motor mechanisms (Uchino 1995c).

Intriguing research programs are underway in Japan on the vibration damping of earthquakes, using piezoelectric actuators (Shimoda *et al.* 1992, Fujita 1994). Active damping of a multilayer piezoactuator has been tested using an actual size H-type steel girder, and has been verified to be effective during earthquakes.

5.3 *Europe*

Ceramic actuator development has begun in Europe with a wide range of research topics. A focus of several major manufacturers is on laboratory equipment products such as laboratory stages and steppers with rather sophisticated structures.

Figure 8 shows a walking piezomotor with four multilayer actuators by Philips (Koster 1994). Two short actuators function as clampers while the longer two provide a proceeding distance in an inchworm mechanism. Physik Instrumente has developed more complicated two-leg type walkers (Gross 1994).

6. Conclusions

The early 1980s saw the start of the intensive development of piezoelectric actuators in Japan, which then spread worldwide. The focus has been shifted to practical device applications. This article has reviewed several reliability issues of multilayer ceramic actuators as well as new actuator structures, and has compared the development of applications in the USA, Japan, and Europe.

The markets in the USA are chiefly limited to military and defense applications, and it is difficult to estimate the amount of sales. Navy needs include smart submarine skins, hydrophone actuators, and propeller noise cancellation. Smart aircraft skins are an air force objective, while the army requires helicopter rotor twisting, aeroservoelastic control, and cabin noise/seat vibration cancellation.

In Japan piezoelectric camera shutters (Minolta) and automatic focusing mechanisms in cameras (Canon), dot matrix printers (NEC), and part-feeders (Sanki) are now commercialized and mass-produced by tens of thousands of pieces per month. During commercialization, new designs and drive-control techniques of ceramic actuators have been developed. A number of patent disclosures have been generated by NEC, TOTO Corporation, Matsushita Electric, Brother Industry, Toyota Motors, Tokin, Hitachi Metal, and Toshiba.

Annual sales in 2005 (expecting the economic recovery in Japan) of ceramic actuator units, camera-related devices, and ultrasonic motors are estimated to reach \$500 million, \$300 million, and \$150 million, respectively. Regarding the final actuator-related products, \$10 billion is a realistic goal (Uchino 1995c).

Future research trends can be divided in two ways: increasing the size of space structures and decreasing that of office equipment. Further decrease in size will also be required in medical diagnostic applications such as blood test kits and surgical catheters.

Key concerns for the future of multilayer ceramic actuators are "miniaturization" and "hybridization." Layers thinner than 10 µm will be introduced in actuator devices replacing 100 µm sheets. Piezoelectric thin films compatible with silicon technology are a focus in microelectromechanical systems. Ultrasonic rotary motors as small as 2 mm in diameter (Flynn *et al.* 1992) and a two-dimensional micro-optical scanner (Goto 1994), both of which are fabricated on a silicon membrane, are good examples.

Nonuniform configurations or heterostructures of different materials, layer thickness, or electrode patterns will be adopted for practical devices. Functionally gradient piezoelectric actuators now being prototyped indicate a new trend (Kim *et al.* 1992).

Bibliography

Abe K, Uchino K, Nomura S 1986 Barium titanate-based actuator with ceramic internal electrodes. *Ferroelectrics* **68**, 215–23

Abe S, Igashira T, Sakakibara Y, Kobayashi F 1994 Development of pilot injection system equipped with piezoelectric actuator for diesel engine. *J. Soc. Automot. Eng. Jpn. Rev.* **15**, 201–8

Aburatani H, Harada S, Uchino K, Furuta A, Fuda Y 1994 Destruction mechanisms in ceramic multilayer actuators. *Jpn. J. Appl. Phys.* **33**, 3091–4

Aburatani H, Uchino K 1996 Stress and fatigue estimation in multilayer ceramic actuators using an internal strain gauge. *Int. Symp. Solid-State Sensors and Actuators*, 191SXIX-37-96. American Ceramic Society

Aburatani H, Uchino K, Furuta A, Fuda Y 1995 Destruction mechanism and destruction detection technique for multilayer ceramic actuators In: *Proc. 9th Int. Symp. Applied Ferroelectrics*, pp. 750–2

Agrawal B N, Bang H 1993 Active vibration control of flexible space structures by using piezoelectric sensors and actuators. *Am. Soc. Mech. Eng. Des. Eng. Div.* **61**, 169–79

Bauer A, Moller F 1995 Piezo actuator special design In: *Proc. 4th Int. Conf. New Actuators* (Actuator '94). Germany, pp. 128–32

Cao H C, Evans A G 1994 Electric-field-induced fatigue crack growth in piezoelectrics. *J. Am. Ceram. Soc.* **77**, 1783–6

von Cieminski J, Beige H 1991 High-signal electrostriction in ferroelectric materials. *J. Phys. D* **24**, 1182–6

Cross L E, Jang S J, Newnham R E, Nomura S, Uchino K 1980 Large electrostrictive effects in relaxor ferroelectrics. *Ferroelectrics* **23** (3), 187

Dibbern U 1995 Piezoelectric actuators in multilayer technique In: *Proc. 4th Int. Conf. New Actuators* (Actuator '94). Germany, pp. 114–8

Dogan A, Fernandez J F, Uchino K, Newnham R E 1994 New piezoelectric composite actuator designs for displacement amplification. *Proc. 4th Euro Ceramics* **5**, 127–32

Flynn A M, Tavrow L S, Bart S F, Brooks R A, Ehrlich D J, Udayakumar K R, Cross L E 1992 Piezoelectric micromotors for microrobots. *IEEE J. Microelectromech. Syst.* **1**, 44–51

Fujita T 1994 Can piezoelectric actuators control vibrations of buildings? *New Ceram.* **7**(12), 52–5

Fukami A, Yano M, Tokuda H, Ohki M, Kizu R 1994 Development of piezo-electric actuators and sensors for electronically controlled suspension. *Int. J. Veh. Des.* **15**, 348–57

Funakubo T, Tsubata T, Taniguchi Y, Kumei K, Fujumura T, Abe C 1995 Ultrasonic linear motor using multilayer piezoelectric actuators. *Jpn. J. Appl. Phys.* **34**, 2756–9

Furuta A, Oh K Y, Uchino K 1992 Shape memory ceramics and their application to latching relays. *Sensors Mater.* **3** (4), 205–15

Furuta K, Uchino K 1986 Electric-field-induced strain in (Pb,La)(Zr,Ti)O_3 ceramics. *Adv. Ceram. Mater.* **1**, 61–3

Furuta A, Uchino K 1993 Dynamic observation of crack propagation in piezoelectric multilayer actuators. *J. Am. Ceram. Soc.* **76**, 1615–7

Furuta A, Uchino K 1994 Destruction mechanism of multilayer ceramic actuators: case of antiferroelectrics. *Ferroelectrics* **160**, 277–85

Goto H 1994 Miniature 2D optical scanner and its application to an optical sensor. *J. Opt. Technol. Contact* **32**, 322–30

Goto H, Imanaka K, Uchino K 1992 Piezoelectric actuators for light beam scanners. *Ultrasonic Technol.* **5**, 48–51

Gross R 1994 A high resolution piezo walk drive In: *Proc. 4th Int. Conf. New Actuators.* Germany, pp. 190–2

Hao T H, Gong X, Suo Z 1996 Fracture mechanics for the design of ceramic multilayer actuators. *J. Mech. Phys. Solids*

Heeg J 1993 Analytical and experimental investigation of flutter suppression by piezoelectric actuation. NASA Technical Paper, NASA-TP-3241, p. 47

Hirose S, Aoyagi M, Tomikawa Y, Takahashi S, Uchino K 1995a High-power characteristics at antiresonance frequency of piezoelectric transducers. *Proc. Ultrasonic Int.* 1–4

Hirose S, Takahashi S, Uchino K, Aoyagi M, Tomikawa Y 1995b Measuring methods for high-power characteristics of piezoelectric materials. *Proc. Mater. Res. Soc.* **360**, 15–20

Kim H S, Choi S C, Lee J K, Jung H J 1992 Fabrication and piezoelectric strain characteristics of PLZT functionally gradient piezoelectric actuator by doctor blade process. *J. Korean Ceram. Soc.* **29**, 695–704

Koster M P 1994 A walking piezo motor In: *Proc. 4th Int. Conf. New Actuators.* Germany. pp. 144–8

Mather G P, Tran B N 1993 Aircraft cabin noise reduction tests using active structural acoustic control. American Institute of Aeronautics and Astronautics paper AIAA-93-4437, p. 7

Nagata K, Kinoshita S 1994 Life time of multilayer ceramic actuator at high temperature. *J. Powder Metall.* **41**, 975–9

Nagata K, Kinoshita S 1995a Effect of humidity on life time of multilayer ceramic actuator. *J. Powder Metall.* **42**, 623–7

Nagata K, Kinoshita S 1995b Relationship between lifetime of multilayer ceramic actuator and temperature. *Jpn. J. Appl. Phys.* **34**, 5266–9

Ohashi J, Fuda Y, Ohno T 1993a Multilayer piezoelectric ceramic actuator with interdigital internal electrodes. *Jpn. J. Appl. Phys.* **32**, 2412–4

Ohashi J, Fuda Y, Ohno T 1993b Multilayer piezoelectric ceramic actuator with interdigital internal electrodes. *Tokin Tech. Rev.* **19**, 55–60

Okada N, Ishikawa K, Murakami K, Nomura T, Ogino M 1993 Improving hysteresis of piezoelectric PZT actuator. *Bull. Shizuoka Univ. Electron. Sci. Grad. School* **14**, 7–13

Onitsuka K, Dogan A, Tressler J F, Xu Q C, Yoshikawa S, Newnham R E 1995 Metal–ceramic composite transducer, the "moonie." *J. Intelligent Mater. Syst. Struct.* **6**, 447–55

Ono Y, Fuda Y 1994 Multilayer piezoelectric ceramic actuator with interdigital internal electrodes for impact dot-matrix printers In: *Proc. 6th Symp. Electro-Magnetic Related Dynamics*. pp. 135–8

Saigoh H, Kawasaki M, Maruko N, Kanayama K 1995 Multilayer piezoelectric motor using the first longitudinal and the second bending vibrations. *Jpn. J. Appl. Phys.* **34**, 2760–4

Sakai T, Ishikiriyama M, Terai Y, Shimazaki R 1992 Improvement in the durability of PZT ceramics for an actuator. *Toyota Tech. Rev.* **42**, 52–9

Sano M, Ohya K, Hamada K, Inoue Y, Kajino Y 1993 Multilayer piezoelectric actuator with strain gauge. *NEC Giho* **46** (10), 100–3

Schneider G A, Rostek A, Zickgraf B, Aldinger F 1994 Crack growth in ferroelectric ceramics under mechanical and electrical loading In: *Proc. 4th Int. Conf. Electronic Ceramics and Applications*. Germany, pp. 1211–6

Shimoda H, Ohmata J, Okamoto F 1992 Seismic response control of a piping system by a semiactive damper with piezoelectric actuators. *J. Precision Eng. Jpn.* **58**, 2111–7

Sugawara Y, Onitsuka K, Yoshikawa S, Xu Q C, Newnham R E, Uchino K 1992 Metal–ceramic composite actuators. *J. Am. Ceram. Soc.* **75**, 996–8

Suo Z 1993 Models for breakdown resistant dielectric and ferroelectric ceramics. *J. Mech. Phys. Solids* **41**, 1155–76

Takada S 1992 Quality improvement of multilayer piezoelectric actuator. *NEC Giho* **45**, 109–13

Takada S, Inoue Y, Oya K, Inagawa M 1994 100 V DC driving type multilayer piezoelectric actuator. *NEC Giho* **47**, 98–102

Takahashi S, Sakaki Y, Hirose S, Uchino K 1995 Stability of $PbZrO_3$–$PbTiO_3$–$Pb(Mn_{1/3}Sb_{2/3})O_3$ piezoelectric ceramics under vibration-level change. *Jpn. J. Appl. Phys.* **34**, 5328–31

Tohda M, Ichikawa S, Uchino K, Kato K 1989 Ultrasonic linear motor using a multilayered piezoelectric actuator. *Ferroelectrics* **93**, 287–94

Tressler J F, Xu Q C, Yoshikawa S, Uchino K, Newnham R E 1994 Composite flextensional transducers for sensing and actuating. *Ferroelectrics* **156**, 67–72

Uchino K 1988 *Ceramic Data Book '88*. Institute of Industrial Manufacturing Technology, Tokyo

Uchino K 1993 Ceramic actuators: principles and applications. *MRS Bull.* **18**, 42–8

Uchino K 1994 New piezoelectric devices for smart actuator/sensor systems In: *Proc. 4th Int. Conf. Electronic Ceramics and Applications*, pp. 179–91

Uchino K 1995a Manufacturing technology of multilayered transducers. *Proc. Am. Ceram. Soc.* 81–93

Uchino K 1995b Materials update: advances in ceramic actuator materials. *Mater. Lett.* **22**, 1–4

Uchino K 1995c Piezoelectric actuators/ultrasonic motors—their developments and markets In: *Proc. 9th Int. Symp. Applied Ferroelectrics*, pp. 319–24

Uchino K 1996 *Piezoelectric Actuators/Ultrasonic Motors*. Kluwer, MA

Uchino K, Aburatani H 1994 Destruction detection techniques for safety piezoelectric actuator systems In: *Proc. 2nd Int. Conf. Intelligent Materials*. pp. 1248–56

Uchino K, Ishii T 1988 Mechanical damper using piezoelectric ceramics. *J. Jpn. Ceram. Soc* **96**, 863–7

Uchino K, Nomura S 1983 Electrostriction in PZT-family antiferroelectrics. *Ferroelectrics* **50** (1), 191–6

Wada B 1993 Summary of precision actuators for space application. JPL Document D-10659

Wang H, Singh R N 1995 Electric field effects on the crack propagation in a electro-strictive PMN-PT ceramic. *Ferroelectrics* **168**, 281–91

Watanabe J, Sometsugu T, Watanabe Y, Johmura S, Kurihara K, Kazama Y 1993 Multilayer piezoelectric ceramic actuator for high temperature use. *Hitachi Metal Giho* **9**, 59–64

Yano T, Fukui I, Sato E, Inui O, Miyazaki Y 1984 Impact dot-matrix printer head using multilayer piezoelectric actuators. *Proc. Electron. Commun. Soc.* **1**, 156

Yonekubo S 1995 Color inkjet printer by multi-layer piezoelectric actuator. *J. Electron. Photo. Soc* **34**, 226–8

Yoshikawa S, Shrout T 1993 Multilayer piezoelectric actuators—structures and reliability. *AIAA/ASME/ASCE/AHS/ASC Struct. Struct. Dyn. Mater. Conf.* **34**, 3581–6

Zhang Q M, Zhao J Z, Uchino K, Zheng J H 1996 Change of the weak field properties of $Pb(Zr,Ti)O_3$ piezoceramics with compressive uniaxial stresses and its links to the effect of dopants on the stability of the polarizations in the materials. *J. Mater. Res.* **12**, 226–34

Zheng J H, Takahashi S, Yoshikawa S, Uchino K 1996 Heat generation in multilayer piezoelectric actuators. *J. Am. Ceram. Soc.* **79**, 3193–8

N

Nanoelectromechanical Systems: Experiments and Modeling

H. D. Espinosa, C. Ke, and N. Pugno

1. *Introduction*

Nanoelectromechanical systems (NEMSs) are systems with characteristic dimensions of a few nanometers. By exploiting nanoscale effects, NEMSs present interesting and unique characteristics, which deviate greatly from their predecessor microelectromechanical systems (MEMSs). For instance, NEMS-based devices can have fundamental frequencies in the microwave range ($\sim 100\,\mathrm{GH}$) (Rueckes *et al.* 2000); mechanical quality factors in the tens of thousands (ultralow energy dissipation); active mass in the femtogram range; force sensitivity at the attonewton level; mass sensitivity up to attogram (Ilic *et al.* 2000) and subattogram (Davis *et al.* 2000) levels; heat capacities far below a "yoctocalorie" (Roukes 1999); power consumption in the order of 10 aw (Roukes 2004); and extreme high integration level, approaching 10^{12} elements per cm^2 (Rueckes *et al.* 2000). All these distinguishing properties of NEMS devices pave the way to applications such as force sensors, chemical sensors, biological sensors, and ultrahigh frequency resonators.

The interesting properties of the NEMS devices typically arise from the behavior of the active parts, which, in most cases, are in the forms of cantilevers or doubly clamped beams with dimensions at the nanometer scale. The materials for those active components include silicon, silicon carbide, carbon nanotubes, gold and platinum, to name a few. Silicon, the basic material employed in integrated circuit (IC) technology and MEMS, has been widely used to build NEMS. However, ultrasmall silicon-based NEMS nanoresonators failed to achieve the much anticipated high-quality factors due to the dominance of surface effects, such as surface oxidation and reconstruction, and thermoelastic damping. Limitations in strength and flexibility also compromise the performance of silicon-based NEMS actuators. Instead, carbon nanotubes appear to be ideal for NEMS given their nearly one-dimensional structures with high aspect ratio and nearly perfect-terminated surfaces and excellent electrical and mechanical properties. Due to significant advances in growth, manipulation, knowledge of electrical and mechanical properties, carbon nanotubes have become the most promising building blocks for the next generation of NEMS. Several carbon-nanotube-based functional NEMS devices have been reported so far (Kim and Lieber 1999, Rueckes *et al.* 2000, Akita *et al.* 2001, Fennimore *et al.* 2003, Kinaret *et al.* 2003, Ke and Espinosa 2004, Sazonova *et al.* 2004). Similarly to carbon nanotubes, nanowires are a type of one-dimensional novel nanostructure well suited for building NEMS because of their size and controllable electrical properties.

The purpose of this article is to provide a review of NEMS devices to date and to summarize the modeling currently being pursued to gain insight into their performance. This article is organized as follows: in the first part, we review carbon-nanotube- and -nanowire-based NEMS. In the second part, we present the multiphysics modeling of NEMS based on continuum theory.

2. *NEMS*

2.1 *Carbon-nanotube-based NEMS Devices*

(*a*) *Nonvolatile random access memory* (*NRAM*). A carbon-nanotube-based nonvolatile random access memory (NRAM) reported by Rueckes *et al.* (2000) is illustrated in Fig. 1. The device is a suspended SWNT crossbar array for both I/O and switchable, bistable device elements with well-defined OFF and ON states. Qualitatively, bistability can be envisioned as arising from the interplay of the elastic energy and the van der Waals energy when the upper nanotube is freestanding or the suspended SWNT is deflected and in contact with the lower nanotube. For a device element, these two states can be read easily by measuring the resistance of the junction. Moreover, it can be switched between OFF and ON states by applying voltage pulses to produce attractive or repulsive electrostatic forces. A key aspect of this device is that the separation between top and bottom conductors must be of the order of 10 nm. In such case, the van der Waals energy overcomes the elastic energy when the junction is actuated (ON state) and remains on this state even if the electrical field is turned off (nonvolatile feature). With multiplexing, this device is envisioned to function as a highly integrated ($\sim 1 \times 10^{12}$ elements per square centimeter), fast ($\sim 200\,\mathrm{GHz}$ operation frequency), and macroscopically addressable nonvolatile random access memory (RAM) structure that could overcome the fundamental limitations of semiconductor RAM in size, speed, and cost. However, small junction gap sizes impose significant challenges in the nanofabrication of parallel device arrays.

(*b*) *Nanotweezers.* There are two types of carbon-nanotube-based nanotweezers reported by Kim and Lieber (1999) and Akita *et al.* (2001), respectively. Both nanotweezers employ multiwalled carbon nanotubes (MWNTs) as tweezers' arms which are actuated by electrostatic forces. The applications of these nanotweezers include the manipulation of nanostructures and two-tip STM or AFM probes (Kim and Lieber 1999). Figures 2(a)–(d) show the motion of the nanotube arms as a function of the applied voltage *V* (Akita *et al.* 2001). In this

Figure 1
Schematics of free-standing nanotube device architecture with multiplex addressing. Reprinted with permission from Rueckes T, *et al.* 2000 Carbon nanotube-based nonvolatile random access memory for molecular computing. *Science* **289**, 94–7. Copyright 2000 AAAS.

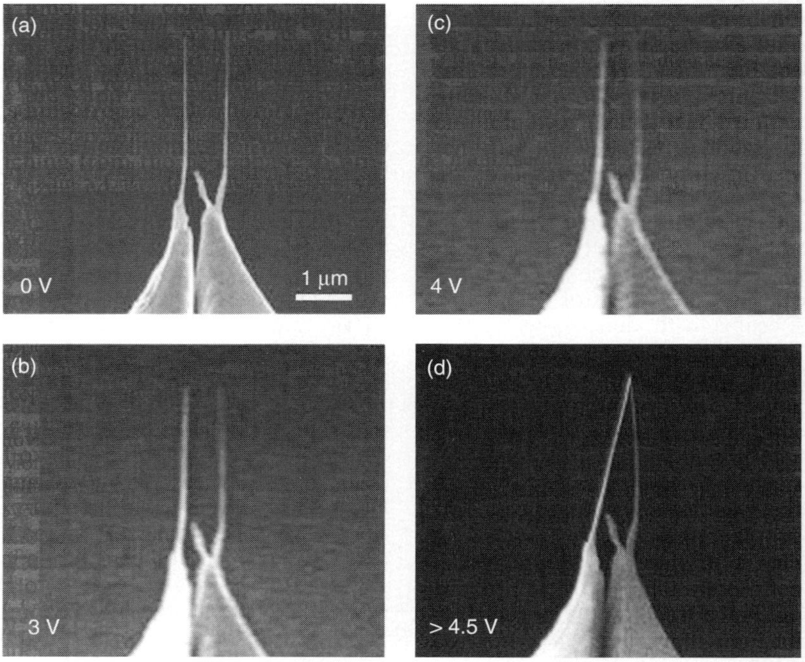

Figure 2
SEM images of the motion of nanotube arms in a pair of nanotweezers as a function of the applied voltage. Reprinted with permission from Akita S, *et al.* 2001 Nanotweezers consisting of carbon nanotubes operating in an atomic force microscope. *Appl. Phys. Lett.* **79**, 1691–3. Copyright 2001, American Institute of Physics.

sequence of images, it is clearly seen that the arms bent and the separation between the tips decreased with increasing applied voltage. The separation became 500 nm at $V = 4$ V and zero at $V > 4.5$ V. The authors noted that the motion shown in Figs. 2(a)–(d) could be repeated many times without any permanent deformation, showing that carbon nanotubes are ideal materials for building NEMSs.

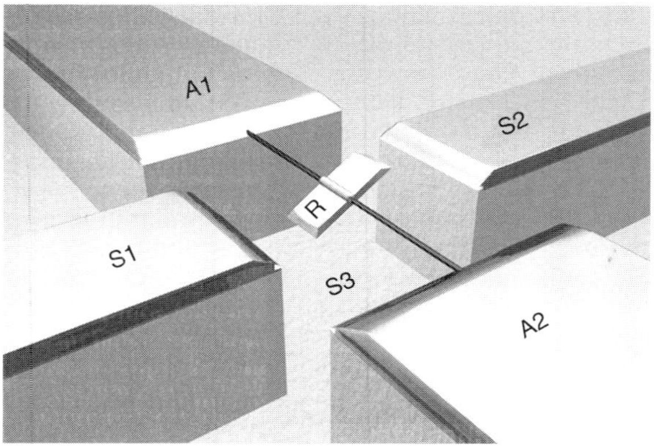

Figure 3
Conceptual drawing of the integrated synthetic NEMS actuator. Reproduced from Fennimore A M, *et al.* 2003 Rotational actuators based on carbon nanotubes. *Nature* **424**, 408–10, with permission from Nature Publishing Group.

(*c*) *Rotational motors.* A carbon-nanotube-based rotational motor reported by Fennimore *et al.* (2003) is conceptually illustrated in Fig. 3. The rotational element (R), a solid rectangular metal plate serving as a rotor, is attached transversely to a suspended support shaft which ends are embedded in electrically conducting anchors (A1, A2) that rest on the oxidized surface of a silicon chip. The rotor plate assembly is surrounded by three fixed stator electrodes (S1, S2, S3). Four independent (DC and/or appropriately phased AC) voltage signals, one to the rotor plate and three to the stators, are applied to control the position, speed, and direction of rotation of the rotor plate. The key component in the assembly is a single MWNT, which serves simultaneously as the rotor plate support shaft and the electrical feed-through to the rotor plate; most importantly, it is also the source of rotational freedom. The experiments show that the MWNT clearly serves as a reliable, presumably wear-free, NEMS element providing rotational freedom. No apparent wear or degradation in performance was observed after many thousands of cycles of rotations. The potential applications of the actuators include ultra-high-density optical sweeping and switching elements.

(*d*) *Tunable oscillators.* A tunable carbon nanotube oscillator was reported by Sazonova *et al.* (2004). It consists of a doubly clamped nanotube, as shown in Fig. 4. Single- or few-walled nanotubes with diameters in the range of 1–4 nm, grown by chemical

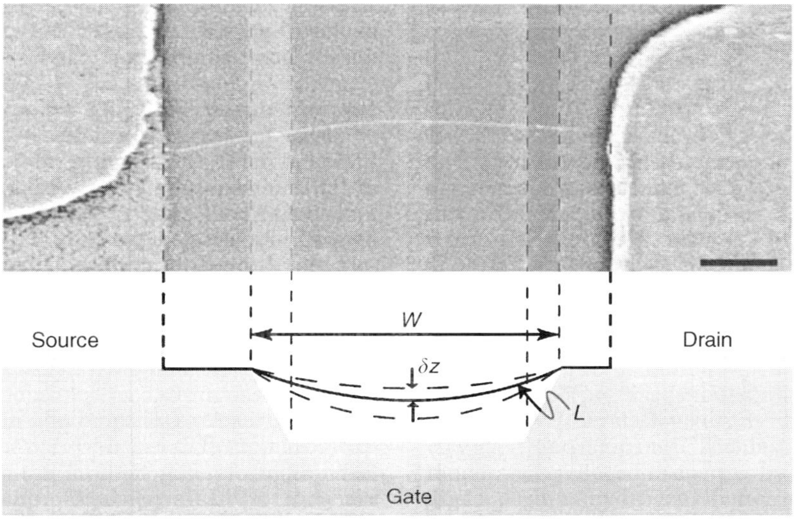

Figure 4
SEM image of a suspended device (top) and a schematic of device geometry (bottom). Scale bar, 300 nm. Reproduced from Sazonova V, *et al.* 2004 A tunable carbon nanotube electromechanical oscillator. *Nature* **431**, 284–7, with permission from Nature Publishing Group.

vapor deposition (CVD) were suspended over a trench between two metal electrodes. The nanotube motion was induced and detected using the electrostatic interaction with the gate electrode underneath the tube. In this device, the gate voltage has both a static (DC) component and a small time-varying (AC) component. The DC voltage at the gate produces a static force on the nanotube that can be used to control its tension, therefore the oscillation frequency. The AC voltage produces a periodic electric force, which sets the nanotube into motion. As the driving frequency approaches the resonance frequency of the tube, the displacement becomes large. The device showed a high force sensitivity (below 5 aN), which made it a small force transducer.

(*e*) *Nanorelays.* Carbon-nanotube-based nanorelays were firstly reported by Kinaret *et al.* (2003) and later experimentally demonstrated by Lee *et al.* (2004). The nanorelay is a three-terminal device including a conducting carbon nanotube placed on a terrace in a silicon substrate and connected to a fixed source electrode (S), as shown in Fig. 5. A gate electrode (G) is positioned underneath the nanotube so that charge can be induced in the nanotube by applying a gate voltage. The resulting electrostatic force brings the tube end into contact with a drain electrode (D) on the lower terrace, thereby closing an electric circuit. Theoretical modeling of the device shows that there is a sharp transition from a nonconducting (OFF) to a conducting (ON) state when the gate voltage is varied at a fixed source-drain voltage. The sharp switching curve allows for amplification of weak signals superimposed on the gate voltage (Kinaret *et al.* 2003).

(*f*) *Feedback-controlled nanocantilevers.* A feedback-controlled carbon-nanotube-based NEMS device reported by Ke and Espinosa (2004), schematically shown in Fig. 6, is made of an MWNT placed as a cantilever over a microfabricated step. The device has two well-defined stable equilibrium positions based on the interaction of the elastic energy, van der Waals energy and the electrostatic energy, as well as the feedback control mechanism through the resistor R in the device circuit. The representative characteristic curve of the device is shown in Fig. 7. At the "upper" equilibrium position, the electrostatic force is balanced by the elastic force from the deflection of the nanotube. There is no current in the circuit—"OFF" state. With the increase in applied voltage, the nanotube cantilever is pulled-in inducing tunneling contact with the bottom electrode. The "lower" equilibrium position is reached with a tunneling gap between the tip and electrode which is determined by the voltage applied on the gap adjusted by the feedback resistor R. There is stable current in the device—"ON" state. These two states can be switched through the pull-in and pull-out processes by controlling the applied voltage U. The potential applications of the device include ultrasonic wave detection for monitoring the health of materials and structures, gap sensing, NEMS switches, memory elements, and logic devices.

The current jump behavior at pull-in has been confirmed by means of *in situ* SEM experiments. In this experiment, a nanotube cantilever, freestanding above an electrode, was actuated by electrostatic forces (Ke *et al.* 2005a, 2005b) as shown in Fig. 8. The measured I–V behavior after the pull-in was found in very good agreement with the theoretical prediction. The parameters used in the theoretical prediction include the length of the nanotube, $L = 3.8\,\mu m$; the diameter of the nanotube, $R_{ext} = 20\,nm$; initial gap between the nanotube cantilever and the electrode, $H = 200\,nm$; resistor resistance, $R = 0.98\,G\Omega$; and contact resistance, $R_0 = 50\,\Omega$ (Ke and Espinosa 2004).

In comparison to nanorelays (Kinaret *et al.* 2003, Lee *et al.* 2004), the device reported in Ke and Espinosa (2004) is a two-terminal device, which provides more flexibility in terms of device realization and control. In comparison to the NRAM described in Rueckes *et al.* (2000), the feedback-controlled

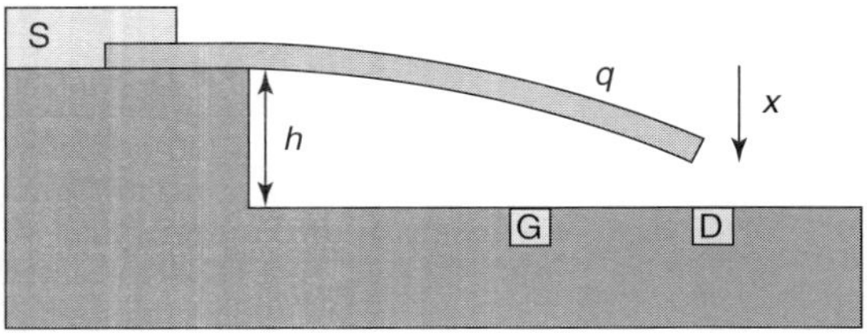

Figure 5
Schematic diagrams of a CNT nanorelay device. Reprinted with permission from Kinaret J, *et al.* 2003 A carbon nanotube based nanorelay. *Appl. Phys. Lett.* **82**, 1287–9. Copyright 2003, American Institute of Physics.

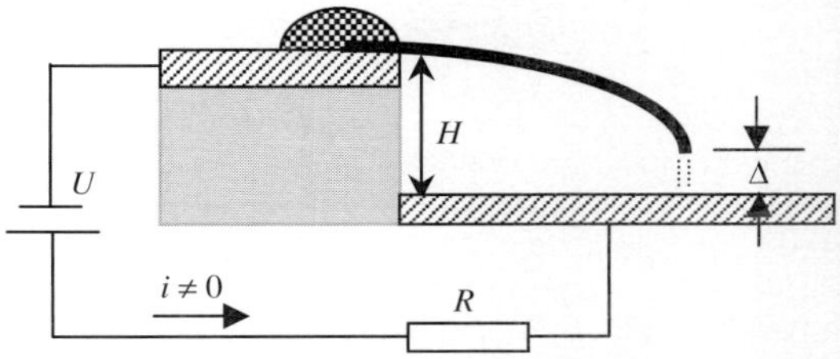

Figure 6
Schematic of nanotube-based device with tunneling contacts. Reprinted with permission from Ke C-H, Espinosa H D 2004 Feedback controlled nanocantilever device. *Appl. Phys. Lett.* **85**, 681–3. Copyright 2004, American Institute of Physics.

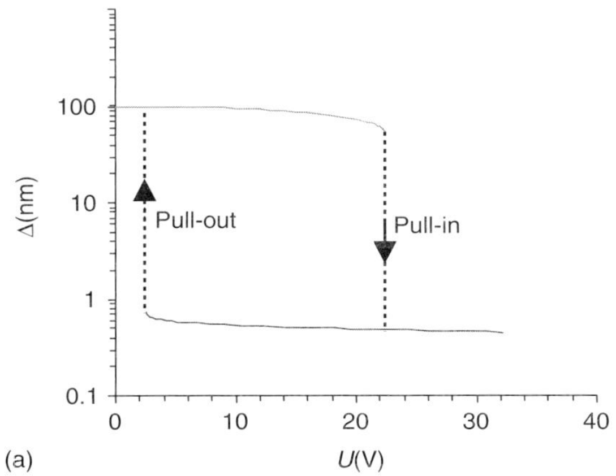

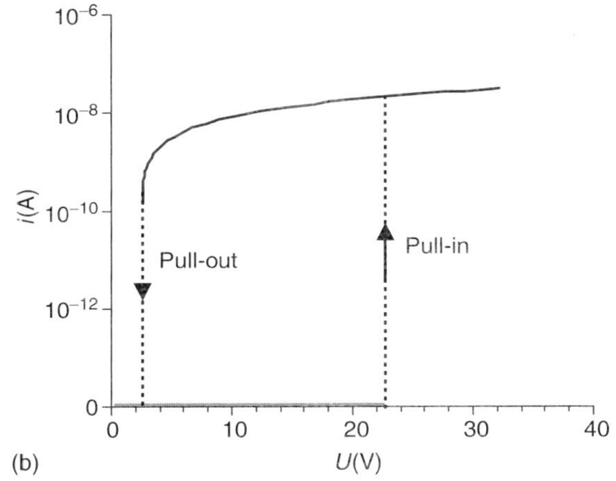

Figure 7
Representative characteristic of pull-in and pull-out processes for the feedback-controlled nanocantilever device. (a) Relationship between the gap Δ and the applied voltage U. (b) Relationship between the current i in the circuit and the applied voltage U. Reprinted with permission from Ke C-H, Espinosa H D 2004 Feedback controlled nanocantilever device. *Appl. Phys. Lett.* **85**, 681–3. Copyright 2004, American Institute of Physics.

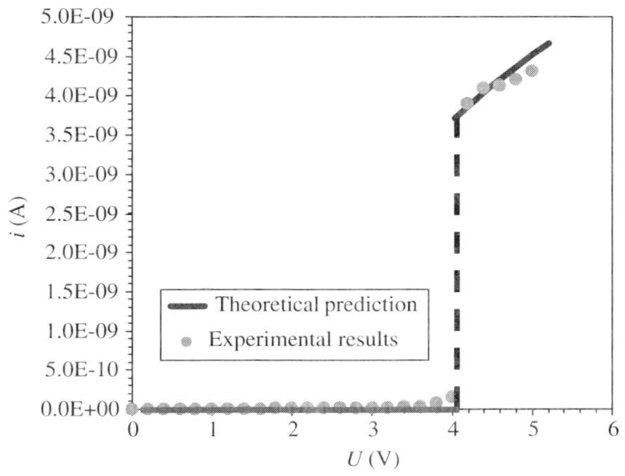

Figure 8
Comparison between theoretical prediction and I–V measurement of an electrostatically actuated free-standing nanotube cantilever with an electronic circuit incorporating a resistor.

NEMS device (Ke and Espinosa 2004) employs an electrical circuit incorporated with a resistor to adjust the electrostatic field and achieve a second stable equilibrium position. This feature reduces the constraints in fabricating devices with nanometer gap control between the freestanding CNTs or NWs and the substrate, and therefore, provides more reliability and tolerance to variability in fabrication parameters. However, the drawback of the feedback-controlled device, in memory applications, is that the memory becomes volatile. Hence, the working principle and the potential applications for these two devices are somewhat complementary.

2.2 Nanowire-based NEMS Devices

(*a*) *Resonators.* Figure 9(a) shows a suspended platinum nanowire resonator, reported by Husain *et al.* (2003), and Fig. 9(b) the circuit used for magnetomotive drive and detection of its motion. The free standing platinum structure with a diameter of 43 nm and length of 1.3 μm was fabricated by random dispersion followed by e-beam lithography. A magnetomotive detection scheme (see Fig. 9(b)), in which an AC current drives a beam in a transverse magnetic field, was used to drive and read out the resonators. The measured quality factor Q was ~8500 and decreased slightly with the increase in magnetic field. It was noted that the response of the beam was in the linear regime. Badzey *et al.* (2004) reported a doubly clamped nanomechanical Si beam working in the nonlinear response region. The nonlinear response of the beam displays notable hysteresis and bistability in the amplitude–frequency space when the frequency sweeps upward and downward. This particular behavior shows that the device can be used as mechanical memory elements.

(*b*) *Nanoelectromechanical programmable read-only memories.* A germanium-nanowire-based nanoelectromechanical programmable read-only memory (NEMPROM) was reported by Ziegler *et al.* (2004). The device has two well-defined states because of the interplay of the electrostatic energy, van der Waals energy, and elastic energy. The electrostatic forces pull in the nanowire to make contact with the electrode ("ON" state) and keep the state even without the electrostatic field because the van der Waals

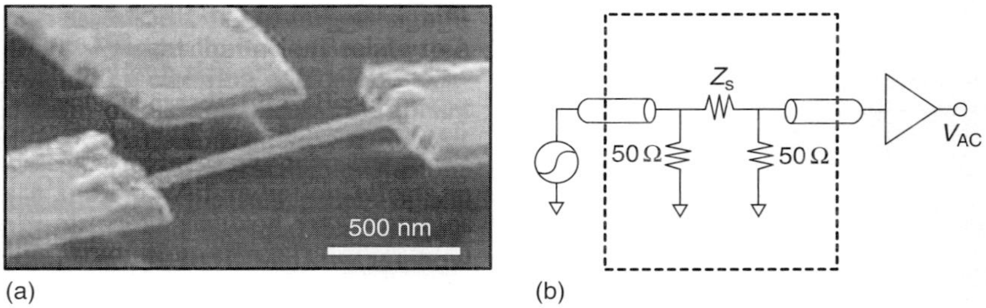

Figure 9
(a) SEM image of the suspended nanowire device, 1.3 μm long and 43 nm in diameter. (b) Measurement circuit used for magnetomotive drive and detection. Reprinted with permission from Husain A, *et al.* 2003 Nanowire-based very-high-frequency electromechanical resonator. *Appl. Phy. Lett.* **83**, 1240–2. Copyright 2003, American Institute of Physics.

force is larger than the electrostatic force. The NEMPROM device can be switched OFF by mechanical motion or by heating the device above the stability limit to overcome the van der Waals attractive forces. The working principle of the NEMPROM is similar to that of the NRAM (Rueckes *et al.* 2000) since both of them employ van der Waals energy to achieve the bistability behavior, although the usage of germanium may provide better size control and electrical behavior than that of carbon nanotube.

3. *Mutiphysics Modeling of NEMS*

The design of NEMS depends on a thorough understanding of the mechanics of the devices themselves and the interactions between the devices and the external forces/fields. With the critical dimension shrinking from micron to nanometer scales, new physics emerges so that the theory typically applied to MEMS does not immediately translate to NEMS. For example, van der Waals forces from atomic interactions play an important role in NEMS, while they can be generally neglected in MEMS. Many NEMS devices can be modeled either as biased cantilever beams as fixed–fixed beams freestanding over a ground substrate. The beams can be carbon nanotubes, nanowires, or small nanofabricated parts. In the following section, we overview the governing equation of equilibrium for both small deformation and finite deformation based on continuum theory.

3.1 Governing Equations

The electromechanical characteristic of nanotube cantilevers or doubly clamped nanotube beams can be determined by coupling the van der Waals, electrostatic, and elastic forces. The governing equation under the small deformation assumption (considering only bending) is given by Desquesnes *et al.* (2002)

$$EI\frac{d^4r}{dx^4} = q_{elec} + q_{vdw} \tag{8b}$$

where r is the gap between the nanotube and the ground plane, x is the position along the tube, E is the Young's modulus (for carbon nanotube $E=$ 1–1.2 TPa), I is the moment of inertia (for nanotubes, $I = \pi/4\left(R_{ext}^4 - R_{int}^4\right)$, R_{ext} and R_{int} are the outer and inner radii of the nanotubes, respectively), q_{elec} is the electrostatic force per unit length force, and q_{vdw} is the van der Waals force per unit length. q_{elec} is given by (Ke *et al.* 2005a, 2005b)

$$q_{elec} = \frac{-\pi\varepsilon_0 V^2}{\sqrt{r(r+2R_{ext})}a\cosh^2(1+(r/R_{ext}))}(1+f_c) \tag{2}$$

where $f_c = 0.85[(H+R_{ext})^2R_{ext}]^{1/3}\delta(x-x_{tip})$ accounts for the concentrated charge at the end of the tube (for doubly clamped tube, $f_c=0$) (Ke and Espinosa 2005), $\varepsilon_0 = 8.854\times10^{-12}\,C^2N^{-1}m^{-2}$ is the permittivity of vacuum, V is the bias voltage, q_{vdw} can be calculated based on the continuum Lennard-Jones potential by assuming that the substrate is layers of graphite sheets (Desquesnes *et al.* 2002, 2004). It is noted that the van der Waals force is only relevant when the gap between nanotube and the substrate is in the order of a few nanometers.

For cantilevers exhibiting large displacements, as shown in Fig. 10(a), the curvature of the deflection should be considered and the governing equation changes into (Ke *et al.* 2005a, 2005b)

$$EI\frac{d^2}{dx^2}\left(\frac{d^2r/dx^2}{\left(1+(dr/dx)^2\right)^{3/2}}\right) = (q_{vdw}+q_{elec})\sqrt{1+\left(\frac{dr}{dx}\right)^2} \tag{3}$$

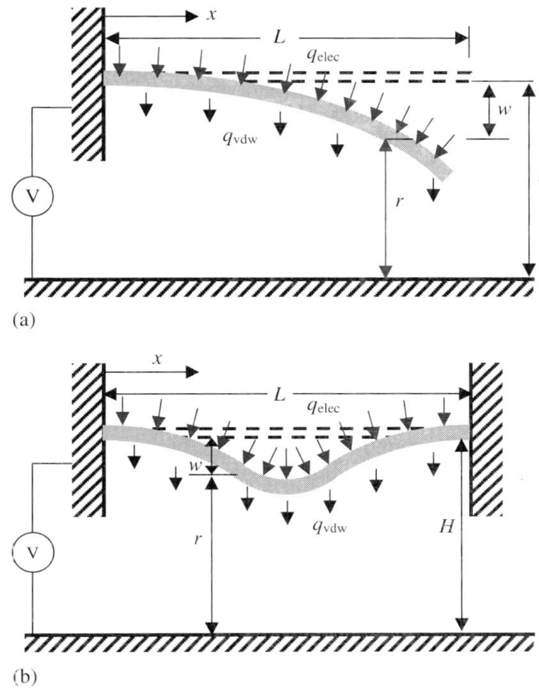

Figure 10
(a) Schematic of finite kinematics configuration of a cantilever nanotube device and (b) doubly clamped nanotube device subjected to electrostatic and van der Waals forces. Reproduced from Ke C-H, *et al.* 2005a Numerical analysis of nanotube based NEMS devices. Part II: Role of finite kinematics, stretching and charge concentrations. *J. Appl. Mech.* **72**, 726–31, with permission from The American Society of Mechanical Engineers.

For doubly clamped structures exhibiting finite kinematics (Fig. 10(b)), stretching becomes significant as a consequence of the rope-like behavior of a doubly clamped nanotube. The corresponding governing equation is expressed as (Desquesnes *et al.* 2004, Ke *et al.* 2005a, 2005b, Pugno *et al.* 2005)

$$EI\frac{d^4r}{dx^4}-\frac{EA}{2L}\int_0^L\left(\frac{dr}{dx}\right)^2 dx\frac{d^2r}{dx^2}=q_{elec}+q_{vdw} \quad (4)$$

where the term $EA/2L\int_0^L(dr/dx)^2\,dx$ is the tension along the axis of the tube due to stretching.

The aforementioned governing equations can be numerically solved by either direct integration or finite difference method. The effect of various factors, such as concentrated charge, finite kinematics, and stretching, on the prediction of pull-in voltages of devices can then be identified.

3.2 Analytical Solutions

In this section, we discuss the electromechanical characteristic of the NEMS devices consisting of both cantilever and double-clamped nanotubes. In particular, the pull-in voltage calculations based on the energy method are reported (Ke *et al.* 2005a, 2005b, Pugno *et al.* 2005).

For nanotube cantilevers (singly clamped), the deflection of the cantilever nanotube can be approximated by the following quadratic function (Ke *et al.* 2005a, 2005b):

$$w(x)\approx\frac{x^2}{L^2}c \quad (5)$$

where L is the length of the nanotube, c is a constant that represents the displacement of the end of the cantilever, and x is the coordinate along the nanotube.

By assuming that the nanotube's (external) radius R_{ext} is much smaller than the distance r between nanotube and ground plane, that is, $R_{ext}/r\ll 1$, the pull-in voltage, considering the nonlinear finite kinematics and the concentrated charges at the free end, is given by Ke *et al.* (2005a, 2005b)

$$V_{S\ PI}\approx k_S\sqrt{\frac{1+K_S^{FK}}{1+K_S^{tip}}\frac{H}{L^2}\ln\left(\frac{2H}{R_{ext}}\right)}\sqrt{\frac{EI}{\varepsilon_0}}$$

$$k_S\approx 0.85,\quad K_S^{FK}\approx\frac{8H^2}{9L^2} \quad (6a)$$

$$K_S^{tip}\approx\frac{2.55\left[R_{ext}(H+R_{ext})^2\right]^{1/3}}{L} \quad (6b)$$

where subscripts S refer to singly clamped boundary conditions for cantilevers, superscript FK refers to finite kinematics, tip refers to the charge concentration being at the free end.

For doubly clamped nanotubes, the deflection $w(x)$ is assumed such that it satisfies the boundary conditions $w(x=0,\ L)=w'(x=0,\ L)=0$, namely (Pugno *et al.* 2005)

$$w(x)\approx 16\left[\left(\frac{x}{L}\right)^2-2\left(\frac{x}{L}\right)^3+\left(\frac{x}{L}\right)^4\right]c \quad (7)$$

where $w(x=L/2)=c$ is here an unknown constant that represents the displacement of the central point. The pull-in voltage can be expressed as (Pugno *et al.* 2005)

$$V_{D\ PI}=k_D\sqrt{1+k_D^{FK}}\,\frac{H+R}{L^2}\ln\left(\frac{2(H+R)}{R}\right)\sqrt{\frac{EI}{\varepsilon_0}} \quad (8a)$$

$$k_{\rm D} = \sqrt{\frac{1024}{5\pi S'(c_{\rm PI})}\left(\frac{c_{\rm PI}}{H+R}\right)}, \quad k_{\rm D}^{\rm FK} = \frac{128}{3003}\left(\frac{c_{\rm PI}}{\rho}\right)^2$$

$$\rho^2 = \frac{I}{A} = \frac{R_{\rm ext}^2 + R_{\rm int}^2}{4} \tag{8b}$$

$$S(c) = \sum_{i=1}^{\infty}\left(\frac{1}{(\ln(2(H+R)/R))^i}\sum_{j=i}^{\infty} a_{ij}\left(\frac{c}{(H+R)}\right)^j\right) \tag{8c}$$

Subscripts D refer to double clamped boundary conditions, $c_{\rm PI}$ is the central deflection of the nanotube at the pull-in, and the $\{a_{ij}\}$ in Eqn. (8c) are known constants (Pugno *et al.* 2005).

The accuracy of the analytical solutions is verified by comparison with both numerical integration of the governing equations (Desquesnes *et al.* 2002, Ke *et al.* 2005a, 2005b) and experimental measurements (see Sect. 3.3) (Ke *et al.* 2005a, 2005b). The comparison between pull-in voltages evaluated numerically and theoretically for doubly (D) and singly (S) clamped nanotube devices is listed in Table 1 (Ke *et al.* 2005a, 2005b). Columns six and seven in Table 1 compare analytical and numerical pull-in voltage predictions under the assumption of small deformations. Columns eight and nine in Table 1 compare analytical and numerical pull-in voltage predictions under the assumption of finite kinematics. The agreement is very good (with a maximum discrepancy of 5%).

3.3 Comparison between Analytical Predictions and Experiments. In this section, a comparison between analytical predictions and experimental data, for both small deformation and finite kinematics regimes, is presented.

(*a*) *Small deformation regime.* The nanotweezers experimental data reported by Akita *et al.* (2001), plotted in Fig. 11, is used to assess the model accuracy under small deformation. In this case, the nanotweezers are equivalent to a nanotube cantilever with length of 2.5 μm freestanding above an electrode with a gap of 390 nm.Symmetry is here exploited. In the same figure, a comparison between the analytically predicted nanotube cantilever deflection and the experimentally measured data is shown (Ke *et al.* 2005a, 2005b). The analytical model includes the van der Waals force and charge concentration at the free end of the nanotube cantilever. Model parameters include Young's modulus, $E = 1$ TPa, external radius $R = R_{\rm ext} = 5.8$ nm, and

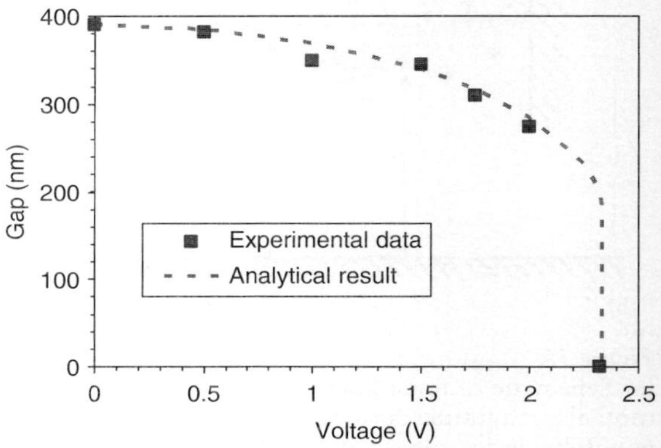

Figure 11
Comparison between experimental data and theoretical prediction in the small deformation regime. Reproduced from Ke C-H *et al.* 2005b Experiments and modeling of nanotube NEMS devices. *J. Mech. Phys. Solids* **53**, 1314–33, with permission from Elsevier.

Table 1
Comparison between pull-in voltages evaluated numerically and theoretically for doubly (D) and singly (S) clamped nanotube devices, respectively. $E = 1$ TPa, $R_{\rm int} = 0$.

Case	BC	H (nm)	L (nm)	$R = R_{\rm ext}$ (nm)	$V_{\rm PI}$ (V) (theo.linear)	$V_{\rm PI}$ (V) (num. linear)	$V_{\rm PI}$ (V) (theo. non-linear)	$V_{\rm PI}$ (V) (num. non-linear)
1	D	100	4000	10	3.20	3.18	9.06	9.54
2	D	100	3000	10	5.69	5.66	16.14	16.95
3	D	100	2000	10	12.81	12.73	36.31	38.14
4	D	150	3000	10	9.45	9.43	38.93	40.92
5	D	200	3000	10	13.53	13.52	73.50	77.09
6	D	100	3000	20	19.21	18.74	31.57	32.16
7	D	100	3000	30	38.57	37.72	51.96	50.63
8	S	100	500	10	27.28 (w)	27.05 (w)	27.52 (w)	27.41 (w)
9	S	100	500	10	27.28 (w)	27.05 (w)	30.87	31.66

For a cantilever nanotube device the symbol (w) denotes that the effect of charge concentration has been included. Reproduced from Ke C-H, *et al.* 2005a Numerical analysis of nanotube based NEMS devices. Part II: Role of finite kinematics, stretching and charge concentrations. *J. Appl. Mech.* **72**, 726–31, with permission from The American Society of Mechanical Engineers.

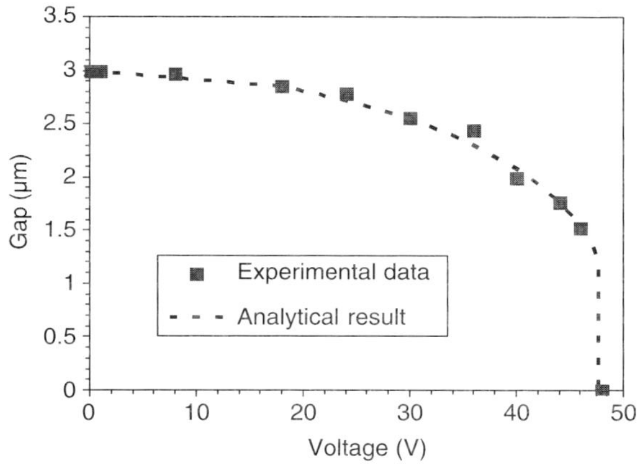

Figure 12
Comparison between experimental data and theoretical prediction in the finite kinematics regime. Reproduced from Ke C-H *et al.* 2005b Experiments and modeling of nanotube NEMS devices. *J. Mech. Phys. Solids* **53**, 1314–33, with permission from Elsevier.

$R_{int}=0$. The pull-in voltage from the analytical model is 2.34 V while the experimentally measured pull-in voltage was 2.33 V. It is clear that the analytical prediction and experimental data for the deflection of the nanotube cantilever, as a function of applied voltage, are in very good agreement.

(*b*) *Finite kinematics regime.* Experimental data corresponding to the deflection of carbon nanotube cantilevers in the finite kinematics regime were recently obtained by *in situ* SEM measurements (Ke *et al.* 2005a, 2005b). An MWNT mounted to a manipulator probe as a cantilever was placed parallel to an electrode and actuated by the electrostatic force. The length of the nanotube is 6.8 μm and the gap between the nanotube and the electrode is 3 μm. Because the ratio between the length of the nanotube and the gap between the nanotube and electrode is 2.3, the deflection of the nanotube can be considered to be in the finite kinematics regime.

The experimentally measured nanotube cantilever deflections, in the finite kinematics regime, are plotted in Fig. 12 (Ke *et al.* 2005a, 2005b). The figure also shows a comparison between analytical prediction and experimental data. The analytical model includes finite kinematics, the van der Waals force, and charge concentration at the free end of the nanotube cantilever. For these predictions, the following parameters were employed: length of the nanotube, $L=6.8$ μm, initial gap between nanotube and electrode, $H=3$ μm, $R=R_{ext}=23.5$ nm, $R_{int}=0$, $E=1$ TPa. The pull-in voltage given by the analytical analysis is 47.8 V, whereas the pull-in voltage experimentally measured was 48 V.

Bibliography

Akita S, Nakayama Y, Mizooka S, Takano Y, Okawa T, Miyatake Y, Yamanaka S, Tsuji M, Nosaka T 2001 Nanotweezers consisting of carbon nanotubes operating in an atomic force microscope. *Appl. Phys. Lett.* **79**, 1691–3

Badzey R L, Zolfagharkhani G, Gaidarzhy A, Mohanty P 2004 A controllable nanomechanical memory element. *Appl. Phys. Lett.* **85**, 3587–9

Davis Z J, Abadal G, Kuhn O, Hansen O, Grey F, Boisen A 2000 Fabrication and characterization of nano-resonating devices for mass detection. *J. Vac. Sci. Technol. B* **18**, 612–6

Desquesnes M, Rotkin S V, Alaru N R 2002 Calculation of pull-in voltages for carbon-nanotube-based nanoelectromechanical switches. *Nanotechnology* **13**, 120–31

Desquesnes M, Tang Z, Aluru N R 2004 Static and dynamic analysis of carbon nanotube-based switches. *J. Eng. Mater. Tech.* **126**, 230–7

Fennimore A M, Yuzvlnsky T D, Han W Q, Fuhrer M S, Cummings J, Zettl A 2003 Rotational actuators based on carbon nanotubes. *Nature* **424**, 408–10

Husain A, Hone J, Postma H W Ch, Huang X M H, Drake T, Barbic M, Scherer A, Roukes M L 2003 Nanowire-based very-high-frequency electromechanical resonator. *Appl. Phys. Lett.* **83**, 1240–2

Ilic B, Craighead H G, Krylov S, Senaratne W, Ober C, Neuzil P 2000 Attogram detection using nanoelectromechanical oscillators. *J. Appl. Phys.* **95**, 3694–703

Ke C-H, Espinosa H D 2004 Feedback controlled nanocantilever device. *Appl. Phys. Lett.* **85**, 681–3

Ke C-H, Espinosa H D 2005 Numerical analysis of nanotube based NEMS devices. Part I Electrostatic charge distribution on multiwalled nanotubes. *J. Appl. Mech.* **72**, 721–5

Ke C-H, Espinosa H D, Pugno N 2005a Numerical analysis of nanotube based NEMS devices. Part II Role of finite kinematics, stretching and charge concentrations. *J. Appl. Mech.* **72**, 726–31

Ke C-H, Pugno N, Peng B, Espinosa H D 2005b Experiments and modeling of nanotube NEMS devices. *J. Mech. Phys. Solids* **53**, 1314–33

Kim P, Lieber C M 1999 Nanotube nanotweezers. *Science* **126**, 2148–50

Kinaret J, Nord T, Viefers S 2003 A carbon nanotube based nanorelay. *Appl. Phys. Lett.* **82**, 1287–9

Lee S, Lee D, Morjan R, Jhang S, Sveningsson M, Nerushev O, Park Y, Campbell E 2004 A three-terminal carbon nanorelay. *Nano Lett.* **4**, 2027–30

Pugno N, Ke C-H, Espinosa H D 2005 Analysis of doubly clamped nanotube devices in the finite deformation regime. *J. Appl. Mech.* **72**, 445–9

Roukes M L 1999 Yoctocalorimetry phonon counting in nanostructures. *Physica B* **263-4**, 1–15

Roukes M L 2004 Nanoelectromechanical Systems, Technical Digest of the 2000 Solid-State Sensor and Actuator Workshop.

Rueckes T, Kim K, Joslevich E, Tseng G Y, Cheung C, Lieber C M 2000 Carbon nanotube-based nonvolatile random access memory for molecular computing. *Science* **289**, 94–7

Sazonova V, Yaish Y, Üstünel H, Roundy D, Arias T, McEuen P 2004 A tunable carbon nanotube electromechanical oscillator. *Nature* **431**, 284–7

Ziegler K J, Lyons D M, Holmes J D, Erts D, Polyakov B, Olin H, Svensson K, Olsson E 2004 Bistable nanoelectromechanical devices. *Appl. Phys. Lett.* **84**, 4074–6

NDT for Non-Nuclear Power Generation: Pressure Vessels, Piping, and Turbines

A. Birring

Boilers, headers, steam lines, turbines, feedwater heaters, and condensers are the main components inspected in a non-nuclear power plant. Unlike nuclear plants where inspections are required for regulatory reasons, non-nuclear plants are inspected for reducing forced outages, condition assessment, and plant safety. Because of a range of equipment and a number of damage mechanisms, a variety of NDT techniques are applied. These include simple magnetic particle testing to advanced ultrasonic phased array systems. In addition to the traditional nondestructive tests, damage assessment methods are used to determine degradation of material properties. These can include sample removal, hardness testing, and *in situ* metallographic replication. Proper selection of NDT technology is necessary for conducting a reliable inspection of these components.

1. Boilers

Boiler tubes are usually the main cause of forced outages in a thermal power plant. Twenty-two failure mechanisms can be held responsible for boiler tube failures.

1.1 Outer Diameter Erosion, Corrosion, and Overheating

Outer diameter (OD) wall loss in a boiler tube is caused by erosion, fireside corrosion, and short-term overheating. OD erosion is measured by ultrasonic thickness measurement. Wall thickness measurements are performed with commercially available ultrasonic digital gauges or portable ultrasonic pulser-receivers using a dual transducer. Calibration is performed on a curved calibration plate to simulate actual boiler tube geometry. In addition, transducer alignment on both the boiler tube and the calibration block should be identical.

1.2 Inner Diameter Corrosion, Hydrogen Damage, Caustic Corrosion, Chemical Attack

Inner diameter (ID) pitting in boiler tubes can be caused by hydrogen damage, caustic corrosion, chemical attack, etc. Because this type of pitting is usually isolated, a careful examination of the whole boiler tube length is required. Digital gauges have severe limitations when measuring tubes with ID pitting. Ultrasonic scattering from ID pits produces an undefined back-surface reflection signal and impairs thickness measurement. When measuring the thickness of a tube with ID surface corrosion, an instrument with a CRT screen display is recommended. The screen presentation identifies the back-wall reflection so as to give reliable thickness measurements.

Hydrogen damage is one of the mechanisms that produces ID corrosion. This damage is produced in the water wall tubes from an imbalance in water chemistry (French 1993). Tube bends, circumferential welds, and tube lengths across the burners are the most susceptible locations for such damage. Hydrogen damage is of serious concern because it not only results in ID wall loss but also in a zone of decarburized material under the corroded area. Ultrasonic thickness scanning is the first step towards detection of corrosion caused by hydrogen damage. Since ID corrosion can be caused by other mechanisms, hydrogen damage should be verified by other NDT methods. Decarburization caused by hydrogen damage reduces the ultrasonic velocity, and so velocity measurement techniques can therefore be applied for verification of such damage.

1.3 Cracking—Corrosion Fatigue, Stress Corrosion, and Thermal Fatigue

OD cracking in a boiler tube can be produced through thermal fatigue, corrosion fatigue, etc. Visual testing (VT), magnetic particle testing (MT), penetrant testing (PT), and radiography testing (RT) are commonly applied for the detection of OD cracking. The depth sizing of such cracks can be performed visually from the crack length and width or by eddy current testing. Special send–receive eddy current surface probes are recommended for crack sizing.

ID cracks with axial orientation may be caused by stress corrosion and corrosion fatigue mechanisms. Refracted ultrasonic shear waves are used to detect these cracks. Inspection is performed by placing a transducer on the OD surface of the tube, with the beam directed towards the area being inspected. A refracted angle that maximizes reflectivity from the crack is selected. Maximum reflectivity from the crack is produced when the incident angle on the crack is 45°. This incident angle is used to calculate the refracted angle of the transducer. The calculated refracted angle is always less than the incident angle.

Dissimilar metal weld (DMW) cracking occurs in welds that join low-alloy steels with stainless steel. These welds are present in high-temperature sections of the boiler, including the superheater and reheater sections. DMW cracking occurs along or near the fusion line between the low-alloy steel and the weld. In addition to cracking, there can be the presence of an oxide notch that is commonly found on the OD surface of the DMW. An oxide notch is initiated because of differences in the creep strength between a weld metal and the low-alloy steel heat affected zone (HAZ). The presence of an oxide notch is not an indication of a crack. Ultrasonic testing (UT) and radiography are two methods to inspect DMW cracks. When properly applied, these methods can resolve DMW cracking from the oxide notch.

1.4 Creep—ID Oxide Scale

ID oxide scale can be produced when tubes in the reheater and superheater experience high temperatures for extended periods of time. The formation of ID scale reduces heat transfer and results in a further increase of tube metal temperature. The increase in ID scale and the associated tube metal temperature increase promotes creep and reduces tube life. The final outcome of excessive scale is a thick lipped, long-term overheat failure. Scale thickness measurements should be taken just upstream of material upgrade and thickness upgrade locations. A history of prior long-term overheat failures should also be used to select tubes for oxide scale inspection. The ultrasonic method for measuring scale thickness is based on transmitting a wave through the tube thickness. The thickness is calculated by measuring the time difference between the signals reflected from the steel/scale interface and the tube ID surface. Because of the extremely small time difference, the application requires the use of high-frequency transducers in the range 15–30 MHz. Knowing the scale thickness and the operating hours, the remaining life of the tube can be calculated.

2. *Headers*

Headers are inspected for cracking in the welds and ligaments. Weld cracking is inspected using ultrasonics and wet fluorescent magnetic particle testing (WFMT). WFMT is used for OD cracking while UT is used for ID or midwall cracks.

Leaks in headers can be caused by ligament cracking. Ligament cracking is produced in the bore holes and the stub tube ID. The cracking occurs due to cyclic events such as startups and shutdowns, transients, and thermal shocks. The hottest areas are the most susceptible to ligament cracking; however, there can be exceptions.

One approach for the detection of ligament cracking is first to remove the stub tube and then perform penetrant inspection. ID grinding is first done to remove the oxide scale. This is followed by the wet fluorescent penetrant method. The inspection determines the length of the crack in the borehole region. The crack length information is then used to make a decision on the disposition of the header.

3. *Steam Lines*

Inspection of steam lines is done to detect cracking in the welds. Bending loads can produce OD cracking in the circumferential welds. WFMT is the recommended approach for such an inspection. The inspection should also be done on the hanger welds to determine their integrity.

High-temperature creep can cause midwall cracking or ID connected cracking in seam welds. Over a long period of time, creep voids can grow to microcracks, interlink, and cause failure of a long seam weld (Viswanathan 1989). The inspection of long seam welds gained significant coverage in the technical literature after the failure of this type of weld caused loss of life. The failure of a steam line can either be a "leak before failure" or a rupture. The type of failure depends on the length of the crack. Cracks longer than the critical length can result in rupture. UT is the recommended approach for such inspection. The inspection is done using refracted shear waves. Either a single transducer or phased arrays can be used. Because of the pipe curvature, proper selection of refracted angles is crucial to this inspection.

4. *Turbines*

There are several components that are inspected in a turbine. These include bore, disk keyway, disk blade attachment area, blades, nozzles, casing, bolts, etc.

4.1 Bore

The mechanism of crack growth in a rotor bore is due to the combined action of creep and fatigue (Viswanathan 1989). Creep is more prevalent in HP rotors that operate close to 540°C. Fatigue is more prevalent in LP rotors that operate at lower temperatures. Sensitivity of the bore examination depends on locations that experience the highest level of stress and temperature. The hoop stress is higher under the disks because of mass loading. The temperature is highest under the control stage. Sensitivity of examination should therefore be highest in the HP rotors at the ID bore surface under the HP disks.

Three methods are commonly used for bore inspection: MT, eddy current testing, and UT. The first two methods are limited to surface defects. MT is performed by applying a circumferential magnetic field at the bore ID. The circumferential field detects axial cracks on the bore surface. Ultrasonic inspection is the only method that can perform a complete volumetric examination. A combination of transducer angles is used to perform the inspection. The transducers are installed on a scanner and the data are recorded on an ultrasonic imaging system. Detection sensitivity is controlled by adjusting the scan step interval. In addition to defect detection, damage assessment is done in the bore to determine any loss of fracture toughness. Indirect measurements such as remote replication and small punch test sample removal are performed to assess material degradation.

4.2 Solid Rotor

The main advantage of a boreless rotor is its lower level of stress compared to a bored rotor. The lower level of stress makes the boreless rotor tolerant to larger flaws. For this reason, inspection procedures for boreless rotors are less stringent. The inspection of a solid rotor is performed using a combination of L- and S-wave transducers. Inspection of boreless rotors requires that the selected angles be able to inspect the entire material volume of interest. As an alternative, phased arrays that sweep a range of angles can be used.

Transverse cracking in LP rotors initiates from corrosion pits and can grow during service by corrosion fatigue. Transverse cracking is easily detected by application of MT on the rotor OD surface. The depth of the crack can be measured using the ultrasonic tip diffraction method.

4.3 Disk Keyway Cracking

The primary cause of disk keyway cracking is stress corrosion. High stress concentration in the keyway region promotes growth of this cracking. Because of the stress corrosion cracking (SCC) mechanism, keyway cracking is mostly observed past the Wilson line (steam transforms from dry to wet). In some cases, SCC may be found before the Wilson line if condensation occurs during standby. Ultrasonic inspection of keyway cracking is performed either using multiple refracted angles or using phased array transducers. Phased array transducers have the capability of sweeping a range of angles electronically. Both pulse-echo and pitch-catch modes are used during inspection. Alignment of the transducers, in the pitch-catch mode, is very critical to assure a reliable inspection.

4.4 Disk Blade Attachment Area

The mechanism of crack initiation and growth in turbine disk blade attachments (steeples) depends on three variables: the operating temperature, stresses, and environment. Creep is the primary mechanism in HP and IP rotors. SCC, combined with fatigue, is the primary mechanism for LP rotors. Initially, cracking in an LP rotor grows slowly by SCC. When the stress intensity, K_I, exceeds K_{th}, crack growth is predominantly due to fatigue. Crack growth rates in this mode are significantly high because of vibratory loads. Generally, failure can be imminent when the threshold for fatigue crack growth, K_{th}, is reached. Therefore it is important that NDE inspections detect cracks before their stress intensity reaches K_{th}.

The inspection methods applied to detect steeple cracking depend on the geometry. Dovetail design (GE turbines) is inspected only by ultrasonics (Ciorau and MacGillivray 2000). Phased array or multiple single-element transducers are used for inspection. Side entry steeples (Westinghouse turbines) allow access for surface inspection. In addition to UT, these disks can be inspected by eddy current testing and MT. However, ultrasonics is the only method that is capable of inspecting the entire length of the side entry steeple under the blade. Once the blades are removed, WFMT is the preferred method for inspecting steeples. The process is slow, but results in a highly sensitive inspection.

4.5 Blades

The failure mechanism of turbine blades is dependent on their temperature, environment, and stress state. Corrosion fatigue is the major failure mechanism of blades in the next-to-last stage of the LP turbine. Creep blade failures are limited to HP turbines. Cracking of blades occurs at the following three locations: blade attachments, airfoil, and tenon. The inspection methods chosen for each of these locations depend on whether the inspection is performed with the blade removed from the disk or not.

Eddy current testing and MT are the two methods used to inspect the blade attachment areas of side entry blades. Eddy current inspection is an attractive method since it can be performed *in situ* without removal of the turbine from the casing. During examination, portholes in the turbine casing are used to gain access to the blades. An eddy current probe along with a fiber optic probe is held on the end of a rod that is inserted in the casing through a porthole for inspection.

Once the turbine has been removed from the casing, the blade attachments become directly accessible for inspection. Either eddy current or MT may be used at this stage. MT is the preferred method because it is faster. WFMT is performed with a.c. coils or a yoke. WFMT is the most commonly used method for inspecting blade length. The inspection is performed by magnetizing the blades with a.c. coils. The a.c. magnetization allows a highly sensitive examination on the surface, while leaving minimal residual magnetism in the blades.

Blade tenons are located at the tip of the blades and hold the shroud. Cracking and failure of the tenons may release the shroud and cause mechanical damage to other blades. The only method available to inspect blade tenons is ultrasonics. An ultrasonic transducer is placed on the tenon; a flat surface on the tenon is required so that contact with the transducer can be achieved. Tenons without a flat face cannot be inspected unless they are ground flat.

4.6 Bolts

Creep rupture and brittle fracture are two primary reasons for bolt failures. The low toughness that leads to brittle fracture is due to the inherent high strength of bolts. The failures are usually initiation controlled. Hence, the failure time is very short after crack initiation. Cracking in bolts occurs only in the threads next to the joint, these threads experiencing the highest level of stress. The stress on the last thread at the end of the bolt is almost zero. Therefore, any inspection should carefully investigate threads next to the joint. UT is the only method that can inspect bolts without removing them from the casing. One of two ultrasonic approaches is used for this inspection: a zero degree examination is performed when the top surface of the bolt is flat, since it allows placement of a normal beam transducer; when the top face of the bolt is not flat, an angle beam examination is performed through the heater holes.

4.7 Retaining Rings

The susceptibility of 18Mn–5Cr steels to SCC produces cracking in retaining rings. Initiation of cracks in the retaining rings occurs when moisture enters and settles on the inner surface. The initiation time of the cracks is quite long. Nevertheless, once the crack has initiated, crack growth can be quite rapid. The high crack growth rates limit the application of NDE for crack detection. No effort is made to size the cracks once they are detected. Repair or replacement actions are initiated once a crack is positively detected.

Four methods are generally used when inspecting the retaining rings: (i) VT, (ii) fluorescent PT, (iii) eddy current testing, and (iv) UT. The first three methods can only be applied after ring removal. Visual inspection is accomplished using borescopes that detect moisture in accessible areas of the ID surface. Evidence of moisture is an indication of possible crack initiation. Penetrant examinations are performed using the fluorescent penetrant method. High-sensitivity lipophilic emulsifiers are used for these inspections. UT is the only method that can be applied without removal of the retaining ring; however, its detection sensitivity is limited. A combination of adverse factors, such as high ultrasonic attenuation and spurious geometrical reflections, results in the low UT detection sensitivity.

5. Condensers

The main reason for inspection of condenser tubes is to determine their condition. The information from the inspection can then be used to make decisions on replacement of all the tubes. The inspection of condensers is normally limited to a 2–5% random sample of tubes. The most common tube materials in a condenser are copper–nickel alloys, brass, titanium, stainless steel, and ferritic stainless steel. Pitting is the most common form of damage in condenser tubing. OD erosion/corrosion is very common in brass tubing. Tubes in the top row are susceptible to OD erosion. Conventional eddy current testing is applied for inspection of condenser tubing. Because of the long length of the tubing, inspection of condenser tubing is done at high speed manually or using pusher–pullers.

6. Feedwater Heaters

Tube failure in feedwater heaters is one of the major causes of forced outages in a fossil fuel power plant. Inspection of HP feedwater heaters produces one of the highest cost benefits of any NDT inspection in a power plant. The most common tube materials in feedwater heaters are carbon steel, stainless steel, brass, and copper–nickel alloys. OD erosion is the most common type of damage in carbon steel tubes. The locations most susceptible to OD wear are the subcooling zone and the desuperheating zone. Pitting can occur in tubes of stainless steel and copper alloys.

The tubing in the feedwater heaters should be periodically inspected to determine its condition. Selection of tubes for inspection is the key to an effective feedwater heater inspection. For a regular inspection the plan should include tubes in the subcooling zone, desuperheating zone, tubes around previously plugged tubes, and some tubes at random. In addition to regular inspections, inspection after a tube failure is highly recommended. A tube failure is an indication of damage to and impending failure in other tubes. During such an inspection, tubes around the leaking tubes should be tested, and tubes with damage above a certain level should be immediately plugged. Such an approach results in effective plugging and avoids future forced outages. Conventional eddy current testing is applied for nonferromagnetic materials. Remote field eddy current testing is quite effective for inspection of carbon steel tubing.

7. Conclusions

The methods described here show the state of the art in NDT. Future developments are expected in areas of damage assessment, data acquisition, analysis, and reliability. Methods to detect and characterize incipient or early stages of damage will be applied on a wider scale. More sophisticated systems will be introduced for inspection using faster, smaller, and economical computer hardware. There will be more emphasis on improving the reliability of NDT methods, especially characterization of defects. This may include upgrading personnel qualifications to improve inspection quality.

Bibliography

Ciorau P, MacGillivray 2000 *In situ* examination of low-pressure steam turbine components using phased array technology In: *Proc. 2nd Int. Conf. on NDE*EPRI, New Orleans, LA

French D N 1993 *Metallurgical Failures in Fossil Fired Boilers.* Wiley, New York

Viswanathan R 1989 *Damage Mechanisms and Life Assessment of High-Temperature Components.* ASM International, Metals Park, OH

NDT of Components in Nuclear Installations

E. Fischer

In-service inspections (ISIs) are a major part of outage activities of nuclear power plants (NPPs) all over the world. Despite many safeguards inherent in component design and provided by plant operating limits, the ISI has been applied to the components of the primary circuit of pressurized water reactors (PWRs) and boiling water reactors (BWRs), especially to the reactor pressure vessel (RPV).

From very early, the first "editions" and "addenda" of the ASME Code, Sect. XI, with detailed descriptions concerning extent of examination, inspection intervals and technical aspects, have been created to lay down the code requirements for the mechanized ultrasonic inspection (UT) with regard to the preservice inspection, performed prior to first hot operation of the NPP and the related ISI after going into operation.

Many other countries, in which NPPs went into operation, followed the ASME code or developed their own code. As a generic feature, the ultrasonic technique is the key element for the ISI of the RPV independent of the specific code. There are some slight differences between the codes. Comparing the German code (KTA 3201.4, describing the in-service inspection of the primary components of NPPs) to ASME, there are more stringent requirements with regard to inspection intervals (e.g., the German code requires RPV inspection at five year intervals, as against 10 year intervals according to the ASME code). There are also differences in detection requirements and angles of incidence (e.g., ASME requires 0°, 45°, 60° plus 70° for the near surface region, while the KTA code requires instead 60°, the tandem technique for components with a wall thickness greater that 100 mm). The main features, including the indication size to be recorded, are listed in Fig. 1. The current development of advanced automated UT systems applied to safety- relevant components of NPPs is oriented towards

(i) UT phased-array technique based on linear arrays;
(ii) modular UT systems, capable of operating simultaneously, conventional and phased array probes, extendable to the simultaneous operation of electromagnetic acoustic transducers (EMAT) and for the purpose of surface inspection by eddy current testing; and
(iii) robotics as a further development step to manipulators that are in use up to now and are capable of operation in areas of high radiation levels.

Many field equipment and service organizations with expertise in the above three areas are available

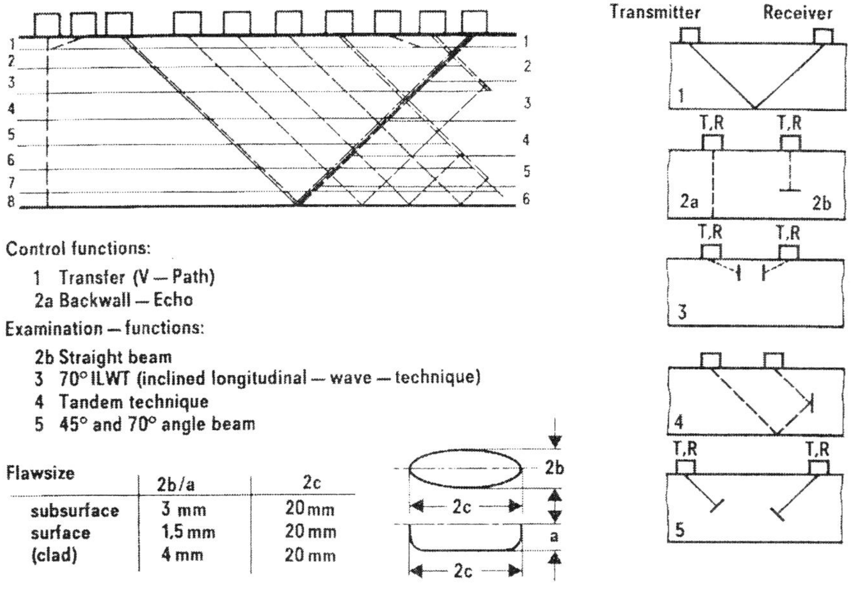

Figure 1
Multiple function UT probe system for cylindrical part of the RPV and guaranteed detectable flaw size.

for carrying out automated UT for ISIs. Based on experience, improvements with regard to successful performance of ISI in NPPs are made based on:

(i) well trained and motivated NDE teams and
(ii) trouble-free and efficient operating equipment, which allows the systems to run with a minimum exposure to radioactivity.

Focussing on the technology the currently applied advanced systems will be presented in more detail.

1. Basic Elements of Automated Inspection Systems

In general one can break down an automated inspection system into the following main elements:

(i) Ultrasonic equipment capable of operating these types of probes simultaneously including EMAT and/or eddy current probes for special applications and also operating with qualified data acquisition and evaluation software;
(ii) Probe system designed for the inspection task complying with the plant specific code requirements (e.g., ASME, KTA (for Germany), RCMR (for France)) based on conventional or phased array probes; and
(iii) Manipulator or robotic system including control unit.

1.1 Ultrasonic Equipment

Nondestructive inspections on reactor components typically have the following characteristics:

(i) Heavy-walled components and extremely large examination volume
(ii) Adapted to complex geometries and materials which are difficult to examine
(iii) Fast processing of measured data and precise analysis of indication.

To comply with these features advanced UT systems are required and have been developed (e.g., Gohlke 1998). These inspection systems with data acquisition and evaluation are the first available systems that combine the conventional inspection technology of standard search units with the advantages of the phased-array technique. The systems provide a considerable number of search unit channels (at least 96 extendable to 488) and, when used with phased-array search units—which can carry out several different inspection tasks through the alternation of transmission and receiving angles—has allowed compact search unit systems to be built which are capable of covering all the necessary functions for simultaneous inspection of longitudinal and transverse defects in heavy-walled components, requiring only a single scanning pass in each area. Compared to previous systems, these new systems require fewer scanning passes and fewer changes of search unit systems, thereby reducing the time during which other activities are blocked out of the reactor well, which impacts the critical path of the outage. This time has been reduced by more than one half during the application of ISI of nuclear power plants in Europe.

These systems are linked by Ethernet LAN (local area network) to the controls of the inspection manipulator as well as to the workstation for data acquisition and evaluation. This allows setting of parameters for subsystems, operation of the equipment and transmission of inspection data and manipulator position data to be carried out while scanning is going on. These functions can also be controlled by a superordinate inspection data management system which has access not only to data from the inspection in progress but also to historical data from previous inspections, allowing rapid and direct comparison of results. Such systems can accommodate simultaneous operation of up to seven phased-array search units each with 16 elements, or 112 conventional search units. Any combination of the two types of search units is also possible.

This type of system, with EMAT and eddy current techniques, offers the following possibilities for data evaluation:

(i) Output coupling images with statistical data
(ii) Geometrically correct top-, side-, and end-projections of amplitudes corresponding to times of flight, incident angles, and scanning directions
(iii) Interactive modification of colour scale as an aid to evaluation
(iv) Definition of indications in the scan presentations using search functions for peaks and defect lengths
(v) Display of amplitude and time histories for specific indications
(vi) Storage of specific indications and documentation of the results in list form
(vii) Display of amplitudes as an image
(viii) Circle reconstruction of indications for analysis purposes.

The software package also contains numerous tools for preparing scans, for securing original data as well as a database for parameter settings and measured data.

Another important feature is the self-check of the system, which encompasses all relevant data acquisition and processing features of the UT instrument. It is performed within 10 minutes, as against five days required for manual inspection. The documentation in its largest extent can serve as an equipment calibration-certification report, at its minimum as system recheck according to code requirements.

These clearly visible steps of implementation of new equipment generations have been accompanied by numerous steps of improved logistics and standardization of performance including calibration and evaluation as well as changes to the equipment which may appear minor in scope, but have been proven to be essential for the increased efficiency to guarantee minimum inspection times.

1.2 Phased-array UT Technique

Starting with the successful blind test for the Oyster Creek plant in 1988 and the subsequent examinations of BWR-RPV-feedwater nozzle from the outer surface, the application of the phased-array techniques for the inspection of reactor components was consequently extended in the 1990s (Engl *et al.* 1998). Up to the beginning of the twenty-first century, a large variety of ISI problem areas were covered by the implementation of phased-array techniques. In most of the cases the driving force to implement this (at least at first view) complex technology was twofold: This technology offered a considerably higher reliability of the inspection in terms of detection, sizing and characterization of defects accompanied by savings in inspection time—mostly in conjunction with a considerable reduction of the radiation exposure to the inspection personnel. The last two criteria can be translated into reduced cost of the examination, even if the price of the UT equipment and of the individual probes for phased array application is considerably higher when compared with the conventional installation. Some recent applications include the following:

(i) The nozzle inner radius of the emergency core cooling system (ECCS) piping into the (ferritic clad) main coolant piping.
(ii) The control rod ligaments in the PWR-RPV top head and the BWR-RPV bottom head.
(iii) Consequently, also the examination of the adjacent circumferential weld(s) between top/bottom head dome and top/bottom flange was enhanced and extended by the use of phased array techniques.

Other examples of the successful application of phased-array technology are BWR-RPV-ISI from inner and outer surfaces, RPV core shroud, RPV bottom ligaments, pressurizer ligaments and nozzle-to-dome welds.

Complicated geometries together with the problem of stainless steel weld inspection in BWR pipe-to-nozzle areas posed a very difficult ISI problem, which could only be solved satisfactorily with the phased-array approach.

A most remarkable milestone for the phased array application was the successful examination of the two RPVs (with Framatome design) in Koeberg, South Africa. The on-vessel-time for the entire examination (10 year scope and entire core region of the shell) was 6.5 days (Pannaye *et al.* 1998)

The phased-array techniques have also been implemented into the ISI of the components of Russian reactors such as the RPV components and steam generator collector flanges. The pioneer plants for these activities are Paks (Hungary) and Bohunice/Mochovce (Slovakia), followed by Rovno (Ukraine) and Balakovo and Novovoronesh (Russia).

Utilizing the advantage of electronic steering of the sound field, the phased-array technique improves dramatically the inspection capability. This has been demonstrated during the recent RPV inspections carried out as per the ASME code. For further clarification of indications, the full ranges of angles of incidence in longitudinal and shear mode are available. This approach for the characterization and sizing of detected defect indications is used and

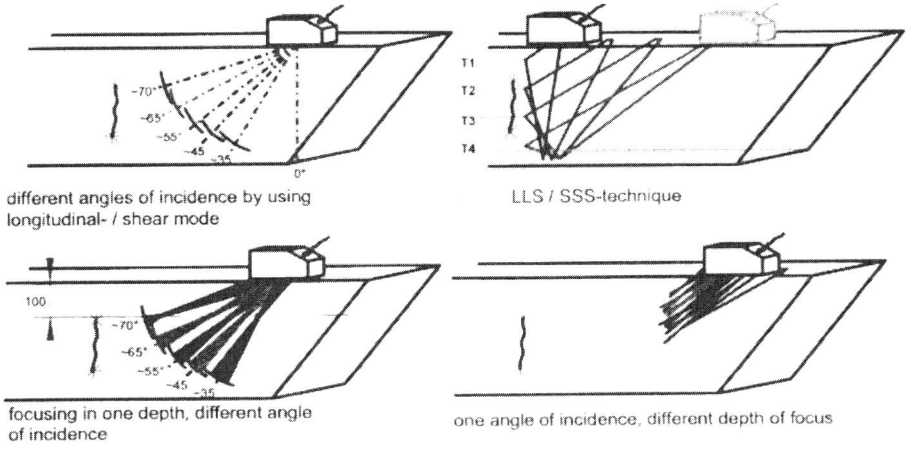

Figure 2
Phased-array features for detection and sizing of defects (L = longitudinal; S = shear wave).

proved to be effective and efficient in view of the close schedule. The flexible programming possibility of UT systems is very helpful in facilitating an immediate re-scan of the area of concern with a different set of angles of incidence and activating extended analysis features in the evaluation software (Fig. 2).

The phased-array system can directly discriminate indications caused by cladding influence and other spurious indications from "true" indications by using the same angles of incidence with longitudinal wave mode for discrimination already based on the detection scan. For the examinations of the nozzle to shell welds, the trimetallic nozzle to safe end welds and the austenitic safe end to pipe welds, phased-array techniques are used as well. For the step of defect characterization and sizing, the same approach is utilized as mentioned above for the shell inspection. It is obvious that the probe parameters have to be adapted to the specific geometry- and material-related-requirements for optimized transducer sizes, frequencies and directions of beam variation (horizontal or vertical).

For the safe end to pipe welds, the x-ray examination previously had been an additional time-consuming item on the ISI program. Based on a demonstration of performance on a full scale test-block including the most stringent trimetallic material conditions as well as the stainless steel weld boundary conditions, with realistic defects, the effectiveness of this UT inspection has been proven to detect all the defects present and to give satisfactory sizing results. Therefore the UT examination is regarded as a viable alternative to RT examination, so that the time-consuming RT examination can be avoided.

2. *Advanced Manipulator and Robotic Technology*

To support reduced inspection time, requirements led to focussing efforts on innovative manipulator and robot system (Dirauf *et al.* 1998). Most of the systems used up to now were developed for a specific application like the central mast manipulator, with which more than 130 PWR RPV inspections have been carried out in most of the reactor types (Fig. 3).

However, this strategy led to a large number of manipulators each for a specific inspection task, therefore each with a very limited usage factor. To cut down investment and maintenance costs, the new generation of manipulators have to be based on a modular robotic system where each robotic axis operates as a self-sufficient module, having integrated a motor, a transmission, a position encoder and a brake including a miniaturized electronic control with intelligent microcontroller. These standardized robotic elements can easily be combined to complex robotic systems consisting of linear and rotating axis depending on the inspection task. Each module has its own digital address and will be operated via the

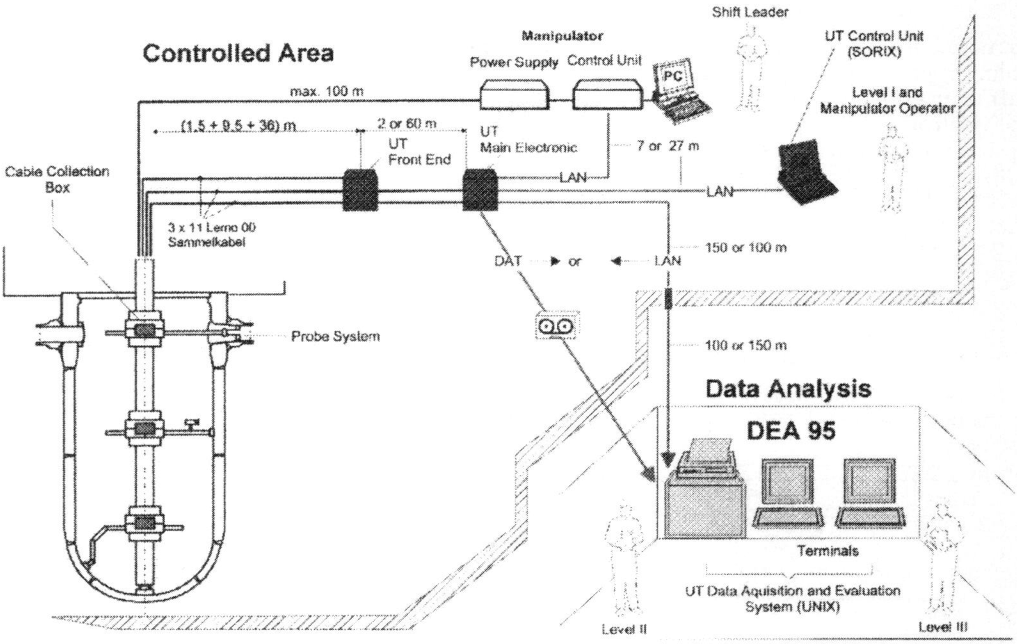

Figure 3
Schematic system set up for the PWR RPV-inspection.

Figure 4
Inspection robot for BWR reactor pressure vessels with four phased-array probes (each with vertical skew) for the weld inspection and one phased-array probe (horizontal skew) for nozzle inner radius inspection.

serial and parallel ports to the central robotic control unit and to an industrial PC. Independent of the number of motion axes only one thin connecting cable is necessary to operate all robotic modules, including the power supply. One of the most advanced systems consists of a six-motion axis for the inspection of BWR RPVs from the outer surface (Fig. 4).

3. *Conclusions*

The technology for the in-service inspection of the components of the primary circuits of the reactor systems was driven by a dynamic process to achieve the high standard of reliability and efficiency of today as a combination of a well trained and experienced inspection team and advanced and reliable technologies.

Several technology leaps have contributed to that level:

(i) Implementation of phased-array probes in most of the inspection tasks
(ii) Major improvements in computer technology to handle high speed large data flows collected by advanced UT systems, and
(iii) New robotic systems based on modular decentralized elements controlled by an easy-to-apply bus system.

Under consideration of current innovative approaches further progress can be expected, e.g., integration of major parts of electronics into the probe head resulting in higher sensitivity and allowing one to achieve the final step of integration by designing phased-array matrix elements, allowing one to generate sound fields in any defect orientation by just one phased-array probe.

However, these innovative approaches can only be realized if the corresponding improvements lead to higher efficiency in a very competitive limited market.

Bibliography

Dirauf F, Kamieth M, Fischer E, Eggers H, Neundorf B 1998 Advanced BWR-RPV O. D. inspection utilizing robotics and phased array technology. *EPRI Conf.* Santa Fe, CA

Engl G, Mohr F, Fischer E, Gremme W 1998 Technical and economical advantages for the inspection of reactor components by application of phased-array techniques. *EPRI Phased Array Inspection Seminar*. Portland, OR

Gohlke B 1998 New generation phased-array technology aims to cut vessel inspection costs,. *Nucl. Eng. Int.* (March), 26–8

Pannaye E, Prince S, Neeson H, Guse G, Engl G 1998 Experience with the application of the phased array technology in the ISI of the Koeberg NPP reactor pressure-vessels. *EPRI Phased Array Inspection Seminar*. Portland, OR

Novel Nitride Phosphors

H. T. Hintzen and Y. Q. Li

Despite intensive research on rare-earth-doped luminescent materials with different structures and varying compositions (Blasse and Grabmaier 1994), the possibility of creating novel promising phosphors based on (oxy)nitrides was only considered a few years ago (Hintzen *et al.* 1996). During the last few years, however, interest in this new type of materials has grown rapidly with the result that novel phosphor materials have been invented with unconventional properties, enabling new or improved applications. In this contribution an overview will be given of the progress obtained so far, as well as some remarks will be made about future prospects. The focus is on rare-earth-doped (oxy)nitride phosphors in which silicon and/or aluminum is present.

1. *Ce- and Tb-doped Rare-earth Silicon Oxynitrides*

Long-wavelength Ce^{3+} $5d \rightarrow 4f$ band emission was observed in Y–Si–O–N materials (van Krevel *et al.* 1998); its exact position depends on the nephelauxetic effect, crystal field splitting of the $5d$-level and Stokes shift. The crystal field splitting decreases when more N^{3-} versus O^{2-} is coordinated to Ce^{3+}, which is ascribed to the higher formal charge of nitrogen versus oxygen (van Krevel *et al.* 1998). The Stokes shift becomes smaller in the series $Y_5(SiO_4)_3N > Y_4Si_2O_7N_2 > YSiO_2N > Y_2Si_3O_3N_4$, related to an increasing rigidity due to more extended cross-linking between SiX_4 ($X = O$, N) tetrahedra in this direction (respectively: single SiX_4 units, Si_2X_7 dimers, Si_3X_9 rings, and $(Si_3X_7)_n$ layers).

The highest emission wavelength was observed for $Y_2Si_3O_3N_4$:Ce (≈ 504 nm), whose position can only be negligibly influenced by replacement of $(SiN)^+$ by $(AlO)^+$ resulting in $Y_2Si_{3-x}Al_xO_{3+x}N_{4-x}$: Ce compounds (van Krevel *et al.* 2000). Despite a high solid solubility up to $x = 0.6$ changing significantly the chemical composition of the host lattice, evidently the local coordination around the Ce^{3+} ion nearly remains the same. This effect was ascribed to the larger ionic size of Ce^{3+} as compared to Y^{3+} resulting in preferential coordination of the dopant ion with oxygen as compared to nitrogen (van Krevel *et al.* 2000).

In a similar way the position of the $5d$ excitation band of Tb^{3+} incorporated in these host lattices can be shifted (van Krevel 2000). For $Y_4Si_2O_7N_2$: 10%Tb the excitation maximum at 255–265 nm is overlapping with the absorption edge of the host lattice, resulting in both a high absorption (90%) and quantum efficiency (85%) for 254 nm excitation (van Krevel 2000). In this way the necessity of sensitization of the Tb emission in commercial phosphors by co-doping with Ce (Blasse and Grabmaier 1994) can be avoided.

2. *Rare-earth-doped α-Sialon Materials*

α-Sialon (for α-Si-AlON) materials originate from α-Si_3N_4 in which Si has been replaced by Al, N has been replaced by O, and for charge compensation a metal M is added, resulting in the general composition $M_{m/val}Si_{12-(m+n)}Al_{(m+n)}O_nN_{16-n}$ ($M =$ Li, Ca, rare earth). These materials are well known because of their high durability under chemical, thermal, and mechanical attack. Based on the criterion of bond directness, these matrices are considered to be promising for rare-earth activation (Grekov and Chernovets 1998).

Doping the α-Sialon matrix with Tb^{3+} gives the characteristic green $^5D_4 \rightarrow {}^7F_5$ emission ~ 545 nm (Xie *et al.* 2002, van Krevel *et al.* 2002), but with low quantum efficiency for 254 nm excitation (van Krevel *et al.* 2002). This is probably due to overlap of the $5d$ Tb^{3+} excitation band with the α-Sialon host-lattice conduction band resulting in radiationless transitions by photo-ionization processes (van Krevel *et al.* 2002).

Pr^{3+}-doped α-Sialon shows red luminescence due to $4f \rightarrow 4f$ transitions from the $^3P_0$ and $^1D_2$ emitting levels (Xie *et al.* 2002).

The emission band of Ce^{3+} in the α-Sialon matrix is at quite a high wavelength (≈ 495 nm (Grekov *et al.* 1999), 515–540 nm (van Krevel *et al.* 2002)). The exact position of the $5d \rightarrow 4f$ emission band depends on the excitation wavelength, indicating that some variation exists in local Ce coordination (van Krevel *et al.* 2002).

For Eu^{2+} doping a similar $5d \rightarrow 4f$ band emission was observed in the range 550–590 nm (Xie *et al.* 2002, van Krevel *et al.* 2002). The emission peak shifts to a longer wavelength with increasing amounts of europium (Xie *et al.* 2002). Despite systematic changes in chemical composition of the host lattice (i.e., m and n), the emission is not strongly influenced. This indicates that the Eu^{2+} ion dominates its local coordination, induced by the difference in size and the available space in the interstitial site (van Krevel *et al.* 2002). The Eu-doped α-Sialon samples are yellow-green colored (Xie *et al.* 2002, van Krevel *et al.* 2002), and therefore can be efficiently excited by visible radiation in the range 400–450 nm.

Recently, long wavelength band emission was also reported for rare-earth ions in Sialon glasses (up to 500 nm for Ce^{3+} (de Graaf *et al.* 2003a), and as long as 640 nm for Eu^{2+} (de Graaf *et al.* 2003b)).

3. *Eu-doped Alkaline-earth Silicon Nitrides*

Deep-red band emission was discovered by van Krevel (2000) for Eu^{2+} incorporated in alkaline-earth silicon nitrides, such as $M_2Si_5N_8$: Eu ($M =$ Ca, Sr,

Ba; $\lambda_{em} = 620–640\,nm$) and $BaSi_7N_{10}$: Eu ($\lambda_{em} \approx 660\,nm$). Emission above 600 nm is very uncommon for Eu^{2+} (Blasse and Grabmaier, 1994), and this is a consequence of the large nephelauxetic effect and large crystal field splitting caused by nitrogen. Due to a concomitant shift of the $4f \rightarrow 5d$ absorption band into the visible region, these compounds are strongly orange-red colored. Moreover, these materials are capable of efficiently converting the absorbed 400–450 nm radiation into red light. Because of the high application potential for light emitting diode (LED) lighting when combined with UV-blue emitting LEDs, in 1999 a patent was filed on such materials (Hintzen *et al.* 2001). In the meantime, also the thermoluminescence behavior of $Ba_2Si_5N_8$: Eu was reported by Höppe *et al.* (2000).

For Eu doped in host lattices with a higher M/Si ratio (i.e., lower degree of cross-linking between SiN_4 tetrahedra), the emission is at lower wavelengths, i.e., blue-green emission at ~515 nm for $MgSiN_2$: Eu (Gaido *et al.* 1974) and red emission at ~620 nm for $CaSiN_2$: Eu (Lee *et al.* 1997). For $MgSiN_2$ also the luminescence of the rare-earth ions Sm and Yb was studied (Dubrovskii *et al.* 1981). While generally excitation takes place with electromagnetic radiation, for $MgSiN_2$: Eu also data have been published for cathode-ray excitation (Bondar *et al.* 2001) and for $CaSiN_2$: Eu also electroluminescence was reported (Lee *et al.* 1997).

The purity of the nitride phosphor can have a strong effect, not only on the luminescence efficiency, but also on the emission characteristics. A nice example is given by $BaSi_7N_{10}$: Eu, for which a drastic influence of the presence of oxygen was established on the position of the Eu band. It has been shown that the dominating factor is not just the oxygen concentration itself but the type of oxygen (Li *et al.* 2003). Charge compensating the oxygen present by aluminum (i.e., replacement of $(SiN)^+ \rightarrow (AlO)^+$) keeps the emission at the same position as found for "oxygen-free" $BaSi_7N_{10}$: Eu (Li *et al.* 2003).

Large amount of oxygen results in silicon-oxynitrides, in which deep red Eu^{2+} emission at 650 nm was reported for $LaEuSi_2N_3O_2$, while green band emission at 549 nm was found for $La_{0.9}Eu_{0.1}Si_3N_{5-x}O_x$ (Uheda *et al.* 2000). The amount of oxygen can be further increased by replacement of $(SiN)^+ \rightarrow (AlO)^+$ and/or $(LaN)^0 \rightarrow (MO)^0$ (M = alkaline earth), resulting in, for example, $SrSiAl_2O_3N_2$ for which blue-green Eu^{2+} emission and blue Ce^{3+} emission were reported at about 487 nm and 466 nm, respectively (Fiedler *et al.* 2002).

4. Eu-doped Aluminum-oxynitrides with β-Alumina Structure

The previously mentioned replacement $(SiN)^+ \rightarrow (AlO)^+$ can be used in the reversed direction for the design of oxynitrides starting from aluminates. Another way to obtain oxynitrides from M-aluminates is to replace $(M^{n+}O)$ by $(M^{(n+1)+}N)$. Examples are given by $LaAl_{12}O_{18}N$ with the magnetoplumbite structure (Wang *et al.* 1988) (derived from $SrAl_{12}O_{19}$ by replacement of $(SrO)^0$ by $(LaN)^0$) and $BaAl_{11}O_{16}N$ with the β-alumina structure (Jansen *et al.* 1997) (derived from $NaAl_{11}O_{17}$ by replacement of $(NaO)^+$ by $(BaN)^+$). Doping this last host lattice with Eu^{2+} gives a blue-emitting phosphor enabling combination of a higher color rendering index with a high lumen output of a tricolor lamp (Jansen *et al.* 1999) as compared to the commercial $BaMgAl_{10}O_{17}$: Eu blue-emitting phosphor (which host lattice can be derived from $NaAl_{11}O_{17}$ by replacement of $(NaAl)^{4+}$ by $(BaMg)^{4+}$). Moreover, the presence of N^{3-} ions prevents the conversion of Eu^{2+} into Eu^{3+}, which is reported to be the main cause for the degradation of $BaMgAl_{10}O_{17}$: Eu (Oshio *et al.* 1998).

5. Outlook

The future of rare-earth-doped (oxy)nitride phosphors is bright, from a scientific as well as an industrial point of view. Despite its short research history, rare-earth-doped (oxy)nitride phosphors have already demonstrated to be promising for lighting applications. So far it is claimed that nitride phosphors improve the efficacy (~5%) of traditional high-pressure mercury vapor lamps (Grekov 2003). More important, future lighting will be revolutionized by LED lamps (Zukauskas *et al.* 2002), for which application of (oxy)nitride phosphors turns out to be very attractive. It goes without saying that many new materials with unconventional or improved properties still remain to be discovered. Work is in progress to study structure–property relationships with the objective to explore the potential of rare-earth activated (oxy)nitride materials for various applications in general and LED lighting in particular.

See also: Rare Earth Phosphors: Fundamentals and Applications; Light Conversion: Organized Lanthanide-containing Molecular Systems; Electroluminescent Phosphors

Bibliography

Blasse G, Grabmaier B C 1994 *Luminescent Materials*. Springer, Berlin

Bondar V, Axelrud L, Davydov V, Felter T 2001 Synthesis, structure and luminescence of $A^2B^4C^5$ nitrides. *Mat. Res. Soc. Symp. Proc.* **639**, G11.36.1–6

de Graaf D, Hintzen H T, de With G 2003a The influence of the composition on the luminescence of Ce(III)–Ln–Si–Al–O–N glasses (Ln = Sc, Y, La, Gd). *J. Lumin.* **104**, 131–6 de Graaf D, Hintzen H T, Hampshire S, de With G 2003b Long wavelength Eu^{2+} emission in Eu-doped Y–Si–Al–O–N glasses. *J. Eur. Ceram. Soc.* **23**, 1093–7

Dubrovskii G P, Zykov A M, Chernovets B V 1981 Luminescence of rare-earth activated $MgSiN_2$. *Izv. Akad. Nauk, SSSR Neorg. Mater.* **17**, 1421–5

Fiedler T, Jermann F, Ellens A, Zachau M 2002 Novel nitride-based phosphors for LED applications. 201st ECS Meeting, Philadelphia, Pennsylvania, USA, abstract 1191

Gaido G K, Dubrovskii G P, Zykov A M 1974 Photoluminescence of $MgSiN_2$ activated by europium. *Izv. Akad. Nauk SSSR, Neorg. Mater.* **10**, 564–6

Grekov F F 2003 Luminescent oxynitrides. http://home.earthlink.net/~grekov/LumMaterials.htm

Grekov F F, Chernovets B V 1998 Criteria for selection of matrices for luminophors activated with lanthanides. *Russ. J. Appl. Chem.* **71**, 194–7

Grekov F F, Feopyentov A V, Chernovets B V, Chernysheva T E 1999 Effect of a complex of coactivators on photoluminescence in silicon nitride. *Russ. J. Appl. Chem.* **72**, 931–6

Hintzen H T, Ronda C R, Metselaar R 1996 Rare-earth activated oxynitrides a novel class of luminescent materials. In: Curien H, Laurent Y, Verdier P (eds.) *International Symposium on Nitrides*. Saint-Malo, France, p. A19

Hintzen H T, van Krevel J W H, Botty G F M 2001 Red-emitting luminescent materials. *Eur. Pat. Appl.* EP 1104799 A1

Höppe H A, Lutz H, Morys P, Schnick W, Seilmeier A 2000 Luminescence in Eu^{2+}-doped $Ba_2Si_5N_8$ fluorescence, thermoluminescence, and upconversion. *J. Phys. Chem. Solid* **61**, 2001–6

Jansen S R, Hintzen H T, Metselaar R 1997 Phase relations in the $BaO–Al_2O_3–AlN$ system materials with the β-alumina structure. *J. Solid State Chem.* **129**, 66–73

Jansen S R, Migchels J, Hintzen H T, Metselaar R 1999 Eu-doped barium aluminium oxynitride with the β-alumina-type structure as a new blue-emitting phosphor. *J. Electrochem. Soc.* **146**, 800–6

Lee S S, Lim S, Sun S S, Wager J F 1997 Photoluminescence and electroluminescence characteristics of $CaSiN_2$: Eu phosphor. In: Hariz A (ed.) *Proceedings of SPIE—The International Society for Optical Engineering SPIE*. Adelaide, Australia, Vol. 3241, pp. 75–83

Li Y Q, Hintzen H T, de With G Effect of oxygen on luminescence properties of Eu^{2+}-doped $BaSi_7N_{10}$ phosphor (to be published)

Oshio S, Matsuoka T, Tanaka S, Kobayashi H 1998 Mechanism of luminance decrease in $BaMgAl_{10}O_{17}$: Eu^{2+} phosphor by oxidation. *J. Electrochem. Soc.* **145**, 3903–7

Uheda K, Takizawa H, Endo T, Yamane H, Shimada M, Wang C M, Mitomo M 2000 Synthesis and luminescent property of Eu^{3+}-doped $LaSi_3N_5$ phosphor. *J. Lumin.* **87–89**, 967–9

van Krevel J W H 2000 On new rare-earth doped M–Si–Al–O–N materials luminescence properties and oxidation resistance. P.hd. thesis, Eindhoven University of Technology.

Van Krevel J W H, Hintzen H T, Metselaar R, Meijerink A 1998 Long-wavelength Ce^{3+} emission in Y–Si–O–N materials. *J. Alloys Comp.* **268**, 272–7

van Krevel J W H, Hintzen H T, Metselaar R 2000 On the Ce^{3+} luminescence in the melilite type oxynitride compound $Y_2Si_{3-x}Al_xO_{3+x}N_{4-x}$. *Mat. Res. Bull.* **35**, 747–54

van Krevel J W H, van Rutten J W T, Hintzen H T, Metselaar R, Mandal H 2002 Luminescence properties of terbium-, cerium-, or europium-doped α-Sialon materials. *J. Solid State Chem.* **165**, 19–24

Wang X H, Lejus A M, Vivien D, Collongues R 1988 Synthesis and characterization of lanthanide aluminum oxynitrides with magnetoplumbite-like structure. *Mater. Res. Bull.* **23**, 43–9

Xie R J, Mitomo M, Uheda K, Xu F F, Akimune Y 2002 Preparation and luminescence spectra of calcium- and rare-earth (R = Eu, Tb, and Pr)-co-doped α-Sialon ceramics. *J. Am. Ceram. Soc.* **85**, 1229–34

Zukauskas A, Shur M S, Gaska R 2002 *Introduction to Solid State Lighting*. Wiley, New York

Nuclear Applications: Zirconium Alloys

S. Banerjee

A combination of several attractive properties such as low absorption cross-section for thermal neutrons, excellent corrosion resistance in water, and good mechanical properties at operating temperatures of nuclear reactors has given zirconium alloys the most distinctive position among all structural materials in nuclear applications. The important characteristics of the metal zirconium can be sketched as follows:

- low thermal neutron absorption cross-section (0.185 barn);
- allotropy, the high temperature b.c.c. (β) phase transforming into the h.c.p. (α) phase at 1135 K;
- anisotropic thermal and mechanical properties leading to unequal thermal expansions along different crystallographic directions and formation of strong crystallographic texture during mechanical working;
- h.c.p. structure of the α phase with a c/a ratio of 1.593, which is less than ideal, making the prismatic slip on $\{10\bar{1}0\}$ planes most predominant;
- high reactivity with oxygen, nitrogen, and carbon and high solubility of these interstitial elements in the α phase, necessitating special care during melting and fabrication processes;
- low solubility, 0.034 ppm at 293 K, of hydrogen in the α phase; excess hydrogen, if present, causing hydride precipitation which may result in serious embrittlement.

These attributes of zirconium play very crucial roles in the in-reactor behavior as well as in the thermomechanical processing of zirconium alloys. Design of zirconium alloys for meeting the service requirements of mechanical and corrosion resistance properties always takes into account the consideration of neutron absorption cross-section. Only a few alloying elements, such as niobium, tin, and oxygen, which have rather low neutron absorption cross-sections, are added to a limited extent in zirconium alloys for applications in nuclear reactors. Chemical compositions of some alloys suitable for nuclear applications are listed in Table 1, which also indicates the upper limits of the impurities such as hafnium, boron, and cadmium, which have very high thermal neutron absorption cross-section.

1. *Physical Metallurgy*

Alloying elements in zirconium alloys can be grouped into α-stabilizers and β-stabilizers. Tin,

Table 1
Chemical compositions of zirconium alloys for nuclear application.*

Alloy element	Zircaloy-2	Zircaloy-4	Zr-1Sn-1Nb	Zr-2.5Nb	Zr-1Nb	Excel	Zr-Nb-Cu	Ozhennite
Sn (wt.%)	1.2–1.7	1.2–1.7	0.9–1.1	–	0.9–1.1	3.5	–	0.2
Fe (wt.%)	0.07–0.2	0.18–0.24	0.1	<650 ppm	<500 ppm	–	–	0.1
Cr (wt.%)	0.05–0.15	0.07–0.13	–	<100 ppm	–	–	–	–
Ni (wt.%)	0.03–0.08	–	–	<35 ppm	<200 ppm	–	–	0.1
O (ppm)	900–1300	900–1400	900–1300	900–1300	<1000	–	900–1300	–
N (ppm)	<80	<65	<65	<65	<60	<60	<65	–
Nb (wt.%)	–	–	0.9–1.0	2.4–2.8	1.0	0.8	2.4–2.8	0.1
Cu (wt.%)	–	–	–	–	–	–	0.3–0.7	–
Mo (wt.%)	–	–	–	–	–	0.8	–	–
H (ppm)	<25	<25	<20	<5	<15	–	<25	–
C (ppm)	<270	150–400	–	<125	<200	–	<270	–
Cl (ppm)	<20	–	–	<0.5	–	–	–	–
P (ppm)	–	–	–	<10	–	–	–	–
	Fuel tube and plug material for BWR, PHWR	Fuel tube and plug material for PWR and PHWR	Experimental fuel tube alloy	Pressure tube material for PHWR	Fuel tube and end plug in VVER	Experimental pressure tube Alloy	Garter spring material for PHWR	

*All the alloys contain Al <75 ppm, B and Cd <0.5 ppm, Co and Mg <20 ppm, and Hf <50–150 ppm.

oxygen, and aluminum belong to α-stabilizers as the addition of these elements raises the α to β transformation temperature. Generally, peritectoid reactions are exhibited in the zirconium-rich end of the binary phase diagrams of alloys with α stabilizing elements. The β phase is stabilized by the addition of iron, chromium, nickel, niobium, molybdenum, and hydrogen; the corresponding binary diagrams usually showing eutectoid or monotectoid reactions in the zirconium-rich end. The wide variety of phase transformations present makes zirconium alloys amenable to microstructural control through suitable combinations of thermal and mechanical treatments.

Two types of athermal transformations can be induced in zirconium alloys, namely, the $\beta \rightarrow \alpha'$ martensitic transformation and the $\beta \rightarrow \omega$ displacive transformation. The Zr–Nb phase diagram is shown in Fig. 1 to illustrate the composition ranges over which the martensitic and the displacive transformations occur on quenching of Zr–Nb alloys from the β-phase field. The corresponding start temperatures are indicated by $M_s(\alpha')$ and $M_s(\omega)$ lines which intersect at niobium level of about 7 wt.% (wt.% and at.% are almost identical in these alloys). While alloys containing <7% Nb transform into h.c.p. α′ martensite, a $\beta + \omega$ structure is produced in alloys containing 7–20% Nb. Full retention of the β phase in a metastable state is possible in alloys richer than 20% Nb.

A strong tendency to phase separation of the β phase in Zr–Nb alloys is reflected in the presence of the miscibility gap $(\beta_1 + \beta_2)$ in the β-phase field and in the occurrence of the monotectoid reaction $\beta_1 \rightarrow \alpha + \beta_2$ at 610 °C. In dilute Zr–Nb alloys which find extensive applications in nuclear reactors, the metastable β_1 phase is usually present along with the predominant α phase.

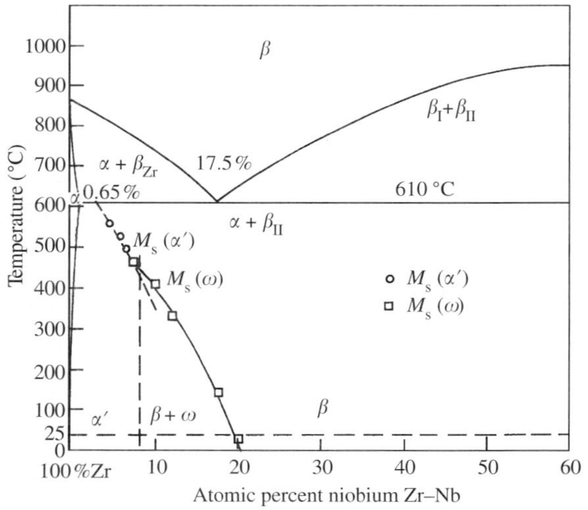

Figure 1
Phase diagram of the Zr-Nb system showing the start temperatures for martensitic [Ms(α′)] and displacive $\beta \rightarrow \omega$ [Ms(ω)] transformations.

A number of β-stabilizing elements, such as iron, nickel, chromium, and molybdenum, have very little solubility in the α phase and they form ordered intermetallic phases which are readily precipitated even in very dilute zirconium alloys resulting in precipitation hardening. The chemical ordering reaction in the Zr–Al system can produce the ordered intermetallic Zr_3Al phase ($L1_2$ structure). A substantially low self-diffusion rate in such ordered phases can be exploited in designing alloys with good creep resistance under irradiation.

A number of zirconium-based binary phase diagrams exhibit eutectoid reactions, with the eutectoid composition very lean in the alloying element. This offers the opportunity of producing a fine lamellar α + intermetallic structure (Fig. 2(a)) in systems like Zr–Cu and Zr–Fe. The possibility of exploiting such a microstructure in strengthening of zirconium alloys has not been explored as yet.

Zircaloys which contain tin, iron, chromium, and in some cases nickel (as shown in Table 1) are most widely used in nuclear reactors. They are essentially α-Zr–Sn solid solution containing fine intermetallic precipitates. Two distinct types of intermetallic particles are observed in zircaloy-2 (Fig. 2(b)), namely, Zr_2(Fe,Ni) and Zr(Cr,Fe)$_2$. The former has a b.c.c. tetragonal structure (space group: I4/mmm, D_{4h}^{17}) while the latter is a Laves phase with a C14 hexagonal structure. Usually precipitate particles formed at the grain boundaries of zircaloy-2 are larger (typical size ≈3 μm) and are of the Zr_2(Fe,Ni) type while intragranular precipitates (typical size ⩽1 μm) are of both Zr_2(Fe,Ni) and Zr(Cr,Fe)$_2$ types. Presence of stacking faults within the latter is the distinguishing feature of the Zr(Cr,Fe)$_2$ particles. Mechanical and corrosion properties of zircaloys (2 and 4) are strongly influenced by the size and distribution of the ordered intermetallic particles which form during the thermal and mechanical processing of zircaloy components.

2. *Deformation Behavior and Strengthening Mechanisms*

The axial ratio (c/a) of the α phase being less than ideal, α-zirconium alloys can be classified as hard h.c.p. material with the $\{10\bar{1}0\}\langle\bar{1}012\rangle$ slip system most predominant. The high ductility of α-Zr alloys cannot be accounted for solely by this prismatic slip which provides only two independent slip systems. Slip on pyramidal planes such as $\{11\bar{2}1\}$ and $\{10\bar{1}1\}$ become operative at high plastic strains and at elevated temperatures.

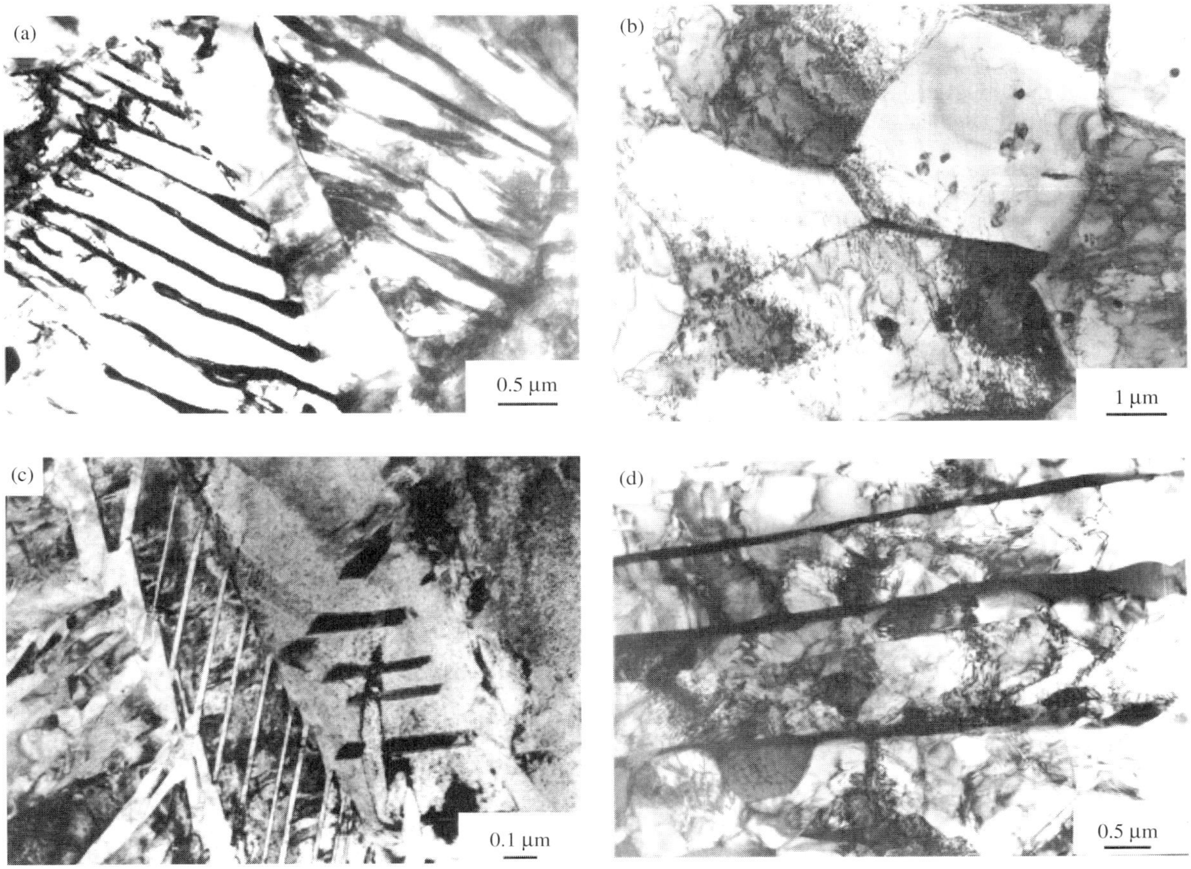

Figure 2
Micrographs of (a) eutectoid structure in Zr-Fe, (b) zircaloy-2 showing fine intermetallic precipitates in equiaxed α grains, (c) martensitic structure in Zr-2.5 Nb alloy, and (d) hot extruded Zr-2.5 Nb with fine lamellar $\alpha + \beta$.

Apart from slip, twinning plays an important role in the plastic deformation of zirconium and its alloys. A number of twinning systems are operative in zirconium alloys depending on the type of loading and the temperature and strain rate of deformation. Under a tensile stress along the c-axis, $\{10\bar{1}2\}\langle\bar{1}011\rangle$ twins are activated at the ambient temperature, while $\{11\bar{2}1\}\langle\bar{1}\bar{1}26\rangle$ twinning operates less frequently. In contrast $\{11\bar{2}2\}\langle\bar{1}\bar{1}23\rangle$ twinning is more dominant during compressive loading. At high temperatures $\{10\bar{1}1\}\langle\bar{1}012\rangle$ twins are observed.

Though the critical resolved shear stress is higher for $\{10\bar{1}2\}\langle\bar{1}011\rangle$ twinning than for prismatic slip at ambient temperatures, twinning is activated before slip in some favorably oriented grains. At the large strains obtained during mechanical processing, steady-state interactions occur between the twin and slip systems that tend to align the basal planes parallel to the direction of rolling and pilgering and perpendicular to the direction of extrusion. Wrought products of α-Zr alloys, therefore, contain a strong crystallographic texture which can be controlled by selection of appropriate conditions of deformation processing.

Pure zirconium containing <100 ppm oxygen in the annealed condition has a very low yield strength (≈ 150 MPa). It is, therefore, essential to strengthen zirconium metal mainly with the incentive of reducing the volume of structural material required in the core of nuclear reactors to minimize parasitic neutron absorption. Oxygen level in the range of 0.11–0.16% is recommended in most of the reactor grade zirconium alloys, lowering of the ductility with increasing oxygen content being the factor deciding the upper limit of oxygen. The solute strengthening contribution due to oxygen comes down with increase in temperature and becomes negligible at temperatures close to 300°C, the service temperature of pressurized water reactors (PWR) and pressurized heavy water reactors (PHWR).

Such a strong temperature dependence of the strengthening contribution of the interstitial oxygen suggests that the barriers introduced by oxygen atoms can be overcome by thermally activated dislocation motion at about 300°C. In contrast, solution strengthening by most of the substitutional solutes contributes towards the athermal component of the flow stress and is, therefore, retained at service temperatures. Nitrogen also acts similarly to oxygen as far as solution strengthening is concerned. But the corrosion resistance is adversely affected if nitrogen level exceeds about 0.07 wt.%.

As stated earlier the neutron economy consideration seriously restricts the selection and the level of alloy additions. Aluminum and tin are two alloying elements which have good solubilities in α-Zr and also possess low thermal neutron absorption cross-sections. While aluminum in α-Zr lattice is not acceptable because of it's deleterious effect on the corrosion properties, tin is extensively used in nuclear grade zirconium alloys for improving both corrosion resistance and strength. The deleterious effect of nitrogen impurity on the oxidation rate of zirconium is mitigated by tin which effectively complexes with nitrogen atoms present in α-zirconium.

Substitutional alloying elements like iron and chromium which possess atomic sizes considerably different from that of zirconium have very low solid solubility and for the same level of solute content the strengthening effects of these solutes are much larger than that of tin. However, the solution-strengthening potentials of these elements cannot be realized in annealed alloys where solutes in excess of the solubility limits are expelled from the α-Zr lattice to form intermetallic precipitates.

Precipitation hardening by a dispersion of hard second phase particles is commonly employed in zirconium alloys which can be supersaturated with solutes by a solution treatment either in the α or in the β field. The solubility of the majority of alloy additions such as niobium, iron, chromium, and nickel being very small in the α phase, a β-solutionizing treatment followed by quenching is required for bringing all these elements into solid solution. In zircaloys, alloying elements except tin are precipitated out during subsequent intermediate annealing operations which are necessary for softening and ductilizing the alloy after imparting a certain level of cold work. The corrosion resistance of zircaloys strongly depends on the size and distribution of intermetallic particles which grow during these annealing stages. In order to restrain the growth of intermetallic particles a control must be exercised on the cumulative annealing operations.

The β-quenching operation also introduces a martensitic structure (Fig. 2(c)) which is associated with both solid solution and substructure strengthening. In general, martensites in dilute zirconium alloys possess adequate ductility and are amenable to further ductilization by subsequent tempering which reduces the level of supersaturation and introduces second-phase precipitates. In fact, the Zr-2.5 Nb alloy in the tempered martensitic condition has tensile strength about 30% higher than that in the cold worked condition. In spite of the higher strength associated with the tempered martensitic condition, the alloy is used in the cold worked state for pressure tube applications in PHWRs because of better in-reactor creep properties of the latter.

Work hardening is another important strengthening mechanism which remains effective at reactor operating temperatures. Zirconium alloys for stressed components such as pressure tubes are therefore used in the cold-worked and stress relieved condition. The level of cold work is chosen to optimize the strength and the resistance against in-reactor creep and growth processes.

Order hardening of zirconium alloys was considered very seriously, mainly to combat the in-reactor creep process. Since the movement of point defects in ordered structures is much slower, there is a higher probability of mutual recombination of vacancies and interstitials produced in a radiation environment. The ordered Zr_3Al alloy ($L1_2$ structure) has high strength, good corrosion resistance, and very low in-reactor creep rate. But finally this alloy could not be accepted for reactor applications due to its tendency for amorphization under irradiation.

Due to the inherently anisotropic behavior of the h.c.p. α-phase which is the major phase in all zirconium alloys used in nuclear reactors, preferred orientations develop in these alloys during fabrication processing and recrystallization steps. The presence of such crystallographic texture is important for two reasons; first, the texture at a given step in the fabrication process will influence the ease with which the next deformation process can be imparted. Second, the texture of the fabricated material controls properties such as the yield strength, thermal and in-reactor creep strengths, distribution of hydride precipitates, and irradiation induced growth behavior. It is feasible not only to predict the life of components made of these textured materials but also to tailor the texture of the finished components by selecting the fabrication route and process parameters in order to meet the design requirements.

3. Consolidation and Fabrication

Zirconium alloys are consolidated by the melting of reactor-grade zirconium sponge with alloying elements by consumable electrode electric arc melting. The process involves cold compaction of zirconium sponge and alloy additions into briquettes which are welded into electrodes. The levels of interstitials such as O, N, C, and H are carefully controlled during the melting stage. Oxygen level is controlled by a proper selection of the quality of zirconium sponge (containing 600–1200 ppm of oxygen) and by appropriate addition of either high purity iodide-refined zirconium (containing as little as 100 ppm oxygen) or ZrO_2 as a source of oxygen. Consumable arc melting operations are repeated for reduction of hydrogen level and for bringing about a better chemical homogeneity of the ingot. Quadruple melting is currently being practiced in order to attain a hydrogen level of about 5 ppm and to reduce trace impurities.

Zirconium alloy components used in nuclear reactors are usually made from tube or plate. The first step of mechanical processing is hot forging, hot rolling, or hot extrusion. The first hot working step breaks down the cast structure and is carried out in the higher temperature range in the α or in the $\alpha+\beta$ phase field. This is often followed by a soaking in the β-phase field and subsequent quenching. This treatment homogenizes the chemical composition, randomizes the crystallographic texture, and produces a martensitic structure. The next hot working operation (extrusion for tubular products and rolling for flat products) introduces the major reduction in the cross-section. The hot working parameters such as temperature and strain rate have a pronounced effect not only on the microstructure of the alloy but also on its capacity to undergo subsequent cold working operations.

Several processes such as recovery, recrystallization, and precipitation occur dynamically during the hot working operation which is usually carried out either in the α or in the $(\alpha+\beta)$ phase field and not in the β-phase field in order to avoid excessive grain growth in the latter case. Soaking in the $\alpha+\beta$ phase field causes partitioning of the alloying elements in accordance with their solubilities in the α and β phases. Beta stabilizing elements, chromium, iron, nickel, and niobium migrate preferentially to the β phase while tin and oxygen remain primarily in the α phase. Hot working in the $\alpha+\beta$ phase field results in the formation of a two-phase microstructure in which the α and the β phase microconstituents are elongated in the form of stringers as seen in the $\alpha+\beta$ hot extruded Zr-2.5%Nb alloy (Fig. 2(d)). While the α-phase contains about 0.5% Nb, the β phase is enriched with 15–20% Nb. Hot extrusion is carried out with the material clad in copper to prevent contact with air. A close control of the dimensions of the hot worked material is necessary to ensure that the finished product fulfills the close dimensional tolerances.

The next stages of fabrication essentially involve a series of cold working steps with intermediate annealing treatments. One of the prime concerns in these operations is the development of a favorable crystallographic texture which is usually represented in terms of the distribution frequency of important crystallographic planes (e.g., basal or prismatic planes) with reference to the axial, radial, and circumferential directions for tubular products and to the rolling, normal, and transverse directions for plate products. The development of texture with increasing degree of cold work is due to the lattice rotations caused by a number of deformation systems. Twinning causes spontaneous lattice rotations through large angles though the resulting strain is small. Consequently at low values of plastic strain, the contribution of twinning is more significant in texture development. After large plastic strains, texture development is mainly controlled by slip deformation which tends to rotate the basal pole by $\approx 20°$ to $40°$ from the radial/ normal direction towards the circumferential/transverse direction, provided the major compressive deformation is along the former.

In the production of zircaloy tubing, tube reduction processes, such as pilger milling involving triaxial stresses and strains, are commonly employed. The reduction in the cross-section (RA), is achieved by reductions in the wall thickness (RW), and in the diameter (RD). Depending on the value of the ratio Q (= RW/RD), different crystallographic textures are developed in pilgered tubes. A high wall thickness to diameter reduction ($Q>1$) produces a texture with the basal pole aligned parallel to the radial direction while a deformation with $Q<1$ leads to a texture with the basal poles aligned parallel to the circumferential direction. Cold rolling of sheets also produces a similar texture with the basal poles being about 20–40° away from the normal direction. Texture development in different types of metal forming operations in zirconium alloys is schematically illustrated in Fig. 3.

The crystallographic texture of zirconium alloys influences the physical, mechanical, and corrosion properties both out-of-pile and in-pile. Under neutron irradiation single crystals of α-zirconium shrink along the c axis and dilate on the basal plane with the volume remaining more or less constant. This means a tube with a predominant circumferential basal pole texture will grow along the axial direction under irradiation. In order to minimize such growth, the fraction of basal poles along the axial direction need to be increased. Similarly, the creep strength along the hoop and axial directions of the cladding can be tailored by appropriately modifying the texture. Anisotropy of plastic deformation and resistance to stress corrosion cracking also depend on the texture. By far the most important application of texture control is in achieving a favorable orientation distribution of brittle hydride platelets in cladding tubings with a strong radial basal pole texture.

The service demands from zirconium alloys for different components in a nuclear reactor are considerably different. Since only a few alloys have found acceptance in nuclear reactors, tailoring their

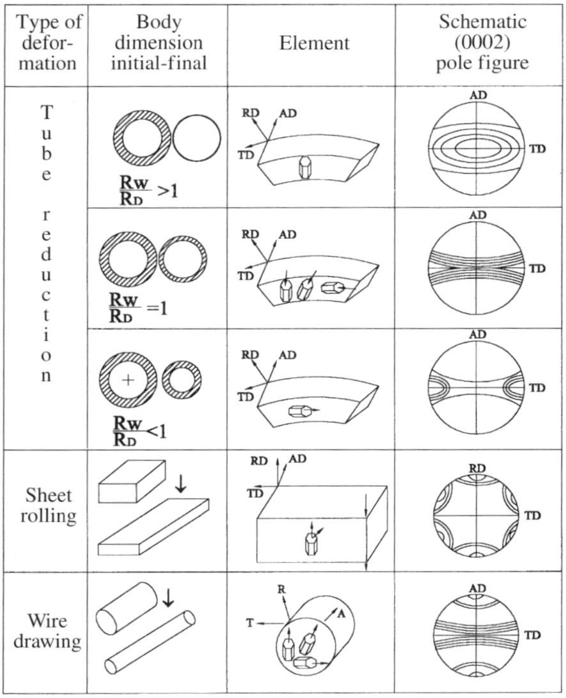

Figure 3
Texture developed due to pilgering, sheet rolling, and wire drawing (cold working) operations.

properties is often achieved through the selection of the right fabrication route. This is illustrated in Fig. 4 showing the important steps in the fabrication flow sheets used for different zirconium alloy components used in PHWRs and boiling water reactors (BWR). Tubes for fuel cladding can be grouped into two classes, namely, *collapsible* and *freestanding*. While the former finds application in PHWRs, the latter is used in PWRs and BWRs. Fuel tubes are exposed to very high fluence of fast neutrons which causes substantial loss of their ductility. In order to maintain integrity of cladding, fuel tubes must retain a minimum level of ductility (at least ~2% total circumferential elongation, TCE) at the end of their service life which is dictated by the desired fuel burn up. Giving the allowance for the drop in TCE during irradiation, a minimum of 15% TCE is specified for collapsible fuel tubes in PHWRs.

Transverse strength and ductility are of prime importance for freestanding cladding tubes used in PWRs and BWRs. Tube burst tests are used widely for assessing the suitability of fuel tubes. The stringent requirements of high strength and high TCE of these tubes can be met by introducing a heavy reduction in wall thickness in the last stages of pilgering. This is followed by annealing at an optimum temperature where the material is just recrystallized to produce a fine-grained structure that is free of residual stress. This fabrication schedule also generates a radial basal pole texture which is essential for orientating hydride plates along the circumferential direction. Calandria tubes in PHWRs experience neither a high stress nor high temperatures during service but they are expected to serve the entire design life of these reactors. The current fabrication practice of calandria tubes includes steps such as sheet rolling followed by tube forming, seam welding, rolling to level the welded seam, and vacuum annealing. The fabrication steps are so designed that the irradiation growth of these tubes remains within a tolerable limit during the design life of a PHWR.

The use of a zirconium alloy component in the pressure boundary of a nuclear reactor was, no doubt, a very bold decision in 1955 when the modular pressure tube design of the PHWR was introduced. Little was known of pressure tubes made of zirconium alloys other than that they were desirable and possible. The requirements for pressure tube alloys differ from those for fuel cladding. This is largely because of the fact that an improvement in the strength of the material can significantly reduce the quantity of parasitic material within the core. In addition, the pressure tube, unlike the fuel cladding, is expected to have a long service life preferably extending up to the design life of a reactor. High strength and good in-reactor creep resistance at the service temperature are, therefore, recognized as the most important properties.

Though zircaloy-2 was first chosen as the pressure tube material because of the confidence gained on this material in the reactor environment, Zr-2.5 Nb soon replaced zircaloy-2 because of a better strength and creep resistance and a lower hydrogen ingress rate in the former. The Zr-2.5 Nb alloy in the tempered martensitic condition has about 62% higher tensile strength than 25% cold-worked zircaloy-2. But after the use of heat-treated Zr-2.5 Nb pressure tubes in only two PHWRs, this alloy is now being used in all subsequent PHWRs in the cold-worked state, the latter exhibiting a lower in-reactor creep rate, the tensile strength being maintained at about 45% higher than cold-worked zircaloy-2.

The fabrication schedule of Zr-2.5 Nb pressure tubes, as largely practiced, involves hot working of the ingot in several stages either by forging or by extrusion followed by a single cold-working step like tube drawing which introduces the required 25% cold work and shapes the final product within the dimensional tolerances specified. The last hot-working step, hot extrusion, is carried out in the $(\alpha+\beta)$ phase field with an accurate control of the temperature, the deformation rate (ram speed), and the extrusion ratio. The volume fractions of the two phases, and consequently their compositions, the aspect ratios of the α and β grains in the $(\alpha+\beta)$ fibrous microstructure (Fig. 2(d)), and the crystallographic texture of the product are essentially determined by the process parameters of the last hot extrusion step. The combination of high strength with a good ductility and toughness (Table 2) of Zr-2.5 Nb pressure tubes is essentially derived from the fine $(\alpha+\beta)$ fibrous microstructure consisting of elongated α-grains containing about 0.5% Nb and the β-phase (15–20% Nb) stringers primarily located at α-grain boundaries. Such a structure cannot remain stable if it is subjected to a cold working—full annealing cycle which tends to coarsen the structure into a distribution of equiaxed α and β grains, resulting in a substantial drop in the ultimate tensile strength.

In recent years, a variation in the fabrication schedule has been adopted, the modified route consisting of cold working by drawing or pilgering in two steps with an optimized intermediate annealing treatment. Though the aspect ratios of the α and β fibers in the tubes manufactured by the modified fabrication route are much smaller, the specified high temperature mechanical properties including creep properties can be met and the in-reactor creep and growth behavior can be improved.

The requirement of the leak-before-break design practice lays emphasis on the improvement of fracture toughness of pressure tube materials. It has been reported that a reduction in the residual chlorine content (from the level of 5 ppm to <0.5 ppm) can significantly improve the fracture toughness value. Quadruple arc melting is being practiced to achieve purity with respect to chlorine (<0.5 ppm), hydrogen (<5 ppm), carbon (<125 ppm), and phosphorus (<10 ppm).

An autoclaving treatment is often introduced as the last step of fabrication of zirconium pressure tubes. Steam autoclaving at around 400°C for an extended period (24–72 h) is quite effective in stress relieving the finished component and in the formation of a thin compact black oxide film. This dense layer of zirconia is rich in the tetragonal allotropic form, a phase normally stable at high pressure and temperature, and provides protection against erosion and corrosion during in-reactor service.

Having discussed the physical metallurgy, deformation behavior, fabrication practices, and the texture evolution of zirconium alloys, the behavior of zirconium alloy components when exposed to a nuclear reactor environment will now be considered. The important processes which are responsible for the in-reactor degradation of material properties and for limiting the component life are (i) corrosion by coolant water and fission products; (ii) hydrogen ingress and consequential effects; (iii) irradiation damages; and (iv) dimensional changes due to creep and growth. The synergy of these different phenomena is responsible for restricting the life of zirconium alloy components.

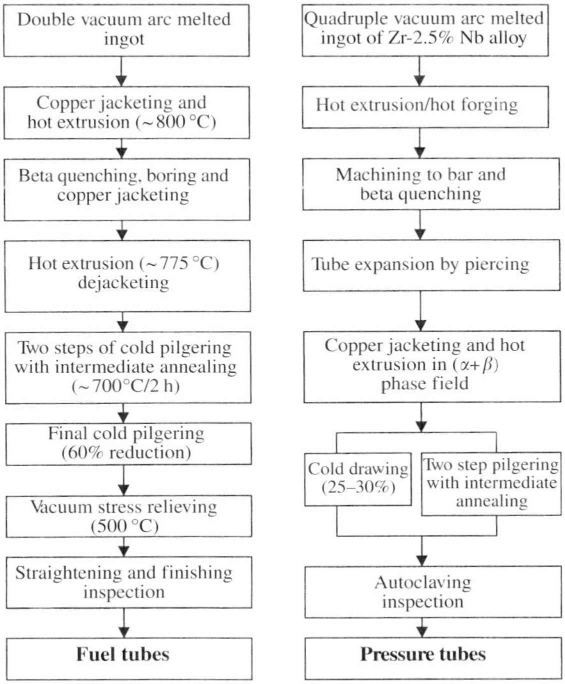

Figure 4
Important steps in fabrication flow sheets of zirconium alloy components for PHWR and BWR.

4. *Corrosion*

Zirconium alloy fuel cladding experiences two types of corrosion; one from the interactions with the coolant water on outer surface, the second with the fuel material and fission products inside the tube. Under most conditions of temperature and environment in a nuclear reactor, corrosion of zirconium alloys by coolant water/steam results in the growth of uniform oxide films, especially in the early stages. Only in two isolated regimes of temperature and environment, namely, in 300°C boiling (or oxygenated) water and in high-temperature, high-pressure steam ($\geqslant$450°C, $\geqslant$5 MPa), a local form of oxidation known as nodular corrosion occurs.

The growth of oxide layer as a result of the corrosion reaction between zirconium alloys and water can be described in terms of two distinct stages, usually known as pre- and post-transition stages. The pretransition stage is characterized by a decreasing rate of weight gain which corresponds to a cubic or quadratic kinetics relation between the weight gain and time (Fig. 5). The deviation from the usual parabolic growth kinetics in this stage is attributed to the fact that the rate-controlling process of the diffusion of oxygen ions through the oxide layer is not by lattice diffusion but by grain boundary diffusion.

Table 2
Mechanical properties of zirconium and zirconium alloys for nuclear application.

Alloy	Temperature	0.2% yield strength (MPa)	Tensile strength (MPa)	Percent elongation (TCE in %)	Fracture toughness	Creep strain rate
Pure Zr	Room temperature	80	220	45	–	At $\sigma = 88$ MPa and 310°C ($\sim 1 \times 10^{-4}\,h^{-1}$)
	300°C	45	120	55	–	
Zircaloy-2 (Fully annealed)	Room temperature	275	435	20 (18)	–	At 300°C, 110 MPa ($\sim 2.4 \times 10^{-8}\,h^{-1}$, out-of-pile)
	343°C	103	200	–	–	($3.0 \times 10^{-7}\,h^{-1}$, in pile)
Zircaloy-4 (Stress relieved)	Room temperature	410	480	20 (10)	–	–
Zr-1Nb (Annealed)	Room temperature	205	272	28	–	At $\sigma = 88$ MPa and 350°C ($\sim 1 \times 10^{-4}\,h^{-1}$)
	380°C	127	145	33	–	
Zircaloy-2 (Cold worked)	Room temperature	420	580	18	$K_{JC} = 55–65$ MPa.m$^{0.5}$	–
	300°C	270	330	12	$K_{JC} = 60–75$ $K_{IH} = 10–17$ MPam$^{0.5}$	–
Zr-2.5 Nb (Cold worked)	Room temperature	600	650	12	$J_{0.2} = 120–160$ kJm^{-2} dJ/da = 120–360 MPa	At $\sigma = 150$ MPa and 350°C ($\sim 1 \times 10^{-4}\,h^{-1}$)
	300°C	325	460	14	$K_{IH} = 6–9$ MPam$^{0.5}$	

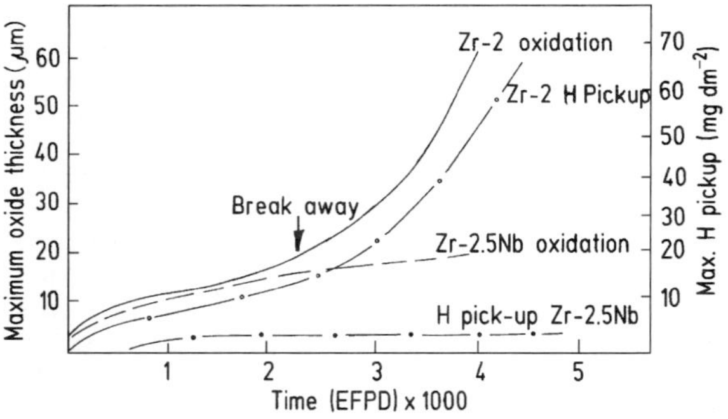

Figure 5
Long-term, in-reactor, oxidation, and hydrogen pick up behavior of zircaloy-2 and Zr-2.5 Nb pressure tubes, showing parabolic and then accelerated linear oxidation and hydrogen pickup kinetics in zircaloy-2. A low and uniform rate of corrosion and hydrogen pick up is seen in the in Zr-2.5 Nb alloy.

The oxide layer grows in the early stages maintaining epitaxy. The ratio of the oxide volume to that of the parent metal (Pilling-Bedworth ratio) is 1.56 for the Zr/ZrO_2 system.

Thus, as the oxide grows, the stress buildup due to the volume expansion accompanying oxidation induces a fibrous texture of the oxide crystals which minimizes the compressive stress in the plane of the oxide layer. The presence of the compressive stress is also a factor responsible for the stabilization of the tetragonal phase in the ZrO_2 layer. An examination of the metal/oxide interface reveals that a thin layer ($\sim$10 nm thick) of amorphous oxide is present right on the metal surface. Fine crystallites of zirconia are distributed in the amorphous matrix and the volume fraction of the crystalline phase and the size of crystallites increase with the distance from the metal surface. The growth of oxide crystals finally results in the formation of a columnar structure. The intercrystalline boundaries provide the diffusion path of O^{2-} ions from the environment to the reaction front at the metal/oxide boundary. The oxidation process and the nature of the oxide layer on a zirconium alloy sample are schematically illustrated in Fig. 6.

As the oxide layer grows, the compressive stress at the outer layer of oxide is not sustained and consequently the tetragonal phase becomes unstable and transforms into the monoclinic phase. Such a transformation causes the formation of a fine interconnected porosity in the oxide film which allows the oxidizing water to come in contact with the metal surface. With the development of an equilibrium pore and crack structure in the oxide layer, the oxidation rate effectively becomes linear, a characteristic

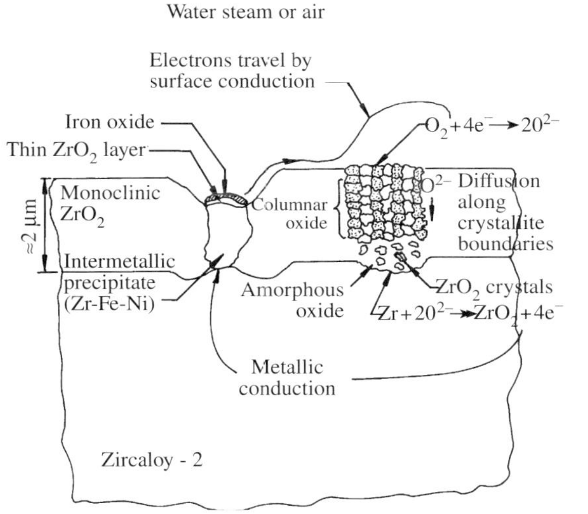

Figure 6
Schematic diagram showing the mechanism of the oxidation process and the oxide film structure on zircaloy.

feature of the post-transition oxidation behavior. Alloying elements, particularly tin, niobium, and iron, present in the α-solid solution strongly influence both the kinetics and the mechanism of oxide growth in zirconium alloys.

Nonuniform oxide formation, usually referred to as *nodular corrosion*, is the limiting design consideration in BWRs. Several mechanisms have been proposed for nodule nucleation which can occur in various sites such as α-Zr grain boundaries and in areas at which the continuous dense oxide layer ruptures in the initial stages due to the presence of large intermetallic precipitates. The fact that the best microstructure for resisting nodular corrosion in a BWR consists of fine intermetallic precipitates (diameter below 0.15 μm) is consistent with the latter possibility. It may be noted here that precipitate size has an opposite impact on corrosion rate in the PWR condition where fine precipitates (<0.1 μm) increase the uniform corrosion rate.

In addition to metallurgical factors, water chemistry has a strong influence on the corrosion process in zirconium alloys. The control of water chemistry as required in different reactor systems include suitable additions of lithium hydroxide, boric acid, hydrogen/deuterium, oxygen, iron, and zinc. In PWRs, boric acid is added to the coolant water for reactivity control. The pH of the coolant is adjusted by addition of lithium hydroxide which renders the coolant slightly alkaline, in order to reduce the corrosion rates of structural materials (stainless steels and Inconels) in the primary heat transport circuit and thereby inhibit deposition of corrosion products on the fuel cladding. Radiolysis of water produces oxidizing species which enhances oxidation rate of zirconium alloys in the radiation environment. The buildup of these can be suppressed by adding hydrogen in the coolant water. Dissolved hydrogen concentration in PWR coolant is maintained at a level of 2.2–4.5 ppm with a view to enhancing recombination with oxygen radicals formed by radiolysis. The PHWR coolant does not contain boric acid and deuterium is added in place of hydrogen. Dissolved oxygen in PHWR coolant is controlled between 10–50 ppb and it has been observed that corrosion in both zircaloy-2 and Zr-2.5 Nb pressure tubes rises to an exceptionally high value at high oxygen concentrations. The BWR coolant usually contains a higher level of oxygen, typically 200–400 ppb. Hydrogen addition in boiling water conditions is not very effective due to the segregation of hydrogen in the steam phase.

The interaction that fuel cladding tubes experience due to the fission products on the inner surface is stress corrosion cracking (SCC). Iodine, caesium, and cadmium are some of the fission products which contribute strongly to the SCC process. Due to the combined effects of thermal expansion of fuel pellets, fuel swelling due to fission gas release, and the coolant pressure on the outer surface of fuel tubes, cladding tubes collapse on the fuel pellets. In PWRs the closure of the gap between the pellet and the clad takes about one full power year while the gap is maintained for a longer period in BWRs. Due to the collapsible clad design of PHWRs, contact between the pellet and the clad is established right from the beginning of the life of the fuel element. Once the pellet/clad gap is closed, any change in pellet dimensions causes stress in the cladding and the synergy between the stress and the corroding medium induces the SCC process. In order to prevent or minimize pellet/clad interactions, both mechanical and chemical, a coating of graphite powder is applied on the inner surface of cladding tubes. The graphite layer provides lubrication to minimize the mechanical interactions, and acts as a barrier against the migration of harmful species to stressed regions of fuel tubes.

5. *Hydrogen Pick up*

The reaction between zirconium components and hot water coolant releases hydrogen as H^+ radicals which are also produced by radiolytic decomposition of water. Since hydrogen solubility for α-Zr is about 60 ppm at 300 °C, hydrogen can be picked up in zirconium alloy components during in-reactor service or even during autoclaving (Fig. 5). The hydrogen pick up process involves chemisorption of hydrogen at the tips of oxide pores, migration of hydrogen from oxide pores to the metal surface, and finally dissolution of hydrogen in α-Zr phase. Only a small fraction of the hydrogen radicals produced is absorbed in α-Zr as the remaining part recombines to form hydrogen molecules which dissolve in the coolant.

The dissociation of hydrogen molecules is catalyzed by nickel which is seen to have a very strong influence on the hydrogen pick up rate. The elimination of nickel has resulted in the reduction of hydrogen uptake rate in zircaloy-4 during service to a level of about 20% of zircaloy-2.

The solubility of hydrogen in α-Zr decreases sharply with temperature. A drop in temperature from 300°C to 30°C results in extensive precipitation of hydride plates in zirconium alloys. Two types of hydride phases are encountered in dilute alloys. They are the equilibrium δ-hydride ($ZrH_{1.66}$) with f.c.c. structure and the metastable γ-hydride (ZrH) with b.c.t. structure. The hydride phase tends to assume a plate shape forming along specific habit planes, $\{10\bar{1}0\}\alpha$ and $\{10\bar{1}7\}\alpha$ for minimizing the strain energy associated with the precipitation. The crystallography of the α to γ-hydride transformation suggests that the transformation process involves a combination of shearing of the α-Zr lattice and interstitial ordering of hydrogen atoms. Precipitate plates of δ-hydride usually cluster in the α-zirconium matrix, the macroscopic growth direction of hydride clusters lying in the direction of maximum tensile stress or in the basal plane if unstressed. Thus the crystallographic texture of zirconium alloy components and the stress distribution within a component control the orientation distribution of hydride precipitates.

Zirconium hydrides, being very brittle, cleave at a very low stress level. A significant reduction in ductility is observed when hydrides are unfavorably oriented with respect to the stress state. For example, a tubular component experiencing a hoop stress can fail at a very low stress if hydride plates are oriented in such a manner that their habit planes are predominantly lying along the radial direction. The primary objective for the control of texture of cladding tubes is to orient the basal planes of α grains predominantly parallel to the circumferential direction along which hydride precipitates form. Such a texture maximizes the hoop stress bearing capability of cladding tubes.

The strong temperature dependence of hydrogen solubility and the high diffusivity of hydrogen in the α phase cause considerable redistribution of hydride precipitates as zirconium alloy components experience temperature variations during start ups, shut downs, and power fluctuations in nuclear reactors. Diffusion of hydrogen up a stress gradient or down a thermal gradient is responsible for several phenomena such as stress-induced reorientation of hydride precipitates, delayed hydride cracking (DHC), and formation of hydride blisters at cold spots. The influence of stress on the distribution of hydride precipitates is demonstrated in Fig. 7(a) which shows reorientation of hydride precipitates in regions where tensile stress exceeded a threshold value. The migration of hydrogen up a stress gradient can cause a serious degradation process known as DHC.

The presence of high tensile stress near stress concentration drives hydrogen to migrate to regions of high tensile stress where hydride phase precipitates out once the solubility limit is exceeded. Hydride plates, not being able to support the concentrated stress at crack tips, undergo fracture resulting in the crack extension. The DHC process comprises dissolution of unfavorably oriented hydrides, hydrogen migration, hydride precipitation and growth, and fracture of hydrides when grown to a critical size leading to crack extension. Repetition of this cycle leads to stable discontinuous crack growth which is referred as the phenomenon of DHC. The crack growth rate due to DHC is higher for Zr-2.5 Nb than for zircaloy-2 owing to a higher strength and to the presence of nearly continuous β-stringers which provide easy hydrogen migration paths in the former. The DHC velocity increases with temperature in accordance with an Arrhenius relation with activation energy of around $70\,kJmol^{-1}$, which corresponds to hydrogen diffusion into triaxial stress field ahead of the crack tip.

Hydride precipitates also concentrate at cold spots where the solubility of hydrogen is low. Such a situation may be encountered in PHWRs when a pressure tube which is at the coolant temperature of about 300°C comes in contact with the calandria tube which is maintained at the moderator temperature of 60–80°C. At the point or the line of contact between a pressure tube and a calandria tube, the hydride volume fraction monotonically increases with time. Since the formation of hydride phase is accompanied by a volume increase, the regions in which the volume fraction of hydride is nearly 100% develop blisters (Fig. 7(b)). A row of blisters can join up to form a crack, which may eventually lead to unstable crack growth. In order to avoid such a catastrophic failure of pressure tubes, a constant vigil is maintained to detect whether contact between pressure tube and calandria tube has taken place.

6. *Irradiation Effects*

Irradiation damage of zirconium alloy components in a nuclear reactor environment results in accumulation of defect clusters, hardening, and accompanying reduction in ductility, dimensional changes brought about by irradiation creep, and growth and changes in the second phase structure. For a collision between a zirconium atom and a 1 MeV neutron, the primary knock-on atom (PKA) has a maximum energy of 44 keV. This PKA loses its energy in secondary collisions with other zirconium atoms in a collision cascade which is a localized region of high displacement density (typically 100–200 atomic

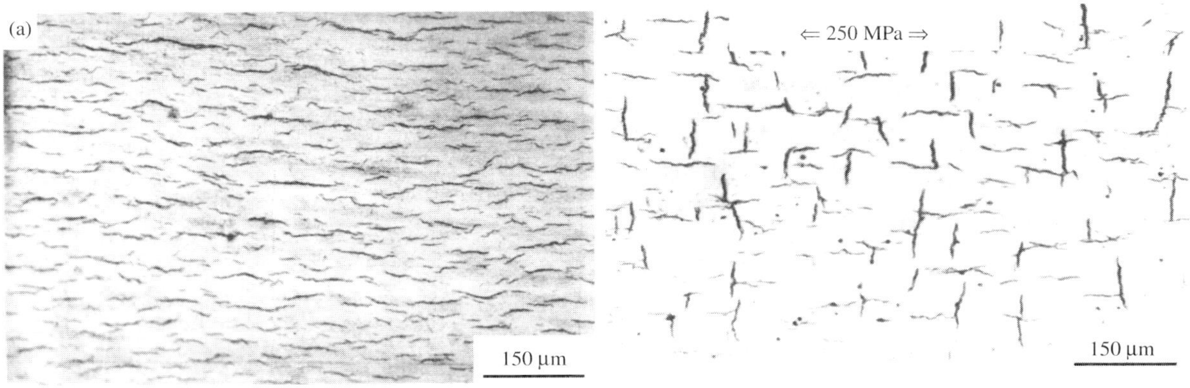

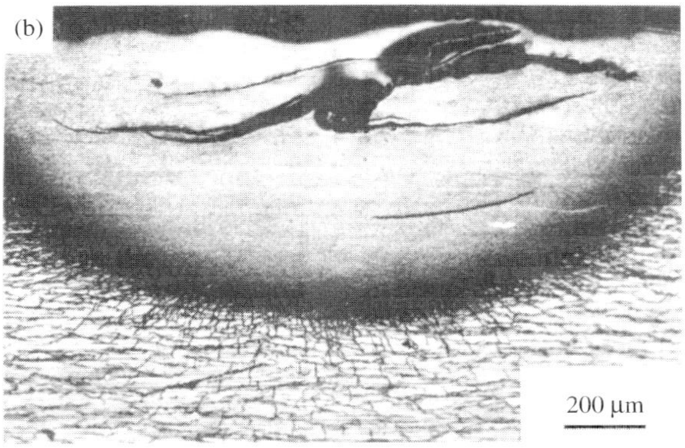

Figure 7
(a) Stress reorientation of circumferential zirconium hydride platelets (left hand-side) at 250 MPa stress level in the direction shown, and (b) a hydride blister in the zirconium alloy pressure tube section.

displacements). Changes in microstructure and in physical and mechanical properties due to radiation damage arise essentially from atomic displacements and the consequent generation of point defects. Radiation damage is, therefore, expressed in terms of displacements per atom (dpa). Typically, zirconium alloy fuel tubes will receive about 20 dpa in a PWR in the 3 years they reside in the reactor, the corresponding damage rate being 1×10^{-7} dpa s^{-1}.

In the case of zirconium alloys both vacancy and interstitial loops are seen in more or less equal number. Most of these dislocation loops are associated with the $\langle a \rangle$ type ($\frac{1}{3}\langle 11\bar{2}0 \rangle$) Burgers vector and they are arranged in layers parallel to the basal plane. The vacancy loops are generally stable up to a temperature of 723 K beyond which thermal emission of vacancies from dislocation loops causes disappearance of some vacancy loops. When the fluence exceeds around 3×10^{25} nm^{-2} at temperatures between 560 and 773 K, $\langle c \rangle$ type dislocation loops start appearing with a corresponding increase in the rate of irradiation growth.

The accumulation of defects under irradiation is manifested in the hardening and growth of zirconium alloys as a function of radiation dose. Both yield strength and ultimate tensile strength increase with fluence and are accompanied by a reduction in ductility and toughness. The effect is less pronounced in cold-worked material when compared with recrystallized material. The density of $\langle a \rangle$ type dislocations, fracture toughness, and strength values saturate when the fluence reaches about 5×10^{25} nm^{-2}. The hardening effect can be rationalized in terms of the requirement of a high critical stress for dislocation glide through planes consisting of an array of dislocation loops; the number density of which saturates with dose due to overlap of absorption volumes for interstitials and vacancies. Apart from this, the strengthening also comes from the defect cloud formation along the length of dislocation which locks them.

Irradiation growth which occurs in the unstressed material has been found to result from the expansion along the $\langle a \rangle$ direction and a corresponding contraction along the $\langle c \rangle$ direction in zirconium single crystals. The growth behavior of polycrystalline zirconium alloys also follows the same trend. For zircaloy cladding tubes in which basal poles are preferentially aligned along the radial direction, there is a tendency for the axial length to increase and the diameter and thickness to decrease due to irradiation growth. In PHWRs calandria tubes are expected to serve for the entire lifetime of the reactor and, therefore, a proper fabrication schedule is selected to produce a crystallographic texture which minimizes irradiation growth.

The present understanding of the irradiation growth phenomenon is based on the partitioning of interstitials and vacancies to different sinks such as dislocations of $\langle a \rangle$, $\langle c \rangle$, and $\langle c+a \rangle$ character and grain boundaries of different orientations. Since these sinks are anisotropically distributed, biased migration of vacancies and interstitials towards these sinks cause preferential growth and contraction in different directions. Irradiation growth is influenced by alloy composition, extent of cold work, residual stress, and irradiation variables such as flux and temperature. It has been observed that the growth strain in cold-worked material is considerably higher than that in recrystallized material and increases linearly with fluence. This linear steady-state behavior corresponds to a sink-dominated regime where most of the irradiation-produced defects are lost to dislocations.

Unlike stainless steels, zirconium alloys do not exhibit void swelling under neutron irradiation. The absence of 3D cavities in irradiated zirconium alloys is believed to be due to the fact that gaseous impurities like oxygen, hydrogen, and nitrogen, which stabilize void nuclei, are highly soluble in α-zirconium. This point has been proved by the experimental observation that swelling occurs in electron-irradiated zirconium which is preinjected with helium. The overall dimensional change of zirconium alloy components in a reactor result from the combination of irradiation growth and irradiation creep. At the operating temperature ($\sim 300\,^\circ C$) of nuclear reactors thermal creep rate in zirconium alloys is not of serious concern. However, in the presence of radiation the creep rate is considerably enhanced. In fact creep deformation can be the life-limiting factor for highly stressed components such as pressure tubes.

The influence of irradiation on creep processes (Fig. 8) can be grouped into three categories, namely irradiation-induced creep, irradiation-enhanced creep, and irradiation-retarded creep. Mechanisms responsible for irradiation-induced creep are those which can operate only in an irradiation environment and do not have any thermal creep counterpart. In contrast, several thermal creep mechanisms are either accelerated or impeded by the presence of point defects in excess of the equilibrium concentration in a radiation environment. These are respectively grouped as radiation-enhanced and radiation-retarded creep. Two main irradiation-induced creep mechanisms are: stress-induced preferred nucleation (SIPN) of dislocation loops and stress-induced preferred absorption (SIPA) of point defects by dislocations. SIPN can be visualized as a process involving preferred nucleation of interstitial loops on atomic planes perpendicular to the stress direction while vacancies migrating to planes parallel to the stress direction or to other sinks. However, SIPA involves preferential absorption of interstitials at dislocation loops oriented perpendicular to the stress axis. In materials containing a high dislocation density, preferential absorption of interstitials at edge dislocations with Burgers vector parallel to the applied stress direction further augments the creep rate by the SIPA mechanism. Irradiation enhances the rate at which a thermal creep mechanism such as jog dragging, dislocation climb/glide, and solute trapping proceeds. All these mechanisms, therefore, can be grouped under radiation-enhanced creep. The defect structure produced by radiation—which include depleted zones, vacancy, and interstitial clusters—all act as obstacles to dislocation motion and are responsible for retarding the creep process. All these different mechanisms can act either in parallel or sequentially to determine the overall creep rate (ε) in the components.

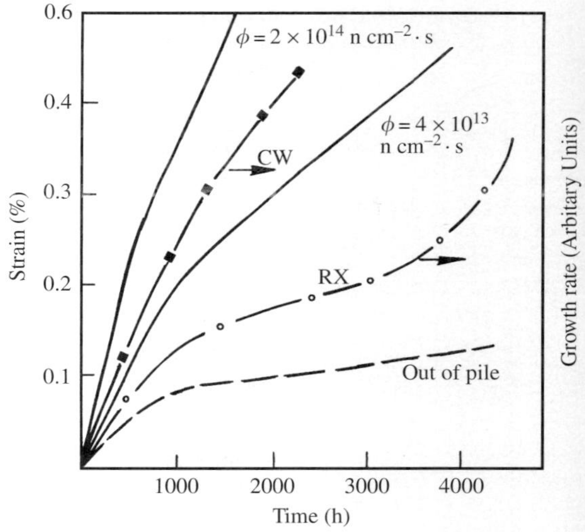

Figure 8
Irradiation creep rate in zircaloy-2 under biaxial loading (150 MPa and 300 °C) and a schematic diagram to show the growth rate of cold-worked (CW) and recrystallized (RX) zircaloy-2.

Since predictions from each of the mechanisms mentioned above are still not quantifiable, the following parametric relation is used, near the reactor operating conditions, for practical purposes.

$$\dot{\varepsilon} = A(\phi t)^m \sigma^n e^{-Q/RT} \quad (1)$$

where the effect of flux (φ), stress (σ), time (t), and temperature (T) can be separated. The usual values for the exponents are $m = 0.6$–1.0, $n \sim 1$, and $Q = 5$–$15\,\mathrm{kJ\,mol^{-1}}$. The influence of microstructural variables on creep rates has not been quantified as yet. To summarize some general trends it may be noted that solutes such as tin, iron, chromium, and nickel significantly increase the irradiation creep and thermal creep strength. Cold working increases the in-reactor creep rate but cold-worked and stress-relieved material is less affected by increased dislocation density due to cold work. Grain size variation within the normally used range of 1–20 µm has marginal effect on the creep rate although the grain shape influences the irradiation growth.

7. Concluding Remarks

Materials science principles as deployed for developing zirconium alloys for nuclear applications have been discussed briefly. The current research in zirconium alloy development is mainly concerned with (i) more quantitative understanding of the mechanisms of creep, growth, and oxidation processes under irradiation, (ii) microstructure and texture evolution in candidate alloys, (iii) development of alloys with longer service life and with a higher level of immunity to damage due to hydrogen, and (iv) development of techniques for joining zirconium and steels.

See also: Nuclear Reactor Materials; Nuclear Reactor Fuel Fabrication (Including Quality Control); Heavy Water Reactor Fuel Design and Performance; Nuclear Reactor Materials: Irradiation Effects; Light Water Reactor Fuel Design and Performance

Bibliography

Britt J C, Linga Murty K 1991 Texture development in zirconium alloys In: *Proc. Symp. on Zirconium Alloys for Reactor Components*. ZARC-91, BARC, Mumbai, India

Conference proceedings published as Special Technical Publications of ASTM, Zirconium in the Nuclear Industry, Vol. 1–10, American Society for Testing of Materials, Philadelphia

Cox B 1990 Pellet-clad interaction (PCI failures of zirconium-alloy fuel cladding: a review. *J. Nucl. Mater* **172**, 249–92

Douglass D L 1971 *The Metallurgy of Zirconium*. Atomic Energy Review, Vienna

Fidleris V 1988 The irradiation creep and growth phenomena. *J. Nucl. Mater* **159**, 22–42

Fleck R G, Price E G, Cheadle B A 1984 Pressure tube development for CANDU reactors. In: Franklin D G, Admason R B (eds.) *Zirconium in the Nuclear Industry, Sixth Int. Symp. ASTM STP 824*. ASTM, Philadelphia, pp. 88–105

Franklin D G, Langs P M 1991 Zirconium-alloy corrosion: a review based on an International Atomic Energy Agency (IAEA) meeting. In: Eucken C M, Garde A M (eds.) *Zirconium in the Nuclear Industry, 9th Int. Symp., ASTM STP 1132*. ASTM, Philadelphia, pp. 3–32

Franklin D G, Lucas G E, Bement A L 1983 *Creep of Zirconium Alloys in Nuclear Reactors*. ASTM STP 815, Philadelphia

Griffiths M, Chow C K, Coleman C E, Holt R A, Sagat S, Urbanic V F 1993 Evolution of microstructure in zircaloy core components in nuclear reactors during service. In: Kumar A S, Gelles D S, Nantad R K, Little E A (eds.) *Effects of Irradiation on Materials: 16th Conf. ASTM STP 1175*. ASTM, Philadelphia, pp. 1077–110

IAEA-TECDOC-684 1993 Corrosion of Zirconium Alloys in Nuclear Power Plants. Int. Atomic Energy Agency

Lemaignan C, Motta A T 1994 Zirconium alloys in nuclear application. In: Cahn R W, Haasen P, Kramer E J (eds.) Materials Science and Technology, Vol. 10 B. Nuclear Materials. VCH, Weinheim, Germany

Nothwood D O, Kosasih U 1983 Hydrides and delayed hydrogen cracking in zirconium and its alloys. *Int. Met. Rev* **28**, 92–121

Srivastava D, Dey G K, Banerjee S 1995 Evolution of microstructure during fabrication of Zr-2.5 wt.%Nb alloy pressure tubes. *Metall. Mater. Trans. A*, **26**, 2707–18

Tenckhoff E 1988 *Deformation Mechanisms, Texture, and Anisotropy in Zirconium and Zircaloy*. ASTM STP 966, Philadelphia

Nuclear Fuel Cycles

P. Rodriguez

Each fuel used for energy production is characterized by a fuel cycle which describes all aspects of obtaining and preparing the fuel, its transportation, using it for energy production, and managing the spent fuel and all effluents and waste products including waste heat. As an example, the typical cycle for a fossil fuel consists of:

(i) exploration
(ii) mining to extract the resource
(iii) processing or refining to remove impurities and otherwise prepare the fuel for use
(iv) energy production
(v) management of waste products such as ash, particulate matter, gaseous and liquid effluents including waste heat as condenser water
(vi) transportation as a step between any of the above and for waste disposal (for example utilization of fly ash as an input for construction material).

The nuclear fuel cycles are more complex, involve additional steps and special aspects of internalities and externalities for the following reasons:

(i) there are only three fissile isotopes by which a nuclear fission chain reaction can be sustained: uranium-235, plutonium-239, and uranium-233
(ii) of these three, only ^{235}U occurs in nature; however, its abundance in natural uranium is only ~0.7%, the predominant isotope in natural uranium being ^{238}U. For most of the fuel cycles, except the heavy water fuel cycle, isotopic enrichment in ^{235}U content becomes necessary and is an extra step in the cycle
(iii) the fissile materials ^{239}Pu and ^{233}U, which do not occur in nature, are produced by transmutation of uranium-238 and thorium-232, respectively; the latter two are, therefore, called *fertile* materials. For fuel cycles using ^{239}Pu or ^{233}U as the fuel, irradiation of ^{238}U and ^{232}Th in the fuel and/or blanket (see *Fast Breeder Reactors: Fuels*) and reprocessing to extract the ^{239}Pu and ^{233}U become important steps in the cycle (see *Purex and Thorex Processes (Aqueous Reprocessing)*).
(iv) the fertile materials and the fissile materials and their fission products are associated with radioactivity, necessitating special safety measures in all the steps of the cycle.
(v) the possibility exists that a neutron chain reaction (criticality) could occur in a fuel material located outside a reactor; this calls for safety measures against a criticality accident during enrichment, fabrication, transportation and storage of the fuel.
(vi) the fission and fusion chain reactions used for nuclear power generation are also used in the respective fission and fusion bombs. This fact brings in new dimensions to the nuclear fuel cycles, the concern for proliferation resistance and the desire to have provision for safeguards, inspection and accounting of fissile materials through all the steps in the cycle.

1. Classification of Nuclear Fuel Cycles

There are many schemes for classification of nuclear fuel cycles. One is to classify them on the basis of the nuclear chain reaction involved; thus there are fission and fusion cycles. A second approach is to name the fuel cycle after the fuel materials (isotopes) used. Thus we have the natural uranium fuel cycle, the ^{233}U cycle, and the plutonium cycle for the fission reactors. The deuterium (D)–tritium (T) reaction is the least difficult fusion reaction to initiate on earth and, therefore, is the most commonly explored fusion reaction; but there are many alternate fuels based on cycles other than D–T that are under consideration. For example, the D–D fusion cycles, proton-based fuel cycles, and D–He_3 fuel cycle (Dobrott 1983). Among the fusion cycles only the D–T cycle will be discussed since the first generation of future fusion reactors will be based on this reaction.

2. Uranium Fuel Cycle

2.1 Mining of Uranium Ore

The uranium content of the natural ore is quite low (0.1–0.3%), i.e., only about 1–3 kg of uranium is obtained from 1 t of mined ore. Concentration of the ore at the mine site is therefore carried out. The steps involved are crushing and grinding, followed by leaching in acid and ion exchange or solvent extraction to separate the uranium which is produced as ammonium diruranate (called *yellow-cake* due to its color). If an enrichment step is involved, the yellow-cake is converted to UF_6; otherwise the conversion is to U_3O_8 and finally to UO_2.

The tailings from uranium mining, which contain most of the radioactive radium, thorium, and actinide elements are considered a general environmental hazard. Since the 1970s there has been an increased awareness of this problem and strict control and adherence to health physics management have led to limiting occupational exposures and effluent disposal to acceptable levels.

2.2 Uranium Enrichment

Eighty five percent of the nuclear electricity (which is ~17% of the total electricity generation in the world today) comes from light water reactors (LWRs) which needs enrichment in U-235 to about 3–5 at.%.

Several uranium enrichment processes have been developed, all of which are based on the difference in mass between the nuclei of ^{235}U and ^{238}U. The gaseous diffusion process using UF_6 gas at ~85°C is the oldest and most widely used method and employs the faster diffusion of the lighter isotope through a porous membrane to achieve an enrichment level of 1.002 times the previous level. A plant producing ~3% enrichment would need ~1400 "stages" of enrichment. A more recent process uses a high-speed centrifuge operating with UF_6 to separate the two uranium isotopes.

Laser-based enrichment processes, which exploit the differences in wavelengths of the light, absorbed by the atoms or molecules of the two isotopes, are under development in many countries. An appropriate combination of lasers selectively excites one of the isotopes. Atomic vapor laser isotope separation (AVLIS) is one such process. The expectation is that if successfully developed on a commercial scale, the laser process can bring in energy savings and a reduction in the number of stages.

2.3 Open and Closed Cycles

The fuel that is discharged from the light water reactors (LWRs) at a burnup of ~30 MWd kg^{-1} (see *Light Water Reactor Fuel Design and Performance*) would have undergone the following approximate composition changes in 100 kg of the fuel.

$$3\,\text{kg}\,^{235}U \rightarrow \sim 1\,\text{kg}\,^{235}U \text{ (2 kg undergo fission)}$$

$$97\,\text{kg}\,^{238}U \rightarrow \sim 95\,\text{kg}\,^{238}U \text{ (2 kg transformed to } ^{239}Pu)$$

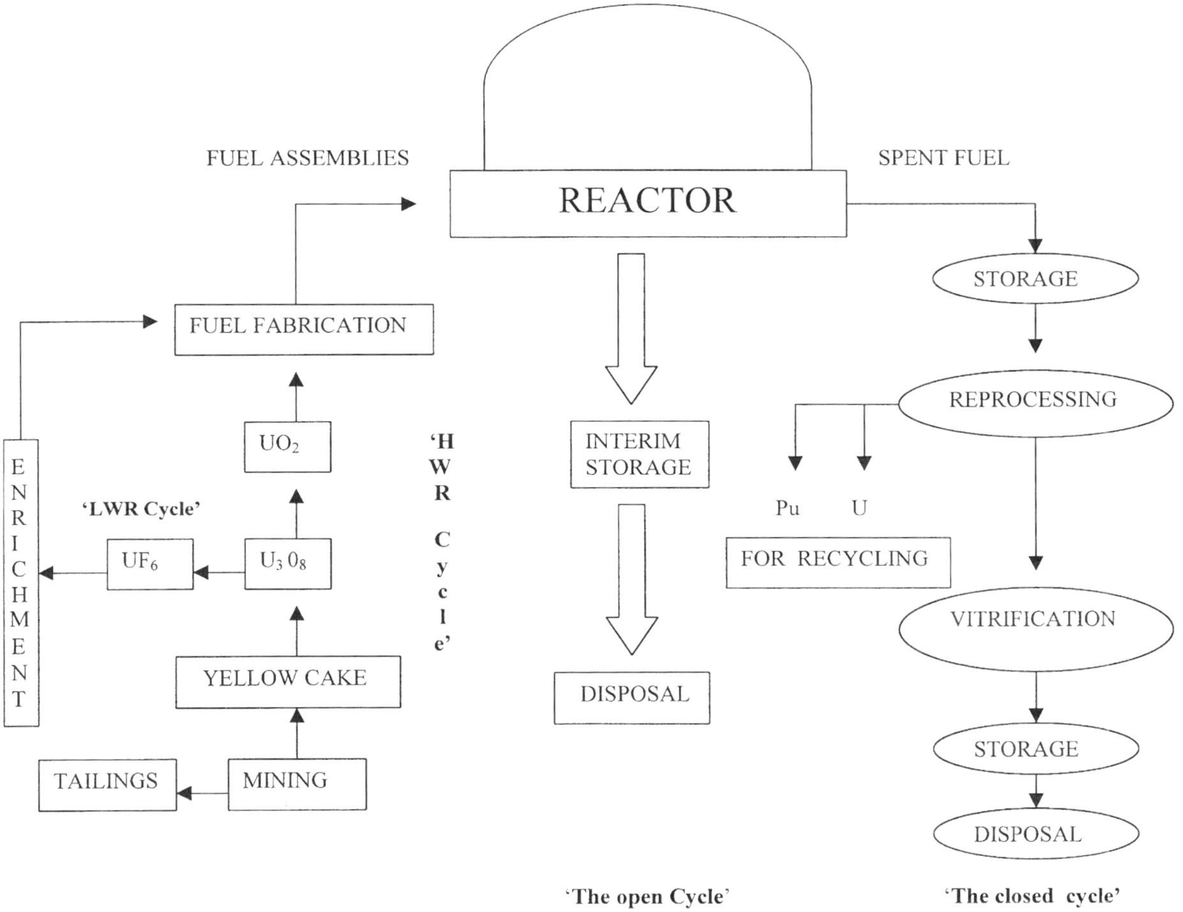

Figure 1
The uranium fuel cycle; from reactor the down path indicates the "open cycle" and the route on the right-hand side the "closed cycle," the UF_6 and enrichment steps are absent in the HWR cycle.

~ 1 kg ^{239}Pu (out of 2 kg above ~ 1 kg undergoes fission)

~ 3 kg of fission products

Heavy water reactors (HWRs) use natural uranium as the fuel and with a burnup of $\sim 8\,\mathrm{MWd\,kg^{-1}}$ the materials balance in HWRs is approximately as follows:

100 kg of natural uranium

~ 0.7 kg ^{235}U $\rightarrow \sim 0.2$ kg ^{235}U

~ 99.3 kg ^{238}U $\rightarrow \sim 98.7$ kg ^{238}U

(0.6 kg transformed to ^{239}Pu) ~ 0.3 kg ^{239}Pu

(~ 0.3 kg of ^{239}Pu undergoes fission)

~ 0.8 kg of fission products

The uranium cycle for the LWRs and HWRs may be implemented as a "closed cycle" in which the spent fuel is reprocessed to separate and recover the plutonium or as an "open cycle" in which the spent fuel is not reprocessed at all. In the latter case, it is stored at an interim storage facility before its final disposal in a safe depository. The "open" and "closed" uranium cycles for the LWR and HWR are illustrated in Fig. 1.

3. *The Plutonium Cycle*

There are two options for the use of the recovered plutonium for nuclear power production. The best usage is as fuel for fast reactors. When operated on the breeding mode, with ^{238}U in the blanket, fast breeder reactors (FBRs) produce more plutonium than they consume. Once an initial inventory of plutonium is built up from LWRs or HWRs to build and operate a FBR it will produce, in a certain time, enough plutonium to fuel another fast breeder reactor; this time is called the *doubling* time. FBRs have the potential for extracting overall 60 times more energy from uranium than the LWRs. Reprocessing of HWR fuel produces plutonium and depleted uranium in which the ^{235}U content is reduced to $\sim 0.2\%$. Depleted uranium is produced in the LWR cycle also in the enrichment process. Depleted uranium from both sources can be used in the FBRs for breeding purpose. FBRs thus have the potential for utilizing virtually all of the uranium produced from the mining operations.

A second use for plutonium is to provide the enrichment needed for the LWR fuel partially or fully through the use of mixed oxide (MOX) fuel.

4. *The Thorium Cycle*

Thorium is about three times more abundant in the earth's crust than uranium. All the long-term strategies and scenarios for nuclear power, envisage the introduction of the thorium cycle at the appropriate stage in the 21st Century for the conversion of the fertile thorium-232 to fissile uranium-233. On the basis of nuclear properties, thermal reactors offer a better potential for the thorium cycle than do fast reactors. Among the various thermal reactors, the molten salt breeder reactor (MSBR), the high temperature gas cooled reactor (HTGR), HWRs, and LWRs rank, in that order, for the application of the thorium cycle (Rodriguez and Sundaram 1981). Ironically, the status of technological development of these reactor systems is roughly in the reverse order to the above ranking.

Though ^{233}U is a better fuel in the thermal spectrum, the breeding of ^{233}U in a fast breeder reactor with ^{232}Th in the blanket, with the use of the bred ^{233}U as fuel in a thermal reactor is a possible strategy. India, which has one of the largest reserves of thorium in its monazite sands, has such a stated policy of a three-stage nuclear power program. In the first stage it is building a series of HWRs using natural uranium as fuel. It is also launching a fast breeder program as the second stage using plutonium as fuel and depleted uranium in the blanket. There are plans to use ^{232}Th also in the blanket to produce ^{233}U which will be used to fuel a series of advanced heavy water reactors (AHWR) in the third stage. Finally, a symbiotic combination of FBRs and AHWRs will lead to sustainable nuclear power for five centuries or more (Basu and Srinivasan 1990, Srinivasan and Kimura 1990, Anantharaman and Sinha 2000).

The European Commission has recently concluded a detailed assessment of the thorium fuel cycles including the status of various national and international programs (Lung 1997).

5. *The Fusion Cycle*

The first generation of commercial fusion reactors will use the D–T reaction for energy production. Both deuterium and tritium are virtually inexhaustible and can provide a total energy equivalent of 1×10^{12} TWYr. Deuterium is obtained from ordinary water (33 g from 1 ton). Tritium is produced from lithium in the blanket through (n, He + T) reactions. Lithium-6 (7.5% in natural lithium, balance 92.5% is lithium-7) is the relevant resource and is available in ore deposits (11 million tons) and in sea water (200 billion tons). Beryllium is used as a neutron multiplier making use of the (n, 2n) reaction.

The environmental effects of the fusion cycle are generally considered to be nonexistent. However, there is a concern about tritium release and nonradioactive toxic materials such as lithium, beryllium, and lead. Another concern is the radioactivity of the structural materials that need to be replaced every five years or so. This latter concern is the reason for the active research towards development of low activation materials for fusion reactors.

6. *Hybrid Cycles*

When a product of one cycle is used to fuel another cycle, it is a "hybrid fuel cycle." The use of ^{233}U as the initial fissile fuel in a uranium cycle or ^{239}Pu as initial fuel in thorium cycle are examples for hybrid fuel cycles which are also called *crossed progeny* cycles.

One hybrid cycle of interest in HWRs is the *dupic fuel cycle*. In Sect. 2.3, it was mentioned that the fuel discharged from the LWRs contains 1% or more of ^{235}U which is higher than the ^{235}U contained in natural uranium. This discharged fuel, if suitably processed and conditioned, can become the fuel in a HWR (see *DUPIC Fuel Cycle*).

Hybrid systems are possible between fusion reactors and fission reactors and accelerators and fission reactor systems. The fusion-fission hybrid system, also called the fusion fissile fuel producing (FFP) reactor, has been evaluated in detail by INFCE (1980). Essentially in the system, 14 MeV fusion neutrons are absorbed in a fertile blanket such as ^{238}U or ^{232}Th to convert it to the fissile nuclide.

Rubbia (1994) has proposed a subcritical fission reactor, driven by an accelerator, based on the thorium fuel cycle, which is a subcritical version of the accelerator-breeder system earlier examined by INFCE (1980). In principle, accelerator-fission hybrids can be based on both the uranium and thorium cycles.

7. *Concluding Remarks*

This article is not written as a comprehensive treatment of the subject of nuclear fuel cycles. The salient features of a fuel cycle have been explained with the uranium cycle in the LWR and HWR as examples. Important aspects of recycling plutonium and the futuristic thorium fuel cycles and the fusion cycles have been highlighted. Fuel fabrication, fuel reprocessing, and waste management are important steps in all nuclear fuel cycles.

See also: Nuclear Reactor Materials; Purex and Thorex Processes (Aqueous Reprocessing); DUPIC Fuel Cycle; Fast Breeder Reactors: Fuels; Nuclear Reactor Fuel Fabrication (Including Quality Control); Mixed Oxide (MOX) Fuel; Inert Matrix Fuels; Heavy Water Reactor Fuel Design and Performance; Light Water Reactor Fuel Design and Performance; Nuclear Waste Management, Materials Aspects of

Bibliography

Anantharaman K and Sinha R K (eds.) 2000 *INSAC Proc. 11th Annual Conf. Indian Nuclear Society: Power from Thorium: Status, Strategies and Directions*. Indian Nuclear Society, Mumbai, India

Basu T K, Srinivasan M 1990 Thorium fuel cycle development activities in India (A Decade of Programme: 1981–1990) Report BARC 1532. Bhabha Atomic Research Centre, Mumbai, India

Carle R 1994 *Nuclear Power*. Presses Universitaires De France, Paris, France

Dobrott T 1983 Alternate fuels in fusion reactors. *Nucl. Technol. Fusion* **4**, 339–47

International Atomic Energy Agency *1977 Proc. Int. Conf. on Nuclear Power and its Fuel Cycle*. IAEA, Vienna

International Atomic Energy Agency *1997 Proc. of the Int. Symp. on Nuclear Fuel Cycle and Reactor Strategies: Adjusting to New Realities*. IAEA, Vienna

International Nuclear Fuel Cycle Evaluation (INFCE) 1980 Report of Working Groups. IAEA, Vienna

Ion S C, Bonser D R 1997 Fuel cycles of the future. *Nuc. Energy* **36**, 127–30

Kazimi M S 1983 Risk constraints for fusion energy. *Nucl. Technol. Fusion* **4**, 527–32

Knief R A 1981 *Nuclear Energy Technology*. Hemisphere, New York

Lung M 1997 *A Present Review of the Thorium Nuclear Fuel Cycles*. European Commission, Luxembourg: Office for Official Publications of the European Communities

Marshall W 1980 The use of plutonium. *Graham Young Memorial Lecture Atom* **282**, 88–103

Marsham T N 1982 Fast reactor fuel cycles. *Atom* **303**, 7–10

Nuclear Energy Agency 1994 The economics of the nuclear fuel cycle, Nuclear Energy Agency. OECD, Paris

Ramanna R, Lee S M 1986 The thorium cycle for fast breeder reactors. Pramāna **27**, 129–37

Rodriguez P, Sundaram C V 1981 Nuclear and materials aspects of the thorium fuel cycle. *J. Nucl. Mater*. **100**, 227–49

Rubbia C 1994 *The Energy Amplifier*. 8th Journées Saturne, Orsay, France

Srinivasan M and Kimura I (eds.) 1990 *Proc. Indo-Japan Seminar on Thorium Utilisation*. Indian Nuclear Society, Mumbai, India

Study Group on Nuclear Fuel Cycles and Waste Management 1978 Report to the APS. *Rev. Mod. Phys*. 50, S1–S186

Uranium Information Centre Ltd 2000 Nuclear Electricity, Melbourne, Australia (http://www.uic.com.au/ne)

Van Oost G, Ongena J P H E 1996 Energy for future centuries: Will fusion be an inexhaustible safe and clean energy source? *Trans. Fusion Technol*. **29**, 9–15

Wymer R G, Vondra B L Jr 1981 *Light Water Reactor Nuclear Fuel Cycle*. CRC, Boca Raton, FL

Nuclear Fuel Reprocessing

J. J. Laidler

When nuclear fuel (typically uranium or a uranium compound such as UO_2) is operated in a nuclear reactor, the uranium atoms can undergo two reactions when impinged upon by neutrons produced in the reactor core. The uranium atom can split (fission), producing energy and releasing additional neutrons for participation in the chain reaction, or it can absorb a neutron. The uranium fission process results in the formation of fission product atoms (generally two), whose combined mass is somewhat less than that of the original uranium atom. These fission products are unstable isotopes that decay by emission of radiation, eventually leading to stable isotopes. Uranium fission normally takes place with the fissile isotope, ^{235}U, which comprises about 0.7% of natural uranium.

Neutron absorption is the more common reaction in the interaction of a neutron with the ^{238}U isotope that makes up 99.3% of natural uranium. The neutron absorption reaction leads to the production of the transuranic elements neptunium, plutonium, americium, and curium.

As these various neutron reactions take place in the uranium fuel, the concentration of fission products and transuranic elements increases. Some of the transuranic isotopes are readily fissionable and contribute in an increasing proportion to the energy released by the fuel as the residence of the fuel in the reactor core continues. Other transuranic isotopes are strong neutron absorbers, as are some of the fission product isotopes. These neutron absorbers tend to slow down the chain reaction until it becomes necessary to shut down the reactor and replace the expended fuel with fresh fuel. The discharged fuel is lethally radioactive and thermally hot due to the energy released in the radioactive decay process; it is commonly stored in a deep water-filled storage pool for a period of a year or more before processing, so that the short-lived fission products can decay away. The unaltered uranium and the transuranics present in the spent fuel can be recovered by a chemical separation process for recycling or for other uses. This has come to be termed "reprocessing." It also describes the recovery of fertile thorium and fissile uranium in operations with the ^{232}Th–^{233}U fuel cycle, but the emphasis here will be on the uranium–plutonium fuel cycle for purposes of simplification.

1. *Historical Perspective*

The technology for reprocessing nuclear fuel was born under a cloak of secrecy during the development of the atomic bomb in the World War II period. It was recognized by workers in those early days that a chemical separation of plutonium was feasible, possibly with greater success than the isotopic separation of fissile uranium. The initial military-related application of reprocessing and the legacy of radioactive wastes produced in the past under the pressures of defense production needs have led to much of the controversy surrounding the use of reprocessing today, even in strictly-controlled peaceful applications.

1.1 Early Processes

Scientists working with minuscule amounts of material found in 1941 that plutonium, like uranium, has several oxidation states and that plutonium in the Pu^{IV} oxidation state can be coprecipitated with lanthanum trifluoride. This led to the development in 1942 of the "bismuth process" that was successfully applied in the wartime years. The bismuth process was based on the coprecipitation of plutonium with a carrier precipitate, due to the very low concentration of plutonium in the discharged fuel. In this process, the fuel discharged from the first production reactors was dissolved in nitric acid and treated with sodium nitrite to bring all the plutonium into the Pu^{IV} state. Then $BiONO_3$ and H_3PO_4 were added to the nitrate solution, precipitating Bi_3PO_4 and $Pu_3(PO_4)_4$; several cycles of the coprecipitation process were performed after centrifuging, washing, and dissolving, to extract as much plutonium as possible. A final coprecipitation with lanthanum trifluoride was followed by dissolution in nitric acid and potassium hydroxide which removed the lanthanum trifluoride. The plutonium was then precipitated as $Pu(NO_3)_4$. Over 7000 tonnes of irradiated fuel were processed at the Hanford site in southeastern Washington using the bismuth process over the period 1944–1956. The bismuth process had some drawbacks that resulted in its eventual replacement: (i) the process did not recover uranium, which was a scarce and valuable commodity in that time period; and (ii) the process generated large amounts of high-level radioactive waste, approximately 40 m^3 per tonne of fuel processed.

The initial replacement for the bismuth precipitation process was the REDOX (REDuction–OXidation) process, developed at the Hanford site beginning in 1947. This process was the first continuous-flow solvent extraction process designed for nuclear fuel reprocessing, and it was the first separation process to recover uranium as well as plutonium. Solvent extraction separation of plutonium and uranium is based on the easy extractability of uranyl nitrate and plutonium(IV) nitrate and the relative inextractability of fission products and plutonium(III) nitrate. The feed to the REDOX process was the clarified solution from the dissolver operation, in which the fuel was dissolved in nitric acid. This feed solution was contacted with hexone (methyl isobutyl ketone), extracting uranium and plutonium into the organic phase and leaving the fission products in the aqueous phase. The organic phase was then contacted with an $Al(NO_3)_3$–HNO_3 solution containing a reducing agent such as ferrous sulfamate ($Fe(H_2SNO_3)_2$) to reduce plutonium to the plutonium(III) state which then precipitated from the organic phase due to extremely limited solubility. The high concentration of salting agents (nitrates) kept the uranium in the organic phase. The organic phase was then scrubbed with a stripping solution of dilute nitric acid to remove uranium; the aqueous phase bearing the uranium was then contacted with hexone to extract the uranium, and that organic phase was then stripped with a nitric acid wash to recover the uranium as uranyl nitrate. The plutonium(III) in the aqueous stream was reoxidized to plutonium(VI), reextracted with hexone, and then stripped with an acid wash. Decontamination factors (defined as the ratio of the concentration of impurity to that of the valued constituent in the feed, divided by the same ratio for these elements in the product) achieved with this process were typically 10^6 for uranium and 10^7 for plutonium. Construction of the REDOX plant at Hanford was completed in late 1951, and the plant reached a throughput capacity of 12 tonnes per day in 1958. The REDOX plant shut down in late 1967 after processing over 19000 tonnes of fuel.

1.2 PUREX Process

The principal drawbacks of the REDOX process appear to have been: (i) added waste generation associated with use of the aluminum nitrate salting agent; (ii) safety hazards posed by the severe flammability of the hexone solvent; and (iii) rather high costs of operation. A replacement process known as PUREX (for Plutonium–URanium EXtraction) was developed in the early 1950s at the Knolls Atomic Power Laboratory in Schenectady, NY. The PUREX process operates with the same feed material as REDOX, but employs a low-flammability organic extractant, tri-*n*-butyl phosphate (TBP) in a kerosene diluent. The TBP extractant is stable in contact with high concentrations of nitric acid, so the acid can be used as a salting agent. The PUREX process was first used at the Savannah River Site in South Carolina in two large plants: F-Canyon became operational in November 1954 and H-Canyon in July 1955. The PUREX plant at the Hanford site became operational in January 1956. These plants, as well as the early bismuth phosphate plants at Hanford, are all arranged in a long, narrow "canyon" configuration with provision for remote operation. The bismuth plants were known as "Queen Marys" because of the resemblance of their shape to the dimensions of the famous ocean liner (311 m in length and 36 m wide); the PUREX canyons have similar dimensions.

The PUREX plant at Hanford was permanently closed in 1992. Operations at the Savannah River F-Canyon were suspended in 1992, but the plant resumed operation in 1995 for the purpose of processing spent nuclear fuels that represented a disposal problem. This plant is expected to remain in operation until about 2003. The Savannah River H-Canyon was shut down in 1992 but restarted in 1997 for conversion of highly enriched uranium stocks to low-enriched uranium. This blend-down of uranium should continue until about 2006, when H-Canyon is expected to close.

A commercial PUREX reprocessing plant was operated in the USA by the Nuclear Fuel Services Company from 1966–1972 at the West Valley site in New York. This plant recycled a substantial quantity of fuel from the Hanford N-Reactor plant. A large plant was built by Allied General Nuclear Services at the Barnwell site in South Carolina, but the plant was never operated for purposes of reprocessing nuclear fuel. A number of other PUREX plants have operated world-wide, as shown in Table 1.

1.3 Melt-refining Process

Work on pyrochemical separations technology began in 1952 when it was recognized that low-decontamination pyrochemical methods were ideally suited to the reprocessing of the metal alloy breeder reactor fuels that were under consideration at that time. The EBR-II reactor began operation with highly enriched metallic uranium fuel in 1964, and a pyrochemical reprocessing method was put into place shortly thereafter for demonstration of closure of the breeder fuel cycle. Because the fuel was limited to rather low burnup at that time due to fuel swelling, a pyrochemical process was developed that permitted the processing of fuel with very short cooling times. This early process involved the mechanical decladding of uranium fuel slugs and melting of the fuel slugs in a zirconium oxide crucible. The more easily oxidized fission products formed a dross that adhered to the crucible walls, while the gaseous fission products and the volatile fission products were released to the hot cell atmosphere. The molten fuel then contained uranium and the small amount of transuranic elements generated, as well as the nonvolatile noble metal fission products (which could be tolerated in the fast neutron environment of the breeder reactor core). The process was termed "melt-refining," and the collection of noble metal fission products recycled with the actinides was called "fissium." The EBR-II reprocessing operation continued until early 1969, when the demonstration of fuel cycle closure was deemed complete. At about the same time, a decision was made by the US government to concentrate on oxide fuels for further breeder reactor development.

Table 1
Commercial PUREX reprocessing plants operated or still operating in 2000.

Country	Location	Facility	Capacity (t yr^{-1})	Operational	Shut down
USA	New York	West Valley	300	1966	1972
	South Carolina	Barnwell	1500	Never operated	
Russia	Chelyabinsk-65	Mayak	600	1978	
UK	BNFL Sellafield	B-205	1500	1964	
	BNFL Sellafield	THORP	700	1994	
France	COGEMA LaHague	UP2	400	1976	1987
	COGEMA LaHague	UP3	800	1990	
	COGEMA LaHague	UP2-800	800	1994	
Germany	Karlsruhe	WAK	35	1971	1990
Belgium	Mol	Eurochemic	100	1966	1974
India	Tarapur	DAE	150	1982	
	Trombay	DAE	50	1985	
	Kalpakkam	DAE	200	1994	
Japan	Tokai-mura	JNC	90	1981	
	Rokkasho-mura	JNFS	800	(2003)	

2. *Contemporary Reprocessing Methods*

2.1 *Aqueous Processes*

(*a*) *PUREX process.* In the PUREX process, spent fuel is initially dissolved with nitric acid in a large dissolver vessel. In the case of aluminum-clad fuel, the dissolution is catalyzed by the addition of mercury, whereas zirconium alloy-clad fuel is dissolved in a mixture of nitric and hydrofluoric acids; the hydrofluoric acid is added to complex the zirconium and prevent the occurrence of highly energetic reactions. Zirconium alloy-clad fuel can be decladded chemically with ammonium fluoride (ZRFLEX process); alternatively, the spent fuel can be chopped into short segments and the fuel material leached out with nitric acid, leaving the cladding undissolved (this is termed the *chop–leach process*). Except for a few fission products that are volatilized during dissolution, all the constituents initially in the fuel are retained in the dissolver solution as soluble nitrate salts or, in the case of some minor constituents, as an insoluble sludge. Generally, the solution must be clarified to accommodate its use as feed to the solvent extraction process.

In the solvent extraction steps, purification of the desired products (generally uranium and plutonium, possibly thorium) is achieved by the selective transfer of these constituents into an organic solution that is immiscible with the original aqueous solution. The most commonly used organic solution consists of the extractant TBP dissolved in purified kerosene or similar materials such as long-chain hydrocarbons (for example, *n*-dodecane). Uranium in aqueous nitrate solutions normally exists only in the U^{IV} and U^{VI} oxidation states. Nitric acid dissolution of spent fuel yields uranium(VI) and nitrogen oxides (NO and NO_2) as products. The uranium(VI) is strongly hydrolyzed, and exists as the uranyl ($UO^{2+}{}_2$) ion. By the same token, plutonium may exist in the Pu^{III} or Pu^{IV} state, with the Pu^{IV} state resulting from nitric acid dissolution. When in the proper oxidation states (($UO^{2+}{}_2$) and Pu^{IV}), the desired products are readily extractable into TBP. Plutonium in the Pu^{III} state has a very low extractability into TBP.

The solvent extraction separation is based on this ready extractability of uranyl nitrate and plutonium(IV) nitrate and the relative inextractability of fission products and plutonium(III) nitrate. The first step of the PUREX solvent extraction process thus involves the coextraction of uranium and plutonium into the organic (TBP + kerosene) phase, leaving most of the fission products in the aqueous phase. The organic solvent is then contacted with a clean aqueous acid solution, and the uranium and plutonium transfer into this clean aqueous phase. The solvent can now be recycled. After reducing the plutonium, typically with ferrous sulfamate, to plutonium(III) it is then a matter of contacting the aqueous phase containing plutonium(III) nitrate and uranyl nitrate with clean organic phase; the uranium transfers to the organic phase and the plutonium remains in the aqueous phase. The uranium–organic phase can then be contacted with another clean aqueous phase to transfer the uranyl nitrate into the aqueous phase.

The products of the PUREX process are separated nitrate solutions of uranium and plutonium. These solutions are treated separately to produce uranium and plutonium oxides. The solvent extraction process yields virtually complete recovery of the uranium and plutonium present in the spent fuel, and both are of high purity. Decontamination factors of 10^7–10^8 are possible.

Because TBP is susceptible to decomposition to dibutyl phosphate and monobutyl phosphate, both of which have significant deleterious effects on plutonium stripping, care must be taken to maintain the purity of the solvent. Otherwise, additional process steps are necessary to recover plutonium, at the cost of increased volume of liquid waste. Limiting the time of exposure of sensitive solvents to high levels of nitric acid and beta/gamma radiation greatly increases their useful lifetime.

Neptunium is present in the irradiated fuel dissolution liquor, commonly in oxidation state Np^{V}. The extractability of neptunium(V) by TBP is quite poor, whereas that for neptunium(VI) is almost as high as it is for uranium(VI) and plutonium(IV). Therefore, in those instances where the separation of neptunium is desirable, it is coextracted with uranium and plutonium in the PUREX first cycle after oxidizing all of the neptunium to the Np^{VI} oxidation state. The coextracted neptunium can thereafter be separated from uranium and plutonium in the second uranium cycle or by introducing specific reagents such as butyraldehydes. There is at present no process for separation of americium and curium.

(*b*) *TRUEX and related processes.* The TRUEX (TRansUranium EXtraction) Process was developed in the 1980s to decontaminate transuranic material from liquid nuclear wastes accumulated in the USA at Department of Energy nuclear material production sites. The active extractant in the TRUEX process is *n*-octylphenyldiisobutylcarbamoylmethylphosphine oxide (CMPO). An effective extraction/stripping process is made possible by the fact that the CMPO extractant exhibits high and low extractability for actinide(III) and lanthanide(III) nitrates at high and low nitric acid concentrations, respectively, in the aqueous phase. In fact, the CMPO is such a strong extractant that it tends to extract many other metallic constituents from the aqueous solution; it also makes stripping of the desired actinides somewhat difficult. Accordingly, CMPO is frequently modified with TBP. A complexing agent such as oxalic acid is often added to the aqueous feed to facilitate the separation of the metallic species from the actinides and lanthanides.

Several other highly specific processes have been developed for application in the extraction of actinides and lanthanides. Brief descriptions follow.

- *DHDECMP process.* This process is similar to the TRUEX process, except the active extractant is dihexyl-*N*, *N*-diethylcarbamoylmethylphosphonate, which is a weaker extractant than CMPO. The transuranic elements and lanthanides are extracted by the solvent and stripped together and report to the end product stream. This process has been successfully tested on a pilot scale with simulated tank waste at the Idaho National Engineering and Environmental Laboratory (INEEL).

- *Phosphine oxide process.* This process uses a phosphine oxide derivative as the active extractant. It was developed jointly between the INEEL and the Khlopin Radium Institute in Russia. The transuranics and lanthanides are extracted by the solvent and stripped together. The process has been successfully demonstrated (>99.5% transuranics removal) with actual tank waste at the INEEL using 2-cm diameter centrifugal contactors in a shielded and remotely maintained hot cell facility.

- *TRPO process.* This was developed as a cooperative program between the Tsinghua University in China and the Institute for Transuranium Elements (ITU) in Karlsruhe, Germany. The process uses trialkylphosphine oxides as the active extractant. It operates most effectively with a waste acidity <2.0 M HNO_3. The transuranics and lanthanides are extracted by the solvent and stripped together. The process has been demonstrated with actual commercial high-level waste (HLW) solution, diluted ten-fold, in centrifugal contactors in China. Greater than 99.97% of the americium was extracted and a decontamination factor of >1400 was obtained for ^{99}Tc.

- *DIDPA processes.* This process was developed by the Japan Atomic Energy Research Institute. It uses diisodecylphosphoric acid as the active extractant. The process has the potential for selective stripping of the actinides from the lanthanides and requires a waste acidity of <0.5 M HNO_3. This low acidity may require dilution which will increase processing time and possibly waste generation. The DIDPA process was successfully tested in Japan with actual commercial HLW (denitrated with formic acid) in mixer-settlers. Greater than 99.99% of the americium was extracted. A closely related process known as TALSPEAK uses HDEHP (di-2-ethylhexylphosphoric acid) as the extractant.

- *DIAMEX process.* The DIAMEX process was first developed at the Fontenay-aux-Roses Research Center of the Commissariat à l'Energie Atomique (CEA, France) and the University of Reading in the UK. The active extractant in the DIAMEX process is dimethyldibutyltetradecylmalonamide. The DIAMEX solvent is fully incinerable and is not adversely affected by radiolytic and hydrolytic degradation. The process requires a waste acidity greater than 3 M. The transuranic elements and lanthanides are extracted by the solvent and stripped together. Technetium is not extracted with this process. This process has been successfully tested with actual waste in France.

Increasing interest is developing in the complete separation of the minor actinide elements (i.e., neptunium, americium, and curium) from the lanthanide fission products. Processes are under development that will accomplish these separations, but the added cost and added waste volume may present a problem. Some processes that are well along in development are:

- *TPTZ*. This process is being developed by the CEA (France) and the University of Reading (UK). The process uses a tripyridyltriazine ligand to extract selectively the transuranic material (oxidation state (III)) from the lanthanides (oxidation state (III)). This process has been successfully tested with simulated waste in France.

- *CYANEX 301*™. This process is being developed in the People's Republic of China and by the ITU in Germany. It uses bis(2,4,4-trimethylpentyl)dithiophosphonic acid to separate selectively transuranic(III) material from the lanthanides(III). The aqueous solution must be adjusted to a pH of 3.5 to 4. This process has been successfully demonstrated at bench scale with radioactive waste in China.

2.2 *Nonaqueous Processes*

Although the PUREX process and its predecessors are highly effective in accomplishing the recovery of plutonium and (in the case of PUREX and REDOX) uranium, these aqueous processes pose a number of issues that must be addressed by countries or commercial entities considering their application. The first issue that must be faced is the extremely low critical mass of plutonium in aqueous solution, as low as 750 g. This requires the careful design of process piping and vessels to ensure against an accidental criticality, and tends to increase the size and hence the cost of a reprocessing facility. Cost considerations favor the construction of plants with rather large capacity, to take advantage of the economy-of-scale principle. A second issue is inherent to all processes involving the chemical alteration of spent nuclear fuel: the potential for separation of materials usable in nuclear weapons. Aqueous processes have been used for the separation of pure plutonium for decades, and they clearly represent a proliferation concern. The third issue is that of radioactive waste generation. As stated previously, the waste volume arising from operation of the early aqueous reprocessing plants was quite large; these wastes have resulted in multibillion dollar remediation programs at the Savannah River, Hanford, and West Valley plants in the USA. More recent commercial plant operations have implemented major efforts to reduce waste generation, with excellent results. These issues of plant costs, proliferation, and waste generation have all contributed in varying degrees to the recent development of nonaqueous processing methods.

(*a*) *American pyrochemical process.* A family of pyrochemical processes has been developed in the USA, starting in 1985, with the express goal of realizing a low-cost, proliferation-resistant spent fuel treatment process that could meet the needs of future nuclear power systems. The initial application was for an inherently safe fast reactor system, the integral fast reactor (IFR). The IFR fuel was to be a metallic alloy, U–20 wt.% Pu–10 wt.% Zr. The fast reactor can be operated as a breeder, producing more fuel (by absorption of neutrons in fertile uranium to produce plutonium) than it consumes, or it can be operated on a break-even basis, producing just enough fuel to sustain operations. In either case reprocessing is required for recovery of the fissile materials remaining in the spent fuel. The fast neutron spectrum of the IFR permits virtually complete fission of all isotopes of the transuranic elements, and the character of the IFR core physics is such that carryover of certain fission products in the recycle fuel can be tolerated without a performance penalty. A decontamination factor of 100 is quite acceptable, compared to that of $>10^6$ possible with aqueous reprocessing. So, it is not necessary to recover the transuranic elements separately, it is not necessary to separate pure plutonium, and it is not necessary to remove fission products down to trace quantities.

The nature of the IFR fuel cycle requirements led to development of a pyrochemical process based on electrorefining. In this process, chopped spent metallic fuel is placed in a basket suspended in a molten salt electrolyte in an electrorefining cell. The basket is made the anode of the cell and a potential is applied between the anode and a suitable cathode. In the specific IFR application, the electrolyte salt is a eutectic mixture of lithium and potassium chlorides, and the cell is operated at a temperature of 500°C. Limitation of the applied voltage enables the operator to control which of the constituents of the spent fuel are dissolved. Those constituents which form stable chlorides will tend to dissolve into the electrolyte salt; this includes the group IA and IIA fission products and some lanthanides (e.g., Cs, Sr, Sm, Eu, etc.). The transition metal fission products tend to remain undissolved in the anode basket. Since the free energy of formation of UCl_3 is lower than that of the transuranic chlorides (e.g., $PuCl_3$), uranium can be selectively electrotransported to the cathode and deposited as a pure metal as long as the ratio of transuranic elements to uranium in the salt remains low. For efficient collection of the transuranic elements, a liquid metal cathode, typically molten cadmium, is used. The transuranics are electrotransported to the liquid metal cathode where they deposit along with about 30% uranium. Due to the absence of water and the presence of a strong neutron absorber, quite large quantities of transuranic elements can be collected in a single cathode crucible without concern for a nuclear criticality incident. After removal of the cathode deposits from the electrorefiner, adhering electrolyte salt is removed by melting the deposit under partial vacuum, whereupon the salt evaporates and is collected for recycle. Processing of the liquid cathode deposits also requires the distillation and recovery of cadmium. Because the cadmium never leaves the process, a waste disposal problem is avoided.

The IFR pyrochemical process, for extraction of uranium only, has been applied on a pilot scale to the treatment of a total of about 1.5 tonnes of sodium-bonded driver fuel and blanket assemblies discharged from the EBR-II reactor. Demonstration of the process with those materials was completed successfully in 1999. Because this particular application was in a waste management role, the plutonium and other transuranics in the fuel and blankets were left in the waste stream. Design studies have shown that it should be possible to realize the technical and economic benefits of the principal features of the pyrochemical process (flowsheet simplicity, compactness of equipment, safety of operation, and resistance to proliferation) in future applications. A number of countries worldwide are exploring the potential of this process.

(*b*) *Russian pyrochemical process.* Russian fast reactor development has focused on the oxide fuel cycle. Oxide fuel is used in the BN-600 reactor and is planned for the BN-800 series. In the late 1970s, Russian scientists began development of a pyrochemical process that would deal with oxide fuel. This process begins with the chopping of spent oxide fuel; the granulated oxide recovered is then chlorinated at 600–650 °C in a molten sodium–potassium chloride salt bath contained in a pyrolytic graphite crucible, to produce uranium ($(UO_2)^{2+}$) oxychloride and Pu^{4+} chloride. An electrowinning step is then used to extract UO_2 and the noble metal fission products; this step is carried out at a voltage that will not codeposit plutonium. The process is repeated until the plutonium concentration in the salt batch increases to 15–20 wt.%. The cathode deposits of UO_2 and fission products are removed from the electrowinning cell, washed with water to remove the adhering salt, and then stored. Next, oxygen is sparged through the salt to precipitate the plutonium as PuO_2, which is recovered with an efficiency of more than 99%. The recovered PuO_2 is directed to a vibropac fuel fabrication process. After the plutonium recovery step, the electrowinning step is repeated to deplete the salt of UO_2; this step also removes most of the remaining fission products. The salt is then treated with Na_3PO_4 to precipitate residual impurities as phosphates, which are incorporated into a phosphate-type waste form for disposal.

A variation of the Russian process involves the codeposition of UO_2 and PuO_2, with the recovered product going directly to the vibropac fuel fabrication process after removal of the electrolyte salt. Both of these processes have been used at rather small scale to treat actual spent fuel from the BOR-60 reactor and recycle it. Neptunium in the spent fuel tends to follow the plutonium and is codeposited with PuO_2 or UO_2/PuO_2. The recovery of americium and curium is made somewhat difficult by their tendency to behave like the rare earth elements, and work is underway to develop means for their extraction.

3. Conclusion

The USA has maintained a national policy forbidding the reprocessing of civilian reactor fuel for plutonium recovery since the mid-1970s. A number of other governments, however, including those of France, the UK, Russia, and Japan, actively support reprocessing as a means for the management of highly-radioactive spent fuel and as a source of fissile material for future nuclear fuel supply.

Obviously, decisions to reprocess are not made on economic grounds only, making it difficult to evaluate the economic viability of reprocessing in various scenarios. In the ideal case, a number of factors must be considered, including: (i) the cost of uranium/U_3O_8, which is currently stable at quite low values due to limited demand and large supplies; (ii) the cost of enrichment, which is also stable due to the large stocks of enriched uranium available at the end of the cold war; (iii) the cost of fabricating plutonium-bearing (MOX) fuels, which is greater than for uranium oxide fuels due to the toxicity of plutonium; (iv) the cost of reprocessing, which is highly dependent upon facility costs; (v) waste disposal cost; and (vi) the fissile content of spent fuel, with high-value spent fuels making reprocessing more profitable.

It is generally accepted that breeder reactors—those that are designed to generate more fuel than is consumed (by absorption of neutrons in blankets comprising fertile isotopes such as ^{238}U or ^{232}Th)—will not be deployed until the latter part of the twenty-first century. With these reactors, reprocessing will be mandatory because it is necessary to recover and recycle the fissile material generated in the blankets. It is reasonable to expect that the economics of reprocessing will be advantageous to the deployment of breeder reactors when the time comes for their introduction into the energy generation repertoire. In the meantime, reprocessing will continue in those countries where it is legislated as a waste management practice or where the recovered plutonium is viewed as an essential fuel resource. The development of improved reprocessing methods is likely to continue, and the approach of the time for retirement of existing PUREX plants may see renewed interest in innovative methods that promise to afford significant technical and economic advantages relative to the predecessor processes.

Bibliography

American Nuclear Society 1993 *Proc. Int. Conf. Technol. Expos. Future Nuclear Systems Emerging Fuel Cycles and Waste Disposal Systems, Global '93*. ANS, LaGrange Park, IL

Benedict M, Pigford T H, Levi H W 1981 *Nuclear Chemical Engineering*, 2nd edn., McGraw-Hill, New York

Hannum W H 1997 The technology of the integral fast reactor and its associated fuel cycle. *Prog. Nuclear Energy* **31**, 1–217

Horwitz E P, Diamond H, Gatrone R C, Nash K L, Rickert P G 1990 *Proc. Int. Solvent Extraction Conf., ISEC '90*, Kyoto, Japan. *Elsevier, New York, pp. 357–62*

International Atomic Energy Agency 1989 *Nuclear Fuel Cycle in the 1990s and Beyond the Century: Some Trends and Foreseeable Problems*. IAEA, Vienna

Long J T 1967 *Engineering for Nuclear Fuel Reprocessing*. Gordon and Breach, New York

National Research Council Committee on Separations Technology and Transmutation Systems 1996 *Nuclear Wastes: Technologies for Separations and Transmutation*. National Academy Press, Washington, DC

Organization for Economic Cooperation and Development Nuclear Energy Agency 1977 *Reprocessing of Spent Nuclear Fuels in OECD Countries*. OECD/NEA, Paris

Organization for Economic Cooperation and Development Nuclear Energy Agency 1999 *Actinide and Fission Product Partitioning and Transmutation*. OECD/NEA, Paris

Peka I, Rak V 1967 *Nonaqueous Reprocessing of Irradiated Fuel*. IAEA, Vienna

Nuclear Fuels: Behavior at High Burnup

I. D. Palmer

A nuclear reactor extracts energy from its fuel via a controlled fission reaction. The amount of energy produced per unit weight of fuel is referred to as the fuel burnup. Increasing burnup can improve the economics of reactor operation and reduce waste. On the other hand, achieving higher burnups involves subjecting the fuel to higher power levels and/or longer dwell times in the reactor. The harsh environment in which the fuel operates produces a variety of chemical, mechanical, and nuclear effects which collectively degrade the fuel materials, eroding operational and safety margins. Achievable burnup is therefore limited by material performance issues, although the limits are gradually being raised as mechanisms are understood and materials and operating practices are improved. This article explains and illustrates some of the main effects which are responsible for constraining fuel burnup in the current generation of commercial light water reactors (LWRs).

The total energy generated by fuel, or alternatively, the integral of power over its lifetime, is termed the fuel burnup. The units used here are the number of MWd (megawatt days) of energy produced divided by the initial fuel weight (in terms of kg of heavy metal). A burnup of 40 MWd kg^{-1} is typical of current LWR fuel duty. In fuel discharged at this level, approximately 4% of the uranium atoms initially present will have been fissioned. The energy produced per unit weight is around five orders of magnitude higher than for conventional combustion processes such as the burning of coal or gas.

Older reactor systems and those relying on natural (unenriched) fuel cycles, including RMBK, Candu, Magnox, and advanced gas-cooled reactors (AGRs), tend to have lower burnup goals. Of the commercial reactors in operation today, it is the LWRs (pressurized water ractors (PWRs) and boiling water reactors (BWRs)) which set the most demanding burnup targets. Average discharge burnups have risen steadily, with many utilities now targeting 50–60 MWd kg^{-1} or higher. However, it is important to emphasize that the improvements in burnup have been achieved at the same time as improvements in fuel reliability. Reliability is the reactor operator's primary requirement. The relatively modest improvements in fuel cycle economics derived from higher burnups would be quickly negated by any lost output resulting from fuel-related problems.

Traditional LWR refueling schemes used annual cycles in which one-third of the fuel was replaced at each outage, producing burnups in the 30–35 MWd kg^{-1} range. Achieving higher burnups involves reusing fuel for a fourth and maybe fifth annual cycle, or alternatively lengthening cycles to 18 or 24 months. Both strategies are now in use, with practices varying around the world depending on the local grid requirements and other factors. Higher burnups lead to more complex core management issues. The difference in fissile content between the highest and lowest burnup assemblies is increased, making it more difficult to achieve uniform power distribution; it is localized power peaking which sets many of the constraints on reactor operation. High burnups would not be achievable without the sophisticated deployment of neutron-absorbing materials to facilitate core management. Gadolinia and erbia doped fuel, and boride-coated pellets, are all now in widespread use. The drive for higher burnup has also produced challenges for all of the component materials used in fuel assemblies, and both materials and designs have had to evolve in response.

The sections below discuss some of the main issues relating to fuel performance and how they are affected by increasing burnup. The focus is on fuel rod behavior under normal operation. Performance in transients and under postulated accident conditions is also affected by burnup, as is the performance of other assembly components (grids, springs, nozzles, etc.), but these are not addressed here. The final section considers possible future developments.

1. *Thermal Performance*

LWR fuel consists of cylindrical UO_2 pellets inside a pressurized cladding tube. Energy, from fission, is generated approximately uniformly throughout the fuel pellets. Heat is extracted at the rod surface by the coolant flow. The temperature at the hottest point, at the pellet centerline, is typically in the range 800–1000 °C, depending on the power level. The metallic cladding is a good conductor, so the temperature rise across the wall is only a few tens of degrees. The helium filling gas is also a reasonably good conductor, but only as long as it remains uncontaminated by fission products. The poorly conducting ceramic fuel pellets are responsible for the majority of the temperature rise. The evolution of the conductivities of the pellet and filling gas determine how the thermal performance develops with burnup.

Conduction of heat in UO_2 arises from two mechanisms: conduction by lattice vibrations (phonons) and conduction by electrons. Electronic conduction makes a significant contribution only at high temperatures (above about 1500 °C), and hence the thermal performance of standard commercial reactor fuel under normal operation is determined by phonon conduction. From the Debye formula for dielectric solids, phonon conductivity depends on the mean free path of the phonons, as governed by the various scattering processes. The dominant effects are phonon–phonon interactions, for which the mean free path is inversely proportional to temperature, together with scattering of phonons by lattice imperfections, for which the mean free path depends on defect concentration. For freshly manufactured fuel, defect concentration is controlled by the inherent properties of the material, and to a lesser extent by fabrication-related parameters such as grain size and porosity distribution.

Under irradiation, additional classes of defect are introduced both by radiation damage, and by the generation of fission products (two fission products for each uranium atom fissioned). While radiation damage effects are rapidly annealed at all but the lowest temperatures (below about 500°C), fission products accumulate throughout the irradiation, becoming increasingly important as burnup increases. For practical purposes, phonon conductivity can be modeled by an expression of the form $(a+bT+c\beta)^{-1}$, where T is the absolute temperature and β is the burnup. The constants a and b are basic properties of the material and can be determined from standard laboratory experiments on fresh fuel.

Measurement of the parameter c presents more difficulty. Laboratory techniques such as laser-flash thermal diffusivity measurements are possible on irradiated fuel, but the fuel microstructure created during irradiation is rapidly annealed at elevated temperature and, therefore, destroyed by the measurement process. Estimates of c are generally derived from in-pile temperature measurements on test rods equipped with centerline thermocouples. Based on such evidence, the phonon conductivity at 1000 K is estimated to be reduced by 30–40% for a burnup of 50 MWd kg^{-1} compared to values for fresh fuel.

The effect of thermal conductivity degradation on the performance of the fuel is illustrated in Fig. 1. This shows the results of calculations in which the power in a rod is artificially held at a constant level. The low burnup behavior is controlled by factors

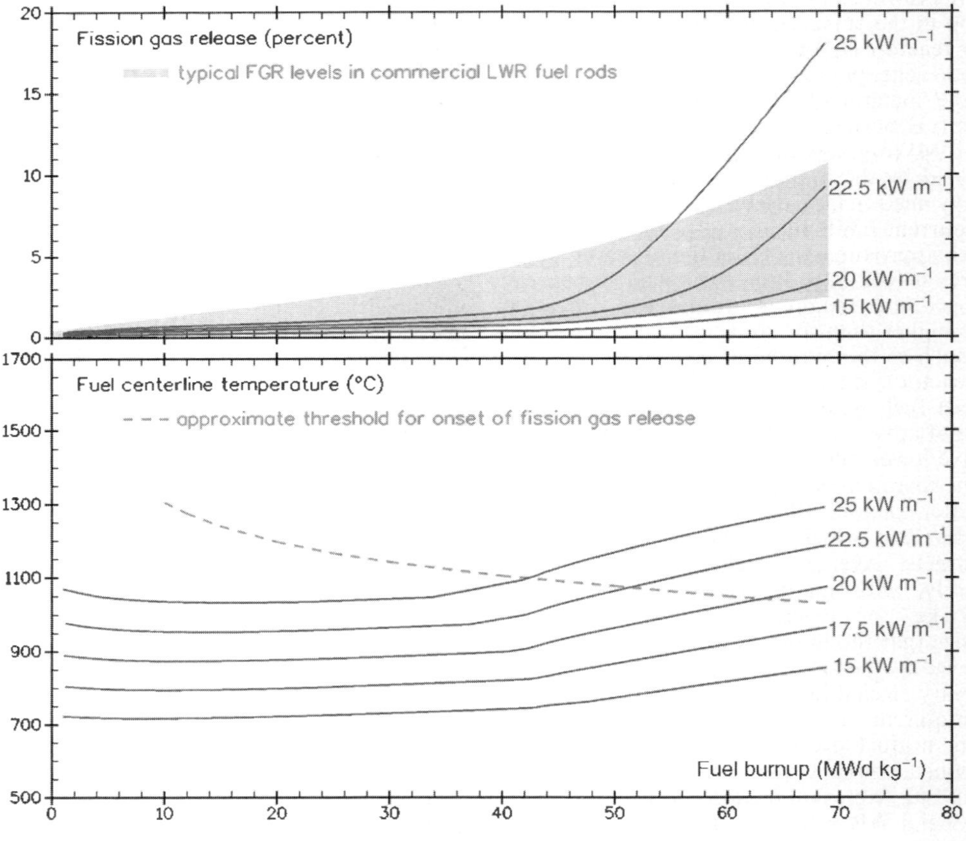

Figure 1
The lower graph illustrates the evolution of fuel centerline temperature with burnup, for power levels in the range 15–25 kW m^{-1}. Fuel thermal conductivity degrades with burnup, causing the temperatures to rise. High burnup fuel may be required to operate above the threshold for the onset of fission gas release. Significant amounts of fission gas can be released, as illustrated in the upper graph. (Courtesy of BNFL.)

such as pellet densification and fuel-clad gap closure, but at higher burnups conductivity degradation becomes increasingly significant, driving the temperatures up to and beyond the level where gaseous and volatile fission products start to be released from the fuel. The consequences of this are considered further in later sections.

2. *Clad Performance*

The combined effects of irradiation and high temperature cause many changes in the pellet and cladding. The processes which dominate the early stages of irradiation are relatively well understood. The cladding tubes are subject to a high compressive stress due to the external coolant pressure. Temperature and neutron flux levels are such that irradiation creep leads to fuel-clad gap closure within 1–2 yr. Meanwhile, the fuel pellets, being brittle, crack due to the thermal gradients, allowing the fragments to relocate. Fission fragments destroy the residual porosity in the fuel, causing the pellets initially to shrink. This is subsequently reversed by the buildup of metallic and oxide fission products in the fuel lattice. Fuel swelling produces an increase in volume of 0.5–1.0% for each $10\,\mathrm{MWd\,kg^{-1}}$ burnup increment. After 3–4 yr of irradiation, the clad is in hard contact with the pellet fragments. Any transient increase in power, causing thermal expansion of the fuel, can induce significant tensile stress in the clad, which may result in a stress corrosion cracking failure. This type of failure has been an issue in the past for BWR, and to a lesser extent PWR, rods up to medium burnup, although it has now been largely eliminated by the use of cladding with a pure zirconium or soft alloy liner. At higher burnups, the magnitude of power increase that fuel can sustain diminishes, and failures of this type become less of a concern.

Oxidation of the cladding occurs at both its inside and outside surfaces. The outside of the tube is exposed to the reactor coolant water. The zirconium is oxidized and a fraction of the liberated hydrogen is absorbed by the cladding. The oxide film acts as a thermal barrier, and hence the reaction rate, which is controlled by the metal–oxide interface temperature, increases as the layer grows. The reaction kinetics change from cubic to linear, and the corrosion layer becomes increasingly porous as it thickens. The absorbed hydrogen produces zirconium hydride platelets which embrittle the cladding. Hydrogen migrates down temperature gradients, producing the largest hydride concentrations nearer to the cooler outer cladding surface. This is illustrated in Fig. 2, which shows a section through moderately oxidized and hydrided cladding. If the oxide layer grows to thicknesses of around 100 μm, it becomes prone to spalling. This produces a local cool spot which can result in failure due to hydride embrittlement. LWR fuels clad in the traditional Zircaloy-2 and Zircaloy-4 alloys have become constrained in burnup by corrosion performance. Some alleviation has been obtained by reducing the tin content of the alloy to the minimum of its specified range, which has been found to improve corrosion resistance. For the highest burnup duty new proprietary alloys such as Zirlo and M5 have been developed to provide even better corrosion performance. These alloys have also been adopted for their reduced growth properties. Alternatively, to limit the impact on other mechanical properties, the corrosion-resistant material can be confined to the outer part of the tube using coextrusion, in so-called duplex cladding. Fuel performance is no longer limited by clad corrosion issues for burnups up to around $60\,\mathrm{MWd\,kg^{-1}}$, and alloy development and testing programs are in place to raise this target further.

Figure 2
Zircaloy cladding in high burnup fuel. The outer 10% of the wall thickness has been oxidized. Circumferentially oriented zirconium hydride stringers are visible in the metal. These are most concentrated in the cooler, outer regions of the tube. (Courtesy of BNFL.)

Excess oxygen from the pellet can also lead to oxidation of the cladding inner surface. The local damage associated with fission product recoils allows high oxygen mobility within a small volume. After fuel–clad contact, this volume encompasses a few microns of the cladding inner surface, allowing oxygen to migrate from pellet to clad. The resulting oxide layer is typically a few microns in thickness, and is generally protective, acting as both a chemical and mechanical barrier. As burnup increases, and the contact force between fuel and clad is raised, the oxide layer becomes more intimately attached to the pellet surface. In very high burnup fuel, the pellet and cladding can become strongly bonded. Figure 3 shows an example of fuel–clad bonding. Although bonding may be beneficial in some respects, for example, reducing the likelihood of clad lift-off, it may also lead to higher clad stresses in some fault scenarios. A number of research programs are currently in place to study the properties of highly bonded fuel.

3. *Fission Gas behavior*

Roughly 25% of all the fission products arising from the fissioning of U^{235} or Pu^{239} are the inert gases xenon and krypton. Their overall rate of generation per MWd of energy produced is roughly 30 cc at standard temperature and pressure. In a high burnup LWR rod producing 100 MWd of energy, the gas generation is large compared to the 10 cc or so of free space available inside the rod. The release of even a small fraction of the fission gas inventory, therefore, has the potential to dramatically raise the rod pressure. Internal pressures significantly above the external reactor coolant pressure cannot easily be tolerated due to the risk of clad lift-off and thermal feedback, threatening rod failure. Alternatively, gas which forms into bubbles inside the fuel can lead to pellet swelling, stressing the clad, and also threatening failure. Accommodation of fission gas is thus one of the main challenges in fuel design, a challenge which becomes increasingly difficult as burnup targets are increased.

Xenon and krypton are insoluble in UO_2. Diffusion occurs as single gas atoms, or strictly as single atoms linked to lattice vacancies. Collections of a few atoms appear as nanometer-sized bubbles, which are present in fuel at densities around 10^{23}–10^{24} m^{-3}, as illustrated in Fig. 4. These intragranular bubbles exist in dynamic equilibrium, being nucleated in the wake of fission tracks, and also destroyed by fission spikes passing nearby (termed irradiation resolution). The presence of intragranular bubbles modifies, but does not prevent, the diffusion of gas atoms to the grain boundaries. Intragranular bubbles play a role in the nucleation of grain boundary bubbles, but in normal operation do not themselves grow in sufficient numbers to produce significant fuel swelling. However, in some transient events, their contribution to swelling may be appreciable.

Gas reaching the fuel grain boundaries forms lenticular shaped bubbles. Figure 5 shows an example from a fuel fracture surface. The bubbles can be up to a few microns in size, on grain faces which are typically of the order of 10 μm in diameter. Unlike the smaller intragranular bubbles, grain face bubbles are relatively stable against irradiation resolution, although a part of their gas content may be returned to the fuel matrix by passing fission spikes. In transient conditions, growth and coalescence of grain face

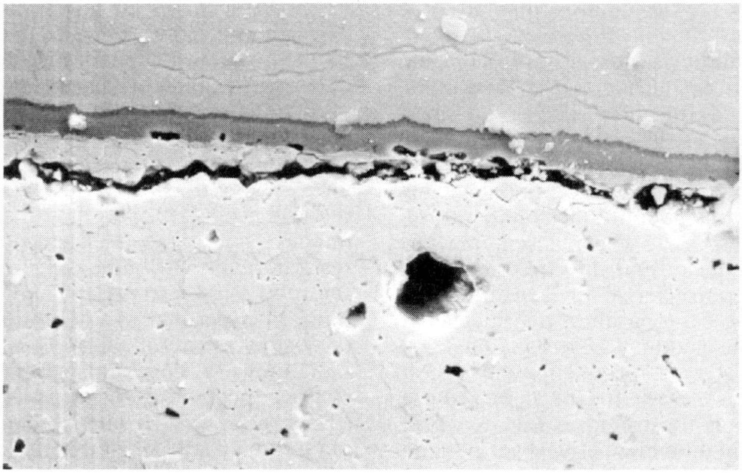

Figure 3
Example of pellet–clad bonding in high burnup fuel. The clad (top) shows hydride stringers and an inner oxide layer a few microns in thickness. The circumferential crack that forms on cooling passes through the pellet rather than the pellet–clad interface, indicating that the fuel is firmly bonded to the oxide layer. (Courtesy of BNFL.)

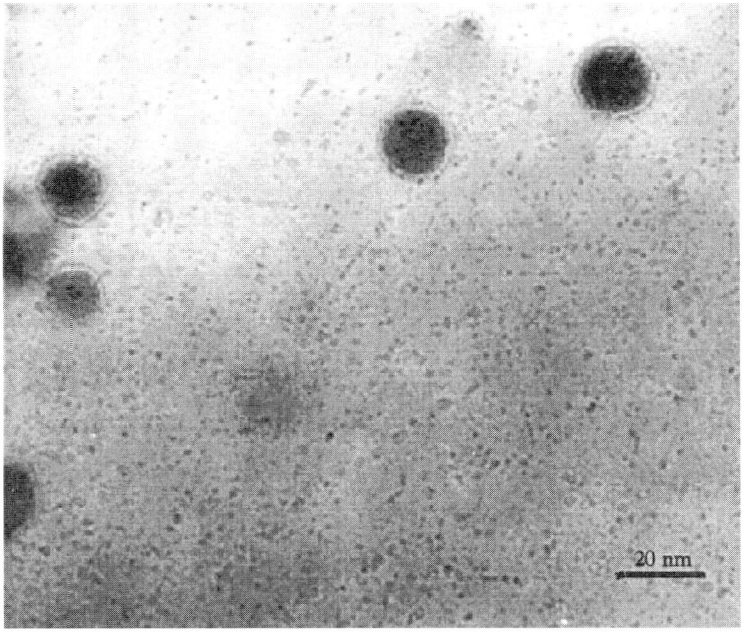

Figure 4
High magnification image of the interior of a fuel grain. A high density of fission gas bubbles in the size range 1–2 nm is present. Some of these bubbles are seen to lie on straight lines, indicating that they are nucleated in the wake of fission fragment tracks. Growth and coalescence produces a secondary population of bubbles in the 10–20 nm range. (Courtesy of BNFL.)

bubbles can occur more rapidly than the diffusive processes responsible for returning them to their equilibrium circular cross-section. This can result in elongated bubbles, as shown in Fig. 6, and to significant fuel swelling.

As more gas reaches the grain boundaries, the grain face bubbles grow, coalesce, and intersect the grain edges (where three grains meet). Interconnected grain edge bubbles can be clearly seen in Fig. 6. Eventually, the grain edge bubbles interlink over a wide area to form a continuous network through which the gas is vented to the rod-free volume. Significant release of fission gas does not occur, even in the hottest central regions of the pellet, until the grain face and grain edge bubble networks have developed along the lines discussed. This hold-up on the grain boundaries is responsible for the incubation effect which is evident in the empirical release threshold shown on the lower part of Fig. 1. The upper part of Fig. 1 shows how the level of gas release can rise sharply with burnup. Commercial LWR rods are discharged with typically a few percent of the fission gases released from the fuel. The conductivity of xenon and krypton is low compared with the helium filling gas, resulting in poorer gap conductance and hence driving pellet temperatures higher still. This positive feedback mechanism is one of the main factors which constrains fuel duty.

4. Evolution of the Fuel Microstructure

In conventional uranic fuels, U^{235} is the dominant source of fissions. Its natural abundance of 0.7% is enhanced by enrichment to values typically in the range 3–5% for commercial reactor use. In mixed oxide (MOX) fuels, recycled plutonium takes the place of U^{235} as the fissile source. In either case the majority of the heavy metal atoms present are U^{238}, the distribution of fissile items is initially homogeneous, and the distribution of power across the radius of a pellet is initially quite uniform.

The fission process is initiated, predominantly, by low-energy (thermal) neutrons, but generates high-energy (fast) neutrons. The fast neutrons must be moderated (reduced in energy) by collisions with light nuclei such as hydrogen, deuterium, oxygen, or carbon before they can initiate a subsequent fission. In the order of 10–20 collisions are involved in reducing neutron energy from the megaelectronvolt range at which they are born down to the fractional electronvolt thermal range. The mean free path at high

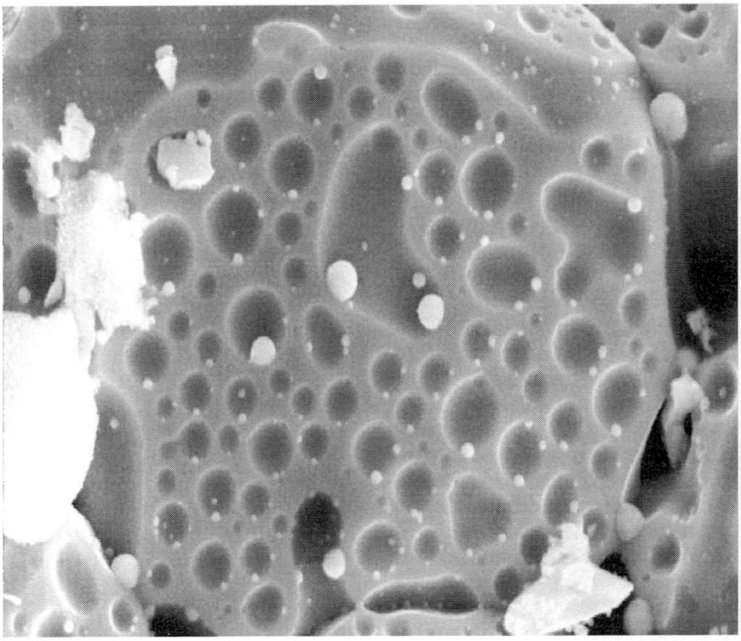

Figure 5
Fuel fracture surface showing lenticular grain face gas bubbles. The smaller particles are metallic precipitates. The larger white particles are resin used in the sample preparation. (Courtesy of BNFL.)

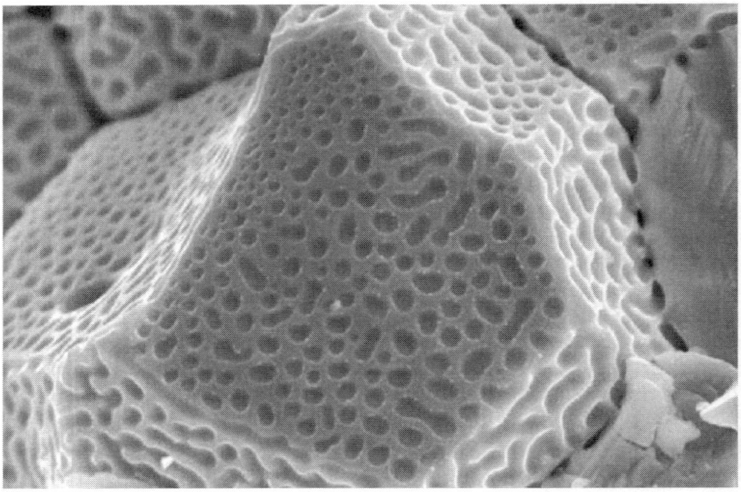

Figure 6
Fuel fracture surface showing the elongated grain face bubble structures that can form under transient conditions. This picture also illustrates the interlinked network of grain edge bubbles through which fission gas is released from the fuel. (Courtesy of BNFL.)

energies is many centimeters. It follows that the source of thermal neutrons is outside a given fuel rod, and hence that thermal neutron flux is highest at the rod periphery and declines with distance into the pellet. This flux depression effect is nevertheless quite small, giving perhaps a 5–10% reduction in power across the radius of a pellet.

Of more significance is that in the intermediate range between thermal and fast energies, U^{238} exhibits a number of resonant capture peaks (principally around 6, 20, and 36 eV). Neutron capture in U^{238} is followed by two short half-life β-decays to produce the fissile isotope Pu^{239}. The resonant peaks have a very high cross-section, meaning that any neutron of the right energy has a high probability of being captured near to the surface of the pellet. Plutonium production is therefore very sharply peaked in the outer few hundred microns of the pellet periphery. The effect grows in significance as irradiation progresses. For typical commercial reactor fuel near the end of its life, the power generation rate at the surface can be more than double the pellet average.

The outer periphery of a fuel pellet operates at relatively low temperatures, typically 400–500°C. At these temperatures the effects of irradiation damage are not rapidly and fully annealed, as occurs in hotter regions. The combination of low temperature and high fission rate and burnup can result in a microstructural transformation referred to as the rim effect. The transformed microstructure, often called the high burnup structure, or HBS, is characterized by small (submicron) grains and a high density of fission gas bubbles. The effect is first visible on the outer surface of pellets that have achieved an average burnup of around 50 MWd kg^{-1}. The effect progresses inwards, covering about 100–200 μm at 60 MWd kg^{-1}, and up to perhaps 1000 μm at 100 MWd kg^{-1}. It is presently unclear whether the rim effect can penetrate further into the hotter regions of a pellet at even higher burnups. However, test irradiations have shown that fuel can be fully transformed to the rim structure if the temperature is maintained at an artificially low level.

At the microscopic level, the grain subdivision associated with the rim effect is observed to begin at the margins of pores and grain boundary gas bubbles. Several factors cause local straining of the lattice, including bubble pressure, high defect concentration, fission product supersaturation, and substitution of plutonium for uranium. Beyond a certain point, dislocation clusters can grow to form dislocation walls, dividing the grain, draining point defects, and purifying the matrix. Eventually, local lattice rotation occurs to relax the stresses and a new subgrain is formed. The local recrystallization causes the xenon and krypton fission gases to be precipitated into bubbles, typically 1 μm or so in size. The subgrains display a distribution of sizes in the range 50–500 nm (compared with about 10 μm in freshly fabricated fuel). A very similar microstructual transformation is seen in the plutonium-rich agglomerates in MOX fuel, for those agglomerates located in the cooler outer regions of a pellet. Rim formation is an active area of research and a full understanding of the mechanisms involved has yet to be achieved. Research work is also in place to develop fuel variants, for example, involving doped materials, which might exhibit more resistance to the rim transformation.

Figure 7 shows a high magnification image of fuel that has begun the rim transformation. The latest evidence indicates that the rim effect does not add significantly to fission gas release (at least for steady-state operation), suggesting that the gases remain confined in bubbles and that the bubbles remain isolated. However other significant effects are known to occur. The porosity formed acts as a barrier to heat flow, reducing fuel conductivity, and hence increasing pellet temperatures. The porosity also adds to pellet swelling. Under fast transient conditions, some experiments have shown a very large contribution to swelling from the rim region, accelerating clad failure. The rim transformation also modifies the mechanical properties of the fuel, for example, increasing fracture toughness, and changes how the fuel and clad interact and bond at the pellet–clad interface.

5. *Possible Future Trends*

Current designs of LWR fuel are able to achieve discharge burnups of around 60 MWd kg^{-1}. Evolution of design and materials, particularly cladding

Figure 7
Example of the characteristic microstructural transformation in the pellet rim region of very high burnup LWR fuel. The formation of a new population of submicron fuel grains begins at the margins of existing pores, as seen on the left of the picture (after Ray *et al.* 1997).

alloy composition, has raised burnup targets significantly since the 1980s. Present limitations relate to a variety of issues, including cladding corrosion resistance, accommodation of fission gases, and pellet microstructure transformation. Evidence from test reactor irradiations suggests that further evolutions of the same basic model—oxide fuel pellets in zirconium alloy cladding—could be capable of achieving discharge burnups up to 100 $MWd\,kg^{-1}$. It now seems unlikely that more radical changes, such as ceramic cladding, will have a significant role to play. As the world's fleet of LWRs ages, investment in revolutionary new fuel designs becomes more difficult to justify.

Fuel discharge burnup influences overall fuel cycle costs, but there are many other parameters involved including enrichment, transportation, and waste disposal costs. Economic modeling of the fuel cycle suggests that currently achievable burnups are already close to the optimum for many reactor operators. The incentive for raising burnups further is likely to be small and only applicable in some scenarios and in certain market conditions. It seems likely that evolutions of existing fuel designs will be pushed to burnup targets of perhaps 70–75 $MWd\,kg^{-1}$ over the medium term, but only in some countries and not in others. What is clearer is that burnups beyond this level would require fuel to be enriched above the 5% limit to which most fabrication, transport, and handling facilities are licensed. Higher burnup levels may, however, be economically favorable in some niche areas, for example, with plutonium-bearing MOX fuel. As a newer, less well understood product, this currently operates to lower burnup targets than standard fuel. But since fabrication costs are greater and enrichment is free, the economically optimum burnup may be at a higher level than for UO_2 fuel. Nevertheless, dramatic increases in burnup are only likely to be seen with the arrival of new reactor systems where fuel cycle infrastructure is designed from the outset to handle appropriately high enrichments.

See also: Light Water Reactor Fuel Design and Performance.

Bibliography

American Nuclear Society 2000 *Proc. 2000 Int. Topical Meeting on LWR Fuel Performance*. American Nuclear Society Inc, La Grange Park, IL

American Society for Testing and Materials 1999 *12th Int. Symp. Zirconium in the Nuclear Industry*, ASTM STP 1354. ASTM, Philadelphia

Bailly H, Menessier D, Prunier C 1999 *The Nuclear Fuel of Pressurised Water Reactors and Fast Reactors: Design and Behaviour*. Lavoisier Publishing, Paris

International Atomic Energy Agency 2001 *Nuclear Fuel Behaviour Modelling at High Burnup and its Experimental Support*. IAEA-TECDOC-1233. IAEA, Vienna

OECD Nuclear Energy Agency 1998 Thermal performance of high burnup LWR fuel. *Proc. Seminar Cadarache, France*. OECD Publications, Paris

OECD Nuclear Energy Agency 2000 Fission gas behaviour in water reactor fuels. *Proc. Seminar Cadarache, France*. OECD Publications, Paris

Ray I L F, Matzke Hj, Thiele H A, Kinoshita M 1997 An electron microscope study of the rim structure of a UO_2 fuel with a high burnup of 7.9% FIMA. *J. Nucl. Mater.* **245**, 115–23

Nuclear Graphite and Radiation Effects

T. D. Burchell

This article provides the reader with an understanding of the physics of neutron damage in nuclear graphite and the effects of neutron damage on the physical and mechanical properties of the material. A background section outlines the role graphite plays as a moderator in nuclear fission reactors and briefly summarizes the manufacturing process for nuclear-grade graphite.

Graphite lattice displacement damage is introduced and the consequences of such damage on graphite single-crystal structure and properties are discussed. The phenomenon of Wigner (stored) energy is introduced and its consequences for reactor operation and design elucidated.

The effect of crystal neutron damage on graphite structure and properties is discussed in the context of neutron irradiation-induced dimensional change. The dependence of dimensional change on both in-crystal changes and structure changes (texture dependent) is discussed. The effect of neutron irradiation on other key properties, including thermal conductivity, coefficient of thermal expansion, electrical resistivity, strength, and elastic modulus is reported. Finally, the significance of radiation creep is discussed.

1. Background

1.1 Nuclear Fission

Nuclear fission has been harnessed as a commercial energy source since the mid-twentieth century. There are numerous graphite-moderated, power-producing fission reactors operating worldwide (Burchell 1996a), the greatest numbers being in the UK (gas cooled) and in the countries of the former Soviet Union (water cooled).

In nuclear fission, energy is derived when a heavy element such as $^{235}{}_{92}U$ fissions or "splits" apart. The fission reaction can be written as

$$^{235}_{92}\mathrm{U} + ^{1}_{0}\mathrm{n} \rightarrow ^{236}_{92}\mathrm{U}^{*} \rightarrow \mathrm{F}_1 + \mathrm{F}_2 + \mathrm{n} + \mathrm{energy} \qquad (1)$$

where F_1 and F_2 are fission fragments. Typically, an impinging neutron initiates the fission reaction, and the reaction yields an average of 2.5 neutrons per fission. The fission fragments and the neutron possess kinetic energy that can be degraded to heat and harnessed to raise steam and hence drive turbine generators. As discussed below, the role of graphite in the fission reactor is to facilitate the nuclear chain reaction by moderation (slowing down) of the high-energy fission neutron.

Nuclear fission reactors produce thermal energy from the fission of heavy isotopes, notably $^{235}_{92}\mathrm{U}$ (Eqn. (1)). The kinetic energy of the fission fragments is converted to thermal energy by successive collisions within the body of the uranium fuel mass in which the fission occurs. The fission neutrons give up their energy within the moderator via the process of elastic collision. The γ-energy given up in the fission reaction is absorbed in the bulk of the reactor outside of the fuel, i.e., moderator, pressure vessel, and shielding. During the fission process, the longer a neutron dwells in the vicinity of a nucleus of a fuel atom the greater the possibility it will be captured and therefore cause the fuel atom to fission.

It is desirable, therefore, to slow the high-energy ($\sim$2 MeV) fission neutron (referred to as a fast neutron) to lower (thermal) energies of $\sim$0.025 eV at room temperature, which corresponds to a velocity of $2.2 \times 10^5\,\mathrm{cm\,s^{-1}}$. The process of thermalization, or slowing down, of fast fission neutrons is referred to as "moderation." The material in a thermal reactor (i.e., a reactor in which fission is caused by neutrons with thermal energies) that is responsible for slowing down the fast fission neutrons is referred to as the "moderator."

1.2 Selection of Moderator Materials

A good moderator material must not react with neutrons, because if they are captured in the moderator they cannot sustain the fission process. Moreover, neutrons should be slowed down over short distances and with few collisions in the moderator. Ideally, the moderator will be inexpensive, yet will be compatible with the other materials used in the reactor core and satisfy the core structural requirements. Finally, it is preferable that the moderator should not undergo any damaging physical or chemical changes when bombarded with neutrons.

A good moderator should efficiently thermalize fast neutrons through the process of elastic collisions. The maximum energy lost per collision occurs when the target nucleus has unit mass and tends to zero for heavy target elements. Low atomic number (Z) is therefore a prime requirement of a good moderator. Thermalization of a neutron takes place over many collisions, the maximum energy being lost in a direct collision, although elastic collisions can take place at many angles. The number of atoms per unit volume (density) of the moderator, and the chance of a scattering collision taking place, must also be considered. A frequently used "figure of merit" applied when assessing moderators is the *slowing-down power*, which accounts for the average energy loss per collision, the number of atoms per unit volume, and the scattering cross-section of the moderator. To gain a full picture of the performance of a material as a nuclear moderator, its slowing-down power (which should be large) should be compared with its tendency to capture neutrons (*capture cross-section*), which should be small. Therefore the ratio of the slowing-down power to the macroscopic absorption (capture) cross-section, or *moderating ratio*, should additionally be considered when evaluating potential moderators. The slowing-down power and the moderating ratio for several candidate moderators are given in Table 1.

In practice, the choice of moderating materials is limited to a few elements with atomic number less than 16. Gases are of little use as moderators because of their low density, but can be used when fixed in chemical compounds such as H_2O and D_2O. The selection of materials that are available rapidly reduces to the four shown in Table 1. Water is relatively unaffected by neutron irradiation, is low in cost, and easily contained. However, the moderating ratio is reduced by neutron absorption in the hydrogen, requiring the use of enriched (in ^{235}U) fuels to maintain the neutron economy. Heavy water is a good moderator because $^{2}_{1}H$ and $^{16}_{8}O$ do not absorb neutrons. The slowing-down power is high and therefore the moderating ratio is very large. Unfortunately, the cost of separating the heavy hydrogen isotope is high. Beryllium or beryllium oxide are good moderators but are expensive, difficult to machine, and suffer from toxicity problems. Finally, graphite is an acceptable moderator. It offers a compromise between nuclear properties, utility as a core structural material, and cost. Unfortunately, the properties of graphite are markedly altered by neutron irradiation, as discussed in a later section of this article.

2. *Graphite Manufacture*

Detailed accounts of the manufacture of polygranular graphite have been published elsewhere (Eatherly and Piper 1962, Ragan and Marsh 1983, Inagaki 2001). The major processing steps in the manufacture of conventional polygranular graphite are summarized in Fig. 1. Polygranular graphite consists of two phases: a filler material and a binder phase. The predominant filler material, particularly in the USA, is a petroleum coke made by the delayed coking process. European nuclear graphites are typically made from a coal tar pitch-derived coke. In the UK, Gilsonite coke derived from a naturally occurring bitumen found in Utah, USA, has been used. The coke is usually calcined (thermally processed) at ~1300°C prior to being crushed and blended. Typically, the binder phase is a coal tar pitch. The binder plasticizes the filler coke particles so that they can be formed. Commonly used forming processes include extrusion, molding, and isostatic pressing. The binder phase is carbonized during the subsequent baking operation (~1000°C). Frequently, engineering graphites are pitch impregnated to densify the carbon artifact, followed by rebaking. Useful increases in density and strength are obtained with up to six impregnations, but two or three are more typical.

The final stage of the manufacturing process is graphitization (2500–3000°C) during which, in simplistic terms, carbon atoms in the baked material migrate to form the thermodynamically more stable graphite lattice. Nuclear graphites require high chemical purity to minimize neutron absorption. Moreover, certain elements catalyze the oxidation of

Table 1
The properties of good moderators (Burchell 1999).

Moderator material	Slowing-down power (cm^{-1})	Moderating ratio
Graphite (C)	0.064	170
Beryllium (Be)	0.176	159
Water (H_2O)	1.530	72
Heavy water (deuterium, D_2O)	0.370	12000

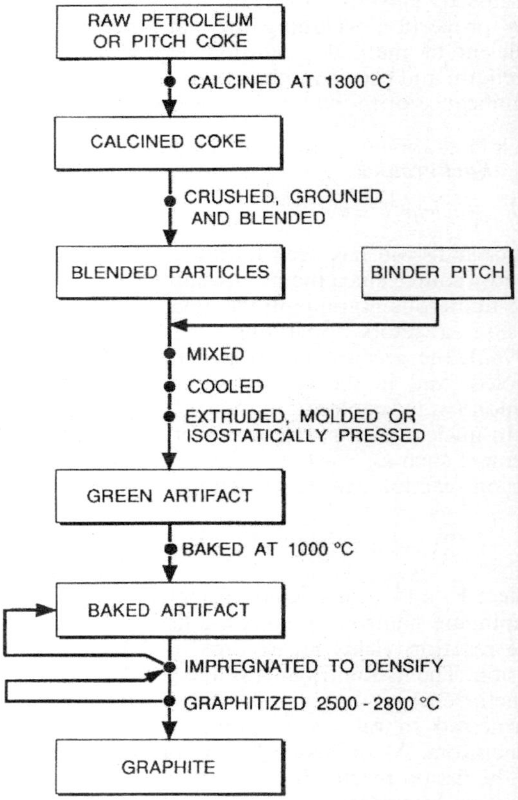

Figure 1
Major processing steps in the manufacture of a polygranular graphite.

graphite and must be reduced to an acceptable level. This is achieved by selecting very pure cokes, using a high graphitization temperature (>2800°C), or by including a halogen purification stage in the manufacture of the cokes or graphite. In its perfect form, the crystal structure of graphite (Fig. 2) consists of tightly bonded (covalent) sheets of carbon atoms in a hexagonal lattice network (Ruland 1968). The sheets are weakly bound with van der Waals-type bonds in an ABAB stacking sequence with a separation of 0.335 nm. The crystals in a manufactured polygranular graphite are less than perfect, with approximately one layer plane in every six constituting a stacking fault. The graphite crystals have two distinct dimensions, the crystallite size L_a measured parallel to the basal plane and the dimension L_c measured perpendicular to the basal planes. In a coke-based nuclear graphite, values of $L_a \sim 80$ nm and $L_c \sim 60$ nm are typical (Kelly 1994).

3. *Graphite-moderated Reactors*

The use of carbons and graphite in fission reactors has been reviewed by Burchell (1999). For the reasons outlined above, the earliest experimental fission reactors used graphite as their moderators. For example, the Chicago Pile No. 1 (CP-1), constructed under the stands at Stagg Field, University of Chicago, USA, by Enrico Fermi in 1942 used grade AGOT graphite manufactured by the National Carbon Company. It was in CP-1 that Fermi and his team achieved the world's first nuclear chain reaction on December 2, 1942. Subsequently, the early weapons materials production reactors constructed in the USA, the UK, France, and the former USSR

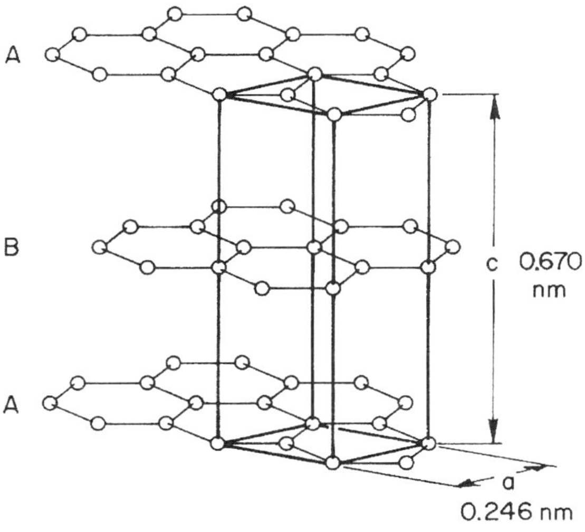

Figure 2
Crystal structure of graphite.

were graphite-moderated reactors. Moreover, the first electric power-generating fission reactors were also graphite moderated.

The essential features of a typical graphite-moderated, gas-cooled, power-producing nuclear fission reactor are shown in Fig. 3. The reactor core comprises stacks of graphite moderator blocks that are usually keyed to one another so that mechanical loads may be transmitted through the core. The graphite moderator is penetrated vertically with a series of channels into which fuel stringers may be placed via the reactor charge face, using the refueling machine. The fuel, which may be natural uranium (a mixture of ^{238}U, ^{235}U, and ^{234}U) or enriched (in ^{235}U) uranium, is typically contained in a metal sheath. The sheath or "cladding" is usually a light alloy such as aluminum or magnesium, or can be stainless steel (requiring enriched fuel) if a higher fuel temperature is desirable (>600°C). The fuel cladding and fuel stringer structural material may be one and the same, as in the UK Magnox design (Burchell 1999), or the fuel may be in the form of steel clad "pins" arranged in graphite fuel sleeves that are joined to one another and form the fuel stringer. The UK Advanced Gas Cooled Reactor (AGR) is an example of this latter configuration (Burchell 1999). The metallic fuel clad retains the gaseous fission products that migrate from the fuel during the fission reaction and prevents direct contact between the coolant fluid and the fuel.

An alternative core layout uses integral fuel/moderator elements in which the uranium fuel is placed directly into cavities in the moderator block and the entire block is discharged from the reactor when the fuel is spent. In this design of fuel element the fuel is typically a ceramic (UO_2 or UC_2), and is overcoated with layers of SiC and pyrolytic graphite (Burchell 1999). The coated fuel particles are usually formed into fuel pucks or fuel rods, but may be consolidated into fuel balls or "pebbles" (Burchell 1999). The SiC and carbon layers act as the fission product barrier and negate the use of a metallic fuel clad, allowing the reactor core to operate at very high temperatures (>1000°C). The US-designed Modular High Temperature Gas Cooled Reactor (MHTGR) and Japanese High Temperature Test Reactor (HTTR) are both examples of gas-cooled reactors with high-temperature ceramic fuel.

Additional vertical channels in the reactor core house the reactor control rods, which regulate the fission reaction by introducing neutron-absorbing materials to the core and therefore reduce the number of neutrons available to sustain the fission process. When the control rods are withdrawn from the core, the self-sustaining fission reaction commences. Heat is generated by the moderation of the fission fragments in the fuel and by the moderation of fast neutrons in the graphite. The heat is removed from the core by a coolant, typically a gas, which flows freely through the core and over the graphite

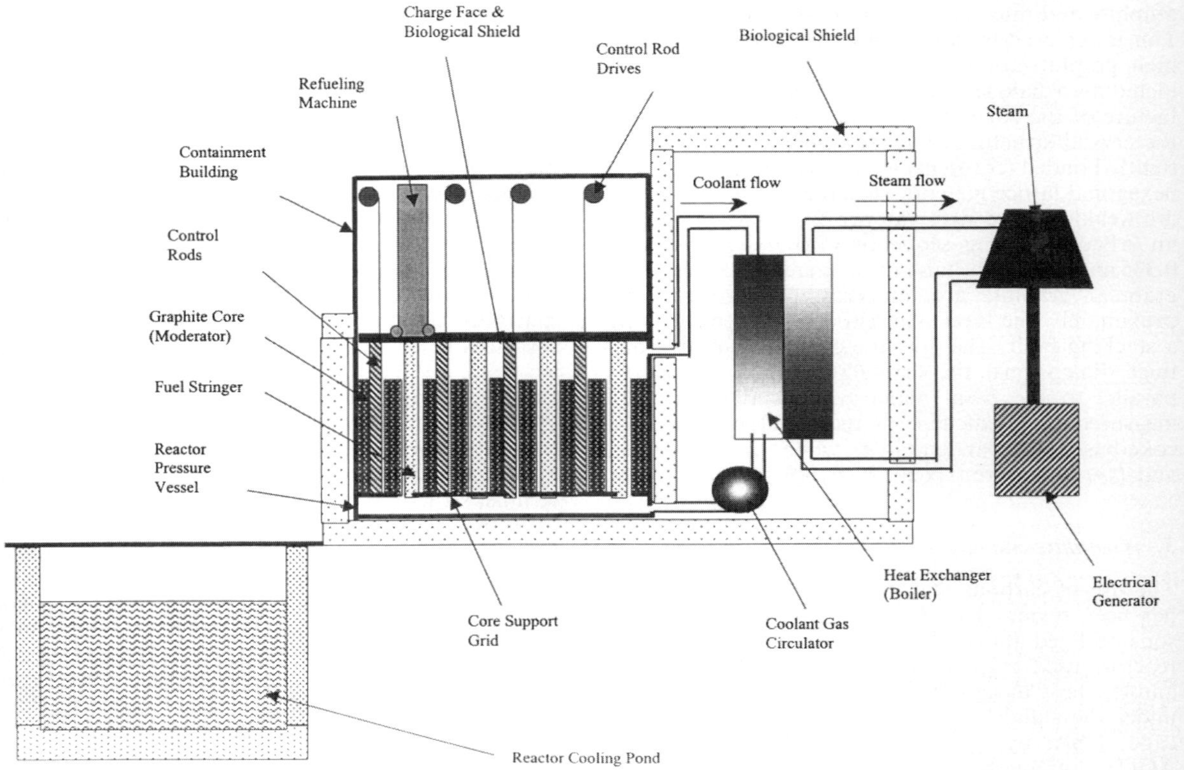

Figure 3
Key features of a graphite-moderated, gas-cooled, power-producing nuclear fission reactor.

moderator. The coolant is forced through the core by a gas circulator and passes into a heat exchanger/boiler (frequently referred to as a steam generator).

The primary coolant loop (i.e., the reactor coolant) is maintained at elevated pressure to improve the heat transfer characteristics of the coolant and therefore a pressure vessel surrounds the reactor core. A secondary coolant (water) loop runs through the heat exchanger and cools the primary coolant so that it may be returned to the reactor core at a reduced temperature. The secondary coolant is raised to steam temperature and is passed to a turbine where it gives up its energy to drive an electric generator. Not all graphite-moderated reactors are gas-cooled ones. In most water-cooled, graphite-moderated reactors the water coolant passes through the core in zirconium alloy tubes at elevated pressure before being fed to a steam generator.

The reactor core and primary coolant loop are enclosed in a concrete biological shield that protects the reactor staff and public from gamma radiation and fission neutrons, and also prevents the escape of radioactive contamination and fission product gasses that originate in the fuel pins/blocks. The charge face, refueling machine, control rod drives, discharge area, and cooling ponds are housed in a containment structure that similarly prevents the spread of any contamination. Additional necessary features of a fission reactor are: (i) a refueling bay, where new fuel stringers or fuel elements are assembled prior to being loaded into the reactor core; and (ii) a discharge area and cooling ponds where spent fuel is placed while the short-lived isotopes are allowed to decay before the fuel can be reprocessed.

4. *Displacement Damage and Induced Structural and Dimensional Changes in Graphite*

The binding energy of a carbon atom in the graphite lattice is about 7 eV (Thrower and Meyer 1978). Impinging energetic particles such as fast neutrons, electrons, or ions can displace carbon atoms from their equilibrium positions. There have been many studies of the energy required to *displace* a carbon atom (E_d), as reviewed by Kelly (1981) and Burchell (1996a). The value of E_d is not well defined but lies between 24 eV and 60 eV. The latter value has gained wide acceptance and use in displacement damage calculations, but a value of 30 eV would be more appropriate.

The primary carbon atom displacement (*primary knock-on atoms*, PKAs) produced by energetic particle collisions produces further carbon atom displacements in a cascade effect. The cascade carbon atoms are referred to as *secondary knock-on atoms* (SKAs). The displaced SKAs tend to be clustered in small groups of 5–10 atoms and for most purposes it is satisfactory to treat the displacements as if they occur randomly. The total number of displaced carbon atoms will depend on the energy of the PKA, which is itself a function of the neutron energy spectrum and the neutron flux. Once displaced, the carbon atoms recoil through the graphite lattice, displacing other carbon atoms and leaving vacant lattice sites. However, not all of the carbon atoms remain displaced. The displaced carbon atoms diffuse between the graphite layer planes in two dimensions and a high proportion will recombine with lattice vacancies. Others will coalesce to form C_2, C_3, or C_4 linear molecules. These in turn may form the nucleus of a dislocation loop—essentially a new graphite plane. Interstitial clusters may, on further irradiation, be destroyed by a fast neutron or carbon knock-on atom (irradiation annealing). Adjacent lattice vacancies in the same graphitic layer are believed to collapse parallel to the layers, thereby forming sinks for other vacancies that are increasingly mobile above 600°C and hence can no longer recombine and annihilate interstitials.

A principal result of carbon atom displacements is crystallite dimensional change. Interstitial defects will cause crystallite growth perpendicular to the layer planes (*c*-axis direction) and relaxation in the plane due to coalescence of vacancies will cause a shrinkage parallel to the layer plane (*a*-axis direction). The damage mechanism and associated dimensional changes are illustrated in Fig. 4. Dimensional changes can be very large, as demonstrated in studies on well-ordered graphite materials, such as pyrolytic graphite, which has frequently been used to study the neutron irradiation-induced dimensional changes of the graphite crystallite (Kelly 1981, Engle and Eatherly 1972). Price (1974) conducted a study of the neutron irradiation-induced dimensional changes in pyrolytic graphite.

Figure 5 shows the crystallite shrinkage in the *a*-direction for neutron doses up to 12 displacements per atom (dpa) for samples graphitized at a temperature of 2200–3300°C prior to being irradiated at 1300–1500°C. The *a*-axis shrinkage increases linearly with dose for all of the samples, but the magnitude of the shrinkage at any given dose decreases with increasing graphitization temperature. Similar trends are noted for the *c*-axis expansion. The significant effect of graphitization temperature on irradiation-induced dimensional change accumulation can be attributed to thermally induced improvements in crystal perfection, thereby reducing the number of vacancy trapping sites in the lattice.

Polygranular graphites possess a polycrystalline structure, usually with significant texture resulting from the method of forming during manufacture. Consequently, structural and dimensional changes in polygranular graphites are a function of the crystallite dimensional changes and the texture of the graphite. In polygranular graphite, thermal shrinkage cracks that occur during manufacture and are preferentially aligned in the crystallographic *a*-direction will initially accommodate the *c*-direction expansion, so mainly *a*-direction contraction will be observed. The graphite thus undergoes a net volume shrinkage. With increasing neutron dose (displacements) the incompatibility of crystallite dimensional changes leads to the generation of new porosity oriented

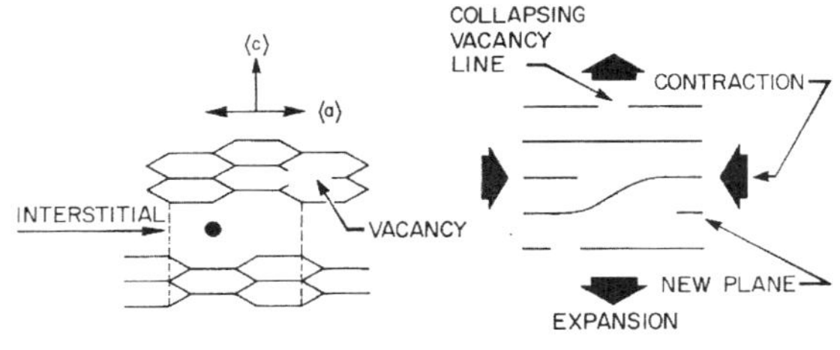

Figure 4
Neutron irradiation damage mechanism illustrating the induced crystal dimensional strains.

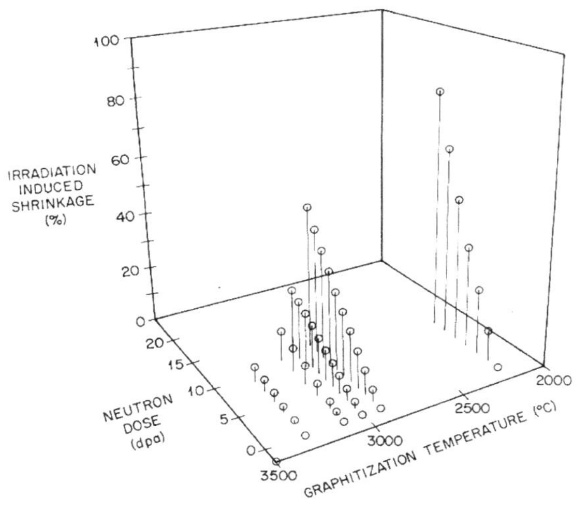

Figure 5
Neutron irradiation-induced *a*-axis shrinkage behavior of pyrolytic graphite showing the effects of graphitization temperature on the magnitude of the dimensional changes (Price 1974).

parallel to the basal planes and the volume shrinkage rate falls, eventually reaching zero. The graphite now begins to swell at an increasing rate with increasing neutron dose. The graphite thus undergoes a volume change "turnaround" into net growth that continues until the generation of cracks and pores in the graphite, owing to differential crystal strain, eventually causes total disintegration of the graphite.

Irradiation-induced dimensional damage data for GraphNOL N3M are shown in Fig. 6. N3M is a molded graphite and thus the filler coke particles are preferentially aligned in the radial direction. Consequently, the crystallographic *a*-direction is predominantly aligned in the radial (perpendicular to forming) direction. Therefore, the *a*-direction irradiation-induced shrinkage is more apparent in the radial direction, as indicated by the radial data (both 600°C and 875°C) in Fig. 6. A general theory of dimensional change in graphite due to Simmons (1965) has been extended by Brocklehurst and Kelly (1993). For a detailed account of the treatment of dimensional changes in graphite the reader is referred to Kelly and Burchell (1994a).

5. Neutron-induced Property Changes

5.1 Stored Energy

The lattice displacement processes described above can cause an excess of energy in the graphite crystallites. The release of this stored energy (or Wigner energy, after the physicist who first postulated its existence) was historically the first problem of radiation damage in graphite to manifest itself. When an interstitial carbon atom and lattice vacancy recombine, their excess energy is given up. If sufficient damage has accumulated in the graphite, the release of this stored energy can result in a rapid rise in temperature. Stored energy accumulation was found to be particularly problematic in the early (air-cooled) graphite-moderated reactors, which operated at relatively low temperatures. Figure 7 shows the rate of release of stored energy with temperature, as a function of temperature, for graphite samples irradiated at 30°C by low doses in the Hanford K reactor. The release rate curves are characterized by a peak occurring at 200°C. The release rate exceeds the specific heat and thus, under adiabatic conditions, the graphite would rise sharply in temperature. For ambient temperature irradiations it is found that the stored energy can attain values up to 2720 J g^{-1}, which if released adiabatically would cause a temperature rise of some 1300°C (Kelly 1994). A simple experiment, whereby samples irradiated at 30°C are placed in a furnace at 200°C and their temperature monitored, shows that when the sample attains a temperature of ~70°C it suddenly increases in temperature to a maximum of about 400°C and then returns to 200°C (Kelly 1994). In order to limit the total amount of stored energy in the early graphite reactors it became necessary periodically to anneal the graphite. The temperature of the graphite was raised sufficiently, by nuclear heating or by the use of inserted electrical heaters, to "trigger" the release of stored energy. The release then self-propagated slowly through the core, raising the graphite moderator temperature and thus partially annealing the graphite core. Indeed, it was during such a reactor anneal that the Windscale (UK) reactor accident occurred in 1957 (Arnold 1992). The accumulation of stored energy in a graphite is both dose- and irradiation temperature-dependent. With increasingly higher irradiation temperatures the total amount of

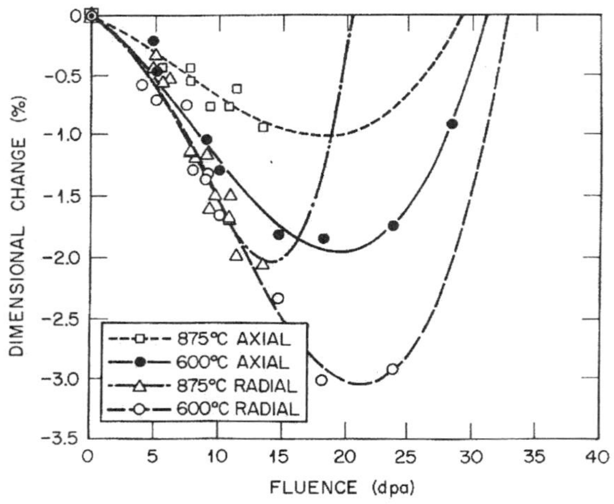

Figure 6
Neutron irradiation-induced dimensional change curves for GraphNOL N3M graphite irradiated at 600°C or 875°C (Burchell and Eatherly 1991).

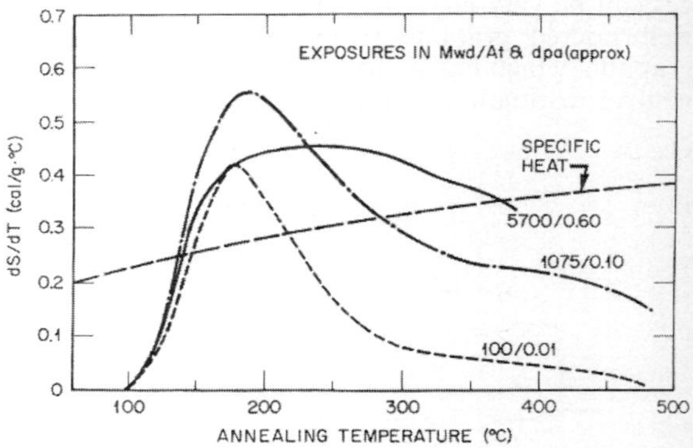

Figure 7
Stored energy release curves for CSF graphite irradiated at ~30°C in the Hanford K reactor cooled test hole (Nightingale 1962).

stored energy and its peak rate of release diminish, such that above an irradiation temperature of ~300°C stored energy ceases to be a problem. Excellent accounts of stored energy in graphite can be found elsewhere (Nightingale 1962, Simmons 1965, Kelly 1994, Burchell 1999).

5.2 *Physical Property Effects*

The physical and mechanical properties of some common nuclear graphites are listed in Table 2. The physical properties of carbon and graphite materials are drastically altered by irradiation damage. Graphite is a phonon conductor of heat. Consequently, improvements in the thermal conductivity can be achieved by thermal annealing (i.e., increased graphitization temperature). These improvements are attributable to increases in crystal perfection and a concomitant increase in the size of the regions of coherent ordering. This minimizes the extent of phonon–defect scattering and results in a larger phonon mean free path. With increasing temperature the dominant phonon interaction becomes phonon–phonon scattering (Umklapp processes). Therefore, there is a reduction of thermal conductivity with increasing temperature (Burchell 2001).

The mechanism of thermal conductivity and the degradation of thermal conductivity have been extensively reviewed (Kelly 1981, Engle and Eatherly 1972, Burchell 1996a, 1997). The increase of thermal resistance owing to irradiation damage has been ascribed by Taylor *et al.* (1969) to the formation of: (i) submicroscopic interstitial clusters, containing 4±2 carbon atoms; (ii) vacant lattice sites, existing as singles, pairs, or small groups; and (iii) vacancy loops, which exist in the graphite crystal basal plane and are too small to have collapsed parallel to the hexagonal axis. The contribution of collapsed lines of vacant lattice sites and interstitial loops to the increased thermal resistance is negligible.

The reduction in thermal conductivity caused by irradiation damage is temperature and dose sensitive. At any irradiation temperature, the decreasing thermal conductivity will reach a "saturation limit." This limit is not exceeded until the graphite undergoes gross structural changes at very high doses. The "saturated" value of conductivity will be attained more rapidly, and will be lower, at lower irradiation temperatures (Snead and Burchell 1995). In graphite, the neutron irradiation-induced degradation of thermal conductivity can be very large, particularly at low temperatures. Bell *et al.* (1962) report that the room temperature thermal conductivity of PGA graphite (the Magnox core graphite) is reduced by more than a factor of 70 when irradiated at 155°C to a dose of ~0.6 dpa.

At an irradiation temperature of 355°C the thermal conductivity of PGA is reduced by less than a factor of 10 at doses twice that obtained at 155°C. Above 600°C the reduction of thermal conductivity is less significant. For example, Kelly (1994) reports the degradation of PGA at a high temperature. At an irradiation temperature of 600°C and a dose of ~13 dpa, the thermal conductivity is only degraded by a factor of ~6. Moreover, at irradiation temperatures of 920°C and 1150°C the degradation is minimal (less than a factor of 4 at ~7 dpa).

The thermal expansion of polygranular graphites is controlled by the thermal closure of aligned internal porosity. Irradiation-induced changes in the pore structure (see above discussion of structural changes) can therefore be expected to modify the thermal expansion behavior of carbon materials. The behavior of GraphNOL N3M (Fig. 8) is typical of many fine-

Table 2
Physical and mechanical properties of some common nuclear graphites (Burchell 1999).

Grade	Source	Forming method[a]	Bulk density ($g\,cm^{-3}$)	Elastic modulus[b] (GP[a])	Strength[b] (MPa) Tensile	Bend	Compression	Thermal conductivity[b] ($W\,m^{-1}\,K^{-1}$)	CTE[b] ($10^{-6}\,K^{-1}$)
PGA	UK	E	1.74	12/5	17/11	19/12	27/27	200/109	0.9/2.8
SM2-24	UK	M	1.7	8/8.5	12	19	47		
Pitch-coke	UK/France	E	1.8	13/10	25/17	32/26	70/63	130/135	4/3.8
IM1-24	UK	M	1.81	11	27.5	23	70	131	4.3
AGOT	USA	E	1.7	10/8	10/9	16/13	41/41	227/138	2.2/3.8
H-451	USA	E	1.75	11/9.6	15/13	20/24	60/60	150/135	3.5/4.5
ASR-1RS	Germany	M	1.78	9.9/9.2	15/14	26/232	67/63	125	4.7/4.9
IG-110	Japan	I	1.75	10	25	34	71	124/138	4/3.6
TSX	USA	E	1.7	14/3.8		25/7			1/4
GR-280	Russia	E	1.72	6.5/5	7.6/6		34/24	103/89	3.2/4.9
GR2-125	Russia	E	1.85	12/8.5	15/8		59/59	160/100	3.9/5.2

[a] E, extruded; M, molded; I, isostatic pressing; [b] Parallel/perpendicular to the forming axis.

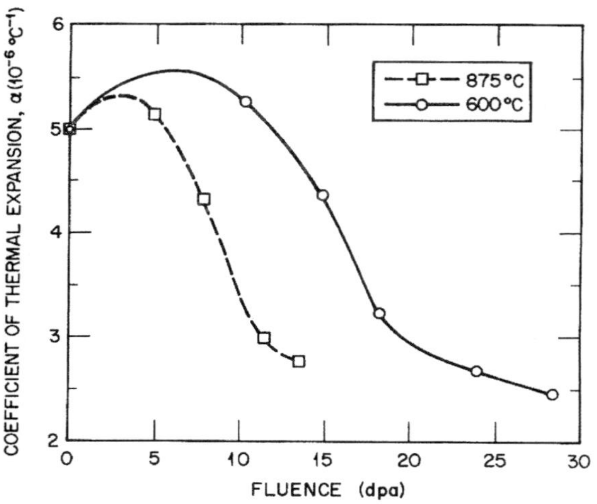

Figure 8
Neutron irradiation-induced changes in the coefficient of thermal expansion of GraphNOL N3M at irradiation temperatures of 600°C and 875°C (Burchell and Eatherly 1991).

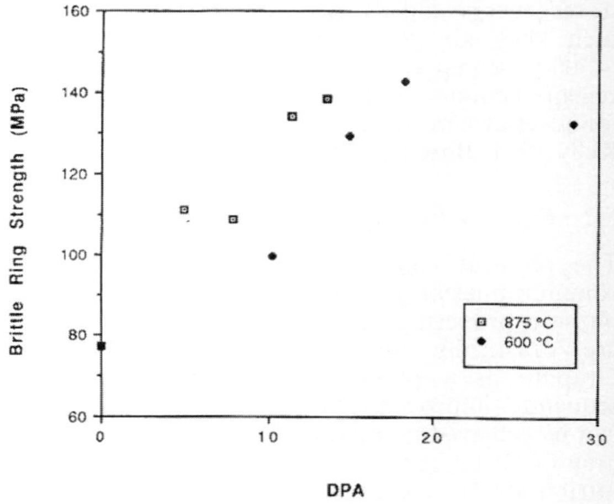

Figure 9
Neutron irradiation-induced strength changes for GraphNOL N3M at irradiation temperatures of 600°C and 875°C (Burchell and Eatherly 1991).

textured graphites, which undergo an initial increase in the coefficient of thermal expansion followed by a steady reduction to a value less than half the unirradiated value of $5\times10^{-6}\ ^{\circ}\mathrm{C}^{-1}$ (Burchell and Eatherly 1991). Similar behavior is reported by Kelly (1994) for the British AGR moderator graphite (grade IM1-24).

The electrical resistivity of graphite will also be affected by radiation damage. The mean free path of the conduction electron in an unirradiated graphite is relatively large, being limited only by crystallite boundary scattering. Neutron irradiation introduces: (i) scattering centers, which reduce charge carrier mobility; (ii) electron traps, which decrease the charge carrier density; and (iii) additional spin resonance. The net effect of these changes is to increase the electrical resistivity on irradiation, initially very rapidly, with little or no subsequent change to relatively high fluence (Burchell 1996a, Burchell and Eatherly 1991). A subsequent decrease at very high neutron doses may be attributed to structural degradation.

5.3 Mechanical Properties

The mechanical properties of graphites are substantially altered by radiation damage. In the unirradiated condition, polygranular graphites behave in a brittle fashion and fail at relatively low strains. The stress–strain curve is nonlinear and the fracture process occurs via the formation of subcritical cracks, which coalesce to produce a critical flaw (Tucker *et al.* 1986, Burchell 1996b). When graphite is irradiated the stress–strain curve becomes more linear, the strain to failure is reduced, and the strength and elastic modulus increased. As shown in Fig. 9, there is a rapid rise in strength attributed to dislocation pinning at irradiation-induced lattice defect sites. This effect has largely saturated at doses >1 dpa. Above ~1 dpa a more gradual increase in strength occurs owing to structural changes within the graphite. For polygranular graphites the dose at which the maximum strength is attained loosely corresponds with the volume change turnaround dose, indicating the importance of pore generation in controlling the high-dose strength behavior.

The strain behavior of polygranular graphite subjected to an externally applied load is largely controlled by shear of the component crystallites. As with strength, irradiation-induced changes in Young's modulus are the combined result of in-crystallite effects, due to low-fluence dislocation pinning, and superimposed structural change external to the crystallite. The effects of these two mechanisms are generally considered separable, and related by

$$(E/E_0)_{\mathrm{irradiated}} = (E/E_0)_{\mathrm{pinning}} \cdot (E/E_0)_{\mathrm{structure}} \quad (2)$$

where E/E_0 is the ratio of the irradiated to unirradiated elastic modulus. The dislocation pinning contribution to the modulus change is due to relatively mobile small defects and is thermally annealable at ~2000°C. Figure 10 shows the irradiation-induced elastic modulus changes for GraphNOL

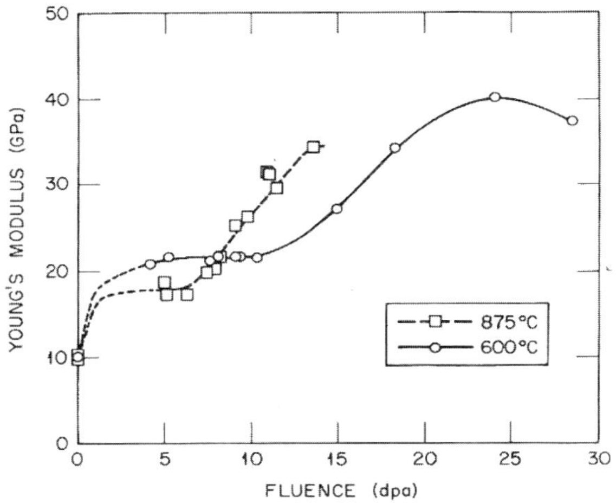

Figure 10
Neutron irradiation-induced Young's modulus changes for GraphNOL N3M at irradiation temperatures of 600°C and 875°C (Burchell and Eatherly 1991).

N3M. The low-dose change owing to dislocation pinning (dashed line) saturates at a dose <1 dpa.

The elastic modulus and strength, σ, are related by a Griffith theory-type relationship:

$$\sigma = (GE/\pi c)^{1/2} \tag{3}$$

where G is the fracture toughness or strain energy release rate (J m^{-2}), E is the elastic modulus (Pa), and c is the flaw size (m). Thus, irradiation-induced changes in σ and E (in the absence of changes in G/c) should follow $\sigma \propto E^{1/2}$. High-dose data reported by Ishiyama *et al.* (1996) show significant deviation from this relationship for grade IG-110 graphite, indicating that changes in G and/or c must occur.

5.4 Radiation Creep

Graphite will creep under neutron irradiation and stress at temperatures where thermal creep is normally negligible. The phenomenon of irradiation creep has been widely studied because of its significance to the operation of graphite-moderated fission reactors. Indeed, if irradiation-induced stresses in graphite moderators could not relax via irradiation creep, rapid core disintegration would result. The observed creep strain has traditionally been separated into a primary reversible component (ε_1) and a secondary irreversible component (ε_2), both proportional to stress and to the appropriate unirradiated elastic compliance (inverse modulus). The total irradiation-induced creep strain (ε_c) is thus

$$\varepsilon_c = \varepsilon_1 + \varepsilon_2 \tag{4}$$

or

$$\varepsilon_c = (\sigma/E_0)[1 - \exp(-b\gamma) + (K/E_0)\sigma\gamma] \tag{5}$$

where E_0 is the unirradiated Young's modulus, b is a constant, γ is the neutron dose, and K is the irradiation creep coefficient. Kelly (1994) has reported that values of -4×10^{-20} for b and 0.23×10^{-20} for K apply to UK data taken over a range of irradiation temperatures (300–650°C). At high fluence, Eqn. (5) must be modified to account for structural changes occurring in the graphite:

$$\varepsilon_c = \frac{\sigma}{E_0} + \left(\frac{d\varepsilon_c}{d\gamma}\right)_0 \cdot \int_S^{\gamma} S^{-1}(\gamma)d\gamma \tag{6}$$

where $(d\varepsilon_c/d\gamma)_0$ is the initial secondary creep rate and $S(\gamma)$ is the "structure factor" normally deduced from Young's modulus changes ascribed to structural effects (i.e., $S(\gamma) = (E/E_p)$, where E is the Young's modulus at fluence γ and E_p is the Young's modulus after the initial increase due to dislocation pinning). Further discussion of irradiation creep in nuclear graphite can be found in Burchell (1999) and Kelly and Burchell (1994b).

6. *Summary and Conclusions*

Nuclear graphite has been a subject of research and development since the mid-twentieth century. The physics of the material and its behavior under neutron irradiation are well understood. The knowledge gained through these many years of work is being used to underwrite the safety cases of graphite-moderated reactors throughout the world. In the early twenty-first century more knowledge will be gained from the new graphite-moderated reactors in Japan and China that operate at higher temperatures. With the realization and acceptance that greenhouse gas emissions from fossil fueled power plants are causing global climate changes, as evidenced by the Kyoto Agreement, the nuclear option may once again become attractive for clean electric power generation. At that time, the inherently safe, graphite-moderated gas-cooled reactors may find popularity and nuclear graphite may once again be elevated to a position of prominence in the carbon and graphite world.

See also: Radiation Damage Theory

Bibliography

Arnold L 1992 *Windscale 1957, Anatomy of a Nuclear Accident*. St. Martin Press, London

Bell J C, Bridge H, Cottrell A H, Greenough G B, Reynold W N, Simmons J W H 1962 Stored energy in the graphite of power producing reactors. *Philos. Trans. R. Soc. London, Ser. A* **245**, 361–95

Brocklehurst J E, Kelly B T 1993 Analysis of the dimensional changes and structural changes in polycrystalline graphite under fast neutron irradiation. *Carbon* **31**, 155–78

Burchell T D 1996a Radiation damage in carbon materials. In: Roth J, Hoffer W O (eds.) *Physical Processes of the Interaction of Fusion Plasmas with Solids*. Academic Press, San Diego, CA

Burchell T D 1996b A microstructurally based fracture model for polygranular graphites. *Carbon* **34**, 297–316

Burchell 1997 Radiation effects in graphite and carbon-based materials. *MRS Bull* **22**, 29–35

Burchell T D 1999 Fission reactor applications of carbon. In: Burchell T D (ed.) *Carbon Materials for Advanced Technologies*. Elsevier, Oxford

Burchell T D 2001 Thermal properties and nuclear energy applications. In: Delhaès P (ed.) *Graphite and Precursors*. Gordon and Breach, The Netherlands

Burchell T D, Eatherly W P 1991 The effects of radiation damage on the properties of GraphNOL N3M. *J. Nucl. Mater* **170–181**, 205–8

Eatherly W P, Piper E L 1962 Manufacture. In: Nightingale R E (ed.) *Nuclear Graphite*. Academic Press, New York

Engle G B, Eatherly W P 1972 Irradiation behavior of graphite at high temperatures. *High Temperatures–High Pressures* **4**, 119–58

Inagaki M 2001 Applications of polycrystalline graphite. In: Delhaès P (ed.) *Graphite and Precursors*. Gordon and Breach, The Netherlands

Ishiyama S, Burchell T D, Strizak J P, Eto M 1996 The effect of high fluence neutron irradiation on the properties of a fine-grained isotropic nuclear graphite. *J. Nucl. Mater* **230**, 1–7

Kelly B T 1981 *Physics of Graphite*. Applied Science Publishers, London

Kelly B T 1994 Nuclear reactor moderator materials. In: Cahn R W, Haasen P, Kramer E J (eds.) *Materials Science and Technology. A Comprehensive Treatment: Vol. 10A. Nuclear Materials*. Wiley VCH, Weinheim, Germany

Kelly B T, Burchell T D 1994a Structure-related property changes in polycrystalline graphite under neutron irradiation. *Carbon* **32**, 499–505

Kelly B T, Burchell T D 1994b The analysis of irradiation creep experiments on nuclear reactor graphite. *Carbon* **32**, 119–25

Nightingale R E 1962 *Nuclear Graphite*. Academic Press, New York

Price R J 1974 High-temperature neutron irradiation of highly oriented carbons and graphites. *Carbon* **12**, 159–69

Ragan S, Marsh H 1983 Review: science and technology of graphite manufacture. *J. Mater. Sci* **18**, 3161–76

Ruland W 1968 X-ray diffraction studies on carbon and graphite. *Chem. Phys. Carbon* **4**, 1

Simmons J W H 1965 *Radiation Damage in Graphite*. Pergamon, Oxford

Snead L L, Burchell T D 1995 Thermal conductivity degradation of graphite due to neutron irradiation at low temperature. *J. Nucl. Mater* **224**, 222–9

Taylor R, Kelly B T, Gilchrist K E 1969 The thermal conductivity of fast neutron irradiated graphite. *J. Phys. Chem. Solids* **30**, 2251–67

Thrower P A, Meyer R M 1978 Review article: point defects and self diffusion in graphite. *Phys. Status Solidi A* **14**, 11–37

Tucker M O, Rose A P G, Burchell T D 1986 The fracture of polygranular graphites. *Carbon* **24**, 581–602

Nuclear Reactor Fuel Fabrication (Including Quality Control)

C. Ganguly

The world's first self-sustaining nuclear fission chain reaction was demonstrated in December 1942 by Fermi and his co-workers in the Chicago "pile" by using some 4 tons (4.1×10^3 kg) of natural uranium fuel in the crude form of lumps of metal and oxide distributed throughout a matrix of 400 tons (4.1×10^5 kg) of graphite. Thereafter, a number of metallic, ceramic, and dispersion fuels, in both pin and plate geometries were designed, fabricated, and irradiated in the 1950s and 1960s in a variety of reactor systems with different neutron energies and with different combinations of coolant, moderator, and cladding materials from which viable fuel systems have emerged for power reactors and research reactors. The nuclear fuels industry is now established in several parts of the world catering for the requirements of more than 300 research reactors that are used for production of radioisotopes, materials irradiation, basic research, training, and education in 65 countries, and some 430 nuclear power reactors with a total installed capacity of 350 GWe in 31 countries.

This article reviews the fabrication of conventional and advanced fuels for the commonly used nuclear research reactors, namely the materials test reactor (MTR) and TRIGA (training, research, isotopes, general atomic); and the popular nuclear power reactors, namely the light water-cooled reactor (LWR), the pressurized heavy water reactor (PHWR), and the liquid metal-cooled fast breeder reactor (LMFBR). LWRs are of two main types: the pressurized water reactor (PWR) and the boiling water reactor (BWR). These are used extensively all over the world and are in operation in the USA, France, Japan, Germany, Sweden, Belgium, Switzerland, Russia, Ukraine, and South Korea. The Russian version of the PWR is known as VVER. The PHWR is the backbone of the nuclear power programs in Canada (where it is known as CANDU) and India. There are very few LMFBRs, but they are likely to play a major role in the coming years.

Uranium and thorium, the basic materials for "nuclear fuels," are radioactive and among the heaviest elements occurring in nature. Uranium occurs natural mainly in two isotopic forms, the "fertile" ^{238}U (99.3%) and the "fissile" ^{235}U (0.7%). Thorium occurs in nature almost entirely as the "fertile" ^{232}Th isotope. ^{235}U is the only fissile isotope occurring in nature. The two synthetic fissile isotopes, ^{239}Pu and ^{233}U, are usually produced by neutron capture reactions of ^{238}U and ^{232}Th, respectively, in nuclear reactors. ^{239}Pu is often associated with ^{241}Pu, which is also a fissile isotope.

Nuclear fuels are made up of one or more fissile isotopes, namely ^{235}U, ^{239}Pu, and ^{233}U, often in combination with a fertile isotope, ^{238}U or ^{232}Th, and are usually in the solid form of metals, alloys, oxides, carbides, nitrides, silicides, etc., of uranium, plutonium, and thorium. The term "fuel" defined as a material for producing heat is in the true sense not applicable to nuclear fuel, since the heat in this case is generated not by burning but by fission or disintegration of the fissile isotopes. Since uranium, plutonium, thorium, and their fission products are radioactive and hazardous to health to varying degrees, nuclear fuels are hermetically sealed inside a nonradioactive structural material, known as cladding. The cladding is an integral part of the nuclear fuel element. It serves as primary containment for radioactive fission products, acts as a barrier avoiding direct contact of the fuel with coolant, and transfers fission heat energy from fuel to coolant. Nuclear fuel elements are usually manufactured in the shape of plates, pins, or rods and are assembled in specific geometric configurations using spacers, end fittings, and other supporting hardware. The package of fuel elements is called a fuel assembly.

1. *Fuels for Research and Power Reactors*

Nuclear research and power reactors operate mainly by the fission of ^{235}U by slow or thermal neutrons. The ^{239}Pu by-product from these thermal reactors is the best fissile material for LMFBRs. However, since the LMFBRs are yet to be deployed commercially, the plutonium is recycled in the interim in several LWRs in France, Germany, Belgium, Switzerland, and Japan. ^{233}U is the ideal fissile material for thermal reactors and it is possible to have a self-sustaining ^{232}Th–^{233}U fuel cycle. ^{233}U fuels have so far been manufactured only on a laboratory scale, primarily for irradiation testing experiments. Table 1 summarizes the conventional and advanced fuels and cladding materials for research and power reactors.

The conventional fuels for LWRs and PHWRs are slightly enriched uranium (1.5–5% ^{235}U) and natural uranium, respectively, in the form of high-density cylindrical UO_2 "pellets," which are stacked inside zirconium alloy cladding tubes and hermetically sealed. Many aspects of uranium oxide fuel are reviewed in Belle and Lustman (1958). In VVER fuel, the uranium oxide pellets have a central hole. For BWRs and PWRs, zircaloy 2 and zircaloy 4 are used as cladding materials, respectively (see *Nuclear Applications: Zirconium Alloys*). The Russian VVER and RBMK reactors use Zr–1% Nb as cladding material. The quaternary Zr–Nb–Sn–Fe alloys are being developed as advanced cladding materials for very high burn-up LWR fuels. Zircaloy 4-clad very high-density ($\geqslant 96\%$ theoretical density (TD)) natural uranium oxide is the reference fuel for PHWRs.

The inner surface of the PHWR cladding tube is coated with graphite. The advanced PHWR fuels aim at improved burn-up. Accordingly, recycled uranium (REU, 0.8–1.1% ^{235}U) oxide and slightly enriched uranium oxide, $(U,Pu)O_2$, $(Th,Pu)O_2$, and $(Th,^{233}U)O_2$, are being developed. The carbides and nitrides of uranium, plutonium, and thorium are not compatible with water and are hence not used as water-cooled reactor fuels.

Austenitic stainless steel (mostly type D-9)-clad mixed uranium–plutonium oxide (MOX) pellets containing between 15% and 25% PuO_2 have been the driver fuel for the first generation of demonstration and prototype LMFBRs in France, the UK, and Japan (see *Fast Reactor Cores*, Austenitic Steels for). The Russian LMFBRs use highly enriched uranium oxide pellets as fuel. Mixed uranium–plutonium monocarbide (MC), mononitride (MN), and U–Pu–Zr are considered as advanced LMFBR fuels on the basis of their superior breeding ratio, higher thermal conductivity, and better chemical compatibility with sodium coolant and stainless steel cladding compared to MOX fuel.

The HT-9 type and oxide-dispersed ferritic steels (see Fast Reactor Cores, FerriticMartensitic Steels for) have emerged as advanced cladding materials for LMFBRs. India has been able to use mixed uranium–plutonium monocarbide driver fuel in their fast breeder test reactor (FBTR) with a hitherto untried plutonium-rich fuel composition (Ganguly and Rodriguez 2000). Metallic fuels, U–Zr and U–Pu–Zr, have been used mostly in the experimental breeder reactor (EBR-II) in the USA and are considered to be the reference fuels for the "inherently safe, integral fast reactor" operating on a proliferation-resistant fuel cycle with co-located facilities for fuel reprocessing and refabrication (Chang 1989, Stevenson 1987).

"Vi-pac" or "sphere-pac" fuel elements containing high-density microspheres of mixed uranium–plutonium oxide, monocarbides, and mononitrides are also considered as advanced LMFBR fuels. In vi-pac pins, the fuel microspheres of two or three size fractions are vibrofilled in a metallic cladding tube and encapsulated.

2. *Challenges in Manufacturing Nuclear Fuels*

The following are the unique features of nuclear fuels, which are taken into consideration during their fabrication.

Table 1
Conventional and advanced fuels for MTRs, LWRs, PHWRs, and LMFBRs.

Reactor	Conventional	Advanced/alternative fuel
Research reactors		
MTR	HEU (>93% ^{235}U): (i) Al–UAl_x (density ≤ 1.7 gU cm^{-3}); (ii) Al–U_3O_8 (density ≤ 1.3 gUcm^{-3})	LEU (<20% ^{235}U): (i) Al–UAl_x (density = 2.3 gUcm^{-3}); (ii) Al–U_3O_8 (density = 3.2 gUcm^{-3}); (iii) Al–U_3Si_2 (density = 4.8 gUcm^{-3}); (iv) Al–U_3Si (density = 6.0 gUcm^{-3}); (v) Al–U–Mo and Al–U–Zr–Nb (density = 7–8 gUcm^{-3})
TRIGA	Aluminum- or stainless steel-clad $UZrH_x$ pin assembly (density = 0.5 gUcm^{-3})	Aluminum- or stainless steel-clad $UZrH_x$ pin assembly (density ≤ 3.7 gUcm^{-3})
LWR: BWR, PWR, and VVER		
Fuel	LEU (≤5% ^{235}U) as UO_2	LEU (5–10% ^{235}U) mixed uranium–plutonium oxide (≤10% PuO_2)
Cladding	Zircaloy 2 (BWR), Zircaloy 4 (PWR)	Zr–Sn–Nb–Fe and Zr– Nb–O alloys
Burn-up	$2\times10^4 - 3\times10^4$ MWD/t	High: ≤ 6×10^4 MWD/t; ultrahigh: ≤ 9×10^4 MWD/t
PHWR		
Fuel	Natural UO_2	REU, SEU in the form of UO_2, $(Th,Pu)O_2$, and $(Th,^{233}U)O_2$, containing ≤ 2% fissile material
Cladding	Zircaloy 4	Zircaloy 4
Burn-up	6700 MWD/t	$1.5\times10^4 - 2\times10^4$ MWD/t
LMFBR		
Fuel	HEU in the form of UO_2 and $(U,Pu)O_2$ (≤25% Pu)	(U,Pu)C, (U,Pu)N, and U–Pu–Zr, (≤25% Pu)
Cladding	Stainless steel D-9	Stainless steel (HT-9 or oxide dispersed)
Burn-up	10^5 MWD/t	≤ 2×10^5 MWD/t
Breeding ratio	≤1.2	≤1.5

2.1 Radioactivity

Uranium, plutonium, and thorium emit alpha particles and are radioactive to varying degrees. Hence, for radiation safety, materials bearing these elements are handled in a well-ventilated special operating area such that the radioactive aerosols are contained within the plant. ^{239}Pu and ^{233}U because of their high radiotoxicity and specific radioactivity are handled inside leak-tight glove boxes or hot cells, maintained at slightly negative pressure (see Hot Cells, Glove Boxes, and Shielded Facilities). Beta–gamma and neutron shielding are not needed for handling pure ^{239}Pu and ^{233}U isotopes. However, the ^{239}Pu obtained by reprocessing spent uranium fuels from power reactors contains significant quantities of other plutonium isotopes, namely ^{238}Pu, ^{240}Pu, ^{241}Pu, and ^{242}Pu, which release spontaneous neutrons (^{238}Pu, ^{240}Pu, and ^{242}Pu) and gamma rays (^{241}Pu decays to ^{241}Am, which is a strong gamma emitter). Likewise, the ^{233}U obtained by reprocessing spent thoria-bearing fuels or blanket materials always contains ^{232}U. Hence, beta–gamma and neutron shielding is necessary for handling plutonium and ^{233}U-bearing materials. In order to keep the radiation exposure of the plant personnel well below the permissible limits, great emphasis is given to remote working, automation, and minimization of radiotoxic dust hazards.

2.2 Pyrophoricity

Uranium and plutonium metal, monocarbide, and mononitride, and uranium silicide are highly susceptible to oxidation and hydrolysis and are pyrophoric in powder form. Hence, these materials are required to be handled in leak-tight enclosures under a high-purity inert cover gas (argon or nitrogen containing <25 ppm each of oxygen and moisture).

2.3 Nonstoichiometry

The dioxide, monocarbide, mononitride, and silicides of uranium and plutonium are nonstoichiometric compounds. Strict control of composition is necessary in order to ensure stability and to avoid chemical interaction between fuel and cladding and fuel and coolant. Hyperstoichiometric $(U,Pu)O_2$ fuel containing a higher percentage of plutonium oxide may cause fuel–clad and fuel–coolant chemical interactions. In addition, in the case of oxide or mixed oxide fuels, there is significant reduction in thermal conductivity even with slight deviation of the oxygen to metal ratio from 2.00. Monocarbides and mononitrides of uranium and plutonium have line compositions.

A slight deviation from the stoichiometric ratio leads to an unwanted metallic second phase for hypostoichiometric compounds and higher carbide or higher nitride for hyperstoichiometric composition. The metallic second phase forms a low-melting eutectic with the iron and nickel of stainless steel cladding and the higher carbides and higher nitrides cause clad carburization or nitridation. Hence, only a controlled amount of the higher carbide second phase is acceptable in monocarbide fuel.

2.4 Criticality

Fissile materials are permitted to be handled only in limited quantities (a few kilograms) in order to avoid an uncontrolled nuclear fission chain reaction or what is known as a "criticality accident."

2.5 Accountability

Precise and real-time accounting of special nuclear materials (SNM: fissile and fertile) is necessary to prevent criticality accidents and also to avoid proliferation of SNM for nonpeaceful purposes. Hence, high security arrangements are needed for SNM.

3. Fabrication of Research Reactor Fuel

The common fuels for water-cooled nuclear research reactors are as follows:

- aluminum-clad, natural uranium metal in the form of rods or clusters of pins;
- aluminum alloy-clad, low-enriched uranium (LEU; 19.75%) as Al–UAl_x, Al–U_3O_8, or Al–U_3Si_2 in the form of plate fuel assemblies for MTRs; and
- aluminum- or stainless steel-clad, low-enriched uranium in the form of uranium zirconium hydride rods for TRIGAs.

3.1 Uranium Metal Fuel

The natural uranium metal fuels used in NRX (Canada) and CIRUS and Dhruva (India) reactors are typical of metallic uranium fuel and are fabricated in two steps (Vijayaraghavan 1989). First, a uranium metal button is prepared by either magnesiothermic or calciothermic reduction of uranium tetrafluoride in a steel mold or bomb.

In the second step, the uranium metal button is subjected to vacuum melting and casting at around 1300 °C in an induction furnace using an yttria-coated graphite crucible and mold. The cast uranium metal ingot is evaluated by ultrasonic testing for intercepting any casting defects and subjected to conventional shaping treatments including extrusion, forging, rolling, swaging, drawing, machining, etc., to obtain the uranium fuel in the desired geometric shape and dimension. Usually, the uranium metal is hot rolled in the alpha phase (orthorhombic structure stable up to 665 °C) and protected from oxidation by using fused salt or inert gas atmosphere. However, since alpha uranium has anisotropic properties, it is susceptible to irradiation growth. In order to avoid this, the uranium metal is often alloyed with molybdenum or zirconium to retain the stable b.c.c. gamma phase structure at ambient temperature. The gamma phase is relatively soft and amenable to cold working.

3.2 Low-enriched Uranium Dispersion Fuels

The MTR was initially developed in the USA in the 1950s and was later used all over the world. Until 1977 high-enriched uranium (HEU; $\geqslant 85\%$ ^{235}U) was used as the fuel in these reactors in the form of aluminum alloy-clad Al–UAl_x and Al–U_3O_8 dispersion fuels with uranium densities of 1.7 $g cm^{-3}$ and 1.3 $g cm^{-3}$, respectively, in both pin and plate geometries. In 1978 Argonne National Laboratory (ANL) in the USA initiated the program on reduced enrichment for research and test reactors (RERTR) in order to avoid diversion of HEU for nonpeaceful purposes. The primary objective of the program was to substitute HEU with low-enriched uranium (LEU; $\leqslant 20\%$ ^{235}U) fuel with high uranium density. Accordingly, several LEU-based high uranium density fuels have emerged as driver fuels for MTRs (RERTR 1996, 1998). The qualified uranium densities in LEU-based fuels extend up to 2.3 $g cm^{-3}$ for Al–UAl_x, 3.2 $g cm^{-3}$ for Al–U_3O_8, and 4.8 $g cm^{-3}$ for Al–U_3Si_2 fuel.

The advanced LEU-based MTR fuels consist of Al–U_3Si with a uranium density in the region of 6.0 $g cm^{-3}$ and Al–U–Mo and Al–U–Zr–Nb with a uranium density in the range 7–8 $g cm^{-3}$. The alloy or dispersion fuel elements are usually manufactured in the form of plates in which the fuel "meat" is sandwiched between aluminum alloy cover plates. The fuel meat consists of UAl_x, U_3O_8, U_3Si_2, U_3Si, U–Mo, or U–Zr–Nb dispersoids, large (44–150 μm) in size compared with the recoiled range of fission products (6–12 μm), which are uniformly dispersed in an aluminum or aluminum alloy matrix. An ideal dispersion "is an array of uniform, spherical fuel particles, sized to permit a reasonable amount of matrix metal between particles to remain undamaged by fission fragments." In order to avoid any overlapping of damaged fission recoil regions, the volume fraction of the fuel particles should be within 70% and the average size of the fuel particles should not be less than 44 μm. The principal steps in the fabrication of aluminum-clad, aluminum matrix plate-type dispersion fuels are the following:

- preparation of aluminum matrix dispersion-type fuel in the form of rectangular plates;
- picture framing and encasing of fuel meat in aluminum or Al–Mg alloy plates to form a sandwich between the top and bottom cover plates of aluminum or Al–Mg alloy;
- hot rolling of the sandwich for metallurgical bonding between mating surfaces of the core and cover plates;
- x-ray radiography for locating the outline of the fuel meat in the roll-bonded sandwich followed by shearing in the region outside the fuel outline to obtain the fuel elements of the desired dimension; and
- assembling several fuel elements in certain geometric configuration to form the fuel assembly.

Aluminum-clad, Al–23% ^{233}U alloy plate fuel has been manufactured in India by the classical melting and casting route (Ganguly *et al.* 1991a). The increasing divergence between the solidus and the liquidus of the hypereutectic Al–U alloys imparts a pronounced tendency for macrosegregation especially during slow cooling of the ingots and leads to the formation of coarse, columnar grains of UAl_4, which are very brittle and are responsible for severe cracking of the ingots during subsequent rolling. This problem is avoided by the addition of a small quantity (1–2%) of zirconium during the Al–U master alloy preparation and by rapid solidification of the cast alloy with the use of a water-cooled metallic mold. Zirconium addition is found to stabilize the relatively softer and denser UAl_3 phase and eliminate the dendritic morphology of UAl_4 platelets in the cast Al–U alloy.

Fuel particles of UAl_x, U_xSi_y and U–Mo alloy in the size range 44–150 μm are fabricated by vacuum induction or arc melting of aluminum and uranium, silicon and uranium, or molybdenum and uranium, followed by casting in the form of buttons, crushing, and sieving. The U_3Si is tough and difficult to grind and is more susceptible to oxidation than U_3Si_2. In addition, the former has higher irradiation swelling. Al–U_3Si_2 fuel with a uranium density ~4.8 $g cm^{-3}$ has evolved as a standard LEU-based fuel for research reactors. With Al–U–Mo or Al–U–Zr–Nb fuel, it is possible to obtain uranium densities in the range 7–8 $g cm^{-3}$.

3.3 Uranium–Zirconium Hydride Fuel

Aluminum- or stainless steel-clad uranium–zirconium hydride ($UZrH_{1.0}$) fuel in the configuration of cylindrical pins is used in some 65 TRIGA reactors worldwide (Razvi 1998). Initially, HEU fuel was used in TRIGA with a uranium density ~0.5 $g cm^{-3}$. After the initiation of the RERTR program, the HEU was replaced by LEU with a uranium density as high as 3.7 $g cm^{-3}$ in $UZrH_x$ fuel (RERTR 1996, 1998). The uranium enrichment is ~20% and zirconium hydride serves as a moderator. Since the hydrogen is intimately mixed with uranium, the U–Zr hydride fuel ensures inherent safety of TRIGA reactors. The TRIGA fuel rods are manufactured in several steps. First, homogenous ingots of uranium–zirconium alloy are prepared by double arc melting zirconium sponge and uranium metal piece. Next, the ingots are hot extruded and machined to the required size. Finally, the alloy is hydrided in a tube furnace under controlled hydrogen atmosphere at around 750°C to form uranium–zirconium hydride rods with hydrogen to zirconium atom ratio of 1:1. The microstructure of the uranium–zirconium hydride fuel thus prepared consists of alpha zirconium, zirconium hydride, and the epsilon phase of zirconium–uranium alloy. The uranium–zirconium hydride rods are assembled into aluminum or stainless steel tubes together with graphite rods and burnable poison disks at both ends and encapsulated to form cylindrical fuel rods.

4. Fabrication of Ceramic Nuclear Fuels for Power Reactors

Figure 1 summarizes the major steps followed in the fabrication of oxide and mixed oxides of uranium, plutonium, and thorium and mixed uranium–plutonium monocarbide and mononitride fuels by powder pellet, vibratory compaction, and sol–gel microsphere pelletization (SGMP) processes. UO_2, PuO_2, and ThO_2 are isostructural (f.c.c., CaF_2 type), completely solid soluble in each other and have very similar thermodynamic and thermophysical properties. Hence, the manufacturing processes for oxide and mixed oxide fuels are similar. The manufacturing processes for mixed uranium–plutonium monocarbide and mononitride fuels are likewise the same as for carbide or nitride, respectively, since UC, PuC, UN, and PuN have the same f.c.c., NaCl-type crystal structure and are completely solid soluble in each other.

The powder pellet route is followed in most countries for fabrication of both low (⩽85% TD) and high (⩾94% TD) density oxide, carbide, and nitride fuel pellets, starting with UO_2, PuO_2, and ThO_2 powders. The mixed uranium–plutonium monocarbide and mononitride powders are usually prepared by carbothermic synthesis of tableted, or granulated oxide plus carbon mixture in vacuum and flowing nitrogen, respectively. The major drawback of the powder pellet process is generation and handling of fine powder or dust particles of the fuel and the problem of radiotoxic dust hazard associated with it. Further, fine powders have poor flowability, which makes remote and automated processing somewhat difficult. In addition, fine powder of carbide and nitride are highly pyrophoric and require a very high purity inert cover gas for handling. The

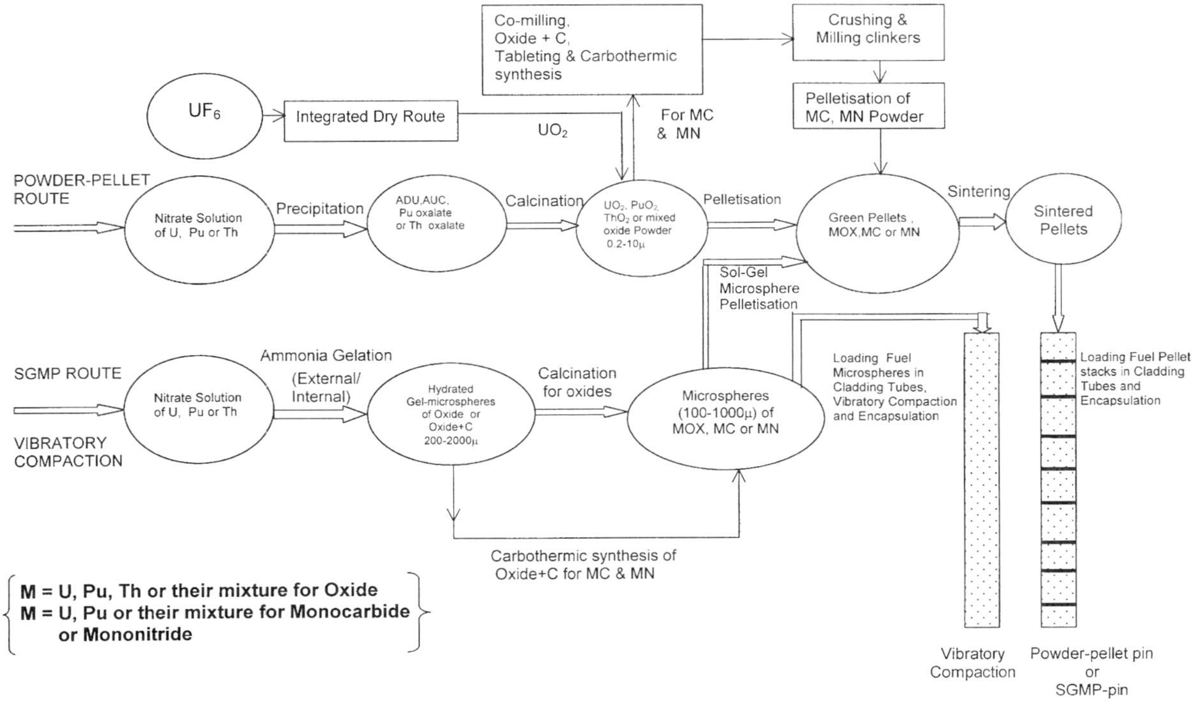

Figure 1
Powder pellet, vibratory compaction, and SGMP processes for fabrication of ceramic nuclear fuels.

objectives of advanced methods of fabrication of ceramic nuclear fuels are safety of plant personnel, minimization of cost, and improved microstructure for high burn-up. The major advanced processes developed so far are as follows:

- advanced powder pellet route: low-temperature oxidative sintering (LTS) of UO_2 and $(U,Pu)O_2$ pellets; admixing TiO_2 dopant for obtaining UO_2 and $(U,Pu)O_2$ pellets of large grain size; low-temperature sintering of ThO_2, $(Th,U)O_2$, and $(Th,Pu)O_2$ pellets by MgO, CaO, or Nb_2O_5 "sintering aid"; pellet impregnation technique for fabrication of high-density $(Th,^{233}U)O_2$ pellets;
- vibro–sol or vibratory compaction process for oxide, carbide, and nitride fuels;
- SGMP of oxide, carbide, and nitride pellets; and
- combined SGMP–LTS process for UO_2 and $(U,Pu)O_2$ pellets.

4.1 Preparation of UO_2 Powder

Sinterable-grade uranium oxide powder is prepared either by the integrated dry route (IDR) starting with UF_6 or by employing the wet chemical routes starting with magnesium diuranate (MDU) or UF_6. The most common wet routes are based on ammonium diuranate (ADU) and ammonium uranyl carbonate (AUC) processes.

The IDR is extensively followed in France, Russia, the UK, and the USA for manufacturing enriched UO_2 powders (1.5–5% ^{235}U) (Chotard *et al.* 1991). The IDR is carried out inside a rotary kiln involving single-stage gas–gas and gas–solid reactions. The vaporized UF_6 is first reacted with superheated dry steam to form uranyl fluoride (UO_2F_2) by hydrolysis. Next, the product is routed through the second part of the kiln, consisting of a rotating tube at 600–700°C, where it is reduced to UO_2 by a countercurrent flow of hydrogen and steam. The overall reaction is as follows:

$$UF_6 + 2H_2O + H_2 = UO_2 + 6HF$$

The process does not generate any liquid effluent. The only by-product is high-purity HF, which could be recovered and sold. The specific surface area of IDR-derived UO_2 powder is generally low, about $2\,m^2g^{-1}$, compared with the powder produced by the wet route (2.5–$6\,m^2g^{-1}$). The powder consists of spherical particles or crystallites with diameters of $\sim 0.2\,\mu m$ aggregated to form brittle particles of $\sim 1\,\mu m$. Despite a low specific surface area, the IDR-derived UO_2 powder has excellent sinterability. It is usually transferred directly into orbital screw blenders for homogenization. If necessary, pore formers could be introduced at the blending stage.

The ADU route is most popular in Canada, India, and Brazil (Hastings 1989, Sejnoha 1999). To prepare natural uranium oxide powder for manufacturing PHWR fuel, it is essential to purify the MDU feed material in order to keep rare earths, boron, and cadmium, which are high neutron absorbers, to less than 1 ppm. The MDU cake is dissolved in nitric acid and purified by solvent extraction with an organic solvent, namely tributyl phosphate (TBP) in kerosene. Uranium is preferentially extracted into the organic phase, leaving the impurities in the aqueous phase. The uranium-laden organic, after a scrubbing treatment, is stripped with water to recover uranium as pure uranyl nitrate aqueous solution. By further steps of ammonia precipitation, filtration, calcination, hydrogen reduction, and stabilization, pure uranium and sinterable UO_2 powder is produced. If UF_6 is used as a starting material, it is subjected to hydrolysis followed by ammonia treatment to obtain ADU precipitate, which after washing and drying is pyrohydrolyzed to remove traces of fluoride and to yield U_3O_8, which after hydrogen reduction and stabilization leads to the formation of sinterable-grade UO_2 powder. The ex-ADU, UO_2 powder is extremely fine ($<1\,\mu m$), and has a specific surface area in the range 3–$5\,m^2g^{-1}$, oxygen to metal ratio between 2.07 and 2.15, and bulk density of 1.8–$2.0\,gcm^{-3}$. The powder is not free flowing and requires additional process steps, namely precompaction and granulation or roll compaction and granulation for producing free-flowing UO_2 granules that are suitable as press-feed material for pelletization in high-speed automatic hydraulic or mechanical presses.

The AUC process was developed in Germany (Assmann and Becker 1979) and is used in South Korea and Argentina. It can use both UF_6 and uranyl nitrate as feed materials and produces free-flowing UO_2 powder suitable for direct pelletization. Unlike ADU, AUC is crystalline, free flowing, and relatively coarse ($40\,\mu m$). Calcination of AUC at around 900 K produces oxide powder that is relatively coarse (10–$20\,\mu m$), free flowing, and with relatively higher specific surface area ($5\,m^2g^{-1}$), suitable for direct pelletization. The AUC process has also been extended to the fabrication of UO_2–Gd_2O_3 fuel pellets for LWRs. The coprecipitated uranium–gadolinium mixed oxide thus produced avoids free Gd_2O_3 in sintered pellets, ensuring excellent microhomogeneity of the gadolinium burnable poison in fuel pellets. The ammonium uranium plutonium carbonate (AuPuC) route is an extension of the AUC process and has been extensively used for manufacturing free-flowing $(U,Pu)O_2$ powders in Germany (Assmann *et al.* 1988, Roepenack *et al.* 1984).

4.2 Preparation of ThO_2 and PuO_2 Powders

ThO_2 and PuO_2 powders are prepared by air calcination of the oxalate precipitates obtained by the addition of oxalic acid to thorium nitrate or plutonium nitrate solution. The calcination temperatures are approximately 850 K and 1050 K, respectively, for plutonium oxalate and thorium oxalate. In order to dope CaO or MgO a sintering aid (0.2–0.4 w/o) in the ThO_2 powder, $Ca(NO_3)_2$ or $Mg(NO_3)_2$, is usually added to the thorium nitrate solution prior to precipitation. The oxalate-derived ThO_2 and PuO_2 powders are extremely fine (<1 μm) and have poor flowability and need to be granulated to produce free-flowing press-feed materials. In addition, the oxalate-derived ThO_2 powder has platelet morphology and requires premilling to break the platelets.

4.3 *Preparation of Carbide and Nitride Powders*

Figure 2 summarizes the different techniques for the preparation of mixed uranium plutonium monocarbide (MC) and mononitride (MN) in the form of powders and microspheres, starting with metals, oxides, or nitrate solutions of uranium and plutonium as feed materials (Sowden *et al.* 1964, Leary and Kittle 1977, IAEA 1984, 1988). Uranium forms three different carbides, namely UC, U_2C_3, and UC_2. Similarly, carbides of plutonium include Pu_3C_2, PuC, PuC_2, and Pu_2C_3. The U–N system is characterized by four compounds, namely UN, α-U_2N_3, β-U_2N_3, and a nitrogen-deficient UN_2. The Pu–N system has only one compound, the mononitride PuN.

The process involving hydriding and dehydriding of massive metal at 450–525 K and 800–1000 K, respectively, for obtaining fine metallic powder of high surface area followed by carburization with methane or propane at 1000–1100 °C for synthesis of monocarbide powder and nitridation with N_2 or N_2+H_2 mixture for preparation of mononitride powder was used in the initial years of carbide and nitride fuel development (Matzke 1986). This method of synthesis of MC and MN has two main advantages. First, the reaction temperatures are low, which, in turn, avoids plutonium volatilization losses. Second, the end products are fine and highly reactive MC and MN powders, which can be pelletized and sintered directly.

For preparation of MC and MN, all techniques using uranium and plutonium metals as starting materials are uneconomical because these metals are prepared from their oxides via expensive steps such as hydrofluorination and calcio- or magnesiothermic reduction. Moreover, storing uranium and plutonium

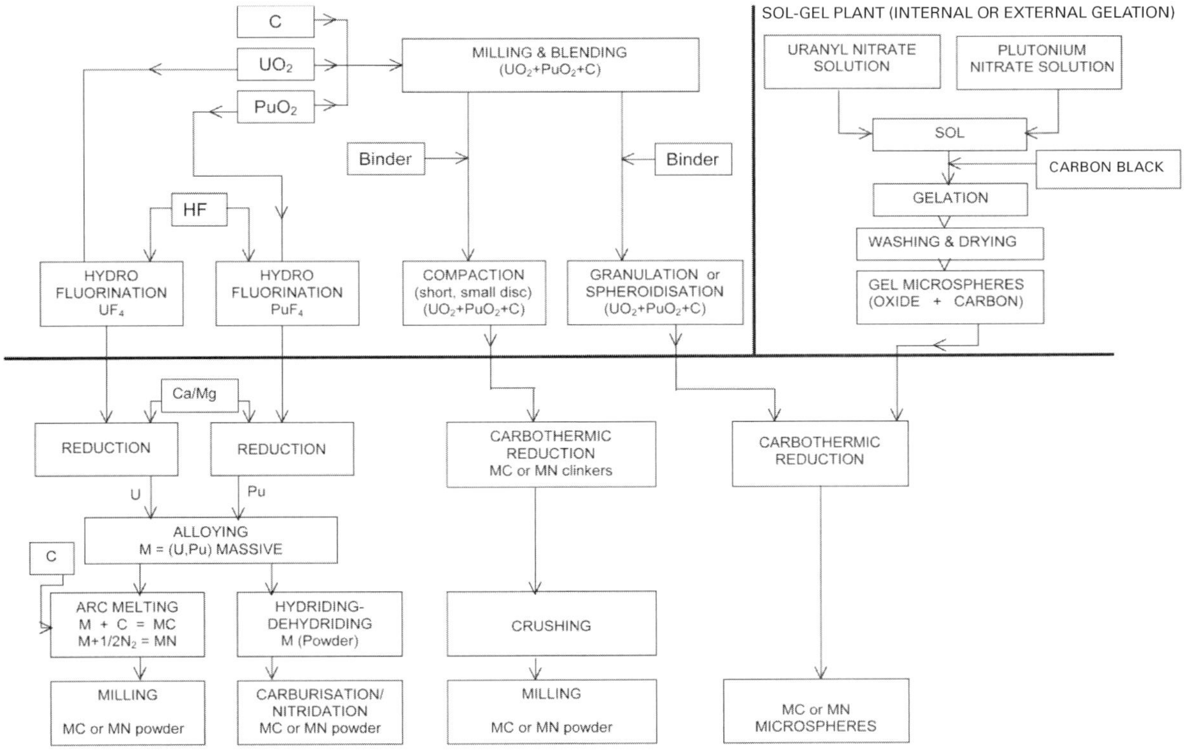

Figure 2
Summary of different methods for preparation of (U,Pu)C and (U,Pu)N powders and microspheres starting with metals, oxides, or nitrate solutions of uranium and plutonium.

in bulk quantities is associated with criticality hazards. Uranium and plutonium powders are pyrophoric and chemically unstable. Oxides have advantages as starting materials mainly because of economic incentives. Further, oxides are chemically stable and are the easiest and safest to store.

Carbothermic reduction of mechanically mixed or coprecipitated oxide is most attractive for large-scale production of MC and MN powders and has, therefore, been studied intensively in all the laboratories associated with the development of MC and MN fuels (IAEA 1988). In this process, a high degree of microhomogeneity of the starting oxide–carbon mixture is necessary. Otherwise, localized deficiencies and excesses of carbon lead to the formation of unwanted second phases. The requisite homogenization is achieved either by prolonged blending of the oxide–carbon powder mixture or alternatively by the wet chemical route, popularly known as the sol–gel process. The homogeneous oxide–carbon powder mixture is compacted into short (2–3 mm long) and small diameter ($\leqslant$25 mm) tablets in order to increase the oxide–carbon contact surface.

(*a*) *Carbothermic synthesis of monocarbide.* The overall simplified chemical equation for preparation of monocarbide from oxide is

$$MO_2 + 3C = MC + 2CO$$

Mixed uranium–plutonium monocarbide powder is usually prepared by single-step carbothermic reduction of a tableted powder mixture of oxide and carbon. Two-stage reduction (Richter 1976) and three-stage reaction sintering routes have also been developed. Thermodynamic calculations show that it is not possible to produce oxygen-free MC by the single-step direct solid-state carbothermic reduction of mechanically mixed oxide without incurring significant plutonium volatilization loss. During carbothermic reduction, control of the partial pressure of CO is very important since the evolution of this gas not only constitutes the primary reduction mechanism, but also controls the kinetics of this reaction. The two-step solid-state synthesis in a static bed aims at the preparation of a single-phase MC with very low oxygen and nitrogen impurities and with practically no loss of plutonium by volatilization.

In the two-step process, carbothermic reduction of the oxide is carried out with excess carbon in order to form M_2C_3 in the first step. M_2C_3, unlike MC, has very little oxygen and nitrogen solubility and can be prepared at relatively low temperatures ($\leqslant$1500 K). In the second step, the M_2C_3 is crushed, milled, and treated with hydrogen at 1125 K in order to reduce it to MC. Any free carbon present is removed as methane. A major drawback of the two-stage carbothermic reduction is that pure H_2 reduction is not recommended for plutonium-bearing material because it is associated with explosion hazards. In a nonexplosive Ar + 8% H_2 gas mixture, the partial pressure of hydrogen is inadequate for complete conversion of M_2C_3 to MC.

In three-stage synthesis (Wedemeyer and Gunther 1975) UC is prepared by carbothermic reduction of UO_2 at 2073 K. In the second stage, plutonium oxycarbide is produced by discontinuing the carbothermic reduction at 1473 K. And finally, in the third stage, uranium carbide and plutonium oxycarbide–carbon powders are blended, compacted, and subjected to reaction sintering. There are three advantages of this method. First, since the primary reduction of PuO_2 is at a low temperature (1473 K), the plutonium volatilization loss is low. Second, a carbon stoichiometric correction is possible before the sintering process because of the separate preparation of uranium carbide and plutonium oxycarbide. Third, the loading of plutonium is a minimum and the UC is produced separately without the penalty of plutonium hazards. A significant disadvantage of this method is the microstructural inhomogeneity of the sintered pellet. In the interior of the pellet a coarsely porous structure is observed while at the edges of the pellet a highly densified zone is seen. Further, because of the substantial release of carbon monoxide during reaction sintering, the maximum achievable sintered density is limited to 94% TD.

(*b*) *Carbothermic synthesis of mononitride.* The carbothermic synthesis of mononitride is represented by the overall reaction

$$MO_2 + 2C + \tfrac{1}{2}N_2 = MN + 2CO$$

Nitrogen plays the dual role of reactant and carrier for the removal of CO. The product of the reaction will have the general formula ($MN_{1-x-y}C_xO_y$) and the relative levels of oxygen and carbon retained in MN will depend on: (i) N_2 flow rate and, in turn, the partial pressure of N_2 and CO, (ii) the oxide to carbon mole ratio, and (iii) use of H_2. Around 10% excess carbon content in the oxide–carbon mixture, a gas atmosphere of N_2 followed by $N_2 + H_2$ followed by argon, a gas flow of around 1000 Lh^{-1}, and a carbothermic synthesis temperature in the range 1825–1925 K are the optimum parameters for obtaining relatively high-purity mixed uranium–plutonium mononitride with minimum residual impurities of oxygen and carbon (Ganguly 1991a).

4.4 Sol–Gel Process for Preparation of Oxide, Carbide, and Nitride Microspheres

The sol–gel process is an alternative wet chemical route whereby "ammonia gelation," dust-free, free-flowing, coarse, and spherical (10–1000 μm in diameter) hydrated gel microspheres of oxide (UO_3, ThO_2, or PuO_2) or mixed oxide and oxide + carbon mixtures of very good microhomogeneity are produced from the heavy metal nitrate feed solutions. The ammonia gelation is achieved either externally via NH_3 gas and NH_4OH or "internally" via an added ammonia generator, namely hexamethylene tetramine (HMTA). The ThO_2 and ThO_2-based mixed oxide microspheres, namely $(Th,U)O_2$ and $(Th,Pu)O_2$, are most efficiently prepared by the external gelation of thorium (EGT) process of Germany. For UO_2 and urania-based mixed oxide, namely $(U,Pu)O_2$, the external gelation of uranium (EGU) process of Germany, the SNAM process of Italy, and the KEMA internal gelation of uranium (IGU) process followed in The Netherlands, Switzerland, and India are popular (IAEA 1974).

The major steps in the sol–gel process are as follows:

- Preparation of sol or broth from nitrate solutions of uranium, thorium, or plutonium.
- Ultrasonic dispersion of fine carbon black particles in the sol or broth for preparation of "porous" oxide microspheres after controlled calcination and carbide and nitride microspheres after carbothermic synthesis (for preparation of "nonporous" oxide microspheres this step is not needed).
- Droplet formation by a vibrating nozzle.
- Droplet gelation in NH_3 gas and NH_4OH bath (for external gelation) or silicon oil bath at 90°C (for internal gelation). In the internal gelation process, microwave heating is also used in place of the silicon oil bath.
- Washing of gel microspheres.
- Drying of gel microspheres.

In the EGU process, the broth is prepared by mixing uranium nitrate solution ($1\,mol\,L^{-1}$) with urea ($4\,mol\,L^{-1}$) and NH_4NO_3 ($2.5\,mol\,L^{-1}$) and boiling the mixture for 30 min. A small amount of poly(vinyl alcohol) ($5\,g\,L^{-1}$) is also added to the mixed solution for mechanical stability of the gel. The same procedure is followed when the EGU process is extended for preparation of mixed uranium–plutonium oxide microspheres starting from nitrate solutions of uranium and plutonium. In the EGT process, the thorium-bearing sols are prepared by passing a controlled amount of ammonia gas through the thorium nitrate solution at 80–100°C, such that 85–88% of pre-neutralization reactions are completed.

The uranium/thorium-bearing broth/sol is dispersed into droplets, which pass through a curtain of NH_3 gas and fall into a 10 w/o aqueous ammonia gelation bath. In the gelation bath, the droplets gel within seconds to rigid gel particles, which are then washed in 1% ammonia solution for removal of NH_4NO_3 and dried on a continuous belt drier at 200–250°C.

In the IGU process, the uranium and plutonium nitrate solutions are mixed with HMTA (ammonia generator) and urea in mole proportions of 1.25, 1.75, and 1.75, respectively, and cooled to 0°C (Vaidya *et al.* 1987). Urea prevents premature gelation of the solution. Droplets of the sol or broth fall into the silicon oil gelation bath at 90 ± 1°C that decomposes HMTA to release ammonia for conversion of the droplets into hydrous gel particles. The gel particles are washed with carbon tetrachloride to remove oil and in a 3 M NH_4OH solution to remove ammonium nitrate.

Ganguly and Basak (1991), Ganguly *et al.* (1986b, 1987), and Hastings (1989 pp. 398–413) made the following modifications to the external and internal gelation processes for production of soft and porous UO_2, $(U,Pu)O_2$ ($\leqslant 30\%$ PuO_2), ThO_2, ThO_2–2% UO_2, and ThO_2–5% PuO_2 microspheres suitable for pelletization and sintering to high-density pellets (the plutonium was simulated with cerium in most cases):

(i) use of a heavy metal nitrate feed solution of lower molarity ($\leqslant 1$);
(ii) addition of carbon black ($30\,g\,mol^{-1}$ heavy metal) pore former to the sol or broth prior to gelation and later removing the carbon black from the gel microspheres by controlled air calcination at 700°C to form porous oxide microspheres; and
(iii) addition of dopants to the heavy metal nitrate feed solutions, which enhance densification and/or grain growth during sintering of oxide pellets.

The hydrated gel microspheres of uranium oxide or mixed oxide thus obtained are subjected to air calcination followed by H_2 reduction and stabilization treatment in order to obtain porous, free-flowing, and soft microspheres with the desired oxygen to metal ratio and specific surface area in the range $10 \pm 5\,m^2g^{-1}$. The porous microspheres crush easily during pelletization and have excellent sinterability. If carbon black pore former is not added, relatively hard and nonporous microspheres are produced.

Mixed uranium–plutonium monocarbide and mononitride microspheres are prepared in two stages. First, hydrated gel microspheres of $UO_3 + PuO_2 + C$ are prepared by the ammonia internal or external gelation process starting with uranyl nitrate and plutonium nitrate solution. Next, the gel microspheres are subjected to carbothermic synthesis at 1450°C in vacuum (1 Pa) for monocarbide and in flowing nitrogen atmosphere (flow rate of $1.2\,m^3\,h^{-1}$) at 1550°C for mononitride. The sol–gel process is particularly attractive for the preparation of monocarbide and mononitride microspheres for the following reasons:

(i) the number of process steps is less than half compared to the conventional powder route;
(ii) generation and handling of fine carbide and nitride powders are avoided, which, in turn, reduces fire and radiotoxic dust hazards;

(iii) high specific surface area and chemical reactivity of gel microspheres containing oxide–carbon mixtures reduce significantly the carbothermic reduction temperature and the soaking period; and

(iv) a high degree of microhomogeneity is achieved in the fuel.

4.5 Cold Pelletization and Sintering

The oxide, carbide, or nitride fuel pellets are universally manufactured by the classical powder pellet route starting with fine powders. The alternative SGMP process is dust-free and has several advantages over the powder pellet route but has not been employed anywhere on an industrial scale.

In the powder pellet route, the fine powders of oxide, carbide, and nitride are granulated by precompaction and granulation, roll compaction and granulation, or extrusion in order to obtain free-flowing, coarse ($\sim$1 mm diameter) granules to ensure constant filling of the dies of the multipunch double-acting hydraulic press or rotary press. UO_2 and $(U,Pu)O_2$ powders obtained by the AUC and AUPuC process, respectively, are free flowing and can be directly pelletized. In the powder pellet route, the pore formers (e.g., poly(vinyl alcohol), methyl cellulose), binders (paraffin wax, naphthalene, etc.), and additives for sintering aid and grain growth are added to the powder during milling or blending prior to granulation.

In case of the SGMP process, the dust-free and free-flowing, oxide, carbide, or nitride microspheres are directly pelletized and sintered. The sintering aid and grain growth additives are added to the solution or sol prior to gelation. The lubricants, if any, are added to the granules or sol–gel particles prior to pelletization.

The sintering of oxide, carbide, or nitride pellets is usually carried out at a high temperature (1600–1750°C) in a reducing atmosphere (H_2, cracked NH_3, N_2+H_2, $Ar+H_2$) in a molybdenum resistance furnace of the batch type or continuous pusher or walking beam type. In the case of carbide and nitride, the sintering is also carried out in vacuum or argon

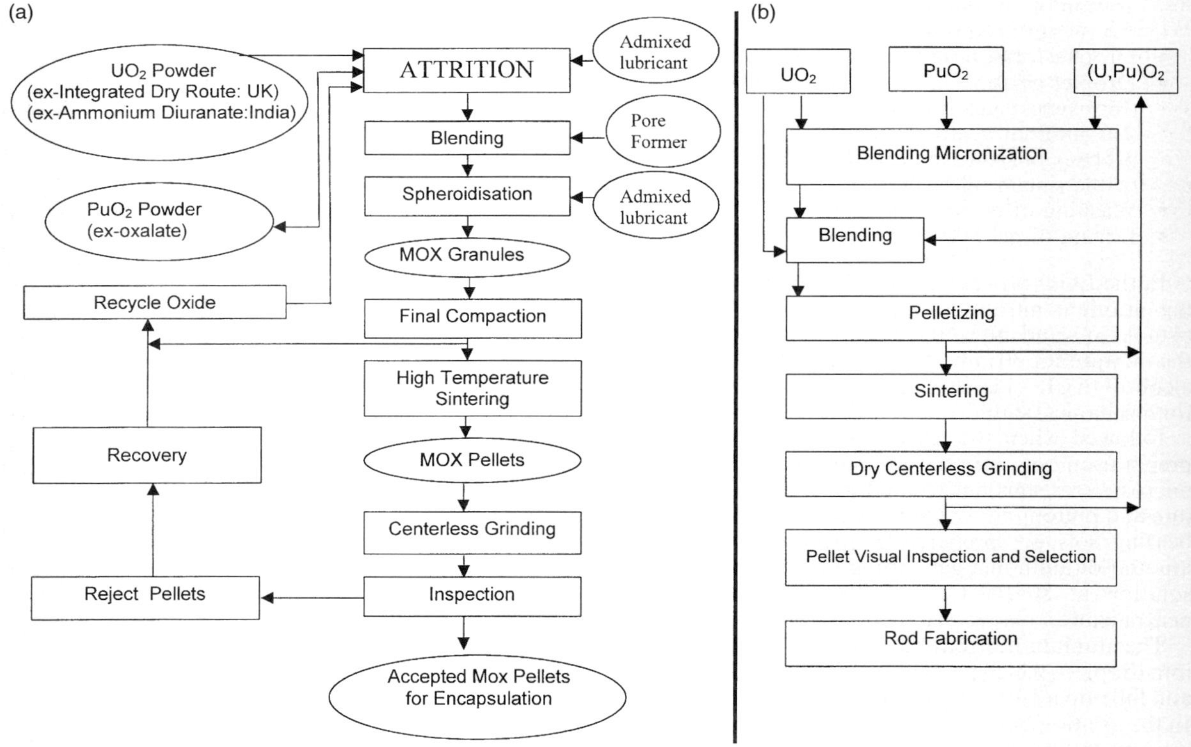

Figure 3
(a) Process flowchart based on the SBR followed in the UK and India for fabrication of mixed uranium–plutonium oxide (MOX) fuel pellets involving attrition of UO_2 and PuO_2 powders. (b) MIMAS blend process followed in Belgium and France for fabrication of mixed uranium–plutonium oxide (MOX) fuel pellets for LWRs and FBRs.

atmosphere. Fabrication of high-density UO_2 and $(U,Pu)O_2$ pellets is possible by LTS ($\leqslant 1200°C$). In LTS and SGMP–LTS, there is significant conservation of electrical energy as a result of which the fuel fabrication cost is reduced and the requirements of molybdenum heating element and charge carriers are avoided.

(*a*) *Sintering of UO_2 and $(U,Pu)O_2$ pellets for water-cooled reactors.* Industrial production of UO_2 fuel pellets for water-cooled reactors is universally carried out by cold pelletization of UO_2 powder followed by sintering at 1700–1750°C in a high-temperature resistance furnace using molybdenum heating elements and molybdenum charge carriers. Industrial-scale production of mixed uranium–plutonium oxide (MOX) fuel for LWRs has so far been carried out mainly in France, Germany, Belgium, and the UK. In addition, several MOX pilot plants are in operation in Japan, India, and Russia (IAEA 1997).

Figures 3(a) and (b) summarize the process flowcharts followed for production of MOX fuel pellets by the micronized master (MIMAS) process and the short binderless route (SBR), respectively. The MIMAS process was developed by Belgonucleaire in the mid-1980s and is being followed in the Dessel plant in Belgium and the MELOX plant in France. These two plants supply MOX reload fuel to several PWRs and BWRs. The MIMAS process produces a master blend of UO_2 and 30% PuO_2 by ball milling, which also ensures micronization of the UO_2 and PuO_2 particles. In the second blending operation, free-flowing UO_2 powders, either of the AUC or the ADU type, is added in order to have 2–8% PuO_2 in the final blend depending on the requirement. The double blending ensures very good isotopic homogeneity of the plutonium in the product, even with plutonium of various origins, forms, and sources. In addition, in the MIMAS process, the rejected pellets, grinding powder, and other scraps are subjected to dry recycling, which is reintroduced at the first or the second blending step. The sintered MOX pellets do not contain any undissolved PuO_2 particles and ensures easy solubility in nitric acid.

The SBR was developed by British Nuclear Fuels Ltd (BNFL) in the early 1990s, based on which the MOX Demonstration Facility (MDF) was set up at Sellafield, UK. The SBR uses a high-energy attritor mill to blend UO_2 and PuO_2 feed powders and a spheroidizer to condition the powder before it is used as a feed to a rotary compaction press. At the milling stage, zinc stearate lubricant and Conpor pore former are added. The MOX pellets are sintered at 1650°C over a cycle time of 24 h. Based on the successful operation of MDF, the Sellafield MOX plant (SMP) has been set up. The use of attritor milling ensures good microhomogeneity of the co-milled powder, which in turn leads to complete solid-solution formation of UO_2 and PuO_2 during sintering. The MOX plant in India also uses an attritor for co-milling ex-ADU UO_2 and oxalate-derived PuO_2 powders. The lubricant is added at the attrition stage after which the co-milled powder is subjected to precompaction and granulation leading to free-flowing oxide powder that is pelletized and sintered. The MOX fuel thus produced would have excellent microhomogeneity and is expected to perform better than UO_2. In addition, such MOX fuel would be easy to reprocess since undissolved PuO_2 particles would be absent. The microwave denitration of mixed nitrate solutions of uranium and plutonium followed in Japan and the AUPuC process of Germany lead to homogeneous MOX powder, which after pelletization and sintering ensures complete solid-solution formation between UO_2 and PuO_2, leaving no undissolved PuO_2 particles.

For high burn-up PHWR and LWR fuels, it is essential to manufacture UO_2 and $(U,Pu)O_2$ pellets of large grain size (from 10–15 μm to ~50 μm) in order to minimize fission gas release from the fuel during operation in the reactor and to avoid pellet–cladding interaction (PCI) failures. This is also true for mixed urania–gadolinia (gadolinium as burnable poison) fuel pins for LWRs. Pellets of large grain size are mostly obtained by addition of dopants such as TiO_2 and Nb_2O_5, which enhance cation diffusion in the UO_2 or $(U,Pu)O_2$ lattice. Large-grained pellets can also be obtained at relatively low temperature (1200°C) by starting with oxide powder having an oxygen to metal ratio in the range 2.25–2.27 by sintering in controlled oxidizing and reducing atmospheres.

The major breakthroughs in the fabrication of UO_2 and $(U,Pu)O_2$ pellets are: (i) LTS, also known as the NIKUSI process developed in Germany (Doerr *et al.* 1986), and (ii) the subsequent SGMP–LTS process developed in India (Ganguly 1993, Ganguly and Basak 1991).

In the LTS process, the sintering of UO_2 or $(U,Pu)O_2$ is carried out at 1100–1200°C for a very short duration (soaking at 1100–1200°C for less than 1 h) in a three-zone continuous sintering furnace maintained at three different atmospheres. The diffusion of uranium cation (D^U) in UO_{2+x} is proportional to x^2. Accordingly, in the LTS furnace, the sintering of hyperstoichiometric UO_2 or $(U,Pu)O_2$ pellets is carried out in the oxidative zone containing a mixture of air and nitrogen with 500–1000 ppm oxygen in order to facilitate the densification process. After densification, the hyperstoichiometric oxide or MOX pellets are reduced to stoichiometric oxide in the reducing zone of pure hydrogen or nitrogen plus hydrogen atmosphere. A nitrogen curtain is used between the oxidizing and reducing zones.

In the SGMP–LTS process, sol–gel-derived dust-free and free-flowing oxide or mixed oxide microspheres are used for pelletization, thus minimizing the radiotoxic dust hazard associated in the powder route and facilitating remote and automated compaction of pellets. High-density UO_2 and $(U,Pu)O_2$ pellets have been manufactured by the SGMP–LTS process using the IGU and EGU processes for preparation of hydrated gel microspheres. In most cases, for preparation of porous microspheres, carbon black pore former is added to the solution prior to gelation and later removed by controlled calcination. The porous microspheres thus prepared are found to have a low crushing strength (2.5 N per microsphere); as a result the microspheres disintegrate easily during pelletization at 350 MPa and yield sintered pellets without microsphere boundaries and with uniformly distributed "closed" spherical pores in the ideal diameter range of 2–5 μm. Such a pore structure contributes to improved thermal conductivity and better retention of fission gas and accommodation of fuel swelling. The high-density oxide and mixed oxide fuel pellets are found to have a single-phase, equiaxed microstructure. The nonporous mixed oxide microspheres retain their individual identity even after pelletization and sintering and lead to a "blackberry" structure with a lot of open porosity (Ganguly and Hegde 1997). Such low-density pellets with open porosity are recommended for LMFBR fuel to facilitate fission gas release and in turn reduce fuel swelling. The mixed oxide pellets manufactured by the SGMP process ensure complete solid-solution formation between UO_2 and PuO_2 and the absence of any undissolved PuO_2. High-density UO_2–5% Gd_2O_3 pellets have also been prepared by the SGMP process, using the EGU process for the preparation of hydrated gel microspheres.

(*b*) *Sintering of* ThO_2*,* $(Th,U)O_2$*, and* $(Th,Pu)O_2$*.* High-density ThO_2, $(Th,U)O_2$ (<4% UO_2), and $(Th,Pu)O_2$ (⩽4% PuO_2) pellets have been fabricated successfully by the classical powder metallurgy route using approximately 0.2–0.4% MgO as a sintering aid (Hastings 1989, pp. 398–413). The MgO is usually added in the form of $Mg(NO_3)_2$ in thorium nitrate solution. The sintering is carried out at 1923 K for 4 h in argon +8% hydrogen atmosphere. High-density ThO_2, $(Th,U)O_2$ (2% UO_2), and $(Th,Ce)O_2$ (4% CeO_2 with cerium as simulator for plutonium) pellets have been successfully fabricated by the SGMP process, using the ammonia EGT process for the preparation of the hydrated gel microspheres (Ganguly *et al.* 1986b). Calcium nitrate (1%) was added to the thorium nitrate solution in order to have around 0.4% CaO sintering aid in the ThO_2 and ThO_2-based mixed oxide pellets. In order to obtain soft and porous microspheres, thorium nitrate solution of low molarity (1–1.2) was used and carbon black was added as a pore former to the sol prior to gelation and later removed by controlled calcination in air.

(*c*) *Pellet impregnation technique for preparation of* $(Th,U)O_2$ *pellets.* The pellet impregnation technique (Hastings 1989 pp. 398–413) is ideally suitable for remote and automated fabrication of highly radiotoxic $(Th,^{233}U)O_2$ fuel containing up to 2% ^{233}U, which is an advanced fuel for PHWRs. In this process, ThO_2 pellets of relatively low density (75–80% TD) are fabricated in an unshielded conventional plant by cold pelletization of ThO_2 powder followed by sintering at relatively low temperature (1373–1573 K). The ThO_2 pellets are then transferred inside a shielded alpha-tight cell and impregnated with acid-deficient uranyl nitrate solution for half an hour in vacuum. After impregnation, the pellets are vacuum dried, calcined, and sintered in vacuum or hydrogen atmosphere at 1973 K to obtain high-density $(Th,^{233}U)O_2$ pellets with single-phase microstructure. Thus, the pellet impregnation route avoids the handling of fine ^{233}U-bearing powder and involves fewer process steps. In addition, only a few parts of the process equipment are required to be exposed to ^{233}U and to be confined in shielded facilities, thereby making plant maintenance easier.

(*d*) *(U,Pu)C and (U,Pu)N pellets for LMFBRs.* Among all the existing methods of consolidation of (U,Pu)C and (U,Pu)N fuels for LMFBRs, the cold pelletization and sintering is the simplest, is well established, and the most widely accepted. MC and MN fuels have been produced on a small scale (batch size ⩽5 kg) mainly for out-of-pile property evaluation and irradiation testing. The cumulative production of MC and MN fuel worldwide will not exceed 1000 kg and 200 kg, respectively. The starting materials consist of either fine powders or sol–gel-derived synthesized microspheres of MC and MN. The MC and MN products synthesized by different methods are subjected to the following treatment for obtaining fine powder:

- crushing and ball milling arc-melted MC and MN buttons;
- crushing and ball milling clinkers, after carbothermic reduction of tableted oxide–carbon powder mixture; and
- milling oxide–carbon spheroids, granules, or microspheres after reaction sintering.

Suitable binders and sintering aids (if any) are added to the milled powder and the milling is continued for proper homogenization. The powders are then compacted into pellets (length to diameter ratio of 1.0–1.5), preferably in a double-action press at 140–350 MPa, followed by sintering in vacuum, argon, or argon–hydrogen mixture in the temperature range 1673–2173 K.

The degree of densification and the microstructural features of sintered MC and MN pellets depend, besides sintering temperature and time, on the average particle size or surface area of the powder, the binding agent (i.e., binder), the sintering aids (if any), and the plutonium and the oxygen contents of the powder. Most of the studies reported on MC and MN pellets pertain to uranium-rich composition with maximum PuC and PuN contents of 20%. Extensive investigations have been carried out in India on the densification of plutonium-rich mixed uranium–plutonium monocarbide and mononitride keeping their potential use as driver fuel in FBTRs (Ganguly *et al.* 1986a, 1991a, 1994). The densification behavior of MC and MN pellets of both uranium- and plutonium-rich compositions are found to be similar.

The powder compaction pressure in the range 140–350 MPa and the sintering atmosphere play only minor roles in controlling the sintered density of MC and MN. For MN pellets, although vacuum sintering is marginally more effective than sintering in argon or nitrogen atmosphere, the decomposition of MN in vacuum limits the maximum permissible sintering temperature to 1873 K. For argon and nitrogen sintering, the MN pellet density increases from about 75% TD at 1773 K to around 95% TD at 2173 K.

The general conclusion regarding the effect of the surface area of UC or (U,Pu)C powders on the sintered density is that powders of surface area of 1.5–2.0 m^2g^{-1} are adequate for achieving high sintered densities ($\sim$96% TD) at temperatures of $\sim$1873 K. UC and (U,Pu)C powders of these surface area values can be obtained by ball milling in a vibratory mill with tungsten carbide balls for 12–16 h.

A small quantity of nickel powder (0.1–0.2 wt.%, freshly hydrogen reduced) when admixed with UC, PuC, or (U,Pu)C powder is found to act as a very effective sintering aid. High-density ($\geqslant$95% TD) pellets are consistently obtained at relatively low temperatures (1723–1823 K) even with considerable variations in the specific surface area, oxygen contents, and sintering conditions of the carbide pellets. However, nickel promotes grain growth and the formation of M_2C_3. Monocarbide pellets with nickel always contain 5–15% M_2C_3 and have a microstructure with angular or rectangular grains and banded structure within the grains.

The densification of UN and (U,Pu)N powder compacts is extremely sensitive to the average particle size (or the surface area) of the nitride powders. A small amount (1–5%) of oxide phase (MO_2) has been found to act as a sintering aid for UN and (U,Pu)N. The effect of oxygen on the densification of MC and MN depends on the form in which the oxygen is present. For MC, the sintered density is not at all affected if the oxygen is dissolved in the monocarbide grains. However, if the oxygen is present as an oxide phase, it acts as a barrier to the interdiffusion of uranium, plutonium, and carbon and inhibits sintering. The sinterability of mononitride is enhanced when a small amount of oxide is present as a second phase.

The mixed carbide and mixed nitride microspheres prepared by the sol–gel process are nonporous and have a relatively high crushing strength (20 N per microsphere). As a result, the microspheres retain their individual identity even after pelletization at high pressure (840 MPa) and sintering at high temperature (1973 K), resulting in a "blackberry structure" with microsphere boundaries and open pores. The blackberry microstructure in sintered pellets is desirable for LMFBR fuel as it facilitates fission gas release, which, in turn, minimizes fuel swelling.

(*e*) *Vibratory compaction.* In this process, high-density microspheres ($\geqslant$98% TD) of oxide, carbide, or nitride fuels of 1, 2, or 3 size fractions in the range 10–1000 μm are directly loaded into one end of a closed fuel cladding tube and subjected to vibratory compaction in order to obtain fuel pins of controlled smear density (weight of the fuel pin divided by internal volume of cladding tube) in the range 60–90% TD. The microspheres are generally prepared by the sol–gel process. In some cases, the microspheres are prepared by the dry route involving spheroidization of powder. For vibratory compaction, various types of pneumatic and electro dynamic vibrators are used to obtain high compaction efficiencies. For smear densities typical of LMFBR fuels (75–80% TD), particles of 2 size fractions, typically 750 μm and 70 μm, are used in the ratio of 3:1. The highest achievable smear density of 90% TD is obtained by mixing the coarse (750–1000 μm), medium (70–100 μm), and fine (7–10 μm) fractions in the ratio 7:2:1. The rates of infiltration and settling of the fine particles in the voids formed by the packing of the coarse particles depend on the frequency, the waveform, and the acceleration of vibration, as well as on the particle shape and size fractions and the internal geometry of the fuel cladding tubes.

The vibratory compaction process avoids generation and handling of fine radioactive powder of uranium-, plutonium-, or thorium-bearing oxide, carbide, or nitrides. The dust-free and free-flowing microspheres minimize radiotoxic dust hazard and the process is amenable to remote and automated fuel fabrication. The process is particularly attractive for carbide and nitride fuels because the number of process steps is less, and it is recommended for fabrication of highly radioactive plutonium- and ^{233}U-bearing fuels. Apart from this, the vibratory compaction route has the maximum flexibility of operation because fuel cladding tubes of any internal dimensions can be easily vibro-filled to a wide range of smeared density (60–90% TD) by packing microspheres of 1, 2, or 3 sizes. However, the maximum achievable smeared density of a vibro-filled fuel pin is 90% TD. Hence, the process is not suitable for manufacturing LWR or PHWR fuels, which have very high ($\geqslant$94% TD) fuel pellet density and high smeared density. The vibro-filled fuel pins are suitable only for use as helium-bonded fuel pins for LMFBRs, which require low smeared density (75–80% TD).

5. *Fabrication of Metallic Fuels for Fast Reactors*

U–Pu–Zr alloys have emerged as safe and commercially viable advanced LMFBR fuels. They are fabricated by vacuum induction melting of uranium, plutonium, and zirconium metals in a yttria/zirconia-coated graphite crucible at around 1450°C, followed by injection casting (by furnace pressurization with 0.05 MPa argon or helium) in a closed-end Vycor (SiO_2) mold coated with yttria, zirconia, or boron nitride (Chang 1989). The zirconium is added to increase the solidus and fuel–clad eutectic temperatures. Further, oxygen and nitrogen impurities in the fuel form ZrO_x and ZrN_x diffusion barriers on the fuel–clad interface and delay eutectic formation with nickel in the stainless steel cladding. The cast pins require no further surface finishing since the metallic fuels use sodium bonding with large gap in the annular space between pellet and cladding.

6. *Conclusion*

The fabrication of nuclear fuels is unique in several ways since uranium, plutonium, and thorium are radioactive and are associated with the problems of criticality (uncontrolled nuclear fission chain reaction), health hazard, security, and pyrophoricity (in some cases). The fuel has to meet stringent dimensional and chemical specifications in order to ensure satisfactory performance in the reactor. The manufacturing technology of both metallic and ceramic fuels for research and power reactors has been well established over the years. The aluminum alloy-clad aluminum matrix dispersion fuel containing U_3Si_2 particles, the reference fuel for MTRs, is fabricated by the picture-frame technique followed by roll bonding. The manufacturing process of aluminum alloy- or stainless steel-clad U–ZrH fuel for TRIGA reactors consists of controlled hydriding of uranium–zirconium alloy rods. For water-cooled nuclear power reactors, zirconium alloy-clad high-density uranium oxide pellets are the reference fuel. Mixed uranium–plutonium oxide containing up to 5% PuO_2 and 25% PuO_2 have been used in PWRs and LMFBRs, respectively. $(Th,U)O_2$ and $(Th,Pu)O_2$ are advanced fuels for PHWRs. The oxide and mixed oxide fuels are mostly manufactured in the form of small cylindrical pellets of controlled density and microstructure by the classical powder pellet route. The mixed uranium–plutonium monocarbide and mononitride advanced fuels for LMFBRs are also fabricated by the powder route starting with UO_2 and PuO_2 powder. The SGMP process is a dust-free advanced method for manufacturing oxide, carbide, and nitride fuels and is best suited for remote and automated fabrication of highly radiotoxic plutonium- and ^{233}U-bearing fuels. The vibratory compacted oxide, carbide, and nitride fuels are suitable for manufacturing LMFBR fuel pins with relatively low packing density (⩽85% TD). The U–Zr and U–Pu–Zr advanced metallic fuels for fast reactors are fabricated by vacuum induction melting followed by injection casting.

See also: DUPIC Fuel Cycle; Fast Breeder Reactors: Fuels; Heavy Water Reactor Fuel Design and Performance; Inert Matrix Fuels; Light Water Reactor Fuel Design and Performance; In-reactor Creep Testing: Techniques; Nuclear Fuel Cycles; Nuclear Reactor Materials

Bibliography

Assmann H, Becker M 1979 Technology of UO_2 fuel fabrication by the AUC process In: *Proc. European Nuclear Conf.* European Nuclear Society, Hamburg, Germany

Assmann H, Peehs M, Roepenack H 1988 Survey of binary oxide fuel manufacture and quality control. *J. Nucl. Mater.* **153**, 115

Belle J, Lustman B 1958 Preparation of UO_2, properties of UO_2 and sintering characteristics. *USAEC Report TID 7546*

Chang Y I 1989 The integral fast reactor. *Nucl. Technol.* **88**, 129

Chotard A, Ledac A, Bernardin M 1991 Fabrication characteristics and in-pile performance of UO_2 pellets prepared from dry route powder. In: Ganguly C, Roy S K, Roy P R (eds.) *Advanced Ceramics, Key Engineering Materials.* Trans Tech, Switzerland, Vols. 56, 57, pp. 471–88

Doerr W, Peehs M, Assmann H 1986 Advantages of the oxidative UO_2 sintering process NIKUSI In: *Proc. 4th Int. ENS/ANS Conf.* European Nuclear Society, Geneva, Switzerland, Vol. 4, p. 233

Ganguly C 1989 In: *Nuclear Fuel Fabrication, Materials Science Forum*, Trans Tech, Switzerland, Vols. 48, 49.

Ganguly C 1993 Sol–gel microsphere pelletization—a "powder-free" advanced process for fabrication of ceramic nuclear fuel pellets. *Bull. Mater. Sci.* **16**, 509–22

Ganguly C, Basak U 1991 Fabrication of high-density UO_2 fuel pellets involving sol–gel microsphere pelletization and low temperature sintering. *J. Nucl. Mater.* **178**, 179–83

Ganguly C, Basak U, Vaidya V N, Sood D D, Roy P R 1989 Sol–gel microsphere pelletization of UO_2 and UO_2–PuO_2 pellets of PHWR fuel specifications using internal gelation process. In: Hastings I J (ed.) *Proc. 2nd Int. Conf. CANDU Fuel.* Canadian Nuclear Society, Pembroke, Canada, pp. 108–24

Ganguly C, Hegde P V 1997 Sol–gel microsphere pelletization process for fabrication of $(U,Pu)O_2$, (U,Pu)C and (U,Pu)N fuel pellets for the prototype fast breeder reactor in India. *J. Sol–Gel Sci. Technol.* **9**, 285

Ganguly C, Hegde P V, Jain G C 1994 Fabrication of $(Pu_{0.55}U_{0.55})C$ fuel pellets for the second core of FBTR. *Nucl. Technol.* **105** (3), 346–54

Ganguly C, Hegde P V, Sengupta A K 1991a Preparation, characterization and out-of-pile property evaluation of (U,Pu)N fuel pellets. *J. Nucl. Mater.* **178**, 234–41

Ganguly C, Jain G C, Hegde P V, Basak U, Mehrotra R S, Majumdar S, Roy P R 1986a Development and fabrication of 70% PuC–30% UC fuel for the Fast Breeder Test Reactor in India. *Nucl. Technol.* **72**, 59–69

Ganguly C, Langen H, Zimmer E, Merz E 1986b Sol–gel microsphere pelletization process for fabrication of high density ThO_2–2% UO_2 fuel for advanced pressurized heavy water reactors. *Nucl. Technol.* **73**, 84–95

Ganguly C, Linke U, Kaiser E 1987 Characterization of $(U,Ce)O_2$ pellets prepared by the sol–gel microsphere pelletization. *Metallography* **20**, 1–14

Ganguly C, Prasad G J, Mahule K N, Ghosh J K, Asari K B J, Chandrasekharan K N, Muralidhar S, Balan T S, Roy P R 1991b Fabrication experience of Al–U^{233} and Al–Pu plate fuel for PURNIMA-III and KAMINI research reactors. *Nucl. Technol.* **96**, 72–83

Ganguly C, Rodriguez P 2000 Indian nuclear fuel cycle: options and strategies In: *Proc. 3rd Meeting Int. Working Group on Nuclear Fuel Cycle Options (IWGNFCO)*. International Atomic Energy Agency, Vienna

Ganguly C, Roy S K, Roy P R 1991c *Advanced Ceramics, Key Engineering Materials*. Trans Tech, Switzerland, Vols. 56, 57

Hastings I J (ed.) 1989 *Proc. 2nd Int. Conf. CANDU Fuel*. Canadian Nuclear Society, Pembroke, Canada

Hastings I J (ed.) 1989 *Proc. 2nd Int. Conf. CANDU Fuel*. Canadian Nuclear Society, Pembroke, Canada

IAEA 1974 IAEA technical report. *Proc. Int. Atomic Energy Agency Panel Meeting on Sol–Gel Processes for Fuel Fabrication*. International Atomic Energy Agency, Vienna

IAEA 1984 IAEA-TECDOC-352. *Proc. IAEA Advisory Group Meeting on Advanced Fuel Technology and Performance*. International Atomic Energy Agency, Vienna

IAEA 1988 IAEA-TECDOC-466. *Proc. IAEA Technical Committee Meeting on Advanced Fuels for Fast Breeder Reactors: Fabrication and Properties and their Optimization*. International Atomic Energy Agency, Vienna

IAEA 1997 IAEA-TECDOC-941. *Proc. IAEA Technical Committee Meeting on Recycling of Plutonium and Uranium in Water Reactor Fuel*. International Atomic Energy Agency, Vienna

Leary J, Kittle H 1977 *Proc. Topical Meeting on Advanced LMFBR Fuels, ERDA 4455*. American Nuclear Society

Mansfield R G 1976 *Preparation of uranium–plutonium mixed carbide*, ORNL–tr–4180, USA, English translation of Wedemeyer H, Guenther E 1975 Report KfK 2238. Karlsruhe, Germany

Matzke Hj 1986 *Science of Advanced LMFBR Fuels*. North Holland, Amsterdam

Razvi J 1998 TRIGA: 40 years of success. *General Atomics Update*, August

RERTR 1996 *Proc. 19th Int. Meeting on Reduced Enrichment for Research and Test Reactors (RERTR)*, RERTR 19. Korea Atomic Energy Research Institute, Seoul, Korea

RERTR 1998 *Proc. 21st Int. Meeting on Reduced Enrichment for Research and Test Reactors (RERTR)*, RERTR 21. Sao Paulo, Brazil

Richter K 1976 Preparation methods for uranium plutonium non-oxide fuels In: *Proc. European Nuclear Conf.* Pergamon, Oxford, Vol. 7.

Roepenack H, Schneider V, Druckenbrodt W G 1984 Experience with AuPuC—Co-conversion process for mixed oxide fuel fabrication. *Am. Ceram. Soc. Bull.* 63

Sejnoha R 1999 In: *Proc. 6th Int. Conf. on CANDU Fuel*. Canadian Nuclear Society, Niagara Falls, Canada, Vols. 1, 2.

Sowden R G, Hodge N, Morten-Smith M S, White D V 1964 *Carbides in Nuclear Energy*. Macmillan, London, Vols. 1, 2, Vols. 1, 2.

Stevenson C E 1987 *The EBR II Fuel Cycle Story*. American Nuclear Society

Vaidya V N, Mukherjee S K, Joshi J K, Kamat R V, Sood D D 1987 A study of chemical parameters of the internal gelation based sol–gel process for uranium dioxide. *J. Nucl. Mater.* **148**, 324

Vijayaraghavan R 1989 Experiences in the fabrication of aluminium clad metallic uranium fuel In: *Materials Science Forum*. Trans Tech, Switzerland, Vols. 48, 49, p. 329

Wedemeyer H, Gunther E 1975 *Germon Report Kfk 2238*

Zimmer E, Ganguly C, Borchardt J, Langen H 1988 SGMP—an advanced method for fabrication of UO_2 and MOX fuel pellets. *J. Nucl. Mater.* **152**, 169–77

Nuclear Reactor Materials

C. K. Gupta

Nuclear fission energy eminently stands today as a relatively well-developed option to produce electric power. Nuclear reactors are most fundamentally classified according to the range of the kinetic energy of those neutrons which are responsible for inducing most fissions in the chain reaction. There are *fast reactors*; fast neutrons induce most fission reactions in them. There are *intermediate*, *epithermal reactors* in which it is the epithermal neutrons that cause most fission reactions. Then there are *thermal reactors*, implying that thermal neutrons cause most fission reactions in these systems. Among the systems cited, it is the thermal reactor which is the most developed. The design, construction, and operation of thermal power reactor systems by nuclear installations all over the world have accounted for the principal share of nuclear energy delivered in the service of man. In addition, based on the functions they perform, the reactors can also be classified into power reactors meant for the production of electric power, research reactors meant to carry out principally nuclear research and materials-testing experiments, and teaching reactors meant to teach students most of the fundamental reactor experiments.

Harnessing of nuclear energy, a primary source of extremely high density, involves a technology that leans very heavily on materials, as far as the economic efficiency and societal impact of this technology are concerned. In the nuclear field, apart from a host of rigorous conventional requirements, such as those pertaining to thermal and mechanical stress, corrosion, and so on, there are novel and unusual nonconventional requirements, such as those pertaining to the resistance to the influences of nuclear radiation, the interaction with neutrons, the production of fission products, and so on. It is, therefore, not surprising for one to perceive the nuclear field not only as including within its realm preoccupation with a large number of new, or less common, materials but also as evolving new standards of references and codes in evaluating and utilizing already known materials. In this respect, it is entirely in order to bring to the fore a specific class of materials which may be called *nuclear reactor materials*, so named as to be synonymously representative of their applications exclusively for the nuclear field.

This article provides a brief account of nuclear reactor materials. It is confined to a selected materials list pertaining only to nuclear fission reactor systems. This materials list, which may also be viewed as a primary component list for nuclear fission reactors, is made up of (i) fissile and fertile fuel materials, (ii) structural materials, (iii) moderator, reflector, and blanket materials, (iv) coolant materials, and (v) control, shielding, and safety system materials. Materials destined to see service in nuclear fission reactors must possess certain properties, and those may conveniently be divided into two categories: general properties (mechanical strength, ductility, toughness, fabricability, etc.), and special properties (neutronic properties, induced radioactivity, irradiation stability, ease of spent fuel processing, etc.).

1. *Nuclear Fissile and Fertile Fuel Materials*

A nuclear fuel must contain fissile nuclei. One can use any of the fissile isotopes of uranium and plutonium. The fissile materials are the odd-A even-Z nuclei (A is the atomic number and Z is the nuclear charge). Important examples are ^{233}U, ^{235}U, ^{239}Pu, and ^{241}Pu. They undergo fission when struck with neutrons with energies in the low and intermediate to fast ranges i.e. at thermal energies (< 1eV) and at energies around 1MeV. These isotopes are called *fissile isotopes*. The even atomic mass isotopes of thorium, uranium, and plutonium are called fertile isotopes. They undergo fission when bombarded only by high-energy neutrons, of the order of 1MeV or more. Thus they do not fission readily and by themselves would not make a good reactor fuel. They can, however, easily absorb a stray neutron in a reactor core, thereby transmuting into odd-A even-Z nuclei, thereby getting converted to a good reactor fuel. Important examples are ^{233}U, ^{239}Pu, and ^{241}Pu. The isotope ^{235}U is the only fissile isotope available in nature, and it amounts to only a small fraction, of the order of 0.7wt.%, of natural uranium; the balance is made up of the more abundant isotope ^{238}U (99.3wt.%). Thorium has only one abundant naturally occurring isotope, ^{232}Th. In contrast, as mentioned, in the case of ^{238}U, naturally occurring uranium is available as a ready made mixture of minor amounts of ^{235}U with ^{238}U. The actual choice of fissile material content in a given fuel is determined by the amount of neutron-capturing material in the particular core arrangement and the amount of energy it is desired to extract before renewing the fuel.

The fissile and fertile materials may now be related to the different reactor systems introduced with different fuel cycles. It is very well known that the number of neutrons released in fission is greater than two. Besides one neutron, which is required to sustain the fission reaction, and the fractions going into nonfission reactions in the fuel (radiative capture), absorption in nonfuel materials (called "parasitic" absorption), and leakage through the surface of the reactor core to the surrounding space (the fraction of neutrons leaking out can be reduced by increasing the reactor core size or the concentration of fissile material), some neutrons are left for absorption in fertile material to produce new fissile material. This process is known as "conversion." Thermal reactors convert relatively little fertile material to fissile material compared with the amount they consume. They are, therefore, called "*burners*." The changes in composition of fuel that occur during irradiation in a general thermal reactor system go to prove the point. The concentration of ^{235}U decreases almost exponentially over the burn-up while ^{239}Pu builds up rapidly at first and then levels off at a value of about 0.6%. An additional fissile material, ^{241}Pu, builds up, through absorption in ^{240}Pu, giving fissile advantage. The concentration of ^{240}Pu, however, is quite low. It is thus seen that although there is some build-up of new fissile nuclei from the fertile material in the core, the dominant process is the burning of existing fissile nuclei, and hence the name "burner." If the ratio of fissile material produced to the fissile material consumed (which is called the *conversion ratio*, CR) is substantial (between 0.5 and 1.0), the reactor system can be called a *converter*. The most interesting case arises when the CR becomes larger than unity. Now more fissile material is produced than consumed, and it is possible, in due course, to build up enough additional fissile material to fuel another reactor. Hence these are called "*breeder reactors*." For breeding to occur it is essential that the number of neutrons released per neutron used up in the fission process is more than 2: one neutron to perpetuate the fission reaction, one to yield a new fissile nucleus (by absorption in a fertile nucleus), and a few extra for various losses. In the thermal energy region, it is only ^{233}U for which the number of neutrons freed in the fission event per neutron absorbed is 2.3. It is for this reason that the thermal breeders conceived so far use ^{233}U as the fuel and ^{232}Th as the fertile isotope.

However, since ^{233}U does not occur naturally, one must start any such reactor with an initial loading of ^{235}U, and later use ^{233}U, when it becomes available in sufficient amounts. It is entirely possible to build breeder reactors by using a fast neutron spectrum, where the neutronic characteristics of ^{239}Pu are most favorable. The number of neutrons released in fission per neutron consumed for this isotope is found to rise markedly above the value of 2 at energies above 10keV, and attains a value of almost 3 at an energy of 1MeV. This is a most desirable feature. The reactor system featuring all that has been talked about so far is known as a *fast breeder reactor*. In fast reactors the bulk of fissions occurs at 100keV, and a true fast breeder reactor is one in which the fissile nuclide produced is the same as that burnt. This involves the use of ^{239}Pu as the fissile material and ^{238}U as the fertile material.

A fast reactor system using ^{235}U as the fissile material and yielding a larger amount of ^{239}Pu (from ^{238}U) is a type of breeder. It, however, does not draw as much attention as a true breeder. The reserves of ^{235}U, the only naturally occurring fissile material, are limited, and a time will come when all the available ^{235}U will be exhausted. Further use of the remaining ^{238}U and of ^{232}Th will then depend on the use of ^{239}Pu or ^{233}U to provide the fission chain. This explains why true breeders, which both generate and utilize these fissile isotopes, are particularly attractive.

Nuclear fissile and fertile materials and related nuclear reactor systems clearly identify two fuel cycles: thorium–uranium and uranium–plutonium. The former differs from the latter in several ways. A significant point of difference is that the neutronics for thorium–^{233}U are promising at thermal neutron energies, whereas plutonium shows superior neutronic properties at high neutron energies. It follows that thermal reactors are best suited as either converters or breeders for the thorium–^{233}U cycle while fast reactors are more favorable for the uranium–plutonium system.

If nuclear physics were the exclusive consideration, then it is obvious that the most common nuclear fuel would be uranium metal. The use of uranium in nuclear power generation is by far the chief use of this element. The metal has a higher "uranium density" than any compound of uranium and does not carry any extraneous atoms to consume neutrons or otherwise meddle with the efficiency of nuclear fission. The route traversed by uranium in various stages, from mining up to its entry into a reactor, is called the "front end of the fuel cycle." It is the uranium-bearing minerals which form the starting point as the ultimate source of nuclear fuel, especially for the light/heavy water reactor systems. Uranium mining and milling activities are extremely important. Two basic mining methods have been widely used: (i) open pit mining and (ii) underground excavation. More recently, solution mining has been added to this list. This last method offers several economic and environmental impact advantages; however, not all uranium sources are amenable to solution mining.

The principal uranium-bearing mineral sources are uraninite (UO_2) and pitchblende (U_3O_8). Going by the recognition of pitchblende as the principal uranium-bearing mineral, uraninite may be regarded as the UO_2 component of pitchblende. Milling operations follow mining and are normally carried out at the mine site to save transportation cost. All processes for recovering uranium from its sources involve leaching and this step is of fundamental importance to the overall flowsheet since it involves a choice between using acidic or alkaline conditions. Of the two minerals, uraninite is completely insoluble in dilute sulfuric acid while pitchblende is partially soluble, which is thought to be due to the presence of uranium in two valence states: the insoluble tetravalent form and the soluble hexavalent form. For this reason, pitchblende is sometimes represented by the chemical formula $UO_2 \cdot 2UO_3$. The usual process of carrying out the sulfuric acid leaching involves the continuous addition of an oxidant, usually manganese dioxide or sodium chlorate. The oxidant plays the important role of oxidizing uranium of those ores containing tetravalent uranium, implying thereby ready dissolution of uraninite and pitchblende, and of preventing reduction of uranium by other ore ingredients or by metallic iron introduced during the ore comminution process. The leaching process is followed by a concentration and purification process which involves either ion exchange or solvent extraction. Milling operations thus extract uranium from ores that normally contain between 0.10% and 0.30% U_3O_8 to produce finally a semirefined product that is referred to as "yellow cake." This product produced at uranium mills is a concentrated product which may contain any of several impure compounds, such as U_3O_8, ammonium diuranate, or magnesium diuranate containing the equivalent of 75–85% U_3O_8; however, it is still far from pure. Refining operations are carried out further, and various uranium compounds prepared which serve as raw materials for different uranium products for nuclear use.

Metallic uranium and its alloys have long been used as fuel in some power reactors and in most teaching and research reactors working at relatively low temperatures. However, it is important to note that although the uranium fuel is always contained in an inert metal cladding, its high chemical reactivity limits the choice of the reactor coolant and, in particular, precludes the use of high-pressure water. It is of interest and also greatly relevant to describe the neutron irradiation-induced (or the primary irradiation) effects on uranium metallic fuel in nuclear engineering applications. These can consist of irradiation growth, thermal-cycling growth, irradiation swelling, and irradiation creep. The metallic uranium fuel is particularly sensitive to these irradiation effects during nuclear reactor operation.

Irradiation and thermal-cycling growth of uranium fuel may cause dimensional and structural instability. The former is caused by basic anisotropy, leads to fuel shape changes, and occurs at relatively low temperatures. The latter occurs when the fuel is subjected to repeated heating and cooling in the orthorhombic phase range. Irradiation swelling is a volumetric instability that alters the fuel volume, and is brought about by the formation of voids and bubbles and agglomeration of inert fission gases, such as helium, krypton, and xenon in uranium fuel. It takes place at high temperatures in the neighborhood of 0.5 times the melting point, and sets a limit to fuel burn-up. Creep is defined as a slow, continuous, plastic deformation under constant load at elevated temperatures, this being commonly known as thermal creep. Irradiation creep is the creep that is induced by neutron irradiation in an intense radiation environment, and is attended by high plasticity.

Any substitution for uranium metal should have enhanced chemical inertness, superior high temperature stability, and, preferably, a cubic crystal structure to minimize irradiation damage and thermal cycling problems. It should also be reasonably easy to produce. The first two of these requisites point toward ceramics rather than metals, and in fact all are met by one particular ceramic material: uranium dioxide (UO_2). Although UO_2 is now the most common nuclear fuel, it does not possess ideal properties in all respects. With a theoretical density of $10.97 gcm^{-3}$, its "uranium density" (i.e. mass of uranium atoms per unit volume) of $9.67 gcm^{-3}$ is only about half that of the metal. Also, like most ceramics, it is brittle and, in common with most oxide ceramics, it possesses a very low thermal conductivity. However, these demerits are not debilitating. The low uranium density can, if required, be made up by slight enrichment of the UO_2 in ^{235}U. Intrinsic brittleness, low strength, and low thermal conductivity cause dense UO_2 to crack under even modest thermal stresses, but in practice it is quite workable to permit extensive cracking inside a structural metal cladding. The low thermal conductivity is outweighed by a high melting point ($\sim 2800°C$), by the absence of phase transformations right up to the melting point, and by generally good and predictable irradiation behavior up to these high temperatures. Another attractive feature of UO_2, significant when alternative fuel cycles are being taken into account, is that it takes its chemically rather similar fissile and fertile relatives—plutonium dioxide (PuO_2) and thoria (ThO_2)—into solid solution across the whole range of compositions. All possess the cubic fluorite structure.

Metal-clad UO_2 is a most successful and established fuel for a variety of reactor types. For UO_2, which, unlike metals, cannot be rolled into thin sheets nor be cheaply bonded to a metallic sheath, a simple rod geometry is by far the most convenient, and has been adopted. A long, thin, straight metal cladding tube filled with short cylindrical UO_2 pellets represent, the most adopted fuel element for the present generation of power reactors. There are several different methods for producing ceramic-grade UO_2 powder, the starting material for the production of consolidated UO_2 fuel. The primary method of fabrication of the UO_2 pellets has been by cold pressing and sintering in an inert (helium or argon) or reducing (hydrogen) atmosphere at temperatures ranging from about 1300°C to 2000°C.

While dealing with UO_2 it is of interest to bring reference to *mixed-oxide* (MOX) fuel which is often written as $(U, Pu)O_2$. This originates from the fact that plutonium extracted from reprocessed thermal reactor fuel can be fed back into thermal reactors in admixture (up to about 7wt.%) with natural or slightly enriched uranium. The in-reactor behavior of $(U, Pu)O_2$ is in many ways similar to that of UO_2 and conceptually, therefore, it seems to be a fairly simple matter to replace UO_2 fuel with $(U, Pu)O_2$. A limited amount of power reactor experience has to date been obtained by selective replacement of standard fuel assemblies in some reactors. It may, however, be mentioned that plutonium recycling in thermal reactors has not yet been introduced on a large scale.

Plutonium, as a fissile fuel, will play a very significant role in the breeder and power reactors. It is a synthetic fuel derived from neutron irradiation of natural or depleted uranium. All power reactors fuelled by uranium create plutonium internally as an inevitable part of their energy production. Some of the plutonium generated in the reactor undergoes fission *in situ*, yielding about a third of the heat produced, and to this extent one can say that plutonium is already in use as a fuel. However, about half of the plutonium resides in the spent fuel.

On the basis of its nuclear properties, plutonium operated in a fast breeder reactor is noticeably superior to uranium fuel. Plutonium alone, however, is not suitable for use as a fast reactor fuel. The principal reasons are: (i) within the short temperature range from room temperature to its melting point, 640 °C, the metal undergoes six phase changes, and (ii) plutonium is so concentrated a fissile material that it must be diluted, e.g. in the form of U–Pu alloy or of mixed oxide ($(U, Pu)O_2$), or of carbide ((U, Pu)C), or of nitride ((U, Pu)N), for use in fast breeder reactors. The mixed-oxide fuel ($(U, Pu)O_2$) provides the best near-term potential for the rapid development of fast breeder reactor fuel elements. The fundamental reasons are: (i) extensive technology and experience already exist for preparative and fabrication aspects of oxide fuel, (ii) the thermal and irradiation features of oxide fuel are known, and (iii) the facility and operating experience of the oxide fuel are well defined.

Thorium is a very well-recognized fertile material in the thorium–uranium fuel cycle in which ^{232}Th absorbs thermal neutrons in a nuclear reactor to become ^{233}Th, which then transmutes by two stages of negative β decay through ^{233}Pa to ^{233}U. The synthetic uranium isotope ^{233}U is a fissile material produced in the thorium–uranium fuel cycle; just ^{239}Pu is produced in the uranium–plutonium fuel cycle. It may be pointed out that while the uranium–plutonium fuel cycle has been well developed and established, the development and technology of the thorium–uranium fuel cycle have been hindered by the occurrence of ^{232}U, a strong γ emitter.

Compared with uranium, thorium has an isotropic structure, higher mechanical strength, and a melting point approximately 600 °C higher. Like uranium dioxide, thorium dioxide has a face-centered cubic structure, its melting point being 500 °C higher than that of UO_2; together they form solid solutions. Its cubic structure gives thorium a greater dimensional stability under irradiation, as compared with uranium.

To increase mechanical and corrosion resistance, several alloying elements have been added to pure metallic thorium. Among metallic thorium alloys, Th–U and Th–Pu alloys provide the possibilities of bringing together the properties of uranium, of being fissile, and of thorium, of being fertile, in a thorium–uranium fuel cycle in thermal power and fast breeder reactors. Among the mixed ceramic thorium compounds, the combinations of mixed oxide, carbide, and nitride fertile and fuel materials may offer a better scope for operating the thorium–uranium fuel cycle for some of the reactor systems. In the thorium–uranium fuel recycle the parasitically produced decay isotopes and their daughter products giving out intense radiation and radioactivity can impose the requirement of heavy shielding for fuel handling, reprocessing, and fabrication of the recycled fertile and fuel materials, adding to the cost.

2. *Structural Materials*

A nuclear reactor is an intricate structure comprising a miscellany of subassemblies that harmonize in sustaining a self-supporting chain fission reaction, and in effecting an exchange of power between the core and outer systems of conversion for utilization of heat of nuclear origin. Each and every material that is a part of such a system has a "structural" function as well. However, a certain class of materials stands out in the sense that their role in the development and control of the nuclear reaction is not of much consequence: these are conventionally described as "structural materials." Structural materials are, for instance, those that go to make the pressure vessel, the pressure tubes through which the coolant is circulated, the impervious cladding hosting the nuclear fuel, the materials used in the devices for driving and steering the fuel elements and control rods, fittings and supports, and so on. Depending on their location in the reactor, structural materials may be classified as in-core and out-of-core materials. This article concentrates on some of the in-core materials.

Beryllium metal is unique in that it has the lowest absorption cross-section for thermal neutrons among all metals. This property, combined with its large scattering cross-section and high melting point, makes it an excellent structural, as well as moderator and reflector, material. For these reasons, beryllium and beryllia (BeO) have been used as structural and reflector materials. A minor application of metallic beryllium as a structural and a reactor material is associated with neutron sources used to initiate a cold nuclear reactor. For the reactor start-up, the materials used in neutron sources, such as ^{239}Pu–^{9}Be, ^{236}Ra–^{9}Be, ^{210}Po–^{9}Be, and ^{123}Sb–^{9}Be, can induce (α, n) or (γ, n) reactions to generate copious neutrons ($^{9}Be + {^{4}He} \rightarrow {^{12}C} + {_0n^1}$; $^{9}Be + \gamma \rightarrow {^{8}Be} + {_0n^1}$). Here, ^{239}Pu and ^{210}Po are α emitters, ^{226}Ra is both an α and a γ emitter, and ^{124}Sb is a γ emitter. As advantages of beryllium as a structural, moderator, and reflector material for nuclear reactors, mention may be made of the following: (i) very low thermal neutron absorption, (ii) high thermal neutron scattering, (iii) high melting point, (iv) high strength-to-weight ratio, (v) good thermal stability and low vapor pressure, and (vi) good corrosion resistance. As disadvantages, mention may be made of (i) poor ductility, (ii) toxicity, (iii) high cost and scarcity, and (iv) fracture rupture.

Magnesium has excellent nuclear properties, with a low thermal neutron absorption cross-section between those of beryllium and aluminum. As a structural material it has good compatibility with uranium fuel and CO_2 coolant in a gas-cooled, graphite-moderated reactor. Magnesium alloys, particularly Magnox A-12 and ZA alloys, have been used as cladding materials on a large scale since the development of the British Calder Hall-type gas-cooled reactors. The melting point of magnesium is 600°C, and its resistance to oxidation from air to CO_2 is reasonable up to 400°C. Therefore, most of the gas-cooled reactors are normally operated at low temperatures and low power density. As advantages of magnesium and its alloys for gas-cooled reactors, mention may be made of (i) their low thermal neutron absorption cross-section, (ii) their compatibility with fuel and coolant, and (iii) that they undergo no significant changes in mechanical strength and ductility under irradiation.

Aluminum and its alloys have been used for nuclear fuel cladding and in the manufacture of the passages traversed by the coolant and the control rods. Aluminum and its alloys are mostly used in reactors using metallic natural or slightly enriched uranium fuel and operating at relatively low temperatures, in certain reactors designed for plutonium production, and in experimental and research reactors including those used for teaching. Their relatively low thermal neutron absorption, high thermal conductivity, high stability under irradiation, good corrosion resistance to air and water, fabricability and weldability, low cost, and abundant availability can be cited as advantages, while their low melting point and low mechanical strength at elevated temperatures can be cited as disadvantages. The choice of aluminum as a structural material is, however, based on the advantages it offers over the disadvantages, because the chief requirement of a teaching or research reactor is a high neutron flux for neutron economy with reliable and safe operation of the fuel elements at a constant power level rather than power generation. Therefore, the operation of teaching and research reactors at low temperatures is preferable when the availability, reliability, and safety of the reactor are concerned. In fact, the mechanical strength, ductility, and thermal conductivity of aluminum and its alloys are adequate for thermal stresses, elongation, and thermal gradients developed during reactor operation. Their resistance to corrosion by water and water vapor is sufficient for coolant temperatures below about 150°C.

Zirconium freed of hafnium to neutronically acceptable levels is commonly known as *nuclear grade zirconium*, and it is such a specified zirconium that is used by the nuclear industry. Ordinary industrial zirconium, however, cannot reach the degree of purity required for suitable corrosion resistance and mechanical characteristics. If special precautions are not taken in the zirconium-making process, certain impurities may be introduced into the material, the most harmful being gases (oxygen, nitrogen, and hydrogen) and carbon. Corrosion resistance is impaired by carbon and nitrogen, hydrogen leads to embrittlement, and oxygen hardens zirconium. In order to prevent these undesirable effects, zirconium must be alloyed. Zirconium alloys containing tin, iron, chromium, and nickel are known as *zircaloys*. The zircaloys have a much better corrosion resistance, greater mechanical strength, sufficient ductility, and acceptable neutronic characteristics compared with unalloyed zirconium. In the family of zirconium alloys, the first manufactured was Zr–2.5% Sn (zircaloy 1). Composition variations around this alloy led to zircaloy 2. Zircaloy 3, a very low-tin variant, was relegated to the back seat in favor of the superior zircaloy 4, a nickel-free variant customized to decrease hydrogen pick-up. Similar experiments with compositional variations in order to improve corrosion resistance and strength led to the development of another family of alloys based on the Zr–Nb binary system. This alloy system offered the possibility of obtaining a fine two-phase structure leading to higher strength. There are some interesting points as regards the corrosion of zircaloy cladding. It can occur either at the water side or at the fuel side. On the water side, zircaloy tends to form a very thin oxide layer, which protects the metal from further corrosion. This is a phenomenon known as passivation. On the fuel side, another form of zircaloy corrosion occurs. The temperature at the inner surface of the tubing is higher. In addition, a hostile environment exists owing to the presence of fission products, the most harmful of which is iodine. When the right combinations in this scenario build up, tube failure occurs by the phenomenon of stress corrosion cracking. To prevent this phenomenon from occurring inside cladding tubes, various strategies have been suggested, mostly depending on a thin barrier between the pellet and cladding to inhibit contact of the cladding with the fission products. These designs are called *barrier fuels*. A scheme involving a graphite coating on the interior surface of the tubes has proved to be quite successful.

Given the high temperature and high neutron fluence imposed on fast breeder cladding candidates, much effort has gone into the selection of special alloys that will give the best overall balance of design.

It is understood that, as in any complex engineering application, optimal properties cannot be consistently obtained from any one material over the full range of parameters desired in nuclear engineering. To strike a general note on the requirements of materials selection for fast breeder reactors, it may in the first instance be mentioned that they operate under a high temperature (about 550°C) and a low system pressure. Moreover, the excellent heat transfer properties of the liquid metal coolant cause fast temperature alternations in the metal structure in the presence of temperature transients (e.g., as occur during reactor shutdown). As a result, there are low primary stresses but high secondary stresses of thermal lineage. It is necessary to have adequate high-temperature (creep) strength and structural stability. The reference materials used for most liquid-metal fast breeder reactor in-core cladding and structural applications are austenitic chromium, nickel steels. They match best the various design requirements for corrosion resistance, high temperature strength, and fabrication. Austenitic chromium, nickel steels of type 304 or type 316 have been used for primary and secondary sodium systems, mainly owing to the availability of data in design codes and ease of weldability compared with types 321/347. In some instances, stabilized grades (type 321) have been selected to bring down the potential risk of carbon loss at high temperatures. More recently, type 316L with a controlled nitrogen content, 316L (N), has drawn attention as a prime candidate.

3. *Moderator, Reflector, and Blanket Materials*

A moderator, as the name implies, is used to moderate or slow down fast fission neutrons from relatively high (kinetic) energies to thermal (ambient temperature) energy levels in a thermal reactor. The process of reducing the energy of the fission-produced neutrons, from an average of 2MeV to a value of 0.025–1eV— which conform to a thermodynamic equilibrium with the environment in the reactor and promote ^{235}U fission in thermal reactors—is called *neutron moderation*, and the materials in the reactor core that are instrumental in achieving this are, accordingly, known as *moderator materials*. The moderation process is implemented by repeated elastic collisions of neutrons with the nuclei of the moderator material. At each collision a neutron transfers part of its energy to the moderator nucleus. The smaller the mass of the target nucleus the larger is the transferred energy, which attains its maximum value when the moderator nucleus has about the same mass as that of the neutron. This situation is obtained only with hydrogen. Apart from lightness, other properties, such as probabilities (called "cross-sections") for scattering and absorption, must also be taken into account in estimating a figure of merit for a moderator. The scattered or leaking neutrons are reflected back into the core in both thermal and fast reactors by *radial and axial reflectors*.

The nuclear requirements for moderators and reflectors are the same in a thermal reactor. The nuclear requirements are a high neutron scattering cross-section, a large energy loss by a neutron per collision, and a low neutron absorption cross-section. The high neutron scattering, or collision, cross-section of the moderators and reflectors means frequent, large-angle collisions and a relatively short mean free path, or distance, over which a fast neutron travels during the slowing-down process, thereby bringing about a reduction in neutron leakage and escape.

The nuclear requirements for blanket materials in a fast reactor are quite different from those for moderators and reflectors in a thermal reactor. The nuclear properties needed for *blankets in fast breeder reactors* are (i) nuclear fertile fuel, (ii) a high neutron absorption cross-section, and (iii) ease of fuel reprocessing.

Over and above the nuclear requirements, solid moderators, reflectors, and blankets must have qualities of structural materials: they must possess adequate mechanical strength, thermal, and irradiation stabilities, fabricability, and corrosion resistance. Reasonably good heat transfer characteristics are also needed.

Among the major moderator and reflector materials, light water and heavy water are used, not only as moderator and reflector materials but also as coolant materials. Eminently abundant and economical, light water has been extensively used as a coolant. As coolants, light water or heavy water can simultaneously perform the function of moderator and reflector in a thermal reactor (for the relative merits of light and heavy water, see the next section). The structural materials beryllium, beryllium oxide, and graphite can also perform the function of moderator and reflector simultaneously. The fertile materials serving as blankets provided in fast reactors can be either depleted natural uranium (^{238}U) or thorium (^{232}Th). Graphite has a long history of proven use as moderator, reflector, and structural material in gas-cooled, graphite-moderated reactors and in high-temperature gas-cooled reactors. The first nuclear fission reactor, CP-1, built in 1942, was a natural uranium–graphite reactor.

Graphite has a low neutron absorption cross-section, a high neutron scattering cross-section, and a low mass number conducive to a good moderating power. In addition, graphite possesses excellent thermal properties, good mechanical strength at high temperatures, and relative ease of machining and manufacturing. Although graphite occurs in appreciable quantities in nature, reactor grade graphite is produced artificially by the graphitization of petroleum coke. Although neutron irradiation enhances the mechanical strength (particularly in compression), hardness, and elastic modulus of graphite, the important factors in relation to reactor moderator and reflector design are: (i) reduction in thermal conductivity at high temperatures, (ii) dimensional changes and instability, (iii) reduction in ductility, and (iv) stored energy.

The blanket of a fast breeder reactor is normally made of an axial and a radial blanket. The former is connected with the upper and lower sections of the fuel elements, and the latter surrounds the fuel elements in the radial direction of the core. In a fast breeder reactor using the plutonium recycling fuel cycle, the blanket material is natural uranium (containing 99.274% ^{238}U) or depleted uranium. The depleted uranium (about 100% ^{238}U) usually comes from either spent fuel reprocessing or from fuel enrichment plants. In the ^{235}U fuel enrichment plant, the blanket material, ^{238}U, is the waste left over from the gaseous diffusion or centrifuge process by which the uranium isotopic enrichment process is carried out. It is, therefore, seen that utilization of the depleted uranium, ^{238}U, as a blanket material can generate a great amount of plutonium for a nuclear energy source in the future.

4. *Coolant Materials*

The thermal energy generated in the fuel elements as a result of nuclear reactions is extracted from the reactor core by means of selected fluids known as coolants. An ideal coolant should have the following qualities: (i) good thermal properties, (ii) low pumping power requirement, (iii) low melting and high boiling points, (iv) thermal stability, (v) irradiation stability, (vi) low neutron absorption and suitable neutron scattering cross-sections, (vii) nontoxic, nonhazardous, and low induced radioactivity, (viii) relative absence of impurities, (ix) low corrosiveness, (x) low viscosity, (xi) compatibility with fuel and structural materials, (xii) easy and safe handling, and (xiii) be abundant and economical. There are gaseous and liquid coolants. The coolants that have been chosen in the reactor systems possess most of the characteristics, but none possesses all of them. For the gas coolants the special requirements list as follows: (i) good chemical and radiochemical stability; (ii) low levels of induced radioactivity with a short half-life of the nuclei generated by neutron-induced reactions; (iii) high density, specific heat, and thermal conductivity; and (iv) good and lasting compatibility with the structural and moderator materials with which they come into contact. For the liquid cooling materials, the following are additionally required: (i) a low melting temperature, a critical temperature as high as possible, and a critical pressure as low as possible; (ii) low viscosity; and (iii) a small neutron capture cross-section. The liquids used for cooling fast reactors must also have a moderating power as weak as possible, to prevent growth of the fraction of low-energy neutrons, which would decrease the efficiency in the generation of fissionable nuclides.

Among the gaseous coolants, particular mention may be made of helium and carbon dioxide. From the standpoint of stability against thermal and irradiation effects, ease of handling, abundance, and economics, the most obvious choice is air as a reactor coolant. It, however, is not a good heat transfer material and requires a high pumping power for a cooling system. At elevated temperatures, the oxygen and nitrogen constituents of the air attack the graphite and structural materials of gas-cooled reactors. The inertness of helium is well known. It possesses good heat transfer properties in comparison with air and carbon dioxide, and needs less pumping power than air and carbon dioxide on account of its low density. Additionally, it has thermal stability, irradiation stability, a very low neutron absorption cross-section, little induced radioactivity, low corrosiveness, suitable compatibility with the fuel and structural materials, and ease of handling. Helium is used as the coolant in high-temperature gas-cooled reactors and advanced gas-cooled reactors. Economically, however, helium is an expensive commodity. It must, therefore, be operated in a closed coolant system or loop. In order to economize on the pumping power, it becomes necessary to deploy a relatively high pressure for the helium cooling system. This introduces another problem in that a requirement arises for a leak-proof piping system and pressurized reactor vessel for a reduction in the leakage and loss of the helium coolant in the reactor system where it is used as a coolant. The use of carbon dioxide as a coolant in Calder Hall-type gas-cooled reactors is well known. It possesses fair heat transfer properties, thermal stability, irradiation stability, low induced radioactivity, low corrosiveness, acceptable compatibility with the fuel, properties as a structural or moderator (graphite) material, ease of handling, and availability. At elevated temperatures this gaseous coolant reacts with the graphite moderator of the reactor system. The reaction has been studied extensively. The high pressure deployed in pumping the coolant can result in a reduction in the CO concentration in the reactor.

The realm of liquid coolants comprises light water, heavy water, and molten metals such as sodium and lithium. Light water holds considerable attraction as a coolant because of its good heat transfer properties, easy and safe handling, abundance and economics. Its neutron absorption cross-section is relatively high, so that a sustained fission reactor with light water as coolant and moderator can be a reality only when an enriched ^{235}U fuel is used. The light water moderated and cooled nuclear power plants are normally situated in locations where large volumes of clear and pure water are available. Heavy water is used as moderator and coolant in heavy water reactors.

The only important difference between heavy and light water with regard to the dual function of moderator and coolant lies in the fact that the thermal neutron absorption cross-section of the former is much less than that of the latter. This permits natural uranium fuel to be the fuel of choice in heavy water reactors. Light or heavy water is deployed in a multitude of applications such as moderator, reflector, coolant or solvent (e.g., to dissolve boric acid (H_3BO_3) in water for chemical shim).

It is known that high power can be secured from relatively small core volumes in fast reactors because there is no moderator and because the liquid metals used as coolants possess excellent heat extraction properties. Some of the primary requirements for a fast reactor coolant are: (i) it must minimize neutron moderation, (ii) it must extract heat adequately from a high power density system, and (iii) it must minimize parasitic neutron absorption. The first of the requirements cited immediately drops water and any organic coolant from the count. The choice goes only to liquid metals among which it is liquid sodium that has drawn the maximum attention as a coolant.

The design concepts of liquid sodium-cooled fast breeder reactor systems are mainly determined by the thermal and nuclear properties of the coolant: good heat transfer, small moderating effects, and low neutron absorption cross-section. The melting point of sodium being 98°C requires preheating of the system for filling or some maintenance operations. The high boiling point (982°C, 1bar) permits high coolant temperatures and very low system pressures (apart from the water-steam circuit) compared with, say, light water reactor systems. Two main design concepts have been deployed up to now: the loop and the pool concepts which are well documented in literature. The compatibility of sodium with structural materials is determined by its purity, temperature and circulation rate. Sodium is quite compatible with stainless steel used at present to make clads and fuel assemblies, and also with oxidic fuel materials (UO_2 and (U, Pu) O_2). The presence of oxygen as an impurity impairs corrosion resistance, so that the contact with oxygen must be kept limited; this necessitates continuous purification of liquid sodium.

5. *Control, Shielding, and Safety System Materials*

A nuclear reactor is always required to be under control and amenable to adjustments within accurate limits in order to have a safe, controlled nuclear (fission or fusion) energy release. The control and safety of nuclear fission reactors are linked with (i) the rate of neutron generation, (ii) the rate of neutron loss by leakage, and (iii) the rate of neutron loss by parasitic absorption in the core. Parasitic absorption refers mainly to fission product generation and buildup (attended by depletion of nuclear fuel). The rates of neutron production and losses are directly related to the influences of reactivity and control of the reactor. The most practical and effective control element is a powerful neutron absorber.

Shielding from the passage and penetration of nuclear radiations to the outside of the reactor is required. It is the shielding materials that reduce all types of radiation to minimum tolerable levels. Neutrons and gamma rays are characterized by their high penetrating power, and they require to be attenuated and absorbed in the shielding. In general, the best shielding for fast and thermal neutrons is provided by light elements or materials of low atomic weight, and that for primary and secondary gamma rays by heavy elements or materials of high atomic weight.

To provide safety systems, primary as well as secondary, in nuclear plants is of major importance. The functions of these systems are (i) to prevent the temperature from shooting up to unacceptably high levels for any reason, (ii) to suppress any contingency of extremely high pressures developing, and (iii) to avert any accidental release of radiation from the reactor core containment. Most hardware of safety systems comprises structural materials (such as carbon steel and stainless steel), pumps, valves, thermal shields, etc.

Over a long period of reactor operation, four important circumstances must be considered in relation to the control elements (rods or plates): (i) reactivity coming down with fuel depletion, (ii) breeding of new fuel through neutron irradiation, (iii) formation and accumulation of fission products, or absorbers, and (iv) increase of burn-up of the control element. These factors need constant adjustment and precise movement of the control elements. The primary requirements for control elements may be listed as (i) very high neutron absorption, (ii) adequate mechanical strength, (iii) light element, hard surface, (iv) high thermal and irradiation stability, (v) good heat transfer characteristics, (vi) high wear and corrosion resistance or hardness and inertness. The control elements are usually in the form of rods and there are three types of control rods in use: (i) shim rods for occasional coarse control and adjustments in neutron density with an intermediate "rod's worth of reactivity," (ii) regulating rods for carrying out fine control and adjustments to maintain the desired power level with a small "rod's worth of reactivity," and (iii) safety rods for shutting down or "scramming" the reactor rapidly in case of emergency. The safety rods have a large "rod's worth of reactivity." In regard to control materials, a representative list of elements may be drawn up as: cadmium, boron, hafnium, and the rare earths. Cadmium is a good neutron absorbent. Its uses as a control material are, however, limited due to its relatively low melting temperature, and to the penetrating γ radiations that are generated during neutron capture. Boron is one of the best neutron absorbents. Its cheapest form is B_4C. This material as such and various products made out of it are accomplished neutron absorber materials. Hafnium possesses a special feature: its higher isotopes formed by neutron capture are found to have neutron-capturing capabilities. Rare earth elements due to their very feeble resistance to corrosion and high propensity to oxidation are used as oxides and dispersions in metallic matrices. Thus, in high-temperature reactors oxides of rare earths are in order, while in other reactor systems dispersions of the rare earth oxides in stainless steel, clad in stainless steel tubes, are preferred. An interesting feature among some rare earths may be accounted for in the present context of neutron absorption. In the case of samarium, for example, the natural nuclide ^{149}Sm makes the greatest contribution to the capture cross-section; when this isotope turns into ^{150}Sm, the cross-section drops significantly, which makes the element a consumable absorbent. The disposition for gadolinium is similar. This is unlike what has been said about hafnium which is non-depleting in character when in use in the duty for neutron absorption.

A nuclear fission reactor is controlled in three most practical ways. These are (i) control rods with B, B_4C (natural or enriched in ^{10}B), Cd, Ag-Cd-In, Hf, Ag-Hf, Ag-Ir-Hf, or rare earth elements as the control material, (ii) chemical shim with boric acid (H_3HO_3) as an absorber or control material, and (iii) a burnable absorber with the oxide of the rare earth elements, Eu_2O_3, Er_2O_3, or Gd_2O_3, as the control material. The control rods carry out the job of primary control, while the chemical shim and the burnable absorber carry out the job of secondary, or supplementary, control. Also, the chemical shim has an interesting characteristic in which solubility of boric acid, dispersed in the reactor core, increases with temperature. In some power reactor systems, a combination of control rods and chemical shim is used, while in others a combination of control rods and burnable absorber is used for reactor control and safety.

The shield design is an integral part of reactor design. Radiation shielding serves three major functions: (i) thermal shielding to protect the pressure vessel, coolant loop, and inner shield from the intense heat generated by the absorption of nuclear radiations, (ii) biological shielding to protect personnel and assure safeguards to public health, and (iii) apparatus and instrument shielding to protect electromagnetic apparatus and electronic instruments that otherwise would not work properly in an intense radiation field. Although the nuclear radiation coming out of a reactor may include α rays, β rays, protons, neutrons, and γ rays, as far as shielding design is concerned only neutrons and γ rays require attention.

The reason is that α rays, β rays and protons, which are charged particles, have very low penetrating power, while neutrons and γ rays, which are neutral, have large penetrating power. The primary requirements for shielding are (i) slowing down fast neutrons, (ii) absorbing slow and thermal neutrons, (iii) attenuating primary γ rays, and (iv) minimizing the production of secondary γ rays. On the basis of their functions, shielding materials used in nuclear reactors can be divided into three principal categories: (i) heavy and moderately heavy elements, (ii) hydrogenous materials, and (iii) light elements.

The heavy and moderately heavy elements are meant to attenuate γ rays and to slow down fast neutrons by inelastic scattering in the high energy region. Among the heavy and moderately heavy elements, lead, bismuth, tantalum, and iron (steels) are often used for attenuating γ rays. Lead is the most common shielding material for γ rays. It is abundant and economical but it has a low melting point. A moderately heavy element that has hitherto been used (after being immersed in water for thermal shielding) is iron in the form of construction steel or stainless steel. The steel can attenuate γ rays and release radiation heat. The water can moderate fast neutrons and remove the radiation heat from the steel plate of the thermal shielding in the reactor core. Ceramics that contain heavy or moderately heavy elements may prove to be useful as shielding materials, and in this context, mention may be made of iron oxides, titanium carbide, silicon carbide, and boron carbide. The primary job of hydrogenous materials, is to moderate and thermalize neutrons by elastic scattering in the intermediate and low energy regions and to absorb slow or thermal neutrons and attenuate γ rays.

The assessment of a hydrogenous material for neutron shielding is based on the following properties: (i) hydrogen content and neutron moderating power, (ii) neutron and γ ray absorption cross-sections, and (iii) abundance and economics. The best overall candidate is light water. A material containing more hydrogen atoms per unit volume than light water is polyethylene, but it is obviously not as abundant and economical as light water. Light water is used not only as an effective neutron shielding material but also for providing an efficient means for removing the heat produced by the absorption of nuclear radiation. Cement concrete containing ceramics, minerals, and water is a general hydrogenous material for shielding of neutrons and γ rays. In addition to the properties required, concrete possesses a high mechanical strength and various adaptabilities and configurations of construction to shield the reactor core. It is used more than any other material for nuclear reactor shielding. Concrete used for radiation shielding can be classified into ordinary concrete and special concrete.

The safety systems of a reactor power plant comprise (i) primary safety systems, and (ii) secondary safety systems. The primary and secondary safety systems may include the emergency core cooling system (ECCS), automatic pressure suppression system (PSS) (or depressurization system), containment isolation system (CIS), containment (or core standby) cooling system, in-core or out-of-reactor instrumentation safety (monitoring) system (ISS), cracking leak control system, and so on. Reactor materials used in the safety systems vary with different components, vessels, valves, pumps, protection apparatus, control equipment, and monitoring instruments involved in the safety analysis and design. The major safety system materials, however, are structural steel, stainless steel, reinforced concrete, light water, borated water, aluminum, and copper.

6. *Summary*

This article is illustrative rather than exhaustive in nature and presents a representative list of materials which are used in the nuclear reactor primary components. The bibliography at the end of the article will be helpful for a comprehensive study of the subject. The article by itself can serve as an introductory text to many other articles appearing in this encyclopedia which have been cited at the end.

Reactors have been classified on the basis of different terms of reference; some of these have been mentioned in the text. Reactor taxonomy can also deal with the kind of moderator, coolant, rated power, etc. For the identification of a reactor system, one often specifies the coolant, the fuel, the moderator and some other features, all defining a type of reactor. Thus there are: (i) LWR (light water reactor)— versions PWR (pressurized water reactor), BWR (boiling water reactor); (ii) HWR (heavy water reactor)—versions PHWR (pressurized heavy water reactor), SGHR (steam-generating heavy water reactor); (iii) GCR (gas-cooled-graphite-moderated reactor) —versions AGR (advanced gas-cooled reactor), HTGR (high temperature gas-cooled reactor), (iv) FBR (fast breeder reactor)—versions LMFBR (liquid metal fast breeder reactor), GCFBR (gas-cooled fast breeder reactor); and (v) MSR (molten salt reactor). Among these, light water reactors for generating electric power contribute the lion's share of nuclear fission energy. Providing a description of this vast array of reactors, with the whole range of materials placed in their respective positions and playing their respective roles, is a very interesting, but arduous task. This purpose can perhaps be best served, if only partially, by picking up just one reactor system as an example. This exercise will demonstrate how the assemblage of a particular set of materials is important for carrying out relevant functions in an operating nuclear reactor.

One could thus consider, a PWR which has the reputation of having the longest reactor years experience in the field of nuclear power generation. The name of the reactor derives from the fact that the primary circuit is maintained at a very high pressure, typically around 15MPa, such that at the maximum prevailing temperature in the system, no boiling can take place. At 15MPa, the boiling temperature of water is about 343°C, well above the maximum temperature of 332°C allowed in the primary circuit water. The reactor system is an avid user of water as coolant and moderator. The reactor vessel is one of the most important components of a PWR in terms of size, weight, cost, and safety significance. It is a massive structure, cylindrical in shape, and containing the reactor core and other related components. The vessel has a removable top head which, when in operation, is held in place by a large number of studs, and is made of an alloy steel of high tensile strength. The base material for the vessel is either carbon steel or a low alloy steel with high toughness.

Typical materials contain 1% Mn, 0.6–0.8% Ni, and 0.5% Mo. All internal surfaces that come in contact with the primary coolant are clad with a layer of stainless steel (or inconel) for minimizing corrosion. The basic fuel component is the fuel rod which consists of a large number of uranium dioxide pellets stacked inside a cladding made of stainless steel or zircaloy. The uranium used is enriched to between 2% and 4% in ^{235}U. A number of assemblies in the core contain control rods. Each control rod contains a neutron-absorbing material (boron carbide or an alloy of 80 Ag-15 In-5 Cd) clad in cold-worked stainless steel (type 304) with stainless steel end pieces welded to the upper and lower ends. Most of the control rods are used to regulate the reactor power level or bring the reactor to a quick shutdown (scram). These are called "shim" safety control rods. In addition to the control rods, a boron solution in the coolant assists in controlling reactivity. It is called "chemical shim."

It may be pointed out that it is difficult to identify any particular line of nuclear energy development as the best available option for all countries. The most suitable option will vary from one country to the other. For example, a country which possesses limited sources of uranium but vast reserves of thorium, and has no established fuel enrichment facility but a well-founded spent-fuel-processing and waste-management program, would pursue natural uranium fueled reactor systems in the first generation and would advance progressively to subsequent generations working on the U-Th fuel cycle.

See also: Fast Reactor Cores; Nuclear Reactors: Shielding Materials; Nuclear Reactors: Coolant Materials; Nuclear Reactors: Moderator and Reflector Materials; Fast Reactor Cores, Ferritic–Martensitic

Steels for; Fast Breeder Reactors: Fuels; Heavy Water Reactor Fuel Design and Performance; Nuclear Reactor Materials: Irradiation Effects; Light Water Reactor Fuel Design and Performance; Mixed Oxide (MOX) Fuel; Nuclear Reactors: Moderator and Reflector Materials; Nuclear Fuel Cycles; Nuclear Reactor Fuel Fabrication (Including Quality Control); Nuclear Reactors: Pressure Vessel Steels; Radiation Effects in Carbon–Carbon Composites; Nuclear Applications: Zirconium Alloys

Bibliography

Cahn R W, Haasen P, Kramer E J 1994 *Material Science and Technology—A Comprehensive Treatment, Vol. 10A. Nuclear Materials, Parts I and II*. VCH, Weiheim, Germany

Gupta C K 1989 In: *Materials in Nuclear Energy Application*, Vols. I and II. CRC Press, Boca Raton

Ma B M 1983 *Nuclear Reactor Materials and Applications*. Van Nostrand Reinhold, New York

Marshall W 1983 *Nuclear Power Technology*. Clarendon Press, Oxford, UK, Vols. 1–3

Rahn F J, Adamantiades A G, Kenton J E, Braun Chain 1984 *A Guide to Nuclear Power Technology—A Source for Decision Making*. Wiley, New York

Ursu I 1985 *Physics and Technology of Nuclear Materials*. Pergamon, Oxford, UK

Walter A E, Reynolds A B 1981 *Fast Breeder Reactors*. Pergamon, Oxford, UK

Wymer R G, Vondra B L 1981 *Light Water Reactor Nuclear Fuel Cycle*. CRC Press, Boca Raton, FL

Nuclear Reactor Materials: Irradiation Effects

P. Rodriguez

The economic viability of nuclear power will increase with an increase in the fuel burn-up (i.e., the percentage of fissile atoms burnt or fissioned before the fuel is removed, also expressed as MWd kg^{-1}, the equivalent energy extracted from 1 kg of fissile material). The burn-up in research reactors is ~ 1 MWd kg^{-1} and that in light water reactors averages about 30–50 MWd kg^{-1}. In the case of fast reactors the target burn-up is 150–200 MWd kg^{-1}. The limitation on the burn-up often comes not from the fuel itself but from the damage caused by the hostile neutron irradiation to which the structural components of the reactor are subjected. The structural components include the thin-walled metallic tubes, called fuel pins, that contain the fuel, the wrapper or can that contains a number of fuel pins as required by the design parameters of the fuel, the vessel in which the core is immersed, and many other components that are contained in it (see *Nuclear Reactor Materials*).

There are two types of primary radiation damage mechanisms, namely atomic displacements and changes at the molecular level. The former is important in metals and the latter in nonmetals; a combination of both mechanisms occurs in semiconductors and insulators. In organic materials, the effects of irradiation include polymerization, cross-linking, and chemical dissociation. Ionizing radiation dissociates water to produce many free radicals including hydrogen, oxygen, and hydrogen peroxide. The focus in this article is on the atomic displacements causing microscopic crystal lattice damage in metals and other crystalline solids.

1. *The Basics of Radiation Damage*

On passage through a crystal lattice, neutrons interact with the lattice atoms and lose some of their energy to these atoms. These processes can be accompanied by damage to the crystal lattice, in general of three types:

(i) production of lattice atoms displaced from their regular lattice positions;
(ii) changes in the chemical composition by capture of neutrons followed by transmutation; and
(iii) excitation of electrons and ionization of atoms (which does not produce permanent damage in metals).

The collisions between the impinging particles and the lattice can be elastic or inelastic. In elastic collisions, the kinetic energy of all the particles is conserved. The kinetic energy, T, transferred from an incident particle of mass m to a lattice atom of mass M is described by

$$T = \frac{4Mm}{(M+m)^2} E \sin^2(\theta/2)$$

where E is the energy of the incident particle and θ is the scattering angle.

The maximum value for T (T_{max}) occurs for a head-on collision when $\theta = 180°$. In inelastic collisions, part of the energy of the incident particle is lost, generally as heat or other forms of radiation. Thus, the energy balance in this case is given by

$$E = T + E' + Q$$

where E is the initial kinetic energy of the particle, T and E' are the kinetic energies of the lattice atom and the particle, respectively, after the collision, and Q is the inelastic loss.

Atoms knocked out of their lattice positions, called *primary knock-on atoms* (PKAs), can have various energies from zero to T_{max}; the energy spectrum of the PKAs depends on the energy and type of incident particles. Differences in the PKA energy spectrum lead to variations in the extent of damage. The formation of a PKA essentially results in a vacancy–interstitial pair (a *Frenkel pair*). The minimum energy required for displacement is called the *threshold energy* (T_d), and its value is between 20 eV and 50 eV for all metals. The lattice atoms can attain this energy in collisions with electrons with energy >1 MeV, neutrons with energy >1 keV, or ions with energy >100 eV. The transfer of energy less than T_d leads to a local increase in the atomic oscillation, i.e., a local rise in temperature that is called the "thermal spike".

The displaced atoms initiate a cascade of displacements that result in the formation of a large number of vacancies and interstitials, as shown schematically in Fig. 1. The higher the energy of the incident particle, the greater is the degree of branching into cascades and subcascades. This leads

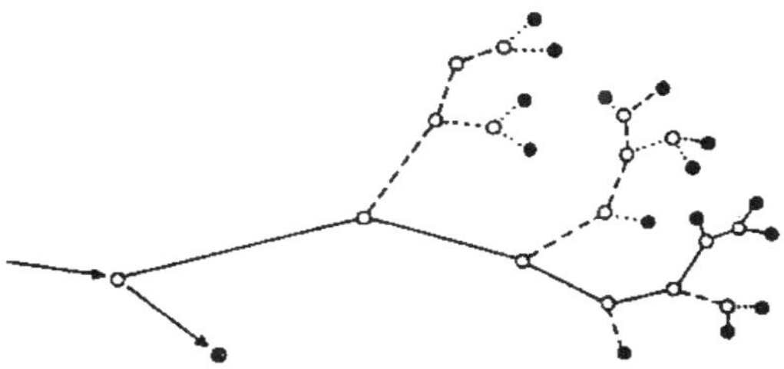

Figure 1
Schematic diagram of a collision cascade caused by particle irradiation on a metallic lattice (○, vacancies; •, interstitials).

to a collective atomic rearrangement in this region called the *displacement spike* with the formation of a random and irregular cluster of vacancies at the center (termed a *depleted zone*), surrounded by an envelope of interstitial atoms. A schematic representation of a displacement spike is shown in Fig. 2. The number of Frenkel pairs, $\nu(T)$, formed by PKAs with energy T is given by the Norgett–Robinson–Torrens (NRT) model (Norgett *et al.* 1975) according to

$$\nu(T) = k\frac{T_{\mathrm{dam}}}{2\langle T_{\mathrm{d}}\rangle}$$

where $T_{\mathrm{dam}} = T - Q$ is that part of the PKA energy corresponding to elastic collision leading to permanent radiation damage, Q is the total energy of inelastic losses in the collision cascade, k is the displacement efficiency dependent on T, and $\langle T_{\mathrm{d}}\rangle$ is the mean of the displacement threshold energy in various crystallographic directions.

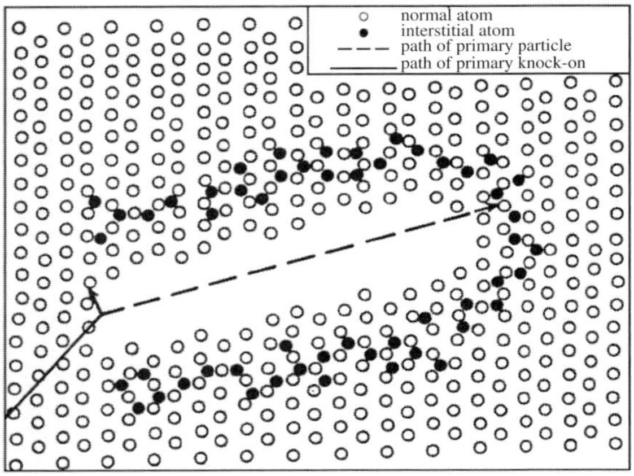

Figure 2
Displacement spike associated with a displacement cascade.

The displacement damage is expressed in terms of *displacements per atom* (dpa) and is calculated from the neutron flux, neutron energy spectrum, and the effective displacement cross-section, which gives the probability of displacement of an atom of the target lattice as a result of collisions with the impinging neutrons (see *Radiation Damage Theory*). The definition and use of "dpa" as the unit of radiation damage started only in the mid-1970s and in literature before then the neutron fluence (n cm^{-2}), the product of the flux (ncm^{-2}s^{-1}) of fast neutrons above a certain energy (usually 0.1 MeV) and irradiation time, was the unit used for damage. The operating conditions in typical nuclear reactors including the flux, fluence, the atomic displacement rate, damage in dpa, along with other parameters of importance are given in Table 1 to enable the reader to appreciate and compare the data in both older and more recent literature. The data given in Table 1 also highlight the fact that there is a quantum increase in radiation damage from research reactors to power reactors and from thermal to fast/fusion reactors.

2. *The Evolution of Defect Structures at Different Irradiation Temperatures*

The defects introduced by irradiation can vary from single vacancies and interstitials of one atom diameter to voids and helium bubbles as large as 3–30 nm in diameter depending on the temperature of irradiation. The primary defects generated are point defects (interstitials and vacancies). At low temperatures, where the mobilities of these defects are limited, they combine to form multiple point defects, e.g., divacancies, trivacancies, etc., or clusters of point defects or complexes with solutes that are a few

Table 1
Typical operating conditions in nuclear reactors.

Parameter	Research reactor	LWR	PHWR	FBR	Fusion reactor (tokamak)
Temperature (K)	<370	~ 575	~ 575	$\sim\cdot 825$	~ 600–1300
Neutron flux[a] (n cm^{-2}s^{-1})	$\sim 10^{12}$	$\sim 10^{14}$	$\sim 10^{14}$	$\sim 5\times 10^{15}$	$\sim 5\times 10^{15}$ (14 MeV)
Neutron fluence (n cm^{-2})	$\sim 10^{18}$	$\sim 10^{22}$	$\sim 10^{22}$	$\sim 3\times 10^{23}$	$\sim 3\times 10^{23}$ (14 MeV)
Atomic displacement rate (dpa s^{-1})	$\sim 10^{-13}$	$\sim 10^{-8}$	$\sim 10^{-8}$	$\sim 10^{-6}$	$\sim 10^{-6}$
Fuel burn-up[b] MWd kg^{-1}	~ 1	30–40	7–10	150–200	
% atoms fissioned	<0.1	~ 3–4	~ 1	~ 15–20	
Radiation damage (dpa)		~ 1	~ 1	~ 100	~ 100
Helium production (appm yr^{-1})	Negligible	~ 1	~ 1	~ 10	

[a]The neutrons in a fusion reactor have energies of ~ 14 MeV. Fission neutrons have an initial energy of ~ 2 MeV; the flux and fluence indicated for fission reactors are for fast neutrons, i.e., with an energy >0.1 MeV. As explained in the text, the dpa unit takes into account the spectrum of neutron energies by integration between the minimum and maximum.
[b]Burn-up is a measure of the extent of fission of the fuel. The unit MWd kg^{-1} measures the fission energy extracted per unit mass of the fissile atom; 10 MWd kg^{-1} is approximately equivalent to 1% fission in the fuel.

atoms in diameter. As the irradiation temperature increases, planar defects like depleted zones that can even collapse into planar vacancy clusters and vacancy loops form. At temperatures between $0.1T_m$ and $0.3T_m$ (where T_m is the melting point in kelvin) the interstitials are mobile and they annihilate by combining with vacancies; annihilation of interstitials at vacancy loops and other sinks also occurs. Interstitials also agglomerate to form clusters of interstitials. At still higher temperatures, 0.3–$0.5T_m$, the vacancies also become mobile, which causes their annihilation by recombination with interstitials; annihilation of both interstitials and vacancies at various sinks like loops, dislocations, etc., also takes place. The interstitial clusters further agglomerate to form interstitial-type dislocation loops. Overall, in this temperature range the concentration of vacancies far exceeds the thermal equilibrium concentration and they coalesce to form voids.

Transmutations change the composition of alloys. Of particular importance is helium generated from (n, α) reactions where the nucleus of the structural atom absorbs a neutron and emits an alpha particle (helium nucleus). Helium is highly insoluble in the metal lattice and precipitates as bubbles that can drastically affect microstructure and properties. Around and above $0.5T_m$ volume diffusion becomes possible and if helium has been released as a transmutation product, helium bubbles form.

A list of the various defect structures (with their sizes) that can be produced in materials at different irradiation temperatures is given in Table 2.

The important phenomena arising due to the defect structures discussed above that determine the dimensional and mechanical stability of core components (fuel element cladding and subassembly wrappers or cans) are irradiation hardening, irradiation embrittlement, void swelling, and irradiation creep.

3. *Macroscopic Effects of Radiation Damage*

3.1 *Irradiation Hardening*

The general effects of neutron irradiation on mechanical behavior are (i) an increase in the yield strength (irradiation hardening); (ii) an increase in the ultimate tensile strength, which is less than the increase in yield strength; (iii) a decrease in the rate of work hardening; and (iv) a reduction in the uniform and total elongations. These effects are illustrated for a typical b.c.c. alloy such as mild steel (which shows a yield point) (Fig. 3) and for a typical f.c.c. metal such as polycrystalline copper (Fig. 4). An interesting aspect is that at very high doses a yield point appears in the stress–strain curve of copper. This implies that

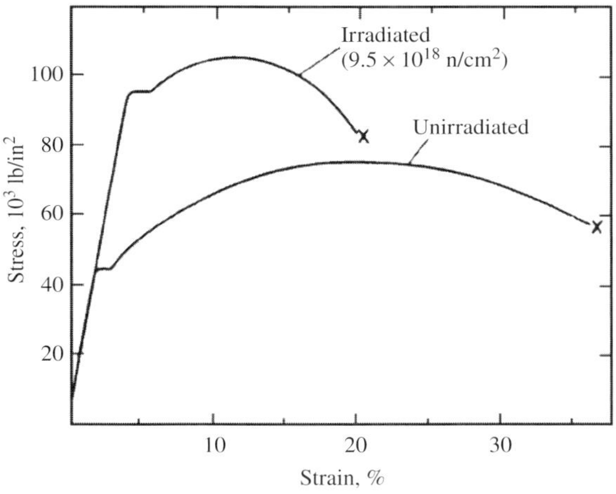

Figure 3
Change in room temperature properties of mild steel produced by neutron irradiation (Little 1976).

Table 2
Summary of defects introduced into the lattice structure during irradiation at different temperatures.

Temperature (T/T_m)	Defect	Size
0	Point defects: vacancies and interstitials	One atomic diameter
0.1	Multiple point defects, clusters of point defects, complexes of vacancies, and interstitials with solutes	A few atomic diameters
↓	Vacancy clusters, loops, and depleted zones	~100 nm length, ~20 nm width
0.3	Interstitial loops	>7 nm
↓	Rafts (agglomerates of clusters and small loops)	6–10 nm thick, 100–200 nm in length and width
	Voids	10–60 nm
0.5	Helium bubbles, transmutation of atoms (produced at all temperatures but agglomerate at $T/T_m > 0.5$)	3–30 nm

the defect structures produced by irradiation cause locking of dislocations by forming atmospheres around them. In the cascade-induced source-hardening model (see Radiation Damage Theory) interstitial loops are postulated to form the atmospheres around dislocation. Linga Murthy (1999b) analyzed radiation hardening in iron and steels also in terms of source hardening and friction hardening and concluded that there is a contribution to both from the irradiation-induced structures.

Rodriguez *et al.* (1984) have analyzed irradiation hardening in terms of thermal and athermal components of flow stress. The barriers to dislocation motion may be divided into two categories, short range (controlling the thermal component of flow stress) and long range (controlling the obstacle athermal component) depending on the distance over which the force exerted by the obstacle extends. Irradiation appears to enhance both short- and long-range dislocation barriers.

3.2 Irradiation Embrittlement

Irradiation embrittlement is a major factor affecting the performance of many reactor components. The loss of ductility depends on irradiation temperature, preirradiation microstructure, and fluence. At low temperatures, the embrittlement is a consequence of irradiation hardening by defects, clustering of solute atoms, and consequent short- and long-range dislocation barrier effects arising from them. These effects manifest themselves as an increase in the ductile–brittle transition temperature in b.c.c. alloys, reduced uniform and total elongations, and a reduced capacity to work harden. At high temperatures ($>0.5T_m$) embrittlement due to helium generated by (n, α) reactions is predominant in f.c.c. alloys.

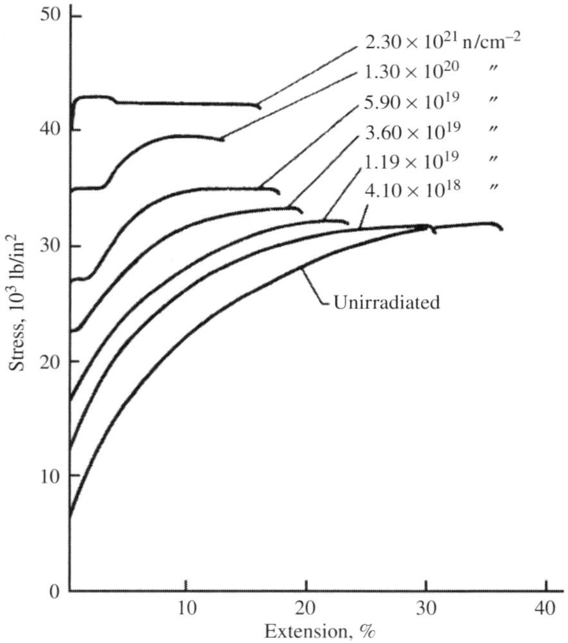

Figure 4
Stress–strain curves for polycrystalline copper tested at 20°C after irradiation at the doses indicated (Sheely 1967).

(a) Irradiation embrittlement in b.c.c. metals and alloys. These alloys undergo a transition from ductile to brittle behavior as the temperature is decreased; below the transition temperature, called the nil-ductility transition temperature (NDTT), fracture occurs by cleavage and the energy required for fracturing a specimen in an impact test is negligible. Irradiation has two effects on b.c.c. metals and alloys: (i) it increases the NDTT and (ii) it brings down the fracture energy of the material even in the ductile region (so-called upper shelf). Irradiation embrittlement and the increase in NDTT are serious engineering problems in nuclear reactors since the materials chosen for pressure vessels and containment vessels of thermal reactor systems are ferritic steels. By code requirements, the operation of the reactor is permissible only if the NDTT of the component is at least ~37 K below its operating temperature.

The ductile–brittle transition in b.c.c. metals and alloys is associated with the strong temperature dependence of the yield strength in these materials. According to the Ludwik–Davidenkov criterion for NDTT illustrated in Fig. 5, the ductile-to-brittle transition occurs at a temperature where the yield

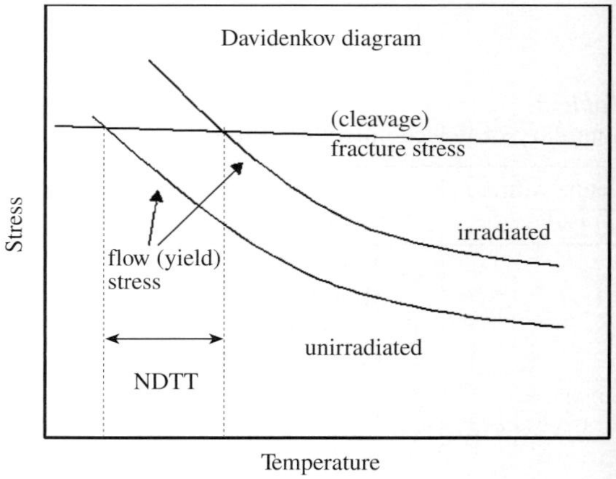

Figure 5
Schematic illustration of the Ludwik–Davidenkov criterion for NDTT and its shift with irradiation.

stress, due to its strong increase with decrease in temperature, reaches the critical stress for cleavage fracture, which is much less temperature dependent. Furthermore, the cleavage fracture stress is relatively independent of microstructural features such as defect clusters introduced by irradiation. The yield stress, however, increases drastically on irradiation, thereby shifting the NDTT to higher values as shown in Fig. 5.

The influence of neutron irradiation on the NDTT of ASTM A 203 grade D steel (a low-alloy ferritic steel containing about 3.5 wt.% nickel, which is the end shield material in the first three Indian pressurized heavy water reactors (PHWRs)) is shown in Fig. 6. Results (Baldev Raj *et al.* 1983) have been obtained from Charpy-V half-size specimens of the steel. Using the criterion of 30 J energy to determine the NDTT, in the unirradiated condition (curve A) the steel has a NDTT of 182 K and an upper shelf energy of about 100 J. Irradiation to a fluence level of 3.5×10^{19} ncm^{-2} (curve B) increases the NDTT to around 423 K while decreasing the upper shelf energy to as low as 40 J. At a lower fluence of 5×10^{18} n cm^{-2} (curve C), the NDTT is shifted to only about 263 K and the upper shelf energy to about 50 J.

Curve D shows the impact energy transition curve for material that has been annealed at 573 K for 15 days following irradiation to a fluence level of 3.5×10^{19} n cm^{-2}, i.e., it shows the recovery in properties that can be obtained by annealing at higher temperatures in specimens that have been irradiated to a brittle condition corresponding to curve B. It can be seen that the NDTT has been decreased to 263 K and the upper shelf energy has increased to about 80 J. This effect of annealing at higher temperatures offers the possibility of using *in situ* annealing of reactor components to remove at least a portion of the

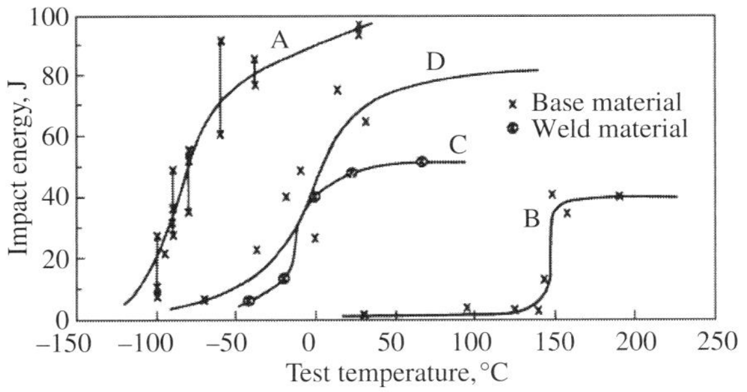

Figure 6
Impact energy vs. temperature curves for ASTM A 203 grade D steel: A = unirradiated; B = irradiated at a fluence of 3.5×10^{19} ncm^{-2}; C = irradiated at a fluence of 5×10^{18} ncm^{-2}; D = annealed at 300°C for 15 days after irradiation at a fluence of 3.5×10^{19} ncm^{-2} (Baldev Raj *et al.* 1983).

irradiation damage and to reduce the NDTT to acceptable levels in those cases where the shift has become excessive. As a permanent solution in subsequent PHWRs, the end shield material has been changed to 304SS, an austenitic steel not prone to a ductile–brittle transition at low temperatures.

Residual elements in ferritic steels are found to have a very significant effect on the extent of irradiation embrittlement. Two elements that have the most deleterious effect are phosphorus and copper. Studies by Brenner *et al.* (1978) using a field ion microscope to determine the structure at an atomic level have shown that copper and phosphorus can stabilize a high concentration of very small microvoids (8×10^{21} voids m^{-3} with a mean diameter of 0.6 nm). These stabilized microvoids would produce the increased flow stress and the associated shift in the NDTT. A prominent example of the successful interaction between basic and applied research in the field of nuclear materials is the development of improved pressure vessel steels for light water reactors. Combining the results of measurements using advanced techniques such as field ion microscopy, high-resolution microscopy, and small-angle neutron scattering, along with mechanical tests on systematically modified alloys, has led to new pressure vessel steels with reduced copper content. These modern steels exhibit a high resistance to radiation embrittlement (see *Nuclear Reactors: Pressure Vessel Steels*).

(*b*) *Irradiation embrittlement in normally ductile metals and alloys.* The effects of irradiation on the shape of the tensile stress–strain curve for copper (Fig. 4) appear to be broadly representative of ductile metals of all the three crystal structures (f.c.c, b.c.c., and h.c.p.) and of pure metals and complex alloys alike. For example, it may be seen that even in a material like mild steel that shows the ductile–brittle transition, the effect on the stress–strain curve at room temperature at which mild steel is ductile (Fig. 3) is quite similar to the effect in f.c.c. copper. The reductions in ductility caused by irradiation in normally ductile materials have been attributed to the phenomenon of dislocation channeling. The radiation-produced defect clusters are apparently removed when slip dislocations are set in motion along the slip plane in the early stages of deformation. The slip process then continues by the motion of subsequent slip dislocations along the same channel in preference to a new band in an adjacent region of the crystal. As a result, deformation becomes considerably more localized than is the case for unirradiated metals. Owing to dislocation channeling, deformation takes place in the glide bands in a catastrophic manner and leads to plastic instability localized in the channels or slip bands. The phenomenon of dislocation channeling is also able to explain the reduction in work hardening rate owing to irradiation in ductile metals. The removal of irradiation-produced obstacles by the moving dislocations will result in a reduction in the enhancement of flow stress with strain and consequently in a reduced work hardening rate. On a macroscopic scale, plastic instability and necking always set in when in a tensile test, within a small strain interval, the increase in stress due to work hardening is too small to compensate for the reduction in cross-section so that the applied load remains constant or even decreases. A general reduction in work hardening will always therefore lead to premature necking and lower values of uniform and total elongations.

(*c*) *Helium embrittlement.* Helium has a marked effect on the elevated temperature fracture behavior of metals and alloys above about $0.4T_m$. The problem of helium embrittlement has come to the fore with the advent of fast breeder reactors, which operate at high temperatures where grain boundary sliding becomes predominant in the structural materials. The most important materials for applications in fast breeder reactors are austenitic stainless steels, and most of the data on the phenomenon of helium embrittlement has been generated for this class of materials. The major source of helium in stainless steels is nickel; helium is also generated from the $^{10}B(n, \alpha)^7Li$ reaction. It is the segregation of helium to the grain boundaries that is the deciding factor in promoting helium embrittlement. Above about 900 K in stainless steels, grain boundary sliding becomes prominent and intergranular fracture occurs by wedge cracking at grain boundary junctions or by the fracture of continuous brittle grain boundary precipitates or by cavitation. Helium produced by the various transmutation reactions in the reactor environment causes intergranular embrittlement by stabilizing the cavities and the wedge cracks: the growth and coalescence of helium bubbles have also been suggested as a possible mechanism for helium embrittlement (see *Fast Reactor Cores, Austenitic Steels for*).

(*d*) *Embrittlement due to other transmutation products.* The mechanical properties of structural materials can be altered by transmutation products other than helium. One example is the case of aluminum (Farrell and King 1973) in which neutron irradiation damage when irradiated to high fluences at about 323 K is found to consist of a fine precipitate of transmutation-produced silicon and the associated dislocation network and very few voids. These defects are responsible for the large increase in the mechanical strength of commercial purity aluminum as shown in Fig. 7. The embrittling effect has been attributed to the production of silicon by the following reaction:

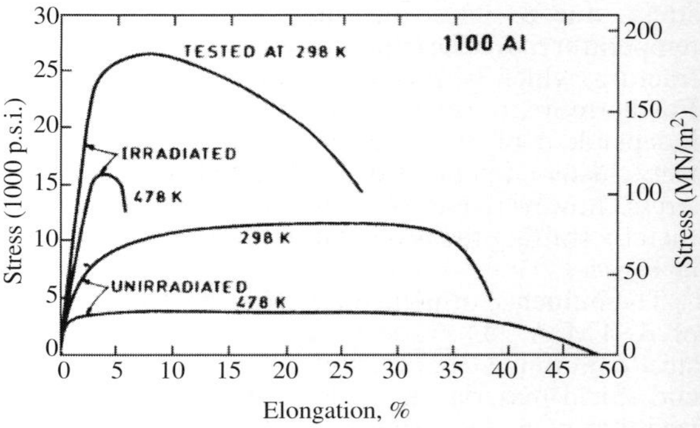

Figure 7
Effect of neutron irradiation (fluence of 2.7×10^{22} ncm^{-2} at 323 K) on the tensile curves of 1100 Al at 298 and 478 K (Farrel and King 1973).

$$^{27}Al(n, \gamma) \rightarrow {}^{28}Al \rightarrow {}^{28}Si + \beta^-$$

Total elongation, which is of the order of 38% in the unirradiated condition at 298 K, decreases to about 27% after irradiation. More important is the reduction in the uniform elongation, which decreases from about 32% to about 7%. It is also seen that the reductions in ductility at 478 K are more severe, with the total elongation decreasing from ~48% to about 6% after irradiation and the uniform elongation becoming as low as 3%. This embrittlement at 478 K has been shown to be due to helium produced by (n, α) reactions of lithium and boron present as impurities as well as the Al(n, α) reaction.

3.3 Void Swelling

Swelling is the volumetric expansion of the structural materials caused by the formation of voids that occurs when these are subjected to fast neutron irradiation. Voids nucleate and grow when the irradiation temperature is in the range ~ 0.3–$0.5T_m$. For stainless steels, these are the temperatures at which core materials in commercial fast breeder reactors operate. In this temperature range irradiation leads to a supersaturation of vacancies, caused either by a bias of dislocations as sinks for interstitials or by a production bias for vacancy clusters surrounded by self-interstitial atoms (see Radiation Damage Theory). Below $0.3T_m$, vacancies are not mobile enough to diffuse and coalesce to voids. Above $0.5T_m$, the thermal vacancy concentration exceeds that introduced by irradiation and there is no supersaturation of vacancies to collapse to voids.

(*a*) *Effect of alloy composition and microstructure on void swelling: austenitic stainless steels.* In austenitic stainless steels, extensive studies have shown that void swelling varies systematically with the chromium and nickel contents. The swelling reduces with increasing nickel content and increases with increasing chromium. The resistance to void swelling can, therefore, be varied considerably by modifying alloy composition with respect to major elements. Further, it is found that the commercial alloys swell less than the simple ternary alloys with the same chromium and nickel contents, suggesting that the minor element additions contribute to the swelling resistance. That impurities or minor elements exert considerable influence on swelling resistance has also been demonstrated by comparing swelling behavior of pure metals with metals containing impurities. Minor elements that have been found particularly effective in increasing the resistance of stainless steels to swelling are titanium and silicon. Among the impurity elements, phosphorus is very effective in providing swelling resistance, particularly in combination with titanium.

Improved resistance to void swelling can also be achieved through certain microstructural modifications. A grain boundary, being a good sink for point defects, depresses the steady-state concentration of radiation-produced vacancies. As a result, voids tend not to form in the immediate vicinity of a boundary. At sufficiently fine grain sizes, void swelling can be inhibited. Although dislocations are slightly more effective in trapping interstitials than vacancies, they are sufficiently good vacancy sinks that the very high dislocation densities in a cold-worked material improve resistance to swelling. Cold-worked stainless steels have been employed in various reactors because of their better resistance to swelling compared to solution-annealed steels. For a detailed discussion on the effects of alloying elements and thermomechanical treatments on void swelling in austenitic steels and of mechanisms thereof, see *Fast Reactor Cores, Austenitic Steels for*.

(*b*) *Effect of alloy composition and microstructure on void swelling: ferritic steels.* In ferritic steels, atomic transport by diffusion is much more rapid than in austenitic materials because of their more open structure. The recombination rates of vacancy–interstitial pairs are much faster and void swelling is much less of a problem than in austenitic materials. The 8–12% chromium martensitic steels have been investigated for use as fuel cladding materials in fast reactors. Typical of these is HT-9 (12Cr–1Mo alloy). These alloys possess a high dislocation density in a fine lath martensite structure, which may also contain fine carbide precipitates. The large density of traps for lattice defects characteristic of these materials also contributes to their swelling resistance. However, significant problems in the use of martensitic steels relate to their poor weldability and tendency to irradiation embrittlement. The increase in NDTT with irradiation is a major consideration in their application. The use of 8–9% Cr–1–2% Mo alloys and reduced-activation versions containing 1–2% tungsten are also being explored. Studies show that 9Cr alloys such as EM10 and those without nickel exhibit a lower shift in NDTT than 12Cr alloys (see *Fast Reactor Cores, Ferritic–Martensitic Steels for*; *Fusion Reactors: Low-activation Structural Materials*).

3.4 Irradiation Creep

Creep is the time-dependent deformation of a material under stress. Knowledge of the creep mechanisms in the absence of irradiation is essential for an understanding of the influence of irradiation on creep. There are at least six distinguishable and independent ways in which a polycrystal can deform and yet remain crystalline. All of them involve the motion through grains, or round their boundaries, of lattice defects. The glide motion of dislocations is the predominant mode of deformation in normal crystals. However, at higher temperatures, where glide becomes quite fast, the climb of dislocations over localized obstacles becomes rate controlling. The third and fourth mechanisms, which occur at very high temperatures and low stresses, are Nabarro–Herring creep and Coble creep, which are two independent kinds of "diffusional flow" resulting from the flow of point defects through the grains and around grain boundaries, respectively. The sliding of grain boundaries is the fifth possible mechanism and may also involve other defects at the boundaries. Finally, twinning provides a sixth mechanism; however, unlike the others it can supply only a limited amount of deformation and is not important under creep conditions.

At low temperatures, below $\sim 0.3T_m$ where dislocation glide is the rate-controlling mechanism, there will be a reduction in creep rate under irradiation arising from irradiation hardening due to the defect clusters. At higher temperatures irradiation will increase the creep rate since climb-controlled glide is the creep mechanism and this can be assisted by the excess concentration of vacancies. One effect of the increased concentration of vacancies due to irradiation is to reduce the temperature range at which creep due to dislocation climb can occur. At very high temperatures, $> 0.6T_m$ where the concentration of vacancies produced by irradiation is small compared to the thermal equilibrium concentration of vacancies themselves, irradiation will not have any significant effect on creep and thermal creep predominates.

There can be new mechanisms of irradiation creep, for example, the mechanisms of stress-induced preferred nucleation (SIPN) or alignment of dislocation loops and stress-induced preferential absorption

(SIPA) of point defects at dislocations. These phenomena have as common features the formation or growth of loops and absorption of defects at the existing dislocations depending on their orientation with respect to the external stress, which causes a macroscopic creep strain (Fig. 8).

3.5 Irradiation Growth

Irradiation growth refers to the dimensional changes at constant volume of an unstressed material under irradiation. Irradiation growth occurs in both single crystals and textured polycrystals of anisotropic metals and alloys by the unequal distribution of interstitial and vacancy loops on different crystallographic planes. Irradiation growth occurs at relatively low temperatures, in the vicinity of $0.2T_m$, and is accompanied by irradiation hardening and embrittlement. Texture, grain size, dislocation density, and alloy content all affect the growth rate, which also increases with temperature. Irradiation growth in zirconium alloys is significant for the design and operation of PHWRs. The texture present in zircaloy cladding tubes causes the axial length to increase and the cladding diameter and thickness to diminish. This has an impact on design and safety, the growth distortions setting operational burn-up limits. Irradiation growth is influenced by microstructural variables such as amount of cold work, residual stresses, alloying additions, and texture, as well as radiation variables such as flux and temperature.

When uranium is used in the metallic form as a fuel, because of its anisotropic orthorhombic crystal structure, it undergoes irradiation growth. Suitable alloying and quenching from the beta phase to produce a fine grain size and random orientation are carried out to produce "adjusted uranium" resistant to irradiation growth.

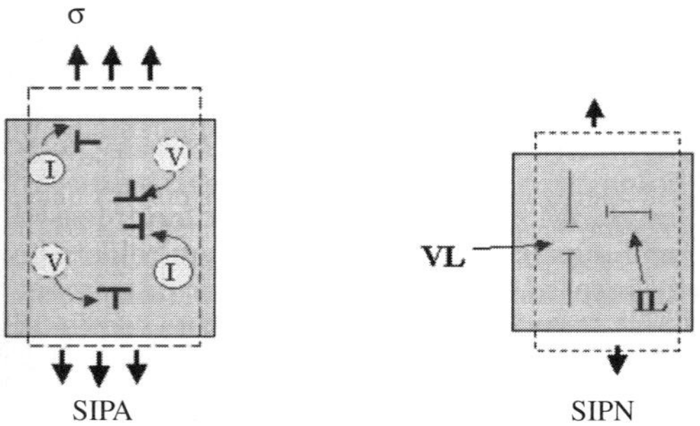

Figure 8
Mechanisms of irradiation creep. Left: stress-induced preferential absorption (SIPA) of vacancies (V) and interstitials (I) to suitably oriented dislocations. Right: stress-induced preferential nucleation (SIPN): V loops parallel to stress and I loops perpendicular to stress axis.

Graphites used as moderators and for shielding are also anisotropic and suffer irradiation growth (see *Radiation Effects in Carbon–Carbon Composites*; *Nuclear Reactors: Moderator and Reflector Materials*; *Nuclear Reactors: Shielding Materials*).

4. Phase Stability under Irradiation

4.1 Radiation-induced Segregation and Precipitation

Point defects migrate to defect sinks such as grain boundaries, other interfaces, and surfaces during irradiation. The persistent flux of point defects to the sinks gives rise to the phenomenon of radiation-induced segregation (RIS), which is the nonequilibrium segregation caused by the preferential coupling of solute atoms to defect fluxes. For a binary alloy AB, the vacancy flux, J_V, to the sink essentially corresponds to the flux of solute atoms, J_A and J_B, away from the sink with $J_V = J_A + J_B$. If both the alloy components A and B diffuse at the same rate, there will be no net enrichment or depletion of any element at the sinks. However, if $J_A > J_B$, A gets depleted at the sink. This phenomenon is also referred to as the *inverse Kirkendall effect*. A similar effect will also be caused by interstitial fluxes; the only difference is that the interstitial flux and the corresponding atomic fluxes will be in the same direction. The alloy component that gets preferentially coupled to interstitials gets enriched at the sinks.

In austenitic stainless steels, austenite grains can transform entirely into ferrite during neutron irradiation, because localized regions get enriched in either chromium or nickel, stabilizing either ferrite or austenite at the respective locations. The segregation arises because elemental diffusivities are different, with $D_{cr} > D_{Fe} > D_{Ni}$. In Fe–Cr–Mn steels it is iron that segregates via the inverse Kirkendall mechanism, manganese being the fastest diffusing element. Maximum RIS is expected at intermediate temperatures where an excess concentration of vacancies and interstitials are produced and they are able to diffuse relatively long distances. At low temperatures, the mobility of the defects is less and at high temperatures the thermal equilibrium concentration of vacancies predominates and rules out any radiation-induced effects. A full understanding of many of these processes is still a subject of active research. However, three factors have been identified that could contribute to these effects: (i) the relative atomic size differences between solvent and solute atoms, (ii) the binding of minor alloying elements with interstitials and vacancies, and (iii) the inverse Kirkendall effect in concentrated alloys with unequal diffusion rates of different alloy components, resulting in segregation.

To get an idea of the size effect in an iron (austenitic) matrix the percentage size differences for different solutes are: aluminum (+12.8), manganese (+4.9), molybdenum (+27.5), niobium (+17.6), nickel (+4.6), phosphorus (−13.2), silicon (−7.9), titanium (+27.5), vanadium (+10.5), and tungsten (+33.0). The elements with a large positive mismatch are relatively more difficult to be transported while elements like silicon and phosphorus with negative misfit volumes are highly mobile and generally migrate towards free surfaces such as grain boundaries and cavities.

In addition to the inverse Kirkendall effect, segregation can also result from the formation of bound defect–solute complexes. These complexes diffuse down their own concentration gradients and tend to sweep solute towards sinks in initially homogeneous alloys providing a new mechanism of solute segregation referred to as "solute drag". All sinks for point defect annihilation will become selectively enriched with elements that have significant binding energies with point defects. Defect–solute complexes are defined as mobile when the following condition is satisfied:

$$E_{\mathrm{B}}^{\mathrm{d-s}} + E_{\mathrm{m}}^{\mathrm{d}} > E_{\mathrm{m}}^{\mathrm{d-s}}$$

i.e., when the sum of the defect–solute binding energy, $E_{\mathrm{B}}^{\mathrm{d-s}}$, and the defect migration energy, $E_{\mathrm{m}}^{\mathrm{d}}$, exceeds the migration energy, $E_{\mathrm{m}}^{\mathrm{d-s}}$, of the complex. Complexes that satisfy this condition may be regarded as distinct entities that can diffuse down their own concentration gradients.

4.2 Radiation-induced Precipitation

Another way in which point defects alter phase stability is through their influence in the nucleation of precipitates. Vacancies and interstitials help to accommodate the strain produced by the formation of a precipitate nucleus. Irradiation-induced chemical disorder plays a major role in precipitate stability; the disordering of small ordered precipitates leads to their dissolution.

One of the most striking manifestations of RIS is the formation and growth of precipitates. Phases that form during irradiation have been classified as (i) radiation induced, (ii) radiation enhanced, and (iii) radiation modified. Radiation-induced phases are those that form as a result of RIS and that would not normally form in the alloy under thermal conditions. In 316 stainless steel, the formation of $Ni_3Si(\gamma')$, where silicon may be replaced by aluminum, niobium, or titanium, the G-phases $M_6Ni_{16}Si_7$ (e.g., M = Cr or Mn), and phosphides are examples of radiation-induced precipitation. The formation of phases rich in nickel and silicon affect not only the mechanical properties of irradiated stainless steels, but also appears to be the major determinant of when the onset of void swelling and accelerated irradiation creep occurs through the removal of nickel and silicon from the matrix.

Radiation-enhanced phases are those that form in the alloy under normal thermal conditions, but during irradiation they are produced more rapidly, more abundantly, and at lower temperatures. The opposite effect, of retardation of a phase, is also possible by radiation; M_6C is generally enhanced and $M_{23}C_6$ is retarded due to irradiation (M = Cr, Fe, Mo, W). Radiation-modified phases are those that form in the alloy under thermal conditions but in which the composition is altered as a result of RIS.

4.3 Radiation-induced Amorphization and Disordering

The defects produced during irradiation give rise to several interesting phenomena that are important from the point of view of microstructure. The increase in point defect concentration can affect the phase stability of materials, the increase in the internal energy due to the accumulation of point defects gives rise to the destabilization of the crystalline phase resulting in amorphization of certain intermetallics. A critical defect density has to build up in the material for the crystalline-to-amorphous transformation to occur. Studies of several intermetallics have shown the existence of a critical dose of irradiation for amorphization. Figure 9 shows a sequence of electron micrographs of Al_6Mn phase irradiated with 30 keV helium ions Nair *et al.* (1993). No amorphization is seen up to a dose of 10^{16} ions cm^{-2}. The onset of amorphization occurs around a dose of 2×10^{16} ions cm^{-2} as evidenced by the faint diffuse ring in the electron diffraction pattern. No point

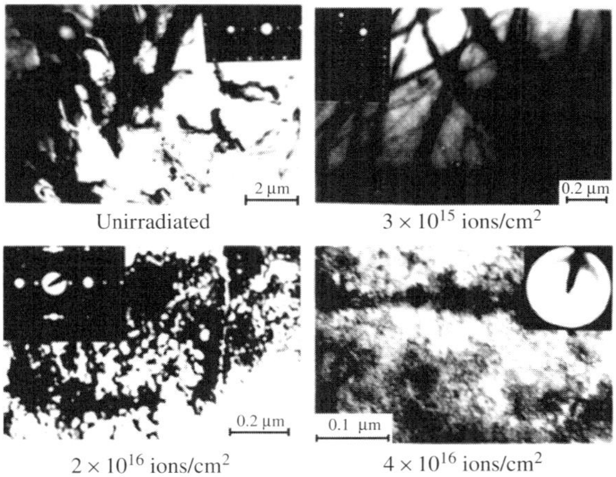

Figure 9
Radiation-induced amorphization of Al_6Mn irradiated with 30 keV helium ions: (a) unirradiated, (b)–(d) irradiated with 30 kV argon ions at various doses indicated. SAD patterns along with the TEM images show various stages of amorphization (Nair *et al.* 1993).

defect clusters like vacancy or interstitial loops are formed in the samples prior to amorphization. This suggests low point defect mobility, which is an essential prerequisite for the buildup of the high defect density needed for the crystalline-to-amorphous transformation. Alloys that show amorphization on irradiation are all ordered intermetallic compounds that retain their ordered structure up to their melting point. In ordered alloys that disorder below their melting temperature, radiation induces disordering.

5. *Practical Consequences of Irradiation Effects*

The operating temperatures of most reactor structural materials range from ambient to about 925 K (corresponding to T/T_m of between 0.15 and 0.55 for steels). The macroscopic radiation damage effects and their implications are given in Table 3.

It can be seen from Table 3 that many of the effects occur concurrently and therefore may also interact among themselves, which complicates prediction of material behavior. The radiation embrittlement of pressure vessel steels in light water reactors (LWRs) is a well-known practical problem. Two other examples of engineering consequences of irradiation effects on the operation of nuclear reactors are discussed, one from PHWRs and the other from fast breeder reactors (FBRs).

5.1 PHWRs

A detailed view of the core of a PHWR is given in Fig. 10. The fuel bundles are placed within pressure tubes surrounded by an annular gap formed by a larger "calandria" tube, which surrounds the pressure tube. Heavy water, which is the moderator, is present outside the calandria tube. The calandria and fuel tubes are made of zirconium alloys. The space between the calandria tubes and fuel tubes is maintained by "garter" springs and is filled with N_2 or CO_2 gas. Several hundred such fuel channels are arranged horizontally in the reactor vessel called a "calandria." One end of the fuel channel tubes is fixed rigidly to the end shields while the other end is free to move on bearings, thus accommodating axial thermal expansion. Irradiation creep and growth of pressure tubes result in the following effects in PHWRs.

- Diametral strain of pressure tubes occurs from irradiation-enhanced creep due to hoop stress in the tube. This creep leads to an increase in the clearance between the fuel bundles and the pressure tube causing an increase in the bypass flow of coolant, affecting proper cooling of the fuel. Ductility limits of the material may also be approached owing to continued creep strain. A maximum limit of 2% on diametral creep strain is imposed on the pressure tubes of zircaloy-2.
- Axial elongation of pressure tubes exhausts the available bearing axial travel and increases the feeder stresses owing to axial movement of the feeder/coolant channel connection.

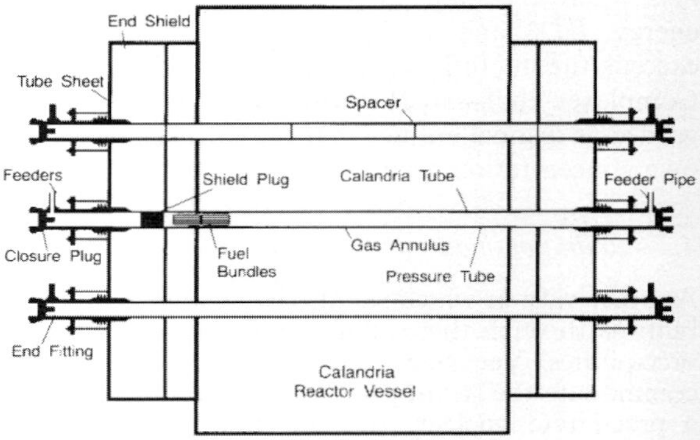

Figure 10
PHWR core showing the calandria and pressure tubes.

Table 3
Macroscopic radiation damage effects.

Effect	Temperature range	Important for
Irradiation-induced segregation and changes in precipitate structure	$T > 0.2T_m$	Corrosion, weldability, and indirect influence on all following effects
Low-temperature embrittlement, increase of NDTT	$T < 0.3T_m$	B.c.c. steels and refractory alloys for pressure vessels
Irradiation creep	$0.2T_m < T < 0.4T_m$	Most in-core structural materials
Irradiation growth	$0.1T_m < T < 0.3T_m$	Anisotropic materials (zirconium alloys, metallic uranium fuel and graphites)
Void swelling	$0.3T_m < T < 0.5T_m$	Austenitic steels for core components in LMFBRs and first wall components in fusion reactors
Helium high-temperature embrittlement	$T > 0.45T_m$	First wall structures in fusion reactors, core components of fast breeder reactors

- Owing to the weight of fuel bundles, coolant, and self-weight, the pressure tubes sag (creep sag). Diametral creep, along with bending creep may cause the pressure tubes to come in contact with cooler calandria tubes. This increases the heat loss to the moderator. Owing to the temperature gradient, hydrogen, which is present in the pressure tube (because of the reaction between zirconium and water), moves towards cooler regions and precipitates there owing to the reduced solubility of hydrogen at low temperatures. Hydride precipitation can lead to brittle failure in the pressure tube.

These effects, particularly the creep sag of the pressure tube and the contact between the pressure tube and the calandria tube, along with high hydrogen pick up, may become unacceptable and a particular channel may have to be "quarantined" (i.e., not used with fuels anymore). When this happens in a significant number of channels, "en masse" retubing of channels becomes a necessity in PHWRs. With zircalloy-2 as the pressure tube material, this retubing becomes necessary after 7–8 years of full-power operation. With the use of cold-worked Zr–2.5 Nb alloy (see *Nuclear Applications: Zirconium Alloys*), the time for retubing may be extended to 15–20 years at full-power operation; i.e., a PHWR still requires one retubing in its lifetime.

5.2 FBRs

In FBRs, void swelling and irradiation creep are important in determining the burn-up limits for the fuel element subassemblies. Void swelling produces increases in the length and diameter of the wrappers and causes the subassemblies to bow owing to gradients in neutron flux and temperature (Fig. 11) and to interact with their neighbors or core restraint structure. Bowing of subassemblies results from differential void swelling on opposite faces of the wrapper and assumes importance for the subassemblies at the edge of the core. This bowing can give rise to difficulties in the unloading/loading of fuel such as unacceptable increase in force required for withdrawal of the assembly. The best solution to this problem obviously lies in the use of materials resistant to void swelling. Further, irradiation creep leads to wrapper dilation under the influence of the internal coolant pressure. The combination of swelling and creep can lead to outward bulging of the wrapper faces, again with the possible interference between adjacent subassemblies. The swelling in the cladding can also limit the achievable burn-up by causing the pins to extend axially to take up the available space within the wrapper and then to buckle, and by increasing the diameters of the pins. These changes in dimensions will not directly produce failure but may constrict coolant flow, resulting in overheating; failure may then ensue by one of several mechanisms, including creep of the cladding under the influence of the fission gas pressure or fuel–clad mechanical interaction and irradiation-induced exhaustion of high-temperature ductility. The importance of helium embrittlement in limiting the performance of fuel pins has already been pointed out. Further, gradients in flux or temperature can cause uneven dimensional changes in fuel pins in the same bundle, which can lead to mechanical problems in fuel handling.

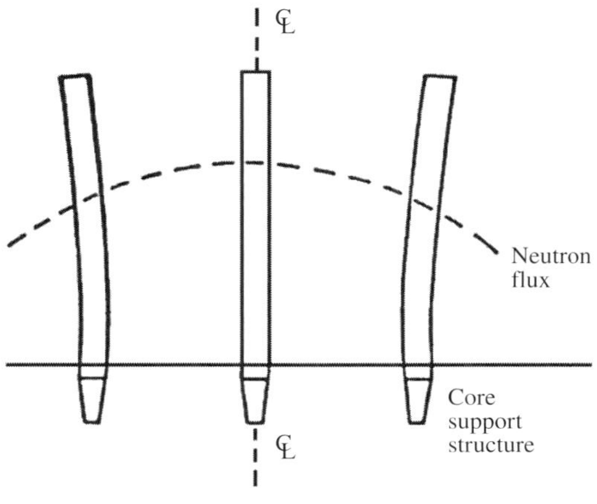

Figure 11
Subassembly bowing in an unrestrained FBR.

These considerations indicate the desirability of selecting or developing wrapper materials that exhibit consistently low void swelling and irradiation creep rates. Furthermore, the fuel pin cladding must retain sufficient in-reactor creep strength and ductility to avoid premature failure. Compatibility of wrapper and cladding tube properties with respect to void swelling and irradiation creep is also important for avoiding interaction between the cladding tubes and wrapper. To accommodate distortions caused by irradiation-induced swelling and creep, additional space or thick ducts and cladding (to reduce stresses) must be provided. This accommodation decreases the fuel volume fraction and degrades the breeding performance. It has already been pointed out that ferritic–martensitic steels, which have superior performance with respect to void swelling, suffer from poorer creep strength than austenitic steels and they present the problem of an increase in NDTT with irradiation. The development of an ideal core structural material for fast/fusion reactors resistant to void swelling, and to high-temperature helium embrittlement, with adequate creep strength at operating temperatures and good toughness (with a safe temperature margin above NDTT) at fuel handling temperatures, continues to be a challenging task for nuclear metallurgists.

6. Conclusion

An attempt has been made in this article to correlate the microscopic aspects of irradiation effects in nuclear structural materials with the macroscopic effects and practical engineering consequences. Figure 12 shows a summary of radiation effects in which the temperature and neutron flux domains over which various irradiation effects take place in steels are illustrated schematically. As the defect structures and microstructural changes introduced by neutron radiation vary with temperature, neutron flux, neutron energy spectrum, and fluence, different effects and phenomena become important for different reactor systems.

From a basic point of view these effects have opened new avenues to the materials scientists by providing defects, defect structures, and microstructural changes that are not normally encountered in alloys under equilibrium conditions. This has led to the naming of the field as "irradiation-enhanced materials science." The excitement of the subject is best summed up by quoting Sir Alan Cottrell from his opening address at the Tenth Anniversary Conference of the British Nuclear Energy Society:

"What a wonderful subject all this is! As I know to my cost, in other places people can agonize for hours trying to decide between basic and applied research. But, here, there is no problem. It is both! We have all the playthings of the basic materials scientist—vacancies, dislocations, grain boundaries, foreign atoms—interacting in the most fascinating of ways, and setting innumerable problems for fundamental science. And yet, at the same time, the performances of the materials are of most profound importance for practical reactor engineering and must be understood and mastered if we are to build good reactors. I doubt there has ever been such a fortunate and intimate combination of fundamental scientific problems and practical needs."

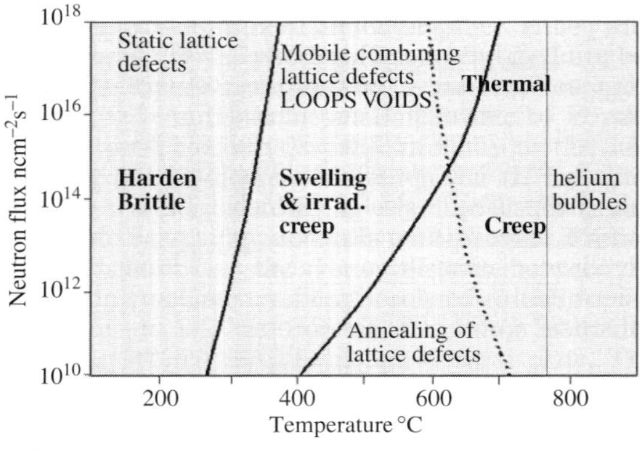

Figure 12
Temperature–neutron flux domains over which various irradiation effects take place in steels.

See also: Fast Reactor Cores, Austenitic Steels for; Fast Reactor Cores, Ferritic–Martensitic Steels for; Heavy Water Reactor Fuel Design and Performance; Light Water Reactor Fuel Design and Performance; Fusion Reactors (Magnetically Confined)–Tokamaks: Materials; Nuclear Reactor Materials; Radiation Effects in Carbon–Carbon Composites; Fusion Conditions; Nuclear Reactors: Shielding Materials; Radiation Damage Theory; Nuclear Applications: Zirconium Alloys

Bibliography

American Society for Testing and Materials 2000 *19th Int. Symp. on Radiation Effects in Materials.* ASTM Special Technical Publication STP1366. All the proceedings of this biennial series of symposia have been published as ASTM special technical publications and are of relevance. Full information is available at http://www.astm.org/search/iatoc

Baldev Raj Kasiviswanathan K V, Rodriguez P 1983 Impact testing of irradiated AST4 A203 grade steel In: *Activity Report of Metallurgy Programme,* 1981 and 1982. Reactor Research Centre, Kalpakkam, India, pp. 67–70

Bement A L Jr. 1972 Irradiation effects on structural materials [in five parts]. *Rev. Roum. Phys* **17**, 361–80; 505–17; 519–25; 607–18; 619–30

Brenner S S, Wagner R, Spitznagel J A 1978 Field-ion microscope detection of ultra-fine defects in neutron-irradiated Fe-0.34 Pct Cu alloy. *Metallurgical Transactions A* **9A**, 1761–4

British Nuclear Energy Society 1972 *Proc. Conf. on Irradiation Embrittlement and Creep in Fuel Cladding and Core Components.* British Nuclear Energy Society, London

Coghlan W A 1986 Review of recent irradiation creep results. *Int. Metals Rev.* **31**, 245–57

Farrell K, King R T 1973 Radiation-induced strengthening and embrittlement in aluminum. *Metallurgical Transactions* **4**, 1223–31

Field G J 1988 Problems caused by irradiation deformation in CANDU reactors. *J. Nuclear Mater.* **159**, 3–11

Frost B R T 1994 Nuclear materials. In: Cahn R W, et al. (eds.) *Materials Science and Technology: A Comprehensive Treatment.* VCH, Weinheim, Germany, Vol. 10.

Garner F A, Gelles D S 1988 Irradiation creep mechanisms: an experimental perspective. *J. Nuclear Mater.* **159**, 286–309

Gilbert E R, Straalsund J L, Wire G L 1977 Irradiation creep data in support of LMFBR core design. *J. Nuclear Mater.* **65**, 266–78

Gittus J 1978 *Irradiation Effects in Crystalline Solids.* Applied Science Publishers, London

Holland J R, Mansur L K, and Potter D I (eds.) 1980 *Proc. Symp. on Phase Stability during Irradiation.* Metallurgical Society of AIME, Pittsburgh, PA

Johnson R A and Orlov A N (eds.) 1986 *Physics of Radiation Effects in Crystals.* North-Holland, Amsterdam

Krishan K, Sundraam C V 1985 Radiation-induced structural changes in alloys. *Pramāna* **24** (1, 2), 361–81

Linga Murthy K 1999a Creep studies for zircaloy life prediction in water reactors. *J. Mater.* 32–9
Linga Murthy K 1999b Role and significance of source hardening in radiation embrittlement of iron and ferritic steels. *J. Nuclear Mater.* **270**, 115–28
Little E A 1976 Neutron-irradiation hardening in irons and ferritic steels. *Int. Metals Rev.* **204**, 25–60
Maziasz P J 1989 Formation and stability of radiation-induced phases in neutron-irradiated austenitic and ferritic steels. *J. Nuclear Mater.* **169**, 95–115
Nair K G M 1995 Non-equilibrium segregation during irradiation. *Metal Mater. Processes* **6**, 217–24
Nair K G M, Vijayalakshmi M, Krishan K 1993 Amorphization of Al6 Mn phase in an Al–14 a/o Mn alloy during low energy helium ion bombardment. *Scr. Metall. Mater.* **29**, 913–7
Norgett M J, Robinson M T, Torrens I M 1975 A proposed method for calculating displacement dose rates. *Nucl. Eng. Des* **33**, 50–4
Proc. Symp. on Irradiation-Enhanced Materials Science and Engineering 1989 Metall. Trans. **20A,** 2577–694
Proc. Symp. on Irradiation-Enhanced Materials Science and Engineering 1990 Metall. Trans. **21A,** 1015–132; 1799–876
Robinson M T, Young Jr. F W (eds.) 1975 *Proc. Int. Conf. on Fundamental Aspects of Radiation Damage in Metals.* US Energy Research and Development Administration, Washington, DC
Rodriguez P 1998 What happens to structural materials in a nuclear reactor: the fascinating radiation-enhanced materials science. G. S. Tendolkar Memorial Lecture. Indian Institute of Technology, Mumbai, Indira Gandhi Centre for Atomic Research, Kalpakkam, India
Rodriguez P, Krishnan R, Sundaram C V 1984 Radiation effects in nuclear reactor materials—correlation with structure. *Bull. Mater. Sci.* **6**, 339–67
Sheely W F (ed.) 1967 *Radiation Effects.* Gordon and Breach, New York
Szenes G (ed.) 1992 Physics of irradiation effects in metals. In: *Materials Science Forum.* Trans Tech, Zurich, Vols. 97–99
Thompson M W 1969 *Defects and Radiation Damage.* Cambridge University Press, Cambridge, UK

Nuclear Reactors: Coolant Materials

S. C. Chetal

The coolant, as the name implies, is used to remove the heat produced in the core and from other parts of nuclear reactors where heat is produced. The choice of coolant for a given type of a reactor is governed by various factors such as nuclear properties, heat transfer and transport properties, physical properties, chemical properties, corrosion behavior, stability, induced radioactivity, and cost. In practice, the coolant is chosen as a compromise between the various requirements. The main nuclear properties are neutron absorption and scattering cross-sections. The coolant should have a low neutron absorption cross-section; the value of the scattering cross-section depends on the reactor type: thermal or fast reactor. In the case of thermal reactors the scattering cross-section should be large, and in the case of fast reactors it should be small. The coolant should have a high heat transfer coefficient and require low pumping power. The boiling point should be high, except in reactors where boiling is desired, and the melting point should be low. The coolant should be compatible with the fuel cladding and the structural materials with which it makes contact. In the case of coolants other than water, the compatibility of coolant with water is an important aspect, in view of possible leaks in the steam generator. The coolant must not give rise to a fire hazard in case of a leak to the atmosphere. The coolant should be stable under operating temperature and radiation environment. The induced activity in the coolant should be a minimum to reduce the need for shielding. Finally, the coolant must be cheap and easily available.

1. *Reactor Coolants*

1.1 *Water*

Water has many desirable characteristics and most reactors for power production use ordinary (light) water as the coolant. In a boiling water reactor (BWR) (Fig. 1), the light water is allowed to boil and the separated steam is fed to the turbine. The condensed steam returns to the reactor through feed water heaters and pumps as in a conventional fossil fuel power station. In a pressurized water reactor (PWR) (Fig. 2), the light water is maintained at a sufficiently high pressure so that the temperature of coolant leaving the reactor is below the saturation temperature. The high-pressure hot water from the reactor transfers heat to the feed water in the steam generators and the steam produced is sent to the turbine. In both BWRs and PWRs, light water is both the moderator as well as the coolant and the fuel is based on either enriched uranium or a mixture of

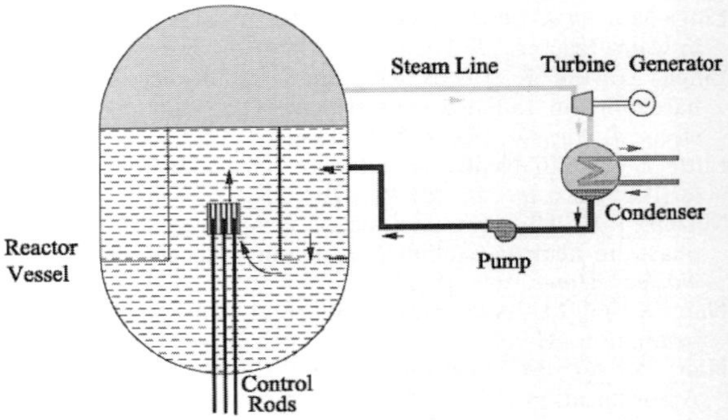

Figure 1
Boiling water reactor.

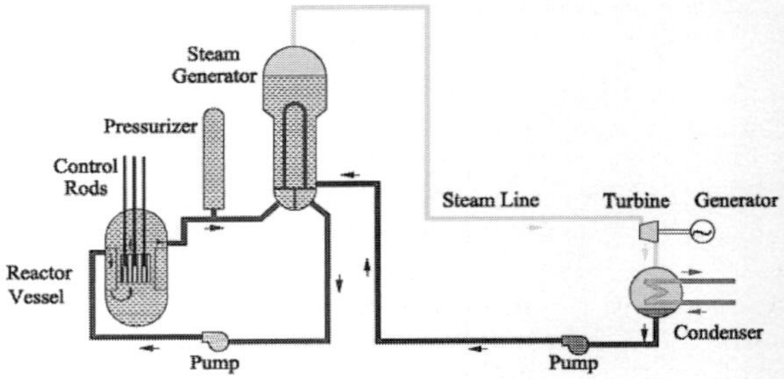

Figure 2
Pressurized water reactor.

natural uranium and plutonium. The principal advantage of heavy-water moderated and cooled reactors is that natural uranium can be used as the fuel. The thermal neutron absorption cross-section of heavy water is much smaller, about 1/500 of that of light water. Except for the very significant advantage of the neutron-moderating property of heavy water in comparison to light water, the difference between them in terms of the thermophysical and transport properties of interest as coolants is small. However, heavy water is very expensive. The pressurized heavy water reactor (PHWR) (Fig. 3) uses the pressure tube concept. Heavy water flows at high pressure through these tubes which contain the fuel. Hot heavy water in turn transfers heat to the feed water in the steam generators and the steam produced is sent to the turbine. There are a few reactors which employ heavy water as the moderator in combination with another material, such as light water, as the coolant. The fuel choice in this particular case is different from natural uranium.

There is no serious neutron activation problem in water-cooled reactors. ^{16}N formed by the (η,p) reaction with ^{16}O has a 7.4 s half-life and therefore the activity is negligible after a few minutes of reactor

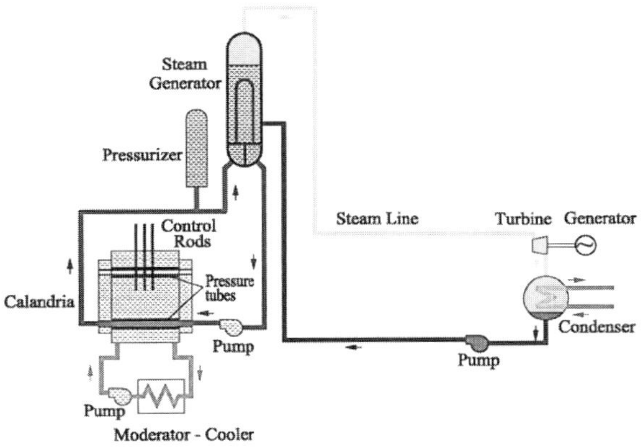

Figure 3
Pressurized heavy water reactor.

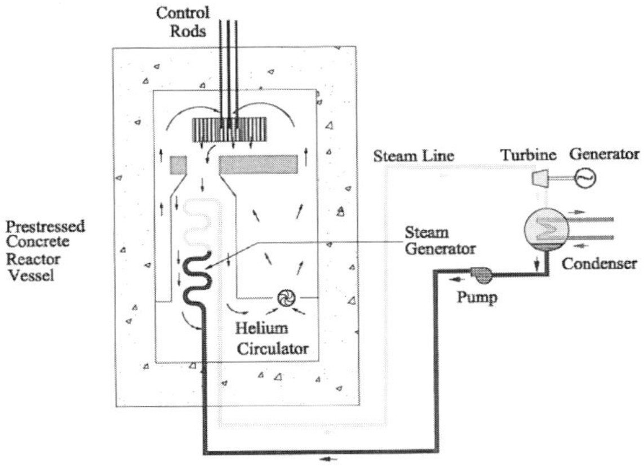

Figure 4
High-temperature gas-cooled reactor.

shutdown. A serious drawback of water as a coolant is its high vapor pressure. The coolant system pressure in BWRs, PWRs, and PHWRs is approximately 7 MPa, 15 MPa, and 10 MPa, respectively. The steam temperature at the turbine inlet in these types of reactors is rather low, approximately 532–573 K, compared to 811 K in modern fossil fuel power stations. The lower steam conditions in these reactors result in lower overall efficiency, of the order of 33% for BWRs and PWRs and 29% for PHWRs, compared to about 42% for fossil fuel power plants.

1.2 Organic Coolants

Coolants based on diphenyl and terphenyl have been used in reactors. Their good moderating properties, high boiling point, noncorrosive nature, and relative cheapness make them attractive as coolants. Their major drawback is decomposition by nuclear radiation. There is virtually no interest in organic-cooled reactors.

1.3 Gases

The heat transfer characteristics of gases are poor in comparison to liquids. To improve heat transfer and reduce pumping power, it is necessary that the gas system be pressurized. Heat removal can be further improved by having fins on the fuel elements. Gases as coolants have the advantages of good radiation and thermal stability, ease of handling, and compatibility with structural materials. Carbon dioxide has been extensively used in graphite-moderated gas-cooled reactors, as in the Magnox and advanced gas-cooled reactors (AGRs) in the UK. AGRs operate with an overall efficiency of about 40% and are comparable in this respect to modern fossil fuel power stations. The heat transfer coefficient of helium is higher than that of carbon dioxide and it also has advantages with respect to lower pumping power and neutron absorption. Carbon dioxide has the additional drawback of interacting with the graphite moderator at high temperatures. Helium is capable of working at extremely high temperatures, and helium-cooled reactors are being employed and considered in some countries for power generation based on both steam turbines (Fig. 4) and direct-cycle gas turbines and to provide process heat in chemical and other industries as well as for domestic heating purposes.

1.4 Liquid Metals

Sodium, sodium–potassium alloy, mercury, lead, and lead–bismuth alloy are the liquid metals that have been used as coolants in nuclear reactors. Sodium is the coolant for all fast reactors in operation. Alkali metals in particular have excellent heat transfer properties and high boiling points. Among the various liquid metals, sodium has found the widest use in fast reactors because of its high heat transfer coefficient, poor moderation, low pumping power, very high boiling point (1155 K), low melting point (371 K), compatibility with structural materials provided oxygen impurity is controlled, and commercial availability at a relatively low price. In a liquid metal-cooled fast breeder reactor, the sodium coolant in the primary circuit picks up the heat produced in the core and transfers it to a secondary sodium circuit, which in turn produces the steam in the steam generators (Fig. 5).

Sodium has drawbacks associated with its intense chemical activity (catching fire) in the case of a leak to the atmosphere. This leads to the incorporation of a double envelope for the primary circuit with an inert interspace. The violent reaction of sodium with water makes the design of sodium-heated steam generators a challenging task. Another drawback of sodium is its activation. ^{24}Na as a result of the (n,γ)

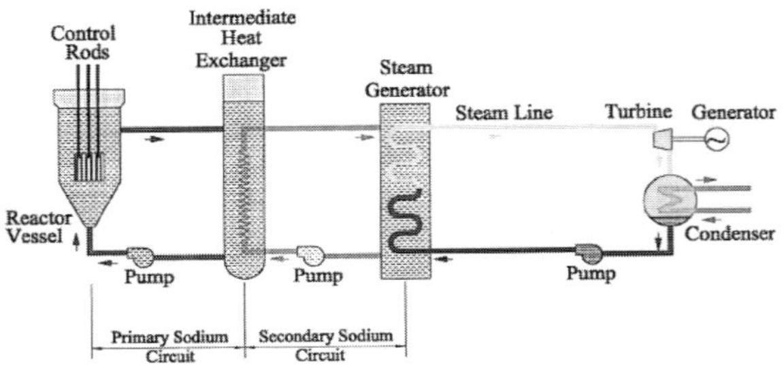

Figure 5
Sodium-cooled fast reactor.

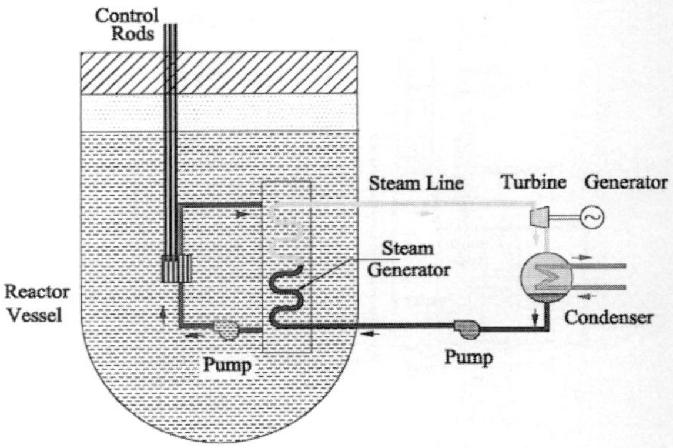

Figure 6
Lead-cooled fast reactor.

reaction and ^{22}Na as a result of the (n,2n) reaction are the main activation products. The half-lives of ^{24}Na and ^{22}Na are 15 h and 2.6 yr, respectively. As a result, sodium-cooled reactors need biological shielding. Also, there is a delay of a few days after reactor shutdown before access can be safely gained to the primary circuit for any maintenance activity. The excellent heat transfer of sodium coupled with its low specific heat leading to higher coolant temperature rise makes the thermal stresses of great significance in the mechanical design of sodium components. The other alkali metals have less favorable properties in comparison to sodium. The eutectic alloy of sodium with potassium (NaK), however, has the advantage that it is in the liquid state at room temperature.

Among the heavy metals there is renewed interest in lead–bismuth and lead as coolants for fast reactors to overcome the serious drawback of the chemical activity of sodium. Using these coolants, the secondary circuit associated with the sodium-cooled reactor is unnecessary. Lead–bismuth and lead have relatively poor heat transfer and transport properties compared to sodium, control of oxygen is needed to minimize corrosion of structural materials, and the melting point of bismuth (398 K) is also higher than sodium. A serious drawback to the use of lead–bismuth alloy is the formation of toxic polonium through neutron capture by bismuth. To overcome the problem of the formation of polonium and the higher cost, lead as a coolant instead of lead–bismuth is under study, particularly in Russia. Figure 6 shows a schematic diagram of a lead-cooled fast reactor. The high melting point (600 K) and the structural material compatibility are the two challenging tasks in the design and operation of a lead-cooled reactor.

Mercury was used in Clementine, the first fast reactor. However, it is not an attractive coolant owing to its relatively poor nuclear and heat transfer properties, toxicity, and high cost.

Liquid metal-cooled reactors, in general, generate steam conditions to provide an overall efficiency of around 40%.

1.5 Molten Salts

Molten salts, especially fluorides, have been used as reactor coolants because they provide advantages associated with a fluid-fuel type of reactor and have favorable high-temperature properties. In one reactor of this type (molten salt reactor experiment), a molten mixture of the fluorides of lithium, beryllium, zirconium, enriched uranium, and thorium is pumped through channels provided in graphite moderator elements. As the molten mixture passes through the core, the system becomes critical and the fission energy is absorbed directly in the fluid. The hot fluid then transfers the heat to a secondary coolant, consisting of a mixture of lithium fluoride and beryllium fluoride, in an intermediate heat exchanger and returns to the reactor. Molten salt reactors have the potential to provide an overall efficiency of about 44% and breeding. However, molten salt reactors have some serious drawbacks, like the highly corrosive nature of the molten salts and a requirement for remotely operated equipment for maintenance and repairs, which do not permit continued interest in their development for power generation.

1.6 Chemistry Control of Nuclear Reactor Coolants

Chemistry control of coolant is of importance in reducing the corrosion of structural materials and in minimizing the man-rem consumption during operation and maintenance of the reactor. The control of pH and oxygen are the two important factors in PWRs and PHWRs. Lithium hydroxide is normally used for pH control of primary water, the pH being between 6.9 and 7.4 at operating temperature. Hydrogen is normally used for oxygen control. BWRs, unlike PWRs, operate at close to neutral pH, as the system is not closed. Oxygen impurity in sodium is the key factor in the corrosion of structural materials in sodium-cooled systems. Oxygen impurity in sodium is kept normally below 5 ppm by purification of sodium in a cold trap. Lead- or lead–bismuth-cooled reactors need a controlled amount of oxygen in the coolant to reduce the corrosion of structural materials.

2. *Future Developments*

Water will continue its dominant position as a coolant for thermal reactors. Helium will receive more attention as a coolant in high-temperature reactors for process heat and power generation purposes. As alternatives to sodium, lead and lead-bismuth are likely to be seriously evaluated as coolants for fast reactors.

See also: Nuclear Reactors: Moderator and Reflector Materials; Nuclear Reactor Materials

Bibliography

Bukousky J, Haack K 1994 *Heavy Water Handbook*. Risø National Laboratory, Roskilde, Denmark

Glasstone S, Sesonske A 1967 *Nuclear Reactor Engineering*. Van Nostrand Reinhold, New York

IAEA 1999 *Status of Liquid Metal Cooled Fast Reactor Technology*, TECDOC-1083. International Atomic Energy Agency, Vienna

Lamarsh J R 1983 *Introduction to Nuclear Engineering*. Addison-Wesley, Reading, MA

Rahn F J, Adamantiades A G, Kenton J E, Braun C 1984 *A Guide to Nuclear Power Technology*. Wiley Interscience, New York

Yevick J G, Amorosi A 1966 *Fast Reactor Technology: Plant Design*. MIT Press, Cambridge, MA

World Nuclear Industry Handbook 1999 Nuclear Engineering International, Wilmington, NJ

Nuclear Reactors: Moderator and Reflector Materials

B. P. Sharma

Moderator and reflector are two important components of a nuclear fission reactor for maintaining neutron economy and sustaining the chain reaction. Depending on the fuel material and the objective (to breed or not), fission is allowed to occur either by fast neutrons (1–2 MeV) or by thermal neutrons (~0.025 eV). Accordingly, reactors are classified as fast or thermal reactors. The neutrons generated in a fission reaction are fast neutrons and thermal reactors call for a moderator material for slowing down the fission neutrons. Fast reactors do not employ moderator and to maintain fast neutron economy, the coolant is also carefully chosen.

The function of a reflector surrounding the reactor core is to scatter back the core neutrons which would otherwise be lost by leakage. A reflector thus directly contributes to the neutron economy and, indirectly, helps to decrease the critical volume of the reactor core.

This article briefly discusses the basic principles of nuclear physics involved in the selection of moderator and reflector materials and describes the different materials employed for these applications, their properties, and methods of manufacture.

1. *Requirement of Properties*

Interaction of a neutron with a nucleus can be broadly classified into two categories, namely, scattering and absorption. Neutron moderation essentially consists of elastic and inelastic scattering of fast neutrons produced in a fission reaction by the nuclei in the surrounding media until their energy is reduced to a level where they are in thermal equilibrium with the media. Fractional loss of energy of a neutron per collision is related to the scattering angle and the mass number of the target nuclei. The lower the mass number of the moderating atom, the greater the fractional loss of energy per collision. A useful quantity to express the slowing down of neutrons is the average logarithmic energy decrement per collision (ξ). ξ can be approximated as:

$$\xi = \frac{2}{A + \frac{2}{3}}$$

where A = mass number and the value of ξ is independent of the initial energy of the neutron. Besides a large ξ, the moderator material must also exhibit a high scattering cross-section (Σ_s) and low absorption cross-section (Σ_a). Thus, an appropriate quantity for comparing the effectiveness of a moderator can be expressed as ($\xi\Sigma_s/\Sigma_a$). This quantity is called the *moderating ratio*. It may be mentioned here that macroscopic cross-sections (Σ) are related to microscopic cross-sections (σ) through Avogadro's number (N_A), density of the material (ρ) and atomic or molecular weight (A), by:

$$\Sigma = \frac{N_A \rho \sigma}{A}$$

The microscopic cross-section (σ) has the unit of area, barn (1 barn $= 1 \times 10^{-24}$ cm^2 (in the early days a researcher called this "as large as a barn door")) and can be considered an indication of the probability of a nuclear reaction. The macroscopic cross-section (Σ) has the unit of inverse length and $1/\Sigma$ is an indication of the mean free path of the neutron in the medium.

The general requirements of a reflector material are high scattering cross-section and low absorption cross-section. The probability of a neutron returning from the reflector to the core increases with decrease in the scattering mean free path. A useful quantity to define this is the diffusion coefficient (D). For a good reflector the $\Sigma_a D$ should be small. The choice of a reflector material is largely governed by the energy distribution of the neutrons in the reactor. For the thermal neutrons, properties required of a moderator and a reflector are essentially same. Thus, for a thermal reactor core containing a moderator, the reflector material may be simply an extension of the moderator beyond the region of fuel loading. In contrast, for fast reactors, the presence of energy moderating material must be avoided. In this case, the reflector material is essentially made of a heavy element, such as iron, bismuth, or lead with a high scattering cross-section for fast neutrons. Usually in a fast reactor a blanket containing fertile materials, such as Th^{232} and U^{238}, is provided to breed new fuels by absorbing escaping neutrons.

Besides the desirable nuclear properties, a good moderator or reflector should also exhibit high thermal conductivity and good thermal and radiation stability.

2. *Moderator and Reflector Materials and their Properties: Effect of Radiation*

Moderator materials essentially consist of elements of low atomic number. Among these, hydrogen, beryllium, and carbon are the most efficient moderating elements because of their high scattering cross-section and low absorption cross-section. Hence, these elements or their appropriate compounds are commonly employed as moderator and reflector materials. Nuclear properties of some of the potential moderator materials are given in Table 1. Deuterium is the naturally occurring heavier isotope (with one extra neutron in its nucleus) of hydrogen and forms

Table 1
Nuclear properties of potential moderator/reflector materials.

	Epithermal region				Thermal region			
Elemental/compound	ξ	Σ_s (m^{-1})	Σ_a (m^{-1})	$\xi\Sigma_s/\Sigma_a$	Σ_s (m^{-1})	Σ_a (m^{-1})	D (m)	$\Sigma_a D$
Light water (H_2O)	0.927	154	2.3	62	345	2.2	1.7×10^{-3}	3.7×10^{-3}
Heavy water (D_2O)	0.510	35	3.7×10^{-3}	4830	45	8.5×10^{-4}	8.5×10^{-3}	7.2×10^{-6}
Beryllium (Be)	0.207	74	1.2×10^{-1}	126	86	1.2×10^{-1}	5.4×10^{-3}	6.5×10^{-4}
Beryllium oxide (BeO)	0.172	71	7.7×10^{-2}	158	50	7.3×10^{-2}	6.6×10^{-3}	4.8×10^{-4}
Graphite (C)	0.158	53	3.8×10^{-2}	216	41	3.2×10^{-2}	9.4×10^{-3}	3.0×10^{-4}

Table 2
Physical properties of ordinary and heavy water.

Properties	H_2O	D_2O
Molecular weight	18.0152	20.02726
Density (gcm^{-3})	1.0	1.105
Melting point (°C)	0.0	3.82
Boiling point (°C)	100.0	101.42

heavy water. Physical properties of ordinary and heavy water differ marginally as shown in Table 2. It may be mentioned here that, though the mass numbers of boron and lithium are low, these are useless as a moderator or reflector because of their high absorption cross-sections.

2.1 *Water*, Heavy Water, *and other Hydrogen Compounds*

It can be seen from Table 1 that ordinary water (H_2O) has a relatively high cross-section for thermal-neutron capture. Therefore, a critical system can be attained, with water as moderator, only if the fuel is enriched (2–4%) in the fissile isotope. However, with enriched material available, the small migration length of thermal neutrons in water makes it possible to design a compact reactor. The very large moderating ratio of heavy water (D_2O) makes it possible to build thermal reactors based on natural uranium (0.7% U^{235}) as fissile material. The combination of natural uranium oxide fuel with heavy water as moderator/reflector (and primary coolant) is used in power reactors of the CANDU (Canadian Deuterium Uranium) type. Reactor systems based on natural uranium fuel should have the highest level of neutron economy built into their design and into the selection of structural and other materials.

The water must be free from impurities since they not only capture neutrons but also may become radioactive as a result of (n, γ) reactions. The decomposition of water by nuclear radiation is also often facilitated by the presence of ionic impurities. Further, water has to be especially free from chloride ions to minimize corrosion. Deuterium undergoes a (γ, n) reaction, thus contributing some neutrons to the chain reaction. A few neutrons which are captured by heavy water in the reactor result in the formation of radioactive tritium. It should be noted that heavy water itself is not radioactive.

One of the main drawbacks to the use of ordinary or heavy water as moderator is their relatively low boiling points. This means that if these are to be utilized in power reactors, even at moderately high temperatures (~330°C), the system must be designed to withstand high pressures.

Another hydrogen bearing compound of particular interest is zirconium hydride (ZrH_2). It is prepared by heating zirconium metal in hydrogen gas at about 350°C. In this compound, the hydrogen atom density approaches that in water. The absorption cross-section of zirconium being small, zirconium hydride exhibits good moderating properties. It serves as the moderator in the TRIGA (Training, Research, Isotopes, General Atomic) reactor in which it is incorporated with enriched uranium metal in the fuel elements.

Hydrogen bearing organic hydrocarbons, such as biphenyl ($C_{12}H_{10}$), polystyrene ($(CH)_n$), or paraffin ($(CH_2)_n$), have also been investigated for use as moderator materials. However, these materials, in general, did not progress beyond the experimental stage because of degradation caused by the combined effect of radiation and temperature.

2.2 *Beryllium and Beryllium Oxide*

Beryllium metal has the lowest thermal neutron absorption cross-section among all metals. This property, combined with a large scattering cross-section, high melting point, and good strength, makes it an excellent moderator and reflector material. However, because of its high cost, beryllium is used only in experimental reactors. Beryllium oxide (BeO, also known as beryllia), owing to its high melting point is a candidate material as the moderator and reflector for high-temperature, gas-cooled reactors (HTGRs).

Some of the properties of beryllium, beryllium oxide and graphite are listed in Table 3. As a result of

Table 3
Properties of beryllium, beryllium oxide, and graphite.

Property	Be	BeO	Graphite
Crystal structure	Hexagonal	Hexagonal	Hexagonal
Density ($kg m^{-3}$)	1850	2850	2250
Melting point (°C)	1283	2524	3650
Vapor pressure at 723°C (mbar)	6×10^{-8}	5×10^{-27}	3×10^{-27}
Thermal conductivity ($W m^{-1} {}^{\circ}C^{-1}$), (at 20°C)	216	248	156
Thermal expansion coefficient ($^{\circ}C^{-1}$) (at 20°C)	11.5×10^{-6}	6.4×10^{-6}	2.0×10^{-6}
Modulus of elasticity (GPa) (at 20°C)	307	386	4.8 (in tension)
Compressive strength (MPa) (at 20°C)	196	798	58

their hexagonal crystal structure, many of the physical and mechanical properties of these materials are anisotropic. As-cast beryllium exhibits a coarse grain structure, micro- and macro-cracks and rather poor strength. The preferred material, for reactors and other applications, is produced by hot pressing the powder. The hot-pressed product has good strength properties which are approximately isotropic. The elongation prior to fracture, however, is small. The brittleness is accentuated by the presence of bubbles of helium gas formed during reactor operation by the (n, α) reaction.

BeO has exceptional resistance to thermal shock for an oxide ceramic. Its thermal conductivity is high, but it decreases markedly with increasing temperature. The oxide has a low vapor pressure in a dry atmosphere. However, in the presence of moisture it vaporizes rapidly, when hot, as a result of the formation of the hydroxide; this is a serious drawback especially in view of the extreme toxicity of beryllium and its compounds.

In the reactor environment, beryllium undergoes following nuclear reactions:

$$Be^9 + n^1 \rightarrow He^6 + \alpha^4 \qquad He^6 \rightarrow Li^6 + \beta^0$$

$$Li^6 + n^1 \rightarrow He^4 + H^3$$

$$Be^9 + \gamma^0 \rightarrow Be^8 + n^1 \qquad Be^8 \rightarrow 2\alpha^4$$

$$Be^9 + n^1 \rightarrow Be^8 + 2n^1$$

Beryllium thus offers an added advantage that (n, 2n) and (γ, n) reactions contribute a few extra neutrons to the chain reaction.

2.3 Graphite

Next to heavy water, graphite has the highest moderating ratio and is extensively used as moderator and reflector for gas cooled thermal reactors. The world's first nuclear fission reactor (CP-1), built in 1942, was based on natural uranium and graphite.

The tensile strength of graphite increases with temperature, attaining a maximum at about 2500°C, when the value is roughly twice that at ordinary temperatures. Above 2500°C, the strength drops very rapidly. The creep rate is very small, below 1500°C. Graphite has excellent resistance to thermal shock.

Radiation exposure affects most of the properties of graphite. Neutron interaction leads to the formation of vacancy-interstitial pairs (see *Nuclear Reactor Materials: Irradiation Effects*). This has implications for reactor safety. The energy of formation of a vacancy-interstitial pair in graphite is about 10 eV. Below about 350°C there is little mobility of carbon atoms and radiation damage causes accumulation of stored energy in the crystal lattice. If this stored energy, also called the *Wigner energy*, is not released in a controlled manner (by annealing at a high temperature) it can have disastrous effects. A point can be reached when the metastable (high-energy) material spontaneously releases the stored energy resulting in a sharp increase in temperature. Such an accident occurred in the Windscale reactor in the UK in 1957. This problem does not exist in reactors operating at high temperature.

3. Production and Processing Techniques

3.1 Heavy Water

Natural water contains 145 ppm of D_2O. Heavy water with a deuterium content of 99.8% is used as moderator in reactors with natural uranium fuel. Because of the close similarity of the physical and chemical properties of H_2O and D_2O, the extent of enrichment required poses a technical challenge and makes D_2O very expensive. Processes that are available for production of heavy water are listed in Table 4.

Currently the bulk of heavy water in the world is produced by the ammonia–hydrogen exchange process. Distillation and electrolysis are highly energy intensive for primary production of heavy water. However, these are quite simple and economic for up-grading of heavy water at reactor sites.

Table 4
Processes for production of heavy water.

Process	Principle	Separation ratio	Remarks
Distillation of water	Difference in boiling point of ordinary (H_2O) and heavy water (D_2O)	1.5 at 50°C 1.026 at 100°C	Large number of equilibrium stages Energy intensive
Ammonia–hydrogen exchange process	Preferential partitioning of deuterium between hydrogen and ammonia	5.2 at −25°C, 350 atmH_2 pressure	Process kinetics is slow Specially designed column for intimate mixing
Girdler sulfide process	Preferential partitioning of deuterium between water and hydrogen sulfide	2.32 at 32°C 1.80 at 138°C	H_2S is highly toxic and corrosive
Electrolysis	Higher mobility and discharge ratio of hydrogen in comparison with deuterium	3.8 at 25°C 13.2 at 15°C	Energy intensive Electrode material and cell design play an important role

3.2 Beryllium and Beryllium Oxide

Beryllium/BeO is extracted from beryl ore ($3BeO.Al_2O_3.6SiO_2$) through a series of hydro-metallurgical steps. Two extraction processes, the sulfate process and the fluoride process, are mainly used.

These processes yield $Be(OH)_2$ which is thermally decomposed to BeO. Since beryllium is a highly reactive metal, reduction of its compounds requires the use of strong reducing metals. Reduction of beryllium fluoride with magnesium is most commonly used for the production of the metal.

Beryllium is also produced by electrolysis of beryllium chloride. BeO is mixed with carbon and chlorinated at 1000°C to obtain beryllium chloride. To reduce its vapor pressure, beryllium chloride is mixed with sodium chloride to obtain an electrolyte which melts at 370°C. Electrolysis is carried out in a nickel stainless steel pot which also acts as the cathode. Graphite is used as the anode. The process yields beryllium flakes which are subsequently purified by vacuum melting.

Powder metallurgy is the route for obtaining a fine-grained structure together with adequate mechanical strength. Beryllium powder is vacuum hot pressed around 1200°C. Hot isostatic pressing is an advanced technique particularly useful for consolidating high-purity material. High-density beryllium oxide is also produced by hot pressing.

3.3 Reactor Grade Graphite

Reactor-grade graphite is made artificially by graphitization of selected petroleum coke (low in neutron absorbers like boron, cadmium, cobalt, etc.). The coke is first heated to drive off volatile matter and then ground and mixed with a coal-tar pitch binder. The mixture is extruded into bars and heated to temperatures up to 1500°C in gas-fired ovens in order to carbonize the pitch and set the binder. These are then impregnated with pitch under vacuum and again heated to increase the bulk density. The resultant carbon is finally graphitized by self-resistance heating. The temperature is raised to ~2700–3000°C for several days, and then about three or four weeks are allowed to cool the material. The physical characteristics of the final product vary with the fineness of the ground coke, the type and amount of pitch used for impregnation, and the temperature and duration of the graphitization treatment.

3.4 Pyrolytic Carbon

Pyrolytic carbon is produced by the thermal decomposition of a hydrocarbon gas, e.g., methane, propane, acetylene, etc. The decomposition temperature may be roughly in the range of 1400–2500°C. The structure and density of the product depend upon the substrate temperature and the nature of the hydrocarbon gas and its flow rate. By changing the conditions, pyrolytic carbon can be deposited with either isotropic or anisotropic structure. It can also be obtained in porous or impervious forms. For use in HTGR, the small fuel particles of uranium and thorium carbides (or oxides) are first coated with a layer of porous anisotropic pyrocarbon at a lower temperature, and then with an impervious layer at a higher temperature. The porous pyrocarbon serves as a reservoir for fission product gases, and the impervious layer provides overall containment like conventional fuel-element cladding.

See also: Nuclear Reactors: Coolant Materials; Nuclear Reactors: Shielding Materials

Bibliography

Dave S M, Sadhukhan H K, Novaro O A 1997 *Heavy Water: Properties, Production and Analysis.* Quest Publication, Mumbai India

Engle G B, Everett M R, Fatherholt W P 1974 Status of Graphite Technology and Requirements for HTGRS. US AEC Report CONF-740501

Everest D A 1964 In: *The Chemistry of Beryllium*. Topics in Inorganic and General Chemistry, Vol. 6. Elsevier, Amsterdam

Frost B R T (ed.) 1993 *Materials Science and Technology, Vol. 10, Nuclear Materials*. VCH, New York

Glasstone S 1955 *Principles of Nuclear Reactor Engineering*, 1st edn., Van Nostrand, New York

Hove J E and Riley W C (eds.) 1965 *Ceramics for Advanced Technologies*. Wiley, New York

Ma B M 1983 *Nuclear Reactor Materials and Applications*. Van Nostrand Reinhold, New York

Nightingale R E 1962 *Nuclear Graphite*. Academic Press, New York

Piccinini N 1975 Coated nuclear fuel particles. *Adv. Nucl. Sci. Technol.* **8**, 255–341

Simmons J H W 1965 *Radiation Damage in Graphite*. Pergamon, Oxford

Tipton C R (ed.) 1960 *Reactor Handbook, Vol. 1, Materials*. Interscience, New York

Nuclear Reactors: Pressure Vessel Steels

G. R. Odette

The majority of the world's nuclear energy is generated by light water reactors. Achieving reasonable thermodynamic efficiency requires a heavy-section steel reactor pressure vessel (RPV) to contain safely coolant water at about 290 °C and pressures ranging from 7 MPa in boiling water reactors (BWRs) to 14 MPa in pressurized water reactors (PWRs).

Regulations require very low RPV failure probabilities both for normal operation and during rare accident events (USNRC 1986, ASME 1986). Designs assume that large cracks are present that reduce the effective stress for catastrophic fast fracture. The minimum fracture stress (σ_f) is governed by the linear elastic fracture toughness (K_{Ic}) of the RPV steel, as $\sigma_f = C_g K_{Ic}/(\pi a)^{1/2}$, where a is the crack length and C_g is a geometric constant.

Vessel integrity assessments require activities ranging from in-service flaw inspections to system-scale thermal-hydraulic stress analysis. However, a basic safety criterion is that the RPV steels are sufficiently tough. This is not an issue for as-fabricated vessels. However, exposure to neutrons from the reactor core degrades fracture toughness. Irradiation embrittlement is usually characterized by the increase in a ductile-to-brittle transition (DBT) temperature. Shifts in the DBT temperature exceed 200 °C in some cases. Hence, embrittlement must be considered in PRV integrity assessments (USNRC 1988) and, if severe, may require either premature plant closure or vessel annealing.

1. *Vessels, Steels, and Service Environments*

Details of RPV design, fabrication, in-service operation, and regulation vary widely. Thus this article primarily reflects RPV technology in the USA, which is reasonably representative of the approaches used in most other countries. RPVs are very large welded structures, typically weighing about 500 tonnes, standing 14 m high by 4.5 m in diameter with a wall thickness of 20 cm or more (Fig. 1(a)). RPV base metals (A302B, A533B plates, and A508 forgings) and welds are quenched and tempered low-alloy steels. Typical compositions are carbon (0.05–0.2%), manganese (0.7–1.6%), molybdenum (0.4–0.6%), nickel (0.2–1.4%), silicon (0.2–0.6%), and chromium (0.05–0.5%). Multiple layer submerged arc welds, made of consumable metal wires, join vessel sections. Weld compositions differ from the base metal, and may vary significantly even within the same weld. In order to permit transportation by rail, the vessels of the VVER reactors of the former Soviet Union are somewhat smaller, and hence use stronger steels alloyed with chromium and vanadium.

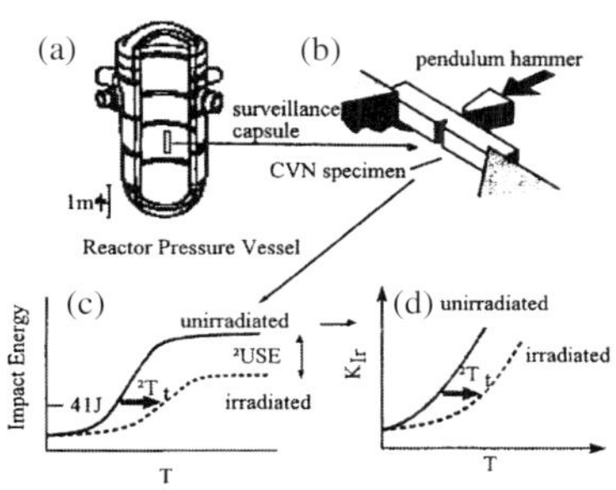

Figure 1
Procedure for establishing the irradiated $K_{Ir}(T)$ curve based on CVN tests: (a) reactor vessel and surveillance capsule location; (b) CVN impact test; (c) unirradiated and irradiated CVN absorbed energy–temperature curves and ΔT_t; (d) unirradiated and irradiated $K_{Ir}(T)$ curves.

Typical RPV stress relief treatments are about 610 °C for 30 h. Common microstructures are tempered bainite, with carbide phases and dislocations. Yield stress (σ_y) values are about 500 ± 50 MPa. Compositions and microstructures vary on both the macro- and microscales. Along with alloy additions of nickel, trace copper and phosphorous impurities increase embrittlement. The copper content is high ($\geqslant$0.4%) in some early US welds and VVER steels have phosphorus contents up to about 0.06%.

Vessels operate at about 290 ± 30 °C (T_i). The neutron flux (ϕ) and fluence (ϕt) are defined as the number of neutrons crossing a unit area per unit time (neutrons $m^{-2}s^{-1}$) and the flux integrated over time (neutrons m^{-2}), respectively. The spectrum of neutron energies ranges from <1 eV to several MeV. High-energy neutrons are the dominant source of embrittlement. A standard unit of neutron exposure is the fluence greater than 1 MeV ($\phi t_{>1}$). The end-of-life $\phi t_{>1}$ for US PWRs is about 10^{23}–3×10^{23} nm^{-2}. For smaller diameter VVER vessels it is of the order of 10^{24} nm^{-2}, while for BWRs it is only about 3×10^{22} nm^{-2}.

Improvements that have reduced the problem of RPV embrittlement include the use of tougher steels, reduction of both trace impurities and neutron fluence, and elimination of beltline welds. However, embrittlement remains a potential problem for a significant number of older vessels, and is an unknown for the extended life of others.

2. *Measures of Irradiation Embrittlement*

Early recognition of the importance of embrittlement led to RPV surveillance programs. Capsules containing representative steels are located on the inside of the RPV where the neutron flux is several times higher than in the vessel. Thus, surveillance data are used to provide early estimates of embrittlement. There also have been numerous accelerated test reactor studies of embrittlement.

Conventional measurements of K_{Ic} require large specimens and sophisticated test procedures. Thus small 10 mm × 10 mm × 55 mm Charpy V-notch (CVN) impact specimens are used in most surveillance programs. The energy consumed in the fracture of a CVN specimen by the impact of a pendulum hammer has a transition between a brittle lower and a ductile upper shelf over a range of test temperatures (Fig. 1(b)). The DBT temperature (T_t) is usually defined at 41 J. Neutron irradiation elevates T_t (ΔT_t) and decreases the upper shelf energy (Fig. 1(c)).

However, CVN tests do not provide direct measures of K_{Ic}. US regulations use a lower bound reference toughness–temperature curve, $K_{Ir}(T' = T - T_{ndt})$, that has a specified shape on a temperature scale relative to the so-called nil ductility transition temperature (T_{ndt}). The unirradiated temperature (T_{ndtu}) is measured for the pertinent vessel steel, and adjusted to account for embrittlement as $T_{ndi} = T_{ndtu} + \Delta T_t$ (Fig. 1(d)). The ΔT_t includes an uncertainty term.

Plant-specific surveillance data are usually not sufficient to predict ΔT_t. More commonly, the ΔT_t values are evaluated using regulatory equations (USNRC 1988). Shifts are controlled by a large number of variables. Statistical fits to the US surveillance database show that ΔT_t depends on T_i, ϕ, $\phi t_{>1}$, copper, nickel, and phosphorus content, and product form (weld, plate, and forging) (Eason *et al.* 1998a). Test reactor studies show that ΔT_t also depends on a number of other variables including manganese content, final heat treatment conditions, and the neutron spectrum (Odette and Lucas 1998). Very strong synergistic interactions between variables, such as copper and nickel content, must also be accounted for.

Postirradiation annealing at temperatures (T_a) well above T_i results in partial to nearly full embrittlement recovery, depending on the copper content, T_i, T_a, ϕ, and annealing time (t_i) (Eason *et al.* 1998b). The rate of re-irradiation embrittlement following annealing is an important issue that is not yet well understood.

Because of the number of variables and variable combinations (e.g., Cu–Ni), coupled with large uncertainties in the surveillance database, purely empirical ΔT_t predictions are not reliable. Fortunately, basic research has provided much improved understanding and physically based models of embrittlement (Odette and Lucas 1998).

3. *Embrittlement Mechanisms and Models*

Steels have a regular b.c.c. crystalline arrangement of the atoms. Defects, including grain boundaries and dislocations, are deviations from a perfect crystal structure, and play a key role in mechanical properties. Alloying elements (carbon, manganese, molybdenum, nickel, etc.) are both dissolved in the iron matrix and located in compositionally, and most often structurally, discrete second-phase precipitates, like carbon–metal compounds (e.g., $Fe_{2.4}Mn_{0.6}C$ and Mo_2C). The overall arrangement of defects and alloy constituents defines the microstructure. The initial microstructure depends on both composition and processing history. Exposure to neutrons results in the formation of nanometer-scale precipitates, as well as clusters of solutes and crystal lattice defects.

The key embrittlement processes, shown in Fig. 2, include (Odette and Lucas 1998):

- Generation of primary lattice defects of single and small clusters of vacancies, self-interstitials, and solutes (Fig. 2(a)).
- Evolution of fine scale defect cluster–solute atom complexes, solute clusters, and distinct phases, primarily copper-rich precipitates (Fig. 2(b)).
- Hardening by these nanofeatures (Fig. 2(c)).
- Hardening-induced ΔT_t shifts (Figs. 2(d) and (e)).

These processes can be combined to model ΔT_t as a function of the key variables (copper, nickel, phosphorus, ... , ϕ, $\phi t_{>1}$, T_i).

3.1 *Hardening-induced Shifts*

Brittle fracture in RPV steels requires a combination of high material strength and a stress concentration. Strength increases with decreasing temperature, and a DBT occurs below a temperature that depends on

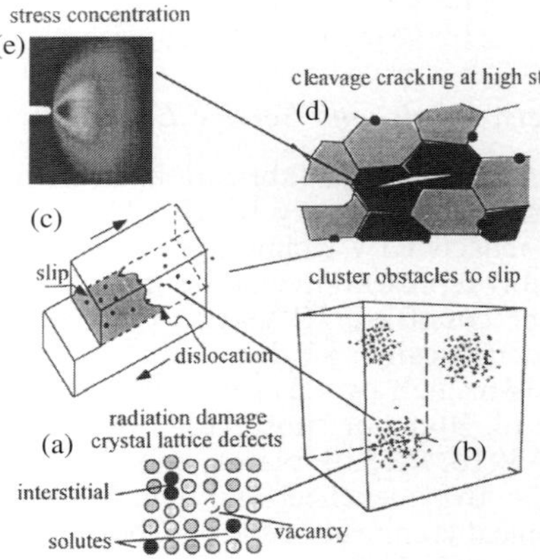

Figure 2
Sequence of basic embrittlement processes: (a) creation of primary radiation damage defects; (b) formation of nanoscale solute and defect clusters (iron atoms not shown); (c) cluster pinning of dislocations and hardening; (d) hardening-enhanced cleavage fracture; (e) stress concentration.

the material and details of the test procedure. As a consequence of irradiation hardening, the DBT temperature increases (Odette *et al.* 1985).

Brittle fracture occurs by propagation of cleavage cracks, with little energy expenditure and plastic deformation at the crack tip (Fig. 3(a)). At higher temperatures, ductile fracture occurs by growth and coalescence of microvoids, requiring significant plastic deformation and large energy expenditure (Fig. 3(b)). Cleavage occurs only if an internal tensile stress (σ_t) exceeds a critical stress (σ^*). The σ_t is roughly proportional to σ_y, which increases with decreasing temperature, while σ^* is relatively constant (Fig. 3(d)).

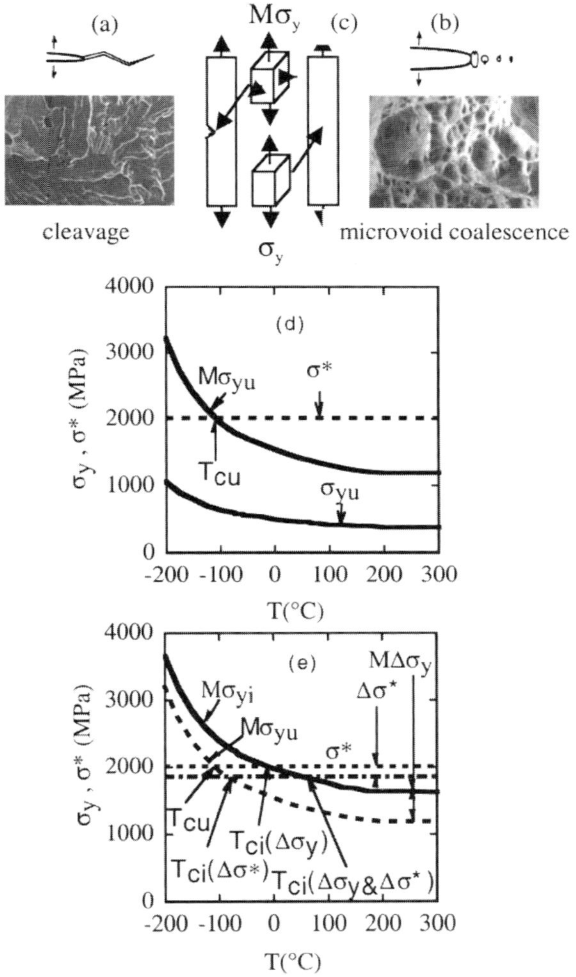

Figure 3
DBT and hardening-induced ΔT_c: (a, b) brittle cleavage vs. ductile microvoid coalescence; (c) stress concentration to a multiple (M) of the yield stress at a notch tip; (d) unirradiated T_{cu}; (e) irradiated T_{ci} owing to $\Delta\sigma_y$, $\Delta\sigma^*$, or both.

Even at very low temperatures, however, σ_t is too low ($\approx \sigma_y \ll \sigma^*$) to cause cleavage in a uniaxial tensile test, where the internal stress acts only in a single direction (Figs. 3(c) and (d)). However, internal stresses are both concentrated and act in all directions in the vicinity of a notch or crack. The corresponding maximum σ_t increases to about $M\sigma_y$, where $M = 2$–5 (Figs. 3(d) and (e)). Thus below a DBT temperature (T_{cu} in Fig. 3(d)), cleavage crack propagation initiates near a notch tip, when $M\sigma_y(T_c) = \sigma^*$. The unirradiated T_c of RPV steels is typically below room temperature.

RPV irradiation conditions do not affect the coarse scale (e.g., carbide) microstructures that govern σ^*. However, irradiation-enhanced phosphorus segregation to grain boundaries can result in a value of σ^* for intergranular fracture that is lower than for cleavage. Thus increases in T_c (ΔT_c) can be caused by irradiation hardening, $\Delta\sigma_y = \sigma_{yi} - \sigma_{yu}$ (where "i" and "u" indicate irradiated and unirradiated conditions, respectively), $\Delta\sigma^*$, or a combination of both. These effects are labeled T_{ci} ($\Delta\sigma_y$, $\Delta\sigma^*$ and $\Delta\sigma_y$, $\Delta\sigma^*$) in Fig. 3(e). Irradiation hardening is the dominant mechanism in RPV steels of Western countries. Since $\Delta\sigma_y$ does not depend on test temperature, $\Delta T_c = T_{ci} - T_{cu}$ is simply defined by the condition $\sigma_{yu}(T_{cu}) = \sigma_{yu}(T_{ci}) + \Delta\sigma_y$ ($= \sigma^*/M$). The actual relationship between $\Delta\sigma_y$ and the CVN ΔT_t at 41 J is more complicated, but can be approximated as $\Delta T_t \approx [0.6 \pm 0.2\,°\mathrm{C\,MPa}^{-1}]\Delta\sigma_y$.

3.2 Irradiation Hardening

The value of σ_y is governed by the shear stresses needed to move dislocations. Dislocations are line defects in a crystalline solid that separate slipped and unslipped regions on a shear plane (Fig. 2(c)). The stress needed to move a dislocation is much less than that needed to produce simultaneous shear of all atoms across a slip plane. Plastic strains are simply the accumulated shear owing to the long-range motion of large numbers of dislocations.

The intrinsic b.c.c. crystal lattice resistance to dislocation motion is high at low temperatures, but is negligible above about room temperature. However, contributions to strength that are relatively insensitive to temperature are provided by grain boundaries, other dislocations, and almost any sizable structural or chemical deviation from the perfect crystal lattice. Second-phase precipitates are very effective in pinning dislocations in this manner (Figs. 2(b) and (c)). The strength contribution of pinning points, $\sigma_{yp} \approx 6 \times 10^{-5}\beta_p\ (N_p d_p)^{1/2}$, where β_p (<1) is the effective pinning strength and d_p and N_p are the diameter and number of pinning points per unit volume, respectively. For example, for $N_p = 10^{24}\ \mathrm{m}^{-3}$, $\beta_p = 0.1$, $d_p = 2\ \mathrm{nm}$, $\sigma_{yp} \approx 270\ \mathrm{MPa}$.

Irradiation produces nanoscale pinning features, including small precipitates and clusters of solute atoms and lattice defects (Fig. 2(b)). Thus $\Delta\sigma_y$ and ΔT_c are controlled by the types, numbers, and sizes of the nanofeatures that form under irradiation.

3.3 Primary Defect Production

Neutrons create defects by displacing atoms from their normal crystal lattice sites. The most basic displacement defects are pairs of lattice sites missing an atom, or vacancies (Fig. 4(d)), and lattice sites shared by two atoms, or self-interstitials (Fig. 4(e)). Displacements are produced in cascades resulting from highly energetic primary recoiling atoms (PRAs) generated by neutron scattering. The interaction of a high-energy neutron with an atomic nucleus results in energy transfer (T) (Fig. 4(c)) (Greenwood 1994). For example, a 1 MeV neutron transfers up to about 70 keV to an iron PRA. The PRA kinetic energy is quickly transferred by secondary, tertiary, and n-subsequent generations of collisional displacements, producing 2^n recoiling atoms at lower energies ($T/2^n$) (Figs. 4(a)–(c)). Some recoil energy is lost to electrons. The process terminates when the kinetic energy of the recoils is below that needed to cause displacements (Robinson 1994).

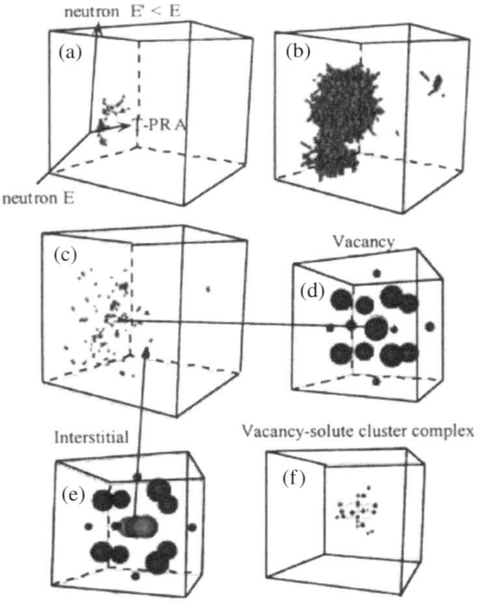

Figure 4
Cascade primary damage production (iron atoms not shown in (a)–(c) and (f)): (a–c) MD simulation snapshots of initial, intermediate, and final dynamic stage of the displacement cascade; (d, e) vacancy and self-interstitial defects; (f) vacancy–cluster solute complex formed after long-term cascade aging.

Understanding of primary damage production is largely based on molecular dynamics (MD) (Bacon and Diaz de la Rubia 1994, Stoller *et al.* 1997), Monte Carlo (MC) computer simulations (Wirth and Odette 1999), as well as from indirect experimental measurements (e.g., electrical resistivity that increases with defect concentration). MD simulations start with a computational lattice of thousands to millions of atoms that interact by prescribed force-separation distance (or interatomic energy potential) laws. All the forces in the crystal exactly balance when the atoms are in their proper, equilibrium positions, including the distortions owing to defects. However, atoms that are out of position move since they feel a net force ($F = ma$). The MD simulations of displacement cascades impart a specified PRA T (and direction) and then integrate the $F = ma$ equations of motion of all the atoms (Figs. 4(a)–(c)) until the kinetic energy is dissipated to heat within about 10–100 ps. On average, a PRA creates $\nu \approx T/D$ displacements, where $D \approx 0.125$ keV. Thus $\nu \approx 160$ in a typical 20 keV cascade.

The total displacement cross-section (σ_{dpa}) is ν times the effective PRA cross-section. For the neutron spectrum at the inside of the vessel, $\sigma_{dpa} \approx 1.5 \times 10^{-25}\,m^2$. The number of displacements per atom (dpa) is $\phi t_{>1}\sigma_{dpa}$. Thus an end-of life $\phi t_{>1} \approx 3 \times 10^{23}\,nm^{-2}$ produces about 0.045 dpa. However, closely spaced interstitial and vacancies recombine and annihilate in periods of less than about 100 ps. As a consequence, only about a third of the initial displacements survive. Typically, this leaves a vacancy-rich cascade core, surrounded by a shell of interstitials (Fig. 4(c)).

The majority of the interstitials quickly form small disk-shaped clusters that are identical to small dislocation loops (Wirth *et al.* 2000). Along with the interstitials, these clusters are very mobile and almost immediately leave the cascade region, unless they are strongly trapped by other defects or solutes. Although they are less mobile than interstitials, vacancies also diffuse by sequences of jumps between lattice sites. This stage of damage evolution can be modeled by Monte Carlo (MC) methods. The MC algorithm exchanges each vacancy (or other defect) with a neighboring site at a rate that decreases exponentially with the height of the energy barrier (E_j) to the jump, $\exp(-E_j/kT)$. E_j depends on the arrangement of the neighboring atoms and vacancies. For example, E_j is higher if the vacancy is bound to a solute or another vacancy. Through a series of jumps, the vacancies and solutes in the cascade evolve to lower energy configurations. Within milliseconds, many of the cascade vacancies jump short distances to form small three-dimensional clusters, while others leave the cascade region (Wirth and Odette 1999). The small clusters are unstable and can dissolve by vacancy emission. However, they also rapidly diffuse as clusters and coalesce with each other, forming larger nanovoids, which persist for much longer times. Solute atoms bind to the vacancies and segregate to clusters (Fig. 4(f)). The vacancy emission rate is lower from vacancy–solute cluster complexes. Small solute clusters left after all the vacancy clusters have finally dissolved.

In summary, cascades produce a range of clusters (defects, solutes, and defect–solute complexes) that directly contribute to irradiation hardening. However, most of the vacancies and interstitials eventually migrate long distances from the cascade region. Long-range diffusion results in additional nanostructural evolutions.

3.4 *Irradiation-induced Nanostructures*

Understanding of embrittlement nanofeatures is based on combinations of sophisticated microstructural and microchemical characterization studies and physical models. Key characterization methods include small-angle neutron and x-ray scattering that depend on the number, size, and chemistry of the nanofeatures (Solt *et al.* 1993, Odette and Lucas 1998, Phythian and English 1993, Williams and Phythian 1998); various types of electron microscopy that provide both high-resolution structural images and compositional information (Phythian and English 1993, Jenkins 1994, Williams and Phythian 1998); and three-dimensional atom probe-field ion microscopy that measures the configuration of the elements within a small sampling volume to near atomic resolution (Miller *et al.* 2000). Hardness recovery during annealing at $<350\,^{\circ}\mathrm{C}$ has also been used to study the matrix features, which have proved very difficult to characterize by other methods. Thermodynamic–kinetic models are used to track the transport and fate of irradiation defects and solutes and to predict the number, size distribution, and composition of the nanofeatures (Odette and Lucas 1998, Liu 1997). While all of these tools have individual limitations, in combination they have provided a reasonable insight into the nanofeatures that can be divided into three broad categories:

- Copper-rich or catalyzed Mn–Ni-rich precipitates (CRPs/MNPs).
- Unstable matrix defects, formed even in steels with little or no copper, which anneal rapidly compared to typical irradiation times.
- Stable matrix features (SMFs) that persist or grow under irradiation even in steels with little or no copper.

The unstable matrix defects play a role in the T_i and ϕ dependence of hardening only in the high ϕ regime, pertinent to accelerated test reactor irradiations. However, they are not important for surveillance or vessel $\phi \ll 10^{16}\,\mathrm{nm^{-2}s^{-1}}$. A large fraction of the vacancies and interstitials annihilate in equal numbers at sinks, like grain boundaries and dislocations. However, clusters formed in cascades may grow owing to a slight bias in the flow of interstitials to dislocations and vacancies to nanovoids. Various solutes also segregate to the loops and nanovoids by long-range diffusion.

An even more important consequence of displacement damage, however, is greatly enhanced rates of solute diffusion. Diffusion normally takes place by solute jumps into a neighboring vacancy. Repetition of this process many times results in long-range solute migration (Fig. 5(a)). Hence, the excess vacancies produced by neutron irradiation greatly increase diffusion rates of elements like copper.

The primary effect of radiation-enhanced diffusion is the formation of fine scale CRPs (Odette and Lucas 1986, 1998). The effective solubility of copper in the iron matrix is about 0.3%. However, the solubility of copper decreases at lower temperatures, to $<0.01\%$ at around $290\,^{\circ}\mathrm{C}$. The excess copper very slowly diffuses and precipitates into a separate phase (Figs. 5(b)–(d)). However, radiation-enhanced diffusion enormously accelerates this process, resulting in the rapid formation of a high number ($\geqslant 10^{23}\,\mathrm{m^{-3}}$) of

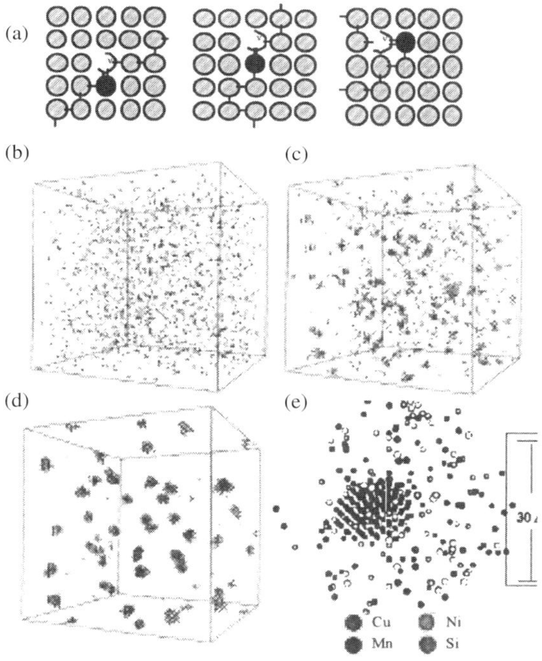

Figure 5
Solute diffusion and precipitation (iron atoms not shown in (b)–(e)): (a) vacancy exchange solute diffusion mechanism; (b–d) MC simulation of copper precipitation; (e) MC simulation of an alloyed (Cu–Mn–Ni–Si) CRP.

small ($\sim$1.5–3 nm diameter) CRPs with a b.c.c. crystal structure that is coherent with the iron matrix.

The CRPs are the dominant hardening feature in sensitive steels that have copper contents greater than about 0.05–0.1%, which is the minimum needed for rapid nucleation. The CRP volume increases with copper up to an effective upper limit around 0.3%, imposed by coarse scale copper pre-precipitation at the final stress relief temperature ($\approx$620 °C). The CRP contribution has a relatively weak T_i dependence and saturates at high ϕt owing to copper depletion from the matrix. At very high ϕ ($>10^{16}\,nm^{-2}s^{-1}$), CRP evolution is delayed by the reduction of irradiation-enhanced diffusion efficiency owing to a large number of unstable matrix defects that act as vacancy–interstitial sinks. At very low ϕ ($<10^{14}\,nm^{-2}s^{-1}$) CRP evolutions may be accelerated owing to a contribution of thermal diffusion to copper precipitation. The corresponding effect of ϕ at intermediate levels is not well established.

The CRPs are enriched in manganese and nickel, as well as smaller amounts of phosphorus and silicon (Fig. 5(e)) (Liu *et al.* 1997, Odette and Lucas 1998). The nickel and manganese strongly bind and amplify the effect of copper by increasing the volume of the precipitates. In some cases this can result in replacement of CRPs with MNPs with a small copper-rich core. The MNPs are promoted by high manganese and nickel content, low copper content (beyond the amount needed for nucleation), and low T_i. This explains the strong interaction between copper and nickel in increasing hardening and embrittlement.

At very low copper levels, distinct MNPs have not been observed up to intermediate ϕt. However, the Cu–Mn–Ni–T_i regime for formation of MNPs at very high ϕt, if any, is not known. The potential for the formation of such late-blooming phases is a concern since, if MNPs do eventually form, rapid embrittlement could occur even in low-copper steels. Other irradiation-enhanced fine scale precipitates include various compounds of molybdenum, vanadium, silicon, phosphorus, nitrogen, and carbon. These features probably play a more important role in VVER steels (Miller *et al.* 2000).

3.5 Annealing and Re-irradiation Embrittlement

Postirradiation annealing (PIA) results in further changes in the nanofeatures, leading to partial to full recovery of the pre-irradiated mechanical properties (Eason *et al.* 1998b). The SMFs dissolve at about 375–400 °C. The CRPs are more stable, but they partially dissolve and become much coarser at 425–450 °C. High T_a PIA quickly redissolves the CRP manganese and nickel. Some copper is also redissolved at short annealing time (t_a), but subsequently flows to the coarsened precipitates. The smaller volume and much lower number of nearly pure copper precipitates result in far less hardening.

Hardening during subsequent re-irradiation is primarily owing to the development of a new population of SMFs. Residual copper in solution above about 0.1% may also precipitate as new CRPs. However, most of the copper can be effectively sequestered in the coarsened precipitates. Thus high T_a is an effective means of persistently reducing embrittlement.

4. Two-feature Engineering Models of Irradiation Embrittlement and Annealing

The physical understanding and detailed models described above provide a very useful basis for developing quantitative engineering predictions of ΔT_t (Odette and Lucas 1986, 1998, Fisher and Buswell 1987, Eason *et al.* 1998a, 1998b). The detailed models have been used to derive simpler, but physically based, equations for $\Delta T_t = f(\text{Cu, Ni}, T_i, \sigma t, T_a, t_a, \ldots)$ that are statistically fit to the engineering database by nonlinear least square regression analysis (Eason *et al.* 1998a). Selecting the best engineering model is guided by consistency with independent data from well-controlled, single-variable test reactor experiments and physical understanding.

A two-feature model (SMF and CRP) provides an excellent fit to the large (609 ΔT_t points) US power reactor surveillance database:

$$\Delta T_t = Af_{\text{smf}}(T_i,\ \phi t,\ \text{P}) + Bf_{\text{crp}}(\text{Cu, Ni}, \phi, \phi t)$$

with a standard error of $\pm$13 °C.

Both the coefficients for the SMF (A) and CRP (B) differ between welds, plates, and forgings. The welds are the most sensitive product form and the forgings the least sensitive. For example, taking Cu = 0.15 and $\phi t = 3 \times 10^{23}$ and the other variables as given below, the total ΔT_t and CRP contributions are 85 and 60 °C (welds); 77 and 49 °C (plates); and 52 and 31 °C (forgings). The lower sensitivity of the forgings is primarily due to their lower manganese content ($\approx$0.8%) compared to $\approx$1.5% for plates and welds. The CRP contribution to ΔT_t is accelerated at $\phi < 10^{14}\,nm^{-2}$. Other representative ΔT_t predictions are shown in Fig. 6. Unless otherwise indicated, the default variables are: welds, $T_i = 290$ °C, $\phi t = 10^{23}\,nm^{-2}$, P = 0.01%, Cu = 0.3%, Ni = 0.8%, and $\phi = 5 \times 10^{14}\,nm^{-2} - s^{-1}$. The CRP term saturates at high ϕt owing to the depletion of matrix copper (Fig. 6(a)). There is a very strong Cu–Ni interaction (Fig. 6(b)). The threshold and the effective maximum copper are about 0.07% and about 0.3% (Fig. 6(c)) owing to the CRP nucleation and pre-precipitation limits, respectively. While detailed refinements are possible, all these trends in the CRP term are in excellent agreement with both independent experiments and understanding of embrittlement mechanisms.

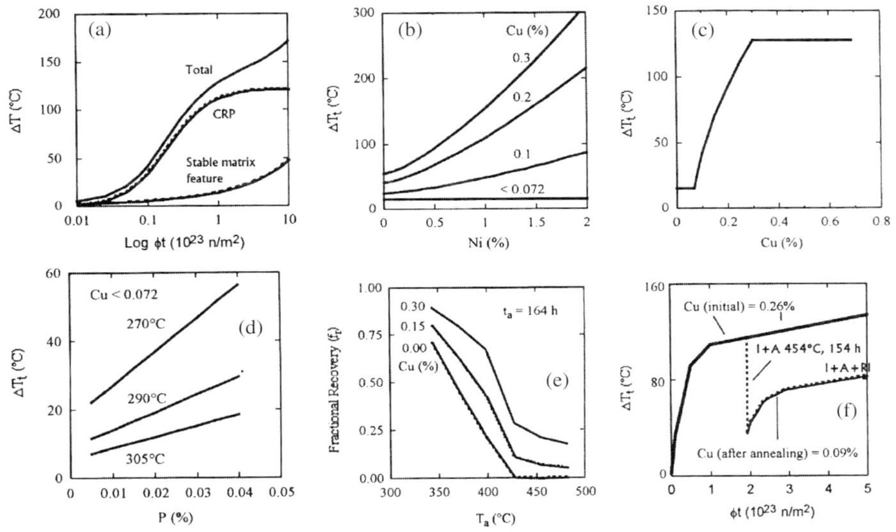

Figure 6
Physically based statistical correlation model predictions of ΔT_t: (a) fluence dependence; (b) copper–nickel interaction dependence; (c) copper dependence; (d) T_i and phosphorus dependence of the SMF; (e) fractional PIA recovery; (f) PIA recovery and re-embrittlement.

The SMF contribution increases roughly with the square root of ϕt (Fig. 6(a)) as well as with increasing phosphorus content and decreasing T_i (Fig. 6(d)). These trends are also consistent with independent sources of information. However, the causes, character, and consequences of the SMFs are not as well understood as in the case of CRPs, and knowledge of them needs to be improved with additional research.

The two-feature model has also been applied to correlating fractional recovery (f_r) PIA data, $\Delta T_{ta} = f_r \Delta T_t$, where ΔT_{ta} is the residual shift after annealing (Eason *et al.* 1998b). f_r primarily depends on T_a, t_a and the copper content (Fig. 6(f)). Below 400 °C, f_r also depends on ϕ owing to a third hardening contribution of the unstable matrix defects in the high-flux test reactor data used in the analysis. Microhardness recovery data from both single-variable test reactor studies and measurements on surveillance specimens provide independent confirmation of the annealing recovery model.

The formulations for embrittlement and PIA can be combined to predict re-embrittlement, assuming that high T_a returns the steel to its original state, except for the redistribution of copper. For example, the results in Fig. 6(g) for a Cu = 0.26% weld annealed at 454 °C for 164 h assume a residual copper content of 0.09%. The re-embrittlement is moderate and PIA provides a persistent ΔT_t advantage relative to the unannealed condition. The residual copper can be estimated from measurements or models and used as a basis to optimize the selection of T_a and t_a.

5. *Summary, Issues, and Opportunities*

Embrittlement is primarily the consequence of radiation defects that both cluster and accelerate solute clustering and precipitation of copper-rich phases. Dislocation pinning by these features leads to hardening and corresponding increases in the DBT temperature. Annealing recovery is the consequence of dissolving and/or coarsening these features. All the major observed trends in embrittlement and PIA can be rationalized by physical models. These models are the basis for the two-feature correlations of engineering databases that provide accurate, quantitative predictions of both embrittlement and recovery by thermal annealing.

However, there are a number of issues that are not fully resolved. One set of issues relates to better quantification of factors such as the effect of ϕ, the maximum effective copper content as a function of details of the steel and vessel thermal history, and the effects of a number of secondary variables and variable combinations that are either not accounted for (e.g., manganese) or are crudely treated (e.g., SMFs and phosphorus). A second class of issues is the potential technical surprises, such as the formation of late blooming MNP phases in low-copper steels, or other new phenomena that occur at long times or very high ϕt, that lie beyond the scope of the database. Perhaps the most difficult issue is that associated with material variability and the inherent uncertainties about the composition and properties of the steels in the RPV itself.

Finally, a proposed master curve procedure provides an important opportunity to replace the indirect and approximate CVN-based method for establishing toughness–temperature curves (ASTM 1998). The master curve method directly determines both mean and statistical lower bound $K_{MC}(T)$ curves based on a relatively small number of tests on specimens potentially as small as precracked CVN bars.

See also: Nuclear Reactor Materials: Irradiation Effects

Bibliography

ASME 1986 Protection against nonductile fracture. *Boiler and Pressure Vessel Code Section III, App. G*. American Society of Mechanical Engineers, New York

ASTM 1998 Standard test method for determination of reference temperature T_o for ferritic steels in the transition range, ASTM E1921-97. *Annual Book of ASTM Standards 03.01*. American Society for Testing and Materials, West Conshohocken, PA, pp. 1068–84

Bacon D J, Diaz de la Rubia T 1994 Molecular dynamics simulations of computer cascades in metals. *J. Nucl. Mater.* **216**, 275–90

Eason E D, Wright J E, Odette G R 1998a *Improved Embrittlement Correlations for Reactor Pressure Vessel Steels, NUREG/CR-6551*. US Nuclear Regulatory Commission, Washington, DC

Eason E D, Wright J E, Nelson E E, Odette G R 1998b Embrittlement recovery due to annealing of reactor pressure vessel steels. *Nucl. Eng. Des.* **179** (3), 257–65

Fisher S B, Buswell J T 1987 A model for PWR pressure vessel embrittlement. *Int. J. Press. Vessels Piping*. **27** (2), 91–135

Greenwood L R 1994 Neutron interactions and atomic recoil spectra. *J. Nucl. Mater.* **216**, 29–44

Jenkins M L 1994 Characterization of radiation damage microstructures by TEM. *J. Nucl. Mater.* **216**, 125–56

Liu C L, Odette G R, Wirth B D, Lucas G E 1997 A lattice Monte Carlo simulation of nanophase compositions and structures in irradiated pressure vessel Fe–Cu–Ni–Mn–Si steels. *Mater. Sci. Eng. A Struct.* **238** (1), 202–9

Miller M K, Pareige P, Burke M G 2000 Understanding pressure vessel steels: an atom probe perspective. *Mater. Charact.* **44**, 235–54

Odette G R, Lombrozo P M, Wullaert R A 1985 The relationship between irradiation hardening and embrittlement of pressure vessel steels. In: Garner F A, Perrin J S (eds.) *Effects of Radiation on Materials—12th Int. Symp., ASTM STP 870*. American Society for Testing and Materials, Philadelphia,, pp. 840–60

Odette G R, Lucas G E 1986 Irradiation embrittlement of reactor pressure vessel steels: mechanisms, models and data correlations. In: Steele L E (ed.) *Radiation Embrittlement of Reactor Pressure Vessel Steels—An International Review, ASTM STP 909*. American Society for Testing and Materials, Philadelphia,, pp. 206–41

Odette G R, Lucas G E 1998 Recent progress in understanding reactor pressure vessel embrittlement. *Radiat. Effects Defects Solids* **144**, 189–231

Odette G R, Wirth B D 1997 A computational microscopy study of nanostructural evolution in irradiated pressure vessel steels. *J. Nucl. Mater.* **251**, 157–71

Phythian W J, English C A 1993 Microstructural evolution in reactor pressure vessel steels. *J. Nucl. Mater.* **205**, 162–77

Robinson M T 1994 Basic physics of radiation damage production. *J. Nucl. Mater.* **216**, 1–28

Solt G F, Frisius F, Waeber W B, Tipping P, 1993 In: Kumar A S *et al.* (eds.) *Effects of Radiation on Materials–16th Int. Symp.*, ASTMSTP 1175. American Society for Testing and Materials, West Conshohocken, PA, pp. 444–462

Stoller R E, Odette G R, Wirth B D 1997 Primary damage formation in BCC iron. *J. Nucl. Mater.* **251**, 49–60

Williams T J, Phythian W J 1996 Electron microscopy and small angle neutron scattering study of precipitates in low-alloy submerged-arc welds. In: Gelles D S, *et al.* (eds.) *Effects of Radiation on Materials—17th Int. Symp., ASTM 1270*. American Society for Testing and Materials, West Conshohocken, PA,, pp. 191–205

USNRC 1986 Rules and Regulations Title 10 Code of Federal Regulations Part 50 Fracture Toughness Requirement. US Nuclear Regulatory Commission, US Government Printing Office, Washington, DC

USNRC 1988 Regulatory Guide 1.99-Rev 2: Radiation Embrittlement to Reactor Pressure Vessel Materials. US Nuclear Regulatory Commission, US Government Printing Office, Washington, DC

Wirth B D, Odette G R 1999 Kinetic lattice Monte Carlo simulations of cascade aging in dilute iron-copper alloys. In: Zinkle S J, *et al.* (eds.) *Microstructural Processes In Irradiated Materials, MRS Symp. Proc. 540*. Materials Research Society, Warrendale, PA,, pp. 637–42

Wirth B D, Odette G R, Maroudas D, Lucas G E 2000 Dislocation loop structure energy and mobility of self-interstitial atom clusters in BCC iron. *J. Nucl. Mater* **276**, 33–40

Nuclear Reactors: Shielding Materials

M. T. Simnad

The term "shielding" in nuclear reactors is generally defined as a system of shielding materials that are assembled for a specific radiation protection purpose. This article presents a short survey of the following subjects related to the design of shields and the selection of shielding materials: (i) factors involved in the design of shielding, (ii) interactions of shield materials with the various types of radiation, (iii) relevant properties of shield materials. The topics addressed include radiation transport within the shield, radioactivation of shield materials, radiation levels in the vicinity of shields and radiation scattering around shields, radiation penetration through holes in the shields, optimization of shield configuration, deposition of heat in shields, radiation protection criteria, and the economics of shield design. The bibliography provides a list of monographs that survey shielding in nuclear reactors in detail. Shielding materials are reviewed in an AEC report (Simnad 1977).

1. *Design Considerations*

The primary purpose of shielding structures is to protect the components of the reactor or a shielded facility such as hot cells (see *Hot Cells, Glove Boxes, and Shielded Facilities*) and people from excessive radiation exposure. The objectives of shielding studies are to: (i) establish the design criteria for radiation shielding, (ii) compile existing data on candidate shielding materials, (iii) establish the basis for selection of configuration and materials, (iv) determine test requirements and conduct the tests, (v) establish the design and materials specifications and (vi) develop procedures and techniques for the fabrication of the shields as required.

Shielding design has to address the various types of radiation that are present in nuclear reactors. These are divided into two types, namely, (i) directly ionizing radiations, which interact very strongly with shielding materials and are readily stopped, and (ii) indirectly ionizing radiations, which are penetrating and require massive shielding. The directly ionizing radiation includes energetic particles carrying electric charge, such as alpha and beta particles and protons. They cause ionization by interacting with electrons in the atoms of the materials through which they pass. The indirectly ionizing radiations, such as neutrons, gamma rays and x rays, involve the emission of energetic secondary charged particles, which cause the ionization. During reactor shutdown and in portions of the coolant circuit, which lie outside the main shield, the types of radiation that are produced by fission product decay, coolant activity, activation of structural materials, and bremsstrahlung must be addressed. The choice of possible shield materials may also be restricted by requirements of reactor safety. For example, during a reactor runaway, melting of the shield material should not occur. Thermal stresses may have to be mitigated by absorbing most of the heat energy in additional thermal shields. Heat energy is generated in shield materials in transfer of the kinetic energy of fast neutrons during slowing down (moderation), absorption of gamma radiation from the core and reflector, and absorption of gamma radiation produced during the neutron capture process in the shield. Also, the chemical compatibility has to be a major concern; for example, a water shield would be a poor choice for a sodium-cooled reactor. The acceptance criteria evolve from analysis of the properties within the bounds of design considerations.

2. *Radiation Interactions with Materials*

Microscopic cross-section is a measure of the probability that a photon or neutron will interact with a nucleus, an atom or an electron. It includes the cross-sections for absorption, scattering, inelastic scattering, and total cross-section, and has the units of area. The unit used is one barn (b), which is equivalent to $10^{-24}\,cm^2$. Macroscopic cross-sections are the product of the microscopic cross-sections and the number of atoms per unit volume of the material. The passage of neutrons and photons through the shield materials includes the following four stages:

(i) interaction with the nuclei or electrons, to produce secondary charged particles;
(ii) the secondary charged particles have kinetic energies related to the energy of the primary particles and the type of interactions;
(iii) the secondary charged particles lose energy through ionization and the emission of photons called bremsstrahlung,
(iv) scattering and other processes by uncharged primary particles may produce additional uncharged particles, which also remove part of the energy of the interaction.

There are two types of primary radiation damage mechanisms, namely, atomic displacements and changes at the molecular level. The former takes place mainly in metals and the latter in nonmetals. A combination of both mechanisms occur in semiconductors and insulators. In organic materials the effects of radiation include polymerization, cross-linking and chemical dissociation. Ionizing radiation dissociates water to produce many free radicals, including hydrogen, oxygen, and hydrogen peroxide.

2.1 *Neutron Interactions with Materials*

Neutron absorption cross-sections for all nuclides are small at all energies except thermal energies. The (n, gamma) reaction is significant only for thermal neutrons and for a few absorption resonances for the heavy elements. In the high energy region, the scattering of fast neutrons is predominant. At energies above about 8 MeV, reactions such as (n, 2n) and (n, n + p) can occur. In light elements interactions such as (n, p) and (n, alpha), which produce charged particles, are possible. Only hydrogen and deuterium exhibit no resonances in their interactions with neutrons. Their cross-sections above 1 eV are almost constant up to the MeV region and then decrease with increasing energy. In practice, an important concern in shield design is with neutrons of energies above 0.5 MeV.

The principal nuclear data required for neutron shielding calculations are the following: (i) high-energy interactions ($1\,\text{eV} < E < 20\,\text{MeV}$), which include elastic and inelastic neutron scattering cross-sections and their angular distributions, gamma ray yields from inelastic neutron scattering, and resonance absorption cross-sections; and (ii) low-energy interactions ($< 1\,\text{eV}$), which include thermal-averaged absorption cross-sections, yield of neutron capture gamma rays, and fission cross-sections, and associated neutron yields and gamma ray yields. Neutron shielding analysis involves attenuation of both the primary neutrons and the secondary particles, as well as the production of gamma rays during neutron inelastic scattering, thermalization of neutrons, capture of thermal neutrons resulting in emission of gamma photons, and in the production of secondary neutrons by fission or (n, 2n) reactions. The analysis may also address the problems of energy deposition and the resulting thermal and radiation damage effects.

2.2 *Gamma Photon Interactions with Materials*

In nuclear reactors attenuations of about a billion are often required for gamma photons. The attenuation of gamma photons is governed primarily by the atomic number and density of the shielding material. The three mechanisms of gamma attenuation are pair production, the photoelectric effect, and Compton scattering. The attenuation proceeds approximately exponentially with distance. The region of gamma photons in the range of 0.5 MeV to 10 MeV is the most important in reactor technology, because it includes the penetrating radiations from fission, capture, and inelastic scattering of neutrons, as well as from the decay of many fission products. It is the region where Compton scattering is the most important and equal masses of all materials are equally effective, with a minimum value of about $60\,\text{g cm}^{-2}$. For heavy elements there is a fairly sharp minimum in the mass absorption cross-section at about 3 MeV, while for light elements this minimum occurs only at very high energies.

3. *Selection of Materials*

Limitation of radiation intensities to prescribed levels in nuclear facilities may be met by many materials and combinations of materials and geometric configurations. However, the specific materials selection process for a particular shield design is governed by many factors. These include radiation attenuation, radiation damage, materials and construction costs, structural strength, weight and space limitations, safety, and corrosion and maintenance considerations. Other important factors include the temperatures and temperature gradients and thermal stresses in the shield, as well as the effects of nuclear irradiation on the properties of the shielding materials. Although the attenuation of neutrons and gamma photons are the primary nuclear properties considered, many shielding materials also produce secondary gamma rays from neutron inelastic scattering and radioactive capture. Thermal protection is often required for inner layers of shielding materials, so that cooling must be provided.

The design objectives for a facility generally influence the selection of shield materials. For example, in mobile systems total weight becomes an important factor and needs to be optimized. The usual approach is to select an effective combination of shield materials compatible with the design rather than to select a single material. In such cases, several component materials are layered so as to furnish the minimum weight by providing the optimum balance between primary and secondary radiation and geometric effects on material volume. For example, the reactor shields of the Russian nuclear icebreaker surface ships are composed of water, steel and heavy concrete. The reactor shield on the US prototype nuclear merchant ship NS *Savannah* was composed of light water, lead, polyethylene, standard concrete and heavy concrete. Other shielding materials combinations include lead plus polyethylene, uranium plus titanium hydride, and tungsten alloys plus lithium hydride. The constraints on shield design include volume limitations, corrosion, dose-rate requirements, and number of shield layers. The streaming of neutrons through gaps and ducts in the shield cannot be mitigated by placing bends in a duct, as may be done with gamma radiation. This requires special care in the shielding design so as to reduce the streaming of neutrons in such channels. Hence, the optimization procedure must address these constraints together within the boundary conditions in order to determine the optimum design parameters.

Installation costs for shielding are influenced primarily by materials and construction costs. Maintenance and salvage value of the materials also influences the economics. However, shield weight and volume often influence other installation costs, such as foundation support and size of the containment structure. Properties of shield materials are discussed in the following sections.

3.1 Water

Both light and heavy water have desirable properties as reactor coolants (see *Nuclear Reactors: Coolant Materials*) and radiation shields. Water is an excellent neutron attenuator because of its large hydrogen content. It is not a good gamma ray attenuator because of its low electron density. In a water shield the dose contributed by neutrons is largely due to fast neutrons. The 330 mb thermal neutron capture cross-section of hydrogen gives only 2.2 MeV gamma rays, so that water is relatively low in secondary gamma-ray production. The oxygen atom in water has no thermal neutron capture. Radiolysis of water occurs in the presence of highly energetic radiation. The water molecules may be ionized, excited, or dissociated by the reactor radiation. This is followed by reactions amongst the species themselves and with normal water molecules.

The relative yields of these reactions depend primarily on the ionization density produced by the radiation. The more diffuse ionization from gamma rays favors the formation of the active radicals H and OH. The overall decomposition process initially produces about 0.4 molecules of hydrogen per 100 eV of energy absorbed in the water for radiations of low ionization density. This is the G-value, which is the number of molecules of hydrogen or deuterium that are produced for every 100 eV of energy liberated by absorption of radiation. For mixed reactor radiations, the G-value is about 1.8 molecule for densely ionizing radiations. In pure water the radicals H and OH combine to form water, hydrogen peroxide and hydrogen. The back reactions serve to reform the water and are facilitated by an increase in hydrogen concentration in the water. Impurities present in the water, such as chlorides, react with the radicals and thus prevent the back reactions. Hence, the water has to be kept very pure by use of ion exchangers. In fact, the hydrogen formed will reach an equilibrium concentration in an enclosed system and the hydrogen pressure builds up until the reverse reaction becomes effective. When water is used as a shield material the problems of purification, containment, temperature, control, and corrosion must be addressed.

Heavy water radiolysis is similar to that of light water, but it is necessary to recover the deuterium released in order to use it to generate heavy water for further use. Also, the reaction $^{2}D \rightarrow (\text{gamma, n}) \rightarrow {}^{1}H$ depletes the deuterium. Isotopic pollution of heavy water results from atmospheric humidity and leakage or condensation of normal water, so that the surface should be covered with a pressurized inert dry gas. The irradiated heavy water has to be subjected to a scrubbing process to remove tritium. Tritium is generated in heavy water by the reaction $^{2}D \rightarrow (\text{n, gamma}) \rightarrow {}^{3}T$, and decays with the emission of beta particles of 0.018 MeV energy and with a 12.3 y half-life.

3.2 Concrete

Concrete is the most widely used shielding material. It can be used to shield against both neutrons and gamma radiation. It is the least expensive material and can readily be formed into complex-shaped shields with good structural properties. Concrete is composed of a mixture of cement (about 13%), water (about 7%), and aggregate (about 80%). Many types of concrete are prepared by varying the nature of the aggregate. Ordinary (portland) concrete is referred to as light and has a density of about 2.2–2.4 $g cm^{-3}$. Heavy concretes have densities ranging from about 3–6 $g cm^{-3}$ and have the advantage of a smaller volume required for a given total attenuation, but are more costly. Scrap iron of small size or iron shots may be dispersed in the concrete to obtain high density. Ferrophosphorus ore aggregate and limonite mixtures have been used because of their greater densities and higher water content. Colemanite aggregate is a hydrated calcium borate with high boron and fixed-water content. The barium sulfate ore barite is also used as an aggregate in heavy concretes. These concretes contain boron for neutron absorption and hydrogen for neutron attenuation. The barite ore is also a good gamma attenuator.

The thermal neutron absorption properties of concrete can be greatly enhanced by the addition of boron compounds. Hydrogen in fixed water (hydrated form) and free water (in the pores) in concrete serves as a good neutron shield. Initially the free water content is about 3% by weight. The water may be lost at ambient temperatures by evaporation, as much as half the water being lost over a 20 to 30-year period. In order to minimize the evaporative loss of water and the thermal stresses, the design criteria limit the temperatures to maximum ambient temperatures of 149 °C for gamma-photon shields and 71 °C for neutron shields. The maximum internal temperature for gamma-photon shields is 171 °C, and 88 °C for neutron shields, 6 °C for maximum ΔT from internal heating, and a 1 °C maximum temperature gradient. Steel reinforcements are normally included in the concrete shielding. Radiation damage to the properties of concrete has not been found to be significant and there is no evidence of loss of hydration by radiation. Over-stressing owing to nuclear heating is a more serious problem. A steel thermal shield of 15–25 cm in thickness is included in most reactors in order to reduce the problems posed by nuclear heating. This also serves to reduce the thickness of the main concrete shield.

3.3 Steel

Steels are present in reactor structures, in thermal shields, and in pressure vessels. These include carbon steels, alloy steels, and austenitic and ferritic stainless steels. Neutron irradiation of the iron in steel produces a number of gamma rays of up to 10 MeV energy, with most of them at 7.6 MeV. The isotope ^{58}Fe is activated by thermal neutrons to produce ^{59}Fe, which emits 1.5 MeV gamma rays and has a half-life of 47 days. The radioactivity may limit access to a reactor after shutdown.

Neutron irradiation affects the mechanical and physical properties of steels, depending on the composition and microstructure and on the neutron fluence and energy spectrum. Steels experience irradiation damage, which is manifested by hardening and embrittlement, loss of ductility, swelling, and enhanced creep. Structural carbon steels and ferritic stainless steels, which have body-centered-cubic crystal structures, exhibit a ductile–brittle transition temperature (NDTT) below which brittle fracture occurs. It may range from −70 to +20 °C. Irradiation increases the NDTT by as much as several hundred degrees at high exposures.

In the USA, the ASME Boiler and Pressure Vessel Code, Section III, Nuclear Power Plant Components, governs design criteria for accounting for permanent steel structural components. These rules generally apply to light water cooled reactors, where the components are designed to operate in the sub-creep range of temperatures. For applications where the components operate in the creep range of temperature regime the ASME Code Case 1592 is applicable. This code provides creep stresses that limit the elastically calculated load-controlled stresses for the stainless steels Types 304 and 316, Alloy 800H, and 2-1/4 Cr-1 Mo ferritic steel. A factor of safety is applied to the analyses for the defects in the materials (Section XI of the ASME Boiler and Pressure Vessel Code). The limits on the service degradation of materials properties are defined in the Code of Federal Regulations (10CFR50) and in the Regulatory Guides issued by the US Nuclear Regulatory Commission (USNRC). They require that the radiation effects be incorporated in structural analyses. The Structural Integrity Plan is used to assess the effects of materials flaws quantitatively in order to preclude failures resulting from flaw instabilities. This plan considers the influence of flaw size and the mechanical properties (particularly fatigue, corrosion fatigue, and stress corrosion) by means of analytical techniques.

A fluence limit based on an allowable 10% residual total elongation (RTE) is adopted for permanent steel internal components. This ductility level is considered acceptable to satisfy ductility requirements implied by the code. In addition, degradation of other mechanical properties seems to be acceptable at fluences required to degrade ductility to 10% RTE. Therefore, when component exposure does not exceed the fluence limit, irradiation damage can be neglected in the mechanical design of the component; that is, unirradiated material properties can be used for the mechanical design. When neutron exposure exceeds the limit to degrade ductility to 10% RTE, elastic or inelastic mechanical design analyses must account for radiation-induced properties changes and damage, such as increased yield and ultimate strength, reduction in fatigue life, enhanced creep, reduction of time to rupture, enhanced susceptibility to crack propagation, and combined swelling–creep deformation. Stainless steel weldments are more susceptible to radiation damage than the base metal and in some cases will limit component exposure.

Pre-stressed concrete reactor vessel liners consist of SA-537 Class B steel in the USA. The liner is treated as a pressure vessel, which is very conservative since pre-stressing of the concrete acts to produce a net compressive stress. The liner is sensitive to radiation embrittlement because of its low operating temperature, which is near 70 °C. The brittle–ductile transition temperature (NDTT) shift becomes limiting. A 33 °C minimum margin is required between the NDTT and the minimum service temperature.

3.4 Lead

Lead is the best material for gamma ray shielding next to uranium, because of its high density and high atomic number. Lead must be constrained in steel containers at temperatures above 250 °C to prevent creep flow and as a container if melting occurs in the event of a reactor runaway. It is generally alloyed with antimony to increase its creep strength. Lead shields can be fabricated by casting, extrusion, rolling, machining, and welding, or it can be used as lead shots as a filler for cavities in plugs and doors in shields. Its corrosion resistance in water is very good. It is used for gamma shielding when limitations on shield thickness are important. Lead has poor neutron attenuation properties and produces 7.4 MeV gamma rays on neutron capture. Commercial lead may include impurities that can result in activation problems.

3.5 Cadmium

Cadmium was used as a neutron absorber in the first nuclear reactor, CP-1, in 1942. It is a very good absorber of neutrons with energies below 0.6 eV. It has been alloyed with lead to improve its mechanical properties, as a 5% dispersion. Cadmium is less costly as an absorber of thermal neutron than boral (see below). However, its neutron capture gamma radiation is much more penetrating than that from boron, but less energetic than the principal iron radiation.

3.6 Tungsten

Tungsten is an excellent material for shielding high energy gamma radiation at elevated temperatures, because of its high density (19.3 gcm^{-3}) and high melting point (3400 °C). Various alloys of tungsten can be fabricated by powder metallurgy techniques, using binder metals such as copper, iron, or nickel, to form very dense (16–18 gcm^{-3}) machineable alloys. However, because of its high cost it is generally used for small shield components near the source. A layered configuration of tungsten, alternating with lithium hydride, provides good neutron and gamma attenuation.

3.7 Uranium

Large quantities of depleted uranium have been produced at the uranium enrichment plants. The ^{238}U shield is the best attenuator of gamma radiation on a mass absorption as well as on a thickness basis. However, depleted uranium does have a residual amount of ^{235}U ($\sim$0.2%) that should be considered in estimating activation, secondary gamma production, and heating. It has a very high density (19 gcm^{-3}) and good strength properties. However, alloying may be required to mitigate its large and anisotropic thermal expansion coefficient if used at high temperatures. Depleted uranium provides relatively poor attenuation for neutrons. Also, activation and transmutation (plutonium production) by neutrons would make it an undesirable shield material for attenuation of neutrons. It can be cast, forged, extruded, and machined. In the machining process protection must be provided against fire and toxicity. Uranium oxidizes in air and moist air will corrode uranium fairly rapidly. In the powder form it is pyrophoric. Protective cladding or alloying (e.g., with 2% molybdenum) is necessary from the corrosion standpoint.

3.8 Boron, Boral, and Boron Carbide

Boron is used primarily to absorb thermal neutrons because of its high absorption cross-section of 3840 barns for the $^{10}B \rightarrow (n, gamma) \rightarrow {}^{7}Li$ reaction by which the majority of the neutron captures result. There is an appreciable cross-section for high-energy neutron captures in the reaction $^{10}B \rightarrow (n, t) \rightarrow 2{}^{4}He$, which also generates tritium. Although the relative amount of radioactivity represented by tritium is quite small, the mobility of this hydrogen isotope in the reactor components has led to concern over its potential release to the environment. The alphas that are emitted are readily absorbed, but gamma heating poses a problem for large incident neutron flux densities. The 0.5 MeV gamma rays that are emitted are of concern in irradiated borated water and polyethylene.

Boron is available in a number of forms, including elemental boron, boric oxide, and as boron carbide. It is also employed as a sandwich material called boral, which consists of a hot-compacted aluminum–boron carbide powder mixture clad with pure aluminum. Boral can be produced in large sheets. It can be cut by shearing. The stability of boral under radiation is adequate.

Boron has been added to steel shields in order to decrease secondary gamma ray production and activation of the iron. In boron-loaded steel there is a reduction in both the number of neutrons giving capture radiation and in the residual activity. Boron can be added to molten steel as ferroboron to a concentration of about 3% by weight. Higher boron contents embrittle the steel. Greater amounts of boron can be added by using a dispersion of boron carbide in steel, in amounts up to about 8% by weight boron, with improved mechanical properties.

Boron carbide has a boron concentration of 85% of that of elemental boron. Natural boron contains 19.8% of the high-cross-section isotope ^{10}B, and the content of ^{10}B in natural boron carbide is 14.7%. The thermal neutron absorption cross-section of ^{10}B is 3840 barns and of natural boron carbide about 600 barns. The neutron absorption cross-section decreases with an increase in the neutron energy, by the $1/V$ relation, for neutron energies below about 100 eV. It remains fairly constant for energies between 100 eV and 0.1 MeV and has several resonances between 0.5 MeV and 5 MeV. The cross-section in a fast breeder spectrum is about 1 barn.

Boron carbide pellets and structures can be produced by cold-pressing and sintering (70–80% density) or by hot pressing to 99% density. Boron carbide structures of large and complex shapes can be fabricated by hot pressing in graphite dies at temperatures of from 2050–2300 °C under a pressure of 1500 psi. The crystal structure of boron carbide is rhombohedral and the composition B_4C is a carbon-saturated solid solution. The vapor pressure of boron carbide is very low below 300 °C, and the vaporization occurs mainly as the individual species boron and carbon. Boron carbide retains its room temperature strength up to about 1800 °C. The thermal conductivity decreases under irradiation. The loss in conductivity saturates rapidly at a level proportional to the temperature. The corrosion resistance of boron carbide in water or steam is very good at temperatures of up to 300 °C. In air, oxidation of the fine powder form becomes significant at about 600 °C, and coarse particles are resistant to oxidation at temperatures of up to about 1000 °C. Studies of the compatibility with stainless steel cladding indicate no significant interaction below 500 °C. Coating the stainless steel with copper restricts the interaction to acceptable levels at temperatures as high as 850 °C. The diametric swelling of neutron-irradiated boron carbide is given by the following equation:

$$\Delta D/D = C[-0.0211 + (62.04/T)]$$

where C = captures/cm^3 × 10 exp^{-20}and T = irradiation temperature in °F.

Helium release from irradiated boron carbide increases with increasing temperature. Irradiation tests of boron carbide pellets indicate that homogeneous microcracking is exhibited throughout the pellet cross-sections in fast neutron irradiated pellets, whereas in thermal neutron irradiated boron carbide the cracks are located near the circumference of the pellets. The difference in behavior is ascribed to the radial burnup gradient and a radial swelling gradient. There is self-shielding in the interior of the pellets in thermal neutron irradiation, but no significant radial neutron flux depression exists in a fast neutron environment and microcracking is consequently uniformly distributed.

3.9 Graphite

Besides its use as a moderator/reflector, graphite is used in thermal shields. It is a good shield material for slowing down fast neutrons and when mixed with boron carbide it is a very effective material for capturing neutrons. Boron-impregnated graphite has a very low residual radioactivity after irradiation. However, it is a very poor gamma-ray attenuator. Its high temperature properties make it suitable for neutron shielding at high temperatures and high flux densities. Graphite is made from petroleum coke and coal tar pitch. Graphite logs in diameters up to 1.25 m are manufactured by a hot extrusion process. It has an anisotropic structure so that its physical and mechanical properties are directional. Irradiation at temperatures below about 250 °C results in the accumulation of strain energy (stored energy), which is released upon annealing at higher temperatures by a self-sustaining energy release during a temperature excursion. This can present operational problems in reactors incorporating graphite at low temperatures, but is of no importance if the operating temperature of the graphite is above 300 °C (see also *Radiation Effects in Carbon–Carbon Composites*).

The most stable commercial-grade graphites are near-isotropic materials, such as H-451, which have been evaluated for the large high temperature gas-cooled reactors. The most important effects of irradiation on the graphites used for shielding are dimensional changes, reduction in thermal conductivity, and enhancement of creep rates. The dimensional change behavior of irradiated nuclear graphite is in the following pattern. In extruded material, there is first a contraction in the direction transverse to the extrusion direction and then a turnaround and rapid expansion. In the direction parallel to extrusion, there is a contraction at an increasing and then a decreasing rate, followed again by a turnaround and expansion. The rapid expansions result from porosity generation between filler particles, and the rate of expansion depends on the graphite microstructure. The isotropic materials expand at a lower rate. The initial shrinkage rate is temperature dependent, decreasing up to 800 °C and then increasing with increasing temperature up to 1200–1400 °C.

The thermal conductivity of the graphite is markedly reduced by irradiation. The rate of reduction declines and the conductivity approaches a saturation level, which increases as the temperature increases. Eventually, when irradiation-induced expansion starts, the conductivity again decreases.

The irradiation-induced creep of graphite has been studied. The transient creep strain and the steady-state creep constant increase with increasing irradiation temperature over the interval 500 °C to 1200 °C. For different graphites, the transient creep strain and the steady-state creep constant are both inversely proportional to Young's modulus. Creep strains up to 2.5% in tension and 5% in compression have been reported. However, there is some indication that in isotropic graphites compressive creep slows down or stops when the strain reaches 2–3%. It is desirable to use different types of graphite materials in the various shielding components. The choice of material is dictated by the fast neutron fluences and the temperatures in the different shield regions. For example, strong graphite would be used in the highest fluence high-temperature regions. The longer residence time would justify the higher cost because of savings accruing from elimination of costly periodic replacement operations.

Boron carbide–graphite bodies are fabricated by either extrusion or warm-pressing and then heat-treated to 2000 °C, while protected from oxidation. Heat treatment is limited to below 2200 °C to prevent migration of boron into the graphite, which would enhance radiation swelling. At temperatures above 2000 °C, the structural stability and strength of the boronated graphite materials degrade. Boron carbide melts between 2140 and 2450 °C and reacts with the graphite matrix. The interaction of boronated graphite bodies with metal cladding has been investigated. The reaction rate was found to be acceptable with Haselloy-X and Incoloy-800 cladding at temperatures up to 800 °C. Boronated graphites containing 23–43% by weight boron as boron carbide were irradiated to very high fast neutron fluences. Irradiation caused an anisotropic dimensional change related to the preferred orientation of the graphite crystallites in the graphite matrix, a decrease in thermal conductivity, and an increase in thermal expansivity. The dimensional changes in boronated graphite (less than 2% change in dimensions at 300–750 °C) were related to the fast neutron fluences and ^{10}B fission damage to the matrix. The damage increased with increasing ^{10}B isotope enrichment of the boron in the boron carbide particles. The use of natural boron in the boronated graphite resulted in dimensional changes independent of the ^{10}B burnup gradient and correlated with the fast neutron fluence. The use of ^{10}B enriched boron carbide led to distortion directly related to the high ^{10}B burnup gradient. The dimensional changes were not influenced by swelling of the boron carbide particles during irradiation.

3.10 *Beryllium and Beryllium Oxide*

Beryllium and beryllium oxide have been used as the moderator, reflector, and shield in a number of nuclear reactors. Beryllium metal has many of the desirable nuclear properties for a moderator and reflector, such as low neutron absorption cross-section, high neutron scattering cross-section, low atomic weight, high melting point, high specific heat, and fairly good corrosion resistance in water. Fast neutron transmutation reactions in beryllium yield helium and tritium, through (n, 2n) and (n, alpha) capture reactions, namely:

$^{9}Be \rightarrow (n, 2n) \rightarrow 2\,^{4}He$, threshold 1.85 MeV, 125 mb cross-section,
$^{9}Be \rightarrow (n, alpha) \rightarrow {}^{6}He$, threshold 0.71 MeV, 25 mb cross-section,
$^{6}He \rightarrow {}^{6}Li$, beta emission, half-life 0.82 s
$^{6}Li \rightarrow (n, alpha) \rightarrow {}^{3}H$, beta emission, 12.3 y half-life, 950 b cross-section
^{3}H–^{3}He

For thermal neutrons, the reactions are: $^{6}Li \rightarrow (n, alpha) \rightarrow {}^{3}H$ and $^{3}He \rightarrow (n, p) \rightarrow {}^{3}H$. The effects of the helium and tritium gases are governed by the fluence and the temperature. The solubility of helium in beryllium is extremely low. Because of its high cross-section, the ^{6}Li soon reaches an equilibrium level, whereas the quantity of ^{3}He will increase with time. Both of these isotopes have high cross-sections. The helium at low temperatures is in enforced solid solution in the beryllium lattice, and no gas bubbles are observed. At high temperatures, the helium migrates and agglomerates into gas bubbles, which results in marked swelling. The gas bubbles themselves can migrate by a surface diffusion mechanism and coalesce.

The amount of swelling in beryllium depends on the fast fluence received. When the neutron exposure rate varies in accordance with the location of the shielding, the difference in swelling in various portions of the shielding is likely to induce internal stresses similar to thermal stress. The combination stress caused by swelling, thermal effects, and possibly mechanically induced stress could cause distortion or even cracking of the beryllium reflector or shield. Considerable swelling occurs when beryllium is irradiated to high fluences at high temperatures or annealed at high temperatures after irradiation at low temperatures. Irradiation affects the mechanical properties of beryllium, resulting in increased hardness, yield and ultimate strength, and decreased ductility, depending on the temperature. The ductility is drastically reduced even at elevated temperatures because of the formation of helium bubbles at the grain boundaries that promote intergranular failure. The stress rupture strength decreases with increasing fluence.

There has been a long-standing interest in the use of beryllium oxide as a reflector and shielding material, particularly at high temperatures. The relatively high cost and deleterious property changes under irradiation have limited the applications of beryllium oxide in nuclear reactors. The properties of BeO bodies are governed largely by the nature of the starting materials and the methods of fabrication. The crystal structure is a distorted, wurtzite-type structure. The important variables in the fabricated pieces include composition, grain size, pore structure, and density. Additives have a pronounced effect on the properties of BeO. Adding 1% MgO ensures densification (>97%) and the material retains the small grain size required for long-term stability at high temperatures. Dimensional changes during low-temperature irradiations result from lattice expansion and the formation of grain boundary cracks. Above ~700 °C, the lattice expansion occurs primarily in the *c*-axis and becomes smaller as the temperature increases. In addition, at higher temperatures helium bubbles and defect clusters contribute to the expansion by forming both at the grain boundaries and inside the grains. Precipitation of helium begins during irradiation at temperatures above 600 °C and peaks at 1200 °C.

3.11 *Metal Hydrides*

The atomic density of hydrogen in many metal hydrides is greater than that in liquid hydrogen or water. Metal hydrides are efficient moderators and neutron shielding materials and are particularly suitable for minimizing the core shield volume. They provide a means to combine in one material both the light and heavy constituents of a reactor shield, with a minimum of gamma radiation. There are two types of metal hydrides, namely, those with an ionic crystal structure, such as lithium hydride, and solid solutions of hydrogen in a heavy metal, such as zirconium or titanium. Metal hydrides have been used in a number of nuclear reactor systems. The TRIGA research reactor (General Atomics) and the SNAP space reactor (Atomics International) used uranium-zirconium-hydride rods as a combination fuel moderator. The sodium-cooled prototype KNK (Interatom and Karlsruhe) contained zirconium hydride as moderator elements for operation at temperatures up to 600 °C. The metal hydrides that merit consideration as reactor shielding materials include lithium hydride, titanium hydride, and zirconium hydride. Consideration has also been given to the use of "mixed hybrid" or "trap" hydrides such as ZrEuH, ZrHfH, EuH, GdH, and mixtures of ZrH–europium oxide and ZrH–boron carbide.

Lithium hydride (LiH) is an ionic compound with a high hydrogen content, about the same as water, and has good thermal and radiation stability. LiH melts at 683.5 °C and has a density of only 0.82 $\mathrm{g\,cm^{-3}}$. It reacts vigorously with water in air and requires containment. It is an excellent neutron attenuator, owing to the ^{6}Li isotope (7.52% in natural lithium), which has a high thermal neutron absorption cross-section of 945 barns and no secondary gamma ray. LiH is difficult to fabricate and is expensive (it has been said that "weight-conscious shield designers like to recommend its use; but mechanical designers shudder at the thought"!). Titanium hydride, with a density of 3.76 $\mathrm{g\,cm^{-3}}$, has a hydrogen density of about 50% greater than water. Its maximum useful temperature is about 600 °C, it is stable in air and will evolve hydrogen at elevated temperatures. Zirconium hydride has a very high hydrogen density, equivalent to that of water at the composition used (ZrH-1.6), is stable at elevated temperatures and has excellent irradiation resistance. It can be used at temperatures as high as 650 °C with a suitable cladding.

4. Conclusion

The shielding designs of nuclear reactors and other nuclear facilities and the selection of shielding materials have been highly successful during the past half century of nuclear reactor operations worldwide. There are now over 400 nuclear power reactors and several hundred research reactors in over 30 countries that are operating within safe levels of radiation in their ambient environment. The progress in analytical capability is increasingly dependent on modern computing capabilities. Future work in shielding materials development will have to address the problems associated with irradiation effects on pressure vessel steels, inner liner steels in prestressed concrete pressure vessels, and the development of "mixed hybrid" or "trap" hydride materials as more efficient shielding materials for special applications. The economic and technical factors involved in the selection of shield materials will also require further studies in order to lower the cost and improve the reliability.

See also: Nuclear Reactors: Coolant Materials; Nuclear Reactor Materials: Irradiation Effects; Nuclear Reactors: Moderator and Reflector Materials; Nuclear Reactors: Pressure Vessel Steels

Bibliography

American Concrete Institute1972 *Concrete for Nuclear Reactors* (3 vols.). American Concrete Institute Special Publication SP-34, Chicago, IL

American Nuclear Society 1996 *Radiation Protection and Shielding*. ANS, La Grange Park, IL

Chilton A B 1984 *Principles of Radiation Shielding*. Prentice-Hall, Englewood, NJ

Goldstein J 1959 *Fundamental Aspects of Reactor Shielding*. Addison-Wesley, Reading, MA

Jaeger T 1965 *Principles of Radiation Protection Engineering*. McGraw-Hill, New York

Kaplan M F 1989 *Concrete Shielding*. Wiley, New York

Price B T, Horton C C, Spinney K T 1957 *Radiation Shielding*. Pergamon, London

Profio E 1979 *Radiation shielding and dosimetry*. Wiley, New York

Roussin R W, Abbot L S, Bartine D E 1977 *Nuclear Reactor Shielding*. Science Press, Princeton, NJ

Shaeffer N M (ed.) 1973 Reactor shielding for nuclear engineers. US Atomic Energy Commission Report TID-25951, US Department of Commerce, Springfield, VA

Shapiro J 1990 *Radiation Protection*. Harvard University Press, Cambridge, MA

Shultis J K, Faw R E 1996 *Radiation Shielding*. Prentice-Hall, Upper Saddle River, NJ

Simnad M T 1977 Shielding and control materials for the gas cooled fast breeder reactor. US Atomic Energy Commission Report GA-14478, NTIS, US Department of Commerce, Springfield, VA

Turner J E 1995 *Atoms, Radiation and Radiation Protection*. Wiley, New York

Nuclear Waste Management, Materials Aspects of

H. Matzke

During the fabrication, use, and consumption of any goods, waste products are formed. These must be disposed of in a safe way. Naturally, this is true for the peaceful use of nuclear energy as well, both for the application of radioactive tracers or of radiation sources in medicine, research, and industry and for the use of nuclear energy for electricity production in nuclear power stations. Because of the long half-life of some radioactive nuclides, the waste has to be suitably treated and stored.

In this article, materials aspects of high-level waste (HLW) and spent nuclear fuel will be treated. Low-level waste and medical waste will not be dealt with. HLW arises mainly from reprocessing of spent nuclear fuel. Nuclear power is based on heat produced by fissioning uranium or plutonium atoms contained in the fuel. Each fission produces two fission products belonging to about 30 elements between germanium and dysprosium. In addition, due to neutron capture by actinide atoms and subsequent β decay, higher (or so-called "minor") actinides are formed: americium and curium in addition to neptunium and plutonium. To separate the fission products and these minor actinides and to recover the unused uranium and the plutonium formed during reactor operation, some nations reprocess the fuel by dissolving it in nitric acid and separating chemically the reusable uranium and plutonium (see *Purex and Thorex Processes (Aqueous Reprocessing)*). The resulting unusable liquid and some insoluble sludges containing a small fraction (<1%) of the uranium and plutonium have to be solidified and incorporated in a stable matrix. About 300 liters of a 2 N acid solution of HLW are produced per tonne of reprocessed fuel. These contain about 26 kg of fission products, ~0.4 kg of neptunium, ~0.3 kg of americium, and ~30 g of curium. In addition, insoluble residues from the zircalloy cladding (~2.2 kg) and some 3 kg of insoluble metallic fission products (ruthenium, rhodium, palladium, molybdenum, and technetium alloys, often called "five-metal particles" in the fuel) have to be handled. Research has been directed to the study of whether the minor actinides and some fission products of long half-life and radiological concern, such as Tc^{99} or J^{129}, can be effectively and economically separated (or "partitioned") from the less long-lived waste. Figure 1 shows the radioactivity and the heat produced by fission products and actinides of a typical light-water reactor (LWR) UO_2 fuel. At the start of the twenty-first century, LWRs are the most common electricity-producing power reactors. Figure 1 clearly shows that after about 500 years storage, only actinides and a few fission products (e.g., Tc^{99} and J^{129}) contribute to the radioactivity. If such partitioning is done, these long-lived radiotoxic nuclides could be solidified separately, or they could be destroyed ("transmuted") by further irradiation in a conventional reactor or in new innovative reactor types (e.g., accelerator-driven systems). For HLW solidification, a stable matrix able to incorporate the complex waste and resistant to radiation and possible interaction with ground water has to be selected. Vitrification in borosilicate glasses or solidification in stable ceramics or ceramic mixtures is the appropriate solution. Extensive reviews have been published on these two options for final storage (IAEA 1985, Matzke 1987, 1988, Lutze and Ewing 1988). For the transmutation option, the matrix additionally has to be stable against the impact of neutrons and fission products, and against the reactor coolant, and it has to have good thermal properties. As an alternative to the reprocessing option, some nations consider the spent UO_2 (or mixed oxide $(U,Pu)O_2$, called "MOX") fuel itself as waste which should be stored directly. It is planned that final storage will be in safe,

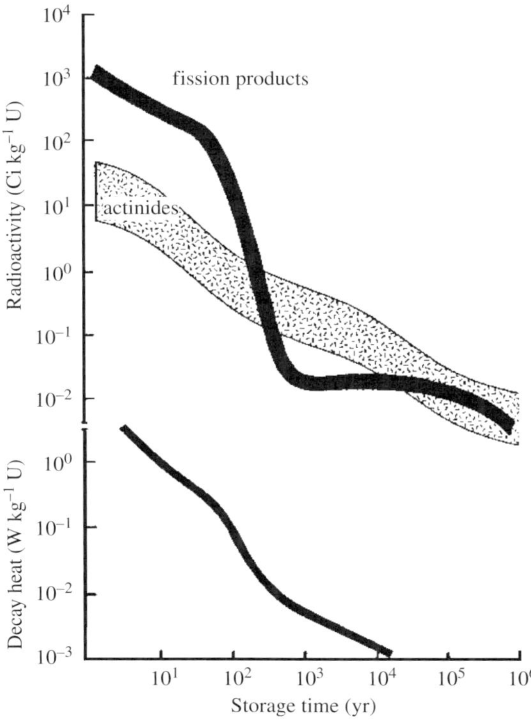

Figure 1
Radioactivity and heat produced by fission products and actinides of an LWR fuel (3.3% enrichment, 34 GWd t^{-1} U burnup, reprocessing 150 d after reactor unloading).

deep geological formations, the most frequently studied alternatives being granite, rock salt, and clay.

1. Vitrification of HLW

The currently favored technology to solidify HLW for long-term storage is vitrification, i.e., the incorporation of HLW into special tailor-made borosilicate glasses. Borosilicate glasses have the capability for incorporation of most of the elements in HLW. Table 1 gives the composition of one of the versions of the French R7T7 glass, i.e., a chemically simulated HLW glass with inactive nuclides replacing the active ones of actual HLW glass. Property measurements are much easier to perform with such simulated waste glasses. Table 2 gives typical properties (see also CEA 1997).

The waste borosilicate glasses can be tailor-made for specific HLW types. Their advantages are ease of incorporation of complex wastes, relatively easy fabrication, and a homogeneous structure (in contrast to that of spent fuel). Glasses of the type given in Table 1 are produced on a large scale in France. The Atelier de Vitrification in Marcoule (AVM) started its operation in 1978, and the two fabrication sites R7 and T7 in La Hague began large-scale vitrification in 1989 and 1991, respectively. Both installations have three vitrification lines with a nominal capacity of 25 kgh^{-1}, and more than 10000 glass containers have been successfully produced.

The HLW in the glass produces decay heat. The thermal power of the glass is about 7 Wkg^{-1} at the time of fabrication, decreasing to about 1 Wkg^{-1} after 50 years and to $\sim$0.05 Wkg^{-1} after 300 years. Recrystallization has to be avoided. The glass composition has to be chosen in such a way that the recrystallization temperature is high ($\sim$750 °C). Other important properties are durability and leaching behavior in case ground water has access to the glass at a later time of storage or if the metal container of the glass block becomes corroded. The kinetics of leaching under different possible conditions (type and composition of leaching solution, pH, temperature, radiation damage, radiolysis, etc.) have therefore been studied and are well known. Granitic and clay ground waters as well as salt brine, representative of the three main repository conditions, were used in these experiments. The inherent resistance of a glass toward aqueous corrosion is characterized by relatively fast initial dissolution followed by much lower leach rates. The sequence of events is an initial hydration process followed by an interdiffusion $H^+ \rightleftharpoons Na^+$, dissolution of the silica network and formation of chemically altered surface layers. Though glass dissolution is nonselective, it becomes incongruent since not all elements go into solution. Finally, redeposition of previously dissolved

Table 1
Composition of the waste glass R7T7.

Component	Composition (wt.%)
SiO_2	45.48
Al_2O_3	4.91
B_2O_3	14.02
Na_2O	9.86
CaO	4.04
ZnO	2.5
Li_2O	1.98
Fe_2O_3	2.91
P_2O_5	0.28
ZrO_2	2.65
NiO	0.42
Cr_2O_3	0.51
SrO	0.33
Y_2O_3	0.20
MoO_3	1.70
MnO_2	0.25
AgO_2	0.03
CdO	0.03
SnO_2	0.02
Sb_2O_3	0.01
TeO_2	0.23
BaO	0.60
La_2O_3	0.56
Ce_2O_3	0.93
Pr_2O_3	0.44
Sm_2O_3	0.31
Gd_2O_3	0.03
Nd_2O_3	1.59
Rb_2O	0.13
Cs_2O	1.29
Ru_2O	0.46
Rh_2O_3	0.12
PdO	0.33
ThO_2[a]	0.33
UO_2	0.52

[a]Simulating the total content of NpO_2, PuO_2, AmO_2, and CmO_2.

Table 2
Relevant properties of the waste glass R7T7.

Property	Value
Density ($t\ m^{-3}$)	2.75
Mean thermal expansion coefficient (300–570 K) (K^{-1})	8.3×10^{-6}
Thermal conductivity at 300 K ($W\ K^{-1} m^{-1}$)	1.1
Viscosity at 1370 K (Pa s)	8
Transformation temperature (K)	775
Young's modulus E at 300 K (GPa)	84
Fracture toughness K_{IC} at 300 K ($MPa\ m^{1/2}$)	0.95

components may occur. Modeling can describe this leaching behavior appropriately (Matzke and Vernaz 1993).

The influence of added near-field engineering barriers, such as steel or copper for the glass container and sand or clay as backfill, have been studied as well as the effects of radiation damage due to α decay of actinides. For the latter, glasses with incorporated short-lived actinides (Cm^{242}, Cm^{244}, Pu^{238}) were used. No significant increases in leach rates were found for equivalent storage times of up to 10^5 years. For the former, the effect of steel is mentioned here: steel corrosion can cause dissolved silicon to become adsorbed (or react with) precipitating iron hydroxides, thus preventing the leaching solution to become saturated in silicon, which delays the above-mentioned decrease in leach rates with time.

A typical initial leach rate in water at 370 K for the glass R7T7 is 0.17 $mgcm^{-2}\,d^{-1}$, increasing by about a factor of 10 if the pH increases from 7 to 10. The temperature dependence is described by an Arrhenius law with an activation enthalpy of $\sim$60 $kJmol^{-1}$. The mass losses decrease therefore to the order of a μg $cm^{-2}\,d^{-1}$ or less for room temperature.

Similar borosilicate glasses such as the French R7T7 product have been developed and tested in other nations, e.g., in Germany, India, Japan, the UK, and the USA. Sintered glasses, lead–iron and other phosphate glasses, and aluminosilicate glasses have also been studied. The final choice was the borosilicate glasses, as the best compromise between fabrication engineering constraints and the physicochemical and thermal properties of the final product, i.e., between requirements of technical feasibility and the highest possible degree of waste containment.

2. *Solidification of HLW in Ceramics*

Radioactive waste can also be solidified in ceramics. Ceramics, in particular titanates, are more leach resistant than glasses. However, they are more difficult to produce, e.g., high-temperature sintering or hot pressing techniques are needed. Also, a given binary or ternary ceramic does not have the same high capability of incorporating into its structure the many elements of HLW as the borosilicate glasses.

Table 3 shows some of the ceramics which were extensively studied as waste forms. The most prestigious of these is the synthetic rock, the SYNROC, suggested by Ringwood (1980). By choosing an assemblage of four titanate minerals (Table 3), most of the many elements occurring in HLW can be incorporated into their crystal structures as a solid solution. Minerals similar to the phases of SYNROC exist in rocks in nature and have survived natural geochemical–geological environments for up to 2000 million years. Several distinct SYNROC formulations exist for a range of radioactive wastes. The principal composition (SYNROC C) is (wt.%): TiO_2, 57; ZrO_2, 5.4; Al_2O_3, 4.3; BaO, 4.4; CaO, 8.9; and HLW, 20. This gives a mineral assemblage of (approximate wt.%) hollandite and zirconolite (30 each), perovskite (20), titanium oxides and calcium–aluminum titanates (15), and alloys (mainly the above "five metals", but also some nickel and iron) (5). Hot pressing is usually employed for fabrication.

Table 3
Some ceramic materials suggested and tested as waste matrices.

Material	Comments
SYNROC	A synthetic rock proposed by Ringwood (1980), Australia, composed of (predominantly) three titanate minerals: perovskite ($CaTiO_3$), Ba-hollandite ($BaAl_2Ti_6O_{16}$), zirconolite ($CaZrTi_2O_7$), and some rutile phase (Ti_nO_{2n-1})
Monazite	Mixture of orthophosphates of the type $CePO_4$
Sphene	$CaTiSiO_5$
Pyrochlore	$(Gd,La)_2Ti_2O_7$
Supercalcine	Produced by adding SiO_2, CaO, Al_2O_3, and SrO to the HLW solution. The reacted product contains crystalline phases of apatite, fluorite, scheelite, pollucite, and spinel structures
Zircon	$ZrSiO_4$
Zirconia	Stabilized ZrO_2, e.g., $(Zr,Y)O_{2-x}$

SYNROC is chemically very stable. Following an initial fast mass or element loss, the leach rates decrease very soon with time. As an example, at 90°C and in demineralized water, mass leach rates are of the order of 0.5 mg $m^{-2}\,d^{-1}$. This is about a factor of 10^3 smaller than the rates for glasses. Similar low leach rates can also be obtained with the other waste ceramics. However, most of these cannot incorporate into their structure equally as many fission product elements as SYNROC does. Also, α-decay radiation damage renders some of the ceramic waste forms amorphous. This is the case for zirconolite and zircon, for example. Amorphization is related to swelling, e.g., by 18% for zircon and 6–7% for zirconolite, and to enhanced leach rates, which can be 10–50 times larger than those for the crystalline phase (Weber *et al.* 1988).

Helpful information on the durability and radiation stability of candidate ceramics has been obtained by studying "natural analogues." These are naturally occurring minerals with a certain content of uranium and/or thorium. These minerals have been subject to significant α-decay damage, and can contain large amounts of helium. Investigating their properties gave very valuable additional information.

As a result of these two lines of research, a good understanding exists of damage formation and its consequences, and reliable models have been developed to describe observed results and predict the behavior of the materials for new parameters of application (Weber *et al.* 1988).

3. Solidification of Actinides for Final Storage or for Transmutation

If only actinides are incorporated into ceramics the products can be used either for final storage or for transmutation by irradiation with neutrons in a "bury or burn" strategy. The actinides can be excess civilian or ex-military Pu^{239} and U^{235}. For "burial," stabilized cubic ZrO_2 is a promising option. For "burning," the matrix must be a good thermal conductor, compatible with cladding and coolant, neutronically favorable and stable not only against neutron and α-decay damage but also against fission damage. Some ceramics are stable against α-decay damage, but turn amorphous under fission product impact. Examples are monazite and spinel. Significant effort has been devoted to the development of uranium-free fuels for transmutation. As a result of these research activities, innovative new fuels have been suggested, such as the incorporation of americium in ZrO_2 (radiation-resistant, but poor thermal conductor) in the form of small (~100 μm) spheres placed in a spinel matrix (good thermal conductor, good stability against neutron damage, but limited stability against fission damage) (Chauvin *et al.* 1999). Such a partitioning–transmutation procedure significantly reduces the long-time radiotoxicity of HLW and decreases the demands on the HLW glass or ceramic waste form for stability against α-decay damage and large amounts of helium formed by α decay.

4. Spent Fuel Direct Storage

If reprocessing is not accepted, the spent fuel has to be stored directly. In contrast to the above-mentioned waste forms, the spent fuel is a heterogeneous matrix for fission products and actinides. Due to the thermal gradients existing during reactor operation, the UO_2 fuel pellets crack, and there may be radial gradients in microstructure and fission product content. In addition, due to resonance neutron capture in the outermost ~100 μm of UO_2 fuel leading to largely increased plutonium formation in this rim, there is a much larger burn-up in this shallow rim zone. If the cross-section average burn-up of the fuel surpasses ~40 GWd t^{-1}M (gigawatt days per tonne of metal), a grain subdivision or polygonization process occurs in this zone, connected with the formation of a large number of micrometer-sized pores. In the case of direct storage and breaching of the cladding, any available water will have first contact with this porous zone with very small grains and high contents of plutonium and fission products. Besides performing well-defined leaching studies, a careful characterization of the high burn-up fuel and knowledge of its properties are needed to assess the risks of spent fuel direct storage. A large data base already exists on such measurements. An additional problem that has attracted much attention is that of radiolysis of the water changing chemistry and thus leach rates. Nuclear energy is a new technology, and there is no "old" fuel available. Since fuel containers are guaranteed for $\sim 10^3$ years, the predominantly existing β–γ radiolysis of "new" fuels will have decayed and α radiolysis will dominate (see Fig. 1). This cannot be studied with high-burn-up fuel but is being investigated with UO_2 doped with short-lived Pu^{238}.

5. Summary

Suitable treatment and safe storage of HLW is a key point in the public acceptance of nuclear energy. The liquid HLW from the reprocessing of spent nuclear fuel is a complex mixture of some 35 elements. Solidification in suitable matrices has posed new and challenging problems for materials research. Remote operation in shielded cells is necessary, and the final product has to withstand large radiation doses for very long times. Borosilicate glasses have been developed as a very suitable waste form, and are produced on a technological scale. Waste ceramics and ceramic assemblages such as SYNROC are alternatives, in particular for specific waste streams. Recent materials research is aimed at reducing the radiotoxicity of HLW by partitioning of the actinides neptunium, americium, and curium, and possibly long-lived fission products, and to transmute them into a uranium-free fuel. Very promising results have been obtained, and much work is ongoing in these fields.

See also: DUPIC Fuel Cycle; Oxide Fuels, Nonaqueous Reprocessing of; Nuclear Fuel Cycles; Purex and Thorex Processes (Aqueous Reprocessing)

Bibliography

Chauvin N, Konings R J M, Matzke Hj 1999 Optimization of inert matrix fuel concepts for americium transmutation. *J. Nucl. Mater.* **274**, 105–11

IAEA 1985 Chemical durability and related properties of solidified high-level waste forms. Technical report series No. 257. International Atomic Energy Agency, Vienna

CEA 1997 Glass – scientific research for high-performance containment. *Proc. Summer University Méjannes-le-Clap*. University Montpellier II, Montpellier, and CEA Valrho, France

Lutze W and Ewing R C (eds.) 1988 *Radioactive Waste Forms for the Future*. North-Holland, Amsterdam

Matzke Hj 1987 Nuclear waste materials. In: Arnold G, Mazzoldi P (eds.) *Ion Beam Modification of Insulators*. Elsevier, Amsterdam, pp. 501–30

Matzke Hj 1988 Nuclear waste materials, a workshop on radiation damage effects in nuclear waste materials. *Nucl. Instrum. Methods Phys. Res. B* **32**, 453–517

Matzke Hj, Vernaz E 1993 Thermal and physicochemical properties important for the long-term behavior of nuclear waste glasses. *J. Nucl. Mater.* **201**, 295–309

Ringwood A E 1980 Safe disposal of high-level nuclear wastes: a new strategy. *Fortschr. Mineral.* **58**, 149

Weber W J, Ewing R C, Catlow C R A, Diaz de la Rubia T, Hobbs L W, Kinoshita C, Matzke Hj, Motta T A, Nastasi M A, Salje E H K, Vance E R, Zinkle S J 1998 Radiation effects in crystalline ceramic phases relevant to the immobilization and disposition of nuclear waste and weapons plutonium. *J. Mater. Res.* **13**, 1434–84

O

Organic Electrolytes and Electrodes for Batteries

R. A. M. Hikmet

Ion conductors are of great interest as they are used in many applications such as electrochromic devices, fuel cells, photoelectric cells, and batteries (Scrosati 1993, Wakihara and Yamamoto 1998). In all these applications electrolytes perform essential functions and facilitate the transport of ions. The recent developments in the field of portable electronic equipment have been the main driving force behind the search for batteries with a high energy density and form flexibility, and hence for new battery materials to satisfy these requirements. Lithium is a very attractive material for high energy density batteries because of its low weight and high potential. The research into lithium batteries which began in the 1950s has led to the current rechargeable lithium batteries (Wakihara and Yamamoto 1998), which are rapidly replacing bulkier nickel–cadmium and nickel–metal hydride cells in portable electronic devices. While aqueous electrolytes are used in nickel–cadmium and nickel–metal hydride batteries, nonaqueous electrolytes need to be used in lithium batteries. In high energy density lithium batteries operating at high voltages the organic electrolyte must show high electrochemical stability. Furthermore, the electrolyte must have a high conductivity in order to enable fast charging/discharging of the battery. Work is now being done to produce thin batteries with form flexibility. Here, the focus will be primarily on lithium ion conductors and flexible electrode materials, which can be used in polymer lithium batteries.

1. Ion Conductivity

Electrolytes are obtained by dissolving salts in organic solvents. The conductivity (σ) of an electrolyte can be expressed as

$$\sigma = \sum_i n_i q_i \mu_i \qquad (1)$$

where n is the number, q is the charge, and μ is the diffusion constant of i-type charge carriers. To have a high conductivity, it is therefore necessary to have a high concentration of fully dissociated ions with a high mobility. Here we shall first consider the solvation mechanism in electrolytes. A solvent will dissolve an inorganic salt if the total energy of the system (Gibbs free energy) is decreased. The change in the Gibbs free energy (ΔG) can be written as

$$\Delta G = \Delta H - T\Delta S \qquad (2)$$

where T is the absolute temperature and ΔH and ΔS are the changes in the enthalpy and entropy, respectively. Dissolution of the salt increases the entropy of the cation and the anion, but coordination of the solvent molecules by the cation decreases the entropy of the solvent. In the case of polymers, cations form pseudo- cross-links and thus reduce the entropy of the system. Therefore, the net change in the entropy upon dissolution of salt molecules in high molecular weight polymers is usually positive, regardless of the type of salt. The enthalpic change is, however, very much dependent on the type of solvent and the salt interactions. One of the approaches used to promote dissolution is to use large salt molecules with low lattice energies. The most important enthalpic interactions between the solvent and the salt are (i) long-range electrostatic forces and (ii) short-range specific interactions.

Electrostatic interactions neglecting all other interactions are described by the Born's model, which relates the solvation energy ΔG to the dielectric constant of the solvent ε as

$$\Delta G = \frac{Nq^2}{2r}\left(\frac{1}{\varepsilon} - 1\right) \qquad (3)$$

where r is the radius of the ion, N is Avogadro's number, and q is the charge of the ion. This equation implies several trends which are also observed in various solvents, for example ions with a small radius dissolve more easily and solvents with a high dielectric constant promote a higher degree of dissociation of the salt, resulting in a higher concentration of free ions. However, the model totally ignores the lattice energy of the salts and the specific interaction with the solvents.

Specific interactions between the solvents and the cations have been considered by coordination chemists as the most important factor determining the solvating ability of solvents. The best-known scheme is that based on donor numbers, which is used to classify solvents according to their ability to donate electron pairs. It is defined as the negative enthalpy of the reaction between the solvent under study and antimony pentachloride ($SbCl_5$). However, the donor number scheme does not take into account other specific interactions such as hydrogen bonding and can hence not be used in the case of solvents forming hydrogen bonds.

Dissolution of the salt molecules does not always lead to the formation of free ions. In poor solvents, even in dilute solutions, extensive ion pairing occurs. With increasing solvent quality, the ion association occurs at higher ion concentrations. In such systems in addition to fully dissociated (free) ions, neutral ion pairs as well as ion clusters with a net charge exist. The state of the ions can be determined using various techniques such as infrared and Raman spectroscopy.

The other factor determining the conductivity is the mobility of the ions. The mobility is dependent on the size of the ion and the free volume within the system. Einstein's relation relates the mobility of an ion with a charge q to the diffusion constant as

$$\mu = \frac{qD}{kT} \tag{4}$$

where k is the Boltzmann constant. In ionic crystals, ions are in an array of energetically equivalent sites, which are partially occupied. If the vacancy is in a potential well with a height ΔH, the ions need to overcome this potential during their thermal motion in order to be able to move to another site and the self-diffusion coefficient can be written as

$$D = D_0 \exp(-\Delta H/kT) \tag{5}$$

This equation implies that a plot of $\log(D)$ against $1/kT$ should be a straight line. However, in the case of organic electrolytes close to their glass transition temperature substantially curved lines have been observed. This indicates that the behavior of the organic electrolytes cannot be described in terms of a thermally activated process with a single activation energy. In liquid electrolytes the motion of the ions is coupled to the movement of the liquid molecules, whereas in polymers the segmental motion of the polymer determines the ionic mobility. It is for this reason that the viscosity of liquids and the storage modulus of polymers have been considered in attempts to explain the ion conductivity in such systems. The Stokes–Einstein relation relates the diffusion constant of an atom with a radius $\boldsymbol{r}$ to the viscosity η of a liquid as

$$D = \frac{kT}{6\pi\eta r} \tag{6}$$

By combining Eqns. (1), (4), and (6), the conductivity can be rewritten as

$$\sigma = \frac{nq^2}{6\pi\eta r} \tag{7}$$

The temperature dependence of the viscosity is in turn described in various ways. According to relationship established by WLF (Williams *et al.* 1955), a shift factor aT can be defined, which relates the viscosity (η) measured at a given temperature T to the viscosity (η_s) measured at a reference temperature T_s as

$$\log aT = \frac{C_1(T - T_s)}{C_2 + (T - T_s)} = \log\left(\frac{\eta_T}{\eta_s}\right) \tag{8}$$

Using this equation, it has been shown that there is a linear relationship between the shift factor aT and the measured temperature at a given temperature. Equation (8) is also described using the free volume theory (Cohen and Turnbull 1959) which relates the viscosity to the free volume v_f as

$$\eta = A\exp\left(\frac{\gamma v_c}{v_f}\right) = A\exp\left(\frac{\gamma}{\alpha(T - T_0)}\right) \tag{9}$$

where A is a constant and γv_c is a dimension corresponding roughly to the volume of the diffusing molecule. T_0 is the temperature at which the free volume disappears. T_0 lies approximately 50 °C below the glass transition temperature ($T_0 \approx T_g - 50$). The free volume dependence of the temperature is then given by $v_f = v_c\alpha(T - T_0)$. It can be shown that Eqn. (9) is in fact the same as the WLF equation. Using the theory of Cohen and Turnbull a modified free volume theory (Hikmet *et al.* 1999) has been established which relates the conductivity to the free volume and the fraction of the components within the system as

$$\sigma = BCX\exp\left(\frac{-\gamma v_c}{v_f}\right) \tag{10}$$

where B is a constant, C is the concentration of salt molecules, and $(1 - X)$ is the fraction of the polymer within the system. This equation is also used for mixed systems containing low molar mass solvents as well as polymer. The free volume is then defined as $v_f = (v - v_g)$, where v is the volume measured at a given temperature and v_g is the average volume of the system at its T_0, at which long-range diffusion within the system extrapolates to zero. This volume is related to the Van der Waals volume (v_w) approximately as $v_g = 1.45\ v_w$. This equation has been successfully used to construct master curves relating the conductivity to the free volume within various systems.

2. *Organic Electrolytes*

Electrolytes enable ion transport and facilitate ion exchange in and out of the anode and cathode materials of batteries. In rocking chair batteries, which will be described below, electrolytes moreover have to form a passivation layer on the graphite anode. This is very important for obtaining a high cycling efficiency. The electrolyte must also show a high electrochemical stability in the range of operation of the battery. Various types of organic electrolytes and their properties that are important for battery applications will be discussed in the following section.

2.1 Lithium Salts

Organic electrolytes are obtained by dissolving salts in organic electrolytes. It is therefore interesting to consider some of the lithium salts used in electrolytes. Of the many lithium salts available only a few have been considered for use in commercial batteries. Important issues regarding the choice of salt are conductivity, safety, and electrochemical stability. Lithium perchlorate ($LiClO_4$) has a high conductivity, but it is not stable and raises some safety issues. Lithium triflate ($LiCF_3SO_3$) is a stable salt, but it has a rather low conductivity. So, salts used in commercial batteries are restricted to $LiBF_4$ and $LiPF_6$, which have good electrochemical stability and show a high conductivity. Recently, perfluoro-alkylsulfomethanides such as $Li(CF_3SO_2)2N$ and perfluoro-alkylsulfomethanides such as $Li(CF_3SO_2)3C$ have also been proposed for use in rechargeable batteries. Lithium hexafluoroarsene ($LiAsF_6$) is not used due to environmental considerations. The conductivity and the electrochemical stability of the salts depend very much on the host solvent. This will be discussed below.

2.2 Liquid Electrolytes

Many molecules have been identified which can be used as host solvents in liquid electrolytes. In battery applications such solvents must meet various requirements: a high thermal and chemical stability, electrochemical stability, the ability to dissociate ions, a low viscosity, and the ability to form a passivation layer to promote high cycling efficiencies in rocking chair batteries. The effect of the salt concentration on the conductivity is shown in Fig. 1. Here the conductivity has been plotted as a function of the salt concentration. It can be seen that the conductivity passes through a maximum. This behavior is a result of two opposing effects: an increase in the concentration of the charge carriers and a decrease in their mobility with an increasing salt concentration. Initially, the effect of the increase in the number of charge carriers is greater than the effect of the increased viscosity, which causes a decrease in the mobility of the charge carriers. However, at higher concentrations of salt the ability of the solvent to dissociate ions decreases and the effect of viscosity prevails. Consequently, the conductivity decreases. The equivalent (molar) conductivity is also shown as a function of concentration in this figure. It can be seen that at low

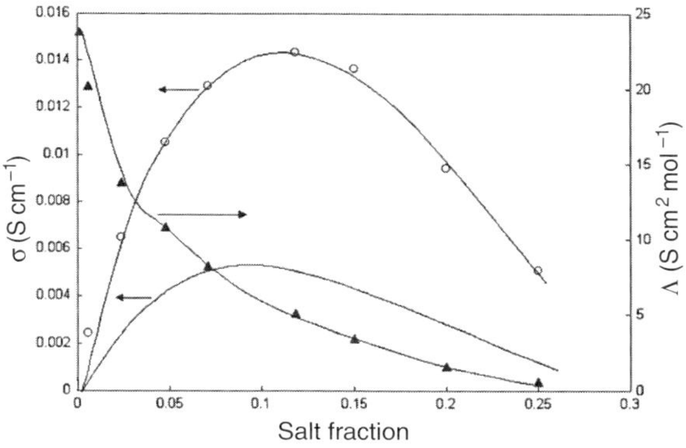

Figure 1
Conductivity measured at ■ = 20 °C and ○ = 60 °C and equivalent conductivity measured at ▲ = 20 °C as a function of the $LiPF_6$ fraction in PC.

concentrations the equivalent ionic conductivity already shows a fast decrease. At low ion concentrations the viscosity is not affected to a large extent. This again indicates that at low salt concentrations, ion association already plays a dominant role in determining ion conductivity.

The effect of the ability of the solvents to dissociate ions and the effect of their viscosity on the ion conductivity are compared in Table 1 which shows various properties. The conductivity should show the order PC > SL > AN > TEGDME > THF on the basis of the dielectric constant of the solvents, and AN > THF > TEGDME > PC > SL on the basis of their viscosity. It is for this reason that mixed solvents have been used to combine the favorable properties of various solvents. Figure 2 shows the ionic conductivity at room temperature of various mixtures of 1 mol $LiPF_6$ solutions in propylene carbonate (PC) and triethyleneglycoldimethylether (TEGDME). It can be seen that the conductivity of the system containing 25% weight fraction TEGDME is the highest. PC with a higher dielectric constant can dissociate ions much better than TEGDME, which on the other hand has a lower viscosity. Therefore, an optimum mixture of the two solvents will show a greater conductivity than the pure components. Not only can the viscosity and dipolar effects of the mixed systems be tuned, but also their microscopic structure can be influenced. This can also have a favorable effect on the conductivity.

2.3 Polymer Electrolytes

The great interest in polymeric ion conductors (Gray 1991, Armand 1994) is attributable to their mechanical stability and low volatility. Polymeric ion conductors are obtained by dissolving a salt in a polymeric matrix. Various types of polymers can be

Table 1
Various physical constants of some organic solvents.

Solvent	m.p. (°C)	b.p. (°C)	ε	η (cP)	Donor number	σ ($LiPF_6$) (mS cm^{-1})
Acetonitrile (AN)	−45	82	38	0.35	14.1	50
Propylene carbonate (PC)	−49	241	64	2.53	15.1	6
Triethyleneglycol dimethylether (TEGDME)	−45	216	8	2.04		3.5
Tetrahydrofuran (THF)	−108	65	7.4	0.46	20	14
Sulfolane (SL)	28	287	42	9.9	19.3	3

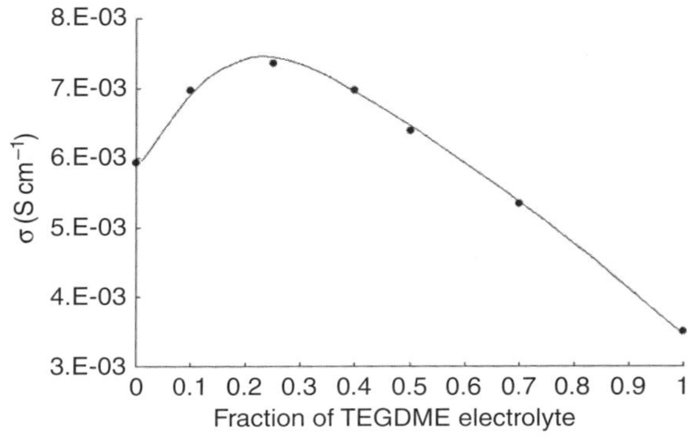

Figure 2
Conductivity as a function of TEGDME in PC for a 1 mol solution of LiPF6.

distinguished, notably: main chain homo polymers, main chain copolymers, side chain polymers, branched polymers, and cross-linked networks.

Main chain polyethers such as poly(ethylene oxide) (PEO) $[-CH_2CH_2O-]_n$ poly(propylene oxide) $[-CH_2CHCH_3O-]_n$, and poly(1,3-dioxolane) $[-CHOCH_2CH_2O-]_n$ are the most commonly used to obtain such ion conductors. Other polymers such as poly(vinyl carpoate) $[-CH_2CHOCOC_5H_{11}-]_n$, poly(ethylene adipate) $[-OCH_2CH_2OOC(CH_2)_4CO-]_n$, have also been studied as ion conductors.

Of the polyethers listed above, PEO, which has the greatest ability to dissociate ions, is the most commonly used to obtain various polymeric ion conductors. However a ten-fold decrease in conductivity is observable below the crystallization temperature of PEO. The copolymers have been synthesized with the aim of suppressing PEO crystallinity. While copolymers of PEO with other polyethers have a good ion solvating ability, PEO block copolymers with nonsolvating blocks such as poly (methyl methacrylate), polyethylene, or polystyrene blocks have better dimensional stability. Such blocks do not have the ability to solvate the ions but they function as physical cross-link points to give the system physical stability. Block copolymers of polyurethanes with polyethers have also been reported. In such rubbery thermoplastic polymers, urethane segments form the cross-links between the soft polyether segments. Experiments have been done with electrolytes of liquid crystalline materials, too, with the hope that order induced by the liquid crystalline units would align the paths of the ions, enabling them to move at higher rates. Examples of such structures are shown below (Lauter *et al.* 1997)

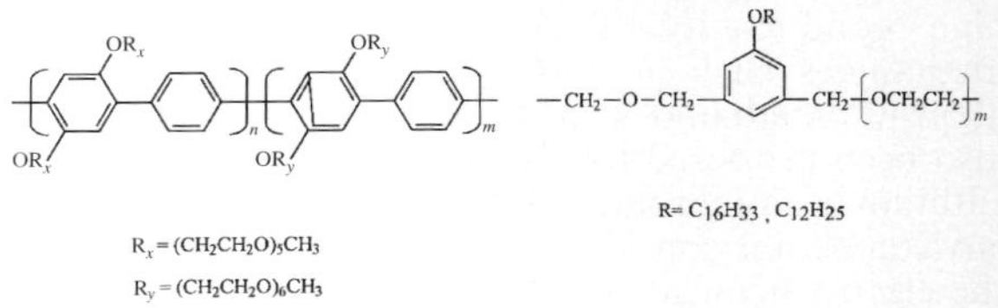

Various side chain polymers containing oligomeric side groups are shown below. Polymers with various main chains such as (meth)acrylate, siloxane, and phosphozane have been synthesized using different synthetic routes. The route followed in grafting ethylene oxide groups to polysiloxanes is shown below

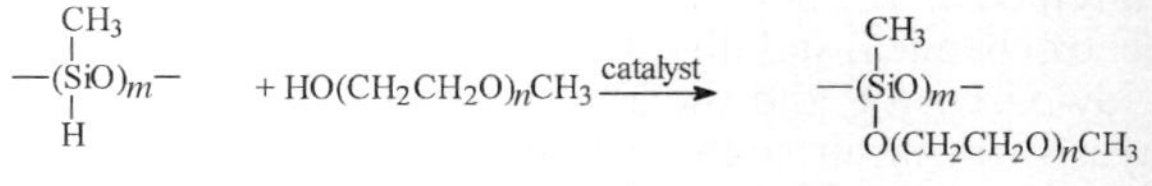

Polyphosphazenes are synthesized as shown below

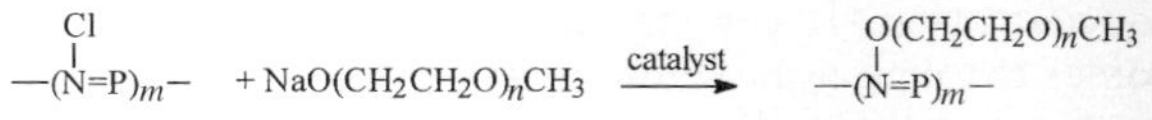

Acrylates are usually polymerized according to the scheme shown below

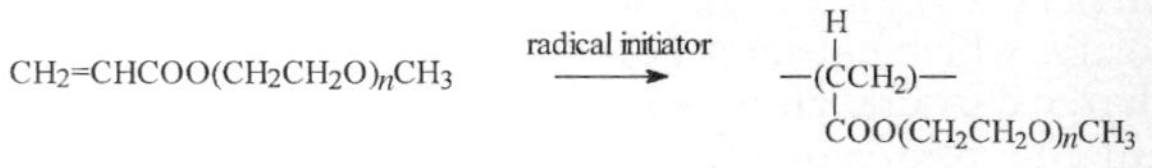

Highly branched polymers with a dendritic structure have also been synthesized. Examples of such polymers are shown below (Allcock *et al.* 1997).

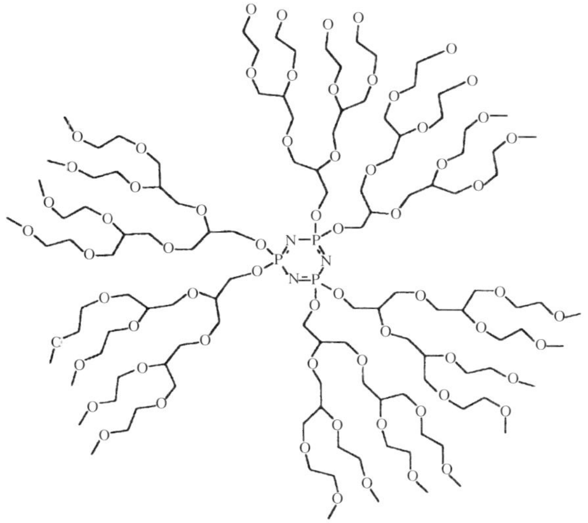

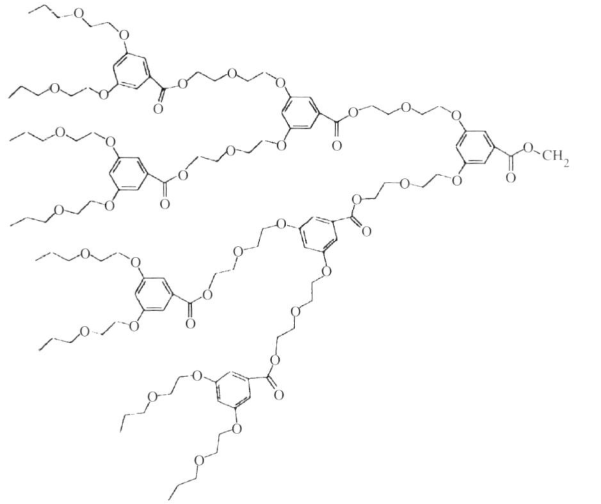

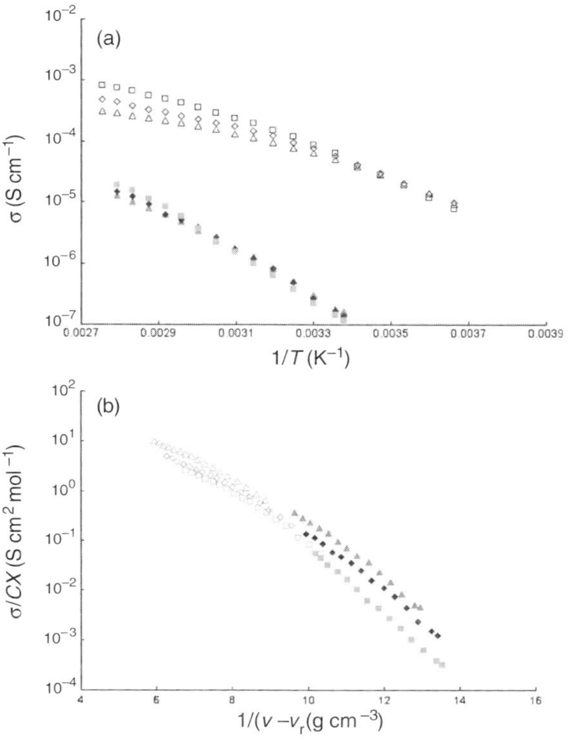

Figure 3
(a) Arrhenius plot of the ionic conductivity obtained for mixtures containing various amounts of $LiSO_3CF_3$ in PEGDA (400); (b) σ/CX as a function of inverse free volume per unit mass for mixtures containing various amounts of $LiSO_3CF_3$ in PEGDA (400). $\Delta = 6.7\%$, $\diamond = 11.6\%$, $\square = 20.7\%$. Open symbols before polymerization, closed symbols after polymerization.

Polymer networks can be obtained following various routes. One of the routes involves the radiation of polyethylene oxide using high-energy radiation such as electrons or gamma rays, which leads to cross-linking of polyethylene oxide chains. Polyurethane-linked polyether networks are also frequently used to obtain cross-linked electrolytes. The reaction scheme for preparation of PEO-polyurethane networks is given below

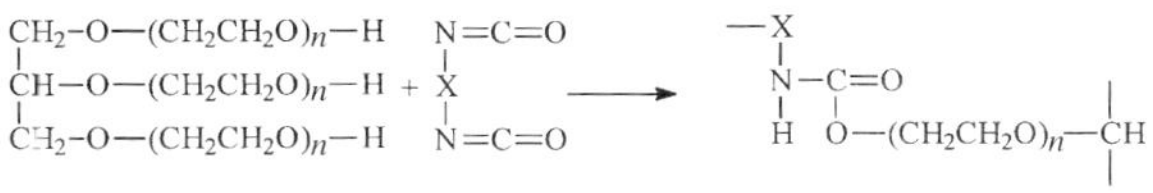

Siloxane networks are obtained when diols are used in the scheme shown above. However, the easiest way of obtaining networks is by using acrylates of PEO with two or more reactive groups. The salt is dissolved in the monomeric PEO acrylate and a photo-initiator is added to the system. The system is then subjected to UV radiation, which results in isothermal vitrification of the system due to cross-linking. The free volume within the polymer network is determined by the cross-link density, which in turn determines the conductivity within the networks. The cross-link density within such networks can be decreased by increasing the length of the PEO groups or by including monoacrylates within the system.

Figure 3(a) shows the conductivity of polyethylene glycol diacrylate (PEGDA, MW = 400) containing various amounts of $LiSO_3CF_3$ before and after polymerization. It can be seen that the conductivity decreases in all cases during polymerization. This decrease in the conductivity can again be explained using the free volume theory described above. Using the Van der Waals volumes and the measured temperature-dependent densities of the samples, σ/XC has been plotted as a function of the inverse free volume per unit mass for mixtures before and after polymerization (Fig. 3(b)). It can be clearly seen that the points obtained for the monomer and the polymer lie on a single line at a given salt concentration. This indeed shows that the reduction in the free volume within the system explains the decrease in the conductivity observed during polymerization. It is also important to point out that the free volume theory does not take into account specific interactions within the system, such as the preferential coordination of Li^+ by polyethers. As the long-distance movement of polyether chains would be prevented by polymerization, the diffusion of Li^+ would be expected to decrease accordingly. At a given composition, the points corresponding to the monomeric mixture and the polymer should hence not lie on a single line. The fact that they do lie on a single line indicates that the diffusion of lithium is retarded already before polymerization and the mobility of $SO_3CF_3^-$ is much greater. This point will be addressed again in the next section, when we discuss the mobility of the components studied by means of PFG-NMR. The other effect observed is a decrease in the molar conductivity with an increasing salt concentration. This effect is also related to the formation of ion pairs and ion clusters at high concentrations of polymer and will be discussed further in the section on ion association.

It has been shown that the conductivity in such polymers depends very much on the free volume available within the polymer. Almost all the polymers with a high molecular weight and sufficient mechanical stability reported in the literature have room temperature conductivities of $<10^{-5}$ S cm^{-1}. The addition of low-mass solvents has therefore been suggested in order to increase the conductivity of such systems. This leads to the formation of various composite electrolytes described below.

2.4 Polymer Blends

As described above, polymer electrolytes show a strong decrease in conductivity upon crystallization. In the molten state polymer electrolytes show an increased conductivity, but their mechanical integrity is then strongly reduced. It has been suggested that polymer blends could be used to obtain polymer electrolytes with good mechanical properties and a high conductivity. Usually an oligomeric polyether electrolyte that is not crystalline is blended with a polymer such as polystyrene or polymethyl methacrylate. Such blends are incompatible, and in the correct composition range the system is made up of two distinct interpenetrating phases. The non-ion-conducting polymer phase provides the system with mechanical integrity while the oligomeric electrolyte is responsible for ion conduction. Using such electrolytes, room-temperature conductivity values of more than 1 mS cm^{-1} have been obtained.

2.5 Gel Ion Conductors

Gel ion conductors are obtained in various ways. In the case of thermo-reversible gels, a polymer is dissolved in a liquid electrolyte at a high temperature. When the system is cooled to room temperature the system phase separates and a gel is formed. In the same way polymer can be dissolved in a liquid electrolyte in the presence of excess solvent. When a film is cast the excess solvent evaporates and a gel forms. Another method is to first dissolve the polymer in a solvent without the salt. The solution is then mixed with a salt solution and the system is annealed. Most commonly used polymers to obtain such gels are poly(acrylonitrile) $(-CH_2CHCN-)_n$ and poly(vinylidene fluoride) $(-CH_2CF_2-)_n$. A disadvantage of these gels is their thermoreversibility, which means that they lose their mechanical integrity at high temperatures. One such successful gel electrolyte is based on a copolymer of vinylidene fluoride and hexafluoropropylene (PVDF-HFP) $([CH_2CF_2]_n[CF_2CFCF_3]$. The copolymer is first dissolved in a solvent. A polymer film is cast and the solvent is extracted. The membrane can be swollen by a low-mass electrolyte. The presence of HFP in PVDF increases the electrolyte uptake of the membrane. This ability is further enhanced by adding inorganic particles to the system.

The easiest way of obtaining gel electrolytes is by using mixtures of liquid electrolytes in acrylates. Polymerization of such a system results in a three-dimensional network containing liquid electrolyte (gel electrolyte). Depending on the solvent and the type of acrylate used, the gels obtained may be homogeneous (single-phase gels) or phase-separated. In the case of homogeneous gels, the polymer network is homogeneously dispersed within the system. In the case of the two-phase gels, however, the system comprises regions which are polymer-rich and regions which contain only the liquid. Homogeneous gels are obtained when polar acrylates such as polyethylene glycol diacrylate (PEGDA) are used. Such polymers interact well with the salt and with many polar solvents, giving rise to single-phase gels. Two-phase gels are formed when an apolar diacrylate such as alkane diacrylate (AD)$(CH_2{=}CH-COO-(CH_2)_n-COO-CH{=}CH_2)$ is used. In the monomeric state, such an acrylate is miscible with the electrolyte.

Polymerization of the system causes the system to phase separate due to the reduced entropy. The effect of the polymer on the conductivity is illustrated in Fig. 4 where the conductivity has been plotted as a function of the inverse temperature for various mixtures containing 40 wt.% acrylates (AD $n=12$

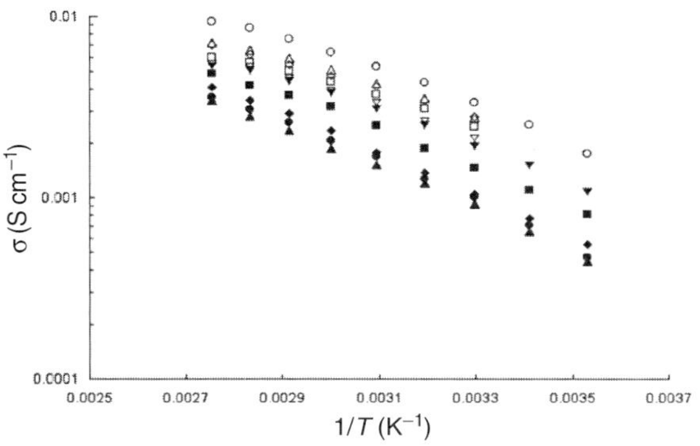

Figure 4
Arrhenius plot of the ionic conductivity obtained for electrolyte (1 mol $LiClO_4$ PC) mixtures containing 40% acrylate of various compositions ○ = 40%PEGDA, Δ = 30%PEGDA,10%DDDA, ◇ = 20%PEGDA, 20%DDDA, □ = 10%PEGDA, 30%DDDA. ∇ = 40% DDDA. Open symbols before polymerization, closed symbols after polymerization.

and PEGDA) of various compositions and 8.2 wt.% $LiClO_4$ in PC. It can be seen that, before polymerization, at a given temperature, the conductivity increases with an increasing PEGDA fraction. The opposite effect is however observed for the polymerized samples (gels). The fact that the conductivity decreases in the case of the gels with increasing PEGDA concentrations indicates the importance of the phase structure on the conductivity. The phase structure of two-phase gels can be observed using scanning electron microscopy. The morphology of the gel depends on the amount of polymer present within the system. Gels containing a small amount of polymer have a rather open structure, which is not desirable for applications in batteries. However, higher polymer concentrations lead to gels with a dense polymer morphology and a good conductivity. Besides the morphology, the specific interaction of the ions with the polymer is also important. This will be discussed in the following sections on IR and NMR spectroscopy.

2.6 Single-ion Conductors

In polymeric ion conductors such as polyethers, the largest contribution to the ionic conductivity is from the anions while higher cation mobility is desired. In order to reduce the mobility of the anions, single-ion conductors in which the anion is directly attached to the polymer have been synthesized. Examples of various single-lithium-ion conductors are shown below

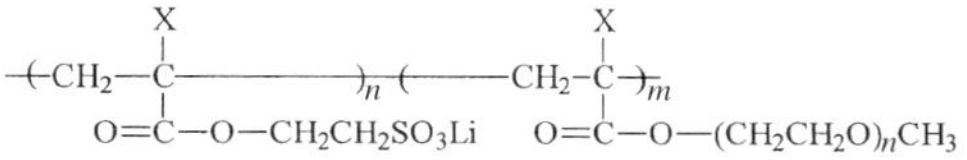

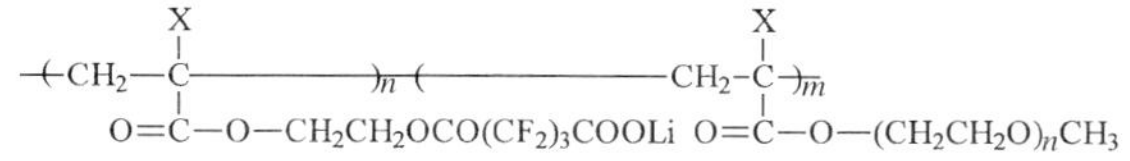

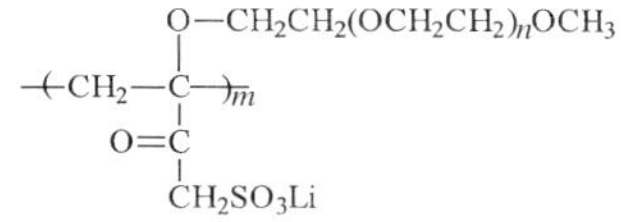

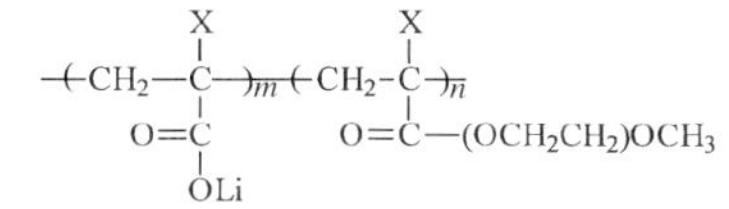

However, because of the high degree of association between the ions only modest values of 10^{-5}–10^{-7} $S\,cm^{-1}$ have been measured for the conductivity in these polymers, at room temperature.

3. Physical Properties

3.1 Diffusion Coefficient

Conductivity measured using impedance spectroscopy is an average measure of the contributions from the anion and the cation. In most applications the most important contribution is from the cation therefore it is important to determine the diffusivity of the individual species making up the ion conductor. The most frequently used technique is the pulse field gradient (PFG)-NMR technique (Johansson *et al.* 1996). This is a versatile technique, which gives information on the mobility of individual components within the system as well as on the occurrence of specific interactions.

Table 2 shows the diffusion constants of Li^+, PF_6^- in PC and in various PEGDA mixtures before and after polymerization. If we first consider the pure liquid electrolyte (1 mol $LiPF_6$ in PC) it can be seen that even though Li^+ is the smallest particle in the system, it has the lowest diffusion coefficient. This is attributed to the increased hydrodynamic radius of the Li^+ ion as a result of association with the solvent molecules which form a shell around the Li^+. If we consider the mixtures of liquid electrolyte in PEGDA it can be seen that the addition of PEGDA causes a decrease in the diffusion coefficient of all the components. However, closer inspection shows that the diffusion constant of Li^+ decreases much more than the other components within the system. It is also evident from this table that the diffusion coefficient of Li^+ decreases with an increasing molecular weight

Table 2
Diffusion constants of various components in various systems produced by using various (a) diacrylates and (b) solvents measured at 25 °C using PFG-NMR. (a).

Polymer fraction	State	D(pol) (10^{-7} cm^2 s^{-1})	D(PC) (10^{-7} cm^2 s^{-1})	D(Li^+) (10^{-7} cm^2 s^{-1})	D(PF_6^-) (10^{-7} cm^2 s^{-1})	t^+
0	mon	N.A.	17	7.5	14.7	0.34
0.2 PEGDA 400	mon	5.8	19	5.1	13.9	0.25
0.2 DDDA	mon	7.9	17.0	6.5	11.2	0.37
0.2 PEGDA 400	pol	0	15.2	0.65	8.1	0.07
0.2 DDDA	pol	0	13.0	5.7	11.1	0.34

(b)

Electrolyte (solvent + salt) fraction	State	D (PEG400) (10^{-7} cm^2 s^{-1})	D (solvent) (10^{-7} cm^2 s^{-1})	D (Li^+) (10^{-7} cm^2 s^{-1})	D (PF_6^-) (10^{-7} cm^2 s^{-1})	t^+
0.8 DMF	mon	17.5	57.6	25.2	42.5	0.37
0.8 TEGDME	mon	7.0	15.8	9.4	14.0	0.4
0.8 DMF	pol	0	48	17.9	37.0	0.33
0.8 TEGDME	pol	0	14.4	5.5	6.7	0.45

of the PEGDA, while the diffusion coefficients of PC and PF_6^- remain almost unchanged. This indicates that the lithium ions are coordinated far more by the PEGDA molecules than by the PC molecules. This is even more evident after polymerization. After polymerization, the long-range diffusion of the polymer stops, and the diffusion coefficient of the lithium becomes much lower. On the other hand, the diffusion coefficient of the PC is almost the same after polymerization whereas that of the PF_6^- is lower. This indicates strong interaction within the PEGDA gels, which influences the motion of the lithium and PF_6^- ions, whereas the diffusion of PC is retarded to a much lesser extent.

It is also interesting to consider the behavior of the liquid electrolytes containing dimethylformamide (DMF) and tri(ethylene glycol) dimethyl ether (TEGDME) instead of PC. Again, in all cases the solvent molecules have the highest diffusion coefficients, followed by PF_6^- and Li^+. The reactive polymer has the lowest diffusion coefficient. However, a major decrease in the diffusion coefficients of all the components is observable upon polymerization. In the gels containing TEGDME and DMF the diffusion coefficients of PF_6^- and Li^+ decreased by about the same factor, which differs from the behavior observed for PEGDA gels containing PC. This indicates that no preferential coordination of lithium by the PEGDA takes place in the gels containing DMF and TEGDME.

The effect of the polarity of the polymer molecules on the diffusion coefficients of the components was also studied using apolar DDDA instead of PEGDA as a polymer. In this case, almost no change in the diffusion coefficients of the ions was observed after polymerization. This is again due to phase separation occurring in the system upon polymerization. Because of the apolar nature of DDDA, no interaction occurs between lithium and the polymer, and the diffusion constant of the ions consequently remains unaltered.

3.2 *Ion Associations*

It has been shown that with an increasing salt concentration the equivalent conductivity decreases. This behavior is associated with the decreased ability of the solvent to dissociate ions. Both ion pairs and ion clusters are consequently formed. The techniques most frequently used to determine the state of ions are infrared and Raman spectroscopy (Miles *et al.* 1969). It is usually the spectral changes in the anion bands which provide information on their environment. Different IR and Raman active bands have been identified for different anions in estimating the populations of various cations. One of the most intensively studied salts is $LiSO_3CF_3$. The IR peak observed at 752 cm^{-1} is associated with the CF_3 deformation band when the ion is totally dissociated and free (free ion). The bands which appear for this salt in polyethers at 757 cm^{-1} and 762 cm^{-1} are associated with the ion pair and ion aggregates, respectively. Figure 5 shows room-temperature IR spectra obtained for $LiSO_3CF_3$ in PEGDA for various salt concentrations. It can clearly be seen that with increasing salt concentration, the relative concentrations of the associated ions (ion pairs and clusters) increase. Studies made as a function of increasing temperature showed that the concentrations

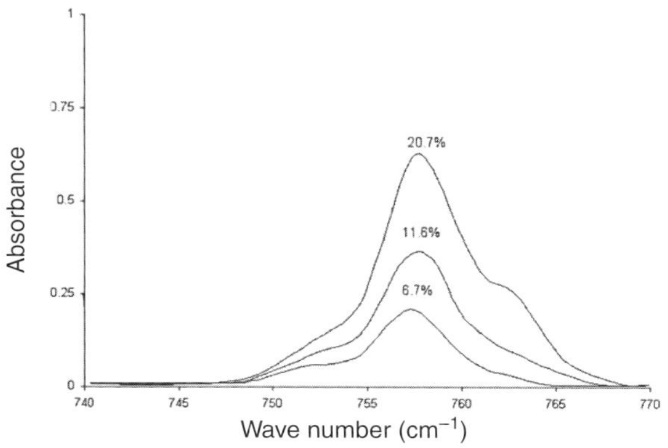

Figure 5
IR spectra obtained for mixtures containing various amounts of $LiSO_3CF_3$ in PEGDA (400).

the free ions and ion clusters increase, while the fraction of ion pairs decreases. It is this kind of behavior that makes it difficult to formulate a comprehensive model describing the ion conductivity of electrolytes.

3.3 Transport Number

Another important requirement in electrolytes in applications such as batteries, besides high conductivity, is a high transport number for cations. The transport number for lithium, defined as $t^+ = D^+/(D^- + D^+)$ can be estimated from the diffusion coefficients D^- and D^+ for PF_6^- and Li^+, respectively. When an ion conductor with a low lithium transport number is used, concentration gradients build up during the charge/discharge of a battery. This will result in an internal potential, which is undesirable. Development of a concentration gradient during the application of a potential has been observed using Schlieren optics. The concentration of the ions decreases at the negative electrode, whereas ion concentration increases at the positive electrode upon the application of voltage. At longer times, a steady-state gradient builds up. This results in a reduction in the potential across the electrolyte, ΔE, given by

$$\Delta E = \frac{RT}{F \ln(C_A/C_C)} \tag{11}$$

where R is the molar gas constant and C_A and C_C are the ion concentrations at the anode and cathode, respectively. The current consequently decreases from the initial value I_0 to the steady-state value I_s. It has been shown that in the absence of interfacial resistance, the transport number is then given by $t^+ = I_s/I_0$.

Transport numbers for various systems estimated on the basis of diffusion coefficients are shown in Table 2. It can be seen that the PEGDA gels containing PC have a low lithium transport number meaning that they are not suitable for practical applications. Two-phase gels obtained using apolar molecules have much higher transport numbers, making them more suitable for use as ion conductors.

3.4 Electrochemical Stability

The other requirement for electrolytes is electrochemical stability. It has been found that, for example, the electrochemical potential μ_A, which is the Fermi energy of the solid reductant, must lie below the LUMO (lowest unoccupied molecular orbital) of a liquid electrolyte to achieve thermodynamic stability against reduction of the electrolyte by the reductant. In the same way the electrochemical potential of the solid cathode must lie above the HOMO (highest unoccupied molecular orbital) of an electrolyte to prevent oxidation of the electrolyte by the oxidant. So the energy gap E_g between HOMO and LUMO of an electrolyte must be larger than the difference between the electrochemical potentials of the anode and the cathode ($E_g > \mu_A - \mu_C$). It has been found that carbonate-based electrolytes containing $LiPF_6$ or $LiBF_4$ have breakdown voltages (vs. Li) of more than 5.1 V. These electrolytes are therefore currently used in lithium ion batteries, which will be discussed in the next section.

4. Polymer Batteries

In a battery, electrochemical energy is converted into electrical power. The battery consists of a negative electrode (the anode) and a positive electrode (the cathode), which are held apart by a porous separator, and the system is saturated by an ion conductor (electrolyte). During the charging of the battery electrons flow outside the battery from the cathode to the anode while the ions move inside the cell. Discharging of the battery occurs when the cell is short-circuited using a load, and the flow directions of the electrons and the ions become reversed.

The open-circuit voltage of a battery is defined by the difference between the electrical potentials of the anode (φ_A) and the cathode (φ_C), given by

$$V_{oc} = \varphi_C - \varphi_A = \frac{\mu_C - \mu_A}{nF} \tag{12}$$

where n is the number of electrons involved in the reaction and μ_A and μ_C are the electrochemical potentials of the anode and the cathode in a certain state.

Intercalating electrodes are used to obtain high cycling efficiency in batteries. In such electrodes, the chemical potential is described by the mean field theory, which makes use of the lattice gas model. The chemical potential is then expressed as

$$\mu = E_0 + \gamma Ux + RT\left(\frac{x}{1-x}\right) \quad (13)$$

where E_0 is the energy associated with the filling of an isolated lattice site, x is the fraction of the available sites occupied by the guest species, and γ is the number of sites coupled to a given site by the interaction energy U. In this equation the first term accounts for the intercalant energy, the second term is the electrostatic interaction term and third term accounts for the entropic contributions. In a phase-separated system ($U<0$) the chemical potentials show a slight dependence on x. In systems where $U>0$ the potential on the contrary drops very rapidly.

One of the parameters defining the characteristics of a battery is the capacity per unit weight (or per unit volume). The energy density of a battery is therefore determined by the amount of charge which can be used during the electrochemical reaction per unit of mass (per unit of volume) times the average open field voltage. Form flexibility of a battery has also become a desired parameter. Here two types of lithium batteries that use polymeric and inorganic composite intercalating electrodes will be described.

4.1 Polymeric Electrodes

In general, two types of polymer electrodes can be defined: (i) conjugated polymers (Killian *et al.* 1996) and (ii) sulfur-containing polymers (Visco *et al.* 1993). In a conjugated polymer, intercalation occurs by oxidation, in which a hole (radical cation) is created and a counterion is inserted into the system to balance the charge for electrical neutrality. This type of doping is known as positive doping or *p*-doping. Several types of conjugated polymers have been studied as intercalation materials for batteries. Some examples of such polymers are polypyrrole $(C_4H_2NH)_n$, polythiophene $(C_4H_2S)_n$, and polyaniline $(C_6H_4NH)_n$. The redox process for these polymers (P) may be written as

$$XP + XC^- \rightleftharpoons (P^+C^-)X + Xe^-$$

where C^- is the cation. Almost all conjugated polymers are *p*-type conductors and are hence suitable for use as cathode materials. They do not exhibit an *n*-doped state. In a cell in which both the anode and the cathode consists of conjugated polymer it is only possible to transfer charge between the neutral and *p*-doped states. Operating in this way it is not possible to access all the stored charge. This problem is solved by incorporating a polymeric anion such as an anion consisting of polystyrenesulfonate. The redox reaction of such a system containing polypyrrole is shown below.

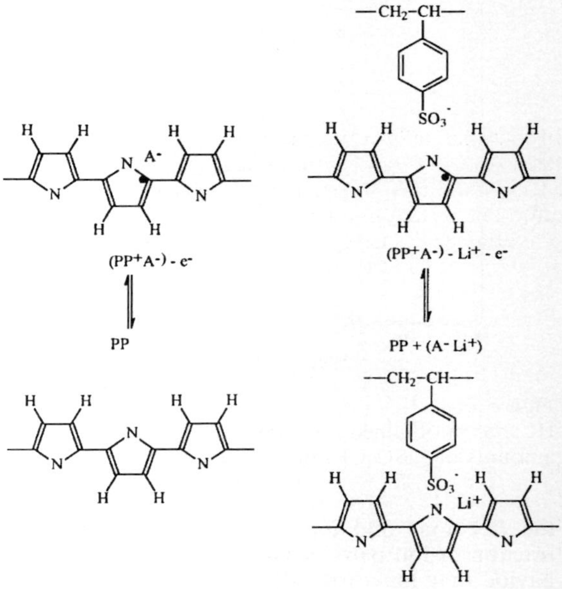

The polymers are obtained through electropolymerization on carbon fibers resulting in a porous structure, which makes the conjugated polymer accessible. The presence of polystyrenesulfonate causes a negative shift of about 350 mV in the redox peaks of polypyrrole. This will turn the polymer into a pseudo-*n*-dopable anode. In this way, polymer batteries with a capacity of 22 mAh g^{-1} operating at an average discharge voltage of 0.4 V are produced.

Polymeric organosulfides (polythiols) have also been suggested for use as electrode materials in batteries. These materials are called solid redox polymerization electrodes. Redox reaction results in the polymerization and depolymerization of these materials as shown below

$$(SRS)_n + 2ne^- + 2nLi^+ \rightleftharpoons n(Li^+\text{-}SRS\text{-}Li)$$

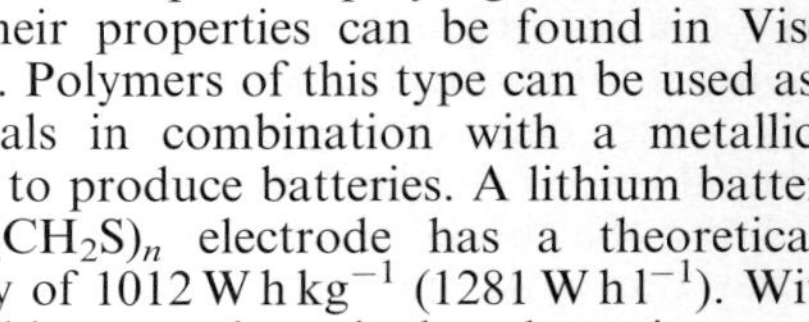

where R is an organic group such as ethyl or phenyl.

Some examples of polyorganodisulfide electrodes and their properties can be found in Visco *et al.* (1993). Polymers of this type can be used as cathode materials in combination with a metallic lithium anode to produce batteries. A lithium battery with a $(SCH_2CH_2S)_n$ electrode has a theoretical energy density of 1012 W h kg^{-1} (1281 W h l^{-1}). With an excess lithium metal anode these batteries can be cycled hundreds of times. However, safety is a big issue for batteries incorporating metallic lithium anodes.

4.2 Composite Electrodes

In commercially available rechargeable lithium batteries, $LiCoO_2$ and graphite are used as cathode and anode materials, respectively. Such a combination of materials results in a battery with a high average Voc potential and energy density. The reaction at the anode side is given by

$$LixC_6 \rightleftharpoons xLi^+ + xe^- + Li_{1-x}C_6$$

and that at the cathode side by

$$LiCoO_2 \rightleftharpoons xLi^+ + xe^- + Li_{1-x}CoO_2$$

For $x = 1$ at the anode side and $x = 0.5$ at the cathode side a battery with a very high cycling efficiency (low loss of capacity during each cycle) is obtained. The high stability of these anode and cathode materials also makes the battery safe. Such a battery is often called a rocking chair battery as lithium ions shuttle between the anode and the cathode, becoming intercalated during the charging and discharging of the battery.

Many attempts have been made to produce batteries with composite anode and cathodes made up of inorganic particles held together by polymers. In this polymer battery concept, anode and cathode materials are laminated together by placing a separator between them. In this way, a single-sheet polymer battery can be produced. Various materials ranging from solid polymer electrolytes to the gel electrolytes described above have been suggested for the production of such batteries. The concept (Tarascon *et al.* 1996) is now being commercialized. In this concept, a copolymer of vinylidene fluoride and hexafluoropropylene (PVDF-HFP) is used. The polymer is dissolved in a solvent and appropriate particles are added to the solution to obtain composite electrodes and the separator. After lamination, the solvent is extracted to produce a porous structure. The laminate is then activated by dipping it in a liquid electrolyte. The battery is subsequently vacuum sealed in laminated aluminum bags for use.

5. Outlook

Various types of lithium ion conductors and their properties have been described above. As long as the demand for high energy density storage devices remains, the market share of lithium batteries and battery materials will grow. This means that the search for stable, nonvolatile electrolyte materials will continue. The new materials are expected to be stable at high voltages and temperatures and expected to make the manufacture of batteries easier and more flexible. The dream of a solid state-polymer battery that can be cut into a desired shape and then put into use is yet to be realized. So much fundamental and applied research into such battery concepts remains to be done.

Bibliography

Allcock H R, Ravikiran R, O'Connor S J M 1997 The effect of oligo(ethyleneoxy)cyclotriphosphazates, tetraglyme, and other small molecules on the ionic conductivity of the poly(bis(methoxyethoxyethoxy)phosphazene) lithium triflate system. *Macromolecules* **30**, 3184–90

Armand M 1994 The history of polymer electrolytes. *Solid State Ionics* **69**, 309–19

Cohen M H, Turnbull D J 1959 Molecular transport in liquids and glasses. *Chem. Phys.* **31**, 1164–9

Gray F M 1991 *Solid Polymer Electrolytes*. VCH, Weinheim, Germany

Hikmet R A M, Peeters M P J, Lub J, Nijssen W 1999 Physical properties and battery performance of novel two-phase ion-conducting gels. *J. Electrochem. Soc.* **146**, 2397–403

Johansson A, Gogoll A, Tegenfeldt J 1996 Diffusion and ionic conductivity in $Li(CF_3SO_3)$PEG10 and $LiN(CF_3SO_2)$2PEG10. *Polymer* **37**, 1387–93

Killian J G, Coffey B M, Gao F, Poehler T O, Searson P C 1996 Polypyrrole composite electrodes in all-polymer battery system. *J. Electrochem. Soc.* **143**, 936–42

Lauter U, Meyer W H, Wegner G 1997 Molecular composites from rigid-rod. Poly(p-phenylene)s with oligo(oxyethylene) side chains as novel polymer electrolytes. *Macromolecules* **30**, 2092–101

Miles M G, Doyle G, Cooney R P, Tobias R S 1969 Raman and infrared spectra and normal coordinates of trifluoromethanesulfonate and trichloromethanesulfonate anions. *Spectrochim. Acta* **25A**, 1515–26

Scrosati B 1993 *Applications of Electroactive Polymers*. Chapman and Hall, London

Tarascon J M, Gozdz A S, Schmutz C, Shokoohi F, Warren P C 1996 Performance of Bellcore's plastic rechargeable Li-ion batteries. *Solid State Ionics* **86-8**, 49–54

Visco S J, Liu M, Doeff M M, Ma Y P 1993 Polyorganodisulfide electrodes for solid-state batteries and electrochromic devices. *Solid State Ionics* **60**, 175–87

Wakihara M, Yamamoto O 1998 *Lithium Ion Batteries*. Wiley-VCH, Tokyo

Williams M L, Landel R F, Ferry J D 1955 The temperature dependence of relaxation mechanisms in amorphous polymers and other glass forming liquids. *J. Am. Chem. Soc.* **77**, 3701–7

Oxide Fuels, Nonaqueous Reprocessing of

O. V. Skiba and A. V. Bygchkov

Nonaqueous methods (NAMs) for the reprocessing of oxide fuel include the processes that are carried out in molten salts (MS) or under the interaction of irradiated fuel with gaseous reagents. NAMs exclude the use of water solutions and organic liquids. These methods can be applied to spent nuclear fuel (SNF) or to high-level wastes (HLW). NAMs have these important differences from aqueous reprocessing: high capacity of fissile material; no neutron moderator needed; minimized HLW volume; and final products are often ready for manufacturing fuel. NAMs provide possible new technologies for a nuclear fuel cycle (NFC).

1. *Reprocessing in Molten Salts*

The following properties of MS are important for reprocessing:

(i) MS are ionic liquids and they are not subject to radiation influence.
(ii) Neutron moderators are absent in MS, therefore criticality problems are solved more easily.
(iii) Both factors allow for working with high radiation loading and with a high content of fissile materials, which results in reduction of equipment volume and HLW amount.
(iv) Strong cation–anion interaction takes place in MS; therefore the vapor pressure of the components is low.
(v) MS with alkaline cations consist of polar ions; therefore voltage of MS decomposition is higher than dissolved components. MS can be restored and impurities can be concentrated by electrochemical processes.
(vi) During reprocessing it is impossible to free the fuel totally from fission products (FPs). This fuel cannot be used for the manufacture of nuclear weapons.

The oxide fuel reprocessing should include three stages:

(i) Dissolution of oxide fuel in halide MS by means of a chlorinating or fluorinating agent. The dissolution in oxy-anion MS could be carried out by direct or dissolving agent treatment.
(ii) The removal of components (actinides) from MS can be carried out by electrolysis. The removal of actinide compounds from MS is possible by change of temperature or by treatment with special gases.
(iii) Many FPs have properties different from actinides in MS; therefore it is easy to carry out their selective extraction. Various methods can be used: electrochemical, precipitation, etc.

Oxide fuel can be reprocessed in a number of salt systems. The main systems are molten chlorides, but alkali fluorides, nitrates, sulfates, molybdates, and tungstates were also studied. Molten alkali and alkaline-earth chlorides are the most studied for fuel reprocessing. There is extensive experience with molten chlorides, and development is being continued in Russia, USA, and Japan. Nowadays, pyroelectrochemical reprocessing is being applied to oxide fuel (RIAR, Russia) and to metal fuel (Argonne National Laboratory (ANL), USA).

The methods which use these salts can be grouped into two sets:

(i) Chloric pyrometallurgy or pyroelectrochemical reprocessing with the use of a chlorinating agent, with direct dissolution of oxide fuel and obtaining of oxide fuel. The processes can be carried out at 500–700°C and include the following operations
(a) fuel chlorination:

$$UO_2 + Cl_2 \rightarrow UO_2Cl_2$$

$$PuO_2 + 2\ Cl_2 \rightarrow PuCl_4 + O_2$$

$$PuCl_4 + O_2 \rightarrow PuO_2Cl_2 + Cl_2$$

(b) electrolysis: cathode:

$$UO_2^{2+} + 2e \rightarrow UO_2$$

$$PuO_2^{2+} + 2e \rightarrow PuO_2$$

anode:

$$2Cl^- - 2e \rightarrow Cl_2$$

(c) precipitation:

$$PuCl_4 + O_2 \rightarrow PuO_2 + 2Cl_2$$

$$ThCl_4 + O_2 \rightarrow ThO_2 + 2Cl_2$$

(ii) Pyrometallurgical processes based on oxide reduction to the metal or nonoxygen-containing cations, followed by electrorefining or purification by exchange reactions and obtaining of metallic actinides. These processes are carried out at 500–700°C and can include the following operations:
(a) reduction in media of liquid alloy (cadmium or others) and salt ($CaCl_2$–CaF_2)

$$UO_2 + Ca(Cd) \rightarrow U(Cd) + CaO$$

$$PuO_2 + Ca(Cd) \rightarrow Pu(Cd) + CaO$$

(CaO is removed from the melt by electrolysis with a carbon anode; lithium also could be used).

(b) electrolysis (in molten LiCl–KCl):

$$\text{anode}: U(Cd) - 3e \rightarrow U^{3+} \quad Pu(Cd) - 3e \rightarrow Pu^{3+}$$

$$\text{cathode}: U^{3+} + 3e \rightarrow U_{(s)} \quad Pu^{3+} + 3e \rightarrow Pu(Cd)_{alloy}$$

During development, most researchers chose molten chlorides only, with alkaline cations in various combinations (for example, 3LiCl–2KCl, or NaCl–KCl). The studies on oxide SNF reprocessing with the production of uranium and plutonium oxides from molten chlorides were carried out in the former Hanford National Laboratory (USA). In the USSR, the work in this direction has developed in RIAR (Dimitrovgrad), where these developments are being continued.

The basic principles of fast breeder reactor nuclear fuel cycles with the application of pyrochemical technology were experimentally demonstrated by RIAR. This included the pyrochemical reprocessing of SNF and waste treatment, the obtaining of dense fuel particles, the manufacturing of fuel pins by vibropacking, and the use of remote-controlled automated equipment. The basic technical problems were solved. The wide reactor tests were carried out and all aspects of such cycle safety were investigated.

Concerning oxide fuel reprocessing technology, the following properties of uranium and plutonium are the most important.

- From an electrochemical point of view, uranium and plutonium oxides behave like metals. During dissolution in the MS, they form MeO_2^{n+} ions that are reduced to oxides on the cathode.
- At temperatures above 400°C, UO_2, and PuO_2 have electric conduction. The lower electric conduction of the oxides in comparison with MS allows the obtaining of compact deposits.
- In molten chlorides, uranium has the stable ions U^{III}, U^{IV}, and U^{VI}. The highest states of plutonium oxidation—Pu^{V} and Pu^{VI}—are stable under a definite ratio of oxidation–reduction potentials of the system. For co-electrodeposition of UO_2 and PuO_2, it is necessary to treat the melt with a chlorine–oxygen gaseous mixture for creation of the PuO_2^{2+} ion.
- Plutonium can be converted into the dioxide from any oxidation state by changing the oxidation–reduction potential of the system.

UO_2 and PuO_2 are reduced in the electropositive area. The majority of FPs are reduced at more negative potentials, so in the process of UO_2 and PuO_2 electrodeposition, their purification from impurities occurs.

Technological processes for production of oxide fuel from the molten chlorides have been developed. The process of UO_2 electrodeposition is carried out from molten NaCl–2CsCl–UO_2Cl_2, prepared by chlorination of uranium oxides. During electrolysis, the formation of dense UO_2 cathodic deposits occurs. The recovery rate of UO_2 is 99.0–99.5%, and it is used for the manufacture of vibropacked fuel. This technology has been developed up to semi-industrial level.

Experiments with irradiated UO_2 from the BOR-60 reactor showed that the pyroelectrochemical method allows the reprocessing of SNF (burn-up: 7.7% h.a.; cooling period: 6 months) with a high recovery rate (99%) and satisfactory decontamination from FPs (DF = 500–1000). Reprocessed UO_2 was irradiated repeatedly in the BOR-60 reactor.

The process of PuO_2 precipitative crystallization is carried out from molten NaCl–KCl after chlorination–dissolution of plutonium. During treatment of the melt by a chlorine–oxygen gas mixture, the formation of crystalline PuO_2 occurs. That process has been developed for the decontamination of PuO_2 from americium, and for conversion of plutonium alloy of military origin into the dioxide.

The process of uranium and plutonium oxides electro-codeposition includes chlorination in molten NaCl–2CsCl, with formation of uranyl chloride and plutonium chlorides and oxidation for the transformation of plutonium into plutonyl chlorides. Electrolysis is carried out with treatment by a chlorine–oxygen gas mixture; thus codeposition of UO_2 and PuO_2 occurs. The cathode deposits obtained are a quasi-homogeneous crystal system of $(U, Pu)O_2$. After removal of captured salts and crushing, the product is used for vibropacking. The fuel recovery rate is 98.5–99.5%. The technology was developed and the semi-industrial facility for production of MOX-fuel was constructed. This facility has been in operation since 1988. It provides manufacture of fuel for the BOR-60 reactor and experimental FAs for the BN-600 reactor.

To perform processes in remote conditions, the necessary equipment was developed. The main device—a chlorinator–electrolyzer—uses pyrolitic graphite as the material for bath-anode, cathode, and gas tubes. The device used nowadays has a working volume of 40 l and is designed for a loading of 30 kg of initial material. Other equipment is used for crushing cathodic deposits, for removal of captured salts from fuel, for drying, and classification.

For pyrochemical reprocessing of irradiated fuel, the following options are under development in RIAR.

(i) Reprocessing of MOX-fuel for only plutonium recycling.
(ii) Reprocessing of MOX-fuel with production of MOX-fuel.

Pyroelectrochemical process of PuO_2 extraction from SNF was developed for demonstration of the closed fuel cycle of the BOR-60 reactor. The thermal or mechanical decladding of fuel is the first stage of the technology. The operations of reprocessing in the chlorinator–electrolyzer are shown in Fig. 1.

(i) Fuel chlorination in molten NaCl–KCl or LiCl–NaCl–KCl–CsCl.
(ii) Electrolysis for partial removal of UO_2 free from plutonium. There are some FPs (Zr, Ru–Rh–Pd, Ag) captured by cathodic deposit.
(iii) Precipitative crystallization of PuO_2 decontaminated from FPs that allows obtaining of PuO_2 ready for vibropacking. Plutonium 99.5–99.9% is collected at the bottom as PuO_2.
(iv) Additional electrolysis for removal of uranium. The main FPs are deposited with UO_2.
(v) MS purification from impurities by the introduction of sodium phosphate into the melt. The formation and precipitation of impurity phosphates occur, which are soluble neither in molten chlorides nor in water. Only caesium, rubidium, and, partially, strontium remain in the MS.

Tests on the reprocessing of irradiated MOX-fuel were carried out. The most important test was carried out with the BOR-60 reactor fuel burned up to 21–24% h.a. Direct recovery of plutonium into the final product was 95.6%. The decontamination of plutonium is quite sufficient from the point of view of fuel nuclear-physical properties (DF was 40–50 for ruthenium, RRE, and more than 3000 for caesium).

The pyrochemical reprocessing of nuclear oxide fuel in molten chlorides is a technology with a high safety level. The safety analysis of the process developed by the ANL for metal fuel has shown a higher level in comparison with aqueous methods. Similar results are given by the analysis for oxide fuel reprocessing.

2. *Fluoride Volatility Method*

An important property of uranium is formation of the hexafluoride, which is stable and gaseous at room temperature. The simple reaction of oxide SNF at 350–400°C with a fluorine–nitrogen mixture:

$$UO_2 + F_2 \rightarrow UO_2F_2, \quad UO_2F_2 + 2F_2 \rightarrow UF_6 + O_2$$

separates uranium from all other actinides, FPs, and impurities.

Developments have been carried out in many countries. The special "FREGAT" facility allowed the carrying out of complex studies on the reprocessing of irradiated UO_2 and MOX-fuel of FBR. The reprocessing flow-sheet for irradiated uranium fuel includes the following stages.

(i) Decladding (and voloxidation).
(ii) Fluorination.
(iii) UF_6 purification from FPs by cooling at 15–30°C.
(iv) FPs sorption of NaF at 400°C, UF_6 sorption on NaF at 100°C.
(v) "Desorption–sorption" cycle for UF_6 purification and condensation at –40 to –60°C.

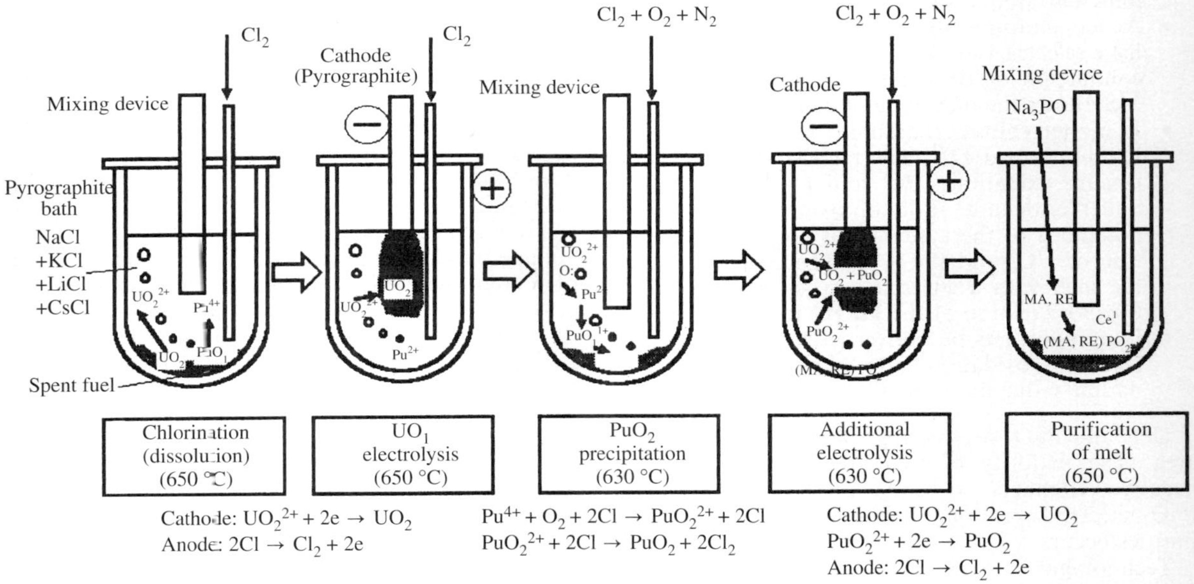

Figure 1
Flow-sheet of pyroelectrochemical reprocessing.

The fuel for tests had burn-up of 10% h.a. and a 3–6 months cooling period. UF_6 had high DF from FPs ($>10^6$). The total extraction of uranium from SNF was 99.4–99.6%, and direct recovery rate into UF_6 was 96.1–98.3%. The losses were 0.4–0.5%. A basic amount of FPs ($\sim 85\%$) is concentrated in the residues of fluorination.

MOX-fuel reprocessing: for PuO_2 fluorination up to PuF_6, a higher temperature is necessary and it was carried out in a special flame fluorinator. Flame temperature is ~ 1300 K. The obtained gas mixture was filtered, FPs fluorides were condensed at 313–323 K, and UF_6 and PuF_6 at 213–233 K. The experiment with the BOR-60 fuel had the following results: (i) DF for UF_6 was 10^7, for PuF_6 was 10^3–10^4; (ii) recovery rate for uranium was $>99\%$, for plutonium 89–91%. The low recovery rate of plutonium is connected to PuF_6 instability.

A big advantage of the method is that UF_6 can be used directly for re-enrichment. Also, hexafluorides can be transformed into dioxides just after reprocessing by pyrohydrolysis.

3. Oxidation–Reduction Method

One property of UO_2 is that it can be oxidized by air at temperatures above 350°C to U_3O_8:

$$3\,UO_2 + O_2 \rightarrow U_3O_8$$

The structure changes and oxide volume is increased. The process named AIROX (Atomics International Reduction Oxidation) is based on this property and includes two stages.

- A spent fuel pin with UO_2 is subjected to oxidation to U_3O_8 by air at 400°C; the volume is increased by 30%, the cladding is broken off, and gaseous FPs removed from the fuel.
- The oxidized product is treated with hydrogen at 600°C.

$$U_3O_8 + 2\,H_2 \rightarrow 3\,UO_2 + 2\,H_2O;$$

It is reduced again to UO_2. The enriched uranium can be added to that product and the final product must be palletized for recycling.

The reactor system "PWR–CANDU" is successful from the point of view of the nuclear–physical properties of fuel. The concept DUPIC (direct use of spent PWR fuel in CANDU reactor) is developed on that basis. The process will include disassembling and decladding of spent LWR fuel, oxidation (with removal of some FPs), reduction, pelletizing, and manufacture of CANDU fuel bundles. KAERI studies are directed to its demonstration and, probably, it will take effect in the future.

See also: Nuclear Fuel Cycles

Bibliography

Bychkov A V, et al. 1993 Pyroelectrochemical reprocessing of irradiated uranium–plutonium oxide fuel for fast reactors In: *Proc. Int. Conf. GLOBAL' 93*. ANS, Seattle, WA, Vol. 2, p. 1351

Bychkov A V, et al. 1997 Pyroelectrochemical reprocessing of irradiated FBR MOX fuel III: experiment on high burn-up fuel of the BOR-60 reactor In: *Proc. Int. Conf. GLOBAL '97*. Vol. 2., p. 912

Demyanovich M A et al. 1982 Fluorination of spent uranium-plutonium oxide fuel in the flame reactor. Preprint RIAR-50(565), Dimitrovgrad, Russia (in Russian)

Grantham L F et al. 1979 AIROX dry pyrochemical processing of oxide fuels. In: Actinide Separation. ACS Symposium Series 117. American Chemical Society, Washington, DC, Chap.16

Hannum W H 1997 The technology of the integral fast reactor and its associated fuel cycle. Progr. Nucl. Energy 31

Harmon K M, Jansen G 1970 The salt cycle process. *Progr. Nucl. Energy* **4**, 429

Kirillovich A P, et al. 1997 Safety analysis of fuel cycle processes based on "dry" pyrochemical fuel reprocessing and vibropac technology In: *Proc. Int. Conf. GLOBAL '97*. Vol. 2, p. 900

Lee J S, Song K S et al. *Proc. Int. Conf. GLOBAL '93*. ANS, Seattle, WA, Vol. 2, p.733

Skiba O V et al. 1976 Test reprocessing of spent uranium fuel of the BOR-60 reactor by fluoride method. Preprint RIAR, P-18(284), Dimitrovgrad, Russia (in Russian).

Skiba O V, et al. 1993 Technology of pyroelectrochemical reprocessing and production of nuclear fuel In: *Proc. Int. Conf. GLOBAL '93*. ANS, Seattle, Vol. 2, p. 1344

P

Petroleum Industry: Corrosion

R. C. John and W. C. Fort III

The petroleum industry has a large capital investment in equipment exposed to a wide range of corrosive environments. Considerable research and engineering effort is expended to reduce corrosion losses and to reduce risks to facilities, personnel, and the environment. The major corrosion problems encountered in the production, transportation, and refining of crude oil and natural gas are discussed below, and typical methods of corrosion control are described.

1. Oil and Gas Production

Internal corrosion and subsequent corrosion fatigue can occur with steel drill pipe in many of the water-based muds used for drilling. Oxygen content of the mud and chloride ion concentration are the major corrosion factors. Corrosion control is achieved by (i) internal coating, (ii) oxygen scavenging (e.g., by sulfite), (iii) inhibition, or (iv) changing to an oil-based mud. If hydrogen sulfide is present, a sulfide scavenger may be added to avoid sulfide stress cracking. Use of fiber-reinforced composite tubulars, which are highly corrosion resistant, is also becoming an option.

For offshore drilling and production structures, the corrosive effects of salt water on external surfaces are countered by use of protective coatings above water and cathodic protection below water, with the splash zone protected with an extra thickness of steel, an alloy wrap, or a polymeric coating.

Internal corrosion of production equipment results from the combination of acid gases (carbon dioxide and/or hydrogen sulfide) and a produced water phase, which may have a high chloride ion or organic acid content. Such internal corrosion is more severe when the industry operates at greater depths where temperatures exceed 260 °C, gas pressures exceed 2×10^8 Pa, carbon dioxide concentrations exceed 1%, and hydrogen sulfide concentrations can exceed 1%. At these depths, the costs of equipment and repair increase, making failures expensive and highly reliable materials more economical. In deep wells, corrosion-resistant alloys such as special Fe–Cr–Ni–Mo stainless steels and Ni–Cr–Mo alloys are being used for tubing and related equipment. Under less severe conditions, corrosion inhibitors can be added continuously or in periodic batches to control corrosion of steel downhole equipment, well head equipment, and flowlines. Chemicals such as methanol or glycol may be used to avoid formation of gas hydrates, which are solid compounds of natural gases and water, at conditions of low temperatures and high pressures. These chemicals may also significantly lower the corrosion rate of alloy steels. Oilfield corrosion inhibitors are proprietary organic compounds, usually with a long hydrocarbon chain and a nitrogenous polar group. They form an oil water-repellent film on the steel surface, and are effective at concentrations of the order of 50–100 ppm. Coatings and cathodic protection are used to protect field tanks and treating vessels where a separated water phase is present. Ingress of oxygen by contamination with air is often a concern.

Air contamination of produced brine being injected into a formation, either for disposal or as means of secondary recovery (water flooding), can result in severe corrosion of steel piping and downhole equipment with oxygen levels as low as 0.01–0.1 ppm. Corrosion control measures include oxygen scavenging with sulfite, inhibition, and use of nonmetallic pipe or use of steel pipe with an organic or cement coating or lining.

Sulfide stress cracking of high-strength steels is a problem in handling sour (hydrogen sulfide containing) crude oil or natural gas. In the USA, NACE International issues a periodically updated standard MR0175 (NACE 2000) for materials selection for such service. It is applied if the hydrogen sulfide content of the gas phase exceeds a partial pressure of 3.5 mbar. For many steels, acceptance for sour service involves controlling the hardness to a maximum of HRC 22.

2. Transportation

External corrosion of pipelines and other buried equipment is controlled by a combination of coatings and cathodic protection used for protecting buried structures. Internal corrosion by crude oil and natural gas resulting from acid gases and water is controlled primarily by reducing the water content as much as possible, with inhibitors being added where necessary. Inhibition is more easily achieved if the velocity of the fluid is controlled such that water cannot collect in low spots in the pipeline. Use of devices called pigs, which travel through pipelines to remove debris and water, can also reduce the potential for corrosion. Pipelines handling refined petroleum products have a potential for corrosion by water and oxygen absorbed from air during prior storage. Corrosion control is by dehydration or by addition of inhibitors soluble in the product.

Tankers handling either crude oil or refined products are subject to internal corrosion due to the presence of oxygenated ballast water. Corrosion control is achieved primarily by application of coatings during construction of the tanker. Some use is made of sacrificial anodes of aluminum or zinc.

3. *Refining*

Corrosion in crude oil distilling and thermal conversion units (including thermal cracking, catalytic cracking, and coking units) results from sulfur compounds, carboxylic acids, and chloride salts in the crude oil feed. Certain sulfur compounds and hydrogen sulfide present in the crude oil or generated by thermal decomposition corrode carbon steel at temperatures above 250 °C. The Cr–Mo steels (5–9% Cr) and Fe–Cr and Fe–Ni–Cr stainless steels offer increasing corrosion resistance to this sulfidic corrosion. Certain hydrocarbon soluble, high molecular weight, carboxylic acids (e.g., naphthenic acids) present in some crude oils corrode carbon steel and Cr–Mo steel equipment operating under high velocity conditions in the 200–400 °C temperature range. Fe–Ni–Cr and Fe–Ni–Cr–Mo stainless steels (especially UNS S31600 and S31700) provide resistance to this naphthenic acid corrosion. Some refiners use proprietary naphthenic acid corrosion inhibitors added to the crude oil feed. Chloride salts dissolved in the crude oil hydrolyze in distilling to form hydrogen chloride, which condenses with water to form hydrochloric acid in distilling column overhead systems. This column overhead corrosion is controlled by a combination of (i) desalting the crude oil charge, (ii) addition of neutralizers and inhibitors in the overhead systems, (iii) water washing, and (iv) specification of certain Ni–Cu, Ni–Cr–Fe, or titanium alloys for specific overhead condensing equipment.

Fluid catalytic cracking units and certain other conversion units circulate and separate large volumes of solid catalyst at temperatures from 480 to 760 °C, which can result in severe erosion of steel and stainless steel components. Erosion is limited primarily by lining vessels and circulating piping and cyclone separators with refractories.

Corrosion and materials degradation in high-pressure hydroprocessing units (including hydrodesulfurizing, hydrotreating, hydrocracking, and catalytic reforming units) results from hydrogen, hydrogen sulfide, ammonia, and hydrogen chloride. High-pressure hydrogen in these units can cause internal decarburization and fissuring of some steels operating above about 220 °C, resulting in loss of mechanical strength. Control of this high-temperature hydrogen attack is achieved by selection of an appropriate Cr–Mo steel (with 1.25–3% Cr), using as a guide the Nelson curves, which are revised periodically by the American Petroleum Institute in its publication RP 941 (API 1983). High-temperature sulfidic corrosion of carbon steel by hydrogen sulfide in hydrogen-rich streams occurs at temperatures above 230 °C.

The Cr–Mo steels and Fe–Cr and Fe–Cr–Ni stainless steels offer increasing resistance to sulfidic corrosion. The thick walled vessels in high-pressure hydroprocessing units are often made from Cr–Mo steels to resist high-temperature hydrogen attack and internally clad or weld overlaid with austenitic stainless steel to resist sulfidic corrosion. Corrosive ammonium bisulfide or ammonium chloride forms from ammonia and hydrogen sulfide or hydrogen chloride in high-pressure reactor effluent streams. They can cause fouling and corrosion, which are normally controlled by combinations of (i) on-stream water washing, (ii) selection of 12% Cr or austenitic alloys, and (iii) addition of proprietary antifoulants or corrosion inhibitors.

Many refining operations produce streams containing sour water (water with hydrogen sulfide with or without ammonia and hydrogen cyanide), which is corrosive to carbon steel. Corrosion charges the steel with hydrogen and can lead to cracking or blistering by a number of mechanisms referred to as wet hydrogen sulfide cracking. The mechanisms include sulfide stress cracking, hydrogen-induced cracking (HIC), and stress-assisted hydrogen-induced cracking (SOHIC). Streams from fluid catalytic cracking and coking units that contain hydrogen cyanide are particularly aggressive. Wet hydrogen sulfide cracking of carbon steel is prevented by (i) selecting specific steel grades, welding practices, and heat treatments to reduce hardness and stresses, (ii) applying internal organic coatings, (iii) using corrosion inhibitors, or (iv) substituting austenitic stainless steels.

Use of austenitic stainless steels and other Fe–Ni–Cr austenitic alloys may introduce secondary corrosion problems resulting from shutdown conditions. Intergranular stress-corrosion cracking can occur if the material has become sensitized and is subsequently exposed to polythionic acids formed from reaction of sulfide scale with water and oxygen in the air. Sensitization occurs as the result of weld fabrication or heat treatment or from in-service exposure in certain temperature ranges, depending on the alloy. Control of this polythionic acid stress-corrosion cracking involves (i) selection of low carbon alloy grades or alloy grades with added titanium or niobium, (ii) control of shutdown conditions to exclude air and/or moisture, or (iii) neutralization of the acids during shutdowns per NACE RP0170 (NACE 1997). Use of austenitic stainless steels in petroleum refining also carries the risk of under-deposit corrosion, pitting, and chloride stress-corrosion cracking where chlorides are present.

Alkylation units manufacture gasoline blending components by treating light hydrocarbons with strong sulfuric or hydrofluoric acid. Carbon steel is used in construction for most equipment. Corrosion is limited in sulfuric acid alkylation units by controlling the water content to low levels, limiting temperature and velocity to acceptably low levels, and specifying Fe–Cr–Ni alloys such as alloy 20 (UNS N08020) for certain components. Corrosion in hydrofluoric acid alkylation units is limited by limiting water content and specifying alloy 400 (UNS N04400) for specific equipment. Both types of alkylation units require caustic neutralization and water washing to remove acid components from the products.

Bibliography

American Petroleum Institute 1983 *Steels for Hydrogen Service at Elevated Temperatures and Pressures in Petroleum Refineries and Petrochemical Plants.* Publication 941. API, Washington, DC

Garverick L (ed.) 1994 *Corrosion in the Petrochemical Industry.* ASM International, Metals Park, OH

Lai G Y 1990 *High-Temperature Corrosion of Engineering Alloys.* ASM International, Metals Park, OH

Nathan C C (ed.) 1973 *Corrosion Inhibitors.* National Association of Corrosion Engineers, Houston, TX

National Association of Corrosion Engineers 1979 *Corrosion Control in Petroleum Production.* NACE International, Houston, TX

National Association of Corrosion Engineers 1994 *Materials and Fabrication Practices for New Pressure Vessels Used in Wet H2S Refinery Services.* Publication 8X194. NACE International, Houston, TX

National Association of Corrosion Engineers 1997 *Protection of Austenitic Stainless Steels and Other Austenitic Alloys from Polythionic Acid Stress Corrosion Cracking During Shutdown of Refinery Equipment*, RP0170. NACE International, Houston, TX

National Association of Corrosion Engineers 2000 *Sulfide Stress Cracking Resistant Metallic Materials for Oilfield Equipment.* NACE International, Houston, TX, NACE MR0175, 2000

White R A 1998 *Materials Selection for Petroleum Refineries and Gathering Facilities.* NACE International, Houston, TX

Photodetectors

P. Fay

Photodetectors play a critical role in many current and emerging technologies. Sensing systems for biotechnology, telecommunications, medicine, atmospheric studies, and the physical sciences are all impacted by the capabilities and limitations of semiconductor-based photodetection technology. In this article, a brief outline of the fundamental physical processes used in common photodetectors is presented, followed by a description of the primary figures of merit for photodetectors. The device structures, operation, and performance of common detector types are described, and the fabrication processes typically used to implement detectors are examined.

1. Physical Fundamentals of Photodetector Operation

Semiconductor photodetectors can be divided into two main categories, based on the fundamental physical processes that give rise to their sensitivity to incident optical radiation. These categories are interband detectors and intersubband detectors. Figure 1 illustrates schematically the fundamental physical mechanisms governing the operation of each of these general device types.

Interband detectors, which constitute the majority of semiconductor photodetectors and include both photoconductors and diode-based detectors, rely on electron–hole pair (EHP) generation in semiconducting materials by photons with energies greater than the bandgap. As depicted in Fig. 1(a), the resulting excess carriers give rise to increased conductivity in the case of photoconductive detectors, or to increased terminal current in diode-based detectors.

Intersubband detectors, which include such devices as quantum well intersubband photodetectors (QWIPs), use engineered band structures to produce bound states and a continuum miniband in the conduction or valence band of the material, and photodetection proceeds via excitation of carriers from the bound to unbound states in the same "main" (i.e., conduction or valence) band. Externally, these structures behave much like conventional photoconductive devices, with the exception that in the case of QWIPs the required photon energy for photocarrier generation is not the bandgap of the absorption layer materials themselves, but rather the energy spacing of the bound-to-unbound transition formed by the quantum well superlattice. The small transition energy permits detection of long-wavelength radiation, while the "engineered" nature of the band structure permits the absorption to be tailored to specific wavelengths and allows for a great degree of flexibility in the design. Tunability of the absorption wavelength through electric field dependencies in the band structure, as well as the ability to fabricate multiple superlattices with different absorption characteristics in a single device structure, allows for multispectral applications of this technology.

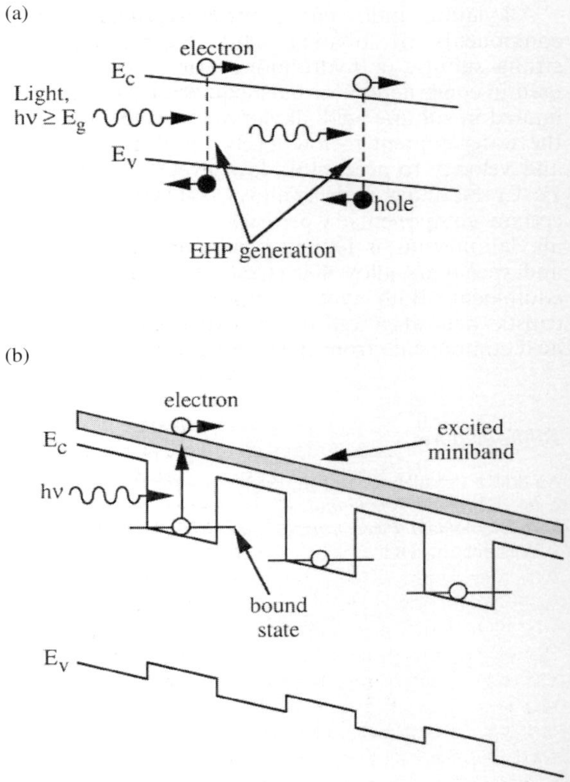

Figure 1
Schematic diagram of (a) interband and (b) intersubband detector operation.

2. Performance Characteristics

Regardless of the particular type of photodetector employed, several device figures of merit help to determine the suitability of a given detector for a specific system application. These figures of merit include the responsivity of the detector, its bandwidth, and the noise added to the signal by the photodetector. Responsivity, $\mathcal{R}$, is a measure of the change in detector current arising from the optical signal; this is conventionally expressed in units of

$A W^{-1}$. For many applications it is desirable to maximize the responsivity in order to maximize the overall signal-to-noise ratio of the system. Several photodetector types can be designed to achieve gain in their responsivity; i.e., the output photocurrent (in electrons s^{-1}) can be larger than the incident absorbed photon flux (in photons s^{-1}). This is possible despite the fact that each incident photon produces only one EHP. For photoconductors, gain is facilitated by using materials with carrier recombination lifetimes longer than the transit time within the device. For photodiodes, gain can be caused instead by impact ionization and avalanche multiplication. Although the presence of internal gain owing to these mechanisms can significantly increase the responsivity of a detector, this may be accompanied by a penalty in the bandwidth and noise performance of the detector. Careful consideration of these effects is required to choose the appropriate responsivity–bandwidth–noise tradeoff for a given application.

The operational speed of a photodetector is another important characteristic. This can be specified either in terms of the frequency- or the time-domain response. For frequency-domain characterization, the bandwidth is specified as the frequency at which the responsivity has fallen off from its low-frequency value by 3 dB. Time-domain approaches characterize the response of the photodetector to very short optical pulses. Most often this response is quoted in terms of the full width at half maximum (FWHM) of the resulting output electrical pulses. For a detector with a Gaussian impulse response, the bandwidth and FWHM are related by the following expression:

$$f_{3\mathrm{dB}} = \frac{2\sqrt{(\ln 2)}}{\pi \cdot \mathrm{FWHM}} \approx \frac{0.533}{\mathrm{FWHM}}$$

In practice, photodetectors do not have purely Gaussian impulse responses, and so caution in the use and interpretation of this expression is required. The generally wide disparity in the transport properties of electrons and holes leads to multicomponent impulse responses that often deviate significantly from simple Gaussian pulses. The consequence is that bandwidth estimates based on the assumption of a Gaussian response often significantly overestimate the actual bandwidth of detectors. This overestimation occurs because the FWHM is dominated by the high-velocity carriers, usually electrons. A more complete treatment of the time response by Fourier transformation of the full impulse response (and thus including the "tail" region of the response owing to slow carriers) results in bandwidths that match those obtained via frequency-domain techniques.

Another important characteristic of detectors is the amount of noise generated. Noise in detectors can be characterized by the r.m.s. noise current, $\langle I^2 \rangle$, the noise equivalent power, NEP, or the specific detectivity, D^*. The noise current spectral density is simply the variance of the photodetector current. When integrated over the desired operating bandwidth, this provides a measure of the r.m.s. noise current. The NEP is the required input optical signal power required to make the desired output signal level (r.m.s.) equal to the r.m.s. noise output in a bandwidth of 1 Hz. The specific detectivity is defined as

$$D* = \frac{\sqrt{(A \cdot \Delta f)}}{\mathrm{NEP}}$$

where A is the detector area and Δf is the bandwidth. D^* is normally specified in a 1 Hz bandwidth. The choice of normalization for D^* is intended to allow direct comparison of devices of differing areas; however, not all detectors follow the $\sqrt{(A)}$ scaling assumed by the definition of D^*.

Generation of noise is fundamental to the operation of photodetectors and arises from several intrinsic physical phenomena. For most types of photodetectors, the noise generated by the detector can be classified into five main categories: thermal noise, shot noise, generation–recombination noise, flicker or $1/f$ noise, and excess noise. Thermal noise arises from the random thermal motion of charge carriers inside the device. Shot noise is caused by the discretization of charge composing an electric current; fluctuations in the arrival statistics of the individual electrons that make up the average current give rise to this type of noise. Generation–recombination noise is owing to the random timing of EHP generation and recombination events. Thermal noise and shot noise have frequency-independent power spectral densities (a "white" spectral distribution), whereas the spectral density of generation–recombination noise is dependent on the device structure and operational details. Flicker noise, often also called $1/f$ noise after the frequency dependence that it exhibits, is present in detectors as well as all other semiconductor devices. This is generally significant for frequencies below a few kilohertz for optimized structures and fabrication processes. Excess noise is any noise contribution of a detector above the limits imposed by the fundamental physical processes of thermal, generation–recombination, and shot noise. Detectors that exhibit gain (e.g., APDs and photoconductors) generally have larger excess noise contributions owing to random processes that participate in producing photocurrent gain. Minimization of excess noise is generally pursued through careful attention to device design and fabrication processing.

3. *Material Selection*

There are several significant issues that govern the suitability or desirability of any given material for photodetector applications. These issues include the optical properties of the material, its electronic properties, and its mechanical properties. Behavioral parameters for materials in all three of these broad categories are important to consider for implementation of high-performance photodetectors.

3.1 Optical Properties

The primary optical properties that must be considered in the design and implementation of a photodetector are the absorption spectrum of the material and its permittivity and associated dispersion. For photodetectors, there is often a need both for materials that absorb radiation at a desired wavelength, as well as for materials that are transparent at the same wavelength. Absorbing materials provide the means to detect optical radiation through EHP generation. Transparent materials are required for the realization of antireflective coatings on detector surfaces, for core and cladding layers in integrated waveguide structures, and as passivation layers to prevent environmental processes from degrading photodetector performance over time. The details of the absorption coefficient and permittivity of a material dictate its suitability for use in each of these applications.

3.2 Electronic Properties

The electronic properties of materials are also important to the photodetector designer. These are intimately related to the optical properties through such fundamental parameters as the bandgap, but also include other properties which do not manifest themselves directly in the optical absorption spectra but nonetheless have a significant impact on photodetector performance. The fundamental electronic properties of interest in photodetectors include the bandgap, the conduction and valence band dispersion relationships (including effective masses and band nonparabolicity), the transport properties, and the density and distribution of trapping states.

The bandgap is perhaps the most fundamental property of semiconducting materials, and strongly dominates both the optical and electronic properties. For intrinsic EHP generation, only incident photons with energies greater than the bandgap energy, E_g, are capable of generating EHPs, and thus absorption occurs only for wavelengths shorter than that corresponding to the bandgap energy. This cutoff wavelength can be expressed as

$$\lambda_c(\mu\mathrm{m}) = \frac{1.24}{E_g(\mathrm{eV})}$$

Table 1 contains a summary of some common semiconducting materials used in photodetectors with their bandgaps and equivalent cutoff wavelengths. The bandgap, in conjunction with other band structure parameters, also contributes to the determination of the intrinsic carrier concentration, n_i, and to the thermal generation rate of EHPs within the material. This strongly influences the conduction and noise properties of photodetectors; a small bandgap leads to a larger thermal generation rate and thus generally to increased dark current, shot noise, and generation–recombination noise.

Table 1
Bandgap and cutoff wavelength for common semiconductor materials.

Material	Bandgap (eV)	Cutoff wavelength (nm)
Si	1.11 (indirect)	1120
Ge	0.67 (indirect)	1850
GaAs	1.43 (direct)	867
InP	1.35 (direct)	919
$In_{0.53}Ga_{0.47}As$	0.75 (direct)	1550
InAs	0.36 (direct)	3440
InSb	0.18 (direct)	6890
GaSb	0.72 (direct)	1720
CdS	2.42 (direct)	512
CdTe	1.5 (direct)	827
HgTe	−0.15 (direct)	
GaN	3.4 (direct)	365
InN	2.4 (direct)	517
AlN	5.9 (indirect)	210

The details of the conduction and valence band dispersion relationships also play a significant role in controlling the behavior of photodetectors. Fundamentally the generation of EHPs by photon absorption is controlled by the availability of filled valence band states and unoccupied conduction band states separated by the incident photon energy under the constraint of conservation of momentum. Thus, the density of states in both the conduction and valence bands, as well as the details of the distributions in energy and in momentum of these states must be considered. Owing to the typically small photon momentum, the conservation of momentum translates to the transition of an electron from a valence band state to a conduction band state with nearly the same momentum. As a consequence, the availability of a "direct" transition is required for efficient absorption. Materials, such as silicon and germanium, which have indirect bandgaps require additional phonon interactions to provide the necessary momentum for photon absorption. This more complicated process results in a generally lower absorption coefficient, and explains the bias towards materials with direct bandgaps for the highest performance photodetector absorption layers. Intrinsic absorption efficiency is increased at shorter wavelengths owing to the increased density of states away from the band edges. Unfortunately, this also causes photon absorption to occur increasingly close to the semiconductor surface, where surface states and defects facilitate rapid recombination. This rapid recombination results in a net decrease in the number of EHPs that can be extracted from the absorption region. This effect, together with the fact that the energy per photon increases with decreasing wavelength, results in a decrease in detector responsivity with increasing photon energy for wavelengths below λ_c.

The presence of trapping levels in the material also leads to optical and electronic effects. States within the bandgap can give rise to absorption of photons with wavelengths larger than the cutoff wavelength. This process, extrinsic EHP generation, proceeds when an incident photon first promotes a charge carrier to a trap state in the gap and then a second photon (or thermal excitation) completes the process by promoting the trapped carrier to the conduction band. Trapping levels with energies near the band edges can also give rise to persistent photoconductivity effects through repeated trapping/emission cycles of photogenerated carriers prior to the ultimate interband recombination of the photocarriers. The electrical performance of photodetectors is also affected by the presence of ionized traps and both charged and neutral defects. Carrier mobility is strongly influenced by defects and charged impurities; high defect densities lead to lowered mobilities and thus larger access resistances, and defects can act as effective recombination centers.

3.3 Lattice Properties

Although fundamentally all of the optical and electronic properties of typical semiconductor materials can be traced back to their origins in the detailed arrangement and chemical properties of their constituent atoms, from the perspective of the device designer the lattice properties that must be considered in the design of photodetectors are dominated by consideration of material compatibility. Owing to the varying lattice constants of potentially useful semiconductor materials for photodetector fabrication, the Matthews–Blakeslee limit governs the maximum (critical) thickness for pseudomorphic growth of lattice-mismatched layers. Growth of layers thicker than this critical thickness is possible, however, and advances in the growth of metamorphic layers permits a degree of flexibility beyond the pseudomorphic film thickness limitations. In addition, more advanced techniques employing alternately strained layers have been demonstrated that also permit more flexibility in the materials and thicknesses that can be used successfully. However, the large density of dislocations and defects that can be expected in thick layers of lattice-mismatched materials exacts a penalty in terms of mobility and noise performance for devices. The ability to grow low defect density materials with different permittivities, absorption coefficients, and bandgaps within the same structure is critical to the formation of guided-wave structures, avalanche photodiodes with separate gain and absorption regions, and quantum well-based devices.

4. Photodetector Architectures

A wide variety of photodetection schemes and device architectures have been demonstrated. A survey of some of the most common types of semiconductor photodetectors is included below.

4.1 Photoconductors

One of the most basic types of photodetectors is the photoconductor. These devices consist of semiconducting material between two ohmic contact electrodes. A simplified schematic of a photoconductor is shown in Fig. 2(a). The exposed semiconductor region between the electrodes is illuminated, and absorption of the incident photons produces EHPs that contribute to an increase in conductivity. Photoconductive detectors can use either intrinsic absorptive processes or extrinsic defect-mediated absorption for long-wavelength detection. The choice of absorption material is dictated by the required spectral response. Materials suitable for use in photoconductive detectors include nearly all semiconductors, including elemental (e.g., silicon, germanium) as well as compound semiconductors (including GaAs, InP, InGaAs, CdS, and HgCdTe).

For the simple device geometry shown in Fig. 2(a), one can derive approximate expressions for the responsivity, bandwidth, and noise characteristics. The reader is referred to a number of excellent texts for a more detailed treatment of this derivation (e.g., Chuang 1995, Bhattacharya 1997, Saleh and Teich 1991, Taguchi 1999). The responsivity is given approximately by

$$\mathscr{R} = \frac{q}{h\nu}\frac{\tau_n \mu_n V}{l^2}\eta_i(1-R)(1-e^{-\alpha d})$$

where τ_n is the electron recombination lifetime, μ_n is the electron mobility, R is the detector surface reflectivity, α is the absorption coefficient, η_i is the intrinsic quantum efficiency, l is the device length, V is the applied voltage, and d is the thickness of the absorption layer. The bandwidth of such a structure, defined to be the frequency at which the electrical output signal generated in response to an amplitude-modulated optical input signal has dropped in magnitude by 3 dB relative to its low-frequency value, can be expressed approximately as

$$f_{3dB} = \frac{1}{2\pi\tau_n}$$

The intrinsic noise generated internally by a photoconductor, which arises from thermal motion of charge carriers, generation–recombination processes, and the finite conductivity of the photoconductive material in the absence of optical excitation, may be expressed as the r.m.s. output noise current, $\langle i^2 \rangle$. This is given by

$$\langle i^2 \rangle = \int_0^{\Delta f} 2q(I_D + I_p) + \frac{4q(I_D + I_p)\tau_n/\tau_t}{1 + (2\pi f \tau_n)^2} + \frac{4k_B T}{R} \, df$$

where q is the electronic charge, I_D is the dark current, I_p is the d.c. photocurrent, $\tau_t = l^2/\mu_n V$ is the electron transit time, T is the temperature, R is the average device resistance, and k_B is Boltzmann's constant.

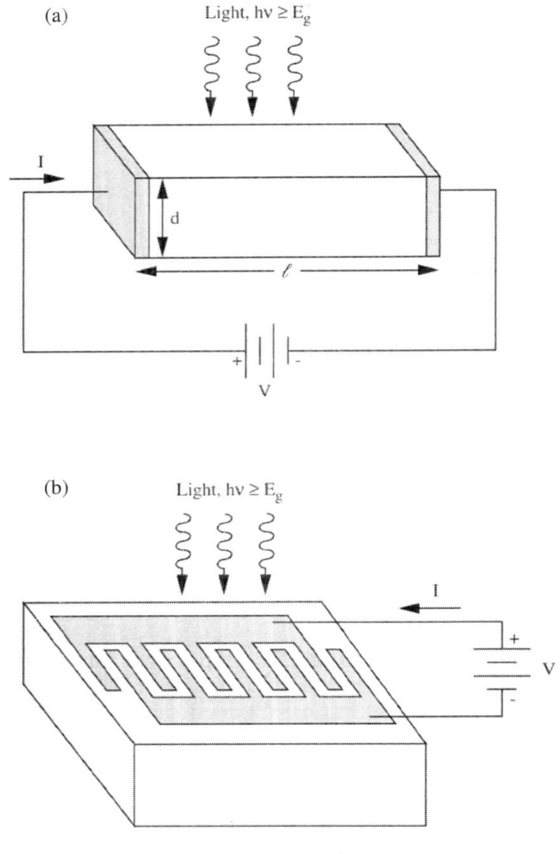

Figure 2
Photoconductor device geometry: (a) simplified geometry for analysis; (b) typical planar geometry.

Although the device geometry shown in Fig. 2(a) is easily amenable to analysis, practical implementations of photoconductors more commonly exhibit a planar geometry of the general form shown schematically in Fig. 2(b). This change in geometry does not significantly alter the fundamental operational principles of the device, and to a good approximation the expressions derived for the simple geometry of Fig. 2(a) can be applied to the planar one, provided that "effective" values of l, d, and R are used to account for the "folding" of the device structure and the shadowing of the absorption area under the contacts by the electrode metallization. As can be seen from the expressions relating device performance to physical parameters, the photoconductor is capable of exhibiting photoconductive gain (i.e., the number of charge carriers that flow through the external circuit in response to an optical signal can exceed the number of EHPs generated by photon absorption), provided $\tau_n > \tau_t$. This process enhances the responsivity of the detector, but exacts a penalty in bandwidth. Since the temporal response of a photoconductive device is directly related to the recombination lifetime of photogenerated EHPs, the maximum achievable bandwidth is limited. The nonzero dark conductivity and thermal generation rate, particularly in small-bandgap materials, also limits the achievable noise performance of photoconductors; for many applications, the noise performance of photoconductors is inadequate.

4.2 *Semiconductor Junction Photodiodes*

A device that overcomes some of the performance deficiencies of photoconductive devices, particularly for high-speed and low-noise applications, is the semiconductor junction photodiode. Photodiodes are generally intrinsic photodetectors, relying on EHPs generated within the depletion region of a p–n junction (and also carriers within a diffusion length on either side of the depletion region) for photodetection. Although junction diodes can in principle be made with any semiconducting material for which both p- and n-type dopants are available, most detectors are made using materials from the silicon, GaAs/AlGaAs, and InGaAs/InAlAs/InP material systems. Devices made from wide-bandgap materials such as GaN and related materials (e.g., InGaN, AlGaN) have been investigated for efficient detection of UV radiation, particularly for solar-blind detectors.

Photodiode detectors have been studied extensively for use in a wide range of applications. As with intrinsic photoconductors, the choice of material for the absorption region of the junction photodiode is dictated by the desired spectral response. Although junction diode-based detectors can be operated in the photovoltaic mode, they are more commonly operated with an external reverse bias voltage applied, and the resulting photocurrent is used for sensing the optical signal. In this operational mode, conventional junction detectors have a maximum external quantum efficiency (i.e., charge carriers collected at the photodetector contacts as a fraction of incident photon flux) of 100%. This is in contrast to the situation with photoconductors, where long recombination lifetimes lead to the possibility of significant photoconductive gain. In practice, nearly 100% efficiency is achievable through the use of optimized designs that include antireflection coatings to minimize reflections from the semiconductor surface and long optical path lengths to increase photon absorption.

Although conventional p–n junctions can be used as photodetectors, devices of this type generally suffer from poor quantum efficiency owing to the small absorption volume defined by the width of the depletion region. In addition, these detectors are generally slow because they have a comparatively large junction capacitance, again owing to the small depletion width. An enhancement to the p–n junction structure that alleviates these two difficulties is the addition of an undoped or lightly doped region between the p- and n-type regions, resulting in a p–i–n structure. The doping level of the intermediate intrinsic region is made low enough so that for relatively small reverse bias voltages the undoped region is completely depleted, and thus this region acts like an extended depletion region. As a consequence, significant enlargement of the absorption volume is possible with structures of this kind, leading to improved quantum efficiency. Furthermore, the junction capacitance is decreased, leading to larger detection bandwidths. An approximate analysis of the p–i–n junction photodiode structure illustrated in Fig. 3(a) yields the following expression for responsivity (Chuang 1995):

$$\mathscr{R} = \frac{q}{h\nu}(1 - R)\eta_{\mathrm{i}}\left(1 - \frac{\mathrm{e}^{-\alpha d}}{1 + \alpha L_{\mathrm{p}}}\right)$$

where d is the thickness of the depletion region and L_{p} is the diffusion length for minority carrier holes.

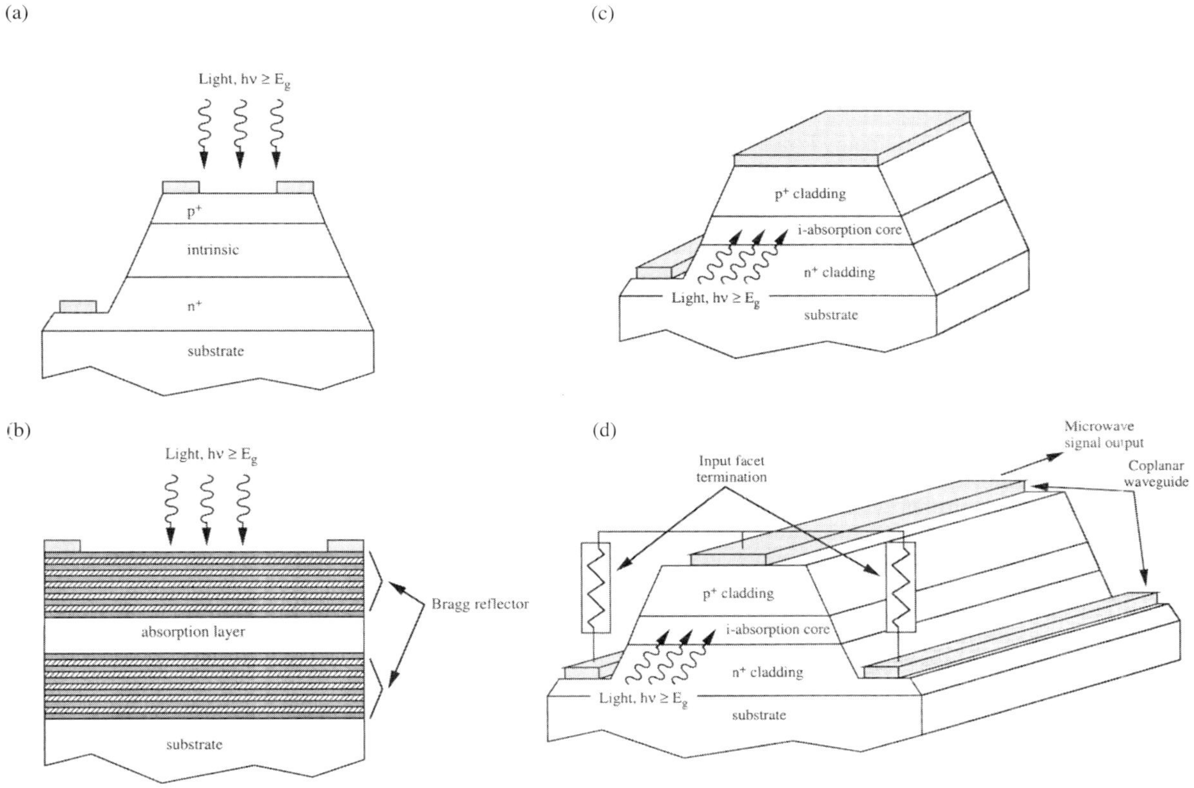

Figure 3
Schematic illustrations of p–i–n photodiode types: (a) conventional normally incident; (b) resonant cavity; (c) waveguide; (d) traveling wave.

This expression is derived assuming that the applied bias voltage is sufficient to fully deplete the intrinsic region and yet is sufficiently small so as to avoid breakdown effects. Similarly, the high-frequency response of these devices can be analyzed, and an approximate expression for the bandwidth is given by

$$f_{3\mathrm{dB}} = \frac{1}{2\pi\sqrt{(R_\mathrm{L} C_\mathrm{j})^2 + \tau^2_\mathrm{t}}}$$

where C_j is the device junction capacitance, R_L is the load resistance, and τ_t is the photocarrier transit time. As can be seen in this expression, junction capacitance plays a significant, but not necessarily dominant, role in determining the bandwidth. The transit time of photogenerated carriers across the depletion region is also a significant factor in determining the frequency response of this type of detector. Both the transit time and the junction capacitance are controlled by the thickness of the intrinsic region; for an optimal design, these contributions are made equal, and thus a unique solution for optimum intrinsic region thickness for a given detector area from bandwidth considerations can be found. The noise performance of p–i–n photodiodes is limited by generation–recombination noise, dark current shot noise, and resistive thermal noise. Considering these effects, the r.m.s. noise current is given by

$$\langle i^2 \rangle = \int_0^{\Delta f} \frac{4k_\mathrm{B}T}{R} + 2q(I_\mathrm{D} + I_\mathrm{p})\mathrm{d}f$$

where the total r.m.s. noise current from the detector is found by integrating the noise current spectral densities from shot and thermal noise over the electrical bandwidth, Δf, of interest.

Design trade-offs are inherent in the design of p–i–n photodiodes. In normally incident device geometries, increases in responsivity require thicker intrinsic absorption regions, but this is diametrically opposed to increasing bandwidth owing to the increase in transit time and subsequent decrease in bandwidth. Several alternative geometries have been explored to attempt to circumvent this constraint. A sampling of the possibilities is shown pictorially in Figs. 3(a)–(d). Figure 3(a) shows the conventional normally incident p–i–n photodiode. Figure 3(b) shows a resonant cavity structure; in this structure, distributed Bragg reflectors are grown above and below the absorption region. These reflectors act as mirrors, bouncing the unabsorbed light back and forth in the absorption region, increasing the effective length of the intrinsic region for the optical field, but keeping the transit length for photogenerated EHPs constant (>Kinsey *et al.* 1998, Knodl *et al.* 1999). In practice, the additional access resistance and heterobarriers that are induced in the current pathway can limit somewhat the achievable improvement in performance. Figure 3(c) illustrates an edge-illuminated waveguide photodetector; in this device a guided-wave structure with an absorptive core is used, and the light is input at the edge of the detector (e.g., Kato *et al.* 1994, Umbach *et al.* 1996, Shishikura *et al.* 1996). As a consequence, the light can be absorbed over a longer optical path length than with the conventional normally incident design. Coupling losses must be carefully controlled, however, since the acceptance angle and cross-sectional area of the photodetector waveguide are often small. This limits the achievable external quantum efficiency, although the internal quantum efficiencies of devices of this type can approach unity. Since in this device the absorption path length and the intrinsic region thickness are decoupled, a short transit time can be maintained while still increasing the absorption length. One limitation of this structure, however, is that the length of the waveguide can be quite long. Although the transit time of the EHPs is kept short, the long device length leads to a large junction capacitance, resulting in limited bandwidth. Figure 3(d) shows a simplified version of a traveling wave photodetector that can overcome this difficulty. In this structure, the top electrode is formed into a coplanar waveguide (CPW), whose propagation velocity is matched to the optical propagation velocity in the integrated optical waveguide.

In this way, a modulated input optical signal and the microwave electrical signal that it generates through EHP generation travel away from the input facet at the same velocity. This circumvents the effects of the excess capacitance in the conventional waveguide detector by distributing it along a CPW, and has been shown to result in extremely wide detection bandwidths (Giboney *et al.* 1995, 1996). The CPW termination near the input facet is used to reduce the occurrence of reflections from the facet end of the CPW. This termination has the undesirable effect of reducing the responsivity (half of the photocurrent from EHP generation is dissipated in the termination, with the remaining half available at the output end of the CPW) as well as contributing additional thermal noise to the output signal, but it significantly reduces pulse distortion in the detector. As with the conventional waveguide detector, input optical coupling losses must be carefully controlled in order to achieve high external quantum efficiency.

4.3 Schottky Diode Photodetectors

A variation on the junction diode detector that finds wide application, especially in high-speed, wide spectral range detectors, is the Schottky diode detector. Schottky diode detectors share much in common with p–i–n detectors; the main structural distinction is that the potential barrier that is induced by the change in doping type and concentration in a p–i–n diode is replaced by a metal–semiconductor Schottky barrier. Since Schottky diodes are majority carrier devices, they are usually implemented using n-type material in order to achieve high speed owing to the generally superior transport properties of electrons relative to holes in common semiconductor materials. Such a detector operates much the same as a p–i–n detector, with incident photons of energy greater than the semiconductor bandgap generating EHPs that are then collected at the contacts as photocurrent. In addition to this dominant process, a less efficient but still useful mode of operation of the Schottky detector allows the detection of longer wavelength photons. In this process, electrons in the metal contact can be excited by an incident photon over the Schottky barrier and injected into the semiconductor. This requires a smaller photon energy than direct band-to-band generation within the semiconductor. The fundamental similarity between the physical operation of p–i–n and Schottky detectors allows the determination of such quantities as responsivity, bandwidth, and noise generation using the same expressions, with suitable modification to the parameter values to account for the presence of the Schottky barrier in place of a p–n junction. One important distinction for high-power detectors is that effects related to diffusion of minority carriers are greatly reduced in Schottky detectors. This is owing to the fact that the major source of these effects in conventional p–i–n detectors, the absorption of photons in the top p^+ contact layer, has been eliminated by replacing the p^+ region with a metal contact.

Variations on the Schottky diode detector include the use of thin metallic Schottky contact layers, so that the incident light can pass through the contact metal (Yuang *et al.* 1995). The use of transparent metal oxide contacts (e.g., indium–tin oxide (ITO) and cadmium–tin oxide (CTO)) as Schottky contacts has also been demonstrated (Wohlmuth *et al.* 1997a, 1997b, Barry *et al.* 1990, Gao *et al.* 1994). For narrow bandgap semiconductors such as $In_{0.53}Ga_{0.47}As$, formation of metal–semiconductor Schottky contacts with sufficiently large barrier heights for low dark current is often difficult. To circumvent this, a heterostructure Schottky detector structure can be used. In this structure, the narrow gap absorption layer is separated from the metallic Schottky contact by a thin layer of a wider bandgap material with a higher Schottky barrier height. For detectors based on InGaAs absorption layers, InAlAs, InAlP, and AlGaAs have been used as Schottky enhancement layers. These layers must be designed to increase the effective barrier height in order to decrease the detector dark current, but at the same time must not significantly hamper the collection of photogenerated EHPs at the contacts. Careful management of the conduction and valence band discontinuities is important in this regard (Wohlmuth *et al.* 1997b).

4.4 Metal–Semiconductor–Metal Photodetectors

The metal–semiconductor–metal photodetector (MSM-PD) is a variation of the Schottky barrier-based photodetector. Unlike conventional Schottky-based detectors in which one contact is a rectifying Schottky contact and the other is an ohmic contact, in MSM-PDs both contacts are rectifying contacts. As a consequence, under bias the device consists of one forward-biased contact and one reverse-biased contact. In the absence of photogenerated EHPs, the saturation current of the reverse-biased contact governs the device current. Under illumination, the photogenerated EHPs drift under the influence of the electric field established by the externally applied bias and produce a photocurrent.

Owing to the symmetry of the contacts in the MSM-PD, typical device layouts resemble those typical of planar geometry photoconductors. Figure 4 shows a typical implementation of an MSM-PD. The interdigitated electrode geometry is almost universally used for MSM-PDs; this geometry allows high electric fields to be established with modest bias voltages, while simultaneously permitting straightforward control of the transit length of photogenerated carriers. A further benefit of the interdigitated electrode geometry is that it results in a very low capacitance. In contrast to the approximately parallel-plate geometry of both the p–i–n and Schottky diode-based detectors, the fringing-dominated electric field distribution associated with an interdigitated electrode geometry exhibits a much

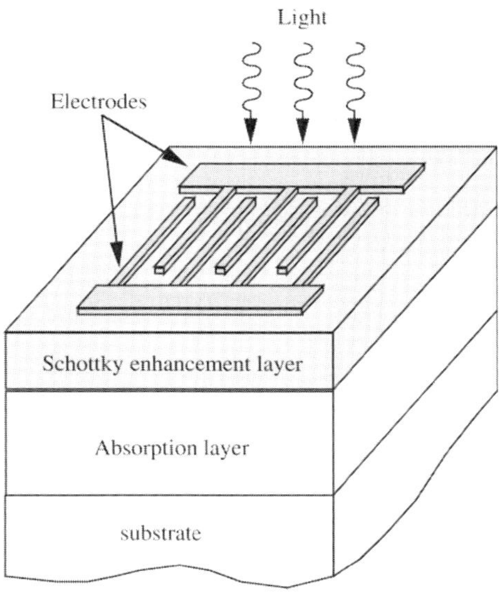

Figure 4
Structure of a typical metal–semiconductor–metal photodetector.

lower capacitance. This provides the possibility of a larger RC-limited temporal response. For the geometry shown in Fig. 4, the detector is generally illuminated from the top surface.

In this configuration, the metal electrodes block a fraction of the light from entering the semiconductor and thus reduce the maximum achievable external quantum efficiency by a factor associated with the "fill factor" of the electrodes, given by

$$\Gamma = \frac{w}{s + w}$$

where s is the spacing between adjacent electrodes and w is the width of the electrodes. As with conventional diode-based detectors, the external quantum efficiency can be improved by backside (through-substrate) illumination to avoid the fill factor quantum efficiency penalty. Unlike p–i–n detectors, however, the nonuniform electric field in the MSM-PD, and particularly the very low electric field present directly beneath the electrodes, contributes to a dramatic reduction in transit time-limited bandwidth for backside-illuminated MSM-PDs. This effect limits the applicability of reverse-illuminated MSM-PDs to those with modest bandwidth requirements.

Owing to the rectifying contacts in MSM-PDs, the re-injection of unrecombined EHPs is precluded, limiting the internal quantum efficiency of the MSM-PDs to unity or less. In practice, however, the high electric fields present along the semiconductor surface in the interdigitated geometry can give rise to surface state-mediated gain mechanisms. Gain has also been shown to arise from image force lowering effects owing to charge accumulation at potential barriers (Burm and Eastman 1996). The dynamics of such defect-assisted mechanisms are generally quite slow, limiting the usefulness of this mechanism for improving responsivity. Neglecting these gain effects and assuming top-side illumination and opaque electrodes, the photocurrent in an MSM-PD under sufficient bias to fully deplete the absorption region is given approximately by

$$\mathscr{R} = \frac{q}{h\nu}(1 - R)\eta_i(1 - e^{-\alpha d})(1 - \Gamma)$$

One advantage of the MSM-PD structure is its relative ease of fabrication. MSM-PDs have been demonstrated with good performance using silicon, GaAs, and InGaAs absorption layers, among others. One difficulty that is encountered in the fabrication of MSM-PDs for low-noise applications is that some of the materials needed for photodetection have very small Schottky barrier heights with most metals. As with conventional Schottky diode-based detectors, this difficulty can be overcome by introducing a Schottky barrier enhancement layer between the small bandgap absorption region and the electrodes. Since the photogenerated carriers must traverse a Schottky barrier to be collected by the electrodes, mitigation of heterobarrier interface effects are especially important in MSM-PDs. The heterobarrier can be reduced by compositional or short-period superlattice grading of the Schottky enhancement layer from the high-barrier contact material to the low-bandgap absorption material. Counter-doping the Schottky layer has also been shown to be effective at minimizing carrier pile-up effects (Chyi *et al.* 1996).

4.5 Avalanche Photodiodes

A device that finds wide application in telecommunication systems is the avalanche photodiode (APD). Based on the p–i–n photodiode structure, APDs offer advantages for high-sensitivity communication systems by providing internal photocurrent gain via avalanche carrier multiplication. This improvement in responsivity is not without cost, however; high-performance APDs are more complicated structures and may suffer from both speed and noise penalties associated with the dynamics of the avalanche gain process. Figure 5 shows a schematic cross-section of two possible structures for an APD. As can be seen, the simplest implementation of an APD is simply a p–i–n photodiode operated under sufficiently high reverse bias so as to cause impact ionization of the photogenerated EHPs. This structure uses impact ionization effects in the absorption region to affect the photocurrent gain. Since the choice of materials that can be used for the absorption region is constrained by the wavelength to be detected, compromise in the gain performance to achieve adequate absorption performance is inevitable. As a consequence, the performance of the impact ionization gain process is limited to that of the particular material for the absorption region. It can be shown that the best avalanche multiplication noise performance is achieved in materials for which avalanche processes are dominated by one carrier type only. In order to afford the device designer the freedom to choose both the optimum absorption material as well as a suitable one-carrier-dominant avalanche region material, device structures of the form shown in Fig. 5(b) can be used. In this structure, the separate absorption/multiplication APD (SAM-APD), the gain and absorption processes occur in separate and distinct layers in the device. Multiplication regions based on InAlAs, InP, and even silicon (through wafer bonding) have been used with $In_{0.53}Ga_{0.47}As$ absorption layers for 1.3–1.55 μm wavelength communication systems (Yu *et al.* 1994, Watanabe *et al.* 1996, Kinsey *et al.* 2000, Levine *et al.* 1997), as well as other material systems for other wavelengths.

Despite the potential limitations on bandwidth and noise performance of APDs, they are frequently used in high-performance telecommunication systems. To

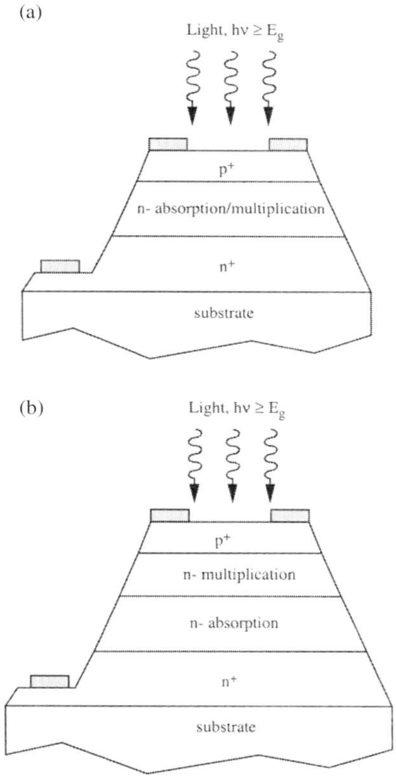

Figure 5
Cross-sectional diagram of avalanche photodiodes: (a) p–i–n diode for avalanche multiplication; (b) separate absorption/multiplication structure.

achieve higher bandwidths, the gain of the APD must generally be reduced; the usual gain–bandwidth product limitations apply to these devices as with many other semiconductor active devices. The noise performance of APDs is also a function of gain; the excess noise factor, F, for SAM-APDs can be expressed as (Chuang 1995)

$$F_n = \frac{\beta_p}{\alpha_n} G_n + \left(1 - \frac{\beta_p}{\alpha_n}\right)\left(2 - \frac{1}{G_n}\right)$$

$$F_p = \frac{\alpha_n}{\beta_p} G_p + \left(1 - \frac{\alpha_n}{\beta_p}\right)\left(2 - \frac{1}{G_p}\right)$$

where F_n is the excess noise for the case when electrons are injected into the multiplication region and F_p corresponds to the injection of holes. G_p and G_n are the avalanche gains, α_n is the impact ionization coefficient for electrons, and β_p is the coefficient for holes. As can be seen in these expressions, the excess noise factor is reduced for materials with large disparities between the impact ionization coefficients for electrons and holes. The noise current spectral density for an APD, including the effects of multiplication noise, is given by

$$\langle i^2(f) \rangle = 2q(I_D + I_p)GF + \frac{4k_B T}{R}$$

where G is the multiplication gain and R is the shunt resistance of the APD. The total noise current is found by integrating this noise current spectral density over the electrical bandwidth of interest.

4.6 QWIPs

One drawback to the intrinsic photodetector types discussed so far is that their absorption spectrum is primarily dictated by the bandgap of the absorption layer material. As a result, it is dictated in practice by the availability and compatibility (in terms of epitaxial growth and processing) of materials with suitable electronic, optical, and lattice properties. To achieve greater flexibility in designing the absorption properties of the detector, a series of engineered quantum wells can be used as the absorption region. Devices based on this concept are QWIPs. A schematic cross-section of this type of structure is shown in Fig. 1(b). Rather than rely on band-to-band generation of EHPs as the origin of the photocurrent, QWIPs use intersubband transitions between different energy states within quantum wells to facilitate detection. The excited state in the quantum well is engineered to be unbound or much less tightly bound than the ground state; thus a modest external electric field causes excited carriers to drift in the excited-state miniband, while carriers in the ground state remain confined. Through engineering of the quantum wells, the energy difference between the ground state and the excited conduction miniband can be well controlled, and thus the absorption spectrum can be directly engineered. This approach can also be thought of as "materials engineering," as the quantum wells and their associated conduction minibands behave like a single material with the desired optical and materials properties. This engineering of absorption spectra and electronic properties has been extended to design of multicolor detectors (Xudong *et al.* 1999, Tidrow *et al.* 1999, Sengupta *et al.* 1998, Chiang *et al.* 1997) and tunable detectors (Tidrow *et al.* 1994, Chiang *et al.* 1996, Lenchyshyn *et al.* 1996). The small intersubband energy spacing achievable with quantum well engineering provides a direct means of detecting long-wavelength far-IR using materials with bandgaps that far exceed the energy of the incident photons, thus providing a possible alternative to traditional small-bandgap materials such as HgCdTe. One side effect of the use of this type of intersubband transition is that the device exhibits significant absorption only for

TM-polarized light. This precludes the direct use of normally incident illumination.

5. Fabrication Processing

Fabrication of photodetectors shares many processes in common with fabrication of other electronic devices. The main distinctions in fabrication processing address the special needs of photodetectors, in particular the need to couple light into the structure, the particular materials required for optical absorption and possibly guided-wave propagation, and the desire to maximize the performance of these structures. An overview of the main processing steps commonly employed in photodetector fabrication and the implications for device performance is presented below.

5.1 Isolation

For nearly all photodetector structures of practical interest it is necessary to employ some type of isolation processing to define the active absorption volume and limit the lateral extent of the detector. Isolation processing can also provide lateral optical confinement, enabling the formation of waveguide photodetectors. Limiting the spatial extent of a detector serves to limit the dark current and thus the noise of the device, since these quantities are related strictly to the size of the device and not to the degree of use of the device. This also improves the bandwidth of the device by reducing the device capacitance, allowing transit time-limited performance to be achieved.

Two main strategies for isolation in common use are mesa isolation and implantation or doping isolation. Despite the difficulty of maintaining adequate wafer planarity in mesa-based processes, mesa isolation is used extensively in the fabrication of devices based on III–V semiconductor materials. Its use requires the availability of suitable wet chemical or dry etchants for the materials comprising the active device layers, as well as suitable passivation techniques for the etched surface. For the majority of materials, suitable etchants are readily available. An enormous amount of work in the processing of silicon and III–V materials for device applications has yielded a large body of knowledge for the processing of these materials. There are many excellent references for etch processing of these materials (e.g., Williams 1990, Hayes 1992, Fay and Adesida 2000). Passivation of the etched surface is especially important in photodetectors. Unpassivated surfaces may contain a high density of surface defect states leading to decreased carrier recombination lifetimes, and the high electric fields typical of high-performance detectors can lead to significant defect-mediated dark current. Doping-based isolation defines the active detector area by either counter-doping the nonactive area or by implanting the nonactive area with a species with a midgap chemical level. The counter-doping approach surrounds the active area with a p–n junction depletion region, while the midgap level implantation pins the Fermi level near midgap and thus produces a semi-insulating film. These approaches are widely used in the fabrication of silicon-based detectors. The sophistication of implant technology developed for silicon device processing and the appreciable bandgap of silicon makes implant isolation viable for silicon-based devices. This approach is less widely used for fabricating III–V devices for several reasons. First, isolation implants into III–V materials are considerably more complex owing to the compound nature of the materials. Promotion of implanted species into substitutional sites is inadequate to ensure electrical activation, and the annealing process can also cause crystal twinning and desorption of volatile semiconductor constituents, leading to stoichiometric variations in the material. Second, in many cases III–V materials are used for their ability to detect long-wavelength photons owing to the comparatively small bandgap of materials like InAs and $In_{0.53}Ga_{0.47}As$. In materials with small bandgaps, implant-based isolation schemes are generally ineffective, since room temperature thermal generation can render the implanted regions conductive, leading to large dark currents and poor noise performance.

5.2 Contacts

Fabrication of the photodetector types discussed above requires the ability to fabricate metal–semiconductor contacts of either an ohmic or rectifying electrical character, depending on the particular device type. Contacts having either of these electrical properties are most often formed by one of two methods. For III–V-based devices, processes based on metal deposition by evaporation followed by patterning by liftoff are common, while for silicon-based devices sputter deposition followed by etching-based definition is nearly universal. Sputter deposition is also used in some III–V processes. The choice of sputtering or evaporation for metal film deposition has implications not only for pattern definition as indicated above, but also for film morphology and step-edge coverage. Sputtered films offer the possibility of superior step-edge coverage compared to evaporated films. To overcome the step-edge coverage limitations of evaporation, planarization techniques based on polymer and dielectric films are employed. These processes result in more nearly planar wafer surfaces, despite the presence of steps in the semiconductor surface owing to mesa isolation processing. Compared to sputter–etch processes, the yield of liftoff processes may be limited by the redeposition of lifted-off metal particles, limiting its usefulness for large-scale integration. This is most often a secondary concern for photodetector fabrication, since the integration level of photodetector structures is small. Evaporation–liftoff processes have the advantage of not requiring a metallization etch. This is particularly helpful in III–V processes where common metal etchants also are effective etchants for the semiconductor material as well. An additional consideration is film composition. For deposition of alloys, evaporation can result in deposited films that differ compositionally from the source material owing to differences in the vapor pressures of the constituent elements. Sputter deposition permits better matching of the source and film composition, and is thus preferable for multiconstituent materials.

(*a*) *Ohmic contacts.* Photodetectors that require ohmic contacts include p–i–n photodiodes, photoconductors, Schottky diode detectors, and APDs. Optimized ohmic contacts are essential to the performance of these devices, since parasitic resistances created by nonideal ohmic contacts can contribute to degradation of the noise and high-frequency performance. For the formation of high-quality ohmic contacts for photodetectors, alloyed or sintered contacts that rely on liquid- or solid-phase reactions between the contact metallization and the semiconductor surface are commonly used.

Alloyed ohmic contacts have been realized using a number of contact metallizations. Most of these consist of a metal (often gold) and a dopant such as germanium for contact to n-type material and beryllium or zinc for contact to p-type material. Nickel is often also incorporated into the contact structure as well, as it has been demonstrated to facilitate diffusion of germanium into GaAs (Anderson *et al.* 1978). Contacts based on AuGe, AuZn, or AuBe are typically realized by evaporation from eutectic. The ohmic contact is formed by thermal annealing in a furnace or rapid thermal processor (RTP) in either an inert ambient (typically argon or helium) or forming gas (15% H_2, 85% N_2). In this approach, the surface doping level attained by the germanium, beryllium, or zinc doping should exceed $2\times10^{19}\,cm^{-3}$ in order to produce low-resistance contact. One limitation of the AuGe and AuGe/Ni metallizations that has been widely observed is that they form a so-called "spiking" contact. This can result in substantial nonuniformity both in terms of the depth of doping and interfacial layer penetration as well as the spatial composition of the contact. In addition to the contact area scaling limits caused by this inhomogeneity, these contacts are not thermally stable and suffer from degradation with subsequent thermal processing or high-temperature device operation (Fatemi and Weizer 1990). One approach to avoid the difficulties associated with these inhomogeneities for ohmic contacts to III–V materials includes replacing the AuGe-based contact metallization with a stack of metals consisting of palladium, germanium, and gold. As with the AuGe-based contacts, after evaporation of the contact material the contact is sintered at 350–450 °C to effect the germanium diffusion. This metallization has proved to be more thermally stable and to produce a smooth surface morphology (Chen *et al.* 1986).

An alternative approach to ohmic contact formation that does not rely on alloying or sintering to achieve ohmic contact is the explicit inclusion of narrow bandgap, highly doped surface layers in the epitaxial heterostructure for the explicit purpose of facilitating ohmic contact formation. On InP, use of n^+ lattice-matched $In_{0.53}Ga_{0.47}As$ has been shown to result in high-quality contacts, with typical contact resistances of 0.07–0.1 Ωmm (Fay *et al.* 1997). The use of such an approach in photodetectors requires care, however, since it may introduce optical absorption in undesired regions of the device. An alternative approach consists of using near-noble transition metal contacts (e.g., titanium, platinum) that are normally associated with Schottky contact formation, but ensuring that the surface doping concentration of the semiconductor is sufficiently high so that a tunneling contact is established. The extremely thin barrier layer that results from the spatially rapid bending of the energy bands produces contacts in which the current is primarily conducted through tunneling (or field emission) through the Schottky barrier rather than thermionic emission over the barrier, thus greatly reducing the degree of rectification. In some processes the contact is further assisted by RTP in order to form a thin interfacial layer (<40 nm) of narrow bandgap intermetallic phases (Katz 1992). Specific contact resistances in the range 10^{-6}–$3\times10^{-8}\,\Omega cm^2$ are obtainable for highly doped material using this method.

(*b*) *Rectifying contacts.* Rectifying contacts based on Schottky barriers are essential to the fabrication of Schottky diode detectors and MSM-PDs. The formation of rectifying Schottky contacts is the natural result of depositing metal on a clean semiconductor surface. Owing to the difference between the metal work function, σ_m, and the semiconductor work function, σ_s, a barrier to charge transfer between the metal and semiconductor is established. Within some very simple approximations, the height of the barrier to electrons traveling from the semiconductor to the metal is $\sigma_m - \sigma_s$, while those electrons traveling from the metal into the semiconductor must overcome a barrier of $\sigma_b = \sigma_m - \sigma$, where σ is the electron affinity of the semiconductor. The asymmetry between these barriers gives rise to the rectifying properties of these contacts. The primary parameters of interest in the formation of Schottky contacts are the Schottky barrier height, σ_b, and the reverse leakage current of the contacts. From the simple expressions above, one would expect that the barrier height could be easily adjusted by judicious selection of contact metallization through selection of an appropriate metal work function. Owing to the presence of surface states that pin the Fermi level at a fixed energy, this behavior is not observed for many important III–V semiconductors.

Typical Schottky contacts consist of metallization stacks of a refractory contact metal and conductive overlayer. Ti/Au, Cr/Au, and Ti/Pt/Au are three common choices. For detectors such as Schottky detectors for which long-wavelength responsivity is desired, or for detectors requiring maximization of responsivity, a transparent contact can be used. Transparent contacts can be implemented in two ways. The first of these approaches is to use a very thin metallization, so that an appreciable fraction of the incident light penetrates through the contact metallization. Thin layers of Ti/Au, with total metal thicknesses of less than 20 nm, have been used in this way (Yuang *et al.* 1995). Such thin metallizations can exhibit high resistances, however, leading to reduced bandwidth and increased thermal noise. An alternative is to use transparent metal electrodes. ITO and CTO are two transparent metallic materials that are widely applied for this purpose. These materials are easily deposited by sputtering, and typically must be annealed after deposition to achieve the best electrical performance. Schottky-based photodetectors with high responsivity and low dark current have been demonstrated using these materials (Wohlmuth *et al.* 1997a, 1997b).

(*c*) *Optical coupling.* Unlike purely electronic semiconductor devices, photodetectors must have not only a low parasitic electrical interconnection to the other components in the system, but must incorporate means for coupling the input optical signal into the absorption region of the device. In many cases of practical interest, the efficiency of this coupling is an important factor that limits the achievable performance, and so must be carefully considered in the photodetector design. The device design and fabrication processing that is appropriate to maximize optical coupling depends strongly on the type of device being considered.

For normal incidence photodetectors such as conventional p–i–n detectors, Schottky detectors, and MSM-PDs, the light impinges on the semiconductor normal to the surface. To minimize reflections from this surface, a transparent dielectric layer with a well-controlled permittivity and thickness is often used. The simplest structure to achieve low reflectivities in most semiconductor-based photodetectors is to use a quarter-wavelength thickness of dielectric, adjusting the dielectric constant to achieve a low reflectivity. SiN_x deposited by plasma enhanced chemical vapor deposition (PECVD) is a common choice because it allows both accurate control of deposited film thickness as well as some control over the stoichiometry, and thus the refractive index, of the deposited film. These films typically have refractive indices, *n*, of 1.7–1.9, which is in the right range to achieve a low surface reflectivity for the typical range of semiconductor permittivities. Such antireflection coatings can also be realized using dielectric evaporation techniques or by reactive sputtering. These processes can be automated to include *in situ* measurement of permittivity and thickness monitoring for optimization of the resulting surface reflectivity. In the case of reactive sputtering, PECVD, and some novel evaporation techniques, the permittivity can be adjusted in real time during deposition to provide enhanced process performance.

For guided-wave detectors, light is incident on the end facet of an integrated waveguide on the wafer surface. Several methods for exposing the active absorption region are commonly employed. Cleaving of the device or waveguide along the desired crystal plane can be used; this provides a very smooth surface for optical coupling. The input facet can also be fabricated by etching, although most wet chemical etchants do not result in acceptable angular profiles. Wet chemical etchants for III–V compound semiconductors most often produce highly crystallographic etch profiles, leading to sloped facets. Owing to the partially polar character of the chemical bonds in III–V compound semiconductors, the etched facet surfaces typically evolve to (111) planes with an angle of 55° with respect to the plane of the surface. Although such angled facets can be used for some applications, in many cases it is more desirable to have more nearly vertical input facets. To achieve vertical facets by etching, dry etching techniques such as reactive ion etching (RIE), inductively coupled plasma RIE (ICP-RIE), chemically assisted ion-beam etching (CAIBE), ion milling, or reactive ion-beam etching (RIBE) are used. Each of these techniques has been demonstrated to be capable of producing vertical or near vertical etch profiles in materials important for photodetector applications. An important consideration in the selection of a processing approach for creating etched vertical facets is the amount of surface damage created by the process. Surface damage contributes to increased surface-state density and increased surface recombination velocity. For photodetectors, this translates into increased dark current (and thus increased levels of shot noise) and decreased responsivity through the decreased recombination lifetime of EHPs photogenerated near the facet. To minimize surface damage, high-density plasma techniques such as ICP-RIE and chemically assisted techniques such as CAIBE and RIBE are often improvements over techniques relying on heavy physical etching components such as ion milling or conventional RIE. As in the case of normal incidence detectors, the deposition of a suitable antireflection coating by CVD or evaporation can be used to minimize reflections from the input facet.

6. Summary

An overview of semiconductor photodetection technology has been undertaken. The fundamental physical principles that govern photodetector operation have been described. The figures of merit for photoresponse, noise, and operational speed have been discussed, and the characteristics of the most common photodetector types been examined. The basic fabrication process technology used in the manufacture of photodetectors has also been described.

Bibliography

Anderson W T, Christou A, Davey J 1978 Development of ohmic contacts for GaAs devices using epitaxial Ge films. *IEEE J. Solid-State Circuits* **13**, 430–5

Barry D M, Snowden C M, Howes M J 1990 Physical modeling of GaAs photodetectors In: *Proc. Colloquium Modelling of Optoelectronic Devices*. IEE, London, pp. 1–7

Bhattacharya P 1997 In: *Semiconductor Optoelectronic Devices*, 2nd edn., Prentice Hall, Upper Saddle River, NJ

Burm J, Eastman L F 1996 Low-frequency gain in MSM photodiodes due to charge accumulation and image force lowering. *IEEE Photon. Technol. Lett.* **8** (1), 1135

Chen C L, Mahoney L J, Finn M C, Brooks R C, Chu A, Mavroides J G 1986 Low resistance Pd/Ge/Au and Ge/Pd/Au ohmic contacts to n-type GaAs. *Appl. Phys. Lett.* **48**, 535–7

Chiang J C, Li S S, Singh A 1997 A two-stack indirect-barrier/triple-coupled quantum well infrared detector for mid-wavelength and long-wavelength infrared dual band detection. *Appl. Phys. Lett.* **71** (24), 3546–8

Chiang J C, Li S S, Tidrow M Z, Ho P, Tsai M, Lee C P 1996 A voltage-tunable multicolor triple-coupled InGaAs/GaAs/AlGaAs quantum-well infrared photodetector for 8–12 μm detection. *Appl. Phys. Lett.* **69** (16), 2412–4

Chuang S L 1995 *Physics of Optoelectronic Devices*. Wiley, New York

Chyi J-I, Chien Y-J, Yuang R-H, Shieh J-L, Pan J-W, Chen J-S 1996 Reduction of hole transit time in GaAs MSM photodetectors by p-type α-doping. *IEEE Photon. Technol. Lett.* **8** (11), 1525–7

Fatemi N S, Weizer V G 1990 Humidity-induced room temperature decomposition of Au contacted indium phosphide. *Appl. Phys. Lett.* **57**, 500–2

Fay P, Adesida I 2000 Processing of epitaxial heterostructure devices. In: Francome M, Wood C E C (eds.) *Heterostructures for High-Performance Devices*. Academic Press, New York

Fay P, Arafa M, Wohlmuth W, Caneau C, Chandrasekhar S, Adesida I 1997 Design, fabrication, and performance of high-speed monolithically integrated InAlAs/InGaAs/InP MSM/HEMT photoreceivers. *IEEE J. Lightwave Technol.* **15**, 1871–9

Gao W, Khan A-S, Berger P R, Hunsperger R G, Zydzik G, O'Bryan H M, Sivco D, Cho A Y 1994 $In_{0.53}Ga_{0.47}As$ metal–semiconductor–metal photodiodes with transparent cadmium tin oxide Schottky contacts. *Appl. Phys. Lett.* **65** (15), 1930–2

Giboney K S, Nagarajan R L, Reynolds T E, Allen S T, Mirin R P, Rodwell M J W, Bowers J E 1995 Travelling-wave photodetectors with 172-GHz bandwidth and 76-GHz bandwidth-efficiency product. *IEEE Photon. Technol. Lett.* **7** (4), 412–4

Giboney K S, Rodwell M J W, Bowers J E 1996 Traveling-wave photodetector design and measurements. *IEEE J. Selected Topics Quantum Electron* **2** (3), 622–9

Hayes T R 1992 Dry etching of In-based semiconductors. In: Katz A (ed.) *Indium Phosphide and Related Materials: Processing, Technology, and Devices*. Artech House, Boston, pp. 277–306

Kato K, Kozen A, Muramoto Y, Itaya Y, Nagatsuma T, Yaita M 1994 110-GHz, 50% efficiency mushroom-mesa waveguide p–i–n photodiode for a 1.55 μm wavelength. *IEEE Photon. Technol. Lett.* **6** (6), 719–21

Katz A 1992 Ohmic contacts to InP and related materials. In: Katz A (ed.) *Indium Phosphide and Related Materials: Processing, Technology, and Devices*. Artech House, Boston, pp. 307–36

Kinsey G S, Hansing C C, Holmes A L Jr., Streetman B G, Campbell J C, Dentai A G 2000 Waveguide $In_{0.53}Ga_{0.47}As$–$In_{0.52}Al_{0.48}As$ avalanche photodiode. *IEEE Photon. Technol. Lett.* **12** (4), 416–8

Kinsey G, Lenox C, Nie H, Campbell J C, Streetman B G 1998 Resonant cavity photodetector with integrated spectral notch filter. *IEEE Photon. Technol. Lett.* **10** (8), 1142–4

Knodl T, Choy H K H, Pan J L, King R, Jager R, Lullo G, Ahadian J F, Ram R J, Fonstad C G Jr., Ebeling K J 1999 RCE photodetectors based on VCSEL structures. *IEEE Photon. Technol. Lett.* **11** (10), 1289–91

Lenchyshyn L C, Liu H C, Buchanan A N, Wasilewski Z R 1996 Voltage tuning in multicolor quantum well infrared photodetector stacks. *J. Appl. Phys.* **79** (10), 8091–7

Levine B F, Hawkins A R, Hiu S, Tseng B J, King C A, Gruezke L A, Johnson R W, Zolnowski D R, Bowers J E 1997 20 GHz high performance planar Si/InGaAs p–i–n photodetector In: *Proc. Int. Conf. Indium Phosphide and Related Materials*. IEEE, Piscataway, NJ, pp. 483–5

Saleh B E A, Teich M C 1991 *Fundamentals of Photonics*. Wiley, New York

Sengupta D K, Gunapala S D, Bandara S V, Pool F, Liu J K, McKelvey M, Luong E, Torezan J, Mumulo J, Hong W, Gill J, Stillman G E, Curtis A P, Kim S, Chou L J, Mares P J, Feng M, Hsieh K C, Chuang S L, Bishop S G, Chang Y C, Liu H C, Wang W I 1998 Monolithically integrated dual-band quantum well infrared photodetector. *Proc. Symp. Mater. Res. Soc.* 205–14

Shishikura M, Nakamura H, Tanaka S, Matsuoka Y, Ono T, Miyazaki T, Tsuji S 1996 High-responsivity, low-dark-current InGaAlAs waveguide photodiode with a symmetric double-core for optical access networks. *Electron. Lett.* **32** (20), 1882–3

Taguchi K 1999 Photodiodes and receivers based on InP materials. In: Wada O, Hasegawa H (eds.) *InP-Based Materials and Devices, Physics and Technology*. Wiley, New York

Tidrow M Z, Choi K K, Farley C W, Chang F 1994 Multicolor infrared detection using a voltage tunable bandpass filter. *Appl. Phys. Lett.* **65** (23), 2996–8

Tidrow M Z, Xudong J, Li S S, Bacher K 1999 A four-color quantum well infrared photodetector. *Appl. Phys. Lett.* **74** (9), 1335–7

Umbach A, Trommer D, Mekonnen G G, Ebert W, Unterborsch G 1996 Waveguide integrated 1.55 μm photodetector with 45 GHz bandwidth. *Electron. Lett.* **32** (23), 2143–5

Watanabe I, Tsuji M, Makita K, Taguchi K 1996 Gain–bandwidth product analysis of InAlGaAs–InAlAs superlattice avalanche photodiodes. *IEEE Photon. Technol. Lett.* **8** (2), 269–71

Williams R 1990 *Modern GaAs Processing Methods*. Artech House, Botson

Wohlmuth W A, Seo J-W, Fay P, Caneau C, Adesida I 1997a A high-speed ITO–InAlAs–InGaAs Schottky-barrier photodetector. *IEEE Photon. Technol. Lett.* **9** (10), 1388–90

Wohlmuth W A, Seo J-W, Fay P, Caneau C, Adesida I 1997b High-speed InGaAs-based vertical Schottky barrier photodetectors In: *Proc. Int. Conf. Indium Phosphide and Related Materials*. IEEE, Piscataway, NJ, pp. 490–3

Xudong J, Li S S, Tidrow M Z 1999 Investigation of a multistack voltage-tunable four-color quantum-well infrared photodetector for mid- and long-wavelength infrared detection. *IEEE J. Quantum Electron* **35** (11), 1685–92

Yu J, Tarof L E, Bruce R, Knight D G, Visvanatha K, Baird T 1994 Noise performance of separate absorption, grading, charge, and multiplication InP/InGaAs avalanche photodiodes. *IEEE Photon. Technol. Lett.* **6** (5), 632–4

Yuang R-H, Chyi J-I, Chan Y-J, Lin W, Tu Y-K 1995 High-responsivity InGaAs MSM photodetectors with semi-transparent Schottky contacts. *IEEE Photon. Technol. Lett.* **7** (11), 1333–5

Photovoltaic Devices from Organic Semiconductors

G. G. Malliaras

Illumination of an organic semiconductor with light of an appropriate wavelength creates free electrons and holes. In photovoltaic devices, a built-in potential forces these carriers to drift towards opposite electrodes. When connected to an appropriate electrical circuit, photovoltaic devices function as photodetectors that generate an electrical current proportional to the incident light intensity, or as solar cells that provide electrical power to an external circuit. The main advantage of organic semiconductors for this type of application is their excellent processability, which allows inexpensive fabrication of large-area films on flat, shaped, or even flexible substrates. As a result, the cost of device fabrication can be significantly reduced compared to their inorganic counterparts. Another advantage of organic materials is the fact that their optoelectronic properties can be tuned over a large range via chemical synthesis and even be tailored to fit specific applications. This unique feature has been exploited to help understand and dramatically improve the properties of photovoltaic devices. As a result, organic photodetectors and solar cells have reached levels of performance that make them technologically interesting.

In the fabrication of these devices, one or more organic semiconductor layers are sandwiched between two metal electrodes, one of which is transparent. The organic layer(s) absorb light, and generate and transport charges, while the metal electrodes create the built-in potential that drives charge separation. Organic semiconductors exhibit distinct characteristics that strongly influence the operation of photovoltaic devices. First, the yield of free charge generation in organic materials is typically very small and highly electric field dependent. Being confined within a molecule or a polymer chain, photogenerated excitons usually undergo geminate recombination instead of dissociation into free carriers. The dissociation probability can be enhanced by an electric field and/or by photoinduced charge transfer at dopant sites and interfaces. Second, charge transport of the free carriers takes place via hopping in a manifold of localized states rather than free propagation in a band. Therefore, the drift mobilities in organic materials are rather low and strongly electric field dependent. The electric field dependence of photogeneration and transport necessitates the use of thin organic films, where the built-in potential causes a large enough field to enhance these two processes.

Based on the principle of operation, organic photovoltaic devices can be classified into three categories: single-layer, bilayer, and blend-based devices. Single-layer devices are discussed first, beginning with the device fabrication and the materials of interest. Subsequently, a device model is presented, the device characteristics are analyzed, and the shortcomings of the single-layer configuration are examined. Finally, bilayer and blend-based devices are described and the article concludes with a discussion of scientific issues that remain to be addressed as well as an overview of the proposed applications and the state of the art.

1. *Single-layer Devices*

The typical configuration for a single-layer device is shown in Fig. 1(a). The first (transparent) electrode is deposited on a transparent substrate (usually glass) by thermal evaporation or sputtering. The organic layer is subsequently deposited by spin coating (polymers) or vacuum deposition (small molecules) and device fabrication is completed with the deposition of the metal electrode. Devices where the top electrode is the transparent one have also been fabricated.

Both hole and electron transport materials, in small molecule and polymer form, have been used as active layers in photovoltaic devices. Examples of hole transport materials include phthalocyanines, porphyrins, merocyanine dyes, oligothiophenes, polythiophenes, and poly(*p*-phenylene vinylene)s. Examples of electron transport molecules include certain porphyrin derivatives, pyrilium dyes, and cyano-substituted poly(*p*-phenylene vinylene)s. Earlier work, including materials of interest, has been reviewed by Wöhrle and Meissner (1991). The thickness of the organic layer is typically of the order of 100 nm, which is thick enough to ensure sufficient light absorption but at the same time thin enough to allow efficient photogeneration and extraction of carriers from the bulk of the organic material (as discussed above, photogeneration and transport are electric field dependent in almost all organic materials). The operation of the device relies on the presence of two metal electrodes with different work functions (ionization potentials): an anode made from a high work function material such as gold, silver, or the transparent conductor indium tin oxide (ITO), and a cathode made from a low work function metal like aluminum, indium, magnesium, or calcium.

The operation of the single-layer device is qualitatively understood in analogy to traditional semiconductor physics (e.g., Sze 1981). The contact potential that is established in a semiconductor adjacent to a metal interface serves as the built-in potential, forcing the photogenerated charges to drift towards opposite electrodes. Since most organic semiconductors have rather low intrinsic carrier

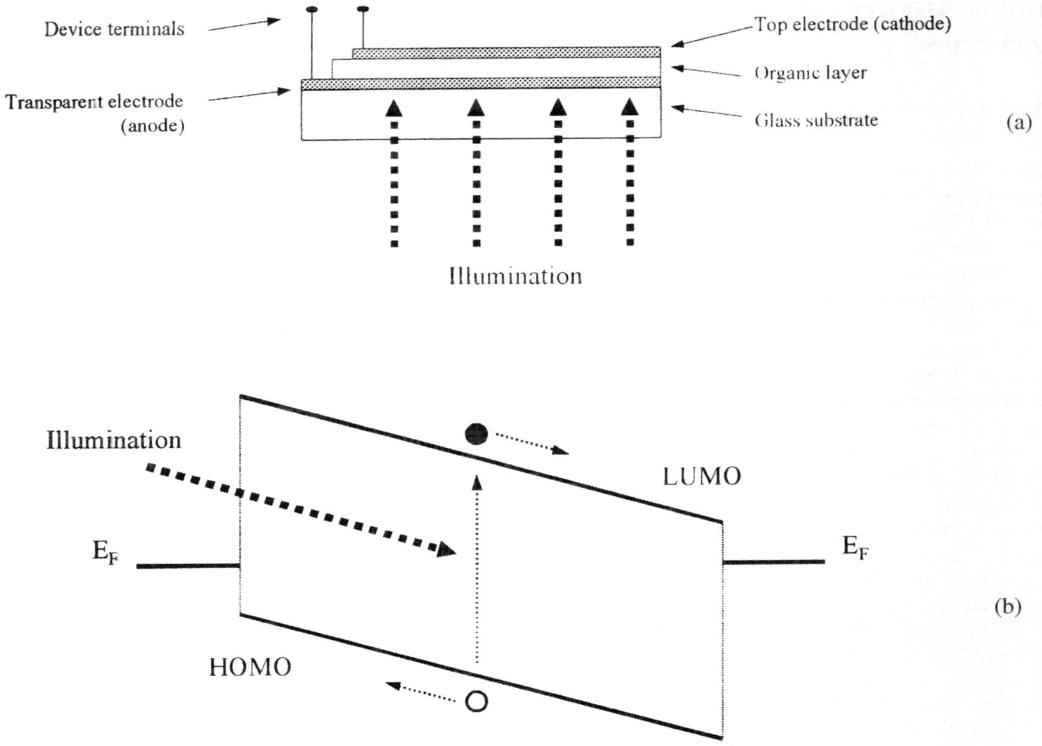

Figure 1
(a) Layout of a single-layer photovoltaic device and (b) the corresponding energy level diagram. Shown are the energy levels of the organic (HOMO and LUMO) and the Fermi levels (E_F) of the anode (left) and the cathode (right). On optical excitation (thick dotted arrow), an electron (•) and a hole (○) are generated and drift towards opposite electrodes.

densities, the built-in potential spans the entire organic layer. A simplified energy level diagram for a single-layer photovoltaic device is shown in Fig. 1(b). This shows the HOMO and LUMO of the organic layer, which correspond to the valence and conduction band of a crystalline semiconductor. On contact with the two metal electrodes, a built-in potential is established across the organic layer (represented in Fig. 1(b) by the tilting of energy levels of the organic). In the absence of interfacial chemical reactions that pin the Fermi level or charge accumulation at the interfaces, the built-in potential is equal to the difference in the work function of the two metals. On illumination with light of appropriate wavelength, absorption in the organic layer leads to the creation of excitons. Some of these excitons dissociate into free electrons and holes, which then drift toward opposite electrodes under the influence of the built-in potential. When the two terminals are connected to each other ($V=0$), a short circuit current (I_{sc}) flows, whereas in the case of an open circuit ($I=0$), an open circuit voltage (V_{oc}) appears. The short circuit current is proportional to the illumination intensity, while the open circuit voltage quickly saturates to a value that is slightly smaller than the built-in potential.

Typical current–voltage characteristics of a single-layer photovoltaic device are shown in Fig. 2. In the dark, the device behaves like a diode: under forward bias (anode positive with respect to the cathode), charge injection from the metal electrodes leads to a sharp rise in the device current (quadrant I in Fig. 2). Injection is not possible under reverse bias and the current remains low (quadrant III in Fig. 2). These current–voltage characteristics change dramatically on illumination. For application as a solar cell, the device drives a load (an external circuit that has a certain resistance) and operates in quadrant IV of Fig. 2. The short circuit current and the open circuit voltage correspond to the extreme cases of zero and infinite load, respectively. For application as a photodetector the device usually operates in quadrant III

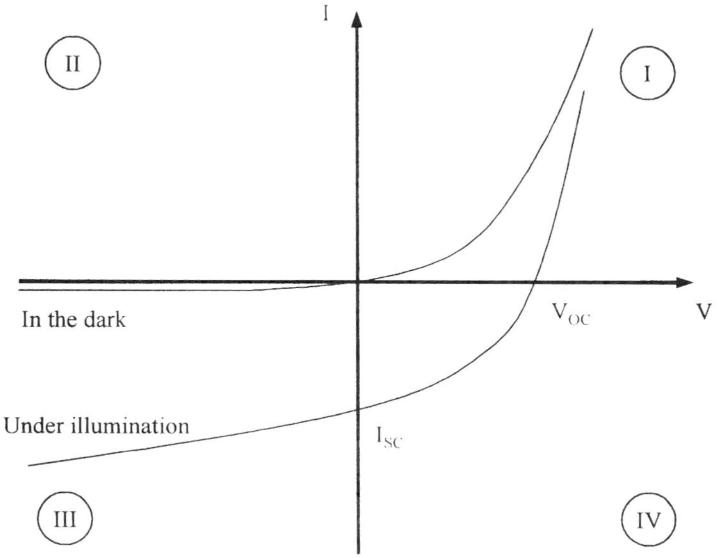

Figure 2
Typical current–voltage characteristics in the dark and under illumination. V_{oc} is the open circuit voltage and I_{sc} is the short circuit current.

of Fig. 2: a reverse bias is applied to enhance the produced photocurrent (note that the reverse bias has the same sign as the built-in potential).

The dependence of the short circuit current on the incident photon energy indicates that only a small part of the organic layer, adjacent to one of the contacts, contributes to the photovoltaic response. For example, in the majority of devices based on hole transport organics, the photocurrent is proportional to the amount of light that reaches the cathode. As a result, illumination through the cathode produces a photocurrent that follows the absorption spectrum (symbatic response), while under illumination through the anode the photocurrent peaks close to the onset of the absorption, rather than at the maximum (antibatic response). This behavior has been attributed to localization of the built-in potential next to one of the contacts, as in the case of traditional extrinsic semiconductors: a p-type semiconductor forms a Schottky contact with a low work function metal (cathode) and an ohmic contact with a high work function metal (anode).

An alternative explanation, that is more appropriate for organic materials in their pristine form, involves charge generation at the interfaces. Since the photogeneration yield in the bulk of the organic is intrinsically rather low, the dominant mechanism of free carrier generation is via electron transfer at the interfaces. As a result, the short circuit current in organics that predominantly transport holes arises from holes photogenerated at the cathode (see Malliaras *et al.* (1998) for a quantitative description).

Although very simple to fabricate, single-layer devices suffer from two main shortcomings. First, the yield of exciton dissociation in organic materials (especially in pristine samples) is typically very small. Therefore, only a very small portion of the absorbed photons contributes to the generation of free carriers. Second, organic materials usually exhibit monoplanar charge transport (only one of the carriers appears to be mobile, indicating that the other one is severely trapped or suffers from an intrinsically low mobility). As a result, the less mobile carrier accumulates in the bulk of the organic creating a space charge field that degrades the device performance.

2. *Bilayer (Heterojunction) Devices*

The main disadvantages of the single-layer device can be overcome to a large degree by bilayer devices, which are based on heterojunctions between hole and electron transport organic layers (Fig. 3(a)). The device fabrication is similar to that for .single-layer devices, using the same organic materials and metal electrodes. For polymer-based devices, the second layer must be coated from a solvent that does not dissolve the first one. Heterojunctions using C_{60}, inorganic semiconductors (CdS, ZnS, TiO_2), and electrolyte solutions have also been used, but their description is beyond the scope of this article (O'Regan and Grätzel (1991) is a good starting point).

A simplified energy level diagram for such a device is shown in Fig. 3(b). The difference between the

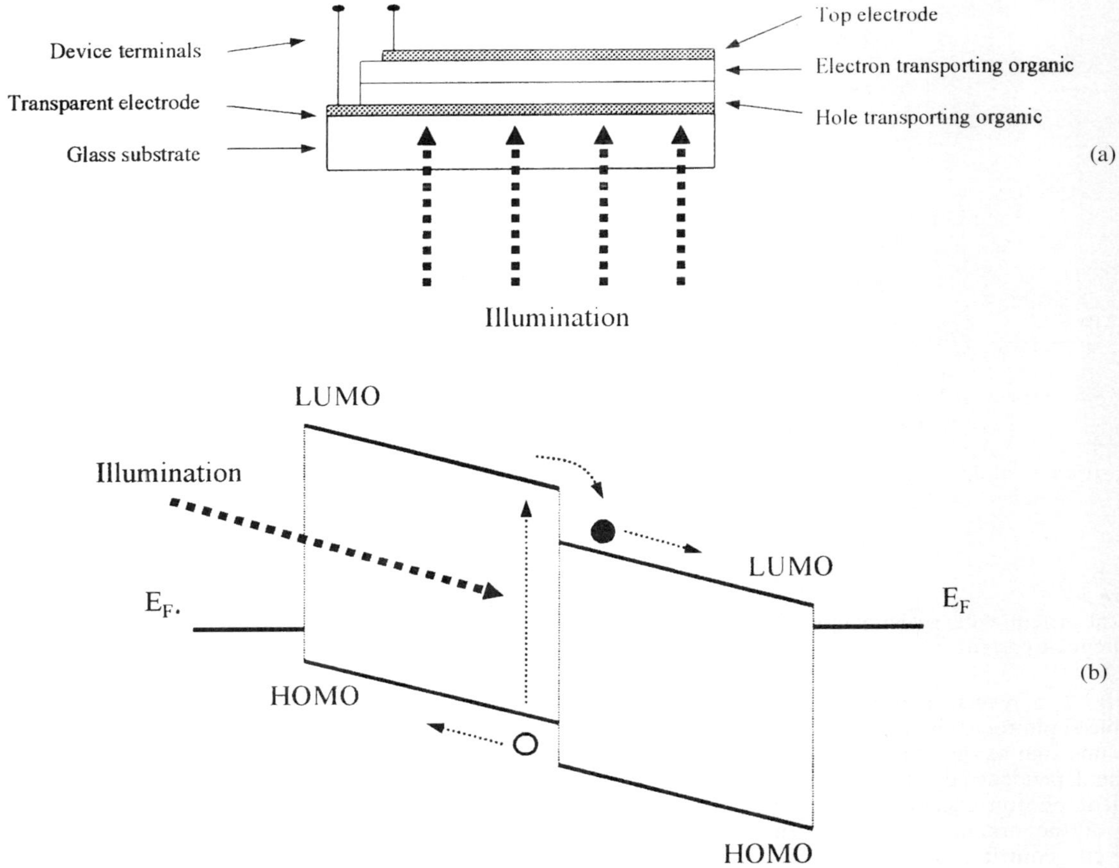

Figure 3
(a) Layout of a heterojunction photovoltaic device and (b) the corresponding energy level diagram. Shown are the energy levels of the two organic layers (HOMO and LUMO) and the Fermi levels (E_F) of the anode (left) and the cathode (right). The hole transport layer is on the left and the electron transport layer on the right. On optical excitation of the hole transport layer (thick dotted arrow), an electron (•) is transferred in the electron transport layer, leaving a hole (○) in the hole transport layer. Both carriers drift towards opposite electrodes.

work function of the two metal electrodes results in a built-in potential. When the hole transport semiconductor is excited, the electron lowers its energy by transferring to the LUMO of the electron transport layer. This process usually takes place via photoinduced electron transfer (PET) and can be up to 100% efficient for excitons that are formed near the interface. In a similar manner, excitons formed in the electron transport layer can dissociate by hole transfer to the hole transport layer. The two carriers, now located at different sides of the interface, drift towards opposite electrodes under the influence of the built-in potential.

The bilayer configuration overcomes the two shortcomings of the single-layer device: exciton dissociation is strongly enhanced at the organic–organic interface and can be up to 100% efficient. Moreover, since the two carriers move into different phases, they are both efficiently extracted from the device. The main disadvantage of bilayer devices is their small active volume: only excitons that are created within the exciton diffusion length from the interface, which is typically of the order of 10 nm, are efficiently dissociated. This drawback is reflected in the spectral dependence of the short circuit current, which is proportional to the amount of light that reaches the heterojunction.

3. *Blend-based Devices*

The blend-based configuration is a novel type of photovoltaic device, which does not have an exact analog in inorganic semiconductors (Yu *et al.* 1995, Halls *et al.* 1995). Its operation is based on a single active layer that is a blend of donor- and acceptor-like moieties. The two moieties are in either small molecule or polymer form. The blends are prepared by mixing two polymers, a polymer and a small molecule, or two small molecules. The first two are prepared alike, by spin coating from a common solution, while the latter is prepared by coevaporation. Blends of conjugated polymers with C_{60} and its derivatives as well as semiconductor or insulator nanoparticles (CdS, TiO_2, SiO_2) have also been developed.

The principle of operation of this type of device is shown in Fig. 4. The blend is sandwiched between two different metal electrodes that create the built-in potential. Under excitation of a donor moiety, photoinduced electron transfer to a neighboring acceptor takes place, which leaves a free hole in the donor and a free electron in the acceptor. The same result is attained on the excitation of an acceptor. Once separated, the two charges drift in different phases thereby minimizing their chance of recombination and allowing the individual optimization of both mobilities. Provided that there are two interpenetrated percolating pathways, both charges can be efficiently collected at the electrodes.

Blend-based devices are the best in terms of performance: As in heterojunctions, they offer efficient exciton dissociation and bipolar charge transport. In addition, the whole bulk of the blend contributes to the charge generation process. Obviously, the device performance is very sensitive to the morphology of the blend. For efficient exciton dissociation, phase separation at the exciton diffusion length scale is required. For efficient bipolar transport, the two phases must each form a percolating network. Morphologies that satisfy these conditions are difficult to achieve without the careful optimization of the preparation conditions of the blend.

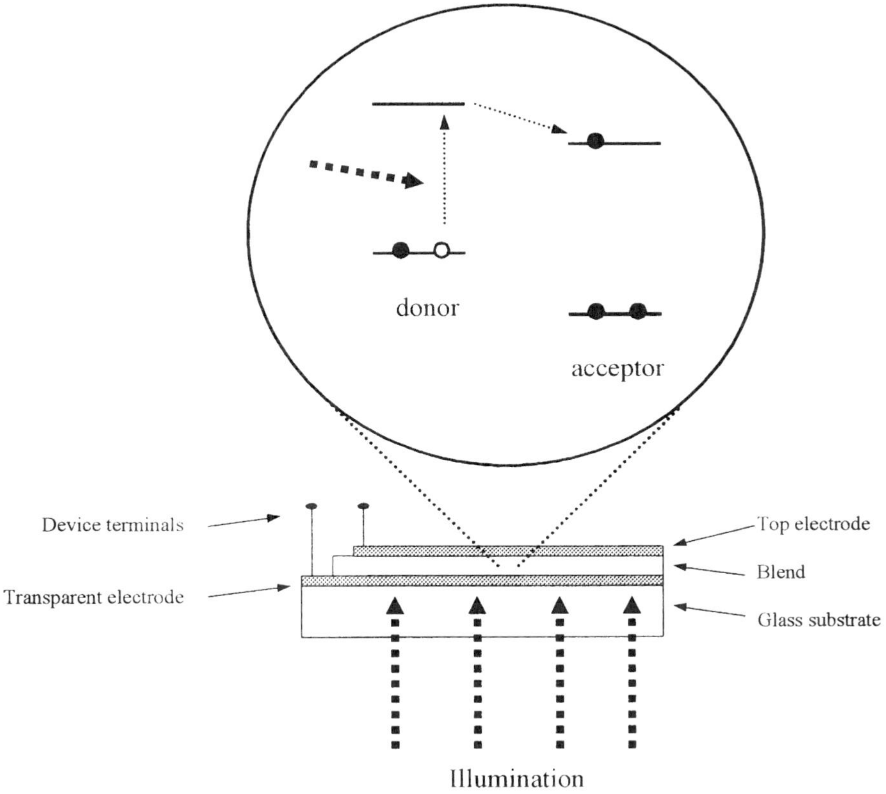

Figure 4
Layout of a blend-based photovoltaic device with showing the detail of the charge generation process. Shown are the energy levels of the two phases (donor and acceptor). On optical excitation (thick dotted arrow), an electron (•) transfers from a donor to an acceptor moiety, leaving a hole (○) in the donor.

4. *Outlook*

The performance of organic photovoltaic devices has improved remarkably. In the solar cell configuration, the power conversion efficiency (electrical energy produced per incident optical energy), which was as low as 10^{-5}% in the early 1970s, has been improved to larger than 2% (Schön *et al.* 2000). In the photodiode configuration, sensitivities (photocurrent produced per incident optical energy, measured in amps per watt) that are larger than in commercially available silicon photodetectors have been achieved (Yu *et al.* 1994). There is plenty of room for further improvement as several issues remain to be addressed. First and foremost, the importance of careful synthesis, purification, and handling of the organic semiconductors is being realized. Accidental impurities, such as oxygen and water, can act as charge traps, charge donors, or photogeneration sites, leading to poor performance and artifacts. On a fundamental level, the energetics at the metal–organic and organic–organic interfaces are not yet well understood. Chemical reactions, dipole layers, and charge accumulation (which have been neglected in the models presented above) will alter the alignment of energy levels at the interface and severely affect the device performance. Understanding and controlling interface energetics should lead to improvements in device efficiency. The process of photoinduced charge transfer is another field that has been attracting a great deal of interest, and thus improvements in the efficiency of free charge generation should be expected. For blend-based devices, a systematic study of morphology vs. performance has not yet been carried out. Materials that can self-assemble to yield interpenetrating morphologies (e.g., block copolymers) could potentially be used as replacements for blends.

Photovoltaic devices from organic semiconductors seem to have a bright future in applications as photodetectors, where they already compete favorably with silicon detectors. For applications as solar cells, a 10-fold improvement in the power conversion efficiency is necessary in order to reach the same levels of performance as amorphous silicon. Their sensitivity in the near-IR must also be improved for this purpose, which necessitates the synthesis of novel materials with extended near-IR sensitivity. To become technologically successful, additional issues such as device stability and reproducibility will have to be addressed. However, since a broader scientific community, including industry, is tackling these problems in connection with the commercialization of organic light emitting diodes, rapid advances are to be expected.

See also: Light Emitting Diodes from Molecular and Polymer Semiconductors

Bibliography

Halls J J M, Walsh C A, Greenham N C, Marseglia E A, Friend R H, Moratti S C, Holmes A B 1995 *Nature* **376**, 498–500

Malliaras G G, Salem J R, Brock P J, Scott J C 1998 *J. Appl. Phys.* **84**, 1583–7

O'Regan B, Grätzel M 1991 *Nature* **353**, 737–40

Schön J H, Kloc Ch, Bucher E, Batlogg B 2000 *Nature* **403**, 408–10

Sze S M 1981 *Physics of Semiconductor Devices.* Wiley, New York

Wöhrle D, Meissner D 1991 *Adv. Mater.* **3**, 129–38

Yu G, Gao J, Hummelen J C, Wudl F, Heeger A J 1995 *Science* **270**, 1789–91

Yu G, Pakbaz K, Heeger A J 1994 *Appl. Phys. Lett.* **64**, 3422–4

Photovoltaics: Advanced Inorganic Materials

A. Slaoui and R. T. Collins

1. Photovoltaic Market

By 2050 world energy consumption is predicted to grow from the present 13 TW yr to 30 TW yr. To meet this huge demand, the global energy sector will be facing two main pressing issues, declining fossil reserves and climate change due to man-made green house gas emissions. To reduce carbon emissions, nuclear power for electricity supply is a potentially important source. The fraction of world needs which can be supplied by nuclear energy will depend upon considerations such as the size of nuclear fuel reserves, the efficiency of old or even new generations of nuclear reactors when using lower-grade uranium ores, and, more importantly, finding satisfactory solutions for nuclear waste disposal or storage. Figure 1 plots the present worldwide noncombustion electricity generation by wind and photovoltaics (PV) compared to nuclear, as well as the projections established by the International Atomic Energy Agency (IAEA). These projections show nuclear flattening out in the near future. They also suggest renewable energy sources have an increasingly important role to play in future sustainable development. While all renewable energy sources need to be considered in meeting the energy challenge, solar energy, which includes biofuels, solar thermal energy, and PV, may arguably be the only renewable source capable of supplying a significant fraction of energy at predicted future levels. This is primarily because of the enormous energy resource represented by the sun which continuously floods the earth with 125 000 TW of power. This sector has only just begun to reveal its enormous potential for growth.

Generating a dramatic increase in PV-produced power is a significant challenge. Present PV production is almost insignificant relative to fossil fuel generation and cost, which is presently almost 10 times higher than coal-produced electricity, is the key issue. At the same time, PV manufacturing is growing exponentially, at approximately 30% per year and this rate is, if anything, increasing (Fig. 2). Thus, the PV industry approached an annual sale of 2.5 GW in

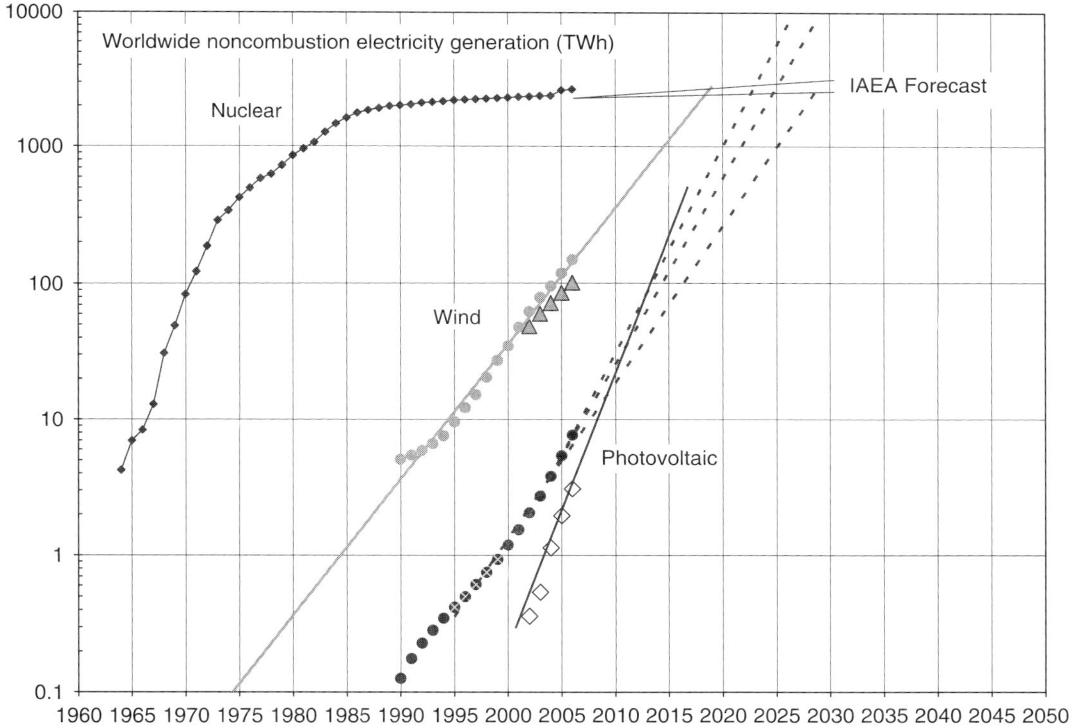

Figure 1
Global share of electricity from nuclear, wind, and photovoltaic. Data source: nuclear—IAEA; PV—EuroObserver and PV news edited by JRC; wind—EWEA and GWA.

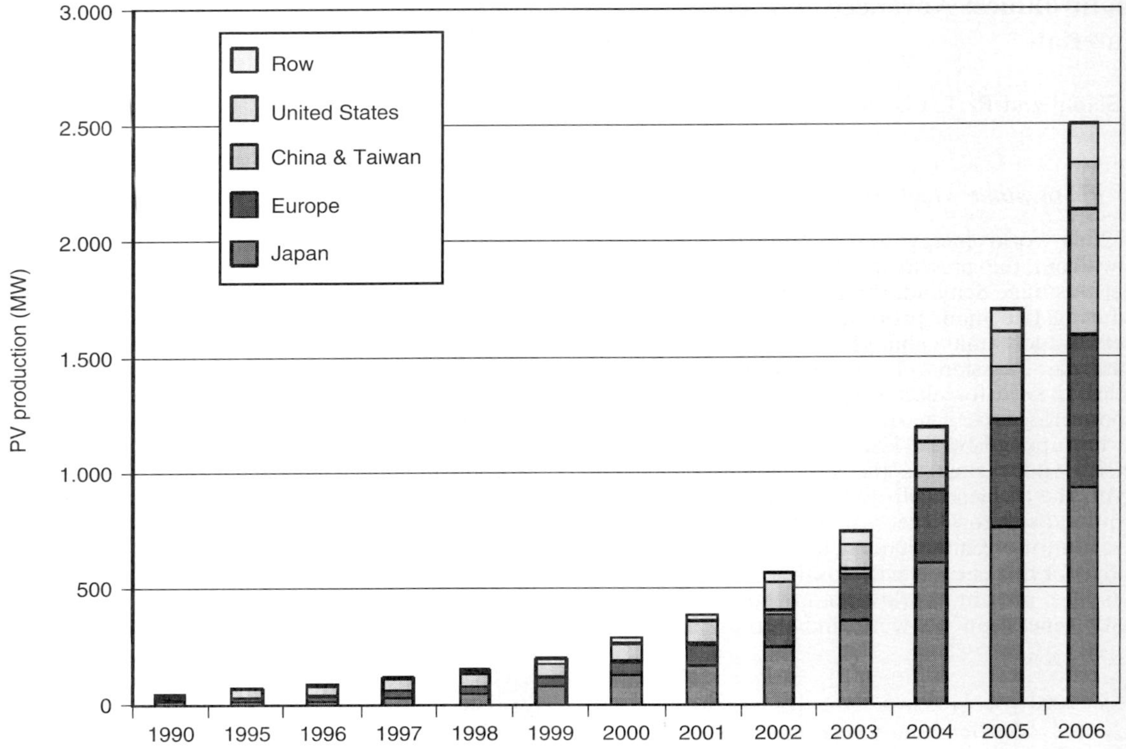

Figure 2
Photovoltaic cell production as reported by the industry. Shipments for Japan, USA, Europe, China, and rest-of-world (ROW) sectors have been identified. Source: PV news, Photon International, IEA PVPS, Photon International Survey.

2006 and more than 10 billion euros. It is true that this growth has been the result of government policies (subsidies or feed-in tariffs), mainly in Japan and Germany. PV production will be further stimulated by the goals for PV set in, for instance, the EU White and Green Paper of 12% of the total energy and 22% of electricity generated by all renewable energies and 3 GWp in Europe for photovoltaics in 2010. Similarly, the Solar America Initiative in the United States calls for investments to make PV power competitive with other forms of renewable energy by 2015 and a growing number of states have mandated renewable energy standards. The impressive progress is also due to the substantial technology improvements in materials quality, devices processing, fabrication and design, and also in characterizations. All these together led to better device performance and reliability, and lowered the systems costs.

Most of the total PV modules shipments in Fig. 2 are based on single-crystal, multicrystalline, and ribbon silicon wafers as shown in Fig. 3. This is typically

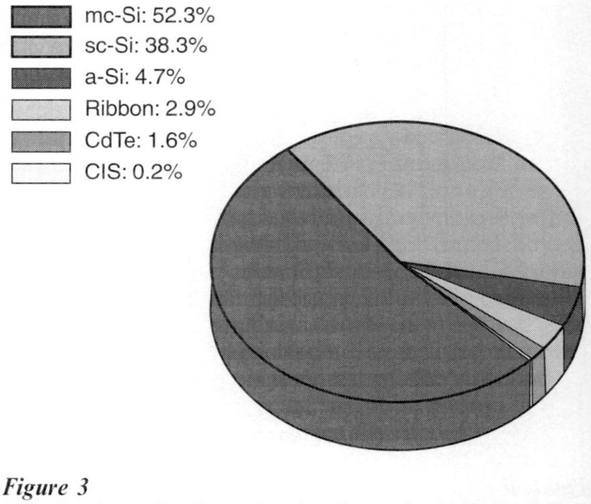

Figure 3
Distribution of cell production by technology in 2006. Source: Photon International, March 2006.

referred to as the first generation of photovoltaics. Fulfilling the projected demand for crystalline silicon-based cells and modules will require innovative research and development to increase the efficiency and reduce the cost. Major activities directed toward such goals will be addressed below. Projections based on historical cost reduction trends suggest that the price for conventional silicon solar cells will not drop to a competitive level with fossil fuel for several decades. For this reason, the second PV generation based on inorganic semiconductor thin films which have the potential for much lower costs will receive special attention here. Finally, third-generation PV concepts aimed at ultrahigh efficiency and/or ultralow cost regimes and involving nanotechnologies, multiple junctions, and bandgap engineering will be considered. These approaches are already on the laboratory research bench and in some instances (e.g., multijunction approaches) nearing market production. Prior to describing these materials and related technologies for PV conversion, a brief discussion of solar cell operation and key figures of merit will be presented.

2. *PV Devices*

A solar cell is a semiconductor device that converts photons from the sun into electricity. Figure 4 shows the generic structure and operation of a single-junction inorganic solar cell. At the heart of the cell is the light-absorbing semiconductor material which converts photons into carriers (electrons and holes) via the photovoltaic effect. The different PV materials discussed in this article are identified by the composition and properties of the absorbing material. After carrier generation, the electrons and holes separate and are collected at the contacts. There are two main modes for charge carrier separation: drift of carriers, driven by the built-in electrostatic field of a p–n junction, and diffusion of carriers from zones of high to low carrier concentration following an electrochemical potential gradient. When drift is important, the structure of the cell can be used to compensate for poor diffusion. When diffusion is used, then improvement in material quality becomes more of the focus. Drift is typical of thin-film inorganic cells which are normally p–n junction devices with relatively high defect densities, while diffusion is more important for high-efficiency single crystal cells and is the driving transport mechanism for dye or organic solar cells. These latter approaches have significant potential for low-cost energy production but will not be discussed here. An in-depth review of organic photovoltaic technologies can be found in an *MRS Bulletin* on organic PVs (Shaheen *et al.* 2005).

The properties of the contacts and window layers are also critical to device performance. At least one contact needs to be both electrically conducting and transparent to photons in the spectral range where

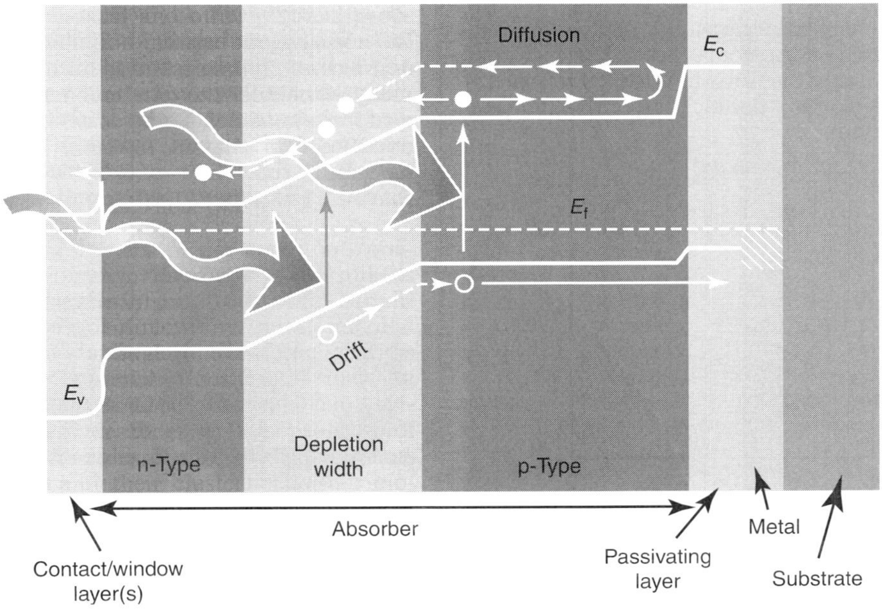

Figure 4
Schematic of the structure and of carrier generation and separation in a p–n junction solar cell. E_V is the valence band energy, E_f is the Fermi level, and E_C is the conduction band energy. Courtesy of Sarah Kurtz.

the absorber creates carriers. Transparent conductive oxides are the material of choice for this purpose. It is also important to point out that the simple structure and the division into well-defined layers with specific functions shown in Fig. 4 is not adequate to describe many solar cells. In a number of designs light enters through the substrate, and contacts can be quite complicated, fabricated from several layers and with significant interdiffusion at metallurgical junctions.

3. *PV Figures of Merit*

Here we discuss some of the basic metrics for characterizing PV devices which are typically quoted by the industry and appear in the rest of this article. Figure 5 shows a typical current–voltage (*J–V*) characteristic for a solar cell in the dark and under illumination. Power is generated when the cell operates in the fourth quadrant of the graph where voltage is positive and current is negative. The fill factor is the power produced at the maximum power point on the illuminated *J–V* curve divided by the product of the open circuit voltage (V_{oc}) and short circuit current (J_{sc}). We can think of it as a measure of how rectangular the *J–V* characteristic is in the fourth quadrant. Another useful measure, quantum efficiency (QE), is the ratio of the number of charge carriers collected by the solar cell to the number of photons at a given illuminating energy. Ideally it is a step function at the bandgap of the absorber. In real devices, however, this is never the case and there is also a roll-off at higher energies.

The most fundamental measure of performance is, of course, the energy conversion efficiency—the percentage of power converted from sunlight to electrical energy. Two decades ago, Shockley and Queisser (S–Q) used a detailed balance approach to estimate the thermodynamic efficiency limit for solar cells with a single absorbing material (a single-junction design, as shown in Fig. 4). Their approach balanced the radiative transfer between the sun, modeled as a black body at 6000 K, and the solar cell assumed to be a 300 K black body which absorbs photons with energy above an energy threshold (the bandgap). They varied the bandgap of the absorber and the solid angle from which the solar cell can collect the sun's radiation (concentration). As the bandgap of the absorber increases, the open circuit voltage increases, but, because only photons with energy above the bandgap are converted into carriers, short circuit current decreases. This leads to a maximum in efficiency as a function of bandgap. For unconcentrated solar irradiance, a maximum efficiency limit of ~31% at a gap near 1.3 eV was found. For full solar concentration (46 300 suns) maximum efficiency was ~41% for a 1.1 eV material. One limiting factor for higher efficiencies is that the excess kinetic energy of hot photogenerated carriers created by the absorption of supra-band-gap photons is lost as heat through phonon emission, and the other limiting factor is that photons less than the bandgap are not absorbed. Designs which attempt to increase cell efficiencies above the S–Q single-junction limit by addressing one or both of these loss mechanisms are often termed third-generation approaches in literature. Several are discussed below.

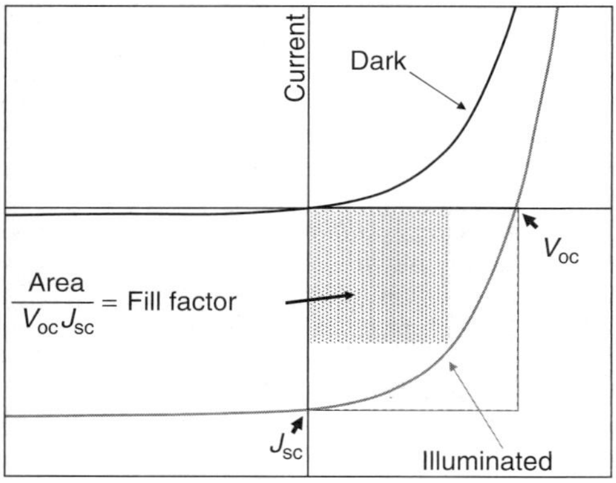

Figure 5
Current voltage (*J–V*) characteristics of a solar cell with and without illumination. The open circuit voltage (V_{OC}), short circuit current (J_{SC}), and fill factor are identified on the figure.

4. *PV Materials*

4.1 *Crystalline Silicon*

As stated above, most of the PV modules being shipped today are single-crystal, multicrystalline, or ribbon-based silicon cells. Thus most efforts at laboratory and industrial scales are directed at improvement of silicon-based solar cell efficiency and at implementing innovative and low-cost processes. Research and development in this area has taken advantage of the wealth of knowledge developed for silicon integrated circuit manufacturing and developed structures that maximize the electron–hole generation and reduce surface and bulk recombination. Thus, cells have reached laboratory performances converting nearly one-fourth of the incident photons into electrical power—reaching about 92% of the reported "theoretical limit" for a silicon absorber. To go further, more complicated designs and structures are sought. Among the advanced structures that allow cells to approach 20% efficiency on multicrystalline silicon and 22% on monocrystalline silicon, there are single-sided and doubled-sided buried contacts (SSBCs and DSBCs), the metal/insulator/n-type/p-type (MINP) bifacial cells, and the heterojunction structure (HIT). Detailed designs for these can be found in the recent proceedings of PV conferences.

Another important research activity dealing with silicon-based cells is how to close the differences between the less perfect materials, namely multicrystalline and ribbon silicon, and single-crystal cells. The roles of defects and impurities—and their interactions and impacts on solar cell performance—remain important. Thus research and development is directed at understanding the nature of the defects resulting from growth and processing and, also, materials treatments to minimize carrier losses at unwanted active sites. These passivation treatments have included processes involving hydrogen, lithium, aluminum, arsenic, and phosphorus for front and back surfaces. The development of plasma and other nitriding processes by the semiconductor electronics industry has led to improvements in single and multicrystalline Si cells and ribbon Si cells in particular. Additionally, the economics of using less material has led the industry to investigate the production and use of thinner wafers, going from standard 300-μm-thick wafers down to 100 μm. This however presents considerable difficulties in handling and processing because of mechanical fragility for these very thin wafer thicknesses and the extremes in temperature and associated thermal stresses they must endure during processing.

Among thinned "wafer" technologies that have emerged in this new century are Sliver® cells composed of silicon strips that are produced by micromachining narrow ~100 μm deep grooves into a Si wafer. These bifacial cells have had efficiencies up to about 18%, with prototype minimodules (560 cm^2 area) above 12% efficiency. The cells are being considered for transparent and concentrating modules, in addition to conventional module use.

4.2 *Thin Films*

Thin-film PVs are now entering the world commercial markets (Fig. 3). Although their market share is still very small, growth is rapid, driven in part by capacity problems with crystalline Si. The arguments favoring thin-film PV are based on low materials utilization, large-scale manufacturing advantages, and better energy economy for production and product recovery. Thin-film technologies also enable other new and desirable properties for specialized applications (flexible modules, semitransparent modules, etc.). In addition, research results with these systems have been very impressive with laboratory device efficiencies for some systems approaching 20%. Despite the very positive potential of these technologies, there is a strong need for basic research to better understand fundamental properties of these materials and devices, especially of interfaces, new multijunction designs, low-cost, high-performance transparent conductor oxide (TCO) materials, alternative module concepts, and of methods for increasing device and module efficiencies. This section begins with a discussion of polycrystalline and amorphous silicon as well as the chalcogenide thin-film cell technologies followed by TCO materials.

(a) From amorphous to polycrystalline silicon films. Amorphous silicon (a-Si:H) is a very attractive material for large-area thin-film electronics, namely as thin-film transistors for flat panel displays, as color sensors, or as the absorbing layer for solar cells. The main advantages of this material are the low deposition temperature that offers the use of glass and flexible substrates, the potential for tuning the energy bandgap (1.7–1.9 eV), the ease of doping by adding dopant gases during film growth, and finally the possibility of stacking layers without severe materials constraints. Several physical or chemical deposition techniques are used to produce hydrogenated amorphous silicon such as magnetron evaporation, plasma-enhanced chemical vapor deposition (PECVD), and, more recently, very high frequency CVD. The main drawbacks when using such disordered material for solar cells are its low charge carrier mobility and diffusion length due to bond angle and bond length distortion, in addition to the strong sensitivity of the silicon–hydrogen bonds to solar radiation. Consequently, while higher open circuit voltages can be achieved on a-Si:H-based solar cells, the photocurrent is limited by insufficient absorption of the solar spectrum and by high defect-related carrier recombination. Despite these limitations, single- and multijunction solar cells and large-area modules based on a-Si:H and related alloys, e.g., a-$Si_{1-x}Ge_x$:H, are now well established in the market. The use of stacked multijunction cells based on a-Si:H and a-SiGe:H has led to record amorphous silicon-based cells. The a-SiGe:H layer is employed as a bottom cell because alloying the silicon with germanium leads to a narrower optical gap and therefore to broader optical absorption. To increase the efficiency further and reduce light-induced degradation of the cells, a combination of amorphous and microcrystalline silicon (μc-Si:H) is presently used (Fig. 6). The μc-Si:H layer consists of an amorphous tissue with coherent regions of crystalline grains having sizes from nanometer scale to micrometer scale. The μc-Si:H can be prepared at low temperatures by the same techniques as amorphous silicon and has absorption spectra similar to crystalline silicon. Furthermore, these materials contain several percent of hydrogen, which assists in neutralizing or passivating grain boundaries.

As for the solar cells, by using a single-junction cell with a homogenous microstructure, a stabilized efficiency of $\eta = 10\%$ on texture-etched ZnO:Al has been obtained. IPV Jülich reported stabilized 12.3% for tandem cells. The highest confirmed efficiency for

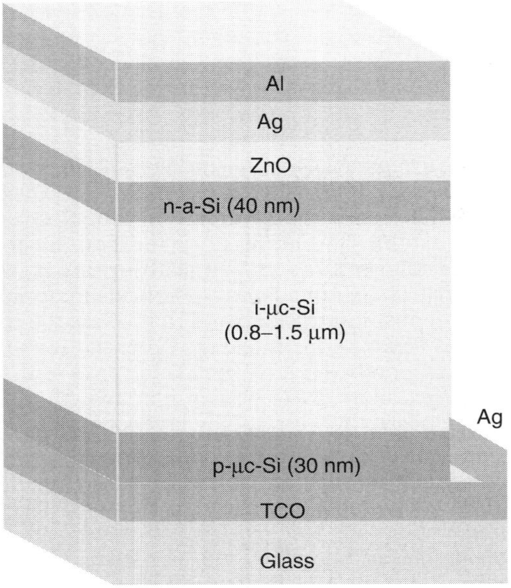

Figure 6
A schematic structure of an amorphous (a-Si)/microcrystalline silicon (μc-Si:H) hybrid solar cell structure. Courtesy of E. Fortunato *et al.*

tandem submodules (Kaneka Co.) based on a-Si/μc-Si is 11.4%. For triple-bandgap a-Si/a-SiGe/μc-Si:H cells, a world record initial efficiency of 15% for thin-film silicon PV technology has been obtained, using optical enhancement techniques, such as textured back reflectors and in some cases also intermediate selectively reflecting layers. Apart from the triple-bandgap a-Si/a-SiGe/μc-Si:H approach, the a-Si/μc-Si/μc-Si structure is also under investigation. United Solar has reported 14.6% initial and 13.0% stable conversion efficiencies using an amorphous silicon-based alloy in a spectrum-splitting, triple-junction structure. Key factors leading to this major advance include improvements made in the low-bandgap amorphous silicon–germanium alloy cell, the p–n tunnel junction between the component cells, and the top conducting oxide. There are still windows for further improvements by better controlling the interfaces and reducing defects in the layers.

As a natural extension of the trend toward higher crystallinity of the active layer, there is an important effort in developing thin-film polycrystalline silicon. The growth of thin crystalline silicon films on various foreign substrates is being explored by a broad range of methods that can be distinguished by the principle of the deposition process: CVD including conventional high temperature CVD (APCVD, LPCVD), CVD on liquid layers (CVDOLL), PECVD, ion-assisted deposition (IAD), liquid-phase epitaxy (LPE), and solid-phase crystallization of amorphous silicon (SPC, aluminum-induced crystallization (AIC)). With the exception of the AIC process, most of these methods result in polycrystalline silicon materials with fine grains ($<10\,\mu m$) and show relatively short minority carrier lifetimes compared to multi- or monocrystalline silicon material. Minority carrier diffusion lengths are usually in the range of only a few micrometers, smaller than or comparable to the film thicknesses but much shorter than the optical absorption length. The analysis of minority carrier properties of poly-Si films is intrinsically complicated by its inhomogeneous nature which arises from a broad grain size distribution and potential fluctuations within individual grain boundaries. As a consequence, the open circuit voltage of such poly-Si thin-film cells is quite small ($<500\,mV$). Efforts to reduce the barrier height and thus charge carrier recombination include doping of the absorber layer, lowering the trap density at the grain boundary by strong hydrogenation and increasing the grain size. The thin-film polycrystalline silicon technology that is closest to industrial implementation is arguably the Crystalline Silicon on Glass (CSG) technology developed by the company previously called Pacific Solar (now CSG Solar). They are using the solid-phase crystallization of amorphous silicon layers, and efficiencies are presently reaching around 10% at mini-module level. Another promising approach is the heterojunction structure (HTJ) that combines an absorbing large grain polysilicon layer formed by the AIC process and thickened by CVD with an emitter made of doped thin-film a-Si:H. Such a combination has resulted, so far, in an efficiency of 8%.

In order to be successful, thin-film polysilicon devices need to improve further and reach 12% module efficiency in a superstrate configuration, with a process that is industrially manufacturable. Considering the fast pace of progress, particularly in the last few years, thin-film polysilicon has potential to emerge as one of the major PV technologies succeeding bulk Si.

(*b*) *Thin-film CdTe, copper indium selenide, its alloys, and related chalcopyrites.* Polycrystalline thin-film solar cells utilizing the II–VI semiconductor cadmium telluride and alloys of the I–III–VI materials, copper indium diselenide ($CuInSe_2$) and copper gallium diselenide ($CuGaSe_2$), are rapidly entering the marketplace. These materials have a number of interesting properties which make them attractive candidates for thin-film PV technologies. Both CdTe and the $Cu(InGa)Se_2$ (CIGS) alloys are direct bandgap compound semiconductors with energy gaps in the 1.1–1.6 eV range which are nearly ideal for single-junction solar cell efficiency (Fig. 7). They have high optical absorption coefficients ($>10^5\,cm^{-1}$), allowing for thin absorber layers. They are also more tolerant of defects than, for example, silicon and III–V compound

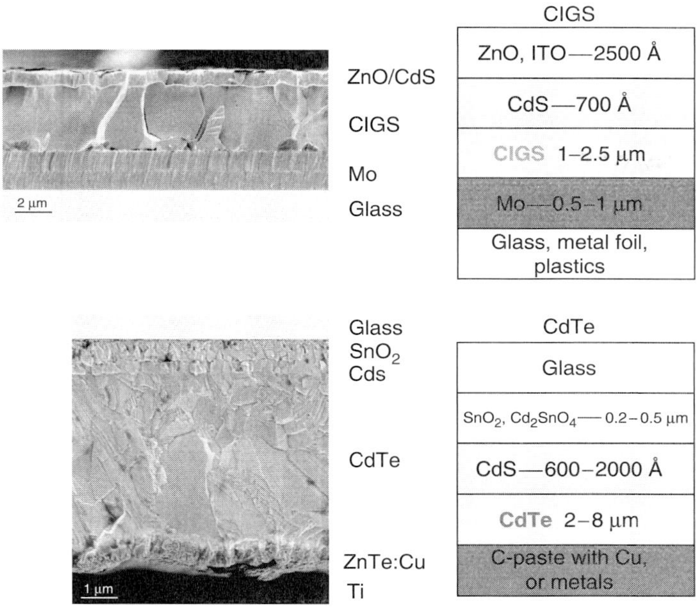

Figure 7
CIGS (top) and CdTe (bottom) PV structures seen in cross section using SEM (left side of panel) and viewed schematically (right side of panel). Courtesy of E. Fortunato *et al.*

semiconductors like GaAs. In spite of the high density of grain boundaries in the films, laboratory solar cells with record efficiencies of ~16.5% (CdTe) and ~19.5% (CIGS) have been demonstrated. There is even evidence that grain boundaries are beneficial, assisting carrier collection. Doping in these materials arises, in fact, from native defect formation during deposition and processing rather than introduction of substitutional impurities. This also means controlled doping, and conductivity is one of the challenges in optimizing cells.

They also have similar cell structures. The junction itself is typically a hetero-interface between n-type CdS and the p-type CIGS or CdTe absorber layer. Light enters through the CdS side of the device and current is carried away by a TCO film which is in contact with the CdS. Since CdS has a bandgap of ~2.4 eV, it absorbs in the blue; however, photons absorbed in it produce little to no photocurrent. For this reason, both CdTe and CIGS cell processing seek to make the CdS layer as thin as possible while still remaining free of pinholes which create shorts between the TCO and absorber layers lowering V_{oc} and cell performance. This is especially problematic for CdTe cells where S diffusion into the CdTe during post-growth anneals further decreasing the CdS layer thickness. The inclusion of thin buffer layers between the TCO and CdS, such as a highly resistive transparent oxide, are found to improve efficiency. The exact role of the buffer layer, whether it simply introduces resistance into shorts or changes the interfacial energetics, is not well understood and optimizing this interface is an important area for materials innovation.

There are also some significant differences in the physical properties and processing of cells from CdTe and CIGS. The highest efficiency CdTe cells operate in a superstrate configuration with light entering through the glass on which the cell is grown. The record-efficiency laboratory devices are prepared by high-temperature absorber growth (near 600 °C). Nearly all preparation schemes follow the CdTe growth with an anneal near 400 °C in an atmosphere of $CdCl_2$ and oxygen that promotes grain regrowth, improves the CdTe carrier lifetime, diffuses the junction region, increases p-type conductivity in CdTe, and forms oxides. Back contacts to CdTe often involve Cu incorporation (e.g., Cu-doped graphite paste or Cu-doped ZnTe). The Cu diffuses rapidly through the device and may be a stability issue. An optimized back contact for CdTe devices, and better understanding of back contact/CdTe interfacial energetics, is a critical need.

In contrast, CIGS thin film cells are grown in a substrate configuration with the TCO contact as the last, rather than the first, layer deposited. The highest-efficiency CIGS cells are grown by physical vapor deposition and the grain microstructure is defined by a complex evolution from Cu-rich to In- and Ga-rich phases during growth. Reproducing this growth path with other more manufacturable deposition options is a major challenge. In optimized devices, the Ga alloy composition is graded by ~5% across the absorber layer. Incorporation of Na ions, which occurs naturally as an aspect of growth on soda lime glass substrates, is also important to optimizing the CIGS device efficiency. Although there is not a definitive explanation for the beneficial effects of Na, Na is believed to reduce the resistivity of the film. Typical absorber deposition temperatures are over 500 °C for CIGS and a post-deposition anneal in a Se atmosphere is recommended. Decreasing deposition and post-growth anneal thermal budgets for both CdTe and $Cu(InGa)Se_2$ are important cost reduction goals. The typical CIGS back contact is a molybdenum film. Device film adhesion issues are often traceable to issues with the molybdenum film. $Cu(InGa)Se_2$ devices where S is substituted for Se are also being explored, allowing broader tunability of electronic and structural properties.

While reasonable stability under operation has been demonstrated for CdTe and CIGS solar cells, the performance of both can degrade in the presence of moisture. For this reason, they are encapsulated between glass sheets to seal the front and back surfaces of the modules, with polymer compounds sealing the edges and bonding the laminate together. Alternative polymer encapsulation schemes are also explored for CIGS devices. Finally, we note that issues of scarce elements (e.g., In) and toxicity (e.g., Cd) are frequently discussed in the context of both technologies. Whether or not either is a truly significant issue depends at least in part upon feasibility, cost, and implementation of recycling and element recovery at end of product life. For example, First Solar, Inc., one of the rapidly growing CdTe panel manufacturers, includes panel recycling as part of product lifecycle management.

(c) TCO-based contacts. Since most PV designs require solar radiation to pass through one of the device contacts, TCOs, which both conduct electricity and transmit visible light, are essential components of many photovoltaic devices. In terms of those discussed in this paper, TCOs form contact layers for a-Si-based devices, crystalline Si HIT cells, CdTe and CIGS-based solar cells, organic solar cells, and are essential components of many third-generation cell approaches discussed below. They are used as electrode elements, structural templates, or diffusion barriers. They are essential to tuning the open circuit device voltage by choosing a TCO with a desired work function. The figures of merit for TCO materials include high optical transmissivity across a wide spectrum and low resistivity. Additionally, TCOs for terrestrial PV applications must use low-cost materials and some may require device-technology-specific properties.

The first TCO, namely CdO, was reported in 1907 and was both optically transparent and electrically conducting. Since then, three oxides have emerged as commercially important transparent conductors: (i) indium oxide, (ii) tin oxide, and (iii) zinc oxide. The properties and crystal chemistry of the transparent conducting indium oxide family of materials are very well documented, since it is the highest-performance, best-understood material in the TCO class. Considerable research is presently directed at finding alternatives to indium-based oxides because of indium's high cost. By volume, the most deposited TCO today is SnO_2 which is used in IR-efficient architectural window applications. ZnO is also, at present, primarily used in window coatings but recent processing-related performance improvements, coupled with Zn's lower cost compared to In, are making ZnO and ZnO alloys attractive replacements for In-based TCOs. Another area under active investigation involves amorphous TCOs. Typically TCOs deposit as fine-grained polycrystalline films. For certain alloy systems (e.g., indium zinc oxide alloys), crystallization is suppressed and amorphous films which still possess high transparency and conductivity can be formed. Potential advantages of amorphous films include less surface roughness, lower processing temperatures, and absence of anisotropy during etching.

4.3 Multijunction Tandem Cells

The multijunction cell approach is the most mature and most successful third-generation technique for improving performance over a single-junction cell. Here solar cells are stacked in order of their bandgap, with solar radiation entering the largest bandgap cell first. Light is automatically filtered as it passes through the stack, with each cell absorbing a narrow energy range of above-bandgap photons, thereby minimizing hot carrier losses. Below-bandgap photons are transmitted. Those with energies above the smallest bandgap in the stack have a high probability of being absorbed, minimizing below-bandgap losses. The result is a multijunction solar cell with an ideal efficiency determined by the number of cells in series. The theoretical efficiency limit is 86.8% for an infinite number of bandgap energies at very high solar concentration levels. If bandgaps are appropriately selected so that each cell in the stack generates nearly the same current, the cells can be interconnected in series making monolithic integration possible.

A multijunction thin-film approach that is now in volume manufacturing utilizes hydrogenated amorphous silicon (a-Si:H) and silicon–germanium (a-SiGe), and microcrystalline silicon (μc-Si:H), as discussed above. While this design represents a materials science *tour de force*, it is difficult to realize the full benefit of multiple junctions when the solar cells are made from polycrystalline or amorphous materials with lower crystal perfection.

Maximizing device efficiency requires long minority carrier lifetimes and, hence, high materials quality. Crystalline group IV (Si and Ge) and III–V compound semiconductors (GaAs, InP, and the numerous III–V alloys) are the best candidates for multijunction cells. Indeed, the highest-efficiency solar cells made are $Ga_{0.5}In_{0.5}P/Ga_{0.99}In_{0.01}As/Ge$ triple-junction devices with concentrated efficiencies of $\sim$40%. The III–V materials, in particular, are ideal for fabricating such multijunction solar cells for two reasons: they can be grown with excellent materials quality using techniques such as molecular beam epitaxy and organometallic CVD and their bandgaps span a wide spectral range (typically between 0.4 eV and 2.4 eV), mostly with direct bandgaps, implying high absorption coefficients. When fabricating these alloys, there are a number of possible ways to mix the endpoint compounds, even within a single-crystal layer, to tailor the optical or electrical properties. In general, lattice-matched materials are the easiest to manufacture with high quality. The lattice-matched nature of the $Ga_{0.5}In_{0.5}P/Ga_{0.99}In_{0.01}As/Ge$ cell is one of the reasons for its strong success. There has been success, however, with lattice-mismatched structures, which increases the range of materials/bandgap combinations that can be considered in optimizing a cell design. Perhaps the most exciting aspect of III–V multijunction cell development is the broad range of phase space yet to be explored in optimizing cell designs. It is known, for example, that the currents in the triple-junction $Ga_{0.5}In_{0.5}P/Ga_{0.99}In_{0.01}As/Ge$ device are not optimally matched, and insertion of an additional junction with a 1.0 eV bandgap directly above the Ge junction would improve efficiency. A GaInAsN alloy which has the desired bandgap and is lattice-matched has been studied extensively for this application but, so far, cannot be integrated with the required materials quality. Modern preparation techniques such as wafer bonding to mitigate the effects of lattice mismatch can also be explored.

The efficiencies already achieved with crystalline multijunction cells are truly astounding, and it seems likely that values near 50% are not that far off. These cells, however, were originally developed for space applications where power-to-weight ratio is the key figure of merit and they are too costly for flat plate terrestrial power production. Integration of the cells with lower-cost solar concentrator designs, however, has the potential for enabling cost-competitive systems. While concentrator designs are not being covered in this article, their development and improvement is equally (or perhaps more) important as improvements in cell efficiency to continuous cost reductions in PV systems based on high-efficiency multijunction cells.

4.4 Emerging Materials and Cells Concepts

To overcome the S–Q efficiency limit mentioned above, there are preliminary reports and experiments on advanced exciting cell concepts which include several nanotechnology and hot-electron approaches such as the quantum dot (QD) cell and relatives using quantum rods and quantum pods. There are several approaches based on semiconductor QDs that could lead to ultrahigh efficiencies through enhanced photocurrent.

(*a*) *Multiple exciton generation.* The formation of multiple excitons from a single absorbed photon occurs when the energy of the photon absorbed is much greater than the semiconductor bandgap. Such a process is well known in bulk semiconductors and is termed impact ionization (II); it is the inverse of the Auger process whereby two electron–hole pairs recombine to produce a single highly energetic electron–hole pair. However, the probability that such a phenomenon will occur in a bulk semiconductor is very low because the excess energy simply dissipates as heat before it can cause other electron–hole pairs to form. But in semiconductor QDs, the rate of energy dissipation is significantly reduced, and the charge carriers are confined within a small volume, thereby increasing their interactions and enhancing the probability for multiple excitons to form (see Fig. 8). Theoretical considerations have reported a thermodynamic limit of about 45% for unconcentrated light. Quite recently, very efficient multiple exciton generation (MEG) has been reported for PbSe, PbS, and PbTe QDs. Up to 3 excitons/photon have been measured in 2.9 nm diameter PbSe QDs when the energy of the photon absorbed is 4 times that of the bandgap, and up to 7 excitons/photon have been reported for excitation at 8 times the bandgap. More importantly, MEG has been reported for silicon QDs that might open larger opportunities for applications of such phenomenon in solar cells. These observations of MEG have effectively occurred in isolated QDs. The enhanced photocurrent possible with MEG has not been observed, to date, in a solar cell structure. The issue is one of finding a device design which incorporates quantum dots or wires and can directly collect the carriers generated within the nanostructure. Several configurations have been proposed including imbedding the QDs in polymers, imbedding them in other semiconductors, applying dots in place of the dye in a dye-sensitized solar cell design, and formation of QD solids from an assembled array of dots. Developing device designs that exhibit and ultimately take advantage of MEG to increase cell efficiency is presently an extremely active area of research.

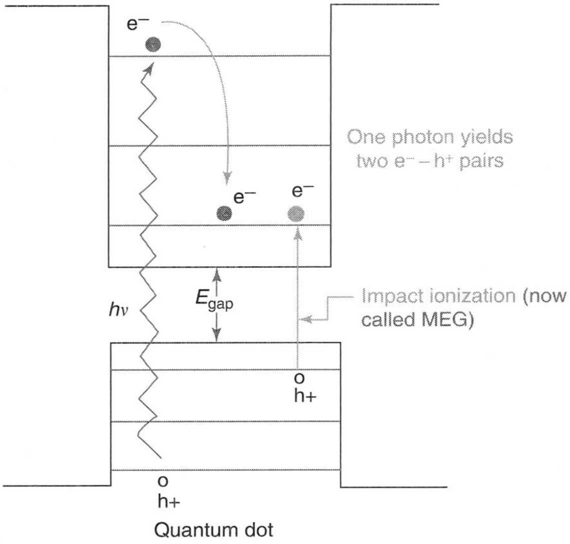

Figure 8
Enhanced electron–hole pair (exciton) multiplication in QDs that could lead to enhanced solar photon conversion efficiency in QD solar cells.

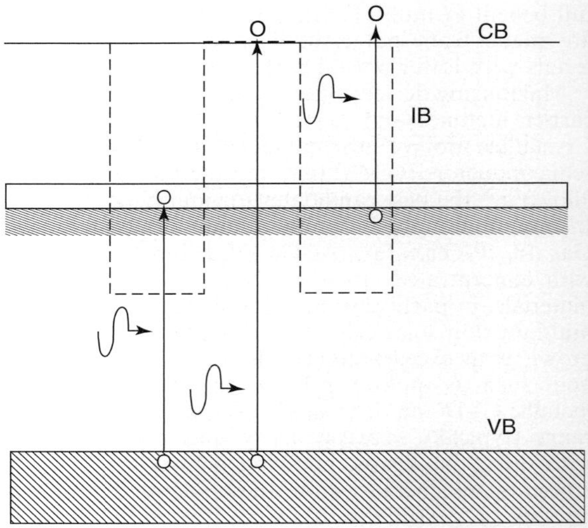

Figure 9
Structure of an intermediate-band material also showing the possible optical transitions.

(*b*) *Intermediate-band approach*. In this approach, quantum dots of binary or ternary semiconductors are used to create an intermediate band (IB) that allows the harvesting of a much larger portion of the available solar spectrum. Because of the IB, below-bandgap energy photons can contribute to the cell photocurrent by pumping electrons from the valence band (VB) to the IB and from the IB to the conduction band (CB; Fig. 9). But most importantly, since the IB is isolated from the CB and the VB by a region with zero density of states, "carrier relaxation" between bands becomes difficult and the carrier statistics in each band is described by its own quasi-Fermi level. The energy states of the QDs forming the intermediate bandgap can be controlled by the size of the dot. It can be shown that, under this assumption, the voltage supplied by the cell, given by the electron and hole quasi-Fermi level split, is still limited by the high bandgap E_G and not by any of the lower bandgaps (E_L or E_H). The study of the limiting PV conversion efficiency of the IB approach reveals a limiting efficiency of ~47% at one sun (~63% at full solar concentration) for this concept as compared to ~43% for a two-junction tandem solar cell at 1 sun (~55% for full solar concentration).

Several groups have been investigating QD-IB structures, but no devices using this approach have yet been validated. Among the prototype cells designs being studied are devices consisting of 10 layers of InAs/GaAs QDs sandwiched by p and n GaAs emitters. The structures are grown by molecular beam epitaxy (MBE) and the dots form in the Stranski–Krastanov growth mode. Such structures still suffer from poor absorption in the QDs and low open circuit voltage due to structural defects induced during the growth.

Another approach considers the addition of nitrogen to III–V materials or inclusion of oxygen in ZnMnTe to form an IB. With the latter, it is reported that the location and the width of the intermediate band is determined by the O content. Calculations based on detailed modeling predict maximum efficiency of more than 55% in alloys with 2% of O. Additional investigations explore epitaxial techniques to grow ZnO_xSe_{1-x} materials.

(*c*) *All-silicon tandem cells approach*. As mentioned above, the tandem cell concept is well established as a way of improving solar cell performance. Obviously there is a strong interest in implementing tandem cells based entirely on silicon thanks to the advantages of QDs. Indeed, the first advantage of QDs is their tunable bandgap. This means that the wavelength at which they will absorb or emit radiation can be adjusted: the smaller the size, the shorter the wavelength of light absorbed, and the greater the output voltage. On the other hand, a lower bandgap results in the capture of more photons including those at the red end of the solar spectrum, resulting in a higher output current but a lower output voltage. Thus, the highest solar-electric energy conversion can be theoretically achieved by using a superlattice based on a mixture of QDs of

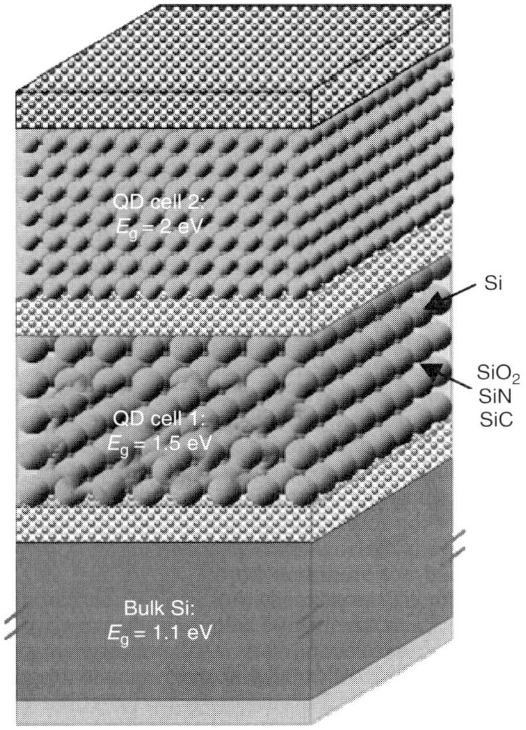

Figure 10
A schematic view of a two-cell "all-silicon" tandem cell using silicon QDs embedded in SiO_2, SiN, or SiC matrix. Courtesy of G. Conibeer, UNSW.

different sizes for harvesting the maximum proportion of the incident light effectively. Ideally, such superlattices can then be used as the higher-bandgap cells in a tandem stack. When applied to silicon (Fig. 10), the efficiency limit is found to reach 42.5% and 47.5% for two-cell and three-cell 'all-silicon' stacks, compared to 29% for a single silicon cell.

To successfully implement the all-silicon tandem concept, the following features need to be demonstrated: control of dot size and therefore bandgap, enhanced optical absorption as size decreases, adequate transport, P–I–N junctions (high V_{oc}), doping, work function control, and interconnected cells. Thanks to the development of nanotechnologies, control of QD size and density in oxide, nitride, and carbide matrices has been demonstrated, as witnessed by high-resolution TEM observations and photoluminescence measurements. However, as is the case with MEG, obtaining the required charge transport properties in such systems is still under investigation in many research institutes, and may be the central issue in harnessing the potential of QDs for solar conversion.

Bibliography

15th International Photovoltaic Science & Engineering Conference (PVSEC-15) 2005 Shanghai,10–15 Oct., 2005, p. 7

Beaucarne G, Slaoui A 2006 Polycrystalline silicon solar cells. In: Poortmans J, Arkhipov V (eds.) *Thin Film Solar Cells: Fabrication, Characterization and Application*. Wiley, New York, pp. 97–132

Ebbessen T W, Lezec H J, Ghaemi H F, Thio T, Wolff P A 1998 Extraordinary optical transmission through subwavelength hole arrays. *Nature* **391**, 667–9

Green M A 2003 Crystalline and thin-film silicon solar cells: state of the art and future potential. *Solar Energy* **74**, 181–92

Green M A, Emery K, King D L, Hisikawa Y, Warta W 2006 Solar efficiency tables (version 27). *Progress in Photovoltaics: Research and Applications* **14**, 45–51

Hamakawa Y 1985 In: Coutts T J, Meakin J D (eds.) *Current Topics in Photovoltaics*. Academic Press, New York, pp. 111–69

Kazmerski L L 2006 Solar photovoltaics R&D at the tipping point: a 2005 technology overview. *J. Electron Spectrosc. Relat. Ph.* **150**, 105–35

Klimov V I, Ivanov S A, Nanda J, Achermann M, Bezel I, Mc Guize J A, Piryatinski A 2007 Single-exciton optical gain in semiconductor nanocrystals. *Nature* **447**, 441–6

Luque A, Hegedus S 2003 *Handbook of Photovoltaic Science and Engineering*. Wiley, New York, online version available at http://www.knovel.com/knovel2/Toc.jsp?BookID = 1081 &VerticalID = 0

Shaheen S E, David S, Ginley D S, Jabbour G E 2005 Organic photovoltaics. *MRS Bull.* **30**, 10–19

Shockley W, Queisser H J 1961 Detailed balance limit of efficiency of p–n junction solar cells. *J. Appl. Phys.* **32**, 510–19

Slaoui A, Poortmans J, Caymax M 2002 High temperature deposition and epitaxy of Si and SiGe. In: Bergmann R (ed.) *Growth, Characterization and Electronic Application of Amorphous and Crystalline Si Thin Films*. Transworld Research Network (TWR), Thiruvananthapuram, pp. 147–81

Slaoui A, Reuben R T 2007 Advanced inorganic materials for photovoltaics. *MRS Bull.* **32**, 211–15

Zeman M, Schropp R E I 1998 *Amorphous and Microcrystalline Silicon Solar Cells: Modeling, Materials and Device Technology*. Kluwer Academic Publishers, Dordrecht

Piezoelectric Thin Films for MEMS

P. Muralt

A significant development in the field of sensors, actuators, and transducers has been the miniaturization enabled by the anisotropic deep etching techniques of silicon combined with the advantages of the well-established manufacturing methods of the semiconductor industry. Systems based on this new technology are commonly termed microelectromechanical systems (MEMS). They are key elements needed in the future for establishing complex robot systems for automotive and military applications, fabrication tools, diagnostics, and processing devices. MEMS are especially required to access nanoscale technology for data storage, processing, and sensing. In general terms, MEMS supply input and output functions to microcomputers. MEMS technology (e.g., Madou 1997) advances in micropatterning processes, in enhancement of functionality, in monolithic integration of sensor and electronics, in packaging, and in integrating new thin-film materials. The latter have led to the use of many kinds of useful physical and chemical effects.

Piezoelectrics are one type of functional materials used in MEMS technology. They provide a transformation between electrical and mechanical energy and can thus be used for mechanical and acoustic input and output devices. Owing to this ability to provide sensor (input) and actuator (output) functions at the same time, piezoelectrics belong to the materials family that is often referred to as "smart materials." The attractive features of piezoelectrics are large achievable output forces, short response times, a high sensitivity, and a low noise level. The main disadvantage is the impossibility to sense directly quasi-static strains and forces, which, however, can be avoided by applying the concept of resonant sensing (e.g., Benes *et al.* 1995). There are good opportunities for piezoelectrics in MEMS, especially in applications where ultrasound frequencies, acoustic waves, combined sensing and actuation (transducer), fast actuation, and resonant sensing are involved or required. Figure 1 shows one of the basic structures, i.e., a micromachined cantilever laminated with a piezoelectric thin film.

1. Principle of Piezoelectric Sensors and Actuators

Parameters needed for the description of the piezoelectric effect in a single crystal are the strain tensor x_i ($i = 1$–6, reduced notation (Nye 1985)), the stress tensor σ_i ($i = 1$–6), the vectors of the electric field E_i, and the electric displacement field D_i ($i = 1$–3). Piezoelectricity means that there is a linear relationship between D-field and strain/stress:

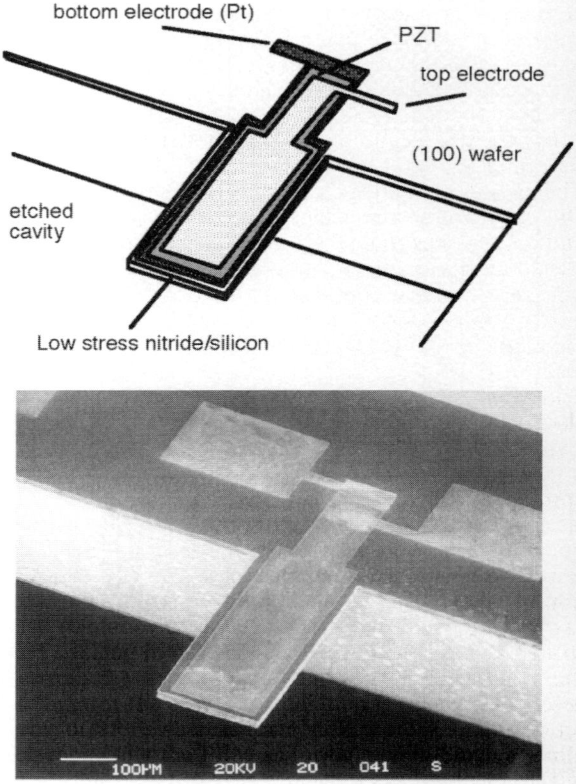

Figure 1
Schematic and SEM image of piezoelectric laminated cantilever structure. The mechanical structure is obtained by wet bulk micromachining techniques combined with dry etching for liberating the cantilever. The sol–gel deposited PZT is sandwiched between top and bottom electrode (courtesy of Ph. Luginbühl).

$$D_i = \sum_k d_{ik}\sigma_k \quad \text{or} \quad D_i = \sum_k e_{ik}x_k \tag{1}$$

These equations describe the so-called direct effect. The piezoelectric coefficients d_{ik} and e_{ik} are the components of a third-rank tensor, which in reduced tensor notation reduces to a 3×6 matrix. The coefficients d and e are related to each other through the stiffness tensor c_{ij}^E:$d_{ik} = e_{ip}c_{pk}^E$. The converse effect is described by the same set of piezoelectric coefficients:

$$x_i = \sum_k d_{ki}E_k \quad \text{or} \quad \sigma_i = -\sum_k e_{ki}E_k \tag{2}$$

Piezoelectricity (see *Stress Coupled Phenomena: Piezoelectric Effect* and Ikeda (1990)) occurs in crystal classes containing no center of inversion (with the exception of crystal class 432). In some of these crystal classes there exists a polar axis. All the piezoelectrics used in thin-film form belong to this material group. (Quartz, which has no polar axis, is only useful in single-crystalline form). It is convention to assign the coordinate index 3 to the polar axis direction, and to define d_{33} as a positive quantity.

In MEMS technology, most of the piezoelectric thin films are, and will be, polycrystalline materials. The piezoelectric effect is averaged over all the grains. One has to note the difference between ferroelectric and nonferroelectric polar materials. The latter do not allow inversion of the polar axis because of a too rigid ion lattice. In this case, the material growth process has to provide for a textured structure that includes the alignment of the polar directions. In ferroelectric materials, the polar axis can be inverted in an electric field by definition. In these materials, an internal electrical polarization, i.e., the spontaneous polarization of the paraelectric-to-ferroelectric phase transition is observed. By means of poling, i.e., the application of a large electric field able to switch polarization, the polarization component parallel to the poling field ends up being positive for all grains. As a result, there is a net piezoelectric effect. It is convention to assign index 3 to this poling direction. For the case of the usually used vertically stacked capacitor structures (see Fig. 1), direction 3 is perpendicular to the film plane. The piezoelectric coefficient d_{33} thus describes the piezoelectric effect in the off-plane direction, as for example, the thickness change of the free film. The directions 1 and 2 are therefore in the plane of the film. In a polycrystalline film, these directions are equivalent. This means that the in-plane strains owing to E_3 are isotropic and d_{31} and d_{32} are the same. The thin-film geometry thus recovers the cylindrical symmetry of poled ceramics, yielding the same non-vanishing elements of the piezoelectric tensor as for tetragonal symmetry (Moulson and Herbert 1990), i.e., d_{33}, d_{31}, d_{15}. In the case of nonferroelectric polycrystalline thin films, where by convention the polar axis is the third one, a (0,0,1) or (0,0, − 1) texture is required. The two possibilities have an opposite piezoelectric response! Uniform texture of either of the two possibilities has thus to be achieved. The in-plane directions are again equivalent owing to averaging over random orientations of in-plane crystallographic directions. The resulting symmetry is thus again the same as for poled ceramics.

The main difference between thin-film materials and bulk materials lies in the fact that thin films are used in a composite structure, where the total elastic properties are often dominated by the other part of the structure. This other part may be a silicon cantilever, or a silicon oxide or nitride membrane, for instance. The interaction with the substrate is very anisotropic. At the interface along the in-plane directions (indices 1, 2), the piezoelectric thin film and the substrate have identical strains. Perpendicular to the film plane, the thin film is free to move, i.e., $\sigma_3 = 0$. As a consequence, there is no deformation mode in which simply one piezoelectric coefficient is involved. The complete equation of state needs to be analyzed. For illustration, a thin film clamped to a much thicker substrate is considered (piezoelectric laminated plate). When a field is applied, the strains in the plane stay at zero ($x_1 = x_2 = 0$), and in-plane stresses and an out-of-plane strain are developed. The equation of state using compliance tensor and d coefficients is (without shear components):

$$\begin{aligned} x_1 &= (s_{11}^E + s_{12}^E)\sigma_1 + d_{31}E_3 = 0 \\ x_3 &= 2s_{13}^E\sigma_1 + d_{33}E_3 \end{aligned} \tag{3}$$

The upper line of Eqn. (3) allows the derivation of the in-plane stresses, i.e., an effective $e_{31,f} = -\sigma_1/E_3$. The lower line leads to the definition of an effective $d_{33,f} = x_3/E_3$:

$$\begin{aligned} e_{31,f} &= \frac{d_{31}}{s_{11}^E + s_{12}^E} \\ d_{33,f} &= d_{33} - \frac{2s_{13}^E}{s_{11}^E + s_{12}^E} d_{31} \end{aligned} \tag{4}$$

Identical expressions can be derived using the e coefficients. Both effective coefficients can be measured directly. It turns out that $d_{33,f}$ is always smaller than d_{33} and that the absolute value of $e_{31,f}$ is always larger than that of e_{31}. The input and output parameters for actuator and sensor applications of piezoelectric laminated plates is shown schematically in Fig. 2. In an actuator, application of a voltage leads to a piezoelectric in-plane stress causing a deflection of the structure, whereas the piezoelectric thin film is strained in the off-plane direction. In the sensor mode, in-plane strains create the piezoelectric charges that record the deformation of the flexible structure. In addition, the film is sensitive to off-plane stress (which also causes in-plane strain!). A cantilever structure, as in Fig. 1, can be modeled by analytical methods (Smits and Choi 1991, Meng *et al.* 1993). In the actuator mode a constant curvature is established at a given voltage. The deflection at the end of the beam is proportional to the square of the beam length. Excursions of 10–20 μm with 500 μm long beams have been obtained (Luginbuhl *et al.* 1996). More complicated structures must be analyzed by means of finite element calculations (Fabula *et al.* 1994).

The piezoelectric coefficients are not the only material parameters of interest. Besides force, deflection, and piezoelectric charge (current), other performance issues are important. In resonating structures the coupling coefficient (k^2) is essential. It describes the effectiveness of energy transformation from electrical to mechanical energy. Part of the electrical energy is dissipated and transformed to heat. The dielectric loss angle (tan δ) is involved in this phenomenon. The dielectric loss also generates a noise current or voltage, which imposes limits to the resolution, or more precisely, to the signal-to-noiseratio of sensors. It depends on application, electronics, and external noise levels as to whether the intrinsic noise owing to tan δ is limiting the performance or not. Table 1 summarizes the figures of merit for the various criteria. Note that the figure of merit for signal-to-noise ratio does not depend on whether voltage or current are detected.

2. *Materials*

Piezoelectric materials exploited or developed for MEMS are polar and exhibit either wurtzite or perovskite structure. The most often applied wurtzite materials are ZnO and AlN. Their polar axis is not invertible. In contrast, the perovskites include many ferroelectrics. The most often applied material of this class is $Pb(Zr,Ti)O_3$ (PZT) (Jaffe *et al.* 1971). In bulk ceramics, the strongest piezoelectric activity is achieved with a Zr/Ti ratio of 53/47, which is close to the morphotropic phase boundary between the titanium-rich tetragonal phase and the zirconium rhombohedral phase.

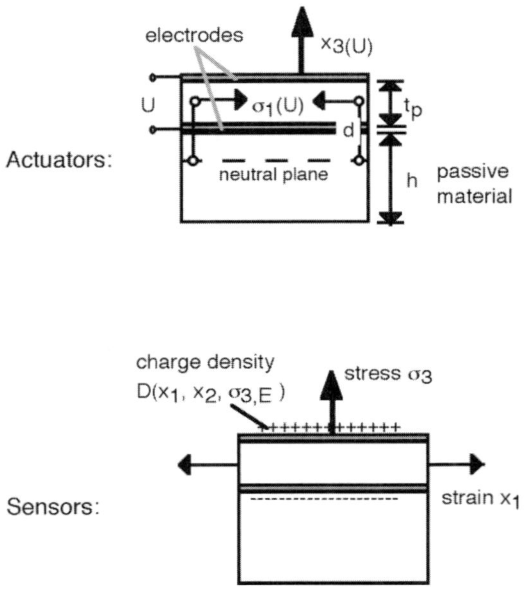

Figure 2
Schematic modes of operation of piezoelectric laminated substrate.

2.1 *Wurtzite Materials*

For MEMS applications, ZnO and AlN are commonly sputter deposited (Hickernell 1976, Rodriguez-Navarro *et al.* 1997). ZnO is preferentially deposited at room temperature in order to obtain a high resistivity. AlN does not crystallize as easily at room temperature. A temperature of 300°C is necessary to achieve good piezoelectric coefficients (Dubois and Muralt 1999). Essential for achieving good piezoelectric coefficients is the homogeneous nucleation of grains with (001) texture of the same polar direction. It appears that a more perfect (001) texture, as determined by x-ray diffraction, leads to a better selectivity between the (0,0,1) and (0,0, – 1) versions. The average polarization, i.e., the $d_{33,\mathrm{f}}$ coefficient, is found to be influenced by substrate as well as deposition conditions. AlN has two major advantages over ZnO. First, it is perfectly compatible with silicon semiconductor technology, whereas zinc is a fast diffuser in silicon. Second, AlN is a large-bandgap (6 eV) material with a large resistivity, whereas ZnO is rather a semiconductor (3.0 eV) with inherently efficient doping by zinc interstitials.

It is in fact difficult to obtain ZnO with a high resistivity. The d.c. conductivity translates into a high dielectric loss at low frequencies, which is especially harmful for sensors and actuators working at frequencies below about 10 kHz. The fact that more MEMS studies have been performed with ZnO than with AlN is explained by the difficulties encountered in reducing mechanical stresses in AlN thin films, and in growing them well textured on electrode films. The two materials are compared in Table 2 on the basis of measured thin-film properties. An accurate method

Table 1
Various figures of merit for materials parameters.

Physical parameter	Figure of merit
Deflection force, piezoelectric charge of deflected piezoelectric laminated structure	$e_{31,\mathrm{f}}$
Piezoelectric voltage in deflected piezoelectric laminated structure	$e_{31,\mathrm{f}}/\varepsilon_0\varepsilon_{33}$
Coupling coefficient for plate wave	$e_{31,\mathrm{f}}^2/\varepsilon_0\varepsilon_{33}$
Coupling coefficient for thickness wave	$e_{33}^2/(\varepsilon_0\varepsilon_{33}c^D{}_{33}) \approx d_{33,f}^2 \cdot c_{33}^E/\varepsilon_0\varepsilon_{33}$
Signal-to-noise ratio	$e_{31,\mathrm{f}}(\varepsilon_0\varepsilon_{33}\tan\delta)^{-1/2}$

Table 2
Typical numerical values for measured and derived materials parameters of thin films. PZT includes various compositions (Muralt 2000).

Coefficients/figures of merit	ZnO	AlN	PZT
$e_{31,f}$ (Cm^{-2})	−0.8	−1.0	−2 to −12
$d_{33,f}$ (pmV^{-1})	7.2	3.9	60–100
ε_{33}	10.9	10.5	300–1300
$e_{31,f}/\varepsilon_0\varepsilon_{33}$ (GVm^{-1})	−8.3	−10.5	−0.7 to −1.8
$e^2_{31,f}/\varepsilon_0\varepsilon_{33}$ (GPa)	6.6	10.5	6–18
tan δ (at 1–10 kHz, $10^5\,Vm^{-1}$)	0.01–0.1	0.003	0.01–0.03
$e_{31,f}/\sqrt{\varepsilon_0\varepsilon_{33}\tan\delta}$ ($10^5\,Pa^{1/2}$)	2–8	18	4–8
c^E_{33} (GPa) (PZT52/48 ceramic)	211	395	98
$d^2_{33,f}\cdot c^E_{33}/\varepsilon_0\varepsilon_{33}$ (%)	11.3	6.3	7–15

to determine $d_{33,f}$ consists of using a double-side laser interferometer, which measures the thickness change unambiguously and avoids errors owing to sample deflections (Kholkin *et al.* 1996). The $e_{31,f}$ coefficient is determined by means of deflecting or deflected structures. The various methods are listed in Shepard *et al.* (1998) and Muralt (2000). Table 2 indicates that AlN is better suited for deflection devices, while ZnO is better suited for bulk wave generation.

2.2 Perovskites

Owing to large permittivities and polarizations, the ferroelectric perovskite materials also exhibit large piezoelectric coefficients. This is especially the case for the strain-related coefficients, d_{ij}, which can be written as

$$D_{3i} = 2Q_{3i}\cdot\varepsilon_0\varepsilon_{33}\cdot P_3 \tag{5}$$

where Q_{3i} is an effective electrostrictive coefficient and P_3 is the spontaneous ferroelectric polarization. The spontaneous polarization is the order parameter of a para-to-ferroelectric phase transition. In the case of the PZT system, the film is grown in the paraelectric phase, which is cubic. When the film is cooled and undergoes the phase transition, several equivalent directions are acquired by the polar axis (6 in tetragonal PZT, 8 in rhombohedral PZT, including 180° flipping). The different solutions are called domains. They are separated by domain walls. The existence of domains complicates the interpretation of the piezoelectric and dielectric response because the observed strain may result in general from lattice distortions combined with domain wall displacements.

In Fig. 3, $d_{33,f}$ evaluated from the strain response to a small a.c. voltage is shown as a function of a superimposed d.c. electric field. The displayed $d_{33,f}$ is shown with a sign, which originates from the phase jump by 180° of the response with respect to the a.c. voltage when the polarization is switching. This type of display reflects the proportionality of polarization and piezoelectric coefficient, as given in Eqn. (5). Two examples with hysterisis loops are shown in Fig. 3: PZT45/55 (100) and PZT53/47 (111). The first has a slightly tetragonal structure and exhibits a larger coercive field than the second that is at the morphotropic phase boundary (MPB). A clear influence of the film texture is observed. (100) textured films show a larger piezoelectric activity. The 53/47 sample exhibits a decrease of $d_{33,f}$ at high fields. This originates from a decrease of the domain wall density when approaching a monodomain state at very high fields. Contributions of domain wall oscillations to the dielectric constant decline to zero. The decrease in

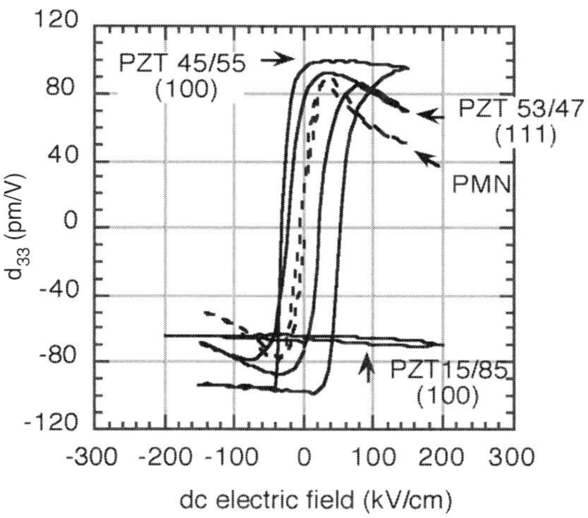

Figure 3
Small-signal piezoelectric $d_{33,f}$ coefficient as a function of a d.c. electric field for $PbZr_xTi_{1-x}O_3$ thin film of two compositions ($x/1-x$) and textures, and for a $PbMg_{1/3}Nb_{2/3}O_3$ thin film (data courtesy of David Taylor, Stephane Hiboux, Zian Kighelman).

dielectric constant, which is indeed observed, reduces $d_{33,f}$ according to Eqn. (5). Furthermore, direct contributions of domain wall oscillations (such as 90° domain wall motions in tetragonal PZT)—if exisiting—decline also to zero at high fields (for a review of these phenomena see Damjanovic (1998)). The effect of high titanium concentrations is illustrated with the sample PZT15/85 (100). The hysteresis is practically nonexistent. The coercive field is very large. Piezoelectric activity is obtained after poling at 150°C.

Films with larger coercive fields yield obviously more stable operating conditions. However, one has to pay with a lower piezoelectric coefficient, and the maximum response is obtained by hot poling only. It is too early to give an exhaustive overview of the figures of merit for the complete PZT solid solution system. The highest piezoelectric coefficients have been reported for compositions near the MPB (45/55 to 53/47). Outside this region, dielectric constant and dielectric loss are lower. Optimal compositions for coupling constant and signal-to-noise ratio are likely to be found in the tetragonal phase of the PZT system. For instance, the low dielectric constant of PZT15/85, combined with the relatively good $d_{33,f}$ make this composition a good candidate as bulk wave resonator material. One of the main differences to bulk ceramics are the high breakdown fields generally found in thin films. This allows one to drive thin films at much higher fields (of the order of $10^7\,\mathrm{Vm}^{-1}$). For stable piezoelectric coefficients, and in order to avoid switching, a d.c. field must be superimposed when operating with large a.c. fields.

The perovskite family also includes relaxor-type materials. They seem to be nearly ferroelectric, and are thus also called incipient ferroelectrics. A characteristic property of these materials is the large polarizability leading to large electrostrictive responses. The electrostrictive strain is, to a good approximation, proportional to the square of the induced polarization. Equation (5) still holds if the spontaneous polarization is replaced by the induced polarization. In Fig. 4, this behavior is represented by the sample of PMN ($PbMg_{1/3}Nb_{2/3}O_3$) (Kighelman *et al.* 1998). The response is zero at zero E-field, increases first linearly with the E-field and reaches saturation because of a saturated induced polarization. The decrease at high fields is because of a decreasing dielectric constant. Relaxor-type materials have the potential to yield higher strain values than PZT, and to avoid hysteresis effects. However, one has always to apply a large d.c. field in the actuator as well as detector mode.

Properties of PZT thin films are compared with those of AlN and ZnO in Table 2. The advantages of PZT are obviously the large output forces and powers, and the good charge response on strain variations. As to coupling coefficients, only the {100} textured PZT films have shown a better value than AlN so far. However, it is too early to state the final performance

Figure 4
TEM dark-field image of a 300 nm thick PZT thin film deposited by the sol–gel technique onto a Pt/TiO_x bottom electrode (oxidized silicon wafer substrate) (courtesy of Laurent Sagalowicz).

of perovskite materials, whereas the data known for ZnO and AlN are likely to be close to the best possible values. The main limitation for applications of perovskite thin films is probably the high processing temperatures. The integration of PZT on silicon devices demands special designs, special electrode materials, and special etching processes. The next section gives an overview on this topic.

3. Processing and Integration of PZT

3.1 Thin-film Deposition

PZT thin films can be grown at much lower temperatures than the typical sintering temperatures of bulk ceramics. This is owing to the much smaller diffusion distances needed with thin-film processing techniques, which provide for a homogeneous, stoichiometric mixture on the molecular level. Most of the existing thin-film deposition techniques have been investigated for the deposition of PZT: ion beam sputtering (Castellano and Feinstein 1979), magnetron sputtering (Okuyama *et al.* 1979, Krupanidhi *et al.* 1983), chemical solution deposition (CSD) by sol–gel reactions (Budd *et al.* 1985, Dey *et al.* 1988), metalorganic decomposition (Spierings *et al.* 1991), metalorganic chemical vapor deposition (MOCVD) (Kojima *et al.* 1983), and pulsed laser deposition (PLD) (Otsubo *et al.* 1990). CSD techniques necessarily need postannealing treatments for the crystallization of the film. All other methods also allow *in situ* growth. The main growth features for PZT are briefly reviewed below. More information can be found in Araujo *et al.* (1996) and Muralt (2000).

The main growth phenomena specific to PZT can be roughly understood in terms of a few key features of the $PbO–ZrO_2–TiO_2$ system, independently of the deposition method.

(i) Nucleation and growth of perovskite require a rather precise stoichiometry. Otherwise competing phases with fluorite and pyrochlore structures nucleate, which are more tolerant in oxygen and lead vacancies.

(ii) Lead ions or PbO molecules that are not incorporated into the perovskite lattice exhibit high diffusivities and volatility above 500 °C. The PbO vapor pressure above PbO is approximately 100 times larger than above PZT. Below 450 °C PbO may oxidize further to give Pb_3O_4 or PbO_2.

(iii) The activation energy for nucleation of perovskite is considerably larger than for its growth. The activation energies increase with increasing zirconium content.

The growth temperature is mainly dictated by the behavior of PbO and is in the range 550–700 °C. *At low temperatures* (<500 °C), the chemical methods are reaction limited. Only physical deposition methods may allow one to grow films below 500 °C. However, the crystallization is not any more tolerant to lead excess, and the oxygen content does not necessarily adjust to stoichiometry by oxygen exchange with the gas phase. Even worse, Pb^{2+} may change its valence to Pb^{4+} owing to further oxidation. Fluorite phases are often obtained. They are very tolerant with respect to composition. *At high temperatures* (>700 °C), the $PbTi_3O_7$ pyrochlore structure is common to all methods. The reason for its formation is heavy lead loss. For this reason, all processes usually use excess lead, in order to compensate for lead loss, which is especially severe before nucleation of the perovskite phase. With *in situ* deposition techniques, one observes a self-stabilization of the lead content at stoichiometry above a critical temperature, even for large quantities of excess lead flux.

Important for optimizing piezoelectric properties is the third key feature mentioned above, i.e., the fact that PZT growth is nucleation controlled (Chen and Mackenzie 1990). This allows for a microstructure consisting of columnar grains nucleated at the bottom electrode (see Fig. 4). Suitable electrode films serving at the same time as seeds enable the nucleation of special film textures (Muralt *et al.* 1998) and reduction of the deposition temperature (Kwok and Desu 1993). With *in situ* sputter deposition, (100) nucleation of $PbTiO_3$ works very well on Pt(111), whereas PZT(100) is more difficult to obtain. However, on a $PbTiO_3$(100) template PZT grows as well in (100) orientation. Selectivity of the seeding by platinum (111) is obtained by an additional 2 nm thick, hot-deposited TiO_2 seed layer, which yields the (111) orientation.

3.2 Integration

High-quality PZT films cannot be grown directly on silicon (Revesz *et al.* 1991). Buffer layers are needed to prevent interdiffusion and oxidation reactions. For most applications, the PZT film has to be grown on an electrode, which obviously should neither oxidize nor become insulating. The most often reported materials include platinum (Al-Shareef *et al.* 1993), and the metal oxides RuO_2 and IrO_2 (rutile structure) (Vijay and Desu 1993), and $SrRuO_3$ and (La,Sr)-CoO_3 (LSCO, perovskite structure) (Ramesh *et al.* 1993). Usually, the chemical barrier function is provided by two or more layers, including the electrode. PZT/Pt/Ti/SiO_2/Si is the most widely applied sequence, where the titanium is needed as an adhesion layer (see Fig. 4). The SiO_2 layer prevents interdiffusion of PZT ions and silicon. This layer may be quite thin. For growth on polysilicon mechanical elements, annealing in O_2 at 950 °C (Kim *et al.* 1998) yields a sufficient oxide thickness. Platinum does not inhibit the diffusion of titanium into PZT, where it reacts with oxygen and serves as nucleation centers for PZT (Willems *et al.* 1997). Pockets of TiO_2 have also been observed between the platinum grains (Maeder *et al.* 1998). There is ample evidence that oxygen migrates along the grain boundaries through the platinum film and reacts with the titanium layer (Sreenivas *et al.* 1994). For stable electrodes the latter has to be preoxidized. A study with titanium, zirconium, and tantalum adhesion layers shows that the most diffusing species is always the one in the corresponding oxide, i.e., titanium in TiO_2, and oxygen in ZrO_2 and Ta_2O_5 (Maeder *et al.* 1998).

Despite this diffusion risk, good-quality films with clean interfaces can be obtained (see Fig. 4). Barrier schemes with insulating films, such as SiO_2 and Si_3N_4, cannot be applied when a direct electrical contact to a silicon or metal substrate is required. In this case, RuO_2 or other oxide electrodes are superior to platinum electrodes. In combination with a second metal (such as chromium, TiN, TiAlN, TaSiN, etc.) that stops the residual oxygen diffusion, RuO_2 turns out to be a good barrier against lead diffusion. RuO_2 allows the growth of PZT on very reactive refractory metals, such as zirconium thin films (Maeder *et al.* 1996). One of the most difficult issues in PZT integration is the film patterning. Platinum and RuO_2 are difficult to etch. No wet etchants are applicable, and only dry etching is possible. Etching of RuO_2 poses somewhat less problems and is thus proposed to replace platinum for this reason (Yoon *et al.* 1998). However, it is commonly observed that the piezoelectric properties of PZT grown on RuO_2 are less good than of PZT grown on platinum.

3.3 Fabrication Example: Ultrasonic Stator for a Micromotor

To give an idea of the process flow the following example is given here. The goal is to fabricate an ultrasonic stator in form of a PZT laminated, thin, clamped circular disk. The first mode with a circular node (B_{10}) has to be excited as efficiently as possible. This is best done by means of an annular electrode. The membrane of the stator is obtained by silicon bulk micromachining in hot KOH (see Fig. 5). A stress compensated bilayer of thermal oxide (SiO_2)

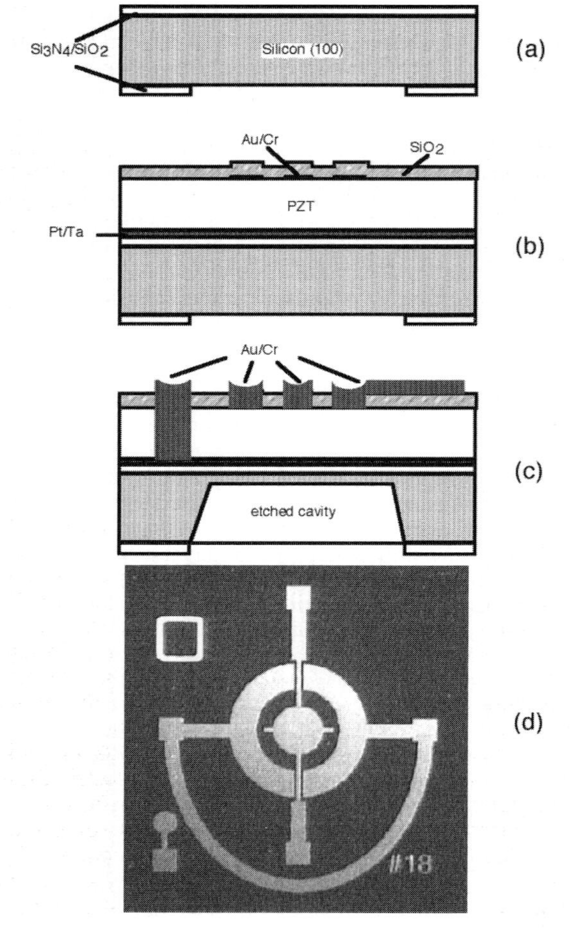

Figure 5
Schematic structure and major fabrication steps of an ultrasonic stator: (a) masking and barrier layers, (b) before opening vias through low permittivity layer to access top electrodes, (c) final cross-section, (d) top view of electrode structure (after Dubois and Muralt 1998).

and LPCVD nitride (Si_3N_4) is applied as masking layer. The first lithographic step consists of opening this masking layer by dry etching (CF_4) for the definition of the membranes. Pt/Ta bottom electrode and PZT deposition follow as the next steps. A sacrificial Au/Cr top electrode pattern serving as etch stop and PZT protection coating is deposited and patterned. A silicon dioxide thin film is then deposited by sputtering. This film reduces the parasitic capacities of conductor lines and contact pads. The silicon dioxide is patterned with dry etching (CF_4) to liberate the top electrode contacts. The sacrificial Au/Cr contacts are needed in this step to avoid deterioration of the PZT surface, i.e., the formation of a low-dielectric fluorinated film. This sacrificial layer is removed and the complete top electrode system, including conductors and pads, is deposited.

4. Devices

Quite a number of thin-film piezoelectric devices have been demonstrated. Usually there is first a version with ZnO, and later a version with PZT. ZnO thin films have been developed for acoustic radio-frequency (RF) applications as, for example, flexural plate wave (or Lamb wave) devices (Uozumi *et al.* 1983), before being applied in MEMS devices (Polla *et al.* 1989, Moroney *et al.* 1991, Mullem *et al.* 1991). The first PZT thin-film devices, still suffering from integration problems, were realized in the early 1990s (Flynn *et al.* 1992, Polla and Francis 1998). The first PZT Lamb wave devices were reported in 1997 (Luginbühl *et al.* 1998), 15 years after the ZnO version. PZT-coated membranes and diaphragms have been applied in peristaltic pumps (Polla and Francis 1998), ultrasonic micromotors (Muralt *et al.* 1995), microphones (Kirby *et al.* 1998), Lamb wave devices for microfluidics (Lugienbühl *et al.* 1998), and transducer arrays for medical applications (Bernstein *et al.* 1997). Cantilever structures have been developed for atomic force microscopy (AFM) (Lee *et al.* 1996, 1999), for accelerometers (Kirby *et al.* 1998), for linear actuators (Polla and Francis 1998), and for micromirror arrays (Hwang *et al.* 1999). PZT grown by a hydrothermal method onto titanium tubes has been used to drive an ultrasonic micromotor (Morita *et al.* 1998). All of these applications are based on the piezoelectric effect in the film plane ($e_{31,f}$). The effect perpendicular to the film plane is applied or proposed much more scarcely. Here, the amplification by laterally extended structures does not take place. Only applications at high frequencies ($>100\,kHz$) are conceivable. These are bulk wave resonators for wireless communications (Maeda *et al.* 1998, Dubois and Muralt 1999), detectors for acoustic emission testing devices (Polla and Francis 1998), and vibration sensors for the ultrasound range (Nemirovsky *et al.* 1996).

The device demonstrators allow a good overall estimation of the performance of piezoelectric thin films. PZT thin-film actuators are outstanding with respect to force and output power at low voltages. Micromotors with comparatively high torques at low speeds have been demonstrated. Torques of almost $1\,\mu Nm$ at $4\,V_{rms}$, or $0.3\,\mu Nm\,V_{rms}^{-1}$, have been achieved (Dubois and Muralt 1998) with millimeter-sized rotors and PZT thin-film stators. This is sufficient for wristwatch applications and much more per voltage than has been demonstrated with electrostatic micromotors (scaled to the same motor size). The low speeds (typically 5 rpm at $2\,V_{rms}$) of ultrasonic micromotors obviate the need for gearboxes, which are very difficult to manufacture in submillimeter dimensions. Piezoelectric thin films exhibit larger breakdown fields than their ceramic counterparts. It is thus possible to operate them at relatively high fields of 100–400 kVcm^{-1}. Such fields yield high energy densities, especially in PZT or related materials. Exhibiting also good coupling coefficients, high mechanical energy densities can be released. This is, for instance, important in flexural plate wave (Lamb wave) devices that can be applied for pumping, filtering, and chemical sensors. Less than 1 μm thick membranes coated with a bottom electrode and a piezoelectric thin film are excited by means of interdigital electrodes placed on top of the piezoelectric film. Asymmetric waves of comparably low sound velocities can be created in this way (see Fig. 6). In contrast to other surface waves (such as Reyleigh- and Love-type waves), which exhibit too high a sound velocity, asymmetric Lamb waves can be used to drag liquids. Another strong point of piezoelectric devices is the inherent transducer property. This property has been applied for medical imaging for many years. It turns out to be useful also in AFM. PZT-coated cantilevers can be used to vibrate the tip, and to detect at the same time changes of the resonance frequency by means of impedance changes. This allows mapping of the forces acting between tip and surface, or gives information on elastic properties of the sample (Lee *et al.* 1999).

In ultrasonic actuation devices, power dissipation is an issue, in spite of the fact that it is small (a few percent) with respect to the reactive power engaged. However, the generated mechanical power in the ultrasonic stator is of the same order of magnitude as the dissipated dielectric loss power, and it still needs to be transformed into useful power at the output (rotor) of the device. This transformation may suffer a lot from frictional loss. For the micromotor mentioned above, the efficiency of this transfer is only about 10%. Such friction problems are in fact a general issue in micromechanical devices and do not exclude ultrasonic actuation.

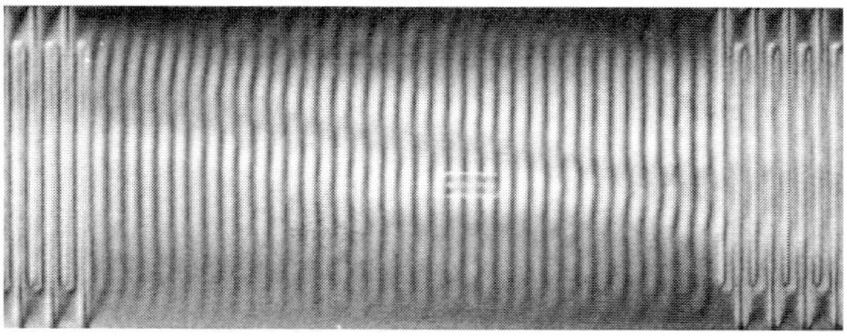

Figure 6
Micrograph of standing Lamb waves. A 75 μm glass bead jumps forward and backwards between two nodal lines (courtesy of Ph. Lugienbühl).

5. *Conclusions and Future Directions*

The art of integrating piezoelectric thin films in MEMS devices is fairly well developed. Good piezoelectric coefficients have been obtained. In the wurtzite materials ZnO and AlN they are close to the theoretical values, in PZT they are in the range 60–80% of values derived from bulk ceramics. In the latter case it is mainly argued that in thin films the domain walls do not contribute as much as in bulk materials. For the future, perovskite thin films have to be tailored or chosen for the requirements of a given application. Ultrasound applications require a material with a high coercive field, a high coupling coefficient, and a low dielectric loss. Linear actuators demand $e_{31,f}$ coefficients as high as possible. In many applications, high breakdown fields are useful. A larger variety of PZT compositions and related materials than in bulk materials will be useful, because thin films can be poled much more easily. Relaxor-type materials avoid problems related to the ferroelectric hysteresis and have the potential to achieve larger strains. The developments in single-crystal piezoelectrics may possibly lead to better thin-film materials as well. However, piezoelectric material properties cannot be considered alone. For MEMS applications, properties have to be evaluated in view of the possibilities in design and processing. For instance, the increase of coupling coefficient or the decrease of frictional losses are very much questions of design and other materials in the device.

Forecasts in specific device applications are not easy to make. The strong points of piezoelectric thin-film devices have been outlined above. The advent of ferroelectric memories will certainly pass some momentum to MEMS activities in order to facilitate the use of PZT. However, for all applications there are competing principles to piezoelectricity. Strains and deflections can also be measured by piezoresistive materials, such as doped polycrystalline silicon, by magnetostrictive materials, or by capacitance changes across an air or vacuum gap. Actuation can be performed also by electrostatic or magnetic forces, by thermal deformation (bi-metal effect), by magnetostriction, or by shape memory alloys. The main criteria for choosing one or the other principle are requirements with regard to strains and forces, frequency domain, stability, linearity, signal-to-noise ratio, power consumption, compatibility with application, and costs. Acoustic applications at high frequencies constitute probably a safe ground for piezoelectric thin films. This will include acoustic imaging at high frequencies, flexural plate wave devices, bulk wave resonators in the microwave range, surface acoustic wave devices based on diamond-coated silicon, and transducer cantilevers for scanning probe surface imaging. Besides high-frequency applications, it is the simplicity and the low power consumption that makes nonresonant piezoelectric cantilever drives very attractive for actuators in low-power, miniaturized robots.

Bibliography

Al-Shareef H N, Gifford K D, Rou S H, Hren P D, Auciello O, Kingon A I 1993 Electrodes for ferroelectric thin films. *Integrated Ferroelectrics* **3**, 321–32

Araujo C P D, Scott J F, and Taylor G W (eds.) 1996 Ferroelectric thin films: synthesis and basic properties. *Ferroelectricity and Related Phenomena*. Gordon and Breach, Reading, UK

Benes E, Gröschl M, Burger W, Schmid M 1995 Sensors based on piezoelectric resonators. *Sensors Actuators A* **48**, 1–21

Bernstein J J, Finberg S L, Houston K, Niles L C, Chen H D, Cross L E, Li K K, Udayakumar K 1997 Micromachined high frequency ferroelectric sonar transducers. *IEEE Trans. UFFC* **44**, 960–9

Budd K D, Dey S K, Payne D A 1985 Br. Ceram. Proc. *Sol–gel processing of PT, PZ, PZT and PLZT thin films* **36**, 107

Castellano R N, Feinstein L G 1979 Ion-beam deposition of thin films of ferroelectric PZT. *J. Appl. Phys* **50**, 4406–11

Chen K C, Mackenzie J D 1990 *Mater. Res. Symp. Proc* **180**, 663–8

Damjanovic D 1998 Ferroelectric, dielectric and piezoelectric properties of ferroelectric thin films and ceramics. *Rep. Prog. Phys.* **61**, 1267–324

Dey S K, Budd K D, Payne D A 1988 Thin film ferroelectrics of PZT by sol–gel processing. *IEEE Trans. UFFC* **35**, 80–1

Dubois M-A, Muralt P 1998 PZT thin film actuated elastic fin micromotor. *IEEE Trans. UFFC* **45**, 1169–77

Dubois M-A, Muralt P 1999 Properties of AlN thin films for piezoelectric transducers and microwave filter applications. *Appl. Phys. Lett.* **74**, 3032–4

Fabula T, Wagner H-J, Schmitt B, Büttgenbach S 1994 Triple beam resonant silicon force sensor based on piezoelectric thin films. *Sensors Actuators A* **41–42**, 375–80

Flynn A M, Tavrow L S, Bart S F, Brooks R A, Ehrlich D J, Udayakumar K R, Cross L E 1992 Piezoelectric micromotors for microrobots. *J. Microelectromechanical Syst.* **1**, 44–51

Hickernell F S 1976 ZnO thin film surface wave transducer. *Proc. IEEE* **64**, 631–5

Hwang K H, Song Y J, Kim S G 1999 Thin film micromirror array for high-brightness projections displays. *Jpn. J. Appl. Phys. Part 1* **37**, 7074–7

Ikeda T 1990 *Fundamentals of Piezoelectricity*. Oxford University Press, Oxford, UK

Jaffe B, Cook W R, Jaffe H 1971 *Piezoelectric Ceramics*. Academic Press, London

Kholkin A L, Wütchrich C, Taylor D V, Setter N 1996 Interferometric measurements of electric field induced displacements in piezoelectric thin films. *Rev. Sci. Instrum.* **67**, 1935–41

Kighelman Z, Damjanovic D, Seifert A, Sagalowicz L, Setler N 1998 Relaxor behavior and electromechanical properties of $Pb(Mg,Nb)O_3$ thin films. *Appl. Phys. Lett.* **73**, 2281–3

Kim J H, Yoon Y S, Polla D L 1998 Characteristics of PZT deposited by MOD on $Pt/Ti/TiO_2/Poly\text{-}Si/Si_3N_4/Si$ for microelectromechanical system devices. *Japn. J. Appl. Phys. Part 1* **37**, 948–50

Kirby P B, Wright R V, Gaucher P, Galtier P, Kojored L, Cullov J O, Von Munch W, Eichner D, Ploss B, Kruger J K 1998 Report on Brite-EuRam project sensors and mechatronic devices using ferroelectric thin films SEMDEFT. *J. Phys. IV* **8**, 161–5

Kojima M, Okuyama M, Nakagawa T, Hamakawa Y 1983 Chemical vapor deposition of $PbTiO_3$ films. *Jpn. J. Appl. Phys.* **22**, 14–7

Krupanidhi S B, Maffei N, Sayer M, El-Assal K 1983 Rf planar magnetron sputtering and characterization of ferroelectric PZT films. *J. Appl. Phys.* **54**, 6601–9

Kwok C K, Desu S B 1993 Low temperature perovskite formation of PZT thin films by a seeding process. *J. Mater. Res.* **8**, 339–44

Lee C K, Itoh T, Suga T 1996 Micromachined piezoelectric force sensors based on PZT thin films. *IEEE Trans. UFFC* **43**, 553–9

Lee C, Itoh T, Suga T 1999 Self-excited piezoelectric PZT microcantilevers for dynamic SFM—with inherent sensing and actuating capabilites. *Sensors Actuators A* **72**, 179–88

Luginbühl P, Collins S D, Racine G A, Gretillat M A, De Rooij N F, Brooks K G, Setler N 1998 Ultrasonic flexural Lamb-wave actuators based on PZT thin film. *Sensors Actuators A* **64**, 41–9

Luginbuhl P, Racine G-A, Lerch P, Romanowicz B, Brooks K G, De Rooij N F, Renaud Ph, Setler N 1996 Piezoelectric cantilever beams actuated by PZT sol–gel film. *Sensors Actuators A* **53**, 530–5

Madou M 1997 *Fundamentals of Microfabrication*. CRC Press, Boca Raton, FL

Maeda C, Yamada A, Uchikawa F, Misa K, Wadaka S, Shinada J 1998 Microwave properties of La-doped lead titanate piezoelectric thin films for bulk acoustic wave resonators. *Proc. IEEE Ultrasonics Symp.* IEEE, New York, pp. 629–32

Maeder T, Muralt P, Sagalovicz L, Reaney I, Kohli M, Kholkin A, Setler N 1996 PZT thin films on Zr membranes for micromechanical applications. *Appl. Phys. Lett.* **68**, 776–8

Maeder T, Sagalowicz L, Muralt P 1998 Stabilized platinum electrodes for PZT thin film deposition using Ti, Zr, and Ta adhesion layers. *Jpn. J. Appl. Phys.* **37**, 2007–12

Meng Q, Mehregany M, Deng K 1993 Modeling of the electromechanical performance of piezoelectric laminated microactuators. *J. Micromech. Microeng.* **3**, 18–23

Morita T, Kurosawa M K, Higuchi T 1998 A cylindrical micro-ultrasonic motor using PZT thin film deposited by hydrothermal method. *IEEE Trans. UFFC* **45**, 1178–87

Moroney R M, White R M, Howe R T 1991 Microtransport induced by ultrasonic Lamb waves. *Appl. Phys. Lett.* **59**, 774–6

Moulson A J, Herbert J M 1990 *Electroceramics*. Chapman & Hall, London

Mullem C J V, Blom F R, Fluitman J H J, Elwenspock M 1991 Piezoelectrically driven silicon beam force sensor. *Sensors Actuators A* **25–27**, 379–83

Muralt P 2000 PZT thin films for microsensors and actuators: Where do we stand? IEEE Trans. UFFC 47 903–15

Muralt P, Kohli M, Maeder T, Kholkin A, Brooks K G, Setler N, Luthier R 1995 Fabrication and characterization of PZT thin-film vibrators for micromotors. *Sensors Actuators A* **48**, 157–65

Muralt P, Maeder T, Sagalowicz L, Hiboux S, Scalese S, Namovic D, Agostino R G, Xanthopoulos N, Mathieu H J, Patthey L, Bullock E L 1998 Texture control of $PbTiO_3$ and PZT thin films with TiO_2 seeding. *J. Appl. Phys.* **83**, 3835–41

Nemirovsky Y, Nemirovsky A, Muralt P, Setler N 1996 Design of a novel thin film piezoelectric accelerometer. *Sensors Actuators A* **56**, 239–49

Nye J F 1985 *Physical Properties of Crystals*. Oxford University Press, Oxford, UK

Okuyama M, Matsui Y, Nakano H, Nakagawa T, Hamakawa Y 1979 Preparation of $PbTiO_3$ ferroelectric thin film by RF sputtering. *Jpn. J. Appl. Phys.* **18, 15**, 1633–4

Otsubo S, Maeda T, Minamikawa T, Yonezawa Y, Morimoto A, Shimizu T 1990 Preparation of PZT films by laser ablation. *Jpn. J. Appl. Phys* **29**, L133–6

Polla D L, Francis L F 1998 Processing and characterization of piezoelectric materials and integration into microelectromechanical systems. *Annu. Rev. Mater. Sci.* **28**, 563–97

Polla D L, Yoon H, Tamagawa T, Voros K 1989 Integration of surface-micromachined zinc oxide sensors in n-well CMOS technology. *Proc. Int. Electron Devices Meeting, IEEE Tech Digest*. IEEE, New York, pp. 495–8

Ramesh R, Gilchrist H, Sands T, Keramidas V G, Haakenaasen R, Fox D K 1993 Ferroelectric LSCO/PZT/LSCO heterostructures on silicon via template growth. *Appl. Phys. Lett.* **63**, 3592–4

Revesz P, Li J, Szabo N, Mayer J W, Caudillo D, Myers E R 1991 PZT interaction with metals and oxides studied by RBS. *Mat. Res. Symp. Proc.* **243**, 101–6

Rodriguez-Navarro A, Otano-Rivera W, Garcia-Ruiz J M, Messier R 1997 Development of preferred orientation in polycrystalline AlN thin films deposited by r.f. sputtering system at low temperatures. *J. Mater. Res.* **12**, 1850–5

Shepard J F, Moses P J, Trolier-McKinstry S 1998 The wafer flexure technique for the determination of the transverse piezoelectric coefficient (d_{31}) of PZT thin films. *Sensors Actuators A* **71**, 133–8

Smits J G, Choi W-S 1991 The constituent equations of piezoelectric heterogeneous bimorphs. *IEEE Trans. UFFC* **38**, 256–70

Spierings G A C M, Ulenaers M J E, Kampschöer G L M, Van Hal H A M, Larsen P K 1991 Preparation and ferroelectric properties of PZT thin films by spin coating and metalorganic decomposition. *J. Appl. Phys.* **70**, 2290–8

Sreenivas K, Reaney I, Maeder T, Setler N, Jagadish C, Elliman R G 1994 Investigation of Pt/Ti bilayer metallization on silicon for ferroelectric thin film integration. *J. Appl. Phys.* **75**, 232–9

Uozumi K, Ohsone K, White R M 1983 Generation and detection of ultrasonic Lamb waves in a thin deposited film using interdigital transducers. *Appl. Phys. Lett.* **43**, 917–9

Vijay D P, Desu S B 1993 Electrodes for PZT ferroelectric thin films. *J. Electrochem. Soc.* **140**, 2640–5

Willems G J, Wouters D J, Maes H E, Nouven R 1997 Nucleation and orientation of sol–gel PZT thin films on Pt electrodes. *Integrated Ferroelectrics* **15**, 19–28

Yoon Y S, Kim J H, Schmidt A M, Polla D L, Wang Q, Gladfelter W L 1998 RuO_2/Ru electrode on Si_3N_4/Si substrate for microelectromechanical systems devices based on PZT film and surface micromachining. *J. Mater. Sci.: Mater. Electron* **9**, 465–71

Polymer-based Electrophosphorescent Devices

M. R. Robinson and G. C. Bazan

The report of a multilayer organic-based light emitting diode (OLED) (Tang and VanSlyke 1987) stimulated the emergence of a new class of materials science and engineering focused on organic semiconductors (Hadziioannou and van Hutten 2000). Further activity was motivated by the demonstration of electroluminescence from a conjugated polymer (Burroughes *et al.* 1990), which contributed substantially to the development of plastic electronics. Research efforts in both academia and industry have resulted in commercially viable polymer light emitting diode (PLED) displays.

PLEDs offer potential key advantages over traditional inorganic LEDs, such as silicon and III–IV semiconductors. For example, PLEDs can be fabricated cheaply by spin casting or printing the electroluminescent polymer from solution. These techniques also allow for easy fabrication of large area displays. Additionally, because the light emitting material is a polymer, the attractive possibility of flexible displays exists—plastic electronics. Indeed, cellular phone and small appliance displays are currently being replaced with OLEDs. Possible future applications include disposable displays and panel lighting.

1. *Organic Electroluminescence: An Overview*

The chemical structure of typical polymers used in PLEDs and a detailed description of PLED fabrication and operation can be found in other articles (see also *Light Emitting Diodes from Molecular and Polymer Semiconductors*).

1.1 Mechanism of Action

A schematic diagram of a working OLED is shown in Fig. 1. The emitting material is oxidized at the anode and reduced at the cathode. The oxidized and reduced molecules correspond to the location of holes and electrons within the film, respectively. Due to the electric field across the film, charges migrate from molecule to molecule in the film, towards the oppositely charged electrode. When a hole and an electron converge, they form an exciton (excited state). It is widely accepted that this exciton is identical to that formed upon photoexcitation. The exciton may radiatively relax to the ground state or follow nonradiative processes, just as in photoluminescence.

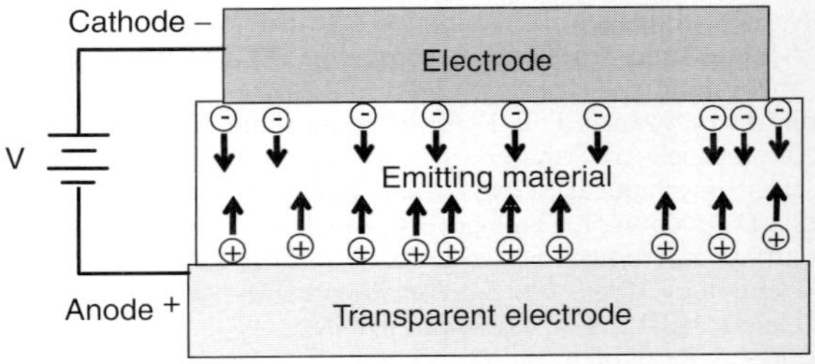

Figure 1
A schematic diagram of a working OLED.

1.2 Polymers and Small Molecules

From a practical perspective, OLEDs made from small molecules differ from those made from conjugated polymers in the morphology of the bulk material and the deposition methods. Although small molecules designed to resist crystallization and to yield amorphous films exist, the majority of them have a tendency towards crystallization, which is detrimental to OLED performance. Conjugated polymers that are soluble in organic solvents form transparent, amorphous films easily by spin casting techniques see Fig. 2. Other deposition methods exist, for example by ink jet printing. In contrast, small molecules suitable for OLEDs do not form good quality films from solution and require vapor deposition (sublimation). Vapor deposition offers more precise control of film thickness than spin casting, however spin casting is a simpler technique, affording cheaper processing options.

Another difference between small molecules and conjugated polymers involves purity. Whereas small molecules can be precisely analyzed and purified by chromatography, crystallization, and sublimation, conjugated polymers have a statistical distribution of molecular weights and are more difficult to purify. Conjugated oligomers of intermediate dimensions with shapes designed to resist crystallization have been developed which combine the benefits of both classes of compounds (Robinson *et al.* 2001).

1.3 Spin Statistics of Charge Recombination

In the vast majority of examples, organic molecules emit light only via a transition from the first singlet excited state to the ground state. Likewise, absorption of light only gives rise to the singlet excited state (Turro 1991). During electroluminescence, taking into account strictly spin statistics, charge recombination yields excited states which have a 75% probability of being triplets (see *Light Emitting Diodes from Molecular and Polymer Semiconductors*). If an organic molecule containing typical light atoms (e.g. C, H, N, O, etc.) finds itself in an excited state of triplet multiplicity, radiative transition is spin forbidden and the molecule will return to the ground state nonradiatively. Thus, using this analysis, in organic molecules, 3/4 of the injected current is

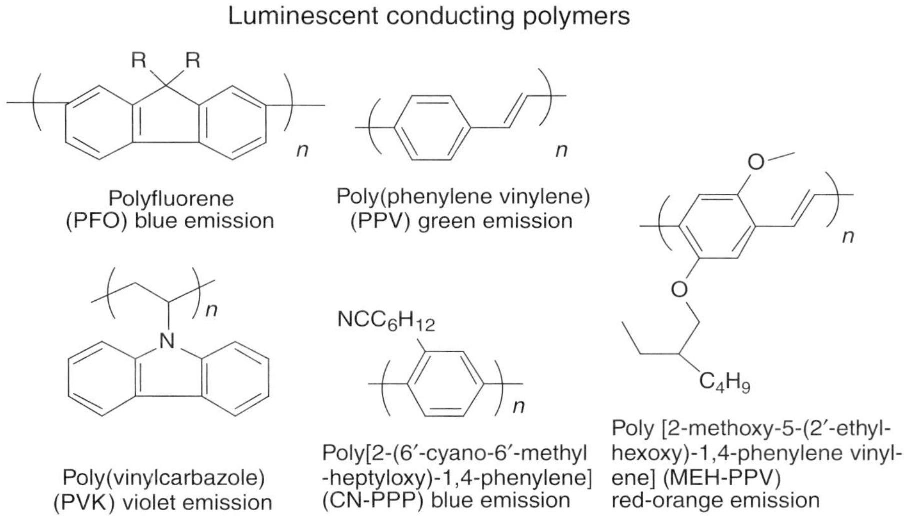

Figure 2
Chemical structures and abbreviated names of the semiconducting polymers mentioned in this article. For more examples (see *Light Emitting Diodes from Molecular and Polymer Semiconductors*) or (Hadziioannou and van Hutten 2000).

wasted by forming non-emissive triplet excitons. Note that the singlet to triplet formation ratio obtained in PLEDs is typically higher than 1:3 (Cao *et al.* 1999, Köhler *et al.* 2002). The mechanism for higher than expected ratios is under debate (Saricifcti 1997). For current work, see Moses *et al.* (2001) and other articles.

2. *Phosphorescent Dopants in Polymer LEDs*

A useful figure of merit for an OLED is the electroluminescence quantum efficiency; defined as the number of emitted photons per injected electron. The formation of non emissive triplets by charge recombination presents a strict efficiency limit that would be desirable to overcome. To make use of the triplet excitons formed in OLEDs one can blend phosphorescent molecules (guest) into the bulk emitting material (host). This host/guest or host/dopant approach retains the injection, mobility, and mechanical properties of the host, while allowing the dopant to exhibit high quantum yield phosphorescence.

With small molecules, phosphorescent dopants are co-sublimed with the host material. The dopant molecules used in conjunction with conjugated polymers are blended in low concentrations (less than 5% by weight) into the polymer solution and are spin cast together with the polymer. This article will focus on phosphorescent guests in conjugated polymers.

Organic phosphorescent dyes, such as benzophenone, have been used as dopants (Hoshino and Suzuki 1996). However they are usually not well suited as the emitting species because their triplet lifetimes can be quite long (>1 ms). Long-lived triplet states lead to emission bottlenecks, because in LEDs made from these materials, charge injection occurs at a faster rate than emission. Metal complexes are better suited as "triplet harvesters" because their triplet lifetimes are considerably shorter than those of organic molecules. Spin-orbit coupling of the electrons in heavier atoms (such as transition metals) facilitates emission from the triplet manifold (Fig. 3). Amorphous metal complexes (no polymer host) can also be employed directly as the emitting layer (Robinson *et al.* 2000, Ostrowski *et al.* 2002). Two distinct types of metal complexes are employed, each with characteristic attributes and drawbacks. These are transition metal based organometallic complexes, and lanthanide (commonly referred to as rare earth) complexes.

3. *Physical Properties and Mechanistic Considerations: Choice of Polymer and Dopant*

An important consideration in determining the host polymer and the guest dopant concerns the energy levels of the host and dopant. Specifically, one must ensure an efficient mechanism of transfer (rate of transfer) and that the energy transfer is thermodynamically feasible. In other words, the spectral overlap between the host and dopant needs to be fine-tuned for optical performance (discussed in detail in the next section). Fine tuning the emission color of

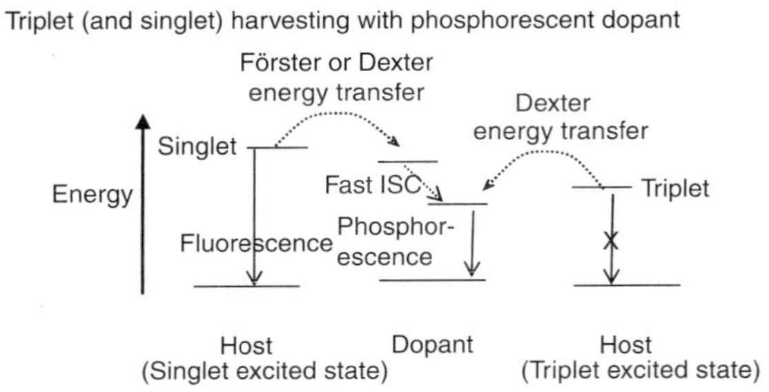

Figure 3
Schematic energy diagram depicting energy transfer from host to phosphorescent dopant (guest). Transfer from the host on the right exemplifies "triplet harvesting."

conjugated polymers can be accomplished by changes in molecular structure. For example PFO is a blue emitter, while MEH-PPV emits red (Fig. 2).

3.1 Mechanisms Leading to Dopant Emission

In the photoluminescence of films with conjugated polymer and phosphorescent dopants, energy transfer quenches the host emission and generates dopant excited states. Two types of energy transfer mechanisms are possible: Förster, which is mediated by dipole–dipole interactions and Dexter, which involves electron exchange via orbital overlap (Fig. 3) (Turro 1991). The distinction is not important for this discussion because the ultimate result is the same; the polymer exciton is transferred to the dopant. An important requirement for efficient energy transfer is that the absorption spectrum of the dopant overlaps the emission spectrum of the host. Förster energy transfer increases efficiency with increasing spectral overlap. For the Dexter mechanism, one needs to establish overlap, but there is little increase in transfer efficiency with increasing overlap.

For energy transfer to be efficient, the distance between the dopant molecule and the nearest host molecule must be sufficiently small. The rate of Förster energy transfer decreases proportionally to the host guest separation (R) to the sixth power (R^6). Dexter energy transfer occurs only at very close distances and the rate decreases exponentially with distance. From a practical standpoint, the average distance between the dopant and polymer is directly related to the doping concentration of the dopant, assuming a homogeneous mixture (McGehee *et al.* 1999).

There are two possible mechanisms in the electroluminescence of dopant/conjugated polymer films. One can envision hole and electron recombination on the polymer to form a polymer-based exciton. At this point, the system evolves as if the exciton was formed optically and energy transfer occurs as described for photoluminescence. The second mechanism involves charge transfer (electron and/or hole) from the polymer to the guest—also known as charge trapping. If this mechanism is dominant, holes and electrons will recombine on the dopant, rather than on the polymer and the emission will come solely from the dopant at concentrations much lower than required for complete photoluminescence quenching (Gong *et al.* 2002).

The contribution of the two mechanisms, exciton transfer from polymer to dopant and charge trapping, depends on the polymer and dopant. For example, energy transfer was determined to be the dominant mechanism leading to dopant emission in the electroluminescence of 2,3,7,8,12,13,17,18-octaethyl-21H,23H-porphyrin platinum(II) (PtOEP in Fig. 4) doped into poly[2,5-bis(cholestanoxy)-1,4-phenylene vinylene] (BCHA-PPV in Fig. 2) (Campbell *et al.* 2002). Changing the host polymer to polyfluorene switches the mechanism to charge trapping (Lane *et al.* 2001).

3.2 Morphology

The topic of conjugated polymer morphology is important since it influences the optical and electronic performance of these materials in the solid state. The emphasis here will be on the interaction of the dopant molecules with the host molecules.

The tendency towards phase separation is prevalent in host/guest systems. Phase separation leads to aggregation of the dopant molecules, which gives rise to self quenching by mechanisms such as excimer formation and/or triplet–triplet annihilation (Hertel *et al.* 2001). Furthermore, aggregation causes the average host/guest separation to vary unpredictably with wt.%, often causing incomplete energy transfer, even at high doping concentrations. Phase separation is often minimized by introducing side groups within the molecular framework of the dopant which are miscible within the polymer matrix.

3.3 Triplet Excitons in the Polymer

To justify using a triplet harvesting phosphorescent dopant, it is important to know the triplet energy level of the host conjugated polymer. If the triplet energy of the polymer lies near or below that of the phosphorescent dopant, then non-radiative relaxation processes from the polymer triplet state will be significant. Under these circumstances a reduction in efficiency would result when singlet excitons from the polymer are converted to triplet excitons on the dopant and subsequently transferred to the non-radiative triplet state of the polymer. Additionally, it is useful to know the ratio of singlet excitons to triplet excitons formed during electroluminescence (Köhler *et al.* 2002). If triplet formation is low, the use of

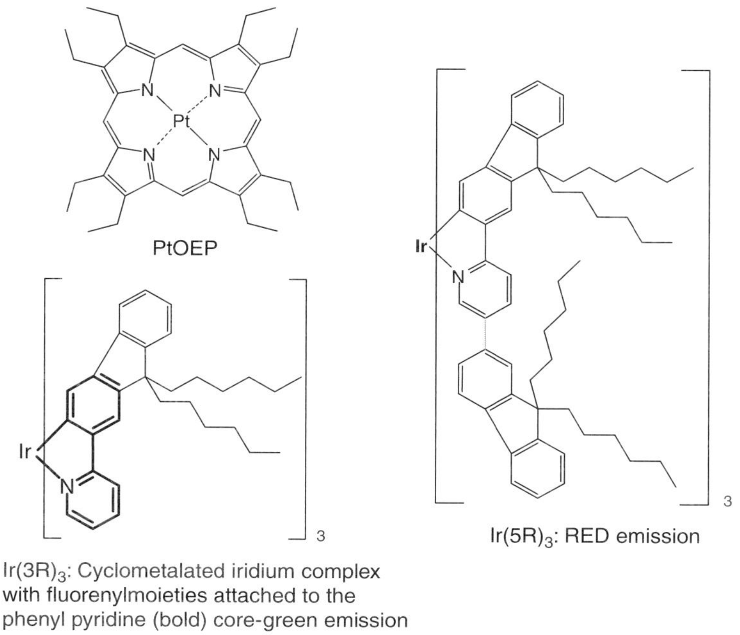

Figure 4
Chemical structures of some phosphorescent dopants. Extending the conjugation of Ir$(3R)_3$ by the addition of a fluorenyl unit to Ir$(5R)_3$ results in a 50 nm red shift in absorption and emission.

dopants would give negligible improvement or introduce unnecessary complications.

4.. *Specific Examples of Phosphorescent Dopants*

4.1 Organic Dyes

The use of phosphorescent dyes to 'harvest' triplet excitons in organic LEDs has been reported (Hoshino and Suzuki 1996, Blumstengel and Dorsinville 1999). However, electrophosphorescence from all-organic molecules will be limited by the long-lived triplet excited states.

4.2 Lanthanide Complexes

Organo-lanthanide complexes have been extensively studied as the emissive component in conjugated polymer blends due to a number of useful properties. Kido and Okamoto (2002) recently published a complete review on the use of lanthanide complexes in OLEDs. A unique feature of the lanthanide (III) complexes is their narrow emission spectra. The emitting state involves atomic transitions of the 4*f* electrons, which are of low oscillator strength and give rise to long-lived excited states (500 μs). Furthermore, the 4*f* electrons are effectively shielded by the filled 5*d* orbitals, so that the emission properties are essentially unperturbed by the surrounding matrix or solvent. It is also significant that the mechanism leading to emission (described below) involves a ligand triplet excited state, thus making lanthanides obvious candidates as dopants for "triplet harvesting" in conjugated polymers.

Direct excitation of the atomic transitions in trivalent lanthanide complexes is parity forbidden. Excitation occurs by energy transfer (or internal conversion) from the triplet state of the organic ligand (Whan and Crosby 1962), provided that this state is higher in energy than the 4*f* electronic transitions on the ion. Intersystem crossing from the ligand singlet to the triplet state is allowed by the heavy lanthanide ion (Fig. 5). Hence, lanthanide complexes offer the possibility of utilizing energy transfer from the normally non-emissive triplet state of the polymer to the triplet state of the ligand on the complex, ultimately resulting in emission from the lanthanide ion.

Emission from europium ions is particularly useful because it fits full color device requirements (612 nm). Doping of various europium complexes into PVK and CN-PPP (see Fig. 2 for polymer structures; see Fig. 6 for europium complex structures) results in red-emitting PLEDs (McGehee *et al.* 1999). The difference in the absorption spectra of the two complexes is shown in Fig. 6. As the conjugation length of the ligand increases, the absorption shifts to lower energy. Consequently, the spectral overlap between the europium complex and the polymer increases and

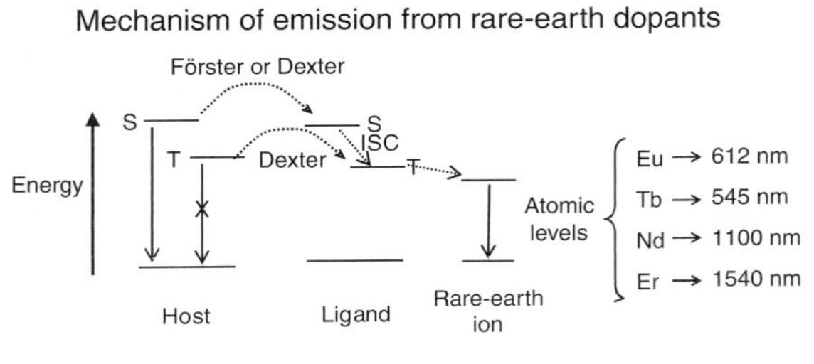

Figure 5
Left: schematic energy diagram depicting energy transfer from host to a lanthanide complex showing that the ligand triplet energy must lie above the lanthanide level. Right: examples of lanthanide ions with their respective emission wavelengths.

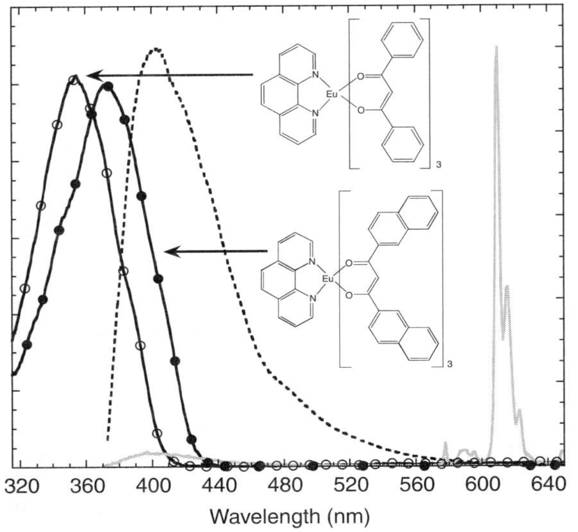

Figure 6
Emission spectrum of CN-PPP (dashed) and absorption spectra of two europium complexes (solid). The overlap between the complex and the polymer is increased by extending the conjugation of the ligand. Also shown is the characteristic narrow emission spectrum (grey) of both europium complexes.

improves the energy transfer efficiency. By modifying the ligand on a lanthanide complex, the optical properties can be tuned to be compatible with the host.

By choosing ligands whose absorption is sufficiently low in energy, stable and efficient host polymers such as MEH-PPV, have been used. The triplet state of such ligands are too low to use with ions emitting visible light. However, narrow infrared emission has been demonstrated from blends doped with porphyrin dopants containing neodymium, erbium and ytterbium complexes (Sloof *et al.* 2001, Harrison *et al.* 2001).

Despite the advantages of using conjugated polymer/lanthanide blends, PLEDs of this type suffer from specific limitations. Although the emitting states in these complexes are not as long-lived as organic dopants, they are sufficiently long-lived to cause bottleneck problems, namely limited brightness and injection limitations. The review by Kido and Okamoto (2002) provides more details.

4.3 Transition Metal Complexes

Certain transition metal complexes are at this stage of development the most promising phosphorescent dopants for fabrication of high efficiency and brightness PLEDs. The metal centers of the complexes used include iridium, platinum, osmium, copper and gold. As in the case with the lanthanide complexes, the metal induces sufficient spin–orbit coupling to allow for efficient phosphorescence. Unlike lanthanide complexes and organic phosphors, transition metal complexes whose excited state is based upon a charge transfer between the metal and the organic ligand, have a relatively short phosphorescence lifetime. Thus, the problems associated with limited brightness and injection bottlenecks are drastically reduced.

Prior to the introduction of transition metal complexes as dopants in conjugated polymers, they were proven useful with vapor deposited small molecules (Baldo *et al.* 1998). OLEDs exhibiting external quantum efficiencies of 19% (Adachi *et al.* 2001) have been achieved using this strategy. Blends of transition metal complexes and conducting polymers have led to high efficiency and high brightness single or double layer PLEDs. The design of the ligands on the metal has proven of critical importance for optimizing properties including HOMO and LUMO energy levels, emission color, and miscibility within the polymer matrix.

Platinum(II) porphyrin complexes (Fig. 4) were among the first transition metal complexes to be used as phosphorescent dopants in polymers (Cleave *et al.* 1999) External quantum efficiencies of 0.6% are reported for blends of 2,3,7,8,12,13,17,18-octaethyl-21H,23H-porphyrin platinum(II) (PtOEP) in PVK (Lee *et al.* 2001). Similar PLEDs using PFO as the host exhibited efficiencies as high as 3.5% (O'Brien *et al.* 2001). However, in both cases the maximum brightness is limited to 240 cd m^{-2}. This limitation is associated with a phosphorescent lifetime ($\sim$100 μs) that is long relative to the fluorescence of organic molecules, albeit shorter than lanthanide and phosphorescent organic molecules.

Cyclometalated iridium complexes are particularly useful because the metal to ligand charge transfer exciton is relatively short lived ($\sim$1 μs) for a triplet state (Garces *et al.* 1988). A number of groups have reported using cyclometalated iridium complexes as a phosphorescent dopant in PVK and in PFO (Gong 2002, Gong *et al.* 2002, Chen *et al.* 2002). Lee *et al.* (2001) and Chen *et al.* (2002) made a direct comparison of PLEDs doped with iridium complexes to PLEDs doped with platinum porphyrins and concluded that iridium complexes indeed lead to brighter and more efficient devices. Osmium(II) complexes have also been examined (Ma *et al.* 1998), as they generally exhibit shorter phosphorescent lifetimes (less than 1 μs) than iridium(III) complexes. However, most recent work has focused on obtaining higher performance from iridium complexes.

The work reported by Gong and co-workers includes the design and synthesis of complexes, such as $Ir(3R)_3$ and $Ir(5R)_3$ in Fig. 4, with molecular structures which closely match that of polyfluorene (Gong 2002, Gong *et al.* 2002). These complexes show good miscibility with host polymers and the aggregation-induced self-quenching is less than that observed with non-functionalized phenylpyridine ligands. Furthermore, the extended conjugation length on the ligands is designed to red shift the absorption frequency and maximize energy transfer from the host in a manner analogous to the lanthanide work. For example, the addition of another fluorenyl in going from $Ir(3R)_3$ to $Ir(5R)_3$ results in a 50 nm red shift in absorption and emission. In contrast, the emission from lanthanide complexes remains invariant, despite a shift in absorption of the ligands. Remarkably high performance PLEDs are fabricated with simple architectures involving a single organic layer comprising PVK or PFO. The best device (1 wt.% $Ir(3R)_3$ in PVK) exhibits green emission with an external quantum efficiency of 11%, and a maximum brightness of over 8000 cd m^{-2}. Red emitting PLEDs with a quantum efficiency of 5% and brightness of over 3000 cd m^{-2} have also been demonstrated in blends of $Ir(5R)_3$ in PVK.

The work by Gong *et al.* (2002) demonstrates that the mechanism leading to iridium electroluminescence is not charge recombination in the host followed by energy transfer to the iridium, but charge trapping within the dopants. This conclusion is drawn from the observation that the amount of iridium complex required to quench the photoluminescence of the host is considerably larger than the amount needed to quench the electroluminescence.

5.. *Conclusions*

Blends of phosphorescent dopants and conjugated polymers can overcome the limit of electroluminescence quantum efficiency imposed on organic molecules by the spin statistics of electron-hole recombination. In some cases this raises the upper limit from 25% to 100%, significantly improving the market potential of organic based light emitting diodes. Lanthanide complexes are under consideration because their emission is narrow and can be chosen at wavelengths ranging from green to infrared, depending on the choice of metal. The emission of cyclometalated iridium(III) complexes can be tuned by controlling conjugation length of the ligand. PLEDs fabricated with the iridium complexes are highly efficient and appear to operate by a charge trapping mechanism. It is highly likely that further advances in molecular design will yield polymer/dopant combinations that will be used in the large scale manufacture of electrophosphorescent PLEDs.

Bibliography

Adachi C, Baldo M A, Thompson M E, Forrest S R 2001 Nearly 100% internal phosphorescence efficiency in an organic light emitting device. *J. Appl. Phys.* **90** (10), 5048–51

Baldo M A, O'Brien D F, You Y, Shoustikov A, Sibley S, Thompson M E, Forrest S R 1998 Highly efficient phosphorescent emission from organic electroluminescent devices. *Nature* **395** (6698), 151–4

Blumstengel S, Dorsinville R 1999 Phosphorescent emission from polymeric light-emitting diodes doped with chrysene-d_{12}. *Jpn. J. Appl. Phys.* **38** (4A), L403–5

Burroughes J H, Bradley D D C, Brown A R, Marks R N, Mackay K, Friend R H, Burn P L, Homes A B 1990 Light emitting diodes based on conjugated polymers. *Nature* **347**, 539–41

Campbell I H, Smith D L, Tretiak S, Martin R L, Neef C J, Ferraris J P 2002 Excitation transfer processes in a phosphor-doped poly(p-phenylene vinylene) light-emitting diode. *Phys. Rev. B* **65**, 085210–218

Cao Y, Parker I D, Yu G, Zhang C, Heeger A J 1999 Improved quantum efficiency for electroluminescence in semiconducting polymers. *Nature* **397** (6718), 414–7

Chen F-C, Yang Y, Thompson M E, Kido J 2002 High-performance light-emitting diodes doped with a red phosphorescent iridium complex. *Appl. Phys. Lett.* **80** (13), 2308–10

Cleave V, Tessler N, Yahioglu G, Le Barny P, Facoetti H, Boueton N, Friend R H 1999 Long lived photo-and electroluminescence from a side-chain polymer/porphyrin blend. *Synth. Met.* **102**, 939–40

Garces F O, King K A, Watts R J 1988 Synthesis, structure, electrochemistry and photophysics of methyl-substituted phenylpyridine ortho-metalated iridium(III) complexes. *Inorg. Chem.* **27**, 3464–71

Gong X 2002 *Appl. Phys. Lett.* (in Press)

Gong X, Robinson M R, Ostrowski J C, Moses D, Bazan G C, Heeger A J 2002 High-efficiency polymer-based electrophosphorescent devices. *Adv. Mater.* **14** (8), 581–5

Hadziioannou G and van Hutten P F (eds.) 2000 *Semiconducting Polymers*. Wiley-VCH, Weinheim, Germany

Harrison B S, Foley T J, Bouguettaya M, Boncella J M, Reynolds J R, Schanze K S, Shim J, Holloway P H, Padmanaban G, Ramakrishnan S 2001 Near-infrared electroluminescence from conjugated polymer/lanthanide porphyrin blends. *Appl. Phys. Lett.* **79** (23), 3770–2

Hertel D, B.ässler H, Guentner R, Scherf U 2001 Triplet–triplet annihilation in a poly(fluorene)-derivative. *J. Chem. Phys.* **115** (21), 10007–13

Hoshino S, Suzuki H 1996 Electroluminescence from triplet excited states of benzophenone. *Appl. Phys. Lett.* **69** (2), 224–6

Jiang X, Jen A K-Y, Carlson B, Dalton L R 2002 Red electrophosphorescence from osmium complexes. *Appl. Phys. Lett.* **80** (5), 713–5

Kido J, Okamoto Y 2002 Organo-lanthanide metal complexes for electroluminescence materials. *Chem. Rev.* **102**, 2357–68

Köhler A, Wilson J S, Friend R H 2002 Fluorescence and phosphorescence in organic materials. *Adv. Mater.* **14** (10), 701–7

Lane P A, Palilis L C, O'Brien D F, Giebeler C, Cadby A J, Lidzey D G, Campbell A J, Blau W, Bradley D D C 2001 Origin of electrophosphorescence from a doped polymer light emitting diode. *Phys. Rev. B* **63**, 235206–208

Lee C-L, Lee K B, Kim J-J 2001 Highly efficient polymer phosphorescent light emitting devices. *Mat. Sci. Eng. B: Solid* **B85**, 228–31

Ma Y, Zhang H, Shen J, Che C 1998 Electroluminescence from triplet metal-ligand charge-transfer excited state of transition metal complexes. *Synth. Met.* **94**, 245–8

Ma Y, Che C-M, Chao H-Y, Zhou X, Chan W-H, Shen J 1999 High luminescence gold(I) and copper(I) complexes with a triplet excited state for use in light-emitting diodes. *Adv. Mater.* **11** (10), 852–7

McGehee M D, Bergstedt T, Zhang C, Staab A P, O'Regan M B, Bazan G C, Srdanov V I, Heeger A J 1999 Narrow bandwidth luminescence from blends with energy transfer from semiconducting polymers to europium complexes. *Adv. Mater.* **11**, 1349–54

Moses D, Soci C, Miranda P, Heeger A J 2001 The role of electron photoemission in the "photoconductivity" of semiconducting polymers. *Chem. Phys. Lett.* **350** (5–6), 531–6

O'Brien D F, Giebeler C, Fletcher R B, Cadby A J, Palilis L C, Lidzey D G, Lane P A, Bradley D D C, Blau W 'Brien et al. 2001 Electrophosphorescence from a doped polymer light emitting diode. *Synth. Met.* **116**, 379–83

Ostrowski J C, Robinson M R, Heeger A J, Bazan G C 2002 Amorphous iridium complexes for electrophosphorescent light emitting devices. *Chem. Comm.* **7**, 784–5

Robinson M R, O'Regan M B, Bazan G C 2000 Synthesis, morphology and optoelectronic properties of tris[(N-ethylcarbazolyl)(3′,5′-hexyloxybenzoyl)methane](phenanthroline)-europium. *Chem. Comm.* **17**, 1645

Robinson M R, Wang S, Heeger A J, Bazan G C 2001 A tetrahedral oligo(phenylenevinylene) molecule of intermediate dimensions: effect of molecular shape on the morphology and electroluminescence of organic glasses. *Adv. Funct. Mater.* **11** (6), 413–9

Saricifcti N S (ed.) 1997 *Primary Photoexcitations in Conjugated Polymers: Molecular Exciton versus Semiconductor Band Model.* World Scientific

Sloof L H, Polman A, Cacialli F, Friend R H, Hebbink G A, van Veggel F C J M, Reinhoudt D N 2001 Near-infrared electroluminescence of polymer light-emitting diodes doped with a lissamine-sensitized Nd^{3+} complex. *Appl. Phys. Lett.* **78** (15), 2122–4

Tang C W, VanSlyke S A 1987 Organic electroluminescent diodes. *Appl. Phys. Lett.* **51**, 913–5

Turro N J 1991 *Modern Molecular Photochemistry*. University Science Books, Sausalito

Whan R E, Crosby G A 1962 Luminescence studies of rare-earth complexes—benzoylacetonate and dibenzoylmethide chelates. *J. Mol. Spectrosc.* **8**, 315

Zhang J, Kan S, Ma Y, Shen J, Chan W, Che C 2001 Energy transfer from singlet to triplet excited states in organic light-emitting devices. *Synth. Met.* **121**, 1723–4

Polymer Light Emitting Diodes, Inkjet Printing of

Y. Yang

Polymer light-emitting diodes (PLEDs) and organic light-emitting diodes (OLEDs) have a potential for extensive applications in multicolor emissive displays, road signs, indicator lights, and logos (Burroughes *et al.* 1990, Gustafsson *et al.* 1992, Nakada and Tohma 1998, Bharathan and Yang 1998). Since some of the conjugated organic materials are solution processable (Braun and Heeger 1991), the traditional method of processing these materials has been spin-coating, a technology incapable of patterning the materials to be deposited. However, to realize the above-mentioned applications, it is necessary to achieve a lateral control during the deposition of the emissive materials. This control can be achieved by using the ink-jet printing (IJP) technology. IJP is a common technology used for desktop publishing. The advantages of this technology over spin-coating are shown in Table 1.

Although IJP is ideal for printing polymer and organic solutions, this technology cannot be used in its original form. This is because the voids in between the solution droplets would subsequently lead to electrical shorting points between the cathode and anode in the devices. This can be clearly understood from Fig. 1. To overcome this drawback, the hybrid ink-jet printing technology (HIJP) has been developed (Yang 1998a, 1998b).

HIJP consists of an ink-jet printed layer in conjunction with another uniform polymer layer (which could be spin-coated). This uniform layer, which is called the buffer layer, serves to seal the pin holes while the ink-jet printed layer consists of the desired pattern, for example the red-green-blue dots for a multicolor display. The use of the polymer buffer layer is demonstrated in Fig. 1. The buffer layer could also work as an ink-absorbing layer to effectively fix the printed materials when a proper solvent for the ink has been chosen. In addition, this buffer layer effectively planarizes the surface roughness produced by the ink-jet printed dots.

The HIJP concept is a unique approach for fabricating polymer and organic electronic devices. It has been used to fabricate polymer light-emitting logos, dual-color PLEDs, multicolor OLEDs, and pixelated PLEDs. These devices and their corresponding ink-jet printing layers are summarized in Fig. 2, but due to the limitation of the space, the individual device structure is not shown. The fabrication of polymer light-emitting logos and multicolor OLEDs are the main focus of this article.

1. *Polymer Light-emitting Logo*

In the polymer light-emitting logo shown in Fig. 2(a), a water-soluble conducting polymer PEDOT (3,4-polyethylenedioxythiophene-polystyrenesulfonate) was used as the ink for the printer and used an emissive polymer MEH-PPV as the buffer layer. To fabricate the polymer light-emitting logos, the desired pattern of PEDOT was printed on the glass/ITO substrates followed by spin-coating of the emissive polymer MEH-PPV. Typical thickness of the MEH-PPV film was around 1000 Å as determined by profilometry. A low

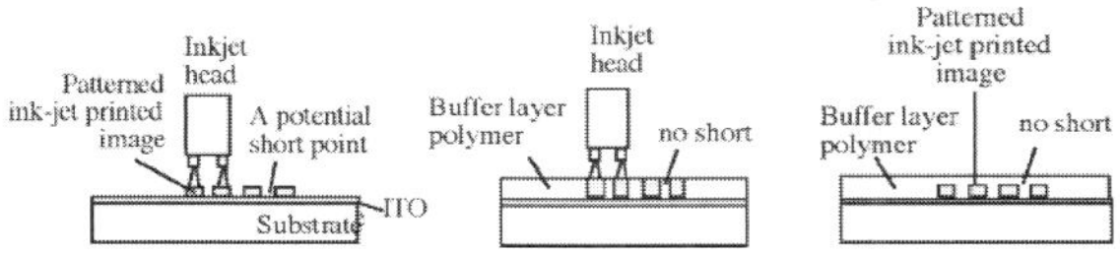

Figure 1
(a) IJP technology in its original form. (b, c) Combination of IJP technology and a buffer layer (the HIJP technology). In (b), the buffer layer is the bottom polymer layer, which absorbs the IJP droplets. In (c), the buffer layer is above the IJP layer and seals the pinholes.

Table 1
Comparison of spin-casting and inkjet printing technologies.

Characteristics	Spin-casting	Ink-jet printing
Patterning capability	No patterning capability	Capable of patterning with micrometer resolution
Large device area capability	Sensitive to dust particles and substrate defects, and not suitable for large area processing	Not sensitive to substrate defects, and a better technology for the fabrication of large area device
Efficiency of using material	More than 99% of the polymer solution is wasted	Less than 2% of the material is wasted
Multicolor display fabrication capability	No multicolor patterning capability	Ideal for multicolor patterning

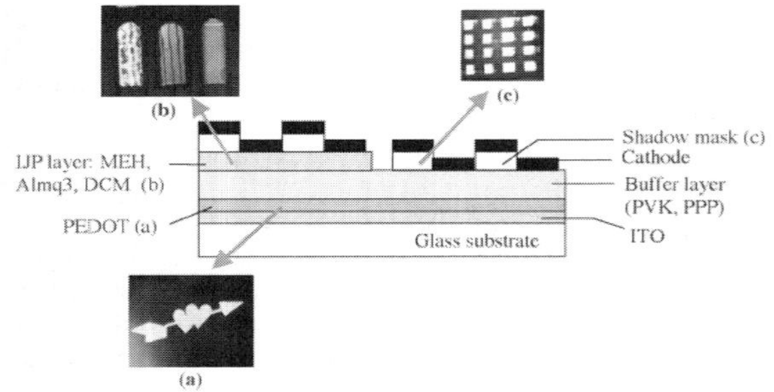

Figure 2
The device structures of three types of polymer LEDs prepared using the HIJP technique. (a) A polymer light-emitting logo, which is achieved by ink-jet patterning the charge injection layer PEDOT. (b) Red, green, and blue OLEDs. (The red and blue LEDs are on the same substrate and the green LED was fabricated on a separated substrate.) These LEDs are patterned by ink-jet printing the red and green organic molecules onto blue PVK buffer layer. (c) Polymer pixelated LEDs. This device is patterned by the built-in shadow mask above the MEH-PPV layer.

work function metal electrode (calcium/aluminum bilayer cathode) was thermally evaporated onto MEH-PPV layer at 10^{-6} torr vacuum. The devices were then encapsulated using a thin epoxy layer and a thin microscope cover glass.

Since the charge injection efficiency of conducting polymer is much better than that of ITO, only the areas covered by the conducting polymer illuminate under the same operating voltage (Yang and Heeger 1994). Therefore the ink-jet printed PEDOT defines the pattern of the logo. For example, as can be seen from Fig. 2(a), at 4 V, the ITO/PEDOT/MEH-PPV/Ca device (the light-emitting area) has a brightness approximately three orders of magnitude higher than the ITO/MEH-PPV/Ca devices (the dark area), giving rise to a very high contrast. Since images from a personal computer can be directly printed to form the light-emitting logo, this technology has very broad applications. For instance, new and complicated emissive logos can be custom-built for a variety of purposes such as greeting cards or other novelty items.

2. *Multicolor Organic Light-emitting Diodes*

In addition to multi-color polymer LEDs (Chang *et al.* 1998, Hebner and Sturm 1998), HIJP can also be applied to fabricate OLEDs (Chang *et al.* 1999). The traditional fabrication of OLEDs involves the thermal sublimation of organic materials in an ultra-high vacuum environment, which is rather time consuming and complicated for patterning fine multicolor pixels. HIJP allows a convenient alternative for the patterning of multicolor pixels through an efficient energy transfer from an appropriate buffer layer (a semiconducting polymer layer) with a wide bandgap to the ink-jet printed materials (dopants) with smaller bandgaps than the buffer layer. Alternatively, the buffer layer can be a continuous polymer layer insoluble in the solvents used by the IJP droplets. In this case, it seals the pin-holes and can also function as the hole-transport layer (HTL). The latter case is ideal for the fabrication of organic LEDs, since the most efficient OLEDs comprise of a bilayer structure of a HTL and an emissive layer (Tang *et al.* 1989).

To fabricate the organic multicolor LED shown in Fig. 2(b) by this technology, a blue-emitting, semiconducting polymer, poly-9-vinylcarbazole (PVK), approximately 1500 Å thick, was used as the buffer layer. The ink-jet printed dopants used were tris(4-methyl-8-quinolinolato)Al (III) ($Almq_3$) (Kido and Iizumi 1998) and DCM (4-(dicyanomethylene)-2-methyl-6-(4-dimethylaminostyryl)-4H-pyran), which were printed on the PVK buffer layer. Based on this principle, multicolor organic LEDs comprising of bilayer structures of PVK/DCM (orange-red emission), PVK/$Almq_3$ (green-blue emission) were fabricated with the blue emission, PVK buffer layer serving as the hole-transport layer.

3. *Built-in Shadow Mask for Pixelated Displays*

In addition to the multicolor patterning capability, another important technical issue for the EL display is the patterning of the anode and the cathode. For the organic EL displays, the easiest and lowest cost way of driving the display is the so-called X-Y passive addressable driving scheme, achieved through the patterning of the anode and the cathode materials into mutually perpendicular row and column electrodes. The patterning of ITO is achieved by regular photolithographic process and the resolution can be very high. On the other hand, the patterning of the cathode is somewhat difficult. The traditional photolithographic process cannot be applied to pattern the cathode, since the solvents that process the photoresist would destroy the underlying organic materials. Therefore, the cathode patterning is usually achieved by using stainless steel shadow masks. However, this technique lacks the required fine resolution. An alternative method could be the use of pre-patterned mushroom shaped photoresist dividers on top of the active organic layers to serve as shadow masks.

Figure 2(c) shows a patterned pixelated polymer LED produced by inkjet printing of an inert. The printed material is an aqueous solution of poly (vinyl alcohol) (PVA). The width of the pixel is about 1 mm. The metal line width will eventually be reduced to the resolution of the inkjet printer, 10 μm (Percin *et al.* 1998). This process will significantly simplify the procedure for fabricating polymeric/organic multicolor, X-Y addressable passive matrix displays.

To summarize, ink-jet printing is compatible with organic materials and also has other significant advantages such as being a low-cost technology and providing a convenient method for processing large area substrates. It is also environmentally friendly since there is minimal wastage of materials. Since this technology is capable of addressing the critical issue of patterning organic materials its applications are widespread. It can be used for the fabrication of logos, indicator lights, and multicolor displays, and in biomedical applications such as biosensors for low-cost diagnostics.

Bibliography

Bharathan J, Yang Y 1998 Polymer electroluminescent devices processed by ink-jet printing: I. Polymer light-emitting logo. *Appl. Phys. Lett.* **72**, 2660–2

Braun D, Heeger A J 1991 Visible light emission from semiconducting polymer diodes. *Appl. Phys. Lett.* **58**, 1982–4

Burroughes J H, Bradley D D C, Brown A R, Marks R N, Mackay K, Friend R H, Burns P L, Holmes A B 1990 Light-emitting diodes based on conjugated polymers. *Nature* **347**, 539–41

Chang S C, Bharathan J, Yang Y, Helgeson R, Wudl F, Ramey M R, Reynolds J R 1998 Dual-color polymer LEDs processed by hybrid inkjet printing technology. *Appl. Phys. Lett.* **73**, 2561–3

Chang S C, Liu J, Bharathan J, Yang Y, Onohara J, Kido J 1999 Multicolor organic light-emitting diodes processed by ink-jet printing. *Adv. Mater.* **11**, 734–6

Gustafsson G, Cao Y, Treacy G M, Klavetter F, Colaneri N, Heeger A J 1992 Flexible light-emitting diodes made from soluble conducting polymers. *Nature* **357**, 477–9

Hebner T R, Sturm J C 1998 Local tuning of organic light-emitting diode color by dye droplet application. *Appl. Phys. Lett.* **73**, 1775–7

Kido J, Iizumi Y 1998 Fabrication of highly efficient organic electroluminescent devices. *Appl. Phys. Lett.* **73**, 2721–3

Nakada H, Tohma T 1998 Organic materials push EL displays into competition with LCDs. *Display Devices* **17**, 29–32

Percin G, Lundgren T S, Khuri-Yakub B T 1998 Controlled ink-jet printing and deposition of organic polymers and solid particles. *Appl. Phys. Lett.* **73**, 2375–7

Tang C W, VanSlyke S A, Chen C H 1989 Electroluminescence of doped organic thin films. *J. Appl. Phys.* **65**, 3610–6

Yang Y 1998a US Patents Pending, Serial Nos. 60/062,294 and 60/072,709

Yang Y, Bharathan J 1998b Polymer light-emitting logos processed by ink-jet printing technology. SPIE Proc. Photonic West Conf. *Light-emitting Diodes*, San Jose, pp. 3278–9

Yang Y, Heeger A J 1994 Polyaniline as a transparent electrode for polymer light-emitting diodes: Lower operating voltage and higher efficiency. *Appl. Phys. Lett.* **64**, 1245–7

Polysilicon and EDF Silicon (Shaped) for Solar Applications

J. C. Lambropoulos

As the efficiency of silicon-based photovoltaic (PV) cells increases, PV energy becomes more attractive as an alternative form of clean, inexpensive energy (Hubbard 1989). Experimental efficiencies in the range 13–23% have been reported.

A commonly used method for the growth from the melt of thin silicon sheets for PV applications is edge-defined film-fed growth (EFG), shown schematically in Fig. 1. The growth of single sheets or tubes eliminates slicing (with kerf loss as high as 50%), grinding, and polishing (to eliminate surface damage) required in bulk-grown silicon. Bell and Kalejs. (1998) compare EFG with other sheet-growing methods.

Potential applications of EFG are in designs requiring large surface areas grown at high productivity. EFG cell efficiencies have increased from 10% in 1980 for 5 cm wide, 400 μm thick, 75 m long single sheets (laser cut to $5 \times 10\,cm^2$ wafers), to 14% in 1995 for 10 cm face width, 300 μm thick, 5 m long octagonal tubes grown at about $2\,cm\,min^{-1}$ (leading to $10 \times 10\,cm^2$ wafers). The price per W has dropped by a factor of more than 20 from 1980 (primarily growth of single sheets) to 1996 (optimized growth of

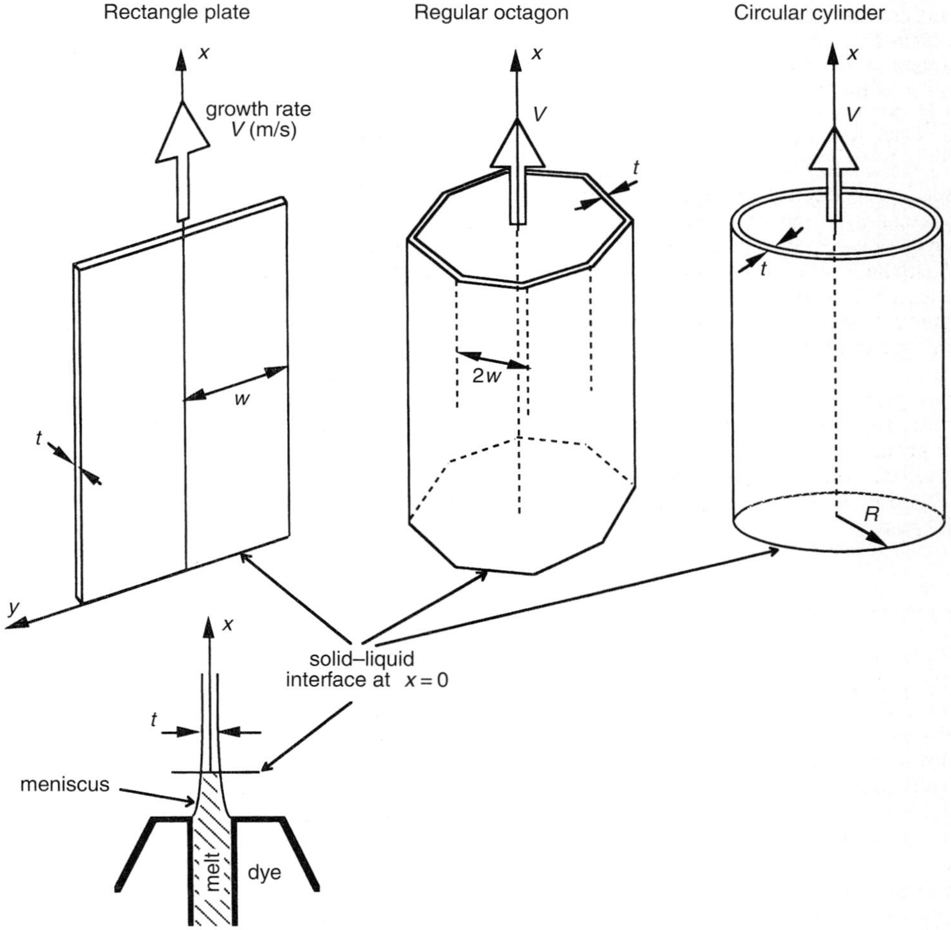

Figure 1
Typical shapes grown by EFG include single sheets (rectangular plate, typically width $2w = 5$ cm, length 80 m), polygonal cylindrical thin tubes (regular octagons or nonagons of 5–10 cm wide faces, length 5 m), or circular cylindrical shells (radius 10–50 cm, length 2 m). Typical thickness t is in range 100–400 μm. Typical growth rates are 2–$4\,cm\,min^{-1}$. The grown sheets may be cut with a laser to produce smaller wafers.

polygonal tubes.) Since 1994 EFG has been in full-scale production.

In EFG silicon plates are grown either directly in the form of rectangular single sheets, as polygonal cylindrical tubes (typically octagons or nonagons), or as polygonal circular cylindrical tubes. Sheets or tubes are subsequently sectioned by laser cutting into rectangular panels.

In EFG the melt passes through a slotted die (typically made of carbon or quartz with a gap of 500 μm) in contact with the melt. Capillary forces raise the liquid to the top of the die. A seed crystal is lowered to make contact with the melt. The liquid extends over the top of the die to the edges. The seed is withdrawn pulling with it the liquid that solidifies, while more liquid enters the die. At typical growth rates, the crystal growth rate at the interface is the same as the pull rate. Heat is primarily removed from the solid–liquid interface by conduction into the solid and the liquid, and by radiation from the meniscus (the solid–gas interface). The meniscus shape is governed by the Young–Laplace equation. Menisci heights range from 0.5 mm to more than several mm (see Bell 1993 for a summary of menisci modeling).

The growth is self-stabilizing because the meniscus height increases as the growth rate increases. For taller meniscus, the meniscus curvature (controlled by gravity and surface tension) produces a smaller crystal thickness, which in turn allows a more rapid heat flux removal and thus a faster growth rate. Different die geometries allow different menisci shapes.

The silicon grown is polycrystalline, with large grains that are elongated in the direction of crystal growth. The grain size is a few mm normal to the growth direction, and several cm parallel to the growth direction. Grains generally extend through the sheet thickness.

1. *Typical Parameters*

Typical widths ($2w$) for flat single sheets are 4–5 cm. More recently, the technology has evolved towards the growth of hollow cylindrical tubes. Besides producing lower thermal stresses, the advantage of tubes is that discontinuous free edges are eliminated. The growth of polygonal tubes increases productivity. The growth of an octagon with 10 cm wide faces is equivalent to a single 80 cm wide sheet. Circular cylinders present the added advantage of circumferential symmetry, so that the crystal can be rotated while growing. The thickness t is in the range of 100–400 μm. Growth rates range from 1–5 cm min^{-1}.

The sheet thickness is a critical parameter in solar cell efficiency. An optimum thickness exists, as determined from fabrication steps and material properties such as surface recombination velocities, minority carrier lifetime, and resistivity of the base material. The optimum thickness ranges from 50–150 μm, leading to efficiencies in the range 14–16 %.

EFG growth requires a large temperature gradient at the solid–liquid interface, e.g., for growing 4 cm wide silicon rectangular plates at 2.5 cm min^{-1}, the crystal thickness is about 300 μm and the thermal gradient at the interface about 1000 K cm^{-1}. For growth at 4 cm min^{-1}, the thickness is 100 μm and the interface gradient is about 800 K cm^{-1}.

The interface thermal gradient may be estimated by a simple thermal balance of a solidifying strip of material. Assuming steady-state crystal growth, and neglecting interface shape, thermal losses by radiation, and thermal gradients in the liquid,

$$k\frac{\mathrm{d}T}{\mathrm{d}x} = L\rho V \tag{1}$$

where k is the silicon thermal conductivity at the melting point $T_{\mathrm{m}} = 1687$ K, L the latent heat (1.8×10^6 J kg^{-1}), ρ the solid density (2.3×10^3 k gm^{-3}), and V the crystal growth rate (cm min^{-1}).

2. *Properties of Silicon Grown by EFG Method*

The properties of the grown silicon sheets are affected by linear defects, such as dislocations, areal defects such as twin boundaries, point defects such as impurities, and mechanical stresses induced by the growth process. Typical resistivity (p-type) is in the range 1–2 Ωcm for EFG silicon.

Twin boundaries are electrically inactive. Dislocations, with densities ranging from 1×10^4 cm^{-2} to 1×10^6 cm^{-2}, are primarily generated by thermal stresses induced by inhomogeneous temperature gradients. In the highest quality EFG silicon grown, the loss in background current is correlated with the dislocation density (rather than grain boundaries), either due to the intrinsic dislocation structure or due to the dislocation-associated solute cloud. In any case, minimizing such dislocations is essential in improving crystal quality. Hydrogen passivation has been found to minimize dislocation recombination.

Typical diffusion lengths are in the range of 50–400 μm. Impurities act as recombination sites, thus reducing diffusion lengths. Typical impurities are carbon (substitutional) and oxygen (interstitial), at levels of typically 1×10^{18} cm^{-3} and 3×10^{17} cm^{-3}, respectively. Notice that the O:C ratio is less than unity, contrary to other methods, such as Czochralski, where the ratio exceeds unity. Although carbon (introduced through the graphite crucible and die) is electrically inactive at lower concentrations, higher concentrations may form extended defects. Oxygen, crucial in maintaining >10 % efficiencies, is introduced via adding CO gas to the growth ambient. Oxygen forms SiO_x precipitates, vacancy oxygen complexes, or thermal donors (Kalejs 1987).

Phosphorus diffusion can be used to get metal impurities with high diffusion coefficients (iron, nickel, copper, typically at concentrations of $1 \times 10^{16}\,\text{cm}^{-3}$). Other precipitates that may form are SiC impurities in the growth of thin sheets. The SiC impurity concentrations are higher near the sheet edges. However, with the growth of closed shapes, such as nonagons or cylindrical shells, this problem has been alleviated.

The effective impurity segregation coefficient $k_{\text{eff}} = C_S/C_0$ (with C_S the concentration in the solid and C_0 that in the liquid) can be quite small. For transition metal impurities, k_{eff} ranges between $k_0 \sim 1 \times 10^{-5}$ (the equilibrium segregation coefficient) and 1×10^{-3}. The effect of the meniscus shape on the segregation coefficient can be found in Ettouney and Brown (1982). A detailed discussion of the relation of EFG process parameters and the silicon sheet quality, including the effects of impurities and defects, is in Kalejs (1987) and Kardauskas *et al.* (1996).

3. Stresses

Stresses are generated in EFG due to inhomogeneous thermal gradients. Stresses and high temperatures both induce the multiplication of dislocations, which are also affected by the crystal growth rate. Alternatively, dislocations relax the stresses. Thus, dislocation multiplication and stress relaxation are coupled processes.

For the growth of silicon sheets two regimes of dislocation multiplication during crystal growth have been identified. In the *temperature-driven* regime, temperature decays slowly as compared to the length scale describing the sheet geometry, thermal stresses are low, and dislocation multiplication is primarily due to the high temperatures prevailing over a long portion of the shell. In the *stress-driven* regime, the temperature decays rapidly, large thermal stresses are induced, which are then primarily responsible for the multiplication of dislocations. In turn, dislocation generation leads to residual stresses when the grown silicon sheet attains room temperature (Bhihe *et al.* 1994).

Compressive residual stresses may lead to buckling of the thin sheets, i.e., out of plane deformations (Tsai and Dillon 1987). Wider, thinner sheets have lower buckling stress thresholds. Stress distributions that are not symmetric through the thickness (yet below the buckling limit) may lead to form errors, causing the grown sheet to deviate from being planar.

Tensile stresses may lead to cracking, e.g., using a fracture toughness of $1\,\text{MPam}^{1/2}$ for silicon at room temperature, a $10\,\mu\text{m}$ surface crack leads to a tensile strength of $\sim 160\,\text{MPa}$, while a $100\,\mu\text{m}$ crack further reduces the strength to $\sim 50\,\text{MPa}$.

Stresses may be estimated by using a simple exponentially decaying temperature profile

$$T(x, z) = T_r + [T_m - T_r + (T_z/t)z]e^{-\gamma x}, \quad |z| \leqslant t/2 \tag{2}$$

where z is the shell normal direction, $z = t/2$ is the outer shell surface, T_r is the room temperature, $T_m = T_r + \Delta T$ the melting point, and T_z is any temperature difference between the outer and inner shell surfaces. It is convenient to use the temperature gradient at the solid–liquid interface $G_0 = \gamma \Delta T$.

3.1 Single Sheets

Finite element calculations of thermoelastic stresses under static conditions (i.e., neglecting dislocation generation and the crystal growth rate) show that the highest stresses (occurring within one sheet width from the interface) increase with (γw). Requiring that the equivalent shear stress $\tau_e = (s_{ij}s_{ij}/2)^{1/2}$ nowhere exceeds the silicon yield stress in shear τ_Y ($\sim 3\,\text{MPa}$ at T_m) leads to the dislocation-free size constraint for EFG single sheets

$$\tau_Y \geqslant 7.5 \times 10^{-2}\,\alpha E G_0 w \tag{3}$$

when the temperature is uniform through the thickness. α is the coefficient of thermal expansion (about $6 \times 10^{-6}\,\text{K}^{-1}$) and E the Young's modulus (160 GPa), e.g., using a width $2w = 4\,\text{cm}$, the interfacial temperature gradient is not to exceed $20\,\text{K cm}^{-1}$. Such a gradient is unrealistic for EFG where interfacial thermal gradients are in the range $500–1000\,\text{K cm}^{-1}$. Thus, EFG of single sheets produces high stresses and high dislocation densities. Some of these issues can be significantly addressed by using long, thin shells of polygonal or circular cross-sections.

3.2 Polygonal Cylindrical Shells

Numerical simulations (Wu and Lambropoulos 1995) show that thermoelastic stresses decrease as the number of sides of the polygonal tube increases (e.g., $n = 8$ for octagons). The stresses are minimized in the limit of the circular cylindrical shell.

3.3 Circular Cylindrical Shells

Because of the simplified and axisymmetric geometry, the stresses can be analytically calculated for the two limiting cases of no stress relaxation or full stress relaxation at the solid–liquid interface.

(*a*) *Static thermoelastic shell, no stress relaxation.* In the first case, the shell is assumed to be thermoelastic and static, i.e., no growth occurs. The temperature profile is superposed on a perfect, stress-free cylinder. The material is linear elastic, and no dislocations relieve any of the stresses.

The stresses lead to forces per unit length N_x, N_θ and moments per unit length M_x, M_θ

$$\sigma_x = \frac{N_x}{t} + 12\frac{M_x z}{t^3}, \quad \sigma_\theta = \frac{N_\theta}{t} + 12\frac{M_\theta z}{t^3} \tag{4}$$

The solid–liquid interface is free of any traction ($Q_x = N_x = M_x = 0$ at $x = 0$.) The results are

$$\frac{N_\theta(x)}{\langle N_0 \rangle} = -\left(\frac{1+\zeta}{1+\frac{\rho^4}{4}}\right)[\rho^2 e^{-\gamma x} + 2e^{-\beta x}((1-\rho)c - s)],$$

$$N_x(x) = 0$$

$$\frac{M_x(x)}{\langle M_0 \rangle} = \left(\frac{1+\zeta}{1+\frac{\rho^4}{4}}\right)[-e^{-\gamma x} + e^{-\beta x}((1-\rho)s + \varepsilon)]$$

$$\frac{M_\theta(x)}{\langle M_0 \rangle} = \left(\frac{v}{1+\frac{\rho^4}{4}}\right)\left[-e^{-\gamma x}\left\{1 + \frac{\zeta}{v}\left(1 + (1-v)\times\frac{\rho^4}{4}\right)\right\}\right.$$
$$\left. + e^{-\beta x}(1+\zeta)((1-\rho)s + c)\right] \tag{5}$$

where

$$\beta = [3(1-v^2)]^{1/4}]/(Rt)^{1/2},$$
$$\langle M_0 \rangle = E\alpha t^3 G^2{}_0/[12(1-v^2)\Delta T],$$
$$\langle N_0 \rangle = E\alpha t^2 R G^2{}_0/[4\Delta T(1-v^2)^{1/2}],$$
$$s = \sin(\beta x), \; c = \cos(\beta x),$$
$$\rho = \gamma/\beta,$$

and

$$\zeta = [(1+v)T_z \Delta T]/[RtG^2{}_0]$$

is the dimensionless effect of the through-thickness temperature gradient. v is the Poisson ratio (0.25 for silicon.) The parameter $\rho = \gamma/\beta$ describes the rate of temperature decay with respect to the geometrical dimensions of the shell. For slow temperature decay $\gamma/\beta \ll 1$. As $x \to \infty$, no residual stresses are present, because no stress relaxation has occurred and no defects have been generated.

Using $R = 0.20$ m, $t = 200\,\mu$m, $G_0 = 800$ K cm^{-1}, $\Delta T = 1300$ K, leads to the values $\beta = 205$ m^{-1}, and $\gamma = 62$ m^{-1}, so that $\gamma/\beta = 0.30$. The variation of τ_e with distance away from the solid–liquid interface is shown in Fig. 2. For these parameters (and with $T_z = 0$), the largest thermally induced stress τ_e, occurring at the shell surface at a distance of about 1 cm above the interface, is ~30 MPa. A thickness of 100 μm would reduce the stress to ~15 MPa. These results clearly identify the vicinity near the solid–liquid interface as the region where all significant stress builds up. For a few mm from the interface, the stress in the middle surface and outer fibers are comparable. Eventually, the stress in the outer and inner shell fibers dominates that in the middle surface, thus showing that the stresses are due to bending.

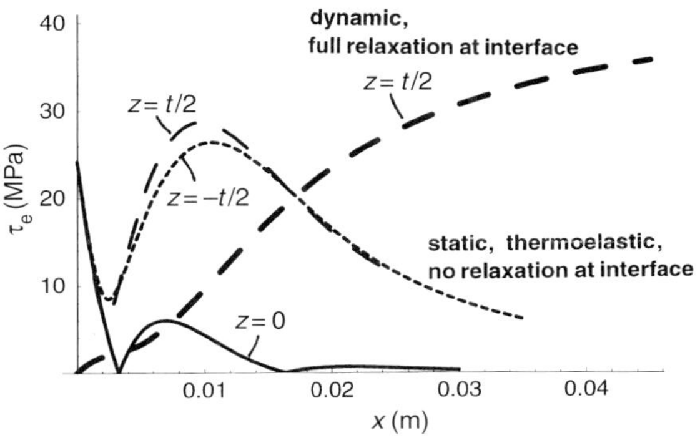

Figure 2
The distribution of the equivalent shear stress τ_e during the growth of thin, circular cylindrical shells for the cases of (a) static, thermoelastic growth, without any stress relaxation at the solid–liquid interface (stresses shown at middle surface, and at inner and outer fibers), and (b) dynamic growth with full stress relaxation at the interface (stress shown at outer fiber). The temperature profile is exponential, decaying with a wave number $\gamma = 62$ m^{-1}. The shell radius is 20 cm, and the thickness 200 μm. The temperature gradient at the solid–liquid interface is 800 K cm^{-1}.

Calculating the largest equivalent shear stress τ_e, and requiring that it nowhere exceeds the yield stress in shear τ_Y now produces the size constraint necessary to avoid any plastic deformation (with no temperature variation through the thickness)

$$\tau_Y \geqslant 0.18\alpha E \frac{G^2{}_0}{\Delta T} Rt \tag{6}$$

Thus, to avoid dislocation generation, thin, small shells must be grown, preferably at low temperature gradients. For $R = 20$ cm, $t = 100\,\mu$m, the temperature gradient G_0 at the interface must not exceed 340 K cm^{-1}. The smaller length scale of $(Rt)^{1/2}$ describing the elastic deformation of circular cylindrical shells essentially diminishes the temperature gradients and thus produces considerably lower stresses than growth of single sheets.

(b) Growing shell, full stress relaxation. With semi-infinite circular cylinder growing through a die the analysis is done in a frame of reference fixed to the die. Transient effects are neglected, and steady-state growth is assumed. The solution is

$$\frac{N_\theta(x)}{\langle N_0 \rangle} = -\left(\frac{1+\zeta}{1+\frac{\rho^4}{4}}\right)[-\rho^2 e^{-\gamma x} + \rho e^{-\beta x}(\rho c + (2-\rho)s)],$$

$$N_x(x) = 0$$

$$\frac{M_\theta(x)}{\langle M_0 \rangle} = \left(\frac{\nu}{1+\frac{\rho^4}{4}}\right)\left[(1-e^{-\gamma x})\left\{1+\frac{\zeta}{\nu}\left(1+(1-\nu)\frac{\rho^4}{4}\right)\right\}.\right.$$
$$\left. -(1+\zeta)\left\{\rho - \frac{\rho^2}{2} + e^{-\beta x}\left(-\left(\rho-\frac{\rho^2}{2}\right)c + \frac{\rho^2}{2}s\right)\right\}\right]$$

$$\frac{M_x(x)}{\langle M_0 \rangle} = \left(\frac{1+\zeta}{1+\frac{\rho^4}{4}}\right)\left[1-\rho+\frac{\rho^2}{2} - e^{-\gamma x}\right.$$
$$\left. + e^{-\beta x}\left(\left(\rho-\frac{\rho^2}{2}\right)c - \frac{\rho^2}{2}s\right)\right] \qquad (7)$$

Now significant residual stresses are present as $x \to \infty$. For the dimensions and parameters of the Sect. 4.3(a), the residual stresses are ~ 35 MPa. These results are shown in Fig. 2, where the stress builds up to a steady-state (residual) value within several centimeters from the solid–liquid interface. The stress in the shell middle surface is much smaller than that in the outer shell fibers, again indicating the importance of bending stresses.

4. *Inelastic Deformation and Dislocation Multiplication in Silicon*

During EFG of silicon from the melt, high temperatures are coupled with thermal stresses to induce the inelastic deformation of silicon once it has solidified. In turn, the multiplication of dislocations relaxes the stresses. The inelastic deformation is described in terms of accumulating strain-rates generated by dislocation multiplication. The high-temperature inelastic deformation of silicon is described by decomposing the strain-rate (denoted by a superposed dot) into elastic, thermal, and creep parts

$$\dot{\varepsilon}_{ij}\,\frac{1+\nu}{E}\left(\dot{\sigma}_{ij} - \frac{\nu}{1+\nu}\dot{\sigma}_{kk}\delta_{ij}\right) + \alpha(T)\dot{T}\delta_{ij} + \dot{\varepsilon}^{cp}_{ij} \qquad (8)$$

Several different models describe the creep strain-rate, including steady-state creep models and dynamic models that treat dislocation density as an evolving internal variable.

4.1 Steady-state Creep Model

During steady-state creep the microstructure of the material is assumed to remain constant, so that the total dislocation density N, as determined by microscopic processes that give rise to dislocation multiplication and annihilation, remains constant (Lambropoulos 1990, Lambropoulos and Wu 1996).

Steady-state creep of silicon may be described by the exponential dependence of creep strain rates on stress via an activation volume (of $\sim 2.7\,\mathrm{nm}^3$ per atom), or by a power-law dependence on the stress via a temperature-dependent stress exponent $n(T)$ ($n = 5$ being typical). The activation energy of about 0.20–0.25 Gb^3 (G being the shear modulus and b the Burgers vector) is close to the value for lattice diffusion.

In crystal growth a given material particle experiences a continuously varying stress temperature and stress field, in contrast to the basic steady-state premise of creep strain at a constant temperature and for a constant applied stress.

4.2 Haasen–Alexander Model

The Haasen–Alexander model for the high-temperature inelastic deformation of semiconductors allows for the explicit generation of dislocations during deformation (Lambropoulos and Wu 1996). The Haasen–Alexander model is based on the combined use of the Orowan equation, an expression for the power-law dependence on stress of the dislocation velocity, and an expression for the rate at which dislocations multiply (typically by cross-slip). The constitutive law is generalized to multiaxial states of stress in the form

$$\dot{\varepsilon}^{cp}_{kl} = \phi b B_0 \exp\left(-\frac{Q}{R_g T}\right)\langle \tau_{eff}\rangle^m N \frac{s_{kl}}{\tau_e},$$

$$N = K B_0 \exp\left(-\frac{Q}{R_g T}\right)\langle \tau_{eff}\rangle^{m+1} N$$

$$\tau_{eff} = \tau_e - A\sqrt{N}, \; \tau_e = \sqrt{s_{ij}s_{ij}}/2 \qquad (9)$$

where τ_{eff} is the effective stress (smaller than the applied equivalent shear stress τ_e due to back stress of existing dislocations), and $\langle\tau_{eff}\rangle = \tau_{eff}$ if $\tau_{eff} > 0$, vanishing otherwise.

The material constants for silicon are $\phi \approx 1/2$, b is the Burgers vector (0.38 nm), exponent $m = 1.1$, and constants $B_0 = 8.6 \times 10^{-4}\,\mathrm{m}^{3.2}\mathrm{N}^{-1.1}\mathrm{s}^{-1}$, $K = 3.1 \times 10^{-4}\,\mathrm{m\,N}^{-1}$, $A = 4\,\mathrm{N\,m}^{-1}$. Activation energy of dislocation motion is $Q = 211\,\mathrm{kJ\,mol}^{-1}$ (lattice or Peierls resistance) and $R_g = 8.314\,\mathrm{J\,mol}^{-1}\mathrm{K}^{-1}$ is the gas constant.

The constitutive model of Haasen–Alexander allows only for dislocation multiplication and for no dislocation annihilation, e.g., via climb and interaction of dislocations with opposite Burgers vectors. Thus, it can not predict a steady-state which requires that dislocations be created and annihilated at the same rate so that the total dislocation density remains constant. A simple model of dislocation climb and annihilation, superposed to dislocation generation by cross-slip, gives the net dislocation generation rate as

$$\dot{N} = K_1 N v \tau_{\text{eff}}^{m+1} - K_2 N^P v \tau_{\text{eff}} \qquad (10)$$

where p is an exponent with $1<p<2$. Since the activation energy for annihilation is higher than that for multiplication, it is expected that only at high temperatures will annihilation become significant.

5. Boundary Conditions at the Solid–Liquid Interface

The finite element formulation of crystal growth problems allows the direct determination of the nodal velocities, from which strain rates and stress rates are then computed. Formulation in terms of rates is obviously needed to account for the incremental nature of the constitutive law and for the growth process. The rates must be integrated along the growth direction x to determine the stresses. For steady-state crystal growth at the rate V, the result is

$$\dot{\sigma}_{ij} = V\frac{\partial \sigma_{ij}}{\partial x}, \text{ or } \sigma_{ij}(x) = \sigma_{ij}^{(0)} + \frac{1}{V}\int_{\xi=0}^{x} \dot{\sigma}_{ij}(\xi)\mathrm{d}\xi \qquad (11)$$

where $\sigma_{ij}^{(0)}$ is the state of stress at the solid–liquid interface. Similarly, the dislocation density is found by integrating the dislocation generation rate

$$N(x) = N^{(0)} + \frac{1}{V}\int_{\xi=0}^{x} \dot{N}(\xi)\mathrm{d}\xi \qquad (12)$$

where $N^{(0)}$ is the initial dislocation density at the interface. $\sigma_{ij}^{(0)}$ and $N^{(0)}$ are possibly dependent on position along the solid–liquid interface, are related to the microscopic process of crystallization, and may vary depending on the material grown or the growth conditions. They are initial conditions that must be independently specified.

It may be argued that at the solid–liquid interface (where $T=T_{\text{m}}$) the material can not sustain any stresses, in which case $\sigma_{ij}^{(0)}$ vanishes and there is thus complete stress relaxation at the interface. Still, at T_{m} the yield stress of silicon is several megapascals. When the interface stress does not vanish, the traction-rate boundary conditions at the interface are

$$\dot{\sigma}_{xy} = -V\frac{\partial \sigma_{yy}}{\partial y} \text{ (single flat sheets)},$$

$$\dot{Q}_x = -V\frac{N_\theta}{R} \text{ (circular cylidrical thin shells)} \qquad (13)$$

These traction rates are entirely driven by the growth process (via the crystal growth rate V), vanishing, as expected, in the absence of any growth.

6. Concluding Remarks

EFG growth of silicon for PV applications has the potential of achieving early in the twenty-first century cell efficiencies of 20% at the full production level. Such levels will require optimum understanding of the EFG process and of the grown material under EFG process conditions.

For the period 2000–2010, the projected direction is the growth of 0.5 m radius hollow cylinders, several meters in length, of thickness 50–100 μm, grown at rates in the range 2–4 cm min^{-1}. Such geometries allow large, single wafers of area up to $160 \times 200\,\text{cm}^2$, such that a module may a single wafer (Kalejs 1997).

Optimization of the die design and the imposed thermal field in the furnace will lead to further reductions of impurities dissolved within the growing material. Increased productivity requires higher growth rates of wider sheets. The growth of polygonal tubes has clearly demonstrated the enhanced productivity. Alternatively, material quality optimization requires the growth of thinner sheets, with fewer defects (impurities or dislocations). Of course, the technical problem of handling thin, large sheets must also be solved, once the growth has been demonstrated.

Computational modeling of thermal, fluid, and solid interactions can help direct the optimization of the EFG process. The details of the thermal field, for example, are critical in guiding the design of more involved temperature profiles including afterheaters, or furnace enclosures. Computational models have identified the physical process of solidification (via the state of stress and defect density at the solid–liquid interface) as a critical input required for predicting the quality of the grown silicon. For such computational models to be effective, though, accurate constitutive models are needed for the high-temperature deformation of silicon, at temperatures up to the melting point, as well as a better understanding of how defects interact with each other (multiplication, annihilation) or with the superimposed thermal and stress fields.

Bibliography

Bell R O 1993 Edge-defined film-fed growth of materials. In: Bloor D, Brook R, Flemings M C, Mahajan S (eds.) *Encyclopedia of Advanced Materials*. Pergamon, Oxford, pp. 671–4

Bell R O, Kalejs J P 1998 Growth of silicon sheets for photovoltaic applications. *J. Mater. Res.* **13**, 2732–9

Bhihe C K, Mataga P A, Hutchinson J W, Rajendran S, Kalejs J P 1994 Residual stresses in crystal growth. *J. Cryst. Growth* **137**, 86–90

Ettouney H M, Brown R A 1982 Effect of heat transfer on melt/solid interface shape and solute segregation in edge-defined film-fed growth: finite element analysis. *J. Cryst. Growth* **58**, 313–29

Hubbard H M 1989 Photovoltaics today and tomorrow. *Science* **244**, 297–304

Kalejs J P 1987 Progress in development of EFG process control in Si ribbon production for photovoltaic applications. In: Khattak C P, Ravi K V (eds.) *Si Processing for Photovoltaics. Vol. II.* Elsevier, Amsterdam, Chap. 4, pp. 185–254

Kalejs J P 1997 Thin self-supporting polycrystalline Si substrates—a new role for EFG cylinders. In: McCommell M (ed.) *1st NREL Conference: Future Generation Photovoltaic Technologies.* American Institute Physics, New York, pp. 207–13

Kardauskas M J, Rosenblum M D, Mackintosh B H, Kalejs J P 1996 The coming of age of a new PV wafer technology—some aspects of EFG polycrystalline Si sheet manufacture In: *25th IEEE Photovoltaic Specialists Conference Record.* Institute of Electrical and Electronic Engineers, New York, pp. 383–8

Lambropoulos J C 1990 High temperature inelastic deformation during shaped crystal growth from the melt. *J. Cryst. Growth* **104**, 1–7

Lambropoulos J C, Wu C-H 1996 Mechanics of shaped crystal growth from the melt. *J. Mater. Res.* **11**, 2163–76

Tsai C T, Dillon O W Jr 1987 Viscoplastic buckling of Si ribbon. *J. Appl. Mech.* **54**, 563–70

Winke W C 1993 Materials for photovoltaics. *Mater. Res. Bulletin* **18**, 18–41

Wu C-H, Lambropoulos J C 1995 Thermoelastic analysis of dislocation generation during EFG of polygonal shells. *J. Cryst. Growth* **155**, 38–46

Power Industry: Corrosion

P. Millet

Corrosion is a leading cause of loss of availability in both fossil and nuclear power plants. The financial impact of corrosion to the power industry is of the order of billions of dollars per year. Corrosion occurs within the steam cycle of nuclear and fossil power plants and on the fire side of fossil-fired power plants. Steam cycle related corrosion has been the most troublesome causing failures to major components such as boilers, steam generators, turbines, feedwater heaters, condensers, and the piping throughout the steam plant. Failures on the fire side of fossil plants include boiler tubes, superheaters, scrubbers, and other pollution-control equipment. Corrosion occurs throughout the steam cycle as illustrated in Fig. 1 (a conventional fossil steam cycle). Multiple forms of corrosion can have an impact on a given component in the system. This is illustrated in Fig. 2 for a steam generator in a pressurized water reactor (PWR)-type nuclear plant.

1. *Methods of Controlling Corrosion*

Corrosion is pervasive in the power industry because of the high operating temperatures. Corrosion processes, like other chemical or electrochemical processes, generally occur faster at elevated temperatures. High temperatures are required to improve the thermodynamic efficiency of the steam cycle and, therefore, limiting the temperature is not the preferred mitigation strategy. Four primary approaches are used to avoid or minimize corrosion in system components, as discussed below.

1.1 Corrosion-resistant Alloys

Stainless steels are used in reactor vessels and piping in nuclear plants as well as high-temperature superheaters in fossil plants. Nickel alloys such as Inconel are widely used for specialized applications such as in the steam generators of PWRs and within subcomponents of nuclear reactors. Other nickel alloys are used, for example, in flue gas desulfurization systems. Titanium alloys are used principally for condenser tubes where resistance to corrosion from salt water and good heat-transfer properties are required. Copper/nickel alloys are still widely used because of their excellent heat-transfer properties for heat exchange applications such as feedwater heating, but owing to the deleterious corrosive effects of copper released from these components on other system components their use is decreasing (see). Carbon steel is widely used for boiler tubes in fossil plants and for piping throughout the steam cycle. Carbon steels with low levels of chromium or other additives are specified for specific corrosion issues such as flow-accelerated corrosion of piping materials.

1.2 Specialized Heat Treatments

Much of the corrosion in power plants, especially of alloys, is localized at grain boundaries. Sensitization of stainless steels and other alloys can be controlled by specialized heat treatments. Heat treatments must be designed to optimize the physical properties of the structural material with its corrosion resistance.

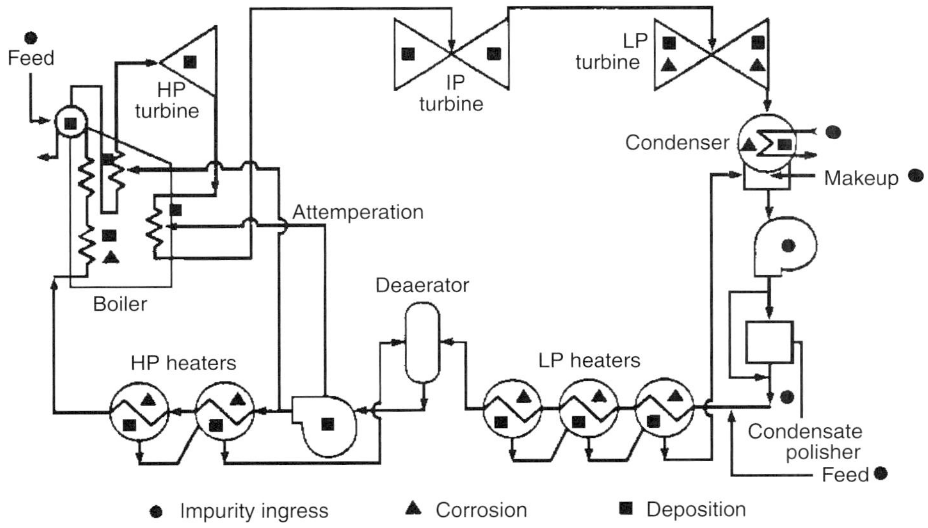

Figure 1
Typical drum boiler fossil plant cycle showing locations of impurity ingress, corrosion, and deposition.

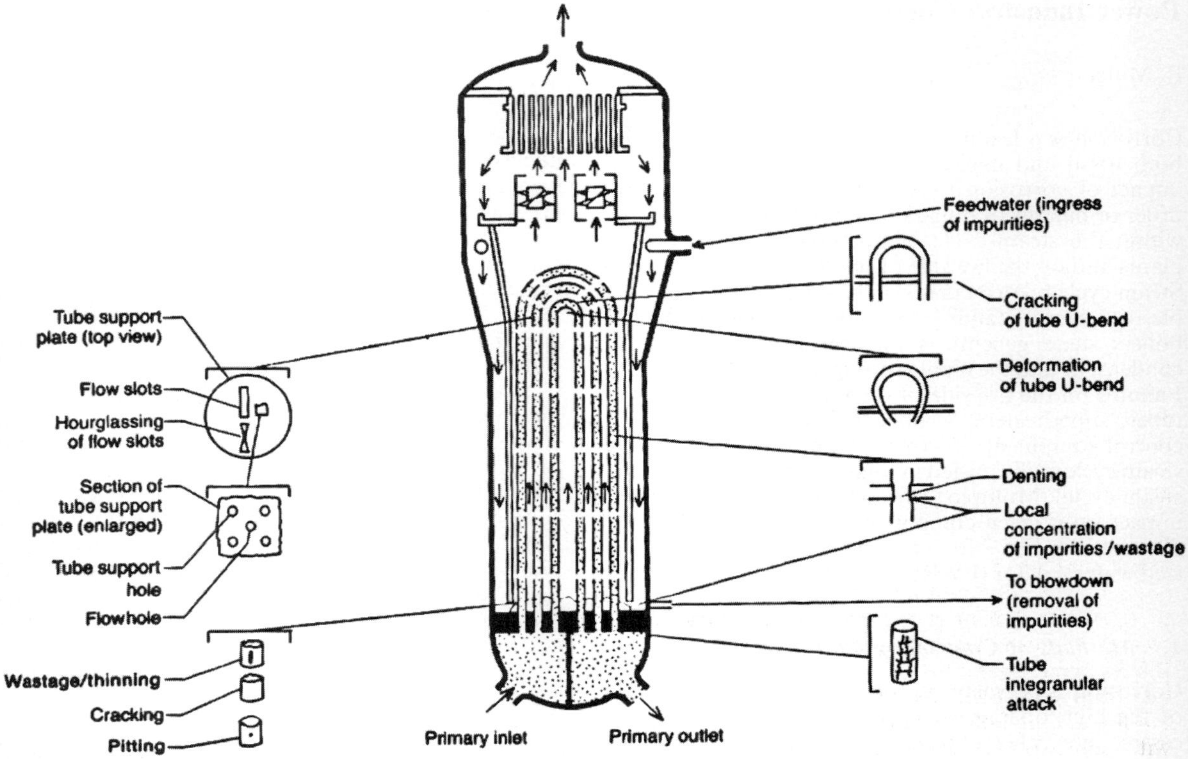

Figure 2
Schematic of recirculating steam generator with various problems indicated.

1.3 Design Features

Stress corrosion cracking (SCC) and crevice corrosion are two of the most serious corrosion problems in the power industry. Reduction of the local and applied stresses both during manufacture and subsequent heat treatment as well as *in situ* stress reduction techniques are important mitigation strategies for SCC. Likewise, designers are careful to eliminate crevices in steam generators, large turbines, and other components. Often crevices are sites where impurities can accumulate by local boiling and electrochemical mechanisms. Concentrated impurities can lead to accelerated corrosion. The crevices between tube support plates and the thousands of heat exchanger tubes in PWR steam generators is one of the most serious corrosion problems facing the nuclear power industry.

1.4 Controlling the Environment

In conventional nuclear and fossil power plants, ultra-pure water is used as the working fluid for producing electricity by the Rankine cycle. The water is purified using ion exchange, filtration, and other processes to remove the impurities before the water is added to the steam cycle. Local crevices and other occluded regions, however, can lead to the accumulation of trace levels of impurities resulting in aggressive environments. Operators of power plants must be diligent in controlling the water chemistry to ensure that any problems that degrade water quality are corrected quickly. Organic and inorganic additives are used to provide further corrosion protection. Oxygen is one of the main corrosion agents in the steam cycle. Adding an oxygen scavenger, such as hydrazine, directly to the condensate water can be used to remove residual oxygen.

Ammonia and other organic amines are added to the feedwater to control the pH. Maintaining alkaline conditions throughout the steam cycle can minimize the general corrosion rate of carbon steels that are widely used in piping systems. Corrosion inhibitors, dispersants, and other chemicals are also used in some power plants to control corrosion and to minimize fouling of heat-transfer surfaces. Sodium phosphate was once used in most nuclear plants and is used in many fossil plants to buffer the pH due to impurity ingress excursions.

2. Commonly Encountered Forms of Corrosion

In fossil plants, boiler tube failures and turbine blade or disk failures are the most serious problems. At least 28 mechanisms have been identified for boiler tube failure (Dooley and Bursik 1995), most of which are related to corrosion. Some of the corrosion-related mechanisms include caustic gouging, hydrogen damage, pitting, fire-side oxidation, corrosion fatigue, and SCC. Turbine availability losses are generally attributed to blade failures by corrosion fatigue and SCC mechanisms.

Nuclear plants have experienced capacity losses due to corrosion damage to a number of components, the most serious of which are SCC of boiling water reactor (BWR) coolant piping and reactor internals and PWR steam generator tubes (Jones 1996). Other corrosion problems include erosion–corrosion of carbon steel steam-system piping, particularly turbine cross-around piping and extraction lines. In the reactor, nuclear fuel cladding failures in BWRs and PWRs can lead to increased operation and maintenance costs, investment losses, reduced cycle efficiency, and increased inspection and reconstitution costs as well as to operational restrictions and outage duration increases. Crud-induced localized corrosion (CILC) in BWRs, which was a major problem in the mid-1970s, has been controlled by reducing copper transport. PWR cladding failures have generally been associated with debris or are due to hydriding. Irradiation-assisted SCC (IASCC) of core internals increases in importance as plants age.

In both fossil and nuclear plants problems with service water systems have included fouling, sedimentation, and various forms of corrosion (including microbial-induced corrosion). These corrosion issues have had a significant impact on maintenance costs at some plants.

3. Economic Considerations

Power plants are large capital investments. The plants are designed to operate nearly continuously for over 40 years. Throughout the world the demand for reliable, inexpensive power has increased the need to operate these systems at full capacity and at as low a cost as possible. Most of the important corrosion issues are due to chronic as opposed to acute causes and may take years or decades to manifest themselves. Operators of power plants must control short-term costs without compromising the long-term integrity of these systems. Replacement of major components is cost prohibitive in many cases. In nuclear plants the contamination of components with radioactive isotopes increases maintenance costs significantly. The industry is committing significant resources towards eliminating or mitigating corrosion in an effort to reduce the cost of power to both the domestic consumer and to industry.

Bibliography

Dooley B, Bursik A 1995 State of the art in fossil plant cycle chemistry. *Proc. 12th Int. Conf. Properties of Water and Steam*

Jones R L 1996 Critical corrosion issues and mitigation strategies impacting the operability of LWRs. *Proc. NACE/96*

Purex and Thorex Processes (Aqueous Reprocessing)

A. Ramanujam

During the operation of nuclear reactors, fission of ^{235}U in uranium fuel generates energy, neutrons, and radioactive fission products. Simultaneously, fertile ^{238}U captures neutrons, yielding plutonium (mainly ^{239}Pu):

$$^{238}_{92}U + ^{1}_{0}n \rightarrow ^{239}_{92}U \xrightarrow{\beta, 23\cdot5\,\text{min}} ^{239}_{93}Np \xrightarrow{\beta,\ 2.3\ \text{d}} ^{239}_{94}Pu \quad (1)$$

The fuel discharged from the reactor, called spent fuel, contains significant quantities of uranium and newly formed plutonium, which can be reused. The process of separating plutonium and uranium from fission products in the spent fuel is termed fuel reprocessing. It differs from conventional chemical processes due to the radioactive nature of the materials processed. The equipment, mostly made of corrosion-resistant SS304L, is installed and operated remotely behind massive concrete shielding with stringent air ventilation and exhaust requirements to protect operating personnel and the environment from radiation and contamination hazards. In the case of equipment failure, the design should provide for duplicate equipment and remote or direct maintenance. Further, the design and chemical flow sheet should preclude the possibility of accidental criticality.

1. *The Purex Process*

The world over, the Purex process, an acronym for "plutonium–uranium redox extraction," is used for the recovery and purification of uranium and plutonium from spent fuels dissolved in nitric acid. It uses tributyl phosphate (TBP) diluted with dodecane as a solvent to extract and separate uranium and plutonium. The essential steps in a generic Purex process are: (i) decladding or dejacketing, (ii) dissolution, (iii) feed preparation, (iv) co-decontamination, (v) a uranium–plutonium partition cycle, (vi) final uranium and plutonium purification cycles, and (vii) reconversion to the respective oxides.

The head-end treatment steps (first three in the list above) are to bring the fuel into an aqueous solution suitable for the Purex process. The aim of the co-decontamination and partitioning steps is to extract and separate uranium and plutonium together from the fission product impurities and from each other. In the subsequent steps, uranium and plutonium are further purified individually before conversion to their oxides. The concentrated fission product waste solution is stored for subsequent treatments such as vitrification and disposal. The general Purex flow scheme is shown in Fig. 1.

The extent of purification of uranium or plutonium achieved is expressed as a decontamination factor (DF), which is the ratio of β or γ activity of fission products associated with unit mass of plutonium or uranium in the feed to that in the product. The larger the DF the greater is the purification efficiency of the process. In general, a DF of 10^6–10^7 for both uranium and plutonium with respect to β and γ activities are aimed at in the process. The plutonium contamination of uranium products should also be reduced similarly.

1.1 *Head-end Process Steps*

(*a*) *Decladding.* Fuel elements use different cladding materials such as aluminum, zircaloy, and stainless steel, depending on the reactor type (see Nuclear Reactor Materials). Based on the cladding material, either chemical or mechanical decladding is selected. In the former, the cladding alone is dissolved whereas the fuel core remains undissolved. Aluminum cladding is preferentially dissolved from irradiated uranium fuels (from research reactors) by a solution of boiling NaOH. Sodium nitrate is added so that the off-gas is ammonia and not hydrogen:

$$20Al + 17NaOH + 21NaNO_3 \rightarrow 20NaAlO_2 + 18NaNO_2 + 4H_2O + 3NH_3 \quad (2)$$

During mechanical decladding, the fuel assemblies are chopped and the fuel core is leached with nitric acid, leaving behind the undissolved cladding material for disposal as a solid waste. This method is adopted for zircaloy-clad fuels from power reactors and stainless steel-clad fast reactor fuels.

(*b*) *Dissolution and feed preparation.* The dissolution of uranium metal and oxide fuels in nitric acid is exothermic and is controlled by optimizing the concentration of the acid as well as the temperature of the reaction. Oxides of nitrogen liberated during the dissolution are reoxidized and returned to the dissolver using down-draft condensers. The off-gases are treated to remove the traces of nitric acid and volatile fission products, cooled, and filtered before being exhausted through tall stacks. Voloxidation methods, where chopped high-burn-up fuel elements are treated with oxygen at 450–500°C before dissolution, are useful in removing and containing the bulk of the iodine and tritium released.

Uranium dioxide dissolves in nitric acid by the net reaction

$$UO_2 + 4H^+ + 4NO_3^- \rightarrow UO_2(NO_3)_2 + 2NO_2 + 2H_2O \quad (3)$$

Using fumeless dissolution techniques, it is possible to dissolve uranium without the evolution of oxides

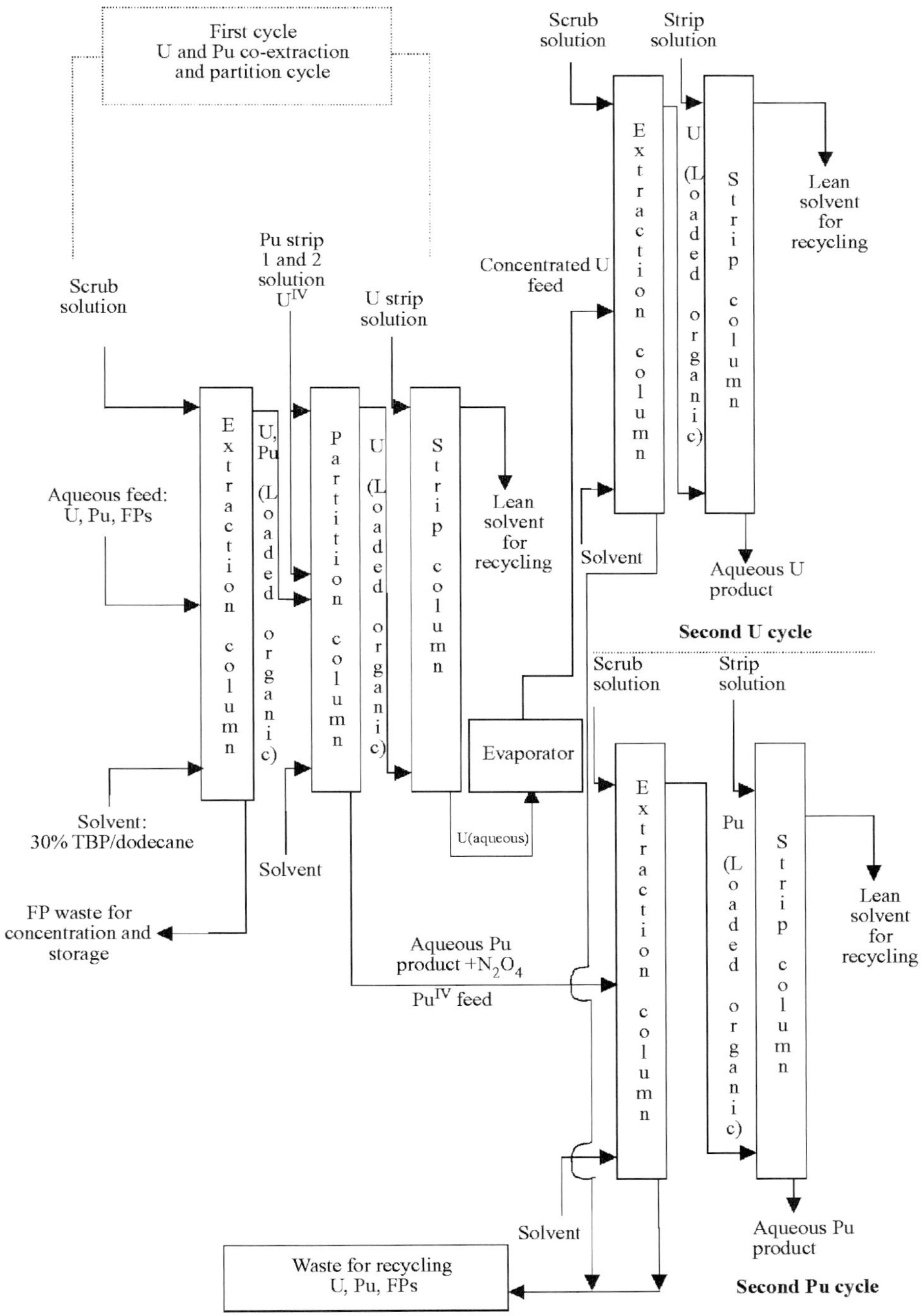

Figure 1
Purex flow scheme (FP, fission product).

of nitrogen by sparging oxygen as the reactant. Fluoride at trace levels may be required for better dissolution of fast reactor fuels.

The enhanced fission product content of high-burn-up fuels leads to the formation of noble metal fission product alloys consisting of ruthenium, rhodium, technetium, molybdenum, and palladium, which are insoluble in boiling nitric acid. After the dissolution of the fuel, these metal particles can be filtered as high-level solid waste (HLSW).

Prior to solvent extraction, the dissolved fuel solution is subjected to one or more of the following operations: (i) feed clarification to avoid choking and emulsification during extraction; (ii) adjustment of feed acidity (2–3M) for the extraction of uranium and plutonium; (iii) and adjustment of salting strength by dilution or evaporation to the desired uranium concentration level for the maximum processing rate.

For good extraction by TBP, conversion of all the plutonium to the Pu^{4+} state is ensured by the addition of $NaNO_2$ and digestion at 50°C:

$$PuO_2^{2+} + NO_2^- + 2H^+ \rightarrow Pu^{4+} + NO_3^- + H_2O \quad (4)$$

$$Pu^{3+} + NO_2^- + 2H^+ \rightarrow Pu^{4+} + NO + H_2O \quad (5)$$

$$6Pu^{3+} + 2NO_2^- + 8H^+ \rightarrow 6Pu^{4+} + N_2 + 4H_2O \quad (6)$$

Use of NO_2 gas for this avoids the introduction of sodium ion and has other advantages.

1.2 Solvent Extraction with TBP: Extraction of Metal Ions

Among the various extractants, TBP is now almost the universal choice as it meets most of the reprocessing requirements. It is highly selective for uranium and plutonium and has very low extractability for fission products. It is cheaply available, easily purifiable, has a high boiling point (266°C) and low mutual solubility. It also has good chemical, thermal, and radiation stability. The drawbacks are its high density (0.973) and viscosity, which are compensated for by dilution with inert diluents such as *n*-dodecane. The diluent for TBP should be nonpolar and unreactive toward HNO_3 and HNO_2 and stable to radiation. In addition, it should have a high flash point and low vapor pressure. The straight-chain aliphatic hydrocarbons containing 12–14 carbon atoms or *n*-paraffin mixtures (boiling point in the range 80–100°C) are preferred. Shell Sol T, Sol trol, etc., are also widely used. Generally 30% (v/v) TBP in *n*-dodecane diluent is used. The plutonium purification cycles often use a lower percentage of TBP.

The good extractability of nitrate complexes of UO_2^{2+} and Pu^{4+} coupled with the inextractability of fission products by TBP forms the main principle of the Purex process. As Pu^{3+} is weakly extracted, further separation of uranium from plutonium is achieved by reducing Pu^{4+} to Pu^{3+}, in which state it is stripped from the organic phase. The species extracted are $MO_2(NO_3)_2 \cdot 2TBP$ for M^{6+} ions, $M(NO_3)_4 \cdot 2TBP$ for M^{4+}, and $M(NO_3)_3 \cdot 3TBP$ for M^{3+}, where M denotes a metal with extractability following the same order. Nitric acid is extracted as $TBP \cdot HNO_3$, and is salted back to the aqueous phase during the loading of TBP with metal nitrates. The extraction equilibria between UO_2^{2+}/Pu^{4+} and TBP are given below:

$$UO_2^{2+} + 2NO_3^- + 2TBP \rightleftharpoons UO_2(NO_3)_2 \cdot 2TBP \quad (7)$$

$$Pu^{4+} + 4NO_3^- + 2TBP \rightleftharpoons Pu(NO_3)_4 \cdot 2TBP \quad (8)$$

The distribution ratio (D) of uranium and plutonium is directly proportional to the square of TBP (free) concentration and to the square (for uranium) and the fourth power (for plutonium) of nitrate ion concentration in the aqueous phase. When nitrates of uranium and plutonium are extracted by TBP, the D values increase sharply with aqueous nitric acid, pass through a maximum and decrease a little thereafter. The initial rise is due to increase in the concentration of NO_3^-, which acts as a salting agent. But at higher acidities the HNO_3 itself competes with uranium and plutonium, resulting in the fall in D values. In the presence of large amounts of extracted species, the available free TBP decreases, retarding further extraction. In some cases, as for thorium and plutonium, the solubility limit of the solvated complex in the diluent is reached, thus splitting the organic phase, leading to third-phase formation, which complicates column operation, and is avoided by controlling the solute concentration, acidity, temperature or modifying the solvent composition. Usually the high-temperature operation is beneficial for scrubbing out fission products and for stripping uranium from TBP.

1.3 Co-decontamination and Partitioning Cycle

After conditioning, the feed from the dissolver is fed to the center of a compound contactor comprising extraction and scrub sections. The solvent, 30% TBP in *n*-paraffin, is fed in at the bottom of the contactor, and 2M or 3M HNO_3 scrub enters from the top. Uranium and plutonium are extracted together by the solvent; the scrub removes the traces of fission products that are extracted. The fission products in the aqueous waste, high-active raffinate waste (HAW), and plutonium and uranium in the solvent stream leave from bottom and top of the column, respectively. An additional scrub column is often provided to improve the decontamination from fission products. While processing long-cooled fuels, ^{90}Sr–^{90}Y, ^{95}Zr–^{95}Nb, ^{106}Ru–^{106}Rh, ^{137}Cs, and ^{144}Ce–^{144}Pr are some of the fission products that affect the product purity. Among these, the behavior of ^{106}Ru–^{106}Rh and ^{95}Zr–^{95}Nb often determines the overall process performance.

During operation, variations in acid profile are possible for better decontamination from zirconium and ruthenium. A high loading of uranium in the solvent phase reduces the extraction of fission products, but too high a loading, on the other hand, will hinder the extraction of plutonium. Hence, an optimum saturation of 60–80% (of the theoretically possible $119 gUl^{-1}$) is generally preferred. Neptunium and americium are the other major actinides present in spent fuel. In Purex plants with appropriate redox conditions, neptunium can be made to follow either the plutonium or uranium stream for further separation. Am^{3+}, like the trivalent rare earths, is poorly extracted by TBP, and thus remains in the HAW. The HAW leaving the column is washed free of entrained TBP by diluent, and concentrated by evaporation to form high-level liquid waste (HLLW). The enhancement in acidity during evaporation is reduced by "killing" the acid using non-salt-forming chemicals such as formaldehyde under carefully controlled conditions.

The solvent loaded with uranium and plutonium from the extraction/scrub column is taken to the partitioning contactor for mutual separation. Back-extraction of plutonium from TBP is achieved by reducing it to the trivalent state with an aqueous solution containing a suitable reductant such as ferrous sulfamate or uranous nitrate stabilized with hydrazine. Use of ferrous sulfamate leads to higher solid content in the waste and corrosion problems due to iron and sulfate. Hence, uranous nitrate stabilized with hydrazine is preferred as the reductant in many plants. The HNO_2 present in the extraction contactors can oxidize U^{4+} and Pu^{3+}. The oxidation of Pu^{3+} is faster than that of U^{4+}. Also, the oxidation of U^{4+} by NO_2^- is autocatalytic. Hence, hydrazine is essential as a stabilizer to destroy NO_2^- in the system. It removes NO_2^- from the 30% TBP phase as well. Acidity of the aqueous phase is maintained at 0.6–2M; a higher acid profile might lead to reoxidation of Pu^{3+}, increasing the possibility of plutonium reflux.

Because of its extractability and poor stability in the TBP phase, U^{4+} is used in excess of stoichiometry for plutonium reduction, and is introduced at the center of the partitioning contactor at or near the plutonium-bearing organic feed point, to reduce and strip the bulk of the plutonium. From here, U^{4+} is carried in both directions by the two streams to all stages of the contactor. An aqueous scrub with U^{4+} and hydrazine in dilute nitric acid removes the residual plutonium from the uranium-bearing organic phase that leaves the column from the top. The Pu^{3+} is further purified from the accompanying U^{4+} and UO_2^{2+} in the bottom scrub section by a fresh TBP stream moving upward, which extracts and removes uranium. After the partitioning, the aqueous plutonium product is sent for final purification. As alternative approaches for uranium/plutonium partitioning, hydroxylamine nitrate at low acidity, a kinetically slow reducing agent, and *in situ* electrolytic reduction of plutonium in specially designed electropulse columns and mixer settlers have also been tested extensively.

The uranium-bearing organic stream is stripped free of uranium and the remaining traces of plutonium in the strip contactor with dilute acid (not lower than 0.01M to avoid plutonium hydrolysis). The organic stream is fed from the bottom of the contactor, and dilute acid strip flows down from the top. The acidity of the incoming organic feed contributes to the acidity of the aqueous phase. The lean organic phase leaving the column is recycled.

TBP is prone to radiolytic and chemical degradation, giving rise to dibutyl phosphate (DBP), monobutyl phosphate (MBP), and butyl alcohol. Nitric acid also reacts to give DBP and butyl nitrate, with complete dealkylation leading to orthophosphoric acid under extreme conditions. The presence of DBP in TBP increases the extraction of uranium, plutonium, and zirconium, and their stripping is difficult, which affects the overall DF. MBP is formed to a lesser extent than DBP, and is water-soluble. The diluent also undergoes degradation, but to a lesser extent. Hence, prior to recycling TBP is given an alkali wash to remove the DBP. However, the diluent degradation products are difficult to wash from TBP.

The uranium product with trace level impurities is concentrated in evaporators. The feeds to all the evaporators are scrubbed with organic diluent to remove the entrained and dissolved organic matter which otherwise will decompose in the evaporators to leave solid residues and heavy organics. These residues have a tendency to absorb and accumulate uranium/plutonium.

1.4 Purification of Uranium and Plutonium

For further purification of uranium, the concentrated feed from the evaporator is adjusted for acidity (1M), and plutonium is reduced to Pu^{3+} and fed to a compound contactor where uranium is extracted by TBP to a higher saturation, scrubbed free of plutonium and fission products and is stripped with 0.01M HNO_3 in a strip column. If required, this cycle is repeated. Finally, the concentrated uranium product is passed through a silica gel column at low feed acidity to further decontaminate it from zirconium and niobium. The product is then precipitated as ammonium diuranate, filtered, and calcined to the oxide.

The plutonium separated in the first cycle is once again passed through a solvent extraction cycle for further purification and concentration. Here, 20% TBP is preferred, as the feed contains only plutonium. The organic-to-aqueous flow ratios are modified in the extraction and strip columns to arrive at the required product concentration. Anion exchange processes have also been employed for this task. Here, Pu^{4+} is preferentially sorbed on the resin as $Pu(NO_3)_6^{2-}$ in the presence of 7.2M HNO_3. After washing at the same acidity, the plutonium-laden resin is eluted with 0.5M HNO_3 to obtain a concentrated product. It is then precipitated as the oxalate, filtered, and calcined to PuO_2.

1.5 Solid, Gaseous, and Liquid Waste Management

During reprocessing, wastes are generated in solid, liquid, and gaseous forms. The cladding material, at the end of the dissolution (in mechanical chop/leach process) is monitored for completion of dissolution and is fixed in a suitable matrix for disposal as HLSW. The HLLW carries the bulk of the fission product activity in a highly concentrated form. It is temporarily stored in stainless steel underground storage tanks with provision for cooling to remove decay heat from fission products, and is finally vitrified for long-term storage in deep geological repositories (see *Nuclear Waste Management, Materials Aspects of*). The medium or intermediate level waste (ILW) is neutralized and stored in carbon steel or stainless steel underground storage tanks, for disposal in a suitable matrix. The low-level waste (LLW) effluents from the plant are treated and released into the sea at levels far below the maximum permissible limits. The medium- and low-level solid wastes are segregated as combustible, noncombustible, α-bearing, non-α-bearing, etc., and treated independently to reduce the volume, and fixed in appropriate matrices for disposal.

TBP–dodecane waste is disposed of by incineration or decomposed by hydrolysis and pyrolysis to separate the mixture into hydrocarbons and active phosphoric acid.

Long-lived fission gases and volatile products such as T_2, THO, ^{85}Kr, Xe, $^{129}I_2$, $^{14}CO_2$, and $^{106}RuO_4$ are liberated during dissolution and other operations. The process equipment such as dissolvers, evaporators, and other vessels, are, therefore, kept under a few centimeters of water vacuum to prevent the escape of activity, and the vessel off-gases are subjected to separate treatment and filtration to remove the fission products and condensable vapors. Finally, the gases along with area ventilation air are released through a tall stack after filtration through high-efficiency particulate activity (HEPA) filters. The environmental releases from reprocessing have decreased considerably over the years due to national and international regulatory controls.

1.6 Partitioning and Transmutation (P & T)

For the sustainable development of nuclear power the environmental impact of the long-term radiotoxicity of HLLW needs to be reduced. As long-term hazards, HLLW contains, in addition to residual uranium and plutonium, several long-lived α-emitting minor actinides such as neptunium, americium, and curium, and fission products such as ^{99}Tc, ^{129}I, ^{93}Zr, and ^{135}Cs. The removal of these long-lived actinides and fission products from the waste through P & T would greatly reduce its long-term radiotoxicity. Removal of short-lived fission products such as ^{90}Sr and ^{137}Cs can reduce the heat generated from these wastes during storage.

The overall approach to partitioning or separation comprises a short-term goal to increase the recovery of plutonium, separate neptunium, and reduce waste volume by improvements in reprocessing facilities and a long-term effort involving separation of all transuranium elements, long-lived fission products, and their transmutation to less hazardous nuclides by recycling in fast-burner reactors or accelerator-driven systems which would need additional facilities. These separation projects have different names in different countries. In the USA, it is termed CURE (clean use of reactor energy) with TRUEX (transuranium extraction) forming a part. In Japan it is known as OMEGA (option for making gain from actinides), while the French use the term SPIN (*separation project incineration nucléaire*), with PURETEX and ACTINEX as its subsets. The need for deep geological repositories for the permanent disposal of HLLW cannot be eliminated by these approaches.

The aqueous separation procedures are based on extractants such as acidic organophosphorus reagents (HDEHP (di-(2-ethylhexyl) phosphoric acid), DIDPA (diisodecyl phosphoric acid)), neutral phosphineoxides (CMPO (octyl (phenyl)-*N*,*N*-diisobutyl carbomoyl methylphosphine oxide), TRPO (trialkyl phosphine oxide)), and diamides (DMDBTDMA; dimethyldibutyl tetradecyl malonamide). An integrated fast reactor approach with a fuel cycle based on nonaqueous pyrochemical processing and recycling is also under consideration as a long-term nonproliferation approach.

2. *The Thorex Process*

The Thorex (thorium uranium extraction) process for treating irradiated thorium has several special features. The ^{233}Pa formed by (n, γ) reactions of ^{232}Th decays with a 27day half-life and its complete decay needs to be ensured prior to reprocessing for ^{233}U and thorium recovery. The (n,2n) reactions encountered during the irradiation of thorium give rise to long-lived ^{231}Pa and shorter-lived ^{232}U (70years) with its hard β/γ-emitting daughter products. The yield of ^{232}U depends on the burn-up and the neutron spectrum in the reactor. The ^{232}U and its daughters in the separated ^{233}U and the ^{229}Th and ^{228}Th in thorium contribute to the radiation dose of these products. The Thorex HLW contains ^{231}Pa and to some extent ^{237}Np as the main long-lived actinides with long-term environmental impact. Once separated, ^{231}Pa should be easier to transmute to the shorter-lived ^{232}U.

2.1 Dissolution of Thorium

As in the case of aluminum-clad uranium, it is advantageous to dissolve the aluminum cladding and thorium separately. After alkali dissolution of aluminum, thorium is dissolved using an HNO_3–HF mixture. The presence of a small amount of the fluoride ion is essential to improve the rate of dissolution of thorium metal or thoria by nitric acid. The corrosion of stainless steel equipment due to fluoride ions is mitigated by the addition of aluminum to complex the excess free fluoride ions. Usually, a mixture of 13M HNO_3 with 0.03–0.05M HF and 0.1M $Al(NO_3)_3$ is used for dissolution. The thorium metal is easier to dissolve than thoria. High-density sintered thoria pellets are difficult to dissolve. If the thoria is clad in zircaloy, then a mechanical chop–leach process is employed for dissolution of the thoria.

2.2 Solvent Extraction and Separation of ^{233}U from Thorium

Here again, TBP is the most widely used extractant, and depending on whether only ^{233}U or both thorium and ^{233}U are to be recovered, the TBP content in the diluent (usually Shell Sol-T, dodecane or *n*-paraffin) varies. Generally, 3–7.5% TBP is used when ^{233}U alone is to be recovered. About 8gl^{-1} thorium is also coextracted in this step when 5% TBP is used. The thorium uptake reduces the extraction of fission products significantly. The coextracted thorium is scrubbed out with 1–2M HNO_3, and the final thorium contamination depends on the number and efficiency of the scrub stages provided. Finally, ^{233}U is stripped from the organic phase with very dilute acid. The thorium remaining in the raffinate can be recovered by a 30–40% TBP flow sheet, either immediately or after allowing for the further decay of ^{228}Th. Highly enriched uranium–aluminum alloy fuels after their dissolution in nitric acid in the presence of small amounts of mercury have been treated similarly.

When thorium is also to be extracted simultaneously with ^{233}U, a higher percentage of TBP, generally 30–42.5% in paraffin, is used. For thorium feeds in 1–3M HNO_3, the organic thorium loading is restricted to about 28–35g l^{-1} in the case of 30% TBP and to about 55gl^{-1} in the case of 42.5% TBP. Acid-deficient thorium feeds have also been used with $Al(NO_3)_3$ (from the cladding dissolved along with thorium) as the salting out agent. The nitric acid salting generates lesser volumes of waste.

For the separation of ^{233}U from thorium extracted together in TBP, a higher acid strip (about 0.3M) is used to remove bulk of the thorium, while very dilute acid is used for ^{233}U. The ^{233}U is further purified by a low-percentage TBP flow sheet or by ion exchange techniques.

2.3 Purification of ^{233}U

The purification of ^{233}U product from thorium is achieved by an anion exchange process, as uranium forms a strong anionic chloride complex at 8M HCl which is sorbed preferentially by the exchanger. As the use of chloride as the process medium has several disadvantages, alternative methods of purification have been proposed: (i) anion exchange in an acetic acid medium for sorption of uranium as an anionic acetate complex; (ii) cation exchange in nitric acid for preferential sorption and separation of Th^{4+}; and (iii) precipitation and separation of thorium from ^{233}U as the oxalate. Of these, the first method has problems similar to that of the chloride. The cation exchange method is amenable for plant operation. The third method is a batch technique useful in laboratory procedures. Conventional routes are followed for the conversion of ^{233}U and thorium products into their respective oxides. As the radiation level of ^{233}U increases with time due to ^{232}U decay, the final product fabrication should be done immediately. Otherwise purification would be required to reduce the radiation level in ^{233}U. This is done by removing ^{228}Th, the longest-lived daughter of ^{232}U, and thus breaking the decay chain (usually by cation exchange technique). The ^{229}Th formed from ^{233}U decay follows ^{228}Th during this step.

3. Recent Developments in Reprocessing

Most of the plutonium available in the world is recovered through the Purex process. Now, the world over, only long-cooled (> 5years) fuels are processed; for these, the Purex process is well entrenched, and is viable in terms of its applicability for processing both the fuels currently encountered and those anticipated in the near future. Developments in chemical engineering technology have had an impact on the Purex process too. One example is the development of the single-cycle Impurex process that depends on computer control of an extraction column with on-line monitoring and automated control of process parameters by sophisticated and sensitive instrumentation systems. Thus, development of single-cycle processes, maintenance-free short-residence contactors, and annular pulse columns that are safe with respect to criticality, near-real-time accountancy of fissile and fertile actinides, use of *in situ* electrochemical, photolytic, and other techniques for plutonium and neptunium reduction and partitioning, and for an ultimate salt free process, minimization of losses to waste, improvements in solvent quality, direct denitration of products to oxides, recovery of useful isotopes such as ^{237}Np, ^{99}Tc, Pd, etc., and also separation of long lived actinides and fission products through P & T with a general reduction in waste volumes are some of the challenging research and development tasks ahead. In processing high-burn-up, short-cooled fast reactor fuels, multiple recycling of plutonium would pose problems in the process and in the handling of the final products due to the α activity arising from increasing amounts of ^{238}Pu, ^{241}Am, and Cm isotopes. The radiolytic degradation of TBP may also be severe, and development of techniques such as pyroelectrochemical processes may become attractive under these circumstances (see *Oxide Fuels, Nonaqueous Reprocessing of*).

In addition to the above-mentioned aspects, the processing of irradiated thorium needs developmental efforts to be focused on the following: better head-end techniques for decladding and dissolution of zircaloy- and stainless steel-clad thoria; corrosion-resistant construction materials; modifications to the Thorex process to deal with ThO_2–PuO_2 fuels; clarification of the behavior of ^{231}Pa in the process and its long-term impact in HLLW; a comparison of the radiological safety in handling $^{233}U/^{232}U$ and that of multirecycled plutonium; and alternative possibilities for on-line solvent-free or nonaqueous process routes for treating high-burn-up fuels for the recovery of either ^{233}Pa or ^{233}U. Finally, state-of-the-art remote handling technology and robotics will have a significant impact on this technology.

See also: Nuclear Fuel Cycles; Oxide Fuels, Nonaqueous Reprocessing of; DUPIC Fuel Cycle

Bibliography

Bruce F R, Fletcher J M, Hyman H H, and Katz J J (eds.) 1956 Second International Conference on the Peaceful Uses of Atomic Energy. Progress in Nuclear Energy Series 3. Process Chemistry. Pergamon, London, Vol. I

Bruce F R, Fletcher J M, and Hyman H H (eds.) 1958 Second International Conference on the Peaceful Uses of Atomic Energy. Progress in Nuclear Energy Series 3. Process Chemistry. Pergamon, London, Vol. II

Bruce F R, Fletcher J M, and Hyman H H (eds.) 1961 Second International Conference on the Peaceful Uses of Atomic Energy. Progress in Nuclear Energy Series 3. Process Chemistry. Pergamon, London, Vol. III

Choppin G, Rydberg J, and Liljenzin J O (eds.) 1995 *Radiochemistry and Nuclear Chemistry*. Butterworth-Heinemann, Oxford, 2nd edn

Edwal B, Detilleux E, Redon A, Centeno J, Drent W, Dewitte R1963 Aqueous Reprocessing Chemistry for Irradiated Fuels, Brussels Symposium. European Nuclear Agency, Paris

Flagg J F (ed.) 1961 *Chemical Processing of Reactor Fuels*. Academic Press, New York

Long J T 1978 *Engineering for Nuclear Fuel Reprocessing*. American Nuclear Society, La Grange Park, IL

Martin F S, Miles G L 1958 *Chemical Processing of Nuclear Fuels*. Butterworth, London

Ryan J L, Wheelwright E J 1959 Recovery, purification and concentration of plutonium by anion exchange in HNO_3. USAEC Report HW 55893

Schulz W W, Burger L L, Navaratil J D, and Bender K P (eds.) 1984 *Science and Technology of Tributylphosphate,* Vol. III. Applications of Tributylphosphate in Fuel Reprocessing. CRC Press, Boca Raton, FL

Sood D D, Patil S K 1996 Chemistry of nuclear fuel reprocessing—current status. *Radioanal. Nucl. Chem. Art* **203** (2), 547

Staller S M and Richards R B (eds.) 1961 *Reactor Hand Book* Vol. 2. Fuel Reprocessing. Interscience, New York

Stevenson C E, Mason E A, and Gresky A T (eds.) 1970 Progress in Nuclear Energy, Series III. Process Chemistry. Pergamon, New York, Vol. 4

Taube M 1974 *Plutonium, A General Survey*. Verlag Chemie, Weinheim

Wick O J (ed.) 1967 *Plutonium Handbook, A Guide to the Technology*. Gordon and Beach, New York, Vols. I and II

R

Radiation Damage Theory

B. N. Singh, H. Trinkaus, and S. I. Golubov

Although not recognized as such at that time, the study of the effects of radiation on the structure and properties of materials started more than a century ago. It was observed in the nineteenth century, for instance, that naturally occurring minerals of an initially crystalline structure became "more-or-less" amorphous if they contained uranium or thorium as impurities. This amorphous state was termed *metamict* (meaning "mixed otherwise" in Latin) by Norwegian mineralogist, Brøgger, in 1893 (see Pabst 1952 for a review). After the discovery of radioactivity by Becquerel in 1896, it was realized that both uranium and thorium were α-particle emitters and this led to the correct explanation by Hamberg (1914) that the metamict state was caused by self-bombardment of these minerals by radioactive decay products.

At this time, however, it was still not known that the accumulation of irradiation-induced atomic displacements were responsible for causing the observed amorphization. It is an interesting historical coincidence that at about the same time Bohr (1913) published the first theoretical study of interactions between energetic projectiles and a solid target. The continuation of theoretical studies along this line finally yielded a full theoretical description of irradiation-induced atomic displacements (see Sect. 1).

It is interesting to note here that it was the development of fission reactors in the 1940s which provided the major driving force for studying the effects of radiation in solids on a large scale. Already in 1946, Wigner had pointed out the possibility of deleterious effects of irradiation on properties of materials exposed to a high flux of fission neutrons. Soon after, his concerns were confirmed by experiments. Hence, the effect of radiation became known as *Wigner disease* (Seitz 1952). This seems to be the reason why it became a tradition to refer to the influence of radiation as "radiation damage." It is important to note that the materials problems related to radiation damage are likely to be even more severe in the case of fusion reactors since the material in the first wall and divertor components of fusion reactors will be exposed to an even higher flux of higher energy neutrons.

The nature and consequences of defect accumulation have been investigated extensively since the 1960s and have identified the following major effects of irradiation on materials properties: (i) hardening and reduction in ductility; (ii) volumetric expansion due to creep, anisotropic growth, and void formation; and (iii) grain boundary embrittlement due to segregation of solid or gaseous impurity atoms. It should be noted that all of these parameters can seriously limit the useful lifetime of materials in service in a fission or a fusion reactor. Thus, for a safe and commercially successful exploitation of fission or fusion reactors as energy sources, it is essential to evaluate the effects of irradiation on properties and performance of materials expected to be employed in the environment of energetic neutrons.

Unfortunately, none of these problems have been solved to the extent where the effects of irradiation with fission or fusion neutrons can be reliably predicted. The main reason for this seems to be the lack of a detailed understanding of the complex physical processes involved between the production of atomic displacements (on a nano scale) and the bulk response of materials employed in components of a device (on a macro scale). Therefore, in order to evaluate the performance and lifetime of materials exposed to different irradiation conditions, it is necessary to understand details of the primary damage production as well as the processes involved in the spatial and temporal evolution of the irradiation-induced microstructure. Furthermore, such understanding is a prerequisite for establishing a realistic framework within which the problems, e.g., if void swelling, irradiation hardening, and plastic instability can be treated in a consistent and cohesive fashion. In this article an attempt is made to identify and highlight the major theoretical issues involved in these problems.

1. Primary Damage Processes

During irradiation, the interaction between energetic projectiles and a solid target is a rather complex process, involving not only interaction with atoms but also the electronic excitations in the target materials. A detailed description of the physics of the collision events and processes leading up to the generation of single or multiple displacements is beyond the scope of this article (see Robinson 1994 for a review). However, since the irradiation-induced lattice defects and their effects on the properties of the target materials originate from displacements, it is of interest to have at least some general idea about their origin as well as dependence on irradiation parameters (e.g., projectile mass and energy) and materials variables (e.g., atomic mass density and electronic structure).

1.1 Collision Events and Atomic Displacements

Historically, the study of interactions between energetic projectiles and a solid target dates back to the original work of Bohr (1913). The continuation of his theoretical studies led to his well-known work *The Penetration of Atomic Particles through Matter*, which was published in 1948. Later, Lindhard joined this systematic study of atomic collisions in solids and, together with his co-workers laid down the foundation of collision physics describing the production of the nascent damage state including cascades and subcascades of displacements (e.g., see Lindhard *et al.* 1968).

During the primary interaction between an energetic projectile particle of initial energy, E, and a solid target, a recoil energy, T, is transferred in less than 1×10^{-15} s to the target atom. In the case of neutrons, this occurs via elastic collisions between neutrons and the target atoms. During these events, when T exceeds the displacement threshold energy, E_d, the struck atom is displaced from its original site, creating a vacancy-interstitial pair (i.e., a Frenkel pair). However, charged particles (e.g., electrons, protons, and heavy ions) at high energy interact primarily inelastically with the electrons in the solid and spend a large fraction of their energy in exciting them. As the projectile slows down, the electronic loss becomes less important and elastic collisions with the target atoms become dominant. Thus, because of electronic losses, including those of the recoiled atoms, not the whole recoil energy, T, but only a part of it, the damage energy (T_{dam})

$$T_{dam}(T) = T - Q(T) \tag{1}$$

is available for further displacements where Q is the electronic energy loss. During charged particle irradiations, target atoms are displaced from their original lattice sites when $T_{dam}(T) > E_d$.

For the sake of completeness, it should be added that γ-photons interact mainly with the electrons of the target atoms. γ-photons may get absorbed in the nuclei which subsequently may emit neutrons via $(\gamma - n)$ reactions. Upon such reactions the atom involved is likely to recoil and may even generate displacement if the damage energy T_{dam} reaches a value above E_d.

Under low recoil energy irradiation conditions (e.g., 1 MeV electrons), practically all displacements yield isolated single vacancies and self-interstitial atoms (SIAs) in the form of Frenkel pairs randomly distributed in space. At primary knock-on atom (PKA) energies substantially higher than the displacement threshold energy, instead of single and isolated displacements, a hierarchy of displacements is produced in a relatively small volume (on nanometer scale) and in a relatively short time ($<1\times10^{-12}$ s). The number and spatial distribution of displaced atoms in a displacement cascade depends on mass, electrical charge, and energy of the projectile particle, as well as on temperature, crystal structure, and mass of target atoms. Kinchin and Pease (1955) were the first to estimate the number of displaced atoms, N_d, in a cascade. Twenty years later this model was modified by Norgett, Robinson, and Torrens (1975). This is commonly known as the NRT model, and takes into account the energy losses in electronic excitation (Eqn. (1)) and considers the fact that the colliding atoms do not behave as hard spheres, thus yielding

$$N_d(T) = 0.8\, T_{dam}/2\, E_d \text{ (for } T_{dam} > 2.5\, E_{Bd}) \tag{2}$$

The integration of the NRT damage function yields the nominal atomic concentration of NRT displacements per atom (dpa). According to ASTM/E 521-96 (1996) the internationally accepted dose unit, dpa, is defined as

$$\text{dpa} = \int_{E_{min}}^{E_{max}} \int_{E_{d,min}}^{T_{max}} \int_0^t N_d(T)\,d\sigma(E,T)\,d\phi(E,t)\,dt \tag{3}$$

where $d\sigma$ is the differential cross-section for the production of PKA_s with energies between T and $T+dT$, $d\phi(E)$ is the flux of irradiating particles with energies between E and $(E+dE)$ and t is the irradiation time. E_{min} and E_{max} are the minimum and maximum values, respectively, of the energy spectrum of irradiation particles. T_{max} is the maximum energy transferable by a projectile of mass, m, and energy, E, to a target atom of mass, M, in a head-on collision and is given by

$$T_{max} = \frac{4Mm}{(M+m)^2}E \tag{4a}$$

and

$$T_{max} = \frac{2m_e}{M}\left(\frac{E}{m_e c^2} + 2\right)E \tag{4b}$$

for nonrelativistic particles and relativistic electrons, respectively where m_e is the mass of electrons and c is the velocity of light.

The NRT model is generally accepted as the international standard for quantifying the number of atomic displacements produced under cascade damage conditions. Unfortunately, the NRT model is based on the elementary theory of isolated binary collisions and cannot be used to determine the amount, spatial distribution, and clustering of lattice defects resulting from the displacements during the collisional phase and surviving at the end of the cooling down phase of cascades (see Sect. 1.2). Thus, the NRT model for atomic *displacements* is not a physically realistic indicator of *defect production* under cascade damage conditions. As discussed below, the lack of this information regarding the transformation of atomic displacements into lattice defects has impeded the adequate theoretical treatment of the damage accumulation and the irradiation-induced changes in physical and mechanical properties of irradiated materials for a number of years.

Finally, the inelastic interactions between the projectile particles and the nuclei of the solid target are considered. The resulting nuclear reactions are particularly important for materials exposed to fission or fusion neutrons since these reactions would not only generate additional displacements but more importantly would produce solid and gaseous (e.g., helium and hydrogen) transmutational impurities. Since helium atoms are practically insoluble in most metals, their continuous generation and accumulation, particularly in the form of bubbles at grain boundaries, may seriously degrade the mechanical performance and lifetime of structural components in fast breeder reactors. This is even more evident in fusion reactors since the generation rate of helium by fusion neutrons is expected to be considerably higher than that by fission neutrons. The generation of solid impurities may also affect the microstructural evolution and the mechanical and other physical properties due to irradiation-induced segregation of impurity atoms at grain boundaries and precipitation of new phases at the boundaries as well as in the grain interior.

1.2 Surviving Defects and Their Clusters in Displacement Cascades

The evolution of damage microstructure, and its impact on the properties of materials under irradiation, is determined not directly by the atomic displacements but rather by lattice defects such as vacancies and self-interstitial atoms (SIAs) surviving in the collision cascades. Over the years many attempts have been made to solve this problem of transformation of cascades of atomic displacements into lattice defects and their clusters by using molecular dynamics (MD) to simulate the evolution of displacement cascades. Although the MD technique was pioneered in the 1960s (e.g., Gibson *et al.* 1960), it is only recently that the availability of more reliable interatomic potentials and more powerful computers have made it possible to simulate displacement cascades induced by recoils with high energies. In the 1990s, a large number of simulation studies (see Diaz de la Rubia 1996, Bacon *et al.* 1997 for reviews) of displacement cascades were carried out in various metals. The properties of the clusters produced in the cascades have been reviewed recently by Osetsky *et al.* (2000a).

High energy recoils generated, for example, by fission and fusion neutrons are likely to create damage in the form of cascades and multiple subcascades (Fig. 1) and can be simulated using the binary collision approximation (BCA) code MARLOWE (e.g., Heinisch and Singh 1993). Although these simulations do not yield information regarding the surviving lattice defects in the cascades, they can be used to determine the number and morphology of subcascades produced at the end of the collisional phase. It has been shown, for example, that the number of

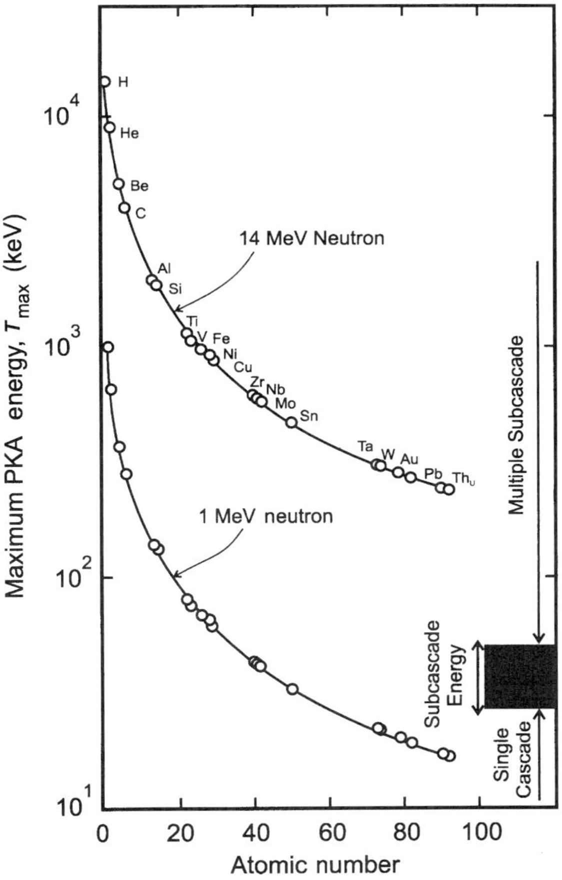

Figure 1
Variation of maximum recoil energies with atomic number of the target materials. Note that the primary damage state may vary significantly not only with varying projectile energy but also with atomic number of target materials (Ishino *et al.* 1989).

subcascades per PKA (or cascade) increases with increasing recoil (or damage) energy. In addition, at a given recoil damage energy, the number of subcascades in different metals decreases with increasing atomic number (e.g., Kiritani *et al.* 1990, Heinisch and Singh 1993). Thus, the binary collision simulations can be used to provide a statistically meaningful description of the initial cascade morphology as a function of recoil energy as well as materials parameters.

The general features of defect production and clustering in displacement cascades as demonstrated by a large number of MD simulations have been summarized recently by Singh (1999a). The following are the crucial aspects of defect evolution in cascades that must be taken into account in the theoretical treatment (see Sect. 2) of defect accumulation under cascade damage conditions:

(i) A substantial decrease in the number of surviving defects compared to the NRT displacements;
(ii) Intracascade clustering of SIAs and vacancies in spatially segregated fashion such that the vacancies cluster i the center and SIAs at the periphery of cascades;
(iii) 1D diffusional transport of SIA clusters; and
(iv) changes in Burgers vector (i.e., direction of motion) of SIA clusters during 1D diffusional transport.

Various fractions of surviving defects are illustrated schematically in Fig. 2. It can be seen that the nature of defect generation under cascade damage conditions is qualitatively different and immensely more complicated than that under single displacement production condition.

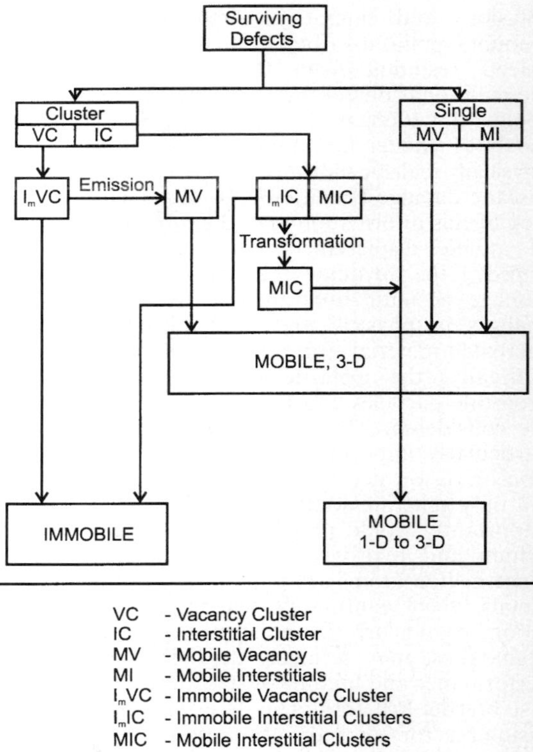

Figure 2
Schematic illustration of various fractions of surviving defects under cascade damage conditions. Obviously, the clustering of SIAs in the cascades and intrinsic properties of these clusters must be incorporated in the treatment of defect kinetics and accumulation under cascade damage conditions (Golubov *et al.* 2000b).

1.3 Properties of Lattice Defects and their Clusters

The intrinsic properties of vacancies and SIAs and their clusters play a crucial role in the evolution of the microstructural features during irradiation. Important parameters of single vacancies and SIAs are their diffusivities and particularly their migration energies which control the temperatures at which they become mobile. The difference in the relaxation volume between vacancies and SIAs gives rise to a preferential attraction of SIAs to dislocations commonly known as dislocation bias (see Greenwood *et al.* 1959). Consequently, more vacancies than SIAs are left in the system. The excess of vacancies thus created leads to void formation and growth. Under Frenkel pair production conditions, the dislocation bias is, in fact, the only driving force for the void swelling. A combination of diffusivity of vacancies and SIAs and the thermal stability of their clusters determines the kinetics and, thereby, the temperature dependence of defect accumulation and the resulting irradiation hardening and void swelling. At irradiation temperatures below the recovery stage III (i.e., when vacancies are immobile), the defect accumulation occurs, for instance, predominantly in the form of clusters of vacancies and SIAs and leads to irradiation hardening. Since vacancies are immobile in this temperature range, no void swelling can be expected.

The defect accumulation under cascade damage conditions is more complicated because of the generation of clusters of vacancies and SIAs directly in the cascades (see Fig. 2). As described already in Sect. 1.2, a fraction of the SIA clusters thus generated are glissile and may change their direction of motion (i.e., Burgers vector) during their transport. As a result, the defect reaction kinetics and, hence, the defect accumulation behavior under cascade damage conditions is expected to change significantly compared to that under Frenkel pair production conditions. Thus, in addition to the thermal stability of these clusters, information about their mobility and propensity for Burgers vector changes is also needed. Finally, it should be recognized that we also need to know about the configurational stability of these clusters produced in the cascades since it affects the ability of a cluster to diffuse one-dimensionally. Investigations have demonstrated, for example, that (i) all stable SIA clusters and loops in b.c.c. iron are glissile whereas only a fraction of them is expected to be glissile in f.c.c. copper, and (ii) that stable vacancy clusters and loops in f.c.c. copper are intrinsically sessile whereas those in b.c.c. iron are likely to perform thermally activated 1D diffusion (Osetsky *et al.* 2000a, 2000b).

2. *Theoretical Treatment of Damage Accumulation*

During the early days of the nuclear technology, a series of experimental observations demonstrated that mechanical and other physical properties of materials could be seriously impaired due to irradiation with fission neutrons. It was shown, for example, that materials exposed to neutrons could suffer from irradiation creep and growth (Konobeevsky *et al.* 1955), radiation hardening (Seeger 1958, Blewitt *et al.* 1960), and void swelling (Cawthorne and Fulton 1967). In addition to intrinsic academic interests in these phenomena, the technological concerns for materials performance in neutron environment provided a strong impetus for establishing a theoretical understanding of the damage accumulation and microstructural evolution during irradiation.

2.1 Basic Considerations

Reactions between primary irradiation-induced defects generated by displacements and between these defects, and the evolving descendant defects such as voids and pre-existing extended defects such as dislocations and grain boundaries result in partial defect annihilation as well as in defect accumulation. A reaction between two defects requires mobility of at least one of them and an attractive interaction between them. At distances large compared to the atomic spacing of the matrix lattice, the interaction between two defects is mainly mediated by lattice strains induced by them and is therefore called *elastic interaction*.

The diffusion, interaction, and reaction of a system of defects in a crystal lattice is a very complicated problem which must be efficiently simplified to make it tractable. For describing the global damage accumulation, the chemical diffusion-reaction theory, or briefly chemical rate theory (Brailsford and Bullough 1972, Wiedersich 1972) has been commonly used. In this mean-field type of approach, the production, distribution, and reaction of a defect, which is assumed to be random in space and time, is approximated by an equivalent homogeneous and continuous production, distribution, and reaction, characterized by a defect production rate density, defect concentration, and reaction rate density, respectively.

Crucial steps in this approach are, (i) the approximation of regions of strong defect interaction by reaction volumes embedded into an effective medium, and (ii) a self-consistent determination of rate constants and associated sink strength for each type of reaction such that (at least in principle) the reaction rates in the effective medium are the same as the average reaction rates in the actual discrete system of interacting defects. It is worth mentioning that the directionality in the diffusion of a defect, with the two extreme cases of strictly 1D and isotropic 3D, plays a crucial role in the dependence of the sink strength for the absorption of the defect on the size and density of the sinks.

Even in the framework of the chemical rate theory, the treatment of defect accumulation under irradiation is still rather complicated because of the generally large number of defects involved in this process. To keep the treatment transparent, it is useful to restrict consideration to the most relevant defects and their reactions. The judgement of the relevance of certain defect reactions depends, of course, on the knowledge of defect production and properties. Accordingly, the history of modeling defect accumulations reflects the level of available knowledge at that time.

2.2 Standard Rate Theory Model

The chemical rate theory approach was employed in the very first attempts to model damage accumulation under irradiation (Brailsford and Bullough 1972, Wiedersich 1972). The following main additional assumptions were made in these early attempts: (i) atomic displacements result in isolated Frenkel pairs, and, correspondingly, single SIAs and vacancies are produced at equal rates, (ii) both SIAs and vacancies perform 3D diffusion when they are mobile, and (iii) the efficiencies for the absorption of SIAs and vacancies are different for different sinks because of differences in the strength of the corresponding elastic interactions (sink bias). The preferential absorption of SIAs by dislocations (single point defect dislocation bias) is the only driving force for the microstructural evolution in this model, referred to in the following as the "Standard Rate Theory" (SRT) model.

Assuming that voids and dislocations form the dominant sinks for mobile SIAs and vacancies (justified for pure metals), the SRT equations for the concentrations of SIAs, C_i, and vacancies, C_v, may be written as

$$\frac{\mathrm{d}C_i}{\mathrm{d}t} = G_{\mathrm{NRT}} - D_iC_i(Z_i^v k_v^2 + Z_i^d \rho) - \mu_r D_i C_i C_v \quad (5a)$$

$$\frac{\mathrm{d}C_v}{\mathrm{d}t} = G_{\mathrm{NRT}} - D_vC_v(Z_v^v k_v^2 + Z_v^d \rho) - \mu_r D_i C_i C_v \quad (5b)$$

Here, G_{NRT} is the NRT Frenkel pair generation rate (Norgett *et al.* 1975), $D_i \gg D_v$ are the diffusion coefficients of SIA and vacancies, respectively. $Z_{i,v}^v k_v^2$ and $Z_{i,v}^d\ \rho$ are the sink strengths of voids and dislocations for the absorption of the 3D diffusing point defects, respectively, where $Z_{i,v}^v \approx 1$, $Z_i^d > Z_v^d \approx 1$, are "bias factors" and $k_v^2 = 4\pi N_v R_v$ (N_v, R_v: number density and mean radius of voids) and μ_r is the recombination coefficient. Note that the total sink strength is equal to the mean reciprocal square of the diffusion range of the defects, $k^2 = \lambda^{-2}$. For the sake of simplicity the thermal emission of vacancies from voids and dislocations is not included in Eqns. (5a) and (5b).

For quasi-steady state where the point defect concentrations follow $\mathrm{d}C_v/\mathrm{d}t \sim \mathrm{d}C_i/\mathrm{d}t \sim 0$, during the slow evolution of the sink strengths, Eqns. (5a) and (5b) yield

$$D_vC_v = \frac{2G_{\mathrm{NRT}}}{(Z_v^v k_v^2 + Z_v^d \rho)} \left(1 + \sqrt{1 + \frac{4G_{\mathrm{NRT}}\mu_r}{D_v(Z_v^v k_v^2 + Z_v^d \rho)(Z_i^v k_v^2 + Z_i^d \rho)}}\right)^{-1} \quad (6a)$$

$$D_iC_i = \frac{(Z_v^v k_v^2 + Z_v^d \rho)}{(Z_i^v k_v^2 + Z_i^d \rho)} D_vC_v \quad (6b)$$

Damage accumulation in the form of voids is controlled by the difference in the fluxes of vacancies and SIAs to voids, determined by the quantity $Z_v^v D_v C_v - Z_i^v D_i C_i$ which represents the effective vacancy supersaturation when normalized to the self-diffusion coefficient. Equations (5a) and (5b) or (6a) and (6b) yield for this quantity

$$Z_v^v D_v C_v - Z_i^v D_i C_i = p_d \frac{Z_v^v Z_v^d \rho}{(Z_i^v k_v^2 + Z_i^d \rho)} D_v C_v, \quad (7a)$$

$$\rightarrow p_d G_{NRT} \frac{Z_v^v Z_v^d \rho}{(Z_i^v k_v^2 + Z_i^d \rho)(Z_v^v k_v^2 + Z_v^d \rho)}, \text{for } \mu_r \rightarrow 0 \quad (7b)$$

where

$$p = \frac{Z_i^d}{Z_v^d} - \frac{Z_i^v}{Z_v^v} \quad (8)$$

is the (single defect) dislocation bias. Equation (7b) represents the simplified explicit version of Eqn. (7a) for negligible recombination. The rate of swelling, $\mathrm{d}S/\mathrm{d}t$, is obtained by multiplying Eqn. (7b) with the sink strength of the voids

$$\frac{\mathrm{d}S}{\mathrm{d}t} = (Z_v^v D_v C_v - Z_i^v D_i C_i) k_v^2 \quad (9)$$

According to Eqns. (7a) and (7b), the most important parameter of the SRT model for damage accumulation in the form of voids is the (single defect) dislocation bias, p_{d}. Equation (9) together with Eqn. (7b) is suited for describing void swelling in pure metals irradiated under the conditions of single Frenkel pair production (e.g., 1 MeV electrons). It should be recognized that at the time the SRT model was proposed, very little was known about the details of intracascade events, even though experimental results showing the formation of vacancy clusters in displacement cascades were reported already in 1960s (see Eyre 1973 for a review). Practically nothing was known at that time regarding SIAs and their clusters in the cascades. In fact, the first indication of intracascade clustering of not only vacancies but also of SIAs emerged in 1973 from the evaluation of diffuse x-ray scattering experiments on copper neutron irradiated at ~4 K (von Guérard and Peisl 1975).

2.3 BEK Model

By the early 1970s a large amount of experimental results had accumulated on cascade collapse behavior in a number of f.c.c. and b.c.c. metals irradiated with heavy ions (mostly self-ions). The most valuable and comprehensive results came from the near-surface cascade collapse studies (see English and Jenkins 1987 for a review). These experiments clearly established that vacancies generated in a cascade collapse into a vacancy loop within the cascade volume. It was also demonstrated that the cascade collapse behavior under fast neutron and heavy ion irradiation is comparable and that the vacancy loops form even at void swelling temperatures. The results showed further that the cascade collapse efficiency (for a given recoil energy) is substantially higher in f.c.c. metals than in b.c.c. metals.

In response to these experimental observations, Bullough *et al.* (1975) modified the SRT model by taking into account the formation of vacancy loops in the cascades and their subsequent annihilation by vacancy emission and excess SIA absorption. This is commonly known as the *BEK model.* It should be emphasized that, as with the SRT model, the BEK model considers the single point defect dislocation bias to be the only driving force for void swelling and does not include considerations of SIA clustering in the cascades.

Assuming that a fraction $\varepsilon_v G_{\mathrm{NRT}}$ of vacancies is produced in cascades in the form of clusters/loops, Eqns. (5a) and (5b) for the point defect concentrations change to

$$\frac{dC_i}{dt} = G_{\rm NRT} - D_iC_i(Z_i^v k_v^2 + Z_i^d \rho + Z_i^{vc} k_{vc}^2) - \mu_r D_i C_i C_v \quad (10a)$$

$$\frac{dC_v}{dt} = (1-\varepsilon_v)G_{\rm NRT} + K_{vc} - D_vC_v(Z_v^v k_v^2 + Z_v^d \rho + Z_v^{vc} k_{vc}^2) - \mu_r D_i C_i C_v, \quad (10b)$$

where K_{vc} is the rate of thermal emission of vacancies from the clusters and $Z_i^{vc}k_{vc}^2$, $Z_v^{vc}k_{vc}^2$ are the sink strengths of the vacancy clusters for absorbing SIAs and vacancies, respectively.

The rate equation for the concentration of vacancies in clusters, $x_{vc}C_{vc}$, where x_{vc} is the average number of vacancies per cluster and C_{vc} is the cluster concentration, may be written in the form

$$x_{vc}\frac{dC_{vc}}{dt} = \varepsilon_v G_{\rm NRT} - K_{vc} - (D_iC_iZ_i^{vc} - D_vC_vZ_v^{vc})k_{vc}^2 \quad (11)$$

At the beginning of irradiation, when C_{vc}, K_{vc} and k_{vc}^2 are still negligibly small, the only modification compared to SRT is a reduction in the production of single vacancies due to vacancy clustering in cascades, resulting in a reduction of the effect of the dislocation bias on the effective vacancy supersaturation (as given by Eqn. (7)).

Alternatively, for quasi-steady-state, elimination of the cluster production term $\varepsilon_v G_{\rm RT}$ from Eqn. (10b) with the aid of Eqn. (11) results in the following balance equations for the point defect concentrations

$$G_{\rm NRT} = D_iC_i(Z_i^v k_v^2 + Z_i^d \rho) + \mu_r D_i C_i C_v + Z_i^{vc} D_i C_i k_{vc}^2 \quad (12a)$$

$$G_{\rm NRT} = D_vC_v(Z_v^v k_v^2 + Z_v^d \rho) + \mu_r D_i C_i C_v + Z_i^{vc} D_i C_i k_{vc}^2 \quad (12b)$$

The last terms on the right hand side of Eqns. (12a) and (12b) describing the effect of vacancy clusters on point defect annihilation in steady-state are equal, in analogy with the recombination term, but in contrast to the first terms describing the biased point defect annihilation at voids and dislocations. This shows clearly that, in steady state, the clusters effectively act as recombination centers by which the point defect concentrations are reduced compared to the ones given by Eqn. (6a) (Golubov 1981). Equations (6b), (7a), and (9) remain valid and in Eqn. (7b) only the term in the second brackets in the denominator has to be augmented by $Z_i^{vc}k_{vc}^2$ which shows the reduction of the effective vacancy supersaturation and the corresponding void swelling in the BEK model compared to the SRT model.

Since the thermal stability and, correspondingly, the sink strength of vacancy clusters, increases with decreasing temperature, the swelling rate decreases with decreasing irradiation temperature faster in the BEK model than that in the SRT model. This is qualitatively in agreement with experimental observations and is the main achievement of the BEK model. However, it must be emphasized that the model cannot give a proper description of defect accumulation under cascade damage conditions since it does not consider the intra-cascade clustering of SIAs and consequences of 1D diffusion of SIA clusters produced in the cascades. It should be mentioned that the so-called *composite* model proposed by Stoller and Odette (1987) is similar to the BEK model except that some additional defect reactions are taken into account and that the vacancy clusters are assumed to be micro-voids instead of vacancy loops.

2.4 Production Bias Model

As outlined in Sects. 2.2 and 2.3, both SRT and BEK models are not suited for treating the kinetics of defect accumulation under irradiation producing clusters not only of vacancies but also of SIAs in the nascent cascades. In fact, it is the clustering of SIAs rather than of vacancies which dominates the damage accumulation behavior under cascade damage conditions at void swelling temperatures. Recognizing the importance of intracascade clustering of SIAs and vacancies, possibly occurring with different clustering efficiencies, and of the difference in thermal stability between SIA and vacancy clusters, Woo and Singh (1990, 1992) proposed the concept of "*production bias*" which was shown to provide a potent driving force for void swelling and to explain qualitatively the temperature dependence of steady-state void swelling.

It was soon recognized (Singh and Foreman 1992), however, that this driving force could be maintained beyond the short transient (i.e., at higher displacement doses) if a substantial fraction of SIA clusters is able to escape to sinks other than voids. The 1D diffusional glide of SIA clusters in the form of small dislocation loops, and their absorption by sinks, was shown to represent an efficient cluster removal mechanism (Trinkaus *et al.* 1992, 1993). Many MD simulations have confirmed the 1D diffusion of planar clusters with migration energies as low as that of single crowdions (e.g., Osetsky *et al.* 2000a, Soneda and Diaz de la Rubia 1998). At void swelling temperatures (i.e., in the presence of vacancy supersaturation), an important effect to be taken into account is the transformation of sessile SIA clusters (produced directly in the cascades or formed by cluster reactions) to glissile SIA clusters by the absorption of vacancies (see Fig. 3).

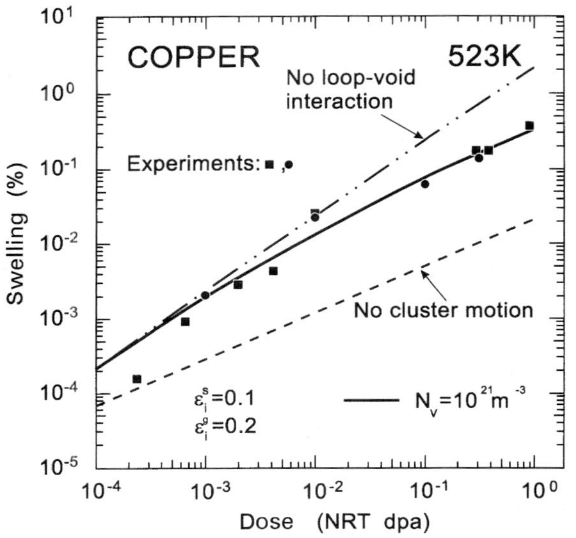

Figure 3
Experimental and theoretical dose dependence of void swelling in neutron-irradiated copper. The void swelling is calculated using the PBM as described in Singh *et al.* (1997). Note that the calculated results agree very well with the experimental measurements when the glissile SIA loops are considered to interact with the voids. ε_i^s and ε_i^g refer to the sessile and glissile fractions of SIA clusters produced in the cascades and N_v is the number density of voids.

Taking into account the reduced defect production efficiency in cascades and intracascade clustering of SIAs as well as vacancies, Eqns. (10a) and (10b) of the BEK model are generalized to

$$\frac{dC_i}{dt} = (1-\varepsilon_i)G + K_{ic} - D_iC_i(Z_i^v k_v^2 + Z_i^d\rho + Z_i^{ic}k_{ic}^2 + Z_i^{vc}k_{vc}^2) - \mu_r D_iC_iC_v \quad (13a)$$

$$\frac{dC_v}{dt} = (1-\varepsilon_v)G + K_{vc} - D_vC_v(Z_v^v k_v^2 + Z_v^d\rho + Z_v^{ic}k_{ic}^2 + Z_v^{vc}k_{vc}^2) - \mu_r D_iC_iC_v \quad (13b)$$

where $G = G_{NRT}(1-\varepsilon_r)$; ε_r is the fraction of Frenkel pairs recombining during the cooling phase of cascades; ε_i is the fraction of SIAs produced in the form of clusters; K_{ic} is the rate of thermal emission of SIAs from clusters; $Z_i^{ic}k_{ic}^2$, $Z_v^{ic}k_{ic}^2$ are the sink strengths of the SIA clusters for absorbing SIAs and vacancies, respectively. An important aspect of the PBM is that, in contrast to vacancy clusters, SIA clusters are thermally stable at void swelling temperatures meaning that K_{ic} may be neglected whereas K_{vc} must be taken into account.

Formally, Eqns. (13a) and (13b) are straightforward extensions of Eqns. (10a) and (10b). Equation (11) for the concentration of vacancies in clusters remains valid. An analogous equation for the concentration of SIAs in clusters would hold if only sessile clusters were produced in cascades. In fact, the effect of cascade-induced sessile SIA clusters on vacancy supersaturation and void swelling is analogous, even though opposite, to that of vacancy clusters. In the transient, such clusters provide a positive contribution to vacancy supersaturation in addition to that provided by the dislocation bias as originally suggested by Woo and Singh (1992). For an assumed quasi-steady state, balance equations analogous to Eqns. (12a) and (12b) indicate that sessile SIA clusters too would act essentially as recombination centers. Real steady state requires, however, that at least part of the SIA clusters are able to escape to sinks.

The whole kinetics of defect accumulation changes qualitatively and, of course, also quantitatively when part of the SIA transport is restricted to 1D, as was discussed in connection with the postulated 1D crowdion diffusion in the so called "two-interstitial model" (Frank and Seeger 1969). Decomposing the total amount of SIAs produced (in cascades) in the form of clusters, ε_i, into sessile (ε_i^s) and 1D diffusing glissile (ε_i^g) cluster fractions, $\varepsilon = \varepsilon_i^s + \varepsilon_i^g$, and assuming the reactions of the latter to be restricted to fixed sinks, the rate equation for the concentration of SIAs in such clusters, $x_i^g C_g$, where x_i^g is the average number of SIAs per cluster and C_g is the cluster concentration, may be written in the form

$$x_i^g \frac{dC_g}{dt} = \varepsilon_i^g G - x_i^g D_g C_g k_g^2 \quad (14)$$

where D_g and k_g^2 are the 1D diffusion coefficients of the glissile clusters and the total sink strength for their absorption, respectively, assumed here for simplicity to be the same for all of them. The transformations of sessile to glissile clusters upon vacancy absorption may be accounted for by substituting ε_i^g by an effective value $\varepsilon_i^{g,eff} > \varepsilon_i^g$ (Singh *et al.* 1997, Golubov *et al.* 2000a, 2000b).

In the present context, the sink strength for the absorption of 1D diffusing defects is of special interest. As in the case of 3D diffusion, it is proportional to the square of the mean reciprocal diffusion range. But differently from the former case, the mean reciprocal diffusion range of 1D diffusing defects itself is, as in collision theory, proportional to the mean defect absorption cross-section (or diameter) and density of the sinks. Hence, the sink strength is quadratic in the sink density establishing according to Eqn. (14), together with the concentration of the mobile defects (clusters), reaction kinetics of 3rd order (Gösele 1984, Trinkaus *et al.* 1993). Because of the quadratic dependence of the 1D sink strength on the sink density, the mean diffusion range (free path) is substantially larger for 1D than for 3D diffusion for low to medium sink densities (Trinkaus *et al.* 1992, 1993). The large spatial range of 1D diffusing clusters (up to several μm) provides an explanation for the heterogeneous and segregated defect accumulation in pure metals under cascade damage conditions. Since, for such large diffusion ranges, grain boundaries may significantly affect damage accumulation within grains, the contribution of grain boundaries to the sink strength must be taken into account. Including the correct numerical factor of two recently derived by Barashev *et al.* (2000), the partial sink strength of a sink of type s for absorbing 1D diffusing defects may be written as

$$k_{gs}^2 = 2\lambda_s^{-1}\lambda^{-1} \tag{15a}$$

$$\lambda^{-1} = \sum_s \sigma_s N_s + \pi\rho d_{abs}/4 + \sqrt{2/[l(2R_g - l)]} \tag{15b}$$

In Eqn. (15a), λ_s^{-1} is one of the contributions in Eqn. (15b) where s ($\neq d$ and g) represents compact sinks, different from dislocations (d) and grain boundaries (g), such as defect clusters and voids, with absorption cross-sections σ_s (for instance $\sigma_{s=v} = \pi R_v^2$ for voids) and number densities N_s; d_{abs} is the effective diameter of dislocations for the absorption of glissile SIA clusters; R_g,l are grain radius and distance from the grain boundary, respectively. Note that the factor two in Eqn. (15a) is missing in Eqn. (2) in the paper by Singh *et al.* (1997). Fortunately, this factor cancels out in the partitioning of glissile cluster over sinks which represents the only relevant effect of these clusters under quasi-steady-state conditions.

In the quasi-steady state, elimination of the cluster production terms $\varepsilon_v G$ and $\varepsilon_i G$ from Eqns. (13a) and (13b) results in the following balance equations for the point defect concentrations

$$\begin{aligned}(1 - \varepsilon_i^g)G = {} & D_i C_i (Z_i^v k_v^2 + Z_i^d \rho) + \mu_r D_i C_i C_v \\ & + D_v C_v Z_v^{ic} k_{ic}^2 + D_i C_i Z_i^{vc} k_{vc}^2 \\ & - \varepsilon_i^g G \lambda \sigma_{ic} N_{ic}\end{aligned} \tag{16a}$$

$$\begin{aligned}G = {} & D_v C_v (Z_v^v k_v^2 + Z_v^d \rho) + \mu_r D_i C_i C_v \\ & + D_v C_v Z_v^{ic} k_{ic}^2 + D_i C_i Z_i^{vc} k_{vc}^2 + \varepsilon_i^g G \lambda \sigma_{vc} N_{vc}\end{aligned} \tag{16b}$$

Equations (16a) and (16b) demonstrate that the effect of glissile SIA clusters is qualitatively different from that of sessile clusters. Whereas the contributions of the sessile clusters are equal in Eqns. (16a) and (16b) and are effective recombination terms, the contributions of the glissile clusters in Eqns. (16a) and (16b) are different, reducing the SIA production term and modifying the production terms of both types of defects by the reaction of glissile with sessile clusters.

Therefore, Eqns. (6b) and (7a) of the SRT model, which were still valid in the BEK model, are no longer valid in the PBM. As a consequence of this, the production of glissile SIA clusters affects the effective vacancy supersaturation and the corresponding void swelling crucially.

In order to illustrate the effect of glissile SIA cluster production on vacancy supersaturation and void swelling, recombination and cluster terms in Eqns. (16a) and (16b) are neglected in the following expressions. With this simplification, Eqns. (16a) and (16b) yield for the effective vacancy supersaturation

$$\begin{aligned}Z_v^v D_v C_v - Z_i^v D_i C_i = {} & p_d \frac{Z_v^v Z_v^d \rho}{(Z_i^v k_v^2 + Z_i^d \rho)(Z_v^v k_v^2 + Z_v^d \rho)} \\ & + \frac{\varepsilon_i^g G}{Z_v^v k_v^2 + Z_v^d \rho}\end{aligned} \tag{17}$$

The first term on the right hand side of Eqn. (17) represents the contribution of the (single defect) dislocation bias already given by Eqn. (7b), the second represents the contribution of glissile SIA clusters produced in cascades ("production bias") which dominates the first one under cascade damage conditions.

Accounting for the flux of glissile SIA clusters to voids in addition to that of point defects, Eqn. (9) for the void swelling rate according to the SRT and BEK models has to be modified as

$$\frac{dS}{dt} = (Z_v^v D_v C_v - Z_i^v D_i C_i)k_v^2 - \varepsilon_i^g G \lambda \sigma_v N_v \tag{18}$$

Using Eqn. (17), the swelling rate can be written in the explicit form as

$$\begin{aligned}\frac{dS}{dt} = {} & p_d G \frac{Z_v^v k_v^2 Z_v^d \rho}{(Z_i^v k_v^2 + Z_i^d \rho)(Z_v^v k_v^2 + Z_v^d \rho)} \\ & + \varepsilon_i^g G \left(\frac{Z_v^v k_v^2}{Z_v^v k_v^2 + Z_v^d \rho} - \lambda \sigma_v N_v \right)\end{aligned} \tag{19}$$

Hence, in the approximations for the vacancy supersaturation and the swelling rate as represented by Eqns. (17) and (19), respectively, the production bias terms depend only on the (effective) production of glissile clusters demonstrating their dominant role in damage accumulation. A characteristic feature of Eqn. (19) is the decrease of the swelling rate. Under appropriate conditions during cascade producing irradiation, the swelling rate may even vanish leading to swelling saturation, as has been occasionally observed under cascade damage conditions. This effect is due to the fact that the absorption of glissile clusters by voids depends more strongly on void size than that of point defects; $\sigma_v \propto R_v^2$ whereas $k_v^2 \propto R_v$ (Trinkaus *et al.* 1992, Golubov *et al.* 2000b).

Recombination and reactions of point defects and glissile clusters with sessile clusters neglected in Eqn.

(19) reduce both the dislocation and production bias contributions to the swelling rate whereas the transformation of sessile to glissile clusters increases the latter contribution (Golubov *et al.* 2000b). In this general case, the swelling rate depends on all relevant cascade damage parameters ε_r, ε_i^s, ε_i^g, x_g, as it should in an adequate model of cascade damage accumulation. This makes it possible to investigate the effect of recoil energy on the damage accumulation which has, in fact, been done recently by combining systematic experiments (Singh *et al.* 2000) with the general form of the PBM (Golubov *et al.* 2000a). The swelling rate depends on temperature and dose mainly via the dislocation density and the densities and sizes of voids and immobile clusters.

It should be noted here, however, that the PBM treatment in its present form as described above is strictly valid only if changes in the diffusion direction of glissile SIA clusters are negligible, as expected to be realized in pure metals, and if the distribution of immobile sinks remains random. An increase in the rate of direction (Burgers vector) changes of mobile clusters, associated with a decrease in the mean 1D diffusion length, has, in fact, been shown recently to cause a continuous change from 1D to 3D reaction kinetics and thus causing a dramatic increase in the reaction rate of such clusters with sinks (Golubov *et al.* 2000b, Barashev *et al.* 2000, Trinkaus *et al.* 2000).

Direction changes are expected to be induced by thermal activation and by the interaction of glissile clusters with other defects. Because of the latter effect, the diffusion reaction kinetics of mobile SIA clusters must be considered to have 3D features in complex technical alloys containing high concentrations of defects such as impurities and precipitates. In addition, it has been shown recently that the 1D diffusion reaction kinetics not only induces ordering of voids (formed initially with random distribution) but upon ordering, the reaction kinetics itself changes significantly because of the ordered distribution of the voids (Barashev *et al.* 2000).

3. *From Dislocation Bias to Production Bias*

Having described the available models for treating the evolution of damage accumulation during irradiation, it is appropriate to examine the scope, applicability, and limitations of these models. Logically, this issue can be addressed by comparing (i) the physical basis of each model against the crucial features of the primary defect production in real irradiation experiments and (ii) the main predictions of each model against relevant experimental observations.

3.1 *Dislocation Bias*

For simple irradiation cases, when defects are generated in the form of single, isolated SIAs and vacancies (e.g., during irradiation with 1–2 MeV electrons), the global accumulation of defects can be treated quite successfully within the framework of the SRT model using the dislocation bias as the only driving force (see Sect. 2.1). The temperature (Bullough *et al.* 1975) and grain size (Singh and Foremann 1974) dependence of void swelling determined experimentally in austenitic stainless steels irradiated in a high voltage electron microscope (HVEM) with 1 MeV electrons have been found to be in good agreement with the theoretical results calculated using the SRT model. Recently, it has been shown that the swelling behavior observed in pure copper irradiated with 2.5 MeV electrons (Singh *et al.* 2000) can be adequately described by the SRT model (Golubov *et al.* 2000a). These calculations yielded a dislocation bias of about 2% which is in good accord with the value of dislocation bias obtained from the swelling results of 1 MeV electron irradiated copper and even pure austenitic stainless steel (see Golubov *et al.* 2000a).

Since both the primary damage process under Frenkel pair production and the description of the damage accumulations (in steady state) are simple (see Eqn. (9)), the combination of the SRT model and appropriate experiments could be used to study the effects of, for example, impurity atoms and crystal structure on dislocation bias. To our knowledge, this kind of investigation unfortunately has not been reported.

Like the SRT model, the BEK model uses the dislocation bias as the only driving force for void swelling. The inclusion of vacancy loops formation by cascade collapse in the BEK model was intended to make the model suitable for describing the damage accumulation under cascade damage conditions. However, the model does not consider the clustering of SIAs in the nascent cascades and therefore does not deal with the properties of SIA clusters and their impact on defect reaction kinetics controlling the damage accumulation. Thus, the BEK model is not suited to study or predict the damage accumulation behavior under cascade damage conditions. Because of the inclusion of vacancy loops in the model, it predicts a part of the temperature dependence of void swelling qualitatively correctly.

The ability of the dislocation bias-based models to explain experimental results obtained under cascade damage conditions, was first questioned in 1981 (Singh and Leffers 1981). Later it was shown by detailed calculations (Leffers *et al.* 1986) that the void formation and growth behavior observed in neutron irradiated copper (Singh *et al.* 1986) could not be explained within the framework of these models. Since then the following experimental observations have been reported that cannot be explained in terms of SRT and BEK models:

(i) Dose dependence of void swelling
(ii) Temperature dependence of void swelling
(iii) Effect of recoil energy on void swelling
(iv) Effect of grain boundaries and grain size on void swelling

These issues have been discussed earlier (Singh and Zinkle 1993, Singh *et al.* 1994, Singh *et al.* 1997, Singh 1999a, 1999b). It is clear, however, that since none of these effects of neutron irradiations can be explained in terms of SRT- and BEK-type models, it would be rather unrealistic to use these models to predict the damage accumulation in materials for nuclear applications.

In summary, the dislocation bias-based models such as SRT- and BEK-type model, neither include in their formulation the fundamental features of the damage production in the form of displacement cascades nor can they explain the results obtained under cascade damage conditions. Historically, it was this situation which made it necessary to formulate a model including the physical features of cascade damage production and which may explain the experimental observations in a consistent manner. Hence, the production bias model.

3.2 Production Bias

As indicated in Sect. 2.4, the concept of production bias originated from the consideration of intracascade clustering of SIAs and vacancies and properties of these clusters. Naturally, the PBM has evolved in sophistication over the years and now incorporates all the physical features of damage production in displacement cascades (see Sect. 1.2 and Fig. 2). Furthermore, it takes into account the properties of the primary clusters and their interactions with other clusters and sinks in the medium. The PBM includes the SRT- and BEK-type models and hence the dislocation bias and its effect. The inclusion of 1D diffusional transport of SIA clusters and changes in reaction kinetics due to changes in the direction of motion (Burgers vector) during 1D transport make the PBM very versatile. These considerations, in fact, provide an appropriate framework for generalizing the PBM so that it can treat the problem of damage accumulation in metals and alloys of different crystal structure in general, including the formation of void superlattices.

In recent years, a considerable amount of effort has been spent on testing the validity of the PBM by comparing its main predictions with available experimental results and in some instances with the results of specific irradiation experiments particularly designed for this purpose. It was shown, for example, that the temperature dependence of void swelling predicted by the PBM is qualitatively consistent with the experimental results (Woo and Singh 1992). The dose dependence of void swelling measured in neutron-irradiated copper agrees quite well with the results calculated using the PBM (Singh *et al.* 1997, see Fig. 3). The enhanced void swelling near grain boundaries (see Fig. 4) (which could not be explained in terms of the SRT- and BEK-type models) has been calculated in terms of the PBM and the results agree quite well with experimental results (e.g., Trinkaus *et al.* 1993).

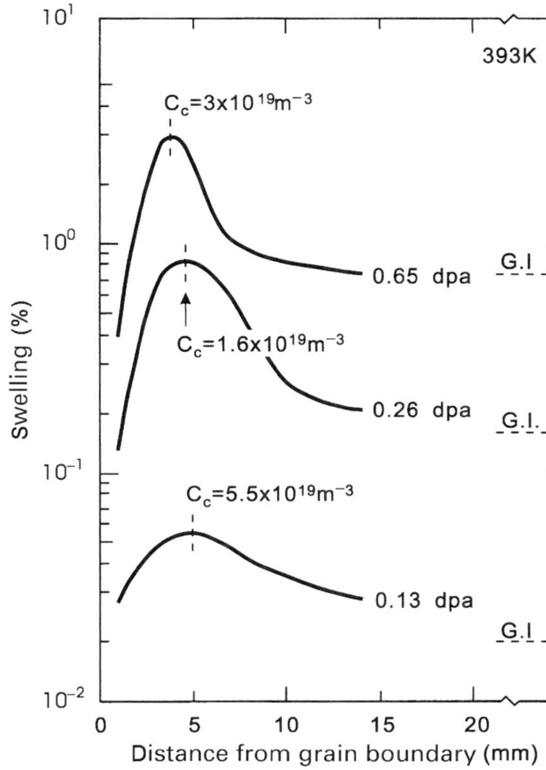

Figure 4
Spatial variation of void swelling with the distance away from grain boundaries in neutron-irradiated aluminum (Foreman *et al.* 1987).

The PBM has been similarly successful in describing the related grain-size dependence of void swelling in neutron irradiated copper (Singh *et al.* 2001). Figure 5 shows void swelling results measured in neutron-irradiated single crystal and polycrystal copper of different grain sizes. Note that the effect of grain size on void swelling is very similar to that observed in pure aluminum (see Singh and Zinkle 1993). For comparison, the results of the grain-size dependence of void swelling in copper, calculated using PBM and SRT, are also shown in Fig. 5. It can be seen that the general trend of the dependence calculated in terms of the PBM is consistent with the trend observed in copper. It is also clear that the experimental results cannot be explained in terms of the SRT (see dashed line in Fig. 5). In fact, the trend predicted by the SRT is similar to the one observed in stainless steel during 1 MeV electron irradiation (see Fig. 6).

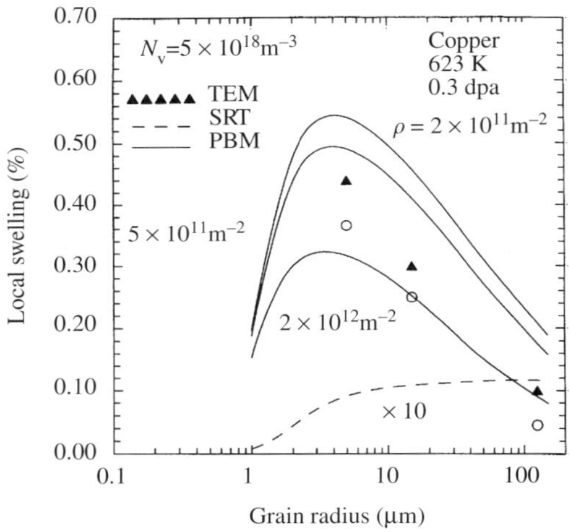

Figure 5
Experimental and theoretical grain size dependence of the local void swelling in neutron irradiated pure copper (Singh *et al.* 2001). The open circles are the swelling values calculated for specific grain sizes using the experimental values of void and dislocation densities. Calculations were carried out using SRT and PBM. Note that the experimental results are qualitatively consistent with the general trend calculated in terms of the PBM and cannot be explained in terms of SRT. The swelling values calculated in terms of the SRT model have been magnified by a factor of 10.

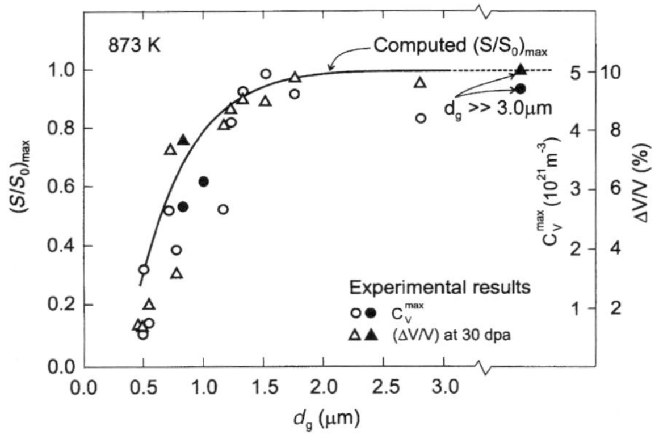

Figure 6
Effect of grain size on the measured void swelling and maximum concentration of voids ($C_v^{\max}$) under 1 MeV electron irradiation and calculated vacancy supersaturation, $(S/S_o)_{\max}$, using SRT in 316 austenitic stainless steel (Singh and Foreman 1974) at 873 K.

The most direct and irrefutable evidence supporting the PBM came from irradiation experiments on pure copper using 2.5 MeV electrons, 3 MeV protons, and fission neutrons specially designed to

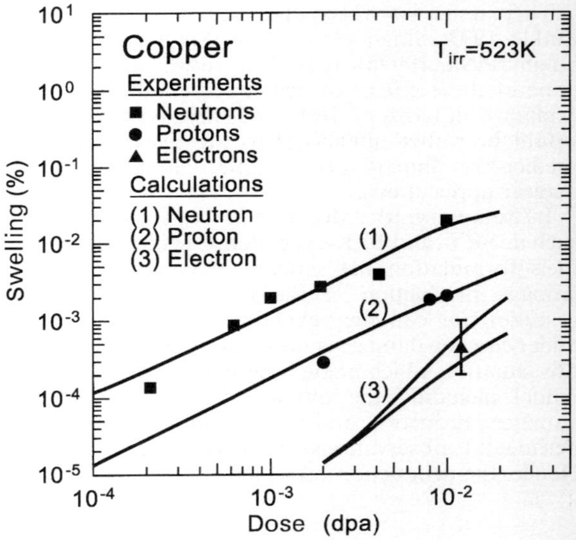

Figure 7
Experimentally measured (Singh *et al.* 2000) and calculated (Golubov *et al.* 2000a) effect of recoil energy on void swelling in pure copper irradiated with 2.5 MeV electrons, 3 MeV protons, and fission neutrons. The void swelling for the electron irradiation is calculated in terms of SRT, whereas for 3 MeV protons and fission neutrons using the PBM as formulated in Singh *et al.* (1997). Note that in both cases the calculated results are in good accord with the measured values.

study the effect of recoil energy on void swelling (Singh *et al.* 2000). The dose dependence of void swelling measured experimentally and calculated using PBM (Golubov *et al.* 2000a) is shown in Fig. 7. The swelling for the electron irradiation is calculated in terms of the SRT. Note that the calculated results agree very well with the experimental values. The results of 3 MeV proton irradiation demonstrate that the production of even small cascades has very significant effects on damage accumulation (see Singh *et al.* 2000, Golubov *et al.* 2000a for details).

3.3 Production Bias and Void Nucleation

Although since the 1970s a considerable amount of effort has been devoted to the problem of void nucleation, a realistic and general theory is still missing. Nonetheless, it is generally accepted that the irradiation-induced vacancy supersaturation is the primary driving force for the void nucleation. The fact that the propensity for void nucleation increases with vacancy supersaturation was demonstrated by correlating calculated vacancy supersaturation with the measured void density (see Fig. 6) under 1 MeV electron irradiation already in 1974 (Singh and

Foreman 1974). According to this, the void nucleation should be sensitive to intracascade clustering of SIAs and vacancies and the properties of these clusters, since, according to the PBM, these features determine the level of vacancy supersaturation (Golubov *et al.* 2000a). This implies that the void nucleation should, for example, be expected to depend on recoil energy.

The experimentally observed recoil energy dependence of damage accumulation in pure copper has indeed demonstrated a very substantial increase in void density with increasing recoil energy (Singh *et al.* 2000). The corresponding vacancy supersaturation for these experiments calculated using PBM was found also to increase very significantly with increasing recoil energy (Golubov *et al.* 2000a).

Another interesting example illustrating the effect of the nature of the primary damage production on vacancy supersaturation and void nucleation is the well-established difference in the void nucleation behavior between f.c.c. and b.c.c. metals irradiated under cascade damage conditions (Singh and Evans 1995, Golubov *et al.* 2000b, Eldrup and Singh 2000). Particularly in the case of f.c.c. copper and b.c.c. iron, the difference in supersaturation and, hence, in void nucleation, can be expected from the facts that (i) in b.c.c. iron the intracascade clustering of vacancies is considerably less efficient than that in f.c.c. copper (English and Jenkins 1987, Phythian *et al.* 1995) and (ii) a higher fraction of SIAs would be transported to sinks (e.g., dislocations and grain boundaries) via 1D diffusional transport of SIA clusters in b.c.c. iron than that in f.c.c. copper since practically all SIA clusters produced in b.c.c. iron are glissile (Osetsky *et al.* 2000b). Experimental results show that the neutron irradiation produces a very high density of voids in iron already at about 0.2 T_m (T_m is the melting temperature) whereas in copper no voids are formed at ~0.27 T_m (Eldrup and Singh 2000). In fact, in copper, voids are first observed at temperatures above 0.35 T_m. These results are consistent with the explanation given above.

In summary, in order to treat the problem of void nucleation under cascade damage conditions, it is necessary to treat the details of the primary damage production in cascades while calculating the vacancy supersaturation which is the main driving force for the nucleation of voids. The PBM provides a proper framework for this kind of calculations.

4. *Decoration of Dislocations and Irradiation Hardening*

As discussed in the preceding sections, the generation of glissile SIA clusters in displacement cascades plays a vital role in the evolution of void swelling. Since these glissile clusters are quite mobile even below the void swelling temperatures, they play a crucial role in the accumulation of defects in the form of defect clusters at lower irradiation temperatures (i.e., below the recovery stage V). Apart from making a

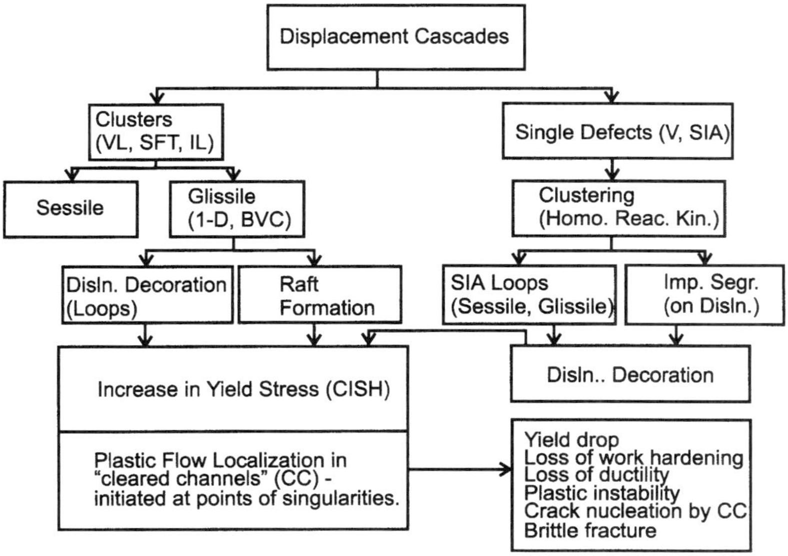

Figure 8
Schematic illustration of production of clusters in displacement cascades, their properties and their role in irradiation hardening and plastic flow localization which may lead to plastic instability and may limit the mechanical performance and lifetime of materials exposed to cascade producing radiation environment (Ghoniem and Singh 1999).

contribution to the global buildup of the cluster density, the glissile SIA clusters tend to segregate in the vicinity of grown-in dislocations, causing at a certain dose level decoration of these dislocations by SIA loops. In fact, the decoration of grown-in dislocations by loops commonly observed during neutron irradiation is perhaps the most striking feature providing evidence for 1D diffusion of SIA clusters (see Trinkaus *et al.* 1997).

Recently, analytical calculations of interactions between the grown-in dislocations and small SIA loops have shown that during neutron irradiation the decoration is likely to occur by trapping of glissile loops (produced in the cascades) in the strain field of grown-in dislocations (Trinkaus *et al.* 1997). The results of these calculations have been confirmed by numerical calculations (using 3D isotropic elasticity) of the interaction energy between the grown-in dislocations and small prismatic loops (Ghoniem and Singh 1999).

The major consequence of dislocation decoration is that the generation of dislocations necessary for the initiation of plastic deformation in the irradiated materials is prevented until a stress level is reached at which the decorated dislocations are unlocked from their atmosphere of SIA loops or fresh dislocations can be generated at sites in the crystal (e.g., grain boundaries, inclusions, and surfaces) where the stress is enhanced. These considerations led to the cascade-induced source hardening (CISH) model (Singh *et al.* 1997, Trinkaus *et al.* 1997).

Processes involved in dislocation decoration and its impact on mechanical properties of materials irradiated under cascade damage conditions are schematically illustrated in Fig. 8. It is shown that the intracascade clustering of SIAs and the 1D glide of SIA clusters play a significant role in dislocation decoration and, hence, in irradiation hardening and irradiation-induced plastic instability. Thus, the concept of production bias is necessary not only to understand void swelling, but also the mechanical performance of irradiated materials.

5. *Concluding Remarks*

In this article various aspects of the production of atomic displacements during irradiation and their subsequent transformation into lattice defects and defect clusters have been described. It has been pointed out that the use of elementary collision physics is not sufficient to describe the efficiency, nature, morphology, and spatial distribution of surviving defects and their clusters generated under cascade damage conditions. Such features can be obtained by MD simulations.

It has been shown that these features of primary damage production in cascades are of crucial importance for the microstructural evolution (e.g., void formation) and irradiation-induced changes in mechanical properties. Theoretical treatments attempting to describe these changes in bulk properties must include, therefore, a proper consideration of these features.

In order to calculate the consequences of these features on damage accumulation, the properties of the SIA clusters such as their thermal stability, ability to diffuse one-dimensionally, and to change their Burgers vector (i.e., direction of motion) must be taken into account. This can be treated only within the framework of specially formulated defect reaction kinetics. The input parameters for such treatments can be obtained by MD simulations.

The PBM is the only theoretical framework available for treating the consequences of all main features of the primary damage production in displacement cascades. The inclusion of 1D diffusion of SIA clusters, occasionally interrupted by changes in the direction of 1D diffusion makes it possible to understand phenomena such as heterogeneous microstructural evolution (in pure metals), the saturation of void swelling, and the formation of void superlattices. In recent years, the PBM has been used to explain experimental observations such as enhanced swelling near grain boundaries, decoration of dislocations with SIA loops, differences in swelling behavior between f.c.c. and b.c.c. metals, and to provide a quantitative treatment of the dependence of void swelling on irradiation temperature, displacement dose, recoil energy, and grain size. Neither the SRT nor the BEK models can rationalize any one of these experimental observations.

Although the PBM in its present form provides an appropriate framework for treating the effects of impurities and alloying elements on damage accumulations, detailed calculations are still missing, primarily due to the lack of information on the effect of impurities on the primary damage production parameters and the properties of defect clusters. Furthermore, a theoretical treatment of void nucleation in terms of the vacancy supersaturation as a primary driving force must be incorporated in the PBM before it can be considered as a general theory for describing damage accumulations under cascade damage conditions. Work is in progress on both of these challenging problems.

In order to explain the irradiation-induced increase in the upper yield stress and sudden yield drop followed by plastic instability, it is necessary to calculate both the global buildup of obstacles in the form of defect clusters, loops, and SFTs and the localized segregation of SIA loops and clusters in the form of dislocation decoration. The PBM provides the appropriate framework for both types of calculations. It has been shown that the considerations of dislocation decoration leads to a credible explanation for the appearance of a yield drop and plastic instability in terms of the cascade-induced source hardening (CISH) model.

See also: Fusion Conditions: Radiation Damage Correlation; Nuclear Reactor Materials: Irradiation Effects

Bibliography

ASTM/E521-96 1996 Standard practice for neutron radiation damage simulation by charged-particle irradiation. Philadelphia, PA: Am. Soc. Test. Mater. 2, Sect. 12

Bacon D J, Calder A F, Gao F 1997 Defect production due to displacement cascades in metals as revealed by computer simulation. *J. Nucl. Mater.* **251**, 1–12

Barashev A V, Golubov S I, Trinkaus H 2000 Reaction kinetics of glissile interstitial clusters in a crystal containing voids and dislocations. *Phil. Mag. A* in press

Blewitt T H, Coltman R R, Jamison R E, Redman J K 1960 Radiation hardening of copper single crystals. *J. Nucl. Mater.* **2**, 277–98

Bohr N 1913 Decrease of speed of electrified particles on passing through matter. *Phil. Mag. A* **25**, 10–31

Bohr N 1948 The penetration of atomic particles through matter. *Kgl. Dan. Vidensk. Selsk. Mat.-Fys. Medd* **18**, 1–62

Brailsford A D, Bullough R 1972 The rate theory of swelling due to void growth in irradiated metals. *J. Nucl. Mater.* **44**, 121–35

Bullough R, Eyre B L, Krishan K 1975 Cascade damage effects on the swelling of irradiated materials. *Proc. Roy. Soc.* **346**, 81–102

Cawthorne C, Fulton E J 1967 Voids in irradiated stainless steel. *Nature* **216**, 575–6

Diaz de la Rubia T 1996 Irradiation-induced defect production in elemental metals and semiconductors: A review of recent molecular dynamics studies. *Annu. Rev. Mater. Sci.* **26**, 613–49

Eldrup M, Singh B N 2000 Study of defect annealing behaviour in neutron irradiated Cu and Fe using positron annihilation and electrical conductivity. *J. Nucl. Mater.* **276**, 269–77

English C A, Jenkins M L 1987 Insight into cascade processes arising from studies of cascade collapse. *Mater. Sci. Forum* **15–18**, 1003–22

Eyre B 1973 Transmission electron microscope studies of point defect clusters in fcc and bcc metals. *J. Phys. F. Met. Phys.* **3**, 422–70

Foreman A J E, Singh B N, Horsewell A 1987 Diffusion mechanisms for enhanced vacancy accumulation near planar sinks. *Mater. Sci. Forum* **15–18**, 895–900

Frank W, Seeger A 1969 The interpretation of electron radiation damage in fcc metals in terms of the conversion—two—interstitial model. *Radiat. Eff.* **1**, 117–31

Ghoniem N M, Singh B N 1999 Dislocation dynamics study of the onset of plastic instabilities in irradiated materials. In: Bilde-Sørensen, Cartensen J V, Hansen N, Jensen D J, Leffers T, Pantleon W, Pederson O B, Winther G (eds.) *Deformation-induced Microstructures: Analysis and Relation to Properties*. Risø National Laboratory, Roskilde, Denmark, pp. 41–60

Gibson J B, Goland A N, Milgram M, Vineyard G H 1960 Dynamics of radiation damage. *Phys. Rev.* **120**, 1229–53

Golubov S I 1981 The influence of vacancy loops on the swelling of irradiated metals. *Phys. Met. Metallogr.* **52**, 86–94

Golubov S I, Singh B N, Trinkaus H 2000a On recoil energy dependent void swelling in pure copper. Part II: Theoretical treatment. Risø Report No. Risø-R-1186(EN), June 2000; *Phil. Mag. A*, in press

Golubov S I, Singh B N, Trinkaus H 2000b Defect accumulation in fcc and bcc metals and alloys under cascade damage conditions—towards a generalisation of the production bias model. *J. Nucl. Mater.* **276**, 78–89

Gösele U 1984 *Prog. Reaction Kinetics* **13**, 63

Greenwood G W, Foreman A J E, Rimmer D E 1959 The role of vacancies and dislocations in the nucleation and growth of gas bubbles in irradiated fissile material. *J. Nucl. Mater.* **1**, 305–24

Hamberg A 1914 Die radioaktiven substanzen und die geologische forschung. *Geol. Fönem Stockholm Föhr* **36**, 31–96

Heinisch H L, Singh B N 1993 On the structure of irradiation-induced collision cascades in metals as a function of recoil energy and crystal structure. *Phil. Mag. A* **67**, 407–24

Ishino S, Rowcliffe A F, Schiller P 1989 Need for and requirements for a neutron irradiation facility for fusion materials testing. *J. Fusion Energy* **8**, 147–55

Kinchin G H, Pease R S 1955 The displacement of atoms in solids. *Rep. Prog. Phys.* **18**, 1–51

Kiritani M, Yoshiie T, Kojima S, Satoh Y 1990 Recoil energy spectrum analysis and impact effect on cascade and subcascade in 14 MeV D-T fusion neutron irradiated fcc metals. *Radiat. Eff. Defects S.* **113**, 75–96

Konobeevsky S T, Pravdynk N F, Kitaitsev V I 1955 Effect of irradiation on structure and properties of fissionable materials. *Proc. 2nd Int. Conf. Peaceful Uses of Atomic Energy* (United Nations, New York 1956) Vol. 7, 433–40

Leffers T, Singh B N, Volobuyev A V, Gann V V 1986 Effects of heterogeneous sink distribution on void swelling. *Phil. Mag. A* **53**, 243–57

Lindhard J, Nielsen V, Scharff M 1968 Approximation method in classical scattering by screened coulomb fields. *Kgl. Dan. Vidensk. Selsk. Mat.-Fys. Medd.* **36**, 1–32

Norgett M J, Robinson M T, Torrens I M 1975 A proposed method for calculating displacement dose rates. *Nucl. Eng. Des.* **33**, 50–4

Osetsky Yu N, Bacon D J, Serra A, Singh B N, Golubov S I 2000a Stability and mobility of defect clusters and dislocation loops in metals. *J. Nucl. Mater.* **276**, 65–77

Osetsky Yu N, Serra A, Singh B N, Golubov S I 2000b Structure and properties of clusters of self-interstitial atoms in fcc copper and bcc iron. *Phil. Mag. A* **80**, 2131–57

Pabst A 1952 The metamict state. *Am. Min.* **37**, 137–57

Phythian W J, Stoller R E, Foreman A J E, Calder A F, Bacon D J 1995 A comparison of displacement cascades in copper and iron by molecular dynamics and its application to microstructural evolution. *J. Nucl. Mater.* **223**, 245–61

Robinson M T 1994 Basic physics of radiation damage production. *J. Nucl. Mater.* **216**, 1–28

Seeger A K 1958 On the theory of radiation damage and radiation hardening. Proc. 2nd Int. Conf. Peaceful Uses of Atomic Energy. Sept. 1958, (United Nations, Geneva 1958) **Vol. 6**, 250–73

Seitz F 1952 Radiation effects in solids. *Phys. Today* **5**, 6–9

Singh B N 1999a Damage production, accumulation and materials performance in radiation environment. *J. Comput. Aided. Mater.* **6**, 195–214

Singh B N 1999b Atomic displacements and defect accumulation during irradiation with energetic particles: anFautobiographical review. *Radiat. Eff. Defects S.* **148**, 383–446

Singh B N, Eldrup M, Horsewell A, Ehrhart P, Dworschak F 2000 On recoil energy dependent void swelling in pure copper Part I: experimental results. *Phil. Mag. A* **80**, 2629–50

Singh B N, Eldrup M, Zinkel S J, Golubov S I 2001 On grain size-dependent void formation in pure copper irradiated with fission neutrons. Risø Report No. Risø-R-1190(EN), March 2001

Singh B N, Evans J H 1995 Significant differences in defect accumulation behaviour between fcc and bcc crystals under cascade damage conditions. *J. Nucl. Mater.* **226**, 277–85

Singh B N, Foreman A J E 1974 Calculated grain size-dependent vacancy supersaturation and its effect on void formation. *Phil. Mag. A* **29**, 847–57

Singh B N, Foreman A J E 1992 Production bias and void swelling in the transient regime under cascade damage conditions. *Phil. Mag. A* **66**, 975–90

Singh B N, Foreman A J E, Trinkaus H 1997 Radiation hardening revisited: role of intracascade clustering. *J. Nucl. Mater.* **249**, 103–15

Singh B N, Golubov S I, Trinkaus H, Serra A, Osetsky Yu N, Barashev A V 1997 Aspects of microstructure evolution under cascade damage conditions. *J. Nucl. Mater.* **251**, 107–22

Singh B N, Leffers T 1981 Some observations on nucleation and growth of voids and dislocation loops in low dose neutron irradiated copper. Voprosy Atomnoi Nauki i Tekhniki, series. *Fizika Radiatsionnykh Povrezhdenii i Radiatsionnoye Materialovedenie* **15**, 3–8

Singh B N, Leffers T, Horsewell A 1986 Dislocation and void segregation in copper during neutron irradiation. *Phil. Mag. A* **53**, 233–42

Singh B N, Trinkaus H, Woo C H 1994 Production bias and cluster annihilation: Why necessary? *J. Nucl. Mater.* **212–215**, 168–74

Singh B N, Zinkle S J 1993 Defect accumulation in pure fcc metals in the transient regime: a review. *J. Nucl. Mater.* **206**, 212–29

Soneda N, Diaz de la Rubia T 1998 Defect production, annealing kinetics and damage evolution in α-Fe: an atomic-scale computer simulation. *Phil. Mag. A* **78**, 995–1019

Stoller R E, Odette G R 1987 A composite model of microstructural evolution in austenitic stainless steel under fast neutron irradiation. In: Garner F A, Packan N H, Kumar A S (eds.) . American Society for Testing and Materials, Philadelphia, pp. 371–92

Trinkaus H, Singh B N, Foreman A J E 1992 Glide of interstitial loops produced under cascade damage conditions: Possible effects on void formation. *J. Nucl. Mater.* **199**, 1–5

Trinkaus H, Singh B N, Foreman A J E 1993 Impact of glissile interstitial loops production in cascades on defect accumulation in the transient. *J. Nucl. Mater.* **206**, 200–11

Trinkaus H, Singh B N, Foreman A J E 1997 Segregation of cascade induced interstitial loops at dislocations: possible effect on initiation of plastic deformation. *J. Nucl. Mater.* **251**, 172–87

Trinkaus H, Singh B N, Golubov S I 2000 Progress in modeling the microstructural evolution in metals under cascade damage conditions. *J. Nucl. Mater.* **283–7**, 89–98

Von Guérard B, Peisl H 1975 Diffuse x-ray scattering from copper after neutron irradiation at 4.6 K. *Radiat. Eff.* **12**, 111–25

Wiedersich H 1972 On the theory of void formation during irradiation. *Radiat. Effects* **12**, 111–25

Wigner E P 1946 Theoretical physics in the metallurgical laboratory of Chicago. *J. Appl. Phys.* **17**, 857–63

Woo C H, Singh B N 1990 The concept of production bias and its possible role in defect accumulation under cascade damage conditions. *Phys. Stat. Sol.* **159**, 609–16

Woo C H, Singh B N 1992 Production bias due to clustering of point defects in irradiation-induced cascades. *Phil. Mag. A* **65**, 889–912

Radiation Effects in Carbon–Carbon Composites

T. D. Burchell

Carbon–carbon (C/C) composite materials are a class of high-temperature materials that consist of a carbon or graphite matrix, reinforced with carbon or graphite fibers, where this class of material is reviewed in detail). C/C materials that are used in nuclear environments are typically manufactured from graphitic fibers, and exhibit graphitic matrices by virtue of the matrix precursors and high heat treatment temperatures employed in their manufacture. Potential nuclear applications for C/C composites include high-temperature control rods for advanced, high-temperature, gas-cooled reactors (Burchell 1999), and the plasma-facing components of a fusion reactor (Snead 1999). Owing to its use as a moderator in nuclear fission reactors, much is known about the irradiation behavior of graphite (see *Nuclear Graphite and Radiation Effects*). In contrast, comparatively little is known about irradiation damage effects in C/C composites (Burchell 1997).

1. *Carbon–Carbon Composites as Fusion Reactor Plasma-facing Materials*

In a fusion reactor, light elements are brought together at enormous temperatures and "fuse" to form a heavier element, energy being released in the process. There are many reactions that could be utilized in a fusion reactor. However, the fusion reaction that has been most widely studied is the deuterium–tritium reaction, which involves two isotopes of hydrogen:

$$_1H^2 + _1H^3 \rightarrow _2He^4 + \text{neutron} + \text{energy}$$

In a fusion reactor, the hot plasma (ionized gas) is magnetically confined within a toroidal-shaped vacuum vessel. The plasma temperature required to sustain the fusion reaction is $> 50 \times 10^6$ K (> 10 keV). If the plasma were allowed to contact the vacuum vessel structural material, the high atomic number elements vaporized by the intense heat of the plasma would enter the plasma and create plasma instability. Consequently, it is necessary to protect or "armor" the inner walls of the vessel with a material of low atomic number, or a highly refractory material. The vessel armor is often referred to as a "plasma-facing component" (PFC) and the materials used are called "plasma-facing materials." The tokamak concept is one of the more popular of the highly specialized designs of fusion reactors (see *Fusion Reactors (Magnetically Confined)–Tokamaks: Materials*). A tokamak utilizes a toroidal magnetic field confinement system in which the plasma ions are confined within, and coupled to, continuous magnetic field lines that travel helically through the toroidal vacuum vessel. Carbon–carbon composites have been widely used as tokamak plasma-facing materials, and are under consideration for use in "next-generation" tokamaks such as ITER (Barabash *et al.* 1998). The interactions that occur between the plasma-facing material and the plasma are particularly severe, and can result in failure due to high thermal stress, melting, sublimation, excessive surface erosion, and neutron damage. The effect of this latter phenomenon on C/C composites is the subject of this article.

2. *Neutron Irradiation Damage Mechanism*

In the graphite crystal, carbon atoms are covalently bound in a hexagonal array with a binding energy in the crystal lattice of about 7 eV. The lattice basal planes are stacked upon one another in an ABAB … stacking sequence with an optimal separation of 0.335 nm. The bonding between the basal planes is of the weak attractive type (Van der Waals). McEnaney (1999) gives a full account of the structure and bonding in carbon materials. Energetic particles, such as electron, ions, or fast neutrons, can displace carbon atoms from their equilibrium lattice positions. The energy required to displace a carbon atom is in the range 24–60 eV (Kelly 1981, Burchell 1996a). The displaced carbon atoms, referred to as primary knock-on carbon atoms (PKAs), may displace further carbon atoms themselves, through secondary collisions with the carbon atoms in the graphite lattice in a cascade effect. The cascade carbon atoms are referred to as secondary knock-on carbon atoms (SKAs). The displaced SKAs tend to be clustered in small groups of 5–10 atoms and are generally treated as if they occur randomly. The total number of displaced carbon atoms produced depends upon the energy of the PKA, which in turn is a function of the neutron energy spectrum and the neutron flux. Displaced carbon atoms recoil through the crystal lattice, displacing other carbon atoms and creating vacant lattice sites. A significant fraction of the PKAs and SKAs do not remain displaced, but diffuse between the graphite crystal layer planes and recombine with lattice vacancies. Others will coalesce to form C_2, C_3, or C_4 linear molecules that in turn form the nucleus of a dislocation loop—essentially a new graphite plane. Adjacent lattice vacancies in the same graphitic layer are believed to collapse parallel to the layers, thus forming sinks for other vacancies, and can no longer recombine and annihilate interstitial defects. Figure 1 is a schematic representation of the neutron damage mechanism in graphite.

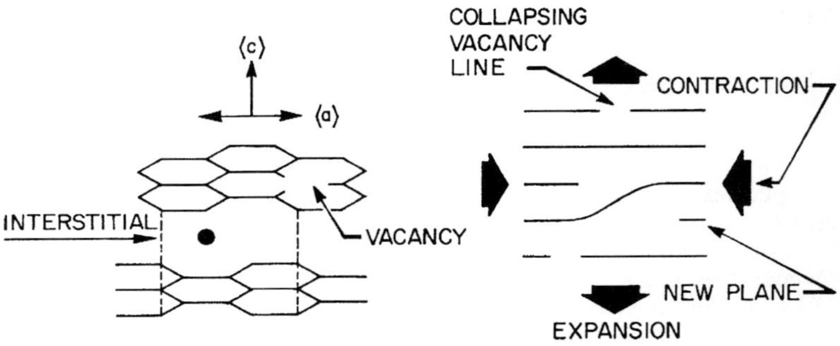

Figure 1
The radiation damage mechanism in graphite showing the induced crystal strains.

3. *Irradiation Induced Dimensional Changes*

Dimensional changes occur in the graphite crystallites because of the carbon atom displacements described above. Interstitial defects will cause the crystallites to expand perpendicularly to the basal planes (**c**-direction in Fig. 1), and relaxation in the plane due to the coalescence of vacancies will cause shrinkage parallel to the basal planes (**a**-direction in Fig. 1). The dimensional changes can be very large, as demonstrated in studies on well-ordered graphite materials, such as pyrolytic graphite (Price 1974). Typically, the dimensional change (**a**-axis shrinkage and **c**-axis growth) in pyrolytic graphite and graphite single crystals increases linearly with dose, but decreases with increasing graphitization temperature. The effect of graphitization temperature can be attributed to thermally induced improvements in crystal perfection, which reduces the number of vacancy trapping sites in the lattice.

Radiation damage in C/C composites has been recently reviewed and materials property data for fusion reactor plasma-facing C/C composites reported (Burchell 1996a, Burchell and Oku 1994). Two-directional (2D) C/C composites have been reported to exhibit excessive swelling perpendicular to the fabric layers (22–37% growth), and shrinkage in directions parallel to the fabric layers (typically 3–19%), on irradiation at 400 °C and to a dose of approximately 4 displacements per atom (dpa) (Burchell *et al.* 1991). C/C composites are an infinitely variable family of materials, and processing and design variables influence their properties and behavior. In order to select a C/C composite for a PFC application in which neutron damage will be significant, we must understand the effect of processing variables on neutron irradiation-induced structure and property changes. Burchell *et al.* (1992) irradiated 1D, 2D, and 3D C/C composites at 600 °C and to damage doses up to 1.5 dpa. Three-directional C/C composites were shown to exhibit more isotropic dimensional changes than either 1D or 2D C/C composite materials. Pitch fiber composites were shown to be more dimensionally stable than polyacrylonitrile (PAN) fiber composites, and high final heat-treatment (graphitization) temperatures were found to be beneficial. Similar trends were reported by Bonal and Wu (1996).

Burchell *et al.* (1996b) reported dimensional, strength, and thermal conductivity changes for so-called 1D, 2D, and 3D C/C composites irradiated to a peak neutron dose of 4.7 dpa at an irradiation temperature of 600 °C. Neutron irradiation-induced dimensional changes for three of the PAN fiber composite materials are summarized in Fig. 2. Specimen length and diameter changes are shown for a unidirectional fiber composite (UFC), a random fiber composite (RFC), and a 3D composite (223). The 1D composite exhibited extremely anisotropic dimensional changes, undergoing rapid shrinkage in the fiber-axis direction (length). In the direction perpendicular to the fiber axis (diameter), the composite shrank at a dose <1 dpa, followed by a reversal to expansion, returning to zero-dimensional change at approximately 2.5 dpa and continuing to expand at an increasing rate. A similar trend was observed for the 2D composites, as typified by the RFC material (Fig. 2), where the fiber axis is in the specimen diameter direction. In the off-axis direction (length), the composite exhibited a slight contraction followed by expansion, returning to the original length at approximately 2 dpa. The diameter (fiber-axis) directions of the RFC specimens exhibited shrinkage, although the magnitude of the shrinkage was much less for the same damage dose than in the case of the 1D composite. In contrast to the 1D and 2D composites, the 3D PAN fiber composite exhibited isotropic behavior at damage doses up to approximately 2 dpa. At doses exceeding 2 dpa, the composite **z**-direction (specimen length) continued to display shrinkage, whereas the fiber **x-y** (specimen diameter) direction exhibited reversal and slight growth.

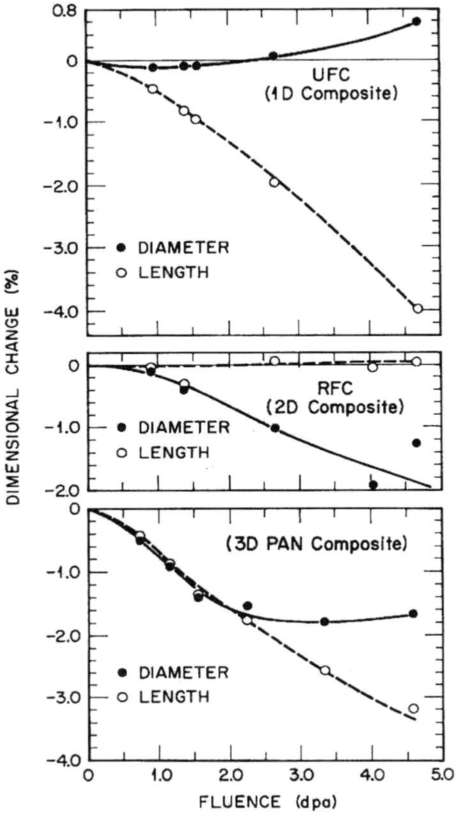

Figure 2
Neutron irradiation-induced dimensional changes of carbon/carbon composites irradiated at 600 °C.

Irradiation-induced dimensional changes in C/C composites has recently been explained in terms of the following (Burchell 1996b):

(i) an understanding of the graphite single crystal dimensional changes;
(ii) a microstructure model of the carbon fibers used in the composite (PAN fibers in all three materials in Fig. 2); and
(iii) a knowledge of the composite architecture and the interactions of the fiber, fiber bundle, matrix, and porosity in the composite.

4. *The Effects of Neutron Irradiation on Thermal Conductivity*

Carbon and graphite are phonon conductors of heat. The thermal conductivity increases with increasing temperature from absolute zero, and peaks at a temperature typically close to room temperature, due to the opposing effects of the increasing specific heat

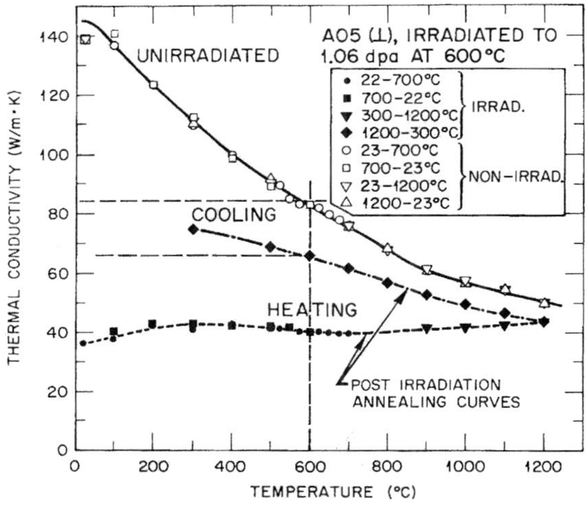

Figure 3
The temperature dependence of thermal conductivity for non-irradiated, irradiated, and annealed carbon/carbon composite material, AO-5.

and decreasing phonon mean free path – resulting from thermally induced phonon self-scattering (Umklapp scattering). The thermal conductivity decreases with further increases in temperature due to the dominating effect Umklapp scattering (Fig. 3). Phonon scattering also occurs from microstructural defects in the materials. Therefore, more graphitic materials exhibit greater thermal conductivity, and the peak in the thermal conductivity–temperature plot is sharper and shifts to lower temperatures. The mechanism of neutron irradiation-induced degradation of thermal conductivity in graphite has been extensively studied and reviewed (Burchell 1996a; Kelly 1981), and data for several C/C composites have been reported by Snead and Burchell (1995), Wu *et al.* (1994) and Eto *et al.* (1998).

The thermal conductivity is rapidly reduced with increasing neutron damage dose, due to the introduction of additional lattice defects, such as interstitial clusters, vacant lattice sites (singles, pairs or small groups of carbon atoms), and basal plane vacancy loops. Figure 3 illustrates the reduction of thermal conductivity with neutron damage dose. The lower curve in Fig. 3 illustrates the temperature dependence of the thermal conductivity of C/C composite A05 after neutron irradiation at 600 °C to ~1 dpa. At the irradiation temperature the thermal conductivity was reduced from $>80\,Wm^{-1}K^{-1}$ to $\sim 40\,Wm^{-1}K^{-1}$. Some fraction of the thermal conductivity may be recovered by thermally annealing the irradiation-induced defects, by heating the material above its irradiation temperature. This phenomenon may be seen in the "cooling" curve of Fig. 3. On cooling from the annealing temperature of

1200 °C to the irradiation temperature (600 °C), the thermal conductivity recovers to 67 $Wm^{-1}K^{-1}$. A greater recovery of thermal conductivity can be expected with increasing annealing temperature. The reduction in thermal conductivity due to irradiation damage is sensitive to dose and temperature. At any given dose and temperature, a saturation limit will be reached that is not exceeded until the composite undergoes gross structural changes at high damage doses. The saturation level of thermal conductivity will be lower, and will be attained more rapidly, at low irradiation temperatures.

5. *Conclusions*

Carbon–carbon composites are currently used in experimental fusion reactors, and are being considered for specialized high temperature gas-cooled fission reactors. In these nuclear applications, the material will be subjected to fast neutrons which result in structural damage and physical property degradation. Although relatively little work has been done to elucidate the effects of neutron damage in C/C composites, a vast body of literature exists for neutron damage in graphite. The data from the earlier studies on graphite has been used successfully to interpret new data on the effects of neutron damage in C/C composites.

See also: Radiation Damage Theory; Nuclear Reactor Materials: Irradiation Effects; Nuclear Reactors: Moderator and Reflector Materials; Nuclear Reactors: Shielding Materials

Bibliography

Barabash V, Akiba M, Bonal J P, Federici G, Matera R, Nakamura K, Pacher H D, Rödig M, Vieider G, Wu C H 1998 Carbon fiber composites application in ITER plasma facing components. *J. Nucl. Mater.* **258–63**, 149–59

Bonal J P, Wu C H 1996 Neutron irradiation effects on the thermal conductivity and the dimensional stability of carbon fiber composites. *Phys. Scr.* **T64**, 26–35

Burchell T D 1996a Radiation damage in carbon materials. In: *Physical Processes of the Interaction of Fusion Plasmas with Solids*. Academic Press, San Diego, CA, pp. 341–84

Burchell T D 1996b Radiation damage in carbon–carbon composites: structure and property effects. *Phys. Scr.* **T64**, 17–25

Burchell T D 1997 Radiation effects in graphite and carbon-based materials. *MRS Bull.* **XXII**, 29–35

Burchell T D 1999 Fission reactor applications of carbon. In: Burchell T D (ed.) *Carbon Materials for Advanced Technologies*. Elsevier, Oxford,, pp. 429–84

Burchell T D, Eatherly W P, Hollenburg G H, Slagle O D, Watson R D 1991 The effects of neutron irradiation on the structure of carbon-carbon composites. In: *Proc. 20th Biennial Conf. Carbon*. American Carbon Society, pp. 598–9

Burchell T D, Eatherly W P, Robbins J M, Strizak J P 1992 The effect of neutron irradiation on the structure and properties of carbon–carbon composite materials. *J. Nucl. Mater.* **191–94**, 295–9

Burchell T D, Oku T 1994 Materials property data for fusion reactor plasma facing carbon–carbon composites. In: Janev R K (ed.) *Nuclear Fusion—Atomic and Plasma-Materials Interactions Data for Fusion*, **5** Supplement to *Nuclear Fusion*. IAEA, Vienna, pp. 77–128

Eto M, Baba S, Ishihara M, Ugachi H 1998 High energy neutron and charged particle irradiation effects on thermomechanical properties of carbon–carbon composites for divertor applications. *J. Nucl. Mater.* **258–63**, 843–7

Kelly B T 1981 *Physics of Graphite*. Applied Science, London

McEnaney B 1999 Structure and bonding in carbon materials. In: Burchell T D (ed.) *Carbon Materials for Advanced Technologies*. Elsevier Science, Oxford, pp. 1–33

Price R J 1974 High-temperature neutron irradiation of highly oriented carbons and graphites. *CARBON* **12**, 159–69

Snead L L 1999 Fusion energy applications. In: Burchell T D (ed.) *Carbon Materials for Advanced Technologies*. Elsevier, Oxford, pp. 389–427

Snead L L, Burchell T D 1995 Thermal conductivity degradation of graphites due to neutron irradiation at low temperature. *J. Nucl. Mater.* **224**, 222–9

Wu C H, Bonal J P, Thiele B 1994 Thermal conductivity changes in graphite and carbon/carbon fiber materials induced by low neutron damages. *J. Nucl. Mater.* **212–15**, 1168–73

Radiation Effects in Polymers

A. Chapiro

1. General Features of Radiation Treatment

The radiation treatment of polymers represents today a multibillion dollar buisiness which includes several different activities: on the one hand, the production of new and improved polymers and on the other, the use of radiation as a biomedical tool, for instance, for the sterilization of medical supplies. In order to understand this rather young technology, a good knowledge of radiation effects in polymers is required. The influence of radiations on polymers was investigated for the first time in the USA, during World War II. The aim was to select those polymers which would best withstand high doses of radiation, without losing their properties, for applications in the vicinity of radiation fields. However, the worldwide interest in radiation processing of polymers followed the results obtained in 1952 by Arthur Charlesby, then at the Atomic Energy Establishment in Harwell, England. Charlesby investigated the influence of the mixed radiation from a nuclear reactor (neutrons and γ-rays) on the properties of polyethylene and found that the irradiated polymer no longer melted at elevated temperatures but exhibited rubber-like elasticity. The linear macromolecules of polyethylene, a paraffin assumed to be chemically unreactive, underwent under irradiation a chemical transformation which linked together the individual macromolecules, giving rise to a tridimensional network. These findings were most unexpected since, at the time, paraffins were only known to react after elaborate initiation at high temperatures, and moreover chemical reactions in solid polymers were not known. This spurred active research and development resulted in today's involvement of industry in this field.

1.1 Radiation Sources

"High-energy radiations" or "ionizing radiations" are generated in nature by radioactive elements which spontaneouly emit α- and β-particles and γ-rays. Manmade machines produce more powerful radiation source, such as x-rays, accelerated electron and heavy particle beams, and nuclear reactor radiation. All these various sources were used in the past for research. Today two types of radiation sources are more widely applied:

(i) γ-rays emitted by an artificial radioactive element: cobalt-60, and
(ii) electron beams generated by accelerators operating over a broad range of energies: 0.3–10 MeV and up to 20 MeV.

These energies are not high enough to initiate nuclear reactions; hence, the irradiated material does not exhibit any radioactivity.

1.2 Radiation Dose—Radiation Units

Radiation dose is the amount of radiation energy absobed per unit mass of material. It is mesured in joules per kg. The unit is the *gray*:

$$1\,\text{Gy} = 1\,\text{J kg}^{-1}$$

The doses used in commercial applications range from 0.1 kGy to 100 kGy. Even higher doses are required for some specific applications. The lowest doses concern those processes which involve chemical chain reactions (polymerization, curing of coatings). Up to the 1980s the accepted unit was the *rad*:

$$1\,\text{rad} = 100\,\text{erg g}^{-1} = 1/100\,\text{Gy}$$

This unit, found in the earlier literature, is rather small. In polymer processing, doses are expressed in terms of megarads:

$$1\,\text{Mrad} = 10\,\text{kGy}$$

1.3 Absorption of Radiation Energy

The following discussion is limited to energy absorption in organic materials which include all natural and artificial polymers (this does not cover inorganic polymers such as silica, asbestos, and the like).

Cobalt-60 γ-rays have quantum energies of $\sim$1.2 MeV. They transfer their energy to the molecules of the medium largely by Compton scattering, ejecting secondary electrons of high velocity ("δ-rays"). These interact with the electrons of the irradiated material, leading to *ionization* and *excitation*. The penetrating power of cobalt-60 γ-rays is very high. The intensity is reduced to 1/10 of its initial value after passing through 43.2 cm of a material of unit density. Cobalt-60 γ-rays are used today in industry for the irradiation of thick blocks of material for cross-linking purposes. Industrial sterilization of food and of biomedical devices also involves γ-ray facilities. However, the dose rates available are limited to ca. 1–10 kGy h^{-1} which requires fairly long exposure times.

Accelerated electrons lose their energy by electrostatic interaction with the electrons of the irradiated medium, generating more electrons, which in turn produce more ionization and excitation. Electron beams are produced at very high dose rates with the powerful accelerators available today. Dose rates of 1–10 kGy s^{-1} are common. However, the penetrating power of electrons is small. For 2 MeV electrons, penetration is limited to 5 mm in a material of unit density. Therefore, such radiation is used for the treatment of moderately thick materials—polymer films, pipes, insulated wires, etc.—which can be processed at very high speeds.

It follows from the above that in all cases the transformation or damage to the irradiated materal is performed by fast electrons, interacting with the electrons of the molecules in the medium. This leads to the following consequences:

(i) For a given amount of absorbed energy, radiation-induced changes are expected to be the same regardless of the type of radiation used: γ-rays or electron beams. This prediction is actually verified in most systems. Differences which may be observed in certain cases are often due to the large differences in dose rates of γ-rays as compared to those available with electron beams.

(ii) Since energy deposition only involves electrostatic interaction of fast moving electrons with the electrons of the medium, it does not depend on the chemical structure of the irradiated substance. This means that all molecules absorb radiation energy, and all substances are subject to radiation damage. There are no "radiation-inert" additives or solvents. This very important feature of ionizing radiation contrasts with the behavior of ultraviolet and visible light which only transfer their energy to substances which contain specific light-absorbing groups and are, therefore, selective in producing active species.

1.4 *Radiation-chemical Yields—"G Values"*

The energy yields of radiation-induced processes are expressed in terms of the number of molecular events arising per 100 eV absorbed in the medium. They are referred to as "*G* values." In polymers, $G(H_2)$ is the number of hydrogen molecules evolved per 100 eV, *G*(c.l.) is the number of cross-links, etc.

2. *Basic Radiation-induced Processes*

2.1 *Ionization–Excitation*

The most obvious change observed in materials subjected to "ionizing" radiation is the appearance of electric charges. These result from the abstraction of electrons from the irradited molecules, leaving behind positive ions. In early work, "ionization chambers" were used to detect and measure the intensity of radiations. For polymers, as for most organic molecules, the ionization potentials, i.e., the binding energies of the most labile electrons, are in the range to 10–15 eV. This defines the lower limit of the energy required for ionization and corresponds to vacuum UV radiation. "High-energy" radiation carries much more energy and therefore ionizations occur with a very high probability. Earlier studies have shown that the energy required for the formation of one ion pair in gases is 25–35 eV on the average (Table 1).

Table 1
Average energy, *W*, required per ion-pair production in various gases.

Gas	*W* (eV)
Hydrogen	36.3
Nitrogen	34.7
Oxygen	30.9
Methane	27.3
Ethane	26.3

It follows that only one-half of the deposited energy is used in ionization, the remainder produces electronic "excitation." This excitation is similar to photochemical excitation produced by visible and UV light. However, the energy levels reached with ionizing radiation are much higher. Highly excited molecules break down with a high probability, giving rise to "free radicals." In solid polymers, ions and free radicals may remain trapped for a considerable length of time (see below).

These basic radiation-induced processes can be summarized as follows:

$$\text{Ionization}: \quad AB \rightsquigarrow AB^{+} + e \qquad (1)$$

$$\text{Excitation}: \quad \mathrm{AB} \rightsquigarrow \mathrm{AB}^{*} \rightarrow \mathrm{A}^{\bullet} + \mathrm{B}^{\bullet} \qquad (2)$$

Rupture of highly excited molecules:

$$\mathrm{AB}^{**} \rightarrow \mathrm{A}^{\bullet} + \mathrm{B}^{\bullet} \qquad (3)$$

Here $\rightsquigarrow$ is the symbol used to designate a radiation-induced event, AB denotes an organic molecule, AB^{+} a positive ion, e an electron, AB^{*} and AB^{**} excited molecules, $A^{\bullet}$ and $B^{\bullet}$ are uncharged fragments (free radicals). These various "active species" are responsible for the changes observed in organic molecules subjected to ionizing radiations. In polymers, some transformations are attributed to ions; however, most reactions can be accounted for by conventional free-radical processes.

2.2 *Charge Recombination*

The electrons absracted from the irradiated molecules are subjected to the strong electric field of positive charges formed earlier. Therefore, charge recombination is a frequent event, either during irradiation or after the end of irradiation, as an "after-effect." In this process the ionization potential

(10–15 eV) is recovered and this generates highly excited molecules, carrying an amount of energy much larger than any bond strength. In organic molecules, the molecular bond strength ranges from 3 eV to 4 eV. As a result, such highly excited molecules will break down into free radicals:

$$AB^{+} + e \rightarrow AB^{**} \rightarrow A^{\bullet} + B^{\bullet} \quad (4)$$

3. *Observed Transformations in Irradiated Polymers*

3.1 Trapped Species

Solid polymers are formed of entangled linear macromolecules which interact with their neighbors by physical and weak physico-chemical forces ("hydrogen bonds") and are more or less free to move. The mobility of polymer "segments" (chain fragments of 10–20 atoms) very much depends on the physical state of the polymer.

In the *vitreous state*, all macromolecules are almost frozen in their positions and segmental motion is strongly limited. The same is true for macromolecules engaged in *crystalline areas*. In contrast, in the *viscoelastic* state, macromolecules vibrate freely and can also migrate, as in a liquid, being only hindered by the high viscosity of the medium. Therefore, "active species" formed in irradiated polymers, i.e., ions and free radicals, which only disappear by mutual interaction in binary reactions with an electron or with another free radical, will remain trapped if their mobility is hindered in the medium. Such trapping is particularly pronounced in vitreous or partially crystalline polymers. These trapped species may interact slowly *after irradiation* as is shown below.

The (low) mobility of trapped ions is responsible for *radiation-induced conductivity*, which is detected if the irradiated sample is placed in an electric field. During irradiation, the intensity of the induced current is a function of the dose rate. After irradiation, the conductivity does not fall off immediately but decays exponentially with time, sometimes over periods of days or even months. Trapped ion recombination is also responsible for the thermoluminescence observed in irradiated polymers. Trapped radicals induce postirradiation damage, such as *oxidative degradation*, by reaction with oxygen (air).

The slow decay of trapped positive charges and trapped radicals is due to recombination with mobile electrons or mobile radicals. This effect carries the radiation-induced process a step further and may result in deleterious consequences. Such *after-effects* are mostly observed in vitreous or partially crystalline polymers, during storage at room temperature after irradiation. They can be avoided, to a large extent, by a postirradiation heat treatment which liberates the trapped species. Alternatively, postirradiation damage is reduced in polymers to which inhibitors or anti-oxidants are added prior to irradiation.

3.2 Influence of Oxygen

The oxygen molecule contains two unpaired electrons which readily react with a free radical, $R^{\bullet}$, resulting in the formation of a peroxy-radical $RO_2^{\bullet}$:

$$R^{\bullet} + O_2 \rightarrow RO_2^{\bullet} \quad (5)$$

Such peroxy-radicals are very reactive and abstract hydrogen atoms, $H^{\bullet}$, from neighboring RH molecules:

$$RO_2^{\bullet} + RH \rightarrow RO_2H + R^{\bullet} \quad (6)$$

A new radical $R^{\bullet}$ is generated in this process which carries the reaction on. This is the basic chain reaction responsible for the oxidation of organic molecules. RO_2H is an unstable hydroperoxide which slowly breaks down at room temperature, producing more radicals and more oxidative damage. This oxidative chain reaction is prevented by the use of proper anti-oxidants.

3.3 Gas Evolution

Gases are generated in all irradiated polymers, as in any organic molecule. Hydrogen accounts for a substantial fraction of the gas. Its formation results from a two-step process. At first a carbon–hydrogen bond is broken by radiation:

$$PH \rightarrow P^{\bullet} + H^{\bullet} \quad (7)$$

Here PH is a hydrogen-containing polymer, $P^{\bullet}$ a polymeric radical and $H^{\bullet}$ a hydrogen atom. This step is followed by hydrogen atom abstraction:

$$H^{\bullet} + PH \rightarrow H_2 + P^{\bullet} \quad (8)$$

In polyethylene, one of the most studied polymers, the radiation chemical yield of hydrogen evolution is $G(H_2) = 4.1$.

Short side-branches are selectively ruptured under irradiation. Thus, methane is a major product in the radiolysis of polypropylene and polyisobutylene. Carbon monoxide and carbon dioxide arise among the gases evolved from polyacrylates and polymethacrylates. Halogenated polymers produce acidic gases: HCl from poly(vinyl chloride) and HF from fluorinated polymers. In vitreous or partially crystalline polymers, the gases generated by irradiation remain trapped and may give rise to strains which produce cracks and may even rupture the irradiated specimen. Here again, a postirradiation heat treatment will favor the diffusion of the gases and release the strains.

3.4 Cross-linking versus Main-chain Scission

Many highly cross-linked polymeric resins, such as epoxy resins for instance, can withstand high doses of radiation without losing their valuable properties. In the case of phenol-formaldehyde resins ("Bakelite"), the aromatic structure further "protects" these materials which can resist extremely high doses up to 1000 kGy or even more. In contrast, the properties of linear (thermoplastic) polymers are strongly affected by radiation which changes the structure of their macromolecules. Two groups of polymers emerge from all preceding studies: polymers undergoing chiefly cross-linking and polymers suffering main-chain scissions. This distinction goes along with the chemical nature of such polymers. Table 2 lists observed yields of cross-linking *G*(c.l.) and Table 3 lists observed yields of main-chain scissions *G*(breaks) in some common polymers of each group.

Cross-linking is the process which follows reactions (7) and (8) in which two polymeric radicals $P^{\bullet}$ arise. These radicals combine to form an intermolecular bond or "cross-link:"

$$P^{\bullet} + P^{\bullet} \rightarrow P - P \qquad (9)$$

This reaction, involving two macromolecular fragments, occurs in solid polymers with unexpectedly high yields. This observation indicates that the mobility of radical sites is large in the solid medium, presumably by transfer of hydrogen atoms from one site to another ("hydrogen hopping"). From a comparison of reactions (7) and (8) with reaction (9), it appears that the formation of one cross-link corresponds to the evolution of one hydrogen molecule. The yield of cross-linking is enhanced in the presence of polyunsaturated compounds, such as polyfunctional vinyl monomers. Polymeric radicals react with the double bonds to form new radicals which carry the reaction on by a short-chain process. Suitable monomers for this process include di- and tetra-acrylates, divinylbenzene, and the like.

Main-chain scission involves the break of a chemical bond in the main polymeric chain. In linear (unbranched) polymers, the resulting two polymeric radicals are trapped in a dose vicinity and readily recombine, restoring the original structure. However, if the polymer contains side branches (methyl groups) in its monomeric unit, particularly if two branches are attached to the same carbon atom (tetra-substituted carbon), the two polymeric radicals interact with transfer of a hydrogen atom and formation of a double bond ("disproportionation"). The scission then becomes permanent. An examination of the data listed in Tables 2 and 3 indeed demonstrate that increased methyl branching changes the fate of a polymer from the cross-linking type to the chain scission type. This appears more clearly wth the polymers listed in Table 4

Reaction (10) illustrates the process of permanent main-chain scission:

$$R'-CH_2-\underset{CH_3}{\overset{CH_3}{C}}-CH_2-R'' \rightarrow R'-CH_2-\underset{CH_3}{\overset{CH_3}{C^{\bullet}}} + {}^{\bullet}CH_2-R''$$
$$\rightarrow R'-CH_2-\underset{CH_3}{\overset{CH_2''}{\overset{\|}{C}}} + CH_3-R'' \qquad (10)$$

3.5 Unsaturation

As is indicated above, double bonds are formed in the interaction of branched free radicals. Moreover, if one compares the radiation chemical yields in polyethylene for hydrogen evolution, $G(H_2) = 4$ and for cross-linking, $G(\text{c.l.}) = 2$ (Table 2), it appears that

Table 2
Yields of cross-linking, *G*(c.l.), in polymers of the cross-linking type.

Polymer	*G*(c.l.)
Polyethylene	2.0
Polypropylene	0.6
Polystyrene	0.05
Polybutadiene	2.0
Poly(methyl acrylate)	0.8
Polydimethylsiloxane	2.5
Polyamides	0.3

Table 3
Yields of main-chain scissions, *G*(breaks), in polymers of the chain-scission type.

Polymer	*G*(breaks)
Polyisobutylene	3.0
Poly(methyl methacrylate)	1.9
Cellulose	10.0

Table 4
Influence of methyl branching on fate of polymers under irradiation.

Cross-linking type	Chain scission type
Polyethylene and polypropylene	Polyisobutylene
Polyacrylates	Polymethacrylates

cross-linking only accounts for one-half of the hydrogen produced. Additional hydrogen is generated in the formation of intra-chain double bonds. This is summarized in reaction (11):

$$R' - CH_2 - CH_2 - R'' \rightarrow R' - CH = CH - R'' + H_2 \quad (11)$$

Subsequent studies confirmed that in polyethylene G(double bonds) = 1.8. Double bonds arise with much higher yields in chlorinated polymers, such as PVC, where HCl is formed by a chain process. As a result, severe corrosion is observed on metal parts located in the vicinity of radiation-treated PVC.

In polymers which initially contain double bonds, such as natural rubber or polybutadiene, a decrease in unsaturation is observed under irradiation. This effect is due to the addition of free radicals onto the double bonds, a process which enhances the yield of cross-linking.

4. *Chain-reactions Initiated by Radiations*

4.1 Polymerization

The free radicals arising under irradiation readily react with most vinyl monomers, resulting in their polymerization. Such radiation-initiated polymerizations were investigated in detail with many conventional monomers. Radiation initiation is temperature independent; thus, the process can be carried out at very low temperatures with significant velocities. Thereby, polymers of more regular structure are obtained. From extensive studies it was concluded that, in most cases, radiation-initiated polymerizations follow normal free-radical kinetics. The ions generated by radiation have very short lifetimes and can only react in polar media where charges become "solvated" and therefore charge recombination is retarded. Under such conditions, ionic polymerizations may arise (Chapiro 1962).

4.2 Radiation Curing

Radiation-initiated polymerization is used commercially, on a very large scale, for curing thin layers of coatings deposited on solid substrates (paper, plastic films, wood panels, and metal foils). In this operation, specifically designed formulations are subjected to low-energy electron beams. The preferred radiation sources are electron accelerators operating in the range of 0.3–0.5 MeV. Such machines require low capital investment for the equipment and for protective shielding. The beam currents are high, allowing very short exposure times and corresponding high production speeds, which are only limited by the heat evolved in the exothermic polymerization. The penetration of the electron beam has to be high enough to cure the deposited layer in depth. In this technology, formulations are used which consist of blends of polymerizable monomers with reactive "prepolymers." The latter are short-chain polymers with reactive double bonds capable of copolymerizing with the monomer. Such blends have the consistency of a paste and can be conveniently deposited on the substrate with a brush or with a knife. Irradiation converts this paste into a hard coating of a cross-linked polymer with excellent strength.

4.3 Graft Copolymerization (Grafting)

A graft copolymer is a polymer which contains, in its individual macromolecules, sequences of two different polymers, say A_n and B_m. As such, a graft copolymer will encompass properties intermediate between those of polymers A_n and B_m. It may also exhibit some of the specific properties of each of these polymers. Graft copolymers can be considered as polymeric alloys. Since most polymers are incompatible, they cannot be easily blended, in contrast to metals. They do not mix, but form brittle microphase separated materials. Binding together two polymers by covalent bonds overcomes chemical repulsion and gives rise to an almost unlimited range of new materials. This accounts for the general interest in graft copolymers.

The most usual route for the preparation of a graft copolymer is to "activate" an initial "trunk polymer" which will then initiate the polymerization of a monomer corresponding to the second polymer. The resulting grafted branches are attached to the trunk polymer by covalent bonds. Radiation is particularly suited for activating polymers since polymeric free radicals arise under irradiation and these are active in initiating polymerization. This is the main advantage of radiation-initiated graft copolymerization.

5. *Gel Formation in Irradiated Polymer Solutions*

Irradiation of polymers dissolved in suitable solvents generates cross-linking with the formation of polymeric gels which are of interest in the biomedical field. Water is a particularly efficient solvent for this process because $H^\bullet$ and $OH^\bullet$ radicals, generated in its radiolysis, abstract hydrogen atoms from the polymer molecule and contribute to gel formation. The structure of the resultant gels is easily controlled. The density of cross-links, which determines the mechanical strength of the gel in the swollen state, is proportional to radiation dose and can be adjusted with precision to specific needs. Poly(*N*-vinylpyrrolodone), a biocompatible polymer, is used for the preparation of gels which find interesting applications as hospital dressings for the treatment of burns. Gels of other water-soluble polymers are used as sensors in different devices.

6. Commercial Applications of Irradiated Polymers

The largest commercial operations involving radiation treatment of polymers are radiation-induced cross-linking and radiation curing of coatings and lacquers, which also includes "drying" of inks. These technologies are at the origin of new and improved products which enjoy acceptance in numerous industries. Radiation sterilization of plastic medical supplies is another large field of application. Here the polymers are not modified in the treatment, but special formulations have to be developed in order to stabilize the polymers and to meet severe requirements to avoid the formation of extractables. Large radiation facilities use cobalt-60 γ-rays or high-energy electron beams. Medical supplies, packed in air-tight bags, are irradiated in their final cardboard boxes, ready for shipping. Smaller-size operations derive from radiation-induced scission of polymer chains which favors molecular weight control. A few graft copolymers are prepared commercially with radiation initiation.

6.1 Molecular Weight Control

Chain scission reduces the molecular weight of the macromolecules of a polymer and this finds some commercial applications in problems concerned with molecular weight control, particularly useful in the reclamation of polymer wastes. Several projects are described in the patent literature. These include the reclaiming of polyisobutylene (butyl rubber) using high doses of radiation which convert wastes of this polymer into valuable lubricating oils. Irradiation of cellulose (wood chips) is another project aimed at improving its uses in the paper and pulp industries. However, these applications appear to be too costly at present for practical purposes and no large-scale operations are in view. Polytetrafluoroethylene (Teflon) is a polymer enjoying a high added value. It is chemically inert and does not decay noticeably under normal natural conditions. Moreover, it does not burn and therefore wastes of Teflon are not disposable by incineration. As a consequence, wastes of this polymer used to accumulate in Teflon production plants. An interesting solution of this problem was found several years ago. Teflon wastes (films, molded parts, etc.) subjected to electron doses of 100–200 kGy crumble into powder and the resulting low-molecular-weight polymer finds useful applications in sprays or in aqueous suspensions used for coating various metals, which thereby acquire improved sliding properties. Several small production units operate along these lines, taking advantage of the very simple technical process involving the irradiation of a solid and the low cost of the raw material.

6.2 Cross-linked Polymers

Cross-linking of linear polymers is the earliest and still one of the major commercial applications of radiations. The most obvious merit of radiation cross-linking of polymeric materials, as opposed to conventional chemical methods of cross-linking, is its ease of operation, the product in its final state being treated at room temperature. Electron beams are the preferred radiation source in this field. The process is carried out at very high speeds, the products (wire and cables, tubings, preformed parts of automobile tyres) passing continuously through the electron beam. The energy of the electrons must be high enough to fully penetrate the product. Therefore, the appropriate machine has to be selected for each particular application. Cross-linked polymers no longer melt at elevated temperatures and exhibit better solvent resistance, improved form stability, and better aging properties. As a result, thinner wall insulators are applicable to wires and cables reducing the weight. These are the main reasons why wires and cables with radiation cross-linked insulators are commonly used in transportation, in aircrafts, automobiles, and space applications. Electrical equipment (motors, transformers, etc.) fitted with radiation cross-linked wire can be operated continuously at more elevated temperatures, thereby improving energy yields. Radiation cross-linked pipes are used in hot water floor heating, where their long lifetime and reliability are of great advantage.

Another large-scale application concerns radiation pre-cross-linked rubber strips which are used in building up automobile tyres. During their manufacture, some of the tire components need to retain a certain strength in order to resist deformation. This "green strength" is improved by irradiating rubber strips with high enough doses to increase natural entanglements of the macromolecules, but without fully cross-linking them. The final structure of the tire is then cross-linked using sulfur and heat (vulcanization).

Numerous sophisticated devices derive from what is sometimes designated as the "memory effect" of cross-linked polymers. When a cross-linked sample of a vitreous or a partially crystalline polymer is heated above its transition temperature (crystalline melting point, glass transition), it does not melt but exhibits rubber-like elasticity. In this state its shape can be altered by streching to another geometric pattern, which is retained upon cooling. When heated again above its transition temperature, the sample recovers its initial shape. This behavior finds interesting applications in "heat shrinkable" packaging materials, tubings, and more sophisticated devices. Heat shrinkable "solder sleeves" are produced in very large numbers by Rachem Corporation, California. Figures 1 and 2 illustrate their uses.

6.3 Radiation curing

Electron beam curing of thin layers of coatings deposited on solid substrates has enjoyed a considerable development in recent years. Numerous monomers and "reactive" prepolymers, especially designed for such applications, are available from suppliers and this makes it possible to "tailor" the coating to exhibit almost any desired properties. Pigments and magnetic powders can be included in the formulation up to the limit of the penetrating power of the electron beam, which for thin layers does not raise any significant problem. Most formulations cure and cross-link at the same time, producing hard coatings with good resistance to wear. Moreover, the migration of solid particles in the bulk of the polymer is strongly reduced by the cross-linked structure of the polymeric matrix. This is of great advantage with magnetic formulations since any imprinted information then remains more stable during long times of storing, particularly at elevated temperatures. The dose required for total cure strongly depends on the composition used. This dose is small in most cases since the process involves a chain reaction. Curing times are therefore reduced to very low values, sometimes to fractions of seconds, which allows large production speeds. The radiation treatment is preferably carried out in an inert atmosphere (nitrogen or burned gases) or after covering the formulation with a thin plastic or aluminum foil in order to avoid inhibition by air and radiation-induced oxidation. Some formulations are specifically designed to cure under normal atmospheric conditions.

Even though radiation curing of coatings was originally initiated by electron beams, it now has to compete with UV cure which requires smaller capital investment and is particularly suited for small production units. UV cure often requires the use of photosensitizers in order to improve the yields. For very large scale productions, UV sources are not powerful enough. Moreover, if the formulation is heavily loaded with pigments or magnetic particles, the penetration of UV light is very low and electron beam curing is of advantage. This technology is widely applied today for curing lacquers and varnishes on wood and metal substrates, in the production of magnetic tapes and disks and for drying inks. Radiation curing of magnetic formulations which at the same time cross-link the polymeric matrix results in the formation of particularly stable products in which migration of magnetic particles is minimized. For all these applications, the production units only need a moderate floor space and the final product is strong enough to permit piling up of the cured layers.

Another promising application of radiation curing is concerned with composite materials. Large carbon fiber-reinforced building elements for aerospace applications are cured at room temperature with radiation. In this technique, deformations observed in heat-cured samples are avoided. Large electron accelerators operating with energies ranging from 10 MeV to 20 MeV are required in order to fully penetrate such pieces.

Figure 1
Top: "solder sleeve" consisting of irradiated heat-shrinkable poly(vinylidene fluoride) sleeve containing a preform of fluxed solder at the center and a thermoplastic sealing ring at each end. Below: three conductors with stripped ends to be soldered. Bottom: the result after heating the assembly with a hot-air torch (courtesy of Raychem Corp.).

6.4 Graft Copolymers Produced by Radiation

Very few graft copolymers are presently used in industry. This is largely due to the difficulty of the manufacture of such materials, which necessarily

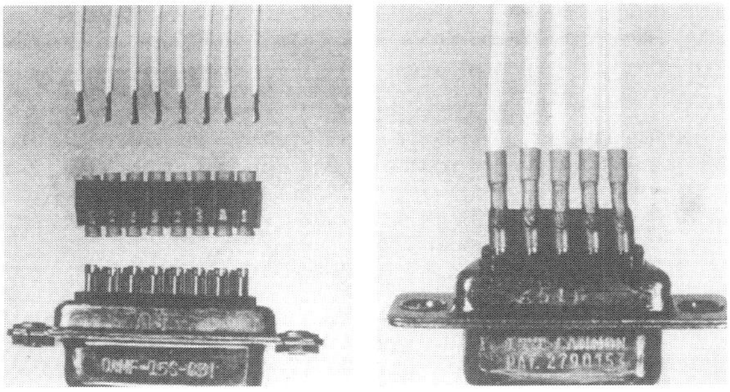

Figure 2
Series of solder sleeves carried on tape used to solder wire to connector pin terminations. Left: before assembly. Right: after heating the assembly with a hot-air torch (courtesy of Raychem Corp.).

increases their production cost. However, in view of their interesting properties, some specific graft copolymers are produced using radiation. The strong adhesive properties of carboxylic acids with respect to metals, plastics, and wood are incorporated by radiation-initiated grafting of acrylic acid into chemically inert polymers such as polyethylene or Teflon. The resulting material then exhibits the combined properties of both constituents, i.e., it combines the chemical inertness of the original polymer with increased wettability and improved adhesion to different surfaces. In a commercial operation, small amounts (10–15%) of acrylic acid are grafted into polyethylene with radiation initiation. The product retains the properties of polyethylene, particularly its ease of handling in molding and in extrusion, but it also exhibits excellent adhesion to metals. This material is used for the manufacture of seals between aluminum capsules and polyethylene bottles. The metal capsules, coated with a thin film of this graft copolymer, are applied under heat and pressure to polyethylene bottles and this produces air-tight seals. Moreover, this material is accepted for food packaging.

Several *permselective membranes* are prepared by grafting water-soluble monomers into the bulk of various polymer films, such as polyethylene, polypropylene, PVC, and Teflon. In this process the membranes are produced directly starting with the polymer film and there is no need for any film-forming treatment which may raise serious problems with such multi-component systems. Teflon-grafted membranes exhibit good resistance in aggressive environments and are candidates for the production of chlorine in electrolytic cells. Radiation-grafted membranes are also used as separators in alkaline batteries. Such operations are presently on stream in various places. One advantage of radiation grafting lies in the fact that the grafted zone can be purposely limited to a surface layer of shaped objects, thereby preserving the physical (mechanical) properties of the trunk polymer. Polyvinylpyrrolidone, a biocompatible polymer, was successfully grafted onto the inner surface of polyurethane and silicone tubings. The resulting samples proved to have improved hemocompatibility in a series of tests on animals (Chapiro *et al.* 1981). Such materials are promising for their use as implants, prostheses, and the like, with improved biological tolerance. The vast domain of radiation-grafted products is still in its infancy and is open to many future developments in numerous fields.

7. *Conclusion*

Radiation treatment of polymers is a fairly recent technology following new applications derived from the nuclear industry. The success of these investigations led to an entirely new industrial activity, the manufacture of electric accelerators to be used as powerful radiation sources. These are now widely spread in industry for radiation processing. The largest productions in the polymer field involve cross-linking of shaped polymer articles and curing of coatings. Sterilization of plastic medical supplies in hospitals is another large-scale application in which the polymers are not modified. This method of sterilization enjoys great popularity for its ease of operation. Smaller-volume applications are found in various fields and involve numerous products. Radiation treatment of polymers is still rapidly expanding, and entirely new applications are expected to come to light in the future.

Bibliography

Chapiro A 1962 *Radiation Chemistry of Polymeric Systems.* Interscience, New York

Chapiro A 2002 Polymer irradiation: past, present and future. *Radiation Phys. Chem.* **63**, 207, presented at the International Meeting on Radiation Processing *XII IMRAP*, Avignon, France, March 2001 (these Meetings, organized every second year, give an overview on the industrial uses of radiation)

Chapiro A, Domurado D, Foëx-Millequant M, Jendry chowska-Bonamour A M 1981 Radiation grafting of N-vinylpyrrolidone into silicone tubings. Synthesis of polymers with improved hemocompatibility and implantation tests in lambs. *Radiation Phys. Chem.* **18**, 1203

Charlesby A 1960 *Atomic Radiation and Polymers.* Pergamon, Oxford

Dole M 1972 *The Radiation Chemistry of Macromolecules.* Academic Press, London

Drobny J G 2003 *Radiation Technology for Polymers.* CRC Press, Boca Raton, FL

Rare Earth Phosphors: Fundamentals and Applications

C. Ronda

Luminescent materials (or phosphors) are mostly solid inorganic materials consisting of a host lattice, usually intentionally doped (Fig. 1). The absorption of energy takes place via either the host lattice or on impurities. In addition, transfer of energy through the lattice can take place. In almost all cases, the emission takes place on intentionally doped impurities, like rare-earth ions, which are present in relatively low concentrations (a few mole percent or less).

Another class of rare-earth-based luminescent materials is governed by complexes of rare-earth ions with organic molecules, e.g., $Eu(thd)_3dpphen$ (Fig. 2). Luminescent materials are applied widely. Major applications are in emissive displays, fluorescent lamps, and x-ray detector systems.

In the last decades, a large number of luminescent materials based on rare-earth ions or rare-earth-based host lattices have been developed. Quite a few of these materials have found their way into applications. In many cases, the application of rare-earth phosphors dramatically improved the performance of the devices in which they were applied.

In the first part of this contribution, the emission properties of rare-earth ions are discussed and compared to the emission properties of other luminescent ions or molecular anion groups. In the second part, the processes leading to emission (absorption and energy transfer) will be elucidated. The third part deals with applications of luminescent materials with emphasis on rare-earth phosphors. This contribution ends with a discussion of new developments in the field of rare-earth phosphors.

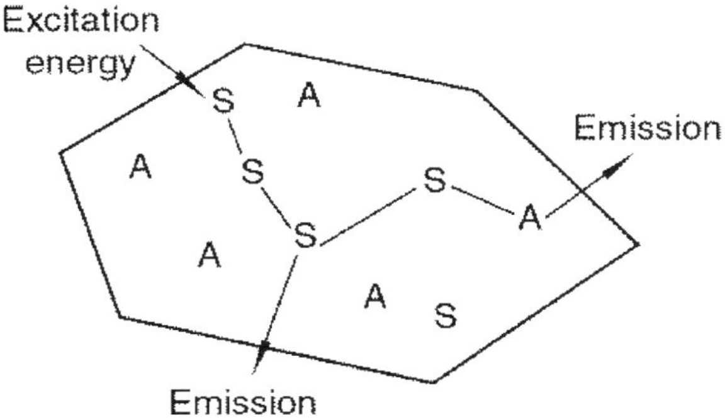

Figure 1
Luminescent material, excited by UV photons or electrons, containing activator ions (A) and sensitizing ions (on which UV absorption takes place, B).

1. *Emission Properties of Rare-earth Ions*

In almost all cases, the emission of rare-earth ions is due to optical transitions within the *f*-electron manifold (e.g., Tb^{3+} ($4f^8$), Gd^{3+} ($4f^7$), and Eu^{3+} ($4f^6$)) which are well shielded from the chemical environment and therefore almost retain their atomic character. For this reason, the *f–f* emission spectra consist of sharp lines. The *f–f* transitions are parity forbidden and many of them also spin forbidden. Therefore, the emission decay times are very long (μs–ms).

For a few rare-earth ions, however, broad emission bands also occur. Examples are Eu^{2+} ($4f^7$) and Ce^{3+} ($4f^1$). In this case, emission is due to *5d–4f* optical transitions. As *d* electrons participate in the chemical bonding, the *d–f* emission spectra consist of broadbands. Furthermore, these *d–f* optical transitions are allowed and are consequently very fast (<μs).

Examples of *f–f* (Y_2O_3:Eu) and *d–f* ($BaMgAl_{10}O_{17}$:Eu) emission spectra are given in Figs. 3 and 4, respectively.

It is interesting to compare the emission properties of rare-earth phosphors with other luminescent centers, to elucidate their unique properties.

Many transition metal ions are known as luminescent center (e.g., Mn^{2+} in Zn_2SiO_4:Mn (Fig. 5)). The $^4T_1$ $(t_{2g})^4$ $(e_g)^1) \rightarrow {}^6A_1$ $(t_{2g})^3$ $(e_g)^2$ transition, leading to green emission in this compound, involves a change in the chemical bonding, due to the transition of an electron from the nonbonding t_{2g} orbital to

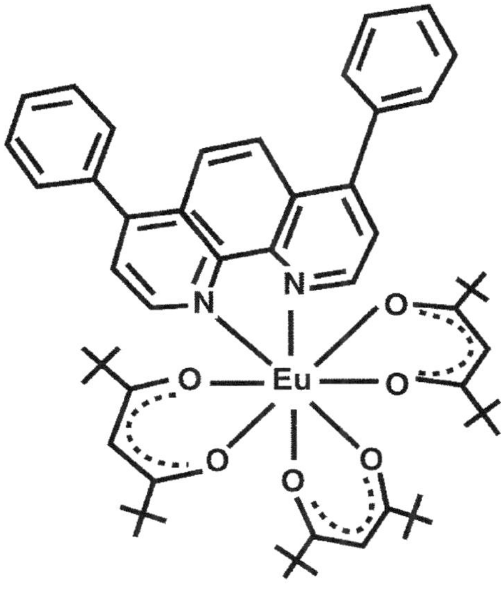

Eu(thd)$_3$dpphen

Figure 2
Luminescent rare-earth ion complex.

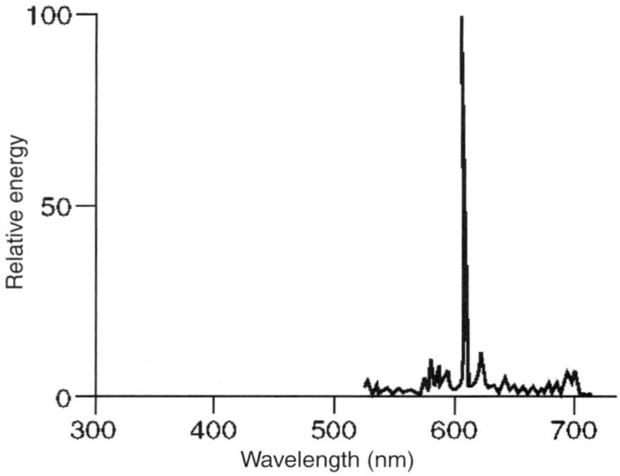

Figure 3
Emission spectrum of Y_2O_3:Eu.

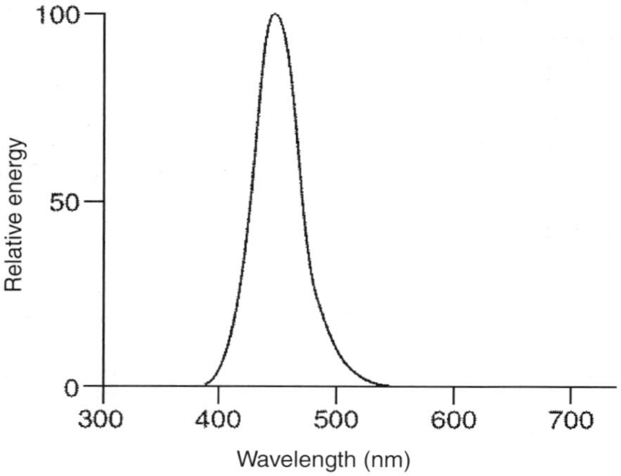

Figure 4
Emission spectrum of $BaMgAl_{10}O_{17}$:Eu.

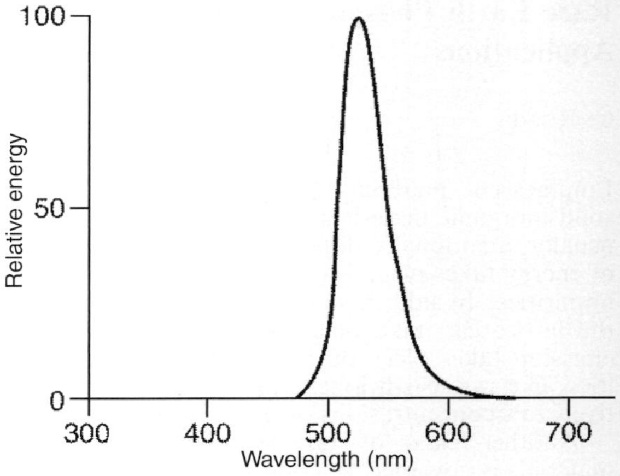

Figure 5
Emission spectrum of Zn_2SiO_4:Mn.

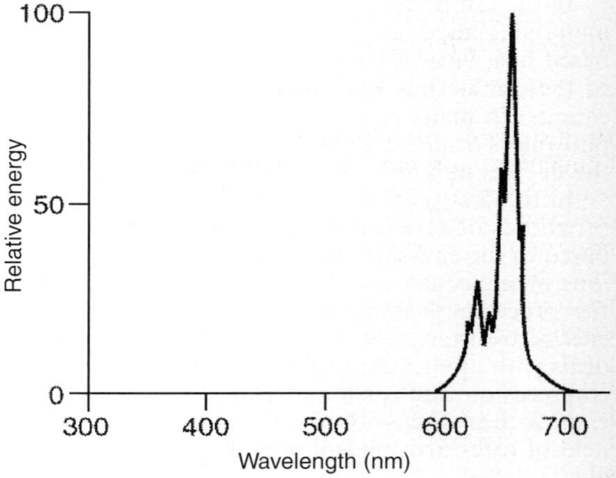

Figure 6
Emission spectrum of $Mg_4GeO_{5.5}F$:Mn.

the antibonding e_g orbital. Therefore, the emission band is broad. The emission spectrum of Mn^{4+} in $Mg_4GeO_{5.5}F$:Mn (Fig. 6) consists of a relatively narrow emission line with some structure due to vibronic interactions. In this case the emission is due to a spin flip transition within the $(t_{2g})^3$ manifold, i.e., the chemical bonding is hardly changed. This latter example shows that also in case of optical transitions involving states which participate in the chemical bonding, relatively sharp emission bands can be observed when the chemical bond strength in ground and excited state does not differ significantly.

Emission, originating from molecular anion groups (e.g., $CaWO_4$, Fig. 7), usually results from a charge transfer transition: a transition involving electronic states of different ions. Here, the chemical bonding will vary strongly, leading to broad emission bands.

This treatment shows that line emission is not a specific property of rare-earth ions and in addition that in case of rare-earth ions, broad emission spectra can be obtained as well. In this respect, rare-earth ions are not unique. The unique properties of rare-earth ions originate from the fact that the spectral position of the emission lines is almost independent of the host lattice, in contrast to line emission generated by transition metal ions. In addition, some of the line-emitting ions (like Tb^{3+} and Eu^{3+}) emit at spectral positions, enabling high lumen efficacies and a very good quality of white light as the emission spectra of these ions are almost optimally adapted to the human eye.

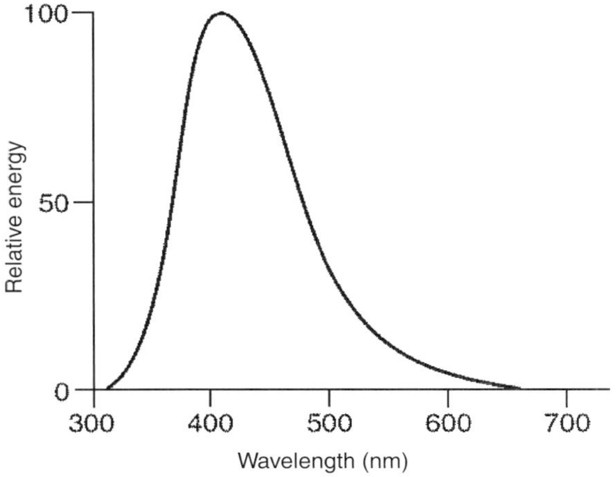

Figure 7
Emission spectrum of $CaWO_4$.

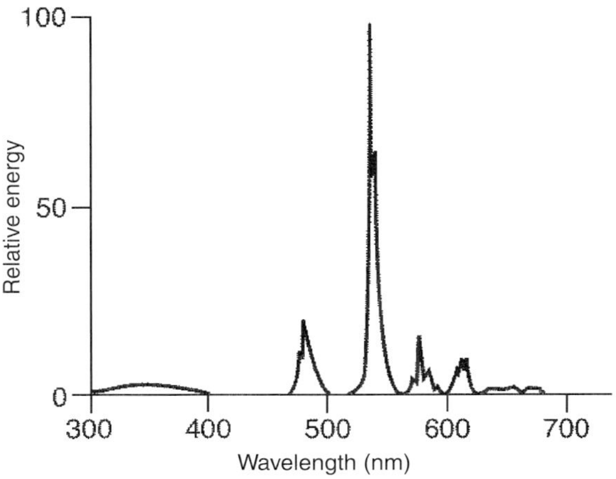

Figure 8
Emission spectrum of $(Ce,Tb)MgAl_{11}O_{19}$.

2. *Excitation and Emission Properties of Applied Rare-earth Phosphors*

2.1 Phosphors for Fluorescent Lamps

In this section, only fluorescent lamps for general illumination purposes, i.e., fluorescent lamps which emit white light, are considered. Apart from the efficiency with which white light is generated, another important property of lamps is their ability to reproduce all colors in a natural way and the color temperature of the light generated. These factors are determined to a large extend by the phosphors applied. The color temperature is equal to the temperature of a black body radiator whose emission spectrum has the same color coordinates. The so-called color rendering index (CRI) is determined by comparing the emission spectrum of the lamp with that of a black body radiator having the same color temperature. A black body radiator (and therefore also an incandescent lamp) has a CRI of 100 by definition.

Until about 20 years ago, mainly $Ca_5(PO_4)_3(F,Cl)$:(Sb,Mn) (halophosphate, CHP) was used in fluorescent lamps for general lighting purposes. In this compound, the UV radiation, generated by the mercury discharge, is absorbed by Sb^{3+} and partly transferred to Mn^{2+}. The combination of a blue (Sb^{3+}) and yellow (Mn^{2+}) emission band results in white light. Lamps based on CHP have a moderate energy efficacy (75 lm W^{-1}) and a relatively low CRI (50–60).

Rare-earth phosphors have changed the situation dramatically. Both the CRI and the energy efficacy of fluorescent lamps based on rare-earth emitters are considerably higher than in the case of lamps based on CHP (e.g., Koedam and Opstelten 1971). In such lamps, CRIs > 80 can be obtained at efficacies of 100 lmW^{-1} (e.g., Ronda 1997). Examples for phosphors used in these high quality fluorescent lamps are given below.

$BaMgAl_{10}O_{17}$ doped with Eu^{2+} (BAM) or $Sr_5(PO_4)_3Cl$:Eu are used as blue-emitting phosphor in high quality fluorescent lamps. In this material the UV photons are absorbed by Eu^{2+} ions, the absorption being due to a $4f \rightarrow 5d$ transition. The emission takes place on Eu^{2+} as well, the emission originates from the $5d \rightarrow 4f$ optical transition.

The red emission in these lamps is generated by Y_2O_3:Eu. Here, excitation involves a charge transfer transition from O^{2-} ions to Eu^{3+}. Emission takes place within the *f*-levels of the Eu^{3+} ions.

In some phosphors, the optical absorption on a center showing the desired emission is not strong enough. In such cases, absorption and emission have to take place on different optical centers. After optical absorption, the energy is transported through the lattice towards the emitting optical center. All green-emitting Tb^{3+} phosphors, applied in high-quality fluorescent lamps, are relying on optical absorption on Ce^{3+} ions, followed by energy transfer to the emitting Tb^{3+} ions. An example is $(Ce,Tb)MgAl_{11}O_{19}$, where the energy is transferred from Ce^{3+} to Tb^{3+}. Its emission spectrum is given in Fig. 8.

Apart from the line emission originating from Tb^{3+}, there is a trace of UV emission originating from Ce^{3+}, indicating that the energy transfer probability is only slightly below unity. A smaller amount of UV radiation is generated by, for example, $(Ce,Gd,Tb)MgB_5O_{10}$, where energy transfer from Ce^{3+} to Tb^{3+} proceeds via the Gd^{3+} ion sublattice.

2.2 Phosphors for CRTs

In standard CRTs, the only rare-earth phosphor applied is the red-emitting Y_2O_2S:Eu. For blue and green, ZnS doped with silver and copper, gold, respectively, are applied. The line emission of the Eu^{3+} ions enables a high luminous efficiency of red emitting Y_2O_2S:Eu despite the relatively low energy efficiency. Especially for red-emitting phosphors being line emitters, this is important in view of the rapid decrease in eye sensitivity for light with wavelength >600 nm.

In CRTs applied in projection television (where the excitation density is much higher), Y_2O_3:Eu is applied as red-emitting phosphor because it shows a better saturation behavior than Y_2O_2S:Eu. Here, the green-emitting phosphor applied is also a rare-earth phosphor; in most cases Y_2SiO_5:Tb is used. Contrary to the ZnS phosphors, the rare-earth activator ion concentrations chosen can be relatively high (of the order of 5%, whereas the activator concentrations in ZnS phosphors are of the order of 100–1000 ppm). No alternatives to ZnS:Ag have been found for blue.

2.3 *Phosphors for Vacuum Ultraviolet Excitation*

In plasma display panels (PDPs) and a new generation of mercury-free fluorescent lamps, a xenon discharge is used to generate photons with high energy. In this case, the photon energies are in the vacuum ultraviolet (VUV) part of the spectrum. Therefore, in many cases excitation of the emission takes place via the phosphor host lattice, followed by transfer of the energy to activator or sensitizer ions.

The performance of a number of phosphors under VUV excitation was reported as early as 1979 (NHK 1979). For display applications, blue-emitting barium magnesium aluminates, doped with Eu^{2+}, have good performance, albeit that some degradation occurs during device operation. The degradation results mainly in a color point shift of the PDP; the contribution to the luminance of the blue-emitting primary is low. The origin of the degradation of BAM is not yet completely clear, however, coatings on $BaMgAl_{10}O_{17}$:Eu have been proven to be a promising way out (Jüstel *et al.* 1998, Bechtel *et al.* 1997, Ronda 1995).

The green primary for PDPs is based on Mn^{2+}; with the emission spectrum of Mn^{2+} in, for example, Zn_2SiO_4:Mn or $(BaO)_x6Al_2O_3$:Mn, being located at the top of the color triangle a very large color gamut can be reproduced. A disadvantage of these materials is the fact that the emission is spin forbidden, resulting in too long a decay time for display applications. The spin selection rule can be lifted by magnetic interaction between magnetic ions (Ronda and Amrein 1996). In this way, the radiative lifetime is shortened, enabling faster decay not originating from loss processes. For lamps, Tb^{3+} emission is preferred in view of the high color quality required.

Red light is generated either by Y_2O_3:Eu or $(Y,Gd)BO_3$:Eu. The VUV absorption of Y_2O_3:Eu is weaker than that of $(Y,Gd)BO_3:Eu^{3+}$, however, the emission spectrum of Y_2O_3:Eu better matches the European Broadcasting Union (EBU) requirements for display applications. The emission spectrum of Y_2O_3:Eu has its maximum at 611 nm, whereas the emission spectrum of the mixed borate has its maximum at 595 nm. This nicely illustrates that, though the emission on Eu^{3+} is due to optical transitions within the f-manifold and, therefore, the spectral position of the individual emission lines is almost fixed, nevertheless the detailed shape of the emission spectrum depends on the crystal host. In this particular case, the emission at 611 nm in Y_2O_3:Eu originates from an electric dipole transition, whereas the emission at 595 nm in $(Y,Gd)BO_3$:Eu is due to a magnetic dipole transition. In $(Y,Gd)BO_3$:Eu, the electric dipole transitions are symmetry forbidden, as the Eu^{3+} ions are incorporated at a site with inversion symmetry. A second consequence is that the Eu^{3+} emission in $(Y,Gd)BO_3$:Eu is rather slow (3 ms).

3. *New Developments*

3.1 *Application of Rare-earth Phosphors in Semiconductor LEDs*

The invention of the blue light emitting diode (LED) based on GaN is a triumph of materials science. Electroluminescence based on III–V nitride semiconductors had already been given up when Nakamura from NICHIA Chemicals in the late 1980s picked up the topic again and developed it to an impressive maturity level within just five years. Very good reviews of the underlying semiconductor photophysics as well as the MOCVD technique used in LED chip production can be found elsewhere (Nakamura and Fasol 1997).

The blue diode is capable of challenging the lighting market as these devices can be used to generate white light from LEDs with efficiencies almost equal to halogen lamps. Three ways of white light generation appear likely:

(i) Blue, green, and red diodes can be combined to yield white light. This is attractive since by extending the GaN technique bright green diodes based on AlInGaN are available as well with high external quantum yields.

(ii) A single white LED can be obtained cheaply by combining a blue LED with green-, yellow-, and/or red-emitting phosphor materials which absorb and convert part of the incident blue light. In comparison to the three LED devices this causes a decrease in the overall efficiency because the quantum deficit of the light conversion from blue to yellow/red has to be taken into account.

(iii) A UV-emitting LED with blue, green, and red-emitting phosphors can be used. This option will have the lowest overall efficiency in view of the increasing quantum deficit.

In the following, the last two options will be discussed with respect to efficiency and light quality. In principle, it is possible to vary the emission wavelength of blue GaN-based LEDs between 370 nm (the bandgap of pure GaN), and 470 nm by increasing the indium content in InGaN devices. Assuming a conversion from the incident light by a phosphor material emitting at 555 nm, the already mentioned quantum deficit decreases in the same direction from about 33% to 15%. At first glance, the obvious option would therefore be a combination of a blue LED emitting at 470 nm and a broadband phosphor which can be excited at that wavelength and re-emitting at about 555 nm, creating white light composed of the blue and yellow-green emissions. Such a device has been realized using $(Y_{1-x}Gd_x)_3(Al_{1-y}Ga_y)_5O_{12}:Ce^{3+}$ (YAG:Ce) as a broadband yellow phosphor (Fig. 9).

Variation of x and y enables a YAG:Ce emission shift between 510 and 570 nm. This allows the adjustment of white light with color temperatures from 8000 K down to 3000 K, but at cost of efficiency to lower color temperatures. The very high color temperature has been chosen for commercialization in the Japanese market, which demands this color property. Due to the broad YAG:Ce emission the CRI is high, reaching 85. Finally, the luminous efficiency is 9 lm W^{-1}. This clearly shows the potential of the blue LED and its possible importance for the future lighting market.

Despite the higher quantum deficit, an LED emitting UV light at 370 nm covered with a rare-earth phosphor mixture would be an attractive option too. In contrast to a phosphor deposited on a blue-emitting LED chip at 470 nm, which only partly absorbs the blue light in order to generate white light, a mixture of blue, green, and red phosphors on a UV LED absorbs all of the incident light emitted by the chip. The white light color coordinates are determined only by the phosphor fractions in the mixture, which can be adjusted rather accurately. Consequently, white LEDs can be produced with color coordinates varying only very little from LED to LED. This is considerably more difficult for an LED of which white light is composed from incident blue emission from the diode and yellow/red light emitted by the phosphor. The additional quantum deficit of about 15% of due to conversion from 370 nm vs. 470 nm can be compensated partially by the very high lumen equivalents of Eu^{3+}- and Tb^{3+}-activated phosphors. In analogy to fluorescent lamps, high color renderings can be obtained at the same time. The search for stable rare-earth phosphors with high absorptions in the near UV/blue spectral region is, therefore, an attractive research task (see also *Light Emitting Diodes*).

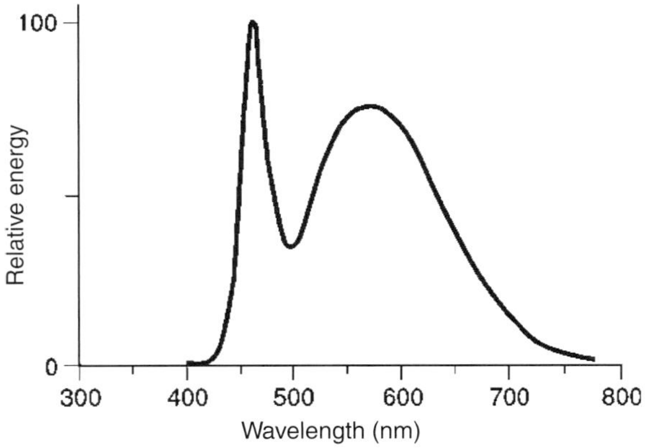

Figure 9
Emission spectrum of a white LED consisting of a blue LED and yellow-emitting YAG:Ce.

3.2 Luminescent Materials with Quantum Yield Larger Than Unity

In 1974, two papers appeared on the generation of more than one visible photon out of one UV photon (Piper *et al.* 1974, Sommerdijk *et al.* 1974). The rare-earth ion performing the quantum cutting or quantum splitting was Pr^{3+} with YF_3 as host lattice. It is quite surprising that with UV photons with wavelength of ~200 nm, blue and red emission, originating from Pr^{3+}, can be generated with a quantum efficiency of about 140%. The optical absorption is due to a $4f \rightarrow 5d$ transition on Pr^{3+} (in this lattice located at about 185 nm). Though a breakthrough, this material has never been applied in fluorescent lamps. An important reason is the fact that the blue line emission, generated by Pr^{3+}, is located at about 410 nm and, therefore, not suitable for lamps with good color rendering, in view of the properties of the human eye. A possible solution is the use of additional phosphors showing a strong absorption in the blue-UV part of the optical spectrum which emit light with longer wavelength, in the same manner as discussed above. In this way, also the very deep blue Pr^{3+} emission could be utilized. The red emission, also generated by the Pr^{3+} in the cascade leading to quantum cutting has very good spectral properties for high quality fluorescent lamps; it is located at ~620 nm.

The observation of the blue line emission of Pr^{3+} in oxidic materials in which the rare-earth ion is coordinated by a large number of oxygen ions has been reported (like in $SrAl_{12}O_{19}$, where the Pr^{3+} is coordinated by 12 oxygen ions) (Srivastava and Beers 1997, Srivastava *et al.* 1996). A prerequisite for the observation of blue emission is that the energy of the $5d$ level is higher than that of the highest $4f$ level, from which the blue emission originates. By choosing a high rare-earth ion coordination number, the mean metal-ligand distance becomes large. The crystal field

(CF) splitting decreases on increasing the metal–ligand distance. As a consequence, in $SrAl_{12}O_{19}$ the Pr^{3+} ion experiences such a weak CF that the energy position of the lowest 5*d* level is higher than that of the highest 4*f* level—a situation until then only observed in fluorides. Again, this can be regarded as a breakthrough; however, the question concerning the blue line emission at 410 nm has not been solved.

In 1997, a luminescent material ($LiGdF_4$:Eu) with internal quantum efficiency of almost 200% (the emission originating almost exclusively from Eu^{3+}) was reported (Wegh *et al.* 1997, 1999). Excitation with VUV radiation is on the Gd^{3+} ion, using radiation with a wavelength of about 200 nm. In the first step, Eu^{3+} ions are excited via cross relaxation between Gd^{3+} and Eu^{3+}. In the second step, Eu^{3+} ions are excited via energy transfer from $Gd^{3+} \rightarrow Eu^{3+}$ (Fig. 10).

This is the first material showing such a high quantum efficiency on excitation with radiation with photon energies accessible with efficient discharges.

3.3 Organic Electroluminescence

Tang and Van Slyke (1987) demonstrated for the first time that light could be generated efficiently by direct recombination of charge carriers in organic materials. A typical organic electroluminescence (EL) device structure is shown in Fig. 11.

A glass substrate is covered with a transparent conductor, in most cases indium tin oxide (ITO). In consecutive evaporation steps, a hole transport layer (HTL), optionally an emissive layer, an electron transport layer (ETL) and finally a cathode are stacked on each other. Hole transport materials are frequently based on aromatic amines known for their low oxidation potential and high hole mobility. A prominent representative which is extensively used as hole transport material in photocopy machines and laser printers is tetraphenylene diaminobiphenylene (TPD) (Fig. 11) (Tang and Van Slyke 1987). The most frequently used electron transport material is a complex of aluminum with 8-hydroxyquinoline (AlQ, Fig. 11). A device with only AlQ and an HTL is already electroluminescent, but only with low efficiency. Stable organic fluorescent dyes doped in small amounts into the ETL (or sandwiched between ETL and HTL) greatly enhance the electroluminescence efficiency and the device stability (Tang *et al.* 1989). Similar to an inorganic phosphor, radiative transitions following energy transfer from the host to the dopant (see below) can be very effective when concentration quenching is avoided. Typically, metals with a low working function like lithium, calcium, magnesium, or their alloys with aluminum or silver are used as cathodes.

The principle of organic EL is explained in Fig. 11. A DC voltage, usually ~5–10 V is applied to the EL device between the ITO and metal cathode layer. Positive charges are injected from the ITO into the HTL, electrons from the metal cathode into the ETL. For this step, it is critical that the Fermi niveaus of the injecting layers and the accepting energy levels of the charge transport layers are adapted to each other in order to avoid injection charges. At the same time, the buildup of space charge has to be avoided which is possible by applying ultra-thin films of the charge

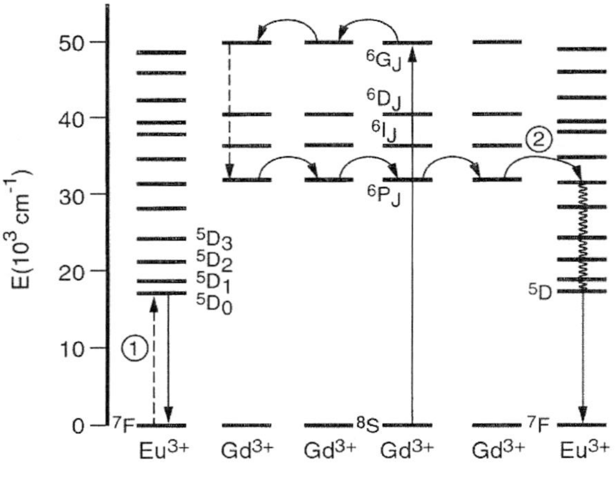

Figure 10
Processes leading to quantum cutting in $LiGdF_4$:Eu.

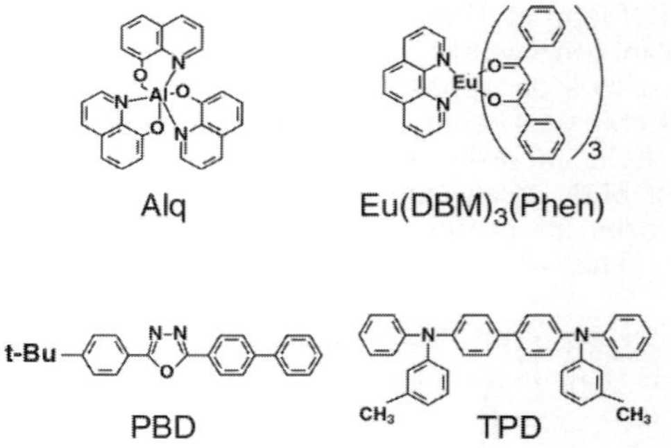

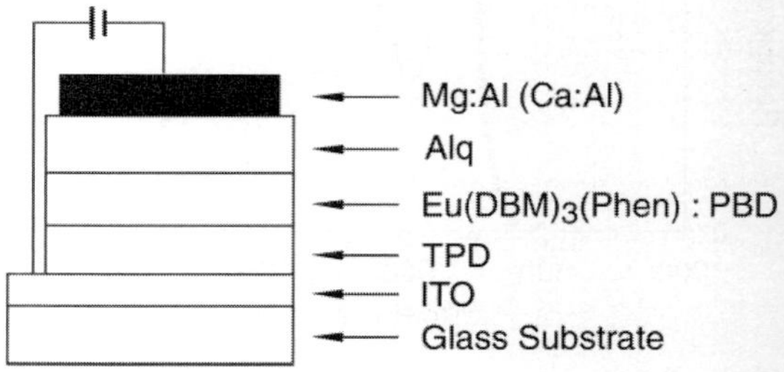

Figure 11
Organic EL device.

transport layers in the order of a few thousand angstroms or less. The HTLs are usually good electron blocking layers while the ETL can efficiently block positive charges. This means that the injected charges will be transported to the HTL–ETL interfacial region where, due to the potential barrier, the probability for recombination is maximized. Charge recombination leads to excitons, 25% of which have singlet and 75% triplet character due to spin statistics. In organic fluorescent dyes only the singlet excitons can populate the ^1S state of the fluorescent dye which consequently decays radiatively, in principle with the same quantum yield as measured for photoluminescence. The triplet excitons are lost in nonradiative transitions or even photochemical side reactions which can be harmful for the device stability. This drawback of organic fluorescent dyes has triggered efforts to use rare-earth ions, such as Eu^{3+} or Tb^{3+} coordinated to organic ligands, as emitters in organic EL (Kido *et al.* 1993, 1994, 1996, Boerner *et al.* 1994, 1997). Here, the organic ligand acts as sensitizer for the otherwise poorly absorbing rare-earth ion. In photoluminescence, UV light is effectively absorbed by the organic ligand leading to an excited singlet state. Due to the close presence of the rare-earth metal ions, intersystem crossing is effective in these compounds populating the ligand triplet state. From there, the energy is efficiently transferred to the f levels of the lanthanide ion. In EL, it appears likely that the ligand triplet levels can be directly populated by energy transfer from the triplet excitons. In theory, internal efficiencies of 100% should thus be possible. Another advantage of rare-earth organic EL is the color purity of the generated emission (see above), i.e., pure green and red for Eu- and Tb-activated devices, respectively.

4. Conclusions

In many applications, the performance of rare-earth phosphors is almost ideal and consequently the use of rare-earth phosphors has pushed the performance of devices based on them to physical limits.

New impetus for research on phosphors is expected from new applications like luminescent materials applied in LEDs, phosphors showing new features e.g., exhibiting quantum efficiencies larger than unity, and organic electroluminescent materials.

Bibliography

Bechtel H, Czarnojan W, Mayr W, Jüstel T, Nikol H, Ronda C R 1997 *German Pat.* 197 27 607.5

Boerner H, Busselt W Jüstel T, Nikol H 1997 *German Pat.* 197 26 472.7

Boerner H, Kynast U, Busselt W, Haase M 1994 *German Pat.* P 4428450.0

Jüstel T, Bechtel H, Nikol H, Ronda C R, Wiechert D U 1998 Improved VUV phosphors for plasma display panels. In: Struck C W, Mishra K C, Di Bartolo B (eds.) *Proc. 7th Int. Symp. Physics and Chemistry of Luminescent Materials.* Electrochemical Society, Pennington, NJ, pp. 103–19

Kido J, Hayase H, Hongawa K, Nai K, Okuyama K 1994 Bright red light-emitting organic electroluminescent devices having a europium complex as an emitter. *Appl. Phys. Lett.* **65**, 2124–6

Kido J, Ikeda W, Kimura M, Nai K 1996 White-light-emitting organic electroluminescent device using lanthanide complexes. *Jpn. J. Appl. Phys.* **35**, L394–6

Kido J, Nagal K, Okamoto Y 1993 Organic electroluminescent devices using lanthanide complexes. *J. Alloys Comp* **192**, 30–3

Koedam M, Opstelten J J 1971 Measurement and computer-aided optimization of spectral power distributions. *Light Res. Technol.* **3**, 205–10

Nakamura S, Fasol G 1997 *The Blue Laser Diode.* Springer, Berlin

NHK Technical Monograph 1979 **28**

Piper W W, DeLuca J A, Ham F S 1974 Cascade fluorescent decay in Pr^{3+}-doped fluorides: achievement of a quantum yield greater than unity for emission of visible light. *J. Lum.* **8**, 344–8

Ronda C R 1995 Recent developments in phosphors for lamps and plasma displays. *Proc. 2nd Int. Display Workshops.* Hamamatsu, Japan, IDW Committee, pp. 69–72

Ronda C R 1997 Recent achievements in research on phosphors for lamps and displays. *J. Lum.* **72–74**, 49–54

Ronda C R, Amrein T 1996 Evidence for exchange induced emission in Zn_2SiO_4:Mn. *J. Lum.* **69**, 245–8

Salbeck J 1996 New low molecular organic compounds with high glass transition temperature as materials for blue electroluminescence. In: Mauch R H, Gumlich H-E (eds.) *Inorganic and Organic Electroluminescence.* W&T Verlag, Berlin, pp. 243–6

Sommerdijk J L, Bril A, de Jager A W 1974 Two photon luminescence with ultraviolet excitation of trivalent praseodymium. *J. Lum.* **8**, 341–4

Srivastava A M, Beers W W 1997 Luminescence of Pr^{3+} in $SrAl_{12}O_{19}$: observation of two photon luminescence in oxide lattice. *J. Lum.* **71**, 285–90

Srivastava A M, Doughty D A, Beers W W 1996 Photon cascade emission of Pr^{3+} in $LaMgB_5O_{10}$. *J. Electrochem. Soc* **143**, 4113–6

Tang C W, Van Slyke S A 1987 Organic electroluminescent diodes. *Appl. Phys. Lett.* **51**, 913–6

Tang C W, Van Slyke S A, Chen C H 1989 Electroluminescence of doped organic films. *J. Appl. Phys.* **65**, 3610–6

Van Slyke S A, Bryan P S, Tang C W 1996 Blue organic light emitting devices. In: Mauch R H, Gumlich H-E (eds.) *Inorganic and Organic Electroluminescence.* W&T Verlag, Berlin, pp. 195–200

Wegh R T Donker H Meijerink A 1997 Vacuum ultraviolet spectroscopy and quantum cutting for trivalent rare-earth ions. In: *Book of Extended Abstracts of the 192nd Meeting of the Electrochemical Society*, Paris, France, Electrochemical Society, p. 1766

Wegh R T, Donker H, Oskam K D, Meijerink A 1999 Visible quantum cutting in LiGdF4:Eu. *Science* **283**, 663–6

Rotary and Linear Machines

R. E. Hanitsch

From the history of electrical machines it is known that the earliest machines were axial-field machines. However, they were replaced by radial-field machines after a short time. It appears that with the high-energy permanent magnets such as NdFeB, the flat shape of axial-field machines is a desirable configuration especially for servo drives. The construction topology of axial-field and radial-field machines can be varied to a large extent. This is also valid for linear machines.

Electrical motor drive systems using electronic conditioning and modern control methods can offer substantial economic advantages over uncontrolled motor drives. The use of digitally controlled electrical drives also enables many applications to be implemented with an accuracy and reliability not otherwise attainable. Significant further advantages in monitoring and multimotor control can arise from the use of digital controllers.

This contribution focuses on the understanding of the design principles using novel soft and hard magnetic materials to achieve efficiency improvements and specific features such as high acceleration. Emphasis will be put on drives in the power range from a few watts to a megawatt.

1. Market Development

In industrialized countries, about one-third of all primary energy is converted into electricity. Of this, about two-thirds are converted back into mechanical energy by motors of different types. The largest part is converted by a.c. induction motors in the power range 1–100 kW/400 V. Motor efficiency is therefore an important economic question. An efficiency increase of only 1% would save around 60 million euro per annum in just one country in northern Europe. The most important aspect of an electric motor is not only the efficiency, it is the cost in combination with reliability and noise.

In the design methodology, the focus is not only put on the electromechanical energy converter, it is on the total drive system including the load and the control. The drive requirements arise from the process needs, the wealth creation objective, and the given environmental requirements. Figure 1 gives an overview of the drive design aspects. The design process itself includes a series of iterations if the requirements don't balance. The trend towards energy conservation and automation of production has led to an increased demand for electronically controlled d.c. and a.c. motors. A wide variety of servomotors is used for automated assembly lines, industrial robots, manipulators, and machine tools.

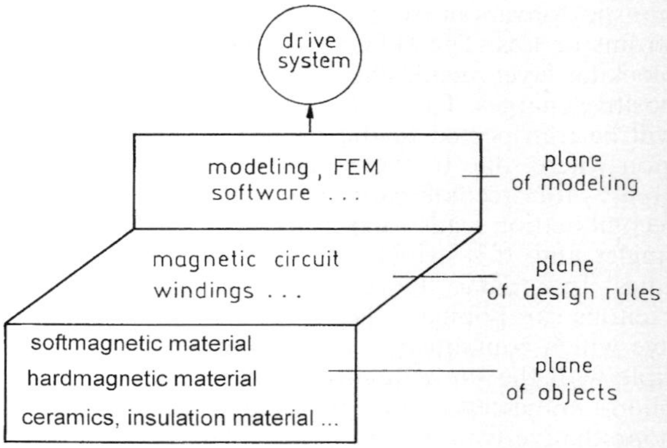

Figure 1
Structure of design process.

d.c. servo motors are still used due to their performance characteristics such as high torque and good dynamic behavior. It is estimated that every year about 15% of the classical brushed d.c. drives are replaced by a.c. drives. The fields of application of motors were analyzed in a recent market survey resulting in a breakdown as: automation 26%, pumps and fans 19%, d.c. drives for automotive subsystems 16%, a.c. vector drives 15%, positioning and CNC machines 14%, and manufacturing 10%.

Although the drive technology is "mature," there is perhaps more development activity today than at any time in the past. This is due to the following factors:

(i) New materials: insulating materials, electrical steels, ceramics, high-energy permanent magnets.
(ii) Precise demands from the "motion-control industry."
(iii) Increased flexibility in drive design resulting from modern power electronic control techniques.
(iv) Improved design techniques based on CAD and finite element analysis.
(v) Improved instrumentation and condition monitoring.

Before illustrating some of these developments, it is necessary to study some interesting motor parameters such as:

(vi) Air gap shear stress (AGSS) in Nm^{-2}, defined as AGSS = current layer × flux density in the air gap.
(vii) Torque per unit rotor volume (TURV) in Nmm^{-3}.

The torque of a motor can be related to an average shear stress at the rotor surface which is related to the product of the magnetic and electric loading. The values quoted in Table 1 relate to continuous ratings. Peak ratings may exceed these values by a factor of two to three, depending on the application and cooling strategy. The TURV can be used to estimate the rotor volume as the starting point of a design procedure.

2. High-efficiency Motors

There are several options in designing motors for higher efficiency. A variety of different activities is often combined:

(i) Improvement of the slot-fill factor.
(ii) Improvement in fan design.
(iii) Better heat transfer.
(iv) Novel soft magnetic materials and/or high-energy permanent magnets.
(v) Refined design methods with finite element analysis/integrated designs.

As a starting point, focus is on asynchronous motors because they are widely used in industry. Large a.c. drives have a specific torque of 3.7–5.0 nmkg^{-1}. Figure 2 shows the power flow for such a motor. The various losses will be studied in detail.

2.1 Copper Losses

A classical way to reduce the copper losses is to increase the copper area using a higher slot-fill factor or by optimizing the slot geometry using finite element analysis. Attention must be paid to the fact that the increased copper area may reduce the iron cross-section and increase the flux density and therefore increase the iron losses. Rapid cooling would help to keep the temperature rise of the winding low, but this will often result in higher fan losses. An alternative might be the water-cooled motor; however, these systems are more complex.

2.2 Iron Losses

These are the sum of hysteresis losses and eddy-current losses and vary with flux density and frequency. Iron losses depend on the composition and the heat treatment of the steel laminations. For low eddy-current losses, the iron laminations should be as thin as possible. Typical laminations have a thickness of 0.23 mm, 0.27 mm, 0.3 mm, 0.35 mm, and 0.5 mm. A further reduction of the eddy-current losses can be achieved by a silicon content of between 1 and 3%. Depending on the application, different types of steel laminations are available:

(i) Grain oriented electrical steels are used especially in the field of transformer construction. They are characterized by low magnetostriction for quiet transformers. A further decrease of the core loss is achieved by refinement of the magnetic domains by laser treatment. Typical loss data are given in Table 2.
(ii) Nonoriented electrical steel grades are used for the magnetically active parts of generators and motors, where the magnetic properties in each direction must be as uniform as possible. The material is finally annealed. For use at high

Table 1
Shear stress and torque per unit rotor volume.

Motor type	AGSS (kNm^{-2})	TURV (kNmm^{-3})
Fractional industrial motors	0.7–2.1	1.5–4
Industrial servos	10–21	20–42
Large liquid-cooled machines	70–105	140–210

Table 2
Examples of grain-oriented electrical steels.

Grade	Nominal thickness (mm)	Maximum core loss, $f = 50$ Hz; 1.0 T; (Wkg^{-1})
080	0.23	1.12
095	0.23	0.93
090 L	0.23	0.85
097	0.30	1.25
100	0.30	0.98

Source: EBG/Bochum.

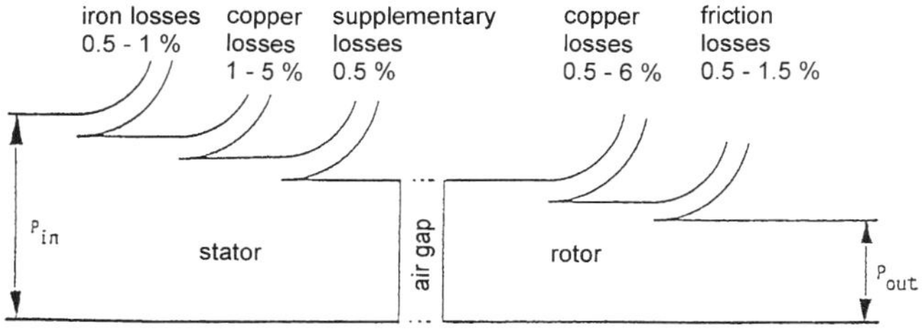

Figure 2
Breakdown of losses for asynchronous motor.

Table 3
Examples of nonoriented steels.

Grade	Nominal thickness (mm)	Maximum core loss, $f = 50$ Hz; 1.0 T; (Wkg^{-1})
235	0.35	2.35
250	0.35	2.50
270	0.50	2.70
330	0.65	3.30

Source: EBG/Bochum.

frequencies, a cold rolled, highly alloyed sheet with a thickness of 0.2 mm is available. The core loss at 400 Hz is $P(1.0) = 13\,\text{Wkg}^{-1}$. Table 3 shows some interesting properties of cold rolled, nonoriented electrical steels, finally annealed.

(iii) Amorphous metal has been used for the core of distribution transformers where the superior magnetic loss performance is exploited. The amorphous alloy is produced in the form of thin ribbons, 30 microns thick, and has several key parameters which have to be allowed for in energy converter design. A laminate, made from six ribbons amalgamated together to form a 0.13 mm thick material can be used as a substitute for silicon–iron laminations. The material has the ability to decrease transformer no-load losses by about 70%. A drawback is the material's low saturation induction of 1.6 T, which dictates a maximum operating induction of 1.35 T: Amorphous materials are a good choice for high-frequency power supplies and high-speed induction motors.

2.3 Supplementary Losses

As these losses are eddy-current losses, they can be reduced by careful material selection and thoughtful design. In designing an "energy efficient" motor, there are obviously several options. While energy saving is always desirable, there are cases where other considerations require a compromise. Guaranteed maximum iron losses will be obtained from the steel producer's curves for the selected grade and thickness of the lamination. These losses are then usually increased by a factor of around 1.15 to allow for the additional losses due to punching. Reviews of the different losses in asynchronous machines have been published by Chalmers (1988) and Nürnberg and Hanitsch (1987). Insulation between the lamination for small motors may be done by steam-bluing the steel surfaces. The larger laminations will employ a phenolic or synthetic resin insulation on one or both sides of the lamination. In the case of single-sided insulation, it is ensured that the laminations are always punched and stacked in one direction, thus avoiding contact between punching burrs. Laminations decrease the amount of magnetic material available to carry flux within a given cross-sectional area. To incorporate this fact in analysis, a stacking factor is defined as:

$$f_{\text{st}} = \frac{\text{cross-section occupied by magnetic material}}{\text{total cross-section}}$$

This factor is important for the calculation of flux densities in laminated magnetic materials. Typical stacking factors range from 0.65 to 0.95. Today, it is also possible to use bonded soft magnetic material to reduce eddy-current loss to a minimum. The soft magnetic material particles are suspended in a nonconductive resin. The small size of the magnetic particles used and their electrical isolation from each other increases the resistivity of the material. The effective permeability of the material, however, is decreased to 500 or 800. In a number of applications employing a three-dimensional flux path, this novel soft magnetic material has demonstrated its benefits. Wire erosion is often the best approach to prepare the magnetic circuit.

2.4 Friction Losses

Depending on the motor power, electric motors will use either rolling or sleeve bearing. For high-speed applications, the friction losses will increase rapidly and then magnetic bearings are a good choice. In the case of large motors, active magnetic bearings are used while for small rotating and linear devices, passive bearings help to overcome the problems with heavy friction losses. Magnetomechanical devices, such as passive bearings, need high-polarization magnets. The polarization must remain constant and the coercive field must be high. Rare earth iron permanent magnets are the best material for this purpose as outlined by Yonnet (1996). For small linear motors, linear passive bearings exist to ensure translational guidance. High-energy permanent magnets for bearings are the essential components of magnetic suspension. The suspension cannot be built completely using permanent-magnet bearings. As a consequence of Earnshaw's theorem, typically a magnetic suspension system can be built with two radial passive bearings and an active axial system.

3. *Selected Designs*

As shown in Fig. 3, the variety of geometries and designs is large. A material-selection decision for these geometries is often disguised as a material substitution decision. An electromechanical device can be characterized as having two components

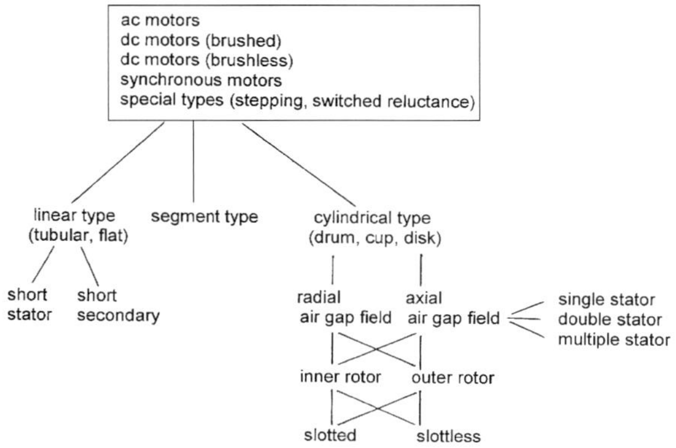

Figure 3
Selected types of electromechanical energy converters.

Table 4
Matrix for materials and designs..

Proven materials, P_M Proven design, P_D (1,1)	Proven materials, P_M Unproven design, U_D (1,2)
Unproven materials, U_M Proven design, P_D (2,1)	Unproven materials, U_M Unproven design, U_D (2,2)

(planes as shown in Fig. 1): material and design. In addition, each plane has just two values: proven or unproven. An electromechanical device has either proven (P_M) or unproven (U_M) materials and the design is again either proven (P_D) or unproven (U_D). This leads to a matrix as shown in Table 4. It is quite common to prefer electromechanical devices to hydraulic, pneumatic, and mechanical devices across a broad spectrum of applications with special demands such as operation at extremely high temperatures, which in some applications can reach 800°C.

3.1 *High-temperature Devices*

In the case of elevated temperatures, special problems arise such as the reduced flux-carrying capability of soft magnetic materials, insulation integrity, and dimensional changes due to thermal expansion. High nickel-content alloys have a Curie temperature around 480°C (e.g., Permalloy) and for 3% silicon–iron the Curie temperature is 720°C. A few grades of cobalt iron exhibit a higher Curie temperature. Focus should be on the 24% and 49% cobalt alloys, which offer a high saturation flux density, and the laminated sheets have low losses. The Curie temperature ranges from 900 to 960°C depending on composition. To move from matrix position (2,1) to (1,1) in Table 4, a new material may be experimentally tested

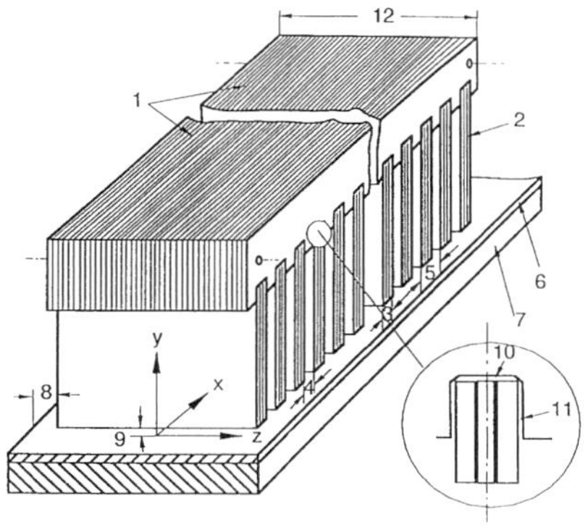

1 stator yoke	5 slot pitch	9 mechanical air gap
2 sheets of tooth	6 reaction plate	10 ground of slot
3 tooth width	7 magnetic return	11 air gap slot/yoke
4 slot width	8 overhang	12 yoke width

Figure 4
Stator construction of asynchronous linear motor (windings are not shown).

for the design specification, e.g., at 750°C, and then validated. The application of permanent magnets is restricted to temperatures of up to 250°C.

3.2 *Linear Motors*

Typical applications for low-speed and low-cost linear asynchronous motors are sliding doors. Instead of punching the lamination with a slot pitch of 10 mm and then using an inverter to operate the linear motor with a frequency below 50 Hz, a novel approach leads to a reduced slot pitch and the linear motor can be supplied from the mains. As shown in Fig. 4, the yoke and the teeth are produced separately. A number of isolated laminations are placed together and form the teeth with a total thickness of 2–3 mm. Later, they are pressed into the slots which are perpendicular to the stator-yoke lamination. The laminations used to create the teeth should have directional magnetic properties to guide the flux.

3.3 *Switched Reluctance Motors*

The switched reluctance (SR) motor is similar in principle to the variable reluctance (VR) stepping motor. But it is designed with rotor-position switching of the phase currents in the stator and optimized for continuous rotation. The advantages over induction and permanent magnet motors are:

(i) The maximum permissible rotor temperature may be higher than in permanent magnet (PM) motors because there are no magnets on the rotor;

(ii) Starting torque can be very high, without the problem of large inrush currents;
(iii) Extremely high speeds are possible;
(iv) The system efficiency of SR drives is 4–7% higher than that of inverter-fed induction machines.

Although there are also some disadvantages, such as the torque ripple and the acoustic noise, SR motors have been built for a number of frame sizes. SiFe alloys with 3% silicon show nearly isotropic magnetic behavior and therefore this material is well suited for the SR geometry to avoid any variation in the inductance profile for the different phases (Miller 1989).

4. *Permanent Magnet (PM) Motors*

In the case of classical d.c. motors and synchronous motors, field windings are used to provide the requested flux density in the air gap. The copper losses in these windings, providing the excitation, can be avoided if the field winding is replaced by an arrangement of permanent magnets. Therefore, in the same frame with the same cooling, the PM motor will have a better efficiency. PM excitation, once thought viable only in small motors, is now also used for industrial drives in the power range up to the megawatt level.

Many different types of PM materials are available today: alnico, ceramic ferrite, samarium-cobalt and neodymium-iron-boron (NdFeB). Most recently developed magnet types are available in both sintered and bonded forms. Bonded magnets are formed by suspending powdered magnet material in a nonconductive, nonmagnetic binder, e.g., resin. Sintered magnets are capable of high performance because the sintering process allows magnets to be formed without a bonding agent.

The alnicos with high remanence of 1.0–1.3 T find limited application in motors because of their nonlinear characteristic. They are prone to irreversible demagnetization by armature reaction fields due to the poor coercivity. Ferrites exhibit generally linear characteristics with a remanence up to 0.43 T. The low cost is the reason for the wide use. The flux concentration technique may be applied to increase the flux density in the airgap.

SmCo magnets offer a high remanence of 0.9–1.1 T and again show linear characteristics. Their low temperature coefficients enable them to operate over a wide range of temperatures. The high cost prevents wide application.

Neodymium-based magnets also have a linear characteristic. The best grades have a remanence of above 1.5 T and an energy product of more than 450 kJm^{-3}. Coercivity, depending on the grade, will vary from 800 kAm^{-1} to 1200 kAm^{-1}. New grades of NdFeB help designers of motors to meet demands for high performance in hostile environments. These grades have enhanced corrosion resistance and are suitable for temperatures of 200°C or more encountered with class F servomotors. Permanent magnets help to improve the energy efficiency of brushed d.c. motors, synchronous motors, and brushless d.c. motors. The high-energy product magnets, such as rare earth cobalt and NdFeB, triggered new brushless motor developments with surface mounted or buried magnets.

4.1 Brushed Types

Brushed d.c. motors with permanent magnets can be found in some fields of drive technology in the power range up to several kilowatts. The permanent d.c. motor is widely accepted, both for general-purpose drives as well as for servo drives.

Servo motors have a linear speed/torque characteristic and a high torque to inertia ratio. The principal application is in automatic position control systems requiring fast response and high power-to-weight ratio. It is evident that with increased flux due to improved magnets, the machine will improve. However, in machines with NdFeB magnets, the effects of magnetic saturation cannot be neglected. The armature magnetomotive force has definite effects on the space distribution of the air gap flux and the magnitude of the net flux per pole. The effect on flux distribution is rather important because the limits of successful commutation are directly influenced (Hanitsch 1996, Hanselman 1994).

4.2 Brushless Types

In a brushless d.c. motor, a multiphase armature winding is located on the stator and connected to an electronic commutation circuit. The rotor is a rotating permanent-magnet assembly. For many applications, a permanent magnet brushless d.c. motor can be designed which is smaller, lighter, and more efficient than an equivalent wound-field machine. The actual achievable size reduction is determined by the choice of the magnet material. The different magnet materials offer a wide range of magnetic, mechanical, and thermal properties (Hanitsch 1996).

Having decided on the type of magnet material to be used, a detailed design work is necessary to find the magnet dimensions required for the desired characteristics. The magnet dimensions, together with the effective airgap length and flux leakage factor determine the open circuit magnetic flux. The magnet dimensions and the layout of the magnetic circuit critically determine the magnet's ability to withstand armature reaction effects on load and particularly under stalled condition. Because of their large coercive force, NdFeB magnets are mostly stable with respect to demagnetization. The classical method of magnet "loadline" calculations is quite helpful in the design procedure. However, for a detailed design, the use of the finite-element method is advantageous.

This method, combined with an interactive CAD/CAM system, will help the design engineer to exploit the materials used in an optimal way. Often a three-dimensional approach is necessary. For proper design of the machine, not only the magnetic characteristics are of interest. Besides thermal characteristics, mechanical characteristics are also needed for a thoughtful design. Corrosion resistance of NdFeB has been improved by the addition of cobalt, copper, and gallium which passivate the reactive neodymium-rich intergranular phase. Handling NdFeB magnets with magnetized poles can be quite difficult.

The outstanding characteristic of the brushless d.c. machine is its versatility. Brushless micromotors have been constructed for medical applications. Another important market segment is the hard disk drive (HDD). To achieve a higher TPI (tracks per inch) along with faster disk rotation, it is necessary to avoid vibrations and acoustic noise from the rotating disk and to keep the head on the track. Almost every HDD today uses a trapezoidal drive waveform to reduce torque ripple. This is possible due to NdFeB magnets.

The replacement of an induction motor rotor by one carrying permanent magnets will produce a synchronous motor, which has a higher efficiency and can run at either leading or lagging power factor. For fixed-speed applications, line-start capability is provided by retaining the rotor cage for startup and embedding the magnets below.

Permanent magnet-excited machines of this type are also used for the exploitation of wind energy. For the low power range, ferrite material is used, while for greater power, NdFeB grades are selected. In addition of the classical designs, toroidal designs are also in use as shown in Fig. 5. Emerging applications of brushless d.c. motors with NdFeB magnets are appearing in hybrid electric vehicles, strip propulsion,

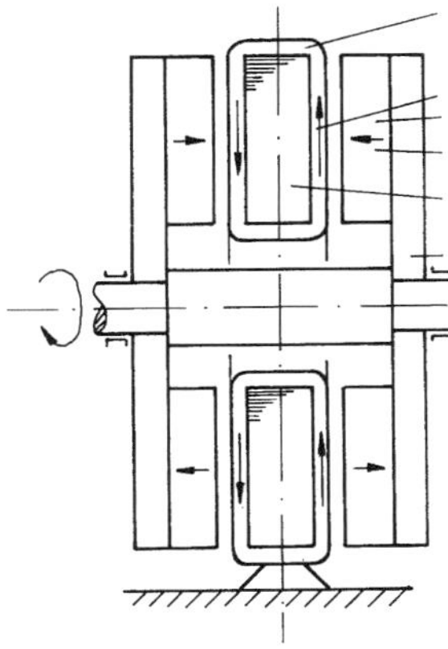

Figure 5
Brushless motor in a toroidal design.

and locomotive systems. The transversal flux motor, recently constructed, has a specific torque of 15–20 Nm kg^{-1} and will be used as a gearless drive.

5. *Conclusions*

As a linear and rotating motion control system consists of subsystems such as: power supply, converter, control system, electric motor, mechanical coupling, and the load, the future trend of the drive technology will be influenced by:

(i) Power electronic components,
(ii) Electric motor technology exploiting novel materials,
(iii) Control strategies (microprocessors and digital signal processor technologies).

Beside the pressure on increased productivity and improved product quality there is always concern for higher energy efficiency. High-energy permanent magnets help to achieve this goal.

With new or improved insulation materials for the windings it will be possible to have "inverter-proof" machines available to withstand the additional stresses created by the voltage pulses and peaks as a result of the fast switching operation of the power electronic circuits.

A motion-control system consisting of a converter and an a.c. motor is a nonlinear multivariable system in which the discrete time effects of the converters introduce additional problems. The control algorithms developed in modern control theory cannot be implemented using only microprocessors, because the hardware limits the sampling frequency. The advent of digital signal processors (DSP) opens new possibilities to realize advanced controllers. The growth of the a.c. drive technology will continue because of the development of high-frequency and high-efficiency power electronic drives. Advanced control algorithms, realized as digital controllers, will contribute to the high performance of efficient electrical drives making use of novel soft and hard magnetic materials.

Bibliography

Chalmers B 1988 *Electric Motor Handbook*. Butterworths, London

Hanitsch R 1996 Permanent magnet motors. In: Coey J M D (ed.) *Rare Earth Iron Permanent Magnets*. Clarendon Press, Oxford, Chap. 10

Hanselman D C 1994 *Brushless Permanent-Magnet Motor Design*. McGraw-Hill, New York

Miller T J E 1989 *Brushless Permanent-Magnet and Reluctance Motor Drives*. Clarendon Press, Oxford

Nürnberg W, Hanitsch R 1987 *Die Prüfung elektrischer Maschinen*. Springer, Berlin

Yonnet J-P 1996 Magnetomechanical devices. In: Coey J M D (ed.) *Rare Earth Iron Permanent Magnets*. Clarendon Press, Oxford, Chap. 9

S

Solid Oxide Fuel Cells

F. Tietz

Introduction

With the sentence "I will ... mention one [experiment] which goes a step further than any hitherto recorded; and affords, I think, an important illustration of the combination of gases by platinum." Sir William Grove started his description of a then-called 'gaseous voltaic battery' (Grove 1839, 1842). Simultaneously, Schönbein delivered the explanation of this polarization phenomenon by a series of galvanic experiments concluding "... that the current in question is caused by the combination of hydrogen with oxygen ... and not by contact" (Schönbein 1839) (referring to a controversially discussed 'contact theory' widely used in the early nineteenth century for the explanation of electrical phenomena). In comparison to batteries with liquid electrolytes, which were also developed during this period (Daniell element), the main disadvantage of the fuel cell is the need of continuous gas supply. Therefore batteries became the more convenient, practical, and compact choice of small-scale power supply until today.

It lasted one century until Baur and Preis constructed the first solid oxide fuel cells (SOFCs), i.e., galvanic cells using a solid oxygen ion conductor as electrolyte (Baur and Preis 1937). In this attempt, they investigated various porcelains, clays, and zirconia-based materials as solid electrolytes. A tubular stack test consisting of eight zircona tubes filled with coke (as fuel and anode) and a kind of powder bath of magnetite used as cathode delivered a power of 0.045 W (i.e., 0.18 Wl^{-1} or about 120 μA cm^{-2}).

Apart from the zirconia, the materials selection of these first SOFCs refers to a description of a "fuel element" (Baur and Treadwell 1916), which is known today as molten carbonate fuel cell. Regarding the applicability of their fuel cells, Baur and Preis concluded that their modified zirconia materials would fulfill all technical requirements but the main drawback is their high price.

Today, stabilized zirconia materials are indispensable compounds for the development of SOFCs. Independent in shape and size, the state-of-the-art SOFCs typically contain 8 mol.% yttria-stabilized zirconia (8YSZ, $Y_{0.15}Zr_{0.85}O_{1.93}$) as electrolyte, a perovskite-based cathode (strontium-substituted modifications of lanthanum manganite like $La_{1-x-y}Sr_xMnO_3$ with $0.15 < x < 0.4$ and $0 < y < 0.05$) and a composite of 8YSZ and metallic nickel as anode.

1. *Description of the Various Types of SOFCs*

Solid oxide fuel cells can be grouped into tubular and planar designs. Both types can consist of one or several single cells per stacking unit, i.e., on a single tube or in a single layer. Depending on the application, the tubular SOFCs have dimensions from needle-like shapes to lengths of about 1.5–2 m for rapid start-up times and large gross power, respectively. The planar designs can be divided into stacks containing metallic or ceramic interconnect material as well as into cells with thick (electrolyte-supported) or thin (electrode-supported) membranes with thicknesses usually of 150–250 μm and 5–20 μm, respectively (Tietz 2003, Blum *et al.* 2005). The size of technologically relevant planar cells varies from 10×10 cm^2 to 25×25 cm^2 or corresponding areas in round shape.

The general tendency today to reduce the operating temperature from about 1000 °C to 500–800 °C favors cell designs with thin electrolytes due to the lower ohmic resistance and hence the higher power density that can be achieved. Therefore, many developers have considered anode-supported cells as the best choice for realizing an SOFC operating at reduced temperature (Tietz 2003, Blum *et al.* 2005).

The term 'anode supported' has been initially introduced to explicitly indicate that in such fuel cells the mechanical stability is transferred from the thick, dense electrolyte layer to an even thicker, porous anode (0.5–1.5 mm). During recent years, other types of SOFCs have also been developed which have the mechanical support on the anode side, though not necessarily the anode substrate. In all these cases, the substrate is always porous to allow gas transport to and from the anode. The materials used for such substrates are metals or alloys, composites like the conventional Ni/8YSZ cermet, or even refractory ceramics. In the case of insulating substrates, they only deliver the fuel gas to the anode and the current path has to be established along the substrate surface instead of passing it perpendicular to the substrate surface for electrically conductive substrates. The thermomechanical compatibility is difficult to achieve with metallic substrates, because on the one hand they often have higher thermal expansion coefficients than the other cell components and on the other hand severe corrosion can occur leading to failure in cell performance.

On the anode substrate, the other cell components are deposited in the sequence anode–electrolyte–cathode. Normally the coatings are realized by screen printing, slurry coating, wet powder spraying, or atmospheric plasma spraying. Even if the anode substrate is made of the anode material, Ni/8YSZ, it has been found beneficial for the electrochemical performance to apply an additional anode layer having a finer microstructure and therefore better electrocatalytic conversion of the fuel gas. The layers typically have thicknesses of 5–20 μm for the anode, 10–20 μm for the electrolyte, and 50–80 μm for the cathode (Fig. 1). Nowadays, the cathode also is composed of two layers, an electrochemically fine-grained composite of the electrolyte material and the electrocatalyst (lanthanum manganite) and a coarse current collection layer supplying air and electrons to the composite layer.

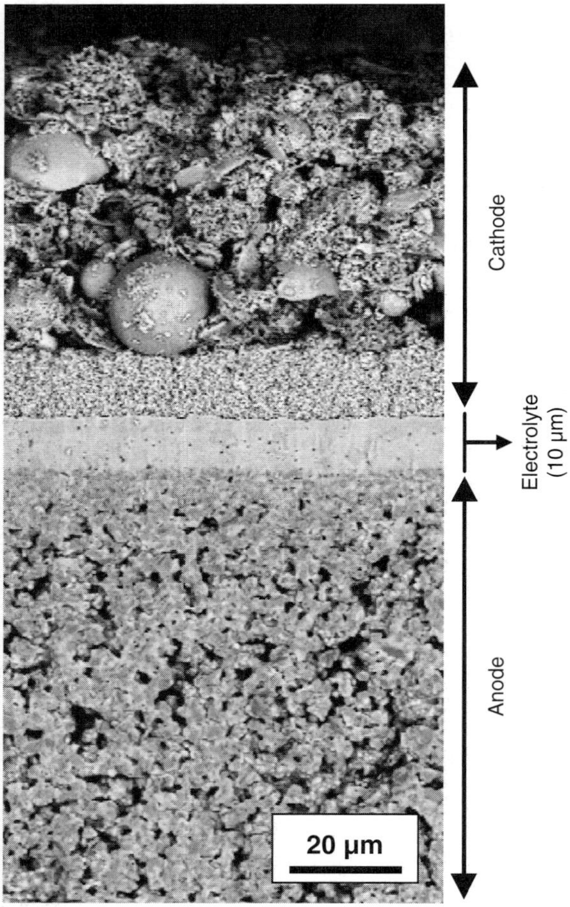

Figure 1
Microstructure of an anode-supported SOFC with a 10 μm thick solid electrolyte. In this magnification, the anode would end up in a thickness of about 1 m. Courtesy of Forschungszentrum Jülich.

2. The Technological Trends during the Recent Years

There have been three main areas of research for making SOFC technology competitive with other electricity generators:

(i) Several approaches have been undertaken to reduce the costs of materials, components, cells, and stacks, e.g., by applying low-cost technical-grade raw materials, by reducing the manufacturing costs (e.g., reduction of sintering temperatures, reduction of sintering steps, application of conventional ceramic or metallurgical processing methods, simplification of production processes), by decreasing the thickness of components, especially the anode substrate, by integration of cheaper interconnect materials.

(ii) Higher performance, power density, and efficiency have been targeted to lower the operating temperature. Simultaneously, this implies the aim to lower degradation processes and costs (due to a smaller number of cells required for a certain gross power). Hence higher performance gives more flexibility in operation and that is why the switch from electrolyte- to anode-supported designs has taken place. Performance has been further increased by using new cathode materials and optimization of their microstructure.

(iii) Another aspect related to costs is the aim to decrease degradation processes and to increase durability and reliability (in operation and in component fabrication). In the latter cases, it has been realized that this aim is easier achieved with small-scale applications in the 1–5 kW class than with larger systems (100–1000 kW). Several degradation phenomena have been addressed like chromium poisoning of the cathode, contact resistances at the interfaces, especially at the interconnect, the long-term stability of electrolytes, anodes, and cathodes, and the influence of fuel gases and contaminants (coking, sulfur poisoning).

3. Progress in Materials Development and Processing

3.1 Electrolytes

As mentioned before, an SOFC uses a solid oxide ion conductor as the electrolyte material. This material must have sufficient ionic conductivity for O^{2-} ions at operating temperature. Due to the aim of lowering operating temperatures, thin electrolyte films on

Table 1
Ionic conductivity (σ_i) of different oxygen ion conductors (in $S\,cm^{-1}$).

Material	σ_i at 800 °C	σ_i at 600 °C
$Zr_{0.85}Y_{0.15}O_{1.93}$	5.0×10^{-2}	6.2×10^{-3}
$Zr_{0.80}Sc_{0.19}Al_{0.02}O_{1.90}$	1.2×10^{-1}	1.1×10^{-2}
$Ce_{0.8}Gd_{0.2}O_{1.9}$	6.5×10^{-2}	1.3×10^{-2}
$La_{0.8}Sr_{0.2}Ga_{0.90}Mg_{0.1}O_{3-x}$	1.1×10^{-1}	1.6×10^{-2}
$La_{10}Si_6O_{27}$	1.7×10^{-2}	8.2×10^{-2}

porous electrode substrates are now preferred. The four main material systems for the electrolyte are partially cation-substituted ZrO_2, CeO_2, and $LaGaO_3$ as well as apatites. The ionic conductivity of these solid electrolytes are listed in Table 1.

Apart from the apatite $La_{10}Si_6O_{27}$, the other solid electrolytes have an ionic conductivity about 2 times higher than that of 8YSZ at 800 °C as well as at lower temperatures. Therefore the improvement can be significant for electrolyte-supported cells but is negligible for electrode-supported cells. Assuming an electrolyte thickness of 10 μm and an ionic conductivity of $0.05\,S\,cm^{-1}$ and $0.12\,S\,cm^{-1}$ for 8YSZ and ScSZ, respectively, at 800 °C (Table 1), this leads to an ohmic loss over the electrolyte layer of about 20 mΩ and 8 mΩ, respectively. These values are much smaller than the overpotentials at the cathode which are in the range of 100–400 mΩ.

Only the apatite, an ion conductor with excess oxygen ions on interstitial sites instead of oxygen vacancies (as in the other materials), has a different temperature dependence and a smaller activation energy of the ionic conductivity. Therefore the value at 800 °C is rather low, but the material surpasses the conductivity of 8YSZ at 600 °C and outperforms also the other materials at and below 500 °C (Nakayama and Sakamoto 1998).

During exposure at 1000 °C it was found in several studies that 8YSZ samples showed a higher decrease of conductivity than samples with a higher Y_2O_3 content. The best conductivity values after annealing were reached by samples with 10 mol.% Y_2O_3. The aging of zirconia stabilized with 7.8 mol.% Sc_2O_3 directly correlated with the decrease in conductivity and temperature. Similar investigations with Sc_2O_3 contents between 8 mol.% and 12 mol.% showed only slight degradation effects after aging at 1000 °C (up to 6000 h). Samples with 11 mol.% Sc_2O_3 remained nearly unchanged with respect to conductivity. Up to the present, however, the lower price and larger availability of yttria lead to its dominant use as a stabilizer in contrast to scandia.

The high ionic conductivity of the gadolinia-substituted ceria changes into a p-type conduction above 600 °C (Blumenthal and Prinz 1967) and the electronic conductivity significantly increases at low oxygen partial pressures, which is attributable to a reduction of Ce^{4+} to Ce^{3+}. However, the problem with ceria is above all its poor sintering properties of commercially available powders, which necessitates high sintering temperatures and causes interactions between CeO_2 and the YSZ layer when it is used as diffusion barrier layer (see Sect. 3.3).

Pure lanthanum gallate is a poor ionic conductor. When replacing La^{3+} or Ga^{3+} ions by Sr^{2+} or Mg^{2+} ions, high ionic conduction is obtained (Ishihara *et al.* 1994). Time-dependent measurements on $La_{0.8}Sr_{0.2}Ga_{0.85}Mg_{0.15}O_{2.825}$ (Huang *et al.* 1998) have shown that the conductivity at 700 °C in air is nearly unchanged, amounting to $0.073\,S\,cm^{-1}$ over a period of 200 h. However, lanthanum gallate reacts with NiO of the anode to form lanthanum–nickel oxide. This leads to a deterioration of the ionic conduction and the anode function. It was possible to suppress this reaction after applying a diffusion barrier layer of $(Ce,Sm)O_{2-x}$.

3.2 Anode Substrates and Anodes

The most commonly applied anode material is a nickel/YSZ cermet; a mixture of nickel oxide and YSZ powder is used as the starting material. In the final structure of the electrode, the nickel oxide is then commonly reduced to nickel during the start-up phase of the cell or stack by the prevailing fuel atmosphere. The nickel serves as a catalyst for the oxidation reaction of the hydrogen and is finely dispersed around the YSZ particles. The function of the zirconia skeleton is to stabilize a porous electrode structure, to create an extended reaction zone by increasing the number of three-phase boundaries, and also to adapt the thermal expansion coefficient between the nickel and the other cell components.

The demands made on the anode substrate such as good electrical conductivity, good gas permeability, and fine-grained homogeneous microstructure can be realized by the powders currently used and were studied in detail for Ni/8YSZ anode substrates. At a stage of development where these demands on the component can be fulfilled reproducibly, consideration can be given to also use technical powder qualities for cost reduction. Therefore, since the anode substrate accounts for more than 90% of the ceramic material expenditure of the ceramic parts, this component is the most promising candidate for saving material. Therefore low-cost nickel oxides of technical powder quality were sought (Tietz *et al.* 2000) to replace the NiO used hitherto which satisfies very high demands on quality.

Because of the introduction of an additional anode functional layer (see Fig. 1), it is possible to divide the two tasks of the Ni/8YSZ anode substrate, i.e., the mechanical stability and the electrocatalytic conversion of the fuel. This allows to separately improve the material properties of the anode functional layer and the anode substrate. Therefore new cermets based on Ni/Al_2O_3 or Ni/TiO_2 were synthesized (Meschke *et al.* 2001) to match the thermal expansion coefficient of the 8YSZ avoiding the thermoelastic bending during cell fabrication (Steinbrech *et al.* 1997) and to reduce the costs of the substrate. However, during cell fabrication, a severe interaction with the anode layer was observed due to liquid-phase sintering, which led to a detrimental decrease of electrochemical performance.

Apart from cermets, other materials also can be used as anode substrate, e.g., metals (ferritic steels) or ceramics with *n*-type conductivity. On the one hand, these alternatives offer solutions for the highly demanded reoxidation stability of anode cermets, because they do not undergo a strong volume expansion during reoxidation. On the other hand, disadvantages like corrosion and chromium diffusion from the steel into the anode need further improvement. Also, cell manufacturing is difficult to realize because steels have to be protected from severe corrosion and highly conductive ceramics may lose their high *n*-type electrical conductivity during sintering in air (Marina *et al.* 2002). Therefore, either the electrolyte has to be deposited by plasma spraying or its sintering has to be performed in reducing atmosphere. Furthermore, the cathode has to be applied without an additional sintering step.

Besides mechanical damage during reoxidation, anode cermets suffer also from sulfur poisoning or coking from the fuel gas, if the methane contains contaminants or insufficient amounts of water for internal reforming, respectively. Although efficient and compact desulfurizers are available and a controlled system operation can solve these problems, there is a continuous search for ceramic anode materials overcoming these drawbacks. In most cases, perovskite materials have been investigated for this purpose (Atkinson *et al.* 2004) such as $(Sr,La)TiO_3$, $(Sr,Y)TiO_3$, $(La,Sr)(Cr,Mn)O_3$, $(La,Sr)CrO_3$, $(La,Sr)(Al,Mn)O_3$, and $(La,Sr)(Ti,Mn)O_3$. As pure ceramics, however, the cell performance has always been significantly lower than for the conventional Ni/8YSZ anode, because these anodes either do not have sufficient electronic or ionic conductivity or are not catalytically active. Actually, there are approaches to improve the performance using ceramic composites with small additions of catalysts, but there is still a lot of development necessary to achieve a similar status as for the SOFCs with conventional anodes.

3.3 Cathode

Perovskite materials of the ABO_3 type are generally used as the cathode material. A frequently used material is lanthanum manganite ($LaMnO_3$) with substitutions of strontium or calcium at the A-site (Minh 1993). Both A- and B-site substitutions have been investigated extensively, so that these ceramics can be selected according to the demands made on their properties such as thermal expansion coefficient, electronic or ionic conductivity, and chemical interaction with the electrolyte.

Lanthanum or strontium cobaltites have attracted much attention because of their high mixed ionic–electronic conduction. Indeed, $(La,Sr)CoO_3$ shows a higher cathodic perfomance than $(La,Sr)MnO_3$, but the cobaltite possesses a very different thermal expansion in comparison to YSZ (Tietz 1999) and it reacts easily with this electrolyte at high temperatures ($\geq 1000\,°C$), forming highly resistive $La_2Zr_2O_7$ or $SrZrO_3$ compounds. These unfavorable solid-state reactions can be avoided, when a ceria-based diffusion barrier layer is applied (Chen *et al.* 1995). As a compromise between thermal expansion mismatch and electrochemical performance, the composition $La_{0.6}Sr_{0.4}Fe_{0.8}Co_{0.2}O_3$ (LSFC) has been established as an alternative to $(La,Sr)MnO_3$ cathodes for reduced operating temperatures since the mid-1990s (Chen *et al.* 1995, Sahibzada *et al.* 1997). SOFCs with cathodes of similar LSCF composition reach current densities of $1.5–2\,A\,cm^{-2}$ and about $1\,A\,cm^{-2}$ at 800 °C and 600 °C, respectively, and at 0.7 V.

The advantage of such mixed ionic–electronic conducting cathodes is their increased reactive catalytic surface area over the whole perovskite grains. Therefore, the charge transfer of adsorbed oxygen atoms takes place directly on the perovskite surface. The increase in oxygen ion conductivity and hence in mixed conductivity is generally correlated with an increasing thermal expansion (Ullmann *et al.* 2000). Therefore, an improvement of the transport properties is always linked to an increasing mismatch in thermal expansion related to the other cell components.

A rough classification of cathode materials regarding their electrocatalytic activity follows the order: $LaCoO_3 > LaMnO_3 > LaFeO_3 > LaCrO_3$. The electrocatalytic activity also varies with alkaline-earth concentration in the perovskite. In general, the electrical and ionic conductivity, the thermal expansion coefficient, and the catalytic properties increase with increasing alkaline-earth content between $0 < x < 0.5$ in $La_{1-x}A_xBO_3$ with A = Ca, Sr, Ba and B = Cr, Fe, Mn, Co.

3.4 Interconnect

The interconnect in SOFC stacks is the component that electrically connects the single cells and in planar systems additionally separates the gas compartments from each other. In an SOFC system, numerous demands are made on the interconnect, like good electrical conductivity, gas tightness, chemical compatibility with the adjacent components of the fuel cell, high corrosion resistance to the reaction gases, matched thermal expansion and last, but not least, reasonable costs. In order to meet these requirements, two classes of materials are commonly used for interconnects, namely ceramic and metallic materials (Anderson and Tietz 2003). Whereas ceramic interconnects played a dominant role in the early SOFC developments, metallic interconnects have also been frequently used in recent developments and are the favored materials worldwide today due to the general trend toward lower operating temperatures.

Practically all ceramic interconnects which have been developed are based on the $LaCrO_3$-type materials because it is one of very few perovskites that do not decompose in fuel gas atmosphere. However, due to the electrical conductivity required, the lanthanum chromites cannot be used at lower temperatures than 800 °C. In addition, lanthanum chromites show undesirable swelling in reducing atmospheres caused by the reduction of Cr^{4+} ions to Cr^{3+} and the formation of oxygen vacancies leading to strong internal stresses and possible cracking during operation.

Apart from these physical handicaps, the materials costs as well as the fabrication and processing costs due to the low sinterability are the most critical issues for ceramic interconnects.

Hence, the advantages of metallic interconnects are obvious: high electrical conductivity, good ability for processing, and lower costs. The disadvantages, however, are corrosion in combination with increasing resistance during operation, chromium evaporation, and low high-temperature strength.

The long-term stability of the metallic interconnect is essentially governed by its corrosion characteristics. The materials used for interconnects are alloys forming chromium oxide, which ensure sufficiently high conductivity for thin oxide scales. Meanwhile several ferritic steels have been developed especially for SOFC application containing only very small amounts of aluminum and silicon to avoid the formation of highly resistive oxide scales and low amounts of manganese for the formation of $(Cr,Mn)_3O_4$ spinels as the outer corrosion scale (Quadakkers *et al.* 2003).

The main focus in interconnect development is still the improvement of the corrosion behavior, but also the reduction of contact resistance of the corrosion scales or in combination with coatings and the reduction of chromium evaporation to avoid a detrimental poisoning of the cathode with chromium species (Hilpert *et al.* 1996). All these phenomena contribute mainly to the observed degradation of stack voltage during operation, but it is not yet clarified to what extent.

3.5 *Protective and Contact Coatings*

Many attempts have been made to reduce the damaging effect of chromium vapors on the cathode side by suitable protective layers. The first approach was plasma-sprayed coatings of lanthanum chromite as a protective method to minimize the evaporation of volatile Cr species. Their function as diffusion barriers against volatile Cr species strongly depends on the quality of the layers (gas tightness, crack density). Additionally, for anode-supported cells, the plasma-sprayed chromite coatings are characterized by high contact resistances and high fabrication costs in comparison to ceramic methods using slurries or pastes. However, the reactivity of lanthanum chromite with metals, either with Cr- or Fe-based alloys, is low and the metal/ceramic interface is very stable under operating conditions.

Chemical interactions between coatings and interconnect material increase when the alkaline earth content in the ceramic is increased. Often perovskite materials have been brought in contact with Cr-based alloys or steels and the formation of chromates (e.g., $CaCrO_4$ or $SrCrO_4$) was observed (Larring and Norby 2000), leading to progressive decomposition of the perovskite material and a brittle interface. As a consequence of these interactions, the application of cobalt oxide or cobalt manganese spinel between perovskite and alloy was investigated (Larring and Norby 2000, Zahid *et al.* 2004) and showed a stable contact resistance with time.

Coating the ferritic steel Crofer22APU by wet powder spraying with a porous Co_3O_4 layer or a metallic Co layer leads to the formation of a new dense Cr-free reactive layer under the SOFC operating conditions (Zahid *et al.* 2004). As a reaction product of Mn from the oxide scale of the steel and the Co_3O_4 layer, this dense layer may act as a barrier against vapor-phase transport of volatile Cr species.

Contact materials are electrically conductive ceramics to improve the contact between interconnect and electrode. Whereas for the anode side metallic nickel is used (as mesh, foam, or paste), on the cathode side often ceramic cathode-like compositions are used, which possess conductivities in the range of 50–500 $S\,cm^{-1}$. Because there are no electrochemical requirements to be obeyed for the contact materials, they can vary significantly from the electrode materials and be optimized with regard to other physical and chemical properties than the electrodes. Apart from the electrical conductivity, the most important properties are the thermal expansion and the sintering behavior at assembling temperatures in the range of 800–1100 °C. Because the contact layer thickness may vary between 30 μm and 200 μm and ceramic materials heat-treated below 1000 °C are usually very porous, the specific conductivity of the material should be high. This is in fact the case for lanthanum cobaltites having specific conductivities up to 1700 $S\,cm^{-1}$. On the contrary, the thermal expansion of these cobaltites has a strong mismatch with the other cell components. For electrically conductive ceramics, therefore, a compromise between acceptable conductivity and tolerable mismatch in thermal expansion has to be made as for the cathode materials.

3.6 Sealings

In planar SOFCs, gastight seals must be realised along the edges of each cell, along the outer edges of the stacking layers, and the gas manifold to separate the two gas compartments and to insulate the stacking layers from each other.

Glass or glass ceramics are well known for joining of metals, especially the joining of steels (Donald 1993). Many of these products are available commercially, but most of the sealing products are produced for low-temperature joining and room-temperature application. However, no commercial product fulfills the requirement of a thermal expansion coefficient in the range of $10–13 \times 10^{-6} K^{-1}$ up to high temperatures. Therefore, many SOFC developers started their own sealing development.

Most studies have been carried out on barium calcium aluminosilicate (BCAS) and barium calcium magnesium aluminosilicate (BCMAS). The comparison of BAS, CAS, and MAS glasses using B_2O_3 as a fluxing agent and TiO_2, ZrO_2, NiO, and Cr_2O_3 as a nucleating agent showed that the barium glass tended to react with chromium steel more than the other glasses, showed a faster crystallization, and an appropriate thermal expansion (Lahl *et al.* 2000, 2002).

BCAS glasses and ferritic steels easily form $BaCrO_4$ in the presence of oxygen. Compatibility studies between steels and glasses clearly showed that the sealing material alone does not lead to a good joining. It is more important to adopt the interconnect/sealant couple.

Under oxidizing conditions, the formed oxide scales of ferritic steels (chromium oxide scale at the inner interface to the steel followed by a scale of manganese chromium spinel at the outer interface toward the glass) are partially dissolved by the glass. This leads, on the one hand, to a good adhesion and sticking between sealant and steel, on the other hand, however, to a destabilization of the steel at the three-phase boundary glass/metal/air. Thus, good sticking and stability against corrosion may become conflicting objectives.

Besides glass-ceramics, compressive seals and brazing systems have also been investigated (Fergus 2005). In the case of compressive seals, metallic gaskets and mica-based materials (vermiculite, phlogopite) were tested. All seals are not hermetically gastight and require high mechanical loads for reaching low leakage rates. To date, no stack has been successfully operated with such sealing systems. Also, the brazing has not yet reached a reliably applied status, but the use of Ag–CuO–Ti brazes in combination with an insulating layer seems to offer technologically interesting alternatives to the glass-ceramic sealants (Weil *et al.* 2005).

4. *Remaining Issues/Technological Hurdles to Be Solved*

During the recent decade, much progress has been made to bring SOFC systems to application and to market. The 100–250 kW systems of Siemens–Westinghouse have demonstrated high reliability and durability with low degradation rates. Other developers have significantly improved the cell performance, allowing the reduction of operating temperature by several hundred degrees, as shown in Fig. 2, applying better processing or materials. Figure 2 also shows the still-not-overcome dilemma: the very durable tubular cells have a low power output, especially at reduced operating temperature and the well-performing anode-supported cells suffer from increasing degradation with increasing current density. Therefore, an SOFC that satisfies long-term stability and high performance (i.e., high efficiency) is not available yet. Hence much effort is devoted to improve the electrochemical performance of electrolyte-supported and tubular cells as well as to the understanding and reduction of the degradation phenomena occurring in anode-supported cells and stacks with metallic interconnects. The most reasonable solution for commercialization today is to use anode-supported SOFCs operated at reduced temperature, which gives the best compromise between performance and long-term stability.

In addition, all types of SOFC systems are still too expensive today and strong cost reduction is required. This issue is as old as the SOFC development

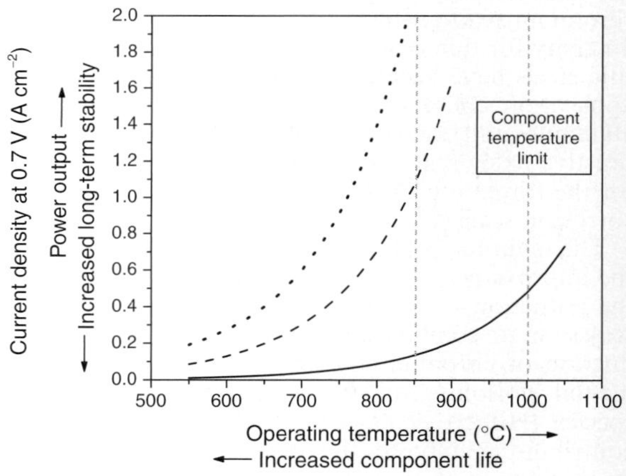

Figure 2
Typical current densities of electrolyte-supported cells and cathode-supported tubular cells (straight line), anode-supported cells with LSM cathode (dashed line), and with LSCF cathode (dotted line). The vertical short-dashed lines indicate the application limit of metallic balance-of-plant components (at 1000 °C) and interconnects made of ferritic steels (at 850 °C).

itself, because already Baur and Preis stated optimistically in 1937: "Meanwhile the unfavourable perspective regarding investment does not need to scare. The ceramic industry will know how to fabricate economically the required components when the demand begins."

See also: Fuel Cells: Electrochemical Reactions; Fuel Cells: Materials; Fuel Cells: Microsystems; Hydrogen–Metal Systems: Applications of Gas–Solid Reactions; Hydrogen–Metal Systems: Technological and Engineering Aspects; Photovoltaics: Advanced Inorganic Materials; Zirconia Electrolyte-based Solid Oxide Fuel Cells.

Bibliography

Anderson H U, Tietz F 2003 Interconnects. In: Singhal S C, Kendall K (eds.) *High-temperature Solid Oxide Fuel Cells: Fundamentals, Design and Applications*, ISBN 1-85617-387-9. Elsevier Advanced Technology, New York, pp. 173–95

Atkinson A, Barnett S, Gorte R J, Irvine J T S, McEvoy A J, Mogensen M, Singhal S C, Vohs J 2004 Advanced anodes for high-temperature fuel cells. *Nature Mater.* **3**, 17–27

Baur E, Preis H 1937 Über Brennstoffzellen-Ketten mit Festleitern. *Z. Elektrochem.* **43**, 727–32

Baur E, Treadwell W E 1916 Brennstoffelement. *German Patent 325783*

Blum L, Meulenberg W A, Nabielek H, Steinberger-Wilckens R 2005 Worldwide SOFC technology overview and benchmark. *Int. J. Appl. Ceram. Techn.* **2**, 482–92

Blumenthal R N, Prinz B A 1967 Nature of the electrical conduction transients observed in CeO_2 and Ca-doped CeO_2. *J. Appl. Phys.* **38**, 2376–83

Chen C C, Nasrallah M M, Anderson H U, Alim M A 1995 Immittance response of $La_{0.6}Sr_{0.4}Co_{0.2}Fe_{0.8}O_3$ based electrochemical cells. *J. Electrochem. Soc.* **142**, 491–6

Donald I W 1993 Preparation, properties and chemistry of glass and glass-ceramic-to-metal seals and coatings. *J. Mater. Sci.* **28**, 2841–86

Fergus J W 2005 Sealants for solid oxide fuel cells. *J. Power Sources* **147**, 46–57

Grove W R 1839 On voltaic series and the combination of gases by platinum. *Phil. Mag., Ser. 3* **14**, 127–30

Grove W R 1842 On a gaseous voltaic battery. *Phil. Mag., Ser. 3* **21**, 417–20

Hilpert K, Das D, Miller M, Peck D H, Weiß R 1996 Chromium vapor species over solid oxide fuel cell interconnect materials and their potential for degradation processes. *J. Electrochem. Soc.* **143**, 3642–7

Huang K, Tichy R S, Goodenough J B 1998 Superior perovskite oxide-ion conductor; strontium- and magnesium-doped $LaGaO_3$: I, phase relationships and electrical properties. *J. Amer. Ceram. Soc.* **81**, 2565–75

Ishihara T, Matsuda H, Takita J 1994 Doped $LaGaO_3$ perovskite type oxide as a new oxide ionic conductor. *J. Am. Chem. Soc.* **116**, 3801–3

Lahl N, Bahadur D, Singh K, Singheiser L, Hilpert K 2002 Chemical interaction between aluminosilicate base sealants and the components on the anode side of solid oxide fuel cells. *J. Electrochem. Soc.* **149**, A607–14

Lahl N, Singh K, Singheiser L, Hilpert K, Bahadur D 2000 Crystallisation kinetics in $AO–Al_2O_3–SiO_2–B_2O_3$ glasses (A = Ba, Ca, Mg). *J. Mater. Sci.* **35**, 3089–96

Larring Y, Norby T 2000 Spinel and perovskite functional layers between Plansee metallic interconnect (Cr-5 wt% Fe-1 wt% Y_2O_3) and ceramic $(La_{0.85}Sr_{0.15})_{0.91}MnO_3$ cathode materials for solid oxide fuel cells. *J. Electrochem. Soc.* **147**, 3251–6

Marina O A, Canfield N L, Stevenson J W 2002 Thermal, electrical and electrocatalytic properties of lanthanum-doped strontium titanate. *Solid State Ionics* **149**, 21–8

Meschke F, Dias F J, Tietz F 2001 Porous Ni/TiO_2 substrates for planar solid oxide fuel cell applications. *J. Mater. Sci.* **36**, 5719–28

Minh N Q 1993 Ceramic fuel cells. *J. Am. Ceram. Soc.* **76**, 563–88

Nakayama S, Sakamoto M 1998 Electrical properties of new type high oxide ionic conductor $RE_{10}Si_6O_{27}$ (RE = La, Pr, Nd, Sm, Gd, Dy). *J. Eur. Ceram. Soc.* **18**, 1413–8

Quadakkers W J, Piron-Abellan J, Shemet V, Singheiser L 2003 Metallic interconnectors for solid oxide fuel cells – a review. *Mater. High Temp.* **20**, 115–27

Sahibzada M, Steele B C H, Zheng K, Rudkin R A, Metcalfe I S 1997 Development of solid oxide fuel cells based on a $Ce(Gd)O_{2-x}$ electrolyte film for intermediate temperature operation. *Catal. Today* **38**, 459–66

Schönbein C F 1839 On the voltaic polarization of certain solids and fluid substances. *Phil. Mag., Ser. 3* **14**, 43–5

Steinbrech R W, Caron A, Blaß G, Dias F J 1997 Influence of sintering characteristics on component curvature of electrolyte-coated anode substrates. In: Stimming U, Singhal S C, Tagawa H, Lehnert W (eds.) *Solid Oxide Fuel Cells (SOFC-V)*. The Electrochemical Society, Pennington, NJ, pp. 727–36

Tietz F 1999 Peculiarities in the thermal expansion behavior of ceramic fuel cell materials. In: Vincenzini P (ed.) *Proc. 9th CIMTEC-World Ceramic Congress and Forum on New Materials*. Techna Publishers S.r.l., Faenza, Italy, Vol. 24: Innovative Materials in Advanced Energy Technologies, pp. 61–70

Tietz F 2003 Materials selection for solid oxide fuel cells. *Mater. Sci. Forum* **426–432**, 4465–70

Tietz F, Dias F J, Simwonis D, Stöver D 2000 Evaluation of commercial nickel oxide powders for components in solid oxide fuel cells. *J. Eur. Ceram. Soc.* **20**, 1023–34

Ullmann H, Trofimenko N, Tietz F, Stöver D, Ahmad-Khanlou A 2000 Correlation between thermal expansion and oxide ion transport in mixed conducting perovskite-type oxides for SOFC cathodes. *Solid State Ionics* **138**, 79–90

Weil K S, Coyle C A, Darsell J T, Xia G G, Hardy J S 2005 Effects of thermal cycling and thermal aging on the hermeticity and strength of silver–copper oxide air-brazed seals. *J. Power Sources* **152**, 97–104

Zahid M, Tietz F, Sebold D, Buchkremer H P 2004 Reactive coatings against chromium evaporation in solid oxide fuel cells. In: Mogensen M (ed.) *Proc. 6th Eur. SOFC Forum*. Lucerne, European Fuel Cell Forum, Oberrohrdorf, Switzerland, Vol. 2, pp. 820–7

Spacecraft Materials

J. Halchak

Materials for space applications fall into two categories: those for structures and those for propulsion systems. Basic requirements are similar to aircraft structures in that high strength, low weight and cost are the most important considerations. However, the space environment can place special demands on structures that are considerably different from aircraft materials. For example, the hard vacuum of space can cause outgassing of polymeric materials, which can both degrade their performance and cause detrimental coating of adjacent optical systems (Paton 1986). So-called "cold" welding (diffusion bonding) of contiguous metallic components has also caused failures of satellite systems. Surface damage from impact with atomic oxygen in low-earth orbit must be taken into account for satellites and space stations (Korb 1987). Consequently, space structural materials require additional testing and evaluation tailored to the unique needs of the space environment; quite different from those needed for aircraft structures.

1. Structures

Materials for vehicle structures generally follow similar design and application practice to that for aircraft structures. An exception to this is the need for thermal protection systems (TPS) on space vehicles and satellites. Vehicles use thermal protection systems to protect the aft portions of the vehicle from the exhaust plume of their rocket engines and, if they are designed to return to earth, from the frictional heating of reentry into the atmosphere. Satellite systems require thermal management coatings and paints on external surfaces to control internal temperature by tailoring of emissivity (Paton 1986).

In the early days of space flight, thermal protection systems usually took the form of ablatives which dissipate heat by burning during reentry. Ablative materials consist of a composite of polymeric materials such as an epoxy filled with either phenolic microballoons or cork. The system is stabilized by some reinforcement such as glass fibers (Korb *et al.* 1981).

More recently, reusable thermal protection systems have been developed which rely upon reradiation of heat to protect the underlying structure. These systems are generally based on either reinforced carbon–carbon (RCC) or silica tiles. Both of these have been used successfully on the Space Shuttle Orbiter. The silica tiles are made from very high-porosity amorphous silica fibers bonded with a slurry and sintered into blocks of appropriate dimension. The tiles are produced in different densities and attached to the substructure with room-temperature vulcanizing (RTV) silicone adhesive through a Nomex strain isolation pad. Tiles are coated with borosilicate glass to provide water resistance and, in high-temperature locations, are coated black by the addition of silicon tetraboride to provide the necessary emittance for reradiation (Korb *et al.* 1981).

RCC is used in TPS areas up to 1650 °C, particularly in the forward sections of reentry vehicles. RCC consists of a carbon matrix reinforced with carbon fibers, coated with silicon carbide to provide oxidation resistance. The material functions extremely well in high-temperature environments, but suffers from susceptibility to impact damage, due to low fracture toughness.

Metallic TPS are also being explored for reusable launch vehicles. These usually take the form of honeycomb panels, made from materials such as nickel-base superalloys. The advantage of metallic-based TPS is greater resistance to mechanical damage and, hence, longer life (Strauss and Hulewicz 1977).

For their primary structures, launch vehicles are generally made from aluminum alloys, although some vehicles, such as Atlas I and II, and Ariane IV, use stainless steel for their propellant tanks (Rycroft 1990). However, conventional aircraft-type, mechanically fastened skin and stringer structures are less frequently used. Instead, integrally stiffened shells are machined from thick aluminum alloy plate, then formed and joined by fusion welding. The Space Shuttle forward fuselage is a combination of fabrication techniques. The inner crew cabin is an all-welded 2219 aluminum pressure vessel, while the outer aeroshell is conventional 2024 sheet metal skin-stringer design (Korb 1987).

Aluminum alloys typically used are 2014, 2219, 2024, and 7075. Alloys such as 2014 and 2219 are used for welded structures, and alloys 7075 and 2024 are used where mechanical joining methods can be employed (Franklin 1987). Care must be taken to use heat treat tempers resistant to stress corrosion cracking, such as T8xx and T73 (Korb 1987). Aluminum–lithium alloys are beginning to be used because of their higher strength-to-weight ratio. The material for the external tank of the Space Shuttle was changed from 2219 to 2195 aluminum–lithium alloy as a weight-saving measure (Malone 1988).

For components such as mounting points and attach fittings, high-strength nickel-base superalloys or titanium alloys are commonly used. More recently, composite structural materials are seeing wider application. Polymer-matrix composites are very attractive because of their light weight. A pioneering application of large polymer composite structures in spacecraft was the use of graphite-epoxy for the cargo bay doors of the Space Shuttle (Korb 1987). The use of carbon-reinforced polymer-matrix composites is now almost universal for the payload fairings of expendable launch vehicles. As improved production techniques have reduced costs, polymer composites are seeing increasing use in launch vehicle primary structures. A good example of this is the interstage structure of the Japanese H-IIA vehicle, in which a graphite-reinforced, polymer-matrix composite structure with integrally molded aluminum alloy stiffeners and attach fittings replaced a fabricated all-aluminum structure with significant savings in both weight and cost (Kobayashi 2000).

Under development are polymer composite fuel tanks and ducting. As liquid hydrogen is increasingly used as the fuel of choice in modern launch vehicles, the extremely low temperature of liquid hydrogen (−252 °C) has required the development of polymers that retain some degree of ductility at cryogenic temperatures (Rogacki *et al.* 2000).

2. *Propulsion Systems*

Space vehicles are propelled by rocket engines, both liquid and solid. Solid rockets are cheaper to manufacturer and offer good value for their cost. Liquid propellant engines offer higher performance; that is, they deliver greater thrust per unit weight of propellant burned. Since liquid engines can be tested several times before flight, they tend to be more reliable, and their ability to be shut down once started provides an extra margin of safety. Liquid propellant engines can also be designed with restart capability to provide orbital maneuvering.

Rocket engine operational factors can be described in terms of extremes: temperatures ranging from that of liquid hydrogen (−252 °C) to 3300 °C; enormous thermal shock (3900 $°C s^{-1}$); large temperature differentials between contiguous components; reactive propellants; extreme acoustic environments; high rotational speeds; extreme power densities, etc. These factors place great demands on materials selection and each must be dealt with while maintaining an engine of the lightest possible weight (Fulton *et al.* 1983, Lewis 1975).

The liquid rocket engines in use today are constructed mostly of metallic alloys. Superalloys based on nickel, cobalt, and iron–nickel systems are used extensively because of their high strength and good toughness over a wide range of temperatures from −252 °C to 1100 °C. Stainless steels, titanium alloys, aluminum alloys, niobium alloys, alloy steels, cobalt-base alloys, and copper alloys also are employed, according to performance demands (Fulton *et al.* 1983, Lewis 1975).

The performance of a rocket engine is a function of the pressure which can be attained in its combustion chamber. Medium-performance engines operate at combustion chamber pressures in the 700 psi to 1500 psi range, whereas pressures in high-performance engines range from 2500 psi to 3500 psi. Obtaining high combustion chamber pressures requires all systems feeding the combustion chamber to operate at high pressures. With pumps, turbines, ducts, and valves operating at high pressures, the use of materials with high strength-to-weight ratios is mandatory in order to keep engine weight from becoming excessive.

High-efficiency, high-capacity turbopumps are required to deliver large amounts of propellant to the combustion chambers of liquid rocket engines. Pump discharge pressures vary from around 2000 psi for lower-performance engines to over 7000 psi in high-performance engines (Lewis 1975). For any given discharge pressure, the higher the rotational speed that the pump impeller can attain, the lighter and more compact the overall pump can be. Maximum rotational speed of a pump impeller is directly proportional to the strength-to-weight of the impeller material. Titanium alloys offer the highest strength-to-weight at cryogenic temperatures.

For this reason, all high-performance hydrogen–oxygen engines utilize titanium alloys for the impellers of their liquid hydrogen fuel pumps (the Japanese LE-7, European Vulcain, Russian RD-0120, and American SSME). As titanium is not compatible with oxygen, the impellers of the corresponding oxygen pumps are made from nickel-base alloys—the Ni-Fe-Cr-Nb-Mo-Al-Ti gamma-prime strengthened alloy 718 is typical of the type of alloy used. Pump housings for both fuel and oxidizer sides are also usually made of nickel-base superalloys. For lower performing engines, pump impellers and housings can be made from aluminum alloys (Lewis 1975, Fujita *et al.* 1988, SEP Brochure No. 16 1989, Chemical Automatics Design Bureau 1990).

Pumps are driven by turbines, which are most often driven by hot gases, either from a separate combustion chamber (gas generator or preburner) or from tapped-off propellants (expander cycle). Turbine materials are most often nickel or cobalt-base superalloys. The efficiency of a turbine can be raised by increasing its operating temperature; however, the compact design of a rocket engine generally precludes cooled turbine blades as are used in turbojet engines. This has led to the examination of ceramic-matrix composites for rocket engine turbines. Most often evaluated are the carbon-fiber-reinforced silicon carbide composites, which can operate at temperatures significantly higher than metal alloys (Effinger *et al.* 2000).

The combustion chambers must be cooled to prevent melting under the high heat fluxes in the combustion zone of a rocket engine. Cooling methods include regenerative, film, transpiration, or ablative, or some combination of these. For high-chamber-pressure liquid engines, copper-alloy, channel-wall, regeneratively cooled combustion chambers are used because of the greater strength and higher heat-transfer rates provided by this design. The copper alloys typically used are Cu-Ag-Zr (Vulcain) and Cu-Cr (LE-7) (Daimler Benz Systems Brochure 1997, Yamada *et al.* 1991). For low- to medium-chamber-pressure engines, brazed tubular-wall designs, composed of stainless steel or nickel-base alloy tubes, are used.

Engine nozzles can be made of brazed or welded tubes (stainless steel, or nickel alloys), brazed channel-wall designs, glass-phenolic ablatives, carbon–carbon, or ceramic-coated sheet metal (alumina or zirconia on a high-temperature alloy). Solid rocket motors typically have ablative nozzles (Franklin 1987, Lewis 1975, Daimler Benz Systems Brochure 1997).

In solid rockets, combustion takes place inside the motor casing itself. Typically, solid rocket motor cases have been welded structures of high-strength, low-alloy steels, such as D6AC (Fe-1.0Cr-1.0Mo-.55Ni-.07V-.75Mn-.45C). More recently, motor cases of carbon filament wound, polymer-matrix composites have become more commonly used. Such composite cases have resulted in significant vehicle weight savings (Franklin 1987).

Environmental factors are a major concern in the selection of materials for a rocket engine. Most of the propellants are reactive fluids. Oxygen, hydrogen, nitrogen tetroxide, hydrazine, red fuming nitric acid—all have potentially detrimental effects on materials. Most organic materials are not compatible with oxygen and can ignite in rocket engine systems under service conditions. Both metallic and nonmetallic materials must be tested for compatibility with oxygen. The primary metallic materials not considered acceptable in oxygen systems are tin, magnesium, and titanium alloys. Their use may result in catastrophic impact ignition (Franklin 1987). Titanium is particularly reactive, and tends to autoignite from pressure alone. Teflon and Kel-F are nominally compatible with liquid and gaseous oxygen, but must be batch tested to assure compatibility. Testing for compatibility with oxygen is performed by either impact or promoted-ignition testing.

Propellant compatibility is also of concern in hydrogen systems, particularly at high pressures (e.g., over 1000 psi). Hydrogen environment embrittlement of metals is a phenomenon encountered in rocket engines for which hydrogen is the fuel. The exact mechanism is not well understood; however, the effect is to drastically reduce notch toughness, ductility, and low-cycle fatigue, and to greatly increase crack growth rates. The phenomenon occurs at room temperature and above, but is not seen at cryogenic temperatures.

Nickel-base alloys tend to be heavily affected. Ferritic and martensitic steels are also adversely affected, whereas austenitic stainless steels tend to be immune or only slightly affected. Aluminum and copper-base alloys are not susceptible to hydrogen environment embrittlement. One technique used to protect alloys susceptible to hydrogen environment embrittlement is to plate their surfaces with copper or gold, which can form an impermeable barrier to hydrogen. Titanium alloys are embrittled by reacting with hydrogen to form hydrides. As this also occurs at room temperature and above, titanium alloys can be used in hydrogen at cryogenic temperatures (Franklin 1987).

Hydrazine and nitrogen tetroxide also present unique compatibility problems. The major concern with hydrazine is decomposition of the hydrazine, which can result in failure due to a large volumetric change. For nitrogen tetroxide systems, most metallic materials are resistant when the N_2O_4 is dry. However, because moisture can easily contaminate such systems (resulting in the formation of nitric acid), the primary materials of construction are those that also have high resistance to nitric acid, such as aluminum alloys, stainless steels, and titanium alloys (Franklin 1987).

Bibliography

Chemical Automatics Design Bureau 1990 Brochure. Chemical Automatics Design Bureau, Voronezh, Russia

Daimler-Benz Systems 1997 Brochure Space and Strategic Systems Group. 14–2863/1

Effinger M R, Genge G G, Kiser J D 2000 Ceramic composite turbine disks for rocket engines. *Adv. Mater. Process* **157** (6), 69–73

Franklin D B 1987 *ASM Handbook*. American Society of Materials, Materials Park, OH, Vol. 13, pp. 1101–5

Fujita T, Aoki H, Okayasu A, Ohta T, Iwama K 1988 Development status of LE-7 fuel turbopump. *Proc. ISTS*. Tokyo, pp. 289–95

Fulton D L, Shoemaker M C, Bashir S 1983 Unique material requirements in the space shuttle main engine. *Proc. 15th National SAMPE Technical Conf.* Azusa, CA

Kobayashi M 2000 Presentation at the SAE Aerospace Manufacturing Technology Conference. Fort Worth, Texas, May, pp. 16–18

Korb L J 1987 *Corrosion in manned spacecraft. ASM Handbook*. American Society of Materials, Materials Park, OH, Vol. 13, pp. 1058–99

Korb L J, Morant C A, Calland R M, Thatcher C S 1981 The shuttle orbiter thermal protection system. *Ceram. Bull.* **60**, 1188–93

Lewis J R 1975 Materials and processes for space shuttle's main engines. *Met. Prog.* 41–51

Malone J 1998 Aluminum–lithium fuel tank for the Space Shuttle. *Adv. Mater. Process.* **153** (4), 7

Paton N E 1986 Space vehicle materials. In: Bever M B (ed.) *Encyclopedia of Materials Science and Engineering*. Pergamon, Oxford, pp. 4532–4

Rogacki J, Ricketts R H, Ronald T 2000 Materials for air and space. *Adv. Mater. Process.* **158** (3), 63–5

Rycroft M 1990 *Cambridge Encyclopedia of Space*. Cambridge University Press, Cambridge

SEP Brochure No. 16 1989 ISSN 0247-3046

Strauss B, Hulewicz J 1977 X-33 advanced metallic thermal protection system. *Adv. Mater. Process.* **151** (5), 55–6

Yamada Y, Nakano T, Watanabe A, Katagi T 1991 Japanese launch vehicle propulsion—status and direction. *AIAA/SAE/ASME 27th Joint Propulsion Conference, June 24–26* Sacramento, CA

Stirling Engine: Micro-CHP System for Residential Application

D. G. Thombarse

Introduction

The Stirling engine micro-combined heat and power system (micro-CHP) is an energy conversion system of primary energy combining electricity generation and heating simultaneously for residential purpose. The energy conversion from fuel energy to electric power is done by the Stirling engine, an external combustion engine, and heating of water and space is done by utilizing waste heat through heat exchanger. The micro-regime is typically designated for a system where power generation is less than 3 kW. Normally natural gas, biogas, wood pallets, or biomass is used as a fuel for such power plants. The heat of combustion of fuel is primarily utilized for mechanical power generation through the Stirling engine and the waste heat is then further utilized for space (air) and water heating for domestic purposes. Combined heating and power (CHP) has proven beneficial in residential as well as many industrial situations due to high overall thermal efficiency, reducing the total power requirement and providing higher-quality power with good reliability. Applying CHP technology to smaller-scale residential and small commercial buildings is an attractive option because of the large potential market. The concept of micro-CHP is having individual heat- and power-producing unit in each home.

1. Benefits of Micro-CHP over Conventional System

(i) Micro-CHP produces heat and power at point on demand which fulfill the domestic electric and heating requirements.

(ii) On-site power production reduces transmission and distribution losses.

(iii) Maximum utilization of primary energy source by reducing waste heat.

(iv) The majority of CO_2 emissions at home is due to space and water heating by conventional methods but micro-CHP offers significant contribution to CO_2 reduction because of external combustion.

2. Stirling Engine Micro-CHP for Domestic Purpose

The concept of the Stirling engine micro-CHP system is illustrated in Fig. 1. A prime mover Stirling engine drives a generator which produces electrical power.

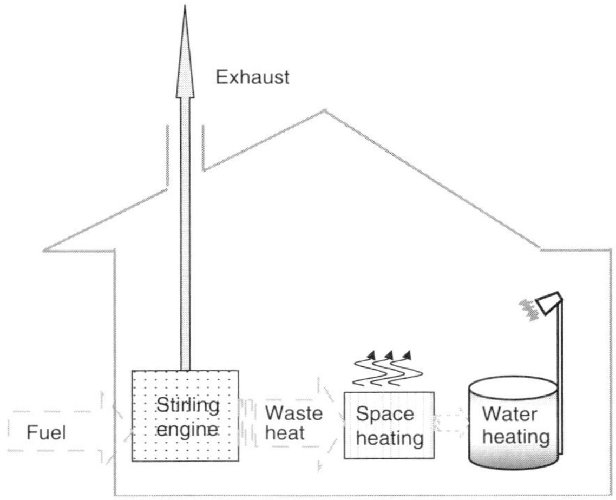

Figure 1
Domestic micro-CHP system.

The waste heat from the prime mover is recovered and used to produce hot water or warm air through the use of heat exchangers.

The micro-CHP system consists of following subsystems, as shown in Fig. 2:

- main burner;
- auxiliary burner;
- gas control valve (for main burner);
- heater heads;
- core engine (Stirling engine);
- exhaust heat exchanger;
- electronic control unit (ECU) and sensors; and
- service plate.

2.1 Main Burner

The function of the main burner is to produce heat by combustion of fuel. The burner assembly is a single-nozzle induced-draught type. Gas is supplied via a multifunctional valve that controls the air/fuel ratio and also incorporates a dual shut-off valve. The gas flow and ignition are managed by an automatic burner control integrated within the engine control unit. Air is supplied to the burner via a balanced flue arrangement and is drawn through the burner by a fan assembly mounted after the exhaust heat exchanger. A flame ionization detector (FID) and an exhaust temperature probe are used to monitor the combustion circuit. Two automatic burner control units manage gas flow and ignition for the main and auxiliary burner.

2.2 Auxiliary Burner

The auxiliary burner is a cylindrical pre-mix gas burner mounted on the heat exchanger. Gas is

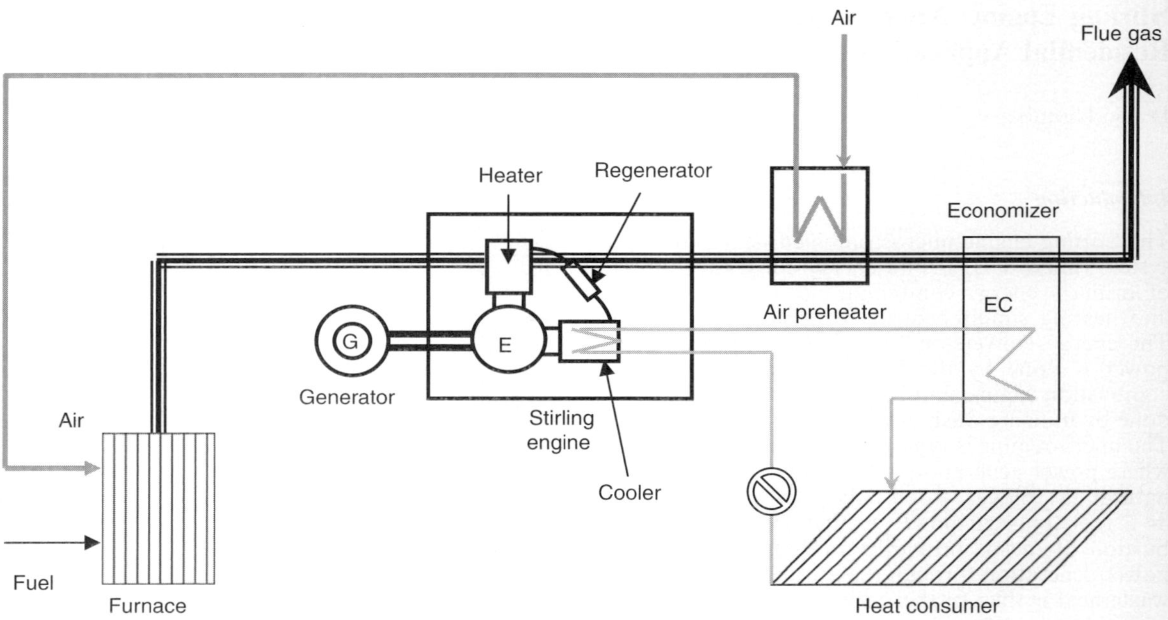

Figure 2
Layout of micro-CHP Stirling system.

supplied to a venturi where it is mixed with the air. The air/fuel ratio is controlled by a multifunctional valve that also incorporates a dual shut-off valve. The gas flow and ignition is managed by an automatic burner control. An FID is used to monitor the combustion circuit. The auxiliary gas burner operates when the heating demand is increased.

2.3 Stirling Engine

The Stirling engine and auxiliary burner are installed in an enclosure that functions as a sealed air box. Air is drawn into the enclosure and circulates through the casing before being drawn into both burners by the fan mounted behind the exhaust heat exchanger.

When a heat demand signal is received, the main burner is started and the heater heads are heated by the hot combustion gases. At a preset temperature, the engine starts running and supplies electricity and heat. The hot exhaust gases leaving the heater of the Stirling engine transfer heat to the air and water in the exhaust heat exchanger, which also contains the auxiliary burner. Depending on the heat load (water temperature), the heat management system can switch on the auxiliary burner to supply more heat. Pressurized nitrogen gas within the engine cylinders is heated in the engine heater and then cooled in the engine cooler. The resulting change in pressure causes the pistons to move up and down. The reciprocating motion is converted in rotary motion depending upon the type of engine configuration used. A special mechanism for phase angle ensures the correct phasing of the pistons and converts the linear motion to rotary motion to drive an electrical generator. Water is pumped through the engine cavities for preliminary heating and then circulated through exhaust heat exchanger. The heat transferred to the water and air is used for central heating and domestic hot water.

2.4 Generator

A 230 V 50 Hz AC power output is generated by a four-pole induction alternator, which is also used as an engine starter. The power output level is controlled by the engine control unit.

2.5 Enclosure

All engine components are contained within a steel enclosure, accessed via a hinged door and covered with a stainless steel front panel. Electronics modules for engine control are fitted to the enclosure door. The enclosure provides protection for the engine and creates a controlled environment for engine operation.

2.6 The Electronic Control Unit

The ECU manages the operation and monitors the electrical output, timing, and central heating system operation. A liquid crystal display (LCD) allows the user to see the system status and clear lockouts.

A passive safety circuit is built into the engine and incorporates a series of switches independent of the ECU. If any switch in the circuit is tripped, the burner controls shut down and the generator set stops operating.

2.7 Heat Exchanger

In the micro-CHP system, the heat exchangers are used for two purposes, one to heat water and another for space or air heating. Normally there are two separate heat exchangers built in the first unit to serve two purposes. For water heating primarily, the water is circulated through the engine cavities which cool the engine components by receiving heat from the engine. For further heating of water, the water is circulated to a heat exchanger where engine exhaust gases are used for water heating. The exhaust gases from the auxiliary burner are directed into the exhaust heat exchanger where heat is transferred to the heating water. These heat exchangers are tube and shell heat exchangers.

3. Operation of the Stirling Engine

The Stirling engine was invented in 1816 by Robert Stirling of Scotland, some 80 years before the invention of the diesel engine. It enjoyed substantial commercial success up to the early 1900s. A Stirling cycle machine is a device which operates on a closed regenerative thermodynamic cycle, with cyclic compression and expansion of the working fluid at different temperature levels. The flow is controlled by volume changes so that there is a net conversion of heat to work. The Stirling engines are frequently called by other names, including hot-air or hot-gas engines, or by one of a number of designations reserved for particular engine arrangements. In the beginning of the nineteenth century, due to the rapid development of internal combustion engines and electrical machines, further development of the Stirling engines was severely hampered. But due to its high heat efficiency, its low noise of operation, and its ability to use many fuels, the demand of an effective use of energy and environmental security today is met. The Stirling engine based on CHP units are considered as being the best among the most effective low-power-range thermal electricity and heat conversion units.

The advantage of the Stirling cycle engine over conventional reciprocating engines is that the heat is not supplied to the cycle by combustion of fuel inside the cylinder, but transferred from the outside through a heat exchanger in the same way as in a steam boiler. Consequently, the combustion system for the Stirling engine can be based on proven furnace technology, thus reducing combustion-related problems typical of solid biomass fuels. The heat input from fuel combustion is transferred to the working gas through a hot heat exchanger at a high temperature, typically between 680 °C and 780 °C. The heat that is not converted to the work on the shaft is rejected to the cooling water in the cold heat exchanger at 25–75 °C.

In order to have a high overall electric efficiency of the CHP plant, the temperature in the heater and the heat exchanger should be as high as possible. Therefore, it is necessary to preheat the combustion air with the flue gas leaving the hot heat exchanger by means of air preheater. Typically the temperature of air is increased to 500–600 °C, resulting in a very high temperature in the combustion chamber. The closed Stirling cycle makes it possible to use a working gas, which is better suited for heat transfer to and from the cycle than is air. Helium or hydrogen are the most suitable gases for Stirling engines.

Following are the advantages of Stirling engines:

(i) The engine works on external combustion principally, so any solid, liquid, or gaseous fuels can be used.
(ii) The engine can use biofuels, so the system is ecofriendly.
(iii) Low emission levels are observed because of controlled combustion.
(iv) Low maintenance of the engine system is needed because of less moving parts in the engine.

4. Thermodynamics of Stirling Cycle Engine

The Stirling engine works on the closed thermodynamic regenerative cycle represented on *PV* and *TS* diagram, as shown in Fig. 3(a). The cycle consists of four processes namely isothermal compression and expansion and isentropic heat addition and rejection processes, in the sequence shown in Fig. 3(a). Consider a cylinder containing two opposed pistons with a regenerator between the pistons as shown in Fig. 3(b). The regenerator is like a thermal sponge alternately absorbing and releasing heat; it is a matrix of finely divided metal in the form of wires or strips. The volume between regenerator and the right-side piston is expansion volume and between regenerator and left-side piston is compression volume. The expansion volume is maintained at high temperature and the compression volume is maintained at low temperature. The temperature gradient $(T_{max} - T_{min})$ between the ends of regenerator is maintained.

To start with a cycle, we assume that the compression space piston is at outer dead point (at extreme right side) and the expansion space piston is at inner dead point close to regenerator. All working fluid is in the cold compression space. The four processes of the thermodynamic cycle are discussed below.

4.1 Process 1–2: Isothermal Compression Process

During the compression process from 1 to 2, the compression piston moves toward the regenerator

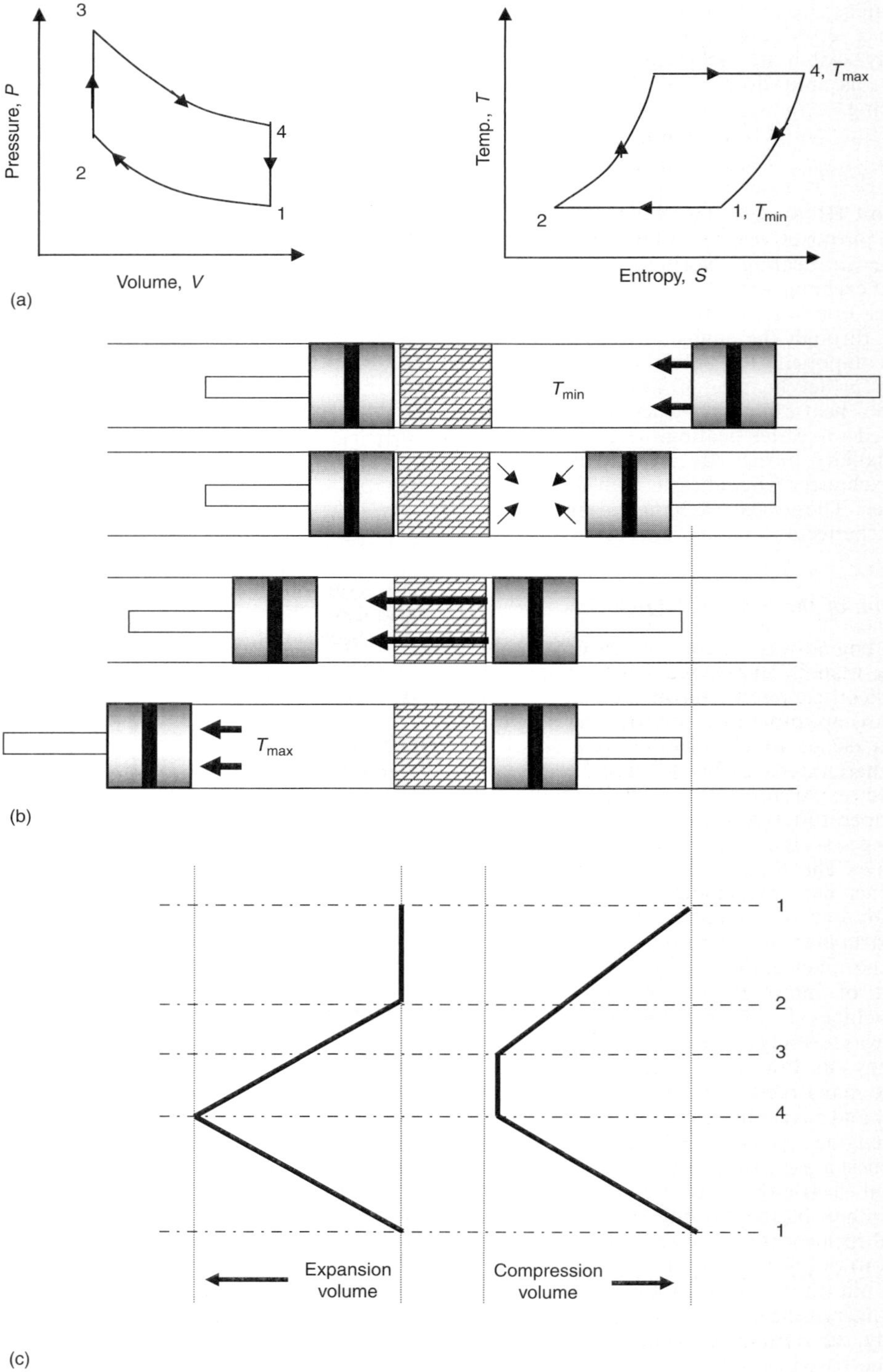

Figure 3
The Stirling cycle. (a) *PV* and *TS* diagram for Stirling engine, (b) engine arrangement and piston, and (c) time–displacement diagram.

while the expansion piston remains stationary. The working fluid is compressed in the compression space and the pressure increases from P_1 to P_2. The temperature is maintained constant due to heat flow from cold space to the surrounding. Work is done on the working fluid equal in magnitude to the heat rejected from the cycle. There is no change in internal energy and there is a decrease in entropy. The isothermal compression of the working fluid involves heat transfer from working fluid to the external dump at T_{min}:

$$P_2 = \frac{P_1 V_1}{V_2} = P_1 r_v \qquad T_1 = T_2 = T_{min}$$

$$\text{Heat transfer } Q = \text{Work done } W$$

$$Q_R = W = P_1 V_1 \ \ln(1/r_v) = mRT_1 \ \ln(1/r_v)$$

$$\text{Change in entropy} = (s_2 - s_1) = R \ln(1/r_v)$$

4.2 Process 2–3: Constant Volume Regenerative Transfer Process

In process 2–3, now, both pistons move simultaneously, i.e., compression piston toward regenerator and expansion piston away from regenerator so that the volume between pistons remains constant. The working fluid is transferred from compression volume to expansion volume through a porous media regenerator. The temperature of the working fluid increased from T_{min} to T_{max} by heat transfer from regenerator matrix to working fluid. The gradual increase in temperature of the working fluid while passing through regenerator causes an increase in pressure. No work is done and there is an increase in the entropy and internal energy of the working fluid. The isochoric (constant volume) heat transfer to working fluid from the regenerator matrix can be characterized as follows:

$$P_3 = \frac{P_2 T_3}{T_2} = \frac{P_2}{\tau}; \qquad V_3 = V_2$$

Here $\tau = T_2/T_3$ is the temperature ratio, as defined in the article of Gustav Schmidt.

$$\text{Heat transfer } Q = C_v\ (T_3 - T_2)$$

$$\text{Work done} = 0$$

$$\text{Change in entropy} = (s_3 - s_2) = C_v \ln(1/\tau)$$

4.3 Process 3–4: Isothermal Expansion Process

In expansion process 3–4, the expansion piston continues to move away from the regenerator toward the outer dead piston while compression piston remains stationary at the inner dead point adjacent to the regenerator. As the expansion proceeds, the pressure decreases as volume increases. The temperature is maintained constant by adding heat to the system from external source at T_{max}. Work is done by the working fluid on the piston equal in magnitude to the heat supplied. There is no change in the internal energy, but an increase in the entropy of the working fluid:

$$P_4 = \frac{P_3 V_3}{V_4} = P_3(1/r_v); \qquad T_4 = T_3 = T_{max}$$

$$\text{Heat transfer} = \text{Work done}$$

$$Q_S = W = P_3 V_3 \ln r_v = mRT_3 \ \ln r_v$$

$$\text{Change in entropy} = (s_3 - s_4) = R \ln r_v$$

4.4 Process 4–1: Constant Volume Regenerative Transfer Process

In process 4–1, both pistons move simultaneously to the transfer working fluid from expansion space to compression space through the regenerator at constant volume. During flow of the working fluid through the regenerator, the heat is transferred from the working fluid to the regenerator matrix, reducing the temperature of the working fluid to T_{min}. No work is done; there is a decrease in the internal energy and the entropy of the working fluid. The isochoric heat rejection follows from

$$P_1 = \frac{P_4 T_4}{T_1} = P_1 \tau; \qquad V_1 = V_4$$

Heat transfer $Q = C_v(T_1 - T_4)$ and change in entropy $= (s_1 - s_4) = C_v \ln \tau$.

And if $r_v = V_4/V_3 = V_1/V_2$, the total heat supplied $= RT_3 \ln r_v$ and the total heat rejected $= RT_1 \ln r_\tau$.

The efficiency can be written as

$$\eta_t = \frac{mRT_3 \ln(r_v) - mRT_1 \ln(r_v)}{mRT_3 \ln(r_v)}$$

The power output of Stirling engine equals $P = 0.015\ p \times f \times V_0$.

5. Maximum Theoretical Obtainable Efficiency of Stirling Cycle Engine

The actual Stirling cycle engine is subjected to heat transfer, internal thermal losses, and mechanical friction losses. To estimate these losses, James Senft has defined some ratios of engine temperatures. The ratio of engine lower to higher operating temperature is defined as $\tau = T_k/T_E$, the ratio of sink to reservoir temperature is defined as $\Gamma = T_C/T_H$, the ratio of reservoir temperature to hot engine temperature is denoted by $\xi = T_E/T_H$, and $\delta = b/a$, $\rho = c/a$, where a and b are heat-transfer coefficients.

So the cycle average power

$$\begin{aligned} P &= Q_s - Q_R \\ &= a(T_H - T_E) - b(T_C - T_K) \\ &= aT_H\{1 + \delta\Gamma - \xi - \xi\delta\tau\} \end{aligned}$$

As per second law, the thermal efficiency is not to exceed the Carnot cycle efficiency:

$$\frac{Q_S - Q_R}{Q_S - Q_T} = \frac{P}{Q_S - Q_T} \le 1 - \frac{T_C}{T_E} = 1 - \tau$$

This condition can be expressed as

$$\xi\{(\delta + 1)\tau - \rho(1 - \tau)^2\} \ge \delta\Gamma + \tau$$

Further, for maximum power condition, the equation can be written as

$$\xi = \frac{(\delta\Gamma + \tau)}{(\delta + 1)\tau - \rho(1 - \tau)^2}$$

Substituting this value, max. power P_t is given as

$$P_t = \frac{aT_H\{\delta(\tau - \Gamma)(1 - \tau) - \rho(1 + \delta\Gamma)(1 - \tau)^2\}}{(\delta + 1)\tau - \rho(1 - \tau)^2}$$

6. *Forms of Cylinder Coupling*

6.1 *Alpha Coupling*

Alpha engines have two pistons in separate cylinders which are connected in series by a heater, regenerator, and cooler, as shown in Fig. 4. The alpha engine is conceptually the simplest Stirling engine configuration; however, it suffers from the disadvantage that both pistons need to have seals to contain the working gas.

In his article, Andy Ross decribes a small air engine with extremely innovative alpha designs, including the classical Ross-yoke drive; more recently, a balanced "Rocker-V" mechanism was introduced. The alpha engine can also be compounded into a compact multiple cylinder configuration, enabling an extremely high specific power output, as is required for an automotive engine. In case of multicylinder arrangement, a number of cylinders are interconnected, so that the expansion space of one cylinder is connected to the compression space of the adjacent cylinder via a series of connected heater, regenerator, and cooler. The pistons are typically driven by a swash plate, resulting in a pure sinusoidal reciprocating motion having a 90° phase difference between the adjacent pistons.

6.2 *Beta Coupling*

Beta engines use displacer–piston arrangements as shown in Fig. 5. The engine construction is such that both displacer and piston are accommodated in same cylinder.

The compression space in this form consists of the space swept by the underside of the displacer and the topside of the power piston. The piston and displacer may or may not physically touch but be connected to the crankshaft by a separate linkage to maintain the required phase angle.

6.3 *Gamma Coupling*

Gamma engines use displacer–piston arrangements similar to the beta engine configuration, with the

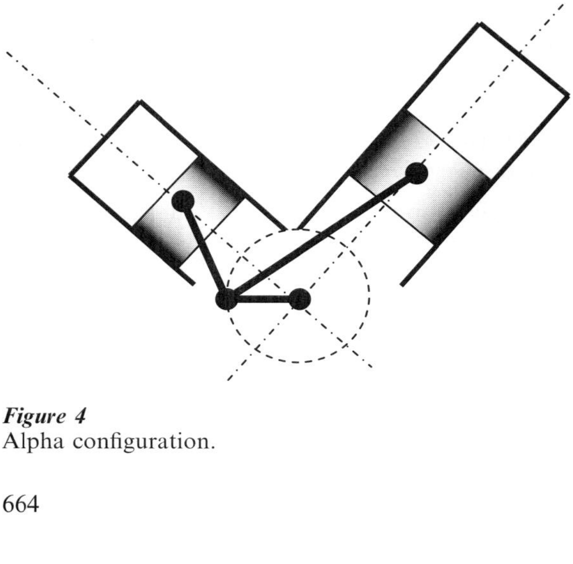

Figure 4
Alpha configuration.

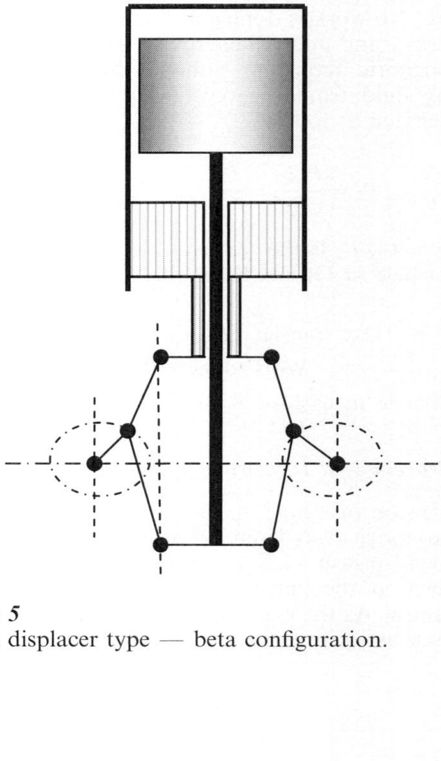

Figure 5
Piston displacer type — beta configuration.

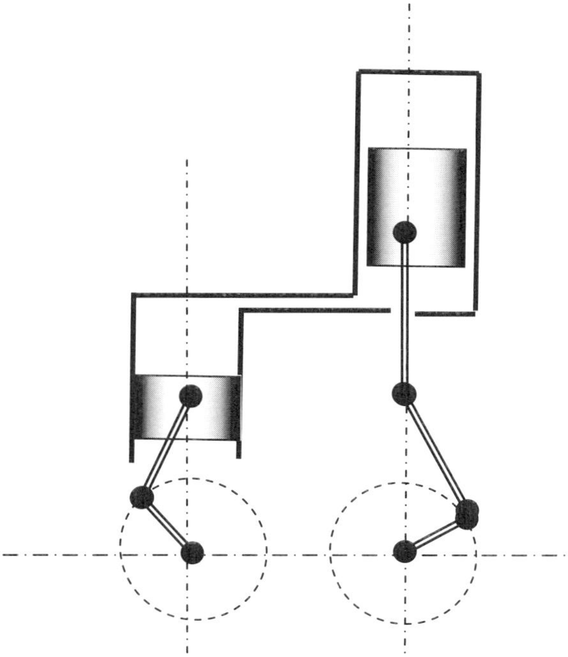

Figure 6
Piston displacer in separate cylinder—beta configuration.

displacer and the piston in separate cylinders as shown in Fig. 6.

In this type of machine, the compression space is split between two cylinders with an interconnecting transfer port. In between the passage from displacer cylinder and compression cylinder, the cooler, heater, and regenerator are connected serially. This arrangement provides advantage of a simple crank mechanism.

7. *Working Fluids for the Stirling Engine*

Any working fluid with high specific heat capacity may be used for the Stirling cycle engine. With few exceptions, the engines in the nineteenth century used air as a working fluid. Most of them operated close to atmospheric pressure. Air was cheap (readily available). The working fluid in a Stirling engine should have the following thermodynamic, heat-transfer, and gas dynamic properties:

(i) high thermal conductivity,
(ii) high specific heat capacity,
(iii) low viscosity, and
(iv) low density.

For better system performance in addition to above ease of availability, cost, safe operation, and storage requirements are also important properties that

Table 1
Relative heat-transfer characteristics for various gases.

Working fluid	Heat transfer	Capability factor
Air	1.0	1.0
Helium	1.42	0.83
Hydrogen	3.42	0.68
Water	1.95	0.39
Sodium–potassium eutectic	32.62	1.32

should not be neglected. The capability of working fluid in terms of specific heat capacity, thermal conductivity, and density is defined by a capability factor which is useful for preliminary selection of the working fluid:

$$\text{Capability factor} = \frac{\text{thermal conductivity}}{\text{specific heat capacity} \times \text{density}}$$

To determine the best working fluid, the whole system performance with different working fluids can be analyzed. The experimental investigation of suitability is difficult and also expensive. Empirical equations do not lend themselves for working fluid assessments, probably because of lack of sufficient experimental data to enable any meaningful correlation to be formed.

A simple approach based on original steady flow analysis can be useful for selection of fluid. By using Reynolds' analogy, a relationship between heat transfer and frictional drag in a flowing stream through duct for a system in terms of heat-transfer ratio and temperature limits is derived. The relation is

$$Q_{\mathrm{wf}} \propto (\rho^2 C_p^2)^{0.5}$$

It is required to simulate engine operation with different working fluids by the available equations so as to select the best working fluid. In Table 1, various fluids are compared using the equations derived above for heat transfer and capability factor at the average temperature and pressure of 800 K and 5 MPa. It may be seen that there is no working fluid that satisfies these two requirements except the NaK eutectic. The feasibility of using this working fluid is currently under investigation at University of California, San Diego. Most of the physical properties involved vary with pressure and temperature and thus heat-transfer (Q_{wf}) and capability factor should be determined under prevailing conditions. NaK seems to be the superior fluid, for the immediate future.

Bibliography

Berggren R W, Moynihan T M 1982 Effect of displacer seal clearance on free-piston Stirling engine performance. In: *17th*

Intersoc. Energy Convers. Eng. Conf., 829313, Aug. 8, 1982. Los Angeles, CA, USA, pp. 1885–90

Bergmann C, Alberto J 1991 Numerical prediction of the instantaneous regenerator and in cylinder heat transfer of a Stirling engine. *Int. J. Energy Res.* **15**, 623–35

Costea M, Feidt M 1998 The effect of the overall heat transfer coefficient variation on the optimal distribution of the heat transfer surface conductance of area in a Stirling engine. *J. Energy Convers. Manag.* **39** (16–18), 1753–61

De Boer P C T 2003 Maximum obtainable performance of Stirling engine and refrigerators. *ASME J. Heat Transfer* **125**, 911–15

Erbay L B, Yavuz H 1997 Analysis of the Stirling heat engine at maximum power conditions. *Int. J. Energy* **22** (7), 645–50

Gu Z, Sato H, Feng X 2001 Using supercritical heat recovery process in Stirling engines for high thermal efficiency. *J. Appl. Thermal Eng.* **21** (16), 1621–30

Hsu S T, Lin F Y, Chiou J S 2003 Heat-transfer aspects of Stirling power generation using incinerator waste energy. *Int. J. Renew. Energy* **28** (1), 59–69

Iwamoto S, Hirata K, Toda F 2001 Performance of Stirling engines. *JSME Int. J.* **44** (1), 140–7

Makhkamov K, Ingham D B 1999 Analysis of the working process and mechanical losses in a Stirling engine for a solar power unit. *ASME J. Solar Energy Eng.* **121** (2), 121–7

Petrescu S, Costea M, Harman C, Florea T 2002 Application of the direct method to irreversible Stirling cycles with finite speed. *Int. J. Energy Res.* **26**, 589–609

Reader G, Hooper C 1983 Stirling Engines. Cambridge University Press, London, pp. 29–35

Rifkin W, Vincent R, Benson O 1980 *Applications of Free Piston Stirling Engines.* American Institute of Aeronautics and Astronautics, New York, 809401, pp. 1982–6

Ross A 1979 Balanced Crankshaft Mechanism of the Two Piston Stirling Engine. *U.S. Patent No.* 4138897

Senft J R 1985 A mathematical model for Ringbom engine operation. *J. Eng. Gas Turbine Power* **107**, 590–5

Senft J R 1998 Theoretical limits on the performance of Stirling engines. *Int. J. Energy Res.* **22**, 991–1000

Senft J R 2000 Extended mechanical efficiency theorem for engines and heat pump. *J. Energy* **24**, 679–93

Senft J R 2001 Optimum Stirling engine geometry. *Int. J. Energy* **26**, 1087–101

Shtrikman S, Urieli I 1982 Linear moving magnet motor/generator for Stirling engine. In: *Proc. 17th Intersoc. Energy Convers. Eng. Conf.* 829310, Aug. 8–12, 1982, pp. 1862–6

Walker G 1973 Stirling Cycle Machines. Clarendon, Oxford, pp. 65–83

Wu F, Chen L, Wu C, Sun F 1998 Optimum performance of irreversible Stirling engine with imperfect regeneration. *J. Energy Conserv. Manag.* **39** (8), 727–32

Zacharias F A 1971 Advanced development of external combustion Stirling engine. In: *Proc. 2nd Symp. Low Pollut. Power Syst. Dev.*, pp. 371–8

Stress Coupled Phenomena: Piezoelectric Effect

P. Muralt

Piezoelectricity is an electromechanical effect resulting from a linear coupling between mechanical stress and strain on the one hand and electric field and displacement field on the other. The name originates from the Greek word *piezos* meaning "pressure" and expresses the observation that electricity is generated on applying pressure to a piezoelectric material. The opposite also occurs: applying an electric field generates a deformation. The first case is called the direct piezoelectric effect, the second the converse piezoelectric effect. Piezoelectricity was discovered by J. Curie and P. Curie in 1880 while studying pyroelectricity in Rochelle salt. Piezoelectricity is found in a variety of electrically insulating materials, including inorganic and organic crystals, ceramics, and polymers, and also in biological materials such as bone and wood. Piezoelectrics find wide application in electrical engineering for use in electromechanical transducers, ultrasonic generators, radio frequency (RF) and microwave filters, sensors, and actuators.

1. Governing Relationships

There are four types of piezoelectric coefficients depending on the pair of independent variables used for describing the equation of state. These can be the electric displacement field, D, or the electric field, E, the strain, ε, or the mechanical stress, σ. For the sake of brevity, only the coefficient d is used in this article, where

$$d = \left(\frac{\partial D}{\partial \sigma}\right)_E = \left(\frac{\partial \varepsilon}{\partial E}\right)_\sigma \tag{1}$$

Since the mechanical and electrical variables are second- and first-rank tensors, respectively, piezoelectric coefficients are components of a third-rank tensor. However, an insight into piezoelectricity can be gained by considering one-dimensional models, which is done below.

Piezoelectricity couples mechanical and electrostatic energy. In the thermodynamic potential there is a bilinear term connecting the two systems. Choosing the intensive variables as E and σ, the Gibbs free energy of a linear piezoelectric body is obtained as (neglecting temperature effects):

$$G(\sigma, E) = -\tfrac{1}{2}\kappa^\sigma E^2 - \tfrac{1}{2}s^E\sigma^2 - d\cdot E\sigma \tag{2}$$

where s^E is the elastic compliance at constant electric field and κ^σ the permittivity at constant stress. The existence of such a thermodynamic potential as a perfect differential also implies that the coefficients for the direct and converse effects are identical (Maxwell relationships), as anticipated in Eqn. (1). The equations of state are obtained from Eqn. (2) by partial differentiation:

$$\varepsilon = -\left(\frac{\partial G}{\partial \sigma}\right)_E = s^E\sigma + d\cdot E;$$
$$D = -\left(\frac{\partial G}{\partial E}\right)_\sigma = \kappa^\sigma E + d\cdot\sigma \tag{3}$$

The coupled nature of the equations of state is best visualized in matrix notation:

$$\begin{pmatrix}\varepsilon\\ D\end{pmatrix} = \begin{pmatrix}s^E & d\\ d & \kappa^\sigma\end{pmatrix}\begin{pmatrix}\sigma\\ E\end{pmatrix} \tag{4}$$

The square of the off-diagonal coupling term divided by the product of the diagonal terms is the so-called coupling coefficient, $k^2 = d^2/s^E\kappa^\sigma$, which is the ratio of the stored mechanical energy and the supplied electrical energy (or vice versa). The coupling coefficient also relates the material constants for different measurement conditions to each other:

$$s^D = s^E(1-k^2); \quad \kappa^\varepsilon = \kappa^\sigma(1-k^2) \tag{5}$$

Considering the material without current flow placed in an open circuit (constant D) to be the more basic state, one may say that the piezoelectricity softens the material when keeping the voltage constant, thus allowing a current flow. This aspect is important for acoustic waves traveling through a piezoelectric material, since they are related to propagating E- and D-fields. In a perfect dielectric material, Maxwell equations require that $D=0$ must hold along the propagation direction, and $E=0$ in the plane perpendicular to this direction. The wave velocity of a purely longitudinal wave is thus determined by the elastic compliance at constant D (σ^D), whereas that for a purely transverse wave is determined by the elastic compliance at constant E (σ^E).

In the open circuit condition ($D=0$ in Eqn. (3)) an electric field is built up to oppose the appearance of true charges on the electrodes. This field is called the depolarizing field (E_{dep}) as it reduces the internal polarization, P, caused by an external stress ($\Delta P = d\cdot\sigma$), a temperature change, or a ferroelectric phase transition:

$$E_{\mathrm{dep}} = -\frac{d}{\kappa^\sigma}\sigma = -\frac{\Delta P}{\kappa^\sigma} \tag{6}$$

Introducing P, the D-field is split into in an ordinary dielectric part, εE, and P, which is because of the polar nature of the structure. P can only be quantified for ferroelectric materials. These exhibit a spontaneous electric polarization, P_s, that can be switched in an external field. In these materials, the polar distortion can also be identified with respect to a nonpolar, high-temperature phase. It makes sense to use the polarization, P, as the essential variable to describe the phenomenon. Strain and polarization are related through

$$\varepsilon = QP^2 \quad (7)$$

where Q is the electrostrictive coefficient. P can be an induced or a spontaneous polarization. In the latter case, the resulting strain is the spontaneous strain related to the ferroelectric phase.

2. *Microscopic Origins*

The microscopic origin of piezoelectricity is based on asymmetric charge distributions caused by anisotropic bonding between different ions. The basic mechanism is visualized in Fig. 1 with a linear chain of positive and negative ions, which are connected by springs with alternating large and small spring constants k_1 and k_2. The spacing between equivalent ions is the lattice period, a. $k_1 \neq k_2$ must hold for a relative atom distance, γ, different from ½. When the chain is stretched by a force F, the two types of springs elongate differently. The center of gravity of the positive charges is displaced by $(k_2^{-1} - k_1^{-1})F$ with respect to the negative ions.

As this happens in all unit cells of the chain with equal polarity, the two ends of the chain exhibit a charge displacement with respect to the unperturbed state. In a three-dimensional body, it is important that all such polar features or "chains" exhibit the same directionality in order to exhibit a net displacement field. In general, the lattice parameter, γ, will also be temperature dependent leading to pyroelectricity. The chain model obviously exhibits a polar axis (the chain direction). For crystals, all point groups allowing for a polar axis, i.e., point groups in which symmetry operations are missing that would invert or rotate the given axis by 180°, are compatible with pyroelectricity. Even fewer restrictions are necessary for piezoelectricity. Only the absence of a center of inversion is required (with the exception of point group 432, which cannot show piezoelectricity).

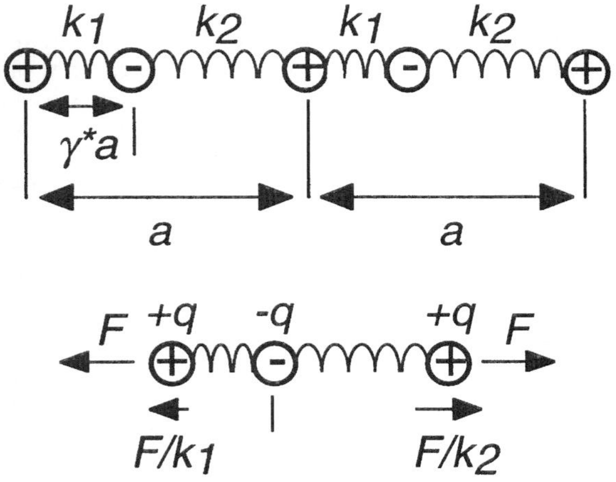

Figure 1
One-dimensional simplified model of piezoelectricity.

In crystals, the full third-rank tensor of d has to be taken into consideration: $D_i = d_{ijk} \cdot \sigma$. Quite often a reduced index notation is applied for the stress tensor: $ii \rightarrow i$; $23 \rightarrow 4$, $13 \rightarrow 5$, $12 \rightarrow 6$. In this notation the nonvanishing elements of the often-encountered point group 4*mm* are d_{33}, d_{31} ($= d_{32}$), and d_{15} ($= d_{14}$). Some numerical values for representative materials are listed in Table 1. The strongest piezoelectricity is found in ionic crystals such as ZnO, $KNbO_3$, and $PbTiO_3$, to name a few. Foils of piezoelectric polymers are very useful for acoustic detectors requiring large surfaces such as hydrophones.

3. *Domain Orientation*

Some piezoelectric materials are ferroelectric and thus can be poled, for instance the PZT mentioned in Table 1. This allows the use of ceramic materials, in contrast to the nonferroelectric materials that have to be singly crystalline or at least textured polycrystalline, such as AlN and ZnO thin films. Ferroelectrics exhibit additional phenomena owing to the existence of domains. In all ferroelectrics there are 180° domain walls that separate regions with opposite polarization. Such walls move when an electric field is applied and enhance the polarization change. In this way, larger piezoelectric strains are obtained as a function of the electric field (see Eqn. (7)). In many ferroelectrics there are also elastic active domains that allow for switching of polarization by a special angle, as defined by the symmetry change at the paraelectric-to-ferroelectric phase transition (quite often 90°).

This switching causes a change of the domain dimensions, which yields a direct contribution to piezoelectric strains. In $Pb(Zr,Ti)O_3$ materials, such domain contributions dominate. The composition is chosen close to a boundary between different ferroelectric phases where the domain walls are especially mobile. In these compounds, 30–50% of the electrical energy can be converted to mechanical energy. Developments in this field include monocrystalline relaxor-type materials (e.g., solid solutions of $Pb(Mg_{1/3}Nb_{2/3})O_3$ and $PbTiO_3$). The crystals are cut to obtain the optimum response in a desired direction. The composition is again close to a phase boundary, the instability of which in the electric field seems to contribute as well to piezoelectricity. Coupling factors, k^2, as high as 80% are observed.

4. *Applications*

The very efficient conversion between electrical and mechanical energy in the very large frequency range

Table 1
Piezoelectric coefficients of some important materials.

Material (point group)	Longitudinal (10^{-12} C/N)	Transverse (10^{-12} C/N)	Shear (10^{-12} C/N)
Quartz (32)	$d_{11}=2.3$	$d_{12}=d_{11}=-2.3$	$d_{14}=-0.67$
$LiNbO_3$ (3*m*)	$d_{33}=6$, $d_{22}=21$	$d_{31}=-1$	$d_{15}=68$
ZnO (6*mm*)	$d_{33}=12.4$	$d_{31}=-5$	$d_{15}=-8.3$
PZT[a] ceramic (as 4*mm*)	$d_{33}=268$	$d_{31}=-119$	$d_{15}=335$
PVDF[b] polymer	$d_{33}=18$	$d_{31}=28$	

[a] $Pb(Zr,Ti)O_3$ solid solution system with additional dopants. [b] Poly(vinylidene fluoride).

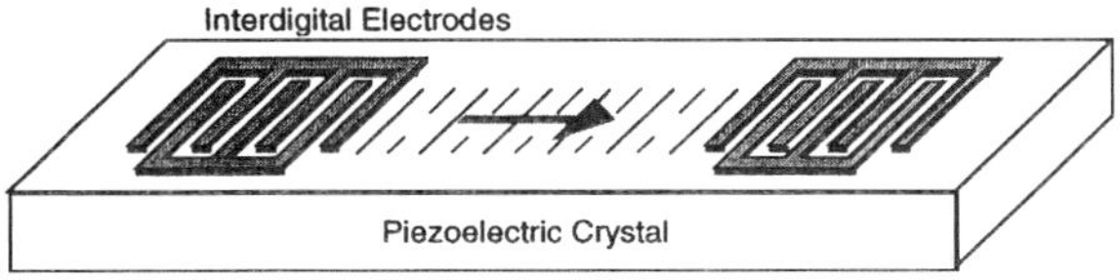

Figure 2
Schematic of a surface acoustic wave (SAW) device.

between audible sound and microwave frequencies allows for unique applications of piezoelectrics. In the audible range there are applications for sonar transducers and hydrophones to locate submarines and fish shoals, or transducers for oil prospecting. Traveling ultrasonic waves in piezoelectric disks (PZT) are used for flat profile motors. Piezoelectric transducers are applied in ultrasonic imaging for medical applications to generate and detect ultrasound waves at megahertz frequencies. Piezoelectrics also supply essential components for signal filtering in mobile telecommunications. For instance, RF filters, which are pass band filters, are needed to suppress signals that do not have the correct carrier frequency (0.9–2.5 GHz). This can be done by surface acoustic wave (SAW) filters based on surface waves (Rayleigh waves) in piezoelectric crystals.

Such waves are excited and detected by means of interdigital electrodes exhibiting the periodicity of the wave. With SAW velocities of around $4000\,\mathrm{m\,s^{-1}}$ the period is typically 4 μm at 1 GHz, thus requiring 1 μm wide electrode fingers (see Fig. 2).

Further applications include linear actuators for scanning probe microscopy (atomic force microscopy, scanning tunneling microscopy, etc.) and further applications in nanotechnology. In these applications one makes use of the high rigidity and high resolution of piezoelectric actuation. Figure 3 shows the schematic setup of a probe scanner as used in the early days of scanning tunneling microscopy. Three orthogonal linear actuators supply the tip excursion for the scanning (x- and y-directions) and the tip-to-sample separation (z). The version shown uses PZT plate elements and exploits the transverse piezoelectric effect (d_{31}). Alternatively, the probe scanner can be realized with a PZT tube whose outer electrode is suitably segmented.

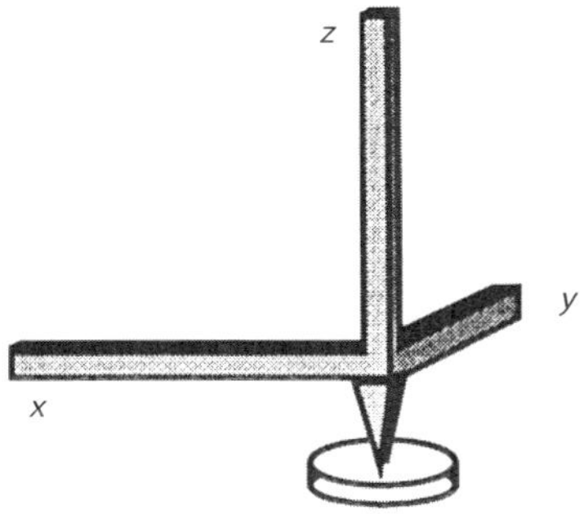

Figure 3
Schematic of a three-dimensional scanning probe microscopy actuator with tip and sample (the cantilever for atomic force microscopy operation is not shown).

Bibliography

Ballato A 1995 Piezoelectricity: old effect, new thrusts. *IEEE Trans. Ultrasonics, Ferroelectrics, and Frequency Control* **42**, 916–26

Beltzer A I 1988 *Acoustics of Solids*. Springer, Heidelberg, Germany

Damjanovic D 1998 Ferroelectric, dielectric and piezoelectric properties of ferroelectric thin films and ceramics. *Rep. Prog. Phys.* **61**, 1267–324

Ikeda T 1990 *Fundamentals of Piezoelectricity*. Oxford University Press, Oxford

Moulson A J, Herbert J M 1990 *Electroceramics*. Chapman and Hall, London

Muralt P 2000 Ferroelectric thin films for microsensors and actuators: a review. *J. Micromech. Microeng.* **10**, 136–46

Park S-E, Shrout T R 1997 Characteristics of relaxor-based piezoelectric single crystals for ultrasonic transducers. *IEEE Trans. Ultrasonics, Ferroelectrics, and Frequency Control* **44**, 1140–7

Strukov B A, Levanyuk A P 1998 *Ferroelectric Phenomena in Crystals*. Springer, Berlin

Tagantsev A K 1991 Electric polarization in crystals and its response to thermal and elastic perturbations. *Phase Transitions* **35**, 119–203

Superconducting Machines: Energy Distribution Components

W. Paul

Most of the potential applications of superconductivity in power systems benefit from zero resistance and high critical current densities in the superconducting state thus reducing losses and size of components, for example the generator, transformer, cable (see *Superconducting Wires and Cables: Low-field Applications*), SMES (superconducting magnetic energy storage) and motor. But, there is also a potential application which utilizes the transition from the superconducting to the normal-conducting state in order to limit fault currents. The device is usually referred to as a superconducting fault current limiter (SCFCL).

1. *Superconducting Transformer*

Transformers are major components of power systems. They allow generation, transmission, and distribution of electrical energy at different voltage levels, leading to economical benefits.

Replacing the copper windings in transformers by a superconductor (SC) offers several opportunities. First it offers reduction in losses, owing to the zero resistance of the conductor. Second it offers reduction of size and weight. Because of higher current densities, the mass of conductor needed can be very small. By applying more winding turns/length in the coils, the amount of iron can be reduced as well. Even iron-free transformers might become feasible. Third, large conventional power transformers are oil-cooled. A superconducting transformer would be more environmentally benign.

In a conventional transformer design the SC will be exposed to alternating magnetic fields up to about 0.3 T. In designs with increased winding turns/length (to reduce iron volume) an even higher magnetic field (B) would be reached. The main challenges for the SC material are high j_c losses, low ac losses at high B, and low cost. Since transformers generate losses not only in the conductor but also in the iron core, it is advisable to leave the core of an SC transformer at room temperature, where cooling is cheaper. Such a design however, poses serious requirements on the cryostat, which would need a warm bore and should ideally be made of nonmagnetic and electrically insulating material in order to avoid additional losses.

1.1 Status of Research

Several SC transformers based on LTS have been built in the past, having a rated power of a few 100 kV A. Commercially available LTS (low temperature superconductor) wires meet technical and economical needs (j_c about $50\,\mathrm{kA\,cm^{-2}}$, a.c. losses $<10^{-5}\,\mathrm{W\,A^{-1}\,m^{-1}}$ at 0.1 T. However, commercialization of LTS transformers has so far been prohibited by the rather high cooling costs, mainly caused by thermal losses through cryostat and current leads.

HTS offers the opportunity to operate SC transformers at 77 K, which could reduce cooling costs to an acceptable level. But, HTS (high temperature superconductor) wires are not nearly as mature as the LTS wires. At present Bi2223 wire is the only HTS available of sufficient length, which could be operated at 77 K. First prototypes based on these materials have been built. A 630 kV A prototype has already been operated for one year under realistic conditions. Even though the technical feasibility has been proven, the present Bi2223 wire will hardly allow a commercial SC transformer to be built.

At present the 77 K field dependence of j_c (j_c(B)), of the wire is too strong and the a.c. losses and wire costs are too high. J_c(B) can be improved by operating the transformer at a lower temperature, which of course increases cooling costs. Bi2223 conductors with reduced a.c. losses are under development. The maximum acceptable losses are in the order of $10^{-4}\,\mathrm{W\,A^{-1}\,m^{-1}}$ at 0.1 T; present values are about 10 times higher.

YBCO ($YBa_2Cu_3O_{7-x}$) tapes, even though only available at length of few meters, offer better j_c(B) and potentially lower cost. The a.c. losses in magnetic fields parallel to the YBCO tape will be negligible. However, it will be a great challenge to reduce the losses due to perpendicular field components (occurring mainly at the end of the coils) to an acceptable value. Nevertheless, considering the status of early twenty-first century technology, the material seems to have the highest potential for transformer applications.

2. *Superconducting Fault Current Limiter*

In electric power systems short circuits might be caused by aged or accidentally damaged insulation or by lightning striking an overhead line. The subsequent fault current is only limited by the impedance of the system between location of fault and power sources. All components of the power system have to be designed to withstand the stresses caused by a short circuit. The higher the fault current anticipated, the higher the costs for the equipment.

Many approaches for realising fault current limiters (FCL) have been based on components with strongly nonlinear current–voltage characteristics (e.g., semiconductors, iron-core reactors, and superconductors). Among the nonlinear materials, superconductors stand out because of their unique transition from zero to finite resistance. FCLs which utilize this transition will be referred to as superconducting fault current limiters (SCFCLs). FCLs which are based on other concepts but might still use SC, e.g., in order to realize low-loss coils, will not be discussed here. There are essentially two concepts in SCFCLs, i.e., the "resistive" and the "shielded core" concepts.

2.1 Resistive SCFCL

The most straightforward concept in an SCFCL is the so called "resistive SCFCL" in which the SC is directly connected in series to the line to be protected. For cooling, the SC might be immersed in a coolant (e.g., liquid helium for LTS or liquid nitrogen for HTS), which is chilled by using a refrigerator.

The cross-section of the SC is determined such that, at nominal current, the SC is operated inside the superconducting state, i.e., at zero resistance, so that its interference with the network is negligible. However, the SC only has true zero impedance for direct current (d.c.)-currents. For the more common a.c. applications the finite length of the SC leads to a finite reactance and the alternating magnetic field generated by the current produces a.c. losses. They both depend very strongly on the geometry of the SC, and can be reduced by optimized conductor architecture, e.g., in the coils with windings in opposite directions or meander structures.

In case of a fault, the increase in j will drive the SC out of the superconducting region into a flux–flow regime. The rapidly increasing resistance limits the current to a value below the unlimited prospective fault current. The actual limitation behavior depends very much on the length of the SC and also on the type of SC material. The three main different limitation behaviors can be realized by just varying the length of the same SC material

First, by using a very long conductor, the electric field $E =$ (line voltage)/(conductor length), to which the SC is exposed during the fault will be very small. Thus, the corresponding current density will exceed the nominal value only by a little. The power density $E.j$ dissipated in the SC will be negligible so that the SC will essentially not warm up (curve 1 in Fig. 1). This "constant temperature" design gives a very good limitation. However, the long length of the conductor will be prohibitively expensive and cause rather high a.c. losses.

The other extreme design would be using a rather short conductor, leading to a relatively large E during the fault. As the fault occurs, the current will increase essentially unlimited by the small resistivity of the flux–flow regime and only limited by the impedance of the power system. However, because of the very high $E.j$ dissipated in the SC, the material will heat up

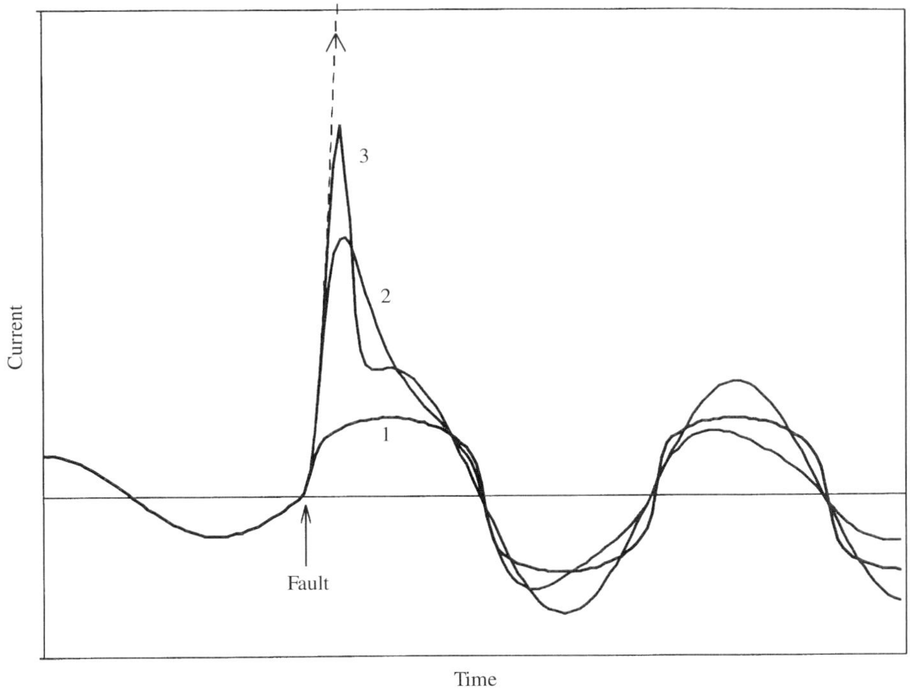

Figure 1
Different designs offer different current limitation behaviors. 1, Constant temperature; 2, slow heating; 3, fast heating.

rapidly and, after a few 100 μs, will become normal conducting and limit the fault current close to or even below the nominal value (curve 3 of Fig. 1). This "fast heating" design needs a rather small amount of SC material. However, the height of the first peak still depends on the unlimited prospective fault current, and instantaneous recovery to normal operation is not possible, since the SC first has to cool down. Furthermore, severe over voltages may arise from the abrupt current reduction, caused by the rapid transition into the normal-conducting state. Last, but not least, the high $E.j$ value during the fault might lead to the SC being damaged.

Using a conductor of intermediate length will lead to an intermediate poor density dissipation in the SC. For the first 10 ms the maximum of the fault current will be limited to several times (e.g., 5–10) that of the nominal current. The SC will keep warming up and after a few tens of milliseconds will enter the normal conducting region ("slow heating" design, curve 2 in Fig. 1).

2.2 Hot Spots

In the above discussion it is assumed that during the limitation process the voltage drop is uniform over the whole length of the conductor. In reality, however, SC tends to develop thermal instabilities called "hot spots," which are due to the strong current and temperature dependence of their resistance. The common measure for reducing the problem is to attach a normal conducting "bypass" or "shunt" in close electrical contact to the SC, which allows the current to bypass the hot spot. Of course, the bypass reduces the total normal resistance of the conductor, which might have to be compensated by increased length. It should be noted that the problem of excessive heating will also limit the maximum possible electric field to which the conductor can be exposed during the fault, i.e., it will also set a lower limit for the length of the SC.

2.3 Shielded Core SCFCL

The second concept in a SCFCL is often referred to as the "shielded core." In this concept the SC is not connected galvanically in the line, but magnetically. The device is essentially a transformer, with its primary normal conducting coil connected in series to the line to be protected, while the secondary side is a superconducting tube (i.e., a one turn coil). In normal operation the iron core sees no magnetic field because it is completely shielded by the SC, thus the name "shielded core." Because of the inductive coupling between the line and the SC, the device is sometimes also referred to as an "inductive SCFCL." The voltage on the secondary is reduced by the number of the turns of the primary coil, while the current is increased by the same factor. The SC on the secondary has to be designed for these values. Assuming an ideal transformer, the SCFCL will behave exactly like a resistive SCFCL.

In order to achieve a certain limitation behavior, both types of SCFCL need the same amount of a certain SC material. Compared to the resistive SCFCL, the shielded core SCFCL shows the disadvantage of being only applicable for a.c. currents and having a much larger size and weight. On the other hand it needs no current leads, which is especially attractive for high current applications, since the losses of the leads are proportional to their current-carrying capability.

2.4 Status of Research

Several SCFCL prototypes with a rated power of several MVA have been realized based on LTS following the resistive concept. Because of the high j_c and low heat capacity of LTS the "fast heating" design was the "natural" choice. However, like in the case of the transformer, commercialization has so far been prohibited by the rather high cooling costs.

To date, the largest HTS SCFCL prototype that has been built utilizes Bi-2212 bulk material. ABB has realized a three-phase 1.2 MV A prototype based on tubes of this material. For ceramic HTS, this geometry is usually more easily realized than long length conductors. The device is of the "shielded core" and "slow heating" type, it was successfully operated for one year under realistic conditions in a Swiss hydropower plant.

Because of their high critical current density (1 MA cm^{-2}) YBCO films are especially suitable for the "fast heating design." Siemens has demonstrated a resistive 300 kV A model based on this material. The films were deposited on planar ceramic substrates, covered with a gold bypass, and patterned into meander.

Bi2223 wires are at present not particularly suitable for SCFCL, because of the low normal resistivity of the Ag-matrix. However, if this resistivity could be increased, the wire would be quite suitable for resistive SCFCL and could even allow for the incorporation of "current limitation" functionality in other devices. Following this approach, a project led by ABB in partnership with American Superconductor and Electricité de France has been launched to develop a current limiting transformer.

Bibliography

Fevrier A, Tavergnier J P, Laumond Y, Bekhaled M 1988 Preliminary test on a superconducting power transformer. *IEEE Trans. Magn.* **24**, 1477–80

Gromoll B, Ries G, Schmidt W, Kraemer H P, Seebacher B, Utz B, Nies R, Neumueller H W, Balzer E, Fissher S, Heismann B 1999 Resistive fault current limiters with YBCO films–100 kV A functional model. *IEEE Trans. Appl. Supercond.* **9**, 656–9

Hoernfeldt S, Albertsson O, Bonmann D, Koenig F 1993 Power transformer with superconducting windings. *IEEE Trans. Magn.* **29**, 3556–8

Paul W, Chen M 1998 Superconducting control for surge currents. *IEEE Spectrum* **May**, 49–54

Paul W, Lakner M, Rhyner J, Unternährer P, Baumann Th, Chen M, Widenhorn L, Guérig A 1997 Test of 1.2 MV A high-T_c superconducting fault current limiter. *Supercond. Sci. Technol.* **10**, 914–18

Schwenterly S W, McConnel B W, Demko J A, Fadnek A, Hsu J, List F A, Walker M S, Hazelton D W, Murray F S, *et al.* (+another 23 authors) 1999 *IEEE Trans. Appl. Supercond.* **9**, 680–4

Seeber B (ed.) 1998 *Handbook of Applied Superconductivity* IOP Publishing, Vol. 2, pp. 1613–26 and 1691–1702.

Verhaege T, Cottevieille C, Estop P, Quemener M, Tavergnier J P, Bekhaled M, Bencharab C, Bonnet P, Laumond Y, Pham V D, Poumarede C, Therond P G 1996 Experiments with a high voltage (40 kV) superconducting fault current limiter. *Cryogenics* **36**, 521–6

Yazawa T, Yoneda E, Nomura S, Tsurunaga K, Urata M, Ohkuma T, Honjo S, Iwata Y, Hara T 1997 Experiments with a 6.6 kV A/1 kA single-phase superconducting fault current limiter. In: Rogalla H, Blank D H A (eds.) *Proceedings of EUCAS 1997, Third European Conference on Applied Superconductivity*. Institute of Physics Conference Series Number 158. Institute of Physics, Bristol, Vol. 2, pp. 1183–6

Zueger H 1998 630 KVA high temperature superconducting transformer. *Cryogenics* **38**, 1169–72

Superconducting Machines: Energy Storage

C. A. Luongo

Superconducting magnetic energy storage (SMES) is an emerging technology offering tremendous benefits to the electric power industry. Many years after its conception (Ferrier 1970), SMES is still not widely implemented and its commercial viability is only slowly being established.

1. *Principles of Operation*

Electric power has always been available on instant demand with a high degree of reliability, flexibility, and control; however, it has never been easy to store in large quantities on a cost-effective and environmentally sound basis.

In SMES, electric energy is stored by circulating a current in a superconducting coil, or inductor. Because no conversion of energy to other forms is involved (e.g., mechanical or chemical), round-trip efficiency can be very high. The energy stored in a d.c. inductor is given by

$$E = \frac{1}{2} LI^2 \tag{1}$$

where E is the stored energy, L is the inductance of the coil (which depends on coil geometry), and I is the circulated current.

As an energy storage device, SMES is a relatively simple concept. It stores electric energy in the magnetic field generated by d.c. current flowing through a coiled wire. If the coils were wound using a conventional wire such as copper, the magnetic energy would be dissipated as heat owing to the resistance of the wire. However, if the wire is superconducting (no resistance), then energy can be stored in a persistent mode, virtually indefinitely, until required.

The SMES coil, a d.c. device, is usually connected to its load (typically an a.c. device) via an a.c./d.c. converter, or "power conditioning system" (PCS). The architecture of the PCS varies depending on the application, but in general is based on standard semiconductor power switches in arrangements to maximize reliability, efficiency, and controllability.

The only loss associated with magnetic storage is the power needed to keep the coil cryogenically cooled as required by the superconductor. Additionally, there are minor losses associated with power transfers through the a.c./d.c. converter. Overall efficiency of a SMES system is heavily dependent on size (stored energy) and power rating, as well as on duty cycle, but in general SMES units can be much more efficient than many other systems for energy storage.

2. *Applications*

In terms of energy storage, SMES competes with other technologies such as batteries, flywheels, and capacitors. The choice of technology is determined by economics (cost) when considering all angles: system performance, capital and operating costs of the unit, reliability and maintainability, environmental and disposal costs, insurance, etc.

SMES is particularly attractive in comparison to other options in cases where a high power/energy ratio is needed (sharp pulses), or when the power pulse repetition rate is high (e.g., tens or even hundreds of pulses per day).

Pulsed power loads such as electromagnetic launch (for military or space applications) could be enabled through the implementation of a SMES system. Buffering pulsating loads in general is a high-payoff application of SMES.

In the electric power sector SMES has two major applications: power quality enhancement, and transmission-related services. Power quality enhancement refers to the ability of SMES systems to sit idle and connected between a load and its utility supply line to "buffer" and "filter" power quality disturbances such as voltage sags and spikes, harmonic distortions, or frequency problems coming through the utility distribution network. Certain applications needing very sensitive loads (such as is required in semiconductor processing plants) can derive great benefit from access to "clean" power and thus pay the premium in price over other storage technologies. At the lower end of stored energy and power (a few megajoules and few megawatts) this is already a commercial application of SMES, with the prospect of capturing a larger market with the development of larger units.

The electric utility industry, being deregulated in most parts of the world, could also be major market for this technology. The applicability of SMES to the utility sector goes well beyond energy storage (daily load leveling). Some of these applications are:

- increased transmission capacity through enhanced line stability;
- spinning reserve;
- leveling of renewable energy sources;
- voltage control;
- frequency control;
- subsynchronous frequency damping;
- wide-area generation control; and
- black start.

Power and energy ratings needed for each of these application vary, from a few hundred megajoules to provide transmission line stabilization, to possibly many gigajoules in the case of spinning reserve for a large network.

In transmission stabilization, a SMES unit is strategically located within a high-voltage network to inject active and/or reactive power on demand to dampen voltage fluctuations that can cause the triggering of protective relays. These fluctuations are usually the result of large generation plants or loads tripping and creating instabilities that cause large-area blackouts. In spinning reserve a SMES unit can act as a infinitely controllable generating unit (with the added benefit of also acting as a "sink") to enhance flexibility of dispatch and to provide that extra cushion need to run a transmission network without the need of having "idling" plants with the concomitant environmental penalty.

In a fully deregulated utility industry (something not yet achieved) SMES will be an integral part of the network.

3. *Technology Development*

The status of SMES in terms of its development and application has been reviewed by Hassenzahl (1989) and by Luongo (1996). The early years of SMES development were mostly devoted to system studies and to determining optimum configurations. During the 1980s the technology saw an increase in activity with the implementation of programs that led to actual hardware being produced. During the 1990s there was somewhat of a retrenchment in terms of development, with relatively few full demonstration programs.

In the early 1980s a 30 MJ SMES unit was installed and tested at the Bonneville Power Authority transmission grid in the USA to demonstrate its ability to dampen subsynchronous resonant frequencies (Rogers *et al.* 1983). Also during the 1980s the US Department of Defense funded the design and development of components needed to build a large-scale SMES units that would serve as pulsed power supplies to military systems. These activities yielded a number of advances in terms of innovative designs and the actual fabrication and testing of major SMES components. Among the advances was the development of a very high-current cable-in-conduit conductor, shown in Fig. 1, rated at 200 kA (at 1.8 K and 5 T), and which was actually tested in short samples under operating condition to its quench current of 300 kA (Peck *et al.* 1994).

In the 1990s most of the development activity took place in Japan with a focus on transmission stability applications as well as the construction and testing of a large SMES model coil (Hamajima 1999).

Previous experience with SMES notwithstanding, it is safe to say that this technology is still in need of an integrated large-scale demonstration under actual operating conditions (e.g., utility environment) before its widespread acceptance in the market.

4. *Outlook*

Despite its great promise and many years of development, SMES is far from being the commercial success once envisioned. Even though the technology is making some inroads into the marketplace (at the lower end of storage/power ratings), it is far from mature or being widely accepted. The main stumbling block for wider dissemination of SMES is not technical, but rather economic. The extremely high cost of superconducting devices does not justify the economic benefits that could be accrued in most applications. The advent of high-temperature superconductor (HTS) materials has done little to advance the deployment of SMES. Major reductions in the price of these superconductors would be needed before they can make an impact on the commercial viability of SMES. Operation at higher temperature alone does not justify adoption of HTS materials, as operating costs are typically a small component of lifecycle cost for SMES units of practical size.

Continued advances in SMES technology are nevertheless likely. One-of-a-kind units for specific applications will give SMES opportunities to demonstrate its benefits. Through this evolutionary chain of development SMES will some day take its promised role in the electric power industry.

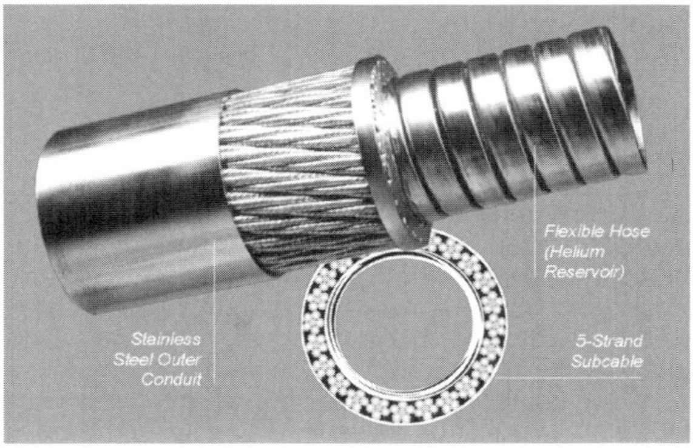

Figure 1

Bibliography

Ferrier M 1970 Stockage d'energie dans un enroulement supraconducteur. *Low Temperature and Electric Power*. Pergamon, Oxford, UK, pp. 425–32

Hamajima T 1999 Test results of the SMES model coil—pulse performance. *Cryogenics* **39**, 351–7

Hassenzahl W V 1989 Superconducting magnetic energy storage. *IEEE Trans. Magnetics* **25**, 750–8

Luongo C A 1996 Superconducting storage systems: an overview. *IEEE Trans. Magnetics* **32**, 2214–23

Peck S D, Zeigler J, Luongo C A 1994 Tests on a 200 kA cable-in-conduit conductor for SMES. *IEEE Trans. Appl. Supercond.* **4**, 199–204

Rogers J D, *et al.* 1983 30-MJ SMES system for electric utility transmission stabilization. *Proc. IEEE* **71**, 1099–107

Superconducting Wires and Cables: High-field Applications

J. H. Schultz

The performance of superconducting wires is fundamentally limited by the magnetic flux density or field, the current density, and the operating temperature, while the magnet applications are limited by current, voltage, and stored energy. Thus the term "high-field applications" is really a generalization for high-performance applications, involving all of these conductor and magnet constraints. As a rule, high field is easiest to obtain in small bores and at low temperatures. This article reviews progress in strands and cables that improves the performance capabilities of large-scale superconducting applications.

All high-current and many high-field superconducting applications use conductors that are made up of cables with multiple strands. A large variety of cable topologies have been used, but two are the most common. High-energy physics multipole (dipole, quadrupole, etc.) magnets generally use Rutherford cables, a high aspect ratio topology with two layers, and a highly regular cabling and crossover pattern. Large energy storage, fusion research, and detector magnets more frequently use cable-in-conduit conductors (CICC), round cables that are built up from smaller round subcables, then compacted into a square or low aspect ratio conduit. The two cable types are illustrated in Fig. 1.

Rutherford cables are favored for multipole magnets, because they can achieve tighter bends with lower conductor strain, while multipole magnets are characterized by modest bore sizes and high length-to-minimum-bend-radius ratios. Cable-in-conduit conductors are favored for the larger bore fusion, detector, and superconducting magnetic energy storage (SMES) magnets, because of their ability to be built up to nearly unlimited current-carrying capabilities, making them most appropriate for very large, high-energy applications. CICC conductors have been built and tested up to 200 kA for large SMES applications (Peck and Michels 1991) and large magnets have been tested to 80 kA (ITER 2001).

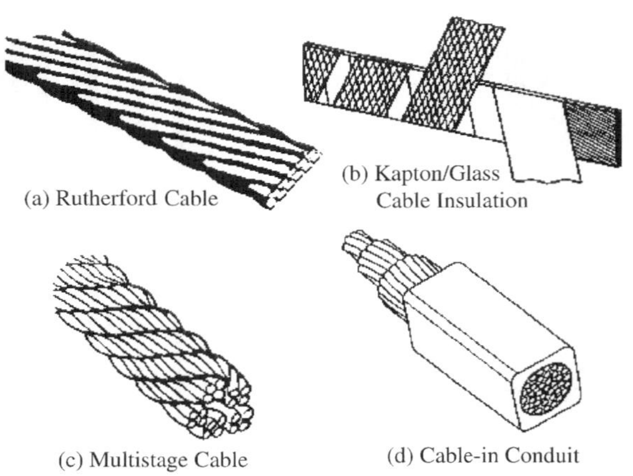

Figure 1
Cable types: (a) simple Rutherford cable, (b) Rutherford cable with kapton/glass insulating wrap, (c) multistage circular cable, (d) cable-in-conduit.

1. Strand

Commercial and scientific applications of superconducting cables are dominated by niobium titanium (NbTi). NbTi, the first practical superconducting material for large-scale applications, is strong, ductile, and relatively inexpensive. The production of NbTi strand is approximately 200 t per year. The largest commercial application is medical magnetic resonance imaging (MRI) magnets, while scientific applications are currently dominated by production of the Large Hadron Collider at CERN (Perin 1991). NbTi magnets have been built with peak flux densities ranging from under 1 T to peaks of 8–9 T at 4.2 K (the boiling temperature of liquid helium) and 9–11 T at 1.8 K (Perin 1991) (near the lambda point of superfluid helium). NbTi is a mature technology, which has slowly been improving, so that it is now possible to purchase large quantities with critical current densities of 3000 A mm^{-2} at 5 T and 4.2 K.

All other magnet applications use Nb_3Sn with an annual production of approximately 2.5 tonnes per year. Nb_3Sn is used in nuclear magnetic resonance and open MRI magnets. Scientific applications include next-generation magnets for high-energy physics and fusion, including 20 t of Nb_3Sn in the KSTAR tokamak being built in South Korea (Lee *et al.* 1999) and an equivalent amount for the recently tested ITER CS Model Coil in Japan (Martovetsky *et al.* 2001). Nb_3Sn cables have been used in large pulsed cables up to 13 T (Tsuji *et al.* 2001), dipole magnets up to 14.7 T (Gourlay *et al.* 2001), small solenoids up to 17 T (Takeo *et al.* 1987), and inserts up to 21 T. Nb_3Sn has superior critical temperature and field, but it is an order of magnitude more expensive than NbTi, requires a long heat treatment at 600–700 °C to become superconducting, and is brittle, requiring great care in handling during fabrication and cooldown to the cryogenic operating temperature. However, the conductor development program for a high-field Very Large Hadron Collider (VLHC) has achieved spectacular progress in critical current density, which has risen over just a few years from 700 A mm^{-2} to 2900 A mm^{-2} at 4.2 K and 12 T. Significant progress in the programme's goal of reducing the cost of Nb_3Sn in large quantities from \$1000 to \$200 kg^{-1} would make several new high-field applications possible.

New materials have been proposed, but have not yet had practical impact on high-field magnet applications. The two most widely studied high-temperature superconductor materials, BSSCO-2223 (bismuth strontium calcium copper oxide) and YBCO (yttrium barium copper oxide), are not available as strands. Artificial pinning center strands have been proposed as an alternative to NbTi, because of superior properties at low field and projected low manufacturing costs (Rudziak *et al.* 1995). Nb_3Al, which has been proposed as an alternative to Nb_3Sn, has similar critical properties, but much lower strain sensitivities. A recently tested Nb_3Al cable has shown performance at 13 T and 46 kA that equals that of previous generation Nb_3Sn and with less degradation over time (Takeuchi 2001). BSSCO-2212 strands have been proposed as high-temperature superconductors, paradoxically because of their superior performance at low temperature and high field. Cables using BSSCO-2212 have achieved 10 kA. None of these alternative strands have demonstrated the existence of more than niche markets and, despite impressive progress, they remain single vendor prototypes. A new strand material that may have a better opportunity to "break out" into commercialization is the recently discovered medium-temperature MgB_2, which benefits from being potentially inexpensive to manufacture, lightweight, and ductile, although its critical temperature is lower than BSSCO or YBCO, while its critical field is lower than Nb_3Sn.

State-of-the-art performance of the different strands is listed in Table 1.

2. *Rutherford Cable*

The Rutherford cable is a two-layer cable of round or square strands, whose high aspect ratio allows winding around the relatively tight radii of constricted end turns in magnets with long straight sections. The simplest form of the Rutherford cable is a two-layer cable with uniform superconducting strand placement and uniform crossover of strands. Several variations of these two forms have been built into a number of historic high-field magnets. Some of these historic cables are illustrated in Fig. 2. Rutherford cables can be compacted for higher strength and packing factor, and "keystoned" so that a winding can fit without gaps on a circumference. As shown in Fig. 2, they can be reinforced by a central core—either steel to add strength and reduce losses, or copper to improve stability and quench protection (Beard *et al.* 1988). Figure 2 also shows how a Rutherford cable can be inserted horizontally or vertically into a copper channel, improving strength, stability, and protection (Beard *et al.* 1988, Schultz *et al.* 1999). It also shows cables that functionally resemble a CICC, by being inserted into a sealed can

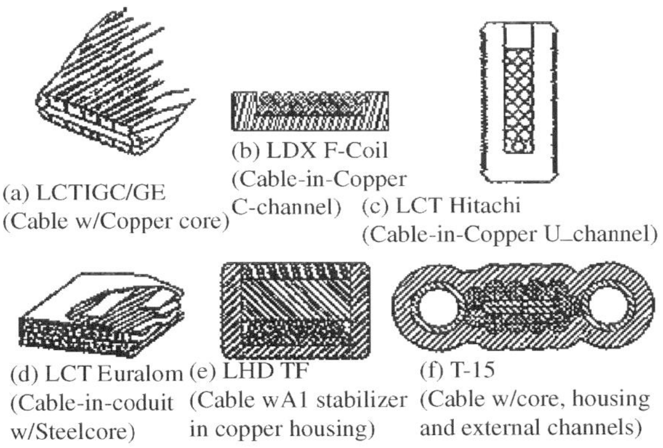

Figure 2
Historic Rutherford cable design options, drawn to scale: (a) large coil task, IGC/General Electric, cable with a copper core; (b) levitated dipole experiment floating coil, cable in a copper C-channel; (c) large coil task, Hitachi, cable in a copper U-channel; (d) large coil task, EURATOM/Vacuumschmelze, cable-in-conduit with an insulated steel core; (e) large helical device, cable with an aluminum stabilizer in a copper housing; (f) T-15 tokamak, cable with core in a housing and external cooling channels.

Table 1
State-of-the-art strands.

Parameter	Units	Value	Name/reference
J_{cnonCu} (4.2 K, 5 T, NbTi)	A mm^{-2}	3800	SSC, OST 14 μm, Furukawa 2.5 μm
J_{cnonCu} (4.2 K, 12 T, Nb_3Sn)	A mm^{-2}	2900	Scanlan
J_{cnonCu} (4.2 K, 20 T, Nb_3Al)	A mm^{-2}	650	Sumption, DOE HEP (Sumption *et al.* 2002)
J_{cnonAg} (4.2 K, 0 T, BSSCO-2212)	A mm^{-2}	2000	Showa, manufacturer's literature
Nstrands/cable (CICC)		1440	ITER Russian PF insert coil (ITER 2000)
Nstrands/cable (Rutherford)		47	LBNL D20
		60	LBNL/Fermilab 2002 (not used in magnet)

with forced-flow helium circulating directly through the conductor or through side-channels (Beard *et al.* 1988, Chernoplekov and Monoszon 1982). The two-layer configuration can also benefit from the ability to achieve a highly regular cabling pattern and near-uniform contact resistances, permitting analysis of cable supercurrents. Cables with as many as 47 strands have been used in high-energy physics dipole magnets (McInturff *et al.* 1997).

Since Rutherford cables are typically used in quasi-steady-state applications, they are usually designed for low contact resistance between strands. The specification for the Large Hadron Collider (Perin 1991) is 10 $\mu\Omega$ per contact. Such a low contact resistance tends to be dominated by the oxide layer on the strand surface. In Tevatron (Cooper and Fisk 1983) and HERA (Meinke 1991), the strands are covered by a thin layer of SnAg (5%) to reduce surface resistance. For applications requiring a significant ramp rate, such as injector magnets, a compromise has to be made between good current transfer for stability and good resistance for low losses. This can usually be accomplished by a thin steel core in the center of the cable, allowing strand–strand current transfer, while eliminating large circulating currents.

3. CICC

Although it has many variants, the cable-in-conduit conductor (CICC) concept has changed little since its invention in the 1960s. Strands are cabled into successive stages, called subcables, in which typically 3–6 subcables are bundled into the next larger cable stage. Cables with as many as 1440 strands have been built up in this manner (ITER 2001). The cable is then jacketed, i.e., inserted into a helium-tight conduit that provides support against winding and Lorentz loads, as well as being a pressure vessel for helium coolant. The most common CICC is a simple circular cable, inserted into a conduit that is squared off with Turks-head rollers, then welded together, either with a continuous seam weld or by orbital welding of discrete sections of pipe. The aspect ratio can be higher than one, when this improves the overall winding packing factor. A circular conduit has the dual advantage of being indifferent to angular position due to any twist during winding or transitions and it also has the lowest stress multiplier for holding internal helium pressure. However, it is incapable of sustaining much compression from radial loads in the winding pack, and has a low packing factor, so that it must be placed in a scalloped structural plate. The cable may have a central coolant hole (Tsuji *et al.* 2001), or it can be placed in a thin inner conduit, leaving four unobstructed cusps in the corners (Steeves *et al.* 1989), so that the hydraulic path will have a larger hydraulic diameter and less friction, allowing the removal of higher heat loads. This philosophy can be extended to where the central hole is much larger than the cable space for very long channels or when needed to allow conduction cooling by superfluid helium (Peck and Michels 1991). Historic examples of these design options are illustrated in Fig. 3.

CICC has several structural, electrical, and conductor cooling advantages. The conduit can be a high-strength superalloy and the winding pack can be completely potted, giving both mechanical rigidity and high electrical integrity. The cable is in direct contact with helium coolant with a high wetted perimeter and can utilize the much higher enthalpy of the helium to give it an energy margin against disturbances that is orders of magnitude higher than a dry magnet. The disadvantages of CICC are the additional cost of the conduit, closure welds, and cable insertion, local stress concentrations in the conduit walls, the need for a forced-flow refrigerator, and the need for electrical isolation from the helium headers. Because of the nearly unlimited ability to build up the cable to high currents and the

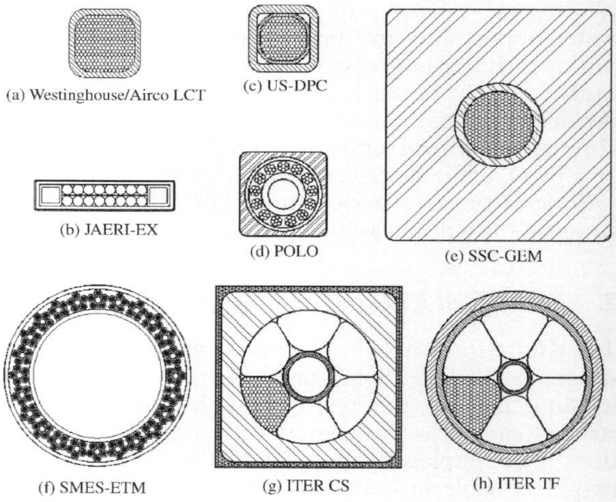

Figure 3
Historic cable-in-conduit design options, drawn to scale: (a) large coil task, Westinghouse/AIRCO, Nb_3Sn; (b) dual purpose coil, Japan Atomic Energy Research Institute, cable-in-conduit with through cable and external cooling channels; (c) US dual purpose coil, central CICC in thick walled conduit and corner channels; (d) POLO, insulated subcable, dual channel CICC; (e) SSC GEM detector, cable in thin steel tube in thick aluminum conduit; (f) superconducting magnetic energy storage-engineering test magnet, large central channel; (g) International Thermonuclear Experimental Reactor (ITER) central solenoid, thick conduit, central cooling channel; (h) ITER toroidal field magnets, thin conduit in plate, central cooling channel.

demonstration of electrical isolation to 25 kV, CICC is favored in large-bore, high-energy storage magnets.

Progress in conduit design and manufacture has included (i) the development of reliable closure welds, (ii) conduit materials and welds with high strength and toughness, even after being subjected to long superconductor heat treatments, and (iii) conduit materials with a good thermal contraction match to the superconductor. Reliable seam welds were demonstrated in thick JBK-75 (Beard *et al.* 1988) and Incoloy (Steeves *et al.* 1989) conduits in the 1980s. The International Thermonuclear Experimental Reactor (ITER), which would have required hundreds of kilometers of high current cable, developed orbital welding techniques with two advantages. The total length of welds in the system was greatly reduced and the welds were fully inspectable, before the insertion of cable. ITER has also demonstrated excellent performance in high-current cables using Incoloy 908 and titanium alloy (Martovetsky *et al.* 2001).

Incoloy 908 has the dual advantages of a good thermal contraction match with Nb_3Sn and higher strength and toughness than stainless steel conduits, after a long heat treatment. Titanium puts the lowest strain in the superconductor of any practical material. Both Incoloy 908 and titanium alloy require tight control of the atmosphere during heat treatment to avoid severe embrittlement during heat treatment. Less-expensive 304 and 316LN stainless steels have been used with NbTi and a low-carbon 316 has been developed to survive the Nb_3Sn heat treatment (Miller *et al.* 1994).

4. *State-of-the-art Magnets and Future Progress*

Several metrics are available for measuring state-of-the-art performance from magnets using strands and cables. These include magnets of solenoidal, toroidal, and multipole topologies, achieving the highest field, the highest current density for the current and voltage, the highest superconducting pinning force (current density-field product), and the highest average stress (e.g., the current density-field-radius product in a solenoid). International experts were surveyed for this article, and the highest historic values of these metrics that were identified are listed in Table 2.

Further improvements in high-field superconducting strands and cables will probably be driven by steady increases in the critical current density, temperature, and field of the strands. It should also be possible to significantly lower the manufacturing costs of all of the advanced strand materials, when they are produced in commercial quantities. The

Table 2
State-of-the-art performance: cable and magnet.

Parameter	Units	Value	Name/reference
B_{max} (solenoid)	T	23.4	NRIM NMR (Kiyoshi *et al.* 2000)
B_{max} (cryocooled solenoid)	T	15.0	INRC, Tohoko U. (Watanabe and Motokawa 2000)
B_{max} (NbTi dipole)	T	10.15	SSC/LBNL D-19
B_{max} (NbTi solenoid)	T	10.1	MIT Hybrid III (Iwasa *et al.* 1993)
B_{max} (NbTi/CICC, 1.8 K)	T	11.1	Euratom LCT (Darweschsad *et al.* 1999)
B_{max} (Nb_3Sn/CICC, 1.8 K)	T	14.1	NHMFL 45 T Outsert (Miller *et al.* 1994)
B_{max} (pulsed coil, > 1 T s^{-1})	T	13	ITER CS model coil
B_{max} (pulsed coil, > 5 T s^{-1})	T	6.7	DPC-EX (Ando *et al.* 1994)
B_{max} (dipole)	T	14.7	LBNL RD-3 (Gourlay *et al.* 2001)
B_{max} (quadrupole)	T	9.63	LHC KEK low-beta quad
B_{grad} (quadrupole)	T m^{-1}	260	LHC MQY Oxford/Cern IR (Strait 2001)
W_m/M_{cold}	kJ kg^{-1}	11.8	CMS solenoid detector (Herve 2000)
I_{cond} (CICC)	kA	200	SMES/ETM (Walker *et al.* 1990)
I_{cond} (dipole)	kA	19.8	U. Twente 11 T (den Ouden *et al.* 1997)
V_{term}	kV	23	POLO (Darweschad *et al.* 1997)
J_{wp} (dipole/quadrupole)	A mm^{-2}	560	LLNL APC HCX prototype (Martovetsky *et al.* 2002)
J_{wp} (solenoid)	A mm^{-2}	415	9.5 T Cryocool outsert (Shen *et al.* 2001)
J_{nonCu} (dipole/quadrupole)	A mm^{-2}	2349	LHC low I insertion quad (Lucas *et al.* 2000)
J_{Cu} (solenoid)	A mm^{-2}	1343	9.5 T Cryocool outsert
J_{Cu} (solenoid/CICC)	A mm^{-2}	550	US-DPC (Painter *et al.* 1992)
J_{Cu} (dipole/quadrupole))	A mm^{-2}	1550	LLNL APC HCX prototype
J_{wp} (dipole)	A mm^{-2}	1627	LBNL D20, outer
$J_{wp}B_{max}R_1$ (solenoid)	MPa	220	NHMFL 45 T outsert (Miller *et al.* 1994)
$J_{wp}B_{max}R_{av}$ (solenoid)	MPa	335	NHMFL 45 T outsert
$R^2_{bore}B^2_o$ (quadrupole)	m^2 T^2	0.73	TJNAF Hall C Q2

same potential for reducing costs with quantity exists for the advanced conduit materials desired for CICCs with heat-treated strands. The copper needed for the protection of CICC magnets should be reduced by the development of internal heaters and sensors with high signal/noise ratios. Advances in the understanding and performance of cables are expected if fully predictive codes become available for calculating supercurrent evolution in each individual strand of a cable, providing reliable predictions of cable losses and individual strand recovery performance.

Bibliography

Ando T, Nakajima H, Sasaki T-Y, Hiyama T, Takahashi Y, Nishi M, Yoshida K, Okuno K, Kato T, Sugimoto M, Isono T, Kawano K, Koizumi N, Hamada K, Oshikiri M, Sasaki T, Wadayma Y, Miyake A, Iwamoto S, Takahashi T, Ebisu H, Hosono T, Yamamoto K, Yasukawa Y, Tsukamoto H, Hanawa H, Seki S, Wakabayashi H, Tsuji H, Takayasu M, Turck B, Wuchuner F 1994 The second test results on the Nb_3Sn demo poloidal coil (DPC-EX). *Adv. Cryogenic Eng.* **39**, 335–41

Beard D S, Klose W, Shimamoto S, Vecsey G, Haubenreich P N, Komarek P, Shimamoto S, Dresner L, Fietz W A, Kato T, Lubell M S, Lue J W, Luton J N, Maurer W, Okuno K, Schwenterly S W, Takahashi Y, Ulbricht A R, Zichy J A 1988 The IEA Large Coil Task: development of superconducting toroidal field magnets for fusion power. *Fusion Eng. Des.* **7**, 1–229 (special issue)

Chernoplekov N A, Monoszon N A 1982 T-15 facility and tests. *IEEE Trans. Magn.* **23**, 826–30

Cooper W E, Fisk H E 1983 Fermilab Tevatron quadrupoles". *IEEE Trans. Magn.* **19**, 1372–5

Darweschsad M, Fink S, Friesinger G, Gruenhagen A, Heller R, Hofmann A, Herz W, Komarek P, Maurer W, Noether G, Rietzschel K, Schleinkofer G, Suesser M, Ulbricht A, Wuechner F, Zahn G 1999 Test of the EURATOM LCT coil (NbTi conductor) with forced flow He II. *Fusion Eng. Des.* **45**, 361–75

Darweschsad M, Friesinger G, Heller R, Irmisch M, Kathol H, Komarek P, Maurer W, Nöther G, Schleinkofer G, Schmidt C, Sihler Ch, Süßer M, Ulbricht A, Wüchner F 1997 Development and test of the poloidal field prototype coil POLO at Forschungszentrum Karlsruhe. *Fusion Eng. Des.* **36**, 227–50

Gourlay S A, Benjegerdes B, Bish P, Byfford D, Caspi S, Chow K, Dietderich D, Gupta R, Hafalia R, Hannaford R, Harnden W, Higley H, Jackson A, Lietzke A, Liggins N, McInturff A, Millos G, O'Neill J, Palmerston E, Sabbi G, Scanlan R, Swanson J 2001 Fabrication and test of Nb_3Sn racetrack coils at high field. *IEEE Trans. Appl. Supercond.* **11** (1), 2164–7

Herve A 2000 The CMS detector magnet. *IEEE Trans. Appl. Supercond.* **10** (1), 389–94

ITER Design Group 2000 *Technical Basis for the ITER FEAT Outline Design.* ITER EDA Documentation Series No. 19. IAEA, Vienna

ITER TFMC Operation and Testing Group 2001 First results of test of the ITER toroidal field model coil in the TOSKA facility of Forschungszentrum Karlsruhe. (in press)

Iwasa Y, Leupold M J, Weggel R J, Williams J E C, Itoh S 1993 Hybrid III: the system, test results, the next step. *IEEE Trans. Appl. Supercond.* **3** (1), 58–65

Kiyoshi T, Kosuge M, Yuyama M, Nagai H, Wada H, Kitaguchi H, Okada M, Tanaka K, Wakuda T, Ohata K, Sato J 2000 Generation of 23.4 T using two Bi-2212 insert coils. *IEEE Trans. Appl. Supercond.* **10** (1), 472–7

Lee G S, Kim J, Hwang S M, Chang C S, Chang H Y, Cho M H 1999 The design of the KSTAR tokamak. *Fusion Eng. Des.* **46**, 405–24

Lucas J, Kirby G A, Ostojic R, Rodrigeuz-Mateos F, Krischel D, Schillo M 2000 Design and construction of a 1 m model of the Low Current Superconducting Quadrupole for the LHC insertions. *IEEE Trans. Appl. Supercond.* **10** (1), 150–3

Martovetsky N, Manahan R, Lietzke A 2002 Development of superconducting focusing quadrupole for heavy ion driver. *Magnet Technology Conf, MT-17.* Geneva, Sept 24–28, 2001

Martovetsky N, Michael P, Minervini J, Radovinsky A, Takayasu M, Thome R, Ando T, Isono T, Kato T, Nakajima H, Nishijima G, Nunoya Y, Sugimoto M, Takahashi Y, Tsuji H, Bessette D, Okuno K, Ricci M 2001 ITER CS model coil and CS insert test results. *IEEE Trans. Appl. Supercond.* **11** (1), 2030–3

McInturff A D, Benjegerdes R, Bish P, Caspi S, Chow K, Dell'Orco D, Dietderich D, Hannaford R, Harnden W, Higley H, Lietzke A, Morrison L, Morrison M, Scanlan R, Smithwick J, Taylor C, van Oort J 1997 Test results for a high field (13 T) Nb3Sn dipole. *Proc. Particle Accelerator Conference.*, Vancouver, Canada, May 12–16, 1997, pp. 3212– 14

Meinke R 1991 Superconducting magnet system for HERA. *IEEE Trans. Magn.* **27**, 1728–31

Miller J R, Bird M D, Bole S, Bonito-Oliva A, Eyssa Y, Kenney W J, Painter T A, Schneider-Muntau H-J, Summers L T, Van Sciver S W, Welton S, Wood R J, Williams J E C, Bobrov E, Iwasa Y, Leupold M, Stejskal V, Weggel R 1994 An overview of the 45-T hybrid magnet system for the new national high magnetic field laboratory. *IEEE Trans. Magn.* **30** (4), 1563–70

den Ouden A, Wessel S, Krooshoop E, ten Kate H 1997 Application of Nb3Sn superconductors in high-field accelerator magnets. *IEEE Trans. Appl. Supercond.* (June), 733–8

Painter T A *et al.* 1992 Test data from the US Demonstration Poloidal Coil Experiment. *MIT Plasma Fusion Center Report,* PFC/RR-91-1, pp. 1–281

Peck S D, Michels P H 1991 Test results from the 200 kA SMES/ETM conductor. *IEEE Trans. Magn.* **27** (2), 1724–7

Perin R 1991 The superconducting magnet system for the LHC. *IEEE Trans. Magn.* **27**, 1735–40

Rudziak M K, Seuntjens J M, Renaud C V, Wong T, Wong J 1995 Development of APC Nb-Ti composite conductors at Supercon, Inc. *IEEE Trans. Appl. Supercond.* **5** (2), 404–7

Schultz J H, Kesner J, Minervini J V, Radovisnky A, Pourrahimi S, Smith B, Thomas P, Wang P W, Zhukovsky A, Myatt R O, Kochan S, Mauel M 1999 The levitated dipole experiment (LDX) magnet system. *IEEE Trans. Appl. Supercond.* **9** (2), 378–81

Shen W, Coffey M, McGhee W 2001 Development of 9.5 T NbTi cryogen-free magnet". *IEEE Trans. Appl. Supercond.* **11** (1), 2619–22

Steeves M M, Hoenig M O, Takayasu M, Randall R N, Tracey J E, Hale J R, Morra M M, Hwang I, Marti P 1989 Progress in the manufacture of the US-DPC test coil. *IEEE Trans. Magn.* **25** (2), 2089–92

Strait J 2001 Very high gradient quadrupoles. *Proc. 2001 Particle Accelerator Conf.*. Chicago, IL, pp. 176–80

Sumption M D, Buta F, Collings E W, Tomsic M, Austen A, Gregory E, Parrell J, Hong S, Zhang Y, Scanlan R M 2002 Nb_3Al strand processing transport properties and cabling. *Proc. Applied Superconductivity Conf.*, Houston, TX, Aug 2002, to be published

Takeo M, Funaki K, Sato S, Yamafuji K, Iwakuma M, Hiramatsu M, Matsumoto K, Hamada M 1987 A 17 tesla superconducting magnet with multifilamentary superconductors. *IEEE Trans. Magn.* **23** (2), 565–8

Takeuchi T 2001 Nb_3Al conductors. *Magnet Technology Conf. MT-17* Geneva, Sept 24–28, 2001

Tsuji H, Okuno K, Thome R, Salpietro E, Egorov S, Martovetsky N, Ricci M, Zanino R, Zahn G, Martinez A, Vecsey G, Arai K, Ishigooka T, Kato T, Ando T, Takahashi Y, Nakajima H, Hiyama T, Sugimoto M, Hosogane N, Matsukawa M, Miura Y, Terakado T, Okano J, Shimada K, Yamashita M, Isono T, Koizumi N, Kawano K, Oshikiri M, Nunoya Y, Matsui K, Tsuchiya Y, Nishijima G, Kubo H, Shimba T, Hara E, Imahashi K, Uno Y, Ohuchi T, Ohtsu K, Okayama J, Kawasaki T, Kawabe M, Seki S, Takano K, Takaya Y, Tajiri F, Tsutsumi F, Nakamura T, Hanawa H, Wakabayashi H, Shimizu T, Kuramochi K, Omine T, Tamiya T, Harada J, Nishii K, Huguet M, Mitchell N, Bessette D, Minervini J, Vieira R, Michael P, Takayasu M, Bevilacqua G, Maix R, Manahan R, Jayakumar R, Savoldi L, Herz W, Ninomiya A 2000 Progress of the ITER Central Solenoid Model Coil Program. *IAEA-CN-77, 18th IAEA Energy Conf.*, Sorrento, Italy, 4-10, IAEA 2001, pp. 354–61

Walker D L 1990 Design of a 200 kA conductor for superconducting magnetic energy storage (SMES). *Adv. Cryogenic Eng.* **35**, 573–9

Watanabe K, Motokawa M 2000 Cryogen-free high field superconducting magnets. *IEEE Trans. Appl. Supercond.* **10** (1), 489–94

Superconducting Wires and Cables: Low-field Applications

L. Gherardi

The discovery in 1986 of high-temperature superconductors (HTSs) with the prospect for operation at liquid nitrogen temperature (77 K) or above provided new impetus for superconducting electric power applications. HTS materials also exhibit higher heat capacity than low-temperature superconductor (LTS) materials that markedly improves their electrical stability, i.e., "normalization" of the superconducting state, during thermal transients. The higher transition temperatures also ease the refrigeration requirements and facilitate cryogenic operation (Blaugher 1996).

1. Cables

Type II superconductors available in the shape of wires and tapes appear as almost "natural" candidates for such electrotechnical applications as power transmission cables. Traditionally, network engineers have relied on increasing voltage to transmit a larger amount of power. Superconducting cables have the capability of carrying much higher currents, so providing the opportunity for reducing the number of circuits, or using lower voltage rating.

In the 1980s a significant research effort, including design studies and experimental activity, was devoted to the assessment of power cables based on low T_c, liquid-helium-cooled materials. Such research led to the conclusion that, although technically attractive, such cables were not economically viable, unless for very high transmitted power, above 2–5 GVA. The main cause for this was identified as the projected cost of refrigeration, which resulted from the high cost and low efficiency of liquid helium, combined with the intrinsically unfavorable geometry (very high surface to volume ratio) of cables. Since 1986 the availability of the new class of high T_c superconducting materials, potentially suitable for (far more effective) cooling with liquid nitrogen, has therefore given a renewed impetus to the development of superconducting power cables.

A range of applications is envisaged. A few instances worth mentioning are the upgrading of existing underground circuits, the matching of the power rating of an overhead line for sections to be relocated underground, and the elimination of intermediate voltage transformations in urban distribution networks.

The expected benefits include, for example, increased transmission efficiency (energy saving, improved stability of high-power circuits), drastic reduction of electromagnetic pollution, an effective solution to the expanded energy demand in congested metropolitan areas, etc.

Development activities are underway in several countries, and include both feasibility studies and experimental work with prototype fabrication and testing. In a number of such experiments, prototype terminations and joints are also being developed. The lengths of such prototypes range from a few meters to more than a hundred meters (Norman *et al.* 1999, Kelley *et al.* 1999).

The first experimental installations of a three-phase superconducting connection in a "real" environment are being realized in the USA. Actually, one link made with cold dielectric (CD) cables (see below) fabricated by SouthWire has already been installed at an industrial plant, and another link, made by Pirelli based on a warm dielectric (WD) design (see below), was planned to be installed in a part of a utility (Detroit Edison) network by the end of 2001.

As type II superconductors show true zero resistance only under d.c. conditions, d.c. cables would appear as the "natural" way of best exploiting HTS materials, by achieving a virtually zero dissipation in the conductor, in addition to negligible dielectric losses. In fact, most power lines transmit a.c. power, and so a.c. designs are also typically assumed for superconducting cables. These are in turn generally classified, according to two main choices that can be made regarding electrical insulation, into "warm" or "cold" dielectric types.

In the WD cable design (Fig. 1), the HTS, made up of one or more layers of current-carrying HTS tapes, is enclosed in a cryogenic environment (the so-called "cryostat"), which in turn is covered by conventional dielectric materials.

This design has several advantages, such as:

- it has similar performance and requires similar installation procedures as a conventional cable.
- it can carry, at the same level of losses, more than twice the power of a conventional cable.
- it uses already developed dielectric materials.
- it needs accessories, both termination and joint, that can be derived in a relatively easy way from conventional ones.
- it requires fewer HTS tapes than the CD coaxial design.

In the CD design, two coaxial HTS conductors, separated by a dielectric at liquid nitrogen temperature, are enclosed in the cryostat (Fig. 2). This approach, which is characterized by the presence of a superconducting "sheath" that acts as a return path for the current and cancels out the external magnetic field, represents the best technical solution for very high-power transmission. It allows one to:

- transmit a large amount of power with reduced losses in "compact" cable systems, eliminating a.c. losses due to external magnetic field of the other phases; and

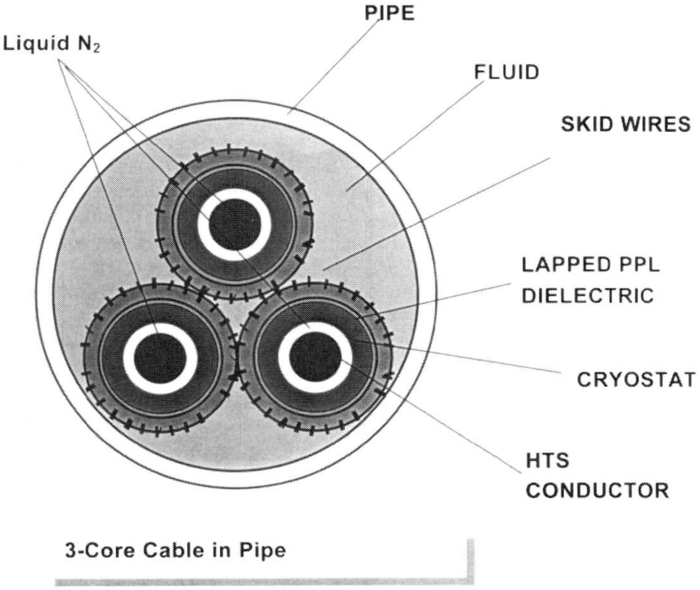

Figure 1
Example of a three-phase WD cable in a pipe.

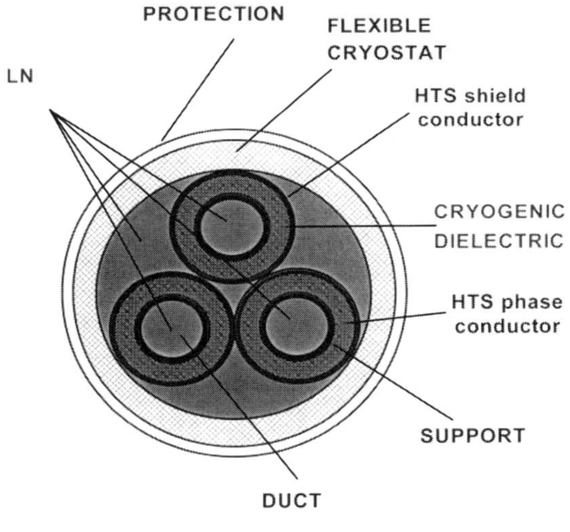

Figure 2
Example of a three-phase CD cable in a pipe.

- eliminate eddy current losses in metallic components, allowing installation close to other cables in service.

However, it requires about twice the amount of superconducting material compared to the WD design, and requires a suitable cryogenic insulation in addition to the development of specific accessories.

Generally, the design of the conductor of a superconducting cable includes a multilayer stranded structure around a suitable former of adequate internal dimension. Optimization of the conductor involves several parameters such as the tape laying angle, the laying tension, the use of binding and/or insulating materials between superconducting layers, the number of tapes and layers, and the former material and dimensions to comply with both electrical and mechanical constraints during fabrication, handling, and service.

Cable manufacturing requires that the superconducting materials are available in the shape of long, flexible wires. Most of the cable prototypes fabricated have been made with BSCCO 2223 tapes typically a few millimeters wide, with a thickness of a few tenths of a millimeter.

For the optimization of the overall electrical and mechanical performance, superconducting tapes are made with a "multifilamentary" structure, with the number of filaments typically ranging from a few tens to over a hundred.

It is worth remarking that the long length required for a wire to be used in cable technology also imposes obvious constraints in terms of performance uniformity. This in turn requires a very good control of the processing parameters (mechanical, "chemical," and thermal) throughout the whole wire fabrication process.

The development of flexible tapes with biaxially textured YBCO, being pursued in the USA, Europe, and Japan with promising results, has not yet reached a stage as to allow the manufacturing of long prototypes. The possibility of replacing BSCCO sintered tapes with YBCO thick films deposited on a flexible substrate is, however, considered extremely interesting for cable applications. In fact, provided that the performances already demonstrated on shorter lengths can be maintained to a significant extent over kilometer-long tapes, the use of YBCO is expected to provide an important improvement in terms of cost-effectiveness, and will therefore possibly give further impetus to the development of superconducting cables.

2. *Fault Current Limiter*

A fault current limiter (FCL) is, in principle, a device that is capable of protecting a power grid from very high surge currents that can arise when a fault (short circuit) occurs anywhere in the grid. As such fault currents are expected to increase with increasing grid interconnection, which is presently a worldwide trend, the need for an effective protection will become more necessary. The problem has been solved partially with the use of nonsuperconducting devices as circuit breakers, fuses, high-impedance transformers, etc, which in general have some drawbacks such as increased operating losses or at least nonideal efficiency. In fact, there is no direct "equivalent" for a superconducting FCL among conventional, nonsuperconducting power components.

A superconducting device can take advantage of the unique property of superconducting materials to undergo a rapid transition to normal state upon a quench, and therefore can in principle offer ideal performance in terms of fast response, fast recovery, and zero impedance throughout normal operation.

As it did for cables, and substantially for the same reasons (insufficient cost-efficiency with low T_c materials), the discovery of HTSs has given new impetus to research activities on FCL development in Europe, the USA, Israel, and Japan. Such activities include design studies as well as demonstrations of small- to full-scale prototypes (Leung 1997).

Various design approaches are considered, namely:

- "Resistive," i.e., the superconducting device is connected directly in series with the power line. The sudden increase of current due to a fault would induce the superconductor-to-normal transition of the material, hence increasing the impedance of the line.
- "Inductive," or "screened core," when the superconductor is *magnetically* (and not galvanically) connected to the line. Such a device is substantially based on a transformer, which has a superconducting shielding tube as the secondary. Under fault conditions the HTS becomes unable to screen the magnetic field generated by the primary coil and the flux penetrates into the iron core thus increasing suddenly the impedance of the device. Different from the resistive type, the HTS component will experience higher current but at lower voltage and does not need current contacts.
- "Hybrid," which is also based on a transformer configuration that includes a resistive superconducting element combined with resistive or inductive shunts.

The different designs are typically based on different superconducting materials. Being studied for FCL applications are BSCCO 2212, BSCCO 2223, and YBCO, in various shapes, namely wires, bulk, and films.

Long tapes with good uniformity are available in the case of a composite tape with BSCCO 2223, which would therefore be a candidate for resistive FCLs. Unfortunately, the composite includes a high-conductivity matrix, which poses severe limits on the efficiency of a type of device based on such tapes. Tapes made with YBCO films deposited on a high-resistivity metallic substrate are available only in short lengths, but, assuming that kilometer-long tapes can be produced, they could represent a good candidate for that purpose. An alternative approach has already been demonstrated with prototypes of resistive FCLs made with YBCO thick films deposited on nonmetallic substrates in a suitable pattern.

YBCO films have also been used as the resistive element in a prototype of a hybrid (with inductive shunt) type of FCL. BSCCO 2212 is more suitable for the manufacture of large pieces, and has been used in prototypes of inductive FCLs in the shape of bulk cylinders.

In general, key material features of superconducting elements for both resistive- and hybrid-type FCLs are essentially their critical current density and normal state resistivity, which determine the required length of the element and the amount of power that will be generated during quench transition, and the sharpness of the superconducting-to-normal transition, which determines the time response of the device.

Other extremely important requirements concern the stability of the superconducting element against the creation of "hot spots" and subsequent thermal instability. From this viewpoint, thin films on nonmetallic substrates imply a different design approach compared to tapes with a metallic matrix or substrate.

As a general consideration, however, it must be noted that there will never be a universal superconducting FCL, since depending on its location inside the network the requirements in terms of recovery, limitation current, and protection coordination are very different. The requirements for the material are also very different. There will therefore probably be opportunities for different materials to be implemented in different types of FCLs in the future.

Bibliography

Blaugher R D 1996 Superconducting electric power applications. *Adv. Cryogenic Eng* **42**, 883–98

Kelley N, Cinquemani P, Belcher P, Wen Y, Nassi M, Corsaro P, Ladiè P, Norman S R, Rahman M M 1999 Applications of HTS cables to power transmission: state of the art and opportunities *Proc. IEEE Transmission and Distribution Conf.* IEEEE, Piscataway, NJ, Vol. 1, pp. 13–19

Leung E 1997 Surge protection for power grids. *IEEE Spectrum* July, 26–30

Norman S R, Ladiè P, Nassi M, Schroth R 1999 Latest development of high temperature superconducting cable systems *Proc. Jicable '99*. Jicable, Versailles, France, paper A3.5, pp. 97–102

Superconducting Wires and Cables: Materials and Processing

P. J. Lee

1. *Introduction*

The phenomenon of superconductivity was first observed in 1912 in the laboratory of Heike Kamerlingh Onnes, at the University of Leiden (Holland). At low temperatures certain materials (~800 superconducting compounds have been recently tabulated by Poole and Farach 2000) suddenly lose their resistance to the flow of electricity. The phenomenon remained a laboratory curiosity until 1954 when G. B. Yntema at the University of Illinois made the first successful superconducting magnet. At that point, magnet performance fell well below that predicted by the properties of the superconductor. The slow progress can be partially attributed to (i) the difficulty in obtaining the extremely low temperatures required and (ii) the lack of understanding of how to create electrically and thermally stable wires from superconductors. The primary limitation, however, was that the superconducting phenomenon was not only limited to low temperatures but also to a restricted range of electrical current density and magnetic field and that the range of these properties was particularly limited in early superconductors. The maximum values of temperature, electrical current, and magnetic field are interdependent and, when plotted in three axes, form a "critical surface" (Fig. 1). Those early superconductors, such as Pb, In, and Hg are classified as type I superconductors. These are superconductors in which magnetic flux is excluded from their bulk and the critical current density is limited to a surface layer of approximately one-tenth of a micrometer. The maximum fields in which these superconductors can operate are usually less that 0.1 T (similar to the flux-density between the poles of a horseshoe permanent magnet). These limitations make type-I superconductors impractical for wire and cable applications. All "technical superconductors," i.e., superconductors that can be readily fabricated into wires and cables for high-current applications, are type-II superconductors. In type-II superconductors, magnetic flux penetrates the bulk of the superconductor to form individual flux quanta. In the most widely used technical superconductors, Nb–Ti and Nb_3Sn, the upper critical fields are 13 T and 27 T, respectively, and current densities $>10^9$ A m^{-2} can be carried (compared to $\sim 10^7$ A m^{-2} for domestic Cu wire). In Fig. 2, the critical current densities at 4.2 K are compared for strands or tapes that are fabricated in lengths of more than 100 m. The temperature range is still limited (the critical temperatures for Nb–Ti and Nb_3Sn are 10 K and 18 K,

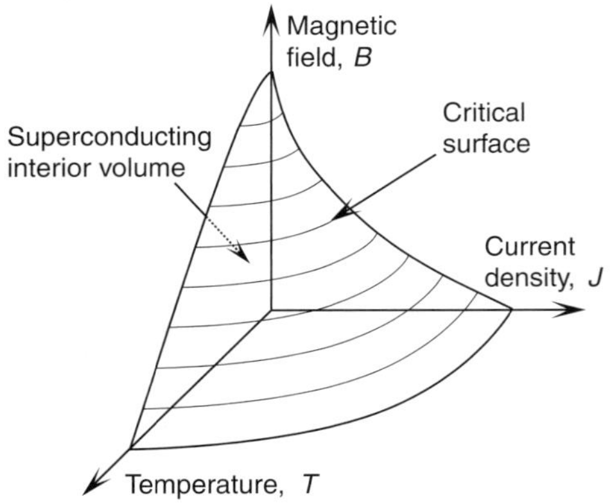

Figure 1
Schematic illustration of the critical surface for a type-II superconductor. The applied field (*B*), the temperature (*T*), and the current density (*J*) must be maintained below the critical surface in order to retain superconductivity.

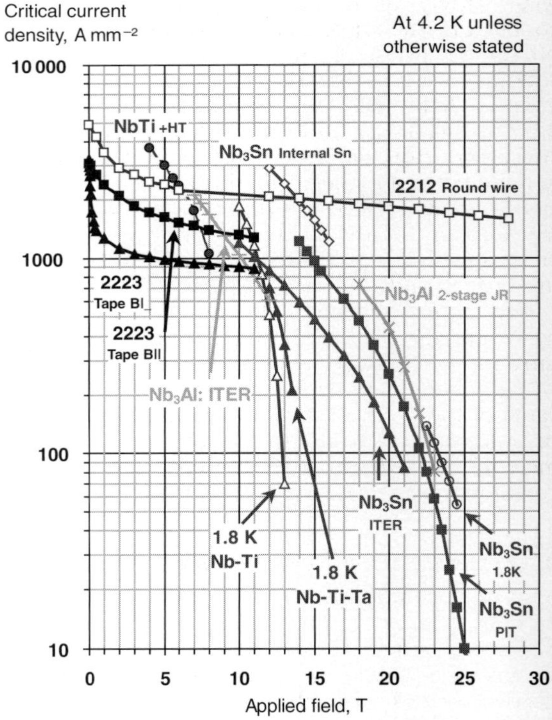

Figure 2
Comparison of critical current densities (at 4.2 K unless otherwise stated) available in strands and tapes of more than 100 m.

respectively), but for many applications the cost of insulation and refrigeration is more than offset by the reduced energy costs resulting from nonresistive current flow. Because the amount of energy lost in current flow through a superconducting magnet is extremely low, it can be operated without a power supply once the magnet has been charged. This mode of operation, termed "persistent" is very desirable for MRI and NMR applications because the current flow (and consequently the field) is very stable. Liquid He is now readily available and provides cooling to 4.2 K at atmospheric pressure and as low as 1.8 K at reduced pressure. Recent advances in insulation and refrigeration technology have also made the application of superconductors more widespread. The discovery of high-temperature superconductivity (HTS) superconductors (ceramic superconductors with critical temperatures above the boiling point of liquid nitrogen (77 K) in 1986) dramatically increased interest in the application of superconductivity. The discovery of superconductivity at 39 K in MgB_2 in 2001 further extended the temperature range for intermetallic superconductors.

Superconductors must be formed into composite wires in order for them to be used in high-current applications and these wires may also be cabled. The wide variety of mechanical properties of superconductors: the range from strong ductile alloys like Nb–Ti to brittle ceramics such as HTS superconductors means that the wire processing routes vary greatly. Of the HTS superconductors only one, Bi-2212, is manufactured in round-wire form and HTS superconductors are more commonly manufactured as multifilamentary tapes. The brittle superconductors cannot undergo wire drawing and are consequently mechanically processed in the form of ductile precursors or fine powders that are reacted to form the final superconductor. For Nb–Ti, the alloy can be cold-worked to a remarkable extent but heat treatments are also applied during processing in order to introduce a second phase into the microstructure that will ultimately pin the fluxoids in the final wire. Because of these variations, the types of processing will be introduced separately and by superconducting material. There are, however, key components that are common to all superconducting strands.

2. General Strand Requirements

2.1 Matrix and Filaments

Superconducting wires are almost always of a multifilamentary design in which individual continuous filaments (<30 μm in diameter) are embedded in a high-conductivity matrix. When conditions in a localized region of a superconductor exceed the critical surface conditions, that area will start to conduct resistively and produce local heating. The resulting heat and current transfer will start a cascade effect and result in a rapid transition of the whole superconductor to the normal state. The ability to transfer heat and current away from localized regions by subdividing the filaments into a high-conductivity matrix was a key step towards producing practical superconductors. The primary components of cryogenic stabilization were not formulated until the mid-1960s (Stekley and Zar 1965). The most desirable matrices are Cu or Al, but Ag is required for the processing of HTS superconductors where oxygen diffusion through the matrix is required during heat treatment. External stabilizer is often added after heat treatment; this is particularly the case for Al, which has thermal and density advantages over Cu, but is difficult to fabricate as the matrix of a composite wire. In addition to stability requirements, fine filaments also serve the purpose of limiting hysteretic loss, not only for a.c. applications but also for d.c. applications. The filament diameter requirements for a.c. applications are quite severe with desired filament diameters of less than 1 μm. For a.c. applications, a Cu–Ni matrix is often used to reduce proximity coupling of the filaments.

2.2 Filament Uniformity

As the critical current per unit cross-sectional area is limited by the material properties, the amount of current that can be carried by an individual filament is reduced if the filament is locally necked to a smaller diameter. Current may be locally transferred to adjacent filaments, but there is a broadening of the critical current transition and a reduction in time that the strand may be operated in the persistent mode (crucial for MRI application). With final filament diameters of less than 20 μm often required, this imposes a very high level of quality control on the processing of the strands. The degree of filament uniformity is reflected in the sharpness of the superconducting transition with increasing current. In the early stages of the resistive transition; the curve of voltage, V, against current, I, can be described by the power law:

$$V \propto I^n \tag{1}$$

The value of n (the transition index) decreases with field; the more uniform the filament cross-sectional areas, the more linear the decrease in n-value and the higher the initial low field value (Warnes and Larbalestier 1986). An electric field, E, is generated along the superconductor carrying a current density, J, that is related to the critical current density of the superconductor J_c at field E_c by the transition index, n:

$$E = E_c(J/J_c)^n \tag{2}$$

For composites based on Nb–Ti strands, the value of n (5 T) can be 50–100, in Nb_3Sn for NMR application the values are in the range of 40–80. The high n-values in Nb–Ti strands correspond to variations in filament cross-sectional area of less than 2% (coefficient of variation) (Lee and Larbalestier 1993). In HTS superconductors the range of available n-values is presently only 10–20 and persistent mode operation is not feasible under these conditions.

2.3 Critical Current Density and Flux Pinning

Pure annealed superconductors do not carry very much current because once current passes through the superconductor, a Lorentz force is induced on the flux lines. When they move, the flux line lattice (FLL) dissipates energy and the superconductor eventually goes "normal". The flux-lines, however, can be held in place by defects in the superconductor, a process termed "flux-pinning." In A15 based superconductors (such as Nb_3Sn), increasing the grain boundary density has been shown to increase critical current density (Scanlan *et al.* 1975, Ochiai *et al.* 1986). In Nb–Ti a linear increase in critical current density is observed with vol.% of α-Ti precipitate (Lee *et al.* 1990). Consequently, the introduction of pinning defects and the refinement of grain size are key elements to the success of technical superconductors.

2.4 Twisting

Twisting of the strand about its drawing axis is typically required to reduce flux-jump instability caused by varying external fields, and to reduce eddy-current losses. The twisting is ideally applied just before a multifilamentary strand has reached final size, so that the twist can be locked into place by a further die-pass. The higher the expected rate of change of field, the tighter the required twist pitch. For the relatively steady-state dipole magnets for the Superconducting Supercollider, ~80 rotations along the drawing axis per meter were required, for a.c. application with a similarly sized strand the number of twists per meter might be 300. Typically, twist pitches cannot be higher than 8 times the wire diameter, and normally a twist pitch of 10–20 times the diameter is used.

2.5 Final Shaping and Cladding

The final cross-section of the strand can be controlled by die shape and size or by using independently adjusted rollers operating along the strand surface. In this way, square or rectangular cross-section filaments can be produced. Alternatively, the strand can by inserted into a channel in a larger form, most commonly a rectangular aluminum external stabilizer.

2.6 Cabling

Individual strands can be cabled or braided together to form a conductor with a higher current carrying capacity. The most common design for Nb–Ti magnets is the Rutherford cable, which consists of a fully transposed, flat cable. Using this approach, high-aspect-ratio cables can be produced with as many as 46 strands (Scanlan *et al.* 1997). Strands may also be drawn into a larger conduit or combined with external stabilizers. A cross-section of a cable-in-conduit-conductor (CICC) is shown in Fig 3. As in the single strand-in-channel aluminum process, multiple strands have been applied to surface channels in circular cross-section aluminum rods.

2.7 Length-Dependent Parameters

Both commercial Nb–Ti and Nb_3Sn strands are made in lengths of over a kilometer and variations can occur over the length of the strand, due to differences in mechanical properties of the components and bonding of the monofilaments that occurred during mechanical reduction. In Nb–Ti variations in critical current, I_c, along the length of strands can be directly correlated with variations in Cu/Nb–Ti ratio along the length as shown by Kanithi *et al.* (1990). In their study of strands developed for the Superconducting Supercollider project, they showed that the variation in I_c (due to variations in the intrinsic superconducting properties, the strand diameter and Cu/Nb–Ti ratio as well as measurement error) was more than twice the variation seen in J_c (due to just the intrinsic superconducting properties and measurement error). As HTS-based strands are still in their production infancy, a different phenomenon is

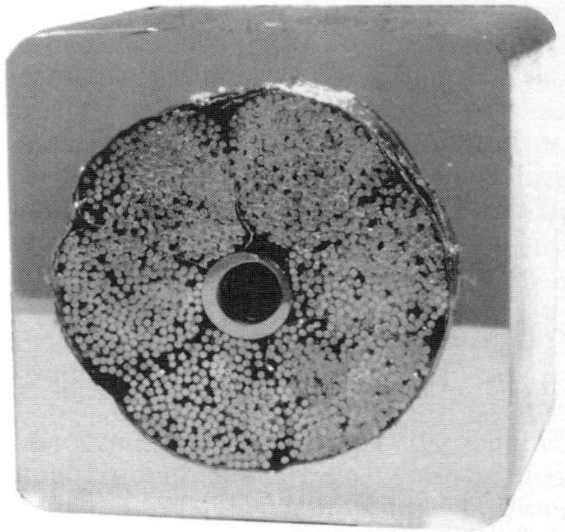

Figure 3
CICC: this 51 mm diameter Incoloy® jacketed cable contains 6 sub-cables; each containing 180 Nb_3Sn based superconducting strands.

observed: as production length is increased, the overall I_c declines because it is very difficult to maintain the ideal conditions (grain alignment, heat treatment temperature, low porosity) necessary for good HTS strand over long lengths.

3. *Superconducting Strand and Cable Manufacture by Material*

3.1 Nb–Ti

(*a*) *Alloy composition.* Nb–Ti has become the dominant commercial superconductor, because it can be economically manufactured in a strong and ductile form with both high current density and high resistive transition index. The starting point for fabrication of Nb–Ti strand is an alloy of between 46.5 wt.% Ti and 50 wt.% Ti. Decreasing Ti over this range increases both the critical temperature, T_c, and the upper critical field, H_{c2} (which at 4.2 K has a maximum value of 11.5 T at 44 wt.% Ti). The increase in T_c and H_{c2}, however, is offset by a decrease in the amount of Ti available for precipitation of the α-Ti phase desired for flux-pinning and high critical density. The composition of the Nb–Ti also affects the precipitate morphology: if the α-Ti is formed at grain boundary intersections, it tends to be uniform in size and distribution and does not significantly affect strand ductility. Higher Ti compositions, however, also produce fine distributions of Widmanstätten-type precipitation in the grain interiors resulting in both a nonuniform microstructure and a significant hardening of the Nb–Ti. As all these factors are sensitive to composition, it is very important to start with a homogeneous alloy. With final filament diameters well below 0.05 mm, the alloy must also be free of unmelted Nb or nonductile inclusions that may result in filament and ultimately strand breakage. Although the large difference in the melting points of Ti and Nb makes this a difficult alloy to manufacture, homogeneity levels in commercial Nb–47 wt.% Ti alloys are now typically much better than ±1.5 wt.% Ti. Impurity content has been traditionally kept low and consistent, but recent work has shown that increasing the allowable content of iron can considerably refine the precipitate size without adversely affecting the precipitate volume (Smathers *et al.* 1996).

(*b*) *Extrusion and billet assembly.* The (as-cast) Nb–Ti ingot is (hot forged) to the required diameter and then annealed in the single phase β region (~2 h at 870°C). For fine filament Cu matrix composites, the Nb–Ti ingot is then sealed inside a high-purity Cu extrusion can, that subsequently (warm-extruded) forms a (well-bonded) monofilament. Monofilament production billets ready for extrusion are shown in Fig. 4. Billet assembly is performed in a clean environment, in order to avoid introducing particles that would not co-deform to the sub-0.05 mm filament size. Where high critical current densities and fine filaments are desired, an Nb diffusion barrier is wrapped around the Nb–Ti rods in order to inhibit the formation of brittle Cu–Ti intermetallics during heat treatment (Wilson 1972). The monofilament is then reduced in diameter by wire drawing, typically finishing with a hexagonal die to produce filaments that can be efficiently (restacked). The thickness of Cu around the Nb–Ti at this stage is important because it determines the spacing between Nb–Ti filaments when Cu/Nb–Ti monofilaments are stacked together. For optimum mechanical stability in Nb–Ti/Cu multifilamentary composites, a filament spacing to filament diameter (*s*/*d*) ratio of 0.15–0.20 is ideal (Gregory *et al.* 1987). Usually additional stabilizer is required and that may be added by increasing the wall thickness of the extrusion can or adding a central core of stabilizer. For d.c. magnet application, it has been established by Ghosh *et al.* (1987) that a minimum Cu thickness of 0.4–0.5 μm is required to reduce the magnetization associated with proximity filament coupling. For *s*/*d* ratios of 0.15–0.20, the minimum Cu thickness requirement limits the minimum filament diameter to ~3 μm when using a pure Cu matrix. If smaller filaments are required, Ni or Mn can be added to the Cu between the filaments (Hlasnik *et al.* 1985, Cave *et al.* 1989, Kreilick *et al.* 1988).

Figure 4
Monofilament Cu-clad Nb–Ti extrusion billets being inspected after welding (image courtesy of Hem Kanithi, Outokumpu Advanced Superconductors Inc.).

The monofilament is then restacked inside another Cu extrusion can to form the final multifilamentary composite. The size of this extrusion can be up to 330 mm in diameter with a weight of 400 kg. As many as 4000 filaments may be assembled in one extrusion. Alternatively, large numbers of fine filaments can be produced by restacking a previously extruded multifilamentary billet. Using this approach, strands with up to 40000 filaments have been manufactured. Where a small number of large filaments are desired (typically for low-field MRI application), Nb–Ti rods can be directly inserted into gun-drilled holes in a solid high-purity Cu extrusion billet. Warm extrusion helps bond the composite, as much of the subsequent processing involves cold work by wire drawing, and the resulting strain introduces into the Nb–Ti an essential element in developing the high critical current microstructure.

(*c*) *Precipitation heat treatment.* Such a treatment is applied after sufficient cold work has been carried out following the multifilamentary extrusion. If too little cold work is applied before heat treatment, Widmanstätten-type precipitation will occur, causing increased hardness (reducing drawability) and producing a non-optimum precipitate size and distribution. Lee *et al.* (1989) found that the amount of (prestrain) required increases linearly with wt.% Ti. Consequently, a higher prestrain is required to suppress Widmanstätten-type precipitation in a chemically inhomogeneous alloy than would be required in a homogeneous alloy. For Nb–47 wt.% Ti, the required prestrain is ~ 5 but this rises steeply to 9 for a 55 wt.% Ti alloy. Heat treatments are typically at 375–420 °C, ranging from 20 h to 80 h in duration. Multiple heat treatment and drawing cycles are required to produce high critical current densities (Cheng-ren *et al.* 1983). Whereas an initial heat treatment will only yield ~ 10 vol.% precipitate, by applying additional cold work strain, the effect of the slow diffusion rate at these temperatures is again overcome and more precipitate is produced. Second and third heat treatments typically yield 15 vol.% and 20 vol.% of α-Ti precipitate respectively. The strain between heat treatments is 0.8–1.5 and at least three heat treatments are usually applied. A linear relationship between the optimized critical current density and the volume of precipitate Nb–47 wt.% Ti alloys has been established by Lee *et al.* (1990) 20% volume of α-Ti precipitate is sufficient to produce a critical current density of $> 3000\ \mathrm{A\ mm^{-2}}$ at 5 T, 4.2 K. Increasing Ti content increases the precipitation rate and amount, but higher pre-strains are required if good ductility and long piece length is to be maintained. Alternatively, the pinning sites can be introduced mechanically by hand-assembling the desired microstructure at a large- size, in the same way as the multifilamentary billets are assembled, and then reducing the dimension to match the fluxoid spacing by multiple restacking and drawing cycles. This approach is called APC for artificial pinning center. With APC strand, no additional heat treatment is necessary and a greater flexibility is possible in the microstructural components. Despite superior low field (< 5 T) performance, the APC approach has not reached commercialization because of the costs associated with the multiple restacking approach.

(*d*) *Final wire drawing.* After final heat treatment the α-Ti precipitates are roughly equiaxed in transverse cross-section, 100–200 nm in diameter with an extension along the wire axis giving an aspect ratio of 5–20. The extensive final wire drawing (an engineering true strain of 5), however, produces a plain strain condition in the b.c.c. β-Nb–Ti, which is supported by (inter-curling) of the Nb–Ti grains. This results in distortion of the α-Ti precipitates into densely folded sheets during final wire drawing (Fig. 5(a)). The folding process rapidly decreases the precipitate thickness and spacing with a dependence

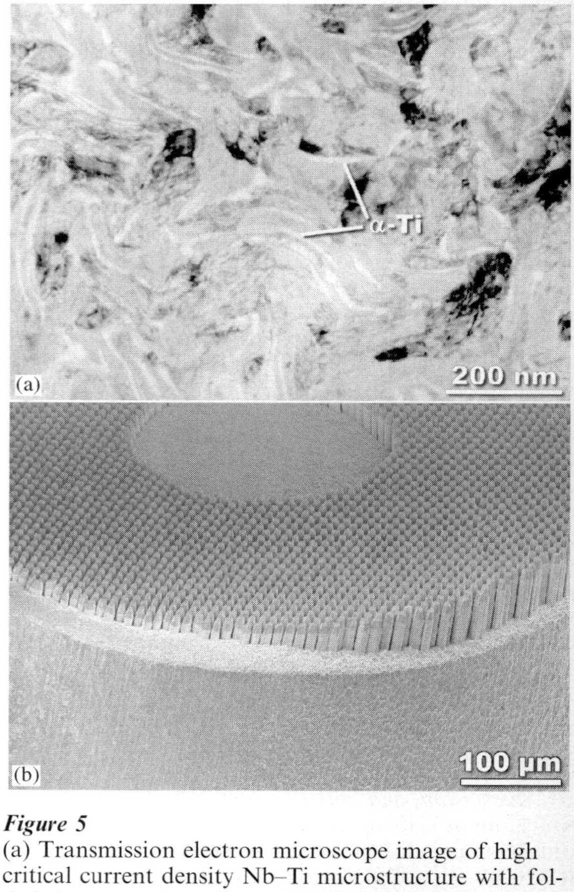

Figure 5
(a) Transmission electron microscope image of high critical current density Nb–Ti microstructure with folded sheets of *alpha*-Ti precipitate. (b) Etched cross-section of Nb–Ti-based superconducting strand fabricated by IGC-AS (now Outokumpu Advanced Superconductors) for the Interaction Region Quadrapole Magnets of the Large Hadron Collider at CERN.

of $d^{1.6}$ (where d is the strand diameter) and increases the precipitate length per area with a dependence of $d^{-1.6}$, as measured by Meingast *et al.* (1989). The critical current density increases as the microstructure is refined until it reaches a peak, after which there is a steady decline. The peak in critical current density for a monofilament or a multifilamentary strand with uniform filaments occurs at a final strain of ~5. If the filaments are nonuniform in cross-section (sausaged), the peak occurs earlier and at a lower critical current density. A strand that has a premature (and lowered) peak in critical current density during final drawing is described as "extrinsically limited" because it has not attained the intrinsic critical current density of the microstructure. The most common source of extrinsic limitation is sausaging of the filaments, due to intermetallic formation or lack of bonding between the components of the composite. A high-performance strand (Fig. 5(b)) has sausaging reduced to a very low level—a coefficient of variation for the filament cross-section of ~2%. With tight quality control, uniform properties and piece lengths exceeding 10 km should be expected.

3.2 Nb_3Sn

Nb_3Sn is the second most widely used superconductor and has a higher upper critical temperature (18.2 K) than Nb–Ti. It is one of several A15 structure compounds that are superconducting. Other A15 superconductors of interest include Nb_3Al (T_c of 19.1 K) and Nb_3Ge (T_c of 23.2). Nb_3Al has been manufactured in long lengths and has a better strain tolerance than Nb_3Sn and better performance at a very high field (> 22 T) making it of important to high-field NMR inserts. Nb_3Al is much more complex and expensive to fabricate than Nb_3Sn because of the mechanical differences between Nb and Al and because the optimum Nb_3Al composition for a A15 structure, can only be formed by high temperature processing (> 1000 °C). Nb_3Sn, on the other hand, is widely manufactured and has shown sufficient strain tolerance in practice to be used successfully in very large solenoids such as the ITER CS model coil which stored 640 MJ of energy (Kato *et al.* 2001) and high-field (> 21 T) inserts for NMR application (Kiyoshi *et al.* 2001). Because A15 materials are brittle, all production routes use ductile precursors that are heat treated to form the A15 at final size by solid-state diffusion. At an early stage it was found that the presence of Cu greatly aids the formation of Nb_3Sn at temperatures below 800 °C. It has also been found that adding Ti (0.7–1 wt.%) or Ta (7.5 wt.%) improves the critical current density at high fields (> 12 T).

The three basic production routes currently used by commercial Nb_3Sn vendors—the bronze process, the internal Sn, and the powder-in-tube (PIT) fabrication techniques—are illustrated in Fig. 6. The internal Sn process may use a jelly roll of Cu and Nb sheets, as illustrated in Fig. 6, or may use Nb (or Nb alloy) rods in a Cu matrix as shown in Fig. 7.

(*a*) *Bronze process.* For MRI application the most common route is the "bronze process" in which Nb alloy rods are first clad in high-purity ductile Cu–Sn bronze and then assembled inside another Cu–Sn tube in a way similar to the production of Nb–Ti filaments in Cu for Nb–Ti strand. As with Nb–Ti alloy, multiple restacks can be used to increase the filament number. Under heat treatment (typically 650–725 °C) the Sn from the bronze reacts with the Nb to form the superconducting A15 compound. As the Cu–Sn alloy does not act as a stabilizer, additional high-purity Cu must be added

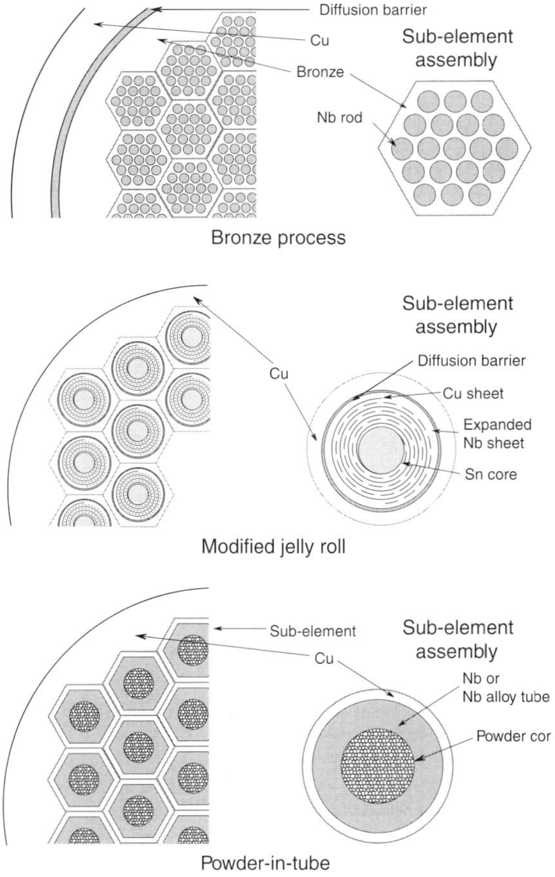

Figure 6
Schematic illustration comparing the bronze process, modified jelly roll (internal Sn), and PIT fabrication techniques for Nb_3Sn-based superconducting strand.

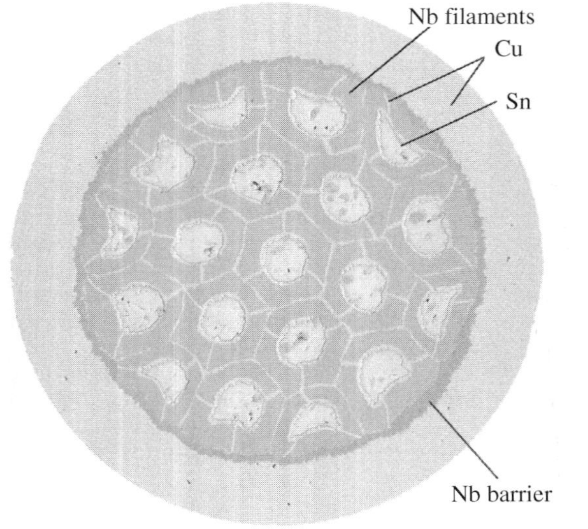

Figure 7
Transverse cross-section of a rod-process internal Sn Nb_3Sn strand prior to diffusion heat treatment. This strand was designed for high critical current density by IGC-AS (now Outokumpu Advanced Superconductors).

outside, and/or inside the bronze–Nb filament composite, with a Ta and/or Nb diffusion barrier to eliminate Sn contamination of the stabilizer. An example of a bronze–process strand is shown in Fig. 8. Vanadium has also been used as a barrier but results in a reduced critical current density. This process can yield a strand with a very high uniformity in filament size and distribution, which makes it an ideal product for high-field NMR and other applications where a minimum of hysteresis loss is desired. The bronze, however, quickly work hardens during wire drawing and must be annealed frequently in order to maintain strand ductility, this extra processing step significantly increasing the cost of the strand. The Sn content in the composite is also limited by the amount of Sn (<15 wt.% Sn) that can be added to bronze while maintaining ductility. The impact of insufficient Sn is most noticeable on critical current density. Recent developments in the bronze process have centered on the use of specially processed high Sn bronzes (up to 16 wt.% Sn) that have good ductility and high homogeneity (Sakamoto *et al.* 2000) (Fig. 9).

(*b*) *Internal Sn process.* Higher Sn content and higher critical current density can be achieved by keeping the Sn separate from the Cu during strand fabrication. This is the basis for the internal Sn process which places an Sn core inside an Nb–

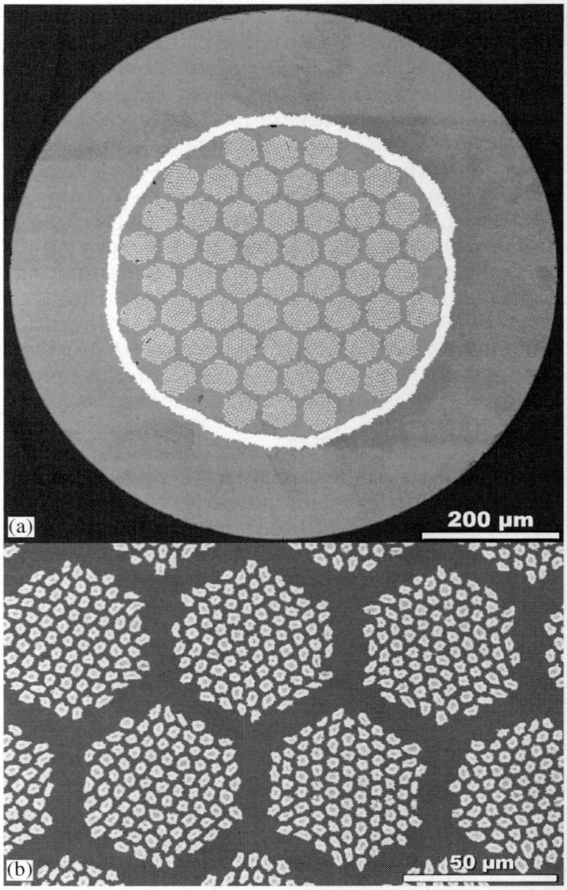

Figure 8
Bronze process Nb_3Sn strand fabricated by Vacuumschmelze for ITER shown in (a) full cross-section and (b) in filament detail. The filaments have a central unreacted core of Nb.

filament/Cu matrix. The Sn core can be alloyed as a carrier for the Nb_3Sn allowing-element or to improve the mechanical properties of the soft Sn. Strands produced by this process have achieved the highest critical current densities for Nb_3Sn (in excess of 2400 A mm^{-2} in the non-stabilizer package). The high critical current densities are achieved, in part, by reducing the amount of Cu in the Cu/filament/Sn package and the resulting reduction in filament coupling results in an undesirable increase in filament coupling. There are two distinct sub-categories of internal Sn process strand based on the form of the Nb filaments. In the "rod-in-tube" (RIT) process internal Sn–Nb filaments are assembled by inserting rods of Nb into pre-drilled holes in a Cu billet. The second internal Sn method is termed modified jelly-roll (MJR) and used an expanded

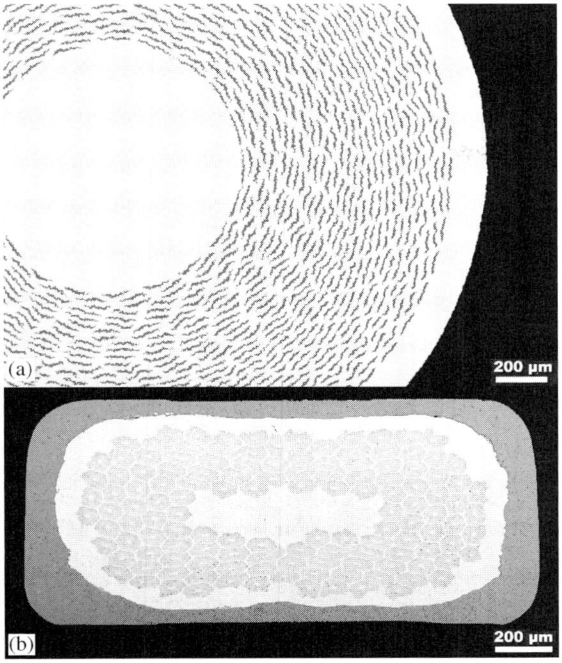

Figure 9
(a) Precursor jelly-roll of Al and Nb fabricated by Oxford Advanced Superconductors. (b) Multifilamentary Nb_3Al tape fabricated by the National Institute for Materials Science in Japan from a rapidly heated and cooled composite of jelly-roll precursors (fabricated by Hitachi Wire and Cable).

metal sheet of Nb co-wrapped with a Cu sheet to create the individual Nb filaments in a Cu matrix. Both of these techniques achieve very high current densities, but limitations to the sizes of the billets (because of the restricted depth of the gun-drilled holes for rod-process and size limitations to deforming the MJR stacks) means that billet sizes are typically only 20 kg and processing costs are subsequently much higher than for Sn. As in the bronze process, the individual Cu–Nb–Sn sub-elements are clad in a diffusion barrier (Ta or Nb or TaNb alloy) to prevent poisioning of the Cu stabilizer by Sn diffusion. Sn diffuses rapidly through Cu at the temperatures used for the A15 forming heat treatment (up to 700°C) and barrier integrity is consequently of great concern. The heat treatment of the strands to form the A15 phase is preceded by multiple lower-temperature steps designed to combine the Sn with the Cu without Sn melting. Both the RIT and MJR technique can also be used to fabricate precursor strand for Nb_3Al production.

(*c*) *PIT process.* Nb_3Sn may also be formed from higher Sn content Nb–Sn phases (Nb_6Sn_5 and $NbSn_2$) by inserting the intermetallics in powder form into tubes of Nb and reacting at 650–700°C. The presence of Cu is also required to allow the formation of the A15 phases in the 650–700°C temperature range in which fine grain size is maintained (Lefranc and Muller 1976). Elemental Sn is also added to improve the Sn-to-Nb ratio. No low-temperature heat treatment is required for these strands and the A15 formation heat treatment takes a relatively short time, 50–75 h (180–270 ks). This process uses the best combination of high critical current densities (>2200 A mm^{-2} at 12 T and 4.2 K in the nonstabilizer volume) and low effective filament diameter (as low as 30 μm in high critical current density strand). The process is used to manufacture small quantities for demanding applications (Lindenhovius *et al.* 2000) and has not received the cost benefits of commercial scale-up. The PIT technique can also be used for MgB_2, Bi-2223, and YBCO, but porosity is an issue for these conductors when fabricated by this technique.

3.3 Nb_3Al

Nb_3Al has not achieved commercial strand status, but it can be described as a technical superconductor as it has been fabricated into high performance and uniform kilometer lengths that have in turn been fabricated into full-scale devices. At magnetic fields above 20 T, critical current densities can be obtained that are comparable to the best Nb_3Sn strands and useful critical current densities (>100 A mm^{-2} non-stabilizer) can be achieved above 24 T. Perhaps the most important feature of Nb_3Al is the high strain tolerance of the Nb_3Al strands compared to Nb_3Sn. One ton (total length of 210 km) of mass-produced "jelly-roll" process strand (Yamada *et al.* 1994) was manufactured for the 13 T–46 kA Nb_3Al insert for the ITER Central Solenoid model coil. The jelly-roll of Nb and Al is created by co-winding sheets of Nb and Al together on an oxygen-free pure copper core. This will eventually become an individual filament in the final multifilamentary strand. The assembled jelly-roll and core are inserted into an oxygen-free pure copper tube to create an extrusion billet. After hydrostatic extrusion, the billet is drawn down to a small enough size that it can be restacked to create a multifilamentary billet. The billet is hydrostatically extruded and drawn for a second time, to the desired strand diameter size. The extrusion and restack process reduces the thickness of the Nb–Al layers in the filaments to less than 1 μm. The production yield rate for the ITER insert strand was 70% for strands longer than 1.5 km. The final strand is heat treated (800°C for 10 h is typical) to produce Nb_3Al. The critical current densities obtained by this technique are ~600 A mm^{-2} at 12 T and 4.2 K (Hosono *et al.* 2002).

The highest critical fields and critical temperatures for Nb_3Al are based around high-temperature heat treatments and rapid cooling. In the rapid-heating, quenching, and transformation (RHQT) (Iijima *et al.* 1997) process Nb/Al composites (MJR or RIT) are resistance heated to 1900–2000°C and then rapidly quenched into a molten Ga bath at 40–50°C producing a metastable b.c.c. supersaturated solid solution, $Nb(Al)_{ss}$, inside an Nb matrix. The heating time is ~0.1 s. The $Nb(Al)_{ss}$ has excellent ductility and the strand can consequently be mechanically clad or cabled with Cu at this stage and the stabilized strand may also be wound into device form. The $Nb(Al)_{ss}$ is then transformed to Nb_3Al at 800°C. Both the critical temperature, T_c, of ~17.8 K and the critical magnetic field, B_{c2}, of ~26 T are 0.7 K and 4 T lower for RHQT composites than the highest values obtained by laser or electron beam irradiation processed Nb_3Al (Takeuchi *et al.* 2000) and this difference has been attributed (Kikuchi *et al.* 2001a) to stacking faults formed in the A15 phase. Two successful modifications to the basic RHQT technique specifically address the stacking fault issue. In the double rapidly heating/quenching technique (DRHQ), the standard RHQT process is followed, but instead of a conventional low temperature (<1000°C) transformation heat treatment, a second RHQ process produces the A15 phase and this is followed by a long-range ordering heat treatment at 800°C for ~12 h (Kikuchi *et al.* 2001b). The T_c for these stacking fault free strands has been measured at 18.4 K, which suggests very good stoichiometry. Furthermore, a critical current density of 135 A mm^{-2} was obtained at 25 T (Kikuchi *et al.* 2002). A disadvantage of this technique is that the brittle A15 phase is formed during high-temperature processing. Alternatively, the transformation-heat-based up-quenching technique (TRUQ), follows the initial RHQ of the RHQT process with a short heating step to 1000°C. At this temperature, the heat released by local transformation from the $Nb(Al)_{ss}$ to the A15 phase produces a rapid local heating and propagation of the reaction along the length of the strand. As the self-heating is localized to the A15 volume, the heat is quickly dissipated after reaction across the Nb and stabilizer volumes, so that external Cu stabilizer applied after the initial RHQ process is not melted (Takeuchi *et al.* 2000).

3.4 *HTS Superconductors*

HTS superconductors have demanded great interest because of their ability to superconduct with relatively inexpensive liquid nitrogen cooling and their impressive high-field performance. HTS wires and tapes have not achieved the same commercial impact as LTS strands, however, because of the relative difficulty of producing long, low-cost lengths of conductor. The primary sources of this difficulty are the anisotropy in their superconducting behavior and the barrier to current flow presented by high-angle grain boundaries in these materials. Consequently, the strands must be fabricated with a remarkable degree of grain alignment. Further complicating the process is a high sensitivity to heat treatment temperature (<1°C control must be exerted across the entire strand length for Bi-2212) and a need for oxygen diffusion into the HTS composite during fabrication. This second requirement effectively means that silver must be used as the superconducting strand stabilizer. Of the HTS superconductors, only Bi-2212 can be readily made into round-wire form strand. All the other HTS superconductors require a tape format. $(Bi,Pb)_2Sr_2Ca_2Cu_3O_x$ based conductors have been shown to be capable of application to superconducting power cables, magnetic energy-storage devices, transformers, fault current limiters and motors, but widespread application will require significant cost reduction in processing. YBCO-based superconductors have the highest recorded critical current densities, but at present there is no viable wire production technique available for YBCO and critical current densities fall rapidly with superconductor thickness above 0.5 μm and with length beyond 1 μm.

(*a*) *Bi-2223.* $(Bi,Pb)_2Sr_2Ca_2Cu_3O_x$ ("Bi-2223") is most readily fabricated using PIT techniques very similar to those used for Nb_3Sn, with Ag replacing the Cu tubes used for Nb_3Sn. In addition to standard wire drawing techniques, groove rolling is also incorporated to bring powder-filled monofilaments to restack size. Hot extrusion can then be used to bond the multifilamentary composite while increasing the powder density and reducing the billet diameter. Extrusion is followed by further drawing and groove rolling, and finally rolling into tape. As a conventionally mechanically processed material, this is readily scalable to large production quantities. The key issue is to increase the connectivity between the grains of the conductor. The prospect of two- to fivefold improvements in the performance of the present process, coupled to cost reductions triggered by significant scale-up and alternative routes such as overpressure processing (Rikel *et al.* 2001) or melt processing (Flükiger *et al.* 2001) could accelerate the early penetration of superconducting technology into the utility network.

Demonstration HTS transmission cable projects have utilized Bi-2223 strand including test installations at Detroit Edison and Southwire in the USA. For the Detroit–Edison project, a conductor was fabricated by American Superconductor that used four layers of Ag-sheathed Bi-2223 tapes (29 km in total), each tape having an average critical current of ~118 A d.c. at 77 K. The tapes utilized oxide-dispersion-strengthened silver matrices as well as stainless steel reinforcement added to both sides of the

HTS tape (Masur *et al.* 2001). For the Southwire project, 30 m cables were produced from PIT Bi-2223 tapes manufactured by IGC. The measured critical current for the cables was 2980 A at 77 K (Sinha *et al.* 2001).

(*b*) *MgB_2*. Since its discovery in early 2001, MgB_2 has generated much interest as it has the highest T_c (39 K) of any intermetallic superconductor and has the lowest cost components. Like Bi-2223, MgB_2 wires or tapes can use essentially the same technology as that used for Nb_3Sn PIT composites. PIT offers a scaleable technology for future production of kilometer-length MgB_2 and has already been exploited for strand fabrication (Jin *et al.* 2001, Grasso *et al.* 2001). The perpendicular irreversibility field is less than that for Nb_3Sn, the most widely used intermetallic superconductor, and this may restrict MgB_2 to applications of lower field.

4. Summary

Nb–Ti and Nb_3Sn superconductors are well established as commercial superconductors with stable markets for NMR and MRI devices as well as high magnetic field testing equipment. Future markets in fusion reactors, accelerator magnets, and energy storage drive further developments. The strain tolerance of Nb_3Al should maintain interest in this material for high-field applications where conductor cost is not an issue. The ability to use liquid nitrogen as a coolant for HTS conductors, such as Bi-2223, is important for applications such as power transmission lines that are not well suited for low temperature refrigeration. Further performance improvement and cost reduction will be required before HTS strands are economically viable. MgB_2 offers the potential of a low-cost superconductor for the future that, while not being able to use liquid nitrogen cooling, is still able to benefit from reduced refrigeration costs due to its high T_c.

Bibliography

Cave J R, Fevrier A, Verhaege T, Lacaze A, Laumond Y 1989 Reduction of AC loss in ultra-fine multifilamentary NbTi wires. *IEEE Trans. Magn.* **25**, 1945–8

Cheng-ren Li, Xiao-zu Wu, Nong Zhou 1983 NbTi superconducting composite with high critical current density. *IEEE Trans. Magn.* **19**, 284–7

Flükiger R, Giannini E, Lomello-Tafin M, Dhallé M, Walker E 2001 Phase formation in Bi,Pb(2223) tapes. *IEEE Trans. Appl. Superconduct.* **11**, 3393–8

Ghosh A K, Sampson W B, Gregory E, Kreilick T S 1987 Anomalous low field magnetization in fine filament NbTi conductors. *IEEE Trans. Magn.* **23**, 1724–7

Grasso G, Malagoli A, Ferdeghini C, Roncallo S, Braccini V, Siri A S, Cimberle M R 2001 Large transport critical currents in unsintered MgB_2 superconducting tapes. *Appl. Phys. Lett.* **79**, 230–2

Gregory E, Kreilick T S, Wong J, Ghosh A K, Sampson W B 1987 Importance of spacing in the development of high current densities in multifilamentary superconductors. *Cryogenics* **27**, 178–82

Hlasnik I, Takacs S, Burjak V P, Majoros M, Krajcik J, Krempasty L, Polak M, Jergei M, Korneeva T A, Mironova O N, Ivan I 1985 Properties of superconducting NbTi superfine filament composites with diameter $0.1 \leq \mu$m. *Cryogenics* **25**, 558–64

Hosono F, Iwaki G, Kikuchi K, Ishida S, Ando T, Kizu K, Miura Y, Sakasai A 2002 Production of a 11 km long jelly roll processed Nb_3Al strand with high copper ratio of 4 for fusion magnets. *IEEE Trans. Appl. Superconduct.* **12**, 1037–40

Iijima Y, Kosuge Y M, Takeuchi T, Inoue K 1997 Nb_3Al multifilamentary wires continuously fabricated by rapid-quenching. *Adv. Cryogenic Eng.* **42**, 1447–54

Jin S, Mavoori H, Bower C, van Dover R B 2001 High critical current in iron-clad superconducting MgB_2 wires. *Nature* **411**, 563–5

Kanithi H, Erdmann M, Valaris P, Krahula F, Lusk R, Gregory E, Zeitlin B 1990 A status report on the development of inner and outer conductors for the SSC dipole and quadrupole magnets. In: McAshan M (ed.) *Supercollider 2*. Plenum, New York, pp. 601–9

Kato T *et al.* 2001 First test results for the ITER central solenoid model coil. *Fusion Eng. Design* **56–57**, 59–70

Kikuchi A, Iijima Y, Inoue K 2001a Microstructures of rapidly-heated/quenched and transformed Nb_3Al multifilamentary superconducting wires. *IEEE Trans. Appl. Superconduct.* **11**, 3615–18

Kikuchi A, Iijima Y, Inoue K 2001b Nb_3Al conductor fabricated by DRHQ (double rapidly-heating/quenching) process. *IEEE Trans. Appl. Superconduct.* **11**, 3968–71

Kikuchi A, Iijima Y, Inoue K, Kosuge M 2002 New Nb_3Al conductor made by DRHQ process. *Adv. Cryogenic Eng.* **48**, 1025–33

Kiyoshi T, Sato A, Takeuchi T, Itoh K, Matsumoto S, Ozaki O, Wada H, Yoshikawa M, Kamikado T, Ito S, Miki T, Hase T, Hamada M, Hayashi S, Kawate Y, Hirose R 2001 Development and operation of superconducting NMR magnet beyond 900 MHz. *IEEE Trans. Appl. Superconduct.* **11**, 2347–50

Kreilick T S, Gregory E, Scanlan R M, Ghosh A K, Sampson W B, Collings E W 1988 Reduction of coupling in fine filamentary Cu/NbTi composites by the addition of manganese to the matrix. *Adv. Cryogenic Eng.* **34**, 895–900

Lee P J, Larbalestier D C 1993 An examination of the properties of SSC Phase II R and D strands. *IEEE Trans. Appl. Superconduct.* **3**, 833–41

Lee P J, McKinnell J C, Larbalestier D C 1989 Progress in the understanding and manipulation in high J_c Nb–Ti alloy composites. In: Murakami Y (ed.) *Proc. New Developments in Applied Superconductivity*. World Scientific Press. pp. 357–62

Lee P J, McKinnell J C, Larbalestier D C 1990 Restricted novel heat treatments for obtaining high J_c in Nb–46.5 wt.%Ti. *Adv. Cryogenic Eng. (Mater.)* **36**, 287–94

Lefranc G, Muller A 1976 Effect of copper additions to superconducting niobium–tin sinter material. *J. Less Common Metals* **45**, 339–42

Lindenhovius J L H, Hornsveld E M, den-Ouden A, Wessel W A J, ten-Kate H H J 2000 Powder-in-tube (PIT) Nb_3Sn

conductors for high-field magnets. *IEEE Trans. Appl. Superconduct.* **10**, 975–8

Masur L, Parker D, Tanner M, Podtburg E, Buczek D, Scudiere J, Caracino P, Spreafico S, Corsaro P, Nassi M 2001 Long length manufacturing of high performance BSCCO–2223 tape for the Detroit Edison Power Cable Project. *IEEE Trans. Appl. Superconduct.* **11**, 3256–60

Meingast C, Lee P J, Larbalestier D C 1989 Quantitative description of a high J_c Nb–Ti superconductor during its final optimization strain: I. Microstructure, T_c, H_{c2} and resistivity. *J. Appl. Phys.* **66**, 5962–70

Ochiai S, Uehara T, Osamura K 1986 Tensile strength and flux pinning force of superconducting Nb_3Sn compounds as a function of grain size. *J. Mat. Sci.* **21**, 1020–6

Patnaik S, Cooley L D, Gurevich A, Polyanskii A A, Jiang J, Cai X Y, Squitieri A A, Naus M T, Lee M K, Choi J H, Belenky L, Bu S D, Letteri J, Song X, Schlom D G, Babcock S E, Eom C B, Hellstrom E E, Larbalestier D C 2001 Electronic anisotropy, magnetic field-temperature phase diagram and their dependence on resistivity in *c*-axis oriented MgB_2 thin films. *Supercond. Sci. Technol.* **14**, 315–9

Poole C P, Farach H A 2000 Tabulations and correlations of transition temperatures of classical superconductors. *J. Superconduct.* **13**, 47–60

Rikel M O, Williams R K, Cai X Y, Polyanskii A A, Jiang J, Wesolowski D, Hellstrom E E, Larbalestier D C, DeMoranville K, Riley G N Jr. 2001 Overpressure processing Bi2223/Ag tapes. *IEEE Trans. Appl. Superconduct.* **11**, 3026–9

Sakamoto H, Higuchi M, Endoh S, Kimura A, Wada K, Meguro S, Ikeda M 2000 Very high critical current density of bronze-processed $(Nb,Ti)_3Sn$ superconducting wire. *IEEE Trans. Appl. Superconduct.* **10**, 971–4

Scanlan R M, Fietz W A, Koch E F 1975 Flux pinning centres in superconducting Nb_3Sn. *J. Appl. Phys.* **46**, 2244–9

Scanlan R, McInturff A D, Taylor C E, Caspi S, Dell'Orco D, Higley H, Gourlay S, Bossert R, Brandt J, Zlobin A V 1997 Design and fabrication of a high aspect ratio cable for a high gradient quadrupole magnet. *IEEE Trans. Appl. Superconduct.* **7**, 936–8

Sinha U K, Lindsay D T, Hughey R L Jr., Stovall J P, Gouge M J, Lue J W, Haldar P, Selvamanickam V, Vo N 2001 Development and test of world's first industrial high temperature superconducting (HTS) power cable. *Power Eng. Soc. Winter Meeting, IEEE* **2**, 442–7

Smathers D B, Leonard D A, Kanithi H C, Hong S, Warnes W H, Lee P J 1996 Improved niobium 47 wt.% titanium composition by iron addition. *Mater. Trans.* (Japanese Institute of Metals) **37**, 519–26

Stekly Z J J, Zar J L 1965 Stable superconducting coils. *IEEE Trans. Nucl. Sci.* **12**, 367–72

Takeuchi T, Banno N, Fukuzaki T, Wada H 2000 Large improvement in high-field critical current densities of Nb_3Al conductors by the transformation-heat-based up-quenching method. *Supercond. Sci. Technol.* **13**, L11–4

Warnes W H, Larbalestier D C 1986 Critical current distributions in superconducting composites. *Cryogenics* **26**, 643–53

Wilson M N 1972 Filamentary composite superconductors for pulsed magnets *Proc. 1972 Appl. Superconduct. Conference, IEEE*. pp. 385–8

Yamada Y, Ayai N, Takahishi K, Sata K, Sugimoto M, Ando T, Takahashi Y, Nishi M 1994 Development of Nb_3Al/Cu Multifilamentary Superconductors. *Adv. Cryogen. Eng.* **40**, 907–14

T

Targets for Inertial Confinement Fusion

A. V. Hamza

1. Introduction

Demonstration of thermonuclear ignition and gain on a laboratory scale is one of science's grand challenges. The National Ignition Facility (NIF) is committed to achieving inertial confinement fusion (ICF) by 2010. Success in this endeavor depends on four elements: the laser driver performance, target design, experimental diagnostics performance, and target fabrication and target materials performance. This article discusses the current state of target fabrication and target materials performance. The first three elements will only be discussed insofar as they relate to target fabrication specifications and target materials performance. Excellent reviews of the physics of ICF are given by Lindl (1998) and Lindl *et al.* (2004).

To achieve conditions under which inertial confinement is sufficient to achieve thermonuclear burn, an imploded fuel capsule is compressed to conditions of high density and temperature. In the laboratory a driver is required to impart energy to the capsule to effect an implosion. There are three drivers currently being considered for ICF in the laboratory: high-powered lasers, accelerated heavy ions, and x-rays resulting from pulsed power machines. Of these, high-powered lasers provide the most symmetric drive and allow the most control over the pulse shape.

Laser drive operates in two configurations. The first is direct drive where the laser energy impinges directly on the ICF capsule and drives the implosion. The second is indirect drive, where the energy from the laser is first absorbed in a high-Z enclosure or hohlraum surrounding the capsule, and the resulting x-rays emitted by the hohlraum material drive the implosion (Z is the atomic number).

Using direct drive the laser beam energy is absorbed by the electrons in the outer corona of the target. The electrons transport the energy to the denser shell region to provide the ablation and the resulting implosion. Ablation due to laser direct drive is generally less efficient and more hydrodynamically unstable than the x-ray driven ablation of indirect drive. The symmetry of the implosion depends sensitively on the balance of the intensity of the individual beams driving the target. Variations in intensity imprint perturbations on the target that are amplified by hydrodynamic instabilities.

Indirect drive is less efficient at coupling energy to a capsule than direct drive because of the conversion of laser energy to x-rays in the hohlraum. However, indirect drive is less sensitive to variations in beam intensity and hydrodynamic instabilities. The ignition threshold for directly driven and indirectly driven targets is about the same. However, the gain is calculated to be about a factor of 2 greater in directly driven targets.

Accelerated heavy ion driven ICF is also an indirect-drive method. The heavy ions impinge on a radiation converter that generates x-rays in a hohlraum. The physics of heavy ion driven ICF is, thus, much the same as for indirect drive.

Z-pinch driven x-ray sources, such as the Z machine (Mehlhorn 2003) at Sandia National Laboratories also can drive ICF capsules. Hammer *et al.* (1999) describe a design for inertial confinement fusion in the laboratory using the Z machine. The Z machine is capable of producing a power pulse of 20 MA and produces 2 MJ of x-rays (290 TW) by passing the current through thin Z-pinch wires. The Z-pinch wires are arranged in a cylindrical array. The axial current produces an azimuthal magnetic field and a radial $\mathbf{j} \times \mathbf{B}$ force that implodes the pinch. The pinch collides with itself at the cylindrical axis and is observed to efficiently convert kinetic and magnetic energy into x-rays. Again, since the x-rays generated drive the capsule, the physics is much the same as for indirect laser-driven capsules.

In order to achieve conditions for ICF, targets have a spherical shell filled with a low density ($\leqslant 1\ \mathrm{mg\,cm^{-3}}$) equimolar mixture of deuterium and tritium (DT) gas. The spherical shell consists of an outer ablator and an inner region of frozen or liquid DT. Energy from the driver is delivered to the ablator, which heats up and expands. As the ablator expands the rest of the shell is forced inward to conserve momentum. The capsule behaves as a spherical, ablation-driven rocket. As the capsule implodes, $P\,\mathrm{d}V$ work heats a central region. Electron conduction and radiative losses act to cool the central region. Fuel convergence ratios of 30–40 and a central fuel temperature of 10 keV ($\sim$110 million K) are required so that α-particle deposition from thermonuclear burn of DT can overcome conduction and radiative losses and a self-sustaining burn wave can be generated. An asymmetric implosion will convert less of the available energy into compression. Assuming the available energy is such that a 25% variation in symmetry is tolerable at peak fuel compression, then $<1\%$ variation in symmetry is acceptable in the precompressed capsule.

Hydrodynamic instabilities, such as the classical Rayleigh–Taylor fluid instability, set the upper limit of the shell compression. Hydrodynamic instabilities set the achievable implosion velocity, which in turn defines the energy required for ignition. In addition, during compression the hydrodynamic instabilities cause mixing of the cold fuel with the central, hot region. A widely used formula for the growth of Rayleigh–Taylor perturbations is given by (Lindl 1983):

$$\gamma = \sqrt{\frac{\text{At} \times k \times a}{1 + \text{At} \times k \times L}} - \beta \times k \times v_a \qquad (1)$$

where γ is the growth rate, At is the Atwood number, k is the mode wave number of the perturbation, a is the acceleration, L is the density gradient length scale in the ablation front, β is a constant between 1 and 3, and v_a is the ablation velocity. The Atwood number is the dimensionless density difference between the two interacting fluids, $\text{At} = (\rho_1 - \rho_2)/(\rho_1 + \rho_2)$. In indirect-drive experiments β is ~ 1 and in direct-drive experiments β is ~ 3. However, since the ablation velocity is approximately a factor-of-10 higher for indirect drive versus direct drive, the Rayleigh–Taylor growth rate is lower for indirect-drive experiments. The growth factors are typically between 200 and 1000. So, for a 100 μm thick ablator, perturbations on the surfaces must be less than 100 nm. As discussed below, since there are more than one interface, the initial perturbation must be significantly less than 100 nm in practice.

2. *Target Design*

The NIF baseline target design is a spherical cryogenic capsule consisting of DT gas, solid DT shell, and a low-Z ablator shell (see Fig. 1). The cryogenic capsule is enclosed in a cylindrical high-Z hohlraum with two laser entrance holes at each end (see Fig. 2).

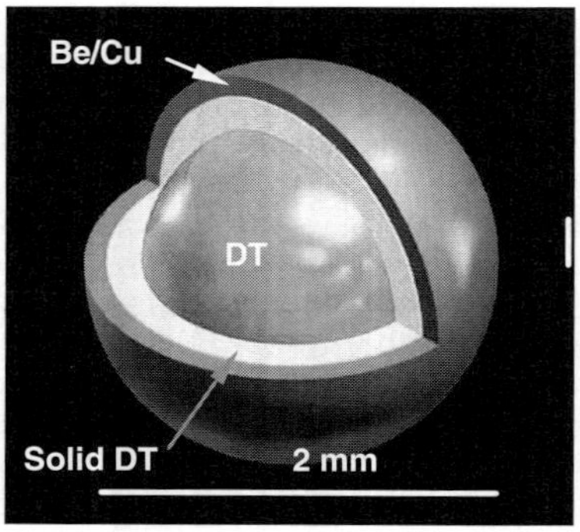

Figure 1
Schematic of cryogenic ICF target.

Three possible ablators are currently under evaluation: Br-doped CH, Cu-doped Be, and polyimide ablators. "CH" is a carbon–hydrogen polymer CH_n with $n \sim 1.3$. The dopants, Br and Cu, are added for x-ray "preheat" control. The high-energy tail of the x-rays generated by the laser drive impinging on the hohlraum walls can penetrate the ablator and "preheat" the fuel. The higher Z dopants can absorb these x-rays. However, varying the dopant concentration radially in the ablator has the potential benefit of increasing the length scale over which the density gradient changes to retard instability growth (denominator in Eqn. (1)).

As mentioned above, the capsules, and hence ablators, must be symmetrically fabricated. A 25% deviation from spherical at peak implosion might be tolerable with sufficient energy absorbed by the capsule to still reach ignition. A 25% deviation at peak implosion (convergence of 30–40) corresponds to a 1% deviation from spherical as fabricated.

Radiation-hydrodynamic simulations of instability growth on all three ablator designs have been performed. While all interfaces contribute to the performance of the capsule, perturbations on the outer surface of the ablator and the inner surface of the DT ice are expected to most sensitively affect the performance. These calculations predict that CH + Br ablator performance will be adequate with a surface root mean square (rms) roughness of 15 nm or less with a smooth DT ice layer ($<0.5\,\mu$m rms roughness). The polyimide and Be + Cu ablator will perform adequately with surface rms roughness of 50 nm or less with a smooth DT ice layer ($<1\,\mu$m rms roughness). Be has a lower albedo (albedo is the fraction of light reflected by a body) than CH and, thus, Be absorbs energy more efficiently leading to higher ablation velocities and hence lower instability growth rates. Because of the low Z of these ablators, the hot coronas do not absorb energy very well, which improves the efficiency of the ablation.

The implosion is very sensitive to nonuniformities in the density and opacity of the ablator. As an estimate of the sensitivity, a 30 nm roughness on the surface for a $\sim 150\,\mu$m thick ablator corresponds to a 1 in 5000 variation in areal density. In addition, the implosion is very sensitive to shock timing during the drive. The fuel entropy should be kept as low as possible to allow for the highest compressibility. Keeping the fuel entropy sufficiently low requires careful timing of four successive shocks to bring the fuel to optimal compression. The timing of the first shock depends on its propagation through the ablator. The sound speed is known to depend on the crystal orientation in Be. Thus, the grain structure of the Be may perturb the first shock. A nonuniform columnar grain structure may adversely affect performance.

All the single-shell ignition targets require that the bulk of the fuel reside in a cryogenic layer on the inside surface of the ablator. For smooth ablators

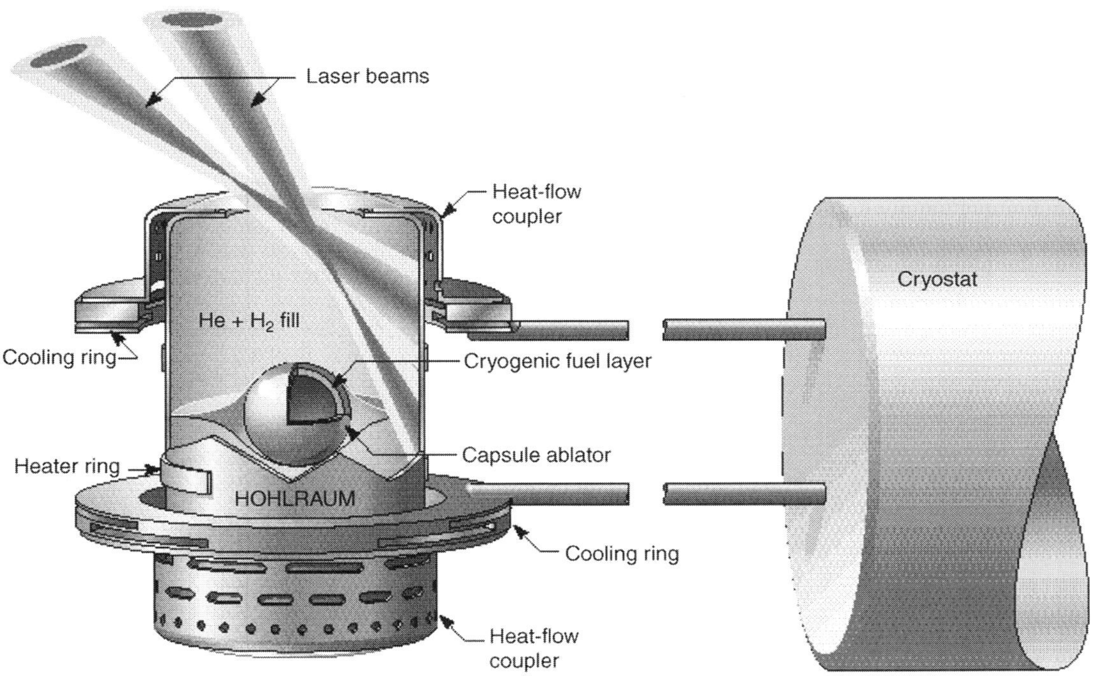

Figure 2
Baseline ICF target inside a cryogenic hohlraum.

(<15 nm rms), simulations predict adequate performance for a CH ablator with the DT ice-roughness $<0.5\,\mu$m rms, a polyimide ablator with the DT ice-roughness $<1\,\mu$m rms, and a Be ablator with the DT ice-roughness $<2\,\mu$m rms.

Bulk heating of the DT by β-decay of the tritium provides a technique to produce smooth DT ice layers. If the ablator is held at a uniform temperature, β-decay will cause thick regions to be at a higher temperature than thinner regions. Sublimation of DT from the hotter regions will produce a DT ice layer of nominally uniform thickness. DT deposits usually as small crystallites. β-layering does not eliminate discontinuities at the boundaries of crystallites, resulting in about $1\,\mu$m rms roughness at the triple point of DT.

β-layering can be improved upon by external heating. Optical heating that couples to rovibrational transitions in DT and radio-frequency heating that couples to free electrons produced by β-decay have been shown to produce smoother DT layers (Collins *et al.*). The optical technique requires transparent ablator shells and the radiofrequency technique requires nonconducting ablator shells. Unfortunately, Be ablators are opaque both in the visible and infrared and conducting.

Low-density foam could be used as a matrix for the cryogenic fuel layer. A foam-filled fuel layer has a higher ignition temperature, which depends on foam density and material composition. Important properties such as DT wicking in nanoporous materials are not known.

The double-shell ICF design is an alternative to the single-shell cryogenic design. Figure 3 shows a double-shell target. The original design advantage of the double shell depended on velocity multiplication. During implosion the massive outer shell collides with the much lighter inner shell and accelerates it to a higher velocity. In the limiting case where the inner-shell mass is negligible compared to the mass of the outer shell and the collision is elastic, the inner shell can reach twice the velocity of the outer shell. Recent simulations have shown, however, that there is not a significant velocity multiplication when the shells collide. There are other advantages that still make the double-shell design interesting. Since volume ignition, used in the double-shell design, requires a lower velocity than "hot spot" ignition in a single-shell design, the double-shell design may ignite at lower absorbed energy than a single-shell design (Amendt *et al.* 2002). The inner shell must be high Z for efficient high-density compression of the fuel contained within it. The high density of the inner shell makes it able to withstand a high-pressure shock and still compress the fuel. The high-density shell inertially confines the fuel for longer time creating a more efficient burn. Also, because the inner shell is high Z, the shell contains the radiation emission from the fuel as it compresses and

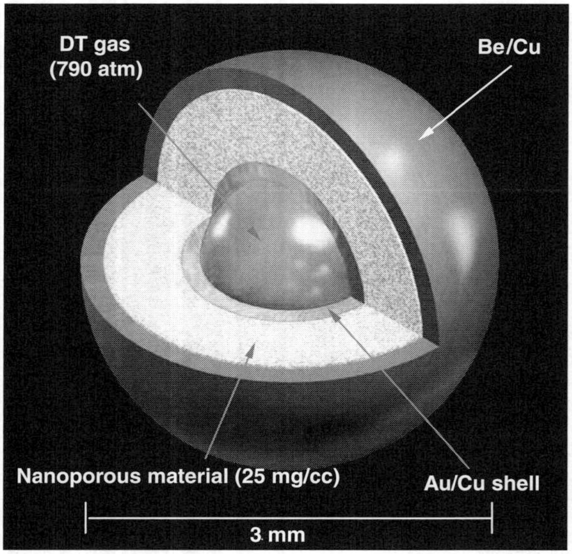

Figure 3
Schematic of double-shell ICF target.

further lowers the ignition threshold. However, the potential advantages of inertial and radiation containment ignore the effects of hydrodynamic instability during the collision of the two shells. The massive, low-density outer shell is pushing on a lighter, high-density inner shell. This is a classical Rayleigh–Taylor unstable situation. Without stringent control of this interface during collision the design entails significant risk. In addition, the smaller amount of fuel contained in the noncryogenic double-shell design makes it a lower gain target.

The inner shell has stringent requirements. In order to keep its mass small compared to the outer shell it must be a thin shell. The thin inner shell must be sufficiently strong to contain ~1000 atmospheres pressure of fuel. For these dimensions (~650 μm diameter and ~50 μm thickness) a tungsten shell has been demonstrated by Los Alamos National Laboratory to have the requisite strength (~700 MPa) (Varnum). As mentioned above, the inner shell must be high Z to contain the radiative emission and have high density to withstand the high-pressure shock during the collision of the shells. Recent designs of double-shell targets have concentrated on control of the hydrodynamic instability that occurs during the collision of the shell. In the double shell the design cannot take advantage of ablative stabilization (the negative term in Eqn. (1)). However, the double-shell design can take advantage of an increase in the length scale over which the density gradient occurs during the collision (the denominator in Eqn. (1)). This requires producing an inner shell with a density gradient from high density to a lower density at the outer surface of the inner shell. The current design calls for an inner shell with pure Au at the inner surface of the inner shell and grading the density by alloying the Au with increasing amounts of Cu until the outer surface of the inner shell is pure Cu.

In order for the double-shell ICF target to perform adequately the two shells must be concentric to within 3–5 μm (Amendt 2002). In order to minimally impede the outer shell collision with the inner shell, there should be as little density as possible between the shells. The material between the shells must be uniform on the length scales that can seed instabilities (~1 to 100 μm). The double-shell design calls for an ultralow density (25–50 mg cm^{-3}) material with sub-500 nm pore sizes with composition to match the outer surface of the inner shell (Cu) and to support the inner shell concentrically within the outer shell. Recently performed experiments on the OMEGA laser at the University of Rochester with double-shell targets measured neutrons from the DD reaction demonstrating efficient compression of the D_2 gas (Amendt *et al.* (in press)). The target materials in this experiment were a Ge-doped CH outer shell, a CH inner shell containing 50 atmospheres of D_2 gas, and a carbonized resorcinol formaldehyde foam with 50 mg cm^{-3} density and sub-100 nm pores. Double-shell capsules with poorly joined outer-shell hemispheres, ultralow density material with greater than 1 μm pore sizes, or stalks or tents to position the inner shells have not demonstrated neutron yield from compression. Some double-shell designs have used formvar polymer "tents" to position the inner shell in the center of the outer shell. Other double-shell designs have used a stalk to hold the inner shell at the center of the outer shell.

Hohlraums for indirect drive experiments are generally cylindrical with laser entrance holes on each side. Gold hohlraums are made in a three-step process. First a mandrel (Cu) is machined to shape, usually by single point diamond turning. The mandrel is coated by electroplating with gold and the mandrel is finally etched away.

The opacity and albedo properties of the hohlraum material are critical to its performance. Any single material will have "holes" (lower opacity regions) in the photon energy region of interest. Thus, cocktail hohlraums have been proposed which use a combination of materials to fill opacity holes and reduce wall energy losses. For example, calculations show Gd opacity complements Au opacity in the 0.1–10 keV photon energy range (Orzechowski *et al.* 1996). The Au/Gd hohlraum can increase indirect-drive efficiency by reducing the wall losses of the hohlraum by 15%.

Hohlraum efficiency can also be improved by reducing wall motion during the radiation of energy in the hohlraum. Energy that goes into ablating material from the hohlraum wall is energy that is not absorbed by the capsule. One way to reduce the fraction of energy that goes into the wall motion is to reduce the density of the holhraum wall, but maintain the atomic number. Thus, a high-Z low-density foam liner could improve the hohlraum coupling efficiency to the capsule. Simulations have shown that a porous gold liner with density of 200–500 mg cm^{-3} could improve the hohlraum coupling by ~15% (Rosen and Hammer (in press)). At 1.8 MJ of laser energy (NIF), a 15% reduction in wall losses corresponds to nearly 100 kJ in energy savings.

3. *Target Materials*

Target materials for ICF consist of two types: materials for shells and nanocellular materials. Microshells are the staple of targets for ICF. Today most capsules are polymer or glass microshells.

Many of the capsules produced for laser-driven ICF experiments are made using a depolymerizable mandrel. The poly(α-methyl-styrene) (PAMS) mandrel is produced by microencapsulation. The immiscibility of oil and water is used to produce liquid shells of polymer material dissolved in the hydrophobic solvent. Upon proper curing and drying, solid polymer shells are formed. Surface tension forces the shells to have excellent symmetry and the immiscibility creates shells with good wall uniformity (<5%) with diameters from 200 μm to 5 mm. A wall uniformity of 1% is required (see Sect. 2). Since many shells are easily produced the best shells can be selected for experiments. Wall thicknesses are typically limited to 10–20% of the diameter size and wall thickness uniformity becomes worse as the wall thickness is increased.

Glow discharge polymer (GDP) or plasma polymer shells are produced by overcoating PAMS shells and then decomposing the PAMS mandrel. Depending on the plasma conditions and process gas(es) used, the GDP coating can be composed of various doped CH-based materials. The materials can be deposited in various combinations to produce shells containing several different layers. Glass capsules can be made by two techniques. The first technique involves a silicon-doped glow discharge polymer coating (Hoppe 2002). The silicon-doped plasma polymer is subsequently oxidized in air, converting the silicon into SiO_2 and removing carbon and hydrogen by their conversion to CO_2 and water. These capsules have good wall uniformity (<5%) and <10 nm rms roughness. The second technique utilizes a heated drop tower.

Polyimide ablators perform intermediate to CH and Be ablators in design simulations, but since they are transparent in the optical and infrared, smoothing of the DT ice layer by coupling to rovibrational transitions can be used with polyimide ablators. An evaporative coating technique for thick polyimide ablator layers has been developed (Letts *et al.* 2004). The polyimide coating technique uses stoichiometrically controlled fluxes from two Knudsen sources containing dianhydride and a diamine to deposit polyamic acid coatings. Heating the polyamic acid coating to 300 °C converts the coating to polyimide. Coated shells are rough owing to particles and damage caused by agitation to ensure a uniform coating. A smoothing process that exposes the initially rough polyamic acid coating to solvent vapor using levitation has been developed. Vapor-smoothed coating of polyimide 160 μm thick with 20 nm rms surface roughness have been produced.

At Los Alamos National Laboratory, the Be capsule development is focused on precision machining of Be/Cu alloy hemishells and subsequent bonding of the hemishells under high-pressure DT gas (Nobile *et al.* 2004). At Lawrence Livermore National Laboratory, Be/Cu capsules are being developed using sputter deposition of Be and Cu onto polymer mandrels (Jankowski *et al.* 2002). It is desirable to produce homogeneous, isotropic, dense, and mechanically strong Be coatings. One way to develop these properties is to refine the grain size to the nanoscale. The compositional effects of Cu, Fe, and B additions on the microstructure have been studied. Fe and B are found to have a dominant effect on grain refinement, capable of producing sub-20 nm grains. Research is also underway to develop DT filling methods for sputter-deposited capsules by laser drilling a small hole (3–5 μm) followed by laser sealing.

Be0.9at.%Cu alloy has been synthesized by blending Be powder (<75 μm particles) and copper flakes (submicron particles) and hot isostatically pressing for 4 h at 1000 °C at 103 MPa (Nobile *et al.* 2004). Equiaxed 10–50 μm grains were produced. Higher atomic number impurity at the grain boundaries is likely oxide, common in powder synthesis techniques. A second technique is arc-melting Be/Cu to eliminate oxide impurities. The as-cast Be/Cu is subsequently extruded by equal channel angular extrusion and annealed to refine grain size. Equiaxed 20 μm grains have been produced (Alexander *et al.* 2004).

Producing smooth thin capsules with sufficient strength to contain ~1000 atmospheres requires novel materials research. The Hall–Petch effect of refining grain size of metals to increase strength is well known. To achieve the strength required this effect will need to be pushed to its limits. For example, a coarse-grained Au/Cu alloy (to be used in double-shell ignition designs) has a yield strength of 60 MPa; for use as capsule material a yield strength of 450 MPa is required. To produce this increase in strength, a grain size between 10–20 nm will be required. At this grain size the empirical Hall–Petch relation breaks down. It is in this regime of grain size where the transition from standard plasticity to new deformation mechanisms occurs. There is significant controversy in the literature over the transition and the new controlling deformation mechanism (van Swygenhoven *et al.* 2003, Yamakov *et al.* 2004). Presently the required increase in strength at the nanometer scale has been demonstrated only for f.c.c. nickel and copper (Budrovic *et al.* 2004). In addition, grain refining also helps the capsules meet the smoothness requirements so as not to seed instabilities during the implosion.

Since materials with cell or pore sizes greater than 1 μm can seed instabilities, only materials with submicron cell or pore sizes are discussed.

A recent review has discussed the chemistry and application of aerogels (Pierre and Pajonk, 2002). Only specific application to ICF will be discussed here. Aerogel preparation will be discussed only briefly.

Resorcinol formaldehyde aerogel is produced by reacting resorcinol and formaldehyde. The resulting polymer gel (sol–gel) is supercritically dried with CO_2 to produce an aerogel with density in the 20–850 mg cm^{-3} range (Pekala 1989). The resorcinol–formaldehyde aerogel can be molded and used or it can be heated to 1000 °C in an inert atmosphere where it becomes an almost pure carbon foam with a pore size of sub-100 nm. Hibbard *et al.* (2004) have used carbonized resorcinol–formaldehyde foam in their double-shell ICF target construction.

Silica aerogels were the first aerogel produced; they were developed for Cherenkov radiation detectors in the 1970s (Bourdinaud *et al.* 1976). They also have sub-100 nm pore size and can be produced in monolithic form at densities from 700 to 1 mg cm^{-3} (the density of air). Silica aerogels can be machined or molded to shape in most cases. Doped silica aerogels have found application as backlighter materials for *in situ* x-ray radiography of laser targets (see below). Silica aerogels have also been used as catchers to recover shrapnel from laser irradiated targets.

Metal nanoporous materials with submicron pore or cell sizes have only been reported since the 1990s. There are two main techniques for their production: dealloying and templating. They have not yet been employed in ICF targets because materials with suitable density, pore size, and mechanical stability have not yet been produced.

Many nanoporous metals have been synthesized by dealloying (Pugh *et al.* 2003). In the dealloying technique, a homogeneous binary mixture of relatively noble metals is etched in, usually, nitric acid solution under an applied potential. The less noble metal is etched completely away over time and a nanoporous more noble metal material remains. The morphology of a typical dealloyed structure is a highly tortuous, completely interconnected porosity with a pore diameter as small as 3 nm. The structures can be coarsened to larger length scales by annealing at elevated temperatures (Li and Sieradzki 1992). Dealloying would be limited to the surface unless a mechanism exists for penetration throughout the alloy. Recently, Erlebacher and Sieradzki (2003) and Erlebacher *et al.* (2001) have discussed the mechanism for porosity to "grow" into the bulk via surface diffusion and capillary action. Figure 4 shows the results of the dealloying process for dealloying of a $Au_{0.3}Ag_{0.7}$ alloy producing nanoporous Au.

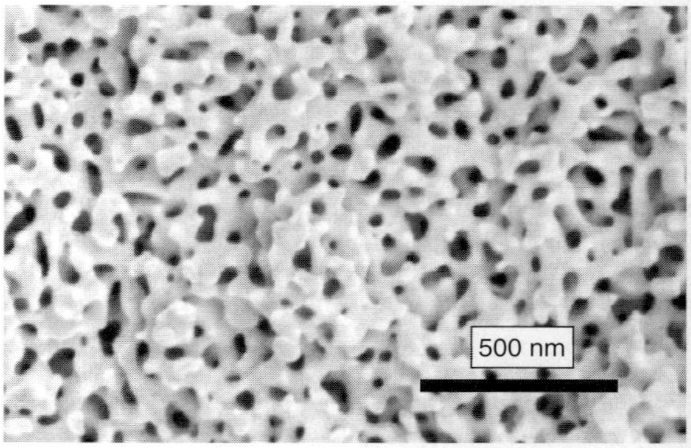

Figure 4
Scanning electron microscope image of nanoporous Au at 30% full density formed by dealloying $Au_{0.3}Ag_{0.7}$. The 500 nm fiducial is shown (image courtesy of Luke Hsuing, LLNL).

Close-packed colloidal crystals are promising precursors for novel materials. Velev *et al.* (1999) have demonstrated the formation of nanoporous Au via colloidal templating. A close-packed crystal of latex spheres was carefully infiltrated with Au nanoparticles in solution. After drying, the Au nanoparticles were sintered and the latex spheres decomposed, leaving behind a nanoporous Au with ~700 nm cell size and low density. Au nanoporous flakes of ~800 μm in diameter and a few microns thick were produced. Because the sintering of Au nanoparticles was used to form the backbone, the density (not characterized) may have been quite low. The size of the sample needs to be increased and the mechanical properties determined to make a useful material for ICF applications, such as hohlraum liners. Figure 5 shows the results of the templating process for producing porous Cu/Ti. The template was produced from 6 μm carbon spheres. The interstitial space was infiltrated with Cu/Ti and the carbon was subsequently burned out. The density and/or pore size still need to be decreased for potential ICF application.

A new class of monolithic, transition metal oxide aerogels have been developed for use in ICF experiments. The approach is straightforward, inexpensive, versatile, and it produces monolithic microporous materials with high surface areas (Gash *et al.* 2001). In this technique epoxides are used as gelation agents for the sol–gel synthesis of transition metal aerogels from simple trivalent transition metal inorganic salts. The results show that rigid monolithic metal oxide aerogels can be prepared from solutions of their respective metal ion salts (Fe^{3+}, Al^{3+}, In^{3+}, Ga^{3+}, Zr^{4+}, Hf^{4+}, Ta^{5+}, Nb^{5+}, and W^{6+}), provided the formal oxidation state of the metal ion is $\geqslant +3$.

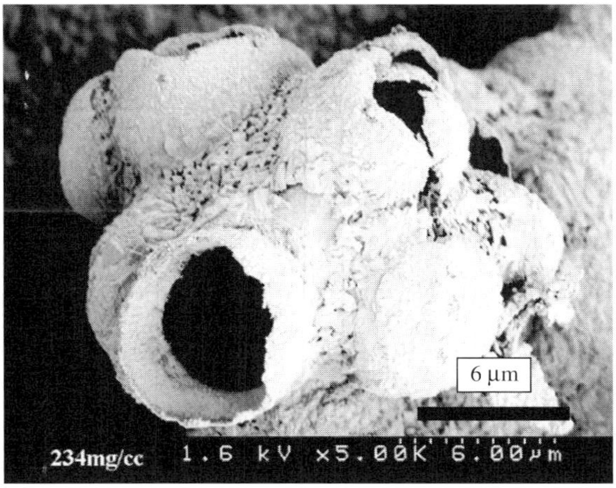

Figure 5
Scanning electron microscope image of porous Cu/Ti at 15% full density formed by templating with carbon spheres. The 6 μm fiducial is shown (image courtesy of Joe Satcher, LLNL).

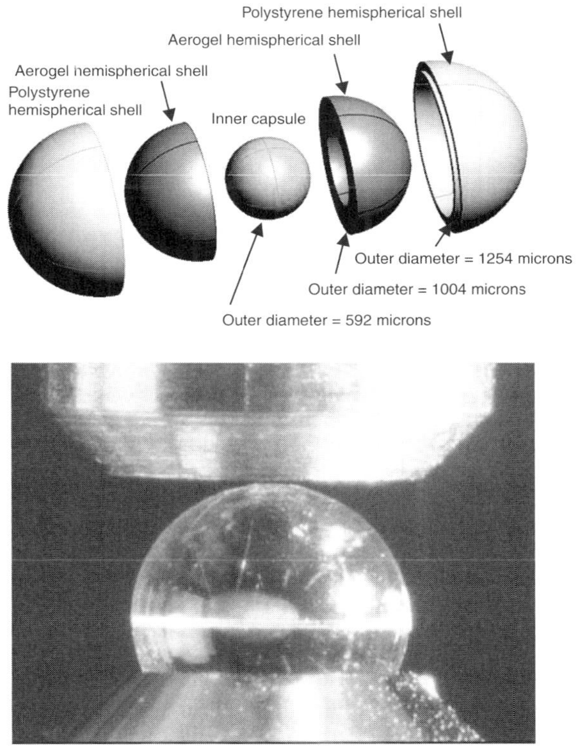

Figure 6
Exploded view of the double-shell target (left) and a completed target in the vacuum chuck holder (image courtesy of Robin Hibbard, LLNL).

Conversely, when divalent transition metal salts are used precipitated solids are the products. These transition metal aerogels can be used in double-shell ICF targets to help control the hydrodynamic instabilities. "X-ray backlighting" for *in situ*, dynamic, radiographic imaging of ICF targets is a staple diagnostic. X-ray backlighting is a method to take snapshots of an experiment. One or more of the laser beams at a facility is delayed compared to the drive beams and aimed at a "backlighter." The laser beam(s) hitting the backlighter generates x-rays that are used to radiograph the target some time, typically nanoseconds, after the drive beams have hit the target or hohlraum. The backlighter is typically a thin metallic film or a "gas bag" filled with a noble gas (Ne, Ar, or Kr).

The radiographic backlit images at useful resolution are almost always photon-starved and generating multi-keV photons for sufficient contrast has required development. Historically, solid targets, usually massive (by comparison to gas bags) disks, have demonstrated efficiencies of only fractions of a percent for multi-keV photon energies. The efficiencies of these massive targets are compromised because much of the laser energy is deposited in a (relatively) low-density region at the critical surface, while the region of the target dominating the radiation output may be a higher-density region that must be conductively heated. One hope to increase the efficiency of the solid targets is to heat volumetrically the whole target by using reduced solid density. Doped aerogels is a promising route. The doped aerogel material is formed by supercritical extraction of the solvent used to mix the dopant and silicon alkoxides. Fournier *et al.* (2004) recently measured the production of $h\nu \sim 4.7$ keV x-rays from low-density Ti-doped (3 at.%) aerogel ($\rho \sim 3$ mg cm^{-3}) targets at the OMEGA laser facility, with the goal of maximizing x-ray output. The laser fully ionizes the target ($n(\mathrm{e})/n(\mathrm{crit}) \leqslant 0.1$), and a laser-bleaching wave excites, supersonically, the high-Z emitter (Ti) ions in the sample. They found between 40 and 260 J of output with $4.67 \leqslant h\nu \leqslant 5.0$ keV.

4. *Target Assembly*

Many targets (ICF single shell, ICF double shell, Rayleigh–Taylor diagnostic targets) require complex assembly to tight tolerances (Balkey *et al.* 2004, Hibbard *et al.* 2004). Assembly of the physics package may involve many micromachining and deposition steps and precision joining of parts with nanometer surface finish and micron features. Figure 6 shows the assembly of an ICF double shell. Generally the capsules must be positioned in the hohlraum with a formvar tent. The hohlraums

Figure 7
Optical microscope image of mounted target package showing backlighter positions and alignment fiducials. The two squares, a transparent one and a black one, are both backlighters. The disk attached to the hohlraum holds one wire fiducial. The second wire is the other alignment fiducial (image courtesy of Russ Wallace, LLNL).

frequently have diagnostic holes in them for mounting backlighters and other diagnostics. The hohlraum is mounted on a stalk, usually with fiducials for laser alignment. Figure 7 is an optical micrograph of an assembled diagnostic target.

Most ICF capsules require filling with either D_2 or DT gas. The standard method for filling is permeation filling at either room or elevated temperature. Capsules that have been produced by coating depolymerizable PAMS shells (e.g., CH shells) are too permeable to hold gas at room temperature. If the α-methyl styrene can diffuse out, the DT gas will also diffuse out rapidly. Poly(vinylalcohol) layers have been used to hold DT gas in the capsules. Aluminum coatings on CH shells can also be used to retain hydrogen gases. SiO_2 glass shells are retentive and can hold more than 60 atmospheres of DT gas with more than 2 week half-life (time for half the hydrogen gas to permeate in or out of the capsule). Permeation filling of high-Z metal inner shells is also possible. The permeation rate for hydrogen through Au/Cu at 500 °C is sufficient to fill a capsule to $\sim$800 atmospheres in about one week (Shelby 1996). Unfortunately, since the strength of the shell to hold such high pressures depends on the microstructure of the metal shell, heating the shell to 500 °C will likely lead to significant grain growth in the metal and weakening of the shell.

A second method for filling is the attachment of a fill tube to the capsule. The use of fill tubes has the advantage that the capsule can be filled and cryogenically cooled and the fuel assembled into a smooth shell after insertion into the target chamber. This technique does not require the extensive cryogenic infrastructure that a prefilled and cryogenically assembled target would require. The fill tube is likely to be a silica capillary with 10 μm outer diameter and 6 μm inner diameter. The tube will have to be precision assembled in the 100–140 μm thick ablator and aligned with the 3–5 μm laser drilled hole in the ablator.

Nonuniformities in bond joints can cause asymmetries in the implosions. Ideally joints will introduce no change in material properties (density, opacity, sound velocity, and surface finish). In the recent double-shell campaign at OMEGA, great care was taken to produce a robust bond joint between the hemispheres of the outer ablator shell (Hibbard *et al.* 2004). The bonding in this case was a two-step process. In the first step, the designed joint required the density of the adhesive match the density of the polystyrene shell to within 10%. A step joint was designed so that the inner half of the step joint is mated and a 2 μm gap is intentionally available for adhesive fill. The adhesive must wick into the 2 μm gap and completely fill it. Otherwise, voids in the joint would adversely affect performance. The adhesive cannot have high-Z impurities or binders that degrade the ablator performance. Further, the adhesive should not wick past the inner step in the joint and into the nanoporous foam (carbonized resorcinol formaldehyde). The additional full density material would also degrade performance. Cyanoacrylate is applied around the perimeter of the joint by placing a droplet on the end of a camel hair. In the second step, epoxy is applied to stabilize the joint during final machining of the outer surface. The epoxy is fully machined away during this machining step. By bonding prior to final machining of the outer ablator surface, precision dispensing of adhesive and crazing of the ablator surface is avoided.

In order to prevent asymmetric shell collision in double-shell ICF targets the two shells must be concentric to within 5 μm. Figure 6 shows the components that are assembled to produce the concentric shells. Great care in the assembly is required. The precision machining produces dimensional errors of <0.3 μm. Since three machined surfaces define the concentricity, concentricity within a micron is possible. However, the fragile nature of the carbonized resorcinol formaldehyde foam introduces the additional difficulty in meeting the 5 μm concentricity requirement. Minimal force during assembly is required so that the foam will not be crushed or deformed. Hence, the mechanical strength of the foam is an important consideration.

5. Target Characterization

A complete ICF experiment requires design, target fabrication and materials performance, and diagnostics. Target fabrication is not complete until the target specifications are characterized and metrologized in the "as built" configuration. Nondestructive evaluation of the assembled target requires "seeing" inside the target. Three probes have been applied to seeing inside the targets: x-rays, protons, and acoustic waves.

X-ray imaging of target assemblies is the most advanced of the characterization technologies and can provide many of the characterization analyses that are required.

Contact radiography involves placing the object to be characterized on the x-ray photographic film and illuminating the object with x-rays. The resolution is determined by the resolution of the film, typically about 1 μm. Figure 8 shows two orthogonal contact radiographs of ICF double-shell targets. The integrity of the bond joint and the concentricity of the target can be measured. Resolution is insufficient to determine surface finish and metrology of the surface must be performed preassembly (see below).

The key strength of x-ray full-field microscopy is the large penetration depth of x-rays into matter, which allows one to image the interior of opaque objects. Combined with tomographic techniques, the three-dimensional inner structure of an object can be reconstructed without the need for difficult and destructive sample preparation (Schroer *et al.* 2004). Point projection x-ray microscopes are currently commercially available. These microscopes can typically perform tomographic imaging and reconstruction of an object. Resolution and field of view change together in these microscopes. Typically a $\sim 2\,\mu$m resolution with 0.5–1 mm field of view can be achieved. Higher resolution (0.5 μm) can be achieved with smaller fields of view. Figure 9 shows a point projection image of a portion of a mock double-shell target.

X-ray projection microscopy which utilizes phase contrast to enhance the quality and information content of images has been developed (Mayo *et al.* 2003). By combining fine x-ray source with an appropriate imaging geometry, phase contrast arising from Fresnel diffraction at edges and boundaries within the sample leads to images with greatly enhanced edge detail. Phase-retrieval algorithms have been developed for quantitative determination of the projected phase shift in the object-plane from either a single image or multiple images acquired at different energies or magnifications. Figure 9 shows a fine defect intentionally machined into the 50 mg cm^{-3} carbonized resorcinol formaldehyde foam. Because of the low density and low Z of the foam, amplitude contrast alone would be insufficient to detect this flaw. However, combining phase contrast and amplitude contrast made this few micron flaw in the foam discernible. This technique is also capable of imaging DT ice structure inside optically opaque ablators (Be).

Structural properties of nanoporous materials can be studied by small-angle x-ray scattering (Pahl *et al.* 1991). The measured SAXS profiles reveal a

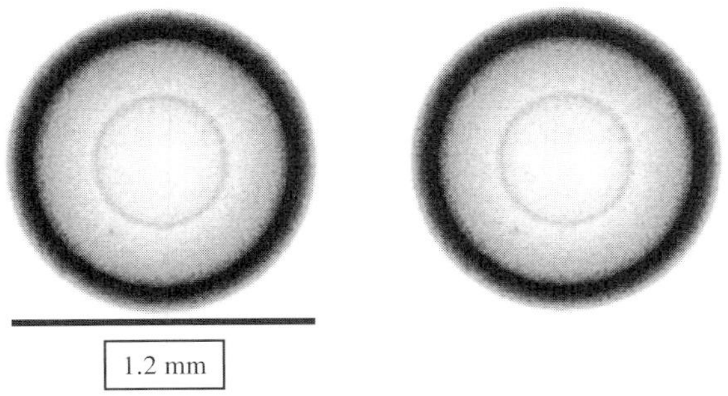

Figure 8
Contact radiograph of a double-shell target. The bond joint is visible in the side view on the left. Top view is shown on the left. Resolution of the image is $\sim 1\,\mu$m (image courtesy of Russell Wallace, LLNL).

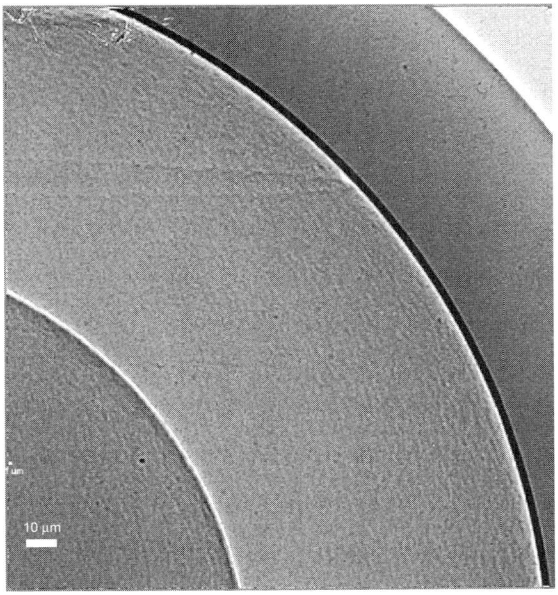

Figure 9
Phase and amplitude contrast image of a mock double-shell target taken with ~ 5 keV x-rays from a point projection microscope (image courtesy of Wenbing Yun, Xradia, Concord, CA).

scattering power depending upon synthetic conditions of the nanoporous material. Structural features such as pore distribution from 2 to 100 nm can be determined from the scattering data. Radii of gyration $R(g)$ are computed and appear to be a measure of the pore (cell) size. Since scattering can be from localized columns (submicron diameters) of material, density gradients, and spatial inhomogeneity of the pores can be measured.

The complementary techniques of ion microtomography (IMT) and particle-induced x-ray emission (PIXE) are used to provide submicron-scale characterization of inertial confinement fusion (ICF) targets for density uniformity, sphericity, and trace-element spatial distributions (Antolak *et al.* 1992). 1% total electron density determinations using IMT with spatial resolution approaching 2 μm can be obtained. Utilizing PIXE, dopant and impurity distributions with elemental detection sensitivities on the order of a few parts per million can be determined.

One potentially valuable method for evaluating the quality of ICF targets is resonant ultrasound spectroscopy (RUS). When applied to simple geometries, such as layered spheres or rectangular parallelepipeds, RUS may yield significant information about alloy homogeneity, elastic constants, cavity geometry, the presence of gross defects such as cracking or hemishell bonding problems, and properties of interior fluids. The strengths of RUS techniques for ICF target characterization include applicability at all temperatures of interest with a single apparatus, high sensitivity in frequency spectral measurements, and the inherent acoustic indifference to optically opaque samples. The acoustic cavity resonance structure can be used to calculate cavity boundary perturbation amplitudes (Asaki *et al.* 1998). Experiments have also been conducted in which the D_2 fill pressure has been determined for several closed millimeter-size aluminum and beryllium shells (Asaki 1999). The vibrational resonance frequency spectrum of the shells was used to calculate the sound velocity of the interior gas. This velocity, along with the equation-of-state, determined the gas pressure and density. The accuracy in determining the fill conditions is within 0.5% in both pressure and density for near critical density ($\rho \geqslant 9\,\mathrm{mol\,l^{-1}}$) gas.

Measuring the surface perturbation of capsules is of considerable importance. Two techniques are available for metrology of the shell surface: interferometry and atomic force profiling.

White light interferometry is used to determine thickness and uniformity of transparent shells (Weinstein 1982). Laser (holographic) interferometry is used to measure cryogenic layer uniformity (Bernat *et al.* 1982). Commercial instruments for capsule measurements are available.

A profilometer based on an atomic force microscope combined with a precision rotary air bearing has been developed (McEachern *et al.* 1995). In this instrument the shell is placed in vacuum chuck and rotated with the precision rotamary air bearing. The AFM tip on its piezo-electric mount profiles the surface to within 1 nm with cantilever deflection (force feedback—contact mode). With reorientations of the shell, a power spectrum (the square of the mode amplitude as a function mode number) for legendre modes (the ratio of the capsule circumference to the wavelength of the perturbation) from 2 to 10 000 for the entire sphere can be calculated. The measured/calculated power spectrum is used as input in implosion simulations.

Acknowledgments

Work was performed under the auspices of the US Department of Energy by Lawrence Livermore National Laboratory under Contract No. W-7405-ENG-48. The author thanks Dr. Peter Amendt and Dr. Bruce Hammel for a critical reading and suggested improvements in this article.

Bibliography

Alexander A J, Cooley J C, Thoma D J, Nobile A 2004 Production of fine grained beryllium-6wt.%Cu for fusion ignition capsules by arc melting and equal channel angular extrusion. *Fusion Sci. Technol.* **45**, 137

Amendt P, Colvin J D, Ramshaw J D, Robey H F, Landen O L 2003 Modified Bell–Plesset effect with compressibility: application to double-shell ignition target designs. *Phys. Plasmas* **10**, 820

Amendt P, Colvin J D, Tipton R E, Hinkel D E, Edwards M J, Landen O L, Ramshaw J D, Suter L J, Varnum W S, Watt R G 2002 Indirect-drive noncryogenic ignition targets for the national ignition facility: design and analysis. *Phys. Plasmas* **9**, 2221

Amendt P, Robey H F, *et al.* (in press).

Antolak A J, Pontau A E, Morse D H, Weirup D L, Heikkinen D W, Cholewa M, Bench G S, Legge G J F 1992 Ion microtomography and particle-induced x-ray-emission analysis of direct drive inertial confinement fusion-targets. *Journal of Vacuum Science & Technology A-Vacuum Surfaces and Films* **10**, 1164–9

Asaki T J 1999 Ultrasonically determined fill pressure and density in closed spherical shells. *Fusion Technology* **35**, 126–30

Asaki T J, Hoffer J K, Sheliak J D 1998 Ultrasonic characterization of inertial confinement fusion targets. *Fusion Technol.* **33**, 171–81

Balkey M M, Day R D, Batha S H, Elliot N E, Pierce T, Sandoval D L, Garrard K P, Sohn A 2004 Production and metrology of cylindrical inertial confinement fusion targets with sinusoidal perturbations. *Fusion Sci. Technol.* **45**, 107

Baumann T F, Satcher J H 2003 Homogeneous incorporation of metal nanoparticles into ordered macroporous carbons. *Chemistry of Materials* **15**, 3745–7

Bernat T P, Darling D H, Sanchez J J 1982 Applications of holographic interferometry to cryogenic ICF target characterization. *J. Vac. Sci. Technol.* **20**, 1362

Bourdinaud M, Cheze J B, Thevenin J C 1976 Use of silica aerogel for Cherenkov radiation counter. *Nucl. Instrum. Methods* **136**, 99–103

Budrovic Z, Van Swygenhoven H, Derlet P M, Van Petegem S, Schmitt B 2004 Plastic deformation with reversible peak broadening in nanocrystalline nickel. *Science* **304**, 273–6

Collins G W, Bernat T P, Mapoles E R, Duriez C *Heat Flux Induced Changes to Multicrystalline D_2 Surfaces*, LLNL Report No. UCRL-JC-124261.

Erlebacher J, Aziz M J, Karma A, Dimitrov N, Sieradzki K 2001 Evolution of nanoporosity in dealloying. *Nature* **410**, 450

Erlebacher J, Sieradzki K 2003 Pattern formation during dealloying. *Scripta Materialia* **49**, 991

Fournier K B, Constantin C, Poco J, Miller M C, Back C A, Suter L J, Satcher J, Davis J, Grun J 2004 Efficient multi-keV x-ray sources from Ti-doped aerogel targets. *Physical Review Letters* **92**, Art. No. 165005

Gash A E, Tillotson T M, Satcher J H, Hrubesh L W, Simpson R L 2001 New sol-gel synthetic route to transition and main-group metal oxide aerogels using inorganic salt precursors. *Journal of Non-Crystalline Solids* **285**, 22–8

Hammer J H, Tabek M, Wilks S C, Lindl J D, Bailey D S, Rambo P W, Roor A, Zimmerman G B, Porter J L Jr. 1999 High yield inertial confinement fusion target design for a *Z*-pinch-driven hohlraum. *Physics of Plasmas* **6**, 2129

Hibbard R L, Bono M J, Amendt P, Bennett D W, Castro C 2004 Precision manufacturing of inertial confinement fusion double shell laser targets for OMEGA. *Fusion Sci. Technol.* **45**, 117

Hoppe M L 2002 Recent developments in making glass shells from silicon doped GDP shells. *Fusion Science and Technology* **41**, 234–7

Jankowski A F, Wall M A, van Buuren A W, Nieh T G, Wadsworth J 2002 From nanocrystalline to amorphous structure in beryllium-based coatings. *Acta Materialia* **50**, 4791

Letts S A, Anthamatten M, Buckley S R, Fearon E, Nissen A E H, Cook R C 2004 Progress toward meeting NIF specifications for vapor deposited polyimide ablator coatings. *Fusion Sci. Technol.* **45**, 180

Li R, Sieradzki K 1992 Ductile-brittle transition in random porous Au. *Phys. Rev. Lett.* **68**, 1168

Lindl J D Hydrodynamic and Plasma Stability Limitations on the Choice of Laser Wavelength for Radiation Driven ICF Implosions, Laser Program Annual Report (1983), Lawrence Livermore National Laboratory, Livermore, CA, UCRL-50055-83, pp. 2–40 to 2–46 (unpublished).

Lindl J D 1998 *Inertial Confinement Fusion*. Springer, New York

Lindl J D, Amendt P, Berger R L, Glendinning S G, Glenzer S H, Haan S W, Kauffman R L, Landen O L, Suter L J 2004 The physics basis for ignition using indirect drive targets on NIF. *Phys. Plasmas* **11**, 339

Mayo S C, Miller P R, Wilkins S W, Davis T J, Gao D, Gureyev T E, Paganin D, Parry D J, Pogany A, Stevenson A W 2003 Applications of phase-contrast x-ray microscopy in an SEM. *Journal de Physique IV* **104**, 543–6

McEachern R L, Moore C E, Wallace R J 1995 Design, performance, and application of an atomic force microscope-based profilometer. *J. Vac. Sci. Technol. A* **13**, 983

Mehlhorn T A, Bailey J E, Bennett G, Chandler G A, Cooper G, Cuneo M E, Golovkin I, Hanson D L, Leeper R J, MacFarlane J J, Mancini R C, Matzen M K, Nash T J, Olson C L, Porter J L, Ruiz C L, Schroen D G, Slutz S A, Varnum W, Vesey R A 2003 Recent experimental results on ICF target implosions by *Z*-pinch radiation sources and their relevance to ICF ignition studies. *Plasma Physics and Controlled Fusion* **45** (Suppl. 12A), A325–34

Nobile A, Dropinski S C, Edwards J M, Rivera G, Margevicius R W, Sebring R J, Olson R E, Tanner D L 2004 Fabrication and characterization of targets for shock propagation and radiation burnthrough measurements on Be0.9At.%Cu alloy. *Fusion Sci. Technol.* **45**, 127

Orzechowski T J, Rosen M D, Kornbulm H N, Porter J L, Suter L J, Thiessen A R, Wallace R J 1996 The Rossland mean opacity of a mixture of gold and gadolinium at high temperature. *Phys. Rev. Lett.* **77**, 3545

Pahl R, Bonse U, Pekala R W, Kinney J H 1991 SAXS investigations on organic aerogels. *Journal of Applied Crystallography* **24**, 771–6

Pekala R W 1989 Organic aerogels from the polycondensation of resorcinol with formaldehyde. *J. Mater. Sci.* **24**, 3221

Pierre A C, Pajonk G M 2002 Chemistry of aerogels and their applications. *Chem. Rev.* **102**, 4243

Pugh D V, Dursun A, Corcoran S G 2003 Formation of nanoporous platinum by selective dissolution of Cu from $Cu_{0.75}Pt_{0.25}$. *J. Mater. Res.* **18**, 216

Rosen M D, Hammer, J. (in press).

Schroer C G, Cloetens P, Rivers M, Snigirev A, Takeuchi A, Yun W B 2004 High-resolution 3D imaging microscopy using hard x-rays. *MRS Bulletin* **29**, 157–65

Shelby J E 1996 *Handbook of Gas diffusion in Solids and Melts*. ASM International, Materials Park, OH

van Swygenhoven H, Derlet P M, Budrovic Z, Hasnaoui A 2003 Unconventional deformation mechanism in nanocrystalline metals? *Zeitschrift Fur Metallkunde* **94**, 1106–10

Varnum W S, Private Communication.

Velev O D, Tessier P M, Lenhoff A M, Kaler E W 1999 Materials - a class of porous metallic nanostructures. *Nature* **401**, 548

Weinstein B W 1982 Physical measurements of inertial confinement fusion targets. *J. Vac. Sci. Technol.* **20**, 1349

Yamakov V, Wolf D, Phillpot S R, Mukherjee A K, Gleiter H 2004 Deformation-mechanism map for nanocrystalline metals by molecular-dynamics simulation. *Nature Materials* **3**, 43–7

Thermoelectric Devices: Refrigeration and Power Generations With No Moving Parts

B. C. Sales

In a solid that conducts both heat and electricity, the thermal and electrical currents are coupled together. This thermoelectric coupling can be used to construct devices that act as refrigerators, power generators, or temperature sensors. The devices work because the electrons or holes in a conducting solid also carry heat as well as electrical charge. Therefore, the electrical carriers are the "working fluid" in a thermoelectric refrigerator. Thermoelectric devices are attractive for many applications as they have no moving parts (except electrons and holes), use no liquid refrigerant, and last indefinitely. The major disadvantage of these devices is poor efficiency.

Thermoelectrics is an old field. In 1823 Thomas Seebeck discovered that a voltage drop appears across a sample that has a temperature gradient. This phenomenon provided the basis for thermocouples used for measuring temperature and for thermoelectric power generators. In 1838 Henrich Lenz placed a drop of water on the junction of metal wires made of bismuth and antimony. Passing an electric current through the junction in one direction caused the water to freeze, and reversing the current caused the ice to quickly melt: thus thermoelectric refrigeration was demonstrated.

These thermoelectric effects remained little more than a curiosity until the 1950s when Abram Ioffe found that doped semiconductors had a much larger thermoelectric effect than did other materials (metals or insulators). He suggested that domestic refrigeration could be accomplished with semiconductors thereby igniting a worldwide frenzy of research and development activity. The early days of semiconductor physics were not driven by the prospect of microelectronics, but were motivated by thermoelectric refrigeration. This focus lasted only a few years, during which virtually every known semiconductor, semimetal, and alloy was investigated. It was found that the best materials at room temperature were Bi_2Te_3–Sb_2Te_3 alloys.

Thermoelectric (also called Peltier) refrigerators with bismuth telluride alloys, however, produce only moderate amounts of cooling, and have poor efficiencies compared to compressor-based refrigerators. As a result thermoelectric refrigerators are used mainly in applications in which reliability or convenience is more important than economy. Thermoelectric coolers can be powered by a car battery for portable beverage storage, or by a power supply to provide active cooling for computer central processing units or for infrared detectors. Thermoelectric generators, usually made with silicon–germanium alloys, are used as power sources for many of NASA's deep-space probes and also are used to convert waste heat into electricity.

This article reviews the basic physics and commercial applications of thermoelectric devices, the properties of state-of-the-art thermoelectric materials, and approaches to the atomic-level design of new thermoelectric materials with improved efficiencies.

1. Basic Thermoelectric Science

The basic thermodynamics of these devices was worked out in the mid-nineteenth century by William Thompson (Lord Kelvin) who showed the mathematical connection between heat and electrical currents in a conducting solid:

$$J = \sigma(E - S\nabla T) \text{ and } Q = (\sigma TS)E - \kappa\nabla T \quad (1)$$

where J is the electrical current density, σ the electrical conductivity, E the electric field, S the Seebeck coefficient, T the temperature, Q the heat current density, and κ the thermal conductivity. Only S causes any conceptual difficulty. It is the average entropy per charge carrier divided by the electron charge. A small temperature difference, ΔT, across any electrical conductor will generate a corresponding voltage, ΔV. If there is no electrical current passing through the sample, the ratio $\Delta V/\Delta T$ is the experimental definition of S. Solving these equations in one dimension, with the appropriate boundary conditions, leads to expressions that describe the efficiency of an idealized device in terms of the electrical and thermal transport coefficients that characterize the thermoelectric material.

The efficiency of a thermoelectric solid is found to depend on material properties through the dimensionless parameter ZT:

$$ZT = TS^2/\kappa\rho \quad (2)$$

where the electrical resistivity is $\rho = 1/\sigma$. Z is defined as the figure of merit and ZT is often referred to as the dimensionless figure of merit. For power generation the efficiency is defined as heat in divided by work out and is given by:

$$\text{Efficiency} = \frac{(T_h - T_c)(\gamma - 1)}{T_c + \gamma T_h} \quad (3)$$

where $T_c(T_h)$ is the temperature of the cold (hot) end and $\gamma = (1 + ZT)^{1/2}$ varies with the average temperature T. For refrigeration, efficiency is defined as the work in divided by the heat pumped out and is called the coefficient of performance (COP). The COP can be greater than unity and is given by:

$$\mathrm{COP} = \frac{\gamma T_c - T_h}{(T_h - T_c)(1+\gamma)} \quad (4)$$

For $ZT \gg 1$, Carnot efficiency is obtained for power generation and the Carnot limit to the COP is obtained for refrigeration. Materials used in thermoelectric devices have ZT values between 0.4 and 1.0.

All thermoelectric materials used in refrigeration or power generation devices are heavily doped semiconductors. Their physical features are illustrated by a calculation of the figure of merit (ZT) as a function of the carrier concentration at room temperature (Fig. 1) using an idealized model of a semiconductor (Goldsmid 1986). One parabolic band is used (i.e., an electron energy proportional to the square of momentum), the carriers have an effective mass of 0.72 times that of the free electron value, a mobility of $625\,\mathrm{cm^2\,V^{-1}s^{-1}}$ at low carrier concentrations, and their scattering is due to acoustic phonons. The lattice contribution to the thermal conductivity is taken to be $1\,\mathrm{Wm^{-1}K^{-1}}$ ($10\,\mathrm{mW\,cm^{-1}\,K^{-1}}$), a value near the lower limit for good thermoelectric materials. Within the framework of this simple model, metals are poor thermoelectric materials because of a low Seebeck coefficient and a large electronic contribution to the thermal conductivity.

Insulators (or lightly doped semiconductors) have large Seebeck coefficients and a small electronic contribution to the thermal conductivity, but have too few carriers, which results in a large electrical resistivity. For this example, the largest value for ZT is obtained midway between these two extremes (metal or insulator) at a carrier concentration of about 10^{19}carriers $\mathrm{cm^{-3}}$. For optimum thermoelectric performance the energy gap of the semiconductor must be at least $10k_BT_{max}$ (T_{max} = maximum operating temperature), otherwise the simultaneous excitation of intrinsic electrons and holes lowers S and ZT.

Another important observation from Fig. 1 is that while it is possible to change some of the electrical and thermal properties of a thermoelectric material by varying the carrier concentration, the ability of the lattice to conduct heat is only slightly affected by changes in the carrier concentration. Hence, if the lattice thermal conductivity is high, ZT will always be low.

2. *Thermoelectric Devices*

All thermoelectric devices are composed of thermoelectric couples such as those illustrated in Fig. 2. Each leg of the couple is a doped semiconductor (or semimetal) with one p-type and one n-type leg. p-Type (n-type) means that the dominant charge carriers are positive (negative). One side of the couple is thermally and electrically shorted together using a metal like copper (dark gray region). Electrical and thermal contact to a heat sink and the battery or load

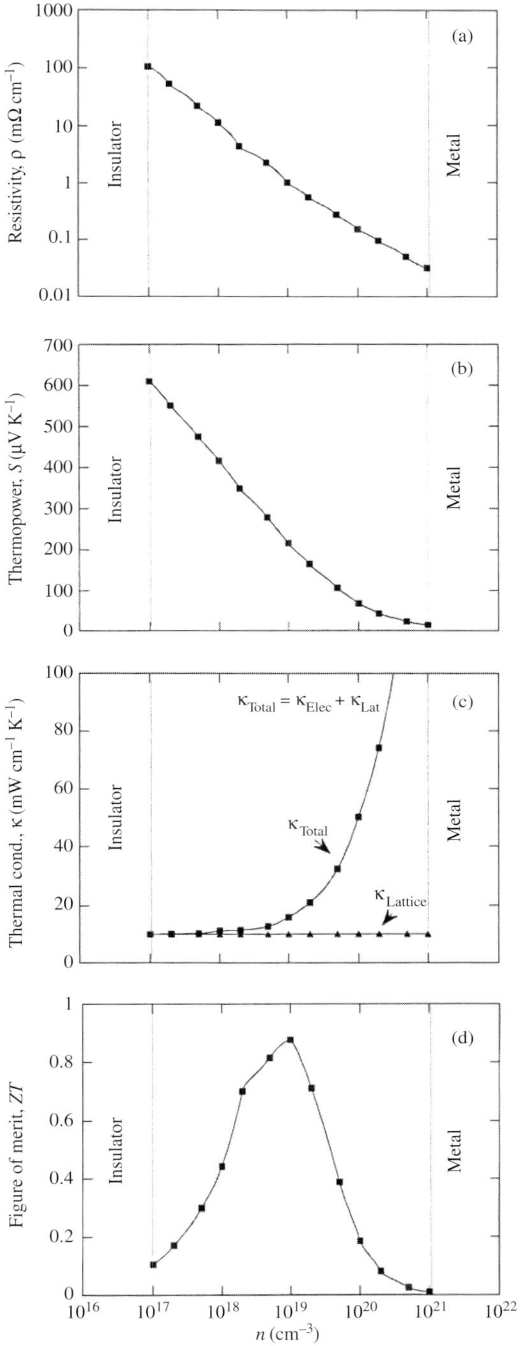

Figure 1
Room temperature transport properties calculated for an idealized semiconductor vs. carrier concentration. For this example, optimum thermoelectric performance is attained at a carrier concentration, n, of about $10^{19}\mathrm{cm^{-3}}$.

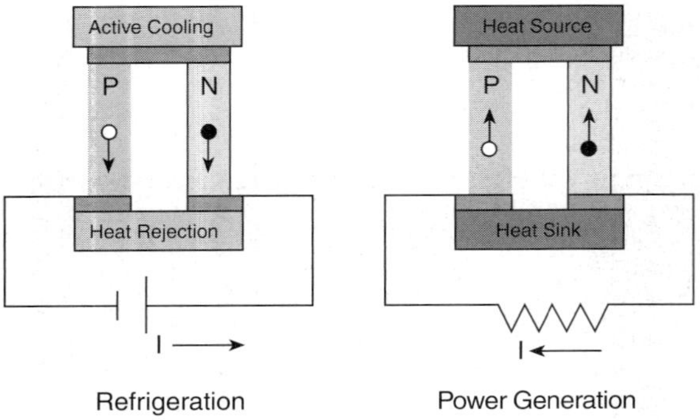

Figure 2
A single thermoelectric couple that has been configured for refrigeration or power generation. "P" (positive) and "N" (negative) refer to the sign of the charge carriers in each leg (open circles correspond to holes and filled circles to electrons). Refrigeration is possible in these devices because electrons (or holes) carry heat, and hot electrons (holes) can be forced away from the cold end of the device by the battery. If a temperature difference is externally imposed on the device, useful power can be extracted.

is made through copper pads at the open end of the couple (dark gray regions). Since the electrons and holes in a material carry heat, a battery can be used to force the hot electrons and holes away from the cold end of the device resulting in cooling of the cold end. If the direction of the current is reversed, the hot and cold ends are also reversed. If a temperature gradient is imposed across the thermoelectric couple, useful power can be extracted. In an actual thermoelectric device, typically several tens of these couples are connected together in series. Figure 3 shows a commercial multistage thermoelectric refrigerator composed of approximately 50 thermoelectric couples of *n*- and *p*-type Bi_2Te_3–Sb_2Te_3. The small copper plate at the top of the refrigerator can reach temperatures as low as 160 K (−113 °C). There are a variety of applications (or proposed applications) that use either thermoelectric refrigerators or thermoelectric generators and a few are listed in Table 1.

NASA has used thermoelectric generators for many of their space probes such as Pioneer, Viking, Galileo, Apollo, and Ulysses. A radioactive module, typically containing ^{238}Pu, is used as the source of heat. Thermoelectric couples are used to convert the heat directly into electricity. The couples are typically made using *p*- and *n*-type PbTe alloys for maximum temperatures of less than 900 K and *p*- and *n*-type GeSi alloys for temperatures less than 1300 K. The generators used by NASA have maximum power outputs between 3 W and 300 W.

Commercial thermoelectric generators have been available since the early 1960s, and they are used in industrial applications all over the world. Over 20 different models of thermoelectric generators are currently available with power outputs ranging from 10 W to 550 W. Most of the generators use natural gas as the heat source, but a few use kerosene or diesel. The main advantage of these generators is that they can operate continuously in remote locations and only need minimal maintenance.

There has been increasing interest in the conversion of low-temperature (< 100–200 °C) "waste" heat sources into useful electrical power. A variety of heat sources are considered such as geothermal, ocean thermal energy, peat deposits, solar ponds, and waste cooling water associated with power stations and other large utilities. Although the efficiency of thermoelectric generators is not very high for weak heat sources, the heat source is free and efficiency is not a main concern.

Most thermoelectric refrigerators are used for spot cooling of electronics or in specialized laboratory equipment. A wide variety of infrared detectors can be reliably cooled using thermoelectric refrigerators without having to continually replace a liquid cryogen. Laser modules of various types are used as the transmitters in fiber optic networks. Thermoelectric refrigerators are used to maintain the laser chip at a constant temperature. If the chip temperature drops too low, the direction of the current through the thermoelectric module is reversed which results in warming of the chip. A constant laser chip temperature is important since both the laser output power and output wavelength are sensitive to small temperature changes. There are a large number of low cooling power (< 100 W) uses for thermoelectric modules in laboratory equipment in the semiconductor, biotechnology, and medical industries. For example, sample tissue can easily be sliced to a thin specimen for microscopy by freezing using a thermoelectric cooling stage. There have been a few uses of thermoelectric coolers in consumer products. The most familiar is the portable picnic box that can keep food and beverages cool using the power available from the cigarette lighter socket in cars. There have been several large-scale thermoelectric air-conditioner units developed for military applications. For example, RCA made a 30 kW air-conditioning unit for the US Navy. There is one company (TECA of Chicago, IL) that manufactures air-conditioning systems based on thermoelectric modules. A typical unit from this company has a cooling power of 400 W. A more complete discussion of the details of thermoelectric devices and their design is given in Rowe (1995).

3. *Thermoelectric Materials*

All thermoelectric generators and refrigerators use materials that were discovered in the early 1960s. For

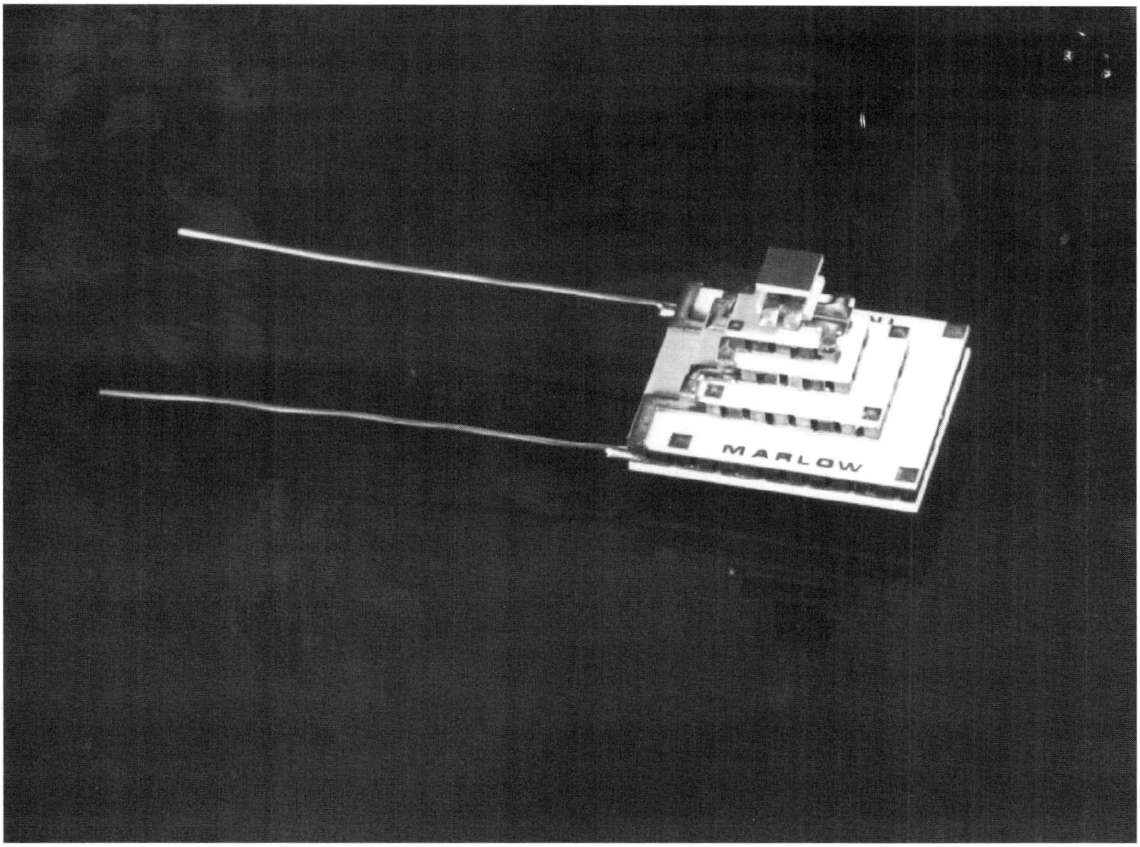

Figure 3
A multistage thermoelectric cooling module capable of reaching temperatures as low as 160K at the small copper pad on the final (highest) stage. The large copper pad at the bottom of the first stage is normally attached to a heat sink. The cooling module is composed of about 50 thermoelectric couples similar to those shown in Fig. 2 (courtesy of Marlow Industries, Dallas, TX).

refrigeration and the conversion of low-temperature waste heat into electricity, materials based on Bi_2Te_3–Sb_2Te_3 alloys are used. Higher temperature thermoelectric generators employ either SiGe or PbTe alloys depending on the temperature range. The major disadvantage of thermoelectric devices containing these materials is their efficiency. Improving the efficiency of these devices requires the development of materials with higher values of the figure of merit, ZT (Eqn. (2)). Although many materials have been investigated since 1965, the maximum value for the dimensionless figure of merit has remained near $ZT = 1$. It has been shown theoretically that there is not an upper limit on the maximum value of ZT (Mahan 1998). Using reasonable extrapolations of the properties of known materials, however, it is unlikely that bulk materials exist with ZT values greater than 4. A ZT value of 3 would make solid-state domestic refrigerators economically competitive with compressor-based refrigeration.

The remainder of this article discusses four different approaches to the development of thermoelectric materials: (i) binary and ternary covalently bonded semiconductors, (ii) semiconductors with "rattling" atoms or molecules, (iii) "correlated" metals or semiconductors, and (iv) superlattices and low-dimensional systems.

3.1 "Normal" Broad-band Semiconductors

This class of materials includes the thermoelectric compounds and alloys used in state-of-the-art thermoelectric power generators and refrigerators. Historically, "normal" semiconductors have received most attention in the search to find more efficient thermoelectric compounds. They are well

Table 1
Current and proposed applications for thermoelectric devices (proposed applications in italics).

Refrigeration
Spot cooling of electronics: infrared detectors, computer CPUs, X-ray detectors, fiber optic laser packages
Picnic coolers (powered by car battery)
Air conditioning in submarines and railway coaches
Water coolers
Superconducting electronics
Domestic refrigerators
Power generation
Power for deep-space probes (Pioneer, Viking, Voyager, Galileo)
Remote weather stations
Remote navigational systems
Submarines
Undersea power generation (for petroleum wellhead valves)
Conversion of waste heat into useful electrical power: large diesel trucks, steel industry, chemical industry

approximated by parabolic conduction and valence bands, with the energy dependence of the carrier scattering time described by a power law of the form $\tau = \tau_0 E^r$. The exponent r is used to parametrize the type of scattering, and typically $r = -1/2$ when the scattering of the carriers is primarily due to acoustic phonons. Within this approximation it can be shown that maximizing ZT is equivalent to maximizing $Nm^{*3/2}\mu/k_{\text{lattice}}$, where m^* is the effective mass of the holes (p-type material) or electrons (n-type material), N is the number of equivalent parabolic bands, and μ is the carrier mobility. It is physically plausible—although by no means proved—that compounds composed of elements that are covalently bonded should have high mobilities. If the elements in a particular compound are ionically bonded it has been argued that the charge fluctuations from atom to atom in the compound will tend to reduce carrier mobility.

All of the known good thermoelectric materials have electronic band structures that are "multivalleyed." This means that the maximum in the valence band and the minimum in the conduction band occur away from the center of the Brillouin zone. The conduction bands of silicon, for example, are composed of six equivalent bands located along $\langle 100 \rangle$ axes. Therefore, the conduction band of silicon is "multivalleyed" with a degeneracy of 6, implying that $N = 6$. It can be shown that the maximum possible value of N depends on the symmetry of the crystal structure, with higher symmetry crystal structures having larger values of N. In the search for new thermoelectric materials, this implies that the crystal structure of the compound should be cubic, hexagonal, or possibly tetragonal. The thermoelectric properties of many broad-band semiconductors (including the ternary compounds that have been investigated) have been reviewed by Wood (1988) and Slack (1995).

Taken together, phenomenological rules suggest that within this class of materials, high ZT semiconducting compounds should be composed of covalently bonded heavy elements, with a large number of atoms in a unit cell of high symmetry. These rules provide some guidelines in the search for high ZT materials and certainly most of the known "good" thermoelectric materials tend to follow these guidelines. For normal broad-band semiconductors these rules suggest compounds made from elements found in the lower right-hand corner of the periodic table (groups 3B to 6B). Using these rules, a ternary thermoelectric compound with the composition Tl_2SnTe_5 has been reported to have thermoelectric properties similar to Bi_2Te_3 (Sharp *et al.* 1999).

It seems unlikely that the next generation of thermoelectric materials will be binary compounds or alloys, most of which have been evaluated. This is much like superconductivity research before high T_c materials were found: binary compositions had been essentially exhausted and ternaries were being evaluated. The simplest high T_c materials discovered are ternary alloys ($La_{2-x}Sr_xCuO_4$) or quaternary compounds ($YBa_2Cu_3O_7$).

3.2 "Rattling" Semiconductors: Phonon Glasses and Electron Crystals

The maximum thermoelectric efficiency is achieved when the lattice contribution to the flow of heat is as low as possible ($Z = S^2/\rho\kappa$, where $\kappa = \kappa_{\text{electronic}} + \kappa_{\text{lattice}}$). It has been suggested that for a particular crystal the minimum lattice thermal conductivity, κ_{min}, corresponds to that of an amorphous solid (or glass) with the same composition. κ_{min} is reached when the mean free path of the heat-carrying phonons is of the order of the phonon wavelength (Slack 1995). In a beautiful set of experiments it has been demonstrated that κ_{min} can be attained in crystalline compounds with specific types of controlled disorder (Cahill *et al.* 1992). It is found that localized vibrational or rotational modes (Einstein oscillators) provide a particularly effective way of reducing the thermal conductivity of a crystalline solid to within 50% of κ_{min} at room temperature. For example, in $(KBr)_{1-x}(KCN)_x$ single crystals $\kappa_{lattice}$ values near κ_{min} are obtained for x values as small as 0.19. The rotation of the CN molecule about its lattice site effectively scatters the heat-carrying phonons at temperatures greater than 10 K. By way of contrast, simple mass fluctuation scattering, such as occurs in $KCl_{1-x}Br_x$ mixed crystals, only reduces $\kappa_{lattice}$ to $4\kappa_{min}$.

The ultimate thermoelectric material should conduct electricity like a crystal but heat like a glass. Crystal structures that contain weakly bound atoms or molecules that "rattle" within an atomic cage will likely have a low lattice thermal conductivity. Since good electrical properties must be maintained, there should be at least three distinct crystallographic sites in the structure, i.e., a true ternary compound. Two of the sites would form the basic framework of the structure, and this framework would dominate the band structure and therefore electronic transport. The third cage-like site would be occupied by the "rattling" atom that scatters the heat-carrying phonons, thus greatly reducing the lattice portion of the thermal conductivity.

These ideas have been demonstrated for two classes of materials, the filled skutterudite antimonides and semiconducting analogues of the ice clathrate structure. The filled skutterudite antimonides are cubic compounds with typical composition $LaFe_4Sb_{12}$. This structure can be described as consisting of square planar rings of four antimony atoms with the rings lying in the (100), (010), or (001) crystallographic planes. The iron atoms form a simple cubic sublattice and the lanthanum atoms are positioned in the two remaining "cages" in the unit cell. X-ray and neutron structure refinements indicate that the lanthanum atoms tend to exhibit exceptionally large thermal parameters corresponding to the "rattling" of these atoms in an oversized atomic cage. The rattling lanthanum atoms result in a large reduction of the lattice thermal conductivity. The values obtained for ZT at elevated temperatures for these materials are comparable to the best values reported for any material ($ZT = 1.4$) (Sales *et al.* 1996). The semiconducting clathrates have typical composition $Sr_8Ga_{16}Ge_{30}$ (Nolas *et al.* 1998). In these compounds the strontium atoms rattle in an oversized atomic cage resulting in a glass-like thermal conductivity. Good electronic transport is achieved through the Ge–Ga tetrahedral network. These materials also appear to have large values of ZT at elevated temperature.

3.3 *Correlated Metals and Semiconductors*

As is shown in Fig. 1, "normal" metals are poor thermoelectrics owing to a low Seebeck coefficient and a large electronic contribution to the total thermal conductivity. The approximations used to generate Fig. 1, however, are not valid for many rare earth intermetallic compounds (usually containing cerium or ytterbium) in which the 4*f* level is near the Fermi energy. For these materials, the 4*f* "bands" are very narrow (resulting in a very high density of states near the Fermi level) and nonparabolic, and the carrier scattering time is not described by a simple power law in the carrier energy but is closer to resonant scattering. Theoretical support for these types of materials has been demonstrated (Mahan and Sofo 1996). In a very general theoretical analysis it was shown that the best electronic structure for thermoelectric materials is a delta function in the transport distribution centered about $2k_BT$ to $3k_BT$ from the Fermi energy. The transport distribution is defined to include the density of states and the energy-dependent scattering time. It is noted, however, that if there is a parallel "normal" channel for carrier conduction (a broad metallic band), then the values of ZT rapidly drop with only a small contribution from this "normal" density of states. Unusual correlated semiconductors (also termed hybridization gap or Kondo insulators) may represent a closer approximation to the ideal electronic structure suggested by Mahan and Sofo (1996). In a correlated semiconductor the electron states near the gap retain much of their free-atom character. Several of these correlated metals and semiconductors have been investigated but ZT values are below 0.25.

A very general and simple analysis shows that for a correlated metal to be a viable thermoelectric material (defined as $ZT = 1$), the Seebeck coefficient must be greater than $156 \mu VK^{-1}$. Recall that $ZT = TS^2/\kappa\rho$, where $\kappa = \kappa_{electronic} + \kappa_{lattice}$. The largest value of ZT occurs when the lattice contribution can be ignored so that $\kappa = \kappa_{electronic}$. ZT is then equal to $TS^2/\rho\kappa_{electronic}$. For a metal, the Wiedemann–Franz law states that $\rho\kappa_{electronic}/T = L_0$, where L_0 is a constant ($2.44 \times 10^{-8} V^{-2} K^{-2}$). The Wiedemann–Franz law is surprisingly general, and theory shows that it even applies for correlated metals. Therefore, $ZT \leqslant S^2/L_0$, which implies that for $ZT = 1$ the minimum value of S is $156 \mu VK^{-1}$. The correlated metal with the highest reported Seebeck coefficient is $CePd_3$, which has a maximum value of $115 \mu VK^{-1}$ at 125 K.

3.4 *Superlattices and Low-dimensional Materials*

Theoretical calculations indicate that semiconductor quantum wells should have an improved figure of merit. This conjecture has been confirmed experimentally (Hicks *et al.* 1996), and it is found that ZT for a single quantum well increases as the well width decreases. The enhancement of ZT arises from an increase of the electronic density of states. The model has been tested using PbTe quantum wells confined by $Pb_{0.927}Eu_{0.073}Te$ barrier layers. The electrons are confined to quantum wells with thicknesses ranging from 17Å to 55Å. An increase by a factor of five is found in $S^2\sigma$ relative to bulk PbTe of the same volume. Using the thermal conductivity of bulk PbTe as a conservative estimate for the quantum well, a value of $ZT=2$ can be estimated within a single well. This novel approach has spurred new activity into these types of devices. Research has extended these ideas to the investigation of quantum wires and other low-dimensional structures as possible advanced thermoelectric materials.

4. Thermionic Refrigeration

A solid-state refrigeration device based on thermionic emission has been proposed (Mahan and Woods 1998). Although technically these are not thermoelectric devices, there are many similarities between a thermoelectric and a thermionic refrigerator—particularly with regard to the material properties needed. Thermionic refrigeration is a solid-state process based on evaporative cooling. A barrier is imposed to the flow of electrons (or holes). Those electrons (holes) that have sufficient kinetic energy to surmount the barrier are swept away by a small voltage. A system cools if its most energetic particles are removed from the system. A practical thermionic refrigerator is envisioned to consist of a metal–semiconductor superlattice. A Schottky barrier at the metal–semiconductor interface of about 0.05eV is required for efficient operation near room temperature. Electrons from the metal with sufficient kinetic energy to surmount the small barrier are ballistically transported across the thin semiconducting layer using a small voltage. As is the case for a thermoelectric refrigerator, the thermal conductivity of the semiconductor material must be low to prevent substantial backflow of heat. Calculations indicate that if these devices can be fabricated, they should have effective ZT values greater than 2. There has been no experimental confirmation of these intriguing calculations.

Bibliography

Cahill D G, Watson S K, Pohl R O 1992 Lower limit to the thermal conductivity of disordered crystals. *Phys. Rev. B* **46**, 6131–40

Goldsmid H J 1986 *Electronic Refrigeration*. Pion, London

Hicks L D, Harmon T C, Sun X, Dresselhaus M S 1996 Experimental study of the effect of quantum-well structures on the thermoelectric figure of merit. *Phys. Rev. B* **53**, 10493–96

Mahan G D 1998 Good thermoelectrics. *Solid State Phys.* **51**, 81–157

Mahan G D, Sofo J O 1996 The best thermoelectric. *Proc. Natl. Acad. Sci. USA* **93**, 7436–9

Mahan G D, Woods L M 1998 Multilayer thermionic refrigeration. *Phys. Rev. Lett.* **80**, 4016–9

Nolas G S, Cohn J L, Slack G A, Schujman S B 1998 Semiconducting Ge clathrates—promising candidates for thermoelectric applications. *Appl. Phys. Lett.* **73**, 178–80

Rowe D M (ed.) 1995 *CRC Handbook of Thermoelectrics*. Chemical Rubber Press, Boca Raton, FL

Sales B C, Mandrus D, Williams R W 1996 Filled skutterudite antimonides: a new class of thermoelectric materials. *Science* **272**, 1325–8

Sharp J W, Sales B C, Chakoumakos B C, Mandrus D 1999 Thermoelectric properties of Tl_2SnTe_5 and Tl_2GeTe_5. *Appl. Phys. Lett.* **74**, 3794–6

Slack G A 1995 New materials and performance limits for thermoelectric cooling. In: Rowe D M (ed.) *CRC Handbook of Thermoelectric Materials*. Chemical Rubber Press, Boca Raton, FL, pp. 407–40

Wood C 1988 Materials for thermoelectric energy conversion. *Rep. Prog. Phys.* **51**, 459–539

Thermoelectric Materials: Principles, Structure, Properties, and Applications

T. M. Tritt

During the 1990s there was a heightened interest in the field of thermoelectrics driven by the need for more efficient materials for electronic refrigeration and power generation (Nolas *et al.* 2002, Tritt *et al.* 1997, Tritt *et al.* 1998, Tritt *et al.* 2000a). Proposed industrial and military applications of thermoelectric (TE) materials are generating increased activity in this field by demanding higher performance, near-room-temperature TE materials than those presently in use. Thermoelectric refrigeration is an environmentally "green" method of small-scale localized cooling in computers, infrared detectors, electronics, and opto-electronics as well as many other applications. However, most of the electronics and opto-electronics technologies typically require only small scale or localized spot cooling of small components which do not impose a large heat load. If significant economical cooling can be achieved the resulting "cold computing" could produce speed gains of 30–200% in some CMOS computer processors. Cooling is perceived by many as the fundamental limit to electronic system performance. Cooling of laser diodes and infrared detectors to temperatures (T), between 100 K and 200 K, would greatly improve performance and sensitivity and thus is extremely important to many technologies. Thus, the potential payoff for the development of low-temperature thermoelectric refrigeration devices is great, and the requirement for compounds with properties optimized over wide temperature ranges has led to a much-expanded interest in new thermoelectric materials.

Utilization of Peltier coolers in relation to refrigeration of biological specimens/samples is an emerging application of thermoelectrics. Power generation applications are being investigated by the automotive industry as a means to develop electrical power from waste engine heat for use in the "next generation vehicle." These uses range from power generation utilizing waste engine heat from the exhaust and radiator cooling system to seat coolers for comfort or electronic component cooling. Of course, the deep space applications of NASA's *Voyager* and *Cassini* missions using radioactive thermoelectric generators (RTGs) are well established. Given the present energy needs experienced in the United States there is even a more pressing need to investigate alternative energy conversion technologies in this country, e.g., the thermal to electrical energy conversion from natural heat gradients that thermoelectric technologies provide. This has already been identified as important in many European and Asian countries. An overview of the state-of-the-art materials, a theoretical and experimental discussion of the basic principles, as well as an overview of some of the recent developments and materials is given in the texts of Tritt and Nolas (Nolas *et al.* 2001a, Tritt 2000a).

1. *Thermoelectric Phenomena*

1.1 *Seebeck and Peltier Effects*

A discussion of thermoelectric effects and devices should start with the most fundamental phenomena, the Seebeck effect or thermopower (Morelli 1997). In the early 1800s Seebeck observed that if two dissimilar materials were joined together and the junctions were held at different temperatures (T and $T+\Delta T$) a voltage difference (ΔV) was developed that was proportional to the temperature difference (ΔT). The ratio of the voltage developed to the temperature gradient ($\Delta V/\Delta T$) is related to an intrinsic property of the materials called the Seebeck coefficient (α) or the thermopower. The Seebeck coefficient is very low for metals (only a few $\mu V\,K^{-1}$) and much larger for semiconductors (typically a few $100\,\mu V\,K^{-1}$). A related effect (the Peltier effect) was discovered a few years later by Peltier, who observed that if an electrical current is passed through the junction of two dissimilar materials, heat is either absorbed or rejected at the junction depending on the direction of the current.

This effect is due to the difference in Fermi energies (discussed later: Sect. 4.1) of the two materials. The Seebeck effect and the Peltier effect are illustrated in Figs. 1 and 2, respectively. These two effects are related to each other, as shown in the definition of the Peltier coefficient, Π;

$$\Pi = \alpha T \tag{1}$$

The rate that the Peltier heat is liberated or rejected at the junction (Q_P) is given by:

$$Q_P = \alpha I T \tag{2}$$

where, I is the current through the junction and T is the temperature in kelvin. There are also a number of thermomagnetic effects such as the Hall, Ettingshausen, and Nernst effects that are beyond the scope of this article. The reader is referred to the text by Nolas *et al.* (2001a) for a discussion of these effects.

1.2 *Thermoelectric Modules—Devices*

The Peltier effect is the basis for many modern day TE refrigeration devices and the Seebeck effect is the basis for TE power generation devices (see *Thermoelectric Devices: Refrigeration and Power Generations With No Moving Parts*). The versatility of TE materials is illustrated in Fig. 3 which shows a diagram of a thermoelectric couple composed of an n-type (negative thermopower and electron carriers) and a

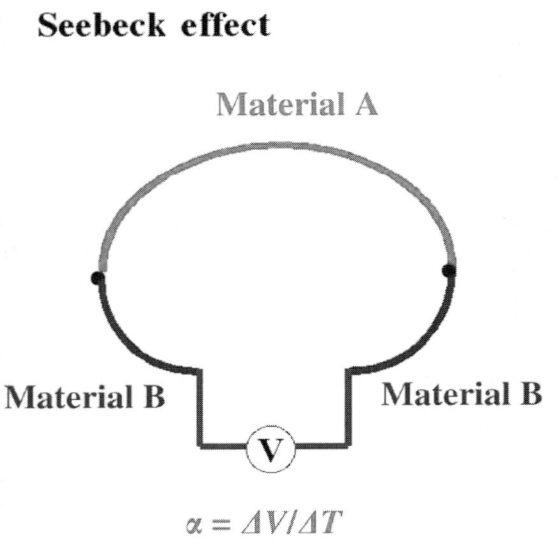

Figure 1
A schematic illustrating the Seebeck effect of a couple of two dissimilar materials, in which a voltage difference (ΔV) is generated that is proportional to the temperature gradient (ΔT) between the ends of the couple. The Seebeck effect is the basis for power generation discussed in the text.

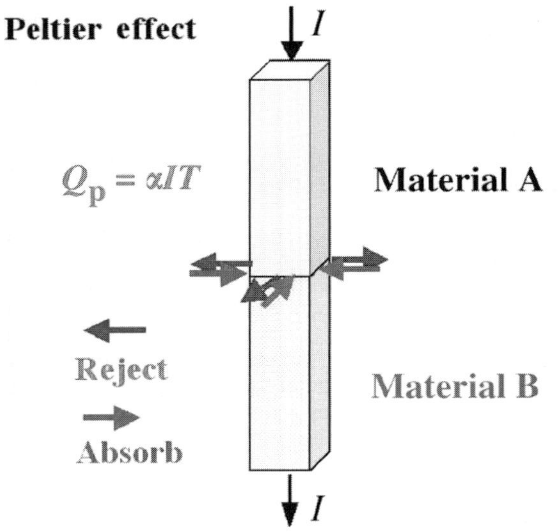

Figure 2
A schematic illustrating the Peltier effect between two dissimilar materials (A and B). As a current (I) flows through the junction of the two materials, heat is either absorbed or rejected depending on the direction of the current through the two materials. The Peltier effect is the basis for electronic refrigeration discussed in the text.

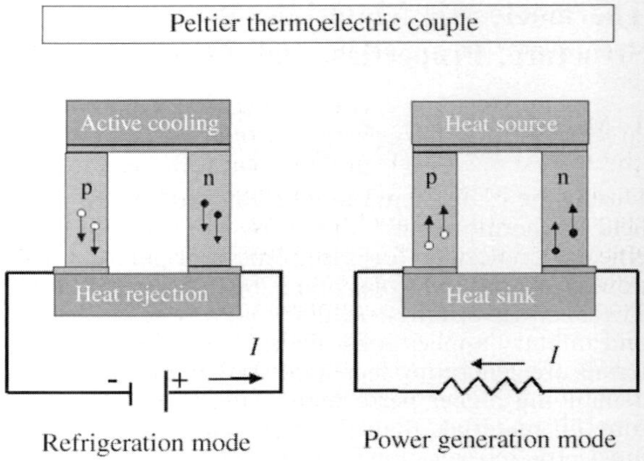

Figure 3
A schematic illustrating a thermoelectric couple or module made of an n-type (electrons, closed circle arrows) material and a p-type (holes, open circle arrows) material. This diagram illustrates both power generation (Seebeck effect) and electronic refrigeration (Peltier effect) capabilities of the thermoelectric module.

p-type (positive thermopower and hole carriers) semiconductor material connected through metallic electrical contact pads. Both refrigeration and power generation may be accomplished using the same module as shown in Fig. 3. A TE module or device is built up of an array of these couples, which are arranged electrically in series and thermally in parallel. Thermoelectric energy conversion utilizes the Peltier heat generated when an electric current is passed through a thermoelectric material to provide a temperature gradient with heat being absorbed on the cold side, transferred through (or pumped by) the thermoelectric materials and rejected at the sink, thus providing a refrigeration capability. The advantages of TE solid state energy conversion are compactness, quietness (no moving parts), and localized heating or cooling. Many of the potential applications and importance of these devices have already been discussed in the background and applications section.

2. *The Figure of Merit*

2.1 *Definition and Description of the Figure of Merit*

The potential of a material for thermoelectric applications is determined in large part to a measure of the material's dimensionless figure of merit,

$$ZT = \frac{\alpha^2 \sigma T}{\kappa} = \frac{\alpha^2 T}{\rho \kappa} \tag{3}$$

where, α is the Seebeck coefficient, σ the electrical conductivity, ρ the electrical resistivity, and κ the total thermal conductivity ($\kappa = \kappa_L + \kappa_E$; the lattice and electronic contributions, respectively). The power factor, $\alpha^2\sigma T$, (or $\alpha^2 T/\rho$) is typically optimized as a function of carrier concentration (typically around 10^{19} carriers cm^{-3}), through doping, to give the largest ZT. High mobility carriers are most desirable in order to have the highest electrical conductivity. Semiconductors have been primarily the materials of choice for thermoelectric applications. The ZT for a single material is somewhat incomplete since an array of TE couples are utilized in a device or module. There are two materials in the TE couple, which is shown in Fig. 3; an n-type and a p-type. The resulting figure of merit for the couple (or a device) is typically given by:

$$ZT = \frac{(\alpha_p - \alpha_n)^2 T}{(\rho_n \kappa_n)^{1/2} + (\rho_p \kappa_p)^{1/2}} \tag{4}$$

The ZT is not a fixed quantity for the couple and can depend on the relative dimensions of the branches. ZT is maximized when the product RK is minimized, where R is the total couple resistance and K is the couple thermal conductance. This is accomplished when:

$$\frac{L_n A_p}{L_p A_n} = \left(\frac{\rho_p \kappa_n}{\rho_n \kappa_p}\right)^{1/2} \tag{5}$$

and the figure of merit for the couple is given by Eqn. (4).

2.2 *Coefficient of Performance and Efficiency*

The coefficient of performance (ϕ) (refrigeration mode) of the thermoelectric couple is determined by the cooling power or net heat pumping rate (Q_C) over the power input (W) and is given below:

$$\varphi = \frac{Q_C}{W} = \frac{(\alpha_p - \alpha_n) I T_C - K\Delta T - \frac{1}{2} I^2 R}{I[(\alpha_p - \alpha_n)\Delta T + IR]} \tag{6}$$

where T_C (T_H) is the cold side (hot side) temperature and $\Delta T = T_H - T_C$. The terms in the numerator are, in order, the Peltier heat, the thermal conductance heat flow (Fourier heat) due to the ΔT, and the Joule heating term due to the total resistance of the couple. The efficiency (η) (power generation mode) of the thermoelectric couple is given by the power input to the load (W) over the net heat flow rate (Q_H), where Q_H is positive for heat flow from the source to the sink and is given below:

$$\eta = \frac{W}{Q_H} = \frac{I[(\alpha_p - \alpha_n)\Delta T - IR]}{K\Delta T + (\alpha_p - \alpha_n) I T_H - \frac{1}{2} I^2 R} \tag{7}$$

Thus, one can see that the efficiency (η) is proportional to $(1 + ZT)^{1/2}$.

The best thermoelectric materials that are currently in devices have a value of $ZT \approx 1$. This value, $ZT \approx 1$, has been a practical upper limit since the 1970s, yet no theoretical or thermodynamic reason exists for a $ZT \approx 1$ as an upper barrier. The value of ZT (Eqn. (3)) can be increased by decreasing κ_L or by increasing either α or σ. However, σ is tied to the electronic thermal conductivity, κ_E, through the Wiedemann–Franz relationship and the ratio is essentially constant at a given temperature. Some of the goals of research efforts in the early part of the twenty-first century are to find new materials which either raise the current efficiency of thermoelectric devices (increase ZT) or have the capability of operating in new and broader temperature regimes, especially at lower temperatures, $T < 250$ K. Since the 1970s, alloys based on the Bi_2Te_3 system $[(Bi_{1-x}Sb_x)_2(Te_{1-x}Se_x)_3]$ and the $Si_{1-x}Ge_x$ system have been extensively studied and optimized for their use as thermoelectric materials to perform a variety of solid state TE refrigeration and power generation applications (Goldsmid 1986, Rowe 1995). Due to the extensive investigations of these traditional thermoelectric materials there appears little room for future improvement, and thus, entirely new classes of compounds will have to be investigated (Tritt 1996). Figure 4 shows the ZT as a function of temperature for the Bi_2Te_3 and $Si_{1-x}Ge_x$ materials as well as many of the more recently developed bulk materials, which are discussed in Sect. 4.

2.3 *Thermopower or Seebeck Coefficient*

The thermoelectric power or Seebeck coefficient (α) can be thought of as the heat per carrier over temperature or more simply the entropy per carrier, $\alpha \approx C/q$ where C is the specific heat and q is the charge of the carrier (Chaiken 1990). For the case of a classical gas each particle has an energy of 3/2 $k_B T$, where k_B is the Boltzmann constant. The thermopower is thus approximately k_B/e, where e is the charge of the electron. Let us estimate an expected magnitude for the thermopower of metals and semiconductors. For metals, the heat per carrier is essentially a product of the electronic specific heat and the temperature divided by the number of carriers (N), i.e., ($\alpha \approx C_{el} T/N$) and then α is approximately:

$$\alpha \approx \frac{C_{el}}{Q} \approx \left(\frac{k_B}{e}\right)\frac{k_B T}{E_F} \tag{8}$$

where E_F is the Fermi energy (related to the chemical potential of the material) which is basically the energy, such that at $T \approx 0$, all states above this energy are vacant, and all the states below are occupied. The quantity, $k_B/e \approx 87 \mu V K^{-1}$, is a constant, which represents the thermopower of a classical electron gas. Metals have thermopower values much less than

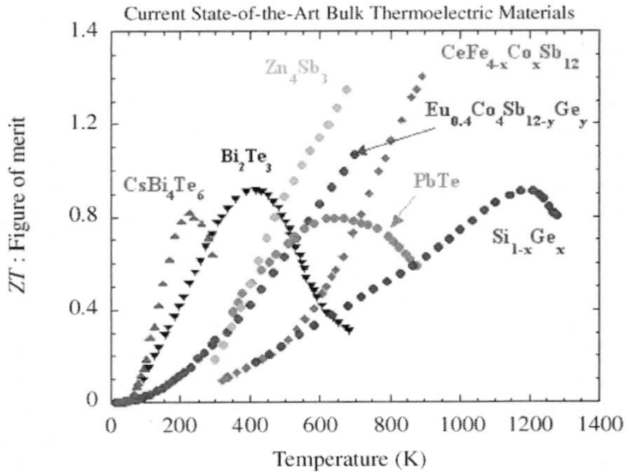

Figure 4
A plot of the figure of merit (ZT) as a function of temperature for several bulk thermoelectric materials as indicated in the figure above. The Bi_2Te_3 material is the current state-of-the-art refrigeration material and the $Si_{1-x}Ge_x$ alloy is the current state-of-the-art power generation material. Several bulk materials are currently under investigation and show some promise, as shown in the figure above.

$87\,\mu V\,K^{-1}$ (on the order of 1–$10\,\mu V\,K^{-1}$) and decrease with decreasing temperature, i.e., $E_F >> k_BT$).

In a semiconductor, a charged particle must first be excited across an energy gap E_G. In this case the thermopower is approximated by:

$$\alpha \approx \frac{C_{el}}{Q} \approx \left(\frac{k_B}{e}\right)\frac{E_G}{k_BT} \tag{9}$$

Thus, the thermopower is larger than the characteristic value $87\,\mu V\,K^{-1}$ and increases with decreasing temperature. Semiconductors can exhibit either electron conduction (negative thermopower) or hole conduction (positive thermopower). The thermopower for different carrier types is given by a weighted average to their electrical conductivity values (σ_e and σ_p)

$$\alpha \approx \frac{\alpha_e\sigma_e + \alpha_p\sigma_p}{\sigma_e + \sigma_p} \tag{10}$$

It is necessary to dope the semiconductors with either donor or acceptor states to allow extrinsic conduction of the appropriate carrier type, electrons, or holes, respectively. It is apparent the total thermopower will be lower than that of either of the individual contributions unless the direct band gap is large enough to effectively minimize minority carrier contributions.

In a measurement of the Seebeck effect, what is actually measured is the difference in the Seebeck coefficients of the two dissimilar materials, α_{AB}, which is called the relative Seebeck of the couple. The relative Seebeck coefficient is given by $\alpha_{AB} = -\Delta V/\Delta T$, where $\alpha_{AB} = \alpha_A - \alpha_B$, which includes both the sample (α_A) and the lead (α_B) contributions. The absolute thermopower of an unknown sample is then given by the measured α_{AB}, and a known lead value, (α_B), such as Pb or some other known standard.

2.4 *Electrical Conductivity*

The electrical conductivity is related to the density of charge carriers (n) and their mobility (μ), typically given by

$$\sigma = ne\mu \tag{11}$$

The mobility is given by:

$$\mu = e\tau/m_e \tag{12}$$

where μ is the mobility of the carriers, m_e is the effective mass, and τ is the mean scattering time between collisions for the carriers. In metals there are many carriers and states available for conduction, typically $n \approx 10^{22}$ carriers cm^{-3}. The electrical conductivity is then very large for metals; on the order of $10^6\,(\Omega\,cm)^{-1}$.

Again for semiconductors, the carriers must be thermally excited across a gap for conduction to occur.

$$\sigma \approx \sigma_0 \exp\left(+\frac{E_G}{k_BT}\right) \tag{13}$$

The conductivity can occur through the contributions of both holes and electrons

$$\sigma = ne\mu_e + pe\mu_h \tag{14}$$

where n and μ_e are the electron concentration and mobility, respectively, and p and μ_h are the hole concentration and mobility, respectively. There are two primary ways to achieve a high conductivity in a semiconductor, either by having a very small gap to excite across ($E_G < k_BT$) or by having very high mobility carriers. Typical values of the electrical conductivity lie between 10^{-4} and $10^4\,(\Omega\,cm)^{-1}$, however these are somewhat arbitrary boundaries.

2.5 *Thermal Conductivity*

The thermal conductivity, κ, is related to the transfer of heat through a material, either by the electrons or via quantized vibrations of the lattice called phonons such that

$$\kappa = \kappa_L + \kappa_E \tag{15}$$

(where κ_L and κ_E are the lattice and electronic contributions, respectively). The electrical conductivity and the thermal conductivity are interrelated, in that σ is tied to κ_E through the Wiedemann–Franz relationship:

$$\kappa_{\mathrm{E}} = L_0 \sigma T \tag{16}$$

The ratio is essentially constant at a given temperature. The Lorentz number is given by

$$L_0 = \frac{\pi^2}{3}\left(\frac{k_{\mathrm{B}}}{e}\right)^2 \tag{17}$$

where, $L_0 = 2.45 \times 10^{-8}\,\mathrm{W\,\Omega\,K^{-2}}$ (or $L_0 = 2.45 \times 10^{-8}$ ($\mathrm{V^2\ K^{-2}}$)). This assumes an elastic scattering mechanism and that the mean scattering time for the electrons and the phonons is the same.

In many areas of research related to new TE materials, attempts are being made to reduce the lattice part of the thermal conductivity essentially to its minimum value, i.e., where a minimum lattice thermal conductivity is achieved (that is, when all the phonons have a mean free path equal to the interatomic spacing of the constituent atoms). This is being attempted by scattering phonons in different frequency ranges utilizing a variety of methods such as mass fluctuation scattering (a mixed crystal, in ternary and quaternary compounds), grain boundary scattering due to the size of the grains and interface scattering in thin films or multilayer systems. In addition, there is the relatively recent idea of "rattling" scattering or inserting new atoms in open structure or cage-like systems to essentially "rattle around inside the cages" and thus resonantly scatter the heat carrying or acoustic phonons. The aspect of thermal conductivity reduction utilizing the "rattling" scattering in the cage structure will become much clearer when this aspect is discussed in more detail later in this article in relation to classes of materials called skutterudites, "filled" skutterudites, and clathrates.

3. *Investigating New Thermoelectric Materials*

3.1 Phonon-glass: Electron-crystal Approach

Slack has described the chemical characteristics of materials that might be candidates for a good thermoelectric material (Slack 1995). The candidates need to be narrow bandgap semiconductors with high-mobility carriers, while the thermal conductivity has to be low. In semiconductors, the Seebeck coefficient and electrical conductivity (both in the numerator of ZT—see Eqns. (3) and (4)) are strong functions of the doping level and chemical composition. These quantities must therefore be optimized for good thermoelectric performance. The thermal conductivity of complex materials can often be modified by chemical substitutions. Understanding these various effects, and selecting optimization strategies can be an exceedingly difficult problem, because in complex materials there are often many possible degrees of freedom. Slack suggested that the best thermoelectric material would behave as a "phonon-glass, electron-crystal" (PGEC) that is, it would have the electrical properties of a crystalline material and the thermal properties of an amorphous or glass-like material. In general, the successful candidate material requires several specific characteristics. The candidate material is typically a narrow bandgap semiconductor ($E_{\mathrm{G}} \approx 10\ k_{\mathrm{B}}T$ or $\approx 0.25\,\mathrm{eV}$ at 300 K). Also, the mobility of the carriers must remain high, ($\mu \approx 2000\ \mathrm{cm^2\ (V\ s)^{-1}}$) while the thermal conductivity must be minimized. This high ratio of mobility to thermal conductivity imparts certain demands on the structure of the material. To maintain a high mobility, the electronegativity difference, $\Delta\chi$, between the elements of the material must be as small as possible, preferably less than approximately $\Delta\chi < 0.5$ per bond on average, using the Allred scale. All these demands insure that the structural elements will be drawn from the heavy metal or metalloid set (p-band electrons) of the periodic table, such as: Bi, Sb, Se, Te, Ge, Sn, etc. It is only these elements which have such small electronegativity differences. In addition, many of the compounds formed by these elements are narrow bandgap semiconductors.

Currently there are no theoretical or thermodynamic limits to the possible values of ZT. Slack estimated that an optimized PGEC material would have values of $ZT \approx 4$ in the 77–300 K temperature range (Slack 1995). This gives encouragement that such materials may be possible. Thus, a systematic search and subsequent thorough investigation of promising materials might eventually yield these much-needed materials for the next-generation TE devices. Whether one is investigating thin film or bulk thermoelectrics materials, the PGEC approach may well be the best strategy to employ.

3.2 Minimum Thermal Conductivity (κ_{min})

The thermal conductivity of complex materials can often be modified by chemical substitutions or alloy scattering of the phonons. While grain boundary scattering can be an important consideration for reducing the lattice thermal conductivity, as discussed, it can also be detrimental to carrier mobility. Understanding the various effects and selecting optimization strategies can be an exceedingly difficult problem due to the fact that in complex materials there are often many possible degrees of freedom. Slack introduced the concept of a minimum lattice thermal conductivity (κ_{min}). The lattice thermal conductivity is given by:

$$\kappa_{\mathrm{L}} \approx v_{\mathrm{S}} C L_{\mathrm{ph}} \tag{18}$$

where, v_{S} is the velocity of sound, C is the heat capacity, and L_{ph} is the mean free path of the phonons.

At high temperatures ($T > \approx 300$ K) the sound velocity and the heat capacity are essentially temperature independent in typical materials.

Therefore, the magnitude and the temperature dependence of κ_L is basically determined by the mean free path of the phonons. Slack defined the minimum thermal conductivity, κ_{min}, as the thermal conductivity when the mean free path is essentially limited by the interatomic distance between the atoms within the crystal (Slack 1979). Typical analysis of κ_{min} results in values of $\kappa_{min} \approx 0.25$–$0.5$ W m^{-1}K^{-1} (Slack 1995). These values will essentially result in a practical limit for new thermoelectric materials, and effectively yields an upper limit of $ZT \approx 2$ at room temperature, $ZT \approx 1.5$ at 100 K, and at high temperatures, a $ZT \approx 4$ may be achievable. This analysis is consistent with estimations of practical limits of ZT by both Slack (1995) and Goldsmid (1986).

Based on the predictions by Slack, the ideal thermoelectric materials will probably be those containing at least three elements, two or more of which form a highly crystalline framework, while the third element occupies a site in which it can create localized disorder (e.g., "rattlers"). More recently this concept has been successfully employed to reduce κ_L in compounds that possess these "rattling" atoms, while maintaining relatively high power factors. One example is the skutterudites, discussed later. The framework should be well ordered for the efficient mobility of charge, while the less ordered element rattles around to inhibit phonon conduction. It was such a recipe, which originally led to the correct prediction that filled skutterudites should have good thermoelectric properties (Slack and Tsoukala 1994).

3.3 *Minimum Thermopower*

There are certain practical limits for each of the parameters used to calculate ZT. These practical limits must be possible in order to achieve a material viable for thermoelectric applications. For example, in Bi_2Te_3, to achieve a $ZT \approx 1$ at $T = 320$ K, $\sigma \approx 1$ m Ω-cm, $\sigma \approx 225\ \mu$V K^{-1}, and $\kappa \approx 1.5$ W m^{-1}K^{-1}. We have already discussed the "ZT barrier" which, in effect is given by minimizing the thermal conductivity. It is practical to investigate materials where the electronic and lattice terms are comparable, on the order of 0.75–1 W m^{-1}K^{-1}. Let us look at the hypothetical situation of a material where the lattice thermal conductivity is zero ($\kappa_L = 0$)! We will also assume the scattering in this system is elastic and that the Wiedemann–Franz relationship, slightly rearranged ($\kappa_E/\sigma = L_0 T$, where L_0 = Lorentz number ($L_0 \approx 2.45 \times 10^{-8}$ V^{2}K^{-2})) is well behaved in this material. Then we can rewrite Eqn. (3):

$$ZT = \frac{\alpha^2 \sigma T}{\kappa_E} = \frac{\alpha^2}{L_0} \qquad (19)$$

Therefore for a material to be a viable TE material it must possess a minimum thermopower which is directly related to the value of ZT and L_0. Given this description, in order to achieve a certain value of ZT the material would require that $\alpha = (L_0)^{0.5} = 157\ \mu$V K^{-1} for a $ZT = 1$ and $\alpha = (2L_0)^{0.5} = 225\ \mu$V K^{-1} for a $ZT = 2$.

3.4 *Bulk Materials Approach*

The discussion thus far indicates that new materials must eventually be able to achieve certain minimum values of important parameters to be considered for a potential thermoelectric material. It does not matter if a material has a $\kappa_L \approx \kappa_{min}$. If it cannot be "tuned" or doped to attain a minimum thermopower of $\alpha \approx 150\ \mu$V K^{-1} then it would not be able to achieve a $ZT \approx 1$. Unless, of course, a material is in serious violation of the Wiedemann–Franz relationship. Some of the materials that are under investigation are highlighted in Sect. 4 and a summary is given as to why these materials are interesting for potential TE applications in Sect. 5. Each of the materials under investigation for thermoelectrics has its own set of challenges that must be overcome.

4. *Recent Results and Promising Materials*

4.1 *Complex Chalcogenides*

One of the important issues relative to the development of low-temperature thermoelectric materials is the identification of mechanisms which might give high thermopower (α) at low temperatures. Possibilities include phonon drag, heavy fermion materials, Kondo systems, and material which exhibit phase transitions, as well as quasi-one-dimensional materials. Low-dimensional systems are known to be susceptible to van Hove singularities (or cusps) in their density of states, $D(E)$, electronic phase transitions, and exotic transport phenomena which can add structure in $D(E)$ near their Fermi energy, E_F. In most materials, at temperatures far from a phase transition, the electrical conductivity and thermopower are related to the electron density of states near the Fermi energy $D(E_F)$. The conductivity is proportional to $D(E_F)$ while α is proportional to

$$\alpha \approx \frac{1}{D}\left[\frac{dD(E)}{dE}\right]_{E=E_F} \qquad (20)$$

evaluated at $E = E_F$. Hence, as D is increased, σ typically increases while α decreases. This is illustrated more precisely in the Mott relation given below:

$$\alpha = \left(\frac{\pi^2}{3}\right)\left(\frac{k_B^2 T}{e}\right)\frac{1}{\sigma}\left(\frac{d\sigma}{dE}\right)_{E=E_F} \qquad (21)$$

Doping can produce very substantial effects in these types of materials and can drastically change their electronic transport. Quantum well systems take advantage of a low-dimensional character through physical confinement in thin film structures to enhance the electronic properties of a given material (Hicks and Dresselhaus 1993).

Results on a new system of materials grown by the Kanatzidis group at Michigan State University exhibit very promising low-temperature thermoelectric properties and have yielded some of the highest ZT values achieved below $T \approx 250$ K (Chung *et al.* 2000). These materials are based on $CsBi_4Te_6$ and exhibit low-dimensional-anisotropic transport behavior. This may be an important aspect of future investigations of new materials. The figure of merit as a function of temperature is shown in Fig. 4 for the $CsBi_4Te_6$ material, with a $ZT \approx 0.8$ at $T \approx 220$ K. Another family of low-dimensional semiconductors or semimetals, called pentatellurides ($HfTe_5$ and $ZrTe_5$), have also shown promise as potential low-temperature thermoelectric materials (Littleton *et al.* 2001). These pentatelluride materials exhibit a thermopower that is relatively large over a broad range at low temperatures, $T < 250$ K. These materials do, in fact, exhibit anisotropic transport properties with the high conductivity axis being the growth axis (a-axis) and have a van der Waals gap between the individual layers (perpendicular to the b-axis), cleaving easily along these planes. This is similar to the Bi_2Te_3 system, which results in interplanar scattering for the phonons. The power factor of the Se-doped pentatellurides exceeds that of the optimally doped Bi_2Te_3 system $[(Bi_{1-x}Sb_x)_2(Te_{1-x}Se_x)_3]$ over the temperature range of measurement. However, investigations of the thermal conductivity indicate it is too high ($\kappa \approx 4$–8 W m^{-1}K^{-1}) for use as a TE material. Another group of complex chalcogenides that are of interest are the Tl_2SnTe_5 and Tl_2GeTe_5 systems that were identified as potential thermoelectric materials by Sharp *et al.* (1999). One aspect of these materials is that they exhibit a very low total thermal conductivity, (≈ 0.5 W m^{-1}K^{-1}) with a lattice thermal conductivity very near that of the estimated minimum thermal conductivity. These materials exhibit a $ZT \approx 0.6$ at $T \approx 300$ K. The practical use of these materials is limited by the extreme toxicity of the oxides of the Tl, however other compounds or derivatives not containing Tl may be possible.

4.2 *Thermoelectric Oxide Materials ($Na_xCo_2O_4$)*

The development of air and water-stable refractory metal oxides as thermoelectric materials has potentially numerous advantages over existing high-temperature TE materials, especially for power generation applications. They would be more rugged and easier to prepare than the conventional materials, and could be expected to have a wide range of operating environments. The class of $Na_xCo_2O_4$ compounds has displayed surprisingly effective thermoelectric properties (Terasaki *et al.* 1997). The materials are members of a large class of layered metal bronzes, wherein the transition metals form edge shared octahedral oxide layers, with the alkali metal ions located within the layers. The compounds contain mixed valence Co ions with itinerant electrons within the layers, and nonstoichiometric amounts of Na^+ ions between the layers. There are many possibilities in this system of materials both in single crystal and polycrystalline form.

4.3 *Zinc Antimonides (β-Zn_4Sb_3)*

β-Zn_4Sb_3 has been re-investigated for thermoelectric applications by the group at the Jet Propulsion Laboratory (Caillat *et al.* 1997). They have measured the thermal conductivity of polycrystalline β-Zn_4Sb_3 samples, and found it is nearly temperature independent between 300 K and 650 K. This is unusual, since typically a $1/T$ dependence is observed in κ_L in this temperature range. β-Zn_4Sb_3 exhibits values of $\kappa_L \approx 0.65$ W m^{-1}K^{-1} at 300 K, nearly two times lower than that of Bi_2Te_3 alloys. These very low values of κ_L along with its unusual temperature dependence can be predominantly attributed to the relatively complex crystal structure and to the presence of vacancies in the lattice. The electronic transport properties are typical of a semimetal with low electrical resistivity increasing with increasing temperature ($d\rho/dT > 0$).

The Seebeck coefficient also increases with increasing temperature and peaks at 675 K with a value of about 200 μV K^{-1}. This relatively high Seebeck value (for a semimetal) is the result of fairly large effective mass. The best ZT obtained to date on polycrystalline samples is about 1.4 at 675 K, as shown in Fig. 4 (Caillat *et al.* 1997). Above 765 K, β-Zn_4Sb_3 transforms into γ-Zn_4Sb_3, which has poorer thermoelectric properties. Band structure calculations (Kim *et al.* 1998) predict a metallic behavior with improved thermoelectric performance for lower doping levels. However, little success has been obtained experimentally in optimizing the doping level of this compound. The compound β-Zn_4Sb_3 forms a full range of solid solution with the isostructural compound Cd_4Sb_3 and further reduction in κ_L can be obtained for $Zn_{4-x}Cd_xSb_3$ mixed crystals, but these crystals appear to be even less temperature stable than β-Zn_4Sb_3 itself. Further optimization of the thermoelectric properties of these materials seems to be limited considering the difficulties with doping them and the restricted compositional variations possible.

4.4 *Half-Heusler Alloys*

Another group of materials, which are under investigation as potential thermoelectric materials are the half-Heusler alloys (Poon 2000). The half-Heusler alloys are intermetallic compounds with the general formula MNiSn where M is a group 4 transition metal (M = Zr, Hf, Ti). Half-Heusler alloys have a MgAgAs-type crystal structure, forming three interpenetrating f.c.c. sublattices with one Ni sublattice vacant. Heusler alloys (e.g., MNi_2Sn) differ from half-Heusler alloys in having the Ni sublattice fully occupied and are also metallic or half-metallic (see *Thermoelectric Devices: Refrigeration and Power Generations With No Moving Parts*) and exhibit interesting magnetic properties. Half-Heusler alloys, on the other hand, are small band gap semiconductors with a gap of ($E_G \approx 0.1$–0.5 eV).

The half-Heusler alloys exhibit a high negative thermopower ($-40\ \mu V\ K^{-1}$ to $-250\ \mu V\ K^{-1}$) and low electrical resistivity values (0.1–8 mΩ cm) both of which are necessary for a potential thermoelectric material. Some of the recent investigations have centered around the effect of Sb doping on the Sn site ($TiNiSn_{1-x}Sb_x$). The Sb doping leads to a relatively large power factor of 0.2–1.0 W $m^{-1}K^{-1}$ at room temperature for small concentrations of Sb. These values are comparable to that of Bi_2Te_3 alloys. The power factor is much larger at $T \approx 700$ K where it is over 4 W $m^{-1}K^{-1}$ making these materials very attractive for potential power generation considerations. Unfortunately the thermal conductivity is relatively high for a thermoelectric material (on the order of 10–20 W $m^{-1}K^{-1}$) and much of the current and future research is related to reducing the thermal conductivity while maintaining the high power factor.

Shen *et al.* (2002) reported the effect of partial substitution of Ni by Pd on the thermoelectric (TE) properties in the ZrNiSn half-Heusler system. This investigation yielded a thermoelectric figure of merit of $ZT = 0.7$ at $T \approx 800$ K in a sample of $Zr_{0.5}Hf_{0.5}Ni_{0.8}Pd_{0.2}Sn_{0.99}Sb_{0.01}$. Shen and co-workers were able to achieve values of $\kappa_L \approx 2$ W $m^{-1}K^{-1}$ at $T \approx 800$ K in a sample of $Zr_{0.5}Hf_{0.5}Ni_{0.5}Pd_{0.5}Sn_{0.99}Sb_{0.01}$ and $\kappa_L \approx 3$ W $m^{-1}K^{-1}$ in the high ZT sample identified above. They further state that possible grain size effects may be able to lower κ_L even further in concert with previous theoretical predictions (Sharp *et al.* 2001). Earlier results presented on reduction of grain boundary scattering on the lattice thermal conductivity appear to have observed these predicted results (Tritt *et al.* 2001).

4.5 Skutterudites

Skutterudites are the next class of materials that have received a lot of attention within the TE community since the 1990s. This is due in large part to the ability to greatly vary κ_L, which is facilitated by the filling of the voids within the structure with small diameter, large mass interstitials such as trivalent rare-earth ions (Nolas *et al.* 1999 and references therein, Uher 2000). Specifically, the smaller more massive ions that are incorporated within the skutterudite voids result in the lowest κ_L. This is shown clearly by Nolas and co-workers in an investigation of the thermal conductivity of $RECo_4Sb_9Ge_3$, with RE = La, Nd, Sm, as compared with the unfilled $IrSb_3$. The skutterudite structure is cubic (with space group Im3) and the unit cell contains eight AB_3 groups where A is the transition metal element Ir, Co, or Rh, and B is the pnictide element such as P, As, and Sb. It contains 32 atoms per unit cell, with the metal atoms on the corners of the eight "cubes" with six (four atom) Sb rings inside the cubes with two "voids" in the remaining "cubes". In context of the doping scheme for thermoelectrics the general crystal structure of these materials can be best described by the following designation, $\square_2CO_8Sb_{24}$.

The open square, □, indicates the presence of two voids in the structure which are typically filled by the rare earth atoms to produce the most pronounced effects on the thermal conductivity reduction. The other metal and pnictide sites are doped to charge compensate for the additional electrons contributed by the rare earth doping and to enhance the electronic properties. For example, a sample of $(Ce_yFe_{1-y})_xCo_{4-x}Sb_{12}$ exhibited a $ZT \approx 1.4$ at $T \approx 1000$ K as shown by the GM, JPL, and Oak Ridge groups, see Fig. 4. The reduction in the thermal conductivity of filled skutterudites is due the dynamic, or "rattling" disorder of the void-filling ions that substantially affect the phonon propagation through the lattice. More loosely bound "rattlers" produce local vibrational modes of lower frequency and are thus more effective in scattering the lower-frequency, heat-carrying phonons. The heavier and smaller the ion in the voids, the larger the disorder that is produced and therefore the larger the reduction in the lattice thermal conductivity, κ_L. This concept, first introduced by Slack, is corroborated by the large atomic displacement parameters (ADPs) that have been observed in alkaline earth and lanthanide-filled skutterudites. ADPs have been found to be very useful in thermoelectric materials research (Sales *et al.* 2000).

Investigations of the TE properties of Yb partially filled skutterudites yielded a figure of merit, $ZT > 1$ at $T \approx 600$ K (Nolas *et al.* 2000a). Similar results have also been observed in a series of Eu-doped Co-based skutterudites, which is shown in Fig. 4 (Lamberton *et al.* 2002). The thermal conductivity of the Yb- and Eu-filled skutterudites is much lower, with a relatively weak temperature dependence, as compared to that of $CoSb_3$. Only a 19% Yb in $CoSb_3$ dramatically decreases κ_L as compared to the "unfilled" compound. The "rattling" motion of the small and massive Yb in the voids of $CoSb_3$ substantially reduces κ_L in these compounds. Note that although the $Yb_{0.19}Co_4Sb_{12}$ specimen has a low thermal conductivity, it is still much higher than the estimated minimum thermal conductivity of $CoSb_3$, $\kappa_{min} \approx 0.3$ W $m^{-1}K^{-1}$. Although these results are encouraging, even higher ZT values may be forthcoming in skutterudite compounds.

4.6 *Clathrates*

Another class of materials that is very interesting for thermoelectrics comprises materials called clathrates (e.g., $Sr_8Ga_{16}Ge_{30}$) (Nolas *et al.* 2001b, Nolas *et al.* 2000b). These materials, like the skutterudites are also cage-like structures, thus allowing for a "rattling" mechanism with which to "tune" the thermal conductivity. Clathrates with frameworks built up from group 14 atoms have been known since the 1970s based on Na_8Si_{46} and Na_xSi_{136} structures. These materials have crystal structures closely related to those of type-I and type-II clathrate hydrates such as $(Cl_2)_8(H_2O)_{46}$ and $(CO_2)_{24}(H_2O)_{136}$. Many different compositions with these two structures are possible. They are of fundamental interest from the perspective of both bonding and their physical properties. Thus far, there has been a substantial amount of work on compounds with the type I crystal structure. A variety of different elements have been encapsulated inside the polyhedra of these structure types, including alkali metal, alkaline earth, and rare earth atoms. The very low glass-like thermal conductivity found in certain type I clathrate compounds has also attracted interest, especially for potential thermoelectric applications.

The bonding scheme in the clathrate hydrates is somewhat similar to that of diamond-structured group 14 elements. A similar framework can therefore be built up from elements such as Si, Ge, or Sn. The type I structure can then be represented by the general formula $X_2Y_6E_{46}$, where X and Y are alkali-metal, alkaline earth, or rare earth metal "guest" atoms that are encapsulated in two different polyhedra and E represents the group 14 elements such as Si, Ge, or Sn. The type II structure is represented by the general formula $X_8Y_{16}E_{136}$. Metallic variants of these compounds are most readily synthesized. Ba containing Si and Ge type I compounds have shown superconducting properties. Ternary type I clathrates, where charge compensation is established by substitutionally doping the tetravalent framework atoms with trivalent atoms such as gallium, result in semiconducting behavior. This was the approach used by Nolas and co-workers in synthesizing polycrystalline semiconducting type I clathrates in order to investigate their transport properties for possible TE applications.

One of the more interesting properties of the type I clathrates is the very distinct thermal conductivity. For example, the semiconductor $Sr_8Ga_{16}Ge_{30}$ exhibits a thermal conductivity of similar magnitude and temperature dependence as that of amorphous materials, where it can be lower than that of vitreous silica and very close to that of amorphous germanium. The low-temperature ($T<1$K) data indicate a T^2 temperature dependence. Higher temperature data show a minimum, or dip, indicative of resonance scattering. It is clear from these results that in the $Sr_8Ga_{16}Ge_{30}$ compound the traditional alloy phonon scattering, which predominantly scatters the highest frequency phonons, has been replaced by one or more much lower frequency scattering mechanisms. The scattering of the low-frequency acoustic phonons by the "rattle" modes of the encapsulated Sr atoms results in the low thermal conductivity. It is suggested that low-frequency vibrational modes of the "guest" atoms inside their oversized polyhedra are responsible for the large reduction in thermal conductivity in these clathrates. Room temperature structural refinements from single crystal and powder neutron scattering and x-ray diffraction reveal large atomic displacement parameters (ADPs) for atoms inside the structure of the type I clathrates. This is an indication of localized disorder within these polyhedra beyond typical thermal vibration. Large single crystals have been synthesized by the Oak Ridge group (Sales and co-workers) and extensive investigations of $X_8Ga_{16}Ge_{30}$ (where X = Eu, Sr, Ba) have been performed (Sales *et al.* 2001). The thermal conductivity properties were consistent with those observed earlier, in that they were low, $\kappa_L<1$ W m^{-1}K^{-1} (at $T\approx300$ K) and the $Eu_8Ga_{16}Ge_{30}$ and $Sr_8Ga_{16}Ge_{30}$ systems exhibited properties of a structural glass while the $Ba_8Ga_{16}Ge_{30}$ system exhibited a temperature dependence consistent with a crystalline material. This study concluded that "tunneling states" as well as "rattling states" are necessary to produce a true glass-like thermal conductivity in these materials. These materials have yet to exhibit a $ZT\approx1$, but these are relatively new TE materials and much work is ahead.

4.7 *Thin Films Materials Approach*

Researchers at the Research Triangle Institute (RTI) have demonstrated a significant enhancement in the thermoelectric figure-of-merit overcoming a major hurdle since the 1960s (Venkatasubramanian *et al.* 2002). These materials exhibited a $ZT\approx2.4$. The enhancement is attributed to creating a "nano-engineered" material that is efficient in thermal insulation while remaining a good electrical conductor, not easily found in nature. The thermal insulation arises from a complex localization-like behavior for phonons while the electron transmission is facilitated by optimal choice of band-offsets in these semiconductor heterostructures. These nano-engineered TE devices would offer a high coefficient of performance in solid-state refrigeration as well as allow high cooling power densities to be achieved, as high as 700 W cm^{-2}. In addition, these thin-film devices are extremely fast acting, within $\sim$10 μs, which is a factor of 23 000 better than state-of-the art bulk TE technology. The combination of performance, power density, and speed can lead to enormous applications in a gamut of different technologies. Also, there have been reports by Harman *et al.* at Lincoln Labs on

PbTe/PbTeSe quantum dot structures that yield a $ZT \approx 0.8$ at $T \approx 300$ K and $ZT \approx 2$ at $T \approx 550$ K (Harman *et al.* 2000). These materials have been grown as thick films, which have been "floated off" the substrate to yield free-standing films, which were measured to yield these results. Conservative estimates of the thermal conductivity were used in these calculations.

5. *Conclusion*

From these discussions it is apparent that there are many strategies towards investigating materials in the hope of identifying the next generation TE materials. In 2002 none of these materials have as yet displaced the current state-of-the-art materials (Bi_2Te_3 and SiGe) which have held that distinction since the 1970s. However, given the many materials that are yet to be investigated, as well as the ongoing research on the most promising new materials, there is certainly much more work ahead. While the thin film results are very exciting they may be most appropriate for small-scale electronic and opto-electronics applications where small heat loads or low levels of power generation are more appropriate. To address large-scale refrigeration (home refrigerators) or power generation (automotive or industrial applications) requirements, then higher performance bulk materials will have to be developed. The existing bulk TE materials are based on binary compounds and the ternary systems are just beginning to be investigated. A number of potential materials have been identified and discussed in Sect. 4. Only a small fraction of these ternary materials have been attempted. Certainly theoretical guidance, in terms of band structure calculations and modeling, will be essential to the identification of the most promising materials and compositions. In addition, rapid yet accurate characterization of materials is also essential to effectively advance this field of research.

See also: Thermoelectric Devices: Refrigeration and Power Generations With No Moving Parts

Bibliography

Caillat T, Fleurial J-P, Borshchevsky A 1997 Preparation and thermoelectric properties of semiconducting Zn_4Sb_3. *J. Phys. Chem. Solids* **58**, 1119–25

Chaiken P M 1990 An introduction to thermopower for those who might want to use it to study organic conductors and superconductors. In: Kresin V Z, Little W A (eds.) *Organic Superconductors*. Plenum Press, New York, pp. 101–5

Chung D Y, Hogan T, Brazis P, Rocci-Lane M, Kannewurf C, Bastea M, Uher C, Kanatzidis M G 2000 $CsBi_4Te_6$; a high-performance thermoelectric material for low-temperature applications. *Science* **287**, 1024–8

Goldsmid H J 1986 *Electronic Refrigeration*. Pion, London

Harman T, Taylor P J, Spears D L, Walsh M P 2000 Thermoelectric quantum-dot superlattices with high ZT. *J. Electron. Mater.* **29**, L1–4

Hicks L D, Dresselhaus M S 1993 Effect of quantum-well structures on the thermoelectric figure of merit. *Phys. Rev. B.* **47**, 12727–31

Kim S-G, Mazin I I, Singh D J 1998 First-principles study of Zn–Sb thermoelectrics. *Phys. Rev. B* **57**, 6199–203

Lamberton G A Jr, Bhattacharya S, Littleton R T IV, Kaeser M A, Tedstrom R H, Tritt T M, Yang J, Nolas G S 2002 High figure of merit in Eu-filled $CoSb_3$-based skutterudites. *Appl. Phys. Lett.* **80**, 598–600

Littleton R T IV, Tritt T M, Korzenski M, Ketchum D, Kolis J W 2001 Effect of Sb substitution on the thermoelectric properties of the Group IV pentatelluride materials $M_{1-X}Y_XTe_5$ (M = Hf, Zr and Ti). *Phys. Rev. B. Rapid Comm.* **64**, 121104–107

Morelli D T 1997 Thermoelectric devices. In: Trigg G L, Immergut E H (eds.) *Encyclopedia of Applied Physics*. Wiley-VCH, New York, Vol. 21, pp. 339–54

Nolas G S, Johnson D, Mandrus D G 2002 Thermoelectric materials and devices In: *Proc. 2002 Materials Research Society*. Warrendale, PA Vol. 691

Nolas G S, Kaeser M, Littleton R T IV, Tritt T M 2000a High figure of merit in ytterbium-filled skutterudite materials. *Appl. Phys. Lett.* **77**, 1855–7

Nolas G S, Morelli D T, Tritt T M 1999 Skutterudites: a phonon-glass-electron-crystal approach to advanced thermoelectrics energy conversion applications. *Annu. Rev. Mater. Sci.* **29**, 89–116

Nolas G S, Sharp J, Goldsmid H J 2001a *Thermoelectrics: Basic Principles and New Materials Developments*. Spinger, New York

Nolas G S, Slack G A 2001b Thermoelectric clathrates. *Am. Sci.* **89**, 136–41

Nolas G S, Slack G A, Schujman S B 2000b Semiconducting clathrates: a phonon electron crystal material with potential for thermoelectric applications glass. In: Tritt T M (ed.) *Semiconductors and Semimetals*. Recent Trends in Thermoelectric Materials Research, Academic Press, San Diego, Vol. 69, Chap. 6, pp. 255–300

Poon S J 2000 Electronic and thermoelectric properties of half-Heusler alloys. In: Tritt T M (ed.) *Semiconductors and Semimetals*. Recent Trends in Thermoelectric Materials Research, Academic Press, San Diego, Vol. 70, Chap. 2, pp. 37–76

Rowe D M (ed.) 1995 *CRC Handbook of Thermoelectrics*. CRC Press, Boca Raton, FL

Sales B C, Chakoumakos B C, Jin R, Thompson J R, Mandrus D 2001 Structural, magnetic, thermal, and transport properties of $X_8Ga_{16}Ge_{30}$ (X = Eu, Sr, Ba) single crystals. *Phys. Rev. B* **63**, 113–21245

Sales B C, Mandrus D G, Chakoumakos B C 2000 Use of atomic displacement parameters in thermoelectric materials research. In: Tritt T M (ed.) *Semiconductors and Semimetals*. Recent Trends in Thermoelectric Materials Research, Academic Press, San Diego, Chap. 1 Vol. 70, pp. 1–36

Sharp J W, Sales B C, Mandrus D G, Chakoumakos B 1999 Thermoelectric properties of Tl_2SnTe_5 and Tl_2GeTe_5. *Appl. Phys. Lett.* **74**, 3794–6

Sharp J W, Poon S J, Goldsmid H J 2001 Boundary scattering of phonons and thermoelectric figure of merit. *Physica Status Solidi (a)* **187**, 507–16

Shen Q, Chen L, Goto T, Hirai T, Yang J, Meissner G P, Uher C 2002 Effects of partial substitution of Ni by Pd on the thermoelectric properties of ZrNiSn-based half-Heusler compounds. *Appl. Phys. Lett.* **79**, 4165–7

Slack G A 1979 The thermal conductivity of non-metallic crystals. In: Seitz F, Turnbull D, Ehrenreich H (eds.) *Solid State Physics*. Academic Press, New York, Vol. 34, pp. 1–71

Slack G A 1995 New materials and performance limits for thermoelectric cooling. In: Rowe D M (ed.) *CRC Handbook of Thermoelectrics*. CRC Press, Boca Raton, FL, pp. 407–40

Slack G A, Tsoukala V G 1994 Some properties of semiconducting $IrSb_3$. *J. Appl. Phys.* **76**, 1665–71

Terasaki I, Sasago Y, Uchinokura K 1997 Large thermoelectric power in $NaCo_2O_4$ single crystals. *Phys. Rev. B* **56**, R82685–689

Tritt T M 1996 Thermoelectric materials beat the heat. *Science* **272**, 1276–7

Tritt T M (ed.) 2000a *Semiconductors and Semimetals, Recent Trends in Thermoelectric Materials Research*. Academic Press, San Diego, **Vol. 69–71**.

Tritt T M, Nolas G S, Mahan G, Mandras D, and Kanatzidis M G (eds.) 2000 *New materials for small-scale thermoelectric refrigeration and power generation applications*. Proc. 2000 Materials Research Society, Warrendale PA, **Vol. 626**.

Tritt T M, Bhattacharya S, Xia Y, Ponnambalam V, Poon S J, Thadhani N 2001 Effect of various grain structures and sizes on the thermal conductivity of Ti-based half-Heusler alloys In: *Proc. 20th Int. Conf. Thermoelectrics*. Beijing, China. ICT 2001, IEEE Press, pp. 7–12

Tritt T M, Kanatzidis M G, Lyon H B, and Mahan G D (eds.) 1997 *Thermoelectric materials—new directions and approaches*. Proc. 1997 Materials Research Society. **Vol. 478,** Warrendale PA

Tritt T M, Kanatzidis M G, Lyon H B, and Mahan G D (eds.) 1998 *New material for small-scale thermoelectric refrigeration and power generation applications*. Proc. 1998 Materials Research Society, Warrendale, PA, **Vol. 545**.

Uher C 2000 Skutterudites: prospective novel thermoelectrics. In: Tritt T M (ed.) *Semiconductors and Semimetals*. Recent Trends in Thermoelectric Materials Research, Academic Press, San Diego

Venkatasubramanian R, Siivola E, Colpitts T, O'Quinn B C 2002 Thin-film thermoelectric devices with high room temperature figures of merit. *Nature* **413**, 597–602

Thin-film Materials for Solid-state Rechargeable Batteries

N. J. Dudney, J. B. Bates, and B. J. Neudecker

There is an extensive research and development effort to develop a compact, lightweight, and reliable rechargeable lithium battery to power a variety of products from micromechanical devices, to consumer electronics, to automobiles. Lithium-based battery chemistries are considered the most promising owing to their high electrochemical potential and lithium's low atomic weight. Within this field is a smaller effort to develop and commercialize thin-film batteries with an inorganic solid electrolyte. Although the ionic conductivity of these solid electrolytes is ~4 orders of magnitude lower than that for liquid electrolytes and ~2 orders of magnitude lower than the best polymer electrolytes, this handicap can be readily overcome by making the electrolyte 1 μm thick. More importantly, an inorganic electrolyte offers superior chemical, mechanical, and thermal stability, which translates into an extended cycle life, shelf-life, and safety for the rechargeable battery. The drawback is the added expense in the fabrication of an all-solid-state thin-film device.

Figure 1 shows a schematic cross-section of such a thin-film battery. Because it is a thin-film device, the structure must be built on a substrate, which in many cases may be part of an integrated device. Typical energy densities are of the order of $3.6\,J\,cm^{-2}$ ($1\,mWh\,cm^{-2}$) so fabrication of batteries with even moderate energies requires large film areas, which may then be rolled, folded, or stacked to form a convenient geometry with the required battery voltage. Thin-film materials 0.1–10 μm thick are most often grown by deposition from the vapor. These synthesis techniques may in some cases yield materials with different compositions or microstructures than can be readily obtained by conventional bulk synthesis.

1. *Thin-film Electrolytes*

The requirements for the battery electrolyte layer are the most stringent. This material must have a good Li^+ conductivity, be electrochemically stable with both the cathode and anode material at high cell potentials, and have a negligible electronic conductivity to minimize the self-discharge of the cell. The electrolyte must tolerate volume changes of the anode and cathode and the high lithium ion flux without developing lithium filaments along cracks or other flaws. Glassy materials are considered preferable to polycrystalline films with internal boundaries that are prone to form leakage paths.

The only single-phase electrolyte currently known to meet all these criteria is an amorphous lithium phosphorus oxynitride, first synthesized by Bates in 1992, with a composition near $Li_{2.9}PO_{3.3}N_{0.46}$. This material, known as "Lipon," is deposited at near ambient temperatures and at rates up to $1\,\mu m\,h^{-1}$ by RF (radio-frequency) magnetron sputtering of a dense Li_3PO_4 ceramic target in a nitrogen plasma. Figure 2 shows a cross-section of a Lipon film on a polymer substrate. The films are dense, $\sim 2.5\,g\,cm^{-3}$, flexible, and form a uniform conformal coating free of grain boundaries and cracks. Electrical measurements by Yu demonstrated that Lipon films are stable in contact with lithium metal at cathode potentials up to 5.5 V. The Li^+ conductivity is about $2\,\mu S\,cm^{-1}$ while that for electronic carriers is $< 10^{-14}\,S\,cm^{-1}$. Compared to sputter-deposited amorphous Li_3PO_4, the addition of nitrogen enhances the Li^+ conductivity by a factor of 40 and increases the electrochemical stability window by 2 V. The role of the nitrogen is not completely understood, but the Li^+ transport properties of Lipon could be associated with the nitrogen cross-linking of two or three $(PO_4)^{3-}$ groups. Cells with the Lipon electrolyte have demonstrated cycle lives of 70000 cycles between 4.3 and 3 V and shelf lives of at least 3 years with no significant self-discharge.

Lipon is an example of a material that cannot be achieved readily by bulk synthesis. With the high lithium content, the Lipon composition is outside of the normal glass-forming region; however, no phase separation occurs in the sputter-deposited films.

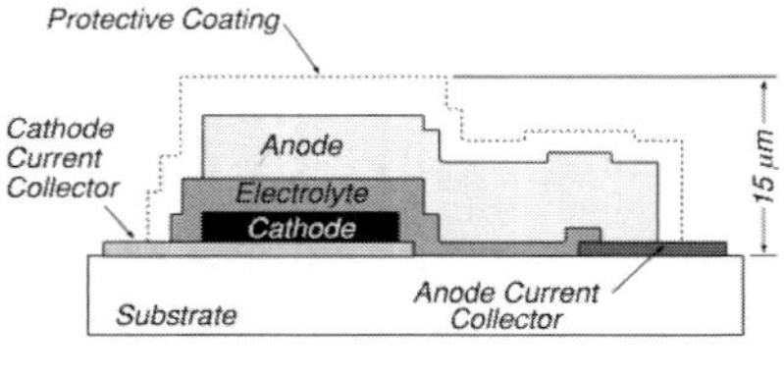

Figure 1
Schematic cross-section of a thin-film battery with a lithium metal anode.

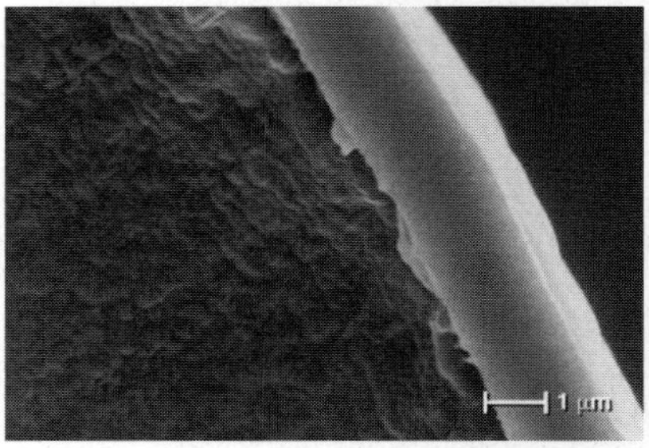

Figure 2
SEM cross-section of a Lipon film deposited on a polypropylene membrane.

Bombardment of the film with energetic ions not only introduces the nitrogen ions, but promotes a dense structure and good adhesion to the substrate material.

As reviewed earlier, others have investigated a number of amorphous lithium electrolyte films prepared by sputtering or thermal evaporation techniques. The most successful of these electrolytes, as demonstrated by extended cycling of a thin-film battery, are $Li_{3.6}Si_{0.6}P_{0.4}O_4$ (Kanehori), $Li_3PO_{2.67}IS_{0.83}$ (Jones and Akridge 1996), $Li_{1.2}B_{0.76}S_{0.31}O_{2.7}$ (Meunier), and $Li_{3.4}V_{0.6}Si_{0.4}O_4$ (Ohtsuka). Of these, the vanadate electrolyte had a significant electronic leakage current and the others reacted with the lithium anode. This reaction was ameliorated by Creus through the addition of a LiI film between the electrolyte and the lithium metal. Recent work on compacted ultrafine particles of lithium-doped BPO_4 appear promising, but although the Li^+ conductivity of this electrolyte is acceptable, the internal resistance of the cell is still very high (Jak *et al.* 1999).

2. *Thin-film Cathodes*

When a lithium cell is discharged, a reduction reaction occurs at the cathode which involves incorporating Li^+ ions from the electrolyte and electrons from the external circuit through the current collector film. The cathode material must therefore be a good mixed Li^+ ion and electronic conductor over a wide composition range and be able to accommodate a large density of lithium ions reversibly at a high cell potential. The cell potential is given by the free energy for the cell reaction

$$\text{cathode} + \text{Li(anode)} \rightarrow \text{Li(cathode)} + \text{anode}$$

Most cathodes are transition metal oxides or sulfides, such as Li_xCoO_2, $Li_xMn_2O_4$, and TiS_2, which can readily accommodate lithium in crystallographic planes or channels with little change in the lattice structure or spacing. Generally a reduction of the transition metal cations is assumed to compensate for the Li^+ addition in these insertion compounds, although recent calculations by Ceder for Li_xCoO_2 found a substantial, $\sim 50\%$, electron transfer to oxygen.

Thin-film cathodes have been deposited by a variety of techniques including RF and d.c. sputtering, electron-beam and thermal evaporation, laser ablation, spray pyrolysis, and chemical vapor deposition onto either ambient temperature or heated substrates. Some depositions are followed by *in situ* or *ex situ* annealing to refine the composition or crystallinity. The details of the processing conditions often result in compositional, crystallographic, and microstructural variations that affect the material's performance in the thin-film battery. As a comprehensive discussion of cathode materials is beyond the scope of this section, two examples have been chosen to illustrate the best performance that can be achieved from crystalline (c) and amorphous (a) materials.

Exceptionally high lithium transport rates have been obtained by Bates *et al.* (2000) for crystalline $(c)Li_xCoO_2$ thin-film cathodes cycled over compositions of $0.5 < x < 1$. This corresponds to cell potentials of 4.2 to 3 V with respect to a lithium anode. Figure 3 shows typical discharge curves for several thin-film batteries including a Li/Lipon/$(c)LiCoO_2$ cell. When discharged at infinitely low current densities, such a curve reflects the equilibrium thermodynamic activity of the lithium in the cathode; at higher current densities, polarization and IR losses suppress the cell potential below the equilibrium value. Batteries with $2.5\,\mu m$ thick films of Li_xCoO_2 will supply $> 50\%$ of the maximum capacity at current densities up to $10\ mA\,cm^{-2}$.

The $(c)LiCoO_2$ cathode films are fabricated by RF magnetron sputter deposition onto unheated substrates. The as-deposited films are amorphous by x-ray diffraction and crystallize upon annealing at temperatures of 400 to 700°C. Figure 4 shows the microstructure of an annealed film. The crystallite grains, $\sim 0.1\,\mu m$ across, are aligned vertically in a fibrous structure. The larger aggregate columns initiate from the grains of the polycrystalline substrate and then shrink during the anneal. The crystalline films have a strong preferential crystallographic orientation which depends on both the film thickness and the substrate temperature (60–300°C) during the film deposition. Crystallite grains in a very thin film are oriented to minimize the surface energy, while the volume strain energy dominates the free energy in thicker polycrystalline films. The observed distribution of grains with different orientations correlate with Hart's calculated volume strain energies.

Cathodes have also been prepared with unannealed films. Elimination of the high-temperature anneal greatly expands the choice of substrates and battery configurations. The lithium diffusivity,

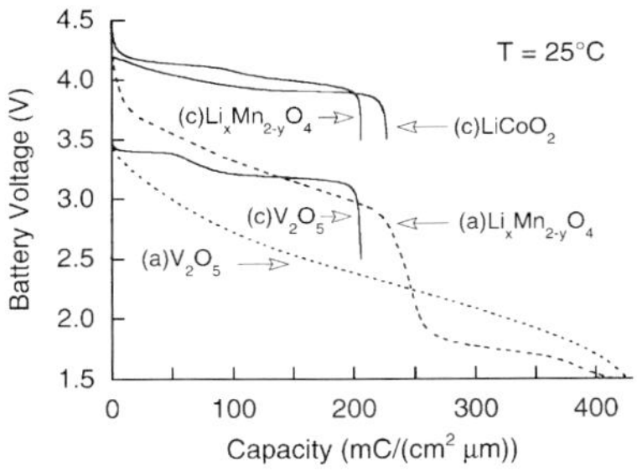

Figure 3
Discharge curves for batteries with a lithium metal anode and the designated cathode. Capacity expressed in mC per unit volume of the cathode film.

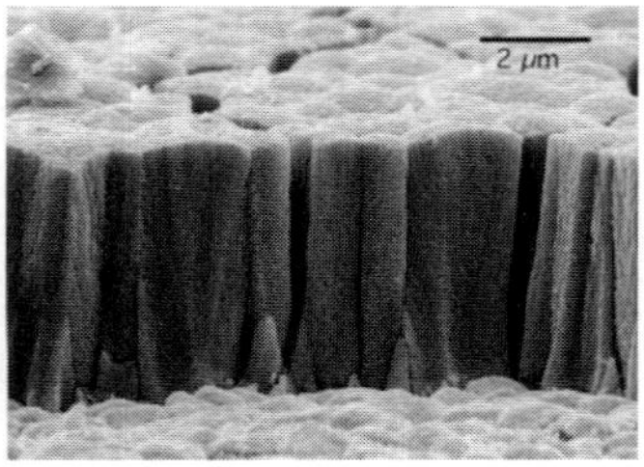

Figure 4
Fracture cross-section of a thick (c)$LiCoO_2$ cathode deposited and crystallized on a platinum-coated polycrystalline Al_2O_3 substrate.

however, is typically 100 times lower than that for more highly crystalline films. This limits the practical discharge current densities to $<50\,\mu A\,cm^{-2}$. Such cathodes have been deposited by RF magnetron sputtering from targets of $LiMn_2O_4$ by Dudney *et al.* (1999). Although the films are amorphous by x-ray diffraction, TEM reveals that the material is at least partially nanocrystalline with electron diffraction rings exhibiting the spacings of the spinel structure. The film composition is invariably manganese deficient, $Li_xMn_{2-y}O_4$, with values of x and y sensitive to the film deposition rate and the process gas composition and pressure. As shown in Fig. 3, these cathodes cycle reversibly between 4.5 and 1.5 V, which corresponds to a change in the manganese oxidation state of Mn^{4+} to Mn^{3+}. These nanocrystalline cathodes can more readily accommodate the phase transformations that occur within this voltage range, which often leads to battery degradation for the more crystalline cathodes.

Other thin-film cathodes that have proven successful in all solid-state batteries for at least modest current densities ($>10\,\mu A\,cm^{-2}$) and cycle numbers (~ 100) include crystalline films of $LiMn_2O_4$ and TiS_2. The (c)$LiMn_2O_4$ films were grown by Park *et al.* (1998) using sputtering plus a rapid anneal, by Ueda and Bates (1999) using sputtering and a furnace anneal, and by Bates (1994) by e-beam evaporation. Crystalline TiS_2 was deposited by Kanehori using CVD and by Jones and Akridge (1996) with RF magnetron sputtering. Several unannealed x-ray amorphous cathodes have been discussed in earlier reviews, including $TiO_{1.5}S_{0.7}$, MnO_{3-x}, and WO_3. More recent reports of cathodes evaluated in solid-state cells include (a)V_2O_5 deposited by Kwak *et al.* (1996) using reactive sputtering and by Lee *et al.* (1999) using thermal evaporation, and MoO_{3-x} of Yamaki *et al.* (1996). Many other thin-film cathodes have been reported including cathodes formed from solution by sol–gel processing (Park *et al.* 1998), and an electrochemical hydrothermal reaction in LiOH (Han *et al.* 1998). These cathodes cycle well in a liquid cell, but may not prove useful with a thin-film solid electrolyte, which does not permeate the open porosity of the cathode.

3. Thin-film Anodes

Most thin-film batteries use a lithium metal anode, which is easily deposited by thermal evaporation and gives the best battery performance. Such "Li-batteries" have the maximum battery voltage, discharge rate capability, and energy density. However, two drawbacks of the lithium anode, the low melting temperature (180°C) and the high reactivity with air, has motivated development of other thin-film anode materials. As for a cathode, the thin-film anode must be a good mixed conductor which can accommodate a large density of Li^+ ions reversibly. The removal of Li^+ must occur at a low electrochemical potential, <1 V with respect to lithium, in order to maximize the battery voltage. Many different anode materials, including lithium alloys, carbon, and oxide insertion compounds, are being developed for thicker batteries known as "Li-ion" or "rocking chair" batteries, but few have as yet been studied as thin-film anodes.

Goldner *et al.* (1996) were the first to report a thin-film lithium-ion battery with a carbon anode deposited by either an anodic arc process or by electron beam evaporation. This anode was used with a $LiCoO_2$ cathode and a sputter-deposited Lipon electrolyte; the complete battery was cycled 400 times at moderate current densities. The charge per cycle corresponded to a maximum lithium content of $x \sim 0.9$ to 1.0 for Li_xC_6.

More recently, following the demonstration of SnO anodes in thick battery assemblies, Neudecker *et al.* (1999) deposited thin films of an amorphous silicon tin oxynitride, "Siton," as a thin-film anode material. The films were deposited by RF magnetron sputtering of a $SnO{:}SiO_2$ target in nitrogen giving a typical film composition of $SiSn_{0.87}O_{1.2}N_{1.72}$. Because Siton contains no lithium, the battery cathode must be synthesized with a fully discharged composition as it is the sole source of cyclable lithium ions. Upon the initial charge of the battery, some of this lithium becomes trapped in the Siton anode, presumably as Li_2O and/or Li_3N, while the additional lithium forms an alloy with silicon and/or tin. The reversible lithium exchange is believed to occur within the alloy of the anode film. Figure 5 shows the discharge curve for such a battery with a (c)$LiCoO_2$ cathode, a Lipon electrolyte, and the Siton anode. Compared with the discharge curve for a similar battery with a lithium metal anode, the lower discharge voltage reflects the decreased lithium activity in the Siton anode and the reduced discharge capacity results from the irreversibly trapped lithium. In addition, Neudecker reports similar behavior for batteries with thin-film sputter deposited zinc and tin nitride films. Each of these

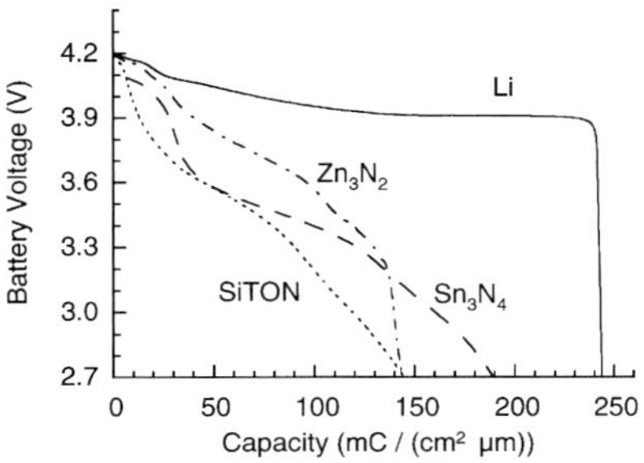

Figure 5
Discharge curves for batteries with a (c)$LiCoO_2$ cathode and the designated anode. Capacity expressed in mC per unit volume of the cathode film.

cells can be discharged at high current densities of $5\,mA\,cm^{-2}$ with good cathode utilization.

4. Protective Films

A protective coating is needed to prevent the loss of lithium, which otherwise would react with gases in the air. This film must form an impermeable barrier to both reactive gases and either Li^+ ions or electrons. Lipon is an ideal protective film, being stable in contact with lithium, free of pores and cracks, and an excellent electronic insulator. No doubt other ceramic insulators would also form good barriers. A multilayer coating ~ 5–$8\,\mu m$ thick of sputtered titanium and poly-*p*-xylylene (Parylene C) used by Bates is the only protective coating that has been evaluated under long-term cycling conditions.

5. Summary

The success of solid state thin-film rechargeable lithium batteries will depend on the efficient and reliable synthesis of thin-film ceramic materials for the electrolyte, cathode and perhaps anode components. Synthesis of materials by vapor deposition techniques present opportunities of forming unique compositions and microstructures which are not amenable to conventional powder synthesis processes.

6. Acknowledgment

Research was sponsored by the US Department of Energy's Divisions of Materials Science and Chemical Science and the Laboratory Technology Transfer Research Program under contract DE-AC05-00OR22725 with the Oak Ridge National Laboratory, managed by UT-Battelle, LLC.

Bibliography

Bates J B 1994 In: Bloor D, Brook R J, Flemings M C, Mahajan S (eds.) *Encyclopedia of Advanced Materials*. Pergamon, Oxford

Bates J B, Dudney N J, Neudecker B J, Hart F X, Jun H P, Hackney S A 2000 Preferred orientation of polycrystalline $LiCoO_2$ films. *J. Electrochem. Soc.* **147**, 59–70

Bates J B, Dudney N J, Neudecker B J, Wang B 2000 In: Osaka T, Datta M (eds.) *New Trends in Electrochemical Technology: Energy Storage Systems in Electronics*. Gordon and Breach, Singapore, pp. 453–85

Bates J B, Lubben D, Dudney N J, Hart F X 1995 Five-volt plateau in $LiMn_2O_4$ thin films. *J. Electrochem. Soc.* **142**, L149–51

Ceder G, Kohan A F, Aydinol M K, Tepesch P D, Van der Ven A 1988 Thermodynamics of oxides with substitutional disorder: a microscopic model and evaluation of important energy contributions. *J. Am. Ceram. Soc.* **81**, 517–25

Dudney N J, Bates J B, Zuhr R A, Young S, Robertson J D, Jun H P, Hackney S A 1999 Nanocrystalline $Li_xMn_{2-y}O_4$ cathodes for solid-state thin-film rechargeable lithium batteries. *J. Electrochem. Soc.* **146**, 2455–64

Goldner R, Lui T Y, Slaven S, Gerouki A, Zerigian P, Haas T E, Arntz F O, Jones S 1996 Development of a thin-film $Li_{1-y}CoO_2/Li_xC_6$ rocking-chair battery. In: Bates J B (ed.) *Thin-film Solid Ionic Devices and Materials*. Electrochemical Society, Boston, 95-22, pp. 173–82

Han K-S, Song S-W, Yoshimura M 1998 Fabrication in a single synthetic step of electrochemically active $LiMO_2$ (M = Ni and Co) thin-film electrodes using soft solution processing at 20–200°C. *Chem. Mater.* **12**, 2183–8

Hart F X, Bates J B 1998 Lattice model calculation of the strain energy density and other properties of crystalline $LiCoO_2$. *J. Appl. Phys.* **83**, 7560–6

Jak M J G, Ooms F G B, Kelder E M, Legerstee W J, Schoonman J, Weisenburger A 1999 Dynamically compacted all-ceramic lithium-ion batteries. *J. Power Sources* **80**, 83–9

Jones S D, Akridge J R 1995 In: Munshi M Z A (ed.) *Handbook of Solid State Batteries and Capacitors*. World Scientific, NJ, Chap. 11, pp. 209–30

Jones S D, Akridge J A 1996 A microfabricated solid-state secondary Li battery. *Solid State Ionics* **86–88**, 1291–4

Julien C 1994 In: Pistoia G (ed.) *Lithium Batteries, New Materials, Developments and Perspectives*. Elsevier, New York, Chap. 5, pp. 167–238

Kwak B S, Bates J B, Hart F X 1996 Capacity fade in thin-film lithium-aV_2O_5 Cells. In: Bates J B (ed.) *Thin-film Solid Ionic Devices and Materials*. Electrochemical Society, Boston, 95-22, pp. 137–47

Lee S-H, Liu P, Tracy C E, Benson D K 1999 All-solid-state rocking chair lithium battery on flexible Al substrate. *Electrochem. Solid-state Lett.* **2**, 425–7

Neudecker B J, Zuhr R A, Bates J B 1999 Lithium silicon tin oxynitride (Li_ySiTON): high-performance anode in thin-film lithium-ion batteries for microelectronics. *J. Power Sources* **81–82**, 27–32

Park Y J, Kim J G, Kim M K, Chung H T, Um W S, Kim M H, Kim H G 1998 Fabrication of $LiMn_2O_4$ thin films by sol–gel method for cathode materials of microbattery. *J. Power Sources* **76**, 41–7

Park Y-S, Lee S-H, Lee B-I, Joo S-K 1999 All-solid-state lithium thin-film rechargeable battery with lithium manganese oxide. *Electrochem. Solid State Lett.* **2**, 58–9

Ueda A, Bates J B 1999 3d metal doped $LiMn_2O_4$ thin films. In: Surampudi S, Marsh R A (eds.) *Lithium Batteries*. Electrochemical Society, Boston, 98-1, pp. 286–95

Yamaki K-I, Ohtsuka H, Shodai T 1996 Rechargeable lithium thin-film cells with inorganic electrolytes. *Solid State Ionics* **86–88**, 1279–84

Yu X, Bates J B, Jellison G E, Sales B C 1997 A stable thin-film lithium electrolyte: lithium phosphorus oxynitride. *J. Electrochem. Soc.* **144**, 524–32

Thorium Fuel Cycles

P. Rodriguez

There are three important isotopes in which a nuclear fission chain reaction can be sustained: uranium-235, plutonium-239, and uranium-233. Of these, only ^{235}U occurs in nature; the other two fissile materials, ^{239}Pu and ^{233}U, are produced by transmutation of ^{238}U and ^{232}Th respectively and are therefore called fertile materials. Owing to the natural abundance of the two fertile materials, (^{238}U is ~140 times more abundant than ^{235}U and ^{232}Th is at least 430 times more abundant than ^{235}U), there has been interest in, and research and development efforts towards establishing plutonium and thorium fuel cycles from the very early days of nuclear technology (see *Nuclear Fuel Cycles*; *Fast Breeder Reactors: Fuels*; *Nuclear Reactor Materials*). This article attempts to summarize and catalogue the activities related to the thorium cycle and also to look at the scenarios emerging for thorium utilization in the twenty-first century.

1. *Salient Physics Characteristics of Thorium*

To understand the nuclear aspects of the thorium cycle, we need to compare the relevant physical characteristics of the fertile isotopes ^{232}Th and ^{238}U and of the fissile isotopes ^{233}U, ^{235}U, and ^{239}Pu. An important aspect of the utilization of thorium, and probably one reason for its slow progress, is the fact that unlike in natural uranium, there is no fissile isotope present. Thus its use in reactors is to be supported by another fuel containing ^{235}U or ^{239}Pu, until the required inventory of ^{233}U is built up by conversion or breeding.

The most important parameter for a fissile isotope is the "number of fission neutrons produced per neutron absorbed in the nuclide," the so-called η (eta) value. Table 1 gives the η values and the relevant radioactivity characteristics of the three fissile isotopes. One of the neutrons produced per fission must be absorbed in another fissile atom to continue the chain reaction. If L is the number of neutrons lost through leakage and absorption in fission products, coolant, moderator, structural and control materials, etc., the number of neutrons available for absorption by any fertile nuclide in the fuel/blanket will be equal to (η–1–L), which is termed the Conversion Ratio (CR)—the number of neutrons available for productive absorption in the fertile material per neutron absorbed in a fissile atom. When CR > 1.0, the reactor system is said to "breed" and the term Breeding Ratio or BR becomes synonymous with CR.

It is clear from Table 1 that with ^{233}U, under favorable conditions, breeding is possible in both thermal and fast reactors. On the other hand with ^{239}Pu, breeding is possible only in a fast reactor.

Even though ^{233}U has an η value much greater than 2.0 in the thermal spectrum, the CR values attained in actual reactor systems are ≤1. There are mainly two factors responsible for this. One is the phenomenon of "resonance," high absorption cross-sections (i.e., reaction probabilities) at discrete neutron energies. The resonance integral of thorium is one third that of ^{238}U and fewer neutrons are absorbed in ^{232}Th. Another factor is the loss of neutrons by absorption in the intermediate nuclide ^{233}Pa in the reaction chain of ^{232}Th to ^{233}U, which is much higher than in the corresponding intermediate nuclide ^{239}Np in the reaction chain of ^{238}U to ^{239}Pu. The net result is that the maximum CR for the thorium cycle postulated in a thermal reactor is 1.07 for the Molten Salt Breeder Reactor (MSBR) and other thermal reactors can at best be optimized to reach a CR value of 1.0.

Notwithstanding the almost equal value of η for ^{233}U and ^{239}Pu in fast reactors, the BR for the Th-232/U-233 fuel cycle is about 20% less than for a $^{238}U/^{239}Pu$ fuel cycle for similar fuel materials (metal/oxide/carbide/nitride). In addition to the two factors mentioned in the last paragraph, there is another factor that contributes in the fast spectrum to lowering the BR in the thorium cycle compared to the plutonium cycle. The fission cross-section of ^{232}Th is much lower than that of ^{238}U, such that only 2% of the fission reactions in the Th/U-233 cycle take place in Th-232 compared to ~15% in U-238 in the plutonium cycle. In spite of this, for the conventional

Table 1
η-value and other characteristics of fissile isotopes.

Isotope	η Thermal	η Fast	Radioactivity	Handling restrictions
U-233	2.28	2.5	Strong γ and α (U-232)[a]	Remote handling, shielding
U-235	2.04	2.2	Weak γ and α (U-234)[a]	Shielding or glove box not required
Pu-239	1.94	2.6	Strong α	Handle in glove box

[a]U-232 is a product in the Th-232 to U-233 conversion reactions and is a problem due to the build-up of highly gamma active daughter products as explained in Section 7 U-234 the corresponding product in the U-238 to U-239 conversion is only weakly radioactive.

oxide fuel, a BR of ~1.1 can be reached for the thorium cycle; the BR further increases to ~1.2 for carbide/nitride and for metallic fuels. The FBR with the use of thorium fuels has two specific advantages, *viz.*, a negative sodium void coefficient of reactivity and a more negative Doppler coefficient, both of which improve reactor safety.

2. *Materials Aspects of the Thorium Cycles*

Fuel fabrication, irradiation performance in the reactor, and refabrication of recycled fuel are the three relevant and important components in evaluating the materials aspects of the thorium cycles and in comparing with the uranium and plutonium cycles. The specific materials problems and consideration in the above context vary with the reactor concept, the specific form (metal oxide, carbide, etc.) in which the fuel is to be used, fuel design, etc. As far as the ceramic fuel compounds are concerned, the respective thorium compounds have the same crystal structures as the uranium/plutonium compounds and they easily form a complete series of solid solutions. The melting points are also high, as for the U/Pu compounds. Thorium metal has a f.c.c. crystal structure, which is isotropic and it retains this crystal form up to 1623 K; this is a distinct advantage compared to uranium and plutonium, which are anisotropic and have several allotropic forms. Rodriguez and Sundaram 1981) have summarized these aspects, including irradiation performance, in a comprehensive review of materials aspects of the thorium cycle.

3. *Reactor Systems for Thorium Cycles*

The different reactor systems in which thorium cycles have been considered are the following: Light Water Reactors (LWR)—both Boiling Water Reactors (BWR) and Pressurized Water Reactors (PWR); High-Temperature Gas-Cooled Reactors (HTGR); Molten Salt Breeder Reactors (MSBR); Pressurized Heavy Water Reactors (PHWR); Aqueous Suspension Heavy Water Reactors (ASHWR); Open Lattice Pressure Vessel Heavy Water Reactors (OLPVHWR); Advanced Heavy Water Reactors (AWHR); Fast Breeder Reactors (FBR); and Source-Driven Reactors (SDR).

For a detailed discussion of the different reactor systems and their relative merits for thorium utilization see Rodriguez and Sundaram (1981), Lung (1997) and International Atomic Energy Agency (IAEA 2000).

4. *Past Experience, Present Status, and Emerging Trends*

In the following, a country-by-country summary of all the work done so far for thorium fuel cycle relating to reactor systems, fuel systems, and irradiation experiments is presented. The present interests and emerging concepts are also discussed. The excellent surveys by Lung (1997) and IAEA (2000) are the sources for the information, unless otherwise indicated specifically.

4.1 USA

Early neutron experiments generating the physics data on thorium and U-233 were carried out in the late 1950s and early 1960s in the Advanced Epithermal Thorium Reactor (AETR) at Oak Ridge. Following this, in the 1960s and 1970s, whole-core demonstrations of thorium–uranium oxide fuels in LWR were carried out with two different approaches:

(i) Mixing ThO_2 with highly enriched UO_2 in a uniform lattice in the BORAX–IV, Indian Point PWR and Elk River BWR reactors.
(ii) Using a heterogeneous arrangement of seed and blanket regions, where the blanket has less uranium with the feasibility of net breeding in the core, in the Shippingport Light Water Breeder Reactor (LWBR).

Of the first type, the 28 MWe Elk River BWR was operated during 1962–1968 and concurrently the 270 MWe Indian Point I PWR was operated during 1962–1965. The Shippingport Power Station, the very first station that demonstrated the PWR concept for commercial power generation, was operated as an experimental LWBR (70 MWe) on the thorium cycle with the seed-blanket configuration from 1977 to 1982. All the above reactor experiments have operated with high availability factor (>65%) and have generated valuable information and data.

In the United States, several reactor types other than LWR were also studied and operated on the thorium cycle. The first gas-cooled, graphite-moderated 40 MWe Peach Bottom prototype HTGR was operated between 1966 and 1974, producing 1400 GWh of electricity with an availability factor of ~70% in about seven years. The industrial follow-up reactor, the 300 MWe Fort St Vrain, was probably too quickly extrapolated as it met with many structural problems and in 12 years (1976–1989) produced only ~6000 GWh of power. The fuel was in microspheres of thorium carbide and $Th/^{235}U$ carbide (4/1) coated with silicon oxide and pyrolytic carbon to retain fission products (see Rodriguez and Sundaram 1981). About 2.5×10^4 kg of thorium fuel was manufactured for the reactor and valuable data on reactor physics data, and fuel performance (which was good) have been generated.

The most celebrated US efforts for the thorium cycle have been the study and development of molten salt reactors at Oak Ridge. These studies started as early as 1947 for reactors for aerospace applications. The potential for civilian applications and thorium cycle was soon recognized and a development program was started in 1956. The Molten Salt Reactor Experiment (MSRE) 8 MWth was operated in the 1960s, fueled with U-235/U-238 in the first two years and with U-233 in the remaining 1.5 years. The program included extensive studies on Molten Salt Breeder Reactor (MSBR) concepts. The program was interrupted in 1973, resumed briefly in 1974, and finally terminated in 1976, mainly because of problems related to structural materials (see Rodriguez and Sundaram 1981).

The above extensive program in the US also involved fuel fabrication of thorium and U–Th fuel elements for "all sorts of fuel and all sorts of processes and matrices and a great experience has been accumulated" (see Lung 1997). The sol–gel process for powder production and vibratory compaction of the powders into the pin cladding was pioneered by Oak Ridge National Laboratory (ORNL) in the context of the thorium cycle. One of the problems in the Th-U-233 cycle is the gamma activity due to the daughter product U-232 which necessitates fuel refabrication in shielded facilities. The KILDROD facility (semi-shielded) was operated at ORNL on a 10 kg/day scale from 1964. A pilot fabrication plant of about 10 kg Heavy Metal (HM)/batch was operated at Babcock Wilcox on the same ORNL process from 1965.

The reprocessing of fuel for U-233 was also piloted by ORNL and the Knolls Atomic Power Laboratory (KAPL). The THOREX flowsheet is very similar to the PUREX process (see *Purex and Thorex Processes (Aqueous Reprocessing)*) and has been adopted for commercial operation in the US. Allis-Chalmers of the US undertook a Uranium–Thorium Fuel Cycle Program (PUCT) in cooperation with CNEN Italy, which included (in addition to a reactor evaluation study), the design, construction, and operation of a remotized facility to reprocess and refabricate thoria–urania fuel at Rotondella in Italy. Campaigns on Elk River BWR fuel elements were also carried out in this plant.

A fully remotized fuel fabrication plant named TURF (Thorium–Uranium Recycle Facility) was developed at ORNL. ORNL also prepared design parameters for large U-233/Th fuel fabricating plants having capacities from 67 kg to 3700 kg of HM/day.

The US has extensive experience also with fabrication, irradiation testing, and pyrochemical reprocessing of metallic fuels containing thorium in the Experimental Breeder Reactor EBR II.

It is estimated that a total of 700 tons of thorium have been irradiated in the various US programs, generating ~5 tons of U-233 of which about 1.6 tons have probably been recovered, until the ban on fuel reprocessing in 1975 during the Carter administration.

It is the greatest irony and tragedy that, in a series of technical meetings held between 1995 to 1997 to review the status of the thorium fuel cycle worldwide, the country statement from the USA stated: "Currently no work is underway in the United States on the thorium cycle nor are there plans to utilize it in the future." (See IAEA 2000.)

Since 1993, General Atomics of USA and the Russian Ministry for Atomic Energy (Miniatom) have been working on a cooperative program to develop a Gas-turbine-modular Helium Reactor (GTMHR). It is a small reactor (600 MWth per module) which is inherently safe due to the negative temperature coefficient of reactivity under all conditions and when coupled with a gas turbine in a direct cycle achieves almost 50% thermodynamic efficiency. The primary emphasis is currently for the burning of surplus weapons grade plutonium in Russia. But the reactor can also be used in thorium-based fuel cycles with a high conversion ratio. The program had a checkered progress, with the US Senate and House of Representatives voting in 1995 to terminate the program. Framatome and Fuji Electric joined as partners in 1996 and 1997 respectively, and in 1999 US Congress renewed funding. In February 2002, General Atomics requested the Nuclear Regulatory Commission of the US to commence the Safety Review of the first 250 MWe GTMHR to be constructed at Seversk (formerly Tomsk-7) in Russia.

4.2 Germany

Germany has carried out pioneering work on the HTGR (pebble bed fuel) for the thorium cycle. The AVR reactor (15 Mwe) at Jülich was operated successfully for 21 years (1967–1988); it saw 1.27×10^5 hours of operation (~70% availability), and approximately 3×10^5 fuel balls containing 1360 kg of Th were used. A maximum fuel burn-up of 1.5×10^5 MWd/t was achieved. The activities included post-irradiation examination of fuel and extensive fuel fabrication, reprocessing, and refabrication. These experiences, gained on a laboratory/pilot scale, gave enough confidence to build a 300 MWe power reactor with fuel fabrication on an industrial scale (NUKEM). The reactor, with 6.75×10^5 pebbles of thorium/highly enriched uranium (HEU) fuel, was operated for only three years between 1985 and 1989. The premature shutdown had two reasons: the first similar to the American experience of hasty extrapolation from 15 MWe to 300 MWe which led to structural problems with the graphite internals and the second, more crucial reason of the political climate against the use of HEU as fuel in reactors.

4.3 India

India, which has one of the largest reservoirs of thorium (~3.6×10^5 Mg) as against only ~6.0×10^4

Mg of uranium, has from the very beginning, emphasized the importance of thorium utilization in its nuclear program (Basu and Srinivasan 1990; Anantharaman and Sinha 2000). A three-stage nuclear power program has been envisaged. The first stage is a series of HWR using natural uranium oxide as the fuel. A total of 12 reactors are in operation, which with the two BWR at Tarapore, give a total capacity ~2720 MWe. The maximum generation capacity possible with the uranium reserves available in the country is limited to $\sim 1.5 \times 10^4$ MWe. Therefore, India has also launched a second stage of FBR based on the reprocessed Pu as fuel and depleted U in the blanket. Th-232 also will be used in the blanket to produce U-233, which will be used in a series of Advanced Heavy Water Reactors (AWHR) in the third stage. Finally, a symbiotic combination of FBR and AHWR will lead to sustainable nuclear power for nearly five centuries or more.

With the objectives stated above, India has continuously implemented a self-sustained program of extraction of thorium, fabrication of thorium-bearing fuels, irradiation of thorium in reactors to produce U-233, reprocessing to obtain U-233, fabrication of U-233 fuels, and building reactors using U-233 as fuel. A policy of thorium irradiation wherever possible has been followed from the beginning. Thorium pellets and rods have been irradiated in both the CIRUS and the DHRUVA research reactors at Trombay. Reprocessing of these to obtain U-233 has been carried out in facilities at both Trombay and Kalpakkam.

The expertise and experience gained by India in reprocessing and fabrication of fuels relevant for the thorium cycle are unique. All aspects of the 'Thorex' process (see *Purex and Thorex Processes (Aqueous Reprocessing)*), including a few alternative extraction options aimed at reducing fission product contamination and criticality risks and at coprocessing of ^{233}U and Th, have been mastered. A myriad form of fuels metal, alloy, oxide, MOX (ThO_2, PuO_2, and ThO_2–UO_2), carbides, and nitrides have been fabricated. Fabrication routes for rods, plates and pellets, including advanced routes such as pellet impregnation techniques and sol–gel microsphere pelletization (SGMP), have been established.

In the PHWR, in order to reduce power peaking at the center and to achieve full power as early as possible, depleted uranium is initially used in the center of the core. Following initial irradiation experiments using four experimental bundles with thoria pellets in one of the reactors of the Madras Atomic Power Station, all the new PHWRs in India from 1993 onwards have used ~500 kg of thorium fuel for power-flattening the initial core.

A U-233 fuelled reactor experiment called PURNIMA-2 with U-233 nitrate solution was carried out in 1984. This was followed by PURNIMA-3, a critical assembly using U-233/Al alloy fabricated into flat plates (U content 20 wt.%) which was a prelude to the KAMINI reactor (30 kW), commissioned in 1996 at Kalpakkam, the only operating reactor in the world today using ^{233}U as fuel.

India has plans to use ThO_2 in the blanket of the Fast Breeder Reactor (FBTR) from the second core onwards. India has also gained experience in the fabrication and irradiation testing of MOX fuels of both (Th + ^{233}U) and (Th + Pu) in the context of adopting new fuelling schemes for its PHWR.

India is presently working on an advanced heavy water reactor (AWHR), which extends the technology of PHWR with many innovative and passive safety features. It is a vertical pressure tube type reactor with boiling light water as the coolant. The fuel design is of the cluster type consisting of 54 fuel rods arranged in three concentric circles. There are 24 (Th + Pu)O_2 fuel rods in the outer circle and 30 (Th + ^{233}U)O_2 fuel rods in the two inner circles distributed as 12 in the innermost and 18 in the second circle. A hollow cylindrical ZrO_2 rod containing absorber material is located at the center of the cluster and also provides, through the holes in it, the paths for emergency core cooling water into the cluster. The fuel and reactor designs have evolved over two decades, and have a slightly negative void coefficient of reactivity as an important passive safety feature; another design feature is that it will be self-sustaining in U-233.

4.4 Russia and Former Soviet States

From the early days of the nuclear era, the Institute of Physics and Power Engineering at Obninsk and the Kurchatov Institute in Moscow have carried out extensive studies. These studies for thorium utilization have concentrated on:

- Physics of VVER (Russian PWR) and BN (Russian FBR) type reactors.
- Development of fuels for the above.
- Development of reprocessing technology.
- Development of thorium-based fuels of the cermet type.
- Thorium fuel elements have been irradiated in the BN-350 fast reactor in Kazakhstan.

4.5 Japan

As a national policy, the uranium–plutonium cycle has been primarily promoted in Japan. However, the significance of thorium cycle was recognized from the early days of nuclear energy; research was very active in the 1960s in government institutions and private companies but has since declined except for the activities of research committees on thorium-fueled reactors of the Atomic Energy Society of Japan, which have been active since 1974 on aspects of nuclear characteristics, fuel and reactor systems, downstream chemistry, and biological effects related to thorium cycle. The main thrust of the studies was on the MSBR and (to a lesser extent) on HTGR and PWR. These studies, carried out in universities with grants from the Government, are continuing as a cooperative program on thorium as an energy source in the twenty-first century.

In recent years, the Japan Atomic Energy Research Institute (JAERI) has been studying reactor systems for "burning" plutonium and higher actinides. The Plutonium rock-like fuel integrated research and development (PROFIT R&D) program is one such. PWR, BWR, HTGR, and FR have been considered, in which Pu will be burnt and U-233 will be bred.

It is also to be recalled that since 1975 JAERI, through its laboratory-scale studies, has built up expertise in fuel-fabrication technologies relevant to the thorium cycle. These included, in addition to fuels for MSBR, the SGMP process for ThO_2 and $(Th–U)O_2$, coated kernels of $(Th–U)O_2$ for HTGR, and a hydride-based nuclear fuel $UThTi_{1.2}Zr_2H_x$. Thorium fuel performance has been studied by irradiating kernels in the Japan Material Test Reactor. Many critical experiments have been conducted in the KUCA facility with metallic Th and enriched U.

4.6 United Kingdom

The interest in thorium was high in the 1950s and 1960s. The 20 MWth Dragon reactor, an OECD project, was located in UK and operated successfully between 1966 and 1973 for 741 full power days, often exceeding the rating by up to 13%. Th/HEU, fuels with Th/U-235 ratios greater than three, were successfully used in carbide and oxide form, reaching irradiations of 1×10^5 MWD/t without problems.

UKAEA had a thorium research program and built a prototype thorium reprocessing plant at Dounreay called D.1203. This plant was used on and off until the 1980s. At the beginning of the twenty-first century the UK has no definite program on thorium cycle.

4.7 Canada

The HWR is a good system for thorium conversion to U-233. This fact naturally led to many studies for adapting the thorium fuel cycle to the CANDU (Canadium deuterium uranium) reactor. Feasibility studies on chemical separation of Th, U, and Pu were successfully carried out at Whiteshell Laboratories. The work was discontinued at the end of the 1970s.

Recently a long-term program was re-established, anticipating interest in the marketplace for CANDU reactors using the thorium cycle.

4.8 France

Feasibility studies on backfitting existing PWRs on a thorium cycle were carried out in France in the 1970s. Limited studies were also carried out on HTGR. Later studies relating to the use of the fast reactor (FR) as a burner of Pu and to the production of U-233 in the blanket for use as fuel in PWR were also carried out. One such FR can produce fuel for five LWRs (of equal reactor power).

4.9 Brazil

Brazil, like India, has rich resources in thorium in its monazite beach sands. During 1979–1988, within a cooperative program between Brazil and Germany, the feasibility of backfitting existing PWR for a thorium cycle was examined and studied in detail. A program of test irradiations of Th/U- and Th-Pu fuel elements in the 900 MWe Angra 1 PWR was prepared (including licensing documentation) but these were not carried out.

4.10 Israel

Studies at Tel Aviv University on a "Non-proliferative Light Water Thorium Reactor" are supported by the Raytheon Corporation of the USA and in cooperation with the Kurchatov Institute and the Brookhaven National Laboratory, had progressed to a theoretical development stage at the beginning of the twenty-first century. The concept is similar to the seed-blanket LWBR of Shippingport, except that breeding is avoided and U-233 is burnt in place as it is formed.

4.11 Netherlands

An aqueous homogeneous suspension reactor project KSTR (1 MWth) was operated successfully during 1974–1977 at Arnhem. The fuel comprised 22.5% ^{235}U, 2.5% ^{238}U, and 75% ThO_2 in the form of 5 μm particles suspended in demineralized water and circulated by a centrifugal pump; BeO + graphite reflectors were used.

5. *Neutron Source Driven Systems*

These are systems in which nonfission neutrons, (neutrons produced in spallation reactions in an accelerator or in thermonuclear fusion) are used for the conversion of fertile into fissile material. This concept is nearly 30 years old, but has been revived in the context of accelerator-driven transmutation for radwaste and similar applications. Such systems can be built around both uranium and thorium; however, thorium systems are in focus at the beginning of the twenty-first century because of the advantage that the system can be maintained in a subcritical state.

5.1 Accelerator Driven Systems (ADS)

Following the celebrated proposal of Rubbia (1994) of an "Energy Amplifier," many countries, notably

France, Germany, and the USA have intensified work on ADS. In an ADS, high-energy neutrons from a spallation reaction (see Rodriguez and Sundaram 1981) are directed to a subcritical reactor containing thorium. ADS can also be used for incineration of long-lived fission products (LLFP).

5.2 Accelerator Breeders

In this concept, neutrons from a spallation reaction in an accelerator are absorbed in a surrounding thorium blanket to breed ^{233}U. A beam of 300 mA protons in a 1GeV accelerator can produce ~ 1000 kg of ^{233}U in a year.

A more recent concept of an accelerator breeder from Jülich uses thermal neutrons and a liquid fuel (lead-actinide), leading to a self-sufficient system breeding its own fuel and producing a surplus of energy.

As a part of the work of the Japan Atomic Energy Society on thorium-fueled reactors cited earlier, the accelerator molten salt breeder (AMSB) concept has evolved over the years. For the AMSB, the expected fissile material doubling time is about four years.

5.3 Fusion–Fission Hybrids

It is now recognized that it would be much easier to design a fusion reactor if it were not expected to be an energy producer. The symbiosis between a fusion reactor, an abundant source of neutrons, and a thorium-fission reactor deficient in neutrons, will be a net energy-producing fissile "factory." Japan is pursuing two possible systems, the inertial-confined fusion hybrid molten salt breeder (IHMSB) and the impact fusion molten salt breeder (IFMSB).

It needs to be emphasized that all the above neutron source driven systems are concepts and no extensive experience exists with industrial prototypes or even with irradiation experiments with thorium. The only exception is the joint experiments carried out by India and Switzerland, in which thoria blanket elements fabricated in India were irradiated in the 14MeV neutron generator of the LOTUS facility at EPFL, Lausanne, Switzerland. Global ^{233}U production in the thoria blanket was in agreement with 2-D calculations (Basu and Srinivasan 1990).

6. Strategies for Thorium Utilization

The main reason for the renewed interest in the thorium cycle is the perception that it is more proliferation-resistant than the uranium–plutonium cycle. The gamma activity due to ^{232}U, which necessitates handling in shielded facilities, makes diversion of the fissile material difficult. It is also possible to denature ^{233}U with natural uranium or depleted uranium, such that expensive and detectable enrichment facilities are needed to produce a critical quantity from such denatured fuel material for weapons production.

Another distinct advantage of the thorium fuel cycle is the significant reduction in the production of long-lived transuranic actinides (TRU) and of LLFP. As a result, the radioactivity of "waste" in the thorium cycle decays out in $\sim 1 \times 10^3$ years in comparison to $> 1 \times 10^5$ years for the U–Pu cycle.

Other potential goals for the thorium cycle have been identified as:

(i) Prolonging the exploitation of LWR and HWR on the basis of ^{233}U fuel.
(ii) Improving the burn-up and fuel utilization in thermal reactors.
(iii) Significantly reducing the ^{235}U enrichment requirements.
(iv) Destruction of accumulated Pu and incineration of TRU and LLFP.
(v) Developing the potential to combine the thorium fuel cycle with ADS or hybrid fusion driven systems, to breed ^{233}U and incinerate long-lived radiotoxic isotopes.

Goals (i) to (iii) can be achieved through a pure Th scenario with no breeding envisaged. Without breeding (and breeding is possible only in MSBR and FBR) the growth potential for nuclear power is severely limited. India for example has evaluated three different cycles applicable to HWR in the thorium cycle:

(i) The self-sustaining equilibrium thorium cycle (SSET).
(ii) The high burn-up, high conversion ratio cycle.
(iii) The once-through thorium cycle (OTT).

For sustainable growth and development, SSET offers the best route among the three but the maximum installed capacity can at best be trebled from what is possible with the HWR using natural uranium as fuel. The uranium–plutonium cycle offers a faster route but in the global scenario today, when the world is awash with plutonium (both civil plutonium and weapons plutonium), and nuclear terrorism is a serious concern, plutonium burning in fast reactors rather than breeding has become important and acceptable.

Combining thorium with plutonium (goal (iv)) is a better option than burning Pu in inert matrix fuels because, with this strategy we are essentially converting Pu into ^{233}U as well as producing energy during conversion. French studies have shown that one fast reactor can supply enough ^{233}U for five LWR loadings. Current Indian strategy, as explained earlier, envisages a role for plutonium breeders mainly from the need to reach a faster power growth, but in the third stage of thorium utilization the FR can be plutonium-burning U-233 producers or

Th–^{233}U breeders. As already pointed out with thorium fuels, FBR safety is higher due to negative sodium void coefficient and a more negative Doppler coefficient.

The strategy of converting Pu into ^{233}U in LWR and HWR is also feasible. Indian studies have shown that a 3.6% initial enrichment with Pu in a PHWR thorium fuel has double advantages: the burn-up is increased fourfold (compared to natural uranium fuel) to 4×10^4 MWd/t, yielding a discharge fuel which contains only 0.2% Pu and $\sim$1.3% ^{233}U. One kilogram of plutonium can thus be converted to 1130 MWd of energy plus 376 g of ^{233}U.

However, such a strategy will not have the advantage of low radiotoxicity in the waste as obtained in pure thorium or fast reactor scenarios, unless the plutonium and thorium are employed in separate driver fuel and blanket regions, and separate streams for reprocessing are also set up. This is, in fact, envisaged for reprocessing AHWR fuel.

Such a need for three separate reprocessing streams will arise whenever hybrid fuels like ^{233}U–^{238}U and Th/Pu are employed and this will add to the fuel cycle costs.

7. *Economics of Thorium Cycles*

The economics of a fuel cycle depends not only on the cost of the power plant, but also on the costs associated with the whole fuel cycle, consisting of mining and extraction, reprocessing, refabrication, waste disposal, and decommissioning. From the experience of five decades with nuclear power plants, we are aware today that the two successful technologies, LWR and HWR, took nearly two decades to traverse the learning curve to reach the maturity in terms of availability and stabilized low cost for electricity generation. Use of these reactors for thorium utilization, therefore, offers the best economic choice. FBR, HTGR, MSBR, and SDS come in that order for decreasing experience with commercial or prototype plants and therefore the cost/economics is likely to increase in the same order. ADS and fusion–fission hybrids are a marriage of two complex technologies (Accelerator/fission reactor in one case and fusion reactor/fission reactor in the other case). Hence the costs are likely to be very high.

It is also to be recognized that there will be an economic penalty in the fuel cycle cost for the thorium cycle, due to the need for shielded and remotized refabrication facilities. The AHWR is expected to have $\sim$200 ppm of ^{232}U in the ^{233}U produced. The ^{233}U bred in the thorium blankets of FBR will have a much higher ^{232}U content of $\sim$5000 ppm. Once the ^{232}U is present in ^{233}U, the gamma activity will build up with time, due to the hard gamma emitters ^{212}Bi and ^{208}Tl (daughter products of ^{232}U). Therefore, it is considered prudent to reduce the ^{232}U content in the ^{233}U to $\sim$5 ppm before refabricating the fuel, so that the need for shielded facilities for fuel refabrication and for remote handling at the reactor site can be avoided at least for a year from the date of purification of ^{233}U. Fuel fabrication can then be carried out in glove boxes similar to those used for (Pu–U) Mox fuel. India is working on a process of separation of the isotopic impurity ^{232}U from ^{233}U by employing a laser-based process. The new process is referred to as the ^{233}U clean-up process in the Indian Nuclear Fuel Cycle. The clean-up step is inevitable for long-term use of thorium cycle since the ^{232}U build up will reach unacceptable levels for refabrication and handling without shielding as the number of cycles increase. This additional step also will have an impact on the fuel cycle costs for the thorium cycle.

The redeeming feature is that since fuel fabrication costs constitute only 7% the electricity generation cost in a nuclear power plant (Ramanna and Lee 1986), the economic penalty may not be significant and will be overshadowed by the myriad benefits and advantages coming from the thorium cycle.

8. *Conclusion*

Considerable experience and expertise have been generated in many countries, notably the USA, Germany, India, and Japan on various aspects of the nuclear cycle. Two nuclear power plant technologies, LWR and HWR, have achieved the maturity to adapt to thorium use in the near future. The FBR, HTGR, MSBR, and SDS rank in that order in terms of increasing need for further development at prototype/commercial scale to prove their economic viability. Fuel cycle costs for the thorium cycle may be higher by a factor of 1.5 to 2 than for the uranium cycle. However, the economic penalty on the unit electricity generation cost may not be significant enough to outweigh the benefits and advantages, particularly with respect to resistance to proliferation, lower radioactivity of the waste, inherent safety features for the reactor systems, higher burn-up, and better fuel utilization, offered by the thorium cycles.

See also: Nuclear Fuel Cycles; Nuclear Reactor Materials; Purex and Thorex Processes (Aqueous Reprocessing); DUPIC Fuel Cycle; Nuclear Reactor Fuel Fabrication (Including Quality Control); Mixed Oxide (MOX) Fuel; Inert Matrix Fuels; Heavy Water Reactor Fuel Design and Performance; Light Water Reactor Fuel Design and Performance.

Bibliography

Anantharaman K, Sinha R K (eds.) 2000 *INSAC—2000, Proc. 11th Annual Conference of Indian Nuclear Society: Power from Thorium: status, strategies, and directions.* Indian Nuclear Society, Mumbai, India

Anon 1996 The role of thorium in nuclear energy. In: *Energy Information Administration/Uranium Industry Annual.* US Department of Energy, Washington, DC, pp. ix–xvii

Basu T K, Srinivasan M 1990 *Thorium fuel cycle development activities in India (A Decade of Programme: 1981–1990). Report BARC 1532.* Bhabha Atomic Research Centre, Mumbai, India

International Atomic Energy Agency 1977 *Proc. Int. Conf. Nuclear Power and its Fuel Cycle.* IAEA, Vienna

International Atomic Energy Agency 1977 *Proc. Int. Symp. Nuclear Fuel Cycle and Reactor Strategies: Adjusting to New Realities.* IAEA, Vienna

International Atomic Energy Agency 2000 *Thorium based fuel options for the generation of electricity: Developments in the 1990s.* IAEA-TEDOC-1155, IAEA, Vienna

International Nuclear Fuel Cycle Evaluation (INFCE) *Report of Working Groups.* IAEA, Vienna

Ion S C, Bonser D R 1997 Fuel cycles of the future. *Nuclear Energy* **36**, 127–30

Lung M 1997 A Present Review of the Thorium Nuclear Fuel Cycles. Office for Official Publications of the European Communities, European Commission, Luxembourg

Nuclear Energy Agency 1994 The economics of the nuclear fuel cycle. Nuclear Energy Agency, OECD, Parishttp://www.nea.fr/html/ndd/reports/efc/

Ramanna R, Lee S M 1986 The thorium cycle for fast breeder reactors. *Pramana* **27**, 129–37

Rodriguez P, Sundaram C V 1981 Nuclear and materials aspects of the thorium fuel cycle. *Journal of Nuclear Materials* **100**, 227–49

Rubbia C 1994 The energy amplifier. *Proc. 8th Journees Saturne.* Orsay, France

Srinivasan M, Kimura I (eds.) 1990 *Proc Indo–Japan Seminar on Thorium Utilisation.* Indian Nuclear Society, Mumbai, India

Uranium Information Centre 2000 Nuclear Electricity, Melbourne, Australia http://www.uic.com.au

World Nuclear Association 2001 Information and Issue Briefs: Thorium. http://www.world-nuclear.org/info/inf62.htm

Twin Cell Technology for Hydrogen Generation

K. G. U. Wijayantha and D. H. Auty

1. *Introduction*

Since photoelectrochemical water splitting is a vast area, it is not our objective to cover all its aspects within this article, but to discuss the twin cell technology based Tandem Cell™ for water splitting. It has long been accepted that world demand for energy will increase steadily for the foreseeable future, but until the last decade more intensive exploitation of conventional sources seemed able to provide an adequate supply. However, today the situation has changed entirely and everyone acknowledges the threat of depletion of fossil fuel. Therefore, nonrenewable fuels need to be conserved. Another major concern associated with the use of fossil fuels is the formation of large amounts of CO_2, SO_2, and NO_x by combustion that adversely contributes to the world climate. Hence, fossil fuels are becoming ecologically unpopular. Since nuclear power presents social and political problems, today searching for a carbon-free renewable energy source has become more important than ever before. It is fair to say that renewable energy sources such as wind and tidal power still face technological challenges.

Hydrogen is a potential fuel with desirable properties. Its energy storage capacity is $120000\,\mathrm{J\,g^{-1}}$ as compared to 40000 for oil and 30000 for coal (Grätzel 1989). Furthermore, hydrogen has the advantage over conventional fossil or nuclear energy sources in that combustion in a thermal engine or fuel cell does not result in pollution of the environment.

Biological photosynthesis is one of the fascinating phenomena of nature in which water and carbon dioxide are converted into oxygen and carbohydrates in the presence of sunlight. In this process a manganese complex, which is in an intricate enzymatic environment, provides an active site for splitting water. Designing a process based on this generic concept to drive a reaction thermodynamically uphill to split water into hydrogen and oxygen (sometimes referred to as artificial photosynthesis) is a highly rewarding task:

$$H_2O \rightarrow 0.5O_2 + H_2$$

$$\Delta G^0_{298} = 237\ \mathrm{kJ\,mol^{-1}} \qquad (1)$$

2. *Historical Background*

Water splitting has been a goal of photoelectrochemistry for nearly a decade. The concept is based on a photosynthetic cell. The photosynthetic cell operates on the principle that there are two redox systems in the electrolyte and upon light absorption and generation of electron and hole pairs in the semiconductor photoanode (photocathode), one redox system reacts with the photogenerated holes (electrons) at the surface of the semiconductor photoanode (photocathode) and the other reacts with the electrons (holes) entering at the counter electrode. A schematic diagram of a water splitting device based on this principle is shown in Fig. 1, where an n-type semiconductor electrode is used as the photoanode to generate oxygen and a suitable counter electrode is used to generate hydrogen. The overall reaction is the cleavage of water by sunlight (Eqn. (5)).

Light absorption and generation of electron/hole pairs at the photoanode,

$$n - SC + h\nu \rightarrow n - SC(e^- + h^+) \qquad (2)$$

Light-driven reaction at the photoanode,

$$2H_2O + 4h^+_{VB} \rightarrow 4H^+ + O_2 \qquad (3)$$

Light-driven reaction at the cathode,

$$2H^+ + 2e^-_{CB} \rightarrow H_2 \qquad (4)$$

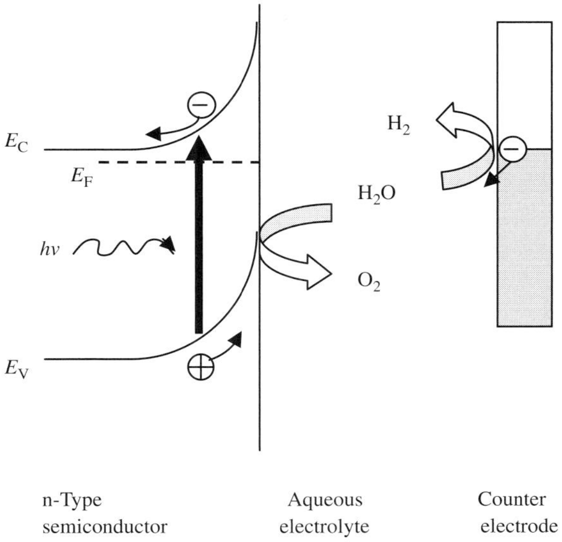

Figure 1
Photolysis of water by using an n-type semiconductor electrode.

The overall reaction is

$$4h\nu + 2H_2O \rightarrow O_2 + 2H_2 \quad (5)$$

Using a titanium dioxide electrode as the photoanode in this type of device, Fujishima and Honda (1972) demonstrated water photolysis for the first time. Having a band gap of 3–3.2 eV, TiO_2 absorbs only 4% of the AM 1.5 solar spectrum (Fig. 2). Therefore, TiO_2 is not the ideal choice for the photoelectrode in water photolysis although its conduction band edge position is favorable. Figure 3 shows the conduction and valence band positions of several semiconductors in contact with aqueous electrolyte at pH 1.

The energy difference between the oxygen evolution potential and the hydrogen evolution potential is 1.229 eV at 25 °C. Therefore, ideally one would like to have a chemically and optically stable semiconductor, in which the conduction and valence bands are positioned favorably so that they straddle the hydrogen and oxygen evolution potentials. To be able to overcome the over-potential, which is mainly associated with oxygen evolution, it is preferred to have a band gap of 1.8–2.0 eV in the prospective semiconductor. Although some chalcogenide semiconductors fulfill these criteria, they are unstable on prolonged exposure to sunlight owing to dissolution and photocorrosion in aqueous electrolytes. Therefore, metal oxide semiconductors are the natural choice for photo-electrochemical water splitting. Generally they show good stability in aqueous electrolytes and are readily available. Considerable efforts have been made to extend the spectral response of semiconductors such as TiO_2 toward the visible region or to develop alternative oxide semiconductors (Asahi *et al.* 2001). However, extensive work on finding a prospective metal oxide semiconductor that meets all the above-mentioned requirements has so far failed.

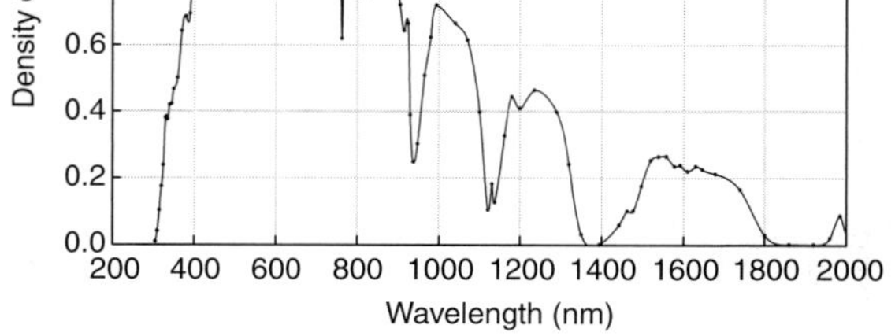

Figure 2
AM 1.5 solar spectrum.

3. Recent Advances of Nanotechnology and the Tandem Cell™

The interest in the use of Tandem cells™ for water cleavage by visible light has reignited in the last two decades (Weber and Dignam 1984, Turner *et al.* 1996). The recent advances of nanotechnology have stimulated photo-electrochemical research, including photo-electrochemical water splitting. The reignited interest in the use of the Tandem Cell™ for water cleavage by visible light and the recent advances in nanotechnology have blended together in the invention of the twin cell technology-based Tandem Cell™ (Grätzel 2001, Grätzel and Augustynski 2002). A three-dimensional schematic representation of a Tandem Cell™ is illustrated in Fig. 4.

The twin cell-based Tandem Cell™ is a low-cost device, which consists of two photosystems connected in series. The cell on the left in Fig. 4 consists of a mesoporous metal oxide semiconductor film (such as WO_3 or Fe_2O_3) that is deposited on a fluorine-doped conducting SnO_2 transparent electrode and a Pt-coated fluorine-doped conducting SnO_2 transparent electrode. Fluorine-doped conducting SnO_2 substrates in both anode and cathode serve as transparent current collectors. The first cell contains an aqueous electrolyte that is subjected to water photolysis. The electrolyte consists of added ions (e.g., H_2SO_4, NaOH) to achieve ionic conductance. The light enters through the Pt-coated fluorine-doped conducting SnO_2 transparent electrode and then through the aqueous electrolyte to the nanocrystalline metal oxide semiconductor electrode where the blue and green part of the solar spectrum is absorbed. If the energy of absorbed photons is higher than the band gap energy of the metal oxide semiconductor, electrons excite into the conduction band leaving holes in the valence band (Eqn. (2)). The conduction band electrons move to the fluorine-doped SnO_2 transparent current collector. The remaining photogenerated holes oxidize water, forming oxygen and protons (Eqn. (3)).

The yellow and red part of the solar spectrum is transmitted through the first photocell and enters the second photocell, which is mounted behind the back wall of the first cell. The second cell is a dye-sensitized nanocrystalline TiO_2 photovoltaic cell. A schematic diagram of a dye-sensitized nanocrystalline TiO_2 photovoltaic cell is shown in Fig. 5. It consists of a nanocrystalline TiO_2 thin film photoanode deposited on a fluorine-doped SnO_2 transparent substrate and a Pt-coated fluorine-doped SnO_2 substrate. In

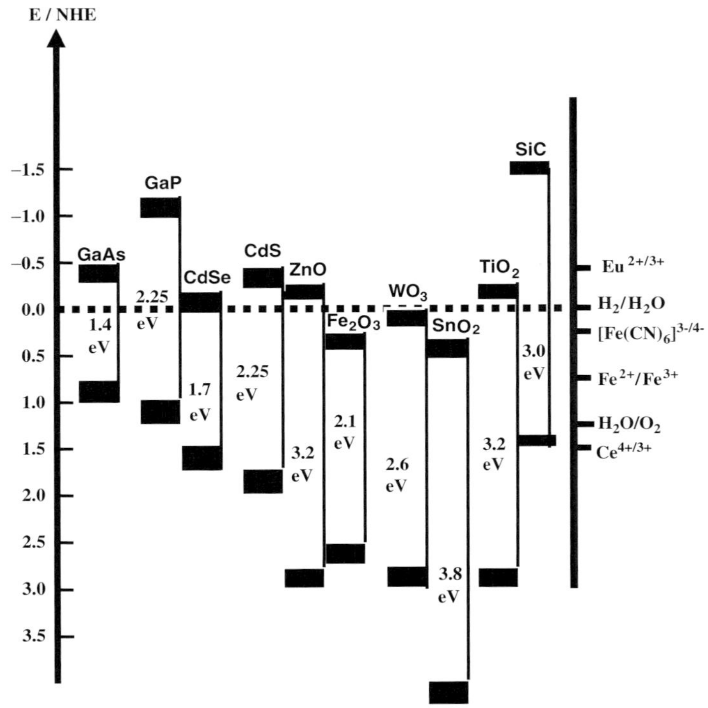

Figure 3
Band positions of several semiconductors in contact with aqueous electrolyte at pH 1.

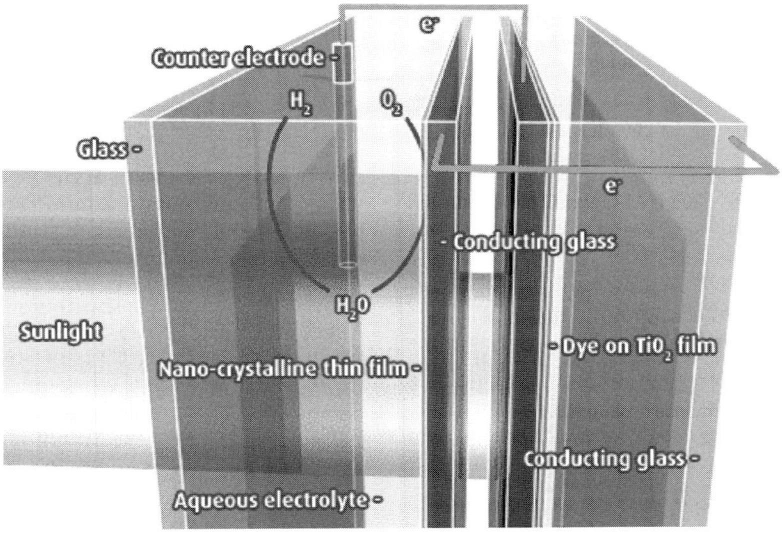

Figure 4
A three-dimensional schematic representation of a Tandem Cell™.

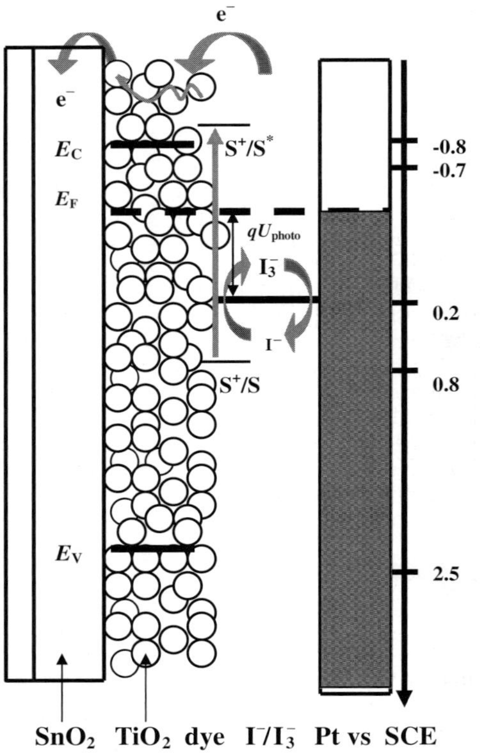

Figure 5
Schematic representation of a dye-sensitized nanocrystalline TiO_2 photovoltaic cell.

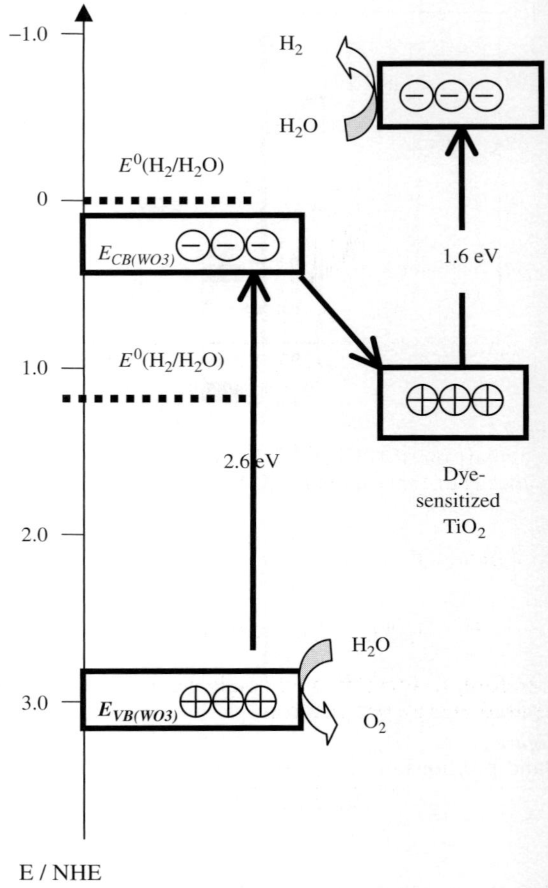

Figure 6
The electron flow diagram for the Tandem Cell™.

nanocrystalline TiO_2 film, each TiO_2 nano-particle is covered with dye at a monolayer coverage. As the yellow and red part of the solar spectrum intercepts many dye-coated TiO_2 nanoparticles, the effective light absorption cross section increased. Optical excitation of the adsorbed dye leads to injection of an electron from the excited dye to the conduction band of TiO_2. The oxidized dye is subsequently reduced back to its native state by an external donor (I^-) present in the non-aqueous electrolyte and the resulting tri-iodide ions (I_3^-) are reduced back to I^- at the Pt-coated counter electrode. Capturing the yellow and red part of the solar spectrum, the dye-sensitized nanocrystalline TiO_2 photovoltaic cell provides the necessary light-driven electric bias. As a result, the electrochemical potential of electrons collected at the fluorine-doped SnO_2 transparent substrate of the photoanode of the first cell increases, enabling them to reduce water to hydrogen (Eqn. (4)). Assuming the photoanode of the first cell is a nanocrystalline WO_3 electrode and the second cell is a dye-sensitized nanocrystalline TiO_2 photovoltaic cell, the electron flow diagram for the twin cell based Tandem Cell™ is shown in Fig. 6.

To describe how each photoelectrode in the first and second cell responds to the solar spectrum, the incident photons-to-current conversion efficiency (IPCE) of each photocell is compared in Fig. 7. The IPCE of the first cell is recorded at 1.0 V against standard calomel electrode. The IPCE of the second cell is recorded under short-circuit conditions. From Fig. 7 it is clear that the blue part of the solar spectrum is used by the WO_3 photoanode and the yellow/red part of the solar spectrum is used by the dye-sensitized nanocrystalline TiO_2 electrode.

However, the second photocell acts merely as a photodriven bias, so that other types of photovoltaic cells such as the Si solar cell or CdS/CdTe solar cell could also be used instead of the dye-sensitized nanocrystalline TiO_2 photovoltaic cell.

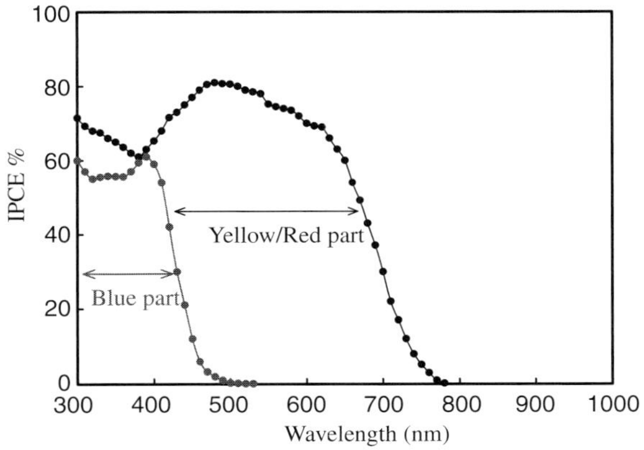

Figure 7
A comparison of IPCEs (%) of the first and second photocells of the Tandem Cell™.

4. *Efficiency*

$$H_2 + 1/2O_2 \rightarrow H_2O \quad 2 \text{ electronreaction}$$

Therefore, to form/break a single water molecule the required energy per electron,

$$= \frac{\text{Enthalpy}}{\text{Faradayconstant}} \times \frac{1}{2}$$
$$= 1.48 \text{ eV}$$

Efficiency = energy out/energy in

$$= I \text{ mA cm}^{-2} \times 1.48 \times 100\%/100 \text{ mW cm}^{-2}$$

Efficiency = $1.48 \times I$ (I in mA cm^{-2})

Because there is a theoretical limit to the feasible photocurrent density, the efficiency calculated in this way cannot exceed 100%.

Therefore, the Tandem Cell™ solar conversion efficiency is given by $1.48 \times$ photocurrent density. Because there is a theoretical limit to the feasible photocurrent density, the efficiency calculated in this way cannot exceed 100%. However, the use of the enthalpy term instead of the free-energy term in the present efficiency calculation is currently being questioned. The twin cell technology based Tandem Cell™ tackles a number of challenges faced by single photo-electrochemical cell based water-splitting. The concept also offers the flexibility of use of a second bias cell (photovoltaic cell). The number of required photovoltaic cells is determined by the photocurrent density of the first cell. Allowing two photocells to absorb complementary parts of the solar spectrum, the Tandem Cell™ represents a low-cost way of achieving direct water cleavage.

Bibliography

Asahi R, Morikawa T, Ohiwaki T, Aoki K, Taga Y 2001 Visible-light photocatalysis in nitrogen-doped titanium oxides. *Science* **293**, 269

Fujishima A, Honda K 1972 Electrochemical photolysis of water at a semiconductor electrode. *Nature* **238**, 37

Grätzel M 1989 *Heterogeneous Photochemical Electron Transfer*. CRC Press, FL

Grätzel M 2001 Photoelectrochemical cells. *Nature* **414**, 338

Grätzel M, Augustynski J 2002 *Eur. Pat.* EP1198621

Turner J A, Arent D, Peterson M, Kocha S 1996 Photoelectrochemical systems for water decomposition. In: Veziroglu T N (ed.) *Hydrogen Energy Progress XI, Proceedings of the World Hydrogen Energy Conference*. Stuttgart, June 23–28, pp. 2749–53

Weber M F, Dignam M J 1984 Efficiency of splitting water with semiconducting photoelectrodes. *J. Electrochem. Soc.* **131**(6), 1258–65

Z

Zirconia Electrolyte-based Solid Oxide Fuel Cells

S. C. Singhal

A solid oxide fuel cell (SOFC) is an all solid-state energy conversion device that produces electricity (and heat) by electrochemically combining fuel and oxidant gases across an ionic conducting oxide. These cells offer a clean, nonpolluting technology to convert chemical energy of a fuel to electrical energy at high efficiencies. They offer several advantages over conventional power generation systems including high efficiency, reliability, modularity, fuel adaptability, and very low emissions of nitrogen and sulfur oxides.

1. *Operating Principle*

A SOFC in essence consists of two porous electrodes separated by a dense oxygen ion-conducting membrane that serves as the electrolyte. The operating principle of a SOFC is shown in Fig. 1. Oxygen supplied at the cathode (also referred to as the air electrode) reacts with electrons from the external circuit to form oxygen ions, which travel through the electrolyte to the anode (also referred to as the fuel electrode) where they react with fuel gases and liberate electrons. The liberated electrons travel through the external circuit to the cathode.

The open circuit voltage is given by the Nernst equation:

$$E^0 = \frac{RT}{4F}\ln\left[\frac{pO_2(\text{oxidant})}{pO_2(\text{fuel})}\right] \quad (1)$$

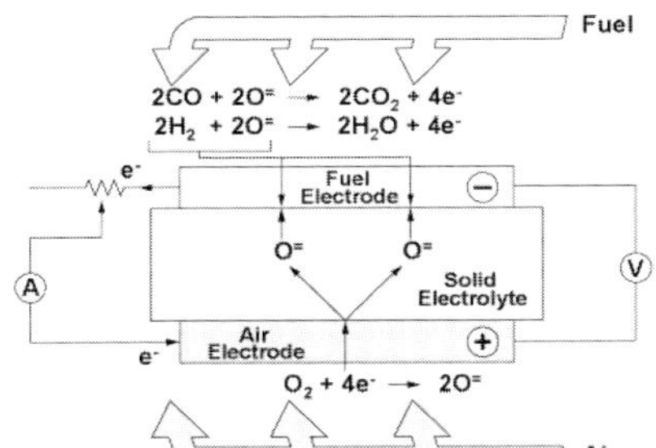

Figure 1
Operating principle of a solid oxide fuel cell.

where E^0 is the open circuit voltage, R is the universal gas constant, T is the cell temperature, F is the Faraday constant, and the pO_2 are the oxygen partial pressures. Under cell operating conditions, i.e., when a current passes through it, the cell voltage (V) is given by

$$V = E^0 - IR - \eta_A - \eta_F \quad (2)$$

where I is the current passing through the cell, R is the resistance of the cell, and η_A and η_F are the voltage losses due to polarization associated with the air electrode and the fuel electrode, respectively. The cell design strategy to obtain high performance is to minimize the resistive losses by using thin electrolyte films and to minimize polarizations at the interfaces by using electrode/electrolyte interlayers.

These fuel cells operate at about 1000°C. Using air as oxidant and natural gas as fuel, the open circuit voltage of a cell is about 1 V. Single cells are connected in series (referred to as a stack) to build voltage. In a stack, a component called a bipolar separator or an interconnection connects the anode of one cell to the cathode of the next cell. The four main components of a SOFC are the electrolyte, the anode, the cathode, and the interconnection. Each component serves one or more functions in the fuel cell and must meet certain property requirements. Each component must be chemically, structurally, morphologically, and dimensionally stable in oxidizing and/or reducing environments. It must have chemical compatibility with other components and should possess proper ionic or electronic conductivity. Furthermore, the components should have similar thermal expansion coefficients in order to minimize the thermal stresses that could develop during cell fabrication and operation. If the components are processed sequentially, the temperature for each successive component fabrication process must be lower than that of the preceding process to avoid altering the microstructure of the preceding components.

2. *Materials for SOFC Components*

The properties of the materials for the four different cell components are discussed below.

2.1 Electrolyte

The most commonly used electrolyte in SOFCs is stabilized zirconia (ZrO_2). In its pure form the ionic conductivity of zirconia is low. At room temperature, zirconia has a monoclinic (m) crystal structure which changes to a tetragonal (t) form above 1170°C and to a cubic fluorite form above 2370°C. However, the addition of aliovalent oxides such as Y_2O_3, CaO, MgO, or Sc_2O_3 stabilizes the cubic fluorite structure of ZrO_2 from room temperature up to its melting point and, at the same time, increases its oxygen vacancy concentration, leading to enhanced ionic conductivity over an extended range of oxygen partial pressures. The creation of oxygen vacancies and stabilization of the cubic phase is accomplished by direct substitution of divalent or trivalent cations of appropriate size for the host lattice cation, Zr^{4+}. For example, the charge compensation equation for stabilization by yttria can be written (using Kroger–Vink notation) as:

$$Y_2O_3 \xrightarrow{ZrO_2} 2Y'_{Zr} + 3O_0^X + V_0^{\cdot\cdot} \quad (3)$$

Oxygen ion conduction takes place by movement of oxygen ions via vacancies. The ionic conductivity of stabilized zirconia strongly depends on the dopant concentration. The isothermal plots of conductivity as a function of dopant concentration generally exhibit a maximum. The decrease in conductivity at higher dopant concentration is believed to be due to defect ordering, vacancy clustering, or electrostatic interactions. For the same dopant concentration, the conductivity increases with decreasing dopant size. This is believed to be due to less strain and consequently fewer associations of the metal cations with oxygen ion vacancies. This leads to a higher oxygen vacancy concentration, and hence, higher ionic conductivity.

The most commonly used electrolyte material in SOFCs is 8–10 mol.% Y_2O_3-stabilized zirconia (YSZ). Although YSZ does not yield the highest conductivity of all dopants, its ease of availability, low cost, and good mechanical and electrical properties make it an attractive choice as the electrolyte. The properties required of the YSZ electrolyte in a SOFC are that it must be free of porosity so that gases cannot permeate from one side of the fuel cell to the other, it should be uniformly thin to minimize the resistive losses in the cell, and the transport number for electronic conduction should be as close to zero as possible to maximize the efficiency of the cell. The electrical conductivity of YSZ follows an Arrhenius-type relationship as a function of temperature. The ionic transport number of YSZ is very close to unity (i.e., negligible electronic conduction) over the range of partial pressures experienced under fuel cell conditions (0.21 atm to about 10^{-20}atm).

The thermal expansion coefficient (TEC) of YSZ ranges over $10–11 \times 10^{-6} K^{-1}$. This TEC value is selected as the baseline and the TECs of other cell components are tailored to be near this value by dopant addition or by varying the porosity to minimize the thermal stresses. It is desirable to have a high fracture toughness in the electrolyte so that it is less sensitive to the presence of flaws and so that it imparts better fracture resistance to the cell during the fabrication process. The fracture toughness of YSZ (8 mol.% yttria) at room temperature is about $3\,MN\ m^{-3/2}$. Limited studies on improving the strength of zirconia indicate that the addition of fine particles of partially stabilized zirconia and alumina increase the fracture toughness of the YSZ electrolyte without significantly affecting its conductivity.

Cells using YSZ electrolyte need to be operated above about 800°C to achieve sufficient ionic conductivity and cell power density. A reduction in the operating temperature would reduce interfacial reactions among various cell components, as well as prolong cell life. Hence, the development of alternate electrolyte materials, possessing sufficient ionic conductivity at reduced temperatures, has attracted interest. Among other oxygen ion-conducting materials, stabilized bismuth oxide (Bi_2O_3) and ceria (CeO_2) exhibit high ionic conductivity at reduced temperatures. However, stabilized Bi_2O_3 is easily reduced and decomposes to bismuth metal under low oxygen partial pressures ($< 10^{-13}$ atm), thus making it unsuitable for use as a SOFC electrolyte. Doped ceria, on the other hand, undergoes a large departure from stoichiometry under elevated temperatures and reducing atmospheres and develops electronic conductivity which is undesirable for its use as an electrolyte. The approaches that have been investigated to overcome this problem include the addition of dopants and the coating of ceria with a layer of zirconia. Several doped perovskites (ABO_3) have also been identified as potential candidates for use as intermediate temperature (600–800°C) electrolytes. The notable ones include doped lanthanum gallates, barium cerates, and strontium zirconates. However, compared with YSZ, there is only a limited understanding of the stability and electrical and mechanical properties of these materials.

2.2 Anode

The properties desired of the anode (fuel electrode) are that it must be stable in the reducing environment of the fuel, should be electronically conducting, and have sufficient porosity to allow the transport of the fuel to and the transport of the products of fuel oxidation away from the electrolyte/fuel electrode interface where the following fuel oxidation reaction takes place:

$$O^{2-}(s) + H_2(g) \rightarrow H_2O(g) + 2e^- \quad (4)$$

The two electrons liberated in the above reaction transport through the outer circuit to the cathode. The reducing conditions present on the fuel side of a SOFC permit the use of a metal as the fuel electrode.

Nickel is the most commonly used metal because of its low cost. However, the thermal expansion of nickel ($13.3 \times 10^{-6} K^{-1}$) is considerably larger than that of YSZ. Nickel can also sinter at the cell's operating temperatures resulting in a decrease in active surface area and reduced conductivity of the anode, which in turn leads to lower cell performance. These problems are minimized by forming a skeleton of YSZ around the nickel particles. The YSZ skeleton prevents sintering of the nickel particles, decreases the fuel electrode thermal expansion coefficient to bring it closer to that of the electrolyte, and provides better adhesion of the fuel electrode with the electrolyte.

The electrical conductivity of the Ni–YSZ cermet electrode is a strong function of its nickel content. The percolation threshold depends on the surface area of the two phases and is usually at about 30 vol.% Ni (corresponding to the mechanism of electronic conduction through the nickel phase).

Other candidate anode materials under investigation include cobalt–YSZ, ruthenium–YSZ, and mixed conducting oxides based on Y_2O_3–ZrO_2–TiO_2 solid solutions.

2.3 *Cathode*

The cathode operates in an oxidizing environment of air or oxygen at $\sim 1000°C$ and participates in the following oxygen reduction reaction:

$$\tfrac{1}{2} O_2(g) + 2e^- \rightarrow O^{2-} \qquad (5)$$

i.e., oxygen in the gas phase is reduced to oxygen ions, consuming two electrons in the process. The requirements of the cathode are high electronic conductivity, chemical and dimensional stability during cell fabrication and cell operation, thermal expansion match with all of the other cell components, and sufficient porosity to facilitate transport of molecular oxygen from the gas phase to the cathode/electrolyte interface.

The high operating temperature of the SOFCs allows the use of only noble metals or electronic conducting oxides as cathode materials. However, the high cost of noble metals such as platinum or palladium prohibits their use. Among the various doped oxides investigated, doped lanthanum manganite ($LaMnO_3$) is the most commonly used. It is a p-type perovskite oxide which can have a range of oxygen stoichiometry depending on processing history, temperature, and the ambient oxygen partial pressure. Although $LaMnO_3$ is stable under oxidizing conditions, it dissociates under highly reducing conditions ($pO_2 \leqslant 10^{-14}$ atm at 1000°C). It has intrinsic p-type conductivity due to the formation of cation vacancies, and electronic conduction arises from a mechanism called small polaron hopping. Essentially, this is due to the hopping of an electron hole between the +3 and +4 valence states of manganese. Doping with a lower valence (divalent) ion such as calcium, strontium, barium, or magnesium enhances this effect and is commonly used to a obtain higher conductivity in $LaMnO_3$. For instance, strontium doping enhances the electronic conductivity by increasing the Mn^{4+} content by substituting La^{3+} with Sr^{2+}.

The temperature dependence of the conductivity of $LaMnO_3$ shows an Arrhenius-type relationship with activation energies of 18.3, 15.4, and 8.7 kJ mol^{-1} for undoped and 10 and 20 mol.% strontium-doped $LaMnO_3$, respectively. The electrical behavior of doped $LaMnO_3$ is significantly affected by oxygen partial pressure. At about 1000°C the conductivity shows little dependence on oxygen partial pressure; however, the range of oxygen partial pressures over which conductivity remains constant becomes smaller as temperature increases. At lower oxygen partial pressures, the conductivity decreases as a function of the fourth root of pO_2 and abruptly decreases at a critical oxygen partial pressure corresponding to the decomposition of the $LaMnO_3$ phase.

The TEC of undoped $LaMnO_3$ is $11.2 \times 10^{-6} K^{-1}$. Doping with strontium increases the TEC of $LaMnO_3$ making it higher than that of the YSZ electrolyte. The lowering of TEC is accomplished by the substitution of lanthanum with smaller cations such as calcium.

In $LaMnO_3$ manganese is known to be the mobile species at elevated temperatures and, if diffused in the electrolyte, can change its electrical properties. A number of studies has been performed on the interactions between $LaMnO_3$ and YSZ and no significant reactions have been observed at temperatures below about 1200°C. Above 1200°C, there is the potential to form an insulating phase $La_2Zr_2O_7$.

Other cathode materials investigated include perovskite-structured oxides such as lanthanum cobaltite, lanthanum ferrites, and lanthanum nickelates, suitably doped with alkali and alkaline earth ions to tailor the conductivity and thermal expansion coefficient.

2.4 *Interconnection*

The requirements for the interconnection are the most severe of all the cell components and include 100% electronic conductivity (in contrast to the electrolyte where 100% ionic conductivity is required), absence of porosity to prevent permeation and mixing of the oxidant and the fuel, a thermal expansion match with the cathode and the electrolyte, and minimal reactivity with the cathode, the electrolyte, and the material for the electrical contacts (usually nickel).

The material that satisfies all of these requirements and is widely used in SOFCs is alkaline earth-doped lanthanum chromite ($LaCrO_3$). This material, like the air electrode material, has the perovskite structure and exhibits p-type conductivity due to the small polaron hopping mechanism. The electrical conduction properties of this material are very similar to those of the air electrode. The important difference between $LaMnO_3$ and $LaCrO_3$ is the ability of the chromite to survive in highly reducing as well as oxidizing atmospheres. Lanthanum chromite is often doped with divalent ions such as strontium, calcium, or magnesium to obtain adequate conductivity for use as an interconnection. The conductivity of doped $LaCrO_3$ can be several orders of magnitude lower in hydrogen than in air. Even though the interconnection is exposed to a dual atmosphere and a conductivity gradient exists across the material, the overall conductivity of doped $LaCrO_3$ is still adequate for interconnection applications.

The TEC of $LaCrO_3$ at room temperature is about $6.7\times10^{-6}\,K^{-1}$ which changes to about $9.2\times10^{-6}\,K^{-1}$ above 240°C due to an orthorhombic–rhombohedral phase transformation. The TEC of $LaCrO_3$ is tailored to match that of the cathode and electrolyte by the addition of aliovalent ions such as aluminum, which decreases the TEC, or calcium and strontium, which increase the TEC. Lanthanum chromite exhibits poor sinterability in air or oxidizing environments. The various approaches investigated to improve the sinterability of $LaCrO_3$ include the use of high-surface-area powders and nonstoichiometric powders and the addition of dopants and sintering aids such as CaO, Al_2O_3, LaF_3, and MgF_2.

3. *Cell Design and Fabrication Processes*

Solid oxide fuel cells are being developed in several designs including tubular and planar (Fig. 2). Materials used to construct the SOFCs in these designs are very similar in nature; however, the power density, the manner of isolating fuel from oxidant, and cell-to-cell electrical connections differ in these designs.

3.1 Tubular Design

Figure 2(a) illustrates the design of the tubular SOFC. A ceramic tube that forms the mechanical support for the cell also serves as the air electrode (cathode). The other components of the cell are deposited as thin layers on the air electrode tube. For cell operation, the oxidant (air) is introduced near the closed end of the cell through a ceramic injector tube where it flows and exits through the annular space between the cell and the injector tube. Fuel flows on the outside of the cell and is electrochemically oxidized while flowing to the open end of the cell.

Individual cells are connected in series or in parallel to form a stack. Porous nickel felt is used to connect the anode of one cell to the cathode of the next cell through the nickel plating on the interconnection for series connection, and the anode of one cell to the anode of an adjacent cell for parallel connection. The major advantage of the tubular design is that it does not require high integrity, high-temperature seals for isolating the fuel from the oxidant. On the other hand, this design has a longer current path, resulting in higher resistive losses.

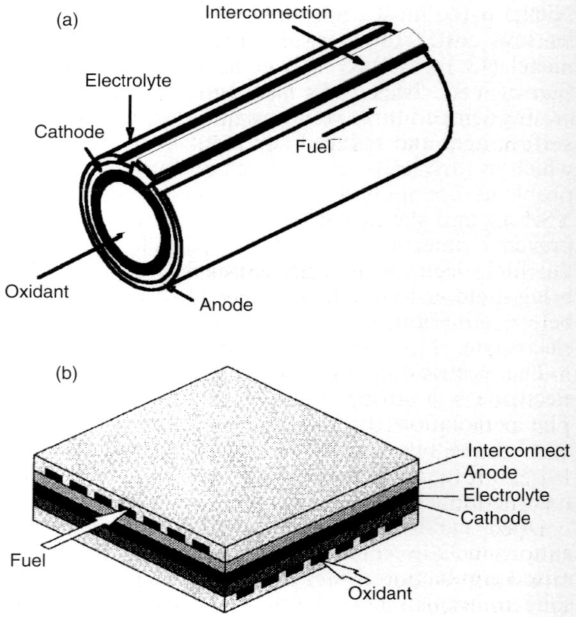

Figure 2
Different designs of SOFCs: (a) tubular; (b) planar.

The fabrication processes for the tubular design commence with formation of the air electrode tube by extrusion, followed by sintering in air at about 1500°C to ~30% porosity. On top of this tube the interconnection is deposited, in the form of an 85 μm thick, 9 mm wide strip on the air electrode, by a plasma spraying process during which the rest of the air electrode is masked to limit the deposition to a thin strip. The masking material is removed during postdeposition cleaning.

The YSZ electrolyte is deposited as a ~40 μm thick film by an electrochemical vapor deposition process (EVD) that involves the reaction of zirconium chloride and yttrium chloride with oxygen at about 1200°C to form YSZ.

The fuel electrode (anode), which is 100–150 μm in thickness, is fabricated in a two-step process. A nickel + YSZ slurry is first applied over the electrolyte followed by sintering at about 1300°C to form a ~40% porous fuel electrode. Finally, nickel is electroplated on to the interconnection to provide electrical contact.

The performance and lifetimes of single cells, cell bundles, and SOFC power generation systems using such cells of tubular design have been extensively investigated. Single cells have been tested for up to ~69000 h and a 100 kW system consisting of 1152 cells has been tested for over 17500 h.

3.2 *Planar Design*

In this design, single cells consisting of anode/electrolyte/cathode and interconnection are formed as thin flat plates (Fig. 2(b)). The interconnection usually has a ribbed structure that allows gas passage and also functions as a bipolar gas separator connecting the anode and cathode of adjacent cells. The planar design offers higher power density compared with the tubular design, and its internal resistance losses are independent of cell area due to in-plane conduction. This permits the use of very thin cell components to minimize electrical resistance. The thin components also offer the possibility of a lower-temperature operation. Further, cell components can be fabricated independently, thus avoiding the difficulties in co-firing different components. However, the planar design requires high-temperature seals at the edges of the plates to isolate fuel from the oxidant. Various seals based on cements, glasses, and glass–ceramics have been proposed but they suffer from a variety of problems including cracking, interfacial reactions, and unacceptably high thermal expansion mismatches between the seals and the cell components.

In the planar design the YSZ electrolyte is usually made by tape casting and sintering to about 50–120 μm thickness. The electrodes are applied by slurry coating, screen printing, or plasma spraying. The interconnection can be either $LaCrO_3$ (fabricated by tape casting or hot pressing) or a high-temperature, oxidation-resistant metallic alloy. Stacks are formed by stacking up the individual cells to the desired thickness. Single cells of planar design have been tested for more than 5000 hours. The operation of multikilowatt stacks of up to 10 kW size has also been demonstrated using the planar design.

4. *Summary*

High-temperature SOFCs provide a high efficiency, nonpolluting power generation technology. Properties of an oxygen ion-conducting zirconia electrolyte have been extensively studied for SOFC applications and are well understood. Performance of SOFCs based on zirconia electrolytes has been demonstrated via successful operation of power generation systems of up to 100 kW in size that incorporate these cells. Among the various cell designs, the tubular design is closest to commercialization. Ongoing research efforts to improve further the efficiency and reduce the cost of SOFCs include studying alternate, lower temperature, and/or higher conductivity electrolytes, and finding lower-cost fabrication processes for the present materials.

Bibliography

Minh N Q, Takahashi T 1995 *Science and Technology of Ceramic Fuel Cells*. Elsevier, Amsterdam

Pal U B, Singhal S C 1990 Electrochemical vapor deposition of yttria stabilized zirconia films. *J. Electrochem. Soc.* **137**, 2937–41

Subbarao E C, Maiti H S 1984 Solid electrolytes with oxygen ion conduction. *Solid State Ionics* **11**, 317–38

Worrell W L, Liou S S 1989 Electrical properties of novel mixed-conducting oxides. *Appl. Phys.* **49**, 25–31

LIST OF CONTRIBUTORS

Contributors are listed in alphabetical order together with their addresses. Titles of articles that they have authored follow in alphabetical order. Where articles are coauthored, this has been indicated by an asterisk preceding the title.

Abe, F.
National Research Institute for Metals
Tsukuba
Japan
Fusion Reactors: Low-activation Structural Materials

Alamo, A.
Commissariat à l'Energie Atomique
Gif-sur-Yvette
France
Fast Reactor Cores, Ferritic–Martensitic Steels for

Auty, D. H.
Twin Cell Technology for Hydrogen Generation

Banerjee, S.
Bhabha Atomic Research Centre
Mumbai
India
Nuclear Applications: Zirconium Alloys

Barber, D. H.
Ontario Power Generation
Toronto, Ontario
Canada
Heavy Water Reactor Fuel Design and Performance

Bates, J. B.
Future Energy Inc.
Marietta, Georgia
USA
Thin-film Materials for Solid-state Rechargeable Batteries

Bazan, G. C.
University of California
Santa Barbara, CA
USA
Polymer-based Electrophosphorescent Devices

Birring, A.
NDE Associates Inc.
Webster, Texas
USA
NDT for Non-Nuclear Power Generation: Pressure Vessels, Piping, and Turbines

Boller, C.
European Aeronautic Defence and Space Company
Muenchen
Germany
Composites for Sensors and Actuators

Bour, D. P.
LumiLeds Lighting
San Jose, California
USA
GaN-based Optoelectronic Devices

Brachet, J. C.
Commissariat à l'Energie Atomique
Gif-sur-Yvette
France
Fast Reactor Cores, Ferritic–Martensitic Steels for

Brandon, N. P.
Imperial College
London
UK
Fuel Cells: Materials

Bünzli, J.-C. G.
University of Lausanne
Switzerland
Light Conversion: Organized Lanthanide-containing Molecular Systems

Burchell, T. D.
Oak Ridge National Laboratory
Oak Ridge, Tennessee
USA
Nuclear Graphite and Radiation Effects
Radiation Effects in Carbon–Carbon Composites

Burkov, A. T.
Russian Academy of Sciences, St Petersburg
Russia
Magnetic Fields: Thermoelectric Power

Bygchkov, A. V.
Research Institute of Atomic Reactors
Dimitrograd
Russia
Oxide Fuels, Nonaqueous Reprocessing of

Chapiro, A.
Grande Condriac
Gif-sur-Yvette
France
Radiation Effects in Polymers

Chauvin, N.
Commissariat à l'Energie Atomique (CEA), Saint Paul Lez Durance
France
Inert Matrix Fuels

Chen, P.
Metal Amides: New Hydrogen Storage Systems

Chetal, S. C.
Indira Gandhi Centre for Atomic Research
Kalpakkam
India
Nuclear Reactors: Coolant Materials

Collins, R. T.
Photovoltaics: Advanced Inorganic Materials

Collocott, S. J.
CSIRO Telecommunications and Industrial Physics
Lindfield
The Netherlands
Magnetic Materials: Domestic Applications

Cugat, O.
Laboratoire d'Electrotechnique INPG B.P. 46
F-38402 Saint Martin d'Heres Grenoble Cedex
France
Magnetic Micro-actuators (MAGMAS)

Delamare, J.
Laboratoire d'Electrotechnique INPG B.P. 46
F-38402 Saint Martin d'Heres Grenoble Cedex
France
Magnetic Micro-actuators (MAGMAS)

Duclot, M.
Laboratoire d'Electrochimie et de Physico-chimie des Matériaux et des Interfaces, Saint Martin d'Hères
France
Batteries: Glassy Electrolytes

Dudney, N. J.
Oak Ridge National Laboratory
Oak Ridge, Tennessee
USA
Thin-film Materials for Solid-state Rechargeable Batteries

Espinosa, H. D.
Nanoelectromechanical Systems: Experiments and Modeling

Fay, P.
University of Notre Dame
Notre Dame, Indiana
USA
Photodetectors

Fischer, E.
Siemens, Erlangen
Germany
NDT of Components in Nuclear Installations

Flandrois, S.
Centre de Recherche Paul-Pascal
Pessac
France
Battery Carbons

Fort, W. C. III
Shell International E & P, Inc.
Houston, Texas
USA
Petroleum Industry: Corrosion

Gadsby, R.
Atomic Energy Canada Limited
Mississauga, Ontario
Canada
DUPIC Fuel Cycle

Ganguly, C.
Nuclear Fuel Complex
Hyderabad
India
Nuclear Reactor Fuel Fabrication (Including Quality Control)

Gao, J.
University of California
Santa Barbara
California
USA
Electrochemical Light Emitting Cells from Semiconducting Polymers

Gherardi, L.
Pirelli Cavi e Sistemi Spa
Milan
The Netherlands
Superconducting Wires and Cables: Low-field Applications

Ghoniem, N. M.
University of California
Los Angeles, California
USA
Fusion Conditions: Radiation Damage Correlation

Golubov, S. I.
Scientific Center of the Russian Federation, Obninsk
Russian Federation
Radiation Damage Theory

Gooch, D. J.
GenExp Ltd.
Swindon
UK
Creep Design and Life Extension of Large-scale Plant

Guha, S.
United Solar Systems Corporation
Troy, Michigan
USA
Amorphous Semiconductor Solar Cells

Gupta, C. K.
Bhabha Atomic Research Centre
Mumbai
India
Nuclear Reactor Materials

Halchak, J.
The Boeing Company
Canoga Park, California
USA
Spacecraft Materials

Hamza, A. V.
Targets for Inertial Confinement Fusion

Hanitsch, R. E.
Technische Universität Berlin
Germany
Rotary and Linear Machines

Hebling, C.
Fuel Cells: Microsystems

Heeger, A. J.
University of California
Santa Barbara, California
USA
Electrochemical Light Emitting Cells from Semiconducting Polymers

Hesketh, K. W.
British Nuclear Fuels Ltd.
Preston
UK
Burnable Poisons in Nuclear Fuels

Hikmet, R. A. M.
Philips Research
Eindhoven
The Netherlands
Organic Electrolytes and Electrodes for Batteries

Hintzen, H. T.
Eindhoven University of Technology
The Netherlands
Novel Nitride Phosphors

Holdsworth, S. R.
ALSTOM Energy Ltd.
Rugby
UK
Creep-resistant Materials for Steam Turbines

Iglesias, F. C.
Ontario Power Generation
Toronto, Ontario
Canada
Heavy Water Reactor Fuel Design and Performance

Inoue, H.
Osaka Prefecture University
Sakai, Osaka
Japan
Hydrogen–Metal Systems: Electrochemical Reactions (Fundamentals and Applications)

Ishino, S.
Tokai University
Kanagawa-ken
Japan
Fusion Reactors (Magnetically Confined)—Tokamaks: Materials

Iwakura, C.
Osaka Prefecture University
Sakai, Osaka
Japan
Hydrogen–Metal Systems: Electrochemical Reactions (Fundamentals and Applications)

John, R. C.
Shell International E & P, Inc.
Houston, Texas
USA
Petroleum Industry: Corrosion

Kamath, H. S.
Bhabha Atomic Research Centre
Trombay
India
Mixed Oxide (MOX) Fuel

Kasiviswanathan, K. V.
Indira Gandhi Center for Atomic Research
Kalpakkam
India
Hot Cells, Glove Boxes, and Shielded Facilities

Ke, C.
Nanoelectromechanical Systems: Experiments and Modeling

Kumar, P. V.
Indira Gandhi Centre for Atomic Research
Kalpakkam
India
In-reactor Creep Testing: Techniques

Laidler, J. J.
Argonne National Laboratory
Illinois
USA
Nuclear Fuel Reprocessing

Lambropoulos, J. C.
University of Rochester
New York
USA
Polysilicon and EDF Silicon (Shaped) for Solar Applications

Lee, P. J.
The University of Wisconsin
Madison, WI
USA
Superconducting Wires and Cables: Materials and Processing

Leskelä, M.
University of Helsinki
Finland
Electroluminescent Phosphors

Li, W.-M.
University of Helsinki
Finland
Electroluminescent Phosphors

Li, Y. Q.
Eindhoven University of Technology
The Netherlands
Novel Nitride Phosphors

Li, Z. P.
Kogakuin University
Tokyo
Japan
Hydrogen–Metal Systems: Fluorinated Metal Hydrides

Liu, B.-H.
Kogakuin University
Tokyo
Japan
Hydrogen–Metal Systems: Hydride Forming Alloys (Properties and Characteristics, Database Information)

Luo, W.
Metal Amides: New Hydrogen Storage Systems

Luongo, C. A.
National High Magnetic Field Laboratory
Tallahassee, Florida
USA
Superconducting Machines: Energy Storage

Malliaras, G. G.
Cornell University, Ithaca
New York
USA
Photovoltaic Devices from Organic Semiconductors

Mannan, S. L.
Indira Gandhi Centre for Atomic Research
Kalpakkam
India
Fast Reactor Cores, Austenitic Steels for

Matzke, H.
European Commission, Joint Research Centre
Institute for Transuranium Elements
Karlsruhe
Germany
Nuclear Waste Management, Materials Aspects of

McCune, R. C.
Ford Motor Company
Dearborn, Michigan
USA
Automotive Engine Materials

Millet, P.
Commissariat à l'Energie Atomique, Cadarache
France and
EPRI
Palo Alto, California
USA
Fast Breeder Reactors: Fuels
Power Industry: Corrosion

Muralt, P.
Swiss Federal Institute of Technology
Lausanne
Switzerland
Piezoelectric Thin Films for MEMS
Stress Coupled Phenomenon: Piezoelectric Effect

Murugan, S.
Indira Gandhi Centre for Atomic Research
Kalpakkam
India
In-reactor Creep Testing: Techniques

Nagamoto, H.
Kogakuin University
Tokyo
Japan
Fuel Cells: Electrochemical Reactions

Neudecker, B. J.
ITN Energy Systems
Littleton, Colorado
USA
Thin-film Materials for Solid-state Rechargeable Batteries

Newman, J.
University of California
Berkeley, California
USA
Advanced Automotive Powertrains: Materials

Noda, T.
National Research Institute for Metals
Tsukuba
Japan
Fusion Reactors: Low-activation Structural Materials

Nohara, S.
Osaka Prefecture University
Sakai, Osaka
Japan
Hydrogen–Metal Systems: Electrochemical Reactions (Fundamentals and Applications)

Odette, G. R.
University of California
Santa Barbara
California
USA
Nuclear Reactors: Pressure Vessel Steels

Olander, D. R.
University of California at Berkeley
California
USA
Light Water Reactor Fuel Design and Performance

Palmer, I. D.
British Nuclear Fuels Limited
Preston
UK
Nuclear Fuels: Behavior at High Burnup

Paul, W.
ABB Corporate Research Ltd
Baden-Daettwil
The Netherlands
Superconducting Machines: Energy Distribution Components

Pelletier, M.
Commissariat à l'Energie Atomique (CEA), Saint Paul Lez Durance
France
Inert Matrix Fuels

Peuzin, J. C.
Laboratoire Louis Néel, CNRS Grenoble
France
Magnetostrictive Materials

Piguet, C.
University of Geneva
Switzerland
Light Conversion: Organized Lanthanide-containing Molecular Systems

Pilgrim, S. M.
NYS College of Ceramics at Alfred University
Alfred
New York
USA
Electrostrictive Ceramics for Sonar Projectors

Pugno, N.
Nanoelectromechanical Systems: Experiments and Modeling

Purushotham, D. S. C.
Bhabha Atomic Research Centre
Trombay
India
Mixed Oxide (MOX) Fuel

Quandt, E.
Center of Advanced European Studies and Research
Bonn
Germany
Actuator Materials for Small-scale Devices

Raehm, L.
Université Louis Pasteur
Strasbourg
France
Molecular Motors

Ramanujam, A.
Bhabha Atomic Research Centre
Mumbai
India
Purex and Thorex Processes (Aqueous Reprocessing)

Reyne, G.
Laboratoire d'Electrotechnique INPG B.P. 46
F-38402 Saint Martin d'Heres Grenoble Cedex
France
Magnetic Micro-actuators (MAGMAS)

Ritala, M.
University of Helsinki
Finland
Electroluminescent Phosphors

Robinson, M. R.
University of California
Santa Barbara, CA
USA
Polymer-based Electrophosphorescent Devices

Rodriguez, P.
Indira Gandhi Centre for Atomic Research
Kalpakkam
India and
Defence Research and Development Organisation
Ministry of Defence, Timarpur
Delhi
India
**In-reactor Creep Testing: Techniques*
Nuclear Fuel Cycles
Nuclear Reactor Materials: Irradiation Effects
Thorium Fuel Cycles

Roedel, R. J.
Arizona State University
Tempe, Arizona
USA
Light Emitting Diodes

Ronda, C.
Philips Research Laboratories
Aachen
Germany
Rare Earth Phosphors: Fundamentals and Applications

Séran, J. L.
Commissariat à l'Energie Atomique
Gif-sur-Yvette
France
Fast Reactor Cores, Ferritic–Martensitic Steels for

Sales, B. C.
Oak Ridge National Laboratory
Oak Ridge, Tennessee
USA
Thermoelectric Devices: Refrigeration and Power Generations With No Moving Parts

Sauvage, J.-P.
Université Louis Pasteur
Strasbourg
France
Molecular Motors

Schafrik, R. E.
Cincinnati
Ohio
USA
Jet Engine Materials

Schultz, J. H.
Massachusetts Institute of Technology
Cambridge, MA
USA
Superconducting Wires and Cables: High-field Applications

Sharma, B. P.
Bhabha Atomic Research Centre
Navi Mumbai
India
Nuclear Reactors: Moderator and Reflector Materials

Simnad, M. T.
University of California
San Diego, California
USA
Nuclear Reactors: Shielding Materials

Singh, B. N.
Risø National Laboratory
Roskilde
Denmark
Radiation Damage Theory

Singhal, S. C.
Siemens Westinghouse Power Corporation
Pittsburgh, Pennsylvania
USA
Zirconia Electrolyte-based Solid Oxide Fuel Cells

Sivaprasad, P. V.
Indira Gandhi Centre for Atomic Research
Kalpakkam
India
Fast Reactor Cores, Austenitic Steels for

Skiba, O. V.
Research Institute of Atomic Reactors
Dimitrograd
Russia
Oxide Fuels, Nonaqueous Reprocessing of

Slaoui, A.
Photovoltaics: Advanced Inorganic Materials

Souquet, J.-L.
Laboratoire d'Electrochimie et de Physico-chimie des Matériaux et des Interfaces, Saint Martin d'Hères
France
Batteries: Glassy Electrolytes

Spearing, S. M.
Massachusetts Institute of Technology
Massachusetts
USA
Micro Devices and Micro Systems, Materials for

Suda, S.
Kogakuin University
Tokyo
Japan
Direct Borohydride Fuel Cell
Hydrogen–Metal Systems: Applications of Gas–Solid Reactions
Hydrogen–Metal Systems: Technological and Engineering Aspects

Tanaka, K.
Nagoya Institute of Technology
Japan
Hydrogen–Metal Systems: Basic Properties (1)

Tessler, N.
Technion Israel Institute of Technology
Haifa
Israel
Light Emitting Diodes from Molecular and Polymer Semiconductors

Thombarse, D. G.
Stirling Engine: Micro-CHP System for Residential Application

Tietz, F.
Solid Oxide Fuel Cells

Trinkaus, H.
Forschungszentrum Jülich
Germany
Radiation Damage Theory

Tritt, T. M.
Clemson University
South Carolina
USA
Thermoelectric Materials: Principles, Structure, Properties, and Applications

Uchino, K.
Pennsylvania State University
University Park, Pennsylvania
USA
Multilayer Ceramic Actuators

Weber, G. A.
Ford Motor Company
Dearborn, Michigan
USA
Automotive Engine Materials

Wijayantha, K. G. U.
Twin Cell Technology for Hydrogen Generation

Williams, J. C.
Cincinnati
Ohio
USA
Jet Engine Materials

Yagasaki, K.
University of the Ryukyus
Okinawa
Japan
Magnetic Fields: Thermoelectric Power

Yang, Myung Seung
Korea Atomic Energy Research Institute
Taejon
Republic of Korea
DUPIC Fuel Cycle

Yang, Y.
University of California
Los Angeles, California
USA
Polymer Light Emitting Diodes, Inkjet Printing of

Yoshinari, O.
Nagoya Institute of Technology
Japan
Hydrogen–Metal Systems: Basic Properties (1)

Yu, H.
Fuel Cells: Microsystems

Zhang, L.
Visteon Corporation
Livonia, Michigan
USA
Batteries, Rechargeable

Zimm, C. B.
Magnetic Refrigerator: Design and Performance

INDEX

The Subject Index has been compiled to assist the reader in locating all references to a particular topic in the Encyclopedia. Index entries are filed in letter-by-letter sequence, and arranged in set-out style with up to three levels of heading. Major coverage of a subject is indicated by ***bold italic*** page numbers. Page numbers followed by *T* (or *F*) refer to Tables (or Figures). *See* references point the user to preferred terms; *see also* references indicate terms of related interest. Every effort has been made to make the index as comprehensive as possible and to standardize the terms used.

A

B

C

D

E

H

I

J

K

L

M

O

P

Q

R

S

T

U

V

W

X

Y